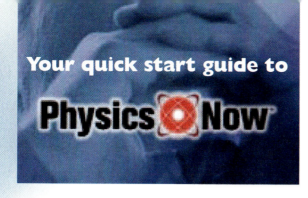

Your quick start guide to Physics Now

Welcome to **PhysicsNow**, your fully integrated system for physics tutorials and self-assessment on the web. To get started, just follow these simple instructions.

Your first visit to PhysicsNow

1. Go to **http://www.cp7e.com** and click the **Register** button.

2. The first time you visit, you will be asked to select your school. Choose your state from the drop-down menu, then type in your school's name in the box provided and click **Search**. A list of schools with names similar to what you entered will show on the right. Find your school and click on it.

3. On the next screen, enter the access code from the card that came with your textbook in the "Content or Course Access Code" box*. Enter your email address in the next box and click **Submit**.

   * PhysicsNow access codes may be purchased separately. Should you need to purchase an access code, go back to **http://www.cp7e.com** and click the **Buy** button.

4. On the next screen, choose a password and click **Submit.**

5. Lastly, fill out the registration form and click **Register and Enter iLrn.** This information will only be used to contact you if there is a problem with your account.

6. You should now see the **PhysicsNow** homepage. Select a chapter and begin!

   Note: Your account information will be sent to the email address that you entered in Step 3, so be sure to enter a valid email address. You will use your email address as your username the next time you login.

Second and later visits

1. Go to **http://www.cp7e.com** and click the **Login** button.

2. Enter your user name (the email address you entered when you registered) and your password and then click **Login.**

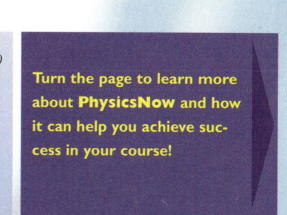

SYSTEM REQUIREMENTS:
(Please see the System Requirements link at www.ilrn.com for complete list.)
PC: Windows 98 or higher, Internet Explorer 5.5 or higher
Mac: OS X or higher, Mozilla browser 1.2.1 or higher

TECHNICAL SUPPORT:
For online help, click on **Technical Support** in the upper right corner of the screen, or contact us at:

 1-800-423-0563 Monday–Friday • 8:30 A.M. to 6:00 P.M. EST
tl.support@thomson.com

Turn the page to learn more about PhysicsNow and how it can help you achieve success in your course!

THOMSON
BROOKS/COLE

PhysicsNow™ Quick Start Guide

iii

What do you need to learn now?

Take charge of your learning with **PhysicsNow**™, a powerful student-learning tool for physics! This interactive resource helps you gauge your unique study needs, then gives you a *Personalized Learning Plan* that will help you focus in on the concepts and problems that will most enhance your understanding. With **PhysicsNow**, you have the resources you need to take charge of your learning!

The access code card included with this new copy of **College Physics** is your ticket to all of the resources in **PhysicsNow.** (See the previous page for login instructions.)

Interact at every turn with the POWER and SIMPLICITY of PhysicsNow!

PhysicsNow combines Serway and Faughn's best-selling *College Physics* with carefully crafted media resources that will help you learn. This dynamic resource and the Seventh Edition of the text were developed in concert, to enhance each other and provide you with a seamless, integrated learning system.

As you work through the text, you will see notes that direct you to the media-enhanced activities in **PhysicsNow.** This precise page-by-page integration means you'll spend less time flipping through pages or navigating websites looking for useful exercises. These multimedia exercises will make all the difference when you're studying and taking exams . . .

after all, it's far easier to understand physics if it's seen in action, and **PhysicsNow** enables you to become a part of the action!

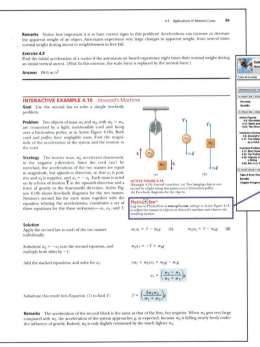

Begin at http://www.cp7e.com and build your own Personalized Learning Plan now!

Log into PhysicsNow at **http://www.cp7e.com** by using the free access code packaged with the text. You'll immediately notice the system's simple, browser-based format. You can build a complete *Personalized Learning Plan* for yourself by taking advantage of all three powerful components found on PhysicsNow:

▶ **What I Know**
▶ **What I Need to Learn**
▶ **What I've Learned**

The best way to maximize the system and optimize your time is to start by taking the *Pre-Test* ▶▶▶

PhysicsNow™ Quick Start Guide

CONVERSION FACTORS

Length

1 m = 39.37 in. = 3.281 ft
1 in. = 2.54 cm
1 km = 0.621 mi
1 mi = 5280 ft = 1.609 km
1 light year (ly) = 9.461×10^{15} m
1 angstrom (Å) = 10^{-10} m

Mass

1 kg = 10^3 g = 6.85×10^{-2} slug
1 slug = 14.59 kg
1 u = 1.66×10^{-27} kg = 931.5 MeV/c^2

Time

1 min = 60 s
1 h = 3600 s
1 day = 8.64×10^4 s
1 yr = 365.242 days = 3.156×10^7 s

Volume

1 L = 1000 cm^3 = 3.531×10^{-2} ft^3
1 ft^3 = 2.832×10^{-2} m^3
1 gal = 3.786 L = 231 in.^{3}

Angle

180° = π rad
1 rad = 57.30°
1° = 60 min = 1.745×10^{-2} rad

Speed

1 km/h = 0.278 m/s = 0.621 mi/h
1 m/s = 2.237 mi/h = 3.281 ft/s
1 mi/h = 1.61 km/h = 0.447 m/s = 1.47 ft/s

Force

1 N = 0.2248 lb = 10^5 dynes
1 lb = 4.448 N
1 dyne = 10^{-5} N = 2.248×10^{-6} lb

Work and energy

1 J = 10^7 erg = 0.738 ft·lb = 0.239 cal
1 cal = 4.186 J
1 ft·lb = 1.356 J
1 Btu = 1.054×10^3 J = 252 cal
1 J = 6.24×10^{18} eV
1 eV = 1.602×10^{-19} J
1 kWh = 3.60×10^6 J

Pressure

1 atm = 1.013×10^5 N/m^2 (or Pa) = 14.70 lb/in.^{2}
1 Pa = 1 N/m^2 = 1.45×10^{-4} lb/in.^{2}
1 lb/in.^{2} = 6.895×10^3 N/m^2

Power

1 hp = 550 ft·lb/s = 0.746 kW
1 W = 1 J/s = 0.738 ft·lb/s
1 Btu/h = 0.293 W

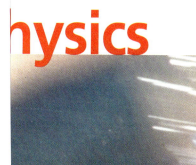

Se...hysics

Chris Vuille
Embry-Riddle Aeronautical University

Charles A. Bennett
The University of North Carolina, Asheville
Contributing Author, Media

THOMSON
BROOKS/COLE

Australia · Canada · Mexico · Singapore · Spain · United Kingdom · United States

THOMSON

★

BROOKS/COLE

Physics Acquisitions Editor: Chris Hall
Publisher: David Harris
Vice President, Editor-in-Chief, Sciences: Michelle Julet
Development Editor: Ed Dodd
Assistant Editor: Annie Mac
Editorial Assistant: Kyra Engelberg
Technology Project Manager: Sam Subity
Marketing Manager: Erik Evans, Julie Conover
Marketing Assistant: Leyla Jowza
Advertising Project Manager: Stacey Purviance
Project Manager, Editorial Production: Teri Hyde
Print/Media Buyer: Barbara Britton

Permissions Editor: Audrey Pettengill
Production Service: Progressive Publishing Alternatives
Text Designer: Patrick Devine
Photo Researcher: Jane Sanders
Copy Editor: Progressive Publishing Alternatives
Illustrator: Rolin Graphics, Progressive Information Technologies
Cover Designer: Patrick Devine
Cover Image: ML Sinibaldi/CORBIS
Cover Printer: Phoenix Color Corp., MD
Compositor: Progressive Information Technologies
Printer: R.R. Donnelley

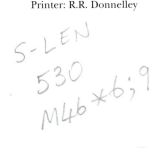

For more information about our products, contact us at:
Thomson Learning Academic Resource Center
1-800-423-0563

For permission to use material from this text or product, submit a request online at **http://www.thomsonrights.com**.
Any additional questions about permissions can be submitted by email to **thomsonrights@thomson.com**.

Thomson Brooks/Cole
10 Davis Drive
Belmont, CA 94002
USA

Library of Congress Control Number: 2004113839

Student Edition: ISBN 0-534-99723-6
Instructor's Edition: ISBN 0-534-99724-4
International Student Edition: ISBN 0-534-49318-1

Asia
Thomson Learning
5 Shenton Way #01-01
UIC Building
Singapore 068808

Australia/New Zealand
Thomson Learning
102 Dodds Street
Southbank, Victoria 3006
Australia

Canada
Nelson
1120 Birchmount Road
Toronto, Ontario M1K 5G4
Canada

Europe/Middle East/Africa
Thomson Learning
High Holborn House
50/51 Bedford Row
London WC1R 4LR
United Kingdom

Latin America
Thomson Learning
Seneca, 53
Colonia Polanco
11560 Mexico D.F.
Mexico

Spain/Portugal
Paraninfo
Calle Magallanes, 25
28015 Madrid, Spain

What I Know

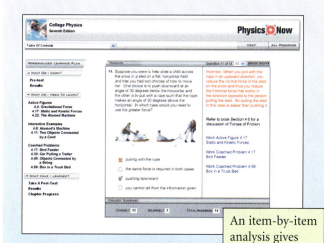

▲ You take a *Pre-Test* to measure your level of comprehension after reading a chapter. Each *Pre-Test* includes approximately 15 questions. The *Pre-Test* is your first step in creating your custom-tailored *Personalized Learning Plan*.

An item-by-item analysis gives you feedback on each of your answers.

▲ Once you've completed the "What I Know" *Pre-Test*, you are presented with a detailed *Personalized Learning Plan*, with text references that outline the elements you need to review in order to master the chapter's most essential concepts. This roadmap to concept mastery guides you to exercises designed to improve skills and to increase your understanding of the basic concepts.

At each stage, the *Personalized Learning Plan* refers to *College Physics* to reinforce the connection between text and technology as a powerful learning tool.

What I Need to Learn

Once you've completed the *Pre-Test*, you're ready to work through tutorials and exercises that will help you master the concepts that are essential to your success in the course.

ACTIVE FIGURES

A remarkable bank of more than 200 animated figures helps you visualize physics in action. Taken straight from illustrations in the text, these *Active Figures* help you master key concepts from the book. By interacting with the animations and accompanying quiz questions, you come to an even greater understanding of the concepts you need to learn from each chapter. ▼

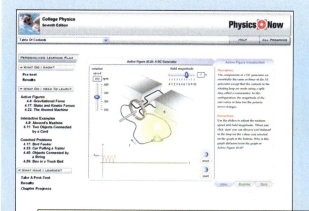

Each figure is titled so you can easily identify the concept you are seeing. The final tab features a *Quiz*. The *Explore* tab guides you through the animation so you understand what you should be seeing and learning.

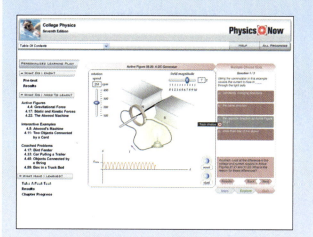

▲ The brief *Quiz* ensures that you mastered the concept played out in the animation—and gives you feedback on each response.

Continued on the next page ▶

COACHED PROBLEMS

Engaging *Coached Problems* reinforce the lessons in the text by taking a step-by-step approach to problem-solving methodology. Each *Coached Problem* gives you the option of breaking down a problem from the text into steps with feedback to 'coach' you toward the solution. There are approximately five *Coached Problems* per chapter.

You can choose to work through the *Coached Problems* by inputting an answer directly or working in steps with the program. If you choose to work in steps, the problem is solved with the same problem-solving methodology used in *College Physics* to reinforce these critical skills. Once you've worked through the problem, you can click **Try Another** to change the variables in the problem for more practice.

▲ Also built into each *Coached Problem* is a link to Brooks/Cole's exclusive **vMentor**™ web-based tutoring service site that lets you interact directly with a live physics tutor. If you're stuck on math, a *MathAssist* link on each *Coached Problem* launches tutorials on math specific to that problem.

INTERACTIVE EXAMPLES

You'll strengthen your problem-solving and visualization skills with *Interactive Examples*. Extending selected examples from the text, *Interactive Examples* utilize the proven and trusted problem-solving methodology presented in *College Physics.* These animated learning modules give you all the tools you need to solve a problem type—you're then asked to apply what you have learned to different scenarios. You will find approximately two *Interactive Examples* for each chapter of the text.▼

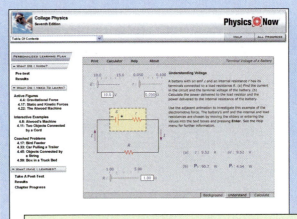

You're guided through the steps to solve the problem and then asked to input an answer in a simulation to see if your result is correct. Feedback is instantaneous.

What I've Learned

▶ After working through the problems highlighted in your *Personalized Learning Plan,* you move on to a *Post Test,* about 15 questions per chapter

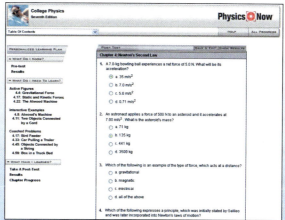

◀ Once you've completed the *Post Test,* you receive your percentage score and specific feedback on each answer. The *Post Test*s give you a new set of questions with each attempt, so you can take them over and over as you continue to build your knowledge and skills and master concepts.

Also available to help you succeed in your course

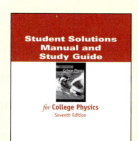

Student Solutions Manual and Study Guide
Volume I (Ch. 1–14) ISBN: 0-534-99920-4
Volume II (Ch. 15–30) ISBN: 0-534-99930-1
These manuals contain detailed solutions to approximately 12 problems per chapter. These problems are indicated in the textbook with boxed problem numbers. Each manual also features a skills section, important notes from key sections of the text, and a list of important equations and concepts.

Core Concepts in College Physics CD-ROM, Version 2.0
ISBN: 0-03-033701-1
Explore the core of physics with this powerful CD-ROM/workbook program! Content screens provide in-depth coverage of abstract and often difficult principles, building connections between physical concepts and mathematics. The presentation contains more than 350 movies—both animated and live video—including laboratory demonstrations, real-world examples, graphic models, and step-by-step explanations of essential mathematics. An accompanying workbook contains practical physics problems directly related to the presentation, along with worked solutions. Package includes three discs and a workbook.

Welcome to your MCAT Test Preparation Guide

The **MCAT Test Preparation Guide** makes your copy of *College Physics,* **Seventh Edition,** the most comprehensive MCAT study tool and classroom resource in introductory physics. The grid, which begins below and continues on the next two pages, outlines twelve concept-based **study courses** for the physics part of your MCAT exam. Use it to prepare for the MCAT, class tests, and your homework assignments.

Vectors

Skill Objectives: To calculate distance, angles between vectors, and magnitudes.

Review Plan:

Distance and Angles:
- Chapter 1, Sections 1.7, 1.8
- Active Figure 1.6
- Chapter Problems 35, 39, 42

Using Vectors:
- Chapter 3, Sections 3.1, 3.2
- Quick Quizzes 3.1–3.3
- Examples 3.1–3.3
- Active Figure 3.3
- Chapter Problems 4, 11, 19

Motion

Skill Objectives: To understand motion in two dimensions, to calculate speed and velocity, centripetal acceleration, and acceleration in free fall problems.

Review Plan:

Motion in 1 Dimension:
- Chapter 2, Sections 2.1–2.6
- Quick Quizzes 2.1–2.8
- Examples 2.1–2.10
- Active Figure 2.15
- Chapter Problems 1, 8, 17, 21, 29, 33, 38, 47, 51

Motion in 2 Dimensions:
- Chapter 3, Sections 3.3, 3.4
- Quick Quizzes 3.4–3.7
- Examples 3.4–3.8
- Active Figures 3.14, 3.15
- Chapter Problems 27, 33

Centripetal Acceleration:
- Chapter 7, Section 7.4
- Quick Quizzes 7.6, 7.7
- Example 7.6

Force

Skill Objectives: To know and understand Newton's Laws, to calculate resultant forces, and weight.

Review Plan:

Newton's Laws:
- Chapter 4, Sections 4.1–4.4
- Quick Quizzes 4.1–4.5
- Examples 4.1–4.4
- Active Figure 4.6
- Chapter Problems 5, 7, 11

Resultant Forces:
- Chapter 4, Section 4.5
- Quick Quizzes 4.6, 4.7
- Examples 4.7, 4.9, 4.10
- Chapter Problems 17, 25, 33

Equilibrium

Skill Objectives: To calculate momentum and impulse, center of gravity, and torque.

Review Plan:

Momentum:
- Chapter 6, Sections 6.1–6.3
- Quick Quizzes 6.2–6.6
- Examples 6.1–6.4, 6.6
- Active Figures 6.7, 6.10, 6.13
- Chapter Problems 7, 16, 21

Torque:
- Chapter 8, Sections 8.1–8.4
- Examples 8.1–8.7
- Chapter Problems 5, 9

Work

Skill Objectives: To calculate friction, work, kinetic energy, potential energy, and power.

Review Plan:

Friction:
- Chapter 4, Section 4.6
- Quick Quizzes 4.8–4.10
- Active Figure 4.19

Work:
- Chapter 5, Section 5.1
- Quick Quiz 5.1
- Example 5.1
- Active Figure 5.5
- Chapter Problems 11, 17

Energy:
- Chapter 5, Sections 5.2, 5.3
- Examples 5.4, 5.5
- Quick Quizzes 5.2, 5.3

Power:
- Chapter 5, Section 5.6
- Examples 5.12, 5.13

Waves

Skill Objectives: To understand interference of waves, to calculate basic properties of waves, properties of springs, and properties of pendulums.

Review Plan:

Wave Properties:
- Chapters 13, Sections 13.1–13.4, 13.7–13.11
- Quick Quizzes 13.1–13.6
- Examples 13.1, 13.6, 13.8–13.10
- Active Figures 13.1, 13.8, 13.12, 13.13, 13.24, 13.26, 13.32, 13.33, 13.34, 13.35
- Chapter Problems 9, 15, 23, 29, 41, 47, 53

Pendulum:
- Chapter 13, Sections 13.5
- Quick Quizzes 13.7–13.9
- Example 13.7
- Active Figures 13.15, 13.16
- Chapter Problem 35

Matter

Skill Objectives: To calculate pressure, density, specific gravity, and flow rates.

Review Plan:

Properties:
- Chapter 9, Sections 9.1–9.3
- Quick Quiz 9.1
- Examples 9.1, 9.3, 9.4
- Active Figure 9.3
- Chapter Problem 7

Pressure:
- Chapter 9, Sections 9.3–9.6
- Quick Quizzes 9.2–9.6
- Examples 9.4–9.9
- Active Figures 9.19, 9.20
- Chapter Problems 13, 23, 39

Flow rates:
- Chapter 9, Sections 9.7, 9.8
- Quick Quiz 9.7
- Examples 9.11–9.14
- Chapter Problem 42

Sound

Skill Objectives: To understand interference of waves, to calculate properties of waves, the speed of sound, Doppler shifts, and intensity.

Review Plan:

Sound Properties:
- Chapter 14, Sections 14.1–14.4, 14.6
- Quick Quizzes 14.1, 14.2
- Examples 14.1, 14.2, 14.4, 14.5
- Active Figures 14.6, 14.11
- Chapter Problems 7, 13,25

Interference/Beats:
- Chapter 14, Sections 14.7, 14.8, 14.11
- Quick Quiz 14.7
- Examples 14.6, 14.11
- Active Figures 14.18, 14.25
- Chapter Problems 33, 37, 51

Light

Skill Objectives: To understand mirrors and lenses, to calculate the angles of reflection, to use the index of refraction, and to find focal lengths.

Review Plan:

Reflection and Refraction:
- Chapter 22, Sections 22.1–22.4
- Quick Quizzes 22.2–22.4
- Examples 22.1–22.4
- Active Figures 22.4, 22.6, 22.7
- Chapter Problems 9, 17, 21, 25

Mirrors and Lenses:
- Chapter 23, Sections 23.1–23.6
- Quick Quizzes 23.1, 23.2, 23.4–23.6
- Examples 23.7, 23.8, 23.9
- Active Figures 23.2, 23.16, 23.25
- Chapter Problems 25, 29, 33, 39

Electrostatics

Skill Objectives: To understand and calculate the electric field, the electrostatic force, and the electric potential.

Review Plan:

Coulomb's Law:
- Chapter 15, Sections 15.1–15.3
- Quick Quiz 15.2
- Examples 15.1–15.3
- Active Figure 15.6
- Chapter Problems 11

Electric Field:
- Chapter 15, Sections 15.4, 15.5
- Quick Quizzes 15.3–15.6
- Examples 15.4, 15.5
- Active Figures 15.11, 15.16
- Chapter Problems 19, 23, 27

Potential:
- Chapter 16, Sections 16.1–16.3
- Quick Quizzes 16.1–16.5
- Examples 16.1, 16.4
- Active Figure 16.7
- Chapter Problems 1, 7, 15

Circuits

Skill Objectives: To understand and calculate current, resistance, voltage, power, and energy, and to use circuit analysis.

Review Plan:

Ohm's Law:
- Chapter 17, Sections 17.1–17.4
- Quick Quizzes 17.1, 17.3
- Examples 17.1, 17.3
- Chapter Problem 15

Power and energy:
- Chapter 17, Section 17.8
- Quick Quizzes 17.6–17.8
- Example 17.6
- Active Figure 17.11
- Chapter Problem 36

Circuits:
- Chapter 18, Sections 18.2, 18.3
- Quick Quizzes 18.1–18.5
- Examples 18.1–18.3
- Active Figures 18.2, 18.6
- Chapter Problems 4, 13

Atoms

Skill Objectives: To calculate half-life, and to understand decay processes and nuclear reactions.

Review Plan:

Atoms:
- Chapter 29, Sections 29.1, 29.2

Radioactive Decay:
- Chapter 29, Sections 29.3–29.5
- Examples 29.3, 29.4, 29.7
- Active Figures 29.6, 29.7
- Chapter Problems 15, 21, 29, 35

Nuclear reactions:
- Chapter 29, Sections 29.6
- Quick Quiz 29.4
- Examples 29.8, 29.9
- Chapter Problems 41, 45

Contents Overview

Contents

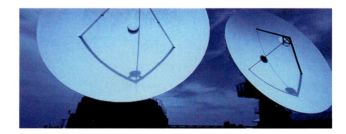

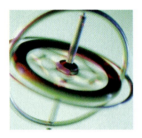

Part 2: Thermodynamics

Part 3: Vibrations and Waves

About the Authors

Raymond A. Serway received his doctorate at Illinois Institute of Technology and is Professor Emeritus at James Madison University. In 1990, he received the Madison Scholar Award at James Madison University, where he taught for 17 years. Dr. Serway began his teaching career at Clarkson University, where he conducted research and taught from 1967 to 1980. He was the recipient of the Distinguished Teaching Award at Clarkson University in 1977 and of the Alumni Achievement Award from Utica College in 1985. As Guest Scientist at the IBM Research Laboratory in Zurich, Switzerland, he worked with K. Alex Müller, 1987 Nobel Prize recipient. Dr. Serway also was a visiting scientist at Argonne National Laboratory, where he collaborated with his mentor and friend, Sam Marshall. In addition to earlier editions of this textbook, Dr. Serway is the co-author of *Physics for Scientists and Engineers,* Sixth Edition; *Principles of Physics,* Fourth Edition; and *Modern Physics,* Third Edition. He also is the author of the high-school textbook *Physics,* published by Holt, Rinehart, & Winston. In addition, Dr. Serway has published more than 40 research papers in the field of condensed matter physics and has given more than 70 presentations at professional meetings. Dr. Serway and his wife Elizabeth enjoy traveling, golfing, and spending quality time with their four children and six grandchildren.

Jerry S. Faughn earned his doctorate at the University of Mississippi. He is Professor Emeritus and former Chair of the Department of Physics and Astronomy at Eastern Kentucky University. Dr. Faughn has also written a microprocessor interfacing text for upper-division physics students. He is co-author of a non-mathematical physics text and a physical science text for general education students, and (with Dr. Serway) the high-school textbook *Physics,* published by Holt, Reinhart, & Winston. He has taught courses ranging from the lower division to the graduate level, but his primary interest is in students just beginning to learn physics. Dr. Faughn has a wide variety of hobbies, among which are reading, travel, genealogy, and old-time radio. His wife Mary Ann is an avid gardener, and he contributes to her efforts by staying out of the way. His daughter Laura is in family practice and his son David is an attorney.

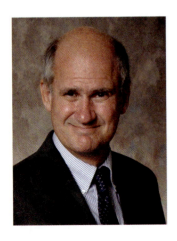

Chris Vuille is an associate professor of physics at Embry-Riddle Aeronautical University, Daytona Beach, Florida, the world's premier institution for aviation higher education. He received his doctorate in physics at the University of Florida in 1989, moving to Daytona after a year at ERAU's Prescott, Arizona campus. While he has taught courses at all levels, including post-graduate, his primary interest has been the delivery of introductory physics. He has received several awards for teaching excellence, including the Senior Class Appreciation Award (three times). He conducts research in general relativity and quantum theory, and was a participant in the JOVE program, a special three-year NASA grant program during which he studied neutron stars. His work has appeared in a number of scientific journals, and he has been a featured science writer in Analog Science Fiction/Science Fact magazine. Dr. Vuille enjoys tennis, lap swimming, guitar and classical piano, and is a former chess champion of St. Petersburg and Atlanta. In his spare time he writes science fiction and goes to the beach. His wife, Dianne Kowing, is an optometrist for a local VA clinic. Teen daughter Kira Vuille-Kowing is an accomplished swimmer, violinist, and mall shopper. He has two sons, twelve-year-old Christopher, a cellist and electronic game expert, and three-year-old James, master of tinkertoys.

Charles A. Bennett received his doctorate at North Carolina State University, and is Professor of Physics at the University of North Carolina at Asheville. His research interests include quantum and physical optics, and laser applications in environmental and fusion energy research. He has collaborated with Oak Ridge National Laboratory since 1983, where he is currently an adjunct research and development associate of the Advanced Laser and Optical Technology and Development group. In addition to his work in optics, Dr. Bennett has a long record of innovation in educational technology, particularly in the integration of active media into on-line homework. He is a past director of the UNCA Center for Teaching and Learning, and has received UNCA's most prestigious recognition for scholarship: the Ruth and Leon Feldman Professorship for 1996–1997. He is an avid hiker, and enjoys spending time with his wife Karen and four children.

Preface

College Physics is written for a one-year course in introductory physics usually taken by students majoring in biology, the health professions, and other disciplines including environmental, earth, and social sciences, and technical fields such as architecture. The mathematical techniques used in this book include algebra, geometry, and trigonometry, but not calculus.

The main objectives of this introductory textbook are twofold: to provide the student with a clear and logical presentation of the basic concepts and principles of physics, and to strengthen an understanding of the concepts and principles through a broad range of interesting applications to the real world. To meet these objectives, we have emphasized sound physical arguments and problem-solving methodology. At the same time, we have attempted to motivate the student through practical examples that demonstrate the role of physics in other disciplines.

This textbook, which covers the standard topics in classical physics and 20th-century physics, is divided into six parts. Part I (Chapters 1–9) deals with Newtonian mechanics and the physics of fluids; Part II (Chapters 10–12) is concerned with heat and thermodynamics; Part III (Chapters 13–14) covers wave motion and sound; Part IV (Chapters 15–21) develops the concepts of electricity and magnetism; Part V (Chapters 22–25) treats the properties of light and the field of geometric and wave optics; and Part VI (Chapters 26–30) provides an introduction to special relativity, quantum physics, atomic physics, and nuclear physics.

CHANGES TO THE SEVENTH EDITION

A number of new features, changes, and improvements have been added to this edition. Based on comments from users of the sixth edition and reviewers' suggestions, a major effort was made to improve clarity of presentation, precision of language, and accuracy throughout. The new pedagogical features added to this edition are based on current trends in science education. The following represent the major changes in the seventh edition.

Pedagogical Changes

- **Examples** All in-text worked examples have been reconstituted for the seventh edition, and now are presented in a two-column format that better aids student learning and helps to better reinforce physical concepts. Special care has been taken in this new edition to present a range of levels within each chapter's collection of worked examples, so that students are better prepared to solve the end-of-chapter problems. The examples are set off from the text for ease of location and are given titles to describe their content. All worked examples now include the following parts:

  (a) **Goal,** which describes the physical concepts being explored within the Worked Example.
  (b) **Problem,** which presents the problem itself.
  (c) **Strategy,** which helps students analyze the problem and create a framework for working out the solution.
  (d) **Solution,** which uses a two-column format that gives the explanation for each step of the solution in the left-hand column, while giving each accompanying mathematical step in the right-hand column. This layout facilitates matching the idea with its execution, and helps students learn how to organize their work. Another benefit: students can easily use this format as a training tool, covering up the solution on the right and solving the problem using the comments on the left as a guide.
  (e) **Remarks,** which follow each Solution and highlight some of the underlying concepts and methodology used in arriving at a correct solution. In addition, the remarks are often used to put the problem into a larger, real world context.
  (f) **Exercise/Answer** By popular demand of users and reviewers, every worked example in the new edition is now followed immediately by an exercise with an answer. These exercises allow the students to reinforce their understanding by working a similar or related problem, with the answers giving them instant feedback. At the option of the instructor, the exercises can also be assigned as homework. Students who work through these exercises on a regular basis will find the end-of-chapter problems less intimidating.

- *PhysicsNow*™ Directly links with *College Physics, Seventh Edition. PhysicsNow*™ and *College Physics, Seventh Edition* were built in concert to enhance each other and provide a seamless, integrated learning system. Throughout the text, readers will see Physics Now™ icons that direct them to media-enhanced activities at the *PhysicsNow*™ Web site (**www.cp7e.com**). The precise page-by-page integration means professors and students will spend less time flipping through pages and navigating Web sites for useful exercises.

■ **Active Figures** Many diagrams from the text have been animated to form Active Figures, part of the new *PhysicsNow*™ integrated Web-based learning system. There are over 120 Active Figures in the seventh edition, available at **www.cp7e.com**. By visualizing phenomena and processes that cannot be fully represented on a static page, students greatly increase their conceptual understanding. Active Figures are easily identified by their red figure legend and Physics⊗Now™ icon, and the caption after the icon describes briefly the nature and contents of the animation. In addition to viewing animations of the figures, students can change variables to see the effects, conduct suggested explorations of the principles involved in the figure, and take and receive feedback on quizzes related to the figure.

■ **Interactive Examples** Thirty-four of the worked examples in the seventh edition have been identified as interactive. As part of the *PhysicsNow*™ Web-based learning system, students can engage in an extension of the problem solved in the example. This often includes elements of both visualization and calculation, and may also involve prediction and intuition-building. Interactive Examples are available at **www.cp7e.com**.

■ **Quick Quizzes** All of the Quick Quizzes in this edition have been cast in an objective format, including multiple choice, true-false, and ranking. Quick Quizzes provide students with opportunities to test their understanding of the physical concepts presented. The questions require students to make decisions on the basis of sound reasoning, and some of them have been written to help students overcome common misconceptions. Answers to all Quick Quiz questions are found at the end of the textbook, while answers with detailed explanations are provided in the *Instructor's Manual*. Many instructors choose to use Quick Quiz questions in a "peer instruction" teaching style.

■ **Summary** The Summary is now organized by each individual chapter heading, for ease of reference.

■ **Problems and Conceptual Questions** All problems have been carefully worded and have been checked for clarity and accuracy, and approximately 15% of the conceptual questions and problems are new to this edition. Solutions to approximately 20% of the end-of-chapter problems are included in the *Student Solutions Manual and Study Guide*. These problems are identified with a box around the problem number. A smaller subset of problems will be available with coached solutions as part of the *PhysicsNow*™ Web-based learning system and will be accessible to students and instructors using *College Physics*. The Physics⊗Now™ icon identifies these problems.

■ **Line-by-Line Revision** The text has been carefully edited to improve clarity of presentation and precision of language. We hope that the result is a book both accurate and enjoyable to read.

CONTENT CHANGES

Although the overall content and organization of the textbook is similar to that of the sixth edition, several changes were implemented.

■ Vectors are now denoted in bold face with arrows over them (for example, $\vec{v}$), making them easier to recognize.

■ In many chapters, one or two examples have been modified so as to combine new concepts with previously studied concepts. This not only reviews prior material, but also integrates different concepts, showing how they work together to enhance our understanding of the physical world.

■ In Chapter 4, on Newton's laws, we have doubled the number of examples, from seven in the sixth edition to fourteen in the seventh, to help aid student understanding.

■ Chapter 5, on work and energy, has been extensively reorganized, with gravitational potential energy and spring potential energy handled in separate sections. The connection between work done by conservative forces and the definition of the potential energy of a system has been emphasized and clarified.

■ The number of examples in Chapter 9, Solids and Fluids, has been increased from thirteen to nineteen.

■ Chapter 12, The Laws of Thermodynamics, now covers the important concepts of degrees of freedom and the equipartition of energy, and relates them to determining molar specific heats. The different kinds of thermodynamic processes have been organized into their own subsections, and the work done during an isothermal process is now presented and used in examples. Generic processes are also briefly addressed. The methods for calculating the quantities appearing in the first law are organized into a table that makes clear the similarities and differences between the processes.

■ The concepts of electric potential and electric potential energy have been clarified in Chapter 16.

■ In Chapter 23, the tables for sign conventions for mirrors and lenses have been reorganized for greater clarity.

■ The methods of handling interference in thin films have been organized into a table in Chapter 24, facilitating student understanding and problem-solving in that area.

- Additional detail on prescribing lenses has been added to the example problems of Chapter 25, Optical Instruments.
- In Chapter 26, the Twin Paradox and concepts of general relativity were clarified and enhanced.

TEXTBOOK FEATURES

Most instructors would agree that the textbook assigned in a course should be the student's primary guide for understanding and learning the subject matter. Furthermore, the textbook should be easily accessible and written in a style that facilitates instruction and learning. With this in mind, we have included many pedagogical features that are intended to enhance the textbook's usefulness to both students and instructors. These features are as follows:

STYLE   To facilitate rapid comprehension, we have attempted to write the book in a style that is clear, logical, relaxed, and engaging. The somewhat informal and relaxed writing style is designed to connect better with students and enhance their reading enjoyment. New terms are carefully defined, and we have tried to avoid the use of jargon.

PREVIEWS   All chapters begin with a brief preview that includes a discussion of the chapter's objectives and content.

UNITS   The international system of units (SI) is used throughout the text. The U.S. customary system of units is used only to a limited extent in the chapters on mechanics and thermodynamics.

MARGINAL NOTES   Comments and notes appearing in the margin can be used to locate important statements, equations, and concepts in the text.

PROBLEM-SOLVING STRATEGIES   A general strategy to be followed by the student is outlined at the end of Chapter 1 and provides students with a structured process for solving problems. In most chapters, more specific strategies and suggestions are included for solving the types of problems featured in both the worked examples and end-of-chapter problems. This feature, highlighted by a surrounding box, is intended to help students identify the essential steps in solving problems and increases their skills as problem solvers.

BIOMEDICAL APPLICATIONS   For biology and pre-med students, icons point the way to various practical and interesting applications of physical principles to biology and medicine. Where possible, an effort was made to include more problems that would be relevant to these disciplines.

APPLYING PHYSICS   provide students with an additional means of reviewing concepts presented in that section. Some Applying Physics examples demonstrate the connection between the concepts presented in that chapter and other scientific disciplines. These examples also serve as models for students when assigned the task of responding to the Conceptual Questions presented at the end of each chapter.

TIPS   are placed in the margins of the text and address common student misconceptions and situations in which students often follow unproductive paths. Approximately 40 Tips are provided in Volume 2 to help students avoid common mistakes and misunderstandings.

IMPORTANT STATEMENTS AND EQUATIONS   Most important statements and definitions are set in **boldface** type or are highlighted with a background screen for added emphasis and ease of review. Similarly, important equations are highlighted with a tan background screen to facilitate location.

ILLUSTRATIONS AND TABLES   The readability and effectiveness of the text material, worked examples, and end-of-chapter conceptual questions and problems are enhanced by the large number of figures, diagrams, photographs, and tables. Full color adds clarity to the artwork and makes illustrations as realistic as possible. Three-dimensional effects are rendered with the use of shaded and lightened areas where appropriate. Vectors are color coded, and curves in graphs are drawn in color. Color photographs have been carefully selected, and their accompanying captions have been written to serve as an added instructional tool. A complete description of the pedagogical use of color appears on the inside front cover.

SIGNIFICANT FIGURES   Significant figures in both worked examples and end-of-chapter problems have been handled with care. Most numerical examples and problems are worked out to either two or three significant figures, depending on the accuracy of the data provided. Intermediate results presented in the examples are rounded to the proper number of significant figures, and only those digits are carried forward.

CONCEPTUAL QUESTIONS   are provided at the end of each chapter (almost 550 questions are included in the seventh edition). The **Applying Physics** examples presented in the text serve as models for students when conceptual questions are assigned, and show how the concepts can be applied to understanding the physical world. The conceptual questions provide the student with a means of self-testing the concepts presented in the chapter. Some conceptual questions are appropriate for initiating classroom discussions. Answers to all odd-numbered conceptual questions are located in the answer section at the end of the book.

End-of-Chapter Problems   An extensive set of problems is included at the end of each chapter (in all, over 2,000 problems are provided in the seventh edition). Answers to odd-numbered problems are given at the end of the book. For the convenience of both the student and instructor, about two thirds of the problems are keyed to specific sections of the chapter. The remaining problems, labeled "Additional Problems," are not keyed to specific sections. There are three levels of problems that are graded according to their difficulty. Straightforward problems are numbered in black, intermediate level problems are numbered in blue, and the most challenging problems are numbered in magenta. The 🧬 icon identifies problems dealing with applications to the life sciences and medicine.

Solutions to approximately 20% of the problems in each chapter are in the *Student Solutions Manual and Study Guide*. Among these, selected problems are identified with *PhysicsNow*™ icons and have coached solutions available at **www.cp7e.com.**

Activities   have been included at the end of each chapter's problem set to encourage students to engage in scientific activities outside of the classroom.

Appendices and Endpapers   Several appendices are provided at the end of the textbook. Most of the appendix material represents a review of mathematical concepts and techniques used in the text, including scientific notation, algebra, geometry, trigonometry, differential calculus, and integral calculus. Reference to these appendices is made as needed throughout the text. Most of the mathematical review sections include worked examples and exercises with answers. In addition to the mathematical review, some appendices contain useful tables that supplement textual information. For easy reference, the front endpapers contain a chart explaining the use of color throughout the book and a list of frequently used conversion factors.

TEACHING OPTIONS

This book contains more than enough material for a one-year course in introductory physics. This serves two purposes. First, it gives the instructor more flexibility in choosing topics for a specific course. Second, the book becomes more useful as a resource for students. On the average, it should be possible to cover about one chapter each week for a class that meets three hours per week. Those sections, examples, and end-of-chapter problems dealing with applications of physics to life sciences are identified with the DNA icon 🧬. We offer the following suggestions for shorter courses for those instructors who choose to move at a slower pace through the year.

Option A: If you choose to place more emphasis on contemporary topics in physics, you should consider omitting all or parts of Chapter 8 (Rotational Equilibrium and Rotational Dynamics), Chapter 21 (Alternating Current Circuits and Electromagnetic Waves), and Chapter 25 (Optical Instruments).

Option B: If you choose to place more emphasis on classical physics, you could omit all or parts of Part VI of the textbook, which deals with special relativity and other topics in 20th-century physics.

The *Instructor's Manual* offers additional suggestions for specific sections and topics that may be omitted without loss of continuity if time presses.

ANCILLARIES

The ancillary package has been updated substantially and streamlined in response to suggestions from users of the sixth edition. The most essential parts of the student package are the two-volume *Student Solutions Manual and Study Guide* with a tight focus on problem-solving and the Web-based *PhysicsNow*™ learning system. Instructors will find increased support for their teaching efforts with new electronic materials.

STUDENT ANCILLARIES

Thomson • Brooks/Cole offers several items to supplement and enhance the classroom experience. These ancillaries will allow instructors to customize the textbook to their students' needs and to their own style of instruction. One or more of these ancillaries may be shrink-wrapped with the text at a reduced price:

Student Solutions Manual and Study Guide   by John R. Gordon, Charles Teague, and Raymond A. Serway. Now offered in two volumes, this manual features detailed solutions to approximately 12 problems per chapter. Boxed numbers identify those problems in the textbook for which complete solutions are found in the manual. The manual also features a skills section, important notes from key sections of the text, and a list of important equations and concepts. Volume 1 contains Chapters 1–14 and Volume 2 contains Chapters 15–30.

Physics⊗Now™   Register for *PhysicsNow*™ at **www.cp7e.com.** The *PhysicsNow*™ system is made up of three interrelated parts:

- How much do you know?
- What do you need to learn?
- What have you learned?

Students maximize their success by starting with the Pre-Test for the relevant chapter. Each Pre-Test is a mix of conceptual and numerical questions. After completing the Pre-Test, each student is presented with a detailed Learning Plan. The Learning Plan outlines elements to review in the text and Web-based media (Active Figures, Interactive Examples, and Coached Problems) in order to master the chapter's most essential concepts. After working through these materials, students move on to a multiple-choice Post-Test presenting them with questions similar to those that might appear on an exam. Results can be e-mailed to instructors.

Physics Laboratory Manual,   2nd edition by David Loyd. This manual supplements the learning of basic physical principles while introducing laboratory procedures and equipment. Each chapter of the manual includes a pre-laboratory assignment, objectives, an equipment list, the theory behind the experiment, experimental procedures, graphs, and questions. A laboratory report is provided for each experiment so the student can record data, calculations, and experimental results. Students are encouraged to apply statistical analysis to their data in order to develop their ability to judge the validity of their results.

WebTutor™ on WebCT and Blackboard   WebTutor™ offers students real-time access to a full array of study tools, including PhysicsNow's Active Figures, chapter outlines, summaries, and learning objectives. New to this edition is a bank of end-of-chapter problems from each chapter of the book coded in multiple choice format and ready to assign as quizzes or homework.

The Brooks/Cole Physics Resource Center   You'll find additional online quizzes, Web links, and animations at **http://physics.brookscole.com.**

INSTRUCTOR ANCILLARIES

The following ancillaries are available to qualified adopters. Please contact your local Thomson • Brooks/Cole sales representative for details.
Ancillaries offered in two volumes are split as follows: Volume 1 contains Chapters 1–14 and Volume 2 contains Chapters 15–30.

Instructor's Solutions Manual   by Jerry Faughn and Charles Teague. Available in two volumes, this manual consists of complete solutions to all the problems in the text, answers to the even-numbered problems and conceptual questions, and full answers with explanations to the Quick Quizzes.

Test Bank   by Ed Oberhofer. Contains approximately 1 750 multiple-choice problems and questions. Answers are provided in a separate key. It is provided in print form (in two volumes) for the instructor who does not have access to a computer, and instructors may duplicate pages for distribution to students. The questions in the *Test Bank* are also available in electronic format with complete answers and solutions in iLrn's Computerized Testing.

Overhead Transparency Acetates   The collection of transparencies in two volumes consists of approximately 200 full-color figures and photographs from the text to enhance lectures. These transparencies feature large print for easy viewing in the classroom.

Instructor's Manual for Physics Laboratory Manual   2nd edition by David Loyd. Each chapter contains a discussion of the experiment, teaching hints, answers to selected questions from the student laboratory manual, and a post-laboratory quiz with short answers and essay questions. The author has also included a list of the suppliers of scientific equipment and a summary of the equipment needed for all the experiments in the manual.

Personal Response System Content   To support your use of personal response systems in the classroom, we've coded the questions from the Quick Quiz questions for use with the system of your choice. Contact your Thomson representative for more details.

Multimedia Manager Instructor's Resource CD   This easy-to-use lecture tool allows you to quickly assemble art, photos, and multimedia with notes to create fluid lectures. The CD-ROM set (Volume 1, Chapters 1–14; Volume 2, Chapters 15–30) includes a database of animations, video clips and digital art from the text as well as PowerPoint lectures and electronic files of the Instructor's Solutions Manual and **Test Bank.** You'll also find additional content from the textbook like the Quick Quizzes and Conceptual Questions.

Physics⊗Now™   *PhysicsNow*™ **Course Management Tools powered by iLrn**   This extension to the student tutorial environment of PhysicsNow™ allows instructors to deliver online assignments in an environment that is familiar to students. This powerful system is your gateway to managing online homework, testing, and course administration all in one shell with the proven content to make your course a success. To see a demonstration of this powerful system, contact your Thomson representative or go to **www.cp7e.com.**

Physics⊗Now™   *PhysicsNow*™ **Homework Management powered by iLrn**   To further enhance PhysicsNow, contributing author Chuck Bennett has created ready-to-use homework assignments which are based on the in-chapter examples. These assignments feature step-by-step guides through algorithmic versions of the in-text examples paired with more challenging examples to extend your students' problem-solving abilities. The problems include hints and feedback to help students pinpoint their errors. The goal of this set of exercises is to bring students' problem solving abilities up to the level they will need to work the end-of-chapter problems.

PhysicsNow™ gives you a rich array of problem types and grading options. Its library of assignable questions includes all of the end-of-chapter problems from the text so that you can select the problems you want to include in your online homework assignments. These well-crafted problems are algorithmically generated so that you can assign the same problem with different variables for each student. A flexible grading tolerance feature allows you to specify a percentage range of correct answers so that your students are not penalized for rounding errors. You can give students the option to work an assignment multiple times and record the highest score or limit the times they are able to attempt it. In addition, you can create your own problems to complement the problems from the text. Results flow automatically to an exportable grade book so that instructors are better able to assess student understanding of the material, even prior to class or to an actual test.

Physics⊗Now™   **iLrn Computerized Testing**   Extend the student experience with iLrn into a testing or quizzing environment. The test item file from the text is included to give you a bank of well-crafted questions that you can deliver online or print out. As with the homework problems, you can use the program's friendly interface to craft your own questions to complement the Serway/Faughn questions. You have complete control over grading, deadlines, and availability and can create multiple tests based on the same material.

WebTutor™ on WebCT and Blackboard   With **WebTutor**™'s text-specific, pre-formatted content and total flexibility, instructors can easily create and manage their own personal Web site. WebTutor™'s course management tool gives instructors the ability to provide virtual office hours, post syllabi, set up threaded discussions, track student progress with the quizzing material, and much more. WebTutor™ also provides robust communication tools, such as a course calendar, asynchronous discussion, real-time chat, a whiteboard, and an integrated e-mail system.

Additional Options for Online Homework

WebAssign: A Web-Based Homework System   WebAssign is the most utilized homework system in physics. Designed by physicists for physicists, this system is a trusted companion to your teaching. An enhanced version of WebAssign is available for College Physics. In addition to selected end-of-chapter problems, this enhanced version includes animations with conceptual questions. We've also added tutorial problems with feedback and hints to guide students' content mastery. Take a look at this new innovation from the most trusted name in physics homework. **www.webassign.net**

LON-CAPA   This Web-based course management system developed at Michigan State University includes selected end of chapter problems from *College Physics*. For more information, visit the LON-CAPA Web site at **www.lon-capa.org.**

UT Austin Homework Service   Details about and a demonstration of this service are available at **http://hw.ph.utexas.edu/hw.html.**

ACKNOWLEDGEMENTS

In preparing the seventh edition of this textbook, we have been guided by the expertise of many people who have reviewed manuscript and/or provided pre-revision suggestions. We wish to acknowledge the following reviewers and express our sincere appreciation for their helpful suggestions, criticism, and encouragement.

Seventh edition reviewers:

Gary B. Adams, *Arizona State University*
Ricardo Alarcon, *Arizona State University*
Natalie Batalha, *San Jose State University*
Ken Bolland, *Ohio State University*
Kapila Clara Castoldi, *Oakland University*
Andrew Cornelius, *University of Nevada, Las Vegas*
Yesim Darici, *Florida International University*
N. John DiNardo, *Drexel University*
Steve Ellis, *University of Kentucky*
Hasan Fakhruddin, *Ball State University/ The Indiana Academy*
Emily Flynn
Lewis Ford, *Texas A & M University*
James R. Goff, *Pima Community College*
Yadin Y. Goldschmidt, *University of Pittsburgh*
Steve Hagen, *University of Florida*
Raymond Hall, *California State University, Fresno*
Patrick Hamill, *San Jose State University*
Joel Handley
Grant W. Hart, *Brigham Young University*
James E. Heath, *Austin Community College*
Grady Hendricks, *Blinn College*
Rhett Herman, *Radford University*
Aleksey Holloway, *University of Nebraska at Omaha*

Joey Huston, *Michigan State University*
Mark James, *Northern Arizona University*
Randall Jones, *Loyola College in Maryland*
Joseph Keane, *St. Thomas Aquinas College*
Dorina Kosztin, *University of Missouri-Columbia*
Martha Lietz, *Niles West High School*
Edwin Lo
Rafael Lopez-Mobilia, *University of Texas at San Antonio*
John A. Milsom, *University of Arizona*
Monty Mola, *Humboldt State University*
Charles W. Myles, *Texas Tech University*
Ed Oberhofer, *Lake Sumter Community College*
Chris Pearson, *University of Michigan-Flint*
Alexey A. Petrov, *Wayne State University*
M. Anthony Reynolds, *Embry-Riddle Aeronautical University*
Dubravka Rupnik, *Louisiana State University*
Surajit Sen, *State University of New York at Buffalo*
Marllin L. Simon, *Auburn University*
Matthew Sirocky
George Strobel, *University of Georgia*
James Wanliss, *Embry-Riddle Aeronautical University*

College Physics, seventh edition was carefully checked for accuracy by Ken Bolland (Ohio State University), Steve Ellis (University of Kentucky), Steve Hagen (University of Florida), Mark James (Northern Arizona University), Randall Jones (Loyola College in Maryland), Edwin Lo, Marllin L. Simon (Auburn University), George Strobel (University of Georgia), M. Anthony Reynolds (Embry-Riddle Aeronautical University), Emily Flynn, Joel Handley and Matthew Sirocky. Though responsibility for any remaining errors rests with us, we thank them for their dedication and vigilance.

We thank the following people for their suggestions and assistance during the preparation of earlier editions of this textbook:

Marilyn Akins, *Broome Community College;* Albert Altman, *University of Lowell;* John Anderson, *University of Pittsburgh;* Lawrence Anderson-Huang, *University of Toledo;* Subhash Antani, *Edgewood College;* Neil W. Ashcroft, *Cornell University;* Charles R. Bacon, *Ferris State University;* Dilip Balamore, *Nassau Community College;* Ralph Barnett, *Florissant Valley Community College;* Lois Barrett, *Western Washington University;* Paul D. Beale, *University of Colorado at Boulder;* Paul Bender, *Washington State University;* David H. Bennum, *University of Nevada at Reno;* Jeffery Braun, *University of Evansville;* John Brennan, *University of Central Florida;* Michael Bretz, *University of Michigan, Ann Arbor;* Michael E. Browne, *University of Idaho;* Joseph Cantazarite, *Cypress College;* Ronald W. Canterna, *University of Wyoming;* Clinton M. Case, *Western Nevada Community College;* Neal M. Cason, *University of Notre Dame;* Roger W. Clapp, *University of South Florida;* Giuseppe Colaccico, *University of South Florida;* Lattie F. Collins, *East Tennessee State University;* Lawrence B. Colman, *University of California, Davis;* Jorge Cossio, *Miami Dade Community College;* Terry T. Crow, *Mississippi State College;* Stephen D. Davis, *University of Arkansas at Little Rock;* John DeFord, *University of Utah;* Chris J. DeMarco, *Jackson Community College;* Michael Dennin, *University of California, Irvine;* N. John DiNardo, *Drexel University;* Robert J. Endorf, *University of Cincinnati;* Paul Feldker, *Florissant Valley Community College;* Leonard X. Finegold, *Drexel University;* Tom French, *Montgomery County Community College;* Albert Thomas Frommhold, Jr., *Auburn University;* Lothar Frommhold, *University of Texas at Austin;* Eric Ganz, *University of Minnesota;* Teymoor Gedayloo, *California Polytechnic State University;* Simon George, *California State University, Long Beach;* John R. Gordon, *James Madison University;* George W. Greenlees, *University of Minnesota;* Wlodzimierz Guryn, *Brookhaven National Laboratory;* James Harmon, *Oklahoma State University;* Grant W. Hart, *Brigham Young University;* Christopher Herbert, *New Jersey City University;* John Ho, *State University of New York at Buffalo;* Murshed Hossain, *Rowan University;* Robert C. Hudson, *Roanoke College;* Fred Inman, *Mankato State University;* Ronald E. Jodoin, *Rochester Institute of Technology;* Drasko Jovanovic, *Fermilab;* George W. Kattawar, *Texas A & M University;* Frank Kolp, *Trenton State University;* Joan P.S. Kowalski, *George Mason University;* Ivan Kramer, *University of Maryland, Baltimore County;* Sol Krasner, *University of Chicago;* Karl F. Kuhn, *Eastern Kentucky University;* David Lamp, *Texas Tech University;* Harvey S. Leff, *California State Polytechnic University;*

Joel Levine, *Orange Coast College;* Michael Lieber, *University of Arkansas;* James Linbald, *Saddleback Community College;* Bill Lochslet, *Pennsylvania State University;* Michael LoPresto, *Henry Ford Community College;* Bo Lou, *Ferris State University;* Jeffrey V. Mallow, *Loyola University of Chicago;* David Markowitz, *University of Connecticut;* Steven McCauley, *California State Polytechnic University, Pomona;* Joe McCauley, Jr., *University of Houston;* Ralph V. McGrew, *Broome Community College;* Bill F. Melton, *University of North Carolina at Charlotte;* H. Kent Moore, *James Madison University;* John Morack, *University of Alaska, Fairbanks;* Steven Morris, *Los Angeles Harbor College;* Carl R. Nave, *Georgia State University;* Martin Nikolo, *Saint Louis University;* Blaine Norum, *University of Virginia;* M. E. Oakes, *University of Texas at Austin;* Lewis J. Oakland, *University of Minnesota;* Ed Oberhofer, *Lake Sumter Community College;* Lewis O'Kelly, *Memphis State University;* David G. Onn, *University of Delaware;* J. Scott Payson, *Wayne State University;* T. A. K. Pillai, *University of Wisconsin, La Crosse;* Lawrence S. Pinsky, *University of Houston;* William D. Ploughe, *Ohio State University;* Patrick Polley, *Beloit College;* Brooke M. Pridmore, *Clayton State University;* Joseph Priest, *Miami University;* James Purcell, *Georgia State University;* W. Steve Quon, *Ventura College;* Michael Ram, *State University of New York at Buffalo;* Kurt Reibel, *Ohio State University;* Barry Robertson, *Queen's University;* Virginia Roundy, *California State University, Fullerton;* Larry Rowan, *University of North Carolina, Chapel Hill;* William R. Savage, *The University of Iowa;* Reinhard A. Schumacher, *Carnegie Mellon University;* John Simon, *University of Toledo;* Donald D. Snyder, *Indiana University at Southbend;* Carey E. Stronach, *Virginia State University;* Thomas W. Taylor, *Cleveland State University;* Perry A. Tompkins, *Samford University;* L. L. Van Zandt, *Purdue University;* Howard G. Voss, *Arizona State University;* Larry Weaver, *Kansas State University;* Donald H. White, *Western Oregon State College;* Bernard Whiting, *University of Florida;* George A. Williams, *The University of Utah;* Jerry H. Wilson, *Metropolitan State College;* Robert M. Wood, *University of Georgia;* Clyde A. Zaidins, *University of Colorado at Denver*

Randall Jones generously contributed many of the new end-of-chapter problems, especially those of interest to the life sciences. Edward F. Redish of the University of Maryland graciously allowed us to list some of his problems from the Activity Based Physics Project as both problems and activities. Some of the remaining activities were written by Robert J. Beichner of North Carolina State University.

We are extremely grateful to the publishing team at the Brooks/Cole Publishing Company for their expertise and outstanding work in all aspects of this project. In particular, we'd like to thank Ed Dodd, who tirelessly coordinated and directed our efforts in preparing the manuscript in its various stages, and Annie Mac, who worked behind the scenes securing text and accuracy reviews, and coordinated and transmitted all the print ancillaries. Jane Sanders, the photo researcher, did a great job finding photos of physical phenomena, Sam Subity coordinated the building of the *PhysicsNow*™ Web site, and Rob Hugel helped translate our rough sketches into accurate, compelling art. Donna King of Progressive Publishing Alternatives managed the difficult task of keeping production moving and on schedule. Chris Hall and Teri Hyde also made numerous valuable contributions. Chris provided just the right amount of guidance and vision throughout the project. We would like to thank David Harris, a great team builder and motivator with loads of enthusiasm and an infectious sense of humor.

Finally, we are deeply indebted to our wives and children for their love, support, and long-term sacrifices.

Raymond A. Serway
St. Petersburg, Florida

Jerry S. Faughn
Richmond, Kentucky

Chris Vuille
Daytona Beach, Florida

Charles A. Bennett
Asheville, North Carolina

Applications

Although physics is relevant to so much in our modern lives, this may not be obvious to students in an introductory course. In this seventh edition of *College Physics,* we continue a design feature begun in the previous edition. This feature makes the relevance of physics to everyday life more obvious by pointing out specific applications in the form of a marginal note. Some of these applications pertain to the life sciences and are marked with the DNA icon ▨. The list below is not intended to be a complete listing of all the applications of the principles of physics found in this textbook. Many other applications are to be found within the text and especially in the worked examples, conceptual questions, and end-of-chapter problems.

To the Student

As a student, it's important that you understand how to use the book most effectively, and how best to go about learning physics. Scanning through the Preface will acquaint you with the various features available, both in the book and online. Awareness of your educational resources and how to use them is essential. Though physics is challenging, it can be mastered with the correct approach.

HOW TO STUDY

Students often ask how best to study physics and prepare for examinations. There's no simple answer to this question, but we'd like to offer some suggestions based on our own experiences in learning and teaching over the years.

First and foremost, maintain a positive attitude toward the subject matter. Like learning a language, physics takes time. Those who keep applying themselves on a daily basis can expect to reach understanding. Keep in mind that physics is the most fundamental of all natural sciences. Other science courses that follow will use the same physical principles, so it is important that you understand and are able to apply the various concepts and theories discussed in the text.

CONCEPTS AND PRINCIPLES

Students often try to do their homework without first studying the basic concepts. It is essential that you understand the basic concepts and principles *before* attempting to solve assigned problems. You can best accomplish this goal by carefully reading the textbook *before* you attend your lecture on the covered material. When reading the text, you should jot down those points that are not clear to you. Also be sure to make a diligent attempt at answering the questions in the Quick Quizzes as you come to them in your reading. We have worked hard to prepare questions that help you judge for yourself how well you understand the material. Pay careful attention to the many Tips throughout the text. These will help you avoid misconceptions, mistakes, and misunderstandings as well as maximize the efficiency of your time by minimizing adventures along fruitless paths. During class, take careful notes and ask questions about those ideas that are unclear to you. Keep in mind that few people are able to absorb the full meaning of scientific material after only one reading.

Be sure to take advantage of the features available in the *PhysicsNow*™ learning system, such as the Active Figures, Interactive Examples, and Coached Problems. Your lectures and laboratory work supplement your textbook and should clarify some of the more difficult material. You should minimize rote memorization of material. Successful memorization of passages from the text, equations, and derivations does not necessarily indicate that you understand the fundamental principles.

Your understanding will be enhanced through a combination of efficient study habits, discussions with other students and with instructors, and your ability to solve the problems presented in the textbook. Ask questions whenever you feel clarification of a concept is necessary.

STUDY SCHEDULE

It is important for you to set up a regular study schedule, preferably a daily one. Make sure you read the syllabus for the course and adhere to the schedule set by your instructor. As a general rule, you should devote about two hours of study time for every hour you are in class. If you are having trouble with the course, seek the advice of the instructor or other students who have taken the course. You may find it necessary to seek further instruction from experienced students. Very often, instructors offer review sessions in addition to regular class periods. It is important that you avoid the practice of delaying study until a day or two before an exam. One hour of study a day for fourteen days is far more effective than fourteen hours the day before the exam. "Cramming" usually produces disastrous results, especially in science. Rather than undertake an all-night study session just before an exam, briefly review the basic concepts and equations and get a good night's rest. If you feel you need additional help in understanding the concepts, in preparing for exams, or in problem-solving, we suggest that you acquire a copy of the *Student Solutions Manual and Study Guide* that accompanies this textbook; this manual should be available at your college bookstore.

USE THE FEATURES

You should make full use of the various features of the text discussed in the preface. For example, marginal notes are useful for locating and describing important equations and concepts, and **boldfaced** type indicates important statements and definitions. Many useful tables are contained in the Appendices, but most tables are incorporated in the text where they are most often referenced. Appendix A is a convenient review of mathematical techniques.

Answers to all Quick Quizzes, as well as odd-numbered conceptual questions and problems, are given at the end of the textbook. Answers to selected end-of-chapter problems are provided in the *Student Solutions Manual and Study Guide*. Problem-Solving Strategies included in selected chapters throughout the text give you additional information about how you should solve problems. The Table of Contents provides an overview of the entire text, while the Index enables you to locate specific material quickly. Footnotes sometimes are used to supplement the text or to cite other references on the subject discussed.

After reading a chapter, you should be able to define any new quantities introduced in that chapter and to discuss the principles and assumptions used to arrive at certain key relations. The chapter summaries and the review sections of the *Student Solutions Manual and Study Guide* should help you in this regard. In some cases, it may be necessary for you to refer to the index of the text to locate certain topics. You should be able to correctly associate with each physical quantity the symbol used to represent that quantity and the unit in which the quantity is specified. Furthermore, you should be able to express each important relation in a concise and accurate prose statement.

PROBLEM-SOLVING

R. P. Feynman, Nobel laureate in physics, once said, "You do not know anything until you have practiced." In keeping with this statement, we strongly advise that you develop the skills necessary to solve a wide range of problems. Your ability to solve problems will be one of the main tests of your knowledge of physics; therefore, you should try to solve as many problems as possible. It is essential that you understand basic concepts and principles before attempting to solve problems. It is good practice to try to find alternate solutions to the same problem. For example, you can solve problems in mechanics using Newton's laws, but very often an alternate method that draws on energy considerations is more direct. You should not deceive yourself into thinking you understand a problem merely because you have seen it solved in class. You must be able to solve the problem and similar problems on your own. We have recast the examples in this book to help you in this regard. After studying an example, see if you can cover up the right-hand side and do it yourself, using only the written descriptions on the left as hints. Once you succeed at that, try solving the example completely on your own. Finally, solve the exercise. Once you have accomplished all this, you will have a good mastery of the problem, its concepts, and mathematical technique.

The approach to solving problems should be carefully planned. A systematic plan is especially important when a problem involves several concepts. First, read the problem several times until you are confident you understand what is being asked. Look for any key words that will help you interpret the problem and perhaps allow you to make certain assumptions. Your ability to interpret a question properly is an integral part of problem-solving. Second, you should acquire the habit of writing down the information given in a problem and those quantities that need to be found; for example, you might construct a table listing both the quantities given and the quantities to be found. This procedure is sometimes used in the worked examples of the textbook. After you have decided on the method you feel is appropriate for a given problem, proceed with your solution. Finally, check your results to see if they are reasonable and consistent with your initial understanding of the problem. General problem-solving strategies of this type are included in the text and are highlighted with a surrounding box. If you follow the steps of this procedure, you will find it easier to come up with a solution and also gain more from your efforts.

Often, students fail to recognize the limitations of certain equations or physical laws in a particular situation. It is very important that you understand and remember the assumptions underlying a particular theory or formalism. For example, certain equations in kinematics apply only to a particle moving with constant acceleration. These equations are not valid for describing motion whose acceleration is not constant, such as the motion of an object connected to a spring or the motion of an object through a fluid.

EXPERIMENTS

Physics is a science based on experimental observations. In view of this fact, we recommend that you try to supplement the text by performing various types of "hands-on" experiments, either at home or in the laboratory. For example, the common Slinky™ toy is excellent for studying traveling waves; a ball swinging on the end of a long string can be used to investigate pendulum motion; various masses attached to the end of a vertical spring or rubber band can be used to determine their elastic nature; an old pair of Polaroid sunglasses and some discarded lenses and a magnifying glass are the components of various experiments in optics; and the approximate measure of the free-fall acceleration can be determined simply by measuring with a stopwatch the time it takes for a ball to drop from a known height. The list of such experiments is endless. When physical models are not available, be imaginative and try to develop models of your own.

Some of the Activities at the end of each chapter can be performed on your own or with the help of a friend and will give you specific guidelines for acquiring "hands-on" experience.

PhysicsNow™ with Active Figures and Interactive Examples

We strongly encourage you to use the *PhysicsNow*™ Web-based learning system that accompanies this textbook. It is far easier to understand physics if you see it in action, and these new materials will enable you to become a part of that action. *PhysicsNow*™ media described in the Preface are accessed at the URL www.cp7e.com, and feature a three-step learning process consisting of a Pre-Test, a personalized learning plan, and a Post-Test.

In addition to the Coached Problems identified with icons, *PhysicsNow*™ includes the following Active Figures and Interactive Examples:

Chapter 1   Active Figures 1.6 and 1.7
Chapter 2   Active Figures 2.2, 2.12, 2.13, and 2.15; Interactive Examples 2.5 and 2.9
Chapter 3   Active Figures 3.3, 3.14, and 3.15; Interactive Examples 3.3 and 3.7
Chapter 4   Active Figures 4.6, 4.18, and 4.19; Interactive Example 4.10
Chapter 5   Active Figures 5.5, 5.15, 5.20, and 5.29; Interactive Example 5.5
Chapter 6   Active Figure 6.7, 6.10, 6.13, and 6.15; Interactive Examples 6.3 and 6.7
Chapter 7   Active Figures 7.5, 7.17, and 7.21; Interactive Example 7.7
Chapter 8   Active Figure 8.25; Interactive Examples 8.8 and 8.11
Chapter 9   Active Figures 9.3, 9.5, 9.6, and 9.19; Interactive Examples 9.7 and 9.13
Chapter 10   Active Figures 10.10, 10.12, and 10.15
Chapter 11   Interactive Example 11.10
Chapter 12   Active Figures 12.1, 12.9, 12.12, 12.15, and 12.16
Chapter 13   Active Figures 13.1, 13.8, 13.12, 13.13, 13.15, 13.16, 13.19, 13.24, 13.26, 13.32, 13.33, 13.34, and 13.35; Interactive Example 13.10
Chapter 14   Active Figures 14.8, 14.10, 14.18, and 14.25; Interactive Examples 14.1, 14.5, and 14.7
Chapter 15   Active Figures 15.6, 15.11, 15.21, and 15.28; Interactive Example 15.2
Chapter 16   Active Figures 16.7, 16.17, and 16.19; Interactive Examples 16.2 and 16.8
Chapter 17   Active Figures 17.4 and 17.11; Interactive Example 17.4
Chapter 18   Active Figures 18.1, 18.6, 18.16, and 18.17; Interactive Examples 18.2 and 18.5
Chapter 19   Active Figures 19.2, 19.17, 19.19, 19.20, 19.23, and 19.28
Chapter 20   Active Figures 20.1, 20.4, 20.13, 20.20, 20.22, 20.27, and 20.28
Chapter 21   Active Figures 21.1, 21.2, 21.4, 21.5, 21.6, 21.7, 21.8, 21.9, and 21.20
Chapter 22   Active Figures 22.4, 22.6, 22.7, 22.19, and 22.25; Interactive Examples 22.1 and 22.4
Chapter 23   Active Figures 23.2, 23.13, 23.16, and 23.25; Interactive Examples 23.2, 23.7, 23.8, and 23.9
Chapter 24   Active Figures 24.1, 24.16, 24.20, 24.21, and 24.26; Interactive Examples 24.1, 24.3, 24.6, and 24.7
Chapter 25   Active Figures 25.7, 25.8, and 25.15
Chapter 26   Active Figures 26.4, 26.8, and 26.11
Chapter 27   Active Figures 27.2, 27.3, 27.4, and 27.5; Interactive Examples 27.3 and 27.6
Chapter 28   Active Figures 28.6, 28.7, 28.17, 28.18, and 28.19; Interactive Example 28.1
Chapter 29   Active Figures 29.1, 29.6, and 29.7; Interactive Example 29.3
Chapter 30   Active Figures 30.2 and 30.10

An Invitation to Physics

It is our hope that you too will find physics an exciting and enjoyable experience and that you will profit from this experience, regardless of your chosen profession. Welcome to the exciting world of physics!

> *To see the World in a Grain of Sand*
> *And a Heaven in a Wild Flower,*
> *Hold infinity in the palm of your hand*
> *And Eternity in an hour.*

William Blake, "Auguries of Innocence"

Stone/Getty Images

Thousands of years ago, people in southern England built Stonehenge, which was used as a calendar. The position of the sun and stars relative to the stones determined seasons for planting or harvesting.

CHAPTER

1

Introduction

The goal of physics is to provide an understanding of the physical world by developing theories based on experiments. A physical theory is essentially a guess, usually expressed mathematically, about how a given physical system works. The theory makes certain predictions about the physical system which can then be checked by observations and experiments. If the predictions turn out to correspond closely to what is actually observed, then the theory stands, although it remains provisional. No theory to date has given a complete description of all physical phenomena, even within a given subdiscipline of physics. Every theory is a work in progress.

The basic laws of physics involve such physical quantities as force, velocity, volume, and acceleration, all of which can be described in terms of more fundamental quantities. In mechanics, the three most fundamental quantities are **length** (L), **mass** (M), and **time** (T); all other physical quantities can be constructed from these three.

1.1 STANDARDS OF LENGTH, MASS, AND TIME

To communicate the result of a measurement of a certain physical quantity, a *unit* for the quantity must be defined. For example, if our fundamental unit of length is defined to be 1.0 meter, and someone familiar with our system of measurement reports that a wall is 2.0 meters high, we know that the height of the wall is twice the fundamental unit of length. Likewise, if our fundamental unit of mass is defined as 1.0 kilogram, and we are told that a person has a mass of 75 kilograms, then that person has a mass 75 times as great as the fundamental unit of mass.

In 1960, an international committee agreed on a standard system of units for the fundamental quantities of science, called **SI** (Système International). Its units of length, mass, and time are the meter, kilogram, and second, respectively.

Physics Now™ Throughout the text, the PhysicsNow icon indicates an opportunity for you to test yourself on key concepts and to explore animations and interactions on the PhysicsNow website at **www.cp7e.com.**

1

Length

In 1799, the legal standard of length in France became the meter, defined as one ten-millionth of the distance from the equator to the North Pole. Until 1960, the official length of the meter was the distance between two lines on a specific bar of platinum-iridium alloy stored under controlled conditions. This standard was abandoned for several reasons, the principal one being that measurements of the separation between the lines are not precise enough. In 1960, the meter was defined as 1 650 763.73 wavelengths of orange-red light emitted from a krypton-86 lamp. In October 1983, this definition was abandoned also, and **the meter was redefined as the distance traveled by light in vacuum during a time interval of 1/299 792 458 second**. This latest definition establishes the speed of light at 299 792 458 meters per second.

◀ Definition of the meter ▶

Mass

◀ Definition of the kilogram ▶

The SI unit of mass, the kilogram, is defined as the mass of a specific platinum-iridium alloy cylinder kept at the International Bureau of Weights and Measures at Sèvres, France (similar to that shown in Figure 1.1a). As we'll see in Chapter 4, mass is a quantity used to measure the resistance to a change in the motion of an object. It's more difficult to cause a change in the motion of an object with a large mass than an object with a small mass.

Time

Before 1960, the time standard was defined in terms of the average length of a solar day in the year 1900. (A solar day is the time between successive appearances of the Sun at the highest point it reaches in the sky each day.) The basic unit of time, the second, was defined to be $(1/60)(1/60)(1/24) = 1/86\,400$ of the average solar day. In 1967, the second was redefined to take advantage of the high precision attainable with an atomic clock, which uses the characteristic frequency of the light emitted from the cesium-133 atom as its "reference clock." **The second is now defined as 9 192 631 700 times the period of oscillation of radiation from the cesium atom**. The newest type of cesium atomic clock is shown in Figure 1.1b.

◀ Definition of the second ▶

(a)

(b)

Figure 1.1 (a) The National Standard Kilogram No. 20, an accurate copy of the International Standard Kilogram kept at Sèvres, France, is housed under a double bell jar in a vault at the National Institute of Standards and Technology. (b) The nation's primary time standard is a cesium fountain atomic clock developed at the National Institute of Standards and Technology laboratories in Boulder, Colorado. This clock will neither gain nor lose a second in 20 million years.

TABLE 1.1

Approximate Values of Some Measured Lengths

	Length (m)
Distance from Earth to most remote known quasar	1×10^{26}
Distance from Earth to most remote known normal galaxies	4×10^{25}
Distance from Earth to nearest large galaxy (M31, the Andromeda galaxy)	2×10^{22}
Distance from Earth to nearest star (Proxima Centauri)	4×10^{16}
One light year	9×10^{15}
Mean orbit radius of Earth about Sun	2×10^{11}
Mean distance from Earth to Moon	4×10^{8}
Mean radius of Earth	6×10^{6}
Typical altitude of satellite orbiting Earth	2×10^{5}
Length of football field	9×10^{1}
Length of housefly	5×10^{-3}
Size of smallest dust particles	1×10^{-4}
Size of cells in most living organisms	1×10^{-5}
Diameter of hydrogen atom	1×10^{-10}
Diameter of atomic nucleus	1×10^{-14}
Diameter of proton	1×10^{-15}

Approximate Values for Length, Mass, and Time Intervals

Approximate values of some lengths, masses, and time intervals are presented in Tables 1.1, 1.2, and 1.3, respectively. Note the wide ranges of values. Study these tables to get a feel for a kilogram of mass (this book has a mass of about 2 kilograms), a time interval of 10^{10} seconds (one century is about 3×10^{9} seconds), or two meters of length (the approximate height of a forward on a basketball team). Appendix A reviews the notation for powers of 10, such as the expression of the number 50 000 in the form 5×10^{4}.

Systems of units commonly used in physics are the Système International, in which the units of length, mass, and time are the meter (m), kilogram (kg), and second (s); the cgs, or Gaussian, system, in which the units of length, mass, and time are the centimeter (cm), gram (g), and second; and the U.S. customary system, in which the units of length, mass, and time are the foot (ft), slug, and second. SI units are almost universally accepted in science and industry, and will be used throughout the book. Limited use will be made of Gaussian and U.S. customary units.

TABLE 1.2

Approximate Values of Some Masses

	Mass (kg)
Observable Universe	1×10^{52}
Milky Way galaxy	7×10^{41}
Sun	2×10^{30}
Earth	6×10^{24}
Moon	7×10^{22}
Shark	1×10^{2}
Human	7×10^{1}
Frog	1×10^{-1}
Mosquito	1×10^{-5}
Bacterium	1×10^{-15}
Hydrogen atom	2×10^{-27}
Electron	9×10^{-31}

TABLE 1.3

Approximate Values of Some Time Intervals

	Time Interval (s)
Age of Universe	5×10^{17}
Age of Earth	1×10^{17}
Average age of college student	6×10^{8}
One year	3×10^{7}
One day (time required for one revolution of Earth about its axis)	9×10^{4}
Time between normal heartbeats	8×10^{-1}
Period[a] of audible sound waves	1×10^{-3}
Period[a] of typical radio waves	1×10^{-6}
Period[a] of vibration of atom in solid	1×10^{-13}
Period[a] of visible light waves	2×10^{-15}
Duration of nuclear collision	1×10^{-22}
Time required for light to travel across a proton	3×10^{-24}

[a]A *period* is defined as the time required for one complete vibration.

TABLE 1.4

Some Prefixes for Powers of Ten Used with "Metric" (SI and cgs) Units

Power	Prefix	Abbreviation
10^{-18}	atto-	a
10^{-15}	femto-	f
10^{-12}	pico-	p
10^{-9}	nano-	n
10^{-6}	micro-	μ
10^{-3}	milli-	m
10^{-2}	centi-	c
10^{-1}	deci-	d
10^{1}	deka-	da
10^{3}	kilo-	k
10^{6}	mega-	M
10^{9}	giga-	G
10^{12}	tera-	T
10^{15}	peta-	P
10^{18}	exa-	E

Some of the most frequently used "metric" (SI and cgs) prefixes representing powers of 10 and their abbreviations are listed in Table 1.4. For example, 10^{-3} m is equivalent to 1 millimeter (mm), and 10^{3} m is 1 kilometer (km). Likewise, 1 kg is equal to 10^{3} g, and 1 megavolt (MV) is 10^{6} volts (V).

1.2   THE BUILDING BLOCKS OF MATTER

A 1-kg ($\approx$ 2-lb) cube of solid gold has a length of about 3.73 cm ($\approx$ 1.5 in.) on a side. Is this cube nothing but wall-to-wall gold, with no empty space? If the cube is cut in half, the two resulting pieces retain their chemical identity as solid gold. But what if the pieces of the cube are cut again and again, indefinitely? Will the smaller and smaller pieces always be the same substance, gold? Questions such as these can be traced back to early Greek philosophers. Two of them—Leucippus and Democritus—couldn't accept the idea that such cutting could go on forever. They speculated that the process ultimately would end when it produced a particle that could no longer be cut. In Greek, *atomos* means "not sliceable." From this term comes our English word *atom*, once believed to be the smallest, ultimate particle of matter, but since found to be a composite of more elementary particles.

The atom can be visualized as a miniature Solar System, with a dense, positively charged nucleus occupying the position of the Sun, with negatively charged electrons orbiting like planets. This model of the atom, first developed by the great Danish physicist Niels Bohr nearly a century ago, led to the understanding of certain properties of the simpler atoms such as hydrogen, but failed to explain many fine details of atomic structure.

Notice the size of a hydrogen atom, listed in Table 1.1, and the size of a proton—the nucleus of a hydrogen atom—one hundred thousand times smaller. If the proton were the size of a Ping Pong ball, the electron would be a tiny speck about the size of a bacterium, orbiting the proton a kilometer away! Other atoms are similarly constructed. So there is a surprising amount of empty space in ordinary matter.

After the discovery of the nucleus in the early 1900s, questions arose concerning its structure. Is the nucleus a single particle or a collection of particles? The exact composition of the nucleus hasn't been defined completely even today, but by the early 1930s a model evolved that helped us understand how the nucleus behaves. Scientists determined that two basic entities—protons and neutrons—occupy the nucleus. The *proton* is nature's fundamental carrier of positive charge (equal in magnitude but opposite in sign to the charge on the electron), and the number of protons in a nucleus determines what the element is. For instance, a nucleus containing only one proton is the nucleus of an atom of hydrogen, regardless of how many neutrons may be present. Extra neutrons correspond to different isotopes of hydrogen—deuterium and tritium—which react chemically in exactly the same way as hydrogen, but are more massive. An atom having two protons in its nucleus, similarly, is always helium, although again, differing numbers of neutrons are possible.

The existence of *neutrons* was verified conclusively in 1932. A neutron has no charge and has a mass about equal to that of a proton. One of its primary purposes is to act as a "glue" to hold the nucleus together. If neutrons were not present, the repulsive electrical force between the positively charged protons would cause the nucleus to fly apart.

The division doesn't stop here; it turns out that protons, neutrons, and a zoo of other exotic particles are now thought to be composed of six particles called **quarks** (rhymes with "forks," though some rhyme it with "sharks"). These particles have been given the names *up, down, strange, charm, bottom,* and *top*. The up, charm, and top quarks each carry a charge equal to $+\frac{2}{3}$ that of the proton, whereas the down, strange, and bottom quarks each carry a charge equal to $-\frac{1}{3}$ the proton charge. The proton consists of two up quarks and one down quark (see Fig. 1.2), giving the correct charge for the proton, $+1$. The neutron is composed of two down quarks and one up quark and has a net charge of zero.

The up and down quarks are sufficient to describe all normal matter, so the existence of the other four quarks indirectly observed in high-energy experiments,

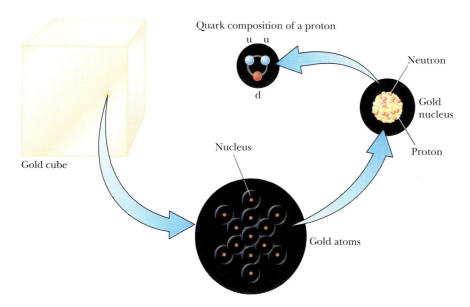

Quark composition of a proton

Gold cube

Nucleus

Gold atoms

Neutron

Gold nucleus

Proton

Figure 1.2 Levels of organization in matter. Ordinary matter consists of atoms, and at the center of each atom is a compact nucleus consisting of protons and neutrons. Protons and neutrons are composed of quarks. The quark composition of a proton is shown.

is something of a mystery. It's also possible that quarks themselves have internal structure. Many physicists believe that the most fundamental particles may be tiny loops of vibrating string.

1.3 DIMENSIONAL ANALYSIS

In physics, the word *dimension* denotes the physical nature of a quantity. The distance between two points, for example, can be measured in feet, meters, or furlongs, which are different ways of expressing the dimension of *length*.

The symbols that we use in this section to specify the dimensions of length, mass, and time are L, M, and T, respectively. Brackets [ ] will often be used to denote the dimensions of a physical quantity. For example, in this notation the dimensions of velocity v are written $[v] = $ L/T, and the dimensions of area A are $[A] = $ L^2. The dimensions of area, volume, velocity, and acceleration are listed in Table 1.5, along with their units in the three common systems. The dimensions of other quantities, such as force and energy, will be described later as they are introduced.

In physics, it's often necessary either to derive a mathematical expression or equation or to check its correctness. A useful procedure for doing this is called **dimensional analysis**, which makes use of the fact that **dimensions can be treated as algebraic quantities**. Such quantities can be added or subtracted only if they have the same dimensions. It follows that the terms on the opposite sides of an equation must have the same dimensions. If they don't, the equation is wrong. If they do, the equation is probably correct, except for a possible constant factor.

To illustrate this procedure, suppose we wish to derive a formula for the distance x traveled by a car in a time t if the car starts from rest and moves with constant acceleration a. The quantity x has the dimension length: $[x] = $ L. Time t, of course, has dimension $[t] = $ T. Acceleration is the change in velocity v with time. Since v has dimensions of length per unit time, or $[v] = $ L/T, acceleration must have dimensions $[a] = $ L/T^2. We organize this information in the form of an equation:

$$[a] = \frac{[v]}{[t]} = \frac{L/T}{T} = \frac{L}{T^2} = \frac{[x]}{[t]^2}$$

TABLE 1.5

Dimensions and Some Units of Area, Volume, Velocity, and Acceleration

System	Area (L^2)	Volume (L^3)	Velocity (L/T)	Acceleration (L/T^2)
SI	m^2	m^3	m/s	m/s^2
cgs	cm^2	cm^3	cm/s	cm/s^2
U.S. customary	ft^2	ft^3	ft/s	ft/s^2

Looking at the left- and right-hand sides of this equation, we might now guess that

$$a = \frac{x}{t^2} \quad \rightarrow \quad x = at^2$$

This is not quite correct, however, because there's a constant of proportionality—a simple numerical factor—that can't be determined solely through dimensional analysis. As will be seen in Chapter 2, it turns out that the correction expression is $x = \frac{1}{2}at^2$.

When we work algebraically with physical quantities, dimensional analysis allows us to check for errors in calculation, which often show up as discrepancies in units. If, for example, the left-hand side of an equation is in meters and the right-hand side is in meters per second, we know immediately that we've made an error.

EXAMPLE 1.1   Analysis of an Equation

Goal   Check an equation using dimensional analysis.

Problem   Show that the expression $v = v_0 + at$, is dimensionally correct, where v and v_0 represent velocities, a is acceleration, and t is a time interval.

Strategy   Analyze each term, finding its dimensions, and then check to see if all the terms agree with each other.

Solution
Find dimensions for v and v_0.

$$[v] = [v_0] = \frac{L}{T}$$

Find the dimensions of at.

$$[at] = \frac{L}{T^2}\,(T) = \frac{L}{T}$$

Remarks   All the terms agree, so the equation is dimensionally correct.

Exercise 1.1
Determine whether the equation $x = vt^2$ is dimensionally correct. If not, provide a correct expression, up to an overall constant of proportionality.

Answer   Incorrect. The expression $x = vt$ is dimensionally correct.

EXAMPLE 1.2   Find an Equation

Goal   Derive an equation by using dimensional analysis.

Problem   Find a relationship between a constant acceleration a, speed v, and distance r from the origin for a particle traveling in a circle.

Strategy   Start with the term having the most dimensionality, a. Find its dimensions, and then rewrite those dimensions in terms of the dimensions of v and r. The dimensions of time will have to be eliminated with v, since that's the only quantity in which the dimension of time appears.

Solution
Write down the dimensions of a:

$$[a] = \frac{L}{T^2}$$

Solve the dimensions of speed for T:

$$[v] = \frac{L}{T} \quad \rightarrow \quad T = \frac{L}{[v]}$$

Substitute the expression for T into the equation for $[a]$:

$$[a] = \frac{L}{T^2} = \frac{L}{(L/[v])^2} = \frac{[v]^2}{L}$$

Substitute L = [r], and guess at the equation:

$$[a] = \frac{[v]^2}{[r]} \quad \rightarrow \quad a = \frac{v^2}{r}$$

Remarks   This is the correct equation for centripetal acceleration—acceleration towards the center of motion—to be discussed in Chapter 7. There isn't any need in this case, to introduce a numerical factor. Such a factor is often displayed explicitly as a constant k in front of the right-hand side—for example, $a = kv^2/r$. As it turns out, $k = 1$ gives the correct expression.

Exercise 1.2
In physics, energy E carries dimensions of mass times length squared, divided by time squared. Use dimensional analysis to derive a relationship for energy in terms of mass m and speed v, up to a constant of proportionality. Set the speed equal to c, the speed of light, and the constant of proportionality equal to 1 to get the most famous equation in physics.

Answer $E = kmv^2 \quad \rightarrow \quad E = mc^2$ when $k = 1$ and $v = $ c.

1.4   UNCERTAINTY IN MEASUREMENT AND SIGNIFICANT FIGURES

Physics is a science in which mathematical laws are tested by experiment. No physical quantity can be determined with complete accuracy, because our senses are physically limited, even when extended with microscopes, cyclotrons, and other gadgets.

Knowing the experimental uncertainties in any measurement is very important. Without this information, little can be said about the final measurement. Using a crude scale, for example, we might find that a gold nugget has a mass of 3 kilograms. A prospective client interested in purchasing the nugget would naturally want to know about the accuracy of the measurement, to ensure paying a fair price. He wouldn't be happy to find that the measurement was good only to within a kilogram, because he might pay for three kilograms and get only two. Of course, he might get four kilograms for the price of three, but most people would be hesitant to gamble that an error would turn out in their favor.

Accuracy of measurement depends on the sensitivity of the apparatus, the skill of the person carrying out the measurement, and the number of times the measurement is repeated. There are many ways of handling uncertainties, and here we'll develop a basic and reliable method of keeping track of them in the measurement itself and in subsequent calculations.

Suppose that in a laboratory experiment we measure the area of a rectangular plate with a meter stick. Let's assume that the accuracy to which we can measure a particular dimension of the plate is ± 0.1 cm. If the length of the plate is measured to be 16.3 cm, we can claim only that it lies somewhere between 16.2 cm and 16.4 cm. In this case, we say that the measured value has three significant figures. Likewise, if the plate's width is measured to be 4.5 cm, the actual value lies between 4.4 cm and 4.6 cm. This measured value has only two significant figures. We could write the measured values as 16.3 ± 0.1 cm and 4.5 ± 0.1 cm. In general, **a significant figure is a reliably known digit** (other than a zero used to locate a decimal point).

Suppose we would like to find the area of the plate by multiplying the two measured values together. The final value can range between (16.3 − 0.1 cm)(4.5 − 0.1 cm) = (16.2 cm)(4.4 cm) = 71.28 cm^2 and (16.3 + 0.1 cm)(4.5 + 0.1 cm) = (16.4 cm)(4.6 cm) = 75.44 cm^2. Claiming to know anything about the hundredths place, or even the tenths place, doesn't make any sense, because it's clear we can't even be certain of the units place, whether it's the 1 in 71, the 5 in 75, or somewhere in between. The tenths and the hundredths places are clearly not significant. We have some information about the units place, so that number is significant. Multiplying the numbers at the middle of the uncertainty ranges gives $(16.3 \text{ cm})(4.5 \text{ cm}) = 73.35 \text{ cm}^2$, which is also in the middle of the area's uncertainty range. Since the hundredths and tenths are not significant, we drop them and take the answer to be 73 cm^2, with an uncertainty of $\pm 2 \text{ cm}^2$. Note that the answer has two significant figures, the same number of figures as the least accurately known quantity being multiplied, the 4.5-cm width.

There are two useful rules of thumb for determining the number of significant figures. The first, concerning multiplication and division, is as follows: **In multiplying (dividing) two or more quantities, the number of significant figures in the final product (quotient) is the same as the number of significant figures in the *least accurate* of the factors being combined, where *least accurate* means *having the lowest number of significant figures*.**

To get the final number of significant figures, it's usually necessary to do some rounding. If the last digit dropped is less than 5, simply drop the digit. If the last digit dropped is greater than or equal to 5, raise the last retained digit by one.

EXAMPLE 1.3 Installing a Carpet

Goal Apply the multiplication rule for significant figures.

Problem A carpet is to be installed in a room of length 12.71 m and width 3.46 m. Find the area of the room, retaining the proper number of significant figures.

Strategy Count the significant figures in each number. The smaller result is the number of significant figures in the answer.

Solution

Count significant figures:

$$12.71 \text{ m} \rightarrow \text{4 significant figures}$$
$$3.46 \text{ m} \rightarrow \text{3 significant figures}$$

Multiply the numbers, keeping only three digits:

$$12.71 \text{ m} \times 3.46 \text{ m} = 43.9766 \text{ m}^2 \rightarrow \boxed{44.0 \text{ m}^2}$$

Remarks In reducing 43.976 6 to three significant figures, we used our rounding rule, adding 1 to the 9, which made 10 and resulted in carrying 1 to the unit's place.

Exercise 1.3
Repeat this problem, but with a room measuring 9.72 m long by 5.3 m wide.

Answer 52 m^2

TIP 1.1 Using Calculators

Calculators were designed by engineers to yield as many digits as the memory of the calculator chip permitted, so be sure to round the final answer down to the correct number of significant figures.

Zeros may or may not be significant figures. Zeros used to position the decimal point in such numbers as 0.03 and 0.007 5 are not significant (but are useful in avoiding errors). Hence, 0.03 has one significant figure, and 0.007 5 has two.

When zeros are placed after other digits in a whole number, there is a possibility of misinterpretation. For example, suppose the mass of an object is given as 1 500 g. This value is ambiguous, because we don't know whether the last two zeros are being used to locate the decimal point or whether they represent significant figures in the measurement.

Using scientific notation to indicate the number of significant figures removes this ambiguity. In this case, we express the mass as 1.5×10^3 g if there are two significant figures in the measured value, 1.50×10^3 g if there are three significant figures, and 1.500×10^3 g if there are four. Likewise, 0.000 15 is expressed in scientific notation as 1.5×10^{-4} if it has two significant figures or as 1.50×10^{-4} if it has three significant figures. The three zeros between the decimal point and the digit 1 in the number 0.000 15 are not counted as significant figures because they only locate the decimal point. In this book, **most of the numerical examples and end-of-chapter problems will yield answers having two or three significant figures**.

For addition and subtraction, it's best to focus on the number of decimal places in the quantities involved rather than on the number of significant figures. **When numbers are added (subtracted), the number of decimal places in the result should equal the smallest number of decimal places of any term in the sum (difference)**. For example, if we wish to compute 123 (zero decimal places) + 5.35 (two decimal places), the answer is 128 (zero decimal places) and not 128.35. If we compute the sum 1.000 1 (four decimal places) + 0.000 3 (four decimal places) = 1.000 4, the result has the correct number of decimal places, namely four. Observe that the rules for multiplying significant figures don't work here because the answer has five significant figures even though one of the terms in the sum, 0.000 3, has only one significant figure. Likewise, if we perform the subtraction 1.002 − 0.998 = 0.004, the result has three decimal places because each term in the subtraction has three decimal places.

To show why this rule should hold, we return to the first example in which we added 123 and 5.35, and rewrite these numbers as 123.*xxx* and 5.35*x*. Digits written with an *x* are completely unknown and can be any digit from 0 to 9. Now we line up 123.*xxx* and 5.35*x* relative to the decimal point and perform the addition, using the rule that an unknown digit added to a known or unknown digit yields an unknown:

$$
\begin{array}{r}
123.xxx \\
+ \quad 5.35x \\
\hline
128.xxx
\end{array}
$$

The answer of 128.*xxx* means that we are justified only in keeping the number 128 because everything after the decimal point in the sum is actually unknown. The example shows that the controlling uncertainty is introduced into an addition or subtraction by the term with the smallest number of decimal places.

In performing any calculation, especially one involving a number of steps, there will always be slight discrepancies introduced by both the rounding process and the algebraic order in which steps are carried out. For example, consider $2.35 \times 5.89/1.57$. This computation can be performed in three different orders. First, we have $2.35 \times 5.89 = 13.842$, which rounds to 13.8, followed by $13.8/1.57 = 8.789\ 8$, rounding to 8.79. Second, $5.89/1.57 = 3.751\ 6$, which rounds to 3.75, resulting in $2.35 \times 3.75 = 8.812\ 5$, rounding to 8.81. Finally, $2.35/1.57 = 1.496\ 8$ rounds to 1.50, and $1.50 \times 5.89 = 8.835$ rounds to 8.84. So three different algebraic orders, following the rules of rounding, lead to answers of 8.79, 8.81, and 8.84, respectively. Such minor discrepancies are to be expected, because the last significant digit is only one representative from a range of possible values, depending on experimental uncertainty. The discrepancies can be reduced by carrying one or more extra digits during the calculation. In our examples, however, intermediate results will be rounded off to the proper number of significant figures, and only those digits will be carried forward. In experimental work, more sophisticated techniques are used to determine the accuracy of an experimental result.

1.5   CONVERSION OF UNITS

Sometimes it's necessary to convert units from one system to another. Conversion factors between the SI and U.S. customary systems for units of length are as follows:

$$1 \text{ mile} = 1\ 609 \text{ m} = 1.609 \text{ km} \qquad 1 \text{ ft} = 0.304\ 8 \text{ m} = 30.48 \text{ cm}$$

$$1 \text{ m} = 39.37 \text{ in.} = 3.281 \text{ ft} \qquad 1 \text{ in.} = 0.025\ 4 \text{ m} = 2.54 \text{ cm}$$

A more extensive list of conversion factors can be found on the inside front cover of this book.

TIP 1.2   No Commas in Numbers with Many Digits

In science, numbers with more than three digits are written in groups of three digits separated by spaces rather than commas; so that 10 000 is the same as the common American notation 10,000. Similarly, $\pi = 3.14159265$ is written as 3.141 592 65.

This road sign near Raleigh, North Carolina, shows distances in miles and kilometers. How accurate are the conversions?

Units can be treated as algebraic quantities that can "cancel" each other. We can make a fraction with the conversion that will cancel the units we don't want, and multiply that fraction by the quantity in question. For example, suppose we want to convert 15.0 in. to centimeters. Because 1 in. = 2.54 cm, we find that

$$15.0 \text{ in.} = 15.0 \text{ in.} \times \left(\frac{2.54 \text{ cm}}{1.00 \text{ in.}}\right) = 38.1 \text{ cm}$$

The next two examples show how to deal with problems involving more than one conversion and with powers.

EXAMPLE 1.4   Pull Over, Buddy!

Goal   Convert units using several conversion factors.

Problem   If a car is traveling at a speed of 28.0 m/s, is it exceeding the speed limit of 55.0 mi/h?

Strategy   Meters must be converted to miles and seconds to hours, using the conversion factors listed on the inside front cover of the book. This requires two or three conversion ratios.

Solution
Convert meters to miles:

$$28.0 \text{ m/s} = \left(28.0 \frac{\text{m}}{\text{s}}\right)\left(\frac{1.00 \text{ mi}}{1\,609 \text{ m}}\right) = 1.74 \times 10^{-2} \text{ mi/s}$$

Convert seconds to hours:

$$1.74 \times 10^{-2} \text{ mi/s}$$
$$= \left(1.74 \times 10^{-2} \frac{\text{mi}}{\text{s}}\right)\left(60.0 \frac{\text{s}}{\text{min}}\right)\left(60.0 \frac{\text{min}}{\text{h}}\right)$$
$$= \boxed{62.6 \text{ mi/h}}$$

Remarks   The driver should slow down because he's exceeding the speed limit. An alternate approach is to use the single conversion relationship 1.00 m/s = 2.24 mi/h:

$$28.0 \text{ m/s} = \left(28.0 \frac{\text{m}}{\text{s}}\right)\left(\frac{2.24 \text{ mi/h}}{1.00 \text{ m/s}}\right) = 62.7 \text{ mi/h}$$

Answers to conversion problems may differ slightly, as here, due to rounding during intermediate steps.

Exercise 1.4
Convert 152 mi/h to m/s.

Answer   68.0 m/s

EXAMPLE 1.5   Press the Pedal to the Metal

Goal   Convert a quantity featuring powers of a unit.

Problem   The traffic light turns green, and the driver of a high-performance car slams the accelerator to the floor. The accelerometer registers 22.0 m/s². Convert this reading to km/min².

Strategy   Here we need one factor to convert meters to kilometers and another two factors to convert seconds squared to minutes squared.

Solution
Insert the necessary factors:

$$\frac{22.0 \text{ m}}{1.00 \text{ s}^2}\left(\frac{1.00 \text{ km}}{1.00 \times 10^3 \text{ m}}\right)\left(\frac{60.0 \text{ s}}{1.00 \text{ min}}\right)^2 = 79.2 \frac{\text{km}}{\text{min}^2}$$

Remarks   Notice that in each conversion factor the numerator equals the denominator when units are taken into account. A common error in dealing with squares is to square the units inside the parentheses while forgetting to square the numbers!

Exercise 1.5
Convert 4.50×10^3 kg/m^3 to g/cm^3.

Answer   4.50 g/cm^3

1.6   ESTIMATES AND ORDER-OF-MAGNITUDE CALCULATIONS

Getting an exact answer to a calculation may often be difficult or impossible, either for mathematical reasons or because limited information is available. In these cases, estimates can yield useful approximate answers that can determine whether a more precise calculation is necessary. Estimates also serve as a partial check if the exact calculations are actually carried out. If a large answer is expected but a small exact answer is obtained, there's an error somewhere.

For many problems, knowing the approximate value of a quantity—within a factor of 10 or so—is sufficient. This approximate value is called an **order-of-magnitude** estimate, and requires finding the power of 10 that is closest to the actual value of the quantity. For example, 75 kg $\sim 10^2$ kg, where the symbol $\sim$ means "is on the order of" or "is approximately." Increasing a quantity by three orders of magnitude means that its value increases by a factor of $10^3 = 1\,000$.

Occasionally, the process of making such estimates results in fairly crude answers, but answers ten times or more too large or small are still useful. For example, suppose you're interested in how many people have contracted a certain disease. Any estimates under ten thousand are small compared with Earth's total population, but a million or more would be alarming. So even relatively imprecise information can provide valuable guidance.

In developing these estimates, you can take considerable liberties with the numbers. For example, $\pi \sim 1$, $27 \sim 10$, and $65 \sim 100$. To get a less crude estimate, it's permissible to use slightly more accurate numbers (e.g., $\pi \sim 3$, $27 \sim 30$, $65 \sim 70$). Better accuracy can also be obtained by systematically underestimating as many numbers as you overestimate. Some quantities may be completely unknown, but it's standard to make reasonable guesses, as the examples show.

EXAMPLE 1.6   How Much Gasoline Do We Use?

Goal   Develop a complex estimate.

Problem   Estimate the number of gallons of gasoline used by all cars in the United States each year.

Strategy   Estimate the number of people in the United States, and then estimate the number of cars per person. Multiply to get the number of cars. Guess at the number of miles per gallon obtained by a typical car and the number of miles driven per year, and from that get the number of gallons each car uses every year. Multiply by the estimated number of cars to get a final answer, the number of gallons of gas used.

Solution

The number of cars equals the number of people times the number of cars per person:

$$\text{number of cars} = (3.00 \times 10^8 \text{ people})$$
$$\times (0.5 \text{ cars/person}) \sim 10^8 \text{ cars}$$

The number of gallons used by one car in a year is the number of miles driven divided by the miles per gallon.

$$\frac{\text{\# gal/yr}}{\text{car}} \approx \frac{\left(\dfrac{10^4 \text{ mi/yr}}{\text{car}}\right)}{10\,\dfrac{\text{mi}}{\text{gal}}} = 10^3\,\frac{\text{gal/yr}}{\text{car}}$$

Multiply these two results together to get an estimate of the number of gallons of gas used per year.

$$\text{\# gal} \sim (10^8 \text{ cars}) \times \left(10^3\,\frac{\text{gal/yr}}{\text{car}}\right) = \boxed{10^{11} \text{ gal/yr}}$$

Remarks   Notice the inexact, and somewhat high, figure for the number of people in the United States, the estimate on the number of cars per person (figuring that every other person has a car of one kind or another), and the truncation of 1.5×10^8 cars to 10^8 cars. A similar estimate was used on the number of miles driven per year by a typical vehicle, and the average fuel economy, 10 mi/gal, looks low. None of this is important, because we are interested only in an order-of-magnitude answer. Few people owning a car would drive just 1 000 miles in a year, and very few would drive 100 000 miles, so 10 000 miles is a good estimate. Similarly, most cars get between 10 and 30 mi/gal, so using 10 is a reasonable estimate, while very few cars would get 100 mi/gal or 1 mi/gal.

In making estimates, it's okay to be cavalier! Feel free to take liberties ordinarily denied.

Exercise 1.6
How many new car tires are purchased in the United States each year? (Use the fact that tires wear out after about 50 000 miles.)

Answer $\sim 10^8$ tires (Individual answers may vary.)

Example 1.7   Stack One-Dollar Bills to the Moon

Goal   Estimate the number of stacked objects required to reach a given height.

Problem   How many one-dollar bills, stacked one on top of the other, would reach the Moon?

Strategy   The distance to the Moon is about 400 000 km. Guess at the number of dollar bills in a millimeter, and multiply the distance by this number, after converting to consistent units.

Solution
We estimate that ten stacked bills form a layer of 1 mm. Convert mm to km:

$$\frac{10 \text{ bills}}{1 \text{ mm}}\left(\frac{10^3 \text{ mm}}{1 \text{ m}}\right)\left(\frac{10^3 \text{ m}}{1 \text{ km}}\right) = \frac{10^7 \text{ bills}}{1 \text{ km}}$$

Multiply this value by the approximate lunar distance:

$$\text{\# of dollar bills} \sim (4 \times 10^5 \text{ km})\left(\frac{10^7 \text{ bills}}{1 \text{ km}}\right)$$

$$= 4 \times 10^{12} \text{ bills}$$

Remarks   That's the same order of magnitude as the U.S. national debt!

Exercise 1.7
How many pieces of cardboard, typically found at the back of a bound pad of paper, would you have to stack up to match the height of the Washington monument, about 170 m tall?

Answer $\sim 10^5$ (Answers may vary.)

EXAMPLE 1.8   Number of Galaxies in the Universe

Goal   Estimate a volume and a number density, and combine.

Problem   Given that astronomers can see about 10 billion light years into space and that there are 14 galaxies in our local group, 2 million light years from the next local group, estimate the number of galaxies in the observable universe. (*Note*: One light year is the distance traveled by light in one year, about 9.5×10^{15} m.) (See Fig. 1.3.)

Strategy   From the known information, we can estimate the number of galaxies per unit volume. The local group of 14 galaxies is contained in a sphere a million light years in radius, with the Andromeda group in a similar sphere, so there are about 10 galaxies within a volume of radius 1 million light years. Multiply that number density by the volume of the observable universe.

Figure 1.3   In this deep-space photograph, there are few stars—just galaxies without end.

Solution

Compute the approximate volume V_{lg} of the local group of galaxies:

$$V_{lg} = \tfrac{4}{3}\pi r^3 \sim (10^6 \text{ ly})^3 = 10^{18} \text{ ly}^3$$

Compute the number of galaxies per cubic light year:

$$\frac{\text{# of galaxies}}{\text{ly}^3} = \frac{\text{# of galaxies}}{V_{lg}}$$

$$\sim \frac{10 \text{ galaxies}}{10^{18} \text{ ly}^3} = 10^{-17}\,\frac{\text{galaxies}}{\text{ly}^3}$$

Compute the approximate volume of the observable universe:

$$V_u = \tfrac{4}{3}\pi r^3 \sim (10^{10} \text{ ly})^3 = 10^{30} \text{ ly}^3$$

Multiply the density of galaxies by V_u:

$$\text{# of galaxies} \sim \left(\frac{\text{# of galaxies}}{\text{ly}^3}\right) V_u$$

$$= \left(10^{-17}\,\frac{\text{galaxies}}{\text{ly}^3}\right)(10^{30} \text{ ly}^3)$$

$$= 10^{13} \text{ galaxies}$$

Remarks   Notice the approximate nature of the computation, which uses $4\pi/3 \sim 1$ on two occasions and $14 \sim 10$ for the number of galaxies in the local group. This is completely justified: Using the actual numbers would be pointless, because the other assumptions in the problem—the size of the observable universe and the idea that the local galaxy density is representative of the density everywhere—are also very rough approximations. Further, there was nothing in the problem that required using volumes of spheres rather than volumes of cubes. Despite all these arbitrary choices, the answer still gives useful information, because it rules out a lot of reasonable possible answers. Before doing the calculation, a guess of a billion galaxies might have seemed plausible.

Exercise 1.8

Given that the nearest star is about 4 light years away and that the galaxy is roughly a disk 100 000 light years across and a thousand light years thick, estimate the number of stars in the Milky Way galaxy.

Answer $\sim 10^{12}$ stars (Estimates will vary. The actual answer is probably close to 4×10^{11} stars.)

1.7   COORDINATE SYSTEMS

Many aspects of physics deal with locations in space, which require the definition of a coordinate system. A point on a line can be located with one coordinate, a point in a plane with two coordinates, and a point in space with three.

A coordinate system used to specify locations in space consists of the following:

- A fixed reference point O, called the *origin*
- A set of specified axes, or directions, with an appropriate scale and labels on the axes
- Instructions on labeling a point in space relative to the origin and axes

One convenient and commonly used coordinate system is the **Cartesian coordinate system**, sometimes called the **rectangular coordinate system**. Such a system in two dimensions is illustrated in Figure 1.4. An arbitrary point in this system is labeled with the coordinates (x, y). For example, the point P in the figure has coordinates $(5, 3)$. If we start at the origin O, we can reach P by moving 5 meters horizontally to the right and then 3 meters vertically upwards. In the same way, the point Q has coordinates $(-3, 4)$, which corresponds to going 3 meters horizontally to the left of the origin and 4 meters vertically upwards from there.

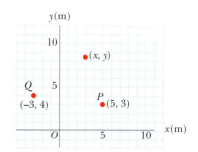

Figure 1.4   Designation of points in a two-dimensional Cartesian coordinate system. Every point is labeled with coordinates (x, y).

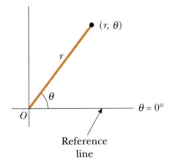

Figure 1.5 A polar coordinate system.

Positive x is usually selected as right of the origin and positive y upward from the origin, but in two dimensions this choice is largely a matter of taste. (In three dimensions, however, there are "right-handed" and "left-handed" coordinates, which lead to minus sign differences in certain operations. These will be addressed as needed.)

Sometimes it's more convenient to locate a point in space by its **plane polar coordinates** (r, θ), as in Figure 1.5. In this coordinate system, an origin O and a reference line are selected as shown. A point is then specified by the distance r from the origin to the point and by the angle θ between the reference line and a line drawn from the origin to the point. The standard reference line is usually selected to be the positive x-axis of a Cartesian coordinate system. The angle θ is considered positive when measured counterclockwise from the reference line and negative when measured clockwise. For example, if a point is specified by the polar coordinates 3 m and 60°, we locate this point by moving out 3 m from the origin at an angle of 60° above (counterclockwise from) the reference line. A point specified by polar coordinates 3 m and $-60°$ is located 3 m out from the origin and 60° below (clockwise from) the reference line.

1.8   TRIGONOMETRY

Consider the right triangle shown in Active Figure 1.6, where side y is opposite the angle θ, side x is adjacent to the angle θ, and side r is the hypotenuse of the triangle. The basic trigonometric functions defined by such a triangle are the ratios of the lengths of the sides of the triangle. These relationships are called the sine (sin), cosine (cos), and tangent (tan) functions. In terms of θ, the basic trigonometric functions are as follows:[1]

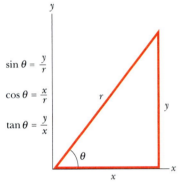

$$\sin \theta = \frac{y}{r}$$
$$\cos \theta = \frac{x}{r}$$
$$\tan \theta = \frac{y}{x}$$

ACTIVE FIGURE 1.6
Certain trigonometric functions of a right triangle.

Physics⊗Now™ Log into PhysicsNow at **www.cp7e.com**, and go to Active Figure 1.6 to move the point and see the changes to the rectangular and polar coordinates and to the sine, cosine, and tangent of angle θ.

$$\sin \theta = \frac{\text{side opposite } \theta}{\text{hypotenuse}} = \frac{y}{r}$$

$$\cos \theta = \frac{\text{side adjacent to } \theta}{\text{hypotenuse}} = \frac{x}{r} \qquad \text{[1.1]}$$

$$\tan \theta = \frac{\text{side opposite } \theta}{\text{side adjacent to } \theta} = \frac{y}{x}$$

For example, if the angle θ is equal to 30°, then the ratio of y to r is always 0.50; that is, $\sin 30° = 0.50$. Note that the sine, cosine, and tangent functions are quantities without units because each represents the ratio of two lengths.

Another important relationship, called the **Pythagorean theorem**, exists between the lengths of the sides of a right triangle:

$$r^2 = x^2 + y^2 \qquad \text{[1.2]}$$

Finally, it will often be necessary to find the values of inverse relationships. For example, suppose you know that the sine of an angle is 0.866, but you need to know the value of the angle itself. The inverse sine function may be expressed as $\sin^{-1}(0.866)$, which is a shorthand way of asking the question "What angle has a sine of 0.866?" Punching a couple of buttons on your calculator reveals that this angle is 60.0°. Try it for yourself and show that $\tan^{-1}(0.400) = 21.8°$. Be sure that your calculator is set for degrees and not radians. In addition, the inverse tangent function can return only values between $-90°$ and $+90°$, so when an angle is in the second or third quadrant, it's necessary to add 180° to the answer in the calculator window.

The definitions of the trigonometric functions and the inverse trigonometric functions, as well as the Pythagorean theorem, can be applied to *any* right triangle, regardless of whether its sides correspond to x- and y-coordinates.

These results from trigonometry are useful in converting from rectangular coordinates to polar coordinates, or vice versa, as the next example shows.

TIP 1.3   Degrees vs. Radians
When calculating trigonometric functions, make sure your calculator setting—degrees or radians—is consistent with the degree measure you're using in a given problem.

---

[1]Many people use the mnemonic *SOHCAHTOA* to remember the basic trigonometric formulas: *Sine* = *Opposite/Hypotenuse, Cosine = Adjacent/Hypotenuse,* and *Tangent = Opposite/Adjacent.* (Thanks go to Professor Don Chodrow for pointing this out.)

EXAMPLE 1.9   Cartesian and Polar Coordinates

Goal   Understand how to convert from plane rectangular coordinates to plane polar coordinates and vice versa.

Problem   (a) The Cartesian coordinates of a point in the xy-plane are $(x, y) = (-3.50, -2.50)$ m, as shown in Active Figure 1.7. Find the polar coordinates of this point. (b) Convert $(r, \theta) = (5.00 \text{ m}, 37.0°)$ to rectangular coordinates.

Strategy   Apply the trigonometric functions and their inverses, together with the Pythagorean theorem.

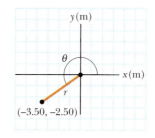

ACTIVE FIGURE 1.7
(Example 1.9) Converting from Cartesian coordinates to polar coordinates.

Physics⊗Now™ Log into PhysicsNow at **www.cp7e.com**, and go to Active Figure 1.7 to move the point in the xy-plane and see how its Cartesian and polar coordinates change.

Solution
(a) Cartesian to Polar

Take the square root of both sides of Equation 1.2 to find the radial coordinate:

$$r = \sqrt{x^2 + y^2} = \sqrt{(-3.50 \text{ m})^2 + (-2.50 \text{ m})^2} = 4.30 \text{ m}$$

Use Equation 1.1 for the tangent function to find the angle with the inverse tangent, adding 180° because the angle is actually in third quadrant:

$$\tan \theta = \frac{y}{x} = \frac{-2.50 \text{ m}}{-3.50 \text{ m}} = 0.714$$

$$\theta = \tan^{-1}(0.714) = 35.5° + 180° = 216°$$

(b) Polar to Cartesian

Use the trigonometric definitions, Equation 1.1.

$$x = r \cos \theta = (5.00 \text{ m}) \cos 37.0° = 3.99 \text{ m}$$

$$y = r \sin \theta = (5.00 \text{ m}) \sin 37.0° = 3.01 \text{ m}$$

Remarks   When we take up vectors in two dimensions in Chapter 3, we will routinely use a similar process to find the direction and magnitude of a given vector from its components, or, conversely, to find the components from the vector's magnitude and direction.

Exercise 1.9
(a) Find the polar coordinates corresponding to $(x, y) = (-3.25, 1.50)$ m. (b) Find the Cartesian coordinates corresponding to $(r, \theta) = (4.00 \text{ m}, 53.0°)$

Answers   (a) $(r, \theta) = (3.58 \text{ m}, 155°)$ (b) $(x, y) = (2.41 \text{ m}, 3.19 \text{ m})$

EXAMPLE 1.10   How High Is the Building?

Goal   Apply basic results of trigonometry.

Problem   A person measures the height of a building by walking out a distance of 46.0 m from its base and shining a flashlight beam toward the top. When the beam is elevated at an angle of 39.0° with respect to the horizontal, as shown in Figure 1.8, the beam just strikes the top of the building. Find the height of the building and the distance the flashlight beam has to travel before it strikes the top of the building.

Strategy   Refer to the right triangle shown in the figure. We know the angle, 39.0°, and the length of the side adjacent to it. Since the height of the building is the side opposite the angle, we can use the tangent function. With the adjacent and opposite sides known, we can then find the hypotenuse with the Pythagorean theorem.

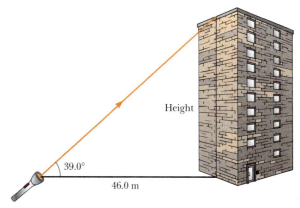

Figure 1.8   (Example 1.10)

Solution

Use the tangent of the given angle:

$$\tan 39.0° = \frac{\text{height}}{46.0 \text{ m}}$$

Solve for the height:

$$\text{Height} = (\tan 39.0°)(46.0 \text{ m}) = (0.810)(46.0 \text{ m})$$
$$= \boxed{37.3 \text{ m}}$$

Find the hypotenuse of the triangle:

$$r = \sqrt{x^2 + y^2} = \sqrt{(37.3 \text{ m})^2 + (46.0 \text{ m})^2} = \boxed{59.2 \text{ m}}$$

Remarks   In a later chapter, right-triangle trigonometry is often used when working with vectors.

Exercise 1.10

High atop a building 50.0 m tall, you spot a friend standing on a street corner. Using a protractor and dangling a plumb bob, you find that the angle between the horizontal and the direction of your friend is 25.0°. Your eyes are located 1.75 m above the top of the building. How far away from the foot of the building is your friend?

Answer   111 m

1.9   PROBLEM-SOLVING STRATEGY

Most courses in general physics require the student to learn the skills used in solving problems, and examinations usually include problems that test such skills. This brief section presents some useful suggestions that will help increase your success in solving problems. An organized approach to problem solving will also enhance your understanding of physical concepts and reduce exam stress. Throughout the book, there will be a number of sections labeled "Problem-Solving Strategy," many of them just a specializing of the list given below (and illustrated in Figure 1.9).

General Problem-Solving Strategy

1. **Read** the problem carefully at least twice. Be sure you understand the nature of the problem before proceeding further.
2. **Draw** a diagram while rereading the problem.
3. **Label** all physical quantities in the diagram, using letters that remind you what the quantity is (e.g., *m* for mass). Choose a coordinate system and label it.
4. **Identify** physical principles, the knowns and unknowns, and list them. Put circles around the unknowns.
5. **Equations**, the relationships between the labeled physical quantities, should be written down next. Naturally, the selected equations should be consistent with the physical principles identified in the previous step.
6. **Solve** the set of equations for the unknown quantities in terms of the known. Do this algebraically, without substituting values until the next step, except where terms are zero.
7. **Substitute** the known values, together with their units. Obtain a numerical value with units for each unknown.
8. **Check** your answer. Do the units match? Is the answer reasonable? Does the plus or minus sign make sense? Is your answer consistent with an order of magnitude estimate?

This same procedure, with minor variations, should be followed throughout the course. The first three steps are extremely important, because they get you mentally oriented. Identifying the proper concepts and physical principles assists you in choosing the correct equations. The equations themselves are essential, because when you understand them, you also understand the relationships between the physical quantities. This understanding comes through a lot of daily practice.

Equations are the tools of physics: To solve problems, you have to have them at hand, like a plumber and his wrenches. Know the equations, and understand what

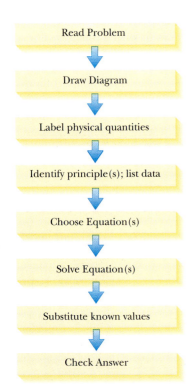

Read Problem
⬇
Draw Diagram
⬇
Label physical quantities
⬇
Identify principle(s); list data
⬇
Choose Equation(s)
⬇
Solve Equation(s)
⬇
Substitute known values
⬇
Check Answer

Figure 1.9   A guide to problem solving.

they mean and how to use them. Just as you can't have a conversation without knowing the local language, you can't solve physics problems without knowing and understanding the equations. This understanding grows as you study and apply the concepts and the equations relating them.

Carrying through the algebra for as long as possible, substituting numbers only at the end, is also important, because it helps you think in terms of the physical quantities involved, not merely the numbers that represent them. Many beginning physics students are eager to substitute, but once numbers are substituted, it's harder to understand relationships and easier to make mistakes.

The physical layout and organization of your work will make the final product more understandable and easier to follow. Although physics is a challenging discipline, your chances of success are excellent if you maintain a positive attitude and keep trying.

EXAMPLE 1.11   A Round Trip by Air

Goal   Illustrate the Problem-Solving Strategy.

Problem   An airplane travels 4.50×10^2 km due east and then travels an unknown distance due north. Finally, it returns to its starting point by traveling a distance of 525 km. How far did the airplane travel in the northerly direction?

Strategy   We've finished reading the problem (step 1), and have drawn a diagram (step 2) in Figure 1.10 and labeled it (step 3). From the diagram, we recognize a right triangle and identify (step 4) the principle involved: the Pythagorean theorem. Side y is the unknown quantity, and the other sides are known.

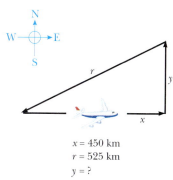

$x = 450$ km
$r = 525$ km
$y = ?$

Figure 1.10   (Example 1.11)

Solution

Write the Pythagorean theorem (step 5):

$$r^2 = x^2 + y^2$$

Solve symbolically for y (step 6):

$$y^2 = r^2 - x^2 \quad \rightarrow \quad y = +\sqrt{r^2 - x^2}$$

Substitute the numbers, with units (step 7):

$$y = \sqrt{(525 \text{ km})^2 - (4.50 \times 10^2 \text{ km})^2} = \boxed{270 \text{ km}}$$

Remarks   Note that the negative solution has been disregarded, because it's not physically meaningful. In checking (step 8), note that the units are correct and that an approximate answer can be obtained by using the easier quantities, 500 km and 400 km. Doing so gives an answer of 300 km, which is approximately the same as our calculated answer of 270 km.

Exercise 1.11

A plane flies 345 km due south, then turns and flies northeast 615 km, until it's due east of its starting point. If the plane now turns and heads for home, how far will it have to go?

Answer   509 km

SUMMARY

1.1   Standards of Length, Mass, and Time

The physical quantities in the study of mechanics can be expressed in terms of three fundamental quantities: length, mass, and time, which have the SI units meters (m), kilograms (kg), and seconds (s), respectively.

1.2   The Building Blocks of Matter

Matter is made of atoms, which in turn are made up of a relatively small nucleus of protons and neutrons within a cloud of electrons. Protons and neutrons are composed of still smaller particles, called quarks.

1.3   Dimensional Analysis

Dimensional analysis can be used to check equations and to assist in deriving them. When the dimensions on both sides of the equation agree, the equation is often correct up to a numerical factor. When the dimensions don't agree, the equation must be wrong.

1.4   Uncertainty in Measurement and Significant Figures

No physical quantity can be determined with complete accuracy. The concept of significant figures affords a basic method of handling these uncertainties. A significant figure is a reliably known digit, other than a zero, used to locate the decimal point. The two rules of significant figures are as follows:

1. When multiplying or dividing using two or more quantities, the result should have the same number of significant figures as the quantity having the fewest significant figures.
2. When quantities are added or subtracted, the number of decimal places in the result should be the same as in the quantity with the fewest decimal places.

Use of scientific notation can avoid ambiguity in significant figures. In rounding, if the last digit dropped is less than 5, simply drop the digit, otherwise raise the last retained digit by one.

1.5   Conversion of Units

Units in physics equations must always be consistent. In solving a physics problem, it's best to start with consistent units, using the table of conversion factors on the inside front cover as necessary.

Converting units is a matter of multiplying the given quantity by a fraction, with one unit in the numerator and its equivalent in the other units in the denominator, arranged so the unwanted units in the given quantity are cancelled out in favor of the desired units.

1.6   Estimates and Order-of-Magnitude Calculations

Sometimes it's useful to find an approximate answer to a question, either because the math is difficult or because

information is incomplete. A quick estimate can also be used to check a more detailed calculation. In an order-of-magnitude calculation, each value is replaced by the closest power of ten, which sometimes must be guessed or estimated when the value is unknown. The computation is then carried out. For quick estimates involving known values, each value can first be rounded to one significant figure.

1.7   Coordinate Systems

The Cartesian coordinate system consists of two perpendicular axes, usually called the x-axis and y-axis, with each axis labeled with all numbers from negative infinity to positive infinity. Points are located by specifying the x- and y-values. Polar coordinates consist of a radial coordinate r which is the distance from the origin, and an angular coordinate θ, which is the angular displacement from the positive x-axis.

1.8   Trigonometry

The three most basic trigonometric functions of a right triangle are the sine, cosine, and tangent, defined as follows:

$$\sin \theta = \frac{\text{side opposite } \theta}{\text{hypotenuse}} = \frac{y}{r}$$

$$\cos \theta = \frac{\text{side adjacent to } \theta}{\text{hypotenuse}} = \frac{x}{r} \qquad [1.1]$$

$$\tan \theta = \frac{\text{side opposite } \theta}{\text{side adjacent to } \theta} = \frac{y}{x}$$

The **Pythagorean theorem** is an important relationship between the lengths of the sides of a right triangle:

$$r^2 = x^2 + y^2 \qquad [1.2]$$

where r is the hypotenuse of the triangle and x and y are the other two sides.

CONCEPTUAL QUESTIONS

1. Estimate the order of magnitude of the length, in meters, of each of the following: (a) a mouse, (b) a pool cue, (c) a basketball court, (d) an elephant, (e) a city block.

2. What types of natural phenomena could serve as time standards?

3. (a) Estimate the number of times your heart beats in a month. (b) Estimate the number of human heartbeats in an average lifetime.

4. An object with a mass of 1 kg weighs approximately 2 lb. Use this information to estimate the mass of the following objects: (a) a baseball; (b) your physics textbook; (c) a pickup truck.

5. Find the order of magnitude of your age in seconds.

6. Estimate the number of atoms in 1 cm^3 of a solid. (Note that the diameter of an atom is about 10^{-10} m.)

7. The height of a horse is sometimes given in units of "hands." Why is this a poor standard of length?

8. How many of the lengths or time intervals given in Tables 1.2 and 1.3 could you verify, using only equipment found in a typical dormitory room?

9. An ancient unit of length called the *cubit* was equal to six palms, where a palm was the width of the four fingers of an open hand. Noah's ark was 300 cubits long, 50 cubits wide, and 30 cubits high. Estimate the volume of the ark in cubic meters. Also, estimate the volume of a typical home in cubic meters, and compare it with the volume of the ark.

10. Do an order-of-magnitude calculation for an everyday situation you encounter. For example, how far do you walk or drive each day?

11. If an equation is dimensionally correct, does this mean that the equation must be true? If an equation is not dimensionally correct, does this mean that the equation can't be true?

12. Figure Q1.12 is a photograph showing unit conversions on the labels of some grocery-store items. Check the accuracy of these conversions. Are the manufacturers using significant figures correctly?

Courtesy of George Semple

Figure Q1.12

PROBLEMS

1, 2, 3 = straightforward, intermediate, challenging   □ = full solution available in *Student Solutions Manual/Study Guide*

Physics⊗Now™ = coached solution with hints available at **www.cp7e.com**   🧬 = biomedical application

Section 1.3 Dimensional Analysis

1. A shape that covers an area A and has a uniform height h has a volume $V = Ah$. (a) Show that $V = Ah$ is dimensionally correct. (b) Show that the volumes of a cylinder and of a rectangular box can be written in the form $V = Ah$, identifying A in each case. (Note that A, sometimes called the "footprint" of the object, can have any shape and that the height can, in general, be replaced by the average thickness of the object.)

2. (a) Suppose that the displacement of an object is related to time according to the expression $x = Bt^2$. What are the dimensions of B? (b) A displacement is related to time as $x = A\sin(2\pi ft)$, where A and f are constants. Find the dimensions of A. (*Hint*: A trigonometric function appearing in an equation must be dimensionless.)

3. The period of a simple pendulum, defined as the time necessary for one complete oscillation, is measured in time units and is given by

$$T = 2\pi\sqrt{\frac{\ell}{g}}$$

where ℓ is the length of the pendulum and g is the acceleration due to gravity, in units of length divided by time squared. Show that this equation is dimensionally consistent. (You might want to check the formula using your keys at the end of a string and a stopwatch.)

4. Each of the following equations was given by a student during an examination:

$$\tfrac{1}{2}mv^2 = \tfrac{1}{2}mv_0^2 + \sqrt{mgh} \qquad v = v_0 + at^2 \qquad ma = v^2$$

Do a dimensional analysis of each equation and explain why the equation can't be correct.

5. Newton's law of universal gravitation is represented by

$$F = G\frac{Mm}{r^2}$$

where F is the gravitational force, M and m are masses, and r is a length. Force has the SI units $kg \cdot m/s^2$. What are the SI units of the proportionality constant G?

6. (a) One of the fundamental laws of motion states that the acceleration of an object is directly proportional to the resultant force on it and inversely proportional to its mass. If the proportionality constant is defined to have no dimensions, determine the dimensions of force. (b) The newton is the SI unit of force. According to the results for (a), how can you express a force having units of newtons by using the fundamental units of mass, length, and time?

Section 1.4 Uncertainty in Measurement and Significant Figures

7. How many significant figures are there in (a) 78.9 ± 0.2, (b) 3.788×10^9, (c) 2.46×10^{-6}, (d) 0.0032?

8. A rectangular plate has a length of (21.3 ± 0.2) cm and a width of (9.8 ± 0.1) cm. Calculate the area of the plate, including its uncertainty.

9. Physics⊗Now™ Carry out the following arithmetic operations: (a) the sum of the measured values 756, 37.2, 0.83, and 2.5; (b) the product 0.0032×356.3; (c) the product $5.620 \times \pi$.

10. The speed of light is now defined to be $2.99\,7924\,58 \times 10^8$ m/s. Express the speed of light to (a) three significant figures, (b) five significant figures, and (c) seven significant figures.

11. A farmer measures the perimeter of a rectangular field. The length of each long side of the rectangle is found to be 38.44 m, and the length of each short side is found to be 19.5 m. What is the perimeter of the field?

12. The radius of a circle is measured to be (10.5 ± 0.2) m. Calculate (a) the area and (b) the circumference of the circle, and give the uncertainty in each value.

13. A fisherman catches two striped bass. The smaller of the two has a measured length of 93.46 cm (two decimal

places, four significant figures), and the larger fish has a measured length of 135.3 cm (one decimal place, four significant figures). What is the total length of fish caught for the day?

14. (a) Using your calculator, find, in scientific notation with appropriate rounding, (a) the value of $(2.437 \times 10^4)(6.5211 \times 10^9)/(5.37 \times 10^4)$ and (b) the value of $(3.14159 \times 10^2)(27.01 \times 10^4)/(1\ 234 \times 10^6)$.

Section 1.5 Conversion of Units

15. A fathom is a unit of length, usually reserved for measuring the depth of water. A fathom is approximately 6 ft in length. Take the distance from Earth to the Moon to be 250 000 miles, and use the given approximation to find the distance in fathoms.

16. Find the height or length of these natural wonders in kilometers, meters, and centimeters: (a) The longest cave system in the world is the Mammoth Cave system in Central Kentucky, with a mapped length of 348 miles. (b) In the United States, the waterfall with the greatest single drop is Ribbon Falls in California, which drops 1 612 ft. (c) At 20 320 feet, Mount McKinley in Alaska is America's highest mountain. (d) The deepest canyon in the United States is King's Canyon in California, with a depth of 8 200 ft.

17. A rectangular building lot measures 100 ft by 150 ft. Determine the area of this lot in square meters (m^2).

18. Suppose your hair grows at the rate of 1/32 inch per day. Find the rate at which it grows in nanometers per second. Since the distance between atoms in a molecule is on the order of 0.1 nm, your answer suggests how rapidly atoms are assembled in this protein synthesis.

19. Using the data in Table 1.1 and the appropriate conversion factors, find the distance to the nearest star, in feet.

20. Using the data in Table 1.3 and the appropriate conversion factors, find the age of Earth in years.

21. The speed of light is about 3.00×10^8 m/s. Convert this figure to miles per hour.

22. A house is 50.0 ft long and 26 ft wide and has 8.0-ft-high ceilings. What is the volume of the interior of the house in cubic meters and in cubic centimeters?

23. The amount of water in reservoirs is often measured in acre-ft. One acre-ft is a volume that covers an area of one acre to a depth of one foot. An acre is 43 560 ft^2. Find the volume in SI units of a reservoir containing 25.0 acre-ft of water.

24. The base of a pyramid covers an area of 13.0 acres (1 acre = 43 560 ft^2) and has a height of 481 ft (Fig. P1.24). If the volume of a pyramid is given by the expression $V = bh/3$, where b is the area of the base and h is the height, find the volume of this pyramid in cubic meters.

Figure P1.24

25. **Physics⊗Now™** A quart container of ice cream is to be made in the form of a cube. What should be the length of a side, in centimeters? (Use the conversion 1 gallon = 3.786 liter.)

26. (a) Find a conversion factor to convert from miles per hour to kilometers per hour. (b) For a while, federal law mandated that the maximum highway speed would be 55 mi/h. Use the conversion factor from part (a) to find the speed in kilometers per hour. (c) The maximum highway speed has been raised to 65 mi/h in some places. In kilometers per hour, how much of an increase is this over the 55-mi/h limit?

27. One cubic centimeter $(1.0\ \text{cm}^3)$ of water has a mass of 1.0×10^{-3} kg. (a) Determine the mass of $1.0\ \text{m}^3$ of water. (b) Assuming that biological substances are 98% water, estimate the masses of a cell with a diameter of 1.0 μm, a human kidney, and a fly. Take a kidney to be roughly a sphere with a radius of 4.0 cm and a fly to be roughly a cylinder 4.0 mm long and 2.0 mm in diameter.

28. A billionaire offers to give you $1 billion if you can count out that sum with only $1 bills. Should you accept her offer? Assume that you can count at an average rate of one bill every second, and be sure to allow for the fact that you need about 8 hours a day for sleeping and eating.

Section 1.6 Estimates and Order-of-Magnitude Calculations

Note: In developing answers to the problems in this section, you should state your important assumptions, including the numerical values assigned to parameters used in the solution.

29. Imagine that you are the equipment manager of a professional baseball team. One of your jobs is to keep baseballs on hand for games. Balls are sometimes lost when players hit them into the stands as either home runs or foul balls. Estimate how many baseballs you have to buy per season in order to make up for such losses. Assume that your team plays an 81-game home schedule in a season.

30. A hamburger chain advertises that it has sold more than 50 billion hamburgers. Estimate how many pounds of hamburger meat must have been used by the chain and how many head of cattle were required to furnish the meat.

31. An automobile tire is rated to last for 50 000 miles. Estimate the number of revolutions the tire will make in its lifetime.

32. Grass grows densely everywhere on a quarter-acre plot of land. What is the order of magnitude of the number of blades of grass? Explain your reasoning. Note that 1 acre = 43 560 ft^2.

33. Estimate the number of Ping-Pong balls that would fit into a typical-size room (without being crushed). In your solution, state the quantities you measure or estimate and the values you take for them.

34. Soft drinks are commonly sold in aluminum containers. To an order of magnitude, how many such containers are thrown away or recycled each year by U.S. consumers? How many tons of aluminum does this represent? In your solution, state the quantities you measure or estimate and the values you take for them.

Section 1.7 Coordinate Systems

35. A point is located in a polar coordinate system by the coordinates $r = 2.5$ m and $\theta = 35°$. Find the x- and

y-coordinates of this point, assuming that the two coordinate systems have the same origin.

36. A certain corner of a room is selected as the origin of a rectangular coordinate system. If a fly is crawling on an adjacent wall at a point having coordinates (2.0, 1.0), where the units are meters, what is the distance of the fly from the corner of the room?

37. Express the location of the fly in Problem 36 in polar coordinates.

38. Two points in a rectangular coordinate system have the coordinates (5.0, 3.0) and (−3.0, 4.0), where the units are centimeters. Determine the distance between these points.

Section 1.8 Trigonometry

39. **Physics⊗Now™** For the triangle shown in Figure P1.39, what are (a) the length of the unknown side, (b) the tangent of θ, and (c) the sine of φ?

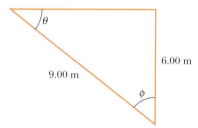

Figure P1.39

40. A ladder 9.00 m long leans against the side of a building. If the ladder is inclined at an angle of 75.0° to the horizontal, what is the horizontal distance from the bottom of the ladder to the building?

41. A high fountain of water is located at the center of a circular pool as shown in Figure P1.41. Not wishing to get his feet wet, a student walks around the pool and measures its circumference to be 15.0 m. Next, the student stands at the edge of the pool and uses a protractor to gauge the angle of elevation at the bottom of the fountain to be 55.0°. How high is the fountain?

Figure P1.41

42. A right triangle has a hypotenuse of length 3.00 m, and one of its angles is 30.0°. What are the lengths of (a) the side opposite the 30.0° angle and (b) the side adjacent to the 30.0° angle?

43. In Figure P1.43, find (a) the side opposite θ, (b) the side adjacent to φ, (c) cos θ, (d) sin φ, and (e) tan φ.

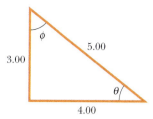

Figure P1.43

44. In a certain right triangle, the two sides that are perpendicular to each other are 5.00 m and 7.00 m long. What is the length of the third side of the triangle?

45. In Problem 44, what is the tangent of the angle for which 5.00 m is the opposite side?

46. A surveyor measures the distance across a straight river by the following method: Starting directly across from a tree on the opposite bank, he walks 100 m along the riverbank to establish a baseline. Then he sights across to the tree. The angle from his baseline to the tree is 35.0°. How wide is the river?

ADDITIONAL PROBLEMS

47. A restaurant offers pizzas in two sizes: small, with a radius of six inches; and large, with a radius of nine inches. A customer argues that if the small one sells for six dollars, the large should sell for nine dollars. Without doing any calculations, is the customer correct? Defend your answer. Calculate the area of each pizza to find out how much pie you are getting in each case. If the small one costs six dollars how much should the large cost?

48. The radius of the planet Saturn is 5.85×10^7 m, and its mass is 5.68×10^{26} kg (Fig. P1.48). (a) Find the density of Saturn (its mass divided by its volume) in grams per cubic centimeter. (The volume of a sphere is given by $(4/3)\pi r^3$.) (b) Find the area of Saturn in square feet. (The surface area of a sphere is given by $4\pi r^2$.)

Figure P1.48 A view of Saturn.

49. The displacement of an object moving under uniform acceleration is some function of time and the acceleration. Suppose we write this displacement as $s = ka^m t^n$, where k is a dimensionless constant. Show by dimensional analysis that this expression is satisfied if $m = 1$ and $n = 2$. Can the analysis give the value of k?

50. Compute the order of magnitude of the mass of (a) a bathtub filled with water and (b) a bathtub filled with pennies. In your solution, list the quantities you estimate and the value you estimate for each.

51. **Physics⊗Now™** You can obtain a rough estimate of the size of a molecule by the following simple experiment: Let a droplet of oil spread out on a smooth surface of water. The resulting oil slick will be approximately one molecule thick. Given an oil droplet of mass 9.00×10^{-7} kg and density 918 kg/m^3 that spreads out into a circle of radius 41.8 cm on the water surface, what is the order of magnitude of the diameter of an oil molecule?

52. In 2003, the U.S. national debt was about $7 trillion. (a) If payments were made at the rate of $1 000 per second, how many years would it take to pay off the debt, assuming that no interest were charged? (b) A dollar bill is about 15.5 cm long. If seven trillion dollar bills were laid end to end around the Earth's equator, how many times would they encircle the planet? Take the radius of the Earth at the equator to be 6 378 km. (*Note:* Before doing any of these calculations, try to guess at the answers. You may be very surprised.)

53. Estimate the number of piano tuners living in New York City. This question was raised by the physicist Enrico Fermi, who was well known for making order-of-magnitude calculations.

54. Sphere 1 has surface area A_1 and volume V_1, and sphere 2 has surface area A_2 and volume V_2. If the radius of sphere 2 is double the radius of sphere 1, what is the ratio of (a) the areas, A_2/A_1 and (b) the volumes, V_2/V_1?

55. (a) How many seconds are there in a year? (b) If one micrometeorite (a sphere with a diameter on the order of 10^{-6} m) struck each square meter of the Moon each second, estimate the number of years it would take to cover the Moon with micrometeorites to a depth of one meter. (*Hint:* Consider a cubic box, 1 m on a side, on the Moon, and find how long it would take to fill the box.)

ACTIVITIES

A.1. Choose a variety of objects that range in length from a few centimeters to a few meters. Try guessing the lengths in a unit appropriate to their size, and then use a meter stick supplied by your instructor to check your guesses. Keep trying until you can estimate consistently to within 20% of an object's actual length.

A.2. Choose a variety of objects that range in mass from a few grams to a few kilograms. Estimate the masses by hefting the objects, then use a balance supplied by your instructor to check your guesses. Keep trying until you can estimate consistently to within 30% of an object's actual mass.

A.3. You know that the measurements of a typical sheet of paper are 8.50 in. by 11.0 in. Convert these measurements to millimeters. Use your results to calculate the length of the diagonal of the sheet of paper by using the Pythagorean theorem. Measure the diagonal with a ruler to see how well you have done. Finally, use a suitable trig function to calculate the angle that the diagonal line makes with the horizontal. Use a protractor to verify your calculation.

Caron/Corbis Sygma

Gravity propels a ski jumper down a straight, snow-covered slope at an acceleration that is approximately constant. The equations of kinematics, studied in this chapter, can give his position and velocity along the slope at any time.

Motion In One Dimension

Life is motion. Our muscles coordinate motion microscopically to enable us to walk and jog. Our hearts pump tirelessly for decades, moving blood through our bodies. Cell wall mechanisms move select atoms and molecules in and out of cells. From the prehistoric chase of antelopes across the savanna to the pursuit of satellites in space, mastery of motion has been critical to our survival and success as a species.

The study of motion and of physical concepts such as force and mass is called **dynamics**. The part of dynamics that describes motion without regard to its causes is called **kinematics**. In this chapter, the focus is on kinematics in one dimension: motion along a straight line. This kind of motion—and, indeed, *any* motion—involves the concepts of displacement, velocity, and acceleration. Here, we use these concepts to study the motion of objects undergoing constant acceleration. In Chapter 3 we will repeat this discussion for objects moving in two dimensions.

The first recorded evidence of the study of mechanics can be traced to the people of ancient Sumeria and Egypt, who were interested primarily in understanding the motions of heavenly bodies. The most systematic and detailed early studies of the heavens were conducted by the Greeks from about 300 B.C. to A.D. 300. Ancient scientists and laypeople regarded the Earth as the center of the Universe. This **geocentric model** was accepted by such notables as Aristotle (384–322 B.C.) and Claudius Ptolemy (about A.D. 140). Largely because of the authority of Aristotle, the geocentric model became the accepted theory of the Universe until the 17th century.

About 250 B.C., the Greek philosopher Aristarchus worked out the details of a model of the Solar System based on a spherical Earth that rotated on its axis and revolved around the Sun. He proposed that the sky appeared to turn westward because the Earth was turning eastward. This model wasn't given much consideration, because it was believed that if the Earth turned, it would set up a great wind as it moved through the air. We know now that the Earth carries the air and everything else with it as it rotates.

SCHWADRON
USA

"Miss Dempswell, send in a frame of reference."

© 2003 Cartoonists & Writers Syndicate

CARTOONISTS & WRITERS SYNDICATE http://CartoonWeb.com

Figure 2.1

The Polish astronomer Nicolaus Copernicus (1473–1543) is credited with initiating the revolution that finally replaced the geocentric model. In his system, called the **heliocentric model**, Earth and the other planets revolve in circular orbits around the Sun.

This early knowledge formed the foundation for the work of Galileo Galilei (1564–1642), who stands out as the dominant facilitator of the entrance of physics into the modern era. In 1609, he became one of the first to make astronomical observations with a telescope. He observed mountains on the Moon, the larger satellites of Jupiter, spots on the Sun, and the phases of Venus. Galileo's observations convinced him of the correctness of the Copernican theory. His quantitative study of motion formed the foundation of Newton's revolutionary work in the next century.

2.1 DISPLACEMENT

Motion involves the displacement of an object from one place in space and time to another. Describing the motion requires some convenient coordinate system and a specified origin. A **frame of reference** is a choice of coordinate axes that defines the starting point for measuring any quantity, an essential first step in solving virtually any problem in mechanics (Fig. 2.1). In Active Figure 2.2a, for example, a car moves along the *x*-axis. The coordinates of the car at any time describe its position in space and, more importantly, its *displacement* at some given time of interest.

Definition of displacement ▶

> The **displacement** Δx of an object is defined as its *change in position*, and is given by
>
> $$\Delta x \equiv x_f - x_i \qquad [2.1]$$
>
> where the initial position of the car is labeled x_i and the final position is x_f. (The indices i and f stand for initial and final, respectively.)
>
> **SI unit: meter (m)**

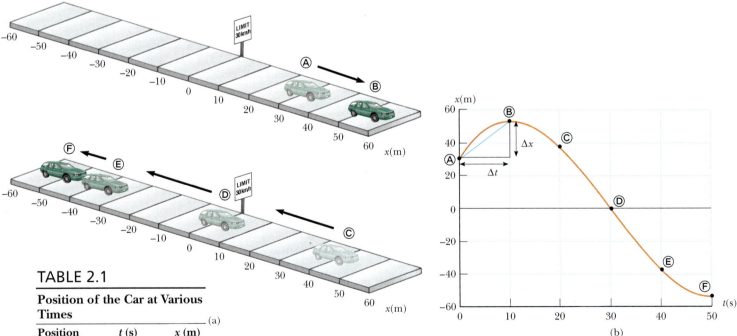

TABLE 2.1

Position of the Car at Various Times

Position	t (s)	x (m)
Ⓐ	0	30
Ⓑ	10	52
Ⓒ	20	38
Ⓓ	30	0
Ⓔ	40	−37
Ⓕ	50	−53

ACTIVE FIGURE 2.2
(a) A car moves back and forth along a straight line taken to be the *x*-axis. Because we are interested only in the car's translational motion, we can model it as a particle. (b) Graph of position vs. time for the motion of the "particle."

Physics ⊗ Now™
Log into PhysicsNow at **www.cp7e.com**, and go to Active Figure 2.2 to move each of the six points Ⓐ through Ⓕ and observe the motion of the car pictorially and graphically as it follows a smooth path through the points.

We will use the Greek letter delta, Δ, to denote a change in any physical quantity. From the definition of displacement, we see that Δx (read "delta ex") is positive if x_f is greater than x_i and negative if x_f is less than x_i. For example, if the car moves from point Ⓐ to point Ⓑ, so that the initial position is $x_i = 30$ m and the final position is $x_f = 52$ m, the displacement is $\Delta x = x_f - x_i = 52$ m $- 30$ m $= +22$ m. However, if the car moves from point Ⓒ to point Ⓕ, then the initial position is $x_i = 38$ m and the final position is $x_f = -53$ m, the displacement is $\Delta x = x_f - x_i = -53$ m $- 38$ m $= -91$ m. A positive answer indicates a displacement in the positive x-direction, whereas a negative answer indicates a displacement in the negative x-direction. Active Figure 2.2b displays the graph of the car's position as a function of time.

Because displacement has both a magnitude (size) and a direction, it's a vector quantity, as are velocity and acceleration. In general, **a vector quantity is characterized by having both a magnitude and a direction**. By contrast, **a scalar quantity has magnitude, but no direction**. Scalar quantities such as mass and temperature are completely specified by a numeric value with appropriate units; no direction is involved.

Vector quantities will be usually denoted in boldface type with an arrow over the top of the letter. For example, $\vec{v}$ represents velocity and $\vec{a}$ denotes an acceleration, both vector quantities. In this chapter, however, it won't be necessary to use that notation, because in one-dimensional motion an object can only move in one of two directions, and these directions are easily specified by plus and minus signs.

TIP 2.1 A Displacement Isn't a Distance!
The displacement of an object is *not* the same as the distance it travels. Toss a tennis ball up and catch it. The ball travels a *distance* equal to twice the maximum height reached, but its *displacement* is zero.

TIP 2.2 Vectors Have Both a Magnitude and a Direction.
Scalars have size. Vectors, too, have size, and they also point in a direction.

2.2 VELOCITY

In day-to-day usage, the terms *speed* and *velocity* are interchangeable. In physics, however, there's a clear distinction between them: Speed is a scalar quantity, having only magnitude, while velocity is a vector, having both magnitude and direction.

Why must velocity be a vector? If you want to get to a town 70 km away in an hour's time, it's not enough to drive at a speed of 70 km/h; you must travel in the correct direction as well. This is obvious, but shows that velocity gives considerably more information than speed, as will be made more precise in the formal definitions.

The **average speed** of an object over a given time interval is defined as the total distance traveled divided by the total time elapsed:

$$\text{Average speed} \equiv \frac{\text{total distance}}{\text{total time}}$$

SI unit: meter per second (m/s)

◀ Definition of average speed

In symbols, this equation might be written $v = d/t$, with the letter v understood in context to be the average speed, and not a velocity. Because total distance and total time are always positive, the average speed will be positive, also. The definition of average speed completely ignores what may happen between the beginning and the end of the motion. For example, you might drive from Atlanta, Georgia, to St. Petersburg, Florida, a distance of about 500 miles, in 10 hours. Your average speed is 500 mi/10 h = 50 mi/h. It doesn't matter if you spent two hours in a traffic jam traveling only 5 mi/h and another hour at a rest stop. For average speed, only the total distance traveled and total elapsed time are important.

EXAMPLE 2.1   The Tortoise and The Hare

Goal Apply the concept of average speed.

Problem A turtle and a rabbit engage in a footrace over a distance of 4.00 km. The rabbit runs 0.500 km and then stops for a 90.0-min nap. Upon awakening, he remembers the race and runs twice as fast. Finishing the course in a total time of 1.75 h, the rabbit wins the race. (a) Calculate the average speed of the rabbit. (b) What was his average speed before he stopped for a nap?

Strategy Finding the overall average speed in part (a) is just a matter of dividing the total distance by the total time. Part (b) requires two equations and two unknowns, the latter turning out to be the two different average speeds: v_1 before the nap and v_2 after the nap. One equation is given in the statement of the problem ($v_2 = 2v_1$), while the other comes from the fact the rabbit ran for only fifteen minutes because he napped for ninety minutes.

Solution
(a) Find the rabbit's overall average speed.

Apply the equation for average speed:
$$\text{Average speed} \equiv \frac{\text{total distance}}{\text{total time}} = \frac{4.00 \text{ km}}{1.75 \text{ h}}$$
$$= 2.29 \text{ km/h}$$

(b) Find the rabbit's average speed before his nap.

Sum the running times, and set the sum equal to 0.25 h: $t_1 + t_2 = 0.250 \text{ h}$

Substitute $t_1 = d_1/v_1$ and $t_2 = d_2/v_2$:
$$\frac{d_1}{v_1} + \frac{d_2}{v_2} = 0.250 \text{ h} \tag{1}$$

Equation (1) and $v_2 = 2v_1$ are the two equations needed, and d_1 and d_2 are known. Solve for v_1 by substitution:
$$\frac{d_1}{v_1} + \frac{d_2}{v_2} = \frac{0.500 \text{ km}}{v_1} + \frac{3.50 \text{ km}}{2v_1} = 0.250 \text{ h}$$
$$v_1 = 9.00 \text{ km/h}$$

Remark As seen in this example, average speed can be calculated regardless of any variation in speed over the given time interval.

Exercise 2.1
Estimate the average speed of the Apollo spacecraft in m/s, given that the craft took five days to reach the Moon from Earth. (The Moon is 3.8×10^8 m from Earth.)

Answer ~900 m/s

Unlike average speed, **average velocity** is a vector quantity, having both a magnitude and a direction. Consider again the car of Figure 2.2, moving along the road (the x-axis). Let the car's position be x_i at some time t_i and x_f at a later time t_f. In the time interval $\Delta t = t_f - t_i$, the displacement of the car is $\Delta x = x_f - x_i$.

Definition of average velocity ▶

The average velocity $\bar{v}$ during a time interval Δt is the displacement Δx divided by Δt:
$$\bar{v} = \frac{\Delta x}{\Delta t} = \frac{x_f - x_i}{t_f - t_i} \tag{2.2}$$

SI Unit: meter per second (m/s)

Unlike the average speed, which is always positive, the average velocity of an object in one dimension can be either positive or negative, depending on the sign of the displacement. (The time interval Δt is always positive.) For example, in Figure 2.2a, the average velocity of the car is positive in the upper illustration, a positive sign indicating motion to the right along the x-axis. Similarly, a negative average velocity for the car in the lower illustration of the figure indicates that it moves to the left along the x-axis.

As an example, we can use the data in Table 2.1 to find the average velocity in the time interval from point Ⓐ to point Ⓑ (assume two digits are significant):

$$\overline{v} = \frac{\Delta x}{\Delta t} = \frac{52 \text{ m} - 30 \text{ m}}{10 \text{ s} - 0 \text{ s}} = 2.2 \text{ m/s}$$

Aside from meters per second, other common units for average velocity are feet per second (ft/s) in the U.S. customary system and centimeters per second (cm/s) in the cgs system.

To further illustrate the distinction between speed and velocity, suppose we're watching a drag race from the Goodyear blimp. In one run we see a car follow the straight-line path from Ⓟ to Ⓠ shown in Figure 2.3 during the time interval Δt, and in a second run a car follows the curved path during the same interval. From the definition in Equation 2.2, the two cars had the same average velocity, because they had the same displacement $\Delta x = x_f - x_i$ during the same time interval Δt. The car taking the curved route, however, traveled a greater distance and had the higher average speed.

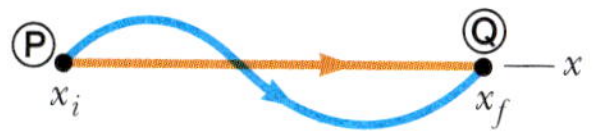

Figure 2.3   A drag race viewed from a blimp. One car follows the red straight-line path from Ⓟ to Ⓠ, and a second car follows the blue curved path.

Quick Quiz 2.1

Figure 2.4 shows the unusual path of a confused football player. After receiving a kickoff at his own goal, he runs downfield to within inches of a touchdown, then reverses direction and races back until he's tackled at the exact location where he first caught the ball. During this run, what is (a) the total distance he travels, (b) his displacement, and (c) his average velocity in the x-direction?

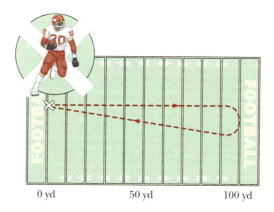

0 yd   50 yd   100 yd

Figure 2.4   (Quick Quiz 2.1) The path followed by a confused football player.

Graphical Interpretation of Velocity

If a car moves along the x-axis from Ⓐ to Ⓑ to Ⓒ, and so forth, we can plot the positions of these points as a function of the time elapsed since the start of the motion. The result is a **position vs. time graph** like those of Figure 2.5. In Figure 2.5a, the graph is a straight line, because the car is moving at constant velocity. The same displacement Δx occurs in each time interval Δt. In this case, the average velocity is always the same and is equal to $\Delta x / \Delta t$. Figure 2.5b is a graph of the data in Table 2.1. Here, the position vs. time graph is not a straight line, because the velocity of the car is changing. Between any two points, however, we can draw a straight line just as in Figure 2.5a, and the slope of that line is the average velocity $\Delta x / \Delta t$ in that time interval. In general, **the average velocity of an object during the time interval Δt is equal to the slope of the straight line joining the initial and final points on a graph of the object's position versus time**.

From the data in Table 2.1 and the graph in Figure 2.5b, we see that the car first moves in the positive x-direction as it travels from Ⓐ to Ⓑ, reaches a position of 52 m at time $t = 10$ s, then reverses direction and heads backwards. In the first 10 s of its motion, as the car travels from Ⓐ to Ⓑ, its average velocity is 2.2 m/s, as previously calculated. In the first 40 seconds, as the car goes from Ⓐ to Ⓔ, its displacement is $\Delta x = -37 \text{ m} - (30 \text{ m}) = -67 \text{ m}$. So the average velocity in this interval, which equals the slope of the blue line in Figure 2.5b from Ⓐ to Ⓔ, is $\overline{v} = \Delta x / \Delta t = (-67 \text{ m})/(40 \text{ s}) = -1.7 \text{ m/s}$. In general, there will be a different average velocity between any distinct pair of points.

Instantaneous Velocity

Average velocity doesn't take into account the details of what happens *during* an interval of time. On a car trip, for example, you may speed up or slow down a

TIP 2.3   Slopes of Graphs

The word *slope* is often used in reference to the graphs of physical data. Regardless of the type of data, the *slope* is given by

$$\text{Slope} = \frac{\text{change in vertical axis}}{\text{change in horizontal axis}}$$

Slope carries units.

TIP 2.4   Average Velocity Versus Average Speed

Average velocity is *not* the same as average speed. If you run from $x = 0$ m to $x = 25$ m and back to your starting point in a time interval of 5 s, the average velocity is zero, while the average speed is 10 m/s.

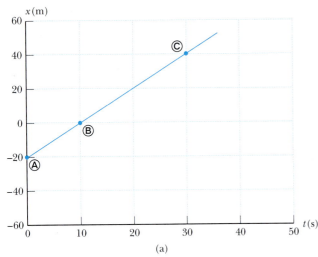

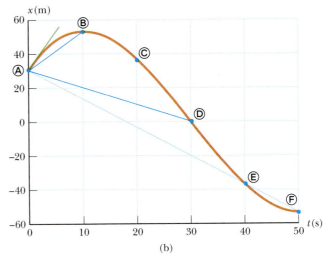

Figure 2.5   (a) Position vs. time graph for the motion of a car moving along the x-axis at constant velocity. (b) Position vs. time graph for the motion of a car with changing velocity, using the data in Table 2.1. The average velocity in the time interval Δt is the slope of the blue straight line connecting Ⓐ and Ⓓ.

number of times in response to the traffic and the condition of the road, and on rare occasions even pull over to chat with a police officer about your speed. What is most important to the police (and to your own safety) is the speed of your car and the direction it was going at a particular instant in time, which together determine the car's **instantaneous velocity**.

So in driving a car between two points, the average velocity must be computed over an interval of time, but the magnitude of instantaneous velocity can be read on the car's speedometer.

Definition of instantaneous velocity ▶

The instantaneous velocity v is the limit of the average velocity as the time interval Δt becomes infinitesimally small:

$$v \equiv \lim_{\Delta t \to 0} \frac{\Delta x}{\Delta t} \qquad [2.3]$$

SI unit: meter per second (m/s)

The notation $\lim_{\Delta t \to 0}$ means that the ratio $\Delta x / \Delta t$ is repeatedly evaluated for smaller and smaller time intervals Δt. As Δt gets extremely close to zero, the ratio $\Delta x / \Delta t$ gets closer and closer to a fixed number, which is defined as the instantaneous velocity.

To better understand the formal definition, consider data obtained on our vehicle via radar (Table 2.2). At $t = 1.00$ s, the car is at $x = 5.00$ m, and at $t = 3.00$ s, it's at $x = 52.5$ m. The average velocity computed for this interval is $\Delta x / \Delta t = (52.5\ \text{m} - 5.00\ \text{m})/(3.00\ \text{s} - 1.00\ \text{s}) = 23.8$ m/s. This result could be used as an estimate for the velocity at $t = 1.00$ s, but it wouldn't be very accurate, because the

TABLE 2.2

Positions of a Car at Specific Instants of Time

t (s)	x (m)
1.00	5.00
1.01	5.47
1.10	9.67
1.20	14.3
1.50	26.3
2.00	34.7
3.00	52.5

TABLE 2.3

Calculated Values of the Time Intervals, Displacements, and Average Velocities for the Car of Table 2.2

Time Interval (s)	Δt (s)	Δx (m)	$\bar{v}$ (m/s)
1.00 to 3.00	2.00	47.5	23.8
1.00 to 2.00	1.00	29.7	29.7
1.00 to 1.50	0.50	21.3	42.6
1.00 to 1.20	0.20	· 9.30	46.5
1.00 to 1.10	0.10	4.67	46.7
1.00 to 1.01	0.01	0.470	47.0

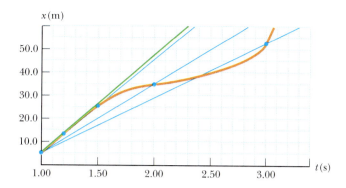

Figure 2.6   Graph representing the motion of the car from the data in Table 2.2. The slope of the blue line represents the average velocity for smaller and smaller time intervals and approaches the slope of the green tangent line.

speed changes considerably in the two-second time interval. Using the rest of the data, we can construct Table 2.3. As the time interval gets smaller, the average velocity more closely approaches the instantaneous velocity. Using the final interval of only 0.010 0 s, we find that the average velocity is $\bar{v} = \Delta x/\Delta t = 0.470 \text{ m}/0.010\ 0 \text{ s} = 47.0 \text{ m/s}$. Since 0.010 0 s is a very short time interval, the actual instantaneous velocity is likely to be very close to this latter average velocity. Finally using the conversion factor on the inside front cover of the book, we see that this is 105 mi/h, a likely violation of the speed limit.

As can be seen in Figure 2.6, the chord formed by the line gradually approaches a tangent line as the time interval becomes smaller. **The slope of the line tangent to the position vs. time curve at A is defined to be the instantaneous velocity at that time**.

The instantaneous speed of an object, which is a scalar quantity, is defined as the magnitude of the instantaneous velocity. Like average speed, instantaneous speed (which we will usually call, simply, "speed") has no direction associated with it and hence carries no algebraic sign. For example, if one object has an instantaneous velocity of $+15$ m/s along a given line and another object has an instantaneous velocity of -15 m/s along the same line, both have an instantaneous speed of 15 m/s.

EXAMPLE 2.2   Slowly Moving Train

Goal   Obtain average and instantaneous velocities from a graph.

Problem   A train moves slowly along a straight portion of track according to the graph of position versus time in Figure 2.7a. Find **(a)** the average velocity for the total trip, **(b)** the average velocity during the first 4.00 s of motion, **(c)** the average velocity during the next 4.00 s of motion, **(d)** the instantaneous velocity at $t = 2.00$ s, and **(e)** the instantaneous velocity at $t = 9.00$ s.

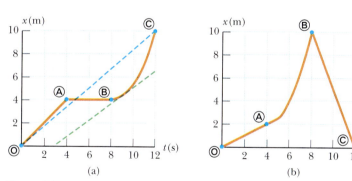

Figure 2.7   (a) (Example 2.2) (b) (Exercise 2.2).

Strategy   The average velocities can be obtained by substituting the data into the definition. The instantaneous velocity at $t = 2.00$ s is the same as the average velocity at that point, because the position vs. time graph is a straight line, indicating constant velocity. Finding the instantaneous velocity when $t = 9.00$ s requires sketching a line tangent to the curve at that point and finding its slope.

Solution

(a) Find the average velocity from ⓞ to ©.

Calculate the slope of the dashed blue line:

$$\bar{v} = \frac{\Delta x}{\Delta t} = \frac{10.0 \text{ m}}{12.0 \text{ s}} = +0.833 \text{ m/s}$$

(b) Find the average velocity during the first 4 seconds of the train's motion.

Again, find the slope:

$$\bar{v} = \frac{\Delta x}{\Delta t} = \frac{4.00 \text{ m}}{4.00 \text{ s}} = +1.00 \text{ m/s}$$

(c) Find the average velocity during the next four seconds.

Here, there is no change in position, so the displacement Δx is zero:

$$\bar{v} = \frac{\Delta x}{\Delta t} = \frac{0 \text{ m}}{4.00 \text{ s}} = 0 \text{ m/s}$$

(d) Find the instantaneous velocity at $t = 2.00$ s.

This is the same as the average velocity found in **(b)**, because the graph is a straight line:

$$v = 1.00 \text{ m/s}$$

(e) Find the instantaneous velocity at $t = 9.00$ s.

The tangent line appears to intercept the x-axis at $(3.0 \text{ s}, 0 \text{ m})$ and graze the curve at $(9.0 \text{ s}, 4.5 \text{ m})$. The instantaneous velocity at $t = 9.00$ s equals the slope of the tangent line through these points.

$$v = \frac{\Delta x}{\Delta t} = \frac{4.5 \text{ m} - 0 \text{ m}}{9.0 \text{ s} - 3.0 \text{ s}} = 0.75 \text{ m/s}$$

Remarks   From the origin to Ⓐ, the train moves at constant speed in the positive x-direction for the first 4.00 s, because the position vs. time curve is rising steadily toward positive values. From Ⓐ to Ⓑ, the train stops at $x = 4.00$ m for 4.00 s. From Ⓑ to Ⓒ, the train travels at increasing speed in the positive x-direction.

Exercise 2.2
Figure 2.7b graphs another run of the train. Find **(a)** the average velocity from Ⓞ to Ⓒ; **(b)** the average and instantaneous velocities from Ⓞ to Ⓐ; **(c)** the approximate instantaneous velocity at $t = 6.0$ s; and **(d)** the average and instantaneous velocity at $t = 9.0$ s.

Answers   (a) 0 m/s   (b) both are +0.5 m/s   (c) 2 m/s   (d) both are −2.5 m/s

---

2.3   ACCELERATION

Going from place to place in your car, you rarely travel long distances at constant velocity. The velocity of the car increases when you step harder on the gas pedal and decreases when you apply the brakes. The velocity also changes when you round a curve, altering your direction of motion. The changing of an object's velocity with time is called **acceleration**.

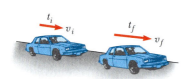

Figure 2.8   A car moving to the right accelerates from a velocity of v_i to a velocity of v_f in the time interval $\Delta t = t_f - t_i$.

Average Acceleration

A car moves along a straight highway as in Figure 2.8. At time t_i it has a velocity of v_i, and at time t_f its velocity is v_f, with $\Delta v = v_f - v_i$ and $\Delta t = t_f - t_i$.

Definition of average acceleration ▶

The average acceleration $\bar{a}$ during the time interval Δt is the change in velocity Δv divided by Δt:

$$\bar{a} \equiv \frac{\Delta v}{\Delta t} = \frac{v_f - v_i}{t_f - t_i}$$ [2.4]

SI unit: meter per second per second (m/s²)

For example, suppose the car shown in Figure 2.8 accelerates from an initial velocity of $v_i = +10$ m/s to a final velocity of $v_f = +20$ m/s in a time interval of 2 s.

(Both velocities are toward the right, selected as the positive direction.) These values can be inserted into Equation 2.4 to find the average acceleration:

$$\bar{a} = \frac{\Delta v}{\Delta t} = \frac{20\ \text{m/s} - 10\ \text{m/s}}{2\ \text{s}} = +5\ \text{m/s}^2$$

Acceleration is a vector quantity having dimensions of length divided by the time squared. Common units of acceleration are meters per second per second ((m/s)/s, which is usually written m/s^2) and feet per second per second (ft/s^2). An average acceleration of $+5\ \text{m/s}^2$ means that, on average, the car increases its velocity by 5 m/s every second in the positive x-direction.

For the case of motion in a straight line, the direction of the velocity of an object and the direction of its acceleration are related as follows: **When the object's velocity and acceleration are in the same direction, the speed of the object increases with time. When the object's velocity and acceleration are in opposite directions, the speed of the object decreases with time.**

To clarify this point, suppose the velocity of a car changes from -10 m/s to -20 m/s in a time interval of 2 s. The minus signs indicate that the velocities of the car are in the negative x-direction; they do *not* mean that the car is slowing down! The average acceleration of the car in this time interval is

$$\bar{a} = \frac{\Delta v}{\Delta t} = \frac{-20\ \text{m/s} - (-10\ \text{m/s})}{2\ \text{s}} = -5\ \text{m/s}^2$$

The minus sign indicates that the acceleration vector is also in the negative x-direction. Because the velocity and acceleration vectors are in the same direction, the speed of the car must increase as the car moves to the left. Positive and negative accelerations specify directions relative to chosen axes, not "speeding up" or "slowing down." The terms "speeding up" or "slowing down" refer to an increase and a decrease in speed, respectively.

TIP 2.5   Negative Acceleration
Negative acceleration doesn't necessarily mean an object is slowing down. If the acceleration is negative and the velocity is also negative, the object is speeding up!

TIP 2.6   Deceleration
The word *deceleration* means a reduction in speed, a slowing down. Some confuse it with a negative acceleration, which can speed something up. (See Tip 2.5.)

Quick Quiz 2.2

True or False? Define east as the negative direction and west as the positive direction. **(a)** If a car is traveling east, its acceleration must be eastward. **(b)** If a car is slowing down, its acceleration may be positive. **(c)** An object with constant nonzero acceleration can never stop and stay stopped.

Instantaneous Acceleration

The value of the average acceleration often differs in different time intervals, so it's useful to define the **instantaneous acceleration**, which is analogous to the instantaneous velocity discussed in Section 2.2.

The instantaneous acceleration a is the limit of the average acceleration as the time interval Δt goes to zero:

$$a \equiv \lim_{\Delta t \to 0} \frac{\Delta v}{\Delta t} \qquad [2.5]$$

SI unit: meter per second per second (m/s^2)

Here again, the notation $\lim_{\Delta t \to 0}$ means that the ratio $\Delta v/\Delta t$ is evaluated for smaller and smaller values of Δt. The closer Δt gets to zero, the closer the ratio gets to a fixed number, which is the instantaneous acceleration.

Figure 2.9, a **velocity vs. time graph**, plots the velocity of an object against time. The graph could represent, for example, the motion of a car along a busy street. The average acceleration of the car between times t_i and t_f can be found by determining the slope of the line joining points Ⓟ and Ⓠ. If we imagine that point Ⓠ is brought closer and closer to point Ⓟ, the line comes closer and closer to becoming tangent at Ⓟ. **The instantaneous acceleration of an object at a given time**

◀ Definition of instantaneous acceleration

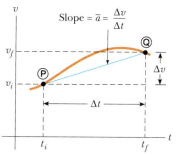

Figure 2.9   Velocity vs. time graph for an object moving in a straight line. The slope of the blue line connecting points Ⓟ and Ⓠ is defined as the average acceleration in the time interval $\Delta t = t_f - t_i$.

Figure 2.10   (Quick Quiz 2.3) Match each velocity vs. time graph to its corresponding acceleration vs. time graph.

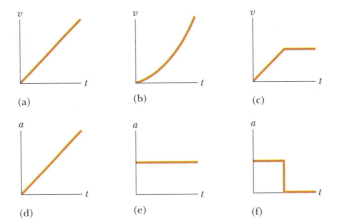

equals the slope of the tangent to the velocity vs. time graph at that time. From now on, we will use the term *acceleration* to mean "instantaneous acceleration."

In the special case where the velocity vs. time graph of an object's motion is a straight line, the instantaneous acceleration of the object at any point is equal to its average acceleration. This also means that the tangent line to the graph overlaps the graph itself. In that case, the object's acceleration is said to be *uniform*, which means that it has a constant value. Constant acceleration problems are important in kinematics and will be studied extensively in this and the next chapter.

Quick Quiz 2.3

Parts (a), (b), and (c) of Figure 2.10 represent three graphs of the velocities of different objects moving in straight-line paths as functions of time. The possible accelerations of each object as functions of time are shown in parts (d), (e), and (f). Match each velocity vs. time graph with the acceleration vs. time graph that best describes the motion.

EXAMPLE 2.3   Catching a Fly Ball

Goal   Apply the definition of instantaneous acceleration.

Problem   A baseball player moves in a straight-line path in order to catch a fly ball hit to the outfield. His velocity as a function of time is shown in Figure 2.11a. Find his instantaneous acceleration at points Ⓐ, Ⓑ, and Ⓒ.

Strategy   At each point, the velocity vs. time graph is a straight line segment, so the instantaneous acceleration will be the slope of that segment. Select two points on each segment and use them to calculate the slope.

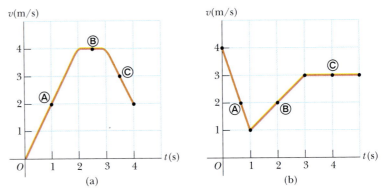

Figure 2.11   (a) (Example 2.3) (b) (Exercise 2.3)

Solution

Acceleration at Ⓐ.

The acceleration equals the slope of the line connecting the points (0 s, 0 m/s) and (2.0 s, 4.0 m/s):

$$a = \frac{\Delta v}{\Delta t} = \frac{4.0 \text{ m/s} - 0}{2.0 \text{ s} - 0} = +2.0 \text{ m/s}^2$$

Acceleration at Ⓑ.

$\Delta v = 0$, because the segment is horizontal:

$$a = \frac{\Delta v}{\Delta t} = \frac{4.0 \text{ m/s} - 4.0 \text{ m/s}}{3.0 \text{ s} - 2.0 \text{ s}} = 0 \text{ m/s}^2$$

Acceleration at Ⓒ.

The acceleration equals the slope of the line connecting the points (3.0 s, 4.0 m/s) and (4.0 s, 2.0 m/s):

$$a = \frac{\Delta v}{\Delta t} = \frac{2.0 \text{ m/s} - 4.0 \text{ m/s}}{4.0 \text{ s} - 3.0 \text{ s}} = -2.0 \text{ m/s}^2$$

Remarks   For the first 2.0 s, the ballplayer moves in the positive x-direction (the velocity is positive) and steadily accelerates (the curve is steadily rising) to a maximum speed of 4.0 m/s. He moves for 1.0 s at a steady speed of 4.0 m/s and then slows down in the last second (the v vs. t curve is falling), still moving in the positive x-direction (v is always positive).

Exercise 2.3
Repeat the problem, using Figure 2.11b.

Answer   The accelerations at Ⓐ, Ⓑ, and Ⓒ are -3.0 m/s², 1.0 m/s², and 0 m/s², respectively.

2.4   MOTION DIAGRAMS

Velocity and acceleration are sometimes confused with each other, but they're very different concepts, as can be illustrated with the help of motion diagrams. A **motion diagram** is a representation of a moving object at successive time intervals, with velocity and acceleration vectors sketched at each position, red for velocity vectors and violet for acceleration vectors, as in Active Figure 2.12. The time intervals between adjacent positions in the motion diagram are assumed equal.

A motion diagram is analogous to images resulting from a stroboscopic photograph of a moving object. Each image is made as the strobe light flashes. Active Figure 2.12 represents three sets of strobe photographs of cars moving along a straight roadway from left to right. The time intervals between flashes of the stroboscope are equal in each diagram.

In Active Figure 2.12a, the images of the car are equally spaced: The car moves the same distance in each time interval. This means that the car moves with *constant positive velocity* and has *zero acceleration*. The red arrows are all the same length (constant velocity) and there are no violet arrows (zero acceleration).

In Active Figure 2.12b, the images of the car become farther apart as time progresses and the velocity vector increases with time, because the car's displacement between adjacent positions increases as time progresses. The car is moving with a *positive velocity* and a constant *positive acceleration*. The red arrows are successively longer in each image, and the violet arrows point to the right.

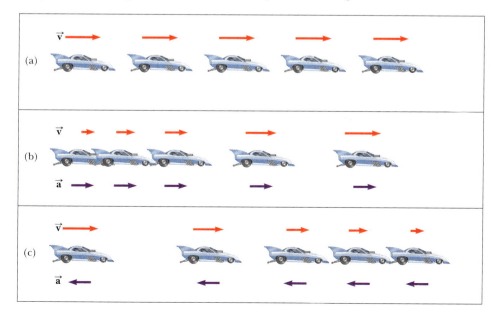

ACTIVE FIGURE 2.12
(a) Motion diagram for a car moving at constant velocity (zero acceleration). (b) Motion diagram for a car undergoing constant acceleration in the direction of its velocity. The velocity vector at each instant is indicated by a red arrow, and the constant acceleration vector by a violet arrow. (c) Motion diagram for a car undergoing constant acceleration in the direction *opposite* the velocity at each instant.

Physics⊗Now™
Log into PhysicsNow at **www.cp7e.com**, and go to Active Figure 2.12, where you can select the constant acceleration and initial velocity of the car and observe pictorial and graphical representations of its motion.

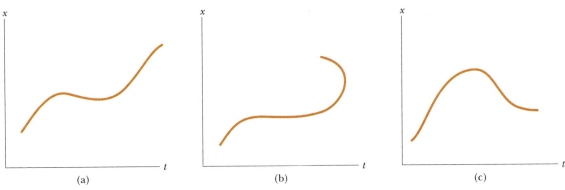

(a) (b) (c)

ACTIVE FIGURE 2.13
(Quick Quiz 2.4) Which position vs. time curve is impossible?

Physics⊗Now™
Log into PhysicsNow at **www.cp7e.com**, and go to Active Figure 2.13, where you can practice matching appropriate velocity vs. time graphs and acceleration vs. time graphs.

In Active Figure 2.12c, the car slows as it moves to the right because its displacement between adjacent positions decreases with time. In this case, the car moves initially to the right with a constant negative acceleration. The velocity vector decreases in time (the red arrows get shorter) and eventually reaches zero, as would happen when the brakes are applied. Note that the acceleration and velocity vectors are *not* in the same direction. The car is moving with a *positive velocity*, but with a *negative acceleration*.

Try constructing your own diagrams for various problems involving kinematics.

Quick Quiz 2.4

The three graphs in Active Figure 2.13 represent the position vs. time for objects moving along the *x*-axis. Which, if any, of these graphs is not physically possible?

Quick Quiz 2.5

Figure 2.14a is a diagram of a multiflash image of an air puck moving to the right on a horizontal surface. The images sketched are separated by equal time intervals, and the first and last images show the puck at rest. (a) In Figure 2.14b, which color graph best shows the puck's position as a function of time? (b) In Figure 2.14c, which color graph best shows the puck's velocity as a function of time? (c) In Figure 2.14d, which color graph best shows the puck's acceleration as a function of time?

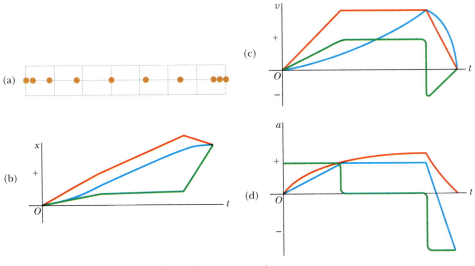

Figure 2.14 (Quick Quiz 2.5) Choose the correct graphs.

2.5 ONE-DIMENSIONAL MOTION WITH CONSTANT ACCELERATION

Many applications of mechanics involve objects moving with *constant acceleration.* This type of motion is important because it applies to numerous objects in nature, such as an object in free fall near Earth's surface (assuming that air resistance can be neglected). A graph of acceleration versus time for motion with constant acceleration is shown in Active Figure 2.15a. **When an object moves with constant acceleration, the instantaneous acceleration at any point in a time interval is equal to the value of the average acceleration over the entire time interval**. Consequently, the velocity increases or decreases at the same rate throughout the motion, and a plot of v versus t gives a straight line with either positive, zero, or negative slope.

Because the average acceleration equals the instantaneous acceleration when a is constant, we can eliminate the bar used to denote average values from our defining equation for acceleration, writing $\bar{a} = a$, so that Equation 2.4 becomes

$$a = \frac{v_f - v_i}{t_f - t_i}$$

The observer timing the motion is always at liberty to choose the initial time, so for convenience, let $t_i = 0$ and t_f be any arbitrary time t. Also, let $v_i = v_0$ (the initial velocity at $t = 0$) and $v_f = v$ (the velocity at any arbitrary time t). With this notation, we can express the acceleration as

$$a = \frac{v - v_0}{t}$$

or

$$v = v_0 + at \qquad \text{(for constant } a\text{)} \qquad \textbf{[2.6]}$$

Equation 2.6 states that the acceleration a steadily changes the initial velocity v_0 by an amount at. For example, if a car starts with a velocity of $+2.0$ m/s to the right and accelerates to the right with $a = +6.0$ m/s^2, it will have a velocity of $+14$ m/s after 2.0 s have elapsed:

$$v = v_0 + at = +2.0 \text{ m/s} + (6.0 \text{ m/s}^2)(2.0 \text{ s}) = +14 \text{ m/s}$$

The graphical interpretation of v is shown in Active Figure 2.15b. The velocity varies linearly with time according to Equation 2.6, as it should for constant acceleration.

Because the velocity is increasing or decreasing *uniformly* with time, we can express the average velocity in any time interval as the arithmetic average of the initial velocity v_0 and the final velocity v:

$$\bar{v} = \frac{v_0 + v}{2} \qquad \text{(for constant } a\text{)} \qquad \textbf{[2.7]}$$

Remember that this expression is valid only when the acceleration is constant, in which case the velocity increases uniformly.

We can now use this result along with the defining equation for average velocity, Equation 2.2, to obtain an expression for the displacement of an object as a function of time. Again, we choose $t_i = 0$ and $t_f = t$, and for convenience, we write $\Delta x = x_f - x_i = x - x_0$. This results in

$$\Delta x = \bar{v}t = \left(\frac{v_0 + v}{2}\right)t$$

$$\Delta x = \tfrac{1}{2}(v_0 + v)t \qquad \text{(for constant } a\text{)} \qquad \textbf{[2.8]}$$

We can obtain another useful expression for displacement by substituting the equation for v (Eq. 2.6) into Equation 2.8:

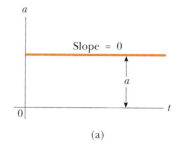

(a)

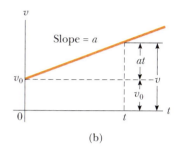

(b)

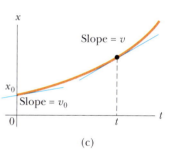

(c)

ACTIVE FIGURE 2.15
A particle moving along the x-axis with constant acceleration a.
(a) the acceleration vs. time graph,
(b) the velocity vs. time graph, and
(c) the position vs. time graph.

Physics Now™
Log into PhysicsNow at **www.cp7e.com**, and go to Active Figure 2.15, where you can adjust the constant acceleration and observe the effect on the position and velocity graphs.

TABLE 2.4

Equations for Motion in a Straight Line Under Constant Acceleration

Equation	Information Given by Equation
$v = v_0 + at$	Velocity as a function of time
$\Delta x = v_0 t + \frac{1}{2}at^2$	Displacement as a function of time
$v^2 = v_0^2 + 2a\Delta x$	Velocity as a function of displacement

Note: Motion is along the x-axis. At $t = 0$, the velocity of the particle is v_0.

$$\Delta x = \tfrac{1}{2}(v_0 + v_0 + at)t$$

$$\Delta x = v_0 t + \tfrac{1}{2}at^2 \qquad \text{(for constant } a) \qquad\qquad [2.9]$$

This equation can also be written in terms of the position x, since $\Delta x = x - x_0$. Active Figure 2.15c shows a plot of x versus t for Equation 2.9, which is related to the graph of velocity vs. time: The area under the curve in Active Figure 2.15b is equal to $v_0 t + \frac{1}{2}at^2$, which is equal to the displacement Δx. In fact, **the area under the graph of v versus t for any object is equal to the displacement Δx of the object.**

Finally, we can obtain an expression that doesn't contain time by solving Equation 2.6 for t and substituting into Equation 2.8, resulting in

$$\Delta x = \tfrac{1}{2}(v + v_0)\left(\frac{v - v_0}{a}\right) = \frac{v^2 - v_0^2}{2a}$$

$$v^2 = v_0^2 + 2a\Delta x \qquad \text{(for constant } a) \qquad\qquad [2.10]$$

Equations 2.6 and 2.9 together can solve any problem in one-dimensional motion with constant acceleration, but Equations 2.7, 2.8, and, especially, 2.10 are sometimes convenient. The three most useful equations—Equations 2.6, 2.9, and 2.10—are listed in Table 2.4.

The best way to gain confidence in the use of these equations is to work a number of problems. There is usually more than one way to solve a given problem, depending on which equations are selected and what quantities are given. The difference lies mainly in the algebra.

Problem-Solving Strategy   Accelerated Motion

The following procedure is recommended for solving problems involving accelerated motion.

1. **Read** the problem.
2. **Draw** a diagram, choosing a coordinate system, labeling initial and final points, and indicating directions of velocities and accelerations with arrows.
3. **Label** all quantities, circling the unknowns. Convert units as needed.
4. **Equations** from Table 2.4 should be selected next. All kinematics problems in this chapter can be solved with the first two equations, and the third is often convenient.
5. **Solve** for the unknowns. Doing so often involves solving two equations for two unknowns. It's usually more convenient to substitute all known values before solving.
6. **Check** your answer, using common sense and estimates.

TIP 2.7   Pigs Don't Fly

After solving a problem, you should think about your answer and decide whether it seems reasonable. If it isn't, look for your mistake!

Most of these problems reduce to writing the kinematic equations from Table 2.4 and then substituting the correct values into the constants a, v_0, and x_0 from the given information. Doing this produces two equations—one linear and one quadratic—for two unknown quantities.

EXAMPLE 2.4   The Daytona 500

Goal   Apply the basic kinematic equations.

Problem   A race car starting from rest accelerates at a constant rate of 5.00 m/s^2. What is the velocity of the car after it has traveled 1.00×10^2 ft?

Strategy   We've read the problem, drawn the diagram in Figure 2.16, and chosen a coordinate system (steps 1 and 2). We'd like to find the velocity v after a certain known displacement Δx. The acceleration a is also known, as is the initial velocity v_0 (step 3, labeling, is complete), so the third equation in Table 2.4 looks most useful. The rest is simple substitution.

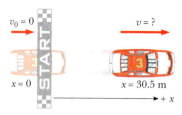

Figure 2.16   (Example 2.4)

Solution

Convert units of Δx to SI, using the information in the inside front cover.

$$1.00 \times 10^2 \text{ ft} = (1.00 \times 10^2 \text{ ft})\left(\frac{1 \text{ m}}{3.28 \text{ ft}}\right) = 30.5 \text{ m}$$

Write the kinematics equation for v^2 (step 4):

$$v^2 = v_0^2 + 2a\,\Delta x$$

Solve for v, taking the positive square root because the car moves to the right (step 5):

$$v = \sqrt{v_0^2 + 2a\,\Delta x}$$

Substitute $v_0 = 0$, $a = 5.00 \text{ m/s}^2$, and $\Delta x = 30.5$ m:

$$v = \sqrt{v_0^2 + 2a\,\Delta x} = \sqrt{(0)^2 + 2(5.00 \text{ m/s}^2)(30.5 \text{ m})}$$

$$= \boxed{17.5 \text{ m/s}}$$

Remarks   The answer is easy to check. An alternate technique is to use $\Delta x = v_0 t + \frac{1}{2}at^2$ to find t and then use the equation $v = v_0 + at$ to find v.

Exercise 2.4

Suppose the driver in this example now slams on the brakes, stopping the car in 4.00 s. Find (a) the acceleration and (b) the distance the car travels, assuming the acceleration is constant.

Answers   (a) $a = -4.38 \text{ m/s}^2$ (b) $d = 35.0$ m

INTERACTIVE EXAMPLE 2.5   Car Chase

Goal   Solve a problem involving two objects, one moving at constant acceleration and the other at constant velocity.

Problem   A car traveling at a constant speed of 24.0 m/s passes a trooper hidden behind a billboard, as in Figure 2.17. One second after the speeding car passes the billboard, the trooper sets off in chase with a constant acceleration of 3.00 m/s^2. **(a)** How long does it take the trooper to overtake the speeding car? **(b)** How fast is the trooper going at that time?

Strategy   Solving this problem involves two simultaneous kinematics equations of position, one for the police motorcycle and the other for the car. Choose $t = 0$ to correspond to the time the trooper takes up the chase, when the car is at $x_{car} = 24.0$ m because of its head start $(24.0 \text{ m/s} \times 1.00 \text{ s})$. The trooper catches up with the car when their positions are the same, which suggests setting $x_{trooper} = x_{car}$ and solving for time, which can then be used to find the trooper's speed in part (b).

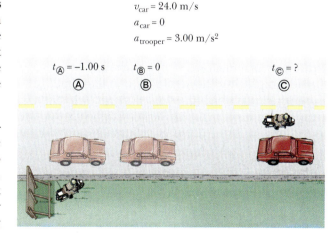

Figure 2.17   (Example 2.5) A speeding car passes a hidden trooper. When does the trooper catch up to the car?

Solution

(a) How long does it take the trooper to overtake the car?

Write the equation for the car's displacement:

$$\Delta x_{car} = x_{car} - x_0 = v_0 t + \tfrac{1}{2}a_{car}t^2$$

Take $x_0 = 24.0$ m, $v_0 = 24.0$ m/s and $a_{car} = 0$. Solve for x_{car}:

$x_{car} = x_0 + vt = 24.0$ m $+ (24.0$ m/s$)t$

Write the equation for the trooper's position, taking $x_0 = 0$, $v_0 = 0$, and $a_{trooper} = 3.00$ m/s^2:

$x_{trooper} = \frac{1}{2}a_{trooper}t^2 = \frac{1}{2}(3.00$ m/s$^2)t^2 = (1.50$ m/s$^2)t^2$

Set $x_{trooper} = x_{car}$, and solve the quadratic equation. (The quadratic formula appears in Appendix A, Equation A.8.) Only the positive root is meaningful.

$(1.50$ m/s$^2)t^2 = 24.0$ m $+ (24.0$ m/s$)t$

$(1.50$ m/s$^2)t^2 - (24.0$ m/s$)t - 24.0$ m $= 0$

$t = \boxed{16.9 \text{ s}}$

(b) Find the trooper's speed at this time.

Substitute the time into the trooper's velocity equation:

$v_{trooper} = v_0 + a_{trooper}t = 0 + (3.00$ m/s$^2)(16.9$ s$)$

$= \boxed{50.7 \text{ m/s}}$

Remarks   The trooper, traveling about twice as fast as the car, must swerve or apply his brakes strongly to avoid a collision! This problem can also be solved graphically, by plotting position versus time for each vehicle on the same graph. The intersection of the two graphs corresponds to the time and position at which the trooper overtakes the car.

Exercise 2.5
A motorist with an expired license tag is traveling at 10.0 m/s down a street, and a policeman on a motorcycle, taking another 5.00 s to finish his donut, gives chase at an acceleration of 2.00 m/s^2. Find (a) the time required to catch the car and (b) the distance the trooper travels while overtaking the motorist.

Answers   (a) 13.7 s   (b) 188 m

Physics⊗Now™ You can study the motion of the car and the trooper for various velocities of the car by logging into PhysicsNow at **www.cp7e.com** and going to Interactive Example 2.5.

EXAMPLE 2.6   The Acela: The Porsche of American Trains

Problem   The sleek high-speed electric train known as the Acela (pronounced ahh-sell-ah) is currently in service on the Washington-New York-Boston run and is shown in Figure 2.18a. The Acela consists of two power cars and six coaches and can carry 304 passengers at speeds up to 170 mi/h. In order to negotiate curves comfortably at high speeds, the train carriages tilt as much as 6° from the vertical, to prevent passengers from being pushed to the side. A velocity vs. time graph for the Acela is shown in Figure 2.18b. **(a)** Describe the motion of the Acela. **(b)** Find the peak acceleration of the Acela in miles per hour per second ((mi/h)/s) as the train speeds up from 45 mi/h to 170 mi/h. **(c)** Find the train's displacement in miles between $t = 0$ and $t = 200$ s. **(d)** Find the average acceleration of the Acela and its displacement in miles in the interval from 200 s to 300 s. (The train has regenerative braking, which means that it feeds energy back into the utility lines each time it stops!) **(e)** Find the total displacement in the interval from 0 to 400 s.

Strategy   Examine the graph in part (a), remembering that the slope of the tangent line at any point of the velocity vs. time graph gives the acceleration at that time. To find the peak acceleration in part (b), study the graph and locate the point at which the slope is steepest. In parts (c)–(e), estimating the area under the curve gives the displacement during a given period, with areas below the time axis, as in part (e), subtracted from the total. The average acceleration in part (d) can be obtained by substituting numbers taken from the graph into the definition of average acceleration, $\bar{a} = \Delta v/\Delta t$.

Solution
(a) Describe the motion.

From about -50 s to 50 s, the Acela cruises at a constant velocity in the $+x$-direction. Then the train accelerates in the $+x$-direction from 50 s to 200 s, reaching a top speed of about 170 mi/h, whereupon it brakes to rest at 350 s and reverses, steadily gaining speed in the $-x$-direction.

(b) Find the peak acceleration.

Calculate the slope of the steepest tangent line, which connects the points (50 s, 50 mi/h) and (100 s, 150 mi/h) (the light blue line in Figure 2.18c):

$a = \text{slope} = \dfrac{\Delta v}{\Delta t} = \dfrac{(1.5 \times 10^2 - 5.0 \times 10^1) \text{ mi/h}}{(1.0 \times 10^2 - 5.0 \times 10^1) \text{ s}}$

$= \boxed{2.0 \text{ (mi/h)/s}}$

(c) Find the displacement between 0 s and 200 s.

Using triangles and rectangles, approximate the area in Figure 2.18d:

$$\Delta x_{0\to200\,s} = \text{area}_1 + \text{area}_2 + \text{area}_3 + \text{area}_4 + \text{area}_5$$
$$\approx (5.0 \times 10^1 \text{ mi/h})(5.0 \times 10^1 \text{ s})$$
$$+ (5.0 \times 10^1 \text{ mi/h})(5.0 \times 10^1 \text{ s})$$
$$+ (1.6 \times 10^2 \text{ mi/h})(1.0 \times 10^2 \text{ s})$$
$$+ \tfrac{1}{2}(5.0 \times 10^1 \text{ s})(1.0 \times 10^2 \text{ mi/h})$$
$$+ \tfrac{1}{2}(1.0 \times 10^2 \text{ s})(1.7 \times 10^2 \text{ mi/h} - 1.6 \times 10^2)$$
$$= 2.4 \times 10^4 (\text{mi/h})\text{s}$$

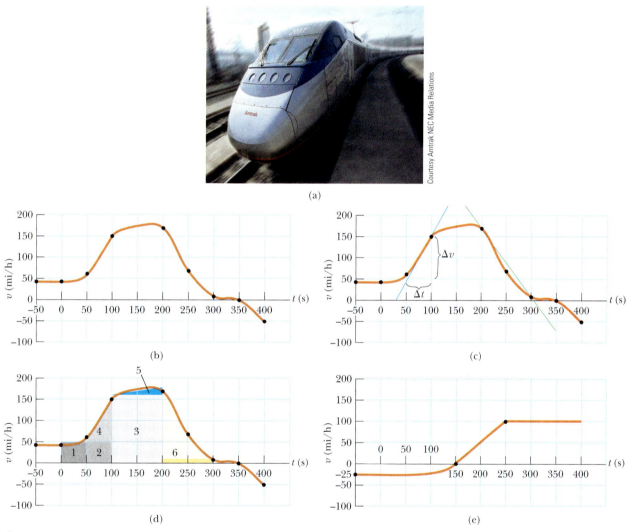

Figure 2.18 (Example 2.6) (a) The Acela, 1 250 000 lb of cold steel thundering along at 170 mi/h. (b) Velocity vs. time graph for the Acela. (c) The slope of the steepest tangent blue line gives the peak acceleration, while the slope of the green line is the average acceleration between 200 s and 300 s. (d) The area under the velocity vs. time graph in some time interval gives the displacement of the Acela in that time interval. (e) (Exercise 2.6).

Convert units to miles by converting hours to seconds:

$$\Delta x_{0\to200\,s} \approx 2.4 \times 10^4 \frac{\text{mi} \cdot \text{s}}{\text{h}} \left(\frac{1 \text{ h}}{3\ 600 \text{ s}} \right) = \boxed{6.7 \text{ mi}}$$

(d) Find the average acceleration from 200 s to 300 s, and find the displacement.

The slope of the green line is the average acceleration from 200 s to 300 s (Fig. 2.18c):

$$\bar{a} = \text{slope} = \frac{\Delta v}{\Delta t} = \frac{(1.0 \times 10^1 - 1.7 \times 10^2) \text{ mi/h}}{1.0 \times 10^2 \text{ s}}$$
$$= \boxed{-1.6 (\text{mi/h})/\text{s}}$$

The displacement from 200 s to 300 s is equal to area$_6$, which is the area of a triangle plus the area of a very narrow rectangle beneath the triangle:

$$\Delta x_{200\to300\,s} \approx \tfrac{1}{2}(1.0\times10^2\,s)(1.7\times10^2 - 1.0\times10^1)\,mi/h$$
$$+ (1.0\times10^1\,mi/h)(1.0\times10^2\,s)$$
$$= 9.0\times10^3(mi/h)(s) = \boxed{2.5\,mi}$$

(e) Find the total displacement from 0 s to 400 s.

The total displacement is the sum of all the individual displacements. We still need to calculate the displacements for the time intervals from 300 s to 350 s and from 350 s to 400 s. The latter is negative, because it's below the time axis.

$$\Delta x_{300\to350\,s} \approx \tfrac{1}{2}(5.0\times10^1\,s)(1.0\times10^1\,mi/h)$$
$$= 2.5\times10^2(mi/h)(s)$$
$$\Delta x_{350\to400\,s} \approx \tfrac{1}{2}(5.0\times10^1\,s)(-5.0\times10^1\,mi/h)$$
$$= -1.3\times10^3(mi/h)(s)$$

Find the total displacement by summing the parts:

$$\Delta x_{0\to400\,s} \approx (2.4\times10^4 + 9.0\times10^3 + 2.5\times10^2$$
$$- 1.3\times10^3)(mi/h)(s) = \boxed{8.9\,mi}$$

Remarks   There are a number of ways of finding the approximate area under a graph. Choice of technique is a personal preference.

Exercise 2.6
Suppose the velocity vs. time graph of another train is given in Figure 2.18e. Find (a) the maximum instantaneous acceleration and (b) the total displacement in the interval from 0 s to 4.00×10^2 s.

Answers   (a) 1.0(mi/h)/s   (b) 4.7 mi

EXAMPLE 2.7   Runway Length

Goal   Apply kinematics to horizontal motion with two phases.

Problem   A typical jetliner lands at a speed of 160 mi/h and decelerates at the rate of (10 mi/h)/s. If the plane travels at a constant speed of 160 mi/h for 1.0 s after landing before applying the brakes, what is the total displacement of the aircraft between touchdown on the runway and coming to rest?

Strategy   See Figure 2.19. First, convert all quantities to SI units. The problem must be solved in two parts, or phases, corresponding to the initial coast after touchdown, followed by braking. Using the kinematic equations, find the displacement during each part and add the two displacements.

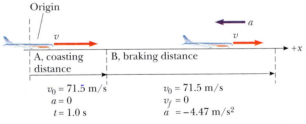

Figure 2.19   (Example 2.7) Coasting and braking distances for a landing jetliner.

Solution
Convert units to SI:

$$v_0 = (160\,mi/h)\left(\frac{0.447\,m/s}{1.00\,mi/h}\right) = 71.5\,m/s$$

$$a = (-10.0\,(mi/h)/s)\left(\frac{0.447\,m/s}{1.00\,mi/h}\right) = -4.47\,m/s^2$$

Taking $a=0$, $v_0=71.5$ m/s, and $t=1.00$ s, find the displacement while the plane is coasting:

$$\Delta x_{coasting} = v_0 t + \tfrac{1}{2}at^2 = (71.5\,m/s)(1.00\,s) + 0 = 71.5\,m$$

Use the time-independent kinematic equation to find the displacement while the plane is braking.

$$v^2 = v_0^2 + 2a\,\Delta x_{braking}$$

Take $a=-4.47$ m/s^2 and $v_0=71.5$ m/s. The negative sign on a means that the plane is slowing down.

$$\Delta x_{braking} = \frac{v^2 - v_0^2}{2a} = \frac{0-(71.5\,m/s)^2}{2.00(-4.47\,m/s^2)} = 572\,m$$

Sum the two results to find the total displacement: $\Delta x_{\text{coasting}} + \Delta x_{\text{braking}} = 72 \text{ m} + 572 \text{ m} = \boxed{644 \text{ m}}$

Remarks To find the displacement while braking, we could have used the two kinematics equations involving time, namely, $\Delta x = v_0 t + \frac{1}{2}at^2$ and $v = v_0 + at$, but because we weren't interested in time, the time-independent equation was easier to use.

Exercise 2.7
A jet lands at 80.0 m/s, applying the brakes 2.00 s after landing. Find the acceleration needed to stop the jet within 5.00×10^2 m.

Answer $a = -9.41 \text{ m/s}^2$

2.6 FREELY FALLING OBJECTS

When air resistance is negligible, all objects dropped under the influence of gravity near Earth's surface fall toward Earth with the same constant acceleration. This idea may seem obvious today, but it wasn't until about 1600 that it was accepted. Prior to that time, the teachings of the great philosopher Aristotle (384–322 B.C.) had held that heavier objects fell faster than lighter ones.

According to legend, Galileo discovered the law of falling objects by observing that two different weights dropped simultaneously from the Leaning Tower of Pisa hit the ground at approximately the same time. Although it's unlikely that this particular experiment was carried out, we know that Galileo performed many systematic experiments with objects moving on inclined planes. In his experiments, he rolled balls down a slight incline and measured the distances they covered in successive time intervals. The purpose of the incline was to reduce the acceleration and enable Galileo to make accurate measurements of the intervals. (Some people refer to this experiment as "diluting gravity.") By gradually increasing the slope of the incline, he was finally able to draw mathematical conclusions about freely falling objects, because a falling ball is equivalent to a ball going down a vertical incline. Galileo's achievements in the science of mechanics paved the way for Newton in his development of the laws of motion, which we will study in Chapter 4.

Try the following experiment: Drop a hammer and a feather simultaneously from the same height. The hammer hits the floor first, because air drag has a greater effect on the much lighter feather. On August 2, 1971, this same experiment was conducted on the Moon by astronaut David Scott, and the hammer and feather fell with exactly the same acceleration, as expected, hitting the lunar surface at the same time. In the idealized case where air resistance is negligible, such motion is called *free fall*.

The expression *freely falling object* doesn't necessarily refer to an object dropped from rest. **A freely falling object is any object moving freely under the influence of gravity alone, regardless of its initial motion**. Objects thrown upward or downward and those released from rest are all considered freely falling.

We denote the magnitude of the **free-fall acceleration** by the symbol g. The value of g decreases with increasing altitude, and varies slightly with latitude, as well. At Earth's surface, the value of g is approximately 9.80 m/s². Unless stated otherwise, we will use this value for g in doing calculations. For quick estimates, use $g \approx 10$ m/s².

If we neglect air resistance and assume that the free-fall acceleration doesn't vary with altitude over short vertical distances, then the motion of a freely falling object is the same as motion in one dimension under constant acceleration. This means that the kinematics equations developed in Section 2.6 can be applied. It's conventional to define "up" as the $+y$-direction and to use y as the position variable. In that case, the acceleration is $a = -g = -9.80$ m/s². In Chapter 7, we study how to deal with variations in g with altitude.

North Wind Archive

GALILEO GALILEI Italian
Physicist and Astronomer
(1564–1642)

Galileo formulated the laws that govern the motion of objects in free fall. He also investigated the motion of an object on an inclined plane, established the concept of relative motion, invented the thermometer, and discovered that the motion of a swinging pendulum could be used to measure time intervals. After designing and constructing his own telescope, he discovered four of Jupiter's moons, found that our own Moon's surface is rough, discovered sunspots and the phases of Venus, and showed that the Milky Way consists of an enormous number of stars. Galileo publicly defended Nicolaus Copernicus's assertion that the Sun is at the center of the Universe (the heliocentric system). He published *Dialogue Concerning Two New World Systems* to support the Copernican model, a view the Church declared to be heretical. After being taken to Rome in 1633 on a charge of heresy, he was sentenced to life imprisonment and later was confined to his villa at Arcetri, near Florence, where he died in 1642.

Quick Quiz 2.6

A tennis player on serve tosses a ball straight up. While the ball is in free fall, does its acceleration (a) increase, (b) decrease, (c) increase and then decrease, (d) decrease and then increase, or (e) remain constant?

Quick Quiz 2.7

As the tennis ball of Quick Quiz 2.6 travels through the air, its speed (a) increases, (b) decreases, (c) decreases and then increases, (d) increases and then decreases, or (e) remains the same.

Quick Quiz 2.8

A skydiver jumps out of a hovering helicopter. A few seconds later, another skydiver jumps out, so they both fall along the same vertical line relative to the helicopter. Both sky divers fall with the same acceleration. Does the vertical distance between them (a) increase, (b) decrease, or (c) stay the same? Does the difference in their velocities (d) increase, (e) decrease, or (f) stay the same? (Assume that g is constant.)

EXAMPLE 2.8   Look Out Below!

Goal   Apply the basic kinematics equations to an object falling from rest under the influence of gravity.

Problem   A golf ball is released from rest at the top of a very tall building. Neglecting air resistance, calculate the position and velocity of the ball after 1.00 s, 2.00 s, and 3.00 s.

Strategy   Make a simple sketch. Because the height of the building isn't given, it's convenient to choose coordinates so that $y = 0$ at the top of the building. Use the velocity and position kinematic equations, substituting known and given values.

Solution

Write the kinematics Equations 2.6 and 2.9:

$$v = at + v_0$$

$$\Delta y = y - y_0 = v_0 t + \tfrac{1}{2}at^2$$

Substitute $y_0 = 0$, $v_0 = 0$, and $a = -g = -9.80 \text{ m/s}^2$ into the preceding two equations:

$$v = at = (-9.80 \text{ m/s}^2)t$$

$$y = \tfrac{1}{2}(-9.80 \text{ m/s}^2)t^2 = -(4.90 \text{ m/s}^2)t^2$$

Substitute in the different times, and create a table.

t (s)	v (m/s)	y (m)
1.00	-9.8	-4.9
2.00	-19.6	-19.6
3.00	-29.4	-44.1

Remarks   The minus signs on v mean that the velocity vectors are directed downward, while the minus signs on y indicates positions below the origin. The velocity of a falling object is directly proportional to the time, and the position is proportional to the time squared, results first proven by Galileo.

Exercise 2.8

Calculate the position and velocity of the ball after 4.00 s has elapsed.

Answer -78.4 m, -39.2 m/s

INTERACTIVE EXAMPLE 2.9
Not a Bad Throw for a Rookie!

Goal Apply the kinematic equations to a freely falling object with a nonzero initial velocity.

Problem A stone is thrown from the top of a building with an initial velocity of 20.0 m/s straight upward, at an initial height of 50.0 m above the ground. The stone just misses the edge of the roof on its way down, as shown in Figure 2.20. Determine **(a)** the time needed for the stone to reach its maximum height, **(b)** the maximum height, **(c)** the time needed for the stone to return to the height from which it was thrown and the velocity of the stone at that instant; **(d)** the time needed for the stone to reach the ground, and **(e)** the velocity and position of the stone at $t = 5.00$ s.

Strategy The diagram in Figure 2.20 establishes a coordinate system with $y_0 = 0$ at the level at which the stone is released from the thrower's hand, with y positive upward. Write the velocity and position kinematic equations for the stone, and substitute the given information. All the answers come from these two equations by using simple algebra or by just substituting the time. In part (a), for example, the stone comes to rest for an instant at its maximum height, so set $v = 0$ at this point and solve for time. Then substitute the time into the displacement equation, obtaining the maximum height.

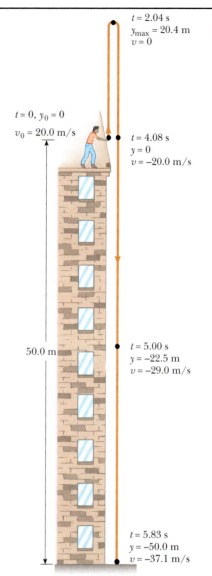

Figure 2.20 (Example 2.9) A freely falling object is thrown upward with an initial velocity of $v_0 = +20.0$ m/s. Positions and velocities are given for several times.

$t = 2.04$ s
$y_{max} = 20.4$ m
$v = 0$

$t = 0, y_0 = 0$
$v_0 = 20.0$ m/s

$t = 4.08$ s
$y = 0$
$v = -20.0$ m/s

50.0 m

$t = 5.00$ s
$y = -22.5$ m
$v = -29.0$ m/s

$t = 5.83$ s
$y = -50.0$ m
$v = -37.1$ m/s

Solution

(a) Find the time when the stone reaches its maximum height.

Write the velocity and position kinematic equations:

$$v = at + v_0$$
$$\Delta y = y - y_0 = v_0 t + \tfrac{1}{2}at^2$$

Substitute $a = -9.80$ m/s^2, $v_0 = 20.0$ m/s, and $y_0 = 0$ into the preceding two equations:

$$v = (-9.80 \text{ m/s}^2)t + 20.0 \text{ m/s} \qquad (1)$$
$$y = (20.0 \text{ m/s})t - (4.90 \text{ m/s}^2)t^2 \qquad (2)$$

Substitute $v = 0$, the velocity at maximum height, into Equation (1) and solve for time:

$$0 = (-9.80 \text{ m/s}^2)t + 20.0 \text{ m/s}$$
$$t = \frac{-20.0 \text{ m/s}}{-9.80 \text{ m/s}^2} = \boxed{2.04 \text{ s}}$$

(b) Determine the stone's maximum height.

Substitute the time $t = 2.04$ s into Equation (2):

$$y_{max} = (20.0 \text{ m/s})(2.04) - (4.90 \text{ m/s}^2)(2.04)^2 = \boxed{20.4 \text{ m}}$$

(c) Find the time the stone takes to return to its initial position, and find the velocity of the stone at that time.

Set $y = 0$ in Equation (2) and solve t:

$$0 = (20.0 \text{ m/s})t - (4.90 \text{ m/s}^2)t^2$$
$$= t(20.0 \text{ m/s} - 4.90 \text{ m/s}^2 t)$$
$$t = \boxed{4.08 \text{ s}}$$

Substitute the time into Equation (1) to get the velocity: $v = 20.0 \text{ m/s} + (-9.80 \text{ m/s}^2)(4.08 \text{ s}) = -20.0 \text{ m/s}$

(d) Find the time required for the stone to reach the ground.

In Equation 2, set $y = -50.0$ m: $-50.0 \text{ m} = (20.0 \text{ m/s})t - (4.90 \text{ m/s}^2)t^2$

Apply the quadratic formula and take the positive root: $t = 5.83 \text{ s}$

(e) Find the velocity and position of the stone at $t = 5.00$ s. Substitute values into Equations (1) and (2):

$$v = (-9.80 \text{ m/s}^2)(5.00 \text{ s}) + 20.0 \text{ m/s} = -29.0 \text{ m/s}$$

$$y = (20.0 \text{ m/s})(5.00 \text{ s}) - (4.90 \text{ m/s}^2)(5.00 \text{ s})^2 = -22.5 \text{ m}$$

Remarks   Notice how everything follows from the two kinematic equations. Once they are written down, and constants correctly identified as in Equations (1) and (2), the rest is relatively easy. If the stone were thrown downward, the initial velocity would have been negative.

Exercise 2.9
A projectile is launched straight up at 60.0 m/s from a height of 80.0 m, at the edge of a sheer cliff. The projectile falls, just missing the cliff and hitting the ground below. Find (a) the maximum height of the projectile above the point of firing, (b) the time it takes to hit the ground at the base of the cliff, and (c) its velocity at impact.

Answers   (a) 184 m   (b) 13.5 s   (c) −72.3 m/s

Physics⊗Now™   You can study the motion of the thrown ball by logging into PhysicsNow at **www.cp7e.com** and going to Interactive Example 2.9.

EXAMPLE 2.10   A Rocket Goes Ballistic

Goal   Solve a problem involving a powered ascent followed by free fall motion.

Problem   A rocket moves straight upward, starting from rest with an acceleration of $+29.4 \text{ m/s}^2$. It runs out of fuel at the end of 4.00 s and continues to coast upward, reaching a maximum height before falling back to Earth. **(a)** Find the rocket's velocity and position at the end of 4.00 s. **(b)** Find the maximum height the rocket reaches. **(c)** Find the velocity the instant before the rocket crashes on the ground.

Strategy   Take $y = 0$ at the launch point and y positive upward, as in Figure 2.21. The problem consists of two phases. In phase 1, the rocket has a net *upward* acceleration of 29.4 m/s², and we can use the kinematic equations with constant acceleration a to find the height and velocity of the rocket at the end of phase 1, when the fuel is burned up. In phase 2, the rocket is in free fall and has an acceleration of −9.80 m/s², with initial velocity and position given by the results of phase 1. Apply the kinematic equations for free fall.

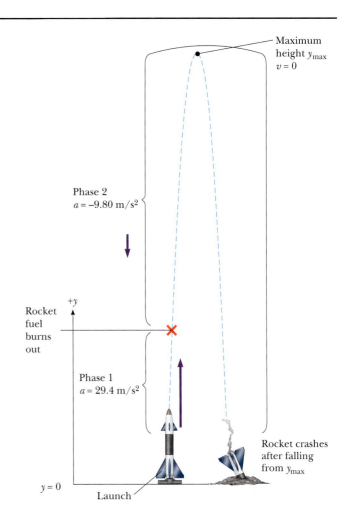

Figure 2.21   (Example 2.10) Two linked phases of motion for a rocket that is launched, uses up its fuel, and crashes.

Solution

(a) Phase 1: Find the rocket's velocity and position after 4.00 s.

Write the velocity and position kinematic equations:

$$v = v_0 + at \tag{1}$$

$$\Delta y = y - y_0 = v_0 t + \tfrac{1}{2}at^2 \tag{2}$$

Adapt these equations to phase 1, substituting $a = 29.4 \text{ m/s}^2$, $v_0 = 0$, and $y_0 = 0$:

$$v = (29.4 \text{ m/s}^2)\,t \tag{3}$$

$$y = \tfrac{1}{2}(29.4 \text{ m/s}^2)\,t^2 = (14.7 \text{ m/s}^2)\,t^2 \tag{4}$$

Substitute $t = 4.00$ s into Equations (3) and (4) to find the rocket's velocity v and position y at the time of burnout. These will be called v_b and y_b, respectively.

$$v_b = 118 \text{ m/s} \quad \text{and} \quad y_b = 235 \text{ m}$$

(b) Phase 2: Find the maximum height the rocket attains.

Adapt Equations (1) and (2) to phase 2, substituting $a = -9.8 \text{ m/s}^2$, $v_0 = v_b = 118 \text{ m/s}$, and $y_0 = y_b = 235 \text{ m}$:

$$v = (-9.8 \text{ m/s}^2)\,t + 118 \text{ m/s} \tag{5}$$

$$y = 235 \text{ m} + (118 \text{ m/s})\,t - (4.90 \text{ m/s}^2)\,t^2 \tag{6}$$

Substitute $v = 0$ (the rocket's velocity at maximum height) in Equation 5 to get the time it takes the rocket to reach its maximum height:

$$0 = (-9.8 \text{ m/s}^2)\,t + 118 \text{ m/s} \rightarrow t = \frac{118 \text{ m/s}}{9.80 \text{ m/s}^2} = 12.0 \text{ s}$$

Substitute $t = 12.0$ s into Equation (6) to find the rocket's maximum height:

$$y_{max} = 235 \text{ m} + (118 \text{ m/s})(12.0 \text{ s}) - (4.90 \text{ m/s}^2)(12.0 \text{ s})^2$$
$$= 945 \text{ m}$$

(c) Phase 2: Find the velocity of the rocket just prior to impact.

Find the time to impact by setting $y = 0$ in Equation (6) and using the quadratic formula:

$$0 = 235 \text{ m} + (118 \text{ m/s})\,t - (4.90 \text{ m/s}^2)\,t^2$$
$$t = 25.9 \text{ s}$$

Substitute this value of t into Equation (5):

$$v = (-9.80 \text{ m/s}^2)(25.9 \text{ s}) + 118 \text{ m/s} = -136 \text{ m/s}$$

Remarks   You may think that it is more natural to break this problem into three phases, with the second phase ending at the maximum height and the third phase a free fall from maximum height to the ground. Although this approach gives the correct answer, it's an unnecessary complication. Two phases are sufficient, one for each different acceleration.

Exercise 2.10
An experimental rocket designed to land upright falls freely from a height of 2.00×10^2 m, starting at rest. At a height of 80.0 m, the rocket's engines start and provide constant upward acceleration until the rocket lands. What acceleration is required if the speed on touchdown is to be zero? (Neglect air resistance.)

Answer 14.7 m/s^2

SUMMARY

Physics✕Now™ Take a practice test by logging into PhysicsNow at **www.cp7e.com** and clicking on the Pre-Test link for this chapter.

2.1 Displacement

The **displacement** of an object moving along the *x*-axis is defined as the change in position of the object,

$$\Delta x \equiv x_f - x_i \tag{2.1}$$

where x_i is the initial position of the object and x_f is its final position.

A **vector** quantity is characterized by both a magnitude and a direction. A **scalar** quantity has a magnitude only.

2.2 Velocity

The **average speed** of an object is given by

$$\text{Average speed} \equiv \frac{\text{total distance}}{\text{total time}}$$

The **average velocity** $\bar{v}$ during a time interval Δt is the displacement Δx divided by Δt.

$$\bar{v} \equiv \frac{\Delta x}{\Delta t} = \frac{x_f - x_i}{t_f - t_i} \qquad \text{[2.2]}$$

The average velocity is equal to the slope of the straight line joining the initial and final points on a graph of the position of the object versus time.

The slope of the line tangent to the position vs. time curve at some point is equal to the **instantaneous velocity** at that time. The **instantaneous speed** of an object is defined as the magnitude of the instantaneous velocity.

2.3 Acceleration

The **average acceleration** $\bar{a}$ of an object undergoing a change in velocity Δv during a time interval Δt is

$$\bar{a} \equiv \frac{\Delta v}{\Delta t} = \frac{v_f - v_i}{t_f - t_i} \qquad \text{[2.4]}$$

The **instantaneous acceleration** of an object at a certain time equals the slope of a velocity vs. time graph at that instant.

2.5 One-Dimensional Motion with Constant Acceleration

The most useful equations that describe the motion of an object moving with constant acceleration along the x axis are as follows:

$$v = v_0 + at \qquad \text{[2.6]}$$

$$\Delta x = v_0 t + \tfrac{1}{2}at^2 \qquad \text{[2.9]}$$

$$v^2 = v_0{}^2 + 2a\Delta x \qquad \text{[2.10]}$$

All problems can be solved with the first two equations alone, the last being convenient when time doesn't explicitly enter the problem. After the constants are properly identified, most problems reduce to one or two equations in as many unknowns.

2.6 Freely Falling Objects

An object falling in the presence of Earth's gravity exhibits a free-fall acceleration directed toward Earth's center. If air friction is neglected and if the altitude of the falling object is small compared with Earth's radius, then we can assume that the free-fall acceleration $g = 9.8 \text{ m/s}^2$ is constant over the range of motion. Equations 2.6, 2.9, and 2.10 apply, with $a = -g$.

CONCEPTUAL QUESTIONS

1. If the velocity of a particle is nonzero, can the particle's acceleration be zero? Explain.

2. If the velocity of a particle is zero, can the particle's acceleration be zero? Explain.

3. If a car is traveling eastward, can its acceleration be westward? Explain.

4. The speed of sound in air is 331 m/s. During the next thunderstorm, try to estimate your distance from a lightning bolt by measuring the time lag between the flash and the thunderclap. You can ignore the time it takes for the light flash to reach you. Why?

5. Can the equations of kinematics be used in a situation where the acceleration varies with time? Can they be used when the acceleration is zero?

6. If the average velocity of an object is zero in some time interval, what can you say about the displacement of the object during that interval?

7. A child throws a marble into the air with an initial speed v_0. Another child drops a ball at the same instant. Compare the accelerations of the two objects while they are in flight.

8. Figure Q2.8 shows strobe photographs taken of a disk moving from left to right under different conditions. The time interval between images is constant. Taking the direction to the right to be positive, describe the motion of the disk in each case. For which case is (a) the acceleration positive? (b) the acceleration negative? (c) the velocity constant?

9. Can the instantaneous velocity of an object at an instant of time ever be greater in magnitude than the average ve-

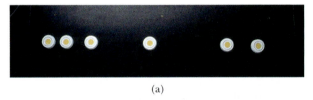

(a)

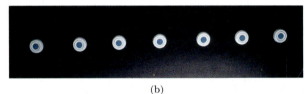

(b)

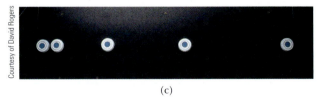

(c)

Figure Q2.8

Courtesy of David Rogers

locity over a time interval containing that instant? Can it ever be less?

10. Car A, traveling from New York to Miami, has a speed of 25 m/s. Car B, traveling from New York to Chicago, also has a speed of 25 m/s. Are their velocities equal? Explain.

11. A ball is thrown vertically upward. (a) What are its velocity and acceleration when it reaches its maximum altitude? (b) What is the acceleration of the ball just before it hits the ground?

12. A rule of thumb for driving is that a separation of one car length for each 10 mi/h of speed should be maintained between moving vehicles. Assuming a constant reaction time, discuss the relevance of this rule for (a) motion with constant velocity and (b) motion with constant acceleration.

13. Two cars are moving in the same direction in parallel lanes along a highway. At some instant, the velocity of car A exceeds the velocity of car B. Does this mean that the acceleration of A is greater than that of B? Explain.

14. Consider the following combinations of signs and values for the velocity and acceleration of a particle with respect to a one-dimensional x-axis:

	Velocity	**Acceleration**
a.	Positive	Positive
b.	Positive	Negative
c.	Positive	Zero
d.	Negative	Positive
e.	Negative	Negative
f.	Negative	Zero
g.	Zero	Positive
h.	Zero	Negative

Describe what the particle is doing in each case, and give a real-life example for an automobile on an east-west one-dimensional axis, with east considered the positive direction.

15. A student at the top of a building of height h throws one ball upward with a speed of v_0 and then throws a second ball downward with the same initial speed, v_0. How do the final velocities compare when the balls reach the ground?

16. A ball is thrown straight upward and moves in free fall. Choose a coordinate system with its origin at the release point of the ball and the positive direction upward. (a) What is the sign of the velocity of the ball just before the ball reaches its maximum height, just after it reaches its maximum height, and at its maximum height. (b) What is the sign of the acceleration of the ball just before the ball reaches its maximum height, just after it reaches its maximum height, and at its maximum height. (c) If the ball takes time t_1 to reach its maximum height, how long will it take to return to ground level? (d) If the ball is thrown upward with a velocity of $+ v_0$, what will be the ball's velocity upon returning to ground level?

17. A pebble is dropped into a water well, and the splash is heard 16 s later, as illustrated in the cartoon strip shown in Figure Q2.17. Estimate the distance from the rim of the well to the water's surface.

18. A ball rolls in a straight line along the horizontal direction. Using motion diagrams (or multiflash photographs), describe the velocity and acceleration of the ball for each of the following situations: (a) The ball moves to the right at a constant speed. (b) The ball moves from right to left and continually slows down. (c) The ball moves from right to left and continually speeds up. (d) The ball moves to the right, first speeding up at a constant rate and then slowing down at a constant rate.

19. You drop a ball from a window on an upper floor of a building. The ball strikes the ground with speed v. You now repeat the drop, but you have a friend down on the street who throws another ball upward at speed v. Your friend throws the ball upward at exactly the same time that you drop yours from the window. At some location, the balls pass each other. Is this location *at* the halfway point between window and ground, *above* that point, or *below* that point?

Figure Q2.17

PROBLEMS

1, 2, 3 = straightforward, intermediate, challenging   ☐ = full solution available in *Student Solutions Manual/Study Guide*
Physics⊗Now™ = coached problem with hints available at **www.cp7e.com**   🐛 = biomedical application

Section 2.1 Displacement
Section 2.2 Velocity

1. A person travels by car from one city to another with different constant speeds between pairs of cities. She drives for 30.0 min at 80.0 km/h, 12.0 min at 100 km/h, and 45.0 min at 40.0 km/h and spends 15.0 min eating lunch and buying gas. (a) Determine the average speed for the trip. (b) Determine the distance between the initial and final cities along the route.

2. (a) Sand dunes on a desert island move as sand is swept up the windward side to settle in the leeward side. Such "walking" dunes have been known to travel 20 feet in a

year and can travel as much as 100 feet per year in particularly windy times. Calculate the average speed in each case in m/s. (b) Fingernails grow at the rate of drifting continents, about 10 mm/yr. Approximately how long did it take for North America to separate from Europe, a distance of about 3 000 mi?

3. Two boats start together and race across a 60-km-wide lake and back. Boat A goes across at 60 km/h and returns at 60 km/h. Boat B goes across at 30 km/h, and its crew, realizing how far behind it is getting, returns at 90 km/h. Turnaround times are negligible, and the boat that completes the round trip first wins. (a) Which boat wins and by how much? (Or is it a tie?) (b) What is the average velocity of the winning boat?

4. The Olympic record for the marathon is 2 h, 9 min, 21 s. The marathon distance is 26 mi, 385 yd. Determine the average speed (in miles per hour) of the record.

5. A motorist drives north for 35.0 minutes at 85.0 km/h and then stops for 15.0 minutes. He then continues north, traveling 130 km in 2.00 h. (a) What is his total displacement? (b) What is his average velocity?

6. A graph of position versus time for a certain particle moving along the x-axis is shown in Figure P2.6. Find the average velocity in the time intervals from (a) 0 to 2.00 s, (b) 0 to 4.00 s, (c) 2.00 s to 4.00 s, (d) 4.00 s to 7.00 s, and (e) 0 to 8.00 s.

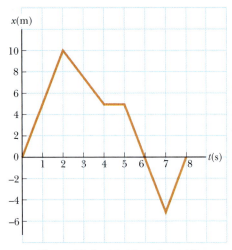

Figure P2.6 (Problems 6 and 15)

7. A tennis player moves in a straight-line path as shown in Figure P2.7. Find her average velocity in the time intervals from (a) 0 to 1.0 s, (b) 0 to 4.0 s, (c) 1.0 s to 5.0 s, and (d) 0 to 5.0 s.

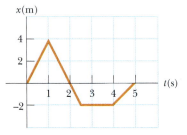

Figure P2.7 (Problems 7 and 17)

8. Two cars travel in the same direction along a straight highway, one at a constant speed of 55 mi/h and the other at 70 mi/h. (a) Assuming that they start at the same point, how much sooner does the faster car arrive at a destination 10 mi away? (b) How far must the faster car travel before it has a 15-min lead on the slower car?

9. An athlete swims the length of a 50.0-m pool in 20.0 s and makes the return trip to the starting position in 22.0 s. Determine her average velocities in (a) the first half of the swim, (b) the second half of the swim, and (c) the round trip.

10. If the average speed of an orbiting space shuttle is 19 800 mi/h, determine the time required for it to circle Earth. Make sure you consider the fact that the shuttle is orbiting about 200 mi above Earth's surface, and assume that Earth's radius is 3 963 miles.

11. **Physics⊗Now™** A person takes a trip, driving with a constant speed of 89.5 km/h, except for a 22.0-min rest stop. If the person's average speed is 77.8 km/h, how much time is spent on the trip and how far does the person travel?

12. A tortoise can run with a speed of 0.10 m/s, and a hare can run 20 times as fast. In a race, they both start at the same time, but the hare stops to rest for 2.0 minutes. The tortoise wins by a shell (20 cm). (a) How long does the race take? (b) What is the length of the race?

13. In order to qualify for the finals in a racing event, a race car must achieve an average speed of 250 km/h on a track with a total length of 1 600 m. If a particular car covers the first half of the track at an average speed of 230 km/h, what minimum average speed must it have in the second half of the event in order to qualify?

14. Runner A is initially 4.0 mi west of a flagpole and is running with a constant velocity of 6.0 mi/h due east. Runner B is initially 3.0 mi east of the flagpole and is running with a constant velocity of 5.0 mi/h due west. How far are the runners from the flagpole when they meet?

15. A graph of position versus time for a certain particle moving along the x-axis is shown in Figure P2.6. Find the instantaneous velocity at the instants (a) $t = 1.00$ s, (b) $t = 3.00$ s, (c) $t = 4.50$ s, and (d) $t = 7.50$ s.

16. A race car moves such that its position fits the relationship

$$x = (5.0 \text{ m/s})t + (0.75 \text{ m/s}^3)t^3$$

where x is measured in meters and t in seconds. (a) Plot a graph of the car's position versus time. (b) Determine the instantaneous velocity of the car at $t = 4.0$ s, using time intervals of 0.40 s, 0.20 s, and 0.10 s. (c) Compare the average velocity during the first 4.0 s with the results of (b).

17. Find the instantaneous velocities of the tennis player of Figure P2.7 at (a) 0.50 s, (b) 2.0 s, (c) 3.0 s, and (d) 4.5 s.

Section 2.3 Acceleration

18. Secretariat ran the Kentucky Derby with times of 25.2 s, 24.0 s, 23.8 s, and 23.0 s for the quarter mile. (a) Find his average speed during each quarter-mile segment. (b) Assuming that Secretariat's instantaneous speed at the finish line was the same as his average speed during the final quarter mile, find his average acceleration for the entire race. (*Hint:* Recall that horses in the Derby start from rest.)

19. A steam catapult launches a jet aircraft from the aircraft carrier *John C. Stennis*, giving it a speed of 175 mi/h in 2.50 s. (a) Find the average acceleration of the plane. (b) Assuming that the acceleration is constant, find the distance the plane moves.

20. A car traveling in a straight line has a velocity of +5.0 m/s at some instant. After 4.0 s, its velocity is +8.0 m/s. What is the car's average acceleration during the 4.0-s time interval?

21. **Physics⊗Now™** A certain car is capable of accelerating at a rate of +0.60 m/s². How long does it take for this car to go from a speed of 55 mi/h to a speed of 60 mi/h?

22. The velocity vs. time graph for an object moving along a straight path is shown in Figure P2.22. (a) Find the average acceleration of the object during the time intervals 0 to 5.0 s, 5.0 s to 15 s, and 0 to 20 s. (b) Find the instantaneous acceleration at 2.0 s, 10 s, and 18 s.

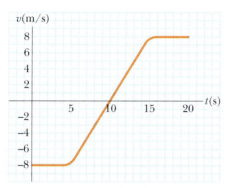

Figure P2.22

23. The engine of a model rocket accelerates the rocket vertically upward for 2.0 s as follows: At $t = 0$, the rocket's speed is zero; at $t = 1.0$ s, its speed is 5.0 m/s; and at $t = 2.0$ s, its speed is 16 m/s. Plot a velocity vs. time graph for this motion, and use the graph to determine (a) the rocket's average acceleration during the 2.0-s interval and (b) the instantaneous acceleration of the rocket at $t = 1.5$ s.

Section 2.5 One-Dimensional Motion with Constant Acceleration

24. A car traveling in a straight-line path has a velocity of +10.0 m/s at some instant. After 3.00 s, its velocity is +6.00 m/s. What is the average acceleration of the car during this time interval?

25. In 1865, Jules Verne proposed sending men to the Moon by firing a space capsule from a 220-m-long cannon with final speed of 10.97 km/s. What would have been the unrealistically large acceleration experienced by the space travelers during their launch? (A human can stand an acceleration of 15*g* for a short time.) Compare your answer with the free-fall acceleration, 9.80 m/s².

26. A truck covers 40.0 m in 8.50 s while smoothly slowing down to a final speed of 2.80 m/s. (a) Find the truck's original speed. (b) Find its acceleration.

27. A speedboat increases its speed uniformly from 20 m/s to 30 m/s in a distance of 200 m. Find (a) the magnitude of its acceleration and (b) the time it takes the boat to travel the 200-m distance.

28. Two cars are traveling along a straight line in the same direction, the lead car at 25.0 m/s and the other car at

30.0 m/s. At the moment the cars are 40.0 m apart, the lead driver applies the brakes, causing his car to have an acceleration of −2.00 m/s². (a) How long does it take for the lead car to stop? (b) Assuming that the chasing car brakes at the same time as the lead car, what must be the chasing car's minimum negative acceleration so as not to hit the lead car? (c) How long does it take for the chasing car to stop?

29. A Cessna aircraft has a lift-off speed of 120 km/h. (a) What minimum constant acceleration does the aircraft require if it is to be airborne after a takeoff run of 240 m? (b) How long does it take the aircraft to become airborne?

30. A truck on a straight road starts from rest and accelerates at 2.0 m/s² until it reaches a speed of 20 m/s. Then the truck travels for 20 s at constant speed until the brakes are applied, stopping the truck in a uniform manner in an additional 5.0 s. (a) How long is the truck in motion? (b) What is the average velocity of the truck during the motion described?

31. A drag racer starts her car from rest and accelerates at 10.0 m/s² for a distance of 400 m ($\frac{1}{4}$ mile). (a) How long did it take the race car to travel this distance? (b) What is the speed of the race car at the end of the run?

32. A jet plane lands with a speed of 100 m/s and can accelerate at a maximum rate of −5.00 m/s² as it comes to rest. (a) From the instant the plane touches the runway, what is the minimum time needed before it can come to rest? (b) Can this plane land on a small tropical island airport where the runway is 0.800 km long?

33. A driver in a car traveling at a speed of 60 mi/h sees a deer 100 m away on the road. Calculate the minimum constant acceleration that is necessary for the car to stop without hitting the deer (assuming that the deer does not move in the meantime).

34. A record of travel along a straight path is as follows:
    1. Start from rest with a constant acceleration of 2.77 m/s² for 15.0 s.
    2. Maintain a constant velocity for the next 2.05 min.
    3. Apply a constant negative acceleration of −9.47 m/s² for 4.39 s.

    (a) What was the total displacement for the trip? (b) What were the average speeds for legs 1, 2, and 3 of the trip, as well as for the complete trip?

35. A train is traveling down a straight track at 20 m/s when the engineer applies the brakes, resulting in an acceleration of −1.0 m/s² as long as the train is in motion. How far does the train move during a 40-s time interval starting at the instant the brakes are applied?

36. A car accelerates uniformly from rest to a speed of 40.0 mi/h in 12.0 s. Find (a) the distance the car travels during this time and (b) the constant acceleration of the car.

37. A car starts from rest and travels for 5.0 s with a uniform acceleration of +1.5 m/s². The driver then applies the brakes, causing a uniform acceleration of −2.0 m/s². If the brakes are applied for 3.0 s, (a) how fast is the car going at the end of the braking period, and (b) how far has the car gone?

38. A train 400 m long is moving on a straight track with a speed of 82.4 km/h. The engineer applies the brakes at a crossing, and later the last car passes the crossing with a

speed of 16.4 km/h. Assuming constant acceleration, determine how long the train blocked the crossing. Disregard the width of the crossing.

39. A hockey player is standing on his skates on a frozen pond when an opposing player, moving with a uniform speed of 12 m/s, skates by with the puck. After 3.0 s, the first player makes up his mind to chase his opponent. If he accelerates uniformly at 4.0 m/s^2, (a) how long does it take him to catch his opponent, and (b) how far has he traveled in that time? (Assume that the player with the puck remains in motion at constant speed.)

40. A glider on an air track carries a flag of length ℓ through a stationary photogate that measures the time interval Δt_d during which the flag blocks a beam of infrared light passing across the gate. The ratio $v_d = \ell/\Delta t_d$ is the average velocity of the glider over this part of its motion. Suppose the glider moves with constant acceleration. (a) Argue for or against the idea that v_d is equal to the instantaneous velocity of the glider when it is halfway through the photogate in terms of distance. (b) Argue for or against the idea that v_d is equal to the instantaneous velocity of the glider when it is halfway through the photogate in terms of time.

41. In the Daytona 500 auto race, a Ford Thunderbird and a Mercedes Benz are moving side by side down a straightaway at 71.5 m/s. The driver of the Thunderbird realizes that she must make a pit stop, and she smoothly slows to a stop over a distance of 250 m. She spends 5.00 s in the pit and then accelerates out, reaching her previous speed of 71.5 m/s after a distance of 350 m. At this point, how far has the Thunderbird fallen behind the Mercedes Benz, which has continued at a constant speed?

42. A certain cable car in San Francisco can stop in 10 s when traveling at maximum speed. On one occasion, the driver sees a dog a distance d m in front of the car and slams on the brakes instantly. The car reaches the dog 8.0 s later, and the dog jumps off the track just in time. If the car travels 4.0 m beyond the position of the dog before coming to a stop, how far was the car from the dog? (*Hint*: You will need three equations.)

Section 2.6 Freely Falling Objects

43. A ball is thrown vertically upward with a speed of 25.0 m/s. (a) How high does it rise? (b) How long does it take to reach its highest point? (c) How long does the ball take to hit the ground after it reaches its highest point? (d) What is its velocity when it returns to the level from which it started?

44. It is possible to shoot an arrow at a speed as high as 100 m/s. (a) If friction is neglected, how high would an arrow launched at this speed rise if shot straight up? (b) How long would the arrow be in the air?

45. A certain freely falling object requires 1.50 s to travel the last 30.0 m before it hits the ground. From what height above the ground did it fall?

46. Traumatic brain injury such as concussion results when the head undergoes a very large acceleration. Generally, an acceleration less than 800 m/s^2 lasting for any length of time will not cause injury, whereas an acceleration greater than 1 000 m/s^2 lasting for at least 1 ms will cause injury. Suppose a small child rolls off a bed that is 0.40 m above the floor. If the floor is hardwood, the child's head

is brought to rest in approximately 2.0 mm. If the floor is carpeted, this stopping distance is increased to about 1.0 cm. Calculate the magnitude and duration of the deceleration in both cases, to determine the risk of injury. Assume that the child remains horizontal during the fall to the floor. Note that a more complicated fall could result in a head velocity greater or less than the speed you calculate.

47. **Physics⊗Now**™ A small mailbag is released from a helicopter that is descending steadily at 1.50 m/s. After 2.00 s, (a) what is the speed of the mailbag, and (b) how far is it below the helicopter? (c) What are your answers to parts (a) and (b) if the helicopter is *rising* steadily at 1.50 m/s?

48. A ball thrown vertically upward is caught by the thrower after 2.00 s. Find (a) the initial velocity of the ball and (b) the maximum height the ball reaches.

49. A model rocket is launched straight upward with an initial speed of 50.0 m/s. It accelerates with a constant upward acceleration of 2.00 m/s^2 until its engines stop at an altitude of 150 m. (a) What is the maximum height reached by the rocket? (b) How long after lift-off does the rocket reach its maximum height? (c) How long is the rocket in the air?

50. A parachutist with a camera descends in free fall at a speed of 10 m/s. The parachutist releases the camera at an altitude of 50 m. (a) How long does it take the camera to reach the ground? (b) What is the velocity of the camera just before it hits the ground?

51. A student throws a set of keys vertically upward to his fraternity brother, who is in a window 4.00 m above. The brother's outstretched hand catches the keys 1.50 s later. (a) With what initial velocity were the keys thrown? (b) What was the velocity of the keys just before they were caught?

52. It has been claimed that an insect called the froghopper (*Philaenus spumarius*) is the best jumper in the animal kingdom. This insect can accelerate at 4 000 m/s^2 over a distance of 2.0 mm as it straightens its specially designed "jumping legs." (a) Assuming a uniform acceleration, what is the velocity of the insect after it has accelerated through this short distance, and how long did it take to reach that velocity? (b) How high would the insect jump if air resistance could be ignored? Note that the actual height obtained is about 0.7 m, so air resistance is important here.

ADDITIONAL PROBLEMS

53. A truck tractor pulls two trailers, one behind the other, at a constant speed of 100 km/h. It takes 0.600 s for the big rig to completely pass onto a bridge 400 m long. For what duration of time is all or part of the truck-trailer combination on the bridge?

54. A speedboat moving at 30.0 m/s approaches a no-wake buoy marker 100 m ahead. The pilot slows the boat with a constant acceleration of -3.50 m/s^2 by reducing the throttle. (a) How long does it take the boat to reach the buoy? (b) What is the velocity of the boat when it reaches the buoy?

55. A bullet is fired through a board 10.0 cm thick in such a way that the bullet's line of motion is perpendicular to the face of the board. If the initial speed of the bullet is 400 m/s and it emerges from the other side of the board with a speed of 300 m/s, find (a) the acceleration of the bullet as it passes through the board and (b) the total time the bullet is in contact with the board.

56. An indestructible bullet 2.00 cm long is fired straight through a board that is 10.0 cm thick. The bullet strikes the board with a speed of 420 m/s and emerges with a speed of 280 m/s. (a) What is the average acceleration of the bullet through the board? (b) What is the total time that the bullet is in contact with the board? (c) What thickness of board (calculated to 0.1 cm) would it take to stop the bullet, assuming that the acceleration through all boards is the same?

57. A ball is thrown upward from the ground with an initial speed of 25 m/s; at the same instant, another ball is dropped from a building 15 m high. After how long will the balls be at the same height?

58. A ranger in a national park is driving at 35.0 mi/h when a deer jumps into the road 200 ft ahead of the vehicle. After a reaction time t, the ranger applies the brakes to produce an acceleration $a = -9.00$ ft/s^2. What is the maximum reaction time allowed if she is to avoid hitting the deer?

59. Two students are on a balcony 19.6 m above the street. One student throws a ball vertically downward at 14.7 m/s; at the same instant, the other student throws a ball vertically upward at the same speed. The second ball just misses the balcony on the way down. (a) What is the difference in the two balls' time in the air? (b) What is the velocity of each ball as it strikes the ground? (c) How far apart are the balls 0.800 s after they are thrown?

60. The driver of a truck slams on the brakes when he sees a tree blocking the road. The truck slows down uniformly with an acceleration of -5.60 m/s^2 for 4.20 s, making skid marks 62.4 m long that end at the tree. With what speed does the truck then strike the tree?

61. A young woman named Kathy Kool buys a sports car that can accelerate at the rate of 4.90 m/s^2. She decides to test the car by drag racing with another speedster, Stan Speedy. Both start from rest, but experienced Stan leaves the starting line 1.00 s before Kathy. If Stan moves with a constant acceleration of 3.50 m/s^2 and Kathy maintains an acceleration of 4.90 m/s^2, find (a) the time it takes Kathy to overtake Stan, (b) the distance she travels before she catches him, and (c) the speeds of both cars at the instant she overtakes him.

62. A mountain climber stands at the top of a 50.0-m cliff that overhangs a calm pool of water. She throws two stones vertically downward 1.00 s apart and observes that they cause a single splash. The first stone had an initial velocity of -2.00 m/s. (a) How long after release of the first stone did the two stones hit the water? (b) What initial velocity must the second stone have had, given that they hit the water simultaneously? (c) What was the velocity of each stone at the instant it hit the water?

63. An ice sled powered by a rocket engine starts from rest on a large frozen lake and accelerates at $+40$ ft/s^2. After some time t_1, the rocket engine is shut down and the sled moves with constant velocity v for a time t_2. If the total distance traveled by the sled is 17 500 ft and the total time is 90 s, find (a) the times t_1 and t_2 and (b) the velocity v. At the 17 500-ft mark, the sled begins to accelerate at -20 ft/s^2. (c) What is the final position of the sled when

it comes to rest? (d) How long does it take to come to rest?

64. In Bosnia, the ultimate test of a young man's courage used to be to jump off a 400-year-old bridge (now destroyed) into the River Neretva, 23 m below the bridge. (a) How long did the jump last? (b) How fast was the jumper traveling upon impact with the river? (c) If the speed of sound in air is 340 m/s, how long after the jumper took off did a spectator on the bridge hear the splash?

65. A person sees a lightning bolt pass close to an airplane that is flying in the distance. The person hears thunder 5.0 s after seeing the bolt and sees the airplane overhead 10 s after hearing the thunder. The speed of sound in air is 1 100 ft/s. (a) Find the distance of the airplane from the person at the instant of the bolt. (Neglect the time it takes the light to travel from the bolt to the eye.) (b) Assuming that the plane travels with a constant speed toward the person, find the velocity of the airplane. (c) Look up the speed of light in air, and defend the approximation used in (a).

66. Another scheme to catch the roadrunner has failed! Now a safe falls from rest from the top of a 25.0-m-high cliff toward Wile E. Coyote, who is standing at the base. Wile first notices the safe after it has fallen 15.0 m. How long does he have to get out of the way?

67. **Physics**⊗**Now**™ A stunt man sitting on a tree limb wishes to drop vertically onto a horse galloping under the tree. The constant speed of the horse is 10.0 m/s, and the man is initially 3.00 m above the level of the saddle. (a) What must be the horizontal distance between the saddle and the limb when the man makes his move? (b) How long is he in the air?

68. A hard rubber ball, released at chest height, falls to the pavement and bounces back to nearly the same height. When the ball is in contact with the pavement, its lower side is temporarily flattened. Before the dent in the ball pops out, suppose that its maximum depth is on the order of 1 cm. Compute an order-of-magnitude estimate for the maximum acceleration of the ball. State your assumptions, the quantities you estimate, and the values you estimate for them.

69. *Vroom—vroom!* As soon as a traffic light turns green, a car speeds up from rest to 50.0 mi/h with a constant acceleration of 9.00 mi/h·s. In the adjoining bike lane, a cyclist speeds up from rest to 20.0 mi/h with a constant acceleration of 13.0 mi/h·s. Each vehicle maintains a constant velocity after reaching its cruising speed. (a) For how long is the bicycle ahead of the car? (b) By what maximum distance does the bicycle lead the car?

70. In order to pass a physical education class at a university, a student must run 1.0 mi in 12 min. After running for 10 min, she still has 500 yd to go. If her maximum acceleration is 0.15 m/s^2, can she make it? If the answer is no, determine what acceleration she would need to be successful.

71. One swimmer in a relay race has a 0.50-s lead and is swimming at a constant speed of 4.0 m/s. He has 50 m to swim before reaching the end of the pool. A second swimmer moves in the same direction as the leader. What constant speed must the second swimmer have in order to catch up to the leader at the end of the pool?

ACTIVITIES

A.1. Estimate a few speeds in metric units, using a stopwatch or a wristwatch. For example, roll a ball across a table and estimate the number of centimeters it moves each second to find its speed. Other speeds you might try are for someone walking across the room, a jogger running, a car moving through some distance, and so forth. To see how well you did, make some actual measurements for those situations in which it is feasible to do so.

A.2. Use what you know about falling objects to measure your reaction time. Hold the index finger and thumb of your dominant hand about 2.5 cm apart, and then have your co-worker hold a ruler vertically in the space between your finger and thumb, as shown in Figure A2.2. Note the position of the ruler relative to your index finger. Your co-worker must release the ruler, and you must catch it (without moving your hand downward) as quickly as you can. The ruler (a freely falling object) falls through a distance $d = \frac{1}{2}gt^2$, where t is the reaction time and

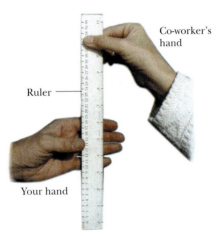

Ruler

Co-worker's hand

Your hand

Figure A2.2

$g = 9.80$ m/s^2. Repeat this measurement of d five times, average your results, and calculate an average value of t. Now measure your co-worker's reaction time, using the same procedure. Compare your results. For most people, the reaction time is at best about 0.2 s. As an extension to this experiment, replace the ruler with a crisp dollar bill. Hold the bill such that your thumb and index finger are just at the level of Washington's face. Unless you are anticipating the time of release, you will not be able to catch the bill when it is released, because the time required for the top to pass out of your hand is less than the typical 0.2-s reaction time.

A.3. Galileo studied accelerated motion by allowing objects to roll down inclined planes so that their motion would be slow enough to make reasonable observations. Try a similar procedure. Make a mark at the top of an inclined plane as the starting point for the motion, and use a metal barrier at the end as a sound cue for stopping a stopwatch. Measure the length of the plane. Record the average time for several trials of a ball rolling down the plane at a measured angle. From the information you obtain, calculate the acceleration. Repeat the experiment for a larger angle of inclination. Do this for several trials, until you can plot a graph of acceleration versus angle. From your graph, can you guess what the acceleration would be if the inclined plane were vertical? Would the results of your experiment be different if you had used a significantly more massive ball? If you are unsure, repeat the experiment to see if there is a difference.

A.4. Perform the activities that follow to verify that all objects fall with the same acceleration. First, try dropping a coffee filter oriented horizontally and also dropping a pencil. Then repeat the experiment with the filter in a loose ball, a tight ball, and, finally, in a compacted wad. You should note that compacting the filter tends to reduce the effects of air resistance and makes the two objects fall more nearly at the same rate.

Legendary motorcycle stuntman Evel Knievel blasts off in his custom rocket-powered Harley-Davidson Skycycle in an attempt to jump the Snake River Canyon in 1974. A parachute prematurely deployed and caused the craft to fall into the canyon, just short of the other side. Knievel survived.

CHAPTER

3

Vectors and Two-Dimensional Motion

In our discussion of one-dimensional motion in Chapter 2, we used the concept of vectors only to a limited extent. In our further study of motion, manipulating vector quantities will become increasingly important, so much of this chapter is devoted to vector techniques. We'll then apply these mathematical tools to two-dimensional motion, especially that of projectiles, and to the understanding of relative motion.

3.1 VECTORS AND THEIR PROPERTIES

Each of the physical quantities we will encounter in this book can be categorized as either a *vector quantity* or a *scalar quantity*. As noted in Chapter 2, a vector has both direction and magnitude (size). A scalar can be completely specified by its magnitude with appropriate units; it has no direction. An example of each kind of quantity is shown in Figure 3.1 (page 54).

As described in Chapter 2, displacement, velocity, and acceleration are vector quantities. Temperature is an example of a scalar quantity. If the temperature of an object is $-5°C$, that information completely specifies the temperature of the object; no direction is required. Masses, time intervals, and volumes are scalars as well. Scalar quantities can be manipulated with the rules of ordinary arithmetic. Vectors can also be added and subtracted from each other, and multiplied, but there are a number of important differences, as will be seen in the following sections.

When a vector quantity is handwritten, it is often represented with an arrow over the letter ($\vec{A}$). As mentioned in Section 2.1, a vector quantity in this book

Figure 3.1 (a) The number of grapes in this bunch ripe for picking is one example of a scalar quantity. Can you think of other examples? (b) This helpful person pointing in the right direction tells us to travel five blocks north to reach the courthouse. A vector is a physical quantity that must be specified by both magnitude and direction.

TIP 3.1 **Vector Addition Versus Scalar Addition**

$\vec{A} + \vec{B} = \vec{C}$ is very different from $A + B = C$. The first is a vector sum, which must be handled graphically or with components, while the second is a simple arithmetic sum of numbers.

will be represented by boldface type with an arrow on top (for example, $\vec{A}$). The magnitude of the vector $\vec{A}$ will be represented by italic type, as A. Italic type will also be used to represent scalars.

Equality of Two Vectors. Two vectors $\vec{A}$ and $\vec{B}$ are equal if they have the same magnitude and the same direction. This property allows us to translate a vector parallel to itself in a diagram without affecting the vector. In fact, for most purposes, any vector can be moved parallel to itself without being affected. (See Fig. 3.2.)

Adding Vectors. When two or more vectors are added, they must all have the same units. For example, it doesn't make sense to add a velocity vector, carrying units of meters per second, to a displacement vector, carrying units of meters. Scalars obey the same rule: It would be similarly meaningless to add temperatures to volumes or masses to time intervals.

Vectors can be added geometrically or algebraically. (The latter is discussed at the end of the next section.) To add vector $\vec{B}$ to vector $\vec{A}$ geometrically, first draw $\vec{A}$ on a piece of graph paper to some scale, such as 1 cm = 1 m, and so that its direction is specified relative a coordinate system. Then draw vector $\vec{B}$ to the same scale with the tail of $\vec{B}$ starting at the tip of $\vec{A}$, as in Active Figure 3.3a. Vector $\vec{B}$ must be drawn along the direction that makes the proper angle relative vector $\vec{A}$. The **resultant vector** $\vec{R} = \vec{A} + \vec{B}$ is the vector drawn from the tail of $\vec{A}$ to the tip of $\vec{B}$. This procedure is known as the **triangle method of addition**.

When two vectors are added, their sum is independent of the order of the addition: $\vec{A} + \vec{B} = \vec{B} + \vec{A}$. This relationship can be seen from the geometric construction in Active Figure 3.3b, and is called the **commutative law of addition**.

This same general approach can also be used to add more than two vectors, as is done in Figure 3.4 for four vectors. The resultant vector sum $\vec{R} = \vec{A} + \vec{B} + \vec{C} + \vec{D}$ is the vector drawn from the tail of the first vector to the tip of the last. Again, the order in which the vectors are added is unimportant.

Negative of a Vector. The negative of the vector $\vec{A}$ is defined as the vector that gives zero when added to $\vec{A}$. This means that $\vec{A}$ and $-\vec{A}$ have the same magnitude but opposite directions.

Figure 3.2 These four vectors are equal because they have equal lengths and point in the same direction.

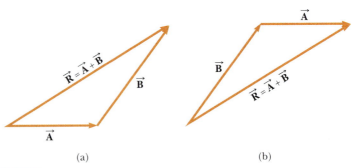

(a)   (b)

ACTIVE FIGURE 3.3
(a) When vector $\vec{B}$ is added to vector $\vec{A}$, the vector sum $\vec{R}$ is the vector that runs from the tail of $\vec{A}$ to the tip of $\vec{B}$. (b) Here the resultant runs from the tail of $\vec{B}$ to the tip of $\vec{A}$. These constructions prove that $\vec{A} + \vec{B} = \vec{B} + \vec{A}$.

Physics⊗Now™
Log into PhysicsNow at **www.cp7e.com**, and go to Active Figure 3.3 to vary $\vec{A}$ and $\vec{B}$ and see the effect on the resultant.

Subtracting Vectors. Vector subtraction makes use of the definition of the negative of a vector. We define the operation $\vec{\mathbf{A}} - \vec{\mathbf{B}}$ as the vector $-\vec{\mathbf{B}}$ added to the vector $\vec{\mathbf{A}}$:

$$\vec{\mathbf{A}} - \vec{\mathbf{B}} = \vec{\mathbf{A}} + (-\vec{\mathbf{B}}) \qquad \text{[3.1]}$$

Vector subtraction is really a special case of vector addition. The geometric construction for subtracting two vectors is shown in Figure 3.5.

Multiplying or Dividing a Vector by a Scalar. Multiplying or dividing a vector by a scalar gives a vector. For example, if vector $\vec{\mathbf{A}}$ is multiplied by the scalar number 3, the result, written $3\vec{\mathbf{A}}$, is a vector with a magnitude three times that of $\vec{\mathbf{A}}$ and pointing in the same direction. If we multiply vector $\vec{\mathbf{A}}$ by the scalar -3, the result is $-3\vec{\mathbf{A}}$, a vector with a magnitude three times that of $\vec{\mathbf{A}}$ and pointing in the opposite direction (because of the negative sign).

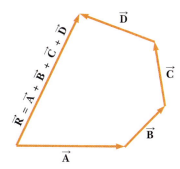

Figure 3.4 A geometric construction for summing four vectors. The resultant vector $\vec{\mathbf{R}}$ is the vector that completes the polygon.

> ### Quick Quiz 3.1
>
> The magnitudes of two vectors $\vec{\mathbf{A}}$ and $\vec{\mathbf{B}}$ are 12 units and 8 units, respectively. What are the largest and smallest possible values for the magnitude of the resultant vector $\vec{\mathbf{R}} = \vec{\mathbf{A}} + \vec{\mathbf{B}}$? (a) 14.4 and 4; (b) 12 and 8; (c) 20 and 4; (d) none of these.

> ### Quick Quiz 3.2
>
> If vector $\vec{\mathbf{B}}$ is added to vector $\vec{\mathbf{A}}$, the resultant vector $\vec{\mathbf{A}} + \vec{\mathbf{B}}$ has magnitude $A + B$ when $\vec{\mathbf{A}}$ and $\vec{\mathbf{B}}$ are (a) perpendicular to each other; (b) oriented in the same direction; (c) oriented in opposite directions; (d) none of these answers.

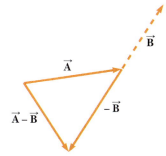

Figure 3.5 This construction shows how to subtract vector $\vec{\mathbf{B}}$ from vector $\vec{\mathbf{A}}$. The vector $-\vec{\mathbf{B}}$ has the same magnitude as the vector $\vec{\mathbf{B}}$, but points in the opposite direction.

EXAMPLE 3.1   Taking a Trip

Goal   Find the sum of two vectors by using a graph.

Problem   A car travels 20.0 km due north and then 35.0 km in a direction 60° west of north, as in Figure 3.6. Using a graph, find the magnitude and direction of a single vector that gives the net effect of the car's trip. This vector is called the car's *resultant displacement*.

Strategy   Draw a graph, and represent the displacement vectors as arrows. Graphically locate the vector resulting from the sum of the two displacement vectors. Measure its length and angle with respect to the vertical.

Solution
Let $\vec{\mathbf{A}}$ represent the first displacement vector, 20.0 km north, $\vec{\mathbf{B}}$ the second displacement vector, extending west of north. Carefully graph the two vectors, drawing a resultant vector $\vec{\mathbf{R}}$ with its base touching the base of $\vec{\mathbf{A}}$ and extending to the tip of $\vec{\mathbf{B}}$. Measure the length of this vector, which turns out to be about 48 km. The angle β, measured with a protractor, is about 39° west of north.

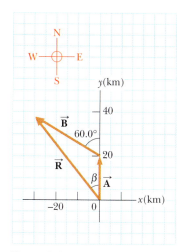

Figure 3.6   (Example 3.1) A graphical method for finding the resultant displacement vector $\vec{\mathbf{R}} = \vec{\mathbf{A}} + \vec{\mathbf{B}}$.

Remarks   Notice that ordinary arithmetic doesn't work here: the correct answer of 48 km is not equal to 20.0 km + 35.0 km = 55.0 km!

Exercise 3.1
Graphically determine the magnitude and direction of the displacement if a man walks 30.0 km 45° north of east and then walks due east 20.0 km.

Answer   46 km, 27° north of east

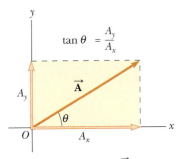

Figure 3.7 Any vector $\vec{A}$ lying in the xy-plane can be represented by its rectangular components A_x and A_y.

TIP 3.2   x- and y-Components

Equation 3.2 for the x- and y-components of a vector associates cosine with the x-component and sine with the y-component, as in Figure 3.8a. This association is due *solely* to the fact that we chose to measure the angle θ with respect to the positive x-axis. If the angle were measured with respect to the y-axis, as in Figure 3.8b, the components would be given by $A_x = A \sin \theta$ and $A_y = A \cos \theta$.

TIP 3.3   Inverse Tangents on Calculators: Right Half the Time

The inverse tangent function on calculators returns an angle between $-90°$ and $+90°$. If the vector lies in the second or third quadrant, the angle, as measured from the positive x-axis, will be the angle returned by your calculator plus 180°.

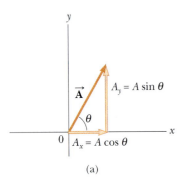

(a)

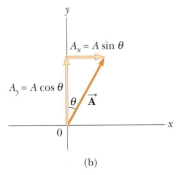

(b)

Figure 3.8 The angle θ need not always be defined from the positive x-axis.

3.2   COMPONENTS OF A VECTOR

One method of adding vectors makes use of the projections of a vector along the axes of a rectangular coordinate system. These projections are called **components**. Any vector can be completely described by its components.

Consider a vector $\vec{A}$ in a rectangular coordinate system, as shown in Figure 3.7. $\vec{A}$ can be expressed as the sum of two vectors: $\vec{A}_x$, parallel to the x-axis; and $\vec{A}_y$, parallel to the y-axis. Mathematically,

$$\vec{A} = \vec{A}_x + \vec{A}_y$$

where $\vec{A}_x$ and $\vec{A}_y$ are the component vectors of $\vec{A}$. The projection of $\vec{A}$ along the x-axis, A_x, is called the x-component of $\vec{A}$, and the projection of $\vec{A}$ along the y-axis, A_y, is called the y-component of $\vec{A}$. These components can be either positive or negative numbers with units. From the definitions of sine and cosine, we see that $\cos \theta = A_x/A$ and $\sin \theta = A_y/A$, so the components of $\vec{A}$ are

$$A_x = A \cos \theta$$
$$A_y = A \sin \theta$$

[3.2]

These components form two sides of a right triangle having a hypotenuse with magnitude A. It follows that $\vec{A}$'s magnitude and direction are related to its components through the Pythagorean theorem and the definition of the tangent:

$$A = \sqrt{A_x^2 + A_y^2}$$

[3.3]

$$\tan \theta = \frac{A_y}{A_x}$$

[3.4]

To solve for the angle θ, which is measured from the positive x-axis by convention, we can write Equation 3.4 in the form

$$\theta = \tan^{-1}\left(\frac{A_y}{A_x}\right)$$

This formula gives the right answer only half the time! The inverse tangent function returns values only from $-90°$ to $+90°$, so the answer in your calculator window will only be correct if the vector happens to lie in first or fourth quadrant. If it lies in second or third quadrant, adding 180° to the number in the calculator window will always give the right answer. The angle in Equations 3.2 and 3.4 must be measured from the positive x-axis. Other choices of reference line are possible, but certain adjustments must then be made. (See Tip 3.2 and Figure 3.8.)

If a coordinate system other than the one shown in Figure 3.7 is chosen, the components of the vector must be modified accordingly. In many applications it's more convenient to express the components of a vector in a coordinate system having axes that are not horizontal and vertical, but are still perpendicular to each other. Suppose a vector $\vec{B}$ makes an angle θ' with the x'-axis defined in Figure 3.9. The rectangular components of $\vec{B}$ along the axes of the figure are given by $B_{x'} = B \cos \theta'$ and $B_{y'} = B \sin \theta'$, as in Equations 3.2. The magnitude and direction of $\vec{B}$ are then obtained from expressions equivalent to Equations 3.3 and 3.4.

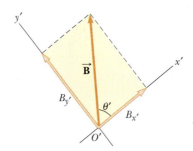

Figure 3.9 The components of vector $\vec{B}$ in a tilted coordinate system.

Quick Quiz 3.3

Figure 3.10 shows two vectors lying in the xy-plane. Determine the signs of the x- and y-components of $\vec{A}$, $\vec{B}$, and $\vec{A} + \vec{B}$, and place your answers in the following table:

Vector	x-component	y-component
$\vec{A}$		
$\vec{B}$		
$\vec{A} + \vec{B}$		

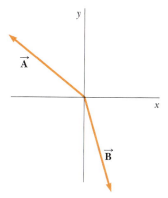

Figure 3.10 (Quick Quiz 3.3)

EXAMPLE 3.2 Help Is on the Way!

Goal Find vector components, given a magnitude and direction, and vice versa.

Problem (a) Find the horizontal and vertical components of the 1.00×10^2 m displacement of a superhero who flies from the top of a tall building along the path shown in Figure 3.11a. (b) Suppose instead the superhero leaps in the other direction along a displacement vector $\vec{B}$, to the top of a flagpole where the displacement components are given by $B_x = -25.0$ m and $B_y = 10.0$ m. Find the magnitude and direction of the displacement vector.

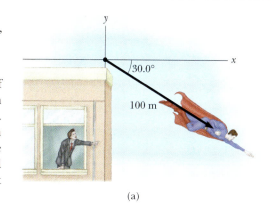

(a)

Strategy (a) The triangle formed by the displacement and its components is shown in Figure 3.11b. Simple trigonometry gives the components relative the standard x-y coordinate system: $A_x = A \cos \theta$ and $A_y = A \sin \theta$ (Equations 3.2). Note that $\theta = -30.0°$, negative because it's measured clockwise from the positive x-axis. (b) Apply Equations 3.3 and 3.4 to find the magnitude and direction of the vector.

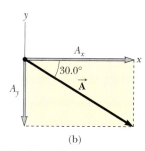

(b)

Figure 3.11 (Example 3.2)

Solution

(a) Find the vector components of $\vec{A}$ from its magnitude and direction.

Use Equations 3.2 to find the components of the displacement vector $\vec{A}$:

$$A_x = A \cos \theta = (1.00 \times 10^2 \text{ m}) \cos(-30.0°) = \boxed{+86.6 \text{ m}}$$

$$A_y = A \sin \theta = (1.00 \times 10^2 \text{ m}) \sin(-30.0°) = \boxed{-50.0 \text{ m}}$$

(b) Find the magnitude and direction of the displacement vector $\vec{B}$ from its components.

Compute the magnitude of $\vec{B}$ from the Pythagorean theorem:

$$B = \sqrt{B_x^2 + B_y^2} = \sqrt{(-25.0 \text{ m})^2 + (10.0 \text{ m})^2} = \boxed{26.9 \text{ m}}$$

Calculate the direction of $\vec{B}$ using the inverse tangent, remembering to add 180° to the answer in your calculator window, because the vector lies in the second quadrant:

$$\theta = \tan^{-1}\left(\frac{B_y}{B_x}\right) = \tan^{-1}\left(\frac{10.0}{-25.0}\right) = -21.8°$$

$$\theta = \boxed{158°}$$

Remarks In part (a), note that $\cos(-\theta) = \cos \theta$; however, $\sin(-\theta) = -\sin \theta$. The negative sign of A_y reflects the fact that displacement in the y-direction is *downward*.

Exercise 3.2

(a) Suppose the superhero had flown 150 m at a 120° angle with respect to the positive x-axis. Find the components of the displacement vector. (b) Suppose instead, the superhero had leaped with a displacement having an x-component of 32.5 m and a y-component of 24.3 m. Find the magnitude and direction of the displacement vector.

Answers   (a) $A_x = -75$ m, $A_y = 130$ m   (b) 40.6 m, 36.8°

Adding Vectors Algebraically

The graphical method of adding vectors is valuable in understanding how vectors can be manipulated, but most of the time vectors are added algebraically in terms of their components. Suppose $\vec{R} = \vec{A} + \vec{B}$. Then the components of the resultant vector $\vec{R}$ are given by

$$R_x = A_x + B_x \qquad\qquad \text{[3.5a]}$$

$$R_y = A_y + B_y \qquad\qquad \text{[3.5b]}$$

So x-components are added only to x-components, and y-components only to y-components. The magnitude and direction of $\vec{R}$ can subsequently be found with Equations 3.3 and 3.4.

Subtracting two vectors works the same way, because it's a matter of adding the negative of one vector to another vector. You should make a rough sketch when adding or subtracting vectors, in order to get an approximate geometric solution as a check.

INTERACTIVE EXAMPLE 3.3   Take a Hike

Goal   Add vectors algebraically and find the resultant vector.

Problem   A hiker begins a trip by first walking 25.0 km southeast from her base camp. On the second day she walks 40.0 km in a direction 60.0° north of east, at which point she discovers a forest ranger's tower. (a) Determine the components of the hiker's displacements in the first and second days. (b) Determine the components of the hiker's total displacement for the trip. (c) Find the magnitude and direction of the displacement from base camp.

Strategy   This is just an application of vector addition using components, Equations 3.5. We denote the displacement vectors on the first and second days by $\vec{A}$ and $\vec{B}$, respectively. Using the camp as the origin of the coordinates, we get the vectors shown in Figure 3.12a. After finding x- and y-components for each vector, we add them "componentwise." Finally, we determine the magnitude and direction of the resultant vector $\vec{R}$, using the Pythagorean theorem and the inverse tangent function.

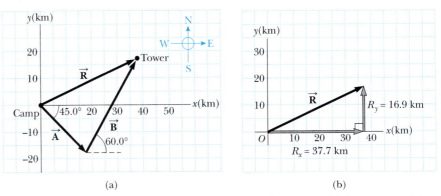

Figure 3.12   (Example 3.3) (a) Hiker's path and the resultant vector. (b) Components of the hiker's total displacement from camp.

Solution

(a) Find the components of $\vec{A}$.

Use Equations 3.2 to find the components of $\vec{A}$:

$$A_x = A \cos(-45.0°) = (25.0 \text{ km})(0.707) = \boxed{17.7 \text{ km}}$$

$$A_y = A \sin(-45.0°) = -(25.0 \text{ km})(0.707) = \boxed{-17.7 \text{ km}}$$

Find the components of $\vec{B}$:

$$B_x = B \cos 60.0° = (40.0 \text{ km})(0.500) = \boxed{20.0 \text{ km}}$$

$$B_y = B \sin 60.0° = (40.0 \text{ km})(0.866) = \boxed{34.6 \text{ km}}$$

(b) Find the components of the resultant vector, $\vec{R} = \vec{A} + \vec{B}$.

To find R_x, add the x-components of $\vec{A}$ and $\vec{B}$:

$$R_x = A_x + B_x = 17.7 \text{ km} + 20.0 \text{ km} = \boxed{37.7 \text{ km}}$$

To find R_y, add the y-components of $\vec{A}$ and $\vec{B}$:

$$R_y = A_y + B_y = -17.7 \text{ km} + 34.6 \text{ km} = \boxed{16.9 \text{ km}}$$

(c) Find the magnitude and direction of $\vec{R}$.

Use the Pythagorean theorem to get the magnitude:

$$R = \sqrt{R_x^2 + R_y^2} = \sqrt{(37.7 \text{ km})^2 + (16.9 \text{ km})^2} = \boxed{41.3 \text{ km}}$$

Calculate the direction of $\vec{R}$ using the inverse tangent function:

$$\theta = \tan^{-1}\left(\frac{16.9 \text{ km}}{37.7 \text{ km}}\right) = \boxed{24.1°}$$

Remarks Figure 3.12b shows a sketch of the components of $\vec{R}$ and their directions in space. The magnitude and direction of the resultant can also be determined from such a sketch.

Exercise 3.3
A cruise ship leaving port, travels 50.0 km 45.0° north of west and then 70.0 km at a heading 30.0° north of east. Find (a) the ship's displacement vector and (b) the displacement vector's magnitude and direction.

Answer (a) $R_x = 25.3$ km, $R_y = 70.4$ km   (b) 74.8 km, 70.2° north of east

Physics⊗Now™ Investigate this problem further by logging into PhysicsNow at **www.cp7e.com** and going to Interactive Example 3.3.

3.3 DISPLACEMENT, VELOCITY, AND ACCELERATION IN TWO DIMENSIONS

In one-dimensional motion, as discussed in Chapter 2, the direction of a vector quantity such as a velocity or acceleration can be taken into account by specifying whether the quantity is positive or negative. The velocity of a rocket, for example, is positive if the rocket is going up and negative if it's going down. This simple solution is no longer available in two or three dimensions. Instead, we must make full use of the vector concept.

Consider an object moving through space as shown in Figure 3.13. When the object is at some point ℗ at time t_i, its position is described by the position vector $\vec{r}_i$, drawn from the origin to ℗. When the object has moved to some other point ℚ at time t_f, its position vector is $\vec{r}_f$. From the vector diagram in Figure 3.13, the final position vector is the sum of the initial position vector and the displacement $\Delta\vec{r}$: $\vec{r}_f = \vec{r}_i + \Delta\vec{r}$. From this relationship, we obtain the following one:

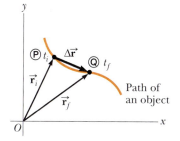

Figure 3.13 An object moving along some curved path between points ℗ and ℚ. The displacement vector $\Delta\vec{r}$ is the difference in the position vectors: $\Delta\vec{r} = \vec{r}_f - \vec{r}_i$.

> An object's **displacement** is defined as the change in its position vector, or
>
> $$\Delta\vec{r} \equiv \vec{r}_f - \vec{r}_i \qquad [3.6]$$
>
> **SI unit: meter (m)**

◀ Displacement vector

We now present several generalizations of the definitions of velocity and acceleration given in Chapter 2.

> An object's **average velocity** during a time interval Δt is its displacement divided by Δt:
>
> $$\vec{v}_{av} \equiv \frac{\Delta\vec{r}}{\Delta t} \qquad [3.7]$$
>
> **SI unit: meter per second (m/s)**

◀ Average velocity

Because the displacement is a vector quantity and the time interval is a scalar quantity, we conclude that the average velocity is a *vector* quantity directed along $\Delta \vec{r}$.

Instantaneous velocity ▶

An object's **instantaneous velocity** $\vec{v}$ is the limit of its average velocity as Δt goes to zero:

$$\vec{v} \equiv \lim_{\Delta t \to 0} \frac{\Delta \vec{r}}{\Delta t} \qquad [3.8]$$

SI unit: meter per second (m/s)

The direction of the instantaneous velocity vector is along a line that is tangent to the object's path and in the direction of its motion.

Average acceleration ▶

An object's **average acceleration** during a time interval Δt is the change in its velocity $\Delta \vec{v}$ divided by Δt, or

$$\vec{a}_{av} \equiv \frac{\Delta \vec{v}}{\Delta t} \qquad [3.9]$$

SI unit: meter per second squared (m/s^2)

Instantaneous acceleration ▶

An object's **instantaneous acceleration** vector $\vec{a}$ is the limit of its average acceleration vector as Δt goes to zero:

$$\vec{a} \equiv \lim_{\Delta t \to 0} \frac{\Delta \vec{v}}{\Delta t} \qquad [3.10]$$

SI unit: meter per second squared (m/s^2)

It's important to recognize that an object can accelerate in several ways. First, the magnitude of the velocity vector (the speed) may change with time. Second, the direction of the velocity vector may change with time, even though the speed is constant, as can happen along a curved path. Third, both the magnitude and the direction of the velocity vector may change at the same time.

Quick Quiz 3.4

Which of the following objects can't be accelerating? (a) An object moving with a constant speed; (b) an object moving with a constant velocity; (c) an object moving along a curve.

Quick Quiz 3.5

Consider the following controls in an automobile: gas pedal, brake, steering wheel. The controls in this list that cause an acceleration of the car are (a) all three controls, (b) the gas pedal and the brake, (c) only the brake, or (d) only the gas pedal.

3.4   MOTION IN TWO DIMENSIONS

In Chapter 2, we studied objects moving along straight-line paths, such as the *x*-axis. In this chapter, we look at objects that move in both the *x*- and *y*-directions simultaneously under constant acceleration. An important special case of this two-

Projectile motion ▶

dimensional motion is called **projectile motion**.

Anyone who has tossed any kind of object into the air has observed projectile motion. If the effects of air resistance and the rotation of Earth are neglected, the

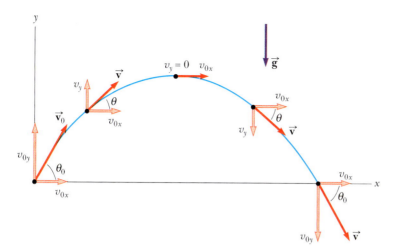

ACTIVE FIGURE 3.14
The parabolic trajectory of a particle that leaves the origin with a velocity of $\vec{v}_0$. Note that $\vec{v}$ changes with time. However, the x-component of the velocity, v_x, remains constant in time. Also, $v_y = 0$ at the peak of the trajectory, but the acceleration is always equal to the free-fall acceleration and acts vertically downward.

Physics⊗Now™
Log into PhysicsNow at **www.cp7e.com**, and go to Active Figure 3.14, where you can change the particle's launch angle and initial speed. You can also observe the changing components of velocity along the trajectory of the projectile.

path of a projectile in Earth's gravity field is curved in the shape of a parabola, as shown in Active Figure 3.14.

The positive x-direction is horizontal and to the right, and the y-direction is vertical and positive upward. The most important experimental fact about projectile motion in two dimensions is that **the horizontal and vertical motions are completely independent of each other**. This means that motion in one direction has no effect on motion in the other direction. If a baseball is tossed in a parabolic path, as in Active Figure 3.14, the motion in the y-direction will look just like a ball tossed straight up under the influence of gravity. Active Figure 3.15 shows the effect of various initial angles; note that complementary angles give the same horizontal range.

In general, the equations of constant acceleration developed in Chapter 2 follow separately for both the x-direction and the y-direction. An important difference is that the initial velocity now has two components, not just one as in that chapter. We assume that at $t = 0$, the projectile leaves the origin with an initial velocity $\vec{v}_0$. If the velocity vector makes an angle θ_0 with the horizontal, where θ_0 is called the *projection angle*, then from the definitions of the cosine and sine functions and Active Figure 3.14, we have

$$v_{0x} = v_0 \cos \theta_0 \qquad \text{and} \qquad v_{0y} = v_0 \sin \theta_0$$

where v_{0x} is the initial velocity (at $t = 0$) in the x-direction and v_{0y} is the initial velocity in the y-direction.

Now, Equations 2.6, 2.9, and 2.10 developed in Chapter 2 for motion with constant acceleration in one dimension carry over to the two-dimensional case; there is one set of three equations for each direction, with the initial velocities modified as just discussed. In the x-direction, with a_x constant, we have

TIP 3.4   Acceleration at the Highest Point

The acceleration in the y-direction is *not* zero at the top of a projectile's trajectory. Only the y-component of the velocity is zero there. If the acceleration were zero, too, the projectile would never come down!

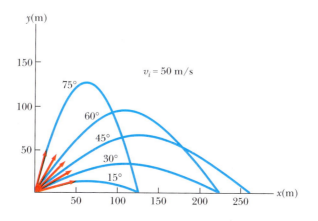

ACTIVE FIGURE 3.15
A projectile launched from the origin with an initial speed of 50 m/s at various angles of projection. Note that complementary values of the initial angle θ result in the same value of R (the range of the projectile).

Physics⊗Now™
Log into PhysicsNow at **www.cp7e.com**, and go to Active Figure 3.15, where you can vary the projection angle to observe the effect on the trajectory and measure the flight time.

$$v_x = v_{0x} + a_x t \qquad \text{[3.11a]}$$

$$\Delta x = v_{0x} t + \tfrac{1}{2} a_x t^2 \qquad \text{[3.11b]}$$

$$v_x^2 = v_{0x}^2 + 2 a_x \Delta x \qquad \text{[3.11c]}$$

where $v_{0x} = v_0 \cos \theta_0$. In the y-direction, we have

$$v_y = v_{0y} + a_y t \qquad \text{[3.12a]}$$

$$\Delta y = v_{0y} t + \tfrac{1}{2} a_y t^2 \qquad \text{[3.12b]}$$

$$v_y^2 = v_{0y}^2 + 2 a_y \Delta y \qquad \text{[3.12c]}$$

where $v_{0y} = v_0 \sin \theta_0$ and a_y is constant. The object's speed v can be calculated from the components of the velocity using the Pythagorean theorem:

$$v = \sqrt{v_x^2 + v_y^2}$$

The angle that the velocity vector makes with the x-axis is given by

$$\theta = \tan^{-1}\left(\frac{v_y}{v_x}\right)$$

This formula for θ, as previously stated, must be used with care, because the inverse tangent function returns values only between $-90°$ and $+90°$. Adding $180°$ is necessary for vectors lying in the second or third quadrant.

The kinematic equations are easily adapted and simplified for projectiles close to the surface of the Earth. In that case, assuming air friction is negligible, the acceleration in the x-direction is 0 (because air resistance is neglected). **This means that $a_x = 0$, and the projectile's velocity component along the x-direction remains constant.** If the initial value of the velocity component in the x-direction is $v_{0x} = v_0 \cos \theta_0$, then this is also the value of v_x at any later time, so

$$v_x = v_{0x} = v_0 \cos \theta_0 = \text{constant} \qquad \text{[3.13a]}$$

while the horizontal displacement is simply

$$\Delta x = v_{0x} t = (v_0 \cos \theta_0) t \qquad \text{[3.13b]}$$

For the motion in the y-direction, we make the substitution $a_y = -g$ and $v_{0y} = v_0 \sin \theta_0$ in Equations 3.12, giving

$$v_y = v_0 \sin \theta_0 - gt \qquad \text{[3.14a]}$$

$$\Delta y = (v_0 \sin \theta_0) t - \tfrac{1}{2} g t^2 \qquad \text{[3.14b]}$$

$$v_y^2 = (v_0 \sin \theta_0)^2 - 2g \Delta y \qquad \text{[3.14c]}$$

The important facts of projectile motion can be summarized as follows:

1. Provided air resistance is negligible, the horizontal component of the velocity v_x remains constant because there is no horizontal component of acceleration.
2. The vertical component of the acceleration is equal to the free fall acceleration $-g$.
3. The vertical component of the velocity v_y and the displacement in the y-direction are identical to those of a freely falling body.
4. Projectile motion can be described as a superposition of two independent motions in the x- and y-directions.

A water fountain. The individual water streams follow parabolic trajectories. The horizontal range and maximum height of a given stream of water depend on the elevation angle of that stream's initial velocity as well as its initial speed.

EXAMPLE 3.4　Projectile Motion with Diagrams

Goal　Approximate answers in projectile motion using a motion diagram.

Problem　A ball is thrown so that its initial vertical and horizontal components of velocity are 40 m/s and 20 m/s, respectively. Use a motion diagram to estimate the ball's total time of flight and the distance it traverses before hitting the ground.

Strategy Use the diagram, estimating the acceleration of gravity as -10 m/s^2. By symmetry, the ball goes up and comes back down to the ground at the same y-velocity as when it left, except with opposite sign. With this fact and the fact that the acceleration of gravity decreases the velocity in the y-direction by 10 m/s every second, we can find the total time of flight and then the horizontal range.

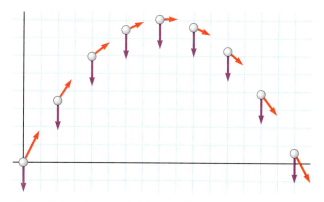

Solution

In the motion diagram shown in Figure 3.16, the acceleration vectors are all the same, pointing downward with magnitude of nearly 10 m/s^2. By symmetry, we know that the ball will hit the ground at the same speed in the y-direction as when it was thrown, so the velocity in the y-direction goes from 40 m/s to -40 m/s in steps of -10 m/s every second; hence, approximately 8 seconds elapse during the motion.

Figure 3.16 (Example 3.4) Motion diagram for a projectile.

The velocity vector constantly changes direction, but the horizontal velocity never changes, because the acceleration in the horizontal direction is zero. Therefore, the displacement of the ball in the x-direction is given by Equation 3.13b, $\Delta x \approx v_{0x}t = (20 \text{ m/s})(8 \text{ s}) = 160$ m.

Remarks This example emphasizes the independence of the x- and y-components in projectile motion problems.

Exercise 3.4

Estimate the maximum height in this same problem.

Answer 80 m

Quick Quiz 3.6

Suppose you are carrying a ball and running at constant speed, and wish to throw the ball and catch it as it comes back down. Should you (a) throw the ball at an angle of about 45° above the horizontal and maintain the same speed, (b) throw the ball straight up in the air and slow down to catch it, or (c) throw the ball straight up in the air and maintain the same speed?

Quick Quiz 3.7

As a projectile moves in its parabolic path, the velocity and acceleration vectors are perpendicular to each other (a) everywhere along the projectile's path, (b) at the peak of its path, (c) nowhere along its path, or (d) not enough information is given.

Problem-Solving Strategy Projectile Motion

1. Select a coordinate system and sketch the path of the projectile, including initial and final positions, velocities, and accelerations.
2. Resolve the initial velocity vector into x- and y-components.
3. Treat the horizontal motion and the vertical motion independently.
4. Follow the techniques for solving problems with constant velocity to analyze the horizontal motion of the projectile.
5. Follow the techniques for solving problems with constant acceleration to analyze the vertical motion of the projectile.

EXAMPLE 3.5   Stranded Explorers

Goal   Solve a two-dimensional projectile motion problem in which an object has an initial horizontal velocity.

Problem   An Alaskan rescue plane drops a package of emergency rations to a stranded hiker, as shown in Figure 3.17. The plane is traveling horizontally at 40.0 m/s at a height of 1.00×10^2 m above the ground. **(a)** Where does the package strike the ground relative to the point at which it was released? **(b)** What are the horizontal and vertical components of the velocity of the package just before it hits the ground?

Strategy   Here, we're just taking Equations 3.13 and 3.14, filling in known quantities, and solving for the remaining unknown quantities. Sketch the problem using a coordinate system as in Figure 3.17. In part (a), set the y-component of the displacement equations equal to -1.00×10^2 m—the ground level where the package lands—and solve for the time it takes the package to reach the ground. Substitute this time into the displacement equation for the x-component to find the range. In part (b), substitute the time found in part (a) into the velocity components. Notice that the initial velocity has only an x-component, which simplifies the math.

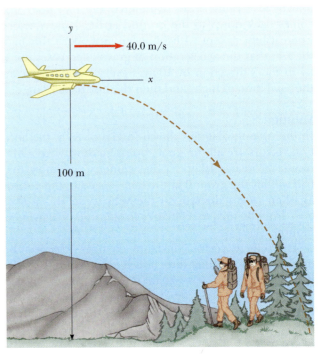

Figure 3.17   (Example 3.5) From the point of view of an observer on the ground, a package released from the rescue plane travels along the path shown.

Solution

(a) Find the range of the package.

Use Equation 3.14b to find the y-displacement:

$$\Delta y = y - y_0 = v_{0y}t - \tfrac{1}{2}gt^2$$

Substitute $y_0 = 0$ and $v_{0y} = 0$, set $y = -1.00 \times 10^2$ m—the final vertical position of the package relative the airplane—and solve for time:

$$y = -(4.90 \text{ m/s}^2)t^2 = -1.00 \times 10^2 \text{ m}$$

$$t = 4.52 \text{ s}$$

Use Equation 3.13b to find the x-displacement:

$$\Delta x = x - x_0 = v_{0x}t$$

Substitute $x_0 = 0$, $v_{0x} = 40.0$ m/s, and the time:

$$x = (40.0 \text{ m/s})(4.52 \text{ s}) = \boxed{181 \text{ m}}$$

(b) Find the components of the package's velocity at impact:

Find the x-component of the velocity at the time of impact:

$$v_x = v_0 \cos \theta = (40.0 \text{ m/s}) \cos 0° = \boxed{40.0 \text{ m/s}}$$

Find the y-component of the velocity at the time of impact:

$$v_y = v_0 \sin \theta - gt = 0 - (9.80 \text{ m/s}^2)(4.52 \text{ s})$$
$$= \boxed{-44.3 \text{ m/s}}$$

Remarks   Notice how motion in the x-direction and motion in the y-direction are handled separately.

Exercise 3.5

A bartender slides a beer mug at 1.50 m/s towards a customer at the end of a frictionless bar that is 1.20 m tall. The customer makes a grab for the mug and misses, and the mug sails off the end of the bar. (a) How far away from the end of the bar does the mug hit the floor? (b) What are the speed and direction of the mug at impact?

Answers   (a) 0.742 m   (b) 5.08 m/s, $\theta = -72.8°$

EXAMPLE 3.6   The Long Jump

Goal   Solve a two-dimensional projectile motion problem involving an object starting and ending at the same height.

Long Jumping

Problem   A long jumper (Fig. 3.18) leaves the ground at an angle of 20.0° to the horizontal and at a speed of 11.0 m/s. **(a)** How long does it take for him to reach maximum height? **(b)** What is the maximum height? **(c)** How far does he jump? (Assume that his motion is equivalent to that of a particle, disregarding the motion of his arms and legs.) **(d)** Find the maximum height he reaches using Equation 3.14c.

Strategy   Again, we take the projectile equations, fill in the known quantities, and solve for the unknowns. At the maximum height, the velocity in the y-direction is zero, so setting Equation 3.14a equal to zero and solving gives the time it takes him to reach his maximum height. By symmetry, given that his trajectory starts and ends at the same height, doubling this time gives the total time of the jump.

Figure 3.18   (Example 3.6) Mike Powell, current holder of the world long-jump record of 8.95 m.

Solution

(a) Find the time t_{max} taken to reach maximum height.

Set $v_y = 0$ in Equation 3.14b and solve for t_{max}:

$$v_y = v_0 \sin \theta_0 - g t_{max} = 0$$

$$t_{max} = \frac{v_0 \sin \theta_0}{g} = \frac{(11.0 \text{ m/s})(\sin 20.0°)}{9.80 \text{ m/s}^2}$$

$$= \boxed{0.384 \text{ s}}$$

(b) Find the maximum height he reaches.

Substitute the time t_{max} into the equation for the y-displacement:

$$y_{max} = (v_0 \sin \theta_0) t_{max} - \tfrac{1}{2} g (t_{max})^2$$

$$y_{max} = (11.0 \text{ m/s})(\sin 20.0°)(0.384 \text{ s})$$
$$- \tfrac{1}{2}(9.80 \text{ m/s}^2)(0.384 \text{ s})^2$$

$$y_{max} = \boxed{0.722 \text{ m}}$$

(c) Find the horizontal distance he jumps.

First find the time for the jump, which is twice t_{max}:

$$t = 2t_{max} = 2(0.384 \text{ s}) = 0.768 \text{ s}.$$

Substitute this result into the equation for the x-displacement:

$$\Delta x = (v_0 \cos \theta_0) t = (11.0 \text{ m/s})(\cos 20.0°)(0.768 \text{ s})$$

$$= \boxed{7.94 \text{ m}}$$

(d) Use an alternate method to find the maximum height.

Use Equation 3.14c, solving for Δy:

$$v_y{}^2 - v_{0y}{}^2 = -2g\Delta y$$

$$\Delta y = \frac{v_y{}^2 - v_{0y}{}^2}{-2g}$$

Substitute $v_y = 0$ at maximum height, and the fact that $v_{0y} = (11.0 \text{ m/s}) \sin 20.0°$:

$$\Delta y = \frac{0 - ((11.0 \text{ m/s}) \sin 20.0°)^2}{-2(9.80 \text{ m/s}^2)} = \boxed{0.722 \text{ m}}$$

Mike Powell/Allsport/Getty Images

Remarks Although modelling the long jumper's motion as that of a projectile is an oversimplification, the values obtained are reasonable.

Exercise 3.6
A grasshopper jumps 1.00 m from rest, with an initial velocity at a 45.0° angle with respect to the horizontal. Find (a) the initial speed of the grasshopper and (b) the maximum height reached.

Answers   (a) 3.13 m/s   (b) 0.250 m

INTERACTIVE EXAMPLE 3.7   That's Quite an Arm

Goal Solve a two-dimensional kinematics problem with a nonhorizontal initial velocity, starting and ending at different heights.

Problem A stone is thrown upward from the top of a building at an angle of 30.0° to the horizontal and with an initial speed of 20.0 m/s, as in Figure 3.19. The point of release is 45.0 m above the ground. **(a)** How long does it take for the stone to hit the ground? **(b)** Find the stone's speed at impact. **(c)** Find the horizontal range of the stone.

Strategy Choose coordinates as in the figure, with the origin at the point of release. **(a)** Fill in the constants of Equation 3.14b for the y-displacement, and set the displacement equal to − 45.0 m, the y-displacement when the stone hits the ground. Using the quadratic formula, solve for the time. To solve **(b)**, substitute the time from part (a) into the components of the velocity, and substitute the same time into the equation for the x-displacement to solve **(c)**.

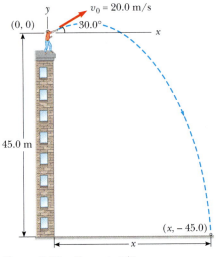

Figure 3.19   (Example 3.7)

Solution
(a) Find the time of flight.

Find the initial x- and y-components of the velocity:

$$v_{0x} = v_0 \cos \theta_0 = (20.0 \text{ m/s}) (\cos 30.0°) = +17.3 \text{ m/s}$$

$$v_{0y} = v_0 \sin \theta_0 = (20.0 \text{ m/s}) (\sin 30.0°) = +10.0 \text{ m/s}$$

Find the y-displacement, taking $y_0 = 0$, $y = -45.0$ m, and $v_{0y} = 10.0$ m/s:

$$\Delta y = y - y_0 = v_{0y}t - \tfrac{1}{2}gt^2$$

$$-45.0 \text{ m} = (10.0 \text{ m/s})t - (4.90 \text{ m/s}^2)t^2$$

Reorganize the equation into standard form and use the quadratic formula (see Appendix A) to find the positive root:

$$t = \boxed{4.22 \text{ s}}$$

(b) Find the speed at impact.

Substitute the value of t found in part (a) into Equation 3.14a to find the y-component of the velocity at impact:

$$v_y = v_{0y} - gt = 10.0 \text{ m/s} - (9.80 \text{ m/s}^2)(4.22 \text{ s})$$
$$= -31.4 \text{ m/s}$$

Use this value of v_y, the Pythagorean theorem, and the fact that $v_x = v_{0x} = 17.3$ m/s to find the speed of the stone at impact:

$$v = \sqrt{v_x^2 + v_y^2} = \sqrt{(17.3 \text{ m/s})^2 + (-31.4 \text{ m/s})^2}$$
$$= \boxed{35.9 \text{ m/s}}$$

(c) Find the horizontal range of the stone.

Substitute the time of flight into the range equation:

$$\Delta x = x - x_0 = (v_0 \cos \theta)t = (20.0 \text{ m/s}) (\cos 30.0°)(4.22 \text{ s})$$
$$= \boxed{73.1 \text{ m}}$$

Exercise 3.7

Suppose the stone is thrown at an angle of 30.0° degrees below the horizontal. If it strikes the ground 57.0 m away, find (a) the time of flight, (b) the initial speed, and (c) the speed and the angle of the velocity vector with respect to the horizontal at impact. (*Hint:* For part (a), use the equation for the *x*-displacement to eliminate $v_0 t$ from the equation for the *y*-displacement.)

Answers   (a) 1.57 s   (b) 41.9 m/s   (c) 51.3 m/s, − 45.0°

Physics⊗Now™ Investigate this problem further by logging into PhysicsNow at **www.cp7e.com** and going to Interactive Example 3.7.

So far we have studied only problems in which an object with an initial velocity follows a trajectory determined by the acceleration of gravity alone. In the more general case, other agents, such as air drag, surface friction, or engines, can cause accelerations. These accelerations, taken together, form a vector quantity with components a_x and a_y. When both components are constant, we can use Equations 3.11 and 3.12 to study the motion, as in the next example.

EXAMPLE 3.8   The Rocket

Goal   Solve a problem involving accelerations in two directions.

Problem   A jet plane traveling horizontally at 1.00×10^2 m/s drops a rocket from a considerable height. (See Figure 3.20.) The rocket immediately fires its engines, accelerating at 20.0 m/s² in the *x*-direction while falling under the influence of gravity in the *y*-direction. When the rocket has fallen 1.00 km, find **(a)** its velocity in the *y*-direction, **(b)** its velocity in the *x*-direction, and **(c)** the magnitude and direction of its velocity. Neglect air drag and aerodynamic lift.

Strategy   Because the rocket maintains a horizontal orientation (say, through gyroscopes), the *x*- and *y*-components of acceleration are independent of each other. Use the time-independent equation for the velocity in the *y*-direction to find the *y*-component of the velocity after the rocket falls 1.00 km. Then calculate the time of the fall, and use that time to find the velocity in the *x*-direction.

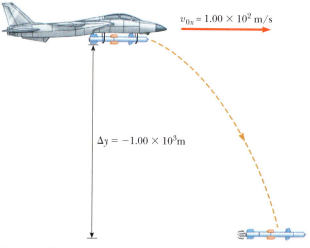

$v_{0x} = 1.00 \times 10^2$ m/s

$\Delta y = -1.00 \times 10^3$ m

Figure 3.20   (Example 3.8)

Solution
(a) Find the velocity in the *y*-direction.

Use Equation 3.14c:

$$v_y{}^2 = v_{0y}{}^2 - 2g\Delta y$$

Substitute $v_{0y} = 0$, $g = -9.80$ m/s², and $\Delta y = -1.00 \times 10^3$ m, and solve:

$$v_y{}^2 - 0 = 2(-9.8 \text{ m/s}^2)(-1.00 \times 10^3 \text{ m})$$

$$v_y = \boxed{-1.40 \times 10^2 \text{ m/s}}$$

(b) Find the velocity in the *x*-direction.

Find the time it takes the rocket to drop 1.00×10^3 m, using the *y*-component of the velocity:

$$v_y = v_{0y} + a_y t$$

$$-1.40 \times 10^2 \text{ m/s} = 0 - (9.80 \text{ m/s}^2)t \quad \rightarrow \quad t = 14.3 \text{ s}$$

Substitute t, v_{0x}, and a_x into Equation 3.11a to find the velocity in the *x*-direction:

$$v_x = v_{0x} + a_x t = 1.00 \times 10^2 \text{ m/s} + (20.0 \text{ m/s}^2)(14.3 \text{ s})$$

$$= \boxed{386 \text{ m/s}}$$

(c) Find the magnitude and direction of the velocity.

Find the magnitude using the Pythagorean theorem and the results of parts (a) and (b):

$$v = \sqrt{v_x{}^2 + v_y{}^2} = \sqrt{(-1.40 \times 10^2 \text{ m/s})^2 + (386 \text{ m/s})^2}$$
$$= \boxed{411 \text{ m/s}}$$

Use the inverse tangent function to find the angle:

$$\theta = \tan^{-1}\left(\frac{v_y}{v_x}\right)\tan^{-1}\left(\frac{-1.40 \times 10^2 \text{ m/s}}{386 \text{ m/s}}\right) = \boxed{-19.9°}$$

Remarks   Notice the symmetry: The kinematic equations for the x- and y-directions are handled in exactly the same way. Having a nonzero acceleration in the x-direction doesn't greatly increase the difficulty of the problem.

Exercise 3.8
Suppose a rocket-propelled motorcycle is fired from rest horizontally across a canyon 1.00 km wide. (a) What minimum constant acceleration in the x-direction must be provided by the engines so the cycle crosses safely if the opposite side is 0.750 km lower than the starting point? (b) At what speed does the motorcycle land if it maintains this constant horizontal component of acceleration? Neglect air drag, but remember that gravity is still acting in the negative y-direction.

Answers   (a) 13.1 m/s^2 (b) 202 m/s

In a stunt similar to that described in Exercise 3.8, motorcycle daredevil Evel Knievel tried to vault across Hells Canyon, part of the Snake River system in Idaho, on his rocket-powered Harley-Davidson X-2 "Skycycle." (See the chapter-opening photo on page 53). He lost consciousness at takeoff and released a lever, prematurely deploying his parachute and falling short of the other side. He landed safely in the canyon.

3.5   RELATIVE VELOCITY

Relative velocity is all about relating the measurements of two different observers, one moving with respect to the other. The measured velocity of an object depends on the velocity of the observer with respect to the object. On highways, for example, cars moving in the same direction are often moving at high speed relative to Earth, but relative each other they hardly move at all. To an observer at rest at the side of the road, a car might be traveling at 60 mi/h, but to an observer in a truck traveling in the same direction at 50 mi/h, the car would appear to be traveling only 10 mi/h.

So measurements of velocity depend on the **reference frame** of the observer. Reference frames are just coordinate systems. Most of the time, we use a **stationary frame of reference** relative to Earth, but occasionally we use a **moving frame of reference** associated with a bus, car, or plane moving with constant velocity relative to Earth.

In two dimensions, relative velocity calculations can be confusing, so a systematic approach is important and useful. Let E be an observer, assumed stationary with respect to Earth. Let two cars be labeled A and B, and introduce the following notation (see Figure 3.21):

$\vec{r}_{AE}$ = the position of Car A as measured by E (in a coordinate system fixed with respect to Earth).

$\vec{r}_{BE}$ = the position of Car B as measured by E.

$\vec{r}_{AB}$ = the position of Car A as measured by an observer in Car B.

According to the preceding notation, the first letter tells us what the vector is pointing at and the second letter tells us where the position vector starts. The position vectors of Car A and Car B relative to E, $\vec{r}_{AE}$ and $\vec{r}_{BE}$, are given in the figure. How do we find $\vec{r}_{AB}$, the position of Car A as measured by an observer in Car B? We simply draw an arrow pointing from Car B to Car A, which can be obtained by subtracting $\vec{r}_{BE}$ from $\vec{r}_{AE}$:

$$\vec{r}_{AB} = \vec{r}_{AE} - \vec{r}_{BE} \qquad [3.15]$$

Now, the rate of change of these quantities with time gives us the relationship

Figure 3.21 The position of Car A relative to Car B can be found by vector subtraction. The rate of change of the resultant vector with respect to time is the relative velocity equation.

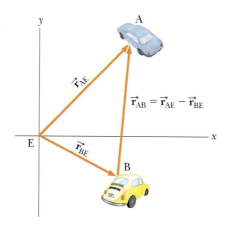

$$\vec{\mathbf{r}}_{AB} = \vec{\mathbf{r}}_{AE} - \vec{\mathbf{r}}_{BE}$$

between the associated velocities:

$$\vec{\mathbf{v}}_{AB} = \vec{\mathbf{v}}_{AE} - \vec{\mathbf{v}}_{BE} \qquad\qquad [3.16]$$

The coordinate system of observer E need not be fixed to Earth, although it often is. Take careful note of the pattern of subscripts; rather than memorize Equation 3.15, it's better to study the short derivation shown in Figure 3.21. Note also that the equation doesn't work for observers traveling a sizeable fraction of the speed of light, when Einstein's theory of special relativity comes into play.

PROBLEM-SOLVING STRATEGY   Relative Velocity

1. Label each object involved (usually three) with a letter that reminds you of what it is (for example, E for Earth).
2. Look through the problem for phrases such as "The velocity of A relative to B," and write the velocities as '$\vec{\mathbf{v}}_{AB}$'. When a velocity is mentioned but it isn't explicitly stated as relative to something, it's almost always relative to Earth.
3. Take the three velocities you've found and assemble them into an equation just like Equation 3.16, with subscripts in an analogous order.
4. There will be two unknown components. Solve for them with the x- and y-components of the equation developed in step 3.

EXAMPLE 3.9   Pitching Practice on the Train

Goal   Solve a one-dimensional relative velocity problem.

Problem   A train is traveling with a speed of 15.0 m/s relative to Earth. A passenger standing at the rear of the train pitches a baseball with a speed of 15.0 m/s relative to the train off the back end, in the direction opposite the motion of the train. What is the velocity of the baseball relative to Earth?

Strategy   Solving these problems involves putting the proper subscripts on the velocities and arranging them as in Equation 3.16. In the first sentence of the problem statement, we are informed that the train travels at "15.0 m/s relative to Earth." This quantity is $\vec{\mathbf{v}}_{TE}$, with T for train and E for Earth. The passenger throws the baseball at "15 m/s relative to the train," so this quantity is $\vec{\mathbf{v}}_{BT}$, where B stands for baseball. The second sentence asks for the velocity of the baseball relative to Earth, $\vec{\mathbf{v}}_{BE}$. The rest of the problem can be solved by identifying the correct components of the known quantities and solving for the unknowns, using an analog of Equation 3.16.

Solution

Write the x-components of the known quantities:

$$(\vec{\mathbf{v}}_{TE})_x = +15 \text{ m/s}$$
$$(\vec{\mathbf{v}}_{BT})_x = -15 \text{ m/s}$$

Follow Equation 3.16:

$$(\vec{\mathbf{v}}_{BT})_x = (\vec{\mathbf{v}}_{BE})_x - (\vec{\mathbf{v}}_{TE})_x$$

Insert the given values, and solve:

$$-15 \text{ m/s} = (\vec{\mathbf{v}}_{BE})_x - 15 \text{ m/s}$$
$$(\vec{\mathbf{v}}_{BE})_x = 0$$

Exercise 3.9

A train is traveling at 27 m/s relative Earth, and a passenger standing in the train throws a ball at 15 m/s relative the train in the same direction as the train's motion. Find the speed of the ball relative to Earth.

Answer   42 m/s

EXAMPLE 3.10   Crossing a River

Goal   Solve a simple two-dimensional relative motion problem.

Problem   The boat in Figure 3.22 is heading due north as it crosses a wide river with a velocity of 10.0 km/h relative to the water. The river has a uniform velocity of 5.00 km/h due east. Determine the velocity of the boat with respect to an observer on the riverbank.

Strategy   Again, we look for key phrases. "The boat (has) . . . a velocity of 10.0 km/h relative to the water" gives $\vec{\mathbf{v}}_{BR}$. "The river has a uniform velocity of 5.00 km/h due east" gives $\vec{\mathbf{v}}_{RE}$, because this implies velocity with respect to Earth. The observer on the riverbank is in a reference frame at rest with respect to Earth. Because we're looking for the velocity of the boat with respect to that observer, this last velocity is designated $\vec{\mathbf{v}}_{BE}$. Take east to be the $+x$-direction, north the $+y$-direction.

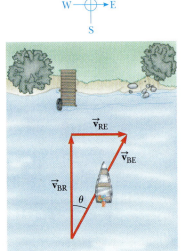

Figure 3.22   (Example 3.10)

Solution

Arrange the three quantities into the proper relative velocity equation:

$$\vec{\mathbf{v}}_{BR} = \vec{\mathbf{v}}_{BE} - \vec{\mathbf{v}}_{RE}$$

Write the velocity vectors in terms of their components. For convenience, these are organized in the following table:

Vector	x-component (km/h)	y-component (km/h)
$\vec{\mathbf{v}}_{BR}$	0	10.0
$\vec{\mathbf{v}}_{BE}$	v_x	v_y
$\vec{\mathbf{v}}_{RE}$	5.00	0

Find the x-component of velocity:

$$0 = v_x - 5.00 \text{ km/h} \quad \rightarrow \quad v_x = 5.00 \text{ km/h}$$

Find the y-component of velocity:

$$10.0 \text{ km/h} = v_y - 0 \quad \rightarrow \quad v_y = 10.0 \text{ km/h}$$

Find the magnitude of $\vec{\mathbf{v}}_{BE}$:

$$v_{BE} = \sqrt{v_x^2 + v_y^2}$$
$$= \sqrt{(5.00 \text{ km/h})^2 + (10.0 \text{ km/h})^2} = \boxed{11.2 \text{ km/h}}$$

Find the direction of $\vec{\mathbf{v}}_{BE}$:

$$\theta = \tan^{-1}\left(\frac{v_x}{v_y}\right) = \tan^{-1}\left(\frac{5.00 \text{ m/s}}{10.0 \text{ m/s}}\right) = \boxed{26.6°}$$

The boat travels at a speed of 11.2 km/h in the direction 26.6° east of north with respect to Earth.

Exercise 3.10

Suppose the river is flowing east at 3.00 m/s and the boat is traveling south at 4.00 m/s with respect to the river. Find the speed and direction of the boat relative to Earth.

Answer   5.00 m/s, 53.1° south of east

EXAMPLE 3.11   Bucking the Current

Goal   Solve a complex two-dimensional relative motion problem.

Problem   If the skipper of the boat of Example 3.10 moves with the same speed of 10.0 km/h relative to the water, but now wants to travel due north, as in Figure 3.23, in what direction should he head? What is the speed of the boat, according to an observer on the shore? The river is flowing east at 5.00 km/h.

Strategy   Proceed as in the previous problem. In this situation, we must find the heading of the boat and its velocity with respect to the water, using the fact that the boat travels due north.

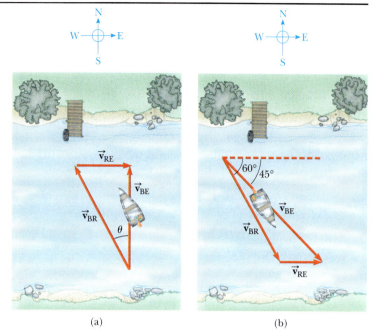

Figure 3.23   (a) (Example 3.11) (b) (Exercise 3.11)

Solution

Arrange the three quantities, as before:

$$\vec{v}_{BR} = \vec{v}_{BE} - \vec{v}_{RE}$$

Organize a table of velocity components:

Vector	x-component (km/h)	y-component (km/h)
$\vec{v}_{BR}$	$-(10.0 \text{ km/h}) \sin \theta$	$(10.0 \text{ km/h}) \cos \theta$
$\vec{v}_{BE}$	0	v
$\vec{v}_{RE}$	5.00 km/h	0

The x-component of the relative velocity equation can be used to find θ:

$$-(10.0 \text{ m/s}) \sin \theta = 0 - 5.00 \text{ km/h}$$

$$\sin \theta = \frac{5.00 \text{ km/h}}{10.0 \text{ km/h}} = \frac{1.00}{2.00}$$

Apply the inverse sine function and find θ, which is the boat's heading, east of north:

$$\theta = \sin^{-1}\left(\frac{1.00}{2.00}\right) = \boxed{30.0°}$$

The y-component of the relative velocity equation can be used to find v:

$$(10.0 \text{ km/h}) \cos \theta = v \quad \rightarrow \quad v = \boxed{8.66 \text{ km/h}}$$

Remarks   From Figure 3.23, we see that this problem can be solved with the Pythagorean theorem, because the problem involves a right triangle: The boat's x-component of velocity exactly cancels the river's velocity. When this is not the case, a more general technique is necessary, as shown in the following exercise. Notice that in the x-component of the relative velocity equation a minus sign had to be included in the term $-(10.0 \text{ km/h}) \sin \theta$ because the x-component of the boat's velocity with respect to the river is negative.

Exercise 3.11

Suppose the river is moving east at 5.00 km/h and the boat is traveling 45.0° south of east with respect to Earth. Find (a) the speed of the boat with respect to Earth and (b) the speed of the boat with respect to the river if the boat's heading in the water is 60.0° south of east. (See Figure 3.23b.) You will have to solve two equations and two unknowns.

Answers   (a) 16.7 km/h   (b) 13.7 km/h

SUMMARY

3.1 Vectors and their Properties

Two vectors $\vec{A}$ and $\vec{B}$ can be added geometrically with the **triangle method**. The two vectors are drawn to scale on graph paper, with the tail of the second vector located at the tip of the first. The **resultant** vector is the vector drawn from the tail of the first vector to the tip of the second.

The negative of a vector $\vec{A}$ is a vector with the same magnitude as $\vec{A}$, but pointing in the opposite direction. A vector can be multiplied by a scalar, changing its magnitude, and its direction if the scalar is negative.

3.2 Components of a Vector

A vector $\vec{A}$ can be split into two components, one pointing in the x-direction and the other in the y-direction. These components form two sides of a right triangle having a hypotenuse with magnitude A and are given by

$$A_x = A \cos \theta \qquad [3.2]$$
$$A_y = A \sin \theta$$

The magnitude and direction of $\vec{A}$ are related to its components through the Pythagorean theorem and the definition of the tangent:

$$A = \sqrt{A_x^2 + A_y^2} \qquad [3.3]$$

$$\tan \theta = \frac{A_y}{A_x} \qquad [3.4]$$

If $\vec{R} = \vec{A} + \vec{B}$, then the components of the resultant vector $\vec{R}$ are

$$R_x = A_x + B_x \qquad [3.5a]$$
$$R_y = A_y + B_y \qquad [3.5b]$$

3.3 Displacement, Velocity, and Acceleration in Two Dimensions

The displacement of an object in two dimensions is defined as the change in the object's position vector:

$$\Delta \vec{r} \equiv \vec{r}_f - \vec{r}_i \qquad [3.6]$$

The average velocity of an object during the time interval Δt is

$$\vec{v}_{av} \equiv \frac{\Delta \vec{r}}{\Delta t} \qquad [3.7]$$

Taking the limit of this expression as Δt gets arbitrarily small gives the instantaneous velocity $\vec{v}$:

$$\vec{v} \equiv \lim_{\Delta t \to 0} \frac{\Delta \vec{r}}{\Delta t} \qquad [3.8]$$

The direction of the instantaneous velocity vector is along a line that is tangent to the path of the object and in the direction of its motion.

The average acceleration of an object with a velocity changing by $\Delta \vec{v}$ in the time interval Δt is

$$\vec{a}_{av} \equiv \frac{\Delta \vec{v}}{\Delta t} \qquad [3.9]$$

Taking the limit of this expression as Δt gets arbitrarily small gives the instantaneous acceleration vector $\vec{a}$:

$$\vec{a} \equiv \lim_{\Delta t \to 0} \frac{\Delta \vec{v}}{\Delta t} \qquad [3.10]$$

3.4 Motion in Two Dimensions

The general kinematic equations in two dimensions for objects with constant acceleration are, for the x-direction,

$$v_x = v_{0x} + a_x t \qquad [3.11a]$$
$$\Delta x = v_{0x} t + \tfrac{1}{2} a_x t^2 \qquad [3.11b]$$
$$v_x^2 = v_{0x}^2 + 2a_x \Delta x \qquad [3.11c]$$

where $v_{0x} = v_0 \cos \theta_0$, and, for the y-direction,

$$v_y = v_{0y} + a_y t \qquad [3.12a]$$
$$\Delta y = v_{0y} t + \tfrac{1}{2} a_y t^2 \qquad [3.12b]$$
$$v_y^2 = v_{0y}^2 + 2a_y \Delta y \qquad [3.12c]$$

where $v_{0y} = v_0 \sin \theta_0$. The speed v of the object at any instant can be calculated from the components of velocity at that instant using the Pythagorean theorem:

$$v = \sqrt{v_x^2 + v_y^2}$$

The angle that the velocity vector makes with the x-axis is given by

$$\theta = \tan^{-1}\left(\frac{v_y}{v_x}\right)$$

The kinematic equations are easily adapted and simplified for projectiles close to the surface of the Earth. The equations for the motion in the horizontal or x-direction are

$$v_x = v_{0x} = v_0 \cos \theta_0 = \text{constant} \qquad [3.13a]$$
$$\Delta x = v_{0x} t = (v_0 \cos \theta_0) t \qquad [3.13b]$$

while the equations for the motion in the vertical or y-direction are

$$v_y = v_0 \sin \theta_0 - gt \qquad [3.14a]$$
$$\Delta y = (v_0 \sin \theta_0) t - \tfrac{1}{2} g t^2 \qquad [3.14b]$$
$$v_y^2 = (v_0 \sin \theta_0)^2 - 2g \Delta y \qquad [3.14c]$$

Problems are solved by algebraically manipulating one or more of these equations, which often reduces the system to two equations and two unknowns.

3.5 Relative Velocity

Let E be an observer, and B a second observer traveling with velocity $\vec{v}_{BE}$ as measured by E. If E measures the velocity of an object A as $\vec{v}_{AE}$, then B will measure A's velocity as

$$\vec{v}_{AB} = \vec{v}_{AE} - \vec{v}_{BE} \qquad [3.16]$$

Solving relative velocity problems involves identifying the velocities properly and labeling them correctly, substituting into Equation 3.16, and then solving for unknown quantities.

CONCEPTUAL QUESTIONS

1. Vector $\vec{A}$ lies in the *xy*-plane. For what orientations of vector $\vec{A}$ will both of its components be negative? When will the components have opposite signs?

2. If $\vec{B}$ is added to $\vec{A}$, under what conditions does the resultant vector have a magnitude equal to $A + B$? Under what conditions is the resultant vector equal to zero?

3. A wrench is dropped from the top of a 10-m mast on a sailing ship while the ship is traveling at 20 knots. Where will the wrench hit the deck? (Galileo posed this problem.)

4. Two vectors have unequal magnitudes. Can their sum be zero? Explain.

5. A projectile is fired on Earth with some initial velocity. Another projectile is fired on the Moon with the same initial velocity. If air resistance is neglected, which projectile has the greater range? Which reaches the greater altitude? (Note that the free-fall acceleration on the Moon is about 1.6 m/s^2.)

6. Can a vector $\vec{A}$ have a component greater than its magnitude *A*?

7. Is it possible to add a vector quantity to a scalar quantity?

8. Under what circumstances would a vector have components that are equal in magnitude?

9. As a projectile moves in its path, is there any point along the path where the velocity and acceleration vectors are (a) perpendicular to each other? (b) parallel to each other?

10. A rock is dropped at the same instant that a ball is thrown horizontally from the same initial elevation. Which will have the greater speed when it reaches ground level?

11. Explain whether the following particles do or do not have an acceleration: (a) a particle moving in a straight line with constant speed and (b) a particle moving around a curve with constant speed.

12. Correct the following statement: "The racing car rounds the turn at a constant velocity of 90 miles per hour."

13. A spacecraft drifts through space at a constant velocity. Suddenly, a gas leak in the side of the spacecraft causes it to constantly accelerate in a direction perpendicular to the initial velocity. The orientation of the spacecraft does not change, so the acceleration remains perpendicular to the original direction of the velocity. What is the shape of the path followed by the spacecraft?

14. A ball is projected horizontally from the top of a building. One second later, another ball is projected horizontally from the same point with the same velocity. At what point in the motion will the balls be closest to each other? Will the first ball always be traveling faster than the second? What will be the time difference between them when the balls hit the ground? Can the horizontal projection velocity of the second ball be changed so that the balls arrive at the ground at the same time?

15. Two projectiles are thrown with the same initial speed, one at an angle θ with respect to the level ground and the other at angle $90° - \theta$. Both projectiles strike the ground at the same distance from the projection point. Are both projectiles in the air for the same length of time?

16. A baseball is thrown such that its initial *x*- and *y*-components of velocity are known. Neglecting air resistance, describe how you would calculate the ball's (a) coordinates, (b) velocity, and (c) acceleration at the instant the ball reaches the top of its trajectory. How would these results change if air resistance were taken into account?

17. A projectile is fired at some angle to the horizontal with some initial speed v_0, and air resistance is neglected. Is the projectile a freely falling body? What is its acceleration in the vertical direction? What is its acceleration in the horizontal direction?

18. Determine which of the following moving objects obey the equations of projectile motion developed in this chapter: (a) A ball is thrown in an arbitrary direction. (b) A jet airplane crosses the sky with its engines thrusting the plane forward. (c) A rocket leaves the launch pad. (d) A rocket moves through the sky after its engines have failed. (e) A stone is thrown under water.

19. How can you throw a projectile so that it has zero speed at the top of its trajectory? So that it has nonzero speed at the top of its trajectory?

20. A ball is thrown upward in the air by a passenger on a train that is moving with constant velocity. (a) Describe the path of the ball as seen by the passenger. Describe the path as seen by a stationary observer outside the train. (b) How would these observations change if the train were accelerating along the track?

PROBLEMS

1, 2, 3 = straightforward, intermediate, challenging   □ = full solution available in *Student Solutions Manual/Study Guide*

Physics⊗Now™ = coached problem with hints available at **www.cp7e.com**   = biomedical application

Section 3.1   Vectors and Their Properties

1. A roller coaster moves 200 ft horizontally and then rises 135 ft at an angle of 30.0° above the horizontal. Next, it travels 135 ft at an angle of 40.0° below the horizontal. Use graphical techniques to find the roller coaster's displacement from its starting point to the end of this movement.

2. An airplane flies 200 km due west from city A to city B and then 300 km in the direction of 30.0° north of west from city B to city C. (a) In straight-line distance, how far is city C from city A? (b) Relative to city A, in what direction is city C?

3. A man lost in a maze makes three consecutive displacements so that at the end of his travel he is right back where he started. The first displacement is 8.00 m westward, and the second is 13.0 m northward. Use the graphical method to find the magnitude and direction of the third displacement.

4. A jogger runs 100 m due west, then changes direction for the second leg of the run. At the end of the run, she is 175 m away from the starting point at an angle of 15.0° north of west. What were the direction and length of her second displacement? Use graphical techniques.

5. A plane flies from base camp to lake A, a distance of 280 km at a direction of 20.0° north of east. After dropping off supplies, the plane flies to lake B, which is 190 km and 30.0° west of north from lake A. Graphically determine the distance and direction from lake B to the base camp.

6. Vector $\vec{A}$ has a magnitude of 8.00 units and makes an angle of 45.0° with the positive x-axis. Vector $\vec{B}$ also has a magnitude of 8.00 units and is directed along the negative x-axis. Using graphical methods, find (a) the vector sum $\vec{A} + \vec{B}$ and (b) the vector difference $\vec{A} - \vec{B}$.

7. Vector $\vec{A}$ is 3.00 units in length and points along the positive x-axis. Vector $\vec{B}$ is 4.00 units in length and points along the negative y-axis. Use graphical methods to find the magnitude and direction of the vectors (a) $\vec{A} + \vec{B}$ and (b) $\vec{A} - \vec{B}$.

8. Each of the displacement vectors $\vec{A}$ and $\vec{B}$ shown in Figure P3.8 has a magnitude of 3.00 m. Graphically find (a) $\vec{A} + \vec{B}$, (b) $\vec{A} - \vec{B}$, (c) $\vec{B} - \vec{A}$, and (d) $\vec{A} - 2\vec{B}$.

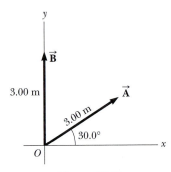

Figure P3.8

Section 3.2   Components of a Vector

9. A golfer takes two putts to get his ball into the hole once he is on the green. The first putt displaces the ball 6.00 m east, the second 5.40 m south. What displacement would have been needed to get the ball into the hole on the first putt?

10. A person walks 25.0° north of east for 3.10 km. How far would the person walk due north and due east to arrive at the same location?

11. **Physics⊗Now™** A girl delivering newspapers covers her route by traveling 3.00 blocks west, 4.00 blocks north, and then 6.00 blocks east. (a) What is her resultant displacement? (b) What is the total distance she travels?

12. While exploring a cave, a spelunker starts at the entrance and moves the following distances: 75.0 m north, 250 m east, 125 m at an angle 30.0° north of east, and 150 m south. Find the resultant displacement from the cave entrance.

13. A vector has an x-component of -25.0 units and a y-component of 40.0 units. Find the magnitude and direction of the vector.

14. A quarterback takes the ball from the line of scrimmage, runs backwards for 10.0 yards, then runs sideways parallel to the line of scrimmage for 15.0 yards. At this point, he throws a 50.0-yard forward pass straight downfield, perpendicular to the line of scrimmage. What is the magnitude of the football's resultant displacement?

15. The eye of a hurricane passes over Grand Bahama Island in a direction 60.0° north of west with a speed of 41.0 km/h. Three hours later, the course of the hurricane suddenly shifts due north, and its speed slows to 25.0 km/h. How far from Grand Bahama is the hurricane 4.50 h after it passes over the island?

16. A small map shows Atlanta to be 730 miles in a direction 5° north of east from Dallas. The same map shows that Chicago is 560 miles in a direction 21° west of north from Atlanta. Assume a flat Earth, and use the given information to find the displacement from Dallas to Chicago.

17. A commuter airplane starts from an airport and takes the route shown in Figure P3.17. The plane first flies to city A, located 175 km away in a direction 30.0° north of east. Next, it flies for 150 km 20.0° west of north, to city B. Finally, the plane flies 190 km due west, to city C. Find the location of city C relative to the location of the starting point.

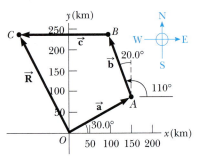

Figure P3.17

18. The helicopter view in Figure P3.18 shows two people pulling on a stubborn mule. Find (a) the single force that is equivalent to the two forces shown and (b) the force that a third person would have to exert on the mule to make the net force equal to zero. The forces are measured in units of newtons (N).

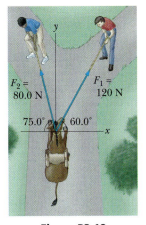

Figure P3.18

19. A man pushing a mop across a floor causes the mop to undergo two displacements. The first has a magnitude of 150 cm and makes an angle of 120° with the positive x-axis. The resultant displacement has a magnitude of 140 cm and is directed at an angle of 35.0° to the positive x-axis. Find the magnitude and direction of the second displacement.

20. An airplane starting from airport A flies 300 km east, then 350 km at 30.0° west of north, and then 150 km north to arrive finally at airport B. (a) The next day, another plane

flies directly from A to B in a straight line. In what direction should the pilot travel in this direct flight? (b) How far will the pilot travel in the flight? Assume there is no wind during either flight.

21. Long John Silver, a pirate, has buried his treasure on an island with five trees located at the following points: A (30.0 m, −20.0 m), B (60.0 m, 80.0 m), C (−10.0 m, −10.0 m), D (40.0 m, −30.0 m), and E (−70.0 m, 60.0 m). All of the points are measured relative to some origin, as in Figure P3.21. Long John's map instructs you to start at A and move toward B, but cover only one-half the distance between A and B. Then move toward C, covering one-third the distance between your current location and C. Then move toward D, covering one-fourth the distance between where you are and D. Finally, move toward E, covering one-fifth the distance between you and E, stop, and dig. (a) What are the coordinates of the point where the pirate's treasure is buried? (b) Rearrange the order of the trees — for instance, B (30 m, −20 m), A (60 m, 80 m), E (−10 m, −10 m), C (40 m, −30 m), and D (−70 m, 60 m) — and repeat the calculation to show that the answer does not depend on the order.

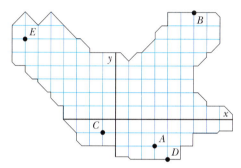

Figure P3.21

Section 3.3 Displacement, Velocity, and Acceleration in Two Dimensions
Section 3.4 Motion in Two Dimensions

22. One of the fastest recorded pitches in major-league baseball, thrown by Billy Wagner in 2003, was clocked at 101.0 mi/h (Fig. P3.22). If a pitch were thrown horizontally with this velocity, how far would the ball fall vertically by the time it reached home plate, 60.5 ft away?

Figure P3.22   Billy Wagner throws a baseball.

23. A peregrine falcon is the fastest bird, flying at a speed of 200 mi/h. Nature has adapted the bird to reach such a speed by placing baffles in its nose to prevent air from rushing in and slowing it down. Also, the bird's eyes adjust their focus faster than the eyes of any other creature, so the falcon can focus quickly on its prey. Assume that a peregrine falcon is moving horizontally at its top speed at a height of 100 m above the ground when it brings its wings into its sides and begins to drop in free fall. How far will the bird fall vertically while traveling horizontally a distance of 100 m?

Figure P3.23   Notice the structure within the peregrine falcon's nostrils.

24. A student stands at the edge of a cliff and throws a stone horizontally over the edge with a speed of 18.0 m/s. The cliff is 50.0 m above a flat, horizontal beach, as shown in Figure P3.24. How long after being released does the stone strike the beach below the cliff? With what speed and angle of impact does the stone land?

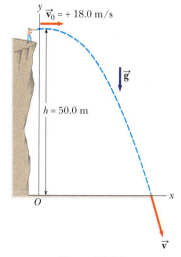

Figure P3.24

25. The best leaper in the animal kingdom is the puma, which can jump to a height of 12 ft when leaving the ground at an angle of 45°. With what speed, in SI units, must the animal leave the ground to reach that height?

26. Tom the cat is chasing Jerry the mouse across the surface of a table 1.5 m above the floor. Jerry steps out of the way at the last second, and Tom slides off the edge of the table at a speed of 5.0 m/s. Where will Tom strike the floor, and what velocity components will he have just before he hits?

27. **Physics⊗Now™** A tennis player standing 12.6 m from the net hits the ball at 3.00° above the horizontal. To clear the net, the ball must rise at least 0.330 m. If the ball just clears the net at the apex of its trajectory, how fast was the ball moving when it left the racquet?

28. An artillery shell is fired with an initial velocity of 300 m/s at 55.0° above the horizontal. To clear an avalanche, it explodes on a mountainside 42.0 s after firing. What are the x- and y-coordinates of the shell where it explodes, relative to its firing point?

29. A brick is thrown upward from the top of a building at an angle of 25° to the horizontal and with an initial speed of 15 m/s. If the brick is in flight for 3.0 s, how tall is the building?

30. A placekicker must kick a football from a point 36.0 m (about 39 yd) from the goal, and the ball must clear the crossbar, which is 3.05 m high. When kicked, the ball leaves the ground with a velocity of 20.0 m/s at an angle of 53° to the horizontal. (a) By how much does the ball clear or fall short of clearing the crossbar? (b) Does the ball approach the crossbar while still rising or while falling?

31. A car is parked on a cliff overlooking the ocean on an incline that makes an angle of 24.0° below the horizontal. The negligent driver leaves the car in neutral, and the emergency brakes are defective. The car rolls from rest down the incline with a constant acceleration of 4.00 m/s^2 for a distance of 50.0 m to the edge of the cliff, which is 30.0 m above the ocean. Find (a) the car's position relative to the base of the cliff when the car lands in the ocean and (b) the length of time the car is in the air.

32. A fireman 50.0 m away from a burning building directs a stream of water from a ground-level fire hose at an angle of 30.0° above the horizontal. If the speed of the stream as it leaves the hose is 40.0 m/s, at what height will the stream of water strike the building?

33. A projectile is launched with an initial speed of 60.0 m/s at an angle of 30.0° above the horizontal. The projectile lands on a hillside 4.00 s later. Neglect air friction. (a) What is the projectile's velocity at the highest point of its trajectory? (b) What is the straight-line distance from where the projectile was launched to where it hits its target?

34. A soccer player kicks a rock horizontally off a 40.0-m-high cliff into a pool of water. If the player hears the sound of the splash 3.00 s later, what was the initial speed given to the rock? Assume the speed of sound in air to be 343 m/s.

Section 3.5   Relative Velocity

35. A jet airliner moving initially at 300 mi/h due east enters a region where the wind is blowing at 100 mi/h in a direction 30.0° north of east. What is the new velocity of the aircraft relative to the ground?

36. A boat moves through the water of a river at 10 m/s relative to the water, regardless of the boat's direction. If the water in the river is flowing at 1.5 m/s, how long does it take the boat to make a round trip consisting of a 300-m displacement downstream followed by a 300-m displacement upstream?

37. A Chinook (King) salmon (Genus *Oncorynchus*) can jump out of water with a speed of 6.26 m/s. (See Problem 4.9, page 109 for an investigation of how the fish can leave the water at a higher speed than it can swim underwater.) If the salmon is in a stream with water speed equal to 1.50 m/s, how high in the air can the fish jump if it leaves the water traveling vertically upwards relative to the Earth?

38. A river flows due east at 1.50 m/s. A boat crosses the river from the south shore to the north shore by maintaining a constant velocity of 10.0 m/s due north relative to the water. (a) What is the velocity of the boat relative to the shore? (b) If the river is 300 m wide, how far downstream has the boat moved by the time it reaches the north shore?

39. A rowboat crosses a river with a velocity of 3.30 mi/h at an angle 62.5° north of west relative to the water. The river is 0.505 mi wide and carries an eastward current of 1.25 mi/h. How far upstream is the boat when it reaches the opposite shore?

40. Suppose a Chinook salmon needs to jump a waterfall that is 1.50 m high. If the fish starts from a distance 1.00 m from the base of the ledge over which the waterfall flows, find the x- and y-components of the initial velocity the salmon would need to just reach the ledge at the top of its trajectory. Can the fish make this jump? (Remember that a Chinook salmon can jump out of the water with a speed of 6.26 m/s.)

41. How long does it take an automobile traveling in the left lane of a highway at 60.0 km/h to overtake (become even with) another car that is traveling in the right lane at 40.0 km/h when the cars' front bumpers are initially 100 m apart?

42. A science student is riding on a flatcar of a train traveling along a straight horizontal track at a constant speed of 10.0 m/s. The student throws a ball along a path that she judges to make an initial angle of 60.0° with the horizontal and to be in line with the track. The student's professor, who is standing on the ground nearby, observes the ball to rise vertically. How high does the ball rise?

ADDITIONAL PROBLEMS

43. A particle undergoes two displacements. The first has a magnitude of 150 cm and makes an angle of 120.0° with the positive x-axis. The resultant of the two displacements is 140 cm, directed at an angle of 35.0° to the positive x-axis. Find the magnitude and direction of the second displacement.

44. Find the sum of these four vector forces: 12.0 N to the right at 35.0° above the horizontal, 31.0 N to the left at 55.0° above the horizontal, 8.40 N to the left at 35.0° below the horizontal, and 24.0 N to the right at 55.0° below the horizontal. (*Hint:* N stands for newton, the SI unit of force. The component method allows the addition of any vectors—forces as well as displacements and velocities. Make a drawing of this situation, and select the best axes for x and y so that you have the least number of components.)

45. A car travels due east with a speed of 50.0 km/h. Rain is falling vertically with respect to Earth. The traces of the rain on the side windows of the car make an angle of 60.0° with the vertical. Find the velocity of the rain with respect to (a) the car and (b) Earth.

46. You can use any coordinate system you like in order to solve a projectile motion problem. To demonstrate the

truth of this statement, consider a ball thrown off the top of a building with a velocity $\vec{v}$ at an angle θ with respect to the horizontal. Let the building be 50.0 m tall, the initial horizontal velocity be 9.00 m/s, and the initial vertical velocity be 12.0 m/s. Choose your coordinates such that the positive y-axis is upward, the x-axis is to the right, and the origin is at the point where the ball is released. (a) With these choices, find the ball's maximum height above the ground, and the time it takes to reach the maximum height. (b) Repeat your calculations choosing the origin at the base of the building.

47. **Physics** **Now**™ Towns A and B in Figure P3.47 are 80.0 km apart. A couple arranges to drive from town A and meet a couple driving from town B at the lake, L. The two couples leave simultaneously and drive for 2.50 h in the directions shown. Car 1 has a speed of 90.0 km/h. If the cars arrive simultaneously at the lake, what is the speed of car 2?

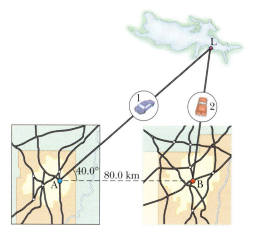

Figure P3.47

48. A Chinook salmon has a maximum underwater speed of 3.58 m/s, but it can jump out of water with a speed of 6.26 m/s. To move upstream past a waterfall, the salmon does not need to jump to the top of the fall, but only to a point in the fall where the water speed is less than 3.58 m/s; it can then swim up the fall for the remaining distance. Because the salmon must make forward progress in the water, let's assume that it can swim to the top if the water speed is 3.00 m/s. If water has a speed of 1.50 m/s as it passes over a ledge, how far below the ledge will the water be moving with a speed of 3.00 m/s? (Note that water undergoes projectile motion once it leaves the ledge.) If the salmon is able to jump vertically upward from the base of the fall, what is the maximum height of waterfall that the salmon can clear?

49. A rocket is launched at an angle of 53.0° above the horizontal with an initial speed of 100 m/s. The rocket moves for 3.00 s along its initial line of motion with an acceleration of 30.0 m/s². At this time, its engines fail and the rocket proceeds to move as a projectile. Find (a) the maximum altitude reached by the rocket, (b) its total time of flight, and (c) its horizontal range.

50. Two canoeists in identical canoes exert the same effort paddling and hence maintain the same speed relative to the water. One paddles directly upstream (and moves upstream), whereas the other paddles directly downstream. With downstream as the positive direction, an observer on shore determines the velocities of the two canoes to be −1.2 m/s and +2.9 m/s, respectively. (a) What is the speed of the water relative to the shore? (b) What is the speed of each canoe relative to the water?

51. If a person can jump a maximum horizontal distance (by using a 45° projection angle) of 3.0 m on Earth, what would be his maximum range on the Moon, where the free-fall acceleration is g/6 and g = 9.80 m/s²? Repeat for Mars, where the acceleration due to gravity is 0.38g.

52. A daredevil decides to jump a canyon. Its walls are equally high and 10 m apart. He takes off by driving a motorcycle up a short ramp sloped at an angle of 15°. What minimum speed must he have in order to clear the canyon?

53. (a) Vector $\vec{A}$ is in the first quadrant of a Cartesian coordinate system. What is the sign of the x-component of $\vec{A}$? What is the sign of the y-component? (b) Vector $\vec{B}$ is in the second quadrant of a Cartesian coordinate system. What is the sign of the x-component of $\vec{B}$? What is the sign of the y-component? (c) The vector sum $\vec{A} + \vec{B}$ _____. Choose the correct fill-in-the-blank answer from (i) must be in either the first or the second quadrant or (ii) could be in any quadrant. (d) Let $\vec{A} = 30$ m at an angle of 30° from the positive x-axis and $\vec{B} = 20$ m at an angle of 40° from the negative x-axis. Test your predictions of parts (a) through (c).

54. A boy and a girl are tossing an apple back and forth between them. Figure P3.54 shows one path the apple follows when watched by an observer looking on from the side. The apple is moving from left to right. Five points are marked on the path. Ignore air resistance. (a) Make a copy of this figure. At each of the marked points, draw an arrow that indicates the magnitude and direction of the apple's velocity when it passes through that point. (b) Make a second copy of the figure. This time, at each marked point, place an arrow indicating the magnitude and direction of any acceleration the apple exhibits at that point.

Figure P3.54

55. A home run is hit in such a way that the baseball just clears a wall 21 m high, located 130 m from home plate. The ball is hit at an angle of 35° to the horizontal, and air resistance is negligible. Find (a) the initial speed of the ball, (b) the time it takes the ball to reach the wall, and (c) the velocity components and the speed of the ball when it reaches the wall. (Assume that the ball is hit at a height of 1.0 m above the ground.)

56. A ball is thrown straight upward and returns to the thrower's hand after 3.00 s in the air. A second ball is thrown at an angle of 30.0° with the horizontal. At what speed must the second ball be thrown so that it reaches the same height as the one thrown vertically?

57. A quarterback throws a football toward a receiver with an initial speed of 20 m/s at an angle of 30° above the horizontal. At that instant, the receiver is 20 m from the quarterback. In what direction and with what constant speed should the receiver run in order to catch the football at the level at which it was thrown?

58. A 2.00-m-tall basketball player is standing on the floor 10.0 m from the basket, as in Figure P3.58. If he shoots the ball at a 40.0° angle with the horizontal, at what initial speed must he throw the basketball so that it goes through the hoop without striking the backboard? The height of the basket is 3.05 m.

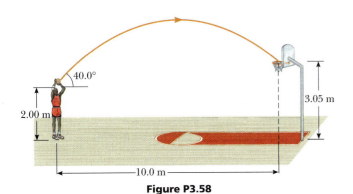

Figure P3.58

59. In a very popular lecture demonstration, a projectile is fired at a falling target as in Figure P3.59. The projectile leaves the gun at the same instant that the target is dropped from rest. Assuming that the gun is initially aimed at the target, show that the projectile will hit the target. (One restriction of this experiment is that the projectile must reach the target before the target strikes the floor.)

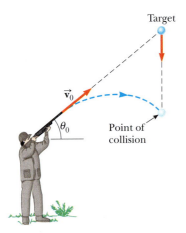

Figure P3.59

60. Figure P3.60 illustrates the difference in proportions between the male (m) and female (f) anatomies. The displacements $\vec{d}_{1m}$ and $\vec{d}_{1f}$ from the bottom of the feet to the navel have magnitudes of 104 cm and 84.0 cm, respectively. The displacements $\vec{d}_{2m}$ and $\vec{d}_{2f}$ have magnitudes of 50.0 cm and 43.0 cm, respectively. (a) Find the vector sum of the displacements $\vec{d}_1$ and $\vec{d}_2$ in each case. (b) The male figure is 180 cm tall, the female 168 cm. Normalize the displacements of each figure to a common height of

200 cm, and re-form the vector sums as in part (a). Then find the vector difference between the two sums.

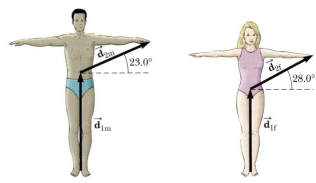

Figure P3.60

61. By throwing a ball at an angle of 45°, a girl can throw the ball a maximum horizontal distance R on a level field. How far can she throw the same ball vertically upward? Assume that her muscles give the ball the same speed in each case. (Is this assumption valid?)

62. A projectile is fired with an initial speed v_0 at an angle θ_0 to the horizontal, as in Figure 3.14. When it reaches its peak, the projectile has (x, y) coordinates given by $(R/2, h)$, and when it strikes the ground, its coordinates are $(R, 0)$, where R is called the *horizontal range*. (a) Show that the projectile reaches a maximum height given by

$$ h = \frac{v_0{}^2 \sin^2 \theta_0}{2g} $$

(b) Show that the horizontal range of the projectile is given by

$$ R = \frac{v_0{}^2 \sin 2\,\theta_0}{g} $$

63. A hunter wishes to cross a river that is 1.5 km wide and flows with a speed of 5.0 km/h parallel to its banks. The hunter uses a small powerboat that moves at a maximum speed of 12 km/h with respect to the water. What is the minimum time necessary for crossing?

64. A water insect maintains an average position on the surface of a stream by darting upstream (against the current) then drifting downstream (with the current) to its original position. The current in the stream is 0.500 m/s relative to the shore, and the insect darts upstream 0.560 m (relative to a spot on shore) in 0.800 s during the first part of its motion. Take upstream as the positive direction. (a) Determine the velocity of the insect relative to the water (i) during its dash upstream and (ii) during its drift downstream. (b) How far upstream relative to the water does the insect move during one cycle of its motion? (c) What is the average velocity of the insect relative to the water?

65. **Physics⊗Now™** A daredevil is shot out of a cannon at 45.0° to the horizontal with an initial speed of 25.0 m/s. A net is positioned a horizontal distance of 50.0 m from the cannon. At what height above the cannon should the net be placed in order to catch the daredevil?

66. Chinook salmon are able to move upstream faster by jumping out of the water periodically; this behavior is called *porpoising*. Suppose a salmon swimming in still water jumps out of the water with a speed of 6.26 m/s at an angle of 45°, sails through the air a distance L before returning to the water, and then swims a distance L underwater at a speed of 3.58 m/s before beginning another porpoising maneuver. Determine the average speed of the fish.

67. A student decides to measure the muzzle velocity of a pellet shot from his gun. He points the gun horizontally. He places a target on a vertical wall a distance x away from the gun. The pellet hits the target a vertical distance y below the gun. (a) Show that the position of the pellet when traveling through the air is given by $y = Ax^2$, where A is a constant. (b) Express the constant A in terms of the initial (muzzle) velocity and the free-fall acceleration. (c) If x = 3.00 m and y = 0.210 m, what is the initial speed of the pellet?

68. A sailboat is heading directly north at a speed of 20 knots (1 knot = 0.514 m/s). The wind is blowing towards the east with a speed of 17 knots. Determine the magnitude and direction of the wind velocity as measured on the boat. What is the component of the wind velocity in the direction parallel to the motion of the boat? (See Problem 4.54 for an explanation of how a sailboat can move "into the wind.")

69. Instructions for finding a buried treasure include the following: Go 75 paces at 240°, turn to 135° and walk 125 paces, and then travel 100 paces at 160°. Determine the resultant displacement from the starting point.

70. When baseball outfielders throw the ball, they usually allow it to take one bounce, on the theory that the ball arrives at its target sooner that way. Suppose that, after the bounce, the ball rebounds at the same angle θ that it had when it was released (as in Fig. P3.70), but loses half its speed. (a) Assuming that the ball is always thrown with the same initial speed, at what angle θ should the ball be thrown in order to go the same d istance D with one bounce as a ball thrown upward at 45.0° with no bounce? (b) Determine the ratio of the times for the one-bounce and no-bounce throws.

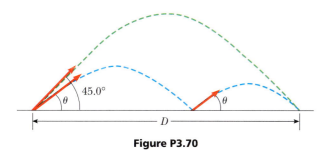

Figure P3.70

71. One strategy in a snowball fight is to throw a snowball at a high angle over level ground. Then, while your opponent is watching that snowball, you throw a second one at a low angle timed to arrive before or at the same time as the first one. Assume that both snowballs are thrown with a speed of 25.0 m/s. The first is thrown at an angle of 70.0° with respect to the horizontal. (a) At what angle should the second snowball be thrown to arrive at the same point

as the first? (b) How many seconds later should the second snowball be thrown after the first in order for both to arrive at the same time?

72. A dart gun is fired while being held horizontally at a height of 1.00 m above ground level and while it is at rest relative to the ground. The dart from the gun travels a horizontal distance of 5.00 m. A college student holds the same gun in a horizontal position while sliding down a 45.0° incline at a constant speed of 2.00 m/s. How far will the dart travel if the student fires the gun when it is 1.00 m above the ground?

73. The determined Wile E. Coyote is out once more to try to capture the elusive roadrunner. The coyote wears a new pair of Acme power roller skates, which provide a constant horizontal acceleration of 15 m/s², as shown in Figure P3.73. The coyote starts off at rest 70 m from the edge of a cliff at the instant the roadrunner zips by in the direction of the cliff. (a) If the roadrunner moves with constant speed, find the minimum speed the roadrunner must have in order to reach the cliff before the coyote. (b) If the cliff is 100 m above the base of a canyon, find where the coyote lands in the canyon. (Assume that his skates are still in operation when he is in "flight" and that his horizontal component of acceleration remains constant at 15 m/².)

Coyote *Roadrunner*
stupidus *delightus*

Figure P3.73

ACTIVITIES

A.1. Take three steps, turn 90°, and then walk four steps. Now count the number of steps it takes to walk in a straight line back to your starting point. Verify your result mathematically.

A.2. For this investigation, you need to be outside with a small ball such as a tennis ball and a wristwatch with a second hand. Throw the ball vertically upward as hard as you can, and find the initial speed of your throw and the approximate maximum height of the ball solely with the use of your wristwatch. What happens when you throw the ball at some angle other than 90°? Does this change the time of flight? Can you still determine the maximum height and initial speed? Give careful explanations for your answers. Your co-worker can eyeball the maximum height by standing at a distance and noting the angle.

A.3. Using as projectiles the drops of water spraying from a garden hose at ground level, test the statement that the maximum range occurs when the angle of projection is 45°. As an additional part of this experiment, hold the hose horizontally above the ground and have your co-worker position a marker at the location where the water strikes the ground. Increase the angle of inclination by about 10°, and record the strike position again. Repeat until you have reached an angle of about 75°. Note the pattern produced. Are there two angles at which the range is the same? Explain the reasoning behind your observations.

A.4. Use a chessboard as a coordinate system, with the intersection of the lines on the board as the positions of the coordinates. Select an origin for your coordinate system and, on index cards, write down several vector displacements that are at right angles to one another. For example, displacement 1 might be a movement of four units to the right, displacement 2 an upward movement of six units, and so forth. Continue this for a total of seven or eight movements until you end up at some particular location on the chessboard. Let your co-worker start at the origin and follow your vector directions to see whether he or she arrives at the expected final location. Now shuffle the cards and repeat the experiment. Does the order of the displacements make any difference as to where you eventually end up?

A.5. Roll a ball off a table. At the very instant the rolling ball leaves the table, drop a second ball from the same height above the floor. (Doing this will require a sharp eye and good reflexes!) Do the two balls hit the floor at the same time? Try varying the speed at which you roll the ball off the table. Does this change affect the time at which the balls strike the floor? Finally, roll one of the balls down an incline, and drop the other ball from the base of the incline at the instant the first ball leaves the slope. Which of these balls hits the floor first in this situation? Explain the reasoning behind your observations.

Forces exerted by Earth, wind, and water, properly channeled by the strength and skill of these windsurfers, combine to create a non-zero net force on their surfboards, driving them forward through the waves.

CHAPTER

4

The Laws of Motion

Classical mechanics describes the relationship between the motion of objects found in our everyday world and the forces acting on them. As long as the system under study doesn't involve objects comparable in size to an atom or traveling close to the speed of light, classical mechanics provides an excellent description of nature.

This chapter introduces Newton's three laws of motion and his law of gravity. The three laws are simple and sensible. The first law states that a force must be applied to an object in order to change its velocity. Changing an object's velocity means accelerating it, which implies a relationship between force and acceleration. This relationship, the second law, states that the net force on an object equals the object's mass times its acceleration. Finally, the third law says that whenever we push on something, it pushes back with equal force in the opposite direction. These are the three laws in a nutshell.

Newton's three laws, together with his invention of calculus, opened avenues of inquiry and discovery that are used routinely today in virtually all areas of mathematics, science, engineering, and technology. Newton's theory of universal gravitation had a similar impact, starting a revolution in celestial mechanics and astronomy that continues to this day. With the advent of this theory, the orbits of all the planets could be calculated to high precision and the tides understood. The theory even led to the prediction of "dark stars," now called black holes, over two centuries before any evidence for their existence was observed.[1] Newton's three laws of motion, together with his law of gravitation, are considered among the greatest achievements of the human mind.

4.1 FORCES

A **force** is commonly imagined as a push or a pull on some object, perhaps rapidly, as when we hit a tennis ball with a racket. (See Figure 4.1.) We can hit the ball at different speeds and direct it into different parts of the opponent's court. This

Figure 4.1 Tennis champion Andy Roddick strikes the ball with his racket, applying a force and directing the ball into the open part of the court.

---

[1]In 1783, John Michell combined Newton's theory of light and theory of gravitation, predicting the existence of "dark stars" from which light itself couldn't escape.

means that we can control the magnitude of the applied force and also its direction, so force is a vector quantity, just like velocity and acceleration.

If you pull on a spring (Fig. 4.2a), the spring stretches. If you pull hard enough on a wagon (Fig. 4.2b), the wagon moves. When you kick a football (Fig. 4.2c), it deforms briefly and is set in motion. These are all examples of **contact forces**, so named because they result from physical contact between two objects.

Another class of forces doesn't involve any direct physical contact. Early scientists, including Newton, were uneasy with the concept of forces that act between two disconnected objects. Nonetheless, Newton used this 'action-at-a-distance' concept in his law of gravity, whereby a mass at one location, such as the Sun, affects the motion of a distant object such as Earth despite no evident physical connection between the two objects. To overcome the conceptual difficulty associated with action at a distance, Michael Faraday (1791–1867) introduced the concept of a *field*. The corresponding forces are called **field forces**. According to this approach, an object of mass *M,* such as the Sun, creates an invisible influence that stretches throughout space. A second object of mass *m,* such as Earth, interacts with the *field* of the Sun, not directly with the Sun itself. So the force of gravitational attraction between two objects, illustrated in Figure 4.2d, is an example of a field force. The force of gravity keeps objects bound to Earth and also gives rise to what we call the *weight* of those objects.

Another common example of a field force is the electric force that one electric charge exerts on another (Fig. 4.2e). A third example is the force exerted by a bar magnet on a piece of iron (Fig. 4.2f).

The known fundamental forces in nature are all field forces. These are, in order of decreasing strength, (1) the strong nuclear force between subatomic particles; (2) the electromagnetic forces between electric charges; (3) the weak nuclear force, which arises in certain radioactive decay processes; and (4) the gravitational force between objects. The strong force keeps the nucleus of an atom from flying apart due to the repulsive electric force of the protons. The weak force

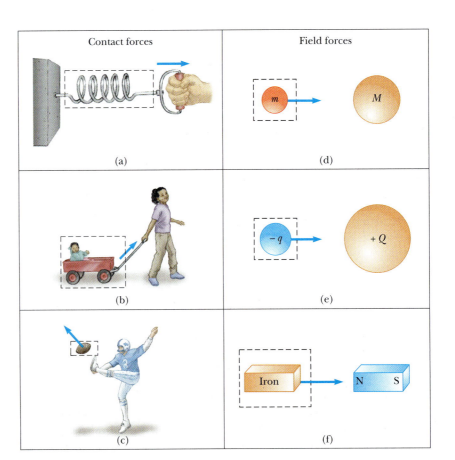

Figure 4.2 Examples of forces applied to various objects. In each case, a force acts on the object surrounded by the dashed lines. Something in the environment external to the boxed area exerts the force.

is involved in most radioactive processes and plays an important role in the nuclear reactions that generate the Sun's energy output. The strong and weak forces operate only on the nuclear scale, with a very short range on the order of 10^{-15} m. Outside this range, they have no influence. Classical physics, however, deals only with gravitational and electromagnetic forces, which have infinite range.

Forces exerted on an object can change the object's shape. For example, striking a tennis ball with a racquet, as in Figure 4.1, deforms the ball to some extent. Even objects we usually consider rigid and inflexible are deformed under the action of external forces. Often the deformations are permanent, as in the case of a collision between automobiles.

4.2   NEWTON'S FIRST LAW

Consider a book lying on a table. Obviously, the book remains at rest if left alone. Now imagine pushing the book with a horizontal force great enough to overcome the force of friction between the book and the table, setting the book in motion. Because the magnitude of the applied force exceeds the magnitude of the friction force, the book accelerates. When the applied force is withdrawn, friction soon slows the book to a stop.

Now imagine pushing the book across a smooth, waxed floor. The book again comes to rest once the force is no longer applied, but not as quickly as before. Finally, if the book is moving on a horizontal frictionless surface, it continues to move in a straight line with constant velocity until it hits a wall or some other obstruction.

Before about 1600, scientists felt that the natural state of matter was the state of rest. Galileo, however, devised thought experiments—such as an object moving on a frictionless surface, as just described—and concluded that **it's not the nature of an object to stop, once set in motion, but rather to continue in its original state of motion**. This approach was later formalized as **Newton's first law of motion**:

> An object moves with a velocity that is constant in magnitude and direction, unless acted on by a nonzero net force.

◄ Newton's first law

The net force on an object is defined as the vector sum of all external forces exerted on the object. External forces come from the object's environment. If an object's velocity isn't changing in either magnitude or direction, then its acceleration and the net force acting on it must both be zero.

Internal forces originate within the object itself and can't change the object's velocity (although they can change the object's rate of rotation, as described in Chapter 8). As a result, internal forces aren't included in Newton's second law. It's not really possible to "pull yourself up by your own bootstraps."

A consequence of the first law is the feasibility of space travel. After just a few moments of powerful thrust, the spacecraft coasts for months or years, its velocity only slowly changing with time under the relatively faint influence of the distant sun and planets.

Mass and Inertia

Imagine hitting a golf ball off a tee with a driver. If you're a good golfer, the ball will sail over two hundred yards down the fairway. Now imagine teeing up a bowling ball and striking it with the same club (an experiment we don't recommend). Your club would probably break, you might sprain your wrist, and the bowling ball, at best, would fall off the tee, take half a roll and come to rest.

From this thought experiment, we conclude that while both balls resist changes in their state of motion, the bowling ball offers much more effective resistance. The tendency of an object to continue in its original state of motion is called **inertia**.

While inertia is the tendency of an object to continue its motion in the absence of a force, **mass** is a measure of the object's resistance to changes in its motion due

Unless acted on by an external force, an object at rest will remain at rest and an object in motion will continue in motion with constant velocity. In this case, the wall of the building did not exert a large enough external force on the moving train to stop it.

Roger Viollet, Mill Valley, CA, University Science Books, 1982

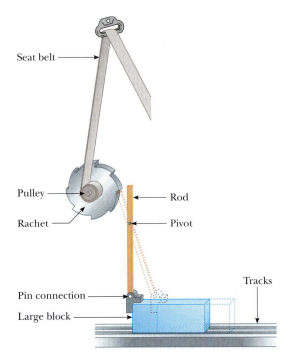

Seat belt

Pulley

Rod

Rachet

Pivot

Tracks

Pin connection

Large block

Figure 4.3 A mechanical arrangement for an automobile seat belt.

to a force. The greater the mass of a body, the less it accelerates under the action of a given applied force. The SI unit of mass is the kilogram. Mass is a scalar quantity that obeys the rules of ordinary arithmetic.

Inertia can be used to explain the operation of one type of seat belt mechanism. In the event of an accident, the purpose of the seat belt is to hold the passenger firmly in place relative to the car, to prevent serious injury. Figure 4.3 illustrates how one type of shoulder harness operates. Under normal conditions, the ratchet turns freely to allow the harness to wind on or unwind from the pulley as the passenger moves. In an accident, the car undergoes a large acceleration and rapidly comes to rest. Because of its inertia, the large block under the seat continues to slide forward along the tracks. The pin connection between the block and the rod causes the rod to pivot about its center and engage the ratchet wheel. At this point, the ratchet wheel locks in place and the harness no longer unwinds.

4.3 NEWTON'S SECOND LAW

Newton's first law explains what happens to an object that has no net force acting on it: The object either remains at rest or continues moving in a straight line with constant speed. Newton's second law answers the question of what happens to an object that *does* have a net force acting on it.

Imagine pushing a block of ice across a frictionless horizontal surface. When you exert some horizontal force on the block, it moves with an acceleration of, say, 2 m/s^2. If you apply a force twice as large, the acceleration doubles to 4 m/s^2. Pushing three times as hard triples the acceleration, and so on. From such observations, we conclude that **the acceleration of an object is directly proportional to the net force acting on it**.

Mass also affects acceleration. Suppose you stack identical blocks of ice on top of each other while pushing the stack with constant force. If the force applied to one block produces an acceleration of 2 m/s^2, then the acceleration drops to half that value, 1 m/s^2, when two blocks are pushed, to one-third the initial value when three blocks are pushed, and so on. We conclude that **the acceleration of an object is inversely proportional to its mass**. These observations are summarized in **Newton's second law:**

TIP 4.1 **Force Causes**
***Changes* in Motion**

Motion can occur even in the absence of forces. Force causes *changes* in motion.

Newton's second law ▶

> The acceleration $\vec{\mathbf{a}}$ of an object is directly proportional to the net force acting on it and inversely proportional to its mass.

The constant of proportionality is equal to one, so in mathematical terms the preceding statement can be written

$$\vec{\mathbf{a}} = \frac{\Sigma \vec{\mathbf{F}}}{m}$$

where $\vec{\mathbf{a}}$ is the acceleration of the object, m is its mass, and $\Sigma\vec{\mathbf{F}}$ is the vector sum of all forces acting on it. Multiplying through by m, we have

$$\Sigma\vec{\mathbf{F}} = m\vec{\mathbf{a}} \qquad\qquad \text{[4.1]}$$

Physicists commonly refer to this equation as '$F = ma$'. The second law is a vector equation, equivalent to the following three component equations:

$$\Sigma F_x = ma_x \qquad \Sigma F_y = ma_y \qquad \Sigma F_z = ma_z \qquad \text{[4.2]}$$

When there is no net force on an object, its acceleration is zero, which means the velocity is constant.

Units of Force and Mass

The SI unit of force is the **newton**. When 1 newton of force acts on an object that has a mass of 1 kg, it produces an acceleration of 1 m/s² in the object. From this definition and Newton's second law, we see that the newton can be expressed in terms of the fundamental units of mass, length, and time as

$$1\ \text{N} \equiv 1\ \text{kg}\cdot\text{m/s}^2 \qquad\qquad \text{[4.3]}$$

In the U.S. customary system, the unit of force is the **pound**. The conversion from newtons to pounds is given by

$$1\ \text{N} = 0.225\ \text{lb} \qquad\qquad \text{[4.4]}$$

The units of mass, acceleration, and force in the SI and U.S. customary systems are summarized in Table 4.1.

Quick Quiz 4.1

True or false? (a) It's possible to have motion in the absence of a force. (b) If an object isn't moving, no external force acts on it.

Quick Quiz 4.2

True or false? (a) If a single force acts on an object, the object accelerates. (b) If an object is accelerating, a force is acting on it. (c) If an object is not accelerating, no external force is acting on it.

Quick Quiz 4.3

True or false? If the net force acting on an object is in the positive x-direction, the object moves only in the positive x-direction.

TABLE 4.1

Units of Mass, Acceleration, and Force

System	Mass	Acceleration	Force
SI	kg	m/s²	N = kg·m/s²
U.S. customary	slug	ft/s²	lb = slug·ft/s²

TIP 4.2 $m\vec{\mathbf{a}}$ Is Not a Force
Equation 4.1 does *not* say that the product $m\vec{\mathbf{a}}$ is a force. All forces exerted on an object are summed as vectors to generate the net force on the left side of the equation. This net force is then equated to the product of the mass and resulting acceleration of the object. Do *not* include an "$m\vec{\mathbf{a}}$ force" in your analysis.

◀ Definition of newton

Giraudon/Art Resource

ISAAC NEWTON English Physicist and Mathematician (1642–1727)

Newton was one of the most brilliant scientists in history. Before he was 30, he formulated the basic concepts and laws of mechanics, discovered the law of universal gravitation, and invented the mathematical methods of the calculus. As a consequence of his theories, Newton was able to explain the motions of the planets, the ebb and flow of the tides, and many special features of the motions of the Moon and Earth. He also interpreted many fundamental observations concerning the nature of light. His contributions to physical theories dominated scientific thought for two centuries and remain important today.

EXAMPLE 4.1   Airboat

Goal Apply Newton's law in one dimension, together with the equations of kinematics.

Problem An airboat with mass 3.50×10^2 kg, including passengers, has an engine that produces a net horizontal force of 7.70×10^2 N, after accounting for forces of resistance. **(a)** Find the acceleration of the airboat. **(b)** Starting from rest, how long does it take the airboat to reach a speed of 12.0 m/s? **(c)** After reaching this speed, the pilot turns off the engine and drifts to a stop over a distance of 50.0 m. Find the resistance force, assuming it's constant.

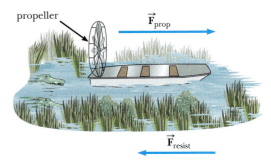

Figure 4.4 (Example 4.1)

Strategy In part **(a)**, apply Newton's second law to find the acceleration, and in part **(b)** use this acceleration in the one-dimensional kinematics equation for the velocity. When the engine is turned off in part **(c)**, only the resistance forces act on the boat, so their net acceleration can be found from $v^2 - v_0^2 = 2a\Delta x$. Then Newton's second law gives the resistance force.

Solution

(a) Find the acceleration of the airboat.

Apply Newton's second law and solve for the acceleration:

$$ma = F_{net} \quad \rightarrow \quad a = \frac{F_{net}}{m} = \frac{7.70 \times 10^2 \text{ N}}{3.50 \times 10^2 \text{ kg}}$$

$$= \boxed{2.20 \text{ m/s}^2}$$

(b) Find the time necessary to reach a speed of 12.0 m/s.

Apply the kinematics velocity equation:

$$v = at + v_0 = (2.20 \text{ m/s}^2)t = 12.0 \text{ m/s} \quad \rightarrow \quad t = \boxed{5.45 \text{ s}}$$

(c) Find the resistance force after the engine is turned off.

Using kinematics, find the net acceleration due to resistance forces:

$$v^2 - v_0^2 = 2a\Delta x$$

$$0 - (12 \text{ m/s})^2 = 2a(50.0 \text{ m}) \quad \rightarrow \quad a = -1.44 \text{ m/s}^2$$

Substitute the acceleration into Newton's second law, finding the resistance force:

$$F_{resist} = ma = (3.50 \times 10^2 \text{ kg})(-1.44 \text{ m/s}^2) = \boxed{-504 \text{ N}}$$

Remarks The negative answer for the acceleration in part **(c)** means that the airboat is slowing down. In this problem, the *components* of the vectors $\vec{a}$ and $\vec{F}$ were used, so they were written in italics without arrows. It's important to bear in mind that $\vec{a}$ and $\vec{F}$ are vectors, not scalars.

TIP 4.3   Newton's Second Law is a *Vector* Equation

In applying Newton's second law, add all of the forces on the object as vectors and then find the resultant vector acceleration by dividing by m. Don't find the individual magnitudes of the forces and add them like scalars.

Exercise 4.1

Suppose the pilot, starting again from rest, opens the throttle partway. At a constant acceleration, the airboat then covers a distance of 60.0 m in 10.0 s. Find the net force acting on the boat.

Answer 4.20×10^2 N

EXAMPLE 4.2   Horses Pulling a Barge

Goal Apply Newton's second law in a two-dimensional problem.

Problem Two horses are pulling a barge with mass 2.00×10^3 kg along a canal, as shown in Figure 4.5. The cable connected to the first horse makes an angle of 30.0° with respect to the direction of the canal, while the cable connected to the second horse makes an angle of 45.0°. Find the initial acceleration of the barge, starting at rest, if each horse exerts a force of magnitude 6.00×10^2 N on the barge. Ignore forces of resistance on the barge.

Strategy   Using trigonometry, find the vector force exerted by each horse on the barge. Add the x-components together to get the x-component of the resultant force, and then do the same with the y-components. Divide by the mass of the barge to get the accelerations in the x- and y-directions.

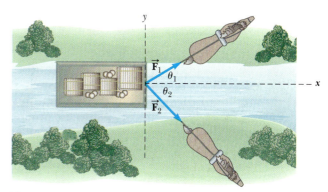

Figure 4.5   (Example 4.2)

Solution

Find the x-components of the forces exerted by the horses.

$$F_{1x} = F_1 \cos\theta_1 = (6.00 \times 10^2\,\text{N})\cos(30.0°) = 5.20 \times 10^2\,\text{N}$$

$$F_{2x} = F_2 \cos\theta_2 = (6.00 \times 10^2\,\text{N})\cos(-45.0°) = 4.24 \times 10^2\,\text{N}$$

Find the total force in the x-direction by adding the x-components:

$$F_x = F_{1x} + F_{2x} = 5.20 \times 10^2\,\text{N} + 4.24 \times 10^2\,\text{N}$$
$$= 9.44 \times 10^2\,\text{N}$$

Find the y-components of the forces exerted by the horses:

$$F_{1y} = F_1 \sin\theta_1 = (6.00 \times 10^2\,\text{N})\sin 30.0° = 3.00 \times 10^2\,\text{N}$$

$$F_{2y} = F_2 \sin\theta_2 = (6.00 \times 10^2\,\text{N})\sin(-45.0°)$$
$$= -4.24 \times 10^2\,\text{N}$$

Find the total force in the y-direction by adding the y-components:

$$F_y = F_{1y} + F_{2y} = 3.00 \times 10^2\,\text{N} - 4.24 \times 10^2\,\text{N}$$
$$= -1.24 \times 10^2\,\text{N}$$

Find the components of the acceleration by dividing the force components by the mass:

$$a_x = \frac{F_x}{m} = \frac{9.44 \times 10^2\,\text{N}}{2.00 \times 10^3\,\text{kg}} = 0.472\,\text{m/s}^2$$

$$a_y = \frac{F_y}{m} = \frac{-1.24 \times 10^2\,\text{N}}{2.00 \times 10^3\,\text{kg}} = -0.0620\,\text{m/s}^2$$

Find the magnitude of the acceleration:

$$a = \sqrt{a_x{}^2 + a_y{}^2} = \sqrt{(0.472\,\text{m/s}^2)^2 + (-0.0620\,\text{m/s}^2)^2}$$
$$= 0.476\,\text{m/s}^2$$

Find the direction of the acceleration.

$$\tan\theta = \frac{a_y}{a_x} = \frac{-0.0620}{0.472} = -0.131$$

$$\theta = \tan^{-1}(-0.131) = -7.46°$$

Exercise 4.2

Repeat Example 4.2, but assume that the upper horse pulls at a 40.0° angle, the lower horse at 20.0°.

Answer $0.520\,\text{m/s}^2$, 10.0°

The Gravitational Force

The **gravitational force** is the mutual force of attraction between any two objects in the Universe. Although the gravitational force can be very strong between very large objects, it's the weakest of the fundamental forces. A good demonstration of how weak it is can be carried out with a small balloon. Rubbing the balloon in your hair gives the balloon a tiny electric charge. Through electric forces, the balloon then adheres to a wall, resisting the gravitational pull of the entire Earth!

In addition to contributing to the understanding of motion, Newton studied gravity extensively. **Newton's law of universal gravitation states that every particle in the Universe attracts every other particle with a force that is directly proportional to the product of the masses of the particles and inversely proportional to the square of the distance between them.** If the particles have masses m_1 and m_2 and are separated by a distance r, as in Active Figure 4.6, the magnitude of the gravitational force, F_g is

◀ Law of universal gravitation

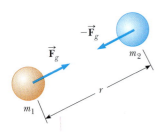

$$F_g = G\frac{m_1 m_2}{r^2} \qquad [4.5]$$

where $G = 6.67 \times 10^{-11}$ N·m²/kg² is the **universal gravitation constant**. We examine the gravitational force in more detail in Chapter 7.

ACTIVE FIGURE 4.6
The gravitational force between two particles is attractive.

Weight

Physics⊗Now™
Log into to PhysicsNow at **www.cp7e.com**, and go to Active Figure 4.6 to change the masses of the particles and the separation between the particles to see the effect on the gravitational force.

The magnitude of the gravitational force acting on an object of mass m near Earth's surface is called the *weight, w,* of the object, given by

$$w = mg \qquad [4.6]$$

where g is the acceleration of gravity.
SI unit: newton (N)

From Equation 4.5, an alternate definition of the weight of an object with mass m can be written as

$$w = G\frac{M_E m}{r^2} \qquad [4.7]$$

where M_E is the mass of Earth and r is the distance from the object to Earth's center. If the object is at rest on Earth's surface, then r is equal to Earth's radius R_E. Since r is in the denominator of Equation 4.7, the weight decreases with increasing r. So the weight of an object on a mountaintop is less than the weight of the same object at sea level.

Comparing Equations 4.6 and 4.7, we see that

$$g = G\frac{M_E}{r^2} \qquad [4.8]$$

Astronaut Edwin E. "Buzz" Aldrin, Jr., walking on the Moon after the *Apollo 11* lunar landing. Aldrin's weight on the Moon is less than it is on Earth, but his mass is the same in both places.

Unlike mass, weight is not an inherent property of an object because it can take different values, depending on the value of g in a given location. If an object has a mass of 70.0 kg, for example, then its weight at a location where $g = 9.80$ m/s² is $mg = 686$ N. In a high-altitude balloon, where g might be 9.76 m/s², the object's weight would be 683 N. The value of g also varies slightly due to the density of matter in a given locality.

Equation 4.8 is a general result that can be used to calculate the acceleration of an object falling near the surface of any massive object if the more massive object's radius and mass are known. Using the values in Table 7.3 (p. 216), you should be able to show that $g_{Sun} = 274$ m/s² and $g_{Moon} = 1.62$ m/s². An important fact is that for spherical bodies, distances are calculated from the centers of the objects, a consequence of Gauss's law (explained in Chapter 15), which holds for both gravitational and electric forces.

Quick Quiz 4.4

A friend, calling from the Moon, tells you she has just won 1 newton of gold in a contest. Excitedly, you tell her that you entered Earth's version of the same contest and also won 1 newton of gold! Who won the prize of greatest value? (a) your friend   (b) you   (c) the values are equal.

EXAMPLE 4.3   May the Force Be with You

Goal   Calculate the magnitude of a gravitational force using Newton's law of gravitation.

Problem   Find the gravitational force exerted by the Sun on a 70.0-kg man located on Earth. The distance from the Sun to the Earth is about 1.50×10^{11} m, and the Sun's mass is 1.99×10^{30} kg.

Strategy   Substitute numbers into Newton's law of gravitation, Equation 4.5, making sure to use the correct units.

Solution

Apply Equation 4.5, substituting values:

$$F_{sun} = G\frac{mM_S}{r^2}$$

$$= (6.67 \times 10^{-11}\,\text{kg}^{-1}\text{m}^3\text{s}^{-2})\,\frac{(70.0\,\text{kg})(1.99 \times 10^{30}\,\text{kg})}{(1.50 \times 10^{11}\,\text{m})^2}$$

$$= \boxed{0.413\,\text{N}}$$

Remarks   The gravitational attraction between the Sun and objects on Earth is easily measurable and has been exploited in experiments to determine whether gravitational attraction depends on the composition of the object. As the exercise shows, the gravitational force on Earth due to the Moon is much weaker than the gravitational force on Earth due to the Sun. Paradoxically, the Moon's effect on the tides is over twice that of the Sun, because the tides depend on *differences* in the gravitational force across the Earth, and those differences are greater for the Moon's gravitational force because it's much closer to Earth than the Sun.

Exercise 4.3

To one significant digit, find the force exerted by the Moon on a 70-kg man on Earth. The Moon has a mass of 7.36×10^{22} kg and is 3.84×10^{8} m from Earth.

Answer $F_{Moon} = 0.002$ N

EXAMPLE 4.4   Weight on Planet X

Goal   Understand the effect of a planet's mass and radius on the weight of an object on the planet's surface.

Problem   An astronaut on a space mission lands on a planet with three times the mass and twice the radius of Earth. What is her weight w_X on this planet as a multiple of her Earth weight w_E?

Strategy   Write M_X and r_X, the mass and radius of the planet, in terms of M_E and R_E, the mass and radius of Earth, respectively, and substitute into the law of gravitation.

Solution

From the statement of the problem, we have the following relationships:

$$M_X = 3M_E \qquad r_X = 2R_E$$

Substitute the preceding expressions into Equation 4.5 and simplify, algebraically associating the terms giving the weight on Earth:

$$w_X = G\frac{M_X m}{r_X^2} = G\frac{3M_E m}{(2R_E)^2} = \frac{3}{4}\,G\frac{M_E m}{R_E^2} = \boxed{\frac{3}{4}w_E}$$

Remarks This problem shows the interplay between a planet's mass and radius in determining the weight of objects on its surface. Because of Earth's much smaller radius, the weight of an object on Jupiter is only 2.64 times its weight on Earth, despite the fact that Jupiter has over 300 times as much mass.

Exercise 4.4
An astronaut lands on Ganymede, a giant moon of Jupiter that is larger than the planet Mercury. Ganymede has one-fortieth the mass of Earth and two-fifths the radius. Find the weight of the astronaut standing on Ganymede in terms of his Earth weight w_E.

Answer $w_G = (5/32)\,w_E$

4.4 NEWTON'S THIRD LAW

In Section 4.1 we found that a force is exerted on an object when it comes into contact with some other object. Consider the task of driving a nail into a block of wood, for example, as illustrated in Figure 4.7a. To accelerate the nail and drive it into the block, the hammer must exert a net force on the nail. Newton recognized, however, that a single isolated force (such as the force exerted by the hammer on the nail), couldn't exist. Instead, **forces in nature always exist in pairs**. According to Newton, as the nail is driven into the block by the force exerted by the hammer, the hammer is slowed down and stopped by the force exerted by the nail.

Newton described such paired forces with his **third law**:

Newton's third law ▶

> If object 1 and object 2 interact, the force $\vec{\mathbf{F}}_{12}$ exerted by object 1 on object 2 is equal in magnitude but opposite in direction to the force $\vec{\mathbf{F}}_{21}$ exerted by object 2 on object 1.

This law, which is illustrated in Figure 4.7b, states that **a single isolated force can't exist**. The force $\vec{\mathbf{F}}_{12}$ exerted by object 1 on object 2 is sometimes called the *action force*, and the force $\vec{\mathbf{F}}_{21}$ exerted by object 2 on object 1 is called the *reaction force*. In reality, either force can be labeled the action or reaction force. **The action force is equal in magnitude to the reaction force and opposite in direction. In all cases, the action and reaction forces act on different objects**. For example, the force acting on a freely falling projectile is the force exerted by Earth on the projectile, $\vec{\mathbf{F}}_g$, and the magnitude of this force is its weight *mg*. The reaction to force $\vec{\mathbf{F}}_g$ is the force exerted by the projectile on Earth, $\vec{\mathbf{F}}_g' = -\vec{\mathbf{F}}_g$. The reaction force $\vec{\mathbf{F}}_g'$ must accelerate the Earth towards the projectile, just as the action force $\vec{\mathbf{F}}_g$ accelerates the projectile towards the Earth.

TIP 4.4 Action-Reaction Pairs
In applying Newton's third law, remember that an action and its reaction force always act on *different* objects. Two external forces acting on the same object, even if they are equal in magnitude and opposite in direction, *can't* be an action-reaction pair.

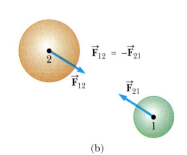

Figure 4.7 Newton's third law. (a) The force exerted by the hammer on the nail is equal in magnitude and opposite in direction to the force exerted by the nail on the hammer. (b) The force $\vec{\mathbf{F}}_{12}$ exerted by object 1 on object 2 is equal in magnitude and opposite in direction to the force $\vec{\mathbf{F}}_{21}$ exerted by object 2 on object 1.

(a)   (b)

Because the Earth has such a large mass, however, its acceleration due to this reaction force is negligibly small.

Newton's third law constantly affects our activities in everyday life. Without it, no locomotion of any kind would be possible, whether on foot, on a bicycle, or in a motorized vehicle. When walking, for example, we exert a frictional force against the ground. The reaction force of the ground against our foot propels us forward. In the same way, the tires on a bicycle exert a frictional force against the ground, and the reaction of the ground pushes the bicycle forward. As we'll see shortly, friction plays a large role in such reaction forces.

For another example of Newton's third law, consider the helicopter. Most helicopters have a large set of blades rotating in a horizontal plane above the body of the vehicle and another, smaller set rotating in a vertical plane at the back. Other helicopters have two large sets of blades above the body rotating in opposite directions. Why do helicopters always have two sets of blades? In the first type of helicopter, the engine applies a force to the blades, causing them to change their rotational motion. According to Newton's third law, however, the blades must exert a force on the engine of equal magnitude and in the opposite direction. This force would cause the body of the helicopter to rotate in the direction opposite the blades. A rotating helicopter would be impossible to control, so a second set of blades is used. The small blades in the back provide a force opposite to that tending to rotate the body of the helicopter, keeping the body oriented in a stable position. In helicopters with two sets of large counterrotating blades, engines apply forces in opposite directions, so there is no net force rotating the helicopter.

As mentioned earlier, the Earth exerts a force $\vec{\mathbf{F}}_g$ on any object. If the object is a TV at rest on a table, as in Figure 4.8a, the reaction force to $\vec{\mathbf{F}}_g$ is the force the TV exerts on the Earth, $\vec{\mathbf{F}}_g'$. The TV doesn't accelerate downward because it's held up by the table. The table, therefore, exerts an upward force $\vec{\mathbf{n}}$, called the **normal force**, on the TV. (*Normal*, a technical term from mathematics, means "perpendicular" in this context.) The normal force is an elastic force arising from the cohesion of matter and is electromagnetic in origin. It balances the gravitational force acting on the TV, preventing the TV from falling through the table, and can have any value needed, up to the point of breaking the table. The reaction to $\vec{\mathbf{n}}$ is the force exerted by the TV on the table, $\vec{\mathbf{n}}'$. Therefore,

$$\vec{\mathbf{F}}_g = -\vec{\mathbf{F}}_g' \qquad \text{and} \qquad \vec{\mathbf{n}} = -\vec{\mathbf{n}}'$$

APPLICATION

Helicopter Flight

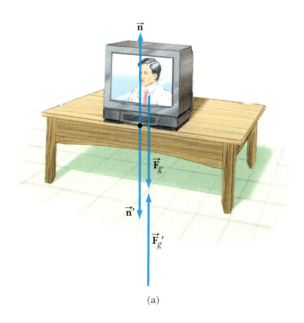

(a)

(b)

Figure 4.8 When a TV set is sitting on a table, the forces acting on the set are the normal force $\vec{\mathbf{n}}$ exerted by the table and the force of gravity, $\vec{\mathbf{F}}_g$, as illustrated in (b). The reaction to $\vec{\mathbf{n}}$ is the force exerted by the TV set on the table, $\vec{\mathbf{n}}'$. The reaction to $\vec{\mathbf{F}}_g$ is the force exerted by the TV set on Earth, $\vec{\mathbf{F}}_g'$.

The forces $\vec{\mathbf{n}}$ and $\vec{\mathbf{n}}'$ both have the same magnitude as $\vec{\mathbf{F}}_g$. Note that the forces acting on the TV are $\vec{\mathbf{F}}_g$ and $\vec{\mathbf{n}}$, as shown in Figure 4.8b. The two reaction forces, $\vec{\mathbf{F}}_g'$ and $\vec{\mathbf{n}}'$, are exerted by the TV on objects other than the TV. Remember, the two forces in an action-reaction pair always act on two different objects.

Because the TV is not accelerating in any direction ($\vec{\mathbf{a}} = 0$), it follows from Newton's second law that $m\vec{\mathbf{a}} = 0 = \vec{\mathbf{F}}_g + \vec{\mathbf{n}}$. However, $F_g = -mg$, so $n = mg$, a useful result.

Quick Quiz 4.5

A small sports car collides head-on with a massive truck. The greater impact force (in magnitude) acts on (a) the car, (b) the truck, (c) neither, the force is the same on both. Which vehicle undergoes the greater magnitude acceleration? (d) the car, (e) the truck, (f) the accelerations are the same.

4.5   APPLICATIONS OF NEWTON'S LAWS

This section applies Newton's laws to objects moving under the influence of constant external forces. We assume that objects behave as particles, so we need not consider the possibility of rotational motion. We also neglect any friction effects and the masses of any ropes or strings involved. With these approximations, the magnitude of the force exerted along a rope, called the **tension**, is the same at all points in the rope. This is illustrated by the rope in Figure 4.9, showing the forces $\vec{\mathbf{T}}$ and $\vec{\mathbf{T}}'$ acting on it. If the rope has mass m, then Newton's second law applied to the rope gives $T - T' = ma$. If the mass m is taken to be negligible, however, as in the upcoming examples, then $T = T'$.

$\vec{\mathbf{T}}' \longleftarrow$ [rope] $\longrightarrow \vec{\mathbf{T}}$

Figure 4.9 Newton's second law applied to a rope gives $T - T' = ma$. However, if $m = 0$, then $T = T'$. Thus, the tension in a massless rope is the same at all points in the rope.

When we apply Newton's law to an object, we are interested only in those forces which act *on the object*. For example, in Figure 4.8b, the only external forces acting on the TV are $\vec{\mathbf{n}}$ and $\vec{\mathbf{F}}_g$. The reactions to these forces, $\vec{\mathbf{n}}'$ and $\vec{\mathbf{F}}_g'$, act on the table and on Earth, respectively, and don't appear in Newton's second law applied to the TV.

Consider a crate being pulled to the right on a frictionless, horizontal surface, as in Figure 4.10a. Suppose you wish to find the acceleration of the crate and the force the surface exerts on it. The horizontal force exerted on the crate acts through the rope. The force that the rope exerts on the crate is denoted by $\vec{\mathbf{T}}$ (because it's a tension force). The magnitude of $\vec{\mathbf{T}}$ is equal to the tension in the rope. What we mean by the words "tension in the rope" is just the force read by a spring scale when the rope in question has been cut and the scale inserted between the cut ends. A dashed circle is drawn around the crate in Figure 4.10a to emphasize the importance of isolating the crate from its surroundings.

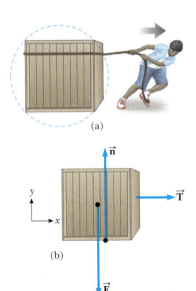

Figure 4.10 (a) A crate being pulled to the right on a frictionless surface. (b) The free-body diagram that represents the forces exerted on the crate.

Because we are interested only in the motion of the crate, we must be able to identify all forces acting on it. These forces are illustrated in Figure 4.10b. In addition to displaying the force $\vec{\mathbf{T}}$, the force diagram for the crate includes the force of gravity $\vec{\mathbf{F}}_g$ exerted by Earth and the normal force $\vec{\mathbf{n}}$ exerted by the floor. Such a force diagram is called a **free-body diagram**, because the environment is replaced by a series of forces on an otherwise free body. The construction of a correct free-body diagram is an essential step in applying Newton's laws. An incorrect diagram will most likely lead to incorrect answers!

The *reactions* to the forces we have listed—namely, the force exerted by the rope on the hand doing the pulling, the force exerted by the crate on Earth, and the force exerted by the crate on the floor—aren't included in the free-body diagram because they act on other objects and not on the crate. Consequently, they don't directly influence the crate's motion. Only forces acting directly on the crate are included.

Now let's apply Newton's second law to the crate. First we choose an appropriate coordinate system. In this case it's convenient to use the one shown in Figure 4.10b, with the x-axis horizontal and the y-axis vertical. We can apply Newton's second law in the x-direction, y-direction, or both, depending on what we're asked to find in a problem. Newton's second law applied to the crate in the x- and y-directions yields the following two equations:

$$ma_x = T \qquad ma_y = n - mg = 0$$

From these equations, we find that the acceleration in the x-direction is constant, given by $a_x = T/m$, and that the normal force is given by $n = mg$. Because the acceleration is constant, the equations of kinematics can be applied to obtain further information about the velocity and displacement of the object.

Problem-Solving Strategy   Newton's Second Law

Problems involving Newton's second law can be very complex. The following protocol breaks the solution process down into smaller, intermediate goals:

1. **Read** the problem carefully at least once.
2. **Draw** a picture of the system, identify the object of primary interest, and indicate forces with arrows.
3. **Label** each force in the picture in a way that will bring to mind what physical quantity the label stands for (e.g., T for tension).
4. **Draw** a free-body diagram of the object of interest, based on the labeled picture. If additional objects are involved, draw separate free-body diagrams for them. Choose convenient coordinates for each object.
5. **Apply Newton's second law**. The x- and y-components of Newton's second law should be taken from the vector equation and written individually. This often results in two equations and two unknowns.
6. **Solve** for the desired unknown quantity, and substitute the numbers.

In the special case of equilibrium, the foregoing process is simplified because the acceleration is zero.

Objects in Equilibrium

Objects that are either at rest or moving with constant velocity are said to be in equilibrium. Because $\vec{a} = 0$, Newton's second law applied to an object in equilibrium gives

$$\sum \vec{F} = 0 \qquad \textbf{[4.9]}$$

This statement signifies that the *vector* sum of all the forces (the net force) acting on an object in equilibrium is zero. Equation 4.9 is equivalent to the set of component equations given by

$$\sum F_x = 0 \quad \text{and} \quad \sum F_y = 0 \qquad \textbf{[4.10]}$$

We won't consider three-dimensional problems in this book, but the extension of Equation 4.10 to a three-dimensional problem can be made by adding a third equation: $\sum F_z = 0$.

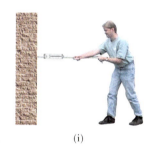

(i)

(ii)

Figure 4.11 (Quick Quiz 4.6) (i) A person pulls with a force of magnitude F on a spring scale attached to a wall. (ii) Two people pull with forces of magnitude F in opposite directions on a spring scale attached between two ropes.

Quick Quiz 4.6

Consider the two situations shown in Figure 4.11, in which there is no acceleration. In both cases, the men pull with a force of magnitude F. Is the reading on the scale in part (i) of the figure (a) greater than, (b) less than, or (c) equal to the reading in part (ii)?

EXAMPLE 4.5   A Traffic Light at Rest

Goal Use the second law in an equilibrium problem requiring two free-body diagrams.

Problem A traffic light weighing 1.00×10^2 N hangs from a vertical cable tied to two other cables that are fastened to a support, as in Figure 4.12a. The upper cables make angles of 37.0° and 53.0° with the horizontal. Find the tension in each of the three cables.

Strategy There are three unknowns, so we need to generate three equations relating them, which can then be solved. One equation can be obtained by applying Newton's second law to the traffic light, which has forces in the y-direction only. Two more equations can be obtained by applying the second law to the knot joining the cables—one equation from the x-component and one equation from the y-component.

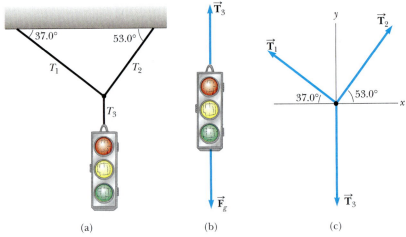

Figure 4.12 (Example 4.5) (a) A traffic light suspended by cables. (b) A free-body diagram for the traffic light. (c) A free-body diagram for the knot joining the cables.

Solution

Find T_3 from Figure 4.12b, using the condition of equilibrium:

$$\sum F_y = 0 \quad \rightarrow \quad T_3 - F_g = 0$$
$$T_3 = F_g = 1.00 \times 10^2 \text{ N}$$

Using Figure 4.12c, resolve all three tension forces into components and construct a table for convenience:

Force	x-component	y-component
$\vec{T}_1$	$-T_1 \cos 37.0°$	$T_1 \sin 37.0°$
$\vec{T}_2$	$T_2 \cos 53.0°$	$T_2 \sin 53.0°$
$\vec{T}_3$	0	-1.00×10^2 N

Apply the conditions for equilibrium to the knot, using the components in the table:

$$\sum F_x = -T_1 \cos 37.0° + T_2 \cos 53.0° = 0 \quad \text{(1)}$$
$$\sum F_y = T_1 \sin 37.0° + T_2 \sin 53.0° - 1.00 \times 10^2 \text{ N} = 0 \text{ (2)}$$

There are two equations and two remaining unknowns. Solve Equation (1) for T_2:

$$T_2 = T_1 \left(\frac{\cos 37.0°}{\cos 53.0°} \right) = T_1 \left(\frac{0.799}{0.602} \right) = 1.33 T_1$$

Substitute the result for T_2 into Equation (2):

$$T_1 \sin 37.0° + (1.33 T_1)(\sin 53.0°) - 1.00 \times 10^2 \text{ N} = 0$$
$$T_1 = \boxed{60.1 \text{ N}}$$
$$T_2 = 1.33 T_1 = 1.33(60.0 \text{ N}) = \boxed{79.9 \text{ N}}$$

Remarks It's very easy to make sign errors in this kind of problem. One way to avoid them is to always measure the angle of a vector from the positive x-direction. The trigonometric functions of the angle will then automatically give the correct signs for the components. For example, $\vec{T}_1$ makes an angle of $180° - 37° = 143°$ with respect to the positive x-axis, and its x-component, $T_1 \cos 143°$, is negative, as it should be.

Exercise 4.5

Suppose the traffic light is hung so that the tensions T_1 and T_2 are both equal to 80.0 N. Find the new angles they make with respect to the x axis. (By symmetry, these angles will be the same.)

Answer Both angles are 38.7°.

EXAMPLE 4.6   Sled on a Frictionless Hill

Goal   Use the second law and the normal force in an equilibrium problem.

Problem   A child holds a sled at rest on a frictionless, snow-covered hill, as shown in Figure 4.13a. If the sled weighs 77.0 N, find the force exerted by the rope on the sled and the magnitude of the force $\vec{\mathbf{n}}$ exerted by the hill on the sled.

Strategy   When an object is on a slope, it's convenient to use tilted coordinates, as in Figure 4.13b, so that the normal force $\vec{\mathbf{n}}$ is in the y-direction and the tension force $\vec{\mathbf{T}}$ is in the x-direction. In the absence of friction, the hill exerts no force on the sled in the x-direction. Because the sled is at rest, the conditions for equilibrium, $\Sigma F_x = 0$ and $\Sigma F_y = 0$, apply, giving two equations for the two unknowns—the tension and the normal force.

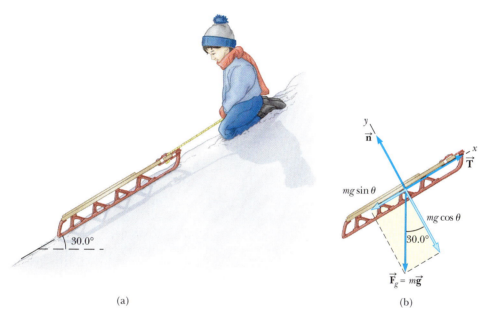

(a)   (b)

Figure 4.13   (Example 4.6) (a) A child holding a sled on a frictionless hill. (b) A free-body diagram for the sled.

Solution

Apply Newton's second law to the sled, with $\vec{\mathbf{a}} = 0$:

$$\sum \vec{\mathbf{F}} = \vec{\mathbf{T}} + \vec{\mathbf{n}} + \vec{\mathbf{F}}_g = 0$$

Extract the x-component from this equation to find T. The x-component of the normal force is zero, and the sled's weight is given by $mg = 77.0$ N.

$$\sum F_x = T + 0 - mg \sin \theta = T - (77.0 \text{ N}) \sin 30.0° = 0$$

$$T = \boxed{38.5 \text{ N}}$$

Write the y-component of Newton's second law. The y-component of the tension is zero, so this equation will give the normal force.

$$\sum F_y - 0 + n - mg \cos \theta = n - (77.0 \text{ N})(\cos 30.0°) = 0$$

$$n = \boxed{66.7 \text{ N}}$$

Remarks   Unlike its value on a horizontal surface, n is less than the weight of the sled when the sled is on the slope. This is because only part of the force of gravity (the x-component) is acting to pull the sled down the slope. The y-component of the force of gravity balances the normal force.

Exercise 4.6

Suppose another child of weight w climbs onto the sled. If the tension force is measured to be 60.0 N, find the weight of the child and the magnitude of the normal force acting on the sled.

Answers $w = 43.0$ N, $n = 104$ N

Quick Quiz 4.7

For the child being pulled forward on the toboggan in Figure 4.14, is the magnitude of the normal force exerted by the ground on the toboggan (a) equal to the total weight of the child plus the toboggan, (b) greater than the total weight, (c) less than the total weight, or (d) possibly greater than or less than the total weight, depending on the size of the weight relative to the tension in the rope?

Figure 4.14   (Quick Quiz 4.7)

Accelerating Objects and Newton's Second Law

When a net force acts on an object, the object accelerates, and we use Newton's second law to analyze the motion.

EXAMPLE 4.7   Moving a Crate

Goal   Use the second law of motion for a system not in equilibrium, together with a kinematics equation.

Problem   The combined weight of the crate and dolly in Figure 4.15 is 3.00×10^2 N. If the man pulls on the rope with a constant force of 20.0 N, what is the acceleration of the system (crate plus dolly), and how far will it move in 2.00 s? Assume that the system starts from rest and that there are no friction forces opposing the motion.

Strategy   We can find the acceleration of the system from Newton's second law. Because the force exerted on the system is constant, its acceleration is constant. Therefore, we can apply a kinematics equation to find the distance traveled in 2.00 s.

Figure 4.15   (Example 4.7)

Solution

Find the mass of the system from the definition of weight, $w = mg$:

$$m = \frac{w}{g} = \frac{3.00 \times 10^2 \text{ N}}{9.80 \text{ m/s}^2} = 30.6 \text{ kg}$$

Find the acceleration of the system from the second law:

$$a_x = \frac{F_x}{m} = \frac{20.0 \text{ N}}{30.6 \text{ kg}} = 0.654 \text{ m/s}^2$$

Use kinematics to find the distance moved in 2.00 s, with $v_0 = 0$:

$$\Delta x = \tfrac{1}{2} a_x t^2 = \tfrac{1}{2} (0.654 \text{ m/s}^2)(2.00 \text{ s})^2 = 1.31 \text{ m}$$

Remarks   Note that the constant applied force of 20.0 N is assumed to act on the system at all times during its motion. If the force were removed at some instant, the system would continue to move with constant velocity and hence zero acceleration. The rollers have an effect that was neglected, here.

Exercise 4.7

A man pulls a 50.0-kg box horizontally from rest while exerting a constant horizontal force, displacing the box 3.00 m in 2.00 s. Find the force the man exerts on the box. (Ignore friction.)

Answer   75.0 N

EXAMPLE 4.8   The Runaway Car

Goal   Apply the second law and kinematic equations to a problem involving a moving object on a slope.

Problem   **(a)** A car of mass m is on an icy driveway inclined at an angle $\theta = 20.0°$, as in Figure 4.16a. Determine the acceleration of the car, assuming that the incline is frictionless. **(b)** If the length of the driveway is 25.0 m and the car starts from rest at the top, how long does it take to travel to the bottom? **(c)** What is the car's speed at the bottom?

Strategy   Choose tilted coordinates as in Figure 4.16b, so that the normal force $\vec{\mathbf{n}}$ is in the positive y-direction, perpendicular to the driveway, and the positive x-axis is down the slope. The force of gravity $\vec{\mathbf{F}}_g$ then has an x-component, $mg \sin \theta$, and a y-component, $-mg \cos \theta$. The components of Newton's second law form a system of two equations and two unknowns for the acceleration down the slope, a_x, and the normal force. Parts (b) and (c) can be solved with the kinematics equations.

(a)

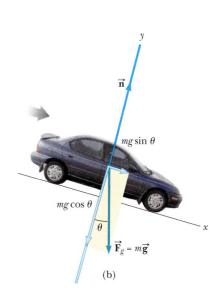

(b)

Figure 4.16   (Example 4.8)

Solution

(a) Find the acceleration of the car.

Apply Newton's second law.

$$m\vec{a} = \sum \vec{\mathbf{F}} = \vec{\mathbf{F}}_g + \vec{\mathbf{n}}$$

Extract the x- and y-components from the second law:

$$ma_x = \sum F_x = mg \sin \theta \qquad (1)$$

$$0 = \sum F_y = -mg \cos \theta + n \qquad (2)$$

Divide Equation (1) by m and substitute the given values:

$$a_x = g \sin \theta = (9.80 \text{ m/s}^2) \sin 20.0° = \boxed{3.35 \text{ m/s}^2}$$

(b) Find the time taken for the car to reach the bottom.

Use Equation 2.9 for displacement, with $v_{0x} = 0$:

$$\Delta x = \tfrac{1}{2} a_x t^2 \quad \rightarrow \quad \tfrac{1}{2} \cdot (3.35 \text{ m/s}^2) t^2 = 25.0 \text{ m}$$

$$t = \boxed{3.86 \text{ s}}$$

(c) Find the speed of the car at the bottom of the driveway.

Use Equation 2.6 for velocity, again with $v_{0x} = 0$:

$$v = at = (3.35 \text{ m/s}^2)(3.86 \text{ s}) = \boxed{12.9 \text{ m/s}}$$

Remarks   Notice that the final answer for the acceleration depends only on g and the angle θ, not the mass. Equation (2), which gives the normal force, isn't useful here, but is essential when friction plays a role.

Exercise 4.8

(a) Suppose a hockey puck slides down a frictionless ramp with an acceleration of 5.00 m/s^2. What angle does the ramp make with respect to the horizontal? **(b)** If the ramp has a length of 6.00 m, how long does it take the puck to reach the bottom? **(c)** Now suppose the mass of the puck is doubled. What's the puck's new acceleration down the ramp?

Answer   **(a)** 30.7° **(b)** 1.55 s **(c)** unchanged, 5.00 m/s^2

EXAMPLE 4.9   Weighing a Fish in an Elevator

Goal   Explore the effect of acceleration on the apparent weight of an object.

Problem   A man weighs a fish with a spring scale attached to the ceiling of an elevator, as shown in Figure 4.17a. While the elevator is at rest, he measures a weight of 40.0 N. **(a)** What weight does the scale read if the elevator accelerates upward at 2.00 m/s^2? **(b)** What does the scale read if the elevator accelerates downward at 2.00 m/s^2? **(c)** If the elevator cable breaks, what does the scale read?

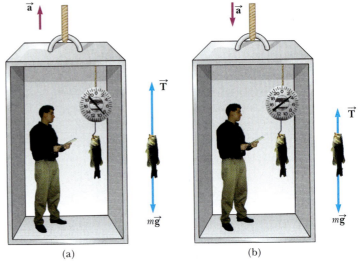

(a)   **(b)**

Figure 4.17   (Example 4.9)

Strategy   Write down Newton's second law for the fish, including the force $\vec{T}$ exerted by the spring scale and the force of gravity, $m\vec{g}$. The scale doesn't measure the true weight, it measures the force T that it exerts on the fish, so in each case solve for this force, which is the apparent weight as measured by the scale.

Solution

(a) Find the scale reading as the elevator accelerates upwards.

Apply Newton's second law to the fish, taking upwards as the positive direction:

$$ma = \Sigma F = T - mg$$

Solve for T:

$$T = ma + mg = m(a + g)$$

Find the mass of the fish from its weight of 40.0 N:

$$m = \frac{w}{g} = \frac{40.0 \text{ N}}{9.80 \text{ m/s}^2} = 4.08 \text{ kg}$$

Compute the value of T, substituting $a = +2.00$ m/s^2:

$$T = m(a + g) = (4.08 \text{ kg})(2.00 \text{ m/s}^2 + 9.80 \text{ m/s}^2)$$
$$= \boxed{48.1 \text{ N}}$$

(b) Find the scale reading as the elevator accelerates downwards.

The analysis is the same, the only change being the acceleration, which is now negative: $a = -2.00$ m/s^2.

$$T = m(a + g) = (4.08 \text{ kg})(-2.00 \text{ m/s}^2 + 9.80 \text{ m/s}^2)$$
$$= \boxed{31.8 \text{ N}}$$

(c) Find the scale reading after the elevator cable breaks.

Now $a = -9.8$ m/s^2, the acceleration due to gravity:

$$T = m(a + g) = (4.08 \text{ kg})(-9.80 \text{ m/s}^2 + 9.80 \text{ m/s}^2)$$
$$= \boxed{0 \text{ N}}$$

Remarks Notice how important it is to have correct signs in this problem! Accelerations can increase or decrease the apparent weight of an object. Astronauts experience very large changes in apparent weight, from several times normal weight during ascent to weightlessness in free fall.

Exercise 4.9

Find the initial acceleration of a rocket if the astronauts on board experience eight times their normal weight during an initial vertical ascent. (*Hint*: In this exercise, the scale force is replaced by the normal force.)

Answer $68.6 \ \text{m/s}^2$

INTERACTIVE EXAMPLE 4.10   Atwood's Machine

Goal Use the second law to solve a simple two-body problem.

Problem Two objects of mass m_1 and m_2, with $m_2 > m_1$, are connected by a light, inextensible cord and hung over a frictionless pulley, as in Active Figure 4.18a. Both cord and pulley have negligible mass. Find the magnitude of the acceleration of the system and the tension in the cord.

Strategy The heavier mass, m_2, accelerates downwards, in the negative y-direction. Since the cord can't be stretched, the accelerations of the two masses are equal in magnitude, but *opposite* in direction, so that a_1 is positive and a_2 is negative, and $a_2 = -a_1$. Each mass is acted on by a force of tension $\vec{\mathbf{T}}$ in the upwards direction and a force of gravity in the downwards direction. Active Figure 4.18b shows free-body diagrams for the two masses. Newton's second law for each mass, together with the equation relating the accelerations, constitutes a set of three equations for the three unknowns— a_1, a_2, and T.

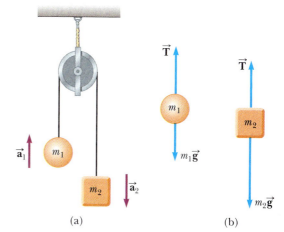

ACTIVE FIGURE 4.18
(Example 4.10) Atwood's machine. (a) Two hanging objects connected by a light string that passes over a frictionless pulley. (b) Free-body diagrams for the objects.

Physics⊗Now™
Log into to PhysicsNow at **www.cp7e.com**, and go to Active Figure 4.18 to adjust the masses of objects on Atwood's machine and observe the resulting motion.

Solution

Apply the second law to each of the two masses individually:

$$m_1 a_1 = T - m_1 g \quad \textbf{(1)} \qquad\qquad m_2 a_2 = T - m_2 g \quad \textbf{(2)}$$

Substitute $a_2 = -a_1$ into the second equation, and multiply both sides by -1:

$$m_2 a_1 = -T + m_2 g$$

Add the stacked equations, and solve for a_1:

$$(m_1 + m_2) a_1 = m_2 g - m_1 g$$

$$a_1 = \left(\frac{m_2 - m_1}{m_1 + m_2} \right) g$$

Substitute this result into Equation (1) to find T:

$$T = \left(\frac{2 m_1 m_2}{m_1 + m_2} \right) g$$

Remarks The acceleration of the second block is the same as that of the first, but negative. When m_2 gets very large compared with m_1, the acceleration of the system approaches g, as expected, because m_2 is falling nearly freely under the influence of gravity. Indeed, m_2 is only slightly restrained by the much lighter m_1.

Exercise 4.10

Suppose that in the same Atwood setup another string is attached to the bottom of m_1 and a constant force f is applied, retarding the upward motion of m_1. If $m_1 = 5.00$ kg and $m_2 = 10.00$ kg, what value of f will reduce the acceleration of the system by 50%?

Answer   24.5 N

Physics⊗Now™ Investigate the response of Atwood's machine by logging into PhysicsNow at **www.cp7e.com** and going to Interactive Example 4.10.

4.6 FORCES OF FRICTION

An object moving on a surface or through a viscous medium such as air or water encounters resistance as it interacts with its surroundings. This resistance is called **friction**. Forces of friction are essential in our everyday lives. Friction makes it possible to grip and hold things, drive a car, walk, and run. Even standing in one spot would be impossible without friction, as the slightest shift would instantly cause you to slip and fall.

Imagine that you've filled a plastic trash can with yard clippings and want to drag the can across the surface of your concrete patio. If you apply an external horizontal force $\vec{F}$ to the can, acting to the right as shown in Active Figure 4.19a, the can remains stationary if $\vec{F}$ is small. The force that counteracts $\vec{F}$ and keeps the can from moving acts to the left, opposite the direction of $\vec{F}$, and is called the **force of static friction, $\vec{f}_s$**. As long as the can isn't moving, $\vec{f}_s = -\vec{F}$. If $\vec{F}$ is increased, $\vec{f}_s$ also increases. Likewise, if $\vec{F}$ decreases, $\vec{f}_s$ decreases. Experiments show that the friction force arises from the nature of the two surfaces: Because of their roughness, contact is made at only a few points, as shown in the magnified view of the surfaces in Active Figure 4.19a.

If we increase the magnitude of $\vec{F}$, as in Active Figure 4.19b, the trash can eventually slips. When the can is on the verge of slipping, f_s is a maximum, as shown in Figure 4.19c. When F exceeds $f_{s,max}$, the can accelerates to the right. When the can is in motion, the friction force is less than $f_{s,max}$ (Fig. 4.19c). We call the friction force for an object in motion the **force of kinetic friction, $\vec{f}_k$**. The net force $F - f_k$ in the x-direction produces an acceleration to the right, according to Newton's second law. If $F = f_k$, the acceleration is zero, and the can moves to the right with constant speed. If the applied force is removed, the friction force acting to the left provides an acceleration of the can in the $-x$-direction and eventually brings it to rest, again consistent with Newton's second law.

Experimentally, to a good approximation, both $f_{s,max}$ and f_k for an object on a surface are proportional to the normal force exerted by the surface on the object. The experimental observations can be summarized as follows:

- The magnitude of the force of static friction between any two surfaces in contact can have the values

$$f_s \leq \mu_s n \qquad [4.11]$$

where the dimensionless constant μ_s is called the **coefficient of static friction** and n is the magnitude of the normal force exerted by one surface on the other. Equation 4.11 also holds for $f_s = f_{s,max} \equiv \mu_s n$ when an object is on the verge of slipping. This situation is called *impending motion*. The strict inequality holds when the component of the applied force parallel to the surfaces is less than $\mu_s n$.

- The magnitude of the force of kinetic friction acting between two surfaces is

$$f_k = \mu_k n \qquad [4.12]$$

where μ_k is the **coefficient of kinetic friction**.

TIP 4.7 Use the Equals Sign in Limited Situations

In Equation 4.11, the equals sign is used *only* when the surfaces are just about to break free and begin sliding. Don't fall into the common trap of using $f_s = \mu_s n$ in *any* static situation.

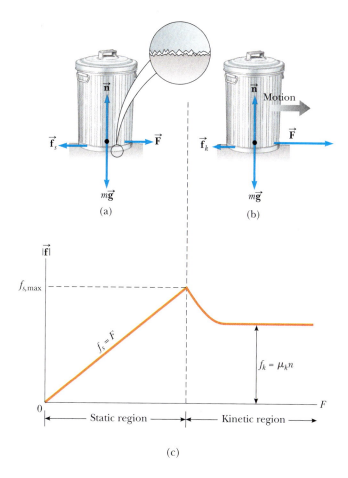

force of friction $\vec{\mathbf{f}}_s$ exerted by a concrete surface on a trash can is directed opposite the force $\vec{\mathbf{F}}$ that you exert on the can. As long as the can is not moving, the magnitude of the force of static friction equals that of the applied force $\vec{\mathbf{F}}$. (b) When the magnitude of $\vec{\mathbf{F}}$ exceeds the magnitude of $\vec{\mathbf{f}}_k$, the force of kinetic friction, the trash can accelerates to the right. (c) A graph of the magnitude of the friction force versus that of the applied force. Note that $f_{s,max} > f_k$.

Physics⊗Now™
Log into PhysicsNow at **www.cp7e.com**, and go to Active Figure 4.19 to vary the applied force on the can and practice sliding it on surfaces of varying roughness. Note the effect on the can's motion and the corresponding behavior of the graph in (c).

- The values of μ_k and μ_s depend on the nature of the surfaces, but μ_k is generally less than μ_s. Table 4.2 lists some reported values.
- The direction of the friction force exerted by a surface on an object is opposite the actual motion (kinetic friction) or the impending motion (static friction) of the object relative to the surface.
- The coefficients of friction are nearly independent of the area of contact between the surfaces.

Although the coefficient of kinetic friction varies with the speed of the object, we will neglect any such variations. The approximate nature of Equations 4.11 and 4.12

TABLE 4.2

Coefficients of Friction[a]

	μ_s	μ_k
Steel on steel	0.74	0.57
Aluminum on steel	0.61	0.47
Copper on steel	0.53	0.36
Rubber on concrete	1.0	0.8
Wood on wood	0.25–0.5	0.2
Glass on glass	0.94	0.4
Waxed wood on wet snow	0.14	0.1
Waxed wood on dry snow	—	0.04
Metal on metal (lubricated)	0.15	0.06
Ice on ice	0.1	0.03
Teflon on Teflon	0.04	0.04
Synovial joints in humans	0.01	0.003

[a]All values are approximate.

Figure 4.20   (Quick Quiz 4.10)

(a)   (b)

is easily demonstrated by trying to get an object to slide down an incline at constant acceleration. Especially at low speeds, the motion is likely to be characterized by alternate stick and slip episodes.

Quick Quiz 4.8

If you press a book flat against a vertical wall with your hand, in what direction is the friction force exerted by the wall on the book?   (a) downward   (b) upward   (c) out from the wall   (d) into the wall.

Quick Quiz 4.9

A crate is sitting in the center of a flatbed truck. As the truck accelerates to the east, the crate moves with it, not sliding on the bed of the truck. In what direction is the friction force exerted by the bed of the truck on the crate?   (a) To the west.   (b) To the east.   (c) There is no friction force, because the crate isn't sliding.

Quick Quiz 4.10

Suppose you're playing with your niece in the snow. She's sitting on a sled and asks you to move her across a flat, horizontal field. You have a choice of (a) pushing her from behind by applying a force downward on her shoulders at 30° below the horizontal (Fig. 4.20a) or (b) attaching a rope to the front of the sled and pulling with a force at 30° above the horizontal (Fig 4.20b). Which option would be easier and why?

EXAMPLE 4.11   A Block on a Ramp

Goal   Apply the concept of static friction to an object resting on an incline.

Problem   Suppose a block with a mass of 2.50 kg is resting on a ramp. If the coefficient of static friction between the block and ramp is 0.350, what maximum angle can the ramp make with the horizontal before the block starts to slip down?

Strategy   This is an application of Newton's second law involving an object in equilibrium. Choose tilted coordinates, as in Figure 4.21. Use the fact that the block is just about to slip when the force of static friction takes its maximum value, $f_s = \mu_s n$.

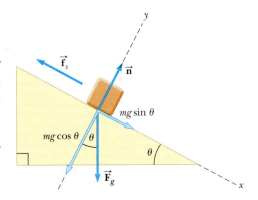

Figure 4.21   (Example 4.11)

Solution

Write Newton's laws for a static system in component form. The gravity force has two components, just as in Examples 4.6 and 4.8.

$$\sum F_x = mg \sin \theta - \mu_s n = 0 \qquad (1)$$
$$\sum F_y = n - mg \cos \theta = 0 \qquad (2)$$

| Rearrange Equation (2) to get an expression for the normal force n: | $n = mg \cos \theta$ |

| Substitute the expression for n into Equation (1) and solve for $\tan \theta$: | $\sum F_x = mg \sin \theta - \mu_s\, mg \cos \theta = 0 \quad \rightarrow \quad \tan \theta = \mu_s$ |

| Apply the inverse tangent function to get the answer: | $\tan \theta = 0.350 \quad \rightarrow \quad \theta = \tan^{-1}(0.350) = \boxed{19.3°}$ |

Remark   It's interesting that the final result depends only on the coefficient of static friction. Notice also how similar Equations (1) and (2) are to the equations developed in Examples 4.6 and 4.8. Recognizing such patterns is key to solving problems successfully.

Exercise 4.11
The ramp in Example 4.11 is roughed up and the experiment repeated. **(a)** What is the new coefficient of static friction if the maximum angle turns out to be 30.0°? **(b)** Find the maximum static friction force that acts on the block.

Answer   **(a)** 0.577 **(b)** 12.2 N

EXAMPLE 4.12   The Sliding Hockey Puck

Goal   Apply the concept of kinetic friction.

Problem   The hockey puck in Figure 4.22, struck by a hockey stick, is given an initial speed of 20.0 m/s on a frozen pond. The puck remains on the ice and slides 1.20×10^2 m, slowing down steadily until it comes to rest. Determine the coefficient of kinetic friction between the puck and the ice.

Strategy   The puck slows "steadily," which means that the acceleration is constant. Consequently, we can use the kinematic equation $v^2 = v_0^2 + 2a\Delta x$ to find a, the acceleration in the x-direction. The x- and y-components of Newton's second law then give two equations and two unknowns for the coefficient of kinetic friction, μ_k, and the normal force n.

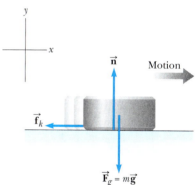

Figure 4.22   (Example 4.12) *After* the puck is given an initial velocity to the right, the external forces acting on it are the force of gravity $\vec{\mathbf{F}}_g$, the normal force $\vec{\mathbf{n}}$, and the force of kinetic friction, $\vec{\mathbf{f}}_k$.

Solution

| Solve the time-independent kinematic equation for the acceleration a: | $v^2 = v_0^2 + 2a\Delta x$

 $a = \dfrac{v^2 - v_0^2}{2\Delta x}$ |

| Substitute $v = 0$, $v_0 = 20.0$ m/s, and $\Delta x = 1.20 \times 10^2$ m. Note the negative sign in the answer: $\vec{\mathbf{a}}$ is opposite $\vec{\mathbf{v}}$. | $a = \dfrac{0 - (20.0 \text{ m/s})^2}{2(1.20 \times 10^2 \text{ m})} = -1.67 \text{ m/s}^2$ |

| Find the normal force from the y-component of the second law: | $\sum F_y = n - F_g = n - mg = 0$

 $n = mg$ |

| Obtain an expression for the force of kinetic friction, and substitute it into the x-component of the second law: | $f_k = \mu_k n = \mu_k mg$

 $ma = \sum F_x = -f_k = -\mu_k mg$ |

| Solve for μ_k and substitute values: | $\mu_k = -\dfrac{a}{g} = \dfrac{1.67 \text{ m/s}^2}{9.80 \text{ m/s}^2} = 0.170$ |

Remarks   Notice how the problem breaks down into three parts: kinematics, Newton's second law in the *y*-direction, and then Newton's law in the *x*-direction.

Exercise 4.12

An experimental rocket plane lands on skids on a dry lake bed. If it's traveling at 80.0 m/s when it touches down, how far does it slide before coming to rest? Assume the coefficient of kinetic friction between the skids and the lake bed is 0.600.

Answer   544 m

Two-body problems can often be treated as single objects and solved with a system approach. When the objects are rigidly connected—say, by a string of negligible mass that doesn't stretch—this approach can greatly simplify the analysis. When the two bodies are considered together, one or more of the forces end up becoming forces that are internal to the system, rather than external forces affecting each of the individual bodies. Both approaches will be used in Example 4.13.

EXAMPLE 4.13   Connected Objects

Goal   Use both the general method and the system approach to solve a connected two-body problem involving gravity and friction.

Problem   (a) A block with mass $m_1 = 4.00$ kg and a ball with mass $m_2 = 7.00$ kg are connected by a light string that passes over a frictionless pulley, as shown in Figure 4.23a. The coefficient of kinetic friction between the block and the surface is 0.300. Find the acceleration of the two objects and the tension in the string. (b) Check the answer for the acceleration by using the system approach.

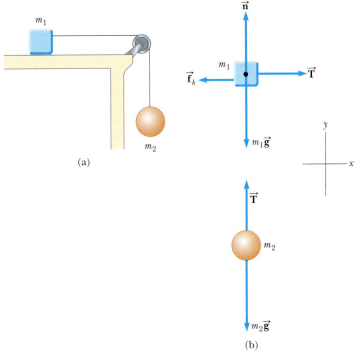

(a)

(b)

Figure 4.23   (Example 4.13) (a) Two objects connected by a light string that passes over a frictionless pulley. (b) Free-body diagrams for the objects.

Strategy   Connected objects are handled by applying Newton's second law separately to each object. The free-body diagrams for the block and the ball are shown in Figure 4.23b, with the + *x*-direction to the right and the + *y*-direction upwards. The magnitude of the acceleration for both objects has the same value, $|a_1| = |a_2| = a$. The block with mass m_1 moves in the positive *x*-direction, and the ball with mass m_2 moves in the negative *y*-direction, so $a_1 = -a_2$. Using Newton's second law, we can develop two equations involving the unknowns T and a that can be solved simultaneously. In part (b), treat the two masses as a single object, with the gravity force on the ball increasing the combined object's speed and the friction force on the block retarding it. The tension forces then become internal and don't appear in the second law.

Solution

(a) Find the acceleration of the objects and the tension in the string.

Write the components of Newton's second law for the cube of mass m_1:

$$\sum F_x = T - f_k = m_1 a_1 \qquad \sum F_y = n - m_1 g = 0$$

The equation for the *y*-component gives $n = m_1 g$. Substitute this value for n and $f_k = \mu_k n$ into the equation for the *x*-component:

$$T - \mu_k m_1 g = m_1 a_1 \qquad\qquad (1)$$

Apply Newton's second law to the ball, recalling that $a_2 = -a_1$:

$$\sum F_y = -m_2 g + T = m_2 a_2 = -m_2 a_1 \qquad (2)$$

Subtract Equation (2) from Equation (1), eliminating T and leaving an equation that can be solved for a_1 (substitution can also be used):

$$m_2 g - \mu_k m_1 g = (m_1 + m_2) a_1$$

$$a_1 = \frac{m_2 g - \mu_k m_1 g}{m_1 + m_2}$$

Substitute the given values to obtain the acceleration.

$$a_1 = \frac{(7.00\ \text{kg})(9.80\ \text{m/s}^2) - (0.300)(4.00\ \text{kg})(9.80\ \text{m/s}^2)}{(4.00\ \text{kg} + 7.00\ \text{kg})}$$

$$= 5.17\ \text{m/s}^2$$

Substitute the value for a_1 into Equation (1) to find the tension T:

$$T = 32.4\ \text{N}$$

(b) Find the acceleration using the system approach, where the system consists of the two blocks.

Apply Newton's second law to the system and solve for a:

$$(m_1 + m_2) a = m_2 g - \mu_k n = m_2 g - \mu_k m_1 g$$

$$a = \frac{m_2 g - \mu_k m_1 g}{m_1 + m_2}$$

Remarks Although the system approach appears quick and easy, it can be applied only in special cases and can't give any information about the internal forces, such as the tension. To find the tension, you must consider the free-body diagram of one of the blocks separately.

Exercise 4.13
What if an additional mass is attached to the ball in Example 4.13? How large must this mass be to increase the downward acceleration by 50%? Why isn't it possible to add enough mass to double the acceleration?

Answer   14.0 kg. Doubling the acceleration to 10.3 m/s² isn't possible simply by suspending more mass, because all objects, regardless of their mass, fall freely at 9.8 m/s² near the Earth's surface.

EXAMPLE 4.14   Two Blocks and a Cord

Goal Apply Newton's second law and static friction in a two-body system.

Problem A block of mass 5.00 kg rides on top of a second block of mass 10.0 kg. A person attaches a string to the bottom block and pulls the system horizontally across a frictionless surface, as in Figure 4.24. Friction between the two blocks keeps the 5.00-kg block from slipping off. If the coefficient of static friction is 0.350, what maximum force can be exerted by the string on the 10.0-kg block without causing the 5.00-kg block to slip?

Strategy Draw a free-body diagram for each block. The static friction force causes the top block to move horizontally, and the maximum such force corresponds to $f_s = \mu_s n$. This same static friction

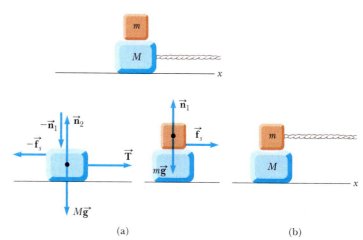

Figure 4.24   (a) (Example 4.14) (b) (Exercise 4.14)

retards the motion of the bottom block. As long as the top block isn't slipping, the acceleration of both blocks is the same. Write Newton's second law for each block, and eliminate the acceleration a by substitution, solving for the tension T.

Chapter 4   The Laws of Motion

Jlution

Write the two components of Newton's second law for the top block:	x-component: $ma = \mu_s n_1$ y-component: $0 = n_1 - mg$
Solve the y-component for n, substitute the result into the x-component, and then solve for a:	$n_1 = mg \rightarrow ma = \mu_s mg \rightarrow a = \mu_s g$
Write the x-component of Newton's second law for the bottom block:	$Ma = -\mu_s mg + T$ (1)
Substitute the expression for $a = \mu_s g$ into Equation (1) and solve for the tension T:	$M\mu_s g = T - \mu_s mg \rightarrow T = (m + M)\mu_s g$
Now evaluate to get the answer:	$T = (5.00 \text{ kg} + 10.0 \text{ kg})(0.350)(9.80 \text{ m/s}^2) = $ 51.5 N

Remarks   Notice that the y-component for the 10.0-kg block wasn't needed, because there was no friction between that block and the underlying surface. It's also interesting to note that the top block was accelerated by the force of static friction.

Exercise 4.14
Suppose instead that the string is attached to the top block in Example 14.4. Find the maximum force that can be exerted by the string on the block without causing the top block to slip.

Answer   25.7 N

Applying Physics 4.1   Cars and Friction

Forces of friction are important in the analysis of the motion of cars and other wheeled vehicles. How do such forces both help and hinder the motion of a car?

Explanation   There are several types of friction forces to consider, the main ones being the force of friction between the tires and the road surface and the retarding force produced by air resistance.

Assuming that the car is a four-wheel-drive vehicle of mass m, as each wheel turns to propel the car forward, the tire exerts a rearward force on the road. The reaction to this rearward force is a forward force $\vec{f}$ exerted by the road on the tire (Fig. 4.25). If we assume that the same forward force $\vec{f}$ is exerted on each tire, the net forward force on the car is $4\vec{f}$, and the car's acceleration is therefore $\vec{a} = 4\vec{f}/m$.

The friction between the moving car's wheels and the road is normally static friction, unless the car is skidding.

When the car is in motion, we must also consider the force of air resistance, $\vec{R}$, which acts in the

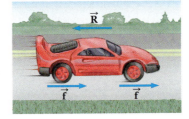

Figure 4.25   (Applying Physics 4.1) The horizontal forces acting on the car are the *forward* forces $\vec{f}$ exerted by the road on each tire and the force of air resistance $\vec{R}$, which acts *opposite* the car's velocity. (The car's tires exert a rearward force on the road, not shown in the diagram.)

direction opposite the velocity of the car. The net force exerted on the car is therefore $4\vec{f} - \vec{R}$, so the car's acceleration is $\vec{a} = (4\vec{f} - \vec{R})/m$. At normal driving speeds, the magnitude of $\vec{R}$ is proportional to the first power of the speed, $R = bv$, where b is a constant, so the force of air resistance increases with increasing speed. When R is equal to $4f$, the acceleration is zero and the car moves at a constant speed. To minimize this resistive force, racing cars often have very low profiles and streamlined contours.

Applying Physics 4.2   Air Drag

Air resistance isn't always undesirable. What are some applications that depend on it?

Explanation   Consider a skydiver plunging through the air, as in Figure 4.26. Despite falling from a height of several thousand meters, she never exceeds a speed of around 120 miles per hour. This is because, aside from the downward force of gravity $m\vec{g}$, there is also an upward force of air resistance, $\vec{R}$. Before she reaches a final constant speed, the magnitude of $\vec{R}$ is less than her weight. As her downward speed increases, the force of air resistance increases. The vector sum of the force of gravity and the force of air resistance gives a total force that decreases with time, so her acceleration decreases. Once the two forces balance each other, the net force is zero, so the acceleration is zero, and she reaches a **terminal speed**.

Terminal speed is generally still high enough to be fatal on impact, although there have been amazing stories of survival. In one case, a man fell flat on his back in a freshly plowed field and survived. (He did, however, break virtually every bone in his body). In another case, a stewardess survived a fall from thirty thousand feet into a snowbank. In neither case would the person have had any chance of surviving without the effects of air drag.

Parachutes and paragliders create a much larger drag force due to their large area and can reduce the terminal speed to a few meters per second. Some

Figure 4.26   (Applying Physics 4.2)

sports enthusiasts have even developed special suits with wings, allowing a long glide to the ground. In each case, a larger cross-sectional area intercepts more air, creating greater air drag, so the terminal speed is lower.

Air drag is also important in space travel. Without it, returning to Earth would require a considerable amount of fuel. Air drag helps slow capsules and spaceships, and aerocapture techniques have been proposed for trips to other planets. These techniques significantly reduce fuel requirements by using air drag to slow the spacecraft down.

SUMMARY

Physics⊗Now™   Take a practice test by logging into PhysicsNow at **www.cp7e.com** and clicking on the Pre-Test link for this chapter.

4.1 Forces

There are four known fundamental forces of nature: (1) the strong nuclear force between subatomic particles; (2) the electromagnetic forces between electric charges; (3) the weak nuclear forces, which arises in certain radioactive decay processes; and (4) the gravitational force between objects. These are collectively called field forces. Classical physics deals only with the gravitational and electromagnetic forces.

Forces such as friction or that characterizing a bat hitting a ball are called contact forces. On a more fundamental level, contact forces have an electromagnetic nature.

4.2 Newton's First Law

Newton's first law states that an object moves at constant velocity unless acted on by a force.

The tendency for an object to maintain its original state of motion is called **inertia**. **Mass** is the physical quantity that measures the resistance of an object to changes in its velocity.

4.3 Newton's Second Law

Newton's second law states that the acceleration of an object is directly proportional to the net force acting on it and inversely proportional to its mass. The net force acting on an object equals the product of its mass and acceleration:

$$\sum \vec{F} = m\vec{a} \qquad [4.1]$$

Newton's universal law of gravitation is

$$F_g = G\frac{m_1 m_2}{r^2} \qquad [4.5]$$

The **weight** w of an object is the magnitude of the force of gravity exerted on that object and is given by

$$w = mg \qquad [4.6]$$

where $g = F_g/m$ is the acceleration of gravity near Earth's surface.

Solving problems with Newton's second law involves finding all the forces acting on a system and writing Equation ₗ for the *x*-component and *y*-component separately. These two equations are then solved algebraically for the unknown quantities.

4.4 Newton's Third Law

Newton's third law states that if two objects interact, the force $\vec{\mathbf{F}}_{12}$ exerted by object 1 on object 2 is equal in magnitude and opposite in direction to the force $\vec{\mathbf{F}}_{21}$ exerted by object 2 on object 1:

$$\vec{\mathbf{F}}_{12} = -\vec{\mathbf{F}}_{21}$$

An isolated force can never occur in nature.

4.5 Applications of Newton's Laws

An **object in equilibrium** has no net external force acting on it, and the second law, in component form, implies that $\Sigma F_x = 0$ and $\Sigma F_y = 0$ for such an object. These two equations are useful for solving problems in statics, in which the object is at rest or moving at constant velocity.

An object under acceleration requires the same two equations, but with the acceleration terms included: $\Sigma F_x = ma_x$ and $\Sigma F_y = ma_y$. When the acceleration is constant, the equations of kinematics can supplement Newton's second law.

4.6 Forces of Friction

The magnitude of the maximum force of static friction, $f_{s,\max}$, between an object and a surface is proportional to the magnitude of the normal force acting on the object. This maximum force occurs when the object is on the verge of slipping. In general,

$$f_s \leq \mu_s n \qquad [4.11]$$

where μ_s is the **coefficient of static friction**. When an object slides over a surface, the direction of the force of kinetic friction, $\vec{\mathbf{f}}_k$, on the object is opposite the direction of the motion of the object relative to the surface, and proportional to the magnitude of the normal force. The magnitude of $\vec{\mathbf{f}}_k$ is

$$f_k = \mu_k n \qquad [4.12]$$

where μ_k is the **coefficient of kinetic friction**. In general, $\mu_k < \mu_s$.

Solving problems that involve friction is a matter of using these two friction forces in Newton's second law. The static friction force must be handled carefully, because it refers to a maximum force, which is not always called upon in a given problem.

CONCEPTUAL QUESTIONS

1. A ball is held in a person's hand. (a) Identify all the external forces acting on the ball and the reaction to each. (b) If the ball is dropped, what force is exerted on it while it is falling? Identify the reaction force in this case. (Neglect air resistance.)

2. If a car is traveling westward with a constant speed of 20 m/s, what is the resultant force acting on it?

3. If a car moves with a constant acceleration, can you conclude that there are no forces acting on it?

4. A rubber ball is dropped onto the floor. What force causes the ball to bounce?

5. If you push on a heavy box that is at rest, you must exert some force to start its motion. However, once the box is sliding, you can apply a smaller force to maintain its motion. Why?

6. If gold were sold by weight, would you rather buy it in Denver or in Death Valley? If it were sold by mass, in which of the two locations would you prefer to buy it? Why?

7. A passenger sitting in the rear of a bus claims that she was injured as the driver slammed on the brakes, causing a suitcase to come flying toward her from the front of the bus. If you were the judge in this case, what disposition would you make? Why?

8. A space explorer is moving through space far from any planet or star. He notices a large rock, taken as a specimen from an alien planet, floating around the cabin of the ship. Should he push it gently or kick it toward the storage compartment? Why?

9. What force causes an automobile to move? A propeller-driven airplane? A rowboat?

10. Analyze the motion of a rock dropped in water in terms of its speed and acceleration as it falls. Assume that a resistive force is acting on the rock that increases as the velocity of the rock increases.

11. In the motion picture *It Happened One Night* (Columbia Pictures, 1934), Clark Gable is standing inside a stationary bus in front of Claudette Colbert, who is seated. The bus suddenly starts moving forward and Clark falls into Claudette's lap. Why did this happen?

12. A weight lifter stands on a bathroom scale. She pumps a barbell up and down. What happens to the reading on the scale? Suppose she is strong enough to actually *throw* the barbell upward. How does the reading on the scale vary now?

13. In a tug-of-war between two athletes, each pulls on the rope with a force of 200 N. What is the tension in the rope? If the rope doesn't move, what horizontal force does each athlete exert against the ground?

14. As a rocket is fired from a launching pad, its speed and acceleration increase with time as its engines continue to operate. Explain why this occurs even though the thrust of the engines remains constant.

15. Identify the action-reaction pairs in the following situations: (a) a man takes a step; (b) a snowball hits a girl in the back; (c) a baseball player catches a ball; (d) a gust of wind strikes a window.

16. The driver of a speeding empty truck slams on the brakes and skids to a stop through a distance *d*. (a) If the truck carried a load that doubled its mass, what would be the truck's "skidding distance"? (b) If the initial speed of the truck were halved, what would be the truck's "skidding distance"?

17. Suppose you are driving a car at a high speed. Why should you avoid slamming on your brakes when you want to stop in the shortest possible distance? (Newer cars have antilock brakes that avoid this problem.)

18. A truck loaded with sand accelerates along a highway. If the driving force on the truck remains constant, what

happens to the truck's acceleration if its trailer leaks sand at a constant rate through a hole in its bottom?

19. A large crate is placed on the bed of a truck, but is not tied down. (a) As the truck accelerates forward, the crate remains at rest relative to it. What force causes the crate to accelerate forward? (b) If the driver slams on the brakes, what could happen to the crate?

20. Draw a free-body diagram for each of the following objects: (a) a projectile in motion in the presence of air resistance, (b) a rocket leaving the launch pad with its engines operating, (c) an athlete running along a horizontal track.

PROBLEMS

1, 2, 3 = straightforward, intermediate, challenging   □ = full solution available in *Student Solutions Manual/Study Guide*

Physics⊗Now™ = coached solution with hints available at **www.cp7e.com**   🧬 = biomedical application

Section 4.1 Forces
Section 4.2 Newton's First Law
Section 4.3 Newton's Second Law
Section 4.4 Newton's Third Law

1. A 6.0-kg object undergoes an acceleration of 2.0 m/s^2. (a) What is the magnitude of the resultant force acting on it? (b) If this same force is applied to a 4.0-kg object, what acceleration is produced?

2. A football punter accelerates a football from rest to a speed of 10 m/s during the time in which his toe is in contact with the ball (about 0.20 s). If the football has a mass of 0.50 kg, what average force does the punter exert on the ball?

3. The heaviest invertebrate is the giant squid, which is estimated to have a weight of about 2 tons spread out over its length of 70 feet. What is its weight in newtons?

4. The heaviest flying bird is the trumpeter swan, which weighs in at about 38 pounds at its heaviest. What is its weight in newtons?

5. A bag of sugar weighs 5.00 lb on Earth. What would it weigh in newtons on the Moon, where the free-fall acceleration is one-sixth that on Earth? Repeat for Jupiter, where *g* is 2.64 times that on Earth. Find the mass of the bag of sugar in kilograms at each of the three locations.

6. A freight train has a mass of 1.5×10^7 kg. If the locomotive can exert a constant pull of 7.5×10^5 N, how long does it take to increase the speed of the train from rest to 80 km/h?

7. The air exerts a forward force of 10 N on the propeller of a 0.20-kg model airplane. If the plane accelerates forward at 2.0 m/s^2, what is the magnitude of the resistive force exerted by the air on the airplane?

8. A 5.0-g bullet leaves the muzzle of a rifle with a speed of 320 m/s. What force (assumed constant) is exerted on the bullet while it is traveling down the 0.82-m-long barrel of the rifle?

9. A Chinook salmon has a maximum underwater speed of 3.0 m/s, and can jump out of the water vertically with a speed of 6.0 m/s. A record salmon has a length of 1.5 m and a mass of 61 kg. When swimming upward at constant speed, and neglecting buoyancy, the fish experiences three forces: an upward force *F* exerted by the tail fin, the downward drag force of the water, and the downward force of gravity. As the fish leaves the surface of the water, however, it experiences a net upward force causing it to accelerate from 3.0 m/s to 6.0 m/s. Assuming that the drag force disappears as soon as the head of the fish breaks the surface

and that *F* is exerted until 2/3 of the fish's length has left the water, determine the magnitude of *F*.

10. Consider a solid metal sphere (S) a few centimeters in diameter and a feather (F). For each quantity in the list that follows, indicate whether the quantity is the same, greater, or lesser in the case of S or in that of F? Explain in each case why you gave the answer you did. Here is the list: (a) the gravitational force; (b) the time it will take to fall a given distance in air; (c) the time it will take to fall a given distance in vacuum; (d) the total force on the object when falling in vacuum.

11. Physics⊗Now™ A boat moves through the water with two forces acting on it. One is a 2 000-N forward push by the water on the propellor, and the other is a 1 800-N resistive force due to the water around the bow. (a) What is the acceleration of the 1000-kg boat? (b) If it starts from rest, how far will the boat move in 10.0 s? (c) What will its velocity be at the end of that time?

12. Two forces are applied to a car in an effort to move it, as shown in Figure P4.12. (a) What is the resultant of these two forces? (b) If the car has a mass of 3 000 kg, what acceleration does it have? Ignore friction.

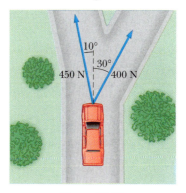

Figure P4.12

13. After falling from rest from a height of 30 m, a 0.50-kg ball rebounds upward, reaching a height of 20 m. If the contact between ball and ground lasted 2.0 ms, what average force was exerted on the ball?

14. The force exerted by the wind on the sails of a sailboat is 390 N north. The water exerts a force of 180 N east. If the boat (including its crew) has a mass of 270 kg, what are the magnitude and direction of its acceleration?

Section 4.5 Applications of Newton's Laws

15. Find the tension in each cable supporting the 600-N cat burglar in Figure P4.15.

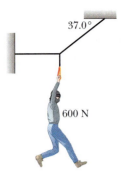

Figure P4.15

16. Find the tension in the two wires that support the 100-N light fixture in Figure P4.16.

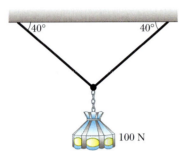

Figure P4.16

17. A 150-N bird feeder is supported by three cables as shown in Figure P4.17. Find the tension in each cable.

Figure P4.17

18. The leg and cast in Figure P4.18 weigh 220 N (w_1). Determine the weight w_2 and the angle α needed so that no force is exerted on the hip joint by the leg plus the cast.

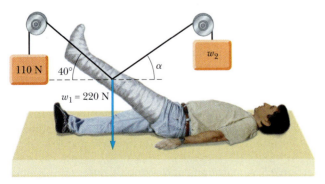

Figure P4.18

19. Two blocks are fastened to the ceiling of an elevator as in Figure P4.19. The elevator accelerates upward at 2.00 m/s². Find the tension in each rope.

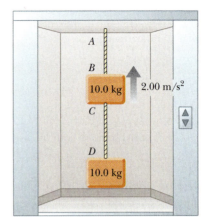

Figure P4.19

20. Two people are pulling a boat through the water as in Figure P4.20. Each exerts a force of 600 N directed at a 30.0° angle relative to the forward motion of the boat. If the boat moves with constant velocity, find the resistive force $\vec{F}$ exerted by the water on the boat.

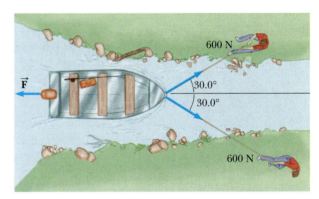

Figure P4.20

21. The distance between two telephone poles is 50.0 m. When a 1.00-kg bird lands on the telephone wire midway between the poles, the wire sags 0.200 m. Draw a free-body diagram of the bird. How much tension does the bird produce in the wire? Ignore the weight of the wire.

22. You are a judge in a children's kite-flying contest, and two children will win prizes for the kites that pull most strongly and least strongly, respectively, on their strings. To measure string tensions, you borrow a weight hanger, some slotted weights, and a protractor from your physics teacher, and use the following protocol, illustrated in Figure P4.22: Wait for a child to get her kite well controlled, hook the hanger onto the kite string about 30 cm from her hand, pile on weight until that section of string is horizontal, record the mass required, and record the angle between the horizontal and the string running up to the kite. (a) Explain how this method works. As you construct your explanation, imagine that the children's parents ask you about the method, that they might make false assumptions about your ability without concrete evidence,

and that your explanation is an opportunity to give them confidence in your evaluation technique. (b) Find the tension in the string if the mass is 132 g and the angle is 46.3°.

Figure P4.22

23. Physics Now™ A 5.0-kg bucket of water is raised from a well by a rope. If the upward acceleration of the bucket is 3.0 m/s², find the force exerted by the rope on the bucket.

24. A shopper in a supermarket pushes a loaded cart with a horizontal force of 10 N. The cart has a mass of 30 kg. (a) How far will it move in 3.0 s, starting from rest? (Ignore friction.) (b) How far will it move in 3.0 s if the shopper places his 30-N child in the cart before he begins to push it?

25. A 2 000-kg car is slowed down uniformly from 20.0 m/s to 5.00 m/s in 4.00 s. (a) What average force acted on the car during that time, and (b) how far did the car travel during that time?

26. Two packing crates of masses 10.0 kg and 5.00 kg are connected by a light string that passes over a frictionless pulley as in Figure P4.26. The 5.00-kg crate lies on a smooth incline of angle 40.0°. Find the acceleration of the 5.00-kg crate and the tension in the string.

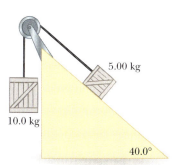

5.00 kg

10.0 kg

40.0°

Figure P4.26

27. Assume that the three blocks portrayed in Figure P4.27 move on a frictionless surface and that a 42-N force acts as shown on the 3.0-kg block. Determine (a) the acceleration given this system, (b) the tension in the cord connecting the 3.0-kg and the 1.0-kg blocks, and (c) the force exerted by the 1.0-kg block on the 2.0-kg block.

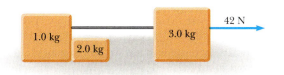

1.0 kg

2.0 kg

3.0 kg   42 N

Figure P4.27

28. An object of mass 2.0 kg starts from rest and slides down an inclined plane 80 cm long in 0.50 s. What *net force* is acting on the object along the incline?

29. A 40.0-kg wagon is towed up a hill inclined at 18.5° with respect to the horizontal. The tow rope is parallel to the incline and has a tension of 140 N. Assume that the wagon starts from rest at the bottom of the hill, and neglect friction. How fast is the wagon going after moving 80.0 m up the hill?

30. An object with mass $m_1 = 5.00$ kg rests on a frictionless horizontal table and is connected to a cable that passes over a pulley and is then fastened to a hanging object with mass $m_2 = 10.0$ kg, as shown in Figure P4.30. Find the acceleration of each object and the tension in the cable.

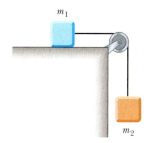

m_1

m_2

Figure P4.30   (Problems 30, 36, and 45)

31. A train has a mass of 5.22×10^6 kg and is moving at 90.0 km/h. The engineer applies the brakes, resulting in a net backward force of 1.87×10^6 N on the train. The brakes are held on for 30.0 s. (a) What is the final speed of the train? (b) How far does the train travel during this period?

32. (a) An elevator of mass m moving upward has two forces acting on it: the upward force of tension in the cable and the downward force due to gravity. When the elevator is accelerating upward, which is greater, T or w? (b) When the elevator is moving at a constant velocity upward, which is greater, T or w? (c) When the elevator is moving upward, but the acceleration is downward, which is greater, T or w? (d) Let the elevator have a mass of 1 500 kg and an upward acceleration of 2.5 m/s². Find T. Is your answer consistent with the answer to part (a)? (e) The elevator of part (d) now moves with a constant upward velocity of 10 m/s. Find T. Is your answer consistent with your answer to part (b)? (f) Having initially moved upward with a constant velocity, the elevator begins to accelerate downward at 1.50 m/s². Find T. Is your answer consistent with your answer to part (c)?

33. A 1 000-kg car is pulling a 300-kg trailer. Together, the car and trailer have an acceleration of 2.15 m/s² in the forward direction. Neglecting frictional forces on the trailer, determine (a) the net force on the car, (b) the net force on the trailer, (c) the force exerted by the trailer on the car, and (d) the resultant force exerted by the car on the road.

34. Two objects with masses of 3.00 kg and 5.00 kg are connected by a light string that passes over a frictionless pulley, as in Figure P4.34. Determine (a) the tension in the string, (b) the acceleration of each object, and (c) the distance each object will move in the first second of motion if both objects start from rest.

Figure P4.34

Section 4.6 Forces of Friction

35. A dockworker loading crates on a ship finds that a 20-kg crate, initially at rest on a horizontal surface, requires a 75-N horizontal force to set it in motion. However, after the crate is in motion, a horizontal force of 60 N is required to keep it moving with a constant speed. Find the coefficients of static and kinetic friction between crate and floor.

36. In Figure P4.30, $m_1 = 10$ kg and $m_2 = 4.0$ kg. The coefficient of static friction between m_1 and the horizontal surface is 0.50, and the coefficient of kinetic friction is 0.30. (a) If the system is released from rest, what will its acceleration be? (b) If the system is set in motion with m_2 moving downward, what will be the acceleration of the system?

37. A 1 000-N crate is being pushed across a level floor at a constant speed by a force $\vec{F}$ of 300 N at an angle of 20.0° below the horizontal, as shown in Figure P4.37a. (a) What is the coefficient of kinetic friction between the crate and the floor? (b) If the 300-N force is instead pulling the block at an angle of 20.0° above the horizontal, as shown in Figure P4.37b, what will be the acceleration of the crate? Assume that the coefficient of friction is the same as that found in (a).

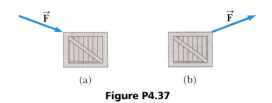

(a)   (b)

Figure P4.37

38. A hockey puck is hit on a frozen lake and starts moving with a speed of 12.0 m/s. Five seconds later, its speed is 6.00 m/s. (a) What is its average acceleration? (b) What is the average value of the coefficient of kinetic friction between puck and ice? (c) How far does the puck travel during the 5.00-s interval?

39. Consider a large truck carrying a heavy load, such as steel beams. A significant hazard for the driver is that the load may slide forward, crushing the cab, if the truck stops suddenly in an accident or even in braking. Assume, for example, that a 10 000-kg load sits on the flat bed of a 20 000-kg truck moving at 12.0 m/s. Assume the load is not tied down to the truck and has a coefficient of static friction of 0.500 with the truck bed. (a) Calculate the minimum stopping distance for which the load will not slide forward relative to the truck. (b) Is any piece of data unnecessary for the solution?

40. A woman at an airport is towing her 20.0-kg suitcase at constant speed by pulling on a strap at an angle θ above the horizontal (Fig. P4.40). She pulls on the strap with a 35.0-N force, and the friction force on the suitcase is 20.0 N. Draw a free-body diagram of the suitcase. (a) What angle does the strap make with the horizontal? (b) What normal force does the ground exert on the suitcase?

Figure P4.40

41. The coefficient of static friction between the 3.00-kg crate and the 35.0° incline of Figure P4.41 is 0.300. What minimum force $\vec{F}$ must be applied to the crate perpendicular to the incline to prevent the crate from sliding down the incline?

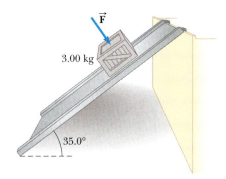

Figure P4.41

42. A box of books weighing 300 N is shoved across the floor of an apartment by a force of 400 N exerted downward at an angle of 35.2° below the horizontal. If the coefficient of kinetic friction between box and floor is 0.570, how long does it take to move the box 4.00 m, starting from rest?

43. An object falling under the pull of gravity is acted upon by a frictional force of air resistance. The magnitude of this force is approximately proportional to the speed of the object, which can be written as $f = bv$. Assume that $b = 15$ kg/s and $m = 50$ kg. (a) What is the terminal speed the object reaches while falling? (b) Does your answer to part (a) depend on the initial speed of the object? Explain.

44. A student decides to move a box of books into her dormitory room by pulling on a rope attached to the box. She pulls with a force of 80.0 N at an angle of 25.0° above the horizontal. The box has a mass of 25.0 kg, and the coefficient of kinetic friction between box and floor is 0.300. (a) Find the acceleration of the box. (b) The student now starts moving the box up a 10.0° incline, keeping her 80.0 N force directed at 25.0° above the line of the incline. If the coefficient of friction is unchanged, what is the new acceleration of the box?

45. **Physics⊗Now™** Objects with masses $m_1 = 10.0$ kg and $m_2 = 5.00$ kg are connected by a light string that passes over a frictionless pulley as in Figure P4.30. If, when the system starts from rest, m_2 falls 1.00 m in 1.20 s, determine the coefficient of kinetic friction between m_1 and the table.

46. A car is traveling at 50.0 km/h on a flat highway. (a) If the coefficient of friction between road and tires on a rainy day is 0.100, what is the minimum distance in which the car will stop? (b) What is the stopping distance when the surface is dry and the coefficient of friction is 0.600?

47. A 3.00-kg block starts from rest at the top of a 30.0° incline and slides 2.00 m down the incline in 1.50 s. Find (a) the acceleration of the block, (b) the coefficient of kinetic friction between the block and the incline, (c) the frictional force acting on the block, and (d) the speed of the block after it has slid 2.00 m.

48. Objects of masses $m_1 = 4.00$ kg and $m_2 = 9.00$ kg are connected by a light string that passes over a frictionless pulley as in Figure P4.48. The object m_1 is held at rest on the floor, and m_2 rests on a fixed incline of $\theta = 40.0°$. The objects are released from rest, and m_2 slides 1.00 m down the incline in 4.00 s. Determine (a) the acceleration of each object, (b) the tension in the string, and (c) the coefficient of kinetic friction between m_2 and the incline.

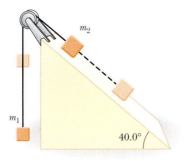

Figure P4.48

49. Find the acceleration reached by each of the two objects shown in Figure P4.49 if the coefficient of kinetic friction between the 7.00-kg object and the plane is 0.250.

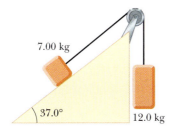

Figure P4.49

50. A 2.00-kg block is held in equilibrium on an incline of angle $\theta = 60.0°$ by a horizontal force $\vec{F}$ applied in the direction shown in Figure P4.50. If the coefficient of static

friction between block and incline is $\mu_s = 0.300$, determine (a) the minimum value of $\vec{F}$ and (b) the normal force exerted by the incline on the block.

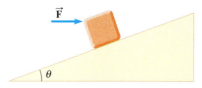

Figure P4.50

51. The person in Figure P4.51 weighs 170 lb. Each crutch makes an angle of 22.0° with the vertical (as seen from the front). Half of the person's weight is supported by the crutches, the other half by the vertical forces exerted by the ground on his feet. Assuming that he is at rest and that the force exerted by the ground on the crutches acts along the crutches, determine (a) the smallest possible coefficient of friction between crutches and ground and (b) the magnitude of the compression force supported by each crutch.

Figure P4.51

52. A block of mass $m = 2.00$ kg rests on the left edge of a block of length $L = 3.00$ m and mass $M = 8.00$ kg. The coefficient of kinetic friction between the two blocks is $\mu_k = 0.300$, and the surface on which the 8.00-kg block rests is frictionless. A constant horizontal force of magnitude $F = 10.0$ N is applied to the 2.00-kg block, setting it

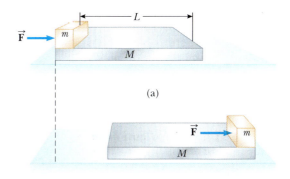

Figure P4.52

in motion as shown in Figure P4.52a. (a) How long will it take before this block makes it to the right side of the 8.00-kg block, as shown in Figure P4.52b? (*Note:* Both blocks are set in motion when the force $\vec{F}$ is applied.) (b) How far does the 8.00-kg block move in the process?

Additional Problems

53. In Figure P4.53, the coefficient of kinetic friction between the two blocks shown is 0.30. The surface of the table and the pulleys are frictionless. (a) Draw a free-body diagram for each block. (b) Determine the acceleration of each block. (c) Find the tension in the strings.

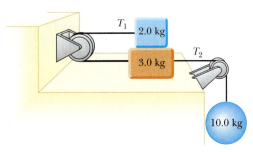

Figure P4.53

54. The force exerted by the wind on a sailboat is approximately perpendicular to the sail and proportional to the component of the wind velocity perpendicular to the sail. For the 800 kg sailboat shown in Figure P4.54, the proportionality constant is

$$F_{sail} = \left(550 \frac{N}{m/s}\right) v_{wind\perp}$$

Water exerts a force along the keel (bottom) of the boat that prevents it from moving sideways, as shown in the figure. Once the boat starts moving forward, water also exerts a drag force backwards on the boat, opposing the forward motion. If a 17-knot wind (1 knot = 0.514 m/s) is blowing to the east, what is the initial acceleration of the sailboat?

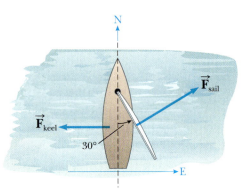

Figure P4.54

55. (a) What is the resultant force exerted by the two cables supporting the traffic light in Figure P4.55? (b) What is the weight of the light?

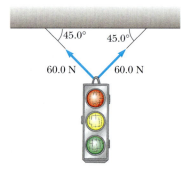

Figure P4.55

56. As a protest against the umpire's calls, a baseball pitcher throws a ball straight up into the air at a speed of 20.0 m/s. In the process, he moves his hand through a distance of 1.50 m. If the ball has a mass of 0.150 kg, find the force he exerts on the ball to give it this upward speed.

57. A boy coasts down a hill on a sled, reaching a level surface at the bottom with a speed of 7.0 m/s. If the coefficient of friction between the sled's runners and the snow is 0.050 and the boy and sled together weigh 600 N, how far does the sled travel on the level surface before coming to rest?

58. (a) What is the minimum force of friction required to hold the system of Figure P4.58 in equilibrium? (b) What coefficient of static friction between the 100-N block and the table ensures equilibrium? (c) If the coefficient of kinetic friction between the 100-N block and the table is 0.250, what hanging weight should replace the 50.0-N weight to allow the system to move at a constant speed once it is set in motion?

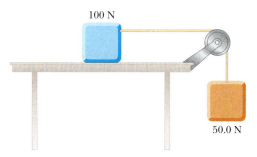

Figure P4.58

59. A box rests on the back of a truck. The coefficient of static friction between the box and the bed of the truck is 0.300. (a) When the truck accelerates forward, what force accelerates the box? (b) Find the maximum acceleration the truck can have before the box slides.

60. A 4.00-kg block is pushed along the ceiling with a constant applied force of 85.0 N that acts at an angle of 55.0°

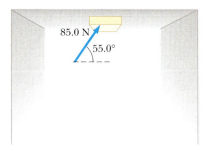

Figure P4.60

with the horizontal, as in Figure P4.60. The block accelerates to the right at 6.00 m/s². Determine the coefficient of kinetic friction between block and ceiling.

61. A frictionless plane is 10.0 m long and inclined at 35.0°. A sled starts at the bottom with an initial speed of 5.00 m/s up the incline. When the sled reaches the point at which it momentarily stops, a second sled is released from the top of the incline with an initial speed v_i. Both sleds reach the bottom of the incline at the same moment. (a) Determine the distance that the first sled traveled up the incline. (b) Determine the initial speed of the second sled.

62. Three objects are connected by light strings as shown in Figure P4.62. The string connecting the 4.00-kg object and the 5.00-kg object passes over a light frictionless pulley. Determine (a) the acceleration of each object and (b) the tension in the two strings.

Figure P4.62

63. A 2.00-kg aluminum block and a 6.00-kg copper block are connected by a light string over a frictionless pulley. The two blocks are allowed to move on a fixed steel block wedge (of angle $\theta = 30.0°$) as shown in Figure P4.63. Making use of Table 4.2, determine (a) the acceleration of the two blocks and (b) the tension in the string.

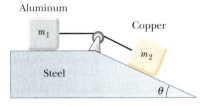

Figure P4.63

64. A 5.0-kg penguin sits on a 10-kg sled, as shown in Figure P4.64. A horizontal force of 45 N is applied to the sled, but the penguin attempts to impede the motion by holding onto a cord attached to a wall. The coefficient of kinetic friction between the sled and the snow, as well as that between the sled and the penguin, is 0.20. (a) Draw a free-body diagram for the penguin and one for the sled, and identify the reaction force for each force you include. Determine (b) the tension in the cord and (c) the acceleration of the sled.

Figure P4.64

65. Two boxes of fruit on a frictionless horizontal surface are connected by a light string as in Figure P4.65, where $m_1 = 10$ kg and $m_2 = 20$ kg. A force of 50 N is applied to the 20-kg box. (a) Determine the acceleration of each box and the tension in the string. (b) Repeat the problem for the case where the coefficient of kinetic friction between each box and the surface is 0.10.

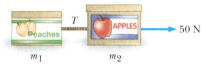

Figure P4.65

66. A high diver of mass 70.0 kg jumps off a board 10.0 m above the water. If her downward motion is stopped 2.00 s after she enters the water, what average upward force did the water exert on her?

67. Two people pull as hard as they can on ropes attached to a 200-kg boat. If they pull in the same direction, the boat has an acceleration of 1.52 m/s² to the right. If they pull in opposite directions, the boat has an acceleration of 0.518 m/s² to the left. What is the force exerted by each person on the boat? (Disregard any other forces on the boat.)

68. A 3.0-kg object hangs at one end of a rope that is attached to a support on a railroad car. When the car accelerates to the right, the rope makes an angle of 4.0° with the vertical, as shown in Figure P4.68. Find the acceleration of the car.

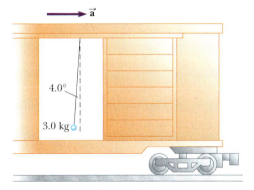

Figure P4.68

69. Three blocks of masses 10.0 kg, 5.00 kg, and 3.00 kg are connected by light strings that pass over frictionless pulleys as shown in Figure P4.69. The acceleration of the 5.00-kg block is 2.00 m/s² to the left, and the surfaces are

rough. Find (a) the tension in each string and (b) the co-efficient of kinetic friction between blocks and surfaces. (Assume the same μ_k for both blocks that are in contact with surfaces.)

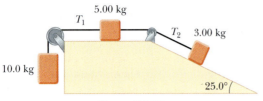

Figure P4.69

70. An inquisitive physics student, wishing to combine pleasure with scientific inquiry, rides on a rollercoaster sitting on a bathroom scale. (Do not try this yourself on a roller coaster that forbids loose heavy packages.) The bottom of the seat in the rollercoaster car is in a plane parallel to the track. The seat has a perpendicular back and a seat belt that fits around the student's chest in a plane parallel to the bottom of the seat. The student lifts his feet from the floor, so that the scale reads his weight, 200 lb, when the car is horizontal. At one point during the ride, the car zooms with negligible friction down a straight slope inclined at 30.0° below the horizontal. What does the scale read at that point?

71. A van accelerates down a hill (Fig. P4.71), going from rest to 30.0 m/s in 6.00 s. During the acceleration, a toy ($m = 0.100$ kg) hangs by a string from the van's ceiling. The acceleration is such that the string remains perpendicular to the ceiling. Determine (a) the angle θ and (b) the tension in the string.

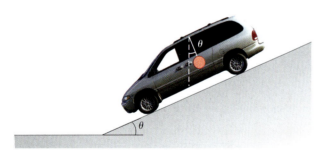

Figure P4.71

72. An 80-kg stuntman jumps from a window of a building situated 30 m above a catching net. Assuming that air resistance exerts a 100-N force on the stuntman as he falls, determine his velocity just before he hits the net.

73. The parachute on a race car of weight 8 820 N opens at the end of a quarter-mile run when the car is traveling at 35 m/s. What total retarding force must be supplied by the parachute to stop the car in a distance of 1 000 m?

74. **Physics⊗Now**™ On an airplane's takeoff, the combined action of the air around the engines and wings of an airplane exerts an 8 000-N force on the plane, directed upward at an angle of 65.0° above the horizontal. The plane rises with constant velocity in the vertical direction while continuing to accelerate in the horizontal direction. (a) What is the weight of the plane? (b) What is its horizontal acceleration?

75. A 72-kg man stands on a spring scale in an elevator. Starting from rest, the elevator ascends, attaining its maximum speed of 1.2 m/s in 0.80 s. The elevator travels with this constant speed for 5.0 s, undergoes a uniform *negative* acceleration for 1.5 s, and then comes to rest. What does the spring scale register (a) before the elevator starts to move? (b) during the first 0.80 s of the elevator's ascent? (c) while the elevator is traveling at constant speed? (d) during the elevator's negative acceleration?

76. A sled weighing 60.0 N is pulled horizontally across snow so that the coefficient of kinetic friction between sled and snow is 0.100. A penguin weighing 70.0 N rides on the sled, as in Figure P4.76. If the coefficient of static friction between penguin and sled is 0.700, find the maximum horizontal force that can be exerted on the sled before the penguin begins to slide off.

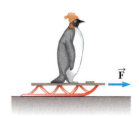

Figure P4.76

77. The board sandwiched between two other boards in Figure P4.77 weighs 95.5 N. If the coefficient of friction between the boards is 0.663, what must be the magnitude of the compression forces (assumed to be horizontal) acting on both sides of the center board to keep it from slipping?

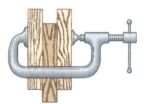

Figure P4.77

78. A magician pulls a tablecloth from under a 200-g mug located 30.0 cm from the edge of the cloth. The cloth exerts a friction force of 0.100 N on the mug and is pulled with a constant acceleration of 3.00 m/s². How far does the mug move relative to the horizontal tabletop before the cloth is completely out from under it? Note that the cloth must move more than 30 cm relative to the tabletop during the process.

79. An inventive child wants to reach an apple in a tree without climbing the tree. Sitting in a chair connected to a rope that passes over a frictionless pulley (Fig. P4.79), the child pulls on the loose end of the rope with such a force that the spring scale reads 250 N. The child's true weight is 320 N, and the chair weighs 160 N. (a) Show that the acceleration of the system is *upward* and find its magnitude. (b) Find the force the child exerts on the chair.

Figure P4.79

80. A fire helicopter carries a 620-kg bucket of water at the end of a 20.0-m-long cable. Flying back from a fire at a constant speed of 40.0 m/s, the cable makes an angle of 40.0° with respect to the vertical. Determine the force exerted by air resistance on the bucket.

81. A bag of cement hangs from three wires as shown in Figure P4.81. Two of the wires make angles θ_1 and θ_2, respectively, with the horizontal. (a) Show that, if the system is in equilibrium, then

$$T_1 = \frac{w \cos \theta_2}{\sin(\theta_1 + \theta_2)}$$

(b) Given that $w = 325$ N, $\theta_1 = 10.0°$, and $\theta_2 = 25.0°$, find the tensions T_1, T_2, and T_3 in the wires.

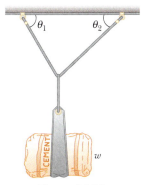

Figure P4.81

Activities

A.1. There is a simple method for measuring the coefficients of static and kinetic friction between an object and some surface. For this investigation, you will need a few coins, your textbook or some other flat surface that can be inclined, a protractor, and some double-stick tape. Place a coin at one edge of the book as it lies on a table, and lift that edge of the book until the coin just slips down the incline, as shown in Figure A4.1. When the coin starts to slip, measure the angle of incline with your protractor. Repeat the measurement five times, and find the average value of this critical angle θ_c. The coefficient of static friction between the coin and book's surface is $\mu_s = \tan \theta_c$. (You should prove this as an exercise.) Calculate the average value of μ_s. To measure the coefficient of kinetic friction, find the angle θ_c' at which the coin moves down the incline with constant speed. This angle should be less than θ_c. Measure this new angle five times, and get its average value. Calculate the average value of μ_k, using the fact that $\mu_k = \tan \theta_c'$, where $\theta_c' < \theta_c$. Repeat these measurements, using two or three stacked coins with double-stick tape between them. You should get the same results as with one coin. Why?

Figure A4.1

A.2. Borrow a spring scale from your instructor and use it to study some of the properties of the force of friction. (1) Attach the scale to a block of wood resting on the surface of a table, and note the force required to start the block moving. You should each take at least five trials and average your results. This measured force is the maximum value of the force of static friction between the block and surface. (2) Now use the spring scale to measure the force required to keep the block moving at constant velocity. Again, perform several trials to find the average value for this force. The force you find is the force of kinetic friction. (3) Turn the block so that a side with a different surface area is in contact with the table. Repeat the preceding experiments to see if the area of contact between the surfaces produces different values for the forces of friction.

A.3. Get a bathroom scale and stand on it while riding on an elevator. Watch carefully what happens to your apparent weight (the reading on the scale) as the elevator moves upward or downward as a function of time. What do the readings on the scale tell you about the acceleration of the elevator during the ride?

An asteroid plunges through Earth's atmosphere while pterodactyls watch. This artist's conception is of a catastrophic event thought to have led to the extinction of dinosaurs. During an impact, an asteroid only a kilometer across releases its awesome energy of motion as heat and light, delivering the explosive equivalent of one hundred million atomic bombs.

NASA

CHAPTER

5

OUTLINE

Energy

Energy is one of the most important concepts in the world of science. In everyday use, energy is associated with the fuel needed for transportation and heating, with electricity for lights and appliances, and with the foods we consume. These associations, however, don't tell us what energy *is*, only what it *does*, and that producing it requires fuel. Our goal in this chapter, therefore, is to develop a better understanding of energy and how to quantify it.

Energy is present in the Universe in a variety of forms, including mechanical, chemical, electromagnetic, and nuclear energy. Even the inert mass of everyday matter contains a very large amount of energy. Although energy can be transformed from one kind to another, all observations and experiments to date suggest that the total amount of energy in the Universe never changes. This is also true for an isolated system—a collection of objects that can exchange energy with each other, but not with the rest of the Universe. If one form of energy in an isolated system decreases, then another form of energy in the system must increase. For example, if the system consists of a motor connected to a battery, the battery converts chemical energy to electrical energy, and the motor converts electrical energy to mechanical energy. Understanding how energy changes from one form to another is essential in all the sciences.

In this chapter the focus is mainly on *mechanical energy*, which is the sum of *kinetic energy*—the energy associated with motion—and *potential energy*—the energy associated with position. Using an energy approach to solve certain problems is often much easier than using forces and Newton's three laws. These two very different approaches are linked through the concept of *work*.

5.1 WORK

Work has a different meaning in physics than it does in everyday usage. In the physics definition, a programmer does very little work typing away at a computer. A mason, by contrast, may do a lot of work laying concrete blocks. In physics, work is done only if an object is moved through some displacement while a force is applied to it. If either the force or displacement is doubled, the work is doubled. Double them both, and the work is quadrupled. Doing work involves applying a force to an object while moving it a given distance.

Figure 5.1 shows a block undergoing a displacement $\Delta \vec{x}$ along a straight line while acted on by a constant force $\vec{F}$ in the same direction. We have the following definition:

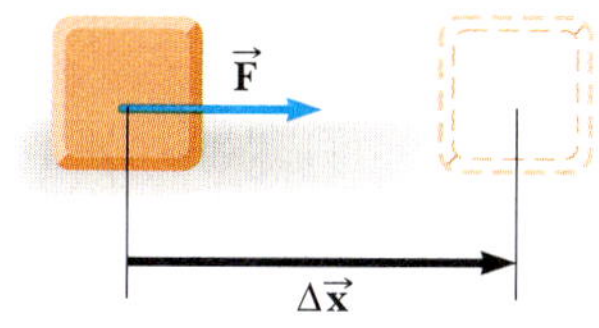

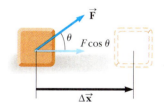

Figure 5.1 A constant force $\vec{F}$ in the same direction as the displacement, $\Delta \vec{x}$, does work $F\Delta x$.

◀ Work by a constant force along the displacement

> The work W done on an object by a constant force $\vec{F}$ is given by
>
> $$W = F\Delta x \qquad \text{[5.1]}$$
>
> where F is the magnitude of the force, Δx is the magnitude of the displacement, and $\vec{F}$ and $\Delta \vec{x}$ point in the same direction.
>
> **SI unit: joule (J) = newton · meter = kg · m²/s²**

It's easy to see the difference between the physics definition and the everyday definition of work. The programmer exerts very little force on the keys of a keyboard, creating only small displacements, so relatively little physics work is done. The mason must exert much larger forces on the concrete blocks and move them significant distances, and so performs a much greater amount of work. Even very tiring tasks, however, may not constitute work according to the physics definition. A truck driver, for example, may drive for several hours, but if he doesn't exert a force, then $F = 0$ in Equation 5.1 and he doesn't do any work. Similarly, a student pressing against a wall for hours in an isometric exercise also does no work, because the displacement in Equation 5.1, Δx, is zero.[1] Atlas, of Greek mythology, bore the world on his shoulders, but that, too, wouldn't qualify as work in the physics definition.

Work is a scalar quantity—a number rather than a vector—and consequently is easier to handle. No direction is associated with it. Further, work doesn't depend explicitly on time, which can be an advantage in problems involving only velocities and positions. Since the units of work are those of force and distance, the SI unit is the **newton-meter** (N · m). Another name for the newton-meter is the **joule (J)** (rhymes with "pool"). The U.S. customary unit of work is the **foot-pound**, because distances are measured in feet and forces in pounds in that system.

Complications in the definition of work occur when the force exerted on an object is not in the same direction as the displacement (Figure 5.2.) The force, however, can always be split into two components—one parallel and the other perpendicular to the direction of displacement. Only the component parallel to the direction of displacement does work on the object. This fact can be expressed in the following more general definition:

TIP 5.1 Work is a Scalar Quantity

Work is a simple number—a scalar, not a vector—so there is no direction associated with it. Energy and energy transfer are also scalars.

Figure 5.2 A constant force $\vec{F}$ exerted at an angle θ with respect to the displacement, $\Delta \vec{x}$, does work $(F\cos \theta)\Delta x$.

> The work W done on an object by a constant force $\vec{F}$ is given by
>
> $$W \equiv (F\cos \theta)\Delta x \qquad \text{[5.2]}$$
>
> where F is the magnitude of the force, Δx is the magnitude of the object's displacement, and θ is the angle between the directions of $\vec{F}$ and $\Delta \vec{x}$.
>
> **SI unit: joule (J)**

◀ Work by a constant force at an angle to the displacement

In Figure 5.3, a man carries a bucket of water horizontally at constant velocity. The upward force exerted by the man's hand on the bucket is perpendicular to

[1]Actually, you do expend energy while doing isometric exercises, because your muscles are continuously contracting and relaxing in the process. This internal muscular movement qualifies as work according to the physics definition.

Figure 5.3 No work is done on a bucket when it is moved horizontally because the applied force $\vec{F}$ is perpendicular to the displacement.

TIP 5.2 **Work is Done** *by Something, on Something Else*

Work doesn't happen by itself. Work is done *by* something in the environment, *on* the object of interest.

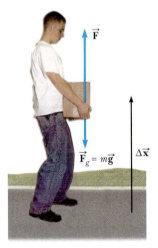

Figure 5.4 The student does positive work when he lifts the box, because the applied force $\vec{F}$ is in the same direction as the displacement. When he lowers the box to the floor, he does negative work.

the direction of motion, so it does no work on the bucket. This can also be seen from Equation 5.2, because the angle between the force exerted by the hand and the direction of motion is 90°, giving cos 90° = 0 and W = 0. Similarly, the force of gravity does no work on the bucket.

Work always requires a system of more than just one object. A nail, for example, can't do work on itself, but a hammer can do work on the nail by driving it into a board. In general, an object may be moving under the influence of several external forces. In that case, the total work done on the object as it undergoes some displacement is just the sum of the amount of work done by each force.

Work can be either positive or negative. In the definition of work in Equation 5.2, F and Δx are magnitudes, which are never negative. Work is therefore positive or negative depending on whether cos θ is positive or negative. This, in turn, depends on the direction of $\vec{F}$ relative the direction of $\Delta\vec{x}$. When these vectors are pointing in the same direction the angle between them is 0°, so cos 0° = +1 and the work is positive. For example, when a student lifts a box as in Figure 5.4, the work he does on the box is positive because the force he exerts on the box is upward, in the same direction as the displacement. In lowering the box slowly back down, however, the student still exerts an upward force on the box, but the motion of the box is downwards. Since the vectors $\vec{F}$ and $\Delta\vec{x}$ are now in opposite directions, the angle between them is 180°, and cos 180° = −1 and the work done by the student is negative. In general, when the part of $\vec{F}$ parallel to $\Delta\vec{x}$ points in the same direction as $\Delta\vec{x}$, the work is positive; and is otherwise negative.

Because Equations 5.1 and 5.2 assume a force constant in both direction and size, they are only special cases of a more general definition of work—that done by a varying force—treated briefly in Section 5.7.

Quick Quiz 5.1

In Active Figure 5.5 (a)–(d), a block moves to the right in the positive *x*-direction through the displacement Δx while under the influence of a force with the same magnitude *F*. Which of the following is the correct order of the amount of work done by the force *F*, from most positive to most negative? (A) d, c, a, b   (B) c, a, b, d   (C) c, a, d, b

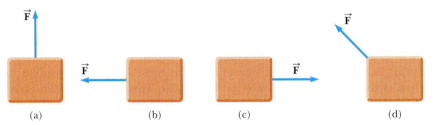

ACTIVE FIGURE 5.5
(Quick Quiz 5.1) A force $\vec{F}$ is exerted on an object that undergoes a displacement to the right. Both the magnitude of the force and the displacement are the same in all four cases.

Physics⊗Now™

Log into PhysicsNow at **www.cp7e.com** and go to Active Figure 5.5, where you can move a block to see how the work done against gravity depends on position.

EXAMPLE 5.1   Sledding through the Yukon

Goal   Apply the basic definitions of work done by a constant force.

Problem   An Eskimo returning from a successful fishing trip pulls a sled loaded with salmon. The total mass of the sled and salmon is 50.0 kg, and the Eskimo exerts a force of 1.20×10^2 N on the sled by pulling on the rope.

(a) How much work does he do on the sled if the rope is horizontal to the ground ($\theta = 0°$ in Figure 5.6) and he pulls the sled 5.00 m? **(b)** How much work does he do on the sled if $\theta = 30.0°$ and he pulls the sled the same distance? (Treat the sled as a point particle, so details such as the point of attachment of the rope make no difference.)

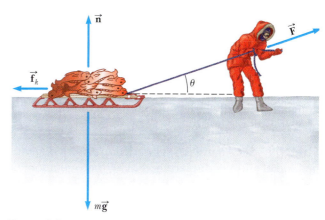

Strategy   Substitute the given values of F and Δx into the basic equations for work, Equations 5.1 and 5.2.

Solution
(a) Find the work done when the force is horizontal.

Figure 5.6   (Examples 5.1 and 5.2) An Eskimo pulling a sled with a rope at an angle θ to the horizontal.

Use Equation 5.1, substituting the given values:

$$W = F\Delta x = (1.20 \times 10^2\,\text{N})(5.00\,\text{m}) = \boxed{6.00 \times 10^2\,\text{J}}$$

(b) Find the work done when the force is exerted at a 30° angle.

Use Equation 5.2, again substituting the given values:

$$W = (F\cos\theta)\Delta x = (1.20 \times 10^2\,\text{N})(\cos 30°)(5.00\,\text{m})$$
$$= \boxed{5.20 \times 10^2\,\text{J}}$$

Remarks   The normal force $\vec{\mathbf{n}}$, the gravitational force $m\vec{\mathbf{g}}$, and the upward component of the applied force do *no* work on the sled, because they're perpendicular to the displacement. The mass of the sled didn't come into play here, but is important when the effects of friction must be calculated, and in the next section, where we introduce the work–energy theorem.

Exercise 5.1
Suppose the Eskimo is pushing the same 50.0-kg sled across level terrain with a force of 50.0 N. (a) If he does $4.00 \times 10^2\,\text{J}$ of work on the sled while exerting the force horizontally, through what distance must he have pushed it? (b) If he exerts the same force at an angle of 45.0° with respect to the horizontal and moves the sled through the same distance, how much work does he do on the sled?

Answers   (a) 8.00 m   (b) 283 J

Work and Dissipative Forces

Frictional work is extremely important in everyday life, because doing almost any other kind of work is impossible without it. The Eskimo in the last example, for instance, depends on surface friction to pull his sled. Otherwise, the rope would slip in his hands and exert no force on the sled, while his feet slid out from underneath him and he fell flat on his face. Cars wouldn't work without friction, nor could conveyor belts, nor even our muscle tissue.

The work done by pushing or pulling an object is the application of a single force. Friction, on the other hand, is a complex process caused by numerous microscopic interactions over the entire area of the surfaces in contact. Consider a metal block sliding over a metal surface. Microscopic "teeth" in the block encounter equally microscopic irregularities in the underlying surface. Pressing against each other, the teeth deform, get hot, and weld to the opposite surface. Work must be done breaking these temporary bonds, and this comes at the expense of the energy of motion of the block, to be discussed in the next section. The energy lost by the block goes into heating both the block and its environment, with some energy converted to sound.

The friction force of two objects in contact and in relative motion to each other always dissipates energy in these relatively complex ways. For our purposes, the phrase "work done by friction" will denote the effect of these processes on mechanical energy alone.

The edge of a razor blade looks smooth to the eye, but under a microscope proves to have numerous irregularities.

EXAMPLE 5.2   More Sledding

Goal   Calculate the work done by friction when an object is acted on by an applied force.

Problem   Suppose that in Example 5.1 the coefficient of kinetic friction between the loaded 50.0-kg sled and snow is 0.200. **(a)** The Eskimo again pulls the sled 5.00 m, exerting a force of 1.20×10^2 N at an angle of 0°. Find the work done on the sled by friction, and the net work. **(b)** Repeat the calculation if the applied force is exerted at an angle of 30.0° with the horizontal.

Strategy   See Figure 5.6. The frictional work depends on the magnitude of the kinetic friction coefficient, the normal force, and the displacement. Use the y-component of Newton's second law to find the normal force $\vec{\mathbf{n}}$, calculate the work done by friction using the definitions, and sum with the result of Example 5.1(a) to obtain the net work on the sled. Part (b) is solved similarly, but the normal force is smaller because it has the help of the applied force $\vec{\mathbf{F}}_{app}$ in supporting the load.

Solution

(a) Find the work done by friction on the sled and the net work, if the applied force is horizontal.

First, find the normal force from the y-component of Newton's second law, which involves only the normal force and the force of gravity:

$$\Sigma F_y = n - mg = 0 \quad \rightarrow \quad n = mg$$

Use the normal force to compute the work done by friction:

$$W_{fric} = -f_k \Delta x = -\mu_k n \Delta x = -\mu_k mg \Delta x$$
$$= -(0.200)(50.0 \text{ kg})(9.80 \text{ m/s}^2)(5.00 \text{ m})$$
$$= -4.90 \times 10^2 \text{ J}$$

Sum the frictional work with the work done by the applied force from Example 5.1 to get the net work (the normal and gravity forces are perpendicular to the displacement, so they don't contribute):

$$W_{net} = W_{app} + W_{fric} + W_n + W_g$$
$$= 6.00 \times 10^2 \text{ J} + (-4.90 \times 10^2 \text{ J}) + 0 + 0$$
$$= 1.10 \times 10^2 \text{ J}$$

(b) Recalculate the frictional work and net work if the applied force is exerted at a 30.0° angle.

Find the normal force from the y-component of Newton's second law:

$$\Sigma F_y = n - mg + F_{app} \sin \theta = 0$$
$$n = mg - F_{app} \sin \theta$$

Use the normal force to calculate the work done by friction:

$$W_{fric} = -f_k \Delta x = -\mu_k n \Delta x = -\mu_k (mg - F_{app} \sin \theta) \, \Delta x$$
$$= -(0.200)(50.0 \text{ kg} \cdot 9.80 \text{ m/s}^2$$
$$- 1.20 \times 10^2 \text{ N} \sin 30.0°)(5.00 \text{ m})$$
$$W_{fric} = -4.30 \times 10^2 \text{ J}$$

Sum this answer with the result of Example 5.1(b) to get the net work (again, the normal and gravity forces don't contribute):

$$W_{net} = W_{app} + W_{fric} + W_n + W_g$$
$$= 5.20 \times 10^2 \text{ J} - 4.30 \times 10^2 \text{ J} + 0 + 0 = 90.0 \text{ J}$$

Remark   The most important thing to notice here is that exerting the applied force at different angles can dramatically affect the work done on the sled. Pulling at the optimal angle (about 15° in this case) will result in the most net work for a given amount of effort.

Exercise 5.2

(a) The Eskimo pushes the same 50.0-kg sled over level ground with a force of 1.75×10^2 N exerted horizontally, moving it a distance of 6.00 m over new terrain. If the net work done on the sled is 1.50×10^2 J, find the coefficient of kinetic friction. **(b)** Repeat the exercise if the applied force is upwards at a 45.0° angle with the horizontal.

Answer   (a) 0.306   (b) 0.270

5.2 KINETIC ENERGY AND THE WORK–ENERGY THEOREM

Solving problems using Newton's second law can be difficult if the forces involved are complicated. An alternative is to relate the speed of an object to the net work done on it by external forces. If the net work can be calculated for a given displacement, the change in the object's speed is easy to evaluate.

Figure 5.7 shows an object of mass m moving to the right under the action of a constant net force $\vec{\mathbf{F}}_{\text{net}}$, also directed to the right. Because the force is constant, we know from Newton's second law that the object moves with constant acceleration $\vec{\mathbf{a}}$. If the object is displaced by Δx, the work done by $\vec{\mathbf{F}}_{\text{net}}$ on the object is

$$W_{\text{net}} = F_{\text{net}}\Delta x = (ma)\Delta x \qquad [5.3]$$

In Chapter 2, we found that the following relationship holds when an object undergoes constant acceleration:

$$v^2 = v_0{}^2 + 2a\Delta x \qquad \text{or} \qquad a\Delta x = \frac{v^2 - v_0{}^2}{2}$$

We can substitute this expression into Equation 5.3 to get

$$W_{\text{net}} = m\left(\frac{v^2 - v_0{}^2}{2}\right)$$

or

$$W_{\text{net}} = \tfrac{1}{2}mv^2 - \tfrac{1}{2}mv_0{}^2 \qquad [5.4]$$

So the net work done on an object equals a change in a quantity of the form $\tfrac{1}{2}mv^2$. Because this term carries units of energy and involves the object's speed, it can be interpreted as energy associated with the object's motion, leading to the following definition:

> The **kinetic energy KE** of an object of mass m moving with a speed v is defined by
>
> $$KE \equiv \tfrac{1}{2}mv^2 \qquad [5.5]$$
>
> **SI unit: joule (J) = kg · m²/s²**

◀ Kinetic energy

Like work, kinetic energy is a scalar quantity. Using this definition in Equation 5.4, we arrive at an important result known as the **work–energy theorem**:

◀ Work–energy theorem

> The net work done on an object is equal to the change in the object's kinetic energy:
>
> $$W_{\text{net}} = KE_f - KE_i = \Delta KE \qquad [5.6]$$
>
> where the change in the kinetic energy is due entirely to the object's change in speed.

The proviso on the speed is necessary because work that deforms or causes the object to warm up invalidates Equation 5.6, although under most circumstances it remains approximately correct. From that equation, a positive net work W_{net} means that the final kinetic energy KE_f is greater than the initial kinetic energy KE_i. This, in turn, means that the object's final speed is greater than its initial speed. So positive net work increases an object's speed, and negative net work decreases its speed.

We can also turn the equation around and think of kinetic energy as the work a moving object can do in coming to rest. For example, suppose a hammer is on the verge of striking a nail, as in Figure 5.8. The moving hammer has kinetic energy and can therefore do work on the nail. The work done on the nail is $F\Delta x$, where F is the average net force exerted on the nail and Δx is the distance the nail is driven

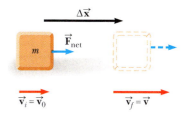

Figure 5.7 An object undergoes a displacement and a change in velocity under the action of a constant net force $\vec{\mathbf{F}}_{\text{net}}$.

Figure 5.8 The moving hammer has kinetic energy and can do work on the nail, driving it into the wall.

into the wall. This work, plus small amounts of energy carried away as heat and sound, is equal to the change in kinetic energy of the hammer, ΔKE.

For convenience, Equation 5.4 was derived under the assumption that the net force acting on the object was constant. A more general derivation, using calculus, would show that Equation 5.4 is valid under all circumstances, including the application of a variable force.

Applying Physics 5.1   Leaving Skid Marks

Suppose a car traveling at a speed v skids a distance d after its brakes lock. Estimate how far it would skid if it were traveling at speed $2v$ when its brakes locked.

Explanation Assume for simplicity that the force of kinetic friction between the car and the road surface is constant and the same at both speeds. From the work–energy theorem, the net force exerted on the car times the displacement of the car, $F_{net}\Delta x$, is equal in magnitude to its initial kinetic energy, $\frac{1}{2}mv^2$. When the speed is doubled, the kinetic energy of the car is quadrupled. So for a given applied friction force, the distance traveled must increase fourfold when the initial speed is doubled, and the estimated distance the car skids is $4d$.

EXAMPLE 5.3   Collision Analysis

Goal Apply the work–energy theorem with a known force.

Problem The driver of a 1.00×10^3 kg car traveling on the interstate at 35.0 m/s (nearly 80.0 mph) slams on his brakes to avoid hitting a second vehicle in front of him, which had come to rest because of congestion ahead. After the brakes are applied, a constant friction force of 8.00×10^3 N acts on the car. Ignore air resistance. **(a)** At what minimum distance should the brakes be applied to avoid a collision with the other vehicle? **(b)** If the distance between the vehicles is initially only 30.0 m, at what speed would the collision occur?

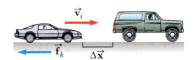

Figure 5.9 (Example 5.3) A braking vehicle just prior to an accident.

Strategy Compute the net work, which involves just the kinetic friction, because the normal and gravity forces are perpendicular to the motion. Then set the net work equal to the change in kinetic energy. To get the minimum distance in part (a), we take the final speed v_f to be zero just as the braking vehicle reaches the rear of the vehicle at rest. Solve for the unknown, Δx. For part (b) proceed similarly, except that the unknown is the final velocity v_f.

Solution
(a) Find the minimum necessary stopping distance.

Apply the work–energy theorem:

$$W_{net} = \tfrac{1}{2}mv_f^2 - \tfrac{1}{2}mv_i^2$$

Substitute an expression for the frictional work and set $v_f = 0$:

$$-f_k\Delta x = 0 - \tfrac{1}{2}mv_i^2$$

Substitute $v_i = 35.0$ m/s, $f_k = 8.00 \times 10^3$ N, and $m = 1.00 \times 10^3$ kg. Solve for Δx:

$$-(8.00 \times 10^3 \text{ N})\Delta x = -\tfrac{1}{2}(1.00 \times 10^3 \text{ kg})(35.0 \text{ m/s})^2$$

$$\Delta x = \boxed{76.6 \text{ m}}$$

(b) At the given distance of 30.0 m, the car is too close to the other vehicle. Find the speed at impact.

Write down the work–energy theorem:

$$W_{net} = W_{fric} = -f_k\Delta x = \tfrac{1}{2}mv_f^2 - \tfrac{1}{2}mv_i^2$$

Multiply by $2/m$ and rearrange terms, solving for the final velocity v_f:

$$v_f^2 = v_i^2 - \frac{2}{m}f_k\Delta x$$

$$v_f^2 = (35.0 \text{ m/s})^2 - \left(\frac{2}{1.00 \times 10^3 \text{ kg}}\right)$$
$$\times (8.00 \times 10^3 \text{ N})(30.0 \text{ m}) = 745 \text{ m}^2/\text{s}^2$$

$$v_f = \boxed{27.3 \text{ m/s}}$$

Remarks   This calculation illustrates how important it is to remain alert on the highway, allowing for an adequate stopping distance at all times. It takes about a second to react to the brake lights of the car in front of you. On a high-speed highway, your car may travel over 30 meters before you can engage the brakes. Bumper-to-bumper traffic at high speed, as often exists on the highways near big cities, is extremely unsafe.

Exercise 5.3
A police investigator measures straight skid marks 27 m long in an accident investigation. Assuming a friction force and car mass the same as in the previous problem, what was the minimum speed of the car when the brakes locked?

Answer   20.8 m/s

Conservative and Nonconservative Forces

It turns out there are two general kinds of forces. The first is called a **conservative force**. Gravity is probably the best example of a conservative force. To understand the origin of the name, think of a diver climbing to the top of a 10-meter platform. The diver has to do work against gravity in making the climb. Once at the top, however, he can recover the work—as kinetic energy—by taking a dive. His speed just before hitting the water will give him a kinetic energy equal to the work he did against gravity in climbing to the top of the platform—minus the effect of some nonconservative forces, such as air drag and internal muscular friction.

A **nonconservative force** is generally dissipative, which means it tends to randomly disperse the energy of bodies on which it acts. This dispersal of energy often takes the form of heat or sound. Kinetic friction and air drag are good examples. Propulsive forces, like the force exerted by a jet engine on a plane or by a propeller on a submarine, are also nonconservative.

Work done against a nonconservative force can't be easily recovered. Dragging objects over a rough surface requires work. When the Eskimo in Example 5.2 dragged the sled across terrain having a non-zero coefficient of friction, the net work was smaller than in the frictionless case. The missing energy went into warming the sled and its environment. As will be seen in the study of thermodynamics, such losses can't be avoided, nor all the energy recovered, so these forces are called nonconservative.

Another way to characterize conservative and nonconservative forces is to measure the work done by a force on an object traveling between two points along different paths. The work done by gravity on someone going down a frictionless slide, as in Figure 5.10, is the same as that done on someone diving into the water from

Figure 5.10   Because the gravity field is conservative, the diver regains as kinetic energy the work she did against gravity in climbing the ladder. Taking the frictionless slide gives the same result.

the same height. This equality doesn't hold for nonconservative forces. For example, sliding a book directly from point Ⓐ to point Ⓓ in Figure 5.11 requires a certain amount of work against friction, but sliding the book along the three other legs of the square, from Ⓐ to Ⓑ, Ⓑ to Ⓒ, and finally Ⓒ to Ⓓ, requires three times as much work. This observation motivates the following definition of a conservative force:

Conservative force ▶

A force is conservative if the work it does moving an object between two points is the same no matter what path is taken.

Nonconservative forces, as we've seen, don't have this property. The work–energy theorem, Equation 5.6, can be rewritten in terms of the work done by conservative forces W_c and the work done by nonconservative forces W_{nc}, because the net work is just the sum of these two:

$$W_{nc} + W_c = \Delta KE \qquad [5.7]$$

It turns out that conservative forces have another useful property: The work they do can be recast as something called **potential energy**, a quantity that depends only on the beginning and end points of a curve, not the path taken.

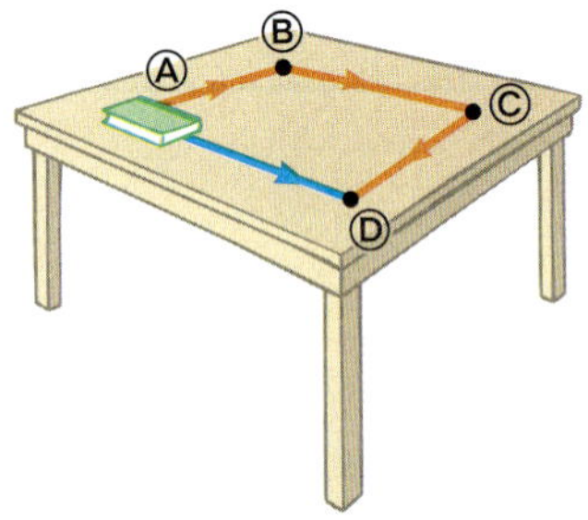

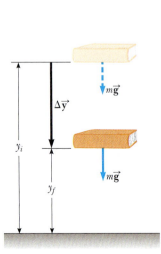

Figure 5.11   Because friction is a nonconservative force, a book pushed along the three segments Ⓐ–Ⓑ, Ⓑ–Ⓒ, and Ⓒ–Ⓓ requires three times the work as pushing the book directly from Ⓐ to Ⓓ.

5.3   GRAVITATIONAL POTENTIAL ENERGY

An object with kinetic energy (energy of motion) can do work on another object, just like a moving hammer can drive a nail into a wall. A brick on a high shelf can also do work: It can fall off the shelf, accelerate downwards, and hit a nail squarely, driving it into the floorboards. The brick is said to have **potential energy** associated with it, because from its location on the shelf it can potentially do work.

Potential energy is a property of a **system**, rather than of a single object, because it's due to a physical position in space relative a center of force, like the falling diver and the Earth of Figure 5.10. In this chapter, we define a system as a collection of objects interacting via forces or other processes that are internal to the system. It turns out that potential energy is another way of looking at the work done by conservative forces.

Gravitational Work and Potential Energy

Using the work–energy theorem in problems involving gravitation requires computing the work done by gravity. For most trajectories—say, for a ball traversing a parabolic arc—finding the gravitational work done on the ball requires sophisticated techniques from calculus. Fortunately, for conservative fields there's a simple alternative: potential energy.

Gravity is a conservative force, and for every conservative force a special expression called a potential energy function can be found. Evaluating that function at any two points in an object's path of motion and finding the difference will give the negative of the work done by that force between those two points. It's also advantageous that potential energy, like work and kinetic energy, is a scalar quantity.

Our first step is to find the work done by gravity on an object when it moves from one position to another. The negative of that work is the change in the gravitational potential energy of the system, and from that expression, we'll be able to identify the potential energy function.

In Figure 5.12, a book of mass m falls from a height y_i to a height y_f, where the positive y-coordinate represents position above the ground. We neglect the force of air friction, so the only force acting on the book is gravitation. How much work is done? The magnitude of the force is mg and that of the displacement is $\Delta y = y_i - y_f$ (a positive number), while both $\vec{F}$ and $\Delta \vec{y}$ are pointing downwards, so the angle between them is zero. We apply the definition of work in Equation 5.2:

$$W_g = F\Delta y \cos \theta = mg(y_i - y_f)\cos 0° = -mg(y_f - y_i) \qquad [5.8]$$

Figure 5.12   The work done by the gravitational force as the book falls from y_i to y_f equals $mgy_i - mgy_f$.

Factoring out the minus sign was deliberate, to clarify the coming connection to potential energy. Equation 5.8 for gravitational work holds for any object, regardless of its trajectory in space, because the gravitational force is conservative. Now, W_g will appear as the work done by gravity in the work–energy theorem. For the rest of this section, assume for simplicity that we are dealing only with systems involving gravity and nonconservative forces. Then Equation 5.7 can be written as

$$W_{\text{net}} = W_{nc} + W_g = \Delta KE$$

where W_{nc} is the work done by the nonconservative forces. Substituting the expression for W_g from Equation 5.8, we obtain

$$W_{nc} - mg(y_f - y_i) = \Delta KE \qquad \text{[5.9a]}$$

Next, we add $mg(y_f - y_i)$ to both sides:

$$W_{nc} = \Delta KE + mg(y_f - y_i) \qquad \text{[5.9b]}$$

Now, by definition, we'll make the connection between gravitational work and gravitational potential energy.

> The gravitational potential energy of a system consisting of the Earth and an object of mass m near the Earth's surface is given by
>
> $$PE \equiv mgy \qquad \text{[5.10]}$$
>
> where g is the acceleration of gravity and y is the vertical position of the mass relative the surface of Earth (or some other reference point).
>
> **SI unit: joule (J)**

◄ Gravitational potential energy

In this definition, $y = 0$ is usually taken to correspond to Earth's surface, but this is not strictly necessary, as discussed in the next subsection. It turns out that only *differences* in potential energy really matter.

So the gravitational potential energy associated with an object located near the surface of the Earth is the object's weight mg times its vertical position y above Earth. From this *definition*, we have the relationship between gravitational work and gravitational potential energy:

$$W_g = -(PE_f - PE_i) = -(mgy_f - mgy_i) \qquad \text{[5.11]}$$

The work done by gravity is one and the same as the negative of the change in gravitational potential energy.

Finally, using the relationship in Equation 5.11 in Equation 5.9b, we obtain an extension of the work–energy theorem:

$$W_{nc} = (KE_f - KE_i) + (PE_f - PE_i) \qquad \text{[5.12]}$$

This equation says that the work done by nonconservative forces, W_{nc}, is equal to the change in the kinetic energy plus the change in the gravitational potential energy.

Equation 5.12 will turn out to be true in general, even when other conservative forces besides gravity are present. The work done by these additional conservative forces will again be recast as changes in potential energy and will appear on the right-hand side along with the expression for gravitational potential energy.

TIP 5.3 Potential Energy Takes Two

Potential energy always takes a system of at least two interacting objects — for example, the Earth and a baseball interacting via the gravitational force.

Reference Levels for Gravitational Potential Energy

In solving problems involving gravitational potential energy, it's important to choose a location at which to set that energy equal to zero. Given the form of Equation 5.10, this is the same as choosing the place where $y = 0$. The choice is completely arbitrary because the important quantity is the *difference* in potential energy, and this difference will be the same regardless of the choice of zero level. However, once this position is chosen, it must remain fixed for a given problem.

Figure 5.13 Any reference level—the desktop, the floor of the room, or the ground outside the building—can be used to represent zero gravitational potential energy in the book–Earth system.

While it's always possible to choose the surface of the Earth as the reference position for zero potential energy, the statement of a problem will usually suggest a convenient position to use. As an example, consider a book at several possible locations, as in Figure 5.13. When the book is at Ⓐ, a natural zero level for potential energy is the surface of the desk. When the book is at Ⓑ, the floor might be a more convenient reference level. Finally, a location such as Ⓒ, where the book is held out a window, would suggest choosing the surface of the Earth as the zero level of potential energy. The choice, however, makes no difference: Any of the three reference levels could be used as the zero level, regardless of whether the book is at Ⓐ, Ⓑ, or Ⓒ. Example 5.4 illustrates this important point.

EXAMPLE 5.4 Wax Your Skis

Goal Calculate the change in gravitational potential energy for different choices of reference level.

Problem A 60.0-kg skier is at the top of a slope, as shown in Figure 5.14. At the initial point Ⓐ, she is 10.0 m vertically above point Ⓑ. (a) Setting the zero level for gravitational potential energy at Ⓑ, find the gravitational potential energy of this system when the skier is at Ⓐ and then at Ⓑ. Finally, find the change in potential energy of the skier–Earth system as the skier goes from point Ⓐ to point Ⓑ. (b) Repeat this problem with the zero level at point Ⓐ. (c) Repeat again, with the zero level 2.00 m higher than point Ⓑ.

Strategy Follow the definition, and be careful with signs. Ⓐ is the initial point, with gravitational potential energy PE_i, and Ⓑ is the final point, with gravitational potential energy PE_f. The location chosen for $y = 0$ is also the zero point for the potential energy, since $PE = mgy$.

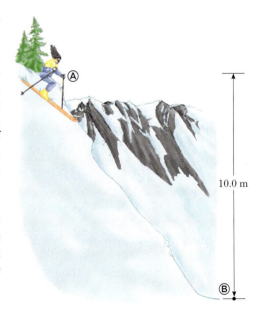

Figure 5.14   (Example 5.4)

Solution

(a) Let $y = 0$ at Ⓑ. Calculate the potential energy at Ⓐ, at Ⓑ, and the change in potential energy.

Find PE_i, the potential energy at Ⓐ, from Equation 5.10:

$$PE_i = mgy_i = (60.0 \text{ kg})(9.80 \text{ m/s}^2)(10.0 \text{ m}) = 5.88 \times 10^3 \text{ J}$$

$PE_f = 0$ at Ⓑ by choice. Find the difference in potential energy between Ⓐ and Ⓑ:

$$PE_f - PE_i = 0 - 5.88 \times 10^3 \text{ J} = \boxed{-5.88 \times 10^3 \text{ J}}$$

(b) Repeat the problem if $y = 0$ at Ⓐ, the new reference point, so that $PE = 0$ at Ⓐ.

Find PE_f, noting that point Ⓑ is now at $y = -10.0$ m:

$$PE_f = mgy_f = (60.0 \text{ kg})(9.80 \text{ m/s}^2)(-10.0 \text{ m})$$
$$= -5.88 \times 10^3 \text{ J}$$
$$PE_f - PE_i = -5.88 \times 10^3 \text{ J} - 0 = \boxed{-5.88 \times 10^3 \text{ J}}$$

(c) Repeat the problem, if $y = 0$ two meters above Ⓑ.

Find PE_i, the potential energy at Ⓐ:

$$PE_i = mgy_i = (60.0 \text{ kg})(9.80 \text{ m/s}^2)(8.00 \text{ m}) = 4.70 \times 10^3 \text{ J}$$

Find PE_f, the potential energy at Ⓑ:

$$PE_f = mgy_f = (60.0 \text{ kg})(9.8 \text{ m/s}^2)(-2.00 \text{ m})$$
$$= -1.18 \times 10^3 \text{ J}$$

Compute the change in potential energy:

$$PE_f - PE_i = -1.18 \times 10^3 \text{ J} - 4.70 \times 10^3 \text{ J}$$
$$= \boxed{-5.88 \times 10^3 \text{ J}}$$

Remarks   These calculations show that the change in the gravitational potential energy when the skier goes from the top of the slope to the bottom is -5.88×10^3 J, *regardless of the zero level selected.*

Exercise 5.4
If the zero level for gravitational potential energy is selected to be midway down the slope, 5.00 m above point Ⓑ, find the initial potential energy, the final potential energy, and the change in potential energy as the skier goes from point Ⓐ to Ⓑ in Figure 5.14.

Answer   2.94 kJ, −2.94 kJ, −5.88 kJ

Gravity and the Conservation of Mechanical Energy

Conservation principles play a very important role in physics. **When a physical quantity is conserved the numeric value of the quantity remains the same throughout the physical process.** Although the form of the quantity may change in some way, **its final value is the same as its initial value.**

The kinetic energy KE of an object falling only under the influence of gravity is constantly changing, as is the gravitational potential energy PE. Obviously, then, these quantities aren't conserved. Because all nonconservative forces are assumed absent, however, we can set $W_{nc} = 0$ in Equation 5.12. Rearranging the equation, we arrive at the following very interesting result:

$$KE_i + PE_i = KE_f + PE_f \qquad [5.13]$$

According to this equation, **the sum of the kinetic energy and the gravitational potential energy remains constant at all times and hence is a conserved quantity.** We denote the total mechanical energy by $E = KE + PE$, and say that **the total mechanical energy is conserved.**

To show how this concept works, think of tossing a rock off a cliff, and ignore the drag forces. As the rock falls, its speed increases, so its kinetic energy increases. As the rock approaches the ground, the potential energy of the rock–Earth system decreases. Whatever potential energy is lost as the rock moves downward appears as kinetic energy, and Equation 5.13 says that in the absence of nonconservative forces like air drag, the trading of energy is exactly even. This is true for all conservative forces, not just gravity.

TIP 5.4   Conservation Principles

There are many conservation laws like the conservation of mechanical energy in isolated systems, as in Equation 5.13. For example, momentum, angular momentum, and electric charge are all conserved quantities, as will be seen later. Conserved quantities may change form during physical interactions, but their sum total for a system never changes.

In any isolated system of objects interacting only through conservative forces, the total mechanical energy $E = KE + PE$, of the system, remains the same at all times.

◀ Conservation of mechanical energy

If the force of gravity is the *only* force doing work within a system, then the principle of conservation of mechanical energy takes the form

$$\tfrac{1}{2}mv_i{}^2 + mgy_i = \tfrac{1}{2}mv_f{}^2 + mgy_f \qquad \text{[5.14]}$$

This form of the equation is particularly useful for solving problems involving only gravity. Further terms have to be added when other conservative forces are present, as we'll soon see.

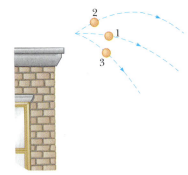

ACTIVE FIGURE 5.15
(Quick Quiz 5.2) Three identical balls are thrown with the same initial speed from the top of a building.

Physics⊗Now™

Log into PhysicsNow at **www.cp7e.com**, and go to Active Figure 5.15 to throw balls at different angles from the top of the building and compare trajectories and speeds as the balls hit the ground.

Quick Quiz 5.2

Three identical balls are thrown from the top of a building, all with the same initial speed. The first ball is thrown horizontally, the second at some angle above the horizontal, and the third at some angle below the horizontal, as in Active Figure 5.15. Neglecting air resistance, rank the speeds of the balls as they reach the ground, from fastest to slowest. (a) 1, 2, 3 (b) 2, 1, 3 (c) 3, 1, 2 (d) all three balls strike the ground at the same speed.

Quick Quiz 5.3

Bob, of mass m, drops from a tree limb at the same time that Esther, also of mass m, begins her descent down a frictionless slide. If they both start at the same height above the ground, which of the following is true about their kinetic energies as they reach the ground?
(a) Bob's kinetic energy is greater than Esther's.
(b) Esther's kinetic energy is greater than Bob's.
(c) They have the same kinetic energy.
(d) The answer depends on the shape of the slide.

Problem-Solving Strategy
Applying Conservation of Mechanical Energy

Take the following steps when applying conservation of mechanical energy to problems involving gravity:
1. Define the system, including all interacting bodies. Choose a location for $y = 0$, the zero point for gravitational potential energy.
2. Select the body of interest and identify two points—one point where you have given information, and the other point where you want to find out something about the body of interest.
3. After verifying the absence of nonconservative forces, write down the conservation of energy equation, Equation 5.14, for the system. Identify the unknown quantity of interest.
4. Solve for the unknown quantity, which is usually either a speed or a position, and substitute known values.

As previously stated, it's usually best to do the algebra with symbols rather than substituting known numbers first, because it's easier to check the symbols for possible errors. The exception is when a quantity is clearly zero, in which case immediate substitution greatly simplifies the ensuing algebra.

INTERACTIVE EXAMPLE 5.5   Platform Diver

Goal   Use conservation of energy to calculate the speed of a body falling straight down in the presence of gravity.

Problem   A diver of mass m drops from a board 10.0 m above the water's surface, as in Figure 5.16. Neglect air resistance. **(a)** Use conservation of mechanical energy to find his speed 5.00 m above the water's surface. **(b)** Find his speed as he hits the water.

Strategy   Refer to the problem-solving strategy. Step 1: The system consists of the diver and the Earth. As the diver falls, only the force of gravity acts on him (neglecting air drag), so the mechanical energy of the system is conserved and we can use conservation of energy for both (a) and (b). Choose $y = 0$ for the water's surface. Step 2: In (a), $y = 10.0$ m and $y = 5.00$ m are the points of interest, while in (b), $y = 10.0$ m and $y = 0$ m are of interest.

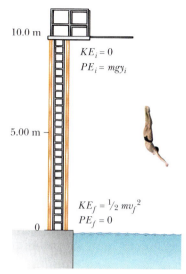

Solution
(a) Find the diver's speed halfway down, at $y = 5.00$ m.

Step 3: we write the energy conservation equation and supply the proper terms:

$$KE_i + PE_i = KE_f + PE_f$$
$$\tfrac{1}{2}mv_i^2 + mgy_i = \tfrac{1}{2}mv_f^2 + mgy_f$$

Step 4: Substitute $v_i = 0$, cancel the mass m and solve for v_f:

$$0 + gy_i = \tfrac{1}{2}v_f^2 + gy_f$$
$$v_f = \sqrt{2g(y_i - y_f)} = \sqrt{2(9.80 \text{ m/s}^2)(10.0\,\text{m} - 5.00\,\text{m})}$$
$$v_f = 9.90 \text{ m/s}$$

Figure 5.16   (Example 5.5) The zero of gravitational potential energy is taken to be at the water's surface.

(b) Find the diver's speed at the water's surface, $y = 0$.

Use the same procedure as in part (a), taking $y_f = 0$:

$$0 + mgy_i = \tfrac{1}{2}mv_f^2 + 0$$
$$v_f = \sqrt{2gy_i} = \sqrt{2(9.80 \text{ m/s}^2)(10.0 \text{ m})} = 14.0 \text{ m/s}$$

Remark   Notice that the speed halfway down is not half the final speed.

Exercise 5.5
Suppose the diver vaults off the springboard, leaving it with an initial speed of 3.50 m/s upwards. Use energy conservation to find his speed when he strikes the water.

Answer   14.4 m/s

Physics⊗Now™ Investigate the conservation of mechanical energy for a dropped ball by logging into PhysicsNow at **www.cp7e.com** and going to Interactive Example 5.5.

EXAMPLE 5.6   The Jumping Bug

Goal   Use conservation of mechanical energy and concepts from ballistics in two dimensions to calculate a speed.

Problem   A powerful grasshopper launches itself at an angle of 45° above the horizontal and rises to a maximum height of 1.00 m during the leap. (See Figure 5.17.) With what speed v_i did it leave the ground? Neglect air resistance.

Strategy   This problem can be solved with conservation of energy and the relation between the initial velocity and

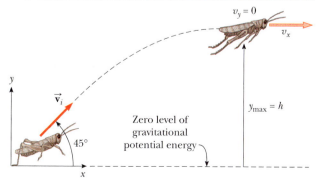

Figure 5.17   (Example 5.6)

its x-component. Aside from the origin, the other point of interest is the maximum height $y = 1.00$ m, where the grasshopper has a velocity v_x in the x-direction only. Energy conservation then gives one equation with two unknowns: the initial speed v_i and speed at maximum height, v_x. Because there are no forces in the x-direction, however, v_x is the same as the x-component of the initial velocity.

Solution

Use energy conservation:

$$\tfrac{1}{2}mv_i^2 + mgy_i = \tfrac{1}{2}mv_f^2 + mgy_f$$

Substitute $y_i = 0$, $v_f = v_x$, and $y_f = h$:

$$\tfrac{1}{2}mv_i^2 = \tfrac{1}{2}mv_x^2 + mgh$$

Multiply each side by $2/m$, obtaining one equation and two unknowns:

$$v_i^2 = v_x^2 + 2gh \tag{1}$$

Eliminate v_x by substituting $v_x = v_i \sin 45°$ into equation (1), solving for v_i, and substituting known values:

$$v_i^2 = (v_i \sin 45°)^2 + 2gh = \tfrac{1}{2}v_i^2 + 2gh$$

$$v_i = 2\sqrt{gh} = 2\sqrt{(9.80 \text{ m/s}^2)(1.00 \text{ m})} = \boxed{6.26 \text{ m/s}}$$

Remarks The final answer is a surprisingly high value and illustrates how strong insects are relative to their size.

Exercise 5.6
A catapult launches a rock at a 30.0° angle with respect to the horizontal. Find the maximum height attained if the speed of the rock at its highest point is 30.0 m/s.

Answer 15.3 m

TIP 5.5 Don't Use Work Done by the Force of Gravity *and* Gravitational Potential Energy!
Gravitational potential energy is just another way of including the work done by the force of gravity in the work–energy theorem. Don't use both of them in the equation at the same time, or you'll count it twice!

Gravity and Nonconservative Forces

When nonconservative forces are involved along with gravitation, the full work–energy theorem must be used, often with techniques from Chapter 4. Solving problems requires the basic procedure of the problem-solving strategy for conservation-of-energy problems in the previous section. The only difference lies in substituting Equation 5.12, the work–energy equation with potential energy, for Equation 5.14.

EXAMPLE 5.7 Der Stuka!

Goal Use the work–energy theorem with gravitational potential energy to calculate the work done by a nonconservative force.

Problem Waterslides are nearly frictionless, hence can provide bored students with high-speed thrills (Fig. 5.18). One such slide, Der Stuka, named for the terrifying German dive bombers of World War II, is 72.0 feet high (21.9 m), found at Six Flags in Dallas, Texas, and at Wet'n Wild in Orlando, Florida. (a) Determine the speed of a 60.0-kg woman at the bottom of such a slide, assuming no friction is present. (b) If the woman is clocked at 18.0 m/s at the bottom of the slide, how much mechanical energy was lost through friction?

Strategy The system consists of the woman, the Earth, and the slide. The normal force, always perpendicular to the displacement, does no work. Let $y = 0$ m represent the bottom of the slide. The two points of interest are $y = 0$ m and $y = 21.9$ m. Without friction, $W_{nc} = 0$, and we can apply conservation of mechanical energy, Equation 5.14. For part (b), use Equation 5.12, substitute two velocities and heights, and solve for W_{nc}.

Figure 5.18 (Example 5.7) If the slide is frictionless, the speed of the thrill seeker at the bottom depends only on the height of the slide, not on the path it takes.

Solution

(a) Find the woman's speed at the bottom of the slide, assuming no friction.

Write down Equation 5.14, for conservation of energy:

$$\tfrac{1}{2}mv_i^2 + mgy_i = \tfrac{1}{2}mv_f^2 + mgy_f$$

Insert the values $v_i = 0$ and $y_f = 0$:

$$0 + mgy_i = \tfrac{1}{2}mv_f^2 + 0$$

Solve for v_f and substitute values for g and y_i:

$$v_f = \sqrt{2gy_i} = \sqrt{2(9.80 \text{ m/s}^2)(21.9 \text{ m})} = \boxed{20.7 \text{ m/s}}$$

(b) Find the mechanical energy lost due to friction if $v_f = 18.0$ m/s < 20.7 m/s.

Write Equation 5.12, substituting expressions for the kinetic and potential energies:

$$W_{nc} = (KE_f - KE_i) + (PE_f - PE_i)$$
$$= (\tfrac{1}{2}mv_f^2 - \tfrac{1}{2}mv_f^2) + (mgy_f - mgy_i)$$

Substitute $m = 60.0$ kg, $v_f = 18.0$ m/s, and $v_i = 0$, and solve for W_{nc}:

$$W_{nc} = [\tfrac{1}{2}\cdot60.0 \text{ kg} \cdot (18.0 \text{ m/s})^2 - 0]$$
$$+ [0 - 60.0 \text{ kg} \cdot (9.80 \text{ m/s}^2)\cdot 21.9 \text{ m}]$$
$$W_{nc} = \boxed{-3.16 \times 10^3 \text{ J}}$$

Remarks The speed found in part (a) is the same as if the thrill seeker fell vertically through a distance of 21.9 m, consistent with our intuition in Quick Quiz 5.3. The result of part (b) is negative because the system loses mechanical energy. Friction transforms part of the mechanical energy into thermal energy and mechanical waves, absorbed partly by the system and partly by the environment.

Exercise 5.7

Suppose a slide similar to Der Stuka is 35.0 meters high, but is a straight slope, inclined at 45.0° with respect to the horizontal. (a) Find the speed of a 60.0-kg thrill seeker at the bottom of the slide, assuming no friction. (b) If the thrill seeker has a speed of 20.0 m/s at the bottom, find the mechanical energy lost due to friction and (c) the magnitude of the force of friction, assumed constant.

Answers   (a) 26.2 m/s   (b) -8.58×10^3 J   (c) 173 N

EXAMPLE 5.8   Hit the Ski Slopes

Goal Combine conservation of mechanical energy with the work–energy theorem involving friction on a horizontal surface.

Problem A skier starts from rest at the top of a frictionless incline of height 20.0 m, as in Figure 5.19. At the bottom of the incline, the skier encounters a horizontal surface where the coefficient of kinetic friction between skis and snow is 0.210. **(a)** Find the skier's speed at the bottom. **(b)** How far does the skier travel on the horizontal surface before coming to rest?

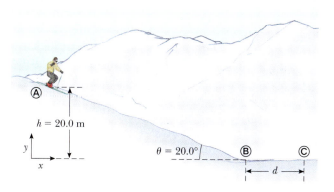

Figure 5.19 (Example 5.8) The skier slides down the slope and onto a level surface, stopping after traveling a distance d from the bottom of the hill.

Strategy Going down the frictionless incline is physically no different than going down the slide of the previous example and is handled the same way, using conservation of mechanical energy to find the speed $v_\circledB$ at the bottom. On the flat, rough surface, use the work–energy theorem, Equation 5.12, with $W_{nc} = W_{\text{fric}} = -f_k d$, where f_k is the magnitude of the force of friction and d is the distance traveled.

Solution

(a) Find the skier's speed at the bottom.

Follow the procedure used in part (a) of the previous example as the skier moves from the top, point Ⓐ, to the bottom, point Ⓑ.

$$v_Ⓑ = \sqrt{2gh} = \sqrt{2(9.80 \text{ m/s}^2)(20.0 \text{ m})} = \boxed{19.8 \text{ m/s}}$$

(b) Find the distance traveled on the horizontal, rough surface.

Apply the work–energy theorem as the skier moves from Ⓑ to Ⓒ:

$$W_{net} = -f_k d = \Delta KE = \tfrac{1}{2}mv_Ⓒ^2 - \tfrac{1}{2}mv_Ⓑ^2$$

Substitute $v_Ⓒ = 0$ and $f_k = \mu_k n = \mu_k mg$:

$$-\mu_k mgd = -\tfrac{1}{2}mv_Ⓑ^2$$

Solve for d:

$$d = \frac{v_Ⓑ^2}{2\mu_k g} = \frac{(19.8 \text{ m/s})^2}{2(0.210)(9.80 \text{ m/s}^2)} = \boxed{95.2 \text{ m}}$$

Remarks   Substituting the symbolic expression $v_Ⓑ = \sqrt{2gh}$ into the equation for the distance d shows that d is linearly proportional to h: Doubling the height doubles the distance traveled.

Exercise 5.8
Find the horizontal distance the skier travels before coming to rest if the incline also has a coefficient of kinetic friction equal to 0.210.

Answer   40.3 m

5.4   SPRING POTENTIAL ENERGY

Springs are important elements in modern technology. They are found in machines of all kinds, in watches, toys, cars, and trains. Springs will be introduced here, then studied in more detail in Chapter 13.

Work done by an applied force in stretching or compressing a spring can be recovered by removing the applied force, so like gravity, the spring force is conservative. This means a potential energy function can be found and used in the work–energy theorem.

Active Figure 5.20a shows a spring in its equilibrium position, where the spring is neither compressed nor stretched. Pushing a block against the spring as in Active Figure 5.20b compresses it a distance x. While x appears to be merely a coordinate, for springs it also represents a *displacement* from the equilibrium position, which for our purposes will always be taken to be at $x = 0$. Experimentally, it turns out that doubling a given displacement requires double the force, while tripling it takes triple the force. This means the force exerted by the spring, F_s, must be proportional to the displacement x, or

$$F_s = -kx \qquad\qquad [5.15]$$

where k is a constant of proportionality, the *spring constant*, carrying units of newtons per meter. Equation 5.15 is called **Hooke's law**, after Sir Robert Hooke, who discovered the relationship. The force F_s is often called a *restoring force*, because the spring always exerts a force in a direction opposite the displacement of its end, tending to restore whatever is attached to the spring to its original position. For positive values of x, the force is negative, pointing back towards equilibrium at $x = 0$, and for negative x, the force is positive, again pointing towards $x = 0$. For a flexible spring, k is a small number (about 100 N/m), whereas for a stiff spring k is large (about 10 000 N/m). The value of the spring constant k is

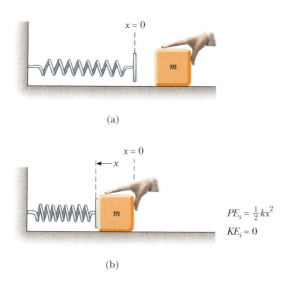

(a)

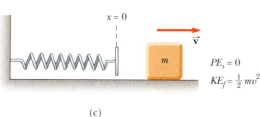

$PE_s = \frac{1}{2}kx^2$

$KE_i = 0$

(b)

$PE_s = 0$

$KE_f = \frac{1}{2}mv^2$

(c)

ACTIVE FIGURE 5.20
(a) A spring at equilibrium, neither compressed nor stretched. (b) A block of mass m on a frictionless surface is pushed against the spring. If x is the compression in the spring, the potential energy stored in the spring is $\frac{1}{2}kx^2$. (c) When the block is released, this energy is transferred to the block in the form of kinetic energy.

Physics⊗Now™
Log into PhysicsNow at **www.cp7e.com**, and go to Active Figure 5.20 to compress the spring by varying amounts and observe the effect on the block's speed.

determined by how the spring was formed, its material composition, and the thickness of the wire. The minus sign ensures that the spring force is always directed back towards the equilibrium point.

As in the case of gravitation, a potential energy, called the **elastic potential energy**, can be associated with the spring force. Elastic potential energy is another way of looking at the work done by a spring during motion, because it is equal to the negative of the work done by the spring. It can also be considered stored energy arising from the work done to compress or stretch the spring.

Consider a horizontal spring and mass at the equilibrium position. We determine the work done by the spring when compressed by an applied force from equilibrium to a displacement x, as in Active Figure 5.20b. The spring force points in the direction opposite the motion, so we expect the work to be negative. When we studied the constant force of gravity near the Earth's surface, we found the work done on an object by multiplying the gravitational force by the vertical displacement of the object. However, this procedure can't be used with a varying force such as the spring force. Instead, we use the average force, $\overline{F}$:

$$\overline{F} = \frac{F_0 + F_1}{2} = \frac{0 - kx}{2} = -\frac{kx}{2}$$

Therefore, the work *done by the spring force* is

$$W_s = \overline{F}x = -\tfrac{1}{2}kx^2$$

In general, when the spring is stretched or compressed from x_i to x_f, the work done by the spring is

$$W_s = -\left(\tfrac{1}{2}kx_f{}^2 - \tfrac{1}{2}kx_i{}^2\right)$$

The work done by a spring can be included in the work–energy theorem. Assume Equation 5.12 now includes the work done by springs on the left-hand side. It then reads

$$W_{nc} - (\tfrac{1}{2}kx_f{}^2 - \tfrac{1}{2}kx_i{}^2) = \Delta KE + \Delta PE_g$$

where PE_g is the gravitational potential energy. We now define the elastic potential energy associated with the spring force, PE_s, by

$$PE_s \equiv \tfrac{1}{2}kx^2 \qquad \text{[5.16]}$$

Inserting this expression into the previous equation and rearranging gives the new form of the work–energy theorem, including both gravitational and elastic potential energy:

$$W_{nc} = (KE_f - KE_i) + (PE_{gf} - PF_{gi}) + (PE_{sf} - PE_{si}) \qquad \text{[5.17]}$$

where W_{nc} is the work done by nonconservative forces, KE is kinetic energy, PE_g is gravitational potential energy, and PE_s is the elastic potential energy. PE, formerly used to denote gravitational potential energy alone, will henceforth denote the total potential energy of a system, including potential energies due to all conservative forces acting on the system.

It's important to remember that the work done by gravity and springs in any given physical system is already included on the right-hand side of Equation 5.17 as potential energy and should not also be included on the left as work.

Active Figure 5.20c shows how the stored elastic potential energy can be recovered. When the block is released, the spring snaps back to its original length and the stored elastic potential energy is converted to kinetic energy of the block. The elastic potential energy stored in the spring is zero when the spring is in the equilibrium position ($x = 0$). As given by Equation 5.16, potential energy is also stored in the spring when it's stretched. Further, the elastic potential energy is a maximum when the spring has reached its maximum compression or extension. Finally, the potential energy is always positive when the spring is not in the equilibrium position, because PE_s is proportional to x^2.

In the absence of nonconservative forces, $W_{nc} = 0$, so the left-hand side of Equation 5.17 is zero, and an extended form for conservation of mechanical energy results:

$$(KE + PE_g + PE_s)_i = (KE + PE_g + PE_s)_f \qquad \text{[5.18]}$$

Problems involving springs, gravity, and other forces are handled in exactly the same way as described in problem-solving strategies 1 and 2, as will be seen in Examples 5.9–5.11.

EXAMPLE 5.9   A Horizontal Spring

Goal   Use conservation of energy to calculate the speed of a block on a horizontal spring with and without friction.

Problem   A block with mass of 5.00 kg is attached to a horizontal spring with spring constant $k = 4.00 \times 10^2$ N/m, as in Figure 5.21. The surface the block rests upon is frictionless. If the block is pulled out to $x_i = 0.0500$ m and released, **(a)** find the speed of the block at the equilibrium point, **(b)** find the speed when $x = 0.0250$ m, and **(c)** repeat part (a) if friction acts on the block, with coefficient $\mu_k = 0.150$.

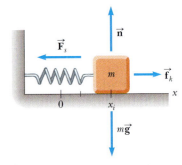

Figure 5.21   (Example 5.9) A mass attached to a spring.

Strategy   In parts (a) and (b) there are no nonconservative forces, so conservation of energy, Equation 5.18, can be applied. In part (c), the definition of work and the work–energy theorem are needed to deal with the loss of mechanical energy due to friction.

Solution

(a) Find the speed of the block at equilibrium point.

Start with Equation 5.18:

$$(KE + PE_g + PE_s)_i = (KE + PE_g + PE_s)_f$$

Substitute expressions for the block's kinetic energy and the potential energy, and set the gravity terms to zero:

$$\tfrac{1}{2}mv_i^2 + \tfrac{1}{2}kx_i^2 = \tfrac{1}{2}mv_f^2 + \tfrac{1}{2}kx_f^2 \qquad \textbf{(1)}$$

Substitute $v_i = 0$ and $x_f = 0$, and multiply by $2/m$:

$$\frac{k}{m}x_i^2 = v_f^2$$

Solve for v_f and substitute the given values:

$$v_f = \sqrt{\frac{k}{m}}\,x_i = \sqrt{\frac{4.00 \times 10^2\,\text{N/m}}{5.00\,\text{kg}}}\,(0.0500\ \text{m})$$

$$= \;0.447\ \text{m/s}$$

(b) Find the speed of the block at the halfway point.

Set $v_i = 0$ in Equation (1), and multiply by $2/m$:

$$\frac{kx_i^2}{m} = v_f^2 + \frac{kx_f^2}{m}$$

Solve for v_f, and substitute the given values:

$$v_f = \sqrt{\frac{k}{m}(x_i^2 - x_f^2)}$$

$$= \sqrt{\frac{4.00 \times 10^2\,\text{N/m}}{5.00\,\text{kg}}((0.050\ \text{m})^2 - (0.025\ \text{m})^2)}$$

$$= \;0.387\ \text{m/s}$$

(c) Repeat part (a), this time with friction.

Apply the work–energy theorem. The work done by the force of gravity and the normal force is zero, because these forces are perpendicular to the motion.

$$W_{\text{fric}} = \tfrac{1}{2}mv_f^2 - \tfrac{1}{2}mv_i^2 + \tfrac{1}{2}kx_f^2 - \tfrac{1}{2}kx_i^2$$

Substitute $v_i = 0$, $x_f = 0$, and $W_{\text{fric}} = -\mu_k n x_i$:

$$-\mu_k n x_i = \tfrac{1}{2}mv_f^2 - \tfrac{1}{2}kx_i^2$$

Set $n = mg$, and solve for v_f:

$$\tfrac{1}{2}mv_f^2 = \tfrac{1}{2}kx_i^2 - \mu_k mg x_i$$

$$v_f = \sqrt{\frac{k}{m}x_i^2 - 2\mu_k g x_i}$$

$$= \sqrt{\frac{4.00 \times 10^2\,\text{N/m}}{5.00\ \text{kg}}(0.05.\ \text{m})^2 - 2(0.150)(9.80\ \text{m/s}^2)(0.050\ \text{m})}$$

$$v_f = \;0.230\ \text{m/s}$$

Remarks   Friction or drag from immersion in a fluid damps the motion of an object attached to a spring, eventually bringing the object to rest.

Exercise 5.9

Suppose the spring system in the last example starts at $x = 0$ and the attached object is given a kick to the right, so it has an initial speed of 0.600 m/s. (a) What distance from the origin does the object travel before coming to rest, assuming the surface is frictionless? (b) How does the answer change if the coefficient of kinetic friction is $\mu_k = 0.150$? (Use the quadratic formula.)

Answer   (a) 0.0671 m   (b) 0.0512 m

EXAMPLE 5.10   Circus Acrobat

Goal   Use conservation of mechanical energy to solve a one-dimensional problem involving gravitational potential energy and spring potential energy.

Problem   A 50.0-kg circus acrobat drops from a height of 2.00 meters straight down onto a springboard with a force constant of 8.00×10^3 N/m, as in Figure 5.22. By what maximum distance does she compress the spring?

Strategy   Nonconservative forces are absent, so conservation of mechanical energy can be applied. At the two points of interest, the acrobat's initial position and the point of maximum spring compression, her velocity is zero, so the kinetic energy terms will be zero. Choose $y = 0$ as the point of maximum compression, so the final gravitational potential energy is zero. This choice also means that the initial position of the acrobat is $y_i = h + d$, where h is the acrobat's initial height above the platform and d is the spring's maximum compression.

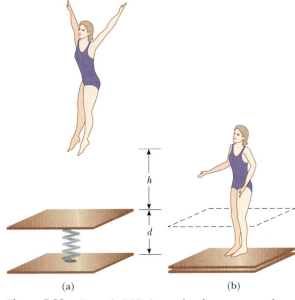

(a)   (b)

Figure 5.22   (Example 5.10) An acrobat drops onto a springboard, causing it to compress.

Solution

Use conservation of mechanical energy:

$$(KE + PE_g + PE_s)_i = (KE + PE_g + PE_s)_f \qquad (1)$$

The only nonzero terms are the initial gravitational potential energy and the final spring potential energy.

$$0 + mg(h + d) + 0 = 0 + 0 + \tfrac{1}{2}kd^2$$

$$mg(h + d) = \tfrac{1}{2}kd^2$$

Substitute the given quantities, and rearrange the equation into standard quadratic form:

$$(50.0 \text{ kg})(9.80 \text{ m/s}^2)(2.00 \text{ m} + d) = \tfrac{1}{2}(8.00 \times 10^3 \text{ N/m})d^2$$

$$d^2 - (0.123 \text{ m})d - 0.245 \text{ m}^2 = 0$$

Solve with the quadratic formula (Equation A.8):

$$d = \boxed{0.560 \text{ m}}$$

Remarks   The other solution, $d = -0.437$ m, can be rejected because d was chosen to be a positive number at the outset. A change in the acrobat's center of mass, say, by crouching as she makes contact with the springboard, also affects the spring's compression, but that effect was neglected. Shock absorbers often involve springs, and this example illustrates how they work. The spring action of a shock absorber turns a dangerous jolt into a smooth deceleration, as excess kinetic energy is converted to spring potential energy.

Exercise 5.10

An 8.00-kg block drops straight down from a height of 1.00 m, striking a platform spring having force constant 1.00×10^3 N/m. Find the maximum compression of the spring.

Answer $d = 0.482$ m

EXAMPLE 5.11   A Block Projected up a Frictionless Incline

Goal   Use conservation of mechanical energy to solve a problem involving gravitational potential energy, spring potential energy, and a ramp.

Problem   A 0.500-kg block rests on a horizontal, frictionless surface as in Figure 5.23. The block is pressed back against a spring having a constant of $k = 625$ N/m, compressing the spring by 10.0 cm to point Ⓐ. Then the block is released. **(a)** Find the maximum distance d the block travels up the frictionless incline if $\theta = 30.0°$. **(b)** How fast is the block going when halfway to its maximum height?

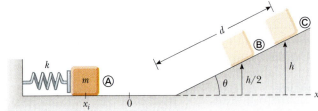

Figure 5.23   (Example 5.11)

Strategy   In the absence of other forces, conservation of mechanical energy applies to parts **(a)** and **(b)**. In part **(a)**, the block starts at rest and is also instantaneously at rest at the top of the ramp, so the kinetic energies at Ⓐ and Ⓒ are both zero. Note that the question asks for a distance d along the ramp, not the height h. In part **(b)**, the system has both kinetic and gravitational potential energy at Ⓑ.

Solution

(a) Find the distance the block travels up the ramp.

Apply conservation of mechanical energy:

$$\tfrac{1}{2}mv_i^2 + mgy_i + \tfrac{1}{2}kx_i^2 = \tfrac{1}{2}mv_f^2 + mgy_f + \tfrac{1}{2}kx_f^2$$

Substitute $v_i = v_f = 0$, $y_i = 0$, $y_f = h = d \sin \theta$, and $x_f = 0$.

$$\tfrac{1}{2}kx_i^2 = mgh = mgd \sin \theta$$

Solve for the distance d and insert the known values:

$$d = \frac{\tfrac{1}{2}kx_i^2}{mg \sin \theta} = \frac{\tfrac{1}{2}(625\ \text{N/m})(-0.100\ \text{m})^2}{(0.500\ \text{kg})(9.80\ \text{m/s}^2)\sin(30.0°)}$$

$$= \boxed{1.28\ \text{m}}$$

(b) Find the velocity at half the height, $h/2$. Note that $h = d \sin \theta = (1.28\ \text{m}) \sin 30.0° = 0.640\ \text{m}$.

Use energy conservation again:

$$\tfrac{1}{2}mv_i^2 + mgy_i + \tfrac{1}{2}kx_i^2 = \tfrac{1}{2}mv_f^2 + mgy_f + \tfrac{1}{2}kx_f^2$$

Take $v_i = 0$, $y_i = 0$, $y_f = \tfrac{1}{2}h$, and $x_f = 0$, yielding

$$\tfrac{1}{2}kx_i^2 = \tfrac{1}{2}mv_f^2 + mg\left(\tfrac{1}{2}h\right)$$

Multiply by $2/m$, and solve for v_f:

$$\frac{k}{m}x_i^2 = v_f^2 + gh$$

$$v_f = \sqrt{\frac{k}{m}x_i^2 - gh}$$

$$= \sqrt{\left(\frac{625\ \text{N/m}}{0.500\ \text{kg}}\right)(-0.100\ \text{m})^2 - (9.80\ \text{m/s}^2)(0.640\ \text{m})}$$

$$v_f = \boxed{2.50\ \text{m/s}}$$

Exercise 5.11

A 1.00-kg block is shot horizontally from a spring, as in the previous example, and travels 0.500 m up along a frictionless ramp before coming to rest and sliding back down. If the ramp makes an angle of 45.0° with respect to the horizontal, and the spring was originally compressed by 0.120 m, find the spring constant.

Answer   481 N/m

Applying Physics 5.2   Accident Reconstruction

Sometimes people involved in automobile accidents make exaggerated claims of chronic pain due to subtle injuries to the neck or spinal column. The likelihood of injury can be determined by finding the change in velocity of a car during the accident. The larger the change in velocity, the more likely it is that the person suffered spinal injury resulting in chronic pain. How can reliable estimates for this change in velocity be found after the fact?

Explanation   The metal and plastic of an automobile acts much like a spring, absorbing the car's kinetic energy by flexing during a collision. When the magnitude of the difference in velocity of the two cars is under five miles per hour, there is usually no visible damage, because bumpers are designed to absorb the impact and return to their original shape at such low speeds. At greater relative speeds there will be permanent damage to the vehicle. Despite the fact the structure of the car may not return to its original shape, a certain force per meter is still required to deform it, just as it takes a certain force per meter to compress a spring. The greater the original kinetic energy, the more the car is compressed during a collision, and the greater the damage. By using data obtained through crash tests, it's possible to obtain effective spring constants for all the different models of cars and determine reliable estimates of the change in velocity of a given vehicle during an accident. Medical research has established the likelihood of spinal injury for a given change in velocity, and the estimated velocity change can be used to help reduce insurance fraud.

5.5 SYSTEMS AND ENERGY CONSERVATION

Recall that the work–energy theorem can be written as

$$W_{nc} + W_c = \Delta KE$$

where W_{nc} represents the work done by nonconservative forces and W_c is the work done by conservative forces in a given physical context. As we have seen, any work done by conservative forces, such as gravity and springs, can be accounted for, by changes in potential energy. The work–energy theorem can therefore be written in the following way:

$$W_{nc} = \Delta KE + \Delta PE = (KE_f - KE_i) + (PE_f - PE_i) \qquad [5.19]$$

where now, as previously stated, *PE* includes all potential energies. This equation is easily rearranged to:

$$W_{nc} = (KE_f + PE_f) - (KE_i + PE_i) \qquad [5.20]$$

Recall, however, that the total mechanical energy is given by $E = KE + PE$. Making this substitution into Equation 5.20, we find that the work done on a system by all nonconservative forces is equal to the change in mechanical energy of that system:

$$W_{nc} = E_f - E_i = \Delta E \qquad [5.21]$$

If the mechanical energy is changing, it has to be going somewhere. The energy either leaves the system and goes into the surrounding environment, or it stays in the system and is converted into a nonmechanical form such as thermal energy.

A simple example is a block sliding along a rough surface. Friction creates thermal energy, absorbed partly by the block and partly by the surrounding environment. When the block warms up, something called *internal energy* increases. The internal energy of a system is related to its temperature, which in turn is a consequence of the activity of its parts, such as the moving atoms of a gas or the vibration of atoms in a solid. (Internal energy will be studied in more detail in Chapter 12.)

Energy can be transferred between a nonisolated system and its environment. If positive work is done on the system, energy is transferred from the environment to the system. If negative work is done on the system, energy is transferred from the system to the environment.

So far, we have encountered three methods of storing energy in a system: kinetic energy, potential energy, and internal energy. On the other hand, we've seen only one way of transferring energy into or out of a system: through work. Other methods will be studied in later chapters, but are summarized here:

- **Work**, in the mechanical sense of this chapter, transfers energy to a system by displacing it with an applied force.
- **Heat** is the process of transferring energy through microscopic collisions between atoms or molecules. For example, a metal spoon resting in a cup of coffee becomes hot because some of the kinetic energy of the molecules in the liquid coffee is transferred to the spoon as internal energy.
- **Mechanical waves** transfer energy by creating a disturbance that propagates through air or another medium. For example, energy in the form of sound leaves your stereo system through the loudspeakers and enters your ears to stimulate the hearing process. Other examples of mechanical waves are seismic waves and ocean waves.
- **Electrical transmission** transfers energy through electric currents. This is how energy enters your stereo system or any other electrical device.
- **Electromagnetic radiation** transfers energy in the form of electromagnetic waves such as light, microwaves, and radio waves. Examples of this method of transfer include cooking a potato in a microwave oven and light energy traveling from the Sun to the Earth through space.

Conservation of Energy in General

The most important feature of the energy approach is the idea that energy is conserved; it can't be created or destroyed, only transferred from one form into another. This is the principle of **conservation of energy**.

The principle of conservation of energy is not confined to physics. In biology, energy transformations take place in myriad ways inside all living organisms. One example is the transformation of chemical energy to mechanical energy that causes flagella to move and propel an organism. Some bacteria use chemical energy to produce light. (See Figure 5.24.) Although the mechanisms that produce these light emissions are not well understood, living creatures often rely on this light for their existence. For example, certain fish have sacs beneath their eyes filled with light-emitting bacteria. The emitted light attracts creatures that become food for the fish.

Quick Quiz 5.4

A book of mass m is projected with a speed v across a horizontal surface. The book slides until it stops due to the friction force between the book and the surface. The surface is now tilted 30°, and the book is projected up the surface with the same initial speed v. When the book has come to rest, how does the decrease in mechanical energy of the book–Earth system compare with that when the book slid over the horizontal surface? (a) It's the same; (b) it's larger on the tilted surface; (c) it's smaller on the tilted surface; (d) more information is needed.

APPLICATION

Flagellar Movement;
Bioluminescence

Figure 5.24   This small plant, found in warm southern waters, exhibits bioluminescence, a process in which chemical energy is converted to light. The red areas are chlorophyll, which glows when excited by blue light.

Applying Physics 5.3   Asteroid Impact!

An asteroid about a kilometer in radius has been blamed for the extinction of the dinosaurs 65 million years ago. How can a relatively small object, which could fit inside a college campus, inflict such injury on the vast biosphere of the Earth?

Explanation   While such an asteroid is comparatively small, it travels at a very high speed relative to the Earth, typically on the order of 40 000 m/s. A roughly spherical asteroid one kilometer in radius and made mainly of rock has a mass of approximately 10 trillion kilograms—a small mountain of matter. The kinetic energy of such an asteroid would be about 10^{22} J, or 10 billion trillion joules. By contrast, the atomic bomb that devastated Hiroshima was equivalent to 15 kilotons of TNT, approximately 6×10^{13} J of energy. On striking the Earth, the asteroid's enormous kinetic energy changes into other forms, such as thermal energy, sound, and light, with a total energy release greater than 100 million Hiroshima explosions! Aside from the devastation in the immediate blast area and fires across a continent, gargantuan tidal waves would scour low-lying regions around the world and dust would block the sun for decades.

For this reason, asteroid impacts represent a threat to life on Earth. Asteroids large enough to cause widespread extinction hit Earth only every 60 million years

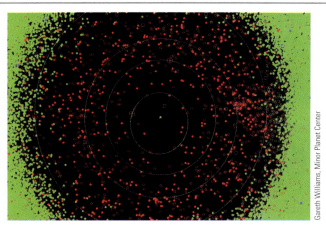

Figure 5.25   Asteroid map of the inner solar system.

or so. Smaller asteroids, of sufficient size to cause serious damage to civilization on a global scale, are thought to strike every five to ten thousand years. There have been several near misses by such asteroids in the last century and even in the last decade. In 1907, a small asteroid or comet fragment struck Tunguska, Siberia, annihilating a region 60 kilometers across. Had it hit northern Europe, millions of people might have perished.

Figure 5.25 is an asteroid map of the inner solar system. More asteroids are being discovered every year.

5.6 POWER

The rate at which energy is transferred is important in the design and use of practical devices, such as electrical appliances and engines of all kinds. The issue is particularly interesting for living creatures, since the maximum work per second, or power output, of an animal varies greatly with output duration. Power is defined as the rate of energy transfer with time:

Average power ▶

If an external force is applied to an object and if the work done by this force is W in the time interval Δt, then the **average power** delivered to the object during this interval is the work done divided by the time interval, or

$$\overline{\mathcal{P}} = \frac{W}{\Delta t} \qquad \qquad [5.22]$$

SI unit: watt (W = J/s)

It's sometimes useful to rewrite Equation 5.22 by substituting $W = F\,\Delta x$ and noticing that $\Delta x/\Delta t$ is the average speed of the object during the time Δt:

$$\overline{\mathcal{P}} = \frac{W}{\Delta t} = \frac{F\,\Delta x}{\Delta t} = F\overline{v} \qquad \qquad [5.23]$$

According to Equation 5.23, average power is a constant force times the average speed. The force F is the component of force in the direction of the average velocity. A more general definition can be written down with a little calculus and has the same form as Equation 5.23:

$$\mathcal{P} = Fv$$

The SI unit of power is the joule/sec, also called the **watt**, named after James Watt:

$$1\ \text{W} = 1\ \text{J/s} = 1\ \text{kg} \cdot \text{m}^2/\text{s}^3 \qquad \qquad [5.24]$$

The unit of power in the U. S. customary system is the horsepower (hp), where

$$1\ \text{hp} \equiv 550\ \frac{\text{ft} \cdot \text{lb}}{\text{s}} = 746\ \text{W} \qquad \qquad [5.25]$$

TIP 5.6   Watts the Difference?
Don't confuse the nonitalic symbol for watts, W, with the italic symbol W for work. A watt is a unit, the same as joules per second. Work is a concept, carrying units of joules.

The horsepower was first defined by Watt, who needed a large power unit to rate the power output of his new invention, the steam engine.

The watt is commonly used in electrical applications, but it can be used in other scientific areas as well. For example, European sports car engines are rated in kilowatts.

In electric power generation, it's customary to use the kilowatt-hour as a measure of energy. One kilowatt-hour (kWh) is the energy transferred in 1 h at the constant rate of 1 kW = 1 000 J/s. Therefore,

$$1\ \text{kWh} = (10^3\ \text{W})(3\ 600\ \text{s}) = (10^3\ \text{J/s})(3\ 600\ \text{s}) = 3.60 \times 10^6\ \text{J}$$

It's important to realize that a kilowatt-hour is a unit of energy, *not* power. When you pay your electric bill, you're buying energy, and that's why your bill lists a charge for electricity of about 10 cents/kWh. The amount of electricity used by an appliance can be calculated by multiplying its power rating (usually expressed in watts and valid only for normal household electrical circuits) by the length of time the appliance is operated. For example, an electric bulb rated at 100 W (= 0.100 kW) "consumes" 3.6×10^5 J of energy in 1 h.

EXAMPLE 5.12 Power Delivered by an Elevator Motor

Goal Apply the force-times-velocity definition of power.

Problem A 1.00×10^3-kg elevator carries a maximum load of 8.00×10^2 kg. A constant frictional force of 4.00×10^3 N retards its motion upward, as in Figure 5.26. What minimum power, in kilowatts and in horsepower, must the motor deliver to lift the fully loaded elevator at a constant speed of 3.00 m/s?

Strategy To solve this problem, we need to determine the force the elevator's motor must deliver through the force of tension in the cable, $\vec{\mathbf{T}}$. Substituting this force together with the given speed v into $\mathcal{P} = Fv$ gives the desired power. The tension in the cable, T, can be found with Newton's second law.

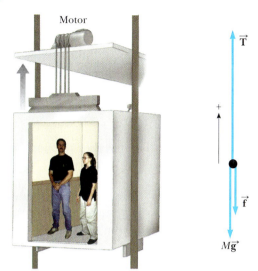

Figure 5.26 (Example 5.12) The motor exerts an upward force $\vec{\mathbf{T}}$ on the elevator. A frictional force $\vec{\mathbf{f}}$ and the force of gravity $M\vec{\mathbf{g}}$ act downwards.)

Solution

Apply Newton's second law to the elevator:

$$\Sigma \vec{\mathbf{F}} = m\vec{\mathbf{a}}$$

The velocity is constant, so the acceleration is zero. The forces acting on the elevator are the force of tension in the cable, $\vec{\mathbf{T}}$, the friction $\vec{\mathbf{f}}$, and gravity $M\vec{\mathbf{g}}$, where M is the mass of the elevator.

$$\vec{\mathbf{T}} + \vec{\mathbf{f}} + M\vec{\mathbf{g}} = 0$$

Write the equation in terms of its components:

$$T - f - Mg = 0$$

Solve this equation for the tension T and evaluate it:

$$T = f + Mg$$
$$= 4.00 \times 10^3 \text{ N} + (1.80 \times 10^3 \text{ kg})(9.80 \text{ m/s}^2)$$
$$T = 2.16 \times 10^4 \text{ N}$$

Substitute this value of T for F in the power equation:

$$\mathcal{P} = Fv = (2.16 \times 10^4 \text{ N})(3.00 \text{ m/s}) = 6.48 \times 10^4 \text{ W}$$
$$\mathcal{P} = 64.8 \text{ kW} = \boxed{86.9 \text{ hp}}$$

Remarks The friction force acts to retard the motion, requiring more power. For a descending elevator, the friction force can actually reduce the power requirement.

Exercise 5.12

Suppose the same elevator with the same load descends at 3.00 m/s. What minimum power is required? (Here, the motor removes energy from the elevator by not allowing it to fall freely.)

Answer 4.09×10^4 W = 54.9 hp

EXAMPLE 5.13 Shamu Sprint

Goal Calculate the power needed to increase an object's kinetic energy.

Problem Killer whales are known to reach 32 ft in length and have a mass of over 8 000 kg. They are also very quick, able to accelerate up to 30 mi/h in a matter of seconds. Disregarding the considerable drag force of water, calculate the average power a killer whale named Shamu with mass 8.00×10^3 kg would need to generate to reach a speed of 12.0 m/s in 6.00 s.

Strategy Find the change in kinetic energy of Shamu and use the work–energy theorem to obtain the minimum work Shamu has to do to effect this change. (Internal and external friction forces increase the necessary amount of energy.) Divide by the elapsed time to get the power.

Solution

Calculate the change in Shamu's kinetic energy. By the work–energy theorem, this equals the minimum work Shamu must do:

$$\Delta KE = \tfrac{1}{2} m v_f{}^2 - \tfrac{1}{2} m v_i{}^2$$
$$= \tfrac{1}{2} \cdot 8.00 \times 10^3 \,\text{kg} \cdot (12.0 \,\text{m/s})^2 - 0$$
$$= 5.76 \times 10^5 \,\text{J}$$

Divide by the elapsed time (Eq. 5.22), noting that $W = \Delta KE$.

$$\mathcal{P} = \frac{W}{\Delta t} = \frac{5.76 \times 10^5 \,\text{J}}{6.00 \,\text{s}} = \boxed{9.60 \times 10^4 \,\text{W}}$$

Remarks This is enough power to run a moderate-sized office building! The actual requirements are larger because of friction in the water and muscular tissues. Something similar can be done with gravitational potential energy, as the exercise illustrates.

Exercise 5.13

What minimum average power must a 35-kg human boy generate climbing up the stairs to the top of the Washington monument? The trip up the nearly 170-m-tall building takes him 10 minutes. Include only work done against gravity, ignoring biological efficiency.

Answer 97 W

EXAMPLE 5.14   Speedboat Power

Goal Combine power, the work–energy theorem and nonconservative forces with one-dimensional kinematics.

Problem How much power would a 1.00×10^3-kg speedboat need to go from rest to 20.0 m/s in 5.00 s, assuming the water exerts a constant drag force of magnitude $f_d = 5.00 \times 10^2$ N and the acceleration is constant.

Strategy The power is provided by the engine, which creates a nonconservative force. Use the work–energy theorem together with the work done by the engine, W_{engine}, and the work done by the drag force, W_{drag}, on the left-hand side. Use one-dimensional kinematics to find the acceleration and then the displacement Δx. Solve the work–energy theorem for W_{engine}, and divide by the elapsed time to get the power.

Solution

Write the work–energy theorem:

$$W_{\text{net}} = \Delta KE = \tfrac{1}{2} m v_f{}^2 - \tfrac{1}{2} m v_i{}^2$$

Fill in the two work terms and take $v_i = 0$:

$$W_{\text{engine}} + W_{\text{drag}} = W_{\text{engine}} - f_d \,\Delta x = \tfrac{1}{2} m v_f{}^2 \qquad (1)$$

To get the displacement Δx, first find the acceleration using the velocity equation of kinematics:

$$v_f = at + v_i \quad \rightarrow \quad v_f = at$$
$$(20.0 \,\text{m/s}) = a(5.00 \,\text{s}) \quad \rightarrow \quad a = 4.00 \,\text{m/s}^2$$

Substitute a into the time-independent kinematics equation, and solve for Δx:

$$v_f{}^2 - v_i{}^2 = 2a \,\Delta x$$
$$(20.0 \,\text{m/s})^2 - 0^2 = 2(4.00 \,\text{m/s}^2)\Delta x$$
$$\Delta x = 50.0 \,\text{m}$$

Now that we know Δx, we can find the mechanical energy lost due to friction.

$$W_{\text{fric}} = -f_d \,\Delta x = -(5.00 \times 10^2 \,\text{N})(50.0 \,\text{m}) = -2.50 \times 10^4 \,\text{J}$$

Solve equation (1) for W_{engine}:

$$W_{\text{engine}} = \tfrac{1}{2} m v_f{}^2 - f_d \,\Delta x$$
$$= \tfrac{1}{2}(1.00 \times 10^3 \,\text{kg})(20.0 \,\text{m/s})^2 - (-2.50 \times 10^4 \,\text{J})$$
$$W_{\text{engine}} = 2.25 \times 10^5 \,\text{J}$$

Compute the power:

$$\mathcal{P} = \frac{W_{\text{engine}}}{\Delta t} = \frac{2.25 \times 10^5 \,\text{J}}{5.00 \,\text{s}} = 4.50 \times 10^4 \,\text{W} = \boxed{60.3 \,\text{hp}}$$

Remarks In fact, drag forces generally get larger with increasing speed.

Exercise 5.14

What power must be supplied to push a 5.00-kg block from rest to 10.0 m/s in 5.00 s when the coefficient of kinetic friction between the block and surface is 0.250? Assume the acceleration is uniform.

Answer   111 W

Energy and Power in a Vertical Jump

The stationary jump consists of two parts: extension and free flight.[2] In the extension phase the person jumps up from a crouch, straightening the legs and throwing up the arms; the free-flight phase occurs when the jumper leaves the ground. Because the body is an extended object and different parts move with different speeds, we describe the motion of the jumper in terms of the position and velocity of the **center of mass (CM)**, which is the point in the body at which all the mass may be considered to be concentrated. Figure 5.27 shows the position and velocity of the CM at different stages of the jump.

Using the principle of the conservation of mechanical energy, we can find H, the maximum increase in height of the CM, in terms of the velocity v_{CM} of the CM at liftoff. Taking PE_i, the gravitational potential energy of the jumper–Earth system just as the jumper lifts off from the ground to be zero, and noting that the kinetic energy KE_f of the jumper at the peak is zero, we have

$$PE_i + KE_i = PE_f + KE_f$$

$$\tfrac{1}{2} m v_{CM}^2 = mgH \quad \text{or} \quad H = \frac{v_{CM}^2}{2g}$$

We can estimate v_{CM} by assuming that the acceleration of the CM is constant during the extension phase. If the depth of the crouch is h and the time for extension is Δt, we find that $v_{CM} = 2\bar{v} = 2h/\Delta t$. Measurements on a group of male college students show typical values of $h = 0.40$ m and $\Delta t = 0.25$ s, the latter value being set by the fixed speed with which muscle can contract. Substituting, we obtain

$$v_{CM} = 2(0.40 \text{ m})/(0.25 \text{ s}) = 3.2 \text{ m/s}$$

and

$$H = \frac{v_{CM}^2}{2g} = \frac{(3.2 \text{ m/s})^2}{2(9.80 \text{ m/s}^2)} = 0.52 \text{ m}$$

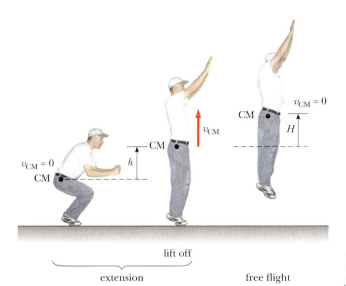

lift off

extension   free flight

Figure 5.27 Extension and free flight in the vertical jump.

---

[2]For more information on this topic, see E. J. Offenbacher, *American Journal of Physics*, **38**, 829 (1969).

TABLE 5.1

Maximum Power Output from Humans over Various Periods

Power	Time
2 hp, or 1 500 W	6 s
1 hp, or 750 W	60 s
0.35 hp, or 260 W	35 min
0.2 hp, or 150 W	5 h
0.1 hp, or 75 W (safe daily level)	8 h

APPLICATION

Diet Versus Exercise in Weight-loss Programs

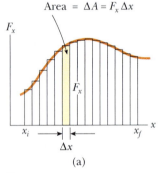

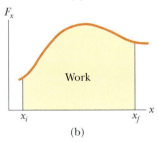

Figure 5.28 (a) The work done by the force component F_x for the small displacement Δx is $F_x\Delta x$, which equals the area of the shaded rectangle. The total work done for the displacement from x_i to x_f is approximately equal to the sum of the areas of all the rectangles. (b) The work done by the component F_x of the varying force as the particle moves from x_i to x_f is exactly equal to the area under the curve shown.

Measurements on this same group of students found that H was between 0.45 m and 0.61 m in all cases, confirming the basic validity of our simple calculation.

In order to relate the abstract concepts of energy, power, and efficiency to humans, it's interesting to calculate these values for the vertical jump. The kinetic energy given to the body in a jump is $KE = \frac{1}{2}mv_{CM}^2$, and for a person of mass 68 kg, the kinetic energy is

$$KE = \tfrac{1}{2}(68 \text{ kg})(3.2 \text{ m/s})^2 = 3.5 \times 10^2 \text{ J}$$

Although this may seem like a large expenditure of energy, we can make a simple calculation to show that jumping and exercise in general are not good ways to lose weight, in spite of their many health benefits. Since the muscles are at most 25% efficient at producing kinetic energy from chemical energy (muscles always produce a lot of internal energy and kinetic energy as well as work—that's why you sweat when you work out), they use up four times the 350 J (about 1 400 J) of chemical energy in one jump. This chemical energy ultimately comes from the food we eat, with energy content given in units of food calories and one food calorie equal to 4 200 J. So the total energy supplied by the body as internal energy and kinetic energy in a vertical jump is only about one-third of a food calorie! You are a lot better off not eating that piece of cheesecake than trying to work it off by jumping.

Finally, it's interesting to calculate the mechanical power that can be generated by the body in strenuous activity for brief periods. Here we find that

$$\mathcal{P} = \frac{KE}{\Delta t} = \frac{3.5 \times 10^2 \text{ J}}{0.25 \text{ s}} = 1.4 \times 10^3 \text{ W}$$

or (1 400 W)(1 hp/746 W) = 1.9 hp. So humans can produce about 2 hp of mechanical power for periods on the order of seconds. Table 5.1 shows the maximum power outputs from humans for various periods while bicycling and rowing, activities in which it is possible to measure power output accurately.

5.7   WORK DONE BY A VARYING FORCE

Suppose an object is displaced along the x-axis under the action of a force F_x that acts in the x-direction and varies with position, as shown in Figure 5.28. The object is displaced in the direction of increasing x from $x = x_i$ to $x = x_f$. In such a situation, we can't use Equation 5.1 to calculate the work done by the force because this relationship applies only when $\vec{F}$ is constant in magnitude and direction. However, if we imagine that the object undergoes the *small* displacement Δx shown in Figure 5.28a, then the x-component F_x of the force is nearly constant over this interval and we can approximate the work done by the force for this small displacement as

$$W_1 \cong F_x\Delta x \qquad [5.26]$$

This quantity is just the area of the shaded rectangle in Figure 5.28a. If we imagine that the curve of F_x versus x is divided into a large number of such intervals, then the total work done for the displacement from x_i to x_f is approximately equal to the sum of the areas of a large number of small rectangles:

$$W \cong F_1\Delta x_1 + F_2\Delta x_2 + F_3\Delta x_3 + \cdots \qquad [5.27]$$

Now imagine going through the same process with twice as many intervals, each half the size of the original Δx. The rectangles then have smaller widths and will better approximate the area under the curve. Continuing the process of increasing the number of intervals while allowing their size to approach zero, the number of terms in the sum increases without limit, but the value of the sum approaches a definite value equal to the area under the curve bounded by F_x and the x-axis in Figure 5.28b. In other words, **the work done by a variable force acting on an object that undergoes a displacement is equal to the area under the graph of F_x versus x.**

A common physical system in which force varies with position consists of a block on a horizontal, frictionless surface connected to a spring, as discussed in Section 5.4. When the spring is stretched or compressed a small distance x from its equilibrium position $x = 0$, it exerts a force on the block given by $F_s = -kx$, where k is the force constant of the spring.

Now let's determine the work done by an *external agent* on the block as the spring is stretched *very slowly* from $x_i = 0$ to $x_f = x_{max}$, as in Active Figure 5.29a. This work can be easily calculated by noting that at any value of the displacement, Newton's third law tells us that the applied force $\vec{F}_{app}$ is equal in magnitude to the spring force $\vec{F}_s$ and acts in the opposite direction, so that $F_{app} = -(-kx) = kx$. A plot of F_{app} versus x is a straight line, as shown in Active Figure 5.29b. Therefore, the work done by this applied force in stretching the spring from $x = 0$ to $x = x_{max}$ is the area under the straight line in that figure, which in this case is the area of the shaded triangle:

$$W_{F_{app}} = \tfrac{1}{2}kx_{max}^2$$

During this same time the spring has done exactly the same amount of work, but that work is negative, because the spring force points in the direction opposite the motion. The potential energy of the system is exactly equal to the work done by the applied force and is the same sign, which is why potential energy is thought of as stored work.

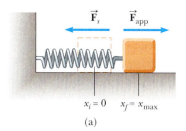

$x_i = 0 \qquad x_f = x_{max}$

(a)

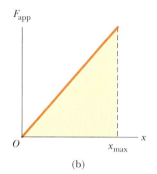

(b)

ACTIVE FIGURE 5.29
(a) A block being pulled from $x_i = 0$ to $x_f = x_{max}$ on a frictionless surface by a force $\vec{F}_{app}$. If the process is carried out very slowly, the applied force is equal in magnitude and opposite in direction to the spring force at all times. (b) A graph of F_{app} versus x.

Physics⊗Now™
Log into PhysicsNow at **www.cp7e.com**, and go to Active Figure 5.29 to observe the block's motion for various maximum displacements and spring constants.

EXAMPLE 5.15 Work Required to Stretch a Spring

Goal Apply the graphical method of finding work.

Problem One end of a horizontal spring ($k = 80.0$ N/m) is held fixed while an external force is applied to the free end, stretching it slowly from $x_{Ⓐ} = 0$ to $x_{Ⓑ} = 4.00$ cm. **(a)** Find the work done by the applied force on the spring. **(b)** Find the additional work done in stretching the spring from $x_{Ⓑ} = 4.00$ cm to $x_{Ⓒ} = 7.00$ cm.

Strategy For part (a), simply find the area of the smaller triangle, using $A = \tfrac{1}{2}bh$, one-half the base times the height. For part (b), the easiest way to find the additional work done from $x_{Ⓑ} = 4.00$ cm to $x_{Ⓒ} = 7.00$ cm is to find the area of the new, larger triangle and subtract the area of the smaller triangle.

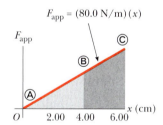

Figure 5.30 (Example 5.15) A graph of the external force required to stretch a spring that obeys Hooke's law versus the elongation of the spring.

Solution
(a) Find the work from $x_{Ⓐ} = 0$ cm to $x_{Ⓑ} = 4.00$ cm.

Compute the area of the smaller triangle:

$W = \tfrac{1}{2}kx_B^2 = \tfrac{1}{2}(80.0 \text{ N/m})(0.040 \text{ m})^2 = \boxed{0.064\,0 \text{ J}}$

(b) Find the work from $x_{Ⓑ} = 4.00$ cm to $x_{Ⓒ} = 7.00$ cm.

Compute the area of the large triangle, and subtract the area of the smaller triangle:

$W = \tfrac{1}{2}kx_C^2 - \tfrac{1}{2}kx_B^2$

$W = \tfrac{1}{2}(80.0 \text{ N/m})(0.070\,0 \text{ m})^2 - 0.064\,0 \text{ J}$

$= 0.196 \text{ J} - 0.064\,0 \text{ J}$

$= \boxed{0.132 \text{ J}}$

Remarks Only simple geometries—rectangles and triangles—can be solved exactly with this method. More complex shapes require calculus or the square-counting technique in the next worked example.

Exercise 5.15
How much work is required to stretch this same spring from $x_i = 5.00$ cm to $x_f = 9.00$ cm?

Answer 0.224 J

EXAMPLE 5.16 Estimating Work by Counting Boxes

Goal Use the graphical method and counting boxes to estimate the work done by a force.

Problem Suppose the force applied to stretch a thick piece of elastic changes with position as indicated in Figure 5.31a. Estimate the work done by the applied force.

Strategy To find the work, simply count the number of boxes underneath the curve, and multiply that number by the area of each box. The curve will pass through the middle of some boxes, in which case only an estimated fractional part should be counted.

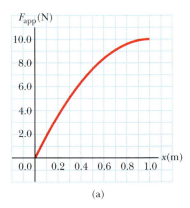

Solution
There are 62 complete or nearly complete boxes under the curve, 6 boxes that are about half under the curve, and a triangular area from $x = 0$ m to $x = 0.10$ m that is equivalent to 1 box, for a total of about 66 boxes. Since the area of each box is 0.10 J, the total work done is approximately 66×0.10 J $= 6.6$ J.

Remarks Mathematically, there are a number of other methods for creating such estimates, all involving adding up regions approximating the area. To get a better estimate, make smaller boxes.

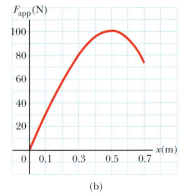

Exercise 5.16
Suppose the applied force necessary to pull the drawstring on a bow is given by Figure 5.31b. Find the approximate work done by counting boxes.

Figure 5.31 (a) (Example 5.16) (b) (Exercise 5.16)

Answer About 50 J. (Individual answers may vary.)

SUMMARY

Physics⊗Now™ Take a practice test by logging into PhysicsNow at **www.cp7e.com** and clicking on the Pre-Test link for this chapter.

5.1 Work

The work done on an object by a constant force is

$$W = (F \cos \theta)\Delta x \qquad [5.2]$$

where F is the magnitude of the force, Δx is the object's displacement, and θ is the angle between the direction of the force $\vec{F}$ and the displacement $\Delta \vec{x}$. Solving simple problems requires substituting values into this equation. More complex problems, such as those involving friction, often require using Newton's second law, $m\vec{a} = \Sigma\vec{F}$, to determine forces.

5.2 Kinetic Energy and the WorK–Energy Theorem

The kinetic energy of a body with mass m and speed v is given by

$$KE \equiv \tfrac{1}{2} mv^2 \qquad [5.5]$$

The work–energy theorem states that the net work done on an object of mass m is equal to the change in its kinetic energy, or

$$W_{\text{net}} = KE_f - KE_i = \Delta KE \qquad [5.6]$$

Work and energy of any kind carry units of joules. Solving problems involves finding the work done by each force acting on the object and summing them up, which is W_{net}, followed by substituting known quantities into Equation 5.6, solving for the unknown quantity.

Conservative forces are special: Work done against them can be recovered—it's conserved. An example is gravity: The work done in lifting an object through a height is effectively stored in the gravity field and can be recovered in the kinetic energy of the object simply by letting it fall. Nonconservative forces, such as surface friction and drag, dissipate energy in a form that can't be readily recovered. To account for such forces, the work–energy theorem can be rewritten as

$$W_{nc} + W_c = \Delta KE \qquad [5.7]$$

where W_{nc} is the work done by nonconservative forces and W_c is the work done by conservative forces.

5.3 Gravitational Potential Energy

The gravitational force is a conservative field. Gravitational potential energy is another way of accounting for gravitational work W_g:

$$W_g = -(PE_f - PE_i) = -(mgy_f - mgy_i) \qquad [5.11]$$

To find the change in gravitational potential energy as an object of mass m moves between two points in a gravitational field, substitute the values of the object's y-coordinates.

The work–energy theorem can be generalized to include gravitational potential energy:

$$W_{nc} = (KE_f - KE_i) + (PE_f - PE_i) \qquad [5.12]$$

Gravitational work and gravitational potential energy should not both appear in the work–energy theorem at the same time, only one or the other, because they're equivalent. Setting the work due to nonconservative forces to zero and substituting the expressions for KE and PE, a form of the conservation of mechanical energy with gravitation can be obtained:

$$\tfrac{1}{2}mv_i^2 + mgy_i = \tfrac{1}{2}mv_f^2 + mgy_f \qquad [5.14]$$

To solve problems with this equation, identify two points in the system—one where information is known and the other where information is desired. Substitute and solve for the unknown quantity.

The work done by other forces, as when frictional forces are present, isn't always zero. In that case, identify two points as before, calculate the work due to all other forces, and solve for the unknown in Equation 5.12.

5.4 Spring Potential Energy

The spring force is conservative, and its potential energy is given by

$$PE_s \equiv \tfrac{1}{2}kx^2 \qquad [5.16]$$

Spring potential energy can be put into the work–energy theorem, which then reads

$$W_{nc} = (KE_f - KE_i) + (PE_{gf} - PE_{gi}) + (PE_{sf} - PE_{si}) \qquad [5.17]$$

When nonconservative forces are absent, $W_{nc} = 0$ and mechanical energy is conserved.

5.5 Systems and Energy Conservation

The principle of the conservation of energy states that energy can't be created or destroyed. It can be transformed, but the total energy content of any isolated system is always constant. The same is true for the universe at large. The work done by all nonconservative forces acting on a system equals the change in the total mechanical energy of the system:

$$W_{nc} = (KE_f + PE_f) - (KE_i + PE_i) = E_f - E_i \qquad [5.20-21]$$

where PE represents all potential energies present.

5.6 Power

Average power is the amount of energy transferred divided by the time taken for the transfer:

$$\overline{\mathcal{P}} = \frac{W}{\Delta t} \qquad [5.22]$$

This expression can also be written

$$\overline{\mathcal{P}} = F\overline{v} \qquad [5.23]$$

where $\overline{v}$ is the object's average speed. The unit of power is the watt ($W = J/s$). To solve simple problems, substitute given quantities into one of these equations. More difficult problems usually require finding the work done on the object using the work–energy theorem or the definition of work.

CONCEPTUAL QUESTIONS

1. Consider a tug-of-war as in Figure Q5.1, in which two teams pulling on a rope are evenly matched, so that no motion takes place. Is work done on the rope? On the pullers? On the ground? Is work done on anything?

2. Discuss whether any work is being done by each of the following agents and, if so, whether the work is positive or negative: (a) a chicken scratching the ground, (b) a person studying, (c) a crane lifting a bucket of concrete, (d) the force of gravity on the bucket in part (c), (e) the leg muscles of a person in the act of sitting down.

3. If the height of a playground slide is kept constant, will the length of the slide or whether it has bumps make any difference in the final speed of children playing on it? Assume that the slide is slick enough to be considered

Figure Q5.1

Arthur Tilley/FPG/Getty Images

frictionless. Repeat this question, assuming that the slide is not frictionless.

4. (a) Can the kinetic energy of a system be negative? (b) Can the gravitational potential energy of a system be negative? Explain.

5. Roads going up mountains are formed into switchbacks, with the road weaving back and forth along the face of the slope such that there is only a gentle rise on any portion of the roadway. Does this configuration require any less work to be done by an automobile climbing the mountain, compared with one traveling on a roadway that is straight up the slope? Why are switchbacks used?

6. (a) If the speed of a particle is doubled, what happens to its kinetic energy? (b) If the net work done on a particle is zero, what can be said about its speed?

7. As a simple pendulum swings back and forth, the forces acting on the suspended object are the force of gravity, the tension in the supporting cord, and air resistance. (a) Which of these forces, if any, does no work on the pendulum? (b) Which of these forces does negative work at all times during the pendulum's motion? (c) Describe the work done by the force of gravity while the pendulum is swinging.

8. A bowling ball is suspended from the ceiling of a lecture hall by a strong cord. The ball is drawn away from its equilibrium position and released from rest at the tip of the demonstrator's nose, as shown in Figure Q5.8. If the demonstrator remains stationary, explain why the ball does not strike her on its return swing. Would this demonstrator be safe if the ball were given a push from its starting position at her nose?

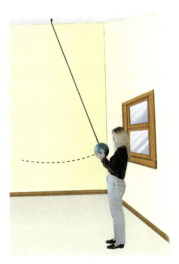

Figure Q5.8

9. An older model car accelerates from 0 to speed v in 10 seconds. A newer, more powerful sports car accelerates from 0 to $2v$ in the same time. What is the ratio of the powers expended by the two cars? Assume the energy coming from the engine appears only as kinetic energy of the cars.

10. During a stress test of the cardiovascular system, a patient walks and runs on a treadmill. (a) Is the energy expended by the patient equivalent to the energy of walking and running on the ground? Explain. (b) What effect, if any, does tilting the treadmill upward have? Discuss.

11. When a punter kicks a football, is he doing any work on the ball while the toe of his foot is in contact with it? Is he doing any work on the ball after it loses contact with his toe? Are any forces doing work on the ball while it is in flight?

12. As a sled moves across a flat, snow-covered field at constant velocity, is any work done? How does air resistance enter into the picture?

13. A weight is connected to a spring that is suspended vertically from the ceiling. If the weight is displaced downward from its equilibrium position and released, it will oscillate up and down. If air resistance is neglected, will the total mechanical energy of the system (weight plus Earth plus spring) be conserved? How many forms of potential energy are there for this situation?

14. The driver of a car slams on her brakes to avoid colliding with a deer crossing the highway. What happens to the car's kinetic energy as it comes to rest?

15. Suppose you are reshelving books in a library. You lift a book from the floor to the top shelf. The kinetic energy of the book on the floor was zero, and the kinetic energy of the book on the top shelf is zero, so there is no change in kinetic energy. Yet you did some work in lifting the book. Is the work–energy theorem violated?

16. The feet of a standing person of mass m exert a force equal to mg on the floor, and the floor exerts an equal and opposite force upwards on the feet, which we call the normal force. During the extension phase of a vertical jump (see page 145), the feet exert a force on the floor that is greater than mg, so the normal force is greater than mg. As you learned in Chapter 4, we can use this result and Newton's second law to calculate the acceleration of the jumper: $a = F_{\text{net}}/m = (n - mg)/m$.

    Using energy ideas, we know that work is performed on the jumper to give him or her kinetic energy. But the normal force can't perform any work here, because the feet don't undergo any displacement. How is energy transferred to the jumper?

17. An Earth satellite is in a circular orbit at an altitude of 500 km. Explain why the work done by the gravitational force acting on the satellite is zero. Using the work–energy theorem, what can you say about the speed of the satellite?

18. In most circumstances, the normal force acting on an object and the force of static friction do no work on the object. However, the reason that the work is zero is different for the two cases. In each case, explain why the work done by the force is zero.

19. In most situations we have encountered in this chapter, frictional forces tend to reduce the kinetic energy of an object. However, frictional forces can sometimes increase an object's kinetic energy. Describe a few situations in which friction causes an increase in kinetic energy.

20. Discuss the energy transformations that occur as a pole vaulter runs at high speeds and attempts to clear a bar that is about 5 m from the ground. In your analysis, you must consider changes in the kinetic energy of the runner, the elastic potential energy of the pole as it bends, and the gravitational potential energy of the vaulter. Ignore rotational motion.

PROBLEMS

1, 2, 3 = straightforward, intermediate, challenging   □ = full solution available in *Student Solutions Manual/Study Guide*

Physics⊗Now™ = coached problem with hints available at **www.cp7e.com**   🧬 = biomedical application

Section 5.1 Work

1. A weight lifter lifts a 350-N set of weights from ground level to a position over his head, a vertical distance of 2.00 m. How much work does the weight lifter do, assuming he moves the weights at constant speed?

2. If a man lifts a 20.0-kg bucket from a well and does 6.00 kJ of work, how deep is the well? Assume that the speed of the bucket remains constant as it is lifted.

3. A tugboat exerts a constant force of 5.00×10^3 N on a ship moving at constant speed through a harbor. How much work does the tugboat do on the ship if each moves a distance of 3.00 km?

4. A shopper in a supermarket pushes a cart with a force of 35 N directed at an angle of 25° downward from the horizontal. Find the work done by the shopper as she moves down a 50-m length of aisle.

5. Starting from rest, a 5.00-kg block slides 2.50 m down a rough 30.0° incline. The coefficient of kinetic friction between the block and the incline is $\mu_k = 0.436$. Determine (a) the work done by the force of gravity, (b) the work done by the friction force between block and incline, and (c) the work done by the normal force.

6. A horizontal force of 150 N is used to push a 40.0-kg packing crate a distance of 6.00 m on a rough horizontal surface. If the crate moves at constant speed, find (a) the work done by the 150-N force and (b) the coefficient of kinetic friction between the crate and surface.

7. A sledge loaded with bricks has a total mass of 18.0 kg and is pulled at constant speed by a rope inclined at 20.0° above the horizontal. The sledge moves a distance of 20.0 m on a horizontal surface. The coefficient of kinetic friction between the sledge and surface is 0.500. (a) What is the tension in the rope? (b) How much work is done by the rope on the sledge? (c) What is the mechanical energy lost due to friction?

8. A block of mass 2.50 kg is pushed 2.20 m along a frictionless horizontal table by a constant 16.0-N force directed 25.0° below the horizontal. Determine the work done by (a) the applied force, (b) the normal force exerted by the table, (c) the force of gravity, and (d) the net force on the block.

Section 5.2 Kinetic Energy and the WorK – Energy Theorem

9. A mechanic pushes a 2.50×10^3-kg car from rest to a speed of v, doing 5 000 J of work in the process. During this time, the car moves 25.0 m. Neglecting friction between car and road, find (a) v and (b) the horizontal force exerted on the car.

10. A 7.00-kg bowling ball moves at 3.00 m/s. How fast must a 2.45-g Ping-Pong ball move so that the two balls have the same kinetic energy?

11. A person doing a chin-up weighs 700 N, exclusive of the arms. During the first 25.0 cm of the lift, each arm exerts an upward force of 355 N on the torso. If the upward movement starts from rest, what is the person's velocity at that point?

12. A crate of mass 10.0 kg is pulled up a rough incline with an initial speed of 1.50 m/s. The pulling force is 100 N parallel to the incline, which makes an angle of 20.0° with the horizontal. The coefficient of kinetic friction is 0.400, and the crate is pulled 5.00 m. (a) How much work is done by gravity? (b) How much mechanical energy is lost due to friction? (c) How much work is done by the 100-N force? (d) What is the change in kinetic energy of the crate? (e) What is the speed of the crate after being pulled 5.00 m?

13. A 70-kg base runner begins his slide into second base when he is moving at a speed of 4.0 m/s. The coefficient of friction between his clothes and Earth is 0.70. He slides so that his speed is zero just as he reaches the base. (a) How much mechanical energy is lost due to friction acting on the runner? (b) How far does he slide?

14. An outfielder throws a 0.150-kg baseball at a speed of 40.0 m/s and an initial angle of 30.0°. What is the kinetic energy of the ball at the highest point of its motion?

15. A 2.0-g bullet leaves the barrel of a gun at a speed of 300 m/s. (a) Find its kinetic energy. (b) Find the average force exerted by the expanding gases on the bullet as it moves the length of the 50-cm-long barrel.

16. A 0.60-kg particle has a speed of 2.0 m/s at point *A* and a kinetic energy of 7.5 J at point *B*. What is (a) its kinetic energy at *A*? (b) its speed at point *B*? (c) the total work done on the particle as it moves from *A* to *B*?

17. **Physics⊗Now**™ A 2 000-kg car moves down a level highway under the actions of two forces: a 1 000-N forward force exerted on the drive wheels by the road and a 950-N resistive force. Use the work–energy theorem to find the speed of the car after it has moved a distance of 20 m, assuming that it starts from rest.

18. On a frozen pond, a 10-kg sled is given a kick that imparts to it an initial speed of $v_0 = 2.0$ m/s. The coefficient of kinetic friction between sled and ice is $\mu_k = 0.10$. Use the work–energy theorem to find the distance the sled moves before coming to rest.

Section 5.3 Gravitational Potential Energy
Section 5.4 Spring Potential Energy

19. Find the height from which you would have to drop a ball so that it would have a speed of 9.0 m/s just before it hits the ground.

20. A flea is able to jump about 0.5 m. It has been said that if a flea were as big as a human, it would be able to jump over a 100-story building! When an animal jumps, it converts work done in contracting muscles into gravitational potential energy (with some steps in between). The maximum force exerted by a muscle is proportional to its cross-sectional area, and the work done by the muscle is this force times the length of contraction. If we magnified a flea by a factor of 1 000, the cross section of its muscle would increase by $1\,000^2$ and the length of contraction would increase by 1 000. How high would this "superflea" be able to jump? (Don't forget that the mass of the "superflea" increases as well.)

21. An athlete on a trampoline leaps straight up into the air with an initial speed of 9.0 m/s. Find (a) the maximum height reached by the athlete relative to the trampoline and (b) the speed of the athlete when she is halfway up to her maximum height.

22. Truck suspensions often have "helper springs" that engage at high loads. One such arrangement is a leaf spring with a helper coil spring mounted on the axle, as shown in Figure P5.22. When the main leaf spring is compressed by distance y_0, the helper spring engages and then helps to support any additional load. Suppose the leaf spring constant is 5.25×10^5 N/m, the helper spring constant is 3.60×10^5 N/m, and $y_0 = 0.500$ m. (a) What is the compression of the leaf spring for a load of 5.00×10^5 N? (b) How much work is done in compressing the springs?

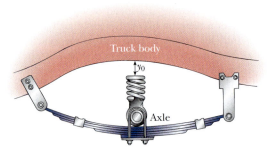

Figure P5.22

23. A daredevil on a motorcycle leaves the end of a ramp with a speed of 35.0 m/s as in Figure P5.23. If his speed is 33.0 m/s when he reaches the peak of the path, what is the maximum height that he reaches? Ignore friction and air resistance.

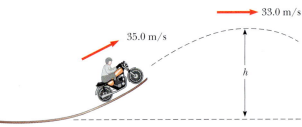

Figure P5.23

24. A softball pitcher rotates a 0.250-kg ball around a vertical circular path of radius 0.600 m before releasing it. The pitcher exerts a 30.0-N force directed parallel to the motion of the ball around the complete circular path. The speed of the ball at the top of the circle is 15.0 m/s. If the ball is released at the bottom of the circle, what is its speed upon release?

25. The chin-up is one exercise that can be used to strengthen the biceps muscle. This muscle can exert a force of approximately 800 N as it contracts a distance of 7.5 cm in a 75-kg male[3]. How much work can the biceps muscles (one in each arm) perform in a single contraction? Compare this amount of work with the energy required to lift a 75-kg person 40 cm in performing a chin-up. Do you think the biceps muscle is the only muscle involved in performing a chin-up?

Section 5.5 Systems and Energy Conservation

26. A 50-kg pole vaulter running at 10 m/s vaults over the bar. Her speed when she is above the bar is 1.0 m/s. Neglect

air resistance, as well as any energy absorbed by the pole, and determine her altitude as she crosses the bar.

27. A child and a sled with a combined mass of 50.0 kg slide down a frictionless slope. If the sled starts from rest and has a speed of 3.00 m/s at the bottom, what is the height of the hill?

28. A 0.400-kg bead slides on a curved wire, starting from rest at point Ⓐ in Figure P5.28. If the wire is frictionless, find the speed of the bead (a) at Ⓑ and (b) at Ⓒ.

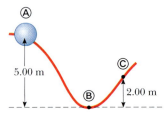

Figure P5.28 (Problems 28 and 36)

29. A 5.00-kg steel ball is dropped onto a copper plate from a height of 10.0 m. If the ball leaves a dent 3.20 mm deep in the plate, what is the average force exerted by the plate on the ball during the impact?

30. A bead of mass $m = 5.00$ kg is released from point Ⓐ and slides on the frictionless track shown in Figure P5.30. Determine (a) the bead's speed at points Ⓑ and Ⓒ and (b) the net work done by the force of gravity in moving the bead from Ⓐ to Ⓒ.

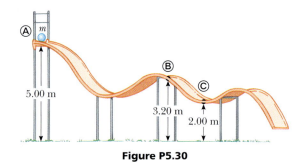

Figure P5.30

31. Tarzan swings on a 30.0-m-long vine initially inclined at an angle of 37.0° with the vertical. What is his speed at the bottom of the swing (a) if he starts from rest? (b) if he pushes off with a speed of 4.00 m/s?

32. Three objects with masses $m_1 = 5.0$ kg, $m_2 = 10$ kg, and $m_3 = 15$ kg, respectively, are attached by strings over frictionless pulleys, as indicated in Figure P5.32. The horizontal surface is frictionless and the system is released from rest. Using energy concepts, find the speed of m_3 after it moves down a distance of 4.0 m.

Figure P5.32 (Problems 32 and 89)

---

[3]G. P. Pappas et.al., "Nonuniform shortening in the biceps brachii during elbow flexion," *Journal of Applied Physiology* **92**, 2381, 2002.

33. **Physics Now™** The launching mechanism of a toy gun consists of a spring of unknown spring constant, as shown in Figure P5.33a. If the spring is compressed a distance of 0.120 m and the gun fired vertically as shown, the gun can launch a 20.0-g projectile from rest to a maximum height of 20.0 m above the starting point of the projectile. Neglecting all resistive forces, determine (a) the spring constant and (b) the speed of the projectile as it moves through the equilibrium position of the spring (where $x = 0$), as shown in Figure P5.33b.

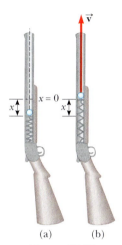

(a)   (b)

Figure P5.33

34. A projectile is launched with a speed of 40 m/s at an angle of 60° above the horizontal. Use conservation of energy to find the maximum height reached by the projectile during its flight.

35. A 0.250-kg block is placed on a light vertical spring ($k = 5.00 \times 10^3$ N/m) and pushed downwards, compressing the spring 0.100 m. After the block is released, it leaves the spring and continues to travel upwards. What height above the point of release will the block reach if air resistance is negligible?

36. The wire in Problem 28 (Fig. P5.28) is frictionless between points Ⓐ and Ⓑ and rough between Ⓑ and Ⓒ. The 0.400-kg bead starts from rest at Ⓐ. (a) Find its speed at Ⓑ. (b) If the bead comes to rest at Ⓒ, find the loss in mechanical energy as it goes from Ⓑ to Ⓒ.

37. (a) A child slides down a water slide at an amusement park from an initial height h. The slide can be considered frictionless because of the water flowing down it. Can the equation for conservation of mechanical energy be used on the child? (b) Is the mass of the child a factor in determining his speed at the bottom of the slide? (c) The child drops straight down rather than following the curved ramp of the slide. In which case will he be traveling faster at ground level? (d) If friction is present, how would the conservation-of-energy equation be modified? (e) Find the maximum speed of the child when the slide is frictionless if the initial height of the slide is 12.0 m.

38. (a) A block with a mass m is pulled along a horizontal surface for a distance x by a constant force $\vec{F}$ at an angle θ with respect to the horizontal. The coefficient of kinetic friction between block and table is μ_k. Is the force exerted by friction equal to $\mu_k mg$? If not, what is the force exerted by friction? (b) How much work is done by the friction force and by $\vec{F}$? (Don't forget the signs.)

(c) Identify all the forces that do no work on the block. (d) Let $m = 2.00$ kg, $x = 4.00$ m, $\theta = 37.0°$, $F = 15.0$ N, and $\mu_k = 0.400$, and find the answers to parts (a) and (b).

39. A 70-kg diver steps off a 10-m tower and drops from rest straight down into the water. If he comes to rest 5.0 m beneath the surface, determine the average resistive force exerted on him by the water.

40. An airplane of mass 1.5×10^4 kg is moving at 60 m/s. The pilot then revs up the engine so that the forward thrust by the air around the propeller becomes 7.5×10^4 N. If the force exerted by air resistance on the body of the airplane has a magnitude of 4.0×10^4 N, find the speed of the airplane after it has traveled 500 m. Assume that the airplane is in level flight throughout this motion.

41. A 2.1×10^3-kg car starts from rest at the top of a 5.0-m-long driveway that is inclined at 20° with the horizontal. If an average friction force of 4.0×10^3 N impedes the motion, find the speed of the car at the bottom of the driveway.

42. A 25.0-kg child on a 2.00-m-long swing is released from rest when the ropes of the swing make an angle of 30.0° with the vertical. (a) Neglecting friction, find the child's speed at the lowest position. (b) If the actual speed of the child at the lowest position is 2.00 m/s, what is the mechanical energy lost due to friction?

43. Starting from rest, a 10.0-kg block slides 3.00 m down to the bottom of a frictionless ramp inclined 30.0° from the floor. The block then slides an additional 5.00 m along the floor before coming to a stop. Determine (a) the speed of the block at the bottom of the ramp, (b) the coefficient of kinetic friction between block and floor, and (c) the mechanical energy lost due to friction.

44. A child slides without friction from a height h along a curved water slide (Fig. P5.44). She is launched from a height $h/5$ into the pool. Determine her maximum airborne height y in terms of h and the launch angle θ.

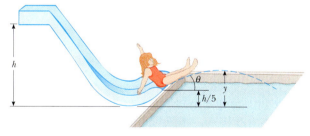

Figure P5.44

45. **Physics Now™** A skier starts from rest at the top of a hill that is inclined 10.5° with respect to the horizontal. The hillside is 200 m long, and the coefficient of friction between snow and skis is 0.075 0. At the bottom of the hill, the snow is level and the coefficient of friction is unchanged. How far does the skier glide along the horizontal portion of the snow before coming to rest?

46. In a circus performance, a monkey is strapped to a sled and both are given an initial speed of 4.0 m/s up a 20° inclined track. The combined mass of monkey and sled is 20 kg, and the coefficient of kinetic friction between sled and incline is 0.20. How far up the incline do the monkey and sled move?

47. An 80.0-kg skydiver jumps out of a balloon at an altitude of 1 000 m and opens the parachute at an altitude of

200.0 m. (a) Assuming that the total retarding force on the diver is constant at 50.0 N with the parachute closed and constant at 3 600 N with the parachute open, what is the speed of the diver when he lands on the ground? (b) Do you think the skydiver will get hurt? Explain. (c) At what height should the parachute be opened so that the final speed of the skydiver when he hits the ground is 5.00 m/s? (d) How realistic is the assumption that the total retarding force is constant? Explain.

Section 5.6 Power

48. A skier of mass 70 kg is pulled up a slope by a motor-driven cable. (a) How much work is required to pull him 60 m up a 30° slope (assumed frictionless) at a constant speed of 2.0 m/s? (b) What power must a motor have to perform this task?

49. Columnist Dave Barry poked fun at the name "The Grand Cities," adopted by Grand Forks, North Dakota, and East Grand Forks, Minnesota. Residents of the prairie towns then named a sewage pumping station for him. At the Dave Barry Lift Station No. 16, untreated sewage is raised vertically by 5.49 m in the amount of 1 890 000 liters each day. With a density of 1 050 kg/m^3, the waste enters and leaves the pump at atmospheric pressure through pipes of equal diameter. (a) Find the output power of the lift station. (b) Assume that a continuously operating electric motor with average power 5.90 kW runs the pump. Find its efficiency. In January 2002, Barry attended the outdoor dedication of the lift station and a festive potluck supper to which the residents of the different Grand Forks sewer districts brought casseroles, Jell-O® salads, and "bars" (desserts).

50. While running, a person dissipates about 0.60 J of mechanical energy per step per kilogram of body mass. If a 60-kg person develops a power of 70 W during a race, how fast is the person running? (Assume a running step is 1.5 m long.)

51. The electric motor of a model train accelerates the train from rest to 0.620 m/s in 21.0 ms. The total mass of the train is 875 g. Find the average power delivered to the train during its acceleration.

52. An electric scooter has a battery capable of supplying 120 Wh of energy. [Note that an energy of 1 Wh = (1 J/s)(3600 s) = 3600 J] If frictional forces and other losses account for 60.0% of the energy usage, what change in altitude can a rider achieve when driving in hilly terrain if the rider and scooter have a combined weight of 890 N?

53. A 1.50 × 10^3-kg car starts from rest and accelerates uniformly to 18.0 m/s in 12.0 s. Assume that air resistance remains constant at 400 N during this time. Find (a) the average power developed by the engine and (b) the instantaneous power output of the engine at t = 12.0 s, just before the car stops accelerating.

54. A 650-kg elevator starts from rest and moves upwards for 3.00 s with constant acceleration until it reaches its cruising speed, 1.75 m/s. (a) What is the average power of the elevator motor during this period? (b) How does this amount of power compare with its power during an upward trip with constant speed?

Section 5.7 Work Done by a Varying Force

55. **Physics**⊗**Now**™ The force acting on a particle varies as in Figure P5.55. Find the work done by the force as the particle moves (a) from x = 0 to x = 8.00 m,

(b) from x = 8.00 m to x = 10.0 m, and (c) from x = 0 to x = 10.0 m.

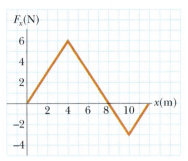

Figure P5.55

56. An object is subject to a force F_x that varies with position as in Figure P5.56. Find the work done by the force on the object as it moves (a) from x = 0 to x = 5.00 m, (b) from x = 5.00 m to x = 10.0 m, and (c) from x = 10.0 m to x = 15.0 m. (d) What is the total work done by the force over the distance x = 0 to x = 15.0 m?

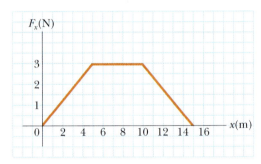

Figure P5.56

57. The force acting on an object is given by F_x = (8x − 16) N, where x is in meters. (a) Make a plot of this force versus x from x = 0 to x = 3.00 m. (b) From your graph, find the net work done by the force as the object moves from x = 0 to x = 3.00 m.

ADDITIONAL PROBLEMS

58. A 2.0-m-long pendulum is released from rest when the support string is at an angle of 25° with the vertical. What is the speed of the bob at the bottom of the swing?

59. An archer pulls her bowstring back 0.400 m by exerting a force that increases uniformly from zero to 230 N. (a) What is the equivalent spring constant of the bow? (b) How much work does the archer do in pulling the bow?

60. A block of mass 12.0 kg slides from rest down a frictionless 35.0° incline and is stopped by a strong spring with k = 3.00 × 10^4 N/m. The block slides 3.00 m from the point of release to the point where it comes to rest against the spring. When the block comes to rest, how far has the spring been compressed?

61. (a) A 75-kg man steps out a window and falls (from rest) 1.0 m to a sidewalk. What is his speed just before his feet strike the pavement? (b) If the man falls with his knees and ankles locked, the only cushion for his fall is an approximately 0.50-cm give in the pads of his feet. Calculate the average force exerted on him by the ground in this sit-

uation. This average force is sufficient to cause damage to cartilage in the joints or to break bones.

62. A toy gun uses a spring to project a 5.3-g soft rubber sphere horizontally. The spring constant is 8.0 N/m, the barrel of the gun is 15 cm long, and a constant frictional force of 0.032 N exists between barrel and projectile. With what speed does the projectile leave the barrel if the spring was compressed 5.0 cm for this launch?

63. Two objects are connected by a light string passing over a light, frictionless pulley as in Figure P5.63. The 5.00-kg object is released from rest at a point 4.00 m above the floor. (a) Determine the speed of each object when the two pass each other. (b) Determine the speed of each object at the moment the 5.00-kg object hits the floor. (c) How much higher does the 3.00-kg object travel after the 5.00-kg object hits the floor?

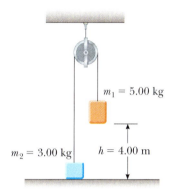

Figure P5.63

64. Two blocks, *A* and *B* (with mass 50 kg and 100 kg, respectively), are connected by a string, as shown in Figure P5.64. The pulley is frictionless and of negligible mass. The coefficient of kinetic friction between block *A* and the incline is $\mu_k = 0.25$. Determine the change in the kinetic energy of block *A* as it moves from Ⓒ to Ⓓ, a distance of 20 m up the incline if the system starts from rest.

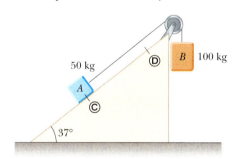

Figure P5.64

65. A 200-g particle is released from rest at point *A* on the inside of a smooth hemispherical bowl of radius $R = 30.0$ cm

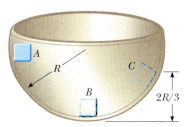

Figure P5.65

(Fig. P5.65). Calculate (a) its gravitational potential energy at *A* relative to *B*, (b) its kinetic energy at *B*, (c) its speed at *B*, (d) its potential energy at *C* relative to *B*, and (e) its kinetic energy at *C*.

66. Energy is conventionally measured in Calories as well as in joules. One Calorie in nutrition is 1 kilocalorie, which we define in Chapter 11 as 1 kcal = 4 186 J. Metabolizing 1 gram of fat can release 9.00 kcal. A student decides to try to lose weight by exercising. She plans to run up and down the stairs in a football stadium as fast as she can and as many times as necessary. Is this in itself a practical way to lose weight? To evaluate the program, suppose she runs up a flight of 80 steps, each 0.150 m high, in 65.0 s. For simplicity, ignore the energy she uses in coming down (which is small). Assume that a typical efficiency for human muscles is 20.0%. This means that when your body converts 100 J from metabolizing fat, 20 J goes into doing mechanical work (here, climbing stairs). The remainder goes into internal energy. Assume the student's mass is 50.0 kg. (a) How many times must she run the flight of stairs to lose 1 pound of fat? (b) What is her average power output, in watts and in horsepower, as she is running up the stairs?

67. In terms of saving energy, bicycling and walking are far more efficient means of transportation than is travel by automobile. For example, when riding at 10.0 mi/h, a cyclist uses food energy at a rate of about 400 kcal/h above what he would use if he were merely sitting still. (In exercise physiology, power is often measured in kcal/h rather than in watts. Here, 1 kcal = 1 nutritionist's Calorie = 4 186 J.) Walking at 3.00 mi/h requires about 220 kcal/h. It is interesting to compare these values with the energy consumption required for travel by car. Gasoline yields about 1.30×10^8 J/gal. Find the fuel economy in equivalent miles per gallon for a person (a) walking and (b) bicycling.

68. An 80.0-N box is pulled 20.0 m up a 30° incline by an applied force of 100 N that points upwards, parallel to the incline. If the coefficient of kinetic friction between box and incline is 0.220, calculate the change in the kinetic energy of the box.

69. A ski jumper starts from rest 50.0 m above the ground on a frictionless track and flies off the track at an angle of 45.0° above the horizontal and at a height of 10.0 m above the level ground. Neglect air resistance. (a) What is her speed when she leaves the track? (b) What is the maximum altitude she attains after leaving the track? (c) Where does she land relative to the end of the track?

70. A 5.0-kg block is pushed 3.0 m up a vertical wall with constant speed by a constant force of magnitude *F* applied at an angle of $\theta = 30°$ with the horizontal, as shown in Figure P5.70. If the coefficient of kinetic friction between

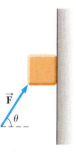

Figure P5.70

block and wall is 0.30, determine the work done by (a) $\vec{F}$, (b) the force of gravity, and (c) the normal force between block and wall. (d) By how much does the gravitational potential energy increase during the block's motion?

71. The ball launcher in a pinball machine has a spring with a force constant of 1.20 N/cm (Fig. P5.71). The surface on which the ball moves is inclined 10.0° with respect to the horizontal. If the spring is initially compressed 5.00 cm, find the launching speed of a 0.100-kg ball when the plunger is released. Friction and the mass of the plunger are negligible.

Figure P5.71

72. The masses of the javelin, discus, and shot are 0.80 kg, 2.0 kg, and 7.2 kg, respectively, and record throws in the corresponding track events are about 98 m, 74 m, and 23 m, respectively. Neglecting air resistance, (a) calculate the minimum initial kinetic energies that would produce these throws, and (b) estimate the average force exerted on each object during the throw, assuming the force acts over a distance of 2.0 m. (c) Do your results suggest that air resistance is an important factor?

73. Jane, whose mass is 50.0 kg, needs to swing across a river filled with crocodiles in order to rescue Tarzan, whose mass is 80.0 kg. However, she must swing into a *constant* horizontal wind force $\vec{F}$ on a vine that is initially at an angle of θ with the vertical. (See Fig. P5.73.) In the figure, $D = 50.0$ m, $F = 110$ N, $L = 40.0$ m, and $\theta = 50.0°$. (a) With what minimum speed must Jane begin her swing in order to just make it to the other side? (*Hint:* First determine the potential energy that can be associated with the wind force. Because the wind force is constant, use an analogy with the constant gravitational force.) (b) Once the rescue is complete, Tarzan and Jane must swing back across the river. With what minimum speed must they begin their swing?

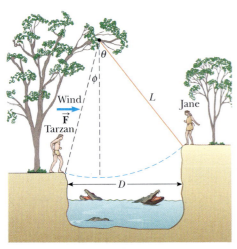

Figure P5.73

74. A hummingbird is able to hover because, as the wings move downwards, they exert a downward force on the air. Newton's third law tells us that the air exerts an equal and opposite force (upwards) on the wings. The average of this force must be equal to the weight of the bird when it hovers. If the wings move through a distance of 3.5 cm with each stroke, and the wings beat 80 times per second, determine the work performed by the wings on the air in 1 minute if the mass of the hummingbird is 3.0 grams.

75. A child's pogo stick (Fig. P5.75) stores energy in a spring $(k = 2.50 \times 10^4$ N/m). At position Ⓐ $(x_1 = -0.100$ m), the spring compression is a maximum and the child is momentarily at rest. At position Ⓑ $(x = 0)$, the spring is relaxed and the child is moving upwards. At position Ⓒ, the child is again momentarily at rest at the top of the jump. Assuming that the combined mass of child and pogo stick is 25.0 kg, (a) calculate the total energy of the system if both potential energies are zero at $x = 0$, (b) determine x_2, (c) calculate the speed of the child at $x = 0$, (d) determine the value of x for which the kinetic energy of the system is a maximum, and (e) obtain the child's maximum upward speed.

Figure P5.75

76. A 2.00-kg block situated on a rough incline is connected to a spring of negligible mass having a spring constant of 100 N/m (Fig. P5.76). The block is released from rest when the spring is unstretched, and the pulley is frictionless. The block moves 20.0 cm down the incline before coming to rest. Find the coefficient of kinetic friction between block and incline.

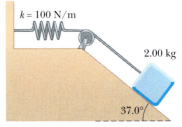

Figure P5.76

77. In the dangerous "sport" of bungee jumping, a daring student jumps from a hot-air balloon with a specially designed elastic cord attached to his waist, as shown in Figure P5.77. The unstretched length of the cord is 25.0 m, the student weighs 700 N, and the balloon is 36.0 m above the surface of a river below. Calculate the required force constant of the cord if the student is to stop safely 4.00 m above the river.

Figure P5.77 Bungee jumping. (Problems 77 and 82)

78. An object of mass m is suspended from the top of a cart by a string of length L as in Figure P5.78a. The cart and object are initially moving to the right at a constant speed v_0. The cart comes to rest after colliding and sticking to a bumper, as in Figure P5.78b, and the suspended object swings through an angle θ. (a) Show that the initial speed is $v_0 = \sqrt{2gL(1 - \cos\theta)}$. (b) If $L = 1.20$ m and $\theta = 35.0°$, find the initial speed of the cart. (*Hint:* The force exerted by the string on the object does no work on the object.)

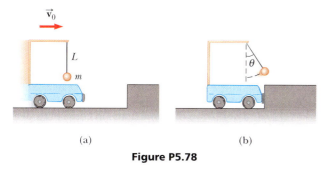

(a)   (b)

Figure P5.78

79. A truck travels uphill with constant velocity on a highway with a 7.0° slope. A 50-kg package sits on the floor of the back of the truck and does not slide, due to a static frictional force. During an interval in which the truck travels 340 m, what is the net work done on the package? What is the work done on the package by the force of gravity, the normal force, and the friction force?

80. As part of a curriculum unit on earthquakes, suppose that 375 000 British schoolchildren stand on their chairs and simultaneously jump down to the floor. Seismographers around the country see whether they can detect the resulting ground tremor. (This experiment was actually based on a suggestion by the children themselves.) (a) Find the energy released in the experiment. Model the children as having average mass 36.0 kg and as stepping from chair seats 38.0 cm above the floor. (b) Most of the energy is converted very rapidly into internal energy within the bodies of the children and the floors of the school buildings. Assume that 1% of the energy is carried away by a seismic wave. The magnitude of an earthquake on the Richter scale is given by

$$M = \frac{\log E - 4.8}{1.5}$$

where E is the seismic wave energy in joules. According to this model, what is the magnitude of the demonstration quake?

81. A loaded ore car has a mass of 950 kg and rolls on rails with negligible friction. It starts from rest and is pulled up a mine shaft by a cable connected to a winch. The shaft is inclined at 30.0° above the horizontal. The car accelerates uniformly to a speed of 2.20 m/s in 12.0 s and then continues at constant speed. (a) What power must the winch motor provide when the car is moving at constant speed? (b) What maximum power must the motor provide? (c) What total energy transfers out of the motor by work by the time the car moves off the end of the track, which is of length 1 250 m?

82. A daredevil wishes to bungee-jump from a hot-air balloon 65.0 m above a carnival midway (Fig. P5.77). He will use a piece of uniform elastic cord tied to a harness around his body to stop his fall at a point 10.0 m above the ground. Model his body as a particle and the cord as having negligible mass and a tension force described by Hooke's force law. In a preliminary test, hanging at rest from a 5.00-m length of the cord, the jumper finds that his body weight stretches it by 1.50 m. He will drop from rest at the point where the top end of a longer section of the cord is attached to the stationary balloon. (a) What length of cord should he use? (b) What maximum acceleration will he experience?

83. The system shown in Figure P5.83 consists of a light, inextensible cord, light frictionless pulleys, and blocks of equal mass. Initially, the blocks are at rest the same height above the ground. The blocks are then released. Find the speed of block A at the moment when the vertical separation of the blocks is h.

Figure P5.83

84. A cafeteria tray dispenser supports a stack of trays on a shelf that hangs from four identical spiral springs under tension, one near each corner of the shelf. Each tray has a mass of 580 g and is rectangular, 45.3 cm by 35.6 cm, and 0.450 cm thick. (a) Show that the top tray in the stack can always be at the same height above the floor, however many trays are in the dispenser. (b) Find the spring constant each spring should have in order for the dispenser to function in this convenient way. Is any piece of data unnecessary for this determination?

85. In bicycling for aerobic exercise, a woman wants her heart rate to be between 136 and 166 beats per minute. Assume that her heart rate is directly proportional to her mechanical power output. Ignore all forces on the woman-plus-bicycle system, except for static friction forward on the drive wheel of the bicycle and an air resistance force proportional to the square of the bicycler's speed. When her speed is 22.0 km/h, her heart rate is 90.0 beats per minute. In what range should her speed be so that her heart rate will be in the range she wants?

86. In a needle biopsy, a narrow strip of tissue is extracted from a patient with a hollow needle. Rather than being pushed by hand, to ensure a clean cut the needle can be fired into the patient's body by a spring. Assume the needle has mass 5.60 g, the light spring has force constant 375 N/m, and the spring is originally compressed 8.10 cm to project the needle horizontally without friction. The tip of the needle then moves through 2.40 cm of skin and soft tissue, which exerts a resistive force of 7.60 N on it. Next, the needle cuts 3.50 cm into an organ, which exerts a backward force of 9.20 N on it. Find (a) the maximum speed of the needle and (b) the speed at which a flange on the back end of the needle runs into a stop, set to limit the penetration to 5.90 cm.

87. The power of sunlight reaching each square meter of the Earth's surface on a clear day in the tropics is close to 1 000 W. On a winter day in Manitoba, the power concentration of sunlight can be 100 W/m². Many human activities are described by a power-per-footprint-area on the order of 10^2 W/m² or less. (a) Consider, for example, a family of four paying $80 to the electric company every 30 days for 600 kWh of energy carried by electric transmission to their house, with floor area 13.0 m by 9.50 m. Compute the power-per-area measure of this energy use. (b) Consider a car 2.10 m wide and 4.90 m long traveling at 55.0 mi/h using gasoline having a "heat of combustion" of 44.0 MJ/kg with fuel economy 25.0 mi/gallon. One gallon of gasoline has a mass of 2.54 kg. Find the power-per-area measure of the car's energy use. It can be similar to that of a steel mill where rocks are melted in blast furnaces. (c) Explain why the direct use of solar energy is not practical for a conventional automobile.

88. In 1887 in Bridgeport, Connecticut, C. J. Belknap built the water slide shown in Figure P5.88. A rider on a small sled, of total mass 80.0 kg, pushed off to start at the top of the slide (point Ⓐ) with a speed of 2.50 m/s. The chute was 9.76 m high at the top, 54.3 m long, and 0.51 m wide. Along its length, 725 wheels made friction negligible. Upon leaving the chute horizontally at its bottom end (point Ⓒ), the rider skimmed across the water of Long Island Sound for as much as 50 m, "skipping along like a flat pebble," before at last coming to rest and swimming ashore, pulling his sled after him. (a) Find the speed of the sled and rider at point Ⓒ. (b) Model the force of water friction as a constant retarding force acting on a particle. Find the work done by water friction in stopping the sled and rider. (c) Find the magnitude of the force the water exerts on the sled. (d) Find the magnitude of the force the chute exerts on the sled at point Ⓑ.

Engraving from *Scientific American*, July 1888.

9.76 m — Ⓐ — Ⓑ — Ⓒ — 54.3 m — 50.0 m

Figure P5.88

89. Three objects with masses $m_1 = 5.0$ kg, $m_2 = 10$ kg, and $m_3 = 15$ kg, respectively, are attached by strings over frictionless pulleys as indicated in Figure P5.32. The horizontal surface exerts a force of friction of 30 N on m_2. If the system is released from rest, use energy concepts to find the speed of m_3 after it moves down 4.0 m.

ACTIVITIES

A.1. Suspend a rubber band from a support and borrow some weights from your instructor to measure the rubber band's spring constant for small extensions. Calculate how much elastic potential energy is stored in the rubber band for a given extension. Use conservation of energy to predict how high a paper wad will go into the air when released from a given extension of the band. Try it to test your prediction.

A.2. Wrap a rubber band tightly around a tennis ball. Now fasten one end of a string through the rubber band and the other end to the top of a doorframe to construct a pendulum. Pull the pendulum to the side at a variety of angles to observe that the energy of the pendulum–Earth system is always conserved as the pendulum swings (almost) to the same height of its arc as the height from which it was released. (The word "almost" in the last sentence applies because some energy is lost to friction at the point of support and to air resistance. You can observe this slight loss of energy by pulling the ball to the side and letting it go from a point about half an inch from your chin. Let it go—but don't push it!—and test your belief in conservation of energy by seeing if you can avoid flinching when the ball swings back toward your chin.)

A.3. While you have your pendulum from the last activity set up, predict what will happen in the following situation and then test your guess: When a pendulum is released from a given height, it swings to the same height at the other end of its arc as you noted above. However, suppose you place a meterstick across the door opening such that the pendulum string strikes the stick about halfway up the string when it moves through the opening. How high will the ball swing in this case? Will it return to the same height as that at which it started, swing to a lower height, or swing to a greater height? Explain your answer.

A.4. Measure your pulse rate while at rest. Now slowly walk up a flight of stairs and measure your pulse rate at the top of the stairs. Repeat this activity, starting with about the same rest pulse rate at the bottom of the stairs. This time, run up the stairs. Based on your pulse rate readings, what can you conclude about the amount of work and power expended in each case? Repeat this experiment with a series of 10 push-ups.

A.5. Many fitness centers have stepper machines that enable a person to climb continuously without actually moving, because the steps move downwards as the person climbs.

The work that the climber performs on the step is determined by the force exerted on the step times the distance the step moves. Since the net force on the climber is zero, the force exerted on the step must equal the climber's weight. A reasonably strenuous workout on this machine is 90 steps per minute, with each step being 8 inches (15.2 cm) high. What is the rate (in watts) at which a 130-lb (60 kg) climber does work on the stair steps? The energy actually expended by the climber is approximately five times the work done. (You may notice that a lot of heat is generated!) The machines usually report this rate in Calories/hour (1 Calorie = 1 kcal = 4186 J). Determine the rate, in Cal/h, at which energy is expended by the 130-lb climber.

Activity: If you have access to a stepper, find the ratio used by the manufacturer to determine the energy expended from the work performed. In some cases, this ratio may vary as the step speed changes. If so, generate a graph of the ratio as a function of step speed.

A small buck from the massive bull transfers a large amount of momentum to the cowboy, resulting in an involuntary dismount.

© Reuters/Corbis

CHAPTER

6

Momentum and Collisions

What happens when two automobiles collide? How does the impact affect the motion of each vehicle, and what basic physical principles determine the likelihood of serious injury? How do rockets work, and what mechanisms can be used to overcome the limitations imposed by exhaust speed? Why do we have to brace ourselves when firing small projectiles at high velocity? Finally, how can we use physics to improve our golf game?

To begin answering such questions, we introduce *momentum*. Intuitively, anyone or anything that has a lot of momentum is going to be hard to stop. In politics, the term is metaphorical. Physically, the more momentum an object has, the more force has to be applied to stop it in a given time. This concept leads to one of the most powerful principles in physics: *conservation of momentum*. Using this law, complex collision problems can be solved without knowing much about the forces involved during contact. We'll also be able to derive information about the average force delivered in an impact. With conservation of momentum, we'll have a better understanding of what choices to make when designing an automobile or a moon rocket, or when addressing a golf ball on a tee.

6.1 MOMENTUM AND IMPULSE

In physics, momentum has a precise definition. A slowly moving brontosaurus has a lot of momentum, but so does a little hot lead shot from the muzzle of a gun. We therefore expect that momentum will depend on an object's mass and velocity.

Linear momentum ▶

The linear momentum $\vec{\mathbf{p}}$ of an object of mass m moving with velocity $\vec{\mathbf{v}}$ is the product of its mass and velocity :

$$\vec{\mathbf{p}} \equiv \vec{\mathbf{v}} \qquad\qquad [6.1]$$

SI unit: kilogram-meter per second (kg · m/s)

Doubling either the mass or the velocity of an object doubles its momentum; doubling both quantities quadruples its momentum. Momentum is a vector quantity with the same direction as the object's velocity. Its components are given in two dimensions by

$$p_x = mv_x \qquad p_y = mv_y$$

where p_x is the momentum of the object in the x-direction and p_y its momentum in the y-direction.

Quick Quiz 6.1

Two objects with masses m_1 and m_2 have equal kinetic energy. How do the magnitudes of their momenta compare? (a) $p_1 < p_2$ (b) $p_1 = p_2$ (c) $p_1 > p_2$ (d) not enough information is given

Changing the momentum of an object requires the application of a force. This is, in fact, how Newton originally stated his second law of motion. Starting from the more common version of the second law, we have

$$\vec{\mathbf{F}}_{net} = m\vec{\mathbf{a}} = m\frac{\Delta \vec{\mathbf{v}}}{\Delta t} = \frac{\Delta(m\vec{\mathbf{v}})}{\Delta t} \qquad [6.2]$$

where the mass m and the forces are assumed constant. The quantity in parentheses is just the momentum, so we have the following result:

The change in an object's momentum $\Delta \vec{\mathbf{p}}$ divided by the elapsed time Δt equals the constant net force $\vec{\mathbf{F}}_{net}$ acting on the object:

$$\frac{\Delta \vec{\mathbf{p}}}{\Delta t} = \frac{\text{change in momentum}}{\text{time interval}} = \vec{\mathbf{F}}_{net} \qquad [6.3]$$

◀ Newton's second law

This equation is also valid when the forces are not constant, provided the limit is taken as Δt becomes infinitesimally small. Equation 6.3 says that if the net force on an object is zero, the object's momentum doesn't change. In other words, the linear momentum of an object is *conserved* when $\vec{\mathbf{F}}_{net} = 0$. Equation 6.3 also tells us that changing an object's momentum requires the continuous application of a force over a period of time Δt, leading to the definition of *impulse:*

If a constant force $\vec{\mathbf{F}}$ acts on an object, the **impulse $\vec{\mathbf{I}}$** delivered to the object over a time interval Δt is given by

$$\vec{\mathbf{I}} \equiv \vec{\mathbf{F}}\Delta t \qquad [6.4]$$

SI unit: kilogram meter per second (kg · m/s)

Impulse is a vector quantity with the same direction as the constant force acting on the object. When a single constant force $\vec{\mathbf{F}}$ acts on an object, Equation 6.3 can be written as

$$\vec{\mathbf{F}}\Delta t = \Delta \vec{\mathbf{p}} = m\vec{\mathbf{v}}_f - m\vec{\mathbf{v}}_i \qquad [6.5]$$

◀ Impulse–momentum theorem

This is a special case of the **impulse–momentum theorem.** Equation 6.5 shows that **the impulse of the force acting on an object equals the change in momentum of that object.** This equality is true even if the force is not constant, as long as the time interval Δt is taken to be arbitrarily small. (The proof of the general case requires concepts from calculus.)

In real-life situations, the force on an object is only rarely constant. For example, when a bat hits a baseball, the force increases sharply, reaches some maximum value, and then decreases just as rapidly. Figure 6.1(a) shows a typical graph of

Figure 6.1 (a) A force acting on an object may vary in time. The impulse is the area under the force vs. time curve. (b) The average force (horizontal dashed line) gives the same impulse to the object in the time interval Δt as the real time-varying force described in (a).

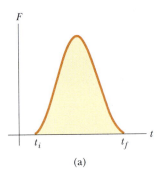

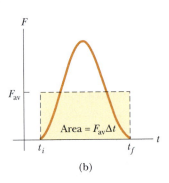

force versus time for such incidents. The force starts out small as the bat comes in contact with the ball, rises to a maximum value when they are firmly in contact, and then drops off as the ball leaves the bat. In order to analyze this rather complex interaction, it's useful to define an **average force** $\vec{\mathbf{F}}_{av}$, shown as the dashed line in Figure 6.1b. This average force is the constant force delivering the same impulse to the object in the time interval Δt as the actual time-varying force. We can then write the impulse–momentum theorem as

$$\vec{\mathbf{F}}_{av}\Delta t = \Delta \vec{\mathbf{p}} \qquad [6.6]$$

The magnitude of the impulse delivered by a force during the time interval Δt is equal to the area under the force vs. time graph as in Figure 6.1a or, equivalently, to $F_{av}\Delta t$ as shown in Figure 6.1b. The brief collision between a bullet and an apple is illustrated in Figure 6.2.

Applying Physics 6.1   Boxing and Brain Injury

In boxing matches of the 19th century, bare fists were used. In modern boxing, fighters wear padded gloves. How do gloves protect the brain of the boxer from injury? Also, why do boxers often "roll with the punch"?

Explanation   The brain is immersed in a cushioning fluid inside the skull. If the head is struck suddenly by a bare fist, the skull accelerates rapidly. The brain matches this acceleration only because of the large impulsive force exerted by the skull on the brain. This large and sudden force (large F_{av} and small Δt) can cause severe brain injury. Padded gloves extend

the time Δt over which the force is applied to the head. For a given impulse $F_{av}\Delta t$, a glove results in a longer time interval than a bare fist, decreasing the average force. Because the average force is decreased, the acceleration of the skull is decreased, reducing (but not eliminating) the chance of brain injury. The same argument can be made for "rolling with the punch": If the head is held steady while being struck, the time interval over which the force is applied is relatively short and the average force is large. If the head is allowed to move in the same direction as the punch, the time interval is lengthened and the average force reduced.

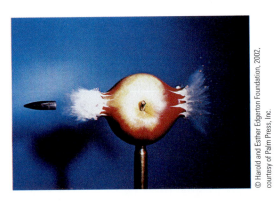

© Harold and Esther Edgerton Foundation, 2002, courtesy of Palm Press, Inc.

Figure 6.2   An apple being pierced by a 30-caliber bullet traveling at a supersonic speed of 900 m/s. This collision was photographed with a microflash stroboscope using an exposure time of 0.33 μs. Shortly after the photograph was taken, the apple disintegrated completely. Note that the points of entry and exit of the bullet are visually explosive.

EXAMPLE 6.1 Teeing Off

Goal Use the impulse–momentum theorem to estimate the average force exerted during an impact.

Problem A golf ball with mass 5.0×10^{-2} kg is struck with a club as in Figure 6.3. The force on the ball varies from zero when contact is made up to some maximum value (when the ball is maximally deformed) and then back to zero when the ball leaves the club, as in the graph of force vs. time in Figure 6.1. Assume that the ball leaves the club face with a velocity of $+44$ m/s. **(a)** Find the magnitude of the impulse due to the collision. **(b)** Estimate the duration of the collision and the average force acting on the ball.

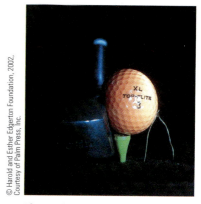

Figure 6.3 (Example 6.1) A golf ball being struck by a club.

© Harold and Esther Edgerton Foundation, 2002, Courtesy of Palm Press, Inc.

Strategy In part (a), use the fact that the impulse is equal to the change in momentum. The mass and the initial and final speeds are known, so this change can be computed. In part (b), the average force is just the change in momentum computed in part (a) divided by an estimate of the duration of the collision. Guess at the distance the ball travels on the face of the club (about 2 cm, roughly the same as the radius of the ball). Divide this distance by the average velocity (half the final velocity) to get an estimate of the time of contact.

Solution

(a) Find the impulse delivered to the ball.

The problem is essentially one dimensional. Note that $v_i = 0$, and calculate the change in momentum, which equals the impulse:

$$I = \Delta p = p_f - p_i = (5.0 \times 10^{-2} \text{ kg})(44 \text{ m/s}) - 0$$

$$= +2.2 \text{ kg} \cdot \text{m/s}$$

(b) Estimate the duration of the collision and the average force acting on the ball.

Estimate the time interval of the collision, Δt, using the approximate displacement (radius of the ball) and its average speed (half the maximum speed):

$$\Delta t = \frac{\Delta x}{v_{av}} = \frac{2.0 \times 10^{-2} \text{ m}}{22 \text{ m/s}} = 9.1 \times 10^{-4} \text{ s}$$

Estimate the average force from Equation 6.6:

$$F_{av} = \frac{\Delta p}{\Delta t} = \frac{2.2 \text{ kg} \cdot \text{m/s}}{9.1 \times 10^{-4} \text{ s}} = +2.4 \times 10^3 \text{ N}$$

Remarks This estimate shows just how large such contact forces can be. A good golfer achieves maximum momentum transfer by shifting weight from the back foot to the front foot, transmitting the body's momentum through the shaft and head of the club. This timing, involving a short movement of the hips, is more effective than a shot powered exclusively by the arms and shoulders. Following through with the swing ensures that the motion isn't slowed at the critical instant of impact.

Exercise 6.1

A 0.150-kg baseball, thrown with a speed of 40.0 m/s, is hit straight back at the pitcher with a speed of 50.0 m/s. **(a)** What is the impulse delivered by the bat to the baseball? **(b)** Find the magnitude of the average force exerted by the bat on the ball if the two are in contact for 2.00×10^{-3} s.

Answer (a) 13.5 kg·m/s (b) 6.75 kN

EXAMPLE 6.2 How Good Are the Bumpers?

Goal Find an impulse and estimate a force in a collision of a moving object with a stationary object.

Problem In a crash test, a car of mass 1.50×10^3 kg collides with a wall and rebounds as in Figure 6.4a. The initial and final velocities of the car are $v_i = -15.0$ m/s and $v_f = 2.60$ m/s, respectively. If the collision lasts for 0.150 s, find

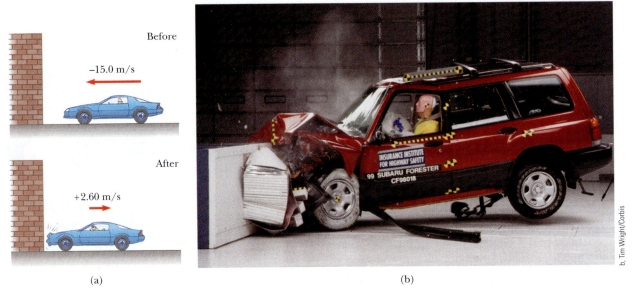

(a)   (b)

Figure 6.4 (Example 6.2) (a) This car's momentum changes as a result of its collision with the wall. (b) In a crash test (an inelastic collision), much of the car's initial kinetic energy is transformed into the energy it took to damage the vehicle.

(a) the impulse delivered to the car due to the collision and **(b)** the size and direction of the average force exerted on the car.

Strategy This problem is similar to the previous example, except that the initial and final momenta are both nonzero. Find the momenta and substitute into the impulse–momentum theorem, Equation 6.6, solving for F_{av}.

Solution

(a) Find the impulse delivered to the car.

Calculate the initial and final momenta of the car:

$$p_i = mv_i = (1.50 \times 10^3 \text{ kg})(-15.0 \text{ m/s})$$
$$= -2.25 \times 10^4 \text{ kg} \cdot \text{m/s}$$

$$p_f = mv_f = (1.50 \times 10^3 \text{ kg})(+2.60 \text{ m/s})$$
$$= +0.390 \times 10^4 \text{ kg} \cdot \text{m/s}$$

The impulse is just the difference between the final and initial momenta:

$$I = p_f - p_i$$
$$= +0.390 \times 10^4 \text{ kg} \cdot \text{m/s} - (-2.25 \times 10^4 \text{ kg} \cdot \text{m/s})$$

$$I = \boxed{2.64 \times 10^4 \text{ kg} \cdot \text{m/s}}$$

(b) Find the average force exerted on the car.

Apply Equation 6.6, the impulse–momentum theorem:

$$F_{av} = \frac{\Delta p}{\Delta t} = \frac{2.64 \times 10^4 \text{ kg} \cdot \text{m/s}}{0.150 \text{ s}} = \boxed{+1.76 \times 10^5 \text{ N}}$$

Remarks When the car doesn't rebound off the wall, the average force exerted on the car is smaller than the value just calculated. With a final momentum of zero, the car undergoes a smaller change in momentum.

Exercise 6.2
Suppose the car doesn't rebound off the wall, but the time interval of the collision remains at 0.150 s. In this case, the final velocity of the car is zero. Find the average force exerted on the car.

Answer $+1.50 \times 10^5$ N

Injury in Automobile Collisions

The main injuries that occur to a person hitting the interior of a car in a crash are brain damage, bone fracture, and trauma to the skin, blood vessels, and internal organs. Here, we compare the rather imprecisely known thresholds for human injury with typical forces and accelerations experienced in a car crash.

A force of about 90 kN (20 000 lb) compressing the tibia can cause fracture. Although the breaking force varies with the bone considered, we may take this value as the threshold force for fracture. It's well known that rapid acceleration of the head, even without skull fracture, can be fatal. Estimates show that head accelerations of $150g$ experienced for about 4 ms or $50g$ for 60 ms are fatal 50% of the time. Such injuries from rapid acceleration often result in nerve damage to the spinal cord where the nerves enter the base of the brain. The threshold for damage to skin, blood vessels, and internal organs may be estimated from whole-body impact data, where the force is uniformly distributed over the entire front surface area of 0.7 m^2 to 0.9 m^2. These data show that if the collision lasts for less than about 70 ms, a person will survive if the whole-body impact pressure (force per unit area) is less than $1.9 \times 10^5 \text{ N/m}^2$ (28 lb/in.^{2}). Death results in 50% of cases in which the whole-body impact pressure reaches $3.4 \times 10^5 \text{ N/m}^2$ (50 lb/in.^{2}).

Armed with the data above, we can estimate the forces and accelerations in a typical car crash and see how seat belts, air bags, and padded interiors can reduce the chance of death or serious injury in a collision. Consider a typical collision involving a 75-kg passenger not wearing a seat belt, traveling at 27 m/s (60 mi/h) who comes to rest in about 0.010 s after striking an unpadded dashboard. Using $F_{av}\Delta t = mv_f - mv_i$, we find that

$$F_{av} = \frac{mv_f - mv_i}{\Delta t} = \frac{0 - (75 \text{ kg})(27 \text{ m/s})}{0.010 \text{ s}} = -2.0 \times 10^5 \text{ N}$$

and

$$a = \left| \frac{\Delta v}{\Delta t} \right| = \frac{27 \text{ m/s}}{0.010 \text{ s}} = 2\ 700 \text{ m/s}^2 = \frac{2\ 700 \text{ m/s}^2}{9.8 \text{ m/s}^2} g = 280g$$

If we assume the passenger crashes into the dashboard and windshield so that the head and chest, with a combined surface area of 0.5 m^2, experience the force, we find a whole-body pressure of

$$\frac{F_{av}}{A} = \frac{2.0 \times 10^5 \text{ N}}{0.5 \text{ m}^2} \cong 4 \times 10^5 \text{ N/m}^2$$

We see that the force, the acceleration, and the whole-body pressure all *exceed* the threshold for fatality or broken bones and that an unprotected collision at 60 mi/h is almost certainly fatal.

What can be done to reduce or eliminate the chance of dying in a car crash? The most important factor is the collision time, or the time it takes the person to come to rest. If this time can be increased by 10 to 100 times the value of 0.01 s for a hard collision, the chances of survival in a car crash are much higher, because the increase in Δt makes the contact force 10 to 100 times smaller. Seat belts restrain people so that they come to rest in about the same amount of time it takes to stop the car, typically about 0.15 s. This increases the effective collision time by an order of magnitude. Figure 6.5 shows the measured force on a car versus time for a car crash.

Air bags also increase the collision time, absorb energy from the body as they rapidly deflate, and spread the contact force over an area of the body of about 0.5 m^2, preventing penetration wounds and fractures. Air bags must deploy very rapidly (in less than 10 ms) in order to stop a human traveling at 27 m/s before he or she comes to rest against the steering column about 0.3 m away. In order to achieve this rapid deployment, accelerometers send a signal to discharge a bank of capacitors (devices that store electric charge), which then ignites an explosive, thereby filling the air bag with gas very quickly. The electrical charge for ignition is

Figure 6.5 Force on a car versus time for a typical collision.

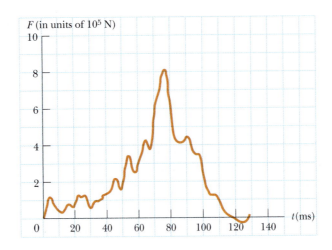

Figure 6.5 Force on a car versus time for a typical collision.

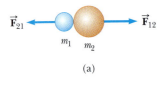

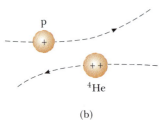

Figure 6.6 (a) A collision between two objects resulting from direct contact. (b) A collision between two charged objects (in this case, a proton and a helium nucleus).

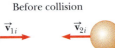

ACTIVE FIGURE 6.7
Before and after a head-on collision between two objects. The momentum of each object changes as a result of the collision, but the total momentum of the system remains constant.

Physics⊗Now™
Log into PhysicsNow at **www.cp7e.com**, and go to Active Figure 6.7 to adjust the masses and velocities of the colliding objects and see the effect on the final velocities.

stored in capacitors to ensure that the air bag continues to operate in the event of damage to the battery or the car's electrical system in a severe collision.

The important reduction in potentially fatal forces, accelerations, and pressures to tolerable levels by the simultaneous use of seat belts and air bags is summarized as follows: If a 75-kg person traveling at 27 m/s is stopped by a seat belt in 0.15 s, the person experiences an average force of 9.8 kN, an average acceleration of $18g$, and a whole-body pressure of 2.8×10^4 N/m^2 for a contact area of 0.5 m^2. These values are about one order of magnitude less than the values estimated earlier for an unprotected person and well below the thresholds for life-threatening injuries.

6.2 CONSERVATION OF MOMENTUM

When a collision occurs in an isolated system, the total momentum of the system doesn't change with the passage of time. Instead, it remains constant both in magnitude and in direction. The momenta of the individual objects in the system may change, but the vector sum of *all* the momenta will not change. The total momentum, therefore, is said to be *conserved*. In this section, we will see how the laws of motion lead us to this important conservation law.

A collision may be the result of physical contact between two objects, as illustrated in Figure 6.6a. This is a common macroscopic event, as when a pair of billiard balls or a baseball and a bat strike each other. By contrast, because contact on a submicroscopic scale is hard to define accurately, the notion of *collision* must be generalized to that scale. Forces between two objects arise from the electrostatic interaction of the electrons in the surface atoms of the objects. As will be discussed in Chapter 15, electric charges are either positive or negative. Charges with the same sign repel each other, while charges with opposite sign attract each other. To understand the distinction between macroscopic and microscopic collisions, consider the collision between two positive charges, as shown in Figure 6.6b. Because the two particles in the figure are both positively charged, they repel each other. During such a microscopic collision, particles need not touch in the normal sense in order to interact and transfer momentum.

Active Figure 6.7 shows an isolated system of two particles before and after they collide. By "isolated," we mean that no external forces, such as the gravitational force or friction, act on the system. Before the collision, the velocities of the two particles are $\vec{\mathbf{v}}_{1i}$ and $\vec{\mathbf{v}}_{2i}$; after the collision, the velocities are $\vec{\mathbf{v}}_{1f}$ and $\vec{\mathbf{v}}_{2f}$. The impulse–momentum theorem applied to m_1 becomes

$$\vec{\mathbf{F}}_{21}\Delta t = m_1\vec{\mathbf{v}}_{1f} - m_1\vec{\mathbf{v}}_{1i}$$

Likewise, for m_2, we have

$$\vec{\mathbf{F}}_{12}\Delta t = m_2\vec{\mathbf{v}}_{2f} - m_2\vec{\mathbf{v}}_{2i}$$

where $\vec{\mathbf{F}}_{21}$ is the average force exerted by m_2 on m_1 during the collision and $\vec{\mathbf{F}}_{12}$ is the average force exerted by m_1 on m_2 during the collision, as in Figure 6.6a.

We use average values for $\vec{\mathbf{F}}_{21}$ and $\vec{\mathbf{F}}_{12}$ even though the actual forces may vary in time in a complicated way, as is the case in Figure 6.8. Newton's third law states that at all times these two forces are equal in magnitude and opposite in direction: $\vec{\mathbf{F}}_{21} = -\vec{\mathbf{F}}_{12}$. In addition, the two forces act over the same time interval. As a result, we have

$$\vec{\mathbf{F}}_{21}\Delta t = -\vec{\mathbf{F}}_{12}\Delta t$$

or

$$m_1\vec{\mathbf{v}}_{1f} - m_1\vec{\mathbf{v}}_{1i} = -(m_2\vec{\mathbf{v}}_{2f} - m_2\vec{\mathbf{v}}_{2i})$$

after substituting the expressions obtained for $\vec{\mathbf{F}}_{21}$ and $\vec{\mathbf{F}}_{12}$. This equation can be rearranged to give the following important result:

$$m_1\vec{\mathbf{v}}_{1i} + m_2\vec{\mathbf{v}}_{2i} = m_1\vec{\mathbf{v}}_{1f} + m_2\vec{\mathbf{v}}_{2f} \qquad \text{[6.7]}$$

This result is a special case of the law of **conservation of momentum** and is true of isolated systems containing any number of interacting objects.

> When no net external force acts on a system, the total momentum of the system remains constant in time.

Defining the isolated system is an important feature of applying this conservation law. A cheerleader jumping upwards from rest might appear to violate conservation of momentum, because initially her momentum is zero and suddenly she's leaving the ground with velocity $\vec{\mathbf{v}}$. The flaw in this reasoning lies in the fact that the cheerleader isn't an isolated system. In jumping, she exerts a downward force on the Earth, changing its momentum. This change in the Earth's momentum isn't noticeable, however, because of the Earth's gargantuan mass compared to the cheerleader's. When we define the system to be *the cheerleader and the Earth*, momentum is conserved.

Action and reaction, together with the accompanying exchange of momentum between two objects, is responsible for the phenomenon known as *recoil*. Everyone knows that throwing a baseball while standing straight up, without bracing your feet against the Earth, is a good way to fall over backwards. This reaction, an example of recoil, also happens when you fire a gun or shoot an arrow. Conservation of momentum provides a straightforward way to calculate such effects, as the next example shows.

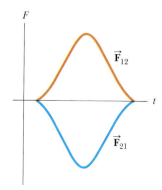

Figure 6.8 Force as a function of time for the two colliding particles in Figures 6.6(a) and 6.7. Note that $\vec{\mathbf{F}}_{21} = -\vec{\mathbf{F}}_{12}$.

◀ Conservation of momentum

TIP 6.1 **Momentum Conservation Applies to a *System*!**

The momentum of an isolated system is conserved, but not necessarily the momentum of one particle within that system, because other particles in the system may be interacting with it. Apply conservation of momentum to an isolated system *only*.

Conservation of momentum is the principle behind these two propulsion systems. (a) The force from a nitrogen-propelled, hand-controlled device allows an astronaut to move about freely in space without restrictive tethers. (b) A squid propels itself by expelling water at a high velocity.

(a)

(b)

Courtesy of NASA

Mike Severns/Stone/Getty Images

INTERACTIVE EXAMPLE 6.3   The Archer

Goal   Calculate recoil velocity using conservation of momentum

Problem   An archer stands at rest on frictionless ice and fires a 0.500-kg arrow horizontally at 50.0 m/s. (See Fig. 6.9.) The combined mass of the archer and bow is 60.0 kg. With what velocity does the archer move across the ice after firing the arrow?

Strategy   Set up the conservation of momentum equation in the horizontal direction and solve for the final velocity of the archer. The system of the archer (including the bow) and the arrow is *not* isolated, because the gravitational and normal forces act on it. These forces, however, are perpendicular to the motion of the system and hence do no work on it.

Figure 6.9   (Interactive Example 6.3) An archer fires an arrow horizontally to the right. Because he is standing on frictionless ice, he will begin to slide to the left across the ice.

Solution

Write the conservation of momentum equation. Let v_{1f} be the archer's velocity and v_{2f} the arrow's velocity.

$$p_i = p_f$$
$$0 = m_1 v_{1f} + m_2 v_{2f}$$

Substitute $m_1 = 60.0$ kg, $m_2 = 0.500$ kg, and $v_{2f} = 50.0$ m/s, and solve for v_{1f}:

$$v_{1f} = -\frac{m_2}{m_1} v_{2f} = -\left(\frac{0.500 \text{ kg}}{60.0 \text{ kg}}\right)(50.0 \text{ m/s})$$
$$= \boxed{-0.417 \text{ m/s}}$$

Remarks   The negative sign on v_{1f} indicates that the archer is moving opposite the direction of motion of the arrow, in accordance with Newton's third law. Because the archer is much more massive than the arrow, his acceleration and consequent velocity are much smaller than the acceleration and velocity of the arrow.

Newton's second law, $\Sigma F = ma$, can't be used in this problem because we have no information about the force on the arrow or its acceleration. An energy approach can't be used either, because we don't know how much work is done in pulling the string back or how much potential energy is stored in the bow. Conservation of momentum, however, readily solves the problem.

Exercise 6.3

A 70.0-kg man and a 55.0-kg woman on ice skates stand facing each other. If the woman pushes the man backwards so that his final speed is 1.50 m/s, at what speed does she recoil?

Answer   1.91 m/s

Physics⊗Now™   You can change the mass of the archer and the mass and speed of the arrow by logging into Physics-Now at **www.cp7e.com** and going to Interactive Example 6.3.

Quick Quiz 6.2

A boy standing at one end of a floating raft that is stationary relative to the shore walks to the opposite end of the raft, away from the shore. As a consequence, the raft (a) remains stationary, (b) moves away from the shore, or (c) moves toward the shore. (*Hint*: Use conservation of momentum.)

6.3   COLLISIONS

We have seen that for any type of collision, the total momentum of the system just before the collision equals the total momentum just after the collision as long as the system may be considered isolated. The total kinetic energy, on the other hand, is generally not conserved in a collision because some of the kinetic energy is converted to internal energy, sound energy, and the work needed to permanently deform the objects involved, such as cars in a car crash. **We define an inelastic collision as a collision in which momentum is conserved, but kinetic energy is not.** The collision of a rubber ball with a hard surface is inelastic, because some of the kinetic energy is lost when the ball is deformed during contact with the surface. **When two objects collide and stick together, the collision is called** *perfectly inelastic.* For example, if two pieces of putty collide, they stick together and move with some common velocity after the collision. If a meteorite collides head on with the Earth, it becomes buried in the Earth and the collision is considered perfectly inelastic. Only in very special circumstances is all the initial kinetic energy lost in a perfectly inelastic collision.

An elastic collision is defined as one in which both momentum and kinetic energy are conserved. Billiard ball collisions and the collisions of air molecules with the walls of a container at ordinary temperatures are highly elastic. Macroscopic collisions such as those between billiard balls are only approximately elastic, because some loss of kinetic energy takes place—for example, in the clicking sound when two balls strike each other. Perfectly elastic collisions do occur, however, between atomic and subatomic particles. Elastic and perfectly inelastic collisions are *limiting* cases; most actual collisions fall into a range in between them.

As a practical application, an inelastic collision is used to detect glaucoma, a disease in which the pressure inside the eye builds up and leads to blindness by damaging the cells of the retina. In this application, medical professionals use a device called a *tonometer* to measure the pressure inside the eye. This device releases a puff of air against the outer surface of the eye and measures the speed of the air after reflection from the eye. At normal pressure, the eye is slightly spongy, and the pulse is reflected at low speed. As the pressure inside the eye increases, the outer surface becomes more rigid, and the speed of the reflected pulse increases. In this way, the speed of the reflected puff of air can measure the internal pressure of the eye.

We can summarize the types of collisions as follows:

- In an elastic collision, both momentum and kinetic energy are conserved.
- In an inelastic collision, momentum is conserved but kinetic energy is not.
- In a *perfectly* inelastic collision, momentum is conserved, kinetic energy is not, and the two objects stick together after the collision, so their final velocities are the same.

In the remainder of this section, we will treat perfectly inelastic collisions and elastic collisions in one dimension.

Quick Quiz 6.3

A car and a large truck traveling at the same speed collide head-on and stick together. Which vehicle experiences the larger change in the magnitude of its momentum? (a) the car   (b) the truck   (c) the change in the magnitude of momentum is the same for both   (d) impossible to determine

Perfectly Inelastic Collisions

Consider two objects having masses m_1 and m_2 moving with known initial velocity components v_{1i} and v_{2i} along a straight line, as in Active Figure 6.10. If the two objects collide head-on, stick together, and move with a common velocity component v_f after the collision, then the collision is perfectly inelastic. Because the total momentum of the two-object isolated system before the collision equals the total momentum of the combined-object system after the collision, we can solve for the final velocity using conservation of momentum alone:

$$m_1v_{1i} + m_2v_{2i} = (m_1 + m_2)v_f \qquad [6.8]$$

TIP 6.2 Momentum and Kinetic Energy in Collisions
The momentum of an isolated system is conserved in all collisions. However, the kinetic energy of an isolated system is conserved only when the collision is elastic.

TIP 6.3 Inelastic vs. Perfectly Inelastic Collisions
If the colliding particles stick together, the collision is perfectly inelastic. If they bounce off each other (and kinetic energy is not conserved), the collision is inelastic.

APPLICATION
Glaucoma Testing

◀ Elastic collision
◀ Inelastic collision

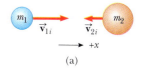
Before collision
(a)

After collision
(b)

ACTIVE FIGURE 6.10
(a) Before and (b) after a perfectly inelastic head-on collision between two objects.

Physics⊗Now™
Log into PhysicsNow at **www.cp7e.com**, and go to Active Figure 6.10 to adjust the masses and velocities of the colliding objects and see the effect on the final velocity.

Final velocity of two objects in a one-dimensional perfectly inelastic collision ▶

$$v_f = \frac{m_1 v_{1i} + m_2 v_{2i}}{m_1 + m_2}$$ [6.9]

It's important to notice that v_{1i}, v_{2i}, and v_f represent the *x*-components of the velocity vectors, so care is needed in entering their known values, particularly with regard to signs. For example, in Active Figure 6.10, v_{1i} would have a positive value (m_1 moving to the right), whereas v_{2i} would have a negative value (m_2 moving to the left). Once these values are entered, Equation 6.9 can be used to find the correct final velocity, as shown in Examples 6.4 and 6.5.

EXAMPLE 6.4   An SUV Versus a Compact

Goal   Apply conservation of momentum to a one-dimensional inelastic collision.

Problem   An SUV with mass 1.80×10^3 kg is traveling eastbound at $+15.0$ m/s, while a compact car with mass 9.00×10^2 kg is traveling westbound at -15.0 m/s. (See Fig. 6.11.) The cars collide head-on, becoming entangled. **(a)** Find the speed of the entangled cars after the collision. **(b)** Find the change in the velocity of each car. **(c)** Find the change in the kinetic energy of the system consisting of both cars.

Strategy   The total momentum of the cars before the collision, p_i, equals the total momentum of the cars after the collision, p_f, if we ignore friction and assume the two cars form an isolated system. (This is called the "impulse approximation.") Solve the momentum conservation equation for the final velocity of the entangled cars. Once the velocities are in hand, the other parts can be solved by substitution.

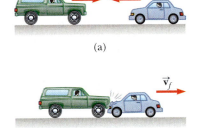

(a)

(b)

Figure 6.11   (Example 6.4)

Solution

(a) Find the final speed after collision.

Let m_1 and v_{1i} represent the mass and initial velocity of the SUV, while m_2 and v_{2i} pertain to the compact. Apply conservation of momentum:

$$p_i = p_f$$
$$m_1 v_{1i} + m_2 v_{2i} = (m_1 + m_2) v_f$$

Substitute the values and solve for the final velocity, v_f:

$$(1.80 \times 10^3 \text{ kg})(15.0 \text{ m/s}) + (9.00 \times 10^2 \text{ kg})(-15.0 \text{ m/s})$$
$$= (1.80 \times 10^3 \text{ kg} + 9.00 \times 10^2 \text{ kg}) v_f$$

$$v_f = +5.00 \text{ m/s}$$

(b) Find the change in velocity for each car.

Change in velocity of the SUV:

$$\Delta v_1 = v_f - v_{1i} = 5.00 \text{ m/s} - 15.0 \text{ m/s} = -10.0 \text{ m/s}$$

Change in velocity of the compact car:

$$\Delta v_2 = v_f - v_{2i} = 5.00 \text{ m/s} - (-15.0 \text{ m/s}) = 20.0 \text{ m/s}$$

(c) Find the change in kinetic energy of the system.

Calculate the initial kinetic energy of the system:

$$KE_i = \tfrac{1}{2} m_1 v_{1i}^2 + \tfrac{1}{2} m_2 v_{2i}^2 = \tfrac{1}{2}(1.80 \times 10^3 \text{ kg})(15.0 \text{ m/s})^2$$
$$+ \tfrac{1}{2}(9.00 \times 10^2 \text{ kg})(-15.0 \text{ m/s})^2$$
$$= 3.04 \times 10^5 \text{ J}$$

Calculate the final kinetic energy of the system and the change in kinetic energy, ΔKE.

$$KE_f = \tfrac{1}{2}(m_1 + m_2) v_f^2$$
$$= \tfrac{1}{2}(1.80 \times 10^3 \text{ kg} + 9.00 \times 10^2 \text{ kg})(5.00 \text{ m/s})^2$$
$$= 3.38 \times 10^4 \text{ J}$$

$$\Delta KE = KE_f - KE_i = -2.70 \times 10^5 \text{ J}$$

Remarks During the collision, the system lost almost 90% of its kinetic energy. The change in velocity of the SUV was only 10.0 m/s, compared to twice that for the compact car. This example underscores perhaps the most important safety feature of any car: its mass. Injury is caused by a change in velocity, and the more massive vehicle undergoes a smaller velocity change in a typical accident.

Exercise 6.4

Suppose the same two vehicles are both traveling eastward, the compact car leading the SUV. The driver of the compact car slams on the brakes suddenly, slowing the vehicle to 6.00 m/s. If the SUV traveling at 18.0 m/s crashes into the compact car, find (a) the speed of the system right after the collision, assuming the two vehicles become entangled, (b) the change in velocity for both vehicles, and (c) the change in kinetic energy of the system, from the instant before impact (when the compact car is traveling at 6.00 m/s) to the instant right after the collision.

Answers (a) 14.0 m/s (b) SUV: $\Delta v_1 = -4.0$ m/s Compact car: $\Delta v_2 = 8.0$ m/s (c) -4.32×10^4 J

EXAMPLE 6.5 The Ballistic Pendulum

Goal Combine the concepts of conservation of energy and conservation of momentum in inelastic collisions.

Problem The ballistic pendulum (Fig. 6.12a) is a device used to measure the speed of a fast-moving projectile such as a bullet. The bullet is fired into a large block of wood suspended from some light wires. The bullet is stopped by the block, and the entire system swings up to a height h. It is possible to obtain the initial speed of the bullet by measuring h and the two masses. As an example of the technique, assume that the mass of the bullet, m_1, is 5.00 g, the mass of the pendulum, m_2, is 1.000 kg, and h is 5.00 cm. Find the initial speed of the bullet, v_{1i}.

Strategy First, use conservation of momentum and the properties of perfectly inelastic collisions to find the initial speed of the bullet, v_{1i}, in terms of the final velocity of the block–bullet system, v_f. Second, use conservation of energy and the height reached by the pendulum to find v_f. Finally, substitute this value of v_f into the previous result to obtain the initial speed of the bullet.

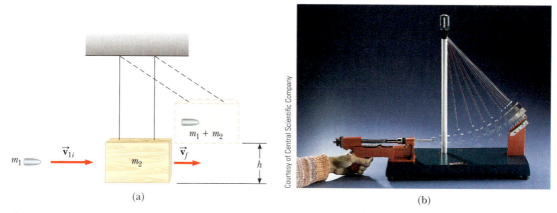

Figure 6.12 (Example 6.5) (a) Diagram of a ballistic pendulum. Note that $\vec{\mathbf{v}}_f$ is the velocity of the system just *after* the perfectly inelastic collision. (b) Multiflash photograph of a laboratory ballistic pendulum.

Courtesy of Central Scientific Company

Solution

Use conservation of momentum, and substitute the known masses. Note that $v_{2i} = 0$ and v_f is the velocity of the system (block + bullet) just after the collision.

$$p_i = p_f$$

$$m_1 v_{1i} + m_2 v_{2i} = (m_1 + m_2) v_f$$

$$(5.00 \times 10^{-3}\text{ kg})v_{1i} + 0 = (1.005\text{ kg})v_f \qquad \textbf{(1)}$$

Apply conservation of energy to the block–bullet system after the collision:

$$(KE + PE)_{\text{after collision}} = (KE + PE)_{\text{top}}$$

Both the potential energy at the bottom and the kinetic energy at the top are zero. Solve for the final velocity of the block–bullet system, v_f:

$$\tfrac{1}{2}(m_1 + m_2)v_f^2 + 0 = 0 + (m_1 + m_2)gh$$

$$v_f^2 = 2gh$$

$$v_f = \sqrt{2gh} = \sqrt{2(9.80 \text{ m/s}^2)(5.00 \times 10^{-2} \text{ m})}$$

$$v_f = 0.990 \text{ m/s}$$

Finally, substitute v_f into Equation 1 to find v_{1i}, the initial speed of the bullet:

$$v_{1i} = \frac{(1.005 \text{ kg})(0.990 \text{ m/s})}{5.00 \times 10^{-3} \text{ kg}} = \boxed{199 \text{ m/s}}$$

Remarks   Because the impact is inelastic, it would be incorrect to equate the initial kinetic energy of the incoming bullet to the final gravitational potential energy associated with the bullet–block combination. The energy isn't conserved!

Exercise 6.5
A bullet with mass 5.00 g is fired horizontally into a 2.000-kg block attached to a horizontal spring. The spring has a constant 6.00×10^2 N/m and reaches a maximum compression of 6.00 cm. (a) Find the initial speed of the bullet–block system. (b) Find the speed of the bullet.

Answer   (a) 1.04 m/s   (b) 417 m/s

Quick Quiz 6.4

An object of mass m moves to the right with a speed v. It collides head-on with an object of mass $3m$ moving with speed $v/3$ in the opposite direction. If the two objects stick together, what is the speed of the combined object, of mass $4m$, after the collision?
(a) 0   (b) $v/2$ (c) v (d) $2v$

Quick Quiz 6.5

A skater is using very low friction rollerblades. A friend throws a Frisbee® at her, on the straight line along which she is coasting. Describe each of the following events as an elastic, an inelastic, or a perfectly inelastic collision between the skater and the Frisbee: (a) She catches the Frisbee and holds it. (b) She tries to catch the Frisbee, but it bounces off her hands and falls to the ground in front of her. (c) She catches the Frisbee and immediately throws it back with the same speed (relative to the ground) to her friend.

Quick Quiz 6.6

In a perfectly inelastic one-dimensional collision between two objects, what condition alone is necessary so that *all* of the original kinetic energy of the system is gone after the collision? (a) The objects must have momenta with the same magnitude but opposite directions. (b) The objects must have the same mass. (c) The objects must have the same velocity. (d) The objects must have the same speed, with velocity vectors in opposite directions.

Elastic Collisions

Now consider two objects that undergo an **elastic head-on collision** (Active Fig. 6.13). In this situation, **both the momentum and the kinetic energy of the system of two objects are conserved**. We can write these conditions as

$$m_1 v_{1i} + m_2 v_{2i} = m_1 v_{1f} + m_2 v_{2f} \qquad [6.10]$$

and

$$\tfrac{1}{2}m_1 v_{1i}^2 + \tfrac{1}{2}m_2 v_{2i}^2 = \tfrac{1}{2}m_1 v_{1f}^2 + \tfrac{1}{2}m_2 v_{2f}^2 \qquad [6.11]$$

where v is positive if an object moves to the right and negative if it moves to the left.

In a typical problem involving elastic collisions, there are two unknown quantities, and Equations 6.10 and 6.11 can be solved simultaneously to find them. These two equations are linear and quadratic, respectively. An alternate approach simplifies the quadratic equation to another linear equation, facilitating solution. Canceling the factor $\frac{1}{2}$ in Equation 6.11, we rewrite the equation as

$$m_1(v_{1i}^2 - v_{1f}^2) = m_2(v_{2f}^2 - v_{2i}^2)$$

Here we have moved the terms containing m_1 to one side of the equation and those containing m_2 to the other. Next, we factor both sides of the equation:

$$m_1(v_{1i} - v_{1f})(v_{1i} + v_{1f}) = m_2(v_{2f} - v_{2i})(v_{2f} + v_{2i}) \qquad \textbf{[6.12]}$$

Now we separate the terms containing m_1 and m_2 in the equation for the conservation of momentum (Equation 6.10) to get

$$m_1(v_{1i} - v_{1f}) = m_2(v_{2f} - v_{2i}) \qquad \textbf{[6.13]}$$

To obtain our final result, we divide Equation 6.12 by Equation 6.13, producing

$$v_{1i} + v_{1f} = v_{2f} + v_{2i}$$

Gathering initial and final values on opposite sides of the equation gives

$$v_{1i} - v_{2i} = -(v_{1f} - v_{2f}) \qquad \textbf{[6.14]}$$

This equation, in combination with Equation 6.10, will be used to solve problems dealing with perfectly elastic head-on collisions. According to Equation 6.14, the relative velocity of the two objects before the collision, $v_{1i} - v_{2i}$, equals the negative of the relative velocity of the two objects after the collision, $-(v_{1f} - v_{2f})$. To better understand the equation, imagine that you are riding along on one of the objects. As you measure the velocity of the other object from your vantage point, you will be measuring the relative velocity of the two objects. In your view of the collision, the other object comes toward you and bounces off, leaving the collision with the same speed, but in the opposite direction. This is just what Equation 6.14 states.

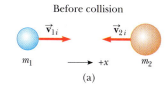

Before collision

(a)

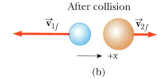

After collision

(b)

ACTIVE FIGURE 6.13
(a) Before and (b) after an elastic head-on collision between two hard spheres.

Physics⊗Now™
Log into PhysicsNow at **www.cp7e.com**, and go to Active Figure 6.13 to adjust the masses and velocities of the colliding objects and see the effect on the final velocities.

Problem-Solving Strategy
One-Dimensional Collisions

The following procedure is recommended for solving one-dimensional problems involving collisions between two objects:

1. **Coordinates**. Choose a coordinate axis that lies along the direction of motion.
2. **Diagram**. Sketch the problem, representing the two objects as blocks and labeling velocity vectors and masses.
3. **Conservation of Momentum**. Write a general expression for the *total* momentum of the system of two objects *before* and *after* the collision, and equate the two, as in Equation 6.10. On the next line, fill in the known values.
4. **Conservation of Energy**. If the collision is elastic, write a general expression for the total energy before and after the collision, and equate the two quantities, as in Equation 6.11 or (preferably) Equation 6.14. Fill in the known values. (*Skip* this step if the collision is *not* perfectly elastic.)
5. **Solve** the equations simultaneously. Equations 6.10 and 6.14 form a system of two linear equations and two unknowns. If you have forgotten Equation 6.14, use Equation 6.11 instead.

Steps 1 and 2 of the problem-solving strategy are generally carried out in the process of sketching and labeling a diagram of the problem. This is clearly the case in our next example, which makes use of Figure 6.13. Other steps are pointed out as they are applied.

EXAMPLE 6.6   Let's Play Pool

Goal   Solve an elastic collision in one dimension.

Problem   Two billiard balls of identical mass move toward each other as in Active Figure 6.13. Assume that the collision between them is perfectly elastic. If the initial velocities of the balls are $+30.0$ cm/s and -20.0 cm/s, what is the velocity of each ball after the collision? Assume friction and rotation are unimportant.

Strategy   Solution of this problem is a matter of solving two equations, the conservation of momentum and conservation of energy equations, for two unknowns, the final velocities of the two balls. Instead of using Equation 6.11 for conservation of energy, use Equation 6.14, which is linear, hence easier to handle.

Solution

Write the conservation of momentum equation. Because $m_1 = m_2$, we can cancel the masses, then substitute $v_{1i} = +30.0$ m/s and $v_{2i} = -20.0$ cm/s (Step 3).

$$m_1 v_{1i} + m_2 v_{2i} = m_1 v_{1f} + m_2 v_{2f}$$
$$30.0 \text{ cm/s} + (-20.0 \text{ cm/s}) = v_{1f} + v_{2f}$$
$$10.0 \text{ cm/s} = v_{1f} + v_{2f} \qquad \textbf{(1)}$$

Next, apply conservation of energy in the form of Equation 6.14 (Step 4):

$$v_{1i} - v_{2i} = -(v_{1f} - v_{2f})$$
$$30.0 \text{ cm/s} - (-20.0 \text{ cm/s}) = v_{2f} - v_{1f}$$
$$50.0 \text{ cm/s} = v_{2f} - v_{1f} \qquad \textbf{(2)}$$

Now solve (1) and (2) simultaneously (Step 5): $v_{1f} = \boxed{-20.0 \text{ cm/s}} \qquad v_{2f} = \boxed{+30.0 \text{ cm/s}}$

Remarks   Notice the balls exchanged velocities—almost as if they'd passed through each other. This is always the case when two objects of equal mass undergo an elastic head-on collision.

Exercise 6.6

Find the final velocity of the two balls if the ball with initial velocity $v_{2i} = -20.0$ cm/s has a mass equal to half that of the ball with initial velocity $v_{1i} = +30.0$ cm/s.

Answer $v_{1f} = -3.33$ cm/s; $v_{2f} = +46.7$ cm/s

INTERACTIVE EXAMPLE 6.7   Two Blocks and a Spring

Goal   Solve an elastic collision involving spring potential energy.

Problem   A block of mass $m_1 = 1.60$ kg, initially moving to the right with a velocity of $+4.00$ m/s on a frictionless horizontal track, collides with a massless spring attached to a second block of mass $m_2 = 2.10$ kg moving to the left with a velocity of -2.50 m/s, as in Figure 6.14a. The spring has a spring constant of 6.00×10^2 N/m. **(a)** Determine the velocity of block 2 at the instant when block 1 is moving to the right with a velocity of $+3.00$ m/s, as in Figure 6.14b. **(b)** Find the compression of the spring.

Strategy   We identify the system as the two blocks and the spring. Write down the conservation of momentum equations, and solve for the final velocity of block 2, v_{2f}. Then use conservation of energy to find the compression of the spring.

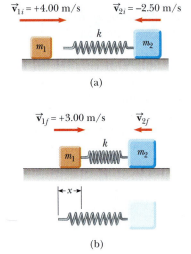

Figure 6.14   (Example 6.7)

Solution
(a) Find the velocity v_{2f} when block 1 has velocity $+3.00$ m/s.

$$m_1 v_{1i} + m_2 v_{2i} = m_1 v_{1f} + m_2 v_{2f}$$

$$v_{2f} = \frac{m_1 v_{1i} + m_2 v_{2i} - m_1 v_{1f}}{m_2}$$

$$= \frac{(1.60 \text{ kg})(4.00 \text{ m/s}) + (2.10 \text{ kg})(-2.50 \text{ m/s}) - (1.60 \text{ kg})(3.00 \text{ m/s})}{2.10 \text{ kg}}$$

Write the conservation of momentum equation for the system and solve for v_{2f}: $v_{2f} = \boxed{-1.74 \text{ m/s}}$

(b) Find the compression of the spring.

Use energy conservation for the system, noticing that potential energy is stored in the spring when it is compressed a distance x:

$$E_i = E_f$$
$$\tfrac{1}{2}m_1v_{1i}^2 + \tfrac{1}{2}m_2v_{2i}^2 + 0 = \tfrac{1}{2}m_1v_{1f}^2 + \tfrac{1}{2}m_2v_{2f}^2 + \tfrac{1}{2}kx^2$$

Substitute the given values and the result of part (a) into the preceding expression, solving for x.

$$x = \;\; \boxed{0.173 \text{ m}}$$

Remarks   The initial velocity component of block 2 is -2.50 m/s because the block is moving to the left. The negative value for v_{2f} means that block 2 is still moving to the left at the instant under consideration.

Exercise 6.7
Find (a) the velocity of block 1 and (b) the compression of the spring at the instant that block 2 is at rest.

Answer   (a) 0.719 m/s to the right   (b) 0.251 m

Physics⊗Now™   You can change the masses and speeds of the blocks and freeze the motion at the maximum compression of the spring by logging into PhysicsNow at **www.cp7e.com** and going to Interactive Example 6.7.

6.4   GLANCING COLLISIONS

In Section 6.2 we showed that the total linear momentum of a system is conserved when the system is isolated (that is, when no external forces act on the system). For a general collision of two objects in three-dimensional space, the conservation of momentum principle implies that the total momentum of the system in each direction is conserved. However, an important subset of collisions takes place in a plane. The game of billiards is a familiar example involving multiple collisions of objects moving on a two-dimensional surface. We restrict our attention to a single two-dimensional collision between two objects that takes place in a plane, and ignore any possible rotation. For such collisions, we obtain two component equations for the conservation of momentum:

$$m_1v_{1ix} + m_2v_{2ix} = m_1v_{1fx} + m_2v_{2fx}$$

$$m_1v_{1iy} + m_2v_{2iy} = m_1v_{1fy} + m_2v_{2fy}$$

We must use three subscripts in this general equation, to represent, respectively, (1) the object in question, and (2) the initial and final values of the components of velocity.

Now, consider a two-dimensional problem in which an object of mass m_1 collides with an object of mass m_2 that is initially at rest, as in Active Figure 6.15. After the collision, object 1 moves at an angle θ with respect to the horizontal, and object 2 moves at an angle ϕ with respect to the horizontal. This is called a *glancing* collision. Applying the law of conservation of momentum in component form, and noting that the initial y-component of momentum is zero, we have

x-component: $m_1v_{1i} + 0 = m_1v_{1f}\cos\theta + m_2v_{2f}\cos\phi$ **[6.15]**

y-component: $0 + 0 = m_1v_{1f}\sin\theta - m_2v_{2f}\sin\phi$ **[6.16]**

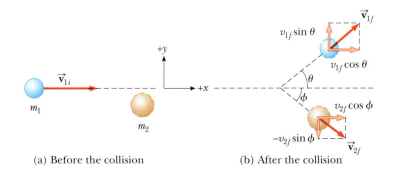

(a) Before the collision   (b) After the collision

ACTIVE FIGURE 6.15
(a) Before and (b) after a glancing collision between two balls.

Physics⊗Now™
Log into PhysicsNow at **www.cp7e.com**, and go to Active Figure 6.15 to adjust the speed and position of the blue particle, adjust the masses of both particles, and see the effects.

If the collision is elastic, we can write a third equation, for conservation of energy, in the form

$$\tfrac{1}{2}m_1v_{1i}^2 = \tfrac{1}{2}m_1v_{1f}^2 + \tfrac{1}{2}m_2v_{2f}^2 \qquad [6.17]$$

If we know the initial velocity v_{1i} and the masses, we are left with four unknowns (v_{1f}, v_{2f}, θ, and ϕ). Because we have only three equations, one of the four remaining quantities must be given in order to determine the motion after the collision from conservation principles alone.

If the collision is inelastic, the kinetic energy of the system is *not* conserved, and Equation 6.17 does *not* apply.

Problem-Solving Strategy
Two-Dimensional Collisions

To solve two-dimensional collisions, follow this procedure:
1. **Coordinate Axes.** Use both *x*- and *y*-coordinates. It's convenient to have either the *x*-axis or the *y*-axis coincide with the direction of one of the initial velocities.
2. **Diagram.** Sketch the problem, labeling velocity vectors and masses.
3. **Conservation of Momentum.** Write a separate conservation of momentum equation for each of the *x*- and *y*-directions. In each case, the total initial momentum in a given direction equals the total final momentum in that direction.
4. **Conservation of Energy.** If the collision is elastic, write a general expression for the total energy before and after the collision, and equate the two expressions, as in Equation 6.11. Fill in the known values. (*Skip* this step if the collision is *not* perfectly elastic.) The energy equation can't be simplified as in the one-dimensional case, so a quadratic expression such as Equation 6.11 or 6.17 must be used when the collision is elastic.
5. **Solve** the equations simultaneously. There are two equations for inelastic collisions and three for elastic collisions.

EXAMPLE 6.8   Collision at an Intersection

Goal   Analyze a two-dimensional inelastic collision.

Problem   A car with mass 1.50×10^3 kg traveling east at a speed of 25.0 m/s collides at an intersection with a 2.50×10^3-kg van traveling north at a speed of 20.0 m/s, as shown in Figure 6.16. Find the magnitude and direction of the velocity of the wreckage after the collision, assuming that the vehicles undergo a perfectly inelastic collision (that is, they stick together) and assuming that friction between the vehicles and the road can be neglected.

Strategy   Use conservation of momentum in two dimensions. (Kinetic energy is *not* conserved.) Choose coordinates as in Figure 6.16. Before the collision, the only object having momentum in the *x*-direction is the car, while the van carries all the momentum in the *y*-direction. After the totally inelastic collision, both vehicles move together at some common speed v_f and angle θ. Solve for these two unknowns, using the two components of the conservation of momentum equation.

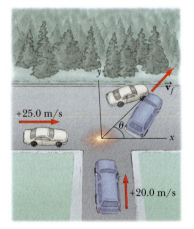

Figure 6.16   (Example 6.8) A top view of a perfectly inelastic collision between a car and a van.

Solution

Find the *x*-components of the initial and final total momenta:

$$\Sigma p_{xi} = m_{car}v_{car} = (1.50 \times 10^3 \text{ kg})(25.0 \text{ m/s})$$
$$= 3.75 \times 10^4 \text{ kg} \cdot \text{m/s}$$

$$\Sigma p_{xf} = (m_{car} + m_{van})v_f \cos\theta = (4.00 \times 10^3 \text{ kg})v_f \cos\theta$$

Set the initial *x*-momentum equal to the final *x*-momentum:

$$3.75 \times 10^4 \text{ kg} \cdot \text{m/s} = (4.00 \times 10^3 \text{ kg})\, v_f \cos\theta \qquad (1)$$

Find the *y*-components of the initial and final total momenta:

$$\Sigma p_{iy} = m_{van} v_{van} = (2.50 \times 10^3 \text{ kg})(20.0 \text{ m/s})$$
$$= 5.00 \times 10^4 \text{ kg} \cdot \text{m/s}$$

$$\Sigma p_{fy} = (m_{car} + m_{van}) v_f \sin \theta = (4.00 \times 10^3 \text{ kg}) v_f \sin \theta$$

Set the initial *y*-momentum equal to the final *y*-momentum:

$$5.00 \times 10^4 \text{ kg} \cdot \text{m/s} = (4.00 \times 10^3 \text{ kg}) v_f \sin \theta \qquad \textbf{(2)}$$

Divide Equation (2) by Equation (1) and solve for θ:

$$\tan \theta = \frac{5.00 \times 10^4 \text{ kg} \cdot \text{m/s}}{3.75 \times 10^4 \text{ kg} \cdot \text{m}} = 1.33$$

$$\theta = \boxed{53.1°}$$

Substitute this angle back into Equation (2) to find v_f:

$$v_f = \frac{5.00 \times 10^4 \text{ kg} \cdot \text{m/s}}{(4.00 \times 10^3 \text{ kg}) \sin 53.1°} = \boxed{15.6 \text{ m/s}}$$

Remark   It's also possible to first find the *x*- and *y*-components v_{fx} and v_{fy} of the resultant velocity. The magnitude and direction of the resultant velocity can then be found with the Pythagorean theorem, $v_f = \sqrt{v_{fx}^2 + v_{fy}^2}$, and the inverse tangent function $\theta = \tan^{-1}(v_{fy}/v_{fx})$. Setting up this alternate approach is a simple matter of substituting $v_{fx} = v_f \cos \theta$ and $v_{fy} = v_f \sin \theta$ in Equations (1) and (2).

Exercise 6.8
A 3.00-kg object initially moving in the positive *x*-direction with a velocity of $+5.00$ m/s collides with and sticks to a 2.00-kg object initially moving in the negative *y*-direction with a velocity of -3.00 m/s. Find the final components of velocity of the composite object.

Answer $v_{fx} = 3.00$ m/s; $v_{fy} = -1.20$ m/s

6.5   ROCKET PROPULSION

When ordinary vehicles such as cars and locomotives move, the driving force of the motion is friction. In the case of the car, this driving force is exerted by the road on the car, a reaction to the force exerted by the wheels against the road. Similarly, a locomotive "pushes" against the tracks; hence, the driving force is the reaction force exerted by the tracks on the locomotive. However, a rocket moving in space has no road or tracks to push against. How can it move forward?

In fact, reaction forces also propel a rocket. (You should review Newton's third law, discussed in Chapter 4.) To illustrate this point, we model our rocket with a spherical chamber containing a combustible gas, as in Figure 6.17a. When an explosion occurs in the chamber, the hot gas expands and presses against all sides of the chamber, as indicated by the arrows. Because the sum of the forces exerted on the rocket is zero, it doesn't move. Now suppose a hole is drilled in the bottom of the chamber, as in Figure 6.17b. When the explosion occurs, the gas presses against the chamber in all directions, but can't press against anything at the hole, where it simply escapes into space. Adding the forces on the spherical chamber now results in a net force upwards. Just as in the case of cars and locomotives, this is a reaction force. A car's wheels press against the ground, and the reaction force of the ground on the car pushes it forward. The wall of the rocket's combustion chamber exerts a force on the gas expanding against it. The reaction force of the gas on the wall then pushes the rocket upward.

In a now infamous article in *The New York Times*, rocket pioneer Robert Goddard was ridiculed for thinking that rockets would work in space, where, according to the *Times*, there was nothing to push against. The *Times* retracted, rather belatedly, during the first Apollo moon landing mission in 1969. The hot gases are not pushing against anything external, but against the rocket itself—and ironically, rockets actually work *better* in a vacuum. In an atmosphere, the gases have to do work against the outside air pressure to escape the combustion chamber, slowing the exhaust velocity and reducing the reaction force.

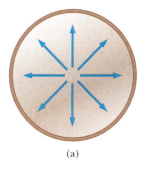

(a)

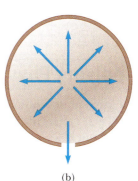

(b)

Figure 6.17   (a) A rocket reaction chamber without a nozzle has reaction forces pushing equally in all directions, so no motion results. (b) An opening at the bottom of the chamber removes the downward reaction force, resulting in a net upward reaction force.

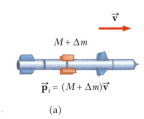

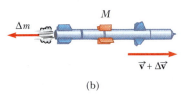

(a)

(b)

Figure 6.18 Rocket propulsion.
(a) The initial mass of the rocket and
fuel is $M + \Delta m$ at a time t, and the
rocket's speed is v. (b) At a time
$t + \Delta t$, the rocket's mass has been
reduced to M, and an amount of fuel
Δm has been ejected. The rocket's
speed increases by an amount Δv.

At the microscopic level, this process is complicated, but it can be simplified by applying conservation of momentum to the rocket and its ejected fuel. In principle, the solution is similar to that in Example 6.3, with the archer representing the rocket and the arrows the exhaust gases.

Suppose that at some time t, the momentum of the rocket plus the fuel is $(M + \Delta m)v$, where Δm is an amount of fuel about to be burned (Fig. 6.18a). This fuel is traveling at a speed v relative to, say, the Earth, just like the rest of the rocket. During a short time interval Δt, the rocket ejects fuel of mass Δm, and the rocket's speed increases to $v + \Delta v$ (Fig. 6.18b). If the fuel is ejected with exhaust speed v_e *relative to the rocket*, the speed of the fuel relative to the Earth is $v - v_e$. Equating the total initial momentum of the system with the total final momentum, we have

$$(M + \Delta m)v = M(v + \Delta v) + \Delta m(v - v_e)$$

Simplifying this expression gives

$$M\Delta v = v_e \Delta m$$

The increase Δm in the mass of the exhaust corresponds to an equal decrease in the mass of the rocket, so that $\Delta m = -\Delta M$. Using this fact, we have

$$M\Delta v = -v_e \Delta M \qquad \text{[6.18]}$$

This result, together with the methods of calculus, can be used to obtain the following equation:

$$v_f - v_i = v_e \ln\left(\frac{M_i}{M_f}\right) \qquad \text{[6.19]}$$

where M_i is the initial mass of the rocket plus fuel and M_f is the final mass of the rocket plus its remaining fuel. This is the basic expression for rocket propulsion; it tells us that the increase in velocity is proportional to the exhaust speed v_e and to the natural logarithm of M_i/M_f. Because the maximum ratio of M_i to M_f for a single-stage rocket is about 10:1, the increase in speed can reach $v_e \ln 10 = 2.3 v_e$ or about twice the exhaust speed! For best results, therefore, the exhaust speed should be as high as possible. Currently, typical rocket exhaust speeds are several kilometers per second.

The **thrust** on the rocket is defined as the force exerted on the rocket by the ejected exhaust gases. We can obtain an expression for the instantaneous thrust by dividing Equation 6.18 by Δt:

Rocket thrust ▶

$$\text{Instantaneous thrust} = Ma = M\frac{\Delta v}{\Delta t} = \left|v_e \frac{\Delta M}{\Delta t}\right| \qquad \text{[6.20]}$$

The absolute value signs are used for clarity: In Equation 6.18, $-\Delta M$ is a positive quantity (as is v_e, a speed). Here we see that the thrust increases as the exhaust velocity increases and as the rate of change of mass $\Delta M/\Delta t$ (the burn rate) increases.

Applying Physics 6.2   Multistage Rockets

The current maximum exhaust speed of $v_e = 4\,500$ m/s can be realized with rocket engines fueled with liquid hydrogen and liquid oxygen. But this means that the maximum speed attainable for a given rocket with a mass ratio of 10 is $v_e \ln 10 \approx 10\,000$ m/s. To reach the moon, however, requires a change in velocity of over $11\,000$ m/s. Further, this change must occur while working against gravity and atmospheric friction. How can that be managed without developing better engines?

Explanation   The answer is the multistage rocket. By dropping stages, the spacecraft becomes lighter, so that fuel burned later in the mission doesn't have to accelerate mass that no longer serves any purpose. Strap-on boosters, as used by the Space Shuttle and a number of other rockets, such as the Titan 4 or Russian Proton, is a similar concept. The boosters are jettisoned after their fuel is exhausted, so the rocket is no longer burdened by their weight.

EXAMPLE 6.9   Single Stage to Orbit (SSTO)

Goal   Apply the velocity and thrust equations of a rocket.

Problem   A rocket has a total mass of 1.00×10^5 kg and a burnout mass of 1.00×10^4 kg, including engines, shell, and payload. The rocket blasts off from Earth and exhausts all its fuel in 4.00 min, burning the fuel at a steady rate with an exhaust velocity of $v_e = 4.50 \times 10^3$ m/s. **(a)** If air friction and gravity are neglected, what is the speed of the rocket at burnout? **(b)** What thrust does the engine develop at liftoff? **(c)** What is the initial acceleration of the rocket if gravity is not neglected? **(d)** Estimate the speed at burnout if gravity isn't neglected.

Strategy   Although it sounds sophisticated, this problem is mainly a matter of substituting values into the appropriate equations. Part (a) requires substituting values into Equation 6.19 for the velocity. For part (b), divide the change in the rocket's mass by the total time, getting $\Delta M/\Delta t$, then substitute into Equation 6.20 to find the thrust. (c) Using Newton's second law, the force of gravity, and the result of (b), we can find the initial acceleration. For part (d), the acceleration of gravity is approximately constant over the few kilometers involved, so the velocity found in (b) will be reduced by roughly $\Delta v_g = -gt$. Add this loss to the result of part (a).

Solution

(a) Calculate the velocity at burnout.

Substitute $v_i = 0$, $v_e = 4.50 \times 10^3$ m/s, $M_i = 1.00 \times 10^5$ kg, and $M_f = 1.00 \times 10^4$ kg into Equation 6.19:

$$v_f = v_i + v_e \ln\left(\frac{M_i}{M_f}\right)$$

$$= 0 + (4.5 \times 10^3 \text{ m/s}) \ln\left(\frac{1.00 \times 10^5 \text{ kg}}{1.00 \times 10^4 \text{ kg}}\right)$$

$$v_f = 1.04 \times 10^4 \text{ m/s}$$

(b) Find the thrust at liftoff.

Compute the change in the rocket's mass:

$$\Delta M = M_f - M_i = 1.00 \times 10^4 \text{ kg} - 1.00 \times 10^5 \text{ kg}$$
$$= -9.00 \times 10^4 \text{ kg}$$

Calculate the rate at which rocket mass changes by dividing the change in mass by the time (4.00 min, converted to seconds):

$$\frac{\Delta M}{\Delta t} = \frac{-9.00 \times 10^4 \text{ kg}}{2.40 \times 10^2 \text{ s}} = -3.75 \times 10^2 \text{ kg/s}$$

Substitute this rate into Equation 6.20, obtaining the thrust:

$$\text{Thrust} = \left| v_e \frac{\Delta M}{\Delta t} \right| = (4.50 \times 10^3 \text{ m/s})(3.75 \times 10^2 \text{ kg/s})$$

$$= 1.69 \times 10^6 \text{ N}$$

(c) Find the initial acceleration.

Write Newton's second law, where T stands for thrust, and solve for the acceleration a:

$$Ma = \Sigma F = T - Mg$$

$$a = \frac{T}{M} - g = \frac{1.69 \times 10^6 \text{ N}}{1.00 \times 10^5 \text{ kg}} - 9.80 \text{ m/s}^2$$

$$= 7.10 \text{ m/s}^2$$

(d) Estimate the speed at burnout when gravity is not neglected.

Find the approximate loss of speed due to gravity:

$$\Delta v_g = -g\Delta t = -(9.80 \text{ m/s}^2)(2.40 \times 10^2 \text{ s})$$
$$= -2.35 \times 10^3 \text{ m/s}$$

Add this loss to the result of part (b):

$$v_f = 1.04 \times 10^4 \text{ m/s} - 2.35 \times 10^3 \text{ m/s}$$

$$= 8.05 \times 10^3 \text{ m/s}$$

Remarks   Even taking gravity into account, the speed is sufficient to attain orbit. Some additional boost may be required to overcome air drag.

Exercise 6.9
A spaceship with a mass of 5.00×10^4 kg is traveling at 6.00×10^3 m/s relative a space station. What mass will the ship have after it fires its engines in order to reach a speed of 8.00×10^3 m/s? Assume an exhaust velocity of 4.50×10^3 m/s.

Answer 3.21×10^4 kg

SUMMARY

Physics⊗Now™ Take a practice test by logging into PhysicsNow at **www.cp7e.com** and clicking on the Pre-Test link for this chapter.

6.1   Momentum and Impulse
The **linear momentum** $\vec{\mathbf{p}}$ of an object of mass m moving with velocity $\vec{\mathbf{v}}$ is defined as

$$\vec{\mathbf{p}} \equiv m\vec{\mathbf{v}} \qquad [6.1]$$

Momentum carries units of kg·m/s. The **impulse** $\vec{\mathbf{I}}$ of a constant force $\vec{\mathbf{F}}$ delivered to an object is equal to the product of the force and the time interval during which the force acts:

$$\vec{\mathbf{I}} \equiv \vec{\mathbf{F}}\Delta t \qquad [6.4]$$

These two concepts are unifed in the **impulse–momentum theorem**, which states that the impulse of a constant force delivered to an object is equal to the change in momentum of the object:

$$\vec{\mathbf{F}}\Delta t = \Delta\vec{\mathbf{p}} \equiv m\vec{\mathbf{v}}_f - m\vec{\mathbf{v}}_i \qquad [6.5]$$

Solving problems with this theorem often involves estimating speeds or contact times (or both), leading to an average force.

6.2   Conservation of Momentum
When no net external force acts on an isolated system, the total momentum of the system is constant. This principle is called **conservation of momentum**. In particular, if the isolated system consists of two objects undergoing a collision, the total momentum of the system is the same before and after the collision. Conservation of momentum can be written mathematically for this case as

$$m_1\vec{\mathbf{v}}_{1i} + m_2\vec{\mathbf{v}}_{2i} = m_1\vec{\mathbf{v}}_{1f} + m_2\vec{\mathbf{v}}_{2f} \qquad [6.7]$$

Collision and recoil problems typically require finding unknown velocities in one or two dimensions. Each vector component gives an equation, and the resulting equations are solved simultaneously.

6.3   Collisions
In an **inelastic collision**, the momentum of the system is conserved, but kinetic energy is not. In a **perfectly inelastic collision**, the colliding objects stick together. In an **elastic collision**, both the momentum and the kinetic energy of the system are conserved.

A one-dimensional **elastic collision** between two objects can be solved by using the conservation of momentum and conservation of energy equations:

$$m_1 v_{1i} + m_2 v_{2i} = m_1 v_{1f} + m_2 v_{2f} \qquad [6.10]$$

$$\tfrac{1}{2}m_1 v_{1i}^2 + \tfrac{1}{2}m_2 v_{2i}^2 = \tfrac{1}{2}m_1 v_{1f}^2 + \tfrac{1}{2}m_2 v_{2f}^2 \qquad [6.11]$$

The following equation, derived from Equations 6.10 and 6.11, is usually more convenient to use than the original conservation of energy equation:

$$v_{1i} - v_{2i} = -(v_{1f} - v_{2f}) \qquad [6.14]$$

These equations can be solved simultaneously for the unknown velocities. Energy is not conserved in **inelastic collisions**, so such problems must be solved with Equation 6.10 alone.

6.4   Glancing Collisions
In glancing collisions, conservation of momentum can be applied along two perpendicular directions: an x-axis and a y-axis. Problems can be solved by using the x- and y-components of Equation 6.7. Elastic two-dimensional collisions will usually require Equation 6.11 as well. (Equation 6.14 doesn't apply to two dimensions.) Generally, one of the two objects is taken to be traveling along the x-axis, undergoing a deflection at some angle θ after the collision. The final velocities and angles can be found with elementary trigonometry.

CONCEPTUAL QUESTIONS

1. A batter bunts a pitched baseball, blocking the ball without swinging. (a) Can the baseball deliver more kinetic energy to the bat and batter than the ball carries initially? (b) Can the baseball deliver more momentum to the bat and batter than the ball carries initially? Explain each of your answers.

2. America will never forget the terrorist attack on September 11, 2001. One commentator remarked that the force of the explosion at the Twin Towers of the World Trade Center was strong enough to blow glass and parts of the steel structure to small fragments. Yet the television coverage showed thousands of sheets of paper floating down, many still intact. Explain how that could be.

3. In perfectly inelastic collisions between two objects, there are events in which all of the original kinetic energy is transformed to forms other than kinetic. Give an example of such an event.

4. If two objects collide and one is initially at rest, is it possible for both to be at rest after the collision? Is it possible for only one to be at rest after the collision? Explain.

5. A ball of clay of mass *m* is thrown with a speed *v* against a brick wall. The clay sticks to the wall and stops. Is the principle of conservation of momentum violated in this example?

6. A skater is standing still on a frictionless ice rink. Her friend throws a Frisbee straight at her. In which of the following cases is the largest momentum transferred to the skater? (a) The skater catches the Frisbee and holds onto it. (b) The skater catches the Frisbee momentarily, but then drops it vertically downward. (c) The skater catches the Frisbee, holds it momentarily, and throws it back to her friend.

7. You are standing perfectly still and then you take a step forward. Before the step your momentum was zero, but afterwards you have some momentum. Is the conservation of momentum violated in this case?

8. If two particles have equal kinetic energies, are their momenta necessarily equal? Explain.

9. A more ordinary example of conservation of momentum than a rocket ship occurs in a kitchen dishwashing machine. In this device, water at high pressure is forced out of small holes on the spray arms. Use conservation of momentum to explain why the arms rotate, directing water to all the dishes.

10. If two automobiles collide, they usually do not stick together. Does this mean the collision is elastic? Explain why a head-on collision is likely to be more dangerous than other types of collisions.

11. An open box slides across a frictionless, icy surface of a frozen lake. What happens to the speed of the box as water from a rain shower collect in it, assuming that the rain falls vertically downward into the box? Explain.

12. Consider a perfectly inelastic collision between a car and a large truck. Which vehicle loses more kinetic energy as a result of the collision?

13. Your physical education teacher throws you a tennis ball at a certain velocity, and you catch it. You are now given the following choice: The teacher can throw you a medicine ball (which is much more massive than the tennis ball) with the same velocity, the same momentum, or the same kinetic energy as the tennis ball. Which option would you choose in order to make the easiest catch, and why?

14. While watching a movie about a superhero, you notice that the superhero hovers in the air and throws a piano at some bad guys while remaining stationary in the air. What's wrong with this scenario?

15. In golf, novice players are often advised to be sure to "follow through" with their swing. Why does this make the ball travel a longer distance? If a shot is taken near the green, very little follow-through is required. Why?

16. An air bag inflates when a collision occurs, protecting a passenger (the dummy in Figure Q6.16) from serious injury. Why does the air bag soften the blow? Discuss the physics involved in this dramatic photograph.

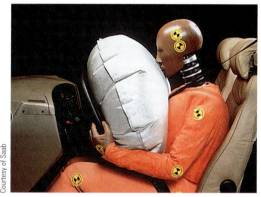

Figure Q6.16

17. A sharpshooter fires a rifle while standing with the butt of the gun against his shoulder. If the forward momentum of a bullet is the same as the backward momentum of the gun, why isn't it as dangerous to be hit by the gun as by the bullet?

18. A large bedsheet is held vertically by two students. A third student, who happens to be the star pitcher on the baseball team, throws a raw egg at the sheet. Explain why the egg doesn't break when it hits the sheet, regardless of its initial speed. (If you try this, make sure the pitcher hits the sheet near its center, and don't allow the egg to fall on the floor after being caught.)

PROBLEMS

1, 2, 3 – straightforward, intermediate, challenging   ☐ = full solution available in *Student Solutions Manual/Study Guide*

Physics⊗Now™ = coached problem with hints available at **www.cp7e.com**   = biomedical application

Section 6.1   Momentum and Impulse

1. A ball of mass 0.150 kg is dropped from rest from a height of 1.25 m. It rebounds from the floor to reach a height of 0.960 m. What impulse was given to the ball by the floor?

2. A tennis player receives a shot with the ball (0.060 0 kg) traveling horizontally at 50.0 m/s and returns the shot with the ball traveling horizontally at 40.0 m/s in the opposite direction. (a) What is the impulse delivered to the ball by the racquet? (b) What work does the racquet do on the ball?

3. Calculate the magnitude of the linear momentum for the following cases: (a) a proton with mass 1.67×10^{-27} kg, moving with a speed of 5.00×10^6 m/s; (b) a 15.0-g bullet moving with a speed of 300 m/s; (c) a 75.0-kg sprinter running with a speed of 10.0 m/s; (d) the Earth (mass = 5.98×10^{24} kg) moving with an orbital speed equal to 2.98×10^4 m/s.

4. A 0.10-kg ball is thrown straight up into the air with an initial speed of 15 m/s. Find the momentum of the ball (a) at its maximum height and (b) halfway to its maximum height.

5. A pitcher claims he can throw a 0.145-kg baseball with as much momentum as a 3.00-g bullet moving with a speed of 1.50×10^3 m/s. (a) What must the baseball's speed be

if the pitcher's claim is valid? (b) Which has greater kinetic energy, the ball or the bullet?

6. A stroboscopic photo of a club hitting a golf ball, such as the photo shown in Figure 6.3, was made by Harold Edgerton in 1933. The ball was initially at rest, and the club was shown to be in contact with the club for about 0.002 0 s. Also, the ball was found to end up with a speed of 2.0×10^2 ft/s. Assuming that the golf ball had a mass of 55 g, find the average force exerted by the club on the ball.

7. **Physics⊗Now™** A professional diver performs a dive from a platform 10 m above the water surface. Estimate the order of magnitude of the average impact force she experiences in her collision with the water. State the quantities you take as data and their values.

8. A 75.0-kg stuntman jumps from a balcony and falls 25.0 m before colliding with a pile of mattresses. If the mattresses are compressed 1.00 m before he is brought to rest, what is the average force exerted by the mattresses on the stuntman?

9. A car is stopped for a traffic signal. When the light turns green, the car accelerates, increasing its speed from 0 to 5.20 m/s in 0.832 s. What are the magnitudes of the linear impulse and the average total force experienced by a 70.0-kg passenger in the car during the time the car accelerates?

10. A 0.500-kg football is thrown toward the east with a speed of 15.0 m/s. A stationary receiver catches the ball and brings it to rest in 0.020 0 s. (a) What is the impulse delivered to the ball as it's caught? (b) What is the average force exerted on the receiver?

11. The force shown in the force vs. time diagram in Figure P6.11 acts on a 1.5-kg object. Find (a) the impulse of the force, (b) the final velocity of the object if it is initially at rest, and (c) the final velocity of the object if it is initially moving along the x-axis with a velocity of -2.0 m/s.

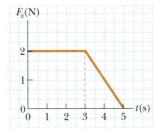

Figure P6.11

12. A force of magnitude F_x acting in the x-direction on a 2.00-kg particle varies in time as shown in Figure P6.12. Find (a) the impulse of the force, (b) the final velocity of

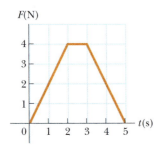

Figure P6.12

the particle if it is initially at rest, and (c) the final velocity of the particle if it is initially moving along the x-axis with a velocity of -2.00 m/s.

13. The forces shown in the force vs. time diagram in Figure P6.13 act on a 1.5-kg particle. Find (a) the impulse for the interval from $t = 0$ to $t = 3.0$ s and (b) the impulse for the interval from $t = 0$ to $t = 5.0$ s. (c) If the forces act on a 1.5-kg particle that is initially at rest, find the particle's speed at $t = 3.0$ s and at $t = 5.0$ s.

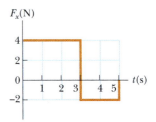

Figure P6.13

14. A 3.00-kg steel ball strikes a massive wall at 10.0 m/s at an angle of 60.0° with the plane of the wall. It bounces off the wall with the same speed and angle (Fig. P6.14). If the ball is in contact with the wall for 0.200 s, what is the average force exerted by the wall on the ball?

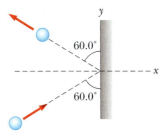

Figure P6.14

15. The front 1.20 m of a 1 400-kg car is designed as a "crumple zone" that collapses to absorb the shock of a collision. If a car traveling 25.0 m/s stops uniformly in 1.20 m, (a) how long does the collision last, (b) what is the magnitude of the average force on the car, and (c) what is the acceleration of the car? Express the acceleration as a multiple of the acceleration of gravity.

16. A pitcher throws a 0.15-kg baseball so that it crosses home plate horizontally with a speed of 20 m/s. The ball is hit straight back at the pitcher with a final speed of 22 m/s. (a) What is the impulse delivered to the ball? (b) Find the average force exerted by the bat on the ball if the two are in contact for 2.0×10^{-3} s.

17. A car of mass 1.6×10^3 kg is traveling east at a speed of 25 m/s along a horizontal roadway. When its brakes are applied, the car stops in 6.0 s. What is the average horizontal force exerted on the car while it is braking?

Section 6.2   Conservation of Momentum

18. A 730-N man stands in the middle of a frozen pond of radius 5.0 m. He is unable to get to the other side because of a lack of friction between his shoes and the ice. To overcome this difficulty, he throws his 1.2-kg physics textbook horizontally toward the north shore at a speed of 5.0 m/s. How long does it take him to reach the south shore?

19. High-speed stroboscopic photographs show that the head of a 200-g golf club is traveling at 55 m/s just before it strikes a 46-g golf ball at rest on a tee. After the collision, the club head travels (in the same direction) at 40 m/s. Find the speed of the golf ball just after impact.

20. A rifle with a weight of 30 N fires a 5.0-g bullet with a speed of 300 m/s. (a) Find the recoil speed of the rifle. (b) If a 700-N man holds the rifle firmly against his shoulder, find the recoil speed of the man and rifle.

21. **Physics⊗Now™** A 45.0-kg girl is standing on a 150-kg plank. The plank, originally at rest, is free to slide on a frozen lake, which is a flat, frictionless surface. The girl begins to walk along the plank at a constant velocity of 1.50 m/s to the right relative to the plank. (a) What is her velocity relative to the surface of the ice? (b) What is the velocity of the plank relative to the surface of the ice?

22. A 65.0-kg person throws a 0.045 0-kg snowball forward with a ground speed of 30.0 m/s. A second person, with a mass of 60.0 kg, catches the snowball. Both people are on skates. The first person is initially moving forward with a speed of 2.50 m/s, and the second person is initially at rest. What are the velocities of the two people after the snowball is exchanged? Disregard friction between the skates and the ice.

23. In Section 6.2, we implied that the kinetic energy of the Earth can be ignored when considering the energy of a system consisting of the Earth and a dropped ball of mass m_b. Verify this statement by first setting up a ratio of the kinetic energy of the Earth to that of the ball as they collide. Then use conservation of momentum to show that

$$\frac{v_E}{v_b} = -\frac{m_b}{m_E} \quad \text{and} \quad \frac{KE_E}{KE_b} = \frac{m_b}{m_E}$$

Find the order of magnitude of the ratio of the kinetic energies, based on data that you specify.

24. Two ice skaters are holding hands at the center of a frozen pond when an argument ensues. Skater A shoves skater B along a horizontal direction. Identify (a) the horizontal forces acting on A and (b) those acting on B. (c) Which force is greater, the force on A or the force on B? (d) Can conservation of momentum be used for the system of A and B? Defend your answer. (e) If A has a mass of 0.900 times that of B, and B begins to move away with a speed of 2.00 m/s, find the speed of A.

Section 6.3   Collisions
Section 6.4   Glancing Collisions

25. An archer shoots an arrow toward a 300-g target that is sliding in her direction at a speed of 2.50 m/s on a smooth, slippery surface. The 22.5-g arrow is shot with a speed of 35.0 m/s and passes through the target, which is stopped by the impact. What is the speed of the arrow after passing through the target?

26. A 75.0-kg ice skater moving at 10.0 m/s crashes into a stationary skater of equal mass. After the collision, the two skaters move as a unit at 5.00 m/s. Suppose the average force a skater can experience without breaking a bone is 4 500 N. If the impact time is 0.100 s, does a bone break?

27. A railroad car of mass 2.00×10^4 kg moving at 3.00 m/s collides and couples with two coupled railroad cars, each of the same mass as the single car and moving in the same direction at 1.20 m/s. (a) What is the speed of the three coupled cars after the collision? (b) How much kinetic energy is lost in the collision?

28. A 7.0-g bullet is fired into a 1.5-kg ballistic pendulum. The bullet emerges from the block with a speed of 200 m/s, and the block rises to a maximum height of 12 cm. Find the initial speed of the bullet.

29. A 0.030-kg bullet is fired vertically at 200 m/s into a 0.15-kg baseball that is initially at rest. How high does the combined bullet and baseball rise after the collision, assuming the bullet embeds itself in the ball?

30. An 8.00-g bullet is fired into a 250-g block that is initially at rest at the edge of a table of height 1.00 m (Fig. P6.30). The bullet remains in the block, and after the impact the block lands 2.00 m from the bottom of the table. Determine the initial speed of the bullet.

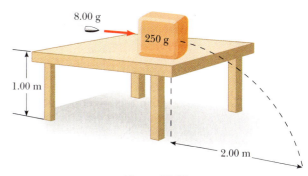

8.00 g

250 g

1.00 m

2.00 m

Figure P6.30

31. Gayle runs at a speed of 4.00 m/s and dives on a sled, initially at rest on the top of a frictionless, snow-covered hill. After she has descended a vertical distance of 5.00 m, her brother, who is initially at rest, hops on her back, and they continue down the hill together. What is their speed at the bottom of the hill if the total vertical drop is 15.0 m? Gayle's mass is 50.0 kg, the sled has a mass of 5.00 kg, and her brother has a mass of 30.0 kg.

32. A 1 200-kg car traveling initially with a speed of 25.0 m/s in an easterly direction crashes into the rear end of a 9 000-kg truck moving in the same direction at 20.0 m/s (Fig. P6.32). The velocity of the car right after the collision is 18.0 m/s to the east. (a) What is the velocity of the truck right after the collision? (b) How much mechanical energy is lost in the collision? Account for this loss in energy.

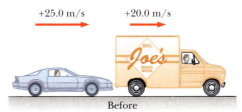

+25.0 m/s   +20.0 m/s

Before

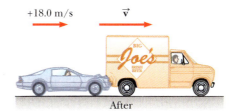

+18.0 m/s $\vec{v}$

After

Figure P6.32

33. A 12.0-g bullet is fired horizontally into a 100-g wooden block that is initially at rest on a frictionless horizontal surface and connected to a spring having spring constant 150 N/m. The bullet becomes embedded in the block. If the bullet–block system compresses the spring by a maximum of 80.0 cm, what was the speed of the bullet at impact with the block?

34. (a) Three carts of masses 4.0 kg, 10 kg, and 3.0 kg move on a frictionless horizontal track with speeds of 5.0 m/s, 3.0 m/s, and 4.0 m/s, as shown in Figure P6.34. The carts stick together after colliding. Find the final velocity of the three carts. (b) Does your answer require that all carts collide and stick together at the same time?

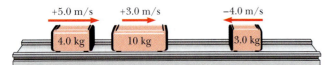

Figure P6.34

35. A 5.00-g object moving to the right at 20.0 cm/s makes an elastic head-on collision with a 10.0-g object that is initially at rest. Find (a) the velocity of each object after the collision and (b) the fraction of the initial kinetic energy transferred to the 10.0-g object.

36. A 10.0-g object moving to the right at 20.0 cm/s makes an elastic head-on collision with a 15.0-g object moving in the opposite direction at 30.0 cm/s. Find the velocity of each object after the collision.

37. A 25.0-g object moving to the right at 20.0 cm/s overtakes and collides elastically with a 10.0-g object moving in the same direction at 15.0 cm/s. Find the velocity of each object after the collision.

38. Four railroad cars, each of mass 2.50×10^4 kg, are coupled together and coasting along horizontal tracks at speed v_i toward the south. A very strong but foolish movie actor riding on the second car uncouples the front car and gives it a big push, increasing its speed to 4.00 m/s south. The remaining three cars continue moving south, now at 2.00 m/s. (a) Find the initial speed of the cars. (b) How much work did the actor do?

39. When fired from a gun into a 1.00-kg block of wood held in a vise, a 7.00-g bullet penetrates the block to a depth of 8.00 cm. The block is then placed on a frictionless, horizontal surface, and a second 7.00-g bullet is fired from the gun into the block. To what depth does the bullet penetrate the block in this case?

40. A billiard ball rolling across a table at 1.50 m/s makes a head-on elastic collision with an identical ball. Find the speed of each ball after the collision (a) when the second ball is initially at rest, (b) when the second ball is moving toward the first at a speed of 1.00 m/s, and (c) when the second ball is moving away from the first at a speed of 1.00 m/s.

41. A 90-kg fullback moving east with a speed of 5.0 m/s is tackled by a 95-kg opponent running north at 3.0 m/s. If the collision is perfectly inelastic, calculate (a) the velocity of the players just after the tackle and (b) the kinetic energy lost as a result of the collision. Can you account for the missing energy?

42. An 8.00-kg object moving east at 15.0 m/s on a frictionless horizontal surface collides with a 10.0-kg object that is initially at rest. After the collision, the 8.00-kg object moves south at 4.00 m/s. (a) What is the velocity of the 10.0-kg object after the collision? (b) What percentage of the initial kinetic energy is lost in the collision?

43. A 2 000-kg car moving east at 10.0 m/s collides with a 3 000-kg car moving north. The cars stick together and move as a unit after the collision, at an angle of 40.0° north of east and a speed of 5.22 m/s. Find the speed of the 3 000-kg car before the collision.

44. Two automobiles of equal mass approach an intersection. One vehicle is traveling with velocity 13.0 m/s toward the east, and the other is traveling north with speed v_{2i}. Neither driver sees the other. The vehicles collide in the intersection and stick together, leaving parallel skid marks at an angle of 55.0° north of east. The speed limit for both roads is 35 mi/h, and the driver of the northward-moving vehicle claims he was within the limit when the collision occurred. Is he telling the truth?

45. **Physics⊗Now™** A billiard ball moving at 5.00 m/s strikes a stationary ball of the same mass. After the collision, the first ball moves at 4.33 m/s at an angle of 30° with respect to the original line of motion. (a) Find the velocity (magnitude and direction) of the second ball after collision. (b) Was the collision inelastic or elastic?

ADDITIONAL PROBLEMS

46. In research in cardiology and exercise physiology, it is often important to know the mass of blood pumped by a person's heart in one stroke. This information can be obtained by means of a *ballistocardiograph*. The instrument works as follows: The subject lies on a horizontal pallet floating on a film of air. Friction on the pallet is negligible. Initially, the momentum of the system is zero. When the heart beats, it expels a mass m of blood into the aorta with speed v, and the body and platform move in the opposite direction with speed V. The speed of the blood can be determined independently (for example, by observing an ultrasound Doppler shift). Assume that the blood's speed is 50.0 cm/s in one typical trial. The mass of the subject plus the pallet is 54.0 kg. The pallet moves 6.00×10^{-5} m in 0.160 s after one heartbeat. Calculate the mass of blood that leaves the heart. Assume that the mass of blood is negligible compared with the total mass of the person. This simplified example illustrates the principle of ballistocardiography, but in practice a more sophisticated model of heart function is used.

47. A 0.50-kg object is at rest at the origin of a coordinate system. A 3.0-N force in the $+x$-direction acts on the object for 1.50 s. (a) What is the velocity at the end of this interval? (b) At the end of the interval, a constant force of 4.0 N is applied in the $-x$-direction for 3.0 s. What is the velocity at the end of the 3.0 s?

48. Consider a frictionless track as shown in Figure P6.48. A block of mass $m_1 = 5.00$ kg is released from Ⓐ. It makes a head-on elastic collision at Ⓑ with a block of mass $m_2 = 10.0$ kg that is initially at rest. Calculate the maximum height to which m_1 rises after the collision.

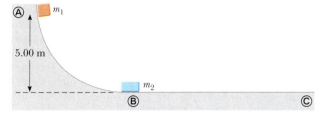

Figure P6.48

49. Most of us know intuitively that in a head-on collision between a large dump truck and a subcompact car, you are better off being in the truck than in the car. Why is this? Many people imagine that the collision force exerted on the car is much greater than that exerted on the truck. To substantiate this view, they point out that the car is crushed, whereas the truck is only dented. This idea of unequal forces, of course, is false; Newton's third law tells us that both objects are acted upon by forces of the same magnitude. The truck suffers less damage because it is made of stronger metal. But what about the two drivers? Do they experience the same forces? To answer this question, suppose that each vehicle is initially moving at 8.00 m/s and that they undergo a perfectly inelastic head-on collision. Each driver has mass 80.0 kg. Including the masses of the drivers, the total masses of the vehicles are 800 kg for the car and 4 000 kg for the truck. If the collision time is 0.120 s, what force does the seat belt exert on each driver?

50. A bullet of mass m and speed v passes completely through a pendulum bob of mass M as shown in Figure P6.50. The bullet emerges with a speed of $v/2$. The pendulum bob is suspended by a stiff rod of length ℓ and negligible mass. What is the minimum value of v such that the bob will barely swing through a complete vertical circle?

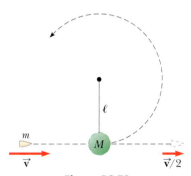

Figure P6.50

51. A 2.0-g particle moving at 8.0 m/s makes a perfectly elastic head-on collision with a resting 1.0-g object. (a) Find the speed of each particle after the collision. (b) Find the speed of each particle after the collision if the stationary particle has a mass of 10 g. (c) Find the final kinetic energy of the incident 2.0-g particle in the situations described in (a) and (b). In which case does the incident particle lose more kinetic energy?

52. A 0.400-kg green bead slides on a curved frictionless wire, starting from rest at point Ⓐ in Figure P6.52. At point Ⓑ, the bead collides elastically with a 0.600-kg blue ball at rest. Find the maximum height the blue ball rises as it moves up the wire.

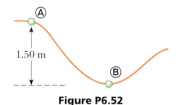

Figure P6.52

53. An 80-kg man standing erect steps off a 3.0-m-high diving platform and begins to fall from rest. The man again comes to rest 2.0 s after reaching the water. What average force did the water exert on him?

54. A 12.0-g bullet is fired horizontally into a 100-g wooden block initially at rest on a horizontal surface. After impact, the block slides 7.5 m before coming to rest. If the coefficient of kinetic friction between block and surface is 0.650, what was the speed of the bullet immediately before impact?

55. A 60.0-kg person running at an initial speed of 4.00 m/s jumps onto a 120-kg cart that is initially at rest (Figure P6.55). The person slides on the cart's top surface and finally comes to rest relative to the cart. The coefficient of kinetic friction between the person and the cart is 0.400. Friction between the cart and ground can be neglected. (a) Find the final speed of the person and cart relative to the ground. (b) Find the frictional force acting on the person while he is sliding across the top surface of the cart. (c) How long does the frictional force act on the person? (d) Find the change in momentum of the person and the change in momentum of the cart. (e) Determine the displacement of the person relative to the ground while he is sliding on the cart. (f) Determine the displacement of the cart relative to the ground while the person is sliding. (g) Find the change in kinetic energy of the person. (h) Find the change in kinetic energy of the cart. (i) Explain why the answers to (g) and (h) differ. (What kind of collision is this, and what accounts for the loss of mechanical energy?)

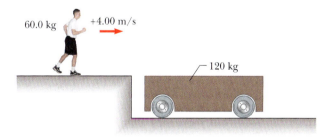

Figure P6.55

56. Two blocks of masses $m_1 = 2.00$ kg and $m_2 = 4.00$ kg are each released from rest at a height of 5.00 m on a frictionless track, as shown in Figure P6.56, and undergo an elastic head-on collision. (a) Determine the velocity of each block just before the collision. (b) Determine the velocity of each block immediately after the collision. (c) Determine the maximum heights to which m_1 and m_2 rise after the collision.

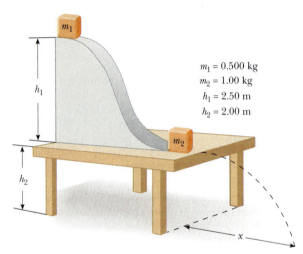

Figure P6.56

57. A 0.500-kg block is released from rest at the top of a frictionless track 2.50 m above the top of a table. It then collides elastically with a 1.00-kg object that is initially at rest on the table, as shown in Figure P6.57. (a) Determine the velocities of the two objects just after the collision. (b) How high up the track does the 0.500-kg object travel back after the collision? (c) How far away from the bottom of the table does the 1.00-kg object land, given that the table is 2.00 m high? (d) How far away from the bottom of the table does the 0.500-kg object eventually land?

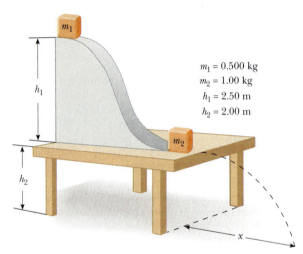

$m_1 = 0.500$ kg
$m_2 = 1.00$ kg
$h_1 = 2.50$ m
$h_2 = 2.00$ m

Figure P6.57

58. Tarzan, whose mass is 80.0 kg, swings from a 3.00-m vine that is horizontal when he starts. At the bottom of his arc, he picks up 60.0-kg Jane in a perfectly inelastic collision. What is the height of the highest tree limb they can reach on their upward swing?

59. A small block of mass $m_1 = 0.500$ kg is released from rest at the top of a curved wedge of mass $m_2 = 3.00$ kg, which sits on a frictionless horizontal surface as in Figure P6.59a. When the block leaves the wedge, its velocity is measured to be 4.00 m/s to the right, as in Figure P6.59b. (a) What is the velocity of the wedge after the block reaches the horizontal surface? (b) What is the height h of the wedge?

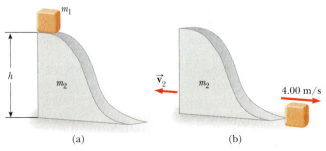

Figure P6.59

60. Two carts of equal mass $m = 0.250$ kg are placed on a frictionless track that has a light spring of force constant $k = 50.0$ N/m attached to one end of it, as in Figure P6.60. The red cart is given an initial velocity of $\vec{v}_0 = 3.00$ m/s to the right, and the blue cart is initially at rest. If the carts collide elastically, find (a) the velocity of the carts just after the first collision and (b) the maximum compression of the spring.

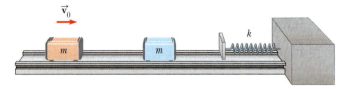

Figure P6.60

61. A cannon is rigidly attached to a carriage, which can move along horizontal rails, but is connected to a post by a large spring, initially unstretched and with force constant $k = 2.00 \times 10^4$ N/m, as in Figure P6.61. The cannon fires a 200-kg projectile at a velocity of 125 m/s directed 45.0° above the horizontal. (a) If the mass of the cannon and its carriage is 5 000 kg, find the recoil speed of the cannon. (b) Determine the maximum extension of the spring. (c) Find the maximum force the spring exerts on the carriage. (d) Consider the system consisting of the cannon, the carriage, and the shell. Is the momentum of this system conserved during the firing? Why or why not?

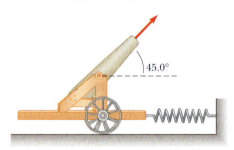

45.0°

Figure P6.61

62. Two objects of masses m and $3m$ are moving toward each other along the x-axis with the same initial speed v_0. The object with mass m is traveling to the left, and the object with mass $3m$ is traveling to the right. They undergo an elastic glancing collision such that m is moving downward after the collision at right angles from its initial direction. (a) Find the final speeds of the two objects. (b) What is the angle θ at which the object with mass $3m$ is scattered?

63. Physics⊗Now™ A neutron in a reactor makes an elastic head-on collision with a carbon atom that is initially at rest. (The mass of the carbon nucleus is about 12 times that of the neutron.) (a) What fraction of the neutron's kinetic energy is transferred to the carbon nucleus? (b) If the neutron's initial kinetic energy is 1.6×10^{-13} J, find its final kinetic energy and the kinetic energy of the carbon nucleus after the collision.

64. A cue ball traveling at 4.00 m/s makes a glancing, elastic collision with a target ball of equal mass that is initially at rest. The cue ball is deflected so that it makes an angle of 30.0° with its original direction of travel. Find (a) the angle between the velocity vectors of the two balls after the collision and (b) the speed of each ball after the collision.

65. A block of mass m lying on a rough horizontal surface is given an initial velocity of $\vec{v}_0$. After traveling a distance d, it makes a head-on elastic collision with a block of mass $2m$. How far does the second block move before coming to rest? (Assume that the coefficient of friction, μ_k, is the same for both blocks.)

66. The "force platform" is a tool that is used to analyze the performance of athletes by measuring the vertical force as a function of time that the athlete exerts on the ground in performing various activities. A simplified force vs. time graph for an athlete performing a standing high jump is shown in Figure P6.66. The athlete started the jump at $t = 0.0$ s. How high did this athlete jump?

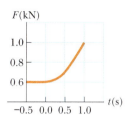

Figure P6.66

67. (a) A car traveling due east strikes a car traveling due north at an intersection, and the two move together as a unit. A property owner on the southeast corner of the intersection claims that his fence was torn down in the collision. Should he be awarded damages by the insurance company? Defend your answer. (b) Let the eastward-moving car have a mass of 1 300 kg and a speed of 30.0 km/h and the northward-moving car a mass of 1 100 kg and a speed of 20.0 km/h. Find the velocity after the collision. Are the results consistent with your answer to part (a)?

68. Two blocks collide on a frictionless surface. After the collision, the blocks stick together. Block A has a mass M and is initially moving to the right at speed v. Block B has a mass $2M$ and is initially at rest. System C is composed of both blocks. (a) Draw a free-body diagram for each block at an instant *during* the collision. (b) Rank the magnitudes of the horizontal forces in your diagram. Explain your reasoning. (c) Calculate the change in momentum of block A, block B, and system C. (d) Is kinetic energy conserved in this collision? Explain your answer. (This problem is courtesy of Edward F. Redish. For more such problems, visit http://www.physics.umd.edu/perg.)

69. A tennis ball of mass 57.0 g is held just above a basketball of mass 590 g. With their centers vertically aligned, both balls are released from rest at the same time, to fall through a distance of 1.20 m, as shown in Figure P6.69. (a) Find the magnitude of the downward velocity with which the basketball reaches the ground. (b) Assume that

an elastic collision with the ground instantaneously reverses the velocity of the basketball while the tennis ball is still moving down. Next, the two balls meet in an elastic collision. (b) To what height does the tennis ball rebound?

70. A 60-kg soccer player jumps vertically upwards and heads the 0.45-kg ball as it is descending vertically with a speed of 25 m/s. If the player was moving upward with a speed of 4.0 m/s just before impact, what will be the speed of the ball immediately after the collision if the ball rebounds vertically upwards and the collision is elastic? If the ball is in contact with the player's head for 20 ms, what is the average acceleration of the ball? (Note that the force of gravity may be ignored during the brief collision time.)

71. Small ice cubes, each of mass 5.00 g, slide down a frictionless ski-jump track in a steady stream, as shown in Figure P6.71. Starting from rest, each cube moves down through a net vertical distance of 1.50 m and leaves the bottom end of the track at an angle of 40.0° above the horizontal. At the highest point of its subsequent trajectory, the cube strikes a vertical wall and rebounds with half the speed it had upon impact. If 10.0 cubes strike the wall per second, what average force is exerted on the wall?

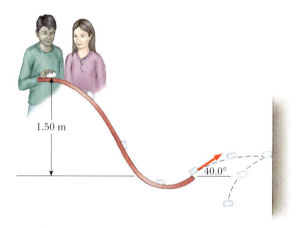

1.50 m

40.0°

Figure P6.71

72. A 0.30-kg puck, initially at rest on a frictionless horizontal surface, is struck by a 0.20-kg puck that is initially moving along the x-axis with a velocity of 2.0 m/s. After the collision, the 0.20-kg puck has a speed of 1.0 m/s at an angle of $\theta = 53°$ to the positive x-axis. (a) Determine the velocity of the 0.30-kg puck after the collision. (b) Find the fraction of kinetic energy lost in the collision.

73. A cannon initially resting on a frictionless surface of mass $m_1 = 800$ kg (when unloaded) is loaded with a "shot" of mass $m_2 = 10.0$ kg. The cannon is aimed at mass $m_3 = 7\,990$ kg, which is connected to a massless spring of force constant $k = 4\,500$ N/m, as in Figure P6.73a. The cannon is then fired, and the shot inelastically collides with mass m_3 and sticks in it, as shown in Figure P6.73b. The combined system compresses the spring a maximum distance of $d = 0.500$ m, as in Figure P6.73c. (a) Determine the speed of m_2 just before it collides with m_3. (You may assume that m_2 travels in a straight line.) (b) Determine the recoil speed of the cannon. (c) The cannon recoils towards the right, and when it passes point A there is friction (with $\mu_k = 0.600$) between the cannon and the ground. How far to the right of A does the cannon slide before coming to rest?

Figure P6.69

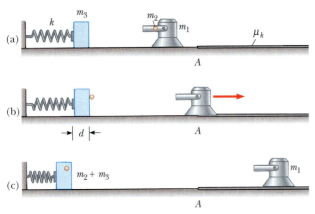

Figure P6.73

74. A flying squid (family Ommastrephidae) is able to "jump" off the surface of the sea by taking water into its body cavity and then ejecting the water vertically downward. A 0.85-kg squid is able to eject 0.30 kg of water with a speed of 20 m/s. (a) What will be the speed of the squid immediately after ejecting the water. (b) How high in the air will the squid rise?

ACTIVITIES

A.1. Tie a string to two tennis balls to make identical pendulums. Support the two such that they are of the same length and just touching. Now model an inelastic collision by attaching a piece of putty or double-stick tape to one of the balls so that when the balls collide they will stick together and move as a unit. Move one pendulum upward to some starting amplitude, and observe the amplitude to which the combination rises after the collision. Discuss and explain the result. (Don't forget that *after* the collision, the energy of the system (the two tennis balls and the Earth) is conserved, and you use this fact to determine the height to which the combination rises.)

A.2. A fun device that illustrates conservation of momentum and conservation of kinetic energy is the so-called executive stress reliever shown in Figure A6.2. It consists of five identical hard balls supported by strings of equal lengths. When one ball is pulled out and released, an almost-elastic collision causes one ball to move out on the opposite side with the same speed as the incoming ball. If two balls are pulled out and released, two balls swing out on the opposite side, and so forth—but how do the balls "know"? For example, is it possible that, on occasion, when one ball is released, two will swing out on the opposite side traveling with half the speed of the incoming ball? If you have access to this toy, try the experiment to see if you can get the latter collision to occur. When you have convinced yourself that it is not going to happen, find the kinetic energy of the system with one incoming ball of mass m and speed v and the kinetic energy of two outgoing balls with mass m and speed $v/2$ to show that kinetic energy is not conserved in such a collision.

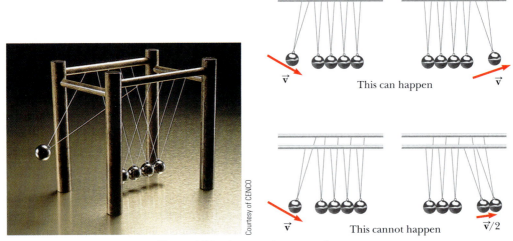

Figure A6.2 An executive stress-reliever

Courtesy NASA

CHAPTER

7

Rotational Motion and the Law of Gravity

Rotational motion is an important part of everyday life. The rotation of the Earth creates the cycle of day and night, the rotation of wheels enables easy vehicular motion, and modern technology depends on circular motion in a variety of contexts, from the tiny gears in a Swiss watch to the operation of lathes and other machinery. The concepts of *angular speed, angular acceleration*, and *centripetal acceleration* are central to understanding the motions of a diverse range of phenomena, from a car moving around a circular race track to clusters of galaxies orbiting a common center.

Rotational motion, when combined with Newton's law of universal gravitation and his laws of motion, can also explain certain facts about space travel and satellite motion, such as where to place a satellite so it will remain fixed in position over the same spot on the Earth. The generalization of gravitational potential energy and energy conservation offers an easy route to such results as planetary escape speed. Finally, we present Kepler's three laws of planetary motion, which formed the foundation of Newton's approach to gravity.

7.1 ANGULAR SPEED AND ANGULAR ACCELERATION

In the study of linear motion, the important concepts are *displacement* Δx, *velocity v,* and *acceleration a.* Each of these concepts has its analog in rotational motion: *angular displacement* $\Delta\theta$, *angular velocity* ω, and *angular acceleration* α.

The *radian*, a unit of angular measure, is essential to the understanding of these concepts. Recall that the distance *s* around a circle is given by $s = 2\pi r$, where *r* is

189

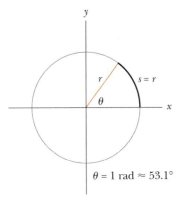

Figure 7.1 For a circle of radius r, one radian is the angle subtended by an arc length equal to r.

TIP 7.1   Remember the Radian

Equation 7.1 defines an angle expressed in *radians*. Angles expressed in terms of degrees must first be converted to radians. Also, be sure to check whether your calculator is in degree or radian mode when solving problems involving rotation.

the radius of the circle. Dividing both sides by r results in $s/r = 2\pi$. This quantity is dimensionless, because both s and r have dimensions of length, but the value 2π corresponds to a displacement around a circle. A half circle would give an answer of π, a quarter circle an answer of $\pi/2$. The numbers 2π, π, and $\pi/2$ correspond to angles of 360°, 180°, and 90°, respectively, so a new unit of angular measure, the **radian**, can be defined as **the arc length s along a circle divided by the radius r**:

$$\theta = \frac{s}{r} \qquad \text{[7.1]}$$

Figure 7.1 illustrates the size of 1 radian, which is approximately 53°. For conversions, we use the fact that 360° = 2π radians (or 180° = π radians). For example, 45° $(2\pi \, \text{rad}/360°) = (\pi/4)$ rad.

Generally, angular quantities in physics must be expressed in radians. Be sure to set your calculator to radian mode; neglecting to do this is a common error.

Armed with the concept of the radian, we can now discuss angular concepts in physics. Consider Figure 7.2a, a top view of a rotating compact disc. Such a disk is an example of a "rigid body," with each part of the body fixed in position relative all other parts of the body. When a rigid body rotates through a given angle, all parts of the body rotate through the same angle at the same time. For the compact disk, the axis of rotation is at the center of the disc, O. A point P on the disc is at a distance r from the origin and moves about O in a circle of radius r. We set up a *fixed* reference line, as shown in Figure 7.2a, and assume that at time $t = 0$ the point P is on that reference line. After a time interval Δt has elapsed, P has advanced to a new position (Fig. 7.2b). In this interval, the line OP has moved through the angle θ with respect to the reference line. The angle θ, measured in radians, is called the **angular position** and is analogous to the linear position variable x. Likewise, P has moved an arc length s measured along the circumference of the circle.

In Figure 7.3, as a point on the rotating disc moves from Ⓐ to Ⓑ in a time Δt, it starts at an angle θ_i and ends at an angle θ_f. The difference $\theta_f - \theta_i$ is called the **angular displacement**.

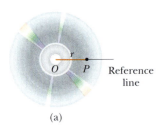

(a)

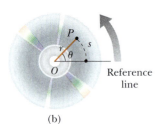

(b)

Figure 7.2 (a) The point P on a rotating compact disc at $t = 0$. (b) As the disc rotates, P moves through an arc length s.

An object's angular displacement, $\Delta\theta$, is the difference in its final and initial angles:

$$\Delta\theta = \theta_f - \theta_i \qquad \text{[7.2]}$$

SI unit: radian (rad)

For example, if a point on a disk is at $\theta_i = 4$ rad and rotates to angular position $\theta_f = 7$ rad, the angular displacement is $\Delta\theta = \theta_f - \theta_i = 7 \, \text{rad} - 4 \, \text{rad} = 3$ rad. Note that we use angular variables to describe the rotating disc because **each point on the disc undergoes the same angular displacement in any given time interval**.

Having defined angular displacements, it's natural to define an angular speed:

Average angular speed ▶

The average angular speed ω_{av} of a rotating rigid object during the time interval Δt is defined as the angular displacement $\Delta\theta$ divided by Δt:

$$\omega_{av} \equiv \frac{\theta_f - \theta_i}{t_f - t_i} = \frac{\Delta\theta}{\Delta t} \qquad \text{[7.3]}$$

SI unit: radian per second (rad/s)

For very short time intervals, the average angular speed approaches the instantaneous angular speed, just as in the linear case.

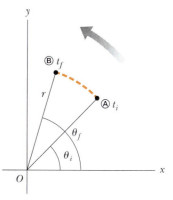

Figure 7.3   As a point on the compact disc moves from Ⓐ to Ⓑ, the disc rotates through the angle $\Delta\theta = \theta_f - \theta_i$.

> The **instantaneous angular speed** ω of a rotating rigid object is defined as the limit of the average speed $\Delta\theta/\Delta t$ as the time interval Δt approaches zero:
>
> $$\omega \equiv \lim_{\Delta t \to 0} \frac{\Delta\theta}{\Delta t} \qquad [7.4]$$
>
> **SI unit: radian per second (rad/s)**

We take ω to be positive when θ is increasing (counterclockwise motion) and negative when θ is decreasing (clockwise motion). When the angular speed is constant, the instantaneous angular speed is equal to the average angular speed.

EXAMPLE 7.1   Whirlybirds

Goal   Convert an angular speed in revolutions per minute to radians per second.

Problem   The rotor on a helicopter turns at an angular speed of 3.20×10^2 revolutions per minute. (In this book, we sometimes use the abbreviation rpm, but in most cases we use rev/min.) Express this angular speed in radians per second.

Strategy   During one revolution, the rotor turns through an angle of 2π radians. Use this relationship as a conversion factor.

Solution

Apply the conversion factors 1 rev = 2π rad and 60 s = 1 min:

$$\omega = 3.20 \times 10^2 \, \frac{\text{rev}}{\text{min}}$$

$$= 3.20 \times 10^2 \, \frac{\text{rev}}{\text{min}} \left(\frac{2\pi \, \text{rad}}{\text{rev}} \right) \left(\frac{1.00 \, \text{min}}{60.0 \, \text{s}} \right)$$

$$= \boxed{33.5 \, \text{rad/s}}$$

Exercise 7.1
A waterwheel turns at 1 500 revolutions per hour. Express this figure in radians per second.

Answer   2.6 rad/s

Quick Quiz 7.1

A rigid body is rotating counterclockwise about a fixed axis. Each of the following pairs of quantities represents an initial angular position and a final angular position of the rigid body. Which of the sets can occur *only* if the rigid body rotates through more than 180°? (a) 3 rad, 6 rad; (b) −1 rad, 1 rad; (c) 1 rad, 5 rad.

Quick Quiz 7.2

Suppose that the change in angular position for each of the pairs of values in Quick Quiz 7.1 occurred in 1 s. Which choice represents the lowest average angular speed?

Figure 7.4 shows a bicycle turned upside down so that a repair technician can work on the rear wheel. The bicycle pedals are turned so that at time t_i the wheel has angular speed ω_i (Fig. 7.4a) and at a later time t_f it has angular speed ω_f

Figure 7.4 An accelerating bicycle wheel rotates with (a) angular speed ω_i at time t_i and (b) angular speed ω_f at time t_f.

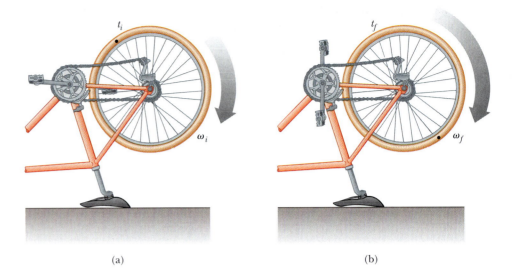

(a)   (b)

(Fig. 7.4b). Just as a changing speed leads to the concept of an acceleration, a changing angular speed leads to the concept of an angular acceleration.

Average angular acceleration ▶

An object's average angular acceleration α_{av} during the time interval Δt is defined as the change in its angular speed $\Delta\omega$ divided by Δt:

$$\alpha_{av} \equiv \frac{\omega_f - \omega_i}{t_f - t_i} = \frac{\Delta\omega}{\Delta t} \qquad [7.5]$$

SI unit: radian per second squared (rad/s^2)

As with angular velocity, positive angular accelerations are in the counterclockwise direction, negative angular accelerations in the clockwise direction. If the angular speed goes from 15 rad/s to 9.0 rad/s in 3.0 s, then the average angular acceleration during that time interval is

$$\alpha_{av} = \frac{\Delta\omega}{\Delta t} = \frac{9.0 \text{ rad/s} - 15 \text{ rad/s}}{3.0 \text{ s}} = -2.0 \text{ rad/s}$$

The negative sign indicates that the angular acceleration is clockwise (though the angular speed, still positive but slowing down, is in the counterclockwise direction). There is also an instantaneous version of angular acceleration:

Instantaneous angular acceleration ▶

The instantaneous angular acceleration α is defined as the limit of the average angular acceleration $\Delta\omega/\Delta t$ as the time interval Δt approaches zero:

$$\alpha \equiv \lim_{\Delta t \to 0} \frac{\Delta\omega}{\Delta t} \qquad [7.6]$$

SI unit: radian per second squared (rad/s^2)

When a rigid object rotates about a fixed axis, as does the bicycle wheel, every portion of the object has the same angular speed and the same angular acceleration. This fact is what makes these variables so useful for describing rotational motion. In contrast, the tangential (linear) speed and acceleration of the object take different values that depend on the distance from a given point to the axis of rotation.

7.2   ROTATIONAL MOTION UNDER CONSTANT ANGULAR ACCELERATION

A number of parallels exist between the equations for rotational motion and those for linear motion. For example, compare the defining equation for the average angular speed,

$$\omega_{av} \equiv \frac{\theta_f - \theta_i}{t_f - t_i} = \frac{\Delta\theta}{\Delta t}$$

with that of the average linear speed,

$$v_{av} \equiv \frac{x_f - x_i}{t_f - t_i} = \frac{\Delta x}{\Delta t}$$

In these equations, ω takes the place of v and θ takes the place of x, so the equations differ only in the names of the variables. In the same way, every linear quantity we have encountered so far has a corresponding "twin" in rotational motion.

The procedure used in Section 2.5 to develop the kinematic equations for linear motion under constant acceleration can be used to derive a similar set of equations for rotational motion under constant angular acceleration. The resulting equations of rotational kinematics, along with the corresponding equations for linear motion, are as follows:

Linear Motion with a Constant (Variables: x and v)	Rotational Motion about a Fixed Axis with α Constant (Variables: θ and ω)	
$v = v_i + at$	$\omega = \omega_i + \alpha t$	[7.7]
$\Delta x = v_i t + \frac{1}{2}at^2$	$\Delta\theta = \omega_i t + \frac{1}{2}\alpha t^2$	[7.8]
$v^2 = v_i^2 + 2a\Delta x$	$\omega^2 = \omega_i^2 + 2\alpha\Delta\theta$	[7.9]

Notice that every term in a given linear equation has a corresponding term in the analogous rotational equation.

Quick Quiz 7.3

Consider again the pairs of angular positions for the rigid object in Quick Quiz 7.1. If the object starts from rest at the initial angular position, moves counterclockwise with constant angular acceleration, and arrives at the final angular position with the same angular speed in all three cases, for which choice is the angular acceleration the highest?

EXAMPLE 7.2   A Rotating Wheel

Goal   Apply the rotational kinematic equations.

Problem   A wheel rotates with a constant angular acceleration of 3.50 rad/s². If the angular speed of the wheel is 2.00 rad/s at $t_i = 0$, **(a)** through what angle does the wheel rotate between $t = 0$ and $t = 2.00$ s? Give your answer in radians and in revolutions. **(b)** What is the angular speed of the wheel at $t = 2.00$ s?

Strategy   The angular acceleration is constant, so this problem just requires substituting given values into Equations 7.7 and 7.8.

Solution

(a) Find the angular displacement after 2.00 s, in both radians and revolutions.

Use Equation 7.8, setting $\omega_i = 2.00$ rad/s, $\alpha = 3.5$ rad/s², and $t = 2.00$ s:

$$\Delta\theta = \omega_i t + \tfrac{1}{2}\alpha t^2$$
$$= (2.00 \text{ rad/s})(2.00 \text{ s}) + \tfrac{1}{2}(3.50 \text{ rad/s}^2)(2.00 \text{ s})^2$$
$$= 11.0 \text{ rad}$$

Convert radians to revolutions. $\Delta\theta = (11.0\ \text{rad})(1.00\ \text{rev}/2\pi\ \text{rad}) = \boxed{1.75\ \text{rev}}$

(b) What is the angular speed of the wheel at $t = 2.00$ s?

Substitute the same values into Equation 7.7: $\omega = \omega_i + \alpha t = 2.00\ \text{rad/s} + (3.50\ \text{rad/s}^2)(2.00\ \text{s})$

$= \boxed{9.00\ \text{rad/s}}$

Remarks   The result of part (b) could also be obtained from Equation 7.9 and the results of part (a).

Exercise 7.2
(a) Find the angle through which the wheel rotates between $t = 2.00$ s and $t = 3.00$ s. (b) Find the angular speed when $t = 3.00$ s.

Answer   (a) 10.8 rad   (b) 12.5 rad/s

EXAMPLE 7.3   Slowing Propellers

Goal   Apply the time-independent rotational kinematic equation.

Problem   An airplane propeller slows from an initial angular speed of 12.5 rev/s to a final angular speed of 5.00 rev/s. During this process, the propeller rotates through 21.0 revolutions. Find the angular acceleration of the propeller in radians per second squared, assuming it's constant.

Strategy   The given quantities are the angular speeds and the displacement, which suggests applying Equation 7.9, the time-independent rotational kinematic equation, to find α.

Solution
First, convert the angular displacement to radians and the angular speeds to rad/s:

$\Delta\theta = (21.0\ \text{rev})(2\pi\ \text{rad/rev}) = 42.0\pi\ \text{rad}$

$\omega_i = (12.5\ \text{rev/s})(2\pi\ \text{rad/rev}) = 25.0\pi\ \text{rad/s}$

$\omega = (5.00\ \text{rev/s})(2\pi\ \text{rad/rev}) = 10.0\pi\ \text{rad/s}$

Substitute these values into Equation 7.9 to find the angular acceleration α:

$\omega^2 = \omega_i^2 + 2\alpha\Delta\theta$

$(10.0\pi\ \text{rad/s})^2 = (25.0\pi\ \text{rad/s})^2 + 2\alpha\,(42\pi\ \text{rad})$

Solve for α: $\alpha = \boxed{-6.25\pi\ \text{rad/s}^2}$

Remark   Waiting until the end to convert revolutions to radians is also possible and requires only one conversion instead of three.

Exercise 7.3
Suppose, instead, the engine speeds up so that the propeller goes through 28.0 revolutions while the angular speed increases uniformly from 5.00 rev/s to 15.0 rev/s. Find the angular acceleration.

Answer $7.14\pi\ \text{rad/s}^2$

7.3   RELATIONS BETWEEN ANGULAR AND LINEAR QUANTITIES

Angular variables are closely related to linear variables. Consider the arbitrarily shaped object in Active Figure 7.5 rotating about the z-axis through the point O. Assume that the object rotates through the angle $\Delta\theta$, and hence point P moves through the arc length Δs, in the interval Δt. We know from the defining equation for radian measure that

$$\Delta\theta = \frac{\Delta s}{r}$$

Dividing both sides of this equation by Δt, the time interval during which the rotation occurs, yields

$$\frac{\Delta\theta}{\Delta t} = \frac{1}{r}\frac{\Delta s}{\Delta t}$$

When Δt is very small, the angle $\Delta\theta$ through which the object rotates is also small and the ratio $\Delta\theta/\Delta t$ is close to the instantaneous angular speed ω. On the other side of the equation, similarly, the ratio $\Delta s/\Delta t$ approaches the instantaneous linear speed v for small values of Δt. Hence, when Δt gets arbitrarily small, the preceding equation is equivalent to

$$\omega = \frac{v}{r}$$

In Active Figure 7.5, the point P traverses a distance Δs along a circular arc during the time interval Δt at a linear speed of v. The direction of P's velocity vector $\vec{\mathbf{v}}$ is *tangent to the circular path*. The magnitude of $\vec{\mathbf{v}}$ is the linear speed $v = v_t$, called the **tangential speed** of a particle moving in a circular path, written

$$v_t = r\omega \qquad\qquad [7.10]$$

◀ Tangential speed

The tangential speed of a point on a rotating object equals the distance of that point from the axis of rotation multiplied by the angular speed. Equation 7.10 shows that the linear speed of a point on a rotating object increases as that point is moved outward from the center of rotation toward the rim, as expected; however, **every point on the rotating object has the same *angular* speed**.

Equation 7.10, derived using the defining equation for radian measure, is valid only when ω is measured in radians per unit time. Other measures of angular speed, such as degrees per second and revolutions per second, shouldn't be used.

To find a second equation relating linear and angular quantities, refer again to Figure 7.5, and suppose the rotating object changes its angular speed by $\Delta\omega$ in the time interval Δt. At the end of this interval, the speed of a point on the object, such as P, has changed by the amount Δv_t. From Equation 7.10 we have

$$\Delta v_t = r\,\Delta\omega$$

Dividing by Δt gives

$$\frac{\Delta v_t}{\Delta t} = r\frac{\Delta\omega}{\Delta t}$$

As the time interval Δt is taken to be arbitrarily small, $\Delta\omega/\Delta t$ approaches the instantaneous angular acceleration. On the left-hand side of the equation, the ratio $\Delta v_t/\Delta t$ tends to the instantaneous linear acceleration, called the tangential acceleration of that point, given by

$$a_t = r\alpha \qquad\qquad [7.11]$$

◀ Tangential acceleration

The tangential acceleration of a point on a rotating object equals the distance of that point from the axis of rotation multiplied by the angular acceleration. Again, radian measure must be used for the angular acceleration term in this equation.

One last equation that relates linear quantities to angular quantities will be derived in the next section.

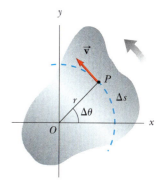

ACTIVE FIGURE 7.5
Rotation of an object about an axis through O (the z-axis) that is perpendicular to the plane of the figure. Note that a point P on the object rotates in a circle of radius r centered at O.

Physics⊗Now™
Log into PhysicsNow at **www.cp7e.com,** and go to Active Figure 7.5 to move point P and observe the tangential velocity as the object rotates.

Quick Quiz 7.4

Andrea and Chuck are riding on a merry-go-round. Andrea rides on a horse at the outer rim of the circular platform, twice as far from the center of the circular platform as Chuck, who rides on an inner horse. When the merry-go-round is rotating at a constant angular speed, Andrea's angular speed is (a) twice Chuck's (b) the same as Chuck's (c) half of Chuck's (d) impossible to determine.

Quick Quiz 7.5

When the merry-go-round of Quick Quiz 7.4 is rotating at a constant angular speed, Andrea's tangential speed is (a) twice Chuck's (b) the same as Chuck's (c) half of Chuck's (d) impossible to determine.

Applying Physics 7.1   ESA Launch Site

Why is the launch area for the European Space Agency in South America and not in Europe?

Explanation   Satellites are boosted into orbit on top of rockets, which provide the large tangential speed necessary to achieve orbit. Due to its rotation, the surface of Earth is already traveling toward the east at a tangential speed of nearly 1 700 m/s at the equator.

This tangential speed is steadily reduced further north, because the distance to the axis of rotation is decreasing. It finally goes to zero at the North Pole. Launching eastward from the equator gives the satellite a starting initial tangential speed of 1 700 m/s, whereas a European launch provides roughly half that speed (depending on the exact latitude).

EXAMPLE 7.4   Compact Discs

Goal   Apply the rotational kinematics equations in tandem with tangential acceleration and speed.

Problem   A compact disc rotates from rest up to an angular speed of 31.4 rad/s in a time of 0.892 s. **(a)** What is the angular acceleration of the disc, assuming the angular acceleration is uniform? **(b)** Through what angle does the disc turn while coming up to speed? **(c)** If the radius of the disc is 4.45 cm, find the final tangential speed of a microbe riding on the rim of the disc. **(d)** What is the magnitude of the tangential acceleration of the microbe at the given time?

Strategy   We can solve parts (a) and (b) by applying the kinematic equations for angular speed and angular displacement (Equations 7.7 and 7.8). Multiplying the radius by the angular acceleration yields the tangential acceleration at the rim, while multiplying the radius by the angular speed gives the tangential speed at that point.

Solution
(a) Find the angular acceleration.

Apply the angular velocity equation $\omega = \omega_i + \alpha t$, taking $\omega_i = 0$ at $t = 0$:

$$\alpha = \frac{\omega}{t} = \frac{31.4 \text{ rad/s}}{0.892 \text{ s}} = \boxed{35.2 \text{ rad/s}^2}$$

(b) Through what angle does the disc turn?

Use Equation 7.8 for angular displacement, with $t = 0.892$ s and $\omega_i = 0$:

$$\Delta\theta = \omega_i t + \tfrac{1}{2}\alpha t^2 = \tfrac{1}{2}(35.2 \text{ rad/s}^2)(0.892 \text{ s})^2 = \boxed{14.0 \text{ rad}}$$

(c) Find the final tangential speed of a microbe at $r = 4.45$ cm.

Substitute into Equation 7.10:

$$v_t = r\omega = (0.0445 \text{ m})(31.4 \text{ rad/s}) = \boxed{1.40 \text{ m/s}}$$

(d) Find the tangential acceleration of the microbe at $r = 4.45$ cm.

Substitute into Equation 7.11:

$$a_t = r\alpha = (0.0445 \text{ m})(35.2 \text{ rad/s}^2) = \boxed{1.57 \text{ m/s}^2}$$

Remarks   Because 2π rad $= 1$ rev, the angular displacement in part (**b**) corresponds to 2.23 rev. In general, dividing the number of radians by six gives a good approximation to the number of revolutions, because $2\pi \sim 6$.

Exercise 7.4
(a) What are the angular speed and angular displacement of the disc 0.300 s after it begins to rotate? (b) Find the tangential speed at the rim at this time.

Answers   (a) 10.6 rad/s; 1.58 rad   (b) 0.472 m/s

Before compact discs became the medium of choice for recorded music, phonographs were popular. There are similarities and differences between the rotational motion of phonograph records and that of compact discs. A phonograph record rotates at a constant angular speed. Popular angular speeds were $33\frac{1}{3}$ rev/min for long-playing albums (hence the nickname "LP"), 45 rev/min for "singles," and 78 rev/min used in very early recordings. At the outer edge of the record, the pickup needle (stylus) moves over the vinyl material at a faster tangential speed than when the needle is close to the center of the record. As a result, the sound information is compressed into a smaller length of track near the center of the record than near the outer edge.

CDs, on the other hand, are designed so that the disc moves under the laser pickup at a constant tangential speed. Because the pickup moves radially as it follows the tracks of information, the angular speed of the compact disc must vary according to the radial position of the laser. Since the tangential speed is fixed, the information density (per length of track) anywhere on the disc is the same. Example 7.5 demonstrates numerical calculations for both compact discs and phonograph records.

APPLICATION

Phonograph Records and Compact Discs

EXAMPLE 7.5   Track Length of a Compact Disc

Goal   Relate angular to linear variables.

Problem   In a compact disc player, as the read head moves out from the center of the disc, the angular speed of the disc changes so that the linear speed at the position of the head remains at a constant value of about 1.3 m/s. (a) Find the angular speed of the compact disc when the read head is at $r = 2.0$ cm and again at $r = 5.6$ cm. (b) An old-fashioned record player rotates at a constant angular speed, so the linear speed of the record groove moving under the detector (the stylus) changes. Find the linear speed of a 45.0-rpm record at points 2.0 and 5.6 cm from the center. (c) In both the CD and phonograph record, information is recorded in a continuous spiral track. Calculate the total length of the track for a CD designed to play for 1.0 h.

Strategy   This problem is just a matter of substituting numbers into the appropriate equations. Part (a) requires relating angular and linear speed with Equation 7.10, $v_t = r\omega$, solving for ω and substituting given values. In part (b), convert from rev/min to rad/s and substitute straight into Equation 7.10 to obtain the linear speeds. In part (c), linear speed multiplied by time gives the total distance.

Solution
(a) Find the angular speed when the read head is at $r = 2.0$ cm and $r = 5.6$ cm.

Solve $v_t = r\omega$ for ω and substitute the numbers. At $r = 2.0$ cm:

$$\omega = \frac{v_t}{r} = \frac{1.3 \text{ m/s}}{2.0 \times 10^{-2} \text{ m}} = \boxed{65 \text{ rad/s}}$$

Likewise, find the angular speed at $r = 5.6$ cm:

$$\omega = \frac{v_t}{r} = \frac{1.3 \text{ m/s}}{5.6 \times 10^{-2} \text{ m}} = \boxed{23 \text{ rad/s}}$$

(b) Find the linear speed in m/s of a 45.0-rpm record at points 2.0 cm and 5.6 cm from the center.

Convert rev/min to rad/s:

$$45.0 \frac{\text{rev}}{\text{min}} = 45.0 \frac{\text{rev}}{\text{min}} \left(\frac{2\pi \text{ rad}}{\text{rev}} \right) \left(\frac{1.00 \text{ min}}{60.0 \text{ s}} \right) = 4.71 \frac{\text{rad}}{\text{s}}$$

Calculate the linear speed at $r = 2.0$ cm:

$$v_t = r\omega = (2.0 \times 10^{-2} \text{ m})(4.71 \text{ rad/s}) = \boxed{0.094 \text{ m/s}}$$

Calculate the linear speed at $r = 5.6$ cm:

$$v_t = r\omega = (5.6 \times 10^{-2} \text{ m})(4.71 \text{ rad/s}) = \boxed{0.26 \text{ m/s}}$$

(c) Calculate the total length of the track for a CD designed to play for 1.0 h.

Multiply the linear speed of the read head by the time in seconds.

$$\Delta x = v_t t = (1.3 \text{ m/s})(3\,600 \text{ s}) = \boxed{4\,700 \text{ m}}$$

Remark   Notice that for the record player in part (b), even though the angular speed is constant at all points along a radial line, the tangential speed steadily increases with increasing r. The calculation for a CD in part (c) is easy only because the linear (tangential) speed is constant. It would be considerably more difficult for a record player, where the tangential speed depends on the distance from the center.

Exercise 7.5
Compute the linear speed of a record playing at $33\frac{1}{3}$ revolutions per minute (a) at $r = 2.00$ cm and (b) at $r = 5.60$ cm.

Answers   (a) 0.069 8 m/s   (b) 0.195 m/s

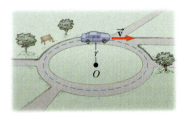

(a)

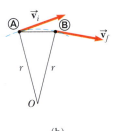

(b)

Figure 7.6   (a) Circular motion of a car moving with constant speed. (b) As the car moves along the circular path from Ⓐ to Ⓑ, the direction of its velocity vector changes, so the car undergoes a centripetal acceleration.

7.4   CENTRIPETAL ACCELERATION

Figure 7.6a shows a car moving in a circular path with *constant linear speed v.* **Even though the car moves at a constant speed, it still has an acceleration.** To understand this, consider the defining equation for average acceleration:

$$\vec{\mathbf{a}}_{av} = \frac{\vec{\mathbf{v}}_f - \vec{\mathbf{v}}_i}{t_f - t_i} \qquad [7.12]$$

The numerator represents the difference between the velocity vectors $\vec{\mathbf{v}}_f$ and $\vec{\mathbf{v}}_i$. These vectors may have the same *magnitude*, corresponding to the same speed, but if they have different *directions*, their difference can't equal zero. The direction of the car's velocity as it moves in the circular path is continually changing, as shown in Figure 7.6b. For circular motion at constant speed, the acceleration vector always points toward the center of the circle. Such an acceleration is called a **centripetal (center-seeking) acceleration**. Its magnitude is given by

$$a_c = \frac{v^2}{r} \qquad [7.13]$$

To derive Equation 7.13, consider Figure 7.7a. An object is first at point Ⓐ with velocity $\vec{\mathbf{v}}_i$ at time t_i and then at point Ⓑ with velocity $\vec{\mathbf{v}}_f$ at a later time t_f. We assume that $\vec{\mathbf{v}}_i$ and $\vec{\mathbf{v}}_f$ differ only in direction; their magnitudes are the same $(v_i = v_f = v)$. To calculate the acceleration, we begin with Equation 7.12,

$$\vec{\mathbf{a}}_{av} = \frac{\vec{\mathbf{v}}_f - \vec{\mathbf{v}}_i}{t_f - t_i} = \frac{\Delta\vec{\mathbf{v}}}{\Delta t} \qquad [7.14]$$

where $\Delta\vec{\mathbf{v}} = \vec{\mathbf{v}}_f - \vec{\mathbf{v}}_i$ is the change in velocity. When Δt is very small, Δs and $\Delta\theta$ are also very small. In Figure 7.7b, $\vec{\mathbf{v}}_f$ is almost parallel to $\vec{\mathbf{v}}_i$, and the vector $\Delta\vec{\mathbf{v}}$ is approximately perpendicular to them, pointing toward the center of the circle. In the limiting case when Δt becomes vanishingly small, $\Delta\vec{\mathbf{v}}$ points exactly toward the center of the circle, and the average acceleration $\vec{\mathbf{a}}_{av}$ becomes the instantaneous acceleration $\vec{\mathbf{a}}$. From Equation 7.14, $\vec{\mathbf{a}}$ and $\Delta\vec{\mathbf{v}}$ point in the same direction (in this limit), so the instantaneous acceleration points to the center of the circle.

The triangle in Figure 7.7a, which has sides Δs and r, is similar to the one formed by the vectors in Figure 7.7b, so the ratios of their sides are equal:

$$\frac{\Delta v}{v} = \frac{\Delta s}{r}$$

or

$$\Delta v = \frac{v}{r} \Delta s \qquad \text{[7.15]}$$

Substituting the result of Equation 7.15 into $a_{av} = \Delta v / \Delta t$ gives

$$a_{av} = \frac{v}{r} \frac{\Delta s}{\Delta t} \qquad \text{[7.16]}$$

But Δs is the distance traveled along the arc of the circle in time Δt, and in the limiting case when Δt becomes very small, $\Delta s / \Delta t$ approaches the instantaneous value of the tangential speed, v. At the same time, the average acceleration a_{av} approaches a_c, the instantaneous centripetal acceleration, so Equation 7.16 reduces to Equation 7.13:

$$a_c = \frac{v^2}{r}$$

Because the tangential speed is related to the angular speed through the relation $v_t = r\omega$ (Eq. 7.10), an alternate form of Equation 7.13 is

$$a_c = \frac{r^2 \omega^2}{r} = r\omega^2 \qquad \text{[7.17]}$$

Dimensionally, $[r] = L$ and $[\omega] = 1/T$, so the units of centripetal acceleration are L/T^2, as they should be. This is a geometric result relating the centripetal acceleration to the angular speed, but physically an acceleration can occur only if some force is present. For example, if a car travels in a circle on flat ground, the force of static friction between the tires and the ground provides the necessary centripetal force.

Note that a_c in Equations 7.13 and 7.17 represents only the *magnitude* of the centripetal acceleration. The acceleration itself is always directed towards the center of rotation.

The foregoing derivations concern circular motion at constant speed. When an object moves in a circle but is speeding up or slowing down, a tangential component of acceleration, $a_t = r\alpha$, is also present. Because the tangential and centripetal components of acceleration are perpendicular to each other, we can find the magnitude of the **total acceleration** with the Pythagorean theorem:

$$a = \sqrt{a_t^2 + a_c^2} \qquad \text{[7.18]}$$

◀ Total acceleration

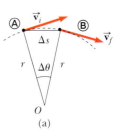

Figure 7.7 (a) As the particle moves from Ⓐ to Ⓑ, the direction of its velocity vector changes from $\vec{v}_i$ to $\vec{v}_f$. (b) The construction for determining the direction of the change in velocity $\Delta \vec{v}$, which is toward the center of the circle.

Quick Quiz 7.6

A race track is constructed such that two arcs of radius 80 m at Ⓐ and 40 m at Ⓑ are joined by two stretches of straight track as in Figure 7.8. In a particular trial run, a driver travels at a constant speed of 50 m/s for one complete lap.

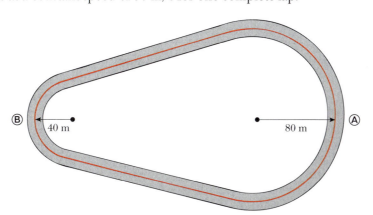

Figure 7.8 (Quick Quiz 7.6)

  1. The ratio of the tangential acceleration at Ⓐ to that at Ⓑ is
       (a) $\frac{1}{2}$ (b) $\frac{1}{4}$ (c) 2 (d) 4 (e) The tangential acceleration is zero at both points.
  2. The ratio of the centripetal acceleration at Ⓐ to that at Ⓑ is
       (a) $\frac{1}{2}$ (b) $\frac{1}{4}$ (c) 2 (d) 4 (e) The centripetal acceleration is zero at both points.
  3. The angular speed is greatest at
       (a) Ⓐ (b) Ⓑ (c) It is equal at both Ⓐ and Ⓑ.

Quick Quiz 7.7

An object moves in a circular path with constant speed v. Which of the following statements is true concerning the object? (a) Its velocity is constant, but its acceleration is changing. (b) Its acceleration is constant, but its velocity is changing. (c) Both its velocity and acceleration are changing. (d) Its velocity and acceleration remain constant.

EXAMPLE 7.6   At the Race Track

Goal   Apply the concepts of centripetal acceleration and tangential speed.

Problem   A race car accelerates uniformly from a speed of 40.0 m/s to a speed of 60.0 m/s in 5.00 s while traveling counterclockwise around a circular track of radius 4.00×10^2 m. When the car reaches a speed of 50.0 m/s, find (a) the magnitude of the car's centripetal acceleration (b) the angular speed (c) the tangential acceleration, and (d) the magnitude of the total acceleration.

Strategy   Substitute values into the definitions of centripetal acceleration (Equation 7.13), tangential speed (Equation 7.10), and total acceleration (Equation 7.18). Dividing the change in linear speed by the time yields the tangential acceleration.

Solution
(a) Find the magnitude of the centripetal acceleration when $v = 50.0$ m/s.

Substitute into Equation 7.13:

$$a_c = \frac{v^2}{r} = \frac{(50.0 \text{ m/s})^2}{(4.00 \times 10^2 \text{ m})} = \boxed{6.25 \text{ m/s}^2}$$

(b) Find the angular speed.

Solve Equation 7.10 for ω and substitute:

$$\omega = \frac{v}{r} = \frac{50.0 \text{ m/s}}{4.00 \times 10^2 \text{ m}} = \boxed{0.125 \text{ rad/s}}$$

(c) Find the tangential acceleration.

Divide the change in linear speed by the time:

$$a_t = \frac{v_f - v_i}{\Delta t} = \frac{60.0 \text{ m/s} - 40.0 \text{ m/s}}{5.00 \text{ s}} = \boxed{4.00 \text{ m/s}^2}$$

(d) Find the magnitude of the total acceleration.

Substitute into Equation 7.18:

$$a = \sqrt{a_t^2 + a_c^2} = \sqrt{(4.00 \text{ m/s}^2)^2 + (6.25 \text{ m/s}^2)^2}$$

$$a = \boxed{7.42 \text{ m/s}^2}$$

Remarks   We can also find the centripetal acceleration by substituting the derived value of ω into Equation 7.17.

Exercise 7.6
Suppose the race car now slows down uniformly from 60.0 m/s to 30.0 m/s in 4.50 s to avoid an accident, while still traversing a circular path 4.00×10^2 m in radius. Find the car's (a) centripetal acceleration, (b) angular speed, (c) tangential acceleration, and (d) total acceleration when the speed is 40.0 m/s.

Answers   (a) 4.00 m/s^2 (b) 0.100 rad/s   (c) -6.67 m/s^2 (d) 7.77 m/s^2

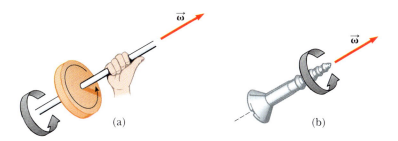

Figure 7.9   (a) The right-hand rule for determining the direction of the angular velocity vector $\vec{\omega}$. (b) The direction of $\vec{\omega}$ is in the direction of advance of a right-handed screw.

Angular Quantities Are Vectors

When we discussed linear motion in Chapter 2, we emphasized the fact that displacement, velocity, and acceleration are all vector quantities. In describing rotational motion, angular displacement, angular velocity, and angular acceleration are also vector quantities.

The direction of the angular velocity vector $\vec{\omega}$ can be found with the **right-hand rule**, as illustrated in Figure 7.9a. Grasp the axis of rotation with your right hand so that your fingers wrap in the direction of rotation. Your extended thumb then points in the direction of $\vec{\omega}$. Figure 7.9b shows that $\vec{\omega}$ is also in the direction of advance of a rotating right-handed screw.

We can apply this rule to a rotating disk viewed along the axis of rotation, as in Figure 7.10. When the disk rotates clockwise (Fig. 7.10a), the right-hand rule shows that the direction of $\vec{\omega}$ is into the page. When the disk rotates counterclockwise (Fig. 7.10b), the direction of $\vec{\omega}$ is out of the page.

Finally, the directions of the angular acceleration $\vec{\alpha}$ and the angular velocity $\vec{\omega}$ are the same if the angular speed ω (the magnitude of $\vec{\omega}$) is increasing with time, and opposite each other if the angular speed is decreasing with time.

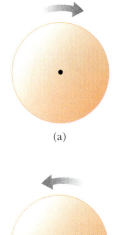

(a)

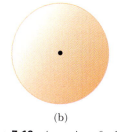

(b)

Figure 7.10   A top view of a disk rotating about an axis through its center perpendicular to the page. (a) When the disk rotates clockwise, $\vec{\omega}$ points into the page. (b) When the disk rotates counterclockwise, $\vec{\omega}$ points out of the page.

Forces Causing Centripetal Acceleration

An object can have a centripetal acceleration *only* if some external force acts on it. For a ball whirling in a circle at the end of a string, that force is the tension in the string. In the case of a car moving on a flat circular track, the force is friction between the car and track. A satellite in circular orbit around Earth has a centripetal acceleration due to the gravitational force between the satellite and Earth.

Some books use the term "centripetal force," which can give the mistaken impression that it is a new force of nature. This is not the case: The adjective "centripetal" in "centripetal force" simply means that the force in question acts toward a center. The gravitational force and the force of tension in the string of a yo-yo whirling in a circle are examples of centripetal forces, as is the force of gravity on a satellite circling the Earth.

Consider a ball of mass m that is tied to a string of length r and is being whirled at constant speed in a horizontal circular path, as illustrated in Figure 7.11. Its weight is supported by a frictionless table. Why does the ball move in a circle? Because of its inertia, the tendency of the ball is to move in a straight line; however, the string prevents motion along a straight line by exerting a radial force on the ball—a tension force—that makes it follow the circular path. The tension is directed along the string toward *the center of the circle, as shown* in the figure.

In general, applying Newton's second law along the radial direction yields the equation relating the net centripetal force F_c—the sum of the radial components of all forces acting on a given object—with the centripetal acceleration:

$$F_c = ma_c = m\frac{v^2}{r}$$ [7.19]

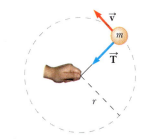

Figure 7.11   A ball attached to a string of length r, rotating in a circular path at constant speed.

TIP 7.2 **Centripetal Force is a** *Type* **of Force, not a Force in Itself!**

"Centripetal force" is a classification that includes forces acting toward a central point, like string tension on a tetherball or gravity on a satellite. A centripetal force must be *supplied* by some actual, physical force.

A net force causing a centripetal acceleration acts toward the center of the circular path and effects a change in the direction of the velocity vector. If that force should vanish, the object would immediately leave its circular path and move along a straight line tangent to the circle at the point where the force vanished.

Applying Physics 7.2 Artificial Gravity

Astronauts spending lengthy periods of time in space experience a number of negative effects due to weightlessness, such as weakening of muscle tissue and loss of calcium in bones. These effects may make it very difficult for them to return to their usual environment on Earth. How could artificial gravity be generated in space to overcome such complications?

Solution A rotating cylindrical space station creates an environment of artificial gravity. The normal force of the rigid walls provides the centripetal force, which keeps the astronauts moving in a circle (Fig. 7.12). To an astronaut, the normal force can't be easily distin-

guished from a gravitational force as long as the radius of the station is large compared with the astronaut's height. (Otherwise there are unpleasant inner ear effects.) This same principle is used in certain amusement park rides in which passengers are pressed against the inside of a rotating cylinder as it tilts in various directions. The visionary physicist Gerard O'Neill proposed creating a giant space colony a kilometer in radius that rotates slowly, creating Earth-normal artificial gravity for the inhabitants in its interior. These inside-out artificial worlds could enable safe transport on a several-thousand-year journey to another star system.

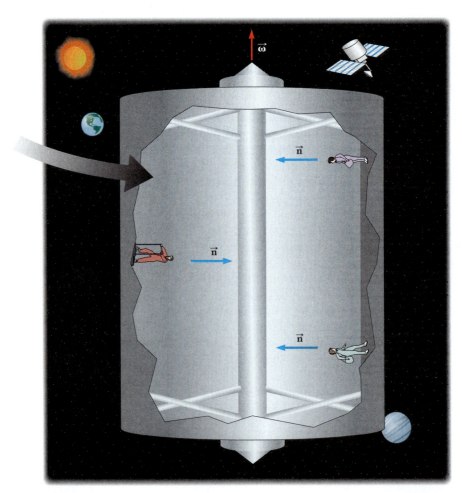

Figure 7.12 Artificial gravity inside a spinning cylinder is provided by the normal force.

Problem Solving Strategy   Forces that Cause Centripetal Acceleration

Use the following steps in dealing with centripetal accelerations and the forces that produce them:

1. **Draw a free-body diagram** of the object under consideration, labeling all forces that act on it.
2. **Choose a coordinate system** that has one axis perpendicular to the circular path followed by the object (the radial direction) and one axis tangent to the circular path (the tangential, or angular, direction). The normal direction, perpendicular to the plane of motion, is also often needed.
3. **Find the net force F_c toward the center** of the circular path, $F_c = \Sigma F_r$, where ΣF_c is the sum of the radial components of the forces. This net radial force causes the centripetal acceleration.
4. **Use Newton's second law for the radial, tangential, and normal directions**, as required, writing $\Sigma F_r = ma_c$, $\Sigma F_t = ma_t$, and $\Sigma F_n = ma_n$. Remember that the magnitude of the centripetal acceleration for uniform circular motion can always be written $a_c = v_t{}^2/r$.
5. **Solve** for the unknown quantities.

INTERACTIVE EXAMPLE 7.7   Buckle Up for Safety

Goal   Calculate the frictional force that causes an object to have a centripetal acceleration.

Problem   A car travels at a constant speed of 30.0 mi/h (13.4 m/s) on a level circular turn of radius 50.0 m, as shown in the bird's-eye view in Figure 7.13a. What minimum coefficient of static friction, μ_s, between the tires and roadway will allow the car to make the circular turn without sliding?

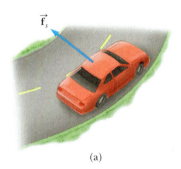

(a)

Strategy   In the car's free-body diagram (Fig. 7.13b) the normal direction is vertical and the tangential direction is into the page (step 2). Use Newton's second law. The net force acting on the car in the radial direction is the force of static friction toward the center of the circular path, which causes the car to have a centripetal acceleration. Calculating the maximum static friction force requires the normal force, obtained from the normal component of the second law.

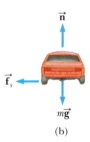

(b)

Figure 7.13   (Example 7.7) (a) Top view of a car on a curved path. (b) A free-body diagram of the car, showing an end view.

Solution

(Step 3, 4) Write the components of Newton's second law. The radial component involves only the maximum static friction force, $f_{s,\text{max}}$:

$$m\frac{v^2}{r} = f_{s,\text{max}} = \mu_s n$$

In the vertical component of the second law, the gravity force and the normal force are in equilibrium:

$$n - mg = 0 \quad \rightarrow \quad n = mg$$

(Step 5) Substitute the expression for n into the first equation and solve for μ_s:

$$m\frac{v^2}{r} = \mu_s mg$$

$$\mu_s = \frac{v^2}{rg} = \frac{(13.4 \text{ m/s})^2}{(50.0 \text{ m})(9.80 \text{ m/s}^2)} = 0.366$$

Remarks   The value of μ_s for rubber on dry concrete is very close to 1, so the car can negotiate the curve with ease. If the road were wet or icy, however, the value for μ_s could be 0.2 or lower. Under such conditions, the radial force provided by static friction wouldn't be great enough to keep the car on the circular path, and it would slide off on a tangent, leaving the roadway.

Exercise 7.7
At what maximum speed can a car negotiate a turn on a wet road with coefficient of static friction 0.230 without sliding out of control? The radius of the turn is 25.0 m.

Answer   7.51 m/s

Physics⊗Now™ Investigate the motion of a car around a level curve by logging into PhysicsNow at **www.cp7e.com** and going to Interactive Example 7.7.

EXAMPLE 7.8   Daytona International Speedway

Goal   Solve a centripetal force problem involving two dimensions.

Problem   The Daytona International Speedway in Daytona Beach, Florida, is famous for its races, especially the Daytona 500, held every spring. Both of its courses feature four-story, 31.0° banked curves, with maximum radius of 316 m. If a car negotiates the curve too slowly, it tends to slip down the incline of the turn, whereas if it's going too fast, it may begin to slide up the incline. **(a)** Find the necessary centripetal acceleration on this banked curve so the car won't slip down or slide up the incline. (Neglect friction.) **(b)** Calculate the speed of the race car.

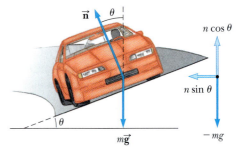

Figure 7.14   (Example 7.8) Front view of a car rounding a banked roadway. Vector components are shown to the right.

Strategy   Two forces act on the race car: the force of gravity and the normal force $\vec{\mathbf{n}}$. (See Fig. 7.14.) Use Newton's second law in the upward and radial directions to find the centripetal acceleration a_c. Solving $a_c = v^2/r$ for v then gives the race car's speed.

Solution
(a) Find the centripetal acceleration.

Write Newton's second law for the car:

$$m\vec{\mathbf{a}} = \sum \vec{\mathbf{F}} = \vec{\mathbf{n}} + m\vec{\mathbf{g}}$$

Use the y-component of Newton's second law to solve for the normal force n:

$$n \cos\theta - mg = 0$$

$$n = \frac{mg}{\cos\theta}$$

Obtain an expression for the horizontal component of $\vec{\mathbf{n}}$, which is the centripetal force F_c in this example:

$$F_c = n \sin\theta = \frac{mg \sin\theta}{\cos\theta} = mg \tan\theta$$

Substitute this expression for F_c into the radial component of Newton's second law and divide by m to get the centripetal acceleration:

$$ma_c = F_c$$

$$a_c = \frac{F_c}{m} = \frac{mg \tan\theta}{m} = g \tan\theta$$

$$a_c = (9.80 \text{ m/s}^2)(\tan 31.0°) = \boxed{5.89 \text{ m/s}^2}$$

(b) Find the speed of the race car.

Apply Equation 7.13:

$$\frac{v^2}{r} = a_c$$

$$v = \sqrt{ra_c} = \sqrt{(316 \text{ m})(5.89 \text{ m/s}^2)} = \boxed{43.1 \text{ m/s}}$$

Remarks   Both banking and friction assist in keeping the race car on the track.

APPLICATION
Banked Roadways

Exercise 7.8
A race track is to have a banked curve with radius of 245 m. What should be the angle of the bank if the normal force alone is to allow safe travel around the curve at 58.0 m/s?

Answer   54.5°

EXAMPLE 7.9   Riding the Tracks

Goal   Combine centripetal force with conservation of energy.

Problem   Figure 7.15a shows a roller-coaster car moving around a circular loop of radius R. **(a)** What speed must the car have so that it will just make it over the top without any assistance from the track? **(b)** What speed will the car subsequently have at the bottom of the loop? **(c)** What will be the normal force on a passenger at the bottom of the loop if the loop has a radius of 10.0 m?

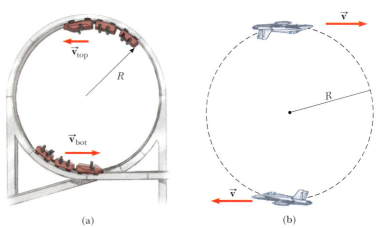

(a)   (b)

Figure 7.15   (a) (Example 7.9) A roller coaster traveling around a nearly circular track. (b) (Exercise 7.9) A jet executing a vertical loop.

Strategy   This problem requires Newton's second law and centripetal acceleration to find an expression for the car's speed at the top of the loop, followed by conservation of energy to find its speed at the bottom. If the car just makes it over the top, the force $\vec{\mathbf{n}}$ must become zero there, so the only force exerted on the car at that point is the force of gravity, $m\vec{\mathbf{g}}$. At the bottom of the loop, the normal force acts up toward the center and the gravity force acts down, away from the center. The difference of these two is the centripetal force. The normal force can then be calculated from Newton's second law.

Solution
(a) Find the speed at the top of the loop.

Write Newton's second law for the car:
$$m\vec{\mathbf{a}}_c = \vec{\mathbf{n}} + m\vec{\mathbf{g}} \tag{1}$$

At the top of the loop, set $n = 0$. The force of gravity acts toward the center and provides the centripetal acceleration $a_c = v^2/R$.
$$m\frac{v_{top}^2}{R} = mg$$

Solve the foregoing equation for v_{top}:
$$v_{top} = \sqrt{gR}$$

(b) Find the speed at the bottom of the loop.

Apply conservation of mechanical energy to find the total mechanical energy at the top of the loop:
$$E_{top} = \tfrac{1}{2}mv_{top}^2 + mgh = \tfrac{1}{2}mgR + mg(2R) = 2.5mgR$$

Find the total mechanical energy at the bottom of the loop:
$$E_{bot} = \tfrac{1}{2}mv_{bot}^2$$

Energy is conserved, so these two energies may be equated and solved for v_{bot}:
$$\tfrac{1}{2}mv_{bot}^2 = 2.5mgR$$
$$v_{bot} = \sqrt{5gR}$$

(c) Find the normal force on a passenger at the bottom. (This is the passenger's perceived weight.)

Use Equation(1). The net centripetal force is $n - mg$:

$$m\frac{v_{bot}^2}{R} = n - mg$$

Solve for n:

$$n = mg + m\frac{v_{bot}^2}{R} = mg + m\frac{5gR}{R} = \boxed{6\,mg}$$

Remarks   The final answer for n shows that the rider experiences a force six times normal weight at the bottom of the loop! Astronauts experience a similar force during space launches.

Exercise 7.9
A jet traveling at a speed of 1.20×10^2 m/s executes a vertical loop with a radius of 5.00×10^2 m. (See Fig. 7.15b.) Find the magnitude of the force of the seat on a 70.0-kg pilot at (a) the top and (b) the bottom of the loop.

Answers   (a) 1.33×10^3 N   (b) 2.70×10^3 N

Fictitious Forces

Anyone who has ridden a merry-go-round as a child (or as a fun-loving grown-up) has experienced what feels like a "center-fleeing" force. Holding onto the railing and moving toward the center feels like a walk up a steep hill.

Actually, this so-called centrifugal force is *fictitious*. In reality, the rider is exerting a centripetal force on his body with his hand and arm muscles. In addition, a smaller centripetal force is exerted by the static friction between his feet and the platform. If the rider's grip slipped, he wouldn't be flung radially away; rather, he would go off on a straight line, tangent to the point in space where he let go of the railing. The rider lands at a point that is further away from the center, but not by "fleeing the center" along a radial line. Instead, he travels perpendicular to a radial line, traversing an angular displacement while increasing his radial displacement. (See Fig. 7.16.)

TIP 7.3   Centrifugal Force
A so-called centrifugal force is most often just the *absence* of an adequate *centripetal force*, arising from measuring phenomena from a noninertial (accelerating) frame of reference such as a merry-go-round.

7.5   NEWTONIAN GRAVITATION

Prior to 1686, a great deal of data had been collected on the motions of the Moon and planets, but no one had a clear understanding of the forces affecting them. In that year, Isaac Newton provided the key that unlocked the secrets of the heavens. He knew from the first law that a net force had to be acting on the Moon. If it were not, the Moon would move in a straight-line path rather than in its almost circular orbit around Earth. Newton reasoned that this force arose as a result of an attractive force between Moon and Earth, called the force of gravity, and that it was the same kind of force that attracted objects—such as apples—close to the surface of the Earth.

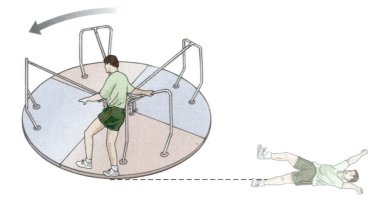

Figure 7.16   A fun-loving student loses his grip and falls along a line tangent to the rim of the merry-go-round.

In 1687, Newton published his work on the law of universal gravitation:

> If two particles with masses m_1 and m_2 are separated by a distance r, then a gravitational force acts along a line joining them, with magnitude given by
>
> $$F = G\frac{m_1 m_2}{r^2} \qquad [7.20]$$
>
> where $G = 6.673 \times 10^{-11}$ kg^{-1} m^3 s^{-2} is a constant of proportionality called the **constant of universal gravitation**. The gravitational force is always attractive.

◀ Law of universal gravitation

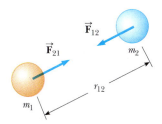

ACTIVE FIGURE 7.17
The gravitational force between two particles is attractive and acts along the line joining the particles. Note that according to Newton's third law, $\vec{F}_{12} = -\vec{F}_{21}$.

Physics⊗Now™
Log into PhysicsNow at **www.cp7e.com**, and go to Active Figure 7.17 to change the masses of the particles and the separation between them to see the effect on the gravitational force.

This force law is an example of an **inverse-square law**, in that it varies as one over the square of the separation of the particles. From Newton's third law, we know that the force exerted by m_1 on m_2, designated $\vec{F}_{12}$ in Active Figure 7.17, is equal in magnitude but opposite in direction to the force $\vec{F}_{21}$ exerted by m_2 on m_1, forming an action–reaction pair.

Another important fact is that **the gravitational force exerted by a uniform sphere on a particle outside the sphere is the same as the force exerted if the entire mass of the sphere were concentrated at its center**. This is called Gauss's law, after the German mathematician and astronomer Karl Friedrich Gauss, and is also true of electric fields, which we will encounter in Chapter 15. Gauss's law is a mathematical result, true because the force falls off as an inverse square of the separation between the particles.

Near the surface of the Earth, the expression $F = mg$ is valid. As shown in Table 7.1, however, the free-fall acceleration g varies considerably with altitude above the Earth.

Quick Quiz 7.8

A ball falls to the ground. Which of the following statements are false? (a) The force that the ball exerts on Earth is equal in magnitude to the force that Earth exerts on the ball. (b) The ball undergoes the same acceleration as Earth. (c) Earth pulls much harder on the ball than the ball pulls on Earth, so the ball falls while Earth remains stationary.

Quick Quiz 7.9

A planet has two moons with identical mass. Moon 1 is in a circular orbit of radius r. Moon 2 is in a circular orbit of radius $2r$. The magnitude of the gravitational force exerted by the planet on Moon 2 is (a) four times as large (b) twice as large (c) the same (d) half as large (e) one-fourth as large as the gravitational force exerted by the planet on Moon 1.

Measurement of the Gravitational Constant

The gravitational constant G in Equation 7.20 was first measured in an important experiment by Henry Cavendish in 1798. His apparatus consisted of two small spheres, each of mass m, fixed to the ends of a light horizontal rod suspended by a thin metal wire, as in Figure 7.18a (see page 208). Two large spheres, each of mass M, were placed near the smaller spheres. The attractive force between the smaller and larger spheres caused the rod to rotate in a horizontal plane and the wire to twist. The angle through which the suspended rod rotated was measured with a light beam reflected from a mirror attached to the vertical suspension. (Such a moving spot of light is an effective technique for amplifying the motion.) The experiment was carefully repeated with different masses at various separations. In addition to providing a value for G, the results showed that the force is attractive, proportional to the product mM, and inversely proportional to the square of the distance r. Modern forms of such experiments are carried out regularly today, in an effort to determine G with greater precision.

TABLE 7.1

Free-Fall Acceleration g at Various Altitudes

Altitude (km)[a]	g (m/s^2)
1 000	7.33
2 000	5.68
3 000	4.53
4 000	3.70
5 000	3.08
6 000	2.60
7 000	2.23
8 000	1.93
9 000	1.69
10 000	1.49
50 000	0.13

[a]All figures are distances above Earth's surface.

Figure 7.18   (a) A schematic diagram of the Cavendish apparatus for measuring G. The smaller spheres of mass *m* are attracted to the large spheres of mass *M*, and the rod rotates through a small angle. A light beam reflected from a mirror on the rotating apparatus measures the angle of rotation. (b) A student Cavendish apparatus.

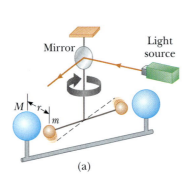

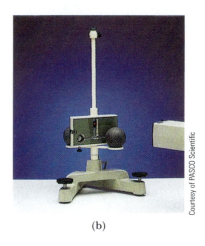

(a)

(b)

EXAMPLE 7.10   Billiards, Anyone?

Goal   Use vectors to find the net gravitational force on an object.

Problem   (a) Three 0.300-kg billiard balls are placed on a table at the corners of a right triangle, as shown from overhead in Figure 7.19. Find the net gravitational force on the cue ball (designated as m_1) resulting from the forces exerted by the other two balls. (b) Find the components of the gravitational force of m_2 on m_3.

Strategy   (a) To find the net gravitational force on the cue ball of mass m_1, we first calculate the force $\vec{F}_{21}$ exerted by m_2 on m_1. This force is the y-component of the net force acting on m_1. Then we find the force $\vec{F}_{31}$ exerted by m_3 on m_1, which is the x-component of the net force acting on m_1. With these two components, we can find the magnitude and direction of the net force on the cue ball. (b) In this case, we must use trigonometry to find the components of the force $\vec{F}_{23}$.

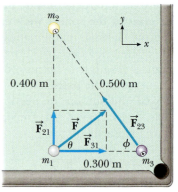

Figure 7.19   (Example 7.10)

Solution

(a) Find the net gravitational force on the cue ball. Find the magnitude of the force $\vec{F}_{21}$ exerted by m_2 on m_1 using the law of gravitation, Equation 7.20:

$$F_{21} = \frac{m_2 m_1}{r_{21}^{\,2}}$$

$$= (6.67 \times 10^{-11}\,\text{N}\cdot\text{m}^2/\text{kg}^2)\frac{(0.300\,\text{kg})\,(0.300\,\text{kg})}{(0.400\,\text{m})^2}$$

$$F_{21} = 3.75 \times 10^{-11}\,\text{N}$$

Find the magnitude of the force $\vec{F}_{31}$ exerted by m_3 on m_1, again using Newton's law of gravity:

$$F_{31} = G\frac{m_3 m_1}{r_{31}^{\,2}}$$

$$= (6.67 \times 10^{-11}\,\text{N}\cdot\text{m}^2/\text{kg}^2)\frac{(0.300\,\text{kg})\,(0.300\,\text{kg})}{(0.300\,\text{m})^2}$$

$$F_{31} = 6.67 \times 10^{-11}\,\text{N}$$

The net force has components $F_x = F_{31}$ and $F_y = F_{21}$. Compute the magnitude of this net force:

$$F = \sqrt{F_x^{\,2} + F_y^{\,2}} = \sqrt{(6.67)^2 + (3.75)^2} \times 10^{-11}\,\text{N}$$

$$= \boxed{7.65 \times 10^{-11}\,\text{N}}$$

Use the inverse tangent to obtain the direction of $\vec{F}$:

$$\theta = \tan^{-1}\left(\frac{F_y}{F_x}\right) = \tan^{-1}\left(\frac{3.75 \times 10^{-11}\,\text{N}}{6.67 \times 10^{-11}\,\text{N}}\right) = \boxed{29.3°}$$

(b) Find the components of the force of m_2 on m_3. First, compute the magnitude of $\vec{F}_{23}$:

$$F_{23} = G\frac{m_2 m_1}{r_{23}^{\,2}}$$

$$= (6.67 \times 10^{-11}\,\text{kg}^{-1}\text{m}^3\text{s}^{-2})\frac{(0.300\,\text{kg})\,(0.300\,\text{kg})}{(0.500\,\text{m})^2}$$

$$= 2.40 \times 10^{-11}\,\text{N}$$

To obtain the x- and y-components of F_{23}, we need $\cos \varphi$ and $\sin \varphi$. Use the sides of the large triangle in Figure 7.19:

$$\cos \varphi = \frac{\text{adj}}{\text{hyp}} = \frac{0.300 \text{ m}}{0.500 \text{ m}} = 0.600$$

$$\sin \varphi = \frac{\text{opp}}{\text{hyp}} = \frac{0.400 \text{ m}}{0.500 \text{ m}} = 0.800$$

Compute the components of $\vec{F}_{23}$. A minus sign must be supplied for the x-component, because it's in the negative x-direction.

$$F_{23x} = -F_{23} \cos \varphi = -(2.40 \times 10^{-11} \text{ N})(0.600)$$
$$= -1.44 \times 10^{-11} \text{ N}$$

$$F_{23y} = F_{23} \sin \varphi = (2.40 \times 10^{-11} \text{ N})(0.800)$$
$$= 1.92 \times 10^{-11} \text{ N}$$

Remarks Notice how small the gravity forces are between everyday objects. Nonetheless, such forces can be measured directly with torsion balances.

Exercise 7.10
Find magnitude and direction of the force exerted by m_1 and m_3 on m_2.

Answers 5.85×10^{-11} N, $-75.8°$

EXAMPLE 7.11   Ceres

Goal Relate Newton's universal law of gravity to mg, and show how g changes with position.

Problem An astronaut standing on the surface of Ceres, the largest asteroid, drops a rock from a height of 10.0 m. It takes 8.06 s to hit the ground. **(a)** Calculate the acceleration of gravity on Ceres. **(b)** Find the mass of Ceres, given that the radius of Ceres is $R_C = 5.10 \times 10^2$ km. **(c)** Calculate the gravitational acceleration 50.0 km from the surface of Ceres.

Strategy Part (a) is a review of one-dimensional kinematics. In part (b), the weight of an object, $w = mg$, is the same as the magnitude of the force given by the universal law of gravity. Solve for the unknown mass of Ceres, after which the answer for (c) can be found by substitution into the universal law of gravity, Equation 7.20.

Solution
(a) Calculate the acceleration of gravity, g_C, on Ceres.

Apply the kinematics displacement equation to the falling rock:

$$\Delta x = \tfrac{1}{2}at^2 + v_0 t$$

Substitute $\Delta x = -10.0$ m, $v_0 = 0$, $a = -g_C$, and $t = 8.06$ s, and solve for the gravitational acceleration on Ceres, g_C:

$$-10.0 \text{ m} = -\tfrac{1}{2}g_C(8.06 \text{ s})^2 \quad \rightarrow \quad g_C = 0.308 \text{ m/s}^2$$

(b) Find the mass of Ceres.

Equate the weight of the rock on Ceres to the gravitational force acting on the rock:

$$mg_C = G\frac{M_C m}{R_C^2}$$

Solve for the mass of Ceres, M_C:

$$M_C = \frac{g_C R_C^2}{G} = 1.20 \times 10^{21} \text{ kg}$$

(c) Calculate the acceleration of gravity at a height of 50.0 km above the surface of Ceres.

Equate the weight at 50.0 km to the gravitational force:

$$mg_C' = G\frac{m M_C}{r^2}$$

Cancel m, and substitute the mass of Ceres and $r = 5.60 \times 10^5$ m:

$$g'_C = G \frac{M_C}{r^2}$$

$$= (6.67 \times 10^{-11} \text{ kg}^{-1}\text{m}^3\text{s}^{-2}) \frac{1.20 \times 10^{21} \text{ kg}}{(5.60 \times 10^5 \text{ m})^2}$$

$$= \boxed{0.255 \text{ m/s}^2}$$

Remarks   This is the standard method of finding the mass of a planetary body: study the motion of a falling (or orbiting) object.

Exercise 7.11
An object takes 2.40 s to fall 5.00 m on a certain planet. (a) Find the acceleration due to gravity on the planet. (b) Find the planet's mass if its radius is 5 250 km.

Solution   (a) 1.74 m/s^2 (b) 7.19×10^{23} kg

Gravitational Potential Energy Revisited

In Chapter 5 we introduced the concept of gravitational potential energy and found that the potential energy associated with an object could be calculated from the equation $PE = mgh$, where h is the height of the object above or below some reference level. This equation, however, is valid *only* when the object is near Earth's surface. For objects high above Earth's surface, such as a satellite, an alternative must be used, because g varies with distance from the surface, as shown in Table 7.1.

General form of gravitational potential energy ▶

The gravitational potential energy associated with an object of mass m at a distance r from the center of Earth is

$$PE = -G \frac{M_E m}{r} \qquad [7.21]$$

where M_E and R_E are the mass and radius of Earth, respectively, with $r > R_E$.
SI units: Joules (J)

As before, gravitational potential energy is a property of a system, in this case the object of mass m and Earth. Equation 7.21 is valid for the special case where the zero level for potential energy is at an infinite distance from the center of Earth. Recall that the gravitational potential energy associated with an object is nothing more than the negative of the work done by the force of gravity in moving the object. If an object falls under the force of gravity from a great distance (effectively infinity), the change in gravitational potential energy is negative, which corresponds to a positive amount of gravitational work done on the system. This positive work is equal to the (also positive) change in kinetic energy, as the next example shows.

EXAMPLE 7.12   A Near-Earth Asteroid

Goal   Use gravitational potential energy to calculate the work done by gravity on a falling object.

Problem   An asteroid with mass $m = 1.00 \times 10^9$ kg comes from deep space, effectively from infinity, and falls toward Earth. **(a)** Find the change in potential energy when it reaches a point 4.00×10^8 m from Earth (just beyond the Moon), assuming it falls from rest at infinity. In addition, find the work done by the force of gravity. **(b)** Calculate the speed of the asteroid at that point. **(c)** How much work would have to be done on the asteroid by some other agent so the asteroid would be traveling at only half the speed found in (b) at the same point?

Strategy Part (a) requires simple substitution into the definition of gravitational potential energy. To find the work done by the force of gravity, recall that the work done on an object by a conservative force is just the negative of the change in potential energy. Part (b) can be solved with conservation of energy, and part (c) is an application of the work–energy theorem.

Solution
(a) Find the change in potential energy and the work done by the force of gravity.

Apply Equation 7.21:

$$\Delta PE = PE_f - PE_i = -\frac{GM_E m}{r_f} - \left(-\frac{GM_E m}{r_i}\right)$$

$$= GM_E m\left(-\frac{1}{r_f} + \frac{1}{r_i}\right)$$

Substitute known quantities. The asteroid's initial position is effectively infinity, so $1/r_i$ is zero.

$$\Delta PE = (6.67 \times 10^{-11}\,\text{kg}^{-1}\text{m}^3/\text{s}^2)(5.98 \times 10^{24}\,\text{kg})$$
$$\times (1.00 \times 10^9\,\text{kg})\left(-\frac{1}{4.00 \times 10^8\,\text{m}} + 0\right)$$

$$\Delta PE = -9.97 \times 10^{14}\,\text{J}$$

Compute the work done by the force of gravity:

$$W_{\text{grav}} = -\Delta PE = 9.97 \times 10^{14}\,\text{J}$$

(b) Find the speed of the asteroid when it reaches $r_f = 4.00 \times 10^8$ m.

Use conservation of energy:

$$\Delta KE + \Delta PE = 0$$
$$(\tfrac{1}{2}mv^2 - 0) - 9.97 \times 10^{14}\,\text{J} = 0$$
$$v = 1.41 \times 10^3\,\text{m/s}$$

(c) Find the work needed to reduce the speed to 7.05×10^2 m/s (half the value just found) at this point.

Apply the work–energy theorem:

$$W = \Delta KE + \Delta PE$$

The change in potential energy remains the same as in part (a), but substitute only half the speed in the kinetic-energy term:

$$W = (\tfrac{1}{2}mv^2 - 0) - 9.97 \times 10^{14}\,\text{J}$$
$$W = \tfrac{1}{2}(1.00 \times 10^9\,\text{kg})(7.05 \times 10^2\,\text{m/s})^2 - 9.97 \times 10^{14}\,\text{J}$$
$$= -7.48 \times 10^{14}\,\text{J}$$

Remark The amount of work calculated in part (c) is negative because an external agent must exert a force against the direction of motion of the asteroid. It would take a thruster with a megawatt of output about 24 years to slow down the asteroid to half its original speed. An asteroid endangering Earth need not be slowed that much: A small change in its speed, if applied early enough, will cause it to miss Earth. Timeliness of the applied thrust, however, is important. By the time you can look over your shoulder and see the Earth, it's already far too late, despite how these scenarios play out in Hollywood. Last minute rescues won't work!

Exercise 7.12
Suppose the asteroid starts from rest at a great distance (effectively infinity), falling toward Earth. How much work would have to be done on the asteroid to slow it to 425 m/s by the time it reached a distance of 2.00×10^8 m from Earth?

Answer -1.90×10^{15} J

Applying Physics 7.3   Why is the Sun Hot?

Explanation The Sun formed when particles in a cloud of gas in coalesced, due to gravitational attraction, into a massive astronomical object. Before this occurred, the particles in the cloud were widely scattered, representing a large amount of gravitational potential energy. As the particles fell closer together, their kinetic energy increased, but the gravitational potential energy of the system decreased, as required by the conservation of energy. With further slow collapse, the cloud became more dense, and the average kinetic energy of the particles increased. This kinetic energy is the internal energy of the cloud, which is proportional to the temperature. If enough particles come together, the temperature can rise to a point at which nuclear fusion occurs, and the ball of gas becomes a star. Otherwise, the temperature may rise, but not enough to ignite fusion reactions, and the object becomes a brown dwarf (a failed star) or a planet.

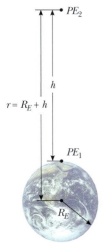

Figure 7.20 Relating the general form of gravitational potential energy to *mgh*.

On inspecting Equation 7.21, some may wonder what happened to *mgh*, the gravitational potential energy expression introduced in Chapter 5. That expression is still valid when *h* is small compared with the Earth's radius. To see this, we write the change in potential energy as an object is raised from the ground to height *h*, using the general form for gravitational potential energy (see Fig. 7.20):

$$PE_2 - PE_1 = -G\frac{M_E m}{(R_E + h)} - \left(-G\frac{M_E m}{R_E}\right)$$

$$= -GM_E m\left[\frac{1}{(R_E + h)} - \frac{1}{R_E}\right]$$

After finding a common denominator and applying some algebra, we obtain

$$PE_2 - PE_1 = \frac{GM_E mh}{R_E(R_E + h)}$$

When the height *h* is very small compared with R_E, *h* can be dropped from the second factor in the denominator, yielding

$$\frac{1}{R_E(R_E + h)} \cong \frac{1}{R_E^2}$$

Substituting this into the previous expression, we have

$$PE_2 - PE_1 \cong \frac{GM_E}{R_E^2}mh$$

Now recall from Chapter 4 that the free-fall acceleration at the surface of Earth is given by $g = GM_E/R_E^2$, giving

$$PE_2 - PE_1 \cong mgh$$

Escape Speed

If an object is projected upward from Earth's surface with a large enough speed, it can soar off into space and never return. This speed is called Earth's **escape speed**. (It is also commonly called the *escape velocity*, but in fact is more properly a speed.)

Earth's escape speed can be found by applying conservation of energy. Suppose an object of mass *m* is projected vertically upward from Earth's surface with an initial speed v_i. The initial mechanical energy (kinetic plus potential energy) of the object–Earth system is given by

$$KE_i + PE_i = \tfrac{1}{2}mv_i^2 - \frac{GM_E m}{R_E}$$

We neglect air resistance and assume that the initial speed is just large enough to allow the object to reach infinity with a speed of zero. This value of v_i is the escape speed v_{esc}. When the object is at an infinite distance from Earth, its kinetic energy is zero, because $v_f = 0$, and the gravitational potential energy is also zero, because $1/r$ goes to zero as r goes to infinity. Hence the total mechanical energy is zero, and the law of conservation of energy gives

$$\frac{1}{2}mv_{esc}^2 - \frac{GM_E m}{R_E} = 0$$

so that

$$v_{esc} = \sqrt{\frac{2GM_E}{R_E}} \qquad [7.22]$$

TABLE 7.2

Escape Speeds for the Planets and the Moon

Planet	v_e (km/s)
Mercury	4.3
Venus	10.3
Earth	11.2
Moon	2.3
Mars	5.0
Jupiter	60.0
Saturn	36.0
Uranus	22.0
Neptune	24.0
Pluto	1.1

The escape speed for Earth is about 11.2 km/s, which corresponds to about 25 000 mi/h. (See Example 7.13.) Note that the expression for v_{esc} doesn't depend on the mass of the object projected from Earth, so a spacecraft has the same escape speed as a molecule. Escape speeds for the planets, the Moon, and the Sun are listed in Table 7.2. Escape speed and temperature determine to a large extent whether a world has an atmosphere and, if so, what the constituents of the atmosphere are. Planets with low escape speeds, such as Mercury, generally don't have atmospheres because the average speed of gas molecules is close to the escape speed. Venus has a very thick atmosphere, but it's almost entirely carbon dioxide, a heavy gas. The atmosphere of Earth has very little hydrogen or helium, but has retained the much heavier nitrogen and oxygen molecules.

EXAMPLE 7.13 From the Earth to the Moon

Goal Apply conservation of energy with the general form of Newton's universal law of gravity.

Problem In Jules Verne's classic novel, *From the Earth to the Moon*, a giant cannon dug into the Earth in Florida fired a spacecraft all the way to the Moon. (a) If the spacecraft leaves the cannon at escape speed, at what speed is it moving when 1.50×10^5 km from the center of Earth? Neglect any friction effects. (b) Approximately what constant acceleration is needed to propel the spacecraft to this speed through a cannon bore a kilometer long?

Strategy For part (a), use conservation of energy and solve for the final speed v_f. Part (b) is an application of the time-independent kinematic equation: solve for the acceleration a.

Solution
(a) Find the speed at $r = 1.50 \times 10^5$ km.

Apply conservation of energy:

$$\frac{1}{2}mv_i^2 - \frac{GM_E m}{R_E} = \frac{1}{2}mv_f^2 - \frac{GM_E m}{r_f}$$

Multiply by $2/m$ and rearrange, solving for v_f^2. Then substitute known values and take the square root.

$$v_f^2 = v_i^2 + \frac{2GM_E}{r_f} - \frac{2GM_E}{R_E} = v_i^2 + 2GM_E\left(\frac{1}{r_f} - \frac{1}{R_E}\right)$$

$$v_f^2 = (1.12 \times 10^4 \text{ m/s})^2 + 2(6.67 \times 10^{-11} \text{ kg}^{-1} \text{ m}^3\text{s}^{-2})$$

$$\times (5.98 \times 10^{24} \text{ kg})\left(\frac{1}{1.50 \times 10^8 \text{ m}} - \frac{1}{6.37 \times 10^6 \text{ m}}\right)$$

$$v_f = \boxed{2.35 \times 10^3 \text{ m/s}}$$

(b) Find the acceleration through the cannon bore, assuming that it's constant.

Use the time-independent kinematics equation:

$$v^2 - v_0^2 = 2a\Delta x$$

$$(1.12 \times 10^4 \text{ m/s})^2 - 0 = 2a(1.00 \times 10^3 \text{ m})$$

$$a = \boxed{6.27 \times 10^4 \text{ m/s}^2}$$

Remark   This result corresponds to an acceleration of over 6 000 times the free-fall acceleration on Earth. Such a huge acceleration is far beyond what the human body can tolerate.

Exercise 7.13
Using the data in Table 7.3 (see page 216), find (a) the escape speed from the surface of Mars and (b) the speed of a space vehicle when it is 1.25×10^7 m from the center of Mars if it leaves the surface at the escape speed.

Answer   (a) 5.04×10^3 m/s   (b) 2.62×10^3 m/s

7.6   KEPLER'S LAWS

The movements of the planets, stars, and other celestial bodies have been observed for thousands of years. In early history, scientists regarded Earth as the center of the Universe. This **geocentric model** was developed extensively by the Greek astronomer Claudius Ptolemy in the second century A.D. and was accepted for the next 1 400 years. In 1543, the Polish astronomer Nicolaus Copernicus (1473–1543) showed that Earth and the other planets revolve in circular orbits around the Sun (the **heliocentric model**).

The Danish astronomer Tycho Brahe (pronounced Brah or BRAH–huh; 1546–1601) made accurate astronomical measurements over a period of 20 years, providing the data for the currently accepted model of the Solar System. Brahe's precise observations of the planets and 777 stars were carried out with nothing more elaborate than a large sextant and compass; the telescope had not yet been invented.

The German astronomer Johannes Kepler, who was Brahe's assistant, acquired Brahe's astronomical data and spent about 16 years trying to deduce a mathematical model for the motions of the planets. After many laborious calculations, he found that Brahe's precise data on the motion of Mars about the Sun provided the answer. Kepler's analysis first showed that the concept of circular orbits about the Sun had to be abandoned. He eventually discovered that the orbit of Mars could be accurately described by an ellipse with the Sun at one focus. He then generalized this analysis to include the motions of all planets. The complete analysis is summarized in three statements known as **Kepler's laws**:

> 1. All planets move in elliptical orbits with the Sun at one of the focal points.
> 2. A line drawn from the Sun to any planet sweeps out equal areas in equal time intervals.
> 3. The square of the orbital period of any planet is proportional to the cube of the average distance from the planet to the Sun.

Newton later demonstrated that these laws are consequences of the gravitational force that exists between any two objects. Newton's law of universal gravitation, together with his laws of motion, provides the basis for a full mathematical description of the motions of planets and satellites.

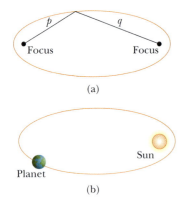

ACTIVE FIGURE 7.21
(a) The sum $p + q$ is the same for every point on the ellipse. (b) In the Solar System, the Sun is at one focus of the elliptical orbit of each planet and the other focus is empty.

Physics⊗Now™
Log into PhysicsNow at **www.cp7e.com**, and go to Active Figure 7.21a to move the focal points, or enter values for a, b, c, and e and see the resulting elliptical shape.

Kepler's First Law

The first law arises as a natural consequence of the inverse-square nature of Newton's law of gravitation. Any object bound to another by a force that varies as $1/r^2$ will move in an elliptical orbit. As shown in Active Figure 7.21a, an ellipse is a curve drawn so that the sum of the distances from any point on the curve to two internal points called *focal points* or *foci* (singular, *focus*) is always the same. The semimajor axis a is half the length of the line that goes across the ellipse and contains both foci. For the Sun–planet configuration (Active Fig. 7.21b), the Sun is at one focus and the other focus is empty. Because the orbit is an ellipse, the distance from the Sun to the planet continuously changes.

Kepler's Second Law

Kepler's second law states that a line drawn from the Sun to any planet sweeps out equal areas in equal time intervals. Consider a planet in an elliptical orbit about the Sun, as in Figure 7.22. In a given period Δt, the planet moves from point Ⓐ to point Ⓑ. The planet moves more slowly on that side of the orbit because it's farther away from the sun. On the opposite side of its orbit, the planet moves from point Ⓒ to point Ⓓ in the same amount of time, Δt, moving faster because it's closer to the sun. Kepler's second law says that any two wedges formed as in Figure 7.22 will always have the same area. As we will see in Chapter 8, Kepler's second law is related to a physical principle known as conservation of angular momentum.

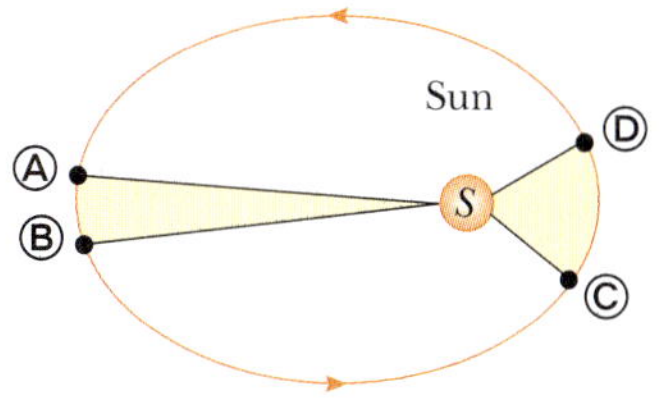

Figure 7.22   The two areas swept out by the planet in its elliptical orbit about the Sun are equal if the time interval between points Ⓐ and Ⓑ is equal to the time interval between points Ⓒ and Ⓓ.

Kepler's Third Law

The derivation of Kepler's third law is simple enough to carry out for the special case of a circular orbit. Consider a planet of mass M_p moving around the Sun, which has a mass of M_S, in a circular orbit. Because the orbit is circular, the planet moves at a constant speed v. Newton's second law, his law of gravitation, and centripetal acceleration then give the following equation:

$$M_p a_c = \frac{M_p v^2}{r} = \frac{GM_S M_p}{r^2}$$

The speed v of the planet in its orbit is equal to the circumference of the orbit divided by the time required for one revolution, T, called the **period** of the planet, so $v = 2\pi r/T$. Substituting, the preceding expression becomes

$$\frac{GM_S}{r^2} = \frac{(2\pi r/T)^2}{r}$$

$$T^2 = \left(\frac{4\pi^2}{GM_S}\right) r^3 = K_S r^3 \qquad \text{[7.23]} \qquad \blacktriangleleft \text{Kepler's third law}$$

where K_S is a constant given by

$$K_S = \frac{4\pi^2}{GM_S} = 2.97 \times 10^{-19} \text{ s}^2/\text{m}^3$$

Equation 7.23 is Kepler's third law for a circular orbit. The orbits of most of the planets are very nearly circular. Comets and asteroids, however, usually have elliptical orbits. For these orbits, the radius r must be replaced with a, the semimajor axis—half the longest distance across the elliptical orbit. (This is also the average distance of the comet or asteroid from the sun.) A more detailed calculation shows that K_S actually depends on the sum of both the mass of a given planet and the Sun's mass. The masses of the planets, however, are negligible compared with the Sun's mass; hence can be neglected, meaning Equation 7.23 is valid for any planet in the Sun's family. If we consider the orbit of a satellite such as the Moon around Earth, then the constant has a different value, with the mass of the Sun replaced by the mass of Earth. In that case, K_E equals $4\pi^2/GM_E$.

The mass of the Sun can be determined from Kepler's third law, because the constant K_S in Equation 7.23 includes the mass of the Sun and the other variables and constants can be easily measured. The value of this constant can be found by substituting the values of a planet's period and orbital radius and solving for K_S. The mass of the Sun is then

$$M_S = \frac{4\pi^2}{GK_S}$$

This same process can be used to calculate the mass of Earth (by considering the period and orbital radius of the Moon) and the mass of other planets in the Solar System that have satellites.

TABLE 7.3

Useful Planetary Data

Body	Mass (kg)	Mean Radius (m)	Period (s)	Mean Distance from Sun (m)	$\dfrac{T^2}{r^3}10^{-19}\left(\dfrac{s^2}{m^3}\right)$
Mercury	3.18×10^{23}	2.43×10^{6}	7.60×10^{6}	5.79×10^{10}	2.97
Venus	4.88×10^{24}	6.06×10^{6}	1.94×10^{7}	1.08×10^{11}	2.99
Earth	5.98×10^{24}	6.38×10^{6}	3.156×10^{7}	1.496×10^{11}	2.97
Mars	6.42×10^{23}	3.37×10^{6}	5.94×10^{7}	2.28×10^{11}	2.98
Jupiter	1.90×10^{27}	6.99×10^{7}	3.74×10^{8}	7.78×10^{11}	2.97
Saturn	5.68×10^{26}	5.85×10^{7}	9.35×10^{8}	1.43×10^{12}	2.99
Uranus	8.68×10^{25}	2.33×10^{7}	2.64×10^{9}	2.87×10^{12}	2.95
Neptune	1.03×10^{26}	2.21×10^{7}	5.22×10^{9}	4.50×10^{12}	2.99
Pluto	1.27×10^{23}	1.14×10^{6}	7.82×10^{9}	5.91×10^{12}	2.96
Moon	7.36×10^{22}	1.74×10^{6}	—	—	—
Sun	1.991×10^{30}	6.96×10^{8}	—	—	—

The last column in Table 7.3 confirms the fact that T^2/r^3 is very nearly constant. When time is measured in Earth years and the semimajor axis in astronomical units (1 AU = the distance from the Earth to the Sun), Kepler's law takes the following simple form:

$$T^2 = a^3$$

This equation can be easily checked: The Earth has a semimajor axis of one astronomical unit (by definition), and it takes one year to circle the sun. This equation, of course, is valid only for the sun and its planets, asteroids, and comets.

Quick Quiz 7.10

Suppose an asteroid has a semimajor axis of 4 AU How long does it take the asteroid to go around the sun? (a) 2 years (b) 4 years (c) 6 years (d) 8 years

---

EXAMPLE 7.14 Geosynchronous Orbit and Telecommunications Satellites

Goal Apply Kepler's third law to an Earth satellite

Problem From a telecommunications point of view, it's advantageous for satellites to remain at the same location relative to a location on the Earth. This can occur only if the satellite's orbital period is the same as the Earth's period of rotation, 24.0 h. **(a)** At what distance from the center of the Earth can this geosynchronous orbit be found? **(b)** What's the orbital speed of the satellite?

Strategy This problem can be solved with the same method that was used to derive a special case of Kepler's third law, with Earth's mass replacing the Sun's mass. There's no need to repeat the analysis, just replace the Sun's mass with Earth's mass in Kepler's third law, substitute the period T (converted to seconds), and solve for r. For part (b), find the circumference of the circular orbit and divide by the elapsed time.

Solution

(a) Find the distance r to geosynchronous orbit.

Apply Kepler's third law:

$$T^2 = \left(\frac{4\pi^2}{GM_E}\right)r^3$$

Substitute the period in seconds, $T = 86\ 400$ s, the gravity constant $G = 6.67 \times 10^{-11}$ kg^{-1} m^3/s^2, and the mass of the Earth, $M_E = 5.98 \times 10^{24}$ kg. Solve for r:

$r = 4.23 \times 10^7$ m

(b) Find the orbital speed:

$$v = \frac{d}{T} = \frac{2\pi r}{T} = \frac{2\pi(4.23 \times 10^7 \text{ m})}{8.64 \times 10^4 \text{ s}} = 3.08 \times 10^3 \text{ m/s}$$

Remarks   Both these results are independent of the mass of the satellite. Notice that Earth's mass could be found by substituting the Moon's distance and period into this form of Kepler's third law.

Exercise 7.14

Mars rotates on its axis once every 1.02 days (almost the same as Earth does). (a) Find the distance from Mars at which a satellite would remain in one spot over the Martian surface. (b) Find the speed of the satellite.

Answer   (a) 2.03×10^7 m   (b) 1.45×10^3 m/s

SUMMARY

Physics⊗Now™ Take a practice test by logging into PhysicsNow at **www.cp7e.com** and clicking on the Pre-Test link for this chapter.

7.1 Angular Speed and Angular Acceleration

The **average angular speed** ω_{av} of a rigid object is defined as the ratio of the angular displacement $\Delta\theta$ to the time interval Δt, or

$$\omega_{av} = \frac{\theta_f - \theta_i}{t_f - t_i} = \frac{\Delta\theta}{\Delta t} \qquad [7.3]$$

where ω_{av} is in radians per second (rad/s).

The **average angular acceleration** α_{av} of a rotating object is defined as the ratio of the change in angular speed $\Delta\omega$ to the time interval Δt, or

$$\alpha_{av} \equiv \frac{\omega_f - \omega_i}{t_f - t_i} = \frac{\Delta\omega}{\Delta t} \qquad [7.5]$$

where α_{av} is in radians per second per second (rad/s^2).

7.2 Rotational Motion under Constant Angular Acceleration

If an object undergoes rotational motion about a fixed axis under a constant angular acceleration α, its motion can be described with the following set of equations:

$$\omega = \omega_i + \alpha t \qquad [7.7]$$
$$\Delta\theta = \omega_i t + \tfrac{1}{2}\alpha t^2 \qquad [7.8]$$
$$\omega^2 = \omega_i^2 + 2\alpha\Delta\theta \qquad [7.9]$$

Problems are solved as in one-dimensional kinematics.

7.3 Relations between Angular and Linear Quantities

When an object rotates about a fixed axis, the angular speed and angular acceleration are related to the tangential speed and tangential acceleration through the relationships

$$v_t = r\omega \qquad [7.10]$$

and

$$a_t = r\alpha \qquad [7.11]$$

7.4 Centripetal Acceleration

Any object moving in a circular path has an acceleration directed toward the center of the circular path, called a **centripetal acceleration**. Its magnitude is given by

$$a_c = \frac{v^2}{r} = r\omega^2 \qquad [7.13, 7.17]$$

Any object moving in a circular path must have a net force exerted on it that is directed toward the center of the path. Some examples of forces that cause centripetal acceleration are the force of gravity (as in the motion of a satellite) and the force of tension in a string.

7.5 Newtonian Gravitation

Newton's law of universal gravitation states that every particle in the Universe attracts every other particle with a force that is directly proportional to the product of their masses and inversely proportional to the square of the distance r between them:

$$F = G\frac{m_1 m_2}{r^2} \qquad [7.20]$$

where $G = 6.673 \times 10^{-11}$ N·m^2/kg^2 is the **constant of universal gravitation**. A general expression for gravitational potential energy is

$$PE = -G\frac{M_E m}{r} \qquad [7.21]$$

This expression reduces to $PE = mgh$ close to the surface of Earth and holds for other worlds through replacement of the mass M_E. Problems such as finding the escape velocity from Earth can be solved by using Equation 7.21 in the conservation of energy equation.

7.6 Kepler's Laws

Kepler derived the following three laws of planetary motion:

1. All planets move in elliptical orbits with the Sun at one of the focal points.
2. A line drawn from the Sun to any planet sweeps out equal areas in equal time intervals.
3. The square of the orbital period of a planet is propor-

tional to the cube of the average distance from the planet to the Sun:

$$T^2 = \left(\frac{4\pi^2}{GM_S} \right) r^3 \qquad [7.23]$$

The third law can be applied to any large body and its system of satellites by replacing the Sun's mass with the body's mass. In particular, it can be used to determine the mass of the central body once the average distance to a satellite and its period are known.

CONCEPTUAL QUESTIONS

1. In a race like the Indianapolis 500, a driver circles the track counterclockwise and feels his head pulled toward one shoulder. To relieve his neck muscles from having to hold his head erect, the driver fastens a strap to one wall of the car and the other to his helmet. The length of the strap is adjusted to keep his head vertical. (a) Which shoulder does his head tend to lean toward? (b) What force or forces produce the centripetal acceleration when there is no strap? (c) What force or forces do so when there is a strap?

2. Two schoolmates, Romeo and Juliet, catch each other's eye across a crowded dance floor at a school dance. Find the order of magnitude of the gravitational attraction that Juliet exerts on Romeo and that Romeo exerts on Juliet. State the quantities you take as data and the values you measure or estimate for them.

3. An object executes circular motion with a constant speed whenever a net force of constant magnitude acts perpendicular to its velocity. What happens to the speed if the force is not perpendicular to the velocity?

4. Explain why Earth is not spherical in shape, but bulges at the equator.

5. If a car's wheels are replaced with wheels of greater diameter, will the reading of the speedometer change? Explain.

6. At night, you are farther away from the Sun than during the day. What's more, the force exerted by the Sun on you is downward into Earth at night, and upward into the sky during the day. If you had a sensitive enough bathroom scale, would you appear to weigh more at night than during the day?

7. Correct the following statement: "The race car rounds the turn at a constant velocity of 90 miles per hour."

8. Why does an astronaut in a spacecraft orbiting Earth experience a feeling of weightlessness?

9. Explain why it's easier to determine the mass of a planet when it has a moon.

10. Because of Earth's rotation about its axis, you weigh slightly less at the equator than at the poles. Why?

11. It has been suggested that rotating cylinders about 10 miles long and 5 miles in diameter be placed in space for colonies. The purpose of their rotation is to simulate gravity for the inhabitants. Explain the concept behind this proposal.

12. Describe the path of a moving object in the event that the object's acceleration is constant in magnitude at all times and (a) perpendicular to its velocity; (b) parallel to its velocity.

13. A pail of water can be whirled in a vertical circular path such that no water is spilled. Why does the water remain in the pail, even when the pail is upside down above your head?

14. The orbital planes of all the planets must pass through the center of the Sun. Why?

15. Is it possible for a car to move in a circular path in such a way that it has a tangential acceleration but no centripetal acceleration?

16. Use Kepler's second law to convince yourself that Earth must move faster in its orbit during the northern-hemisphere winter, when it is closest to the Sun, than during the summer, when it is farthest from the Sun.

17. If the mass of Earth were doubled at the same time its radius were doubled, the free-fall acceleration would (a) increase, (b) decrease, or (c) stay the same?

18. A satellite in orbit is not truly traveling through a vacuum—it's moving through very thin air. Does the resulting air friction cause the satellite to slow down?

PROBLEMS

1, 2, 3 = straightforward, intermediate, challenging   ☐ = full solution available in *Student Solutions Manual/Study Guide*
Physics⊗Now™ = coached problem with hints available at **www.cp7e.com**   🧬 = biomedical application

Section 7.1 Angular Speed and Angular Acceleration

1. The tires on a new compact car have a diameter of 2.0 ft and are warranted for 60 000 miles. (a) Determine the angle (in radians) through which one of these tires will rotate during the warranty period. (b) How many revolutions of the tire are equivalent to your answer in (a)?

2. A wheel has a radius of 4.1 m. How far (path length) does a point on the circumference travel if the wheel is rotated through angles of 30°, 30 rad, and 30 rev, respectively?

3. Find the angular speed of Earth about the Sun in radians per second and degrees per day.

4. A potter's wheel moves from rest to an angular speed of 0.20 rev/s in 30 s. Find its angular acceleration in radians per second per second.

Section 7.2 Rotational Motion under Constant Angular Acceleration
Section 7.3 Relations between Angular and Linear Quantities

5. A dentist's drill starts from rest. After 3.20 s of constant angular acceleration, it turns at a rate of 2.51×10^4 rev/min. (a) Find the drill's angular acceleration. (b) Determine the angle (in radians) through which the drill rotates during this period.

6. A centrifuge in a medical laboratory rotates at an angular speed of 3 600 rev/min. When switched off, it rotates through 50.0 revolutions before coming to rest. Find the constant angular acceleration of the centrifuge.

7. A machine part rotates at an angular speed of 0.60 rad/s; its speed is then increased to 2.2 rad/s at an angular acceleration of 0.70 rad/s². Find the angle through which the part rotates before reaching this final speed.

8. A tire placed on a balancing machine in a service station starts from rest and turns through 4.7 revolutions in 1.2 s before reaching its final angular speed. Calculate its angular acceleration.

9. Physics⊗Now™ The diameters of the main rotor and tail rotor of a single-engine helicopter are 7.60 m and 1.02 m, respectively. The respective rotational speeds are 450 rev/min and 4 138 rev/min. Calculate the speeds of the tips of both rotors. Compare these speeds with the speed of sound, 343 m/s.

10. The tub of a washer goes into its spin-dry cycle, starting from rest and reaching an angular speed of 5.0 rev/s in 8.0 s. At this point, the person doing the laundry opens the lid, and a safety switch turns off the washer. The tub slows to rest in 12.0 s. Through how many revolutions does the tub turn during the entire 20-s interval? Assume constant angular acceleration while it is starting and stopping.

11. A standard cassette tape is placed in a standard cassette player. Each side lasts for 30 minutes. The two tape wheels of the cassette fit onto two spindles in the player. Suppose that a motor drives one spindle at constant angular velocity of approximately 1 rad/s and the other spindle is free to rotate at any angular speed. Find the order of magnitude of the tape's thickness. Specify any other quantities you estimate and the values you take for them.

12. A coin with a diameter of 2.40 cm is dropped on edge onto a horizontal surface. The coin starts out with an initial angular speed of 18.0 rad/s and rolls in a straight line without slipping. If the rotation slows with an angular acceleration of magnitude 1.90 rad/s², how far does the coin roll before coming to rest?

13. A rotating wheel requires 3.00 s to rotate 37.0 revolutions. Its angular velocity at the end of the 3.00-s interval is 98.0 rad/s. What is the constant angular acceleration of the wheel?

Section 7.4 Centripetal Acceleration

14. It has been suggested that rotating cylinders about 10 mi long and 5.0 mi in diameter be placed in space and used as colonies. What angular speed must such a cylinder have

so that the centripetal acceleration at its surface equals the free-fall acceleration on Earth?

15. Find the centripetal accelerations of (a) a point on the equator of Earth and (b) the North Pole, due to the rotation of Earth about its axis.

16. A tire 2.00 ft in diameter is placed on a balancing machine, where it is spun so that its tread is moving at a constant speed of 60.0 mi/h. A small stone is stuck in the tread of the tire. What is the acceleration of the stone as the tire is being balanced?

17. (a) What is the tangential acceleration of a bug on the rim of a 10-in.-diameter disk if the disk moves from rest to an angular speed of 78 rev/min in 3.0 s? (b) When the disk is at its final speed, what is the tangential velocity of the bug? (c) One second after the bug starts from rest, what are its tangential acceleration, centripetal acceleration, and total acceleration?

18. A race car starts from rest on a circular track of radius 400 m. The car's speed increases at the constant rate of 0.500 m/s². At the point where the magnitudes of the centripetal and tangential accelerations are equal, determine (a) the speed of the race car, (b) the distance traveled, and (c) the elapsed time.

19. A 55.0-kg ice-skater is moving at 4.00 m/s when she grabs the loose end of a rope, the opposite end of which is tied to a pole. She then moves in a circle of radius 0.800 m around the pole. (a) Determine the force exerted by the horizontal rope on her arms. (b) Compare this force with her weight.

20. A sample of blood is placed in a centrifuge of radius 15.0 cm. The mass of a red blood cell is 3.0×10^{-16} kg, and the magnitude of the force acting on it as it settles out of the plasma is 4.0×10^{-11} N. At how many revolutions per second should the centrifuge be operated?

21. A certain light truck can go around a flat curve having a radius of 150 m with a maximum speed of 32.0 m/s. With what maximum speed can it go around a curve having a radius of 75.0 m?

22. The *cornering performance* of an automobile is evaluated on a skid pad, where the maximum speed that a car can maintain around a circular path on a dry, flat surface is measured. Then the centripetal acceleration, also called the lateral acceleration, is calculated as a multiple of the free-fall acceleration *g*. The main factors affecting the performance of the car are its tire characteristics and suspension system. A Dodge Viper GTS can negotiate a skid pad of radius 61.0 m at 86.5 km/h. Calculate its maximum lateral acceleration.

23. Physics⊗Now™ A 50.0-kg child stands at the rim of a merry-go-round of radius 2.00 m, rotating with an angular speed of 3.00 rad/s. (a) What is the child's centripetal acceleration? (b) What is the minimum force between her feet and the floor of the carousel that is required to keep her in the circular path? (c) What minimum coefficient of static friction is required? Is the answer you found reasonable? In other words, is she likely to stay on the merry-go-round?

24. An engineer wishes to design a curved exit ramp for a toll road in such a way that a car will not have to rely on friction to round the curve without skidding. He does so by banking the road in such a way that the force causing the

centripetal acceleration will be supplied by the component of the normal force toward the center of the circular path. (a) Show that, for a given speed v and a radius r, the curve must be banked at the angle θ such that $\tan\theta = v^2/rg$. (b) Find the angle at which the curve should be banked if a typical car rounds it at a 50.0-m radius and a speed of 13.4 m/s.

25. An air puck of mass 0.25 kg is tied to a string and allowed to revolve in a circle of radius 1.0 m on a frictionless horizontal table. The other end of the string passes through a hole in the center of the table, and a mass of 1.0 kg is tied to it (Fig. P7.25). The suspended mass remains in equilibrium while the puck on the tabletop revolves. (a) What is the tension in the string? (b) What is the horizontal force acting on the puck? (c) What is the speed of the puck?

Figure P7.25

26. Tarzan ($m = 85$ kg) tries to cross a river by swinging from a 10-m-long vine. His speed at the bottom of the swing (as he just clears the water) is 8.0 m/s. Tarzan doesn't know that the vine has a breaking strength of 1 000 N. Does he make it safely across the river? Justify your answer.

27. A 40.0-kg child takes a ride on a Ferris wheel that rotates four times each minute and has a diameter of 18.0 m. (a) What is the centripetal acceleration of the child? (b) What force (magnitude and direction) does the seat exert on the child at the lowest point of the ride? (c) What force does the seat exert on the child at the highest point of the ride? (d) What force does the seat exert on the child when the child is halfway between the top and bottom?

28. A roller-coaster vehicle has a mass of 500 kg when fully loaded with passengers (Fig. P7.28). (a) If the vehicle has a speed of 20.0 m/s at point Ⓐ, what is the force of the track on the vehicle at this point? (b) What is the maximum speed the vehicle can have at point Ⓑ in order for gravity to hold it on the track?

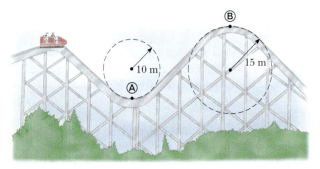

Figure P7.28

Section 7.5 Newtonian Gravitation

29. The average distance separating Earth and the Moon is 384 000 km. Use the data in Table 7.3 to find the net gravitational force exerted by Earth and the Moon on a 3.00×10^4-kg spaceship located halfway between them.

30. During a solar eclipse, the Moon, Earth, and Sun all lie on the same line, with the Moon between Earth and the Sun. (a) What force is exerted by the Sun on the Moon? (b) What force is exerted by Earth on the Moon? (c) What force is exerted by the Sun on Earth? (See Table 7.3 and Problem 29.)

31. A coordinate system (in meters) is constructed on the surface of a pool table, and three objects are placed on the table as follows: a 2.0-kg object at the origin of the coordinate system, a 3.0-kg object at (0, 2.0), and a 4.0-kg object at (4.0, 0). Find the resultant gravitational force exerted by the other two objects on the object at the origin.

32. Use the data of Table 7.3 to find the point between Earth and the Sun at which an object can be placed so that the net gravitational force exerted by Earth and the Sun on that object is zero.

33. Objects with masses of 200 kg and 500 kg are separated by 0.400 m. (a) Find the net gravitational force exerted by these objects on a 50.0-kg object placed midway between them. (b) At what position (other than infinitely remote ones) can the 50.0-kg object be placed so as to experience a net force of zero?

34. Two objects attract each other with a gravitational force of magnitude 1.00×10^{-8} N when separated by 20.0 cm. If the total mass of the objects is 5.00 kg, what is the mass of each?

Section 7.6 Kepler's Laws

35. **Physics⊗Now™** A satellite moves in a circular orbit around Earth at a speed of 5 000 m/s. Determine (a) the satellite's altitude above the surface of Earth and (b) the period of the satellite's orbit.

36. A 600-kg satellite is in a circular orbit about Earth at a height above Earth equal to Earth's mean radius. Find (a) the satellite's orbital speed, (b) the period of its revolution, and (c) the gravitational force acting on it.

37. Io, a satellite of Jupiter, has an orbital period of 1.77 days and an orbital radius of 4.22×10^5 km. From these data, determine the mass of Jupiter.

38. A satellite has a mass of 100 kg and is located at 2.00×10^6 m above the surface of Earth. (a) What is the potential energy associated with the satellite at this location? (b) What is the magnitude of the gravitational force on the satellite?

39. A satellite of mass 200 kg is launched from a site on Earth's equator into an orbit 200 km above the surface of Earth. (a) Assuming a circular orbit, what is the orbital period of this satellite? (b) What is the satellite's speed in it's orbit? (c) What is the minimum energy necessary to place the satellite in orbit, assuming no air friction?

ADDITIONAL PROBLEMS

40. Neutron stars are extremely dense objects that are formed from the remnants of supernova explosions. Many rotate very rapidly. Suppose that the mass of a certain spherical

neutron star is twice the mass of the sun and its radius is 10.0 km. Determine the greatest possible angular speed the neutron star can have so that the matter at its surface on the equator is just held in orbit by the gravitational force.

41. One method of pitching a softball is called the "windmill" delivery method, in which the pitcher's arm rotates through approximately 360° in a vertical plane before the 198-gram ball is released at the lowest point of the circular motion. An experienced pitcher can throw a ball with a speed of 98.0 mi/h. Assume that the angular acceleration is uniform throughout the pitching motion, and take the distance between the softball and the shoulder joint to be 74.2 cm. (a) Determine the angular speed of the arm in rev/s at the instant of release. (b) Find the value of the angular acceleration in rev/s² and the radial and tangential acceleration of the ball just before it is released. (c) Determine the force exerted on the ball by the pitcher's hand (both radial and tangential components) just before it is released.

42. The Mars probe *Pathfinder* is designed to drop the vehicle's instrument package from a height of 20 meters above the Martian surface, after the speed of the probe has been brought to zero by a combination parachute–rocket system at that height. To cushion the landing, giant air bags surround the package. The mass of Mars is 0.107 4 times that of Earth, and the radius of Mars is 0.528 2 that of Earth. Find (a) the acceleration due to gravity at the surface of Mars and (b) how long it takes for the instrument package to fall the last 20 meters.

43. An athlete swings a 5.00-kg ball horizontally on the end of a rope. The ball moves in a circle of radius 0.800 m at an angular speed of 0.500 rev/s. What are (a) the tangential speed of the ball and (b) its centripetal acceleration? (c) If the maximum tension the rope can withstand before breaking is 100 N, what is the maximum tangential speed the ball can have?

44. A digital audio compact disc carries data along a continuous spiral track from the inner circumference of the disc to the outside edge. Each bit occupies 0.6 μm of the track. A CD player turns the disc to carry the track counterclockwise above a lens at a constant speed of 1.30 m/s. Find the required angular speed (a) at the beginning of the recording, where the spiral has a radius of 2.30 cm, and (b) at the end of the recording, where the spiral has a radius of 5.80 cm. (c) A full-length recording lasts for 74 min, 33 s. Find the average angular acceleration of the disc. (d) Assuming that the acceleration is constant, find the total angular displacement of the disc as it plays. (e) Find the total length of the track.

45. The Solar Maximum Mission Satellite was placed in a circular orbit about 150 mi above Earth. Determine (a) the orbital speed of the satellite and (b) the time required for one complete revolution.

46. A car rounds a banked curve where the radius of curvature of the road is R, the banking angle is θ, and the coefficient of static friction is μ. (a) Determine the range of speeds the car can have without slipping up or down the road. (b) What is the range of speeds possible if $R = 100$ m, $\theta = 10°$, and $\mu = 0.10$ (slippery conditions)?

47. **Physics⊗Now**™ A car moves at speed v across a bridge made in the shape of a circular arc of radius r. (a) Find an expression for the normal force acting on the car when it is at the top of the arc. (b) At what minimum speed will the normal force become zero (causing the occupants of the car to seem weightless) if $r = 30.0$ m?

48. A 0.400-kg pendulum bob passes through the lowest part of its path at a speed of 3.00 m/s. (a) What is the tension in the pendulum cable at this point if the pendulum is 80.0 cm long? (b) When the pendulum reaches its highest point, what angle does the cable make with the vertical? (c) What is the tension in the pendulum cable when the pendulum reaches its highest point?

49. Because of Earth's rotation about its axis, a point on the equator has a centripetal acceleration of 0.034 0 m/s², while a point at the poles has no centripetal acceleration. (a) Show that, at the equator, the gravitational force on an object (the object's true weight) must exceed the object's apparent weight. (b) What are the apparent weights of a 75.0-kg person at the equator and at the poles? (Assume Earth is a uniform sphere, and take $g = 9.800$ m/s².)

50. A stuntman whose mass is 70 kg swings from the end of a 4.0-m-long rope along the arc of a vertical circle. Assuming that he starts from rest when the rope is horizontal, find the tensions in the rope that are required to make him follow his circular path (a) at the beginning of his motion, (b) at a height of 1.5 m above the bottom of the circular arc, and (c) at the bottom of the arc.

51. In a popular amusement park ride, a rotating cylinder of radius 3.00 m is set in rotation at an angular speed of 5.00 rad/s, as in Figure P7.51. The floor then drops away, leaving the riders suspended against the wall in a vertical position. What minimum coefficient of friction between a rider's clothing and the wall is needed to keep the rider from slipping? (*Hint:* Recall that the magnitude of the maximum force of static friction is equal to μn, where n is the normal force — in this case, the force causing the centripetal acceleration.)

Figure P7.51

52. A 0.50-kg ball that is tied to the end of a 1.5-m light cord is revolved in a horizontal plane, with the cord making a 30° angle with the vertical. (See Fig. P7.52.) (a) Determine the ball's speed. (b) If, instead, the ball is revolved so that its speed is 4.0 m/s, what angle does the cord make with the vertical? (c) If the cord can withstand a maximum tension of 9.8 N, what is the highest speed at which the ball can move?

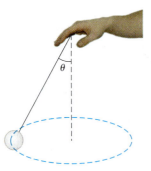

Figure P7.52

53. A skier starts at rest at the top of a large hemispherical hill (Fig. P7.53). Neglecting friction, show that the skier will leave the hill and become airborne at a distance $h = R/3$ below the top of the hill. (*Hint:* At this point, the normal force goes to zero.)

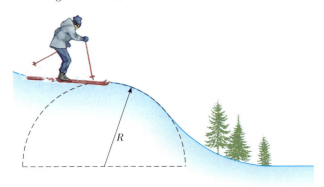

Figure P7.53

54. After consuming all of its nuclear fuel, a massive star can collapse to form a black hole, which is an immensely dense object whose escape speed is greater than the speed of light. Newton's law of universal gravitation still describes the force that a black hole exerts on objects outside it. A spacecraft in the shape of a long cylinder has a length of 100 m, and its mass with occupants is 1 000 kg. It has strayed too close to a black hole having a mass 100 times that of the Sun (Figure P7.54). If the nose of the spacecraft points toward the center of the black hole, and if the distance between the nose of the spacecraft and the black hole's center is 10 km, (a) determine the total force on the spacecraft. (b) What is the difference in the force per kilogram of mass felt by the occupants in the nose of the ship versus those in the rear of the ship farthest from the black hole?

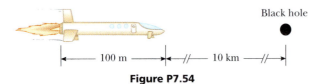

Figure P7.54

55. In Robert Heinlein's *The Moon Is a Harsh Mistress*, the colonial inhabitants of the Moon threaten to launch rocks down onto Earth if they are not given independence (or at least representation). Assuming that a gun could launch a rock of mass m at twice the lunar escape speed, calculate the speed of the rock as it enters Earth's atmosphere.

56. Show that the escape speed from the surface of a planet of uniform density is directly proportional to the radius of the planet.

57. A massless spring of constant $k = 78.4$ N/m is fixed on the left side of a level track. A block of mass $m = 0.50$ kg is pressed against the spring and compresses it a distance d, as in Figure P7.57. The block (initially at rest) is then released and travels toward a circular loop-the-loop of radius $R = 1.5$ m. The entire track and the loop-the-loop are frictionless, except for the section of track between points A and B. Given that the coefficient of kinetic friction between the block and the track along AB is $\mu_k = 0.30$, and that the length of AB is 2.5 m, determine the minimum compression d of the spring that enables the block to just make it through the loop-the-loop at point C. (*Hint:* The force exerted by the track on the block will be zero if the block barely makes it through the loop-the-loop.)

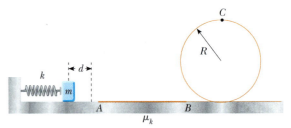

Figure P7.57

58. A small block of mass $m = 0.50$ kg is fired with an initial speed of $v_0 = 4.0$ m/s along a horizontal section of frictionless track, as shown in the top portion of Figure P7.58. The block then moves along the frictionless, semicircular, *vertical* tracks of radius $R = 1.5$ m. (a) Determine the force exerted by the track on the block at points Ⓐ and Ⓑ. (b) The bottom of the track consists of a section ($L = 0.40$ m) with friction. Determine the coefficient of kinetic friction between the block and that portion of the bottom track if the block just makes it to point Ⓒ on the first trip. (*Hint:* If the block just makes it to point Ⓒ, the force of contact exerted by the track on the block at that point is zero.)

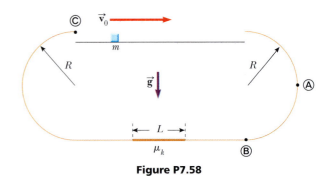

Figure P7.58

59. A frictionless roller coaster is given an initial speed v_0 at height h, as in Figure P7.59. The radius of curvature of the track at point Ⓐ is R. (a) Find the maximum value of v_0 so that the roller coaster stays on the track at Ⓐ solely because of gravity. (b) Using the value of v_0 calculated in (a), determine the value of h' that is necessary if the roller coaster just makes it to point Ⓑ.

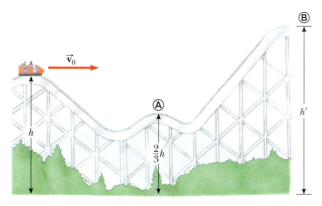

Figure P7.59

60. A roller coaster travels in a circular path. (a) Identify the forces on a passenger at the top of the circular loop that cause centripetal acceleration. Show the direction of all forces in a sketch. (b) Identify the forces on the passenger at the bottom of the loop that produce centripetal acceleration. Show these in a sketch. (c) Based on your answers to (a) and (b), at what point, top or bottom, should the seat exert the greatest force on the passenger? (d) Assume the speed of the roller coaster is 4.00 m/s at the top of the loop of radius 8.00 m. Find the force exerted by the seat on a 70.0-kg passenger at the top of the loop. Then, assume the speed remains the same at the bottom of the loop, and find the force exerted by the seat on the passenger at this point. Are your answers consistent with your choice of answers for (a) and (b)?

61. Assume that you are agile enough to run across a horizontal surface at 8.50 m/s, independently of the value of the gravitational field. What would be (a) the radius and (b) the mass of an airless spherical asteroid of uniform density 1.10×10^3 kg/m³ on which you could launch yourself into orbit by running? (c) What would be your period?

62. Figure P7.62 shows the elliptical orbit of a spacecraft around Earth. Take the origin of your coordinate system to be at the center of Earth.

Figure P7.62

(a) On a copy of the figure (enlarged if necessary), draw vectors representing
(i) the position of the spacecraft when it is at Ⓐ and Ⓑ;
(ii) the velocity of the spacecraft when it is at Ⓐ and Ⓑ;
(iii) the acceleration of the spacecraft when it is at Ⓐ and Ⓑ.
Make sure that each type of vector can be distinguished. Provide a legend that shows how each type is represented.

(b) Have you drawn the velocity vector at Ⓐ longer than, shorter than, or the same length as the one at Ⓑ? Explain. Have you drawn the acceleration vector at Ⓐ longer than, shorter than, or the same length as the one at Ⓑ? Explain. (Problem 62 is courtesy of E. F. Redish. For more problems of this type, visit www.physics.umd.edu/perg/)

63. In a home laundry drier, a cylindrical tub containing wet clothes rotates steadily about a horizontal axis, as in Figure P7.63. The clothes are made to tumble so that they will dry uniformly. The rate of rotation of the smooth-walled tub is chosen so that a small piece of cloth loses contact with the tub when the cloth is at an angle of 68.0° above the horizontal. If the radius of the tub is 0.330 m, what rate of revolution is needed?

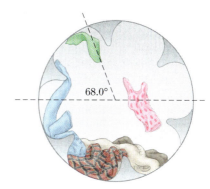

Figure P7.63

64. Casting of molten metal is important in many industrial processes. *Centrifugal casting* is used for manufacturing pipes, bearings, and many other structures. A cylindrical enclosure is rotated rapidly and steadily about a horizontal axis, as in Figure P7.64. Molten metal is poured into the rotating cylinder and then cooled, forming the finished product. Turning the cylinder at a high rotation rate forces the solidifying metal strongly to the outside. Any bubbles are displaced toward the axis so that unwanted voids will not be present in the casting.

Suppose that a copper sleeve of inner radius 2.10 cm and outer radius 2.20 cm is to be cast. To eliminate bubbles and give high structural integrity, the centripetal acceleration of each bit of metal should be 100g. What rate of rotation is required? State the answer in revolutions per minute.

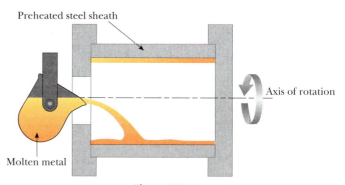

Figure P7.64

65. Suppose that a 1 800-kg car passes over a bump in a roadway that follows the arc of a circle of radius 20.4 m, as in Figure P7.65. (a) What force does the road exert on the car as the car passes the highest point of the bump if the car travels at 8.94 m/s? (b) What is the maximum speed the car can have without losing contact with the road as it passes this highest point?

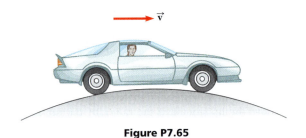

Figure P7.65

66. One popular design of a household juice machine is a conical, perforated stainless-steel basket 3.30 cm high, with a closed bottom of diameter 8.00 cm and an open top of diameter 13.70 cm, that spins at 20 000 revolutions per minute about a vertical axis (Fig. P7.66). Solid pieces of fruit or vegetables are chopped into granules by cutters on the bottom of the spinning cone. The dry pulp is ejected from the top of the cone. After passing through the perforations on the surface of the cone, the juice is collected in an enclosure immediately surrounding the cone. (a) What centripetal acceleration does a bit of fruit experience as it spins with the basket at a point midway between the top and bottom of the basket? Express the answer as a multiple of *g*. (b) Observe that the weight of the fruit is a negligible force. What is the normal force on 2.00 g of fruit at the midway point? (c) If the coefficient of kinetic friction between the fruit and the cone is 0.600, with what acceleration relative to the cone will the bit of fruit start to slide up the wall of the cone at that point, after being temporarily stuck?

Figure P7.66

67. (a) A luggage carousel at an airport has the form of a section of a large cone, steadily rotating about its vertical axis. Its metallic surface slopes downward toward the outside, making an angle of 20.0° with the horizontal. A 30.0-kg piece of luggage is placed on the carousel, 7.46 m from the axis of rotation. The travel bag goes around once in 38.0 s. Calculate the force of static friction between the bag and the carousel. (b) The drive motor is shifted to turn the carousel at a higher constant rate of rotation, and the piece of luggage is bumped to a position 7.94 m from the axis of rotation. The bag is on the verge

of slipping as it goes around once every 34.0 s. Calculate the coefficient of static friction between the bag and the carousel.

68. A merry-go-round is stationary. A dog is running on the ground just outside its circumference, moving with a constant angular speed of 0.750 rad/s. The dog does not change his pace when he sees a bone resting on the edge of the merry-go-round one-third of a revolution in front of him. At the instant the dog sees the bone, the merry-go-round begins to move in the direction the dog is running, with a constant angular acceleration of 0.015 0 rad/s². (a) After what time will the dog reach the bone? (b) The confused dog keeps running and passes the bone. How long after the merry-go-round starts to turn do the dog and the bone draw even with each other for the second time?

69. Figure P7.69 shows the drive train of a bicycle that has wheels 67.3 cm in diameter and pedal cranks 17.5 cm long. The cyclist pedals at a steady angular rate of 76.0 rev/min. The chain engages with a front sprocket 15.2 cm in diameter and a rear sprocket 7.00 cm in diameter. (a) Calculate the speed of a link of the chain relative to the bicycle frame. (b) Calculate the angular speed of the bicycle wheels. (c) Calculate the speed of the bicycle relative to the road. (d) What pieces of data, if any, are not necessary for the calculations?

Figure P7.69

70. The maximum lift force on a bat is proportional to the square of its flying speed *v*. For the hoary bat (*Lasiurus cinereus*), the magnitude of the lift force is given by

$$F_L \le (0.018 \text{ N} \cdot \text{s}^2/\text{m}^2) v^2$$

The bat can fly in a horizontal circle by "banking" its wings at an angle θ, as shown in Figure P7.70. In this situation, the magnitude of the vertical component of the lift force must equal the bat's weight. The horizontal component of the force provides the centripetal acceleration. (a) What is the minimum speed that the bat can have if its mass is 0.031 kg? (b) If the maximum speed of the bat is 10 m/s, what is the maximum banking angle that allows the bat to stay in a horizontal plane? (c) What is the radius of the circle of its flight when the bat flies at its maximum speed? (d) Can the bat turn with a smaller radius by flying more slowly?

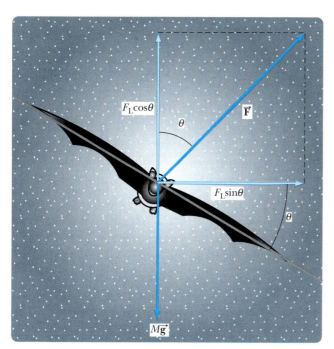

$F_L \cos\theta$

$\vec{\mathbf{F}}$

θ

$F_L \sin\theta$

θ

$M\vec{g}$

Figure P7.70

71. (a) Find the acceleration due to gravity at the surface of a neutron star of mass 1.5 solar masses and having a radius of 10.0 km. (b) Find the weight of a 0.120-kg baseball on this star. (c) Assume the equation $PE = mgh$ applies, and calculate the energy that a 70.0-kg person would expend climbing a 1.00-cm-tall mountain on the neutron star.

ACTIVITIES

A.1. Experiment with a bicycle wheel like the one in Figure 7.4. Wrap tape around the middle of one of the spokes so that you can easily see it, and draw a chalk mark on a point on the rim at the end of the spoke. Make the tire rotate by gently and steadily turning the crank by hand. Compare the linear speed of the tape with that of the chalk mark. Why do these two points move at different speeds? What happens to the speed of the tape as you slide it closer to the center of the wheel?

With the tape at a given distance from the center, measure the time it takes for the tape to make five rotations. From this time, find the angular speed and the linear speed of the wheel. Repeat at least five times and average your results. Move the tape to a different point and repeat the observations.

A.2. Tie a tennis ball to a string having a length of about 1.0 m. At a safe distance from other students, whirl the ball in a horizontal circle. Note the increasing tension in the string as you whirl it faster. Now whirl the ball in a vertical circle, and observe the difference in tension at the top and at the bottom of the path. Why is there such a difference? While whirling the ball, release it and observe that it flies off in a direction tangent to its circular path.

A.3. On a clear night, spend a few hours outside observing the stars. Choose a pattern of stars toward the north and others toward the east, the south, and the west. As time passes, notice how these patterns move. Obtain a star chart of the constellations from your instructor and also a list of the planets that are visible on your date of observation. Locate as many of these constellations and stars as possible.

Rotational motion is key in the harnessing of energy for power and propulsion, as illustrated by a steamboat. The rotating paddlewheel, driven by a steam engine, pushes water backwards, and the reaction force of the water thrusts the boat forward.

© Kevin Fleming/Corbis

CHAPTER

8

Rotational Equilibrium and Rotational Dynamics

In the study of linear motion, objects were treated as point particles without structure. It didn't matter *where* a force was applied, only *whether* it was applied or not.

The reality is that the point of application of a force *does* matter. In football, for example, if the ball carrier is tackled near his midriff, he might carry the tackler several yards before falling. If tackled well below the waistline, however, his center of mass rotates toward the ground, and he can be brought down immediately. Tennis provides another good example. If a tennis ball is struck with a strong horizontal force acting through its center of mass, it may travel a long distance before hitting the ground, far out of bounds. Instead, the same force applied in an upward, glancing stroke will impart topspin to the ball, which can cause it to land in the opponent's court.

The concepts of rotational equilibrium and rotational dynamics are also important in other disciplines. For example, students of architecture benefit from understanding the forces that act on buildings and biology students should understand the forces at work in muscles and on bones and joints. These forces create torques, which tell us how the forces affect an object's equilibrium and rate of rotation.

We will find that an object remains in a state of uniform rotational motion unless acted on by a net torque. This principle is the equivalent of Newton's first law. Further, the angular acceleration of an object is proportional to the net torque acting on it, which is the analog of Newton's second law. A net torque acting on an object causes a change in its rotational energy.

Finally, torques applied to an object through a given time interval can change the object's angular momentum. In the absence of external torques, angular momentum is conserved, a property that explains some of the mysterious and formidable properties of pulsars—remnants of supernova explosions that rotate at equatorial speeds approaching that of light.

8.1   TORQUE

Forces cause accelerations; *torques* cause angular accelerations. There is a definite relationship, however, between the two concepts.

Figure 8.1 depicts an overhead view of a door hinged at point O. From this viewpoint, the door is free to rotate around an axis perpendicular to the page and passing through O. If a force $\vec{F}$ is applied to the door, there are three factors that determine the effectiveness of the force in opening the door: the *magnitude* of the force, the *position* of application of the force, and the *angle* at which it is applied.

For simplicity, we restrict our discussion to position and force vectors lying in a plane. When the applied force $\vec{F}$ is perpendicular to the outer edge of the door, as in Figure 8.1, the door rotates counterclockwise with constant angular acceleration. The same perpendicular force applied at a point nearer the hinge results in a smaller angular acceleration. In general, a larger radial distance r between the applied force and the axis of rotation results in a larger angular acceleration. Similarly, a larger applied force will also result in a larger angular acceleration. These considerations motivate the basic definition of **torque** for the special case of forces perpendicular to the position vector:

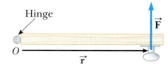

Figure 8.1   A bird's-eye view of a door hinged at O, with a force applied perpendicular to the door.

Let $\vec{F}$ be a force acting on an object, and let $\vec{r}$ be a position vector from a chosen point O to the point of application of the force, with $\vec{F}$ perpendicular to $\vec{r}$. The magnitude of the torque $\vec{\tau}$ exerted by the force $\vec{F}$ is given by

$$\tau = rF \qquad\qquad \textbf{[8.1]}$$

where r is the length of the position vector and F is the magnitude of the force.

SI unit: Newton-meter (N·m)

◄ Basic definition of torque

The vectors $\vec{r}$ and $\vec{F}$ lie in a plane. As discussed in detail shortly in conjunction with Figure 8.4, the torque $\vec{\tau}$ is then perpendicular to this plane. The point O is usually chosen to coincide with the axis the object is rotating around, such as the hinge of a door or hub of a merry-go-round. (Other choices are possible as well.) In addition, we consider only forces acting in the plane perpendicular to the axis of rotation. This criterion excludes, for example, a force with upward component on a merry-go-round railing, which cannot affect the merry-go-round's rotation.

Under these conditions, an object can rotate around the chosen axis in one of two directions. By convention, counterclockwise is taken to be the positive direction, clockwise the negative direction. When an applied force causes an object to rotate counterclockwise, the torque on the object is positive. When the force causes the object to rotate clockwise, the torque on the object is negative. When two or more torques act on an object at rest, the torques are added. If the net torque isn't zero, the object starts rotating at an ever-increasing rate. If the net torque is zero, the object's rate of rotation doesn't change. These considerations lead to the rotational analog of the first law: **the rate of rotation of an object doesn't change, unless the object is acted on by a net torque.**

EXAMPLE 8.1   Battle of the Revolving Door

Goal   Apply the basic definition of torque.

Problem   Two disgruntled businessmen are trying to use a revolving door, as in Figure 8.2. The door has a diameter of 2.60 m. The businessman on the left exerts a force of 625 N perpendicular to the door and 1.20 m from the hub's center, while the one on the right exerts a force of 8.50×10^2 N perpendicular to the door and 0.800 m from the hub's center. Find the net torque on the revolving door.

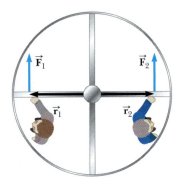

Figure 8.2   (Example 8.1)

Strategy   Calculate the individual torques on the door using the definition of torque, Equation 8.1, and then sum to find the net torque on the door. The first businessman exerts a negative torque, the second a positive torque. Their positions of application also differ.

Solution

Calculate the torque exerted by the first businessman. A negative sign must be supplied, because $\vec{F}_1$, if unopposed, would cause a clockwise rotation.

$$\tau_1 = -r_1 F_1 = -(1.20 \text{ m})(625 \text{ N}) = -7.50 \times 10^2 \text{ N} \cdot \text{m}$$

Calculate the torque exerted by the second businessman. The torque is positive because $\vec{F}_2$, if unopposed, would cause a counterclockwise rotation.

$$\tau_2 = r_2 F_2 = (0.800 \text{ m})(8.50 \times 10^2 \text{ N})$$
$$= 6.80 \times 10^2 \text{ N} \cdot \text{m}$$

Sum the torques to find the net torque on the door:

$$\tau_{\text{net}} = \tau_1 + \tau_2 = \boxed{-7.0 \times 10^1 \text{ N} \cdot \text{m}}$$

Remark   The negative result here means the net torque will produce a clockwise rotation.

Exercise 8.1

A businessman enters the same revolving door on the right, pushing with 576 N of force directed perpendicular to the door and 0.700 m from the hub, while a boy exerts a force of 365 N perpendicular to the door, 1.25 m to the left of the hub. Find (a) the torques exerted by each person and (b) the net torque on the door.

Answers   (a) $\tau_{\text{boy}} = -456 \text{ N} \cdot \text{m}$, $\tau_{\text{man}} = 403 \text{ N} \cdot \text{m}$ (b) $\tau_{\text{net}} = -53 \text{ N} \cdot \text{m}$

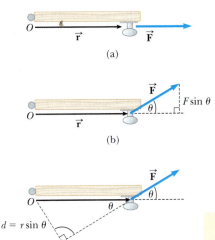

Figure 8.3   (a) A force $\vec{F}$ acting at an angle $\theta = 180°$ exerts zero torque about the pivot O. (b) The part of the force perpendicular to the door, $F \sin \theta$, exerts torque $rF \sin \theta$ about O. (c) An alternate interpretation of torque in terms of a *lever arm* $d = r \sin \theta$.

The applied force isn't always perpendicular to the position vector $\vec{r}$. Suppose the force $\vec{F}$ exerted on a door is directed away from the axis, as in Figure 8.3a, say, by someone's grasping the doorknob and pushing to the right. Exerting the force in this direction couldn't possibly open the door. However, if the applied force acts at an angle to the door as in Figure 8.3b, the component of the force *perpendicular* to the door will cause it to rotate. This figure shows that the component of the force perpendicular to the door is $F \sin \theta$, where θ is the angle between the position vector $\vec{r}$ and the force $\vec{F}$. When the force is directed away from the axis, $\theta = 0°$, $\sin (0°) = 0$, and $F \sin (0°) = 0$. When the force is directed toward the axis, $\theta = 180°$ and $F \sin (180°) = 0$. The maximum absolute value of $F \sin \theta$ is attained only when $\vec{F}$ is perpendicular to $\vec{r}$—that is, when $\theta = 90°$ or $\theta = 270°$. These considerations motivate a more general definition of torque:

Let $\vec{F}$ be a force acting on an object, and let $\vec{r}$ be a position vector from a chosen point O to the point of application of the force. The magnitude of the torque $\vec{\tau}$ exerted by the force $\vec{F}$ is

$$\tau = rF \sin \theta \qquad \text{[8.2]}$$

where r is the length of the position vector, F the magnitude of the force, and θ the angle between $\vec{r}$ and $\vec{F}$.

SI unit: Newton-meter (N $\cdot$ m)

Again the vectors $\vec{r}$ and $\vec{F}$ lie in a plane and for our purposes the chosen point O will usually correspond to an axis of rotation perpendicular to the plane.

A second way of understanding the $\sin\theta$ factor is to associate it with the magnitude r of the position vector $\vec{r}$. The quantity $d = r\sin\theta$ is called the **lever arm**, which is the perpendicular distance from the axis of rotation to a line drawn along the direction of the force. This alternate interpretation is illustrated in Figure 8.3c.

It's important to remember that **the value of τ depends on the chosen axis of rotation**. Torques can be computed around any axis, regardless of whether there is some actual, physical rotation axis present. Once the point is chosen, however, it must be used consistently throughout a given problem.

Torque is a vector perpendicular to the plane determined by the position and force vectors, as illustrated in Figure 8.4. The direction can be determined by the *right-hand rule*:

1. Point the fingers of your right hand in the direction of $\vec{r}$.
2. Curl your fingers toward the direction of vector $\vec{F}$.
3. Your thumb then points approximately in the direction of the torque, in this case out of the page.

Problems used in this book will be confined to objects rotating around an axis perpendicular to the plane containing $\vec{r}$ and $\vec{F}$, so if these vectors are in the plane of the page, the torque will always point either into or out of the page, parallel to the axis of rotation. If your right thumb is pointed in the direction of a torque, your fingers curl naturally in the direction of rotation that the torque would produce on an object at rest.

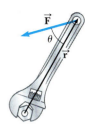

Figure 8.4 The right-hand rule: Point the fingers of your right hand along $\vec{r}$ and curl them in the direction of $\vec{F}$. Your thumb then points in the direction of the torque (out of the page, in this case).

EXAMPLE 8.2   The Swinging Door

Goal Apply the more general definition of torque.

Problem (a) A man applies a force of $F = 3.00 \times 10^2$ N at an angle of 60.0° to the door of Figure 8.5a, 2.00 m from the hinges. Find the torque on the door, choosing the position of the hinges as the axis of rotation. (b) Suppose a wedge is placed 1.50 m from the hinges on the other side of the door. What minimum force must the wedge exert so that the force applied in part (a) won't open the door?

Strategy Part (a) can be solved by substitution into the general torque equation. In part (b), the hinges, the wedge, and the applied force all exert torques on the door. The door doesn't open, so the sum of these torques must be zero, a condition that can be used to find the wedge force.

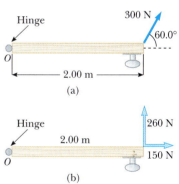

Figure 8.5 (Example 8.2a) (a) Top view of a door being pushed by a 300-N force. (b) The components of the 300-N force.

Solution

(a) Compute the torque due to the applied force exerted at 60.0°.

Substitute into the general torque equation:

$$\tau_F = rF\sin\theta = (2.00\text{ m})(3.00\times10^2\text{ N})\sin 60.0°$$
$$= (2.00\text{ m})(2.60\times10^2\text{ N}) = \boxed{5.20\times10^2\text{ N}\cdot\text{m}}$$

(b) Calculate the force exerted by the wedge on the other side of the door.

Set the sum of the torques equal to zero:

$$\tau_{\text{hinge}} + \tau_{\text{wedge}} + \tau_F = 0$$

The hinge force provides no torque because it acts at the axis ($r = 0$). The wedge force acts at an angle of $-90.0°$, opposite F_y.

$$0 + F_{\text{wedge}}(1.50\text{ m})\sin(-90.0°) + 5.20\times10^2\text{ N}\cdot\text{m} = 0$$

$$F_{\text{wedge}} = \boxed{347\text{ N}}$$

Remark Notice that the angle from the position vector to the wedge force is $-90°$. This is because, starting at the position vector, it's necessary to go 90° clockwise (the negative angular direction) to get to the force vector. Measuring the angle in this way automatically supplies the correct sign for the torque term and is consistent with the right-hand rule. Alternately, the magnitude of the torque can be found and the correct sign chosen based on physical intuition.

Exercise 8.2

A man ties one end of a strong rope 8.00 m long to the bumper of his truck, 0.500 m from the ground, and the other end to a vertical tree trunk at a height of 3.00 m. He uses the truck to create a tension of 8.00×10^2 N in the rope. Compute the magnitude of the torque on the tree due to the tension in the rope, with the base of the tree acting as the reference point.

Answer 2.28×10^3 N·m

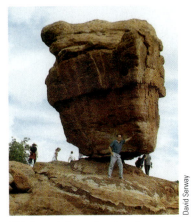

This large balanced rock at the Garden of the Gods in Colorado, Springs, Colorado, is in mechanical equilibrium.

8.2 TORQUE AND THE TWO CONDITIONS FOR EQUILIBRIUM

An object in mechanical equilibrium must satisfy the following two conditions:

1. The net external force must be zero: $\Sigma \vec{\mathbf{F}} = 0$

2. The net external torque must be zero: $\Sigma \vec{\tau} = 0$

The first condition is a statement of translational equilibrium: The sum of all forces acting on the object must be zero, so the object has no translational acceleration, $\vec{\mathbf{a}} = 0$. The second condition is a statement of rotational equilibrium: The sum of all torques on the object must be zero, so the object has no angular acceleration, $\vec{\boldsymbol{\alpha}} = 0$. For an object to be in equilibrium, it must both translate and rotate at a constant rate.

Because we can choose any location for calculating torques, it's usually best to select an axis that will make at least one torque equal to zero, just to simplify the net torque equation.

EXAMPLE 8.3 Balancing Act

Goal Apply the conditions of equilibrium and illustrate the use of different axes for calculating the net torque on an object.

Problem A woman of mass $m = 55.0$ kg sits on the left end of a seesaw—a plank of length $L = 4.00$ m, pivoted in the middle as in Figure 8.6. **(a)** First compute the torques on the seesaw about an axis that passes through the pivot point. Where should a man of mass $M = 75.0$ kg sit if the system (seesaw plus man and woman) is to be balanced? **(b)** Find the normal force exerted by the pivot if the plank has a mass of $m_{pl} = 12.0$ kg. **(c)** Repeat part (b), but this time compute the torques about an axis through the left end of the plank.

Strategy In part (a), apply the second condition of equilibrium, $\Sigma \tau = 0$, computing torques around the pivot point. The mass of the plank forming the seesaw is distributed evenly on either side of the pivot point, so the torque exerted by gravity on the plank, $\tau_{gravity}$, can be computed as if all the plank's mass is concentrated at the pivot point. Then

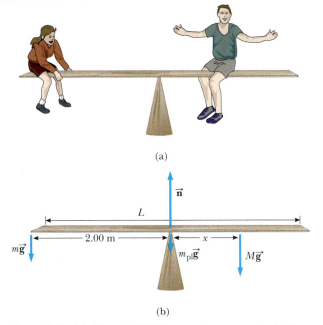

Figure 8.6 (a) (Example 8.3) Two people on a see-saw. (b) Free body diagram for the plank.

$\tau_{gravity}$ is zero, as is the torque exerted by the pivot, because their moment arms are zero. In part (b), the first condition of equilibrium, $\Sigma \vec{F} = 0$, must be applied. Part (c) is a repeat of part (a) showing that choice of a different axis yields the same answer.

Solution

(a) Where should the man sit to balance the seesaw?

Apply the second condition of equilibrium to the plank by setting the sum of the torques equal to zero:

$$\tau_{pivot} + \tau_{gravity} + \tau_{man} + \tau_{woman} = 0$$

The first two torques are zero. Let x represent the man's distance from the pivot. The woman is at a distance $L/2$ from the pivot.

$$0 + 0 - Mgx + mg(L/2) = 0$$

Solve this equation for x and evaluate it:

$$x = \frac{m(L/2)}{M} = \frac{(55.0\text{ kg})(2.00\text{ m})}{75.0\text{ kg}} = \boxed{1.47\text{ m}}$$

(b) Find the normal force n exerted by the pivot on the seesaw.

Apply for first condition of equilibrium to the plank, solving the resulting equation for the unknown normal force, n:

$$-Mg - mg - m_{pl}g + n = 0$$
$$n = (M + m + m_{pl})g$$
$$= (75.0\text{ kg} + 55.0\text{ kg} + 12.0\text{ kg})(9.80\text{ m/s}^2)$$
$$n = \boxed{1.39 \times 10^3\text{ N}}$$

(c) Repeat part (a), choosing a new axis through the left end of the plank.

Compute the torques using this axis, and set their sum equal to zero. Now the pivot and gravity forces on the plank result in nonzero torques.

$$\tau_{man} + \tau_{woman} + \tau_{plank} + \tau_{pivot} = 0$$
$$-Mg(L/2 + x) + mg(0) - m_{pl}g(L/2) + n(L/2) = 0$$

Substitute all known quantities:

$$-(75.0\text{ kg})(9.80\text{ m/s}^2)(2.00\text{ m} + x) + 0$$
$$- (12.0\text{ kg})(9.80\text{ m/s}^2)(2.00\text{ m}) + n(2.00\text{ m}) = 0$$
$$-(1.47 \times 10^3\text{ N}\cdot\text{m}) - (735\text{ N})x - (235\text{ N}\cdot\text{m})$$
$$+ (2.00\text{ m})n = 0$$

Solve for x, substituting the normal force found in part (b):

$$x = \boxed{1.46\text{ m}}$$

Remarks   The answers for x in parts (a) and (c) agree except for a small round-off discrepancy. This illustrates how choosing a different axis leads to the same solution.

Exercise 8.3

Suppose a 30.0-kg child sits 1.50 m to the left of center on the same seesaw. A second child sits at the end on the opposite side, and the system is balanced. (a) Find the mass of the second child. (b) Find the normal force acting at the pivot point.

Answers   (a) 22.5 kg   (b) 632 N

8.3 THE CENTER OF GRAVITY

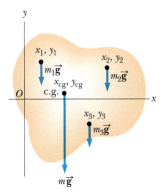

Figure 8.7 The net gravitational torque on an object is zero if computed around the center of gravity. The object will balance if supported at that point (or at any point along a vertical line above or below that point).

In the example of the seesaw in the previous section, we guessed that the torque due to the force of gravity on the plank was the same as if all the plank's weight were concentrated at its center. This is a general procedure: To compute the torque on a rigid body due to the force of gravity, the body's entire weight can be thought of as concentrated at a single point. The problem then reduces to finding the location of that point. If the body is homogeneous (its mass is distributed evenly) and symmetric, it's usually possible to guess the location of that point, as in Example 8.3. Otherwise, it's necessary to calculate the point's location, as explained in this section.

Consider an object of arbitrary shape lying in the xy-plane, as in Figure 8.7. The object is divided into a large number of very small particles of weight m_1g, m_2g, m_3g, . . . having coordinates (x_1, y_1), (x_2, y_2), (x_3, y_3), ... If the object is free to rotate around the origin, each particle contributes a torque about the origin that is equal to its weight multiplied by its lever arm. For example, the torque due to the weight m_1g is m_1gx_1, and so forth.

We wish to locate the point of application of the single force of magnitude $w = F_g = Mg$ (the total weight of the object), where the effect on the rotation of the object is the same as that of the individual particles. This point is called the object's **center of gravity**. Equating the torque exerted by w at the center of gravity to the sum of the torques acting on the individual particles gives

$$(m_1g + m_2g + m_3g + \cdots)x_{cg} = m_1gx_1 + m_2gx_2 + m_3gx_3 + \cdots$$

We assume that g is the same everywhere in the object (which is true for all objects we will encounter). Then the g factors in the preceding equation cancel, resulting in

$$x_{cg} = \frac{m_1x_1 + m_2x_2 + m_3x_3 + \cdots}{m_1 + m_2 + m_3 + \cdots} = \frac{\Sigma m_i x_i}{\Sigma m_i} \qquad \text{[8.3a]}$$

where x_{cg} is the x-coordinate of the center of gravity. Similarly, the y-coordinate and z-coordinate of the center of gravity of the system can be found from

$$y_{cg} = \frac{\Sigma m_i y_i}{\Sigma m_i} \qquad \text{[8.3b]}$$

and

$$z_{cg} = \frac{\Sigma m_i z_i}{\Sigma m_i} \qquad \text{[8.3c]}$$

These three equations are identical to the equations for a similar concept called **center of mass**. The center of mass and center of gravity of an object are exactly the same when g doesn't vary significantly over the object.

It's often possible to guess the location of the center of gravity. **The center of gravity of a homogeneous, symmetric body must lie on the axis of symmetry.** For example, the center of gravity of a homogeneous rod lies midway between the ends of the rod, and the center of gravity of a homogeneous sphere or a homogeneous cube lies at the geometric center of the object. The center of gravity of an irregularly shaped object, such as a wrench, can be determined experimentally by suspending the wrench from two different arbitrary points (Fig. 8.8). The wrench is first hung from point A, and a vertical line AB (which can be established with a plumb bob) is drawn when the wrench is in equilibrium. The wrench is then hung from point C, and a second vertical line CD is drawn. The center of gravity coincides with the intersection of these two lines. In fact, if the wrench is hung freely from any point, the center of gravity always lies straight below the point of support, so the vertical line through that point must pass through the center of gravity.

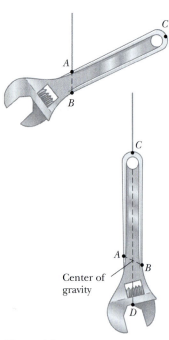

Figure 8.8 An experimental technique for determining the center of gravity of a wrench. The wrench is hung freely from two different pivots, A and C. The intersection of the two vertical lines, AB and CD, locates the center of gravity.

Several examples in Section 8.4 involve homogeneous, symmetric objects where the centers of gravity coincide with their geometric centers. A rigid object in a uniform gravitational field can be balanced by a single force equal in magnitude to the weight of the object, as long as the force is directed upward through the object's center of gravity.

EXAMPLE 8.4   Where Is the Center of Gravity?

Goal   Find the center of gravity of a system of particles.

Problem   Three particles are located in a coordinate system as shown in Figure 8.9. Find the center of gravity.

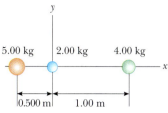

Strategy   The y-coordinate and z-coordinate of the center of gravity are both zero because all the particles are on the x-axis. We can find the x-coordinate of the center of gravity using Equation 8.3a.

Figure 8.9   (Example 8.4) Locating the center of gravity of a system of three particles.

Solution

Apply Equation 8.3a to the system of three particles:

$$x_{cg} = \frac{\Sigma m_i x_i}{\Sigma m_i} = \frac{m_1 x_1 + m_2 x_2 + m_3 x_3}{m_1 + m_2 + m_3} \quad \textbf{(1)}$$

Compute the numerator of Equation (1):

$$\Sigma m_i x_i = m_1 x_1 + m_2 x_2 + m_3 x_3$$
$$= (5.00 \text{ kg})(-0.500 \text{ m}) + (2.00 \text{ kg})(0 \text{ m})$$
$$+ (4.00 \text{ kg})(1.00 \text{ m})$$
$$= 1.50 \text{ kg} \cdot \text{m}$$

Substitute the denominator, $\Sigma m_i = 11.0$ kg, and the numerator into Equation (1).

$$x_{cg} = \frac{1.50 \text{ kg} \cdot \text{m}}{11.0 \text{ kg}} = \boxed{0.136 \text{ m}}$$

Exercise 8.4

If a fourth particle of mass 2.00 kg is placed at $x = 0$, $y = 0.250$ m, find the x-and y-coordinates of the center of gravity for this system of four particles.

Answer $x_{cg} = 0.115$ m; $y_{cg} = 0.0385$ m

EXAMPLE 8.5   Locating Your Lab Partner's Center of Gravity

Goal   Use torque to find a center of gravity.

Problem   In this example, we show how to find the location of a person's center of gravity. Suppose your lab partner has a height L of 173 cm (5 ft, 8 in) and a weight w of 715 N (160 lb). You can determine the position of his center of gravity by having him stretch out on a uniform board supported at one end by a scale, as shown in Figure 8.10. If the board's weight w_b is 49 N and the scale reading F is 3.50×10^2 N, find the distance of your lab partner's center of gravity from the left end of the board.

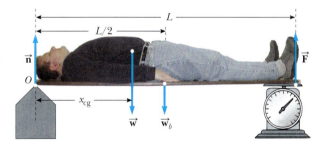

Figure 8.10   (Example 8.5) Determining your lab partner's center of gravity

Strategy   To find the position x_{cg} of the center of gravity, compute the torques using an axis through O. Set the sum of the torques equal to zero and solve for x_{cg}.

Solution

Apply the second condition of equilibrium. There is no torque due to the normal force $\vec{\mathbf{n}}$ because its moment arm is zero.

$$\Sigma \tau_O = 0$$
$$-w x_{cg} - w_b(L/2) + FL = 0$$

Solve for x_{cg} and substitute known values:

$$x_{cg} = \frac{FL - w_b(L/2)}{w}$$

$$= \frac{(350 \text{ N})(173 \text{ cm}) - (49 \text{ N})(86.5 \text{ cm})}{715 \text{ N}} = \boxed{79 \text{ cm}}$$

Remarks The given information is sufficient only to determine the *x*-coordinate of the center of gravity. The other two coordinates can be estimated, based on the body's symmetry.

Exercise 8.5
Suppose a 416-kg alligator of length 3.5 m is stretched out on a board of the same length weighing 65 N. If the board is supported on the ends as in Figure 8.10, and the scale reads 1 880 N, find the *x*-component of the alligator's center of gravity.

Answer 1.59 m

8.4   EXAMPLES OF OBJECTS IN EQUILIBRIUM

Recall from Chapter 4 that when an object is treated as a geometric point, equilibrium requires only that the net force on the object is zero. In this chapter, we have shown that for extended objects a second condition for equilibrium must also be satisfied: The net torque on the object must be zero. The following general procedure is recommended for solving problems that involve objects in equilibrium.

> **Problem-Solving Strategy**   Objects in Equilibrium
>
> 1. **Diagram the system.** Include coordinates and choose a convenient rotation axis for computing the net torque on the object.
> 2. **Draw a free-body diagram** of the object of interest, showing all external forces acting on it. For systems with more than one object, draw a *separate* diagram for each object. (Most problems will have a single object of interest.)
> 3. **Apply $\Sigma \tau_i = 0$, the second condition of equilibrium**. This condition yields a single equation for each object of interest. If the axis of rotation has been carefully chosen, the equation often has only one unknown and can be solved immediately.
> 4. **Apply $\Sigma F_x = 0$ and $\Sigma F_y = 0$, the first condition of equilibrium**. This yields two more equations per object of interest.
> 5. **Solve the system of equations.** For each object, the two conditions of equilibrium yield three equations, usually with three unknowns. Solve by substitution.

EXAMPLE 8.6   A Weighted Forearm

Goal Apply the equilibrium conditions to the human body.

Problem A 50.0-N (11-lb) weight is held in a person's hand with the forearm horizontal, as in Figure 8.11. The biceps muscle is attached 0.030 0 m from the joint, and the weight is 0.350 m from the joint. Find the upward force $\vec{\mathbf{F}}$ exerted by the biceps on the forearm (the ulna) and the downward force $\vec{\mathbf{R}}$ exerted by the humerus on the forearm, acting at the joint. Neglect the weight of the forearm.

Strategy The forces acting on the forearm are equivalent to those acting on a bar of length 0.350 m, as shown in Figure 8.11b. Choose the usual *x*- and *y*-coordinates as shown and the axis at *O* on the left end. (This completes Steps 1 and 2.) Use the conditions of equilibrium to generate equations for the unknowns, and solve.

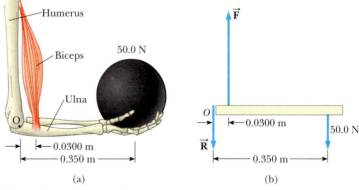

Figure 8.11 (Example 8.6) (a) A weight held with the forearm horizontal. (b) The mechanical model for the system.

Solution

Apply the second condition for equilibrium (step 3):

$$\Sigma \tau_i = \tau_R + \tau_F + \tau_{BB} = 0$$

$$R(0) + F(0.0300 \text{ m}) - (50.0 \text{ N})(0.350 \text{ m}) = 0$$

$$F = \boxed{583 \text{ N } (131 \text{ lb})}$$

Apply the first condition for equilibrium (step 4):

$$\Sigma F_y = F - R - 50.0 \text{ N} = 0$$

$$R = F - 50.0 \text{ N} = 583 \text{ N} - 50 \text{ N} = \boxed{533 \text{ N } (120 \text{ lb})}$$

Exercise 8.6

Suppose you wanted to limit the force acting on your joint to a maximum value of 8.00×10^2 N. (a) Under these circumstances, what maximum weight would you attempt to lift? (b) What force would your biceps apply while lifting this weight?

Answers   (a) 75.0 N   (b) 875 N

EXAMPLE 8.7   Walking a Horizontal Beam

Goal   Solve an equilibrium problem with nonperpendicular torques.

Problem   A uniform horizontal beam 5.00 m long and weighing 3.00×10^2 N is attached to a wall by a pin connection that allows the beam to rotate. Its far end is supported by a cable that makes an angle of $53.0°$ with the horizontal (Fig. 8.12a). If a person weighing 6.00×10^2 N stands 1.50 m from the wall, find the magnitude of the tension $\vec{T}$ in the cable and the force $\vec{R}$ exerted by the wall on the beam.

Strategy

The second condition of equilibrium, $\Sigma \tau_i = 0$, with torques computed around the pin, can be solved for the tension T in the cable. The first condition of equilibrium, $\Sigma \vec{F}_i = 0$, gives two equations and two unknowns for the two components of the force exerted by the wall, R_x and R_y.

Solution

From Figure 8.12, the forces causing torques are the wall force $\vec{R}$, the gravity forces on the beam and the man, w_B and w_M, and the tension force $\vec{T}$. Apply the condition of rotational equilibrium:

$$\Sigma \tau_i = \tau_R + \tau_B + \tau_M + \tau_T = 0$$

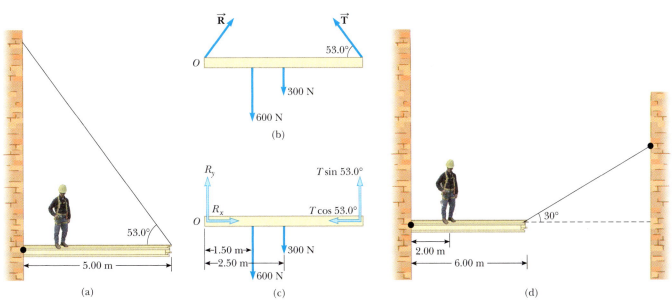

Figure 8.12   (Example 8.7) (a) A uniform beam attached to a wall and supported by a cable. (b) A free-body diagram for the beam. (c) The component form of the free-body diagram. (d) (Exercise 8.7)

Compute torques around the pin at O, so $\tau_R = 0$ (zero moment arm). The torque due to the beam's weight acts at the beam's center of gravity.

$$\Sigma \tau_i = 0 - w_B(L/2) - w_M(1.50 \text{ m}) + TL \sin(53°) = 0$$

Substitute $L = 5.00$ m and the weights, solving for T:

$$- (3.00 \times 10^2 \text{ N})(2.50 \text{ m})$$
$$- (6.00 \times 10^2 \text{ N})(1.50 \text{ m})$$
$$+ (T \sin 53.0°)(5.00 \text{ m}) = 0$$

$$T = \boxed{413 \text{ N}}$$

Now apply the first condition of equilibrium to the beam:

$$\Sigma F_x = R_x - T \cos 53.0° = 0 \tag{1}$$
$$\Sigma F_y = R_y - w_B - w_M + T \sin 53.0° = 0 \tag{2}$$

Substituting the value of T found in the previous step and the weights, obtain the components of $\vec{R}$:

$$R_x = \boxed{249 \text{ N}} \qquad R_y = \boxed{5.70 \times 10^2 \text{ N}}$$

Remarks Even if we selected some other axis for the torque equation, the solution would be the same. For example, if the axis were to pass through the center of gravity of the beam, the torque equation would involve both T and R_y. Together with equations (1) and (2), however, the unknowns could still be found—a good exercise.

Exercise 8.7
A person with mass 55.0 kg stands 2.00 m away from the wall on a 6.00-m beam, as shown in Figure 8.12d. The mass of the beam is 40.0 kg. Find the hinge force components and the tension in the wire.

Answers $T = 751$ N, $R_x = -6.50 \times 10^2$ N, $R_y = 556$ N

INTERACTIVE EXAMPLE 8.8 Don't Climb the Ladder

Goal Apply the two conditions of equilibrium.

Problem A uniform ladder 10.0 m long and weighing 50.0 N rests against a smooth vertical wall as in Figure 8.13a. If the ladder is just on the verge of slipping when it makes a 50.0° angle with the ground, find the coefficient of static friction between the ladder and ground.

Strategy Figure 8.13b is the free-body diagram for the ladder. The first condition of equilibrium, $\Sigma \vec{F}_i = 0$, gives two equations for three unknowns: the magnitudes of the static friction force f and the normal force n, both acting on the base of the ladder, and the magnitude of the force of the wall, P, acting on the top of the ladder. The second condition of equilibrium, $\Sigma \tau_i = 0$, gives a third equation (for P), so all three quantities can be found. The definition of static friction then allows computation of the coefficient of static friction.

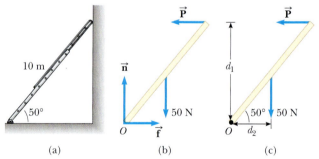

Figure 8.13 (Interactive Example 8.8) (a) A ladder leaning against a frictionless wall. (b) A free-body diagram of the ladder. (c) Lever arms for the force of gravity and $\vec{P}$.

Solution
Apply the first condition of equilibrium to the ladder:

$$\Sigma F_x = f - P = 0 \quad \rightarrow \quad f = P \tag{1}$$
$$\Sigma F_y = n - 50.0 \text{ N} = 0 \quad \rightarrow \quad n = 50.0 \text{ N} \tag{2}$$

Apply the second condition of equilibrium, computing torques around the base of the ladder, with τ_{grav} standing for the torque due to the ladder's 50.0-N weight:

$$\Sigma \tau_i = \tau_f + \tau_n + \tau_{grav} + \tau_P = 0$$

The torques due to friction and the normal force are zero about O because their moment arms are zero. (Moment arms can be found from Figure 8.13c.)

$$0 + 0 - (50.0 \text{ N})(5.00 \text{ m}) \sin 40.0°$$
$$+ P(10.0 \text{ m}) \sin 50.0° = 0$$

$$P = 21.0 \text{ N}$$

From Equation (1), we now have $f = P = 21.0$ N. The ladder is on the verge of slipping, so write an expression for the maximum force of static friction and solve for μ_s:

$$21.0 \text{ N} = f = f_{s,\text{max}} = \mu_s n = \mu_s (50.0 \text{ N})$$

$$\mu_s = \frac{21.0 \text{ N}}{50.0 \text{ N}} = \boxed{0.420}$$

Remarks Note that torques were computed around an axis through the bottom of the ladder so that only $\vec{\mathbf{P}}$ and the force of gravity contributed nonzero torques. This choice of axis reduces the complexity of the torque equation, often resulting in an equation with only one unknown.

Exercise 8.8
If the coefficient of static friction is 0.360, and the same ladder makes a 60.0° angle with respect to the horizontal, how far along the length of the ladder can a 70.0-kg painter climb before the ladder begins to slip?

Answer 6.33 m

Physics⊗Now™ You can adjust the angle of the ladder and watch what happens when it is released by logging into PhysicsNow at **www.cp7e.com** and going to Interactive Example 8.8.

8.5 RELATIONSHIP BETWEEN TORQUE AND ANGULAR ACCELERATION

When a rigid object is subject to a net torque, it undergoes an angular acceleration that is directly proportional to the net torque. This result, which is analogous to Newton's second law, is derived as follows.

The system shown in Figure 8.14 consists of an object of mass m connected to a very light rod of length r. The rod is pivoted at the point O, and its movement is confined to rotation on a frictionless *horizontal* table. Assume that a force F_t acts perpendicular to the rod and hence is tangent to the circular path of the object. Because there is no force to oppose this tangential force, the object undergoes a tangential acceleration a_t in accordance with Newton's second law:

$$F_t = ma_t$$

Multiply both sides of this equation by r:

$$F_t r = mra_t$$

Substituting the equation $a_t = r\alpha$ relating tangential and angular acceleration into the above expression gives

$$F_t r = mr^2 \alpha \qquad \text{[8.4]}$$

The left side of Equation 8.4 is the torque acting on the object about its axis of rotation, so we can rewrite it as

$$\tau = mr^2 \alpha \qquad \text{[8.5]}$$

Equation 8.5 shows that the torque on the object is proportional to the angular acceleration of the object, where the constant of proportionality mr^2 is called the **moment of inertia** of the object of mass m. (Because the rod is very light, its moment of inertia can be neglected.)

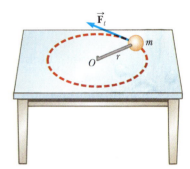

Figure 8.14 An object of mass m attached to a light rod of length r moves in a circular path on a frictionless horizontal surface while a tangential force $\vec{\mathbf{F}}_t$ acts on it.

Quick Quiz 8.1

Using a screwdriver, you try to remove a screw from a piece of furniture, but can't get it to turn. To increase the chances of success, you should use a screwdriver that (a) is longer, (b) is shorter, (c) has a narrower handle, or (d) has a wider handle.

Torque on a Rotating Object

Consider a solid disk rotating about its axis as in Figure 8.15a. The disk consists of many particles at various distances from the axis of rotation. (See Fig. 8.15b.) The

Figure 8.15 (a) A solid disk rotating about its axis. (b) The disk consists of many particles, all with the same angular acceleration.

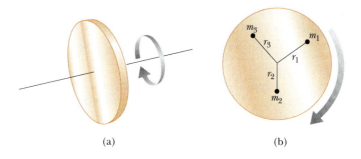

(a)   (b)

torque on each one of these particles is given by Equation 8.5. The *net* torque on the disk is given by the sum of the individual torques on all the particles:

$$\Sigma\tau = (\Sigma mr^2)\alpha \qquad [8.6]$$

Because the disk is rigid, all of its particles have the *same* angular acceleration, so α is not involved in the sum. If the masses and distances of the particles are labeled with subscripts as in Figure 8.15b, then

$$\Sigma mr^2 = m_1 r_1^2 + m_2 r_2^2 + m_3 r_3^2 + \cdots$$

This quantity is the moment of inertia, I, of the whole body:

Moment of inertia ▶

$$I \equiv \sum mr^2 \qquad [8.7]$$

The moment of inertia has the SI units kg·m². Using this result in Equation 8.6, we see that the net torque on a rigid body rotating about a fixed axis is given by

Rotational analog of Newton's second law ▶

$$\sum \tau = I\alpha \qquad [8.8]$$

Equation 8.8 says that **the angular acceleration of an extended rigid object is proportional to the net torque acting on it**. This equation is the rotational analog of Newton's second law of motion, with torque replacing force, moment of inertia replacing mass, and angular acceleration replacing linear acceleration. Although the moment of inertia of an object is related to its mass, there is an important difference between them. The mass m depends only on the quantity of matter in an object while the moment of inertia, I, depends on both the quantity of matter and its distribution (through the r^2 term in $I = \Sigma mr^2$) in the rigid object.

Quick Quiz 8.2

A constant net torque is applied to an object. Which one of the following will *not* be constant? (a) angular acceleration, (b) angular velocity, (c) moment of inertia, or (d) center of gravity.

Quick Quiz 8.3

The two rigid objects shown in Figure 8.16 have the same mass, radius, and angular speed. If the same braking torque is applied to each, which takes longer to stop? (a) A (b) B (c) more information is needed

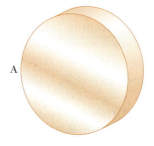

A

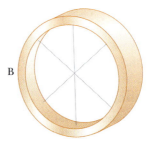

B

Figure 8.16 (Quick Quiz 8.3)

The gear system on a bicycle provides an easily visible example of the relationship between torque and angular acceleration. Consider first a five-speed gear system in which the drive chain can be adjusted to wrap around any of five gears attached to the back wheel (Fig. 8.17). The gears, with different radii, are concentric with the wheel hub. When the cyclist begins pedaling from rest, the chain is attached to the largest gear. Because it has the largest radius, this gear provides the largest torque to the drive wheel. A large torque is required initially, because the bicycle starts from rest. As the bicycle rolls faster, the tangential speed of the chain increases, eventually becoming too fast for the cyclist to maintain by pushing the pedals. The chain is then moved to a gear with a smaller radius, so the chain has a smaller tangential speed that the cyclist can more easily maintain. This gear doesn't provide as much torque as the first, but the cyclist needs to accelerate only to a somewhat higher speed. This process continues as the bicycle moves faster and faster and the cyclist shifts through all five gears. The fifth gear supplies the lowest torque, but now the main function of that torque is to counter the frictional torque from the rolling tires, which tends to reduce the speed of the bicycle. The small radius of the fifth gear allows the cyclist to keep up with the chain's movement by pushing the pedals.

A 15-speed bicycle has the same gear structure on the drive wheel, but has three gears on the sprocket connected to the pedals. By combining different positions of the chain on the rear gears and the sprocket gears, 15 different torques are available.

APPLICATION

Bicycle Gears

Figure 8.17   The drive wheel and gears of a bicycle.

More on the Moment of Inertia

As we have seen, a small object (or a particle) has a moment of inertia equal to mr^2 about some axis. The moment of inertia of a *composite* object about some axis is just the sum of the moments of inertia of the object's components. For example, suppose a majorette twirls a baton as in Figure 8.18. Assume that the baton can be modeled as a very light rod of length 2ℓ with a heavy object at each end. (The rod of a real baton has a significant mass relative to its ends.) Because we are neglecting the mass of the rod, the moment of inertia of the baton about an axis through its center and perpendicular to its length is given by Equation 8.7:

$$I = \Sigma mr^2$$

Because this system consists of two objects with equal masses equidistant from the axis of rotation, $r = \ell$ for each object, and the sum is

$$I = \Sigma mr^2 = m\ell^2 + m\ell^2 = 2m\ell^2$$

If the mass of the rod were not neglected, we would have to include its moment of inertia to find the total moment of inertia of the baton.

We pointed out earlier that I is the rotational counterpart of m. However, there are some important distinctions between the two. For example, mass is an intrinsic property of an object that doesn't change, whereas **the moment of inertia of a system depends on how the mass is distributed and on the location of the axis of rotation**. Example 8.9 illustrates this point.

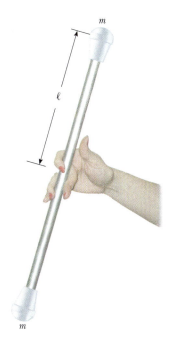

Figure 8.18   A baton of length 2ℓ and mass $2m$. (The mass of the connecting rod is neglected.) The moment of inertia about the axis through the baton's center and perpendicular to its length is $2m\ell^2$.

EXAMPLE 8.9   The Baton Twirler

Goal   Calculate a moment of inertia.

Problem   In an effort to be the star of the half-time show, a majorette twirls an unusual baton made up of four spheres fastened to the ends of very light rods (Fig. 8.19). Each rod is 1.0 m long. **(a)** Find the moment of inertia of the baton about an axis perpendicular to the page and passing through the point where the rods cross. **(b)** The majorette tries spinning her strange baton about the axis OO', as shown in Figure 8.20. Calculate the moment of inertia of the baton about this axis.

Strategy In Figure 8.19, all four balls contribute to the moment of inertia, whereas in Figure 8.20, with the new axis, only the two balls on the left and right contribute. Technically, the balls on the top and bottom still make a small contribution because they're not really point particles. However, their moment of inertia can be neglected, because the radius of the sphere is much smaller than the radius formed by the rods.

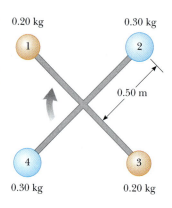

Figure 8.19 (Example 8.9a) Four objects connected to light rods rotating in the plane of the page.

Solution

(a) Calculate the moment of inertia of the baton when oriented as in Figure 8.19.

Apply Equation 8.7, neglecting the mass of the connecting rods:

$$I = \Sigma mr^2 = m_1 r_1^2 + m_2 r_2^2 + m_3 r_3^2 + m_4 r_4^2$$
$$= (0.20 \text{ kg})(0.50 \text{ m})^2 + (0.30 \text{ kg})(0.50 \text{ m})^2$$
$$+ (0.20 \text{ kg})(0.50 \text{ m})^2 + (0.30 \text{ kg})(0.50 \text{ m})^2$$
$$I = \boxed{0.25 \text{ kg} \cdot \text{m}^2}$$

(b) Calculate the moment of inertia of the baton when oriented as in Figure 8.20.

Apply Equation 8.7 again, neglecting the radii of the 0.20-kg spheres.

$$I = \Sigma mr^2 = m_1 r_1^2 + m_2 r_2^2 + m_3 r_3^2 + m_4 r_4^2$$
$$= (0.20 \text{ kg})(0)^2 + (0.30 \text{ kg})(0.50 \text{ m})^2$$
$$+ (0.20 \text{ kg})(0)^2 + (0.30 \text{ kg})(0.50 \text{ m})^2$$
$$I = \boxed{0.15 \text{ kg} \cdot \text{m}^2}$$

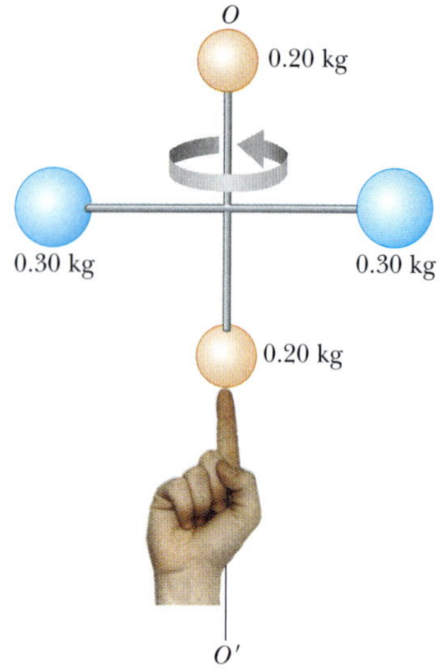

Figure 8.20 (Example 8.9b) A double baton rotating about the axis OO'.

Remarks The moment of inertia is smaller in part (b) because in this configuration the 0.20-kg spheres are essentially located on the axis of rotation.

Exercise 8.9

Yet another bizarre baton is created by taking four identical balls, each with mass 0.300 kg, and fixing them as before, except that one of the rods has a length of 1.00 m and the other has a length of 1.50 m. Calculate the moment of inertia of this baton (a) when oriented as in Figure 8.19; (b) when oriented as in Figure 8.20, with the shorter rod vertical; and (c) when oriented as in Figure 8.20, but with longer rod vertical.

Answers (a) $0.488 \text{ kg} \cdot \text{m}^2$ (b) $0.338 \text{ kg} \cdot \text{m}^2$ (c) $0.150 \text{ kg} \cdot \text{m}^2$

Calculation of Moments of Inertia for Extended Objects

The method used for calculating moments of inertia in Example 8.9 is simple when only a few small objects rotate about an axis. When the object is an extended one, such as a sphere, a cylinder, or a cone, techniques of calculus are often required, unless some simplifying symmetry is present. One such extended object amenable to a simple solution is a hoop rotating about an axis perpendicular to its plane and passing through its center, as shown in Figure 8.21. (A bicycle tire, for example, would approximately fit into this category.)

To evaluate the moment of inertia of the hoop, we can still use the equation $I = \Sigma mr^2$ and imagine that the mass of the hoop M is divided into n small segments having masses m_1, m_2, m_3, . . . , m_n, as in Figure 8.21, with $M = m_1 + m_2 + m_3 + . . . + m_n$. This approach is just an extension of the baton problem described in the preceding examples, except that now we have a large number of small masses in rotation instead of only four.

We can express the sum for I as

$$I = \Sigma mr^2 = m_1r_1^2 + m_2r_2^2 + m_3r_3^2 + \cdots + m_nr_n^2$$

All of the segments around the hoop are at the *same distance R* from the axis of rotation, so we can drop the subscripts on the distances and factor out R^2 to obtain

$$I = (m_1 + m_2 + m_3 + \cdots + m_n)R^2 = MR^2 \qquad [8.9]$$

This expression can be used for the moment of inertia of any ring-shaped object rotating about an axis through its center and perpendicular to its plane. Note that the result is strictly valid only if the thickness of the ring is small relative to its inner radius.

The hoop we selected as an example is unique in that we were able to find an expression for its moment of inertia by using only simple algebra. Unfortunately, for most extended objects the calculation is much more difficult because the mass elements are not all located at the same distance from the axis, so the methods of integral calculus are required. The moments of inertia for some other common shapes are given without proof in Table 8.1. You can use this table as needed to determine the moment of inertia of a body having any one of the listed shapes.

If mass elements in an object are redistributed parallel to the axis of rotation, the moment of inertia of the object doesn't change. Consequently, the expression $I = MR^2$ can be used equally well to find the axial moment of inertia of an embroidery hoop or of a long sewer pipe. Likewise, a door turning on its hinges is described by the same moment-of-inertia expression as that tabulated for a long thin rod rotating about an axis through its end.

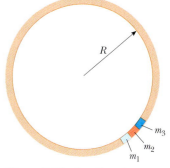

Figure 8.21 A uniform hoop can be divided into a large number of small segments that are equidistant from the center of the hoop.

TIP 8.3   No Single Moment of Inertia

Moment of inertia is analogous to mass, but there are major differences. Mass is an inherent property of an object. The moment of inertia of an object depends on the shape of the object, its mass, and the choice of rotation axis.

TABLE 8.1

Moments of Inertia for Various Rigid Objects of Uniform Composition

Hoop or thin cylindrical shell $I = MR^2$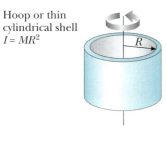

Solid sphere $I = \frac{2}{5} MR^2$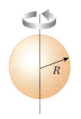

Solid cylinder or disk $I = \frac{1}{2} MR^2$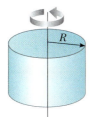

Thin spherical shell $I = \frac{2}{3} MR^2$

Long thin rod with rotation axis through center $I = \frac{1}{12} ML^2$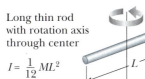

Long thin rod with rotation axis through end $I = \frac{1}{3} ML^2$

EXAMPLE 8.10   Warming Up

Goal Find a moment of inertia and apply the rotational analog of Newton's second law.

Problem A baseball player loosening up his arm before a game tosses a 0.150-kg baseball, using only the rotation of his forearm to accelerate the ball (Fig. 8.22). The forearm has a mass of 1.50 kg and a length of 0.350 m. The ball starts at rest and is released with a speed of 30.0 m/s in 0.300 s. **(a)** Find the constant angular acceleration of the arm and ball. **(b)** Calculate the moment of inertia of the system consisting of the forearm and ball. **(c)** Find the torque exerted on the ball to give it the angular acceleration found in part (a).

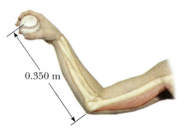
0.350 m

Figure 8.22 (Example 8.10) A ball being tossed by a pitcher. The forearm is used to accelerate the ball.

Strategy The angular acceleration can be found with rotational kinematic equations, while the moment of inertia of the system can be obtained by summing the separate moments of inertia of the ball and forearm. Multiplying these two results together gives the torque.

Solution
(a) Find the angular acceleration of the ball.

The angular acceleration is constant, so use the angular velocity kinematic equation with $\omega_i = 0$.

$$\omega = \omega_i + \alpha t \;\;\rightarrow\;\; \alpha = \frac{\omega}{t}$$

The ball accelerates along a circular arc with radius given by the length of the forearm. Solve $v = r\omega$ for ω and substitute:

$$\alpha = \frac{\omega}{t} = \frac{v}{rt} = \frac{30.0 \text{ m/s}}{(0.350 \text{ m})(0.300 \text{ s})} = \boxed{286 \text{ rad/s}^2}$$

(b) Find the moment of inertia of the system (forearm plus ball).

Find the moment of inertia of the ball about an axis that passes through the elbow, perpendicular to the arm:

$$I_{ball} = mr^2 = (0.150 \text{ kg})(0.350 \text{ m})^2$$
$$= 1.84 \times 10^{-2} \text{ kg} \cdot \text{m}^2$$

Obtain the moment of inertia of the forearm, modeled as a rod, by consulting Table 8.1:

$$I_{forearm} = \tfrac{1}{3} ML^2 = \tfrac{1}{3}(1.50 \text{ kg})(0.350 \text{ m})^2$$
$$= 6.13 \times 10^{-2} \text{ kg} \cdot \text{m}^2$$

Sum the individual moments of inertia to obtain the moment of inertia of the system (ball plus forearm):

$$I_{system} = I_{ball} + I_{forearm} = \boxed{7.97 \times 10^{-2} \text{ kg} \cdot \text{m}^2}$$

(c) Find the torque exerted on the ball.

Apply Equation 8.8, using the results of parts (a) and (b):

$$\tau = I_{system}\alpha = (7.97 \times 10^{-2} \text{ kg} \cdot \text{m}^2)(286 \text{ rad/s}^2)$$
$$= \boxed{22.8 \text{ N} \cdot \text{m}}$$

Remarks Notice that having a long forearm can greatly increase the torque and hence the acceleration of the ball. This is one reason it's advantageous for a pitcher to be tall—the pitching arm is proportionately longer. A similar advantage holds in tennis, where taller players can usually deliver faster serves.

Exercise 8.10
A catapult with a radial arm 4.00 m long accelerates a ball of mass 20.0 kg through a quarter circle. The ball leaves the apparatus at 45.0 m/s. If the mass of the arm is 25.0 kg and the acceleration is uniform, find **(a)** the angular acceleration, **(b)** the moment of inertia of the arm and ball, and **(c)** the net torque exerted on the ball and arm. *Hint:* Use the time-independent rotational kinematics equation to find the angular acceleration, rather than the angular velocity equation.

Answers (a) 40.3 rad/s² (b) 453 kg·m² (c) 1.83×10^4 N·m

INTERACTIVE EXAMPLE 8.11   The Falling Bucket

Goal   Combine Newton's second law with its rotational analog.

Problem   A solid, frictionless cylindrical reel of mass $M = 3.00$ kg and radius $R = 0.400$ m is used to draw water from a well (Fig. 8.23a). A bucket of mass $m = 2.00$ kg is attached to a cord that is wrapped around the cylinder. **(a)** Find the tension T in the cord and acceleration a of the bucket. **(b)** If the bucket starts from rest at the top of the well and falls for 3.00 s before hitting the water, how far does it fall?

Strategy   This problem involves three equations and three unknowns. The three equations are Newton's second law applied to the bucket, $ma = \Sigma F_i$; the rotational version of the second law applied to the cylinder, $I\alpha = \Sigma \tau_i$; and the relationship between linear and angular acceleration, $a = r\alpha$, which connects the dynamics of the bucket and cylinder. The three unknowns are the acceleration a of the bucket, the angular acceleration a of the cylinder, and the tension T in the rope. Assemble the terms of the three equations and solve for the three unknowns by substitution. Part (b) is a review of kinematics.

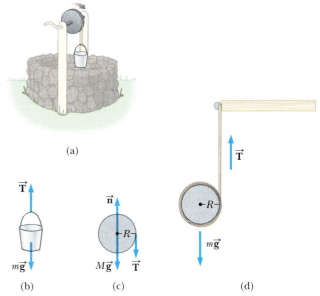

Figure 8.23   (Interactive Example 8.11) (a) A water bucket attached to a rope passing over a frictionless reel. (b) A free-body diagram for the bucket. (c) The tension produces a torque on the cylinder about its axis of rotation. (d) A falling cylinder (Exercise 8.11).

Solution

(a) Find the tension in the cord and the acceleration of the bucket.

Apply Newton's second law to the bucket in Figure 8.23b. There are two forces: the tension $\vec{\mathbf{T}}$ acting upwards and gravity $m\vec{\mathbf{g}}$ acting downwards.

$$ma = -mg + T \tag{1}$$

Apply $\tau = I\alpha$ to the cylinder in Figure 8.23c:

$$\Sigma \tau = I\alpha = \tfrac{1}{2}MR^2\alpha \quad \text{(solid cylinder)}$$

Notice the angular acceleration is clockwise, so the torque is negative. The normal and gravity forces have zero moment arm, and don't contribute any torque.

$$-TR = \tfrac{1}{2}MR^2\alpha \tag{2}$$

Solve for T and substitute $\alpha = a/R$ (notice that both α and a are negative):

$$T = -\tfrac{1}{2}MR\alpha = -\tfrac{1}{2}Ma \tag{3}$$

Substitute the expression for T in Equation (3) into Equation (1), and solve for the acceleration:

$$ma = -mg - \tfrac{1}{2}Ma \quad \rightarrow \quad a = -\frac{mg}{m + \tfrac{1}{2}M}$$

Substitute the values for m, M, and g, getting a, then substitute a into Equation (3) to get T.

$$a = \boxed{-5.60 \text{ m/s}^2} \qquad T = \boxed{8.40 \text{ N}}$$

(b) Find the distance the bucket falls in 3.00 s.
Apply the displacement kinematic equation for constant acceleration, with $t = 3.00$ s and $v_0 = 0$:

$$\Delta y = v_0 t + \tfrac{1}{2}at^2 = -\tfrac{1}{2}(5.60 \text{ m/s}^2)(3.00 \text{ s})^2 = \boxed{-25.2 \text{ m}}$$

Remarks   Proper handling of signs is very important in these problems. All such signs should be chosen initially and checked mathematically and physically. In this problem, for example, both the angular acceleration α and the acceleration a are negative, so $\alpha = a/R$ applies. If the rope had been wound the other way on the cylinder, causing counterclockwise rotation, the torque would have been positive, and the relationship would have been $\alpha = -a/R$, with the double negative making the right-hand side positive, just like the left-hand side.

Exercise 8.11

A hollow cylinder of mass 0.100 kg and radius 4.00 cm has a string wrapped several times around it, as in Figure 8.23d. If the string is attached to a rigid support and the cylinder allowed to drop from rest, find (a) the acceleration of the cylinder and (b) the speed of the cylinder when a meter of string has unwound off of it.

Answers   (a) -4.90 m/s^2 (b) 3.13 m/s

Physics⊗Now™ You can change the mass of the object and the mass and radius of the wheel to see the effect on how the system moves by logging into PhysicsNow at **www.cp7e.com** and going to Interactive Example 8.11.

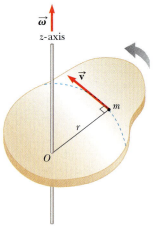

Figure 8.24 A rigid plate rotating about the *z*-axis with angular speed ω. The kinetic energy of a particle of mass m is $\frac{1}{2}mv^2$. The total kinetic energy of the plate is $\frac{1}{2}I\omega^2$.

8.6 ROTATIONAL KINETIC ENERGY

In Chapter 5 we defined the kinetic energy of a particle moving through space with a speed v as the quantity $\frac{1}{2}mv^2$. Analogously, **an object rotating about some axis with an angular speed ω has rotational kinetic energy given by $\frac{1}{2}I\omega^2$**. To prove this, consider an object in the shape of a thin rigid plate rotating around some axis perpendicular to its plane, as in Figure 8.24. The plate consists of many small particles, each of mass m. All these particles rotate in circular paths around the axis. If r is the distance of one of the particles from the axis of rotation, the speed of that particle is $v = r\omega$. Because the *total* kinetic energy of the plate's rotation is the sum of all the kinetic energies associated with its particles, we have

$$KE_r = \Sigma(\tfrac{1}{2}mv^2) = \Sigma(\tfrac{1}{2}mr^2\omega^2) = \tfrac{1}{2}(\Sigma mr^2)\omega^2$$

In the last step, the ω^2 term is factored out because it's the same for every particle. Now, the quantity in parentheses on the right is the moment of inertia of the plate in the limit as the particles become vanishingly small, so

$$KE_r = \tfrac{1}{2}I\omega^2 \qquad\qquad \text{[8.10]}$$

where $I = \Sigma mr^2$ is the moment of inertia of the plate.

A system such as a bowling ball rolling down a ramp is described by three types of energy: **gravitational potential energy PE_g, translational kinetic energy KE_t,** and **rotational kinetic energy KE_r.** All these forms of energy, plus the potential energies of any other conservative forces, must be included in our equation for the conservation of mechanical energy of an isolated system:

Conservation of mechanical energy ▶

$$(KE_t + KE_r + PE)_i = (KE_t + KE_r + PE)_f \qquad \text{[8.11]}$$

where i and f refer to initial and final values, respectively, and PE includes the potential energies of all conservative forces in a given problem. This relation is true *only* if we ignore dissipative forces such as friction. In that case, it's necessary to resort to a generalization of the work–energy theorem:

Work–energy theorem including rotational energy ▶

$$W_{nc} = \Delta KE_t + \Delta KE_r + \Delta PE \qquad \text{[8.12]}$$

Problem-Solving Strategy   Energy Methods and Rotation

1. **Choose two points of interest**, one where all necessary information is known, and the other where information is desired.
2. **Identify** the conservative and nonconservative forces acting on the system being analyzed.
3. **Write the general work–energy theorem**, Equation 8.12, or Equation 8.11 if all forces are conservative.
4. **Substitute general expressions** for the terms in the equation.
5. **Use $v = r\omega$** to eliminate either ω or v from the equation.
6. **Solve** for the unknown.

EXAMPLE 8.12   A Ball Rolling Down an Incline

Goal   Combine gravitational, translational, and rotational energy.

Problem   A ball of mass M and radius R starts from rest at a height of 2.00 m and rolls down a 30.0° slope, as in Active Figure 8.25. What is the linear speed of the ball when it leaves the incline? Assume that the ball rolls without slipping.

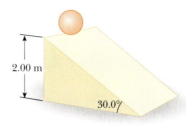

2.00 m

30.0°

ACTIVE FIGURE 8.25
(Example 8.12) A ball starts from rest at the top of an incline and rolls to the bottom without slipping.

Physics⊗Now™
Log into PhysicsNow at **www.cp7e.com**, and go to Active Figure 8.25 to roll several objects down a hill and see how the final speed depends on the shape of the object.

Strategy   The two points of interest are the top and bottom of the incline, with the bottom acting as the zero point of gravitational potential energy. The force of static friction converts translational kinetic energy to rotational kinetic energy without dissipating any energy, so mechanical energy is conserved and Equation 8.11 can be applied.

Solution

Apply conservation of energy with $PE = PE_g$, the potential energy associated with gravity.

$$(KE_t + KE_r + PE_g)_i = (KE_t + KE_r + PE_g)_f$$

Substitute the appropriate general expressions, noting that $(KE_t)_i = (KE_r)_i = 0$ and $(PE_g)_f = 0$ (obtain the moment of inertia of a ball from Table 8.1):

$$0 + 0 + Mgh = \tfrac{1}{2}Mv^2 + \tfrac{1}{2}(\tfrac{2}{5}MR^2)\omega^2 + 0$$

The ball rolls without slipping, so $R\omega = v$, the "no-slip condition," can be applied:

$$Mgh = \tfrac{1}{2}Mv^2 + \tfrac{1}{5}Mv^2 = \tfrac{7}{10}Mv^2$$

Solve for v, noting that M cancels.

$$v = \sqrt{\frac{10gh}{7}} = \sqrt{\frac{10(9.80 \text{ m/s}^2)(2.00 \text{ m})}{7}} = \boxed{5.29 \text{ m/s}}$$

Exercise 8.12

Repeat this example for a solid cylinder of the same mass and radius as the ball and released from the same height. In a race between the two objects on the incline, which one would win?

Answer $v = \sqrt{4gh/3} = 5.11$ m/s; the ball would win.

Quick Quiz 8.4

Two spheres, one hollow and one solid, are rotating with the same angular speed around an axis through their centers. Both spheres have the same mass and radius. Which sphere, if either, has the higher rotational kinetic energy? (a) The hollow sphere. (b) The solid sphere. (c) They have the same kinetic energy.

Quick Quiz 8.5

Which arrives at the bottom first, (a) a ball rolling without sliding down a certain incline A, (b) a solid cylinder rolling without sliding down incline A, or (c) a box of the same mass as the ball sliding down a frictionless incline B having the same dimensions as A? Assume that each object is released from rest at the top of its incline.

EXAMPLE 8.13   Blocks and Pulley

Goal   Solve a system requiring rotation concepts and the work–energy theorem.

Problem   Two blocks with masses $m_1 = 5.00$ kg and $m_2 = 7.00$ kg are attached by a string as in Figure 8.26a, over a pulley with mass $M = 2.00$ kg. The pulley, which turns on a frictionless axle, is a hollow cylinder with radius 0.050 0 m over which the string moves without slipping. The horizontal surface has coefficient of kinetic friction 0.350. Find the speed of the system when the block of mass m_2 has dropped 2.00 m.

Strategy   This problem can be solved with the extension of the work–energy theorem, Equation 8.12. If the block of mass m_2 falls from height h to 0, then the block of mass m_1 moves the same distance, $\Delta x = h$. Apply the work–energy theorem, solve for v, and substitute. Kinetic friction the sole nonconservative force.

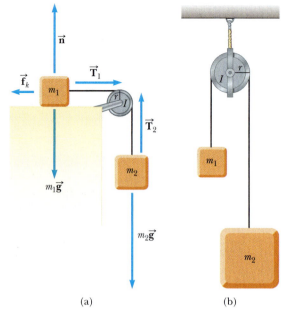

(a)   (b)

Figure 8.26   (a) (Example 8.13) (b) (Exercise 8.13) In both cases, $\vec{T}_1$ and $\vec{T}_2$ exert torques on the pulley.

Solution

Apply the work–energy theorem, with $PE = PE_g$, the potential energy associated with gravity.

$$W_{nc} = \Delta KE_t + \Delta KE_r + \Delta PE_g$$

Substitute the frictional work for W_{nc}, kinetic energy changes for the two blocks, the rotational kinetic energy change for the pulley, and the potential energy change for the second block.

$$-\mu_k n \Delta x = -\mu_k(m_1 g)\Delta x = (\tfrac{1}{2}m_1 v^2 - 0) + (\tfrac{1}{2}m_2 v^2 - 0)$$
$$+ (\tfrac{1}{2}I\omega^2 - 0) + (0 - m_2 gh)$$

Substitute $\Delta x = h$, and write I as $(I/r^2)r^2$:

$$-\mu_k(m_1 g)h = \tfrac{1}{2}m_1 v^2 + \tfrac{1}{2}m_2 v^2 + \tfrac{1}{2}\left(\frac{I}{r^2}\right)r^2\omega^2 - m_2 gh$$

For a hoop, $I = Mr^2$ so $(I/r^2) = M$. Substitute this quantity and $v = r\omega$:

$$-\mu_k(m_1 g)h = \tfrac{1}{2}m_1 v^2 + \tfrac{1}{2}m_2 v^2 + \tfrac{1}{2}Mv^2 - m_2 gh$$

Solve for v:

$$m_2 gh - \mu_k(m_1 g)h = \tfrac{1}{2}m_1 v^2 + \tfrac{1}{2}m_2 v^2 + \tfrac{1}{2}Mv^2$$
$$= \tfrac{1}{2}(m_1 + m_2 + M)v^2$$
$$v = \sqrt{\frac{2gh(m_2 - \mu_k m_1)}{m_1 + m_2 + M}}$$

Substitute $m_1 = 5.00$ kg, $m_2 = 7.00$ kg, $M = 2.00$ kg, $g = 9.80$ m/s^2, $h = 2.00$ m, and $\mu_k = 0.350$:

$$v = \boxed{3.83 \text{ m/s}}$$

Remarks   In the expression for the speed v, the mass m_1 of the first block and the mass M of the pulley all appear in the denominator, reducing the speed, as they should. In the numerator, m_2 is positive while the friction term is negative. Both assertions are reasonable, because the force of gravity on m_2 increases the speed of the system while the force of friction on m_1 slows it down. This problem can also be solved with Newton's second law together with $\tau = I\alpha$, a difficult exercise (though it can be facilitated with a system approach).

Exercise 8.13

Two blocks with masses $m_1 = 2.00$ kg and $m_2 = 9.00$ kg are attached over a pulley with mass $M = 3.00$ kg, hanging straight down as in Atwood's machine (Fig. 8.26b). The pulley is a solid cylinder with radius 0.050 0 m, and there is some friction in the axle. The system is released from rest, and the string moves without slipping over the pulley. If the larger mass is traveling at a speed of 2.50 m/s when it has dropped 1.00 m, how much mechanical energy was lost due to friction in the pulley's axle?

[*Hint:* This exercise is slightly easier than the associated example because the friction force need not be determined.]

Answer   29.5 J

8.7 ANGULAR MOMENTUM

In Figure 8.27, an object of mass m rotates in a circular path of radius r, acted on by a net force, $\vec{F}_{net}$. The resulting net torque on the object increases its angular speed from the value ω_0 to the value ω in a time interval Δt. Therefore, we can write

$$\Sigma\tau = I\alpha = I\frac{\Delta\omega}{\Delta t} = I\left(\frac{\omega - \omega_0}{\Delta t}\right) = \frac{I\omega - I\omega_0}{\Delta t}$$

If we define the product

$$L \equiv I\omega \qquad \text{[8.13]}$$

◄ Definition of angular momentum

as the **angular momentum** of the object, then we can write

$$\Sigma\tau = \frac{\text{change in angular momentum}}{\text{time interval}} = \frac{\Delta L}{\Delta t} \qquad \text{[8.14]}$$

Equation 8.14 is the rotational analog of Newton's second law in the form $F = \Delta p/\Delta t$ and states that **the net torque acting on an object is equal to the time rate of change of the object's angular momentum**. Recall that this equation also parallels the impulse–momentum theorem.

When the net external torque ($\Sigma\tau$) acting on a system is zero, Equation 8.14 gives that $\Delta L/\Delta t = 0$, which says that the time rate of change of the system's angular momentum is zero. We then have the following important result:

Let L_i and L_f be the angular momenta of a system at two different times, and suppose there is no net external torque, so $\Sigma\tau = 0$. Then

$$L_i = L_f \qquad \text{[8.15]}$$

and angular momentum is said to be *conserved*.

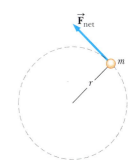

Figure 8.27 An object of mass m rotating in a circular path under the action of a constant torque.

◄ Conservation of angular momentum

Equation 8.15 gives us a third conservation law to add to our list: **conservation of angular momentum**. We can now state that **the mechanical energy, linear momentum, and angular momentum of an isolated system all remain constant**.

If the moment of inertia of an isolated rotating system changes, the system's angular speed will change. Conservation of angular momentum then requires that

$$I_i\omega_i = I_f\omega_f \qquad \text{if} \qquad \Sigma\tau = 0 \qquad \text{[8.16]}$$

Note that conservation of angular momentum applies to macroscopic objects such as planets and people, as well as to atoms and molecules. There are many examples of conservation of angular momentum; one of the most dramatic is that of a figure skater spinning in the finale of her act. In Figure 8.28a, the skater has pulled her arms and legs close to her body, reducing their distance from her axis of rotation and hence also reducing her moment of inertia. By conservation of angular momentum, a reduction in her moment of inertia must increase her angular velocity. Coming out of the spin in Figure 8.28b, she needs to reduce her angular velocity, so she extends her arms and legs again, increasing her moment of inertia and thereby slowing her rotation.

APPLICATION

Figure Skating

Figure 8.28 Michelle Kwan controls her moment of inertia. (a) By pulling in her arms and legs, she reduces her moment of inertia and increases her angular velocity (rate of spin). (b) Upon landing, extending her arms and legs increases her moment of inertia and helps slow her spin.

(a)

(b)

APPLICATION

Aerial Somersaults

Tightly curling her body, a diver decreases her moment of inertia, increasing her angular velocity.

Similarly, when a diver or an acrobat wishes to make several somersaults, she pulls her hands and feet close to the trunk of her body in order to rotate at a greater angular speed. In this case, the external force due to gravity acts through her center of gravity and hence exerts no torque about her axis of rotation, so the angular momentum about her center of gravity is conserved. For example, when a diver wishes to double her angular speed, she must reduce her moment of inertia to half its initial value.

An interesting astrophysical example of conservation of angular momentum occurs when a massive star, at the end of its lifetime, uses up all its fuel and collapses under the influence of gravitational forces, causing a gigantic outburst of energy called a supernova. The best-studied example of a remnant of a supernova explosion is the Crab Nebula, a chaotic, expanding mass of gas (Fig. 8.29). In a supernova, part of the star's mass is ejected into space, where it eventually condenses into new stars and planets. Most of what is left behind typically collapses into a **neutron star**—an extremely dense sphere of matter with a diameter of about 10 km, greatly reduced from the 10^6-km diameter of the original star and containing a large fraction of the star's original mass. In a neutron star, pressures become so great that atomic electrons combine with protons, becoming neutrons. As the moment of inertia of the system decreases

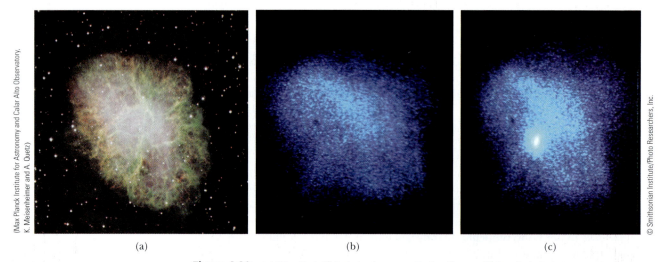

(a)

(b)

(c)

Figure 8.29 (a) The Crab Nebula in the constellation Taurus. This nebula is the remnant of a supernova seen on Earth in A.D. 1054. It is located some 6 300 lightyears away and is approximately 6 lightyears in diameter, still expanding outward. A pulsar deep inside the nebula flashes 30 times every second. (b) Pulsar off. (c) Pulsar on.

during the collapse, the star's rotational speed increases. More than 700 rapidly rotating neutron stars have been identified since their first discovery in 1967, with periods of rotation ranging from a millisecond to several seconds. The neutron star is an amazing system—an object with a mass greater than the Sun, fitting comfortably within the space of a small county and rotating so fast that the tangential speed of the surface approaches a sizeable fraction of the speed of light!

APPLICATION

Rotating Neutron Stars

Quick Quiz 8.6

A horizontal disk with moment of inertia I_1 rotates with angular speed ω_1 about a vertical frictionless axle. A second horizontal disk, with moment of inertia I_2 and initially not rotating, drops onto the first. Because their surfaces are rough, the two eventually reach the same angular speed ω. The ratio ω/ω_1 is equal to (a) I_1/I_2 (b) I_2/I_1 (c) $I_1/(I_1 + I_2)$ (d) $I_2/(I_1 + I_2)$

Quick Quiz 8.7

If global warming continues, it's likely that some ice from the polar ice caps of the Earth will melt and the water will be distributed closer to the Equator. If this occurs, would the length of the day (one revolution) (a) increase, (b) decrease, or (c) remain the same?

EXAMPLE 8.14   The Spinning Stool

Goal Apply conservation of angular momentum to a simple system.

Problem A student sits on a pivoted stool while holding a pair of weights. (See Fig. 8.30.) The stool is free to rotate about a vertical axis with negligible friction. The moment of inertia of student, weights, and stool is $2.25 \text{ kg} \cdot \text{m}^2$. The student is set in rotation with arms outstretched, making one complete turn every 1.26 s, arms outstretched. **(a)** What is the initial angular speed of the system? **(b)** As he rotates, he pulls the weights inward so that the new moment of inertia of the system (student, objects, and stool) becomes $1.80 \text{ kg} \cdot \text{m}^2$. What is the new angular speed of the system? **(c)** Find the work done by the student on the system while pulling in the weights. (Ignore energy lost through dissipation in his muscles.)

Figure 8.30 (Example 8.14) (a) The student is given an initial angular speed while holding two weights out. (b) The angular speed increases as the student draws the weights inwards.

Strategy (a) The angular frequency can be obtained from the frequency, which is the inverse of the period. (b) There are no external torques acting on the system, so the new angular speed can be found with the principle of conservation of angular momentum. (c) The work done on the system during this process is the same as the system's change in rotational kinetic energy.

Solution
(a) Find the initial angular speed of the system.

Invert the period to get the frequency, and multiply by 2π: $\omega_i = 2\pi f = 2\pi/T = \boxed{4.99 \text{ rad/s}}$

(b) After he pulls the weights in, what's the system's new angular speed?

Equate the initial and final angular momenta of the system:

$$L_i = L_f \quad \rightarrow \quad I_i\omega_i = I_f\omega_f \tag{1}$$

Substitute and solve for the final angular speed ω_f:

$$(2.25 \text{ kg} \cdot \text{m}^2)(4.99 \text{ rad/s}) = (1.80 \text{ kg} \cdot \text{m}^2)\omega_f \qquad (2)$$

$$\omega_f = \boxed{6.24 \text{ rad/s}}$$

(c) Find the work the student does on the system.

Apply the work–energy theorem:

$$W_{student} = \Delta K_r = \tfrac{1}{2}I_f\omega_f^2 - \tfrac{1}{2}I_i\omega_i^2$$
$$= \tfrac{1}{2}(1.80 \text{ kg} \cdot \text{m}^2)(6.24 \text{ rad/s})^2$$
$$- \tfrac{1}{2}(2.25 \text{ kg} \cdot \text{m}^2)(4.99 \text{ rad/s})^2$$

$$W_{student} = \boxed{7.03 \text{ J}}$$

Remarks   Although the angular momentum of the system is conserved, mechanical energy is not conserved because the student does work on the system.

Exercise 8.14
A star with an initial radius of 1.0×10^8 m and period of 30.0 days collapses suddenly to a radius of 1.0×10^4 m. (a) Find the period of rotation after collapse. (b) Find the work done by gravity during the collapse if the mass of the star is 2.0×10^{30} kg. (c) What is the speed of an indestructible person standing on the equator of the collapsed star? (Neglect any relativistic or thermal effects, and assume the star is spherical before and after it collapses.)

Answers   (a) 2.6×10^{-2} s   (b) 2.3×10^{42} J   (c) 2.4×10^6 m/s

---

EXAMPLE 8.15   The Merry-Go-Round

Goal   Apply conservation of angular momentum while combining two moments of inertia.

Problem   A merry-go-round modeled as a disk of mass $M = 1.00 \times 10^2$ kg and radius $R = 2.00$ m is rotating in a horizontal plane about a frictionless vertical axle (Fig. 8.31). **(a)** After a student with mass $m = 60.0$ kg jumps onto the merry-go-round, the system's angular speed decreases to 2.00 rad/s. If the student walks slowly from the edge toward the center, find the angular speed of the system when she reaches a point 0.500 m from the center. **(b)** Find the change in the system's rotational kinetic energy caused by her movement to the center. **(c)** Find the work done on the student as she walks to $r = 0.500$ m.

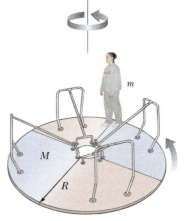

Figure 8.31   (Example 8.15) As the student walks toward the center of the rotating platform, the moment of inertia of the system (student plus platform) decreases. Because angular momentum is conserved, the angular speed of the system must increase.

Strategy   This problem can be solved with conservation of angular momentum by equating the system's initial angular momentum when the student stands at the rim to the angular momentum when the student has reached $r = 0.500$ m. The key is to find the different moments of inertia.

Solution
(a) Find the angular speed when the student reaches a point 0.500 m from the center.

Calculate the moment of inertia of the disk, I_D:

$$I_D = \tfrac{1}{2}MR^2 = \tfrac{1}{2}(1.00 \times 10^2 \text{ kg})(2.00 \text{ m})^2$$
$$= 2.00 \times 10^2 \text{ kg} \cdot \text{m}^2$$

Calculate the initial moment of inertia of the student. This is the same as the moment of inertia of a mass a distance R from the axis:

$$I_S = mR^2 = (60.0 \text{ kg})(2.00 \text{ m})^2 = 2.40 \times 10^2 \text{ kg} \cdot \text{m}^2$$

Sum the two moments of inertia and multiply by the initial angular speed to find L_i, the initial angular momentum of the system:

$$L_i = (I_D + I_S)\omega_i$$
$$= (2.00 \times 10^2 \text{ kg} \cdot \text{m}^2 + 2.40 \times 10^2 \text{ kg} \cdot \text{m}^2)(2.00 \text{ rad/s})$$
$$= 8.80 \times 10^2 \text{ kg} \cdot \text{m}^2/\text{s}$$

Calculate the student's final moment of inertia, I_{Sf}, when she is 0.500 m from the center:

$$I_{Sf} = mr_f^2 = (60.0\ \text{kg})(0.50\ \text{m})^2 = 15.0\ \text{kg} \cdot \text{m}^2$$

The moment of inertia of the platform is unchanged. Add it to the student's final moment of inertia, and multiply by the unknown final angular speed to find L_f:

$$L_f = (I_D + I_{Sf})\omega_f = (2.00 \times 10^2\ \text{kg} \cdot \text{m}^2 + 15.0\ \text{kg} \cdot \text{m}^2)\omega_f$$

Equate the initial and final angular momenta and solve for the final angular speed of the system:

$$L_i = L_f$$

$$(8.80 \times 10^2\ \text{kg} \cdot \text{m}^2/\text{s}) = (2.15 \times 10^2\ \text{kg} \cdot \text{m}^2)\omega_f$$

$$\omega_f = \boxed{4.09\ \text{rad/s}}$$

(b) Find the change in the rotational kinetic energy of the system.

Calculate the initial kinetic energy of the system:

$$KE_i = \tfrac{1}{2}I_i\omega_i^2 = \tfrac{1}{2}(4.40 \times 10^2\ \text{kg} \cdot \text{m}^2)(2.00\ \text{rad/s})^2$$
$$= 8.80 \times 10^2\ \text{J}$$

Calculate the final kinetic energy of the system:

$$KE_f = \tfrac{1}{2}I_f\omega_f^2 = \tfrac{1}{2}(215\ \text{kg} \cdot \text{m}^2)(4.09\ \text{rad/s})^2 = 1.80 \times 10^3\ \text{J}$$

Calculate the change in kinetic energy of the system.

$$KE_f - KE_i = \boxed{920\ \text{J}}$$

(c) Find the work done on the student.

The student undergoes a change in kinetic energy that equals the work done on her. Apply the work–energy theorem:

$$W = \Delta KE_{\text{student}} = \tfrac{1}{2}I_{Sf}\omega_f^2 - \tfrac{1}{2}I_S\omega_i^2$$
$$= \tfrac{1}{2}(15.0\ \text{kg} \cdot \text{m}^2)(4.09\ \text{rad/s})^2$$
$$- \tfrac{1}{2}(2.40 \times 10^2\ \text{kg} \cdot \text{m}^2)(2.00\ \text{rad/s})^2$$

$$W = \boxed{-355\ \text{J}}$$

Remarks   The angular momentum is unchanged by internal forces; however, the kinetic energy increases, because the student must perform positive work in order to walk toward the center of the platform.

Exercise 8.15
(a) Find the angular speed of the merry-go-round before the student jumped on, assuming the student didn't transfer any momentum or energy as she jumped on the merry-go-round. (b) By how much did the kinetic energy of the system change when the student jumped on? Notice that energy is lost in this process, as should be expected, since it is essentially a perfectly inelastic collision.

Answers   (a) 4.4 rad/s   (b) $KE_f - KE_i = -1.06 \times 10^3$ J.

SUMMARY

Physics⊗Now™ Take a practice test by logging into Physics-Now at **www.cp7e.com** and clicking on the Pre-Test link for this chapter.

8.1 Torque

Let $\vec{F}$ be a force acting on an object, and let $\vec{r}$ be a position vector from a chosen point O to the point of application of the force. Then the magnitude of the torque $\vec{\tau}$ of the force $\vec{F}$ is given by

$$\tau = rF \sin \theta \qquad \text{[8.2]}$$

where r is the length of the position vector, F the magnitude of the force, and θ the angle between $\vec{F}$ and $\vec{r}$.

The quantity $d = r \sin \theta$ is called the *lever arm* of the force.

8.2 Torque and the Two Conditions for Equilibrium

An object in mechanical equilibrium must satisfy the following two conditions:

1. The net external force must be zero: $\Sigma \vec{F} = 0$.
2. The net external torque must be zero: $\Sigma \vec{\tau} = 0$.

These two conditions, used in solving problems involving rotation in a plane—result in three equations and three unknowns—two from the first condition (corresponding to the x- and y-components of the force) and one from the second condition, on torques. These equations must be solved simultaneously.

8.5 Relationship Between Torque and Angular Acceleration

The **moment of inertia** of a group of particles is

$$I \equiv \Sigma mr^2 \qquad \text{[8.7]}$$

If a rigid object free to rotate about a fixed axis has a net external torque $\Sigma\tau$ acting on it, then the object undergoes an angular acceleration α, where

$$\Sigma\tau = I\alpha \qquad \text{[8.8]}$$

This equation is the rotational equivalent of the second law of motion.

Problems are solved by using Equation 8.8 together with Newton's second law and solving the resulting equations simultaneously. The relation $a = r\alpha$ is often key in relating the translational equations to the rotational equations.

8.6 Rotational Kinetic Energy

If a rigid object rotates about a fixed axis with angular speed ω, its **rotational kinetic energy** is

$$KE_r \equiv \tfrac{1}{2}I\omega^2 \qquad \text{[8.10]}$$

where I is the moment of inertia of the object around the axis of rotation.

A system involving rotation is described by three types of energy: potential energy PE, translational kinetic energy KE_t, and rotational kinetic energy KE_r. All these forms of energy must be included in the equation for conservation of mechanical energy for an isolated system:

$$(KE_t + KE_r + PE)_i = (KE_t + KE_r + PE)_f \qquad \text{[8.11]}$$

where i and f refer to initial and final values, respectively. When non-conservative forces are present, it's necessary to use a generalization of the work–energy theorem:

$$W_{nc} = \Delta KE_t + \Delta KE_r + \Delta PE \qquad \text{[8.12]}$$

8.7 Angular Momentum

The **angular momentum** of a rotating object is given by

$$L \equiv I\omega \qquad \text{[8.13]}$$

Angular momentum is related to torque in the following equation:

$$\Sigma\tau = \frac{\text{change in angular momentum}}{\text{time interval}} = \frac{\Delta L}{\Delta t} \qquad \text{[8.14]}$$

If the net external torque acting on a system is zero, then the total angular momentum of the system is constant,

$$L_i = L_f \qquad \text{[8.15]}$$

and is said to be conserved. Solving problems usually involves substituting into the expression

$$I_i\omega_i = I_f\omega_f \qquad \text{[8.16]}$$

and solving for the unknown.

CONCEPTUAL QUESTIONS

1. Why can't you put your heels firmly against a wall and then bend over without falling?

2. Why does a tall athlete have an advantage over a smaller one when the two are competing in the high jump?

3. Both torque and work are products of force and distance. How are they different? Do they have the same units?

4. Is it possible to calculate the torque acting on a rigid object without specifying an origin? Is the torque independent of the location of the origin?

5. Can an object be in equilibrium when only one force acts on it? If you believe the answer is yes, give an example to support your conclusion.

6. The polar ice caps contain about 2.3×10^{19} kg of ice. This mass contributes almost nothing to the moment of inertia of the Earth because it is located at the poles, close to the Earth's axis of rotation. *Estimate* the change in the length of the day that would be expected if the polar ice caps were to melt and the water were distributed uniformly over the surface of the Earth. (Note that the moment of inertia of a thin spherical shell of radius r and mass m is $2mr^2/3$.) (Question 6 is courtesy of Edward F. Redish. For more questions of this type, see www.physics.umd.edu/perg/.)

7. In some motorcycle races, the riders drive over small hills, and the motorcycle becomes airborne for a short time. If the motorcycle racer keeps the throttle open while leav-ing the hill and going into the air, the motorcycle's nose tends to rise upwards. Why does this happen?

8. In the movie *Jurassic Park*, there is a scene in which some members of the visiting group are trapped in the kitchen with dinosaurs outside. The paleontologist is pressing against the center of the door, trying to keep out the dinosaurs on the other side. The botanist throws herself against the door at the edge near the hinge. A pivotal point in the film is that she cannot reach a gun on the floor because she is trying to hold the door closed. If the paleontologist is pressing at the center of the door, and the botanist is pressing at the edge about 8 cm from the hinge, estimate how far the paleontologist would have to relocate in order to have a greater effect on keeping the door closed than both of them pushing together have in their original positions. (Question 8 is courtesy of Edward F. Redish. For more questions of this type, see www.physics.umd.edu/perg/.)

9. Suppose you are designing a car for a coasting race—a race in which the cars have no engines, but simply coast downhill. Do you want large wheels or small wheels? Do you want solid, disklike wheels or hooplike wheels? Should the wheels be heavy or light?

10. If you toss a textbook into the air, rotating it each time about one of the three axes perpendicular to it, you will find that it will not rotate smoothly about one of those axes. (Try placing a strong rubber band around the book

before the toss so that it will stay closed.) The book's rotation is stable about those axes having the largest and smallest moments of inertia, but unstable about the axis of intermediate moment. Try this on your own to find the axis that has this intermediate moment of inertia.

11. Stars originate as large bodies of slowly rotating gas. Because of gravity, these clumps of gas slowly decrease in size. What happens to the angular speed of a star as it shrinks? Explain.

12. If a high jumper positions his body correctly when going over the bar, the center of gravity of the athlete may actually pass under the bar. (See Fig. Q8.12.) Explain how this is possible.

Figure Q8.12

13. In a tape recorder, the tape is pulled past the read–write heads at a constant speed by the drive mechanism. Consider the reel from which the tape is pulled: As the tape is pulled off, the radius of the roll of remaining tape decreases. How does the torque on the reel change with time? How does the angular speed of the reel change with time? If the tape mechanism is suddenly turned on so that the tape is quickly pulled with a large force, is the tape more likely to break when pulled from a nearly full reel or from a nearly empty reel?

14. (a) Give an example in which the net force acting on an object is zero, yet the net torque is nonzero. (b) Give an example in which the net torque acting on an object is zero, yet the net force is nonzero.

15. A mouse is initially at rest on a horizontal turntable mounted on a frictionless vertical axle. If the mouse begins to walk clockwise around the perimeter of the table, what happens to the turntable? Explain.

16. A cat usually lands on its feet regardless of the position from which it is dropped. A slow-motion film of a cat falling shows that the upper half of its body twists in one direction while the lower half twists in the opposite direction. (See Fig. Q8.16.) Why does this type of rotation occur?

Figure Q8.16   A falling, twisting cat.

17. A ladder rests inclined against a wall. Would you feel safer climbing up the ladder if you were told that the floor was frictionless, but the wall was rough, or that the wall was frictionless, but the floor was rough? Justify your answer.

18. Two solid spheres—a large, massive sphere and a small sphere with low mass—are rolled down a hill. Which one reaches the bottom of the hill first? Next, we roll a large, low-density sphere and a small, high-density sphere, both with the same mass. Which one wins the race?

PROBLEMS

1, 2, 3 = straightforward, intermediate, challenging   ☐ = full solution available in *Student Solutions Manual/Study Guide*
Physics⊗Now™ = coached problem with hints available at **www.cp7e.com**   🐝 = biomedical application

Section 8.1 Torque

1. If the torque required to loosen a nut that is holding a flat tire in place on a car has a magnitude of 40.0 N · m, what *minimum* force must be exerted by the mechanic at the end of a 30.0-cm lug wrench to accomplish the task?

2. A steel band exerts a horizontal force of 80.0 N on a tooth at point *B* in Figure P8.2. What is the torque on the root of the tooth about point *A*?

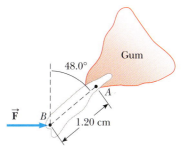

Figure P8.2

3. Calculate the net torque (magnitude and direction) on the beam in Figure P8.3 about (a) an axis through O perpendicular to the page and (b) an axis through C perpendicular to the page.

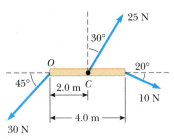

Figure P8.3

4. Write the necessary equations of equilibrium of the object shown in Figure P8.4. Take the origin of the torque equation about an axis perpendicular to the page through the point O.

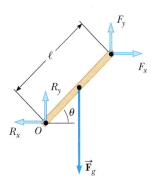

Figure P8.4

5. A simple pendulum consists of a small object of mass 3.0 kg hanging at the end of a 2.0-m-long light string that is connected to a pivot point. Calculate the magnitude of the torque (due to the force of gravity) about this pivot point when the string makes a 5.0° angle with the vertical.

6. A fishing pole is 2.00 m long and inclined to the horizontal at an angle of 20.0° (Fig. P8.6). What is the torque exerted by the fish about an axis perpendicular to the page and passing through the hand of the person holding the pole?

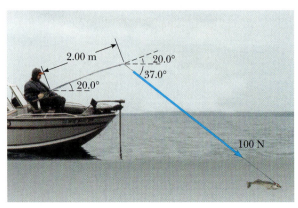

Figure P8.6

Section 8.2 Torque and the Two Conditions for Equilibrium
Section 8.3 The Center of Gravity
Section 8.4 Examples of Objects in Equilibrium

7. The arm in Figure P8.7 weighs 41.5 N. The force of gravity acting on the arm acts through point A. Determine the magnitudes of the tension force $\vec{F}_t$ in the deltoid muscle and the force $\vec{F}_s$ exerted by the shoulder on the humerus (upper-arm bone) to hold the arm in the position shown.

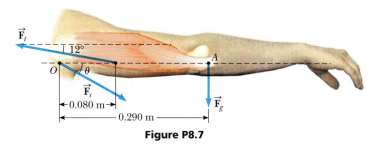

Figure P8.7

8. A water molecule consists of an oxygen atom with two hydrogen atoms bound to it as shown in Figure P8.8. The bonds are 0.100 nm in length, and the angle between the two bonds is 106°. Use the coordinate axes shown, and determine the location of the center of gravity of the molecule. Take the mass of an oxygen atom to be 16 times the mass of a hydrogen atom.

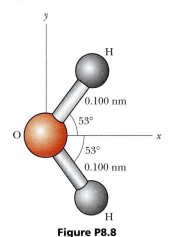

Figure P8.8

9. **Physics⊗Now™** A cook holds a 2.00-kg carton of milk at arm's length (Fig. P8.9). What force $\vec{F}_B$ must be exerted by the biceps muscle? (Ignore the weight of the forearm.)

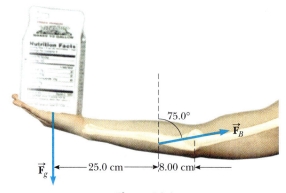

Figure P8.9

10. A meterstick is found to balance at the 49.7-cm mark when placed on a fulcrum. When a 50.0-gram mass is attached at the 10.0-cm mark, the fulcrum must be moved to the 39.2-cm mark for balance. What is the mass of the meter stick?

11. Find the x- and y-coordinates of the center of gravity of a 4.00-ft by 8.00-ft uniform sheet of plywood with the upper right quadrant removed as shown in Figure P8.11.

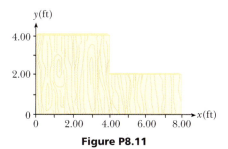

Figure P8.11

12. Consider the following mass distribution, where x- and y-coordinates are given in meters: 5.0 kg at (0.0, 0.0) m, 3.0 kg at (0.0, 4.0) m, and 4.0 kg at (3.0, 0.0) m. Where should a fourth object of 8.0 kg be placed so that the center of gravity of the four-object arrangement will be at (0.0, 0.0) m?

13. Many of the elements in horizontal-bar exercises can be modeled by representing the gymnast by four segments consisting of arms, torso (including the head), thighs, and lower legs, as shown in Figure P8.13a. Inertial parameters for a particular gymnast are as follows:

Segment	Mass (kg)	Length (m)	r_{cg} (m)	I (kg-m^2)
Arms	6.87	0.548	0.239	0.205
Torso	33.57	0.601	0.337	1.610
Thighs	14.07	0.374	0.151	0.173
Legs	7.54	—	0.227	0.164

Note that in Figure P8.13a r_{cg} is the distance to the center of gravity measured from the joint closest to the bar and the masses for the arms, thighs, and legs include both appendages. I is the moment of inertia of each segment about its center of gravity. Determine the distance from the bar to the center of gravity of the gymnast for the two positions shown in Figures P8.13b and P8.13c.

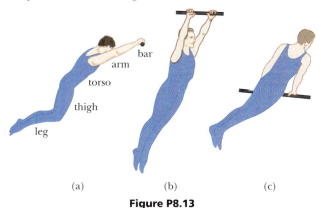

Figure P8.13

14. Using the data given in Problem 13 and the coordinate system shown in Figure P8.14b, calculate the position of the center of gravity of the gymnast shown in Figure P8.14a. Pay close attention to the definition of r_{cg} in the table.

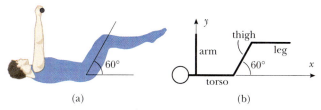

Figure P8.14

15. A person bending forward to lift a load "with his back" (Fig. P8.15a) rather than "with his knees" can be injured by large forces exerted on the muscles and vertebrae. The spine pivots mainly at the fifth lumbar vertebra, with the principal supporting force provided by the erector spinalis muscle in the back. To see the magnitude of the forces involved, and to understand why back problems are common among humans, consider the model shown in Fig. P8.15b of a person bending forward to lift a 200-N object. The spine and upper body are represented as a uniform horizontal rod of weight 350 N, pivoted at the base of the spine. The erector spinalis muscle, attached at a point two-thirds of the way up the spine, maintains the position of the back. The angle between the spine and this muscle is 12.0°. Find the tension in the back muscle and the compressional force in the spine.

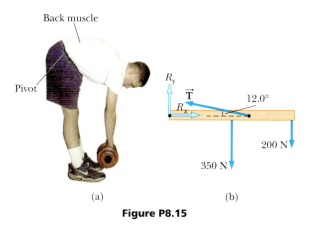

Figure P8.15

16. When a person stands on tiptoe (a strenuous position), the position of the foot is as shown in Figure P8.16a. The total gravitational force on the body, $\vec{F}_g$, is supported by the force $\vec{n}$ exerted by the floor on the toes of one foot. A mechanical model of the situation is shown in Figure P8.16b, where $\vec{T}$ is the force exerted by the Achilles tendon on the foot and $\vec{R}$ is the force exerted by the tibia on the foot. Find the values of T, R, and θ when $F_g = 700$ N.

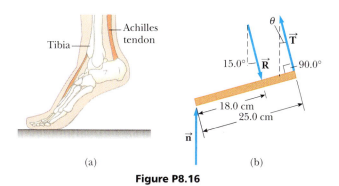

Figure P8.16

17. A 500-N uniform rectangular sign 4.00 m wide and 3.00 m high is suspended from a horizontal, 6.00-m-long, uniform, 100-N rod as indicated in Figure P8.17. The left end of the rod is supported by a hinge, and the right end is supported by a thin cable making a 30.0° angle with the vertical. (a) Find the tension T in the cable. (b) Find the horizontal and vertical components of force exerted on the left end of the rod by the hinge.

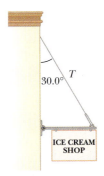

Figure P8.17

18. A window washer is standing on a scaffold supported by a vertical rope at each end. The scaffold weighs 200 N and is 3.00 m long. What is the tension in each rope when the 700-N worker stands 1.00 m from one end?

19. The chewing muscle, the masseter, is one of the strongest in the human body. It is attached to the mandible (lower jawbone) as shown in Figure P8.19a. The jawbone is pivoted about a socket just in front of the auditory canal. The forces acting on the jawbone are equivalent to those acting on the curved bar in Figure P8.19b: $\vec{F}_c$ is the force exerted by the food being chewed against the jawbone, $\vec{T}$ is the force of tension in the masseter, and $\vec{R}$ is the force exerted by the socket on the mandible. Find $\vec{T}$ and $\vec{R}$ for a person who bites down on a piece of steak with a force of 50.0 N.

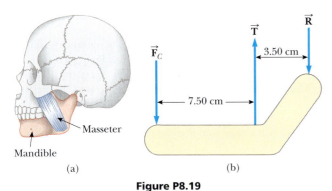

Figure P8.19

20. A hungry 700-N bear walks out on a beam in an attempt to retrieve some "goodies" hanging at the end (Fig. P8.20). The beam is uniform, weighs 200 N, and is 6.00 m long; the goodies weigh 80.0 N. (a) Draw a free-body diagram of the beam. (b) When the bear is at $x = 1.00$ m, find the tension in the wire and the components of the reaction force at the hinge. (c) If the wire can withstand a maximum tension of 900 N, what is the maximum distance the bear can walk before the wire breaks?

Figure P8.20

21. A uniform semicircular sign 1.00 m in diameter and of weight w is supported by two wires as shown in Figure P8.21. What is the tension in each of the wires supporting the sign?

Figure P8.21

22. A 20.0-kg floodlight in a park is supported at the end of a horizontal beam of negligible mass that is hinged to a pole, as shown in Figure P8.22. A cable at an angle of 30.0° with the beam helps to support the light. Find (a) the tension in the cable and (b) the horizontal and vertical forces exerted on the beam by the pole.

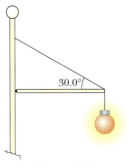

Figure P8.22

23. A uniform plank of length 2.00 m and mass 30.0 kg is supported by three ropes, as indicated by the blue vectors in Figure P8.23. Find the tension in each rope when a 700-N person is 0.500 m from the left end.

24. A 15.0-m, 500-N uniform ladder rests against a frictionless wall, making an angle of 60.0° with the horizontal. (a) Find the horizontal and vertical forces exerted on the base of the ladder by the Earth when an 800-N firefighter is 4.00 m from the bottom. (b) If the ladder is just on the

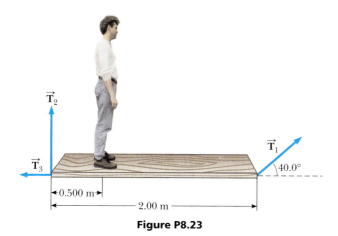

Figure P8.23

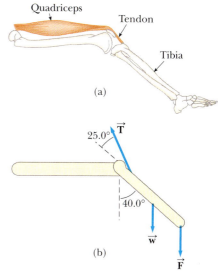

Figure P8.27

verge of slipping when the firefighter is 9.00 m up, what is the coefficient of static friction between ladder and ground?

25. An 8.00-m, 200-N uniform ladder rests against a smooth wall. The coefficient of static friction between the ladder and the ground is 0.600, and the ladder makes a 50.0° angle with the ground. How far up the ladder can an 800-N person climb before the ladder begins to slip?

26. A 1 200-N uniform boom is supported by a cable perpendicular to the boom as in Figure P8.26. The boom is hinged at the bottom, and a 2 000-N weight hangs from its top. Find the tension in the supporting cable and the components of the reaction force exerted on the boom by the hinge.

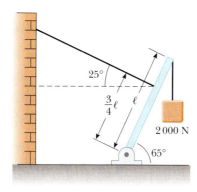

Figure P8.26

27. The large quadriceps muscle in the upper leg terminates at its lower end in a tendon attached to the upper end of the tibia (Fig. P8.27a). The forces on the lower leg when the leg is extended are modeled as in Figure P8.27b, where $\vec{T}$ is the force of tension in the tendon, $\vec{w}$ is the force of gravity acting on the lower leg, and $\vec{F}$ is the force of gravity acting on the foot. Find $\vec{T}$ when the tendon is at an angle of 25.0° with the tibia, assuming that $w = 30.0$ N, $F = 12.5$ N, and the leg is extended at an angle θ of 40.0° with the vertical. Assume that the center of gravity of the lower leg is at its center and that the tendon attaches to the lower leg at a point one-fifth of the way down the leg.

28. One end of a uniform 4.0-m-long rod of weight w is supported by a cable. The other end rests against a wall, where it is held by friction. (See Fig. P8.28.) The coefficient of static friction between the wall and the rod is

$\mu_s = 0.50$. Determine the minimum distance x from point A at which an additional weight w (the same as the weight of the rod) can be hung without causing the rod to slip at point A.

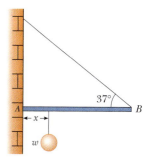

Figure P8.28

Section 8.5 Relationship Between Torque and Angular Acceleration

29. Four objects are held in position at the corners of a rectangle by light rods as shown in Figure P8.29. Find the moment of inertia of the system about (a) the x-axis, (b) the y-axis, and (c) an axis through O and perpendicular to the page.

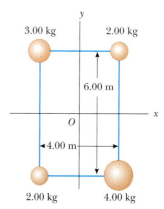

Figure P8.29 (Problems 29 and 30)

30. If the system shown in Figure P8.29 is set in rotation about each of the axes mentioned in Problem 29, find the torque that will produce an angular acceleration of 1.50 rad/s^2 in each case.

31. A model airplane with mass 0.750 kg is tethered by a wire so that it flies in a circle 30.0 m in radius. The airplane engine provides a net thrust of 0.800 N perpendicular to the tethering wire. (a) Find the torque the net thrust produces about the center of the circle. (b) Find the angular acceleration of the airplane when it is in level flight. (c) Find the linear acceleration of the airplane tangent to its flight path.

32. A potter's wheel having a radius of 0.50 m and a moment of inertia of 12 kg·m^2 is rotating freely at 50 rev/min. The potter can stop the wheel in 6.0 s by pressing a wet rag against the rim and exerting a radially inward force of 70 N. Find the effective coefficient of kinetic friction between the wheel and the wet rag.

33. A cylindrical fishing reel has a moment of inertia $I = 6.8 \times 10^{-4}$ kg·m^2 and a radius of 4.0 cm. A friction clutch in the reel exerts a restraining torque of 1.3 N·m if a fish pulls on the line. The fisherman gets a bite, and the reel begins to spin with an angular acceleration of 66 rad/s^2. (a) What is the force exerted by the fish on the line? (b) How much line unwinds in 0.50 s?

34. A bicycle wheel has a diameter of 64.0 cm and a mass of 1.80 kg. Assume that the wheel is a hoop with all the mass concentrated on the outside radius. The bicycle is placed on a stationary stand, and a resistive force of 120 N is applied tangent to the rim of the tire. (a) What force must be applied by a chain passing over a 9.00-cm-diameter sprocket in order to give the wheel an acceleration of 4.50 rad/s^2? (b) What force is required if you shift to a 5.60-cm-diameter sprocket?

35. A 150-kg merry-go-round in the shape of a uniform, solid, horizontal disk of radius 1.50 m is set in motion by wrapping a rope about the rim of the disk and pulling on the rope. What constant force must be exerted on the rope to bring the merry-go-round from rest to an angular speed of 0.500 rev/s in 2.00 s?

36. A 5.00-kg cylindrical reel with a radius of 0.600 m and a frictionless axle starts from rest and speeds up uniformly as a 3.00-kg bucket falls into a well, making a light rope unwind from the reel (Fig. P8.36). The bucket starts from rest and falls for 4.00 s. (a) What is the linear acceleration of the falling bucket? (b) How far does it drop? (c) What is the angular acceleration of the reel?

37. An airliner lands with a speed of 50.0 m/s. Each wheel of the plane has a radius of 1.25 m and a moment of inertia of 110 kg·m^2. At touchdown, the wheels begin to spin under the action of friction. Each wheel supports a weight of 1.40×10^4 N, and the wheels attain their angular speed in 0.480 s while rolling without slipping. What is the coefficient of kinetic friction between the wheels and the runway? Assume that the speed of the plane is constant.

Section 8.6 Rotational Kinetic Energy

38. A constant torque of 25.0 N·m is applied to a grindstone whose moment of inertia is 0.130 kg·m^2. Using energy principles and neglecting friction, find the angular speed after the grindstone has made 15.0 revolutions. [*Hint:* The angular equivalent of $W_{net} = F\Delta x = \frac{1}{2}mv_f{}^2 - \frac{1}{2}mv_i{}^2$ is

$W_{net} = \tau\Delta\theta = \frac{1}{2}I\omega_f{}^2 - \frac{1}{2}I\omega_i{}^2$. You should convince yourself that this relationship is correct.)

39. Physics⊗Now™ A 10.0-kg cylinder rolls without slipping on a rough surface. At an instant when its center of gravity has a speed of 10.0 m/s, determine (a) the translational kinetic energy of its center of gravity, (b) the rotational kinetic energy about its center of gravity, and (c) its total kinetic energy.

40. Use conservation of energy to determine the angular speed of the spool shown in Figure P8.36 after the 3.00-kg bucket has fallen 4.00 m, starting from rest. The light string attached to the bucket is wrapped around the spool and does not slip as it unwinds.

5.00 kg

0.600 m

3.00 kg

Figure P8.36   (Problems 36 and 40)

41. A horizontal 800-N merry-go-round of radius 1.50 m is started from rest by a constant horizontal force of 50.0 N applied tangentially to the merry-go-round. Find the kinetic energy of the merry-go-round after 3.00 s. (Assume it is a solid cylinder.)

42. A car is designed to get its energy from a rotating flywheel with a radius of 2.00 m and a mass of 500 kg. Before a trip, the flywheel is attached to an electric motor, which brings the flywheel's rotational speed up to 5 000 rev/min. (a) Find the kinetic energy stored in the flywheel. (b) If the flywheel is to supply energy to the car as a 10.0-hp motor would, find the length of time the car could run before the flywheel would have to be brought back up to speed.

43. The top in Figure P8.43 has a moment of inertia of 4.00×10^{-4} kg·m^2 and is initially at rest. It is free to rotate about a stationary axis AA'. A string wrapped around a peg along the axis of the top is pulled in such a manner as to maintain a constant tension of 5.57 N in the string. If the string does not slip while wound around the peg, what

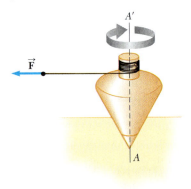

A'

$\vec{F}$

A

Figure P8.43

is the angular speed of the top after 80.0 cm of string has been pulled off the peg? [*Hint:* Consider the work that is done.]

44. A 240-N sphere 0.20 m in radius rolls without slipping 6.0 m down a ramp that is inclined at 37° with the horizontal. What is the angular speed of the sphere at the bottom of the slope if it starts from rest?

Section 8.7 Angular Momentum

45. A light rigid rod 1.00 m in length rotates about an axis perpendicular to its length and through its center, as shown in Figure P8.45. Two particles of masses 4.00 kg and 3.00 kg are connected to the ends of the rod. What is the angular momentum of the system if the speed of each particle is 5.00 m/s? (Neglect the rod's mass.)

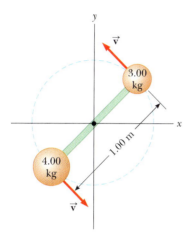

Figure P8.45

46. Halley's comet moves about the Sun in an elliptical orbit, with its closest approach to the Sun being 0.59 A.U. and its greatest distance being 35 A.U. (1 A.U. is the Earth–Sun distance). If the comet's speed at closest approach is 54 km/s, what is its speed when it is farthest from the Sun? You may neglect any change in the comet's mass and assume that its angular momentum about the Sun is conserved.

47. The system of small objects shown in Figure P8.47 is rotating at an angular speed of 2.0 rev/s. The objects are connected by light, flexible spokes that can be lengthened or shortened. What is the new angular speed if the spokes are shortened to 0.50 m? (An effect similar to that illustrated in this problem occurred in the early stages of the formation of our galaxy. As the massive cloud of dust and gas that was the source of the stars and planets contracted, an initially small angular speed increased with time.)

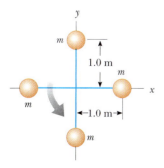

Figure P8.47

48. A playground merry-go-round of radius 2.00 m has a moment of inertia $I = 275$ kg·m² and is rotating about a frictionless vertical axle. As a child of mass 25.0 kg stands at a distance of 1.00 m from the axle, the system (merry-go-round and child) rotates at the rate of 14.0 rev/min. The child then proceeds to walk toward the edge of the merry-go-round. What is the angular speed of the system when the child reaches the edge?

49. A solid, horizontal cylinder of mass 10.0 kg and radius 1.00 m rotates with an angular speed of 7.00 rad/s about a fixed vertical axis through its center. A 0.250-kg piece of putty is dropped vertically onto the cylinder at a point 0.900 m from the center of rotation and sticks to the cylinder. Determine the final angular speed of the system.

50. A student sits on a rotating stool holding two 3.0-kg objects. When his arms are extended horizontally, the objects are 1.0 m from the axis of rotation and he rotates with an angular speed of 0.75 rad/s. The moment of inertia of the student plus stool is 3.0 kg·m² and is assumed to be constant. The student then pulls in the objects horizontally to 0.30 m from the rotation axis. (a) Find the new angular speed of the student. (b) Find the kinetic energy of the student before and after the objects are pulled in.

51. The puck in Figure P8.51 has a mass of 0.120 kg. Its original distance from the center of rotation is 40.0 cm, and it moves with a speed of 80.0 cm/s. The string is pulled downward 15.0 cm through the hole in the frictionless table. Determine the work done on the puck. [*Hint:* Consider the change in kinetic energy of the puck.]

Figure P8.51

52. A merry-go-round rotates at the rate of 0.20 rev/s with an 80-kg man standing at a point 2.0 m from the axis of rotation. (a) What is the new angular speed when the man walks to a point 1.0 m from the center? Assume that the merry-go-round is a solid 25-kg cylinder of radius 2.0 m. (b) Calculate the change in kinetic energy due to the man's movement. How do you account for this change in kinetic energy?

53. A 60.0-kg woman stands at the rim of a horizontal turntable having a moment of inertia of 500 kg·m² and a radius of 2.00 m. The turntable is initially at rest and is free to rotate about a frictionless, vertical axle through its center. The woman then starts walking around the rim clockwise (as viewed from above the system) at a constant speed of 1.50 m/s relative to the Earth. (a) In what direction and with what angular speed does the turntable rotate? (b) How much work does the woman do to set herself and the turntable into motion?

54. A space station shaped like a giant wheel has a radius of 100 m and a moment of inertia of $5.00 \times 10^8 \ \text{kg} \cdot \text{m}^2$. A crew of 150 lives on the rim, and the station is rotating so that the crew experiences an apparent acceleration of $1g$ (Fig. P8.54). When 100 people move to the center of the station for a union meeting, the angular speed changes. What apparent acceleration is experienced by the managers remaining at the rim? Assume the average mass of a crew member is 65.0 kg.

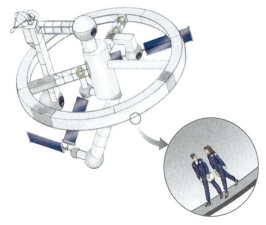

Figure P8.54

ADDITIONAL PROBLEMS

55. A cylinder with moment of inertia I_1 rotates with angular velocity ω_0 about a frictionless vertical axle. A second cylinder, with moment of inertia I_2, initially not rotating, drops onto the first cylinder (Fig. P8.55). Since the surfaces are rough, the two cylinders eventually reach the same angular speed ω. (a) Calculate ω. (b) Show that kinetic energy is lost in this situation, and calculate the ratio of the final to the initial kinetic energy.

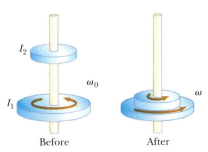

Before   After

Figure P8.55

56. A new General Electric stove has a mass of 68.0 kg and the dimensions shown in Figure P8.56. The stove comes with a warning that it can tip forward if a person stands or sits on the oven door when it is open. What can you conclude about the weight of such a person? Could it be a child? List the assumptions you make in solving this problem. (The stove is supplied with a wall bracket to prevent the accident.)

57. A 40.0-kg child stands at one end of a 70.0-kg boat that is 4.00 m long (Fig. P8.57). The boat is initially 3.00 m from the pier. The child notices a turtle on a rock beyond the far end of the boat and proceeds to walk to that end to catch the turtle. (a) Neglecting friction between the boat

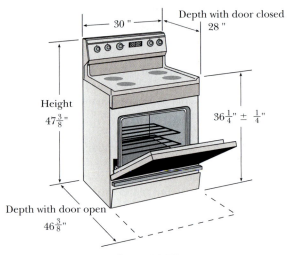

Figure P8.56

and water, describe the motion of the system (child plus boat). (b) Where will the child be relative to the pier when he reaches the far end of the boat? (c) Will he catch the turtle? (Assume that he can reach out 1.00 m from the end of the boat.)

Figure P8.57

58. Figure P8.58 shows a clawhammer as it is being used to pull a nail out of a horizontal board. If a force of magnitude 150 N is exerted horizontally as shown, find (a) the

Figure P8.58

force exerted by the hammer claws on the nail and (b) the force exerted by the surface at the point of contact with the hammer head. Assume that the force the hammer exerts on the nail is parallel to the nail.

59. The pulley in Figure P8.59 has a moment of inertia of $5.0 \text{ kg} \cdot \text{m}^2$ and a radius of 0.50 m. The cord supporting the masses m_1 and m_2 does not slip, and the axle is frictionless. (a) Find the acceleration of each mass when $m_1 = 2.0$ kg and $m_2 = 5.0$ kg. (b) Find the tension in the cable supporting m_1 and the tension in the cable supporting m_2. [*Note:* The two tensions are different].

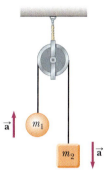

Figure P8.59

60. A 12.0-kg object is attached to a cord that is wrapped around a wheel of radius $r = 10.0$ cm (Fig. P8.60). The acceleration of the object down the frictionless incline is measured to be 2.00 m/s^2. Assuming the axle of the wheel to be frictionless, determine (a) the tension in the rope, (b) the moment of inertia of the wheel, and (c) the angular speed of the wheel 2.00 s after it begins rotating, starting from rest.

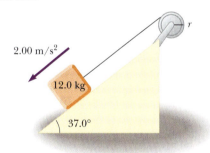

Figure P8.60

61. A uniform ladder of length L and weight w is leaning against a vertical wall. The coefficient of static friction between the ladder and the floor is the same as that between the ladder and the wall. If this coefficient of static friction is $\mu_s = 0.500$, determine the smallest angle the ladder can make with the floor without slipping.

62. A uniform 10.0-N picture frame is supported as shown in Figure P8.62. Find the tension in the cords and the magnitude of the horizontal force at P that are required to hold the frame in the position shown.

63. **Physics⊗Now™** A solid 2.0-kg ball of radius 0.50 m starts at a height of 3.0 m above the surface of the Earth and *rolls* down a 20° slope. A solid disk and a ring start at the same time and the same height. The ring and disk each have the same mass and radius as the ball. Which of the three wins the race to the bottom if all roll without slipping?

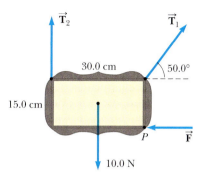

Figure P8.62

64. A common physics demonstration (Fig. P8.64) consists of a ball resting at the end of a board of length ℓ that is elevated at an angle θ with the horizontal. A light cup is attached to the board at r_c so that it will catch the ball when the support stick is suddenly removed. Show that (a) the ball will lag behind the falling board when $\theta < 35.3°$ and (b) the ball will fall into the cup when the board is supported at this limiting angle and the cup is placed at

$$r_c = \frac{2\ell}{3 \cos \theta}$$

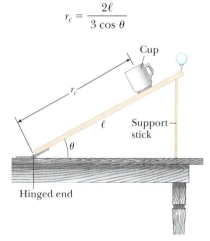

Figure P8.64

65. In Figure P8.65, the sliding block has a mass of 0.850 kg, the counterweight has a mass of 0.420 kg, and the pulley is a uniform solid cylinder with a mass of 0.350 kg and an outer radius of 0.0300 m. The coefficient of kinetic friction between the block and the horizontal surface is 0.250. The pulley turns without friction on its axle. The light cord does not stretch and does not slip on the pulley. The block has a

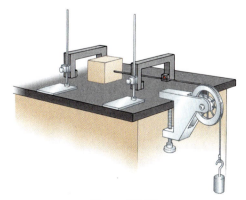

Figure P8.65

velocity of 0.820 m/s toward the pulley when it passes through a photogate. (a) Use energy methods to predict the speed of the block after it has moved to a second photogate 0.700 m away. (b) Find the angular speed of the pulley at the same moment.

66. (a) Without the wheels, a bicycle frame has a mass of 8.44 kg. Each of the wheels can be roughly modeled as a uniform solid disk with a mass of 0.820 kg and a radius of 0.343 m. Find the kinetic energy of the whole bicycle when it is moving forward at 3.35 m/s. (b) Before the invention of a wheel turning on an axle, ancient people moved heavy loads by placing rollers under them. (Modern people use rollers, too: Any hardware store will sell you a roller bearing for a lazy Susan.) A stone block of mass 844 kg moves forward at 0.335 m/s, supported by two uniform cylindrical tree trunks, each of mass 82.0 kg and radius 0.343 m. There is no slipping between the block and the rollers or between the rollers and the ground. Find the total kinetic energy of the moving objects.

67. In exercise physiology studies, it is sometimes important to determine the location of a person's center of gravity. This can be done with the arrangement shown in Figure P8.67. A light plank rests on two scales that read $F_{g1} = 380$ N and $F_{g2} = 320$ N. The scales are separated by a distance of 2.00 m. How far from the woman's feet is her center of gravity?

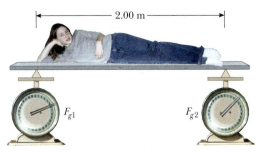

Figure P8.67

68. Two astronauts (Fig. P8.68), each having a mass of 75.0 kg, are connected by a 10.0-m rope of negligible mass. They are isolated in space, moving in circles around the point halfway between them at a speed of 5.00 m/s. Treating the astronauts as particles, calculate (a) the magnitude of the angular momentum and (b) the rotational energy of the system. By pulling on the rope, the astronauts shorten the distance between them to 5.00 m. (c) What is the new angular momentum of the system?

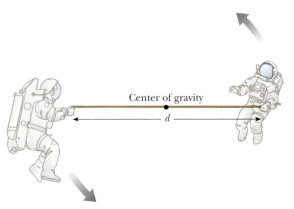

Figure P8.68 (Problems 68 and 69)

(d) What are their new speeds? (e) What is the new rotational energy of the system? (f) How much work is done by the astronauts in shortening the rope?

69. Two astronauts (Fig. P8.68), each having a mass M, are connected by a rope of length d having negligible mass. They are isolated in space, moving in circles around the point halfway between them at a speed v. (a) Calculate the magnitude of the angular momentum of the system by treating the astronauts as particles. (b) Calculate the rotational energy of the system. By pulling on the rope, the astronauts shorten the distance between them to $d/2$. (c) What is the new angular momentum of the system? (d) What are their new speeds? (e) What is the new rotational energy of the system? (f) How much work is done by the astronauts in shortening the rope?

70. Two window washers, Bob and Joe, are on a 3.00-m-long, 345-N scaffold supported by two cables attached to its ends. Bob weighs 750 N and stands 1.00 m from the left end, as shown in Figure P8.70. Two meters from the left end is the 500-N washing equipment. Joe is 0.500 m from the right end and weighs 1 000 N. Given that the scaffold is in rotational and translational equilibrium, what are the forces on each cable?

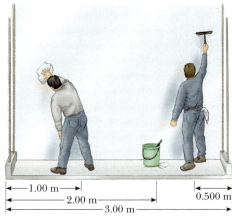

Figure P8.70

71. We have all complained that there aren't enough hours in a day. In an attempt to change that, suppose that all the people in the world lined up at the equator and started running east at 2.5 m/s relative to the surface of the Earth. By how much would the length of a day increase? (Assume that there are 5.5×10^9 people in the world with an average mass of 70 kg each and that the Earth is a solid, homogeneous sphere. In addition, you may use the result $1/(1 - x) \approx 1 + x$ for small x.)

72. In a circus performance, a large 5.0-kg hoop of radius 3.0 m rolls without slipping. If the hoop is given an angular speed of 3.0 rad/s while rolling on the horizontal ground and is then allowed to roll up a ramp inclined at 20° with the horizontal, how far along the incline does the hoop roll?

73. **Physics⊗Now™** A uniform solid cylinder of mass M and radius R rotates on a frictionless horizontal axle (Fig. P8.73). Two objects with equal masses m hang from light cords wrapped around the cylinder. If the system is released from rest, find (a) the tension in each cord and (b) the acceleration of each object after the objects have descended a distance h.

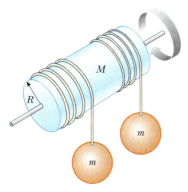

Figure P8.73

74. Figure P8.74 shows a vertical force applied tangentially to a uniform cylinder of weight *w*. The coefficient of static friction between the cylinder and all surfaces is 0.500. Find, in terms of *w*, the maximum force $\vec{F}$ that can be applied without causing the cylinder to rotate. [*Hint:* When the cylinder is on the verge of slipping, both friction forces are at their maximum values. Why?]

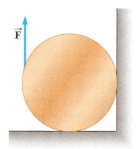

Figure P8.74

75. Due to a gravitational torque exerted by the Moon on the Earth, our planet's period of rotation slows at a rate on the order of 1 ms/century. (a) Determine the order of magnitude of Earth's angular acceleration. (b) Find the order of magnitude of the torque. (c) Find the order of magnitude of the size of the wrench an ordinary person would need to exert such a torque, as in Figure P8.75. Assume the person can brace his feet against a solid firmament.

Figure P8.75

76. A uniform pole is propped between the floor and the ceiling of a room. The height of the room is 7.80 ft, and the

coefficient of static friction between the pole and the ceiling is 0.576. The coefficient of static friction between the pole and the floor is greater than that. What is the length of the longest pole that can be propped between the floor and the ceiling?

77. A *war-wolf*, or *trebuchet*, is a device used during the Middle Ages to throw rocks at castles and now sometimes used to fling pumpkins and pianos. A simple trebuchet is shown in Figure P8.77. Model it as a stiff rod of negligible mass 3.00 m long and joining particles of mass 60.0 kg and 0.120 kg at its ends. It can turn on a frictionless horizontal axle perpendicular to the rod and 14.0 cm from the particle of larger mass. The rod is released from rest in a horizontal orientation. Find the maximum speed that the object of smaller mass attains.

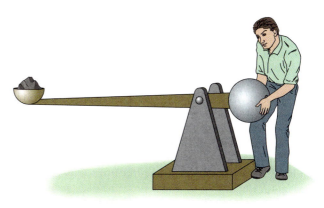

Figure P8.77

78. A painter climbs a ladder leaning against a smooth wall. At a certain height, the ladder is on the verge of slipping. (a) Explain why the force exerted by the vertical wall on the ladder is horizontal. (b) If the ladder of length *L* leans at an angle θ with the horizontal, what is the lever arm for this horizontal force with the axis of rotation taken at the base of the ladder? (c) If the ladder is uniform, what is the lever arm for the force of gravity acting on the ladder? (d) Let the mass of the painter be 80 kg, *L* = 4.0 m, the ladder's mass be 30 kg, θ = 53°, and the coefficient of friction between ground and ladder be 0.45. Find the maximum distance the painter can climb up the ladder.

79. A 4.00-kg mass is connected by a light cord to a 3.00-kg mass on a smooth surface (Fig. P8.79). The pulley rotates about a frictionless axle and has a moment of inertia of 0.500 kg·m² and a radius of 0.300 m. Assuming that the cord does not slip on the pulley, find (a) the acceleration of the two masses and (b) the tensions T_1 and T_2.

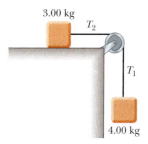

3.00 kg

T_2

T_1

4.00 kg

Figure P8.79

80. A string is wrapped around a uniform cylinder of mass M and radius R. The cylinder is released from rest with the string vertical and its top end tied to a fixed bar (Fig. P8.80). Show that (a) the tension in the string is one-third the weight of the cylinder, (b) the magnitude of the acceleration of the center of gravity is $2g/3$, and (c) the speed of the center of gravity is $(4gh/3)^{1/2}$ after the cylinder has descended through distance h. Verify your answer to (c) with the energy approach.

Figure P8.80

81. A person in a wheelchair wishes to roll up over a sidewalk curb by exerting a horizontal force $\vec{F}$ to the top of each of the wheelchair's main wheels (Fig. P8.81a). The main wheels have radius r and come in contact with a curb of height h (Fig. P8.81b). (a) Assume that each main wheel supports half of the total load, and show that the magnitude of the minimum force necessary to raise the wheelchair from the street is given by

$$F = \frac{mg\sqrt{2rh - h^2}}{2(2r - h)}$$

where mg is the combined weight of the wheelchair and person. (b) Estimate the value of F, taking $mg = 1\,400$ N, $r = 30$ cm, and $h = 10$ cm.

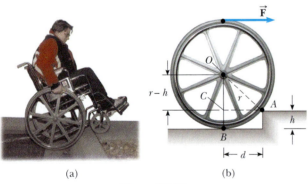

(a)   (b)

Figure P8.81

82. The truss structure in Figure P8.82 represents part of a bridge. Assume that the structural components are connected by pin joints and that the entire structure is free to slide horizontally at each end. Assume furthermore that the mass of the structure is negligible compared with the load it must support. In this situation, the force exerted by each of the bars (struts) on the pin joints is either a force of tension or one of compression and must be along the length of the bar. Calculate the force in each strut when the bridge supports a 7 200-N load at its center.

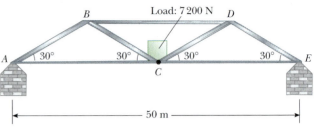

Figure P8.82

83. The Iron Cross When a gymnast weighing 750 N executes the iron cross as in Figure P8.83a, the primary muscles involved in supporting this position are the latissimus dorsi ("lats") and the pectoralis major ("pecs"). The rings exert an upward force on the arms and support the weight of the gymnast. The force exerted by the shoulder joint on the arm is labeled $\vec{F}_s$ while the two muscles exert a total force $\vec{F}_m$ on the arm. Estimate the magnitude of the force $\vec{F}_m$. Note that one ring supports half the weight of the gymnast, which is 375 N as indicated in Figure P8.83b. Assume that the force $\vec{F}_m$ acts at an angle of 45° below the horizontal at a distance of 4.0 cm from the shoulder joint. In your estimate, take the distance from the shoulder joint to the hand to be 70 cm and ignore the weight of the arm.

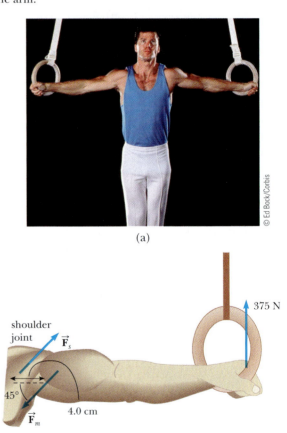

(a)

(b)

Figure P8.83

84. Swinging on a high bar The gymnast shown in Figure P8.84 is performing a backwards giant swing on the high bar. Starting from rest in a near-vertical orientation, he rotates around the bar in a counterclockwise direction, keeping his body and arms straight. Friction between the bar and the gymnast's hands exerts a constant torque opposing the rotational motion. If the angular velocity of the gymnast at position 2 is measured to be 4.0 rad/s, determine his angular velocity at position 3. (Note that this maneuver is called a backwards giant swing, even though the motion of the gymnast would seem to be forwards.)

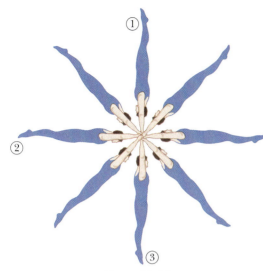

Figure P8.84

ACTIVITIES

A.1. Compare the motion of an empty soup can and a filled soup can down the same incline, such as a tilted table. If they are released from rest at the same height on the incline, which one reaches the bottom first? Repeat your observations with different kinds of soup (tomato, chicken noodle, etc.) and a can of beans. Compare their motions, and try to explain your observations. Finally, compare the motion of a filled soup can with a tennis ball, and explain your results to a friend.

A.2. Before attempting this exercise, review Example 8.14, dealing with the spinning stool. The techniques used here are similar to those used there. (a) First, make an estimate of the moment of inertia of your body. One way to do this would be to model your body as a solid cylinder and find I from $I = \frac{1}{2}MR^2$. You would have to determine your mass and estimate your average "radius" for this approach. Can you think of an alternative way to estimate I? (b) Now use the approach of Example 8.13 to measure I: Sit on a rotating stool, hold two weights (say, two books), and determine the angular speed of rotation with the books extended and after they are pulled in. The angular speed can be found by estimating the time taken for a given number of rotations. Use conservation of angular momentum to determine I. Do this five times to determine an average value for I. How well do your results for (a) and (b) compare, and if they differ greatly, what might cause the discrepancy?

A.3. This experiment demonstrates a simple way to find the center of gravity of an irregularly shaped object. Cut out an irregular shape from a piece of cardboard, and punch three to five holes around the edge of the shape. Put a pushpin through one of the holes, and tack the shape to a corkboard so that the shape can rotate freely. Now tie a weight to one end of a string, and hang the other end of the string from the pushpin. When the string stops moving, trace a line on the cardboard that follows the string. Repeat this for each of the holes in the cardboard. You will find that there is a point where all the lines intersect. This point is the center of gravity of the object.

In the Dead Sea, a lake between Jordan and Israel, the high percentage of salt dissolved in the water raises the fluid's density, dramatically increasing the buoyant force. Bathers can kick back and enjoy a good read, dispensing with the floating lounge chairs.

© Alison Wright/Corbis

CHAPTER

9

Solids and Fluids

There are four known states of matter: solids, liquids, gases, and plasmas. In the universe at large, plasmas—systems of charged particles interacting electromagnetically—are the most common. In our environment on Earth, solids, liquids, and gases predominate.

An understanding of the fundamental properties of these different states of matter is important in all the sciences, in engineering, and in medicine. Forces put stresses on solids, and stresses can strain, deform, and break those solids, whether they are steel beams or bones. Fluids under pressure can perform work, or they can carry nutrients and essential solutes, like the blood flowing through our arteries and veins. Flowing gases cause pressure differences that can lift a massive cargo plane or the roof off a house in a hurricane. High-temperature plasmas created in fusion reactors may someday allow humankind to harness the energy source of the sun.

The study of any one of these states of matter is itself a vast discipline. Here, we'll introduce basic properties of solids and liquids, the latter including some properties of gases. In addition, we'll take a brief look at surface tension, viscosity, osmosis, and diffusion.

9.1 STATES OF MATTER

Matter is normally classified as being in one of three states: **solid**, **liquid**, or **gas**. Often this classification system is extended to include a fourth state of matter, called a **plasma**.

Everyday experience tells us that a solid has a definite volume and shape. A brick, for example, maintains its familiar shape and size day in and day out. A liquid has a definite volume but no definite shape. When you fill the tank on a lawn mower, the gasoline changes its shape from that of the original container to that of the tank on the mower, but the original volume is unchanged. A gas differs from solids and liquids in that it has neither definite volume nor definite shape. Because gas can flow, however, it shares many properties with liquids.

All matter consists of some distribution of atoms or molecules. The atoms in a solid, held together by forces that are mainly electrical, are located at specific positions with respect to one another and vibrate about those positions. At low temperatures, the vibrating motion is slight and the atoms can be considered essentially fixed. As energy is added to the material, the amplitude of the vibrations increases. A vibrating atom can be viewed as being bound in its equilibrium position by springs attached to neighboring atoms. A collection of such atoms and imaginary springs is shown in Figure 9.1. We can picture applied external forces as compressing these tiny internal springs. When the external forces are removed, the solid tends to return to its original shape and size. Consequently, a solid is said to have *elasticity*.

Solids can be classified as either crystalline or amorphous. In a **crystalline solid** the atoms have an ordered structure. For example, in the sodium chloride crystal (common table salt), sodium and chlorine atoms occupy alternate corners of a cube, as in Figure 9.2a. In an **amorphous solid**, such as glass, the atoms are arranged almost randomly, as in Figure 9.2b.

For any given substance, the liquid state exists at a higher temperature than the solid state. The intermolecular forces in a liquid aren't strong enough to keep the molecules in fixed positions, and they wander through the liquid in random fashion (Fig. 9.2c). Solids and liquids both have the property that when an attempt is made to compress them, strong repulsive atomic forces act internally to resist the compression.

In the gaseous state, molecules are in constant random motion and exert only weak forces on each other. The average distance between the molecules of a gas is quite large compared with the size of the molecules. Occasionally the molecules collide with each other, but most of the time they move as nearly free, noninteracting particles. As a result, unlike solids and liquids, gases can be easily compressed. We'll say more about gases in subsequent chapters.

When a gas is heated to high temperature, many of the electrons surrounding each atom are freed from the nucleus. The resulting system is a collection of free, electrically charged particles—negatively charged electrons and positively charged ions. Such a highly ionized state of matter containing equal amounts of positive and negative charges is called a **plasma**. Unlike a neutral gas, the long-range electric and magnetic forces allow the constituents of a plasma to interact with each other. Plasmas are found inside stars and in accretion disks around black holes, for example, and are far more common than the solid, liquid, and gaseous states because there are far more stars around than any other form of celestial matter,

Crystals of natural quartz (SiO_2), one of the most common minerals on Earth. Quartz crystals are used to make special lenses and prisms and are employed in certain electronic applications.

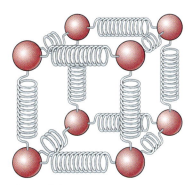

Figure 9.1   A model of a portion of a solid. The atoms (spheres) are imagined as being attached to each other by springs, which represent the elastic nature of the interatomic forces. A solid consists of trillions of segments like this, with springs connecting all of them.

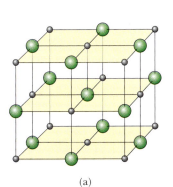

(a)

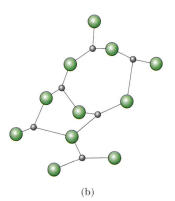

(b)

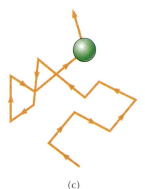

(c)

Figure 9.2   (a) The NaCl structure, with the Na^+ (gray) and Cl^- (green) ions at alternate corners of a cube. (b) In an amorphous solid, the atoms are arranged randomly. (c) Erratic motion of a molecule in a liquid.

except possibly **dark matter**. Dark matter, inferred by observations of the motion of stars around the galaxy, makes up about 90% of the matter in the universe and is of unknown composition. In this chapter, however, we ignore plasmas and dark matter and concentrate on the more familiar solid, liquid, and gaseous forms that make up the environment of our planet.

9.2   THE DEFORMATION OF SOLIDS

While a solid may be thought of as having a definite shape and volume, it's possible to change its shape and volume by applying external forces. A sufficiently large force will permanently deform or break an object, but otherwise, when the external forces are removed, the object tends to return to its original shape and size. This is called *elastic behavior*.

The elastic properties of solids are discussed in terms of stress and strain. **Stress** is the force per unit area causing a deformation; **strain** is a measure of the amount of the deformation. For sufficiently small stresses, **stress is proportional to strain**, with the constant of proportionality depending on the material being deformed and on the nature of the deformation. We call this proportionality constant the **elastic modulus**:

$$\text{stress} = \text{elastic modulus} \times \text{strain} \qquad [9.1]$$

The elastic modulus is analogous to a spring constant. It can be taken as the stiffness of a material: A material having a large elastic modulus is very stiff and difficult to deform. There are three relationships having the form of Equation 9.1, corresponding to tensile, shear, and bulk deformation, and all of them satisfy an equation similar to Hooke's law for springs:

$$F = k\Delta x \qquad [9.2]$$

where F is the applied force, k is the spring constant, and Δx is the amount by which the spring is compressed.

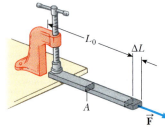

ACTIVE FIGURE 9.3
A long bar clamped at one end is stretched by the amount ΔL under the action of a force $\vec{\mathbf{F}}$.

Physics⊗Now™
Log into PhysicsNow at **www.cp7e.com**, and go to Active Figure 9.3 to adjust the values of the applied force and Young's modulus and observe the change in length of the bar.

Young's Modulus: Elasticity in Length

Consider a long bar of cross-sectional area A and length L_0, clamped at one end (Active Fig. 9.3). When an external force $\vec{\mathbf{F}}$ is applied along the bar, perpendicular to the cross section, internal forces in the bar resist the distortion ("stretching") that $\vec{\mathbf{F}}$ tends to produce. Nevertheless, the bar attains an equilibrium in which (1) its length is greater than L_0 and (2) the external force is balanced by internal forces. Under these circumstances, the bar is said to be *stressed*. We define the **tensile stress** as the ratio of the magnitude of the external force F to the cross-sectional area A. The word "tensile" has the same root as the word "tension" and is used because the bar is under tension. The SI unit of stress is the newton per square meter (N/m^2), called the **pascal** (Pa):

The pascal ▶

$$1 \text{ Pa} \equiv 1 \text{ N/m}^2$$

The **tensile strain** in this case is defined as the ratio of the change in length ΔL to the original length L_0 and is therefore a dimensionless quantity. Using Equation 9.1, we can write an equation relating tensile stress to tensile strain:

$$\frac{F}{A} = Y\frac{\Delta L}{L_0} \qquad [9.3]$$

In this equation, Y is the constant of proportionality, called **Young's modulus**. Notice that Equation 9.3 could be solved for F and put in the form $F = k\Delta L$, where $k = YA/L_0$, making it look just like Hooke's law, Equation 9.2.

A material having a large Young's modulus is difficult to stretch or compress. This quantity is typically used to characterize a rod or wire stressed under either tension or compression. Because strain is a dimensionless quantity, Y is in pascals.

TABLE 9.1

Typical Values for the Elastic Modulus

Substance	Young's Modulus (Pa)	Shear Modulus (Pa)	Bulk Modulus (Pa)
Aluminum	7.0×10^{10}	2.5×10^{10}	7.0×10^{10}
Bone	1.8×10^{10}	8.0×10^{10}	—
Brass	9.1×10^{10}	3.5×10^{10}	6.1×10^{10}
Copper	11×10^{10}	4.2×10^{10}	14×10^{10}
Steel	20×10^{10}	8.4×10^{10}	16×10^{10}
Tungsten	35×10^{10}	14×10^{10}	20×10^{10}
Glass	$6.5–7.8 \times 10^{10}$	$2.6–3.2 \times 10^{10}$	$5.0–5.5 \times 10^{10}$
Quartz	5.6×10^{10}	2.6×10^{10}	2.7×10^{10}
Rib Cartilage	1.2×10^{7}	—	—
Rubber	0.1×10^{7}	—	—
Tendon	2×10^{7}	—	—
Water	—	—	0.21×10^{10}
Mercury	—	—	2.8×10^{10}

Typical values are given in Table 9.1. Experiments show that (1) the change in length for a fixed external force is proportional to the original length and (2) the force necessary to produce a given strain is proportional to the cross-sectional area. The value of Young's modulus for a given material depends on whether the material is stretched or compressed. A human femur, for example, is stronger under tension than compression.

It's possible to exceed the **elastic limit** of a substance by applying a sufficiently great stress (Fig. 9.4). At the elastic limit, the stress–strain curve departs from a straight line. A material subjected to a stress beyond this limit ordinarily doesn't return to its original length when the external force is removed. As the stress is increased further, it surpasses the **ultimate strength**: the greatest stress the substance can withstand without breaking. The **breaking point** for brittle materials is just beyond the ultimate strength. For ductile metals like copper and gold, after passing the point of ultimate strength, the metal thins and stretches at a lower stress level before breaking.

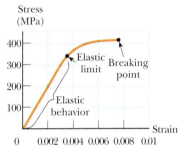

Figure 9.4   Stress-versus-strain curve for an elastic solid.

Shear Modulus: Elasticity of Shape

Another type of deformation occurs when an object is subjected to a force $\vec{\mathbf{F}}$ *parallel* to one of its faces while the opposite face is held fixed by a second force (Active Fig. 9.5a). If the object is originally a rectangular block, such a parallel force results in a shape with the cross section of a parallelogram. This kind of stress is called a **shear stress**. A book pushed sideways, as in Active Figure 9.5b, is being subjected to a shear stress. There is no change in volume with this kind of deformation. It's important to remember that in shear stress, the applied force is *parallel* to the cross-sectional area, whereas in tensile stress the force is *perpendicular* to the cross-sectional area. We define **the shear stress as F/A, the ratio of the magnitude of the**

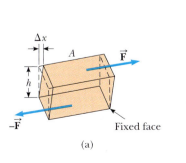

(a)

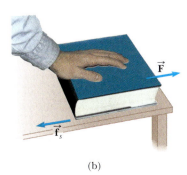

(b)

ACTIVE FIGURE 9.5
(a) A shear deformation in which a rectangular block is distorted by forces applied tangent to two of its faces. (b) A book under shear stress.

Physics⊗Now™
Log into PhysicsNow at **www.cp7e.com**, and go to Active Figure 9.5 to adjust the values of the applied force and the shear modulus and observe the change in shape of the block in part (a).

Fixed face

ACTIVE FIGURE 9.6
When a solid is under uniform pressure, it undergoes a change in volume, but no change in shape. This cube is compressed on all sides by forces normal to its six faces.

Physics⊗Now™
Log into PhysicsNow at **www.cp7e.com**, and go to Active Figure 9.6 to adjust the values of the applied force and the bulk modulus and observe the change in volume of the cube.

Bulk modulus ▶

parallel force to the area A of the face being sheared. The shear strain is the ratio $\Delta x/h$, where Δx is the horizontal distance the sheared face moves and h is the height of the object. The shear stress is related to the shear strain according to

$$\frac{F}{A} = S\frac{\Delta x}{h} \qquad [9.4]$$

where S is the **shear modulus** of the material, with units of pascals (force per unit area). Once again, notice the similarity to Hooke's law.

A material having a large shear modulus is difficult to bend. Shear moduli for some representative materials are listed in Table 9.1.

Bulk Modulus: Volume Elasticity

The bulk modulus characterizes the response of a substance to uniform squeezing. Suppose that the external forces acting on an object are all perpendicular to the surface on which the force acts and are distributed uniformly over the surface of the object (Active Fig. 9.6). This occurs when an object is immersed in a fluid. An object subject to this type of deformation undergoes a change in volume but no change in shape. **The volume stress ΔP is defined as the ratio of the magnitude of the change in the applied force ΔF to the surface area A.** (In dealing with fluids, we'll refer to the quantity F/A as the **pressure**, to be defined and discussed more formally in the next section.) The volume strain is equal to the change in volume ΔV divided by the original volume V. Again using Equation 9.1, we can relate a volume stress to a volume strain by the formula

$$\Delta P = -B\frac{\Delta V}{V} \qquad [9.5]$$

A material having a large bulk modulus doesn't compress easily. Note that a negative sign is included in this defining equation so that B is always positive. An increase in pressure (positive ΔP) causes a decrease in volume (negative ΔV) and vice versa.

Table 9.1 lists bulk modulus values for some materials. If you look up such values in a different source, you may find that the reciprocal of the bulk modulus, called the **compressibility** of the material, is listed. Note from the table that both solids and liquids have bulk moduli. There is neither a Young's modulus nor shear modulus for liquids, however, because liquids simply flow when subjected to a tensile or shearing stress.

EXAMPLE 9.1   Built to Last

Goal   Calculate a compression due to tensile stress, and maximum load.

Problem   A vertical steel beam in a building supports a load of 6.0×10^4 N. **(a)** If the length of the beam is 4.0 m and its cross-sectional area is 8.0×10^{-3} m², find the distance the beam is compressed along its length. **(b)** What maximum load in newtons could the steel beam support before failing?

Strategy   Equation 9.3 pertains to compressive stress and strain and can be solved for ΔL, followed by substitution of known values. For part (b), set the compressive stress equal to the ultimate strength of steel from Table 9.2. Solve for the magnitude of the force, which is the total weight the structure can support.

Solution
(a) Find the amount of compression in the beam.

Solve Equation 9.3 for ΔL and substitute, using the value of Young's modulus from Table 9.1:

$$\frac{F}{A} = Y\frac{\Delta L}{L_0}$$

$$\Delta L = \frac{FL_0}{YA} = \frac{(6.0 \times 10^4\ \text{N})(4.0\ \text{m})}{(2.0 \times 10^{11}\ \text{Pa})(8.0 \times 10^{-3}\ \text{m}^2)}$$

$$= 1.5 \times 10^{-4}\ \text{m}$$

(b) Find the maximum load that the beam can support.

Set the compressive stress equal to the ultimate
compressive strength from Table 9.2, and solve for F:

$$\frac{F}{A} = \frac{F}{8.0 \times 10^{-3}\ \text{m}^2} = 5.0 \times 10^8\ \text{Pa}$$

$$F = 4.0 \times 10^6\ \text{N}$$

Remarks   In designing load-bearing structures of any kind, it's always necessary to build in a safety factor. No one would
drive a car over a bridge that had been designed to supply the minimum necessary strength to keep it from collapsing.

Exercise 9.1
A cable used to lift heavy materials like steel I-beams must be strong enough to resist breaking even under a load of
1.0×10^6 N. For safety, the cable must support twice that load. (a) What cross-sectional area should the cable have if
it's to be made of steel? (b) By how much will an 8.0-m length of this cable stretch when subject to the 1.0×10^6-N
load?

Answers   (a) $4.0 \times 10^{-3}\ \text{m}^2$ (b) $1.0 \times 10^{-2}\ \text{m}$

TABLE 9.2

Ultimate Strength of Materials

Material	Tensile Strength (N/m²)	Compressive Strength (N/m²)
Iron	1.7×10^8	5.5×10^8
Steel	5.0×10^8	5.0×10^8
Aluminum	2.0×10^8	2.0×10^8
Bone	1.2×10^8	1.5×10^8
Marble	—	8.0×10^7
Brick	1×10^6	3.5×10^7
Concrete	2×10^6	2×10^7

EXAMPLE 9.2   Explosive Bolts

Goal   Calculate the maximum shear stress supported by a set of bolts.

Problem   Until launch, rockets are generally held to the launch pad by explosive
bolts. Such bolts are also used in escape hatches and to secure different stages of
the rocket, external tanks, and strap-on boosters, allowing rapid release when a
part needs to be separated from the rest of the vehicle. Suppose a rocket has a
strap-on booster supported by eight horizontal steel bolts, each 9.00 cm in diameter and oriented horizontally. (Bolts similar to these would be used in rockets like
the Titan IV, shown at right.) **(a)** What maximum load can be placed on these
bolts before they are sheared off? Assume the load is shared equally by the eight
bolts. The ultimate shear strength of steel is 2.50×10^8 Pa. **(b)** If the booster has a
mass of 3.00×10^5 kg, calculate the shear deformation of one of the bolts if the
length of the bolt between rocket and booster is 8.00 cm.

NASA

The Titan IV launch vehicle, with its
two solid rocket boosters, is capable
of placing large satellites in geosynchronous orbit.

Strategy   (a) The total force required to shear off the bolts increases with the
number of bolts, but the necessary shear stress does not. Set the ultimate shear
strength equal to the shear stress and solve for the force, multiplying the answer
by eight to find the total shear force that can be applied. Part (b) can be solved by
substituting values into Equation 9.4, which relates shear stress to shear strain.

Solution
(a) Find the maximum load the bolts can support.

Set the shear stress for one bolt equal to its ultimate
shear strength:

$$\frac{F_1}{A} = \frac{F_1}{\pi r^2} = 2.50 \times 10^8\ \text{N/m}^2$$

Solve for the force needed to shear off one bolt:

$$F_1 = \pi r^2 (2.50 \times 10^8 \text{ Pa})$$
$$= \pi (4.50 \times 10^{-2} \text{ m})^2 (2.50 \times 10^8 \text{ Pa})$$
$$= 1.59 \times 10^6 \text{ N}$$

Multiply this result by eight to get the total shear force that the eight bolts can support:

$$F_{tot} = 1.27 \times 10^7 \text{ N}$$

(b) Calculate the shear deformation of one bolt at half the maximum load.

The bolts must support the booster against the force of gravity. The reaction to this force is the force exerted by the booster on the bolts, equal in magnitude to the weight of the booster.

$$F_{tot} = mg = (3.00 \times 10^5 \text{ kg})(9.80 \text{ m/s}^2) = 2.94 \times 10^6 \text{ N}$$

This force is shared among eight bolts. Use Equation 9.4, taking $F = F_{tot}/8$ and solving for Δx:

$$\frac{F}{A} = S \frac{\Delta x}{h}$$

$$\Delta x = \frac{(F_{tot}/8)h}{AS} = \frac{(3.68 \times 10^5 \text{ N})(0.080 \text{ m})}{\pi (4.50 \times 10^{-2} \text{ m})^2 (8.40 \times 10^{10} \text{ Pa})}$$

$$= 5.51 \times 10^{-5} \text{ m}$$

Remarks   The Titan IV launch vehicle, which is capable of putting large satellites into geosynchronous orbit, has a pair of strap-on boosters similar to this one. Notice that the bolts in this example are capable of supporting about four times as much weight as needed. Because all materials either have or develop microscopic defects, a safety factor has to be built in, so structures are designed to tolerate several times the maximum stresses they are expected to undergo.

Exercise 9.2
Calculate the diameter of a single steel horizontal bolt if it is expected to support a maximum load having a mass of 2.00×10^3 kg, but for safety reasons must be designed to support three times that load.

Answer   1.73 cm

EXAMPLE 9.3   Stressing a Lead Ball

Goal   Apply the concepts of bulk stress and strain.

Problem   A solid lead sphere of volume 0.50 m³, dropped in the ocean, sinks to a depth of 2.0×10^3 m (about 1 mile), where the pressure increases by 2.0×10^7 Pa. Lead has a bulk modulus of 4.2×10^{10} Pa. What is the change in volume of the sphere?

Strategy   Solve Equation 9.5 for ΔV and substitute the given quantities.

Solution
Start with the definition of bulk modulus:

$$B = -\frac{\Delta P}{\Delta V/V}$$

Solve for ΔV:

$$\Delta V = -\frac{V \Delta P}{B}$$

Substitute the known values:

$$\Delta V = -\frac{(0.50 \text{ m}^3)(2.0 \times 10^7 \text{ Pa})}{4.2 \times 10^{10} \text{ Pa}} = -2.4 \times 10^{-4} \text{ m}^3$$

Remarks   The negative sign indicates a *decrease* in volume. The following exercise shows that even water can be compressed, though not by much, despite the depth.

Exercise 9.3

(a) By what percentage does a similar globe of water shrink at that same depth? (b) What is the ratio of the new radius to the initial radius?

Answer (a) 0.95% (b) 0.997

Arches and the Ultimate Strength of Materials

As we have seen, the ultimate strength of a material is the maximum force per unit area the material can withstand before it breaks or fractures. Such values are of great importance, particularly in the construction of buildings, bridges, and roads. Table 9.2 gives the ultimate strength of a variety of materials under both tension and compression. Note that bone and a variety of building materials (concrete, brick, and marble) are stronger under compression than under tension. The greater ability of brick and stone to resist compression is the basis of the semicircular arch, developed and used extensively by the Romans in everything from memorial arches to expansive temples and aqueduct supports.

Before the development of the arch, the principal method of spanning a space was the simple post-and-beam construction (Fig. 9.7a), in which a horizontal beam is supported by two columns. This type of construction was used to build the great Greek temples. The columns of these temples were closely spaced because of the limited length of available stones and the low ultimate tensile strength of a sagging stone beam.

The semicircular arch (Fig. 9.7b) developed by the Romans was a great technological achievement in architectural design. It effectively allowed the heavy load of a wide roof span to be channeled into horizontal and vertical forces on narrow supporting columns. The stability of this arch depends on the compression between its wedge-shaped stones. The stones are forced to squeeze against each other by the uniform loading, as shown in the figure. This compression results in horizontal outward forces at the base of the arch where it starts curving away from the vertical. These forces must then be balanced by the stone walls shown on the sides of the arch. It's common to use very heavy walls (buttresses) on either side of the arch to provide horizontal stability. If the foundation of the arch should move, the compressive forces between the wedge-shaped stones may decrease to the extent that the arch collapses. The stone surfaces used in the arches constructed by the Romans were cut to make very tight joints; mortar was usually not used. The resistance to slipping between stones was provided by the compression force and the friction between the stone faces.

APPLICATION

Arch Structures in Buildings

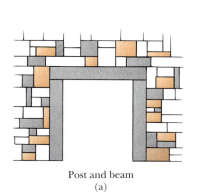

Post and beam
(a)

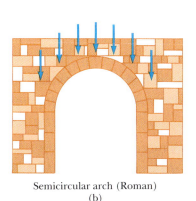

Semicircular arch (Roman)
(b)

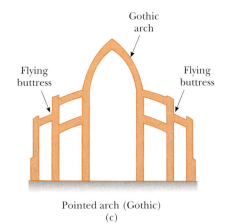

Pointed arch (Gothic)
(c)

Figure 9.7 (a) A simple post-and-beam structure. (b) The semicircular arch developed by the Romans. (c) Gothic arch with flying buttresses to provide lateral support.

Another important architectural innovation was the pointed Gothic arch, shown in Figure 9.7c. This type of structure was first used in Europe beginning in the 12th century, followed by the construction of several magnificent Gothic cathedrals in France in the 13th century. One of the most striking features of these cathedrals is their extreme height. For example, the cathedral at Chartres rises to 118 ft, and the one at Reims has a height of 137 ft. Such magnificent buildings evolved over a very short time, without the benefit of any mathematical theory of structures. However, Gothic arches required flying buttresses to prevent the spreading of the arch supported by the tall, narrow columns.

9.3 DENSITY AND PRESSURE

Equal masses of aluminum and gold have an important physical difference: The aluminum takes up over seven times as much space as the gold. While the reasons for the difference lie at the atomic and nuclear levels, a simple measure of this difference is the concept of *density*.

Density ▶

> The **density** ρ of an object having uniform composition is defined as its mass M divided by its volume V:
>
> $$\rho \equiv \frac{M}{V} \qquad\qquad [9.6]$$
>
> **SI unit: kilogram per meter cubed (kg/m³)**

The most common units used for density are kilograms per cubic meter in the SI system and grams per cubic centimeter in the cgs system. Table 9.3 lists the densities of some substances. The densities of most liquids and solids vary slightly with changes in temperature and pressure; the densities of gases vary greatly with such changes. Under normal conditions, the densities of solids and liquids are about 1 000 times greater than the densities of gases. This difference implies that the average spacing between molecules in a gas under such conditions is about ten times greater than in a solid or liquid.

The **specific gravity** of a substance is the ratio of its density to the density of water at 4°C, which is 1.0×10^3 kg/m³. (The size of the kilogram was originally defined to make the density of water 1.0×10^3 kg/m³ at 4°C.) By definition, specific gravity is a dimensionless quantity. For example, if the specific gravity of a substance is 3.0, its density is $3.0(1.0 \times 10^3 \text{ kg/m}^3) = 3.0 \times 10^3$ kg/m³.

TABLE 9.3

Densities of Some Common Substances

Substance	$\rho(\text{kg/m}^3)^{\text{a}}$	Substance	$\rho(\text{kg/m}^3)^{\text{a}}$
Ice	0.917×10^3	Water	1.00×10^3
Aluminum	2.70×10^3	Glycerin	1.26×10^3
Iron	7.86×10^3	Ethyl alcohol	0.806×10^3
Copper	8.92×10^3	Benzene	0.879×10^3
Silver	10.5×10^3	Mercury	13.6×10^3
Lead	11.3×10^3	Air	1.29
Gold	19.3×10^3	Oxygen	1.43
Platinum	21.4×10^3	Hydrogen	8.99×10^{-2}
Uranium	18.7×10^3	Helium	1.79×10^{-1}

ᵃAll values are at standard atmospheric temperature and pressure (STP), defined as 0°C (273 K) and 1 atm $(1.013 \times 10^5$ Pa). To convert to grams per cubic centimeter, multiply by 10^{-3}.

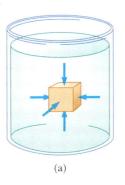

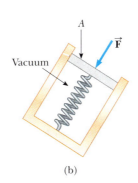

FIGURE 9.8 (a) The force exerted by a fluid on a submerged object at any point is perpendicular to the surface of the object. The force exerted by the fluid on the walls of the container is perpendicular to the walls at all points and increases with depth. (b) A simple device for measuring pressure in a fluid.

(a) (b)

Quick Quiz 9.1

Suppose you have one cubic meter of gold, two cubic meters of silver, and six cubic meters of aluminum. Rank them by mass, from smallest to largest. (a) gold, aluminum, silver (b) gold, silver, aluminum (c) aluminum, gold, silver (d) silver, aluminum, gold

Fluids don't sustain shearing stresses, so the only stress that a fluid can exert on a submerged object is one that tends to compress it, which is bulk stress. The force exerted by the fluid on the object is always perpendicular to the surfaces of the object, as shown in Figure 9.8a.

The pressure at a specific point in a fluid can be measured with the device pictured in Figure 9.8b: an evacuated cylinder enclosing a light piston connected to a spring that has been previously calibrated with known weights. As the device is submerged in a fluid, the fluid presses down on the top of the piston and compresses the spring until the inward force exerted by the fluid is balanced by the outward force exerted by the spring. Let F be the magnitude of the force on the piston and A the area of the top surface of the piston. Notice that the force that compresses the spring is spread out over the entire area, motivating our formal definition of pressure:

> If F is the magnitude of a force exerted perpendicular to a given surface of area A, then the pressure P is the force divided by the area:
>
> $$P \equiv \frac{F}{A} \qquad [9.7]$$
>
> **SI unit: pascal (Pa)**

Because pressure is defined as force per unit area, it has units of pascals (newtons per square meter). The English customary unit for pressure is the pound per inch squared. Atmospheric pressure at sea level is 14.7 lb/in², which in SI units is 1.01×10^5 Pa.

As we see from Equation 9.7, the effect of a given force depends critically on the area to which it's applied. A 700-N man can stand on a vinyl-covered floor in regular street shoes without damaging the surface, but if he wears golf shoes, the metal cleats protruding from the soles can do considerable damage to the floor. With the cleats, the same force is concentrated into a smaller area, greatly elevating the pressure in those areas, resulting in a greater likelihood of exceeding the ultimate strength of the floor material.

Snowshoes use the same principle (Fig. 9.9). The snow exerts an upward normal force on the shoes to support the person's weight. According to Newton's third law, this upward force is accompanied by a downward force exerted by the shoes on the snow. If the person is wearing snowshoes, that force is distributed over the very large area of each snowshoe, so that the pressure at any given point is relatively low and the person doesn't penetrate very deeply into the snow.

TIP 9.1 Force and Pressure

Equation 9.7 makes a clear distinction between force and pressure. Another important distinction is that *force is a vector* and *pressure is a scalar*. There is no direction associated with pressure, but the direction of the force associated with the pressure is perpendicular to the surface of interest.

◄ Average pressure

© Royalty-Free/Corbis

Figure 9.9 Snowshoes prevent the person from sinking into the soft snow because the force on the snow is spread over a larger area, reducing the pressure on the snow's surface.

Applying Physics 9.1   Bed of Nails Trick

After an exciting but exhausting lecture, a physics professor stretches out for a nap on a bed of nails, as in Figure 9.10, suffering no injury and only moderate discomfort. How is this possible?

Explanation   If you try to support your entire weight on a single nail, the pressure on your body is your weight divided by the very small area of the end of the nail. The resulting pressure is large enough to penetrate the skin. If you distribute your weight over several hundred nails, however, as demonstrated by the professor, the pressure is considerably reduced because the area that supports your weight is the total area of all nails in contact with your body. (Why is lying on a bed of nails more comfortable than sitting on the same bed? Extend the logic to show that it

would be more uncomfortable yet to stand on a bed of nails without shoes.)

Figure 9.10   (Applying Physics 9.1) Does anyone have a pillow?

EXAMPLE 9.4   The Water Bed

Goal   Calculate a density and a pressure from a weight.

Problem   A water bed is 2.00 m on a side and 30.0 cm deep. **(a)** Find its weight. **(b)** Find the pressure that the water bed exerts on the floor. Assume that the entire lower surface of the bed makes contact with the floor.

Strategy   Density is mass per unit volume: first, find the volume of the bed and multiply it by the density of water to get the bed's mass. Multiplying by the acceleration of gravity then gives the weight of the bed. The weight divided by the area of floor the bed rests upon gives the pressure exerted on the floor.

Solution
(a) Find the weight of the water bed.

First, find the volume of the bed:

$$V = lwh = (2.00 \text{ m})(2.00 \text{ m})(0.300 \text{ m}) = 1.20 \text{ m}^3$$

Solve the density equation for the mass and substitute, then multiply the result by g to get the weight:

$$\rho = \frac{M}{V}$$

$$M = \rho V = (1.00 \times 10^3 \text{ kg/m}^3)(1.20 \text{ m}^3) = 1.20 \times 10^3 \text{ kg}$$

$$w = Mg = (1.20 \times 10^3 \text{ kg})(9.80 \text{ m/s}^2) = \boxed{1.18 \times 10^4 \text{ N}}$$

(b) Find the pressure that the bed exerts on the floor.

Use the cross-sectional area $A = 4.00 \text{ m}^2$ and the value of w from part (a) to get the pressure:

$$P = \frac{F}{A} = \frac{w}{A} = \frac{1.18 \times 10^4 \text{ N}}{4.00 \text{ m}^2} = \boxed{2.95 \times 10^3 \text{ Pa}}$$

Remarks   Notice that the answer to part (b) is far less than atmospheric pressure. Water is heavier than air for a given volume, but the air is stacked up considerably higher (100 km!). The total pressure exerted on the floor would include the pressure of the atmosphere.

Exercise 9.4
Calculate the pressure exerted by the water bed on the floor if the bed rests on its side.

Answer $1.97 \times 10^4 \text{ Pa}$

9.4   VARIATION OF PRESSURE WITH DEPTH

When a fluid is at rest in a container, **all portions of the fluid must be in static equilibrium**—at rest with respect to the observer. Furthermore, **all points at the same depth must be at the same pressure**. If this were not the case, fluid would flow from the higher pressure region to the lower pressure region. For example, consider the small block of fluid shown in Figure 9.11a. If the pressure were greater on the left side of the block than on the right, $\vec{F}_1$ would be greater than $\vec{F}_2$, and the block would accelerate to the right and thus would not be in equilibrium.

Next, let's examine the fluid contained within the volume indicated by the darker region in Figure 9.11b. This region has cross-sectional area A and extends from position y_1 to position y_2 below the surface of the liquid. Three external forces act on this volume of fluid: the force of gravity, Mg; the upward force P_2A exerted by the liquid below it; and a downward force P_1A exerted by the fluid above it. Because the given volume of fluid is in equilibrium, these forces must add to zero, so we get

$$P_2A - P_1A - Mg = 0 \qquad \text{[9.8]}$$

From the definition of density, we have

$$M = \rho V = \rho A(y_1 - y_2) \qquad \text{[9.9]}$$

Substituting Equation 9.9 into Equation 9.8, canceling the area A, and rearranging terms, we get

$$P_2 = P_1 + \rho g(y_1 - y_2) \qquad \text{[9.10]}$$

Notice that $(y_1 - y_2)$ is positive, because $y_2 < y_1$. The force P_2A is greater than the force P_1A by exactly the weight of water between the two points. This is the same principle experienced by the person at the bottom of a pileup in football or rugby.

Atmospheric pressure is also caused by a piling up of fluid—in this case, the fluid is the gas of the atmosphere. The weight of all the air from sea level to the edge of space results in an atmospheric pressure of $P_0 = 1.013 \times 10^5$ Pa (equivalent to 14.7 lb/in.²) at sea level. This result can be adapted to find the pressure P at any depth $h = (y_1 - y_2) = (0 - y_2)$ below the surface of the water:

$$P = P_0 + \rho g h \qquad \text{[9.11]}$$

According to Equation 9.11, **the pressure P at a depth h below the surface of a liquid open to the atmosphere is greater than atmospheric pressure by the amount $\rho g h$**. Moreover, the pressure isn't affected by the shape of the vessel, as shown in Figure 9.12.

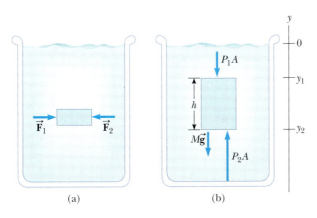

(a)   (b)

Figure 9.11   (a) If the block of fluid is to be in equilibrium, the force $\vec{F}_1$ must balance the force $\vec{F}_2$. (b) The net force on the volume of liquid within the darker region must be zero.

Courtesy of Central Scientific Company

Figure 9.12   This photograph illustrates the fact that the pressure in a liquid is the same at all points lying at the same elevation. For example, the pressure is the same at points A, B, C, and D. Note that the shape of the vessel does not affect the pressure.

The pressure at the bottom of a glass filled with water (ρ = 1 000 kg/m³) is P. The water is poured out and the glass is filled with ethyl alcohol (ρ = 806 kg/m³). The pressure at the bottom of the glass is now (a) smaller than P (b) equal to P (c) larger than P (d) indeterminate.

EXAMPLE 9.5   Oil and Water

Goal   Calculate pressures created by layers of different fluids.

Problem   In a huge oil tanker, salt water has flooded an oil tank to a depth of 5.00 m. On top of the water is a layer of oil 8.00 m deep, as in the cross-sectional view of the tank in Figure 9.13. The oil has a density of 0.700 g/cm³. Find the pressure at the bottom of the tank. (Take 1 025 kg/m³ as the density of salt water.)

Strategy   Equation 9.11 must be used twice. First, use it to calculate the pressure P_1 at the bottom of the oil layer. Then use this pressure in place of P_0 in Equation 9.11 and calculate the pressure P_{bot} at the bottom of the water layer.

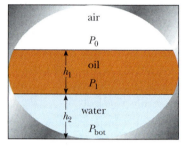

Figure 9.13   (Example 9.5)

Solution

Use Equation 9.11 to calculate the pressure at the bottom of the oil layer:

$$P_1 = P_0 + \rho g h_1 \tag{1}$$
$$= 1.01 \times 10^5 \text{ Pa}$$
$$+ (7.00 \times 10^2 \text{ kg/m}^3)(9.80 \text{ m/s}^2)(8.00 \text{ m})$$
$$P_1 = 1.56 \times 10^5 \text{ Pa}$$

Now adapt Equation 9.11 to the new starting pressure, and use it to calculate the pressure at the bottom of the water layer:

$$P_{bot} = P_1 + \rho g h_2 \tag{2}$$
$$= 1.56 \times 10^5 \text{ Pa}$$
$$+ (1.025 \times 10^3 \text{ kg/m}^3)(9.80 \text{ m/s}^2)(5.00 \text{ m})$$
$$P_{bot} = \boxed{2.06 \times 10^5 \text{ Pa}}$$

Remark   The weight of the atmosphere results in P_0 at the surface of the oil layer. Then the weight of the oil and the weight of the water combine to create the pressure at the bottom.

Exercise 9.5

Calculate the pressure on the top lid of a chest buried under 4.00 meters of mud with density 1.75×10^3 kg/m³ at the bottom of a 10.0-m-deep lake.

Answer 2.68×10^5 Pa

EXAMPLE 9.6   A Pain in the Ear

Goal   Calculate a pressure difference at a given depth, and estimate a force.

Problem   Estimate the net force exerted on your eardrum due to the water above when you are swimming at the bottom of a pool that is 5.0 m deep.

Strategy   Use Equation 9.11 to find the pressure difference across the eardrum at the given depth. The air inside the ear is generally at atmospheric pressure. Estimate the eardrum's surface area, then use the definition of pressure to get the net force exerted on the eardrum.

Solution

Use Equation 9.11 to calculate the difference between the water pressure at the depth h and the pressure inside the ear:

$$\Delta P = P - P_0 = \rho g h$$
$$= (1.00 \times 10^3 \text{ kg/m}^3)(9.80 \text{ m/s}^2)(5.0 \text{ m})$$
$$= 4.9 \times 10^4 \text{ Pa}$$

Mutliply by area A to get the net force on the eardrum associated with this pressure difference, estimating the area of the eardrum as 1 cm².

$$F_{net} = A\Delta P \approx (1 \times 10^{-4} \text{ m}^2)(4.9 \times 10^4 \text{ Pa}) \approx \boxed{5 \text{ N}}$$

Remarks Because a force on the eardrum of this magnitude is uncomfortable, swimmers often "pop their ears" by swallowing or expanding their jaws while underwater, an action that pushes air from the lungs into the middle ear. Using this technique equalizes the pressure on the two sides of the eardrum and relieves the discomfort.

Exercise 9.6
An airplane takes off at sea level and climbs to a height of 425 m. Estimate the net outward force on a passenger's eardrum assuming the density of air is approximately constant at 1.3 kg/m^3 and that the inner ear pressure hasn't been equalized.

Answer   0.54 N

In view of the fact that the pressure in a fluid depends on depth and on the value of P_0, any increase in pressure at the surface must be transmitted to every point in the fluid. This was first recognized by the French scientist Blaise Pascal (1623–1662) and is called **Pascal's principle**:

> A change in pressure applied to an enclosed fluid is transmitted undiminished to every point of the fluid and to the walls of the container.

An important application of Pascal's principle is the hydraulic press (Fig. 9.14a). A downward force $\vec{F}_1$ is applied to a small piston of area A_1. The pressure is transmitted through a fluid to a larger piston of area A_2. As the pistons move and the fluids in the left and right cylinders change their relative heights, there are slight differences in the pressures at the input and output pistons. Neglecting these small differences, the fluid pressure on each of the pistons may be taken to be the same; $P_1 = P_2$. From the definition of pressure, it then follows that $F_1/A_1 = F_2/A_2$. Therefore, the magnitude of the force $\vec{F}_2$ is larger than the magnitude of $\vec{F}_1$ by the factor A_2/A_1. That's why a large load, such as a car, can be moved on the large piston by a much smaller force on the smaller piston. Hydraulic brakes, car lifts, hydraulic jacks, forklifts, and other machines make use of this principle.

APPLICATION

Hydraulic Lifts

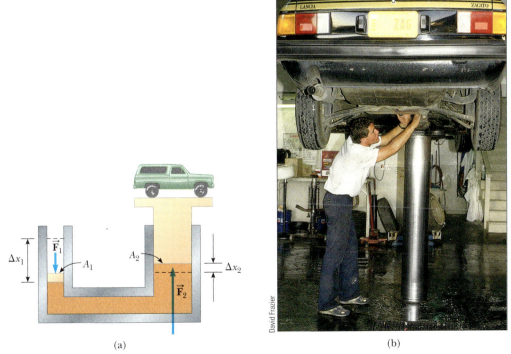

(a)

(b)

Figure 9.14   (a) Diagram of a hydraulic press (Example 9.7). Because the pressure is the same at the left and right sides, a small force $\vec{F}_1$ at the left produces a much larger force $\vec{F}_2$ at the right. (b) A vehicle under repair is supported by a hydraulic lift in a garage.

INTERACTIVE EXAMPLE 9.7   The Car Lift

Goal   Apply Pascal's principle to a car lift, and show that the input work is the same as the output work.

Problem   In a car lift used in a service station, compressed air exerts a force on a small piston of circular cross section having a radius of $r_1 = 5.00$ cm. This pressure is transmitted by an incompressible liquid to a second piston of radius $r_2 = 15.0$ cm. **(a)** What force must the compressed air exert on the small piston in order to lift a car weighing 13 300 N? Neglect the weights of the pistons. **(b)** What air pressure will produce a force of that magnitude? **(c)** Show that the work done by the input and output pistons is the same.

Strategy   Substitute into Pascal's principle in part (a), while recognizing that the magnitude of the output force, F_2, must be equal to the car's weight in order to support it. Use the definition of pressure in part (b). In part (c), use $W = F\Delta x$ to find the ratio W_1/W_2, showing that it must equal 1. This requires combining Pascal's principle with the fact that the input and output pistons move through the same volume.

Solution
(a) Find the necessary force on the small piston.

Substitute known values into Pascal's principle, using $A = \pi r^2$ for the area.

$$F_1 = \left(\frac{A_1}{A_2}\right)F_2 = \frac{\pi r_1^2}{\pi r_2^2} F_2$$

$$= \frac{\pi(5.00 \times 10^{-2}\,\text{m})^2}{\pi(15.0 \times 10^{-2}\,\text{m})^2}\,(1.33 \times 10^4\,\text{N})$$

$$= \boxed{1.48 \times 10^3\,\text{N}}$$

(b) Find the air pressure producing F_1.

Substitute into the definition of pressure:

$$P = \frac{F_1}{A_1} = \frac{1.48 \times 10^3\,\text{N}}{\pi(5.00 \times 10^{-2}\,\text{m})^2} = \boxed{1.88 \times 10^5\,\text{Pa}}$$

(c) Show that the work done by the input and output pistons is the same.

First equate the volumes, and solve for the ratio of A_2 to A_1:

$$V_1 = V_2 \quad \rightarrow \quad A_1\Delta x_1 = A_2\Delta x_2$$

$$\frac{A_2}{A_1} = \frac{\Delta x_1}{\Delta x_2}$$

Now use Pascal's principle to get a relationship for F_1/F_2:

$$\frac{F_1}{A_1} = \frac{F_2}{A_2} \quad \rightarrow \quad \frac{F_1}{F_2} = \frac{A_1}{A_2}$$

Evaluate the work ratio, substituting the preceding two results:

$$\frac{W_1}{W_2} = \frac{F_1\Delta x_1}{F_2\Delta x_2} = \left(\frac{F_1}{F_2}\right)\left(\frac{\Delta x_1}{\Delta x_2}\right) = \left(\frac{A_1}{A_2}\right)\left(\frac{A_2}{A_1}\right) = 1$$

$$W_1 = W_2$$

Remark   In this problem, we didn't address the effect of possible differences in the heights of the pistons. If the column of fluid is higher in the small piston, the fluid weight assists in supporting the car, reducing the necessary applied force. If the column of fluid is higher in the large piston, both the car and the extra fluid must be supported, so additional applied force is required.

Exercise 9.7
A hydraulic lift has pistons with diameters 8.00 cm and 36.0 cm, respectively. If a force of 825 N is exerted at the input piston, what maximum mass can be lifted by the output piston?

Answer 1.70×10^3 kg

Physics⊗Now™   You can adjust the weight of the truck in Figure 9.14a by logging into PhysicsNow at **www.cp7e.com** and going to Interactive Example 9.7.

Applying Physics 9.2   Building the Pyramids

A corollary to the statement that pressure in a fluid increases with depth is that water always seeks its own level. This means that if a vessel is filled with water, then regardless of the vessel's shape the surface of the water is perfectly flat and at the same height at all points. The ancient Egyptians used this fact to make the pyramids level. Devise a scheme showing how this could be done.

Explanation There are many ways it could be done, but Figure 9.15 shows the scheme used by the Egyptians. The builders cut grooves in the base of the pyramid as in (a) and partially filled the grooves with

water. The height of the water was marked as in (b), and the rock was chiseled down to the mark, as in (c). Finally, the groove was filled with crushed rock and gravel, as in (d).

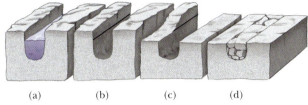

(a) (b) (c) (d)

Figure 9.15 (Applying Physics 9.2)

9.5 PRESSURE MEASUREMENTS

A simple device for measuring pressure is the open-tube manometer (Fig. 9.16a). One end of a U-shaped tube containing a liquid is open to the atmosphere, and the other end is connected to a system of unknown pressure P. The pressure at point B equals $P_0 + \rho g h$, where ρ is the density of the fluid. The pressure at B, however, equals the pressure at A, which is also the unknown pressure P. We conclude that $P = P_0 + \rho g h$.

The pressure P is called the **absolute pressure**, and $P - P_0$ is called the **gauge pressure**. If P in the system is greater than atmospheric pressure, h is positive. If P is less than atmospheric pressure (a partial vacuum), h is negative, meaning that the right-hand column in Figure 9.16a is lower than the left-hand column.

Another instrument used to measure pressure is the **barometer** (Fig. 9.16b), invented by Evangelista Torricelli (1608–1647). A long tube closed at one end is filled with mercury and then inverted into a dish of mercury. The closed end of the tube is nearly a vacuum, so its pressure can be taken to be zero. It follows that $P_0 = \rho g h$, where ρ is the density of the mercury and h is the height of the mercury column. Note that the barometer measures the pressure of the atmosphere, whereas the manometer measures pressure in an enclosed fluid.

One atmosphere of pressure is defined to be the pressure equivalent of a column of mercury that is exactly 0.76 m in height at 0°C with $g = 9.806\ 65\ \text{m/s}^2$. At this temperature, mercury has a density of $13.595 \times 10^3\ \text{kg/m}^3$; therefore,

$$P_0 = \rho g h = (13.595 \times 10^3\ \text{kg/m}^3)(9.806\ 65\ \text{m/s}^2)(0.760\ 0\ \text{m})$$
$$= 1.013 \times 10^5\ \text{Pa} = 1\ \text{atm}$$

It is interesting to note that the force of the atmosphere on our bodies (assuming a body area of 2 000 in²) is extremely large, on the order of 30 000 lb! If it were not for the fluids permeating our tissues and body cavities, our bodies would collapse. The fluids provide equal and opposite forces. In the upper atmosphere or in space, sudden decompression can lead to serious injury and death. Air retained in the lungs can damage the tiny alveolar sacs, and intestinal gas can even rupture internal organs.

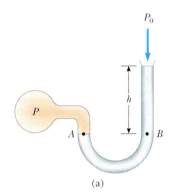

(a)

(b)

Figure 9.16 Two devices for measuring pressure: (a) an open-tube manometer and (b) a mercury barometer.

Quick Quiz 9.3

Several common barometers are built using a variety of fluids. For which fluid will the column of fluid in the barometer be the highest? (Refer to Table 9.3.) (a) mercury (b) water (c) ethyl alcohol (d) benzene.

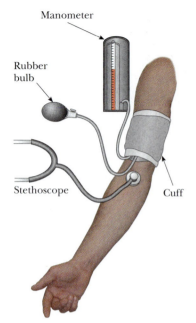

Manometer

Rubber
bulb

Stethoscope   Cuff

Figure 9.17 A sphygmomanometer can be used to measure blood pressure.

Blood Pressure Measurements

A specialized manometer (called a sphygmomanometer) is often used to measure blood pressure. In this application, a rubber bulb forces air into a cuff wrapped tightly around the upper arm and simultaneously into a manometer, as in Figure 9.17. The pressure in the cuff is increased until the flow of blood through the brachial artery in the arm is stopped. A valve on the bulb is then opened, and the measurer listens with a stethoscope to the artery at a point just below the cuff. When the pressure in the cuff and brachial artery is just below the maximum value produced by the heart (the systolic pressure), the artery opens momentarily on each beat of the heart. At this point, the velocity of the blood is high and turbulent, and the flow is noisy and can be heard with the stethoscope. The manometer is calibrated to read the pressure in millimeters of mercury, and the value obtained is about 120 mm for a normal heart. Values of 130 mm or above are considered high, and medication to lower the blood pressure is often prescribed for such patients. As the pressure in the cuff is lowered further, intermittent sounds are still heard until the pressure falls just below the minimum heart pressure (the diastolic pressure). At this point, continuous sounds are heard. In the normal heart, this transition occurs at about 80 mm of mercury, and values above 90 require medical intervention. Blood pressure readings are usually expressed as the ratio of the systolic pressure to the diastolic pressure, which is 120/80 for a healthy heart.

Quick Quiz 9.4

Blood pressure is normally measured with the cuff of the sphygmomanometer around the arm. Suppose that the blood pressure is measured with the cuff around the calf of the leg of a standing person. Would the reading of the blood pressure be (a) the same here as it is for the arm? (b) greater than it is for the arm? or (c) less than it is for the arm?

Applying Physics 9.3   Ballpoint Pens

In a ballpoint pen, ink moves down a tube to the tip, where it is spread on a sheet of paper by a rolling stainless steel ball. Near the top of the ink cartridge, there is a small hole open to the atmosphere. If you seal this hole, you will find that the pen no longer functions. Use your knowledge of how a barometer works to explain this behavior.

Explanation If the hole is sealed, or if it were not present, the pressure of the air above the ink would decrease as the ink gets used. Consequently, atmospheric pressure exerted against the ink at the bottom of the cartridge would prevent some of the ink from flowing out. The hole allows the pressure above the ink to remain at atmospheric pressure. Why does a ballpoint pen seem to run out of ink when you write on a vertical surface?

9.6 BUOYANT FORCES AND ARCHIMEDES'S PRINCIPLE

A fundamental principle affecting objects submerged in fluids was discovered by the Greek mathematician and natural philosopher Archimedes. **Archimedes's principle** can be stated as follows:

Archimedes's principle ▶

> Any object completely or partially submerged in a fluid is buoyed up by a force with magnitude equal to the weight of the fluid displaced by the object.

Many historians attribute the concept of buoyancy to Archimedes's "bathtub epiphany," when he noticed an apparent change in his weight upon lowering himself into a tub of water. As will be seen in Example 9.8, buoyancy yields a method of determining density.

Everyone has experienced Archimedes's principle. It's relatively easy, for example, to lift someone if you're both standing in a swimming pool, whereas lifting that same individual on dry land may be a difficult task. Water provides partial support to any object placed in it. We often say that an object placed in a fluid is buoyed up by the fluid, so we call this upward force the **buoyant force**.

The buoyant force is *not* a mysterious new force that arises in fluids. In fact, the physical cause of the buoyant force is the pressure difference between the upper and lower sides of the object, which can be shown to be equal to the weight of the displaced fluid. In Figure 9.18a, the fluid inside the indicated sphere, colored darker blue, is pressed on all sides by the surrounding fluid. Arrows indicate the forces arising from the pressure. Because pressure increases with depth, the arrows on the underside are larger than those on top. Adding them all up, the horizontal components cancel, but there is a net force upwards. This force, due to differences in pressure, is the buoyant force $\vec{\mathbf{B}}$. The sphere of water neither rises nor falls, so the vector sum of the buoyant force and the force of gravity on the sphere of fluid must be zero, and it follows that $B = Mg$, where M is the mass of the fluid.

Replacing the shaded fluid with a bowling ball of the same volume, as in Figure 9.18b, changes only the mass on which the pressure acts, so the buoyant force is the same: $B = Mg$, where M is the mass of the displaced fluid, *not* the mass of the bowling ball. The force of gravity on the heavier ball is greater than it was on the fluid, so the bowling ball sinks.

Archimedes's principle can also be obtained from Equation 9.8, relating pressure and depth, using Figure 9.11b. Horizontal forces from the pressure cancel, but in the vertical direction $P_2 A$ acts upwards on the bottom of the block of fluid and $P_1 A$ and the gravity force on the fluid, Mg, act downwards, giving

$$B = P_2 A - P_1 A = Mg \qquad \text{[9.12a]}$$

where the buoyancy force has been identified as a difference in pressure equal in magnitude to the weight of the displaced fluid. This buoyancy force remains the same regardless of the material occupying the volume in question because it's due to the *surrounding* fluid. Using the definition of density, Equation 9.12a becomes

$$B = \rho_{\text{fluid}} V_{\text{fluid}} g \qquad \text{[9.12b]}$$

where ρ_{fluid} is the density of the fluid and V_{fluid} is the volume of the displaced fluid. This result applies equally to all shapes, because any irregular shape can be approximated by a large number of infinitesimal cubes.

It's instructive to compare the forces on a totally submerged object with those on a floating object.

Case I: A Totally Submerged Object.   When an object is *totally* submerged in a fluid of density ρ_{fluid}, the upward buoyant force acting on the object has a magnitude of $B = \rho_{\text{fluid}} V_{\text{obj}} g$, where V_{obj} is the volume of the object. If the object has density ρ_{obj}, the downward gravitational force acting on the object has a magni-

North Wind Picture Archives

ARCHIMEDES: Greek Mathematician, Physicist, and Engineer (287–212 B.C.)

Archimedes was perhaps the greatest scientist of antiquity. He is well known for discovering the nature of the buoyant force and was a gifted inventor. According to legend, Archimedes was asked by King Hieron to determine whether the king's crown was made of pure gold or merely a gold alloy. The task was to be performed without damaging the crown. Archimedes allegedly arrived at a solution while taking a bath, noting a partial loss of weight after submerging his arms and legs in the water. As the story goes, he was so excited about his great discovery that he ran naked through the streets of Syracuse, shouting, "Eureka!" which is Greek for "I have found it."

TIP 9.2   Buoyant Force is Exerted by the Fluid

The buoyant force on an object is exerted by the fluid and is the same, regardless of the density of the object. Objects more dense than the fluid sink; objects less dense rise.

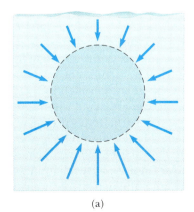

(a)   (b)

Figure 9.18   (a) The arrows indicate forces on the sphere of fluid due to pressure, larger on the underside because pressure increases with depth. The net upward force is the buoyant force. (b) The buoyant force, which is caused by the *surrounding* fluid, is the same on any object of the same volume, including this bowling ball. The magnitude of the buoyant force is equal to the weight of the displaced fluid.

Hot-air balloons. Because hot air is less dense than cold air, there is a net upward force on the balloons.

ACTIVE FIGURE 9.20
An object floating on the surface of a fluid is acted upon by two forces: the gravitational force $\vec{F}_g$ and the buoyant force $\vec{B}$. These two forces are equal in magnitude and opposite in direction.

Physics⊗Now™
Log into PhysicsNow at **www.cp7e.com**, and go to Active Figure 9.20 to change the densities of the object and the fluid and see the results.

APPLICATION
Cerebrospinal Fluid

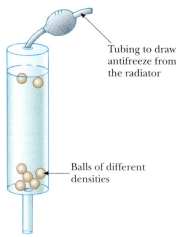

Tubing to draw antifreeze from the radiator

Balls of different densities

Figure 9.21 The number of balls that float in this device is a measure of the density of the antifreeze solution in a vehicle's radiator and, consequently, a measure of the temperature at which freezing will occur.

ACTIVE FIGURE 9.19
(a) A totally submerged object that is less dense than the fluid in which it is submerged is acted upon by a net upward force. (b) A totally submerged object that is denser than the fluid sinks.

Physics⊗Now™
Log into PhysicsNow at **www.cp7e.com**, and go to Active Figure 9.19 to move the object to new positions, as well as change the density of the object, and see the results.

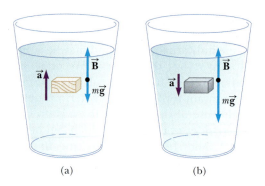

(a)   (b)

tude equal to $w = mg = \rho_{obj}V_{obj}g$, and the net force on it is $B - w = (\rho_{fluid} - \rho_{obj})V_{obj}g$. Therefore, if the density of the object is *less* than the density of the fluid, as in Active Figure 9.19a, the net force exerted on the object is *positive* (upward) and the object accelerates *upward*. If the density of the object is *greater* than the density of the fluid, as in Active Figure 9.19b, the net force is *negative* and the object accelerates *downwards*.

Case II: A Floating Object.   Now consider a partially submerged object in static equilibrium floating in a fluid, as in Active Figure 9.20. In this case, the upward buoyant force is balanced by the downward force of gravity acting on the object. If V_{fluid} is the volume of the fluid displaced by the object (which corresponds to the volume of the part of the object beneath the fluid level), then the magnitude of the buoyant force is given by $B = \rho_{fluid}V_{fluid}g$. Because the weight of the object is $w = mg = \rho_{obj}V_{obj}g$, and because $w = B$, it follows that $\rho_{fluid}V_{fluid}g = \rho_{obj}V_{obj}g$, or

$$\frac{\rho_{obj}}{\rho_{fluid}} = \frac{V_{fluid}}{V_{obj}} \qquad \text{[9.13]}$$

Equation 9.13 neglects the buoyant force of the air, which is slight, because the density of air is only 1.29 kg/m³ at sea level.

Under normal circumstances, the average density of a fish is slightly greater than the density of water, so it would sink if it didn't have a mechanism for adjusting its density. By changing the size of an internal swim bladder, fish maintain neutral buoyancy as they swim to various depths.

The human brain is immersed in a fluid (the cerebrospinal fluid) of density 1 007 kg/m³, which is slightly less than the average density of the brain, 1 040 kg/m³. Consequently, most of the weight of the brain is supported by the buoyant force of the surrounding fluid. In some clinical procedures, a portion of this fluid must be removed for diagnostic purposes. During such procedures, the nerves and blood vessels in the brain are placed under great strain, which in turn can cause extreme discomfort and pain. Great care must be exercised with such patients until the initial volume of brain fluid has been restored by the body.

When service station attendants check the antifreeze in your car or the condition of your battery, they often use devices that apply Archimedes's principle. Figure 9.21 shows a common device that is used to check the antifreeze in a car radiator. The small balls in the enclosed tube vary in density, so that all of them float when the tube is filled with pure water, none float in pure antifreeze, one floats in a 5% mixture, two in a 10% mixture, and so forth. The number of balls that float is a measure of the percentage of antifreeze in the mixture, which in turn is used to determine the lowest temperature the mixture can withstand without freezing.

Similarly, the degree of charge in some car batteries can be determined with a so-called magic-dot process that is built into the battery (Fig. 9.22). Inside a viewing port in the top of the battery, the appearance of an orange dot indicates that the battery is sufficiently charged; a black dot indicates that the battery has lost its charge. If the battery has sufficient charge, the density of the battery fluid is high enough to cause the orange ball to float. As the battery loses its charge, the density

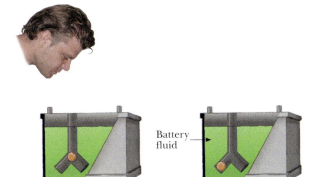

Figure 9.22   The orange ball in the plastic tube inside the battery serves as an indicator of whether the battery is (a) charged or (b) discharged. As the battery loses its charge, the density of the battery fluid decreases, and the ball sinks out of sight.

APPLICATION

Checking the Battery Charge

Charged battery   Discharged battery

Battery fluid

of the battery fluid decreases and the ball sinks beneath the surface of the fluid, leaving the dot to appear black.

Quick Quiz 9.5

Atmospheric pressure varies from day to day. The level of a floating ship on a high-pressure day is (a) higher (b) lower, or (c) no different than on a low-pressure day.

Quick Quiz 9.6

The density of lead is greater than iron, and both metals are denser than water. Is the buoyant force on a solid lead object (a) greater than, (b) equal to, or (c) less than the buoyant force acting on a solid iron object of the same dimensions?

Most of the volume of this iceberg is beneath the water. Can you determine what fraction of the total volume is under water?

EXAMPLE 9.8   A Red-Tag Special on Crowns

Goal   Apply Archimedes's principle to a submerged object.

Problem   A bargain hunter purchases a "gold" crown at a flea market. After she gets home, she hangs it from a scale and finds its weight to be 7.84 N (Fig. 9.23a). She then weighs the crown while it is immersed in water, as in Figure 9.23b, and now the scale reads 6.86 N. Is the crown made of pure gold?

Strategy   The goal is to find the density of the crown and compare it to the density of gold. We already have the weight of the crown in air, so we can get the mass by dividing by the acceleration of gravity. If we can find the volume of the crown, we can obtain the desired density by dividing the mass by this volume.

When the crown is fully immersed, the displaced water is equal to the volume of the crown. This same volume is used in calculating the buoyant force. So our strategy is as follows: (1) Apply Newton's second law to the crown, both in the water and in the air to find the buoyant force. (2) Use the buoyant force to find the crown's volume. (3) Divide the crown's scale weight in air by the acceleration of gravity to get the mass, then by the volume to get the density.

Figure 9.23   (Example 9.8) (a) When the crown is suspended in air, the scale reads $T_{air} = mg$, the crown's true weight. (b) When the crown is immersed in water, the buoyant force $\vec{B}$ reduces the scale reading by the magnitude of the buoyant force, $T_{water} = mg - B$.

Solution

Apply Newton's second law to the crown when it's weighed in air. There are two forces on the crown—gravity $m\vec{g}$ and $\vec{T}_{air}$, the force exerted by the scale on the crown, with magnitude equal to the reading on the scale.

$$T_{air} - mg = 0 \qquad (1)$$

When the crown is immersed in water, the scale force is $\vec{T}_{\text{water}}$, with magnitude equal to the scale reading, and there is an upward buoyant force $\vec{B}$ and the force of gravity.

$$T_{\text{water}} - mg + B = 0 \qquad (2)$$

Solve Equation (1) for mg, substitute into Equation (2), and solve for the buoyant force, which equals the difference in scale readings:

$$T_{\text{water}} - T_{\text{air}} + B = 0$$

$$B = T_{\text{air}} - T_{\text{water}} = 7.84 \text{ N} - 6.86 \text{ N} = 0.980 \text{ N}$$

Find the volume of the displaced water, using the fact that the magnitude of the buoyant force equals the weight of the displaced water:

$$B = \rho_{\text{water}} g V_{\text{water}} = 0.980 \text{ N}$$

$$V_{\text{water}} = \frac{0.980 \text{ N}}{g \rho_{\text{water}}} = \frac{0.980 \text{ N}}{(9.80 \text{ m/s}^2)(1.00 \times 10^3 \text{ kg/m}^3)}$$

$$= 1.00 \times 10^{-4} \text{ m}^3$$

The crown is totally submerged, so $V_{\text{crown}} = V_{\text{water}}$. From Equation (1), the mass is the crown's weight in air, T_{air}, divided by g:

$$m = \frac{T_{\text{air}}}{g} = \frac{7.84 \text{ N}}{9.80 \text{ m/s}^2} = 0.800 \text{ kg}$$

Find the density of the crown:

$$\rho_{\text{crown}} = \frac{m}{V_{\text{crown}}} = \frac{0.800 \text{ kg}}{1.00 \times 10^{-4} \text{ m}^3} = \boxed{8.00 \times 10^3 \text{ kg/m}^3}$$

Remarks   Because the density of gold is $19.3 \times 10^3 \text{ kg/m}^3$, the crown is either hollow, made of an alloy, or both. Despite the mathematical complexity, it is certainly conceivable that this was the method that occurred to Archimedes. Conceptually, it's a matter of realizing (or guessing) that equal weights of gold and a silver–gold alloy would have different scale readings when immersed in water, because their densities and hence their volumes are different, leading to differing buoyant forces.

Exercise 9.8
The weight of a metal bracelet is measured to be 0.100 N in air and 0.092 N when immersed in water. Find its density.

Answer $1.25 \times 10^4 \text{ kg/m}^3$

EXAMPLE 9.9   Floating down the River

Goal   Apply Archimedes's principle to a partially submerged object.

Problem   A raft is constructed of wood having a density of $6.00 \times 10^2 \text{ kg/m}^3$. Its surface area is 5.70 m^2, and its volume is 0.60 m^3. When the raft is placed in fresh water as in Figure 9.24, to what depth h is the bottom of the raft submerged?

Strategy   There are two forces acting on the raft: the buoyant force of magnitude B, acting upwards, and the force of gravity, acting downwards. Because the raft is in equilibrium, the sum of these forces is zero. The buoyant force depends on the submerged volume $V_{\text{water}} = Ah$. Set up Newton's second law and solve for h, the depth reached by the bottom of the raft.

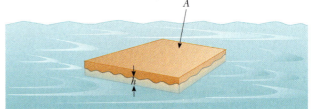

Figure 9.24   (Example 9.9) A raft partially submerged in water.

Solution
Apply Newton's second law to the raft, which is in equilibrium:

$$B - m_{\text{raft}} g = 0 \quad \rightarrow \quad B = m_{\text{raft}} g$$

The volume of the raft submerged in water is given by $V_{\text{water}} = Ah$. The magnitude of the buoyant force is equal to the weight of this displaced volume of water:

$$B = m_{\text{water}} g = (\rho_{\text{water}} V_{\text{water}}) g = (\rho_{\text{water}} Ah) g$$

Now rewrite the gravity force on the raft using the raft's density and volume:

$$m_{\text{raft}} g = (\rho_{\text{raft}} V_{\text{raft}}) g$$

Substitute these two expressions into Newton's second law, $B = m_{raft}g$, and solve for h (note that g cancels):

$$(\rho_{water}Ah)\cancel{g} = (\rho_{raft}V_{raft})\cancel{g}$$

$$h = \frac{\rho_{raft}V_{raft}}{\rho_{water}A}$$

$$= \frac{(6.00 \times 10^2 \text{ kg/m}^3)(0.600 \text{ m}^3)}{(1.00 \times 10^3 \text{ kg/m}^3)(5.70 \text{ m}^2)}$$

$$= \boxed{0.063\ 2 \text{ m}}$$

Remarks   How low the raft rides in the water depends on the density of the raft. The same is true of the human body: Fat is less dense than muscle and bone, so those with a higher percentage of body fat float better.

Exercise 9.9
Calculate how much of an iceberg is beneath the surface of the ocean, given that the density of ice is 917 kg/m^3, and salt water has density 1 025 kg/m^3.

Answer   89.5%

EXAMPLE 9.10   Floating in Two Fluids

Goal   Apply Archimedes's principle to an object floating in a fluid having two layers with different densities.

Problem   A 1.00×10^3-kg cube of aluminum is placed in a tank. Water is then added to the tank until half the cube is immersed. **(a)** What is the normal force on the cube? (See Fig. 9.25a.) **(b)** Mercury is now slowly poured into the tank until the normal force on the cube goes to zero. (See Fig. 9.25b.) How deep is the layer of mercury?

Strategy   Both parts of this problem involve applications of Newton's second law for a body in equilibrium, together with the concept of a buoyant force. In part (a), the normal, gravitational, and buoyant force of water act on the cube. In part (b), there is an additional buoyant force of mercury, while the normal force goes to zero. Using $V_{Hg} = Ah$, solve for the height of mercury, h.

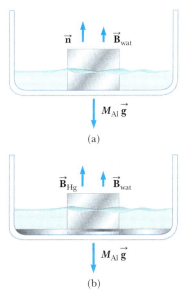

Figure 9.25   (Example 9.10)

Solution
(a) Find the normal force on the cube when half-immersed in water.

Calculate the volume V of the cube and the length d of one side, for future reference (both quantities will be needed for what follows):

$$V_{Al} = \frac{M_{Al}}{\rho_{Al}} = \frac{1.00 \times 10^3 \text{ kg}}{2.70 \times 10^3 \text{ kg/m}^3} = 0.370 \text{ m}^3$$

$$d = V_{Al}^{1/3} = 0.718 \text{ m}$$

Write Newton's second law for the cube, and solve for the normal force. The buoyant force is equal to the weight of the displaced water (half the volume of the cube).

$$n - M_{Al}g + B_{wat} = 0$$

$$n = M_{Al}g - B_{wat} = M_{Al}g - \rho_{wat}(V/2)g$$

$$= (1.00 \times 10^3 \text{ kg})(9.80 \text{ m/s}^2)$$

$$- (1.00 \times 10^3 \text{ kg/m}^3)(0.370 \text{ m}^3/2.00)(9.80 \text{ m/s}^2)$$

$$n = 9.80 \times 10^3 \text{ N} - 1.81 \times 10^3 \text{ N} = \boxed{7.99 \times 10^3 \text{ N}}$$

(b) Calculate the level h of added mercury.

Apply Newton's second law to the cube:

$$n - M_{Al}g + B_{wat} + B_{Hg} = 0$$

Set $n = 0$ and solve for the buoyant force of mercury:

$$B_{Hg} = (\rho_{Hg} Ah)g = M_{Al}g - B_{wat} = 7.99 \times 10^3 \text{ N}$$

Solve for h, noting that $A = d^2$:

$$h = \frac{M_{Al}g - B_{wat}}{\rho_{Hg}Ag} = \frac{7.99 \times 10^3 \text{ N}}{(13.6 \times 10^3 \text{ kg/m}^3)(0.718 \text{ m})^2(9.80 \text{ m/s}^2)}$$

$$h = 0.116 \text{ m}$$

Remarks   Notice that the buoyant force of mercury calculated in part (b) is the same as the normal force in part (a). This is naturally the case, because enough mercury was added to exactly cancel out the normal force. We could have used this fact to take a shortcut, simply writing $B_{Hg} = 7.99 \times 10^3$ N immediately, solving for h, and avoiding a second use of Newton's law. Most of the time, however, we won't be so lucky! Try calculating the normal force when the level of mercury is 4.00 cm.

Exercise 9.10
A cube of aluminum 1.00 m on a side is immersed one-third in water and two-thirds in glycerin. What is the normal force on the cube?

Answer 1.50×10^4 N

9.7   FLUIDS IN MOTION

When a fluid is in motion, its flow can be characterized in one of two ways. The flow is said to be **streamline**, or **laminar**, if every particle that passes a particular point moves along exactly the same smooth path followed by previous particles passing that point. This path is called a *streamline* (Fig. 9.26). Different streamlines can't cross each other under this steady-flow condition, and the streamline at any point coincides with the direction of the velocity of the fluid at that point.

In contrast, the flow of a fluid becomes irregular, or **turbulent**, above a certain velocity or under any conditions that can cause abrupt changes in velocity. Irregular motions of the fluid, called *eddy currents*, are characteristic in turbulent flow, as shown in Figure 9.27.

In discussions of fluid flow, the term **viscosity** is used for the degree of internal friction in the fluid. This internal friction is associated with the resistance between

Figure 9.26   An illustration of streamline flow around an automobile in a test wind tunnel. The streamlines in the airflow are made visible by smoke particles.

two adjacent layers of the fluid moving relative to each other. A fluid such as kerosene has a lower viscosity than does crude oil or molasses.

Many features of fluid motion can be understood by considering the behavior of an **ideal fluid**, which satisfies the following conditions:

1. **The fluid is nonviscous**, which means there is no internal friction force between adjacent layers.
2. **The fluid is incompressible**, which means its density is constant.
3. **The fluid motion is steady**, meaning that the velocity, density, and pressure at each point in the fluid don't change with time.
4. **The fluid moves without turbulence**. This implies that each element of the fluid has zero angular velocity about its center, so there can't be any eddy currents present in the moving fluid. A small wheel placed in the fluid would translate but not rotate.

Equation of Continuity

Figure 9.28a represents a fluid flowing through a pipe of nonuniform size. The particles in the fluid move along the streamlines in steady-state flow. In a small time interval Δt, the fluid entering the bottom end of the pipe moves a distance $\Delta x_1 = v_1 \Delta t$, where v_1 is the speed of the fluid at that location. If A_1 is the cross-sectional area in this region, then the mass contained in the bottom blue region is $\Delta M_1 = \rho_1 A_1 \Delta x_1 = \rho_1 A_1 v_1 \Delta t$, where ρ_1 is the density of the fluid at A_1. Similarly, the fluid that moves out of the upper end of the pipe in the same time interval Δt has a mass of $\Delta M_2 = \rho_2 A_2 v_2 \Delta t$. However, **because mass is conserved and because the flow is steady**, the mass that flows into the bottom of the pipe through A_1 in the time Δt must equal the mass that flows out through A_2 in the same interval. Therefore, $\Delta M_1 = \Delta M_2$, or

$$\rho_1 A_1 v_1 = \rho_2 A_2 v_2 \qquad \text{[9.14]}$$

For the case of an incompressible fluid, $\rho_1 = \rho_2$ and Equation 9.14 reduces to

$$A_1 v_1 = A_2 v_2 \qquad \text{[9.15]}$$

◀ Equation of continuity

This expression is called the **equation of continuity**. From this result, we see that **the product of the cross-sectional area of the pipe and the fluid speed at that cross section is a constant**. Therefore, the speed is high where the tube is constricted and low where the tube has a larger diameter. The product Av, which has dimensions of volume per unit time, is called the **flow rate**. **The condition Av = constant is equivalent to the fact that the volume of fluid that enters one end of the tube in a given time interval equals the volume of fluid leaving the tube in the same interval, assuming that the fluid is incompressible and there are no leaks.** Figure 9.28b

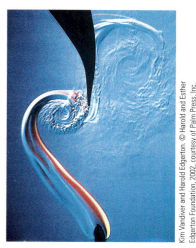

Figure 9.27 Turbulent flow: The tip of a rotating blade (the dark region at the top) forms a vortex in air that is being heated by an alcohol lamp. (The wick is at the bottom.) Note the air turbulence on both sides of the blade.

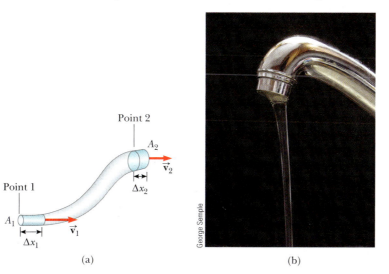

(a)   (b)

Figure 9.28 (a) A fluid moving with streamline flow through a pipe of varying cross-sectional area. The volume of fluid flowing through A_1 in a time interval Δt must equal the volume flowing through A_2 in the same time interval. Therefore, $A_1 v_1 = A_2 v_2$. (b) Water flowing slowly out of a faucet. The width of the stream narrows as the water falls and speeds up in accord with the continuity equation.

TIP 9.3 **Continuity Equations**
The rate of flow of fluid into a system equals the rate of flow out of the system. The incoming fluid occupies a certain volume and can enter the system only if the fluid already inside goes out, thereby making room.

is an example of an application of the equation of continuity: As the stream of water flows continuously from a faucet, the width of the stream narrows as it falls and speeds up.

There are many instances in everyday experience that involve the equation of continuity. Reducing the cross-sectional area of a garden hose by putting a thumb over the open end makes the water spray out with greater speed; hence the stream goes farther. Similar reasoning explains why smoke from a smoldering piece of wood first rises in a streamline pattern, getting thinner with height, eventually breaking up into a swirling, turbulent pattern. The smoke rises because it's less dense than air and the buoyant force of the air accelerates it upward. As the speed of the smoke stream increases, the cross-sectional area of the stream decreases, in accordance with the equation of continuity. The stream soon reaches a speed so great that streamline flow is not possible. We will study the relationship between speed of fluid flow and turbulence in a later discussion on the Reynolds number.

EXAMPLE 9.11   Niagara Falls

Goal   Apply the equation of continuity.

Problem   Each second, 5 525 m³ of water flows over the 670-m-wide cliff of the Horseshoe Falls portion of Niagara Falls. The water is approximately 2 m deep as it reaches the cliff. Estimate its speed at that instant?

Strategy   This is an estimate, so only one significant figure will be retained in the answer. The volume flow rate is given, and according to the equation of continuity, is a constant equal to Av. Find the cross-sectional area, substitute, and solve for the speed.

Solution

Calculate the cross-sectional area of the water as it reaches the edge of the cliff:

$$A = (670 \text{ m})(2 \text{ m}) = 1\ 340 \text{ m}^2$$

Multiply this result by the speed and set it equal to the flow rate. Then solve for v.

$$Av = \text{volume flow rate}$$

$$(1340 \text{ m}^2)v = 5\ 525 \text{ m}^3/\text{s} \quad \rightarrow \quad v \approx \boxed{4 \text{ m/s}}$$

Exercise 9.11

The Garfield Thomas water tunnel at Pennsylvania State University has a circular cross section that constricts from a diameter of 3.6 m to the test section, which is 1.2 m in diameter. If the speed of flow is 3.0 m/s in the larger-diameter pipe, determine the speed of flow in the test section.

Answer   27 m/s

EXAMPLE 9.12   Watering a Garden

Goal   Combine the equation of continuity with concepts of flow rate and kinematics.

Problem   A water hose 2.50 cm in diameter is used by a gardener to fill a 30.0-liter bucket. (One liter = 1 000 cm³.) The gardener notices that it takes 1.00 min to fill the bucket. A nozzle with an opening of cross-sectional area 0.500 cm² is then attached to the hose. The nozzle is held so that water is projected horizontally from a point 1.00 m above the ground. Over what horizontal distance can the water be projected?

Strategy   We can find the volume flow rate through the hose by dividing the volume of the bucket by the time it takes to fill it. After finding the flow rate, apply the equation of continuity to find the speed at which the water shoots horizontally out the nozzle. The rest of the problem is an application of two-dimensional kinematics. The answer obtained is the same as would be found for a ball having the same initial velocity and height.

Solution

Calculate the volume flow rate into the bucket, and convert to m^3/s:

$$\text{volume flow rate} =$$

$$= \frac{30.0 \text{ L}}{1.00 \text{ min}}\left(\frac{1.00 \times 10^3 \text{ cm}^3}{1.00 \text{ L}}\right)\left(\frac{1.00 \text{ m}}{100.0 \text{ cm}}\right)^3\left(\frac{1.00 \text{ min}}{60.0 \text{ s}}\right)$$

$$= 5.00 \times 10^{-4} \text{ m}^3/\text{s}$$

Solve the equation of continuity for v_{0x}, the x-component of the initial velocity of the stream exiting the hose:

$$A_1v_1 = A_2v_2 = A_2v_{0x}$$

$$v_{0x} = \frac{A_1v_1}{A_2} = \frac{5.00 \times 10^{-4} \text{ m}^3/\text{s}}{0.500 \times 10^{-4} \text{ m}^2} = 10.0 \text{ m/s}$$

Calculate the time for the stream to fall 1.00 m, using kinematics. Initially, the stream is horizontal, so v_{0y} is zero

$$\Delta y = v_{0y}t - \tfrac{1}{2}gt^2$$

Set $v_{0y} = 0$ in the preceding equation and solve for t noting that $\Delta y = -1.00$ m:

$$t = \sqrt{\frac{-2\Delta y}{g}} = \sqrt{\frac{-2(-1.00 \text{ m})}{9.80 \text{ m/s}^2}} = 0.452 \text{ s}$$

Find the horizontal distance the stream travels:

$$x = v_{0x}t = (10.0 \text{ m/s})(0.452 \text{ s}) = \boxed{4.52 \text{ m}}$$

Remark It's interesting that the motion of fluids can be treated with the same kinematics equations as individual objects.

Exercise 9.12

The nozzle is replaced with a Y-shaped fitting that splits the flow in half. Garden hoses are connected to each end of the Y, with each hose having a 0.400 cm^2 nozzle. (a) How fast does the water come out of one of the nozzles? (b) How far would one of the nozzles squirt water if both were operated simultaneously and held horizontally 1.00 m off the ground? [*Hint:* Find the volume flow rate through each 0.400-cm^2 nozzle, then follow the same steps as before.]

Answer (a) 6.25 m/s (b) 2.82 m

Bernoulli's Equation

As a fluid moves through a pipe of varying cross section and elevation, the pressure changes along the pipe. In 1738 the Swiss physicist Daniel Bernoulli (1700–1782) derived an expression that relates the pressure of a fluid to its speed and elevation. Bernoulli's equation is not a freestanding law of physics; rather, it's **a consequence of energy conservation as applied to an ideal fluid**.

In deriving Bernoulli's equation, we again assume that the fluid is incompressible, nonviscous, and flows in a nonturbulent, steady-state manner. Consider the flow through a nonuniform pipe in the time Δt, as in Figure 9.29. The force on the lower end of the fluid is P_1A_1, where P_1 is the pressure at the lower end. The work done on the lower end of the fluid by the fluid behind it is

$$W_1 = F_1\,\Delta x_1 = P_1A_1\,\Delta x_1 = P_1V$$

where V is the volume of the lower blue region in the figure. In a similar manner, the work done on the fluid on the upper portion in the time Δt is

$$W_2 = -P_2A_2\Delta x_2 = -P_2V$$

The volume is the same because by the equation of continuity, the volume of fluid that passes through A_1 in the time Δt equals the volume that passes through A_2 in the same interval. The work W_2 is negative because the force on the fluid at the top is opposite its displacement. The net work done by these forces in the time Δt is

$$W_{\text{fluid}} = P_1V - P_2V$$

Part of this work goes into changing the fluid's kinetic energy, and part goes into changing the gravitational potential energy of the fluid–Earth system. If m is the

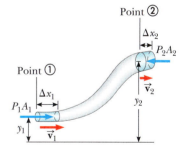

Figure 9.29 A fluid flowing through a constricted pipe with streamline flow. The fluid in the section with a length of Δx_1 moves to the section with a length of Δx_2. The volumes of fluid in the two sections are equal.

DANIEL BERNOULLI, Swiss Physicist and Mathematician (1700–1782)

In his most famous work, *Hydrodynamica*, Bernoulli showed that, as the velocity of fluid flow increases, its pressure decreases. In this same publication, Bernoulli also attempted the first explanation of the behavior of gases with changing pressure and temperature; this was the beginning of the kinetic theory of gases.

Bernoulli's equation ▶

TIP 9.4 **Bernoulli's Principle for Gases**

Equation 9.16 isn't strictly true for gases because they aren't incompressible. The qualitative behavior is the same, however: As the speed of the gas increases, its pressure decreases.

mass of the fluid passing through the pipe in the time interval Δt, then the change in kinetic energy of the volume of fluid is

$$\Delta KE = \tfrac{1}{2}mv_2{}^2 - \tfrac{1}{2}mv_1{}^2$$

The change in the gravitational potential energy is

$$\Delta PE = mgy_2 - mgy_1$$

Because the net work done by the fluid on the segment of fluid shown in Figure 9.29 changes the kinetic energy and the potential energy of the nonisolated system, we have

$$W_{\text{fluid}} = \Delta KE + \Delta PE$$

The three terms in this equation are those we have just evaluated. Substituting expressions for each of the terms gives

$$P_1V - P_2V = \tfrac{1}{2}mv_2{}^2 - \tfrac{1}{2}mv_1{}^2 + mgy_2 - mgy_1$$

If we divide each term by V and recall that $\rho = m/V$, this expression becomes

$$P_1 - P_2 = \tfrac{1}{2}\rho v_2{}^2 - \tfrac{1}{2}\rho v_1{}^2 + \rho gy_2 - \rho gy_1$$

Rearrange the terms as follows:

$$P_1 + \tfrac{1}{2}\rho v_1{}^2 + \rho gy_1 = P_2 + \tfrac{1}{2}\rho v_2{}^2 + \rho gy_2 \qquad \text{[9.16]}$$

This is **Bernoulli's equation**, often expressed as

$$P + \tfrac{1}{2}\rho v^2 + \rho gy = \text{constant} \qquad \text{[9.17]}$$

Bernoulli's equation states that the sum of the pressure P, the kinetic energy per unit volume, $\tfrac{1}{2}\rho v^2$, and the potential energy per unit volume, ρgy, has the same value at all points along a streamline.

An important consequence of Bernoulli's equation can be demonstrated by considering Figure 9.30, which shows water flowing through a horizontal constricted pipe from a region of large cross-sectional area into a region of smaller cross-sectional area. This device, called a **Venturi tube**, can be used to measure the speed of fluid flow. Because the pipe is horizontal, $y_1 = y_2$, and Equation 9.16 applied to points 1 and 2 gives

$$P_1 + \tfrac{1}{2}\rho v_1{}^2 = P_2 + \tfrac{1}{2}\rho v_2{}^2 \qquad \text{[9.18]}$$

Because the water is not backing up in the pipe, its speed v_2 in the constricted region must be greater than its speed v_1 in the region of greater diameter. From Equation 9.18, we see that P_2 must be less than P_1 because $v_2 > v_1$. This result is often expressed by the statement that **swiftly moving fluids exert less pressure than do slowly moving fluids**. This important fact enables us to understand a wide range of everyday phenomena.

Figure 9.30 (a) The pressure P_1 is greater than the pressure P_2, because $v_1 < v_2$. This device can be used to measure the speed of fluid flow. (b) A Venturi tube, located at the top of the photograph. The higher level of fluid in the middle column shows that the pressure at the top of the column, which is in the constricted region of the Venturi tube, is lower than the pressure elsewhere in the column.

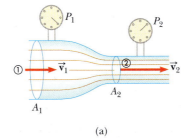

(a)

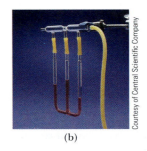

(b)

Quick Quiz 9.7

You observe two helium balloons floating next to each other at the ends of strings secured to a table. The facing surfaces of the balloons are separated by 1–2 cm. You blow through the opening between the balloons. What happens to the balloons? (a) They move toward each other; (b) they move away from each other; (c) they are unaffected.

INTERACTIVE EXAMPLE 9.13   Shoot-Out at the Old Water Tank

Goal   Apply Bernoulli's equation to find the speed of a fluid.

Problem   A nearsighted sheriff fires at a cattle rustler with his trusty six-shooter. Fortunately for the rustler, the bullet misses him and penetrates the town water tank, causing a leak (Fig. 9.31). **(a)** If the top of the tank is open to the atmosphere, determine the speed at which the water leaves the hole when the water level is 0.500 m above the hole. **(b)** Where does the stream hit the ground if the hole is 3.00 m above the ground?

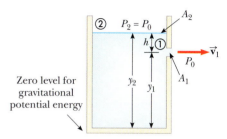

Figure 9.31   (Example 9.13) The water speed v_1 from the hole in the side of the container is given by $v_1 = \sqrt{2gh}$.

Strategy   (a) Assume the tank's cross-sectional area is large compared to the hole's ($A_2 \gg A_1$), so the water level drops very slowly and $v_2 \approx 0$. Apply Bernoulli's equation to points ① and ② in Figure 9.31, noting that P_1 equals atmospheric pressure P_0 at the hole and is approximately the same at the top of the water tank. Part (b) can be solved with kinematics, just as if the water were a ball thrown horizontally.

Solution

(a) Find the speed of the water leaving the hole.

Substitute $P_1 = P_2 = P_0$ and $v_2 \approx 0$ into Bernoulli's equation, and solve for v_1:

$$P_0 + \tfrac{1}{2}\rho v_1^2 + \rho g y_1 = P_0 + \rho g y_2$$

$$v_1 = \sqrt{2g(y_2 - y_1)} = \sqrt{2gh}$$

$$v_1 = \sqrt{2(9.80 \text{ m/s}^2)(0.500 \text{ m})} = \boxed{3.13 \text{ m/s}}$$

(b) Find where the stream hits the ground.

Use the displacement equation to find the time of the fall, noting that the stream is initially horizontal, so $v_{0y} = 0$.

$$\Delta y = -\tfrac{1}{2}g t^2 + v_{0y} t$$

$$-3.00 \text{ m} = -(4.90 \text{ m/s}^2) t^2$$

$$t = 0.782 \text{ s}$$

Compute the horizontal distance the stream travels in this time:

$$x = v_{0x} t = (3.13 \text{ m/s})(0.782 \text{ s}) = \boxed{2.45 \text{ m}}$$

Remarks   As the analysis of part (a) shows, the speed of the water emerging from the hole is equal to the speed acquired by an object falling freely through the vertical distance h. This is known as **Torricelli's law**.

Exercise 9.13

Suppose, in a similar situation, the water hits the ground 4.20 m from the hole in the tank. If the hole is 2.00 m above the ground, how far above the hole is the water level?

Answer   2.20 m above the hole

Physics⊗Now™   You can move the hole vertically and see where the water lands by logging into PhysicsNow at **www.cp7e.com** and going to Interactive Example 9.13.

Example 9.14   Fluid Flow in a Pipe

Goal   Solve a problem combining Bernoulli's equation and the equation of continuity.

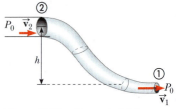

Problem   A large pipe with a cross-sectional area of 1.00 m^2 descends 5.00 m and narrows to 0.500 m^2, where it terminates in a valve (Fig. 9.32). If the pressure at point ② is atmospheric pressure, and the valve is opened wide and water allowed to flow freely, find the speed of the water leaving the pipe.

Figure 9.32   (Example 9.14)

Strategy   The equation of continuity, together with Bernoulli's equation, constitute two equations in two unknowns: the speeds v_1 and v_2. Eliminate v_2 from Bernoulli's equation with the equation of continuity, and solve for v_1.

Solution

Bernoulli's equation is

$$P_1 + \tfrac{1}{2}\rho v_1^2 + \rho g y_1 = P_2 + \tfrac{1}{2}\rho v_2^2 + \rho g y_2 \qquad (1)$$

Solve the equation of continuity for v_2:

$$A_2 v_2 = A_1 v_1$$

$$v_2 = \frac{A_1}{A_2} v_1 \qquad (2)$$

In Equation 1, set $P_1 = P_2 = P_0$, and substitute the expression for v_2. Then solve for v_1.

$$P_0 + \tfrac{1}{2}\rho v_1^2 + \rho g y_1 = P_0 + \tfrac{1}{2}\rho \left(\frac{A_1}{A_2} v_1\right)^2 + \rho g y_2$$

$$v_1^2 \left(1 - \left(\frac{A_1}{A_2}\right)^2\right) = 2g(y_2 - y_1) = 2gh$$

$$v_1 = \frac{\sqrt{2gh}}{\sqrt{1 - (A_1/A_2)^2}}$$

Substitute the given values:

$$v_1 = \boxed{11.4 \text{ m/s}}$$

Remarks   This speed is slightly higher than the speed predicted by Torricelli's law, because the narrowing pipe squeezes the fluid.

Exercise 9.14

Water flowing in a horizontal pipe is at a pressure of 1.4×10^5 Pa at a point where its cross-sectional area is 1.00 m^2. When the pipe narrows to 0.400 m^2, the pressure drops to 1.16×10^5 Pa. Find the water's speed (a) in the wider pipe and (b) in the narrower pipe.

Answer   (a) 3.02 m/s   (b) 7.56 m/s

9.8   OTHER APPLICATIONS OF FLUID DYNAMICS

In this section we describe some common phenomena that can be explained, at least in part, by Bernoulli's equation.

   In general, an object moving through a fluid is acted upon by a net upward force as the result of any effect that causes the fluid to change its direction as it flows past the object. For example, a golf ball struck with a club is given a rapid backspin, as shown in Figure 9.33. The dimples on the ball help entrain the air along the curve of the ball's surface. The figure shows a thin layer of air wrapping partway around the ball and being deflected downward as a result. Because the

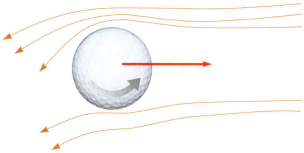

FIGURE 9.33   A spinning golf ball is acted upon by a lifting force that allows it to travel much further than it would if it were not spinning.

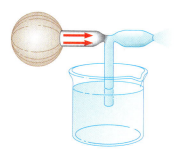

Figure 9.34   A stream of air passing over a tube dipped in a liquid causes the liquid to rise in the tube. This effect is used in perfume bottles and paint sprayers.

ball pushes the air down, by Newton's third law the air must push up on the ball and cause it to rise. Without the dimples, the air isn't as well entrained, so the golf ball doesn't travel as far. A tennis ball's fuzz performs a similar function, though the desired result is ball placement rather than greater distance.

Many devices operate in the manner illustrated in Figure 9.34. A stream of air passing over an open tube reduces the pressure above the tube, causing the liquid to rise into the airstream. The liquid is then dispersed into a fine spray of droplets. You might recognize that this so-called atomizer is used in perfume bottles and paint sprayers. The same principle is used in the carburetor of a gasoline engine. In that case, the low-pressure region in the carburetor is produced by air drawn in by the piston through the air filter. The gasoline vaporizes, mixes with the air, and enters the cylinder of the engine for combustion.

In a person with advanced arteriosclerosis, the Bernoulli effect produces a symptom called **vascular flutter**. In this condition, the artery is constricted as a result of accumulated plaque on its inner walls, as shown in Figure 9.35. To maintain a constant flow rate, the blood must travel faster than normal through the constriction. If the speed of the blood is sufficiently high in the constricted region, the blood pressure is low, and the artery may collapse under external pressure, causing a momentary interruption in blood flow. During the collapse there is no Bernoulli effect, so the vessel reopens under arterial pressure. As the blood rushes through the constricted artery, the internal pressure drops and the artery closes again. Such variations in blood flow can be heard with a stethoscope. If the plaque becomes dislodged and ends up in a smaller vessel that delivers blood to the heart, it can cause a heart attack.

An **aneurysm** is a weakened spot on an artery where the artery walls have ballooned outward. Blood flows more slowly though this region, as can be seen from the equation of continuity, resulting in an increase in pressure in the vicinity of the aneurysm relative to the pressure in other parts of the artery. This condition is dangerous because the excess pressure can cause the artery to rupture.

The lift on an aircraft wing can also be explained in part by the Bernoulli effect. Airplane wings are designed so that the air speed above the wing is greater than the speed below. As a result, the air pressure above the wing is less than the pressure below, and there is a net upward force on the wing, called the *lift*. (There is also a horizontal component called the *drag*.) Another factor influencing the lift on a wing, shown in Figure 9.36, is the slight upward tilt of the wing. This causes air molecules striking the bottom to be deflected downward, producing a reaction force upward by Newton's third law. Finally, turbulence also has an effect. If the wing is tilted too much, the flow of air across the upper surface becomes turbulent, and the pressure difference across the wing is not as great as that predicted by the Bernoulli effect. In an extreme case, this turbulence may cause the aircraft to stall.

APPLICATION

"Atomizers" in Perfume Bottles and Paint Sprayers

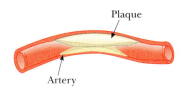

Figure 9.35   Blood must travel faster than normal through a constricted region of an artery.

APPLICATION

Vascular Flutter and Aneurysms

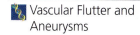

Figure 9.36   Streamline flow around an airplane wing. The pressure above is less than the pressure below, and there is a dynamic upward lift force.

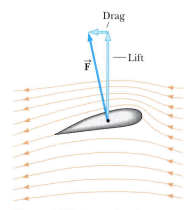

APPLICATION

Lift on Aircraft Wings

EXAMPLE 9.15   Lift on an Airfoil

Goal   Use Bernoulli's equation to calculate the lift on an airplane wing.

Problem   An airplane has wings, each with area 4.00 m^2, designed so that air flows over the top of the wing at 245 m/s and underneath the wing at 222 m/s. Find the mass of the airplane such that the lift on the plane will support its weight, assuming the force from the pressure difference across the wings is directed straight upwards.

Strategy   This problem can be solved by substituting values into Bernoulli's equation to find the pressure difference between the air under the wing and the air over the wing, followed by applying Newton's second law to find the mass the airplane can lift.

Solution

Apply Bernoulli's equation to the air flowing under the wing (point 1) and over the wing (point 2). Gravitational potential energy terms are small compared with the other terms, and can be neglected.

$$P_1 + \tfrac{1}{2}\rho v_1^2 = P_2 + \tfrac{1}{2}\rho v_2^2$$

Solve this equation for the pressure difference:

$$\Delta P = P_1 - P_2 = \tfrac{1}{2}\rho v_2^2 - \tfrac{1}{2}\rho v_1^2 = \tfrac{1}{2}\rho(v_2^2 - v_1^2)$$

Substitute the given speeds and $\rho = 1.29 \text{ kg/m}^3$, the density of air:

$$\Delta P = \tfrac{1}{2}(1.29 \text{ kg/m}^3)(245^2 \text{ m}^2/\text{s}^2 - 222^2 \text{ m}^2/\text{s}^2)$$

$$\Delta P = 6.93 \times 10^3 \text{ Pa}$$

Apply Newton's second law. To support the plane's weight, the sum of the lift and gravity forces must equal zero. Solve for the mass m of the plane.

$$2A\Delta P - mg = 0 \quad \rightarrow \quad m = \boxed{5.66 \times 10^3 \text{ kg}}$$

Remarks   Note the factor of two in the last equation, needed because the airplane has two wings. The density of the atmosphere drops steadily with increasing height, reducing the lift. As a result, all aircraft have a maximum operating altitude.

Exercise 9.15

Approximately what size wings would an aircraft need on Mars if its engine generates the same differences in speed as in the example and the total mass of the craft is 400 kg? The density of air on the surface of Mars is approximately one percent Earth's density at sea level, and the acceleration of gravity on the surface of Mars is about 3.8 m/s^2.

Answer   Rounding to one significant digit, each wing would have to have an area of about 10 m^2. There have been proposals for solar-powered robotic Mars aircraft, which would have to be gossamer-light with large wings.

Applying Physics 9.4   Sailing Upwind

How can a sailboat accomplish the seemingly impossible task of sailing into the wind?

Explanation   As shown in Figure 9.37, the wind blowing in the direction of the arrow causes the sail to billow out and take on a shape similar to that of an airplane wing. By Bernoulli's equation, just as for an airplane wing, there is a force on the sail in the direction shown. The component of force perpendicular to the boat tends to make the boat move sideways in the water, but the keel prevents this sideways motion. The component of the force in the forward direction drives the boat almost against the wind. The word *almost* is used because a sailboat can move forward only

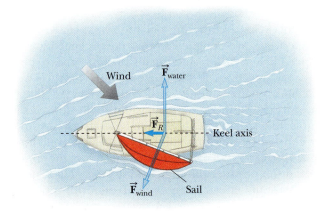

Figure 9.37   (Applying Physics 9.4)

when the wind direction is about 10 to 15° with re-spect to the forward direction. This means that in order to sail directly against the wind, a boat must follow a zigzag path, a procedure called *tacking*, so that the wind is always at some angle with respect to the direction of travel.

Applying Physics 9.5   Home Plumbing

Consider the portion of a home plumbing system shown in Figure 9.38. The water trap in the pipe below the sink captures a plug of water that prevents sewer gas from finding its way from the sewer pipe, up the sink drain, and into the home. Suppose the dishwasher is draining, so that water is moving to the left in the sewer pipe. What is the purpose of the vent, which is open to the air above the roof of the house? In which direction is air moving at the opening of the vent, upwards or downwards?

Explanation   Imagine that the vent isn't present, so that the drainpipe for the sink is simply connected through the trap to the sewer pipe. As water from the dishwasher moves to the left in the sewer pipe, the pressure in the sewer pipe is reduced below atmospheric pressure, in accordance with Bernoulli's principle. The pressure at the drain in the sink is still at atmospheric pressure. This pressure difference can push the plug of water in the water trap of the sink down the drainpipe and into the sewer pipe, removing it as a barrier to sewer gas. With the addition of the vent to the roof, the reduced pressure of the dishwasher water will result in air entering the vent pipe at the roof. This inflow of air will keep the pressure in the vent pipe and the right-hand side of the sink drainpipe close to atmospheric pressure, so that the plug of water in the water trap will remain in place.

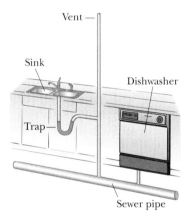

Figure 9.38   (Applying Physics 9.5)

The exhaust speed of a rocket engine can also be understood qualitatively with Bernoulli's equation, though, in actual practice a large number of additional variables need to be taken into account. Rockets actually work better in vacuum than in the atmosphere, contrary to an early *New York Times* article criticizing rocket pioneer Robert Goddard, which held that they wouldn't work at all, having no air to push against. The pressure inside the combustion chamber is P, and the pressure just outside the nozzle is the ambient atmospheric pressure, P_{atm}. Differences in height between the combustion chamber and the end of the nozzle result in negligible contributions of gravitational potential energy. In addition, the gases inside the chamber flow at negligible speed compared to gases going through the nozzle. The exhaust speed can be found from Bernoulli's equation,

$$v_{ex} = \sqrt{\frac{2(P - P_{atm})}{\rho}}$$

APPLICATION

Rocket Engines

This equation shows that the exhaust speed is reduced in the atmosphere, so rockets are actually more effective in the vacuum of space. Also of interest is the appearance of the density ρ in the denominator. A lower density working fluid or gas will give a higher exhaust speed, which partly explains why liquid hydrogen, which has a very low density, is a fuel of choice.

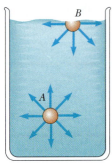

Figure 9.39 The net force on a molecule at A is zero because such a molecule is completely surrounded by other molecules. The net force on a surface molecule at B is downward because it isn't completely surrounded by other molecules.

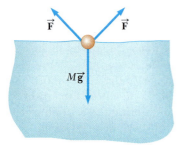

Figure 9.40 End view of a needle resting on the surface of water. The components of surface tension balance the force of gravity.

9.9 SURFACE TENSION, CAPILLARY ACTION, AND VISCOUS FLUID FLOW

If you look closely at a dewdrop sparkling in the morning sunlight, you will find that the drop is spherical. The drop takes this shape because of a property of liquid surfaces called **surface tension**. In order to understand the origin of surface tension, consider a molecule at point A in a container of water, as in Figure 9.39. Although nearby molecules exert forces on this molecule, the net force on it is zero because it's completely surrounded by other molecules and hence is attracted equally in all directions. The molecule at B, however, is not attracted equally in all directions. Because there are no molecules above it to exert upward forces, the molecule at B is pulled toward the interior of the liquid. The contraction at the surface of the liquid ceases when the inward pull exerted on the surface molecules is balanced by the outward repulsive forces that arise from collisions with molecules in the interior of the liquid. **The net effect of this pull on all the surface molecules is to make the surface of the liquid contract and, consequently, to make the surface area of the liquid as small as possible.** Drops of water take on a spherical shape because a sphere has the smallest surface area for a given volume.

If you place a sewing needle very carefully on the surface of a bowl of water, you will find that the needle floats even though the density of steel is about eight times that of water. This phenomenon can also be explained by surface tension. A close examination of the needle shows that it actually rests in a depression in the liquid surface as shown in Figure 9.40. The water surface acts like an elastic membrane under tension. The weight of the needle produces a depression, increasing the surface area of the film. Molecular forces now act at all points along the depression, tending to restore the surface to its original horizontal position. The vertical components of these forces act to balance the force of gravity on the needle. The floating needle can be sunk by adding a little detergent to the water, which reduces the surface tension.

The **surface tension** γ in a film of liquid is defined as the magnitude of the surface tension force F divided by the length L along which the force acts:

$$\gamma \equiv \frac{F}{L} \qquad \text{[9.19]}$$

The SI unit of surface tension is the newton per meter, and values for a few representative materials are given in Table 9.4.

Surface tension can be thought of as the energy content of the fluid at its surface per unit surface area. To see that this is reasonable, we can manipulate the units of surface tension γ as follows:

$$\frac{\text{N}}{\text{m}} = \frac{\text{N} \cdot \text{m}}{\text{m}^2} = \frac{\text{J}}{\text{m}^2}$$

In general, in **any equilibrium configuration of an object, the energy is a minimum**. Consequently, a fluid will take on a shape such that its surface area is as small as possible. For a given volume, a spherical shape has the smallest surface area; therefore, a drop of water takes on a spherical shape.

TABLE 9.4

Surface Tensions for Various Liquids

Liquid	T (°C)	Surface Tension (N/m)
Ethyl alcohol	20	0.022
Mercury	20	0.465
Soapy water	20	0.025
Water	20	0.073
Water	100	0.059

An apparatus used to measure the surface tension of liquids is shown in Figure 9.41. A circular wire with a circumference L is lifted from a body of liquid. The surface film clings to the inside and outside edges of the wire, holding back the wire and causing the spring to stretch. If the spring is calibrated, the force required to overcome the surface tension of the liquid can be measured. In this case, the surface tension is given by

$$\gamma = \frac{F}{2L}$$

We use $2L$ for the length because the surface film exerts forces on both the inside and outside of the ring.

The surface tension of liquids decreases with increasing temperature, because the faster moving molecules of a hot liquid aren't bound together as strongly as are those in a cooler liquid. In addition, certain ingredients called surfactants decrease surface tension when added to liquids. For example, soap or detergent decreases the surface tension of water, making it easier for soapy water to penetrate the cracks and crevices of your clothes to clean them better than plain water does. A similar effect occurs in the lungs. The surface tissue of the air sacs in the lungs contains a fluid that has a surface tension of about 0.050 N/m. A liquid with a surface tension this high would make it very difficult for the lungs to expand during inhalation. However, as the area of the lungs increases with inhalation, the body secretes into the tissue a substance that gradually reduces the surface tension of the liquid. At full expansion, the surface tension of the lung fluid can drop to as low as 0.005 N/m.

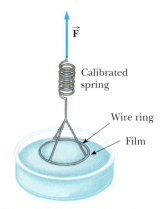

Figure 9.41 An apparatus for measuring the surface tension of liquids. The force on the wire ring is measured just before the ring breaks free of the liquid.

APPLICATION
Air Sac Surface Tension

EXAMPLE 9.16   Walking on Water

Goal   Apply the surface tension equation.

Problem   Many insects can literally walk on water, using surface tension for their support. To show this is feasible, assume that the insect's "foot" is spherical. When the insect steps onto the water with all six legs, a depression is formed in the water around each foot, as shown in Figure 9.42a. The surface tension of the water produces upward forces on the water that tend to restore the water surface to its normally flat shape. If the insect has a mass of 2.0×10^{-5} kg and if the radius of each foot is 1.5×10^{-4} m, find the angle θ.

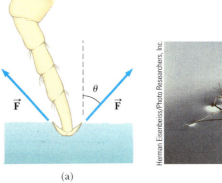
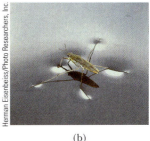

(a)   (b)

Figure 9.42   (Example 9.16) (a) One foot of an insect resting on the surface of water. (b) This water strider resting on the surface of a lake remains on the surface, rather than sinking, because an upward surface tension force acts on each leg, balancing the force of gravity on the insect.

Strategy   Find an expression for the magnitude of the net force F directed tangentially to the depressed part of the water surface, and obtain the part that is acting vertically, in opposition to the downward force of gravity. Assume that the radius of depression is the same as the radius of the insect's foot. Because the insect has six legs, one-sixth of the insect's weight must be supported by one of the legs, assuming the weight is distributed evenly. The length L is just the distance around a circle. Using Newton's second law for a body in equilibrium (zero acceleration), solve for θ.

Solution

Start with the surface tension equation:

$$F = \gamma L$$

Focus on one circular foot, substituting $L = 2\pi r$. Multiply by cos θ to get the vertical component F_v:

$$F_v = \gamma(2\pi r)\cos\theta$$

Write Newton's second law for the insect's one foot, which supports one-sixth of the insect's weight:

$$\Sigma\, F = F_v - F_{grav} = 0$$

$$\gamma 2\pi r \cos\theta - \tfrac{1}{6}mg = 0$$

Solve for cos θ and substitute:

$$\cos \theta = \frac{mg}{12\pi r\gamma}$$

$$= \frac{(2.0 \times 10^{-5}\ \text{kg})(9.80\ \text{m/s}^2)}{12\pi(1.5 \times 10^{-4}\ \text{m})(0.073\ \text{N/m})} = 0.47 \quad \textbf{(1)}$$

Take the inverse cosine of both sides to find the angle θ: $\theta = \cos^{-1}(0.47) = \boxed{62°}$

Remarks   If the weight of the insect were great enough to make the right side of Equation (1) greater than one, a solution for θ would be impossible because the cosine of an angle can never be greater than one. In this circumstance, the insect would sink.

Exercise 9.16
A typical sewing needle floats on water. Create an estimate for the needle's maximum possible mass. Assume the sewing needle is two inches long. [*Hint:* The cosine of an angle is never larger than 1.]

Answer   0.8 g

The Surface of Liquid

If you have ever closely examined the surface of water in a glass container, you may have noticed that the surface of the liquid near the walls of the glass curves upwards as you move from the center to the edge, as shown in Figure 9.43a. However, if mercury is placed in a glass container, the mercury surface curves downwards, as in Figure 9.43b. These surface effects can be explained by considering the forces between molecules. In particular, we must consider the forces that the molecules of the liquid exert on one another and the forces that the molecules of the glass surface exert on those of the liquid. In general terms, forces between like molecules, such as the forces between water molecules, are called **cohesive forces**, and forces between unlike molecules, such as those exerted by glass on water, are called **adhesive forces**.

Water tends to cling to the walls of the glass because the adhesive forces between the molecules of water and the glass molecules are *greater* than the cohesive forces between the water molecules. In effect, the water molecules cling to the surface of the glass rather than fall back into the bulk of the liquid. When this condition prevails, the liquid is said to "wet" the glass surface. The surface of the mercury curves downward near the walls of the container because the cohesive forces between the mercury atoms are greater than the adhesive forces between mercury and glass. A mercury atom near the surface is pulled more strongly toward other mercury atoms than toward the glass surface, so mercury doesn't wet the glass surface.

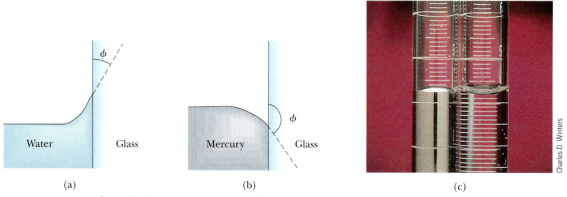

Figure 9.43   A liquid in contact with a solid surface. (a) For water, the adhesive force is greater than the cohesive force. (b) For mercury, the adhesive force is less than the cohesive force. (c) The surface of mercury (*left*) curves downwards in a glass container, whereas the surface of water (*right*) curves upwards, as you move from the center to the edge.

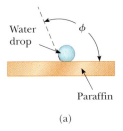

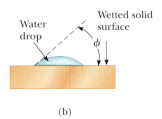

(a)   (b)

Figure 9.44 (a) The contact angle between water and paraffin is about 107°. In this case, the cohesive force is greater than the adhesive force. (b) When a chemical called a wetting agent is added to the water, it wets the paraffin surface, and $f < 90°$. In this case, the adhesive force is greater than the cohesive force.

The angle ϕ between the solid surface and a line drawn tangent to the liquid at the surface is called the **contact angle** (Fig. 9.44). The angle ϕ is less than 90° for any substance in which adhesive forces are stronger than cohesive forces and greater than 90° if cohesive forces predominate. For example, if a drop of water is placed on paraffin, the contact angle is approximately 107° (Fig. 9.44a). If certain chemicals, called wetting agents or detergents, are added to the water, the contact angle becomes less than 90°, as shown in Figure 9.44b. The addition of such substances to water ensures that the water makes intimate contact with a surface and penetrates it. For this reason, detergents are added to water to wash clothes or dishes.

On the other hand, it is sometimes necessary to *keep* water from making intimate contact with a surface, as in waterproof clothing, where a situation somewhat the reverse of that shown in Figure 9.44 is called for. The clothing is sprayed with a waterproofing agent, which changes ϕ from less than 90° to greater than 90°. The water beads up on the surface and doesn't easily penetrate the clothing.

APPLICATION

Detergents and Waterproofing Agents

Capillary Action

In capillary tubes the diameter of the opening is very small, on the order of a hundredth of a centimeter. In fact, the word *capillary* means "hairlike." If such a tube is inserted into a fluid for which adhesive forces dominate over cohesive forces, the liquid rises into the tube, as shown in Figure 9.45. The rising of the liquid in the tube can be explained in terms of the shape of the liquid's surface and surface tension effects. At the point of contact between liquid and solid, the upward force of surface tension is directed as shown in the figure. From Equation 9.19, the magnitude of this force is

$$F = \gamma L = \gamma(2\pi r)$$

(We use $L = 2\pi r$ here because the liquid is in contact with the surface of the tube at all points around its circumference.) The vertical component of this force due to surface tension is

$$F_v = \gamma(2\pi r)(\cos\phi) \qquad [9.20]$$

In order for the liquid in the capillary tube to be in equilibrium, this upward force must be equal to the weight of the cylinder of water of height h inside the capillary tube. The weight of this water is

$$w = Mg = \rho V g = \rho g \pi r^2 h \qquad [9.21]$$

Equating F_v in Equation 9.20 to w in Equation 9.21 (applying Newton's second law for equilibrium), we have

$$\gamma(2\pi r)(\cos\phi) = \rho g \pi r^2 h$$

Solving for h gives the height to which water is drawn into the tube:

$$h = \frac{2\gamma}{\rho g r}\cos\phi \qquad [9.22]$$

If a capillary tube is inserted into a liquid in which cohesive forces dominate over adhesive forces, the level of the liquid in the capillary tube will be below the surface of the surrounding fluid, as shown in Figure 9.46. An analysis similar to the above would show that the distance h to the depressed surface is given by Equation 9.22.

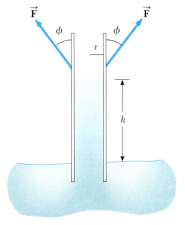

Figure 9.45 A liquid rises in a narrow tube because of capillary action, a result of surface tension and adhesive forces.

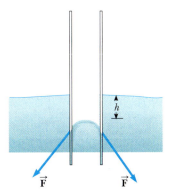

Figure 9.46 When cohesive forces between molecules of a liquid exceed adhesive forces, the level of the liquid in the capillary tube is below the surface of the surrounding fluid.

Capillary tubes are often used to draw small samples of blood from a needle prick in the skin. Capillary action must also be considered in the construction of concrete-block buildings, because water seepage through capillary pores in the blocks or the mortar may cause damage to the inside of the building. To prevent such damage, the blocks are usually coated with a waterproofing agent either outside or inside the building. Water seepage through a wall is an undesirable effect of capillary action, but there are many useful effects. Plants depend on capillary action to transport water and nutrients, and sponges and paper towels use capillary action to absorb spilled fluids.

EXAMPLE 9.17   Rising Water

Goal   Apply surface tension to capillary action.

Problem   Find the height to which water would rise in a capillary tube with a radius equal to 5.0×10^{-5} m. Assume that the contact angle between the water and the material of the tube is small enough to be considered zero.

Strategy   This problem requires substituting values into Equation 9.22.

Solution

Substitute the known values into Equation 9.22:

$$h = \frac{2\gamma \cos 0°}{\rho g r}$$

$$= \frac{2(0.073 \text{ N/m})}{(1.00 \times 10^3 \text{ kg/m}^3)(9.80 \text{ m/s}^2)(5.0 \times 10^{-5} \text{ m})}$$

$$= \boxed{0.30 \text{ m}}$$

Exercise 9.17

Suppose ethyl alcohol rises 0.250 m in a thin tube. Estimate the radius of the tube, assuming the contact angle is approximately zero.

Answer 2.23×10^{-5} m

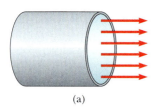

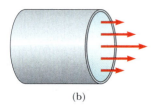

Figure 9.47   (a) The particles in an ideal (nonviscous) fluid all move through the pipe with the same velocity. (b) In a viscous fluid, the velocity of the fluid particles is zero at the surface of the pipe and increases to a maximum value at the center of the pipe.

Viscous Fluid Flow

It is considerably easier to pour water out of a container than to pour honey. This is because honey has a higher viscosity than water. In a general sense, **viscosity refers to the internal friction of a fluid**. It's very difficult for layers of a viscous fluid to slide past one another. Likewise, it's difficult for one solid surface to slide past another if there is a highly viscous fluid, such as soft tar, between them.

When an ideal (nonviscous) fluid flows through a pipe, the fluid layers slide past one another with no resistance. If the pipe has a uniform cross section, each layer has the same velocity, as shown in Figure 9.47a. In contrast, the layers of a viscous fluid have different velocities, as Figure 9.47b indicates. The fluid has the greatest velocity at the center of the pipe, whereas the layer next to the wall doesn't move because of adhesive forces between molecules and the wall surface.

To better understand the concept of viscosity, consider a layer of liquid between two solid surfaces, as in Figure 9.48. The lower surface is fixed in position, and the top surface moves to the right with a velocity $\vec{\mathbf{v}}$ under the action of an external force $\vec{\mathbf{F}}$. Because of this motion, a portion of the liquid is distorted from its original shape, $ABCD$, at one instant to the shape $AEFD$ a moment later. The force required to move the upper plate and distort the liquid is proportional to both the area A in contact with the fluid and the speed v of the fluid. Furthermore, the

TABLE 9.5

Viscosities of Various Fluids

Fluid	$T\,(°C)$	Viscosity $\eta\ (\mathrm{N \cdot s/m^2})$
Water	20	1.0×10^{-3}
Water	100	0.3×10^{-3}
Whole blood	37	2.7×10^{-3}
Glycerin	20	1500×10^{-3}
10-wt motor oil	30	250×10^{-3}

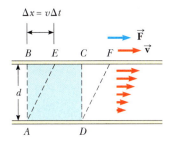

Figure 9.48 A layer of liquid between two solid surfaces in which the lower surface is fixed and the upper surface moves to the right with a velocity $\vec{v}$.

force is inversely proportional to the distance d between the two plates. We can express these proportionalities as $F \propto Av/d$. The force required to move the upper plate at a fixed speed v is therefore

$$F = \eta\,\frac{Av}{d} \qquad [9.23]$$

where η (the lowercase Greek letter *eta*) is the **coefficient of viscosity** of the fluid.

◀ Coefficient of viscosity

The SI units of viscosity are $\mathrm{N \cdot s/m^2}$. The units of viscosity in many reference sources are often expressed in $\mathrm{dyne \cdot s/cm^2}$, called 1 **poise**, in honor of the French scientist J. L. Poiseuille (1799–1869). The relationship between the SI unit of viscosity and the poise is

$$1 \text{ poise} = 10^{-1}\ \mathrm{N \cdot s/m^2} \qquad [9.24]$$

Small viscosities are often expressed in centipoise (cp), where 1 cp = 10^{-2} poise. The coefficients of viscosity for some common substances are listed in Table 9.5.

Poiseuille's Law

Figure 9.49 shows a section of a tube of length L and radius R containing a fluid under a pressure P_1 at the left end and a pressure P_2 at the right. Because of this pressure difference, the fluid flows through the tube. The rate of flow (volume per unit time) depends on the pressure difference $(P_1 - P_2)$, the dimensions of the tube, and the viscosity of the fluid. The result, known as **Poiseuille's law**, is

$$\text{Rate of flow} = \frac{\Delta V}{\Delta t} = \frac{\pi R^4 (P_1 - P_2)}{8 \eta L} \qquad [9.25]$$

◀ Poiseuille's law

where η is the coefficient of viscosity of the fluid. We won't attempt to derive this equation here, because the methods of integral calculus are required. However, it is reasonable that the rate of flow should increase if the pressure difference across the tube or the tube radius increases. Likewise, the flow rate should decrease if the viscosity of the fluid or the length of the tube increases. So the presence of R and the pressure difference in the numerator of Equation 9.25 and of L and η in the denominator makes sense.

From Poiseuille's law, we see that in order to maintain a constant flow rate, the pressure difference across the tube has to increase if the viscosity of the fluid increases. This fact is important in understanding the flow of blood through the circulatory system. The viscosity of blood increases as the number of red blood cells rises. Blood with a high concentration of red blood cells requires greater pumping pressure from the heart to keep it circulating than does blood of lower red blood cell concentration.

Note that the flow rate varies as the radius of the tube raised to the fourth power. Consequently, if a constriction occurs in a vein or artery, the heart will have to work considerably harder in order to produce a higher pressure drop and hence maintain the required flow rate.

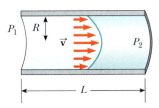

Figure 9.49 Velocity profile of a fluid flowing through a uniform pipe of circular cross section. The rate of flow is given by Poiseuille's law. Note that the fluid velocity is greatest at the middle of the pipe.

EXAMPLE 9.18 A Blood Transfusion

Goal Apply Poiseuille's law.

Problem A patient receives a blood transfusion through a needle of radius 0.20 mm and length 2.0 cm. The density of blood is 1 050 kg/m^3. The bottle supplying the blood is 0.50 m above the patient's arm. What is the rate of flow through the needle?

Strategy Find the pressure difference between the level of the blood and the patient's arm. Substitute into Poiseuille's law, using the value for the viscosity of whole blood in Table 9.5.

Solution

Calculate the pressure difference:

$$P_1 - P_2 = \rho g h = (1\,050 \text{ kg/m}^3)(9.80 \text{ m/s}^2)(0.50 \text{ m})$$
$$= 5.15 \times 10^3 \text{ Pa}$$

Substitute into Poiseuille's law:

$$\frac{\Delta V}{\Delta t} = \frac{\pi R^4 (P_1 - P_2)}{8\eta L}$$

$$= \frac{\pi (2.0 \times 10^{-4} \text{ m})^4 (5.15 \times 10^3 \text{ Pa})}{8(2.7 \times 10^{-3} \text{ N} \cdot \text{s/m}^2)(2.0 \times 10^{-2} \text{ m})}$$

$$= 6.0 \times 10^{-8} \text{ m}^3/\text{s}$$

Remarks Compare this to the volume flow rate in the absence of any viscosity. Using Bernoulli's equation, the calculated volume flow rate is approximately five times as great. As expected, viscosity greatly reduces flow rate.

Exercise 9.18

A pipe carrying water from a tank 20.0 m tall must cross 3.00×10^2 km of wilderness to reach a remote town. Find the radius of pipe so that the volume flow rate is at least 0.0500 m^3/s. (Use the viscosity of water at 20°C.)

Answer 0.118 m

Reynolds Number

At sufficiently high velocities, fluid flow changes from simple streamline flow to turbulent flow, characterized by a highly irregular motion of the fluid. Experimentally, the onset of turbulence in a tube is determined by a dimensionless factor called the **Reynolds number**, RN, given by

Reynolds number ▶

$$RN = \frac{\rho v d}{\eta}$$

[9.26]

where ρ is the density of the fluid, v is the average speed of the fluid along the direction of flow, d is the diameter of the tube, and η is the viscosity of the fluid. If RN is below about 2 000, the flow of fluid through a tube is streamline; turbulence occurs if RN is above 3 000. In the region between 2 000 and 3 000, the flow is unstable, meaning that the fluid can move in streamline flow, but any small disturbance will cause its motion to change to turbulent flow.

EXAMPLE 9.19 Turbulent Flow of Blood

Goal Use the Reynolds number to determine a speed associated with the onset of turbulence.

Problem Determine the speed at which blood flowing through an artery of diameter 0.20 cm will become turbulent.

Strategy The solution requires only the substitution of values into Equation 9.26 giving the Reynolds number and then solving it for the speed v.

Solution

Solve Equation 9.26 for v, and substitute the viscosity and density of blood from Example 9.18, the diameter d of the artery, and a Reynolds number of 3.00×10^3:

$$v = \frac{\eta(RN)}{\rho d} = \frac{(2.7 \times 10^{-3}\,\text{N}\cdot\text{s/m}^2)(3.00 \times 10^3)}{(1.05 \times 10^3\,\text{kg/m}^3)(0.20 \times 10^{-2}\,\text{m})}$$

$$v = \boxed{3.9\,\text{m/s}}$$

Remark The exercise shows that rapid ingestion of soda through a straw may create a turbulent state.

Exercise 9.19

Determine the speed v at which water at $20°$ sucked up a straw would become turbulent. The straw has a diameter of 0.0060 m.

Answer $v = 0.50\,\text{m/s}$

9.10   TRANSPORT PHENOMENA

When a fluid flows through a tube, the basic mechanism that produces the flow is a difference in pressure across the ends of the tube. This pressure difference is responsible for the transport of a mass of fluid from one location to another. The fluid may also move from place to place because of a second mechanism—one that depends on a difference in *concentration* between two points in the fluid, as opposed to a pressure difference. When the concentration (the number of molecules per unit volume) is higher at one location than at another, molecules will flow from the point where the concentration is high to the point where it is lower. The two fundamental processes involved in fluid transport resulting from concentration differences are called *diffusion* and *osmosis*.

Diffusion

In a diffusion process, molecules move from a region where their concentration is high to a region where their concentration is lower. To understand why diffusion occurs, consider Figure 9.50, which depicts a container in which a high concentration of molecules has been introduced into the left side. The dashed line in the figure represents an imaginary barrier separating the two regions. Because the molecules are moving with high speeds in random directions, many of them will cross the imaginary barrier moving from left to right. Very few molecules will pass through moving from right to left, simply because there are very few of them on the right side of the container at any instant. As a result, there will always be a *net* movement from the region with many molecules to the region with fewer molecules. For this reason, the concentration on the left side of the container will decrease, and that on the right side will increase with time. Once a concentration equilibrium has been reached, there will be no *net* movement across the cross-sectional area: The rate of movement of molecules from left to right will equal the rate from right to left.

The basic equation for diffusion is **Fick's law**,

$$\text{Diffusion rate} = \frac{\text{mass}}{\text{time}} = \frac{\Delta M}{\Delta t} = DA\left(\frac{C_2 - C_1}{L}\right) \qquad \textbf{[9.27]}$$

◀ Fick's law

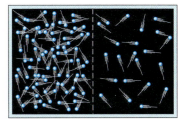

Figure 9.50 When the concentration of gas molecules on the left side of the container exceeds the concentration on the right side, there will be a net motion (diffusion) of molecules from left to right.

where D is a constant of proportionality. The left side of this equation is called the *diffusion rate* and is a measure of the mass being transported per unit time. The equation says that the rate of diffusion is proportional to the cross-sectional area A and to the change in concentration per unit distance, $(C_2 - C_1)/L$, which is called the *concentration gradient*. The concentrations C_1 and C_2 are measured in kilograms per cubic meter. The proportionality constant D is called the **diffusion coefficient** and has units of square meters per second. Table 9.6 (page 306) lists diffusion coefficients for a few substances.

TABLE 9.6

Diffusion Coefficients of Various Substances at 20°C

Substance	D (m²/s)
Oxygen through air	6.4×10^{-5}
Oxygen through tissue	1×10^{-11}
Oxygen through water	1×10^{-9}
Sucrose through water	5×10^{-10}
Hemoglobin through water	76×10^{-11}

APPLICATION

 Effect of Osmosis on Living Cells

APPLICATION

 Kidney Function and Dialysis

The Size of Cells and Osmosis

Diffusion through cell membranes is vital in carrying oxygen to the cells of the body and in removing carbon dioxide and other waste products from them. Cells require oxygen for those metabolic processes in which substances are either synthesized or broken down. In such processes, the cell uses up oxygen and produces carbon dioxide as a by-product. A fresh supply of oxygen diffuses from the blood, where its concentration is high, into the cell, where its concentration is low. Likewise, carbon dioxide diffuses from the cell into the blood, where it is in lower concentration. Water, ions, and other nutrients also pass into and out of cells by diffusion.

A cell can function properly only if it can transport nutrients and waste products rapidly across the cell membrane. The surface area of the cell should be large enough so that the exposed membrane area can exchange materials effectively while the volume should be small enough so that materials can reach or leave particular locations rapidly. This requires a large surface-area-to-volume ratio.

Model a cell as a cube, each side with length L. The total surface area is $6L^2$ and the volume is L^3. The surface area to volume is then

$$\frac{\text{surface area}}{\text{volume}} = \frac{6L^2}{L^3} = \frac{6}{L}$$

Because L is in the denominator, a smaller L means a larger ratio. This shows that the smaller the size of a body, the more efficiently it can transport nutrients and waste products across the cell membrane. Cells range in size from a millionth of a meter to several millionths, so a good estimate of a typical cell's surface-to-volume ratio is 10^6.

The diffusion of material through a membrane is partially determined by the size of the pores (holes) in the membrane wall. Small molecules, such as water, may pass through the pores easily, while larger molecules, such as sugar, may pass through only with difficulty or not at all. A membrane that allows passage of some molecules but not others is called a **selectively permeable** membrane.

Osmosis is the diffusion of water across a selectively permeable membrane from a high water concentration to a low water concentration. As in the case of diffusion, osmosis continues until the concentrations on the two sides of the membrane are equal.

To understand the effect of osmosis on living cells, consider a particular cell in the body that contains a sugar concentration of 1%. (A 1% solution is 1 g of sugar dissolved in enough water to make 100 ml of solution; "ml" is the abbreviation for milliliters, so 10^{-3} L = 1 cm³.) Assume this cell is immersed in a 5% sugar solution (5 g of sugar dissolved in enough water to make 100 ml). Compared to the 1% solution, there are five times as many sugar molecules per unit volume in the 5% sugar solution, so there must be fewer water molecules. Accordingly, water will diffuse from inside the cell, where its concentration is higher, across the cell membrane to the outside solution, where the concentration of water is lower. This loss of water from the cell would cause it to shrink and perhaps become damaged through dehydration. If the concentrations were reversed, water would diffuse *into* the cell, causing it to swell and perhaps burst. If solutions are introduced into the body intravenously, care must be taken to ensure that they don't disturb the osmotic balance of the body, else cell damage can occur. For example, if a 9% saline solution surrounds a red blood cell, the cell will shrink. By contrast, if the solution is about 1%, the cell will eventually burst.

In the body, blood is cleansed of impurities by osmosis as it flows through the kidneys. (See Fig. 9.51a.) Arterial blood first passes through a bundle of capillaries known as a *glomerulus*, where most of the waste products and some essential salts and minerals are removed. From the glomerulus, a narrow tube emerges that is in intimate contact with other capillaries throughout its length. As blood passes through the tubules, most of the essential elements are returned to it; waste products are not allowed to reenter and are eventually removed in urine.

If the kidneys fail, an artificial kidney or a dialysis machine can filter the blood. Figure 9.51b shows how this is done. Blood from an artery in the arm is mixed

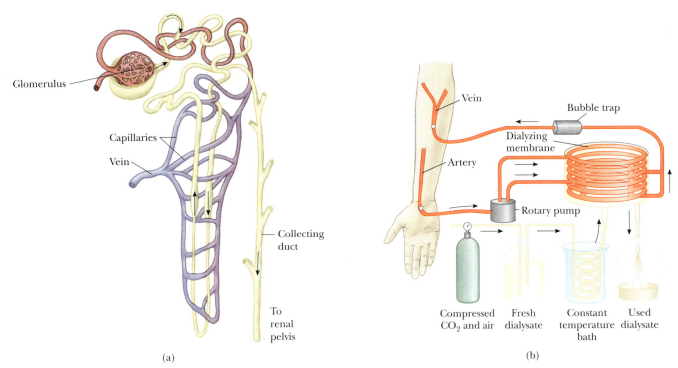

Glomerulus

Capillaries

Vein

Collecting
duct

To
renal
pelvis

(a)

Vein

Bubble trap

Dialyzing
membrane

Artery

Rotary pump

Compressed   Fresh   Constant   Used
CO_2 and air dialysate temperature dialysate
                           bath

(b)

Figure 9.51   (a) Diagram of a single nephron in the human excretory system. (b) An artificial kidney.

with heparin, a blood thinner, and allowed to pass through a tube covered with a semipermeable membrane. The tubing is immersed in a bath of a dialysate fluid with the same chemical composition as purified blood. Waste products from the blood enter the dialysate by diffusion through the membrane. The filtered blood is then returned to a vein.

Motion through a Viscous Medium

When an object falls through air, its motion is impeded by the force of air resistance. In general, this force is dependent on the shape of the falling object and on its velocity. The force of air resistance acts on all falling objects, but the exact details of the motion can be calculated only for a few cases in which the object has a simple shape, such as a sphere. In this section, we will examine the motion of a tiny spherical object falling slowly through a viscous medium.

In 1845 a scientist named George Stokes found that the magnitude of the resistive force on a very small spherical object of radius r falling slowly through a fluid of viscosity η with speed v is given by

$$F_r = 6\pi\eta rv \qquad \text{[9.28]}$$

This equation, called **Stokes's law**, has many important applications. For example, it describes the sedimentation of particulate matter in blood samples. It was used by Robert Millikan (1886–1953) to calculate the radius of charged oil droplets falling through air. From this, Millikan was ultimately able to determine the charge on the electron, and was awarded the Nobel prize in 1923 for his pioneering work on elemental charges.

As a sphere falls through a viscous medium, three forces act on it, as shown in Figure 9.52: $\vec{F}_r$, the force of friction; $\vec{B}$, the buoyant force of the fluid; and $\vec{w}$, the force of gravity acting on the sphere. The magnitude of $\vec{w}$ is given by

$$w = \rho g V = \rho g \left(\frac{4}{3}\pi r^3\right)$$

where ρ is the density of the sphere and $\frac{4}{3}\pi r^3$ is its volume. According to Archimedes's principle, the magnitude of the buoyant force is equal to the weight

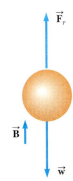

Figure 9.52   A sphere falling through a viscous medium. The forces acting on the sphere are the resistive frictional force $\vec{F}_r$, the buoyant force $\vec{B}$, and the force of gravity $\vec{w}$ acting on the sphere.

of the fluid displaced by the sphere,

$$B = \rho_f g V = \rho_f g \left(\frac{4}{3} \pi r^3 \right)$$

where ρ_f is the density of the fluid.

At the instant the sphere begins to fall, the force of friction is zero because the speed of the sphere is zero. As the sphere accelerates, its speed increases and so does $\vec{F}_r$. Finally, **at a speed called the terminal speed v_t, the net force goes to zero.** This occurs when the net upward force balances the downward force of gravity. Therefore, the sphere reaches terminal speed when

$$F_r + B = w$$

or

$$6 \pi \eta r v_t + \rho_f g \left(\frac{4}{3} \pi r^3 \right) = \rho g \left(\frac{4}{3} \pi r^3 \right)$$

When this equation is solved for v_t, we get

Terminal speed ▶

$$v_t = \frac{2 r^2 g}{9 \eta} (\rho - \rho_f) \qquad \text{[9.29]}$$

Sedimentation and Centrifugation

If an object isn't spherical, we can still use the basic approach just described to determine its terminal speed. The only difference is that we can't use Stokes's law for the resistive force. Instead, we assume that the resistive force has a magnitude given by $F_r = kv$, where k is a coefficient that must be determined experimentally. As discussed previously, the object reaches its terminal speed when the downward force of gravity is balanced by the net upward force, or

$$w = B + F_r \qquad \text{[9.30]}$$

where $B = \rho_f g V$ is the buoyant force. The volume V of the displaced fluid is related to the density ρ of the falling object by $V = m/\rho$. Hence, we can express the buoyant force as

$$B = \frac{\rho_f}{\rho} mg$$

We substitute this expression for B and $F_r = kv_t$ into Equation 9.30 (terminal speed condition):

$$mg = \frac{\rho_f}{\rho} mg + kv_t$$

or

$$v_t = \frac{mg}{k} \left(1 - \frac{\rho_f}{\rho} \right) \qquad \text{[9.31]}$$

The terminal speed for particles in biological samples is usually quite small. For example, the terminal speed for blood cells falling through plasma is about 5 cm/h in the gravitational field of the Earth. The terminal speeds for the molecules that make up a cell are many orders of magnitude smaller than this because of their much smaller mass. The speed at which materials fall through a fluid is called the **sedimentation rate** and is important in clinical analysis.

The sedimentation rate in a fluid can be increased by increasing the effective acceleration g that appears in Equation 9.31. A fluid containing various biological molecules is placed in a centrifuge and whirled at very high angular speeds (Fig. 9.53). Under these conditions, the particles gain a large radial acceleration

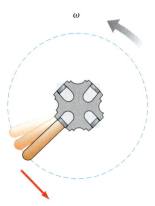

Figure 9.53 Simplified diagram of a centrifuge (top view).

$a_c = v^2/r = \omega^2 r$ that is much greater than the free-fall acceleration, so we can replace g in Equation 9.31 by $\omega^2 r$ and obtain

$$v_t = \frac{m\omega^2 r}{k}\left(1 - \frac{\rho_f}{\rho}\right) \qquad [9.32]$$

This equation indicates that the sedimentation rate is enormously speeded up in a centrifuge ($\omega^2 r \gg g$) and that those particles with the greatest mass will have the largest terminal speed. Consequently the most massive particles will settle out on the bottom of a test tube first.

SUMMARY

Physics⊗Now™ Take a practice test by logging into PhysicsNow at **www.cp7e.com** and clicking on the Pre-Test link for this chapter.

9.1 States of Matter

Matter is normally classified as being in one of three states: solid, liquid, or gaseous. The fourth state of matter is called a plasma, which consists of a neutral system of charged particles interacting electromagnetically.

9.2 The Deformation of Solids

The elastic properties of a solid can be described using the concepts of stress and strain. **Stress** is related to the force per unit area producing a deformation; **strain** is a measure of the amount of deformation. Stress is proportional to strain, and the constant of proportionality is the **elastic modulus**:

$$\text{Stress} = \text{Elastic modulus} \times \text{strain} \qquad [9.1]$$

Three common types of deformation are (1) the resistance of a solid to elongation or compression, characterized by **Young's modulus** Y; (2) the resistance to displacement of the faces of a solid sliding past each other, characterized by the shear modulus S; and (3) the resistance of a solid or liquid to a change in volume, characterized by the bulk modulus B.

All three types of deformation obey laws similar to Hooke's law for springs. Solving problems is usually a matter of identifying the given physical variables and solving for the unknown variable.

9.3 Density and Pressure

The **density** ρ of a substance of uniform composition is its mass per unit volume—kilograms per cubic meter (kg/m^3) in the SI system:

$$\rho \equiv \frac{M}{V} \qquad [9.6]$$

The **pressure** P in a fluid, measured in pascals (Pa), is the force per unit area that the fluid exerts on an object immersed in it:

$$P \equiv \frac{F}{A} \qquad [9.7]$$

9.4 Variation of Pressure with Depth

The pressure in an incompressible fluid varies with depth h according to the expression

$$P = P_0 + \rho g h \qquad [9.11]$$

where P_0 is atmospheric pressure (1.013×10^5 Pa) and ρ is the density of the fluid.

Pascal's principle states that when pressure is applied to an enclosed fluid, the pressure is transmitted undiminished to every point of the fluid and to the walls of the containing vessel.

9.6 Buoyant Forces and Archimedes's Principle

When an object is partially or fully submerged in a fluid, the fluid exerts an upward force, called the **buoyant force**, on the object. This force is, in fact, just the net difference in pressure between the top and bottom of the object. It can be shown that the magnitude of the buoyant force B is equal to the weight of the fluid displaced by the object, or

$$B = \rho_{\text{fluid}} V_{\text{fluid}} g \qquad [9.12b]$$

Equation 9.12b is known as **Archimedes's principle**.

Solving a buoyancy problem usually involves putting the buoyant force into Newton's second law and then proceeding as in Chapter 4.

9.7 Fluids in Motion

Certain aspects of a fluid in motion can be understood by assuming that the fluid is nonviscous and incompressible and that its motion is in a steady state with no turbulence:

1. The flow rate through the pipe is a constant, which is equivalent to stating that the product of the cross-sectional area A and the speed v at any point is constant. At any two points, therefore, we have

$$A_1 v_1 = A_2 v_2 \qquad [9.15]$$

This relation is referred to as the **equation of continuity**.

2. The sum of the pressure, the kinetic energy per unit volume, and the potential energy per unit volume is the same at any two points along a streamline:

$$P_1 + \tfrac{1}{2}\rho v_1{}^2 + \rho g y_1 = P_2 + \tfrac{1}{2}\rho v_2{}^2 + \rho g y_2 \qquad [9.16]$$

Equation 9.16 is known as **Bernoulli's equation**. Solving problems with Bernoulli's equation is similar to solving problems with the work–energy theorem, whereby two points are chosen, one point where a quantity is unknown and another where all quantities are known. Equation 9.16 is then solved for the unknown quantity.

CONCEPTUAL QUESTIONS

1. Baseball home-run hitters like to play in Denver, but curveball pitchers do not. Why?

2. The density of air is 1.3 kg/m^3 at sea level. From your knowledge of air pressure at ground level, estimate the height of the atmosphere. As a simplifying assumption, take the atmosphere to be of uniform density up to some height, after which the density rapidly falls to zero. (In reality, the density of the atmosphere decreases as we go up.) (Question 2 is courtesy of Edward F. Redish. For more questions of this type, see http://www.physics.umd.edu/perg/.)

3. A woman wearing high-heeled shoes is invited into a home in which the kitchen has vinyl floor covering. Why should the homeowner be concerned?

4. Figure Q9.4 shows aerial views from directly above two dams. Both dams are equally long (the vertical dimension in the diagram) and equally deep (into the page in the diagram). The dam on the left holds back a very large lake, while the dam on the right holds back a narrow river. Which dam has to be built more strongly?

 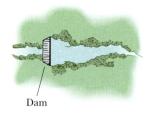

Dam Dam

Figure Q9.4

5. A typical silo on a farm has many bands wrapped around its perimeter, as shown in Figure Q9.5. Why is the spacing between successive bands smaller at the lower portions of the silo?

6. During inhalation, the pressure in the lungs is slightly less than external pressure and the muscles controlling exhalation are relaxed. Under water, the body equalizes internal and external pressures. Discuss the condition of the muscles if a person under water is breathing through a snorkel. Would a snorkel work in deep water?

7. Colloquially, we say a nurse uses a syringe to "draw" blood. Why is it more accurate to say that a syringe is used to "accept" blood?

8. Many people believe that a vacuum created inside a vacuum cleaner causes particles of dirt to be drawn in. Actually, the dirt is pushed in. Explain.

9. Suppose a damaged ship just barely floats in the ocean after a hole in its hull has been sealed. It is pulled by a tugboat toward shore and into a river, heading toward a dry dock for repair. As the boat is pulled up the river, it sinks. Why?

10. Will an ice cube float higher in water or in an alcoholic beverage?

11. A pound of Styrofoam and a pound of lead have the same weight. If they are placed on a sensitive equal-arm balance, will the scales balance?

12. An ice cube is placed in a glass of water. What happens to the level of the water as the ice melts?

13. Place two cans of soft drinks, one regular and one diet, in a container of water. You will find that the diet drink floats while the regular one sinks. Use Archimedes' principle to devise an explanation. [*Broad Hint:* The artificial sweetener used in diet drinks is less dense than sugar.]

14. Prairie dogs live in underground burrows with at least two entrances. They ventilate their burrows by building a mound over one entrance, as shown in Figure Q9.14. This entrance is open to a stream of air when a breeze blows from any direction. A second entrance at ground level is open to almost stagnant air. How does this construction create an airflow through the burrow?

Figure Q9.5

Figure Q9.14

15. When you are driving a small car on the freeway and a truck passes you at high speed, why do you feel pulled toward the truck?

16. A barge is carrying a load of gravel along a river. It approaches a low bridge, and the captain realizes that the top of the pile of gravel is not going to make it under the bridge. The captain orders the crew to quickly shovel gravel from the pile into the water. Is this a good decision?

17. Tornadoes and hurricanes often lift the roofs of houses. Use the Bernoulli effect to explain why. Why should you keep your windows open under these conditions?

18. Water is poured to the same level in each of the three vessels shown in Figure Q9.18. Each vessel has the same base area. Because the water fills each to the same depth, each vessel will have the same pressure at the bottom. Because the area and pressure at each base is the same, each liquid should exert the same force on the base of the vessel. Yet, if the vessels are weighed, three different values are obtained. (The one in the center clearly holds less liquid than the one at the left, so it will weigh less.) How can you resolve this apparent contradiction? (Question 18 is courtesy of Edward F. Redish. For more questions of this type, see http://www.physics.umd.edu/perg/.)

Figure Q9.18

PROBLEMS

1, 2, 3 = straightforward, intermediate, challenging   ☐ = full solution available in *Student Solutions Manual/Study Guide*
Physics⊗Now™ = coached problem with hints available at **www.cp7e.com**   🧬 = biomedical application

Section 9.1 States of Matter
Section 9.2 The Deformation of Solids

1. If the elastic limit of steel is 5.0×10^8 Pa, determine the minimum diameter a steel wire can have if it is to support a 70-kg circus performer without its elastic limit being exceeded.

2. If the shear stress in steel exceeds about 4.00×10^8 N/m^2, the steel ruptures. Determine the shearing force necessary to (a) shear a steel bolt 1.00 cm in diameter and (b) punch a 1.00-cm-diameter hole in a steel plate 0.500 cm thick.

3. The heels on a pair of women's shoes have radii of 0.50 cm at the bottom. If 30% of the weight of a woman weighing 480 N is supported by each heel, find the stress on each heel.

4. When water freezes, it expands about 9.00%. What would be the pressure increase inside your automobile engine block if the water in it froze? The bulk modulus of ice is 2.00×10^9 N/m^2.

5. For safety in climbing, a mountaineer uses a nylon rope that is 50 m long and 1.0 cm in diameter. When supporting a 90-kg climber, the rope elongates 1.6 m. Find its Young's modulus.

6. A stainless-steel orthodontic wire is applied to a tooth, as in Figure P9.6. The wire has an unstretched length of 3.1 cm and a diameter of 0.22 mm. If the wire is stretched 0.10 mm, find the magnitude and direction of the force on the tooth. Disregard the width of the tooth, and assume that Young's modulus for stainless steel is 18×10^{10} Pa.

Figure P9.6

7. Bone has a Young's modulus of about 18×10^9 Pa. Under compression, it can withstand a stress of about 160×10^6 Pa before breaking. Assume that a femur (thighbone) is 0.50 m long, and calculate the amount of compression this bone can withstand before breaking.

8. The distortion of the Earth's crustal plates is an example of shear on a large scale. A particular crustal rock has a shear modulus of 1.5×10^{10} Pa. What shear stress is involved when a 10-km layer of this rock is sheared through a distance of 5.0 m?

9. A child slides across a floor in a pair of rubber-soled shoes. The friction force acting on each foot is 20 N, the footprint area of each foot is 14 cm^2, and the thickness of the soles is 5.0 mm. Find the horizontal distance traveled by the sheared face of the sole. The shear modulus of the rubber is 3.0×10^6 Pa.

10. A high-speed lifting mechanism supports an 800-kg object with a steel cable that is 25.0 m long and 4.00 cm^2 in cross-sectional area. (a) Determine the elongation of the cable. (b) By what additional amount does the cable increase in length if the object is accelerated upwards at a rate of 3.0 m/s^2? (c) What is the greatest mass that can be accelerated upwards at 3.0 m/s^2 if the stress in the cable is not to exceed the elastic limit of the cable, which is 2.2×10^8 Pa?

11. Determine the elongation of the rod in Figure P9.11 if it is under a tension of 5.8×10^3 N.

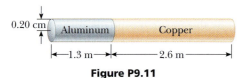

0.20 cm

Aluminum   Copper

←—1.3 m—→|←——2.6 m——→|

Figure P9.11

12. The total cross-sectional area of the load-bearing calcified portion of the two forearm bones (radius and ulna) is approximately 2.4 cm². During a car crash, the forearm is slammed against the dashboard. The arm comes to rest from an initial speed of 80 km/h in 5.0 ms. If the arm has an effective mass of 3.0 kg and bone material can withstand a maximum compressional stress of 16×10^7 Pa, is the arm likely to withstand the crash?

Section 9.3 Density and Pressure

13. A 50.0-kg ballet dancer stands on her toes during a performance with four square inches (26.0 cm²) in contact with the floor. What is the pressure exerted by the floor over the area of contact (a) if the dancer is stationary and (b) if the dancer is leaping upwards with an acceleration of 4.00 m/s²?

14. The four tires of an automobile are inflated to a gauge pressure of 2.0×10^5 Pa. Each tire has an area of 0.024 m² in contact with the ground. Determine the weight of the automobile.

15. Air is trapped above liquid ethyl alcohol in a rigid container, as shown in Figure P9.15. If the air pressure above the liquid is 1.10 atm, determine the pressure inside a bubble 4.0 m below the surface of the liquid.

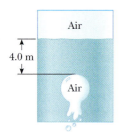

Air

4.0 m

Air

Figure P9.15

16. A 70-kg man in a 5.0-kg chair tilts back so that all the weight is balanced on two legs of the chair. Assume that each leg makes contact with the floor over a circular area with a radius of 1.0 cm, and find the pressure exerted by each leg on the floor.

17. If 1.0 m³ of concrete weighs 5.0×10^4 N, what is the height of the tallest cylindrical concrete pillar that will not collapse under its own weight? The compression strength

of concrete (the maximum pressure that can be exerted on the base of the structure) is 1.7×10^7 Pa.

Section 9.3 Density and Pressure

Section 9.4 Variation of Pressure with Depth

Section 9.5 Pressure Measurements

18. The deepest point in the ocean is in the Mariana Trench, about 11 km deep. The pressure at the ocean floor is huge, about 1.13×10^8 N/m². (a) Calculate the change in volume of 1.00 m³ of water carried from the surface to the bottom of the Pacific. (b) The density of water at the surface is 1.03×10^3 kg/m³. Find its density at the bottom. (c) Is it a good approximation to think of water as incompressible?

19. A collapsible plastic bag (Figure P9.19) contains a glucose solution. If the average gauge pressure in the vein is 1.33×10^3 Pa, what must be the minimum height h of the bag in order to infuse glucose into the vein? Assume that the specific gravity of the solution is 1.02.

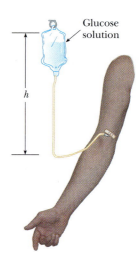

Glucose solution

h

Figure P9.19

20. (a) A very powerful vacuum cleaner has a hose 2.86 cm in diameter. With no nozzle on the hose, what is the weight of the heaviest brick it can lift? (b) A very powerful octopus uses one sucker, of diameter 2.86 cm, on each of the two shells of a clam, in an attempt to pull the shells apart. Find the greatest force the octopus can exert in salt water 32.3 m deep.

21. For the cellar of a new house, a hole is dug in the ground, with vertical sides going down 2.40 m. A concrete foundation wall is built all the way across the 9.60-m width of the excavation. The foundation wall is 0.183 m away from the front of the cellar hole. During a rainstorm, drainage from the street fills up the space in front of the concrete wall, but not the cellar behind the wall. The water does not soak into the clay soil. Find the force the water

exerts on the foundation wall. For comparison, the weight of the water is $2.40 \text{ m} \times 9.60 \text{ m} \times 0.183 \text{ m} \times 1\,000 \text{ kg/m}^3 \times 9.80 \text{ m/s}^2 = 41.3 \text{ kN}$.

22. Blaise Pascal duplicated Torricelli's barometer using a red Bordeaux wine of density 984 kg/m^3 as the working liquid (Fig. P9.22). What was the height h of the wine column for normal atmospheric pressure? Would you expect the vacuum above the column to be as good as for mercury?

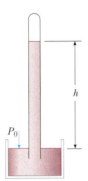

Figure P9.22

23. **Physics ⊗ Now™** A container is filled to a depth of 20.0 cm with water. On top of the water floats a 30.0-cm-thick layer of oil with specific gravity 0.700. What is the absolute pressure at the bottom of the container?

24. Piston ① in Figure P9.24 has a diameter of 0.25 in.; piston ② has a diameter of 1.5 in. In the absence of friction, determine the force $\vec{F}$ necessary to support the 500-lb weight.

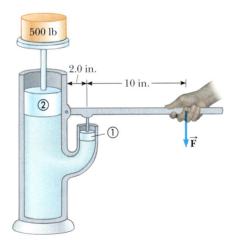

Figure P9.24

25. Figure P9.25 shows the essential parts of a hydraulic brake system. The area of the piston in the master cylinder is 6.4 cm^2, and that of the piston in the brake cylinder is 1.8 cm^2. The coefficient of friction between shoe and

wheel drum is 0.50. If the wheel has a radius of 34 cm, determine the frictional torque about the axle when a force of 44 N is exerted on the brake pedal.

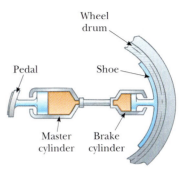

Figure P9.25

Section 9.6 Buoyant Forces and Archimedes' Principle

26. A frog in a hemispherical pod finds that he just floats without sinking in a fluid of density 1.35 g/cm^3. If the pod has a radius of 6.00 cm and negligible mass, what is the mass of the frog? (See Fig. P9.26.)

Figure P9.26

27. A small ferryboat is 4.00 m wide and 6.00 m long. When a loaded truck pulls onto it, the boat sinks an additional 4.00 cm into the river. What is the weight of the truck?

28. The density of ice is 920 kg/m^3, and that of sea water is $1\,030 \text{ kg/m}^3$. What fraction of the total volume of an iceberg is exposed?

29. As a first approximation, the Earth's continents may be thought of as granite blocks floating in a denser rock (called peridotite) in the same way that ice floats in water. (a) Show that a formula describing this phenomenon is

$$\rho_g t = \rho_p d$$

where ρ_g is the density of granite ($2.8 \times 10^3 \text{ kg/m}^3$), ρ_p is the density of peridotite ($3.3 \times 10^3 \text{ kg/m}^3$), t is the thickness of a continent, and d is the depth to which a continent floats in the peridotite. (b) If a continent sinks 5.0 km into the peridotite layer (this surface may be thought of as the ocean floor), what is the thickness of the continent?

30. A 10.0-kg block of metal is suspended from a scale and immersed in water, as in Figure P9.30. The dimensions of the block are 12.0 cm × 10.0 cm × 10.0 cm. The 12.0-cm dimension is vertical, and the top of the block is 5.00 cm below the surface of the water. (a) What are the forces exerted by the water on the top and bottom of the block? (Take $P_0 = 1.013\ 0 \times 10^5\ \text{N/m}^2$.) (b) What is the reading of the spring scale? (c) Show that the buoyant force equals the difference between the forces at the top and bottom of the block.

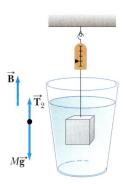

Figure P9.30

31. A bathysphere used for deep sea exploration has a radius of 1.50 m and a mass of 1.20×10^4 kg. In order to dive, the sphere takes on mass in the form of sea water. Determine the mass the bathysphere must take on so that it can descend at a constant speed of 1.20 m/s when the resistive force on it is 1 100 N upward. The density of sea water is $1.03 \times 10^3\ \text{kg/m}^3$.

32. The United States possesses the eight largest warships in the world—aircraft carriers of the *Nimitz* class—and is building one more. Suppose that, at a location where $g = 9.78\ \text{m/s}^2$, one of the ships bobs up to float 11.0 cm higher in the water when 50 fighters take off from it in 25 minutes. Bristling with bombs and missiles, each plane has an average mass of 29 000 kg. Find the horizontal area enclosed by the waterline of the $4-billion ship. By comparison, its flight deck has area of 18 000 m². Below decks are passageways hundreds of meters long, so narrow that two large men cannot pass each other.

33. An empty rubber balloon has a mass of 0.012 0 kg. The balloon is filled with helium at a density of 0.181 kg/m³. At this density, the balloon has a radius of 0.500 m. If the filled balloon is fastened to a vertical line, what is the tension in the line?

34. A light spring of force constant $k = 160$ N/m rests vertically on the bottom of a large beaker of water (Fig. P9.34a). A 5.00-kg block of wood (density = 650 kg/m³) is connected to the spring, and the block–spring system is allowed to come to static equilibrium (Fig. P9.34b). What is the elongation ΔL of the spring?

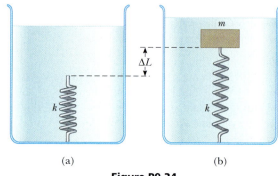

(a)   (b)

Figure P9.34

35. A sample of an unknown material appears to weigh 300 N in air and 200 N when immersed in alcohol of specific gravity 0.700. What are (a) the volume and (b) the density of the material?

36. An object weighing 300 N in air is immersed in water after being tied to a string connected to a balance. The scale now reads 265 N. Immersed in oil, the object appears to weigh 275 N. Find (a) the density of the object and (b) the density of the oil.

37. A thin spherical shell of mass 0.400 kg and diameter 0.200 m is filled with alcohol ($\rho = 806\ \text{kg/m}^3$). It is then released from rest at the bottom of a pool of water. Find the acceleration of the alcohol-filled shell as it starts to rise toward the surface of the water.

38. A rectangular air mattress is 2.0 m long, 0.50 m wide, and 0.08 m thick. If it has a mass of 2.0 kg, what additional mass can it support in water?

39. **Physics ⊗ Now™** A 1.00-kg beaker containing 2.00 kg of oil (density = 916 kg/m³) rests on a scale. A 2.00-kg block of iron is suspended from a spring scale and is completely submerged in the oil (Fig. P9.39). Find the equilibrium readings of both scales.

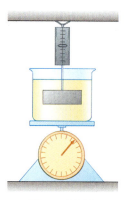

Figure P9.39

Section 9.7 Fluids in Motion

Section 9.8 Other Applications of Fluid Dynamics

40. Water is pumped into a storage tank from a well delivering 20.0 gallons of water in 30.0 seconds through a pipe of 1.00-in.² cross-sectional area. What is the average velocity of the water in the pipe as the water is pumped from the well?

41. (a) Calculate the mass flow rate (in grams per second) of blood ($\rho = 1.0 \text{ g/cm}^3$) in an aorta with a cross-sectional area of 2.0 cm^2 if the flow speed is 40 cm/s. (b) Assume that the aorta branches to form a large number of capillaries with a combined cross-sectional area of $3.0 \times 10^3 \text{ cm}^2$. What is the flow speed in the capillaries?

42. A liquid ($\rho = 1.65 \text{ g/cm}^3$) flows through two horizontal sections of tubing joined end to end. In the first section, the cross-sectional area is 10.0 cm^2, the flow speed is 275 cm/s, and the pressure is $1.20 \times 10^5 \text{ Pa}$. In the second section, the cross-sectional area is 2.50 cm^2. Calculate the smaller section's (a) flow speed and (b) pressure.

43. A hypodermic syringe contains a medicine with the density of water (Fig. P9.43). The barrel of the syringe has a cross-sectional area of $2.50 \times 10^{-5} \text{ m}^2$. In the absence of a force on the plunger, the pressure everywhere is 1.00 atm. A force $\vec{\mathbf{F}}$ of magnitude 2.00 N is exerted on the plunger, making medicine squirt from the needle. Determine the medicine's flow speed through the needle. Assume that the pressure in the needle remains equal to 1.00 atm and that the syringe is horizontal.

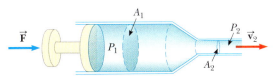

Figure P9.43

44. When a person inhales, air moves down the bronchus (windpipe) at 15 cm/s. The average flow speed of the air doubles through a constriction in the bronchus. Assuming incompressible flow, determine the pressure drop in the constriction.

45. **Physics⊗Now™** A jet of water squirts out horizontally from a hole near the bottom of the tank shown in Figure P9.45. If the hole has a diameter of 3.50 mm, what is the height h of the water level in the tank?

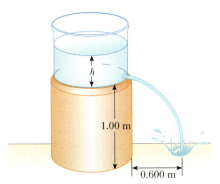

Figure P9.45

46. A large storage tank, open to the atmosphere at the top and filled with water, develops a small hole in its side at a point 16.0 m below the water level. If the rate of flow from the leak is $2.50 \times 10^{-3} \text{ m}^3/\text{min}$, determine (a) the speed at which the water leaves the hole and (b) the diameter of the hole.

47. The inside diameters of the larger portions of the horizontal pipe depicted in Figure P9.47 are 2.50 cm. Water flows to the right at a rate of $1.80 \times 10^{-4} \text{ m}^3/\text{s}$. Determine the inside diameter of the constriction.

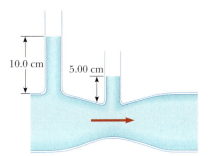

Figure P9.47

48. Water is pumped through a pipe of diameter 15.0 cm from the Colorado River up to Grand Canyon Village, on the rim of the canyon. The river is at 564 m elevation and the village is at 2 096 m. (a) At what minimum pressure must the water be pumped to arrive at the village? (b) If $4\,500 \text{ m}^3$ are pumped per day, what is the speed of the water in the pipe? (c) What additional pressure is necessary to deliver this flow? [*Note:* You may assume that the free-fall acceleration and the density of air are constant over the given range of elevations.]

49. Old Faithful geyser in Yellowstone Park erupts at approximately 1-hour intervals, and the height of the fountain reaches 40.0 m. (a) Consider the rising stream as a series of separate drops. Analyze the free-fall motion of one of the drops to determine the speed at which the water leaves the ground. (b) Treat the rising stream as an ideal fluid in streamline flow. Use Bernoulli's equation to determine the speed of the water as it leaves ground level. (c) What is the pressure (above atmospheric pressure) in the heated underground chamber 175 m below the vent? You may assume that the chamber is large compared with the geyser vent.

50. An airplane is cruising at an altitude of 10 km. The pressure outside the craft is 0.287 atm; within the passenger compartment, the pressure is 1.00 atm and the temperature is 20°C. The density of air is 1.20 kg/m^3 at 20°C and 1 atm of pressure. A small leak forms in one of the window seals in the passenger compartment. Model the air as an ideal fluid to find the speed of the stream of air flowing through the leak.

51. A siphon is a device that allows a fluid to seemingly defy gravity (Fig. P9.51). The flow must be initiated by a partial vacuum in the tube, as in a drinking straw. (a) Show that the speed at which the water emerges from the siphon is given by $v = \sqrt{2gh}$. (b) For what values of y will the siphon work?

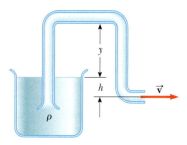

Figure P9.51

Section 9.9 Surface Tension, Capillary Action, and Viscous Fluid Flow

52. In order to lift a wire ring of radius 1.75 cm from the surface of a container of blood plasma, a vertical force of 1.61×10^{-2} N greater than the weight of the ring is required. Calculate the surface tension of blood plasma from this information.

53. A square metal sheet 3.0 cm on a side and of negligible thickness is attached to a balance and inserted into a container of fluid. The contact angle is found to be zero, as shown in Figure P9.53a, and the balance to which the metal sheet is attached reads 0.40 N. A thin veneer of oil is then spread over the sheet, and the contact angle becomes 180°, as shown in Figure P9.53b. The balance now reads 0.39 N. What is the surface tension of the fluid?

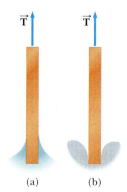

(a)   (b)

Figure P9.53

54. Whole blood has a surface tension of 0.058 N/m and a density of 1 050 kg/m³. To what height can whole blood rise in a capillary blood vessel that has a radius of 2.0×10^{-6} m if the contact angle is zero?

55. A certain fluid has a density of 1 080 kg/m³ and is observed to rise to a height of 2.1 cm in a 1.0-mm-diameter tube. The contact angle between the wall and the fluid is zero. Calculate the surface tension of the fluid.

56. A staining solution used in a microbiology laboratory has a surface tension of 0.088 N/m and a density 1.035 times the density of water. What must be the diameter of a capillary tube so that this solution will rise to a height of 5 cm? (Assume a contact angle of zero.)

57. The block of ice (temperature 0°C) shown in Figure P9.57 is drawn over a level surface lubricated by a layer of

water 0.10 mm thick. Determine the magnitude of the force $\vec{F}$ needed to pull the block with a constant speed of 0.50 m/s. At 0°C, the viscosity of water has the value $\eta = 1.79 \times 10^{-3}$ N·s/m².

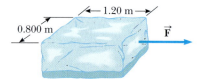

Figure P9.57

58. A thin 1.5-mm coating of glycerine has been placed between two microscope slides of width 1.0 cm and length 4.0 cm. Find the force required to pull one of the microscope slides at a constant speed of 0.30 m/s relative to the other slide.

59. A straight horizontal pipe with a diameter of 1.0 cm and a length of 50 m carries oil with a coefficient of viscosity of 0.12 N·s/m². At the output of the pipe, the flow rate is 8.6×10^{-5} m³/s and the pressure is 1.0 atm. Find the gauge pressure at the pipe input.

60. The pulmonary artery, which connects the heart to the lungs, has an inner radius of 2.6 mm and is 8.4 cm long. If the pressure drop between the heart and lungs is 400 Pa, what is the average speed of blood in the pulmonary artery?

61. Spherical particles of a protein of density 1.8 g/cm³ are shaken up in a solution of 20°C water. The solution is allowed to stand for 1.0 h. If the depth of water in the tube is 5.0 cm, find the radius of the largest particles that remain in solution at the end of the hour.

62. A hypodermic needle is 3.0 cm in length and 0.30 mm in diameter. What excess pressure is required along the needle so that the flow rate of water through it will be 1 g/s? (Use 1.0×10^{-3} Pa·s as the viscosity of water.)

63. What diameter needle should be used to inject a volume of 500 cm³ of a solution into a patient in 30 min? Assume that the length of the needle is 2.5 cm and that the solution is elevated 1.0 m above the point of injection. Furthermore, assume the viscosity and density of the solution are those of pure water, and assume that the pressure inside the vein is atmospheric.

64. Water is forced out of a fire extinguisher by air pressure, as shown in Figure P9.64. What gauge air pressure in the tank (above atmospheric pressure) is required for the water to have a jet speed of 30.0 m/s when the water level in the tank is 0.500 m below the nozzle?

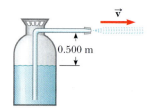

Figure P9.64

65. The aorta in humans has a diameter of about 2.0 cm, and at certain times the blood speed through it is about 55 cm/s. Is the blood flow turbulent? The density of whole blood is 1 050 kg/m³, and its coefficient of viscosity is 2.7×10^{-3} N·s/m².

66. A pipe carrying 20°C water has a diameter of 2.5 cm. Estimate the maximum flow speed if the flow must be streamline.

Section 9.10 Transport Phenomena

67. Sucrose is allowed to diffuse along a 10-cm length of tubing filled with water. The tube is 6.0 cm² in cross-sectional area. The diffusion coefficient is equal to 5.0×10^{-10} m²/s, and 8.0×10^{-14} kg is transported along the tube in 15 s. What is the difference in the concentration levels of sucrose at the two ends of the tube?

68. Glycerin in water diffuses along a horizontal column that has a cross-sectional area of 2.0 cm². The concentration gradient is 3.0×10^{-2} kg/m⁴, and the diffusion rate is found to be 5.7×10^{-15} kg/s. Determine the diffusion coefficient.

69. The viscous force on an oil drop is measured to be 3.0×10^{-13} N when the drop is falling through air with a speed of 4.5×10^{-4} m/s. If the radius of the drop is 2.5×10^{-6} m, what is the viscosity of air?

70. Small spheres of diameter 1.00 mm fall through 20°C water with a terminal speed of 1.10 cm/s. Calculate the density of the spheres.

ADDITIONAL PROBLEMS

71. An iron block of volume 0.20 m³ is suspended from a spring scale and immersed in a flask of water. Then the iron block is removed, and an aluminum block of the same volume replaces it. (a) In which case is the buoyant force the greatest, for the iron block or the aluminum block? (b) In which case does the spring scale read the largest value? (c) Use the known densities of these materials to calculate the quantities requested in parts (a) and (b). Are your calculations consistent with your previous answers to part (a) and (b)?

72. Take the density of blood to be ρ and the distance between the feet and the heart to be h_H. Ignore the flow of blood. (a) Show that the difference in blood pressure between the feet and the heart is given by $P_F - P_H = \rho g h_H$. (b) Take the density of blood to be 1.05×10^3 kg/m³ and the distance between the heart and the feet to be 1.20 m. Find the difference in blood pressure between these two points. This problem indicates that pumping blood from the extremities is very difficult for the heart. The veins in the legs have valves in them that open when blood is pumped toward the heart and close when blood flows away from the heart. Also, pumping action produced by physical activities such as walking and breathing assists the heart.

73. **Physics⊗Now**™ The approximate inside diameter of the aorta is 0.50 cm; that of a capillary is 10 μm. The approximate average blood flow speed is 1.0 m/s in the aorta and 1.0 cm/s in the capillaries. If all the blood in the aorta eventually flows through the capillaries, estimate the number of capillaries in the circulatory system.

74. Superman attempts to drink water through a very long vertical straw. With his great strength, he achieves maximum possible suction. The walls of the straw don't collapse. (a) Find the maximum height through which he can lift the water? (b) Still thirsty, the Man of Steel repeats his attempt on the Moon, which has no atmosphere. Find the difference between the water levels inside and outside the straw.

75. The human brain and spinal cord are immersed in the cerebrospinal fluid. The fluid is normally continuous between the cranial and spinal cavities and exerts a pressure of 100 to 200 mm of H_2O above the prevailing atmospheric pressure. In medical work, pressures are often measured in units of mm of H_2O because body fluids, including the cerebrospinal fluid, typically have nearly the same density as water. The pressure of the cerebrospinal fluid can be measured by means of a *spinal tap*. A hollow tube is inserted into the spinal column, and the height to which the fluid rises is observed, as shown in Figure P9.75. If the fluid rises to a height of 160 mm, we write its gauge pressure as 160 mm H_2O. (a) Express this pressure in pascals, in atmospheres, and in millimeters of mercury. (b) Sometimes it is necessary to determine whether an accident victim has suffered a crushed vertebra that is blocking the flow of cerebrospinal fluid in the spinal column. In other cases, a physician may suspect that a tumor or other growth is blocking the spinal column and inhibiting the flow of cerebrospinal fluid. Such conditions can be investigated by means of the *Queckensted test*. In this procedure, the veins in the patient's neck are compressed, to make the blood pressure rise in the brain. The increase in pressure in the blood vessels is transmitted to the cerebrospinal fluid. What should be the normal effect on the height of the fluid in the spinal tap? (c) Suppose that compressing the veins had no effect on the level of the fluid. What might account for this phenomenon?

Figure P9.75

76. Determining the density of a fluid has many important applications. A car battery contains sulfuric acid, and the battery will not function properly if the acid density is too low. Similarly, the effectiveness of antifreeze in your car's engine coolant depends on the density of the mixture (usually ethylene glycol and water). When you donate blood to a blood bank, its screening includes a determination of the density of the blood, since higher density correlates with higher hemoglobin content. A *hydrometer* is an instrument used to determine the density of a liquid. A simple one is sketched in Figure P9.76. The bulb of a syringe is squeezed and released to lift a sample of the liquid of interest into a tube containing a calibrated rod of known density. (Assume the rod is cylindrical.) The rod, of length L and average density ρ_0, floats partially immersed in the liquid of density ρ. A length h of the rod protrudes above the surface of the liquid. Show that the density of the liquid is given by

$$\rho = \frac{\rho_0 L}{L - h}$$

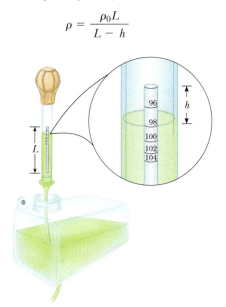

Figure P9.76

77. A 600-kg weather balloon is designed to lift a 4 000-kg package. What volume should the balloon have after being inflated with helium at standard temperature and pressure (see Table 9.3) in order that the total load can be lifted?

78. A helium-filled balloon is tied to a 2.0-m-long, 0.050-kg string. The balloon is spherical with a radius of 0.40 m. When released, it lifts a length h of the string and then remains in equilibrium, as in Figure P9.78. Determine the value of h. When deflated, the balloon has a mass of 0.25 kg. [*Hint:* Only that part of the string above the floor contributes to the load being held up by the balloon.]

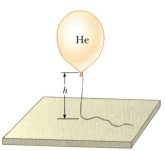

Figure P9.78

79. A block of wood weighs 50.0 N in air. A sinker is attached to the block, and the weight of the wood–sinker combination is 200 N when the sinker alone is immersed in water. Finally, the wood–sinker combination is completely immersed, and the weight is measured to be 140 N. Find the density of the block.

80. A U-tube open at both ends is partially filled with water (Fig. P9.80a). Oil ($\rho = 750$ kg/m³) is then poured into the right arm and forms a column $L = 5.00$ cm high (Fig. P9.80b). (a) Determine the difference h in the heights of the two liquid surfaces. (b) The right arm is then shielded from any air motion while air is blown across the top of the left arm until the surfaces of the two liquids are at the same height (Fig. 9.80c). Determine the speed of the air being blown across the left arm. Assume that the density of air is 1.29 kg/m³.

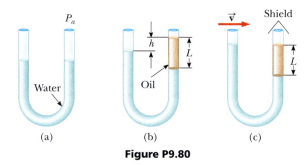

Figure P9.80

81. Figure P9.81 shows a water tank with a valve. If the valve is opened, what is the maximum height attained by the stream of water coming out of the right side of the tank? Assume that $h = 10.0$ m, $L = 2.00$ m, and $\theta = 30.0°$. Assume also that the cross-sectional area at A is very large compared with that at B.

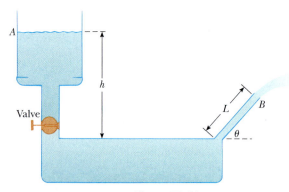

Figure P9.81

82. A solid copper ball with a diameter of 3.00 m at sea level is placed on the bottom of the ocean, at a depth of 10.0 km. If the density of sea water is 1 030 kg/m³, how much does the diameter of the ball decrease when it reaches bottom?

83. A 1.0-kg hollow ball with a radius of 0.10 m and filled with air is released from rest at the bottom of a 2.0-m-deep

pool of water. How high above the water does the ball shoot upward? Neglect all frictional effects, and neglect changes in the ball's motion when it is only partially submerged.

84. In about 1657, Otto von Guericke, inventor of the air pump, evacuated a sphere made of two brass hemispheres (Fig. P9.84). Two teams of eight horses each could pull the hemispheres apart only on some trials and even then with the greatest difficulty. (a) Show that the force required to pull the evacuated hemispheres apart is $\pi R^2 (P_0 - P)$, where R is the radius of the hemispheres and P is the pressure inside the sphere, which is much less than atmospheric pressure P_0. (b) Determine the required force if $P = 0.10\ P_0$ and $R = 0.30$ m.

Figure P9.84 The colored engraving, dated 1672, illustrates Otto von Guericke's demonstration of the force due to air pressure, as performed before Emporer Ferdinand III in 1657.

85. A 2.0-cm-thick bar of soap is floating on a water surface so that 1.5 cm of the bar is under water. Bath oil of specific gravity 0.60 is poured into the water and floats on top of it. What is the depth of the oil layer when the top of the soap is just level with the upper surface of the oil?

86. A cube of ice whose edge is 20.0 mm is floating in a glass of ice-cold water with one of its faces parallel to the water's surface. (a) How far below the water surface is the bottom face of the block? (b) Ice-cold ethyl alcohol is gently poured onto the water surface to form a layer 5.00 mm thick above the water. When the ice cube attains hydrostatic equilibrium again, what will be the distance from the top of the water to the bottom face of the

block? (c) Additional cold ethyl alcohol is poured onto the water surface until the top surface of the alcohol coincides with the top surface of the ice cube (in hydrostatic equilibrium). How thick is the required layer of ethyl alcohol?

87. A water tank open to the atmosphere at the top has two small holes punched in its side, one above the other. The holes are 5.00 cm and 12.0 cm above the floor. How high does water stand in the tank if the two streams of water hit the floor at the same place?

88. Oil having a density of 930 kg/m³ floats on water. A rectangular block of wood 4.00 cm high and with a density of 960 kg/m³ floats partly in the oil and partly in the water. The oil completely covers the block. How far below the interface between the two liquids is the bottom of the block?

89. A hollow object with an average density of 900 kg/m³ floats in a pan containing 500 cm³ of water. Ethanol is added to the water and mixed into it until the object is just on the verge of sinking. What volume of ethanol has been added? (Disregard the loss of volume caused by mixing.)

90. A walkway suspended across a hotel lobby is supported at numerous points along its edges by a vertical cable above each point and a vertical column underneath. The steel cable is 1.27 cm in diameter and is 5.75 m long before loading. The aluminum column is a hollow cylinder with an inside diameter of 16.14 cm, an outside diameter of 16.24 cm, and an unloaded length of 3.25 m. When the walkway exerts a load force of 8 500 N on one of the support points, through what distance does the point move down?

ACTIVITIES

A.1. You will need a large spring scale, a clear container partially filled with water, and a few cylinders of the same size, but made of different materials. Your instructor may be able to supply these items. Measure the volume of each cylinder. Hang a cylinder on the scale, record the reading, and start lowering the cylinder into the water. What happens to the reading on the scale as more of the cylinder is submerged? Why does the reading behave as it does? Is the effect independent of the type of material used for the cylinder? Record the reading for a cylinder when it is not submerged and when it is completely immersed. How can you use these two readings to verify Archimedes's principle?

A.2. Place an egg at the bottom of a container of fresh water. Now use a funnel to slowly add a salt solution to the water. You will observe that the egg begins to rise to the surface. Use Archimedes's principle to explain your observation.

A.3. Suppose you have the following collection of objects: a pencil, a coin, an empty plastic box for a tape cassette with its edges taped shut, a needle, an unopened can of soft drink, and an empty can of soft drink. Which of these objects do you expect will float and which will sink in water? Will it make a difference if you carefully place the object with its largest surface on the surface of the water? In which cases? Explain your reasoning. After you have written your answer, perform the experiments and compare the results with your predictions. (Activity 3 is courtesy of Edward F. Redish. For more problems of this type, see http://www.physics.umd.edu/perg/.)

High temperatures inside a volcano turn water into a high pressure steam. Unless the steam and other gases vent into the atmosphere, pressure can build until a catastrophic explosion results.

Thermal Physics

How can trapped water blow off the top of a volcano in a giant explosion? What causes a sidewalk or road to fracture and buckle spontaneously when the temperature changes? How can thermal energy be harnessed to do work, running the engines that make everything in modern living possible?

Answering these and related questions is the domain of **thermal physics**, the study of temperature, heat, and how they affect matter. Quantitative descriptions of thermal phenomena require careful definitions of the concepts of temperature, heat, and internal energy. Heat leads to changes in internal energy and thus to changes in temperature, which cause the expansion or contraction of matter. Such changes can damage roadways and buildings, create stress fractures in metal, and render flexible materials stiff and brittle, the latter resulting in compromised O-rings and the *Challenger* disaster. Changes in internal energy can also be harnessed for transportation, construction, and food preservation.

Gases are critical in the harnessing of thermal energy to do work. Within normal temperature ranges, a gas acts like a large collection of non-interacting point particles, called an ideal gas. Such gases can be studied on either a macroscopic or microscopic scale. On the macroscopic scale, the pressure, volume, temperature, and number of particles associated with a gas can be related in a single equation known as the ideal gas law. On the microscopic scale, a model called the kinetic theory of gases pictures the components of a gas as small particles. This model will enable us to understand how processes on the atomic scale affect macroscopic properties like pressure, temperature, and internal energy.

10.1 TEMPERATURE AND THE ZEROTH LAW OF THERMODYNAMICS

Temperature is commonly associated with how hot or cold an object feels when we touch it. While our senses provide us with qualitative indications of temperature, they are unreliable and often misleading. A metal ice tray feels colder to the hand, for example, than a package of frozen vegetables at the same temperature, due to the fact metals conduct thermal energy more rapidly than a cardboard package. What we need is a reliable and reproducible method of making quantitative measurements that establish the relative "hotness" or "coldness" of objects—a method related solely to temperature. Scientists have developed a variety of thermometers for making such measurements.

When placed in contact with each other, two objects at different initial temperatures will eventually reach a common intermediate temperature. If a cup of hot coffee is cooled with an ice cube, for example, the ice rises in temperature and eventually melts while the temperature of the coffee decreases.

Understanding the concept of temperature requires understanding *thermal contact* and *thermal equilibrium*. Two objects are in **thermal contact** if energy can be exchanged between them. Two objects are in **thermal equilibrium** if they are in thermal contact and there is no net exchange of energy.

The exchange of energy between two objects because of differences in their temperatures is called **heat**, a concept examined in more detail in Chapter 11.

Using these ideas, we can develop a formal definition of temperature. Consider two objects A and B that are not in thermal contact with each other, and a third object C that acts as a **thermometer**—a device calibrated to measure the temperature of an object. We wish to determine whether A and B would be in thermal equilibrium if they were placed in thermal contact. The thermometer (object C) is first placed in thermal contact with A until thermal equilibrium is reached, as in Figure 10.1a, whereupon the reading of the thermometer is recorded. The thermometer is then placed in thermal contact with B, and its reading is again recorded at equilibrium (Fig. 10.1b). If the two readings are the same, then A and B are in thermal equilibrium with each other. If A and B are placed in thermal contact with each other, as in Figure 10.1c, there is no net transfer of energy between them.

We can summarize these results in a statement known as the **zeroth law of thermodynamics (the law of equilibrium)**:

Zeroth law of thermodynamics ▶

> If objects A and B are separately in thermal equilibrium with a third object C, then A and B are in thermal equilibrium with each other.

This statement is important because it makes it possible to define **temperature**. We can think of temperature as the property that determines whether or not an object is in thermal equilibrium with other objects. **Two objects in thermal equilibrium with each other are at the same temperature**.

Figure 10.1 The zeroth law of thermodynamics. (a) and (b): If the temperatures of A and B are found to be the same as measured by object C (a thermometer), no energy will be exchanged between them when they are placed in thermal contact with each other, as in (c).

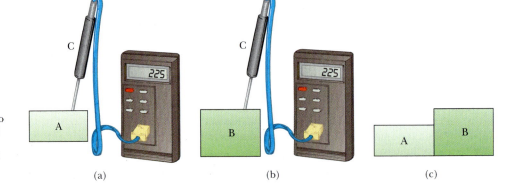

(a)   (b)   (c)

Two objects with different sizes, masses, and temperatures are placed in thermal contact. Choose the best answer: Energy travels (a) from the larger object to the smaller object (b) from the object with more mass to the one with less mass (c) from the object at higher temperature to the object at lower temperature.

10.2   THERMOMETERS AND TEMPERATURE SCALES

Thermometers are devices used to measure the temperature of an object or a system. When a thermometer is in thermal contact with a system, energy is exchanged until the thermometer and the system are in thermal equilibrium with each other. For accurate readings, the thermometer must be much smaller than the system, so that the energy the thermometer gains or loses doesn't significantly alter the energy content of the system. All thermometers make use of some physical property that changes with temperature and can be calibrated to make the temperature measurable. Some of the physical properties used are (1) the volume of a liquid, (2) the length of a solid, (3) the pressure of a gas held at constant volume, (4) the volume of a gas held at constant pressure, (5) the electric resistance of a conductor, and (6) the color of a very hot object.

One common thermometer in everyday use consists of a mass of liquid—usually mercury or alcohol—that expands into a glass capillary tube when its temperature rises (Fig. 10.2). In this case the physical property that changes is the volume of a liquid. To serve as an effective thermometer, the change in volume of the liquid with change in temperature must be very nearly constant over the temperature ranges of interest. When the cross-sectional area of the capillary tube is constant as well, the change in volume of the liquid varies linearly with its length along the tube. We can then define a temperature in terms of the length of the liquid column. The thermometer can be calibrated by placing it in thermal contact with environments that remain at constant temperature. One such environment is a mixture of water and ice in thermal equilibrium at atmospheric pressure. Another commonly used system is a mixture of water and steam in thermal equilibrium at atmospheric pressure.

Once we have marked the ends of the liquid column for our chosen environment on our thermometer, we need to define a scale of numbers associated with various temperatures. An example of such a scale is the **Celsius temperature scale**, formerly called the centigrade scale. On the Celsius scale, the temperature of the ice–water mixture is defined to be zero degrees Celsius, written 0°C and called the **ice point** or **freezing point** of water. The temperature of the water–steam mixture is defined as 100°C, called the **steam point** or **boiling point** of water. Once the ends of the liquid column in the thermometer have been marked at these two points, the distance between marks is divided into 100 equal segments, each corresponding to a change in temperature of one degree Celsius.

Thermometers calibrated in this way present problems when extremely accurate readings are needed. For example, an alcohol thermometer calibrated at the ice and steam points of water might agree with a mercury thermometer only at the calibration points. Because mercury and alcohol have different thermal expansion properties, when one indicates a temperature of 50°C, say, the other may indicate a slightly different temperature. The discrepancies between different types of thermometers are especially large when the temperatures to be measured are far from the calibration points.

The Constant-Volume Gas Thermometer and the Kelvin Scale

We can construct practical thermometers such as the mercury thermometer, but these types of thermometers don't define temperature in a fundamental way. One thermometer, however, *is* more fundamental, and offers a way to define temperature and relate it directly to internal energy: the **gas thermometer**. In a gas thermometer, the temperature readings are nearly independent of the substance used in the thermometer. One type of gas thermometer is the constant-volume unit shown in Figure 10.3.

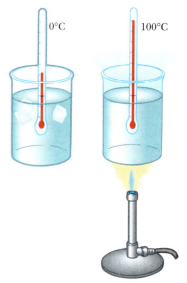

Figure 10.2   Schematic diagram of a mercury thermometer. Because of thermal expansion, the level of the mercury rises as the temperature of the mercury changes from 0°C (the ice point) to 100°C (the steam point).

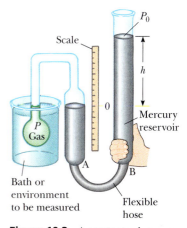

Figure 10.3   A constant-volume gas thermometer measures the pressure of the gas contained in the flask immersed in the bath. The volume of gas in the flask is kept constant by raising or lowering reservoir B to keep the mercury level constant.

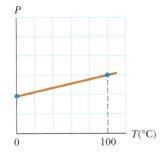

Figure 10.4 A typical graph of pressure versus temperature taken with a constant-volume gas thermometer. The dots represent known reference temperatures (the ice and the steam points of water).

The behavior observed in this device is the variation of pressure with temperature of a fixed volume of gas. When the constant-volume gas thermometer was developed, it was calibrated using the ice and steam points of water as follows (a different calibration procedure, to be discussed shortly, is now used): The gas flask is inserted into an ice–water bath, and mercury reservoir B is raised or lowered until the volume of the confined gas is at some value, indicated by the zero point on the scale. The height h, the difference between the levels in the reservoir and column A, indicates the pressure in the flask at 0°C. The flask is inserted into water at the steam point, and reservoir B is readjusted until the height in column A is again brought to zero on the scale, ensuring that the gas volume is the same as it had been in the ice bath (hence the designation "constant-volume"). A measure of the new value for h gives a value for the pressure at 100°C. These pressure and temperature values are then plotted on a graph, as in Figure 10.4. The line connecting the two points serves as a calibration curve for measuring unknown temperatures. If we want to measure the temperature of a substance, we place the gas flask in thermal contact with the substance and adjust the column of mercury until the level in column A returns to zero. The height of the mercury column tells us the pressure of the gas, and we could then find the temperature of the substance from the calibration curve.

Now suppose that temperatures are measured with various gas thermometers containing different gases. Experiments show that the thermometer readings are nearly independent of the type of gas used, as long as the gas pressure is low and the temperature is well above the point at which the gas liquifies.

We can also perform the temperature measurements with the gas in the flask at different starting pressures at 0°C. As long as the pressure is low, we will generate straight-line calibration curves for each starting pressure, as shown for three experimental trials (solid lines) in Figure 10.5.

If the curves in Figure 10.5 are extended back toward negative temperatures, we find a startling result: In every case, regardless of the type of gas or the value of the low starting pressure, **the pressure extrapolates to zero when the temperature is −273.15°C**. This fact suggests that this particular temperature is universal in its importance, because it doesn't depend on the substance used in the thermometer. In addition, because the lowest possible pressure is $P = 0$, a perfect vacuum, the temperature −273.15°C must represent a lower bound for physical processes. We define this temperature as **absolute zero**.

Absolute zero is used as the basis for the **Kelvin temperature scale**, which sets −273.15°C as its zero point (0 K). The size of a "degree" on the Kelvin scale is chosen to be identical to the size of a degree on the Celsius scale. The relationship between these two temperature scales is

$$T_C = T - 273.15 \qquad [10.1]$$

where T_C is the Celsius temperature and T is the Kelvin temperature (sometimes called the **absolute temperature**).

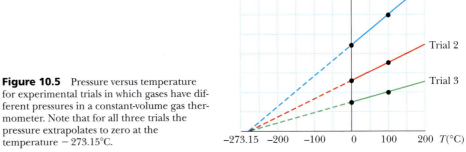

Figure 10.5 Pressure versus temperature for experimental trials in which gases have different pressures in a constant-volume gas thermometer. Note that for all three trials the pressure extrapolates to zero at the temperature −273.15°C.

Technically, Equation 10.1 should have units on the right-hand side so that it reads $T_C = T \,°\text{C}/\text{K} - 273.15 \,°\text{C}$. The units are rather cumbersome in this context, so we will usually suppress them in such calculations except in the final answer. (This will also be the case when discussing the Celsius and Fahrenheit scales).

Early gas thermometers made use of ice and steam points according to the procedure just described. These points are experimentally difficult to duplicate, however, because they are pressure-sensitive. Consequently, a procedure based on two new points was adopted in 1954 by the International Committee on Weights and Measures. The first point is absolute zero. The second point is **the triple point of water, which is the single temperature and pressure at which water, water vapor, and ice can coexist in equilibrium**. This point is a convenient and reproducible reference temperature for the Kelvin scale; it occurs at a temperature of 0.01°C and a pressure of 4.58 mm of mercury. The temperature at the triple point of water on the Kelvin scale occurs at 273.16 K. Therefore, **the SI unit of temperature, the kelvin, is defined as 1/273.16 of the temperature of the triple point of water**. Figure 10.6 shows the Kelvin temperatures for various physical processes and structures. Absolute zero has been closely approached but never achieved.

◀ The kelvin

What would happen to a substance if its temperature could reach 0 K? As Figure 10.5 indicates, the substance would exert zero pressure on the walls of its container (assuming that the gas doesn't liquefy or solidify on the way to absolute zero). In Section 10.5 we show that the pressure of a gas is proportional to the kinetic energy of the molecules of that gas. According to classical physics, therefore, the kinetic energy of the gas would go to zero, and there would be no motion at all of the individual components of the gas. According to quantum theory, however (to be discussed in Chapter 27), the gas would always retain some residual energy, called the *zero-point energy*, at that low temperature.

The Celsius, Kelvin, and Fahrenheit Temperature Scales

Equation 10.1 shows that the Celsius temperature T_C is shifted from the absolute (Kelvin) temperature T by 273.15. Because the size of a Celsius degree is the same as a kelvin, a temperature difference of 5°C is equal to a temperature difference of 5 K. The two scales differ only in the choice of zero point. The ice point (273.15 K) corresponds to 0.00°C, and the steam point (373.15 K) is equivalent to 100.00°C.

The most common temperature scale in use in the United States is the Fahrenheit scale. It sets the temperature of the ice point at 32°F and the temperature of the steam point at 212°F. The relationship between the Celsius and Fahrenheit temperature scales is

$$T_F = \tfrac{9}{5} T_C + 32 \qquad\qquad \textbf{[10.2a]}$$

For example, a temperature of 50.0°F corresponds to a Celsius temperature of 10.0°C and an absolute temperature of 283 K.

Equation 10.2 can be inverted to give Celsius temperatures in terms of Fahrenheit temperatures:

$$T_C = \tfrac{5}{9}(T_F - 32) \qquad\qquad \textbf{[10.2b]}$$

Equation 10.2 can also be used to find a relationship between changes in temperature on the Celsius and Fahrenheit scales. In a problem at the end of the chapter you will be asked to show that if the Celsius temperature changes by ΔT_C, the Fahrenheit temperature changes by the amount

$$\Delta T_F = \tfrac{9}{5} \Delta T_C \qquad\qquad \textbf{[10.3]}$$

Figure 10.7 (page 326) compares the three temperature scales we have discussed.

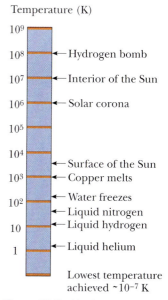

Temperature (K)

10^9
10^8 ← Hydrogen bomb
10^7 ← Interior of the Sun
10^6 ← Solar corona
10^5
10^4 ← Surface of the Sun
10^3 ← Copper melts
10^2 ← Water freezes
← Liquid nitrogen
10 ← Liquid hydrogen
1 ← Liquid helium

Lowest temperature achieved ~10^{-7} K

Figure 10.6 Absolute temperatures at which various selected physical processes take place. Note that the scale is logarithmic.

Figure 10.7 A comparison of the Celsius, Fahrenheit, and Kelvin temperature scales.

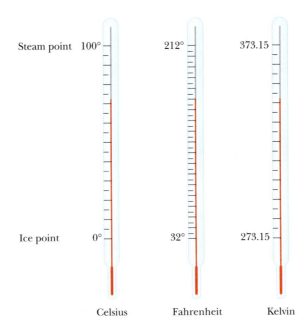

EXAMPLE 10.1   Skin Temperature

Goal Apply the temperature conversion formulas.

Problem The temperature gradient between the skin and the air is regulated by cutaneous (skin) blood flow. If the cutaneous blood vessels are constricted, the skin temperature and the temperature of the environment will be about the same. When the vessels are dilated, more blood is brought to the surface. Suppose that during dilation the skin warms from 72.0°F to 84.0°F. **(a)** Convert these temperatures to Celsius and find the difference. **(b)** Convert the temperatures to Kelvin, again finding the difference.

Strategy This is a matter of applying the conversion formulas, Equations 10.1 and 10.2. For part (b) it's easiest to use the answers for Celsius, rather than develop another set of conversion equations.

Solution
(a) Convert the temperatures from Fahrenheit to Celsius and find the difference.

Convert the lower temperature, using Equation 10.2b.

$T_C = \frac{5}{9}(T_F - 32) = \frac{5}{9}(72 - 32) = \boxed{22°C}$

Convert the upper temperature:

$T_C = \frac{5}{9}(T_F - 32) = \frac{5}{9}(84 - 32) = \boxed{29°C}$

Find the difference of the two temperatures:

$\Delta T_C = 29°C - 22°C = \boxed{7°C}$

(b) Convert the temperatures from Fahrenheit to Kelvin and find their difference.

Convert the lower temperature, using the answers for Celsius found in part (a):

$T_C = T - 273.15 \rightarrow T = T_C + 273.15$
$T = 22 + 273.15 = \boxed{285\ K}$

Convert the upper temperature:

$T = 29 + 273.15 = \boxed{292\ K}$

Find the difference of the two temperatures:

$\Delta T = 292\ K - 285\ K = \boxed{7\ K}$

Remark The change in temperature in Kelvin and Celsius is the same, as it should be.

Exercise 10.1
Core body temperature can rise from 98.6°F to 107°F during extreme exercise, such as a marathon run. Such elevated temperatures can also be caused by viral or bacterial infections or tumors and are dangerous if sustained. (a) Convert the given temperatures to Celsius and find the difference. (b) Convert the temperatures to Kelvin, again finding the difference.

Answer   (a) 37.0°C, 41.7°C, 4.7°C   (b) 310.2 K, 314.9 K, 4.7 K

EXAMPLE 10.2   Extraterrestrial Temperature Scale

Goal   Understand how to relate different temperature scales.

Problem   An extraterrestrial scientist invents a temperature scale such that water freezes at $-75°E$ and boils at $325°E$, where E stands for an extraterrestrial scale. Find an equation that relates temperature in °E to temperature in °C.

Strategy   Using the given data, find the ratio of the number of °E between the two temperatures to the number of °C. This ratio will be the same as a similar ratio for any other such process—say, from the freezing point to an unknown temperature—corresponding to T_E and T_C. Setting the two ratios equal and solving for T_E in terms of T_C yields the desired relationship.

Solution

Find the change in temperature in °E between the freezing and boiling points of water:

$$\Delta T_E = 325°E - (-75°E) = 400°E$$

Find the change in temperature in °C between the freezing and boiling points of water:

$$\Delta T_C = 100°C - 0°C = 100°C$$

Form the ratio of these two quantities.

$$\frac{\Delta T_E}{\Delta T_C} = \frac{400°E}{100°C} = 4\frac{°E}{°C}$$

This ratio is the same between any other two temperatures —say, from the freezing point to an unknown final temperature. Set the two ratios equal to each other:

$$\frac{\Delta T_E}{\Delta T_C} = \frac{T_E - (-75°E)}{T_C - 0°C} = 4\frac{°E}{°C}$$

Solve for T_E:

$$T_E - (-75°E) = 4(°E/°C)(T_C - 0°C)$$

$$T_E = \boxed{4T_C - 75}$$

Remark   The relationship between any other two temperatures scales can be derived in the same way.

Exercise 10.2
Find the equation converting °F to °E.

Answer $T_E = \frac{20}{9}T_F - 146$

10.3   THERMAL EXPANSION OF SOLIDS AND LIQUIDS

Our discussion of the liquid thermometer made use of one of the best-known changes that occur in most substances: As temperature of the substance increases, its volume increases. This phenomenon, known as **thermal expansion**, plays an important role in numerous applications. Thermal expansion joints, for example,

(a)

George Sample

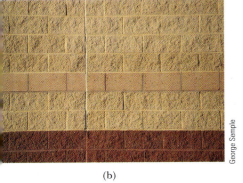

(b)

George Sample

Figure 10.8 (a) Thermal expansion joints are used to separate sections of roadways on bridges. Without these joints, the surfaces would buckle due to thermal expansion on very hot days or crack due to contraction on very cold days. (b) The long, vertical joint is filled with a soft material that allows the wall to expand and contract as the temperature of the bricks changes.

TIP 10.1 Coefficients of Expansion are not Constants

The coefficients of expansion can vary somewhat with temperature, so the given coefficients are actually averages.

must be included in buildings, concrete highways, and bridges to compensate for changes in dimensions with variations in temperature (Fig. 10.8).

The overall thermal expansion of an object is a consequence of the change in the average separation between its constituent atoms or molecules. To understand this idea, consider how the atoms in a solid substance behave. These atoms are located at fixed equilibrium positions; if an atom is pulled away from its position, a restoring force pulls it back. We can imagine that the atoms are particles connected by springs to their neighboring atoms. (See Fig. 9.1 in the previous chapter.) If an atom is pulled away from its equilibrium position, the distortion of the springs provides a restoring force.

At ordinary temperatures, the atoms vibrate around their equilibrium positions with an amplitude (maximum distance from the center of vibration) of about 10^{-11} m, with an average spacing between the atoms of about 10^{-10} m. As the temperature of the solid increases, the atoms vibrate with greater amplitudes and the average separation between them increases. Consequently, the solid as a whole expands.

If the thermal expansion of an object is sufficiently small compared with the object's initial dimensions, then the change in any dimension is, to a good approximation, proportional to the first power of the temperature change. Suppose an object has an initial length L_0 along some direction at some temperature T_0. Then the length increases by ΔL for a change in temperature ΔT. So for small changes in temperature,

$$\Delta L = \alpha L_0 \, \Delta T \qquad [10.4]$$

or

$$L - L_0 = \alpha L_0 (T - T_0)$$

where L is the object's final length T is its final temperature, and the proportionality constant α is called the **coefficient of linear expansion** for a given material and has units of $(°C)^{-1}$.

Table 10.1 lists the coefficients of linear expansion for various materials. Note that for these materials α is positive, indicating an increase in length with increasing temperature.

Thermal expansion affects the choice of glassware used in kitchens and laboratories. If hot liquid is poured into a cold container made of ordinary glass, the container may well break due to thermal stress. The inside surface of the glass becomes hot and expands, while the outside surface is at room temperature, and ordinary glass may not withstand the difference in expansion without breaking. Pyrex® glass has a coefficient of linear expansion of about one-third that of ordinary glass, so the thermal stresses are smaller. Kitchen measuring cups and laboratory beakers are often made of Pyrex so they can be used with hot liquids.

TABLE 10.1

Average Coefficients of Expansion for Some Materials Near Room Temperature

Material	Average Coefficient of Linear Expansion $[(°C)^{-1}]$	Material	Average Coefficient of Volume Expansion $[(°C)^{-1}]$
Aluminum	24×10^{-6}	Ethyl alcohol	1.12×10^{-4}
Brass and bronze	19×10^{-6}	Benzene	1.24×10^{-4}
Copper	17×10^{-6}	Acetone	1.5×10^{-4}
Glass (ordinary)	9×10^{-6}	Glycerin	4.85×10^{-4}
Glass (Pyrex®)	3.2×10^{-6}	Mercury	1.82×10^{-4}
Lead	29×10^{-6}	Turpentine	9.0×10^{-4}
Steel	11×10^{-6}	Gasoline	9.6×10^{-4}
Invar (Ni-Fe alloy)	0.9×10^{-6}	Air	3.67×10^{-3}
Concrete	12×10^{-6}	Helium	3.665×10^{-3}

EXAMPLE 10.3   Expansion of a Railroad Track

Goal Apply the concept of linear expansion and relate it to stress.

Problem (a) A steel railroad track has a length of 30.000 m when the temperature is 0°C. What is its length on a hot day when the temperature is 40.0°C? (b) Suppose the track is nailed down so that it can't expand. What stress results in the track due to the temperature change?

Strategy (a) Apply the linear expansion equation, using Table 10.1 and Equation 10.4. (b) A track that cannot expand by ΔL due to external constraints is equivalent to compressing the track by ΔL, creating a stress in the track. Using the equation relating tensile stress to tensile strain together with the linear expansion equation, the amount of (compressional) stress can be calculated using Equation 9.3.

(Example 10.3) Thermal expansion: The extreme heat of a July day in Asbury Park, New Jersey, caused these railroad tracks to buckle.

AP/Wide World Photos

Solution

(a) Find the length of the track at 40.0°C.

Substitute given quantities into Equation 10.4, finding the change in length:

$$\Delta L = \alpha L_0 \Delta T = [11 \times 10^{-6}(°C)^{-1}](30.000 \text{ m})(40.0°C)$$
$$= 0.013 \text{ m}$$

Add the change to the original length to find the final length:

$$L = L_0 + \Delta L = \boxed{30.013 \text{ m}}$$

(b) Find the stress if the track cannot expand.

Substitute into Equation 9.3 to find the stress:

$$\frac{F}{A} = Y\frac{\Delta L}{L} = (2.00 \times 10^{11} \text{ Pa})\left(\frac{0.013 \text{ m}}{30.0 \text{ m}}\right)$$
$$= \boxed{8.67 \times 10^7 \text{ Pa}}$$

Remarks Repeated heating and cooling is an important part of the weathering process that gradually wears things out, weakening structures over time.

Exercise 10.3

What is the length of the same railroad track on a cold winter day when the temperature is 0°F?

Answer 29.994 m

Applying Physics 10.1   Bimetallic Strips and Thermostats

How can different coefficients of expansion for metals be used as a temperature gauge and control electronic devices such as air conditioners?

Explanation When the temperatures of a brass rod and a steel rod of equal length are raised by the same amount from some common initial value, the brass rod expands more than the steel rod because brass has a larger coefficient of expansion than steel. A simple device that uses this principle is a **bimetallic strip**. Such strips can be found in the thermostats of certain home heating systems. The strip is made by securely bonding two different metals together. As the temperature of the strip increases, the two metals expand by different amounts and the strip bends, as in Figure 10.9 (page 330). The change in shape can make or break an electrical connection.

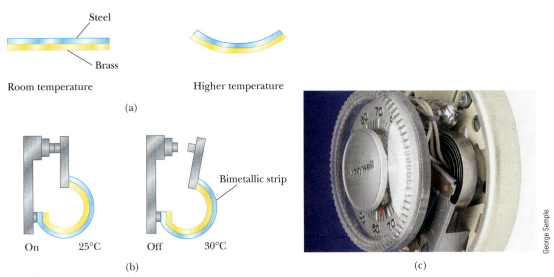

Figure 10.9   (Applying Physics 10.1) (a) A bimetallic strip bends as the temperature changes because the two metals have different coefficients of expansion. (b) A bimetallic strip used in a thermostat to break or make electrical contact. (c) The interior of a thermostat, showing the coiled bimetallic strip. Why do you suppose the strip is coiled?

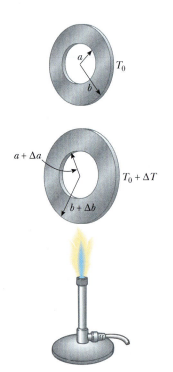

ACTIVE FIGURE 10.10
Thermal expansion of a homogeneous metal washer. As the washer is heated, all dimensions increase. (Note that the expansion is exaggerated in this figure.)

Physics ⊗ Now™
Log into PhysicsNow at **www.cp7e.com**, and go to Active Figure 10.10 to compare expansions for various temperatures of the burner and various materials from which the washer is made.

It may be helpful to picture a thermal expansion as a magnification or a photographic enlargement. For example, as the temperature of a metal washer increases (Active Fig. 10.10), all dimensions, including the radius of the hole, increase according to Equation 10.4.

One practical application of thermal expansion is the common technique of using hot water to loosen a metal lid stuck on a glass jar. This works because the circumference of the lid expands more than the rim of the jar.

Because the linear dimensions of an object change due to variations in temperature, it follows that surface area and volume of the object also change. Consider a square of material having an initial length L_0 on a side and therefore an initial area $A_0 = L_0^2$. As the temperature is increased, the length of each side increases to

$$L = L_0 + \alpha L_0 \, \Delta T$$

The new area A is

$$A = L^2 = (L_0 + \alpha L_0 \Delta T)(L_0 + \alpha L_0 \Delta T) = L_0^2 + 2\alpha L_0^2 \, \Delta T + \alpha^2 L_0^2 (\Delta T)^2$$

The last term in this expression contains the quantity $\alpha \Delta T$ raised to the second power. Because $\alpha \Delta T$ is much less than one, squaring it makes it even smaller. Consequently, we can neglect this term to get a simpler expression:

$$A = L_0^2 + 2\alpha L_0^2 \, \Delta T$$

$$A = A_0 + 2\alpha A_0 \, \Delta T$$

so that

$$\Delta A = A - A_0 = \gamma A_0 \, \Delta T \qquad [10.5]$$

where $\gamma = 2\alpha$. The quantity γ (Greek letter gamma) is called the **coefficient of area expansion**.

EXAMPLE 10.4   Rings and Rods

Goal   Apply the equation of area expansion.

Problem   **(a)** A circular copper ring at 20.0°C has a hole with an area of 9.98 cm^2. What minimum temperature must it have so that it can be slipped onto a steel metal rod having a cross-sectional area of 10.0 cm^2? **(b)** Suppose the ring and the rod are heated simultaneously. What change in temperature of both will allow the ring to be slipped onto the end of the rod? (Assume no significant change in the coefficients of linear expansion over this temperature range.)

Strategy   In part (a), finding the necessary temperature change is just a matter of substituting given values into Equation 10.5, the equation of area expansion. Remember that $\gamma = 2\alpha$. Part (b) is a little harder, because now the rod is also expanding. If the ring is to slip onto the rod, however, the final cross-sectional areas of both ring and rod must be equal. Write this condition in mathematical terms, using Equation 10.5 on both sides of the equation, and solve for ΔT.

Solution

(a) Find the temperature of the ring that will allow it to slip onto the rod.

Substitute the desired change in area into Equation 10.5, finding the necessary change in temperature:

$$\Delta A = \gamma A_0 \, \Delta T = [34 \times 10^{-6} \, (°C)^{-1}](9.98 \text{ cm}^2)(\Delta T)$$
$$= 0.02 \text{ cm}^2$$

Solve for ΔT, then add this change to the initial temperature to get the final temperature:

$$\Delta T = 58.9°C$$
$$T = T_0 + \Delta T = 20.0°C + 58.9°C = \boxed{78.9°C}$$

(b) Increase the temperature of both ring and rod to remove the ring.

Set the final areas of the copper ring and steel rod equal to each other:

$$A_C + \Delta A_C = A_S + \Delta A_S$$

Substitute for each change in area, ΔA:

$$A_C + \gamma_C A_C \Delta T = A_S + \gamma_S A_S \Delta T$$

Rearrange terms to get ΔT on one side only, factor it out and solve:

$$\gamma_C A_C \Delta T - \gamma_S A_S \Delta T = A_S - A_C$$
$$(\gamma_C A_C - \gamma_S A_S) \, \Delta T = A_S - A_C$$
$$\Delta T = \frac{A_S - A_C}{\gamma_C A_C - \gamma_S A_S}$$
$$= \frac{10.0 \text{ cm}^2 - 9.98 \text{ cm}^2}{(34 \times 10^{-6} \, °C^{-1})(9.98 \text{ cm}^2) - (22 \times 10^{-6} \, °C^{-1})(10.0 \text{ cm}^2)}$$
$$\Delta T = \boxed{168°C}$$

Exercise 10.4

A steel ring with a hole having area of 3.99 cm^2 is to be placed on an aluminum rod with cross-sectional area of 4.00 cm^2. Both rod and ring are initially at a temperature of 35.0°C. At what common temperature can the steel ring be slipped onto one end of the aluminum rod?

Answer $-61°C$

We can also show that the *increase in volume* of an object accompanying a change in temperature is

$$\Delta V = \beta V_0 \, \Delta T \qquad\qquad \text{[10.6]}$$

where β, the **coefficient of volume expansion**, is equal to 3α. (Note that $\gamma = 2\alpha$ and $\beta = 3\alpha$ only if the coefficient of linear expansion of the object is the same in all directions.) The proof of Equation 10.6 is similar to the proof of Equation 10.5.

As Table 10.1 indicates, each substance has its own characteristic coefficients of expansion.

The thermal expansion of water has a profound influence on rising ocean levels. At current rates of global warming, scientists predict that about one-half of the expected rise in sea level will be caused by thermal expansion; the remainder will be due to the melting of polar ice.

APPLICATION

Rising Sea Levels

Quick Quiz 10.2

If you quickly plunge a room-temperature mercury thermometer into very hot water, the mercury level will (a) go up briefly before reaching a final reading, (b) go down briefly before reaching a final reading, or (c) not change.

Quick Quiz 10.3

If you are asked to make a very sensitive glass thermometer, which of the following working fluids would you choose? (a) mercury (b) alcohol (c) gasoline (d) glycerin

Quick Quiz 10.4

Two spheres are made of the same metal and have the same radius, but one is hollow and the other is solid. The spheres are taken through the same temperature increase. Which sphere expands more? (a) solid sphere, (b) hollow sphere, (c) they expand by the same amount, or (d) not enough information to say.

EXAMPLE 10.5   Global Warming and Coastal Flooding

Goal   Apply the volume expansion equation together with linear expansion.

Problem   (a) Estimate the fractional change in the volume of Earth's oceans due to an average temperature change of 1°C. (b) Use the fact that the average depth of the ocean is 4.00×10^3 m to estimate the change in depth. Note that $\beta_{water} = 2.07 \times 10^{-4}(°C)^{-1}$

Strategy   In part (a), solve the volume expansion expression, Equation 10.6, for $\Delta V/V$. For part (b), use linear expansion to estimate the increase in depth. Neglect the expansion of landmasses, which would reduce the rise in sea level only slightly.

Solution

(a) Find the fractional change in volume.

Divide the volume expansion equation by V_0 and substitute:

$$\Delta V = \beta V_0 \, \Delta T$$

$$\left(\frac{\Delta V}{V_0}\right) = \beta \Delta T = (2.07 \times 10^{-4}(°C)^{-1} \cdot (1°C) = \boxed{2 \times 10^{-4}}$$

(b) Find the approximate increase in depth.

Use the linear expansion equation. Divide the volume expansion coefficient of water by three to get the equivalent linear expansion coefficient:

$$\Delta L = \alpha L_0 \, \Delta T = \left(\frac{\beta}{3}\right) L_0 \, \Delta T$$

$$\Delta L = (6.90 \times 10^{-5}(°C)^{-1}(4\,000 \text{ m})(1°C) \approx \boxed{0.3 \text{ m}}$$

Remarks   Three-tenths of a meter may not seem significant, but combined with increased melting of the polar ice caps, some coastal areas could experience flooding. An increase of several degrees increases the value of ΔL several times and could significantly reduce the value of waterfront property.

Exercise 10.5

A 1.00-liter aluminum cylinder at 5.00°C is filled to the brim with gasoline at the same temperature. If the aluminum and gasoline are warmed to 65.0°C, how much of the gasoline spills out? [*Hint:* Be sure to account for the expansion of the container. Also, ignore the possibility of evaporation, and assume the volume coefficients are good to three digits.]

Answer   The volume spilled is 53.3 cm³. Forgetting to take into account the expansion of the cylinder results in a (wrong) answer of 57.6 cm³.

The Unusual Behavior of Water

Liquids generally increase in volume with increasing temperature and have volume expansion coefficients about ten times greater than those of solids. Over a small temperature range, water is an exception to this rule, as we can see from its density-versus-temperature curve in Figure 10.11. As the temperature increases from 0°C to 4°C, water contracts, so its density increases. Above 4°C, water exhibits the expected expansion with increasing temperature. The density of water reaches its maximum value of 1 000 kg/m³ at 4°C.

We can use this unusual thermal expansion behavior of water to explain why a pond freezes slowly from the top down. When the atmospheric temperature drops from 7°C to 6°C, say, the water at the surface of the pond also cools and consequently decreases in volume. This means the surface water is more dense than the water below it, which has not yet cooled nor decreased in volume. As a result, the surface water sinks and warmer water from below is forced to the surface to be cooled, a process called *upwelling*. When the atmospheric temperature is between 4°C and 0°C, however, the surface water expands as it cools, becoming less dense than the water below it. The sinking process stops, and eventually the surface water freezes. As the water freezes, the ice remains on the surface because ice is less dense than water. The ice continues to build up on the surface, and water near the bottom of the pool remains at 4°C. Further, the ice forms an insulating layer that slows heat loss from the underlying water, offering thermal protection for marine life.

Without buoyancy and the expansion of water upon freezing, life on Earth may not have been possible. If ice had been more dense than water, it would have sunk to the bottom of the ocean and built up over time. This could have led to a freezing of the oceans, turning the Earth into an icebound world similar to Hoth in the Star Wars epic *The Empire Strikes Back*.

The same peculiar thermal expansion properties of water sometimes cause pipes to burst in winter. As energy leaves the water through the pipe by heat and is transferred to the outside cold air, the outer layers of water in the pipe freeze first.

APPLICATION

Bursting Pipes in Winter

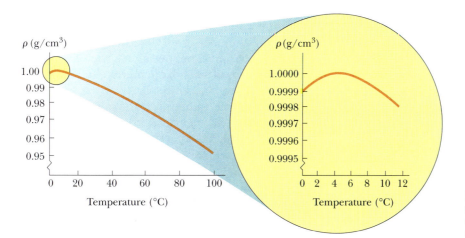

Figure 10.11   The density of water as a function of temperature. The inset at the right shows that the maximum density of water occurs at 4°C.

The continuing energy transfer causes ice to form ever closer to the center of the pipe. As long as there is still an opening through the ice, the water can expand as its temperature approaches 0°C or as it freezes into more ice, pushing itself into another part of the pipe. Eventually, however, the ice will freeze to the center somewhere along the pipe's length, forming a plug of ice at that point. If there is still liquid water between this plug and some other obstruction, such as another ice plug or a spigot, then no additional volume is available for further expansion and freezing. The pressure in the pipe builds and can rupture the pipe.

10.4 MACROSCOPIC DESCRIPTION OF AN IDEAL GAS

The properties of gases are important in a number of thermodynamic processes. Our weather is a good example of the types of processes that depend on the behavior of gases.

If we introduce a gas into a container, it expands to fill the container uniformly, with its pressure depending on the size of the container, the temperature, and the amount of gas. A larger container results in a lower pressure, while higher temperatures or larger amounts of gas result in a higher pressure. The pressure P, volume V, temperature T, and amount n of gas in a container are related to each other by an *equation of state.*

The equation of state can be very complicated, but is found experimentally to be relatively simple if the gas is maintained at a low pressure (or a low density). Such a low-density gas approximates what is called an **ideal gas**. Most gases at room temperature and atmospheric pressure behave approximately as ideal gases. **An ideal gas is a collection of atoms or molecules that move randomly and exert no long-range forces on each other. Each particle of the ideal gas is individually point-like, occupying a negligible volume**.

A gas usually consists of a very large number of particles, so it's convenient to express the amount of gas in a given volume in terms of the number of **moles**, n. A mole is a number. The same number of particles is found in a mole of helium as in a mole of iron or aluminum. This number is known as *Avogadro's number* and is given by

Avogadro's number ▶

$$N_A = 6.02 \times 10^{23} \text{ particles/mole}$$

Avogadro's number and the definition of a mole are fundamental to chemistry and related branches of physics. The number of moles of a substance is related to its mass m by the expression

$$n = \frac{m}{\text{molar mass}} \qquad [10.7]$$

where the molar mass of the substance is defined as the mass of one mole of that substance, usually expressed in grams per mole.

There are lots of atoms in the world, so it's natural and convenient to choose a very large number like Avogadro's number when describing collections of atoms. At the same time, Avogadro's number must be special in some way, because otherwise why not just count things in terms of some large power of ten, like 10^{24}?

It turns out that Avogadro's number was chosen so that the mass in grams of one Avogadro's number of an element is numerically the same as the mass of one atom of the element, expressed in atomic mass units (u).

This relationship is very convenient. Looking at the periodic table of the elements in the back of the book, we find that carbon has an atomic mass of 12 u, so 12 g of carbon consists of exactly 6.02×10^{23} atoms of carbon. The atomic mass of oxygen is 16 u, so in 16 g of oxygen there are again 6.02×10^{23} atoms of oxygen. The same holds true for molecules: The molecular mass of molecular hydrogen, H_2, is 2 u, and there is an Avogadro's number of molecules in 2 g of molecular hydrogen.

The technical definition of a mole is as follows: **One mole (mol) of any substance is that amount of the substance that contains as many particles (atoms, molecules, or other particles) as there are atoms in 12 g of the isotope carbon-12**.

Taking carbon-12 as a test case, let's find the mass of an Avogadro's number of carbon-12 atoms. A carbon-12 atom has an atomic mass of 12 u, or 12 atomic mass units. One atomic mass unit is equal to 1.66×10^{-24} g, about the same as the mass of a neutron or proton—particles that make up atomic nuclei. The mass m of an Avogadro's number of carbon-12 atoms is then given by

$$m = N_A(12 \text{ u}) = 6.02 \times 10^{23}(12 \text{ u})\left(\frac{1.66 \times 10^{-24} \text{ g}}{\text{u}}\right) = 12.0 \text{ g}$$

So we see that Avogadro's number is deliberately chosen to be the inverse of the number of grams in an atomic mass unit. In this way, the atomic mass of an atom expressed in atomic mass units is numerically the same as the mass of an Avogadro's number of that kind of atom expressed in grams. Because there are 6.02×10^{23} particles in one mole of *any* element, the mass per atom for a given element is

$$m_{\text{atom}} = \frac{\text{molar mass}}{N_A}$$

For example, the mass of a helium atom is

$$m_{\text{He}} = \frac{4.00 \text{ g/mol}}{6.02 \times 10^{23} \text{ atoms/mol}} = 6.64 \times 10^{-24} \text{ g/atom}$$

Now suppose an ideal gas is confined to a cylindrical container with a volume that can be changed by moving a piston, as in Active Figure 10.12. Assume that the cylinder doesn't leak, so the number of moles remains constant. Experiments yield the following observations: First, when the gas is kept at a constant temperature, its pressure is inversely proportional to its volume (Boyle's law). Second, when the pressure of the gas is kept constant, the volume of the gas is directly proportional to the temperature (Charles's law.) Third, when the volume of the gas is held constant, the pressure is directly proportional to the temperature (Gay-Lussac's law). These observations can be summarized by the following equation of state, known as the **ideal gas law**:

$$PV = nRT \qquad [10.8]$$

◀ Equation of state for an ideal gas

In this equation, R is a constant for a specific gas that must be determined from experiments, while T is the temperature in kelvins. Each point on a P versus V diagram would represent a different state of the system. Experiments on several gases show that, as the pressure approaches zero, the quantity PV/nT approaches the same value of R for all gases. For this reason, R is called the **universal gas constant**. In SI units, where pressure is expressed in pascals and volume in cubic meters,

$$R = 8.31 \text{ J/mol·K} \qquad [10.9]$$

◀ The universal gas constant

If the pressure is expressed in atmospheres and the volume is given in liters (recall that 1 L = 10^3 cm^3 = 10^{-3} m^3), then

$$R = 0.0821 \text{ L·atm/mol·K}$$

Using this value of R and Equation 10.8, the volume occupied by 1 mol of any ideal gas at atmospheric pressure and at 0°C (273 K) is 22.4 L.

ACTIVE FIGURE 10.12
A gas confined to a cylinder whose volume can be varied with a movable piston.

Physics Now™
Log into PhysicsNow at **www.cp7e.com**, and go to Active Figure 10.12, where you can choose to keep either the temperature or the pressure constant and verify the ideal gas law.

TIP 10.2 Only Kelvin Works!
Temperatures used in the ideal gas law must always be in kelvins.

TIP 10.3 Standard Temperature and Pressure
Chemists often define standard temperature and pressure (STP) to be 20°C and 1.0 atm. We choose STP to be 0°C and 1.0 atm. (See Table 9.3.)

Quick Quiz 10.5

A helium-filled balloon is released into the atmosphere. Assuming constant temperature, as the balloon rises, it (a) expands, (b) contracts, or (c) remains unchanged in size.

EXAMPLE 10.6   An Expanding Gas

Goal   Use the ideal gas law to analyze a system of gas.

Problem   An ideal gas at 20.0°C and a pressure of 1.50×10^5 Pa is in a container having a volume of 1.00 L. **(a)** Determine the number of moles of gas in the container. **(b)** The gas pushes against a piston, expanding to twice its original volume, while the pressure falls to atmospheric pressure. Find the final temperature.

Strategy   **(a)** Solve the ideal gas equation of state for the number of moles, n, and substitute the known quantities. Be sure to convert the temperature from Celsius to Kelvin! **(b)** When comparing two states of a gas, it's often most convenient to divide the ideal gas equation of the final state by the equation of the initial state. Then quantities that don't change can immediately be cancelled, simplifying the algebra.

Solution
(a) Find the number of moles of gas.

Convert the temperature to kelvins:

$$T = T_C + 273 = 20.0 + 273 = 293 \text{ K}$$

Solve the ideal gas law for n and substitute:

$$PV = nRT$$

$$n = \frac{PV}{RT} = \frac{(1.50 \times 10^5 \text{ Pa})(1.00 \times 10^{-3} \text{ m}^3)}{(8.31 \text{ J/mol·K})(293 \text{ K})}$$

$$= 6.16 \times 10^{-2} \text{ mol}$$

(b) Find the temperature after the gas expands to 2.00 L.

Divide the ideal gas law for the final state by the ideal gas law for the initial state:

$$\frac{P_f V_f}{P_i V_i} = \frac{nRT_f}{nRT_i}$$

Cancel the number of moles n and the gas constant R, and solve for T_f:

$$\frac{P_f V_f}{P_i V_i} = \frac{T_f}{T_i}$$

$$T_f = \frac{P_f V_f}{P_i V_i} T_i = \frac{(1.01 \times 10^5 \text{ Pa})(2.00 \text{ L})}{(1.50 \times 10^5 \text{ Pa})(1.00 \text{ L})} (293 \text{ K})$$

$$= 395 \text{ K}$$

Remark   Remember the trick used in part (b), it's often useful in ideal gas problems. Notice that it wasn't necessary to convert units from liters to cubic meters, since the units were going to cancel anyway.

Exercise 10.6
Suppose the temperature of 4.50 L of ideal gas drops from 375 K to 275 K. **(a)** If the volume remains constant and the initial pressure is atmospheric pressure, find the final pressure. **(b)** Find the number of moles of gas.

Answer   (a) 7.41×10^4 Pa   (b) 0.146 mol

EXAMPLE 10.7   Message in a Bottle

Goal   Apply the ideal gas law in tandem with Newton's second law.

Problem   A beachcomber finds a corked bottle containing a message. The air in the bottle is at atmospheric pressure and a temperature of 30.0°C. The cork has a cross-sectional area of 2.30 cm². The beachcomber places the bottle over a fire, figuring the increased pressure will push out the cork. At a temperature of 99°C the cork is ejected from the bottle. **(a)** What was the pressure in the bottle just before the cork left it? **(b)** What force of friction held the cork in place? Neglect any change in volume of the bottle.

Strategy (a) The number of moles of air in the bottle remains the same as it warms over the fire. Take the ideal gas equation for the final state and divide by the ideal gas equation for the initial state. Solve for the final pressure. (b) There are three forces acting on the cork: a friction force, the exterior force of the atmosphere pushing in, and the force of the air inside the bottle pushing out. Apply Newton's second law. Just before the cork begins to move, the three forces are in equilibrium and the static friction force has its maximum value.

Solution
(a) Find the final pressure.

Divide the ideal gas law at the final point by the ideal gas law at the initial point:

$$\frac{P_f V_f}{P_i V_i} = \frac{nRT_f}{nRT_i} \qquad (1)$$

Cancel n, R, and V, which don't change, and solve for P_f:

$$\frac{P_f}{P_i} = \frac{T_f}{T_i} \quad \rightarrow \quad P_f = P_i \frac{T_f}{T_i}$$

Substitute known values, obtaining the final pressure:

$$P_f = (1.01 \times 10^5 \text{ Pa}) \frac{372 \text{ K}}{303 \text{ K}} = \boxed{1.24 \times 10^5 \text{ Pa}}$$

(b) Find the magnitude of the friction force acting on the cork.

Apply Newton's second law to the cork just before it leaves the bottle. P_{in} is the pressure inside the bottle, P_{out} the pressure outside.

$$\Sigma F = 0 \quad \rightarrow \quad P_{in} A - P_{out} A - F_{friction} = 0$$

$$F_{friction} = P_{in} A - P_{out} A = (P_{in} - P_{out}) A$$

$$= (1.24 \times 10^5 \text{ Pa} - 1.01 \times 10^5 \text{ Pa})$$

$$\times (2.30 \times 10^{-4} \text{ m}^2)$$

$$F_{friction} = \boxed{5.29 \text{ N}}$$

Remark Notice the use, once again, of the ideal gas law in Equation (1). Whenever comparing the state of a gas at two different points, this is the best way to do the math. One other point: heating the gas blasted the cork out of the bottle, which meant the gas did work on the cork. The work done by an expanding gas—driving pistons and generators—is one of the foundations of modern technology and will be studied extensively in Chapter 12.

Exercise 10.7
A tire contains air at a gauge pressure of 5.00×10^4 Pa at a temperature of 30.0°C. After nightfall, the temperature drops to -10.0°C. Find the new gauge pressure in the tire. (Recall that gauge pressure is absolute pressure minus atmospheric pressure. Assume constant volume.)

Answer 3.01×10^4 Pa

EXAMPLE 10.8 Submerging a Balloon

Goal Combine the ideal gas law with the equation of hydrostatic equilibrium and buoyancy.

Problem A sturdy balloon with volume 0.500 m³ is attached to a 2.50×10^2-kg iron weight and tossed overboard into a freshwater lake. The balloon is made of a light material of negligible mass and elasticity (though it can be compressed). The air in the balloon is initially at atmospheric pressure. The system fails to sink and there are no more weights, so a skin diver decides to drag it deep enough so that the balloon will remain submerged. (a) Find the volume of the balloon at the point where the system will remain submerged, in equilibrium. (b) What's the balloon's pressure at that point? (c) Assuming constant temperature, to what minimum depth must the balloon be dragged?

Strategy As the balloon and weight are dragged deeper into the lake, the air in the balloon is compressed and the volume is reduced along with the buoyancy. At some depth h the total buoyant force acting on the balloon and weight, $B_{bal} + B_{Fe}$, will equal the total weight, $w_{bal} + w_{Fe}$, and the balloon will remain at that depth. Substitute these forces into Newton's second law and solve for the unknown volume of the balloon, answering part (a). Then use the ideal gas law to find the pressure, and the equation of hydrostatic equilibrium to find the depth.

Solution

(a) Find the volume of the balloon at the equilibrium point.

Find the volume of the iron, V_{Fe}:

$$V_{Fe} = \frac{m_{Fe}}{\rho_{Fe}} = \frac{2.50 \times 10^2 \text{ kg}}{7.86 \times 10^3 \text{ kg/m}^3} = 0.0318 \text{ m}^3$$

Find the mass of the balloon, which is equal to the mass of the air if we neglect the mass of the balloon's material:

$$m_{bal} = \rho_{air} V_{bal} = (1.29 \text{ kg/m}^3)(0.500 \text{ m}^3) = 0.65 \text{ kg}$$

Apply Newton's second law to the system when it's in equilibrium:

$$B_{Fe} - w_{Fe} + B_{bal} - w_{bal} = 0$$

Substitute the appropriate expression for each term:

$$\rho_{wat} V_{Fe} g - m_{Fe} g + \rho_{wat} V_{bal} g - m_{bal} g = 0$$

Cancel the g's and solve for the volume of the balloon, V_{bal}:

$$V_{bal} = \frac{m_{bal} + m_{Fe} - \rho_{wat} V_{Fe}}{\rho_{wat}}$$

$$= \frac{0.65 \text{ kg} + 2.50 \times 10^2 \text{ kg} - (1.00 \times 10^3 \text{ kg/m}^3)(0.0318 \text{ m}^3)}{1.00 \times 10^3 \text{ kg/m}^3}$$

$$V_{bal} = 0.219 \text{ m}^3$$

(b) What's the balloon's pressure at the equilibrium point?

Now use the ideal gas law to find the pressure, assuming constant temperature, so that $T_i = T_f$.

$$\frac{P_f V_f}{P_i V_i} = \frac{nRT_f}{nRT_i} = 1$$

$$P_f = \frac{V_i}{V_f} P_i = \frac{0.500 \text{ m}^3}{0.219 \text{ m}^3} (1.01 \times 10^5 \text{ Pa})$$

$$= 2.31 \times 10^5 \text{ Pa}$$

(c) To what minimum depth must the balloon be dragged?

Use the equation of hydrostatic equilibrium to find the depth:

$$P_f = P_{atm} + \rho g h$$

$$h = \frac{P_f - P_{atm}}{\rho g} = \frac{2.31 \times 10^5 \text{ Pa} - 1.01 \times 10^5 \text{ Pa}}{(1.00 \times 10^3 \text{ kg/m}^3)(9.80 \text{ m/s}^2)}$$

$$= 13.3 \text{ m}$$

Remark   Once again, the ideal gas law was used to good effect. This problem shows how even answering a fairly simple question can require the application of several different physical concepts: density, buoyancy, the ideal gas law, and hydrostatic equilibrium.

Exercise 10.8

A boy takes a 30.0-cm^3 balloon holding air at 1.00 atm at the surface of a freshwater lake down to a depth of 4.00 m. Find the volume of the balloon at this depth. Assume the balloon is made of light material of little elasticity (though it can be compressed), and that the temperature of the trapped air remains constant.

Answer   21.6 cm^3

As previously stated, the number of molecules contained in one mole of any gas is Avogadro's number, $N_A = 6.02 \times 10^{23}$ particles/mol, so

$$n = \frac{N}{N_A} \qquad\qquad \text{[10.10]}$$

where n is the number of moles and N is the number of molecules in the gas. With Equation 10.10, we can rewrite the ideal gas law in terms of the total number of molecules as

$$PV = nRT = \frac{N}{N_A} RT$$

or

$$PV = Nk_B T \qquad\qquad \text{[10.11]} \qquad \blacktriangleleft \text{Ideal gas law}$$

where

$$k_B = \frac{R}{N_A} = 1.38 \times 10^{-23}\,\text{J/K} \qquad\qquad \text{[10.12]} \qquad \blacktriangleleft \text{Boltzmann's constant}$$

is **Boltzmann's constant**. This reformulation of the ideal gas law will be used in the next section to relate the temperature of a gas to the average kinetic energy of particles in the gas.

10.5   THE KINETIC THEORY OF GASES

In Section 10.4, we discussed the macroscopic properties of an ideal gas, including pressure, volume, number of moles, and temperature. In this section we consider the ideal gas model from the microscopic point of view. We will show that the macroscopic properties can be understood on the basis of what is happening on the atomic scale. In addition, we reexamine the ideal gas law in terms of the behavior of the individual molecules that make up the gas.

Using the model of an ideal gas, we will describe the **kinetic theory of gases**. With this theory we can interpret the pressure and temperature of an ideal gas in terms of microscopic variables. The kinetic theory of gases model makes the following assumptions:

1. **The number of molecules in the gas is large, and the average separation between them is large compared with their dimensions.** The fact that the number of molecules is large allows us to analyze their behavior statistically. The large separation between molecules means that the molecules occupy a negligible volume in the container. This assumption is consistent with the ideal gas model, in which we imagine the molecules to be pointlike.   ◀ Assumptions of kinetic theory for an ideal gas
2. **The molecules obey Newton's laws of motion, but as a whole they move randomly.** By "randomly" we mean that any molecule can move in any direction with equal probability, with a wide distribution of speeds.
3. **The molecules interact only through short-range forces during elastic collisions.** This assumption is consistent with the ideal gas model, in which the molecules exert no long-range forces on each other.
4. **The molecules make elastic collisions with the walls.**
5. **All molecules in the gas are identical.**

Although we often picture an ideal gas as consisting of single atoms, *molecular* gases exhibit ideal behavior at low pressures. On average, effects associated with molecular structure have no effect on the motions considered, so we can apply the results of the following development to molecular gases as well as to monatomic gases.

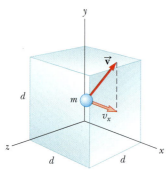

Figure 10.13 A cubical box with sides of length d containing an ideal gas. The molecule shown moves with velocity $\vec{v}$.

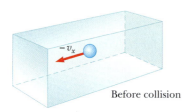

Before collision

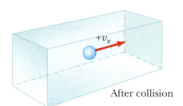

After collision

Figure 10.14 A molecule moving along the x-axis makes elastic collisions with the walls of the container. In colliding with a wall, the molecule's momentum is reversed, and the molecule exerts a force on the wall.

The glass vessel contains dry ice (solid carbon dioxide). The white cloud is carbon dioxide vapor, which is denser than air and hence falls from the vessel as shown.

Molecular Model for the Pressure of an Ideal Gas

As a first application of kinetic theory, we derive an expression for the pressure of an ideal gas in a container in terms of microscopic quantities. The pressure of the gas is the result of collisions between the gas molecules and the walls of the container. During these collisions, the gas molecules undergo a change of momentum as a result of the force exerted on them by the walls.

We now derive an expression for the pressure of an ideal gas consisting of N molecules in a container of volume V. In this section, we use m to represent the mass of one molecule. The container is a cube with edges of length d (Fig. 10.13). Consider the collision of one molecule moving with a velocity $-v_x$ toward the left-hand face of the box (Fig. 10.14). After colliding elastically with the wall, the molecule moves in the positive x-direction with a velocity $+v_x$. Because the momentum of the molecule is $-mv_x$ before the collision and $+mv_x$ afterward, the change in its momentum is

$$\Delta p_x = mv_x - (-mv_x) = 2mv_x$$

If F_1 is the magnitude of the average force exerted by a molecule on the *wall* in the time Δt, then applying Newton's second law to the wall gives

$$F_1 = \frac{\Delta p_x}{\Delta t} = \frac{2mv_x}{\Delta t}$$

In order for the molecule to make two collisions with the same wall, it must travel a distance $2d$ along the x-direction in a time Δt. Therefore, the time interval between two collisions with the same wall is $\Delta t = 2d/v_x$, and the force imparted to the wall by a single molecule is

$$F_1 = \frac{2mv_x}{\Delta t} = \frac{2mv_x}{2d/v_x} = \frac{mv_x^2}{d}$$

The total force F exerted by all the molecules on the wall is found by adding the forces exerted by the individual molecules:

$$F = \frac{m}{d}(v_{1x}^2 + v_{2x}^2 + \cdots)$$

In this equation, v_{1x} is the x-component of velocity of molecule 1, v_{2x} is the x-component of velocity of molecule 2, and so on. The summation terminates when we reach N molecules because there are N molecules in the container.

Note that the average value of the square of the velocity in the x-direction for N molecules is

$$\overline{v_x^2} = \frac{v_{1x}^2 + v_{2x}^2 + \cdots + v_{Nx}^2}{N}$$

where $\overline{v_x^2}$ is the average value of v_x^2. The total force on the wall can then be written

$$F = \frac{Nm}{d}\overline{v_x^2}$$

Now we focus on one molecule in the container traveling in some arbitrary direction with velocity $\vec{v}$ and having components v_x, v_y, and v_z. In this case, we must express the total force on the wall in terms of the speed of the molecules rather than just a single component. The Pythagorean theorem relates the square of the speed to the square of these components according to the expression $v^2 = v_x^2 + v_y^2 + v_z^2$. Hence, the average value of v^2 for all the molecules in the container is related to the average values $\overline{v_x^2}$, $\overline{v_y^2}$, and $\overline{v_z^2}$ according to the expression $\overline{v^2} = \overline{v_x^2} + \overline{v_y^2} + \overline{v_z^2}$. Because the motion is completely random, the average values $\overline{v_x^2}$, $\overline{v_y^2}$, and $\overline{v_z^2}$ are equal to each other. Using this fact and the earlier equation for $\overline{v_x^2}$, we find that

$$\overline{v_x^2} = \tfrac{1}{3}\overline{v^2}$$

The total force on the wall, then, is

$$F = \frac{N}{3}\left(\frac{m\overline{v^2}}{d}\right)$$

This expression allows us to find the total pressure exerted on the wall by dividing by the force by the area:

$$P = \frac{F}{A} = \frac{F}{d^2} = \tfrac{1}{3}\left(\frac{N}{d^3}\,m\overline{v^2}\right) = \tfrac{1}{3}\left(\frac{N}{V}\right)m\overline{v^2}$$

$$P = \tfrac{2}{3}\left(\frac{N}{V}\right)\left(\tfrac{1}{2}m\overline{v^2}\right) \qquad\qquad \text{[10.13]}$$ ◀ Pressure of an ideal gas

Equation 10.13 says that **the pressure is proportional to the number of molecules per unit volume and to the average translational kinetic energy of a molecule, $\tfrac{1}{2}m\overline{v^2}$.** With this simplified model of an ideal gas, we have arrived at an important result that relates the large-scale quantity of pressure to an atomic quantity—the average value of the square of the molecular speed. This relationship provides a key link between the atomic world and the large-scale world.

Equation 10.13 captures some familiar features of pressure. One way to increase the pressure inside a container is to increase the number of molecules per unit volume in the container. You do this when you add air to a tire. The pressure in the tire can also be increased by increasing the average translational kinetic energy of the molecules in the tire. As we will see shortly, this can be accomplished by increasing the temperature of the gas inside the tire. That's why the pressure inside a tire increases as the tire warms up during long trips. The continuous flexing of the tires as they move along the road transfers energy to the air inside them, increasing the air's temperature, which in turn raises the pressure.

Molecular Interpretation of Temperature

Having related the pressure of a gas to the average kinetic energy of the gas molecules, we now relate temperature to a microscopic description of the gas. We can obtain some insight into the meaning of temperature by multiplying Equation 10.13 by the volume:

$$PV = \tfrac{2}{3}N\left(\tfrac{1}{2}m\overline{v^2}\right)$$

Comparing this equation with the equation of state for an ideal gas in the form of Equation 10.11, $PV = Nk_BT$, we note that the left-hand sides of the two equations are identical. Equating the right-hand sides, we obtain

$$T = \frac{2}{3k_B}\left(\tfrac{1}{2}m\overline{v^2}\right) \qquad\qquad \text{[10.14]}$$ ◀ Temperature is proportional to average kinetic energy

This means that **the temperature of a gas is a direct measure of the average molecular kinetic energy of the gas**. As the temperature of a gas increases, the molecules move with higher average kinetic energy.

Rearranging Equation 10.14, we can relate the translational molecular kinetic energy to the temperature:

$$\tfrac{1}{2}m\overline{v^2} = \tfrac{3}{2}k_BT \qquad\qquad \text{[10.15]}$$ ◀ Average kinetic energy per molecule

So the average translational kinetic energy per molecule is $\tfrac{3}{2}k_BT$. The total translational kinetic energy of N molecules of gas is simply N times the average energy per molecule,

$$KE_{\text{total}} = N\left(\tfrac{1}{2}m\overline{v^2}\right) = \tfrac{3}{2}Nk_BT = \tfrac{3}{2}nRT \qquad\qquad \text{[10.16]}$$ ◀ Total kinetic energy of N molecules

where we have used $k_B = R/N_A$ for Boltzmann's constant and $n = N/N_A$ for the number of moles of gas. From this result, we see that **the total translational kinetic energy of a system of molecules is proportional to the absolute temperature of the system**.

For a monatomic gas, translational kinetic energy is the only type of energy the molecules can have, so Equation 10.16 gives the **internal energy U for a monatomic gas**:

$$U = \tfrac{3}{2} nRT \quad \text{(monatomic gas)} \qquad \textbf{[10.17]}$$

For diatomic and polyatomic molecules, additional possibilities for energy storage are available in the vibration and rotation of the molecule.

The square root of $\overline{v^2}$ is called the **root-mean-square (rms) speed** of the molecules. From Equation 10.15, we get, for the rms speed,

◀ Root-mean-square speed

$$v_{\text{rms}} = \sqrt{\overline{v^2}} = \sqrt{\frac{3k_B T}{m}} = \sqrt{\frac{3RT}{M}} \qquad \textbf{[10.18]}$$

where M is the molar mass in *kilograms per mole*, if R is given in SI units. Equation 10.18 shows that, at a given temperature, lighter molecules tend to move faster than heavier molecules. For example, if gas in a vessel consists of a mixture of hydrogen and oxygen, the hydrogen (H_2) molecules, with a molar mass of 2.0×10^{-3} kg/mol, move four times faster than the oxygen (O_2) molecules, with molar mass 32×10^{-3} kg/mol. If we calculate the rms speed for hydrogen at room temperature (~ 300 K), we find

TIP 10.4 Kilograms, not Grams Per Mole

In the equation for the rms speed, the units of molar mass M must be consistent with the units of the gas constant R. In particular, if R is in SI units, M must be expressed in kilograms per mole, not grams per mole.

$$v_{\text{rms}} = \sqrt{\frac{3RT}{M}} = \sqrt{\frac{3(8.31 \text{ J/mol} \cdot \text{K})(300 \text{ K})}{2.0 \times 10^{-3} \text{ kg/mol}}} = 1.9 \times 10^3 \text{ m/s}$$

This speed is about 17% of the escape speed for Earth, as calculated in Chapter 7. Because it is an average speed, a large number of molecules have much higher speeds and can therefore escape from Earth's atmosphere. This is why Earth's atmosphere doesn't currently contain hydrogen—it has all bled off into space.

Table 10.2 lists the rms speeds for various molecules at 20°C. A system of gas at a given temperature will exhibit a variety of speeds. This distribution of speeds is known as the *Maxwell velocity distribution*. An example of such a distribution for nitrogen gas at two different temperatures is given in Active Figure 10.15. The horizontal axis is speed, and the vertical axis is the number of molecules per unit speed. Notice that three speeds are of special interest: the most probable speed, corresponding to the peak in the graph; the average speed, which is found by averaging over all the possible speeds; and the rms speed. For every gas, $v_{\text{mp}} < v_{\text{av}} < v_{\text{rms}}$. As the temperature rises, these three speeds shift to the right.

TABLE 10.2

Some rms Speeds

Gas	Molar Mass (kg/mol)	v_{rms} at 20°C (m/s)
H_2	2.02×10^{-3}	1 902
He	4.0×10^{-3}	1 352
H_2O	18×10^{-3}	637
Ne	20.2×10^{-3}	602
N_2 and CO	28.0×10^{-3}	511
NO	30.0×10^{-3}	494
O_2	32.0×10^{-3}	478
CO_2	44.0×10^{-3}	408
SO_2	64.1×10^{-3}	338

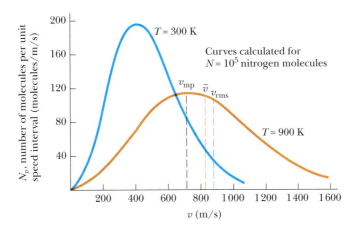

ACTIVE FIGURE 10.15
The Maxwell speed distribution for 10^5 nitrogen molecules at 300 K and 900 K. The total area under either curve equals the total number of molecules. The most probable speed v_{mp}, the average speed v_{av}, and the root-mean-square speed v_{rms} are indicated for the 900-K curve.

Physics⊗Now™
Log into PhysicsNow at **www.cp7e.com**, and go to Active Figure 10.15 to set the desired temperature and see the effect on the distribution curve.

Quick Quiz 10.6

One container is filled with argon gas and another with helium gas. Both containers are at the same temperature. Which atoms have the higher rms speed? (a) argon, (b) helium, (c) they have the same speed, or (d) not enough information to say.

Applying Physics 10.2   Expansion and Temperature

Imagine a gas in an insulated cylinder with a movable piston. The piston has been pushed inward, compressing the gas, and is now released. As the molecules of the gas strike the piston, they move it outward. Explain, from the point of view of the kinetic theory, how the expansion of this gas causes its temperature to drop.

Explanation   From the point of view of kinetic theory, a molecule colliding with the piston causes the piston to move with some velocity. According to the conservation of momentum, the molecule must rebound with less speed than it had before the collision. As these collisions occur, therefore, the average speed of the collection of molecules is reduced. Because temperature is related to the average speed of the molecules, the temperature of the gas drops.

EXAMPLE 10.9   A Cylinder of Helium

Goal   Calculate the internal energy of a system and the average kinetic energy per molecule.

Problem   A cylinder contains 2.00 mol of helium gas at 20.0°C. Assume that the helium behaves like an ideal gas. **(a)** Find the total internal energy of the system. **(b)** What is the average kinetic energy per molecule? **(c)** How much energy would have to be added to the system to double the rms speed? The molar mass of helium is 4.00×10^{-3} kg/mol.

Strategy   This problem requires substitution of given information into the appropriate equations: Equation 10.17 for part (a) and Equation 10.15 for part (b). In part (c), use the equations for the rms speed and internal energy together. A *change* in the internal energy must be computed.

Solution
(a) Find the total internal energy of the system.

Substitute values into Equation 10.17 with $n = 2.00$ and $T = 293$ K:

$$U = \tfrac{3}{2}(2.00 \text{ mol})(8.31 \text{ J/mol}\cdot\text{K})(293 \text{ K}) = \boxed{7.30 \times 10^3 \text{ J}}$$

(b) What is the average kinetic energy per molecule?

Substitute given values into Equation 10.15:

$$\tfrac{1}{2}m\overline{v^2} = \tfrac{3}{2}k_B T = \tfrac{3}{2}(1.38 \times 10^{-23} \text{ J/K})(293 \text{ K})$$
$$= \boxed{6.07 \times 10^{-21} \text{ J}}$$

(c) How much energy must be added to double the rms speed?

From Equation 10.18, doubling the rms speed requires quadrupling T. Calculate the required change of internal energy, which is the energy that must be put into the system:

$$\Delta U = U_f - U_i = \tfrac{3}{2}nRT_f - \tfrac{3}{2}nRT_i = \tfrac{3}{2}nR(T_f - T_i)$$

$$\Delta U = \tfrac{3}{2}(2.00 \text{ mol})(8.31 \text{ J/mol} \cdot \text{K})((4.00 \times 293 \text{ K})$$

$$- 293 \text{ K}) = \boxed{2.19 \times 10^4 \text{ J}}$$

Remark   Computing changes in internal energy will be important in understanding engine cycles in Chapter 12.

Exercise 10.9
The temperature of 5.00 moles of argon gas is lowered from 3.00×10^2 K to 2.40×10^2 K. (a) Find the change in the internal energy, ΔU, of the gas. (b) Find the change in the average kinetic energy per atom.

Answer   (a) $\Delta U = -3.74 \times 10^3$ J   (b) -1.24×10^{-21} J

SUMMARY

Physics⊗Now™ Take a practice test by logging into Physics-Now at **www.cp7e.com** and clicking on the Pre-Test link for this chapter.

10.1 Temperature and the Zeroth Law of Thermodynamics

Two systems are in **thermal contact** if energy can be exchanged between them, and in **thermal equilibrium** if they're in contact and there is no net exchange of energy. The exchange of energy between two objects because of differences in their temperatures is called **heat**.

The **zeroth law of thermodynamics** states that if two objects A and B are separately in thermal equilibrium with a third object, then A and B are in thermal equilibrium with each other. Equivalently, if the third object is a **thermometer**, then the **temperature** it measures for A and B will be the same. Two objects in thermal equilibrium are at the same temperature.

10.2 Thermometers and Temperature Scales

Thermometers measure temperature and are based on physical properties, such as the temperature-dependent expansion or contraction of a solid, liquid, or gas. These changes in volume are related to a linear scale, the most common being the **Fahrenheit, Celsius,** and **Kelvin scales**. The Kelvin temperature scale takes its zero point as **absolute zero** (0 K = $-273.15°$C), the point at which, by extrapolation, the pressure of all gases falls to zero.

The relationship between the Celsius temperature T_C and the Kelvin (absolute) temperature T is

$$T_C = T - 273.15 \qquad \text{[10.1]}$$

The relationship between the Fahrenheit and Celsius temperatures is

$$T_F = \tfrac{9}{5}T_C + 32 \qquad \text{[10.2a]}$$

10.3 Thermal Expansion of Solids and Liquids

Ordinarily a substance expands when heated. If an object has an initial length L_0 at some temperature and undergoes a change in temperature ΔT, its linear dimension changes by the amount ΔL, which is proportional to the object's initial length and the temperature change:

$$\Delta L = \alpha L_0 \Delta T \qquad \text{[10.4]}$$

The parameter α is called the **coefficient of linear expansion**. The change in area of a substance with change in temperature is given by

$$\Delta A = \gamma A_0 \Delta T \qquad \text{[10.5]}$$

where $\gamma = 2\alpha$ is the **coefficient of area expansion**. Similarly, the change in volume with temperature of most substances is proportional to the initial volume V_0 and the temperature change ΔT;

$$\Delta V = \beta V_0 \Delta T \qquad \text{[10.6]}$$

where $\beta = 3\alpha$ is the **coefficient of volume expansion**.

The expansion and contraction of material due to changes in temperature creates stresses and strains, sometimes sufficient to cause fracturing.

10.4 Macroscopic Description of an Ideal Gas

Avogadro's number is $N_A = 6.02 \times 10^{23}$ particles/mol. A mole of anything, by definition, consists of an Avogadro's number of particles. The number is defined so that one mole of carbon-12 atoms has a mass of exactly 12 g. The mass of one mole of a pure substance in grams is the same, numerically, as that substance's atomic (or molecular) mass.

An **ideal gas** obeys the equation

$$PV = nRT \qquad \text{[10.8]}$$

where P is the pressure of the gas, V is its volume, n is the number of moles of gas, R is the universal gas constant

(8.31 J/mol·K), and T is the absolute temperature in kelvins. A real gas at very low pressures behaves approximately as an ideal gas.

Solving problems usually entails comparing two different states of the same system of gas, dividing the ideal gas equation for the final state by the ideal gas equation for the initial state, canceling factors that don't change and solving for the unknown quantity.

10.5   The Kinetic Theory of Gases

The **pressure** of N molecules of an ideal gas contained in a volume V is given by

$$P = \tfrac{2}{3}\left(\frac{N}{V}\right)\left(\tfrac{1}{2}m\overline{v^2}\right) \qquad \textbf{[10.13]}$$

where $\tfrac{1}{2}m\overline{v^2}$ is the **average kinetic energy per molecule**.

The average kinetic energy of the molecules of a gas is directly proportional to the absolute temperature of the gas:

$$\tfrac{1}{2}m\overline{v^2} = \tfrac{3}{2}k_B T \qquad \textbf{[10.15]}$$

The quantity k_B is **Boltzmann's constant** (1.38×10^{-23} J/K).

The internal energy of n moles of a monatomic ideal gas is

$$U = \tfrac{3}{2}nRT \qquad \textbf{[10.17]}$$

The **root-mean-square (rms) speed** of the molecules of a gas is

$$v_{\text{rms}} = \sqrt{\frac{3k_B T}{m}} = \sqrt{\frac{3RT}{M}} \qquad \textbf{[10.18]}$$

CONCEPTUAL QUESTIONS

1. Why does an ordinary glass dish usually break when placed on a hot stove? Dishes made of Pyrex glass don't break so easily. What characteristic of Pyrex prevents breakage?

2. The balance wheel of a mechanical watch governs the frequency with which the watch ticks. A wheel cut from a single piece of metal will expand upon heating, thus increasing its moment of inertia. Will this expansion cause the watch to speed up or slow down?

3. In an astronomy class, the temperature at the core of a star is given by the teacher as 1.5×10^7 degrees. A student asks if this is Kelvins or degrees Celsius. How would you respond?

4. When a car engine overheats, you are warned not to remove the radiator cap to add cold water until there is time for the engine to cool down. Is this good advice? Why or why not?

5. Common thermometers are made of a mercury column in a glass tube. Based on the operation of these common thermometers, which has the larger coefficient of linear expansion—glass or mercury? (Don't answer this question by looking in a table.)

6. A steel wheel bearing is 1 mm smaller in diameter than an axle. How can the bearing be fit onto the axle without removing any material from the axle?

7. Objects deep beneath the surface of the ocean are subjected to extremely high pressures, as we saw in Chapter 9.

Some bacteria in these environments have adapted to pressures as much as a thousand times atmospheric pressure. How might such bacteria be affected if they were rapidly moved to the surface of the ocean?

8. Why is a power line more likely to break in winter than in summer, even if it is loaded with the same weight?

9. Although the average speed of gas molecules in thermal equilibrium at some temperature is greater than zero, the average velocity is zero. Explain.

10. After food is cooked in a pressure cooker, why is it very important to cool the container with cold water before attempting to remove the lid?

11. Some picnickers stop at a convenience store to buy food, including bags of potato chips. They then drive up into the mountains to their picnic site. When they unload the food, they notice that the bags of chips are puffed up like balloons. Why did this happen?

12. Markings to indicate length are placed on a steel tape in a room that is at a temperature of 22°C. Measurements are then made with the same tape on a day when the temperature is 27°C. Are the measurements too long, too short, or accurate?

13. Why do vapor bubbles in a pot of boiling water get larger as they approach the surface?

14. Why do small planets tend to have little or no atmosphere?

PROBLEMS

1, 2, 3 = straightforward, intermediate, challenging   □ = full solution available in *Student Solutions Manual/Study Guide*

Physics⊗Now™ = coached problem with hints available at **www.cp7e.com**   = biomedical application

Section 10.1 Temperature and the Zeroth Law of Thermodynamics

Section 10.2 Thermometers and Temperature Scales

1. For each of the following temperatures, find the equivalent temperature on the indicated scale: (a) $-273.15°C$ on the Fahrenheit scale, (b) $98.6°F$ on the Celsius scale, and (c) 100 K on the Fahrenheit scale.

2. The pressure in a constant-volume gas thermometer is 0.700 atm at 100°C and 0.512 atm at 0°C. (a) What is the temperature when the pressure is 0.0400 atm? (b) What is the pressure at 450°C?

3. Convert the following temperatures to their values on the Fahrenheit and Kelvin scales: (a) the boiling point of liquid hydrogen, $-252.87°C$; (b) the temperature of a room at 20°C.

4. Death Valley holds the record for the highest recorded temperature in the United States. On July 10, 1913, at a place called Furnace Creek Ranch, the temperature rose to 134°F. The lowest U.S. temperature ever recorded occurred at Prospect Creek Camp in Alaska on January 23, 1971, when the temperature plummeted to $-79.8°$ F. Convert these temperatures to the Celsius scale.

5. Show that the temperature $-40°$ is unique in that it has the same numerical value on the Celsius and Fahrenheit scales.

6. A constant-volume gas thermometer is calibrated in dry ice ($-80.0°C$) and in boiling ethyl alcohol (78.0°C). The respective pressures are 0.900 atm and 1.635 atm. (a) What value of absolute zero does the calibration yield? (b) What pressures would be found at the freezing and boiling points of water? (Note that we have the linear relationship $P = A + BT$, where A and B are constants.)

7. Show that if the temperature on the Celsius scale changes by ΔT_C, the Fahrenheit temperature changes by $\Delta T_F = (9/5)\Delta T_C$.

8. The temperature difference between the inside and the outside of an automobile engine is 450°C. Express this difference on (a) the Fahrenheit scale and (b) the Kelvin scale.

9. The melting point of gold is 1 064°C, and the boiling point is 2 660°C. (a) Express these temperatures in Kelvins. (b) Compute the difference of the two temperatures in Celsius degrees and in Kelvins.

Section 10.3 Thermal Expansion of Solids and Liquids

10. A cylindrical brass sleeve is to be shrink-fitted over a brass shaft whose diameter is 3.212 cm at 0°C. The diameter of the sleeve is 3.196 cm at 0°C. (a) To what temperature must the sleeve be heated before it will slip over the shaft? (b) Alternatively, to what temperature must the shaft be cooled before it will slip into the sleeve?

11. The New River Gorge bridge in West Virginia is a 518-m-long steel arch. How much will its length change between temperature extremes of $-20°C$ and 35°C?

12. A grandfather clock is controlled by a swinging brass pendulum that is 1.3 m long at a temperature of 20°C. (a) What is the length of the pendulum rod when the temperature drops to 0.0°C? (b) If a pendulum's period is given by $T = 2\pi\sqrt{L/g}$, where L is its length, does the change in length of the rod cause the clock to run fast or slow?

13. A pair of eyeglass frames are made of epoxy plastic (coefficient of linear expansion $= 1.30 \times 10^{-4}$ °C^{-1}). At room temperature (20.0°C), the frames have circular lens holes 2.20 cm in radius. To what temperature must the frames be heated if lenses 2.21 cm in radius are to be inserted into them?

14. A cube of solid aluminum has a volume of 1.00 m^3 at 20°C. What temperature change is required to produce a 100-cm^3 increase in the volume of the cube?

15. **Physics⊗Now**™ A brass ring of diameter 10.00 cm at 20.0°C is heated and slipped over an aluminum rod of diameter 10.01 cm at 20.0°C. Assuming the average coefficients of linear expansion are constant, (a) to what temperature must the combination be cooled to separate the two metals? Is that temperature attainable? (b) What if the aluminum rod were 10.02 cm in diameter?

16. Show that the coefficient of volume expansion, β, is related to the coefficient of linear expansion, α, through the expression $\beta = 3\alpha$.

17. A gold ring has an inner diameter of 2.168 cm at a temperature of 15.0°C. Determine its inner diameter at 100°C ($\alpha_{gold} = 1.42 \times 10^{-5}$ °C^{-1}).

18. A construction worker uses a steel tape to measure the length of an aluminum support column. If the measured length is 18.700 m when the temperature is 21.2°C, what is the measured length when the temperature rises to 29.4°C? (*Note:* Don't neglect the expansion of the tape.)

19. The band in Figure P10.19 is stainless steel (coefficient of linear expansion $= 17.3 \times 10^{-6}$ °C^{-1}; Young's modulus $= 18 \times 10^{10}$ N/m^2). It is essentially circular with an initial mean radius of 5.0 mm, a height of 4.0 mm, and a thickness of 0.50 mm. If the band just fits snugly over the tooth when heated to a temperature of 80°C, what is the tension in the band when it cools to a temperature of 37°C?

Figure P10.19

Figure P10.25

20. The Trans-Alaskan pipeline is 1 300 km long, reaching from Prudhoe Bay to the port of Valdez, and is subject to temperatures ranging from $-73°C$ to $+35°C$. How much does the steel pipeline expand due to the difference in temperature? How can this expansion be compensated for?

21. An automobile fuel tank is filled to the brim with 45 L (12 gal) of gasoline at $10°C$. Immediately afterward, the vehicle is parked in the sunlight, where the temperature is $35°C$. How much gasoline overflows from the tank as a result of the expansion? (Neglect the expansion of the tank.)

22. When the hot water in a certain upstairs bathroom is turned on, a series of 18 "ticks" is heard as the copper hot-water pipe slowly heats up and increases in length. The pipe runs vertically from the hot-water heater in the basement, through a hole in the floor 5.0 m above the water heater. The "ticks" are caused by the pipe sticking in the hole in the floor until the tension in the expanding pipe is great enough to unstick the pipe, enabling it to jump a short distance through the hole. If the hot-water temperature is $46°C$ and room temperature is $20°C$, determine (a) the distance the pipe moves with each "tick" and (b) the force required to unstick the pipe if the cross-sectional area of the copper in the pipe is 3.55×10^{-5} m^2.

23. The average coefficient of volume expansion for carbon tetrachloride is 5.81×10^{-4} $(°C)^{-1}$. If a 50.0-gal steel container is filled completely with carbon tetrachloride when the temperature is $10.0°C$, how much will spill over when the temperature rises to $30.0°C$?

24. On a day when the temperature is $20.0°C$, a concrete walk is poured in such a way that its ends are unable to move. (a) What is the stress in the cement when its temperature is $50.0°C$ on a hot, sunny day? (b) Does the concrete fracture? Take Young's modulus for concrete to be 7.00×10^9 N/m^2 and the compressive strength to be 2.00×10^7 N/m^2.

25. Figure P10.25 shows a circular steel casting with a gap. If the casting is heated, (a) does the width of the gap increase or decrease? (b) The gap width is 1.600 cm when the temperature is $30.0°C$. Determine the gap width when the temperature is $190°C$.

26. A hollow aluminum cylinder 20.0 cm deep has an internal capacity of 2.000 L at $20.0°C$. It is completely filled with turpentine and then warmed to $80.0°C$. (a) How much turpentine overflows? (b) If it is then cooled back to $20.0°C$, how far below the surface of the cylinder's rim is the turpentine surface?

Section 10.4 Macroscopic Description of an Ideal Gas

27. **Physics⊗Now**™ One mole of oxygen gas is at a pressure of 6.00 atm and a temperature of $27.0°C$. (a) If the gas is heated at constant volume until the pressure triples, what is the final temperature? (b) If the gas is heated so that both the pressure and volume are doubled, what is the final temperature?

28. Gas is contained in an 8.0-L vessel at a temperature of $20°C$ and a pressure of 9.0 atm. (a) Determine the number of moles of gas in the vessel. (b) How many molecules are in the vessel?

29. (a) An ideal gas occupies a volume of 1.0 cm^3 at $20°C$ and atmospheric pressure. Determine the number of molecules of gas in the container. (b) If the pressure of the 1.0-cm^3 volume is reduced to 1.0×10^{-11} Pa (an extremely good vacuum) while the temperature remains constant, how many moles of gas remain in the container?

30. A tank having a volume of 0.100 m^3 contains helium gas at 150 atm. How many balloons can the tank blow up if each filled balloon is a sphere 0.300 m in diameter at an absolute pressure of 1.20 atm?

31. A cylinder with a movable piston contains gas at a temperature of $27.0°C$, a volume of 1.50 m^3, and an absolute pressure of 0.200×10^5 Pa. What will be its final temperature if the gas is compressed to 0.700 m^3 and the absolute pressure increases to 0.800×10^5 Pa?

32. The density of helium gas at $T = 0°C$ is $\rho_0 = 0.179$ kg/m^3. The temperature is then raised to $T = 100°C$, but the pressure is kept constant. Assuming that the helium is an ideal gas, calculate the new density ρ_f of the gas.

33. A weather balloon is designed to expand to a maximum radius of 20 m at its working altitude, where the air pressure is 0.030 atm and the temperature is 200 K. If the balloon is filled at atmospheric pressure and 300 K, what is its radius at liftoff?

34. A cylindrical diving bell 3.00 m in diameter and 4.00 m tall with an open bottom is submerged to a depth of 220 m in the ocean. The surface temperature is 25.0°C, and the temperature 220 m down is 5.00°C. The density of seawater is 1 025 kg/m³. How high does the seawater rise in the bell when it is submerged?

35. An air bubble has a volume of 1.50 cm³ when it is released by a submarine 100 m below the surface of a lake. What is the volume of the bubble when it reaches the surface? Assume that the temperature and the number of air molecules in the bubble remain constant during its ascent.

Section 10.5 The Kinetic Theory of Gases

36. A sealed cubical container 20.0 cm on a side contains three times Avogadro's number of molecules at a temperature of 20.0°C. Find the force exerted by the gas on one of the walls of the container.

37. What is the average kinetic energy of a molecule of oxygen at a temperature of 300 K?

38. (a) What is the total random kinetic energy of all the molecules in 1 mole of hydrogen at a temperature of 300 K? (b) With what speed would a mole of hydrogen have to move so that the kinetic energy of the mass as a whole would be equal to the total random kinetic energy of its molecules?

39. Use Avogadro's number to find the mass of a helium atom.

40. The temperature near the top of the atmosphere on Venus is 240 K. (a) Find the rms speed of hydrogen (H_2) at that point in Venus's atmosphere. (b) Repeat for carbon dioxide (CO_2). (c) It has been found that if the rms speed exceeds one-sixth of the planet's escape velocity, the gas eventually leaks out of the atmosphere and into outer space. If the escape velocity on Venus is 10.3 km/s, does hydrogen escape? Does carbon dioxide?

41. A cylinder contains a mixture of helium and argon gas in equilibrium at a temperature of 150°C. (a) What is the average kinetic energy of each type of molecule? (b) What is the rms speed of each type of molecule?

42. Three moles of nitrogen gas, N_2, at 27.0°C are contained in a 22.4-L cylinder. Find the pressure the gas exerts on the cylinder walls.

43. **Physics**⊗**Now**™ Superman leaps in front of Lois Lane to save her from a volley of bullets. In a 1-minute interval, an automatic weapon fires 150 bullets, each of mass 8.0 g, at 400 m/s. The bullets strike his mighty chest, which has an area of 0.75 m². Find the average force exerted on Superman's chest if the bullets bounce back after an elastic, head-on collision.

44. In a period of 1.0 s, 5.0×10^{23} nitrogen molecules strike a wall of area 8.0 cm². If the molecules move at 300 m/s and strike the wall head on in a perfectly elastic collision, find the pressure exerted on the wall. (The mass of one N_2 molecule is 4.68×10^{-26} kg.)

ADDITIONAL PROBLEMS

45. Inside the wall of a house, an L-shaped section of hot-water pipe consists of a straight horizontal piece 28.0 cm long, an elbow, and a straight vertical piece 134 cm long (Fig. P10.45). A stud and a second-story floorboard hold the ends of this section of copper pipe stationary. Find the magnitude and direction of the displacement of the pipe elbow when the water flow is turned on, raising the temperature of the pipe from 18.0°C to 46.5°C.

Figure P10.45

46. The active element of a certain laser is an ordinary glass rod 20 cm long and 1.0 cm in diameter. If the temperature of the rod increases by 75°C, find its increases in (a) length, (b) diameter, and (c) volume.

47. A popular brand of cola contains 6.50 g of carbon dioxide dissolved in 1.00 L of soft drink. If the evaporating carbon dioxide is trapped in a cylinder at 1.00 atm and 20.0°C, what volume does the gas occupy?

48. A 1.5-m-long glass tube that is closed at one end is weighted and lowered to the bottom of a freshwater lake. When the tube is recovered, an indicator mark shows that water rose to within 0.40 m of the closed end. Determine the depth of the lake. Assume constant temperature.

49. Long-term space missions require reclamation of the oxygen in the carbon dioxide exhaled by the crew. In one method of reclamation, 1.00 mol of carbon dioxide produces 1.00 mol of oxygen, with 1.00 mol of methane as a by-product. The methane is stored in a tank under pressure and is available to control the attitude of the spacecraft by controlled venting. A single astronaut exhales 1.09 kg of carbon dioxide each day. If the methane generated in the recycling of three astronauts' respiration during one week of flight is stored in an originally empty 150-L tank at $-45.0°C$, what is the final pressure in the tank?

50. A vertical cylinder of cross-sectional area 0.050 m² is fitted with a tight-fitting, frictionless piston of mass 5.0 kg (Fig. P10.50). If there are 3.0 mol of an ideal gas in the cylinder at 500 K, determine the height h at which the piston will be in equilibrium under its own weight.

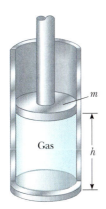

Figure P10.50

51. A liquid with a coefficient of volume expansion of β just fills a spherical flask of volume V_0 at temperature T (Fig. P10.51). The flask is made of a material that has a coefficient of linear expansion of α. The liquid is free to expand into a capillary of cross-sectional area A at the top. (a) Show that if the temperature increases by ΔT, the liquid rises in the capillary by the amount $\Delta h = (V_0/A)(\beta - 3\alpha)\Delta T$. (b) For a typical system, such as a mercury thermometer, why is it a good approximation to neglect the expansion of the flask?

52. A hollow aluminum cylinder is to be fitted over a steel piston. At 20°C, the inside diameter of the cylinder is 99% of the outside diameter of the piston. To what common temperature should the two pieces be heated in order that the cylinder just fit over the piston?

53. A steel measuring tape was designed to read correctly at 20°C. A parent uses the tape to measure the height of a

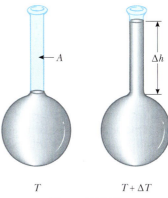

Figure P10.51

1.1-m-tall child. If the measurement is made on a day when the temperature is 25°C, is the tape reading longer or shorter than the actual height, and by how much?

54. Before beginning a long trip on a hot day, a driver inflates an automobile tire to a gauge pressure of 1.80 atm at 300 K. At the end of the trip, the gauge pressure has increased to 2.20 atm. (a) Assuming that the volume has remained constant, what is the temperature of the air inside the tire? (b) What percentage of the original mass of air in the tire should be released so the pressure returns to its original value? Assume that the temperature remains at the value found in (a) and the volume of the tire remains constant as air is released.

55. Two concrete spans of a 250-m-long bridge are placed end to end so that no room is allowed for expansion (Fig. P10.55a). If the temperature increases by 20.0°C, what is the height y to which the spans rise when they buckle (Fig. P10.55b)?

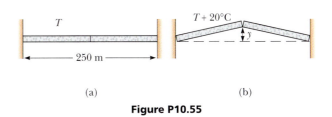

(a)   (b)

Figure P10.55

56. A copper rod and a steel rod are heated. At 0°C, the copper rod has a length L_C and the steel one has a length L_S. When the rods are being heated or cooled, a difference of 5.00 cm is maintained between their lengths. Determine the values of L_C and L_S.

57. If 9.00 g of water is placed in a 2.00-L pressure cooker and heated to 500°C, what is the pressure inside the container?

58. An expandable cylinder has its top connected to a spring with force constant 2.00×10^3 N/m. (See Fig. P10.58.) The cylinder is filled with 5.00 L of gas with the spring relaxed at a pressure of 1.00 atm and a temperature of 20.0°C. (a) If the lid has a cross-sectional area of 0.010 0 m^2 and negligible mass, how high will the lid rise when the temperature is raised to 250°C? (b) What is the pressure of the gas at 250°C?

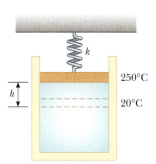

Figure P10.58

59. A swimmer has 0.820 L of dry air in his lungs when he dives into a lake. Assuming the pressure of the dry air is 95% of the external pressure at all times, what is the volume of the dry air at a depth of 10.0 m? Assume that atmospheric pressure at the surface is 1.013×10^5 Pa.

60. Two small containers, each with a volume of 100 cm^3, contain helium gas at 0°C and 1.00 atm pressure. The two containers are joined by a small open tube of negligible volume, allowing gas to flow from one container to the other. What common pressure will exist in the two containers if the temperature of one container is raised to 100°C while the other container is kept at 0°C?

61. **Physics⊗Now**™ A bimetallic bar is made of two thin strips of dissimilar metals bonded together. As they are heated, the one with the larger average coefficient of expansion expands more than the other, forcing the bar into an arc, with the outer strip having both a larger radius and a larger circumference. (See Fig. P10.61.) (a) Derive an expression for the angle of bending, θ, as a function of the initial length of the strips, their average coefficients of linear expansion, the change in temperature, and the separation of the centers of the strips ($\Delta r = r_2 - r_1$). (b) Show that the angle of bending goes to zero when ΔT goes to zero or when the two coefficients

of expansion become equal. (c) What happens if the bar is cooled?

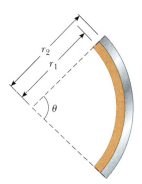

Figure P10.61

62. A 250-m-long bridge is improperly designed so that it cannot expand with temperature. It is made of concrete with $\alpha = 12 \times 10^{-6}$ °C^{-1}. (a) Assuming that the maximum change in temperature at the site is expected to be 20°C, find the change in length the span would undergo if it were free to expand. (b) Show that the stress on an object with Young's modulus Y when raised by ΔT with its ends firmly fixed is given by $\alpha Y \Delta T$. (c) If the maximum stress the bridge can withstand without crumbling is 2.0×10^7 Pa, will it crumble because of this temperature increase? Young's modulus for concrete is about 2.0×10^{10} Pa.

63. The density of gasoline is 730 kg/m^3 at 0°C. Its volume expansion coefficient is 9.6×10^{-4} °C^{-1}. If 1.00 gal of gasoline occupies 0.003 8 m^3, how many extra kilograms of gasoline are obtained when 10 gallons of gasoline are bought at 0°C rather than at 20°C?

ACTIVITIES

A.1. Fill one basin with hot tap water (not to exceed about 100°F). Fill another basin with cold tap water, and add ice until about one-third of the mixture is ice. Fill the third basin with an equal mixture of hot and cold tap water. Place your left hand in the hot water and your right hand in the cold water for about 15 s. Then place both hands in the basin of lukewarm water for 15 s. Describe whether the water feels hot or cold to either of your hands and why this effect occurs.

A.2. Tape two plastic straws tightly together along their entire length, but with a 2-cm offset. Hold them in a stream of very hot water from a faucet so that the water pours through one straw but not the other. Quickly hold the straws up and sight along their length. You should be able to see a very slight curvature in the tape. The effect is small, so look closely. Running cold water through the same straw and again sighting along the length will help you see the small change in shape more clearly. Explain these observations.

A.3. You can study the thermal expansion of air with a Florence flask and a balloon. To do so, boil a small amount of water in the flask. Then quickly remove the flask from the heat source and place the mouth of the balloon over the mouth of the flask. Observe what happens to the balloon when you place the flask in a basin of cold water. Explain your observation. Reheat the water in the flask with the balloon still in place, and explain your observations.

Glacier fragments fall into the sea. Global warming could melt enough ice to swell the oceans and threaten coastal cities around the world.

CHAPTER

11

O U T L I N E

Steve Bly/Getty Images

Energy in Thermal Processes

When two objects with different temperatures are placed in thermal contact, the temperature of the warmer object decreases while the temperature of the cooler object increases. With time, they reach a common equilibrium temperature somewhere in between their initial temperatures. During this process, we say that energy is transferred from the warmer object to the cooler one.

Until about 1850, the subjects of thermodynamics and mechanics were considered two distinct branches of science, and the principle of conservation of energy seemed to describe only certain kinds of mechanical systems. Experiments performed by the English physicist James Joule (1818–1889) and others showed that the decrease in mechanical energy (kinetic plus potential) of an isolated system was equal to the increase in internal energy of the system. Today, internal energy is treated as a form of energy that can be transformed into mechanical energy and vice versa. Once the concept of energy was broadened to include internal energy, the law of conservation of energy emerged as a universal law of nature.

This chapter focuses on some of the processes of energy transfer between a system and its surroundings.

11.1   HEAT AND INTERNAL ENERGY

A major distinction must be made between internal energy and heat. These terms are not interchangeable—heat involves a *transfer* of internal energy from one location to another. The following formal definitions will make the distinction precise.

Internal energy U is the energy associated with the microscopic components of a system—the atoms and molecules of the system. The internal energy includes kinetic and potential energy associated with the random translational, rotational, and vibrational motion of the particles that make up the system, and any potential energy bonding the particles together.

◀ Internal energy

In Chapter 10 we showed that the internal energy of a monatomic ideal gas is associated with the translational motion of its atoms. In this special case, the internal energy is the total translational kinetic energy of the atoms; the higher the temperature of the gas, the greater the kinetic energy of the atoms and the greater the internal energy of the gas. For more complicated diatomic and polyatomic gases, internal energy includes other forms of molecular energy, such as rotational kinetic energy and the kinetic and potential energy associated with molecular vibrations. Internal energy is also associated with the intermolecular potential energy ("bond energy") between molecules in a liquid or solid.

Heat was introduced in Chapter 5 as one possible method of transferring energy between a system and its environment, and we provide a formal definition here:

Heat is the transfer of energy between a system and its environment due to a temperature difference between them.

The symbol Q is used to represent the amount of energy transferred by heat between a system and its environment. For brevity, we will often use the phrase "the energy Q transferred to a system . . ." rather than "the energy Q transferred by heat to a system . . ."

If a pan of water is heated on the burner of a stove, it's incorrect to say more heat is in the water. Heat is the *transfer* of thermal energy, just as work is the transfer of mechanical energy. When an object is pushed, it doesn't have more work; rather, it has more mechanical energy transferred *by* work. Similarly, the pan of water has more thermal energy transferred by heat.

JAMES PRESCOTT JOULE,
British physicist (1818–1889)

Joule received some formal education in mathematics, philosophy, and chemistry from John Dalton, but was in large part self-educated. Joule's most active research period, from 1837 through 1847, led to the establishment of the principle of conservation of energy and the relationship between heat and other forms of energy transfer. His study of the quantitative relationship among electrical, mechanical, and chemical effects of heat culminated in his announcement in 1843 of the amount of work required to produce a unit of internal energy.

Units of Heat

Early in the development of thermodynamics, before scientists realized the connection between thermodynamics and mechanics, heat was defined in terms of the temperature changes it produced in an object, and a separate unit of energy, the **calorie**, was used for heat. The calorie (cal) is defined as **the energy necessary to raise the temperature of 1 g of water from 14.5° to 15.5°C.** (The "Calorie," with a capital "C," used in describing the energy content of foods, is actually a kilocalorie.) Likewise, the unit of heat in the U.S. customary system, the **British thermal unit** (Btu), was defined as **the energy required to raise the temperature of 1 lb of water from 63°F to 64°F.**

In 1948, scientists agreed that because heat (like work) is a measure of the transfer of energy, its SI unit should be the joule. The calorie is now defined to be exactly 4.186 J:

◀ Definition of the calorie

$$1 \text{ cal} \equiv 4.186 \text{ J} \qquad [11.1]$$

◀ The mechanical equivalent of heat

This definition makes no reference to raising the temperature of water. The calorie is a general energy unit, introduced here for historical reasons, though we will make little use of it. The definition in Equation 11.1 is known, from the historical background we have discussed, as the **mechanical equivalent of heat**.

EXAMPLE 11.1   Working Off Breakfast

Goal   Relate caloric energy to mechanical energy.

Problem   A student eats a breakfast consisting of two bowls of cereal and milk, containing a total of 3.20×10^2 Calories of energy. He wishes to do an equivalent amount of work in the gymnasium by doing curls with a 25.0-kg barbell (Fig. 11.1). How many times must he raise the weight to expend that much energy? Assume that he raises it through a vertical displacement of 0.400 m each time, the distance from his lap to his upper chest.

Figure 11.1   (Example 11.1)

Strategy   Convert the energy in Calories to joules, then equate that energy to the work necessary to do n repetitions of the barbell exercise. The work he does lifting the barbell can be found from the work–energy theorem and the change in potential energy of the barbell. He does negative work on the barbell going down, to keep it from speeding up. The net work on the barbell during one repetition is zero, but his muscles expend the same energy both in raising and lowering.

Solution

Convert his breakfast Calories, E, to joules:

$$E = (3.20 \times 10^2 \text{ Cal}) \left(\frac{1.00 \times 10^3 \text{ cal}}{1.00 \text{ Cal}} \right) \left(\frac{4.186 \text{ J}}{\text{cal}} \right)$$

$$= 1.34 \times 10^6 \text{ J}$$

Use the work–energy theorem to find the work necessary to lift the barbell up to its maximum height.

$$W = \Delta KE + \Delta PE = (0 - 0) + (mgh - 0) = mgh$$

The student must expend the same amount of energy lowering the barbell, making $2mgh$ per repetition. Multiply this amount by n repetitions and set it equal to the food energy E:

$$n(2mgh) = E$$

Solve for n, substituting the food energy for E:

$$n = \frac{E}{2mgh} = \frac{1.34 \times 10^6 \text{ J}}{2(25.0 \text{ kg})(9.80 \text{ m/s}^2)(0.400 \text{ m})}$$

$$= 6.84 \times 10^3 \text{ times}$$

Remarks   If the student does one repetition every five seconds, it will take him 9.5 hours to work off his breakfast! In exercising, a large fraction of energy is lost through heat, however, due to the inefficiency of the body in doing work. This transfer of energy dramatically reduces the exercise requirement by at least three-quarters, a little over two hours. All the same, it might be best to forego that second bowl of cereal!

Exercise 11.1

How many sprints from rest to a speed of 5.0 m/s would a 65-kg woman have to complete in order to burn off 5.0×10^2 Calories? (Assume 100% efficiency in converting food energy to mechanical energy).

Answer 2.6×10^3 sprints

APPLICATION

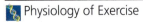 Physiology of Exercise

Getting proper exercise is an important part of staying healthy and keeping weight under control. As seen in the preceding example, the body expends energy when doing mechanical work, and these losses are augmented by the inefficiency of converting the body's internal stores of energy into useful work, with three-quarters or more leaving the body through heat. In addition, exercise tends to elevate the body's general metabolic rate, which persists even after the exercise is over. The increase in metabolic rate due to exercise, more so than the exercise itself, is helpful in weight reduction.

11.2 SPECIFIC HEAT

The historical definition of the calorie is the amount of energy necessary to raise the temperature of one gram of a specific substance—water—by one degree. That amount is 4.186 J. Raising the temperature of one kilogram of water by 1° requires 4 186 J of energy. The amount of energy required to raise the temperature of one kilogram of an arbitrary substance by 1° varies with the substance. For example, the energy required to raise the temperature of one kilogram of copper by 1.0°C is 387 J. Every substance requires a unique amount of energy per unit mass to change the temperature of that substance by 1.0°C.

If a quantity of energy Q is transferred to a substance of mass m, changing its temperature by $\Delta T = T_f - T_i$, the **specific heat** c of the substance is defined by

$$c \equiv \frac{Q}{m\,\Delta T} \qquad \text{[11.2]}$$

SI unit: Joule per kilogram-degree Celsius (J/kg·°C)

Table 11.1 lists specific heats for several substances. From the definition of the calorie, the specific heat of water is 4 186 J/kg · °C.

From the definition of specific heat, we can express the energy Q needed to raise the temperature of a system of mass m by ΔT as

$$Q = mc\,\Delta T \qquad \text{[11.3]}$$

The energy required to raise the temperature of 0.500 kg of water by 3.00°C, for example, is $Q = (0.500 \text{ kg})(4\,186 \text{ J/kg} \cdot °\text{C})(3.00°\text{C}) = 6.28 \times 10^3$ J. Note that when the temperature increases, ΔT and Q are *positive*, corresponding to energy flowing *into* the system. When the temperature decreases, ΔT and Q are *negative*, and energy flows *out* of the system.

Table 11.1 shows that water has the highest specific heat relative to most other common substances. This high specific heat is responsible for the moderate temperatures found in regions near large bodies of water. As the temperature of a body of water decreases during winter, the water transfers energy to the air, which carries the energy landward when prevailing winds are toward the land. Off the western coast of the United States, the energy liberated by the Pacific Ocean is carried to the east, keeping coastal areas much warmer than they would otherwise be. Winters are generally colder in the eastern coastal states, because the prevailing winds tend to carry the energy away from land.

The fact that the specific heat of water is higher than the specific heat of sand is responsible for the pattern of airflow at a beach. During the day, the Sun adds roughly equal amounts of energy to the beach and the water, but the lower specific heat of sand causes the beach to reach a higher temperature than the water. As a result, the air above the land reaches a higher temperature than the air above the water. The denser cold air pushes the less dense hot air upward (due to Archimedes's principle), resulting in a breeze from ocean to land during the day. Because the hot air gradually cools as it rises, it subsequently sinks, setting up the circulation pattern shown in Figure 11.2.

A similar effect produces rising layers of air called *thermals* that can help eagles soar higher and hang gliders stay in flight longer. A thermal is created when a portion of the Earth reaches a higher temperature than neighboring regions. This often happens to plowed fields, which are warmed by the Sun to higher temperatures

TABLE 11.1

Specific Heats of Some Materials at Atmospheric Pressure

Substance	J/kg·°C	cal/g·°C
Aluminum	900	0.215
Beryllium	1 820	0.436
Cadmium	230	0.055
Copper	387	0.0924
Germanium	322	0.077
Glass	837	0.200
Gold	129	0.0308
Ice	2 090	0.500
Iron	448	0.107
Lead	128	0.0305
Mercury	138	0.033
Silicon	703	0.168
Silver	234	0.056
Steam	2 010	0.480
Water	4 186	1.00

TIP 11.1 Finding ΔT

In Equation 11.3, be sure to remember that ΔT is *always* the final temperature minus the initial temperature: $\Delta T = T_f - T_i$.

Figure 11.2 Circulation of air at the beach. On a hot day, the air above the sand warms faster than the air above the cooler water. The warmer air floats upward due to Archimedes's principle, resulting in the movement of cooler air toward the beach.

APPLICATION

Sea Breezes and Thermals

than nearby fields shaded by vegetation. The cooler, denser air over the vegetation-covered fields pushes the expanding air over the plowed field upwards, and a thermal is formed.

Quick Quiz 11.1

Suppose you have 1 kg each of iron, glass, and water, and all three samples are at 10°C. (a) Rank the samples from lowest to highest temperature after 100 J of energy is added to each by heat. (b) Rank them from least to greatest amount of energy transferred by heat if enough energy is transferred so that each increases in temperature by 20°C.

EXAMPLE 11.2   Stressing a Strut

Goal   Use the energy transfer equation in the context of linear expansion and compressional stress.

Problem   A steel strut near a ship's furnace is 2.00 m long, with a mass of 1.57 kg and cross-sectional area of 1.00×10^{-4} m². During operation of the furnace, the strut absorbs thermal energy in a net amount of 2.50×10^5 J. **(a)** Find the change in temperature of the strut. **(b)** Find the increase in length of the strut. **(c)** If the strut is not allowed to expand because it's bolted at each end, find the compressional stress developed in the strut.

Strategy   This problem can be solved by substituting given quantities into three different equations. In part (a), the change in temperature can be computed by substituting into Equation 11.3, which relates temperature change to the energy transferred by heat. In part (b), substituting the result of part (a) into the linear expansion equation yields the change in length. If that change of length is thwarted by poor design, as in part (c), the result is compressional stress, found with the compressional stress–strain equation.

Solution

(a) Find the change in temperature.

Solve Equation 11.3 for the change in temperature and substitute:

$$Q = m_s c_s \Delta T \quad \rightarrow \quad \Delta T = \frac{Q}{m_s c_s}$$

$$\Delta T = \frac{(2.50 \times 10^5 \text{ J})}{(1.57 \text{ kg})(448 \text{ J/kg} \cdot °\text{C})} = \boxed{355°C}$$

(b) Find the change in length of the strut if it's allowed to expand.

Substitute into the linear expansion equation:

$$\Delta L = \alpha L_0 \Delta T = (11 \times 10^{-6} \, °\text{C}^{-1})(2.00 \text{ m})(355°C)$$
$$= \boxed{7.8 \times 10^{-3} \text{ m}}$$

(c) Find the compressional stress in the strut if it is not allowed to expand.

Substitute into the compressional stress–strain equation:

$$\frac{F}{A} = Y \frac{\Delta L}{L_0} = (2.00 \times 10^{11} \text{ Pa}) \frac{7.8 \times 10^{-3} \text{ m}}{2.01 \text{ m}}$$
$$= \boxed{7.8 \times 10^8 \text{ Pa}}$$

Remarks   Notice the use of 2.01 m in the denominator of the last calculation, rather than 2.00 m. This is because, in effect, the strut was compressed back to the original length from the length to which it would have expanded. (The difference is negligible, however.) The answer exceeds the ultimate compressive strength of steel and underscores the importance of allowing for thermal expansion. Of course, it's likely the strut would bend, relieving some of the stress (creating some shear stress in the process). Finally, if the strut is attached at both ends by bolts, thermal expansion and contraction would exert sheer stresses on the bolts, possibly weakening or loosening them over time.

Exercise 11.2
Suppose a steel strut with cross-sectional area 5.00×10^{-4} m^2 and length 2.50 m is bolted between two rigid bulk-heads in the engine room of a submarine. (a) Calculate the change in temperature of the strut if it absorbs an energy of 3.00×10^5 J. (b) Calculate the compressional stress in the strut.

Answers (a) 68.2°C (b) 1.50×10^8 Pa

11.3 CALORIMETRY

One technique for measuring the specific heat of a solid or liquid is to raise the temperature of the substance to some value, place it into a vessel containing cold water of known mass and temperature, and measure the temperature of the combination after equilibrium is reached. Define the system as the substance and the water. If the vessel is assumed to be a good insulator, so that energy doesn't leave the system, then we can assume the system is isolated. Vessels having this property are called **calorimeters**, and analysis performed using such vessels is called **calorimetry**.

The principle of conservation of energy for this isolated system requires that the net result of all energy transfers is zero. If one part of the system loses energy, another part has to gain the energy, because the system is isolated and the energy has nowhere else to go. When a warm object is placed in the cooler water of a calorimeter, the warm object becomes cooler while the water becomes warmer. This principle can be written

$$Q_{cold} = -Q_{hot} \qquad [11.4]$$

Q_{cold} is positive because energy is flowing into cooler objects, and Q_{hot} is negative because energy is leaving the hot object. The negative sign on the right-hand side of Equation 11.4 ensures that the right-hand side is a positive number, consistent with the left-hand side. The equation is valid only when the system it describes is isolated.

Calorimetry problems involve solving Equation 11.4 for an unknown quantity, usually either a specific heat or a temperature.

EXAMPLE 11.3 Finding a Specific Heat

Goal Solve a calorimetry problem involving only two substances.

Problem A 125-g block of an unknown substance with a temperature of 90.0°C is placed in a Styrofoam cup containing 0.326 kg of water at 20.0°C. The system reaches an equilibrium temperature of 22.4°C. What is the specific heat, c_x, of the unknown substance if the heat capacity of the cup is neglected?

Strategy The water gains thermal energy Q_{cold}, while the block loses thermal energy Q_{hot}. Using Equation 11.3, substitute expressions into Equation 11.4 and solve for the unknown specific heat, c_x.

Solution
Let T be the final temperature, and let T_w and T_x be the initial temperatures of the water and block, respectively. Apply Equations 11.3 and 11.4:

$$Q_{cold} = -Q_{hot}$$
$$m_w c_w (T - T_w) = -m_x c_x (T - T_x)$$

Solve for c_x and substitute numerical values:

$$c_x = \frac{m_w c_w (T - T_w)}{m_x (T_x - T)}$$
$$= \frac{(0.326 \text{ kg})(4\,190 \text{ J/kg} \cdot °C)(22.4°C - 20.0°C)}{(0.125 \text{ kg})(90.0°C - 22.4°C)}$$

$$c_x = 388 \text{ J/kg} \cdot °C$$

Remarks   Comparing our results to values given in Table 11.1, the unknown substance is probably copper.

Exercise 11.3

A 255-g block of gold at 85.0°C is immersed in 155 g of water at 25.0°C. Find the equilibrium temperature, assuming the system is isolated and the heat capacity of the cup can be neglected.

Answer   27.9°C

TIP 11.2   Celsius versus Kelvin

In equations in which T appears, such as the ideal gas law, the Kelvin temperature *must* be used. In equations involving ΔT, such as calorimetry equations, it's possible to use either Celsius or Kelvin temperatures, because a change in temperature is the same on both scales. When in doubt, use Kelvin.

As long as there are no more than two substances involved, Equation 11.4 can be used to solve elementary calorimetry problems. Sometimes, however, there may be three (or more) substances exchanging thermal energy, each at a different temperature. If the problem requires finding the final temperature, it may not be clear whether the substance with the middle temperature gains or loses thermal energy. In such cases, Equation 11.4 can't be used reliably.

For example, suppose we want to calculate the final temperature of a system consisting initially of a glass beaker at 25°C, hot water at 40°C, and a block of aluminum at 37°C. We know that after the three are combined, the glass beaker warms up and the hot water cools, but we don't know for sure whether the aluminum block gains or loses energy because the final temperature is unknown.

Fortunately, we can still solve such a problem as long as it's set up correctly. With an unknown final temperature T_f, the expression $Q = mc(T_f - T_i)$ will be positive if $T_f > T_i$ and negative if $T_f < T_i$. Equation 11.4 can be written as

$$\Sigma Q_k = 0 \qquad\qquad [11.5]$$

where Q_k is the energy change in the kth object. Equation 11.5 says that the sum of all the gains and losses of thermal energy must add up to zero, as required by the conservation of energy for an isolated system. Each term in Equation 11.5 will have the correct sign automatically. Applying Equation 11.5 to the water, aluminum, and glass problem, we get

$$Q_w + Q_{al} + Q_g = 0$$

There's no need to decide in advance whether a substance in the system is gaining or losing energy. This equation is similar in style to the conservation of mechanical energy equation, where the gains and losses of kinetic and potential energies sum to zero for an isolated system: $\Delta K + \Delta PE = 0$. As will be seen, changes in thermal energy can be included on the left-hand side of this equation.

When more than two substances exchange thermal energy, it's easy to make errors substituting numbers, so it's a good idea to construct a table to organize and assemble all the data. This strategy is illustrated in the next example.

EXAMPLE 11.4   Calculate an Equilibrium Temperature

Goal   Solve a calorimetry problem involving three substances at three different temperatures.

Problem   Suppose 0.400 kg of water initially at 40.0°C is poured into a 0.300-kg glass beaker having a temperature of 25.0°C. A 0.500-kg block of aluminum at 37.0°C is placed in the water, and the system insulated. Calculate the final equilibrium temperature of the system.

Strategy   The energy transfer for the water, aluminum, and glass will be designated Q_w, Q_{al}, and Q_g, respectively. The sum of these transfers must equal zero, by conservation of energy. Construct a table, assemble the three terms from the given data, and solve for the final equilibrium temperature, T.

Solution

Apply Equation 11.5 to the system:

$$Q_w + Q_{al} + Q_g = 0$$
$$m_w c_w(T - T_w) + m_{al}c_{al}(T - T_{al}) + m_g c_g(T - T_g) = 0 \quad (1)$$

Construct a data table:

Q (J)	m (kg)	c (J/kg °C)	T_f	T_i
Q_w	0.400	4 190	T	40.0°C
Q_{al}	0.500	9.00×10^2	T	37.0°C
Q_g	0.300	837	T	25.0°C

Using the table, substitute into Equation 1:

$$(1.68 \times 10^3 \text{ J/°C})(T - 40.0°C)$$
$$+ (4.50 \times 10^2 \text{ J/°C})(T - 37.0°C)$$
$$+ (2.51 \times 10^2 \text{ J/°C})(T - 25.0°C) = 0$$
$$(1.68 \times 10^3 \text{ J/°C} + 4.50 \times 10^2 \text{ J/°C} + 2.51 \times 10^2 \text{ J/°C})T$$
$$= 9.01 \times 10^4 \text{ J}$$

$$T = 37.9°C$$

Remarks   The answer turned out to be very close to the aluminum's initial temperature, so it would have been impossible to guess in advance whether the aluminum would lose or gain energy. Notice the way the table was organized, mirroring the order of factors in the different terms. This kind of organization helps prevent substitution errors, which are common in these problems.

Exercise 11.4

A 20.0-kg gold bar at 35.0°C is placed in a large, insulated 0.800-kg glass container at 15.0°C and 2.00 kg of water at 25.0°C. Calculate the final equilibrium temperature.

Answer   26.6°C

11.4   LATENT HEAT AND PHASE CHANGE

A substance usually undergoes a change in temperature when energy is transferred between the substance and its environment. In some cases, however, the transfer of energy doesn't result in a change in temperature. This can occur when the physical characteristics of the substance change from one form to another, commonly referred to as a **phase change**. Some common phase changes are solid to liquid (melting), liquid to gas (boiling), and a change in the crystalline structure of a solid. Any such phase change involves a change in the internal energy, but *no change in the temperature.*

The energy Q needed to change the phase of a given pure substance is

$$Q = \pm mL \qquad \text{[11.6]}$$

where L, called the **latent heat** of the substance, depends on the nature of the phase change as well as on the substance.

◀ Latent heat

The unit of latent heat is the joule per kilogram (J/kg). The word *latent* means "lying hidden within a person or thing." The positive sign in Equation 11.6 is chosen when energy is absorbed by a substance, as when ice is melting. The negative sign is chosen when energy is removed from a substance, as when steam condenses to water.

The **latent heat of fusion** L_f is used when a phase change occurs during melting or freezing, while the **latent heat of vaporization** L_v is used when a phase change occurs during boiling or condensing.[1] For example, at atmospheric pressure the

TIP 11.3   Signs are Critical

For phase changes, use the correct explicit sign in Equation 11.6, positive if you are adding energy to the substance, negative if you're taking it away.

[1] When a gas cools, it eventually returns to the liquid phase, or *condenses.* The energy per unit mass given up during the process is called the *heat of condensation,* and it equals the heat of vaporization. When a liquid cools, it eventually solidifies, and the *heat of solidification* equals the heat of fusion.

TABLE 11.2

Latent Heats of Fusion and Vaporization

Substance	Melting Point (°C)	Latent Heat of Fusion (J/kg)	(cal/g)	Boiling Point (°C)	Latent Heat of Vaporization (J/kg)	(cal/g)
Helium	−269.65	5.23×10^3	(1.25)	−268.93	2.09×10^4	(4.99)
Nitrogen	−209.97	2.55×10^4	(6.09)	−195.81	2.01×10^5	(48.0)
Oxygen	−218.79	1.38×10^4	(3.30)	−182.97	2.13×10^5	(50.9)
Ethyl alcohol	−114	1.04×10^5	(24.9)	78	8.54×10^5	(204)
Water	0.00	3.33×10^5	(79.7)	100.00	2.26×10^6	(540)
Sulfur	119	3.81×10^4	(9.10)	444.60	3.26×10^5	(77.9)
Lead	327.3	2.45×10^4	(5.85)	1 750	8.70×10^5	(208)
Aluminum	660	3.97×10^5	(94.8)	2 450	1.14×10^7	(2 720)
Silver	960.80	8.82×10^4	(21.1)	2 193	2.33×10^6	(558)
Gold	1 063.00	6.44×10^4	(15.4)	2 660	1.58×10^6	(377)
Copper	1 083	1.34×10^5	(32.0)	1 187	5.06×10^6	(1 210)

latent heat of fusion for water is 3.33×10^5 J/kg, and the latent heat of vaporization for water is 2.26×10^6 J/kg. The latent heats of different substances vary considerably, as can be seen in Table 11.2.

Another process, sublimation, is the passage from the solid to the gaseous phase without going through a liquid phase. The fuming of dry ice (frozen carbon dioxide) illustrates this process, which has its own latent heat associated with it—the heat of sublimation.

EXAMPLE 11.5   Boiling Liquid Helium

Goal   Apply the concept of latent heat of vaporization to liquid helium.

Problem   Liquid helium has a very low boiling point, 4.2 K, as well as a low latent heat of vaporization, 2.09×10^4 J/kg. If energy is transferred to a container of liquid helium at the boiling point from an immersed electric heater at a rate of 10.0 W, how long does it take to boil away 2.00 kg of the liquid?

Strategy   Because $L_v = 2.09 \times 10^4$ J/kg, boiling away each kilogram of liquid helium requires 2.09×10^4 J of energy. Joules divided by watts is time, so find the total energy needed and divide by the power to find the time.

Solution

Find the energy needed to vaporize 2.00 kg of liquid helium at its boiling point:

$$Q = mL_v = (2.00 \text{ kg})(2.09 \times 10^4 \text{ J/kg}) = 4.18 \times 10^4 \text{ J}$$

Divide this result by the power to find the time:

$$\Delta t = \frac{Q}{\mathcal{P}} = \frac{mL_v}{\mathcal{P}} = \frac{4.18 \times 10^4 \text{ J}}{10.0 \text{ W}}$$

$$\Delta t = 4.18 \times 10^3 \text{ s} = \boxed{69.7 \text{ min}}$$

Remark   Notice that no change of temperature was involved. During such processes, the transferred energy goes into changing the state of the substance involved.

Exercise 11.5

If 10.0 W of power is supplied to 2.00 kg of water at 1.00×10^2°C, how long will it take for the water to completely boil away?

Answer   126 h

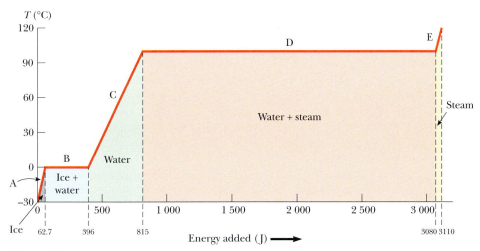

Figure 11.3   A plot of temperature versus energy added when 1.00 g of ice, initially at $-30.0°C$, is converted to steam at 120°C.

To better understand the physics of phase changes, consider the addition of energy to a 1.00-g cube of ice at $-30.0°C$ in a container held at constant pressure. Suppose this input of energy turns the ice to steam (water vapor) at 120.0°C. Figure 11.3 is a plot of the experimental measurement of temperature as energy is added to the system. We examine each portion of the curve separately.

Part A   During this portion of the curve, the temperature of the ice changes from $-30.0°C$ to 0.0°C. Because the specific heat of ice is 2 090 J/kg · °C, we can calculate the amount of energy added from Equation 11.3:

$$Q = mc_{ice}\,\Delta T = (1.00 \times 10^{-3}\,\text{kg})(2\,090\,\text{J/kg} \cdot °\text{C})(30.0°\text{C}) = 62.7\,\text{J}$$

Part B   When the ice reaches 0°C, the ice–water mixture remains at that temperature—even though energy is being added—until all the ice melts to become water at 0°C. According to Equation 11.6, the energy required to melt 1.00 g of ice at 0°C is

$$Q = mL_f = (1.00 \times 10^{-3}\,\text{kg})(3.33 \times 10^5\,\text{J/kg}) = 333\,\text{J}$$

Part C   Between 0°C and 100°C, no phase change occurs. The energy added to the water is used to increase its temperature, as in part A. The amount of energy necessary to increase the temperature from 0°C to 100°C is

$$Q = mc_{water}\,\Delta T = (1.00 \times 10^{-3}\,\text{kg})(4.19 \times 10^3\,\text{J/kg} \cdot °\text{C})(1.00 \times 10^2\,°\text{C})$$

$$Q = 4.19 \times 10^2\,\text{J}$$

Part D   At 100°C, another phase change occurs as the water changes to steam at 100°C. As in Part B, the water–steam mixture remains at constant temperature, this time at 100°C—even though energy is being added—until all the liquid has been converted to steam. The energy required to convert 1.00 g of water at 100°C to steam at 100°C is

$$Q = mL_v = (1.00 \times 10^{-3}\,\text{kg})(2.26 \times 10^6\,\text{J/kg}) = 2.26 \times 10^3\,\text{J}$$

Part E   During this portion of the curve, as in parts A and C, no phase change occurs, so all the added energy goes into increasing the temperature of the steam. The energy that must be added to raise the temperature of the steam to 120.0°C is

$$Q = mc_{steam}\,\Delta T = (1.00 \times 10^{-3}\,\text{kg})(2.01 \times 10^3\,\text{J/kg} \cdot °\text{C})(20.0°\text{C}) = 40.2\,\text{J}$$

The total amount of energy that must be added to change 1.00 g of ice at $-30.0°C$ to steam at 120.0°C is the sum of the results from all five parts of the curve—3.11×10^3 J. Conversely, to cool 1.00 g of steam at 120.0°C down to the point at which it becomes ice at -30.0°C, 3.11×10^3 J of energy must be removed.

Phase changes can be described in terms of rearrangements of molecules when energy is added to or removed from a substance. Consider first the liquid-to-gas phase change. The molecules in a liquid are close together, and the forces between them are stronger than the forces between the more widely separated molecules of a gas. Work must therefore be done on the liquid against these attractive molecular forces in order to separate the molecules. The latent heat of vaporization is the amount of energy that must be added to the one kilogram of liquid to accomplish this separation.

Similarly, at the melting point of a solid, the amplitude of vibration of the atoms about their equilibrium positions becomes great enough to allow the atoms to pass the barriers of adjacent atoms and move to their new positions. On average, these new positions are less symmetrical than the old ones and therefore have higher energy. The latent heat of fusion is equal to the work required at the molecular level to transform the mass from the ordered solid phase to the disordered liquid phase.

The average distance between atoms is much greater in the gas phase than in either the liquid or the solid phase. Each atom or molecule is removed from its neighbors, overcoming the attractive forces of nearby neighbors. Therefore, more work is required at the molecular level to vaporize a given mass of a substance than to melt it, so in general the latent heat of vaporization is much greater than the latent heat of fusion (Table 11.2).

Quick Quiz 11.2

Calculate the slopes for the A, C, and E portions of Figure 11.3. Rank the slopes from least to greatest and explain what your ranking means. (a) A, C, E (b) C, A, E (c) E, A, C (d) E, C, A

Problem-Solving Strategy
Calorimetry with Phase Changes

1. **Make a table for all data**. Include separate rows for different phases and for any transition between phases. Include columns for each quantity used and a final column for the combination of the quantities. Transfers of thermal energy in this last column are given by $Q = mc\,\Delta T$, while phase changes are given by $Q = \pm mL_f$ for changes between liquid and solid, and by $Q = \pm mL_v$ for changes between liquid and gas.
2. **Apply conservation of energy**. If the system is isolated, use $\Sigma Q_k = 0$ (Eq. 11.5). For a nonisolated system, the net energy change should replace the zero on the right-hand side of that equation. ΣQ_k is just the sum of all the terms in the last column of the table.
3. **Solve** for the unknown quantity.

EXAMPLE 11.6 Ice Water

Goal Solve a problem involving heat transfer and a phase change from solid to liquid.

Problem At a party, 6.00 kg of ice at -5.00°C is added to a cooler holding 30 liters of water at 20.0°C. What is the temperature of the water when it comes to equilibrium?

Strategy In this problem, it's best to make a table. With the addition of thermal energy Q_{ice}, the ice will warm to 0°C; then melt at 0°C with the addition of energy Q_{melt}. Next, the melted ice will warm to some final temperature T by absorbing energy $Q_{ice-water}$, obtained from the energy change of the original liquid water, Q_{water}. By conservation of energy, these quantities must sum to zero.

Solution

Calculate the mass of liquid water:

$$m_{water} = \rho_{water}V$$

$$= (1.00 \times 10^3 \, kg/m^3)(30.0 \, L)\frac{1.00 \, m^3}{1.00 \times 10^3 \, L}$$

$$= 30.0 \, kg$$

Write the equation of thermal equilibrium:

$$Q_{ice} + Q_{melt} + Q_{ice-water} + Q_{water} = 0 \qquad (1)$$

Construct a comprehensive table:

Q	m (kg)	c (J/kg·°C)	L (J/kg)	T_f (°C)	T_i (°C)	Expression
Q_{ice}	6.00	2 090		0	−5.00	$m_{ice}c_{ice}(T_f - T_i)$
Q_{melt}	6.00		3.33×10^5	0	0	$m_{ice}L_f$
$Q_{ice-water}$	6.00	4 190		T	0	$m_{ice}c_{wat}(T_f - T_i)$
Q_{water}	30.0	4 190		T	20.0	$m_{wat}c_{wat}(T_f - T_i)$

Substitute all quantities in the second through sixth columns into the last column and sum (which is the evaluation of Equation 1), and solve for T:

$$6.27 \times 10^4 \, J + 2.00 \times 10^6 \, J$$
$$+ (2.51 \times 10^4 \, J/°C)(T - 0°C)$$
$$+ (1.26 \times 10^5 \, J/°C)(T - 20.0°C) = 0$$

$$T = \boxed{3.03°C}$$

Remarks   Making a table is optional. However, simple substitution errors are extremely common, and the table makes such errors less likely.

Exercise 11.6

What mass of ice at −10.0°C is needed to cool a whale's water tank, holding $1.20 \times 10^3 \, m^3$ of water, from 20.0°C down to a more comfortable 10.0°C?

Answer 1.27×10^5 kg

EXAMPLE 11.7   Partial Melting

Goal   Understand how to handle an incomplete phase change.

Problem   A 5.00-kg block of ice at 0°C is added to an insulated container partially filled with 10.0 kg of water at 15.0°C. **(a)** Find the final temperature, neglecting the heat capacity of the container. **(b)** Find the mass of the ice that was melted.

Strategy   Part (a) is tricky, because the ice does not entirely melt in this example. When there is any doubt concerning whether there will be a complete phase change, some preliminary calculations are necessary. First, find the total energy required to melt the ice, Q_{melt}, and then find Q_{water}, the maximum energy that can be delivered by the water above 0°C. If the energy delivered by the water is high enough, all the ice melts. If not, there will usually be a final mixture of ice and water at 0°C, unless the ice starts at a temperature far below 0°C, in which case all the liquid water freezes.

Solution
(a) Find the equilibrium temperature.

First, compute the amount of energy necessary to completely melt the ice:

$$Q_{melt} = m_{ice}L_f = (5.00 \, kg)(3.33 \times 10^5 \, J/kg)$$
$$= 1.67 \times 10^6 \, J$$

Next, calculate the maximum energy that can be lost by the initial mass of liquid water without freezing it:

$$Q_{water} = m_{water}c\Delta T$$
$$= (10.0 \, kg)(4 \, 190 \, J/kg·°C)(0°C - 15.0°C)$$
$$= -6.29 \times 10^5 \, J$$

This is less than half the energy necessary to melt all the ice, so the final state of the system is a mixture of water and ice at the freezing point:

$T = \boxed{0°C}$

(b) Compute the mass of ice melted.

Set the total available energy equal to the heat of fusion of m grams of ice, mL_f:

$6.29 \times 10^5 \, J = mL_f = m(3.33 \times 10^5 \, J/kg)$

$m = \boxed{1.89 \, kg}$

Remarks If this problem is solved assuming (wrongly) that all the ice melts, a final temperature of $T = -16.5°C$ is obtained. The only way that could happen is if the system were not isolated, contrary to the statement of the problem. In the following exercise, you must also compute the thermal energy needed to warm the ice to its melting point.

Exercise 11.7
If 8.00 kg of ice at $-5.00°$ is added to 12.0 kg of water at $20.0°$, compute the final temperature. How much ice remains, if any?

Answer $T = 0°C$, 5.22 kg

Sometimes problems involve changes in mechanical energy. During a collision, for example, some kinetic energy can be transformed to the internal energy of the colliding objects. This kind of transformation is illustrated in Example 11.8, involving a possible impact of a comet on Earth. In this example, a number of liberties will be taken in order to estimate the magnitude of the destructive power of such a catastrophic event. The specific heats depend on temperature and pressure, for example, but that will be ignored. Also, the ideal gas law doesn't apply at the temperatures and pressures attained, and the result of the collision wouldn't be superheated steam, but a plasma of charged particles. Despite all these simplifications, the example yields good order-of-magnitude results.

EXAMPLE 11.8   Armageddon!

Goal   Link mechanical energy to thermal energy, phase changes, and the ideal gas law to create an estimate.

Problem   A comet half a kilometer in radius consisting of ice at 273 K hits Earth at a speed of 4.00×10^4 m/s. For simplicity, assume that all the kinetic energy converts to thermal energy on impact and that all the thermal energy goes into warming the comet. **(a)** Calculate the volume and mass of the ice. **(b)** Use conservation of energy to find the final temperature of the comet material. Assume, contrary to fact, that the result is superheated steam and that the usual specific heats are valid, though in fact they depend on both temperature and pressure. **(c)** Assuming the steam retains a spherical shape and has the same initial volume as the comet, calculate the pressure of the steam using the ideal gas law. This law actually doesn't apply to a system at such high pressure and temperature, but can be used to get an estimate.

Strategy   Part (a) requires the volume formula for a sphere and the definition of density. In part (b), conservation of energy can be applied. There are four processes involved: (1) melting the ice, (2) warming the ice water to the boiling point, (3) converting the boiling water to steam, and (4) warming the steam. The energy needed for these processes will be designated Q_{melt}, Q_{water}, Q_{vapor}, and Q_{steam}, respectively. These quantities plus the change in kinetic energy ΔK sum to zero because they are assumed to be internal to the system. In this case, the first three Q's can be neglected compared to the (extremely large) kinetic energy term. Solve for the unknown temperature, and substitute it into the ideal gas law in part (c).

Solution
(a) Find the volume and mass of the ice.

Apply the volume formula for a sphere:

$V = \dfrac{4}{3}\pi r^3 = \dfrac{4}{3}(3.14)(5.00 \times 10^2 \, m)^3$

$= \boxed{5.23 \times 10^8 \, m^3}$

Apply the density formula to find the mass of the ice:
$$m = \rho V = (917 \text{ kg/m}^3)(5.23 \times 10^8 \text{ m}^3)$$
$$= 4.80 \times 10^{11} \text{ kg}$$

(b) Find the final temperature of the cometary material.

Use conservation of energy:
$$Q_{\text{melt}} + Q_{\text{water}} + Q_{\text{vapor}} + Q_{\text{steam}} + \Delta K = 0 \tag{1}$$
$$mL_f + mc_{\text{water}}\,\Delta T_{\text{water}} + mL_v + mc_{\text{steam}}\,\Delta T_{\text{steam}}$$
$$+ (0 - \tfrac{1}{2}mv^2) = 0 \tag{2}$$

The first three terms are negligible compared to the kinetic energy. The steam term involves the unknown final temperature, so retain only it and the kinetic energy, canceling the mass and solving for T:

$$mc_{\text{steam}}\,(T - 373 \text{ K}) - \tfrac{1}{2}mv^2 = 0$$
$$T = \frac{\tfrac{1}{2}v^2}{c_{\text{steam}}} + 373 \text{ K} = \frac{\tfrac{1}{2}(4.00 \times 10^4\,\text{m/s})^2}{2\,010 \text{ J/kg}\cdot\text{K}} + 373 \text{ K}$$
$$T = 3.98 \times 10^5 \text{ K}$$

(c) Estimate the pressure of the gas, using the ideal gas law.

First, compute the number of moles of steam:
$$n = (4.80 \times 10^{11}\,\text{kg})\left(\frac{1 \text{ mol}}{0.018 \text{ kg}}\right) = 2.67 \times 10^{13} \text{ mol}$$

Solve for the pressure, using $PV = nRT$:
$$P = \frac{nRT}{V}$$
$$= \frac{(2.67 \times 10^{13}\,\text{mol})(8.31 \text{ J/mol}\cdot\text{K})(3.98 \times 10^5\,\text{K})}{5.23 \times 10^8\,\text{m}^3}$$
$$P = 1.69 \times 10^{11} \text{ Pa}$$

Remarks   The estimated pressure is several hundred times greater than the ultimate shear stress of steel! This high-pressure region would expand rapidly, destroying everything within a very large radius. Fires would ignite across a continent-sized region, and tidal waves would wrap around the world, wiping out coastal regions everywhere. The sun would be obscured for at least a decade, and numerous species, possibly including *Homo sapiens*, would become extinct. Such extinction events are rare, but in the long run represent a significant threat to life on Earth.

Exercise 11.8
Suppose a lead bullet with mass 5.00 g and an initial temperature of 65.0°C hits a wall and completely liquifies. What minimum speed did it have before impact? (*Hint:* The minimum speed corresponds to the case where all the kinetic energy becomes internal energy of the lead and the final temperature of the lead is at its melting point. Don't neglect any terms here!)

Answer   341 m/s

11.5   ENERGY TRANSFER

For some applications it's necessary to know the rate at which energy is transferred between a system and its surroundings and the mechanisms responsible for the transfer. This is particularly important in weatherproofing buildings or in medical applications, such as human survival time when exposed to the elements.

Earlier in this chapter we defined heat as a transfer of energy between a system and its surroundings due to a temperature difference between them. In this section, we take a closer look at heat as a means of energy transfer and consider the processes of thermal conduction, convection, and radiation.

Figure 11.4 Conduction makes the metal handle of a cooking pan hot.

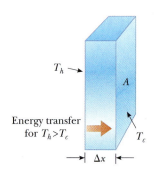

Figure 11.5 Energy transfer through a conducting slab of cross-sectional area A and thickness L. The opposite faces are at different temperatures T_c and T_h.

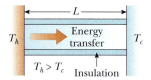

Figure 11.6 Conduction of energy through a uniform, insulated rod of length L. The opposite ends are in thermal contact with energy reservoirs at different temperatures.

TIP 11.4   Blankets and Coats in Cold Weather

When you sleep under a blanket in the winter or wear a warm coat outside, the blanket or coat serves as a layer of material with low thermal conductivity in order to reduce the transfer of energy away from your body by heat. The primary insulating medium is the air trapped in small pockets within the material.

Thermal Conduction

The energy transfer process most closely associated with a temperature difference is called **thermal conduction** or simply **conduction**. In this process, the transfer can be viewed on an atomic scale as an exchange of kinetic energy between microscopic particles—molecules, atoms, and electrons—with less energetic particles gaining energy as they collide with more energetic particles. An inexpensive pot, as in Figure 11.4, may have a metal handle with no surrounding insulation. As the pot is warmed, the temperature of the metal handle increases, and the cook must hold it with a cloth potholder to avoid being burned.

The way the handle warms up can be understood by looking at what happens to the microscopic particles in the metal. Before the pot is placed on the stove, the particles are vibrating about their equilibrium positions. As the stove coil warms up, those particles in contact with it begin to vibrate with larger amplitudes. These particles collide with their neighbors and transfer some of their energy in the collisions. Metal atoms and electrons farther and farther from the flame gradually increase the amplitude of their vibrations, until eventually those in the metal near your hand are affected. This increased vibration represents an increase in temperature of the metal (and possibly a burned hand!).

Although the transfer of energy through a substance can be partly explained by atomic vibrations, the rate of conduction depends on the properties of the substance. For example, it's possible to hold a piece of asbestos in a flame indefinitely. This fact implies that very little energy is conducted through the asbestos. In general, metals are good thermal conductors because they contain large numbers of electrons that are relatively free to move through the metal and can transport energy from one region to another. In a good conductor such as copper, conduction takes place via the vibration of atoms and the motion of free electrons. Materials such as asbestos, cork, paper, and fiberglass are poor thermal conductors. Gases are also poor thermal conductors because of the large distance between their molecules.

Conduction occurs only if there is a difference in temperature between two parts of the conducting medium. The temperature difference drives the flow of energy. Consider a slab of material of thickness Δx and cross-sectional area A with its opposite faces at different temperatures T_c and T_h, where $T_h > T_c$ (Fig. 11.5). The slab allows energy to transfer from the region of higher temperature to the region of lower temperature by thermal conduction. The rate of energy transfer, $\mathcal{P} = Q/\Delta t$, is proportional to the cross-sectional area of the slab and the temperature difference and is inversely proportional to the thickness of the slab:

$$\mathcal{P} = \frac{Q}{\Delta t} \propto A \frac{\Delta T}{\Delta x}$$

Note that $\mathcal{P}$ has units of watts when Q is in joules and Δt is in seconds.

Suppose a substance is in the shape of a long, uniform rod of length L, as in Figure 11.6. We assume the rod is insulated, so thermal energy can't escape by conduction from its surface except at the ends. One end is in thermal contact with an energy reservoir at temperature T_c and the other end is in thermal contact with a reservoir at temperature $T_h > T_c$. When a steady state is reached, the temperature at each point along the rod is constant in time. In this case, $\Delta T = T_h - T_c$ and $\Delta x = L$, so

$$\frac{\Delta T}{\Delta x} = \frac{T_h - T_c}{L}$$

The rate of energy transfer by conduction through the rod is given by

$$\mathcal{P} = kA \frac{(T_h - T_c)}{L} \qquad \text{[11.7]}$$

where k, a proportionality constant that depends on the material, is called the **thermal conductivity**. Substances that are good conductors have large thermal conductivities, whereas good insulators have low thermal conductivities. Table 11.3 lists the thermal conductivities for various substances.

Quick Quiz 11.3

Will an ice cube wrapped in a wool blanket remain frozen for (a) less time, (b) the same length of time, or (c) a longer time than an identical ice cube exposed to air at room temperature?

Quick Quiz 11.4

Two rods of the same length and diameter are made from different materials. The rods are to connect two regions of different temperature so that energy will transfer through the rods by heat. They can be connected in series, as in Figure 11.7a, or in parallel, as in Figure 11.7b. In which case is the rate of energy transfer by heat larger? (a) When the rods are in series (b) When the rods are in parallel (c) The rate is the same in both cases.

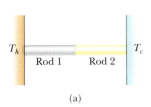

(a)

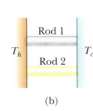

(b)

Figure 11.7 (Quick Quiz 11.4) In which case is the rate of energy transfer larger?

TABLE 11.3

Thermal Conductivities

Substance	Thermal Conductivity $(\text{J}/\text{s}\cdot\text{m}\cdot°\text{C})$
Metals (at 25°C)	
Aluminum	238
Copper	397
Gold	314
Iron	79.5
Lead	34.7
Silver	427
Gases (at 20°C)	
Air	0.023 4
Helium	0.138
Hydrogen	0.172
Nitrogen	0.023 4
Oxygen	0.023 8
Nonmetals	
Asbestos	0.25
Concrete	1.3
Glass	0.84
Ice	1.6
Rubber	0.2
Water	0.60
Wood	0.10

EXAMPLE 11.9   Energy Transfer through a Concrete Wall

Goal   Apply the equation of heat conduction.

Problem   Find the energy transferred in 1.00 h by conduction through a concrete wall 2.0 m high, 3.65 m long, and 0.20 m thick if one side of the wall is held at 20°C and the other side is at 5°C.

Strategy   Equation 11.7 gives the rate of energy transfer by conduction in joules per second. Multiply by the time and substitute given values to find the total thermal energy transferred.

Solution

Multiply Equation 11.7 by Δt to find an expression for the total energy Q transferred through the wall:

$$Q = \mathcal{P}\,\Delta t = kA\left(\frac{T_h - T_c}{L}\right)\Delta t$$

Substitute the numerical values to obtain Q, consulting Table 11.3 for k:

$$Q = (1.3\,\text{J}/\text{s}\cdot\text{m}\cdot°\text{C})(7.3\,\text{m}^2)\left(\frac{15°\text{C}}{0.20\,\text{m}}\right)(3\,600\,\text{s})$$

$$= 2.6 \times 10^6\,\text{J}$$

Remarks   Early houses were insulated with thick masonry walls, which restrict energy loss by conduction because k is relatively low. The large thickness L also decreases energy loss by conduction, as shown by Equation 11.7. There are much better insulating materials, however, and layering is also helpful. Despite the low thermal conductivity of masonry, the amount of energy lost is still rather large—enough to raise the temperature of 600 kg of water by more than 1°C. There are better insulating materials than masonry.

Exercise 11.9

A wooden shelter has walls constructed of wooden planks 1.00 cm thick. If the exterior temperature is $-20.0°C$ and the interior is $5.00°C$, find the rate of energy loss through a wall that has dimensions 2.00 m by 2.00 m.

Answer 1.00×10^3 W

Home Insulation

To determine whether to add insulation to a ceiling or some other part of a building, the preceding discussion of conduction must be extended, for two reasons:

1. The insulating properties of materials used in buildings are usually expressed in engineering (U.S. customary) rather than SI units. Measurements stamped on a package of fiberglass insulating board will be in units such as British thermal units, feet, and degrees Fahrenheit.
2. In dealing with the insulation of a building, conduction through a compound slab must be considered, with each portion of the slab having a certain thickness and a specific thermal conductivity. A typical wall in a house consists of an array of materials, such as wood paneling, drywall, insulation, sheathing, and wood siding.

The rate of energy transfer by conduction through a compound slab is

$$\frac{Q}{\Delta t} = \frac{A(T_h - T_c)}{\sum_i L_i/k_i} \quad \text{[11.8]}$$

where T_h and T_c are the temperatures of the *outer extremities* of the slab and the summation is over all portions of the slab. This formula can be derived algebraically, using the facts that the temperature at the interface between two insulating materials must be the same and that the rate of energy transfer through one insulator must be the same as through all the other insulators. If the slab consists of three different materials, the denominator is the sum of three terms. In engineering practice, the term L/k for a particular substance is referred to as the

TABLE 11.4

R Values for Some Common Building Materials

Material	R value $(\text{ft}^2 \cdot °\text{F} \cdot \text{h/Btu})$
Hardwood siding (1.0 in. thick)	0.91
Wood shingles (lapped)	0.87
Brick (4.0 in. thick)	4.00
Concrete block (filled cores)	1.93
Styrofoam (1.0 in. thick)	5.0
Fiber glass batting (3.5 in. thick)	10.90
Fiber glass batting (6.0 in. thick)	18.80
Fiber glass board (1.0 in. thick)	4.35
Cellulose fiber (1.0 in. thick)	3.70
Flat glass (0.125 in. thick)	0.89
Insulating glass (0.25-in. space)	1.54
Vertical air space (3.5 in. thick)	1.01
Stagnant layer of air	0.17
Dry wall (0.50 in. thick)	0.45
Sheathing (0.50 in. thick)	1.32

R value of the material, so Equation 11.8 reduces to

$$\frac{Q}{\Delta t} = \frac{A(T_h - T_c)}{\sum\limits_i R_i}$$ **[11.9]**

The R values for a few common building materials are listed in Table 11.4. Note the unit of R and the fact that the R values are defined for specific thicknesses.

Next to any vertical outside surface is a very thin, stagnant layer of air that must be considered when the total R value for a wall is computed. The thickness of this stagnant layer depends on the speed of the wind. As a result, energy loss by conduction from a house on a day when the wind is blowing is greater than energy loss on a day when the wind speed is zero. A representative R value for a stagnant air layer is given in Table 11.4.

INTERACTIVE EXAMPLE 11.10   The R Value of a Typical Wall

Goal   Calculate the R value of a wall consisting of several layers of insulating material.

Problem   Calculate the total R value for a wall constructed as shown in Figure 11.8a. Starting outside the house (to the left in the figure) and moving inward, the wall consists of 4.0 in. brick, 0.50 in. sheathing, an air space 3.5 in. thick, and 0.50 in. drywall.

Strategy   Add all the R values together, remembering the stagnant air layers inside and outside the house.

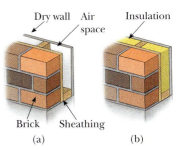

Figure 11.8   (Example 11.10) A cross-sectional view of an exterior wall containing (a) an air space and (b) insulation.

Solution
Refer to Table 11.4, and sum. All quantities are in units of $ft^2 \cdot °F \cdot h/Btu$.

$R_{total} = R_{outside\ air\ layer} + R_{brick} + R_{sheath} + R_{air\ space}$
$+\ R_{drywall} + R_{inside\ air\ layer} = (0.17 + 4.00 + 1.32 + 1.01$
$+\ 0.45 + 0.17)\,ft^2 \cdot °F \cdot h/Btu$

$R_{total} = \boxed{7.12\ ft^2 \cdot °F \cdot h/Btu}$

Exercise 11.10
If a layer of fiber glass insulation 3.5 in. thick is placed inside the wall to replace the air space, as in Figure 11.8b, what is the new total R value? By what factor is the energy loss reduced?

Answer $R = 17\ ft^2 \cdot °F \cdot h/Btu$; 2.4

Physics Now™   Study the R values of various types of common building materials by logging into PhysicsNow at **www.cp7e.com** and going to Interactive Example 11.10.

EXAMPLE 11.11   Staying Warm in the Arctic

Goal   Combine two layers of insulation.

Problem   An arctic explorer builds a wooden shelter out of wooden planks that are 1.0 cm thick. To improve the insulation, he covers the shelter with a layer of ice 3.2 cm thick. **(a)** Compute the R factors for the wooden planks and the ice. **(b)** If the temperature outside the shelter is −20.0°C and the temperature inside is 5.00°C, find the rate of energy loss through one of the walls, if the wall has dimensions 2.00 m by 2.00 m. **(c)** Find the temperature at the interface between the wood and the ice.

Strategy   After finding the R values, substitute into Equation 11.9 to get the rate of energy transfer. To answer part (c), use Equation 11.7 for one of the layers, setting it equal to the rate found in part (b), solving for the temperature.

Solution

(a) Compute the R values using the data in Table 11.3.

Find the R value for the wooden wall:

$$R_{wood} = \frac{L_{wood}}{k_{wood}} = \frac{0.01 \text{ m}}{0.10 \text{ J/s} \cdot \text{m} \cdot {}^\circ\text{C}} = \boxed{0.10 \text{ m}^2 \cdot \text{s} \cdot {}^\circ\text{C/J}}$$

Find the R-value for the ice layer:

$$R_{ice} = \frac{L_{ice}}{k_{ice}} = \frac{0.032 \text{ m}}{1.6 \text{ J/s} \cdot \text{m} \cdot {}^\circ\text{C}} = \boxed{0.020 \text{ m}^2 \cdot \text{s} \cdot {}^\circ\text{C/J}}$$

(b) Find the rate of heat loss.

Apply Equation 11.9:

$$\mathcal{P} = \frac{Q}{\Delta t} = \frac{A(T_h - T_c)}{\sum_i R_i}$$

$$= \frac{(4.00 \text{ m}^2)(5.00{}^\circ\text{C} - (-20.0{}^\circ\text{C}))}{0.12 \text{ m}^2 \cdot \text{s} \cdot {}^\circ\text{C/J}}$$

$$\mathcal{P} = \boxed{830 \text{ W}}$$

(c) Find the temperature in between the ice and wood.

Apply the equation of heat conduction to the wood:

$$\frac{k_{wood} A(T_h - T_c)}{L} = \mathcal{P}$$

$$\frac{(0.10 \text{ J/s} \cdot \text{m} \cdot {}^\circ\text{C})(4.00 \text{ m}^2)(5.00{}^\circ\text{C} - T)}{0.010 \text{ m}} = 830 \text{ W}$$

Solve for the unknown temperature:

$$T = \boxed{-16{}^\circ\text{C}}$$

Remarks The outer side of the wooden wall and the inner surface of the ice must have the same temperature, and the rate of energy transfer through the ice must be the same as through the wooden wall. Using Equation 11.7 for ice instead of wood gives the same answer. This rate of energy transfer is only a modest improvement over the thousand-watt rate in Exercise 11.9. The choice of insulating material is important!

Exercise 11.11
Rather than use ice to cover the wooden shelter, the explorer glues pressed cork with thickness 0.500 cm to the outside of his wooden shelter. Find the new rate of energy loss through the same wall. (Note that $k_{cork} = 0.046 \text{ J/s} \cdot \text{m} \cdot {}^\circ\text{C}$.)

Answer 480 W

Convection

When you warm your hands over an open flame, as illustrated in Figure 11.9, the air directly above the flame, being warmed, expands. As a result, the density of this air decreases and the air rises, warming your hands as it flows by. **The transfer of energy by the movement of a substance is called convection.** When the movement results from differences in density, as with air around a fire, it's referred to as *natural convection*. Airflow at a beach is an example of natural convection, as is the mixing that occurs as surface water in a lake cools and sinks. When the substance is forced to move by a fan or pump, as in some hot air and hot water heating systems, the process is called *forced convection*.

Convection currents assist in the boiling of water. In a teakettle on a hot stovetop, the lower layers of water are warmed first. The warmed water has a lower density and rises to the top, while the denser, cool water at the surface sinks to the bottom of the kettle and is warmed.

Figure 11.9 Warming a hand by convection.

The same process occurs when a radiator raises the temperature of a room. The hot radiator warms the air in the lower regions of the room. The warm air expands and, because of its lower density, rises to the ceiling. The denser cooler air from above sinks, setting up the continuous air current pattern shown in Figure 11.10.

An automobile engine is maintained at a safe operating temperature by a combination of conduction and forced convection. Water (actually, a mixture of water and antifreeze) circulates in the interior of the engine. As the metal of the engine block increases in temperature, energy passes from the hot metal to the cooler water by thermal conduction. The water pump forces water out of the engine and into the radiator, carrying energy along with it (by forced convection). In the radiator, the hot water passes through metal pipes that are in contact with the cooler outside air, and energy passes into the air by conduction. The cooled water is then returned to the engine by the water pump to absorb more energy. The process of air being pulled past the radiator by the fan is also forced convection.

The algal blooms often seen in temperate lakes and ponds during the spring or fall are caused by convection currents in the water. To understand this process, consider Figure 11.11. During the summer, bodies of water develop temperature gradients, with an upper, warm layer of water separated from a lower, cold layer by a buffer zone called a thermocline. In the spring or fall, temperature changes in the water break down this thermocline, setting up convection currents that mix the water. The mixing process transports nutrients from the bottom to the surface. The nutrient-rich water forming at the surface can cause a rapid, temporary increase in the algae population.

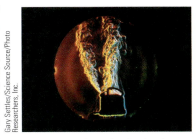

Photograph of a teakettle, showing steam and turbulent convection air currents.

APPLICATION

Cooling Automobile Engines

APPLICATION

Algal Blooms in Ponds and Lakes

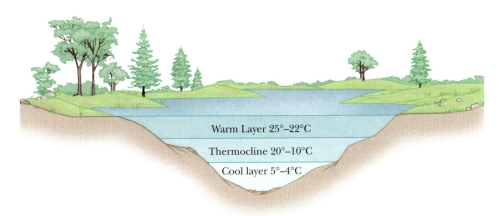

(a) Summer layering of water

Warm Layer 25°–22°C

Thermocline 20°–10°C

Cool layer 5°–4°C

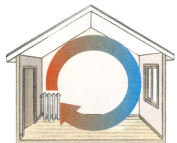

Figure 11.10 Convection currents are set up in a room warmed by a radiator.

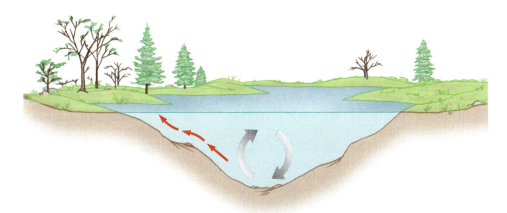

(b) Fall and spring upwelling

Figure 11.11 (a) During the summer, a warm upper layer of water is separated from a cooler lower layer by a thermocline. (b) Convection currents during the spring or fall mix the water and can cause algal blooms.

Applying Physics 11.1   Body Temperature

The body temperature of mammals ranges from about 35°C to 38°C, while that of birds ranges from about 40°C to 43°C. How can these narrow ranges of body temperature be maintained in cold weather?

Explanation   A natural method of maintaining body temperature is via layers of fat beneath the skin. Fat protects against both conduction and convection because of its low thermal conductivity and because there are few blood vessels in fat to carry blood to the surface, where energy losses by convection can occur. Birds ruffle their feathers in cold weather in order to trap a layer of air with a low thermal conductivity between the feathers and the skin. Bristling the fur produces the same effect in fur-bearing animals.

Humans keep warm with wool sweaters and down jackets that trap the warmer air in regions close to their bodies, reducing energy loss by convection and conduction.

Figure 11.12 Warming hands by radiation.

Radiation

Another process of transferring energy is through **radiation**. Figure 11.12 shows how your hands can be warmed at an open flame through radiation. Because your hands aren't in physical contact with the flame and the conductivity of air is very low, conduction can't account for the energy transfer. Nor can convection be responsible for any transfer of energy, because your hands aren't above the flame in the path of convection currents. The warmth felt in your hands, therefore, must come from the transfer of energy by radiation.

All objects radiate energy continuously in the form of electromagnetic waves due to thermal vibrations of their molecules. These vibrations create the orange glow of an electric stove burner, an electric space heater, and the coils of a toaster.

The rate at which an object radiates energy is proportional to the fourth power of its absolute temperature. This is known as **Stefan's law**, expressed in equation form as

Stefan's law ▶

$$\mathcal{P} = \sigma A e T^4 \qquad\qquad [11.10]$$

where $\mathcal{P}$ is the power in watts (or joules per second) radiated by the object, σ is the Stefan–Boltzmann constant, equal to $5.669\ 6 \times 10^{-8}\ \mathrm{W/m^2 \cdot K^4}$, A is the surface area of the object in square meters, e is a constant called the **emissivity** of the object, and T is the object's Kelvin temperature. The value of e can vary between zero and one, depending on the properties of the object's surface.

Approximately 1 340 J of electromagnetic radiation from the Sun passes through each square meter at the top of the Earth's atmosphere every second. This radiation is primarily visible light, accompanied by significant amounts of infrared and ultraviolet. We will study these types of radiation in detail in Chapter 21. Some of this energy is reflected back into space, and some is absorbed by the atmosphere, but enough arrives at the surface of the Earth each day to supply all our energy needs hundreds of times over—if it could be captured and used efficiently. The growth in the number of solar houses in the United States is one example of an attempt to make use of this abundant energy. Radiant energy from the Sun affects our day-to-day existence in a number of ways, influencing Earth's average temperature, ocean currents, agriculture, and rain patterns. It can also affect behavior.

As another example of the effects of energy transfer by radiation, consider what happens to the atmospheric temperature at night. If there is a cloud cover above Earth, the water vapor in the clouds absorbs part of the infrared radiation emitted by Earth and re-emits it back to the surface. Consequently, the temperature at the surface remains at moderate levels. In the absence of cloud cover, there is nothing to prevent the radiation from escaping into space, so the temperature drops more on a clear night than when it's cloudy.

As an object radiates energy at a rate given by Equation 11.10, it also absorbs radiation. If it didn't, the object would eventually radiate all its energy and its temperature would reach absolute zero. The energy an object absorbs comes from its environment, which consists of other bodies that radiate energy. If an object is at a temperature T, and its surroundings are at a temperature T_0, the net energy gained or lost each second by the object as a result of radiation is

$$\mathscr{P}_{\text{net}} = \sigma A e (T^4 - T_0^4)$$ **[11.11]**

When an object is in equilibrium with its surroundings, it radiates and absorbs energy at the same rate, so its temperature remains constant. When an object is hotter than its surroundings, it radiates more energy than it absorbs and so cools.

An **ideal absorber** is an object that absorbs all the light radiation incident on it, including invisible infrared and ultraviolet light. Such an object is called a **black body** because a room-temperature black body would look black. Since a black body doesn't reflect radiation at any wavelength, any light coming from it is due to atomic and molecular vibrations alone. A perfect black body has emissivity $e = 1$. An ideal absorber is also an ideal radiator of energy. The Sun, for example, is nearly a perfect black body. This statement may seem contradictory, because the Sun is bright, not dark; however, the light that comes from the Sun is emitted, not reflected. Black bodies are perfect absorbers that look black at room temperature because they don't reflect any light. All black bodies, except those at absolute zero, emit light that has a characteristic spectrum, to be discussed in Chapter 27. In contrast to black bodies, an object for which $e = 0$ absorbs none of the energy incident on it, reflecting it all. Such a body is an **ideal reflector**.

White clothing is more comfortable to wear in the summer than black clothing. Black fabric acts as a good absorber of incoming sunlight and as a good emitter of this absorbed energy. About half of the emitted energy, however, travels toward the body, causing the person wearing the garment to feel uncomfortably warm. White or light-colored clothing reflects away much of the incoming energy.

The amount of energy radiated by an object can be measured with temperature-sensitive recording equipment via a technique called **thermography**. An image of the pattern formed by varying radiation levels, called a **thermogram**, is brightest in the warmest areas. Figure 11.13 reproduces a thermogram of a house. More energy escapes in the lighter regions, such as the door and windows. The owners of this house could conserve energy and reduce their heating costs by adding insulation to the attic area and by installing thermal draperies over the windows. Thermograms have also been used to image injured or diseased tissue in medicine, since such areas are often at a different temperature than surrounding healthy tissue, though many radiologists consider thermograms inadequate as a diagnostic tool.

Figure 11.14 shows a recently developed radiation thermometer that has removed most of the risk of taking the temperature of young children or the aged with a rectal thermometer—risks such as bowel perforation or bacterial contamination. The instrument measures the intensity of the infrared radiation leaving

APPLICATION
Light-Colored Summer Clothing

APPLICATION
Thermography

APPLICATION
 Radiation Thermometers for Measuring Body Temperature

Figure 11.13 This thermogram of a house, made during cold weather, shows colors ranging from white and yellow (areas of greatest energy loss) to blue and purple (areas of least energy loss).

Figure 11.14 A radiation thermometer measures a patient's temperature by monitoring the intensity of infrared radiation leaving the ear.

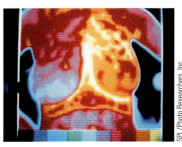

Thermogram of a woman's breasts. Her left breast is diseased (red and orange) and her right breast (blue) is healthy.

the eardrum and surrounding tissues and converts this information to a standard numerical reading. The eardrum is a particularly good location to measure body temperature because it's near the hypothalamus—the body's temperature control center.

Quick Quiz 11.5

Stars A and B have the same temperature, but star A has twice the radius of star B. (a) What is the ratio of star A's power output to star B's output due to electromagnetic radiation? The emissivity of both stars can assumed to be 1. (b) Repeat the question if the stars have the same radius, but star A has twice the absolute temperature of star B. (c) What's the ratio if star A has both twice the radius and twice the absolute temperature of star B?

Applying Physics 11.2   Thermal Radiation and Night Vision

How can thermal radiation be used to see objects in near total darkness?

Explanation   There are two methods of night vision, one enhancing a combination of very faint visible light and infrared light, and another using infrared light only. The latter is valuable for creating images in absolute darkness. Because all objects above absolute zero emit thermal radiation due to the vibrations of

their atoms, the infrared (invisible) light can be focused by a special lens and scanned by an array of infrared detector elements. These elements create a thermogram. The information from thousands of separate points in the field of view is converted to electrical impulses and translated by a microchip into a form suitable for display. Different temperature areas are assigned different colors, which can then be easily discerned on the display.

EXAMPLE 11.12   Polar Bear Club

Goal   Apply Stefan's law.

Problem   A member of the Polar Bear Club, dressed only in bathing trunks of negligible size, prepares to plunge into the Baltic Sea from the beach in St. Petersburg, Russia. The air is calm, with a temperature of 5°C. If the swimmer's surface body temperature is 25°C, compute the net rate of energy loss from his skin due to radiation. How much energy is lost in 10.0 min? Assume his emissivity is 0.900, and his surface area is 1.50 m^2.

Strategy   Use Equation 11.11, the thermal radiation equation, substituting the given information. Remember to convert temperatures to Kelvin by adding 273 to each value in degrees Celsius!

Solution

Convert temperatures from Celsius to Kelvin:

$$T_{5°C} = T_C + 273 = 5 + 273 = 278 \text{ K}$$

$$T_{25°C} = T_C + 273 = 25 + 273 = 298 \text{ K}$$

Compute the net rate of energy loss, using Equation 11.11:

$$\mathcal{P}_{net} = \sigma A e (T^4 - T_0^4)$$

$$= (5.67 \times 10^{-8} \text{ W/m}^2 \cdot \text{K}^4)(1.50 \text{ m}^2)$$

$$\times (0.90)[(298 \text{ K})^4 - (278 \text{ K})^4]$$

$$\mathcal{P}_{net} = \boxed{146 \text{ W}}$$

Multiply the preceding result by the time, 10 minutes, to get the energy lost in that time due to radiation:

$$Q = \mathcal{P}_{net} \times \Delta t = (146)(6.00 \times 10^2 \text{ s}) = \boxed{8.76 \times 10^4 \text{ J}}$$

Remarks   Energy is also lost from the body through convection and conduction. Clothing traps layers of air next to the skin, which are warmed by radiation and conduction. In still air these warm layers are more readily retained. Even a Polar Bear Club member enjoys some benefit from the still air, better retaining a stagnant air layer next to the surface of his skin.

Exercise 11.12
Repeat the calculation when the man is standing in his bedroom, with an ambient temperature of 20.0°C. Assume his body surface temperature is 27.0°C, with emissivity of 0.900.

Answer   55.9 W, 3.35×10^4 J

The Dewar Flask

The Thermos bottle, also called a **Dewar flask** (after its inventor), is designed to minimize energy transfer by conduction, convection, and radiation. The thermos can store either cold or hot liquids for long periods. The standard vessel (Fig. 11.15) is a double-walled Pyrex glass with silvered walls. The space between the walls is evacuated to minimize energy transfer by conduction and convection. The silvered surface minimizes energy transfer by radiation because silver is a very good reflector and has very low emissivity. A further reduction in energy loss is achieved by reducing the size of the neck. Dewar flasks are commonly used to store liquid nitrogen (boiling point 77 K) and liquid oxygen (boiling point 90 K).

To confine liquid helium (boiling point 4.2 K), which has a very low heat of vaporization, it's often necessary to use a double Dewar system in which the Dewar flask containing the liquid is surrounded by a second Dewar flask. The space between the two flasks is filled with liquid nitrogen.

Some of the principles of the Thermos bottle are used in the protection of sensitive electronic instruments in orbiting space satellites. In half of its orbit around the Earth a satellite is exposed to intense radiation from the Sun, and in the other half it lies in the Earth's cold shadow. Without protection, its interior would be subjected to tremendous extremes of temperature. The interior of the satellite is wrapped with blankets of highly reflective aluminum foil. The foil's shiny surface reflects away much of the Sun's radiation while the satellite is in the unshaded part of the orbit and helps retain interior energy while the satellite is in the Earth's shadow.

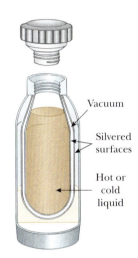

Figure 11.15   A cross-sectional view of a Thermos bottle designed to store hot or cold liquids.

APPLICATION

Thermos Bottles

11.6   GLOBAL WARMING AND GREENHOUSE GASES

Many of the principles of energy transfer, and opposition to it, can be understood by studying the operation of a glass greenhouse. During the day, sunlight passes into the greenhouse and is absorbed by the walls, soil, plants, and so on. This absorbed visible light is subsequently reradiated as infrared radiation, causing the temperature of the interior to rise.

In addition, convection currents are inhibited in a greenhouse. As a result, warmed air can't rapidly pass over the surfaces of the greenhouse that are exposed to the outside air and thereby cause an energy loss by conduction through those surfaces. Most experts now consider this restriction to be a more important warming effect than the trapping of infrared radiation. In fact, experiments have shown that when the glass over a greenhouse is replaced by a special glass known to transmit infrared light, the temperature inside is lowered only slightly. On the basis of this evidence, the primary mechanism that raises the temperature of a greenhouse is not the trapping of infrared radiation, but the inhibition of airflow that occurs under any roof (in an attic, for example).

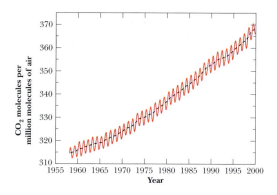

Figure 11.16 The concentration of atmospheric carbon dioxide in parts per million (ppm) of dry air as a function of time during the latter part of the 20th century. These data were recorded at Mauna Loa Observatory in Hawaii. The yearly variations (red curve) coincide with growing seasons, because vegetation absorbs carbon dioxide from the air. The steady increase (black curve) is of concern to scientists.

A phenomenon commonly known as the **greenhouse effect** can also play a major role in determining the Earth's temperature. First, note that the Earth's atmosphere is a good transmitter (and hence a poor absorber) of visible radiation and a good absorber of infrared radiation. The visible light that reaches the Earth's surface is absorbed and reradiated as infrared light, which in turn is absorbed (trapped) by the Earth's atmosphere. An extreme case is the warmest planet, Venus, which has a carbon dioxide (CO_2) atmosphere and temperatures approaching 850°F.

As fossil fuels (coal, oil, and natural gas) are burned, large amounts of carbon dioxide are released into the atmosphere, causing it to retain more energy. This is of great concern to scientists and governments throughout the world. Many scientists are convinced that the 10% increase in the amount of atmospheric carbon dioxide in the past 30 years could lead to drastic changes in world climate. The increase in concentration of atmospheric carbon dioxide in the latter part of the 20th century is shown in Figure 11.16. According to one estimate, doubling the carbon dioxide content in the atmosphere will cause temperatures to increase by 2°C. In temperate regions, such as Europe and the United States, a 2°C temperature rise would save billions of dollars per year in fuel costs. Unfortunately, it would also melt a large amount of ice from the polar ice caps, which could cause flooding and destroy many coastal areas. A 2°C rise would also increase the frequency of droughts, and consequently decrease already low crop yields in tropical and subtropical countries. Even slightly higher average temperatures might make it impossible for certain plants and animals to survive in their customary ranges.

At present, about 3.5×10^{11} tons of CO_2 are released into the atmosphere each year. Most of this gas results from human activities such as the burning of fossil fuels, the cutting of forests, and manufacturing processes. Another greenhouse gas is methane (CH_4), which is released in the digestive process of cows and other ruminants. This gas originates from that part of the animal's stomach called the *rumen*, where cellulose is digested. Termites are also major producers of this gas. Finally, greenhouse gases such as nitrous oxide (N_2O) and sulfur dioxide (SO_2) are increasing due to automobile and industrial pollution.

Whether the increasing greenhouse gases are responsible or not, there is convincing evidence that global warming is underway. The evidence comes from the melting of ice in Antarctica and the retreat of glaciers at widely scattered sites throughout the world (see Fig. 11.17). For example, satellite images of Antarctica show James Ross Island completely surrounded by water for the first time since maps were made, about 100 years ago. Previously, the island was connected to the mainland by an ice bridge. In addition, at various places across the continent, ice shelves are retreating, some at a rapid rate.

Perhaps at no place in the world are glaciers monitored with greater interest than in Switzerland. There, it is found that the Alps have lost about 50% of their glacial ice compared to 130 years ago. The retreat of glaciers on high-altitude peaks in the tropics is even more severe than in Switzerland. The Lewis glacier on Mount Kenya and the snows of Kilimanjaro are two examples. However, in certain regions of the planet where glaciers are near large bodies of water and are fed by

(a)   (b)

Figure 11.17 Glacier melting in Alaska. (a) Hiker on a ridge above the Muir Glacier in Glacier Bay National Park, August 26, 1978. (b) Hiker on the same ridge, June 27, 1993.

large and frequent snows, glaciers continue to advance, so the overall picture of a catastrophic global-warming scenario may be premature. In about 50 years, however, the amount of carbon dioxide in the atmosphere is expected to be about twice what it was in the preindustrial era. Because of the possible catastrophic consequences, most scientists voice the concern that reductions in greenhouse gas emissions need to be made now.

SUMMARY

Physics⊗Now™ Take a practice test by logging into Physics-Now at **www.cp7e.com** and clicking on the Pre-Test link for this chapter.

11.1 Heat and Internal Energy

Internal energy is associated with a system's microscopic components. Internal energy includes the kinetic energy of translation, rotation, and vibration of molecules, as well as potential energy.

Heat is the transfer of energy across the boundary of a system resulting from a temperature difference between the system and its surroundings. The symbol Q represents the amount of energy transferred.

The **calorie** is the amount of energy necessary to raise the temperature of 1 g of water from 14.5°C to 15.5°C. The **mechanical equivalent of heat** is 4.186 J/cal.

11.2 Specific Heat
11.3 Calorimetry

The energy required to change the temperature of a substance of mass m by an amount ΔT is

$$Q = mc\,\Delta T \qquad [11.3]$$

where c is the **specific heat** of the substance. In calorimetry problems, the specific heat of a substance can be determined by placing it in water of known temperature, isolating the system, and measuring the temperature at equilibrium. The sum of all energy gains and losses for all the objects in an isolated system is given by

$$\Sigma Q_k = 0 \qquad [11.5]$$

where Q_k is the energy change in the kth object in the system. This equation can be solved for the unknown specific heat, or used to determine an equilibrium temperature.

11.4 Latent Heat and Phase Change

The energy required to change the phase of a pure substance of mass m is

$$Q = \pm\, mL \qquad [11.6]$$

where L is the **latent heat** of the substance. The latent heat of fusion, L_f, describes an energy transfer during a change from a solid phase to a liquid phase (or vice-versa), while the latent heat of vaporizaion, L_v, describes an energy transfer during a change from a liquid phase to a gaseous phase (or vice-versa). Calorimetry problems involving phase changes are handled with Equation 11.5, with latent heat terms added to the specific heat terms.

11.5 Energy Transfer

Energy can be transferred by several different processes, including work, discussed in Chapter 5, and by conduction, convection, and radiation. **Conduction** can be viewed as an exchange of kinetic energy between colliding molecules or electrons. The rate at which energy transfers by conduction through a slab of area A and thickness L is

$$\mathscr{P} = kA\frac{(T_h - T_c)}{L} \qquad [11.7]$$

where k is the **thermal conductivity** of the material making up the slab.

Energy is transferred by **convection** as a substance moves from one place to another.

All objects emit **radiation** from their surfaces in the form of electromagnetic waves at a net rate of

$$\mathscr{P}_{\text{net}} = \sigma A e (T^4 - T_0{}^4) \qquad [11.11]$$

where T is the temperature of the object and T_0 is the temperature of the surroundings. An object that is hotter than its surroundings radiates more energy than it absorbs, whereas a body that is cooler than its surroundings absorbs more energy than it radiates.

CONCEPTUAL QUESTIONS

1. Rub the palm of your hand on a metal surface for 30–45 seconds. Place the palm of your other hand on an unrubbed portion of the surface and then the rubbed portion. The rubbed portion will feel warmer. Now repeat this process on a wooden surface. Why does the temperature difference between the rubbed and unrubbed portions of the wood surface seem larger than for the metal surface?

2. Pioneers stored fruits and vegetables in underground cellars. Discuss fully this choice for a storage site.

3. In usually warm climates that experience an occasional hard freeze, fruit growers will spray the fruit trees with water, hoping that a layer of ice will form on the fruit. Why would such a layer be advantageous?

4. In winter, why did the pioneers (mentioned in Question 2) store an open barrel of water alongside their produce?

5. Cups of water for coffee or tea can be warmed with a coil that is immersed in the water and raised to a high temperature by means of electricity. Why do the instructions warn users not to operate the coils in the absence of water? Can the immersion coil be used to warm up a cup of stew?

6. The U.S. penny is now made of copper-coated zinc. Can a calorimetric experiment be devised to test for the metal content in a collection of pennies? If so, describe the procedure.

7. On a clear, cold night, why does frost tend to form on the tops, rather than the sides, of mailboxes and cars?

8. A warning sign often seen on highways just before a bridge is "Caution—Bridge Surface Freezes before Road Surface." Of the three energy transfer processes discussed in Sections 11.5 to 11.7, which is most important in causing a bridge surface to freeze before the road surface on very cold days?

9. A tile floor may feel uncomfortably cold to your bare feet, but a carpeted floor in an adjoining room at the same temperature feels warm. Why?

10. On a very hot day, it's possible to cook an egg on the hood of a car. Would you select a black car or a white car on which to cook your egg? Why?

11. Concrete has a higher specific heat than does soil. Use this fact to explain (partially) why a city has a higher average temperature than the surrounding countryside. Would you expect evening breezes to blow from city to country or from country to city? Explain.

12. You need to pick up a very hot cooking pot in your kitchen. You have a pair of hot pads. Should you soak them in cold water or keep them dry in order to pick up the pot most comfortably?

13. In a daring demonstration, a professor dips her wetted fingers into molten lead (327°C) and withdraws them quickly without getting burned. How is this possible?

14. The air temperature above coastal areas is profoundly influenced by the large specific heat of water. One reason is that the energy released when 1 cubic meter of water cools by 1.0°C will raise the temperature of an enormously larger volume of air by 1.0°C. Estimate that volume of air. The specific heat of air is approximately 1.0 kJ/kg · °C. Take the density of air to be 1.3 kg/m^3.

15. Ethyl alcohol has about one-half the specific heat of water. Compare the temperature increases of equal masses of alcohol and water in separate beakers that are supplied with the same amount of energy.

16. Energy is added to ice, raising its temperature from −10°C to −5°C. A larger amount of energy is added to the same mass of liquid water, raising its temperature from 15°C to 20°C. From these results, we can conclude that (a) overcoming the latent heat of fusion of ice requires an input of energy (b) the latent heat of fusion of ice delivers some energy to the system (c) the specific heat of ice is less than that of water (d) the specific heat of ice is greater than that of water.

17. The specific heat of substance A is greater than the specific heat of substance B. Both A and B are at the same initial temperature when equal amounts of energy are added to them. Assuming no melting, freezing, or evaporation occurs, which of the following can be concluded about the final temperature T_A of substance A and the final temperature T_B of substance B? (a) $T_A > T_B$ (b) $T_A < T_B$ (c) $T_A = T_B$ (d) more information is needed.

PROBLEMS

1, 2, 3 = straightforward, intermediate, challenging   □ = full solution available in *Student Solutions Manual/Study Guide*

Physics⊗Now™ = coached problem with hints available at **www.cp7e.com**   🎵 = biomedical application

Section 11.1 Heat and Internal Energy
Section 11.2 Specific Heat

1. Water at the top of Niagara Falls has a temperature of 10.0°C. If it falls a distance of 50.0 m and all of its potential energy goes into heating the water, calculate the temperature of the water at the bottom of the falls.

2. A 50.0-g piece of cadmium is at 20°C. If 400 cal of energy is transferred to the cadmium, what is its final temperature?

3. Lake Erie contains roughly 4.00 × 10^{11} m^3 of water. (a) How much energy is required to raise the temperature of that volume of water from 11.0°C to 12.0°C? (b) How many years would it take to supply this amount of energy by

using the 1 000-MW exhaust energy of an electric power plant?

4. An aluminum rod is 20.0 cm long at 20°C and has a mass of 350 g. If 10 000 J of energy is added to the rod by heat, what is the change in length of the rod?

5. How many joules of energy are required to raise the temperature of 100 g of gold from 20.0°C to 100°C?

6. As part of an exercise routine, a 50.0-kg person climbs 10.0 meters up a vertical rope. How many (food) Calories are expended in a single climb up the rope? (1 food Calorie = 10^3 calories)

7. **Physics**⊗**Now**™ A 75.0-kg weight watcher wishes to climb a mountain to work off the equivalent of a large piece of chocolate cake rated at 500 (food) Calories. How high must the person climb? (1 food Calorie = 10^3 calories)

8. The apparatus shown in Figure P11.8 was used by Joule to measure the mechanical equivalent of heat. Work is done on the water by a rotating paddle wheel, which is driven by two blocks falling at a constant speed. The tem-

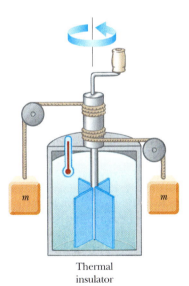

Figure P11.8 The falling weights rotate the paddles, causing the temperature of the water to increase.

perature of the stirred water increases due to the friction between the water and the paddles. If the energy lost in the bearings and through the walls is neglected, then the loss in potential energy associated with the blocks equals the work done by the paddle wheel on the water. If each block has a mass of 1.50 kg and the insulated tank is filled with 200 g of water, what is the increase in tempera-

ture of the water after the blocks fall through a distance of 3.00 m?

9. A 5.00-g lead bullet traveling at 300 m/s is stopped by a large tree. If half the kinetic energy of the bullet is transformed into internal energy and remains with the bullet while the other half is transmitted to the tree, what is the increase in temperature of the bullet?

10. A 1.5-kg copper block is given an initial speed of 3.0 m/s on a rough horizontal surface. Because of friction, the block finally comes to rest. (a) If the block absorbs 85% of its initial kinetic energy as internal energy, calculate its increase in temperature. (b) What happens to the remaining energy?

11. A 200-g aluminum cup contains 800 g of water in thermal equilibrium with the cup at 80°C. The combination of cup and water is cooled uniformly so that the temperature decreases by 1.5°C per minute. At what rate is energy being removed? Express your answer in watts.

Section 11.3 Calorimetry

12. Lead pellets, each of mass 1.00 g, are heated to 200°C. How many pellets must be added to 500 g of water that is initially at 20.0°C to make the equilibrium temperature 25.0°C? Neglect any energy transfer to or from the container.

13. What mass of water at 25.0°C must be allowed to come to thermal equilibrium with a 3.00-kg gold bar at 100°C in order to lower the temperature of the bar to 50.0°C?

14. In a showdown on the streets of Laredo, the good guy drops a 5.0-g silver bullet at a temperature of 20°C into a 100-cm³ cup of water at 90°C. Simultaneously, the bad guy drops a 5.0-g copper bullet at the same initial temperature into an identical cup of water. Which one ends the showdown with the coolest cup of water in the west? Neglect any energy transfer into or away from the container.

15. **Physics**⊗**Now**™ An aluminum cup contains 225 g of water and a 40-g copper stirrer, all at 27°C. A 400-g sample of silver at an initial temperature of 87°C is placed in the water. The stirrer is used to stir the mixture until it reaches its final equilibrium temperature of 32°C. Calculate the mass of the aluminum cup.

16. It is desired to cool iron parts from 500°F to 100°F by dropping them into water that is initially at 75°F. Assuming that all the heat from the iron is transferred to the water and that none of the water evaporates, how many kilograms of water are needed per kilogram of iron?

17. A 100-g aluminum calorimeter contains 250 g of water. The two substances are in thermal equilibrium at 10°C. Two metallic blocks are placed in the water. One is a 50-g piece of copper at 80°C. The other sample has a mass of 70 g and is originally at a temperature of 100°C.

The entire system stabilizes at a final temperature of 20°C. Determine the specific heat of the unknown second sample.

18. When a driver brakes an automobile, the friction between the brake drums and the brake shoes converts the car's kinetic energy to thermal energy. If a 1 500-kg automobile traveling at 30 m/s comes to a halt, how much does the temperature rise in each of the four 8.0-kg iron brake drums? (The specific heat of iron is 448 J/kg·°C.)

19. A student drops two metallic objects into a 120-g steel container holding 150 g of water at 25°C. One object is a 200-g cube of copper that is initially at 85°C, and the other is a chunk of aluminum that is initially at 5.0°C. To the surprise of the student, the water reaches a final temperature of 25°C, precisely where it started. What is the mass of the aluminum chunk?

Section 11.4 Latent Heat and Phase Change

20. A 50-g ice cube at 0°C is heated until 45 g has become water at 100°C and 5.0 g has become steam at 100°C. How much energy was added to accomplish the transformation?

21. A 100-g cube of ice at 0°C is dropped into 1.0 kg of water that was originally at 80°C. What is the final temperature of the water after the ice has melted?

22. How much energy is required to change a 40-g ice cube from ice at −10°C to steam at 110°C?

23. What mass of steam that is initially at 120°C is needed to warm 350 g of water and its 300-g aluminum container from 20°C to 50°C?

24. A resting adult of average size converts chemical energy in food into internal energy at the rate of 120 W, called her *basal metabolic rate*. To stay at a constant temperature, energy must be transferred out of the body at the same rate. Several processes exhaust energy from your body. Usually the most important is thermal conduction into the air in contact with your exposed skin. If you are not wearing a hat, a convection current of warm air rises vertically from your head like a plume from a smokestack. Your body also loses energy by electromagnetic radiation, by your exhaling warm air, and by the evaporation of perspiration. Now consider still another pathway for energy loss: moisture in exhaled breath. Suppose you breathe out 22.0 breaths per minute, each with a volume of 0.600 L. Suppose also that you inhale dry air and exhale air at 37°C containing water vapor with a vapor pressure of 3.20 kPa. The vapor comes from the evaporation of liquid water in your body. Model the water vapor as an ideal gas. Assume its latent heat of evaporation at 37°C is the same as its heat of vaporization at 100°C. Calculate the rate at which you lose energy by exhaling humid air.

25. A 75-kg cross-country skier glides over snow as in Figure P11.25. The coefficient of friction between skis and snow is 0.20. Assume all the snow beneath his skis is at 0°C and that all the internal energy generated by friction is added to snow, which sticks to his skis until it melts. How far would he have to ski to melt 1.0 kg of snow?

Figure P11.25   A cross-country skier.

26. When you jog, most of the food energy you burn above your basal metabolic rate (BMR) ends up as internal energy that would raise your body temperature if it were not eliminated. The evaporation of perspiration is the primary mechanism for eliminating this energy. Determine the amount of water you lose to evaporation when running for 30 minutes at a rate that uses 400 kcal/h above your BMR. (That amount is often considered to be the "maximum fat-burning" energy output.) The metabolism of 1 gram of fat generates approximately 9.0 kcal of energy and produces approximately 1 gram of water. (The hydrogen atoms in the fat molecule are transferred to oxygen to form water.) What fraction of your need for water will be provided by fat metabolism? (The latent heat of vaporization of water at room temperature is 2.5×10^6 J/kg).

27. A 40-g block of ice is cooled to −78°C and is then added to 560 g of water in an 80-g copper calorimeter at a temperature of 25°C. Determine the final temperature of the system consisting of the ice, water, and calorimeter. (If not all the ice melts, determine how much ice is left.) Remember that the ice must first warm to 0°C, melt, and then continue warming as water. The specific heat of ice is 0.500 cal/g·°C = 2090 J/kg·°C.

28. A 60.0-kg runner expends 300 W of power while running a marathon. Assuming that 10.0% of the energy is delivered to the muscle tissue and that the excess energy is

removed from the body primarily by sweating, determine the volume of bodily fluid (assume it is water) lost per hour. (At 37.0°C, the latent heat of vaporization of water is 2.41×10^6 J/kg.)

Figure P11.28 Timothy Cherigat of Kenya, winner of the Boston Marathon in 2004.

29. A high-end gas stove usually has at least one burner rated at 14 000 Btu/h. If you place a 0.25-kg aluminum pot containing 2.0 liters of water at 20°C on this burner, how long will it take to bring the water to a boil, assuming all of the heat from the burner goes into the pot? How long will it take to boil all of the water out of the pot?

30. A beaker of water sits in the sun until it reaches an equilibrium temperature of 30°C. The beaker is made of 100 g of aluminum and contains 180 g of water. In an attempt to cool this system, 100 g of ice at 0°C is added to the water. (a) Determine the final temperature of the system. If $T_f = 0$°C, determine how much ice remains. (b) Repeat your calculations for 50 g of ice.

31. Steam at 100°C is added to ice at 0°C. (a) Find the amount of ice melted and the final temperature when the mass of steam is 10 g and the mass of ice is 50 g. (b) Repeat with steam of mass 1.0 g and ice of mass 50 g.

Section 11.5 Energy Transfer

32. The average thermal conductivity of the walls (including windows) and roof of a house in Figure P11.32 is 4.8×10^{-4} kW/m·°C, and their average thickness is 21.0 cm. The house is heated with natural gas, with a heat of combustion (energy released per cubic meter of gas burned) of 9300 kcal/m^3. How many cubic meters of gas must be burned each day to maintain an inside tempera-

ture of 25.0°C if the outside temperature is 0.0°C? Disregard radiation and energy loss by heat through the ground.

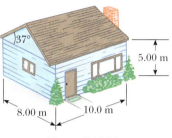

Figure P11.32

33. (a) Find the rate of energy flow through a copper block of cross-sectional area 15 cm^2 and length 8.0 cm when a temperature difference of 30°C is established across the block. Repeat the calculation, assuming that the material is (b) a block of stagnant air with the given dimensions; (c) a block of wood with the given dimensions.

34. A window has a glass surface area of 1.6×10^3 cm^2 and a thickness of 3.0 mm. (a) Find the rate of energy transfer by conduction through the window when the temperature of the inside surface of the glass is 70°F and the outside temperature is 90°F. (b) Repeat for the same inside temperature and an outside temperature of 0°F.

35. **Physics⊗Now™** A steam pipe is covered with 1.50-cm-thick insulating material of thermal conductivity 0.200 cal/cm·°C·s. How much energy is lost every second when the steam is at 200°C and the surrounding air is at 20.0°C? The pipe has a circumference of 800 cm and a length of 50.0 m. Neglect losses through the ends of the pipe.

36. A box with a total surface area of 1.20 m^2 and a wall thickness of 4.00 cm is made of an insulating material. A 10.0-W electric heater inside the box maintains the inside temperature at 15.0°C above the outside temperature. Find the thermal conductivity k of the insulating material.

37. Determine the R value for a wall constructed as follows: The outside of the house consists of lapped wood shingles placed over 0.50-in.-thick sheathing, over 3.0 in. of cellulose fiber, over 0.50 in. of drywall.

38. A thermopane window consists of two glass panes, each 0.50 cm thick, with a 1.0-cm-thick sealed layer of air in between. If the inside surface temperature is 23°C and the outside surface temperature is 0.0°C, determine the rate

of energy transfer through 1.0 m² of the window. Compare your answer with the rate of energy transfer through 1.0 m² of a single 1.0-cm-thick pane of glass.

39. A copper rod and an aluminum rod of equal diameter are joined end to end in good thermal contact. The temperature of the free end of the copper rod is held constant at 100°C, and that of the far end of the aluminum rod is held at 0°C. If the copper rod is 0.15 m long, what must be the length of the aluminum rod so that the temperature at the junction is 50°C?

40. A Styrofoam box has a surface area of 0.80 m² and a wall thickness of 2.0 cm. The temperature of the inner surface is 5.0°C, and the outside temperature is 25°C. If it takes 8.0 h for 5.0 kg of ice to melt in the container, determine the thermal conductivity of the Styrofoam.

41. A sphere that is a perfect blackbody radiator has a radius of 0.060 m and is at 200°C in a room where the temperature is 22°C. Calculate the net rate at which the sphere radiates energy.

42. The surface temperature of the Sun is about 5 800 K. Taking the radius of the Sun to be 6.96×10^8 m, calculate the total energy radiated by the Sun each second. (Assume $e = 0.965$.)

43. A large, hot pizza 70 cm in diameter and 2.0 cm thick, at a temperature of 100°C, floats in outer space. Assume its emissivity is 0.8. What is the order of magnitude of its rate of energy loss?

44. Calculate the temperature at which a tungsten filament that has an emissivity of 0.90 and a surface area of 2.5×10^{-5} m² will radiate energy at the rate of 25 W in a room where the temperature is 22°C.

45. Measurements on two stars indicate that Star X has a surface temperature of 5 727°C and Star Y has a surface temperature of 11 727°C. If both stars have the same radius, what is the ratio of the luminosity (total power output) of Star Y to the luminosity of Star X? Both stars can be considered to have an emissivity of 1.0.

46. At high noon, the Sun delivers 1.00 kW to each square meter of a blacktop road. If the hot asphalt loses energy only by radiation, what is its equilibrium temperature?

ADDITIONAL PROBLEMS

47. The bottom of a copper kettle has a 10-cm radius and is 2.0 mm thick. The temperature of the outside surface is 102°C, and the water inside the kettle is boiling at 1 atm of pressure. Find the rate at which energy is being transferred through the bottom of the kettle.

48. A family comes home from a long vacation with laundry to do and showers to take. The water heater has been turned off during the vacation. If the heater has a capacity of 50.0 gallons and a 4 800-W heating element, how much time is required to raise the temperature of the water from 20.0°C to 60.0°C? Assume that the heater is well insulated and no water is withdrawn from the tank during that time.

49. Solar energy can be the primary source of winter space heating for a typical house (with floor area 130 m² = 1 400 ft²) in the north central United States. If the house has very good insulation, you may model it as losing energy by heat steadily at the rate of 1 000 W during the winter, when the average exterior temperature is −5°C. The passive solar-energy collector can consist simply of large windows facing south. Sunlight shining in during the daytime is absorbed by the floor, interior walls, and objects in the house, raising their temperature to 30°C. As the sun goes down, insulating draperies or shutters are closed over the windows. During the period between 4 PM and 8 AM, the temperature of the house will drop, and a sufficiently large "thermal mass" is required to keep it from dropping too far. The thermal mass can be a large quantity of stone (with specific heat 800 J/kg · °C) in the floor and the interior walls exposed to sunlight. What mass of stone is required if the temperature is not to drop below 18°C overnight?

50. A water heater is operated by solar power. If the solar collector has an area of 6.00 m², and the intensity delivered by sunlight is 550 W/m², how long does it take to increase the temperature of 1.00 m³ of water from 20.0°C to 60.0°C?

51. A 40-g ice cube floats in 200 g of water in a 100-g copper cup; all are at a temperature of 0°C. A piece of lead at 98°C is dropped into the cup, and the final equilibrium temperature is 12°C. What is the mass of the lead?

52. The evaporation of perspiration is the primary mechanism for cooling the human body. Estimate the amount of water you will lose when you bake in the sun on the beach for an hour. Use a value of 1 000 W/m² for the intensity of sunlight, and note that the energy required to evaporate a liquid at a particular temperature is approximately equal to the sum of the energy required to raise its temperature to the boiling point and the latent heat of vaporization (determined at the boiling point).

53. A 200-g block of copper at a temperature of 90°C is dropped into 400 g of water at 27°C. The water is contained in a 300-g glass container. What is the final temperature of the mixture?

54. A class of 10 students taking an exam has a power output per student of about 200 W. Assume that the initial temperature of the room is 20°C and that its dimensions are

6.0 m by 15.0 m by 3.0 m. What is the temperature of the room at the end of 1.0 h if all the energy remains in the air in the room and none is added by an outside source? The specific heat of air is 837 J/kg·°C, and its density is about 1.3×10^{-3} g/cm^3.

55. The human body must maintain its core temperature inside a rather narrow range around 37°C. Metabolic processes (notably, muscular exertion) convert chemical energy into internal energy deep in the interior. From the interior, energy must flow out to the skin or lungs, to be lost by heat to the environment. During moderate exercise, an 80-kg man can metabolize food energy at the rate of 300 kcal/h, do 60 kcal/h of mechanical work, and put out the remaining 240 kcal/h of energy by heat. Most of the energy is carried from the interior of the body out to the skin by "forced convection" (as a plumber would say): Blood is warmed in the interior and then cooled at the skin, which is a few degrees cooler than the body core. Without blood flow, living tissue is a good thermal insulator, with a thermal conductivity about 0.210 W/m·°C. Show that blood flow is essential to keeping the body cool by calculating the rate of energy conduction, in kcal/h, through the tissue layer under the skin. Assume that its area is 1.40 m^2, its thickness is 2.50 cm, and it is maintained at 37.0°C on one side and at 34.0°C on the other side.

56. An aluminum rod and an iron rod are joined end to end in good thermal contact. The two rods have equal lengths and radii. The free end of the aluminum rod is maintained at a temperature of 100°C, and the free end of the iron rod is maintained at 0°C. (a) Determine the temperature of the interface between the two rods. (b) If each rod is 15 cm long and each has a cross-sectional area of 5.0 cm^2, what quantity of energy is conducted across the combination in 30 min?

57. Water is being boiled in an open kettle that has a 0.500-cm-thick circular aluminum bottom with a radius of 12.0 cm. If the water boils away at a rate of 0.500 kg/min, what is the temperature of the lower surface of the bottom of the kettle? Assume that the top surface of the bottom of the kettle is at 100°C.

58. A 3.00-g copper penny at 25.0°C drops 50.0 m to the ground. (a) If 60.0% of the initial potential energy associated with the penny goes into increasing its internal energy, determine the final temperature of the penny. (b) Does the result depend on the mass of the coin? Explain.

59. A bar of gold (Au) is in thermal contact with a bar of silver (Ag) of the same length and area (Fig. P11.59). One end of the compound bar is maintained at 80.0°C, while the opposite end is at 30.0°C. Find the temperature at the junction when the energy flow reaches a steady state.

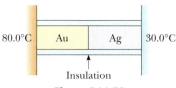

Figure P11.59

60. An iron plate is held against an iron wheel so that a sliding frictional force of 50 N acts between the two pieces of metal. The relative speed at which the two surfaces slide over each other is 40 m/s. (a) Calculate the rate at which mechanical energy is converted to internal energy. (b) The plate and the wheel have masses of 5.0 kg each, and each receives 50% of the internal energy. If the system is run as described for 10 s, and each object is then allowed to reach a uniform internal temperature, what is the resultant temperature increase?

61. An automobile has a mass of 1 500 kg, and its aluminum brakes have an overall mass of 6.0 kg. (a) Assuming that all of the internal energy transformed by friction when the car stops is deposited in the brakes, and neglecting energy transfer, how many times could the car be braked to rest starting from 25 m/s (56 mi/h) before the brakes would begin to melt? (Assume an initial temperature of 20°C.) (b) Identify some effects that are neglected in part (a), but are likely to be important in a more realistic assessment of the temperature increase of the brakes.

62. A 1.0-m-long aluminum rod of cross-sectional area 2.0 cm^2 is inserted vertically into a thermally insulated vessel containing liquid helium at 4.2 K. The rod is initially at 300 K. If half of the rod is inserted into the helium, how many liters of helium boil off in the very short time while the inserted half cools to 4.2 K? The density of liquid helium at 4.2 K is 122 kg/m^3.

63. Physics⊗Now™ A *flow calorimeter* is an apparatus used to measure the specific heat of a liquid. The technique is to measure the temperature difference between the input and output points of a flowing stream of the liquid while adding energy at a known rate. (a) Start with the equations $Q = mc(\Delta T)$ and $m = \rho V$, and show that the rate at which energy is added to the liquid is given by the expression $\Delta Q/\Delta t = \rho c(\Delta T)(\Delta V/\Delta t)$. (b) In a particular experiment, a liquid of density 0.72 g/cm^3 flows through the calorimeter at the rate of 3.5 cm^3/s. At steady state, a temperature difference of 5.8°C is established between the input and output points when energy is supplied at the rate of 40 J/s. What is the specific heat of the liquid?

64. Three liquids are at temperatures of 10°C, 20°C, and 30°C, respectively. Equal masses of the first two liquids are mixed, and the equilibrium temperature is 17°C. Equal masses of the second and third are then mixed, and the equilibrium temperature is 28°C. Find the equilibrium temperature when equal masses of the first and third are mixed.

65. At time $t = 0$, a vessel contains a mixture of 10 kg of water and an unknown mass of ice in equilibrium at 0°C. The temperature of the mixture is measured over a period of an hour, with the following results: During the first 50 min, the mixture remains at 0°C; from 50 min to 60 min, the temperature increases steadily from 0°C to 2°C. Neglecting the heat capacity of the vessel, determine the mass of ice that was initially placed in it. Assume a constant power input to the container.

66. A wood stove is used to heat a single room. The stove is cylindrical in shape, with a diameter of 40.0 cm and a length of 50.0 cm and operates at a temperature of 400°F. (a) If the temperature of the room is 70.0°F determine the amount of radiant energy delivered to the room by the stove each second if the emissivity is 0.920. (b) If the room is a square with walls that are 8.00 ft high and 25.0 ft wide, determine the R value needed in the walls and ceiling to maintain the inside temperature at 70.0°F if the outside temperature is 32.0°F. Note that we are ignoring any heat conveyed by the stove via convection and any energy lost through the walls (and windows!) via convection or radiation.

67. A "solar cooker" consists of a curved reflecting mirror that focuses sunlight onto the object to be heated (Fig. P11.67). The solar power per unit area reaching the Earth at the location of a 0.50-m-diameter solar cooker is 600 W/m². Assuming that 50% of the incident energy is converted to thermal energy, how long would it take to boil away 1.0 L of water initially at 20°C? (Neglect the specific heat of the container.)

Figure P11.67

68. For bacteriological testing of water supplies and in medical clinics, samples must routinely be incubated for 24 h at 37°C. A standard constant temperature bath with electric heating and thermostatic control is not suitable in developing nations without continuously operating electric power lines. Peace Corps volunteer and MIT engineer Amy Smith invented a low cost, low maintenance incubator to fill the need. The device consists of a foam-insulated box containing several packets of a waxy material that melts at 37.0°C, interspersed among tubes, dishes, or bottles containing the test samples and growth medium (food for bacteria). Outside the box, the waxy material is first melted by a stove or solar energy collector. Then it is put into the box to keep the test samples warm as it solidifies. The heat of fusion of the phase-change material is 205 kJ/kg. Model the insulation as a panel with surface area 0.490 m², thickness 9.50 cm, and conductivity 0.012 0 W/m°C. Assume the exterior temperature is 23.0°C for 12.0 h and 16.0°C for 12.0 h. (a) What mass of the waxy material is required to conduct the bacteriological test? (b) Explain why your calculation can be done without knowing the mass of the test samples or of the insulation.

69. What mass of steam initially at 130°C is needed to warm 200 g of water in a 100-g glass container from 20.0°C to 50.0°C?

ACTIVITIES

1. A plot of the decreasing temperature of a substance over time is called a cooling curve and has the same shape and basic explanation as the curve shown in Figure 11.3. You can plot such a curve by observing some water in a container in the freezer compartment of a refrigerator. Place a thermometer in the liquid, and record the reading of the thermometer every minute until about five minutes after the liquid has frozen completely. Explain your observations. A material that is a little easier to work with is naphthalene (mothballs). You can plot the cooling curve in this case without a freezer. Melt a small amount of naphthalene in a container, and plot a graph of temperature versus time as before. Again, explain your observations.

2. You have probably heard someone say that hot water freezes faster than cold water. Is this an urban legend or is it true? To test this hypothesis, fill one container with hot water, at about 200°F, and another with cooler water, at about 70°F. Place the two containers in the freezer compartment of a refrigerator, and find out for yourself. There are a number of variables that you need to attempt to control in such an experiment: (1) The two containers need to be placed at similar locations in the freezer compartment. That is, one should not be near the door while the other is in the back of the compartment. (2) The two containers should not be placed close together, or an un-

wanted exchange of energy will take place between them. (3) If the freezer is one of the old types that forms frost on its walls, the hot container should not be allowed to melt through the frost and make intimate contact with the cold walls of the freezer. Can you list any more variables that you need to control?

3. You may have heard that you can greatly reduce the baking time for potatoes in a *conventional* oven by inserting a nail through each potato. Are there any scientific reasons for believing that this hypothesis is true? Test it with a couple of similar-sized potatoes—but don't bake them in a microwave oven!

Lance Armstrong is an engine: he requires fuel and oxygen to burn it, and the result is work that drives him up the mountainside as his excess, waste energy is expelled in his evaporating sweat.

© Tim De Waele/Corbis

The Laws of Thermodynamics

According to the first law of thermodynamics, the internal energy of a system can be increased either by adding energy to the system or by doing work on it. This means the internal energy of a system, which is just the sum of the molecular kinetic and potential energies, can change as a result of two separate types of energy transfer across the boundary of the system. Although the first law imposes conservation of energy for both energy added by heat and work done on a system, it doesn't predict which of several possible energy-conserving processes actually occur in nature.

The second law of thermodynamics constrains the first law by establishing which processes allowed by the first law actually occur. For example, the second law tells us that energy never flows by heat spontaneously from a cold object to a hot object. One important application of this law is in the study of heat engines (such as the internal combustion engine) and the principles that limit their efficiency.

12.1 WORK IN THERMODYNAMIC PROCESSES

Energy can be transferred to a system by heat and by work done on the system. In most cases of interest treated here, the system is a volume of gas, which is important in understanding engines. All such systems of gas will be assumed to be in thermodynamic equilibrium, so that every part of the gas is at the same temperature and pressure. If that were not the case, the ideal gas law wouldn't apply and most of the results presented here wouldn't be valid. Consider a gas contained by a cylinder fitted with a movable piston (Active Fig. 12.1a) and in equilibrium. The gas occupies a volume V and exerts a uniform pressure P on the cylinder walls and the piston. The gas is compressed slowly enough so the system remains essentially in thermodynamic equilibrium at all times. As the piston is pushed downward by an external force F through a distance Δy, the work done on the gas is

$$W = -F\,\Delta y = -PA\,\Delta y$$

where we have set the magnitude F of the external force equal to PA, possible because the pressure is the same everywhere in the system (by the assumption of equilibrium). Note that if the piston is pushed downward, $\Delta y = y_f - y_i$ is negative, so we need an explicit negative sign in the expression for W to make the work positive. The change in volume of the gas is $\Delta V = A\,\Delta y$, which leads to the following definition:

> The **work W done on a gas** at constant pressure is given by
>
> $$W = -P\,\Delta V \qquad [12.1]$$
>
> where P is the pressure throughout the gas and ΔV is the change in volume of the gas during the process.

If the gas is compressed as in Active Figure 12.1b, ΔV is negative and the work done on the gas is positive. If the gas expands, ΔV is positive and the work done on the gas is negative. The work done by the gas on its environment, W_{env}, is simply the negative of the work done on the gas. In the absence of a change in volume, the work is zero.

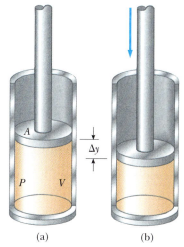

ACTIVE FIGURE 12.1
(a) A gas in a cylinder occupying a volume V at a pressure P. (b) Pushing the piston down compresses the gas.

Physics ⊗ Now™
Log into PhysicsNow at
www.cp7e.com, and go to Active Figure 12.1 to move the piston and see the resulting work done on the gas.

EXAMPLE 12.1   Work Done by an Expanding Gas

Goal   Apply the definition of work at constant pressure.

Problem   In a system similar to that shown in Active Figure 12.1, the gas in the cylinder is at a pressure of 1.01×10^5 Pa and the piston has an area of 0.100 m². As energy is slowly added to the gas by heat, the piston is pushed up a distance of 4.00 cm. Calculate the work done by the expanding gas on the surroundings, W_{env}, assuming the pressure remains constant.

Strategy   The work done on the environment is the negative of the work done on the gas given in Equation 12.1. Compute the change in volume and multiply by the pressure.

Solution

Find the change in volume of the gas, ΔV, which is the cross-sectional area times the displacement:

$$\Delta V = A\,\Delta y = (0.100\ \text{m}^2)(4.00 \times 10^{-2}\ \text{m})$$
$$= 4.00 \times 10^{-3}\ \text{m}^3$$

Multiply this result by the pressure, getting the work the gas does on the environment, W_{env}:

$$W_{\text{env}} = P\,\Delta V = (1.01 \times 10^5\ \text{Pa})(4.00 \times 10^{-3}\ \text{m}^3)$$
$$= \boxed{404\ \text{J}}$$

Remark   The volume of the gas increases, so the work done on the environment is positive. The work done on the system during this process is $W = -404$ J. The energy required to perform positive work on the environment must come from the energy of the gas. (See the next section for more details.)

Exercise 12.1

Gas in a cylinder similar to Figure 12.1 moves a piston with area 0.20 m² as energy is slowly added to the system. If 2.00×10^3 J of work is done on the environment and the pressure of the gas in the cylinder remains constant at 1.01×10^5 Pa, find the displacement of the piston.

Answer 9.90×10^{-2} m

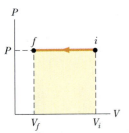

Figure 12.2 The *PV* diagram for a gas being compressed at constant pressure. The shaded area represents the work done on the gas.

Equation 12.1 can be used to calculate the work done on the system *only* when the pressure of the gas remains constant during the expansion or compression. A process in which the pressure remains constant is called an **isobaric process**. The pressure vs. volume graph, or ***PV* diagram**, of an isobaric process is shown in Figure 12.2. The curve on such a graph is called the *path* taken between the initial and final states, with the arrow indicating the direction the process is going, in this case from smaller to larger volume. The area under the graph is

$$\text{Area} = P(V_f - V_i) = P\Delta V$$

The area under the graph in a *PV* diagram is equal in magnitude to the work done on the gas.

This is true in general, whether or not the process proceeds at constant pressure. Just draw the *PV* diagram of the process, find the area underneath the graph (and above the horizontal axis), and that area will be the equal to the magnitude of the work done on the gas. If the arrow on the graph points toward larger volumes, the work done on the gas is negative. If the arrow on the graph points toward smaller volumes, the work done on the gas is positive.

Whenever negative work is done on a system, positive work is done by the system on its environment. The negative work done on the system represents a loss of energy from the system—the cost of doing positive work on the environment.

Quick Quiz 12.1

By visual inspection, order the *PV* diagrams shown in Figure 12.3 from the most negative work done on the system to the most positive work done on the system. (a) a,b,c,d (b) a,c,b,d (c) d,b,c,a (d) d,a,c,b

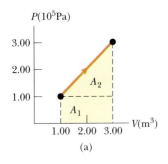
(a)

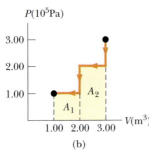
(b)

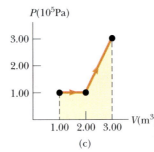
(c)

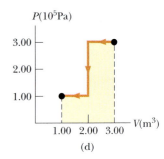
(d)

Figure 12.3   (Quick Quiz 12.1 and Example 12.2)

Notice that the graphs in Figure 12.3 all have the same endpoints, but the areas beneath the curves are different. The work done on a system depends on the path taken in the *PV* diagram.

EXAMPLE 12.2   Work and *PV* Diagrams

Goal   Calculate work from a *PV* diagram.

Problem   Find the numeric value of the work done on the gas in (a) Figure 12.3a and (b) Figure 12.3b.

Strategy   The regions in question are composed of rectangles and triangles. Use basic geometric formulas to find the area underneath each curve. Check the direction of the arrow to determine signs.

Solution
(a) Find the work done on the gas in Figure 12.3a.

Compute the areas A_1 and A_2 in Figure 12.3a. A_1 is a rectangle and A_2 is a triangle.

$$A_1 = \text{height} \times \text{width} = (1.00 \times 10^5 \text{ Pa})(2.00 \text{ m}^3)$$
$$= 2.00 \times 10^5 \text{ J}$$

$$A_2 = \text{one-half base} \times \text{height}$$
$$= \tfrac{1}{2}(2.00 \text{ m}^3)(2.00 \times 10^5 \text{ Pa}) = 2.00 \times 10^5 \text{ J}$$

Sum the areas (the arrows point to increasing volume, so the work done on the gas is negative):

$$\text{Area} = A_1 + A_2 = 4.00 \times 10^5 \text{ J} \quad \rightarrow$$

$$W = -4.00 \times 10^5 \text{ J}$$

(b) Find the work done on the gas in Figure 12.3b.

Compute the areas of the two rectangular regions:

$$A_1 = \text{height} \times \text{width} = (1.00 \times 10^5 \text{ Pa})(1.00 \text{ m}^3)$$
$$= 1.00 \times 10^5 \text{ J}$$

$$A_2 = \text{height} \times \text{width} = (2.00 \times 10^5 \text{ Pa})(1.00 \text{ m}^3)$$
$$= 2.00 \times 10^5 \text{ J}$$

Sum the areas (the arrows point to decreasing volume, so the work done on the gas is positive):

$$\text{Area} = A_1 + A_2 = 3.00 \times 10^5 \text{ J} \quad \rightarrow$$

$$W = +3.00 \times 10^5 \text{ J}$$

Remarks   Notice that in both cases the paths in the *PV* diagrams start and end at the same points, but the answers are different.

Exercise 12.2
Compute the work done on the system in Figures 12.3c and 12.3d.

Answers -3.00×10^5 J, $+4.00 \times 10^5$ J

12.2   THE FIRST LAW OF THERMODYNAMICS

The **first law of thermodynamics** is another energy conservation law that relates changes in internal energy—the energy associated with the position and jiggling of all the molecules of a system—to energy transfers due to heat and work. The first law is universally valid, applicable to all kinds of processes, providing a connection between the microscopic and macroscopic worlds.

There are two ways energy can be transferred between a system and its surroundings: by doing work, which requires a macroscopic displacement of an object through the application of a force; and by heat, which occurs through random molecular collisions. Both mechanisms result in a *change in internal energy*, ΔU, of the system and therefore in measurable changes in the macroscopic variables of the system, such as the pressure, temperature, and volume. This change in the internal energy can be summarized in the **first law of thermodynamics**:

> If a system undergoes a change from an initial state to a final state, where Q is the energy transferred to the system by heat and W is the work done on the system, the change in the internal energy of the system, ΔU, is given by
>
> $$\Delta U = U_f - U_i = Q + W \qquad \text{[12.2]}$$

◀ First law of thermodynamics

The quantity Q is positive when energy is transferred into the system by heat and negative when energy is transferred out of the system by heat. The quantity W is positive when work is done on the system and negative when the system does work on its environment. All quantities in the first law, Equation 12.2, must have the same energy units. Any change in the internal energy of a system—the positions and vibrations of the molecules—is due to the transfer of energy by heat or work (or both).

From Equation 12.2, we also see that the internal energy of any isolated system must remain constant, so that $\Delta U = 0$. Even when a system isn't isolated, the change in internal energy will be zero if the system goes through a cyclic process in which all the thermodynamic variables—pressure, volume, temperature, and moles of gas—return to their original values.

It's important to remember that the quantities in Equation 12.2 concern a *system*, not the effect on the system's environment through work. If the system is hot steam expanding against a piston, for example, the system work W is *negative*, because the piston can only expand at the expense of the internal energy of the gas. The work W_{env} done by the hot steam on the *environment*—in this case, moving a piston which moves the train—is positive, but that's not the work W in Equation 12.2. This way of defining work in the first law makes it consistent with the concept of work defined in Chapter 5. There, positive work done on a system (for example, a block) increased its mechanical energy, while negative work decreased its energy. In this chapter, positive work done on a system (typically, a volume of gas) increases its internal energy, and negative work decreases that internal energy. In both the mechanical and thermal cases, the effect on the system is the same: positive work increases the system's energy, and negative work decreases the system's energy.

Some textbooks identify W as the work done by the gas on its environment. This is an equivalent formulation, but it means that W must carry a minus sign in the first law. That convention isn't consistent with previous discussions of the energy of a system, because when W is positive the system *loses* energy, whereas in Chapter 5 positive W means the system *gains* energy. For that reason, the old convention is not used in this book.

EXAMPLE 12.3   Heating a Gas

Goal   Combine the first law of thermodynamics with work done during a constant pressure process.

Problem   An ideal gas absorbs 5.00×10^3 J of energy while doing 2.00×10^3 J of work on the environment during a constant pressure process. **(a)** Compute the change in the internal energy of the gas. **(b)** If the internal energy now drops by 4.50×10^3 J and 2.00×10^3 J is expelled from the system, find the change in volume, assuming a constant pressure process at 1.01×10^5 Pa.

Strategy   Part (a) requires substitution of the given information into the first law, Equation 12.2. Notice, however, that the given work is done on the *environment*. The negative of this amount is the work done on the *system*, representing a loss of internal energy. Part (b) is a matter of substituting the equation for work at constant pressure into the first law and solving for the change in volume.

Solution
(a) Compute the change in internal energy.

Substitute values into the first law, noting that the work done on the gas is negative:

$$\Delta U = Q + W = 5.00 \times 10^3 \text{ J} - 2.00 \times 10^3 \text{ J}$$
$$= 3.00 \times 10^3 \text{ J}$$

(b) Find the change in volume, noting that ΔU and Q are both negative in this case.

Substitute the equation for work done at constant pressure into the first law:

$$\Delta U = Q + W = Q - P\Delta V$$
$$-4.50 \times 10^3 \text{ J} = -2.00 \times 10^3 \text{ J} - (1.01 \times 10^5 \text{ J})\Delta V$$

Solve for the change in volume, ΔV:

$$\Delta V = 2.48 \times 10^{-2} \text{ m}^3$$

Remarks   The change in volume is positive, so the system expands, doing positive work on the environment, while the work W on the system is negative.

Exercise 12.3
Suppose the internal energy of an ideal gas drops by 3.00×10^3 J at a constant pressure of 1.00×10^5 Pa, while the system gains 5.00×10^2 J of energy by heat. Find the change in volume of the system.

Answer 3.50×10^{-2} m^3

Recall that an expression for the internal energy of an ideal gas is

$$U = \tfrac{3}{2}nRT \qquad\qquad \text{[12.3a]}$$

This expression is valid only for a *monatomic* ideal gas, which means the particles of the gas consist of single atoms. The change in the internal energy, ΔU, for such a gas is given by

$$\Delta U = \tfrac{3}{2}nR\Delta T \qquad\qquad \text{[12.3b]}$$

The **molar specific heat at constant volume** of a monatomic ideal gas, C_v, is defined by

$$C_v \equiv \tfrac{3}{2}R \qquad\qquad \text{[12.4]}$$

The change in internal energy of an ideal gas can then be written

$$\Delta U = nC_v\Delta T \qquad\qquad \text{[12.5]}$$

For ideal gases, this expression is always valid, even when the volume isn't constant. The value of the molar specific heat, however, depends on the gas and can vary under different conditions of temperature and pressure.

A gas with a larger molar specific heat requires more energy to realize a given temperature change. The size of the molar specific heat depends on the structure of the gas molecule and how many different ways it can store energy. A monatomic gas such as helium can store energy as motion in three different directions. A gas such as hydrogen, on the other hand, is diatomic in normal temperature ranges, and aside from moving in three directions, it can also tumble, rotating in two different directions. So hydrogen molecules can store energy in the form of translational motion, and in addition can store energy through tumbling. Further, molecules can also store energy in the vibrations of their constituent atoms. A gas composed of molecules with more ways to store energy will have a larger molar specific heat.

Each different way a gas molecule can store energy is called a *degree of freedom*. Each degree of freedom contributes $\tfrac{1}{2}R$ to the molar specific heat. Because an atomic ideal gas can move in three directions, it has a molar specific heat capacity $C_v = 3(\tfrac{1}{2}R) = \tfrac{3}{2}R$. A diatomic gas like molecular oxygen, O_2, can also tumble in two different directions. This adds $2 \times \tfrac{1}{2}R = R$ to the molar heat specific heat, so $C_v = \tfrac{5}{2}R$ for diatomic gases. The spinning about the long axis connecting the two atoms is generally negligible. Vibration of the atoms in a molecule can also contribute to the heat capacity. A full analysis of a given system is often complex, so in general, molar specific heats must be determined by experiment. Some representative values of C_v can be found in Table 12.1 (page 392).

There are four basic types of thermal processes, which will be studied and illustrated by their effect on an ideal gas.

Isobaric Processes

Recall from Section 12.1 that in an isobaric process the pressure remains constant as the gas expands or is compressed. An expanding gas does work on its environment, given by $W_{\text{env}} = P\Delta V$. The *PV* diagram of an isobaric expansion is given in Figure 12.2. As previously discussed, the magnitude of the work done on the gas is just the area under the path in its *PV* diagram: height times length, or $P\Delta V$. The negative of this quantity, $W = -P\Delta V$, is the energy lost by the gas because the gas does work as it expands. This is the quantity that should be substituted into the first law.

The work done by the gas on its environment must come at the expense of the change in its internal energy, ΔU. Because the change in the internal energy of an ideal gas is given by $\Delta U = nC_v\Delta T$, the temperature of an expanding gas must decrease as the internal energy decreases. Expanding volume and decreasing temperature means the pressure must also decrease, in conformity with the ideal gas law, $PV = nRT$. Consequently, the only way such a process can remain at constant

TABLE 12.1

Molar Specific Heats of Various Gases

Gas	Molar Specific Heat $(J/mol \cdot K)^a$			
	C_P	C_V	$C_P - C_V$	$\gamma = C_P/C_V$
Monatomic Gases				
He	20.8	12.5	8.33	1.67
Ar	20.8	12.5	8.33	1.67
Ne	20.8	12.7	8.12	1.64
Kr	20.8	12.3	8.49	1.69
Diatomic Gases				
H_2	28.8	20.4	8.33	1.41
N_2	29.1	20.8	8.33	1.40
O_2	29.4	21.1	8.33	1.40
CO	29.3	21.0	8.33	1.40
Cl_2	34.7	25.7	8.96	1.35
Polyatomic Gases				
CO_2	37.0	28.5	8.50	1.30
SO_2	40.4	31.4	9.00	1.29
H_2O	35.4	27.0	8.37	1.30
CH_4	35.5	27.1	8.41	1.31

[a]All values except that for water were obtained at 300 K.

pressure is if thermal energy Q is transferred into the gas by heat. Rearranging the first law, we obtain

$$Q = \Delta U - W = \Delta U + P\Delta V$$

Now we can substitute the expression in Equation 12.3b for ΔU and use the ideal gas law to substitute $P\Delta V = nR\Delta T$:

$$Q = \tfrac{3}{2}nR\Delta T + nR\Delta T = \tfrac{5}{2}nR\Delta T$$

Another way to express this transfer by heat is

$$Q = nC_p\Delta T \qquad [12.6]$$

where $C_p = \tfrac{5}{2}R$. For ideal gases, the molar heat capacity at constant pressure, C_p, is the sum of the molar heat capacity at constant volume, C_v, and the gas constant R:

$$C_p = C_v + R \qquad [12.7]$$

This can be seen in the fourth column of Table 12.1, where $C_p - C_v$ is calculated for a number of different gases. The difference works out to be approximately R in virtually every case.

EXAMPLE 12.4 Expanding Gas

Goal Use molar specific heats and the first law in a constant pressure process.

Problem Suppose a system of monatomic ideal gas at 2.00×10^5 Pa and an initial temperature of 293 K slowly expands at constant pressure from a volume of 1.00 L to 2.50 L. **(a)** Find the work done on the environment. **(b)** Find the change in internal energy of the gas. **(c)** Use the first law of thermodynamics to obtain the thermal energy absorbed by the gas during the process. **(d)** Use the molar heat capacity at constant pressure to find the thermal energy absorbed. **(e)** How would the answers change for a diatomic ideal gas?

Strategy This problem mainly involves substituting into the appropriate equations. Substitute into the equation for work at constant pressure to obtain the answer to part (a). In part (b), use the ideal gas law twice, to find the temperature when $V = 2.00$ L and to find the number of moles of the gas. These quantities can then be used to obtain the change in internal energy, ΔU. Part (c) can then be solved by substituting into the first law, yielding Q, the answer checked in part (d) with Equation 12.6. Repeat these steps for part (e) after increasing the molar specific heats by R because of the extra two degrees of freedom associated with a diatomic gas.

Solution

(a) Find the work done on the environment.

Apply the definition of work at constant pressure:

$$W_{env} = P\Delta V = (2.00 \times 10^5\,\text{Pa})(2.50 \times 10^{-3}\,\text{m}^3 - 1.00 \times 10^{-3}\,\text{m}^3)$$

$$W_{env} = \boxed{3.00 \times 10^2\,\text{J}}$$

(b) Find the change in the internal energy of the gas.

First, obtain the final temperature, using the ideal gas law, noting that $P_i = P_f$:

$$\frac{P_f V_f}{P_i V_i} = \frac{T_f}{T_i} \;\rightarrow\; T_f = T_i\frac{V_f}{V_i} = (293\,\text{K})\frac{(2.50 \times 10^{-3}\,\text{m}^3)}{(1.00 \times 10^{-3}\,\text{m}^3)}$$

$$T_f = 733\,\text{K}$$

Again using the ideal gas law, obtain the number of moles of gas:

$$n = \frac{P_i V_i}{R T_i} = \frac{(2.00 \times 10^5\,\text{Pa})(1.00 \times 10^{-3}\,\text{m}^3)}{(8.31\,\text{J/K}\cdot\text{mol})(293\,\text{K})}$$

$$= 8.21 \times 10^{-2}\,\text{mol}$$

Use these results and given quantities to calculate the change in internal energy, ΔU:

$$\Delta U = nC_v\Delta T = \tfrac{3}{2}nR\Delta T$$

$$= \tfrac{3}{2}(8.21 \times 10^{-2}\,\text{mol})(8.31\,\text{J/K}\cdot\text{mol})(733\,\text{K} - 293\,\text{K})$$

$$\Delta U = \boxed{4.50 \times 10^2\,\text{J}}$$

(c) Use the first law to obtain the energy transferred by heat.

Solve the first law for Q, and substitute ΔU and $W = -W_{env} = -3.00 \times 10^2\,\text{J}$:

$$\Delta U = Q + W \;\rightarrow\; Q = \Delta U - W$$

$$Q = 4.50 \times 10^2\,\text{J} - (-3.00 \times 10^2\,\text{J}) = \boxed{7.50 \times 10^2\,\text{J}}$$

(d) Use the molar heat capacity at constant pressure to obtain Q:

Substitute values into Equation 12.6:

$$Q = nC_p\Delta T = \tfrac{5}{2}nR\Delta T$$

$$= \tfrac{5}{2}(8.21 \times 10^{-2}\,\text{mol})(8.31\,\text{J/K}\cdot\text{mol})(733\,\text{K} - 293\,\text{K})$$

$$= \boxed{7.50 \times 10^2\,\text{J}}$$

(e) How would the answers change for a diatomic gas?

Obtain the new change in internal energy, ΔU, noting that $C_v = \tfrac{5}{2}R$ for a diatomic gas:

$$\Delta U = nC_v\Delta T = \left(\tfrac{3}{2} + 1\right)nR\Delta T$$

$$= \tfrac{5}{2}(8.21 \times 10^{-2}\,\text{mol})(8.31\,\text{J/K}\cdot\text{mol})(733\,\text{K} - 293\,\text{K})$$

$$\Delta U = \boxed{7.50 \times 10^2\,\text{J}}$$

Obtain the new energy transferred by heat, Q:

$$Q = nC_p\Delta T = \left(\tfrac{5}{2} + 1\right)nR\Delta T$$

$$= \tfrac{7}{2}(8.21 \times 10^{-2}\,\text{mol})(8.31\,\text{J/K}\cdot\text{mol})(733\,\text{K} - 293\,\text{K})$$

$$Q = \boxed{1.05 \times 10^3\,\text{J}}$$

Remarks   Notice that problems involving diatomic gases are no harder than those with monatomic gases. It's just a matter of adjusting the molar specific heats.

Exercise 12.4

Suppose an ideal monatomic gas at an initial temperature of 475 K is compressed from 3.00 L to 2.00 L while its pressure remains constant at 1.00×10^5 Pa. Find (a) the work done on the gas, (b) the change in internal energy, and (c) the energy lost by heat, Q.

Answers   (a) $1.00 \times 10^2\,\text{J}$ (b) $-150\,\text{J}$ (c) $-250\,\text{J}$

Adiabatic Processes

In an adiabatic process, no energy enters or leaves the system by heat. Such a system is insulated—thermally isolated from its environment. In general, however, the system isn't mechanically isolated, so it can still do work. A sufficiently rapid process may be considered approximately adiabatic because there isn't time for any significant transfer of energy by heat.

For adiabatic processes $Q = 0$, so the first law becomes

$$\Delta U = W$$

The work done during an adiabatic process can be calculated by finding the change in the internal energy. Alternately, the work can be computed from a PV diagram. For an ideal gas undergoing an adiabatic process, it can be shown that

$$PV^\gamma = \text{constant} \qquad\qquad \text{[12.8a]}$$

where

$$\gamma = \frac{C_p}{C_v} \qquad\qquad \text{[12.8b]}$$

is called the *adiabatic index* of the gas. Values of the adiabatic index for several different gases are given in Table 12.1. After computing the constant on the right-hand side of Equation 12.8a and solving for the pressure P, the area under the curve in the PV diagram can be found by counting boxes, yielding the work.

If a hot gas is allowed to expand so quickly that there is no time for energy to enter or leave the system by heat, the work done on the gas is negative and the internal energy decreases. This decrease occurs because kinetic energy is transferred from the gas molecules to the moving piston. Such an adiabatic expansion is of practical importance and is nearly realized in an internal combustion engine when a gasoline–air mixture is ignited and expands rapidly against a piston. The following example illustrates this process.

EXAMPLE 12.5   Work and an Engine Cylinder

Goal   Use the first law to find the work done in an adiabatic expansion.

Problem   In a car engine operating at 1.80×10^3 rev/min, the expansion of hot, high-pressure gas against a piston occurs in about 10 ms. Because energy transfer by heat typically takes a time on the order of minutes or hours, it's safe to assume that little energy leaves the hot gas during the expansion. Estimate the work done by the gas on the piston during this adiabatic expansion by assuming the engine cylinder contains 0.100 moles of an ideal monatomic gas which goes from 1.200×10^3 K to 4.00×10^2 K, typical engine temperatures, during the expansion.

Strategy   Find the change in internal energy using the given temperatures. For an adiabatic process, this equals the work done on the gas, which is the negative of the work done on the environment—in this case, the piston.

Solution

Start with the first law, taking $Q = 0$.

$$W = \Delta U - Q = \Delta U - 0 = \Delta U$$

Find ΔU from the expression for the internal energy of an ideal monatomic gas.

$$\Delta U = U_f - U_i = \tfrac{3}{2}nR(T_f - T_i)$$
$$= \tfrac{3}{2}(0.100 \text{ mol})(8.31 \text{ J/mol} \cdot \text{K})(4.00 \times 10^2 \text{ K}$$
$$- 1.20 \times 10^3 \text{ K})$$

$$\Delta U = -9.97 \times 10^2 \text{ J}$$

The change in internal energy equals the work done on the system, which is the negative of the work done on the piston.

$$W_{\text{piston}} = -W = -\Delta U = \boxed{9.97 \times 10^2 \text{ J}}$$

Remarks   The work done on the piston comes at the expense of the internal energy of the gas. In an ideal adiabatic expansion, the loss of internal energy is completely converted into useful work. In a real engine, there are always losses.

Exercise 12.5

A monatomic ideal gas with volume 0.200 L is rapidly compressed, so the process can be considered adiabatic. If the gas is initially at 1.01×10^5 Pa and 3.00×10^2 K and the final temperature is 477 K, find the work done by the gas on the environment, W_{env}.

Answer   17.9 J

EXAMPLE 12.6   An Adiabatic Expansion

Goal   Use the adiabatic pressure vs. volume relation to find a change in pressure and the work done on a gas.

Problem   A monatomic ideal gas at a pressure 1.01×10^5 Pa expands adiabatically from an initial volume of 1.50 m^3, doubling its volume. **(a)** Find the new pressure. **(b)** Sketch the *PV* diagram and estimate the work done on the gas.

Strategy   There isn't enough information to solve this problem with the ideal gas law. Instead, use Equation 12.8 and the given information to find the adiabatic index and the constant *C* for the process. For part (b), sketch the *PV* diagram and count boxes to estimate the area under the graph, which gives the work.

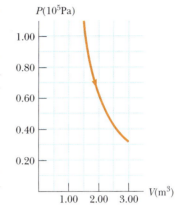

Figure 12.4   (Example 12.6) The *PV* diagram of an adiabatic expansion: the graph of $P = CV^{-\gamma}$, where *C* is a constant and $\gamma = C_p/C_v$.

Solution
(a) Find the new pressure.

First, calculate the adiabatic index:

$$\gamma = \frac{C_p}{C_v} = \frac{\frac{5}{2}R}{\frac{3}{2}R} = \frac{5}{3}$$

Use Equation 12.8a to find the constant *C*:

$$C = P_1 V_1^\gamma = (1.01 \times 10^5 \text{ Pa})(1.50 \text{ m}^3)^{5/3}$$
$$= 1.99 \times 10^5 \text{ Pa} \cdot \text{m}^5$$

The constant *C* is fixed for the entire process and can be used to find P_2:

$$C = P_2 V_2^\gamma = P_2(3.00 \text{ m}^3)^{5/3}$$

$$1.99 \times 10^5 \text{ Pa} \cdot \text{m}^5 = P_2 (6.24 \text{ m}^5)$$

$$P_2 = \boxed{3.19 \times 10^4 \text{ Pa}}$$

(b) Estimate the work done on the gas from a *PV* diagram.

Count the boxes between $V_1 = 1.50$ m^3 and $V_2 = 3.00$ m^3 in the graph of $P = (1.99 \times 10^5 \text{ Pa} \cdot \text{m}^5) V^{-5/3}$ in the *PV* diagram shown in Figure 12.4:

number of boxes ≈ 17

Each box has 'area' 5.00×10^3 J.

$$W \approx -17 \cdot 5.00 \times 10^3 \text{ J} = \boxed{-8.5 \times 10^4 \text{ J}}$$

Remarks   The exact answer, obtained with calculus, is -8.43×10^4 J, so our result is a very good estimate. The answer is negative because the gas is expanding, doing positive work on the environment, thereby reducing its own internal energy.

Exercise 12.6

Repeat the preceding calculations for an ideal diatomic gas expanding adiabatically from an initial volume of 0.500 m^3 to a final volume of 1.25 m^3, starting at a pressure of $P_1 = 1.01 \times 10^5$ Pa. (You must sketch the curve to find the work.)

Answers $P_2 = 2.80 \times 10^4$ Pa, $W \approx -4 \times 10^4$ J

Isovolumetric Processes

An **isovolumetric process**, sometimes called an *isochoric* process (which is harder to remember), proceeds at constant volume, corresponding to vertical lines in a *PV* diagram. If the volume doesn't change, no work is done on or by the system, so $W = 0$, and the first law of thermodynamics reads

$$\Delta U = Q \quad \text{(isovolumetric process)}$$

This result tells us that **in an isovolumetric process, the change in internal energy of a system equals the energy transferred to the system by heat**. From Equation 12.5, the energy transferred by heat in constant volume processes is given by

$$Q = nC_v\Delta T \qquad \text{[12.9]}$$

EXAMPLE 12.7   An Isovolumetric Process

Goal   Apply the first law to a constant-volume process.

Problem   How much thermal energy must be added to 5.00 moles of monatomic ideal gas at 3.00×10^2 K and with a constant volume of 1.50 L in order to raise the temperature of the gas by 3.80×10^2 K?

Strategy   The energy transferred by heat is equal to the change in the internal energy of the gas, which can be calculated by substitution into Equation 12.9.

Solution

Apply Equation 12.9, using the fact that $C_v = 3R/2$ for an ideal monatomic gas:

$$Q = \Delta U = nC_v\Delta T = \tfrac{3}{2}nR\Delta T$$
$$= \tfrac{3}{2}(5.00 \text{ mol})(8.31 \text{ J/K} \cdot \text{mol})(80.0° \text{ K})$$

$$Q = \boxed{4.99 \times 10^3 \text{ J}}$$

Remark   Constant volume processes are the simplest to handle, and include such processes as heating a solid or liquid, in which the work of expansion is negligible.

Exercise 12.7

Find the change in temperature of 22.0 mol of a monatomic ideal gas if it absorbs 9 750 J at constant volume.

Answer   35.6 K

Isothermal Processes

During an isothermal process, the temperature of a system doesn't change. In an ideal gas the internal energy U depends only on the temperature, so it follows that $\Delta U = 0$ because $\Delta T = 0$. In this case, the first law of thermodynamics gives

$$W = -Q \quad \text{(isothermal process)}$$

We see that if the system is an ideal gas undergoing an isothermal process, the work done on the system is equal to the negative of the thermal energy transferred to the system. Such a process can be visualized in Figure 12.5. A cylinder filled with gas is in contact with a large energy reservoir that can exchange energy with the gas without changing its temperature. For a constant temperature ideal gas,

$$P = \frac{nRT}{V}$$

where the numerator on the right-hand side is constant. The *PV* diagram of a typical isothermal process is graphed in Figure 12.6, contrasted with an adiabatic process. When the process is adiabatic, the pressure falls off more rapidly.

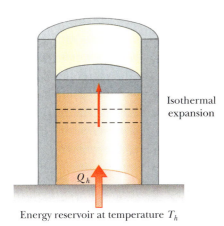

Figure 12.5 The gas in the cylinder expands iso-thermally while in contact with a reservoir at temperature T_h.

Isothermal expansion

Q_h

Energy reservoir at temperature T_h

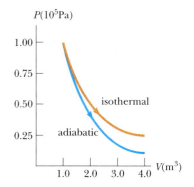

Figure 12.6 The PV diagram of an isothermal expansion, graph of $P = CV^{-1}$, where C is a constant. Contrasted with an adiabatic expansion, $P = C_A V^{-\gamma}$. C_A is a constant equal in magnitude to C in this case, but carrying different units.

Using methods of calculus, it can be shown that the work done on the environment during an isothermal process is given by

$$W_{env} = nRT \ln\left(\frac{V_f}{V_i}\right)$$ [12.10]

The symbol "ln" in Equation 12.10 is an abbreviation for the natural logarithm, discussed in Appendix A. The work W done on the gas is just the negative of W_{env}.

EXAMPLE 12.8   An Isothermally Expanding Balloon

Goal Find the work done during an isothermal expansion.

Problem A balloon contains 5.00 moles of a monatomic ideal gas. As energy is added to the system by heat (say, by absorption from the Sun), the volume increases by 25% at a constant temperature of 27.0°C. Find the work W_{env} done by the gas in expanding the balloon, the thermal energy Q transferred to the gas, and the work W done on the gas.

Strategy Be sure to convert temperatures to kelvins. Use the equation for isothermal work to find the work done on the balloon, which is the work done on the environment. The latter is equal to the thermal energy Q transferred to the gas, and the negative of this quantity is the work done on the gas.

Solution
Substitute into Equation 12.10, finding the work done during the isothermal expansion. Note that $T = 27.0°C = 3.00 \times 10^2$ K.

$$W_{env} = nRT \ln\left(\frac{V_f}{V_i}\right)$$
$$= (5.00 \text{ mol})(8.31 \text{ J/K·mol})(3.00 \times 10^2 \text{ K})$$
$$\times \ln\left(\frac{1.25 V_0}{V_0}\right)$$
$$W_{env} = \boxed{2.78 \times 10^3 \text{ J}}$$
$$Q = W_{env} = \boxed{2.78 \times 10^3 \text{ J}}$$

The negative of this amount is the work done on the gas:

$$W = -W_{env} = \boxed{-2.78 \times 10^3 \text{ J}}$$

Remarks Notice the relationship between the work done on the gas, the work done on the environment, and the energy transferred. These relationships are true of all isothermal processes.

Exercise 12.8
Suppose that subsequent to this heating, 1.50×10^4 J of thermal energy is removed from the gas isothermally. Find the final volume in terms of the initial volume of the example, V_0. (*Hint:* Follow the same steps as in the example, but in reverse. Also note that the initial volume in this exercise is $1.25 V_0$.)

Answer $0.375 V_0$

General Case

When a process follows none of the four given models, it's still possible to use the first law to get information about it. The work can be computed from the area under the curve of the *PV* diagram, and if the temperatures at the endpoints can be found, ΔU follows from Equation 12.5, as illustrated in the following example.

EXAMPLE 12.9   A General Process

Goal   Find thermodynamic quantities for a process that doesn't fall into any of the four previously discussed categories.

Problem   A quantity of 4.00 moles of a monatomic ideal gas expands from an initial volume of 0.100 m³ to a final volume of 0.300 m³ and pressure of 2.5×10^5 Pa (Fig. 12.7a). Compute **(a)** the work done on the gas, **(b)** the change in internal energy of the gas, and **(c)** the thermal energy transferred to the gas.

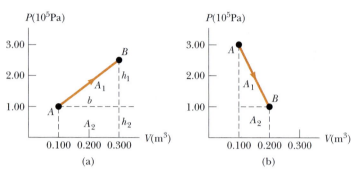

Figure 12.7   (a) (Example 12.9) (b) (Exercise 12.9)

Strategy   The work done on the gas is just equal to the negative of the area under the curve in the *PV* diagram. Use the ideal gas law to get the temperature change and, subsequently, the change in internal energy. Finally, the first law gives the thermal energy transferred by heat.

Solution

(a) Find the work done on the gas by computing the area under the curve in Figure 12.7a.

Find A_1, the area of the triangle:

$$A_1 = \tfrac{1}{2} b h_1 = \tfrac{1}{2}(0.200 \text{ m}^3)(1.50 \times 10^5 \text{ Pa}) = 1.50 \times 10^4 \text{ J}$$

Find A_2, the area of the rectangle:

$$A_2 = b h_2 = (0.200 \text{ m}^3)(1.00 \times 10^5 \text{ Pa}) = 2.00 \times 10^4 \text{ J}$$

Sum the two areas (the gas is expanding, so the work done on the gas is negative and a minus sign must be supplied):

$$W = -(A_1 + A_2) = \boxed{-3.50 \times 10^4 \text{ J}}$$

(b) Find the change in the internal energy during the process.

Compute the temperature at points A and B with the ideal gas law:

$$T_A = \frac{P_A V_A}{nR} = \frac{(1.00 \times 10^5 \text{ Pa})(0.100 \text{ m}^3)}{(4.00 \text{ mol})(8.31 \text{ J/K·mol})} = 301 \text{ K}$$

$$T_B = \frac{P_B V_B}{nR} = \frac{(2.50 \times 10^5 \text{ Pa})(0.300 \text{ m}^3)}{(4.00 \text{ mol})(8.31 \text{ J/K·mol})} = 2.26 \times 10^3 \text{ K}$$

Compute the change in internal energy:

$$\Delta U = \tfrac{3}{2} nR \Delta T$$
$$= \tfrac{3}{2}(4.00 \text{ mol})(8.31 \text{ J/K·mol})(2.26 \times 10^3 \text{ K} - 301 \text{ K})$$
$$\Delta U = \boxed{9.77 \times 10^4 \text{ J}}$$

(c) Compute Q with the first law:

$$Q = \Delta U - W = 9.77 \times 10^4 \text{ J} - (-3.50 \times 10^4 \text{ J})$$
$$= \boxed{1.33 \times 10^5 \text{ J}}$$

Remarks   As long as it's possible to compute the work, cycles involving these more exotic processes can be completely analyzed. Usually, however, it's necessary to use calculus.

Exercise 12.9

Figure 12.7b represents a process involving 3.00 moles of a monatomic ideal gas expanding from 0.100 m³ to 0.200 m³. Find the work done on the system, the change in the internal energy of the system, and the thermal energy transferred in the process.

Answers $W = -2.00 \times 10^4 \text{ J}$, $\Delta U = -1.50 \times 10^4 \text{ J}$, $Q = 5.00 \times 10^3 \text{ J}$

TABLE 12.2

**The First Law and Thermodynamic Processes
(Ideal Gases)**

Process	ΔU	Q	W
Isobaric	$nC_v\Delta T$	$nC_p\Delta T$	$-P\Delta V$
Adiabatic	$nC_v\Delta T$	0	ΔU
Isovolumetric	$nC_v\Delta T$	ΔU	0
Isothermal	0	$-W$	$-nRT\ln\left(\dfrac{V_f}{V_i}\right)$
General	$nC_v\Delta T$	$\Delta U - W$	$(PV\,\text{Area})$

Given all the different processes and formulae, it's easy to become confused when approaching one of these ideal gas problems, though most of the time only substitution into the correct formula is required. The essential facts and formulas are compiled in Table 12.2, both for easy reference and also to display the similarities and differences between the processes.

Quick Quiz 12.2

Identify the paths A, B, C, and D in Figure 12.8 as isobaric, isothermal, isovolumetric, or adiabatic. For path B, $Q = 0$.

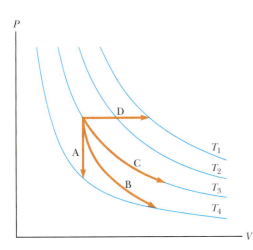

Figure 12.8   (Quick Quiz 12.2) Identify the nature of paths A, B, C, and D.

12.3   HEAT ENGINES AND THE SECOND LAW OF THERMODYNAMICS

A **heat engine** takes in energy by heat and partially converts it to other forms, such as electrical and mechanical energy. In a typical process for producing electricity in a power plant, for instance, coal or some other fuel is burned, and the resulting internal energy is used to convert water to steam. The steam is then directed at the blades of a turbine, setting it rotating. Finally, the mechanical energy associated with this rotation is used to drive an electric generator. In another heat engine—the internal combustion engine in an automobile—energy enters the engine as fuel is injected into the cylinder and combusted, and a fraction of this energy is converted to mechanical energy.

In general, a heat engine carries some working substance through a **cyclic process**[1] during which (1) energy is transferred by heat from a source at a high

◀ Cyclic process

---

[1]Strictly speaking, the internal combustion engine is not a heat engine according to the description of the cyclic process, because the air–fuel mixture undergoes only one cycle and is then expelled through the exhaust system.

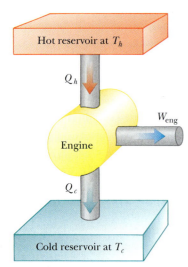

ACTIVE FIGURE 12.9
A schematic representation of a heat engine. The engine receives energy Q_h from the hot reservoir, expels energy Q_c to the cold reservoir, and does work W.

Physics⊗Now™
Log into PhysicsNow at **www.cp7e.com** and go to Active Figure 12.9 to select the efficiency of the engine and observe the transfer of energy.

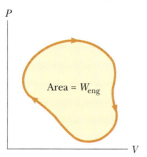

Figure 12.10 The PV diagram for an arbitrary cyclic process. The area enclosed by the curve equals the net work done.

temperature, (2) work is done by the engine, and (3) energy is expelled by the engine by heat to a source at lower temperature. As an example, consider the operation of a steam engine in which the working substance is water. The water in the engine is carried through a cycle in which it first evaporates into steam in a boiler and then expands against a piston. After the steam is condensed with cooling water, it returns to the boiler, and the process is repeated.

It's useful to draw a heat engine schematically, as in Active Figure 12.9. The engine absorbs energy Q_h from the hot reservoir, does work W_{eng}, then gives up energy Q_c to the cold reservoir. (Note that *negative* work is done *on* the engine, so that $W = -W_{eng}$.) Because the working substance goes through a cycle, always returning to its initial thermodynamic state, its initial and final internal energies are equal, so $\Delta U = 0$. From the first law of thermodynamics, therefore,

$$\Delta U = 0 = Q + W \quad \rightarrow \quad Q_{net} = -W = W_{eng}$$

The last equation shows that **the work W_{eng} done by a heat engine equals the net energy absorbed by the engine**. As we can see from Active Figure 12.9, $Q_{net} = |Q_h| - |Q_c|$. Therefore,

$$W_{eng} = |Q_h| - |Q_c| \qquad \text{[12.11]}$$

Ordinarily, a transfer of thermal energy Q can be either positive or negative, so the use of absolute value signs makes the signs of Q_h and Q_c explicit.

If the working substance is a gas, then **the work done by the engine for a cyclic process is the area enclosed by the curve representing the process on a PV diagram**. This area is shown for an arbitrary cyclic process in Figure 12.10.

The **thermal efficiency** e of a heat engine is defined as the work done by the engine, W_{eng}, divided by the energy absorbed during one cycle:

$$e \equiv \frac{W_{eng}}{|Q_h|} = \frac{|Q_h| - |Q_c|}{|Q_h|} = 1 - \frac{|Q_c|}{|Q_h|} \qquad \text{[12.12]}$$

We can think of thermal efficiency as the ratio of the benefit received (work) to the cost incurred (energy transfer at the higher temperature). Equation 12.12 shows that a heat engine has 100% efficiency ($e = 1$) only if $Q_c = 0$—meaning no energy is expelled to the cold reservoir. In other words, a heat engine with perfect efficiency would have to expel all the input energy by doing mechanical work. This isn't possible, as will be seen in Section 12.4.

EXAMPLE 12.10 The Efficiency of an Engine

Goal Apply the efficiency formula to a heat engine.

Problem During one cycle, an engine extracts 2.00×10^3 J of energy from a hot reservoir and transfers 1.50×10^3 J to a cold reservoir. **(a)** Find the thermal efficiency of the engine. **(b)** How much work does this engine do in one cycle? **(c)** How much power does the engine generate if it goes through four cycles in 2.50 s?

Strategy Apply Equation 12.12 to obtain the thermal efficiency, then use the first law, adapted to engines (Equation 12.11), to find the work done in one cycle. To obtain the power generated, just divide the work done in four cycles by the time it takes to run those cycles.

Solution
(a) Find the engine's thermal efficiency.

Substitute Q_c and Q_h into Equation 12.12:

$$e = 1 - \frac{|Q_c|}{|Q_h|} = 1 - \frac{1.50 \times 10^3 \text{ J}}{2.00 \times 10^3 \text{ J}} = \boxed{0.250, \text{ or } 25.0\%}$$

(b) How much work does this engine do in one cycle?

Apply the first law in the form of Equation 12.11 to find the work done by the engine:

$$W_{eng} = |Q_h| - |Q_c| = 2.00 \times 10^3 \, J - 1.50 \times 10^3 \, J$$
$$= 5.00 \times 10^2 \, J$$

(c) Find the power output of the engine.

Multiply the answer of part (b) by four and divide by time:

$$\mathcal{P} = \frac{W}{\Delta t} = \frac{4.00 \times (5.00 \times 10^2 \, J)}{2.50 \, s} = 8.00 \times 10^2 \, W$$

Remark   Problems like this usually reduce to solving two equations and two unknowns, as here, where the two equations are the efficiency equation and the first law and the unknowns are the efficiency and the work done by the engine.

Exercise 12.10
The energy absorbed by an engine is three times as great as the work it performs. **(a)** What is its thermal efficiency? **(b)** What fraction of the energy absorbed is expelled to the cold reservoir? **(c)** What is the power output of the engine if the energy input is 1 650 J each cycle and it goes through two cycles every 3 seconds?

Answer   (a) 1/3   (b) 2/3   (c) 367 W

EXAMPLE 12.11   Analyzing an Engine Cycle

Goal   Combine several concepts to analyze an engine cycle.

Problem   A heat engine contains an ideal monatomic gas confined to a cylinder by a movable piston. The gas starts at A, where $T = 3.00 \times 10^2$ K. (See Fig. 12.11a.) $B \rightarrow C$ is an isothermal expansion. **(a)** Find the number n of moles of gas and the temperature at B. **(b)** Find ΔU, Q, and W for the isovolumetric process $A \rightarrow B$. **(c)** Repeat for the isothermal process $B \rightarrow C$. **(d)** Repeat for the isobaric process $C \rightarrow A$. **(e)** Find the net change in the internal energy for the complete cycle. **(f)** Find the thermal energy Q_h transferred into the system, the thermal energy rejected, Q_c, the thermal efficiency, and net work on the environment performed by the engine.

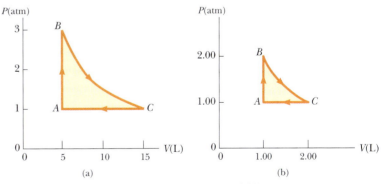

Figure 12.11   (a) (Example 12.11) (b) (Exercise 12.11)

Strategy   In part (a) n, T, and V can be found from the ideal gas law, which connects the equilibrium values of P, V, and T. Once the temperature T is known at the points A, B, and C, the change in internal energy, ΔU, can be computed from the formula in Table 12.2 for each process. Q and W can be similarly computed, or deduced from the first law, using the techniques applied in the single process examples.

Solution
(a) Find n and T_B with the ideal gas law:

$$n = \frac{P_A V_A}{R T_A} = \frac{(1.00 \, atm)(5.00 \, L)}{(0.0821 \, L \cdot atm/mol \cdot K)(3.00 \times 10^2 \, K)}$$
$$= 0.203 \, mol$$

$$T_B = \frac{P_B V_B}{nR} = \frac{(3.00 \, atm)(5.00 \, L)}{(0.203 \, mol)(0.0821 \, L \cdot atm/mol \cdot K)}$$
$$= 9.00 \times 10^2 \, K$$

(b) Find ΔU_{AB}, Q_{AB}, and W_{AB} for the constant volume process $A \rightarrow B$.

Compute ΔU_{AB}, noting that $C_v = \frac{3}{2}R = 12.5 \text{ J/mol} \cdot \text{K}$:

$$\Delta U_{AB} = nC_v\Delta T = (0.203 \text{ mol})(12.5 \text{ J/mol} \cdot \text{K})$$
$$\times (9.00 \times 10^2 \text{ K} - 3.00 \times 10^2 \text{ K})$$

$$\Delta U_{AB} = \boxed{1.52 \times 10^3 \text{ J}}$$

$\Delta V = 0$ for isovolumetric processes, so no work is done:

$$W_{AB} = \boxed{0}$$

We can find Q_{AB} from the first law:

$$Q_{AB} = \Delta U_{AB} = \boxed{1.52 \times 10^3 \text{ J}}$$

(c) Find ΔU_{BC}, Q_{BC}, and W_{BC} for the isothermal process $B \rightarrow C$.

This process is isothermal, so the temperature doesn't change, and the change in internal energy is zero:

$$\Delta U_{BC} = nC_v\Delta T = \boxed{0}$$

Compute the work done on the system, using the negative of Equation 12.10:

$$W_{BC} = -nRT \ln\left(\frac{V_C}{V_B}\right)$$
$$= -(0.203 \text{ mol})(8.31 \text{ J/mol} \cdot \text{K})(9.00 \times 10^2 \text{ K})$$
$$\times \ln\left(\frac{1.50 \times 10^{-2} \text{ m}^3}{5.00 \times 10^{-3} \text{ m}^3}\right)$$

$$W_{BC} = \boxed{-1.67 \times 10^3 \text{ Pa}}$$

Compute Q_{BC} from the first law:

$$0 = Q_{BC} + W_{BC} \quad \rightarrow \quad Q_{BC} = -W_{BC} = \boxed{1.67 \times 10^3 \text{ J}}$$

(d) Find ΔU_{CA}, Q_{CA}, and W_{CA} for the isobaric process $C \rightarrow A$.

Compute the work on the system, with pressure constant:

$$W_{CA} = -P\Delta V = -(1.01 \times 10^5 \text{ Pa})(5.00 \times 10^{-3} \text{ m}^3$$
$$- 1.50 \times 10^{-2} \text{ m}^3)$$

$$W_{CA} = \boxed{1.01 \times 10^3 \text{ J}}$$

Find the change in internal energy, ΔU_{CA}:

$$\Delta U_{CA} = \frac{3}{2}nRT = \frac{3}{2}(0.203 \text{ mol})(8.31 \text{ J/K} \cdot \text{mol})$$
$$\times (3.00 \times 10^2 \text{ K} - 9.00 \times 10^2 \text{ K})$$

$$\Delta U_{CA} = \boxed{-1.52 \times 10^3 \text{ J}}$$

Compute the thermal energy, Q_{CA}, from the first law:

$$Q_{CA} = \Delta U_{CA} - W_{CA} = -1.52 \times 10^3 \text{ J} - 1.01 \times 10^3 \text{ J}$$
$$= \boxed{-2.53 \times 10^3 \text{ J}}$$

(e) Find the net change in internal energy, ΔU_{net}, for the cycle:

$$\Delta U_{net} = \Delta U_{AB} + \Delta U_{BC} + \Delta U_{CA}$$
$$= 1.52 \times 10^3 \text{ J} + 0 - 1.52 \times 10^3 \text{ J} = \boxed{0}$$

(f) Find the energy input, Q_h; the energy rejected, Q_c; the thermal efficiency; and the net work performed by the engine:

Sum all the positive contributions to find Q_h:

$$Q_h = Q_{AB} + Q_{BC} = 1.52 \times 10^3 \text{ J} + 1.67 \times 10^3 \text{ J}$$
$$= \boxed{3.19 \times 10^3 \text{ J}}$$

Sum any negative contributions (in this case, there is only one):

$$Q_c = \boxed{-2.53 \times 10^3 \text{ J}}$$

Find the engine efficiency and the net work done by the engine:

$$e = 1 - \frac{|Q_c|}{|Q_h|} = 1 - \frac{2.53 \times 10^3\,\text{J}}{3.19 \times 10^3\,\text{J}} = 0.207$$

$$\begin{aligned} W_{\text{eng}} &= -(W_{AB} + W_{BC} + W_{CA}) \\ &= -(0 - 1.67 \times 10^3\,\text{J} + 1.01 \times 10^3\,\text{J}) \\ &= 6.60 \times 10^2\,\text{J} \end{aligned}$$

Remarks   Cyclic problems are rather lengthy; however, the individual steps are often short substitutions. Notice that the change in internal energy for the cycle is zero and that the net work done on the environment is identical to the net thermal energy transferred, both as they should be.

Exercise 12.11
4.05×10^{-2} mol of monatomic ideal gas goes through the process shown in Figure 12.11b. The temperature at point A is 3.00×10^2 K and is 6.00×10^2 K during the isothermal process $B \rightarrow C$. (a) Find Q, ΔU, and W for the constant volume process $A \rightarrow B$. (b) Do the same for the isothermal process $B \rightarrow C$. (c) Repeat, for the constant pressure process $C \rightarrow A$. (d) Find Q_h, Q_c, and the efficiency. (e) Find W_{eng}.

Answers   (a) $Q_{AB} = \Delta U_{AB} = 151$ J, $W_{AB} = 0$ (b) $\Delta U_{BC} = 0$, $Q_{BC} = -W_{BC} = 1.40 \times 10^2$ J   (c) $Q_{CA} = -252$ J, $\Delta U_{CA} = -151$ J, $W_{CA} = 101$ J   (d) $Q_h = 291$ J, $Q_c = -252$ J, $e = 0.134$ (e) $W_{\text{eng}} = 39.0$ J

Refrigerators and Heat Pumps

Heat engines can operate in reverse. In this case, energy is injected into the engine, modeled as work W in Active Figure 12.12, resulting in energy being extracted from the cold reservoir and transferred to the hot reservoir. The system now operates as a heat pump, a common example being a refrigerator (Fig. 12.13). Energy Q_c is extracted from the interior of the refrigerator and delivered as energy Q_h to the warmer air in the kitchen. The work is done in the compressor unit of the refrigerator, compressing a refrigerant such as freon, causing its temperature to increase.

A household air conditioner is another example of a heat pump. Some homes are both heated and cooled by heat pumps. In the winter, the heat pump extracts energy Q_c from the cool outside air and delivers energy Q_h to the warmer air inside. In summer, energy Q_c is removed from the cool inside air, while energy Q_h is ejected to the warm air outside.

For a refrigerator or an air conditioner—a heat pump operating in cooling mode—work W is what you pay for, in terms of electrical energy running the compressor, while Q_c is the desired benefit. The most efficient refrigerator or air conditioner is one that removes the greatest amount of energy from the cold reservoir in exchange for the least amount of work.

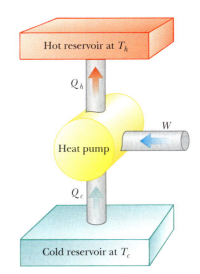

ACTIVE FIGURE 12.12
Schematic diagram of a heat pump, which takes in energy $Q_c > 0$ from a cold reservoir and expels energy $Q_h < 0$ to a hot reservoir. Work W is done *on* the heat pump. A refrigerator works the same way.

Physics⊗Now™
Log into PhysicsNow at
www.cp7e.com, and go to Active Figure 12.12 to select the coefficient of performance (COP) of the heat pump and observe the transfer of energy.

The coefficient of performance for a refrigerator or an air conditioner is the magnitude of the energy extracted from the cold reservoir, $|Q_c|$, divided by the work W performed by the device:

$$\text{COP(cooling mode)} = \frac{|Q_c|}{W} \qquad \text{[12.13]}$$

SI unit: dimensionless

The larger this ratio, the better the performance, since more energy is being removed for a given amount of work. A good refrigerator or air conditioner will have a COP of 5 or 6.

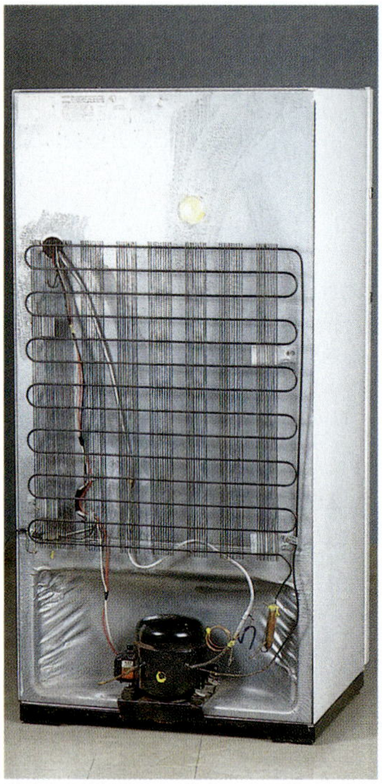

Figure 12.13 The coils on the back of a refrigerator transfer energy by heat to the air.

A heat pump operating in heating mode warms the inside of a house in winter by extracting energy from the colder outdoor air. This may seem paradoxical, but recall that this is equivalent to a refrigerator removing energy from its interior and ejecting it into the kitchen.

The coefficient of performance of a heat pump operating in the heating mode is the magnitude of the energy rejected to the hot reservoir, $|Q_h|$, divided by the work W done by the pump:

$$\text{COP(heating mode)} = \frac{|Q_h|}{W} \qquad \text{[12.14]}$$

SI unit: dimensionless

In effect, the COP of a heat pump in the heating mode is the ratio of what you gain (energy delivered to the interior of your home) to what you give (work input). Typical values for this COP are greater than one, because $|Q_h|$ is usually greater than W.

In a groundwater heat pump, energy is extracted in the winter from water deep in the ground rather than from the outside air, while energy is delivered to that water in the summer. This strategy increases the year-round efficiency of the heating and cooling unit, because the groundwater is at a higher temperature than the air in winter and at a cooler temperature than the air in summer.

EXAMPLE 12.12   Cooling the Leftovers

Goal   Apply the coefficient of performance of a refrigerator.

Problem   2.00 L of leftover soup at a temperature of 323 K is placed in a refrigerator. Assume the specific heat of the soup is the same as that of water and the density is 1.25×10^3 kg/m³. The refrigerator cools the soup to 283 K. **(a)** If the COP of the refrigerator is 5.00, find the energy needed, in the form of work, to cool the soup. **(b)** If the compressor has a power rating of 0.250 hp, for what minimum length of time must it operate to cool the soup to 283 K? (The minimum time assumes the soup cools at the same rate that the heat pump ejects thermal energy from the refrigerator.)

Strategy   The solution to this problem requires three steps. First, find the total mass m of the soup. Second, using $Q = mc\Delta T$, where $Q = Q_c$, find the energy transfer required to cool the soup. Third, substitute Q_c and the COP into Equation 12.13, solving for W. Divide the work by the power to get an estimate of the time required to cool the soup.

Solution
(a) Find the work needed to cool the soup.

Calculate the mass of the soup:

$$m = \rho V = (1.25 \times 10^3 \text{ kg/m}^3)(2.00 \times 10^{-3} \text{ m}^3) = 2.50 \text{ kg}$$

Find the energy transfer required to cool the soup:

$$Q_c = Q = mc\Delta T$$
$$= (2.50 \text{ kg})(4\,190 \text{ J/kg} \cdot \text{K})(283 \text{ K} - 323 \text{ K})$$
$$= -4.19 \times 10^5 \text{ J}$$

Substitute Q_c and the COP into Equation 12.13:

$$\text{COP} = \frac{|Q_c|}{W} = \frac{4.19 \times 10^5 \text{ J}}{W} = 5.00$$

$$W = \boxed{8.38 \times 10^4 \text{ J}}$$

(b) Find the time needed to cool the food.

Convert horsepower to watts:

$$\mathcal{P} = (0.250 \text{ hp})(746 \text{ W/1 hp}) = 187 \text{ W}$$

Divide the work by the power to find the elapsed time:

$$\Delta t = \frac{W}{\mathcal{P}} = \frac{8.38 \times 10^4 \, J}{187 \, W} = \boxed{448 \, s}$$

Remarks   This example illustrates how cooling different substances requires differing amounts of work, due to differences in specific heats. The problem doesn't take into account the insulating properties of the soup container and of the soup itself, which retard the cooling process.

Exercise 12.12

(a) How much work must a heat pump with a COP of 2.50 do in order to extract 1.00 MJ of thermal energy from the outdoors (the cold reservoir)? (b) If the unit operates at 0.500 hp, how long will the process take? (Be sure to use the correct COP!)

Answers   (a) $6.67 \times 10^5 \, J$ (b) $1.79 \times 10^3 \, s$

The Second Law of Thermodynamics

There are limits to the efficiency of heat engines. The ideal engine would convert all input energy into useful work, but it turns out that such an engine is impossible to construct. The Kelvin–Planck formulation of the **second law of thermodynamics** can be stated as follows:

> No heat engine operating in a cycle can absorb energy from a reservoir and use it entirely for the performance of an equal amount of work.

This form of the second law means that the efficiency $e = W_{eng}/|Q_h|$ of engines must always be less than one. Some energy Q_c must always be lost to the environment. In other words, it's theoretically impossible to construct a heat engine with an efficiency of 100%.

To summarize, the first law says **we can't get a greater amount of energy out of a cyclic process than we put in**, and the second law says **we can't break even**.

Reversible and Irreversible Processes

No engine can operate with 100% efficiency, but different designs yield different efficiencies, and it turns out one design in particular delivers the maximum possible efficiency. This design is the Carnot cycle, discussed in the next subsection. Understanding it requires the concepts of reversible and irreversible processes. In a **reversible** process, every state along the path is an equilibrium state, so the system can return to its initial conditions by going along the same path in the reverse direction. A process that doesn't satisfy this requirement is **irreversible**.

Most natural processes are known to be irreversible—the reversible process is an idealization. Although real processes are always irreversible, some are *almost* reversible. If a real process occurs so slowly that the system is virtually always in equilibrium, the process can be considered reversible. Imagine compressing a gas very slowly by dropping grains of sand onto a frictionless piston, as in Figure 12.14. The temperature can be kept constant by placing the gas in thermal contact with an energy reservoir. The pressure, volume, and temperature of the gas are well defined during this isothermal compression. Each added grain of sand represents a change to a new equilibrium state. The process can be reversed by slowly removing grains of sand from the piston.

The Carnot Engine

In 1824, in an effort to understand the efficiency of real engines, a French engineer named Sadi Carnot (1796–1832) described a theoretical engine now called a *Carnot engine* that is of great importance from both a practical and a theoretical

LORD KELVIN, British Physicist and Mathematician (1824–1907)

Born William Thomson in Belfast, Kelvin was the first to propose the use of an absolute scale of temperature. His study of Carnot's theory led to the idea that energy cannot pass spontaneously from a colder object to a hotter object; this principle is known as the second law of thermodynamics.

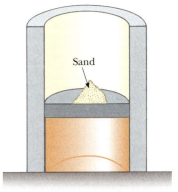

Sand

Energy reservoir

Figure 12.14   A gas in thermal contact with an energy reservoir is compressed slowly by grains of sand dropped onto a piston. The compression is isothermal and reversible.

J-L. Charmet/SPL/Photo Researchers, Inc.

SADI CARNOT, French Engineer (1796–1832)

Carnot is considered to be the founder of the science of thermodynamics. Some of his notes found after his death indicate that he was the first to recognize the relationship between work and heat.

viewpoint. He showed that a heat engine operating in an ideal, reversible cycle—now called a **Carnot cycle**—between two energy reservoirs is the most efficient engine possible. Such an engine establishes an upper limit on the efficiencies of all real engines. **Carnot's theorem** can be stated as follows:

No real engine operating between two energy reservoirs can be more efficient than a Carnot engine operating between the same two reservoirs.

In a Carnot cycle, an ideal gas is contained in a cylinder with a movable piston at one end. The temperature of the gas varies between T_c and T_h. The cylinder walls and the piston are thermally nonconducting. Active Figure 12.15 shows the four stages of the Carnot cycle, and Active Figure 12.16 is the PV diagram for the cycle. The cycle consists of two adiabatic and two isothermal processes, all reversible:

1. The process $A \rightarrow B$ is an isothermal expansion at temperature T_h in which the gas is placed in thermal contact with a hot reservoir (a large oven, for example) at temperature T_h (Active Fig. 12.15a). During the process, the gas absorbs energy Q_h from the reservoir and does work W_{AB} in raising the piston.
2. In the process $B \rightarrow C$, the base of the cylinder is replaced by a thermally non-conducting wall and the gas expands adiabatically, so no energy enters or

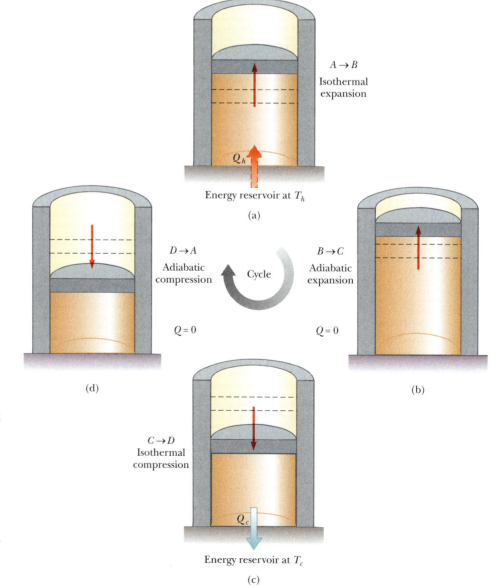

ACTIVE FIGURE 12.15
The Carnot cycle. In process $A \rightarrow B$, the gas expands isothermally while in contact with a reservoir at T_h. In process $B \rightarrow C$, the gas expands adiabatically ($Q = 0$). In process $C \rightarrow D$, the gas is compressed isothermally while in contact with a reservoir at $T_c < T_h$. In process $D \rightarrow A$, the gas is compressed adiabatically. The upward arrows on the piston indicate the removal of sand during the expansions, and the downward arrows indicate the addition of sand during the compressions.

Physics⊗Now™
Log into PhysicsNow at
www.cp7e.com, and go to Active Figure 12.15 to observe the motion of the piston in the Carnot cycle while you also observe the cycle on the PV diagram of Active Figure 12.16.

leaves the system by heat (Active Fig. 12.15b). During the process, the temperature falls from T_h to T_c and the gas does work W_{BC} in raising the piston.

3. In the process $C \rightarrow D$, the gas is placed in thermal contact with a cold reservoir at temperature T_c (Active Fig. 12.15c) and is compressed isothermally at temperature T_c. During this time, the gas expels energy Q_c to the reservoir and the work done on the gas is W_{CD}.

4. In the final process, $D \rightarrow A$, the base of the cylinder is again replaced by a thermally nonconducting wall (Active Fig. 12.15d) and the gas is compressed adiabatically. The temperature of the gas increases to T_h, and the work done on the gas is W_{DA}.

For a Carnot engine, the following relationship between the thermal energy transfers and the absolute temperatures can be derived:

$$\frac{|Q_c|}{|Q_h|} = \frac{T_c}{T_h} \qquad [12.15]$$

Substituting this expression into Equation 12.12, we find that the thermal efficiency of a Carnot engine is

$$e_C = 1 - \frac{T_c}{T_h} \qquad [12.16]$$

where T must be in kelvins. From this result, we see that **all Carnot engines operating reversibly between the same two temperatures have the same efficiency**.

Equation 12.16 can be applied to any working substance operating in a Carnot cycle between two energy reservoirs. According to that equation, the efficiency is zero if $T_c = T_h$. The efficiency increases as T_c is lowered and as T_h is increased. The efficiency can be one (100%), however, only if $T_c = 0$ K. According to the **third law of thermodynamics**, it's impossible to lower the temperature of a system to absolute zero in a finite number of steps, so such reservoirs are not available, and the maximum efficiency is always less than one. In most practical cases, the cold reservoir is near room temperature, about 300 K, so increasing the efficiency requires raising the temperature of the hot reservoir. **All real engines operate irreversibly, due to friction and the brevity of their cycles, and are therefore *less* efficient than the Carnot engine.**

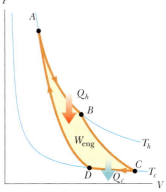

ACTIVE FIGURE 12.16
The PV diagram for the Carnot cycle. The net work done, W_{eng}, equals the net energy received by heat in one cycle, $|Q_h| - |Q_c|$.

Physics⊗Now™
Log into PhysicsNow at **www.cp7e.com**, and go to Active Figure 12.15 to observe the Carnot cycle on the PV diagram while you also observe the motion of the piston in Active Figure 12.15.

TIP 12.2   Don't Shop for a Carnot Engine

The Carnot engine is only an idealization. If a Carnot engine were developed in an effort to maximize efficiency, it would have zero power output, because, in order for all of the processes to be reversible, the engine would have to run infinitely slowly.

Quick Quiz 12.3

Three engines operate between reservoirs separated in temperature by 300 K. The reservoir temperatures are as follows:

Engine A: $T_h = 1\,000$ K, $T_c = 700$ K

Engine B: $T_h = 800$ K, $T_c = 500$ K

Engine C: $T_h = 600$ K, $T_c = 300$ K

Rank the engines in order of their theoretically possible efficiency, from highest to lowest. (a) A, B, C (b) B, C, A (c) C, B, A (d) C, A, B

EXAMPLE 12.13   The Steam Engine

Goal   Apply the equations of an ideal (Carnot) engine.

Problem   A steam engine has a boiler that operates at 5.00×10^2 K. The energy from the boiler changes water to steam, which drives the piston. The temperature of the exhaust is that of the outside air, 3.00×10^2 K. **(a)** What is the engine's efficiency if it's an ideal engine? **(b)** If the 3.50×10^3 J of energy is supplied from the boiler, find the work done by the engine on its environment.

Strategy   This problem requires substitution into Equations 12.15 and 12.16, both applicable to a Carnot engine. The first equation relates the ratio Q_c/Q_h to the ratio T_c/T_h, and the second gives the Carnot engine efficiency.

Solution
(a) Find the engine's efficiency, assuming it's ideal.

Substitute into Equation 12.16, the equation for the efficiency of a Carnot engine:

$$e_C = 1 - \frac{T_c}{T_h} = 1 - \frac{3.00 \times 10^2 \text{ K}}{5.00 \times 10^2 \text{ K}} = 0.400, \text{ or } 40\%$$

(b) Find the work done on the environment if 3.50×10^3 J is delivered to the engine during one cycle.

The ratio of energies equals the ratio of temperatures:

$$\frac{|Q_c|}{|Q_h|} = \frac{T_c}{T_h} \quad \rightarrow \quad |Q_c| = |Q_h|\frac{T_c}{T_h}$$

$$|Q_c| = (3.50 \times 10^3 \text{ J})\left(\frac{3.00 \times 10^2 \text{ K}}{5.00 \times 10^2 \text{ K}}\right) = 2.10 \times 10^3 \text{ J}$$

Use Equation 12.11 to find W:

$$W_{\text{eng}} = |Q_h| - |Q_c| = 3.50 \times 10^3 \text{ J} - 2.10 \times 10^3 \text{ J}$$
$$= 1.40 \times 10^3 \text{ J}$$

Remarks   This problem differs from the earlier examples on work and efficiency because we used the special Carnot relationships, Equations 12.15 and 12.16. Remember that these equations can only be used when the cycle is identified as ideal or a Carnot.

Exercise 12.13
The highest theoretical efficiency of a gasoline engine based on the Carnot cycle, is 0.300, or 30.0%. (a) If this engine expels its gases into the atmosphere, which has a temperature of 3.00×10^2 K, what is the temperature in the cylinder immediately after combustion? (b) If the heat engine absorbs 837 J of energy from the hot reservoir during each cycle, how much work can it perform in each cycle?

Answers   (a) 429 K   (b) 251 J

12.4   ENTROPY

Temperature and internal energy, associated with the zeroth and first laws of thermodynamics, respectively, are both state variables, meaning they can be used to describe the thermodynamic state of a system. A state variable called the **entropy** S is related to the second law of thermodynamics. We define entropy on a macroscopic scale as the German physicist Rudolf Clausius (1822–1888) first expressed it in 1865:

Entropy ▶

Let Q_r be the energy absorbed or expelled during a reversible, constant temperature process between two equilibrium states. Then the change in entropy during any constant temperature process connecting the two equilibrium states is defined as

$$\Delta S \equiv \frac{Q_r}{T} \qquad \qquad \text{[12.17]}$$

SI unit: joules/kelvin (J/K)

TIP 12.3   Entropy ≠ Energy
Don't confuse energy and entropy—though the names sound similar, they are different concepts.

A similar formula holds when the temperature isn't constant, but its derivation entails calculus and won't be considered here. Calculating the change in entropy, ΔS, during a transition between two equilibrium states requires finding a reversible path that connects the states. The entropy change calculated on that reversible path is taken to be ΔS for the *actual* path. This approach is necessary because quantities such as the temperature of a system can be defined only for systems in equilibrium, and a reversible path consists of a sequence of equilibrium states. The subscript r on the term Q_r emphasizes that the path chosen for the calculation must be reversible. The change in entropy ΔS, like changes in internal energy ΔU and changes in potential energy, depends only on the endpoints, and not on the path connecting them.

The concept of entropy gained wide acceptance in part because it provided another variable to describe the state of a system, along with pressure, volume, and temperature. Its significance was enhanced when it was found that **the entropy of the Universe increases in all natural processes**. This is yet another way of stating the second law of thermodynamics.

Although the entropy of the *Universe* increases in all natural processes, the entropy of a *system* can decrease. For example, if system A transfers energy Q to system B by heat, the entropy of system A decreases. This transfer, however, can only occur if the temperature of system B is less than the temperature of system A. Because temperature appears in the denominator in the definition of entropy, system B's increase in entropy will be greater than system A's decrease, so taken together, the entropy of the Universe increases.

For centuries, individuals have attempted to build perpetual motion machines that operate continuously without any input of energy or increase in entropy. The laws of thermodynamics preclude the invention of any such machines.

The concept of entropy is satisfying because it enables us to present the second law of thermodynamics in the form of a mathematical statement. In the next section, we find that entropy can also be interpreted in terms of probabilities, a relationship that has profound implications.

RUDOLF CLAUSIUS, German Physicist (1822–1888)

Born with the name Rudolf Gottlieb, he adopted the classical name of Clausius, which was a popular thing to do in his time. "I propose . . . to call S the entropy of a body, after the Greek word 'transformation.' I have designedly coined the word 'entropy' to be similar to energy, for these two quantities are so analogous in their physical significance, that an analogy of denominations seems to be helpful."

Quick Quiz 12.4

Which of the following is true for the entropy change of a system that undergoes a reversible, adiabatic process? (a) $\Delta S < 0$ (b) $\Delta S = 0$ (c) $\Delta S > 0$

EXAMPLE 12.14   Melting a Piece of Lead

Goal   Calculate the change in entropy due to a phase change.

Problem   (a) Find the change in entropy of 3.00×10^2 g of lead when it melts at 327°C. Lead has a latent heat of fusion of 2.45×10^4 J/kg. (b) Suppose the same amount of energy is used to melt part of a piece of silver, which is already at its melting point of 961°C. Find the change in the entropy of the silver.

Strategy   This problem can be solved by substitution into Equation 12.17. Be sure to use the Kelvin temperature scale.

Solution
(a) Find the entropy change of the lead.

Find the energy necessary to melt the lead:

$Q = mL_f = (0.300 \text{ kg})(2.45 \times 10^4 \text{ J/kg}) = 7.35 \times 10^3 \text{ J}$

Convert the temperature in degrees Celsius to kelvins:

$T = T_C + 273 = 327 + 273 = 6.00 \times 10^2 \text{ K}$

Substitute the quantities found into the entropy equation:

$\Delta S = \dfrac{Q}{T} = \dfrac{7.35 \times 10^3 \text{ J}}{6.00 \times 10^2 \text{ K}} = \boxed{12.3 \text{ J/K}}$

(b) Find the entropy change of the silver.

The added energy is the same as in part (a), by supposition. Substitute into the entropy equation, after first converting the melting point of silver to kelvins:

$T = T_C + 273 = 961 + 273 = 1.234 \times 10^2 \text{ K}$

$\Delta S = \dfrac{Q}{T} = \dfrac{7.35 \times 10^3 \text{ J}}{1.234 \times 10^3 \text{ K}} = \boxed{5.96 \text{ J/K}}$

Remarks   This example shows that adding a given amount of energy to a system increases its disorder, but adding the same amount of energy to another system at higher temperature results in a smaller increase in disorder. This is because the change in entropy is inversely proportional to the temperature.

Exercise 12.14
Find the change in entropy of a 2.00-kg block of gold at 1 063°C when it melts to become liquid gold at 1 063°C.

Answer   96.4 J/K

EXAMPLE 12.15   Ice, Steam, and the Entropy of the Universe

Goal   Calculate the change in entropy for a system and its environment.

Problem   A block of ice at 273 K is put in thermal contact with a container of steam at 373 K, converting 25.0 g of ice to water at 273 K while condensing some of the steam to water at 373 K. **(a)** Find the change in entropy of the ice. **(b)** Find the change in entropy of the steam. **(c)** Find the change in entropy of the Universe.

Strategy   First, calculate the energy transfer necessary to melt the ice. The amount of energy gained by the ice is lost by the steam. Compute the entropy change for each process, and sum to get the entropy change of the universe.

Solution
(a) Find the change in entropy of the ice.

Use the latent heat of fusion, L_f, to compute the thermal energy needed to melt 25.0 g of ice:

$$Q_{ice} = mL_f = (0.025 \text{ kg})(3.33 \times 10^5 \text{ J}) = 8.33 \times 10^3 \text{ J}$$

Calculate the change in entropy of the ice:

$$\Delta S_{ice} = \frac{Q_{ice}}{T_{ice}} = \frac{8.33 \times 10^3 \text{ J}}{273 \text{ K}} = \boxed{30.5 \text{ J/K}}$$

(b) Find the change in entropy of the steam.

By supposition, the thermal energy lost by the steam is equal to the thermal energy gained by the ice:

$$\Delta S_{steam} = \frac{Q_{steam}}{T_{steam}} = \frac{-8.33 \times 10^3 \text{ J}}{373 \text{ K}} = \boxed{-22.3 \text{ J/K}}$$

(c) Find the change in entropy of the Universe.

Sum the two changes in entropy:

$$\Delta S_{universe} = \Delta S_{ice} + \Delta S_{steam} = 30.5 \text{ J/k} - 22.3 \text{ J/K}$$
$$= \boxed{+8.2 \text{ J/K}}$$

Remark   Notice that the entropy of the Universe increases, as it must in all natural processes.

Exercise 12.15
A 4.00-kg block of ice at 273 K encased in a thin plastic shell of negligible mass melts in a large lake at 293 K. At the instant the ice has completely melted in the shell and is still at 273 K, calculate the change in entropy of (a) the ice (b) the lake (which essentially remains at 293 K), and (c) the universe.

Answers   (a) 4.88×10^3 J/K   (b) -4.55×10^3 J/K   (c) $+3.30 \times 10^2$ J/K

EXAMPLE 12.16   A Falling Boulder

Goal   Combine mechanical energy and entropy.

Problem   A chunk of rock of mass 1.00×10^3 kg at 293 K falls from a cliff of height 125 m into a large lake, also at 293 K. Find the change in entropy of the lake, assuming that all of the rock's kinetic energy upon entering the lake converts to thermal energy absorbed by the lake.

Strategy   Gravitational potential energy when the rock is at the top of the cliff converts to kinetic energy of the rock before it enters the lake, and then is transferred to the lake as thermal energy. The change in the lake's temperature is negligible (due to its mass). Divide the mechanical energy of the rock by the temperature of the lake to estimate the lake's change in entropy.

Solution

Calculate the gravitational potential energy associated with the rock at the top of the cliff:

$$PE = mgh = (1.00 \times 10^3 \text{ kg})(9.80 \text{ m/s}^2)(125 \text{ m})$$
$$= 1.23 \times 10^6 \text{ J}$$

This energy is transferred to the lake as thermal energy, resulting in an entropy increase of the lake:

$$\Delta S = \frac{Q}{T} = \frac{1.23 \times 10^6 \text{ J}}{293 \text{ K}} = 4.20 \times 10^3 \text{ J/K}$$

Remarks   This example shows how even simple mechanical processes can bring about increases in the Universe's entropy.

Exercise 12.16

Estimate the change in entropy of a tree trunk at 15.0°C when a bullet of mass 5.00 g traveling at 1.00×10^3 m/s embeds itself in it. (Assume the kinetic energy of the bullet transforms to thermal energy, all of which is absorbed by the tree.)

Answer   8.68 J/K

Entropy and Disorder

A large element of chance is inherent in natural processes. The spacing between trees in a natural forest, for example, is random; if you discovered a forest where all the trees were equally spaced, you would conclude that it had been planted. Likewise, leaves fall to the ground with random arrangements. It would be highly unlikely to find the leaves laid out in perfectly straight rows. We can express the results of such observations by saying that **a disorderly arrangement is much more probable than an orderly one if the laws of nature are allowed to act without interference**.

Entropy originally found its place in thermodynamics, but its importance grew tremendously as the field of statistical mechanics developed. This analytical approach employs an alternate interpretation of entropy. In statistical mechanics, the behavior of a substance is described by the statistical behavior of the atoms and molecules contained in it. One of the main conclusions of the statistical mechanical approach is that **isolated systems tend toward greater disorder, and entropy is a measure of that disorder**.

In light of this new view of entropy, Boltzmann found another method for calculating entropy through use of the relation

$$S = k_B \ln W \qquad \text{[12.18]}$$

where $k_B = 1.38 \times 10^{-23}$ J/K is Boltzmann's constant and W is a number proportional to the probability that the system has a particular configuration. The symbol "ln" again stands for natural logarithm, discussed in Appendix A.

Equation 12.18 could be applied to a bag of marbles. Imagine that you have 100 marbles—50 red and 50 green—stored in a bag. You are allowed to draw four marbles from the bag according to the following rules: Draw one marble, record its color, return it to the bag, and draw again. Continue this process until four marbles have been drawn. Note that because each marble is returned to the bag before the next one is drawn, the probability of drawing a red marble is always the same as the probability of drawing a green one.

The results of all possible drawing sequences are shown in Table 12.3. For example, the result RRGR means that a red marble was drawn first, a red one second, a green one third, and a red one fourth. The table indicates that there is only one possible way to draw four red marbles. There are four possible sequences that produce one green and three red marbles, six sequences that produce two green

TIP 12.4   Don't Confuse the *W*'s

The symbol W used here is a *probability*, not to be confused with the same symbol used for work.

TABLE 12.3

Possible Results of Drawing Four Marbles from a Bag

End Result	Possible Draws	Total Number of Same Results
All R	RRRR	1
1G, 3R	RRRG, RRGR, RGRR, GRRR	4
2G, 2R	RRGG, RGRG, GRRG, RGGR, GRGR, GGRR	6
3G, 1R	GGGR, GGRG, GRGG, RGGG	4
All G	GGGG	1

and two red, four sequences that produce three green and one red, and one sequence that produces all green. From Equation 12.18, we see that the state with the greatest disorder (two red and two green marbles) has the highest entropy, because it is most probable. In contrast, the most ordered states (all red marbles and all green marbles) are least likely to occur and are states of lowest entropy.

The outcome of the draw can range between these highly ordered (lowest-entropy) and highly disordered (highest-entropy) states. Entropy can be regarded as an index of how far a system has progressed from an ordered to a disordered state.

The second law of thermodynamics is really a statement of what is most probable rather than of what must be. Imagine placing an ice cube in contact with a hot piece of pizza. There is nothing in nature that absolutely forbids the transfer of energy by heat from the ice to the much warmer pizza. Statistically, it's possible for a slow-moving molecule in the ice to collide with a faster-moving molecule in the pizza so that the slow one transfers some of its energy to the faster one. However, when the great number of molecules present in the ice and pizza are considered, the odds are overwhelmingly in favor of the transfer of energy from the faster-moving molecules to the slower-moving molecules. Furthermore, this example demonstrates that a system naturally tends to move from a state of order to a state of disorder. The initial state, in which all the pizza molecules have high kinetic energy and all the ice molecules have lower kinetic energy, is much more ordered than the final state after energy transfer has taken place and the ice has melted.

Even more generally, the second law of thermodynamics defines the direction of time for all events as the direction in which the entropy of the universe increases. Although conservation of energy isn't violated if energy flows spontaneously from a cold object (the ice cube) to a hot object (the pizza slice), that event violates the second law because it represents a spontaneous increase in order of course, such an event also violates everyday experience. If the melting ice cube is filmed and the film speeded up, the difference between running the film in forward and reverse directions would be obvious to an audience. The same would be true of filming any event involving a large number of particles, such as a dish dropping to the floor and shattering.

As another example, suppose you were able to measure the velocities of all the air molecules in a room at some instant. It's very unlikely that you would find all molecules moving in the same direction with the same speed—that would be a highly ordered state, indeed. The most probable situation is a system of molecules moving haphazardly in all directions with a wide distribution of speeds—a highly disordered state. This physical situation can be compared to the drawing of marbles from a bag: If a container held 10^{23} molecules of a gas, the probability of finding all of the molecules moving in the same direction with the same speed at some instant would be similar to that of drawing a marble from the bag 10^{23} times and getting a red marble on every draw—clearly an unlikely set of events.

The tendency of nature to move toward a state of disorder affects the ability of a system to do work. Consider a ball thrown toward a wall. The ball has kinetic energy, and its state is an ordered one, which means all of the atoms and molecules of the ball move in unison at the same speed and in the same direction (apart from their random internal motions). When the ball hits the wall, however, part of the ball's kinetic energy is transformed into the random, disordered, internal motion of the molecules in the ball and the wall, and the temperatures of the ball and the wall both increase slightly. Before the collision, the ball was capable of doing work. It could drive a nail into the wall, for example. With the transformation of part of the ordered energy into disordered internal energy, this capability of doing work is reduced. The ball rebounds with less kinetic energy than it originally had, because the collision is inelastic.

Various forms of energy can be converted to internal energy, as in the collision between the ball and the wall, but the reverse transformation is never complete. In general, given two kinds of energy, A and B, if A can be completely converted to B and vice versa, we say that A and B are of the *same grade*. However, if A can be completely converted to B and the reverse is never complete, then A is of a *higher grade* of energy than B. In the case of a ball hitting a wall, the kinetic energy of the ball is of a higher grade than the internal energy contained in the ball and the wall after the collision. When high-grade energy is converted to internal energy, it can never be fully recovered as high-grade energy.

This conversion of high-grade energy to internal energy is referred to as the **degradation of energy**. The energy is said to be degraded because it takes on a form that is less useful for doing work. In other words, **in all real processes, the energy available for doing work decreases**.

Finally, note once again that the statement that entropy must increase in all natural processes is true only for isolated systems. There are instances in which the entropy of some system decreases, but with a corresponding net increase in entropy for some other system. When all systems are taken together to form the Universe, **the entropy of the Universe always increases**.

Ultimately, the entropy of the Universe should reach a maximum. When it does, the Universe will be in a state of uniform temperature and density. All physical, chemical, and biological processes will cease, because a state of perfect disorder implies no available energy for doing work. This gloomy state of affairs is sometimes referred to as the ultimate "heat death" of the Universe.

Quick Quiz 12.5

Suppose you are throwing two dice in a friendly game of craps. For any given throw, the two numbers that are face up can have a sum of 2, 3, 4, 5, 6, 7, 8, 9, 10, 11, or 12. Which outcome is most probable? Which is least probable?

12.5   HUMAN METABOLISM

Animals do work and give off energy by heat, and this lead us to believe the first law of thermodynamics can be applied to living organisms to describe them in a general way. The internal energy stored in humans goes into other forms needed for maintaining and repairing the major body organs and is transferred out of the body by work as a person walks or lifts a heavy object, and by heat when the body is warmer than its surroundings. Because the rates of change of internal energy, energy loss by heat, and energy loss by work vary widely with the intensity and duration of human activity, it's best to measure the time rates of change of ΔU, Q, and W. Rewriting the first law, these time rates of change are related by

$$\frac{\Delta U}{\Delta t} = \frac{Q}{\Delta t} + \frac{W}{\Delta t}$$ **[12.19]**

On average, energy Q flows *out* of the body, and work is done *by* the body on its surroundings, so both $Q/\Delta t$ and $W/\Delta t$ are negative. This means that $\Delta U/\Delta t$ would be negative and the internal energy and body temperature would decrease with time if a human were a closed system with no way of ingesting matter or replenishing internal energy stores. Because all animals are actually open systems, they acquire internal energy (chemical potential energy) by eating and breathing, so their internal energy and temperature are kept constant. Overall, the energy from the

(a)

(b)

A full house is a very good hand in the game of poker. Can you calculate the probability of being dealt a full house (a pair and three of a kind) from a standard deck of 52 cards? (a) A royal flush is a highly ordered poker hand with a low probability of occurrence. (b) A disordered and worthless poker hand. The probability of this *particular* hand occurring is the same as that of the royal flush. There are so many worthless hands, however, that the probability of being dealt a worthless hand is much higher than that of being dealt a royal flush.

oxidation of food ultimately supplies the work done by the body and energy lost from the body by heat, and this is the interpretation we give Equation 12.19. That is, $\Delta U/\Delta t$ is the rate at which internal energy is added to our bodies by food, and this term just balances the rate of energy loss by heat, $Q/\Delta t$, and by work, $W/\Delta t$. Finally, if we have a way of measuring $\Delta U/\Delta t$ and $W/\Delta t$ for a human, we can calculate $Q/\Delta t$ from Equation 12.19 and gain useful information on the efficiency of the body as a machine.

Measuring the Metabolic Rate $\Delta U/\Delta t$

The value of $W/\Delta t$, the work done by a person per unit time, can easily be determined by measuring the power output supplied by the person (in pedaling a bike, for example). **The metabolic rate $\Delta U/\Delta t$ is the rate at which chemical potential energy in food and oxygen are transformed into internal energy to just balance the body losses of internal energy by work and heat**. Although the mechanisms of food oxidation and energy release in the body are complicated, involving many intermediate reactions and enzymes (organic compounds that speed up the chemical reactions taking place at "low" body temperatures), an amazingly simple rule summarizes these processes: **The metabolic rate is directly proportional to the rate of oxygen consumption by volume**. It is found that for an average diet, the consumption of one liter of oxygen releases 4.8 kcal, or 20 kJ, of energy. We may write this important summary rule as

$$\frac{\Delta U}{\Delta t} = 4.8 \frac{\Delta V_{O_2}}{\Delta t} \qquad \text{[12.20]}$$

where the metabolic rate $\Delta U/\Delta t$ is measured in kcal/s and $\Delta V_{O_2}/\Delta t$, the volume rate of oxygen consumption, is in L/s. Measuring the rate of oxygen consumption during various activities ranging from sleep to intense bicycle racing effectively measures the variation of metabolic rate or the variation in the total power the body generates. A simultaneous measurement of the work per unit time done by a person along with the metabolic rate allows the efficiency of the body as a machine to be determined. Figure 12.17 shows a person monitored for oxygen consumption while riding a bike attached to a dynamometer, a device for measuring power output.

Figure 12.17 This bike rider is being monitored for oxygen consumption.

Metabolic Rate, Activity, and Weight Gain

Table 12.4 shows the measured rate of oxygen consumption in milliliters per minute per kilogram of body mass and the calculated metabolic rate for a 65-kg male engaged in various activities. A sleeping person uses about 80 W of power, the **basal metabolic rate**, just to maintain and run different body organs—heart, lungs, liver, kidneys, brain, and skeletal muscles. More intense activity increases the metabolic rate to a maximum of about 1 600 W for a superb racing cyclist, although such a high rate can only be maintained for periods of a few seconds.

TABLE 12.4

Oxygen Consumption and Metabolic Rates for Various Activities for a 65-kg Male[a]

Activity	O$_2$ Use Rate (mL/min · kg)	Metabolic Rate (kcal/h)	Metabolic Rate (W)
Sleeping	3.5	70	80
Light activity (dressing, walking slowly, desk work)	10	200	230
Moderate activity (walking briskly)	20	400	465
Heavy activity basketball, swimming a fast breaststroke)	30	600	700
Extreme activity (bicycle racing)	70	1400	1600

[a]*Source: A Companion to Medical Studies*, 2/e, R. Passmore, Philadelphia, F. A. Davis, 1968.

When we sit watching a riveting film, we give off about as much energy by heat as a bright (250-W) lightbulb.

Regardless of level of activity, the daily food intake should just balance the loss in internal energy if a person is not to gain weight. Further, exercise is a poor substitute for dieting as a method of losing weight. For example, the loss of one pound of body fat requires the muscles to expend 4 100 kcal of energy. If the goal is to lose a pound of fat in 35 days, a jogger could run an extra mile a day, because a 65-kg jogger uses about 120 kcal to jog one mile (35 days × 120 kcal/day = 4 200 kcal). An easier way to lose the pound of fat would be to diet and eat two fewer slices of bread per day for 35 days, because bread has a calorie content of 60 kcal/slice (35 days × 2 slices/day × 60 kcal/slice = 4 200 kcal).

EXAMPLE 12.17   Fighting Fat

Goal   Estimate human energy usage during a typical day.

Problem   In the course of 24 hours, a 65-kg person spends 8 h at a desk, 2 h puttering around the house, 1 h jogging 5 miles, 5 h in moderate activity, and 8 h sleeping. What is the change in her internal energy during this period?

Strategy   The time rate of energy usage—or power—multiplied by time gives the amount of energy used during a given activity. Use Table 12.4 to find the power $\mathscr{P}_i$ needed for each activity, multiply each by the time, and sum them all up.

Solution

$$\Delta U = -\Sigma \mathscr{P}_i \Delta t_i = -(\mathscr{P}_1 \Delta t_1 + \mathscr{P}_2 \Delta t_2 + \ldots + \mathscr{P}_n \Delta t_n)$$

$$= -(200 \text{ kcal/h})(10 \text{ h}) - (5 \text{ mi/h})(120 \text{ kcal/mi})(1 \text{ h}) - (400 \text{ kcal/h})(5 \text{ h}) - (70 \text{ kcal/h})(8 \text{ h})$$

$$\Delta U = \boxed{-5\ 000 \text{ kcal}}$$

Remarks   If this is a typical day in the woman's life, she will have to consume less than 5 000 kilocalories on a daily basis in order to lose weight. A complication lies in the fact that human metabolism tends to drop when food intake is reduced.

Exercise 12.17
If a 60.0-kg woman ingests 3 000 kcal a day and spends 6 h sleeping, 4 h walking briskly, 8 h sitting at a desk job, 1 h swimming a fast breaststroke, and 5 h watching action movies on TV, about how much weight will the woman gain or lose every day? (*Note:* Recall that using about 4 100 kcal of energy will burn off a pound of fat.)

Answer   She'll lose a little over half a pound of fat a day.

Physical Fitness and Efficiency of the Human Body as a Machine

One measure of a person's physical fitness is his or her maximum capacity to use or consume oxygen. This "aerobic" fitness can be increased and maintained with regular exercise, but falls when training stops. Typical maximum rates of oxygen consumption and corresponding fitness levels are shown in Table 12.5; we see that the maximum oxygen consumption rate varies from 28 mL/min · kg of body mass for poorly conditioned subjects to 70 mL/min · kg for superb athletes.

We have already pointed out that the first law of thermodynamics can be rewritten to relate the metabolic rate $\Delta U/\Delta t$ to the rate at which energy leaves the body by work and by heat:

$$\frac{\Delta U}{\Delta t} = \frac{Q}{\Delta t} + \frac{W}{\Delta t}$$

Now consider the body as a machine capable of supplying mechanical power to the outside world, and ask for its efficiency. The body's efficiency e is defined as

TABLE 12.5
Physical Fitness and Maximum Oxygen Consumption Rate[a]

Fitness Level	Maximum Oxygen Consumption Rate (mL/min · kg)
Very poor	28
Poor	34
Fair	42
Good	52
Excellent	70

[a]*Source: Aerobics*, K. H. Cooper, Bantam Books, New York, 1968.

TABLE 12.6

Metabolic Rate, Power Output, and Efficiency for Different Activities[a]

Activity	Metabolic Rate $\dfrac{\Delta U}{\Delta t}$ (watts)	Power Output $\dfrac{W}{\Delta t}$ (watts)	Efficiency e
Cycling	505	96	0.19
Pushing loaded coal cars in a mine	525	90	0.17
Shoveling	570	17.5	0.03

[a]*Source:* "Inter- and Intra-Individual Differences in Energy Expenditure and Mechanical Efficiency," C. H. Wyndham et al., *Ergonomics* **9**, 17 (1966).

the ratio of the mechanical power supplied by a human to the metabolic rate or the total power input to the body:

$$e = \text{body's efficiency} = \frac{\left|\dfrac{W}{\Delta t}\right|}{\left|\dfrac{\Delta U}{\Delta t}\right|} \qquad [12.21]$$

In this definition, absolute-value signs are used to show that e is a positive number and to avoid explicitly using minus signs required by our definitions of W and Q in the first law. Table 12.6 shows the efficiency of workers engaged in different activities for several hours. These values were obtained by measuring the power output and simultaneous oxygen consumption of mine workers and calculating the metabolic rate from their oxygen consumption. The table shows that a person can steadily supply mechanical power for several hours at about 100 W with an efficiency of about 17%. It also shows the dependence of efficiency on activity, and that e can drop to values as low as 3% for highly inefficient activities like shoveling, which involves many starts and stops. Finally, it is interesting in comparison to the average results of Table 12.6 that a superbly-conditioned athlete efficiently coupled to a mechanical device for extracting power (a bike!) can supply a power of around 300 W for about 30 minutes at a peak efficiency of 22%.

SUMMARY

Physics⊗Now™ Take a practice test by logging into Physics-Now at **www.cp7e.com** and clicking on the Pre-Test link for this chapter.

12.1 Work in Thermodynamic Processes

The work done on a gas at a constant pressure is

$$W = -P\,\Delta V \qquad [12.1]$$

The work done on the gas is positive if the gas is compressed (ΔV is negative) and negative if the gas expands (ΔV is positive). In general, the work done on a gas that takes it from some initial state to some final state is the negative of the area under the curve on a PV diagram.

12.2 The First Law of Thermodynamics

According to the first law of thermodynamics, when a system undergoes a change from one state to another, the **change in its internal energy** ΔU is

$$\Delta U = U_f - U_i = Q + W \qquad [12.2]$$

where Q is the energy transferred into the system by heat and W is the work done on the system. Q is positive when

energy enters the system by heat and negative when the system loses energy. W is positive when work is done on the system (for example, by compression) and negative when the system does positive work on its environment.

The change of the internal energy, ΔU, of an ideal gas is given by

$$\Delta U = nC_v\Delta T \qquad [12.5]$$

where C_v is the molar specific heat at constant volume.

An **isobaric process** is one that occurs at constant pressure. The work done on the system in such a process is $-P\,\Delta V$, while the thermal energy transferred by heat is given by

$$Q = nC_p\Delta T \qquad [12.6]$$

with the molar heat capacity at constant pressure given by $C_p = C_v + R$.

In an **adiabatic process** no energy is transferred by heat between the system and its surroundings ($Q = 0$). In this case, the first law gives $\Delta U = W$, which means the internal energy changes solely as a consequence of work being done on the system. The pressure and volume in adiabatic processes are related by

$$PV^\gamma = \text{constant} \qquad \textbf{[12.8a]}$$

where $\gamma = C_p/C_v$ is the adiabatic index.

In an **isovolumetric process** the volume doesn't change and no work is done. For such processes, the first law gives $\Delta U = Q$.

An **isothermal process** occurs at constant temperature. The work done by an ideal gas on the environment is

$$W_{\text{env}} = nRT \ln\left(\frac{V_f}{V_i}\right) \qquad \textbf{[12.10]}$$

12.3   Heat Engines and the Second Law of Thermodynamics

In a **cyclic process** (in which the system returns to its initial state), $\Delta U = 0$ and therefore $Q = W_{\text{eng}}$, meaning the energy transferred into the system by heat equals the work done on the system during the cycle.

A **heat engine** takes in energy by heat and partially converts it to other forms of energy, such as mechanical and electrical energy. The work W_{eng} done by a heat engine in carrying a working substance through a cyclic process ($\Delta U = 0$) is

$$W_{\text{eng}} = |Q_h| - |Q_c| \qquad \textbf{[12.11]}$$

where Q_h is the energy absorbed from a hot reservoir and Q_c is the energy expelled to a cold reservoir.

The **thermal efficiency** of a heat engine is defined as the ratio of the work done by the engine to the energy transferred into the engine per cycle:

$$e = \frac{W_{\text{eng}}}{|Q_h|} = 1 - \frac{|Q_c|}{|Q_h|} \qquad \textbf{[12.12]}$$

Heat pumps are heat engines in reverse. In a refrigerator, the heat pump removes thermal energy from inside the refrigerator. Heat pumps operating in cooling mode have coefficient of performance given by

$$\text{COP(cooling mode)} = \frac{|Q_c|}{W} \qquad \textbf{[12.13]}$$

A heat pump in heating mode has coefficient of performance

$$\text{COP(heating mode)} = \frac{|Q_h|}{W} \qquad \textbf{[12.14]}$$

Real processes proceed in an order governed by the **second law of thermodynamics**, which can be stated in two ways:

1. Energy will not flow spontaneously by heat from a cold object to a hot object.
2. No heat engine operating in a cycle can absorb energy from a reservoir and perform an equal amount of work.

No real heat engine operating between the Kelvin temperatures T_h and T_c can exceed the efficiency of an engine operating between the same two temperatures in a **Carnot cycle**, given by

$$e_C = 1 - \frac{T_c}{T_h} \qquad \textbf{[12.16]}$$

Perfect efficiency of a Carnot engine requires a cold reservoir of 0 K, absolute zero. According to the **third law of thermodynamics**, however, it is impossible to lower the temperature of a system to absolute zero in a finite number of steps.

12.4   Entropy

The second law can also be stated in terms of a quantity called **entropy** (S). The **change in entropy** of a system is equal to the energy flowing by heat into (or out of) the system as the system changes from one state (A) to another (B) by a reversible process, divided by the absolute temperature:

$$\Delta S \equiv \frac{Q_r}{T} \qquad \textbf{[12.17]}$$

One of the primary findings of statistical mechanics is that systems tend toward disorder and that entropy is a measure of that disorder. An alternate statement of the second law is that the entropy of the Universe increases in all natural processes.

CONCEPTUAL QUESTIONS

1. What are some factors that affect the efficiency of automobile engines?

2. "Energy is the mistress of the Universe and entropy is her shadow." Writing for an audience of general readers, argue for the truth of this statement with examples. Alternately, argue for the view that entropy is like a decisive hands-on executive instantly determining what will happen, while energy is like a wretched back-office bookkeeper telling us how little we can afford.

3. For an ideal gas in an isothermal process, there is no change in internal energy. Suppose the gas does work W during such a process. How much energy was transferred by heat?

4. If you shake a jar full of jelly beans of different sizes, the larger beans tend to appear near the top and the smaller ones tend to fall to the bottom. Why does this occur? Does this process violate the second law of thermodynamics?

5. Consider the human body performing a strenuous exercise, such as lifting weights or riding a bicycle. Work is being done by the body, and energy is leaving by conduction from the skin into the surrounding air. According to the first law of thermodynamics, the temperature of the body should be steadily decreasing during the exercise. This isn't what happens, however. Is the first law invalid for this situation? Explain.

6. Clearly distinguish among temperature, heat, and internal energy.

7. What is wrong with the following statement? "Given any two objects, the one with the higher temperature contains more heat."

8. A steam-driven turbine is one major component of an electric power plant. Why is it advantageous to increase the temperature of the steam as much as possible?

9. When a sealed Thermos bottle full of hot coffee is shaken, what changes, if any, take place in (a) the temperature of the coffee and (b) its internal energy?

10. In solar ponds constructed in Israel, the Sun's energy is concentrated near the bottom of a salty pond. With the proper layering of salt in the water, convection is prevented, and temperatures of 100°C may be reached. Can you guess the maximum efficiency with which useful mechanical work can be extracted from the pond?

11. Is it possible to construct a heat engine that creates no thermal pollution?

12. Suppose your roommate is "Mr. Clean" and tidies up your messy room after a big party. Because more order is being created by your roommate, does this tidying up represent a violation of the second law of thermodynamics?

13. A thermodynamic process occurs in which the entropy of a system changes by −8.0 J/K. According to the second law of thermodynamics, what can you conclude about the entropy change of the environment?

14. If a supersaturated sugar solution is allowed to evaporate slowly, sugar crystals form in the container. Hence, sugar molecules go from a disordered form (in solution) to a highly ordered, crystalline form. Does this process violate the second law of thermodynamics? Explain.

15. The first law of thermodynamics says we can't get more out of a process than we put in, but the second law says that we can't break even. Explain this statement.

16. Give some examples of irreversible processes that occur in nature. Give an example of a process in nature that is nearly reversible.

17. Imagine a gas in an insulated cylinder with a movable piston. The piston has been pushed inward, compressing the gas, and is now released. As the molecules of the gas strike the piston, they move it outward. From the point of view of energy principles, explain how this expansion causes the temperature of the gas to drop.

PROBLEMS

1, 2, 3 = straightforward, intermediate, challenging   □ = full solution available in *Student Solutions Manual/Study Guide*

Physics⊗Now™ = coached problem with hints available at **www.cp7e.com**   🧬 = biomedical application

Section 12.1 Work in Thermodynamic Processes

1. Physics⊗Now™ The only form of energy possessed by molecules of a monatomic ideal gas is translational kinetic energy. Using the results from the discussion of kinetic theory in Section 10.5, show that the internal energy of a monatomic ideal gas at pressure P and occupying volume V may be written as $U = \frac{3}{2}PV$.

2. Sketch a PV diagram and find the work done by the gas during the following stages: (a) A gas is expanded from a volume of 1.0 L to 3.0 L at a constant pressure of 3.0 atm. (b) The gas is then cooled at constant volume until the pressure falls to 2.0 atm. (c) The gas is then compressed at a constant pressure of 2.0 atm from a volume of 3.0 L to 1.0 L. (*Note:* Be careful of signs.) (d) The gas is heated until its pressure increases from 2.0 atm to 3.0 atm at a constant volume. (e) Find the net work done during the complete cycle.

3. A container of volume 0.40 m³ contains 3.0 mol of argon gas at 30°C. Assuming argon behaves as an ideal gas, find the total internal energy of the gas. (*Hint:* See Problem 1.)

4. A 40.0-g projectile is launched by the expansion of hot gas in an arrangement shown in Figure P12.4a. The cross-sectional area of the launch tube is 1.0 cm², and the length that the projectile travels down the tube after starting from rest is 32 cm. As the gas expands, the pressure varies as shown in Figure P12.4b. The values for the initial pressure and volume are $P_i = 11 \times 10^5$ Pa and $V_i = 8.0$ cm³ while the final values are $P_f = 1.0 \times 10^5$ Pa and $V_f = 40.0$ cm³. Friction between the projectile and the launch tube is negligible. (a) If the projectile is launched into a vacuum, what is the speed of the projectile as it leaves the launch tube? (b) If instead the projectile is launched into air at a pressure of 1.0×10^5 Pa, what fraction of the work done by the expanding gas in the tube is spent by the projectile pushing air out of the way as it proceeds down the tube?

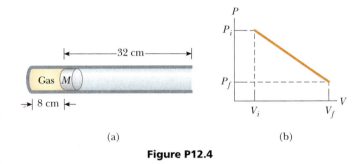

(a)   (b)

Figure P12.4

5. A gas expands from I to F along the three paths indicated in Figure P12.5. Calculate the work done *on* the gas along paths (a) *IAF*, (b) *IF*, and (c) *IBF*.

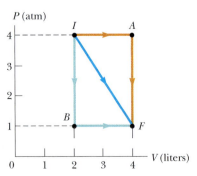

Figure P12.5 (Problems 5 and 15)

6. Sketch a *PV* diagram of the following processes: (a) A gas expands at constant pressure P_1 from volume V_1 to volume V_2. It is then kept at constant volume while the pressure is reduced to P_2. (b) A gas is reduced in pressure from P_1 to P_2 while its volume is held constant at V_1. It is then expanded at constant pressure P_2 to a final volume V_2. (c) In which of the processes is more work done *by* the gas? Why?

7. Gas in a container is at a pressure of 1.5 atm and a volume of 4.0 m^3. What is the work done *on* the gas (a) if it expands at constant pressure to twice its initial volume? (b) if it is compressed at constant pressure to one-quarter its initial volume?

8. A movable piston having a mass of 8.00 kg and a cross-sectional area of 5.00 cm^2 traps 0.200 moles of an ideal gas in a vertical cylinder. If the piston slides without friction in the cylinder, how much work is done *on* the gas when its temperature is increased from 20°C to 300°C?

9. One mole of an ideal gas initially at a temperature of $T_i = 0$°C undergoes an expansion at a constant pressure of 1.00 atm to four times its original volume. (a) Calculate the new temperature T_f of the gas. (b) Calculate the work done *on* the gas during the expansion.

10. (a) Determine the work done *on* a fluid that expands from *i* to *f* as indicated in Figure P12.10. (b) How much work is done *on* the fluid if it is compressed from *f* to *i* along the same path?

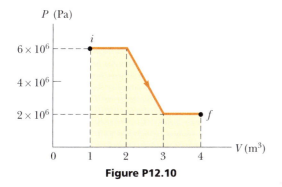

Figure P12.10

Section 12.2 The First Law of Thermodynamics

11. A container is placed in a water bath and held at constant volume as a mixture of fuel and oxygen is burned inside it. The temperature of the water is observed to rise during the burning. (The water is also held at constant volume. (a) Consider the burning mixture to be the system. What are the signs of Q, ΔU, and W? (b) What are the signs of these quantities if the water bath is considered to be the system?

12. A quantity of a monatomic ideal gas undergoes a process in which both its pressure and volume are doubled as shown in Figure P12.12. What is the energy absorbed by heat into the gas during this process? (*Hint:* See Problem 1.)

13. A gas is compressed at a constant pressure of 0.800 atm from 9.00 L to 2.00 L. In the process, 400 J of energy leaves the gas by heat. (a) What is the work done *on* the gas? (b) What is the change in its internal energy?

14. A monatomic ideal gas undergoes the thermodynamic process shown in the *PV* diagram of Figure P12.14. Determine whether each of the values ΔU, Q, and W for the gas is positive, negative, or zero. (*Hint:* See Problem 1.)

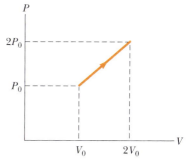

Figure P12.12

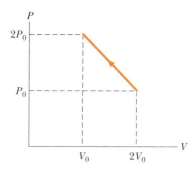

Figure P12.14

15. **Physics⊗Now™** A gas expands from *I* to *F* in Figure P12.5. The energy added to the gas by heat is 418 J when the gas goes from *I* to *F* along the diagonal path. (a) What is the change in internal energy of the gas? (b) How much energy must be added to the gas by heat for the indirect path *IAF* to give the same change in internal energy?

16. A gas is taken through the cyclic process described by Figure P12.16. (a) Find the net energy transferred to the system by heat during one complete cycle. (b) If the cycle is reversed—that is, the process follows the path *ACBA*—what is the net energy transferred by heat per cycle?

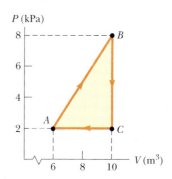

Figure P12.16   (Problems 16 and 18)

17. A gas is enclosed in a container fitted with a piston of cross-sectional area 0.150 m^2. The pressure of the gas is maintained at 6 000 Pa as the piston moves inward 20.0 cm. (a) Calculate the work done by the gas. (b) If the internal energy of the gas decreases by 8.00 J, find the amount of heat removed from the system by heat during the compression.

18. Consider the cyclic process described by Figure P12.16. If Q is negative for the process *BC* and ΔU is negative for the process *CA*, determine the signs of Q, W, and ΔU associated with each process.

19. One gram of water changes to ice at a constant pressure of 1.00 atm and a constant temperature of 0°C. In the process, the volume changes from 1.00 cm^3 to 1.09 cm^3. (a) Find the work done *on* the water and (b) the change in the internal energy of the water.

20. A thermodynamic system undergoes a process in which its internal energy decreases by 500 J. If at the same time 220 J of work is done on the system, find the energy transferred to or from it by heat.

21. A 5.0-kg block of aluminum is heated from 20°C to 90°C at atmospheric pressure. Find (a) the work done by the aluminum, (b) the amount of energy transferred to it by heat, and (c) the increase in its internal energy.

22. One mole of gas initially at a pressure of 2.00 atm and a volume of 0.300 L has an internal energy equal to 91.0 J. In its final state, the gas is at a pressure of 1.50 atm and a volume of 0.800 L, and its internal energy equals 180 J. For the paths *IAF*, *IBF*, and *IF* in Figure P12.22, calculate (a) the work done *on* the gas and (b) the net energy transferred to the gas by heat in the process.

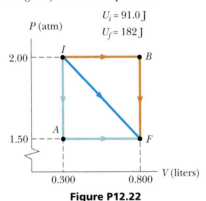

$U_i = 91.0$ J
$U_f = 182$ J

Figure P12.22

Section 12.3 Heat Engines and the Second Law of Thermodynamics

23. A heat engine operates between two reservoirs at temperatures of 20°C and 300°C. What is the maximum efficiency possible for this engine?

24. A steam engine has a boiler that operates at 300°F, and the temperature of the exhaust is 150°F. Find the maximum efficiency of this engine.

25. The energy absorbed by an engine is three times greater than the work it performs. (a) What is its thermal efficiency? (b) What fraction of the energy absorbed is expelled to the cold reservoir?

26. A particular engine has a power output of 5.00 kW and an efficiency of 25.0%. If the engine expels 8 000 J of energy in each cycle, find (a) the energy absorbed in each cycle and (b) the time required to complete each cycle.

27. One of the most efficient engines ever built is a coal-fired steam turbine engine in the Ohio River valley, driving an electric generator as it operates between 1 870°C and 430°C. (a) What is its maximum theoretical efficiency? (b) Its actual efficiency is 42.0%. How much mechanical power does it deliver if it absorbs 1.40×10^5 J of energy each second from the hot reservoir?

28. A gun is a heat engine. In particular, it is an internal combustion piston engine that does not operate in a cycle, but comes apart during its adiabatic expansion process. A certain gun consists of 1.80 kg of iron. It fires one 2.40-g bullet at 320 m/s with an energy efficiency of 1.10%. Assume that the body of the gun absorbs all of the energy exhaust and increases uniformly in temperature for a short time before it loses any energy by heat into the environment. Find its temperature increase.

29. An engine absorbs 1 700 J from a hot reservoir and expels 1 200 J to a cold reservoir in each cycle. (a) What is the engine's efficiency? (b) How much work is done in each cycle? (c) What is the power output of the engine if each cycle lasts for 0.300 s?

30. A power plant has been proposed that would make use of the temperature gradient in the ocean. The system is to operate between 20.0°C (surface water temperature) and 5.00°C (water temperature at a depth of about 1 km). (a) What is the maximum efficiency of such a system? (b) If the useful power output of the plant is 75.0 MW, how much energy is absorbed per hour? (c) In view of your answer to (a), do you think such a system is worthwhile (considering that there is no charge for fuel)?

31. In one cycle, a heat engine absorbs 500 J from a high-temperature reservoir and expels 300 J to a low-temperature reservoir. If the efficiency of this engine is 60% of the efficiency of a Carnot engine, what is the ratio of the low temperature to the high temperature in the Carnot engine?

32. A heat engine operates in a Carnot cycle between 80.0°C and 350°C. It absorbs 21 000 J of energy per cycle from the hot reservoir. The duration of each cycle is 1.00 s. (a) What is the mechanical power output of this engine? (b) How much energy does it expel in each cycle by heat?

33. A nuclear power plant has an electrical power output of 1 000 MW and operates with an efficiency of 33%. If excess energy is carried away from the plant by a river with a flow rate of 1.0×10^6 kg/s, what is the rise in temperature of the flowing water?

34. A 20.0%-efficient real engine is used to speed up a train from rest to 5.00 m/s. It is known that an ideal (Carnot) engine using the same cold and hot reservoirs would accelerate the same train from rest to a speed of 6.50 m/s using the same amount of fuel. If the engines use air at 300 K as a cold reservoir, find the temperature of the steam serving as the hot reservoir.

Section 12.4 Entropy

35. **Physics⊗Now**™ A freezer is used to freeze 1.0 L of water completely into ice. The water and the freezer remain at a constant temperature of $T = 0°C$. Determine (a) the change in the entropy of the water and (b) the change in the entropy of the freezer.

36. What is the change in entropy of 1.00 kg of liquid water at 100°C as it changes to steam at 100°C?

37. A 70-kg log falls from a height of 25 m into a lake. If the log, the lake, and the air are all at 300 K, find the change in entropy of the Universe during this process.

38. Two 2 000-kg cars, both traveling at 20 m/s, undergo a head-on collision and stick together. Find the change in entropy of the Universe resulting from the collision if the temperature is 23°C.

39. The surface of the Sun is approximately at 5 700 K, and the temperature of the Earth's surface is approximately 290 K. What entropy change occurs when 1 000 J of energy is transferred by heat from the Sun to the Earth?

40. Repeat the procedure used to construct Table 12.3 (a) for the case in which you draw three marbles rather than four from your bag and (b) for the case in which you draw five rather than four.

41. Prepare a table like Table 12.3 for the following occurrence: You toss four coins into the air simultaneously and record all the possible results of the toss in terms of the numbers of heads and tails that can result. (For example, HHTH and HTHH are two possible ways in which three heads and one tail can be achieved.) (a) On the basis of your table, what is the most probable result of a toss? In terms of entropy, (b) what is the most ordered state, and (c) what is the most disordered?

42. Consider a standard deck of 52 playing cards that has been thoroughly shuffled. (a) What is the probability of drawing the ace of spades in one draw? (b) What is the probability of drawing any ace? (c) What is the probability of drawing any spade?

ADDITIONAL PROBLEMS

43. A student claims that she has constructed a heat engine that operates between the temperatures of 200 K and 100 K with 60% efficiency. The professor does not give her credit for the project. Why not?

44. A Carnot engine operates between the temperatures $T_h = 100°C$ and $T_c = 20°C$. By what factor does the theoretical efficiency increase if the temperature of the hot reservoir is increased to 550°C?

45. A Carnot heat engine extracts energy Q_h from a hot reservoir at constant temperature T_h and rejects energy Q_c to a cold reservoir at constant temperature T_c. Find the entropy changes of (a) the hot reservoir, (b) the cold reservoir, (c) the engine, and (d) the complete system.

46. One end of a copper rod is in thermal contact with a hot reservoir at $T = 500$ K, and the other end is in thermal contact with a cooler reservoir at $T = 300$ K. If 8 000 J of energy is transferred from one end of the rod to the other, with no change in the temperature distribution, find the entropy change of each reservoir and the total entropy change of the Universe.

47. Find the change in temperature of a river due to the exhausted energy from a nuclear power plant. Assume that the input power to the boiler in the plant is 25×10^8 W, the efficiency of use of this power is 30%, and the river flow rate is 9.0×10^6 kg/min.

48. A Carnot engine operates between 100°C and 20°C. How much ice can the engine melt from its exhaust after it has done 5.0×10^4 J of work?

49. A 1500-kW heat engine operates at 25% efficiency. The heat energy expelled at the low temperature is absorbed by a stream of water that enters the cooling coils at 20°C. If 60 L flows across the coils per second, determine the increase in temperature of the water.

50. When a gas follows path 123 on the PV diagram in Figure P12.50, 418 J of energy flows into the system by heat and -167 J of work is done *on the gas*. (a) What is the change in the internal energy of the system? (b) How much energy Q flows into the system if the gas follows path 143? The work done on the gas along this path is -63.0 J. What net work would be done on or by the system if the system followed (c) path 12341? (d) path 14321? (e) What

is the change in internal energy of the system in the processes described in parts (c) and (d)?

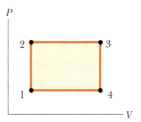

Figure P12.50

51. A substance undergoes the cyclic process shown in Figure P12.51. Work output occurs along path *AB* while work input is required along path *BC*, and no work is involved in the constant volume process *CA*. Energy transfers by heat occur during each process involved in the cycle. (a) What is the work output during process *AB*? (b) How much work input is required during process *BC*? (c) What is the net energy input Q during this cycle?

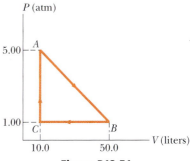

Figure P12.51

52. A power plant having a Carnot efficiency produces 1 000 MW of electrical power from turbines that take in steam at 500 K and eject water at 300 K into a flowing river. The water downstream is 6.00 K warmer due to the output of the plant. Determine the flow rate of the river.

53. A 100-kg steel support rod in a building has a length of 2.0 m at a temperature of 20°C. The rod supports a hanging load of 6 000 kg. Find (a) the work done *on* the rod as the temperature increases to 40°C, (b) the energy Q added to the rod (assume the specific heat of steel is the same as that for iron), and (c) the change in internal energy of the rod.

54. An ideal gas initially at pressure P_0, volume V_0, and temperature T_0 is taken through the cycle described in Figure P12.54. (a) Find the net work done *by* the gas per cycle in

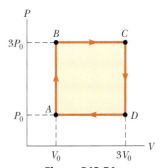

Figure P12.54

terms of P_0 and V_0. (b) What is the net energy Q added to the system per cycle? (c) Obtain a numerical value for the net work done per cycle for 1.00 mol of gas initially at 0°C. (*Hint:* Recall that the work done by the system equals the area under a PV curve.)

55. One mole of neon gas is heated from 300 K to 420 K at constant pressure. Calculate (a) the energy Q transferred to the gas, (b) the change in the internal energy of the gas, and (c) the work done *on* the gas. Note that neon has a molar specific heat of $c = 20.79$ J/mol $\cdot$ K for a constant-pressure process.

56. A 1.0-kg block of aluminum is heated at atmospheric pressure so that its temperature increases from 22°C to 40°C. Find (a) the work done *on* the aluminum, (b) the energy Q added to the aluminum, and (c) the change in internal energy of the aluminum.

57. **Physics⊗Now™** Suppose a heat engine is connected to two energy reservoirs, one a pool of molten aluminum at 660°C and the other a block of solid mercury at −38.9°C. The engine runs by freezing 1.00 g of aluminum and melting 15.0 g of mercury during each cycle. The latent heat of fusion of aluminum is 3.97×10^5 J/kg, and that of mercury is 1.18×10^4 J/kg. (a) What is the efficiency of this engine? (b) How does the efficiency compare with that of a Carnot engine?

58. One mole of an ideal gas is taken through the cycle shown in Figure P12.58. At point A, the pressure, volume, and temperature are P_0, V_0, and T_0. In terms of R and T_0, find (a) the total energy entering the system by heat per cycle, (b) the total energy leaving the system by heat per cycle, (c) the efficiency of an engine operating in this cycle, and (d) the efficiency of an engine operating in a Carnot cycle between the temperature extremes for this process. (*Hint:* Recall that work done on the gas is the negative of the area under a PV curve.)

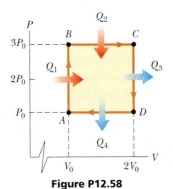

Figure P12.58

59. An electrical power plant has an overall efficiency of 15%. The plant is to deliver 150 MW of electrical power to a city, and its turbines use coal as fuel. The burning coal produces steam at 190°C, which drives the turbines. The steam is condensed into water at 25°C by passing through coils that are in contact with river water. (a) How many metric tons of coal does the plant consume each day (1 metric ton = 1×10^3 kg)? (b) What is the total cost of the fuel per year if the delivery price is $8 per metric ton? (c) If the river water is delivered at 20°C, at what minimum rate must it flow over the cooling coils in order

that its temperature not exceed 25°C? (*Note:* The heat of combustion of coal is 7.8×10^3 cal/g.)

60. At atmospheric pressure $(1.013 \times 10^5$ Pa) and 20.0°C, 1.00 g of water occupies a volume of 1.00 cm^3. (a) Find the change in internal energy when the water is heated to the boiling point. (b) When the water is boiled, it becomes 1 671 cm^3 of steam. Calculate the change in internal energy for this process. Assume the steam vapor doesn't mix with the surrounding air and that it expands at atmospheric pressure.

61. A gas is enclosed in a container fitted with a piston of cross-sectional area 0.10 m^2. The pressure of the gas is maintained at 8 000 Pa while energy is slowly added by heat; as a result, the piston is pushed up a distance of 4.0 cm. (Recall that any process in which the pressure remains constant is an isobaric process.) (a) If 42 J of energy is added to the system by heat during the expansion, what is the change in internal energy of the system? (b) If 42 J of energy is added by heat to the system with the piston clamped in a *fixed* position, what is the work done by the gas? What is the change in its internal energy?

62. Hydrothermal vents deep on the ocean floor spout water at temperatures as high as 570°C. This temperature is below the boiling point of water because of the immense pressure at that depth. Since the surrounding ocean temperature is at 4.0°C, an organism could use the temperature gradient as a source of energy. (a) Assuming the specific heat of water under these conditions is 1.0 cal/g $\cdot$ °C, how much energy is released when 1.0 liter of water is cooled from 570°C to 4.0°C? (b) What is the maximum useable energy an organism can extract from this energy source? (Assume that the organism has some internal type of heat engine acting between the two temperature extremes.) (c) Water from these vents contains hydrogen sulfide (H_2S) at a concentration of 0.90 mmole/liter. Oxidation of one mole of H_2S produces 310 kJ of energy. How much energy is available through H_2S oxidation of 1.0 L of water?

63. Suppose you spend 30.0 minutes on a stair-climbing machine, climbing at a rate of 90.0 steps per minute, with each step 8.00 inches high. If you weigh 150 lb and the machine reports that 600 kcal have been burned at the end of the workout, what efficiency is the machine using in obtaining this result? If your actual efficiency is 0.18, how many kcal did you really burn?

ACTIVITIES

1. Bend a paper clip back and forth several times and touch it to your lip. Why does it warm up? Repeat the process with a rubber band after it is stretched several times. Notice that it also warms up after stretching. Now take the stretched rubber band from your lip and wait for it to come to equilibrium with the surrounding air. Touch the stretched band to your lip, let it relax to its unstretched length, and note that it becomes cooler. Explain how the second law of thermodynamics helps you understand these results.

2. You should be able to locate a helium tank, a nitrogen tank, or a CO_2 tank in the chemistry or physics department of your university. Ask to use one of these for a short period for the following observation: Learn to open and close the valve or valves correctly. Touch the tanks and

nozzle to verify that they are at room temperature. Open a tank for 2–3 seconds and then close it. Touch the nozzle again, and you will find that it is very cold. Why? (*Hint:* The expansion of the gas is nearly adiabatic.)

3. You can study order and disorder and verify the answer to Quick Quiz 12.5 by rolling a pair of dice 100 times and recording the number of spots appearing on the dice for each throw. Which total comes up most frequently?

4. Another way to study order and disorder is to toss pennies. Predict in advance which is the more likely occurrence when tossing three pennies: having the pennies ordered with all the same orientation (all heads or all tails) or having them disordered (not all the same). Show that the ordered result includes two out of eight possibilities, while the disordered result includes six out of eight. Toss the three pennies 100 times to see how close you come to this prediction. Repeat the process for four pennies (The all-the-same result should be expected two times in sixteen throws.)

CHAPTER

13

© Rick Doyle/Corbis

Vibrations and Waves

Periodic motion, from masses on springs to vibrations of atoms, is one of the most important kinds of physical behavior. In this chapter we take a more detailed look at Hooke's law, where the force is proportional to the displacement, tending to restore objects to some equilibrium position. A large number of physical systems can be successfully modeled with this simple idea, including the vibrations of strings, the swinging of a pendulum, and the propagation of waves of all kinds. All these physical phenomena involve periodic motion.

Periodic vibrations can cause disturbances that move through a medium in the form of waves. Many kinds of waves occur in nature, such as sound waves, water waves, seismic waves, and electromagnetic waves. These very different physical phenomena are described by common terms and concepts introduced here.

13.1 HOOKE'S LAW

One of the simplest types of vibrational motion is that of an object attached to a spring, previously discussed in the context of energy in Chapter 5. We assume that the object moves on a frictionless horizontal surface. If the spring is stretched or compressed a small distance x from its unstretched or equilibrium position and then released, it exerts a force on the object as shown in Active Figure 13.1. From experiment this spring force is found to obey the equation

Hooke's law ▶

$$F_s = -kx$$

[13.1]

where x is the displacement of the object from its equilibrium position ($x = 0$) and k is a positive constant called the **spring constant**. This force law for springs was discovered by Robert Hooke in 1678 and is known as **Hooke's law**. The value of k is a measure of the stiffness of the spring. Stiff springs have large k values, and soft springs have small k values.

The negative sign in Equation 13.1 means that the force exerted by the spring is always directed *opposite* the displacement of the object. When the object is to the right of the equilibrium position, as in Active Figure 13.1a, x is positive and F_s is negative. This means that the force is in the negative direction, to the left. When the object is to the left of the equilibrium position, as in Active Figure 13.1c, x is negative and F_s is positive, indicating that the direction of the force is to the right. Of course, when $x = 0$, as in Active Figure 13.1b, the spring is unstretched and $F_s = 0$. Because the spring force always acts toward the equilibrium position, it is sometimes called a restoring force. **A restoring force always pushes or pulls the object toward the equilibrium position**.

Suppose the object is initially pulled a distance A to the right and released from rest. The force exerted by the spring on the object pulls it back toward the equilibrium position. As the object moves toward $x = 0$, the magnitude of the force decreases (because x decreases) and reaches zero at $x = 0$. However, the object gains speed as it moves toward the equilibrium position, reaching its maximum speed when $x = 0$. The momentum gained by the object causes it to overshoot the equilibrium position and compress the spring. As the object moves to the left of the equilibrium position (negative x-values), the spring force acts on it to the right, steadily increasing in strength, and the speed of the object decreases. The object finally comes briefly to rest at $x = -A$ before accelerating back towards $x = 0$ and ultimately returning to the original position at $x = A$. The process is then repeated, and the object continues to oscillate back and forth over the same path. This type of motion is called **simple harmonic motion. Simple harmonic motion occurs when the net force along the direction of motion obeys Hooke's law— when the net force is proportional to the displacement from the equilibrium point and is always directed toward the equilibrium point**.

Not all periodic motions over the same path can be classified as simple harmonic motion. A ball being tossed back and forth between a parent and a child moves repetitively, but the motion isn't simple harmonic motion, because the force acting on the ball doesn't take the form of Hooke's law, Equation 13.1.

The motion of an object suspended from a vertical spring is also simple harmonic. In this case, the force of gravity acting on the attached object stretches the spring until equilibrium is reached and the object is suspended at rest. By definition the equilibrium position of the object is $x = 0$. When the object is moved away from equilibrium by a distance x and released, a net force acts toward the equilibrium position. Because the net force is proportional to x, the motion is simple harmonic.

The following three concepts are important in discussing any kind of periodic motion:

- The **amplitude** A is the maximum distance of the object from its equilibrium position. In the absence of friction, an object in simple harmonic motion oscillates between the positions $x = -A$ and $x = +A$.
- The **period** T is the time it takes the object to move through one complete cycle of motion, from $x = A$ to $x = -A$ and back to $x = A$.
- The **frequency** f is the number of complete cycles or vibrations per unit of time, and is the reciprocal of the period ($f = 1/T$).

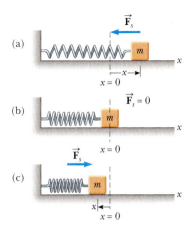

ACTIVE FIGURE 13.1
The force exerted by a spring on an object varies with the displacement of the object from the equilibrium position, $x = 0$. (a) When x is positive (the spring is stretched), the spring force is to the left. (b) When x is zero (the spring is unstretched), the spring force is zero. (c) When x is negative (the spring is compressed), the spring force is to the right.

Physics⊗Now™
Log into PhysicsNow at **www.cp7e.com**, and go to Active Figure 13.1 to choose the spring constant and the initial position and velocity of the block and see the resulting simple harmonic motion.

EXAMPLE 13.1   Measuring the Spring Constant

Goal   Use Newton's second law together with Hooke's law to calculate a spring constant.

Problem   A common technique used to evaluate a spring constant is illustrated in Figure 13.2. A spring is hung vertically (Fig. 13.2a), and an object of mass m is attached to the lower end of the spring and slowly lowered a distance d to the equilibrium point (Fig. 13.2b). Find the value of the spring constant if the spring is displaced by 2.00 cm and the mass is 0.550 kg.

Figure 13.2 (Example 13.1) Determining the spring constant. The elongation d of the spring is due to the suspended weight mg. Because the upward spring force balances the weight when the system is in equilibrium, it follows that $k = mg/d$.

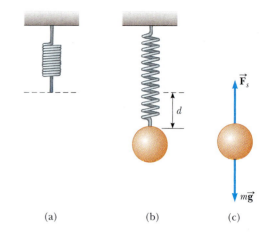

(a)   (b)   (c)

Strategy This is an application of Newton's second law. The spring is stretched by a distance d from its initial position under the action of the load mg. The spring force is upward, balancing the downward force of gravity mg *when the system is in equilibrium.* (See Fig. 13.2c.) The suspended mass is in equilibrium, so set the sum of the forces equal to zero.

Solution

Apply the second law (with $a = 0$) and solve for the spring constant k:

$$\Sigma F = F_g + F_s = -mg + kd = 0$$

$$k = \frac{mg}{d} = \frac{(0.550 \text{ kg})(9.80 \text{ m/s}^2)}{2.00 \times 10^{-2} \text{ m}} = 2.70 \times 10^2 \text{ N/m}$$

Remarks In this case the spring force is positive, because it's directed upward. Once the mass is pulled down from the equilibrium position and released, it oscillates around the equilibrium position, just like the horizontal spring.

Exercise 13.1

A spring with constant $k = 475$ N/m stretches 4.50 cm when an object of mass 25.0 kg is attached to the end of the spring. Find the acceleration of gravity in this location.

Answer 0.855 m/s^2 (The location is evidently an asteroid or small moon.)

The acceleration of an object moving with simple harmonic motion can be found by using Hooke's law in the equation for Newton's second law, $F = ma$. This gives

$$ma = F = -kx$$

Acceleration in simple harmonic motion ▶

$$a = -\frac{k}{m}x \qquad [13.2]$$

Equation 13.2, an example of a *harmonic oscillator equation*, gives the acceleration as a function of position. Because the maximum value of x is defined to be the amplitude A, the acceleration ranges over the values $-kA/m$ to $+kA/m$. In the next section we will find equations for velocity as a function of position and for position as a function of time.

TIP 13.1 Constant-Acceleration Equations Don't Apply

The acceleration a of a particle in simple harmonic motion is *not* constant; it changes, varying with x, so we can't apply the constant acceleration kinematic equations of Chapter 2.

Quick Quiz 13.1

A block on the end of a spring is pulled to position $x = A$ and released. Through what total distance does it travel in one full cycle of its motion? (a) $A/2$ (b) A (c) $2A$ (d) $4A$

Quick Quiz 13.2

For a simple harmonic oscillator, which of the following pairs of vector quantities can't both point in the same direction? (The position vector is the displacement from equilibrium.) (a) position and velocity (b) velocity and acceleration (c) position and acceleration

EXAMPLE 13.2 Simple Harmonic Motion on a Frictionless Surface

Goal Calculate forces and accelerations for a horizontal spring system.

Problem A 0.350-kg object attached to a spring of force constant 1.30×10^2 N/m is free to move on a frictionless horizontal surface, as in Active Figure 13.1. If the object is released from rest at $x = 0.100$ m, find the force on it and its acceleration at $x = 0.100$ m, $x = 0.050\ 0$ m, $x = 0$ m, $x = -0.050\ 0$ m, and $x = -0.100$ m.

Strategy Substitute given quantities into Hooke's law to find the forces, then calculate the accelerations with Newton's second law. The amplitude A is the same as the point of release from rest, $x = 0.100$ m.

Solution

Write Hooke's force law:

$$F_s = -kx$$

Substitute the value for k, and take $x = A = 0.100$ m, finding the force at that point:

$$F_{max} = -kA = -(1.30 \times 10^2 \text{ N/m})(0.100 \text{ m})$$
$$= -13.0 \text{ N}$$

Solve Newton's second law for a and substitute to find the acceleration at $x = A$:

$$ma = F_{max}$$
$$a = \frac{F_{max}}{m} = \frac{-13.0 \text{ N}}{0.350 \text{ kg}} = -37.1 \text{ m/s}^2$$

Repeat the same process for the other four points, assembling a table:

Position (m)	Force (N)	Acceleration (m/s^2)
0.100	− 13.0	− 37.1
0.050	− 6.50	− 18.6
0	0	0
− 0.050	+ 6.50	+ 18.6
− 0.100	+ 13.0	+ 37.1

Remarks Notice that when the initial position is halved, the force and acceleration are also halved. Further, positive values of x give negative values of the force and acceleration, while negative values of x give positive values of the force and acceleration. As the object moves to the left and passes the equilibrium point, the spring force becomes positive (for negative values of x), slowing the object down.

Exercise 13.2

For the same spring and mass system, find the force exerted by the spring and the position x when the object's acceleration is $+ 9.00$ m/s^2.

Answers 3.15 N, − 2.42 cm

13.2 ELASTIC POTENTIAL ENERGY

In this section we review the material covered in Section 4 of Chapter 5.

A system of interacting objects has potential energy associated with the configuration of the system. A compressed spring has potential energy that, when allowed to expand, can do work on an object, transforming spring potential energy into the object's kinetic energy. As an example, Figure 13.3 (page 428) shows a ball being projected from a spring-loaded toy gun, where the spring is compressed a distance x. As the gun is fired, the compressed spring does work on the ball and imparts kinetic energy to it.

Recall that the energy stored in a stretched or compressed spring or some other elastic material is called elastic potential energy, PE_s, given by

$$PE_s \equiv \tfrac{1}{2}kx^2 \qquad\qquad \text{[13.3]} \qquad \blacktriangleleft \text{Elastic potential energy}$$

Figure 13.3 A ball projected from a spring-loaded gun. The elastic potential energy stored in the spring is transformed into the kinetic energy of the ball.

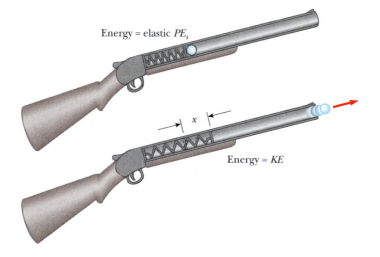

Recall also that the law of conservation of energy, including both gravitational and spring potential energy, is given by

$$(KE + PE_g + PE_s)_i = (KE + PE_g + PE_s)_f \qquad \text{[13.4]}$$

If nonconservative forces such as friction are present, then the change in mechanical energy must equal the work done by the nonconservative forces:

$$W_{nc} = (KE + PE_g + PE_s)_f - (KE + PE_g + PE_s)_i \qquad \text{[13.5]}$$

Rotational kinetic energy must be included in both Equation 13.4 and Equation 13.5 for systems involving torques.

As an example of the energy conversions that take place when a spring is included in a system, consider Figure 13.4. A block of mass m slides on a frictionless horizontal surface with constant velocity $\vec{v}_i$ and collides with a coiled spring. The description that follows is greatly simplified by assuming that the spring is very light and therefore has negligible kinetic energy. As the spring is compressed, it exerts a force to the left on the block. At maximum compression, the block comes to rest for just an instant (Fig. 13.4c). The initial total energy in the system (block plus spring) before the collision is the kinetic energy of the block. After the block

Figure 13.4 A block sliding on a frictionless horizontal surface collides with a light spring. (a) Initially, the mechanical energy is entirely the kinetic energy of the block. (b) The mechanical energy at some arbitrary position is the sum of the kinetic energy of the block and the elastic potential energy stored in the spring. (c) When the block comes to rest, the mechanical energy is entirely elastic potential energy stored in the compressed spring. (d) When the block leaves the spring, the mechanical energy is equal to the block's kinetic energy. The total energy remains constant.

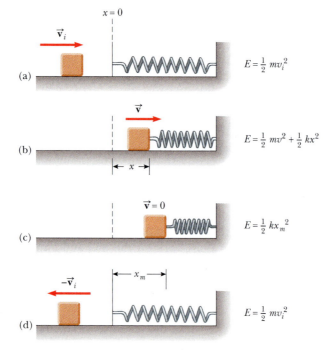

collides with the spring and the spring is partially compressed, as in Figure 13.4b, the block has kinetic energy $\frac{1}{2}mv^2$ (where $v < v_i$) and the spring has potential energy $\frac{1}{2}kx^2$. When the block stops for an instant at the point of maximum compression, the kinetic energy is zero. Because the spring force is conservative and because there are no external forces that can do work on the system, **the total mechanical energy of the system consisting of the block and spring remains constant.** Energy is transformed from the kinetic energy of the block to the potential energy stored in the spring. As the spring expands, the block moves in the opposite direction and regains all of its initial kinetic energy, as in Figure 13.4d.

When an archer pulls back on a bowstring, elastic potential energy is stored in both the bent bow and stretched bowstring (Fig. 13.5). When the arrow is released, the potential energy stored in the system is transformed into the kinetic energy of the arrow. Devices such as crossbows and slingshots work the same way.

Figure 13.5   Elastic potential energy is stored in this drawn bow.

APPLICATION

Archery

Quick Quiz 13.3

When an object moving in simple harmonic motion is at its maximum displacement from equilibrium, which of the following is at a maximum? (a) velocity, (b) acceleration, or (c) kinetic energy.

EXAMPLE 13.3   Stop That Car!

Goal   Apply conservation of energy and the work–energy theorem with spring and gravitational potential energy.

Problem   A 13 000-N car starts at rest and rolls down a hill from a height of 10.0 m (Fig. 13.6). It then moves across a level surface and collides with a light spring-loaded guardrail. **(a)** Neglecting any losses due to friction, find the maximum distance the spring is compressed. Assume a spring constant of 1.0×10^6 N/m. **(b)** Calculate the maximum acceleration of the car after contact with the spring, assuming no frictional losses. **(c)** If the spring is compressed by only 0.30 m, find the energy lost through friction.

Figure 13.6   (Example 13.3) A car starts from rest on a hill at the position shown. When the car reaches the bottom of the hill, it collides with a spring-loaded guardrail.

Strategy   Because friction losses are neglected, use conservation of energy in the form of Equation 13.4 to solve for the spring displacement in part (a). The initial and final values of the car's kinetic energy are zero, so the initial potential energy of the car–spring–Earth system is completely converted to elastic potential energy in the spring at the end of the ride. In part (b), apply Newton's second law, substituting the answer to part (a) for x because the maximum compression will give the maximum acceleration. In part (c) friction is no longer neglected, so use the work–energy theorem, Equation 13.5. The change in mechanical energy must equal the mechanical energy lost due to friction.

Solution

(a) Find the maximum spring compression, assuming no energy losses due to friction.

Apply conservation of mechanical energy. Initially, there is only gravitational potential energy, and at maximum compression of the guardrail, there is only spring potential energy.

$$(KE + PE_g + PE_s)_i = (KE + PE_g + PE_s)_f$$

$$0 + mgh + 0 = 0 + 0 + \tfrac{1}{2}kx^2$$

Solve for x:

$$x = \sqrt{\frac{2mgh}{k}} = \sqrt{\frac{2(13\,000\text{ N})(10.0\text{ m})}{1.0 \times 10^6\text{ N/m}}} = \boxed{0.51\text{ m}}$$

(b) Calculate the maximum acceleration of the car by the spring, neglecting friction.

Apply Newton's second law:

$$ma = -kx \quad \rightarrow \quad a = -\frac{kx}{m} = -\frac{kxg}{mg} = -\frac{kxg}{w}$$

Substitute values:

$$a = -\frac{(1.0 \times 10^6 \text{ N/m})(0.51 \text{ m})(9.8 \text{ m/s}^2)}{13\ 000 \text{ N}}$$

$$= -380 \text{ m/s}^2$$

(c) If the compression of the guardrail is only 0.30 m, find the mechanical energy lost due to friction.

Use the work–energy theorem:

$$\begin{aligned} W_{nc} &= (KE + PE_g + PE_s)_f - (KE + PE_g + PE_s)_i \\ &= (0 + 0 + \tfrac{1}{2}kx^2) - (0 + mgh + 0) \\ &= \tfrac{1}{2}(1.0 \times 10^6 \text{ N/m})(0.30)^2 - (13\ 000 \text{ N})(10.0 \text{ m}) \end{aligned}$$

$$W_{nc} = \boxed{-8.5 \times 10^4 \text{ J}}$$

Remarks   The answer to part (b) is about 40 times greater than the acceleration of gravity, so we'd better be wearing our seat belts. Note that the solution didn't require calculation of the velocity of the car.

Exercise 13.3

A spring-loaded gun fires a 0.100-kg puck along a tabletop. The puck slides up a curved ramp and flies straight up into the air. If the spring is displaced 12.0 cm from equilibrium and the spring constant is 875 N/m, how high does the puck rise, neglecting friction? (b) If instead it only rises to a height of 5.00 m because of friction, what is the change in mechanical energy?

Answer   (a) 6.43 m   (b) −1.40 J

In addition to studying the preceding example, it's a good idea to review those given in Section 5.4.

Velocity as a Function of Position

Conservation of energy provides a simple method of deriving an expression for the velocity of an object undergoing periodic motion as a function of position. The object in question is initially at its maximum extension A (Fig. 13.7a) and is then released from rest. The initial energy of the system is entirely elastic potential energy stored in the spring, $\tfrac{1}{2}kA^2$. As the object moves toward the origin to some new position x (Fig. 13.7b), part of this energy is transformed into kinetic energy, and the potential energy stored in the spring is reduced to $\tfrac{1}{2}kx^2$. Because the total energy of the system is equal to $\tfrac{1}{2}kA^2$ (the initial energy stored in the spring), we can equate this quantity to the sum of the kinetic and potential energies at the position x:

$$\tfrac{1}{2}kA^2 = \tfrac{1}{2}mv^2 + \tfrac{1}{2}kx^2$$

Solving for v, we get

$$v = \pm\sqrt{\frac{k}{m}(A^2 - x^2)} \qquad [13.6]$$

This expression shows that the object's speed is a maximum at $x = 0$ and is zero at the extreme positions $x = \pm A$.

The right side of Equation 13.6 is preceded by the $\pm$ sign because the square root of a number can be either positive or negative. If the object in Figure 13.7 is moving to the right, v is positive; if the object is moving to the left, v is negative.

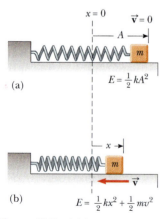

(a)

$$x = 0$$
$$\vec{v} = 0$$
$$A$$
$$m$$
$$E = \tfrac{1}{2}kA^2$$

(b)

$$x$$
$$m$$
$$\vec{v}$$
$$E = \tfrac{1}{2}kx^2 + \tfrac{1}{2}mv^2$$

Figure 13.7   (a) An object attached to a spring on a frictionless surface is released from rest with the spring extended a distance A. Just before the object is released, the total energy is the elastic potential energy $kA^2/2$. (b) When the object reaches position x, it has kinetic energy $mv^2/2$ and the elastic potential energy has decreased to $kx^2/2$.

EXAMPLE 13.4   The Object–Spring System Revisited

Goal   Apply the time-independent velocity expression, Equation 13.6.

Problem   A 0.500-kg object connected to a light spring with a spring constant of 20.0 N/m oscillates on a frictionless horizontal surface. **(a)** Calculate the total energy of the system and the maximum speed of the object if the amplitude of the motion is 3.00 cm. **(b)** What is the velocity of the object when the displacement is 2.00 cm? **(c)** Compute the kinetic and potential energies of the system when the displacement is 2.00 cm.

Strategy   The total energy of the system can be found most easily at the amplitude $x = A$, where the kinetic energy is zero. There, the potential energy alone is equal to the total energy. Conservation of energy then yields the speed at $x = 0$. For part (b), obtain the velocity by substituting the given value of x into the time-independent velocity equation. Using this result, the kinetic energy asked for in part (c) can be found by substitution, and the potential energy from substitution into Equation 13.3.

Solution

(a) Calculate the total energy and maximum speed if the amplitude is 3.00 cm.

Substitute $x = A = 3.00$ cm and $k = 20.0$ N/m into the equation for the total mechanical energy E:

$$E = KE + PE_g + PE_s$$
$$= 0 + 0 + \tfrac{1}{2}kA^2 = \tfrac{1}{2}(20.0 \text{ N/m})(3.00 \times 10^{-2} \text{ m})^2$$
$$= 9.00 \times 10^{-3} \text{ J}$$

Use conservation of energy with $x_i = A$ and $x_f = 0$ to compute the speed of the object at the origin:

$$(KE + PE_g + PE_s)_i = (KE + PE_g + PE_s)_f$$
$$0 + 0 + \tfrac{1}{2}kA^2 = \tfrac{1}{2}mv_{max}^2 + 0 + 0$$
$$\tfrac{1}{2}mv_{max}^2 = 9.00 \times 10^{-3} \text{ J}$$
$$v_{max} = \sqrt{\frac{18.0 \times 10^{-3} \text{ J}}{0.500 \text{ kg}}} = 0.190 \text{ m/s}$$

(b) Compute the velocity of the object when the displacement is 2.00 cm.

Substitute known values directly into Equation 13.6:

$$v = \pm \sqrt{\frac{k}{m}(A^2 - x^2)}$$
$$= \pm \sqrt{\frac{20.0 \text{ N/m}}{0.500 \text{ kg}}((0.030\,0 \text{ m})^2 - (0.020\,0 \text{ m})^2)}$$
$$= \pm 0.141 \text{ m/s}$$

(c) Compute the kinetic and potential energies when the displacement is 2.00 cm.

Substitute into the equation for kinetic energy:

$$KE = \tfrac{1}{2}mv^2 = \tfrac{1}{2}(0.500 \text{ kg})(0.141 \text{ m/s})^2 = 4.97 \times 10^{-3} \text{ J}$$

Substitute into the equation for spring potential energy:

$$PE_s = \tfrac{1}{2}kx^2 = \tfrac{1}{2}(20.0 \text{ N/m})(2.00 \times 10^{-2} \text{ m})^2$$
$$= 4.00 \times 10^{-3} \text{ J}$$

Remark   With the given information, it is impossible to choose between the positive and negative solutions in part (b). Notice that the sum $KE + PE_s$ in part (c) equals the total energy E found in part (a), as it should (except for a small discrepancy due to rounding).

Exercise 13.4

For what values of x is the speed of the object 0.10 m/s?

Answer ± 2.55 cm

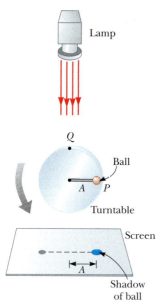

ACTIVE FIGURE 13.8
An experimental setup for demonstrating the connection between simple harmonic motion and uniform circular motion. As the ball rotates on the turntable with constant angular speed, its shadow on the screen moves back and forth with simple harmonic motion.

Physics⊗Now™
Log into PhysicsNow at **www.cp7e.com**, and go to Active Figure 13.8 to adjust the frequency and radial position of the ball and see the resulting simple harmonic motion of the shadow.

APPLICATION
Pistons and Drive Wheels

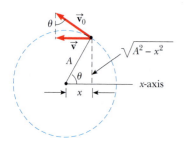

Figure 13.9   The ball rotates with constant speed v_0. The x-component of the ball's velocity equals the projection of $\vec{v}_0$ on the x-axis.

13.3   COMPARING SIMPLE HARMONIC MOTION WITH UNIFORM CIRCULAR MOTION

We can better understand and visualize many aspects of simple harmonic motion along a straight line by looking at its relationship to uniform circular motion. Active Figure 13.8 is a top view of an experimental arrangement that is useful for this purpose. A ball is attached to the rim of a turntable of radius A, illuminated from the side by a lamp. We find that **as the turntable rotates with constant angular speed, the shadow of the ball moves back and forth with simple harmonic motion**.

This fact can be understood from Equation 13.6, which says that the velocity of an object moving with simple harmonic motion is related to the displacement by

$$v = C\sqrt{A^2 - x^2}$$

where C is a constant. To see that the shadow also obeys this relation, consider Figure 13.9, which shows the ball moving with a constant speed v_0 in a direction tangent to the circular path. At this instant, the velocity of the ball in the x-direction is given by $v = v_0 \sin \theta$, or

$$\sin \theta = \frac{v}{v_0}$$

From the larger triangle in the figure we can obtain a second expression for $\sin \theta$:

$$\sin \theta = \frac{\sqrt{A^2 - x^2}}{A}$$

Equating the right-hand sides of the two expressions for $\sin \theta$, we find the following relationship between the velocity v and the displacement x:

$$\frac{v}{v_0} = \frac{\sqrt{A^2 - x^2}}{A}$$

or

$$v = \frac{v_0}{A}\sqrt{A^2 - x^2} = C\sqrt{A^2 - x^2}$$

The velocity of the ball in the x-direction is related to the displacement x in exactly the same way as the velocity of an object undergoing simple harmonic motion. The shadow therefore moves with simple harmonic motion.

A valuable example of the relationship between simple harmonic motion and circular motion can be seen in vehicles and machines that use the back-and-forth motion of a piston to create rotational motion in a wheel. Consider the drive wheel of a locomotive. In Figure 13.10, the curved housing at the left contains a piston that moves back and forth in simple harmonic motion. The piston is connected to an arrangement of rods that transforms its back-and-forth motion into rotational motion of the wheels. A similar mechanism in an automobile engine transforms the back-and-forth motion of the pistons to rotational motion of the crankshaft.

Period and Frequency

The period T of the shadow in Active Figure 13.8, which represents the time required for one complete trip back and forth, is also the time it takes the ball to make one complete circular trip on the turntable. Because the ball moves through the distance $2\pi A$ (the circumference of the circle) in the time T, the speed v_0 of the ball around the circular path is

$$v_0 = \frac{2\pi A}{T}$$

and the period is

$$T = \frac{2\pi A}{v_0} \qquad [13.7]$$

Imagine that the ball moves from P to Q, a quarter of a revolution, in Active Figure 13.8. The motion of the shadow is equivalent to the horizontal motion of an object on the end of a spring. For this reason, the radius A of the circular motion is the same as the amplitude A of the simple harmonic motion of the shadow. During the quarter of a cycle shown, the shadow moves from a point where the energy of the system (ball and spring) is solely elastic potential energy to a point where the energy is solely kinetic energy. By conservation of energy, we have

$$\tfrac{1}{2}kA^2 = \tfrac{1}{2}mv_0^2$$

which can be solved for A/v_0:

$$\frac{A}{v_0} = \sqrt{\frac{m}{k}}$$

Substituting this expression for A/v_0 in Equation 13.7, we find that the period is

$$T = 2\pi\sqrt{\frac{m}{k}} \qquad [13.8]$$

◀ The period of an object–spring system moving with simple harmonic motion

Equation 13.8 represents the time required for an object of mass m attached to a spring with spring constant k to complete one cycle of its motion. The square root of the mass is in the numerator, so a large mass will mean a large period, in agreement with intuition. The square root of the spring constant k is in the denominator, so a large spring constant will yield a small period, again agreeing with intuition. It's also interesting that the period doesn't depend on the amplitude A.

The inverse of the period is the frequency of the motion:

$$f = \frac{1}{T} \qquad [13.9]$$

Therefore, the frequency of the periodic motion of a mass on a spring is

$$f = \frac{1}{2\pi}\sqrt{\frac{k}{m}} \qquad [13.10]$$

◀ Frequency of an object–spring system

The units of frequency are cycles per second (s^{-1}), or **hertz** (Hz). The **angular frequency** ω is

$$\omega = 2\pi f = \sqrt{\frac{k}{m}} \qquad [13.11]$$

◀ Angular frequency of an object–spring system

The frequency and angular frequency are actually closely related concepts. The unit of frequency is cycles per second, where a cycle may be thought of as a unit of angular measure corresponding to 2π radians, or $360°$. Viewed in this way, angular frequency is just a unit conversion of frequency. Radian measure is used for angles mainly because it provides a convenient and natural link between linear and angular quantities.

Although an ideal mass–spring system has a period proportional to the square root of the object's mass m, experiments show that a graph of T^2 versus m doesn't pass through the origin. This is because the spring itself has a mass. The coils of the spring oscillate just like the object, except the amplitudes are smaller for all coils but the last. For a cylindrical spring, energy arguments can be used to show that the *effective* additional mass of a light spring is one-third the mass of the spring. The square of the period is proportional to the total oscillating mass, so a graph of T^2 versus total mass (the mass hung on the spring plus the effective oscillating mass of the spring) would pass through the origin.

Figure 13.10 The drive wheel mechanism of an old locomotive.

© Link/Visuals Unlimited

TIP 13.2 Twin Frequencies

The *frequency* gives the number of cycles per second, while the *angular frequency* gives the number of radians per second. These two physical concepts are nearly identical—linked by the conversion factor 2π rad/cycle.

Quick Quiz 13.4

An object of mass m is attached to a horizontal spring, stretched to a displacement A from equilibrium and released, undergoing harmonic oscillations on a frictionless surface with period T_0. The experiment is then repeated with a mass of $4m$. What's the new period of oscillation? (a) $2T_0$ (b) T_0 (c) $T_0/2$ (d) $T_0/4$.

Quick Quiz 13.5

Consider the situation in Quick Quiz 13.4. The subsequent total mechanical energy of the object with mass $4m$ is (a) greater than, (b) less than, or (c) equal to the original total mechanical energy.

Applying Physics 13.1   Bungee Jumping

A bungee cord can be modeled as a spring. If you go bungee jumping, you will bounce up and down at the end of the elastic cord after your dive off a bridge (Fig. 13.11). Suppose you perform a dive and measure the frequency of your bouncing. You then move to another bridge, but find that the bungee cord is too long for dives off this bridge. What possible solutions might be applied? In terms of the original frequency, what is the frequency of vibration associated with the solution?

Explanation   There are two possible solutions: Make the bungee cord smaller or fold it in half. The latter would be the safer of the two choices, as we'll see. The force exerted by the bungee cord, modeled as a spring, is proportional to the separation of the coils as the spring is extended. First, we extend the spring by a given distance and measure the distance between coils. We then cut the spring in half. If one of the half-springs is now extended by the same distance, the coils will be twice as far apart as they were for the complete spring. Therefore, it takes twice as much force to stretch the half-spring through the same displacement, so the half-spring has a spring constant twice that of the complete spring. The folded bungee cord can then be modeled as two half-springs in parallel. Each half has a spring constant that is twice the original spring constant of the bungee cord. In addition, an object hanging on the folded bungee cord will experience two forces—one from each half-spring. As a result, the required force for a given extension will be

Figure 13.11   (Applying Physics 13.1) Bungee jumping from a bridge.

four times as much as for the original bungee cord. The effective spring constant of the folded bungee cord is therefore four times as large as the original spring constant. Because the frequency of oscillation is proportional to the square root of the spring constant, your bouncing frequency on the folded cord will be twice what it was on the original cord.

This discussion neglects the fact that the coils of a spring have an initial separation. It's also important to remember that a shorter coil may lose elasticity more readily, possibly even going beyond the elastic limit for the material, with disastrous results. Bungee jumping is dangerous—discretion is advised!

EXAMPLE 13.5   That Car Needs Shock Absorbers!

Goal   Understand the relationships between period, frequency, and angular frequency.

Problem   A 1.30×10^3-kg car is constructed on a frame supported by four springs. Each spring has a spring constant of 2.00×10^4 N/m. If two people riding in the car have a combined mass of 1.60×10^2 kg, find the frequency of vibration of the car when it is driven over a pothole in the road. Find also the period and the angular frequency. Assume the weight is evenly distributed.

Strategy   Because the weight is evenly distributed, each spring supports one-fourth of the mass. Substitute this value and the spring constant into Equation 13.10 to get the frequency. The reciprocal is the period, and multiplying the frequency by 2π gives the angular frequency.

Solution

Compute one-quarter of the total mass:

$$m = \tfrac{1}{4}(m_{car} + m_{pass}) = \tfrac{1}{4}(1.30 \times 10^3 \text{ kg} + 1.60 \times 10^2 \text{ kg})$$

$$= 365 \text{ kg}$$

Substitute into Equation 13.10 to find the frequency:

$$f = \frac{1}{2\pi}\sqrt{\frac{k}{m}} = \frac{1}{2\pi}\sqrt{\frac{2.00 \times 10^4 \text{ N/m}}{365 \text{ kg}}} = \boxed{1.18 \text{ Hz}}$$

Invert the frequency to get the period:

$$T = \frac{1}{f} = \frac{1}{1.18 \text{ Hz}} = \boxed{0.847 \text{ s}}$$

Multiply the frequency by 2π to get the angular frequency:

$$\omega = 2\pi f = 2\pi(1.18 \text{ Hz}) = \boxed{7.41 \text{ rad/s}}$$

Remark   Solving this problem didn't require any knowledge of the size of the pothole, because the frequency doesn't depend on the amplitude of the motion.

Exercise 13.5

A 45.0-kg boy jumps on a 5.00-kg pogo stick with spring constant 3 650 N/m. Find (a) the angular frequency, (b) the frequency, and (c) the period of the boy's motion.

Answers   (a) 8.54 rad/s   (b) 1.36 Hz   (c) 0.735 s

13.4   POSITION, VELOCITY, AND ACCELERATION AS A FUNCTION OF TIME

We can obtain an expression for the position of an object moving with simple harmonic motion as a function of time by returning to the relationship between simple harmonic motion and uniform circular motion. Again, consider a ball on the rim of a rotating turntable of radius A, as in Active Figure 13.12. We refer to the circle made by the ball as the *reference circle* for the motion. We assume that the turntable revolves at a *constant* angular speed ω. As the ball rotates on the reference circle, the angle θ made by the line OP with the x-axis changes with time. Meanwhile, the projection of P on the x-axis, labeled point Q, moves back and forth along the axis with simple harmonic motion.

From the right triangle OPQ, we see that $\cos\theta = x/A$. Therefore, the x-coordinate of the ball is

$$x = A\cos\theta$$

Because the ball rotates with constant angular speed, it follows that $\theta = \omega t$ (see Chapter 7), so we have

$$x = A\cos(\omega t) \qquad \textbf{[13.12]}$$

In one complete revolution, the ball rotates through an angle of 2π rad in a time equal to the period T. In other words, the motion repeats itself every T seconds. Therefore,

$$\omega = \frac{\Delta\theta}{\Delta t} = \frac{2\pi}{T} = 2\pi f \qquad \textbf{[13.13]}$$

where f is the frequency of the motion. The angular speed of the ball as it moves around the reference circle is the same as the angular frequency of the projected simple harmonic motion. Consequently, Equation 13.12 can be written

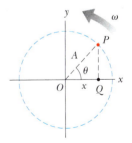

ACTIVE FIGURE 13.12
A reference circle. As the ball at P rotates in a circle with uniform angular speed, its projection Q along the x-axis moves with simple harmonic motion.

Physics⊗Now™
Log into PhysicsNow at **www.cp7e.com**, and go to Active Figure 13.12 to compare the oscillations of two blocks starting from different initial positions and to verify that the frequency is independent of the amplitude.

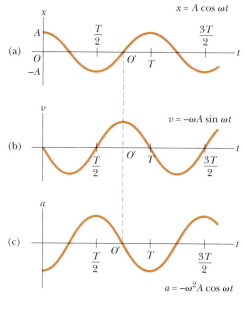

$$x = A \cos(2\pi ft) \qquad \text{[13.14a]}$$

This cosine function represents the position of an object moving with simple harmonic motion as a function of time, and is graphed in Active Figure 13.13a. Because the cosine function varies between 1 and -1, x varies between A and $-A$. The shape of the graph is called *sinusoidal*.

Active Figures 13.13b and 13.13c represent curves for velocity and acceleration as a function of time. To find the equation for the velocity, use Equations 13.6 and 13.14a together with the identity $\cos^2\theta + \sin^2\theta = 1$, obtaining

$$v = -A\omega \sin(2\pi ft) \qquad \text{[13.14b]}$$

where we have used the fact that $\omega = \sqrt{k/m}$. The $\pm$ sign is no longer needed, because sine can take both positive and negative values. Deriving an expression for the acceleration involves substituting Equation 13.14a into Equation 13.2, Newton's second law for springs:

$$a = -A\omega^2 \cos(2\pi ft) \qquad \text{[13.14c]}$$

The detailed steps of these derivations are left as an exercise for the student. Notice that when the displacement x is at a maximum, at $x = A$ or $x = -A$, the velocity is zero, and when x is zero, the magnitude of the velocity is a maximum. Further, when $x = +A$, its most positive value, the acceleration is a maximum but in the negative x-direction, and when x is at its most negative position, $x = -A$, the acceleration has its maximum value in the positive x-direction. These facts are consistent with our earlier discussion of the points at which v and a reach their maximum, minimum, and zero values.

The maximum values of the position, velocity, and acceleration are always equal to the magnitude of the expression in front of the trigonometric function in each equation, because the largest value of either cosine or sine is 1.

Figure 13.14 illustrates one experimental arrangement that demonstrates the sinusoidal nature of simple harmonic motion. An object connected to a spring has a marking pen attached to it. While the object vibrates vertically, a sheet of paper is moved horizontally with constant speed. The pen traces out a sinusoidal pattern.

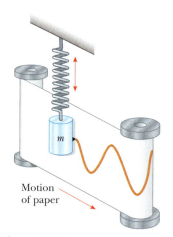

Figure 13.14
An experimental apparatus for
demonstrating simple harmonic motion. A pen attached to the oscillating
object traces out a sinusoidal wave on
the moving chart paper.

Quick Quiz 13.6

If the amplitude of a system moving in simple harmonic motion is doubled, which of the following quantities *doesn't* change? (a) total energy, (b) maximum speed, (c) maximum acceleration, (d) period.

EXAMPLE 13.6   The Vibrating Object–Spring System

Goal   Identify the physical parameters of a harmonic oscillator from its mathematical description.

Problem   (a) Find the amplitude, frequency, and period of motion for an object vibrating at the end of a horizontal spring if the equation for its position as a function of time is

$$x = (0.250 \text{ m}) \cos\left(\frac{\pi}{8.00} t\right)$$

(b) Find the maximum magnitude of the velocity and acceleration. **(c)** What is the position, velocity, and acceleration of the object after 1.00 s has elapsed?

Strategy   In part (a), the amplitude and frequency can be found by comparing the given equation with the standard form in Equation 13.14a, matching up the numerical values with the corresponding terms in the standard form. (b) The maximum speed will occur when the sine function in Equation 13.14b equals 1 or − 1, the extreme values of the sine function (and similarly for the acceleration and the cosine function). In each case, find the magnitude of the expression in front of the trigonometric function. Part (c) is just a matter of substituting values into Equations 13.14a, b, and c.

Solution
(a) Find the amplitude, frequency, and period.

Write the standard form given by Equation 13.14a, and underneath it write the given equation:

$$x = A \cos(2\pi ft) \tag{1}$$

$$x = (0.250 \text{ m}) \cos\left(\frac{\pi}{8.00} t\right) \tag{2}$$

Equate the factors in front of the cosine functions to find the amplitude:

$$A = 0.250 \text{ m}$$

The angular frequency ω is the factor in front of t in equations (1) and (2). Equate these factors:

$$\omega = 2\pi f = \frac{\pi}{8.00} \text{ rad/s} = 0.393 \text{ rad/s}$$

Divide ω by 2π to get the frequency f:

$$f = \frac{\omega}{2\pi} = 0.062\ 5 \text{ Hz}$$

The period T is the reciprocal of the frequency:

$$T = \frac{1}{f} = 16.0 \text{ s}$$

(b) Find the maximum magnitudes of the velocity and the acceleration.

Calculate the maximum speed from the factor in front of the sine function in Equation 13.14b:

$$v_{max} = A\omega = (0.250 \text{ m})(0.393 \text{ rad/s}) = 0.098\ 3 \text{ m/s}$$

Calculate the maximum acceleration from the factor in front of the cosine function in Equation 13.14c:

$$a_{max} = A\omega^2 = (0.250 \text{ m})(0.393 \text{ rad/s})^2 = 0.038\ 6 \text{ m/s}^2$$

(c) Find the position, velocity, and acceleration of the object after 1.00 s.

Substitute $t = 1.00$ s in the given equation:

$$x = (0.250 \text{ m}) \cos(0.393 \text{ rad}) = 0.231 \text{ m}$$

Substitute values into the velocity equation:

$$v = -A\omega \sin(\omega t)$$
$$= -(0.250 \text{ m})(0.393 \text{ rad/s}) \sin(0.393 \text{ rad/s} \cdot 1.00 \text{ s})$$
$$v = -0.037\ 6 \text{ m/s}$$

Substitute values into the acceleration equation:

$$a = -A\omega^2 \cos(\omega t)$$
$$= -(0.250 \text{ m})(0.393 \text{ rad/s}^2)^2 \cos(0.393 \text{ rad/s} \cdot 1.00 \text{ s})$$
$$a = -0.035\ 7 \text{ m/s}^2$$

Remarks   In evaluating the cosine function, the angle is in radians, so you should either set your calculator to evaluate trigonometric functions based on radian measure or convert from radians to degrees.

Exercise 13.6
If the object–spring system is described by $x = (0.330 \text{ m}) \cos(1.50t)$, find (a) the amplitude, the angular frequency, the frequency, and the period, (b) the maximum magnitudes of the velocity and acceleration, and (c) the position, velocity, and acceleration when $t = 0.250$ s.

Answers   (a) $A = 0.330$ m, $\omega = 1.50$ rad/s, $f = 0.239$ Hz , $T = 4.19$ s;   (b) $v_{max} = 0.495$ m/s, $a_{max} = 0.743$ m/s^2; (c) $x = 0.307$ m, $v = -0.181$ m/s, $a = -0.691$ m/s^2

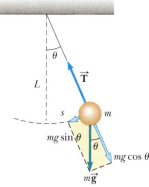

ACTIVE FIGURE 13.15

A simple pendulum consists of a bob of mass m suspended by a light string of length L. (L is the distance from the pivot to the center of mass of the bob.) The restoring force that causes the pendulum to undergo simple harmonic motion is the component of gravitational force tangent to the path of motion, $mg \sin \theta$.

Physics⊗Now™

Log into PhysicsNow at **www.cp7e.com**, and go to Active Figure 13.15 to adjust the mass of the bob, the length of the string, and the initial angle and see the resulting oscillation of the pendulum. The period is slightly larger for larger initial angles.

TIP 13.3   Pendulum Motion is not Harmonic

Remember that the pendulum *does not* exhibit true simple harmonic motion for *any* angle. If the angle is less than about 15°, the motion can be *modeled* as approximately simple harmonic.

13.5   MOTION OF A PENDULUM

A simple pendulum is another mechanical system that exhibits periodic motion. It consists of a small bob of mass m suspended by a light string of length L fixed at its upper end, as in Active Figure 13.15. (By a light string, we mean that the string's mass is assumed to be very small compared with the mass of the bob and hence can be ignored.) When released, the bob swings to and fro over the same path; but is its motion simple harmonic?

Answering this question requires examining the restoring force—the force of gravity—that acts on the pendulum. The pendulum bob moves along a circular arc, rather than back and forth in a straight line. When the oscillations are small, however, the motion of the bob is nearly straight, so Hooke's law may apply approximately.

In Active Figure 13.15, s is the displacement of the bob from equilibrium along the arc. Hooke's law is $F = -kx$, so we are looking for a similar expression involving s, $F_t = -ks$, where F_t is the force acting in a direction tangent to the circular arc. From the figure, the restoring force is

$$F_t = -mg \sin \theta$$

Since $s = L\theta$, the equation for F_t can be written as

$$F_t = -mg \sin\left(\frac{s}{L}\right)$$

This expression isn't of the form $F_t = -ks$, so in general, the motion of a pendulum is *not* simple harmonic. For small angles less than about 15 degrees, however, the angle θ measured in radians and the sine of the angle are approximately equal. For example, $\theta = 10.0° = 0.175$ rad, and $\sin(10.0°) = 0.174$. Therefore, if we restrict the motion to *small* angles, the approximation $\sin \theta \approx \theta$ is valid, and the restoring force can be written

$$F_t = -mg \sin \theta \approx -mg \theta$$

Substituting $\theta = s/L$, we obtain

$$F_t = -\left(\frac{mg}{L}\right) s$$

This equation follows the general form of Hooke's force law $F_t = -ks$, with $k = mg/L$. We are justified in saying that a pendulum undergoes simple harmonic motion only when it swings back and forth at small amplitudes (or, in this case, small values of θ, so that $\sin \theta \cong \theta$).

Recall that for the object–spring system, the angular frequency is given by Equation 13.11:

$$\omega = 2\pi f = \sqrt{\frac{k}{m}}$$

Substituting the expression of k for a pendulum, we obtain

$$\omega = \sqrt{\frac{mg/L}{m}} = \sqrt{\frac{g}{L}}$$

This angular frequency can be substituted into Equation 13.12, which then mathematically describes the motion of a pendulum. The frequency is just the angular frequency divided by 2π, while the period is the reciprocal of the frequency, or

The period of a simple pendulum depends only on L and g ▶

$$T = 2\pi \sqrt{\frac{L}{g}}$$

[13.15]

This equation reveals the somewhat surprising result that the period of a simple pendulum doesn't depend on the mass, but only on the pendulum's length and on the free-fall acceleration. Furthermore, the amplitude of the motion isn't a

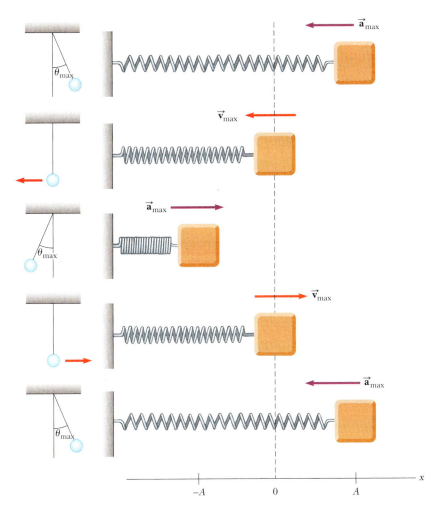

ACTIVE FIGURE 13.16
Simple harmonic motion for an
object–spring system, and its analogy,
the motion of a simple pendulum.

Physics⊗Now™
Log into PhysicsNow at
www.cp7e.com, and go to Active
Figure 13.16 to set the initial position
of the block and see the
block–spring motion and the
analogous pendulum motion.

factor as long as it's relatively small. The analogy between the motion of a simple
pendulum and the object–spring system is illustrated in Active Figure 13.16.

Galileo first noted that the period of a pendulum was independent of its amplitude. He supposedly observed this while attending church services at the cathedral
in Pisa. The pendulum he studied was a swinging chandelier that was set in motion
when someone bumped it while lighting candles. Galileo was able to measure its
period by timing the swings with his pulse.

The dependence of the period of a pendulum on its length and on the free-fall
acceleration allows us to use a pendulum as a timekeeper for a clock. A number of
clock designs employ a pendulum, with the length adjusted so that its period
serves as the basis for the rate at which the clock's hands turn. Of course, these
clocks are used at different locations on the Earth, so there will be some variation
of the free-fall acceleration. To compensate for this variation, the pendulum of a
clock should have some movable mass so that the effective length can be adjusted.

Geologists often make use of the simple pendulum and Equation 13.15 when
prospecting for oil or minerals. Deposits beneath the Earth's surface can produce
irregularities in the free-fall acceleration over the region being studied. A specially
designed pendulum of known length is used to measure the period, which in turn
is used to calculate g. Although such a measurement in itself is inconclusive, it's an
important tool for geological surveys.

APPLICATION
Pendulum Clocks

APPLICATION
Use of Pendulum
in Prospecting

Quick Quiz 13.7

A simple pendulum is suspended from the ceiling of a stationary elevator, and the
period is measured. If the elevator moves with constant velocity, does the period
(a) increase, (b) decrease, or (c) remain the same? If the elevator accelerates upward, does the period (a) increase, (b) decrease, or (c) remain the same?

Quick Quiz 13.8

A pendulum clock depends on the period of a pendulum to keep correct time. Suppose a pendulum clock is keeping correct time and then Dennis the Menace slides the bob of the pendulum downward on the oscillating rod. Does the clock run (a) slow, (b) fast, or (c) correctly?

Quick Quiz 13.9

The period of a simple pendulum is measured to be T on Earth. If the same pendulum were set in motion on the Moon, would its period be (a) less than T, (b) greater than T, or (c) equal to T?

EXAMPLE 13.7   Measuring the Value of g

Goal   Determine g from pendulum motion.

Problem   Using a small pendulum of length 0.171 m, a geologist counts 72.0 complete swings in a time of 60.0 s. What is the value of g in this location?

Strategy   First calculate the period of the pendulum by dividing the total time by the number of complete swings. Solve Equation 13.15 for g and substitute values.

Solution

Calculate the period by dividing the total elapsed time by the number of complete oscillations:

$$T = \frac{\text{time}}{\text{\# of oscillations}} = \frac{60.0 \text{ s}}{72.0} = 0.833 \text{ s}$$

Solve Equation 13.15 for g and substitute values:

$$T = 2\pi \sqrt{\frac{L}{g}} \quad \rightarrow \quad T^2 = 4\pi^2 \frac{L}{g}$$

$$g = \frac{4\pi^2 L}{T^2} = \frac{(39.5)(0.171 \text{ m})}{(0.833 \text{ s})^2} = \boxed{9.73 \text{ m/s}^2}$$

Exercise 13.7

What would be the period of the same pendulum on the Moon, where the acceleration of gravity is 1.62 m/s^2?

Answer   2.04 s

The Physical Pendulum

The simple pendulum discussed thus far consists of a mass attached to a string. A pendulum, however, can be made from an object of any shape. The general case is called the *physical pendulum*.

In Figure 13.17, a rigid object is pivoted at point O, which is a distance L from the object's center of mass. The center of mass oscillates along a circular arc, just like the simple pendulum. The period of a physical pendulum is given by

$$T = 2\pi \sqrt{\frac{I}{mgL}} \qquad [13.16]$$

where I is the object's moment of inertia and m is the object's mass. As a check, notice that in the special case of a simple pendulum with an arm of length L and negligible mass, the moment of inertia is $I = mL^2$. Substituting into Equation 13.16 results in

$$T = 2\pi \sqrt{\frac{mL^2}{mgL}} = 2\pi \sqrt{\frac{L}{g}}$$

which is the correct period for a simple pendulum.

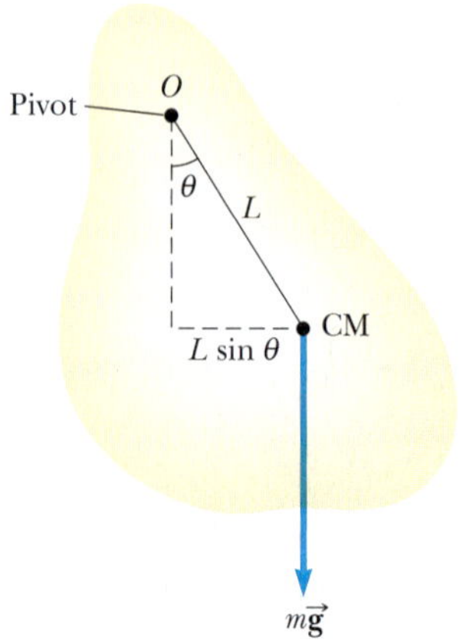

Figure 13.17   A physical pendulum pivoted at O.

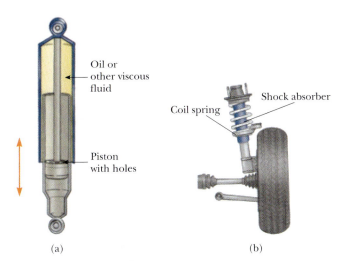

Figure 13.18 (a) A shock absorber consists of a piston oscillating in a chamber filled with oil. As the piston oscillates, the oil is squeezed through holes between the piston and the chamber, causing a damping of the piston's oscillations. (b) One type of automotive suspension system, in which a shock absorber is placed inside a coil spring at each wheel.

13.6   DAMPED OSCILLATIONS

The vibrating motions we have discussed so far have taken place in ideal systems that *oscillate indefinitely* under the action of a linear restoring force. In all real mechanical systems, forces of friction retard the motion, so the systems don't oscillate indefinitely. The friction reduces the mechanical energy of the system as time passes, and the motion is said to be **damped**.

Shock absorbers in automobiles (Fig. 13.18) are one practical application of damped motion. A shock absorber consists of a piston moving through a liquid such as oil. The upper part of the shock absorber is firmly attached to the body of the car. When the car travels over a bump in the road, holes in the piston allow it to move up and down in the fluid in a damped fashion.

Damped motion varies with the fluid used. For example, if the fluid has a relatively low viscosity, the vibrating motion is preserved but the amplitude of vibration decreases in time and the motion ultimately ceases. This is known as *underdamped* oscillation. The position vs. time curve for an object undergoing such oscillation appears in Active Figure 13.19. Figure 13.20 compares three types of damped motion, with curve (a) representing underdamped oscillation. If the fluid viscosity is increased, the object returns rapidly to equilibrium after it's released and doesn't oscillate. In this case, the system is said to be *critically damped*, and is shown as curve (b) in Figure 13.20. The piston returns to the equilibrium position in the shortest time possible without once overshooting the equilibrium position. If the viscosity is made greater still, the system is said to be *overdamped*. In this case, the piston returns to equilibrium without ever passing through the equilibrium point, but the time required to reach equilibrium is greater than in critical damping, as illustrated by curve (c) in Figure 13.20.

To make automobiles more comfortable to ride in, shock absorbers are designed to be slightly underdamped. This can be demonstrated by a sharp downward push on the hood of a car: After the applied force is removed, the body of the car oscillates a few times about the equilibrium position before returning to its fixed position.

13.7   WAVES

The world is full of waves: sound waves, waves on a string, seismic waves, and electromagnetic waves, such as visible light, radio waves, television signals, and *x*-rays. All of these waves have as their source a vibrating object, so we can apply the concepts of simple harmonic motion in describing them.

In the case of sound waves, the vibrations that produce waves arise from sources such as a person's vocal chords or a plucked guitar string. The vibrations of electrons in an antenna produce radio or television waves, and the simple up-and-down motion

APPLICATION

Shock Absorbers

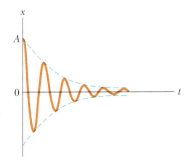

ACTIVE FIGURE 13.19

A graph of displacement versus time for an underdamped oscillator. Note the decrease in amplitude with time.

Physics⊗Now™

Log into PhysicsNow at **www.cp7e.com**, and go to Active Figure 13.19 to adjust the spring constant, the mass of the object, and the damping constant and see the resulting damped oscillation of the object.

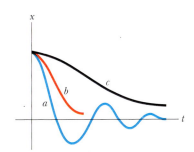

Figure 13.20 Plots of displacement versus time for (a) an underdamped oscillator, (b) a critically damped oscillator, and (c) an overdamped oscillator.

of a hand can produce a wave on a string. Certain concepts are common to all waves, regardless of their nature. In the remainder of this chapter, we focus our attention on the general properties of waves. In later chapters we will study specific types of waves, such as sound waves and electromagnetic waves.

What Is a Wave?

When you drop a pebble into a pool of water, the disturbance produces water waves, which move away from the point where the pebble entered the water. A leaf floating near the disturbance moves up and down and back and forth about its original position, but doesn't undergo any net displacement attributable to the disturbance. This means that the water wave (or disturbance) moves from one place to another, *but the water isn't carried with it.*

When we observe a water wave, we see a rearrangement of the water's surface. Without the water, there wouldn't be a wave. Similarly, a wave traveling on a string wouldn't exist without the string. Sound waves travel through air as a result of pressure variations from point to point. Therefore, we can consider a wave to be *the motion of a disturbance.* In a later chapter we will discuss electromagnetic waves, which don't require a medium.

The mechanical waves discussed in this chapter require (1) some source of disturbance, (2) a medium that can be disturbed, and (3) some physical connection or mechanism through which adjacent portions of the medium can influence each other. All waves carry energy and momentum. The amount of energy transmitted through a medium and the mechanism responsible for the transport of energy differ from case to case. The energy carried by ocean waves during a storm, for example, is much greater than the energy carried by a sound wave generated by a single human voice.

Applying Physics 13.2   Burying Bond

At one point in *On Her Majesty's Secret Service,* a James Bond film from the 1960s, Bond was escaping on skis. He had a good lead and was a hard-to-hit moving target. There was no point in wasting bullets shooting at him, so why did the bad guys open fire?

Explanation   These misguided gentlemen had a good understanding of the physics of waves. An impulsive sound, like a gunshot, can cause an

acoustical disturbance that propagates through the air. If it impacts a ledge of snow that is ready to break free, an avalanche can result. Such a disaster occurred in 1916 during World War I when Austrian soldiers in the Alps were smothered by an avalanche caused by cannon fire. So the bad guys, who have never been able to hit Bond with a bullet, decided to use the sound of gunfire to start an avalanche.

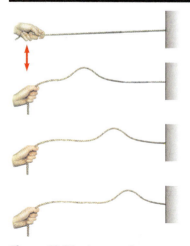

Figure 13.21   A wave pulse traveling along a stretched rope. The shape of the pulse is approximately unchanged as it travels.

Types of Waves

One of the simplest ways to demonstrate wave motion is to flip one end of a long rope that is under tension and has its opposite end fixed, as in Figure 13.21. The bump (called a pulse) travels to the right with a definite speed. A disturbance of this type is called a **traveling wave**. The figure shows the shape of the rope at three closely spaced times.

As such a wave pulse travels along the rope, **each segment of the rope that is disturbed moves in a direction perpendicular to the wave motion**. Figure 13.22 illustrates this point for a particular tiny segment P. The rope never moves in the direction of the wave. A traveling wave in which the particles of the disturbed medium move in a direction perpendicular to the wave velocity is called a **transverse wave**. Figure 13.23a illustrates the formation of transverse waves on a long spring.

In another class of waves, called **longitudinal waves, the elements of the medium undergo displacements parallel to the direction of wave motion**. Sound

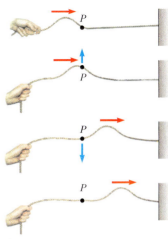

Figure 13.22   A pulse traveling on a stretched rope is a transverse wave. Any element P on the rope moves (blue arrows) in a direction perpendicular to the direction of propagation of the wave motion (red arrows).

waves in air are longitudinal. Their disturbance corresponds to a series of high- and low-pressure regions that may travel through air or through any material medium with a certain speed. A longitudinal pulse can easily be produced in a stretched spring, as in Figure 13.23b. The free end is pumped back and forth along the length of the spring. This action produces compressed and stretched regions of the coil that travel along the spring, parallel to the wave motion.

Waves need not be purely transverse or purely longitudinal: ocean waves exhibit a superposition of both types. When an ocean wave encounters a cork, the cork executes a circular motion, going up and down while going forward and back.

Another type of wave, called a **soliton**, consists of a solitary wave front that propagates in isolation. Ordinary water waves generally spread out and dissipate, but solitons tend to maintain their form. The study of solitons began in 1849, when the Scottish engineer John Scott Russell noticed a solitary wave leaving the turbulence in front of a barge and propagating forward all on its own. The wave maintained its shape and traveled down a canal at about 10 mi/h. Russell chased the wave two miles on horseback before losing it. Only in the 1960s did scientists take solitons seriously; they are now widely used to model physical phenomena, from elementary particles to the Giant Red Spot of Jupiter.

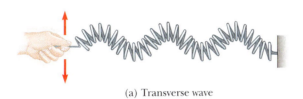

(a) Transverse wave

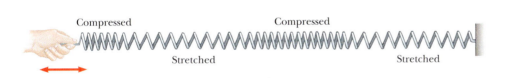

(b) Longitudinal wave

Figure 13.23   (a) A transverse wave is set up in a spring by moving one end of the spring perpendicular to its length. (b) A longitudinal pulse along a stretched spring. The displacement of the coils is in the direction of the wave motion. For the starting motion described in the text, the compressed region is followed by a stretched region.

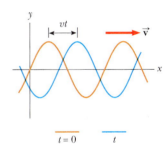

ACTIVE FIGURE 13.24
A one-dimensional sinusoidal wave traveling to the right with a speed v. The brown curve is a snapshot of the wave at $t = 0$, and the blue curve is another snapshot at some later time t.

Physics❋Now™
Log into PhysicsNow at **www.cp7e.com**, and go to Active Figure 13.24 to watch the wave move and to take snapshots of it at various times.

Picture of a Wave

Active Figure 13.24 shows the curved shape of a vibrating string. This pattern is a sinusoidal curve, the same as in simple harmonic motion. The brown curve can be thought of as a snapshot of a traveling wave taken at some instant of time, say, $t = 0$; the blue curve is a snapshot of the same traveling wave at a later time. This picture can also be used to represent a wave on water. In such a case, a high point would correspond to the *crest* of the wave and a low point to the *trough* of the wave.

The same waveform can be used to describe a longitudinal wave, even though no up-and-down motion is taking place. Consider a longitudinal wave traveling on a spring. Figure 13.25a is a snapshot of this wave at some instant, and Figure 13.25b shows the sinusoidal curve that represents the wave. Points where the coils of the spring are compressed correspond to the crests of the waveform, and stretched regions correspond to troughs.

The type of wave represented by the curve in Figure 13.25b is often called a *density wave* or *pressure wave*, because the crests, where the spring coils are compressed, are regions of high density, and the troughs, where the coils are stretched, are regions of low density. Sound waves are longitudinal waves, propagating as a series of high- and low-density regions.

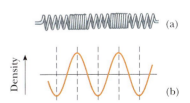

Figure 13.25   (a) A longitudinal wave on a spring. (b) The crests of the waveform correspond to compressed regions of the spring, and the troughs correspond to stretched regions of the spring.

13.8 FREQUENCY, AMPLITUDE, AND WAVELENGTH

Active Figure 13.26 illustrates a method of producing a continuous wave or a steady stream of pulses on a very long string. One end of the string is connected to a blade that is set vibrating. As the blade oscillates vertically with simple harmonic motion, a traveling wave moving to the right is set up in the string. Active Figure 13.26 consists of views of the wave at intervals of one-quarter of a period. Note that **each small segment of the string, such as P, oscillates vertically in the y-direction with simple harmonic motion.** This must be the case, because each segment follows the simple harmonic motion of the blade. Every segment of the string can therefore be treated as a simple harmonic oscillator vibrating with the same frequency as the blade that drives the string.

The frequencies of the waves studied in this course will range from rather low values for waves on strings and waves on water, to values for sound waves between 20 Hz and 20 000 Hz (recall that $1\text{ Hz} = 1\text{ s}^{-1}$), to much higher frequencies for electromagnetic waves. These waves have different physical sources, but can be described with the same concepts.

The horizontal dashed line in Active Figure 13.26 represents the position of the string when no wave is present. The maximum distance the string moves above or below this equilibrium value is called the **amplitude** A of the wave. For the waves we work with, the amplitudes at the crest and the trough will be identical.

Active Figure 13.26b illustrates another characteristic of a wave. The horizontal arrows show the distance between two successive points that behave identically. This distance is called the **wavelength** λ (the Greek letter lambda).

We can use these definitions to derive an expression for the speed of a wave. We start with the defining equation for the **wave speed** v:

$$v = \frac{\Delta x}{\Delta t}$$

The wave speed is the speed at which a particular part of the wave—say, a crest—moves through the medium.

A wave advances a distance of one wavelength in a time interval equal to one period of the vibration. Taking $\Delta x = \lambda$ and $\Delta t = T$, we see that

$$v = \frac{\lambda}{T}$$

Because the frequency is the reciprocal of the period, we have

Wave speed ▶

$$v = f\lambda \qquad\qquad \textbf{[13.17]}$$

This important general equation applies to many different types of waves, such as sound waves and electromagnetic waves.

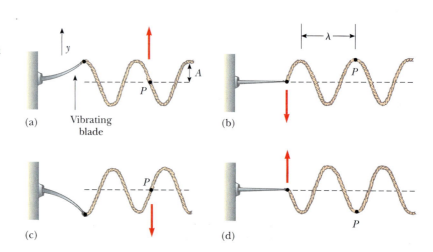

(a) Vibrating blade

(b)

(c)

(d)

EXAMPLE 13.8   A Traveling Wave

Goal   Obtain information about a wave directly from its graph.

Problem   A wave traveling in the positive x-direction is pictured in Figure 13.27a. Find the amplitude, wavelength, speed, and period of the wave if it has a frequency of 8.00 Hz. In Figure 13.27a, $\Delta x = 40.0$ cm and $\Delta y = 15.0$ cm.

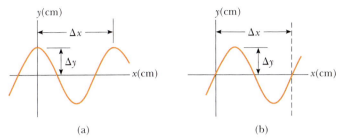

Figure 13.27   (a) (Example 13.8) (b) (Exercise 13.8)

Strategy   The amplitude and wavelength can be read directly from the figure: The maximum vertical displacement is the amplitude, and the distance from one crest to the next is the wavelength. Multiplying the wavelength by the frequency gives the speed, while the period is just the reciprocal of the frequency.

Solution

The maximum wave displacement is the amplitude A:

$$A = \Delta y = 15.0 \text{ cm} = \boxed{0.150 \text{ m}}$$

The distance from crest to crest is the wavelength:

$$\lambda = \Delta x = 40.0 \text{ cm} = \boxed{0.400 \text{ m}}$$

Multiply the wavelength by the frequency to get the speed of the wave.

$$v = f\lambda = (8.00 \text{ Hz})(0.400 \text{ m}) = \boxed{3.20 \text{ m/s}}$$

Take the reciprocal of the frequency to get the period:

$$T = \frac{1}{f} = \frac{1}{8.00} \text{ s} = \boxed{0.125 \text{ s}}$$

Exercise 13.8

A wave traveling in the positive x-direction is pictured in Figure 13.27b. Find the amplitude, wavelength, speed, and period of the wave if it has a frequency of 15.0 Hz. In the figure, $\Delta x = 72.0$ cm and $\Delta y = 25.0$ cm.

Answers $A = 0.25$ m, $\lambda = 0.720$ m, $v = 10.8$ m/s, $T = 0.066\ 7$ s

EXAMPLE 13.9   Sound and Light

Goal   Perform elementary calculations using speed, wavelength, and frequency.

Problem   A wave has a wavelength of 3.00 m. Calculate the frequency of the wave if it is **(a)** a sound wave and **(b)** a light wave. Take the speed of sound as 343 m/s and the speed of light as 3.00×10^8 m/s.

Solution

(a) Find the frequency of a sound wave with $\lambda = 3.00$ m.

Solve Equation 3.17 for the frequency and substitute:

$$f = \frac{v}{\lambda} = \frac{343 \text{ m/s}}{3.00 \text{ m}} = \boxed{114 \text{ Hz}} \qquad (1)$$

(b) Find the frequency of a light wave with $\lambda = 3.00$ m.

Substitute into Equation (1), using the speed of light for c:

$$f = \frac{c}{\lambda} = \frac{3.00 \times 10^8 \text{ m/s}}{3.00 \text{ m}} = \boxed{1.00 \times 10^8 \text{ Hz}}$$

Remark   The same equation can be used to find the frequency in each case, despite the great difference between the physical phenomena. Notice how much larger frequencies of light waves are than frequencies of sound waves.

Exercise 13.9

(a) Find the wavelength of an electromagnetic wave with frequency 9.00 GHz = 9.00×10^9 Hz (G = giga = 10^9), which is in the microwave range. (b) Find the speed of a sound wave in an unknown fluid medium if a frequency of 567 Hz has a wavelength of 2.50 m.

Answers   (a) 0.0333 m   (b) 1.42×10^3 m/s

13.9   THE SPEED OF WAVES ON STRINGS

In this section we focus our attention on the speed of a transverse wave on a stretched string.

For a vibrating string, there are two speeds to consider. One is the speed of the physical string that vibrates up and down, transverse to the string, in the y-direction. The other is the *wave speed*, which is the rate at which the disturbance propagates along the length of the string in the x-direction. We wish to find an expression for the wave speed.

If a horizontal string under tension is pulled vertically and released, it starts at its maximum displacement, $y = A$, and takes a certain amount of time to go to $y = -A$ and back to A again. This amount of time is the period of the wave, and is the same as the time needed for the wave to advance *horizontally* by one wavelength. Dividing the wavelength by the period of one transverse oscillation gives the wave speed.

For a fixed wavelength, a string under greater tension F has a greater wave speed because the period of vibration is shorter, and the wave advances one wavelength during one period. It also makes sense that a string with greater mass per unit length, μ, vibrates more slowly, leading to a longer period and a slower wave speed. The wave speed is given by

$$v = \sqrt{\frac{F}{\mu}}$$ [13.18]

where F is the tension in the string and μ is the mass of the string per unit length, called the *linear density*. From Equation 13.18, it's clear that a larger tension F results in a larger wave speed, while a larger linear density μ gives a slower wave speed, as expected.

According to Equation 13.18, the propagation speed of a mechanical wave, such as a wave on a string, depends only on the properties of the string through which the disturbance travels. It doesn't depend on the amplitude of the vibration. This turns out to be generally true of waves in various media.

A proof of Equation 13.18 requires calculus, but dimensional analysis can easily verify that the expression is dimensionally correct. The dimensions of F are ML/T^2, and the dimensions of μ are M/L. The dimensions of F/μ are therefore L^2/T^2, so those of $\sqrt{F/\mu}$ are L/T, giving the dimensions of speed. No other combination of F and μ is dimensionally correct, so in the case where the tension and mass density are the only relevant physical factors, we have verified Equation 13.18 up to an overall constant.

APPLICATION

Bass Guitar Strings

According to Equation 13.18, we can increase the speed of a wave on a stretched string by increasing the tension in the string. Increasing the mass per unit length, on the other hand, decreases the wave speed. These physical facts lie behind the metallic windings on the bass strings of pianos and guitars. The windings increase the mass per unit length, μ, decreasing the wave speed and hence the frequency, resulting in a lower tone. Tuning a string to a desired frequency is a simple matter of changing the tension in the string.

INTERACTIVE EXAMPLE 13.10   A Pulse Traveling on a String

Goal   Calculate the speed of a wave on a string.

Problem   A uniform string has a mass M of 0.030 0 kg and a length L of 6.00 m. Tension is maintained in the string by suspending a block of mass $m = 2.00$ kg from one end (Fig. 13.28). **(a)** Find the speed of a transverse wave pulse on this string. **(b)** Find the time it takes the pulse to travel from the wall to the pulley.

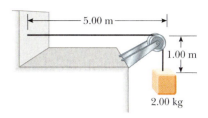

Figure 13.28 (Interactive Example 13.10) The tension F in the string is maintained by the suspended block. The wave speed is given by the expression $v = \sqrt{F/\mu}$.

Strategy   The tension F can be obtained from Newton's second law for equilibrium applied to the block, and the mass per unit length of the string is $\mu = M/L$. With these quantities, the speed of the transverse pulse can be found by substitution into Equation 13.18. Part (b) requires the formula $d = vt$.

Solution

(a) Find the speed of the wave pulse.

Apply the second law to the block: the tension F is equal and opposite to the force of gravity.

$$\Sigma F = F - mg = 0 \quad \rightarrow \quad F = mg$$

Substitute expressions for F and μ into Equation 13.18:

$$v = \sqrt{\frac{F}{\mu}} = \sqrt{\frac{mg}{M/L}}$$

$$= \sqrt{\frac{(2.00 \text{ kg})(9.80 \text{ m/s}^2)}{(0.030\ 0 \text{ kg})/(6.00 \text{ m})}} = \sqrt{\frac{19.6 \text{ N}}{0.005\ 00 \text{ kg/m}}}$$

$$= \boxed{62.6 \text{ m/s}}$$

(b) Find the time it takes the pulse to travel from the wall to the pulley.

Solve the distance formula for time:

$$t = \frac{d}{v} = \frac{5.00 \text{ m}}{62.6 \text{ m/s}} = 0.0799 \text{ s}$$

Exercise 13.10

To what tension must a string with mass 0.010 0 kg and length 2.50 m be tightened so that waves will travel on it at a speed of 125 m/s?

Answer   62.5 N

Physics⊗Now™ Investigate the transmission of such pulses by logging into PhysicsNow at **www.cp7e.com** and going to Interactive Example 13.10.

13.10   INTERFERENCE OF WAVES

Many interesting wave phenomena in nature require two or more waves passing through the same region of space at the same time. **Two traveling waves can meet and pass through each other without being destroyed or even altered**. For instance, when two pebbles are thrown into a pond, the expanding circular waves don't destroy each other. In fact, the ripples pass through each other. Likewise, when sound waves from two sources move through air, they pass through each other. In the region of overlap, the resultant wave is found by adding the displacements of the individual waves. For such analyses, the **superposition principle** applies:

When two or more traveling waves encounter each other while moving through a medium, the resultant wave is found by adding together the displacements of the individual waves point by point.

◀ Superposition principle

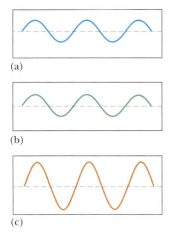

Figure 13.29 Constructive interference. If two waves having the same frequency and amplitude are in phase, as in (a) and (b), the resultant wave when they combine (c) has the same frequency as the individual waves, but twice their amplitude.

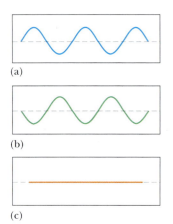

Figure 13.30 Destructive interference. When two waves with the same frequency and amplitude are 180° out of phase, as in (a) and (b), the result when they combine (c) is complete cancellation.

Figure 13.31 Interference patterns produced by outward-spreading waves from many drops of liquid falling into a body of water.

Experiments show that the superposition principle is valid only when the individual waves have small amplitudes of displacement—an assumption we make in all our examples.

Figures 13.29a and 13.29b show two waves of the same amplitude and frequency. If at some instant of time these two waves were traveling through the same region of space, the resultant wave at that instant would have a shape like that shown in Figure 13.29c. For example, suppose the waves are water waves of amplitude 1 m. At the instant they overlap so that crest meets crest and trough meets trough, the resultant wave has an amplitude of 2 m. Waves coming together like this are said to be *in phase* and to exhibit **constructive interference**.

Figures 13.30a and 13.30b show two similar waves. In this case, however, the crest of one coincides with the trough of the other, so one wave is *inverted* relative to the other. The resultant wave, shown in Figure 13.30c, is seen to be a state of complete cancellation. If these were water waves coming together, one of the waves would exert an upward force on an individual drop of water at the same instant the other wave was exerting a downward force. The result would be no motion of the water at all. In such a situation, the two waves are said to be 180° out of phase and to exhibit **destructive interference**. Figure 13.31 illustrates the interference of water waves produced by drops of water falling into a pond.

Active Figure 13.32 shows constructive interference in two pulses moving toward each other along a stretched string; Active Figure 13.33 shows destructive interference in two pulses. Notice in both figures that when the two pulses separate, their shapes are unchanged, as if they had never met!

13.11 REFLECTION OF WAVES

In our discussion so far, we have assumed that waves could travel indefinitely without striking anything. Often, such conditions are not realized in practice. Whenever a traveling wave reaches a boundary, part or all of the wave is reflected. For example, consider a pulse traveling on a string that is fixed at one end (Active Fig. 13.34). When the pulse reaches the wall, it is reflected.

Note that the reflected pulse is inverted. This can be explained as follows: When the pulse meets the wall, the string exerts an upward force on the wall. According to Newton's third law, the wall must exert an equal and opposite (downward) reaction force on the string. This downward force causes the pulse to invert on reflection.

Now suppose the pulse arrives at the string's end, and the end is attached to a ring of negligible mass that is free to slide along the post without friction (Active Fig. 13.35). Again the pulse is reflected, but this time it is not inverted. On reaching

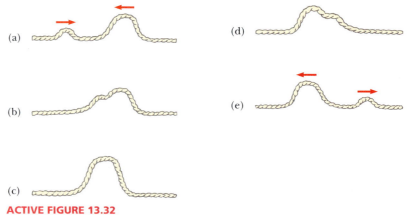

ACTIVE FIGURE 13.32

Two wave pulses traveling on a stretched string in opposite directions pass through each other. When the pulses overlap, as in (b), (c), and (d), the net displacement of the string equals the sum of the displacements produced by each pulse.

Physics⊗Now™

Log into PhysicsNow at **www.cp7e.com**, and go to Active Figure 13.32 to choose the amplitude and orientation of each of the pulses and study the interference between them as they pass each other.

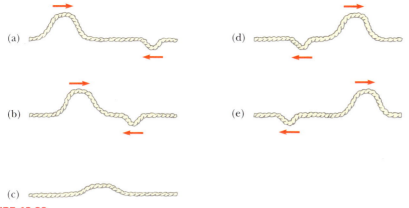

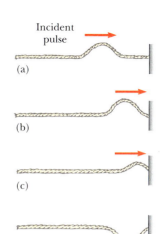

ACTIVE FIGURE 13.33
Two wave pulses traveling in opposite directions with displacements that are inverted relative to each other. When the two overlap, as in (c), their displacements subtract from each other.

Physics⊗Now™
Log into PhysicsNow at **www.cp7e.com**, and go to Active Figure 13.33 to choose the amplitude and orientation of each of the pulses and study the interference between them as they pass each other.

the post, the pulse exerts a force on the ring, causing it to accelerate upward. The ring is then returned to its original position by the downward component of the tension in the string.

An alternate method of showing that a pulse is reflected without inversion when it strikes a free end of a string is to send the pulse down a string hanging vertically. When the pulse hits the free end, it's reflected without inversion, just as is the pulse in Active Figure 13.35.

Finally, when a pulse reaches a boundary, it's partly reflected and partly transmitted past the boundary into the new medium. This effect is easy to observe in the case of two ropes of different density joined at some boundary.

ACTIVE FIGURE 13.34
The reflection of a traveling wave at the fixed end of a stretched string. Note that the reflected pulse is inverted, but its shape remains the same.

Physics⊗Now™
Log into PhysicsNow at **www.cp7e.com**, and go to Active Figure 13.34 to adjust the linear mass density of the string and the transverse direction of the initial pulse.

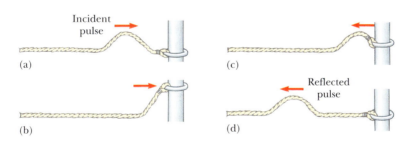

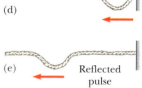

ACTIVE FIGURE 13.35
The reflection of a traveling wave at the free end of a stretched string. In this case, the reflected pulse is not inverted.

Physics⊗Now™
Log into PhysicsNow at **www.cp7e.com**, and go to Active Figure 13.35 to adjust the linear mass density of the string and the transverse direction of the initial pulse.

SUMMARY

Physics⊗Now™ Take a practice test by logging into PhysicsNow at **www.cp7e.com** and clicking on the Pre-Test link for this chapter.

13.1 Hooke's Law
Simple harmonic motion occurs when the net force on an object along the direction of motion is proportional to the object's displacement and in the opposite direction:

$$F_s = -kx \quad [13.1]$$

This is called Hooke's law. The time required for one complete vibration is called the **period** of the motion. The reciprocal of the period is the **frequency** of the motion, which is the number of oscillations per second.

When an object moves with simple harmonic motion, its **acceleration** as a function of position is

$$a = -\frac{k}{m}x \quad [13.2]$$

13.2 Elastic Potential Energy
The energy stored in a stretched or compressed spring or in some other elastic material is called **elastic potential energy**:

$$PE_s \equiv \tfrac{1}{2}kx^2 \quad [13.3]$$

The **velocity** of an object as a function of position, when the object is moving with simple harmonic motion, is

$$v = \pm\sqrt{\frac{k}{m}(A^2 - x^2)} \quad [13.6]$$

13.4 Position, Velocity, and Acceleration as a Function of Time

The **period** of an object of mass m moving with simple harmonic motion while attached to a spring of spring constant k is

$$T = 2\pi\sqrt{\frac{m}{k}} \qquad \text{[13.8]}$$

where T is independent of the amplitude A.

The **frequency** of an object–spring system is $f = 1/T$. The **angular frequency** ω of the system in rad/s is

$$\omega = 2\pi f = \sqrt{\frac{k}{m}} \qquad \text{[13.11]}$$

When an object is moving with simple harmonic motion, the **position**, **velocity**, and **acceleration** of the object as a function of time are given by

$$x = A\cos(2\pi ft) \qquad \text{[13.14a]}$$

$$v = -A\omega\sin(2\pi ft) \qquad \text{[13.14b]}$$

$$a = -A\omega^2\cos(2\pi ft) \qquad \text{[13.14c]}$$

13.5 Motion of a Pendulum

A **simple pendulum** of length L moves with simple harmonic motion for small angular displacements from the vertical, with a period of

$$T = 2\pi\sqrt{\frac{L}{g}} \qquad \text{[13.15]}$$

13.7 Waves

In a **transverse wave** the elements of the medium move in a direction perpendicular to the direction of the wave. An example is a wave on a stretched string.

In a **longitudinal wave** the elements of the medium move parallel to the direction of the wave velocity. An example is a sound wave.

13.8 Frequency, Amplitude, and Wavelength

The relationship of the speed, wavelength, and frequency of a wave is

$$v = f\lambda \qquad \text{[13.17]}$$

This relationship holds for a wide variety of different waves.

13.9 The Speed of Waves on Strings

The speed of a wave traveling on a stretched string of mass per unit length μ and under tension F is

$$v = \sqrt{\frac{F}{\mu}} \qquad \text{[13.18]}$$

13.10 Interference of Waves

The **superposition principle** states that if two or more traveling waves are moving through a medium, the resultant wave is found by adding the individual waves together point by point. When waves meet crest to crest and trough to trough, they undergo **constructive interference**. When crest meets trough, the waves undergo **destructive interference**.

13.11 Reflection of Waves

When a wave pulse reflects from a rigid boundary, the pulse is inverted. When the boundary is free, the reflected pulse is not inverted.

CONCEPTUAL QUESTIONS

1. If one end of a heavy rope is attached to one end of a light rope, the speed of a wave will change as the wave goes from the heavy rope to the light one. Will the speed increase or decrease? What happens to the frequency? To the wavelength?

2. If a spring is cut in half, what happens to its spring constant?

3. An object–spring system undergoes simple harmonic motion with an amplitude A. Does the total energy change if the mass is doubled but the amplitude isn't changed? Are the kinetic and potential energies at a given point in its motion affected by the change in mass? Explain.

4. The speed of sound in air as given in this chapter (343 m/s) is an enormous speed compared to the speed of common objects. Yet the speed of sound is of the same order of magnitude as the rms speed of air molecules at 1 atmosphere and 20°C as given by the kinetic theory of gases. Is this just a remarkable coincidence? Explain.

5. An object is hung on a spring, and the frequency of oscillation of the system, f, is measured. The object, a second identical object, and the spring are carried to space in the Space Shuttle. The two objects are attached to the ends of the spring, and the system is taken out into space on a space walk. The spring is extended, and the system is released to oscillate while floating in space. The coils of the spring don't bump into one another. What is the frequency of oscillation for this system, in terms of f?

6. If an object–spring system is hung vertically and set into oscillation, why does the motion eventually stop?

7. Is a bouncing ball an example of simple harmonic motion? Is the daily movement of a student from home to school and back simple harmonic motion?

8. If a pendulum clock keeps perfect time at the base of a mountain, will it also keep perfect time when it is moved to the top of the mountain? Explain.

9. A pendulum bob is made from a sphere filled with water. What would happen to the frequency of vibration of this pendulum if the sphere had a hole in it that allowed the water to leak out slowly?

10. If a grandfather clock were running slow, how could we adjust the length of the pendulum to correct the time?

11. A grandfather clock depends on the period of a pendulum to keep correct time. Suppose such a clock is calibrated

correctly and then the temperature of the room in which it resides increases. Does the clock run slow, fast, or correctly? [*Hint:* A material expands when its temperature increases.]

12. If you stretch a rubber hose and pluck it, you can observe a pulse traveling up and down the hose. What happens to the speed of the pulse if you stretch the hose more tightly? What happens to the speed if you fill the hose with water?

13. In a long line of people waiting to buy tickets at a movie theater, when the first person leaves, a pulse of motion occurs as people step forward to fill in the gap. The gap moves through the line of people. What determines the speed of this pulse? Is it transverse or longitudinal? How about the "wave" at a baseball game, where people in the stands stand up and shout as the wave arrives at their location (Fig. Q13.13) and the pulse moves around the stadium—what determines the speed of this pulse? Is it transverse or longitudinal?

Gregg Adams/Stone/Getty Images

Figure Q13.13

14. As part of a physics open house, a department sets up a bungee jump from the top of the physics building. Assume that one end of the elastic band will be firmly attached to the top of the building and the other to the waist of a courageous participant. The participant will step off the edge of the building, to be slowed down and brought back up by the elastic band before hitting the ground. Estimate the length and spring constant of the elastic you would recommend. (Question 14 is courtesy of Edward F. Redish. For more questions of this type, see http://www.physics.umd.edu/perg/.)

15. In mechanics, massless strings are often assumed. Why is this not a good assumption when discussing waves on strings?

16. What happens to the wavelength of a wave on a string when the frequency is doubled? Assume that the tension in the string remains the same.

17. Explain why the kinetic and potential energies of an object–spring system can never be negative.

18. What happens to the speed of a wave on a string when the frequency is doubled? Assume that the tension in the string remains the same.

19. By what factor would you have to multiply the tension in a stretched spring in order to double the wave speed?

20. The left end of a spring is attached to a wall, and its right end is attached to a cart lying on a frictionless horizontal surface. An experimenter pulls the cart away from the wall and holds it there.
(a) What forces are acting on the spring? What is the total force on the spring?
(b) What forces are acting on the cart? What is the total force on the cart?
(c) If the cart is released, describe its motion.

PROBLEMS

1, 2, 3 = straightforward, intermediate, challenging   □ = full solution available in *Student Solutions Manual/Study Guide*
Physics⊗Now™ = coached problem with hints available at **www.cp7e.com**   🧬 = biomedical application

Section 13.1 Hooke's Law

1. A 0.40-kg object is attached to a spring with force constant 160 N/m so that the object is allowed to move on a horizontal frictionless surface. The object is released from rest when the spring is compressed 0.15 m. Find (a) the force on the object and (b) its acceleration at that instant.

2. A load of 50 N attached to a spring hanging vertically stretches the spring 5.0 cm. The spring is now placed horizontally on a table and stretched 11 cm. (a) What force is required to stretch the spring by that amount? (b) Plot a graph of force (on the *y*-axis) versus spring displacement from the equilibrium position along the *x*-axis.

3. **Physics⊗Now™** A ball dropped from a height of 4.00 m makes a perfectly elastic collision with the ground. Assuming that no mechanical energy is lost due to air resistance, (a) show that the motion is periodic and (b) determine the period of the motion. (c) Is the motion simple harmonic? Explain.

4. A small ball is set in horizontal motion by rolling it with a speed of 3.00 m/s across a room 12.0 m long between two walls. Assume that the collisions made with each wall are perfectly elastic and that the motion is perpendicular to the two walls. (a) Show that the motion is periodic and determine its period. (b) Is the motion simple harmonic? Explain.

5. A spring is hung from a ceiling, and an object attached to its lower end stretches the spring by a distance of 5.00 cm from its unstretched position when the system is in equilibrium. If the spring constant is 47.5 N/m, determine the mass of the object.

6. An archer must exert a force of 375 N on the bowstring shown in Figure P13.6a (page 452) such that the string makes an angle of $\theta = 35.0°$ with the vertical. (a) Determine the tension in the bowstring. (b) If the applied force is replaced by a stretched spring as in Figure P13.6b, and the spring is stretched 30.0 cm from its unstretched length, what is the spring constant?

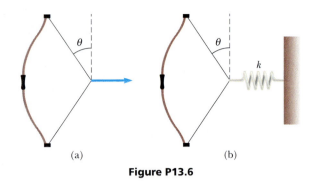

(a)   (b)

Figure P13.6

Section 13.2 Elastic Potential Energy

7. A slingshot consists of a light leather cup containing a stone. The cup is pulled back against two parallel rubber bands. It takes a force of 15 N to stretch either one of these bands 1.0 cm. (a) What is the potential energy stored in the two bands together when a 50-g stone is placed in the cup and pulled back 0.20 m from the equilibrium position? (b) With what speed does the stone leave the slingshot?

8. An archer pulls her bowstring back 0.400 m by exerting a force that increases uniformly from zero to 230 N. (a) What is the equivalent spring constant of the bow? (b) How much work is done in pulling the bow?

9. A child's toy consists of a piece of plastic attached to a spring (Fig. P13.9). The spring is compressed against the floor a distance of 2.00 cm, and the toy is released. If the toy has a mass of 100 g and rises to a maximum height of 60.0 cm, estimate the force constant of the spring.

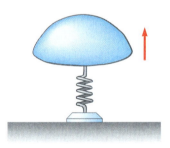

Figure P13.9

10. An automobile having a mass of 1 000 kg is driven into a brick wall in a safety test. The bumper behaves like a spring with constant 5.00×10^6 N/m and is compressed 3.16 cm as the car is brought to rest. What was the speed of the car before impact, assuming that no energy is lost in the collision with the wall?

11. A simple harmonic oscillator has a total energy E. (a) Determine the kinetic and potential energies when the displacement is one-half the amplitude. (b) For what value of the displacement does the kinetic energy equal the potential energy?

12. A 1.50-kg block at rest on a tabletop is attached to a horizontal spring having constant 19.6 N/m, as in Figure P13.12. The spring is initially unstretched. A constant 20.0-N horizontal force is applied to the object, causing the spring to stretch. (a) Determine the speed of the block after it has moved 0.300 m from equilibrium if the surface between the block and tabletop is frictionless.

(b) Answer part (a) if the coefficient of kinetic friction between block and tabletop is 0.200.

Figure P13.12

13. A 10.0-g bullet is fired into, and embeds itself in, a 2.00-kg block attached to a spring with a force constant of 19.6 N/m and whose mass is negligible. How far is the spring compressed if the bullet has a speed of 300 m/s just before it strikes the block and the block slides on a frictionless surface? [*Note:* You must use conservation of momentum in this problem. Why?]

14. A 1.5-kg block is attached to a spring with a spring constant of 2 000 N/m. The spring is then stretched a distance of 0.30 cm and the block is released from rest. (a) Calculate the speed of the block as it passes through the equilibrium position if no friction is present. (b) Calculate the speed of the block as it passes through the equilibrium position if a constant frictional force of 2.0 N retards its motion. (c) What would be the strength of the frictional force if the block reached the equilibrium position the first time with zero velocity?

Section 13.3 Comparing Simple Harmonic Motion with Uniform Circular Motion
Section 13.4 Position, Velocity, and Acceleration as a Function of Time

15. A 0.40-kg object connected to a light spring with a force constant of 19.6 N/m oscillates on a frictionless horizontal surface. If the spring is compressed 4.0 cm and released from rest, determine (a) the maximum speed of the object, (b) the speed of the object when the spring is compressed 1.5 cm, and (c) the speed of the object when the spring is stretched 1.5 cm. (d) For what value of x does the speed equal one-half the maximum speed?

16. An object–spring system oscillates with an amplitude of 3.5 cm. If the spring constant is 250 N/m and the object has a mass of 0.50 kg, determine (a) the mechanical energy of the system, (b) the maximum speed of the object, and (c) the maximum acceleration of the object.

17. At an outdoor market, a bunch of bananas is set into oscillatory motion with an amplitude of 20.0 cm on a spring with a force constant of 16.0 N/m. It is observed that the maximum speed of the bunch of bananas is 40.0 cm/s. What is the weight of the bananas in newtons?

18. A 50.0-g object is attached to a horizontal spring with a force constant of 10.0 N/m and released from rest with an amplitude of 25.0 cm. What is the velocity of the object when it is halfway to the equilibrium position if the surface is frictionless?

19. While riding behind a car traveling at 3.00 m/s, you notice that one of the car's tires has a small hemispherical bump on its rim, as in Figure P13.19. (a) Explain why the bump, from your viewpoint behind the car, executes simple harmonic motion. (b) If the radius of the car's tires is 0.30 m, what is the bump's period of oscillation?

Figure P13.19

20. An object moves uniformly around a circular path of radius 20.0 cm, making one complete revolution every 2.00 s. What are (a) the translational speed of the object, (b) the frequency of motion in hertz, and (c) the angular speed of the object?

21. Consider the simplified single-piston engine in Figure P13.21. If the wheel rotates at a constant angular speed ω, explain why the piston rod oscillates in simple harmonic motion.

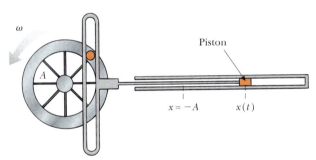

Figure P13.21

22. The frequency of vibration of an object–spring system is 5.00 Hz when a 4.00-g mass is attached to the spring. What is the force constant of the spring?

23. A spring stretches 3.9 cm when a 10-g object is hung from it. The object is replaced with a block of mass 25 g that oscillates in simple harmonic motion. Calculate the period of motion.

24. When four people with a combined mass of 320 kg sit down in a car, they find that the car drops 0.80 cm lower on its springs. Then they get out of the car and bounce it up and down. What is the frequency of the car's vibration if its mass (when it is empty) is 2.0×10^3 kg?

25. A cart of mass 250 g is placed on a frictionless horizontal air track. A spring having a spring constant of 9.5 N/m is attached between the cart and the left end of the track. When in equilibrium, the cart is located 12 cm from the left end of the track. If the cart is displaced 4.5 cm from its equilibrium position, find (a) the period at which it oscillates, (b) its maximum speed, and (c) its speed when it is 14 cm from the left end of the track.

26. The motion of an object is described by the equation

$$x = (0.30 \text{ m}) \cos\left(\frac{\pi t}{3}\right)$$

Find (a) the position of the object at $t = 0$ and $t = 0.60$ s, (b) the amplitude of the motion, (c) the frequency of the motion, and (d) the period of the motion.

27. A 2.00-kg object on a frictionless horizontal track is attached to the end of a horizontal spring whose force constant is 5.00 N/m. The object is displaced 3.00 m to the right from its equilibrium position and then released, initiating simple harmonic motion. (a) What is the force (magnitude and direction) acting on the object 3.50 s after it is released? (b) How many times does the object oscillate in 3.50 s?

28. A spring of negligible mass stretches 3.00 cm from its relaxed length when a force of 7.50 N is applied. A 0.500-kg particle rests on a frictionless horizontal surface and is attached to the free end of the spring. The particle is pulled horizontally so that it stretches the spring 5.00 cm and is then released from rest at $t = 0$. (a) What is the force constant of the spring? (b) What are the angular frequency ω, the frequency, and the period of the motion? (c) What is the total energy of the system? (d) What is the amplitude of the motion? (e) What are the maximum velocity and the maximum acceleration of the particle? (f) Determine the displacement x of the particle from the equilibrium position at $t = 0.500$ s.

29. **Physics ⊗ Now™** Given that $x = A \cos (\omega t)$ is a sinusoidal function of time, show that v (velocity) and a (acceleration) are also sinusoidal functions of time. [*Hint:* Use Equations 13.6 and 13.2.]

Section 13.5 Motion of a Pendulum

30. A man enters a tall tower, needing to know its height. He notes that a long pendulum extends from the ceiling almost to the floor and that its period is 15.5 s. (a) How tall is the tower? (b) If this pendulum is taken to the Moon, where the free-fall acceleration is 1.67 m/s², what is the period there?

31. A simple 2.00-m-long pendulum oscillates at a location where $g = 9.80$ m/s². How many complete oscillations does it make in 5.00 min?

32. An aluminum clock pendulum having a period of 1.00 s keeps perfect time at 20.0°C. (a) When placed in a room at a temperature of −5.0°C, will it gain time or lose time? (b) How much time will it gain or lose every hour? [*Hint:* See Chapter 10.]

33. A pendulum clock that works perfectly on Earth is taken to the Moon. (a) Does it run fast or slow there? (b) If the clock is started at 12:00 midnight, what will it read after one Earth day (24.0 h)? Assume that the free-fall acceleration on the Moon is 1.63 m/s².

34. A simple pendulum is 5.00 m long. (a) What is the period of simple harmonic motion for this pendulum if it is located in an elevator accelerating upward at 5.00 m/s²? (b) What is its period if the elevator is accelerating downward at 5.00 m/s²? (c) What is the period of simple harmonic motion for the pendulum if it is placed in a truck that is accelerating horizontally at 5.00 m/s²?

35. The free-fall acceleration on Mars is 3.7 m/s². (a) What length of pendulum has a period of 1 s on Earth? What length of pendulum would have a 1-s period on Mars? (b) An object is suspended from a spring with force constant 10 N/m. Find the mass suspended from this spring that would result in a period of 1 s on Earth and on Mars.

Section 13.6 Damped Oscillations
Section 13.7 Waves
Section 13.8 Frequency, Amplitude, and Wavelength

36. A cork on the surface of a pond bobs up and down two times per second on ripples having a wavelength of

8.50 cm. If the cork is 10.0 m from shore, how long does it take a ripple passing the cork to reach the shore?

37. A wave traveling in the positive *x*-direction has a frequency of 25.0 Hz, as in Figure P13.37. Find the (a) amplitude, (b) wavelength, (c) period, and (d) speed of the wave.

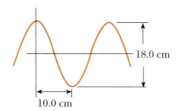

18.0 cm

10.0 cm

Figure P13.37

38. A bat can detect small objects, such as an insect, whose size is approximately equal to one wavelength of the sound the bat makes. If bats emit a chirp at a frequency of 60.0 kHz, and if the speed of sound in air is 340 m/s, what is the smallest insect a bat can detect?

39. If the frequency of oscillation of the wave emitted by an FM radio station is 88.0 MHz, determine (a) the wave's period of vibration and (b) its wavelength. (Radio waves travel at the speed of light, 3.00×10^8 m/s.)

40. The distance between two successive maxima of a transverse wave is 1.20 m. Eight crests, or maxima, pass a given point along the direction of travel every 12.0 s. Calculate the wave speed.

41. A harmonic wave is traveling along a rope. It is observed that the oscillator that generates the wave completes 40.0 vibrations in 30.0 s. Also, a given maximum travels 425 cm along the rope in 10.0 s. What is the wavelength?

42. Ocean waves are traveling to the east at 4.0 m/s with a distance of 20 m between crests. With what frequency do the waves hit the front of a boat (a) when the boat is at anchor and (b) when the boat is moving westward at 1.0 m/s?

Section 13.9 The Speed of Waves on Strings

43. A phone cord is 4.00 m long and has a mass of 0.200 kg. A transverse wave pulse is produced by plucking one end of the taut cord. The pulse makes four trips down and back along the cord in 0.800 s. What is the tension in the cord?

44. A circus performer stretches a tightrope between two towers. He strikes one end of the rope and sends a wave along it toward the other tower. He notes that it takes the wave 0.800 s to reach the opposite tower, 20.0 m away. If a 1-m length of the rope has a mass of 0.350 kg, find the tension in the tightrope.

45. Transverse waves with a speed of 50.0 m/s are to be produced on a stretched string. A 5.00-m length of string with a total mass of 0.060 0 kg is used. (a) What is the required tension in the string? (b) Calculate the wave speed in the string if the tension is 8.00 N.

46. An astronaut on the Moon wishes to measure the local value of *g* by timing pulses traveling down a wire that has a large object suspended from it. Assume a wire of mass 4.00 g is 1.60 m long and has a 3.00-kg object suspended from it. A pulse requires 36.1 ms to traverse the length of the wire. Calculate g_{Moon} from these data. (You may neglect the mass of the wire when calculating the tension in it.)

47. **Physics ⓧ Now™** A simple pendulum consists of a ball of mass 5.00 kg hanging from a uniform string of mass 0.0600 kg and length *L*. If the period of oscillation of the pendulum is 2.00 s, determine the speed of a transverse wave in the string when the pendulum hangs vertically.

48. A string is 50.0 cm long and has a mass of 3.00 g. A wave travels at 5.00 m/s along this string. A second string has the same length, but half the mass of the first. If the two strings are under the same tension, what is the speed of a wave along the second string?

49. Tension is maintained in a string as in Figure P13.49. The observed wave speed is 24 m/s when the suspended mass is 3.0 kg. (a) What is the mass per unit length of the string? (b) What is the wave speed when the suspended mass is 2.0 kg?

3.0 kg

Figure P13.49

50. The elastic limit of a piece of steel wire is 2.70×10^9 Pa. What is the maximum speed at which transverse wave pulses can propagate along the wire without exceeding its elastic limit? (The density of steel is 7.86×10^3 kg/m³.)

51. Transverse waves travel at 20.0 m/s on a string that is under a tension of 6.00 N. What tension is required for a wave speed of 30.0 m/s in the string?

Section 13.10 Interference of Waves
Section 13.11 Reflection of Waves

52. A series of pulses of amplitude 0.15 m is sent down a string that is attached to a post at one end. The pulses are reflected at the post and travel back along the string without loss of amplitude. What is the amplitude at a point on the string where two pulses are crossing (a) if the string is rigidly attached to the post? (b) if the end at which reflection occurs is free to slide up and down?

53. A wave of amplitude 0.30 m interferes with a second wave of amplitude 0.20 m traveling in the same direction. What are (a) the largest and (b) the smallest resultant amplitudes that can occur, and under what conditions will these maxima and minima arise?

ADDITIONAL PROBLEMS

54. The position of a 0.30-kg object attached to a spring is described by

$$x = (0.25 \text{ m}) \cos(0.4\pi t)$$

Find (a) the amplitude of the motion, (b) the spring constant, (c) the position of the object at *t* = 0.30 s, and (d) the object's speed at *t* = 0.30 s.

55. A large block *P* executes horizontal simple harmonic motion as it slides across a frictionless surface with a frequency $f = 1.50$ Hz. Block B rests on it, as shown in Figure P13.55, and the coefficient of static friction between the two is $\mu_s = 0.600$. What maximum amplitude of oscillation can the system have if block B is not to slip?

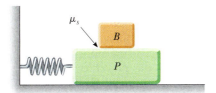

Figure P13.55

56. A 500-g block is released from rest and slides down a frictionless track that begins 2.00 m above the horizontal, as shown in Figure P13.56. At the bottom of the track, where the surface is horizontal, the block strikes and sticks to a light spring with a spring constant of 20.0 N/m. Find the maximum distance the spring is compressed.

Figure P13.56

57. A 3.00-kg object is fastened to a light spring, with the intervening cord passing over a pulley (Fig. P13.57). The pulley is frictionless, and its inertia may be neglected. The object is released from rest when the spring is unstretched. If the object drops 10.0 cm before stopping, find (a) the spring constant of the spring and (b) the speed of the object when it is 5.00 cm below its starting point.

Figure P13.57

58. A 5.00-g bullet moving with an initial speed of 400 m/s is fired into and passes through a 1.00-kg block, as in Figure P13.58. The block, initially at rest on a frictionless horizontal surface, is connected to a spring with a spring constant of 900 N/m. If the block moves 5.00 cm to the right after impact, find (a) the speed at which the bullet emerges from the block and (b) the mechanical energy lost in the collision.

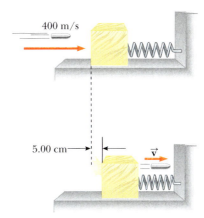

Figure P13.58

59. A 25-kg block is connected to a 30-kg block by a light string that passes over a frictionless pulley. The 30-kg block is connected to a light spring of force constant 200 N/m, as in Figure P13.59. The spring is unstretched when the system is as shown in the figure, and the incline is smooth. The 25-kg block is pulled 20 cm down the incline (so that the 30-kg block is 40 cm above the floor) and is released from rest. Find the speed of each block when the 30-kg block is 20 cm above the floor (that is, when the spring is unstretched).

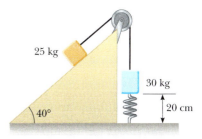

Figure P13.59

60. A spring in a toy gun has a spring constant of 9.80 N/m and can be compressed 20.0 cm beyond the equilibrium position. A 1.00-g pellet resting against the spring is propelled forward when the spring is released. (a) Find the muzzle speed of the pellet. (b) If the pellet is fired horizontally from a height of 1.00 m above the floor, what is its range?

61. A 2.00-kg block hangs without vibrating at the end of a spring ($k = 500$ N/m) that is attached to the ceiling of an elevator car. The car is rising with an upward acceleration of $g/3$ when the acceleration suddenly ceases (at $t = 0$). (a) What is the angular frequency of oscillation of the block after the acceleration ceases? (b) By what amount is the spring stretched during the time that the elevator car is accelerating? This distance will be the amplitude of the ensuing oscillation of the block.

62. An object of mass m is connected to two rubber bands of length L, each under tension F, as in Figure P13.62. The object is displaced vertically by a small distance y. Assuming the tension does not change, show that (a) the restoring force is $-(2F/L)y$ and (b) the system exhibits simple harmonic motion with an angular frequency $\omega = \sqrt{2F/mL}$.

Figure P13.62

63. A light balloon filled with helium of density 0.180 kg/m^3 is tied to a light string of length $L = 3.00$ m. The string is tied to the ground, forming an "inverted" simple pendulum (Fig. P13.63a). If the balloon is displaced slightly from equilibrium, as in Figure P13.63b, show that the motion is simple harmonic, and determine the period of the motion. Take the density of air to be 1.29 kg/m^3. [*Hint:* Use an analogy with the simple pendulum discussed in the text, and see Chapter 9.]

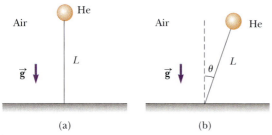

Figure P13.63

64. A light string of mass 10.0 g and length $L = 3.00$ m has its ends tied to two walls that are separated by the distance $D = 2.00$ m. Two objects, each of mass $M = 2.00$ kg, are suspended from the string as in Figure P13.64. If a wave pulse is sent from point A, how long does it take to travel to point B?

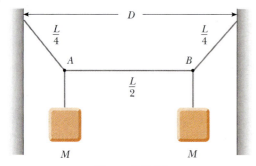

Figure P13.64

65. Assume that a hole is drilled through the center of the Earth. It can be shown that an object of mass m at a distance r from the center of the Earth is pulled toward the center only by the material in the shaded portion of Figure P13.65.

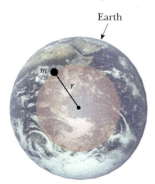

Figure P13.65

Assume Earth has a uniform density ρ. Write down Newton's law of gravitation for an object at a distance r from the center of the Earth, and show that the force on it is of the form of Hooke's law, $F = -kr$, with an effective force constant of $k = (\frac{4}{3})\pi\rho Gm$, where G is the gravitational constant.

66. A 60.0-kg firefighter slides down a pole while a constant frictional force of 300 N retards his motion. A horizontal 20.0-kg platform is supported by a spring at the bottom of the pole to cushion the fall. The firefighter starts from rest 5.00 m above the platform, and the spring constant is 2500 N/m. Find (a) the firefighter's speed just before he collides with the platform and (b) the maximum distance the spring is compressed. Assume that the frictional force acts during the entire motion. [*Hint:* The collision between the firefighter and the platform is perfectly inelastic.]

67. **Physics⊗Now**™ An object of mass $m_1 = 9.0$ kg is in equilibrium while connected to a light spring of constant $k = 100$ N/m that is fastened to a wall, as in Figure P13.67a. A second object, of mass $m_2 = 7.0$ kg, is slowly pushed up against m_1, compressing the spring by the amount $A = 0.20$ m, as shown in Figure P13.67b. The system is then released, causing both objects to start moving to the right on the frictionless surface. (a) When m_1 reaches the equilibrium point, m_2 loses contact with m_1 (Fig. P13.67c) and moves to the right with velocity $\vec{v}$. Determine the magnitude of $\vec{v}$. (b) How far apart are the objects when the spring is fully stretched for the first time (Fig. P13.67d)? [*Hint:* First determine the period of oscillation and the amplitude of the m_1–spring system after m_2 loses contact with m_1.]

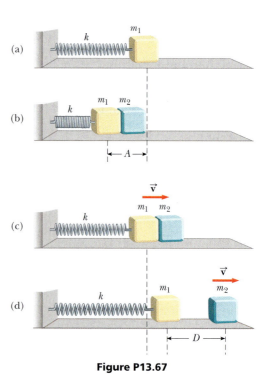

Figure P13.67

68. An 8.00-kg block travels on a rough horizontal surface and collides with a spring. The speed of the block just before the collision is 4.00 m/s. As it rebounds to the left with the spring uncompressed, the block travels at 3.00 m/s. If the coefficient of kinetic friction between the block and the surface is 0.400, determine (a) the loss in

mechanical energy due to friction while the block is in contact with the spring and (b) the maximum distance the spring is compressed.

69. Two points, *A* and *B*, on Earth are at the same longitude and 60.0° apart in latitude. An earthquake at point *A* sends two waves toward *B*. A transverse wave travels along the surface of Earth at 4.50 km/s, and a longitudinal wave travels through Earth at 7.80 km/s. (a) Which wave arrives at *B* first? (b) What is the time difference between the arrivals of the two waves at *B*? Take the radius of Earth to be 6.37×10^6 m.

70. Figure P13.70 shows a crude model of an insect wing. The mass *m* represents the entire mass of the wing, which pivots about the fulcrum *F*. The spring represents the surrounding connective tissue. Motion of the wing corresponds to vibration of the spring. Suppose the mass of the wing is 0.30 g and the effective spring constant of the tissue is 4.7×10^{-4} N/m. If the mass *m* moves up and down a distance of 2.0 mm from its position of equilibrium, what is the maximum speed of the outer tip of the wing?

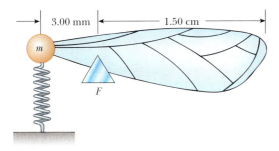

3.00 mm   1.50 cm

m

F

Figure P13.70

71. A 1.6-kg block on a horizontal surface is attached to a spring with a force constant of 1.0×10^3 N/m, as in Active Figure 13.1. The spring is compressed a distance of 2.0 cm, and the block is released from rest. (a) Calculate the speed of the block as it passes through the equilibrium position, $x = 0$, if the surface is frictionless. (b) Calculate the speed of the block as it passes through the equilibrium position if a constant frictional force of 4.0 N retards its motion. (c) How far does the block travel before coming to rest in part (b)?

ACTIVITIES

A.1. Construct a simple pendulum by tying a metal bolt to one end of a string and taping the other end to the top of a doorframe. Adjust the length of the pendulum to be about 1.0 m, and measure it precisely. Obtain the period by timing 25 complete oscillations, making sure the string always makes small angles with the vertical. Repeat the measurements for precisely measured pendulum lengths ranging from 0.4 m to 1.6 m in increments of 0.2 m. Plot the square of the period versus the length of the pendulum and measure the slope of the line best fitting your data points. Does your slope agree with that predicted by $T^2 = (4\pi^2/g)L$? What value of *g* do you obtain from your data?

While you have your pendulum in position, use a procedure similar to the preceding to verify that the period is also independent of the amplitude for small angles and that the period is also independent of the mass.

A.2. Attach one end of a rope (or a spring such as a Slinky™) to a wall, stretch it taut, and use the system to study the following aspects of wave motion:
(a) Send a pulse down the rope by striking it sharply from the side. An observer watching from the side can measure the time elapsed during 3–5 trips of the pulse from one end to the other. Dividing the total distance traveled by the elapsed time yields the wave speed. To get reliable results, take a number of readings and average them.
(b) Use the same setup as in part (a) to test whether the initial amplitude of the pulse changes the wave speed.
(c) Have two people hold opposite ends of the rope (or spring). Then, at the same instant, have both people hit the rope sharply from the side. Observe what happens when the two pulses meet. The superposition lasts only for the short time that the pulses overlap, and you must look carefully to see the effect.
(d) Using the same setup, devise a way to check whether pulses traveling toward one another pass through when they collide or reflect off each other.
(e) Tie one end of a rope to a doorknob and send a pulse down it. Observe what happens when the pulse reflects from the door. Does it return on the same side of the rope, or does it invert?

Patrick Ward/CORBIS

The characteristic sound of any instrument is referred to as the quality of that sound. What is it about the sound from the tuba that allows us to distinguish between it and the sound from a flute?

CHAPTER

14

Sound

Sound waves are the most important example of longitudinal waves. In this chapter we discuss the characteristics of sound waves: how they are produced, what they are, and how they travel through matter. We then investigate what happens when sound waves interfere with each other. The insights gained in this chapter will help you understand how we hear.

14.1 PRODUCING A SOUND WAVE

Whether it conveys the shrill whine of a jet engine or the soft melodies of a crooner, any sound wave has its source in a vibrating object. Musical instruments produce sounds in a variety of ways. The sound of a clarinet is produced by a vibrating reed, the sound of a drum by the vibration of the taut drumhead, the sound of a piano by vibrating strings, and the sound from a singer by vibrating vocal cords.

Sound waves are longitudinal waves traveling through a medium, such as air. In order to investigate how sound waves are produced, we focus our attention on the tuning fork, a common device for producing pure musical notes. A tuning fork consists of two metal prongs, or tines, that vibrate when struck. Their vibration disturbs the air near them, as shown in Figure 14.1. (The amplitude of vibration of the tine shown in the figure has been greatly exaggerated for clarity.) When a tine swings to the right, as in Figure 14.1a, the molecules in an element of air in front of its movement are forced closer together than normal. Such a region of high molecular density and high air pressure is called a **compression**. This compression moves away from the fork like a ripple on a pond. When the tine swings to the left, as in Figure 14.1b, the molecules in an element of air to the right of the tine spread apart, and the density and air pressure in this region are then lower than normal. Such a region of reduced density is called a **rarefaction** (pronounced "rare a fak' shun"). Molecules to the right of the rarefaction in the figure move to the left. The rarefaction itself therefore moves to the right, following the previously produced compression.

As the tuning fork continues to vibrate, a succession of compressions and rarefactions forms and spreads out from it. The resultant pattern in the air is somewhat like that pictured in Figure 14.2a. We can use a sinusoidal curve to represent a sound wave, as in Figure 14.2b. Notice that there are crests in the sinusoidal wave at the points where the sound wave has compressions and troughs where the sound wave has rarefactions. The compressions and rarefactions of the sound waves are superposed on the random thermal motion of the atoms and molecules of the air (discussed in Chapter 10), so sound waves in gases travel at about the molecular rms speed.

14.2 CHARACTERISTICS OF SOUND WAVES

As already noted, the general motion of elements of air near a vibrating object is back and forth between regions of compression and rarefaction. This back-and-forth motion of elements of the medium in the direction of the disturbance is characteristic of a longitudinal wave. **The motion of the elements of the medium in a longitudinal sound wave is back and forth along the direction in which the wave travels.** By contrast, **in a transverse wave, the vibrations of the elements of the medium are at right angles to the direction of travel of the wave**.

Categories of Sound Waves

Sound waves fall into three categories covering different ranges of frequencies. **Audible waves** are longitudinal waves that lie within the range of sensitivity of the human ear, approximately 20 to 20 000 Hz. **Infrasonic waves** are longitudinal waves with frequencies below the audible range. Earthquake waves are an example. **Ultrasonic waves** are longitudinal waves with frequencies above the audible range for humans and are produced by certain types of whistles. Animals such as dogs can hear the waves emitted by these whistles.

Applications of Ultrasound

Ultrasonic waves are sound waves with frequencies greater than 20 kHz. Because of their high frequency and corresponding short wavelengths, ultrasonic waves can be used to produce images of small objects and are currently in wide use in medical applications, both as a diagnostic tool and in certain treatments. Internal organs can be examined via the images produced by the reflection and absorption of ultrasonic waves. Although ultrasonic waves are far safer than x-rays, their images don't always have as much detail. Certain organs, however, such as the liver and the spleen, are invisible to x-rays but can be imaged with ultrasonic waves.

High-density region

(a)

Low-density region

(b)

Royalty-Free/Corbis

(c)

Figure 14.1 A vibrating tuning fork. (a) As the right tine of the fork moves to the right, a high-density region (compression) of air is formed in front of its movement. (b) As the right tine moves to the left, a low-density region (rarefaction) of air is formed behind it. (c) A tuning fork.

APPLICATION

Medical Uses of Ultrasound

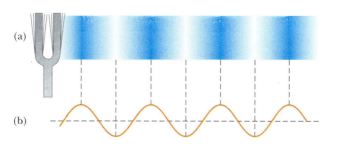

Figure 14.2 (a) As the tuning fork vibrates, a series of compressions and rarefactions moves outward, away from the fork. (b) The crests of the wave correspond to compressions, the troughs to rarefactions.

Direction of vibration

Electrical connections

Crystal

Figure 14.3 An alternating voltage applied to the faces of a piezoelectric crystal causes the crystal to vibrate.

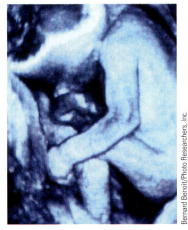

An ultrasound image of a human fetus in the womb.

Bernard Benoit/Photo Researchers, Inc.

APPLICATION

Ultrasonic Ranging Unit for Cameras

Medical workers can measure the speed of the blood flow in the body with a device called an ultrasonic flow meter, which makes use of the Doppler effect (discussed in Section 14.6). The flow speed is found by comparing the frequency of the waves scattered by the flowing blood with the incident frequency.

Figure 14.3 illustrates the technique that produces ultrasonic waves for clinical use. Electrical contacts are made to the opposite faces of a crystal, such as quartz or strontium titanate. If an alternating voltage of high frequency is applied to these contacts, the crystal vibrates at the same frequency as the applied voltage, emitting a beam of ultrasonic waves. At one time, a technique like this was used to produce sound in nearly all headphones. This method of transforming electrical energy into mechanical energy, called the **piezoelectric effect**, is reversible: If some external source causes the crystal to vibrate, an alternating voltage is produced across it. A single crystal can therefore be used to both generate and receive ultrasonic waves.

The primary physical principle that makes ultrasound imaging possible is the fact that a sound wave is partially reflected whenever it is incident on a boundary between two materials having different densities. If a sound wave is traveling in a material of density ρ_i and strikes a material of density ρ_t, the percentage of the incident sound wave intensity reflected, PR, is given by

$$PR = \left(\frac{\rho_i - \rho_t}{\rho_i + \rho_t} \right)^2 \times 100$$

This equation assumes that the direction of the incident sound wave is perpendicular to the boundary and that the speed of sound is approximately the same in the two materials. The latter assumption holds very well for the human body because the speed of sound doesn't vary much in the organs of the body.

Physicians commonly use ultrasonic waves to observe fetuses. This technique presents far less risk than do x-rays, which deposit more energy in cells and can produce birth defects. First the abdomen of the mother is coated with a liquid, such as mineral oil. If this were not done, most of the incident ultrasonic waves from the piezoelectric source would be reflected at the boundary between the air and the mother's skin. Mineral oil has a density similar to that of skin, and a very small fraction of the incident ultrasonic wave is reflected when $\rho_i \approx \rho_t$. The ultrasound energy is emitted in pulses rather than as a continuous wave, so the same crystal can be used as a detector as well as a transmitter. An image of the fetus is obtained by using an array of transducers placed on the abdomen. The reflected sound waves picked up by the transducers are converted to an electric signal, which is used to form an image on a fluorescent screen. Difficulties such as the likelihood of spontaneous abortion or of breech birth are easily detected with this technique. Fetal abnormalities such as spina bifida and water on the brain are also readily observed.

A relatively new medical application of ultrasonics is the *cavitron ultrasonic surgical aspirator* (CUSA). This device has made it possible to surgically remove brain tumors that were previously inoperable. The probe of the CUSA emits ultrasonic waves (at about 23 kHz) at its tip. When the tip touches a tumor, the part of the tumor near the probe is shattered and the residue can be sucked up (aspirated) through the hollow probe. Using this technique, neurosurgeons are able to remove brain tumors without causing serious damage to healthy surrounding tissue.

Ultrasound is also used to break up kidney stones that are otherwise too large to pass. Previously, invasive surgery was more often required.

Another interesting application of ultrasound is the ultrasonic ranging unit used in some cameras to provide an almost instantaneous measurement of the distance between the camera and the object to be photographed. The principal component of this device is a crystal that acts as both a loudspeaker and a microphone. A pulse of ultrasonic waves is transmitted from the transducer to the object, which then reflects part of the signal, producing an echo that is detected by the device. The time interval between the outgoing pulse and the detected echo is electronically converted to a distance, because the speed of sound is a known quantity.

14.3   THE SPEED OF SOUND

The speed of a sound wave in a fluid depends on the fluid's compressibility and inertia. If the fluid has a bulk modulus B and an equilibrium density ρ, the speed of sound in it is

$$v = \sqrt{\frac{B}{\rho}}$$ [14.1]

◀ Speed of sound in a fluid

Equation 14.1 also holds true for a gas. Recall from Chapter 9 that the bulk modulus is defined as the ratio of the change in pressure, ΔP, to the resulting fractional change in volume, $\Delta V/V$:

$$B \equiv -\frac{\Delta P}{\Delta V/V}$$ [14.2]

B is always positive because an increase in pressure (positive ΔP) results in a decrease in volume. Hence, the ratio $\Delta P/\Delta V$ is always negative.

It's interesting to compare Equation 14.1 with Equation 13.18 for the speed of transverse waves on a string, $v = \sqrt{F/\mu}$, discussed in Chapter 13. In both cases, the wave speed depends on an elastic property of the medium (B or F) and on an inertial property of the medium (ρ or μ). In fact, the speed of all mechanical waves follows an expression of the general form

$$v = \sqrt{\frac{\text{elastic property}}{\text{inertial property}}}$$

Another example of this general form is the **speed of a longitudinal wave in a solid rod**, which is

$$v = \sqrt{\frac{Y}{\rho}}$$ [14.3]

where Y is the Young's modulus of the solid (see Eqn. 9.3), and ρ is its density. This expression is valid only for a thin, solid rod.

Table 14.1 lists the speeds of sound in various media. The speed of sound is much higher in solids than in gases, because the molecules in a solid interact more strongly with each other than do molecules in a gas. Striking a long steel rail with a hammer, for example, produces two sound waves, one moving through the rail and a slower wave moving through the air. A student with an ear pressed against the rail first hears the faster sound moving through the rail, then the sound moving through air. In general, sound travels faster through solids than liquids and faster through liquids than gases, although there are exceptions.

The speed of sound also depends on the temperature of the medium. For sound traveling through air, the relationship between the speed of sound and temperature is

$$v = (331 \text{ m/s}) \sqrt{\frac{T}{273 \text{ K}}}$$ [14.4]

where 331 m/s is the speed of sound in air at 0°C and T is the absolute (Kelvin) temperature. Using this equation the speed of sound in air at 293 K (a typical room temperature) is approximately 343 m/s.

Quick Quiz 14.1

Which of the following actions will increase the speed of sound in air? (a) decreasing the air temperature (b) increasing the frequency of the sound (c) increasing the air temperature (d) increasing the amplitude of the sound wave (e) reducing the pressure of the air.

TABLE 14.1

Speeds of Sound in Various Media

Medium	v (m/s)
Gases	
Air (0°C)	331
Air (100°C)	386
Hydrogen (0°C)	1 290
Oxygen (0°C)	317
Helium (0°C)	972
Liquids at 25°C	
Water	1 490
Methyl alcohol	1 140
Sea water	1 530
Solids	
Aluminum	5 100
Copper	3 560
Iron	5 130
Lead	1 320
Vulcanized rubber	54

Applying Physics 14.1   The Sounds Heard During a Storm

How does lightning produce thunder, and what causes the extended rumble?

Explanation   Assume that you're at ground level, and neglect ground reflections. When lightning strikes, a channel of ionized air carries a large electric current from a cloud to the ground. This results in a rapid temperature increase of the air in the channel as the current moves through it, causing a similarly rapid expansion of the air. The expansion is so sudden and so intense that a tremendous disturbance is produced in the air—thunder. The entire length of the channel produces the sound at essentially the same instant of time. Sound produced at the bottom of the channel reaches you first, because that's the point closest to you. Sounds from progressively higher portions of the channel reach you at later times, resulting in an extended roar. If the lightning channel were a perfectly straight line, the roar might be steady, but the zigzag shape of the path results in the rumbling variation in loudness, with different quantities of sound energy from different segments arriving at any given instant.

INTERACTIVE EXAMPLE 14.1   Sound Waves in Various Media

Goal   Calculate and compare the speeds of sound in different media.

Problem   (a) If a solid bar of aluminum 1.00 m long is struck at one end with a hammer, a longitudinal pulse propagates down the bar. Find the speed of sound in the bar, which has a Young's modulus of 7.0×10^{10} Pa and a density of 2.7×10^3 kg/m³. (b) Calculate the speed of sound in ethyl alcohol, which has a density of 806 kg/m³ and bulk modulus of 1.0×10^9 Pa. (c) Compute the speed of sound in air at 35.0°C.

Strategy   Substitute the given values into the appropriate equations.

Solution
(a) Compute the speed of sound in an aluminum bar.

Substitute values into Equation 14.3:

$$v_{Al} = \sqrt{\frac{Y}{\rho}} = \sqrt{\frac{7.0 \times 10^{10} \text{ Pa}}{2.7 \times 10^3 \text{ kg/m}^3}}$$

$$= 5\ 100 \text{ m/s, or about 11 000 mi/h !}$$

(b) Compute the speed of sound in ethyl alcohol.

Substitute values into Equation 14.1:

$$v = \sqrt{\frac{B}{\rho}} = \sqrt{\frac{1.0 \times 10^9 \text{ Pa}}{806 \text{ kg/m}^3}} = 1.1 \times 10^3 \text{ m/s}$$

(c) Compute the speed of sound in air at 35.0°C.

Substitute values into Equation 14.4:

$$v = (331 \text{ m/s}) \sqrt{\frac{(273 \text{ K} + 35.0 \text{ K})}{273 \text{ K}}} = 352 \text{ m/s}$$

Remark   The speed of sound in aluminum is dramatically higher than in either liquid alcohol or air.

Exercise 14.1

Compute the speed of sound in the following substances at 273 K: (a) lead ($Y = 1.6 \times 10^{10}$ Pa), (b) mercury ($B = 2.8 \times 10^{10}$ Pa), and (c) air at −15.0°C.

Answers   (a) 1.2×10^3 m/s   (b) 1.4×10^3 m/s   (c) 322 m/s

Physics⊗Now™

You can compare the speeds of sound through various media by logging into PhysicsNow at **www.cp7e.com** and going to Interactive Example 14.1.

14.4 ENERGY AND INTENSITY OF SOUND WAVES

As the tines of a tuning fork move back and forth through the air, they exert a force on a layer of air and cause it to move. In other words, the tines do work on the layer of air. The fact that the fork pours sound energy into the air is one of the reasons the vibration of the fork slowly dies out. (Other factors, such as the energy lost to friction as the tines bend, are also responsible for the lessening of movement.)

> The average **intensity** I of a wave on a given surface is defined as the rate at which energy flows through the surface, $\Delta E/\Delta t$, divided by the surface area A:
>
> $$I \equiv \frac{1}{A}\frac{\Delta E}{\Delta t} \qquad\qquad [14.5]$$
>
> where the direction of energy flow is perpendicular to the surface at every point.
>
> **SI unit: watt per meter squared (W/m^2)**

A rate of energy transfer is power, so Equation 14.5 can be written in the alternate form

$$I \equiv \frac{\text{power}}{\text{area}} = \frac{\mathcal{P}}{A} \qquad\qquad [14.6] \qquad \blacktriangleleft \text{Intensity of a wave}$$

where $\mathcal{P}$ is the sound power passing through the surface, measured in watts, and the intensity again has units of watts per square meter.

The faintest sounds the human ear can detect at a frequency of 1 000 Hz have an intensity of about 1×10^{-12} W/m^2. This intensity is called the **threshold of hearing**. The loudest sounds the ear can tolerate have an intensity of about 1 W/m^2 (the **threshold of pain**). At the threshold of hearing, the increase in pressure in the ear is approximately 3×10^{-5} Pa over normal atmospheric pressure. Because atmospheric pressure is about 1×10^5 Pa, this means the ear can detect pressure fluctuations as small as about 3 parts in 10^{10}! The maximum displacement of an air molecule at the threshold of hearing is about 1×10^{-11} m—a remarkably small number! If we compare this displacement with the diameter of a molecule (about 10^{-10} m), we see that the ear is an extremely sensitive detector of sound waves.

The loudest sounds the human ear can tolerate at 1 kHz correspond to a pressure variation of about 29 Pa away from normal atmospheric pressure, with a maximum displacement of air molecules of 1×10^{-5} m.

Intensity Level in Decibels

The loudest tolerable sounds have intensities about 1.0×10^{12} times greater than the faintest detectable sounds. The most intense sound, however, isn't perceived as being 1.0×10^{12} times louder than the faintest sound, because the sensation of loudness is approximately logarithmic in the human ear. (For a review of logarithms, see Section A.3, heading G, in Appendix A.) The relative intensity of a sound is called the **intensity level** or **decibel level**, defined by

$$\beta \equiv 10 \log\left(\frac{I}{I_0}\right) \qquad\qquad [14.7] \qquad \blacktriangleleft \text{Intensity level}$$

The constant $I_0 = 1.0 \times 10^{-12}$ W/m^2 is the reference intensity, the sound intensity at the threshold of hearing—I is the intensity, and β is the corresponding intensity

word *decibel*, which is one-tenth of a *bel*,
...tor of the telephone, Alexander Graham Bell

...ous decibel levels, we can substitute a few representative
...tion 14.7, starting with $I = 1.0 \times 10^{-12}$ W/m²:

$$\beta = 10 \log \left(\frac{1.0 \times 10^{-12} \text{ W/m}^2}{1.0 \times 10^{-12} \text{ W/m}^2} \right) = 10 \log(1) = 0 \text{ dB}$$

From this result, we see that the lower threshold of human hearing has been chosen to be zero on the decibel scale. Progressing upward by powers of ten yields

$$\beta = 10 \log \left(\frac{1.0 \times 10^{-11} \text{ W/m}^2}{1.0 \times 10^{-12} \text{ W/m}^2} \right) = 10 \log(10) = 10 \text{ dB}$$

$$\beta = 10 \log \left(\frac{1.0 \times 10^{-10} \text{ W/m}^2}{1.0 \times 10^{-12} \text{ W/m}^2} \right) = 10 \log(100) = 20 \text{ dB}$$

Notice the pattern: *Multiplying* a given intensity by ten *adds* 10 db to the intensity level. This pattern holds throughout the decibel scale. For example, a 50-dB sound is 10 times as intense as a 40-dB sound, while a 60-dB sound is 100 times as intense as a 40-dB sound.

On this scale, the threshold of pain ($I = 1.0$ W/m²) corresponds to an intensity level of $\beta = 10 \log(1/1 \times 10^{-12}) = 10 \log(10^{12}) = 120$ dB. Nearby jet airplanes can create intensity levels of 150 dB, and subways and riveting machines have levels of 90 to 100 dB. The electronically amplified sound heard at rock concerts can attain levels of up to 120 dB, the threshold of pain. Exposure to such high intensity levels can seriously damage the ear. Earplugs are recommended whenever prolonged intensity levels exceed 90 dB. Recent evidence suggests that noise pollution, which is common in most large cities and in some industrial environments, may be a contributing factor to high blood pressure, anxiety, and nervousness. Table 14.2 gives the approximate intensity levels of various sounds.

TABLE 14.2

Intensity Levels in Decibels for Different Sources

Source of Sound	β(dB)
Nearby jet airplane	150
Jackhammer, machine gun	130
Siren, rock concert	120
Subway, power mower	100
Busy traffic	80
Vacuum cleaner	70
Normal conversation	50
Mosquito buzzing	40
Whisper	30
Rustling leaves	10
Threshold of hearing	0

EXAMPLE 14.2 A Noisy Grinding Machine

Goal Work with watts and decibels.

Problem A noisy grinding machine in a factory produces a sound intensity of 1.00×10^{-5} W/m². Calculate **(a)** the decibel level of this machine, and **(b)** the new intensity level when a second, identical machine is added to the factory. **(c)** A certain number of additional such machines are put into operation alongside these two. When all the machines are running at the same time the decibel level is 77.0 dB. Find the sound intensity.

Strategy Parts (a) and (b) require substituting into the decibel formula, Equation 14.7, with the intensity in part (b) twice the intensity in part (a). In part (c), the intensity level in decibels is given, and it's necessary to work backwards, using the inverse of the logarithm function, to get the intensity in watts per meter squared.

Solution
(a) Calculate the intensity level of the single grinder.

Substitute the intensity into the decibel formula:

$$\beta = 10 \log \left(\frac{1.00 \times 10^{-5} \text{ W/m}^2}{1.00 \times 10^{-12} \text{ W/m}^2} \right) = 10 \log(10^7)$$

$$= \boxed{70.0 \text{ dB}}$$

(b) Calculate the new intensity level when an additional machine is added.

Substitute twice the intensity of part (a) into the decibel formula:

$$\beta = 10 \log \left(\frac{2.00 \times 10^{-5} \text{ W/m}^2}{1.00 \times 10^{-12} \text{ W/m}^2} \right) = \boxed{73.0 \text{ dB}}$$

(c) Find the intensity corresponding to an intensity level of 77.0 dB.

Substitute 77.0 dB into the decibel formula and divide both sides by 10:

$$\beta = 77.0 \text{ dB} = 10 \log \left(\frac{I}{I_0} \right)$$

$$7.70 = \log \left(\frac{I}{10^{-12} \text{ W/m}^2} \right)$$

Make each side the exponent of 10. On the right-hand side, $10^{\log u} = u$, by definition of base ten logarithms.

$$10^{7.70} = 5.01 \times 10^7 = \frac{I}{1.00 \times 10^{-12} \text{ W/m}^2}$$

$$I = \boxed{5.01 \times 10^{-5} \text{ W/m}^2}$$

Remark The answer is five times the intensity of the single grinder, so in part (c) there are five such machines operating simultaneously. Because of the logarithmic definition of intensity level, large changes in intensity correspond to small changes in intensity level.

Exercise 14.2
Suppose a manufacturing plant has an average sound intensity level of 97.0 dB created by 25 identical machines. (a) Find the total intensity created by all the machines. (b) Find the sound intensity created by one such machine. (c) What's the sound intensity level if five such machines are running?

Answers (a) $5.01 \times 10^{-3} \text{ W/m}^2$ (b) $2.00 \times 10^{-4} \text{ W/m}^2$ (c) 90.0 dB

Federal regulations now demand that no office or factory worker be exposed to noise levels that average more than 90 dB over an 8-h day. From a management point of view, here's the good news: one machine in the factory may produce a noise level of 70 dB, but a second machine, while doubling the total intensity, increases the noise level by only 3 dB. Because of the logarithmic nature of intensity levels, doubling the intensity doesn't double the intensity level; in fact, it alters it by a surprisingly small amount. This means that equipment can be added to the factory without appreciably altering the intensity level of the environment.

Now here's the bad news: as you remove noisy machinery, the intensity level isn't lowered appreciably. In Exercise 14.2, reducing the intensity level by 7 dB would require the removal of 20 of the 25 machines! To lower the level another 7 dB would require removing 80% of the remaining machines, in which case only one machine would remain.

APPLICATION
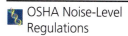
OSHA Noise-Level Regulations

14.5 SPHERICAL AND PLANE WAVES

If a small spherical object oscillates so that its radius changes periodically with time, a spherical sound wave is produced (Fig. 14.4, page 466). The wave moves outward from the source at a constant speed.

Because all points on the vibrating sphere behave in the same way, we conclude that the energy in a spherical wave propagates equally in all directions. This means that no one direction is preferred over any other. If $\mathcal{P}_{av}$ is the average power emitted by the source, then at any distance r from the source, this power must be distributed over a spherical surface of area $4\pi r^2$, assuming no absorption in the medium. (Recall that $4\pi r^2$ is the surface area of a sphere.) Hence, the **intensity** of the sound at a distance r from the source is

$$I = \frac{\text{average power}}{\text{area}} = \frac{\mathcal{P}_{av}}{A} = \frac{\mathcal{P}_{av}}{4\pi r^2} \qquad \textbf{[14.8]}$$

This equation shows that the intensity of a wave decreases with increasing distance from its source, as you might expect. The fact that I varies as $1/r^2$ is a result of the

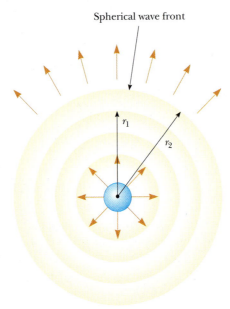

Spherical wave front

Figure 14.4 A spherical wave propagating radially outward from an oscillating sphere. The intensity of the wave varies as $1/r^2$.

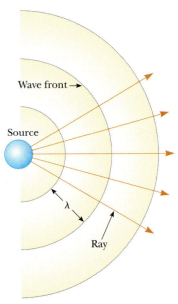

Wave front →

Source

λ

Ray

Figure 14.5 Spherical waves emitted by a point source. The circular arcs represent the spherical wave fronts concentric with the source. The rays are radial lines pointing outward from the source, perpendicular to the wavefronts.

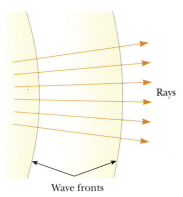

Rays

Wave fronts

Figure 14.6 Far away from a point source, the wave fronts are nearly parallel planes and the rays are nearly parallel lines perpendicular to the planes. Hence, a small segment of a spherical wavefront is approximately a plane wave.

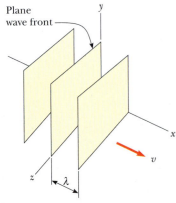

Plane wave front

λ

Figure 14.7 A representation of a plane wave moving in the positive x-direction with a speed v. The wavefronts are planes parallel to the yz-plane.

assumption that the small source (sometimes called a **point source**) emits a spherical wave. (In fact, light waves also obey this so-called inverse-square relationship.) Because the average power is the same through any spherical surface centered at the source, we see that the intensities at distances r_1 and r_2 (Fig. 14.4) from the center of the source are

$$I_1 = \frac{\mathcal{P}_{av}}{4\pi r_1{}^2} \qquad I_2 = \frac{\mathcal{P}_{av}}{4\pi r_2{}^2}$$

The ratio of the intensities at these two spherical surfaces is

$$\frac{I_1}{I_2} = \frac{r_2{}^2}{r_1{}^2} \qquad\qquad [14.9]$$

It's useful to represent spherical waves graphically with a series of circular arcs (lines of maximum intensity) concentric with the source representing part of a spherical surface, as in Figure 14.5. We call such an arc a **wave front**. The distance between adjacent wave fronts equals the wavelength λ. The radial lines pointing outward from the source and perpendicular to the arcs are called **rays**.

Now consider a small portion of a wave front that is at a *great* distance (relative to λ) from the source, as in Figure 14.6. In this case, the rays are nearly parallel to each other and the wave fronts are very close to being planes. At distances from the source that are great relative to the wavelength, therefore, we can approximate the wave front with parallel planes, called **plane waves**. Any small portion of a spherical wave that is far from the source can be considered a plane wave. Figure 14.7 illustrates a plane wave propagating along the x-axis. If the positive x-direction is taken to be the direction of the wave motion (or ray) in this figure, then the wave fronts are parallel to the plane containing the y- and z-axes.

EXAMPLE 14.3 Intensity Variations of a Point Source

Goal Relate sound intensities and their distances from a point source.

Problem A small source emits sound waves with a power output of 80.0 W. **(a)** Find the intensity 3.00 m from the source. **(b)** At what distance would the intensity be one-fourth as much as it is at $r = 3.00$ m? **(c)** Find the distance at which the sound level is 40.0 dB.

Strategy The source is small, so the emitted waves are spherical and the intensity in part (a) can be found by substituting values into Equation 14.8. Part (b) involves solving for r in Equation 14.8 followed by substitution (though Equation 14.9 can be used instead). In part (c), convert from the sound intensity level to the intensity in W/m², using Equation 14.7. Then substitute into Equation 14.9 (though 14.8 could be used, instead) and solve for r_2.

Solution
(a) Find the intensity 3.00 m from the source.

Substitute $\mathscr{P}_{av} = 80.0$ W and $r = 3.00$ m into Equation 14.8:

$$I = \frac{\mathscr{P}_{av}}{4\pi r^2} = \frac{80.0 \text{ W}}{4\pi(3.00 \text{ m})^2} = 0.707 \text{ W/m}^2$$

(b) At what distance would the intensity be one-fourth as much as it is at $r = 3.00$ m?

Take $I = (0.707 \text{ W/m}^2)/4$, and solve for r in Equation 14.8:

$$r = \left(\frac{\mathscr{P}_{av}}{4\pi I}\right)^{1/2} = \left(\frac{80.0 \text{ W}}{4\pi(0.707 \text{ W/m}^2)/4.0}\right)^{1/2} = 6.00 \text{ m}$$

(c) Find the distance at which the sound level is 40.0 dB.

Convert the intensity level of 40.0 dB to an intensity in W/m² by solving Equation 14.7 for I:

$$40.0 = 10 \log\left(\frac{I}{I_0}\right) \rightarrow 4.00 = \log\left(\frac{I}{I_0}\right)$$

$$10^{4.00} = \frac{I}{I_0} \rightarrow I = 10^{4.00}I_0 = 1.00 \times 10^{-8} \text{ W/m}^2$$

Solve Equation 14.9 for r_2^2, substitute the intensity and the result of part (a), and take the square root:

$$\frac{I_1}{I_2} = \frac{r_2^2}{r_1^2} \rightarrow r_2^2 = r_1^2 \frac{I_1}{I_2}$$

$$r_2^2 = (3.00 \text{ m})^2 \left(\frac{0.707 \text{ W/m}^2}{1.00 \times 10^{-8} \text{ W/m}^2}\right)$$

$$r_2 = 2.52 \times 10^4 \text{ m}$$

Remarks Once the intensity is known at one position a certain distance away from the source, it's easier to use Equation 14.9 rather than Equation 14.8 to find the intensity at any other location. This is particularly true for part (b), where, using Equation 14.9, we can see right away that doubling the distance reduces the intensity to one-quarter its previous value.

Exercise 14.3
Suppose a certain jet plane creates an intensity level of 125 dB at a distance of 5.00 m. What intensity level does it create on the ground directly underneath it when flying at an altitude of 2.00 km?

Answer 73.0 dB

14.6 THE DOPPLER EFFECT

If a car or truck is moving while its horn is blowing, the frequency of the sound you hear is higher as the vehicle approaches you and lower as it moves away from you. This is one example of the *Doppler effect*, named for the Austrian physicist Christian Doppler (1803–1853), who discovered it. The same effect is heard if you're on a motorcycle and the horn is stationary: the frequency is higher as you approach the source and lower as you move away.

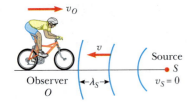

ACTIVE FIGURE 14.8
An observer moving with a speed v_O *toward* a stationary point source (S) hears a frequency f_O that is *greater* than the source frequency f_S.

Physics⊗Now™

Log into PhysicsNow at **www.cp7e.com** and go to Active Figure 14.8 to adjust the speed of the observer.

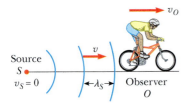

Figure 14.9 An observer moving with a speed of v_O *away* from a stationary source hears a frequency f_O that is *lower* than the source frequency f_S.

TIP 14.2   Doppler Effect Doesn't Depend on Distance

The sound from a source approaching at constant speed will increase in intensity, but the observed (elevated) frequency will remain unchanged. The Doppler effect doesn't depend on distance.

Although the Doppler effect is most often associated with sound, it's common to all waves, including light.

In deriving the Doppler effect, we assume that the air is stationary and that all speed measurements are made relative to this stationary medium. The speed v_O is the speed of the observer, v_S is the speed of the source, and v is the speed of sound.

Case 1: The Observer Is Moving Relative to a Stationary Source

In Active Figure 14.8 an observer is moving with a speed of v_O toward the source (considered a point source), which is at rest ($v_S = 0$).

We take the frequency of the source to be f_S, the wavelength of the source to be λ_S, and the speed of sound in air to be v. If both observer and source are stationary, the observer detects f_S wave fronts per second. (That is, when $v_O = 0$ and $v_S = 0$, the observed frequency f_O equals the source frequency f_S.) When moving toward the source, the observer moves a distance of $v_O t$ in t seconds. During this interval, **the observer detects an additional number of wave fronts**. The number of extra wave fronts is equal to the distance traveled, $v_O t$, divided by the wavelength λ_S:

$$\text{Additional wave fronts detected} = \frac{v_O t}{\lambda_S}$$

Divide this equation by the time t to get the number of additional wave fronts detected *per second*, v_O/λ_S. Hence, the frequency heard by the observer is *increased* to

$$f_O = f_S + \frac{v_O}{\lambda_S}$$

Substituting $\lambda_S = v/f_S$ into this expression for f_O we obtain

$$f_O = f_S \left(\frac{v + v_O}{v} \right) \qquad \text{[14.10]}$$

When the observer is *moving away* from a stationary source (Fig. 14.9), the observed frequency decreases. A derivation yields the same result as Equation 14.10, but with $v - v_O$ in the numerator. Therefore, when the observer is moving away from the source, substitute $-v_O$ for v_O in Equation 14.10.

Case 2: The Source Is Moving Relative to a Stationary Observer

Now consider a source moving toward an observer at rest, as in Active Figure 14.10. Here, the wave fronts passing observer A are closer together because the source is moving in the direction of the outgoing wave. As a result, the wavelength λ_O measured by observer A is shorter than the wavelength λ_S of the source at rest. During each vibration, which lasts for an interval T (the period), the source moves a distance $v_S T = v_S/f_S$ and **the wavelength is shortened by that amount**. The observed wavelength is therefore given by

$$\lambda_O = \lambda_S - \frac{v_S}{f_S}$$

Because $\lambda_S = v/f_S$, the frequency observed by A is

$$f_O = \frac{v}{\lambda_O} = \frac{v}{\lambda_S - \dfrac{v_S}{f_S}} = \frac{v}{\dfrac{v}{f_S} - \dfrac{v_S}{f_S}}$$

or

$$f_O = f_S \left(\frac{v}{v - v_S} \right) \qquad \text{[14.11]}$$

As expected, **the observed frequency increases when the source is moving toward the observer**. When the source is *moving away* from an observer at rest, the minus sign in the denominator must be replaced with a plus sign, so the factor becomes $(v + v_S)$.

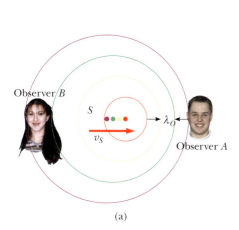

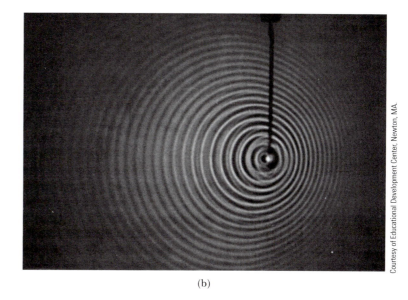

(a)   (b)

ACTIVE FIGURE 14.10

(a) A source S moving with speed v_S toward stationary observer A and away from stationary observer B. Observer A hears an *increased* frequency, and observer B hears a *decreased* frequency. (b) The Doppler effect in water, observed in a ripple tank. The source producing the water waves is moving to the right.

Physics ⊗ Now™

Log into PhysicsNow at **www.cp7e.com**, and go to Active Figure 14.10 to adjust the speed of the source.

General Case

When both the source and the observer are in motion relative to Earth, Equations 14.10 and 14.11 can be combined to give

$$f_O = f_S \left(\frac{v + v_O}{v - v_S} \right)$$ [14.12]

◄ Doppler shift equation—observer and source in motion

In this expression, the signs for the values substituted for v_O and v_S depend on the direction of the velocity. When the observer moves *toward* the source, a *positive* speed is substituted for v_O; when the observer moves *away from* the source, a *negative* speed is substituted for v_O. Similarly, a *positive* speed is substituted for v_S when the source moves *toward* the observer, a *negative* speed when the source moves away from the observer.

Choosing incorrect signs is the most common mistake made in working a Doppler effect problem. The following rules may be helpful: The word *toward* is associated with an *increase* in the observed frequency; the words *away from* are associated with a *decrease* in the observed frequency.

These two rules derive from the physical insight that when the observer is moving toward the source (or the source toward the observer), there is a smaller observed period between wave crests, hence a larger frequency, with the reverse holding—a smaller observed frequency—when the observer is moving away from the source (or the source away from the observer). Keep the physical insight in mind whenever you're in doubt about the signs in Equation 14.12: Adjust them as necessary to get the correct physical result.

The second most common mistake made in applying Equation 14.12 is to accidentally reverse numerator and denominator. Some find it helpful to remember the equation in the following form:

$$\frac{f_O}{v + v_O} = \frac{f_S}{v - v_S}$$

The advantage of this form is its symmetry: both sides are very nearly the same, with O's on the left and S's on the right. Forgetting which side has the plus sign and which has the minus sign is not a serious problem, as long as physical insight is used to check the answer and make adjustments as necessary.

Quick Quiz 14.2

Suppose you're on a hot air balloon ride, carrying a buzzer that emits a sound of frequency f. If you accidentally drop the buzzer over the side while the balloon is rising at constant speed, what can you conclude about the sound you hear as the buzzer falls toward the ground? (a) the frequency and intensity increase, (b) the frequency decreases and the intensity increases, (c) the frequency decreases and the intensity decreases, or (d) the frequency remains the same, but the intensity decreases.

Applying Physics 14.2   Out-of-Tune Speakers

Suppose you place your stereo speakers far apart and run past them from right to left or left to right. If you run rapidly enough and have excellent pitch discrimination, you may notice that the music playing seems to be out of tune when you're between the speakers. Why?

Explanation When you are between the speakers, you are running away from one of them and toward the other, so there is a Doppler shift downward for the sound from the speaker behind you and a Doppler shift upward for the sound from the speaker ahead of you. As a result, the sound from the two speakers will not be in tune. A calculation shows that a world-class sprinter could run fast enough to generate about a semitone difference in the sound from the two speakers.

EXAMPLE 14.4   Listen, but Don't Stand on the Track

Goal Solve a Doppler shift problem when only the source is moving.

Problem A train moving at a speed of 40.0 m/s sounds its whistle, which has a frequency of 5.00×10^2 Hz. Determine the frequency heard by a stationary observer as the train *approaches* the observer. The ambient temperature is 24.0°C.

Strategy Use Equation 14.4 to get the speed of sound at the ambient temperature, then substitute values into Equation 14.12 for the Doppler shift. Because the train approaches the observer, the observed frequency will be larger. Choose the sign of v_S to reflect this fact.

Solution

Use Equation 14.4 to calculate the speed of sound in air at $T = 24.0°C$:

$$v = (331 \text{ m/s})\sqrt{\frac{T}{273 \text{ K}}}$$

$$= (331 \text{ m/s})\sqrt{\frac{(273 + 24.0)\text{ K}}{273 \text{ K}}} = \boxed{345 \text{ m/s}}$$

The observer is stationary, so $v_O = 0$. The train is moving *toward* the observer, so $v_S = +40.0$ m/s (*positive*). Substitute these values and the speed of sound into the Doppler shift equation:

$$f_O = f_S \left(\frac{v + v_O}{v - v_S}\right)$$

$$= (5.00 \times 10^2 \text{ Hz})\left(\frac{345 \text{ m/s}}{345 \text{ m/s} - 40.0 \text{ m/s}}\right)$$

$$= \boxed{566 \text{ Hz}}$$

Remark If the train were going away from the observer, $v_S = -40.0$ m/s would have been chosen instead.

Exercise 14.4

Determine the frequency heard by the stationary observer as the train *recedes* from the observer.

Answer 448 Hz

INTERACTIVE EXAMPLE 14.5   The Noisy Siren

Goal   Solve a Doppler shift problem when both the source and observer are moving.

Problem   An ambulance travels down a highway at a speed of 75.0 mi/h, its siren emitting sound at a frequency of 4.00×10^2 Hz. What frequency is heard by a passenger in a car traveling at 55.0 mi/h in the opposite direction as the car and ambulance **(a)** *approach* each other and **(b)** pass and *move away* from each other? Take the speed of sound in air to be $v = 345$ m/s.

Strategy   Aside from converting from mi/h to m/s, this problem only requires substitution into the Doppler formula, but two signs must be chosen correctly in each part. In part (a), the observer moves toward the source and the source moves toward the observer, so both v_O and v_S should be chosen to be positive. Switch signs after they pass each other.

Solution

Convert the speeds from mi/h to m/s:

$$v_S = (75.0 \text{ mi/h}) \left(\frac{0.447 \text{ m/s}}{1.00 \text{ mi/h}} \right) = 33.5 \text{ m/s}$$

$$v_O = (55.0 \text{ mi/h}) \left(\frac{0.447 \text{ m/s}}{1.00 \text{ mi/h}} \right) = 24.6 \text{ m/s}$$

(a) Compute the observed frequency as the ambulance and car approach each other.

Each vehicle goes toward the other, so substitute $v_O = +24.6$ m/s and $v_S = +33.5$ m/s into the Doppler shift formula:

$$f_O = f_S \left(\frac{v + v_O}{v - v_S} \right)$$

$$= (4.00 \times 10^2 \text{ Hz}) \left(\frac{345 \text{ m/s} + 24.6 \text{ m/s}}{345 \text{ m/s} - 33.5 \text{ m/s}} \right) = \boxed{475 \text{ Hz}}$$

(b) Compute the observed frequency as the ambulance and car recede from each other.

Each vehicle goes away from the other, so substitute $v_O = -24.6$ m/s and $v_S = -33.5$ m/s into the Doppler shift formula:

$$f_O = f_S \left(\frac{v + v_O}{v - v_S} \right)$$

$$= (4.00 \times 10^2 \text{ Hz}) \left(\frac{345 \text{ m/s} + (-24.6 \text{ m/s})}{345 \text{ m/s} - (-33.5 \text{ m/s})} \right)$$

$$= \boxed{339 \text{ Hz}}$$

Remarks   Notice how the signs were handled: In part (b), the negative signs were required on the speeds because both observer and source were moving away from each other. Sometimes, of course, one of the speeds is negative and the other is positive.

Exercise 14.5

Repeat this problem, but assume the ambulance and car are going the same direction, with the ambulance initially behind the car. The speeds and the frequency of the siren are the same as in the example. Find the frequency heard by the observer in the car **(a)** before and **(b)** after the ambulance passes the car. [*Note:* The highway patrol subsequently gives the driver of the car a ticket for not pulling over for an emergency vehicle!]

Answers   (a) 411 Hz   (b) 391 Hz

Physics⊗Now™

You can alter the relative speeds of two submarines and observe the Doppler-shifted frequency by logging into PhysicsNow at **www.cp7e.com** and going to Interactive Example 14.5.

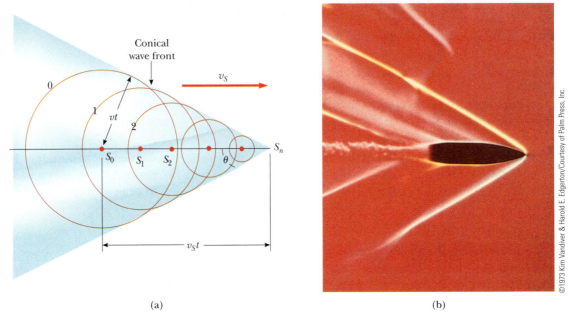

Conical
wave front

v_S

S_n

vt

S_0 S_1 S_2 θ

$v_S t$

(a)

(b)

©1973 Kim Vandiver & Harold E. Edgerton/Courtesy of Palm Press, Inc.

Figure 14.11 (a) A representation of a shock wave, produced when a source moves from S_0 to S_n with a speed v_s that is *greater* than the wave speed v in that medium. The envelope of the wave fronts forms a cone with half-angle of $\sin \theta = v/v_s$. (b) A stroboscopic photograph of a bullet moving at supersonic speed through the hot air above a candle.

Shock Waves

What happens when the source speed v_S *exceeds* the wave velocity v? Figure 14.11a describes this situation graphically. The circles represent spherical wave fronts emitted by the source at various times during its motion. At $t = 0$, the source is at point S_0, and at some later time t, the source is at point S_n. In the interval t, the wave front centered at S_0 reaches a radius of vt. In this same interval, the source travels to S_n, a distance of $v_s t$. At the instant the source is at S_n, the waves just beginning to be generated at this point have wave fronts of zero radius. The line drawn from S_n to the wave front centered on S_0 is tangent to all other wave fronts generated at intermediate times. All such tangent lines lie on the surface of a cone. The angle θ between one of these tangent lines and the direction of travel is given by

$$\sin \theta = \frac{v}{v_s}$$

© 1994, Comstock

Figure 14.12 The V-shaped bow wave of a boat is formed because the boat travels at a speed greater than the speed of the water waves. A bow wave is analogous to a shock wave formed by an airplane traveling faster than sound.

The ratio v_s/v is called the **Mach number**. The conical wave front produced when $v_s > v$ (supersonic speeds) is known as a **shock wave**. Figure 14.11b is a photograph of a bullet traveling at supersonic speed through the hot air rising above a candle. Notice the shock waves in the vicinity of the bullet. Another interesting example of a shock wave is the V-shaped wave front produced by a boat (the bow wave) when the boat's speed exceeds the speed of the water waves (Fig. 14.12).

Jet aircraft and space shuttles traveling at supersonic speeds produce shock waves that are responsible for the loud explosion, or sonic boom, heard on the ground. A shock wave carries a great deal of energy concentrated on the surface of the cone, with correspondingly great pressure variations. Shock waves are unpleasant to hear and can damage buildings when aircraft fly supersonically at low altitudes. In fact, an airplane flying at supersonic speeds produces a double boom, because two shock waves are formed—one from the nose of the plane and one from the tail (Fig. 14.13).

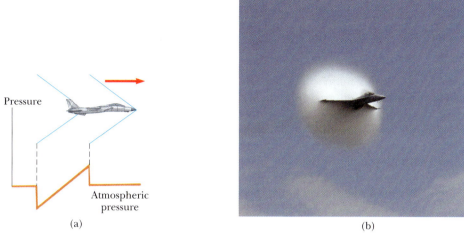

© Keith Lawson/Bettmann/Corbis

(a)   (b)

Figure 14.13   (a) The two shock waves produced by the nose and tail of a jet airplane traveling at supersonic speed. (b) A shock wave due to a jet traveling at the speed of sound is made visible as a fog of water vapor. The large pressure variation in the shock wave causes the water in the air to condense into water droplets.

Quick Quiz 14.3

As an airplane flying with constant velocity moves from a cold air mass into a warm air mass, does the Mach number (a) increase, (b) decrease, or (c) remain the same?

14.7   INTERFERENCE OF SOUND WAVES

Sound waves can be made to interfere with each other, a phenomenon that can be demonstrated with the device shown in Figure 14.14. Sound from a loudspeaker at S is sent into a tube at P, where there is a T-shaped junction. The sound splits and follows two separate pathways, indicated by the red arrows. Half of the sound travels upward, half downward. Finally, the two sounds merge at an opening where a listener places her ear. If the two paths r_1 and r_2 have the same length, waves that enter the junction will separate into two halves, travel the two paths, and then combine again at the ear. This reuniting of the two waves produces *constructive interference*, so the listener hears a loud sound. If the upper path is adjusted to be one full wavelength longer than the lower path, constructive interference of the two waves occurs again, and a loud sound is detected at the receiver. We have the following result: **If the path difference $r_2 - r_1$ is zero or some integer multiple of wavelengths, then constructive interference occurs and**

$$r_2 - r_1 = n\lambda \quad (n = 0, 1, 2, \ldots) \qquad [14.13]$$

◀ Condition for constructive interference

Suppose, however, that one of the path lengths, r_2, is adjusted so that the upper path is half a wavelength *longer* than the lower path r_1. In this case, an entering sound wave splits and travels the two paths as before, but now the wave along the upper path must travel a distance equivalent to half a wavelength farther than the wave traveling along the lower path. As a result, the crest of one wave meets the trough of the other when they merge at the receiver, causing the two waves to

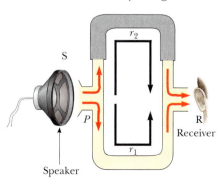

Figure 14.14   An acoustical system for demonstrating interference of sound waves. Sound from the speaker enters the tube and splits into two parts at P. The two waves combine at the opposite side and are detected at R. The upper path length is varied by the sliding section.

cancel each other. This phenomenon is called *totally destructive interference*, and no sound is detected at the receiver. In general, **if the path difference $r_2 - r_1$ is $\frac{1}{2}, 1\frac{1}{2}, 2\frac{1}{2} \ldots$ wavelengths, destructive interference occurs** and

$$r_2 - r_1 = (n + \tfrac{1}{2})\lambda \ (n = 0, 1, 2, \ldots)\qquad \text{[14.14]}$$

Nature provides many other examples of interference phenomena, most notably in connection with light waves, described in Chapter 24.

In connecting the wires between your stereo system and loudspeakers, you may notice that the wires are usually color coded and that the speakers have positive and negative signs on the connections. The reason for this is that the speakers need to be connected with the same "polarity." If they aren't, then the same electrical signal fed to both speakers will result in one speaker cone moving outward at the same time that the other speaker cone is moving inward. In this case, the sound leaving the two speakers will be 180° out of phase with each other. If you are sitting midway between the speakers, the sounds from both speakers travel the same distance and preserve the phase difference they had when they left. In an ideal situation, for a 180° phase difference, you would get complete destructive interference and no sound! In reality, the cancellation is not complete and is much more significant for bass notes (which have a long wavelength) than for the shorter wavelength treble notes. Nevertheless, to avoid a significant reduction in the intensity of bass notes, the color-coded wires and the signs on the speaker connections should be carefully noted.

APPLICATION

Connecting Your Stereo Speakers

TIP 14.3   Do Waves Really Interfere?

In popular usage, to *interfere* means "to come into conflict with" or "to intervene to affect an outcome." This differs from its use in physics, where waves pass through each other and interfere, but don't affect each other in any way.

EXAMPLE 14.6   Two Speakers Driven by the Same Source

Goal   Use the concept of interference to compute a frequency.

Problem   Two speakers placed 3.00 m apart are driven by the same oscillator (Fig. 14.15). A listener is originally at point *O*, which is located 8.00 m from the center of the line connecting the two speakers. The listener then walks to point *P*, which is a perpendicular distance 0.350 m from *O*, before reaching the *first minimum* in sound intensity. What is the frequency of the oscillator? Take the speed of sound in air to be $v_s = 343$ m/s.

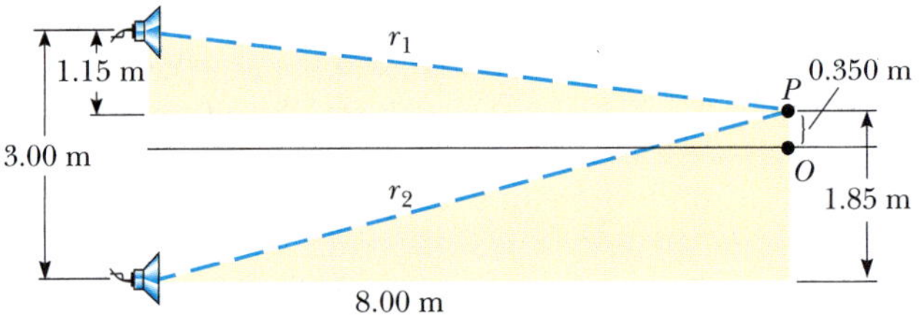

Strategy   The position of the first minimum in sound intensity is given, which is a point of destructive interference. We can find the path lengths r_1 and r_2 with the Pythagorean theorem and then use Equation 14.14 for destructive interference to find the wavelength λ. Using $v = f\lambda$ then yields the frequency.

Figure 14.15   (Example 14.6) Two loudspeakers driven by the same source can produce interference.

Solution
Use the Pythagorean theorem to find the path lengths r_1 and r_2:

$$r_1 = \sqrt{(8.00 \text{ m})^2 + (1.15 \text{ m})^2} = 8.08 \text{ m}$$
$$r_2 = \sqrt{(8.00 \text{ m})^2 + (1.85 \text{ m})^2} = 8.21 \text{ m}$$

Substitute these values and $n = 0$ into Equation 14.14, solving for the wavelength:

$$r_2 - r_1 = (n + \tfrac{1}{2})\lambda$$
$$8.21 \text{ m} - 8.08 \text{ m} = 0.13 \text{ m} = \lambda/2 \rightarrow \lambda = 0.26 \text{ m}$$

Solve $v = \lambda f$ for the frequency f and substitute the speed of sound and the wavelength:

$$f = \frac{v}{\lambda} = \frac{343 \text{ m/s}}{0.26 \text{ m}} = 1.3 \text{ kHz}$$

Remark   For problems involving constructive interference, the only difference is that Equation 14.13, $r_2 - r_1 = n\lambda$, would be used instead of Equation 14.14.

Exercise 14.6
If the oscillator frequency is adjusted so that the location of the first minimum is at a distance of 0.750 m from *O*, what is the new frequency?

Answer   0.642 kHz

14.8 STANDING WAVES

Standing waves can be set up in a stretched string by connecting one end of the string to a stationary clamp and connecting the other end to a vibrating object, such as the end of a tuning fork, or by shaking the hand holding the string up and down at a steady rate (Fig. 14.16). Traveling waves then reflect from the ends and move in both directions on the string. The incident and reflected waves combine according to the **superposition principle**. (See Section 13.10.) If the string vibrates at exactly the right frequency, the wave appears to stand still—hence its name, **standing wave**. A **node** occurs where the two traveling waves always have the same magnitude of displacement but the opposite sign, so the net displacement is zero at that point. There is no motion in the string at the nodes, but midway between two adjacent nodes, at an **antinode**, the string vibrates with the largest amplitude.

Figure 14.17 shows snapshots of the oscillation of a standing wave during half of a cycle. The pink arrows indicate the direction of motion of different parts of the string. Notice that **all points on the string oscillate together vertically with the same frequency, but different points have different amplitudes of motion**. The points of attachment to the wall and all other stationary points on the string are called nodes, labeled N in Figure 14.17a. From the figure, observe that the distance between adjacent nodes is one-half the wavelength of the wave:

$$d_{NN} = \tfrac{1}{2}\lambda$$

Consider a string of length L that is fixed at both ends, as in Active Figure 14.18. For a string, we can set up standing-wave patterns at many frequencies—the more loops, the higher the frequency. Three such patterns are shown in Active Figures 14.18b, 14.18c, and 14.18d. Each has a characteristic frequency, which we will now calculate.

First, **the ends of the string must be nodes, because these points are fixed**. If the string is displaced at its midpoint and released, the vibration shown in Active Figure 14.18b can be produced, in which case the center of the string is an antinode, labeled A. Note that from end to end, the pattern is N–A–N. The distance from a node to its adjacent antinode, N–A, is always equal to a quarter wavelength, $\lambda_1/4$. There are two such segments, N–A and A–N, so $L = 2(\lambda_1/4) = \lambda_1/2$, and $\lambda_1 = 2L$. The frequency of this vibration is therefore

$$f_1 = \frac{v}{\lambda_1} = \frac{v}{2L} \qquad [14.15]$$

Recall that the speed of a wave on a string is $v = \sqrt{F/\mu}$, where F is the tension in the string and μ is its mass per unit length (Chapter 13). Substituting into Equation 14.15, we obtain

$$f_1 = \frac{1}{2L}\sqrt{\frac{F}{\mu}} \qquad [14.16]$$

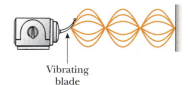

Vibrating
blade

Figure 14.16 Standing waves can be set up in a stretched string by connecting one end of the string to a vibrating blade. When the blade vibrates at one of the natural frequencies of the string, large-amplitude standing waves are created.

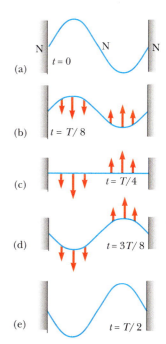

Figure 14.17 A standing-wave pattern in a stretched string, shown by snapshots of the string during one-half of a cycle. In part (a), N denotes a node.

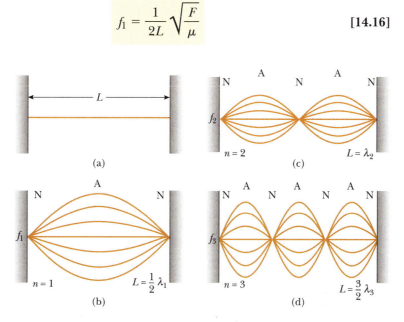

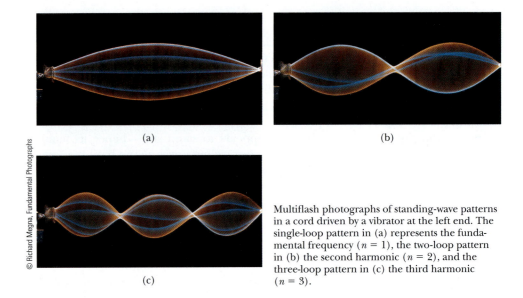

(a)

(b)

(c)

Multiflash photographs of standing-wave patterns in a cord driven by a vibrator at the left end. The single-loop pattern in (a) represents the fundamental frequency ($n = 1$), the two-loop pattern in (b) the second harmonic ($n = 2$), and the three-loop pattern in (c) the third harmonic ($n = 3$).

© Richard Megna, Fundamental Photographs

This lowest frequency of vibration is called the **fundamental frequency** of the vibrating string, or the **first harmonic**.

The first harmonic has nodes only at the ends—the points of attachment, with node–antinode pattern of N–A–N. The next harmonic, called the **second harmonic** (also called the **first overtone**) can be constructed by inserting an additional node–antinode segment between the endpoints. This makes the pattern N–A–N–A–N, as in Active Figure 14.18c. We count the node–antinode pairs: N–A, A–N, N–A, and A–N, four segments in all, each representing a quarter wavelength. We then have $L = 4(\lambda_2/4) = \lambda_2$, and the second harmonic (first overtone) is

$$f_2 = \frac{v}{\lambda_2} = \frac{v}{L} = 2\left(\frac{v}{2L}\right) = 2f_1$$

This frequency is equal to *twice* the fundamental frequency. The **third harmonic** (**second overtone**) is constructed similarly. Inserting one more N–A segment, we obtain Active Figure 14.18c, the pattern of nodes reading N–A–N–A–N–A–N. There are six node–antinode segments, so $L = 6(\lambda_3/4) = 3(\lambda_3/2)$, which means that $\lambda_3 = 2L/3$, giving

$$f_3 = \frac{v}{\lambda_3} = \frac{3v}{2L} = 3f_1$$

All the higher harmonics, it turns out, are positive integer multiples of the fundamental:

$$f_n = nf_1 = \frac{n}{2L}\sqrt{\frac{F}{\mu}} \qquad n = 1, 2, 3 \ldots \qquad \text{[14.17]}$$

The frequencies f_1, $2f_1$, $3f_1$, and so on form a **harmonic series**.

Quick Quiz 14.4

Which of the following frequencies are higher harmonics of a string with fundamental frequency of 150 Hz? (a) 200 Hz (b) 300 Hz (c) 400 Hz (d) 500 Hz (e) 600 Hz.

When a stretched string is distorted to a shape that corresponds to any one of its harmonics, after being released it vibrates only at the frequency of that harmonic. If the string is struck or bowed, however, the resulting vibration includes different amounts of various harmonics, including the fundamental frequency. Waves not in the harmonic series are quickly damped out on a string fixed at both ends. In effect, when disturbed, the string "selects" the standing-wave frequencies. As we'll see later, the presence of several harmonics on a string gives stringed instruments their characteristic sound, which enables us to distinguish one from another even when they are producing identical fundamental frequencies.

The frequency of a string on a musical instrument can be changed by varying either the tension or the length. The tension in guitar and violin strings is varied by turning pegs on the neck of the instrument. As the tension is increased, the frequency of the harmonic series increases according to Equation 14.17. Once the instrument is tuned, the musician varies the frequency by pressing the strings against the neck at a variety of positions, thereby changing the effective lengths of the vibrating portions of the strings. As the length is reduced, the frequency again increases, as follows from Equation 14.17.

Finally, Equation 14.17 shows that a string of fixed length can be made to vibrate at a lower fundamental frequency by increasing its mass per unit length. This is achieved in the bass strings of guitars and pianos by wrapping them with metal windings.

APPLICATION
Tuning a Musical Instrument

INTERACTIVE EXAMPLE 14.7 Guitar Fundamentals

Goal Apply standing-wave concepts to a stringed instrument.

Problem The high E string on a certain guitar measures 64.0 cm in length and has a fundamental frequency of 329 Hz. When a guitarist presses down so that the string is in contact with the first fret (Fig. 14.19a), the string is shortened so that it plays an F note that has a frequency of 349 Hz. **(a)** How far is the fret from the nut? **(b)** Overtones can be produced on a guitar string by gently placing the index finger in the location of a node of a higher harmonic. The string should be touched, but not depressed against a fret. (Given the width of a finger, pressing too hard will damp out higher harmonics as well.) The fundamental frequency is thereby suppressed, making it possible to hear overtones. Where on the guitar string relative to the nut should the finger be lightly placed so as to hear the second harmonic? The fourth harmonic? (This is equivalent to finding the location of the nodes in each case.)

Strategy For part (a) use Equation 14.15, corresponding to the fundamental frequency, to find the speed of waves on the string. Shortening the string by playing a higher note doesn't affect the wave speed, which depends only on the tension and linear density of the string (which are unchanged). Solve Equation 14.15 for the new length L, using the new fundamental frequency, and subtract this length from the original length to find the distance from the nut to the first fret. In part (b), remember that the distance from node to node is half a wavelength. Calculate the wavelength, divide it in two, and locate the nodes, which are integral numbers of half-wavelengths from the nut. [*Note:* The nut is a small piece of wood or ebony at the top of the fret board. The distance from the nut to the bridge (below the sound hole) is the length of the string. (See Fig. 14.19b.)]

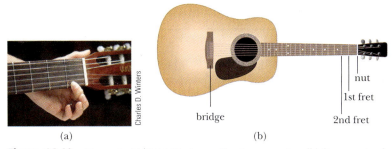

Figure 14.19 (Example 14.7) (a) Playing an F note on a guitar. (b) Some parts of a guitar.

Solution
(a) Find the distance from the nut to the first fret.

Substitute $L_0 = 0.640$ m and $f_1 = 329$ Hz into Equation 14.15, finding the wave speed on the string:

$$f_1 = \frac{v}{2L_0}$$

$$v = 2L_0 f_1 = 2(0.640 \text{ m})(329 \text{ Hz}) = 421 \text{ m/s}$$

Solve Equation 14.15 for the length L, and substitute the wave speed and the frequency of F.

$$L = \frac{v}{2f} = \frac{421 \text{ m/s}}{2(349 \text{ Hz})} = 0.603 \text{ m} = 60.3 \text{ cm}$$

Subtract this length from the original length L_0 to find the distance from the nut to the first fret:

$$\Delta x = L_0 - L = 64.0 \text{ cm} - 60.3 \text{ cm} = \boxed{3.7 \text{ cm}}$$

(b) Find the locations of nodes for the second and fourth harmonics.

The second harmonic has a wavelength $\lambda_2 = L_0 = 64.0$ cm. The distance from nut to node corresponds to half a wavelength.

$$\Delta x = \tfrac{1}{2}\lambda_2 = \tfrac{1}{2}L_0 = 32.0 \text{ cm}$$

The fourth harmonic, of wavelength $\lambda_4 = \tfrac{1}{2}L_0 = 32.0$ cm, has three nodes between the endpoints:

$$\Delta x = \tfrac{1}{2}\lambda_4 = \boxed{16.0 \text{ cm}}, \quad \Delta x = 2(\lambda_4/2) = \boxed{32.0 \text{ cm}},$$

$$\Delta x = 3(\lambda_4/2) = \boxed{48.0 \text{ cm}}$$

Remarks   Placing a finger at the position $\Delta x = 32.0$ cm damps out the fundamental and odd harmonics, but not all the higher even harmonics. The second harmonic dominates, however, because the rest of the string is free to vibrate. Placing the finger at $\Delta x = 16.0$ cm or 48.0 cm damps out the first through third harmonics, allowing the fourth harmonic to be heard.

Exercise 14.7
Pressing the E-string down on the fret board just above the second fret pinches the string firmly against the fret, giving an F sharp, which has frequency 3.70×10^2 Hz. (a) Where should the second fret be located? (b) Find two locations where you could touch the open E-string and hear the third harmonic.

Answer   (a) 7.1 cm from the nut and 3.4 cm from the first fret. Note that the distance from the first to the second fret isn't the same as from the nut to the first fret. (b) 21.3 cm and 42.7 cm from the nut.

Physics⊗Now™
Explore this situation by logging into PhysicsNow at **www.cp7e.com** and going to Interactive Example 14.7.

EXAMPLE 14.8   Harmonics of a Stretched Wire
Goal   Calculate string harmonics, relate them to sound and combine them with tensile stress.

Problem   (a) Find the frequencies of the fundamental, second, and third harmonics of a steel wire 1.00 m long with a mass per unit length of 2.00×10^{-3} kg/m and under a tension of 80.0 N. (b) Find the wavelengths of the sound waves created by the vibrating wire for all three modes. Assume the speed of sound in air is 345 m/s. (c) Suppose the wire is carbon steel with a density of 7.80×10^3 kg/m³, a cross-sectional area $A = 2.56 \times 10^{-7}$ m², and an elastic limit of 2.80×10^8 Pa. Find the fundamental frequency if the wire is tightened to the elastic limit. Neglect any stretching of the wire (which would slightly reduce the mass per unit length).

Strategy   (a) It's easiest to find the speed of waves on the wire then substitute into Equation 14.15 to find the first harmonic. The next two are multiples of the first, given by Equation 14.17. (b) The frequencies of the sound waves are the same as the frequencies of the vibrating wire, but the wavelengths are different. Use $v_s = f\lambda$, where v_s is the speed of sound in air, to find the wavelengths in air. (c) Find the force corresponding to the elastic limit, and substitute it into Equation 14.16.

Solution
(a) Find the first three harmonics at the given tension.

Use Equation 13.17 to calculate the speed of the wave on the wire:

$$v = \sqrt{\frac{F}{\mu}} = \sqrt{\frac{80.0 \text{ N}}{2.00 \times 10^{-3} \text{ kg/m}}} = 2.00 \times 10^2 \text{ m/s}$$

Find the wire's fundamental frequency from Equation 14.15:

$$f_1 = \frac{v}{2L} = \frac{2.00 \times 10^2 \text{ m/s}}{2(1.00 \text{ m})} = 1.00 \times 10^2 \text{ Hz}$$

Find the next two harmonics by multiplication:

$$f_2 = 2f_1 = \boxed{2.00 \times 10^2 \text{ Hz}}, \quad f_3 = 3f_1 = \boxed{3.00 \times 10^2 \text{ Hz}}$$

(b) Find the wavelength of the sound waves produced.

Solve $v_s = f\lambda$ for the wavelength and substitute the frequencies.

$$\lambda_1 = v_s/f_1 = (345 \text{ m/s})/(1.00 \times 10^2 \text{ Hz}) = \boxed{3.45 \text{ m}}$$

$$\lambda_2 = v_s/f_2 = (345 \text{ m/s})/(2.00 \times 10^2 \text{ Hz}) = \boxed{1.73 \text{ m}}$$

$$\lambda_3 = v_s/f_3 = (345 \text{ m/s})/(3.00 \times 10^2 \text{ Hz}) = \boxed{1.15 \text{ m}}$$

(c) Find the fundamental frequency corresponding to the elastic limit.

Calculate the tension in the wire from the elastic limit:

$$\frac{F}{A} = \text{elastic limit} \quad \rightarrow \quad F = (\text{elastic limit})A$$

$$F = (2.80 \times 10^8 \text{ Pa})(2.56 \times 10^{-7} \text{ m}^2) = 71.7 \text{ N}$$

Substitute the values of F, μ, and L into Equation 14.16:

$$f_1 = \frac{1}{2L} \sqrt{\frac{F}{\mu}}$$

$$f_1 = \frac{1}{2(1.00 \text{ m})} \sqrt{\frac{71.7 \text{ N}}{2.00 \times 10^{-3} \text{ kg/m}}} = \boxed{94.7 \text{ Hz}}$$

Remarks From the answer to part (c), it appears we need to choose a thicker wire or use a better grade of steel with a higher elastic limit. The frequency corresponding to the elastic limit is smaller than the fundamental!

Exercise 14.8
(a) Find the fundamental frequency and second harmonic if the tension in the wire is increased to 115 N. (Assume the wire doesn't stretch or break.) **(b)** Using a sound speed of 345 m/s, find the wavelengths of the sound waves produced.

Answer (a) 1.20×10^2 Hz, 2.40×10^2 Hz (b) 2.88 m, 1.44 m

14.9 FORCED VIBRATIONS AND RESONANCE

In Chapter 13 we learned that the energy of a damped oscillator decreases over time because of friction. It's possible to compensate for this energy loss by applying an external force that does positive work on the system.

For example, suppose an object–spring system having some natural frequency of vibration f_0 is pushed back and forth by a periodic force with frequency f. The system vibrates at the frequency f of the driving force. This type of motion is referred to as a **forced vibration**. Its amplitude reaches a maximum when the frequency of the driving force equals the natural frequency of the system f_0, called the **resonant frequency** of the system. Under this condition, the system is said to be in **resonance**.

In Section 14.8 we learned that a stretched string can vibrate in one or more of its natural modes. Here again, if a periodic force is applied to the string, the amplitude of vibration increases as the frequency of the applied force approaches one of the string's natural frequencies of vibration.

Resonance vibrations occur in a wide variety of circumstances. Figure 14.20 illustrates one experiment that demonstrates a resonance condition. Several pendulums of different lengths are suspended from a flexible beam. If one of them, such as A, is set in motion, the others begin to oscillate because of vibrations in the flexible beam. Pendulum C, the same length as A, oscillates with the greatest amplitude because its natural frequency matches that of pendulum A (the driving force).

Another simple example of resonance is a child being pushed on a swing, which is essentially a pendulum with a natural frequency that depends on its length. The

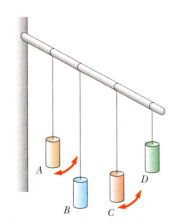

Figure 14.20 Resonance. If pendulum A is set in oscillation, only pendulum C, with a length matching that of A, will eventually oscillate with a large amplitude, or resonate. The arrows indicate motion perpendicular to the page.

Figure 14.21 (*Top*) Standing-wave
pattern in a vibrating wineglass. The
glass will shatter if the amplitude of
vibration becomes too large. (*Bottom*)
A wineglass shattered by the
amplified sound of a human voice.

swing is kept in motion by a series of appropriately timed pushes. For its amplitude
to increase, the swing must be pushed each time it returns to the person's hands.
This corresponds to a frequency equal to the natural frequency of the swing. If the
energy put into the system per cycle of motion equals the energy lost due to friction, the amplitude remains constant.

Opera singers have been known to set crystal goblets in audible vibration with
their powerful voices, as shown in Figure 14.21. This is yet another example of resonance: The sound waves emitted by the singer can set up large-amplitude vibrations in the glass. If a highly amplified sound wave has the right frequency, the amplitude of forced vibrations in the glass increases to the point where the glass
becomes heavily strained and shatters.

The classic example of structural resonance occurred in 1940, when the Tacoma
Narrows bridge in the state of Washington was set in oscillation by the wind (Fig.
14.22). The amplitude of the oscillations increased rapidly and reached a high
value until the bridge ultimately collapsed (probably because of metal fatigue). In
recent years, however, a number of researchers have called this explanation into
question. Gusts of wind, in general, don't provide the periodic force necessary for a
sustained resonance condition, and the bridge exhibited large twisting oscillations,
rather than the simple up-and-down oscillations expected of resonance.

A more recent example of destruction by structural resonance occurred during
the Loma Prieta earthquake near Oakland, California, in 1989. In a mile-long section of the double-decker Nimitz Freeway, the upper deck collapsed onto the
lower deck, killing several people. The collapse occurred because that particular
section was built on mud fill while other parts were built on bedrock. As seismic
waves pass through mud fill or other loose soil, their speed decreases and their
amplitude increases. The section of the freeway that collapsed oscillated at the
same frequency as other sections, but at a much larger amplitude.

14.10 STANDING WAVES IN AIR COLUMNS

Standing longitudinal waves can be set up in a tube of air, such as an organ pipe, as
the result of interference between sound waves traveling in opposite directions. The
relationship between the incident wave and the reflected wave depends on whether
the reflecting end of the tube is open or closed. A portion of the sound wave is reflected back into the tube even at an open end. **If one end is closed, a node must exist at that end because the movement of air is restricted. If the end is open, the elements of air have complete freedom of motion, and an antinode exists**.

Figure 14.23a shows the first three modes of vibration of a pipe open at both
ends. When air is directed against an edge at the left, longitudinal standing waves
are formed and the pipe vibrates at its natural frequencies. Note that, from end to
end, the pattern is A–N–A, the same pattern as in the vibrating string, except
node and antinode have exchanged positions. As before, an antinode and its adjacent node, A–N, represent a quarter-wavelength, and there are two, A–N and
N–A, so $L = 2(\lambda_1/4) = \lambda_1/2$ and $\lambda_1 = 2L$. The fundamental frequency of the
pipe open at both ends is then $f_1 = v/\lambda_1 = v/2L$. The next harmonic has an addi-

Figure 14.22 The collapse of the
Tacoma Narrows suspension bridge
in 1940 has been cited as a demonstration of mechanical resonance.
High winds set up standing waves in
the bridge, causing it to oscillate at
one of its natural frequencies. Once
established, the resonance may have
led to the bridge's collapse (although
this interpretation is currently being
challenged by mathematicians and
physical scientists).

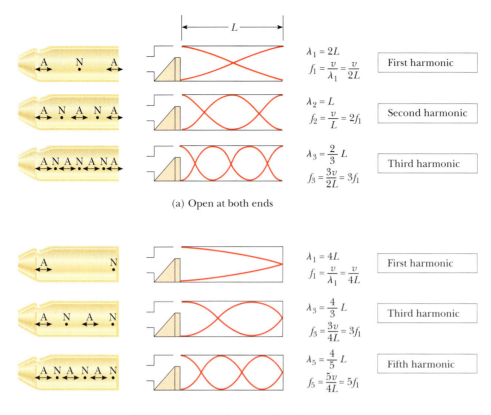

Figure 14.23   (a) Standing longitudinal waves in an organ pipe open at both ends. The natural frequencies $f_1, 2f_1, 3f_1 \ldots$ form a harmonic series. (b) Standing longitudinal waves in an organ pipe closed at one end. Only *odd* harmonics are present, and the natural frequencies are $f_1, 3f_1, 5f_1$, and so on.

TIP 14.4   Sound Waves Are Not Transverse

The standing longitudinal waves in Figure 14.23 are drawn as transverse waves only because it's difficult to draw longitudinal displacements—they're in the same direction as the wave propagation. In the figure, the vertical axis represents either pressure or horizontal displacement of the elements of the medium.

tional node and antinode between the ends, creating the pattern A–N–A–N–A. We count the pairs: A–N, N–A, A–N, and N–A, making four segments, each with length $\lambda_2/4$. We have $L = 4(\lambda_2/4) = \lambda_2$, and the second harmonic (first overtone) is $f_2 = v/\lambda_2 = v/L = 2(v/2L) = 2f_1$. All higher harmonics, it turns out, are positive integer multiples of the fundamental:

$$f_n = n\frac{v}{2L} = nf_1 \quad n = 1, 2, 3, \ldots \qquad [14.18]$$

◀ Pipe open at both ends; all harmonics are present

where v is the speed of sound in air. Notice the similarity to Equation 14.17, which also involves multiples of the fundamental.

   If a pipe is open at one end and closed at the other, the open end is an antinode while the closed end is a node (Fig. 14.23b). In such a pipe, the fundamental frequency consists of a single antinode–node pair, A–N, so $L = \lambda_1/4$ and $\lambda_1 = 4L$. The fundamental harmonic for a pipe closed at one end is then $f_1 = v/\lambda_1 = v/4L$. The first overtone has another node and antinode between the open end and closed end, making the pattern A–N–A–N. There are three antinode–node segments in this pattern (A–N, N–A, and A–N), so $L = 3(\lambda_3/4)$ and $\lambda_3 = 4L/3$. The first overtone, therefore, has frequency $f_3 = v/\lambda_3 = 3v/4L = 3f_1$. Similarly, $f_5 = 5f_1$. In contrast to the pipe open at both ends, **there are no even multiples of the fundamental harmonic**. The odd harmonics for a pipe open at one end only are given by

$$f_n = n\frac{v}{4L} = nf_1 \quad n = 1, 3, 5, \ldots \qquad [14.19]$$

◀ Pipe closed at one end; only odd harmonics are present

Quick Quiz 14.5

A pipe open at both ends resonates at a fundamental frequency f_{open}. When one end is covered and the pipe is again made to resonate, the fundamental frequency is f_{closed}. Which of the following expressions describes how these two resonant frequencies compare? (a) $f_{closed} = f_{open}$, (b) $f_{closed} = \frac{3}{2}f_{open}$, (c) $f_{closed} = 2 f_{open}$, (d) $f_{closed} = \frac{1}{2}f_{open}$, (e) none of these.

Quick Quiz 14.6

Balboa Park in San Diego has an outdoor organ. When the air temperature increases, the fundamental frequency of one of the organ pipes (a) increases (b) decreases (c) stays the same (d) impossible to determine. (The thermal expansion of the pipe is negligible.)

Applying Physics 14.3   Oscillations in a Harbor

Why do passing ocean waves sometimes cause the water in a harbor to undergo very large oscillations, called a *seiche* (pronounced *saysh*)?

Explanation Water in a harbor is enclosed and possesses a natural frequency based on the size of the harbor. This is similar to the natural frequency of the enclosed air in a bottle, which can be excited by blowing across the edge of the opening. Ocean waves pass by the opening of the harbor at a certain frequency. If this frequency matches that of the enclosed harbor, then a large standing wave can be set up in the water by resonance. This situation can be simulated by carrying a fish tank filled with water. If your walking frequency matches the natural frequency of the water as it sloshes back and forth, a large standing wave develops in the fish tank.

Applying Physics 14.4   Why Are Instruments Warmed Up?

Why do the strings go flat and the wind instruments go sharp during a performance if an orchestra doesn't warm up beforehand?

Explanation Without warming up, all the instruments will be at room temperature at the beginning of the concert. As the wind instruments are played, they fill with warm air from the player's exhalation. The increase in temperature of the air in the instruments causes an increase in the speed of sound, which raises the resonance frequencies of the air columns. As a result, the instruments go sharp. The strings on the stringed instruments also increase in temperature due to the friction of rubbing with the bow. This results in thermal expansion, which causes a decrease in tension in the strings. With the decrease in tension, the wave speed on the strings drops, and the fundamental frequencies decrease, so the stringed instruments go flat.

Applying Physics 14.5   How Do Bugles Work?

A bugle has no valves, keys, slides, or finger holes. How can it be used to play a song?

Explanation Songs for the bugle are limited to harmonics of the fundamental frequency, because there is no control over frequencies without valves, keys, slides, or finger holes. The player obtains different notes by changing the tension in the lips as the bugle is played, in order to excite different harmonics. The normal playing range of a bugle is among the third, fourth, fifth, and sixth harmonics of the fundamental. "Reveille," for example, is played with just the three notes G, C, and F. And "Taps" is played with these three notes and the G one octave above the lower G.

EXAMPLE 14.9   Harmonics of a Pipe

Goal Find frequencies of open and closed pipes.

Problem A pipe is 2.46 m long. **(a)** Determine the frequencies of the first three harmonics if the pipe is open at both ends. Take 343 m/s as the speed of sound in air. **(b)** How many harmonic frequencies of this pipe lie in the audible range, from 20 Hz to 20 000 Hz? **(c)** What are the three lowest possible frequencies if the pipe is closed at one end and open at the other?

Strategy Substitute into Equation 14.18 for part (a) and Equation 14.19 for part (c). All harmonics, $n = 1, 2,$ $3 . . .$ are available for the pipe open at both ends, but only the harmonics with $n = 1, 3, 5, . . .$ for the pipe closed at one end. For part (b), set the frequency in Equation 14.18 equal to 2.00×10^4 Hz.

Solution

(a) Find the frequencies if the pipe is open at both ends.

Substitute into Equation 14.18, with $n = 1$:

$$f_1 = \frac{v}{2L} = \frac{343 \text{ m/s}}{2(2.46 \text{ m})} = \boxed{69.7 \text{ Hz}}$$

Multiply to find the second and third harmonics:

$$f_2 = 2f_1 = \boxed{139 \text{ Hz}} \qquad f_3 = 3f_1 = \boxed{209 \text{ Hz}}$$

(b) How many harmonics lie between 20 Hz and 20 000 Hz for this pipe?

Set the frequency in Equation 14.18 equal to 2.00×10^4 and solve for n:

$$f_n = n\,\frac{v}{2L} = n\,\frac{343 \text{ m/s}}{2 \cdot 2.46 \text{ m}} = 2.00 \times 10^4 \text{ Hz}$$

This works out to $n = 286.88$, which must be truncated down ($n = 287$ gives a frequency over 2.00×10^4 Hz).

$$\boxed{n = 286}$$

(c) Find the frequencies for the pipe closed at one end.

Apply Equation 14.19 with $n = 1$:

$$f_1 = \frac{v}{4L} = \frac{343 \text{ m/s}}{4(2.46 \text{ m})} = \boxed{34.9 \text{ Hz}}$$

The next two harmonics are odd multiples of the first:

$$f_3 = 3f_1 = \boxed{105 \text{ Hz}} \qquad f_5 = 5f_1 = \boxed{175 \text{ Hz}}$$

Exercise 14.9

(a) What length pipe open at both ends has a fundamental frequency of 3.70×10^2 Hz? Find the first overtone. **(b)** If the one end of this pipe is now closed, what is the new fundamental frequency? Find the first overtone. **(c)** If the pipe is open at one end only, how many harmonics are possible in the normal hearing range from 20 to 20 000 Hz?

Answer   (a) 0.464 m, 7.40×10^2 Hz   (b) 185 Hz, 555 Hz   (c) 54

EXAMPLE 14.10   Resonance in a Tube of Variable Length

Goal   Understand resonance in tubes and perform elementary calculations.

Problem   Figure 14.24a shows a simple apparatus for demonstrating resonance in a tube. A long tube open at both ends is partially submerged in a beaker of water, and a vibrating tuning fork of unknown frequency is placed near the top of the tube. The length of the air column, L, is adjusted by moving the tube vertically. The sound waves generated by the fork are reinforced when the length of the air column corresponds to one of the resonant frequencies of the tube. Suppose the smallest value of L for which a peak occurs in the sound intensity is 9.00 cm. **(a)** With this measurement, determine the frequency of the tuning fork. **(b)** Find the wavelength and the next two air-column lengths giving resonance. Take the speed of sound to be 345 m/s.

Strategy   Once the tube is in the water, the setup is the same as a pipe closed at one end. For part (a), substitute values for v and L into Equation 14.19 with $n = 1$ and find the frequency of the tuning fork. **(b)** The next resonance maximum occurs when the water level is low enough to allow a second node, which is another half-wavelength in distance. The third resonance occurs when the third node is reached, requiring yet another half-wavelength of distance. The frequency in each case is the same, because it's generated by the tuning fork.

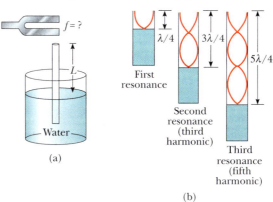

Figure 14.24   (Example 14.10) (a) Apparatus for demonstrating the resonance of sound waves in a tube closed at one end. The length L of the air column is varied by moving the tube vertically while it is partially submerged in water. (b) The first three resonances of the system.

Solution

(a) Find the frequency of the tuning fork.

Substitute $n = 1$, $v = 345$ m/s, and $L_1 = 9.00 \times 10^{-2}$ m into Equation 14.19:

$$f_1 = \frac{v}{4L_1} = \frac{345 \text{ m/s}}{4(9.00 \times 10^{-2} \text{ m})} = \boxed{958 \text{ Hz}}$$

(b) Find the wavelength and the next two water levels giving resonance.

Calculate the wavelength, using the fact that, for a tube open at one end, $\lambda = 4L$ for the fundamental.

$$\lambda = 4L_1 = 4(9.00 \times 10^{-2} \text{ m}) = \boxed{0.360 \text{ m}}$$

Add a half-wavelength of distance to L_1 to get the next resonance position:

$$L_2 = L_1 + \lambda/2 = 0.0900 \text{ m} + 0.180 \text{ m} = \boxed{0.270 \text{ m}}$$

Add another half-wavelength to L_2 to obtain the third resonance position:

$$L_3 = L_2 + \lambda/2 = 0.270 \text{ m} + 0.180 \text{ m} = \boxed{0.450 \text{ m}}$$

Remark This experimental arrangement is often used to measure the speed of sound, in which case the frequency of the tuning fork must be known in advance.

Exercise 14.10

An unknown gas is introduced into the aforementioned apparatus using the same tuning fork, and the first resonance occurs when the air column is 5.84 cm long. Find the speed of sound in the gas.

Answer 224 m/s

14.11 BEATS

The interference phenomena we have been discussing so far have involved the superposition of two or more waves with the same frequency, traveling in opposite directions. Another type of interference effect results from the superposition of two waves with slightly different frequencies. In such a situation, the waves at some fixed point are periodically in and out of phase, corresponding to an alternation in time between constructive and destructive interference. In order to understand this phenomenon, consider Active Figure 14.25. The two waves shown in Active Figure 14.25a were emitted by two tuning forks having slightly different frequencies; Active Figure 14.25b shows the superposition of these waves. At some time t_a the waves are in phase and constructive interference occurs, as demonstrated by the resultant curve in Active Figure 14.25b. At some later time, however, the vibrations of the two forks move out of step with each other. At time t_b, one fork emits a compression while the other emits a rarefaction, and destructive interference occurs, as demonstrated by the curve shown. As time passes, the vibrations of the two forks move out of phase, then into phase again, and so on. As a consequence, a listener at some fixed point hears an alternation in loudness, known as **beats**. The number of beats per second, or the *beat frequency*, equals the difference in frequency between the two sources:

$$f_b = |f_2 - f_1| \qquad \text{[14.20]}$$

where f_b is the beat frequency and f_1 and f_2 are the two frequencies. The absolute value is used because the beat frequency is a positive quantity and will occur regardless of the order of subtraction.

APPLICATION

Using Beats to Tune a Musical Instrument

A stringed instrument such as a piano can be tuned by beating a note on the instrument against a note of known frequency. The string can then be tuned to the desired frequency by adjusting the tension until no beats are heard.

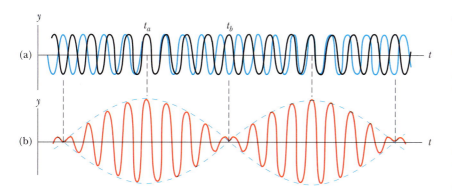

ACTIVE FIGURE 14.25
Beats are formed by the combination
of two waves of slightly different
frequencies traveling in the same
direction. (a) The individual waves
heard by an observer at a fixed point
in space. (b) The combined wave has
an amplitude (dashed line) that
oscillates in time.

Physics⊗Now™
Log into PhysicsNow at
www.cp7e.com, and go to Active
Figure 14.25 to choose two
frequencies and see the
corresponding beats.

Quick Quiz 14.7

You are tuning a guitar by comparing the sound of the string with that of a standard tuning fork. You notice a beat frequency of 5 Hz when both sounds are present. As you tighten the guitar string, the beat frequency rises steadily to 8 Hz. In order to tune the string exactly to the tuning fork, you should (a) continue to tighten the string (b) loosen the string (c) impossible to determine from the given information.

EXAMPLE 14.11   Sour Notes

Goal   Apply the beat frequency concept.

Problem   A certain piano string is supposed to vibrate at a frequency of 4.40×10^2 Hz. In order to check its frequency, a tuning fork known to vibrate at a frequency of 4.40×10^2 Hz is sounded at the same time the piano key is struck, and a beat frequency of 4 beats per second is heard. **(a)** Find the two possible frequencies at which the string could be vibrating. **(b)** Suppose the piano tuner runs toward the piano, holding the vibrating tuning fork while his assistant plays the note, which is at 436 Hz. At his maximum speed, the piano tuner notices the beat frequency drops from 4 Hz to 2 Hz (without going through a beat frequency of zero). How fast is he moving? Use a sound speed of 343 m/s. **(c)** While the piano tuner is running, what beat frequency is observed by the assistant? [*Note:* Assume all numbers are accurate to two decimal places, necessary for this last calculation.]

Strategy   **(a)** The beat frequency is equal to the absolute value of the difference in frequency between the two sources of sound and occurs if the piano string is tuned either too high or too low. Solve Equation 14.20 for these two possible frequencies. **(b)** Moving toward the piano raises the observed piano string frequency. Solve the Doppler shift formula, Equation 14.12, for the speed of the observer. **(c)** The assistant observes a Doppler shift for the tuning fork. Apply Equation 14.12.

Solution
(a) Find the two possible frequencies.

Case 1: $f_2 - f_1$ is already positive, so just drop the absolute-value signs:

$$f_b = f_2 - f_1 \quad \rightarrow \quad 4 \text{ Hz} = f_2 - 4.40 \times 10^2 \text{ Hz}$$

$$f_2 = \boxed{444 \text{ Hz}}$$

Case 2: $f_2 - f_1$ is negative, so drop the absolute-value signs, but apply an overall negative sign:

$$f_b = -(f_2 - f_1) \quad \rightarrow \quad 4 \text{ Hz} = -(f_2 - 4.40 \times 10^2 \text{ Hz})$$

$$f_2 = \boxed{436 \text{ Hz}}$$

(b) Find the speed of the observer if running toward the piano results in a beat frequency of 2 Hz.

Apply the Doppler shift to the case where frequency of the piano string heard by the running observer is $f_O = 438$ Hz:

$$f_O = f_S \left(\frac{v + v_O}{v - v_S} \right)$$

$$438 \text{ Hz} = (436 \text{ Hz}) \left(\frac{343 \text{ m/s} + v_O}{343 \text{ m/s}} \right)$$

$$v_O = \left(\frac{438 \text{ Hz} - 436 \text{ Hz}}{436 \text{ Hz}} \right) (343 \text{ m/s}) = \boxed{1.57 \text{ m/s}}$$

(c) What beat frequency does the assistant observe?

Apply Equation 14.12. Now the source is the tuning
fork, so $f_S = 4.40 \times 10^2$ Hz.

$$f_O = f_S \left(\frac{v + v_O}{v - v_S} \right)$$

$$= (4.40 \times 10^2 \text{ Hz}) \left(\frac{343 \text{ m/s}}{343 \text{ m/s} - 1.57 \text{ m/s}} \right) = 442 \text{ Hz}$$

Compute the beat frequency.

$$f_b = f_2 - f_1 = 442 \text{ Hz} - 436 \text{ Hz} = \boxed{6 \text{ Hz}}$$

Remarks   The assistant on the piano bench and the tuner running with the fork observe different beat frequencies. Many physical observations depend on the state of motion of the observer, a subject discussed more fully in Chapter 26, on relativity.

Exercise 14.11
The assistant adjusts the tension in the same piano string, and a beat frequency of 2.00 Hz is heard when the note and the tuning fork are struck at the same time. (a) Find the two possible frequencies of the string. (b) Assume the actual string frequency is the higher frequency. If the piano tuner runs away from the piano at 4.00 m/s while holding the vibrating tuning fork, what beat frequency does he hear? (c) What beat frequency does the assistant on the bench hear? Use 343 m/s for the speed of sound.

Answers   (a) 438 Hz, 442 Hz   (b) 3 Hz   (c) 7 Hz

14.12   QUALITY OF SOUND

The sound-wave patterns produced by most musical instruments are complex. Figure 14.26 shows characteristic waveforms (pressure is plotted on the vertical axis, time on the horizontal axis) produced by a tuning fork, a flute, and a clarinet, each playing the same steady note. Although each instrument has its own characteristic pattern, the figure reveals that each of the waveforms is periodic. Note that the tuning fork produces only one harmonic (the fundamental frequency), but the two instruments emit mixtures of harmonics. Figure 14.27 graphs the harmonics of the waveforms of Figure 14.26. When the note is played on the flute (Fig. 14.26b), part of the sound consists of a vibration at the fundamental frequency, an even higher intensity is contributed by the second harmonic, the fourth harmonic produces about the same intensity as the fundamental, and so on. These sounds add together according to the principle of superposition to give the complex waveform shown. The clarinet emits a certain intensity at a frequency of the first harmonic, about half as much intensity at the frequency of the second harmonic, and so forth. The resultant superposition of these frequencies produces the pattern shown in Figure 14.26c. The tuning fork (Figs. 14.26a and 14.27a) emits sound only at the frequency of the first harmonic.

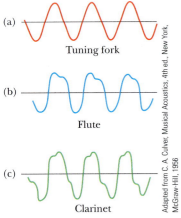

Figure 14.26   Waveforms produced by (a) a tuning fork, (b) a flute, and (c) a clarinet, all at approximately the same frequency. Pressure is plotted vertically, time horizontally.

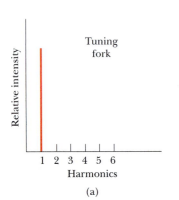

(a)

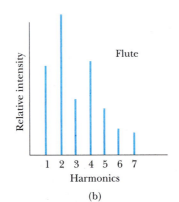

(b)

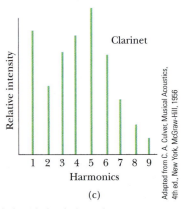

(c)

Figure 14.27   Harmonics of the waveforms in Figure 14.26. Note their variation in intensity.

In music, the characteristic sound of any instrument is referred to as the *quality*, or *timbre*, of the sound. The quality depends on the mixture of harmonics in the sound. We say that the note C on a flute differs in quality from the same C on a clarinet. Instruments such as the bugle, trumpet, violin, and tuba are rich in harmonics. A musician playing a wind instrument can emphasize one or another of these harmonics by changing the configuration of the lips, thereby playing different musical notes with the same valve openings.

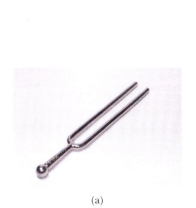

(a)   (b)   (c)

a – c, Royalty-Free/Corbis

Each musical instrument has its own characteristic sound and mixture of harmonics. (See Figures 14.26 and 14.27.) Instruments shown are (a) the tuning fork, (b) the flute, and (c) the clarinet.

Applying Physics 14.6   Why Does the Professor Sound Like Donald Duck?

A professor performs a demonstration in which he breathes helium and then speaks with a comical voice. One student explains, "The velocity of sound in helium is higher than in air, so the fundamental frequency of the standing waves in the mouth is increased." Another student says, "No, the fundamental frequency is determined by the vocal folds and cannot be changed. Only the quality of the voice has changed." Which student is correct?

Explanation   The second student is correct. The fundamental frequency of the complex tone from the voice is determined by the vibration of the vocal folds and is not changed by substituting a different gas in the mouth. The introduction of the helium into the mouth results in harmonics of higher frequencies being excited more than in the normal voice, but the fundamental frequency of the voice is the same—only the quality has changed. The unusual inclusion of the higher frequency harmonics results in a common description of this effect as a "high-pitched" voice, but that description is incorrect. (It is really a "quacky" timbre.)

14.13   THE EAR

The human ear is divided into three regions: the outer ear, the middle ear, and the inner ear (Fig. 14.28, page 488). The *outer ear* consists of the ear canal (which is open to the atmosphere), terminating at the eardrum (tympanum). Sound waves travel down the ear canal to the eardrum, which vibrates in and out in phase with the pushes and pulls caused by the alternating high and low pressures of the waves. Behind the eardrum are three small bones of the *middle ear*, called the hammer, the anvil, and the stirrup because of their shapes. These bones transmit the vibration to the *inner ear*, which contains the cochlea, a snail-shaped tube about 2 cm long. The cochlea makes contact with the stirrup at the oval window and is divided along its length by the basilar membrane, which consists of small hairs (cilia) and nerve fibers. This membrane varies in mass per unit length and in tension along its length, and different portions of it resonate at different frequencies. (Recall that the natural frequency of a string depends on its mass per unit length

Figure 14.28 The structure of the human ear. The three tiny bones (ossicles) that connect the eardrum to the window of the cochlea act as a double-lever system to decrease the amplitude of vibration and hence increase the pressure on the fluid in the cochlea.

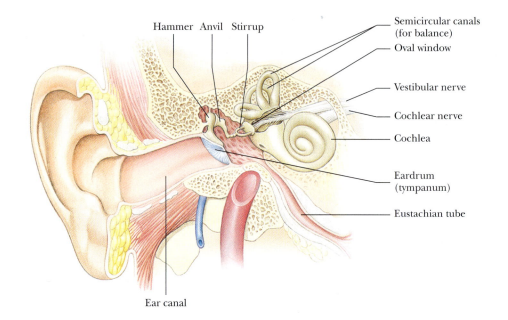

and on the tension in it.) Along the basilar membrane are numerous nerve endings, which sense the vibration of the membrane and in turn transmit impulses to the brain. The brain interprets the impulses as sounds of varying frequency, depending on the locations along the basilar membrane of the impulse-transmitting nerves and on the rates at which the impulses are transmitted.

Figure 14.29 shows the frequency response curves of an average human ear for sounds of equal loudness, ranging from 0 to 120 dB. To interpret this series of graphs, take the bottom curve as the threshold of hearing. Compare the intensity level on the vertical axis for the two frequencies 100 Hz and 1 000 Hz. The vertical axis shows that the 100-Hz sound must be about 38 dB greater than the 1 000-Hz sound to be at the threshold of hearing, which means that the threshold of hearing is very strongly dependent on frequency. The easiest frequencies to hear are around 3 300 Hz; those above 12 000 Hz or below about 50 Hz must be relatively intense to be heard.

Now consider the curve labeled 80. This curve uses a 1 000-Hz tone at an intensity level of 80 dB as its reference. The curve shows that a tone of frequency 100 Hz would have to be about 4 dB louder than the 80-dB, 1 000-Hz tone in order to sound as loud. Notice that the curves flatten out as the intensities levels of the sounds increase, so when sounds are loud, all frequencies can be heard equally well.

Figure 14.29 Curves of intensity level versus frequency for sounds that are perceived to be of equal loudness. Note that the ear is most sensitive at a frequency of about 3 300 Hz. The lowest curve corresponds to the threshold of hearing for only about 1% of the population.

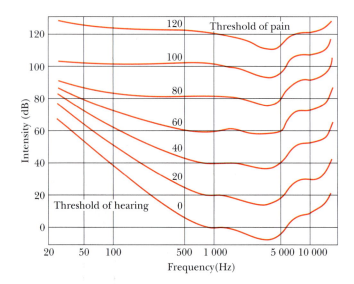

The small bones in the middle ear represent an intricate lever system that increases the force on the oval window. The pressure is greatly magnified because the surface area of the eardrum is about 20 times that of the oval window (in analogy with a hydraulic press). The middle ear, together with the eardrum and oval window, in effect acts as a matching network between the air in the outer ear and the liquid in the inner ear. The overall energy transfer between the outer ear and the inner ear is highly efficient, with pressure amplification factors of several thousand. In other words, pressure variations in the inner ear are much greater than those in the outer ear.

The ear has its own built-in protection against loud sounds. The muscles connecting the three middle-ear bones to the walls control the volume of the sound by changing the tension on the bones as sound builds up, thus hindering their ability to transmit vibrations. In addition, the eardrum becomes stiffer as the sound intensity increases. These two events make the ear less sensitive to loud incoming sounds. There is a time delay between the onset of a loud sound and the ear's protective reaction, however, so a very sudden loud sound can still damage the ear.

The complex structure of the human ear is believed to be related to the fact that mammals evolved from seagoing creatures. In comparison, insect ears are considerably simpler in design, because insects have always been land residents. A typical insect ear consists of an eardrum exposed directly to the air on one side and to an air-filled cavity on the other side. Nerve cells communicate directly with the cavity and the brain, without the need for the complex intermediary of an inner and middle ear. This simple design allows the ear to be placed virtually anywhere on the body. For example, a grasshopper has its ears on its legs. One advantage of the simple insect ear is that the distance and orientation of the ears can be varied so that it is easier to locate sources of sound, such as other insects.

One of the most amazing medical advances in recent decades is the cochlear implant, allowing the deaf to hear. Deafness can occur when the hairlike sensors (cilia) in the cochlea break off over a lifetime or sometimes because of prolonged exposure to loud sounds. Because the cilia don't grow back, the ear loses sensitivity to certain frequencies of sound. The cochlear implant stimulates the nerves in the ear electronically to restore hearing loss that is due to damaged or absent cilia.

SUMMARY

Physics⊗Now™ Take a practice test by logging into PhysicsNow at **www.cp7e.com** and clicking on the Pre-Test link for this chapter.

14.2 Characteristics of Sound Waves

Sound waves are longitudinal waves. **Audible waves** are sound waves with frequencies between 20 and 20 000 Hz. **Infrasonic waves** have frequencies below the audible range, and **ultrasonic waves** have frequencies above the audible range.

14.3 The Speed of Sound

The speed of sound in a medium of bulk modulus B and density ρ is

$$v = \sqrt{\frac{B}{\rho}} \qquad [14.1]$$

The speed of sound also depends on the temperature of the medium. The relationship between temperature and the speed of sound in air is

$$v = (331 \text{ m/s}) \sqrt{\frac{T}{273 \text{ K}}} \qquad [14.4]$$

where T is the absolute (Kelvin) temperature and 331 m/s is the speed of sound in air at 0°C.

14.4 Energy and Intensity of Sound Waves

The **average intensity** of sound incident on a surface is defined by

$$I \equiv \frac{\text{power}}{\text{area}} = \frac{\mathcal{P}}{A} \qquad [14.6]$$

where the power $\mathcal{P}$ is the energy per unit time flowing through the surface, which has area A. The **intensity level** of a sound wave is given by

$$\beta \equiv 10 \log\left(\frac{I}{I_0}\right) \qquad [14.7]$$

The constant $I_0 = 1.0 \times 10^{-12}$ W/m² is a reference intensity, usually taken to be at the threshold of hearing, and I is the intensity at level β, measured in **decibels** (dB).

14.5 Spherical and Plane Waves

The **intensity** of a *spherical wave* produced by a point source is proportional to the average power emitted and inversely proportional to the square of the distance from the source:

$$I = \frac{\mathcal{P}_{av}}{4\pi r^2} \qquad [14.8]$$

14.6   The Doppler Effect

The change in frequency heard by an observer whenever there is relative motion between a source of sound and the observer is called the **Doppler effect**. If the observer is moving with speed v_O and the source is moving with speed v_S, the observed frequency is

$$f_O = f_S \left(\frac{v + v_O}{v - v_S} \right) \qquad [14.12]$$

where v is the speed of sound. A positive speed is substituted for v_O when the observer moves toward the source, a negative speed when the observer moves away from the source. Similarly, a positive speed is substituted for v_S when the sources moves toward the observer, a negative speed when the source moves away.

14.7   Interference of Sound Waves

When waves interfere, the resultant wave is found by adding the individual waves together point by point. When crest meets crest and trough meets trough, the waves undergo **constructive interference**, with path length difference

$$r_2 - r_1 = n\lambda \quad (n = 0, 1, 2, \ldots) \qquad [14.13]$$

When crest meets trough, **destructive interference** occurs, with path length difference

$$r_2 - r_1 = (n + \tfrac{1}{2})\lambda \quad (n = 0, 1, 2, \ldots) \qquad [14.14]$$

14.8   Standing Waves

Standing waves are formed when two waves having the same frequency, amplitude, and wavelength travel in opposite directions through a medium. The natural frequencies of vibration of a stretched string of length L, fixed at both ends, are

$$f_n = nf_1 = \frac{n}{2L} \sqrt{\frac{F}{\mu}} \quad n = 1, 2, 3, \ldots \qquad [14.17]$$

where F is the tension in the string and μ is its mass per unit length.

14.9   Forced Vibrations and Resonance

A system capable of oscillating is said to be in **resonance** with some driving force whenever the frequency of the driving force matches one of the natural frequencies of the system. When the system is resonating, it oscillates with maximum amplitude.

14.10   Standing Waves in Air Columns

Standing waves can be produced in a tube of air. If the reflecting end of the tube is *open*, all harmonics are present and the natural frequencies of vibration are

$$f_n = n \frac{v}{2L} = nf_1 \quad n = 1, 2, 3, \ldots \qquad [14.18]$$

If the tube is *closed* at the reflecting end, only the *odd* harmonics are present and the natural frequencies of vibration are

$$f_n = n \frac{v}{4L} = nf_1 \quad n = 1, 3, 5, \ldots \qquad [14.19]$$

14.11   Beats

The phenomenon of **beats** is an interference effect that occurs when two waves with slightly different frequencies combine at a fixed point in space. For sound waves, the intensity of the resultant sound changes periodically with time. The *beat frequency* is

$$f_b = |f_2 - f_1| \qquad [14.20]$$

where f_2 and f_1 are the two source frequencies.

CONCEPTUAL QUESTIONS

1. (a) You are driving down the highway in your car when a police car sounding its siren overtakes you and passes you. If its frequency at rest is f_0, is the frequency you hear while the car is catching up to you higher or lower than f_0? (b) What about the frequency you hear after the car has passed you?

2. A crude model of the human throat is that of a pipe open at both ends with a vibrating source to introduce the sound into the pipe at one end. Assuming the vibrating source produces a range of frequencies, discuss the effect of changing the pipe's length.

3. An autofocus camera sends out a pulse of sound and measures the time taken for the pulse to reach an object, reflect off of it, and return to be detected. Can the temperature affect the camera's focus?

4. To keep animals away from their cars, some people mount short, thin pipes on the fenders. The pipes give out a high-pitched wail when the cars are moving. How do they create the sound?

5. Secret agents in the movies always want to get to a secure phone with a voice scrambler. How do these devices work?

6. When a bell is rung, standing waves are set up around its circumference. What boundary conditions must be satisfied by the resonant wavelengths? How does a crack in the bell, such as in the Liberty Bell, affect the satisfying of the boundary conditions and the sound emanating from the bell?

7. How does air temperature affect the tuning of a wind instrument?

8. Explain how the distance to a lightning bolt can be determined by counting the seconds between the flash and the sound of thunder.

9. You are driving toward a cliff and you honk your horn. Is there a Doppler shift of the sound when you hear the echo? Is it like a moving source or moving observer? What if the reflection occurs not from a cliff, but from the forward edge of a huge alien spacecraft moving toward you as you drive?

10. Of the following sounds, state which is most likely to have an intensity level of 60 dB: a rock concert, the turning of a page in this text, a normal conversation, a cheering crowd at a football game, and background noise at a church?

11. Guitarists sometimes play a "harmonic" by lightly touching a string at its exact center and plucking the string. The result is a clear note one octave higher than the fundamental frequency of the string, even though the string is not pressed to the fingerboard. Why does this happen?

12. Will two separate 50-dB sounds together constitute a 100-dB sound? Explain.

13. An archer shoots an arrow from a bow. Does the string of the bow exhibit standing waves after the arrow leaves? If so, and if the bow is perfectly symmetric so that the arrow leaves from the center of the string, what harmonics are excited?

14. The radar systems used by police to detect speeders are sensitive to the Doppler shift of a pulse of radio waves. Discuss how this sensitivity can be used to measure the speed of a car.

15. As oppositely moving pulses of the same shape (one upward, one downward) on a string pass through each other, there is one instant at which the string shows no displacement from the equilibrium position at any point. Has the energy carried by the pulses disappeared at this instant of time? If not, where is it?

16. A soft drink bottle resonates as air is blown across its top. What happens to the resonant frequency as the level of fluid in the bottle decreases?

17. A blowing whistle is attached to the roof of a car that moves around a circular race track. Assuming you're standing near the outside of the track, explain the nature of the sound you hear as the whistle comes by each time.

18. Despite a reasonably steady hand, a person often spills his coffee when carrying it to his seat. Discuss resonance as a possible cause of this difficulty, and devise a means for solving the problem.

19. An airplane mechanic notices that the sound from a twin-engine aircraft varies rapidly in loudness when both engines are running. What could be causing this variation from loud to soft?

20. Why does a vibrating guitar string sound louder when placed on the instrument than it would if allowed to vibrate in the air while off the instrument?

PROBLEMS

1, 2, 3 = straightforward, intermediate, challenging   □ = full solution available in *Student Solutions Manual/Study Guide*
Physics⊗Now™ = coached problem with hints available at **www.cp7e.com**   🧬 = biomedical application

Section 14.2 Characteristics of Sound Waves
Section 14.3 The Speed of Sound
Unless otherwise stated, use 345 m/s as the speed of sound in air.

1. Suppose that you hear a clap of thunder 16.2 s after seeing the associated lightning stroke. The speed of sound waves in air is 343 m/s and the speed of light in air is 3.00×10^8 m/s. How far are you from the lightning stroke?

2. A dolphin located in sea water at a temperature of 25°C emits a sound directed toward the bottom of the ocean 150 m below. How much time passes before it hears an echo?

3. A sound wave has a frequency of 700 Hz in air and a wavelength of 0.50 m. What is the temperature of the air?

4. The range of human hearing extends from approximately 20 Hz to 20 000 Hz. Find the wavelengths of these extremes at a temperature of 27°C.

5. A group of hikers hears an echo 3.00 s after shouting. If the temperature is 22.0°C, how far away is the mountain that reflected the sound wave?

6. A stone is dropped from rest into a well. The sound of the splash is heard exactly 2.00 s later. Find the depth of the well if the air temperature is 10.0°C.

7. You are watching a pier being constructed on the far shore of a saltwater inlet when some blasting occurs. You hear the sound in the water 4.50 s before it reaches you through the air. How wide is the inlet? [*Hint:* See Table 14.1. Assume the air temperature is 20°C.]

8. The speed of sound in a column of air is measured to be 356 m/s. What is the temperature of the air?

Section 14.4 Energy and Intensity of Sound Waves

9. The toadfish makes use of resonance in a closed tube to produce very loud sounds. The tube is its swim bladder, used as an amplifier. The sound level of this creature has been measured as high as 100 dB. (a) Calculate the intensity of the sound wave emitted. (b) What is the intensity level if three of these fish try to imitate three frogs by saying "Budweiser" at the same time.

10. The area of a typical eardrum is about 5.0×10^{-5} m². Calculate the sound power (the energy per second) incident on an eardrum at (a) the threshold of hearing and (b) the threshold of pain.

11. There is evidence that elephants communicate via infrasound, generating rumbling vocalizations as low as 14 hz that can travel up to 10 km. The intensity level of these sounds can reach 103 dB, measured a distance of 5.0 m from the source. Determine the intensity level of the infrasound 10 km from the source, assuming the sound energy radiates uniformly in all directions.

12. Two sounds have measured intensities of $I_1 = 100$ W/m² and $I_2 = 200$ W/m². By how many decibels is the level of sound 1 lower than that of sound 2?

13. A noisy machine in a factory produces sound with a level of 80 dB. How many identical machines could you add to the factory without exceeding the 90-dB limit?

14. A family ice show is held at an enclosed arena. The skaters perform to music playing at a level of 80.0 dB. This intensity level is too loud for your baby, who yells at 75.0 dB. (a) What total sound intensity engulfs you? (b) What is the combined sound level?

15. Calculate the sound level in decibels of a sound wave that has an intensity of 4.00 $\mu W/m^2$.

Section 14.5 Spherical and Plane Waves

16. An outside loudspeaker (considered a small source) emits sound waves with a power output of 100 W. (a) Find the intensity 10.0 m from the source. (b) Find the intensity level in decibels at that distance. (c) At what distance would you experience the sound at the threshold of pain, 120 dB?

17. **Physics⊗Now™** A train sounds its horn as it approaches an intersection. The horn can just be heard at a level of 50 dB by an observer 10 km away. (a) What is the average power generated by the horn? (b) What intensity level of the horn's sound is observed by someone waiting at an intersection 50 m from the train? Treat the horn as a point source and neglect any absorption of sound by the air.

18. A skyrocket explodes 100 m above the ground (Fig. P14.18). Three observers are spaced 100 m apart, with the first (A) directly under the explosion. (a) What is the ratio of the sound intensity heard by observer A to that heard by observer B? (b) What is the ratio of the intensity heard by observer A to that heard by observer C?

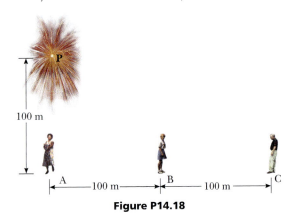

100 m

|← A —100 m—→ B —100 m—→ C|

Figure P14.18

19. Show that the difference in decibel levels β_1 and β_2 of a sound source is related to the ratio of its distances r_1 and r_2 from the receivers by the formula

$$\beta_2 - \beta_1 = 20 \log \left(\frac{r_1}{r_2} \right)$$

Section 14.6 The Doppler Effect

20. An airplane traveling at half the speed of sound ($v = 172$ m/s) emits a sound of frequency 5.00 kHz. At what frequency does a stationary listener hear the sound (a) as the plane approaches? (b) after it passes?

21. A commuter train passes a passenger platform at a constant speed of 40.0 m/s. The train horn is sounded at its characteristic frequency of 320 Hz. (a) What overall change in frequency is detected by a person on the platform as the train moves from approaching to receding? (b) What wavelength is detected by a person on the platform as the train approaches?

22. At rest, a car's horn sounds the note A (440 Hz). The horn is sounded while the car is moving down the street. A bicyclist moving in the same direction with one-third the car's speed hears a frequency of 415 Hz. What is the speed of the car? Is the cyclist ahead of or behind the car?

23. Two trains on separate tracks move towards one another. Train 1 has a speed of 130 km/h, train 2 a speed of 90.0 km/h. Train 2 blows its horn, emitting a frequency of 500 Hz. What is the frequency heard by the engineer on train 1?

24. A bat flying at 5.0 m/s emits a chirp at 40 kHz. If this sound pulse is reflected by a wall, what is the frequency of the echo received by the bat?

25. An alert physics student stands beside the tracks as a train rolls slowly past. He notes that the frequency of the train whistle is 442 Hz when the train is approaching him and 441 Hz when the train is receding from him. Using these frequencies, he calculates the speed of the train. What value does he find?

26. Expectant parents are thrilled to hear their unborn baby's heartbeat, revealed by an ultrasonic motion detector. Suppose the fetus's ventricular wall moves in simple harmonic motion with amplitude 1.80 mm and frequency 115 per minute. (a) Find the maximum linear speed of the heart wall. Suppose the motion detector in contact with the mother's abdomen produces sound at precisely 2 MHz, which travels through tissue at 1.50 km/s. (b) Find the maximum frequency at which sound arrives at the wall of the baby's heart. (c) Find the maximum frequency at which reflected sound is received by the motion detector. (By electronically "listening" for echoes at a frequency different from the broadcast frequency, the motion detector can produce beeps of audible sound in synchrony with the fetal heartbeat.)

27. A tuning fork vibrating at 512 Hz falls from rest and accelerates at 9.80 m/s². How far below the point of release is the tuning fork when waves of frequency 485 Hz reach the release point? Take the speed of sound in air to be 340 m/s.

28. A supersonic jet traveling at Mach 3 at an altitude of 20 000 m is directly overhead at time $t = 0$, as in Figure P14.28. (a) How long will it be before the ground observer encounters the shock wave? (b) Where will the plane be when it is finally heard? (Assume an average value of 330 m/s for the speed of sound in air.)

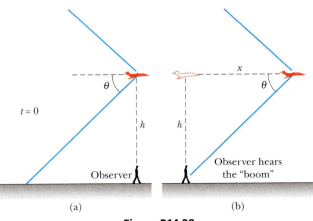

Figure P14.28

29. The now-discontinued *Concorde* flew at Mach 1.5, which meant the speed of the plane was 1.5 times the speed of sound in air. What was the angle between the direction of propagation of the shock wave and the direction of the plane's velocity?

Section 14.7 Interference of Sound Waves

30. The acoustical system shown in Figure 14.14 is driven by a speaker emitting a 400-Hz note. If *destructive* interference occurs at a particular instant, how much must the path length in the U-shaped tube be increased in order to hear (a) constructive interference and (b) destructive interference once again?

31. The ship in Figure P14.31 travels along a straight line parallel to the shore and 600 m from it. The ship's radio receives simultaneous signals of the same frequency from antennas *A* and *B*. The signals interfere constructively at point *C*, which is equidistant from *A* and *B*. The signal goes through the first minimum at point *D*. Determine the wavelength of the radio waves.

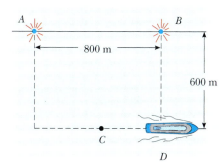

Figure P14.31

32. Two loudspeakers are placed above and below one another, as in Figure 14.15, and are driven by the same source at a frequency of 500 Hz. (a) What minimum distance should the top speaker be moved back in order to create destructive interference between the speakers? (b) If the top speaker is moved back twice the distance calculated in part (a), will there be constructive or destructive interference?

33. A pair of speakers separated by 0.700 m are driven by the same oscillator at a frequency of 690 Hz. An observer originally positioned at one of the speakers begins to walk along a line perpendicular to the line joining the speakers. (a) How far must the observer walk before reaching a relative maximum in intensity? (b) How far will the observer be from the speaker when the first relative minimum is detected in the intensity?

Section 14.8 Standing Waves

34. A steel wire in a piano has a length of 0.700 0 m and a mass of 4.300×10^{-3} kg. To what tension must this wire be stretched in order that the fundamental vibration correspond to middle C ($f_C = 261.6$ Hz on the chromatic musical scale)?

35. A stretched string fixed at each end has a mass of 40.0 g and a length of 8.00 m. The tension in the string is 49.0 N. (a) Determine the positions of the nodes and antinodes for the third harmonic. (b) What is the vibration frequency for this harmonic?

36. Resonance of sound waves can be produced within an aluminum rod by holding the rod at its midpoint and stroking it with an alcohol-saturated paper towel. In this resonance mode, the middle of the rod is a node while the ends are antinodes; no other nodes or antinodes are present. What is the frequency of the resonance if the rod is 1.00 m long?

37. **Physics⊗Now™** Two speakers are driven by a common oscillator at 800 Hz and face each other at a distance of 1.25 m. Locate the points along a line joining the speakers where relative minima of the amplitude of the pressure would be expected. (Use $v = 343$ m/s.)

38. Two pieces of steel wire with identical cross sections have lengths of L and $2L$. The wires are each fixed at both ends and stretched so that the tension in the longer wire is four times greater than in the shorter wire. If the fundamental frequency in the shorter wire is 60 Hz, what is the frequency of the second harmonic in the longer wire?

39. A 12-kg object hangs in equilibrium from a string of total length $L = 5.0$ m and linear mass density $\mu = 0.001\ 0$ kg/m. The string is wrapped around two light, frictionless pulleys that are separated by the distance $d = 2.0$ m (Fig. P14.39a). (a) Determine the tension in the string. (b) At what frequency must the string between the pulleys vibrate in order to form the standing-wave pattern shown in Figure P14.39b?

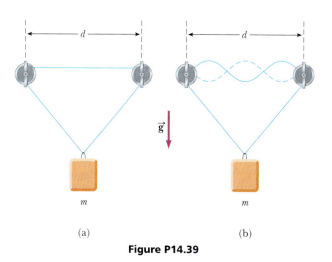

(a)   (b)

Figure P14.39

40. In the arrangement shown in Figure P14.40, an object of mass $m = 5.0$ kg hangs from a cord around a light pulley. The length of the cord between point P and the pulley is $L = 2.0$ m. (a) When the vibrator is set to a frequency of 150 Hz, a standing wave with six loops is formed. What must be the linear mass density of the cord? (b) How many loops (if any) will result if m is changed to 45 kg? (c) How many loops (if any) will result if m is changed to 10 kg?

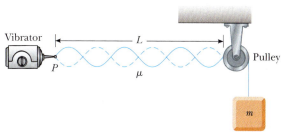

Figure P14.40

41. A 60.00-cm guitar string under a tension of 50.000 N has a mass per unit length of 0.100 00 g/cm. What is the highest resonant frequency that can be heard by a person capable of hearing frequencies up to 20 000 Hz?

Section 14.9 Forced Vibrations and Resonance

42. Standing-wave vibrations are set up in a crystal goblet with four nodes and four antinodes equally spaced around the 20.0-cm circumference of its rim. If transverse waves move around the glass at 900 m/s, an opera singer would have to produce a high harmonic with what frequency in order to shatter the glass with a resonant vibration?

Section 14.10 Standing Waves in Air Columns

43. The windpipe of a typical whooping crane is about 5.0 feet long. What is the lowest resonant frequency of this pipe, assuming that it is closed at one end? Assume a temperature of 37°C.

44. The overall length of a piccolo is 32.0 cm. The resonating air column vibrates as in a pipe that is open at both ends. (a) Find the frequency of the lowest note a piccolo can play, assuming the speed of sound in air is 340 m/s. (b) Opening holes in the side effectively shortens the length of the resonant column. If the highest note a piccolo can sound is 4 000 Hz, find the distance between adjacent antinodes for this mode of vibration.

45. The human ear canal is about 2.8 cm long. If it is regarded as a tube that is open at one end and closed at the eardrum, what is the fundamental frequency around which we would expect hearing to be most sensitive? Take the speed of sound to be 340 m/s.

46. A shower stall measures 86.0 cm × 86.0 cm × 210 cm. When you sing in the shower, which frequencies will sound the richest (because of resonance)? Assume the stall acts as a pipe closed at both ends, with nodes at opposite sides. Assume also that the voices of various singers range from 130 Hz to 2 000 Hz. Let the speed of sound in the hot shower stall be 355 m/s.

47. Physics⊗Now™ A pipe open at both ends has a fundamental frequency of 300 Hz when the temperature is 0°C. (a) What is the length of the pipe? (b) What is the fundamental frequency at a temperature of 30°C?

48. A 2.00-m-long air column is open at both ends. The frequency of a certain harmonic is 410 Hz, and the frequency of the next-higher harmonic is 492 Hz. Determine the speed of sound in the air column.

Section 14.11 Beats

49. Two identical mandolin strings under 200 N of tension are sounding tones with frequencies of 523 Hz. The peg of one string slips slightly, and the tension in it drops to 196 N. How many beats per second are heard?

50. The G string on a violin has a fundamental frequency of 196 Hz. It is 30.0 cm long and has a mass of 0.500 g. While this string is sounding, a nearby violinist effectively shortens the G string on her identical violin (by sliding her finger down the string) until a beat frequency of 2.00 Hz is heard between the two strings. When this occurs, what is the effective length of her string?

51. Two train whistles have identical frequencies of 180 Hz. When one train is at rest in the station, sounding its whistle, a beat frequency of 2 Hz is heard from a moving train.

What two possible speeds and directions can the moving train have?

52. Two pipes of equal length are each open at one end. Each has a fundamental frequency of 480 Hz at 300 K. In one pipe, the air temperature is increased to 305 K. If the two pipes are sounded together, what beat frequency results?

53. A student holds a tuning fork oscillating at 256 Hz. He walks toward a wall at a constant speed of 1.33 m/s. (a) What beat frequency does he observe between the tuning fork and its echo? (b) How fast must he walk away from the wall to observe a beat frequency of 5.00 Hz?

Section 14.13 The Ear

54. If a human ear canal can be thought of as resembling an organ pipe, closed at one end, that resonates at a fundamental frequency of 3 000 Hz, what is the length of the canal? Use a normal body temperature of 37°C for your determination of the speed of sound in the canal.

55. Some studies suggest that the upper frequency limit of hearing is determined by the diameter of the eardrum. The wavelength of the sound wave and the diameter of the eardrum are approximately equal at this upper limit. If the relationship holds exactly, what is the diameter of the eardrum of a person capable of hearing 20 000 Hz? (Assume a body temperature of 37°C.)

ADDITIONAL PROBLEMS

56. A commuter train blows its horn as it passes a passenger platform at a constant speed of 40.0 m/s. The horn sounds at a frequency of 320 Hz when the train is at rest. What is the frequency observed by a person on the platform (a) as the train approaches and (b) as the train recedes from him? (c) What wavelength does the observer find in each case?

57. A quartz watch contains a crystal oscillator in the form of a block of quartz that vibrates by contracting and expanding. Two opposite faces of the block, 7.05 mm apart, are antinodes, moving alternately towards and away from each other. The plane halfway between these two faces is a node of the vibration. The speed of sound in quartz is 3.70 km/s. Find the frequency of the vibration. An oscillating electric voltage accompanies the mechanical oscillation, so the quartz is described as *piezoelectric*. An electric circuit feeds in energy to maintain the oscillation and also counts the voltage pulses to keep time.

58. A flowerpot is knocked off a balcony 20.0 m above the sidewalk and falls toward an unsuspecting 1.75-m-tall man who is standing below. How close to the sidewalk can the flowerpot fall before it is too late for a warning shouted from the balcony to reach the man in time? Assume that the man below requires 0.300 s to respond to the warning.

59. Physics⊗Now™ On a workday, the average decibel level of a busy street is 70 dB, with 100 cars passing a given point every minute. If the number of cars is reduced to 25 every minute on a weekend, what is the decibel level of the street?

60. A variable-length air column is placed just below a vibrating wire that is fixed at both ends. The length of the column, open at one end, is gradually increased from zero until the first position of resonance is observed at

$L = 34.0$ cm. The wire is 120 cm long and is vibrating in its third harmonic. If the speed of sound in air is 340 m/s, what is the speed of transverse waves in the wire?

61. A block with a speaker bolted to it is connected to a spring having spring constant $k = 20.0$ N/m, as shown in Figure P14.61. The total mass of the block and speaker is 5.00 kg, and the amplitude of the unit's motion is 0.500 m. If the speaker emits sound waves of frequency 440 Hz, determine the lowest and highest frequencies heard by the person to the right of the speaker.

Figure P14.61

62. A flute is designed so that it plays a frequency of 261.6 Hz, middle C, when all the holes are covered and the temperature is 20.0°C. (a) Consider the flute to be a pipe open at both ends, and find its length, assuming that the middle-C frequency is the fundamental frequency. (b) A second player, nearby in a colder room, also attempts to play middle C on an identical flute. A beat frequency of 3.00 beats/s is heard. What is the temperature of the room?

63. When at rest, two trains have sirens that emit a frequency of 300 Hz. The trains travel toward one another and toward an observer stationed between them. One of the trains moves at 30.0 m/s, and the observer hears a beat frequency of 3.0 beats per second. What is the speed of the second train, which travels faster than 30.0 m/s?

64. Many artists sing very high notes in *ad lib* ornaments and cadenzas. The highest note written for a singer in a published score was F-sharp above high C, 1.480 kHz, sung by Zerbinetta in the original version of Richard Strauss's opera *Ariadne auf Naxos*. (a) Find the wavelength of this sound in air. (b) In response to complaints, Strauss later transposed the note down to F above high C, 1.397 kHz. By what increment did the wavelength change?

65. A speaker at the front of a room and an identical speaker at the rear of the room are being driven at 456 Hz by the same sound source. A student walks at a uniform rate of 1.50 m/s away from one speaker and towards the other. How many beats does the student hear per second?

66. Two identical speakers separated by 10.0 m are driven by the same oscillator with a frequency of $f = 21.5$ Hz (Fig. P14.66). Explain why a receiver at A records a minimum in sound intensity from the two speakers. (b) If the receiver is moved in the plane of the speakers, what path should it take so that the intensity remains at a minimum? That is, determine the relationship between x and y (the coordinates of the receiver) such that the receiver will record a minimum in sound intensity. Take the speed of sound to be 344 m/s.

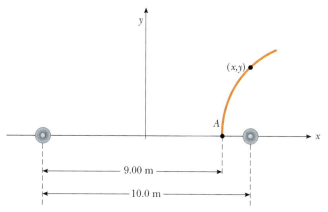

Figure P14.66

67. By proper excitation, it is possible to produce both longitudinal and transverse waves in a long metal rod. In a particular case, the rod is 150 cm long and 0.200 cm in radius and has a mass of 50.9 g. Young's modulus for the material is 6.80×10^{10} Pa. Determine the required tension in the rod so that the ratio of the speed of longitudinal waves to the speed of transverse waves is 8.

68. A student stands several meters in front of a smooth reflecting wall, holding a board on which a wire is fixed at each end. The wire, vibrating in its third harmonic, is 75.0 cm long, has a mass of 2.25 g, and is under a tension of 400 N. A second student, moving towards the wall, hears 8.30 beats per second. What is the speed of the student approaching the wall? Use 340 m/s as the speed of sound in air.

69. Two ships are moving along a line due east. The trailing vessel has a speed of 64.0 km/h relative to a land-based observation point, and the leading ship has a speed of 45.0 km/h relative to the same station. The trailing ship transmits a sonar signal at a frequency of 1 200 Hz. What frequency is monitored by the leading ship? (Use 1 520 m/s as the speed of sound in ocean water.)

70. The Doppler equation presented in the text is valid when the motion between the observer and the source occurs on a straight line, so that the source and observer are moving either directly toward or directly away from each other. If this restriction is relaxed, one must use the more general Doppler equation

$$f_O = \left[\frac{v + v_O \cos(\theta_O)}{v - v_S \cos(\theta_S)} \right] f_S$$

where θ_O and θ_S are defined in Figure P14.70a. (a) If both observer and source are moving away from each other along a straight line, show that the preceding equation yields the same result as Equation 14.12 in the text.

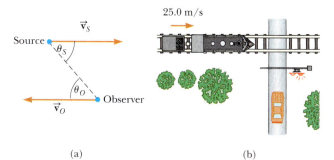

(a)   (b)

Figure P14.70

(b) Use the preceding equation to solve the following problem: A train moves at a constant speed of 25.0 m/s toward the intersection shown in Figure P14.70b. A car is stopped near the intersection, 30.0 m from the tracks. If the train's horn emits a frequency of 500 Hz, what is the frequency heard by the passengers in the car when the train is 40.0 m to the left of the intersection? Take the speed of sound to be 343 m/s.

71. A rescue plane flies horizontally at a constant speed, searching for a disabled boat. When the plane is directly above the boat, the boat's crew blows a loud horn. By the time the plane's sound detector perceives the horn's sound, the plane has traveled a distance equal to one-half its altitude above the ocean. If the sound takes 2.00 s to reach the plane, determine (a) the plane's altitude and (b) its speed.

72. In order to determine her speed, a skydiver carries a tone generator. A friend on the ground at the landing site has equipment for receiving and analyzing sound waves. While the skydiver is falling at terminal speed, her tone generator emits a steady tone of 1.80 kHz. (Assume that the air is calm, that the speed of sound is 343 m/s, independent of altitude.) (a) If her friend on the ground (directly beneath the skydiver) receives waves of frequency 2.15 kHz, what is the skydiver's speed of descent? (b) If the skydiver were also carrying sound-receiving equipment sensitive enough to detect waves reflected from the ground, what frequency of waves would she receive?

ACTIVITIES

A.1. Use an empty 1-liter soft-drink container, blow over the open end, and listen to the sound that is produced. Add some water to the container to change the height of the air column, and repeat the procedure. How does the frequency that you hear change with the height of the air column?

   If you want to investigate this phenomenon in more detail, construct a musical instrument made up of several soft-drink bottles with different amounts of water in each. You can play your instrument as a wind instrument by blowing over the mouths of the bottles.

A.2. Beats can easily be heard on a guitar. When a finger is placed at the fifth fret of the second string, the note produced when the string is plucked should be identical to the note from the first string when it is played without fingering. With your finger in position on the second string, pluck the two strings simultaneously. If one of the strings is slightly out of tune, a very pronounced beat frequency will be heard. What happens to the beat frequency as the string tension is changed in small increments from too low for the intended tuning to too high?

A.3. Attach a rope to a door and shake the other end to see how many of the standing-wave patterns in Figure 14.18 you can produce. When a pattern is formed, note that the amplitude of the rope's vibration is much larger than the movement of your hand.

A.4. Snip off the corners of one end of a paper straw so that the end tapers to a point, as shown in Figure A14.4. Chew on this end to flatten it, and you have created a double-reed instrument. Put your lips around the tapered end of the straw, press them together slightly, and blow through the straw. When you hear a steady tone, slowly snip off a piece of the straw at the other end. Be careful to keep about the same amount of pressure with your lips. How does the frequency of the sound change as the straw becomes shorter? Why does this change occur? You may be able to produce more than one tone for any given length of the straw. Why?

Figure A14.4

A.5. Inflate a balloon just enough to form a small sphere. Measure its diameter. Use a marker to color in a 1-cm square on its surface. Now continue inflating the balloon until it reaches twice the original diameter. Measure the size of the square now. Note how the color of the marked area has changed. Use the information in Section 14.5 to explain these results.

"I love hearing that lonesome wail of the train whistle as the magnitude of the frequency of the wave changes due to the Doppler effect."

This nighttime view of multiple bolts of lightning was photographed in Tucson, Arizona. During a thunderstorm, a high concentration of electrical charge in a thundercloud creates a higher-than-normal electric field between the thundercloud and the negatively charged Earth's surface. This strong electric field creates an electric discharge between the charged cloud and the ground—an enormous spark. Other discharges that are observed in the sky include cloud-to-cloud discharges and the more frequent intracloud discharges.

CHAPTER

15

Electric Forces and Electric Fields

Electricity is the lifeblood of technological civilization and modern society. Without it, we revert to the mid-nineteenth century: no telephones, no television, none of the household appliances that we take for granted. Modern medicine would be a fantasy, and due to the lack of sophisticated experimental equipment and fast computers—and especially the slow dissemination of information—science and technology would grow at a glacial pace.

Instead, with the discovery and harnessing of electric forces and fields, we can view arrangements of atoms, probe the inner workings of the cell, and send spacecraft beyond the limits of the solar system. All this has become possible in just the last few generations of human life, a blink of the eye compared to the million years our kind spent foraging the savannahs of Africa.

Around 700 B.C. the ancient Greeks conducted the earliest known study of electricity. It all began when someone noticed that a fossil material called amber would attract small objects after being rubbed with wool. Since then we have learned that this phenomenon is not restricted to amber and wool, but occurs (to some degree) when almost any two nonconducting substances are rubbed together.

In the current chapter we use the effect of charging by friction to begin an investigation of electric forces. We then discuss Coulomb's law, which is the fundamental law of force between any two stationary charged particles. The concept of an electric field associated with charges is introduced and its effects on other charged particles described. We end with discussions of the Van de Graaff generator and Gauss's law.

15.1 PROPERTIES OF ELECTRIC CHARGES

After running a plastic comb through your hair, you will find that the comb attracts bits of paper. The attractive force is often strong enough to suspend the

Figure 15.1 (a) A negatively charged rubber rod, suspended by a thread, is attracted to a positively charged glass rod. (b) A negatively charged rubber rod is repelled by another negatively charged rubber rod.

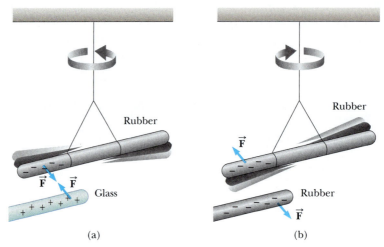

(a)   (b)

BENJAMIN FRANKLIN (1706–1790)

Franklin was a printer, author, physical scientist, inventor, diplomat, and a founding father of the United States. His work on electricity in the late 1740s changed a jumbled, unrelated set of observations into a coherent science.

paper from the comb, defying the gravitational pull of the entire Earth. The same effect occurs with other rubbed materials, such as glass and hard rubber.

Another simple experiment is to rub an inflated balloon against wool (or across your hair). On a dry day, the rubbed balloon will then stick to the wall of a room, often for hours. These materials have become **electrically charged**. You can give your body an electric charge by vigorously rubbing your shoes on a wool rug or by sliding across a car seat. You can then surprise and annoy a friend or co-worker with a light touch on the arm, delivering a slight shock to both yourself and your victim. (If the co-worker is your boss, don't expect a promotion!) These experiments work best on a dry day because excessive moisture can facilitate a leaking away of the charge.

Experiments also demonstrate that there are two kinds of electric charge, which Benjamin Franklin (1706–1790) named **positive** and **negative**. Figure 15.1 illustrates the interaction of the two charges. A hard rubber (or plastic) rod that has been rubbed with fur is suspended by a piece of string. When a glass rod that has been rubbed with silk is brought near the rubber rod, the rubber rod is attracted toward the glass rod (Fig. 15.1a). If two charged rubber rods (or two charged glass rods) are brought near each other, as in Figure 15.1b, the force between them is repulsive. These observations may be explained by assuming that the rubber and glass rods have acquired different kinds of excess charge. We use the convention suggested by Franklin, where the excess electric charge on the glass rod is called positive and that on the rubber rod is called negative. On the basis of observations such as these, we conclude that **like charges repel one another and unlike charges attract one another**. Objects usually contain equal amounts of positive and negative charge—electrical forces between objects arise when those objects have net negative or positive charges.

Nature's basic carriers of positive charge are protons, which, along with neutrons, are located in the nuclei of atoms. The nucleus, about 10^{-15} m in radius, is surrounded by a cloud of negatively charged electrons about ten thousand times larger in extent. An electron has the same magnitude charge as a proton, but the opposite sign. In a gram of matter there are approximately 10^{23} positively charged protons and just as many negatively charged electrons, so the net charge is zero. Because the nucleus of an atom is held firmly in place inside a solid, protons never move from one material to another. Electrons are far lighter than protons and hence more easily accelerated by forces. Furthermore, they occupy the outer regions of the atom. Consequently, objects become charged by gaining or losing electrons.

Charge transfers readily from one type of material to another. Rubbing the two materials together serves to increase the area of contact, facilitating the transfer process.

Like charges repel
Unlike charges attract ▶

An important characteristic of charge is that **electric charge is always conserved**. Charge isn't *created* when two neutral objects are rubbed together; rather, the objects become charged because **negative charge is transferred from one object to the other**. One object gains a negative charge while the other loses an equal amount of negative charge and hence is left with a net positive charge. When a glass rod is rubbed with silk, as in Figure 15.2, electrons are transferred from the rod to the silk. As a result, the glass rod carries a net positive charge, the silk a net negative charge. Likewise, when rubber is rubbed with fur, electrons are transferred from the fur to the rubber.

In 1909 Robert Millikan (1886–1953) discovered that if an object is charged, its charge is always a multiple of a fundamental unit of charge, designated by the symbol *e*. In modern terms, the charge is said to be **quantized**, meaning that charge occurs in discrete chunks that can't be further subdivided. An object may have a charge of $\pm e$, $\pm 2e$, $\pm 3e$, and so on, but never[1] a fractional charge of $\pm 0.5e$ or $\pm 0.22e$. Other experiments in Millikan's time showed that the electron has a charge of $-e$ and the proton has an equal and opposite charge of $+e$. Some particles, such as a neutron, have no net charge. A neutral atom (an atom with no net charge) contains as many protons as electrons. The value of *e* is now known to be $1.602\ 19 \times 10^{-19}$ C. (The SI unit of electric charge is the **coulomb** [C].)

◄ Charge is conserved; charge is quantized

Figure 15.2 When a glass rod is rubbed with silk, electrons are transferred from the glass to the silk. Because of conservation of charge, each electron adds negative charge to the silk, and an equal positive charge is left behind on the rod. Also, because the charges are transferred in discrete bundles, the charges on the two objects are $\pm e$, $\pm 2e$, $\pm 3e$, and so on.

15.2 INSULATORS AND CONDUCTORS

Substances can be classified in terms of their ability to conduct electric charge.

> In **conductors**, electric charges move freely in response to an electric force. All other materials are called **insulators**.

Glass and rubber are insulators. When such materials are charged by rubbing, only the rubbed area becomes charged, and there is no tendency for the charge to move into other regions of the material. In contrast, materials such as copper, aluminum, and silver are good conductors. When such materials are charged in some small region, the charge readily distributes itself over the entire surface of the material. If you hold a copper rod in your hand and rub the rod with wool or fur, it will not attract a piece of paper. This might suggest that a metal can't be charged. However, if you hold the copper rod with an insulator and then rub it with wool or fur, the rod remains charged and attracts the paper. In the first case, the electric charges produced by rubbing readily move from the copper through your body and finally to ground. In the second case, the insulating handle prevents the flow of charge to ground.

Semiconductors are a third class of materials, and their electrical properties are somewhere between those of insulators and those of conductors. Silicon and germanium are well-known semiconductors that are widely used in the fabrication of a variety of electronic devices.

Charging by Conduction

Consider a negatively charged rubber rod brought into contact with an insulated neutral conducting sphere. The excess electrons on the rod repel electrons on the sphere, creating local positive charges on the neutral sphere. On contact, some electrons on the rod are now able to move onto the sphere, as in Figure 15.3, neutralizing the positive charges. When the rod is removed, the sphere is left with a net negative charge. This process is referred to as charging by **conduction.** The object being charged in such a process (the sphere) is always left with a charge having the same sign as the object doing the charging (the rubber rod).

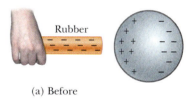

(a) Before

(b) Contact

(c) After breaking contact

Figure 15.3 Charging a metallic object by conduction. (a) Just before contact, the negative rod repels the sphere's electrons, inducing a localized positive charge. (b) After contact, electrons from the rod flow onto the sphere, neutralizing the local positive charges. (c) When the rod is removed, the sphere is left with a negative charge.

---

[1]There is strong evidence for the existence of fundamental particles called **quarks** that have charges of $\pm e/3$ or $\pm 2e/3$. The charge is *still* quantized, but in units of $\pm e/3$ rather than $\pm e$. A more complete discussion of quarks and their properties is presented in Chapter 30.

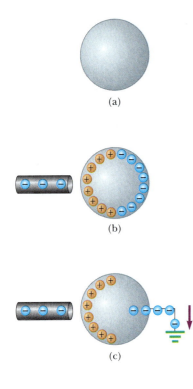

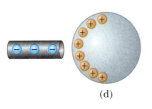

Figure 15.4   Charging a metallic object by *induction*. (a) A neutral metallic sphere with equal numbers of positive and negative charges. (b) The charge on a neutral metal sphere is redistributed when a charged rubber rod is placed near the sphere. (c) When the sphere is grounded, some of the electrons leave it through the ground wire. (d) When the ground connection is removed, the nonuniformly charged sphere is left with excess positive charge. (e) When the rubber rod is moved away, the charges on the sphere redistribute themselves until the sphere's surface becomes uniformly charged.

Charging by Induction

An object connected to a conducting wire or copper pipe buried in the Earth is said to be **grounded**. The Earth can be considered an infinite reservoir for electrons; in effect, it can accept or supply an unlimited number of electrons. With this idea in mind, we can understand the charging of a conductor by a process known as **induction**.

Consider a negatively charged rubber rod brought near a neutral (uncharged) conducting sphere that is insulated, so there is no conducting path to ground (Fig. 15.4). Initially the sphere is electrically neutral (Fig. 15.4a). When the negatively charged rod is brought close to the sphere, the repulsive force between the electrons in the rod and those in the sphere causes some electrons to move to the side of the sphere farthest away from the rod (Fig. 15.4b). The region of the sphere nearest the negatively charged rod has an excess of positive charge because of the migration of electrons away from that location. If a grounded conducting wire is then connected to the sphere, as in Figure 15.4c, some of the electrons leave the sphere and travel to ground. If the wire to ground is then removed (Fig. 15.4d), the conducting sphere is left with an excess of induced positive charge. Finally, when the rubber rod is removed from the vicinity of the sphere (Fig. 15.4e), the induced positive charge remains on the ungrounded sphere. Even though the positively charged atomic nuclei remain fixed, this excess positive charge becomes uniformly distributed over the surface of the ungrounded sphere because of the repulsive forces among the like charges and the high mobility of electrons in a metal.

In the process of inducing a charge on the sphere, the charged rubber rod doesn't lose any of its negative charge because it never comes in contact with the sphere. Furthermore, the sphere is left with a charge opposite that of the rubber rod. **Charging an object by induction requires no contact with the object inducing the charge**.

A process similar to charging by induction in conductors also takes place in insulators. In most neutral atoms or molecules, the center of positive charge coincides with the center of negative charge. However, in the presence of a charged object, these centers may separate slightly, resulting in more positive charge on one side of the molecule than on the other. This effect is known as **polarization**. The realignment of charge within individual molecules produces an induced charge on the surface of the insulator, as shown in Figure 15.5a. This explains why a balloon charged through rubbing will stick to an electrically neutral wall, or the comb you just used on your hair attracts tiny bits of neutral paper.

Quick Quiz 15.1

A suspended object *A* is attracted to a neutral wall. It's also attracted to a positively charged object *B*. Which of the following is true about object *A*? (a) It is uncharged. (b) It has a negative charge. (c) It has a positive charge. (d) It may be either charged or uncharged.

15.3   COULOMB'S LAW

In 1785 Charles Coulomb (1736–1806) experimentally established the fundamental law of electric force between two stationary charged particles.

An **electric force** has the following properties:

1. It is directed along a line joining the two particles and is inversely proportional to the square of the separation distance r, between them.
2. It is proportional to the product of the magnitudes of the charges, $|q_1|$ and $|q_2|$, of the two particles.
3. It is attractive if the charges are of opposite sign and repulsive if the charges have the same sign.

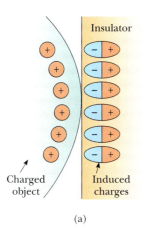

Charged object / Induced charges

(a)   (b)

Figure 15.5 (a) The charged object on the left induces charges on the surface of an insulator. (b) A charged comb attracts bits of paper because charges are displaced in the paper.

From these observations, Coulomb proposed the following mathematical form for the electric force between two charges:

> The magnitude of the electric force F between charges q_1 and q_2 separated by a distance r is given by
>
> $$F = k_e \frac{|q_1||q_2|}{r^2} \qquad [15.1]$$
>
> where k_e is a constant called the *Coulomb constant*.

◄ Coulomb's law

Equation 15.1, known as **Coulomb's law**, applies exactly only to point charges and to spherical distributions of charges, in which case r is the distance between the two centers of charge. Electric forces between unmoving charges are called *electrostatic* forces. Moving charges, in addition, create magnetic forces, studied in Chapter 19.

The value of the Coulomb constant in Equation 15.1 depends on the choice of units. The SI unit of charge is the **coulomb** (C). From experiment, we know that the **Coulomb constant** in SI units has the value

$$k_e = 8.9875 \times 10^9 \, \text{N} \cdot \text{m}^2/\text{C}^2 \qquad [15.2]$$

This number can be rounded, depending on the accuracy of other quantities in a given problem. We'll use either two or three significant digits, as usual.

The charge on the proton has a magnitude of $e = 1.6 \times 10^{-19}$ C. Therefore, it would take $1/e = 6.3 \times 10^{18}$ protons to create a total charge of $+1.0$ C. Likewise, 6.3×10^{18} electrons would have a total charge of -1.0 C. Compare this with the number of free electrons in 1 cm^3 of copper, which is on the order of 10^{23}. Even so, 1.0 C is a very large amount of charge. In typical electrostatic experiments in which a rubber or glass rod is charged by friction, there is a net charge on the order of 10^{-6} C ($= 1 \, \mu$C). Only a very small fraction of the total available charge is transferred between the rod and the rubbing material. Table 15.1 lists the charges and masses of the electron, proton, and neutron.

TABLE 15.1

Charge and Mass of the Electron, Proton, and Neutron

Particle	Charge (C)	Mass (kg)
Electron	-1.60×10^{-19}	9.11×10^{-31}
Proton	$+1.60 \times 10^{-19}$	1.67×10^{-27}
Neutron	0	1.67×10^{-27}

CHARLES COULOMB (1736–1806)

Coulomb's major contribution to science was in the field of electrostatics and magnetism. During his lifetime, he also investigated the strengths of materials and identified the forces that affect objects on beams, thereby contributing to the field of structural mechanics.

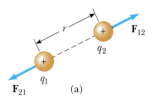

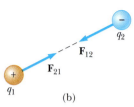

ACTIVE FIGURE 15.6
Two point charges separated by a distance r exert a force on each other given by Coulomb's law. The force on q_1 is equal in magnitude and opposite in direction to the force on q_2. (a) When the charges are of the same sign, the force is repulsive. (b) When the charges are of opposite sign, the force is attractive.

Physics⊗Now™

Log into PhysicsNow at **www.cp7e.com** and go to Active Figure 15.6, where you can move the charges to any position in two-dimensional space and observe the electric forces acting on them.

When using Coulomb's force law, remember that force is a vector quantity and must be treated accordingly. Active Figure 15.6a shows the electric force of repulsion between two positively charged particles. Like other forces, electric forces obey Newton's third law; hence, the forces $\vec{F}_{12}$ and $\vec{F}_{21}$ are equal in magnitude but opposite in direction. (The notation $\vec{F}_{12}$ denotes the force exerted by particle 1 on particle 2; likewise, $\vec{F}_{21}$ is the force exerted by particle 2 on particle 1.) From Newton's third law, F_{12} and F_{21} are always equal regardless of whether q_1 and q_2 have the same magnitude.

Quick Quiz 15.2

Object A has a charge of $+2\ \mu C$, and object B has a charge of $+6\ \mu C$. Which statement is true?

(a) $\vec{F}_{AB} = -3\vec{F}_{BA}$ (b) $\vec{F}_{AB} = -\vec{F}_{BA}$ (c) $3\vec{F}_{AB} = -\vec{F}_{BA}$

The Coulomb force is similar to the gravitational force. Both act at a distance without direct contact. Both are inversely proportional to the distance squared, with the force directed along a line connecting the two bodies. The mathematical form is the same, with the masses m_1 and m_2 in Newton's law replaced by q_1 and q_2 in Coulomb's law and with Newton's constant G replaced by Coulomb's constant k_e. There are two important differences: (1) electric forces can be either attractive or repulsive, but gravitational forces are always attractive, and (2) the electric force between charged elementary particles is far stronger than the gravitational force between the same particles, as the next example shows.

EXAMPLE 15.1   Forces in a Hydrogen Atom

Goal   Contrast the magnitudes of an electric force and a gravitational force.

Problem   The electron and proton of a hydrogen atom are separated (on the average) by a distance of about 5.3×10^{-11} m. Find the magnitudes of the electric force and the gravitational force that each particle exerts on the other, and the ratio of the electric force F_e to the gravitational force F_g.

Strategy   Solving this problem is just a matter of substituting known quantities into the two force laws and then finding the ratio.

Solution

Substitute $|q_1| = |q_2| = e$ and the distance into Coulomb's law to find the electric force:

$$F_e = k_e \frac{|e|^2}{r^2} = \left(8.99 \times 10^9\ \frac{\text{N} \cdot \text{m}^2}{\text{C}^2}\right) \frac{(1.6 \times 10^{-19}\ \text{C})^2}{(5.3 \times 10^{-11}\ \text{m})^2}$$

$$= 8.2 \times 10^{-8}\ \text{N}$$

Substitute the masses and distance into Newton's law of gravity to find the gravitational force:

$$F_g = G \frac{m_e m_p}{r^2}$$

$$= \left(6.67 \times 10^{-11}\ \frac{\text{N} \cdot \text{m}^2}{\text{kg}^2}\right) \frac{(9.11 \times 10^{-31}\ \text{kg})(1.67 \times 10^{-27}\ \text{kg})}{(5.3 \times 10^{-11}\ \text{m})^2}$$

$$= 3.6 \times 10^{-47}\ \text{N}$$

Find the ratio of the two forces:

$$\frac{F_e}{F_g} = 2.27 \times 10^{39}$$

Remarks   The gravitational force between the charged constituents of the atom is negligible compared with the electric force between them. The electric force is so strong, however, that any net charge on an object quickly attracts nearby opposite charges, neutralizing the object. As a result, gravity plays a greater role in the mechanics of moving objects in everyday life.

Exercise 15.1

Find the magnitude of the electric force between two protons separated by one femtometer $(10^{-15}\,\text{m})$, approximately the distance between two protons in the nucleus of a helium atom. The answer may not appear large, but if not for the strong nuclear force, the two protons would fly apart at an initial acceleration of nearly $7 \times 10^{28}\,\text{m/s}^2$!

Answers $2.30 \times 10^2\,\text{N}$

The Superposition Principle

When a number of separate charges act on the charge of interest, each exerts an electric force. These electric forces can all be computed separately, one at a time, then added as vectors. This is another example of the **superposition principle**. The following example illustrates this procedure in one dimension.

INTERACTIVE EXAMPLE 15.2   May the Force Be Zero

Goal   Apply Coulomb's law in one dimension.

Problem   Three charges lie along the x-axis as in Figure 15.7. The positive charge $q_1 = 15\,\mu\text{C}$ is at $x = 2.0$ m, and the positive charge $q_2 = 6.0\,\mu\text{C}$ is at the origin. Where must a *negative* charge q_3 be placed on the x-axis so that the resultant electric force on it is zero?

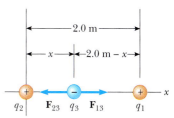

Strategy   If q_3 is to the right or left of the other two charges, then the net force on q_3 can't be zero, because then $\vec{\mathbf{F}}_{13}$ and $\vec{\mathbf{F}}_{23}$ act in the same direction. Consequently, q_3 must lie between the two other charges. Write $\vec{\mathbf{F}}_{13}$ and $\vec{\mathbf{F}}_{23}$ in terms of the unknown coordinate position x, sum them and set them equal to zero, solving for the unknown. The solution can be obtained with the quadratic formula.

Figure 15.7   (Example 15.2) Three point charges are placed along the x-axis. The charge q_3 is negative, whereas q_1 and q_2 are positive. If the resultant force on q_3 is zero, then the force $\vec{\mathbf{F}}_{13}$ exerted by q_1 on q_3 must be equal in magnitude and opposite the force $\vec{\mathbf{F}}_{23}$ exerted by q_2 on q_3.

Solution

Write the x-component of $\vec{\mathbf{F}}_{13}$:

$$F_{13x} = + k_e \frac{(15 \times 10^{-6}\,\text{C})|q_3|}{(2.0\,\text{m} - x)^2}$$

Write the x-component of $\vec{\mathbf{F}}_{23}$:

$$F_{23x} = - k_e \frac{(6.0 \times 10^{-6}\,\text{C})|q_3|}{x^2}$$

Set the sum equal to zero:

$$k_e \frac{(15 \times 10^{-6}\,\text{C})|q_3|}{(2.0\,\text{m} - x)^2} - k_e \frac{(6.0 \times 10^{-6}\,\text{C})|q_3|}{x^2} = 0$$

Cancel k_e, 10^{-6} and q_3 from the equation, and rearrange terms (explicit significant figures and units are temporarily suspended for clarity):

$$6(2 - x)^2 = 15x^2$$

Put this equation into standard quadratic form, $ax^2 + bx + c = 0$:

$$6(4 - 4x + x^2) = 15x^2 \;\;\rightarrow\;\; 2(4 - 4x + x^2) = 5x^2$$
$$3x^2 + 8x - 8 = 0$$

Apply the quadratic formula:

$$x = \frac{-8 \pm \sqrt{64 - (4)(3)(-8)}}{2 \cdot 3} = \frac{-4 \pm 2\sqrt{10}}{3}$$

Only the positive root makes sense:

$$x = \boxed{0.77\,\text{m}}$$

Remarks   Notice that it was necessary to use physical reasoning to choose between the two possible answers for x. This is nearly always the case when quadratic equations are involved.

Exercise 15.2

Three charges lie along the x-axis. The positive charge $q_1 = 10.0 \ \mu C$ is at $x = 1.00$ m, and the *negative charge* $q_2 = -2.00 \ \mu C$ is at the origin. Where must a *positive* charge q_3 be placed on the x-axis so that the resultant force on it is zero?

Answer $x = -0.809$ m

Physics⊗Now™ You can predict where on the x-axis the electric force is zero for random values of of q_1 and q_2 by logging into PhysicsNow at **www.cp7e.com** and going to Interactive Example 15.2.

EXAMPLE 15.3 A Charge Triangle

Goal Apply Coulomb's law in two dimensions.

Problem Consider three point charges at the corners of a triangle, as shown in Figure 15.8, where $q_1 = 6.00 \times 10^{-9}$ C, $q_2 = -2.00 \times 10^{-9}$ C, and $q_3 = 5.00 \times 10^{-9}$ C. **(a)** Find the components of the force $\vec{F}_{23}$ exerted by q_2 on q_3. **(b)** Find the components of the force $\vec{F}_{13}$ exerted by q_1 on q_3. **(c)** Find the resultant force on q_3, in terms of components and also in terms of magnitude and direction.

Strategy Coulomb's law gives the magnitude of each force, which can be split with right-triangle trigonometry into x- and y-components. Sum the vectors component-wise, and then find the magnitude and direction of the resultant vector.

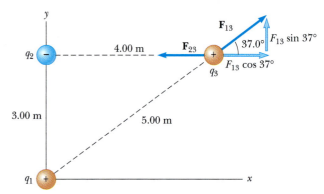

Figure 15.8 (Example 15.3) The force exerted by q_1 on q_3 is $\vec{F}_{13}$. The force exerted by q_2 on q_3 is $\vec{F}_{23}$. The *resultant force* $\vec{F}_3$ exerted on q_3 is the *vector* sum $\vec{F}_{13} + \vec{F}_{23}$.

Solution

(a) Find the components of the force exerted by q_2 on q_3.

Find the magnitude of $\vec{F}_{23}$ with Coulomb's law:

$$F_{23} = k_e \frac{|q_2||q_3|}{r^2}$$

$$= (8.99 \times 10^9 \ \text{N·m}^2/\text{C}^2) \frac{(2.00 \times 10^{-9} \ \text{C})(5.00 \times 10^{-9} \ \text{C})}{(4.00 \ \text{m})^2}$$

$$F_{23} = 5.62 \times 10^{-9} \ \text{N}$$

Because $\vec{F}_{23}$ is horizontal and points in the negative x-direction, the negative of the magnitude gives the x-component, and the y-component is zero:

$$F_{23x} = \boxed{-5.62 \times 10^{-9} \ \text{N}}$$

$$F_{23y} = \boxed{0}$$

(b) Find the components of the force exerted by q_1 on q_3.

Find the magnitude of $\vec{F}_{13}$:

$$F_{13} = k_e \frac{|q_1||q_3|}{r^3}$$

$$= (8.99 \times 10^9 \ \text{N·m}^2/\text{C}^2) \frac{(6.00 \times 10^{-9} \ \text{C})(5.00 \times 10^{-9} \ \text{C})}{(5.00 \ \text{m})^2}$$

$$F_{13} = 1.08 \times 10^{-8} \ \text{N}$$

Use the given triangle to find the components:

$$F_{13x} = F_{13} \cos\theta = (1.08 \times 10^{-8} \ \text{N})\cos(37°)$$
$$= \boxed{8.63 \times 10^{-9} \ \text{N}}$$

$$F_{13y} = F_{13} \sin\theta = (1.08 \times 10^{-8} \ \text{N})\sin(37°)$$
$$= \boxed{6.50 \times 10^{-9} \ \text{N}}$$

(c) Find the components of the resultant vector.

Sum the x-components to find the resultant F_x:

$$F_x = -5.62 \times 10^{-9}\,\text{N} + 8.63 \times 10^{-9}\,\text{N}$$
$$= 3.01 \times 10^{-9}\,\text{N}$$

Sum the y-components to find the resultant F_y:

$$F_y = 0 + 6.50 \times 10^{-9}\,\text{N} = 6.50 \times 10^{-9}\,\text{N}$$

Find the magnitude of the resultant force on the charge q_3, using the Pythagorean theorem:

$$|\vec{\mathbf{F}}| = \sqrt{F_x^{\,2} + F_y^{\,2}}$$
$$= \sqrt{(3.01 \times 10^{-9}\,\text{N})^2 + (6.50 \times 10^{-9}\,\text{N})^2}$$
$$= 7.16 \times 10^{-9}\,\text{N}$$

Find the angle the force vector makes with respect to the positive x-axis:

$$\theta = \tan^{-1}\left(\frac{F_y}{F_x}\right) = \tan^{-1}\left(\frac{6.50 \times 10^{-9}\,\text{N}}{3.01 \times 10^{-9}\,\text{N}}\right) = 65.2^\circ$$

Remarks   The methods used here are just like those used with Newton's law of gravity in two dimensions.

Exercise 15.3
Using the same triangle, find the vector components of the electric force on q_1 and the vector's magnitude and direction.

Answers $F_x = -8.63 \times 10^{-9}\,\text{N}$, $F_y = 5.50 \times 10^{-9}\,\text{N}$, $F = 1.02 \times 10^{-8}\,\text{N}$, $\theta = 147^\circ$

15.4   THE ELECTRIC FIELD

The gravitational force and the electrostatic force are both capable of acting through space, producing an effect even when there isn't any physical contact between the objects involved. Field forces can be discussed in a variety of ways, but an approach developed by Michael Faraday (1791–1867) is the most practical. In this approach, an **electric field** is said to exist in the region of space around a charged object. The electric field exerts an electric force on any other charged object within the field. This differs from the Coulomb's law concept of a force exerted at a distance, in that the force is now exerted by something—the field—that is in the same location as the charged object.

Figure 15.9 shows an object with a small positive charge q_0 placed near a second object with a much larger positive charge Q.

The electric field $\vec{\mathbf{E}}$ produced by a charge Q at the location of a small "test" charge q_0 is defined as the electric force $\vec{\mathbf{F}}$ exerted by Q on q_0, divided by the test charge q_0:

$$\vec{\mathbf{E}} \equiv \frac{\vec{\mathbf{F}}}{q_0} \qquad \text{[15.4]}$$

SI Unit: newton per coulomb (N/C)

Conceptually and experimentally, the test charge q_0 is required to be very small (arbitrarily small, in fact), so it doesn't cause any significant rearrangement of the charge creating the electric field $\vec{\mathbf{E}}$. Mathematically, however, the size of the test charge makes no difference: the calculation comes out the same, regardless. In view of this, using $q_0 = 1$ C in Equation 15.4 can be convenient if not rigorous.

When a positive test charge is used, the electric field always has the same direction as the electric force on the test charge. This follows from Equation 15.4. Hence in Figure 15.9, the direction of the electric field is horizontal and to the

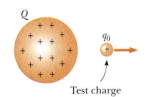

Figure 15.9   A small object with a positive charge q_0 placed near an object with a larger positive charge Q is subject to an electric field $\vec{\mathbf{E}}$ directed as shown. The magnitude of the electric field at the location of q_0 is defined as the electric force on q_0 divided by the charge q_0.

Figure 15.10 (a) The electric field at A due to the negatively charged sphere is downward, toward the negative charge. (b) The electric field at P due to the positively charged conducting sphere is upward, away from the positive charge. (c) A test charge q_0 placed at P will cause a rearrangement of charge on the sphere, unless q_0 is very small compared with the charge on the sphere.

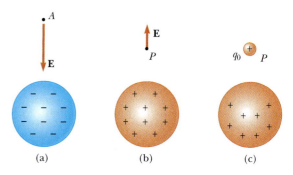

(a)   (b)   (c)

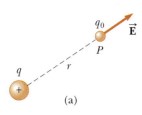

(a)

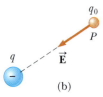

(b)

ACTIVE FIGURE 15.11

A test charge q_0 at P is a distance r from a point charge q. (a) If q is positive, the electric field at P points radially *outwards* from q. (b) If q is negative, the electric field at P points radially *inwards* toward q.

Physics⊗Now™

Log into PhysicsNow at **www.cp7e.com** and go to Active Figure 15.11, where you can move point P to any position in two-dimensional space and observe the electric field due to q.

right. The electric field at point A in Figure 15.10a is vertical and downward because at that point a positive test charge would be attracted toward the negatively charged sphere.

Once the electric field due to a given arrangement of charges is known at some point, the force on *any* particle with charge q placed at that point can be calculated from a rearrangement of Equation 15.4:

$$\vec{F} = q\vec{E} \qquad [15.5]$$

Here q_0 has been replaced by q, which need not be a mere test charge.

As shown in Active Figure 15.11, the direction of $\vec{E}$ is the direction of the force that acts on a positive test charge q_0 placed in the field. We say that **an electric field exists at a point if a test charge at that point is subject to an electric force there**.

Consider a point charge q located a distance r from a test charge q_0. According to Coulomb's law, the *magnitude* of the electric force of the charge q on the test charge is

$$F = k_e \frac{|q||q_0|}{r^2}$$

Because the magnitude of the electric field at the position of the test charge is defined as $E = F/q_0$, we see that the *magnitude* of the electric field due to the charge q at the position of q_0 is

$$E = k_e \frac{|q|}{r^2} \qquad [15.6]$$

Equation 15.6 points out an important property of electric fields that makes them useful quantities for describing electrical phenomena. As the equation indicates, an electric field at a given point depends only on the charge q on the object setting up the field and the distance r from that object to a specific point in space. As a result, we can say that an electric field exists at point P in Active Figure 15.11 whether or not there is a test charge at P.

The principle of superposition holds when the electric field due to a group of point charges is calculated. We first use Equation 15.6 to calculate the electric field produced by each charge individually at a point and then add the electric fields together as vectors.

It's also important to exploit any symmetry of the charge distribution. For example, if equal charges are placed at $x = a$ and at $x = -a$, the electric field is zero at the origin, by symmetry. Similarly, if the x-axis has a uniform distribution of positive charge, it can be guessed by symmetry that the electric field points away from the x-axis and is zero parallel to that axis.

Quick Quiz 15.3

A test charge of $+3\ \mu C$ is at a point P where the electric field due to other charges is directed to the right and has a magnitude of 4×10^6 N/C. If the test charge is replaced with a charge of $-3\ \mu C$, the electric field at P (a) has the same magnitude as before, but changes direction, (b) increases in magnitude and changes direction, (c) remains the same, or (d) decreases in magnitude and changes direction.

Quick Quiz 15.4

A circular ring of charge of radius b has a total charge q uniformly distributed around it. The magnitude of the electric field at the center of the ring is

(a) 0   (b) $k_e q/b^2$ (c) $k_e q^2/b^2$ (d) $k_e q^2/b$ (e) none of these.

Quick Quiz 15.5

A "free" electron and a "free" proton are placed in an identical electric field. Which of the following statements are true? (a) Each particle is acted upon by the same electric force and has the same acceleration. (b) The electric force on the proton is greater in magnitude than the force on the electron, but in the opposite direction. (c) The electric force on the proton is equal in magnitude to the force on the electron, but in the opposite direction. (d) The magnitude of the acceleration of the electron is greater than that of the proton. (e) Both particles have the same acceleration.

EXAMPLE 15.4   Electrified Oil

Goal   Use electric forces and fields together with Newton's second law in a one-dimensional problem.

Problem   Tiny droplets of oil acquire a small negative charge while dropping through a vacuum (pressure $= 0$) in an experiment. An electric field of magnitude 5.92×10^4 N/C points straight down. **(a)** One particular droplet is observed to remain suspended against gravity. If the mass of the droplet is 2.93×10^{-15} kg, find the charge carried by the droplet. **(b)** Another droplet of the same mass falls 10.3 cm from rest in 0.250 s, again moving through a vacuum. Find the charge carried by the droplet.

Strategy   We use Newton's second law with both gravitational and electric forces. In both parts, the electric field $\vec{E}$ is pointing down, taken as the negative direction, as usual. In part (a), the acceleration is equal to zero. In part (b), the acceleration is uniform, so the kinematic equations yield the acceleration. Newton's law can then be solved for q.

Solution
(a) Find the charge on the suspended droplet.

Apply Newton's second law to the droplet in the vertical direction:

$$(1) \quad ma = \Sigma F = -mg + Eq$$

E points downward, hence is negative.
Set $a = 0$ and solve for q:

$$q = \frac{mg}{E} = \frac{(2.93 \times 10^{-15}\,\text{kg})\,(9.80\,\text{m/s}^2)}{-5.92 \times 10^4\,\text{N/C}}$$

$$= \boxed{-4.85 \times 10^{-19}\,\text{C}}$$

(b) Find the charge on the falling droplet.

Use the kinematic displacement equation to find the acceleration:

$$\Delta y = \tfrac{1}{2} a t^2 + v_0 t$$

Substitute $\Delta y = -0.103$ m, $t = 0.250$ s, and $v_0 = 0$:

$$-0.103\,\text{m} = \tfrac{1}{2} a(0.250\,\text{s})^2 \quad \rightarrow \quad a = -3.30\,\text{m/s}^2$$

Solve Equation 1 for q and substitute:

$$q = \frac{m(a + g)}{E}$$

$$= \frac{(2.93 \times 10^{-15}\,\text{kg})(-3.30\,\text{m/s}^2 + 9.80\,\text{m/s}^2)}{-5.92 \times 10^4\,\text{N/C}}$$

$$= \boxed{-3.22 \times 10^{-19}\,\text{C}}$$

Remark This example exhibits features similar to the Millikan Oil-Drop experiment discussed in Section 15.7, which determined the value of the fundamental electric charge e. Notice that in both parts of the example, the charge is very nearly a multiple of e.

Exercise 15.4

Suppose a droplet of unknown mass remains suspended against gravity when $E = -2.70 \times 10^5$ N/C. What is the minimum mass of the droplet?

Answer 4.41×10^{-15} kg

Problem-Solving Strategy
Calculating Electric Forces and Fields

The following procedure is used to calculate electric forces (the same procedure can be used to calculate an electric field, a simple matter of replacing the charge of interest, q, with a convenient test charge and dividing by the test charge at the end):

1. **Draw** a diagram of the charges in the problem.
2. **Identify** the charge of interest, q, and circle it.
3. **Convert all units** to SI, with charges in coulombs and distances in meters, so as to be consistent with the SI value of the Coulomb constant k_e.
4. **Apply Coulomb's law.** For each charge Q, find the electric force on the charge of interest, q. The magnitude of the force can be found using Coulomb's law. The vector direction of the electric force is along the line of the two charges, directed away from Q if the charges have the same sign, toward Q if the charges have the opposite sign. Find the angle θ this vector makes with the positive x-axis. The x-component of the electric force exerted by Q on q will be $F\cos\theta$, and the y-component will be $F\sin\theta$.
5. **Sum all the x-components,** getting the x-component of the resultant electric force.
6. **Sum all the y-components,** getting the y-component of the resultant electric force.
7. **Use the Pythagorean theorem and trigonometry** to find the magnitude and direction of the resultant force if desired.

EXAMPLE 15.5   Electric Field Due to Two Point Charges

Goal   Use the superposition principle to calculate the electric field due to two point charges.

Problem   Charge $q_1 = 7.00\ \mu C$ is at the origin, and charge $q_2 = -5.00\ \mu C$ is on the x-axis, 0.300 m from the origin (Fig. 15.12). **(a)** Find the magnitude and direction of the electric field at point P, which has coordinates $(0, 0.400)$ m. **(b)** Find the force on a charge of 2.00×10^{-8} C placed at P.

Strategy   Follow the problem-solving strategy, finding the electric field at point P due to each individual charge in terms of x- and y-components, then adding the components of each type to get the x- and y-components of the resultant electric field at P. The magnitude of the force in part (b) can be found by simply multiplying the magnitude of the electric field by the charge.

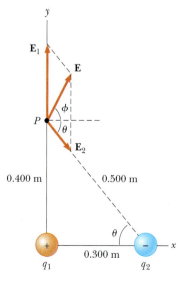

Figure 15.12   (Example 15.5) The resultant electric field $\vec{E}$ at P equals the vector sum $\vec{E}_1 + \vec{E}_2$, where $\vec{E}_1$ is the field due to the positive charge q_1 and $\vec{E}_2$ is the field due to the negative charge q_2.

Solution

(a) Calculate the electric field at P.

Find the magnitude of $\vec{E}_1$ with Equation 15.6:

$$E_1 = k_e \frac{|q_1|}{r_1^2} = (8.99 \times 10^9 \, \text{N·m}^2/\text{C}^2) \frac{(7.00 \times 10^{-6} \, \text{C})}{(0.400 \, \text{m})^2}$$
$$= 3.93 \times 10^5 \, \text{N/C}$$

The vector $\vec{E}_1$ is vertical, making an angle of 90° with respect to the positive x-axis. Use this fact to find its components:

$$E_{1x} = E_1 \cos(90°) = 0$$
$$E_{1y} = E_1 \sin(90°) = 3.93 \times 10^5 \, \text{N/C}$$

Next, find the magnitude of $\vec{E}_2$, again with Equation 15.6:

$$E_2 = k_e \frac{|q_2|}{r_2^2} = (8.99 \times 10^9 \, \text{N·m}^2/\text{C}^2) \frac{(5.00 \times 10^{-6} \, \text{C})}{(0.500 \, \text{m})^2}$$
$$= 1.80 \times 10^5 \, \text{N/C}$$

Obtain the x-component of $\vec{E}_2$, using the triangle in Figure 15.12 to find $\cos \theta$:

$$\cos \theta = \frac{\text{adj}}{\text{hyp}} = \frac{0.300}{0.500} = 0.600$$
$$E_{2x} = E_2 \cos \theta = (1.80 \times 10^5 \, \text{N/C})(0.600)$$
$$= 1.08 \times 10^5 \, \text{N/C}$$

Obtain the y-component in the same way, but a minus sign has to be provided for $\sin \theta$ because this component is directed downwards:

$$\sin \theta = \frac{\text{opp}}{\text{hyp}} = \frac{0.400}{0.500} = 0.800$$
$$E_{2y} = E_2 \sin \theta = (1.80 \times 10^5 \, \text{N/C})(-0.800)$$
$$= -1.44 \times 10^5 \, \text{N/C}$$

Sum the x-components to get the x-component of the resultant vector:

$$E_x = E_{1x} + E_{2x} = 0 + 1.08 \times 10^5 \, \text{N/C} = 1.08 \times 10^5 \, \text{N/C}$$

Sum the y-components to get the y-component of the resultant vector:

$$E_y = E_{1y} + E_{2y} = 0 + 3.93 \times 10^5 \, \text{N/C} - 1.44 \times 10^5 \, \text{N/C}$$
$$E_y = 2.49 \times 10^5 \, \text{N/C}$$

Use the Pythagorean theorem to find the magnitude of the resultant vector:

$$E = \sqrt{E_x^2 + E_y^2} = 2.71 \times 10^5 \, \text{N/C}$$

The inverse tangent function yields the direction of the resultant vector:

$$\phi = \tan^{-1}\left(\frac{E_y}{E_x}\right) = \tan^{-1}\left(\frac{2.49 \times 10^5 \, \text{N/C}}{1.08 \times 10^5 \, \text{N/C}}\right) = 66.6°$$

(b) Find the force on a charge of 2.00×10^{-8} C placed at P.

Calculate the magnitude of the force (the direction is the same as that of $\vec{E}$ because the charge is positive):

$$F = Eq = (2.71 \times 10^5 \, \text{N/C})(2.00 \times 10^{-8} \, \text{C})$$
$$= 5.42 \times 10^{-3} \, \text{N}$$

Remarks   There were numerous steps to this problem, but each was very short. When attacking such problems, it's important to focus on one small step at a time. The solution comes not from a leap of genius, but from the assembly of a number of relatively easy parts.

Exercise 15.5
(a) Place a charge of $-7.00 \, \mu\text{C}$ at point P and find the magnitude and direction of the electric field at the location of q_2. (b) Find the magnitude and direction of the force on q_2.

Answer   (a) $5.84 \times 10^5 \, \text{N/C}$, $\phi = 20.2°$ (b) $F = 2.92 \, \text{N}$, $\phi = 200.°$

15.5 ELECTRIC FIELD LINES

A convenient aid for visualizing electric field patterns is to draw lines pointing in the direction of the electric field vector at any point. These lines, introduced by Michael Faraday and called **electric field lines**, are related to the electric field in any region of space in the following way:

1. The electric field vector $\vec{E}$ is tangent to the electric field lines at each point.
2. The number of lines per unit area through a surface perpendicular to the lines is proportional to the strength of the electric field in a given region.

Note that $\vec{E}$ is large when the field lines are close together and small when the lines are far apart.

Figure 15.13a shows some representative electric field lines for a single positive point charge. This two-dimensional drawing contains only the field lines that lie in the plane containing the point charge. The lines are actually directed radially outward from the charge in *all* directions, somewhat like the quills of an angry porcupine. Because a positive test charge placed in this field would be repelled by the charge q, the lines are directed radially away from the positive charge. The electric field lines for a single negative point charge are directed toward the charge (Fig. 15.13b), because a positive test charge is attracted by a negative charge. In either case, the lines are radial and extend all the way to infinity. Note that the lines are closer together as they get near the charge, indicating that the strength of the field is increasing. Equation 15.6 verifies that this is indeed the case.

The rules for drawing electric field lines for any charge distribution follow directly from the relationship between electric field lines and electric field vectors:

1. The lines for a group of point charges must begin on positive charges and end on negative charges. In the case of an excess of charge, some lines will begin or end infinitely far away.
2. The number of lines drawn leaving a positive charge or ending on a negative charge is proportional to the magnitude of the charge.
3. No two field lines can cross each other.

Figure 15.14 shows the beautifully symmetric electric field lines for two point charges of equal magnitude but opposite sign. This charge configuration is called an **electric dipole**. Note that the number of lines that begin at the positive charge must equal the number that terminate at the negative charge. At points very near either charge, the lines are nearly radial. The high density of lines between the charges indicates a strong electric field in this region.

TIP 15.1   Electric Field Lines Aren't Paths of Particles

Electric field lines are *not* material objects. They are used only as a pictorial representation of the electric field at various locations. Except in special cases, they *do not* represent the path of a charged particle released in an electric field.

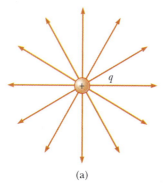
(a)

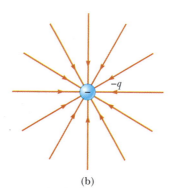
(b)

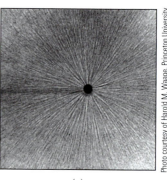
(c)

Figure 15.13   The electric field lines for a point charge. (a) For a positive point charge, the lines radiate outward. (b) For a negative point charge, the lines converge inward. Note that the figures show only those field lines which lie in the plane containing the charge. (c) The dark lines are small pieces of thread suspended in oil, which align with the electric field produced by a small charged conductor at the center.

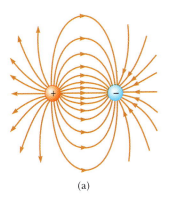

(a)

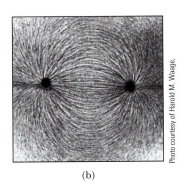

(b)

Photo courtesy of Harold M. Waage, Princeton University

Figure 15.14   (a) The electric field lines for two equal and opposite point charges (an electric dipole). Note that the number of lines leaving the positive charge equals the number terminating at the negative charge. (b) The dark lines are small pieces of thread suspended in oil, which align with the electric field produced by two charged conductors.

Figure 15.15 shows the electric field lines in the vicinity of two equal positive point charges. Again, close to either charge the lines are nearly radial. The same number of lines emerges from each charge because the charges are equal in magnitude. At great distances from the charges, the field is approximately equal to that of a single point charge of magnitude $2q$. The bulging out of the electric field lines between the charges reflects the repulsive nature of the electric force between like charges. Also, the low density of field lines between the charges indicates a weak field in this region, unlike the dipole.

Finally, Active Figure 15.16 is a sketch of the electric field lines associated with the positive charge $+2q$ and the negative charge $-q$. In this case, the number of lines leaving charge $+2q$ is twice the number terminating on charge $-q$. Hence, only half of the lines that leave the positive charge end at the negative charge. The remaining half terminate on negative charges that we assume to be located at infinity. At great distances from the charges (great compared with the charge separation), the electric field lines are equivalent to those of a single charge $+q$.

Quick Quiz 15.6

Rank the magnitudes of the electric field at points *A*, *B*, and *C* in Figure 15.15, with the largest magnitude first.
(a) *A*, *B*, *C* (b) *A*, *C*, *B* (c) *C*, *A*, *B* (d) Can't be determined by visual inspection

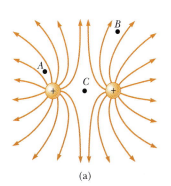

(a)

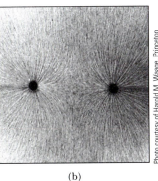

(b)

Photo courtesy of Harold M. Waage, Princeton University

Figure 15.15   (a) The electric field lines for two positive point charges. The points *A*, *B*, and *C* will be discussed in Quick Quiz 15.6. (b) The dark lines are small pieces of thread suspended in oil, which align with the electric field produced by two charged conductors.

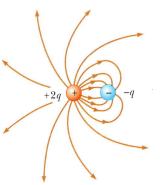

ACTIVE FIGURE 15.16
The electric field lines for a point charge of $+2q$ and a second point charge of $-q$. Note that two lines leave the charge $+2q$ for every line that terminates on $-q$.

Physics⊗Now™
Log into PhysicsNow at
www.cp7e.com and go to Active Figure 15.16, where you can choose the values and signs for the two charges and observe the resulting electric field lines.

Applying Physics 15.1   Measuring Atmospheric Electric Fields

The electric field near the surface of the Earth in fair weather is about 100 N/C downward. Under a thundercloud, the electric field can be very large, on the order of 20 000 N/C. How are these electric fields measured?

Explanation A device for measuring these fields is called the *field mill*. Figure 15.17 shows the fundamental components of a field mill: two metal plates parallel to the ground. Each plate is connected to ground with a wire, with an ammeter (a low-resistance device for measuring the flow of charge, to be discussed in Section 19.6) in one path. Consider first just the lower plate. Because it's connected to ground and the ground carries a negative charge, the plate is negatively charged. The electric field lines, therefore, are directed downward, ending on the plate as in

Figure 15.17a. Now imagine that the upper plate is suddenly moved over the lower plate, as in Figure 15.17b. This plate is also connected to ground and is also negatively charged, so the field lines now end on the upper plate. The negative charges in the lower plate are repelled by those on the upper plate and must pass through the ammeter, registering a flow of charge. The amount of charge that was on the lower plate is related to the strength of the electric field. In this way, the flow of charge through the ammeter can be calibrated to measure the electric field. The plates are normally designed like the blades of a fan, with the upper plate rotating so that the lower plate is alternately covered and uncovered. As a result, charges flow back and forth continually through the ammeter, and the reading can be related to the electric field strength.

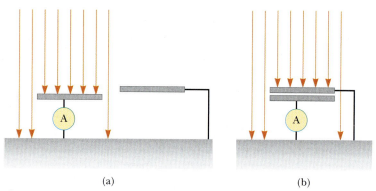

(a)   (b)

Figure 15.17 (Applying Physics 15.1) In (a), electric field lines end on negative charges on the lower plate. In (b), the second plate is moved above the lower plate. Electric field lines now end on the upper plate, and the negative charges in the lower plate are repelled through the ammeter.

15.6 CONDUCTORS IN ELECTROSTATIC EQUILIBRIUM

A good electric conductor like copper, though electrically neutral, contains charges (electrons) that aren't bound to any atom and are free to move about within the material. When no net motion of charge occurs within a conductor, the conductor is said to be in **electrostatic equilibrium**. An isolated conductor (one that is insulated from ground) has the following properties:

Properties of an isolated conductor ▶

1. The electric field is zero everywhere inside the conducting material.
2. Any excess charge on an isolated conductor resides entirely on its surface.
3. The electric field just outside a charged conductor is perpendicular to the conductor's surface.
4. On an irregularly shaped conductor, the charge accumulates at sharp points, where the radius of curvature of the surface is smallest.

The first property can be understood by examining what would happen if it were *not* true. If there were an electric field inside a conductor, the free charge

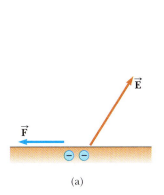

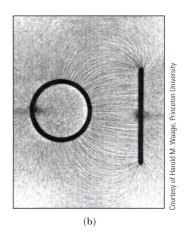

(a)   (b)

Figure 15.18   (a) Negative charges at the surface of a conductor. If the electric field were at an angle to the surface, as shown, an electric force would be exerted on the charges along the surface and they would move to the left. Because the conductor is assumed to be in electrostatic equilibrium, $\vec{\mathbf{E}}$ cannot have a component along the surface and hence must be perpendicular to it. (b) The electric field pattern of a charged conducting plate near an oppositely charged conducting cylinder. Small pieces of thread suspended in oil align with the electric field lines. Note that (1) the electric field lines are perpendicular to the conductors and (2) there are no lines inside the cylinder ($\vec{\mathbf{E}} = 0$).

there would move and a flow of charge, or current, would be created. However, if there were a net movement of charge, the conductor would no longer be in electrostatic equilibrium.

Property 2 is a direct result of the $1/r^2$ repulsion between like charges described by Coulomb's law. If by some means an excess of charge is placed inside a conductor, the repulsive forces between the like charges push them as far apart as possible, causing them to quickly migrate to the surface. (We won't prove it here, but the excess charge resides on the surface due to the fact that Coulomb's law is an inverse-square law. With any other power law, an excess of charge would exist on the surface, but there would be a distribution of charge, of either the same or opposite sign, inside the conductor.)

Property 3 can be understood by again considering what would happen if it were not true. If the electric field in Figure 15.18a were not perpendicular to the surface, it would have a component along the surface, which would cause the free charges of the conductor to move (to the left in the figure). If the charges moved, however, a current would be created and the conductor would no longer be in electrostatic equilibrium. Therefore, $\vec{\mathbf{E}}$ must be perpendicular to the surface.

To see why property 4 must be true, consider Figure 15.19a, which shows a conductor that is fairly flat at one end and relatively pointed at the other. Any excess charge placed on the object moves to its surface. Figure 15.19b shows the forces between two such charges at the flatter end of the object. These forces are predominantly directed parallel to the surface, so the charges move apart until repulsive forces from other nearby charges establish an equilibrium. At the sharp end, however, the forces of repulsion between two charges are directed predominantly away from the surface, as in Figure 15.19c. As a result, there is less tendency for the charges to move apart along the surface here, and the amount of charge per unit area is greater than at the flat end. The cumulative effect of many such outward forces from nearby charges at the sharp end produces a large resultant force directed away from the surface that can be great enough to cause charges to leap from the surface into the surrounding air.

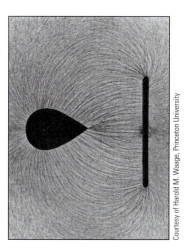

Electric field pattern of a charged conducting plate near an oppositely charged pointed conductor. Small pieces of thread suspended in oil align with the electric field lines. Note that the electric field is most intense near the pointed part of the conductor, where the radius of curvature is the smallest. Also, the lines are perpendicular to the conductors.

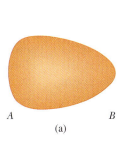

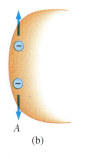

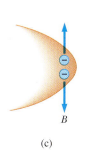

A   B   A   B
    (a)   (b)   (c)

Figure 15.19   (a) A conductor with a flatter end A and a relatively sharp end B. Excess charge placed on this conductor resides entirely at its surface and is distributed so that (b) there is less charge per unit area on the flatter end and (c) there is a large charge per unit area on the sharper end.

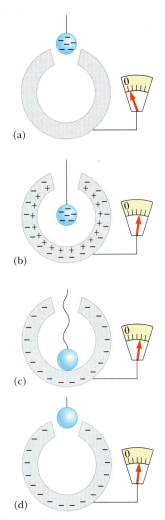

Figure 15.20 An experiment showing that any charge transferred to a conductor resides on its surface in electrostatic equilibrium. The hollow conductor is insulated from ground, and the small metal ball is supported by an insulating thread.

APPLICATION

Lightning Rods

Many experiments have shown that the net charge on a conductor resides on its surface. One such experiment was first performed by Michael Faraday, and is referred to as *Faraday's ice-pail experiment*. Faraday lowered a metal ball having a negative charge at the end of a silk thread (an insulator) into an uncharged hollow conductor insulated from ground, a metal ice-pail as in Figure 15.20a. As the ball entered the pail, the needle on an electrometer attached to the outer surface of the pail was observed to deflect. (An electrometer is a device used to measure charge.) The needle deflected because the charged ball induced a positive charge on the inner wall of the pail, which left an equal negative charge on the outer wall (Fig. 15.20b).

Faraday next touched the inner surface of the pail with the ball and noted that the deflection of the needle did not change, either when the ball touched the inner surface of the pail (Fig. 15.20c) or when it was removed (Fig. 15.20d). Further, he found that the ball was now uncharged, because when it touched the inside of the pail, the excess negative charge on the ball had been drawn off, neutralizing the induced positive charge on the inner surface of the pail. In this way, Faraday discovered the useful result that *all* the excess charge on an object can be transferred to an already charged metal shell if the object is touched to the *inside* of the shell. As we will see, this is the principle of operation of the Van de Graaff generator.

Faraday concluded that because the deflection of the needle in the electrometer didn't change when the charged ball touched the inside of the pail, the positive charge induced on the inside surface of the pail was just enough to neutralize the negative charge on the ball. As a result of his investigations, he concluded that a charged object suspended inside a metal container rearranged the charge on the container so that the sign of the charge on its inside surface was *opposite* the sign of the charge on the suspended object. This produced a charge on the outside surface of the container of the same sign as that on the suspended object.

Faraday also found that if the electrometer was connected to the inside surface of the pail after the experiment had been run, the needle showed no deflection. Thus, the *excess* charge acquired by the pail when contact was made between ball and pail appeared on the outer surface of the pail.

If a metal rod having sharp points is attached to a house, most of any charge on the house passes through these points, eliminating the induced charge on the house produced by storm clouds. In addition, a lightning discharge striking the house passes through the metal rod and is safely carried to the ground through wires leading from the rod to the Earth. Lightning rods using this principle were first developed by Benjamin Franklin. Some European countries couldn't accept the fact that such a worthwhile idea could have originated in the New World, so they "improved" the design by eliminating the sharp points!

Applying Physics 15.2   Conductors and Field Lines

Suppose a point charge $+Q$ is in empty space. Wearing rubber gloves, you proceed to surround the charge with a concentric spherical conducting shell. What effect does this have on the field lines from the charge?

Explanation When the spherical shell is placed around the charge, the charges in the shell rearrange to satisfy the rules for a conductor in equilibrium. A net charge of $-Q$ moves to the interior surface of the conductor, so that the electric field inside the conductor becomes zero. This means the field lines originating on the $+Q$ charge now terminate on the negative charges. The movement of the negative charges to the inner surface of the sphere leaves a net charge of $+Q$ on the outer surface of the sphere. Then the field lines outside the sphere look just as before: the only change, overall, is the absence of field lines within the conductor.

Applying Physics 15.3   Driver Safety During Electrical Storms

Why is it safe to stay inside an automobile during a lightning storm?

Explanation Many people believe that staying inside the car is safe because of the insulating characteristics of the rubber tires, but in fact this isn't true. Lightning can travel through several kilometers of air, so it can certainly penetrate a centimeter of rubber. The safety of remaining in the car is due to the fact that charges on the metal shell of the car will reside on the outer surface of the car, as noted in property 2 discussed earlier. As a result, an occupant in the automobile touching the inner surfaces is not in danger.

15.7   THE MILLIKAN OIL-DROP EXPERIMENT

From 1909 to 1913, Robert Andrews Millikan (1868–1953) performed a brilliant set of experiments at the University of Chicago in which he measured the elementary charge e of the electron and demonstrated the quantized nature of the electronic charge. The apparatus he used, diagrammed in Active Figure 15.21, contains two parallel metal plates. Oil droplets that have been charged by friction in an atomizer are allowed to pass through a small hole in the upper plate. A horizontal light beam is used to illuminate the droplets, which are viewed by a telescope with axis at right angles to the beam. The droplets then appear as shining stars against a dark background, and the rate of fall of individual drops can be determined.

We assume that a single drop having a mass of m and carrying a charge of q is being viewed and that its charge is negative. If no electric field is present between the plates, the two forces acting on the charge are the force of gravity, $m\vec{g}$, acting downward, and an upward viscous drag force $\vec{D}$ (Fig. 15.22a). The drag force is proportional to the speed of the drop. When the drop reaches its terminal speed, v, the two forces balance each other ($mg = D$).

Now suppose that an electric field is set up between the plates by a battery connected so that the upper plate is positively charged. In this case, a third force, $q\vec{E}$, acts on the charged drop. Because q is negative and $\vec{E}$ is downward, the electric

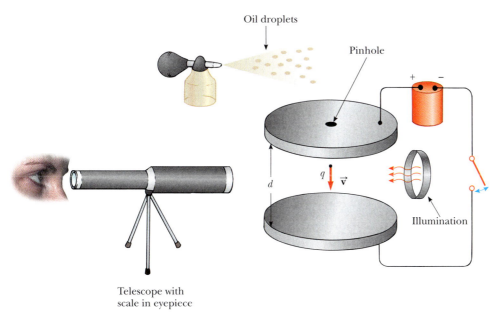

ACTIVE FIGURE 15.21
A schematic view of Millikan's oil-drop apparatus.

Physics⊗Now™
Log into PhysicsNow at **www.cp7e.com** and go to Active Figure 15.21, where you can do a simulation of the experiment, taking data on a number of oil drops and determining the elementary charge from your data.

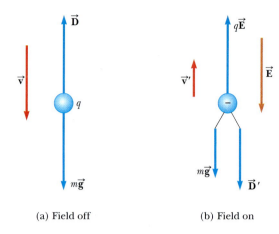

Figure 15.22 The forces on a charged oil droplet in Millikan's experiment.

(a) Field off

(b) Field on

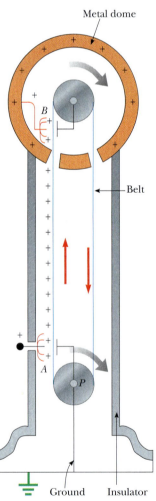

Figure 15.23 A diagram of a Van de Graaff generator. Charge is transferred to the dome by means of a rotating belt. The charge is deposited on the belt at point A and transferred to the dome at point B.

force is *upward* as in Figure 15.22b. If this force is great enough, the drop moves upward and the drag force $\vec{D}'$ acts downward. When the upward electric force, $q\vec{E}$, balances the sum of the force of gravity and the drag force, both acting downward, the drop reaches a new terminal speed v'.

With the field turned on, a drop moves slowly upward, typically at a rate of *hundredths* of a centimeter per second. The rate of fall in the absence of a field is comparable. Hence, a single droplet with constant mass and radius can be followed for hours as it alternately rises and falls, simply by turning the electric field on and off.

After making measurements on thousands of droplets, Millikan and his coworkers found that, to within about 1% precision, every drop had a charge equal to some positive or negative integer multiple of the elementary charge e,

$$q = ne \qquad n = 0, \pm 1, \pm 2, \pm 3, \ldots \qquad \text{[15.7]}$$

where $e = 1.60 \times 10^{-19}$ C. It was later established that positive integer multiples of e would arise when an oil droplet had lost one or more electrons. Likewise, negative integer multiples of e would arise when a drop had gained one or more electrons. Gains or losses in integral numbers provide conclusive evidence that charge is quantized. In 1923, Millikan was awarded the Nobel prize in physics for this work.

15.8 THE VAN DE GRAAFF GENERATOR

In 1929 Robert J. Van de Graaff (1901–1967) designed and built an electrostatic generator that has been used extensively in nuclear physics research. The principles of its operation can be understood with knowledge of the properties of electric fields and charges already presented in this chapter. Figure 15.23 shows the basic construction of this device. A motor-driven pulley P moves a belt past positively charged comb-like metallic needles positioned at A. Negative charges are attracted to these needles from the belt, leaving the left side of the belt with a net positive charge. The positive charges attract electrons onto the belt as it moves past a second comb of needles at B, increasing the excess positive charge on the dome. Because the electric field inside the metal dome is negligible, the positive charge on it can easily be increased regardless of how much charge is already present. The result is that the dome is left with a large amount of positive charge.

This accumulation of charge on the dome can't continue indefinitely. As more and more charge appears on the surface of the dome, the magnitude of the electric field at that surface is also increasing. Finally, the strength of the field becomes great enough to partially ionize the air near the surface, increasing the conductivity of the air. Charges on the dome now have a pathway to leak off into the air, producing some spectacular "lightning bolts" as the discharge occurs. As noted earlier, charges find it easier to leap off a surface at points where the curvature is great. As a result, one way to inhibit the electric discharge, and to increase the amount of charge that can be stored on the dome, is to increase its radius.

Another method for inhibiting discharge is to place the entire system in a container filled with a high-pressure gas, which is significantly more difficult to ionize than air at atmospheric pressure.

If protons (or other charged particles) are introduced into a tube attached to the dome, the large electric field of the dome exerts a repulsive force on the protons, causing them to accelerate to energies high enough to initiate nuclear reactions between the protons and various target nuclei.

15.9   ELECTRIC FLUX AND GAUSS'S LAW

Gauss's law is essentially a technique for calculating the average electric field on a closed surface, developed by Karl Friedrich Gauss (1777–1855). When the electric field, because of its symmetry, is constant everywhere on that surface and perpendicular to it, the exact electric field can be found. In such special cases, Gauss's law is far easier to apply than Coulomb's law.

Gauss's law relates the electric flux through a closed surface and the total charge inside that surface. A *closed surface* has an inside and an outside: an example is a sphere. *Electric flux* is a measure of how much the electric field vectors penetrate through a given surface. If the electric field vectors are tangent to the surface at all points, for example, then they don't penetrate the surface and the electric flux through the surface is zero. These concepts will be discussed more fully in the next two subsections. As we'll see, Gauss's law states that the electric flux through a closed surface is proportional to the charge contained *inside* the surface.

Electric Flux

Consider an electric field that is uniform in both magnitude and direction, as in Figure 15.24. The electric field lines penetrate a surface of area A, which is perpendicular to the field. The technique used for drawing a figure such as Figure 15.24 is that the number of lines per unit area, N/A, is proportional to the magnitude of the electric field, or $E \propto N/A$. We can rewrite this as $N \propto EA$, which means that the number of field lines is proportional to the *product* of E and A, called the **electric flux** and represented by the symbol Φ_E:

$$\Phi_E = EA \qquad [15.8]$$

Note that Φ_E has SI units of $\text{N} \cdot \text{m}^2/\text{C}$ and is proportional to the number of field lines that pass through some area A oriented perpendicular to the field. (It's called flux by analogy with the term *flux* in fluid flow, which is just the volume of liquid flowing through a perpendicular area per second.). If the surface under consideration is not perpendicular to the field, as in Figure 15.25, the expression for the electric flux is

$$\Phi_E = EA \cos \theta \qquad [15.9]$$

where a vector perpendicular to the area A is at an angle θ with respect to the field. This vector is often said to be *normal* to the surface, and we will refer to it as

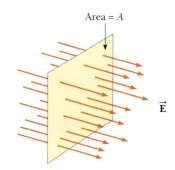

Area = A

$\vec{\mathbf{E}}$

Figure 15.24   Field lines of a uniform electric field penetrating a plane of area A perpendicular to the field. The electric flux Φ_E through this area is equal to EA.

◄ Electric flux

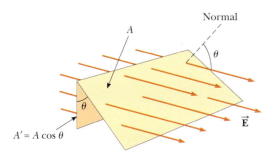

Normal

A

θ

$\vec{\mathbf{E}}$

$A' = A \cos \theta$

Figure 15.25   Field lines for a uniform electric field through an area A that is at an angle of $(90° - \theta)$ to the field. Because the number of lines that go through the shaded area A' is the same as the number that go through A, we conclude that the flux through A' is equal to the flux through A and is given by $\Phi_E = EA \cos \theta$.

"the normal vector to the surface." The number of lines that cross this area is equal to the number that cross the projected area A', which is perpendicular to the field. We see that the two areas are related by $A' = A \cos \theta$. From Equation 15.9, we see that the flux through a surface of fixed area has the maximum value EA when the surface is perpendicular to the field (when $\theta = 0°$) and that the flux is zero when the surface is parallel to the field (when $\theta = 90°$). **By convention, for a closed surface, the flux lines passing into the interior of the volume are negative and those passing out of the interior of the volume are positive.** This convention is equivalent to requiring the normal vector of the surface to point outward when computing the flux through a closed surface.

Quick Quiz 15.7

Calculate the magnitude of the flux of a constant electric field of 5.00 N/C in the z-direction through a rectangle with area 4.00 m^2 in the xy-plane. (a) 0 (b) 10.0 $\text{N} \cdot \text{m}^2/\text{C}$ (c) 20.0 $\text{N} \cdot \text{m}^2/\text{C}$ (d) more information is needed

Quick Quiz 15.8

Suppose the electric field of Quick Quiz 15.7 is tilted 60° away from the positive z-direction. Calculate the magnitude of the flux through the same area. (a) 0 (b) 10.0 $\text{N} \cdot \text{m}^2/\text{C}$ (c) 20.0 $\text{N} \cdot \text{m}^2/\text{C}$ (d) more information is needed

---

EXAMPLE 15.6   Flux Through a Cube

Goal   Calculate the electric flux through a closed surface.

Problem   Consider a uniform electric field oriented in the x-direction. Find the electric flux through each surface of a cube with edges L oriented as shown in Figure 15.26, and the net flux.

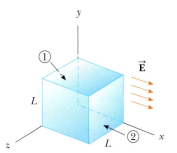

Strategy   This problem involves substituting into the definition of electric flux given by Equation 15.9. E and $A = L^2$ are the same in each case; the only difference is the angle θ that the electric field makes with respect to a vector perpendicular to a given surface and pointing outward (the normal vector to the surface). The angles can be determined by inspection. The flux through a surface parallel to the xy-plane will be labeled Φ_{xy} and further designated by position (front, back); others will be labeled similarly: Φ_{xz} top or bottom, and Φ_{yz} left or right.

Figure 15.26   (Example 15.6) A hypothetical surface in the shape of a cube in a uniform electric field parallel to the x-axis. The net flux through the surface is zero when the net charge inside the cube is zero.

Solution

The normal vector to the xy-plane points in the negative z-direction. This, in turn, is perpendicular to $\vec{E}$, so $\theta = 90°$. (The opposite side works similarly.)

$$\Phi_{xy} = EA \cos(90°) = \boxed{0} \quad \text{(back and front)}$$

The normal vector to the xz-plane points in the negative y-direction. This, in turn, is perpendicular to $\vec{E}$, so again $\theta = 90°$. (The opposite side works similarly.)

$$\Phi_{xz} = EA \cos(90°) = \boxed{0} \quad \text{(top and bottom)}$$

The normal vector to surface ① (the yz-plane) points in the negative x-direction. This is antiparallel to $\vec{E}$, so $\theta = 180°$.

$$\Phi_{yz} = EA \cos(180°) = \boxed{-EL^2} \quad \text{(surface ①)}$$

Surface ② has normal vector pointing in the positive x-direction, so $\theta = 0°$.

$$\Phi_{yz} = EA \cos(0°) = \boxed{EL^2} \quad \text{(surface ②)}$$

We calculate the net flux by summing:

$$\Phi_{net} = 0 + 0 + 0 + 0 - EL^2 + EL^2 = \boxed{0}$$

Remarks   In doing this calculation, it is necessary to remember that the angle in the definition of flux is measured from the normal vector to the surface and that this vector must point outwards for a closed surface. As a result, the normal vector for the yz-plane on the left points in the negative x-direction, and the normal vector to the plane parallel to the yz-plane on the right points in the positive x-direction. Notice that there aren't any charges in the box. The net electric flux is always zero for closed surfaces that don't contain net charge.

Exercise 15.6
Suppose the constant electric field in Example 15.6 points in the positive y-direction instead. Calculate the flux through the xz-plane and the surface parallel to it. What's the net electric flux through the surface of the cube?

Answers $\Phi_{xz} = -EL^2$ (bottom), $\Phi_{xz} = +EL^2$ (top). The net flux is still zero.

Gauss's Law

Consider a point charge q surrounded by a spherical surface of radius r centered on the charge, as in Figure 15.27a. The magnitude of the electric field everywhere on the surface of the sphere is

$$E = k_e \frac{q}{r^2}$$

Note that the electric field is perpendicular to the spherical surface at all points on the surface. The electric flux through the surface is therefore EA, where $A = 4\pi r^2$ is the surface area of the sphere:

$$\Phi_E = EA = k_e \frac{q}{r^2} (4\pi r^2) = 4\pi k_e q$$

It's sometimes convenient to express k_e in terms of another constant, ϵ_0, as $k_e = 1/(4\pi\epsilon_0)$. The constant ϵ_0 is called the **permittivity of free space** and has the value

$$\epsilon_0 = \frac{1}{4\pi k_e} = 8.85 \times 10^{-12} \ \mathrm{C^2/N \cdot m^2} \qquad \textbf{[15.10]}$$

The use of k_e or ϵ_0 is strictly a matter of taste. The electric flux through the closed spherical surface that surrounds the charge q can now be expressed as

$$\Phi_E = 4\pi k_e q = \frac{q}{\epsilon_0}$$

This result says that the electric flux through a sphere that surrounds a charge q is equal to the charge divided by the constant ϵ_0. Using calculus, this result can be proven for *any* closed surface that surrounds the charge q. For example, if the surface surrounding q is irregular, as in Figure 15.27b, the flux through that surface is also q/ϵ_0. This leads to the following general result, known as Gauss's Law:

> The electric flux Φ_E through any closed surface is equal to the net charge inside the surface, Q_{inside}, divided by ϵ_0:
>
> $$\Phi_E = \frac{Q_{\text{inside}}}{\epsilon_0} \qquad \textbf{[15.11]}$$

Though it's not obvious, Gauss's law describes how charges create electric fields. In principle, it can always be used to calculate the electric field of a system of charges or a continuous distribution of charge. In practice, the technique is useful only in a limited number of cases in which there is a high degree of symmetry, such as spheres, cylinders, or planes. With the symmetry of these special shapes, the charges can be surrounded by an imaginary surface, called a Gaussian surface. This imaginary surface is used strictly for mathematical calculation, and need not be an actual, physical surface. If the imaginary surface is chosen so that the

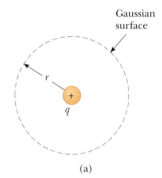

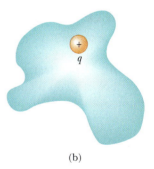

Figure 15.27   (a) The flux through a spherical surface of radius r surrounding a point charge q is $\Phi_E = q/\epsilon_0$. (b) The flux through any arbitrary surface surrounding the charge is also equal to q/ϵ_0.

◄ Gauss's Law

![pencil] **TIP 15.2 Gaussian Surfaces Aren't Real**

A Gaussian surface is an imaginary surface, created solely to facilitate a mathematical calculation. It doesn't necessarily coincide with the surface of a physical object.

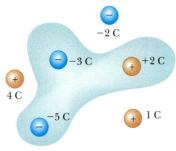

Physics⊗Now™
Log into PhysicsNow at
www.cp7e.com and go to Active
Figure 15.28, where you can change
the size and shape of a closed surface
and see the effect of surrounding
combinations of charge on the electric flux going through that surface.

electric field is constant everywhere on it, then the electric field can be computed
with

$$EA = \Phi_E = \frac{Q_{inside}}{\epsilon_0}$$ [15.12]

as will be seen in the examples. Though Gauss's law in this form can be used to obtain the electric field only for problems with a lot of symmetry, it can *always* be
used to obtain the *average* electric field on *any* surface.

Quick Quiz 15.9

Find the electric flux through the surface in Active Figure 15.28. (a) $-(3\,C)/\epsilon_0$
(b) $(3\,C)/\epsilon_0$ (c) 0   (d) $-(6\,C)/\epsilon_0$

Quick Quiz 15.10

For a closed surface through which the net flux is zero, each of the following four
statements *could* be true. Which of the statements *must* be true? (There may be
more than one.) (a) There are no charges inside the surface. (b) The net charge
inside the surface is zero. (c) The electric field is zero everywhere on the surface.
(d) The number of electric field lines entering the surface equals the number
leaving the surface.

EXAMPLE 15.7   The Electric Field of a Charged Spherical Shell

Goal   Use Gauss's law to determine electric fields when the symmetry is spherical.

Problem   A spherical conducting shell of inner radius a and outer radius b carries a total charge $+Q$ distributed on
the surface of a conducting shell (Fig. 15.29a). The quantity Q is taken to be positive. **(a)** Find the electric field in the
interior of the conducting shell, for $r < a$, and **(b)** the electric field outside the shell, for $r > b$. **(c)** If an additional
charge of $-Q$ is placed at the center, find the electric field for $r > b$.

Strategy   For each part, draw a spherical Gaussian surface in the region of interest. Add up the charge inside the
Gaussian surface, substitute it and the area into Gauss's law, and solve for the electric field.

Figure 15.29   (Example 15.7)
(a) The electric field inside a uniformly charged spherical shell is *zero*.
It is also zero for the conducting material in the region $a < r < b$. The
field outside is the same as that of a
point charge having a total charge Q
located at the center of the shell.
(b) The construction of a Gaussian
surface for calculating the electric
field *inside* a spherical shell. (c) The
construction of a Gaussian surface
for calculating the electric field
outside a spherical shell.

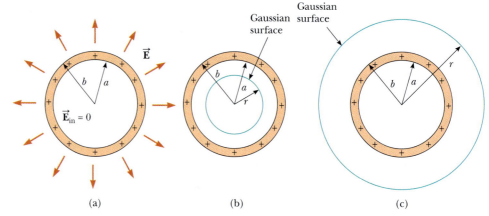

Solution

(a) Find the electric field for $r < a$.

Apply Gauss's law, Equation 15.12, to the Gaussian surface illustrated in Figure 15.29b (note that there isn't
any charge inside this surface):

$$EA = E(4\pi r^2) = \frac{Q_{inside}}{\epsilon_0} = 0 \rightarrow \boxed{E = 0}$$

(b) Find the electric field for $r > b$.

Apply Gauss's law, Equation 15.12, to the Gaussian surface illustrated in Figure 15.29c:

$$EA = E(4\pi r^2) = \frac{Q_{inside}}{\epsilon_0} = \frac{Q}{\epsilon_0}$$

Divide by the area:

$$E = \frac{Q}{4\pi\epsilon_0 r^2}$$

(c) Now an additional charge of $-Q$ is placed at the center of the sphere. Compute the new electric field outside the sphere, for $r > b$.

Apply Gauss's law as in part (b), including the new charge in Q_{inside}:

$$EA = E(4\pi r^2) = \frac{Q_{inside}}{\epsilon_0} = \frac{+Q-Q}{\epsilon_0} = 0$$

$$E = 0$$

Remarks   The important thing to notice is that in each case, the charge is spread out over a region with spherical symmetry or is located at the exact center. This is what allows the computation of a value for the electric field.

Exercise 15.7
Suppose the charge at the center is now increased to $-2Q$, while the surface of the conductor still retains a charge of $+Q$. (a) Find the electric field exterior to the sphere, for $r > b$. (b) What's the electric field inside the conductor, for $a < r < b$?

Answers   (a) $E = -Q/4\pi\epsilon_0 r^2$ (b) $E = 0$, which is always the case when charges are not moving in a conductor.

In Example 15.7, not much was said about the distribution of charge on the conductor. Whenever there is a net nonzero charge, the individual charges will try to get as far away from each other as possible. Hence, charge will reside either on the inside surface or on the outside surface. Because the electric field in the conductor is zero, there will always be enough charge on the inner surface to cancel whatever charge is at the center. In part (b), there is no charge on the inner surface and a charge of $+Q$ on the outer surface. In part (c), with a $-Q$ charge at the center, $+Q$ is on the inner surface, 0 C on the outer surface. Finally, in the exercise, with $-2Q$ in the center, there must be $+2Q$ on the inner surface and $-Q$ on the outer surface. In each case, the total charge on the conductor remains the same, $+Q$; it's just arranged differently.

Problems like Example 15.7 are often said to have "thin nonconducting shells" carrying a uniformly distributed charge. In these cases, no distinction need be made between the outer surface and inner surface of the shell. The next example makes that implicit assumption.

EXAMPLE 15.8   A Nonconducting Plane Sheet of Charge

Goal   Apply Gauss's law to a problem with plane symmetry.

Problem   Find the electric field above and below a nonconducting infinite plane sheet of charge with uniform positive charge per unit area σ (Figure 15.30a).

Strategy   By symmetry, the electric field must be perpendicular to the plane and directed away from it on either side, as shown in Figure 15.30b. For the Gaussian surface, choose a small cylinder with axis perpendicular to the plane, each end having area A_0. No electric field lines pass through the curved surface of the cylinder, only through the two ends, which have total area $2A_0$. Apply Gauss's law, using Figure 15.30b.

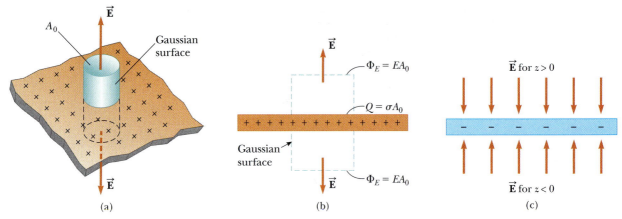

Figure 15.30 (Example 15.8) (a) A cylindrical Gaussian surface penetrating an infinite sheet of charge. (b) A cross section of the same Gaussian cylinder. The flux through each end of the Gaussian surface is EA_0. There is no flux through the cylindrical surface. (c) (Exercise 15.8).

Solution

(a) Find the electric field above and below a plane of uniform charge.

Apply Gauss's law, Equation 15.12:

$$EA = \frac{Q_{inside}}{\epsilon_0}$$

The total charge inside the Gaussian cylinder is the charge density times the cross-sectional area:

$$Q_{inside} = \sigma A_0$$

The electric flux comes entirely from the two ends, each having area A_0. Substitute $A = 2A_0$ and Q_{inside} and solve for E.

$$E = \frac{\sigma A_0}{(2A_0)\,\epsilon_0} = \boxed{\frac{\sigma}{2\epsilon_0}}$$

This is the *magnitude* of the electric field. Find the z-component of the field above and below the plane. The electric field points away from the plane, so it's positive above the plane and negative below the plane.

$$E_z = \frac{\sigma}{2\epsilon_0} \qquad z > 0$$

$$E_z = -\frac{\sigma}{2\epsilon_0} \qquad z < 0$$

Remarks Notice here that the plate was taken to be a thin nonconducting shell. If it's made of metal, of course, the electric field inside it is zero, with half the charge on the upper surface and half on the lower surface.

Exercise 15.8

Suppose an infinite non-conducting plane of charge as in Example 15.8 has a uniform negative charge density of $-\sigma$. Find the electric field above and below the plate. Sketch the field.

Answers

$$E_z = \frac{-\sigma}{2\epsilon_0} \quad z > 0; \quad E_z = \frac{\sigma}{2\epsilon_0} \quad z < 0$$

See Figure 15.30c for the sketch.

An important circuit element that will be studied extensively in the next chapter is the parallel plate capacitor. The device consists of a plate of positive charge, as in Example 15.8, with the negative plate of Exercise 15.8 placed above it. The sum of these two fields is illustrated in Figure 15.31. The result is an electric field with double the magnitude in between the two plates:

$$E = \frac{\sigma}{\epsilon_0} \qquad [15.13]$$

Outside the plates, the electric fields cancel.

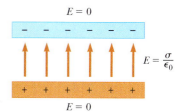

Figure 15.31 Cross section of an idealized parallel-plate capacitor. Electric field vector contributions sum together in between the plates, but cancel outside.

SUMMARY

Physics⊗Now™ Take a practice test by logging into Physics-Now at **www.cp7e.com** and clicking on the Pre-Test link for this chapter.

15.1 Properties of Electric Charges

Electric charges have the following properties:

1. Unlike charges attract one another and like charges repel one another.
2. Electric charge is always conserved.
3. Charge comes in discrete packets that are integral multiples of the basic electric charge $e = 1.6 \times 10^{-19}$ C.
4. The force between two charged particles is proportional to the inverse square of the distance between them.

15.2 Insulators and Conductors

Conductors are materials in which charges move freely in response to an electric field. All other materials are called **insulators**.

15.3 Coulomb's Law

Coulomb's law states that the electric force between two stationary charged particles separated by a distance r has the magnitude

$$F = k_e \frac{|q_1||q_2|}{r^2} \qquad [15.1]$$

where $|q_1|$ and $|q_2|$ are the magnitudes of the charges on the particles in coulombs and

$$k_e \approx 8.99 \times 10^9 \, \text{N} \cdot \text{m}^2/\text{C}^2 \qquad [15.2]$$

is the **Coulomb constant**.

15.4 The Electric Field

An electric field $\vec{E}$ exists at some point in space if a small test charge q_0 placed at that point is acted upon by an electric force $\vec{F}$. The electric field is defined as

$$\vec{E} \equiv \frac{\vec{F}}{q_0} \qquad [15.4]$$

The **direction** of the electric field at a point in space is defined to be the direction of the electric force that would be exerted on a small positive charge placed at that point.

The magnitude of the electric field due to a *point charge* q at a distance r from the point charge is

$$E = k_e \frac{|q|}{r^2} \qquad [15.6]$$

15.5 Electric Field Lines

Electric field lines are useful for visualizing the electric field in any region of space. The electric field vector $\vec{E}$ is tangent to the electric field lines at every point. Furthermore, the number of electric field lines per unit area through a surface perpendicular to the lines is proportional to the strength of the electric field at that surface.

15.6 Conductors in Electrostatic Equilibrium

A **conductor in electrostatic equilibrium** has the following properties:

1. The electric field is zero everywhere inside the conducting material.
2. Any excess charge on an isolated conductor must reside entirely on its surface.
3. The electric field just outside a charged conductor is perpendicular to the conductor's surface.
4. On an irregularly shaped conductor, charge accumulates where the radius of curvature of the surface is smallest, at sharp points.

15.9 Electric Flux and Gauss's Law

Gauss's law states that the electric flux through any closed surface is equal to the net charge Q inside the surface divided by the permittivity of free space, ϵ_0:

$$EA = \Phi_E = \frac{Q_{\text{inside}}}{\epsilon_0} \qquad [15.12]$$

For highly symmetric distributions of charge, Gauss's law can be used to calculate electric fields.

CONCEPTUAL QUESTIONS

1. A glass object is charged to $+3$ nC by rubbing it with a silk cloth. In the rubbing process, have protons been added to the object or have electrons been removed from it?

2. Why must hospital personnel wear special conducting shoes while working around oxygen in an operating room. What might happen if the personnel wore shoes with rubber soles?

3. Two insulated rods are oppositely charged on their ends. They are mounted at the centers so that they are free to rotate, and then held in the position shown in Figure Q15.3 in a view from above. The rods rotate in the plane of the paper. Will the rods stay in those positions when released? If not, into what position(s) will they move? Will the final configuration(s) be stable?

Figure Q15.3

4. Explain from an atomic viewpoint why charge is usually transferred by electrons.

5. Explain how a positively charged object can be used to leave another metallic object with a net negative charge. Discuss the motion of charges during the process.

6. If a suspended object A is attracted to a charged object B, can we conclude that A is charged? Explain.

7. If a metal object receives a positive charge, does its mass increase, decrease, or stay the same? What happens to its mass if the object receives a negative charge?

8. When defining the electric field, why is it necessary to specify that the magnitude of the test charge be very small?

9. In fair weather, there is an electric field at the surface of the Earth, pointing down into the ground. What is the electric charge on the ground in this situation?

10. A student stands on a thick piece of insulating material, places her hand on top of a Van de Graaff generator, and then turns on the generator. Does she receive a shock?

11. An uncharged, metallic-coated Styrofoam ball is suspended in the region between two vertical metal plates. If the two plates are charged, one positively and one negatively, describe the motion of the ball after it is brought into contact with one of the plates.

12. Is it possible for an electric field to exist in empty space? Explain.

13. There are great similarities between electric and gravitational fields. A room can be electrically shielded so that there are no electric fields in the room by surrounding it with a conductor. Can a room be gravitationally shielded?

Why or why not? [*Hint:* There are two kinds of charge in nature, but only one kind of mass.]

14. Would life be different if the electron were positively charged and the proton were negatively charged? Does the choice of signs have any bearing on physical and chemical interactions? Explain.

15. Explain why Gauss's law cannot be used to calculate the electric field of (a) a polar molecule consisting of a positive and a negative charge separated by a very small distance, (b) a charged disk, and (c) three point charges at the corner of a triangle.

16. Why should a ground wire be connected to the metal support rod for a television antenna?

17. A balloon negatively charged by rubbing clings to a wall. Does this mean that the wall is positively charged? Why does the balloon eventually fall?

18. A spherical surface surrounds a point charge q. Describe what happens to the total flux through the surface if (a) the charge is tripled, (b) the volume of the sphere is doubled, (c) the surface is changed to a cube, (d) the charge is moved to another location inside the surface, and (e) the charge is moved outside the surface.

19. A charged comb often attracts small bits of dry paper that then fly away when they touch the comb. Explain.

20. Answer the given questions with one of the statements that follow and defend your answer. The answer to each part is either $+Q$, 0, or $-Q$. A spherical conducting object A with a charge of $+Q$ is lowered through a hole into a metal container B that is initially uncharged.
    (a) When A is at the center of B, but not touching it, the charge on the inner surface of B is _____.
    (b) The charge on the outer surface of B is _____.
    (c) Object A is now allowed to touch the inner surface of B. The charge on A is now _____.
    (d) The charge on the inner surface of B is now _____.
    (e) The charge on the outer surface of B is now _____.

21. A positively charged ball hanging from a nonconducting string is brought near a nonconducting object. Based on the behavior of the ball–string combination, the ball is seen to be attracted to the object. From this experiment, it is not possible to determine whether the object is negatively charged or neutral. Why not? What additional experiment would help you decide between these two possibilities?

22. An electron moving horizontally passes between two horizontal plates, the upper charged negatively, the lower positively. A uniform, upward-directed electric field exists in the region between the plates, and this field exerts an electric force downward on the electron. Describe the movement of the electron in this region.

PROBLEMS

1, **2**, 3 = straightforward, intermediate, challenging   ☐ = full solution available in *Student Solutions Manual/Study Guide*

Physics⊗Now™ = coached solution with hints available at **www.cp7e.com**   = biomedical application

Section 15.3 Coulomb's Law

1. A charge of 4.5×10^{-9} C is located 3.2 m from a charge of -2.8×10^{-9} C. Find the electrostatic force exerted by one charge on the other.

2. The Moon and Earth are bound together by gravity. If, instead, the force of attraction were the result of each having a charge of the same magnitude but opposite in

sign, find the quantity of charge that would have to be placed on each to produce the required force.

3. An alpha particle (charge $= +2.0e$) is sent at high speed toward a gold nucleus (charge $= +79e$). What is the electrical force acting on the alpha particle when it is 2.0×10^{-14} m from the gold nucleus?

4. Four point charges are situated at the corners of a square with sides of length a, as in Figure P15.4. Find the expression for the resultant force on the positive charge q.

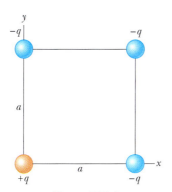

Figure P15.4

5. **Physics⊗Now™** The nucleus of ^{8}Be, which consists of 4 protons and 4 neutrons, is very unstable and spontaneously breaks into two alpha particles (helium nuclei, each consisting of 2 protons and 2 neutrons). (a) What is the force between the two alpha particles when they are 5.00×10^{-15} m apart, and (b) what will be the magnitude of the acceleration of the alpha particles due to this force? Note that the mass of an alpha particle is 4.0026 u.

6. A molecule of DNA (deoxyribonucleic acid) is 2.17 μm long. The ends of the molecule become singly ionized — negative on one end, positive on the other. The helical molecule acts like a spring and compresses 1.00% upon becoming charged. Determine the effective spring constant of the molecule.

7. Suppose that 1.00 g of hydrogen is separated into electrons and protons. Suppose also that the protons are placed at the Earth's North Pole and the electrons are placed at the South Pole. What is the resulting compressional force on the Earth?

8. An electron is released a short distance above the surface of the Earth. A second electron directly below it exerts an electrostatic force on the first electron just great enough to cancel the gravitational force on it. How far below the first electron is the second?

9. Two identical conducting spheres are placed with their centers 0.30 m apart. One is given a charge of 12×10^{-9} C, the other a charge of -18×10^{-9} C. (a) Find the electrostatic force exerted on one sphere by the other. (b) The spheres are connected by a conducting wire. Find the electrostatic force between the two after equilibrium is reached.

10. Calculate the magnitude and direction of the Coulomb force on each of the three charges shown in Figure P15.10.

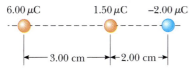

Figure P15.10   (Problems 10 and 18)

11. Three charges are arranged as shown in Figure P15.11. Find the magnitude and direction of the electrostatic force on the charge at the origin.

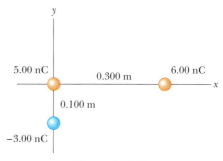

Figure P15.11

12. Three charges are arranged as shown in Figure P15.12. Find the magnitude and direction of the electrostatic force on the 6.00-nC charge.

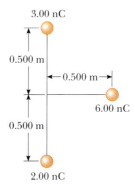

Figure P15.12

13. Three point charges are located at the corners of an equilateral triangle as in Figure P15.13. Calculate the net electric force on the 7.00-μC charge.

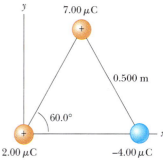

Figure P15.13

14. Two small beads having positive charges $3q$ and q are fixed at the opposite ends of a horizontal insulating rod, extending from the origin to the point $x = d$. As shown in Figure P15.14, a third small charged bead is free to slide on the rod. At what position is the third bead in equilibrium? Can it be in stable equilibrium?

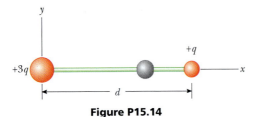

Figure P15.14

15. Two small metallic spheres, each of mass 0.20 g, are suspended as pendulums by light strings from a common point as shown in Figure P15.15. The spheres are given the same electric charge, and it is found that they come to equilibrium when each string is at an angle of 5.0° with the vertical. If each string is 30.0 cm long, what is the magnitude of the charge on each sphere?

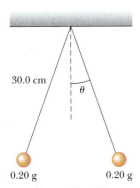

Figure P15.15

16. A charge of 6.00×10^{-9} C and a charge of -3.00×10^{-9} C are separated by a distance of 60.0 cm. Find the position at which a third charge, of 12.0×10^{-9} C, can be placed so that the net electrostatic force on it is zero.

Section 15.4 The Electric Field

17. An object with a net charge of 24 μC is placed in a uniform electric field of 610 N/C, directed vertically. What is the mass of the object if it "floats" in the electric field?

18. (a) Determine the electric field strength at a point 1.00 cm to the left of the middle charge shown in Figure P15.10. (b) If a charge of -2.00 μC is placed at this point, what are the magnitude and direction of the force on it?

19. **Physics⊗Now™** An airplane is flying through a thundercloud at a height of 2 000 m. (This is a very dangerous thing to do because of updrafts, turbulence, and the possibility of electric discharge.) If there are charge concentrations of $+40.0$ C at a height of 3 000 m within the cloud and -40.0 C at a height of 1 000 m, what is the electric field $\vec{E}$ at the aircraft?

20. An electron is accelerated by a constant electric field of magnitude 300 N/C. (a) Find the acceleration of the electron. (b) Use the equations of motion with constant acceleration to find the electron's speed after 1.00×10^{-8} s, assuming it starts from rest.

21. A Styrofoam® ball covered with a conducting paint has a mass of 5.0×10^{-3} kg and has a charge of 4.0 μC. What electric field directed upward will produce an electric force on the ball that will balance its weight?

22. Each of the protons in a particle beam has a kinetic energy of 3.25×10^{-15} J. What are the magnitude and direction of the electric field that will stop these protons in a distance of 1.25 m?

23. A proton accelerates from rest in a uniform electric field of 640 N/C. At some later time, its speed is 1.20×10^6 m/s. (a) Find the magnitude of the acceleration of the proton. (b) How long does it take the proton to reach this speed? (c) How far has it moved in that interval? (d) What is its kinetic energy at the later time?

24. Three charges are at the corners of an equilateral triangle, as shown in Figure P15.24. Calculate the electric field at a point midway between the two charges on the x-axis.

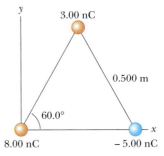

Figure P15.24

25. Three identical charges ($q = -5.0$ μC) lie along a circle of radius 2.0 m at angles of 30°, 150°, and 270°, as shown in Figure P15.25. What is the resultant electric field at the center of the circle?

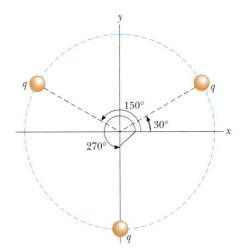

Figure P15.25

26. Two point charges lie along the y-axis. A charge of $q_1 = -9.0$ μC is at $y = 6.0$ m, and a charge of $q_2 = -8.0$ μC is at $y = -4.0$ m. Locate the point (other than infinity) at which the total electric field is zero.

27. In Figure P15.27, determine the point (other than infinity) at which the total electric field is zero.

Figure P15.27

Section 15.5 Electric Field Lines
Section 15.6 Conductors in Electrostatic Equilibrium

28. Figure P15.28 shows the electric field lines for two point charges separated by a small distance. (a) Determine the ratio q_1/q_2. (b) What are the signs of q_1 and q_2?

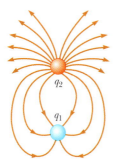

Figure P15.28

29. (a) Sketch the electric field lines around an isolated point charge $q > 0$. (b) Sketch the electric field pattern around an isolated negative point charge of magnitude $-2q$.

30. (a) Sketch the electric field pattern around two positive point charges of magnitude $1\,\mu C$ placed close together. (b) Sketch the electric field pattern around two negative point charges of $-2\,\mu C$, placed close together. (c) Sketch the pattern around two point charges of $+1\,\mu C$ and $-2\,\mu C$, placed close together.

31. Two point charges are a small distance apart. (a) Sketch the electric field lines for the two if one has a charge four times that of the other and both charges are positive. (b) Repeat for the case in which both charges are negative.

32. (a) Sketch the electric field pattern set up by a positively charged hollow sphere. Include regions inside and regions outside the sphere. (b) A conducting cube is given a positive charge. Sketch the electric field pattern both inside and outside the cube.

33. Refer to Figure 15.20. The charge lowered into the center of the hollow conductor has a magnitude of $5\,\mu C$. Find the magnitude and sign of the charge on the inside and outside of the hollow conductor when the charge is as shown in (a) Figure 15.20a; and (b) Figure 15.20b; (c) Figure 15.20c; and (d) Figure 15.20d.

Section 15.8 The Van de Graaff Generator

34. The dome of a Van de Graaff generator receives a charge of 2.0×10^{-4} C. Find the strength of the electric field (a) inside the dome; (b) at the surface of the dome, assuming it has a radius of 1.0 m; and (c) 4.0 m from the center of the dome. [*Hint:* See Section 15.6 to review properties of conductors in electrostatic equilibrium. Also, use the fact that the points on the surface are outside a spherically symmetric charge distribution; the total charge may be considered to be located at the center of the sphere.]

35. **Physics⊗Now™** If the electric field strength in air exceeds 3.0×10^6 N/C, the air becomes a conductor. Using this fact, determine the maximum amount of charge that can be carried by a metal sphere 2.0 m in radius. (See the hint in Problem 34.)

36. In the Millikan oil drop experiment, an atomizer (a sprayer with a fine nozzle) is used to introduce many tiny droplets of oil between two oppositely charged parallel metal plates. Some of the droplets pick up one or more excess electrons. The charge on the plates is adjusted so that the electric force on the excess electrons exactly balances the weight of the droplet. The idea is to look for a droplet that has the smallest electric force and assume that it has only one excess electron. This strategy lets the observer measure the charge on the electron. Suppose we are using an electric field of 3×10^4 N/C. The charge on one electron is about 1.6×10^{-19} C. Estimate the radius of an oil drop of density $858\,\text{kg/m}^3$ for which its weight could be balanced by the electric force of this field on one electron. (Problem 36 is courtesy of E.F. Redish. For more problems of this type, visit http://www.physics.umd.edu/perg/.)

37. A Van de Graaff generator is charged so that the electric field at its surface is 3.0×10^4 N/C. Find (a) the electric force exerted on a proton released at its surface and (b) the acceleration of the proton at that instant of time.

Section 15.9 Electric Flux and Gauss's Law

38. A flat surface having an area of $3.2\,\text{m}^2$ is rotated in a uniform electric field of magnitude $E = 6.2 \times 10^5$ N/C. Determine the electric flux through this area (a) when the electric field is perpendicular to the surface and (b) when the electric field is parallel to the surface.

39. An electric field of intensity 3.50 kN/C is applied along the x-axis. Calculate the electric flux through a rectangular plane 0.350 m wide and 0.700 m long if (a) the plane is parallel to the yz-plane; (b) the plane is parallel to the xy-plane; and (c) the plane contains the y-axis, and its normal makes an angle of $40.0°$ with the x-axis.

40. The electric field everywhere on the surface of a thin spherical shell of radius 0.750 m is measured to be equal to 890 N/C and points radially toward the center of the sphere. (a) What is the net charge within the sphere's surface? (b) What can you conclude about the nature and distribution of the charge inside the spherical shell?

41. A 40-cm-diameter loop is rotated in a uniform electric field until the position of maximum electric flux is found. The flux in that position is measured to be $5.2 \times 10^5\,\text{N}\cdot\text{m}^2/\text{C}$. Calculate the electric field strength in this region.

42. A point charge of $+5.00\,\mu C$ is located at the center of a sphere with a radius of 12.0 cm. Determine the electric flux through the surface of the sphere.

43. A point charge q is located at the center of a spherical shell of radius a that has a charge $-q$ uniformly distributed on its surface. Find the electric field (a) for all points outside the spherical shell and (b) for a point inside the shell a distance r from the center.

44. Use Gauss's law and the fact that the electric field inside any closed conductor in electrostatic equilibrium is zero to show that any excess charge placed on the conductor must reside on its surface.

45. An infinite plane conductor has charge spread out on its surface as shown in Figure P15.45. Use Gauss's law to show that the electric field at any point outside the conductor is given by $E = \sigma/\epsilon_0$, where σ is the charge per unit area on the conductor. [*Hint:* Choose a Gaussian surface in the shape of a cylinder with one end inside the conductor and one end outside the conductor.]

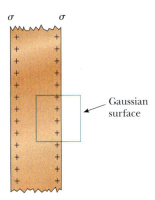

Figure P15.45

46. Show that the electric field just outside the surface of a good conductor of any shape is given by $E = \sigma/\epsilon_0$, where σ is the charge per unit area on the conductor. [*Hint:* The electric field just outside the surface of a charged conductor is perpendicular to its surface.]

ADDITIONAL PROBLEMS

47. Two protons in an atomic nucleus are typically separated by a distance of 2×10^{-15} m. The electric repulsion force between the protons is huge, but the attractive nuclear force is even stronger and keeps the nucleus from bursting apart. What is the magnitude of the electrical force between two protons separated by 2.00×10^{-15} m?

48. In the Bohr theory of the hydrogen atom, an electron moves in a circular orbit about a proton. The radius of the orbit is 0.53×10^{-10} m. (a) Find the electrostatic force acting on each particle. (b) If this force causes the centripetal acceleration of the electron, what is the speed of the electron?

49. Three point charges are aligned along the x-axis as shown in Figure P15.49. Find the electric field at the position $x = +2.0$ m, $y = 0$.

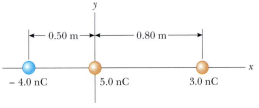

Figure P15.49

50. A small 2.00-g plastic ball is suspended by a 20.0-cm-long string in a uniform electric field, as shown in Figure P15.50. If the ball is in equilibrium when the string makes a 15.0° angle with the vertical as indicated, what is the net charge on the ball?

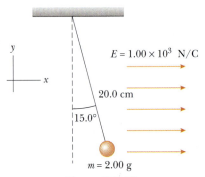

Figure P15.50

51. (a) Two identical point charges $+q$ are located on the y-axis at $y = +a$ and $y = -a$. What is the electric field along the x-axis at $x = b$? (b) A circular ring of charge of radius a has a total positive charge Q distributed uniformly around it. The ring is in the $x = 0$ plane with its center at the origin. What is the electric field along the x-axis at $x = b$ due to the ring of charge? [*Hint:* Consider the charge Q to consist of many pairs of identical point charges positioned at the ends of diameters of the ring.]

52. A positively charged bead having a mass of 1.00 g falls from rest in a vacuum from a height of 5.00 m in a uniform vertical electric field with a magnitude of 1.00×10^4 N/C. The bead hits the ground at a speed of 21.0 m/s. Determine (a) the direction of the electric field (upward or downward), and (b) the charge on the bead.

53. A solid conducting sphere of radius 2.00 cm has a charge of 8.00 μC. A conducting spherical shell of inner radius 4.00 cm and outer radius 5.00 cm is concentric with the solid sphere and has a charge of -4.00 μC. Find the electric field at (a) $r = 1.00$ cm, (b) $r = 3.00$ cm, (c) $r = 4.50$ cm, and (d) $r = 7.00$ cm from the center of this charge configuration.

54. Two small silver spheres, each with a mass of 100 g, are separated by 1.00 m. Calculate the fraction of the electrons in one sphere that must be transferred to the other in order to produce an attractive force of 1.00×10^4 N (about a ton) between the spheres. (The number of electrons per atom of silver is 47, and the number of atoms per gram is Avogadro's number divided by the molar mass of silver, 107.87 g/mol.)

55. A vertical electric field of magnitude 2.00×10^4 N/C exists above the Earth's surface on a day when a thunderstorm is brewing. A car with a rectangular size of 6.00 m by 3.00 m is traveling along a roadway sloping downward at 10.0°. Determine the electric flux through the bottom of the car.

56. A 2.00-μC charged 1.00-g cork ball is suspended vertically on a 0.500-m-long light string in the presence of a uniform downward-directed electric field of magnitude $E = 1.00 \times 10^5$ N/C. If the ball is displaced slightly from the vertical, it oscillates like a simple pendulum. (a) Determine the period of the ball's oscillation. (b) Should gravity be included in the calculation for part (a)? Explain.

57. **Physics⊗Now**™ Two 2.0-g spheres are suspended by 10.0-cm-long light strings (Fig. P15.57). A uniform electric field is applied in the x-direction. If the spheres have charges of -5.0×10^{-8} C and $+5.0 \times 10^{-8}$ C, determine the electric field intensity that enables the spheres to be in equilibrium at $\theta = 10°$.

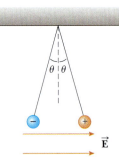

Figure P15.57

58. Two point charges like those in Figure P15.58 are called an electric dipole. Show that the electric field at a distant point along the x-axis is given by $E_x = 4k_e qa/x^3$.

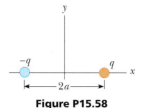

Figure P15.58

59. A charged cork ball of mass 1.00 g is suspended on a light string in the presence of a uniform electric field as in Figure P15.59. When the electric field has an x-component of 3.00×10^5 N/C and a y-component of 5.00×10^5 N/C, the ball is in equilibrium at $\theta = 37.0°$. Find (a) the charge on the ball and (b) the tension in the string.

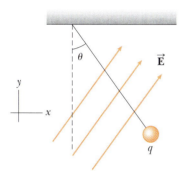

Figure P15.59

60. A point charge of magnitude 5.00 μC is at the origin of a coordinate system, and a charge of -4.00 μC is at the point $x = 1.00$ m. There is a point on the x-axis, at x less than infinity, where the electric field goes to zero. (a) Show by conceptual arguments that this point cannot be located between the charges. (b) Show by conceptual arguments that the point cannot be at any location between $x = 0$ and negative infinity. (c) Show by conceptual arguments that the point must be between $x = 1.00$ m and $x =$ positive infinity. (d) Use the values given to find the point and show that it is consistent with your conceptual argument.

61. Two hard rubber spheres of mass 15 g are rubbed vigorously with fur on a dry day and are then suspended from a rod with two insulating strings of length 5.0 cm. They are observed to hang at equilibrium as shown in Figure P15.61, each at an angle of 10° with the vertical. Estimate the amount of charge that is found on each sphere. (Problem 61 is courtesy of E.F. Redish. For more problems of this type, visit http://www.physics.umd.edu/perg/.)

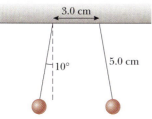

Figure P15.61

62. Three identical point charges, each of mass $m = 0.100$ kg, hang from three strings, as shown in Figure P15.62. If the lengths of the left and right strings are each $L = 30.0$ cm, and if the angle θ is 45.0°, determine the value of q.

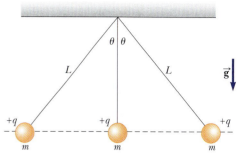

Figure P15.62

63. Each of the electrons in a particle beam has a kinetic energy of 1.60×10^{-17} J. (a) What is the magnitude of the uniform electric field (pointing in the direction of the electrons' movement) that will stop these electrons in a distance of 10.0 cm? (b) How long will it take to stop the electrons? (c) After the electrons stop, what will they do? Explain.

64. Protons are projected with an initial speed $v_0 = 9\,550$ m/s into a region where a uniform electric field $E = 720$ N/C is present (Fig. P15.64). The protons are to hit a target that lies a horizontal distance of 1.27 mm from the point where the protons are launched. Find (a) the two projection angles θ that will result in a hit and (b) the total duration of flight for each of the two trajectories.

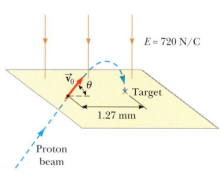

Figure P15.64

ACTIVITIES

1. The following are a number of experiments that you can perform to investigate static electricity.

   (a) Attach two inflated balloons to the ends of a light string having a length of about 2 m. Tape the center of the string to the top of an open doorway as in Figure A15.1. Note that the balloons touch each other as they hang freely. Now rub each balloon several times with a wool cloth, and let them hang freely once again. Why are the balloons no longer touching each other, but are now separated?

   (b) Rub an inflated balloon with a piece of wool and press it against a wall. Note that the balloon adheres to the wall. Why?

(c) Rub nylon hose with a plastic dry-cleaner bag and observe how the hose expands like a balloon. Why does it do so?

(d) Tear some paper into very small pieces. Comb your hair and then bring the comb close to the pieces of paper. Notice that they accelerate toward the comb. How does the magnitude of the electric force compare with the magnitude of the gravitational force exerted on the paper? Keep watching, and you might see a few pieces jump away from the comb. They do not fall away; they are clearly repelled. What causes this repulsion?

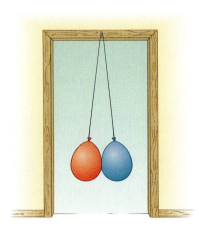

Figure A15.1

2. For this experiment, you will need two 20-cm strips of transparent tape. (The mass of each is about 65 mg.) Fold about 1 cm of tape over at one end of each strip to create a handle. Press both pieces of tape side by side onto a tabletop, rubbing your finger back and forth across the strips. Quickly pull the strips off the surface so that they become charged. Hold the tape handles together, and the strips will repel each other, forming an inverted "V" shape. Measure the angle between the pieces and estimate the excess charge on each strip. Assume that the charges act as if they were located at the center of mass of each strip.

3. Rub an inflated balloon with a piece of wool cloth and place the balloon near a fine stream of water falling from a faucet. The stream of water will deflect toward the balloon. Why? Vary the distance between the balloon and the stream, and observe the displacement of the water stream for different distances. What is the relationship between the displacement of the stream and the distance of separation?

4. If you have access to a Van de Graaff generator, here are a few interesting things to try: (a) Stack several aluminum pie plates on top of the generator and then turn the generator on. What do you think is going to happen? Why? Try it. (b) Tape one pie plate on top of the generator and pour in some puffed rice or pieces of paper. What do you think will happen when you turn the generator on? Why? Try it and see. (c) Tape a glass beaker to the top of the generator and pour in some puffed rice. What do you think will happen when you turn the generator on? Try it. (d) Bring a fluorescent bulb close to a charged generator and observe what happens. Why does it happen?

(Courtesy of Resonance Research Corporation)

Everything in the foreground of this picture is at the same electrical potential of many kilovolts. With no differences in potential, no charge is moving and no one gets a shock.

CHAPTER

16

Electrical Energy and Capacitance

The concept of potential energy was first introduced in Chapter 5 in connection with the conservative forces of gravity and springs. By using the principle of conservation of energy, we were often able to avoid working directly with forces when solving problems. Here we learn that the potential energy concept is also useful in the study of electricity. Because the Coulomb force is conservative, we can define an electric potential energy corresponding to that force. In addition, we define an electric potential—the potential energy per unit charge—corresponding to the electric field.

With the concept of electric potential in hand, we can begin to understand electric circuits, starting with an investigation of a common circuit element called a capacitor. These simple devices store electrical energy and have found uses virtually everywhere, from etched circuits on a microchip to the creation of enormous bursts of power in fusion experiments.

16.1 POTENTIAL DIFFERENCE AND ELECTRIC POTENTIAL

Electric potential energy and electric potential are closely related concepts. The electric potential turns out to be just the electric potential energy per unit charge. This is similar to the relationship between electric force and the electric field, which is the electric force per unit charge.

Work and Electric Potential Energy

Recall from Chapter 5 that the work done by a conservative force $\vec{F}$ on an object depends only on the initial and final positions of the object and not on the path taken between those two points. This, in turn, means that a potential energy

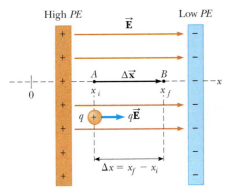

Figure 16.1 When a charge q moves in a uniform electric field $\vec{E}$ from point A to point B, the work done on the charge by the electric force is $qE_x\Delta x$.

function PE exists. As we have seen, potential energy is a scalar quantity with change equal to the negative of the work done by the conservative force: $\Delta PE = PE_f - PE_i = - W_F$.

Both the Coulomb force law and the universal law of gravity are proportional to $1/r^2$. Because they have the same mathematical form, and because the gravity force is conservative, it follows that **the Coulomb force is also conservative**. As with gravity, an electrical potential energy function can be associated with this force.

To make these ideas more quantitative, imagine a small positive charge placed at point A in a *uniform* electric field $\vec{E}$, as in Figure 16.1. For simplicity, we first consider only constant electric fields and charges which move parallel to that field in one dimension (taken to be the x-axis). The electric field between equally and oppositely charged parallel plates is an example of a field that is approximately constant. (See Chapter 15.) As the charge moves from point A to point B under the influence of the electric field $\vec{E}$, the work done on the charge by the electric field is equal to the part of the electric force $q\vec{E}$ acting parallel to the displacement, times the displacement $\Delta x = x_f - x_i$:

$$W_{AB} = F_x\Delta x = qE_x(x_f - x_i)$$

In this expression, q is the charge and E_x is the vector component of $\vec{E}$ in the x-direction (*not* the magnitude of $\vec{E}$). Unlike the magnitude of $\vec{E}$, the component E_x can be positive or negative depending on the direction of $\vec{E}$, though in Figure 16.1 E_x is positive. Finally, note that the displacement, like q and E_x, can also be either positive or negative, depending on the direction of the displacement.

The preceding expression for the work done by an electric field on a charge moving in one dimension is valid for both positive and negative charges and for constant electric fields pointing in *any* direction. When numbers are substituted with correct signs, the overall correct sign automatically results. In some books the expression $W = qEd$ is used, instead, where E is the magnitude of the electric field and d the distance the particle travels. The weakness of this formulation is that it doesn't allow, mathematically, for negative electric work on positive charges, nor for positive electric work on negative charges! Nonetheless, the expression is easy to remember and useful for finding magnitudes: the magnitude of the work done by a constant electric field on a charge moving parallel to the field is always given by $|W| = |q|Ed$.

We can substitute our definition of electric work into the work-energy theorem (assume other forces are absent):

$$W = qE_x\Delta x = \Delta KE$$

The electric force is conservative, so the electric work depends only on the end-points of the path, A and B, not on the path taken. Therefore, as the charge accelerates to the right in Figure 16.1, it gains kinetic energy, and loses an equal amount of potential energy. Recall from Chapter 5 that **the work done by a conservative force can be reinterpreted as the negative of the change in a potential energy associated with that force.** This motivates the definition of the change in electric potential energy:

Change in electric potential energy ▶

The change in the electric potential energy, ΔPE, of a system consisting of an object of charge q moving through a displacement Δx in a constant electric field $\vec{E}$ is given by

$$\Delta PE = - W_{AB} = - qE_x\Delta x \qquad \text{[16.1]}$$

where E_x is the x-component of the electric field and $\Delta x = x_f - x_i$ is the displacement of the charge along the x-axis.

SI Unit: joule (J)

Although potential energy can be defined for any electric field, **Equation 16.1 is valid only for the case of a uniform (i.e., constant) electric field, for a particle that**

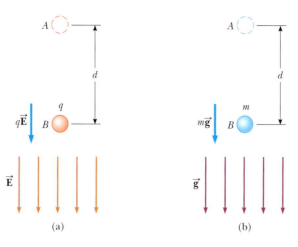

Figure 16.2   (a) When the electric field $\vec{\mathbf{E}}$ is directed downward, point B is at a lower electric potential than point A. As a positive test charge moves from A to B, the electric potential energy decreases. (b) An object of mass m moves in the direction of the gravitational field $\vec{\mathbf{g}}$, the gravitational potential energy decreases.

undergoes a displacement along a given axis (here called the x-axis). Because the electric field is conservative, the change in potential energy doesn't depend on the path. Consequently, it's unimportant whether or not the charge remains on the axis at all times during the displacement: the change in potential energy will be the same. In subsequent sections we will examine situations in which the electric field is not uniform.

Electric and gravitational potential energy can be compared in Figure 16.2. In this figure, the electric and gravitational fields are both directed downwards. We see that positive charge in an electric field acts very much like mass in a gravity field: a positive charge at point A falls in the direction of the electric field, just as a positive mass falls in the direction of the gravity field. Let point B be the zero point for potential energy in both Figure 16.2a and Figure 16.2b. From conservation of energy, in falling from point A to point B the positive charge gains kinetic energy equal in magnitude to the loss of electric potential energy:

$$\Delta KE + \Delta PE_{el} = \Delta KE + (0 - |q|Ed) \quad \rightarrow \quad \Delta KE = |q|Ed$$

The absolute value signs on q are there only to make explicit that the charge is positive in this case. Similarly, the object in Figure 16.2b gains kinetic energy equal in magnitude to the loss of gravitational potential energy:

$$\Delta KE + \Delta PE_{g} = \Delta KE + (0 - mgd) \quad \rightarrow \quad \Delta KE = mgd$$

So for positive charges, electric potential energy works very much like gravitational potential energy. In both cases, moving an object opposite the direction of the field results in a gain of potential energy, and upon release, the potential energy is converted to the object's kinetic energy.

Electric potential energy differs significantly from gravitational potential energy, however, in that there are two kinds of electrical charge—positive and negative—whereas gravity has only positive "gravitational charge" (i.e. mass). A negatively charged particle at rest at point A in Figure 16.2a would have to be *pushed* down to point B. To see this, apply the work–energy theorem to a negative charge at rest at point A and assumed to have some speed v on arriving at point B:

$$W = \Delta KE + \Delta PE_{el} = (\tfrac{1}{2}mv^2 - 0) + (0 - (-|q|Ed))$$

$$W = \tfrac{1}{2}mv^2 + |q|Ed$$

Notice that the negative charge, $-|q|$, unlike the positive charge, had a positive change in electric potential energy in moving from point A to point B. If the negative charge has any speed at point B, the kinetic energy corresponding to that speed is also positive. Because both terms on the right-hand side of the work–energy equation are positive, there is no way of getting the negative charge from point A to point B without doing positive work W on it. In fact, if the negative charge is simply released at point A, it will "fall" upwards against the direction of the field!

EXAMPLE 16.1 Potential Energy Differences in an Electric Field

Goal Illustrate the concept of electric potential energy.

Problem A proton is released from rest at $x = -2.00$ cm in a constant electric field with magnitude 1.50×10^3 N/C, pointing in the positive x-direction. **(a)** Calculate the change in the electric potential energy associated with the proton when it reaches $x = 5.00$ cm. **(b)** An electron is now fired in the same direction from the same position. What is its change in electric potential energy associated with the electron if it reaches $x = 12.0$ cm? **(c)** If the direction of the electric field is reversed and an electron is released from rest at $x = 3.00$ cm, by how much has the electric potential energy changed when the electron reaches $x = 7.00$ cm?

Strategy This problem requires a straightforward substitution of given values into the definition of electric potential energy, Equation 16.1.

Solution
(a) Calculate the change in the electric potential energy associated with the proton.

Apply Equation 16.1:

$$\Delta PE = -qE_x \Delta x = -qE_x (x_f - x_i)$$
$$= -(1.60 \times 10^{-19}\,\text{C})(1.50 \times 10^3\,\text{N/C})$$
$$x[0.050\,0\,\text{m} - (-0.020\,0\,\text{m})]$$
$$= -1.68 \times 10^{-17}\,\text{J}$$

(b) Find the change in electric potential energy associated with an electron fired from $x = -0.0200$ m and reaching $x = 0.120$ m.

Apply Equation 16.1, but in this case note that the electric charge q is negative:

$$\Delta PE = -qE_x \Delta x = -qE_x (x_f - x_i)$$
$$= -(-1.60 \times 10^{-19}\,\text{C})(1.50 \times 10^3\,\text{N/C})$$
$$x[(0.120\,\text{m} - (-0.020\,0\,\text{m})]$$
$$= +3.36 \times 10^{-17}\,\text{J}$$

(c) Find the change in potential energy associated with an electron traveling from $x = 3.00$ cm to $x = 7.00$ cm if the direction of the electric field is reversed.

Substitute, but now the electric field points in the negative x-direction, hence carries a minus sign:

$$\Delta PE = -qE_x \Delta x = -qE_x (x_f - x_i)$$
$$= -(-1.60 \times 10^{-19}\,\text{C})(-1.50 \times 10^3\,\text{N/C})$$
$$\times (0.070\,\text{m} - 0.030\,\text{m})$$
$$= -9.60 \times 10^{-18}\,\text{J}$$

Remarks Notice that the proton (actually the proton–field system) lost potential energy when it moved in the positive x-direction, while the electron gained potential energy when it moved in the same direction. Finding changes in potential energy with the field reversed was just a matter of supplying a minus sign, bringing the total number in this case to three! It's important not to drop any of the signs.

Exercise 16.1
Find the change in electric potential energy associated with the electron in part (b) as it goes on from $x = 0.120$ m to $x = -0.180$ m. (Note that the electron must turn around and go back at some point. The location of the turning point is unimportant, because changes in potential energy depend only on the end points of the path.)

Answer -7.20×10^{-17} J

INTERACTIVE EXAMPLE 16.2   Dynamics of Charged Particles

Goal   Use electric potential energy in conservation of energy problems.

Problem   (a) Find the speed of the proton at $x = 0.050\ 0$ m in part (a) of Example 16.1. (b) Find the initial speed of the electron (at $x = -2.00$ cm) in part (b) of Example 16.1, given that its speed has fallen by half when it reaches $x = 0.120$ m.

Strategy   Apply conservation of energy, solving for the unknown speeds. Part (b) involves two equations: the conservation equation, and the condition $v_f = \frac{1}{2}v_i$ for the unknown initial and final speeds. The changes in electric potential energy have already been calculated in Example 16.1.

Solution
(a) Calculate the proton's speed at $x = 0.050$ m.

Use conservation of energy, with an initial speed of zero:

$$\Delta KE + \Delta PE = 0 \quad \rightarrow \quad (\tfrac{1}{2}mv^2 - 0) + \Delta PE = 0$$

Solve for v, and substitute the change in potential energy found in Example 16.1a:

$$v^2 = -\frac{2}{m}\Delta PE$$

$$v = \sqrt{-\frac{2}{m}\Delta PE}$$

$$= \sqrt{-\frac{2}{(1.67 \times 10^{-27}\ \text{kg})}(-1.68 \times 10^{-17}\ \text{J})}$$

$$v = \boxed{1.42 \times 10^5\ \text{m/s}}$$

(b) Find the electron's initial speed, given that its speed has fallen by half at $x = 0.120$ m.

Apply conservation of energy once again, substituting expressions for the initial and final kinetic energies:

$$\Delta KE + \Delta PE = 0$$

$$(\tfrac{1}{2}mv_f^2 - \tfrac{1}{2}mv_i^2) + \Delta PE = 0$$

Substitute the condition $v_f = \frac{1}{2}v_i$, and subtract the change in potential energy from both sides:

$$\tfrac{1}{2}m(\tfrac{1}{2}v_i)^2 - \tfrac{1}{2}mv_i^2 = -\Delta PE$$

Combine terms and solve for v_i, the initial speed, and substitute the change in potential energy found in Example 16.1b:

$$-\tfrac{3}{8}mv_i^2 = -\Delta PE$$

$$v_i = \sqrt{\frac{8\Delta PE}{3m}} = \sqrt{\frac{8(3.36 \times 10^{-17}\ \text{J})}{3(9.11 \times 10^{-31}\ \text{kg})}}$$

$$= \boxed{9.92 \times 10^6\ \text{m/s}}$$

Remarks   While the changes in potential energy associated with the proton and electron were similar in magnitude, the effect on their speeds differed dramatically. The change in potential energy had a proportionately much greater effect on the much lighter electron than on the proton.

Exercise 16.2
Refer to Exercise 16.1. Find the electron's speed at $x = -0.180$ m. The answer is 4.5% of the speed of light.

Answer 1.35×10^7 m/s

Physics⊗Now™   You can predict and observe the speed of the proton as it arrives at the negative plate, for random values of the electric field, by logging into PhysicsNow at **www.cp7e.com** and going to Interactive Example 16.2.

Electric Potential

In Chapter 15, it was convenient to define an electric field $\vec{E}$ related to the electric force $\vec{F} = q\vec{E}$. In this way, the properties of fixed collections of charges could be easily studied, and the force on any particle in the electric field could be obtained simply by multiplying by the particle's charge q. For the same reasons, it's useful to define an *electric potential difference* ΔV related to the potential energy by $\Delta PE = q\Delta V$:

Potential difference between two points ▶

> The electric potential difference ΔV between points A and B is the change in electric potential energy as a charge q moves from A to B, divided by the charge q:
>
> $$\Delta V = V_B - V_A = \frac{\Delta PE}{q} \qquad [16.2]$$
>
> **SI Unit: joule per coulomb, or volt (J/C, or V)**

This definition is completely general, although in many cases calculus would be required to compute the change in potential energy of the system. Because electric potential energy is a scalar quantity, **electric potential is also a scalar quantity**. From Equation 16.2, we see that electric potential difference is a measure of the change in electric potential energy per unit charge. Alternately, the electric potential difference is the work per unit charge that would have to be done by some force to move a charge from point A to point B in the electric field. The SI unit of electric potential is the joule per coulomb, called the volt (V). From the definition of that unit, 1 J of work must be done to move a 1-C charge between two points that are at a potential difference of 1 V. In the process of moving through a potential difference of 1 V, the 1-C charge gains 1 J of energy.

For the special case of a uniform electric field such as that between charged parallel plates, dividing Equation 16.1 by q gives

$$\frac{\Delta PE}{q} = -E_x \Delta x$$

Comparing this equation with Equation 16.2, we find that

$$\Delta V = -E_x \Delta x \qquad [16.3]$$

Equation 16.3 shows that potential difference also has units of electric field times distance. From this it follows that the SI unit of the electric field, the newton per coulomb, can also be expressed as volts per meter:

$$1 \text{ N/C} = 1 \text{ V/m}$$

Because Equation 16.3 is directly related to Equation 16.1, remember that it's valid only for the system consisting of a uniform electric field and a charge moving in one dimension.

Released from rest, positive charges accelerate spontaneously from regions of high potential to low potential. If a positive charge is given some initial velocity in the direction of high potential, it can move in that direction, but will slow and finally turn around, just like a ball tossed upwards in a gravity field. Negative charges do exactly the opposite: Released from rest, they accelerate from regions of low potential toward regions of high potential. Work must be done on negative charges to make them go in the direction of lower electric potential.

TIP 16.1 Potential and Potential Energy

Electric potential is characteristic of the field only, independent of a test charge that may be placed in that field. On the other hand, potential energy is a characteristic of the charge–field system due to an interaction between the field and a charge placed in the field.

Quick Quiz 16.2

Figure 16.3 is a graph of an electric potential as a function of position. If a positively charged particle is placed at point A, what will its subsequent motion be? It will (a) go to the right (b) go to the left (c) remain at point A (d) oscillate around point B.

Now, if a negatively charged particle is placed at point B and given a very small kick to the right, what will its subsequent motion be? It will (a) go to the right, and not return (b) go to the left (c) remain at point B (d) oscillate around point B.

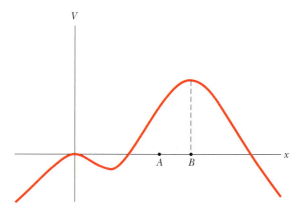

Figure 16.3   (Quick Quiz 16.2)

An application of potential difference is the 12-V battery found in an automobile. Such a battery maintains a potential difference across its terminals, with the positive terminal 12 V higher in potential than the negative terminal. In practice, the negative terminal is usually connected to the metal body of the car, which can be considered to be at a potential of zero volts. The battery provides the electrical current necessary to operate headlights, a radio, power windows, motors, and so forth. Now consider a charge of $+1$ C, to be moved around a circuit that contains the battery connected to some of these external devices. As the charge is moved inside the battery from the negative terminal (at 0 V) to the positive terminal (at 12 V), the work done on the charge by the battery is 12 J. Every coulomb of positive charge that leaves the positive terminal of the battery carries an energy of 12 J. As the charge moves through the external circuit toward the negative terminal, it gives up its 12 J of electrical energy to the external devices. When the charge reaches the negative terminal, its electrical energy is zero again. At this point, the battery takes over and restores 12 J of energy to the charge as it is moved from the negative to the positive terminal, enabling it to make another transit of the circuit. The actual amount of charge that leaves the battery each second and traverses the circuit depends on the properties of the external devices, as seen in the next chapter.

APPLICATION

Automobile Batteries

EXAMPLE 16.3   TV Tubes and Atom Smashers

Goal   Relate electric potential to an electric field and conservation of energy.

Problem   In atom smashers (also known as cyclotrons and linear accelerators), charged particles are accelerated in much the same way they are accelerated in TV tubes: through potential differences. Suppose a proton is injected at a speed of 1.00×10^6 m/s between two plates 5.00 cm apart, as shown in Figure 16.4. The proton subsequently accelerates across the gap and exits through the opening. **(a)** What must the electric potential difference be if the exit speed is to be 3.00×10^6 m/s? **(b)** What is the magnitude of the electric field between the plates?

Strategy   Use conservation of energy, writing the change in potential energy in terms of the change in electric potential, ΔV, and solve for ΔV. For part (b), solve Equation 16.3 for the electric field.

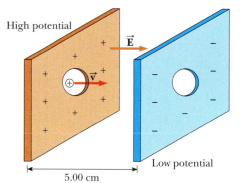

Figure 16.4   (Example 16.3) A proton enters a cavity and accelerates from one charged plate toward the other in an electric field $\vec{E}$.

Solution

(a) Find the electric potential yielding the desired exit speed of the proton.

Apply conservation of energy, writing the potential energy in terms of the electric potential:

$$\Delta KE + \Delta PE = \Delta KE + q\Delta V = 0$$

Solve the energy equation for the change in potential:

$$\Delta V = -\frac{\Delta KE}{q} = -\frac{\frac{1}{2}mv_f^2 - \frac{1}{2}mv_i^2}{q} = -\frac{m}{2q}(v_f^2 - v_i^2)$$

Substitute the given values, obtaining the necessary potential difference:

$$\Delta V = -\frac{(1.67 \times 10^{-27}\,\text{kg})}{2(1.60 \times 10^{-19}\,\text{C})}[(3.00 \times 10^6\,\text{m/s})^2$$
$$- (1.00 \times 10^6\,\text{m/s})^2]$$
$$\Delta V = \boxed{-4.18 \times 10^4\,\text{V}}$$

(b) What electric field must exist between the plates?

Solve Equation 16.3 for the electric field, and substitute:

$$E = -\frac{\Delta V}{\Delta x} = \frac{4.18 \times 10^4\,\text{V}}{0.0500\,\text{m}} = \boxed{8.36 \times 10^5\,\text{N/C}}$$

Remarks   Systems of such cavities, consisting of alternating positive and negative plates, are used to accelerate charged particles to high speed before smashing them into targets. To prevent a slowing of, say, a positively-charged particle after it passes through the negative plate of one cavity and enters the next, the charges on the plates are reversed. Otherwise, the particle would be traveling from the negative plate to a positive plate in the second cavity, and the kinetic energy gained in the previous cavity would be lost in the second.

Exercise 16.3
Suppose electrons in a TV tube are accelerated through a potential difference of 2.00×10^4 V from the heated cathode(negative electrode), where they are produced, towards the screen, which also serves as the anode (positive electrode), 25.0 cm away. (a) At what speed would the electrons impact the phosphors on the screen? Assume they accelerate from rest, and ignore relativistic effects (Chapter 26). (b) What's the magnitude of the electric field, if it is assumed constant?

Answers   (a) 8.38×10^7 m/s   (b) 8.00×10^4 V/m

16.2   ELECTRIC POTENTIAL AND POTENTIAL ENERGY DUE TO POINT CHARGES

In electric circuits, a point of zero electric potential is often defined by grounding (connecting to Earth) some point in the circuit. For example, if the negative terminal of a 12-V battery were connected to ground, it would be considered to have a potential of zero, while the positive terminal would have a potential of + 12 V. The potential difference created by the battery, however, is only locally defined. In this section we describe the electric potential of a point charge, which is defined throughout space.

The electric field of a point charge extends throughout space, so its electric potential does, also. The zero point of electric potential could be taken anywhere, but is usually taken to be an infinite distance from the charge, far from its influence and the influence of any other charges. With this choice, the methods of calculus can be used to show that the electric potential created by a point charge q at any distance r from the charge is given by

Electric potential created
by a point charge ▶

$$V = k_e\frac{q}{r} \qquad\qquad \text{[16.4]}$$

Equation 16.4 shows that the electric potential, or work per unit charge, required to move a test charge in from infinity to a distance r from a positive point charge q increases as the positive test charge moves closer to q. A plot of Equation 16.4 in Figure 16.5 shows that the potential associated with a point charge decreases as $1/r$ with increasing r, in contrast to the magnitude of the charge's electric field, which decreases as $1/r^2$.

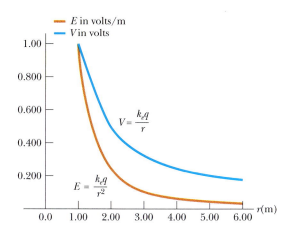

Figure 16.5   Electric field and electric potential versus distance from a point charge of 1.11×10^{-10} C. Note that V is proportional to $1/r$, while E proportional to $1/r^2$.

The electric potential of two or more charges is obtained by applying the **superposition principle: the total electric potential at some point *P* due to several point charges is the algebraic sum of the electric potentials due to the individual charges.** This is similar to the method used in Chapter 15 to find the resultant electric field at a point in space. Unlike electric field superposition, which involves a sum of vectors, the superposition of electric potentials requires evaluating a sum of scalars. As a result, it's much easier to evaluate the electric potential at some point due to several charges than to evaluate the electric field, which is a vector quantity.

◀ Superposition principle

Figure 16.6 is a computer-generated plot of the electric potential associated with an electric dipole, which consists of two charges of equal magnitude but opposite in sign. The charges lie in a horizontal plane at the center of the potential spikes. The value of the potential is plotted in the vertical dimension. The computer program has added the potential of each charge to arrive at total values of the potential.

Just as in the case of constant electric fields, there is a relationship between electric potential and electric potential energy. If V_1 is the electric potential due to charge q_1 at a point P (Active Figure 16.7a, page 540), then the work required to bring charge q_2 from infinity to P without acceleration is q_2V_1. By definition, this work equals the potential energy PE of the two-particle system when the particles are separated by a distance r (Active Fig. 16.7b).

We can therefore express the electrical potential energy of the *pair* of charges as

$$PE = q_2 V_1 = k_e \frac{q_1 q_2}{r} \qquad [16.5]$$

◀ Potential energy of a pair of charges

If the charges are of the *same* sign, PE is positive. This is consistent with the fact that like charges repel, so positive work must be done on the system by an external agent to force the two charges near one another. Conversely, if the charges are of

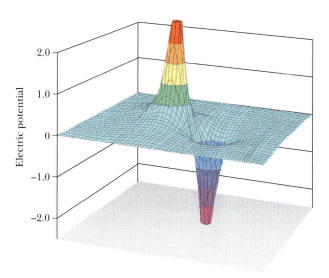

Figure 16.6   The electric potential (in arbitrary units) in the plane containing an electric dipole. Potential is plotted in the vertical dimension.

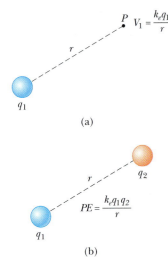

(a)

(b)

ACTIVE FIGURE 16.7
(a) The electric potential V_1 at P due to the point charge q_1 is $V_1 = k_e q_1/r$. (b) If a second charge, q_2, is brought from infinity to P, the potential energy of the pair is $PE = k_e q_1 q_2/r$.

Physics❋Now™
Log into PhysicsNow at
www.cp7e.com and go to Active
Figure 16.7, where you can move one
of the charges and see the effect on
the electric potential energy, or move
a point P to see the effect on the
electric potential at P.

opposite sign, the force is attractive and *PE* is negative. This means that negative work must be done to prevent unlike charges from accelerating toward each other as they are brought close together.

Quick Quiz 16.3

Consider a collection of charges in a given region, and suppose all other charges are distant and have a negligible effect. Further, the electric potential is taken to be zero at infinity. If the electric potential at a given point in the region is zero, which of the following statements must be true? (a) The electric field is zero at that point. (b) The electric potential energy is a minimum at that point. (c) There is no net charge in the region. (d) Some charges in the region are positive and some are negative. (e) The charges have the same sign and are symmetrically arranged around the given point.

Quick Quiz 16.4

A spherical balloon contains a positively charged particle at its center. As the balloon is inflated to a larger volume while the charged particle remains at the center, which of the following are true? (a) The electric potential at the surface of the balloon increases. (b) The magnitude of the electric field at the surface of the balloon increases. (c) The electric flux through the balloon remains the same. (d) none of these.

Problem-Solving Strategy   Electric Potential

1. Draw a diagram of all charges, and circle the point of interest.
2. Calculate the distance from each charge to the point of interest, labeling it on the diagram.
3. For each charge q, calculate the scalar quantity $V = \dfrac{k_e q}{r}$. *The sign of each charge must be included in your calculations!*
4. Sum all the numbers found in the previous step, obtaining the electric potential at the point of interest.

EXAMPLE 16.4   Finding the Electric Potential

Goal   Calculate the electric potential due to a collection of point charges.

Problem   A 5.00-μC point charge is at the origin, and a point charge $q_2 = -2.00\ \mu$C is on the x-axis at $(3.00, 0)$ m, as in Figure 16.8. **(a)** If the electric potential is taken to be zero at infinity, find the total electric potential due to these charges at point P with coordinates $(0, 4.00)$ m. **(b)** How much work is required to bring a third point charge of 4.00 μC from infinity to P?

Strategy   (a) The electric potential at P due to each charge can be calculated from $V = k_e q/r$. The total electric potential at P is the sum of these two numbers. (b) Use the work–energy theorem, together with Equation 16.5, recalling that the potential at infinity is taken to be zero.

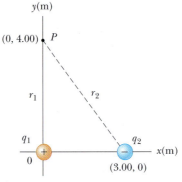

Figure 16.8   (Example 16.4) The electric potential at point P due to the point charges q_1 and q_2 is the algebraic sum of the potentials due to the individual charges.

Solution
(a) Find the electric potential at point P.

Calculate the electric potential at P due to the 5.00-μC charge:

$$V_1 = k_e \frac{q_1}{r_1} = \left(8.99 \times 10^9\ \frac{\text{N} \cdot \text{m}^2}{\text{C}^2}\right)\left(\frac{5.00 \times 10^{-6}\ \text{C}}{4.00\ \text{m}}\right)$$

$$= 1.12 \times 10^4\ \text{V}$$

Find the electric potential at P due to the $-2.00\text{-}\mu C$ charge:

$$V_2 = k_e \frac{q_2}{r_2} = \left(8.99 \times 10^9 \frac{\text{N} \cdot \text{m}^2}{\text{C}^2}\right)\left(\frac{-2.00 \times 10^{-6}\,\text{C}}{5.00\,\text{m}}\right)$$

$$= -0.360 \times 10^4\,\text{V}$$

Sum the two numbers to find the total electric potential at P:

$$V_P = V_1 + V_2 = 1.12 \times 10^4\,\text{V} + (-0.360 \times 10^4\,\text{V})$$

$$= \boxed{7.60 \times 10^3\,\text{V}}$$

(b) Find the work needed to bring the 4.00-μC charge from infinity to P.

Apply the work–energy theorem, with Equation 16.5:

$$W = \Delta PE = q_3 \Delta V = q_3(V_P - V_\infty)$$

$$= (4.00 \times 10^{-6}\,\text{C})(7.60 \times 10^3\,\text{V} - 0)$$

$$W = \boxed{3.04 \times 10^{-2}\,\text{J}}$$

Exercise 16.4

Suppose a charge of $-2.00\ \mu C$ is at the origin and a charge of $3.00\ \mu C$ is at the point $(0, 3.00)$ m. (a) Find the electric potential at $(4.00, 0)$ m, assuming the electric potential is zero at infinity, and (b) find the work necessary to bring a $4.00\ \mu C$ charge from infinity to the point $(4.00, 0)$ m.

Answers   (a) $8.99 \times 10^2\,\text{V}$ (b) $3.60 \times 10^{-3}\,\text{J}$

16.3   POTENTIALS AND CHARGED CONDUCTORS

The electric potential at all points on a charged conductor can be determined by combining Equations 16.1 and 16.2. From Equation 16.1, we see that the work done on a charge by electric forces is related to the change in electrical potential energy of the charge by

$$W = -\Delta PE$$

From Equation 16.2, we see that the change in electric potential energy between two points A and B is related to the potential difference between those points by

$$\Delta PE = q(V_B - V_A)$$

Combining these two equations, we find that

$$\boxed{W = -q(V_B - V_A)} \qquad [16.6]$$

Using this equation, we obtain the following general result: **No net work is required to move a charge between two points that are at the same electric potential.** In mathematical terms, this says that $W = 0$ whenever $V_B = V_A$.

In Chapter 15 we found that when a conductor is in electrostatic equilibrium, a net charge placed on it resides entirely on its surface. Further, we showed that the electric field just outside the surface of a charged conductor in electrostatic equilibrium is perpendicular to the surface and that the field inside the conductor is zero. We now show that **all points on the surface of a charged conductor in electrostatic equilibrium are at the same potential.**

Consider a surface path connecting any points A and B on a charged conductor, as in Figure 16.9. The charges on the conductor are assumed to be in equilibrium with each other, so none are moving. In this case, the electric field $\vec{E}$ is always perpendicular to the displacement along this path. This must be so, for otherwise the part of the electric field tangent to the surface would move the charges. Because $\vec{E}$ is perpendicular to the path, no work is done if a charge is moved between the given two points. From Equation 16.6 we see that if the work done is zero, the difference in electric potential, $V_B - V_A$, is also zero. It

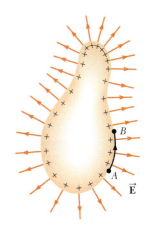

Figure 16.9   An arbitrarily shaped conductor with an excess positive charge. When the conductor is in electrostatic equilibrium, all of the charge resides at the surface, $\vec{E} = 0$ inside the conductor, and the electric field just outside the conductor is perpendicular to the surface. The potential is constant inside the conductor and is equal to the potential at the surface.

follows that **the electric potential is a constant everywhere on the surface of a charged conductor in equilibrium**. Further, because the electric field inside a conductor is zero, no work is required to move a charge between two points inside the conductor. Again, Equation 16.6 shows that if the work done is zero, the difference in electric potential between any two points inside a conductor must also be zero. We conclude that the electric potential is constant everywhere inside a conductor.

Finally, because one of the points inside the conductor could be arbitrarily close to the surface of the conductor, we conclude that **the electric potential is constant everywhere inside a conductor and equal to that same value at the surface**. As a consequence, no work is required to move a charge from the interior of a charged conductor to its surface. (It's important to realize that the potential inside a conductor is not necessarily zero, even though the interior electric field is zero.)

The Electron Volt

An appropriately-sized unit of energy commonly used in atomic and nuclear physics is the electron volt (eV). For example, electrons in normal atoms typically have energies of tens of eV's, excited electrons in atoms emitting x-rays have energies of thousands of eV's, and high energy gamma rays (electromagnetic waves) emitted by the nucleus have energies of millions of eV's.

Definition of the electron volt ▶

> The **electron volt** is defined as the kinetic energy that an electron gains when accelerated through a potential difference of 1 V.

Because $1\,\text{V} = 1\,\text{J/C}$ and because the magnitude of the charge on the electron is 1.60×10^{-19} C, we see that the electron volt is related to the joule by

$$1\,\text{eV} = 1.60 \times 10^{-19}\,\text{C} \cdot \text{V} = 1.60 \times 10^{-19}\,\text{J} \qquad \text{[16.7]}$$

Quick Quiz 16.5

An electron initially at rest accelerates through a potential difference of 1 V, gaining kinetic energy KE_e, while a proton, also initially at rest, accelerates through a potential difference of -1 V, gaining kinetic energy KE_p. Which of the following relationships holds? (a) $KE_e = KE_p$ (b) $KE_e < KE_p$. (c) $KE_e > KE_p$ (d) cannot be determined from the given information.

16.4 EQUIPOTENTIAL SURFACES

A surface on which all points are at the same potential is called an **equipotential surface**. The potential difference between any two points on an equipotential surface is zero. Hence, **no work is required to move a charge at constant speed on an equipotential surface**.

Equipotential surfaces have a simple relationship to the electric field: **The electric field at every point of an equipotential surface is perpendicular to the surface**. If the electric field $\vec{E}$ had a component parallel to the surface, that component would produce an electric force on a charge placed on the surface. This force would do work on the charge as it moved from one point to another, in contradiction to the definition of an equipotential surface.

Equipotential surfaces can be represented on a diagram by drawing equipotential contours, which are two-dimensional views of the intersections of the equipotential surfaces with the plane of the drawing. These equipotential contours are generally referred to simply as **equipotentials**. Figure 16.10a shows the equipotentials (in blue) associated with a positive point charge. Note that the equipotentials are perpendicular to the electric field lines (in red) at all points. Recall that the electric potential created by a point charge q is given by $V = k_e q/r$. This relation shows that, for a single point charge, the potential is constant on any surface on which r is constant. It follows that the equipotentials of a point charge are a family

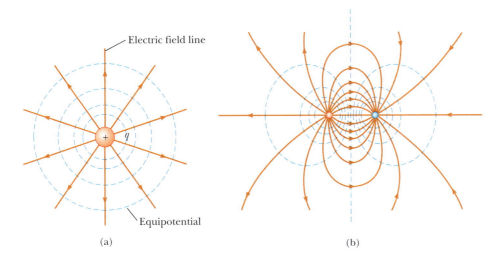

Figure 16.10 Equipotentials (dashed blue lines) and electric field lines (red lines) for (a) a positive point charge and (b) two point charges of equal magnitude and opposite sign. In all cases, the equipotentials are *perpendicular* to the electric field lines at every point.

of spheres centered on the point charge. Figure 16.10b shows the equipotentials associated with two charges of equal magnitude but opposite sign.

16.5   APPLICATIONS

The Electrostatic Precipitator

One important application of electric discharge in gases is a device called an *electrostatic precipitator*. This device removes particulate matter from combustion gases, thereby reducing air pollution. It's especially useful in coal-burning power plants and in industrial operations that generate large quantities of smoke. Systems currently in use can eliminate approximately 90% by mass of the ash and dust from the smoke. Unfortunately, a very high percentage of the lighter particles still escape, and these contribute significantly to smog and haze.

Figure 16.11 illustrates the basic idea of the electrostatic precipitator. A high voltage (typically 40 kV to 100 kV) is maintained between a wire running down the center of a duct and the outer wall, which is grounded. The wire is maintained at a negative electric potential with respect to the wall, so the electric field is directed toward the wire. The electric field near the wire reaches a high enough value to cause a discharge around the wire and the formation of positive ions, electrons, and negative ions, such as O_2^-. As the electrons and negative ions are accelerated

APPLICATION

The Electrostatic Precipitator

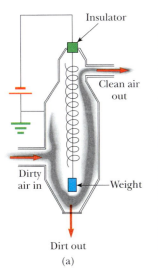

Insulator

Clean air out

Dirty air in

Weight

Dirt out

(a)   (b)   (c)

b. Riei O'Harra/Black Star/PNI; c. Greig Cranna/Stock, Boston/PNI

Figure 16.11   (a) A schematic diagram of an electrostatic precipitator. The high voltage maintained on the central wires creates an electric discharge in the vicinity of the wire. Compare the air pollution when the precipitator is (b) operating, and (c) turned off.

toward the outer wall by the nonuniform electric field, the dirt particles in the streaming gas become charged by collisions and ion capture. Because most of the charged dirt particles are negative, they are also drawn to the outer wall by the electric field. When the duct is shaken, the particles fall loose and are collected at the bottom.

In addition to reducing the amounts of harmful gases and particulate matter in the atmosphere, the electrostatic precipitator recovers valuable metal oxides from the stack.

APPLICATION

The Electrostatic Air Cleaner

A similar device called an *electrostatic air cleaner* is used in homes to relieve the discomfort of allergy sufferers. Air laden with dust and pollen is drawn into the device across a positively charged mesh screen. The airborne particles become positively charged when they make contact with the screen, and then pass through a second, negatively charged mesh screen. The electrostatic force of attraction between the positively charged particles in the air and the negatively charged screen causes the particles to precipitate out on the surface of the screen, removing a very high percentage of contaminants from the air stream.

APPLICATION

Xerographic Copiers

Xerography and Laser Printers

Xerography is widely used to make photocopies of printed materials. The basic idea behind the process was developed by Chester Carlson, who was granted a patent for his invention in 1940. In 1947 the Xerox Corporation launched a full-scale program to develop automated duplicating machines using Carlson's process. The huge success of that development is evident: today, practically all offices and libraries have one or more duplicating machines, and the capabilities of these machines continue to evolve.

Some features of the xerographic process involve simple concepts from electrostatics and optics. However, the one idea that makes the process unique is the use of photoconductive material to form an image. A photoconductor is a material that is a poor conductor of electricity in the dark, but a reasonably good conductor when exposed to light.

Figure 16.12 illustrates the steps in the xerographic process. First, the surface of a plate or drum is coated with a thin film of the photoconductive material (usually selenium or some compound of selenium), and the photoconductive surface is given a positive electrostatic charge in the dark (Fig. 16.12a). The page to be copied is then projected onto the charged surface (Fig. 16.12b). The photoconducting surface becomes conducting only in areas where light strikes; there the light produces charge carriers in the photoconductor which neutralize the positively charged surface. The charges remain on those areas of the photoconductor not exposed to light, however, leaving a hidden image of the object in the form of a positive distribution of surface charge.

Next, a negatively charged powder called a *toner* is dusted onto the photoconducting surface (Fig. 16.12c). The charged powder adheres only to the areas that contain the positively charged image. At this point, the image becomes visible. It is then transferred to the surface of a sheet of positively charged paper. Finally, the toner is "fixed" to the surface of the paper by heat (Fig. 16.12d), resulting in a permanent copy of the original.

APPLICATION

Laser Printers

The steps for producing a document on a laser printer are similar to those used in a photocopy machine, in that parts (a), (c), and (d) of Figure 16.12 remain essentially the same. The difference between the two techniques lies in the way the image is formed on the selenium-coated drum. In a laser printer, the command to print the letter O, for instance, is sent to a laser from the memory of a computer. A rotating mirror inside the printer causes the beam of the laser to sweep across the selenium-coated drum in an interlaced pattern (Fig. 16.12e). Electrical signals generated by the printer turn the laser beam on and off in a pattern that traces out the letter "O" in the form of positive charges on the selenium. Toner is then applied to the drum, and the transfer to paper is accomplished as in a photocopy machine.

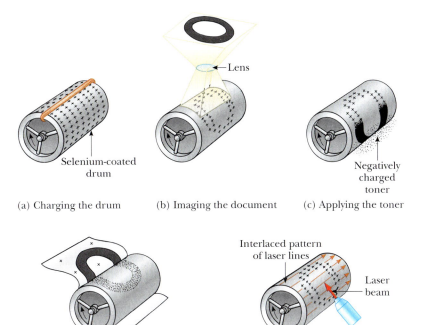

(a) Charging the drum   (b) Imaging the document   (c) Applying the toner

(d) Transferring the
toner to the paper

(e) Laser printer drum

Selenium-coated
drum

Lens

Negatively
charged
toner

Interlaced pattern
of laser lines

Laser
beam

Figure 16.12   The xerographic process. (a) The photoconductive surface is positively charged. (b) Through the use of a light source and a lens, a hidden image is formed on the charged surface in the form of positive charges. (c) The surface containing the image is covered with a negatively charged powder, which adheres only to the image area. (d) A piece of paper is placed over the surface and given a charge. This transfers the image to the paper, which is then heated to "fix" the powder to the paper. (e) The image on the drum of a laser printer is produced by turning a laser beam on and off as it sweeps across the selenium-coated drum.

16.6   CAPACITANCE

A **capacitor** is a device used in a variety of electric circuits—for example, to tune the frequency of radio receivers, eliminate sparking in automobile ignition systems, or store short-term energy for rapid release in electronic flash units. Figure 16.13 shows a typical design for a capacitor. It consists of two parallel metal plates separated by a distance d. Used in an electric circuit, the plates are connected to the positive and negative terminals of a battery or some other voltage source. When this connection is made, electrons are pulled off one of the plates, leaving it with a charge of $+Q$, and are transferred through the battery to the other plate, leaving it with a charge of $-Q$, as shown in the figure. The transfer of charge stops when the potential difference across the plates equals the potential difference of the battery. A charged capacitor is a device that stores energy that can be reclaimed when needed for a specific application.

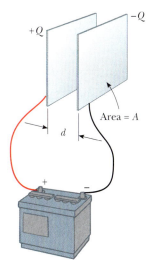

Figure 16.13   A parallel-plate capacitor consists of two parallel plates, each of area A, separated by a distance d. The plates carry equal and opposite charges.

◀ Capacitance of a pair of conductors

The capacitance C of a capacitor is the ratio of the magnitude of the charge on either conductor (plate) to the magnitude of the potential difference between the conductors (plates):

$$C \equiv \frac{Q}{\Delta V} \qquad [16.8]$$

SI Unit: farad (F) = coulomb per volt (C/V)

The quantities Q and ΔV are always taken to be positive when used in Equation 16.8. For example, if a 3.0-μF capacitor is connected to a 12-V battery, the magnitude of the charge on each plate of the capacitor is

$$Q = C\Delta V = (3.0 \times 10^{-6} \text{ F})(12 \text{ V}) = 36 \ \mu\text{C}$$

From Equation 16.8, we see that a large capacitance is needed to store a large amount of charge for a given applied voltage. The farad is a very large unit of capacitance. In practice, most typical capacitors have capacitances ranging from microfarads (1 μF = 1 × 10^{-6} F) to picofarads (1 pF = 1 × 10^{-12} F).

TIP 16.2   Potential Difference Is ΔV, Not V

Use the symbol ΔV for the potential difference across a circuit element or a device (many other books use simply V for potential difference.) The dual use of V to represent potential in one place and a potential difference in another can lead to unnecessary confusion.

16.7 THE PARALLEL-PLATE CAPACITOR

The capacitance of a device depends on the geometric arrangement of the conductors. The capacitance of a parallel-plate capacitor with plates separated by air (see Fig. 16.13) can be easily calculated from three facts. First, recall from Chapter 15 that the magnitude of the electric field between two plates is given by $E = \sigma/\epsilon_0$, where σ is the magnitude of the charge per unit area on each plate. Second, we found earlier in this chapter that the potential difference between two plates is $\Delta V = Ed$, where d is the distance between the plates. Third, the charge on one plate is given by $q = \sigma A$, where A is the area of the plate. Substituting these three facts into the definition of capacitance gives the desired result:

$$C = \frac{q}{\Delta V} = \frac{\sigma A}{Ed} = \frac{\sigma A}{(\sigma/\epsilon_0)\,d}$$

Canceling the charge per unit area, σ, yields

Capacitance of a parallel-plate capacitor ▶

$$C = \epsilon_0 \frac{A}{d}$$ [16.9]

where A is the area of one of the plates, d is the distance between the plates, and ϵ_0 is the permittivity of free space.

From Equation 16.9, we see that plates with larger area can store more charge. The same is true for a small plate separation d, because then the positive charges on one plate exert a stronger force on the negative charges on the other plate, allowing more charge to be held on the plates.

Figure 16.14 shows the electric field lines of a more realistic parallel-plate capacitor. The electric field is very nearly constant in the center between the plates, but becomes less so when approaching the edges. For most purposes, however, the field may be taken as constant throughout the region between the plates.

APPLICATION

Camera Flash Attachments

One practical device that uses a capacitor is the flash attachment on a camera. A battery is used to charge the capacitor, and the stored charge is then released when the shutter-release button is pressed to take a picture. The stored charge is delivered to a flash tube very quickly, illuminating the subject at the instant more light is needed.

APPLICATION

Computer Keyboards

Computers make use of capacitors in many ways. For example, one type of computer keyboard has capacitors at the bases of its keys, as in Figure 16.15. Each key is connected to a movable plate, which represents one side of the capacitor; the fixed plate on the bottom of the keyboard represents the other side of the capacitor. When a key is pressed, the capacitor spacing decreases, causing an increase in capacitance. External electronic circuits recognize each key by the *change* in its capacitance when it is pressed.

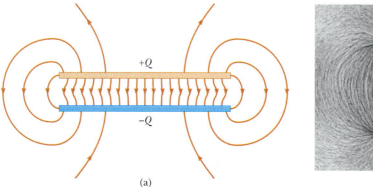

+Q

−Q

(a)

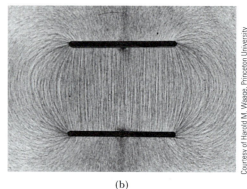

(b)

Figure 16.14   (a) The electric field between the plates of a parallel-plate capacitor is uniform near the center, but nonuniform near the edges. (b) Electric field pattern of two oppositely charged conducting parallel plates. Small pieces of thread on an oil surface align with the electric field.

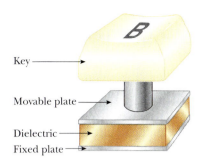

Key ⟶

Movable plate ⟶

Dielectric ⟶

Fixed plate ⟶

Figure 16.15   When the key of one type of keyboard is pressed, the capacitance of a parallel-plate capacitor increases as the plate spacing decreases. The substance labeled "dielectric" is an insulating material, as described in Section 16.10.

Capacitors are useful for storing a large amount of charge that needs to be delivered quickly. A good example on the forefront of fusion research is electrostatic confinement. In this role, capacitors discharge their electrons through a grid. The negatively charged electrons in the grid draw positively charged particles to them and therefore to each other, causing some particles to fuse and release energy in the process.

APPLICATION

Electrostatic Confinement

EXAMPLE 16.5   A Parallel-Plate Capacitor

Goal   Calculate fundamental physical properties of a parallel-plate capacitor.

Problem   A parallel-plate capacitor has an area $A = 2.00 \times 10^{-4} \, \text{m}^2$ and a plate separation $d = 1.00 \times 10^{-3} \, \text{m}$. **(a)** Find its capacitance. **(b)** How much charge is on the positive plate if the capacitor is connected to a 3.00-V battery? **(c)** Calculate the charge density on the positive plate, assuming the density is uniform, and **(d)** the magnitude of the electric field between the plates.

Strategy   Parts (a) and (b) can be solved by substituting into the basic equations for capacitance. In part (c), use the fact that the voltage difference equals the electric field times the distance.

Solution
(a) Find the capacitance.

Substitute into Equation 16.9:

$$C = \epsilon_0 \frac{A}{d} = (8.85 \times 10^{-12} \, \text{C}^2/\text{N} \cdot \text{m}^2)\left(\frac{2.00 \times 10^{-4} \, \text{m}^2}{1.00 \times 10^{-3} \, \text{m}}\right)$$

$$C = \boxed{1.77 \times 10^{-12} \, \text{F} = 1.77 \, \text{pF}}$$

(b) Find the charge on the positive plate after the capacitor is connected to a 3.00-V battery.

Substitute into Equation 16.8:

$$C = \frac{Q}{\Delta V} \quad \rightarrow \quad Q = C\Delta V = (1.77 \times 10^{-12} \, \text{F})(3.00 \, \text{V})$$

$$= 5.31 \times 10^{-12} \, \text{C}$$

(c) Calculate the charge density on the positive plate.

Charge density is charge divided by area:

$$\sigma = \frac{Q}{A} = \frac{5.31 \times 10^{-12} \, \text{C}}{2.00 \times 10^{-4} \, \text{m}^2} = \boxed{2.66 \times 10^{-8} \, \text{C}/\text{m}^2}$$

(d) Calculate the magnitude of the electric field between the plates.

Apply Equation 15.13:

$$E = \frac{\sigma}{\epsilon_0} = \frac{2.66 \times 10^{-8} \, \text{C}/\text{m}^2}{8.85 \times 10^{-12} \, \text{C}^2/\text{N} \cdot \text{m}^2} = \boxed{3.01 \times 10^3 \, \text{N}/\text{C}}$$

Remarks   The answer to part (d) could also have been obtained from the electric potential, which is $\Delta V = Ed$ for a parallel-plate capacitor.

Exercise 16.5

Two plates, each of area $3.00 \times 10^{-4} \, \text{m}^2$, are used to construct a parallel-plate capacitor with capacitance 1.00 pF. (a) Find the necessary separation distance. (b) If the positive plate is to hold a charge of 5.00×10^{-12} C, find the charge density. (c) Find the electric field between the plates. (d) What voltage battery should be attached to the plate to obtain the preceding results?

Answers   (a) 2.66×10^{-3} m   (b) 1.67×10^{-8} C/m^2 (c) 1.89×10^3 N/C   (d) 5.00 V

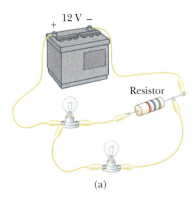

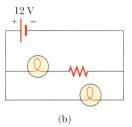

Figure 16.16   (a) A real circuit and (b) its equivalent circuit diagram.

Symbols for Circuit Elements and Circuits

The symbol that is commonly used to represent a capacitor in a circuit is, —||— or sometimes —|(—. Don't confuse either of these symbols with the circuit symbol, —+||−— which is used to designate a battery (or any other source of direct current). The positive terminal of the battery is at the higher potential and is represented by the longer vertical line in the battery symbol. In the next chapter we discuss another circuit element, called a resistor, represented by the symbol —Ʌ/Ʌ—. When wires in a circuit don't have appreciable resistance compared with the resistance of other elements in the circuit, the wires are represented by straight lines.

It's important to realize that a circuit is a collection of real objects, usually containing a source of electrical energy (such as a battery) connected to elements that convert electrical energy to other forms (light, heat, sound) or store the energy in electric or magnetic fields for later retrieval. A real circuit and its schematic diagram are sketched side by side in Figure 16.16. The circuit symbol for a lightbulb shown in Figure 16.16b is ——(Ɩ)——.

If you are not familiar with circuit diagrams, trace the path of the real circuit with your finger to see that it is equivalent to the geometrically regular schematic diagram.

16.8 COMBINATIONS OF CAPACITORS

Two or more capacitors can be combined in circuits in several ways, but most reduce to two simple configurations, called *parallel* and *series*. The idea, then, is to find the single equivalent capacitance due to a combination of several different capacitors that are in parallel or in series with each other. Capacitors are manufactured with a number of different standard capacitances, and by combining them in different ways, any desired value of the capacitance can be obtained.

Capacitors in Parallel

Two capacitors connected as shown in Active Figure 16.17a are said to be in *parallel*. The left plate of each capacitor is connected to the positive terminal of the battery by a conducting wire, so the left plates are at the same potential. In the same way, the right plates, both connected to the negative terminal of the battery, are also at the same potential. This means that **capacitors in parallel both have the same potential difference ΔV across them**. Capacitors in parallel are illustrated in Active Figure 16.17b.

When the capacitors are first connected in the circuit, electrons are transferred from the left plates through the battery to the right plates, leaving the left plates positively charged and the right plates negatively charged. The energy source for this transfer of charge is the internal chemical energy stored in the

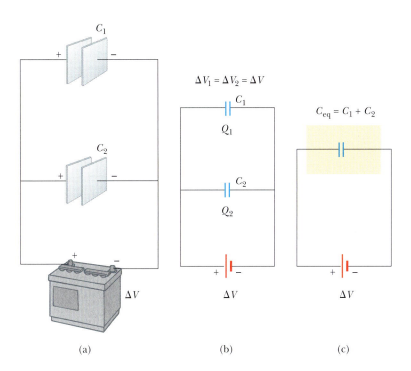

$\Delta V_1 = \Delta V_2 = \Delta V$

$C_{eq} = C_1 + C_2$

(a)   (b)   (c)

ACTIVE FIGURE 16.17
(a) A parallel connection of two
capacitors. (b) The circuit diagram
for the parallel combination. (c) The
potential differences across the capacitors are the same, and the equivalent
capacitance is $C_{eq} = C_1 + C_2$.

Physics⊗Now™
Log into PhysicsNow at
www.cp7e.com and go to Active
Figure 16.17, where you can adjust
the battery voltage and the individual
capacitances to see the resulting
charges and voltages on the
capacitors. You can combine up to
four capacitors in parallel.

battery, which is converted to electrical energy. The flow of charge stops when
the voltage across the capacitors equals the voltage of the battery, at which time
the capacitors have their maximum charges. If the maximum charges on the two
capacitors are Q_1 and Q_2, respectively, then the *total charge*, Q, stored by the two
capacitors is

$$Q = Q_1 + Q_2 \qquad \textbf{[16.10]}$$

We can replace these two capacitors with one equivalent capacitor having a
capacitance of C_{eq}. This equivalent capacitor must have exactly the same external
effect on the circuit as the original two, so it must store Q units of charge and have
the same potential difference across it. The respective charges on each capacitor
are

$$Q_1 = C_1 \Delta V \qquad \text{and} \qquad Q_2 = C_2 \Delta V$$

The charge on the equivalent capacitor is

$$Q = C_{eq} \Delta V$$

Substituting these relationships into Equation 16.10 gives

$$C_{eq} \Delta V = C_1 \Delta V + C_2 \Delta V$$

or

$$C_{eq} = C_1 + C_2 \qquad \begin{pmatrix} \text{parallel} \\ \text{combination} \end{pmatrix} \qquad \textbf{[16.11]}$$

If we extend this treatment to three or more capacitors connected in parallel,
the equivalent capacitance is found to be

$$C_{eq} = C_1 + C_2 + C_3 + \cdots \qquad \begin{pmatrix} \text{parallel} \\ \text{combination} \end{pmatrix} \qquad \textbf{[16.12]}$$

We see that **the equivalent capacitance of a parallel combination of capacitors is
greater than any of the individual capacitances**.

**TIP 16.3   Voltage is the Same
as Potential Difference**

A voltage *across* a device, such as a
capacitor, has the same meaning as
the potential difference across the
device. For example, if we say that the
voltage across a capacitor is 12 V, we
mean that the potential difference
between its plates is 12 V.

EXAMPLE 16.6   Four Capacitors Connected in Parallel

Goal   Analyze a circuit with several capacitors in parallel.

Problem   **(a)** Determine the capacitance of the single capacitor that is equivalent to the parallel combination of capacitors shown in Figure 16.18, and **(b)** find the charge on the 12.0-μF capacitor.

Strategy   (a) Add the individual capacitances. (b) Apply the formula $C = Q/\Delta V$ to the 12.0-μF capacitor. The voltage difference is the same as the difference across the battery.

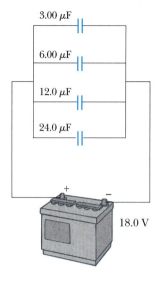

3.00 μF

6.00 μF

12.0 μF

24.0 μF

+   −

18.0 V

Figure 16.18   (Example 16.6) Four capacitors connected in parallel.

Solution

(a) Find the equivalent capacitance.

Apply Equation 16.12:

$$C_{eq} = C_1 + C_2 + C_3 + C_4$$
$$= 3.00~\mu F + 6.00~\mu F + 12.0~\mu F + 24.0~\mu F$$
$$= \boxed{45.0~\mu F}$$

(b) Find the charge on the 12-μF capacitor.

Solve the capacitance equation for Q and substitute:

$$Q = C\Delta V = (12.0 \times 10^{-6}~F)(18.0~V) = 216 \times 10^{-6}~C$$
$$= \boxed{216~\mu C}$$

Remarks   The charge on any one of the parallel capacitors can be found as in part (b), since the potential difference is the same.

Exercise 16.6
Find the charge on the 24.0-μF capacitor.

Answer   432 μC

Capacitors in Series

Q is the same for all capacitors connected in series ▶

Now consider two capacitors connected in *series*, as illustrated in Active Figure 16.19a. **For a series combination of capacitors, the magnitude of the charge must be the same on all the plates.** To understand this principle, consider the charge transfer process in some detail. When a battery is connected to the circuit, electrons with total charge $-Q$ are transferred from the left plate of C_1 to the right plate of C_2 through the battery, leaving the left plate of C_1 with a charge of $+Q$. As a consequence, the magnitudes of the charges on the left plate of C_1 and the right plate of C_2 must be the same. Now consider the right plate of C_1 and the left plate of C_2, in the middle. These plates are not connected to the battery (because of the gap across the plates) and, taken together, are electrically neutral. The charge of $+Q$ on the left plate of C_1, however, attracts negative charges to the right plate of C_1. These charges will continue to accumulate until the left and right plates of C_1, taken together, become electrically neutral, which means the charge on the right plate of C_1 is $-Q$. This negative charge could only have come from the left plate of C_2, so C_2 has a charge of $+Q$.

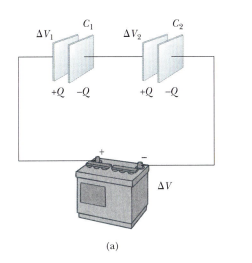
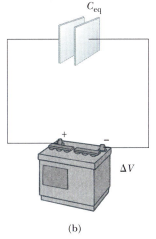

(a)   (b)

ACTIVE FIGURE 16.19
A series combination of two capacitors. The charges on the capacitors are the same, and the equivalent capacitance can be calculated from the reciprocal relationship $1/C_{eq} = (1/C_1) + (1/C_2)$.

Physics⊗Now™
Log into PhysicsNow at **www.cp7e.com** and go to Active Figure 16.19, where you can adjust the battery voltage and the individual capacitances to see the resulting charges and voltages on the capacitors. You can combine up to four capacitors in series.

Therefore, regardless of how many capacitors are in series or what their capacitances are, **all of the right plates gain charges of $-Q$ and all the left plates have charges of $+Q$.** (This is a consequence of the conservation of charge.)

After an equivalent capacitor for a series of capacitors is fully charged, **the equivalent capacitor must end up with a charge of $-Q$ on its right plate and a charge of $+Q$ on its left plate.** Applying the definition of capacitance to the circuit in Active Figure 16.19b, we have

$$\Delta V = \frac{Q}{C_{eq}}$$

where ΔV is the potential difference between the terminals of the battery and C_{eq} is the equivalent capacitance. Because $Q = C\Delta V$ can be applied to each capacitor, the potential differences across them are given by

$$\Delta V_1 = \frac{Q}{C_1} \qquad \Delta V_2 = \frac{Q}{C_2}$$

From Active Figure 16.19a, we see that

$$\Delta V = \Delta V_1 + \Delta V_2 \qquad \textbf{[16.13]}$$

where ΔV_1 and ΔV_2 are the potential differences across capacitors C_1 and C_2. (This is a consequence of the conservation of energy.)

The potential difference across any number of capacitors (or other circuit elements) in series equals the sum of the potential differences across the individual capacitors. Substituting these expressions into Equation 16.13, and noting that $\Delta V = Q/C_{eq}$, we have

$$\frac{Q}{C_{eq}} = \frac{Q}{C_1} + \frac{Q}{C_2}$$

Canceling Q, we arrive at the following relationship:

$$\frac{1}{C_{eq}} = \frac{1}{C_1} + \frac{1}{C_2} \qquad \left(\begin{array}{c}\text{series} \\ \text{combination}\end{array}\right) \qquad \textbf{[16.14]}$$

If this analysis is applied to three or more capacitors connected in series, the equivalent capacitance is found to be

$$\frac{1}{C_{eq}} = \frac{1}{C_1} + \frac{1}{C_2} + \frac{1}{C_3} + \cdots \qquad \left(\begin{array}{c}\text{series} \\ \text{combination}\end{array}\right) \qquad \textbf{[16.15]}$$

As we will show in Example 16.7, Equation 16.15 implies that **the equivalent capacitance of a series combination is always less than any individual capacitance in the combination.**

A capacitor is designed so that one plate is large and the other is small. If the plates are connected to a battery, (a) the large plate has a greater charge than the small plate, (b) the large plate has less charge than the small plate, or (c) the plates have equal, but opposite, charge.

EXAMPLE 16.7   Four Capacitors Connected in Series

Goal   Find an equivalent capacitance of capacitors in series, and the charge and voltage on each capacitor.

Problem   Four capacitors are connected in series with a battery, as in Figure 16.20. **(a)** Calculate the capacitance of the equivalent capacitor. **(b)** Compute the charge on the 12-μF capacitor. **(c)** Find the voltage drop across the 12-μF capacitor.

Strategy   Combine all the capacitors into a single, equivalent capacitor using Equation 16.15. Find the charge on this equivalent capacitor using $C = Q/\Delta V$. This charge is the same as on the individual capacitors. Use this same equation again to find the voltage drop across the 12-μF capacitor.

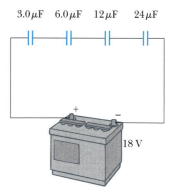

Figure 16.20   (Example 16.7) Four capacitors connected in series.

Solution

(a) Calculate the equivalent capacitance of the series.

Apply Equation 16.15:

$$\frac{1}{C_{eq}} = \frac{1}{3.0\ \mu F} + \frac{1}{6.0\ \mu F} + \frac{1}{12\ \mu F} + \frac{1}{24\ \mu F}$$

$$C_{eq} = \boxed{1.6\ \mu F}$$

(b) Compute the charge on the 12-μF capacitor.

The desired charge equals the charge on the equivalent capacitor:

$$Q = C_{eq}\Delta V = (1.6 \times 10^{-6}\ \text{F})(18\ \text{V}) = \boxed{29\ \mu C}$$

(c) Find the voltage drop across the 12-μF capacitor.

Apply the basic capacitance equation:

$$C = \frac{Q}{\Delta V} \quad \rightarrow \quad \Delta V = \frac{Q}{C} = \frac{29\ \mu C}{12\ \mu F} = \boxed{2.4\ \text{V}}$$

Remarks   Notice that the equivalent capacitance is less than that of any of the individual capacitors. The relationship $C = Q/\Delta V$ can be used to find the voltage drops on the other capacitors, just as in part (c).

Exercise 16.7

The 24-μF capacitor is removed from the circuit, leaving only three capacitors in series. Find (a) the equivalent capacitance, (b) the charge on the 6-μF capacitor, and (c) the voltage drop across the 6-μF capacitor.

Answers   (a) 1.7 μF   (b) 31 μC   (c) 5.2 V

Problem-Solving Strategy
Complex Capacitor Combinations

1. **Combine** capacitors that are in series or in parallel, following the derived formulas.
2. **Redraw** the circuit after every combination.
3. **Repeat** the first two steps until there is only a single equivalent capacitor.
4. **Find the charge** on the single equivalent capacitor, using $C = Q/\Delta V$.

5. **Work backwards** through the diagrams to the original one, finding the charge and voltage drop across each capacitor along the way. To do this, use the following collection of facts:
   A. The capacitor equation: $C = Q/\Delta V$
   B. Capacitors in parallel: $C_{eq} = C_1 + C_2$
   C. Capacitors in parallel all have the same voltage difference, ΔV, as does their equivalent capacitor.
   D. Capacitors in series: $\dfrac{1}{C_{eq}} = \dfrac{1}{C_1} + \dfrac{1}{C_2}$
   E. Capacitors in series all have the same charge, Q, as does their equivalent capacitor.

INTERACTIVE EXAMPLE 16.8   Equivalent Capacitance

Goal   Solve a complex combination of series and parallel capacitors.

Problem   (a) Calculate the equivalent capacitance between a and b for the combination of capacitors shown in Figure 16.21a. All capacitances are in microfarads. **(b)** If a 12-V battery is connected across the system between points a and b, find the charge on the 4.0-μF capacitor in the first diagram and the voltage drop across it.

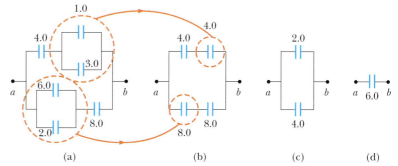

Strategy   (a) Using Equations 16.12 and 16.15, we reduce the combination step by step, as indicated in the figure. (b) To find the charge on the 4.0-μF capacitor, start with Figure 16.21c, finding the charge on the 2.0-μF capacitor. This same charge is on

Figure 16.21   (Example 16.8) To find the equivalent capacitance of the circuit in (a), use the series and parallel rules described in the text to successively reduce the circuit as indicated in (b), (c), and (d).

each of the 4.0-μF capacitors in the second diagram, by fact 5E of the Problem-Solving Strategy. One of these 4.0-μF capacitors in the second diagram is just the original 4.0-μF capacitor in the first diagram.

Solution

(a) Calculate the equivalent capacitance.

Find the equivalent capacitance of the parallel 1.0-μF and 3.0-μF capacitors in Figure 16.21a:

$$C_{eq} = C_1 + C_2 = 1.0\ \mu F + 3.0\ \mu F = 4.0\ \mu F$$

Find the equivalent capacitance of the parallel 2.0-μF and 6.0-μF capacitors in Figure 16.21a:

$$C_{eq} = C_1 + C_2 = 2.0\ \mu F + 6.0\ \mu F = 8.0\ \mu F$$

Combine the two series 4.0-μF capacitors in Figure 16.21b:

$$\frac{1}{C_{eq}} = \frac{1}{C_1} + \frac{1}{C_2} = \frac{1}{4.0\ \mu F} + \frac{1}{4.0\ \mu F}$$

$$= \frac{1}{2.0\ \mu F} \quad \rightarrow \quad C_{eq} = 2.0\ \mu F$$

Combine the two series 8.0-μF capacitors in Figure 16.21b:

$$\frac{1}{C_{eq}} = \frac{1}{C_1} + \frac{1}{C_2} = \frac{1}{8.0\ \mu F} + \frac{1}{8.0\ \mu F}$$

$$= \frac{1}{4.0\ \mu F} \quad \rightarrow \quad C_{eq} = 4.0\ \mu F$$

Finally, combine the two parallel capacitors in Figure 16.21c to find the equivalent capacitance between a and b:

$$C_{eq} = C_1 + C_2 = 2.0\ \mu F + 4.0\ \mu F = \boxed{6.0\ \mu F}$$

(b) Find the charge on the 4.0-μF capacitor and the voltage drop across it.

Compute the charge on the 2.0-μF capacitor in Figure 16.21c, which is the same as the charge on the 4.0-μF capacitor in Figure 16.21a:

$$C = \frac{Q}{\Delta V} \quad \rightarrow \quad Q = C\Delta V = (2.0\ \mu F)(12\ V) = \boxed{24\ \mu C}$$

Use the basic capacitance equation to find the voltage drop across the 4.0-μF capacitor in Figure 16.21a:

$$C = \frac{Q}{\Delta V} \quad \rightarrow \quad \Delta V = \frac{Q}{C} = \frac{24\ \mu C}{4.0\ \mu F} = \boxed{6.0\ V}$$

Remarks To find the rest of the charges and voltage drops, it's just a matter of using $C = Q/\Delta V$ repeatedly, together with facts 5C and 5E in the Problem-Solving Strategy. The voltage drop across the 4.0-μF capacitor could also have been found by noticing, in Figure 16.21b, that both capacitors had the same value and so by symmetry would split the total drop of 12 volts between them.

Exercise 16.8
(a) In Example 16.8, find the charge on the 8.0-μF capacitor in Figure 16.21a and the voltage drop across it. (b) Do the same for the 6.0-μF capacitor in Figure 16.21a.

Answers   (a) 48 μC, 6.0 V   (b) 36 μC, 6.0 V

Physics⊗Now™ You can practice reducing a combination of capacitors to a single equivalent capacitor by logging into PhysicsNow at **www.cp7e.com** and going to Interactive Example 16.8.

16.9   ENERGY STORED IN A CHARGED CAPACITOR

Almost everyone who works with electronic equipment has at some time verified that a capacitor can store energy. If the plates of a charged capacitor are connected by a conductor such as a wire, charge transfers from one plate to the other until the two are uncharged. The discharge can often be observed as a visible spark. If you accidentally touched the opposite plates of a charged capacitor, your fingers would act as a pathway by which the capacitor could discharge, inflicting an electric shock. The degree of shock would depend on the capacitance and voltage applied to the capacitor. *Where high voltages and large quantities of charge are present, as in the power supply of a television set, such a shock can be fatal.*

Capacitors store electrical energy, and that energy is the same as the work required to move charge onto the plates. If a capacitor is initially uncharged (both plates are neutral), so that the plates are at the same potential, very little work is required to transfer a small amount of charge ΔQ from one plate to the other. However, once this charge has been transferred, a small potential difference $\Delta V = \Delta Q/C$ appears between the plates, so work must be done to transfer additional charge against this potential difference. From Equation 16.6, if the potential difference at any instant during the charging process is ΔV, then the work ΔW required to move more charge ΔQ through this potential difference is given by

$$\Delta W = \Delta V \Delta Q$$

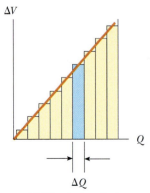

Figure 16.22 A plot of voltage versus charge for a capacitor is a straight line with slope $1/C$. The work required to move a charge of ΔQ through a potential difference of ΔV across the capacitor plates is $\Delta W = \Delta V \Delta Q$, which equals the area of the blue rectangle. The *total work* required to charge the capacitor to a final charge of Q is the area under the straight line, which equals $Q\Delta V/2$.

We know that $\Delta V = Q/C$ for a capacitor that has a total charge of Q. Therefore, a plot of voltage versus total charge gives a straight line with a slope of $1/C$, as shown in Figure 16.22. The work ΔW, for a particular ΔV, is the area of the shaded rectangle. Adding up all the rectangles gives an approximation of the total work needed to fill the capacitor. In the limit as ΔQ is taken to be infinitesimally small, the total work needed to charge the capacitor to a final charge Q and voltage ΔV is the area under the line. This is just the area of a triangle, one-half the base times the height, so it follows that

$$W = \tfrac{1}{2} Q \Delta V \qquad\qquad \text{[16.16]}$$

As previously stated, W is also the energy stored in the capacitor. From the definition of capacitance, we have $Q = C\Delta V$; hence, we can express the energy stored three different ways:

$$\text{Energy stored} = \tfrac{1}{2}Q\Delta V = \tfrac{1}{2}C(\Delta V)^2 = \frac{Q^2}{2C} \qquad \textbf{[16.17]}$$

For example, the amount of energy stored in a 5.0-μF capacitor when it is connected across a 120-V battery is

$$\text{Energy stored} = \tfrac{1}{2}C(\Delta V)^2 = \tfrac{1}{2}(5.0 \times 10^{-6}\,\text{F})(120\,\text{V})^2 = 3.6 \times 10^{-2}\,\text{J}$$

In practice, there is a limit to the maximum energy (or charge) that can be stored in a capacitor. At some point, the Coulomb forces between the charges on the plates become so strong that electrons jump across the gap, discharging the capacitor. For this reason, capacitors are usually labeled with a maximum operating voltage. (This physical fact can actually be exploited to yield a circuit with a regularly blinking light).

Large capacitors can store enough electrical energy to cause severe burns or even death if they are discharged so that the flow of charge can pass through the heart. Under the proper conditions, however, they can be used to *sustain* life by stopping cardiac fibrillation in heart attack victims. When fibrillation occurs, the heart produces a rapid, irregular pattern of beats. A fast discharge of electrical energy through the heart can return the organ to its normal beat pattern. Emergency medical teams use portable defibrillators that contain batteries capable of charging a capacitor to a high voltage. (The circuitry actually permits the capacitor to be charged to a much higher voltage than the battery.) In this case and others (camera flash units and lasers used for fusion experiments), capacitors serve as energy reservoirs that can be slowly charged and then quickly discharged to provide large amounts of energy in a short pulse. The stored electrical energy is released through the heart by conducting electrodes, called paddles, placed on both sides of the victim's chest. The paramedics must wait between applications of electrical energy due to the time it takes for the capacitors to become fully charged. The high voltage on the capacitor can be obtained from a low-voltage battery in a portable machine through the phenomenon of *electromagnetic induction*, to be studied in Chapter 20.

APPLICATION
Defibrillators

EXAMPLE 16.9   Typical Voltage, Energy, and Discharge Time for a Defibrillator

Goal   Apply energy and power concepts to a capacitor.

Problem   A fully charged defibrillator contains 1.20 kJ of energy stored in a 1.10×10^{-4}-F capacitor. In a discharge through a patient, 6.00×10^2 J of electrical energy are delivered in 2.50 ms. **(a)** Find the voltage needed to store 1.20 kJ in the unit. **(b)** What average power is delivered to the patient?

Strategy   (a) Because we know the energy stored and the capacitance, we can use Equation 16.17 to find the required voltage. (b) Dividing the energy delivered by the time gives the average power.

Solution
(a) Find the voltage needed to store 1.20 kJ in the unit.

Solve Equation 16.17 for ΔV:

$$\text{Energy stored} = \tfrac{1}{2}C\Delta V^2$$

$$\Delta V = \sqrt{\frac{2 \times (\text{Energy stored})}{C}}$$

$$= \sqrt{\frac{2(1.20 \times 10^3\,\text{J})}{1.10 \times 10^{-4}\,\text{F}}}$$

$$= 4.67 \times 10^3\,\text{V}$$

(b) What average power is delivered to the patient?

Divide the energy delivered by the time:

$$\mathscr{P}_{av} = \frac{\text{Energy delivered}}{\Delta t} = \frac{6.00 \times 10^2 \text{ J}}{2.50 \times 10^{-3} \text{ s}}$$

$$= 2.40 \times 10^5 \text{ W}$$

Remarks The power delivered by a draining capacitor isn't constant, as we'll find in the study of RC circuits in Chapter 18. For that reason, we were able to find only an average power. Capacitors are necessary in defibrillators because they can deliver energy far more quickly than batteries. Batteries provide current through relatively slow chemical reactions, whereas capacitors release charge that has already been produced and stored.

Exercise 16.9
(a) Find the energy contained in a 2.50×10^{-5}-F parallel-plate capacitor if it holds 1.75×10^{-3} C of charge. (b) What's the voltage between the plates? (c) What new voltage will result in a doubling of the stored energy?

Answers (a) 6.13×10^{-2} J (b) 70.0 V (c) 99.0 V

Applying Physics 16.1 Maximum Energy Design

How should three capacitors and two batteries be connected so that the capacitors will store the maximum possible energy?

Explanation The energy stored in the capacitor is proportional to the capacitance and the square of the

potential difference, so we would like to maximize each of these quantities. If the three capacitors are put in parallel, their capacitances add, and if the batteries are in series, their potential differences, similarly, also add together.

Quick Quiz 16.7

A parallel-plate capacitor is disconnected from a battery, and the plates are pulled a small distance further apart. Do the following quantities increase, decrease, or stay the same?
(a) C (b) Q (c) E between the plates (d) ΔV (e) energy stored in the capacitor

16.10 CAPACITORS WITH DIELECTRICS

A **dielectric** is an insulating material, such as rubber, plastic, or waxed paper. When a dielectric is inserted between the plates of a capacitor, the capacitance increases. If the dielectric completely fills the space between the plates, the capacitance is multiplied by the factor κ, called the **dielectric constant**.

The following experiment illustrates the effect of a dielectric in a capacitor. Consider a parallel-plate capacitor of charge Q_0 and capacitance C_0 in the absence of a dielectric. The potential difference across the capacitor plates can be measured, and is given by $\Delta V_0 = Q_0/C_0$ (Fig. 16.23a). Because the capacitor is not connected to an external circuit, there is no pathway for charge to leave or be added to the plates. If a dielectric is now inserted between the plates as in Figure 16.23b, the voltage across the plates is *reduced* by the factor κ to the value

$$\Delta V = \frac{\Delta V_0}{\kappa}$$

Because $\kappa > 1$, ΔV is less than ΔV_0. Because the charge Q_0 on the capacitor doesn't change, we conclude that the capacitance in the presence of the dielectric must change to the value

$$C = \frac{Q_0}{\Delta V} = \frac{Q_0}{\Delta V_0/\kappa} = \frac{\kappa Q_0}{\Delta V_0}$$

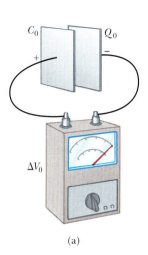
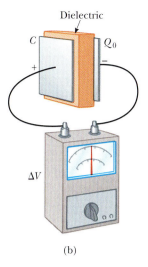

Figure 16.23 (a) With air between the plates, the voltage across the capacitor is ΔV_0, the capacitance is C_0, and the charge is Q_0. (b) With a dielectric between the plates, the charge remains at Q_0, but the voltage and capacitance change.

or

$$C = \kappa C_0 \qquad [16.18]$$

According to this result, the capacitance is *multiplied* by the factor κ when the dielectric fills the region between the plates. For a parallel-plate capacitor, where the capacitance in the absence of a dielectric is $C_0 = \epsilon_0 A/d$, we can express the capacitance in the presence of a dielectric as

$$C = \kappa \epsilon_0 \frac{A}{d} \qquad [16.19]$$

From this result, it appears that the capacitance could be made very large by decreasing d, the separation between the plates. In practice, the lowest value of d is limited by the electric discharge that can occur through the dielectric material separating the plates. For any given plate separation, there is a maximum electric field that can be produced in the dielectric before it breaks down and begins to conduct. This maximum electric field is called the **dielectric strength**, and for air its value is about 3×10^6 V/m. Most insulating materials have dielectric strengths greater than that of air, as indicated by the values listed in Table 16.1. Figure 16.24 shows an instance of dielectric breakdown in air.

Figure 16.24 Dielectric breakdown in air. Sparks are produced when a large alternating voltage is applied across the wires by a high-voltage induction coil power supply.

TABLE 16.1

Dielectric Constants and Dielectric Strengths of Various Materials at Room Temperature

Material	Dielectric Constant κ	Dielectric Strength (V/m)
Vacuum	1.000 00	—
Air	1.000 59	3×10^6
Bakelite®	4.9	24×10^6
Fused quartz	3.78	8×10^6
Pyrex® glass	5.6	14×10^6
Polystyrene	2.56	24×10^6
Teflon®	2.1	60×10^6
Neoprene rubber	6.7	12×10^6
Nylon	3.4	14×10^6
Paper	3.7	16×10^6
Strontium titanate	233	8×10^6
Water	80	—
Silicone oil	2.5	15×10^6

Figure 16.25 Three commercial capacitor designs. (a) A tubular capacitor whose plates are separated by paper and then rolled into a cylinder. (b) A high-voltage capacitor consisting of many parallel plates separated by oil. (c) An electrolytic capacitor.

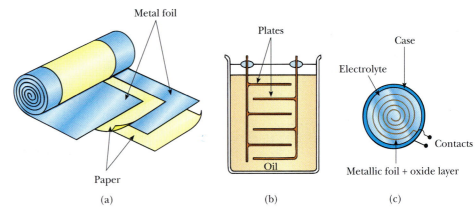

Metal foil

Plates

Case

Electrolyte

Contacts

Metallic foil + oxide layer

Oil

Paper

(a)   (b)   (c)

(a)

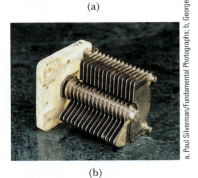

(b)

Figure 16.26 (a) A collection of capacitors used in a variety of applications. (b) A variable capacitor. When one set of metal plates is rotated so as to lie between a fixed set of plates, the capacitance of the device changes.

a, Paul Silverman/Fundamental Photographs; b, George Semple

Commercial capacitors are often made by using metal foil interlaced with thin sheets of paraffin-impregnated paper or Mylar®, which serves as the dielectric material. These alternate layers of metal foil and dielectric are rolled into a small cylinder (Fig. 16.25a). One type of a high-voltage capacitor consists of a number of interwoven metal plates immersed in silicone oil (Fig. 16.25b). Small capacitors are often constructed from ceramic materials. Variable capacitors (typically 10 pF to 500 pF) usually consist of two interwoven sets of metal plates, one fixed and the other movable, with air as the dielectric.

An electrolytic capacitor (Fig. 16.25c) is often used to store large amounts of charge at relatively low voltages. It consists of a metal foil in contact with an electrolyte—a solution that conducts charge by virtue of the motion of the ions contained in it. When a voltage is applied between the foil and the electrolyte, a thin layer of metal oxide (an insulator) is formed on the foil, and this layer serves as the dielectric. Enormous capacitances can be attained because the dielectric layer is very thin.

Figure 16.26 shows a variety of commercially available capacitors. Variable capacitors are used in radios to adjust the frequency.

When electrolytic capacitors are used in circuits, the polarity (the plus and minus signs on the device) must be observed. If the polarity of the applied voltage is opposite that intended, the oxide layer will be removed and the capacitor will conduct rather than store charge. Further, reversing the polarity can result in such a large current that the capacitor may either burn or produce steam and explode.

Applying Physics 16.2   Stud Finders

If you have ever tried to hang a picture on a wall securely, you know that it can be difficult to locate a wooden stud in which to anchor your nail or screw. The principles discussed in this section can be used to detect a stud electronically. The primary element of an electronic stud finder is a capacitor with its plates arranged side by side instead of facing one another, as in Figure 16.27. How does this device work?

Explanation As the detector is moved along a wall, its capacitance changes when it passes across a stud because the dielectric constant of the material "between" the plates changes. The change in capacitance can be used to cause a light to come on, signaling the presence of the stud.

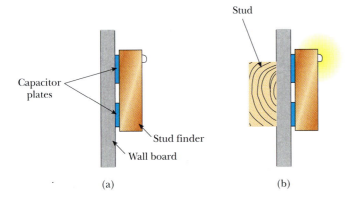

Stud

Capacitor plates

Stud finder

Wall board

(a)   (b)

Figure 16.27 (Applying Physics 16.2) A stud finder. (a) The materials between the plates of the capacitor are the drywall and the air behind it. (b) The materials become drywall and wood when the detector moves across a stud in the wall. The change in the dielectric constant causes a signal light to illuminate.

A fully charged parallel-plate capacitor remains connected to a battery while you slide a dielectric between the plates. Do the following quantities increase, decrease, or stay the same? (a) C (b) Q (c) E between the plates (d) ΔV (e) energy stored in the capacitor.

EXAMPLE 16.10   A Paper-Filled Capacitor

Goal   Calculate fundamental physical properties of a parallel-plate capacitor with a dielectric.

Problem   A parallel-plate capacitor has plates 2.0 cm by 3.0 cm. The plates are separated by a 1.0-mm thickness of paper. Find **(a)** the capacitance of this device, and **(b)** the maximum charge that can be placed on the capacitor.

Strategy   (a) Obtain the dielectric constant for paper from Table 16.1, and substitute, with other given quantities, into Equation 16.19. (b) Table 16.1 also gives the dielectric strength of paper, which is the maximum electric field that can be applied before electrical breakdown occurs. Use Equation 16.3, $\Delta V = Ed$, to obtain the maximum voltage, and substitute into the basic capacitance equation.

Solution
(a) Find the capacitance of this device.

Substitute into Equation 16.19:

$$C = \kappa \epsilon_0 \frac{A}{d}$$

$$= 3.7 \left(8.85 \times 10^{-12} \frac{C^2}{N \cdot m^2} \right) \left(\frac{6.0 \times 10^{-4} \, m^2}{1.0 \times 10^{-3} \, m} \right)$$

$$= 2.0 \times 10^{-11} \, F$$

(b) Find the maximum charge that can be placed on the capacitor.

Calculate the maximum applied voltage, using the dielectric strength of paper, E_{max}.

$$\Delta V_{max} = E_{max} d = (16 \times 10^6 \, V/m)(1.0 \times 10^{-3} \, m)$$
$$= 1.6 \times 10^4 \, V$$

Solve the basic capacitance equation for Q_{max}, and substitute ΔV_{max} and C:

$$Q_{max} = C \Delta V_{max} = (2.0 \times 10^{-11} \, F)(1.6 \times 10^4 \, V)$$
$$= 0.32 \, \mu C$$

Remarks   Dielectrics allow κ times as much charge to be stored on a capacitor for a given voltage. They also allow an increase in the applied voltage by increasing the threshold of electrical breakdown.

Exercise 16.10
A parallel-plate capacitor has plate area of $2.50 \times 10^{-3} \, m^2$ and distance between the plates of 2.00 mm. (a) Find the maximum charge that can be placed on the capacitor if air is between the plates. (b) Find the maximum charge if the air is replaced by polystyrene.

Answers   (a) $7 \times 10^{-8} \, C$ (b) $1.4 \times 10^{-6} \, C$

An Atomic Description of Dielectrics

The explanation of why a dielectric increases the capacitance of a capacitor is based on an atomic description of the material, which in turn involves a property of some molecules called **polarization**. A molecule is said to be polarized when there is a separation between the average positions of its negative charge and its positive charge. In some molecules, such as water, this condition is always present. To see why, consider the geometry of a water molecule (Fig. 16.28, page 560).

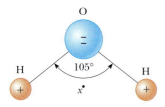

Figure 16.28 The water molecule, H_2O, has a permanent polarization resulting from its bent geometry. The point labeled x is the center of positive charge.

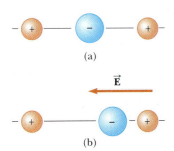

Figure 16.29 (a) A symmetric molecule has no permanent polarization. (b) An external electric field induces a polarization in the molecule.

The molecule is arranged so that the negative oxygen atom is bonded to the positively charged hydrogen atoms with a 105° angle between the two bonds. The center of negative charge is at the oxygen atom, and the center of positive charge lies at a point midway along the line joining the hydrogen atoms (point x in the diagram). Materials composed of molecules that are permanently polarized in this way have large dielectric constants, and indeed, Table 16.1 shows that the dielectric constant of water is large ($\kappa = 80$) compared to other common substances.

A symmetric molecule (Fig. 16.29a) can have no permanent polarization, but a polarization can be induced in it by an external electric field. A field directed to the left, as in Figure 16.29b, would cause the center of positive charge to shift to the left from its initial position and the center of negative charge to shift to the right. This *induced polarization* is the effect that predominates in most materials used as dielectrics in capacitors.

To understand why the polarization of a dielectric can affect capacitance, consider the slab of dielectric shown in Figure 16.30. Before placing the slab in between the plates of the capacitor, the polar molecules are randomly oriented (Fig. 16.30a). The polar molecules are dipoles, and each creates a dipole electric field, but because of their random orientation, this field averages to zero.

After insertion of the dielectric slab into the electric field $\vec{E}_0$ between the plates (Fig. 16.30b), the positive plate attracts the negative ends of the dipoles, while the negative plate attracts the positive ends of the dipoles. These forces exert a torque on the molecules making up the dielectric, reorienting them so that on the average the negative pole is more inclined toward the positive plate and the positive pole more aligned toward the negative plate. The positive and negative charges in the middle still cancel each other, but there is a net accumulation of negative charge in the dielectric next to the positive plate and a net accumulation of positive charge next to the negative plate. This configuration can be modeled as an additional pair of charged plates, as in Figure 16.30c, creating an induced electric field $\vec{E}_{ind}$ that partly cancels the original electric field $\vec{E}_0$. If the battery is not connected when the dielectric is inserted, the potential difference ΔV_0 across the plates is reduced to $\Delta V_0/\kappa$.

If the capacitor is still connected to the battery, however, the negative poles push more electrons off the positive plate, making it more positive. Meanwhile, the positive poles attract more electrons onto the negative plate. This situation continues until the potential difference across the battery reaches its original magnitude, equal to the potential gain across the battery. The net effect is an increase in the amount of charge stored on the capacitor. Because the plates can store more charge for a given voltage, it follows from $C = Q\Delta V$ that the capacitance must increase.

Quick Quiz 16.9

Consider a parallel-plate capacitor with a dielectric material between the plates. If the temperature of the dielectric increases, the capacitance (a) decreases (b) increases (c) remains the same.

Figure 16.30 (a) In the absence of an external electric field, polar molecules are randomly oriented. (b) When an external electric field is applied, the molecules partially align with the field. (c) The charged edges of the dielectric can be modeled as an additional pair of parallel plates establishing an electric field $\vec{E}_{ind}$ in the direction opposite to $\vec{E}_0$.

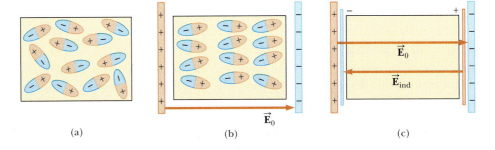

SUMMARY

Physics ⊗ Now™ Take a practice test by logging into PhysicsNow at www.cp7e.com and clicking on the Pre-Test link for this chapter.

16.1 Potential Difference and Electric Potential

The change in the electric potential energy of a system consisting of an object of charge q moving through a displacement Δx in a constant electric field $\vec{E}$ is given by

$$\Delta PE = -W_{AB} = -qE_x\Delta x \qquad \text{[16.1]}$$

where E_x is the component of the electric field in the x-direction and $\Delta x = x_f - x_i$. The **difference in electric potential** between two points A and B is

$$\Delta V = V_B - V_A \equiv \frac{\Delta PE}{q} \qquad \text{[16.2]}$$

where ΔPE is the *change* in electrical potential energy as a charge q moves between A and B. The units of potential difference are joules per coulomb, or **volts**; $1\ \text{J/C} = 1\ \text{V}$.

The **electric potential difference** between two points A and B in a *uniform* electric field $\vec{E}$ is

$$\Delta V = -E_x\Delta x \qquad \text{[16.3]}$$

where $\Delta x = x_f - x_i$ is the displacement between A and B and E_x is the x-component of the electric field in that region.

16.2 Electric Potential and Potential Energy Due to Point Charges

The **electric potential** due to a point charge q at distance r from the point charge is

$$V = k_e\frac{q}{r} \qquad \text{[16.4]}$$

The **electric potential energy** of a pair of point charges separated by distance r is

$$PE = k_e\frac{q_1q_2}{r} \qquad \text{[16.5]}$$

These equations can be used in the solution of conservation of energy problems and in the work–energy theorem.

16.3 Potentials and Charged Conductors
16.4 Equipotential Surfaces

Every point on the surface of a charged conductor in electrostatic equilibrium is at the same potential. Further, the potential is constant everywhere inside the conductor and equals its value on the surface.

The **electron volt** is defined as the energy that an electron (or proton) gains when accelerated through a potential difference of 1 V. The conversion between electron volts and joules is

$$1\ \text{eV} = 1.60 \times 10^{-19}\ \text{C} \cdot \text{V} = 1.60 \times 10^{-19}\ \text{J} \qquad \text{[16.7]}$$

Any surface on which the potential is the same at every point is called an equipotential surface. The electric field is always oriented perpendicular to an equipotential surface.

16.6 Capacitance

A capacitor consists of two metal plates with charges that are equal in magnitude but opposite in sign. The capacitance C of any capacitor is the ratio of the magnitude of the charge Q on either plate to the magnitude of potential difference ΔV between them:

$$C \equiv \frac{Q}{\Delta V} \qquad \text{[16.8]}$$

Capacitance has the units coulombs per volt, or farads; $1\ \text{C/V} = 1\ \text{F}$.

16.7 The Parallel-Plate Capacitor

The capacitance of two parallel metal plates of area A separated by distance d is

$$C = \epsilon_0\frac{A}{d} \qquad \text{[16.9]}$$

where $\epsilon_0 = 8.85 \times 10^{-12}\ \text{C}^2/\text{N}\cdot\text{m}^2$ is a constant called the **permittivity of free space**.

16.8 Combinations of Capacitors

The **equivalent capacitance of a parallel combination** of capacitors is

$$C_{eq} = C_1 + C_2 + C_3 + \cdots \qquad \text{[16.12]}$$

If two or more capacitors are connected in series, the **equivalent capacitance of the series combination** is

$$\frac{1}{C_{eq}} = \frac{1}{C_1} + \frac{1}{C_2} + \frac{1}{C_3} + \cdots \qquad \text{[16.15]}$$

Problems involving a combination of capacitors can be solved by applying Equations 16.12 and 16.13 repeatedly to a circuit diagram, simplifying it as much as possible. This is followed by working backwards to the original diagram, applying $C = Q/\Delta V$, the fact that parallel capacitors have the same voltage drop, and the fact that series capacitors have the same charge.

16.9 Energy Stored in a Charged Capacitor

Three equivalent expressions for calculating the **energy stored** in a charged capacitor are

$$\text{Energy stored} = \tfrac{1}{2}Q\Delta V = \tfrac{1}{2}C(\Delta V)^2 = \frac{Q^2}{2C} \qquad \text{[16.17]}$$

16.10 Capacitors with Dielectrics

When a nonconducting material, called a **dielectric**, is placed between the plates of a capacitor, the capacitance is multiplied by the factor κ, which is called the **dielectric constant**, a property of the dielectric material. The capacitance of a parallel-plate capacitor filled with a dielectric is

$$C = \kappa\epsilon_0\frac{A}{d} \qquad \text{[16.19]}$$

CONCEPTUAL QUESTIONS

1. (a) Describe the motion of a proton after it is released from rest in a uniform electric field. (b) Describe the changes (if any) in its kinetic energy and the electric potential energy associated with the proton.

2. Describe how you can increase the maximum operating voltage of a parallel-plate capacitor for a fixed plate separation.

3. A parallel-plate capacitor is charged by a battery, and the battery is then disconnected from the capacitor. Because the charges on the capacitor plates are opposite in sign, they attract each other. Hence, it takes positive work to increase the plate separation. Show that the external work done when the plate separation is increased leads to an increase in the energy stored in the capacitor.

4. Distinguish between electric potential and electrical potential energy.

5. Suppose you are sitting in a car and a 20-kV power line drops across the car. Should you stay in the car or get out? The power line potential is 20 kV compared to the potential of the ground.

6. Why is it important to avoid sharp edges or points on conductors used in high-voltage equipment?

7. Explain why, under static conditions, all points in a conductor must be at the same electric potential.

8. If you are given three different capacitors C_1, C_2, and C_3, how many different combinations of capacitance can you produce, using all capacitors in your circuits?

9. Why is it dangerous to touch the terminals of a high-voltage capacitor even after the voltage source that charged the battery is disconnected from the capacitor? What can be done to make the capacitor safe to handle after the voltage source has been removed?

10. The plates of a capacitor are connected to a battery. What happens to the charge on the plates if the connecting wires are removed from the battery? What happens to the charge if the wires are removed from the battery and connected to each other?

11. Can electric field lines ever cross? Why or why not? Can equipotentials ever cross? Why or why not?

12. Is it always possible to reduce a combination of capacitors to one equivalent capacitor with the rules developed in this chapter? Explain.

13. If you were asked to design a capacitor for which a small size and a large capacitance were required, what factors would be important in your design?

14. Explain why a dielectric increases the maximum operating voltage of a capacitor even though the physical size of the capacitor doesn't change.

15. (a) Capacitors connected in parallel all have the same (i) charge on them, (ii) potential difference across them, or (iii) neither of the above. (b) Capacitors connected in series all have the same (i) charge on them, (ii) potential difference across them, or (iii) neither of the above.

16. (a) The equivalent capacitance for a group of capacitors connected in parallel is (i) greater than the capacitance of any of the capacitors in the group, (ii) less than the capacitance of any of the capacitors in the group, or (iii) neither of the above. (b) The equivalent capacitance for a group of capacitors connected in series is (i) greater than the capacitance of any of the capacitors in the group, (ii) less than the capacitance of any of the capacitors in the group, or (iii) neither of the above.

17. Suppose scientists had chosen to measure small energies in proton volts rather than electron volts. What difference would this make?

18. (a) Under what conditions can the equation $V_B - V_A = -E_x \Delta x$ be used? (b) Can the equation be used to find the difference in potential between two points in an electric field set up by a point charge? (c) Can the equation be used to find the difference in potential between two points in the electric field between the plates of a charged parallel-plate capacitor?

PROBLEMS

1, 2, 3 = straightforward, intermediate, challenging   ☐ = full solution available in *Student Solutions Manual/Study Guide*

Physics⊗Now™ = coached problem with hints available at **www.cp7e.com**   🐝 = biomedical application

Section 16.1 Potential Difference and Electric Potential

1. A proton moves 2.00 cm parallel to a uniform electric field of $E = 200$ N/C. (a) How much work is done by the field on the proton? (b) What change occurs in the potential energy of the proton? (c) What potential difference did the proton move through?

2. A uniform electric field of magnitude 250 V/m is directed in the positive x-direction. A 12-μC charge moves from the origin to the point $(x, y) = (20$ cm, 50 cm$)$. (a) What was the change in the potential energy of this charge? (b) Through what potential difference did the charge move?

3. A potential difference of 90 mV exists between the inner and outer surfaces of a cell membrane. The inner surface is negative relative to the outer surface. How much work is required to eject a positive sodium ion (Na^+) from the interior of the cell?

4. An ion accelerated through a potential difference of 60.0 V has its potential energy decreased by 1.92×10^{-17} J. Calculate the charge on the ion.

5. The potential difference between the accelerating plates of a TV set is about 25 kV. If the distance between the plates is 1.5 cm, find the magnitude of the uniform electric field in the region between the plates.

6. To recharge a 12-V battery, a battery charger must move 3.6×10^5 C of charge from the negative terminal to the positive terminal. How much work is done by the charger? Express your answer in joules.

7. Physics⊗Now™ Oppositely charged parallel plates are separated by 5.33 mm. A potential difference of 600 V exists between the plates. (a) What is the magnitude of the electric field between the plates? (b) What is the magnitude of the force on an electron between the plates? (c) How much work must be done on the electron to move it to the negative plate if it is initially positioned 2.90 mm from the positive plate?

8. Calculate the speed of a proton that is accelerated from rest through a potential difference of 120 V. (b) Calculate the speed of an electron that is accelerated through the same potential difference.

9. A 4.00-kg block carrying a charge $Q = 50.0 \ \mu C$ is connected to a spring for which $k = 100$ N/m. The block lies on a frictionless horizontal track, and the system is immersed in a uniform electric field of magnitude $E = 5.00 \times 10^5$ V/m directed as in Figure P16.9. (a) If the block is released at rest when the spring is unstretched (at $x = 0$), by what maximum amount does the spring expand? (b) What is the equilibrium position of the block?

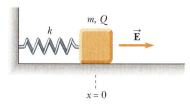

Figure P16.9

10. On planet Tehar, the free-fall acceleration is the same as that on Earth, but there is also a strong downward electric field that is uniform close to the planet's surface. A 2.00-kg ball having a charge of $5.00 \ \mu C$ is thrown upward at a speed of 20.1 m/s. It hits the ground after an interval of 4.10 s. What is the potential difference between the starting point and the top point of the trajectory?

Section 16.2 Electric Potential and Potential Energy Due to Point Charges
Section 16.3 Potentials and Charged Conductors
Section 16.4 Equipotential Surfaces

11. (a) Find the electric potential 1.00 cm from a proton. (b) What is the electric potential difference between two points that are 1.00 cm and 2.00 cm from a proton?

12. Two point charges are on the y-axis, one of magnitude 3.0×10^{-9} C at the origin and a second of magnitude 6.0×10^{-9} C at the point $y = 30$ cm. Calculate the potential at $y = 60$ cm.

13. (a) Find the electric potential, taking zero at infinity, at the upper right corner (the corner without a charge) of the rectangle in Figure P16.13. (b) Repeat if the 2.00-μC charge is replaced with a charge of $- 2.00 \ \mu C$.

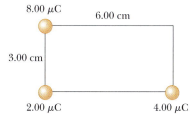

Figure P16.13 (Problems 13 and 14)

14. Three charges are situated at corners of a rectangle as in Figure P16.13. How much energy would be expended in moving the 8.00-μC charge to infinity?

15. Two point charges $Q_1 = + 5.00$ nC and $Q_2 = - 3.00$ nC are separated by 35.0 cm. (a) What is the electric potential at a point midway between the charges? (b) What is the potential energy of the pair of charges? What is the significance of the algebraic sign of your answer?

16. A point charge of 9.00×10^{-9} C is located at the origin. How much work is required to bring a positive charge of 3.00×10^{-9} C from infinity to the location $x = 30.0$ cm?

17. The three charges in Figure P16.17 are at the vertices of an isosceles triangle. Let $q = 7.00$ nC, and calculate the electric potential at the midpoint of the base.

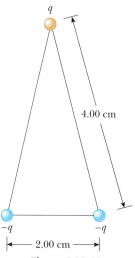

Figure P16.17

18. An electron starts from rest 3.00 cm from the center of a uniformly charged sphere of radius 2.00 cm. If the sphere carries a total charge of 1.00×10^{-9} C, how fast will the electron be moving when it reaches the surface of the sphere?

19. Physics⊗Now™ In Rutherford's famous scattering experiments that led to the planetary model of the atom, alpha particles (having charges of $+ 2e$ and masses of 6.64×10^{-27} kg) were fired toward a gold nucleus with charge $+ 79e$. An alpha particle, initially very far from the gold nucleus, is fired at 2.00×10^7 m/s directly toward the nucleus, as in Figure P16.19. How close does the alpha particle get to the gold nucleus before turning around? Assume the gold nucleus remains stationary.

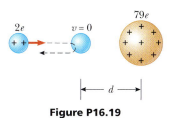

Figure P16.19

20. Starting with the definition of work, prove that the surface must be perpendicular to the local electric field at every point on an equipotential surface.

21. A small spherical object carries a charge of 8.00 nC. At what distance from the center of the object is the potential equal to 100 V? 50.0 V? 25.0 V? Is the spacing of the equipotentials proportional to the change in voltage?

Section 16.6 Capacitance
Section 16.7 The Parallel-Plate Capacitor

22. (a) How much charge is on each plate of a 4.00-μF capacitor when it is connected to a 12.0-V battery? (b) If this same capacitor is connected to a 1.50-V battery, what charge is stored?

23. Consider the Earth and a cloud layer 800 m above the planet to be the plates of a parallel-plate capacitor. (a) If the cloud layer has an area of $1.0 \text{ km}^2 = 1.0 \times 10^6 \text{ m}^2$, what is the capacitance? (b) If an electric field strength greater than 3.0×10^6 N/C causes the air to break down and conduct charge (lightning), what is the maximum charge the cloud can hold?

24. The potential difference between a pair of oppositely charged parallel plates is 400 V. (a) If the spacing between the plates is doubled without altering the charge on the plates, what is the new potential difference between the plates? (b) If the plate spacing is doubled while the potential difference between the plates is kept constant, what is the ratio of the final charge on one of the plates to the original charge?

25. An air-filled capacitor consists of two parallel plates, each with an area of 7.60 cm^2 and separated by a distance of 1.80 mm. If a 20.0-V potential difference is applied to these plates, calculate (a) the electric field between the plates, (b) the capacitance, and (c) the charge on each plate.

26. A 1-megabit computer memory chip contains many 60.0×10^{-15}-F capacitors. Each capacitor has a plate area of $21.0 \times 10^{-12} \text{ m}^2$. Determine the plate separation of such a capacitor. (Assume a parallel-plate configuration). The diameter of an atom is on the order of 10^{-10} m = 1 Å. Express the plate separation in angstroms.

27. A parallel-plate capacitor has an area of 5.00 cm^2, and the plates are separated by 1.00 mm with air between them. The capacitor stores a charge of 400 pC. (a) What is the potential difference across the plates of the capacitor? (b) What is the magnitude of the uniform electric field in the region between the plates?

28. A small object with a mass of 350 mg carries a charge of 30.0 nC and is suspended by a thread between the vertical plates of a parallel-plate capacitor. The plates are separated by 4.00 cm. If the thread makes an angle of 15.0° with the vertical, what is the potential difference between the plates?

Section 16.8 Combinations of Capacitors

29. A series circuit consists of a 0.050-μF capacitor, a 0.100-μF capacitor, and a 400-V battery. Find the charge (a) on each of the capacitors and (b) on each of the capacitors if they are reconnected in parallel across the battery.

30. Three capacitors, $C_1 = 5.00$ μF, $C_2 = 4.00$ μF, and $C_3 = 9.00$ μF, are connected together. Find the effective capacitance of the group (a) if they are all in parallel, and (b) if they are all in series.

31. (a) Find the equivalent capacitance of the capacitors in Figure P16.31. (b) Find the charge on each capacitor and the potential difference across it.

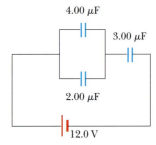

Figure P16.31

32. Two capacitors give an equivalent capacitance of 9.00 pF when connected in parallel and an equivalent capacitance of 2.00 pF when connected in series. What is the capacitance of each capacitor?

33. Four capacitors are connected as shown in Figure P16.33. (a) Find the equivalent capacitance between points a and b. (b) Calculate the charge on each capacitor if a 15.0-V battery is connected across points a and b.

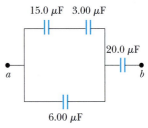

Figure P16.33

34. Consider the combination of capacitors in Figure P16.34. (a) What is the equivalent capacitance of the group? (b) Determine the charge on each capacitor.

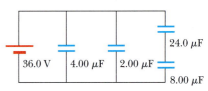

Figure P16.34

35. **Physics⊗Now™** Find the charge on each of the capacitors in Figure P16.35.

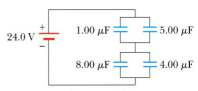

Figure P16.35

36. To repair a power supply for a stereo amplifier, an electronics technician needs a 100-μF capacitor capable of withstanding a potential difference of 90 V between its plates. The only available supply is a box of five 100-μF capacitors, each having a maximum voltage capability of 50 V. Can the technician substitute a combination of these capacitors that has the proper electrical characteristics, and if so, what will be the maximum voltage across any of the capacitors used? [*Hint:* The technician may not have to use all the capacitors in the box.]

37. A 25.0-μF capacitor and a 40.0-μF capacitor are charged by being connected across separate 50.0-V batteries. (a) Determine the resulting charge on each capacitor. (b) The capacitors are then disconnected from their batteries and connected to each other, with each negative plate connected to the other positive plate. What is the final charge of each capacitor, and what is the final potential difference across the 40.0-μF capacitor?

38. A 10.0-μF capacitor is fully charged across a 12.0-V battery. The capacitor is then disconnected from the battery and connected across an initially uncharged capacitor with capacitance C. The resulting voltage across each capacitor is 3.00 V. What is the value of C?

39. A 1.00-μF capacitor is charged by being connected across a 10.0-V battery. It is then disconnected from the battery and connected across an uncharged 2.00-μF capacitor. Determine the resulting charge on each capacitor.

40. Find the equivalent capacitance between points a and b for the group of capacitors connected as shown in Figure P16.40 if $C_1 = 5.00$ μF, $C_2 = 10.0$ μF, and $C_3 = 2.00$ μF.

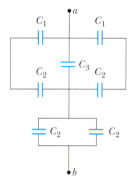

Figure P16.40 (Problems 40 and 41)

41. For the network described in the previous problem and in Figure P16.40, if the potential between points a and b is 60.0 V, what charge is stored on C_3?

42. Find the equivalent capacitance between points a and b in the combination of capacitors shown in Figure P16.42.

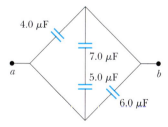

Figure P16.42

Section 16.9 Energy Stored in a Charged Capacitor

43. A parallel-plate capacitor has 2.00-cm^2 plates that are separated by 5.00 mm with air between them. If a 12.0-V battery is connected to this capacitor, how much energy does it store?

44. Two capacitors, $C_1 = 25.0$ μF and $C_2 = 5.00$ μF, are connected in parallel and charged with a 100-V power supply. (a) Calculate the total energy stored in the two capacitors. (b) What potential difference would be required across the same two capacitors connected in *series*

in order that the combination store the same energy as in (a)?

45. Consider the parallel-plate capacitor formed by the Earth and a cloud layer as described in Problem 16.23. Assume this capacitor will discharge (i.e., produce lightning) when the electric field strength between the plates reaches 3.0×10^6 N/C. What is the energy released if the capacitor discharges completely during a lightning strike?

46. A certain storm cloud has a potential difference of 1.00×10^8 V relative to a tree. If, during a lightning storm, 50.0 C of charge is transferred through this potential difference and 1.00% of the energy is absorbed by the tree, how much water (sap in the tree) initially at 30.0°C can be boiled away? Water has a specific heat of 4 186 J/kg°C, a boiling point of 100°C, and a heat of vaporization of 2.26×10^6 J/kg.

Section 16.10 Capacitors with Dielectrics

47. A capacitor with air between its plates is charged to 100 V and then disconnected from the battery. When a piece of glass is placed between the plates, the voltage across the capacitor drops to 25 V. What is the dielectric constant of the glass? (Assume the glass completely fills the space between the plates.)

48. Two parallel plates, each of area 2.00 cm^2, are separated by 2.00 mm with purified nonconducting water between them. A voltage of 6.00 V is applied between the plates. Calculate (a) the magnitude of the electric field between the plates, (b) the charge stored on each plate, and (c) the charge stored on each plate if the water is removed and replaced with air.

49. **Physics ⊗ Now™** Determine (a) the capacitance and (b) the maximum voltage that can be applied to a Teflon®-filled parallel-plate capacitor having a plate area of 175 cm^2 and an insulation thickness of 0.040 0 mm.

50. A commercial capacitor is constructed as in Figure 16.25a. This particular capacitor is made from a strip of aluminum foil separated by two strips of paraffin-coated paper. Each strip of foil and paper is 7.00 cm wide. The foil is 0.004 00 mm thick, and the paper is 0.025 0 mm thick and has a dielectric constant of 3.70. What length should the strips be if a capacitance of 9.50×10^{-8} F is desired before the capacitor is rolled up? (Use the parallel-plate formula. Adding a second strip of paper and rolling up the capacitor doubles its capacitance by allowing both surfaces of each strip of foil to store charge.)

51. A model of a red blood cell portrays the cell as a spherical capacitor—a positively charged liquid sphere of surface area A separated from the surrounding negatively charged fluid by a membrane of thickness t. Tiny electrodes introduced into the interior of the cell show a potential difference of 100 mV across the membrane. The membrane's thickness is estimated to be 100 nm and whose dielectric constant is 5.00. (a) If an average red blood cell has a mass of 1.00×10^{-12} kg, estimate the volume of the cell and thus find its surface area. The density of blood is 1 100 kg/m^3. (b) Estimate the capacitance of the cell. (c) Calculate the charge on the surface of the membrane. How many electronic charges does the surface charge represent?

ADDITIONAL PROBLEMS

52. Three parallel-plate capacitors are constructed, each having the same plate spacing d and with C_1 having plate area A_1, C_2 having area A_2, and C_3 having area A_3. Show that the total capacitance C of the three capacitors connected in parallel is the same as that of a capacitor having plate spacing d and plate area $A = A_1 + A_2 + A_3$.

53. Three parallel-plate capacitors are constructed, each having the same plate area A and with C_1 having plate spacing d_1, C_2 having plate spacing d_2, and C_3 having plate spacing d_3. Show that the total capacitance C of the three capacitors connected in series is the same as a capacitor of plate area A and with plate spacing $d = d_1 + d_2 + d_3$.

54. Two capacitors give an equivalent capacitance of C_p when connected in parallel and an equivalent capacitance of C_s when connected in series. What is the capacitance of each capacitor?

55. An isolated capacitor of unknown capacitance has been charged to a potential difference of 100 V. When the charged capacitor is disconnected from the battery and then connected in parallel to an uncharged 10.0-μF capacitor, the voltage across the combination is measured to be 30.0 V. Calculate the unknown capacitance.

56. Two charges of 1.0 μC and -2.0 μC are 0.50 m apart at two vertices of an equilateral triangle as in Figure P16.56. (a) What is the electric potential due to the 1.0-μC charge at the third vertex, point P? (b) What is the electric potential due to the -2.0-μC charge at P? (c) Find the total electric potential at P. (d) What is the work required to move a 3.0-μC charge from infinity to P.

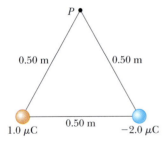

Figure P16.56

57. Find the equivalent capacitance of the group of capacitors shown in Figure P16.57.

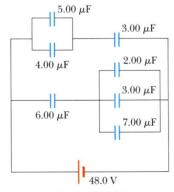

Figure P16.57

58. A spherical capacitor consists of a spherical conducting shell of radius b and charge $-Q$ concentric with a smaller conducting sphere of radius a and charge Q. (a) Find the capacitance of this device. (b) Show that as the radius b

of the outer sphere approaches infinity, the capacitance approaches the value $a/k_e = 4\pi\epsilon_0 a$.

59. The immediate cause of many deaths is ventricular fibrillation, an uncoordinated quivering of the heart, as opposed to proper beating. An electric shock to the chest can cause momentary paralysis of the heart muscle, after which the heart will sometimes start organized beating again. A *defibrillator* is a device that applies a strong electric shock to the chest over a time of a few milliseconds. The device contains a capacitor of a few microfarads, charged to several thousand volts. Electrodes called paddles, about 8 cm across and coated with conducting paste, are held against the chest on both sides of the heart. Their handles are insulated to prevent injury to the operator, who calls, "Clear!" and pushes a button on one paddle to discharge the capacitor through the patient's chest. Assume that an energy of 300 W·s is to be delivered from a 30.0-μF capacitor. To what potential difference must it be charged?

60. When a certain air-filled parallel-plate capacitor is connected across a battery, it acquires a charge of 150 μC on each plate. While the battery connection is maintained, a dielectric slab is inserted into, and fills, the region between the plates. This results in the accumulation of an additional charge of 200 μC on each plate. What is the dielectric constant of the slab?

61. Capacitors $C_1 = 6.0$ μF and $C_2 = 2.0$ μF are charged as a parallel combination across a 250-V battery. The capacitors are disconnected from the battery and from each other. They are then connected positive plate to negative plate and negative plate to positive plate. Calculate the resulting charge on each capacitor.

62. Capacitors $C_1 = 4.0$ μF and $C_2 = 2.0$ μF are charged as a series combination across a 100-V battery. The two capacitors are disconnected from the battery and from each other. They are then connected positive plate to positive plate and negative plate to negative plate. Calculate the resulting charge on each capacitor.

63. The charge distribution shown in Figure P16.63 is referred to as a *linear quadrupole*. (a) Show that the electric potential at a point on the x-axis where $x > d$ is

$$V = \frac{2k_e Q d^2}{x^3 - x d^2}$$

(b) Show that the expression obtained in (a) when $x \gg d$ reduces to

$$V = \frac{2k_e Q d^2}{x^3}$$

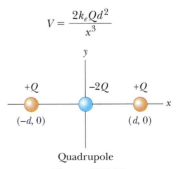

Quadrupole

Figure P16.63

64. The energy stored in a 52.0-μF capacitor is used to melt a 6.00-mg sample of lead. To what voltage must the capacitor be initially charged, assuming that the initial temperature of the lead is 20.0°C? Lead has a specific heat of 128 J/kg·°C, a melting point of 327.3°C, and a latent heat of fusion of 24.5 kJ/kg.

65. Consider a parallel-plate capacitor with charge Q and area A, filled with dielectric material having dielectric constant κ. It can be shown that the magnitude of the attractive force exerted on each plate by the other is $F = Q^2/(2\kappa\epsilon_0 A)$. When a potential difference of 100 V exists between the plates of an air-filled 20-μF parallel-plate capacitor, what force does each plate exert on the other if they are separated by 2.0 mm?

66. An electron is fired at a speed $v_0 = 5.6 \times 10^6$ m/s and at an angle $\theta_0 = -45°$ between two parallel conducting plates that are $D = 2.0$ mm apart, as in Figure P16.66. If the voltage difference between the plates is $\Delta V = 100$ V, determine (a) how close, d, the electron will get to the bottom plate and (b) where the electron will strike the top plate.

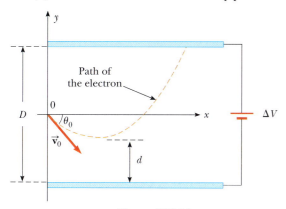

Figure P16.66

ACTIVITIES

1. It takes an electric field of about 30 kV/cm to cause a spark in dry air. Shuffle across a rug and reach toward a doorknob. By estimating the length of the spark, determine the electric potential difference that existed between your finger and the doorknob just before you touched the knob. Try this experiment again on a very humid day, and you will find that the spark is much shorter or is imperceptible. Why?

2. Suppose you are given a battery, a capacitor, two switches, a lightbulb, and several pieces of connecting wire. On a sheet of paper, design a circuit that will do the following: (1) When switch 1 is closed and switch 2 is open, the capacitor charges, but no current moves through the lightbulb. (2) Then, when switch 1 is opened and switch 2 closed, the lightbulb is connected to the capacitor, but not to the battery. Describe the motion of charge in the circuit when switch 1 is closed and switch 2 is open. Is energy being stored in the capacitor? What measurements would you have to make to determine how much energy, if any, is stored? What happens to the lightbulb when switch 1 is opened after the capacitor has charged and switch 2 is then closed? Will the bulb light and stay lit? What happens to the charge on the capacitor when switch 2 is closed in this way?

These power lines transfer energy from the power company to homes and businesses. The energy is transferred at a very high voltage, possibly hundreds of thousands of volts in some cases. The high voltage results in less loss of power due to resistance in the wires, so it is used despite the fact that it makes power lines very dangerous.

Telegraph Colour Library/FPG/Getty Images

CHAPTER

17

Current and Resistance

Many practical applications and devices are based on the principles of static electricity, but electricity was destined to become an inseparable part of our daily lives when scientists learned how to produce a continuous flow of charge for relatively long periods of time using batteries. The battery or voltaic cell was invented in 1800 by the Italian physicist Alessandro Volta. Batteries supplied a continuous flow of charge at low potential, in contrast to earlier electrostatic devices that produced a tiny flow of charge at high potential for brief periods. This steady source of electric current allowed scientists to perform experiments to learn how to control the flow of electric charges in circuits. Today, electric currents power our lights, radios, television sets, air conditioners, computers, and refrigerators. They ignite the gasoline in automobile engines, travel through miniature components making up the chips of microcomputers, and provide the power for countless other invaluable tasks.

In this chapter we define current and discuss some of the factors that contribute to the resistance to the flow of charge in conductors. We also discuss energy transformations in electric circuits. These topics will be the foundation for additional work with circuits in later chapters.

17.1 ELECTRIC CURRENT

In Figure 17.1, charges move in a direction perpendicular to a surface of area A. (The area could be the cross-sectional area of a wire, for example.) **The current is the rate at which charge flows through this surface**.

Suppose ΔQ is the amount of charge that flows through an area A in a time interval Δt and that the direction of flow is perpendicular to the area. Then the current I is equal to the amount of charge divided by the time interval:

$$I \equiv \frac{\Delta Q}{\Delta t} \qquad \text{[17.1]}$$

SI unit: coulomb/second (C/s), or the ampere (A).

One ampere of current is equivalent to one coulomb of charge passing through the cross-sectional area in a time interval of 1 s.

When charges flow through a surface as in Figure 17.1, they can be positive, negative, or both. **The direction of conventional current used in this book is the direction positive charges flow.** (This historical convention originated about 200 years ago, when the ideas of positive and negative charges were introduced.) In a common conductor such as copper, the current is due to the motion of negatively charged electrons, so the direction of the current is opposite the direction of motion of the electrons. On the other hand, for a beam of positively-charged protons in an accelerator, the current is in the same direction as the motion of the protons. In some cases—gases and electrolytes, for example—the current is the result of the flows of both positive and negative charges. Moving charges, whether positive or negative, are referred to as *charge carriers*. In a metal, for example, the charge carriers are electrons.

In electrostatics, where charges are stationary, the electric potential is the same everywhere in a conductor. This is no longer true for conductors carrying current: as charges move along a wire, the electric potential is continually decreasing (except in the special case of superconductors).

TIP 17.1 *Current Flow* is Redundant

The phrases *flow of current* and *current flow* are commonly used, but here the word *flow* is redundant because current is already defined as a flow (of charge). Avoid this construction!

◀ Direction of current

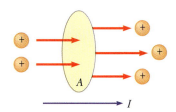

Figure 17.1 Charges in motion through an area A. The time rate of flow of charge through the area is defined as the current I. The direction of the current is the direction of flow of positive charges.

EXAMPLE 17.1 Turn on the Light

Goal Apply the concept of current.

Problem The amount of charge that passes through the filament of a certain lightbulb in 2.00 s is 1.67 C. Find **(a)** the current in the bulb and **(b)** the number of electrons that pass through the filament in 5.00 s.

Strategy Substitute into Equation 17.1 for part (a), then multiply the answer by the time given in part (b) to get the total charge that passes in that time. The total charge equals the number N of electrons going through the circuit times the charge per electron.

Solution
(a) Compute the current in the lightbulb.

Substitute the charge and time into Equation 17.1:

$$I = \frac{\Delta Q}{\Delta t} = \frac{1.67 \text{ C}}{2.00 \text{ s}} = \boxed{0.835 \text{ A}}$$

(b) Find the number of electrons passing through the filament in 5.00 s.

The total number N of electrons times the charge per electron equals the total charge, $I \Delta t$:

$$(1) \quad Nq = I \Delta t$$

Substitute and solve for N:

$$N(1.60 \times 10^{-19} \text{ C/electron}) = (0.835 \text{ A})(5.00 \text{ s})$$

$$N = \boxed{2.61 \times 10^{19} \text{ electrons}}$$

Remarks In developing the solution, it was important to use units to ensure the correctness of equations such as Equation (1). Notice the enormous number of electrons passing through a given point in a typical circuit.

Exercise 17.1
Suppose 6.40×10^{21} electrons pass through a wire in 2.00 min. Find the current.

Answer 8.53 A

Quick Quiz 17.1

Consider positive and negative charges moving horizontally through the four regions in Figure 17.2. Rank the magnitudes of the currents in these four regions from lowest to highest. (I_a is the current in Figure 17.2a, I_b the current in Figure 17.2b, etc.) (a) I_d, I_a, I_c, I_b (b) I_a, I_c, I_b, I_d (c) I_c, I_a, I_d, I_b (d) I_d, I_b, I_c, I_a (e) I_a, I_b, I_c, I_d (f) none of these

Figure 17.2 (Quick Quiz 17.1)

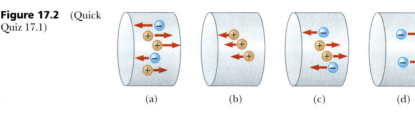

(a)　　(b)　　(c)　　(d)

Figure 17.3 A section of a uniform conductor of cross-sectional area A. The charge carriers move with a speed v_d, and the distance they travel in time Δt is given by $\Delta x = v_d \Delta t$. The number of mobile charge carriers in the section of length Δx is given by $nAv_d\Delta t$, where n is the number of mobile carriers per unit volume.

17.2 A MICROSCOPIC VIEW: CURRENT AND DRIFT SPEED

Macroscopic currents can be related to the motion of the microscopic charge carriers making up the current. It turns out that current depends on the average speed of the charge carriers in the direction of the current, the number of charge carriers per unit volume, and the size of the charge carried by each.

Consider identically charged particles moving in a conductor of cross-sectional area A (Fig. 17.3). The volume of an element of length Δx of the conductor is $A \Delta x$. If n represents the number of mobile charge carriers per unit volume, then the number of carriers in the volume element is $nA \Delta x$. The mobile charge ΔQ in this element is therefore

$$\Delta Q = \text{number of carriers} \times \text{charge per carrier} = (nA \Delta x)q$$

where q is the charge on each carrier. If the carriers move with a constant average speed called the **drift speed** v_d, the distance they move in the time interval Δt is $\Delta x = v_d \Delta t$. We can therefore write

$$\Delta Q = (nAv_d \Delta t)q$$

If we divide both sides of this equation by Δt, we see that the current in the conductor is

$$I = \frac{\Delta Q}{\Delta t} = nqv_dA \qquad \textbf{[17.2]}$$

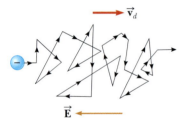

ACTIVE FIGURE 17.4

A schematic representation of the zigzag motion of a charge carrier in a conductor. The sharp changes in direction are due to collisions with atoms in the conductor. Note that the net motion of electrons is opposite the direction of the electric field.

Physics⊗Now™

Log into PhysicsNow at **www.cp7e.com** and go to Active Figure 17.4, where you can observe the random zigzag motion of a charge carrier and see how the motion is affected by an electric field.

TIP 17.2 Electrons are Everywhere in the Circuit

Electrons don't have to travel from the light switch to the light for the light to operate. Electrons already in the filament of the lightbulb move in response to the electric field set up by the battery. Also, the battery does *not* provide electrons to the circuit; it provides *energy* to the existing electrons.

To understand the meaning of drift speed, consider a conductor in which the charge carriers are free electrons. If the conductor is isolated, these electrons undergo random motion similar to the motion of the molecules of a gas. The drift speed is normally much smaller than the free electrons' average speed between collisions with the fixed atoms of the conductor. When a potential difference is applied between the ends of the conductor (say, with a battery), an electric field is set up in the conductor, creating an electric force on the electrons and hence a current. In reality, the electrons don't simply move in straight lines along the conductor. Instead, they undergo repeated collisions with the atoms of the metal, and the result is a complicated zigzag motion with only a small average drift speed along the wire (Active Fig. 17.4). The energy transferred from the electrons to the metal atoms during a collision increases the vibrational energy of the atoms and causes a corresponding increase in the temperature of the conductor. Despite the collisions, however, the electrons move slowly along the conductor in a direction opposite $\vec{\mathbf{E}}$ with the drift velocity $\vec{\mathbf{v}}_d$.

EXAMPLE 17.2   Drift Speed of Electrons

Goal   Calculate a drift speed and compare it with the rms speed of an electron gas.

Problem   A copper wire of cross-sectional area 3.00×10^{-6} m^2 carries a current of 10.0 A. **(a)** Assuming that each copper atom contributes one free electron to the metal, find the drift speed of the electrons in this wire. **(b)** Use the ideal gas model to compare the drift speed with the random rms speed an electron would have at 20.0°C. The density of copper is 8.92 g/cm^3, and its atomic mass is 63.5 u.

Strategy   All the variables in Equation 17.2 are known except for n, the number of free charge carriers per unit volume. We can find n by recalling that one mole of copper contains an Avogadro's number (6.02×10^{23}) of atoms and each atom contributes one charge carrier to the metal. The volume of one mole can be found from copper's known density and atomic mass. The atomic mass is the same, numerically, as the number of grams in a mole of the substance.

Solution

(a) Find the drift speed of the electrons.

Calculate the volume of one mole of copper from its density and its atomic mass:

$$V = \frac{m}{\rho} = \frac{63.5 \text{ g}}{8.92 \text{ g/cm}^3} = 7.12 \text{ cm}^3$$

Convert the volume from cm^3 to m^3:

$$7.12 \text{ cm}^3 \left(\frac{1 \text{ m}}{10^2 \text{ cm}} \right)^3 = 7.12 \times 10^{-6} \text{ m}^3$$

Divide Avogadro's number (the number of electrons in one mole) by the volume per mole to obtain the number density:

$$n = \frac{6.02 \times 10^{23} \text{ electrons/mole}}{7.12 \times 10^{-6} \text{ m}^3/\text{mole}}$$
$$= 8.46 \times 10^{28} \text{ electrons/m}^3$$

Solve Equation 17.2 for the drift speed, and substitute:

$$v_d = \frac{I}{nqA}$$

$$= \frac{10.0 \text{ C/s}}{(8.46 \times 10^{28} \text{ electrons/m}^3)(1.60 \times 10^{-19} \text{ C})(3.00 \times 10^{-6} \text{ m}^2)}$$

$$v_d = \boxed{2.46 \times 10^{-4} \text{ m/s}}$$

(b) Find the rms speed of a gas of electrons at 20.0°C.

Apply Equation 10.18:

$$v_{rms} = \sqrt{\frac{3k_B T}{m_e}}$$

Convert the temperature to the Kelvin scale, and substitute values:

$$v_{rms} = \sqrt{\frac{3(1.38 \times 10^{-23} \text{ J/K})(293 \text{ K})}{9.11 \times 10^{-31} \text{ kg}}}$$

$$= \boxed{1.15 \times 10^5 \text{ m/s}}$$

Remarks   The drift speed of an electron in a wire is very small—only about one-billionth of its random thermal speed.

Exercise 17.2

What current in a copper wire with a cross-sectional area of 7.50×10^{-7} m^2 would result in a drift speed of 5.00×10^{-4} m/s?

Answer   5.08 A

Example 17.2 shows that drift speeds are typically very small. In fact, the drift speed is much smaller than the average speed between collisions. Electrons traveling at 2.46×10^{-4} m/s, as in the example, would take about 68 min to travel 1 m!

In view of this low speed, you might wonder why a light turns on almost instantaneously when a switch is thrown. Think of the flow of water through a pipe. If a drop of water is forced into one end of a pipe that is already filled with water, a drop must be pushed out the other end of the pipe. Although it may take an individual drop a long time to make it through the pipe, a flow initiated at one end produces a similar flow at the other end very quickly. Another familiar analogy is the motion of a bicycle chain. When the sprocket moves one link, the other links all move more or less immediately, even though it takes a given link some time to make a complete rotation. In a conductor, the electric field driving the free electrons travels at a speed close to that of light, so when you flip a light switch, the message for the electrons to start moving through the wire (the electric field) reaches them at a speed on the order of 10^8 m/s!

Quick Quiz 17.2

Suppose a current-carrying wire has a cross-sectional area that gradually becomes smaller along the wire, so that the wire has the shape of a very long cone. How does the drift speed vary along the wire? (a) It slows down as the cross section becomes smaller. (b) It speeds up as the cross section becomes smaller. (c) It doesn't change. (d) More information is needed.

17.3 CURRENT AND VOLTAGE MEASUREMENTS IN CIRCUITS

To study electric current in circuits, we need to understand how to measure currents and voltages.

The circuit shown in Figure 17.5a is a drawing of the actual circuit necessary for measuring the current in Example 17.1. Figure 17.5b shows a stylized figure called a circuit diagram which represents the actual circuit of Figure 17.5a. This circuit consists of only a battery and a lightbulb. The word "circuit" means "a closed loop

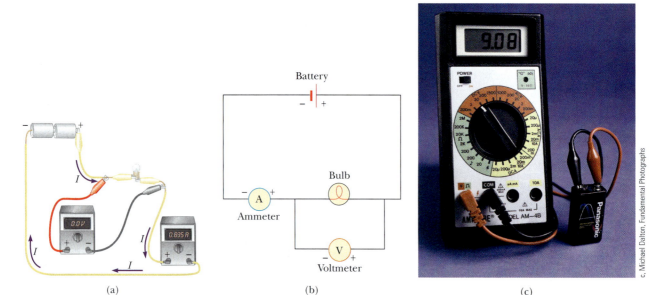

Figure 17.5 (a) A sketch of an actual circuit used to measure the current in a flashlight bulb and the potential difference across it. (b) A schematic diagram of the circuit shown in part (a). (c) A digital multimeter can be used to measure both currents and potential differences. Here, the meter is measuring the potential difference across a 9-V battery.

of some sort around which current circulates." The battery pumps charge through the bulb and around the loop. No charge would flow without a complete conducting path from the positive terminal of the battery into one side of the bulb, out the other side, and through the copper conducting wires back to the negative terminal of the battery. The most important quantities that characterize how the bulb works in different situations are the current I in the bulb and the potential difference ΔV across the bulb. To measure the current in the bulb, we place an ammeter, the device for measuring current, in line with the bulb so there is no path for the current to bypass the meter; all of the charge passing through the bulb must also pass through the ammeter. The voltmeter measures the potential difference, or voltage, between the two ends of the bulb's filament. If we use two meters simultaneously as in Figure 17.5a, we can remove the voltmeter and see if its presence affects the current reading. Figure 17.5c shows a digital multimeter—a convenient device, with a digital readout, that can be used to measure voltage, current, or resistance. An advantage of using a digital multimeter as a voltmeter is that it will usually not affect the current, since a digital meter has enormous resistance to the flow of charge in the voltmeter mode.

At this point, you can measure the current as a function of voltage (an I–ΔV curve) of various devices in the lab. All you need is a variable voltage supply (an adjustable battery) capable of supplying potential differences from about -5 V to $+5$ V, a bulb, a resistor, some wires and alligator clips, and a couple of multimeters. Be sure to always start your measurements using the highest multimeter scales (say, 10 A and 1 000 V), and increase the sensitivity one scale at a time to obtain the highest accuracy without overloading the meters. (Increasing the sensitivity means lowering the maximum current or voltage that the scale reads.) Note that the meters must be connected with the proper polarity with respect to the voltage supply, as shown in Figure 17.5b. Finally, follow your instructor's directions carefully to avoid damaging the meters and incurring a soaring lab fee.

Quick Quiz 17.3

Look at the four "circuits" shown in Figure 17.6 and select those that will light the bulb.

(a)   (b)   (c)   (d)

Figure 17.6   (Quick Quiz 17.3)

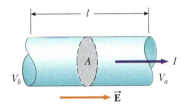

Figure 17.7   A uniform conductor of length l and cross-sectional area A. The current I in the conductor is proportional to the applied voltage $\Delta V = V_b - V_a$. The electric field $\vec{E}$ set up in the conductor is also proportional to the current.

17.4   RESISTANCE AND OHM'S LAW

When a voltage (potential difference) ΔV is applied across the ends of a metallic conductor as in Figure 17.7, the current in the conductor is found to be proportional to the applied voltage; $I \propto \Delta V$. If the proportionality holds, we can write $\Delta V = IR$, where the proportionality constant R is called the *resistance* of the conductor. In fact, we define the **resistance** as the ratio of the voltage across the conductor to the current it carries:

$$R \equiv \frac{\Delta V}{I}$$ [17.3]

◀ Resistance

GEORG SIMON OHM
(1787–1854)

A high school teacher in Cologne and later a professor at Munich, Ohm formulated the concept of resistance and discovered the proportionalities expressed in Equation 17.5.

Resistance has SI units of volts per ampere, called **ohms** (Ω). If a potential difference of 1 V across a conductor produces a current of 1 A, the resistance of the conductor is 1 Ω. For example, if an electrical appliance connected to a 120-V source carries a current of 6 A, its resistance is 20 Ω.

The concepts of electric current, voltage, and resistance can be compared with the flow of water in a river. As water flows downhill in a river of constant width and depth, the flow rate (water current) depends on the steepness of descent of the river and the effects of rocks, the riverbank, and other obstructions. The voltage difference is analogous to the steepness, and the resistance to the obstructions. Based on this analogy, it seems reasonable that increasing the voltage applied to a circuit should increase the current in the circuit, just as increasing the steepness of descent increases the water current. Also, increasing the obstructions in the river's path will reduce the water current, just as increasing the resistance in a circuit will lower the electric current. Resistance in a circuit arises due to collisions between the electrons carrying the current with fixed atoms inside the conductor. These collisions inhibit the movement of charges in much the same way as would a force of friction. For many materials, including most metals, experiments show that **the resistance remains constant over a wide range of applied voltages or currents**. This statement is known as **Ohm's law**, after Georg Simon Ohm (1789–1854), who was the first to conduct a systematic study of electrical resistance.

Ohm's law is given by

Ohm's law ▶

$$\Delta V = IR \qquad \text{[17.4]}$$

where R is understood to be independent of ΔV, the potential drop across the resistor, and I, the current in the resistor. We will continue to use this traditional form of Ohm's law when discussing electrical circuits. A **resistor** is a conductor that provides a specified resistance in an electric circuit. The symbol for a resistor in circuit diagrams is a zigzag line: ———W———.

Ohm's law is an empirical relationship valid only for certain materials. Materials that obey Ohm's law, and hence have a constant resistance over a wide range of voltages, are said to be **ohmic**. Materials having resistance that changes with voltage or current are **nonohmic**. Ohmic materials have a linear current–voltage relationship over a large range of applied voltages (Fig. 17.8a). Nonohmic materials have a nonlinear current–voltage relationship (Fig. 17.8b). One common semiconducting device that is nonohmic is the *diode*, a circuit element that acts like a one-way valve for current. Its resistance is small for currents in one direction (positive ΔV) and large for currents in the reverse direction (negative ΔV). Most modern electronic devices, such as transistors, have nonlinear current–voltage relationships; their operation depends on the particular ways in which they violate Ohm's law.

Figure 17.8 (a) The current–voltage curve for an ohmic material. The curve is linear, and the slope gives the resistance of the conductor. (b) A nonlinear current–voltage curve for a semiconducting diode. This device doesn't obey Ohm's law.

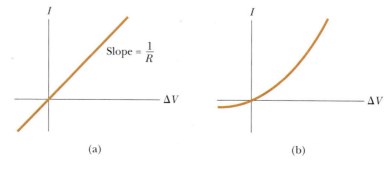

Quick Quiz 17.4

In Figure 17.8b, does the resistance of the diode (a) increase or (b) decrease as the positive voltage ΔV increases?

An assortment of resistors used for a variety of applications in electronic circuits.

EXAMPLE 17.3 Resistance of a Steam Iron

Goal Use Ohm's law to calculate a resistance.

Problem All electric devices are required to have identifying plates that specify their electrical characteristics. The plate on a certain steam iron states that the iron carries a current of 6.40 A when connected to a source of 1.20×10^2 V. What is the resistance of the steam iron?

Strategy Substitute into Ohm's law.

Solution
Apply Equation 17.3:

$$R = \frac{\Delta V}{I} = \frac{1.20 \times 10^2 \text{ V}}{6.40 \text{ A}} = \boxed{18.8 \; \Omega}$$

Exercise 17.3
The resistance of a hot plate is 48.0 Ω. How much current does the plate carry when connected to a 1.20×10^2-V source?

Answer 2.50 A

17.5 RESISTIVITY

Electrons don't move in straight-line paths through a conductor. Instead, they undergo repeated collisions with the metal atoms. Consider a conductor with a voltage applied across its ends. An electron gains speed as the electric force associated with the internal electric field accelerates it, giving it a velocity in the direction opposite that of the electric field. A collision with an atom randomizes the electron's velocity, reducing it in the direction opposite the field. The process then repeats itself. Together, these collisions affect the electron somewhat as a force of internal friction would. This is the origin of a material's resistance.

The resistance of an ohmic conductor increases with length, which makes sense because the electrons going through it must undergo more collisions in a longer conductor. A smaller cross-sectional area also increases the resistance of a conductor, just as a smaller pipe slows the fluid moving through it. The resistance, then, is proportional to the conductor's length l and inversely proportional to its cross-sectional area A,

$$R = \rho \frac{l}{A} \qquad \text{[17.5]}$$

where the constant of proportionality, ρ, is called the **resistivity** of the material.[1] Every material has a characteristic resistivity that depends on its electronic structure and on temperature. Good electric conductors have very low resistivities, and good insulators have very high resistivities. Table 17.1 lists the resistivities of various materials at 20°C. Because resistance values are in ohms, resistivity values must be in ohm-meters ($\Omega \cdot$ m).

---

[1]The symbol ρ used for resistivity shouldn't be confused with the same symbol used earlier in the book for density. Often, a single symbol is used to represent different quantities.

TABLE 17.1

Resistivities and Temperature Coefficients of Resistivity for Various Materials (at 20°C)

Material	Resistivity ($\Omega \cdot m$)	Temperature Coefficient of Resistivity $[(°C)^{-1}]$
Silver	1.59×10^{-8}	3.8×10^{-3}
Copper	1.7×10^{-8}	3.9×10^{-3}
Gold	2.44×10^{-8}	3.4×10^{-3}
Aluminum	2.82×10^{-8}	3.9×10^{-3}
Tungsten	5.6×10^{-8}	4.5×10^{-3}
Iron	10.0×10^{-8}	5.0×10^{-3}
Platinum	11×10^{-8}	3.92×10^{-3}
Lead	22×10^{-8}	3.9×10^{-3}
Nichrome[a]	150×10^{-8}	0.4×10^{-3}
Carbon	3.5×10^{5}	-0.5×10^{-3}
Germanium	0.46	-48×10^{-3}
Silicon	640	-75×10^{-3}
Glass	$10^{10}-10^{14}$	
Hard rubber	$\approx 10^{13}$	
Sulfur	10^{15}	
Quartz (fused)	75×10^{16}	

[a]A nickel–chromium alloy commonly used in heating elements.

Applying Physics 17.1 Dimming of Aging Lightbulbs

As a lightbulb ages, why does it gives off less light than when new?

Explanation There are two reasons for the lightbulb's behavior, one electrical and one optical, but both are related to the same phenomenon occurring within the bulb. The filament of an old lightbulb is made of a tungsten wire that has been kept at a high temperature for many hours. High temperatures evaporate tungsten from the filament, decreasing its radius. From $R = \rho l/A$, we see that a decreased cross-sectional area leads to an increase in the resistance of the filament.

This increasing resistance with age means that the filament will carry less current for the same applied voltage. With less current in the filament, there is less light output, and the filament glows more dimly.

At the high operating temperature of the filament, tungsten atoms leave its surface, much as water molecules evaporate from a puddle of water. The atoms are carried away by convection currents in the gas in the bulb and are deposited on the inner surface of the glass. In time, the glass becomes less transparent because of the tungsten coating, which decreases the amount of light that passes through the glass.

INTERACTIVE EXAMPLE 17.4   The Resistance of Nichrome Wire

Goal Combine the concept of resistivity with Ohm's law.

Problem (a) Calculate the resistance per unit length of a 22-gauge nichrome wire of radius 0.321 mm. (b) If a potential difference of 10.0 V is maintained across a 1.00-m length of the nichrome wire, what is the current in the wire? (c) The wire is melted down and recast with twice its original length. Find the new resistance R_N as a multiple of the old resistance R_O.

Strategy Part (a) requires substitution into Equation 17.5, after calculating the cross-sectional area, while part (b) is a matter of substitution unto Ohm's law. Part (c) requires some algebra. The idea is to take the expression for the new resistance and substitute expressions for l_N and A_N, the new length and cross-sectional area, in terms of the old length and cross-section. For the area substitution, use the fact that the volumes of the old and new wires are the same.

Solution

(a) Calculate the resistance per unit length.

Find the cross-sectional area of the wire:

$$A = \pi r^2 = \pi(0.321 \times 10^{-3}\text{ m})^2 = 3.24 \times 10^{-7}\text{ m}^2$$

Obtain the resistivity of nichrome from Table 17.1, solve Equation 17.5 for R/l, and substitute:

$$\frac{R}{l} = \frac{\rho}{A} = \frac{1.5 \times 10^{-6}\,\Omega\cdot\text{m}}{3.24 \times 10^{-7}\text{ m}^2} = \boxed{4.6\ \Omega/\text{m}}$$

(b) Find the current in a 1.00-m segment of the wire if the potential difference across it is 10.0 V.

Substitute given values into Ohm's law:

$$I = \frac{\Delta V}{R} = \frac{10.0\text{ V}}{4.6\ \Omega} = \boxed{2.2\text{ A}}$$

(c) If the wire is melted down and recast with twice its original length, find the new resistance as a multiple of the old.

Find the new area A_N in terms of the old area A_O, using the fact the volume doesn't change and $l_N = 2l_O$:

$$V_N = V_O \;\rightarrow\; A_N l_N = A_O l_O \;\rightarrow\; A_N = A_O(l_O/l_N)$$
$$A_N = A_O(l_O/2l_O) = A_O/2$$

Substitute into Equation 17.5:

$$R_N = \frac{\rho l_N}{A_N} = \frac{\rho(2l_O)}{(A_O/2)} = 4\,\frac{\rho l_O}{A_O} = \boxed{4R_O}$$

Remarks   From Table 17.1, the resistivity of nichrome is about 100 times that of copper, a typical good conductor. Therefore, a copper wire of the same radius would have a resistance per unit length of only 0.052 $\Omega/$m, and a 1.00-m length of copper wire of the same radius would carry the same current (2.2 A) with an applied voltage of only 0.115 V.

   Because of its resistance to oxidation, nichrome is often used for heating elements in toasters, irons, and electric heaters.

Exercise 17.4

What is the resistance of a 6.0-m length of nichrome wire that has a radius 0.321 mm? How much current does it carry when connected to a 120-V source?

Answer   28 Ω; 4.3 A

Physics⊗Now™ You can explore the resistance of different materials by logging into PhysicsNow at **www.cp7e.com** and going to Interactive Example 17.4.

Quick Quiz 17.5

Suppose an electrical wire is replaced with one having every linear dimension doubled (i.e. the length and radius have twice their original values). Does the wire now have (a) more resistance than before, (b) less resistance, or (c) the same resistance?

17.6   TEMPERATURE VARIATION OF RESISTANCE

The resistivity ρ, and hence the resistance, of a conductor depends on a number of factors. One of the most important is the temperature of the metal. For most metals, resistivity increases with increasing temperature. This correlation can be understood as follows: as the temperature of the material increases, its constituent atoms vibrate with greater amplitudes. As a result, the electrons find it more diffi-cult to get by those atoms, just as it is more difficult to weave through a crowded

In an old-fashioned carbon filament incandescent lamp, the electrical resistance is typically 10 Ω, but changes with temperature.

Courtesy of Central Scientific Company

room when the people are in motion than when they are standing still. The increased electron scattering with increasing temperature results in increased resistivity. Technically, thermal expansion also affects resistance; however, this is a very small effect.

Over a limited temperature range, the resistivity of most metals increases linearly with increasing temperature according to the expression

$$\rho = \rho_0[1 + \alpha(T - T_0)] \qquad [17.6]$$

where ρ is the resistivity at some temperature T (in Celsius degrees), ρ_0 is the resistivity at some reference temperature T_0 (usually taken to be 20°C), and α is a parameter called the **temperature coefficient of resistivity**. Temperature coefficients for various materials are provided in Table 17.1. The interesting negative values of α for semiconductors arise because these materials possess weakly bound charge carriers that become free to move and contribute to the current as the temperature rises.

Because the resistance of a conductor with a uniform cross section is proportional to the resistivity according to Equation 17.5 ($R = \rho l/A$), the temperature variation of resistance can be written

$$R = R_0[1 + \alpha(T - T_0)] \qquad [17.7]$$

Precise temperature measurements are often made using this property, as shown by the following example.

EXAMPLE 17.5 A Platinum Resistance Thermometer

Goal Apply the temperature dependence of resistance.

Problem A resistance thermometer, which measures temperature by measuring the change in resistance of a conductor, is made of platinum and has a resistance of 50.0 Ω at 20.0°C. **(a)** When the device is immersed in a vessel containing melting indium, its resistance increases to 76.8 Ω. From this information, find the melting point of indium. **(b)** The indium is heated further until it reaches a temperature of 235°C. What is the ratio of the new current in the platinum to the current I_{mp} at the melting point?

Strategy In part (a), solve Equation 17.7 for $T - T_0$ and get α for platinum from Table 17.1, substituting known quantities. For part (b), use Ohm's law in Equation 17.7.

Solution
(a) Find the melting point of indium.

Solve Equation 17.7 for $T - T_0$:

$$T - T_0 = \frac{R - R_0}{\alpha R_0} = \frac{76.8\ \Omega - 50.0\ \Omega}{[3.92 \times 10^{-3}\ (°C)^{-1}][50.0\ \Omega]}$$
$$= 137°C$$

Substitute $T_0 = 20.0°C$ and obtain the melting point of indium:

$$T = \boxed{157°C}$$

(b) Find the ratio of the new current to the old when the temperature rises from 157°C to 235°C.

Write Equation 17.7, with R_0 and T_0 replaced by R_{mp} and T_{mp}, the resistance and temperature at the melting point.

$$R = R_{mp}[1 + \alpha(T - T_{mp})]$$

According to Ohm's law, $R = \Delta V/I$ and $R_{mp} = \Delta V/I_{mp}$. Substitute these expressions into Equation 17.7:

$$\frac{\Delta V}{I} = \frac{\Delta V}{I_{mp}}[1 + \alpha(T - T_{mp})]$$

Cancel the voltage differences, invert the two expressions, and then divide both sides by I_{mp}:

$$\frac{I}{I_{mp}} = \frac{1}{1 + \alpha(T - T_{mp})}$$

Substitute $T = 235°C$, $T_{mp} = 157°C$, and the value for α, obtaining the desired ratio:

$$\frac{I}{I_{mp}} = \boxed{0.766}$$

Exercise 17.5

Suppose a wire made of an unknown alloy and having a temperature of 20.0°C carries a current of 0.450 A. At 52.0°C the current is 0.370 A for the same potential difference. Find the temperature coefficient of resistivity of the alloy.

Answer 6.76×10^{-3} $(°C)^{-1}$

17.7 SUPERCONDUCTORS

There is a class of metals and compounds with resistances that fall virtually to *zero* below a certain temperature T_c called the *critical temperature*. These materials are known as **superconductors**. The resistance–temperature graph for a superconductor follows that of a normal metal at temperatures above T_c (Fig. 17.9). When the temperature is at or below T_c, however, the resistance suddenly drops to zero. This phenomenon was discovered in 1911 by the Dutch physicist H. Kamerlingh Onnes as he and a graduate student worked with mercury, which is a superconductor below 4.1 K. Recent measurements have shown that the resistivities of superconductors below T_c are less than 4×10^{-25} $\Omega \cdot m$—around 10^{17} times smaller than the resistivity of copper and in practice considered to be zero.

Today thousands of superconductors are known, including such common metals as aluminum, tin, lead, zinc, and indium. Table 17.2 lists the critical temperatures of several superconductors. The value of T_c is sensitive to chemical composition, pressure, and crystalline structure. Interestingly, copper, silver, and gold, which are excellent conductors, don't exhibit superconductivity.

One of the truly remarkable features of superconductors is the fact that once a current is set up in them, it persists *without any applied voltage* (because $R = 0$). In fact, steady currents in superconducting loops have been observed to persist for years with no apparent decay!

An important development in physics that created much excitement in the scientific community was the discovery of high-temperature copper-oxide-based superconductors. The excitement began with a 1986 publication by J. Georg Bednorz and K. Alex Müller, scientists at the IBM Zurich Research Laboratory in Switzerland, in which they reported evidence for superconductivity at a temperature near 30 K in an oxide of barium, lanthanum, and copper. Bednorz and Müller were awarded the Nobel Prize for physics in 1987 for their important discovery. The discovery was remarkable in view of the fact that the critical temperature was significantly higher than those of any previously known superconductors. Shortly thereafter a new family of compounds was investigated, and research activity in the field of superconductivity proceeded vigorously. In early 1987, groups at the University of Alabama at Huntsville and the University of Houston announced the discovery of superconductivity at about 92 K in an oxide of yttrium, barium, and copper ($YBa_2Cu_3O_7$), shown as the gray disk in Figure 17.10. Late in 1987, teams of scientists from Japan and the United States reported superconductivity at 105 K in an oxide of bismuth, strontium, calcium, and copper. More recently, scientists have reported superconductivity at temperatures as high as 150 K in an oxide containing mercury. The search for novel superconducting materials continues, with the hope of someday obtaining a room-temperature superconducting material. This research is important both for scientific reasons and for practical applications.

An important and useful application is the construction of superconducting magnets in which the magnetic field intensities are about ten times greater than those of the best normal electromagnets. Such magnets are being considered as a means of storing energy. The idea of using superconducting power lines to transmit

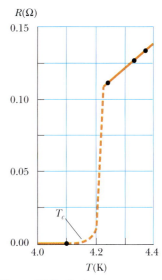

$R(\Omega)$

Figure 17.9 Resistance versus temperature for a sample of mercury (Hg). The graph follows that of a normal metal above the critical temperature T_c. The resistance drops to zero at the critical temperature, which is 4.2 K for mercury, and remains at zero for lower temperatures.

TABLE 17.2

Critical Temperatures for Various Superconductors

Material	T_c (K)
Zn	0.88
Al	1.19
Sn	3.72
Hg	4.15
Pb	7.18
Nb	9.46
Nb_3Sn	18.05
Nb_3Ge	23.2
$YBa_2Cu_3O_7$	90
Bi–Sr–Ca–Cu–O	105
Tl–Ba–Ca–Cu–O	125
$HgBa_2Ca_2Cu_3O_8$	134

Figure 17.10 A small permanent magnet floats freely above a ceramic disk made of the superconductor $YBa_2Cu_3O_7$, cooled by liquid nitrogen at 77 K. The superconductor has zero electric resistance at temperatures below 92 K and expels any applied magnetic field.

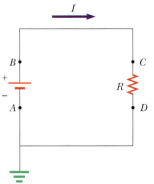

ACTIVE FIGURE 17.11
A circuit consisting of a battery and a resistance R. Positive charge flows clockwise from the positive to the negative terminal of the battery. Point A is grounded.

Physics⊗Now™

Log into PhysicsNow at **www.cp7e.com** and go to Active Figure 17.11, where you can adjust the battery voltage and the resistance, and see the resulting current in the circuit and the power dissipated as heat by the resistor.

power efficiently is also receiving serious consideration. Modern superconducting electronic devices consisting of two thin-film superconductors separated by a thin insulator have been constructed. Among these devices are magnetometers (magnetic-field measuring devices) and various microwave devices.

17.8 ELECTRICAL ENERGY AND POWER

If a battery is used to establish an electric current in a conductor, chemical energy stored in the battery is continuously transformed into kinetic energy of the charge carriers. This kinetic energy is quickly lost as a result of collisions between the charge carriers and fixed atoms in the conductor, causing an increase in the temperature of the conductor. In this way, the chemical energy stored in the battery is continuously transformed into thermal energy.

In order to understand the process of energy transfer in a simple circuit, consider a battery with terminals connected to a resistor (Active Fig. 17.11; remember that the positive terminal of the battery is always at the higher potential). Now imagine following a quantity of positive charge ΔQ around the circuit from point A, through the battery and resistor, and back to A. Point A is a reference point that is grounded (the ground symbol is ⏚), and its potential is taken to be zero. As the charge ΔQ moves from A to B through the battery, the electrical potential energy of the system increases by the amount $\Delta Q\,\Delta V$, and the chemical potential energy in the battery decreases by the same amount. (Recall from Chapter 16 that $\Delta PE = q\,\Delta V$.) However, as the charge moves from C to D through the resistor, it loses this electrical potential energy during collisions with atoms in the resistor. In the process, the energy is transformed to internal energy corresponding to increased vibrational motion of those atoms. Because we can ignore the very small resistance of the interconnecting wires, no energy transformation occurs for paths BC and DA. When the charge returns to point A, the net result is that some of the chemical energy in the battery has been delivered to the resistor and has caused its temperature to rise.

The charge ΔQ loses energy $\Delta Q\,\Delta V$ as it passes through the resistor. If Δt is the time it takes the charge to pass through the resistor, then the rate at which it loses electric potential energy is

$$\frac{\Delta Q}{\Delta t}\,\Delta V = I\Delta V$$

where I is the current in the resistor and ΔV is the potential difference across it. Of course, the charge regains this energy when it passes through the battery, at the expense of chemical energy in the battery. The rate at which the system loses potential energy as the charge passes through the resistor is equal to the rate at which the system gains internal energy in the resistor. Therefore, the power $\mathcal{P}$, representing the rate at which energy is delivered to the resistor, is

Power ▶

$$\mathcal{P} = I\Delta V \qquad\qquad [17.8]$$

While this result was developed by considering a battery delivering energy to a resistor, Equation 17.8 can be used to determine the power transferred from a voltage source to *any* device carrying a current I and having a potential difference ΔV between its terminals.

Using Equation 17.8 and the fact that $\Delta V = IR$ for a resistor, we can express the power delivered to the resistor in the alternate forms

Power delivered to a resistor ▶

$$\mathcal{P} = I^2R = \frac{\Delta V^2}{R} \qquad\qquad [17.9]$$

When I is in amperes, ΔV in volts, and R in ohms, the SI unit of power is the watt (introduced in Chapter 5). The power delivered to a conductor of resistance R is often referred to as an I^2R *loss*. Note that Equation 17.9 applies only to resistors and not to nonohmic devices such as lightbulbs and diodes.

Regardless of the ways in which you use electrical energy in your home, you ultimately must pay for it or risk having your power turned off. The unit of energy used by electric companies to calculate consumption, the **kilowatt-hour**, is defined in terms of the unit of power and the amount of time it's supplied. One kilowatt-hour (kWh) is the energy converted or consumed in 1 h at the constant rate of 1 kW. It has the numerical value

$$1 \text{ kWh} = (10^3 \text{ W})(3600 \text{ s}) = 3.60 \times 10^6 \text{ J} \qquad \text{[17.10]}$$

On an electric bill, the amount of electricity used in a given period is usually stated in multiples of kilowatt-hours.

Applying Physics 17.2   Lightbulb Failures

Why do lightbulbs fail so often right after they're turned on?

Explanation Once the switch is closed, the line voltage is applied across the bulb. As the voltage is applied across the cold filament when the bulb is first turned on, the resistance of the filament is low, the current is high, and a relatively large amount of power is delivered to the bulb. This current spike at the beginning of operation is the reason why lightbulbs often fail just after they are turned on. As the filament warms, its resistance rises and the current decreases. As a result, the power delivered to the bulb decreases, and the bulb is less likely to burn out.

Quick Quiz 17.6

A voltage ΔV is applied across the ends of a nichrome heater wire having a cross-sectional area A and length L. The same voltage is applied across the ends of a second heater wire having a cross-sectional area A and length $2L$. Which wire gets hotter? (a) the shorter wire, (b) the longer wire, or (c) more information is needed.

Quick Quiz 17.7

For the two resistors shown in Figure 17.12, rank the currents at points a through f from largest to smallest.

(a) $I_a = I_b > I_e = I_f > I_c = I_d$
(b) $I_a = I_b > I_c = I_d > I_e = I_f$
(c) $I_e = I_f > I_c = I_d > I_a = I_b$

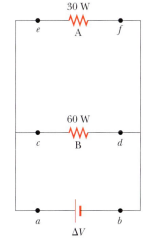

Figure 17.12 (Quick Quiz 17.7)

Quick Quiz 17.8

Two resistors, A and B, are connected in a series circuit with a battery. The resistance of A is twice that of B. Which resistor dissipates more power? (a) resistor A (b) resistor B (c) More information is needed.

Example 17.6   The Cost of Lighting Up Your Life

Goal   Apply the electric power concept, and calculate the cost of power usage using kilowatt-hours.

Problem   A circuit provides a maximum current of 20.0 A at an operating voltage of 1.20×10^2 V. **(a)** How many 75 W bulbs can operate with this voltage source? **(b)** At \$0.120 per kilowatt-hour, how much does it cost to operate these bulbs for 8.00 h?

Strategy   Find the necessary power with $\mathcal{P} = I\Delta V$, then divide by 75.0 W per bulb to get the total number of bulbs. To find the cost, convert power to kilowatts and multiply by the number of hours, then multiply by the cost per kilowatt-hour.

Solution

(a) Find the number of bulbs that can be lighted.

Substitute into Equation 17.8 to get the total power:

$$\mathcal{P}_{total} = I\Delta V = (20.0 \text{ A})(1.20 \times 10^2 \text{ V}) = 2.40 \times 10^3 \text{ W}$$

Divide the total power by the power per bulb to get the number of bulbs.

$$\text{Number of bulbs} = \frac{\mathcal{P}_{total}}{\mathcal{P}_{bulb}} = \frac{2.40 \times 10^3 \text{ W}}{75.0 \text{ W}} = \boxed{32.0}$$

(b) Calculate the cost of this electricity for an 8.00-h day.

Find the energy in kilowatt-hours:

$$\text{Energy} = \mathcal{P}t = (2.40 \times 10^3 \text{ W})\left(\frac{1.00 \text{ kW}}{1.00 \times 10^3 \text{ W}}\right)(8.00 \text{ h})$$

$$= 19.2 \text{ kWh}$$

Multiply by the cost per kilowatt-hour:

$$\text{Cost} = (19.2 \text{ kWh})(\$0.12/\text{kWh}) = \boxed{\$2.30}$$

Remarks   This amount of energy might correspond to what a small office uses in a working day, taking into account all power requirements (not just lighting). In general, resistive devices can have variable power output, depending on how the circuit is wired. Here, power outputs were specified, so such considerations were unnecessary.

Exercise 17.6

(a) How many Christmas tree lights drawing 5.00 W of power each could be run on a circuit operating at 1.20×10^2 V and providing 15.0 A of current? **(b)** Find the cost to operate one such string 24.0 h per day for the Christmas season (two weeks), using the rate \$0.12/kWh.

Answers   **(a)** 3.60×10^2 bulbs   **(b)** \$72.60

EXAMPLE 17.7   The Power Converted by an Electric Heater

Goal   Calculate an electrical power output, and link to its effect on the environment through the first law of thermodynamics.

Problem   An electric heater is operated by applying a potential difference of 50.0 V to a nichrome wire of total resistance 8.00 Ω. **(a)** Find the current carried by the wire and the power rating of the heater. **(b)** Using this heater, how long would it take to heat 2.50×10^3 moles of diatomic gas (e.g., a mixture of oxygen and nitrogen—air) from a chilly 10.0°C to 25.0°C? Take the molar specific heat at constant volume of air to be $\frac{5}{2}R$.

Strategy   For part (a), find the current with Ohm's law and substitute into the expression for power. Part (b) is an isovolumetric process, so the thermal energy provided by the heater all goes into the change in internal energy, ΔU. Calculate this quantity using the first law of thermodynamics, and divide by the power to get the time.

Solution

(a) Compute the current and power output.

Apply Ohm's law to get the current:

$$I = \frac{\Delta V}{R} = \frac{50.0 \text{ V}}{8.00 \text{ Ω}} = \boxed{6.25 \text{ A}}$$

Substitute into Equation 17.9 to find the power: $\mathcal{P} = I^2R = (6.25 \text{ A})^2(8.00 \text{ }\Omega) = \boxed{313 \text{ W}}$

(b) How long does it take to heat the gas?

Calculate the thermal energy transfer from the first law. Note that $W = 0$ because the volume doesn't change.

$$Q = \Delta U = nC_v\Delta T$$
$$= (2.50 \times 10^3 \text{ mol})(\tfrac{5}{2}\cdot 8.31 \text{ J/mol}\cdot\text{K})(298 \text{ K} - 283 \text{ K})$$
$$= 7.79 \times 10^5 \text{ J}$$

Divide the thermal energy by the power, to get the time: $t = \dfrac{Q}{\mathcal{P}} = \dfrac{7.79 \times 10^5 \text{ J}}{313 \text{ W}} = \boxed{2.49 \times 10^3 \text{ s}}$

Remarks   The number of moles of gas given here is approximately what would be found in a bedroom. Warming the air with this space heater requires only about forty minutes. However, the calculation doesn't take into account conduction losses. Recall that a 20-cm-thick concrete wall, as calculated in Chapter 11, permitted the loss of over two megajoules an hour by conduction!

Exercise 17.7
A hot-water heater is rated at 4.50×10^3 W and operates at 2.40×10^2 V. (a) Find the resistance in the heating element, and the current. (b) How long does it take to heat 125 L of water from 20.0°C to 50.0°C, neglecting conduction and other losses?

Answers   (a) 12.8 Ω, 18.8 A   (b) 3.49×10^3 s

17.9   ELECTRICAL ACTIVITY IN THE HEART
Electrocardiograms

Every action involving the body's muscles is initiated by electrical activity. The voltages produced by muscular action in the heart are particularly important to physicians. Voltage pulses cause the heart to beat, and the waves of electrical excitation that sweep across the heart associated with the heartbeat are conducted through the body via the body fluids. These voltage pulses are large enough to be detected by suitable monitoring equipment attached to the skin. A sensitive voltmeter making good electrical contact with the skin by means of contacts attached with conducting paste can be used to measure heart pulses, which are typically of the order of 1 mV at the surface of the body. The voltage pulses can be recorded on an instrument called an **electrocardiograph**, and the pattern recorded by this instrument is called an **electrocardiogram** (EKG). In order to understand the information contained in an EKG pattern, it is necessary first to describe the underlying principles concerning electrical activity in the heart.

The right atrium of the heart contains a specialized set of muscle fibers called the SA (sinoatrial) node that initiates the heartbeat (Fig. 17.13). Electric impulses that originate in these fibers gradually spread from cell to cell throughout the right and left atrial muscles, causing them to contract. The pulse that passes through the muscle cells is often called a *depolarization wave* because of its effect on individual cells. If an individual muscle cell were examined in its resting state, a double-layer electric charge distribution would be found on its surface, as shown in Figure 17.14a (page 584). The impulse generated by the SA node momentarily and locally allows positive charge on the outside of the cell to flow in and neutralize the negative charge on the inside layer. This effect changes the cell's charge distribution to that shown in Figure 17.14b. Once the depolarization wave has passed through an individual heart muscle cell, the cell recovers the resting-state charge distribution (positive out, negative in) shown in Figure 17.14a in about 250 ms. When the impulse reaches the atrioventricular (AV) node (Fig. 17.13), the muscles of the atria begin to relax, and the pulse is directed to the ventricular

APPLICATION
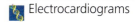 Electrocardiograms

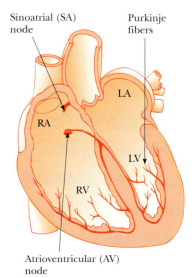

Sinoatrial (SA) node

Purkinje fibers

LA

RA

LV

RV

Atrioventricular (AV) node

Figure 17.13   The electrical conduction system of the human heart. (RA: right atrium; LA: left atrium; RV: right ventricle; LV: left ventricle.)

Figure 17.14   (a) Charge distribution of a muscle cell in the atrium before a depolarization wave has passed through the cell. (b) Charge distribution as the wave passes.

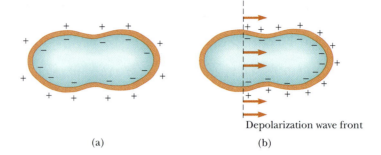

Depolarization wave front

(a)   (b)

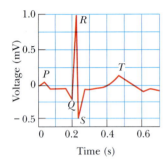

Figure 17.15   An EKG response for a normal heart.

APPLICATION

 Cardiac Pacemakers

muscles by the AV node. The muscles of the ventricles contract as the depolarization wave spreads through the ventricles along a group of fibers called the *Purkinje fibers*. The ventricles then relax after the pulse has passed through. At this point, the SA node is again triggered and the cycle is repeated.

A sketch of the electrical activity registered on an EKG for one beat of a normal heart is shown in Figure 17.15. The pulse indicated by *P* occurs just before the atria begin to contract. The *QRS* pulse occurs in the ventricles just before they contract, and the *T* pulse occurs when the cells in the ventricles begin to recover. EKGs for an abnormal heart are shown in Figure 17.16. The *QRS* portion of the pattern shown in Figure 17.16a is wider than normal, indicating that the patient may have an enlarged heart. (Why?) Figure 17.16b indicates that there is no constant relationship between the *P* pulse and the *QRS* pulse. This suggests a blockage in the electrical conduction path between the SA and AV nodes which results in the atria and ventricles beating independently and inefficient heart pumping. Finally, Figure 17.16c shows a situation in which there is no *P* pulse and an irregular spacing between the *QRS* pulses. This is symptomatic of irregular atrial contraction, which is called *fibrillation*. In this condition, the atrial and ventricular contractions are irregular.

As noted previously, the sinoatrial node directs the heart to beat at the appropriate rate, usually about 72 beats per minute. However, disease or the aging process can damage the heart and slow its beating, and a medical assist may be necessary in the form of a *cardiac pacemaker* attached to the heart. This matchbox-sized electrical device implanted under the skin has a lead that is connected to the wall of the right ventricle. Pulses from this lead stimulate the heart to maintain its proper rhythm. In general, a pacemaker is designed to produce pulses at a rate of about 60 per minute, slightly slower than the normal number of beats per minute,

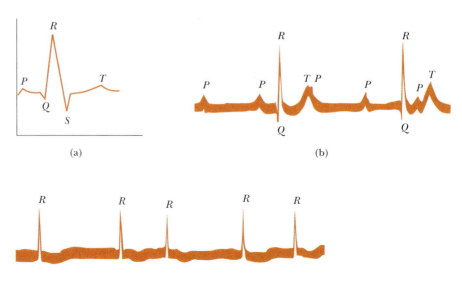

Figure 17.16   Abnormal EKGs.

but sufficient to maintain life. The circuitry basically consists of a capacitor charging up to a certain voltage from a lithium battery and then discharging. The design of the circuit is such that, if the heart is beating normally, the capacitor is not allowed to charge completely and send pulses to the heart.

An Emergency Room in Your Chest

In June 2001, an operation on Vice President Dick Cheney focused attention on the progress in treating heart problems with tiny implanted electrical devices. Aptly termed "an emergency room in your chest" by Cheney's attending physician, devices called *I*mplanted *C*ardioverter *D*efibrillators (**ICD's**) can monitor, record, and logically process heart signals and then supply different corrective signals to hearts beating too slowly, too rapidly, or irregularly. ICD's can even monitor and send signals to the atria and ventricles independently! Figure 17.17a shows a sketch of an ICD with conducting leads that are implanted in the heart. Figure 17.17b shows an actual titanium-encapsulated dual-chamber ICD.

The latest ICD's are sophisticated devices capable of a number of functions:

1. monitoring both atrial and ventricular chambers to differentiate between atrial and potentially fatal ventricular arrhythmias, which require prompt regulation;
2. storing about a half hour of heart signals that can easily be read out by a physician;
3. being easily reprogrammed with an external magnetic wand;
4. performing complicated signal analysis and comparison;
5. supplying either 0.25- to 10-V repetitive pacing signals to speed up or slow down a malfunctioning heart, or a high-voltage pulse of about 800 V to halt the potentially fatal condition of ventricular fibrillation, in which the heart quivers rapidly rather than beats (people who have experienced such a high-voltage jolt say that it feels like a kick or a bomb going off in the chest);
6. automatically adjusting the number of pacing pulses per minute to match the patient's activity.

ICD's are powered by lithium batteries and have implanted lifetimes of 4–6 years. Some basic properties of these adjustable ICD's are given in Table 17.3 (page 586). In the table, *tachycardia* means "rapid heartbeat" and *bradycardia*

APPLICATION

Implanted Cardioverter Defibrillators

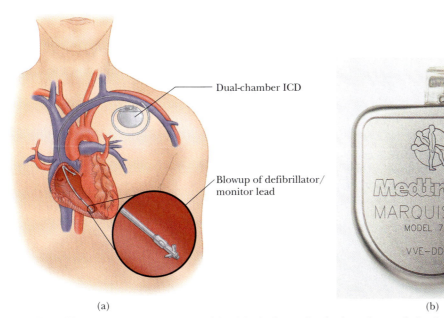

(a)

(b)

Courtesy of Medtronic, Inc.

FIGURE 17.17 (a) A dual-chamber ICD with leads in the heart. One lead monitors and stimulates the right atrium, and the other monitors and stimulates the right ventricle. (b) Medtronic Dual Chamber ICD.

TABLE 17.3

Properties of Implanted Cardioverter Defibrillators[a]

Physical Specifications	
Mass (g)	85
Size (cm)	7.3 × 6.2 × 1.3 (about five stacked silver dollars)
Antitachycardia Pacing	ICD delivers a burst of critically timed low-energy pulses
Number of Bursts	1–15
Burst Cycle Length (ms)	200–552
Number of Pulses per Burst	2–20
Pulse Amplitude (V)	7.5 or 10
Pulse Width (ms)	1.0 or 1.9
High-Voltage Defibrillation	
Pulse energy (J)	37 stored/33 delivered
Pulse Amplitude (V)	801
Bradycardia Pacing	A dual-chamber ICD can steadily deliver repetitive pulses to both the atrium and the ventricle
Base Frequency (beats/minute)	40–100
Pulse Amplitude (V)	0.25–7.5
Pulse Width (ms)	0.05, 0.1–1.5, 1.9

[a]For more information see **www.photonicd.com/specs.html**.

means "slow heartbeat." A key factor in developing tiny electrical implants that serve as defibrillators is the development of capacitors with relatively large capacitance (125 μf) and small physical size.

SUMMARY

Physics⊗Now™ Take a practice test by logging into Physics-Now at **www.cp7e.com** and clicking on the Pre-Test link for this chapter.

17.1 Electric Current

The **electric current** I in a conductor is defined as

$$I \equiv \frac{\Delta Q}{\Delta t} \qquad [17.1]$$

where ΔQ is the charge that passes through a cross section of the conductor in time Δt. The SI unit of current is the **ampere** (A); 1 A = 1 C/s. By convention, the direction of current is the direction of flow of positive charge.

17.2 A Microscopic View: Current and Drift Speed

The current in a conductor is related to the motion of the charge carriers by

$$I = nqv_dA \qquad [17.2]$$

where n is the number of mobile charge carriers per unit volume, q is the charge on each carrier, v_d is the drift speed

of the charges, and A is the cross-sectional area of the conductor.

17.4 Resistance and Ohm's Law

The **resistance** R of a conductor is defined as the ratio of the potential difference across the conductor to the current in it:

$$R \equiv \frac{\Delta V}{I} \qquad [17.3]$$

The SI units of resistance are volts per ampere, or **ohms** (Ω); 1 Ω = 1 V/A.

Ohm's law describes many conductors, for which the applied voltage is directly proportional to the current it causes. The proportionality constant is the resistance:

$$\Delta V = IR \qquad [17.4]$$

17.5 Resistivity

If a conductor has length l and cross-sectional area A, its **resistance** is

$$R = \rho \frac{l}{A} \qquad [17.5]$$

where ρ, is an intrinsic property of the conductor called the **electrical resistivity**. The SI unit of resistivity is the **ohm-meter** $(\Omega \cdot m)$.

17.6 Temperature Variation of Resistance

Over a limited temperature range, the resistivity of a conductor varies with temperature according to the expression

$$\rho = \rho_0[1 + \alpha(T - T_0)] \qquad \textbf{[17.6]}$$

where α is the **temperature coefficient of resistivity** and ρ_0 is the resistivity at some reference temperature T_0 (usually taken to be $20°C$).

The resistance of a conductor varies with temperature according to the expression

$$R = R_0[1 + \alpha(T - T_0)] \qquad \textbf{[17.7]}$$

17.8 Electrical Energy and Power

If a potential difference ΔV is maintained across an electrical device, the **power**, or rate at which energy is supplied to the device, is

$$\mathscr{P} = I \, \Delta V \qquad \textbf{[17.8]}$$

Because the potential difference across a resistor is $\Delta V = IR$, the **power delivered to a resistor** can be expressed as

$$\mathscr{P} = I^2 R = \frac{\Delta V^2}{R} \qquad \textbf{[17.9]}$$

A **kilowatt-hour** is the amount of energy converted or consumed in one hour by a device supplied with power at the rate of 1 kW. This is equivalent to

$$1 \text{ kWh} = 3.60 \times 10^6 \text{ J} \qquad \textbf{[17.10]}$$

CONCEPTUAL QUESTIONS

1. Car batteries are often rated in ampere-hours. Does this unit designate the amount of current, power, energy, or charge that can be drawn from the battery?

2. We have seen that an electric field must exist inside a conductor that carries a current. How is that possible in view of the fact that in electrostatics we concluded that the electric field must be zero inside a conductor?

3. Why don't the free electrons in a metal fall to the bottom of the metal due to gravity? And charges in a conductor are supposed to reside on the surface—why don't the free electrons all go to the surface?

4. In an analogy between traffic flow and electrical current, what would correspond to the charge Q? What would correspond to the current I?

5. Newspaper articles often have statements such as "10 000 volts of electricity surged *through* the victim's body." What is wrong with this statement?

6. Two lightbulbs are each connected to a voltage of 120 V. One has a power of 25 W, the other 100 W. Which bulb has the higher resistance? Which bulb carries more current?

7. When the voltage across a certain conductor is doubled, the current is observed to triple. What can you conclude about the conductor?

8. There is an old admonition given to experimenters to "keep one hand in the pocket" when working around high voltages. Why is this warning a good idea?

9. What factors affect the resistance of a conductor?

10. Some homes have light dimmers that are operated by rotating a knob. What is being changed in the electric circuit when the knob is rotated?

11. Two wires A and B with circular cross section are made of the same metal and have equal lengths, but the resistance of wire A is three times greater than that of wire B. What is the ratio of their cross-sectional areas? How do the radii compare?

12. What single experimental requirement makes superconducting devices expensive to operate? In principle, can this limitation be overcome?

13. What could happen to the drift velocity of the electrons in a wire and to the current in the wire if the electrons could move through it freely without resistance?

14. Use the atomic theory of matter to explain why the resistance of a material should increase as its temperature increases.

15. When is more power delivered to a lightbulb, just after it is turned on and the glow of the filament is increasing or after it has been on for a few seconds and the glow is steady?

PROBLEMS

1, 2, 3 = straightforward, intermediate, challenging   □ = full solution available in *Student Solutions Manual/Study Guide*

Physics⊗Now™ = coached problem with hints available at **www.cp7e.com**   = biomedical application

Section 17.1 Electric Current
Section 17.2 A Microscopic View: Current and Drift Speed

1. If a current of 80.0 mA exists in a metal wire, how many electrons flow past a given cross section of the wire in 10.0 min? Sketch the direction of the current and the direction of the electrons' motion.

2. A certain conductor has 7.50×10^{28} free electrons per cubic meter, a cross-sectional area of $4.00 \times 10^{-6} \text{ m}^2$, and carries a current of 2.50 A. Find the drift speed of the electrons in the conductor.

3. A 1.00-V potential difference is maintained across a $10.0\text{-}\Omega$ resistor for a period of 20.0 s. What total charge passes through the wire in this time interval?

4. In a particular television picture tube, the measured beam current is 60.0 μA. How many electrons strike the screen every second?

5. In the Bohr model of the hydrogen atom, an electron in the lowest energy state moves at a speed of 2.19×10^6 m/s in a circular path having a radius of 5.29×10^{-11} m. What is the effective current associated with this orbiting electron?

6. If 3.25×10^{-3} kg of gold is deposited on the negative electrode of an electrolytic cell in a period of 2.78 h, what is the current in the cell during that period? Assume that the gold ions carry one elementary unit of positive charge.

7. Physics ⊗ Now™ A 200-km-long high-voltage transmission line 2.0 cm in diameter carries a steady current of 1 000 A. If the conductor is copper with a free charge density of 8.5×10^{28} electrons per cubic meter, how many years does it take one electron to travel the full length of the cable?

8. An aluminum wire carrying a current of 5.0 A has a cross-sectional area of 4.0×10^{-6} m^2. Find the drift speed of the electrons in the wire. The density of aluminum is 2.7 g/cm^3. (Assume that one electron is supplied by each atom.)

9. If the current carried by a conductor is doubled, what happens to (a) the charge carrier density? (b) the electron drift velocity?

Section 17.4 Resistance and Ohm's Law
Section 17.5 Resistivity

10. A lightbulb has a resistance of 240 Ω when operating at a voltage of 120 V. What is the current in the bulb?

11. A person notices a mild shock if the current along a path through the thumb and index finger exceeds 80 μA. Compare the maximum possible voltage without shock across the thumb and index finger with a dry-skin resistance of 4.0×10^5 Ω and a wet-skin resistance of 2 000 Ω.

12. Suppose that you wish to fabricate a uniform wire out of 1.00 g of copper. If the wire is to have a resistance $R = 0.500$ Ω, and if all of the copper is to be used, what will be (a) the length and (b) the diameter of the wire?

13. Calculate the diameter of a 2.0-cm length of tungsten filament in a small lightbulb if its resistance is 0.050 Ω.

14. Eighteen-gauge wire has a diameter of 1.024 mm. Calculate the resistance of 15 m of 18-gauge copper wire at 20°C.

15. A potential difference of 12 V is found to produce a current of 0.40 A in a 3.2-m length of wire with a uniform radius of 0.40 cm. What is (a) the resistance of the wire? (b) the resistivity of the wire?

16. Make an order-of-magnitude estimate of the cost of one person's routine use of a hair dryer for 1 yr. If you do not use a blow dryer yourself, observe or interview someone who does. State the quantities you estimate and their values.

17. A wire 50.0 m long and 2.00 mm in diameter is connected to a source with a potential difference of 9.11 V, and the current is found to be 36.0 A. Assume a temperature of 20°C, and, using Table 17.1, identify the metal out of which the wire is made.

18. A rectangular block of copper has sides of length 10 cm, 20 cm, and 40 cm. If the block is connected to a 6.0-V source across two of its opposite faces, what are (a) the

maximum current and (b) the minimum current that the block can carry?

19. A wire of initial length L_0 and radius r_0 has a measured resistance of 1.0 Ω. The wire is drawn under tensile stress to a new uniform radius of $r = 0.25r_0$. What is the new resistance of the wire?

Section 17.6 Temperature Variation of Resistance

20. A certain lightbulb has a tungsten filament with a resistance of 19 Ω when cold and 140 Ω when hot. Assume that Equation 17.7 can be used over the large temperature range involved here, and find the temperature of the filament when it is hot. Assume an initial temperature of 20°C.

21. While taking photographs in Death Valley on a day when the temperature is 58.0°C, Bill Hiker finds that a certain voltage applied to a copper wire produces a current of 1.000 A. Bill then travels to Antarctica and applies the same voltage to the same wire. What current does he register there if the temperature is −88.0°C? Assume that no change occurs in the wire's shape and size.

22. A metal wire has a resistance of 10.00 Ω at a temperature of 20°C. If the same wire has a resistance of 10.55 Ω at 90°C, what is the resistance of the wire when its temperature is −20°C?

23. At 20°C, the carbon resistor in an electric circuit connected to a 5.0-V battery has a resistance of 200 Ω. What is the current in the circuit when the temperature of the carbon rises to 80°C?

24. A wire 3.00 m long and 0.450 mm^2 in cross-sectional area has a resistance of 41.0 Ω at 20°C. If its resistance increases to 41.4 Ω at 29.0°C, what is the temperature coefficient of resistivity?

25. The copper wire used in a house has a cross-sectional area of 3.00 mm^2. If 10.0 m of this wire is used to wire a circuit in the house at 20.0°C, find the resistance of the wire at temperatures of (a) 30.0°C and (b) 10.0°C.

26. A 100-cm-long copper wire of radius 0.50 cm has a potential difference across it sufficient to produce a current of 3.0 A at 20°C. (a) What is the potential difference? (b) If the temperature of the wire is increased to 200°C, what potential difference is now required to produce a current of 3.0 A?

27. Physics ⊗ Now™ (a) A 34.5-m length of copper wire at 20.0°C has a radius of 0.25 mm. If a potential difference of 9.0 V is applied across the length of the wire, determine the current in the wire. (b) If the wire is heated to 30.0°C while the 9.0-V potential difference is maintained, what is the resulting current in the wire?

28. A toaster rated at 1 050 W operates on a 120-V household circuit and a 4.00-m length of nichrome wire as its heating element. The operating temperature of this element is 320°C. What is the cross-sectional area of the wire?

29. In one form of plethysmograph (a device for measuring volume), a rubber capillary tube with an inside diameter of 1.00 mm is filled with mercury at 20°C. The resistance of the mercury is measured with the aid of electrodes sealed into the ends of the tube. If 100.00 cm of the tube is wound in a spiral around a patient's upper arm, the blood flow during a heartbeat causes the arm to expand, stretching the tube to a length of 100.04 cm. From this observation, and

assuming cylindrical symmetry, you can find the change in volume of the arm, which gives an indication of blood flow. (a) Calculate the resistance of the mercury. (b) Calculate the fractional change in resistance during the heartbeat. [*Hint:* The fraction by which the cross-sectional area of the mercury thread decreases is the fraction by which the length increases, since the volume of mercury is constant.] Take $\rho_{Hg} = 9.4 \times 10^{-7} \, \Omega \cdot m$.

30. A platinum resistance thermometer has resistances of 200.0 Ω when placed in a 0°C ice bath and 253.8 Ω when immersed in a crucible containing melting potassium. What is the melting point of potassium? [*Hint:* First determine the resistance of the platinum resistance thermometer at room temperature, 20°C.]

Section 17.8 Electrical Energy and Power

31. A toaster is rated at 600 W when connected to a 120-V source. What current does the toaster carry, and what is its resistance?

32. If electrical energy costs 12 cents, or $0.12, per kilowatt-hour, how much does it cost to (a) burn a 100-W lightbulb for 24 h? (b) operate an electric oven for 5.0 h if it carries a current of 20.0 A at 220 V?

33. How many 100-W lightbulbs can you use in a 120-V circuit without tripping a 15-A circuit breaker? (The bulbs are connected in parallel, which means that the potential difference across each lightbulb is 120 V.)

34. A high-voltage transmission line with a resistance of 0.31 Ω/km carries a current of 1 000 A. The line is at a potential of 700 kV at the power station and carries the current to a city located 160 km from the station. (a) What is the power loss due to resistance in the line? (b) What fraction of the transmitted power does this loss represent?

35. The heating element of a coffeemaker operates at 120 V and carries a current of 2.00 A. Assuming that the water absorbs all of the energy converted by the resistor, calculate how long it takes to heat 0.500 kg of water from room temperature (23.0°C) to the boiling point.

36. The power supplied to a typical black-and-white television set is 90 W when the set is connected to 120 V. (a) How much electrical energy does this set consume in 1 hour? (b) A color television set draws about 2.5 A when connected to 120 V. How much time is required for it to consume the same energy as the black-and-white model consumes in 1 hour?

37. What is the required resistance of an immersion heater that will increase the temperature of 1.50 kg of water from 10.0°C to 50.0°C in 10.0 min while operating at 120 V?

38. A certain toaster has a heating element made of Nichrome resistance wire. When the toaster is first connected to a 120-V source of potential difference (and the wire is at a temperature of 20.0°C), the initial current is 1.80 A. However, the current begins to decrease as the resistive element warms up. When the toaster reaches its final operating temperature, the current has dropped to 1.53 A. (a) Find the power the toaster converts when it is at its operating temperature. (b) What is the final temperature of the heating element?

39. Physics⊗Now™ A copper cable is designed to carry a current of 300 A with a power loss of 2.00 W/m. What is the required radius of this cable?

40. A small motor draws a current of 1.75 A from a 120-V line. The output power of the motor is 0.20 hp. (a) At a rate of $0.060/kWh, what is the cost of operating the motor for 4.0 h? (b) What is the efficiency of the motor?

41. It has been estimated that there are 270 million plug-in electric clocks in the United States, approximately one clock for each person. The clocks convert energy at the average rate of 2.50 W. To supply this energy, how many metric tons of coal are burned per hour in coal-fired electric generating plants that are, on average, 25.0% efficient? The heat of combustion for coal is 33.0 MJ/kg.

42. The cost of electricity varies widely throughout the United States; $0.120/kWh is a typical value. At this unit price, calculate the cost of (a) leaving a 40.0-W porch light on for 2 weeks while you are on vacation, (b) making a piece of dark toast in 3.00 min with a 970-W toaster, and (c) drying a load of clothes in 40.0 min in a 5 200-W dryer.

43. How much does it cost to watch a complete 21-hour-long World Series on a 180-W television set? Assume that electricity costs $0.070/kWh.

44. A house is heated by a 24.0-kW electric furnace that uses resistance heating. The rate for electrical energy is $0.080/kWh. If the heating bill for January is $200, how long must the furnace have been running on an average January day?

45. An 11-W energy-efficient fluorescent lamp is designed to produce the same illumination as a conventional 40-W lamp. How much does the energy-efficient lamp save during 100 hours of use? Assume a cost of $0.080/kWh for electrical energy.

46. An office worker uses an immersion heater to warm 250 g of water in a light, covered, insulated cup from 20°C to 100°C in 4.00 minutes. The heater is a Nichrome resistance wire connected to a 120-V power supply. Assume that the wire is at 100°C throughout the 4.00-min time interval. Specify a diameter and a length that the wire can have. Can it be made from less than 0.5 cm³ of Nichrome?

47. The heating coil of a hot-water heater has a resistance of 20 Ω and operates at 210 V. If electrical energy costs $0.080/kWh, what does it cost to raise the 200 kg of water in the tank from 15°C to 80°C? (See Chapter 11.)

ADDITIONAL PROBLEMS

48. One lightbulb is marked "25 W 120 V," and another "100 W 120 V"; this means that each converts its respective power when plugged into a constant 120-V potential difference. (a) Find the resistance of each bulb. (b) How long does it take for 1.00 C to pass through the dim bulb? How is this charge different upon its exit from, versus its entry into, the bulb? (c) How long does it take for 1.00 J to pass through the dim bulb? How is this energy different upon its exit from, versus its entry into, the bulb? (d) Find the cost of running the dim bulb continuously for 30.0 days if the electric company sells its product at $0.070 0 per kWh. What physical quantity *does* the electric company sell? What is its price for one SI unit of this quantity?

49. A particular wire has a resistivity of $3.0 \times 10^{-8} \, \Omega \cdot m$ and a cross-sectional area of $4.0 \times 10^{-6} \, m^2$. A length of this wire is to be used as a resistor that will develop 48 W of power when connected across a 20-V battery. What length of wire is required?

50. A steam iron draws 6.0 A from a 120-V line. (a) How many joules of internal energy are produced in 20 min? (b) How much does it cost, at $0.080/kWh, to run the steam iron for 20 min?

51. An experiment is conducted to measure the electrical resistivity of Nichrome in the form of wires with different lengths and cross-sectional areas. For one set of measurements, a student uses 30-gauge wire, which has a cross-sectional area of 7.30×10^{-8} m². The student measures the potential difference across the wire and the current in the wire with a voltmeter and an ammeter, respectively. For each of the measurements given in the following table taken on wires of three different lengths, calculate the resistance of the wires and the corresponding value of the resistivity:

L (m)	ΔV (V)	I (A)	R (Ω)	ρ ($\Omega \cdot$ m)
0.540	5.22	0.500		
1.028	5.82	0.276		
1.543	5.94	0.187		

What is the average value of the resistivity, and how does this value compare with the value given in Table 17.1?

52. Birds resting on high-voltage power lines are a common sight. The copper wire on which a bird stands is 2.2 cm in diameter and carries a current of 50 A. If the bird's feet are 4.0 cm apart, calculate the potential difference across its body.

53. You are cooking breakfast for yourself and a friend using a 1 200-W waffle iron and a 500-W coffeepot. Usually, you operate these appliances from a 110-V outlet for 0.500 h each day. (a) At 12 cents per kWh, how much do you spend to cook breakfast during a 30.0 day period? (b) You find yourself addicted to waffles and would like to upgrade to a 2 400-W waffle iron that will enable you to cook twice as many waffles during a half-hour period, but you know that the circuit breaker in your kitchen is a 20-A breaker. Can you do the upgrade?

54. The current in a conductor varies in time as shown in Figure P17.54. (a) How many coulombs of charge pass through a cross section of the conductor in the interval from $t = 0$ to $t = 5.0$ s? (b) What constant current would transport the same total charge during the 5.0-s interval as does the actual current?

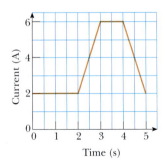

Figure P17.54

55. **Physics✕Now™** An electric car is designed to run off a bank of 12.0-V batteries with a total energy storage of 2.00×10^7 J. (a) If the electric motor draws 8.00 kW, what

is the current delivered to the motor? (b) If the electric motor draws 8.00 kW as the car moves at a steady speed of 20.0 m/s, how far will the car travel before it is "out of juice"?

56. (a) A 115-g mass of aluminum is formed into a right circular cylinder, shaped so that its diameter equals its height. Calculate the resistance between the top and bottom faces of the cylinder at 20°C. (b) Calculate the resistance between opposite faces if the same mass of aluminum is formed into a cube.

57. A length of metal wire has a radius of 5.00×10^{-3} m and a resistance of 0.100 Ω. When the potential difference across the wire is 15.0 V, the electron drift speed is found to be 3.17×10^{-4} m/s. On the basis of these data, calculate the density of free electrons in the wire.

58. A carbon wire and a Nichrome wire are connected one after the other. If the combination has a total resistance of 10.0 kΩ at 20°C, what is the resistance of each wire at 20°C so that the resistance of the combination does not change with temperature?

59. (a) Determine the resistance of a lightbulb marked 100 W @ 120 V. (b) Assuming that the filament is tungsten and has a cross-sectional area of 0.010 mm², determine the length of the wire inside the bulb when the bulb is operating. (c) Why do you think the wire inside the bulb is tightly coiled? (d) If the temperature of the tungsten wire is 2 600°C when the bulb is operating, what is the length of the wire after the bulb is turned off and has cooled to 20°C? (See Chapter 10, and use 4.5×10^{-6}/°C as the coefficient of linear expansion for tungsten.)

60. In a certain stereo system, each speaker has a resistance of 4.00 Ω. The system is rated at 60.0 W in each channel. Each speaker circuit includes a fuse rated at a maximum current of 4.00 A. Is this system adequately protected against overload?

61. A resistor is constructed by forming a material of resistivity 3.5×10^5 $\Omega \cdot$ m into the shape of a hollow cylinder of length 4.0 cm and inner and outer radii 0.50 cm and 1.2 cm, respectively. In use, a potential difference is applied between the ends of the cylinder, producing a current parallel to the length of the cylinder. Find the resistance of the cylinder.

62. The graph in Figure P17.62a shows the current I in a diode as a function of the potential difference ΔV across the diode. Figure P17.62b shows the circuit used to make the measurements. The symbol ▶| represents the diode. (a) Using Equation 17.3, make a table of the resistance of the diode for different values of ΔV in the range from -1.5 V to $+1.0$ V. (b) Based on your results, what amazing electrical property does a diode possess?

63. An x-ray tube used for cancer therapy operates at 4.0 MV, with a beam current of 25 mA striking the metal target. Nearly all the power in the beam is transferred to a stream of water flowing through holes drilled in the target. What rate of flow, in kilograms per second, is needed if the rise in temperature (ΔT) of the water is not to exceed 50°C?

64. A 50.0-g sample of a conducting material is all that is available. The resistivity of the material is measured to be 11×10^{-8} $\Omega \cdot$ m, and the density is 7.86 g/cm³. The material is to be shaped into a solid cylindrical wire that has a total resistance of 1.5 Ω. (a) What length of wire is required? (b) What must be the diameter of the wire?

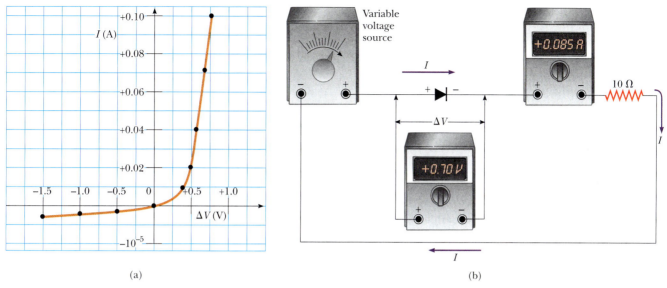

(a)   (b)

Figure P17.62

65. (a) A sheet of copper ($\rho = 1.7 \times 10^{-8}\ \Omega \cdot m$) is 2.0 mm thick and has surface dimensions of 8.0 cm × 24 cm. If the long edges are joined to form a tube 24 cm in length, what is the resistance between the ends? (b) What mass of copper is required to manufacture a 1 500-m-long spool of copper cable with a total resistance of 4.5 Ω?

66. When a straight wire is heated, its resistance changes according to the equation

$$R = R_0[1 + \alpha(T - T_0)]$$

(Eq. 17.7), where α is the temperature coefficient of resistivity. (a) Show that a more precise result, which includes the fact that the length and area of a wire change when it is heated, is

$$R = \frac{R_0[1 + \alpha(T - T_0)][1 + \alpha'(T - T_0)]}{[1 + 2\alpha'(T - T_0)]}$$

where α' is the coefficient of linear expansion. (See Chapter 10.) (b) Compare the two results for a 2.00-m-long copper wire of radius 0.100 mm, starting at 20.0°C and heated to 100.0°C.

67. A man wishes to vacuum his car with a canister vacuum cleaner marked 535 W at 120 V. The car is parked far from the building, so he uses an extension cord 15.0 m long to plug the cleaner into a 120-V source. Assume that the cleaner has constant resistance. (a) If the resistance of each of the two conductors of the extension cord is 0.900 Ω, what is the actual power delivered to the cleaner? (b) If, instead, the power is to be at least 525 W, what must be the diameter of each of two identical copper conductors in the cord the young man buys? (c) Repeat part (b) if the power is to be at least 532 W. [*Suggestion*: A symbolic solution can simplify the calculations.]

ACTIVITIES

1. Connect one terminal of a D-cell battery to the base of a flashlight bulb with insulated wire, tape a second wire to the other battery terminal, and tape a third wire to the center conductor of the bulb, as in Figure A17.1. Make

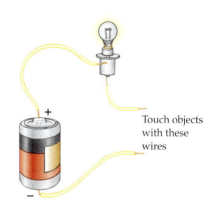

Figure A17.1

sure to remove about 1 cm of insulation from the ends of all wires before making the connections. Now bridge the gap between the open wires with different objects, such as a plastic pen, an aluminum can, a penny, a rubber band, and a spoon. Which objects make the bulb light up? Explain your observations.

2. When the lightbulbs in your home are turned on, they are always connected across the same potential difference. Which do you believe has a filament with the highest resistance when cool, a 60-W bulb or a 100-W bulb? To check your prediction, ask your instructor to lend you a device called an ohmmeter and to instruct you in its use. A resistor must always be disconnected from a circuit when its resistance is measured with an ohmmeter.

3. Examine the labels on several appliances, such as a toaster, a television set, a lamp, a stereo system, an air conditioner, and a clock. From each label, determine the power rating of the device in watts. Check the billing statement from your electric utility company to find the cost of electrical energy per kilowatt-hour. (Prices usually range from about a nickel to 20 cents.) Calculate the cost of running each appliance for 1 h. Estimate how many hours per day each appliance is used. Then, on the basis of your daily estimate, calculate the monthly cost of using each appliance.

Direct-Current Circuits

Batteries, resistors, and capacitors can be used in various combinations to construct electric circuits, which direct and control the flow of electricity and the energy it conveys. Such circuits make possible all the modern conveniences in a home—electric lights, electric stove tops and ovens, washing machines, and a host of other appliances and tools. Electric circuits are also found in our cars, in tractors that increase farming productivity, and in all types of medical equipment that saves so many lives every day.

In this chapter, we study and analyze a number of simple direct-current circuits. The analysis is simplified by the use of two rules known as Kirchhoff's rules, which follow from the principle of conservation of energy and the law of conservation of charge. Most of the circuits are assumed to be in *steady state*, which means that the currents are constant in magnitude and direction. We close the chapter with a discussion of circuits containing resistors and capacitors; in which current varies with time.

18.1 SOURCES OF EMF

A current is maintained in a closed circuit by a source of emf.[1] Among such sources are any devices (for example, batteries and generators) that increase the potential energy of the circulating charges. A source of emf can be thought of as a "charge pump" that forces electrons to move in a direction opposite the electrostatic field inside the source. The emf $\mathcal{E}$ of a source is the work done per unit charge; hence the SI unit of emf is the volt.

Consider the circuit in Active Figure 18.1a consisting of a battery connected to a resistor. We assume that the connecting wires have no resistance. If we neglect the internal resistance of the battery, the potential drop across the battery (the terminal voltage) equals the emf of the battery. Because a real battery always has some

---

[1]The term was originally an abbreviation for *electromotive force*, but emf is not really a force, so the long form is discouraged.

internal resistance r, however, the terminal voltage is not equal to the emf. The circuit of Active Figure 18.1a can be described schematically by the diagram in Active Figure 18.1b. The battery, represented by the dashed rectangle, consists of a source of emf $\mathcal{E}$ in series with an internal resistance r. Now imagine a positive charge moving through the battery from a to b in the figure. As the charge passes from the negative to the positive terminal of the battery, the potential of the charge increases by $\mathcal{E}$. As the charge moves through the resistance r, however, its potential decreases by the amount Ir, where I is the current in the circuit. The terminal voltage of the battery, $\Delta V = V_b - V_a$, is therefore given by

$$\Delta V = \mathcal{E} - Ir \qquad \text{[18.1]}$$

From this expression, we see that $\mathcal{E}$ **is equal to the terminal voltage when the current is zero**, called the **open-circuit voltage**. By inspecting Figure 18.1b, we find that the terminal voltage ΔV must also equal the potential difference across the external resistance R, often called the **load resistance**; that is, $\Delta V = IR$. Combining this relationship with Equation 18.1, we arrive at

$$\mathcal{E} = IR + Ir \qquad \text{[18.2]}$$

Solving for the current gives

$$I = \frac{\mathcal{E}}{R + r}$$

The preceding equation shows that the current in this simple circuit depends on both the resistance external to the battery and the internal resistance of the battery. If R is much greater than r, we can neglect r in our analysis (an option we usually select).

If we multiply Equation 18.2 by the current I, we get

$$I\mathcal{E} = I^2R + I^2r$$

This equation tells us that the total power output $I\mathcal{E}$ of the source of emf is converted at the rate I^2R at which energy is delivered to the load resistance, *plus* the rate I^2r at which energy is delivered to the internal resistance. Again, if $r << R$, most of the power delivered by the battery is transferred to the load resistance.

Unless otherwise stated, we will assume in our examples and end-of-chapter problems that the internal resistance of a battery in a circuit is negligible.

18.2   RESISTORS IN SERIES

When two or more resistors are connected end to end as in Active Figure 18.2, they are said to be in *series*. The resistors could be simple devices, such as lightbulbs or heating elements. When two resistors R_1 and R_2 are connected to a battery as in Active Figure 18.2, **the current is the same in the two resistors, because any charge that flows through R_1 must also flow through R_2**. This is analogous to water flowing through a pipe with two constrictions, corresponding to R_1 and R_2. Whatever volume of water flows in one end in a given time interval must exit the opposite end.

Because the potential difference between a and b in Active Figure 18.2b equals IR_1 and the potential difference between b and c equals IR_2, the potential difference between a and c is

$$\Delta V = IR_1 + IR_2 = I(R_1 + R_2)$$

Regardless of how many resistors we have in series, the sum of the potential differences across the resistors is equal to the total potential difference across the combination. As we will show later, this is a consequence of the conservation of energy. Active Figure 18.2c shows an equivalent resistor R_{eq} that can replace the two resistors of the original circuit. The equivalent resistor has the same effect on the circuit because it results in the same current in the circuit as the two resistors. Applying Ohm's law to this equivalent resistor, we have

$$\Delta V = IR_{eq}$$

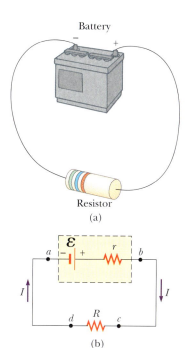

ACTIVE FIGURE 18.1
(a) A circuit consisting of a resistor connected to the terminals of a battery. (b) A circuit diagram of a source of emf $\mathcal{E}$ having internal resistance r connected to an external resistor R.

An assortment of batteries.

TIP 18.1   What's Constant in a Battery?
Equation 18.2 shows that the current in a circuit depends on the resistance of the battery, so a battery can't be considered a source of constant current. Even the terminal voltage of a battery given by Equation 18.1 can't be considered constant, because the internal resistance can change (due to warming, for example, during the operation of the battery). A battery is, however, a source of constant emf.

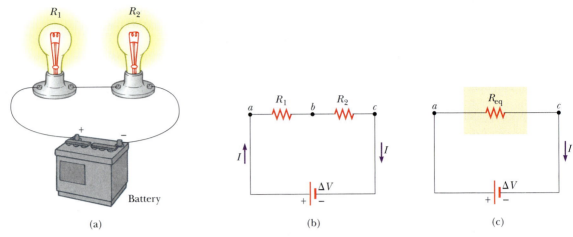

(a)   (b)   (c)

ACTIVE FIGURE 18.2
A series connection of two resistors, R_1 and R_2. The currents in the resistors are the same, and the equivalent resistance of the combination is given by $R_{eq} = R_1 + R_2$.

Physics⊗Now™
Log into PhysicsNow at **www.cp7e.com** and go to Active Figure 18.2, where you can adjust the battery voltage and resistances R_1 and R_2, observing the effect on the current and voltages of the individual resistors.

Equating the preceding two expressions, we have

$$IR_{eq} = I(R_1 + R_2)$$

or

$$R_{eq} = R_1 + R_2 \quad \text{(series combination)} \qquad [18.3]$$

An extension of the preceding analysis shows that the equivalent resistance of three or more resistors connected in series is

◀ Equivalent resistance of a series combination of resistors

$$R_{eq} = R_1 + R_2 + R_3 + \cdots \qquad [18.4]$$

Therefore, **the equivalent resistance of a series combination of resistors is the algebraic sum of the individual resistances and is always greater than any individual resistance**.

Note that if the filament of one lightbulb in Active Figure 18.2 were to fail, the circuit would no longer be complete (an open-circuit condition would exist) and the second bulb would also go out.

Applying Physics 18.1   Christmas Lights in Series

A new design for Christmas tree lights allows them to be connected in series. A failed bulb in such a string would result in an open circuit, and all of the bulbs would go out. How can the bulbs be redesigned to prevent this from happening?

Explanation   If the string of lights contained the usual kind of bulbs, a failed bulb would be hard to locate. Each bulb would have to be replaced with a good bulb, one by one, until the failed bulb was found. If there happened to be two or more failed bulbs in the string of lights, finding them would be a lengthy and annoying task.

Filament

Jumper

Glass insulator

Figure 18.3   (Applying Physics 18.1) Diagram of a modern miniature holiday lightbulb, with a jumper connection to provide a current if the filament breaks.

Christmas lights use special bulbs that have an insulated loop of wire (a jumper) across the conducting supports to the bulb filaments (Fig. 18.3). If the filament breaks and the bulb fails, the bulb's resistance increases dramatically. As a result, most of the applied voltage appears across the loop of wire. This voltage causes the insulation around the loop of wire to burn, causing the metal wire to make electrical contact with the supports. This produces a conducting path through the bulb, so the other bulbs remain lit.

Quick Quiz 18.1

When a piece of wire is used to connect points b and c in Figure 18.2b, the brightness of bulb R_1 (a) increases, (b) decreases but remains lit, (c) stays the same, (d) goes out. The brightness of bulb R_2 (a) increases, (b) decreases but remains lit, (c) stays the same, (d) goes out. (Assume connecting wires have no resistance.)

Quick Quiz 18.2

In Figure 18.4a the current is measured with the ammeter at the right side of the circuit. When the switch is opened as in Figure 18.4b, the reading on the ammeter (a) increases (b) decreases (c) doesn't change.

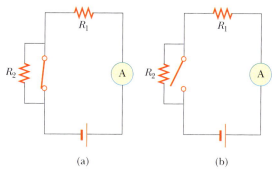

(a)   (b)

Figure 18.4   (Quick Quiz 18.2)

EXAMPLE 18.1   Four Resistors in Series

Goal   Analyze several resistors connected in series.

Problem   Four resistors are arranged as shown in Figure 18.5a. Find **(a)** the equivalent resistance of the circuit and **(b)** the current in the circuit if the emf of the battery is 6.0 V.

Strategy   Because the resistors are connected in series, summing their resistances gives the equivalent resistance. Ohm's law can then be used to find the current.

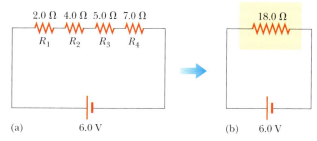

Figure 18.5   (Example 18.1) (a) Four resistors connected in series. (b) The equivalent resistance of the circuit in (a).

Solution

(a) Find the equivalent resistance of the circuit.

Apply Equation 18.4, summing the resistances:

$$R_{eq} = R_1 + R_2 + R_3 + R_4 = 2.0\ \Omega + 4.0\ \Omega + 5.0\ \Omega + 7.0\ \Omega$$
$$= \boxed{18.0\ \Omega}$$

(b) Find the current in the circuit.

Apply Ohm's law to the equivalent resistor in Figure 18.5b, solving for the current:

$$I = \frac{\Delta V}{R_{eq}} = \frac{6.0\ \text{V}}{18.0\ \Omega} = \boxed{\tfrac{1}{3}\ \text{A}}$$

Exercise 18.1

Because the current in the equivalent resistor is $\frac{1}{3}$ A, this must also be the current in each resistor of the original circuit. Find the voltage drop across each resistor.

Answers $\Delta V_{2\Omega} = \frac{2}{3}$ V; $\Delta V_{4\Omega} = \frac{4}{3}$ V; $\Delta V_{5\Omega} = \frac{5}{3}$ V; $\Delta V_{7\Omega} = \frac{7}{3}$ V.

18.3   RESISTORS IN PARALLEL

Now consider two resistors connected in parallel, as in Active Figure 18.6. In this case, **the potential differences across the resistors are the same because each is connected directly across the battery terminals**. The currents are generally not the same. When charges reach point a (called a junction) in Active Figure 18.6b, the current splits into two parts: I_1, flowing through R_1; and I_2, flowing through R_2. If R_1 is greater than R_2, then I_1 is less than I_2. In general, more charge travels through the path with less resistance. **Because charge is conserved, the current I that enters point a must equal the total current $I_1 + I_2$ leaving that point.** Mathematically, this is written

$$I = I_1 + I_2$$

The potential drop must be the same for the two resistors and must also equal the potential drop across the battery. Ohm's law applied to each resistor yields

$$I_1 = \frac{\Delta V}{R_1} \qquad I_2 = \frac{\Delta V}{R_2}$$

Ohm's law applied to the equivalent resistor in Active Figure 18.6c gives

$$I = \frac{\Delta V}{R_{eq}}$$

When these expressions for the currents are substituted into the equation $I = I_1 + I_2$ and the ΔV's are cancelled, we obtain

$$\frac{1}{R_{eq}} = \frac{1}{R_1} + \frac{1}{R_2} \qquad \text{(parallel combination)} \qquad \text{[18.5]}$$

ACTIVE FIGURE 18.6
(a) A parallel connection of two lightbulbs with resistances R_1 and R_2. (b) Circuit diagram for the two-resistor circuit. The potential differences across R_1 and R_2 are the same. (c) The equivalent resistance of the combination is given by the reciprocal relationship $1/R_{eq} = 1/R_1 + 1/R_2$.

Physics⊗Now™

Log into PhysicsNow at **www.cp7e.com**, and go to Active Figure 18.6, where you can adjust the battery voltage and resistances R_1 and R_2 and see the effect on the currents and voltages in the individual resistors.

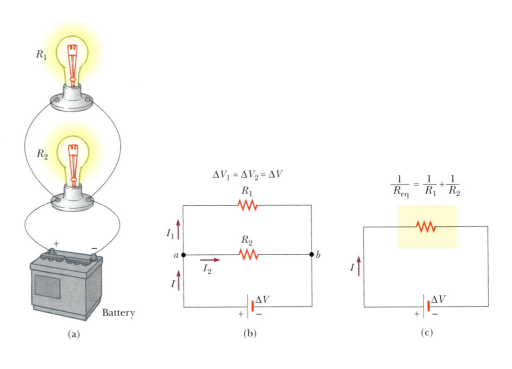

(a)   (b)   (c)

Battery

An extension of this analysis to three or more resistors in parallel produces the following general expression for the equivalent resistance:

$$\frac{1}{R_{eq}} = \frac{1}{R_1} + \frac{1}{R_2} + \frac{1}{R_3} + \cdots$$

[18.6]   ◀ Equivalent resistance of a parallel combination of resistors

From this expression, we see that **the inverse of the equivalent resistance of two or more resistors connected in parallel is the sum of the inverses of the individual resistances and is always less than the smallest resistance in the group**.

INTERACTIVE EXAMPLE 18.2   Three Resistors in Parallel

Goal   Analyze a circuit having resistors connected in parallel.

Problem   Three resistors are connected in parallel as in Figure 18.7. A potential difference of 18 V is maintained between points a and b. **(a)** Find the current in each resistor. **(b)** Calculate the power delivered to each resistor and the total power. **(c)** Find the equivalent resistance of the circuit. **(d)** Find the total power delivered to the equivalent resistance.

Strategy   We can use Ohm's law and the fact that the voltage drops across parallel resistors are all the same to get the current in each resistor. The rest of the problem just requires substitution into the equation for power delivered to a resistor, $\mathcal{P} = I^2R$, and the reciprocal-sum law for parallel resistors.

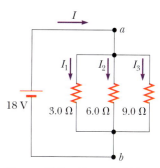

Figure 18.7   (Example 18.2) Three resistors connected in parallel. The voltage across each resistor is 18 V.

Solution

(a) Find the current in each resistor.

Apply Ohm's law, solved for the current I delivered by the battery to find the current in each resistor:

$$I_1 = \frac{\Delta V}{R_1} = \frac{18\ \text{V}}{3.0\ \Omega} = \boxed{6.0\ \text{A}}$$

$$I_2 = \frac{\Delta V}{R_2} = \frac{18\ \text{V}}{6.0\ \Omega} = \boxed{3.0\ \text{A}}$$

$$I_3 = \frac{\Delta V}{R_3} = \frac{18\ \text{V}}{9.0\ \Omega} = \boxed{2.0\ \text{A}}$$

(b) Calculate the power delivered to each resistor and the total power.

Apply $\mathcal{P} = I^2R$ to each resistor, substituting the results from part (a).

$3\ \Omega$: $\mathcal{P}_1 = I_1^2 R_1 = (6.0\ \text{A})^2(3.0\ \Omega) = \boxed{110\ \text{W}}$

$6\ \Omega$: $\mathcal{P}_2 = I_2^2 R_2 = (3.0\ \text{A})^2(6.0\ \Omega) = \boxed{54\ \text{W}}$

$9\ \Omega$: $\mathcal{P}_3 = I_3^2 R_3 = (2.0\ \text{A})^2(9.0\ \Omega) = \boxed{36\ \text{W}}$

Sum to get the total power:

$\mathcal{P}_{tot} = 110\ \text{W} + 54\ \text{W} + 36\ \text{W} = \boxed{2.0 \times 10^2\ \text{W}}$

(c) Find the equivalent resistance of the circuit.

Apply the reciprocal-sum rule, Equation 18.6:

$$\frac{1}{R_{eq}} = \frac{1}{R_1} + \frac{1}{R_2} + \frac{1}{R_3}$$

$$\frac{1}{R_{eq}} = \frac{1}{3.0\ \Omega} + \frac{1}{6.0\ \Omega} + \frac{1}{9.0\ \Omega} = \frac{11}{18\ \Omega}$$

$$R_{eq} = \frac{18}{11}\ \Omega = \boxed{1.6\ \Omega}$$

(d) Compute the power dissipated by the equivalent resistance.

Use the alternate power equation:

$$\mathcal{P} = \frac{(\Delta V)^2}{R_{eq}} = \frac{(18\ V)^2}{(1.6\ \Omega)} = 2.0 \times 10^2\ W$$

Remarks   There's something important to notice in part (a): the smallest 3.0 Ω resistor carries the largest current, while the other, larger resistors of 6.0 Ω and 9.0 Ω carry smaller currents. The largest current is always found in the path of least resistance. In part (b), the power could also be found with $\mathcal{P} = (\Delta V)^2 / R$. Note that $\mathcal{P}_1 = 108\ W$, but is rounded to 110 W because there are only two significant figures. Finally, notice that the total power dissipated in the equivalent resistor is the same as the sum of the power dissipated in the individual resistors, as it should be.

TIP 18.2   Don't Forget to Flip It!

The most common mistake in calculating the equivalent resistance for resistors in parallel is to forget to invert the answer after summing the reciprocals. Don't forget to flip it!

Exercise 18.2
Suppose the resistances in the example are 1.0 Ω, 2.0 Ω, and 3.0 Ω, respectively, and a new voltage source is provided. If the current measured in the 3.0-Ω resistor is 2.0 A, find (a) the potential difference provided by the new battery, and the currents in each of the remaining resistors, (b) the power delivered to each resistor, and the total power, (c) the equivalent resistance, and (d) the total current, and the power dissipated by the equivalent resistor.

Answers   (a) $\mathcal{E} = 6.0\ V$, $I_1 = 6.0\ A$, $I_2 = 3.0\ A$ (b) $\mathcal{P}_1 = 36\ W$, $\mathcal{P}_2 = 18\ W$, $\mathcal{P}_3 = 12\ W$, $\mathcal{P}_{tot} = 66\ W$
(c) $\frac{6}{11}\ \Omega$ (d) $I = 11\ A$, $\mathcal{P}_{eq} = 66\ W$

Physics⊗Now™ Explore different configurations of the battery and resistors by logging into PhysicsNow at www.cp7e.com and going to Interactive Example 18.2.

Household circuits are always wired so that the electrical devices are connected in parallel, as in Active Figure 18.6a. In this way, each device operates independently of the others, so that if one is switched off, the others remain on. For example, if one of the lightbulbs in Active Figure 18.6 were removed from its socket, the other would continue to operate. Equally important, each device operates at the same voltage. If the devices were connected in series, the voltage across any one device would depend on how many devices were in the combination and on their individual resistances.

In many household circuits, circuit breakers are used in series with other circuit elements for safety purposes. A circuit breaker is designed to switch off and open the circuit at some maximum value of the current (typically 15 A or 20 A) that depends on the nature of the circuit. If a circuit breaker were not used, excessive currents caused by operating several devices simultaneously could result in excessive wire temperatures, perhaps causing a fire. In older home construction, fuses were used in place of circuit breakers. When the current in a circuit exceeded some value, the conductor in a fuse melted and opened the circuit. The disadvantage of fuses is that they are destroyed in the process of opening the circuit, whereas circuit breakers can be reset.

APPLICATION

Circuit Breakers

Applying Physics 18.2   Lightbulb Combinations

Compare the brightness of the four identical bulbs shown in Figure 18.8. What happens if bulb A fails, so that it cannot conduct current? What if C fails? What if D fails?

Explanation   Bulbs A and B are connected in series across the emf of the battery, whereas bulb C is connected by itself across the battery. This means the

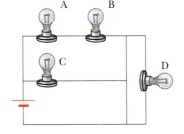

Figure 18.8   (Applying Physics 18.2)

voltage drop across C has the same magnitude as the battery emf, whereas this same emf is split between bulbs A and B. As a result, bulb C will glow more brightly than either of bulbs A and B, which will glow equally brightly. Bulb D has a wire connected across it—a short circuit—so the potential difference across bulb D is zero and it doesn't glow. If bulb A fails, B goes out, but C stays lit. If C fails, there is no effect on the other bulbs. If D fails, the event is undetectable, because D was not glowing initially.

Applying Physics 18.3   Three-Way Lightbulbs

Figure 18.9 illustrates how a three-way lightbulb is constructed to provide three levels of light intensity. The socket of the lamp is equipped with a three-way switch for selecting different light intensities. The bulb contains two filaments. Why are the filaments connected in parallel? Explain how the two filaments are used to provide the three different light intensities.

Explanation   If the filaments were connected in series and one of them were to fail, there would be no current in the bulb, and the bulb would not glow, regardless of the position of the switch. However, when the filaments are connected in parallel and one of them (say, the 75-W filament) fails, the bulb will still operate in one of the switch positions because there is current in the other (100-W) filament. The three light intensities are made possible by selecting one of three values of filament resistance, using a single value of 120 V for the applied voltage. The 75-W filament offers one value of resistance, the 100-W filament offers a second value, and the third resistance is obtained by combining the two filaments in parallel. When switch S_1 is closed and switch S_2 is opened, only the 75-W filament carries current. When switch S_1 is open and switch S_2 is closed, only the 100-W filament carries current. When both switches are closed, both filaments carry current and a total illumination corresponding to 175 W is obtained.

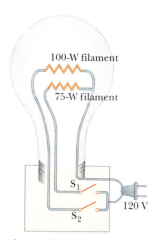

Figure 18.9 (Applying Physics 18.3)

Quick Quiz 18.3

In Figure 18.10a the current is measured with the ammeter on the right side of the circuit diagram. When the switch is closed, the reading on the ammeter (a) increases, (b) decreases, or (c) remains the same.

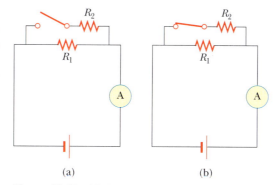

(a)   (b)

Figure 18.10 (Quick Quiz 18.3)

Quick Quiz 18.4

Suppose you have three identical lightbulbs, some wire, and a battery. You connect one lightbulb to the battery and take note of its brightness. You add a second lightbulb, connecting it in parallel with the previous bulbs, again taking note of the brightness. Repeat the process with the third bulb, connecting it in parallel with the other two. As the lightbulbs are added, what happens to (a) the brightness of the bulbs? (b) the individual currents in the bulbs? (c) the power delivered by the battery? (d) the lifetime of the battery? (Neglect the battery's internal resistance)

If the lightbulbs in Quick Quiz 18.4 are connected, one by one, in series instead of in parallel, what happens to (a) the brightness of the bulbs? (b) the individual currents in the bulbs? (c) the power delivered by the battery? (d) the lifetime of the battery? (Again, neglect the battery's internal resistance)

Problem-Solving Strategy   Simplifying Circuits with Resistors

1. **Combine all resistors in series** by summing the individual resistances, and draw the new, simplified circuit diagram.
   Useful facts: $R_{eq} = R_1 + R_2 + R_3 + \cdots$
   The current in each resistor is the same.

2. **Combine all resistors in parallel** by summing the reciprocals of the resistances and then taking the reciprocal of the result. Draw the new, simplified circuit diagram.
   Useful facts: $\dfrac{1}{R_{eq}} = \dfrac{1}{R_1} + \dfrac{1}{R_2} + \dfrac{1}{R_3} + \cdots$
   The potential difference across each resistor is the same.

3. **Repeat** the first two steps as necessary, until no further combinations can be made. If there is only a single battery in the circuit, this will usually result in a single equivalent resistor in series with the battery.

4. **Use Ohm's Law,** $\Delta V = IR$, to determine the current in the equivalent resistor. Then work backwards through the diagrams, applying the useful facts listed in step 1 or step 2 to find the currents in the other resistors. (In more complex circuits, Kirchhoff's rules will be needed, as described in the next section).

EXAMPLE 18.3   Equivalent Resistance

Goal   Solve a problem involving both series and parallel resistors.

Problem   Four resistors are connected as shown in Figure 18.11a. **(a)** Find the equivalent resistance between points a and c. **(b)** What is the current in each resistor if a 42-V battery is connected between a and c?

Strategy   Reduce the circuit in steps, as shown in Figures 18.11b and 18.11c, using the sum rule for resistors in series and the reciprocal-sum rule for resistors in parallel. Finding the currents is a matter of applying Ohm's law while working backwards through the diagrams.

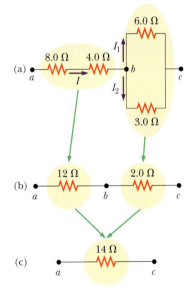

Figure 18.11   (Example 18.3) The four resistors shown in (a) can be reduced in steps to an equivalent 14-Ω resistor.

Solution

(a) Find the equivalent resistance of the circuit.

The 8.0-Ω and 4.0-Ω resistors are in series, so use the sum rule to find the equivalent resistance between a and b:

$$R_{eq} = R_1 + R_2 = 8.0\ \Omega + 4.0\ \Omega = 12\ \Omega$$

The 6.0-Ω and 3.0-Ω resistors are in parallel, so use the reciprocal-sum rule to find the equivalent resistance between b and c (don't forget to invert!):

$$\frac{1}{R_{eq}} = \frac{1}{R_1} + \frac{1}{R_2} = \frac{1}{6.0\ \Omega} + \frac{1}{3.0\ \Omega} = \frac{1}{2.0\ \Omega}$$

$$R_{eq} = 2.0\ \Omega$$

In the new diagram, 18.11b, there are now two resistors in series. Combine them with the sum rule to find the equivalent resistance of the circuit:

$$R_{eq} = R_1 + R_2 = 12\ \Omega + 2.0\ \Omega = \boxed{14\ \Omega}$$

(b) Find the current in each resistor if a 42-V battery is connected between points a and c.

Find the current in the equivalent resistor in Figure 8.11c, which is the total current. Resistors in series all carry the same current, so this is the current in the 12-Ω resistor in Figure 8.11b, and also in the 8.0-Ω and 4.0-Ω resistors in Figure 8.11a.

$$I = \frac{\Delta V_{ac}}{R_{eq}} = \frac{42\ \text{V}}{14\ \Omega} = \boxed{3.0\ \text{A}}$$

Apply the junction rule to point b:

$$(1) \quad I = I_1 + I_2$$

The 6.0-Ω and 3.0-Ω resistors are in parallel, so the voltage drops across them are the same:

$$\Delta V_{6\Omega} = \Delta V_{3\Omega} \quad \rightarrow \quad (6.0\ \Omega)I_1 = (3.0\ \Omega)I_2 \quad \rightarrow$$

$$2.0I_1 = I_2$$

Substitute this result into Equation (1), with $I = 3.0$ A:

$$3.0\ \text{A} = I_1 + 2I_1 = 3I_1 \quad \rightarrow \quad I_1 = \boxed{1.0\ \text{A}}$$

$$I_2 = \boxed{2.0\ \text{A}}$$

Remarks   As a final check, note that $\Delta V_{bc} = (6.0\ \Omega)I_1 = (3.0\ \Omega)I_2 = 6.0$ V and $\Delta V_{ab} = (12\ \Omega)I_1 = 36$ V; therefore, $\Delta V_{ac} = \Delta V_{ab} + \Delta V_{bc} = 42$ V, as expected.

Exercise 18.3
Suppose the series resistors in Example 18.3 are now 6.00 Ω and 3.00 Ω while the parallel resistors are 8.00 Ω (top) and 4.00 Ω (bottom), and the battery provides an emf of 27.0 V. Find (a) the equivalent resistance and (b) the currents I, I_1, and I_2.

Answers   (a) 11.7 Ω (b) $I = 2.31$ A, $I_1 = 0.770$ A, $I_2 = 1.54$ A

18.4   KIRCHHOFF'S RULES AND COMPLEX DC CIRCUITS

As demonstrated in the preceding section, we can analyze simple circuits using Ohm's law and the rules for series and parallel combinations of resistors. However, there are many ways in which resistors can be connected so that the circuits formed can't be reduced to a single equivalent resistor. The procedure for analyzing more complex circuits can be facilitated by the use of two simple rules called **Kirchhoff's rules**:

1. The sum of the currents entering any junction must equal the sum of the currents leaving that junction. (This rule is often referred to as the **junction rule**.)
2. The sum of the potential differences across all the elements around any closed circuit loop must be zero. (This rule is usually called the **loop rule**.)

The junction rule is a statement of *conservation of charge*. Whatever current enters a given point in a circuit must leave that point because charge can't build up or disappear at a point. If we apply this rule to the junction in Figure 18.12a, we get

$$I_1 = I_2 + I_3$$

Figure 18.12b represents a mechanical analog of the circuit shown in Figure 18.12a. In this analog, water flows through a branched pipe with no leaks. The flow rate into the pipe equals the total flow rate out of the two branches.

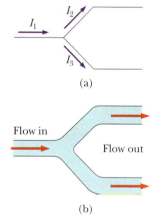

Figure 18.12   (a) A schematic diagram illustrating Kirchhoff's junction rule. Conservation of charge requires that whatever current enters a junction must leave that junction. In this case, therefore, $I_1 = I_2 + I_3$. (b) A mechanical analog of the junction rule: the net flow in must equal the net flow out.

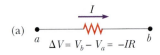

(a) $\Delta V = V_b - V_a = -IR$

(b) $\Delta V = V_b - V_a = +IR$

(c) $\Delta V = V_b - V_a = +\mathcal{E}$

(d) $\Delta V = V_b - V_a = -\mathcal{E}$

Figure 18.13 Rules for determining the potential differences across a resistor and a battery, assuming the battery has no internal resistance.

The loop rule is equivalent to the principle of *conservation of energy*. Any charge that moves around any closed loop in a circuit (starting and ending at the same point) must gain as much energy as it loses. It gains energy as it is pumped through a source of emf. Its energy may decrease in the form of a potential drop $-IR$ across a resistor or as a result of flowing backward through a source of emf, from the positive to the negative terminal inside the battery. In the latter case, electrical energy is converted to chemical energy as the battery is charged.

When applying Kirchhoff's rules, you must make two decisions at the beginning of the problem:

1. Assign symbols and directions to the currents in all branches of the circuit. Don't worry about guessing the direction of a current incorrectly; the resulting answer will be negative, but *its magnitude will be correct.* (This is because the equations are *linear* in the currents—all currents are to the first power.)
2. When applying the loop rule, you must choose a direction for traversing the loop, and be consistent in going either clockwise or counterclockwise. As you traverse the loop, record voltage drops and rises according to the following rules (summarized in Figure 18.13, where it is assumed that movement is from point *a* toward point *b*):

   (a) If a resistor is traversed in the direction of the current, the change in electric potential across the resistor is $-IR$ (Fig. 18.13a).
   (b) If a resistor is traversed in the direction opposite the current, the change in electric potential across the resistor is $+IR$ (Fig. 18.13b).
   (c) If a source of emf is traversed in the direction of the emf (from $-$ to $+$ on the terminals), the change in electric potential is $+\mathcal{E}$ (Fig. 18.13c).
   (d) If a source of emf is traversed in the direction opposite the emf (from $+$ to $-$ on the terminals), the change in electric potential is $-\mathcal{E}$ (Fig. 18.13d).

There are limits to the number of times the junction rule and the loop rule can be used. You can use the junction rule as often as needed as long as, each time you write an equation, you include in it a current that has not been used in a previous junction-rule equation. (If this procedure isn't followed, the new equation will just be a combination of two other equations that you already have.) In general, the number of times the junction rule can be used is one fewer than the number of junction points in the circuit. The loop rule can also be used as often as needed, so long as a new circuit element (resistor or battery) or a new current appears in each new equation. **To solve a particular circuit problem, you need as many independent equations as you have unknowns.**

GUSTAV KIRCHHOFF, German Physicist (1824–1887)

Together with German chemist Robert Bunsen, Kirchhoff, a professor at Heidelberg, invented the spectroscopy that we study in Chapter 28. He also formulated another rule that states, "A cool substance will absorb light of the same wavelengths that it emits when hot."

Problem-Solving Strategy   Applying Kirchhoff's Rules to a Circuit

1. **Assign labels and symbols** to all the known and unknown quantities.
2. **Assign *directions* to the currents** in each part of the circuit. Although the assignment of current directions is arbitrary, you must stick with your original choices throughout the problem as you apply Kirchhoff's rules.
3. **Apply the junction rule** to any junction in the circuit. The rule may be applied as many times as a new current (one not used in a previously found equation) appears in the resulting equation.
4. **Apply Kirchhoff's loop rule** to as many loops in the circuit as are needed to solve for the unknowns. In order to apply this rule, you must correctly identify the change in electric potential as you cross each element in traversing the closed loop. Watch out for signs!
5. **Solve the equations** simultaneously for the unknown quantities, using substitution or any other method familiar to the student.
6. **Check your answers** by substituting them into the original equations.

EXAMPLE 18.4 Applying Kirchhoff's Rules

Goal Use Kirchhoff's rules to find currents in a circuit with three currents and one battery.

Problem Find the currents in the circuit shown in Figure 18.14 by using Kirchhoff's rules.

Strategy There are three unknown currents in this circuit, so we must obtain three independent equations, which then can be solved by substitution. We can find the equations with one application of the junction rule and two applications of the loop rule. We choose junction c. (Junction d gives the same equation.) For the loops, we choose the bottom loop and the top loop, both shown by blue arrows, which indicate the direction we are going to traverse the circuit mathematically (not necessarily the direction of the current). The third loop gives an equation that can be obtained by a linear combination of the other two, so it provides no additional information.

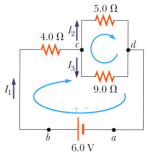

Figure 18.14 (Example 18.4)

Solution

Apply the junction rule to point c. I_1 is directed into the junction, I_2 and I_3 are directed out of the junction.

$$I_1 = I_2 + I_3$$

Select the bottom loop, and traverse it clockwise starting at point a, generating an equation with the loop rule:

$$\Sigma \Delta V = \Delta V_{bat} + \Delta V_{4.0\Omega} + \Delta V_{9.0\Omega} = 0$$
$$6.0 \text{ V} - (4.0 \text{ }\Omega)I_1 - (9.0 \text{ }\Omega)I_3 = 0$$

Select the top loop, and traverse it clockwise from point c. Notice the gain across the 9.0-Ω resistor, because it is traversed *against* the direction of the current!

$$\Sigma \Delta V = \Delta V_{5.0\Omega} + \Delta V_{9.0\Omega} = 0$$
$$- (5.0 \text{ }\Omega)I_2 + (9.0 \text{ }\Omega)I_3 = 0$$

Rewrite the three equations, rearranging terms and dropping units for the moment, for convenience:

(1) $\quad I_1 = I_2 + I_3$

(2) $\quad 4.0I_1 + 9.0I_3 = 6.0$

(3) $\quad -5.0I_2 + 9.0I_3 = 0$

Solve Equation 3 for I_2 and substitute into Equation 1:

$$I_2 = 1.8I_3$$
$$I_1 = I_2 + I_3 = 1.8I_3 + I_3 = 2.8I_3$$

Substitute the latter expression into Equation 2 and solve for I_3:

$$4.0(2.8I_3) + 9.0I_3 = 6.0 \quad \rightarrow \quad I_3 = \boxed{0.30 \text{ A}}$$

Substitute I_3 back into Equation 3 to get I_2:

$$-5.0I_2 + 9.0(0.30 \text{ A}) = 0 \quad \rightarrow \quad I_2 = \boxed{0.54 \text{ A}}$$

Substitute I_3 into Equation 2 to get I_1:

$$4.0I_1 + 9.0(0.30 \text{ A}) = 6.0 \quad \rightarrow \quad I_1 = \boxed{0.83 \text{ A}}$$

Remarks Substituting these values back into the original equations verifies that they are correct, with any small discrepancies due to rounding. The problem can also be solved by first combining resistors.

Exercise 18.4

Suppose the 6.0-V battery is replaced by a battery of unknown emf, and an ammeter measures $I_1 = 1.5$ A. Find the other two currents and the emf of the battery.

Answers $I_2 = 0.96$ A, $I_3 = 0.54$ A, $\mathcal{E} = 11$ V

TIP 18.3 More Current Goes in the Path of Less Resistance

You may have heard the statement "Current takes the path of least resistance." For a parallel combination of resistors, this statement is inaccurate, because current actually follows all paths. The most current, however, travels in the path of least resistance.

INTERACTIVE EXAMPLE 18.5 Another Application of Kirchhoff's Rules

Goal Find the currents in a circuit with three currents and two batteries when some current directions are chosen wrongly.

Problem Find I_1, I_2, and I_3 in Figure 18.15a on page 604.

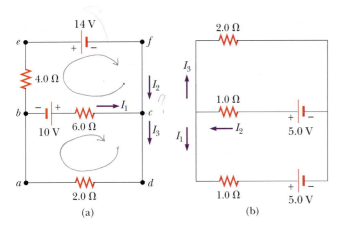

Figure 18.15 (a) (Example 18.5) (b) (Exercise 18.5)

Strategy Use Kirchhoff's two rules, the junction rule once and the loop rule twice, to develop three equations for the three unknown currents. Solve the equations simultaneously.

Solution

Apply Kirchhoff's junction rule to junction c. Because of the chosen current directions, I_1 and I_2 are directed into the junction and I_3 is directed out of the junction.

$$(1) \qquad I_3 = I_1 + I_2$$

Apply Kirchhoff's loop rule to the loops $abcda$ and $befcb$. (Loop $aefda$ gives no new information.) In loop $befcb$, a positive sign is obtained when the 6.0-Ω resistor is traversed, because the direction of the path is opposite the direction of the current I_1.

(2) Loop $abcda$: $\qquad 10\,\text{V} - (6.0\ \Omega)I_1 - (2.0\ \Omega)I_3 = 0$

(3) Loop $befcb$:
$$-14\,\text{V} + (6.0\ \Omega)I_1 - 10\,\text{V} - (4.0\ \Omega)I_2 = 0$$

Using Equation (1), eliminate I_3 from Equation (2) (ignore units for the moment):

$$10 - 6.0I_1 - 2.0(I_1 + I_2) = 0$$
$$(4) \qquad 10 = 8.0I_1 + 2.0I_2$$

Divide each term in Equation (3) by 2 and rearrange the equation so that the currents are on the right side:

$$(5) \qquad -12 = -3.0I_1 + 2.0I_2$$

Subtracting Equation (5) from Equation (4) eliminates I_2 and gives I_1:

$$22 = 11I_1 \quad \rightarrow \quad I_1 = \boxed{2.0\,\text{A}}$$

Substituting this value of I_1 into Equation (5) gives I_2:

$$2.0I_2 = 3.0I_1 - 12 = 3.0(2.0) - 12 = -6.0\,\text{A}$$
$$I_2 = \boxed{-3.0\,\text{A}}$$

Finally, substitute the values found for I_1 and I_2 into Equation (1) to obtain I_3:

$$I_3 = I_1 + I_2 = 2.0\,\text{A} - 3.0\,\text{A} = \boxed{-1.0\,\text{A}}$$

Remarks The fact that I_2 and I_3 are both negative indicates that the wrong directions were chosen for these currents. Nonetheless, the magnitudes are correct. Choosing the right directions of the currents at the outset is unimportant because the equations are linear, and wrong choices result only in a minus sign in the answer.

Exercise 18.5 Find the three currents in Figure 18.15b. (Note that the direction of one current was deliberately chosen wrongly!)

Answers $I_1 = -1.0\,\text{A}$, $I_2 = 1.0\,\text{A}$, $I_3 = 2.0\,\text{A}$

Physics⊗Now™ Practice applying Kirchhoff's rules for different values of resistance and voltage by logging into PhysicsNow at **www.cp7e.com** and going to Interactive Example 18.5.

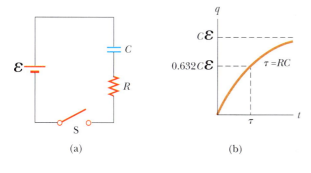

ACTIVE FIGURE 18.16
(a) A capacitor in series with a resistor, a battery, and a switch. (b) A plot of the charge on the capacitor versus time after the switch on the circuit is closed. After one time constant τ, the charge is 63% of the maximum value, $C\mathcal{E}$. The charge approaches its maximum value as t approaches infinity.

Physics⊗Now™
Log into PhysicsNow at **www.cp7e.com** and go to Active Figure 18.16, where you can adjust the values of R and C and observe the effect on the charging of the capacitor.

18.5 *RC* CIRCUITS

So far, we have been concerned with circuits with constant currents. We now consider direct-current circuits containing capacitors, in which the currents vary with time. Consider the series circuit in Active Figure 18.16. We assume that the capacitor is initially uncharged with the switch opened. After the switch is closed, the battery begins to charge the plates of the capacitor and the charge passes through the resistor. As the capacitor is being charged, the circuit carries a changing current. The charging process continues until the capacitor is charged to its maximum equilibrium value, $Q = C\mathcal{E}$, where $\mathcal{E}$ is the maximum voltage across the capacitor. Once the capacitor is fully charged, the current in the circuit is zero. If we assume that the capacitor is uncharged before the switch is closed, and if the switch is closed at $t = 0$, we find that the charge on the capacitor varies with time according to the equation

$$q = Q(1 - e^{-t/RC}) \qquad \textbf{[18.7]}$$

where $e = 2.718 \ldots$ is Euler's constant, the base of the natural logarithms. Active Figure 18.16b is a graph of this equation. The charge is zero at $t = 0$ and approaches its maximum value, Q, as t approaches infinity. The voltage ΔV across the capacitor at any time is obtained by dividing the charge by the capacitance: $\Delta V = q/C$.

As you can see from Equation 18.7, it would take an infinite amount of time, in this model, for the capacitor to become fully charged. The reason for this is mathematical: in obtaining that equation, charges are assumed to be infinitely small, whereas in reality the smallest charge is that of an electron, with a magnitude of 1.60×10^{-19} C. For all practical purposes, the capacitor is fully charged after a finite amount of time. The term RC that appears in Equation 18.7 is called the **time constant** τ (Greek letter tau), so

$$\tau = RC \qquad \textbf{[18.8]}$$

◀ Time constant of an *RC* circuit

The time constant represents the time required for the charge to increase from zero to 63.2% of its maximum equilibrium value. This means that in a period of time equal to one time constant, the charge on the capacitor increases from zero to $0.632Q$. This can be seen by substituting $t = \tau = RC$ into Equation 18.7 and solving for q. (Note that $1/e = 0.632$.) It's important to note that a capacitor charges very slowly in a circuit with a long time constant, whereas it charges very rapidly in a circuit with a short time constant. After a time equal to ten time constants, the capacitor is over 99.99% charged.

Now consider the circuit in Active Figure 18.17a, consisting of a capacitor with an initial charge Q, a resistor, and a switch. Before the switch is closed, the potential difference across the charged capacitor is Q/C. Once the switch is closed, the charge begins to flow through the resistor from one capacitor plate to the other until the capacitor is fully discharged. If the switch is closed at $t = 0$, it can be shown that the charge q on the capacitor varies with time according to the equation

$$q = Qe^{-t/RC} \qquad \textbf{[18.9]}$$

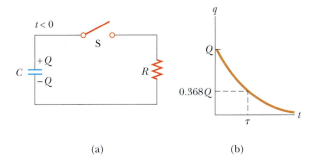

ACTIVE FIGURE 18.17
(a) A charged capacitor connected to a resistor and a switch. (b) A graph of the charge on the capacitor versus time after the switch is closed.

Physics☒Now™
Log into PhysicsNow at **www.cp7e.com** and go to Active Figure 18.17, where you can adjust the values of R and C and see the effect on the discharging of the capacitor.

(a)   (b)

The charge decreases exponentially with time, as shown in Active Figure 18.17b. In the interval $t = \tau = RC$, the charge decreases from its initial value Q to $0.368Q$. In other words, in a time equal to one time constant, the capacitor loses 63.2% of its initial charge. Because $\Delta V = q/C$, the voltage across the capacitor also decreases exponentially with time according to the equation $\Delta V = \mathcal{E}e^{-t/RC}$, where $\mathcal{E}$ (which equals Q/C) is the initial voltage across the fully charged capacitor.

Applying Physics 18.4 Timed Windshield Wipers

Many automobiles are equipped with windshield wipers that can be used intermittently during a light rainfall. How does the operation of this feature depend on the charging and discharging of a capacitor?

Explanation The wipers are part of an RC circuit with time constant that can be varied by selecting different values of R through a multiposition switch. The brief time that the wipers remain on and the time they are off are determined by the value of the time constant of the circuit.

Applying Physics 18.5 Bacterial Growth

In biological applications concerned with population growth, an equation is used that is similar to the exponential equations encountered in the analysis of RC circuits. Applied to a number of bacteria, this equation is

$$N_f = N_i 2^n$$

where N_f is the number of bacteria present after n doubling times, N_i is the number present initially, and n is the number of growth cycles or doubling times. Doubling times vary according to the organism. The doubling time for the bacteria responsible for leprosy is about 30 days, and that for the salmonella bacteria

responsible for food poisoning is about 20 minutes. Suppose only 10 salmonella bacteria find their way onto a turkey leg after your Thanksgiving meal. Four hours later you come back for a midnight snack. How many bacteria are present now?

Explanation The number of doubling times is 240 min/20 min = 12. Thus, we have

$$N_f = N_i 2^n = (10 \text{ bacteria})(2^{12}) = 40\,960 \text{ bacteria}.$$

So your system will have to deal with an invading host of about 41 000 bacteria, which are going to continue to double in a very promising environment.

Applying Physics 18.6 Roadway Flashers

Many roadway construction sites have flashing yellow lights to warn motorists of possible dangers. What causes the lights to flash?

Explanation A typical circuit for such a flasher is shown in Figure 18.18. The lamp L is a gas-filled lamp that acts as an open circuit until a large potential difference causes a discharge, which gives off a bright light. During this discharge, charge flows through the gas between the electrodes of the lamp. When the switch is closed, the battery charges the capacitor. At the beginning, the current is high and the charge on the capacitor is low, so that most of the potential difference appears across the resistance *R*. As the capacitor charges, more potential difference appears across it, reflecting the lower current and lower potential difference across the resistor. Eventually, the potential difference across the capacitor reaches a value at which the lamp will conduct, causing a flash. This discharges the capacitor through the lamp, and the process of charging begins again. The period between flashes can be adjusted by changing the time constant of the *RC* circuit.

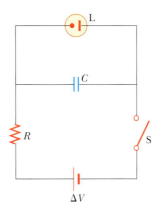

Figure 18.18 (Applying Physics 18.6)

Quick Quiz 18.6

The switch is closed in Figure 18.19. After a long time compared to the time constant of the capacitor, what will the current be in the 2-Ω resistor? (a) 4 A (b) 3 A (c) 2 A (d) 1 A (e) more information is needed

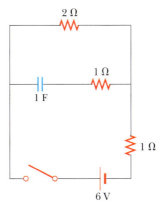

Figure 18.19 (Quick Quiz 18.6)

EXAMPLE 18.6   Charging a Capacitor in an *RC* Circuit

Goal Calculate elementary properties of a simple *RC* circuit.

Problem An uncharged capacitor and a resistor are connected in series to a battery, as in Active Figure 18.16a. If $\mathcal{E}$ = 12.0 V, C = 5.00 μF, and R = 8.00 × 10^5 Ω, find **(a)** the time constant of the circuit, **(b)** the maximum charge on the capacitor, **(c)** the charge on the capacitor after 6.00 s, **(d)** the potential difference across the resistor after 6.00 s, and **(e)** the current in the resistor at that time.

Strategy Finding the time constant in part (a) requires substitution into Equation 18.8. For part (b), the maximum charge occurs after a long time, when the current has dropped to zero. By Ohm's law, $\Delta V = IR$, the potential difference across the resistor is also zero at that time, and Kirchhoff's loop rule then gives the maximum charge. Finding the charge at some particular time, as in part (c), is a matter of substituting into Equation 18.7. Kirchhoff's loop rule and the capacitance equation can be used to indirectly find the potential drop across the resistor in part (d), and then Ohm's law yields the current.

Solution

(a) Find the time constant of the circuit.

Use the definition of the time constant, Equation 18.8: $\tau = RC = (8.00 \times 10^5 \ \Omega)(5.00 \times 10^{-6} \ \text{F}) = \boxed{4.00 \ \text{s}}$

(b) Calculate the maximum charge on the capacitor.

Apply Kirchhoff's loop rule to the RC circuit, going clockwise, which means that the voltage difference across the battery is positive and the differences across the capacitor and resistor are negative.

$$(1) \qquad \Delta V_{bat} + \Delta V_C + \Delta V_R = 0$$

From the definition of capacitance (Equation 16.8) and Ohm's law, we have $\Delta V_C = -q/C$ and $\Delta V_R = -IR$. These are voltage drops, so they're negative. Also, $\Delta V_{bat} = +\mathcal{E} = 12.0$ V.

$$(2) \qquad \mathcal{E} - \frac{q}{c} - IR = 0$$

When the maximum charge $q = Q$ is reached, $I = 0$. Solve Equation (2) for the charge:

$$\mathcal{E} - \frac{Q}{C} = 0 \quad \rightarrow \quad Q = C\mathcal{E}$$

Substitute to find the maximum charge:

$$Q = (5.00 \times 10^{-6}\,\text{F})(12.0\,\text{V}) = \boxed{60.0\ \mu\text{C}}$$

(c) Find the charge on the capacitor after 6.00 s.

Substitute into Equation 18.7:

$$q = Q(1 - e^{-t/\tau}) = (60.0\ \mu\text{C})(1 - e^{-6.00\,\text{s}/4.00\,\text{s}})$$
$$= \boxed{46.6\ \mu\text{C}}$$

(d) Compute the potential difference across the resistor after 6.00 s.

Compute the voltage drop ΔV_C across the capacitor at that time:

$$\Delta V_C = -\frac{q}{C} = \frac{-46.6\ \mu\text{C}}{5.00\ \mu\text{F}} = -9.32\,\text{V}$$

Solve Equation 1 for ΔV_R, and substitute:

$$\Delta V_R = -\Delta V_{bat} - \Delta V_C = -12.0 - (-9.32\,\text{V})$$
$$= \boxed{-2.68\,\text{V}}$$

(e) Find the current in the resistor after 6.00 s.

Apply Ohm's law, using the results of part (d) (remember that $\Delta V_R = -IR$ here):

$$I = \frac{-\Delta V_R}{R} = -\frac{(-2.68\,\text{V})}{(8.0 \times 10^5\ \Omega)}$$
$$= \boxed{3.4 \times 10^{-6}\,\text{A}}$$

Remark   In solving this problem, we paid scrupulous attention to signs. These signs must always be chosen when applying Kirchhoff's loop rule, and must remain consistent throughout the problem. Alternately, magnitudes can be used and the signs chosen by physical intuition. For example, the magnitude of the potential difference across the resistor must equal the magnitude of the potential difference across the battery minus the magnitude of the potential difference across the capacitor.

Exercise 18.6
Find (a) the charge on the capacitor after 2.00 s have elapsed, (b) the magnitude of the potential difference across the capacitor after 2.00 s, and (c) the magnitude of the potential difference across the resistor at that same time.

Answers   (a) 23.6 μC   (b) 4.72 V   (c) 7.28 V

EXAMPLE 18.7   Discharging a Capacitor in an *RC* Circuit

Goal   Calculate some elementary properties of a discharging capacitor in an *RC* circuit.

Problem   Consider a capacitor C being discharged through a resistor R as in Figure 18.17a, page 606. **(a)** How long does it take for the charge on the capacitor to drop to one-fourth of its initial value? **(b)** Compute the initial charge and time constant. **(c)** How long does it take to discharge all but the last quantum of charge, 1.60×10^{-19} C, if the initial potential difference across the capacitor is 12.0 V, the capacitance is 3.50×10^{-6} F, and the resistance is 2.00 Ω? (Assume an exponential decrease during the entire discharge process.)

Strategy This problem requires substituting given values into various equations, as well as a couple algebraic manipulations involving the natural logarithm. In part (a), set $q = \frac{1}{4}Q$ in Equation 18.9 for a discharging capacitor, where Q is the initial charge, and solve for time t. For part (b), substitute into Equations 16.8 and 18.8 to find the initial capacitor charge and time constant, respectively. In part (c), substitute the results of part (b) and $q = 1.60 \times 10^{-19}$ C into the discharging-capacitor equation, again solving for time.

Solution

(a) How long does it take for the capacitor to drain to one-fourth its initial value?

Apply Equation 18.9:

$$q(t) = Qe^{-t/RC}$$

Substitute $q(t) = Q/4$ into the preceding equation and cancel Q:

$$\tfrac{1}{4}Q = Qe^{-t/RC} \quad \rightarrow \quad \tfrac{1}{4} = e^{-t/RC}$$

Take natural logarithms of both sides and solve for the time t:

$$\ln\left(\tfrac{1}{4}\right) = -t/RC$$

$$t = -RC\ln\left(\tfrac{1}{4}\right) = 1.39RC = \boxed{1.39\tau}$$

(b) Compute the initial charge and time constant from the given data.

Use the capacitance equation to find the initial charge:

$$C = \frac{Q}{\Delta V} \quad \rightarrow \quad Q = C\Delta V = (3.50 \times 10^{-6}\text{ F})(12.0\text{ V})$$

$$Q = \boxed{4.20 \times 10^{-5}\text{ C}}$$

Now calculate the time constant:

$$\tau = RC = (2.00\text{ Ω})(3.50 \times 10^{-6}\text{ F}) = \boxed{7.00 \times 10^{-6}\text{ s}}$$

(c) How long does it take to drain all but the last quantum of charge?

Apply Equation 18.9, divide by Q, and take natural logarithms of both sides:

$$q(t) = Qe^{-t/\tau} \quad \rightarrow \quad e^{-t/\tau} = \frac{q}{Q}$$

Take the natural logs of both sides:

$$-t/\tau = \ln\left(\frac{q}{Q}\right) \quad \rightarrow \quad t = -\tau\ln\left(\frac{q}{Q}\right)$$

Substitute $q = 1.60 \times 10^{-19}$ C and the values for Q and τ found in part (b):

$$t = -(7.00 \times 10^{-6}\text{ s})\ln\left(\frac{1.60 \times 10^{-19}\text{ C}}{4.20 \times 10^{-5}\text{ C}}\right)$$

$$= \boxed{2.32 \times 10^{-4}\text{ s}}$$

Remarks Part (a) shows how useful information can often be obtained even when no details concerning capacitances, resistances, or voltages are known. Part (c) demonstrates that capacitors can be rapidly discharged (or conversely, charged), despite the mathematical form of Equations 18.7 and 18.9, which indicate an infinite time would be required.

Exercise 18.7
Suppose the same type of series circuit has $R = 8.00 \times 10^4$ Ω, $C = 5.00$ μF, and an initial voltage across the capacitor of 6.0 V. (a) How long does it take the capacitor to lose half its initial charge? (b) How long does it take to lose all but the last 10 electrons on the negative plate?

Answers (a) 0.277 s (b) 12.2 s

18.6 HOUSEHOLD CIRCUITS

Household circuits are a practical application of some of the ideas presented in this chapter. In a typical installation, the utility company distributes electric power to individual houses with a pair of wires, or power lines. Electrical devices in a

APPLICATION

Fuses and Circuit Breakers

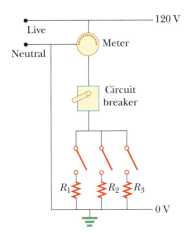

Figure 18.20 A wiring diagram for a household circuit. The resistances R_1, R_2, and R_3 represent appliances or other electrical devices that operate at an applied voltage of 120 V.

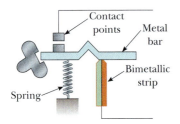

Figure 18.21 A circuit breaker that uses a bimetallic strip for its operation.

house are then connected in parallel to these lines, as shown in Figure 18.20. The potential difference between the two wires is about 120 V. (These currents and voltages are actually alternating currents and voltages, but for the present discussion we will assume that they are direct currents and voltages.) One of the wires is connected to ground, and the other wire, sometimes called the "hot" wire, is at a potential of 120 V. A meter and a circuit breaker (or a fuse) are connected in series with the wire entering the house, as indicated in the figure.

In modern homes, circuit breakers are used in place of fuses. When the current in a circuit exceeds some value (typically 15 A), the circuit breaker acts as a switch and opens the circuit. Figure 18.21 shows one design for a circuit breaker. Current passes through a bimetallic strip, the top of which bends to the left when excessive current heats it. If the strip bends far enough to the left, it settles into a groove in the spring-loaded metal bar. When this occurs, the bar drops enough to open the circuit at the contact point. The bar also flips a switch which indicates that the circuit breaker is not operational. (After the overload is removed, the switch can be flipped back on.) Circuit breakers based on this design have the disadvantage that some time is required for the heating of the strip, so the circuit may not be opened rapidly enough when it is overloaded. Because of this, many circuit breakers are now designed to use electromagnets (discussed in Chapter 19).

The wire and circuit breaker are carefully selected to meet the current demands of a circuit. If the circuit is to carry currents as large as 30 A, a heavy-duty wire and an appropriate circuit breaker must be used. Household circuits that are normally used to power lamps and small appliances often require only 20 A. Each circuit has its own circuit breaker to accommodate its maximum safe load.

As an example, consider a circuit that powers a toaster, a microwave oven, and a heater (represented by R_1, R_2, and R_3 in Fig. 18.20). Using the equation $\mathcal{P} = I \, \Delta V$, we can calculate the current carried by each appliance. The toaster, rated at 1 000 W, draws a current of $1\,000/120 = 8.33$ A. The microwave oven, rated at 800 W, draws a current of 6.67 A, and the heater, rated at 1 300 W, draws a current of 10.8 A. If the three appliances are operated simultaneously, they draw a total current of 25.8 A. Therefore, the breaker should be able to handle at least this much current, or else it will be tripped. As an alternative, the toaster and microwave oven could operate on one 20-A circuit and the heater on a separate 20-A circuit.

Many heavy-duty appliances, such as electric ranges and clothes dryers, require 240 V to operate. The power company supplies this voltage by providing, in addition to a live wire that is 120 V above ground potential, another wire, also considered live, that is 120 V below ground potential (Fig. 18.22). Therefore, the potential drop across the two live wires is 240 V. An appliance operating from a 240-V line requires half the current of one operating from a 120-V line; consequently, smaller wires can be used in the higher-voltage circuit without becoming overheated.

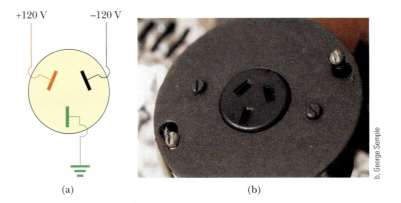

Figure 18.22 Power connections for a 240-V appliance.

(a)

(b)

b, George Semple

18.7 ELECTRICAL SAFETY

A person can be electrocuted by touching a live wire (which commonly is live be-cause of a frayed cord and exposed conductors) while in contact with ground. The ground contact might be made by touching a water pipe (which is normally at ground potential) or by standing on the ground with wet feet, because impure water is a good conductor. Obviously, such situations should be avoided at all costs.

Electric shock can result in fatal burns, or it can cause the muscles of vital or-gans, such as the heart, to malfunction. The degree of damage to the body depends on the magnitude of the current, the length of time it acts, and the part of the body through which it passes. Currents of 5 mA or less can cause a sensation of shock, but ordinarily do little or no damage. If the current is larger than about 10 mA, the hand muscles contract and the person may be unable to let go of the live wire. If a current of about 100 mA passes through the body for just a few seconds, it can be fatal. Such large currents paralyze the respiratory muscles. In some cases, currents of about 1 A through the body produce serious (and sometimes fatal) burns.

As an additional safety feature for consumers, electrical equipment manufactur-ers now use electrical cords that have a third wire, called a *case ground*. To under-stand how this works, consider the drill being used in Figure 18.23. A two-wire device that has one wire, called the "hot" wire, is connected to the high-potential (120-V) side of the input power line, and the second wire is connected to ground (0 V). If the high-voltage wire comes in contact with the case of the drill (Fig 18.23a), a short circuit occurs. In this undesirable circumstance, the pathway for the current is from the high-voltage wire through the person holding the drill and to Earth—a pathway that can be fatal. Protection is provided by a third wire, con-nected to the case of the drill (Fig. 18.23b). In this case, if a short occurs, the path of least resistance for the current is from the high-voltage wire through the case

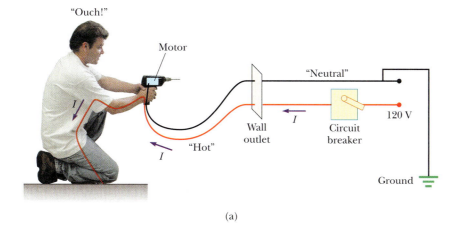

(a)

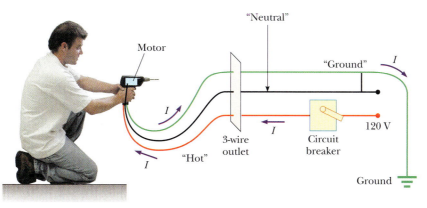

(b)

Figure 18.23 The "hot" (or "live") wire, at 120 V, always includes a cir-cuit breaker for safety. (a) When the drill is operated with two wires, the normal current path is from the "hot" wire, through the motor connections, and back to ground through the "neutral" wire. However, here the high-voltage side has come in contact with the drill case, so that the person holding the drill receives an electrical shock. (b) Shock can be prevented by a third wire running from the drill case to the ground.

and back to ground through the third wire. The resulting high current produced will blow a fuse or trip a circuit breaker before the consumer is injured.

Special power outlets called ground-fault interrupters (GFIs) are now being used in kitchens, bathrooms, basements, and other hazardous areas of new homes. They are designed to protect people from electrical shock by sensing small currents—approximately 5 mA and greater—leaking to ground. When current above this level is detected, the device shuts off (interrupts) the current in less than a millisecond. (Ground-fault interrupters are discussed in Chapter 19.)

18.8 CONDUCTION OF ELECTRICAL SIGNALS BY NEURONS[2]

The most remarkable use of electrical phenomena in living organisms is found in the nervous system of animals. Specialized cells in the body called **neurons** form a complex network that receives, processes, and transmits information from one part of the body to another. The center of this network is located in the brain, which has the ability to store and analyze information. On the basis of this information, the nervous system controls parts of the body.

The nervous system is highly complex and consists of about 10^{10} interconnected neurons. Some aspects of the nervous system are well known. Over the past 45 years, the method of signal propagation through the nervous system has been established. The messages transmitted by neurons are voltage pulses called *action potentials*. When a neuron receives a strong enough stimulus, it produces identical voltage pulses that are actively propagated along its structure. The strength of the stimulus is conveyed by the number of pulses produced. When the pulses reach the end of the neuron, they activate either muscle cells or other neurons. There is a "firing threshold" for neurons: action potentials propagate along a neuron only if the stimulus is sufficiently strong.

Neurons can be divided into three classes: sensory neurons, motor neurons, and interneurons. The sensory neurons receive stimuli from sensory organs that monitor the external and internal environment of the body. Depending on their specialized functions, the sensory neurons convey messages about factors such as light, temperature, pressure, muscle tension, and odor to higher centers in the nervous system. The motor neurons carry messages that control the muscle cells. The messages are based on the information provided by the sensory neurons and by the brain. The interneurons transmit information from one neuron to another.

Each neuron consists of a cell body to which are attached input ends called **dendrites** and a long tail called the **axon**, which transmits the signal away from the cell (Fig. 18.24). The far end of the axon branches into nerve endings that transmit the signal across small gaps to other neurons or to muscle cells. A simple sensorimotor neuron circuit is shown in Figure 18.25. A stimulus from a muscle produces nerve impulses that travel to the spine. Here the signal is transmitted to a motor neuron, which in turn sends impulses to control the muscle. Figure 18.26 shows an electron microscope image of neurons in the brain.

The axon, which is an extension of the neuron cell, conducts electric impulses away from the cell body. Some axons are extremely long. In humans, for example, the axons connecting the spine with the fingers and toes are more than 1 m long. The neuron can transmit messages because of the special active electrical characteristics of the axon. (The axon acts as an **active** source of energy like a battery, rather than like a **passive** stretch of resistive wire.) Much of the information about the electrical and chemical properties of the axon is obtained by inserting small needlelike probes into it. Figure 18.27 shows an experimental setup.

Note that the outside of the axon is grounded, so that all measured voltages are with respect to a zero potential on the outside. With these probes, it is possible to inject current into the axon, measure the resulting action potential as a function of time at a fixed point, and sample the cell's chemical composition. Such experiments

[2]This section is based upon an essay by Paul Davidovits of Boston College.

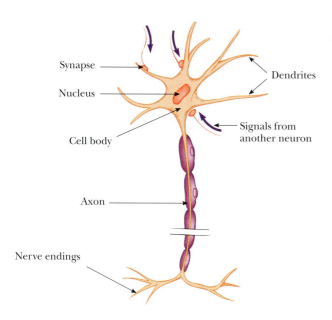

Figure 18.24   Diagram of a neuron.

are usually difficult to run because the diameter of most axons is very small. Even the largest axons in the human nervous system have a diameter of only about 20×10^{-4} cm. The giant squid, however, has an axon with a diameter of about 0.5 mm, which is large enough for the convenient insertion of probes. Much of the information about signal transmission in the nervous system has come from experiments with the squid axon.

In the aqueous environment of the body, salts and other molecules dissociate into positive and negative ions. As a result, body fluids are relatively good conductors of electricity. The inside of the axon is filled with an ionic fluid that is separated from the surrounding body fluid by a thin membrane that is only about 5 nm to 10 nm thick. The resistivities of the internal and external fluids are about the same, but their chemical compositions are substantially different. The external fluid is similar to seawater: Its ionic solutes are mostly positive sodium ions and negative chloride ions. Inside the axon, the positive ions are mostly potassium ions and the negative ions are mostly large organic ions.

Ordinarily, the concentrations of sodium and potassium ions inside and outside the axon would be equalized by diffusion. The axon, however, is a living cell with an energy supply and can change the permeability of its membranes on a time scale of milliseconds.

When the axon is not conducting an electric pulse, the axon membrane is highly permeable to potassium ions, slightly permeable to sodium ions, and impermeable to large organic ions. Consequently, although sodium ions cannot easily enter the axon, potassium ions can leave it. As the potassium ions leave the axon, however, they leave behind large negative organic ions, which cannot follow them through the membrane. As a result, a negative potential builds up inside the axon

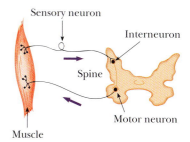

Figure 18.25   A simple neural circuit.

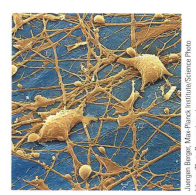

Juergen Berger, Max-Planck Institute/Science Photo Library/Photo Researchers, Inc.

Figure 18.26   Stellate neuron from human cortex.

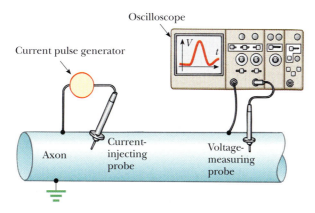

Figure 18.27   An axon stimulated electrically. The left probe injects a short pulse of current, and the right probe measures the resulting action potential as a function of time.

Figure 18.28 (a) Typical action potential as a function of time. (b) Current in the axon membrane wall as a function of time.

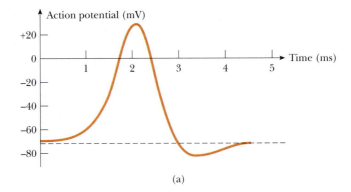

(a)

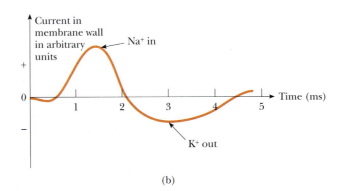

(b)

with respect to the outside. The final negative potential reached, which has been measured at about -70 mV, holds back the outflow of potassium ions so that at equilibrium, the concentration of ions is as we have stated.

The mechanism for the production of an electric signal by the neuron is conceptually simple, but was experimentally difficult to sort out. When a neuron changes its resting potential because of an appropriate stimulus, the properties of its membrane change locally. As a result, there is a sudden flow of sodium ions into the cell that lasts for about two milliseconds. This produces the $+30$ mV peak in the action potential shown in Figure 18.28a. Immediately after, there is an increase in potassium ion flow out of the cell which restores the resting action potential of -70 mV in an additional 3 ms. Both the Na^+ and K^+ ion flows have been measured by using radioactive Na and K tracers. The nerve signal has been measured to travel along the axon at speeds of 50 m/s to about 150 m/s. This flow of charged particles (or signal transmission) in a nerve axon is *unlike* signal transmission in a metal wire. In an axon, charges move in a direction perpendicular to the direction of travel of the nerve signal, and the nerve signal moves much more slowly than a voltage pulse traveling along a metallic wire.

Although the axon is a highly complex structure, and much of how Na^+ and K^+ ion channels open and close is not understood, standard electric circuit concepts of current and capacitance can be used to analyze axons. It is left as a problem (Problem 41) to show that the axon, having equal and opposite charges separated by a thin dielectric membrane, acts like a capacitor.

SUMMARY

18.1 Sources of emf

Any device, such as a battery or generator, that increases the electric potential energy of charges in an electric circuit is called a **source of emf**. Batteries convert chemical energy

into electrical potential energy, and generators convert mechanical energy into electrical potential energy.

The terminal voltage ΔV of a battery is given by

$$\Delta V = \mathcal{E} - Ir \qquad [18.1]$$

where $\mathcal{E}$ is the emf of the battery, I is the current, and r is the internal resistance of the battery. Generally, the internal resistance is small enough to be neglected.

18.2 Resistors in Series

The **equivalent resistance** of a set of resistors connected in **series** is

$$R_{eq} = R_1 + R_2 + R_3 + \cdots \qquad [18.4]$$

The current remains at a constant value as it passes through a series of resistors. The potential difference across any two resistors in series is different, unless the resistors have the same resistance.

18.3 Resistors in Parallel

The **equivalent resistance** of a set of resistors connected in **parallel** is

$$\frac{1}{R_{eq}} = \frac{1}{R_1} + \frac{1}{R_2} + \frac{1}{R_3} + \cdots \qquad [18.6]$$

The potential difference across any two parallel resistors is the same; however, the current in each resistor will be different unless the two resistances are equal.

18.4 Kirchhoff's Rules and Complex DC Circuits

Complex circuits can be analyzed using **Kirchhoff's rules**:

1. The sum of the currents entering any junction must equal the sum of the currents leaving that junction.
2. The sum of the potential differences across all the elements around any closed circuit loop must be zero.

The first rule, called the junction rule, is a statement of **conservation of charge**. The second rule, called the loop rule, is a statement of **conservation of energy**. Solving problems involves using these rules to generate as many equations as there are unknown currents. The equations can then be solved simultaneously.

18.5 *RC* Circuits

In a simple *RC* circuit with a battery, a resistor, and a capacitor in series, the charge on the capacitor increases according to the equation

$$q = Q(1 - e^{-t/RC}) \qquad [18.7]$$

The term *RC* in Equation 18.7 is called the **time constant** τ (Greek letter tau), so

$$\tau = RC \qquad [18.8]$$

The time constant represents the time required for the charge to increase from zero to 63.2% of its maximum equilibrium value.

A simple *RC* circuit consisting of a charged capacitor in series with a resistor discharges according to the expression

$$q = Qe^{-t/RC} \qquad [18.9]$$

Problems can be solved by substituting into these equations. The voltage ΔV across the capacitor at any time is obtained by dividing the charge by the capacitance: $\Delta V = q/C$. Using Kirchhoff's loop rule yields the potential difference across the resistor. Ohm's law applied to the resistor then gives the current.

CONCEPTUAL QUESTIONS

1. Is the direction of current in a battery always from the negative terminal to the positive one? Explain.

2. Given three lightbulbs and a battery, sketch as many different circuits as you can.

3. Suppose the energy transferred to a dead battery during charging is *W*. The recharged battery is then used until fully discharged again. Is the total energy transferred out of the battery during use also *W*?

4. (a) A group of resistors connected in parallel have the same (i) current in them, (ii) potential difference across them, or (iii) neither of the above. (b) A group of resistors connected in series have the same (i) current in them, (ii) potential difference across them, or (iii) neither of the above. Justify your answers by considering a circuit consisting of a 3.0-Ω resistor and a 5.0-Ω resistor connected across a 12-V battery.

5. If you have your headlights on while you start your car, why do they dim while the car is starting?

6. (a) The equivalent resistance of a group of resistors connected in parallel is (i) greater than any of the resistors in the group, (ii) less than any of the resistors in the group, or (iii) neither of the above. (b) The equivalent resistance of a group of resistors connected in series is (i) greater than any of the resistors in the group, (ii) less than any of the resistors in the group, or (iii) neither of the above. Justify your answers by considering a circuit consisting of a

3.0-Ω resistor and a 5.0-Ω resistor connected across a 12-V battery.

7. Electrical devices are often rated with a voltage and a current—for example, 120 V, 5 A. Batteries, however, are rated only with a voltage—for example, 1.5 V. Why?

8. A short circuit is a circuit containing a path of very low resistance in parallel with some other part of the circuit. Discuss the effect of a short circuit on the portion of the circuit it parallels. Use a lamp with a frayed line cord as an example.

9. Connecting batteries in series increases the emf applied to a circuit. What advantage might there be to connecting them in parallel?

10. If electrical power is transmitted over long distances, the resistance of the wires becomes significant. Why? Which mode of transmission would result in less energy loss—high current and low voltage or low current and high voltage? Discuss.

11. Describe what happens to the lightbulb in Figure Q18.11 after the switch is closed. Assume the capacitor has a large capacitance and is initially uncharged. Assume also that the bulb lights up when connected directly across the battery terminals.

12. Two sets of Christmas tree lights are available. For set A, when one bulb is removed, the remaining bulbs remain illuminated. For set B, when one bulb is removed, the

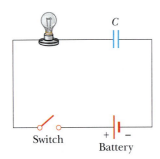

Figure Q18.11

remaining bulbs do not operate. Explain the difference in wiring for the two sets.

13. Why is it possible for a bird to sit on a high-voltage wire without being electrocuted? (See Fig. Q18.13.)

Figure Q18.13   Birds on a high-voltage wire.

14. (a) Two resistors are connected in series across a battery. The power delivered to each resistor is (i) the same or (ii) not necessarily the same. (b) Two resistors are connected in parallel across a battery. The power delivered to each resistor is (i) the same or (ii) not necessarily the same.

15. Embodied in Kirchhoff's rules are two conservation laws. What are they?

16. A ski resort consists of a few chairlifts and several interconnected downhill runs on the side of a mountain, with a lodge at the bottom. The lifts are analogous to batteries and the runs are analogous to resistors. Describe how two runs can be in series. Describe how three runs can be in parallel. Sketch a junction of one lift and two runs. One of the skiers is carrying an altimeter. State Kirchhoff's junction rule and Kirchhoff's loop rule for ski resorts.

17. Suppose you are flying a kite when it strikes a high-voltage wire (a very dangerous situation). What factors determine how great a shock you will receive?

18. Why is it dangerous to turn on a light when you are in a bathtub?

19. Suppose a parachutist lands on a high-voltage wire and grabs the wire as she prepares to be rescued. Will she be electrocuted? If the wire then breaks, should she continue to hold onto the wire as she falls to the ground?

20. Would a fuse or circuit breaker work successfully if it were placed in parallel with the device it was supposed to protect?

21. A series circuit consists of three identical lamps connected to a battery as in Figure Q18.21. When the switch S is closed, what happens (a) to the intensities of lamps A and B, (b) to the intensity of lamp C, (c) to the current in the circuit, and (d) to the voltage drop across the three lamps? (e) Does the power dissipated in the circuit increase, decrease, or remain the same?

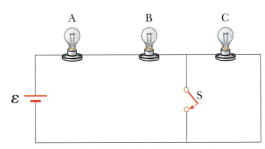

Figure Q18.21

22. Figure Q18.22 shows a series connection of three lamps, all rated at 120 V, with power ratings of 60 W, 75 W, and 200 W, respectively. Why do the intensities of the lamps differ? Which lamp has the greatest resistance? How would their intensities differ if they were connected in parallel?

Figure Q18.22

23. A student claims that, of two lightbulbs connected, in series the second is less bright than the first, because the first bulb uses up some of the current. How would you respond to this statement?

24. Two identical, parallel copper wires are placed underground between two points 1.00 mile apart. These wires are usually not connected, but a construction accident shorts the wires together at some point. To try to isolate the spot so that repair can be initiated, a technician goes to end A of the lines and finds that a 12.0-V battery connected across the wires at that end produces a current of 1.00 A. Doing the same at the other end, B, produces a current of 0.20 A. Is the break closer to end A or to end B?

PROBLEMS

1, 2, 3 = straightforward, intermediate, challenging   ☐ = full solution available in *Student Solutions Manual/Study Guide*
Physics⊗Now™ = coached problem with hints available at **www.cp7e.com**   = biomedical application

Section 18.1 Sources of emf
Section 18.2 Resistors in Series
Section 18.3 Resistors in Parallel

1. A battery having an emf of 9.00 V delivers 117 mA when connected to a 72.0-Ω load. Determine the internal resistance of the battery.

2. A 4.0-Ω resistor, an 8.0-Ω resistor, and a 12-Ω resistor are connected in series with a 24-V battery. What are (a) the equivalent resistance and (b) the current in each resistor? (c) Repeat for the case in which all three resistors are connected in parallel across the battery.

3. A lightbulb marked "75 W [at] 120 V" is screwed into a socket at one end of a long extension cord in which each of the two conductors has a resistance of 0.800 Ω. The other end of the extension cord is plugged into a 120-V outlet. Draw a circuit diagram, and find the actual power of the bulb in the circuit described.

4. A 9.0-Ω resistor and a 6.0-Ω resistor are connected in series with a power supply. (a) The voltage drop across the 6.0-Ω resistor is measured to be 12 V. Find the voltage output of the power supply. (b) The two resistors are connected in parallel across a power supply, and the current through the 9.0-Ω resistor is found to be 0.25 A. Find the voltage setting of the power supply.

5. Physics⊗Now™ (a) Find the equivalent resistance between points *a* and *b* in Figure P18.5. (b) Calculate the current in each resistor if a potential difference of 34.0 V is applied between points *a* and *b*.

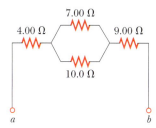

Figure P18.5

6. Find the equivalent resistance of the circuit in Figure P18.6.

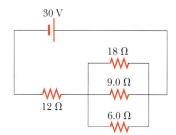

Figure P18.6

7. What is the equivalent resistance of the combination of resistors between points *a* and *b* in Figure P18.7? Note that one end of the vertical resistor is left free.

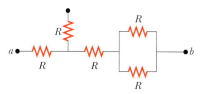

Figure P18.7

8. (a) Find the equivalent resistance of the circuit in Figure P18.8. (b) If the total power supplied to the circuit is 4.00 W, find the emf of the battery.

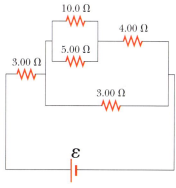

Figure P18.8

9. Consider the circuit shown in Figure P18.9. Find (a) the current in the 20.0-Ω resistor and (b) the potential difference between points *a* and *b*.

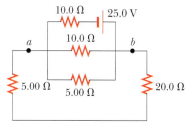

Figure P18.9

10. Two resistors, *A* and *B*, are connected in parallel across a 6.0-V battery. The current through *B* is found to be 2.0 A. When the two resistors are connected in series to the 6.0-V battery, a voltmeter connected across resistor *A* measures a voltage of 4.0 V. Find the resistances of *A* and *B*.

11. The resistance between terminals *a* and *b* in Figure P18.11 is 75 Ω. If the resistors labeled *R* have the same value, determine *R*.

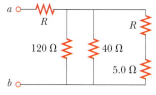

Figure P18.11

12. Three 100-Ω resistors are connected as shown in Figure P18.12. The maximum power that can safely be delivered to any one resistor is 25.0 W. (a) What is the maximum voltage that can be applied to the terminals a and b? (b) For the voltage determined in part (a), what is the power delivered to each resistor? What is the total power delivered?

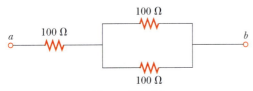

Figure P18.12

13. Find the current in the 12-Ω resistor in Figure P18.13.

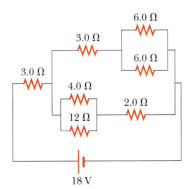

Figure P18.13

14. Calculate the power delivered to each resistor in the circuit shown in Figure P18.14.

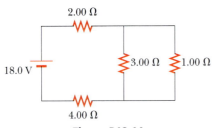

Figure P18.14

15. (a) You need a 45-Ω resistor, but the stockroom has only 20-Ω and 50-Ω resistors. How can the desired resistance be achieved under these circumstances? (b) What can you do if you need a 35-Ω resistor?

Section 18.4 Kirchhoff's Rules and Complex DC Circuits

Note: For some circuits, the currents are not necessarily in the direction shown.

16. The ammeter shown in Figure P18.16 reads 2.00 A. Find I_1, I_2, and $\mathcal{E}$.

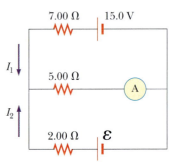

Figure P18.16

17. Determine the current in each branch of the circuit shown in Figure P18.17.

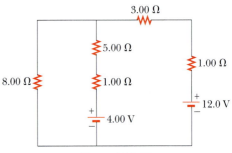

Figure P18.17

18. Determine the potential difference ΔV_{ab} for the circuit in Figure P18.18.

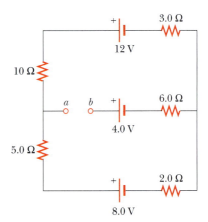

Figure P18.18

19. Figure P18.19 shows a circuit diagram. Determine (a) the current, (b) the potential of wire A relative to ground, and (c) the voltage drop across the 1 500-Ω resistor.

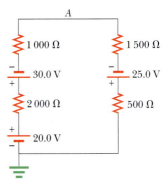

Figure P18.19

20. In the circuit of Figure P18.20, the current I_1 is 3.0 A while the values of $\mathcal{E}$ and R are unknown. What are the currents I_2 and I_3?

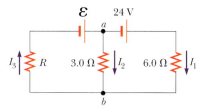

Figure P18.20

21. What is the emf $\mathcal{E}$ of the battery in the circuit of Figure P18.21?

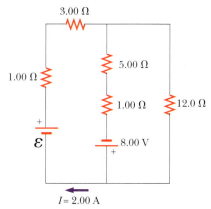

Figure P18.21

22. Four resistors are connected to a battery with a terminal voltage of 12 V, as shown in Figure P18.22. Determine the power delivered to the 50-Ω resistor.

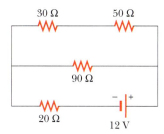

Figure P18.22

23. Using Kirchhoff's rules, (a) find the current in each resistor shown in Figure P18.23 and (b) find the potential difference between points c and f.

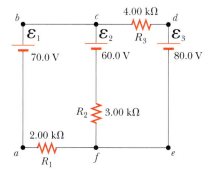

Figure P18.23

24. Two 1.50-V batteries—with their positive terminals in the same direction—are inserted in series into the barrel of a flashlight. One battery has an internal resistance of $0.255\ \Omega$, the other an internal resistance of $0.153\ \Omega$. When the switch is closed, a current of 0.600 A passes through the lamp. (a) What is the lamp's resistance? (b) What fraction of the power dissipated is dissipated in the batteries?

25. Calculate each of the unknown currents I_1, I_2, and I_3 for the circuit of Figure P18.25.

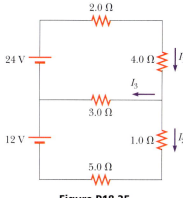

Figure P18.25

26. A dead battery is charged by connecting it to the live battery of another car with jumper cables (Fig. P18.26). Determine the current in the starter and in the dead battery.

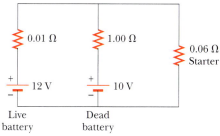

Figure P18.26

27. Find the current in each resistor in Figure P18.27.

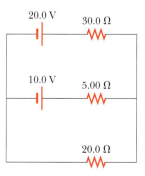

Figure P18.27

28. (a) Determine the potential difference ΔV_{ab} for the circuit in Figure P18.28. Note that each battery has an internal resistance as indicated in the figure. (b) If points a and b are connected by a 7.0-Ω resistor, what is the current in this resistor?

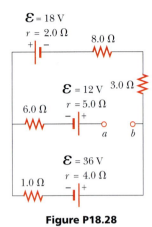

Figure P18.28

29. **Physics⊗Now**™ Find the potential difference across each resistor in Figure P18.29.

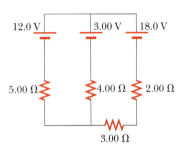

Figure P18.29

Section 18.5 *RC* Circuits

30. Show that $\tau = RC$ has units of time.

31. Consider a series *RC* circuit for which $C = 6.0\ \mu F$, $R = 2.0 \times 10^6\ \Omega$, and $\mathcal{E} = 20\ V$. Find (a) the time constant of the circuit and (b) the maximum charge on the capacitor after a switch in the circuit is closed.

32. An uncharged capacitor and a resistor are connected in series to a source of emf. If $\mathcal{E} = 9.00\ V$, $C = 20.0\ \mu F$, and $R = 100\ \Omega$, find (a) the time constant of the circuit, (b) the maximum charge on the capacitor, and (c) the charge on the capacitor after one time constant.

33. Consider a series *RC* circuit for which $R = 1.0\ M\Omega$, $C = 5.0\ \mu F$, and $\mathcal{E} = 30\ V$. Find the charge on the capacitor 10 s after the switch is closed.

34. A series combination of a 12-kΩ resistor and an unknown capacitor is connected to a 12-V battery. One second after the circuit is completed, the voltage across the capacitor is 10 V. Determine the capacitance of the capacitor.

35. A capacitor in an *RC* circuit is charged to 60.0% of its maximum value in 0.900 s. What is the time constant of the circuit.

36. A series *RC* circuit has a time constant of 0.960 s. The battery has an emf of 48.0 V, and the maximum current in the circuit is 0.500 mA. What are (a) the value of the capacitance and (b) the charge stored in the capacitor 1.92 s after the switch is closed?

Section 18.6 Household Circuits

37. An electric heater is rated at 1 300 W, a toaster at 1 000 W, and an electric grill at 1 500 W. The three appliances are connected in parallel to a common 120-V circuit. (a) How

much current does each appliance draw? (b) Is a 30.0-A circuit breaker sufficient in this situation? Explain.

38. A lamp ($R = 150\ \Omega$), an electric heater ($R = 25\ \Omega$), and a fan ($R = 50\ \Omega$) are connected in parallel across a 120-V line. (a) What total current is supplied to the circuit? (b) What is the voltage across the fan? (c) What is the current in the lamp? (d) What power is expended in the heater?

39. **Physics⊗Now**™ A heating element in a stove is designed to dissipate 3 000 W when connected to 240 V. (a) Assuming that the resistance is constant, calculate the current in the heating element if it is connected to 120 V. (b) Calculate the power it dissipates at that voltage.

40. Your toaster oven and coffeemaker each dissipate 1 200 W of power. Can you operate them together if the 120-V line that feeds them has a circuit breaker rated at 15 A? Explain.

Section 18.8 Conduction of Electrical Signals by Neurons

41. Assume that a length of axon membrane of about 10 cm is excited by an action potential (length excited = nerve speed × pulse duration = 50 m/s × 2.0 ms = 10 cm). In the resting state, the outer surface of the axon wall is charged positively with K^+ ions and the inner wall has an equal and opposite charge of negative organic ions, as shown in Figure P18.41. Model the axon as a parallel-plate capacitor, and take $C = \kappa \varepsilon_0 A/d$ and $Q = C\Delta V$ to investigate the charge as follows: Use typical values for a cylindrical axon of cell wall thickness $d = 1.0 \times 10^{-8}$ m, axon radius $r = 10\ \mu m$, and cell-wall dielectric constant $\kappa = 3.0$. (a) Calculate the positive charge on the outside of a 10-cm piece of axon when it is not conducting an electric pulse. How many K^+ ions are on the outside of the axon? Is this a large charge per unit area? [*Hint:* Calculate the charge per unit area in terms of the number of angstroms (Å^2) per electronic charge. An atom has a cross section of about $1\ \text{Å}^2$ ($1\ \text{Å} = 10^{-10}$ m)] (b) How much positive charge must flow through the cell membrane to reach the excited state of $+30$ mV from the resting state of -70 mV? How many sodium ions (Na^+) is this? (c) If it takes 2.0 ms for the Na^+ ions to enter the axon, what is the average current in the axon wall in this process? (d) How much energy does it take to raise the potential of the inner axon wall to $+30$ mV, starting from the resting potential of -70 mV?

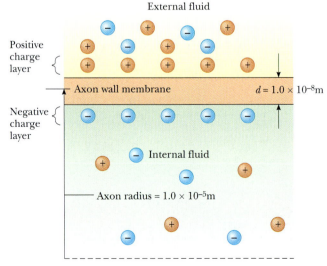

Figure P18.41   (Problems 41 and 42)

42. Consider the model of the axon as a capacitor from Problem 41 and Figure P18.41. (a) How much energy does it take to restore the inner wall of the axon to -70 mV, starting from $+30$ mV? (b) Find the average current in the axon wall during this process.

43. Using Figure 18.28b and the results of Problems 18.41d and 18.42a, find the power supplied by the axon per action potential.

ADDITIONAL PROBLEMS

44. Consider an RC circuit in which the capacitor is being charged by a battery connected in the circuit. After a time equal to two time constants, what percentage of the *final* charge is present on the capacitor?

45. Find the equivalent resistance between points a and b in Figure P18.45.

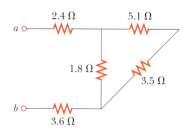

Figure P18.45

46. For the circuit in Figure P18.46, calculate (a) the equivalent resistance of the circuit and (b) the power dissipated by the entire circuit. (c) Find the current in the 5.0-Ω resistor.

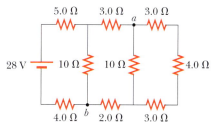

Figure P18.46

47. Find (a) the equivalent resistance of the circuit in Figure P18.47, (b) each current in the circuit, (c) the potential difference across each resistor, and (d) the power dissipated by each resistor.

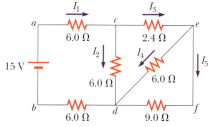

Figure P18.47

48. Three 60.0-W, 120-V lightbulbs are connected across a 120-V power source, as shown in Figure P18.48. Find (a) the total power delivered to the three bulbs and (b) the potential difference across each. Assume that the resistance of each bulb is constant (even though, in reality, the resistance increases markedly with current).

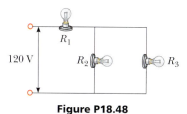

Figure P18.48

49. An automobile battery has an emf of 12.6 V and an internal resistance of 0.080 Ω. The headlights have a total resistance of 5.00 Ω (assumed constant). What is the potential difference across the headlight bulbs (a) when they are the only load on the battery and (b) when the starter motor is operated, taking an additional 35.0 A from the battery?

50. In Figure P18.50, suppose that the switch has been closed for a length of time sufficiently long for the capacitor to become fully charged. Find (a) the steady-state current in each resistor and (b) the charge on the capacitor.

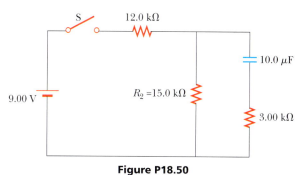

Figure P18.50

51. Find the values of I_1, I_2, and I_3 for the circuit in Figure P18.51.

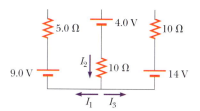

Figure P18.51

52. The resistance between points a and b in Figure P18.52 drops to one-half its original value when switch S is closed. Determine the value of R.

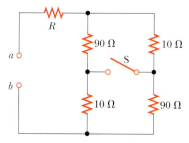

Figure P18.52

53. **Physics⊗Now™** A generator has a terminal voltage of 110 V when it delivers 10.0 A and 106 V when it delivers 30.0 A. Calculate the emf and the internal resistance of the generator.

54. An emf of 10 V is connected to a series RC circuit consisting of a resistor of $2.0 \times 10^6 \ \Omega$ and a capacitor of $3.0 \ \mu F$. Find the time required for the charge on the capacitor to reach 90% of its final value.

55. The student engineer of a campus radio station wishes to verify the effectiveness of the lightning rod on the antenna mast (Fig. P18.55). The unknown resistance R_x is between points C and E. Point E is a "true ground," but is inaccessible for direct measurement, since the stratum in which it is located is several meters below the Earth's surface. Two identical rods are driven into the ground at A and B, introducing an unknown resistance R_y. The procedure for finding the unknown resistance R_x is as follows: Measure resistance R_1 between points A and B. Then connect A and B with a heavy conducting wire, and measure resistance R_2 between points A and C. (a) Derive a formula for R_x in terms of the observable resistances R_1 and R_2. (b) A satisfactory ground resistance would be $R_x < 2.0 \ \Omega$. Is the grounding of the station adequate if measurements give $R_1 = 13 \ \Omega$ and $R_2 = 6.0 \ \Omega$?

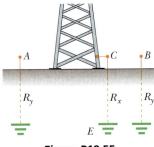

Figure P18.55

56. The resistor R in Figure P18.56 dissipates 20 W of power. Determine the value of R.

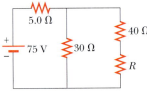

Figure P18.56

57. A voltage ΔV is applied to a series configuration of n resistors, each of resistance R. The circuit components are reconnected in a parallel configuration, and voltage ΔV is again applied. Show that the power consumed by the series configuration is $1/n^2$ times the power consumed by the parallel configuration.

58. For the network in Figure P18.58, show that the resistance between points a and b is $R_{ab} = \frac{27}{17} \ \Omega$. [*Hint*: Connect a battery with emf $\mathcal{E}$ across points a and b and determine $\mathcal{E}/I$, where I is the current in the battery.]

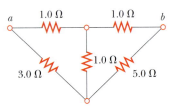

Figure P18.58

59. A battery with an internal resistance of $10.0 \ \Omega$ produces an open-circuit voltage of 12.0 V. A variable load resistance with a range from 0 to $30.0 \ \Omega$ is connected across the battery. (*Note*: A battery has a resistance that depends on the condition of its chemicals and that increases as the battery ages. This internal resistance can be represented in a simple circuit diagram as a resistor in series with the battery.) (a) Graph the power dissipated in the load resistor as a function of the load resistance. (b) With your graph, demonstrate the following important theorem: *The power delivered to a load is a maximum if the load resistance equals the internal resistance of the source.*

60. The circuit in Figure P18.60 contains two resistors, $R_1 = 2.0 \ k\Omega$ and $R_2 = 3.0 \ k\Omega$, and two capacitors, $C_1 = 2.0 \ \mu F$ and $C_2 = 3.0 \ \mu F$, connected to a battery with emf $\mathcal{E} = 120$ V. If there are no charges on the capacitors before switch S is closed, determine the charges q_1 and q_2 on capacitors C_1 and C_2, respectively, as functions of time, after the switch is closed. [*Hint*: First reconstruct the circuit so that it becomes a simple RC circuit containing a single resistor and single capacitor in series, connected to the battery, and then determine the total charge q stored in the circuit.]

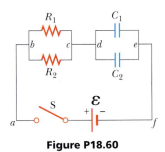

Figure P18.60

61. Consider the circuit shown in Figure P18.61. Find (a) the potential difference between points a and b and (b) the current in the 20.0-Ω resistor.

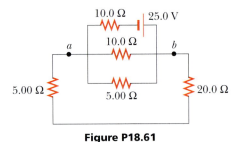

Figure P18.61

62. In Figure P18.62, $R_1 = 0.100 \ \Omega$, $R_2 = 1.00 \ \Omega$, and $R_3 = 10.0 \ \Omega$. Find the equivalent resistance of the circuit and the current in each resistor when a 5.00-V power supply is connected between (a) points A and B, (b) points A and C, and (c) points A and D.

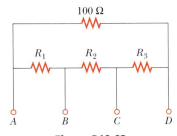

Figure P18.62

63. What are the expected readings of the ammeter and voltmeter for the circuit in Figure P18.63?

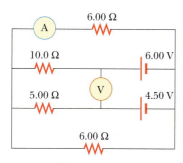

Figure P18.63

64. Consider the two arrangements of batteries and bulbs shown in Figure P18.64. The two bulbs are identical and have resistance R, and the two batteries are identical with output voltage ΔV. (a) In case 1, with the two bulbs in series, compare the brightness of each bulb, the current in each bulb, and the power delivered to each bulb? (b) In case 2, with the two bulbs in parallel, compare the brightness of each bulb, the current in each bulb, and the power supplied to each bulb. (c) Which bulbs are brighter, those in case 1 or those in case 2? (d) In each case, if one bulb fails, will the other go out as well? If the other bulb doesn't fail, will it get brighter or stay the same?

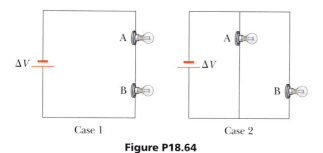

Figure P18.64

(Problem 64 is courtesy of E.F. Redish. For other problems of this type, visit http://www.physics.umd.edu/perg/.)

ACTIVITIES

1. Using insulated wire, connect one terminal of a D-cell battery to the base of a flashlight bulb, tape a second wire to the other battery terminal, and tape a third wire to the center conductor of the bulb, as shown in the Figure A18.1a. Be sure to remove about 1 cm of insulation from the ends of all wires before making connections. Connect the two open wires together to complete the circuit, and note the illumination of the bulb. Now add a second D-cell battery to the circuit as in Figure A18.1b to give a total voltage of 3.0 V, connect the two open wires together to complete the circuit, and note the illumination of the bulb again. Why does the bulb grow brighter in this case?

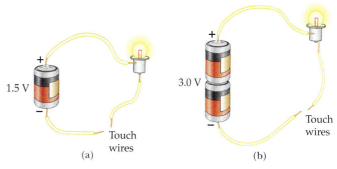

Figure A18.1

2. Use the basic equipment of activity 1 plus a few more items to test some additional features of circuits. First, note the brightness of a single bulb connected to the battery. Now connect two bulbs in series with each other and the battery. Predict whether the bulbs will be dimmer or brighter than when operated separately. Try it and see. Continue for three bulbs in series.

3. Repeat activity 2, but this time connect the bulbs in parallel with each other. Predict how the brightness will change as you add more bulbs in parallel. Explain.

4. Continue your experimentation with circuits by connecting the battery to one bulb, followed in the same circuit by two bulbs in parallel and then back to the battery. Predict the brightness of each bulb in this situation before you connect the circuit. Explain the results you obtain.

Aurora borealis, the Northern Lights. Displays such as this one are caused by cosmic ray particles trapped in the magnetic field of Earth. When the particles collide with atoms in the atmosphere, they cause the atoms to emit visible light.

Johnny Johnson/Stone/Getty

CHAPTER

19

Magnetism

In terms of applications, magnetism is one of the most important fields in physics. Large electromagnets are used to pick up heavy loads. Magnets are used in such devices as meters, motors, and loudspeakers. Magnetic tapes and disks are used routinely in sound- and video-recording equipment and to store computer data. Intense magnetic fields are used in magnetic resonance imaging (MRI) devices to explore the human body with better resolution and greater safety than x-rays can provide. Giant superconducting magnets are used in the cyclotrons that guide particles into targets at nearly the speed of light, and magnetic bottles hold antimatter, possibly the key to future space propulsion systems.

Magnetism is closely linked with electricity. Magnetic fields affect moving charges, and moving charges produce magnetic fields. Changing magnetic fields can even create electric fields. These phenomena signify an underlying unity of electricity and magnetism, which James Clerk Maxwell first described in the 19th century. The ultimate source of any magnetic field is electric current.

19.1 MAGNETS

Most people have had experience with some form of magnet. You are most likely familiar with the common iron horseshoe magnet that can pick up iron-containing objects such as paper clips and nails. Several commercially available magnets are shown in Figure 19.1. In the discussion that follows, we assume the magnet has the shape of a bar. Iron objects are most strongly attracted to either end of such a bar magnet, called its **poles**. One end is called the **north pole** and the other the **south pole**. The names come from the behavior of a magnet in the presence of Earth's magnetic field. If a bar magnet is suspended from its midpoint by a piece of string so that it can swing freely in a horizontal plane, it will rotate until its north pole points to the north and its south pole points to the south. The same idea is used to construct a simple compass. Magnetic poles also exert attractive or repulsive forces on each other similar to the electrical forces between charged objects. In fact, simple

Figure 19.1 An assortment of commercially available magnets. The four red magnets and the large black magnet on the left are made of an alloy of iron, aluminum, and cobalt. The six horseshoe magnets on the right are made of different nickel–steel alloys. The rectangular magnets on the lower right are ceramics made of iron, nickel, and beryllium oxides.

Courtesy of Central Scientific Company

experiments with two bar magnets show that **like poles repel each other and unlike poles attract each other**.

Although the force between opposite magnetic poles is similar to the force between positive and negative electric charges, there is an important difference: positive and negative electric charges can exist in isolation of each other; north and south poles don't. No matter how many times a permanent magnet is cut, each piece always has a north pole and a south pole. There is some theoretical basis for the speculation that magnetic monopoles (isolated north or south poles) exist in nature, and the attempt to detect them is currently an active experimental field of investigation.

An unmagnetized piece of iron can be magnetized by stroking it with a magnet. Magnetism can also be induced in iron (and other materials) by other means. For example, if a piece of unmagnetized iron is placed near a strong permanent magnet, the piece of iron eventually becomes magnetized. The process can be accelerated by heating and then cooling the iron.

Naturally occurring magnetic materials such as magnetite are magnetized in this way because they have been subjected to Earth's magnetic field for long periods of time. The extent to which a piece of material retains its magnetism depends on whether it is classified as magnetically hard or soft. **Soft** magnetic materials, such as iron, are easily magnetized, but also tend to lose their magnetism easily. In contrast, **hard** magnetic materials, such as cobalt and nickel, are difficult to magnetize, but tend to retain their magnetism.

In earlier chapters we described the interaction between charged objects in terms of electric fields. Recall that an electric field surrounds any stationary electric charge. The region of space surrounding a *moving* charge includes a magnetic field as well. A magnetic field also surrounds a properly magnetized magnetic material.

To describe any type of vector field, we must define its magnitude, or strength, and its direction. The direction of a magnetic field $\vec{B}$ at any location is the direction in which the north pole of a compass needle points at that location. Active Figure 19.2a shows how the magnetic field of a bar magnet can be traced with the aid of a compass, defining a **magnetic field line**. Several magnetic field

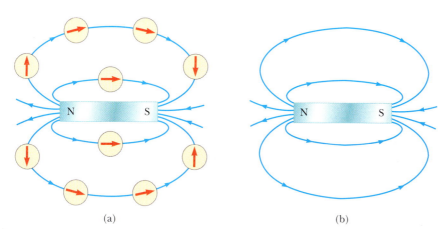

(a)   (b)

ACTIVE FIGURE 19.2

(a) Tracing the magnetic field of a bar magnet. (b) Several magnetic field lines of a bar magnet.

Physics⊗Now™

Log into PhysicsNow at **www.cp7e.com** and go to Active Figure 19.2, where you can move the compass around and trace the magnetic field for yourself.

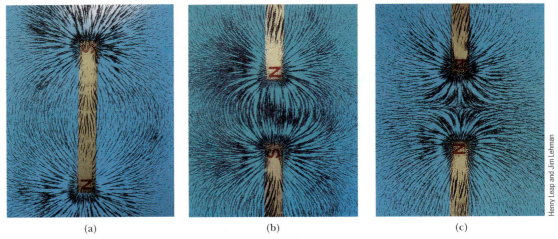

(a)   (b)   (c)

Henry Leap and Jim Lehman

Figure 19.3 (a) The magnetic field pattern of a bar magnet, as displayed with iron filings on a sheet of paper. (b) The magnetic field pattern between *unlike* poles of two bar magnets, as displayed with iron filings. (c) The magnetic field pattern between two *like* poles.

lines of a bar magnet traced out in this way appear in the two-dimensional representation in Active Figure 19.2b. Magnetic field patterns can be displayed by placing small iron filings in the vicinity of a magnet, as in Figure 19.3.

Forensic scientists use a technique similar to that shown in Figure 19.3 to find fingerprints at a crime scene. One way to find latent, or invisible, prints is by sprinkling a powder of iron dust on a surface. The iron adheres to any perspiration or body oils that are present and can be spread around on the surface with a magnetic brush that never comes into contact with the powder or the surface.

APPLICATION

Dusting for Fingerprints

TIP 19.1   The Geographic North Pole is the Magnetic South Pole

The north pole of a magnet in a compass points north because it's attracted to the Earth's *magnetic* south pole—located near the Earth's *geographic* north pole.

19.2   EARTH'S MAGNETIC FIELD

A small bar magnet is said to have north and south poles, but it's more accurate to say it has a "north-seeking" pole and a "south-seeking" pole. By these expressions, we mean that if such a magnet is used as a compass, one end will "seek," or point to, the geographic North Pole of Earth and the other end will "seek," or point to, the geographic South Pole of Earth. We conclude that **the geographic North Pole of Earth corresponds to a magnetic south pole, and the geographic South Pole of Earth corresponds to a magnetic north pole.** In fact, the configuration of Earth's magnetic field, pictured in Figure 19.4, very much resembles

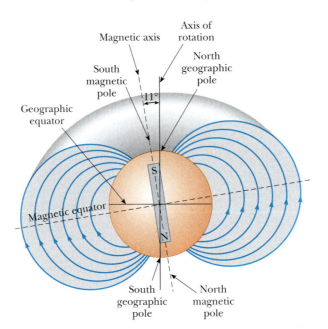

Figure 19.4 Earth's magnetic field lines. Note that magnetic south is at the north geographic pole and magnetic north is at the south geographic pole.

what would be observed if a huge bar magnet were buried deep in the Earth's interior.

If a compass needle is suspended in bearings that allow it to rotate in the vertical plane as well as in the horizontal plane, the needle is horizontal with respect to Earth's surface only near the equator. As the device is moved northward, the needle rotates so that it points more and more toward the surface of Earth. The angle between the direction of the magnetic field and the horizontal is called the **dip angle**. Finally, at a point just north of Hudson Bay in Canada, the north pole of the needle points directly downward, with a dip angle of 90°. This site, first found in 1832, is considered to be the location of the south magnetic pole of Earth. It is approximately 1 300 miles from Earth's geographic North Pole and varies with time. Similarly, Earth's magnetic north pole is about 1 200 miles from its geographic South Pole. This means that compass needles point only approximately north. The difference between true north, defined as the geographic North Pole, and north indicated by a compass varies from point to point on Earth, a difference referred to as *magnetic declination*. For example, along a line through South Carolina and the Great Lakes a compass indicates true north, whereas in Washington state it aligns 25° east of true north (Fig. 19.5).

Although the magnetic field pattern of Earth is similar to the pattern that would be set up by a bar magnet placed at its center the source of Earth's field can't consist of large masses of permanently magnetized material. Earth does have large deposits of iron ore deep beneath its surface, but the high temperatures in the core prevent the iron from retaining any permanent magnetization. It's considered more likely that the true source of Earth's magnetic field is electric current in the liquid part of its core. This current, which is not well understood, may be driven by an interaction between the planet's rotation and convection in the hot liquid core. There is some evidence that the strength of a planet's magnetic field is related to the planet's rate of rotation. For example, Jupiter rotates faster than Earth, and recent space probes indicate that Jupiter's magnetic field is stronger than Earth's, even though Jupiter lacks an iron core. Venus, on the other hand, rotates more slowly than Earth, and its magnetic field is weaker. Investigation into the cause of Earth's magnetism continues.

An interesting fact concerning Earth's magnetic field is that its direction reverses every few million years. Evidence for this phenomenon is provided by basalt (an iron-containing rock) that is sometimes spewed forth by volcanic activity on the ocean floor. As the lava cools, it solidifies and retains a picture of the direction of Earth's magnetic field. When the basalt deposits are dated, they provide evidence for periodic reversals of the magnetic field. The cause of these field reversals is still not understood.

It has long been speculated that some animals, such as birds, use the magnetic field of Earth to guide their migrations. Studies have shown that a type of anaerobic bacterium that lives in swamps has a magnetized chain of magnetite as part of its internal structure. (The term *anaerobic* means that these bacteria live and grow

APPLICATION
Magnetic Bacteria

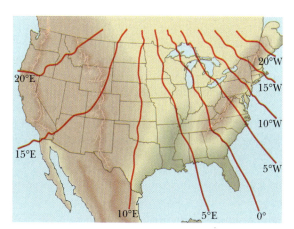

Figure 19.5   A map of the continental United States showing the declination of a compass from true north.

without oxygen; in fact, oxygen is toxic to them.) The magnetized chain acts as a compass needle that enables the bacteria to align with Earth's magnetic field. When they find themselves out of the mud on the bottom of the swamp, they return to their oxygen-free environment by following the magnetic field lines of Earth. Further evidence for their magnetic sensing ability is the fact that bacteria found in the Northern Hemisphere have internal magnetite chains that are opposite in polarity to those of similar bacteria in the Southern Hemisphere. This is consistent with the fact that in the Northern Hemisphere Earth's field has a downward component, whereas in the Southern Hemisphere it has an upward component. Recently, a meteorite originating on Mars has been found to contain a chain of magnetite. NASA scientists believe it may be a fossil of ancient Martian bacterial life.

APPLICATION

Labeling Airport Runways

The magnetic field of Earth is used to label runways at airports according to their direction. A large number is painted on the end of the runway so that it can be read by the pilot of an incoming airplane. This number describes the direction in which the airplane is traveling, expressed as the magnetic heading, in degrees measured clockwise from magnetic north divided by 10. A runway marked 9 would be directed toward the east (90° divided by 10), while a runway marked 18 would be directed toward magnetic south.

Applying Physics 19.1   Compasses Down Under

On a business trip to Australia, you take along your American-made compass that you may have used on a camping trip. Does this compass work correctly in Australia?

Explanation   There's no problem with using the compass in Australia. The north pole of the magnet in the compass will be attracted to the south magnetic pole near the geographic North Pole, just as it was in the United States. The only difference in the magnetic field lines is that they have an upward component in Australia, whereas they have a downward component in the United States. Held in a horizontal plane, your compass can't detect this, however—it only displays the direction of the horizontal component of the magnetic field.

19.3   MAGNETIC FIELDS

Experiments show that a stationary charged particle doesn't interact with a static magnetic field. **When a charged particle is *moving* through a magnetic field, however, a magnetic force acts on it.** This force has its maximum value when the charge moves in a direction perpendicular to the magnetic field lines, decreases in value at other angles, and becomes zero when the particle moves along the field lines. This is quite different from the electric force, which exerts a force on a charged particle whether it's moving or at rest. Further, the electric force is directed parallel to the electric field while the magnetic force on a moving charge is directed perpendicular to the magnetic field.

In our discussion of electricity, the electric field at some point in space was defined as the electric force per unit charge acting on some test charge placed at that point. In a similar way, we can describe the properties of the magnetic field $\vec{\mathbf{B}}$ at some point in terms of the magnetic force exerted on a test charge at that point. Our test object is a charge q moving with velocity $\vec{\mathbf{v}}$. It is found experimentally that the strength of the magnetic force on the particle is proportional to the magnitude of the charge q, the magnitude of the velocity $\vec{\mathbf{v}}$, the strength of the external magnetic field $\vec{\mathbf{B}}$, and the sine of the angle θ between the direction of $\vec{\mathbf{v}}$ and the direction of $\vec{\mathbf{B}}$. These observations can be summarized by writing the magnitude of the magnetic force as

$$F = qvB \sin \theta \qquad [19.1]$$

This expression is used to define the magnitude of the magnetic field as

$$B \equiv \frac{F}{qv \sin\theta} \qquad \text{[19.2]}$$

If F is in newtons, q in coulombs, and v in meters per second, then the SI unit of magnetic field is the **tesla** (T), also called the **weber** (Wb) **per square meter** (1 T = 1 Wb/m^2). If a 1-C charge moves in a direction perpendicular to a magnetic field of magnitude 1 T with a speed of 1 m/s, the magnetic force exerted on the charge is 1 N. We can express the units of $\vec{B}$ as

$$[B] = T = \frac{\text{Wb}}{\text{m}^2} = \frac{\text{N}}{\text{C} \cdot \text{m/s}} = \frac{\text{N}}{\text{A} \cdot \text{m}} \qquad \text{[19.3]}$$

In practice, the cgs unit for magnetic field, the **gauss** (G), is often used. The gauss is related to the tesla through the conversion

$$1 \text{ T} = 10^4 \text{ G}$$

Conventional laboratory magnets can produce magnetic fields as large as about 25 000 G, or 2.5 T. Superconducting magnets that can generate magnetic fields as great as 3×10^5 G, or 30 T, have been constructed. These values can be compared with the value of Earth's magnetic field near its surface, which is about 0.5 G, or 0.5×10^{-4} T.

From Equation 19.1 we see that the force on a charged particle moving in a magnetic field has its maximum value when the particle's motion is *perpendicular* to the magnetic field, corresponding to $\theta = 90°$, so that $\sin \theta = 1$. The magnitude of this maximum force has the value

$$F_{\max} = qvB \qquad \text{[19.4]}$$

Also from Equation 19.1, F is zero when $\vec{v}$ is parallel to $\vec{B}$ (corresponding to $\theta = 0°$ or 180°), so no magnetic force is exerted on a charged particle when it moves in the direction of the magnetic field or opposite the field.

Experiments show that the direction of the magnetic force is always perpendicular to both $\vec{v}$ and $\vec{B}$, as shown in Figure 19.6 for a positively charged particle. To determine the direction of the force, we employ the **right-hand rule number 1**:

1. Point the fingers of your right hand in the direction of the velocity $\vec{v}$.
2. Curl the fingers in the direction of the magnetic field $\vec{B}$, moving through the smallest angle (as in Fig. 19.7).
3. Your thumb is now pointing in the direction of the magnetic force $\vec{F}$ exerted on a positive charge.

If the charge is negative rather than positive, the force $\vec{F}$ is directed *opposite* that shown in Figures 19.6 and 19.7. So if q is negative, simply use the right-hand rule to find the direction for positive q, and then reverse that direction for the negative charge.

Quick Quiz 19.1

A charged particle moves in a straight line through a region of space. Which of the following answers *must* be true? (Assume any other fields are negligible.) The magnetic field (a) has a magnitude of zero (b) has a zero component perpendicular to the particle's velocity (c) has a zero component parallel to the particle's velocity in that region.

Quick Quiz 19.2

The north-pole end of a bar magnet is held near a stationary positively charged piece of plastic. Is the plastic (a) attracted, (b) repelled, or (c) unaffected by the magnet?

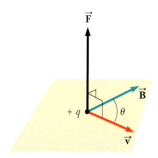

Figure 19.6   The direction of the magnetic force on a positively charged particle moving with a velocity $\vec{v}$ in the presence of a magnetic field. When $\vec{v}$ is at an angle θ with respect to $\vec{B}$, the magnetic force is perpendicular to both $\vec{v}$ and $\vec{B}$.

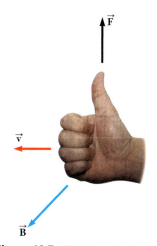

Figure 19.7   Right-hand rule number 1 for determining the direction of the magnetic force on a positive charge moving with a velocity $\vec{v}$ in a magnetic field $\vec{B}$. Point the fingers of your right hand in the direction of $\vec{v}$, and then curl them in the direction of $\vec{B}$. The magnetic force $\vec{F}$ points in the direction of your right thumb.

EXAMPLE 19.1 A Proton Traveling in Earth's Magnetic Field

Goal Calculate the magnitude and direction of a magnetic force.

Problem A proton moves with a speed of 1.00×10^5 m/s through Earth's magnetic field, which has a value of $55.0\ \mu$T at a particular location. When the proton moves eastward, the magnetic force acting on it is directed straight upward, and when it moves northward, no magnetic force acts on it. **(a)** What is the direction of the magnetic field, and **(b)** what is the strength of the magnetic force when the proton moves eastward? **(c)** Calculate the gravitational force on the proton and compare it with the magnetic force. Compare it also with the electric force if there were an electric field with magnitude $E = 1.50 \times 10^2$ N/C at that location, a common value at Earth's surface. Note that the mass of the proton is 1.67×10^{-27} kg.

Strategy The direction of the magnetic field can be found from an application of the right-hand rule, together with the fact that no force is exerted on the proton when it's traveling north. Substituting into Equation 19.1 yields the magnitude of the magnetic field.

Solution

(a) Find the direction of the magnetic field.

No magnetic force acts on the proton when it's going north, so the angle such a proton makes with the magnetic field direction must be either 0° or 180°. Therefore, the magnetic field $\vec{B}$ must point either north or south. Now apply the right-hand rule. When the particle travels east, the magnetic force is directed upward. Point your thumb in the direction of the force and your fingers in the direction of the velocity eastward. When you curl your fingers, they point north, which must therefore be the direction of the magnetic field.

(b) Find the magnitude of the magnetic force.

Substitute the given values and the charge of a proton into Equation 19.1. From part (a), the angle between the velocity $\vec{v}$ of the proton and the magnetic field $\vec{B}$ is 90.0°.

$$F = qvB \sin\theta = (1.60 \times 10^{-19}\text{ C})(1.00 \times 10^5\text{ m/s})$$
$$\times (55.0 \times 10^{-6}\text{ T})\sin(90.0°)$$
$$= 8.80 \times 10^{-19}\text{ N}$$

(c) Calculate the gravitational force on the proton and compare it with the magnetic force, and also with the electric force if $E = 1.50 \times 10^2$ N/C:

$$F_{grav} = mg = (1.67 \times 10^{-27}\text{ kg})(9.80\text{ m/s}^2)$$
$$= 1.64 \times 10^{-26}\text{ N}$$

$$F_{elec} = qE = (1.60 \times 10^{-19}\text{ C})(1.50 \times 10^2\text{ N/C})$$
$$= 2.40 \times 10^{-17}\text{ N}$$

Remarks The information regarding a proton moving north was necessary to fix the direction of the magnetic field. Otherwise, an upward magnetic force on an eastward-moving proton could be caused by a magnetic field pointing anywhere northeast or northwest. Notice in part (c) the relative strengths of the forces, with the electric force larger than the magnetic force and both much larger than the gravitational force, all for typical field values found in nature.

Exercise 19.1

Suppose an electron is moving due west in the same magnetic field as in Example 19.1 at a speed of 2.50×10^5 m/s. Find the magnitude and direction of the magnetic force on the electron.

Answer 2.20×10^{-18} N, straight up. (Don't forget, the electron is negatively charged!)

EXAMPLE 19.2 A Proton Moving in a Magnetic Field

Goal Calculate the magnetic force and acceleration when a particle moves at an angle other than ninety degrees to the field.

Problem A proton moves at 8.00×10^6 m/s along the x-axis. It enters a region in which there is a magnetic field of magnitude 2.50 T, directed at an angle of 60.0° with the x-axis and lying in the xy-plane (Fig. 19.8).

(a) Find the initial magnitude and direction of the magnetic force on the proton. **(b)** Calculate the proton's initial acceleration.

Strategy   Finding the magnitude and direction of the magnetic force requires substituting values into the equation for magnetic force, Equation 19.1, and using the right-hand rule. Applying Newton's second law solves part (b).

Solution
(a) Find the magnitude and direction of the magnetic force on the proton.

Substitute $v = 8.00 \times 10^6$ m/s, the magnetic field strength $B = 2.50$ T, the angle, and the charge of a proton into Equation 19.1:

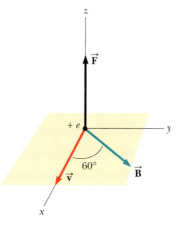

Figure 19.8   (Example 19.2) The magnetic force $\vec{F}$ on a proton is in the positive z-direction when $\vec{v}$ and $\vec{B}$ lie in the xy-plane.

$$F = qvB \sin \theta$$
$$= (1.60 \times 10^{-19}\text{ C})(8.00 \times 10^6\text{ m/s})(2.50\text{ T})(\sin 60°)$$
$$F = 2.77 \times 10^{-12}\text{ N}$$

Apply the right-hand rule number 1 to find the initial direction of the magnetic force:

Point the fingers of the right hand in the x-direction (the direction of $\vec{v}$), and then curl them towards $\vec{B}$. The thumb points upwards, in the positive z-direction.

(b) Calculate the proton's initial acceleration.

Substitute the force and the mass of a proton into Newton's second law:

$$ma = F \quad \rightarrow \quad (1.67 \times 10^{-27}\text{ kg})a = 2.77 \times 10^{-12}\text{ N}$$
$$a = 1.66 \times 10^{15}\text{ m/s}^2$$

Remarks   The initial acceleration is also in the positive z-direction. Because the direction of $\vec{v}$ changes, however, the subsequent direction of the magnetic force also changes. In applying right-hand rule number 1 to find the direction, it was important to take into consideration the charge. A negatively charged particle accelerates in the opposite direction.

Exercise 19.2
Calculate the acceleration of an electron that moves through the same magnetic field as in Example 19.2, at the same velocity as the proton. The mass of an electron is 9.11×10^{-31} kg.

Answer 3.04×10^{18} m/s² in the negative z-direction

19.4   MAGNETIC FORCE ON A CURRENT-CARRYING CONDUCTOR

If a magnetic field exerts a force on a single charged particle when it moves through a magnetic field, it should be no surprise that magnetic forces are exerted on a current-carrying wire, as well (see Fig. 19.9). This follows from the fact that the current is a collection of many charged particles in motion; hence, the resultant force on the wire is due to the sum of the individual forces on the charged particles. The force on the particles is transmitted to the "bulk" of the wire through collisions with the atoms making up the wire.

Some explanation is in order concerning notation in many of the figures. To indicate the direction of $\vec{B}$, we use the following conventions:

If $\vec{B}$ is directed into the page, as in Figure 19.10 (page 632), we use a series of blue crosses, representing the tails of arrows. If $\vec{B}$ is directed out of the page, we use a series of blue dots, representing the tips of arrows. If $\vec{B}$ lies in the plane of the page, we use a series of blue field lines with arrowheads.

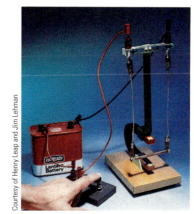

Figure 19.9   This apparatus demonstrates the force on a current-carrying conductor in an external magnetic field. Why does the bar swing *away* from the magnet after the switch is closed?

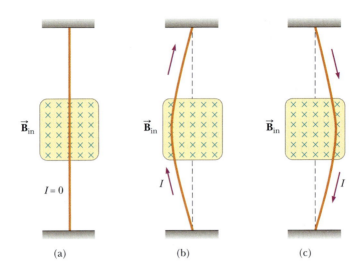

Figure 19.10 A segment of a flexible vertical wire partially stretched between the poles of a magnet, with the field (blue crosses) directed into the page. (a) When there is no current in the wire, it remains vertical. (b) When the current is upward, the wire deflects to the left. (c) When the current is downward, the wire deflects to the right.

(a)   (b)   (c)

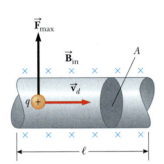

Figure 19.11 A section of a wire containing moving charges in an external magnetic field $\vec{\mathbf{B}}$.

TIP 19.2 The Origin of the Magnetic Force on a Wire

When a magnetic field is applied at some angle to a wire carrying a current, a magnetic force is exerted on each moving charge in the wire. The total magnetic force on the wire is the sum of all the magnetic forces on the individual charges producing the current.

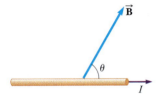

Figure 19.12 A wire carrying a current I in the presence of an external magnetic field $\vec{\mathbf{B}}$ that makes an angle θ with the wire.

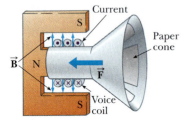

Figure 19.13 A diagram of a loudspeaker.

The force on a current-carrying conductor can be demonstrated by hanging a wire between the poles of a magnet, as in Figure 19.10. In this figure, the magnetic field is directed into the page and covers the region within the shaded area. The wire deflects to the right or left when it carries a current.

We can quantify this discussion by considering a straight segment of wire of length ℓ and cross-sectional area A carrying current I in a uniform external magnetic field $\vec{\mathbf{B}}$, as in Figure 19.11. We assume that the magnetic field is perpendicular to the wire and is directed into the page. A force of magnitude $F_{max} = qv_dB$ is exerted on each charge carrier in the wire, where v_d is the drift velocity of the charge. To find the total force on the wire, we multiply the force on one charge carrier by the number of carriers in the segment. Because the volume of the segment is $A\ell$, the number of carriers is $nA\ell$, where n is the number of carriers per unit volume. Hence, the magnitude of the total magnetic force on the wire of length ℓ is as follows:

Total force = force on each charge carrier × total number of carriers

$$F_{max} = (qv_dB)(nA\ell)$$

From Chapter 17, however, we know that the current in the wire is given by $I = nqv_dA$, so

$$F_{max} = BI\ell \qquad [19.5]$$

This equation can be used only when the current and the magnetic field are at right angles to each other.

If the wire is not perpendicular to the field, but is at some arbitrary angle, as in Figure 19.12, the magnitude of the magnetic force on the wire is

$$F = BI\ell \sin\theta \qquad [19.6]$$

where θ is the angle between $\vec{\mathbf{B}}$ and the direction of the current. The direction of this force can be obtained by the use of right-hand rule number 1. However, in this case you must place your fingers in the direction of the positive current I, rather than in the direction of $\vec{\mathbf{v}}$. The current, naturally, is made up of charges moving at some velocity, so this really isn't a separate rule. In Figure 19.12, the direction of the magnetic force on the wire is out of the page.

Finally, when the current is either in the direction of the field or opposite the direction of the field, the magnetic force on the wire is zero.

The fact that a magnetic force acts on a current-carrying wire in a magnetic field is the operating principle of most speakers in sound systems. One speaker design, shown in Figure 19.13, consists of a coil of wire called the voice coil, a flexible paper cone that acts as the speaker, and a permanent magnet. The coil of wire surrounding the north pole of the magnet is shaped so that the magnetic field lines are directed radially outward from the coil's axis. When an electrical signal is sent to the coil, producing a current in the coil as in Figure 19.13, a magnetic force to the left acts on the coil. (This can be seen by applying right-hand rule number 1 to each turn of wire.) When the current reverses direction, as it would for a current that varied sinusoidally, the magnetic force on the coil also reverses direction, and the cone, which is attached to the coil, accelerates to the right. An alternating current through the coil causes an alternating force on the coil, which results in vibrations of the cone. The vibrating cone creates sound waves as it pushes and pulls on the air in front of it. In this way, a 1-kHz electrical signal is converted to a 1-kHz sound wave.

An unusual application of the force on a current-carrying conductor is illustrated by the electromagnetic pump shown in Figure 19.14. Artificial hearts require a pump to keep the blood flowing, and kidney dialysis machines also require a pump to assist the heart in pumping blood that is to be cleansed. Ordinary mechanical pumps create problems because they damage the blood cells as they move through the pump. The mechanism shown in the figure has demonstrated some promise in such applications. A magnetic field is established across a segment of the tube containing the blood, flowing in the direction of the velocity $\vec{v}$. An electric current passing through the fluid in the direction shown has a magnetic force acting on it in the direction of $\vec{v}$, as applying the right-hand rule shows. This force helps to keep the blood in motion.

APPLICATION
Loudspeaker Operation

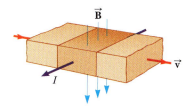

Figure 19.14   A simple electromagnetic pump has no moving parts to damage a conducting fluid, such as blood, passing through. Application of the right-hand rule #1 (right fingers in the direction of the current I, curl them in the direction of $\vec{B}$, thumb points in the direction of the force) shows that the force on the current-carrying segment of the fluid is in the direction of the velocity.

APPLICATION
Electromagnetic Pumps for Artificial Hearts and Kidneys

Applying Physics 19.2   Lightning Strikes

In a lightning strike there is a rapid movement of negative charge from a cloud to the ground. In what direction is a lightning strike deflected by Earth's magnetic field?

Explanation   The downward flow of negative charge in a lightning stroke is equivalent to a current moving upward. Consequently, we have an upward-moving current in a northward-directed magnetic field. According to right-hand rule number 1, the lightning strike would be deflected toward the west.

EXAMPLE 19.3   A Current-Carrying Wire in Earth's Magnetic Field

Goal   Compare the magnetic force on a current-carrying wire with the gravitational force exerted on the wire.

Problem   A wire carries a current of 22.0 A from west to east. Assume that at this location the magnetic field of Earth is horizontal and directed from south to north and that it has a magnitude of 0.500×10^{-4} T. **(a)** Find the magnitude and direction of the magnetic force on a 36.0-m length of wire. **(b)** Calculate the gravitational force on the same length of wire if it's made of copper and has a cross-sectional area of 2.50×10^{-6} m².

Solution
(a) Calculate the magnetic force on the wire.

Substitute into Equation 19.6, using the fact that the magnetic field and the current are at right angles to each other:

$F = BI\ell \sin \theta = (0.500 \times 10^{-4} \text{ T})(22.0 \text{ A})(36.0 \text{ m}) \sin 90.0°$
$= 3.96 \times 10^{-2}$ N

Apply right-hand rule number 1 to find the direction of the magnetic force:

With the fingers of your right hand pointing west to east in the direction of the current, curl them north in the direction of the magnetic field. Your thumb points upward.

(b) Calculate the gravitational force on the wire segment.

First, obtain the mass of the wire from the density of copper, the length, and cross-sectional area of the wire:

$m = \rho V = \rho(A\ell)$
$= (8.92 \times 10^3 \text{ kg/m}^3)(2.50 \times 10^{-6} \text{ m}^2 \cdot 36.0 \text{ m})$
$= 0.803$ kg

To get the gravitational force, multiply the mass by the acceleration of gravity:

$$F_{grav} = mg = \boxed{7.87 \text{ N}}$$

Remarks   This calculation demonstrates that under normal circumstances, the gravitational force on a current-carrying conductor is much greater than the magnetic force due to the Earth's magnetic field.

Exercise 19.3

What current would make the magnetic force in the example equal in magnitude to the gravitational force?

Answer 4.37×10^3 A, a large current that would very rapidly melt the wire.

19.5   TORQUE ON A CURRENT LOOP AND ELECTRIC MOTORS

In the preceding section we showed how a magnetic force is exerted on a current-carrying conductor when the conductor is placed in an external magnetic field. With this as a starting point, we now show that a torque is exerted on a current loop placed in a magnetic field. The results of this analysis will be of great practical value when we discuss generators and motors in Chapter 20.

Consider a rectangular loop carrying current I in the presence of an external uniform magnetic field in the plane of the loop, as shown in Figure 19.15a. The forces on the sides of length a are zero because these wires are parallel to the field. The magnitudes of the magnetic forces on the sides of length b, however, are

$$F_1 = F_2 = BIb$$

The direction of $\vec{F}_1$, the force on the left side of the loop, is out of the page, and that of $\vec{F}_2$, the force on the right side of the loop, is into the page. If we view the loop from the side, as in Figure 19.15b, the forces are directed as shown. If we assume that the loop is pivoted so that it can rotate about point O, we see that these two forces produce a torque about O that rotates the loop clockwise. The magnitude of this torque, τ_{max}, is

$$\tau_{max} = F_1\frac{a}{2} + F_2\frac{a}{2} = (BIb)\frac{a}{2} + (BIb)\frac{a}{2} = BIab$$

where the moment arm about O is $a/2$ for both forces. Because the area of the loop is $A = ab$, the torque can be expressed as

$$\tau_{max} = BIA \qquad [19.7]$$

This result is valid only when the magnetic field is *parallel* to the plane of the loop, as in Figure 19.15b. If the field makes an angle θ with a line perpendicular to the plane of the loop, as in Figure 19.15c, the moment arm for each force is given by $(a/2)\sin\theta$. An analysis similar to the previous gives, for the magnitude of the torque,

Figure 19.15   (a) Top view of a rectangular loop in a uniform magnetic field $\vec{B}$. No magnetic forces act on the sides of length a parallel to $\vec{B}$, but forces do act on the sides of length b. (b) A side view of the rectangular loop shows that the forces $\vec{F}_1$ and $\vec{F}_2$ on the sides of length b create a torque that tends to twist the loop clockwise. (c) If $\vec{B}$ is at an angle θ with a line perpendicular to the plane of the loop, the torque is given by $BIA \sin\theta$.

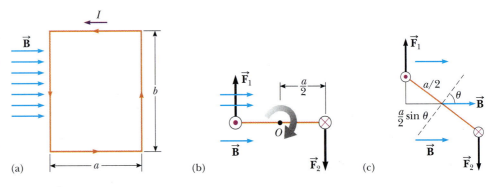

$$\tau = BIA \sin \theta \qquad\qquad \textbf{[19.8]}$$

This result shows that the torque has the *maximum* value BIA when the field is parallel to the plane of the loop ($\theta = 90°$) and is *zero* when the field is perpendicular to the plane of the loop ($\theta = 0$). As seen in Figure 19.15c, the loop tends to rotate to smaller values of θ (so that the normal to the plane of the loop rotates toward the direction of the magnetic field).

Although the foregoing analysis was for a rectangular loop, a more general derivation shows that Equation 19.8 applies regardless of the shape of the loop. Further, the torque on a coil with N turns is

$$\tau = BIAN \sin \theta \qquad\qquad \textbf{[19.9a]}$$

The quantity $\mu = IAN$ is defined as the magnitude of a vector $\vec{\mu}$ called the *magnetic moment* of the coil. The vector $\vec{\mu}$ always points perpendicular to the plane of the loop(s). The angle θ in Equations 19.8 and 19.9 lies between the directions of the magnetic moment $\vec{\mu}$ and the magnetic field $\vec{B}$. The magnetic torque can then be written

$$\tau = \mu B \sin \theta \qquad\qquad \textbf{[19.9b]}$$

Quick Quiz 19.3

A square and a circular loop with the same area lie in the *xy*-plane, where there is a uniform magnetic field $\vec{B}$ pointing at some angle θ with respect to the positive *z*-direction. Each loop carries the same current, in the same direction. Which magnetic torque is larger? (a) the torque on the square loop (b) the torque on the circular loop (c) the torques are the same (d) more information is needed

---

EXAMPLE 19.4   The Torque on a Circular Loop in a Magnetic Field

Goal   Calculate a magnetic torque on a loop of current.

Problem   A circular wire loop of radius 1.00 m is placed in a magnetic field of magnitude 0.500 T. The normal to the plane of the loop makes an angle of 30.0° with the magnetic field (Fig. 19.16a). The current in the loop is 2.00 A in the direction shown. **(a)** Find the magnetic moment of the loop and the magnitude of the torque at this instant. **(b)** The same current is carried by the rectangular 2.00-m by 3.00-m coil with three loops shown in Figure 19.16b. Find the magnetic moment of the coil and the magnitude of the torque acting on the coil at that instant.

Strategy   For each part, we just have to calculate the area, use it in the calculation of the magnetic moment, and multiply the result by $B \sin \theta$. Altogether, this amounts to substituting values into Equation 19.9.

Solution

(a) Find the magnetic moment of the circular loop and the magnetic torque exerted on it.

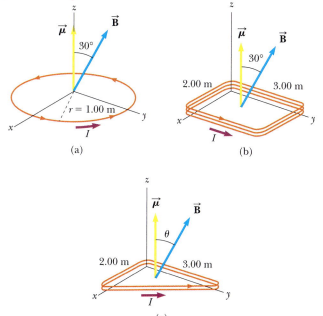

Figure 19.16   (Example 19.4) (a) A circular current loop in an external magnetic field $\vec{B}$. (b) A rectangular current loop in the same field. (c) (Exercise 19.4)

First, calculate the enclosed area of the circular loop:

$$A = \pi r^2 = \pi (1.00 \text{ m})^2 = 3.14 \text{ m}^2$$

Calculate the magnetic moment of the loop:

$$\mu = IAN = (2.00 \text{ A})(3.14 \text{ m}^2)(1) = \boxed{6.28 \text{ A} \cdot \text{m}^2}$$

Now substitute values for the magnetic moment, magnetic field, and θ into Equation 19.9b:

$$\tau = \mu B \sin \theta = (6.28 \text{ A} \cdot \text{m}^2)(0.500 \text{ T})(\sin 30.0°)$$

$$= \boxed{1.57 \text{ N} \cdot \text{m}}$$

(b) Find the magnetic moment of the rectangular coil and the magnetic torque exerted on it.

Calculate the area of the coil:

$$A = L \times H = (2.00 \text{ m})(3.00 \text{ m}) = 6.00 \text{ m}^2$$

Calculate the magnetic moment of the coil:

$$\mu = IAN = (2.00 \text{ A})(6.00 \text{ m}^2)(3) = \boxed{36.0 \text{ A} \cdot \text{m}^2}$$

Substitute values into Equation 19.9b:

$$\tau = \mu B \sin \theta = (0.500 \text{ T})(36.0 \text{ A} \cdot \text{m}^2)(\sin 30.0°)$$

$$= \boxed{9.00 \text{ N} \cdot \text{m}}$$

Remarks In calculating a magnetic torque, it's not strictly necessary to calculate the magnetic moment. Instead, Equation 19.9a can be used directly.

Exercise 19.4
Suppose a right triangular coil with base of 2.00 m and height 3.00 m having two loops carries a current of 2.00 A as shown in Figure 19.16c. Find the magnetic moment and the torque on the coil. The magnetic field is again 0.500 T and makes an angle of 30.0° with respect to the normal direction.

Answer $\mu = 12.0 \text{ A} \cdot \text{m}^2$, $\tau = 3.00 \text{ N} \cdot \text{m}$

Electric Motors

APPLICATION

Electric Motors

It's hard to imagine life in the 21st century without electric motors. Some appliances that contain motors include computer disk drives, CD players, VCR and DVD players, food processors and blenders, car starters, furnaces, and air conditioners. The motors convert electrical energy to kinetic energy of rotation, and consist basically of a rigid current-carrying loop that rotates when placed in the field of a magnet.

As we have just seen (Fig. 19.15), the torque on such a loop rotates the loop to smaller values of θ until the torque becomes zero, when the magnetic field is perpendicular to the plane of the loop and $\theta = 0$. If the loop turns past this angle and the current remains in the direction shown in the figure, the torque reverses direction and turns the loop in the opposite direction—that is, counterclockwise. To overcome this difficulty and provide continuous rotation in one direction, the current in the loop must periodically reverse direction. In alternating current (AC) motors, such a reversal occurs naturally 120 times each second. In direct current (DC) motors, the reversal is accomplished mechanically with split-ring contacts (commutators) and brushes, as shown in Active Figure 19.17.

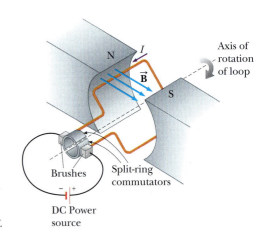

ACTIVE FIGURE 19.17
Simplified sketch of a DC electric motor.

Physics⊗Now™
Log into PhysicsNow at **www.cp7e.com** and go to Active Figure 19.17, where you can adjust the speed of rotation and the strength of the field and see the effects on the generated emf.

Figure 19.18 The Honda Insight combines a three-cylinder gasoline automobile engine with a thin electric motor for improved efficiency and added power when needed. The electric motor (circled) also acts as a generator during braking or coasting downhill to recharge the batteries, with the result that they never need to be recharged by the owner.

Although actual motors contain many current loops and commutators, for simplicity Active Figure 19.17 shows only a single loop and a single set of split-ring contacts rigidly attached to and rotating with the loop. Electrical stationary contacts called *brushes* are maintained in electrical contact with the rotating split ring. These brushes are usually made of graphite, because graphite is a good electrical conductor as well as a good lubricant. Just as the loop becomes perpendicular to the magnetic field and the torque becomes zero, inertia carries the loop forward in the clockwise direction and the brushes cross the gaps in the ring, causing the loop current to reverse its direction. This provides another pulse of torque in the clockwise direction for another 180°, the current reverses, and the process repeats itself. Figure 19.18 shows a modern motor used to power a hybrid gas–electric car.

19.6 MOTION OF A CHARGED PARTICLE IN A MAGNETIC FIELD

Consider the case of a positively charged particle moving in a uniform magnetic field so that the direction of the particle's velocity is *perpendicular to the field*, as in Active Figure 19.19. The label $\vec{\mathbf{B}}_{in}$ and the crosses indicate that $\vec{\mathbf{B}}$ is directed into the page. Application of the right-hand rule at point P shows that the direction of the magnetic force $\vec{\mathbf{F}}$ at that location is upward. The force causes the particle to alter its direction of travel and to follow a curved path. Application of the right-hand rule at any point shows that **the magnetic force is always directed toward the center of the circular path**; therefore, the magnetic force causes a centripetal acceleration, which changes only the direction of $\vec{\mathbf{v}}$ and not its magnitude. Because $\vec{\mathbf{F}}$ produces the centripetal acceleration, we can equate its magnitude, qvB in this case, to the mass of the particle multiplied by the centripetal acceleration v^2/r. From Newton's second law, we find that

$$F = qvB = \frac{mv^2}{r}$$

which gives

$$r = \frac{mv}{qB}$$ [19.10]

This equation says that the radius of the path is proportional to the momentum mv of the particle and is inversely proportional to the charge and the magnetic field. Equation 19.10 is often called the *cyclotron equation*, because it's used in the design of these instruments (popularly known as atom smashers).

If the initial direction of the velocity of the charged particle is not perpendicular to the magnetic field, as shown in Active Figure 19.20, the path followed by the particle is a spiral (called a helix) along the magnetic field lines.

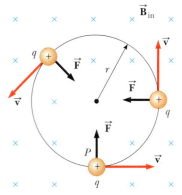

ACTIVE FIGURE 19.19
When the velocity of a charged particle is perpendicular to a uniform magnetic field, the particle moves in a circle whose plane is perpendicular to $\vec{\mathbf{B}}$, which is directed into the page. (The crosses represent the tails of the magnetic field vectors.) The magnetic force $\vec{\mathbf{F}}$ on the charge is always directed toward the center of the circle.

Physics⊗Now™
Log into PhysicsNow at **www.cp7e.com** and go to Active Figure 19.19, where you can adjust the mass, speed, charge of the particle, and the magnitude of the magnetic field, and observe the resulting circular motion.

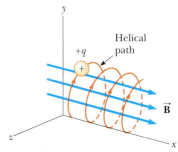

ACTIVE FIGURE 19.20
A charged particle having a velocity directed at an angle with a uniform magnetic field moves in a helical path.

Physics⊗Now™
Log into PhysicsNow at **www.cp7e.com** and go to Active Figure 19.20, where you can adjust the *x*-component of the velocity of the particle and observe the resulting helical motion.

Applying Physics 19.3 Trapping Charges

Storing charged particles is important for a variety of applications. Suppose a uniform magnetic field exists in a finite region of space. Can a charged particle be injected into this region from the outside and remain trapped in the region by magnetic force alone?

Explanation It's best to consider separately the components of the particle velocity parallel and perpendicular to the field lines in the region. There is no magnetic force on the particle associated with the velocity component parallel to the field lines, so that velocity component remains unchanged.

Now consider the component of velocity perpendicular to the field lines. This component will result in a magnetic force that is perpendicular to both the field lines and the velocity component itself. The path of a particle for which the force is always perpendicular to the velocity is a circle. The particle therefore follows a circular arc and exits the field on the other side of the circle, as shown in Figure 19.21 for a particle with constant kinetic energy. On the other hand, a particle can become trapped if it loses some kinetic energy in a collision after entering the field, as in Active Figure 19.20.

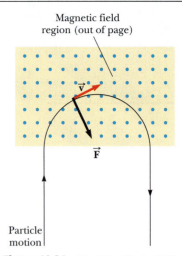

Figure 19.21 (Applying Physics 19.3)

Particles *can* be injected and contained if, in addition to the magnetic field, electrostatic fields are involved. These fields are used in the *Penning trap*. With these devices, it's possible to store charged particles for extended periods. Such traps are useful, for example, in the storage of antimatter, which disintegrates completely on contact with ordinary matter.

Quick Quiz 19.4

As a charged particle moves freely in a circular path in the presence of a constant magnetic field applied perpendicular to the particle's velocity, its kinetic energy (a) remains constant, (b) increases, or (c) decreases.

EXAMPLE 19.5 The Mass Spectrometer: Identifying Particles

Goal Use the cyclotron equation to identify a particle.

Problem A charged particle enters the magnetic field of a mass spectrometer at a speed of 1.79×10^6 m/s. It subsequently moves in a circular orbit with a radius of 16.0 cm in a uniform magnetic field of magnitude 0.350 T having a direction perpendicular to the particle's velocity. Find the particle's mass-to-charge ratio, and identify it based on the table on page 639.

Strategy After finding the mass-to-charge ratio with Equation 19.10, compare it to the values in the table, identifying the particle.

Solution

Write the cyclotron equation:

$$r = \frac{mv}{qB}$$

Solve this equation for the mass divided by the charge, m/q and substitute values:

$$\frac{m}{q} = \frac{rB}{v} = \frac{(0.160 \text{ m})(0.350 \text{ T})}{1.79 \times 10^6 \text{ m/s}} = \boxed{3.13 \times 10^{-8} \frac{\text{kg}}{\text{C}}}$$

Identify the particle from the table. All particles are completely ionized.

particle	m (kg)	q (C)	m/q (kg/C)
Hydrogen	1.67×10^{-27}	1.60×10^{-19}	1.04×10^{-8}
Deuterium	3.35×10^{-27}	1.60×10^{-19}	2.09×10^{-8}
Tritium	5.01×10^{-27}	1.60×10^{-19}	3.13×10^{-8}
Helium-3	5.01×10^{-27}	3.20×10^{-19}	1.57×10^{-8}

The particle is tritium.

Remarks   The mass spectrometer is an important tool in both chemistry and physics. Unknown chemicals can be heated and ionized, and the resulting particles passed through the mass spectrometer and subsequently identified.

Exercise 19.5
Suppose a second charged particle enters the mass spectrometer at the same speed as the particle in Example 19.5. If it travels in a circle with radius 10.7 cm, find the mass-to-charge ratio and identify the particle from the table above.

Answer 2.09×10^{-8} kg/C; deuterium

EXAMPLE 19.6   The Mass Spectrometer: Separating Isotopes

Goal   Apply the cyclotron equation to the process of separating isotopes.

APPLICATION
Mass Spectrometers

Problem   Two singly ionized atoms move out of a slit at point S in Figure 19.22 and into a magnetic field of magnitude 0.100 T pointing into the page. Each has a speed of 1.00×10^6 m/s. The nucleus of the first atom contains one proton and has a mass of 1.67×10^{-27} kg, while the nucleus of the second atom contains a proton and a neutron and has a mass of 3.34×10^{-27} kg. Atoms with the same number of protons in the nucleus but different masses are called isotopes. The two isotopes here are hydrogen and deuterium. Find their distance of separation when they strike a photographic plate at P.

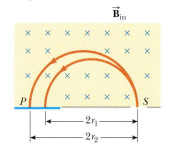

Figure 19.22   (Example 19.6) Two isotopes leave the slit at point S and travel in different circular paths before striking a photographic plate at P.

Strategy   Apply the cyclotron equation to each atom, finding the radius of the path of each. Double the radii to find the path diameters, and then find their difference.

Solution

Use Equation 19.10 to find the radius of the circular path followed by the lighter isotope, hydrogen.

$$r_1 = \frac{m_1 v}{qB} = \frac{(1.67 \times 10^{-27} \text{ kg})(1.00 \times 10^6 \text{ m/s})}{(1.60 \times 10^{-19} \text{ C})(0.100 \text{ T})}$$
$$= 0.104 \text{ m}$$

Use the same equation to calculate the radius of the path of deuterium, the heavier isotope:

$$r_2 = \frac{m_2 v}{qB} = \frac{(3.34 \times 10^{-27} \text{ kg})(1.00 \times 10^6 \text{ m/s})}{(1.60 \times 10^{-19} \text{ C})(0.100 \text{ T})}$$
$$= 0.209 \text{ m}$$

Multiply the radii by 2 to find the diameters, and take the difference, getting the separation x between the isotopes:

$$x = 2r_2 - 2r_1 = 0.210 \text{ m}$$

Remarks   During World War II, mass spectrometers were used to separate the highly radioactive uranium isotope U-235 from its far more common isotope, U-238.

Exercise 19.6
Use the same mass spectrometer as in Example 19.6 to find the separation between two isotopes of helium: normal helium-4, which has a nucleus consisting of two protons and two neutrons, and helium-3, which has two protons and a single neutron. Assume both nuclei, doubly ionized (having a charge of $2e = 3.20 \times 10^{-19}$ C), enter the field at 1.00×10^6 m/s. The helium-4 nucleus has a mass of 6.64×10^{-27} kg, and the helium-3 nucleus has a mass of 5.01×10^{-27} kg.

Answer   0.102 m

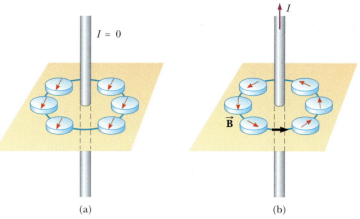

ACTIVE FIGURE 19.23
(a) When there is no current in the vertical wire, all compass needles point in the same direction.
(b) When the wire carries a strong current, the compass needles deflect in directions tangent to the circle, pointing in the direction of $\vec{\mathbf{B}}$, due to the current.

Physics⊗Now™
Log into PhysicsNow at **www.cp7e.com** and go to Active Figure 19.23, where you can change the value of the current and see the effect on the compasses.

HANS CHRISTIAN OERSTED
(1777–1851), Danish Physicist and Chemist

Oersted is best known for observing that a compass needle deflects when placed near a wire carrying a current. This important discovery was the first evidence of the connection between electric and magnetic phenomena. Oersted was also the first to prepare pure aluminum.

TIP 19.3   Raise Your Right Hand!

We have introduced two right-hand rules in this chapter. Be sure to use *only* your right hand when applying these rules.

19.7   MAGNETIC FIELD OF A LONG, STRAIGHT WIRE AND AMPÈRE'S LAW

During a lecture demonstration in 1819, the Danish scientist Hans Oersted (1777–1851) found that an electric current in a wire deflected a nearby compass needle. This momentous discovery, linking a magnetic field with an electric current for the first time, was the beginning of our understanding of the origin of magnetism.

A simple experiment first carried out by Oersted in 1820 clearly demonstrates that a current-carrying conductor produces a magnetic field. In this experiment, several compass needles are placed in a horizontal plane near a long vertical wire, as in Active Figure 19.23a. When there is no current in the wire, all needles point in the same direction (that of Earth's field), as one would expect. However, when the wire carries a strong, steady current, the needles all deflect in directions tangent to the circle, as in Active Figure 19.23b. These observations show that the direction of $\vec{\mathbf{B}}$ is consistent with the following convenient rule, **right-hand rule number 2**:

> Point the thumb of your right hand along a wire in the direction of positive current, as in Figure 19.24a. Your fingers then naturally curl in the direction of the magnetic field $\vec{\mathbf{B}}$.

When the current is reversed, the filings in Figure 19.24b also reverse.

Figure 19.24 (a) Right-hand rule number 2 for determining the direction of the magnetic field due to a long, straight wire carrying a current. Note that the magnetic field lines form circles around the wire. (b) Circular magnetic field lines surrounding a current-carrying wire, displayed by iron filings.

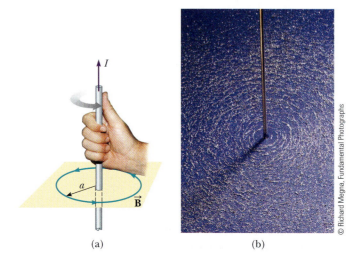

Because the filings point in the direction of $\vec{\mathbf{B}}$, we conclude that the lines of $\vec{\mathbf{B}}$ form circles about the wire. By symmetry, the magnitude of $\vec{\mathbf{B}}$ is the same everywhere on a circular path centered on the wire and lying in a plane perpendicular to the wire. By varying the current and distance from the wire, it can be experimentally determined that $\vec{\mathbf{B}}$ is proportional to the current and inversely proportional to the distance from the wire. These observations lead to a mathematical expression for the strength of the magnetic field due to the current I in a long, straight wire:

$$B = \frac{\mu_0 I}{2\pi r} \qquad \qquad \text{[19.11]}$$

◀ Magnetic field due to a long, straight wire

The proportionality constant μ_0, called the **permeability of free space**, has the value

$$\mu_0 \equiv 4\pi \times 10^{-7}\,\text{T}\cdot\text{m/A} \qquad \qquad \text{[19.12]}$$

Ampère's Law and a Long, Straight Wire

Equation 19.11 enables us to calculate the magnetic field due to a long, straight wire carrying a current. A general procedure for deriving such equations was proposed by the French scientist André-Marie Ampère (1775–1836); it provides a relation between the current in an arbitrarily shaped wire and the magnetic field produced by the wire.

Consider an arbitrary closed path surrounding a current as in Figure 19.25. The path consists of many short segments, each of length $\Delta \ell$. Multiply one of these lengths by the component of the magnetic field parallel to that segment, where the product is labeled $B_\parallel \Delta \ell$. According to Ampère, the sum of all such products over the closed path is equal to μ_0 times the net current I that passes through the surface bounded by the closed path. This statement, known as **Ampère's circuital law**, can be written

$$\Sigma B_\parallel \Delta \ell = \mu_0 I \qquad \qquad \text{[19.13]}$$

where $B_\parallel$ is the component of $\vec{\mathbf{B}}$ parallel to the segment of length $\Delta \ell$ and $\Sigma B_\parallel \Delta \ell$ means that we take the sum over all the products $B_\parallel \Delta \ell$ around the closed path. Ampère's law is the fundamental law describing how electric currents create magnetic fields in the surrounding empty space.

We can use Ampère's circuital law to derive the magnetic field due to a long, straight wire carrying a current I. As discussed earlier, each of the magnetic field lines of this configuration forms a circle with the wire at its centers. The magnetic field is tangent to this circle at every point, and its magnitude has the same value B over the entire circumference of a circle of radius r, so that $B_\parallel = B$, as shown in Figure 19.26 (page 642). In calculating the sum $\Sigma B_\parallel \Delta \ell$ over the circular path, notice that $B_\parallel$ can be removed from the sum (because it has the same value B for each element on the circle). Equation 19.13 then gives

$$\Sigma B_\parallel \Delta \ell = B_\parallel \sum \Delta \ell = B(2\pi r) = \mu_0 I$$

Dividing both sides by the circumference $2\pi r$, we obtain

$$B = \frac{\mu_0 I}{2\pi r}$$

This is identical to Equation 19.11, which is the magnetic field due to the current I in a long, straight wire.

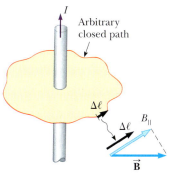

Figure 19.25   An arbitrary closed path around a current is used to calculate the magnetic field of the current by the use of Ampère's rule.

ANDRÉ-MARIE AMPÈRE
(1775–1836)

Ampère, a Frenchman, is credited with the discovery of electromagnetism—the relationship between electric currents and magnetic fields. Ampère's genius, particularly in mathematics, became evident by the age of 12, but his personal life was filled with tragedy. His father, a wealthy city official, was guillotined during the French Revolution, and his wife died young, in 1803. Ampère died of pneumonia at the age of 61. His judgment of his life is clear from the epitaph he chose for his gravestone: *Tandem felix* (Happy at last).

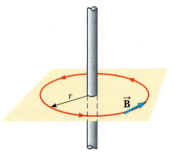

Figure 19.26 A closed circular path of radius *r* around a long, straight current-carrying wire is used to calculate the magnetic field set up by the wire.

Ampère's circuital law provides an elegant and simple method for calculating the magnetic fields of highly symmetric current configurations. However, it can't easily be used to calculate magnetic fields for complex current configurations that lack symmetry. In addition, Ampère's circuital law in this form is valid only when the currents and fields don't change with time.

EXAMPLE 19.7   The Magnetic Field of a Long Wire

Goal   Calculate the magnetic field of a long, straight wire and the force that the field exerts on a particle.

Problem   A long, straight wire carries a current of 5.00 A. At one instant, a proton, 4.00 mm from the wire, travels at 1.50×10^3 m/s parallel to the wire and in the same direction as the current (Fig. 19.27). **(a)** Find the magnitude and direction of the magnetic field created by the wire. **(b)** Find the magnitude and direction of the magnetic force the wire's magnetic field exerts on the proton.

Strategy   First use Equation 19.11 to find the magnitude of the magnetic field at the given point. Use right-hand rule number 2 to find the direction of the field. Finally, substitute into Equation 19.1, computing the magnetic force on the proton.

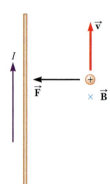

Figure 19.27   (Example 19.7) The magnetic field due to the current is into the page at the location of the proton, and the magnetic force on the proton is to the left.

Solution

(a) Find the magnitude and direction of the wire's magnetic field.

Use Equation 19.11 to calculate the magnitude of the magnetic field 4.00 mm from the wire:

$$B = \frac{\mu_0 I}{2\pi r} = \frac{(4\pi \times 10^{-7}\,\text{T}\cdot\text{m/A})(5.00\,\text{A})}{2\pi(4.00 \times 10^{-3}\,\text{m})}$$

$$= \boxed{2.50 \times 10^{-4}\,\text{T}}$$

Apply right-hand rule number 2 to find the direction of the magnetic field $\vec{\mathbf{B}}$:

With the right thumb pointing in the direction of the current in Figure 19.27, the fingers curl into the page at the location of the proton. The angle θ between $\vec{\mathbf{v}}$ and $\vec{\mathbf{B}}$ is therefore 90°.

(b) Compute the magnetic force exerted by the wire on the proton.

Substitute into Equation 19.1, which gives the magnitude of the magnetic force on a charged particle:

$$F = qvB\sin\theta = (1.60 \times 10^{-19}\,\text{C})(1.50 \times 10^3\,\text{m/s})$$
$$\times (2.50 \times 10^{-4}\,\text{T})(\sin 90°)$$

$$= \boxed{6.00 \times 10^{-20}\,\text{N}}$$

Find the direction of the magnetic force with right-hand rule number 1:

Point your right fingers in the direction of $\vec{\mathbf{v}}$, curling them into the page toward $\vec{\mathbf{B}}$. Your thumb points to the left, which is the direction of the magnetic force.

Remarks   The location of the proton is important. On the left-hand side, the wire's magnetic field points outward, and the magnetic force on the proton is to the right.

Exercise 19.7

Find (a) the magnetic field created by the wire and (b) the magnetic force on a helium-3 nucleus located 7.50 mm to the left of the wire in Figure 19.27, traveling 2.50×10^3 m/s opposite the direction of the current. (See the data table presented in Example 19.5 on page 639).

Answers   (a) 1.33×10^{-4} T   (b) 1.07×10^{-19} N, directed to the left in Figure 19.27.

19.8   MAGNETIC FORCE BETWEEN TWO PARALLEL CONDUCTORS

As we have seen, a magnetic force acts on a current-carrying conductor when the conductor is placed in an external magnetic field. Because a conductor carrying a current creates a magnetic field around itself, it is easy to understand that two current-carrying wires placed close together exert magnetic forces on each other. Consider two long, straight, parallel wires separated by the distance d and carrying currents I_1 and I_2 in the same direction, as shown in Active Figure 19.28. Wire 1 is directly above wire 2. What's the magnetic force on one wire due to a magnetic field set up by the other wire?

In this calculation, we are finding the force on wire 1 due to the magnetic field of wire 2. The current I_2, sets up magnetic field $\vec{B}_2$ at wire 1. The direction of $\vec{B}_2$ is perpendicular to the wire, as shown in the figure. Using Equation 19.11, we find that the magnitude of this magnetic field is

$$B_2 = \frac{\mu_0 I_2}{2\pi d}$$

According to Equation 19.5, the magnitude of the magnetic force on wire 1 in the presence of field $\vec{B}_2$ due to I_2 is

$$F_1 = B_2 I_1 \ell = \left(\frac{\mu_0 I_2}{2\pi d}\right) I_1 \ell = \frac{\mu_0 I_1 I_2 \ell}{2\pi d}$$

We can rewrite this relationship in terms of the force per unit length:

$$\frac{F_1}{\ell} = \frac{\mu_0 I_1 I_2}{2\pi d} \qquad\qquad \textbf{[19.14]}$$

The direction of $\vec{F}_1$ is downward, toward wire 2, as indicated by right-hand rule number 1. This calculation is completely symmetric, which means that the force $\vec{F}_2$ on wire 2 is equal to and opposite $\vec{F}_1$, as expected from Newton's third law of action–reaction.

We have shown that parallel conductors carrying currents in the same direction *attract* each other. You should use the approach indicated by Figure 19.28 and the steps leading to Equation 19.14 to show that parallel conductors carrying currents in opposite directions *repel* each other.

The force between two parallel wires carrying a current is used to define the SI unit of current, the **ampere** (A), as follows:

If two long, parallel wires 1 m apart carry the same current, and the magnetic force per unit length on each wire is 2×10^{-7} N/m, then the current is defined to be 1 A.

◀ Definition of the ampere

The SI unit of charge, the **coulomb** (C), can now be defined in terms of the ampere as follows:

If a conductor carries a steady current of 1 A, then the quantity of charge that flows through any cross section in 1 s is 1 C.

◀ Definition of the coulomb

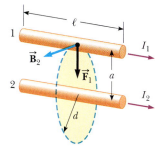

ACTIVE FIGURE 19.28
Two parallel wires, oriented vertically, carry steady currents and exert forces on each other. The field $\vec{B}_2$ at wire 1 due to wire 2 produces a force on wire 1 given by $F_1 = B_2 I_1 \ell$. The force is attractive if the currents have the same direction, as shown, and repulsive if the two currents have opposite directions.

Physics⊗Now™
Log into PhysicsNow at **www.cp7e.com** and go to Active Figure 19.28, where you can adjust the currents in the wires and the distance between them, and see the effect on the force.

If, in Figure 19.28, $I_1 = 2$ A and $I_2 = 6$ A, which of the following is true? (a) $F_1 = 3F_2$ (b) $F_1 = F_2$ or (c) $F_1 = F_2/3$

EXAMPLE 19.8   Levitating a Wire

Goal   Calculate the magnetic force of one current-carrying wire on a parallel current-carrying wire.

Problem   Two wires, each having a weight per unit length of 1.00×10^{-4} N/m, are parallel with one directly above the other. Assume that the wires carry currents that are equal in magnitude and opposite in direction. The wires are 0.10 m apart, and the sum of the magnetic force and gravitational force on the upper wire is zero. Find the current in the wires. (Neglect Earth's magnetic field.)

Strategy   The upper wire must be in equilibrium under the forces of magnetic repulsion and gravity. Set the sum of the forces equal to zero and solve for the unknown current, I.

Solution

Set the sum of the forces equal to zero, and substitute the appropriate expressions. Notice that the magnetic force between the wires is repulsive.

$$\vec{F}_{grav} + \vec{F}_{mag} = 0$$

$$-mg + \frac{\mu_0 I_1 I_2}{2\pi d}\ell = 0$$

The currents are equal, so $I_1 = I_2 = I$. Make these substitutions and solve for I^2.

$$\frac{\mu_0 I^2}{2\pi d}\ell = mg \quad \rightarrow \quad I^2 = \frac{(2\pi d)(mg/\ell)}{\mu_0}$$

Substitute given values, finding I^2, then take the square root. Notice that the weight per unit length, mg/ℓ, is given.

$$I^2 = \frac{(2\pi \cdot 0.100 \text{ m})(1.00 \times 10^{-4} \text{ N/m})}{(4\pi \times 10^{-7} \text{ T} \cdot \text{m})} = 50.0 \text{ A}^2$$

$$I = 7.07 \text{ A}$$

Remark   Exercise 19.3 showed that using the Earth's magnetic field to levitate a wire required extremely large currents. Currents in wires can create much stronger magnetic fields than Earth's in regions near the wire.

Exercise 19.8

If the current in each wire is doubled, how far apart should the wires be placed if the magnitudes of the gravitational and magnetic forces on the upper wire are to be equal?

Answer   0.400 m

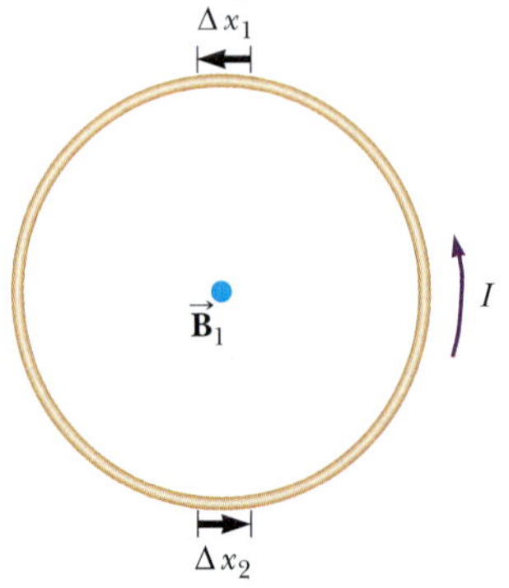

19.9 MAGNETIC FIELDS OF CURRENT LOOPS AND SOLENOIDS

The strength of the magnetic field set up by a piece of wire carrying a current can be enhanced at a specific location if the wire is formed into a loop. You can understand this by considering the effect of several small segments of the current loop, as in Figure 19.29. The small segment at the top of the loop, labeled Δx_1, produces a magnetic field of magnitude B_1 at the loop's center, directed out of the page. The direction of $\vec{B}$ can be verified using right-hand rule number 2 for a long, straight wire. Imagine holding the wire with your right hand, with your thumb pointing in the direction of the current. Your fingers then curl around in the direction of $\vec{B}$.

A segment of length Δx_2 at the bottom of the loop also contributes to the field at the center, increasing its strength. The field produced at the center by the segment Δx_2 has the same magnitude as B_1 and is also directed out of the page. Similarly, all other such segments of the current loop contribute to the field. The net effect is a magnetic field for the current loop as pictured in Figure 19.30a.

Figure 19.29   All segments of the current loop produce a magnetic field at the center of the loop, directed *out of the page*.

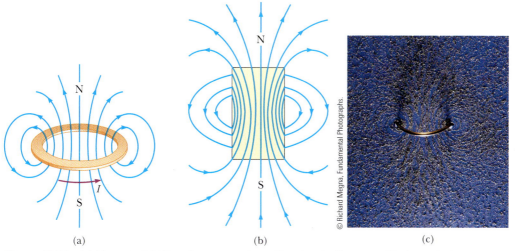

Figure 19.30 (a) Magnetic field lines for a current loop. Note that the lines resemble those of a bar magnet. (b) The magnetic field of a bar magnet is similar to that of a current loop. (c) Field lines of a current loop, displayed by iron filings.

Notice in Figure 19.30a that the magnetic field lines enter at the bottom of the current loop and exit at the top. Compare this to Figure 19.30b, illustrating the field of a bar magnet. The two fields are similar. One side of the loop acts as though it were the north pole of a magnet, and the other acts as a south pole. The similarity of these two fields will be used to discuss magnetism in matter in an upcoming section.

Applying Physics 19.4   Twisted Wires

In electrical circuits, it is often the case that insulated wires carrying currents in opposite directions are twisted together. What is the advantage of doing this?

Explanation If the wires are not twisted together, the combination of the two wires forms a current loop, which produces a relatively strong magnetic field. This magnetic field generated by the loop could be strong enough to affect adjacent circuits or components. When the wires are twisted together, their magnetic fields tend to cancel.

The magnitude of the magnetic field at the center of a circular loop carrying current I as in Figure 19.30a is given by

$$B = \frac{\mu_0 I}{2R}$$

This must be derived with calculus. However, it can be shown to be reasonable by calculating the field at the center of four long wires, each carrying current I and forming a square, as in Figure 19.31, with a circle of radius R inscribed within it. Intuitively, this arrangement should give a magnetic field at the center that is similar in magnitude to the field produced by the circular loop. The current in the circular wire is closer to the center, so that wire would have a magnetic field somewhat stronger than just the four legs of the rectangle, but the lengths of the straight wires beyond the rectangle compensate for this. Each wire contributes the same magnetic field at the exact center, so the total field is given by

$$B = 4 \times \frac{\mu_0 I}{2\pi R} = \frac{4}{\pi}\left(\frac{\mu_0 I}{2R}\right) = (1.27)\left(\frac{\mu_0 I}{2R}\right)$$

This is *approximately* the same as the field produced by the circular loop of current.

When the coil has N loops, each carrying current I, the magnetic field at the

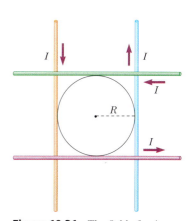

Figure 19.31 The field of a circular loop carrying current I can be approximated by the field due to four straight wires, each carrying current I.

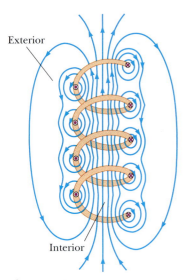

Figure 19.32 The magnetic field lines for a loosely wound solenoid.

The magnetic field inside a solenoid ▶

center is given by

$$B = N\frac{\mu_0 I}{2R} \qquad [19.15]$$

Magnetic Field of a Solenoid

If a long, straight wire is bent into a coil of several closely spaced loops, the resulting device is a **solenoid**, often called an **electromagnet**. This device is important in many applications because it acts as a magnet only when it carries a current. The magnetic field inside a solenoid increases with the current and is proportional to the number of coils per unit length.

Figure 19.32 shows the magnetic field lines of a loosely wound solenoid of length ℓ and total number of turns N. Notice that the field lines inside the solenoid are nearly parallel, uniformly spaced, and close together. As a result, the field inside the solenoid is strong and approximately uniform. The exterior field at the sides of the solenoid is nonuniform, much weaker than the interior field, and *opposite in direction* to the field inside the solenoid.

If the turns are closely spaced, the field lines are as shown in Figure 19.33a, entering at one end of the solenoid and emerging at the other. This means that one end of the solenoid acts as a north pole and the other end acts as a south pole. If the length of the solenoid is much greater than its radius, the lines that leave the north end of the solenoid spread out over a wide region before returning to enter the south end. The more widely separated the field lines are, the weaker the field. This is in contrast to a much stronger field *inside* the solenoid, where the lines are close together. Also, the field inside the solenoid has a constant magnitude at all points far from its ends. As will be shown subsequently, these considerations allow the application of Ampere's law to the solenoid, giving a result of

$$B = \mu_0 n I \qquad [19.16]$$

for the field inside the solenoid, where $n = N/\ell$ is the number of turns per unit length of the solenoid.

So-called steering magnets placed along the neck of the picture tube in a televi-

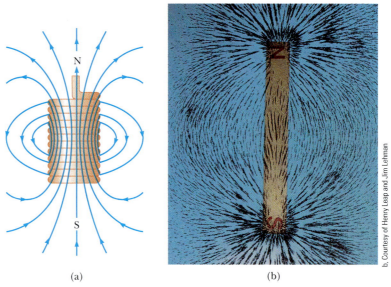

(a) (b)

Figure 19.33 (a) Magnetic field lines for a tightly wound solenoid of finite length carrying a steady current. The field inside the solenoid is nearly uniform and strong. Note that the field lines resemble those of a bar magnet, so the solenoid effectively has north and south poles. (b) The magnetic field pattern of a bar magnet, displayed by small iron filings on a sheet of paper.

EXAMPLE 19.9 The Magnetic Field inside a Solenoid

Goal Calculate the magnetic field of a solenoid from given data and the momentum of a charged particle in this field.

Problem A certain solenoid consists of 100 turns of wire and has a length of 10.0 cm. **(a)** Find the magnitude of the magnetic field inside the solenoid when it carries a current of 0.500 A. **(b)** What is the momentum of a proton orbiting inside the solenoid in a circle with a radius of 0.020 m? The axis of the solenoid is perpendicular to the plane of the orbit. **(c)** Approximately how much wire would be needed to build this solenoid? Assume the solenoid's radius is 5.00 cm.

Strategy In part (a), calculate the number of turns per meter and substitute that and given information into Equation 19.16, getting the magnitude of the magnetic field. Part (b) is an application of Newton's second law.

Solution
(a) Find the magnitude of the magnetic field inside the solenoid when it carries a current of 0.500 A.

Calculate the number of turns per unit length:

$$n = \frac{N}{\ell} = \frac{100 \text{ turns}}{0.100 \text{ m}} = 1.00 \times 10^3 \text{ turns/m}$$

Substitute n and I into Equation 19.16 to find the magnitude of the magnetic field:

$$B = \mu_0 n I$$
$$= (4\pi \times 10^{-7} \text{ T} \cdot \text{m/A})(1.00 \times 10^3 \text{ turns/m})(0.500 \text{ A})$$
$$= 6.28 \times 10^{-4} \text{ T}$$

(b) Find the momentum of a proton orbiting in a circle of radius 0.020 m near the center of the solenoid.

Write Newton's second law for the proton:

$$ma = F = qvB$$

Substitute the centripetal acceleration $a = v^2/r$:

$$m\frac{v^2}{r} = qvB$$

Cancel one factor of v on both sides and multiply by r, getting the momentum mv:

$$mv = rqB = (0.020 \text{ m})(1.60 \times 10^{-19} \text{ C})(6.28 \times 10^{-4} \text{ T})$$
$$p = mv = \boxed{2.01 \times 10^{-24} \text{ kg} \cdot \text{m/s}}$$

(c) Approximately how much wire would be needed to build this solenoid?

Multiply the number of turns by the circumference of one loop:

$$\text{length of wire} \approx (\text{number of turns})(2\pi r)$$
$$= (1.00 \times 10^2 \text{ turns})(2\pi \cdot 0.0500 \text{ m})$$
$$= \boxed{31.4 \text{ m}}$$

Remarks An electron in part (b) would have the same momentum as the proton, but a much higher speed. It would also orbit in the opposite direction. The length of wire in part (c) is only an estimate, because the wire has a certain thickness, slightly increasing the size of each loop. In addition the wire loops aren't perfect circles, because they wind slowly up along the solenoid.

Exercise 19.9
Suppose you have a 32.0-m length of copper wire. If the wire is wrapped into a solenoid 0.240 m long and having a radius of 0.0400 m, how strong is the resulting magnetic field in its center when the current is 12.0 A?

Answer $8.00 \times 10^{-3} \text{ T}$

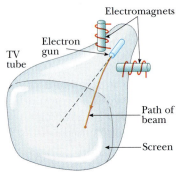

Figure 19.34 Electromagnets are used to deflect electrons to desired positions on the screen of a television tube.

sion set, as in Figure 19.34, are used to make the electron beam move to the desired locations on the screen, tracing out the images. The rate at which the electron beam sweeps over the screen is so fast that to the eye it looks like a picture rather than a sequence of dots.

Ampère's Law Applied to a Solenoid

We can use Ampère's law to obtain the expression for the magnetic field inside a solenoid carrying a current I. A cross section taken along the length of part of our solenoid is shown in Figure 19.35. $\vec{\mathbf{B}}$ inside the solenoid is uniform and parallel to the axis, and $\vec{\mathbf{B}}$ outside is approximately zero. Consider a rectangular path of length L and width w, as shown in the figure. We can apply Ampère's law to this path by evaluating the sum of $B_{\parallel} \Delta\ell$ over each side of the rectangle. The contribution along side 3 is clearly zero, because $\vec{\mathbf{B}} = 0$ in this region. The contributions from sides 2 and 4 are both zero, because $\vec{\mathbf{B}}$ is perpendicular to $\Delta\ell$ along these paths. Side 1 of length L gives a contribution BL to the sum, because $\vec{\mathbf{B}}$ is uniform along this path, and parallel to $\Delta\ell$. Therefore, the sum over the closed rectangular path has the value

$$\Sigma\, B_{\parallel} \Delta\ell = BL$$

The right side of Ampère's law involves the total current that passes through the area bounded by the path chosen. In this case, the total current through the rectangular path equals the current through each turn of the solenoid, multiplied by the number of turns. If N is the number of turns in the length L, then the total current through the rectangular path equals NI. Ampère's law applied to this path therefore gives

$$\Sigma\, B_{\parallel} \Delta\ell = BL = \mu_0 NI$$

or

$$B = \mu_0 \frac{N}{L} I = \mu_0 nI$$

where $n = N/L$ is the number of turns per unit length.

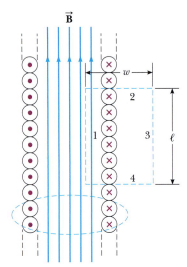

Figure 19.35 A cross-sectional view of a tightly wound solenoid. If the solenoid is long relative to its radius, we can assume that the magnetic field inside is uniform and the field outside is zero. Ampère's law applied to the blue dashed rectangular path can then be used to calculate the field inside the solenoid.

19.10 MAGNETIC DOMAINS

The magnetic field produced by a current in a coil of wire gives us a hint as to what might cause certain materials to exhibit strong magnetic properties. A single coil like that in Figure 19.30a has a north pole and a south pole, but if this is true for a coil of wire, it should also be true for any current confined to a circular path. In particular, *an individual atom should act as a magnet because of the motion of the electrons about the nucleus.* Each electron, with its charge of 1.6×10^{-19} C, circles the atom once in about 10^{-16} s. If we divide the electric charge by this time interval, we see that the orbiting electron is equivalent to a current of 1.6×10^{-3} A. Such a current produces a magnetic field on the order of 20 T at the center of the circular path. From this we see that a very strong magnetic field would be produced if several of these atomic magnets could be aligned inside a material. This doesn't occur, however, because the simple model we have described is not the complete story. A thorough analysis of atomic structure shows that the magnetic field produced by one electron in an atom is often canceled by an oppositely revolving electron in the same atom. The net result is that **the magnetic effect produced by the electrons orbiting the nucleus is either zero or very small for most materials.**

The magnetic properties of many materials can be explained by the fact that an electron not only circles in an orbit, but also spins on its axis like a top, with spin magnetic moment as shown (Fig. 19.36). (This classical description should not be taken too literally. The property of electron *spin* can be understood only in the context of quantum mechanics, which we will not discuss here.) The spinning elec-

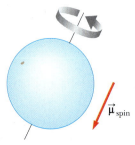

Figure 19.36 Classical model of a spinning electron.

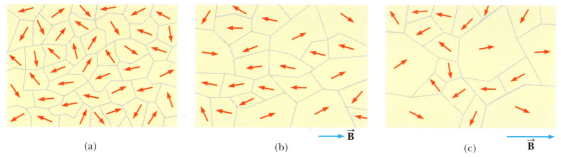

Figure 19.37 (a) Random orientation of domains in an unmagnetized substance. (b) When an external magnetic field $\vec{\mathbf{B}}$ is applied, the domains tend to align with the magnetic field. (c) As the field is made even stronger, the domains not aligned with the external field become very small.

tron represents a charge in motion that produces a magnetic field. The field due to the spinning is generally stronger than the field due to the orbital motion. In atoms containing many electrons, the electrons usually pair up with their spins opposite each other, so that their fields cancel each other. That is why most substances are not magnets. However, in certain strongly magnetic materials, such as iron, cobalt, and nickel, the magnetic fields produced by the electron spins don't cancel completely. Such materials are said to be **ferromagnetic**. In ferromagnetic materials, strong coupling occurs between neighboring atoms, forming large groups of atoms with spins that are aligned. Called **domains**, the sizes of these groups typically range from about 10^{-4} cm to 0.1 cm. In an unmagnetized substance the domains are randomly oriented, as shown in Figure 19.37a. When an external field is applied, as in Figure 19.37b, the magnetic field of each domain tends to come nearer to alignment with the external field, resulting in magnetization.

In what are called hard magnetic materials, domains remain aligned even after the external field is removed; the result is a **permanent magnet**. In soft magnetic materials, such as iron, once the external field is removed, thermal agitation produces motion of the domains and the material quickly returns to an unmagnetized state.

The alignment of domains explains why the strength of an electromagnet is increased dramatically by the insertion of an iron core into the magnet's center. The magnetic field produced by the current in the loops causes the domains to align, thus producing a large net external field. The use of iron as a core is also advantageous because it is a soft magnetic material that loses its magnetism almost instantaneously after the current in the coils is turned off.

TIP 19.4   The Electron Spins—but Doesn't!

Even though we use the word *spin*, the electron, unlike a child's top, isn't physically spinning in this sense. The electron has an intrinsic angular momentum that causes it to act *as if it were spinning*, but the concept of spin angular momentum is actually a relativistic quantum effect.

SUMMARY

Physics⊗Now™ Take a practice test by logging into Physics-Now at **www.cp7e.com** and clicking on the Pre-Test link for this chapter.

19.3 Magnetic Fields

The **magnetic force** that acts on a charge q moving with velocity $\vec{\mathbf{v}}$ in a magnetic field $\vec{\mathbf{B}}$ has magnitude

$$F = qvB \sin \theta \qquad [19.1]$$

where θ is the angle between $\vec{\mathbf{v}}$ and $\vec{\mathbf{B}}$.

To find the direction of this force, use **right-hand rule number 1**: point the fingers of your open right hand in the direction of $\vec{\mathbf{v}}$ and then curl them in the direction of $\vec{\mathbf{B}}$. Your thumb then points in the direction of the magnetic force $\vec{\mathbf{F}}$.

If the charge is *negative* rather than positive, the force is directed opposite the force given by the right-hand rule.

The SI unit of the magnetic field is the **tesla** (T), or weber per square meter (Wb/m^2). An additional commonly used unit for the magnetic field is the **gauss** (G); $1\ \text{T} = 10^4\ \text{G}$.

19.4 Magnetic Force on a Current-Carrying Conductor

If a straight conductor of length ℓ carries current I, the magnetic force on that conductor when it is placed in a uniform external magnetic field $\vec{\mathbf{B}}$ is

$$F = BI\ell \sin \theta \qquad [19.6]$$

where θ is the angle between the direction of the current and the direction of the magnetic field.

Right-hand rule number 1 also gives the direction of the magnetic force on the conductor. In this case, however, you must point your fingers in the direction of the current rather than in the direction of $\vec{\mathbf{v}}$.

19.5 Torque on a Current Loop and Electric Motors

The torque $\vec{\tau}$ on a current-carrying loop of wire in a magnetic field $\vec{B}$ has magnitude

$$\tau = BIA \sin \theta \qquad \text{[19.8]}$$

where I is the current in the loop and A is its cross-sectional area. The magnitude of the magnetic moment of a current-carrying coil is defined by $\mu = IAN$, where N is the number of loops. The magnetic moment is considered a vector, $\vec{\mu}$, that is perpendicular to the plane of the loop. The angle between $\vec{B}$ and $\vec{\mu}$ is θ.

19.6 Motion of a Charged Particle in a Magnetic Field

If a charged particle moves in a uniform magnetic field so that its initial velocity is perpendicular to the field, it will move in a circular path in a plane perpendicular to the magnetic field. The radius r of the circular path can be found from Newton's second law and centripetal acceleration, and is given by

$$r = \frac{mv}{qB} \qquad \text{[19.10]}$$

where m is the mass of the particle and q is its charge.

19.7 Magnetic Field of a Long, Straight Wire and Ampère's Law

The magnetic field at distance r from a **long, straight wire** carrying current I has the magnitude

$$B = \frac{\mu_0 I}{2\pi r} \qquad \text{[19.11]}$$

where $\mu_0 = 4\pi \times 10^{-7}\ \text{T}\cdot\text{m/A}$ is the **permeability of free space**. The magnetic field lines around a long, straight wire are circles concentric with the wire.

Ampère's law can be used to find the magnetic field around certain simple current-carrying conductors. It can be written

$$\Sigma B_{\parallel} \Delta \ell = \mu_0 I \qquad \text{[19.13]}$$

where $B_{\parallel}$ is the component of $\vec{B}$ tangent to a small current element of length $\Delta \ell$ that is part of a closed path and I is the total current that penetrates the closed path.

19.8 Magnetic Force between Two Parallel Conductors

The force per unit length on each of two parallel wires separated by the distance d and carrying currents I_1 and I_2 has the magnitude

$$\frac{F}{\ell} = \frac{\mu_0 I_1 I_2}{2\pi d} \qquad \text{[19.14]}$$

The forces are attractive if the currents are in the same direction and repulsive if they are in opposite directions.

19.9 Magnetic Field of Current Loops and Solenoids

The magnetic field at the center of a coil of N circular loops of radius R, each carrying current I, is given by

$$B = N\frac{\mu_0 I}{2R} \qquad \text{[19.15]}$$

The magnetic field inside a solenoid has the magnitude

$$B = \mu_0 n I \qquad \text{[19.16]}$$

where $n = N/\ell$ is the number of turns of wire per unit length.

CONCEPTUAL QUESTIONS

1. In your home television set, a beam of electrons moves from the back of the picture tube to the screen, where it strikes a fluorescent dot that glows with a particular color when hit. The Earth's magnetic field at the location of the television set is horizontal and toward the north. In which direction(s) should the set be oriented so that the beam undergoes the largest deflection?

2. Can a constant magnetic field set a proton at rest into motion? Explain your answer.

3. A proton moving horizontally enters a region where a uniform magnetic field is directed perpendicular to the proton's velocity, as shown in Figure Q19.3. Describe the subsequent motion of the proton. How would an electron behave under the same circumstances?

4. No magnetic force acts upon a current-carrying conductor when it is placed in a certain manner in a uniform magnetic field. Explain.

5. How can the motion of a charged particle be used to distinguish between a magnetic field and an electric field in a certain region?

6. Which way would a compass point if you were at Earth's north magnetic pole?

7. Why does the picture on a television screen become distorted when a magnet is brought near the screen as in Figure Q19.7? [*Caution:* You should not do this at home on a color television set, because it may permanently affect the picture quality.]

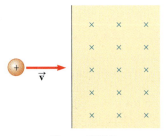

Figure Q19.3

Figure Q19.7

8. A magnet attracts a piece of iron. The iron can then attract another piece of iron. On the basis of domain alignment, explain what happens in each piece of iron.

9. A Hindu ruler once suggested that he be entombed in a magnetic coffin with the polarity arranged so that he could be forever suspended between heaven and Earth. Is such magnetic levitation possible? Discuss.

10. Will a nail be attracted to either pole of a magnet? Explain what is happening inside the nail when it is placed near the magnet.

11. Suppose you move along a wire at the same speed as the drift speed of the electrons in the wire. Do you now measure a magnetic field of zero?

12. Describe the change in the magnetic field in the space enclosed by a solenoid carrying a steady current I if (a) the length of the solenoid is doubled, but the number of turns remains the same, and (b) the number of turns is doubled, but the length remains the same.

13. Can you use a compass to detect the currents in wires in the walls near light switches in your home?

14. Why do charged particles from outer space, called cosmic rays, strike Earth more frequently at the poles than at the equator?

15. Two wires carry currents in opposite directions and are oriented parallel, with one above the other. The wires repel each other. Is the upper wire in a stable levitation over the lower wire? Suppose the current in one wire is reversed, so that the wires now attract. Is the lower wire hanging in a stable attraction to the upper wire?

16. How can a current loop be used to determine the presence of a magnetic field in a given region of space?

17. A hanging Slinky® toy is attached to a powerful battery and a switch. When the switch is closed so that the toy now carries current, does the Slinky® compress or expand?

18. Is it possible to orient a current loop in a uniform magnetic field such that the loop will not tend to rotate?

19. Parallel wires exert magnetic forces on each other. What about perpendicular wires? Imagine two wires oriented perpendicular to each other and almost touching. Each wire carries a current. Is there a force between the wires?

20. Is the magnetic field created by a current loop uniform? Explain.

21. The electron beam in Figure Q19.21 is projected to the right. The beam deflects downward in the presence of a magnetic field produced by a pair of current-carrying coils. (a) What is the direction of the magnetic field? (b) What would happen to the beam if the magnetic field were reversed in direction?

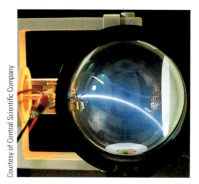

Figure Q19.21

22. Figure Q19.22 shows two permanent magnets, each having a hole through its center. Note that the upper magnet is levitated above the lower one. (a) How does this occur? (b) What purpose does the pencil serve? (c) What can you say about the poles of the magnets from this observation? (d) If the upper magnet were inverted, what do you suppose would happen?

Figure Q19.22

PROBLEMS

1, 2, 3 = straightforward, intermediate, challenging   □ = full solution available in *Student Solutions Manual/Study Guide*

Physics⊗Now™ = coached problem with hints available at **www.cp7e.com**   = biomedical application

Section 19.3 Magnetic Fields

1. An electron gun fires electrons into a magnetic field directed straight downward. Find the direction of the force exerted by the field on an electron for each of the following directions of the electron's velocity: (a) horizontal and due north; (b) horizontal and 30° west of north; (c) due north, but at 30° below the horizontal; (d) straight upward. (Remember that an electron has a negative charge.)

2. (a) Find the direction of the force on a proton (a positively charged particle) moving through the magnetic fields in Figure P19.2 (page 652), as shown. (b) Repeat part (a), assuming the moving particle is an electron.

3. Find the direction of the magnetic field acting on the positively charged particle moving in the various situations shown in Figure P19.3 (page 652) if the direction of the magnetic force acting on it is as indicated.

4. Determine the initial direction of the deflection of charged particles as they enter the magnetic fields, as shown in Figure P19.4 (page 652).

5. At the equator, near the surface of Earth, the magnetic field is approximately 50.0 μT northward, and the electric field is about 100 N/C downward in fair weather. Find the gravitational, electric, and magnetic forces on an electron with an instantaneous velocity of 6.00×10^6 m/s directed to the east in this environment.

6. The magnetic field of the Earth at a certain location is directed vertically downward and has a magnitude of 50.0 μT. A proton is moving horizontally toward the west

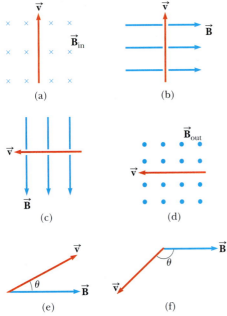

Figure P19.2 (Problems 2 and 12) For Problem 12, replace the velocity vector with a current in that direction.

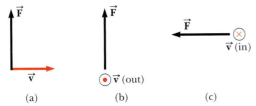

Figure P19.3 (Problems 3 and 13) For Problem 13, replace the velocity vector with a current in that direction.

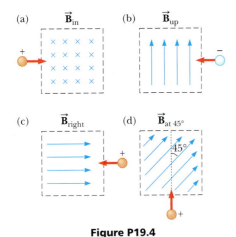

Figure P19.4

in this field with a speed of 6.20×10^6 m/s. What are the direction and magnitude of the magnetic force the field exerts on the proton?

7. **Physics⊗Now**™ What velocity would a proton need to circle Earth 1 000 km above the magnetic equator, where Earth's magnetic field is directed horizontally north and has a magnitude of 4.00×10^{-8} T?

8. An electron is accelerated through 2 400 V from rest and then enters a region where there is a uniform 1.70-T magnetic field. What are (a) the maximum and (b) the minimum magnitudes of the magnetic force acting on this electron?

9. A proton moves perpendicularly to a uniform magnetic field $\vec{B}$ at 1.0×10^7 m/s and exhibits an acceleration of 2.0×10^{13} m/s² in the $+x$-direction when its velocity is in the $+z$-direction. Determine the magnitude and direction of the field.

10. Sodium ions (Na⁺) move at 0.851 m/s through a bloodstream in the arm of a person standing near a large magnet. The magnetic field has a strength of 0.254 T and makes an angle of 51.0° with the motion of the sodium ions. The arm contains 100 cm³ of blood with 3.00×10^{20} Na⁺ ions per cubic centimeter. If no other ions were present in the arm, what would be the magnetic force on the arm?

Section 19.4 Magnetic Force on a Current-Carrying Conductor

11. A current $I = 15$ A is directed along the positive x-axis and perpendicular to a magnetic field. A magnetic force per unit length of 0.12 N/m acts on the conductor in the negative y-direction. Calculate the magnitude and direction of the magnetic field in the region through which the current passes.

12. In Figure P19.2, assume that in each case the velocity vector shown is replaced with a wire carrying a current in the direction of the velocity vector. For each case, find the direction of the magnetic force acting on the wire.

13. In Figure P19.3, assume that in each case the velocity vector shown is replaced with a wire carrying a current in the direction of the velocity vector. For each case, find the direction of the magnetic field that will produce the magnetic force shown.

14. A wire having a mass per unit length of 0.500 g/cm carries a 2.00-A current horizontally to the south. What are the direction and magnitude of the minimum magnetic field needed to lift this wire vertically upward?

15. A wire carries a current of 10.0 A in a direction that makes an angle of 30.0° with the direction of a magnetic field of strength 0.300 T. Find the magnetic force on a 5.00-m length of the wire.

16. At a certain location, Earth has a magnetic field of 0.60×10^{-4} T, pointing 75° below the horizontal in a north–south plane. A 10.0-m-long straight wire carries a 15-A current. (a) If the current is directed horizontally toward the east, what are the magnitude and direction of the magnetic force on the wire? (b) What are the magnitude and direction of the force if the current is directed vertically upward?

17. A wire with a mass of 1.00 g/cm is placed on a horizontal surface with a coefficient of friction of 0.200. The wire carries a current of 1.50 A eastward and moves horizontally to the north. What are the magnitude and the direction of the *smallest* vertical magnetic field that enables the wire to move in this fashion?

18. A conductor suspended by two flexible wires as shown in Figure P19.18 has a mass per unit length of 0.040 0 kg/m. What current must exist in the conductor for the tension in the supporting wires to be zero when the magnetic field is 3.60 T into the page? What is the required direction for the current?

19. An unusual message delivery system is pictured in Figure P19.19. A 15-cm length of conductor that is free to move is held in place between two thin conductors. When a 5.0-A

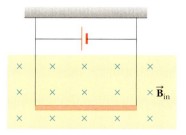

Figure P19.18

current is directed as shown in the figure, the wire segment moves upward at a constant velocity. If the mass of the wire is 15 g, find the magnitude and direction of the minimum magnetic field that is required to move the wire. (The wire slides without friction on the two vertical conductors.)

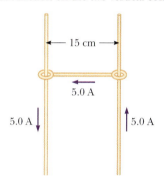

Figure P19.19

20. A wire 2.80 m in length carries a current of 5.00 A in a region where a uniform magnetic field has a magnitude of 0.390 T. Calculate the magnitude of the magnetic force on the wire, assuming the angle between the magnetic field and the current is (a) $60.0°$, (b) $90.0°$, (c) $120°$.

21. In Figure P19.21, the cube is 40.0 cm on each edge. Four straight segments of wire—ab, bc, cd, and da—form a closed loop that carries a current $I = 5.00$ A in the direction shown. A uniform magnetic field of magnitude $B = 0.020\ 0$ T is in the positive y-direction. Determine the magnitude and direction of the magnetic force on each segment.

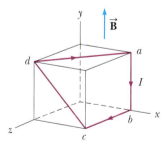

Figure P19.21

Section 19.5 Torque on a Current Loop and Electric Motors

22. A current of 17.0 mA is maintained in a single circular loop with a circumference of 2.00 m. A magnetic field of 0.800 T is directed parallel to the plane of the loop. What is the magnitude of the torque exerted by the magnetic field on the loop?

23. **Physics⊗Now™** An eight-turn coil encloses an elliptical area having a major axis of 40.0 cm and a minor axis of 30.0 cm (Fig. P19.23). The coil lies in the plane of the page and has a 6.00-A current flowing clockwise around

it. If the coil is in a uniform magnetic field of 2.00×10^{-4} T directed toward the left of the page, what is the magnitude of the torque on the coil? [*Hint*: The area of an ellipse is $A = \pi ab$, where a and b are, respectively, the semimajor and semiminor axes of the ellipse.]

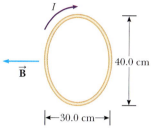

Figure P19.23

24. A rectangular loop consists of 100 closely wrapped turns and has dimensions 0.40 m by 0.30 m. The loop is hinged along the y-axis, and the plane of the coil makes an angle of $30.0°$ with the x-axis (Fig. P19.24). What is the magnitude of the torque exerted on the loop by a uniform magnetic field of 0.80 T directed along the x-axis when the current in the windings has a value of 1.2 A in the direction shown? What is the expected direction of rotation of the loop?

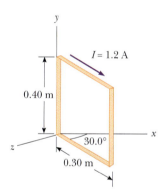

Figure P19.24

25. A long piece of wire with a mass of 0.100 kg and a total length of 4.00 m is used to make a square coil with a side of 0.100 m. The coil is hinged along a horizontal side, carries a 3.40-A current, and is placed in a vertical magnetic field with a magnitude of 0.010 0 T. (a) Determine the angle that the plane of the coil makes with the vertical when the coil is in equilibrium. (b) Find the torque acting on the coil due to the magnetic force at equilibrium.

26. A copper wire is 8.00 m long and has a cross-sectional area of 1.00×10^{-4} m². The wire forms a one-turn loop in the shape of square and is then connected to a battery that applies a potential difference of 0.100 V. If the loop is placed in a uniform magnetic field of magnitude 0.400 T, what is the maximum torque that can act on it? The resistivity of copper is 1.70×10^{-8} $\Omega \cdot$m.

Section 19.6 Motion of a Charged Particle in a Magnetic Field

27. A proton moving freely in a circular path perpendicular to a constant magnetic field takes 1.00 μs to complete one revolution. Determine the magnitude of the magnetic field.

28. A cosmic-ray proton in interstellar space has an energy of 10.0 MeV and executes a circular orbit having a radius equal to that of Mercury's orbit around the Sun, which is 5.80×10^{10} m. What is the magnetic field in that region of space?

29. Figure P19.29a is a diagram of a device called a velocity selector, in which particles of a specific velocity pass through undeflected while those with greater or lesser velocities are deflected either upwards or downwards. An electric field is directed perpendicular to a magnetic field, producing an electric force and a magnetic force on the charged particle that can be equal in magnitude and opposite in direction (Fig. P19.29b) and hence cancel. Show that particles with a speed of $v = E/B$ will pass through the velocity selector undeflected.

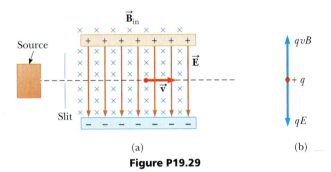

(a)

(b)

Figure P19.29

30. Consider the mass spectrometer shown schematically in Figure P19.30. The electric field between the plates of the velocity selector is 950 V/m, and the magnetic fields in both the velocity selector and the deflection chamber have magnitudes of 0.930 T. Calculate the radius of the path in the system for a singly charged ion with mass $m = 2.18 \times 10^{-26}$ kg. [*Hint:* See Problem 29.]

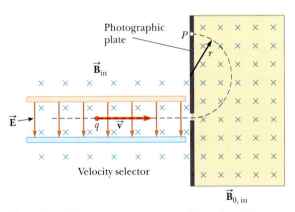

Figure P19.30 A mass spectrometer. Charged particles are first sent through a velocity selector. They then enter a region where a magnetic field $\vec{B}_0$ (directed inward) causes positive ions to move in a semicircular path and strike a photographic film at P.

31. A singly charged positive ion has a mass of 2.50×10^{-26} kg. After being accelerated through a potential difference of 250 V, the ion enters a magnetic field of 0.500 T, in a direction perpendicular to the field. Calculate the radius of the path of the ion in the field.

32. A mass spectrometer is used to examine the isotopes of uranium. Ions in the beam emerge from the velocity selector at a speed of 3.00×10^5 m/s and enter a uniform

magnetic field of 0.600 T directed perpendicularly to the velocity of the ions. What is the distance between the impact points formed on the photographic plate by singly charged ions of ^{235}U and ^{238}U?

33. A proton is at rest at the plane vertical boundary of a region containing a uniform vertical magnetic field B. An alpha particle moving horizontally makes a head-on elastic collision with the proton. Immediately after the collision, both particles enter the magnetic field, moving perpendicular to the direction of the field. The radius of the proton's trajectory is R. Find the radius of the alpha particle's trajectory. The mass of the alpha particle is four times that of the proton, and its charge is twice that of the proton.

Section 19.7 Magnetic Field of a Long, Straight Wire and Ampère's Law

34. In each of parts (a), (b), and (c) of Figure P19.34, find the direction of the current in the wire that would produce a magnetic field directed as shown.

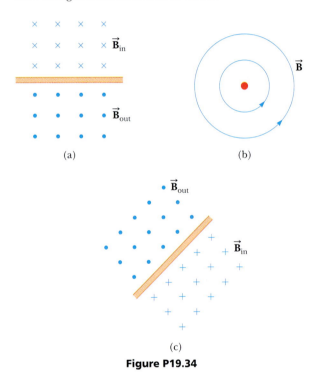

(a)

(b)

(c)

Figure P19.34

35. A lightning bolt may carry a current of 1.00×10^4 A for a short time. What is the resulting magnetic field 100 m from the bolt? Suppose that the bolt extends far above and below the point of observation.

36. In 1962, measurements of the magnetic field of a large tornado were made at the Geophysical Observatory in Tulsa, Oklahoma. If the magnitude of the tornado's field was $B = 1.50 \times 10^{-8}$ T pointing north when the tornado was 9.00 km east of the observatory, what current was carried up or down the funnel of the tornado? Model the vortex as a long, straight wire carrying a current.

37. A cardiac pacemaker can be affected by a static magnetic field as small as 1.7 mT. How close can a pacemaker wearer come to a long, straight wire carrying 20 A?

38. The two wires shown in Figure P19.38 carry currents of 5.00 A in opposite directions and are separated by 10.0 cm. Find the direction and magnitude of the net magnetic field (a) at a point midway between the wires; (b) at point P_1, 10.0 cm to the right of the wire on the right, and (c) at point P_2, 20.0 cm to the left of the wire on the left.

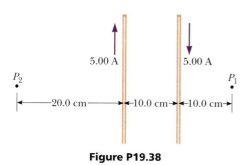

Figure P19.38

39. Four long, parallel conductors carry equal currents of $I = 5.00$ A. Figure P19.39 is an end view of the conductors. The direction of the current is into the page at points A and B (indicated by the crosses) and out of the page at C and D (indicated by the dots). Calculate the magnitude and direction of the magnetic field at point P, located at the center of the square with edge of length 0.200 m.

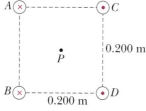

Figure P19.39

40. The two wires in Figure P19.40 carry currents of 3.00 A and 5.00 A in the direction indicated. (a) Find the direction and magnitude of the magnetic field at a point midway between the wires. (b) Find the magnitude and direction of the magnetic field at point P, located 20.0 cm above the wire carrying the 5.00-A current.

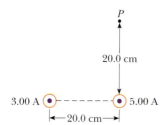

Figure P19.40

41. **Physics⊗Now™** A wire carries a 7.00-A current along the x-axis, and another wire carries a 6.00-A current along the y-axis, as shown in Figure P19.41. What is the magnetic field at point P, located at $x = 4.00$ m, $y = 3.00$ m?

42. A long, straight wire lies on a horizontal table and carries a current of $1.20~\mu$A. In a vacuum, a proton moves parallel to the wire (opposite the direction of the current) with a constant velocity of 2.30×10^4 m/s at a constant distance d above the wire. Determine the value of d. (You may ignore the magnetic field due to Earth.)

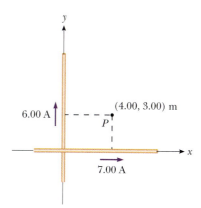

Figure P19.41

43. The magnetic field 40.0 cm away from a long, straight wire carrying current 2.00 A is 1.00 μT. (a) At what distance is it 0.100 μT? (b) At one instant, the two conductors in a long household extension cord carry equal 2.00-A currents in opposite directions. The two wires are 3.00 mm apart. Find the magnetic field 40.0 cm away from the middle of the straight cord, in the plane of the two wires. (c) At what distance is it one-tenth as large? (d) The center wire in a coaxial cable carries current 2.00 A in one direction, and the sheath around it carries current 2.00 A in the opposite direction. What magnetic field does the cable create at points outside?

Section 19.8 Magnetic Force between Two Parallel Conductors

44. Two parallel wires are 10.0 cm apart, and each carries a current of 10.0 A. (a) If the currents are in the same direction, find the force per unit length exerted on one of the wires by the other. Are the wires attracted to or repelled by each other? (b) Repeat the problem with the currents in opposite directions.

45. A wire with a weight per unit length of 0.080 N/m is suspended directly above a second wire. The top wire carries a current of 30.0 A and the bottom wire carries a current of 60.0 A. Find the distance of separation between the wires so that the top wire will be held in place by magnetic repulsion.

46. In Figure P19.46, the current in the long, straight wire is $I_1 = 5.00$ A, and the wire lies in the plane of the rectangular loop, which carries 10.0 A. The dimensions shown are $c = 0.100$ m, $a = 0.150$ m, and $\ell = 0.450$ m. Find the magnitude and direction of the net force exerted by the magnetic field due to the straight wire on the loop.

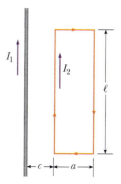

Figure P19.46

Section 19.9 Magnetic Fields of Current Loops and Solenoids

47. What current is required in the windings of a long solenoid that has 1 000 turns uniformly distributed over a length of 0.400 m in order to produce a magnetic field of magnitude 1.00×10^{-4} T at the center of the solenoid?

48. It is desired to construct a solenoid that will have a resistance of 5.00 Ω (at 20°C) and produce a magnetic field of 4.00×10^{-2} T at its center when it carries a current of 4.00 A. The solenoid is to be constructed from copper wire having a diameter of 0.500 mm. If the radius of the solenoid is to be 1.00 cm, determine (a) the number of turns of wire needed and (b) the length the solenoid should have.

49. **Physics⊗Now™** A single-turn square loop of wire 2.00 cm on a side carries a counterclockwise current of 0.200 A. The loop is inside a solenoid, with the plane of the loop perpendicular to the magnetic field of the solenoid. The solenoid has 30 turns per centimeter and carries a counterclockwise current of 15.0 A. Find the force on each side of the loop and the torque acting on the loop.

50. An electron is moving at a speed of 1.0×10^4 m/s in a circular path of radius of 2.0 cm inside a solenoid. The magnetic field of the solenoid is perpendicular to the plane of the electron's path. Find (a) the strength of the magnetic field inside the solenoid and (b) the current in the solenoid if it has 25 turns per centimeter.

ADDITIONAL PROBLEMS

51. A circular coil consisting of a single loop of wire has a radius of 30.0 cm and carries a current of 25 A. It is placed in an external magnetic field of 0.30 T. Find the torque on the wire when the plane of the coil makes an angle of 35° with the direction of the field.

52. An electron enters a region of magnetic field of magnitude 0.010 0 T, traveling perpendicular to the linear boundary of the region. The direction of the field is perpendicular to the velocity of the electron. (a) Determine the time it takes for the electron to leave the "field-filled" region, noting that its path is a semicircle. (b) Find the kinetic energy of the electron if the radius of its semicircular path is 2.00 cm.

53. Two long, straight wires cross each other at right angles, as shown in Figure P19.53. (a) Find the direction and magnitude of the magnetic field at point P, which is in the same plane as the two wires. (b) Find the magnetic field at a point 30.0 cm above the point of intersection (30.0 cm out of the page, toward you).

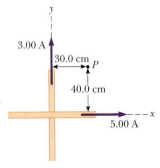

Figure P19.53

54. A 0.200-kg metal rod carrying a current of 10.0 A glides on two horizontal rails 0.500 m apart. What vertical magnetic field is required to keep the rod moving at a constant speed if the coefficient of kinetic friction between the rod and rails is 0.100?

55. Two species of singly charged positive ions of masses 20.0×10^{-27} kg and 23.4×10^{-27} kg enter a magnetic field at the same location with a speed of 1.00×10^5 m/s. If the strength of the field is 0.200 T, and the ions move perpendicularly to the field, find their distance of separation after they complete one-half of their circular path.

56. Two parallel conductors carry currents in opposite directions, as shown in Figure P19.56. One conductor carries a current of 10.0 A. Point A is the midpoint between the wires, and point C is 5.00 cm to the right of the 10.0-A current. I is adjusted so that the magnetic field at C is zero. Find (a) the value of the current I and (b) the value of the magnetic field at A.

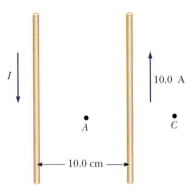

Figure P19.56

57. Using an electromagnetic flowmeter (Fig. P19.57), a heart surgeon monitors the flow rate of blood through an artery. Electrodes A and B make contact with the outer surface of the blood vessel, which has interior diameter 3.00 mm. (a) For a magnetic field magnitude of 0.040 0 T, a potential difference of 160 μV appears between the electrodes. Calculate the speed of the blood. (b) Verify that electrode A is positive, as shown. Does the sign of the emf depend on whether the mobile ions in the blood are predominantly positively or negatively charged? Explain.

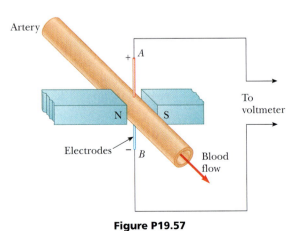

Figure P19.57

58. Two circular loops are parallel, coaxial, and almost in contact 1.00 mm apart (Fig. P19.58). Each loop is 10.0 cm in radius. The top loop carries a clockwise current of 140 A. The bottom loop carries a counterclockwise current of 140 A. (a) Calculate the magnetic force that the bottom loop exerts on the top loop. (b) The upper loop has a mass of 0.021 0 kg. Calculate its acceleration, assuming that the only forces acting on it are the force in part (a) and its weight. [*Hint*: The distance between the loops is small in comparison to their radius of curvature, so the loops may be treated as long, straight parallel wires.]

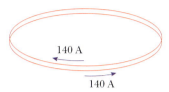

Figure P19.58

59. A 1.00-kg ball having net charge $Q = 5.00 \ \mu C$ is thrown out of a window horizontally at a speed $v = 20.0$ m/s. The window is at a height $h = 20.0$ m above the ground. A uniform horizontal magnetic field of magnitude $B = 0.010 \ 0$ T is perpendicular to the plane of the ball's trajectory. Find the magnitude of the magnetic force acting on the ball just before it hits the ground. [*Hint*: Ignore magnetic forces in finding the ball's final velocity.]

60. The idea that static magnetic fields might have a therapeutic value has been around for centuries. A currently available rare-Earth magnet that is advertised to relieve joint pain is shown in Figure P19.60. It is 1.0 mm thick and has a field strength of 5.0×10^{-2} T at the center of the flat surface. The magnetic field strength at points away from the center of the disk is inversely proportional to h^3, where h is the distance from the midplane of the disk. How far from the surface of the disk will the field strength be reduced to that of Earth (5.0×10^{-5} T)?

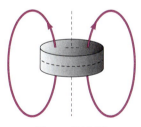

Figure P19.60

61. Two long parallel conductors carry currents $I_1 = 3.00$ A and $I_2 = 3.00$ A, both directed into the page in Figure P19.61. Determine the magnitude and direction of the resultant magnetic field at P.

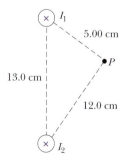

Figure P19.61

62. A uniform horizontal wire with a linear mass density of 0.50 g/m carries a 2.0-A current. It is placed in a constant magnetic field with a strength of 4.0×10^{-3} T. The field is horizontal and perpendicular to the wire. As the wire moves upward starting from rest, (a) what is its acceleration and (b) how long does it take to rise 50 cm? Neglect the magnetic field of Earth.

63. Three long parallel conductors carry currents of $I = 2.0$ A. Figure P19.63 is an end view of the conductors, with each current coming out of the page. Given that $a = 1.0$ cm, determine the magnitude and direction of the magnetic field at points A, B, and C.

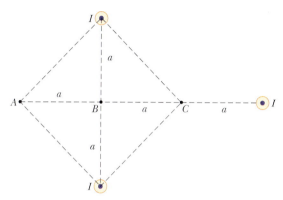

Figure P19.63

64. Two long parallel wires, each with a mass per unit length of 40 g/m, are supported in a horizontal plane by 6.0-cm-long strings, as shown in Figure P19.64. Each wire carries the same current I, causing the wires to repel each other so that the angle θ between the supporting strings is 16°. (a) Are the currents in the same or opposite directions? (b) Determine the magnitude of each current.

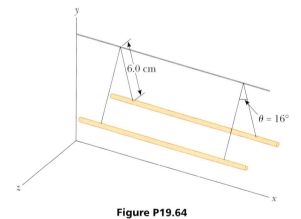

Figure P19.64

65. Protons having a kinetic energy of 5.00 MeV are moving in the positive x-direction and enter a magnetic field of 0.050 0 T in the z-direction, out of the plane of the page, and extending from $x = 0$ to $x = 1.00$ m as in Figure P19.65 (page 658). (a) Calculate the y-component of the protons' momentum as they leave the magnetic field. (b) Find the angle α between the initial velocity vector of the proton beam and the velocity vector after the beam emerges from the field. [*Hint*: Neglect relativistic effects and note that $1 \text{ eV} = 1.60 \times 10^{-19}$ J.]

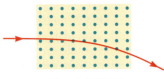

Figure P19.65

66. A straight wire of mass 10.0 g and length 5.0 cm is suspended from two identical springs that, in turn, form a closed circuit (Fig. P19.66). The springs stretch a distance of 0.50 cm under the weight of the wire. The circuit has a total resistance of 12 Ω. When a magnetic field directed out of the page (indicated by the dots in the figure is turned on, the springs are observed to stretch an additional 0.30 cm. What is the strength of the magnetic field? (The upper portion of the circuit is fixed.)

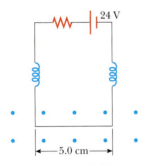

Figure P19.66

67. A solenoid 10.0 cm in diameter and 75.0 cm long is made from copper wire of diameter 0.100 cm with very thin insulation. The wire is wound onto a cardboard tube in a single layer, with adjacent turns touching each other. To produce a field of magnitude 20.0 mT at the center of the solenoid, what power must be delivered to the solenoid?

68. Assume that the region to the right of a certain vertical plane contains a vertical magnetic field of magnitude 1.00 mT and that the field is zero in the region to the left of the plane. An electron, originally traveling perpendicular to the boundary plane, passes into the region of the field. (a) Noting that the path of the electron is a semicircle, determine the time interval required for the electron to leave the "field-filled" region. (b) Find the kinetic energy of the electron if the maximum depth of penetration into the field is 2.00 cm.

69. Three long wires (wire 1, wire 2, and wire 3) are coplanar and hang vertically. The distance between wire 1 and wire 2 is 20.0 cm. On the left, wire 1 carries an upward current of 1.50 A. To the right, wire 2 carries a downward current of 4.00 A. Wire 3 is located such that when it carries a certain current, no net force acts upon any of the wires. Find (a) the position of wire 3 and (b) the magnitude and direction of the current in wire 3.

70. Two long parallel conductors separated by 10.0 cm carry currents in the same direction. The first wire carries a current $I_1 = 5.00$ A and the second carries $I_2 = 8.00$ A. (a) What is the magnitude of the magnetic field created by I_1 at the location of I_2? (b) What is the force per unit length exerted by I_1 on I_2? (c) What is the magnitude of the magnetic field created by I_2 at the

location of I_1? (d) What is the force per length exerted by I_2 on I_1?

ACTIVITIES

1. For this activity, you will need a small bar magnet, a small plastic container, and a bowl of water. Tape the magnet to the bottom of the container, and float the container and magnet on the surface of the bowl as in Figure A19.1. The magnet and the container should rotate and come to equilibrium, with the magnet pointing along a north–south line. The compass you have constructed is similar to the type used by early sailing vessels. How can you determine which direction is north and which is south?

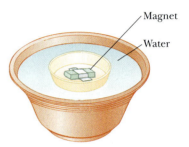

Figure A19.01

2. In the Northern Hemisphere, the direction of Earth's magnetic field becomes more and more nearly vertical the farther north one goes. To find the variation from the horizontal of the magnetic field in your locale, try the following: press an unmagnetized needle through a Ping-Pong® ball, and balance the structure between two drinking glasses that are lined up along an east–west line. Next, press a magnetized needle through the ball at right angles to the unmagnetized needle so that the needle points north. The magnetized needle can now rotate in the vertical direction and will point in the direction of Earth's magnetic field, which is at some angle below the horizontal. Take several measurements of this dip angle and obtain an average value.

3. Construct an electromagnet by wrapping about 1 meter of small-diameter insulated wire around a steel nail. Tape the ends of the wires to a D-cell battery as in Figure A19.3. How many staples or paper clips can you pick up with your electromagnet? How would you increase the magnetic field set up by the nail? Disconnect the wires from

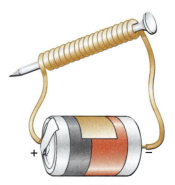

Figure A19.03

the battery and test how much magnetism is retained by the nail by seeing how many staples it can pick up. A convenient way to test the strength of a magnet is to attach a paper clip to a rubber band. Note how far the rubber band is stretched before the clip comes free of the magnet. Test your electromagnet in this way. Where is the magnetic field of the electromagnet strongest, at the ends of the nail or near its center? When you have your nail magnetized, bang it against a table or the floor and then check its magnetism. Why does the nail lose its magnetism by this procedure?

4. You can trace out the field pattern of a magnet with iron filings. Any machine shop will supply the filings, which should be soaked in a soap solution to remove grit and oil and then dried. Scatter them lightly over the surface of a paper covering the magnet, and then tap the paper gently to jar the filings into alignment. Explain why the filings form their pattern. Examine the field pattern set up in the following situations: (a) Arrange two bar magnets about 4 cm apart, aligned with opposite poles facing each other. (b) Use two bar magnets about 4 cm apart, aligned with like poles facing each other. (c) Use a horseshoe magnet.

PhotoDisc/Getty Images

The vibrating strings induce a voltage in pickup coils that detect and amplify the musical sounds being produced. The details of how this phenomenon works are discussed in this chapter.

Induced Voltages and Inductance

In 1819, Hans Christian Oersted discovered that an electric current exerted a force on a magnetic compass. Although there had long been speculation that such a relationship existed, Oersted's finding was the first evidence of a link between electricity and magnetism. Because nature is often symmetric, the discovery that electric currents produce magnetic fields led scientists to suspect that magnetic fields could produce electric currents. Indeed, experiments conducted by Michael Faraday in England and independently by Joseph Henry in the United States in 1831 showed that a changing magnetic field could induce an electric current in a circuit. The results of these experiments led to a basic and important law known as Faraday's law. In this chapter we discuss Faraday's law and several practical applications, one of which is the production of electrical energy in power generation plants throughout the world.

20.1 INDUCED EMF AND MAGNETIC FLUX

An experiment first conducted by Faraday demonstrated that a current can be produced by a changing magnetic field. The apparatus shown in Active Figure 20.1 (page 661) consists of a coil connected to a switch and a battery. We will refer to this coil as the *primary coil* and to the corresponding circuit as the primary circuit. The coil is wrapped around an iron ring to intensify the magnetic field produced by the current in the coil. A second coil, at the right, is wrapped around the iron ring and is connected to an ammeter. This is called the *secondary coil*, and the corresponding circuit is called the secondary circuit. It's important to notice that **there is no battery in the secondary circuit**.

At first glance, you might guess that no current would ever be detected in the secondary circuit. However, when the switch in the primary circuit in Active Figure 20.1 is suddenly closed, something amazing happens: the ammeter

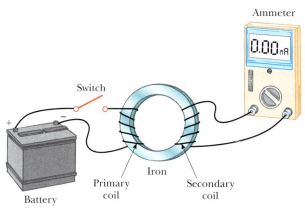

Ammeter

Switch

Primary coil

Iron

Secondary coil

Battery

Physics⊗Now™
Log into PhysicsNow at
www.cp7e.com and go to Active
Figure 20.1, where you can open and
close the switch and observe the
current in the ammeter.

ACTIVE FIGURE 20.1
Faraday's experiment. When the switch in the primary circuit at the left is closed, the ammeter in the secondary circuit at the right measures a momentary current. The emf in the secondary circuit is induced by the changing magnetic field through the coil in that circuit.

measures a current in the secondary circuit and then returns to zero! When the switch is opened again, the ammeter reads a current in the opposite direction and again returns to zero. Finally, whenever there is a steady current in the primary circuit, the ammeter reads zero.

From observations such as these, Faraday concluded that an electric current could be produced by a *changing* magnetic field. (A steady magnetic field doesn't produce a current, unless the coil is moving, as explained below.) The current produced in the secondary circuit occurs only for an instant while the magnetic field through the secondary coil is changing. In effect, the secondary circuit behaves as though a source of emf were connected to it for a short time. It's customary to say that **an induced emf is produced in the secondary circuit by the changing magnetic field**.

Magnetic Flux

In order to evaluate induced emfs quantitatively, we need to understand what factors affect the phenomenon. While changing magnetic fields always induce electric fields, there are also situations in which the magnetic field remains constant, yet an induced electric field is still produced. The best example of this is an electric generator: A loop of conductor rotating in a constant magnetic field creates an electric current.

The physical quantity associated with magnetism that creates an electric field is a **changing magnetic flux**. Magnetic flux is defined in the same way as electric flux (Section 15.9) and is proportional to both the strength of the magnetic field passing through the plane of a loop of wire and the area of the loop.

The **magnetic flux Φ_B** through a loop of wire with area A is defined by

$$\Phi_B \equiv B_\perp A = BA \cos \theta \qquad \textbf{[20.11]}$$

where $B_\perp$ is the component of $\vec{\mathbf{B}}$ perpendicular to the plane of the loop, as in Figure 20.2a, and θ is the angle between $\vec{\mathbf{B}}$ and the normal (perpendicular) to the plane of the loop.
SI unit: weber (Wb)

◀ Magnetic flux

MICHAEL FARADAY, British physicist and chemist (1791–1867)

Faraday is often regarded as the greatest experimental scientist of the 1800s. His many contributions to the study of electricity include the invention of the electric motor, electric generator, and transformer, as well as the discovery of electromagnetic induction and the laws of electrolysis. Greatly influenced by religion, he refused to work on military poison gas for the British government.

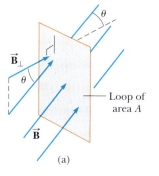

θ

$\vec{\mathbf{B}}_\perp$

θ

Loop of area A

$\vec{\mathbf{B}}$

(a)

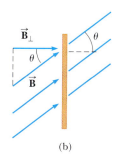

$\vec{\mathbf{B}}_\perp$

θ

θ

$\vec{\mathbf{B}}$

(b)

Figure 20.2   (a) A uniform magnetic field $\vec{\mathbf{B}}$ making an angle θ with a direction normal to the plane of a wire loop of area A. (b) An edge view of the loop.

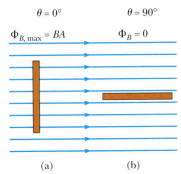

$\theta = 0°$ $\theta = 90°$

$\Phi_{B,\,max} = BA$ $\Phi_B = 0$

(a)   (b)

Figure 20.3 An edge view of a loop in a uniform magnetic field. (a) When the field lines are perpendicular to the plane of the loop, the magnetic flux through the loop is a maximum and equal to $\Phi_B = BA$. (b) When the field lines are parallel to the plane of the loop, the magnetic flux through the loop is zero.

From Equation 20.1, it follows that $B_\perp = B \cos \theta$. The magnetic flux, in other words, is the magnitude of the part of $\vec{B}$ that is perpendicular to the plane of the loop times the area of the loop. Figure 20.2b is an edge view of the loop and the penetrating magnetic field lines. When the field is perpendicular to the plane of the loop as in Figure 20.3a, $\theta = 0$ and Φ_B has a maximum value, $\Phi_{B,\,max} = BA$. When the plane of the loop is parallel to $\vec{B}$ as in Figure 20.3b, $\theta = 90°$ and $\Phi_B = 0$. The flux can also be negative. For example, when $\theta = 180°$, the flux is equal to $-BA$. Because the SI unit of B is the tesla, or weber per square meter, the unit of flux is $T \cdot m^2$, or weber (Wb).

We can emphasize the qualitative meaning of Equation 20.1 by first drawing magnetic field lines, as in Figure 20.3. The number of lines per unit area increases as the field strength increases. **The value of the magnetic flux is proportional to the total number of lines passing through the loop.** We see that the most lines pass through the loop when its plane is perpendicular to the field, as in Figure 20.3a, so the flux has its maximum value at that time. As Figure 20.3b shows, no lines pass through the loop when its plane is parallel to the field, so in that case $\Phi_B = 0$.

Applying Physics 20.1   Flux Compared

Argentina has more land area (2.8×10^6 km^2) than Greenland (2.2×10^6 km^2). Why is the magnetic flux of the Earth's magnetic field larger through Greenland than through Argentina?

Explanation Greenland (latitude 60° north to 80° north) is closer to a magnetic pole than Argentina (latitude 20°

south to 50° south), so the magnetic field is stronger there. That in itself isn't sufficient to conclude that the magnetic flux is greater, but Greenland's proximity to a pole also means the angle magnetic field lines make with the vertical is smaller than in Argentina. As a result, more field lines penetrate the surface in Greenland, despite Argentina's slightly larger area.

EXAMPLE 20.1   Magnetic Flux

Goal   Calculate magnetic flux and a change in flux.

Problem   A conducting circular loop of radius 0.250 m is placed in the xy-plane in a uniform magnetic field of 0.360 T that points in the positive z-direction, the same direction as the normal to the plane. **(a)** Calculate the magnetic flux through the loop. **(b)** Suppose the loop is rotated clockwise around the x-axis, so the normal direction now points at a 45.0° angle with respect to the z-axis. Recalculate the magnetic flux through the loop. **(c)** What is the change in flux due to the rotation of the loop?

Strategy   After finding the area, substitute values into the equation for magnetic flux for each part.

Solution

(a) Calculate the initial magnetic flux through the loop.

First, calculate the area of the loop:

$$A = \pi r^2 = \pi (0.250 \text{ m})^2 = 0.196 \text{ m}^2$$

Substitute A, B, and $\theta = 0°$ into Equation 20.1 to find the initial magnetic flux:

$$\Phi_B = AB \cos \theta = (0.196 \text{ m}^2)(0.360 \text{ T}) \cos (0°)$$
$$= 0.070\,6 \text{ T} \cdot m^2 = \boxed{0.070\,6 \text{ Wb}}$$

(b) Calculate the magnetic flux through the loop after it has rotated 45.0° around the x-axis.

Make the same substitutions as in part (a), except the angle between $\vec{B}$ and the normal is now $\theta = 45.0°$:

$$\Phi_B = AB \cos \theta = (0.196 \text{ m}^2)(0.360 \text{ T}) \cos (45.0°)$$
$$= 0.049\,9 \text{ T} \cdot m^2 = \boxed{0.049\,9 \text{ Wb}}$$

(c) Find the change in the magnetic flux due to the rotation of the loop.

Subtract the result of part (a) from the result of part (b): $\Delta\Phi_B = 0.049\ 9\ \text{Wb} - 0.070\ 6\ \text{Wb} = \boxed{-0.020\ 7\ \text{Wb}}$

Remarks   Notice that the rotation of the loop, not any change in the magnetic field, is responsible for the change in flux. This changing magnetic flux is essential in the functioning of electric motors and generators.

Exercise 20.1

The loop, having rotated by 45°, rotates clockwise another 30°, so the normal to the plane points at an angle of 75° with respect to the direction of the magnetic field. Find (a) the magnetic flux through the loop when $\theta = 75°$ and (b) the change in magnetic flux during the rotation from 45° to 75°.

Answers   (a) 0.018 3 Wb   (b) $-$ 0.031 6 Wb

20.2 FARADAY'S LAW OF INDUCTION

The usefulness of the concept of magnetic flux can be made obvious by another simple experiment that demonstrates the basic idea of electromagnetic induction. Consider a wire loop connected to an ammeter as in Active Figure 20.4. If a magnet is moved toward the loop, the ammeter reads a current in one direction, as in Active Figure 20.4a. When the magnet is held stationary, as in Active Figure 20.4b, the ammeter reads zero current. If the magnet is moved away from the loop, the ammeter reads a current in the opposite direction, as in Active Figure 20.4c. If the magnet is held stationary and the loop is moved either toward or away from the magnet, the ammeter also reads a current. From these observations, it can be concluded that **a current is set up in the circuit as long as there is relative motion between the magnet and the loop**. The same experimental results are found whether the loop moves or the magnet moves. We call such a current an **induced current**, because it is produced by an **induced emf**.

This experiment is similar to the Faraday experiment discussed in Section 20.1. In each case, an emf is induced in a circuit when the magnetic flux through the circuit changes with time. It turns out that the instantaneous emf induced in a circuit equals the negative of the rate of change of magnetic flux with respect to time through the circuit. This is **Faraday's law of magnetic induction**.

TIP 20.1 Induced Current Requires a Change in Magnetic Flux

The existence of magnetic flux through an area is not sufficient to create an induced emf. A *change* in the magnetic flux over some time interval Δt must occur for an emf to be induced.

If a circuit contains N tightly wound loops and the magnetic flux through each loop changes by the amount $\Delta\Phi_B$ during the interval Δt, the average emf induced in the circuit during time Δt is

$$\varepsilon = -N\frac{\Delta\Phi_B}{\Delta t} \qquad [20.2]$$

◀ Faraday's law

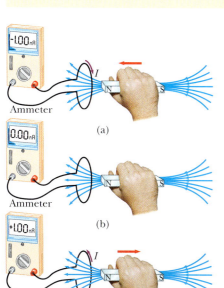

ACTIVE FIGURE 20.4
(a) When a magnet is moved toward a wire loop connected to an ammeter, the ammeter reads a current as shown, indicating that a current I is induced in the loop. (b) When the magnet is held stationary, no current is induced in the loop, even when the magnet is inside the loop. (c) When the magnet is moved away from the loop, the ammeter reads a current in the opposite direction, indicating an induced current going opposite the direction of the current in part (a).

Physics⊗Now™
Log into PhysicsNow at **www.cp7e.com** and go to Active Figure 20.4, where you can move the magnet and observe the current in the ammeter.

Because $\Phi_B = BA \cos \theta$, a change of any of the factors B, A, or θ with time produces an emf. We explore the effect of a change in each of these factors in the following sections. The minus sign in Equation 20.2 is included to indicate the polarity of the induced emf. This polarity simply determines which of two different directions current will flow in a loop, a direction given by **Lenz's law**:

Lenz's law ▶

> The current caused by the induced emf travels in the direction that creates a magnetic field with flux opposing the change in the original flux through the circuit.

Lenz's law says that if the magnetic flux through a loop is becoming more positive, say, then the induced emf creates a current and associated magnetic field that produces negative magnetic flux. Some mistakenly think this "counter magnetic field" created by the induced current, called $\vec{B}_{ind}$ ("ind" for induced) will always point in a direction opposite the applied magnetic field $\vec{B}$, but this is only true half the time! Figure 20.5 shows a field penetrating a loop. The graph in Figure 20.5b shows that the magnitude of the magnetic field $\vec{B}$ shrinks with time. This means the flux of $\vec{B}$ is shrinking with time, so the induced field $\vec{B}_{ind}$ will actually be in the same direction as $\vec{B}$. In effect, $\vec{B}_{ind}$ "shores up" the field $\vec{B}$, slowing the loss of flux through the loop.

The direction of the current in Figure 20.5 can be determined by right-hand rule number 2: Point your right thumb in the direction that will cause the fingers on your right hand to curl in the direction of the induced field $\vec{B}_{ind}$. In this case, that direction is counterclockwise: with the right thumb pointed in the direction of the current, your fingers curl down outside the loop and around and **up through the inside of the loop**. Remember, inside the loop is where it's important for the induced magnetic field to be pointing up.

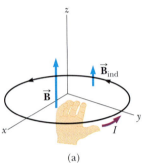

(a)

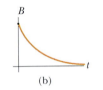

(b)

Figure 20.5 (a) The magnetic field $\vec{B}$ becomes smaller with time, reducing the flux, so current is induced in a direction that creates an induced magnetic field $\vec{B}_{ind}$ opposing the change in magnetic flux. (b) Graph of the magnitude of the magnetic field as a function of time.

Quick Quiz 20.1

Figure 20.6 is a graph of the magnitude B versus time for a magnetic field that passes through a fixed loop and is oriented perpendicular to the plane of the loop. Rank the magnitudes of the emf generated in the loop at the three instants indicated, from largest to smallest.

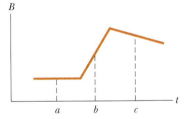

Figure 20.6 (Quick Quiz 20.1)

EXAMPLE 20.2   Faraday and Lenz to the Rescue

Goal Calculate an induced emf and current with Faraday's law, and apply Lenz's law, when the magnetic field changes with time.

Problem A coil with 25 turns of wire is wrapped on a frame with a square cross-section 1.80 cm on a side. Each turn has the same area, equal to that of the frame, and the total resistance of the coil is 0.350 Ω. An applied uniform magnetic field is perpendicular to the plane of the coil, as in Figure 20.7. **(a)** If the field changes uniformly from 0.00 T to 0.500 T in 0.800 s, find the induced emf in the coil while the field is changing. Find **(b)** the magnitude and **(c)** the direction of the induced current in the coil while the field is changing.

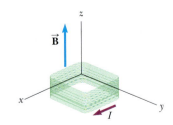

Figure 20.7 (Example 20.2)

Strategy Part (a) requires substituting into Faraday's law, Equation 20.2. The necessary information is given, except for $\Delta\Phi_B$, the change in the magnetic flux during the elapsed time. Compute the initial and final magnetic fluxes with Equation 20.1, find the difference, and assemble all terms in Faraday's law. The current can then be found with Ohm's law, and its direction with Lenz's law.

Solution

(a) Find the induced emf in the coil.

To compute the flux, the area of the coil is needed:

$$A = L^2 = (0.018\ 0\ \text{m})^2 = 3.24 \times 10^{-4}\ \text{m}^2$$

The magnetic flux $\Phi_{B,i}$ through the coil at $t = 0$ is zero because $B = 0$. Calculate the flux at $t = 0.800$ s:

$$\Phi_{B,f} = BA \cos \theta = (0.500\ \text{T})(3.24 \times 10^{-4}\ \text{m}^2) \cos (0°)$$
$$= 1.62 \times 10^{-4}\ \text{Wb}$$

Compute the change in the magnetic flux through the cross-section of the coil over the 0.800-s interval:

$$\Delta\Phi_B = \Phi_{B,f} - \Phi_{B,i} = 1.62 \times 10^{-4}\ \text{Wb}$$

Substitute into Faraday's law of induction to find the induced emf in the coil:

$$\mathcal{E} = -N\frac{\Delta\Phi_B}{\Delta t} = (25\ \text{turns})\left(\frac{1.62 \times 10^{-4}\ \text{Wb}}{0.800\ \text{s}}\right)$$
$$= -5.06 \times 10^{-3}\ \text{V}$$

(b) Find the magnitude of the induced current in the coil.

Substitute the voltage difference and the resistance into Ohm's law:

$$I = \frac{\Delta V}{R} = \frac{5.06 \times 10^{-3}\ \text{V}}{0.350\ \Omega} = 1.45 \times 10^{-2}\ \text{A}$$

(c) Find the direction of the induced current in the coil.

The magnetic field is increasing up through the loop, in the same direction as the normal to the plane; hence the flux is positive and increasing, also. A downward-pointing induced magnetic field will create negative flux, opposing the change. If you point your right thumb in the clockwise direction along the loop, your fingers curl down through the loop—the correct direction for the counter magnetic field.

Remark   Lenz's law can best be handled by sketching a diagram, first.

Exercise 20.2

Suppose the magnetic field changes uniformly from 0.500 T to 0.200 T in the next 0.600 s. Compute (a) the induced emf in the coil and (b) the magnitude and direction of the induced current.

Answers   (a) 4.05×10^{-3} V   (b) 1.16×10^{-2} A, counterclockwise

---

The ground fault interrupter (GFI) is an interesting safety device that protects people against electric shock when they touch appliances and power tools. Its operation makes use of Faraday's law. Figure 20.8 shows the essential parts of a ground fault interrupter. Wire 1 leads from the wall outlet to the appliance to be protected, and wire 2 leads from the appliance back to the wall outlet. An iron ring surrounds the two wires to confine the magnetic field set up by each wire. A sensing coil, which can activate a circuit breaker when changes in magnetic flux occur, is wrapped around part of the iron ring. Because the currents in the wires are in opposite directions, the net magnetic field through the sensing coil due to the currents is zero. However, if a short circuit occurs in the appliance so that there is no returning current, the net magnetic field through the sensing coil is no longer zero. This can happen if, for example, one of the wires loses its insulation, providing a path through you to ground if you happen to be touching the appliance and are grounded as in Figure 18.23a. Because the current is alternating, the magnetic flux through the sensing coil changes with time, producing an induced voltage in the coil. This induced voltage is used to trigger a circuit breaker, stopping the

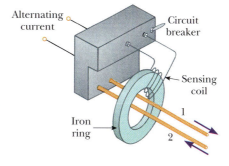

Figure 20.8   Essential components of a ground fault interrupter (contents of the gray box in Fig. 20.9a). In newer homes, such devices are built directly into wall outlets. The purpose of the sensing coil and circuit breaker is to cut off the current before damage is done.

APPLICATION

Ground Fault Interrupters

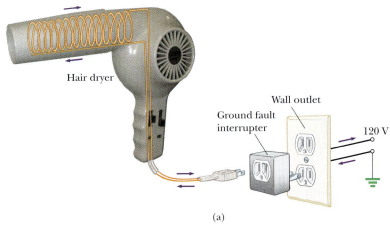

b. George Semple

(a)

Figure 20.9 (a) This hair dryer has been plugged into a ground fault interrupter that is in turn plugged into an unprotected wall outlet. (b) You likely have seen this kind of ground fault interrupter in a hotel bathroom, where hair dryers and electric shavers are often used by people just out of the shower or who might touch a water pipe, providing a ready path to ground in the event of a short circuit.

current quickly in about a millisecond before it reaches a level that might be harmful to the person using the appliance. A ground fault interrupter provides faster and more complete protection than even the case-ground-and-circuit-breaker combination shown in Figure 18.23b. For this reason, ground fault interrupters are commonly found in bathrooms, where electricity poses a hazard to people. (See Fig. 20.9.)

Another interesting application of Faraday's law is the production of sound in an electric guitar. A vibrating string induces an emf in a coil (Fig. 20.10). The pickup coil is placed near the vibrating guitar string, which is made of a metal that can be magnetized. The permanent magnet inside the coil magnetizes the portion of the string nearest the coil. When the guitar string vibrates at some frequency, its magnetized segment produces a changing magnetic flux through the pickup coil. The changing flux induces a voltage in the coil; the voltage is fed to an amplifier. The output of the amplifier is sent to the loudspeakers, producing the sound waves that we hear.

Sudden infant death syndrome (SIDS) is a devastating affliction in which a baby suddenly stops breathing during sleep without an apparent cause. One type of monitoring device, called an apnea monitor, is sometimes used to alert caregivers of the cessation of breathing. The device uses induced currents, as shown in Figure 20.11. A coil of wire attached to one side of the chest carries an alternating current. The varying magnetic flux produced by this current passes through a pickup coil attached to the opposite side of the chest. Expansion and contraction of the chest caused by breathing or movement changes the strength of the voltage induced in the pickup coil. However, if breathing stops, the pattern of the induced voltage stabilizes, and external circuits monitoring the voltage sound an alarm to the caregivers after a momentary pause to ensure that a problem actually does exist.

APPLICATION

Electric Guitar Pickups

APPLICATION

Apnea Monitors

Figure 20.10 (a) In an electric guitar, a vibrating string induces a voltage in the pickup coil. (b) Several pickups allow the vibration to be detected from different portions of the string.

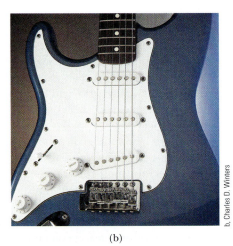

b, Charles D. Winters

(a)   (b)

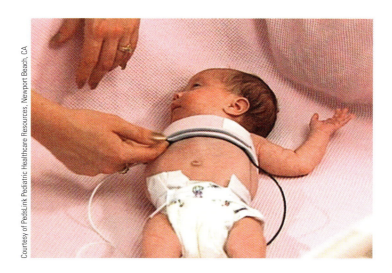

Courtesy of PedsLink Pediatric Healthcare Resources, Newport Beach, CA

Figure 20.11 This infant is wearing a monitor designed to alert caregivers if breathing stops. Note the two wires attached to opposite sides of the chest.

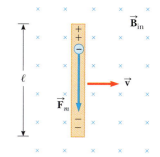

Figure 20.12 A straight conductor of length ℓ moving with velocity $\vec{v}$ through a uniform magnetic field $\vec{B}$ directed perpendicular to $\vec{v}$. The vector $\vec{F}_m$ is the magnetic force on an electron in the conductor. An emf of $B\ell v$ is induced between the ends of the bar.

20.3 MOTIONAL emf

In Section 20.2, we considered emfs induced in a circuit when the magnetic field changes with time. In this section we describe a particular application of Faraday's law in which a so-called **motional emf** is produced. This is the emf induced in a conductor moving through a magnetic field.

First consider a straight conductor of length ℓ moving with constant velocity through a uniform magnetic field directed into the paper, as in Figure 20.12. For simplicity, we assume that the conductor moves in a direction perpendicular to the field. A magnetic force of magnitude $F_m = qvB$, directed downward, acts on the electrons in the conductor. Because of this magnetic force, the free electrons move to the lower end of the conductor and accumulate there, leaving a net positive charge at the upper end. As a result of this charge separation, an electric field is produced in the conductor. The charge at the ends builds up until the downward magnetic force qvB is balanced by the upward electric force qE. At this point, charge stops flowing and the condition for equilibrium requires that

$$qE = qvB \quad \text{or} \quad E = vB$$

Because the electric field is uniform, the field produced in the conductor is related to the potential difference across the ends by $\Delta V = E\ell$, giving

$$\Delta V = E\ell = B\ell v \qquad \textbf{[20.3]}$$

Because there is an excess of positive charge at the upper end of the conductor and an excess of negative charge at the lower end, the upper end is at a higher potential than the lower end. There is a potential difference across a conductor as long as it moves through a field. If the motion is reversed, the polarity of the potential difference is also reversed.

A more interesting situation occurs if the moving conductor is part of a closed conducting path. This situation is particularly useful for illustrating how a changing loop area induces a current in a closed circuit described by Faraday's law. Consider a circuit consisting of a conducting bar of length ℓ, sliding along two fixed parallel conducting rails, as in Active Figure 20.13a. For simplicity, assume that the moving bar has zero resistance and that the stationary part of the circuit has constant resistance R. A uniform and constant magnetic field $\vec{B}$ is applied perpendicular to the plane of the circuit. As the bar is pulled to the right with velocity $\vec{v}$ under the influence of an applied force $\vec{F}_{app}$, a magnetic force along the length of the bar acts on the free charges in the bar. This force in turn sets up an induced current because the charges are free to move in a closed conducting path. In this case, the changing magnetic flux through the loop and the corresponding induced emf across the moving bar arise from the *change in area of the loop* as the bar moves through the magnetic field.

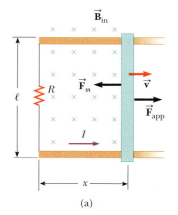

(a)

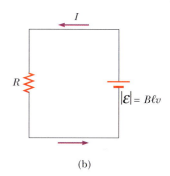

(b)

ACTIVE FIGURE 20.13
(a) A conducting bar sliding with velocity $\vec{v}$ along two conducting rails under the action of an applied force $\vec{F}_{app}$. The magnetic force $\vec{F}_m$ opposes the motion, and a counterclockwise current is induced in the loop. (b) The equivalent circuit of that in (a).

Physics⊗Now™
Log into PhysicsNow at **www.cp7e.com** and go to Active Figure 20.13, where you can adjust the applied force, the magnetic field, and the resistance, and observe the effects on the motion of the bar.

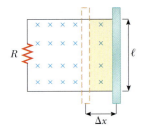

Figure 20.14 As the bar moves to the right, the area of the loop increases by the amount $\ell\Delta x$ and the magnetic flux through the loop increases by $B\ell\Delta x$.

Assume that the bar moves a distance Δx in time Δt, as shown in Figure 20.14. The increase in flux $\Delta\Phi_B$ through the loop in that time is the amount of flux that now passes through the portion of the circuit that has area $\ell\Delta x$:

$$\Delta\Phi_B = BA = B\ell\,\Delta x$$

Using Faraday's law and noting that there is one loop ($N = 1$), we find that the magnitude of the induced emf is

$$|\mathcal{E}| = \frac{\Delta\Phi_B}{\Delta t} = B\ell\frac{\Delta x}{\Delta t} = B\ell v \qquad [20.4]$$

This induced emf is often called a **motional emf** because it arises from the motion of a conductor through a magnetic field.

Further, if the resistance of the circuit is R, the magnitude of the induced current in the circuit is

$$I = \frac{|\mathcal{E}|}{R} = \frac{B\ell v}{R} \qquad [20.5]$$

Active Figure 20.13b shows the equivalent circuit diagram for this example.

Applying Physics 20.2   Space Catapult

Applying a force on the bar will result in an induced emf in the circuit shown in Active Figure 20.13. Suppose we remove the external magnetic field in the diagram and replace the resistor with a high-voltage source and a switch, as in Figure 20.15. What will happen when the switch is closed? Will the bar move, and does it matter which way we connect the high-voltage source?

Explanation Suppose the source is capable of establishing high current. Then the two horizontal conducting rods will create a strong magnetic field in the area between them, directed into the page. (The

movable bar also creates a magnetic field, but this field can't exert force on the bar itself.) Because the moving bar carries a downward current, a magnetic force is exerted on the bar, directed to the right. Hence, the bar accelerates along the rails away from the power supply. If the polarity of the power were reversed, the magnetic field would be out of the page, the current in the bar would be upward, and the force on the bar would still be directed to the right. The $BI\ell$ force exerted by a magnetic field according to Equation 19.6 causes the bar to accelerate away from the voltage source. Studies have shown that it's possible to launch payloads into space with this technology. (This is the working principle of a rail gun.) Very large accelerations can be obtained with currently available technology, with payloads being accelerated to a speed of several kilometers per second in a fraction of a second. This is a larger acceleration than humans can tolerate.

Rail guns have been proposed as propulsion systems for moving asteroids into more useful orbits. The material of the asteroid could be mined and launched off the surface by a rail gun, which would act like a rocket engine, modifying the velocity and hence the orbit of the asteroid. Some asteroids contain trillions of dollars worth of valuable metals.

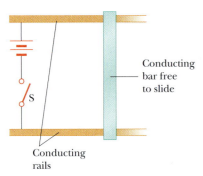

Conducting bar free to slide

Conducting rails

Figure 20.15 (Applying Physics 20.2)

Quick Quiz 20.2

A horizontal metal bar oriented east–west drops straight down in a location where the Earth's magnetic field is due north. As a result, an emf develops between the ends. Which end is positively charged? (a) the east end (b) the west end (c) neither end carries a charge

You intend to move a rectangular loop of wire into a region of uniform magnetic field at a given speed so as to induce an emf in the loop. The plane of the loop must remain perpendicular to the magnetic field lines. In which orientation should you hold the loop while you move it into the region with the magnetic field in order to generate the largest emf? (a) With the long dimension of the loop parallel to the velocity vector, (b) with the short dimension of the loop parallel to the velocity vector, or (c) either way—the emf is the same regardless of orientation.

EXAMPLE 20.3   The Electrified Airplane Wing

Goal   Find the emf induced by motion through a magnetic field.

Problem   An airplane with a wingspan of 30.0 m flies due north at a location where the downward component of the Earth's magnetic field is 0.600×10^{-4} T. There is also a component pointing due north which has a magnitude of 0.470×10^{-4} T. **(a)** Find the difference in potential between the wingtips when the speed of the plane is 2.50×10^2 m/s. **(b)** Which wingtip is positive?

Strategy   Because the plane is flying north, the northern component of the magnetic field won't have any effect on the induced emf. The induced emf across the wing is caused solely by the downward component of the Earth's magnetic field. Substitute the given quantities into Equation 20.4. Use right-hand rule number 1 to find the direction positive charges would be propelled by the magnetic force.

Solution
(a) Calculate the difference in potential across the wingtips.

Write the motional emf equation and substitute the given quantities:

$$\mathcal{E} = B\ell v = (0.600 \times 10^{-4}\,\text{T})(30.0\,\text{m})(2.50 \times 10^2\,\text{m/s})$$
$$= \boxed{0.450\,\text{V}}$$

(b) Which wingtip is positive?

Apply right hand rule number 1:

Point your right fingers north, in the direction of the velocity, curl them down, in the direction of the magnetic field. Your thumb points   west.

Remark   An induced emf such as this can cause problems on an aircraft.

Exercise 20.3
Suppose the magnetic field in a given region of space is parallel to the Earth's surface, points north, and has magnitude 1.80×10^{-4} T. A metal cable attached to a space station stretches radially outwards 2.50 km. (a) Estimate the potential difference that develops between the ends of the cable if it's traveling eastward around Earth at 7.70×10^3 m/s. (b) Which end of the cable is positive, the lower end or the upper end?

Answer   (a) 3.47×10^3 V   (b) The upper end is positive.

EXAMPLE 20.4   Where Is the Energy Source?

Goal   Use motional emf to find an induced emf and a current.

Problem   **(a)** The sliding bar in Figure 20.13a has a length of 0.500 m and moves at 2.00 m/s in a magnetic field of magnitude 0.250 T. Using the concept of motional emf, find the induced voltage in the moving rod. **(b)** If the resistance in the circuit is 0.500 Ω, find the current in the circuit and the power delivered to the resistor. (*Note*: The current, in this case, goes counterclockwise around the loop.) **(c)** Calculate the magnetic force on the bar. **(d)** Use the concepts of work and power to calculate the applied force.

Strategy For part (a), substitute into Equation 20.4 for the motional emf. Once the emf is found, substitution into Ohm's law gives the current. In part (c), use Equation 19.6 for the magnetic force on a current-carrying conductor. In part (d), use the fact that the power dissipated by the resistor multiplied by the elapsed time must equal the work done by the applied force.

Solution

(a) Find the induced emf with the concept of motional emf.

Substitute into Equation 20.4 to find the induced emf:

$$\mathcal{E} = B\ell v = (0.250 \text{ T})(0.500 \text{ m})(2.00 \text{ m/s}) = \boxed{0.250 \text{ V}}$$

(b) Find the induced current in the circuit and the power dissipated by the resistor.

Substitute the emf and the resistance into Ohm's law to find the induced current:

$$I = \frac{\mathcal{E}}{R} = \frac{0.250 \text{ V}}{0.500 \, \Omega} = \boxed{0.500 \text{ A}}$$

Substitute I and $\mathcal{E} = 0.250$ V into Equation 17.8 to find the power dissipated by the 0.500-Ω resistor:

$$\mathcal{P} = I\Delta V = (0.500 \text{ A})(0.250 \text{ V}) = \boxed{0.125 \text{ W}}$$

(c) Calculate the magnitude and direction of the magnetic force on the bar.

Substitute values for I, B, and ℓ into Equation 19.6 (with $\sin \theta = \sin (90°) = 1$) to find the magnitude of the force:

$$F_m = IB\ell = (0.500 \text{ A})(0.250 \text{ T})(0.500 \text{ m}) = \boxed{6.25 \times 10^{-2} \text{ N}}$$

Apply right hand rule number 2 to find the direction of the force:

Point the fingers of your right hand in the direction of the positive current, then curl them in the direction of the magnetic field. Your thumb points in the negative x-direction.

(d) Find the value of F_{app}, the applied force.

Set the work done by the applied force equal to the dissipated power times the elapsed time:

$$W_{app} = F_{app}d = \mathcal{P}\Delta t$$

Solve for F_{app} and substitute $d = v\Delta t$:

$$F_{app} = \frac{\mathcal{P}\Delta t}{d} = \frac{\mathcal{P}\Delta t}{v\Delta t} = \frac{\mathcal{P}}{v} = \frac{0.125 \text{ W}}{2.00 \text{ m/s}} = \boxed{6.25 \times 10^{-2} \text{ N}}$$

Remarks Part (d) could be solved by using Newton's second law for an object in equilibrium: two forces act horizontally on the bar, and the acceleration of the bar is zero, so the forces must be equal in magnitude and opposite in direction. Notice the agreement between the answers for F_m and F_{app}, despite the very different concepts used.

Exercise 20.4

Suppose the current suddenly increases to 1.25 A in the same direction as before, due to an increase in speed of the bar. Find (a) the emf induced in the rod, (b) the new speed of the rod.

Answers (a) 0.625 V (b) 5.00 m/s

20.4 LENZ'S LAW REVISITED (The Minus Sign in Faraday's Law)

To reach a better understanding of Lenz's law, consider the example of a bar moving to the right on two parallel rails in the presence of a uniform magnetic field directed into the paper (Fig. 20.16a). As the bar moves to the right, the magnetic flux through the circuit increases with time because the area of the loop increases. Lenz's law says that the induced current must be in a direction such that the flux *it* produces opposes the change in the external magnetic flux. Because the flux due to the external field is increasing *into* the paper, the induced current, to oppose the change, must produce a flux *out* of the paper. Hence, the induced current must be counterclockwise when the bar moves to the right. (Use right-hand rule number 2 from Chapter 19 to verify this direction.) On the other hand, if the bar is moving to the left, as in Figure 20.16b, the magnetic flux through the loop decreases with time. Because the flux is into the paper, the induced current has to be clockwise to produce its own flux into the paper (which opposes the decrease in the external flux). In either case, the induced current tends to maintain the original flux through the circuit.

Now we examine this situation from the viewpoint of energy conservation. Suppose that the bar is given a slight push to the right. In the preceding analysis, we found that this motion led to a counterclockwise current in the loop. Let's see what would happen if we assume that the current is clockwise, opposite the direction required by Lenz's law. For a clockwise current I, the direction of the magnetic force $BI\ell$ on the sliding bar is to the right. This force accelerates the rod and increases its velocity. This, in turn, causes the area of the loop to increase more rapidly, thereby increasing the induced current, which increases the force, which increases the current, and so forth. In effect, the system acquires energy with zero input energy. This is inconsistent with all experience and with the law of conservation of energy, so we're forced to conclude that the current must be counterclockwise.

Consider another situation. A bar magnet is moved to the right toward a stationary loop of wire, as in Figure 20.17a. As the magnet moves, the magnetic flux through the loop increases with time. To counteract this rise in flux, the induced current produces a flux to the left, as in Figure 20.17b; hence, the induced current is in the direction shown. Note that the magnetic field lines associated with the induced current oppose the motion of the magnet. The left face of the current loop is therefore a north pole and the right face is a south pole.

On the other hand, if the magnet were moving to the left, as in Figure 20.17c, its flux through the loop, which is toward the right, would decrease with time. Under these circumstances, the induced current in the loop would be in a direction to set up a field directed from left to right through the loop, in an effort to maintain a constant number of flux lines. Hence, the induced current in the loop would be as shown in Figure 20.17d. In this case, the left face of the loop would be a south pole and the right face would be a north pole.

As another example, consider a coil of wire placed near an electromagnet, as in Figure 20.18a (page 672). We wish to find the direction of the induced current in the coil at various times: at the instant the switch is closed, after the switch has been closed for several seconds, and when the switch is opened.

When the switch is closed, the situation changes from a condition in which no lines of flux pass through the coil to one in which lines of flux pass through in the

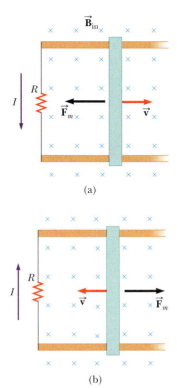

Figure 20.16   (a) As the conducting bar slides on the two fixed conducting rails, the magnetic flux through the loop increases with time. By Lenz's law, the induced current must be *counterclockwise* so as to produce a counteracting flux *out of the paper*. (b) When the bar moves to the left, the induced current must be *clockwise*. Why?

TIP 20.2   There are Two Magnetic Fields to Consider

When applying Lenz's law, there are *two* magnetic fields to consider. The first is the external changing magnetic field that induces the current in a conducting loop. The second is the magnetic field produced by the induced current in the loop.

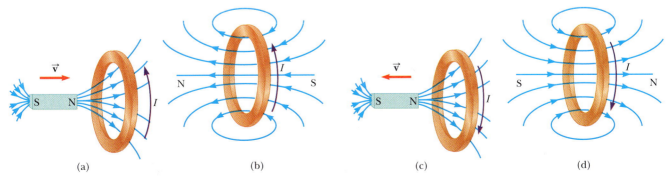

Figure 20.17   (a) When the magnet is moved toward the stationary conducting loop, a current is induced in the direction shown. (b) This induced current produces its own flux to the left to counteract the increasing external flux to the right. (c) When the magnet is moved away from the stationary conducting loop, a current is induced in the direction shown. (d) This induced current produces its own flux to the right to counteract the decreasing external flux to the right.

Figure 20.18 An example of Lenz's law.

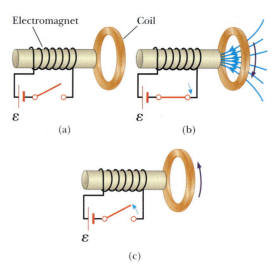

direction shown in Figure 20.18b. To counteract this change in the number of lines, the coil must set up a field from left to right in the figure. This requires a current directed as shown in Figure 20.18b.

After the switch has been closed for several seconds, there is no change in the number of lines through the loop; hence, the induced current is zero.

Opening the switch causes the magnetic field to change from a condition in which flux lines thread through the coil from right to left to a condition of zero flux. The induced current must then be as shown in Figure 20.18c, so as to set up its own field from right to left.

Quick Quiz 20.4

A bar magnet is falling through a loop of wire with constant velocity with the north pole entering first. Viewed from the same side of the loop as the magnet, as the north pole approaches the loop, what is the direction of the induced current? (a) clockwise (b) zero (c) counterclockwise (d) along the length of the magnet

Tape Recorders

One common practical use of induced currents and emfs is associated with the tape recorder. Many different types of tape recorders are made, but the basic principles are the same for all. A magnetic tape moves past a recording and playback head, as in Figure 20.19a. The tape is a plastic ribbon coated with iron oxide or

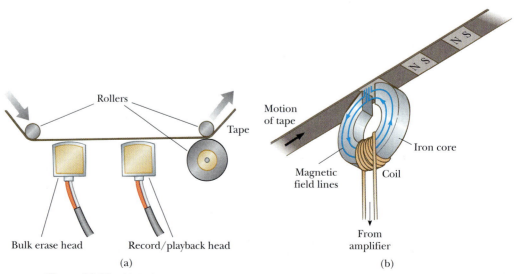

Figure 20.19 (a) Major parts of a magnetic tape recorder. If a new recording is to be made, the bulk erase head wipes the tape clean of signals before recording. (b) The fringing magnetic field magnetizes the tape during recording.

chromium oxide. The hard drives in computers work on the same principle, but use a coated disk instead of tape, allowing for faster access.

The recording process uses the fact that a current in an electromagnet produces a magnetic field. Figure 20.19b illustrates the steps in the process. A sound wave sent into a microphone is transformed into an electric current, amplified, and allowed to pass through a wire coiled around a doughnut-shaped piece of iron, which functions as the recording head. The iron ring and the wire constitute an electromagnet, in which the lines of the magnetic field are contained completely inside the iron except at the point where a slot is cut in the ring. Here the magnetic field fringes out of the iron and magnetizes the small pieces of iron oxide embedded in the tape. As the tape moves past the slot, it becomes magnetized in a pattern that reproduces both the frequency and the intensity of the sound signal entering the microphone.

To reconstruct the sound signal, the previously magnetized tape is allowed to pass through a recorder head operating in the playback mode. A second wire-wound doughnut-shaped piece of iron with a slot in it passes close to the tape, so that the varying magnetic fields on the tape produce changing field lines through the wire coil. The changing flux induces a current in the coil which corresponds to the current in the recording head that originally produced the tape. This changing electric current can be amplified and used to drive a speaker. Playback is thus an example of induction of a current by a moving magnet.

20.5   GENERATORS

Generators and motors are important practical devices that operate on the principle of electromagnetic induction. First, consider the **alternating-current** (AC) **generator**, a device that converts mechanical energy to electrical energy. In its simplest form, the AC generator consists of a wire loop rotated in a magnetic field by some external means (Active Fig. 20.20a). In commercial power plants, the energy required to rotate the loop can be derived from a variety of sources. For example, in a hydroelectric plant, falling water directed against the blades of a turbine produces the rotary motion; in a coal-fired plant, heat produced by burning coal is used to convert water to steam, and this steam is directed against the turbine blades. As the loop rotates, the magnetic flux through it changes with time, inducing an emf and a current in an external circuit. The ends of the loop are connected to slip rings that rotate with the loop.

APPLICATION

Alternating-Current Generators

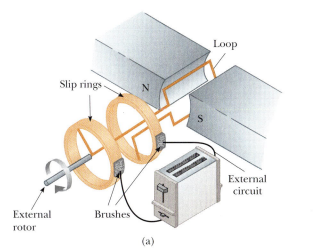

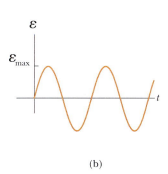

(a)   (b)

ACTIVE FIGURE 20.20
(a) A schematic diagram of an AC generator. An emf is induced in a coil, which rotates by some external means in a magnetic field. (b) A plot of the alternating emf induced in the loop versus time.

Physics⊗Now™

Log into PhysicsNow at **www.cp7e.com** and go to Active Figure 20.20, where you can adjust the speed of rotation and the strength of the field to see the effects on the generated emf.

Figure 20.21 (a) A loop rotating at a constant angular velocity in an external magnetic field. The emf induced in the loop varies sinusoidally with time. (b) An edge view of the rotating loop.

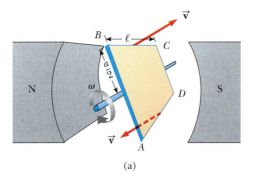

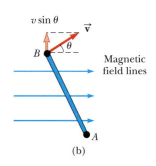

(a)

(b)

Connections to the external circuit are made by stationary brushes in contact with the slip rings.

We can derive an expression for the emf generated in the rotating loop by making use of the equation for motional emf, $\mathcal{E} = B\ell v$. Figure 20.21a shows a loop of wire rotating clockwise in a uniform magnetic field directed to the right. The magnetic force (qvB) on the charges in wires AB and CD is not along the lengths of the wires. (The force on the electrons in these wires is perpendicular to the wires.) Hence, an emf is generated only in wires BC and AD. At any instant, wire BC has velocity $\vec{v}$ at an angle θ with the magnetic field, as shown in Figure 20.21b. (Note that the component of velocity parallel to the field has no effect on the charges in the wire, whereas the component of velocity perpendicular to the field produces a magnetic force on the charges that moves electrons from C to B.) The emf generated in wire BC equals $B\ell v_{\perp}$, where ℓ is the length of the wire and $v_{\perp}$ is the component of velocity perpendicular to the field. An emf of $B\ell v_{\perp}$ is also generated in wire DA, and the sense of this emf is the same as that in wire BC. Because $v_{\perp} = v \sin \theta$, the total induced emf is

$$\mathcal{E} = 2B\ell v_{\perp} = 2B\ell v \sin \theta \qquad [20.6]$$

If the loop rotates with a constant angular speed ω, we can use the relation $\theta = \omega t$ in Equation 20.6. Furthermore, because every point on the wires BC and DA rotates in a circle about the axis of rotation with the same angular speed ω, we have $v = r\omega = (a/2)\omega$, where a is the length of sides AB and CD. Equation 20.6 therefore reduces to

$$\mathcal{E} = 2B\ell \left(\frac{a}{2}\right) \omega \sin \omega t = B\ell a\omega \sin \omega t$$

If a coil has N turns, the emf is N times as large because each loop has the same emf induced in it. Further, because the area of the loop is $A = \ell a$, the total emf is

$$\mathcal{E} = NBA\omega \sin \omega t \qquad [20.7]$$

This result shows that the emf varies sinusoidally with time, as plotted in Active Figure 20.20b. Note that the maximum emf has the value

$$\mathcal{E}_{\text{max}} = NBA\omega \qquad [20.8]$$

which occurs when $\omega t = 90°$ or $270°$. In other words, $\mathcal{E} = \mathcal{E}_{\text{max}}$ when the plane of the loop is parallel to the magnetic field. Further, the emf is zero when $\omega t = 0$ or $180°$, which happens whenever the magnetic field is perpendicular to the plane of the loop. In the United States and Canada the frequency of rotation for commercial generators is 60 Hz, whereas in some European countries 50 Hz is used. (Recall that $\omega = 2\pi f$, where f is the frequency in hertz.)

The **direct current** (DC) **generator** is illustrated in Active Figure 20.22a. The components are essentially the same as those of the AC generator, except that the contacts to the rotating loop are made by a split ring, or commutator. In this design, the output voltage always has the same polarity and the current is a pulsating direct current, as in Active Figure 20.22b. This can be understood by noting

Turbines turn electric generators at a hydroelectric power plant.

APPLICATION

Direct Current Generators

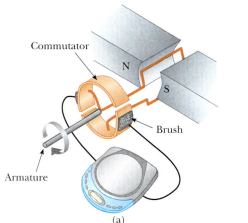

Commutator

N

S

Brush

Armature

(a)

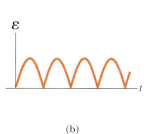

(b)

ACTIVE FIGURE 20.22
(a) A schematic diagram of a DC generator. (b) The emf fluctuates in magnitude, but always has the same polarity.

Physics ⊗ Now™
Log into PhysicsNow at **www.cp7e.com** and go to Active Figure 20.22, where you can adjust the speed of rotation and the strength of the field, observing the effects on the generated emf.

that the contacts to the split ring reverse their roles every half cycle. At the same time, the polarity of the induced emf reverses. Hence, the polarity of the split ring remains the same.

A pulsating DC current is not suitable for most applications. To produce a steady DC current, commercial DC generators use many loops and commutators distributed around the axis of rotation so that the sinusoidal pulses from the loops overlap in phase. When these pulses are superimposed, the DC output is almost free of fluctuations.

EXAMPLE 20.5   Emf Induced in an AC Generator

Goal   Understand physical aspects of an AC generator.

Problem   An AC generator consists of eight turns of wire, each having area $A = 0.090\ 0\ \text{m}^2$, with a total resistance of $12.0\ \Omega$. The loop rotates in a magnetic field of 0.500 T at a constant frequency of 60.0 Hz. **(a)** Find the maximum induced emf. **(b)** What is the maximum induced current? **(c)** Determine the induced emf and current as functions of time. **(d)** What maximum torque must be applied to keep the coil turning?

Strategy   From the given frequency, calculate the angular frequency ω and substitute it, together with given quantities, into Equation 20.8. As functions of time, the emf and current have the form $A \sin \omega t$, where A is the maximum emf or current, respectively. For part (d), calculate the magnetic torque on the coil when the current is at a maximum. (See Chapter 19.) The applied torque must do work against this magnetic torque to keep the coil turning.

Solution
(a) Find the maximum induced emf.

First, calculate the angular frequency of the rotational motion:

$$\omega = 2\pi f = 2\pi(60.0\ \text{Hz}) = 377\ \text{rad/s}.$$

Substitute the values for N, A, B, and ω into Equation 20.8, obtaining the maximum induced emf:

$$\mathcal{E}_{\text{max}} = NAB\omega = 8(0.090\ 0\ \text{m}^2)(0.500\ \text{T})(377\ \text{rad/s})$$
$$= \boxed{136\ \text{V}}$$

(b) What is the maximum induced current?

Substitute the maximum induced emf $\mathcal{E}_{\text{max}}$ and the resistance R into Ohm's law to find the maximum induced current:

$$I_{\text{max}} = \frac{\mathcal{E}_{\text{max}}}{R} = \frac{136\ \text{V}}{12.0\ \Omega} = \boxed{11.3\ \text{A}}$$

(c) Determine the induced emf and the current as functions of time.

Substitute $\mathcal{E}_{\text{max}}$ and ω into Equation 20.7 to obtain the variation of $\mathcal{E}$ with time t in seconds:

$$\mathcal{E} = \mathcal{E}_{\text{max}} \sin \omega t = \boxed{(136\ \text{V}) \sin 377t}$$

The time variation of the current looks just like this, except with the maximum current out in front:

$$I = (11.3 \text{ A}) \sin 377t$$

(d) Calculate the maximum applied torque necessary to keep the coil turning.

Write the equation for magnetic torque:

$$\tau = \mu B \sin \theta$$

Calculate the maximum magnetic moment of the coil, μ:

$$\mu = I_{max}AN = (11.3 \text{ A})(0.090 \text{ m}^2)(8) = 8.14 \text{ A} \cdot \text{m}^2$$

Substitute into the magnetic torque equation, with $\theta = 90°$ to find the maximum applied torque:

$$\tau_{max} = (8.14 \text{ A} \cdot \text{m}^2)(0.500 \text{ T}) \sin 90° = 4.07 \text{ N} \cdot \text{m}$$

Remarks The number of loops, N, can't be arbitrary, because there must be a force strong enough to turn the coil.

Exercise 20.5

An AC generator is to have a maximum output of 301 V. Each coil has an area of 0.100 m² and a resistance of 16.0 Ω and rotates in a magnetic field of 0.600 T with a frequency of 40.0 Hz. (a) How many turns of wire should the coil have to produce the desired emf? (b) Find the maximum current induced in the coil. (c) Determine the induced emf as a function of time.

Answers   (a) 20 turns   (b) 18.8 A   (c) $\mathcal{E} = (301 \text{ V}) \sin 251t$

Motors and Back emf

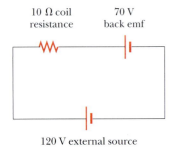

Figure 20.23   A motor can be represented as a resistance plus a back emf.

Motors are devices that convert electrical energy to mechanical energy. Essentially, **a motor is a generator run in reverse**: instead of a current being generated by a rotating loop, a current is supplied to the loop by a source of emf, and the magnetic torque on the current-carrying loop causes it to rotate.

A motor can perform useful mechanical work when a shaft connected to its rotating coil is attached to some external device. As the coil in the motor rotates, however, the changing magnetic flux through it induces an emf which acts to reduce the current in the coil. If it *increased* the current, Lenz's law would be violated. The phrase **back emf** is used for an emf that tends to reduce the applied current. The back emf increases in magnitude as the rotational speed of the coil increases. We can picture this state of affairs as the equivalent circuit in Figure 20.23. For illustrative purposes, assume that the external power source supplying current in the coil of the motor has a voltage of 120 V, that the coil has a resistance of 10 Ω, and that the back emf induced in the coil at this instant is 70 V. The voltage available to supply current equals the difference between the applied voltage and the back emf, 50 V in this case. The current is always reduced by the back emf.

When a motor is turned on, there is no back emf initially, and the current is very large because it's limited only by the resistance of the coil. As the coil begins to rotate, the induced back emf opposes the applied voltage and the current in the coil is reduced. If the mechanical load increases, the motor slows down, which decreases the back emf. This reduction in the back emf increases the current in the coil and therefore also increases the power needed from the external voltage source. As a result, the power requirements for starting a motor and for running it under heavy loads are greater than those for running the motor under average loads. If the motor is allowed to run under no mechanical load, the back emf reduces the current to a value just large enough to balance energy losses by heat and friction.

EXAMPLE 20.6   Induced Current in a Motor

Goal   Apply the concept of a back emf in calculating the induced current in a motor.

Problem   A motor has coils with a resistance of 10.0 Ω and is supplied by a voltage of $\Delta V = 1.20 \times 10^2$ V. When the motor is running at its maximum speed, the back emf is 70.0 V. Find the current in the coils **(a)** when the motor is first turned on and **(b)** when the motor has reached its maximum rotation rate.

Strategy   For each part, find the net voltage, which is the applied voltage minus the induced emf. Divide the net voltage by the resistance to get the current.

Solution

(a) Find the initial current, when the motor is first turned on.

If the coil isn't rotating, the back emf is zero and the current has its maximum value. Calculate the difference between the emf and the initial back emf and divide by the resistance R, obtaining the initial current:

$$I = \frac{\mathcal{E} - \mathcal{E}_{back}}{R} = \frac{1.20 \times 10^2 \, V - 0}{10.0 \, \Omega} = \boxed{12.0 \, A}$$

(b) Find the current when the motor is rotating at its maximum rate.

Repeat the calculation, using the maximum value of the back emf:

$$I = \frac{\mathcal{E} - \mathcal{E}_{back}}{R} = \frac{1.20 \times 10^2 \, V - 70.0 \, V}{10.0 \, \Omega} = \frac{50.0 \, V}{10.0 \, \Omega}$$

$$= \boxed{5.00 \, A}$$

Remark   The phenomenon of back emf is one way in which the rotation rate of electric motors is limited.

Exercise 20.6

If the current in the motor is 8.00 A at some instant, what is the back emf at that time?

Answer   40.0 V

20.6   SELF-INDUCTANCE

Consider a circuit consisting of a switch, a resistor, and a source of emf, as in Figure 20.24. When the switch is closed, the current doesn't immediately change from zero to its maximum value, $\mathcal{E}/R$. The law of electromagnetic induction—Faraday's law—prevents this. What happens instead is the following: as the current increases with time, the magnetic flux through the loop due to this current also increases. The increasing flux induces an emf in the circuit that opposes the change in magnetic flux. By Lenz's law, the induced emf is in the direction indicated by the dashed battery in the figure. The net potential difference across the resistor is the emf of the battery minus the opposing induced emf. As the magnitude of the current increases, the *rate* of increase lessens and hence the induced emf decreases. This opposing emf results in a gradual increase in the current. For the same reason, when the switch is opened, the current doesn't immediately fall to zero. This effect is called **self-induction** because the changing flux through the circuit arises from the circuit itself. The emf that is set up in the circuit is called a **self-induced emf**.

As a second example of self-inductance, consider Figure 20.25 (page 678), which shows a coil wound on a cylindrical iron core. (A practical device would have several hundred turns.) Assume that the current changes with time. When the current is in the direction shown, a magnetic field is set up inside the coil, directed from right to left. As a result, some lines of magnetic flux pass through

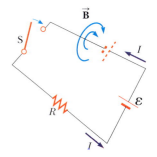

Figure 20.24   After the switch in the circuit is closed, the current produces its own magnetic flux through the loop. As the current increases towards its equilibrium value, the flux changes in time and induces an emf in the loop. The battery drawn with dashed lines is a symbol for the self-induced emf.

Figure 20.25 (a) A current in the coil produces a magnetic field directed to the left. (b) If the current increases, the coil acts as a source of emf directed as shown by the dashed battery. (c) The induced emf in the coil changes its polarity if the current decreases.

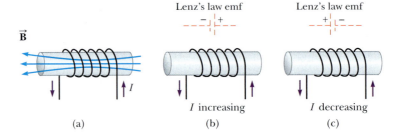

the cross-sectional area of the coil. As the current changes with time, the flux through the coil changes and induces an emf in the coil. Lenz's law shows that this induced emf has a direction so as to oppose the change in the current. If the current is increasing, the induced emf is as pictured in Figure 20.25b, and if the current is decreasing, the induced emf is as shown in Figure 20.25c.

To evaluate self-inductance quantitatively, first note that, according to Faraday's law, the induced emf is given by Equation 20.2:

$$\mathcal{E} = -N\frac{\Delta\Phi_B}{\Delta t}$$

The magnetic flux is proportional to the magnetic field, which is proportional to the current in the coil. Therefore, **the self-induced emf must be proportional to the rate of change of the current with time**, or

$$\mathcal{E} \equiv -L\frac{\Delta I}{\Delta t} \qquad [20.9]$$

JOSEPH HENRY, American physicist (1797–1878)

Henry became the first director of the Smithsonian Institution and first president of the Academy of Natural Science. He was the first to produce an electric current with a magnetic field, but he failed to publish his results as early as Faraday because of his heavy teaching duties at the Albany Academy in New York State. He improved the design of the electromagnet and constructed one of the first motors. He also discovered the phenomenon of self-induction. The unit of inductance, the henry, is named in his honor.

where L is a proportionality constant called the **inductance** of the device. The negative sign indicates that a changing current induces an emf in opposition to the change. This means that if the current is increasing (ΔI positive), the induced emf is negative, indicating opposition to the increase in current. Likewise, if the current is decreasing (ΔI negative), the sign of the induced emf is positive, indicating that the emf is acting to oppose the decrease.

The inductance of a coil depends on the cross-sectional area of the coil and other quantities, all of which can be grouped under the general heading of geometric factors. The SI unit of inductance is the **henry** (H), which, from Equation 20.9, is equal to 1 volt-second per ampere:

$$1\ H = 1\ V\cdot s/A$$

In the process of calculating self-inductance, it is often convenient to equate Equations 20.2 and 20.9 to find an expression for L:

$$N\frac{\Delta\Phi_B}{\Delta t} = L\frac{\Delta I}{\Delta t}$$

Inductance ▶

$$L = N\frac{\Delta\Phi_B}{\Delta I} = \frac{N\Phi_B}{I} \qquad [20.10]$$

Applying Physics 20.3 Making Sparks Fly

In some circuits, a spark occurs between the poles of a switch when the switch is opened. Why isn't there a spark when the switch for this circuit is closed?

Explanation According to Lenz's law, the direction of induced emfs is such that the induced magnetic field opposes change in the original magnetic flux. When

the switch is opened, the sudden drop in the magnetic field in the circuit induces an emf in a direction that opposes change in the original current. This induced emf can cause a spark as the current bridges the air gap between the poles of the switch. The spark doesn't occur when the switch is closed, because the original current is zero and the induced emf opposes any change in that current.

In general, determining the inductance of a given current element can be challenging. Finding an expression for the inductance of a common solenoid, however, is straightforward. Let the solenoid have N turns and length ℓ. Assume that ℓ is large compared with the radius and that the core of the solenoid is air. We take the interior magnetic field to be uniform and given by Equation 19.16,

$$B = \mu_0 nI = \mu_0 \frac{N}{\ell} I$$

where $n = N/\ell$ is the number of turns per unit length. The magnetic flux through each turn is therefore

$$\Phi_B = BA = \mu_0 \frac{N}{\ell} AI$$

where A is the cross-sectional area of the solenoid. From this expression and Equation 20.10, we find that

$$L = \frac{N\Phi_B}{I} = \frac{\mu_0 N^2 A}{\ell} \qquad \text{[20.11a]}$$

This shows that L depends on the geometric factors ℓ and A and on μ_0 and is proportional to the square of the number of turns. Because $N = n\ell$, we can also express the result in the form

$$L = \mu_0 \frac{(n\ell)^2}{\ell} A = \mu_0 n^2 A\ell = \mu_0 n^2 V \qquad \text{[20.11b]}$$

where $V = A\ell$ is the volume of the solenoid.

EXAMPLE 20.7   Inductance, Self-Induced emf, and Solenoids

Goal   Calculate the inductance and self-induced emf of a solenoid.

Problem   (a) Calculate the inductance of a solenoid containing 300 turns if the length of the solenoid is 25.0 cm and its cross-sectional area is $4.00 \times 10^{-4} \text{ m}^2$. (b) Calculate the self-induced emf in the solenoid described in (a) if the current in the solenoid decreases at the rate of 50.0 A/s.

Strategy   Substituting given quantities into Equation 20.11a gives the inductance L. For part (b), substitute the result of part (a) and $\Delta I/\Delta t = -50.0$ A/s into Equation 20.9 to get the self-induced emf.

Solution
(a) Calculate the inductance of the solenoid.

Substitute the number N of turns, the area A, and the length ℓ into Equation 20.11a to find the inductance:

$$L = \frac{\mu_0 N^2 A}{\ell}$$

$$= (4\pi \times 10^{-7} \text{ T} \cdot \text{m/A}) \frac{(300)^2 (4.00 \times 10^{-4} \text{ m}^2)}{25.0 \times 10^{-2} \text{ m}}$$

$$= 1.81 \times 10^{-4} \text{ T} \cdot \text{m}^2/\text{A} = \boxed{0.181 \text{ mH}}$$

(b) Calculate the self-induced emf in the solenoid.

Substitute L and $\Delta I/\Delta t = -50.0$ A/s into Equation 20.9, finding the self-induced emf:

$$\mathcal{E} = -L \frac{\Delta I}{\Delta t} = -(1.81 \times 10^{-4} \text{ H})(-50.0 \text{ A/s})$$

$$= \boxed{9.05 \text{ mV}}$$

Remark   Notice that $\Delta I/\Delta t$ is negative because the current is decreasing with time.

Exercise 20.7

A solenoid is to have an inductance of 0.285 mH, a cross-sectional area of 6.00×10^{-4} m², and a length of 36.0 cm. (a) How many turns per unit length should it have? (b) If the self-induced emf is -12.5 mV at a given time, at what rate is the current changing at that instant?

Answers   (a) 1 025 turns/m   (b) 43.9 A/s

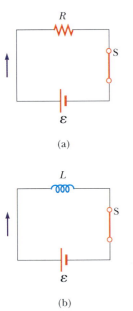

(a)

(b)

Figure 20.26 A comparison of the effect of a resistor with that of an inductor in a simple circuit.

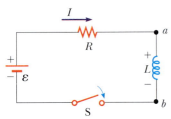

ACTIVE FIGURE 20.27
A series *RL* circuit. As the current increases towards its maximum value, the inductor produces an emf that opposes the increasing current.

Physics⊗Now™
Log into PhysicsNow at **www.cp7e.com** and go to Active Figure 20.27, where you can adjust the values of *R* and *L* and observe the effect on current. A graphical display as in Active Figure 20.28 is available.

20.7 *RL* CIRCUITS

A circuit element that has a large inductance, such as a closely wrapped coil of many turns, is called an **inductor**. The circuit symbol for an inductor is ⸺ꞁꞁꞁꞁ⸺ . We will always assume that the self-inductance of the remainder of the circuit is negligible compared with that of the inductor in the circuit.

To gain some insight into the effect of an inductor in a circuit, consider the two circuits in Figure 20.26. Figure 20.26a shows a resistor connected to the terminals of a battery. For this circuit, Kirchhoff's loop rule is $\mathcal{E} - IR = 0$. The voltage drop across the resistor is

$$\Delta V_R = -IR \qquad [20.12]$$

In this case, **we interpret resistance as a measure of opposition to the current**. Now consider the circuit in Figure 20.26b, consisting of an inductor connected to the terminals of a battery. At the instant the switch in this circuit is closed, because $IR = 0$, the emf of the battery equals the back emf generated in the coil. Hence, we have

$$\mathcal{E}_L = -L\frac{\Delta I}{\Delta t} \qquad [20.13]$$

From this expression, **we can interpret L as a measure of opposition to the rate of change of current**.

Active Figure 20.27 shows a circuit consisting of a resistor, an inductor, and a battery. Suppose the switch is closed at $t = 0$. The current begins to increase, but the inductor produces an emf that opposes the increasing current. As a result, the current can't change from zero to its maximum value of $\mathcal{E}/R$ instantaneously. Equation 20.13 shows that the induced emf is a maximum when the current is changing most rapidly, which occurs when the switch is first closed. As the current approaches its steady-state value, the back emf of the coil falls off because the current is changing more slowly. Finally, when the current reaches its steady-state value, the rate of change is zero and the back emf is also zero. Active Figure 20.28 plots current in the circuit as a function of time.[1] This plot is similar to that of the charge on a capacitor as a function of time, discussed in Chapter 18. In that case, we found it convenient to introduce a quantity called the *time constant of the circuit*, which told us something about the time required for the capacitor to approach its steady-state charge. In the same way, time constants are defined for circuits containing resistors and inductors. The **time constant τ for an *RL* circuit** is the time required for the current in the circuit to reach 63.2% of its final value $\mathcal{E}/R$; the time constant of an *RL* circuit is given by

Time constant for an *RL* circuit ▶

$$\tau = \frac{L}{R} \qquad [20.14]$$

---

[1]The equation for the current in the circuit as a function of time may be obtained from calculus and is

$$I = \frac{\mathcal{E}}{R}(1 - e^{-Rt/L})$$

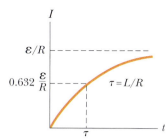

ACTIVE FIGURE 20.28
A plot of current versus time for the *RL* circuit shown in Figure 20.27. The switch is closed at *t* = 0, and the current increases towards its maximum value $\mathcal{E}/R$. The time constant τ is the time it takes the current to reach 63.2% of its maximum value.

Physics⊗Now™
Log into PhysicsNow at **www.cp7e.com** and go to Active Figure 20.28, where you can observe this graph develop after the switch in Active Figure 20.27 is closed.

Quick Quiz 20.5

The switch in the circuit shown in Figure 20.29 is closed and the lightbulb glows steadily. The inductor is a simple air-core solenoid. An iron rod is inserted into the interior of the solenoid, increasing the magnitude of the magnetic field in the solenoid. As the rod is inserted, the brightness of the lightbulb (a) increases, (b) decreases, or (c) remains the same.

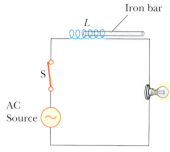

Figure 20.29 (Quick Quiz 20.5)

EXAMPLE 20.8 An *RL* Circuit

Goal Calculate a time constant and relate it to current in an *RL* circuit.

Problem A 12.6-V battery is in a circuit with a 30.0-mH inductor and a 0.150-Ω resistor, as in Active Figure 20.27. The switch is closed at *t* = 0. **(a)** Find the time constant of the circuit. **(b)** Find the current after one time constant has elapsed. **(c)** Find the voltage drops across the resistance when *t* = 0 and *t* = one time constant. **(d)** What's the rate of change of the current after one time constant?

Solution
(a) What's the time constant of the circuit?

Substitute the inductance *L* and resistance *R* into Equation 20.14, finding the time constant:

$$\tau = \frac{L}{R} = \frac{30.0 \times 10^{-3}\,\text{H}}{0.150\,\Omega} = \boxed{0.200\,\text{s}}$$

(b) Find the current after one time constant has elapsed.

First, use Ohm's law to compute the final value of the current after many time constants have elapsed:

$$I_{max} = \frac{\mathcal{E}}{R} = \frac{12.6\,\text{V}}{0.150\,\Omega} = 84.0\,\text{A}$$

After one time constant, the current rises to 63.2% of its final value:

$$I_{1\tau} = (0.632)I_{max} = (0.632)(84.0\,\text{A}) = \boxed{53.1\,\text{A}}$$

(c) Find the voltage drops across the resistance when *t* = 0 and *t* = one time constant.

Initially, the current in the circuit is zero, so, from Ohm's law, the voltage across the resistor is zero:

$$\Delta V_R = IR$$
$$\Delta V_R\,(t = 0\,\text{s}) = (0\,\text{A})(0.150\,\Omega) = \boxed{0}$$

Next, using Ohm's law, find the magnitude of the voltage drop across the resistor after one time constant:

$$\Delta V_R\,(t = 0.200\,\text{s}) = (53.1\,\text{A})(0.150\,\Omega) = \boxed{7.97\,\text{V}}$$

(d) What's the rate of change of the current after one time constant?

Using Kirchhoff's voltage rule, calculate the voltage drop across the inductor at that time:

$$\mathcal{E} + \Delta V_R + \Delta V_L = 0$$

Solve for ΔV_L:

$$\Delta V_L = -\mathcal{E} - \Delta V_R = -12.6 \text{ V} - (-7.97 \text{ V}) = -4.6 \text{ V}$$

Now solve Equation 20.13 for $\Delta I/\Delta t$ and substitute:

$$\Delta V_L = -L\frac{\Delta I}{\Delta t}$$

$$\frac{\Delta I}{\Delta t} = -\frac{\Delta V_L}{L} = -\frac{-4.6 \text{ V}}{30.0 \times 10^{-3} \text{ H}} = \boxed{150 \text{ A/s}}$$

Remarks   The values used in this problem were taken from actual components salvaged from the starter system of a car. Because the current in such an RL circuit is initially zero, inductors are sometimes referred to as "chokes," since they temporarily choke off the current. In solving part (d), we traversed the circuit in the direction of positive current, so the voltage difference across the battery was positive and the differences across the resistor and inductor were negative.

Exercise 20.8
A 12.6-V battery is in series with a resistance of 0.350 Ω and an inductor. (a) After a long time, what is the current in the circuit? (b) What is the current after one time constant? (c) What's the voltage drop across the inductor at this time? (d) Find the inductance if the time constant is 0.130 s.

Answers   (a) 36.0 A   (b) 22.8 A   (c) 4.62 V   (d) 4.55×10^{-2} H

20.8   ENERGY STORED IN A MAGNETIC FIELD

The emf induced by an inductor prevents a battery from establishing an instantaneous current in a circuit. The battery has to do work to produce a current. We can think of this needed work as energy stored in the inductor in its magnetic field. In a manner similar to that used in Section 16.9 to find the energy stored in a capacitor, we find that the energy stored by an inductor is

Energy stored in an inductor ▶

$$PE_L = \tfrac{1}{2}LI^2 \qquad\qquad \textbf{[20.15]}$$

Note that the result is similar in form to the expression for the energy stored in a charged capacitor (Equation 16.18):

Energy stored in a capacitor ▶

$$PE_C = \tfrac{1}{2}C(\Delta V)^2$$

EXAMPLE 20.9   Magnetic Energy

Goal   Relate the storage of magnetic energy to currents in an RL circuit.

Problem   A 12.0-V battery is connected in series to a 25.0-Ω resistor and a 5.00-H inductor. (a) Find the maximum current in the circuit. (b) Find the energy stored in the inductor at this time. (c) How much energy is stored in the inductor when the current is changing at a rate of 1.50 A/s?

Strategy   In part (a), Ohm's law and Kirchhoff's voltage rule yield the maximum current, because the voltage across the inductor is zero when the current is maximal. Substituting the current into Equation 20.15 gives the energy stored in the inductor. In part (c), the given rate of change of the current can be used to calculate the voltage drop across the inductor at the specified time. Kirchhoff's voltage rule and Ohm's law then give the current I at that time, which can be used to find the energy stored in the inductor.

Solution
(a) Find the maximum current in the circuit.

Apply Kirchhoff's voltage rule to the circuit:

$$\Delta V_{\text{batt}} + \Delta V_R + \Delta V_L = 0$$

$$\mathcal{E} - IR - L\frac{\Delta I}{\Delta t} = 0$$

When the maximum current is reached, $\Delta I/\Delta t$ is zero, so the voltage drop across the inductor is zero. Solve for the maximum current I_{max}:

$$I_{max} = \frac{\varepsilon}{R} = \frac{12.0\ \text{V}}{25.0\ \Omega} = \boxed{0.480\ \text{A}}$$

(b) Find the energy stored in the inductor at this time.

Substitute known values into Equation 20.15:

$$PE_L = \tfrac{1}{2}LI_{max}^2 = \tfrac{1}{2}(5.00\ \text{H})(0.480\ \text{A})^2 = \boxed{0.576\ \text{J}}$$

(c) Find the energy in the inductor when the current changes at a rate of 1.50 A/s.

Apply Kirchhoff's voltage rule to the circuit, once again:

$$\varepsilon - IR - L\frac{\Delta I}{\Delta t} = 0$$

Solve this equation for the current I and substitute:

$$I = \frac{1}{R}\left(\varepsilon - L\frac{\Delta I}{\Delta t}\right)$$

$$= \frac{1}{25.0\ \Omega}[12.0\ \text{V} - (5.00\ \text{H})(1.50\ \text{A/s})] = 0.180\ \text{A}$$

Finally, substitute the value for the current into Equation 20.15, finding the energy stored in the inductor:

$$PE_L = \tfrac{1}{2}LI^2 = \tfrac{1}{2}(5.00\ \text{H})(0.180\ \text{A})^2 = \boxed{0.081\ 0\ \text{J}}$$

Remark   Notice how important it is to combine concepts from previous chapters. Here, Ohm's law and Kirchhoff's loop rule were essential to the solution of the problem.

Exercise 20.9
For the same circuit, find the energy stored in the inductor when the rate of change of the current is 1.00 A/s.

Answer   0.196 J

SUMMARY

Physics⊗Now™ Take a practice test by logging into PhysicsNow at **www.cp7e.com** and clicking on the Pre-Test link for this chapter.

20.1   Induced emf and Magnetic Flux
The magnetic flux Φ_B through a closed loop is defined as

$$\Phi_B \equiv BA\cos\theta \qquad \text{[20.1]}$$

where B is the strength of the uniform magnetic field, A is the cross-sectional area of the loop, and θ is the angle between $\vec{B}$ and a direction perpendicular to the plane of the loop.

20.2   Faraday's Law of Induction
Faraday's law of induction states that the instantaneous emf induced in a circuit equals the negative of the rate of change of magnetic flux through the circuit,

$$\varepsilon = -N\frac{\Delta\Phi_B}{\Delta t} \qquad \text{[20.2]}$$

where N is the number of loops in the circuit. The magnetic flux Φ_B can change with time whenever the magnetic field $\vec{B}$, the area A, or the angle θ changes with time.

Lenz's law states that the current from the induced emf creates a magnetic field with flux opposing the *change* in magnetic flux through a circuit.

20.3   Motional emf
If a conducting bar of length ℓ moves through a magnetic field with a speed v so that $\vec{B}$ is perpendicular to the bar, then the emf induced in the bar, often called a **motional emf**, is

$$|\varepsilon| = B\ell v \qquad \text{[20.4]}$$

20.5   Generators
When a coil of wire with N turns, each of area A, rotates with constant angular speed ω in a uniform magnetic field $\vec{B}$, the emf induced in the coil is

$$\varepsilon = NAB\omega\sin\omega t \qquad \text{[20.7]}$$

Such generators naturally produce alternating current (AC), which changes direction with frequency $\omega/2\pi$. The AC current can be transformed to direct current.

20.7   *RL* Circuits
When the current in a coil changes with time, an emf is induced in the coil according to Faraday's law. This

self-induced emf is defined by the expression

$$\mathcal{E} \equiv -L\frac{\Delta I}{\Delta t} \qquad \text{[20.9]}$$

where L is the inductance of the coil. The SI unit for inductance is the henry (H); $1\ \text{H} = 1\ \text{V} \cdot \text{s/A}$.

The **inductance** of a coil can be found from the expression

$$L = \frac{N\Phi_B}{I} \qquad \text{[20.10]}$$

where N is the number of turns on the coil, I is the current in the coil, and Φ_B is the magnetic flux through the coil produced by that current. For a solenoid, the inductance is given by

$$L = \frac{\mu_0 N^2 A}{\ell} \qquad \text{[20.11]}$$

If a resistor and inductor are connected in series to a battery and a switch is closed at $t = 0$, the current in the circuit doesn't rise instantly to its maximum value. After one **time constant** $\tau = L/R$, the current in the circuit is 63.2% of its final value $\mathcal{E}/R$. As the current approaches its final, maximum value, the voltage drop across the inductor approaches zero.

20.8   Energy Stored in a Magnetic Field

The **energy stored** in the magnetic field of an inductor carrying current I is

$$PE_L = \tfrac{1}{2}LI^2 \qquad \text{[20.15]}$$

As the current in an RL circuit approaches its maximum value, the stored energy also approaches a maximum value.

CONCEPTUAL QUESTIONS

1. A circular loop is located in a uniform and constant magnetic field. Describe how an emf can be induced in the loop in this situation.

2. Does dropping a magnet down a copper tube produce a current in the tube? Explain.

3. A spacecraft orbiting the Earth has a coil of wire in it. An astronaut measures a small current in the coil, although there is no battery connected to it and there are no magnets in the spacecraft. What is causing the current?

4. A loop of wire is placed in a uniform magnetic field. For what orientation of the loop is the magnetic flux a maximum? For what orientation is the flux zero?

5. As the conducting bar in Figure Q20.5 moves to the right, an electric field directed downward is set up. If the bar were moving to the left, explain why the electric field would be upward.

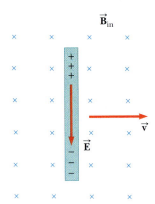

Figure Q20.5   (Conceptual Questions 5 and 6)

6. As the bar in Figure Q20.5 moves perpendicular to the field, is an external force required to keep it moving with constant speed?

7. Wearing a metal bracelet in a region of strong magnetic field could be hazardous. Discuss this statement.

8. How is electrical energy produced in dams? (That is, how is the energy of motion of the water converted to AC electricity?)

9. Eddy currents are induced currents set up in a piece of metal when it moves through a nonuniform magnetic field. For example, consider the flat metal plate swinging at the end of a bar as a pendulum, as shown in Figure Q20.9. At position 1, the pendulum is moving from a region where there is no magnetic field into a region where the field $\vec{\mathbf{B}}_{in}$ is directed into the paper. Show that at position 1 the direction of the eddy current is counterclockwise. Also, at position 2 the pendulum is moving out of the field into a region of zero field. Show that the direction of the eddy current is clockwise in this case. Use right-hand rule number 2 to show that these eddy currents lead to a magnetic force on the plate directed as shown in the figure. Because the induced eddy current always produces a retarding force when the plate enters or leaves the field, the swinging plate quickly comes to rest.

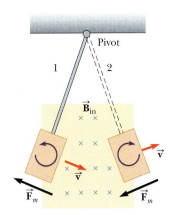

Figure Q20.9

10. Suppose you would like to steal power for your home from the electric company by placing a loop of wire near a transmission cable in order to induce an emf in the loop (Don't do this; it's illegal.) Should you locate the loop so that the transmission cable passes through your loop or simply place your loop near the transmission cable? Does the orientation of the loop matter?

11. A piece of aluminum is dropped vertically downward between the poles of an electromagnet. Does the magnetic

field affect the velocity of the aluminum? [*Hint*: See Conceptual Question 9.]

12. A bar magnet is dropped toward a conducting ring lying on the floor. As the magnet falls toward the ring, does it move as a freely falling object?

13. If the current in an inductor is doubled, by what factor does the stored energy change?

14. Is it possible to induce a constant emf for an infinite amount of time?

15. Why is the induced emf that appears in an inductor called a back (counter) emf?

16. A magneto is used to cause the spark in a spark plug in many lawn mowers today. A magneto consists of a permanent magnet mounted on a flywheel so that it spins past a fixed coil. Explain how this arrangement generates a large enough potential difference to cause the spark.

17. A ramp runs from the bed of a truck down to the level ground. The ramp holds two parallel conducting rails connected at its base. A metal bar slides on the rails without friction. A magnet supplies an external magnetic field directed toward the ground. It is found that the bar slides down the ramp at a constant speed. (a) What is the direction of the induced current in the bar as viewed from above? (b) What can you conclude about the forces exerted on the bar?

18. A bar magnet is held above the center of a wire loop in a horizontal plane, as shown in Figure Q20.18. The south end of the magnet is toward the loop. The magnet is dropped. Find the direction of the current in the resistor

as viewed from above (a) while the magnet is falling toward the loop and (b) after the magnet has passed through the loop and moved away from it.

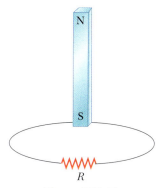

Figure Q20.18

19. What is the direction of the current induced in the resistor when the current in the long, straight wire in Figure Q20.19 decreases rapidly to zero?

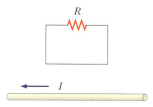

Figure Q20.19

PROBLEMS

1, 2, 3 = straightforward, intermediate, challenging   □ = full solution available in *Student Solutions Manual/Study Guide*
Physics⊗Now™ = coached problem with hints available at **www.cp7e.com**   🧬 = biomedical application

Section 20.1 Induced emf and Magnetic Flux

1. A magnetic field of strength 0.30 T is directed perpendicular to a plane circular loop of wire of radius 25 cm. Find the magnetic flux through the area enclosed by this loop.

2. Find the flux of the Earth's magnetic field of magnitude 5.00×10^{-5} T through a square loop of area 20.0 cm^2 (a) when the field is perpendicular to the plane of the loop, (b) when the field makes a 30.0° angle with the normal to the plane of the loop, and (c) when the field makes a 90.0° angle with the normal to the plane.

3. A square loop 2.00 m on a side is placed in a magnetic field of magnitude 0.300 T. If the field makes an angle of 50.0° with the normal to the plane of the loop, find the magnetic flux through the loop.

4. A long, straight wire carrying a current of 2.00 A is placed along the axis of a cylinder of radius 0.500 m and a length of 3.00 m. Determine the total magnetic flux through the cylinder.

5. A long, straight wire lies in the plane of a circular coil with a radius of 0.010 m. The wire carries a current of 2.0 A and is placed along a diameter of the coil. (a) What is the net flux through the coil? (b) If the wire passes through the center of the coil and is perpendicular to the plane of the coil, find the net flux through the coil.

6. A solenoid 4.00 cm in diameter and 20.0 cm long has 250 turns and carries a current of 15.0 A. Calculate the magnetic flux through the circular cross-sectional area of the solenoid.

7. A cube of edge length $\ell = 2.5$ cm is positioned as shown in Figure P20.7. There is a uniform magnetic field throughout the region with components $B_x = +5.0$ T, $B_y = +4.0$ T, and $B_z = +3.0$ T. (a) Calculate the flux through the shaded face of the cube. (b) What is the total flux

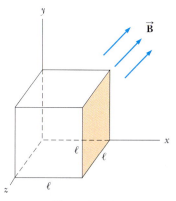

Figure P20.7

emerging from the volume enclosed by the cube (i.e., the total flux through all six faces)?

Section 20.2 Faraday's Law of Induction

8. Transcranial magnetic stimulation (TMS) is a noninvasive technique used to stimulate regions of the human brain. A small coil is placed on the scalp, and a brief burst of current in the coil produces a rapidly changing magnetic field inside the brain. The induced emf can be sufficient to stimulate neuronal activity. One such device generates a magnetic field within the brain that rises from zero to 1.5 T in 120 ms. Determine the induced emf within a circle of tissue of radius 1.6 mm and that is perpendicular to the direction of the field.

9. A square, single-turn coil 0.20 m on a side is placed with its plane perpendicular to a constant magnetic field. An emf of 18 mV is induced in the coil winding when the area of the coil decreases at the rate of 0.10 m^2/s. What is the magnitude of the magnetic field?

10. The flexible loop in Figure P20.10 has a radius of 12 cm and is in a magnetic field of strength 0.15 T. The loop is grasped at points A and B and stretched until its area is nearly zero. If it takes 0.20 s to close the loop, find the magnitude of the average induced emf in it during this time.

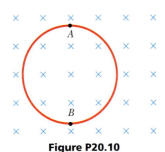

Figure P20.10

11. A wire loop of radius 0.30 m lies so that an external magnetic field of magnitude 0.30 T is perpendicular to the loop. The field reverses its direction, and its magnitude changes to 0.20 T in 1.5 s. Find the magnitude of the average induced emf in the loop during this time.

12. A 500-turn circular-loop coil 15.0 cm in diameter is initially aligned so that its axis is parallel to the Earth's magnetic field. In 2.77 ms, the coil is flipped so that its axis is perpendicular to the Earth's magnetic field. If an average voltage of 0.166 V is thereby induced in the coil, what is the value of the Earth's magnetic field at that location?

13. The plane of a rectangular coil, 5.0 cm by 8.0 cm, is perpendicular to the direction of a magnetic field $\vec{B}$. If the coil has 75 turns and a total resistance of 8.0 Ω, at what rate must the magnitude of $\vec{B}$ change to induce a current of 0.10 A in the windings of the coil?

14. A square, single-turn wire loop 1.00 cm on a side is placed inside a solenoid that has a circular cross section of radius 3.00 cm, as shown in Figure P20.14. The solenoid is 20.0 cm long and wound with 100 turns of wire. (a) If the current in the solenoid is 3.00 A, find the flux through the loop. (b) If the current in the solenoid is reduced to zero in 3.00 s, find the magnitude of the average induced emf in the loop.

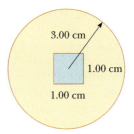

Figure P20.14

15. A 300-turn solenoid with a length of 20 cm and a radius of 1.5 cm carries a current of 2.0 A. A second coil of four turns is wrapped tightly about this solenoid so that it can be considered to have the same radius as the solenoid. Find (a) the change in the magnetic flux through the coil and (b) the magnitude of the average induced emf in the coil when the current in the solenoid increases to 5.0 A in a period of 0.90 s.

16. A circular coil enclosing an area of 100 cm^2 is made of 200 turns of copper wire. The wire making up the coil has resistance of 5.0 Ω, and the ends of the wire are connected to form a closed circuit. Initially, a 1.1-T uniform magnetic field points perpendicularly upward through the plane of the coil. The direction of the field then reverses so that the final magnetic field has a magnitude of 1.1 T and points downward through the coil. If the time required for the field to reverse directions is 0.10 s, what average current flows through the coil during that time?

17. To monitor the breathing of a hospital patient, a thin belt is girded around the patient's chest as in Figure P20.17. The belt is a 200-turn coil. When the patient inhales, the area encircled by the coil increases by 39.0 cm^2. The magnitude of the Earth's magnetic field is 50.0 μT and makes an angle of 28.0° with the plane of the coil. Assuming a patient takes 1.80 s to inhale, find the magnitude of the average induced emf in the coil during that time.

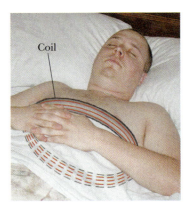

Figure P20.17

Section 20.3 Motional emf

18. Consider the arrangement shown in Figure P20.18. Assume that $R = 6.00$ Ω and $\ell = 1.20$ m, and that a uniform 2.50-T magnetic field is directed *into* the page. At what speed should the bar be moved to produce a current of 0.500 A in the resistor?

19. A Boeing 747 jet with a wingspan of 60.0 m is flying horizontally at a speed of 300 m/s over Phoenix, Arizona, at a location where the Earth's magnetic field is 50.0 μT at

58.0° below the horizontal. What voltage is generated between the wingtips?

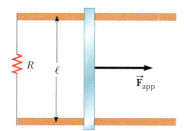

Figure P20.18   (Problems 18 and 57)

20. A 12.0-m-long steel beam is accidentally dropped by a construction crane from a height of 9.00 m. The horizontal component of the Earth's magnetic field over the region is 18.0 μT. What is the induced emf in the beam just before impact with the Earth? Assume the long dimension of the beam remains in a horizontal plane, oriented perpendicular to the horizontal component of the Earth's magnetic field.

21. **Physics ⊗ Now**™ An automobile has a vertical radio antenna 1.20 m long. The automobile travels at 65.0 km/h on a horizontal road where the Earth's magnetic field is 50.0 μT, directed toward the north and downwards at an angle of 65.0° below the horizontal. (a) Specify the direction the automobile should move in order to generate the maximum motional emf in the antenna, with the top of the antenna positive relative to the bottom. (b) Calculate the magnitude of this induced emf.

22. A helicopter has blades of length 3.0 m, rotating at 2.0 rev/s about a central hub. If the vertical component of Earth's magnetic field is 5.0×10^{-5} T, what is the emf induced between the blade tip and the central hub?

Section 20.4 Lenz's Law Revisited (the Minus Sign in Faraday's Law)

23. A bar magnet is positioned near a coil of wire as shown in Figure P20.23. What is the direction of the current in the resistor when the magnet is moved (a) to the left? (b) to the right?

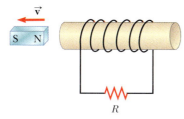

Figure P20.23

24. A conducting rectangular loop of mass M, resistance R, and dimensions w by ℓ falls from rest into a magnetic field $\vec{B}$ as shown in Figure P20.24. During the time interval before the top edge of the loop reaches the field, the loop approaches a terminal speed v_T. (a) Show that

$$v_T = \frac{MgR}{B^2 w^2}$$

(b) Why is v_T proportional to R? (c) Why is it inversely proportional to B^2?

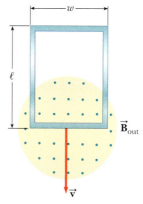

Figure P20.24

25. A rectangular coil with resistance R has N turns, each of length ℓ and width w as shown in Figure P20.25. The coil moves into a uniform magnetic field $\vec{B}$ with constant velocity $\vec{v}$. What are the magnitude and direction of the total magnetic force on the coil (a) as it enters the magnetic field, (b) as it moves within the field, and (c) as it leaves the field?

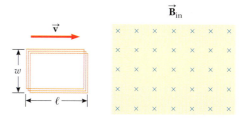

Figure P20.25

26. In Figure P20.26, what is the direction of the current induced in the resistor at the instant the switch is closed?

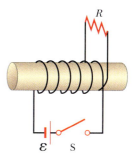

Figure P20.26

27. A copper bar is moved to the right while its axis is maintained in a direction perpendicular to a magnetic field, as shown in Figure P20.27. If the top of the bar becomes

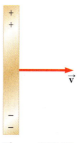

Figure P20.27

positive relative to the bottom, what is the direction of the magnetic field?

28. Find the direction of the current in the resistor shown in Figure P20.28 (a) at the instant the switch is closed, (b) after the switch has been closed for several minutes, and (c) at the instant the switch is opened.

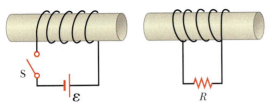

Figure P20.28

29. Find the direction of the current in the resistor R shown in Figure P20.29 after each of the following steps (taken in the order given): (a) The switch is closed. (b) The variable resistance in series with the battery is decreased. (c) The circuit containing resistor R is moved to the left. (d) The switch is opened.

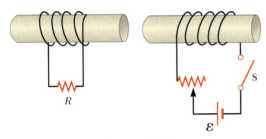

Figure P20.29

Section 20.5 Generators

30. A 100-turn square wire coil of area 0.040 m² rotates about a vertical axis at 1 500 rev/min, as indicated in Figure P20.30. The horizontal component of the Earth's magnetic field at the location of the loop is 2.0×10^{-5} T. Calculate the maximum emf induced in the coil by the Earth's field.

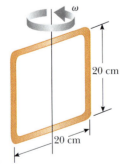

Figure P20.30

31. Considerable scientific work is currently underway to determine whether weak oscillating magnetic fields such as those found near outdoor electric power lines can effect human health. One study indicated that a magnetic field of magnitude 1.0×10^{-3} T, oscillating at 60 Hz, might stimulate red blood cells to become cancerous. If the diameter of a red blood cell is 8.0 μm, determine the

maximum emf that can be generated around the perimeter of the cell.

32. A motor has coils with a resistance of 30 Ω and operates from a voltage of 240 V. When the motor is operating at its maximum speed, the back emf is 145 V. Find the current in the coils (a) when the motor is first turned on and (b) when the motor has reached maximum speed. (c) If the current in the motor is 6.0 A at some instant, what is the back emf at that time?

33. A coil of 10.0 turns is in the shape of an ellipse having a major axis of 10.0 cm and a minor axis of 4.00 cm. The coil rotates at 100 rpm in a region in which the magnitude of the Earth's magnetic field is 55 μT. What is the maximum voltage induced in the coil if the axis of rotation of the coil is along its major axis and is aligned (a) perpendicular to the Earth's magnetic field and (b) parallel to the Earth's magnetic field? (Note that the area of an ellipse is given by $A = \pi ab$, where a is the length of the *semi*major axis and b is the length of the *semi*minor axis.)

34. A flat coil enclosing an area of 0.10 m² is rotating at 60 rev/s, with its axis of rotation perpendicular to a 0.20-T magnetic field. (a) If there are 1 000 turns on the coil, what is the maximum voltage induced in the coil? (b) When the maximum induced voltage occurs, what is the orientation of the coil with respect to the magnetic field?

35. **Physics⊗Now™** In a model AC generator, a 500-turn rectangular coil 8.0 cm by 20 cm rotates at 120 rev/min in a uniform magnetic field of 0.60 T. (a) What is the maximum emf induced in the coil? (b) What is the instantaneous value of the emf in the coil at $t = (\pi/32)$ s? Assume that the emf is zero at $t = 0$. (c) What is the smallest value of t for which the emf will have its maximum value?

Section 20.6 Self-Inductance

36. A coiled telephone cord forms a spiral with 70.0 turns, a diameter of 1.30 cm, and an unstretched length of 60.0 cm. Determine the self-inductance of one conductor in the unstretched cord.

37. A coil has an inductance of 3.0 mH, and the current in it changes from 0.20 A to 1.5 A in 0.20 s. Find the magnitude of the average induced emf in the coil during this period.

38. Show that the two expressions for inductance given by

$$L = \frac{N\Phi_B}{I} \quad \text{and} \quad L = \frac{-\mathcal{E}}{\Delta I/\Delta t}$$

have the same units.

39. A solenoid of radius 2.5 cm has 400 turns and a length of 20 cm. Find (a) its inductance and (b) the rate at which current must change through it to produce an emf of 75 mV.

40. An emf of 24.0 mV is induced in a 500-turn coil when the current is changing at a rate of 10.0 A/s. What is the magnetic flux through each turn of the coil at an instant when the current is 4.00 A?

Section 20.7 *RL* Circuits

41. Show that the SI units for the inductive time constant $\tau = L/R$ are seconds.

42. An *RL* circuit with $L = 3.00$ H and an *RC* circuit with $C = 3.00$ μF have the same time constant. If the two circuits have the same resistance *R*, (a) what is the value of *R* and (b) what is this common time constant?

43. A 6.0-V battery is connected in series with a resistor and an inductor. The series circuit has a time constant of 600 μs, and the maximum current is 300 mA. What is the value of the inductance?

44. A 25-mH inductor, an 8.0-Ω resistor, and a 6.0-V battery are connected in series. The switch is closed at $t = 0$. Find the voltage drop across the resistor (a) at $t = 0$ and (b) after one time constant has passed. Also, find the voltage drop across the inductor (c) at $t = 0$ and (d) after one time constant has elapsed.

45. **Physics Now™** Calculate the resistance in an *RL* circuit in which $L = 2.50$ H and the current increases to 90.0% of its final value in 3.00 s.

46. Consider the circuit shown in Figure P20.46. Take $\mathcal{E} = 6.00$ V, $L = 8.00$ mH, and $R = 4.00$ Ω. (a) What is the inductive time constant of the circuit? (b) Calculate the current in the circuit 250 μs after the switch is closed. (c) What is the value of the final steady-state current? (d) How long does it take the current to reach 80.0% of its maximum value?

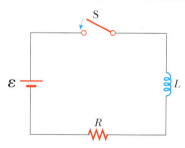

Figure P20.46

20.8 Energy Stored in a Magnetic Field

47. How much energy is stored in a 70.0-mH inductor at an instant when the current is 2.00 A?

48. A 300-turn solenoid has a radius of 5.00 cm and a length of 20.0 cm. Find (a) the inductance of the solenoid and (b) the energy stored in the solenoid when the current in its windings is 0.500 A.

49. **Physics Now™** A 24-V battery is connected in series with a resistor and an inductor, with $R = 8.0$ Ω and $L = 4.0$ H, respectively. Find the energy stored in the inductor (a) when the current reaches its maximum value and (b) one time constant after the switch is closed.

ADDITIONAL PROBLEMS

50. What is the time constant for (a) the circuit shown in Figure P20.50a and (b) the circuit shown in Figure P20.50b?

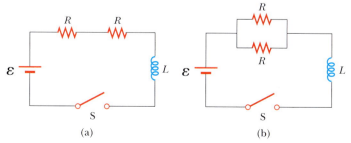

(a)   (b)

Figure P20.50

51. In Figure P20.51, the bar magnet is being moved toward the loop. Is $(V_a - V_b)$ positive, negative, or zero during this motion? Explain.

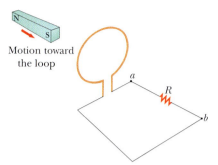

Figure P20.51

52. Your physics teacher asks you to help her set up a demonstration of Faraday's law for the class. The apparatus consists of a strong permanent magnet that has a field of 0.10 T, a small 10-turn coil of radius 2.0 cm cemented on a wood frame with a handle, some flexible connecting wires, and an ammeter, as in Figure P20.52. The idea is to pull the coil out of the center of the magnetic field as quickly as possible and read the average current registered on the meter. The combined resistance of the coil, leads, and meter is 2.0 Ω, and you must flip the coil out of the field in about 0.20 s. The ammeter you must use has a full-scale sensitivity of 1 000 μA. Will this meter be sensitive enough to show the induced current clearly?

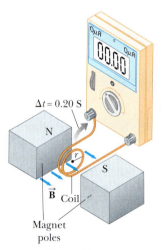

Figure P20.52

53. An 820-turn wire coil of resistance 24.0 Ω is placed on top of a 12 500-turn, 7.00-cm-long solenoid, as in Figure P20.53 (page 690). Both coil and solenoid have cross-sectional areas of 1.00×10^{-4} m². (a) How long does it take the solenoid current to reach 0.632 times its maximum value? (b) Determine the average back emf caused by the self-inductance of the solenoid during this interval. The magnetic field produced by the solenoid at the location of the coil is one-half as strong as the field at the center of the solenoid. (c) Determine the average rate of change in magnetic flux through each turn of the coil during the stated interval. (d) Find the magnitude of the average induced current in the coil.

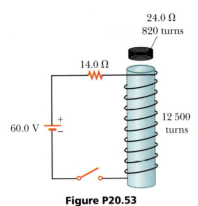

24.0 Ω
820 turns

14.0 Ω

60.0 V

12 500 turns

Figure P20.53

54. Figure P20.54 is a graph of induced emf versus time for a coil of N turns rotating with angular speed ω in a uniform magnetic field directed perpendicular to the axis of rotation of the coil. Copy this sketch (increasing the scale), and, on the same set of axes, show the graph of emf versus t when (a) the number of turns in the coil is doubled, (b) the angular speed is doubled, and (c) the angular speed is doubled while the number of turns in the coil is halved.

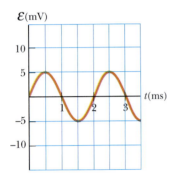

$\mathcal{E}$(mV)

Figure P20.54

55. The plane of a square loop of wire with edge length $a = 0.200$ m is perpendicular to the Earth's magnetic field at a point where $B = 15.0\ \mu$T, as in Figure P20.55. The total resistance of the loop and the wires connecting it to the ammeter is $0.500\ \Omega$. If the loop is suddenly collapsed by horizontal forces as shown, what total charge passes through the ammeter?

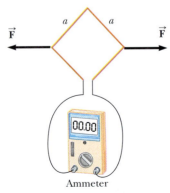

$\vec{F}$ a a $\vec{F}$

Ammeter

Figure P20.55

56. A novel method of storing electrical energy has been proposed. A huge underground superconducting coil

1.00 km in diameter would be fabricated. It would carry a maximum current of 50.0 kA through each winding of a 150-turn Nb₃Sn solenoid. (a) If the inductance of this huge coil were 50.0 H, what is the total stored energy? (b) What is the compressive force per meter acting between two adjacent windings 0.250 m apart? [*Hint*: Because the radius of the coil is so large, the magnetic field created by one winding and acting on an adjacent turn can be considered to be that of a long, straight wire.]

57. A conducting rod of length ℓ moves on two horizontal frictionless rails, as in Figure P20.18. A constant force of magnitude 1.00 N moves the bar at a uniform speed of 2.00 m/s through a magnetic field $\vec{B}$ that is directed into the page. (a) What is the current in an 8.00-Ω resistor R? (b) What is the rate of energy dissipation in the resistor? (c) What is the mechanical power delivered by the constant force?

58. The square loop in Figure P20.58 is made of wires with a total series resistance of $10.0\ \Omega$. It is placed in a uniform 0.100-T magnetic field directed perpendicular into the plane of the paper. The loop, which is hinged at each corner, is pulled as shown until the separation between points A and B is 3.00 m. If this process takes 0.100 s, what is the average current generated in the loop? What is the direction of the current?

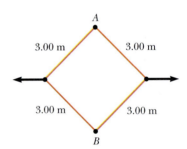

A
3.00 m   3.00 m
3.00 m   3.00 m
B

Figure P20.58

59. The bolt of lightning depicted in Figure P20.59 passes 200 m from a 100-turn coil oriented as shown. If the current in the lightning bolt falls from 6.02×10^6 A to zero in 10.5 μs, what is the average voltage induced in the coil? Assume that the distance to the center of the coil determines the average magnetic field at the coil's position. Treat the lightning bolt as a long, vertical wire.

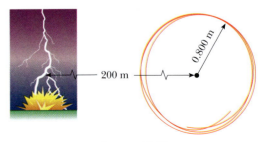

200 m   0.800 m

Figure P20.59

60. The wire shown in Figure P20.60 is bent in the shape of a "tent" with $\theta = 60°$ and $L = 1.5$ m, and is placed in a uniform magnetic field of 0.30 T directed perpendicular to the tabletop. The wire is "hinged" at points a and b. If the tent is flattened out on the table in 0.10 s,

what is the average induced emf in the wire during this time?

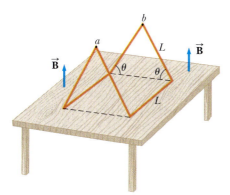

Figure P20.60

61. The magnetic field shown in Figure P20.61 has a uniform magnitude of 25.0 mT directed into the paper. The initial diameter of the kink is 2.00 cm. (a) The wire is quickly pulled taut, and the kink shrinks to a diameter of zero in 50.0 ms. Determine the average voltage induced between endpoints A and B. Include the polarity. (b) Suppose the kink is undisturbed, but the magnetic field increases to 100 mT in 4.00×10^{-3} s. Determine the average voltage across terminals A and B, including polarity, during this period.

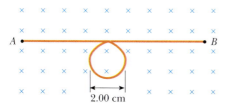

Figure P20.61

62. An aluminum ring of radius 5.00 cm and resistance 3.00×10^{-4} Ω is placed around the top of a long air-core solenoid with 1 000 turns per meter and a smaller radius of 3.00 cm, as in Figure P20.62. If the current in the solenoid is increasing at a constant rate of 270 A/s, what is the induced current in the ring? Assume that the magnetic field produced by the solenoid over the area at the end of the solenoid is one-half as strong as the field at the center of the solenoid. Assume also that the solenoid produces a negligible field outside its cross-sectional area.

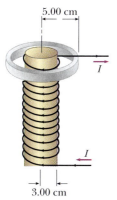

Figure P20.62

63. In Figure P20.63, the rolling axle, 1.50 m long, is pushed along horizontal rails at a constant speed $v = 3.00$ m/s. A resistor $R = 0.400$ Ω is connected to the rails at points a and b, directly opposite each other. (The wheels make good electrical contact with the rails, so the axle, rails, and R form a closed-loop circuit. The only significant resistance in the circuit is R.) A uniform magnetic field $B = 0.800$ T is directed vertically downwards. (a) Find the induced current I in the resistor. (b) What horizontal force $\vec{F}$ is required to keep the axle rolling at constant speed? (c) Which end of the resistor, a or b, is at the higher electric potential? (d) After the axle rolls past the resistor, does the current in R reverse direction?

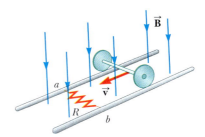

Figure P20.63

64. In 1832, Faraday proposed that the apparatus shown in Figure P20.64 could be used to generate electric current from the water flowing in the Thames River.[2] Two conducting plates of length a and width b are placed facing one another on opposite sides of the river, a distance w apart and immersed entirely. The flow velocity of the river is $\vec{v}$, and the vertical component of Earth's magnetic field is B. Show that the current in the load resistor R is

$$I = \frac{abvB}{\rho + abR/w}$$

where ρ is the resistivity of the water. (b) Calculate the short-circuit current $(R = 0)$ if $a = 100$ m, $b = 5.00$ m, $v = 3.00$ m/s, $B = 50.0$ μT, and $\rho = 100$ $\Omega \cdot$m.

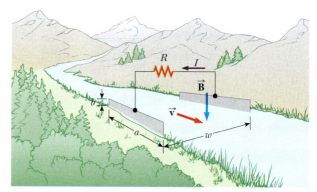

Figure P20.64

65. A horizontal wire is free to slide on the vertical rails of a conducting frame, as in Figure P20.65 (page 692). The wire has mass m and length ℓ, and the resistance of the circuit is R. If a uniform magnetic field is directed perpendicular to

[2]The idea for this problem and Figure P20.64 is from Oleg D. Jefimenko, *Electricity and Magnetism: An Introduction to the Theory of Electric and Magnetic Fields* (Star City, WV, Electret Scientific Co., 1989).

the frame, what is the terminal speed of the wire as it falls under the force of gravity? (Neglect friction.)

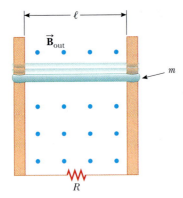

Figure P20.65

66. A one-turn coil of wire of area 0.20 m² and resistance 0.25 Ω is in a magnetic field that varies with time as shown in Figure P20.66a. The magnetic flux through the coil at $t = 0$ is as shown in Figure P20.66b. (a) When is the induced current the largest? (b) When is it zero? (c) Is the induced current always in the same direction? (d) Find the direction (clockwise or counterclockwise) and magnitude of the current between times $t = 0$ and $t = 2.0$ s, between $t = 2.0$ s and $t = 4.0$ s, and between $t = 4.0$ s and $t = 6.0$ s.

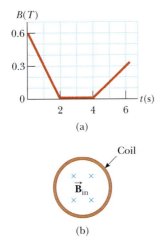

Figure P20.66

ACTIVITIES

1. Experimenting with induced currents is not easy. For small magnets and small coils of wire, the resulting induced currents are so small that they are difficult to detect. Thus, you may have to try this exercise several times before you are satisfied with your results. Wind a coil of wire on a cardboard mailing tube. Use insulated wire with as small a diameter as possible, because you need as many turns as possible on the coil. Connect the coil to a flashlight bulb, and see if you can get it to light by moving a bar magnet into and out of the coil in rapid succession. Why does the speed of movement make a difference? If you are unsuccessful, place two magnets side by side and repeat the experiment.

   After you have finished experimenting with the bulb, ask your instructor to let you use a galvanometer as a current detector. These devices are capable of measuring very small currents, and they have the added advantage of detecting the direction of the current in a circuit. Use your equipment to observe or test the following: (a) Does the magnitude of the induced current depend on the speed of movement of the magnet? (b) Can you induce a current by holding the magnet still and moving the coil over it? (c) Does the direction of the current depend on whether the magnet is pushed in or pulled out of the coil? (d) Does the direction of the current depend on whether the inserted pole of the magnet is the north pole or the south pole? (e) Can you predict the direction of the current by using Lenz's law? (f) Replace your bar magnet with the electromagnet you constructed in the last chapter, and repeat the preceding observations.

2. As explained in the text, a cassette tape is made up of tiny particles of metal oxide attached to a long plastic strip. Pull a tape out of a cassette that you do not mind destroying, and see if it is repelled or attracted by a refrigerator magnet. Also, try this with an expendable floppy computer disk.

3. This experiment takes steady hands, a dime, and a strong magnet. After verifying that a dime is not attracted to the magnet, carefully balance the coin on its edge. (This will not work with other coins, because they require too much force to topple them.) Hold one pole of the magnet within a millimeter of the face of the dime, but do not make contact with it. Now very rapidly pull the magnet straight back away from the coin. Which way does the dime tip? Does the coin fall the same way most of the time? Explain what is going on in terms of Lenz's law.

Arecibo, a large radio telescope in Puerto Rico, gathers electromagnetic radiation in the form of radio waves. These long wavelengths pass through obscuring dust clouds, allowing astronomers to create images of the core region of the Milky Way Galaxy, which can't be observed in the visible spectrum.

Alternating Current Circuits and Electromagnetic Waves

Every time we turn on a television set, a stereo system, or any of a multitude of other electric appliances, we call on alternating currents (AC) to provide the power to operate them. We begin our study of AC circuits by examining the characteristics of a circuit containing a source of emf and one other circuit element: a resistor, a capacitor, or an inductor. Then we examine what happens when these elements are connected in combination with each other. Our discussion is limited to simple series configurations of the three kinds of elements.

We conclude this chapter with a discussion of **electromagnetic waves**, which are composed of fluctuating electric and magnetic fields. Electromagnetic waves in the form of visible light enable us to view the world around us; infrared waves warm our environment; radio-frequency waves carry our television and radio programs, as well as information about processes in the core of our galaxy. X-rays allow us to perceive structures hidden inside our bodies, and study properties of distant, collapsed stars. Light is key to our understanding of the universe.

21.1   RESISTORS IN AN AC CIRCUIT

An AC circuit consists of combinations of circuit elements and an AC generator or an AC source, which provides the alternating current. We have seen that the output of an AC generator is sinusoidal and varies with time according to

$$\Delta v = \Delta V_{max} \sin 2\pi f t \qquad \text{[21.1]}$$

where Δv is the instantaneous voltage, ΔV_{max} is the maximum voltage of the AC generator, and f is the frequency at which the voltage changes, measured in hertz (Hz). (Compare Equations 20.7 and 20.8 with Equation 21.1.) We first consider a simple

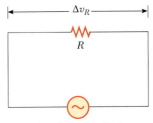

$$\Delta v = \Delta V_{max} \sin 2\pi ft$$

ACTIVE FIGURE 21.1

A series circuit consisting of a resistor R connected to an AC generator, designated by the symbol

.

Physics⊗Now™

Log into PhysicsNow at **www.cp7e.com** and go to Active Figure 21.1, where you can adjust the resistance, the frequency, and the maximum voltage of the circuit shown. The results can be studied with the graph and phasor diagram in Active Figure 21.2.

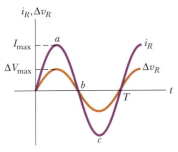

ACTIVE FIGURE 21.2

A plot of current and voltage across a resistor versus time.

Physics⊗Now™

Log into PhysicsNow at **www.cp7e.com** and go to Active Figure 21.2, where you can adjust the resistance, the frequency, and the maximum voltage of the circuit in Active Figure 21.1. The results can be studied with the graph and phasor diagram in Active Figure 21.20.

circuit consisting of a resistor and an AC source (designated by the symbol ⊸~⊸), as in Active Figure 21.1. The current and the voltage across the resistor are shown in Active Figure 21.2.

To explain the concept of alternating current, we begin by discussing the current-versus-time curve in Active Figure 21.2. At point *a* on the curve, the current has a maximum value in one direction, arbitrarily called the positive direction. Between points *a* and *b*, the current is decreasing in magnitude but is still in the positive direction. At point *b*, the current is momentarily zero; it then begins to increase in the opposite (negative) direction between points *b* and *c*. At point *c*, the current has reached its maximum value in the negative direction.

The current and voltage are in step with each other because they vary identically with time. **Because the current and the voltage reach their maximum values at the same time, they are said to be in phase.** Notice that **the average value of the current over one cycle is zero.** This is because the current is maintained in one direction (the positive direction) for the same amount of time and at the same magnitude as it is in the opposite direction (the negative direction). However, the direction of the current has no effect on the behavior of the resistor in the circuit: the collisions between electrons and the fixed atoms of the resistor result in an increase in the resistor's temperature regardless of the direction of the current.

We can quantify this discussion by recalling that the rate at which electrical energy is dissipated in a resistor, the power $\mathcal{P}$, is

$$\mathcal{P} = i^2 R$$

where i is the *instantaneous* current in the resistor. Because the heating effect of a current is proportional to the *square* of the current, it makes no difference whether the sign associated with the current is positive or negative. However, the heating effect produced by an alternating current with a maximum value of I_{max} *is not the same* as that produced by a direct current of the same value. The reason is that the alternating current has this maximum value for only an instant of time during a cycle. The important quantity in an AC circuit is a special kind of average value of current, called the **rms current**—the direct current that dissipates the same amount of energy in a resistor that is dissipated by the actual alternating current. To find the rms current, we first square the current, Then find its average value, and finally take the square root of this average value. Hence, the rms current is the square *root* of the average (*mean*) of the *square* of the current. Because i^2 varies as $\sin^2 2\pi ft$, the average value of i^2 is $\frac{1}{2} I^2_{max}$ (Fig. 21.3b).[1] Therefore, the rms current I_{rms} is related to the maximum value of the alternating current I_{max} by

$$I_{rms} = \frac{I_{max}}{\sqrt{2}} = 0.707 I_{max} \qquad [21.2]$$

This equation says that an alternating current with a maximum value of 3 A produces the same heating effect in a resistor as a direct current of $(3/\sqrt{2})$ A. We can therefore say that the average power dissipated in a resistor that carries alternating current I is

$$\mathcal{P}_{av} = I^2_{rms} R.$$

---

[1] The fact that $(i^2)_{av} = I^2_{max}/2$ can be shown as follows: The current in the circuit varies with time according to the expression $i = I_{max} \sin 2\pi ft$, so $i^2 = I^2_{max} \sin^2 2\pi ft$. Therefore, we can find the average value of i^2 by calculating the average value of $\sin^2 2\pi ft$. Note that a graph of $\cos^2 2\pi ft$ versus time is identical to a graph of $\sin^2 2\pi ft$ versus time, except that the points are shifted on the time axis. Thus, the time average of $\sin^2 2\pi ft$ is equal to the time average of $\cos^2 2\pi ft$, taken over one or more cycles. That is,

$$(\sin^2 2\pi ft)_{av} = (\cos^2 2\pi ft)_{av}$$

With this fact and the trigonometric identity $\sin^2 \theta + \cos^2 \theta = 1$, we get

$$(\sin^2 2\pi ft)_{av} + (\cos^2 2\pi ft)_{av} = 2(\sin^2 2\pi ft)_{av} = 1$$

$$(\sin^2 2\pi ft)_{av} = \frac{1}{2}$$

When this result is substituted into the expression $i^2 = I^2_{max} \sin^2 2\pi ft$, we get $(i^2)_{av} = I^2_{rms} = I^2_{max}/2$, or $I_{rms} = I_{max}/\sqrt{2}$, where I_{rms} is the rms current.

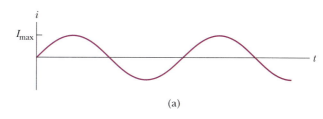

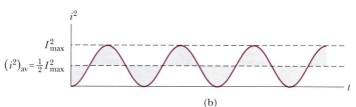

Figure 21.3   (a) Plot of the current in a resistor as a function of time. (b) Plot of the square of the current in a resistor as a function of time. Notice that the gray shaded regions *under* the curve and *above* the dashed line for $I_{max}^2/2$ have the same area as the gray shaded regions *above* the curve and *below* the dashed line for $I_{max}^2/2$. Thus, the average value of I^2 is $I_{max}^2/2$.

Alternating voltages are also best discussed in terms of rms voltages, with a relationship identical to the preceding one,

$$\Delta V_{rms} = \frac{\Delta V_{max}}{\sqrt{2}} = 0.707\,\Delta V_{max} \qquad \text{[21.3]}$$

◀ rms voltage

where ΔV_{rms} is the rms voltage and ΔV_{max} is the maximum value of the alternating voltage.

When we speak of measuring an AC voltage of 120 V from an electric outlet, we really mean an rms voltage of 120 V. A quick calculation using Equation 21.3 shows that such an AC voltage actually has a peak value of about 170 V. In this chapter we use rms values when discussing alternating currents and voltages. One reason is that AC ammeters and voltmeters are designed to read rms values. Further, if we use rms values, many of the equations for alternating current will have the same form as those used in the study of direct-current (DC) circuits. Table 21.1 summarizes the notations used throughout the chapter.

Consider the series circuit in Figure 21.1, consisting of a resistor connected to an AC generator. A resistor impedes the current in an AC circuit, just as it does in a DC circuit. Ohm's law is therefore valid for an AC circuit, and we have

$$\Delta V_{R,rms} = I_{rms}R \qquad \text{[21.4a]}$$

The rms voltage across a resistor is equal to the rms current in the circuit times the resistance. This equation is also true if maximum values of current and voltage are used:

$$\Delta V_{R,max} = I_{max}R \qquad \text{[21.4b]}$$

TABLE 21.1

Notation Used in This Chapter

	Voltage	Current
Instantaneous value	Δv	i
Maximum value	ΔV_{max}	I_{max}
rms value	ΔV_{rms}	I_{rms}

Quick Quiz 21.1

Which of the following statements can be true for a resistor connected in a simple series circuit to an operating AC generator? (a) $\mathcal{P}_{av} = 0$ and $i_{av} = 0$ (b) $\mathcal{P}_{av} = 0$ and $i_{av} > 0$ (c) $\mathcal{P}_{av} > 0$ and $i_{av} = 0$ (d) $\mathcal{P}_{av} > 0$ and $i_{av} > 0$

EXAMPLE 21.1   What Is the rms Current?

Goal   Perform basic AC circuit calculations for a purely resistive circuit.

Problem   An AC voltage source has an output of $\Delta v = (2.00 \times 10^2\,\text{V})\sin 2\pi ft$. This source is connected to a 1.00×10^2-Ω resistor as in Figure 21.1. Find the rms voltage and rms current in the resistor.

Strategy   Compare the expression for the voltage output just given with the general form, $\Delta v = \Delta V_{max}\sin 2\pi ft$, finding the maximum voltage. Substitute this result into the expression for the rms voltage.

Solution

Obtain the maximum voltage by comparison of the given expression for the output with the general expression:

$$\Delta v = (2.00 \times 10^2 \text{ V}) \sin 2\pi ft \qquad \Delta v = \Delta V_{max} \sin 2\pi ft$$
$$\rightarrow \quad \Delta V_{max} = 2.00 \times 10^2 \text{ V}$$

Next, substitute into Equation 21.3 to find the rms voltage of the source:

$$\Delta V_{rms} = \frac{\Delta V_{max}}{\sqrt{2}} = \frac{2.00 \times 10^2 \text{ V}}{\sqrt{2}} = \boxed{141 \text{ V}}$$

Substitute this result into Ohm's law to find the rms current:

$$I_{rms} = \frac{\Delta V_{rms}}{R} = \frac{141 \text{ V}}{1.00 \times 10^2 \text{ }\Omega} = \boxed{1.41 \text{ A}}$$

Remarks   Notice how the concept of rms values allows the handling of an AC circuit quantitatively in much the same way as a DC circuit.

Exercise 21.1

Find the maximum current in the circuit and the average power delivered to the circuit.

Answer   2.00 A; 2.00×10^2 W

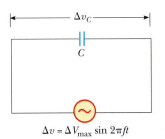

Figure 21.4   A series circuit consisting of a capacitor C connected to an AC generator.

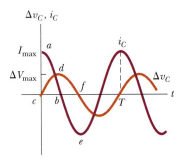

Figure 21.5   Plots of current and voltage across a capacitor versus time in an AC circuit. The voltage lags the current by 90°.

The voltage across a capacitor lags the current by 90° ▶

21.2   CAPACITORS IN AN AC CIRCUIT

To understand the effect of a capacitor on the behavior of a circuit containing an AC voltage source, we first review what happens when a capacitor is placed in a circuit containing a DC source, such as a battery. When the switch is closed in a series circuit containing a battery, a resistor, and a capacitor, the initial charge on the plates of the capacitor is zero. The motion of charge through the circuit is therefore relatively free, and there is a large current in the circuit. As more charge accumulates on the capacitor, the voltage across it increases, opposing the current. After some time interval, which depends on the time constant RC, the current approaches zero. Consequently, a capacitor in a DC circuit limits or impedes the current so that it approaches zero after a brief time.

Now consider the simple series circuit in Figure 21.4, consisting of a capacitor connected to an AC generator. We sketch curves of current versus time and voltage versus time, and then attempt to make the graphs seem reasonable. The curves are shown in Figure 21.5. First, note that the segment of the current curve from a to b indicates that the current starts out at a rather large value. This can be understood by recognizing that there is no charge on the capacitor at $t = 0$; as a consequence, there is nothing in the circuit except the resistance of the wires to hinder the flow of charge at this instant. However, the current decreases as the voltage across the capacitor increases from c to d on the voltage curve. When the voltage is at point d, the current reverses and begins to increase in the opposite direction (from b to e on the current curve). During this time, the voltage across the capacitor decreases from d to f because the plates are now losing the charge they accumulated earlier. The remainder of the cycle for both voltage and current is a repeat of what happened during the first half of the cycle. The current reaches a maximum value in the opposite direction at point e on the current curve and then decreases as the voltage across the capacitor builds up.

In a purely resistive circuit, the current and voltage are always in step with each other. This isn't the case when a capacitor is in the circuit. In Figure 21.5, when an alternating voltage is applied across a capacitor, the voltage reaches its maximum value one-quarter of a cycle after the current reaches its maximum value. We say that **the voltage across a capacitor always lags the current by 90°**.

The impeding effect of a capacitor on the current in an AC circuit is expressed in terms of a factor called the **capacitive reactance** X_C, defined as

Capacitive reactance ▶

$$X_C \equiv \frac{1}{2\pi fC}$$ [21.5]

When C is in farads and f is in hertz, the unit of X_C is the ohm. Notice that $2\pi f = \omega$, the angular frequency.

From Equation 21.5, as the frequency f of the voltage source increases, the capacitive reactance X_C (the impeding effect of the capacitor) decreases, so the current increases. At high frequency, there is less time available to charge the capacitor, so less charge and voltage accumulate on the capacitor, which translates into less opposition to the flow of charge and, consequently, a higher current. The analogy between capacitive reactance and resistance means that we can write an equation of the same form as Ohm's law to describe AC circuits containing capacitors. This equation relates the rms voltage and rms current in the circuit to the capacitive reactance:

$$\Delta V_{C,\text{rms}} = I_{\text{rms}} X_C \qquad [21.6]$$

EXAMPLE 21.2   A Purely Capacitive AC Circuit

Goal   Perform basic AC circuit calculations for a capacitive circuit.

Problem   An 8.00-μF capacitor is connected to the terminals of an AC generator with an rms voltage of 1.50×10^2 V and a frequency of 60.0 Hz. Find the capacitive reactance and the rms current in the circuit.

Strategy   Substitute values into Equations 21.5 and 21.6.

Solution

Substitute the values of f and C into Equation 21.5:

$$X_C = \frac{1}{2\pi f C} = \frac{1}{2\pi \,(60.0 \text{ Hz})\,(8.00 \times 10^{-6} \text{ F})} = \boxed{332 \;\Omega}$$

Solve Equation 21.6 for the current, and substitute X_C and the rms voltage to find the rms current:

$$I_{\text{rms}} = \frac{\Delta V_{C,\text{rms}}}{X_C} = \frac{1.50 \times 10^2 \text{ V}}{332 \;\Omega} = \boxed{0.452 \text{ A}}$$

Remark   Again, notice how similar the technique is to that of analyzing a DC circuit with a resistor.

Exercise 21.2

If the frequency is doubled, what happens to the capacitive reactance and the rms current?

Answer X_C is halved, and I_{rms} is doubled.

21.3   INDUCTORS IN AN AC CIRCUIT

Now consider an AC circuit consisting only of an inductor connected to the terminals of an AC source, as in Active Figure 21.6. (In any real circuit, there is some resistance in the wire forming the inductive coil, but we ignore this for now.) The changing current output of the generator produces a back emf that impedes the current in the circuit. The magnitude of this back emf is

$$\Delta v_L = L \frac{\Delta I}{\Delta t} \qquad [21.7]$$

The effective resistance of the coil in an AC circuit is measured by a quantity called the **inductive reactance**, X_L:

$$X_L \equiv 2\pi f L \qquad [21.8]$$

When f is in hertz and L is in henries, the unit of X_L is the ohm. The inductive reactance *increases* with increasing frequency and increasing inductance. Contrast these facts with capacitors, where increasing frequency or capacitance *decreases* the capacitive reactance.

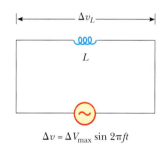

$$\Delta v = \Delta V_{\text{max}} \sin 2\pi f t$$

ACTIVE FIGURE 21.6
A series circuit consisting of an inductor L connected to an AC generator.

Physics⊗Now™
Log into PhysicsNow at **www.cp7e.com** and go to Active Figure 21.6, where you can adjust the inductance, the frequency, and the maximum voltage. The results can be studied with the graph and phasor diagram in Active Figure 21.7.

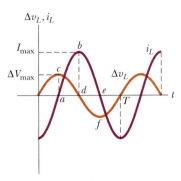

ACTIVE FIGURE 21.7

Plots of current and voltage across an inductor versus time in an AC circuit. The voltage leads the current by 90°.

Physics⊗Now™

Log into PhysicsNow at **www.cp7e.com** and go to Active Figure 21.6, where you can adjust the inductance, the frequency, and the maximum voltage. The results can be studied with the graph and phasor diagram in Active Figure 21.7.

To understand the meaning of inductive reactance, compare Equation 21.8 with Equation 21.7. First, note from Equation 21.8 that the inductive reactance depends on the inductance L. This is reasonable, because the back emf (Eq. 21.7) is large for large values of L. Second, note that the inductive reactance depends on the frequency f. This, too, is reasonable, because the back emf depends on $\Delta I/\Delta t$, a quantity that is large when the current changes rapidly, as it would for high frequencies.

With inductive reactance defined in this way, we can write an equation of the same form as Ohm's law for the voltage across the coil or inductor:

$$\Delta V_{L,\text{rms}} = I_{\text{rms}} X_L \qquad [21.9]$$

where $\Delta V_{L,\text{rms}}$ is the rms voltage across the coil and I_{rms} is the rms current in the coil.

Active Figure 21.7 shows the instantaneous voltage and instantaneous current across the coil as functions of time. When a sinusoidal voltage is applied across an inductor, the voltage reaches its maximum value one-quarter of an oscillation period before the current reaches its maximum value. In this situation, we say that **the voltage across an inductor always leads the current by 90°.**

To see why there is a phase relationship between voltage and current, we examine a few points on the curves of Active Figure 21.7. At point a on the current curve, the current is beginning to increase in the positive direction. At this instant, the rate of change of current, $\Delta I/\Delta t$ (the slope of the current curve), is at a maximum, and we see from Equation 21.7 that the voltage across the inductor is consequently also at a maximum. As the current rises between points a and b on the curve, $\Delta I/\Delta t$ gradually decreases until it reaches zero at point b. As a result, the voltage across the inductor is decreasing during this same time interval, as the segment between c and d on the voltage curve indicates. Immediately after point b, the current begins to decrease, although it still has the same direction it had during the previous quarter cycle. As the current decreases to zero (from b to e on the curve), a voltage is again induced in the coil (from d to f), but the polarity of this voltage is opposite the polarity of the voltage induced between c and d. This occurs because back emfs always oppose the change in the current.

We could continue to examine other segments of the curves, but no new information would be gained because the current and voltage variations are repetitive.

EXAMPLE 21.3   A Purely Inductive AC Circuit

Goal   Perform basic AC circuit calculations for an inductive circuit.

Problem   In a purely inductive AC circuit (see Active Fig. 21.6), $L = 25.0$ mH and the rms voltage is 1.50×10^2 V. Find the inductive reactance and rms current in the circuit if the frequency is 60.0 Hz.

Solution

Substitute L and f into Equation 21.8 to get the inductive reactance:

$$X_L = 2\pi f L = 2\pi (60.0 \text{ s}^{-1})(25.0 \times 10^{-3} \text{ H}) = \boxed{9.42 \ \Omega}$$

Solve Equation 21.9 for the rms current and substitute:

$$I_{\text{rms}} = \frac{\Delta V_{L,\text{rms}}}{X_L} = \frac{1.50 \times 10^2 \text{ V}}{9.42 \ \Omega} = \boxed{15.9 \text{ A}}$$

Remark   The analogy with DC circuits is even closer than in the capacitive case, because in the inductive equivalent of Ohm's law, the voltage across an inductor is *proportional* to the inductance L, just as the voltage across a resistor is proportional to R in Ohm's law.

Exercise 21.3

Calculate the inductive reactance and rms current in a similar circuit if the frequency is again 60.0 Hz, but the rms voltage is 85.0 V and the inductance is 47.0 mH.

Answers $X_L = 17.7 \ \Omega$; $I = 4.80$ A

21.4   THE *RLC* SERIES CIRCUIT

In the foregoing sections, we examined the effects of an inductor, a capacitor, and a resistor when they are connected separately across an AC voltage source. We now consider what happens when these devices are combined.

Active Figure 21.8 shows a circuit containing a resistor, an inductor, and a capacitor connected in series across an AC source that supplies a total voltage Δv at some instant. The current in the circuit is the same at all points in the circuit at any instant and varies sinusoidally with time, as indicated in Active Figure 21.9a. This fact can be expressed mathematically as

$$i = I_{max} \sin 2\pi ft$$

Earlier, we learned that the voltage across each element may or may not be in phase with the current. The instantaneous voltages across the three elements, shown in Active Figure 21.9, have the following phase relations to the instantaneous current:

1. The instantaneous voltage Δv_R across the resistor is *in phase* with the instantaneous current. (See Active Fig. 21.9b.)
2. The instantaneous voltage Δv_L across the inductor *leads* the current by 90°. (See Active Fig. 21.9c.)
3. The instantaneous voltage Δv_C across the capacitor *lags* the current by 90°. (See Active Fig. 21.9d.)

The net instantaneous voltage Δv supplied by the AC source equals the sum of the instantaneous voltages across the separate elements: $\Delta v = \Delta v_R + \Delta v_C + \Delta v_L$. This doesn't mean, however, that the voltages measured with an AC voltmeter across R, C, and L sum to the measured source voltage! In fact, the measured voltages *don't* sum to the measured source voltage, because the voltages across R, C, and L all have different phases.

To account for the different phases of the voltage drops, we use a technique involving vectors. We represent the voltage across each element with a rotating vector, as in Figure 21.10. The rotating vectors are referred to as **phasors**, and the diagram is called a **phasor diagram**. This particular diagram represents the circuit voltage given by the expression $\Delta v = \Delta V_{max} \sin(2\pi ft + \phi)$, where ΔV_{max} is the maximum voltage (the magnitude or length of the rotating vector or phasor) and ϕ is the angle between the phasor and the $+$ x-axis when $t = 0$. The phasor can be viewed as a vector of magnitude ΔV_{max} rotating at a constant frequency f so that its projection along the y-axis is the instantaneous voltage in the circuit. Because ϕ is the phase angle between the voltage and current in the circuit, the phasor for the current (not shown in Fig. 21.10) lies along the positive x-axis when $t = 0$ and is expressed by the relation $i = I_{max} \sin(2\pi ft)$.

The phasor diagrams in Figure 21.11 (page 700) are useful for analyzing the *series RLC* circuit. Voltages in phase with the current are represented by vectors along the positive x-axis, and voltages out of phase with the current lie along other directions. ΔV_R is horizontal and to the right because it's in phase with the current. Likewise, ΔV_L is represented by a phasor along the positive y-axis because it leads the current by 90°. Finally, ΔV_C is along the negative y-axis because it lags the current[2] by 90°. If the phasors are added as vector quantities in order to account for the different phases of the voltages across R, L, and C, Figure 21.11a shows that the only x-component for the voltages is ΔV_R and the net y-component is $\Delta V_L - \Delta V_C$. We now add the phasors vectorially to find the phasor ΔV_{max} (Fig. 21.11b), which represents the maximum voltage. The right triangle in Figure 21.11b gives the following equations for the maximum voltage and the phase angle ϕ between the maximum voltage and the current:

---

[2]A mnemonic to help you remember the phase relationships in *RLC* circuits is " *ELI* the *ICE* man." *E* represents the voltage ε, *I* the current, *L* the inductance, and *C* the capacitance. Thus, the name *ELI* means that, in an inductive circuit, the voltage ε leads the current *I*. In a capacitive circuit, *ICE* means that the current leads the voltage.

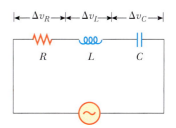

ACTIVE FIGURE 21.8
A series circuit consisting of a resistor, an inductor, and a capacitor connected to an AC generator.

Physics ⊗ Now™
Log into PhysicsNow at **www.cp7e.com** and go to Active Figure 21.8, where you can adjust the resistance, the inductance, and the capacitance. The results can be studied with the graph in Active Figure 21.9 and the phasor diagram in Figure 21.10.

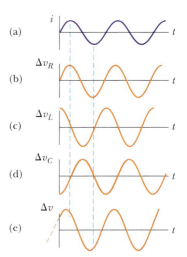

ACTIVE FIGURE 21.9
Phase relations in the series *RLC* circuit shown in Figure 21.8.

Physics ⊗ Now™
Log into PhysicsNow at **www.cp7e.com** and go to Active Figure 21.9, where you can adjust the resistance, the inductance, and the capacitance in Active Figure 21.8. The results can be studied with the graph in this figure and the phasor diagram in Figure

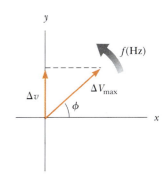

Figure 21.10   A phasor diagram for the voltage in an AC circuit, where ϕ is the phase angle between the voltage and the current and Δv is the instantaneous voltage.

Figure 21.11 (a) A phasor diagram for the *RLC* circuit. (b) Addition of the phasors as vectors gives $\Delta V_{\text{max}} = \sqrt{\Delta V_R{}^2 + (\Delta V_L - \Delta V_C)^2}$. (c) The reactance triangle that gives the impedance relation $Z = \sqrt{R^2 + (X_L - X_C)^2}$.

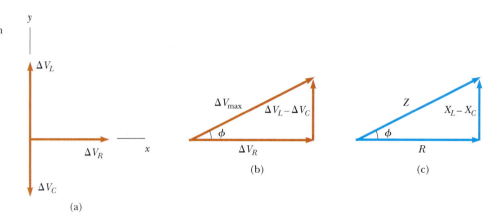

$$\Delta V_{\text{max}} = \sqrt{\Delta V_R{}^2 + (\Delta V_L - \Delta V_C)^2} \qquad [21.10]$$

$$\tan \phi = \frac{\Delta V_L - \Delta V_C}{\Delta V_R} \qquad [21.11]$$

In these equations, all voltages are maximum values. Although we choose to use maximum voltages in our analysis, the preceding equations apply equally well to rms voltages, because the two quantities are related to each other by the same factor for all circuit elements. The result for the maximum voltage ΔV_{max} given by Equation 21.10 reinforces the fact that **the voltages across the resistor, capacitor, and inductor are not in phase, so one cannot simply add them to get the voltage across the combination of element, or the source voltage.**

Quick Quiz 21.2

For the circuit of Figure 21.8, is the instantaneous voltage of the source equal to (a) the sum of the maximum voltages across the elements, (b) the sum of the instantaneous voltages across the elements, or (c) the sum of the rms voltages across the elements?

We can write Equation 21.10 in the form of Ohm's law, using the relations $\Delta V_R = I_{\text{max}}R$, $\Delta V_L = I_{\text{max}}X_L$, and $\Delta V_C = I_{\text{max}}X_C$, where I_{max} is the maximum current in the circuit:

$$\Delta V_{\text{max}} = I_{\text{max}}\sqrt{R^2 + (X_L - X_C)^2} \qquad [21.12]$$

It's convenient to define a parameter called the **impedance** Z of the circuit as

Impedance ▶

$$Z \equiv \sqrt{R^2 + (X_L - X_C)^2} \qquad [21.13]$$

so that Equation 21.12 becomes

$$\Delta V_{\text{max}} = I_{\text{max}}Z \qquad [21.14]$$

Equation 21.14 is in the form of Ohm's law, $\Delta V = IR$, with R replaced by the impedance in ohms. Indeed, Equation 21.14 can be regarded as a generalized form of Ohm's law applied to a series AC circuit. Both the impedance and, therefore, the current in an AC circuit depend on the resistance, the inductance, the capacitance, *and* the frequency (because the reactances are frequency dependent).

It's useful to represent the impedance Z with a vector diagram such as the one depicted in Figure 21.11c. A right triangle is constructed with right side $X_L - X_C$, base R, and hypotenuse Z. Applying the Pythagorean theorem to this triangle, we see that

$$Z - \sqrt{R^2 + (X_L - X_C)^2}$$

which is Equation 21.13. Furthermore, we see from the vector diagram in Figure 21.11c that the phase angle ϕ between the current and the voltage obeys the

TABLE 21.2

Impedance Values and Phase Angles for Various Combinations of Circuit Elements[a]

Circuit Elements	Impedance Z	Phase Angle ϕ
R	R	$0°$
C	X_C	$-90°$
L	X_L	$+90°$
R C	$\sqrt{R^2 + X_C^2}$	Negative, between $-90°$ and $0°$
R L	$\sqrt{R^2 + X_L^2}$	Positive, between $0°$ and $90°$
R L C	$\sqrt{R^2 + (X_L - X_C)^2}$	Negative if $X_C > X_L$ Positive if $X_C < X_L$

[a] In each case, an AC voltage (not shown) is applied across the combination of elements (that is, across the dots).

relationship

$$\tan \phi = \frac{X_L - X_C}{R} \qquad [21.15]$$

◀ Phase angle ϕ

The physical significance of the phase angle will become apparent in Section 21.5.

Table 21.2 provides impedance values and phase angles for some series circuits containing different combinations of circuit elements.

Parallel alternating current circuits are also useful in everyday applications. We won't discuss them here, however, because their analysis is beyond the scope of this book.

Quick Quiz 21.3

The switch in the circuit shown in Figure 21.12 is closed and the lightbulb glows steadily. The inductor is a simple air-core solenoid. As an iron rod is being inserted into the interior of the solenoid, the brightness of the lightbulb (a) increases, (b) decreases, or (c) remains the same.

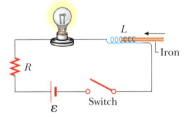

Figure 21.12 (Quick Quiz 21.3)

NIKOLA TESLA (1856–1943)

Tesla was born in Croatia, but spent most of his professional life as an inventor in the United States. He was a key figure in the development of alternating-current electricity, high-voltage transformers, and the transport of electrical power via AC transmission lines. Tesla's viewpoint was at odds with the ideas of Edison, who committed himself to the use of direct current in power transmission. Tesla's AC approach won out.

Problem-Solving Strategy   Alternating Current

The following procedure is recommended for solving alternating-current problems:
1. Calculate as many of the unknown quantities, such as X_L and X_C, as possible.
2. Apply the equation $\Delta V_{max} = I_{max} Z$ to the portion of the circuit of interest. For example, if you want to know the voltage drop across the combination of an inductor and a resistor, the equation for the voltage drop reduces to $\Delta V_{max} = I_{max}\sqrt{R^2 + X_L^2}$.

EXAMPLE 21.4   An *RLC* Circuit

Goal Analyze a series *RLC* AC circuit and find the phase angle.

Problem A series *RLC* AC circuit has resistance $R = 2.50 \times 10^2 \, \Omega$, inductance $L = 0.600$ H, capacitance $C = 3.50 \, \mu$F, frequency $f = 60.0$ Hz, and maximum voltage $\Delta V_{max} = 1.50 \times 10^2$ V. Find **(a)** the impedance, **(b)** the maximum current in the circuit, **(c)** the phase angle, and **(d)** the maximum voltages across the elements.

Strategy   Calculate the inductive and capacitive reactances, then substitute them and given quantities into the appropriate equations.

Solution

(a) Find the impedance of the circuit.

First, calculate the inductive and capacitive reactances:

$X_L = 2\pi f L = 226\ \Omega \qquad X_C = 1/2\pi f C = 758\ \Omega$

Substitute these results and the resistance R into Equation 21.13 to obtain the impedance of the circuit:

$Z = \sqrt{R^2 + (X_L - X_C)^2}$

$= \sqrt{(2.50 \times 10^2\ \Omega)^2 + (226\ \Omega - 758\ \Omega)^2} = \boxed{588\ \Omega}$

(b) Find the maximum current.

Use Equation 21.12, the equivalent of Ohm's law, to find the maximum current:

$I_{max} = \dfrac{\Delta V_{max}}{Z} = \dfrac{1.50 \times 10^2\ \text{V}}{588\ \Omega} = \boxed{0.255\ \text{A}}$

(c) Find the phase angle.

Calculate the phase angle between the current and the voltage with Equation 21.15:

$\phi = \tan^{-1}\dfrac{X_L - X_C}{R} = \tan^{-1}\left(\dfrac{226\ \Omega - 758\ \Omega}{2.50 \times 10^2\ \Omega}\right) = \boxed{-64.8°}$

(d) Find the maximum voltages across the elements.

Substitute into the "Ohm's law" expressions for each individual type of current element:

$\Delta V_{R,\,max} = I_{max}R = (0.255\ \text{A})(2.50 \times 10^2\ \Omega) = \boxed{63.8\ \text{V}}$

$\Delta V_{L,\,max} = I_{max}X_L = (0.255\ \text{A})(2.26 \times 10^2\ \Omega) = \boxed{57.6\ \text{V}}$

$\Delta V_{C,\,max} = I_{max}X_C = (0.255\ \text{A})(7.58 \times 10^2\ \Omega) = \boxed{193\ \text{V}}$

Remarks   Because the circuit is more capacitive than inductive ($X_C > X_L$), ϕ is negative. A negative phase angle means that the current leads the applied voltage. Notice also that the sum of the maximum voltages across the elements is $\Delta V_R + \Delta V_L + \Delta V_C = 314$ V, which is much greater than the maximum voltage of the generator, 150 V. As we saw in Quick Quiz 21.2, the sum of the maximum voltages is a meaningless quantity because when alternating voltages are added, *both their amplitudes and their phases* must be taken into account. We know that the maximum voltages across the various elements occur at different times, so it doesn't make sense to add all the maximum values. The correct way to "add" the voltages is through Equation 21.10.

Exercise 21.4

Analyze a series RLC AC circuit for which $R = 175\ \Omega$, $L = 0.500$ H, $C = 22.5\ \mu$F, $f = 60.0$ Hz, and $\Delta V_{max} = 325$ V. Find (a) the impedance, (b) the maximum current, (c) the phase angle, and (d) the maximum voltages across the elements.

Answers   (a) 189 Ω (b) 1.72 A   (c) 22.0°   (d) $\Delta V_{R,max} = 301$ V, $\Delta V_{L,max} = 324$ V, $\Delta V_{C,max} = 203$ V

21.5   POWER IN AN AC CIRCUIT

No power losses are associated with pure capacitors and pure inductors in an AC circuit. A pure capacitor, by definition, has no resistance or inductance, while a pure inductor has no resistance or capacitance. (These are idealizations: in a real capacitor, for example, inductive effects could become important at high frequencies.) We begin by analyzing the power dissipated in an AC circuit that contains only a generator and a capacitor.

When the current increases in one direction in an AC circuit, charge accumulates on the capacitor and a voltage drop appears across it. When the voltage reaches its maximum value, the energy stored in the capacitor is

$$PE_C = \tfrac{1}{2}C(\Delta V_{max})^2$$

However, this energy storage is only momentary: When the current reverses direction, the charge leaves the capacitor plates and returns to the voltage source. During one-half of each cycle the capacitor is being charged, and during the other half the charge is being returned to the voltage source. Therefore, the average power supplied by the source is zero. In other words, **no power losses occur in a capacitor in an AC circuit**.

Similarly, the source must do work against the back emf of an inductor that is carrying a current. When the current reaches its maximum value, the energy stored in the inductor is a maximum and is given by

$$PE_L = \tfrac{1}{2}LI^2_{max}$$

When the current begins to decrease in the circuit, this stored energy is returned to the source as the inductor attempts to maintain the current in the circuit. The average power delivered to a resistor in an RLC circuit is

$$\mathcal{P}_{av} = I^2_{rms}R \qquad\qquad \textbf{[21.16]}$$

The average power delivered by the generator is converted to internal energy in the resistor. No power loss occurs in an ideal capacitor or inductor.

An alternate equation for the average power loss in an AC circuit can be found by substituting (from Ohm's law) $R = \Delta V_R/I_{rms}$ into Equation 21.16:

$$\mathcal{P}_{av} = I_{rms}\Delta V_R$$

It's convenient to refer to a voltage triangle that shows the relationship among ΔV_{rms}, ΔV_R, and $\Delta V_L - \Delta V_C$, such as Figure 21.11b. (Remember that Fig. 21.11 applies to *both* maximum and rms voltages.) From this figure, we see that the voltage drop across a resistor can be written in terms of the voltage of the source, ΔV_{rms}:

$$\Delta V_R = \Delta V_{rms} \cos \phi$$

Hence, the average power delivered by a generator in an AC circuit is

$$\mathcal{P}_{av} = I_{rms}\Delta V_{rms} \cos \phi \qquad\qquad \textbf{[21.17]}$$ ◀ Average power

where the quantity $\cos \phi$ is called the **power factor**.

Equation 21.17 shows that the power delivered by an AC source to any circuit depends on the phase difference between the source voltage and the resulting current. This fact has many interesting applications. For example, factories often use devices such as large motors in machines, generators, and transformers that have a large inductive load due to all the windings. To deliver greater power to such devices without using excessively high voltages, factory technicians introduce capacitance in the circuits to shift the phase.

APPLICATION

Shifting Phase to Deliver More Power

EXAMPLE 21.5   Average Power in an *RLC* Series Circuit

Goal   Understand power in *RLC* series circuits.

Problem   Calculate the average power delivered to the series *RLC* circuit described in Example 21.4.

Strategy   After finding the rms current and rms voltage with Equations 21.2 and 21.3, substitute into Equation 21.17, using the phase angle found in Example 21.4.

Solution

First, use Equations 21.2 and 21.3 to calculate the rms current and rms voltage:

$$I_{rms} = \frac{I_{max}}{\sqrt{2}} = \frac{0.255\ \text{A}}{\sqrt{2}} = 0.180\ \text{A}$$

$$\Delta V_{rms} = \frac{\Delta V_{max}}{\sqrt{2}} = \frac{1.50 \times 10^2\ \text{V}}{\sqrt{2}} = 106\ \text{V}$$

Substitute these results and the phase angle $\phi = -64.8°$ into Equation 21.17 to find the average power:

$$\mathcal{P}_{av} = I_{rms}\Delta V_{rms} \cos \phi = (0.180\ \text{A})(106\ \text{V})\cos(-64.8°)$$

$$= \boxed{8.12\ \text{W}}$$

Remark The same result can be obtained from Equation 21.16, $\mathcal{P}_{av} = I_{rms}^2 R$.

Exercise 21.5

Repeat this problem, using the system described in Exercise 21.4.

Answer 259 W

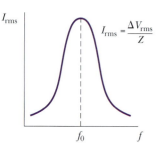

Figure 21.13 A plot of current amplitude in a series *RLC* circuit versus frequency of the generator voltage. Note that the current reaches its maximum value at the resonance frequency f_0.

Resonance frequency ▶

21.6 RESONANCE IN A SERIES *RLC* CIRCUIT

In general, the rms current in a series *RLC* circuit can be written

$$I_{rms} = \frac{\Delta V_{rms}}{Z} = \frac{\Delta V_{rms}}{\sqrt{R^2 + (X_L - X_C)^2}} \qquad [21.18]$$

From this equation, we see that if the frequency is varied, the current has its *maximum* value when the impedance has its *minimum* value. This occurs when $X_L = X_C$. In such a circumstance, the impedance of the circuit reduces to $Z = R$. The frequency f_0 at which this happens is called the **resonance frequency** of the circuit. To find f_0, we set $X_L = X_C$, which gives, from Equations 21.5 and 21.8,

$$2\pi f_0 L = \frac{1}{2\pi f_0 C}$$

$$f_0 = \frac{1}{2\pi\sqrt{LC}} \qquad [21.19]$$

Figure 21.13 is a plot of current as a function of frequency for a circuit containing a fixed value for both the capacitance and the inductance. From Equation 21.18, it must be concluded that the current would become infinite at resonance when $R = 0$. Although Equation 21.18 predicts this result, real circuits always have some resistance, which limits the value of the current.

The tuning circuit of a radio is an important application of a series resonance circuit. The radio is tuned to a particular station (which transmits a specific radio-frequency signal) by varying a capacitor, which changes the resonance frequency of the tuning circuit. When this resonance frequency matches that of the incoming radio wave, the current in the tuning circuit increases.

APPLICATION

Tuning Your Radio

Applying Physics 21.1 Metal Detectors in Airports

When you walk through the doorway of an airport metal detector, as the person in Figure 21.14 is doing, you are really walking through a coil of many turns. How might the metal detector work?

Explanation The metal detector is essentially a resonant circuit. The portal you step through is an inductor (a large loop of conducting wire) that is part of the circuit. The frequency of the circuit is tuned to the resonant frequency of the circuit when there is no metal in the inductor. When you walk through with metal in your pocket, you change the effective inductance of the resonance circuit, resulting in a change in the current in the circuit. This change in current is detected, and an electronic circuit causes a sound to be emitted as an alarm.

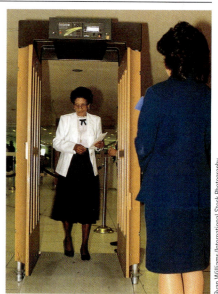

Figure 21.14 (Applying Physics 21.1) An airport metal detector.

EXAMPLE 21.6 A Circuit in Resonance

Goal Understand resonance frequency and its relation to inductance, capacitance, and the rms current.

Problem Consider a series RLC circuit for which $R = 1.50 \times 10^2\ \Omega$, $L = 20.0$ mH, $\Delta V_{rms} = 20.0$ V, and $f = 796\ \text{s}^{-1}$. (a) Determine the value of the capacitance for which the rms current is a maximum. (b) Find the maximum rms current in the circuit.

Strategy The current is a maximum at the resonance frequency f_0, which should be set equal to the driving frequency, $796\ \text{s}^{-1}$. The resulting equation can be solved for C. For part (b), substitute into Equation 21.18 to get the maximum rms current.

Solution
(a) Find the capacitance giving the maximum current in the circuit (the resonance condition).

Solve the resonance frequency for the capacitance:

$$f_0 = \frac{1}{2\pi\sqrt{LC}} \rightarrow \sqrt{LC} = \frac{1}{2\pi f_0} \rightarrow LC = \frac{1}{4\pi^2 f_0^2}$$

$$C = \frac{1}{4\pi^2 f_0^2 L}$$

Insert the given values, substituting the source frequency for the resonance frequency, f_0:

$$C = \frac{1}{4\pi^2(796\ \text{Hz})^2(20.0 \times 10^{-3}\ \text{H})} = 2.00 \times 10^{-6}\ \text{F}$$

(b) Find the maximum rms current in the circuit.

The capacitive and inductive reactances are equal, so $Z = R = 1.50 \times 10^2\ \Omega$. Substitute into Equation 21.18 to find the rms current:

$$I_{rms} = \frac{\Delta V_{rms}}{Z} = \frac{20.0\ \text{V}}{1.50 \times 10^2\ \Omega} = 0.133\ \text{A}$$

Remark Because the impedance Z is in the denominator of Equation 21.18, the maximum current will always occur when $X_L = X_C$, since that yields the minimum value of Z.

Exercise 21.6
Consider a series RLC circuit for which $R = 1.20 \times 10^2\ \Omega$, $C = 3.10 \times 10^{-5}$ F, $\Delta V_{rms} = 35.0$ V, and $f = 60.0\ \text{s}^{-1}$. (a) Determine the value of the inductance for which the rms current is a maximum. (b) Find the maximum rms current in the circuit.

Answers (a) 0.227 H (b) 0.292 A

21.7 THE TRANSFORMER

It's often necessary to change a small AC voltage to a larger one or vice versa. Such changes are effected with a device called a transformer.

In its simplest form, the **AC transformer** consists of two coils of wire wound around a core of soft iron, as shown in Figure 21.15. The coil on the left, which is connected to the input AC voltage source and has N_1 turns, is called the primary winding, or the *primary*. The coil on the right, which is connected to a resistor R and consists of N_2 turns, is the *secondary*. The purpose of the common iron core is to increase the magnetic flux and to provide a medium in which nearly all the flux through one coil passes through the other.

When an input AC voltage ΔV_1 is applied to the primary, the induced voltage across it is given by

$$\Delta V_1 = -N_1 \frac{\Delta \Phi_B}{\Delta t} \qquad [21.20]$$

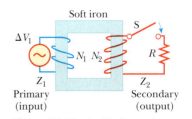

Figure 21.15 An ideal transformer consists of two coils wound on the same soft iron core. An AC voltage ΔV_1 is applied to the primary coil, and the output voltage ΔV_2 is observed across the load resistance R after the switch is closed.

where Φ_B is the magnetic flux through each turn. If we assume that no flux leaks from the iron core, then the flux through each turn of the primary equals the flux through each turn of the secondary. Hence, the voltage across the secondary coil is

$$\Delta V_2 = -N_2 \frac{\Delta \Phi_B}{\Delta t} \qquad \text{[21.21]}$$

The term $\Delta \Phi_B / \Delta t$ is common to Equations 21.20 and 21.21 and can be algebraically eliminated, giving

$$\Delta V_2 = \frac{N_2}{N_1} \Delta V_1 \qquad \text{[21.22]}$$

When N_2 is greater than N_1, ΔV_2 exceeds ΔV_1 and the transformer is referred to as a *step-up transformer*. When N_2 is less than N_1, making ΔV_2 less than ΔV_1, we have a *step-down transformer*.

By Faraday's law, a voltage is generated across the secondary only when there is a *change* in the number of flux lines passing through the secondary. The input current in the primary must therefore change with time, which is what happens when an alternating current is used. When the input at the primary is a direct current, however, a voltage output occurs at the secondary only at the instant a switch in the primary circuit is opened or closed. Once the current in the primary reaches a steady value, the output voltage at the secondary is zero.

It may seem that a transformer is a device in which it is possible to get something for nothing. For example, a step-up transformer can change an input voltage from, say, 10 V to 100 V. This means that each coulomb of charge leaving the secondary has 100 J of energy, whereas each coulomb of charge entering the primary has only 10 J of energy. That is not the case, however, because **the power input to the primary equals the power output at the secondary**:

◀ In an ideal transformer, the input power equals the output power.

$$I_1 \Delta V_1 = I_2 \Delta V_2 \qquad \text{[21.23]}$$

While the *voltage* at the secondary may be, say, ten times greater than the voltage at the primary, the *current* in the secondary will be smaller than the primary's current by a factor of ten. Equation 21.23 assumes an **ideal transformer**, in which there are no power losses between the primary and the secondary. Real transformers typically have power efficiencies ranging from 90% to 99%. Power losses occur because of such factors as eddy currents induced in the iron core of the transformer, which dissipate energy in the form of I^2R losses.

APPLICATION

Long-Distance Electric Power Transmission

When electric power is transmitted over large distances, it's economical to use a high voltage and a low current because the power lost via resistive heating in the transmission lines varies as I^2R. This means that if a utility company can reduce the current by a factor of ten, for example, the power loss is reduced by a factor of one hundred. In practice, the voltage is stepped up to around 230 000 V at the generating station, then stepped down to around 20 000 V at a distribution station, and finally stepped down to 120 V at the customer's utility pole.

EXAMPLE 21.7   Distributing Power to a City

Goal   Understand transformers and their role in reducing power loss.

Problem   A generator at a utility company produces 1.00×10^2 A of current at 4.00×10^3 V. The voltage is stepped up to 2.40×10^5 V by a transformer before being sent on a high-voltage transmission line across a rural area to a city. Assume that the effective resistance of the power line is 30.0 Ω and that the transformers are ideal. **(a)** Determine the percentage of power lost in the transmission line. **(b)** What percentage of the original power would be lost in the transmission line if the voltage were not stepped up?

Strategy   Solving this problem is just a matter of substitution into the equation for transformers and the equation for power loss. To obtain the fraction of power lost, it's also necessary to compute the power output of the generator—the current times the potential difference created by the generator.

Solution

(a) Determine the percentage of power lost in the line.

Substitute into Equation 21.23 to find the current in the transmission line:

$$I_2 = \frac{I_1 \Delta V_1}{\Delta V_2} = \frac{(1.00 \times 10^2 \text{ A})(4.00 \times 10^3 \text{ V})}{2.40 \times 10^5 \text{ V}} = 1.67 \text{ A}$$

Now use Equation 21.16 to find the power lost in the transmission line:

$$(1) \quad \mathscr{P}_{\text{lost}} = I_2{}^2 R = (1.67 \text{ A})^2 (30.0 \ \Omega) = 83.7 \text{ W}$$

Calculate the power output of the generator:

$$\mathscr{P} = I_1 \Delta V_1 = (1.00 \times 10^2 \text{ A})(4.00 \times 10^3 \text{ V}) = 4.00 \times 10^5 \text{ W}$$

Finally, divide $\mathscr{P}_{\text{lost}}$ by the power output and multiply by 100 to find the percentage of power lost:

$$\% \text{ power lost} = \left(\frac{83.7 \text{ W}}{4.00 \times 10^5 \text{ W}} \right) \times 100 = \boxed{0.020\ 9\%}$$

(b) What percentage of the original power would be lost in the transmission line if the voltage were not stepped up?

Replace the stepped-up current in equation (1) by the original current of 1.00×10^2 A.

$$\mathscr{P}_{\text{lost}} = I^2 R = (1.00 \times 10^2 \text{ A})^2 (30.0 \ \Omega) = 3.00 \times 10^5 \text{ W}$$

Calculate the percentage loss, as before:

$$\% \text{ power lost} = \left(\frac{3.00 \times 10^5 \text{ W}}{4.00 \times 10^5 \text{ W}} \right) \times 100 = \boxed{75\%}$$

Remarks   This example illustrates the advantage of high-voltage transmission lines. At the city, a transformer at a substation steps the voltage back down to about 4 000 V, and this voltage is maintained across utility lines throughout the city. When the power is to be used at a home or business, a transformer on a utility pole near the establishment reduces the voltage to 240 V or 120 V.

This cylindrical step-down transformer drops the voltage from 4 000 V to 220 V for delivery to a group of residences.

George Semple

Exercise 21.7

Suppose the same generator has the voltage stepped up to only 7.50×10^4 V and the resistance of the line is 85.0 Ω. Find the percentage of power lost in this case.

Answer   0.604%

21.8   MAXWELL'S PREDICTIONS

During the early stages of their study and development, electric and magnetic phenomena were thought to be unrelated. In 1865, however, James Clerk Maxwell (1831–1879) provided a mathematical theory that showed a close relationship between all electric and magnetic phenomena. In addition to unifying the formerly separate fields of electricity and magnetism, his brilliant theory predicted that electric and magnetic fields can move through space as waves. The theory he developed is based on the following four pieces of information:

1. Electric field lines originate on positive charges and terminate on negative charges.
2. Magnetic field lines always form closed loops—they don't begin or end anywhere.
3. A varying magnetic field induces an emf and hence an electric field. This is a statement of Faraday's law (Chapter 20).
4. Magnetic fields are generated by moving charges (or currents), as summarized in Ampère's law (Chapter 19).

North Wind Photo Archives

JAMES CLERK MAXWELL,
Scottish Theoretical Physicist
(1831–1879)

Maxwell developed the electromagnetic theory of light, the kinetic theory of gases, and explained the nature of Saturn's rings and color vision. Maxwell's successful interpretation of the electromagnetic field resulted in the equations that bear his name. Formidable mathematical ability combined with great insight enabled him to lead the way in the study of electromagnetism and kinetic theory.

The first statement is a consequence of the nature of the electrostatic force between charged particles, given by Coulomb's law. It embodies the fact that **free charges (electric monopoles) exist in nature**.

The second statement—that magnetic fields form continuous loops—is exemplified by the magnetic field lines around a long, straight wire, which are closed circles, and the magnetic field lines of a bar magnet, which form closed loops. It says, in contrast to the first statement, that **free magnetic charges (magnetic monopoles) don't exist in nature**.

The third statement is equivalent to Faraday's law of induction, and the fourth is equivalent to Ampère's law.

In one of the greatest theoretical developments of the 19th century, Maxwell used these four statements within a corresponding mathematical framework to prove that electric and magnetic fields play symmetric roles in nature. It was already known from experiments that a changing magnetic field produced an electric field according to Faraday's law. Maxwell believed that nature was symmetric, and he therefore hypothesized that a changing electric field should produce a magnetic field. This hypothesis could not be proven experimentally at the time it was developed, because the magnetic fields generated by changing electric fields are generally very weak and therefore difficult to detect.

To justify his hypothesis, Maxwell searched for other phenomena that might be explained by it. He turned his attention to the motion of rapidly oscillating (accelerating) charges, such as those in a conducting rod connected to an alternating voltage. Such charges are accelerated and, according to Maxwell's predictions, generate changing electric and magnetic fields. The changing fields cause electromagnetic disturbances that travel through space as waves, similar to the spreading water waves created by a pebble thrown into a pool. The waves sent out by the oscillating charges are fluctuating electric and magnetic fields, so they are called *electromagnetic waves*. From Faraday's law and from Maxwell's own generalization of Ampère's law, Maxwell calculated the speed of the waves to be equal to the speed of light, $c = 3 \times 10^8$ m/s. He concluded that visible light and other electromagnetic waves consist of fluctuating electric and magnetic fields traveling through empty space, with each varying field inducing the other! This was truly one of the greatest discoveries of science, on a par with Newton's discovery of the laws of motion. Like Newton's laws, it had a profound influence on later scientific developments.

21.9   HERTZ'S CONFIRMATION OF MAXWELL'S PREDICTIONS

Bettmann/Corbis

HEINRICH RUDOLF HERTZ,
German Physicist (1857–1894)

Hertz made his most important discovery of radio waves in 1887. After finding that the speed of a radio wave was the same as that of light, Hertz showed that radio waves, like light waves, could be reflected, refracted, and diffracted. Hertz died of blood poisoning at the age of 36. During his short life, he made many contributions to science. The hertz, equal to one complete vibration or cycle per second, is named after him.

In 1887, after Maxwell's death, Heinrich Hertz (1857–1894) was the first to generate and detect electromagnetic waves in a laboratory setting, using *LC* circuits. In such a circuit, a charged capacitor is connected to an inductor, as in Figure 21.16. When the switch is closed, oscillations occur in the current in the circuit and in the charge on the capacitor. If the resistance of the circuit is neglected, no energy is dissipated and the oscillations continue.

In the following analysis, we neglect the resistance in the circuit. We assume that the capacitor has an initial charge of Q_{max} and that the switch is closed at $t = 0$. When the capacitor is fully charged, the total energy in the circuit is stored in the electric field of the capacitor and is equal to $Q_{max}^2/2C$. At this time, the current is zero, so no energy is stored in the inductor. As the capacitor begins to discharge, the energy stored in its electric field decreases. At the same time, the current increases and energy equal to $LI^2/2$ is now stored in the magnetic field of the inductor. Thus, energy is transferred from the electric field of the capacitor to the magnetic field of the inductor. When the capacitor is fully discharged, it stores no energy. At this time, the current reaches its maximum value and all of the energy is stored in the inductor. The process then repeats in the reverse direction. The energy continues to transfer between the inductor and the capacitor, corresponding to oscillations in the current and charge.

As we saw in Section 21.6, the frequency of oscillation of an LC circuit is called the *resonance frequency* of the circuit and is given by

$$f_0 = \frac{1}{2\pi\sqrt{LC}}$$

The circuit Hertz used in his investigations of electromagnetic waves is similar to that just discussed and is shown schematically in Figure 21.17. An induction coil (a large coil of wire) is connected to two metal spheres with a narrow gap between them to form a capacitor. Oscillations are initiated in the circuit by short voltage pulses sent via the coil to the spheres, charging one positive, the other negative. Because L and C are quite small in this circuit, the frequency of oscillation is quite high, $f \approx 100$ MHz. This circuit is called a transmitter because it produces electromagnetic waves.

Several meters from the transmitter circuit, Hertz placed a second circuit, the receiver, which consisted of a single loop of wire connected to two spheres. It had its own effective inductance, capacitance, and natural frequency of oscillation. Hertz found that energy was being sent from the transmitter to the receiver when the resonance frequency of the receiver was adjusted to match that of the transmitter. The energy transfer was detected when the voltage across the spheres in the receiver circuit became high enough to produce ionization in the air, which caused sparks to appear in the air gap separating the spheres. Hertz's experiment is analogous to the mechanical phenomenon in which a tuning fork picks up the vibrations from another, identical tuning fork.

Hertz hypothesized that the energy transferred from the transmitter to the receiver is carried in the form of waves, now recognized as electromagnetic waves. In a series of experiments, he also showed that the radiation generated by the transmitter exhibits wave properties: interference, diffraction, reflection, refraction, and polarization. As you will see shortly, all of these properties are exhibited by light. It became evident that Hertz's electromagnetic waves had the same known properties of light waves and differed only in frequency and wavelength. Hertz effectively confirmed Maxwell's theory by showing that Maxwell's mysterious electromagnetic waves existed and had all the properties of light waves.

Perhaps the most convincing experiment Hertz performed was the measurement of the speed of waves from the transmitter, accomplished as follows: waves of known frequency from the transmitter were reflected from a metal sheet so that an interference pattern was set up, much like the standing-wave pattern on a stretched string. As we learned in our discussion of standing waves, the distance between nodes is $\lambda/2$, so Hertz was able to determine the wavelength λ. Using the relationship $v = \lambda f$, he found that v was close to 3×10^8 m/s, the known speed of visible light. Hertz's experiments thus provided the first evidence in support of Maxwell's theory.

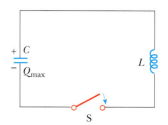

Figure 21.16 A simple LC circuit. The capacitor has an initial charge of Q_{max} and the switch is closed at $t = 0$.

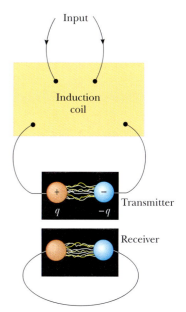

Figure 21.17 A schematic diagram of Hertz's apparatus for generating and detecting electromagnetic waves. The transmitter consists of two spherical electrodes connected to an induction coil, which provides short voltage surges to the spheres, setting up oscillations in the discharge. The receiver is a nearby single loop of wire containing a second spark gap.

21.10 PRODUCTION OF ELECTROMAGNETIC WAVES BY AN ANTENNA

In the previous section, we found that the energy stored in an LC circuit is continually transferred between the electric field of the capacitor and the magnetic field of the inductor. However, this energy transfer continues for prolonged periods of time only when the changes occur slowly. If the current alternates rapidly, the circuit loses some of its energy in the form of electromagnetic waves. In fact, electromagnetic waves are radiated by *any* circuit carrying an alternating current. The fundamental mechanism responsible for this radiation is the acceleration of a charged particle. **Whenever a charged particle accelerates it radiates energy**.

An alternating voltage applied to the wires of an antenna forces electric charges in the antenna to oscillate. This is a common technique for accelerating charged particles and is the source of the radio waves emitted by the broadcast antenna of a radio station.

APPLICATION

Radio-Wave Transmission

Figure 21.18 An electric field set up by oscillating charges in an antenna. The field moves away from the antenna at the speed of light.

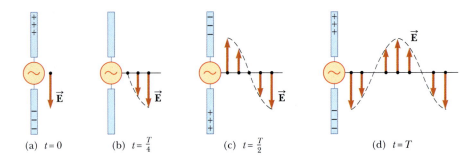

(a) $t = 0$ (b) $t = \frac{T}{4}$ (c) $t = \frac{T}{2}$ (d) $t = T$

TIP 21.1   Accelerated Charges Produce Electromagnetic Waves

Stationary charges produce only electric fields, while charges in uniform motion (i.e., constant velocity) produce electric and magnetic fields, but no electromagnetic waves. In contrast, accelerated charges produce electromagnetic waves as well as electric and magnetic fields. An accelerating charge also radiates energy.

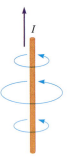

Figure 21.19 Magnetic field lines around an antenna carrying a changing current.

Figure 21.18 illustrates the production of an electromagnetic wave by oscillating electric charges in an antenna. Two metal rods are connected to an AC source, which causes charges to oscillate between the rods. The output voltage of the generator is sinusoidal. At $t = 0$, the upper rod is given a maximum positive charge and the bottom rod an equal negative charge, as in Figure 21.18a. The electric field near the antenna at this instant is also shown in the figure. As the charges oscillate, the rods become less charged, the field near the rods decreases in strength, and the downward-directed maximum electric field produced at $t = 0$ moves away from the rod. When the charges are neutralized, as in Figure 21.18b, the electric field has dropped to zero, after an interval equal to one-quarter of the period of oscillation. Continuing in this fashion, the upper rod soon obtains a maximum negative charge and the lower rod becomes positive, as in Figure 21.18c, resulting in an electric field directed upward. This occurs after an interval equal to one-half the period of oscillation. The oscillations continue as indicated in Figure 21.18d. Note that the electric field near the antenna oscillates in phase with the charge distribution: the field points down when the upper rod is positive and up when the upper rod is negative. Further, the magnitude of the field at any instant depends on the amount of charge on the rods at that instant.

As the charges continue to oscillate (and accelerate) between the rods, the electric field set up by the charges moves away from the antenna in all directions at the speed of light. Figure 21.18 shows the electric field pattern on one side of the antenna at certain times during the oscillation cycle. As you can see, one cycle of charge oscillation produces one full wavelength in the electric field pattern.

Because the oscillating charges create a current in the rods, a magnetic field is also generated when the current in the rods is upward, as shown in Figure 21.19. The magnetic field lines circle the antenna (recall right-hand rule number 2) and are perpendicular to the electric field at all points. As the current changes with time, the magnetic field lines spread out from the antenna. At great distances from the antenna, the strengths of the electric and magnetic fields become very weak. At these distances, however, it is necessary to take into account the facts that (1) a changing magnetic field produces an electric field and (2) a changing electric field produces a magnetic field, as predicted by Maxwell. These induced electric and magnetic fields are in phase: at any point, the two fields reach their maximum values at the same instant. This synchrony is illustrated at one instant of time in Active Figure 21.20. Note that (1) the $\vec{\mathbf{E}}$ and $\vec{\mathbf{B}}$ fields are perpendicular to each other, and (2) both fields are perpendicular to the direction of motion of the wave. This second property is characteristic of transverse waves. Hence, we see that **an electromagnetic wave is a transverse wave**.

21.11   PROPERTIES OF ELECTROMAGNETIC WAVES

We have seen that Maxwell's detailed analysis predicted the existence and properties of electromagnetic waves. In this section we summarize what we know about electromagnetic waves thus far and consider some additional properties. In our discussion here and in future sections, we will often make reference to a type of wave called a **plane wave**. A plane electromagnetic wave is a wave traveling from a very distant source. Active Figure 21.20 pictures such a wave at a given instant of time. In

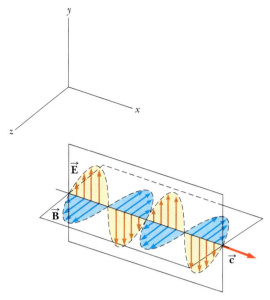

ACTIVE FIGURE 21.20

An electromagnetic wave sent out by oscillating charges in an antenna, represented at one instant of time and far from the antenna, moving in the positive *x*-direction with speed *c*. Note that the electric field is perpendicular to the magnetic field, and both are perpendicular to the direction of wave propagation. The variations of *E* and *B* with time are sinusoidal.

Physics⊗Now™

Log into PhysicsNow at **www.cp7e.com** and go to Active Figure 21.20, where you can observe the wave and the variations of the fields. In addition, you can take a "snapshot" of the wave at an instant of time and investigate the electric and magnetic fields at that instant.

this case, the oscillations of the electric and magnetic fields take place in planes perpendicular to the *x*-axis and are therefore perpendicular to the direction of travel of the wave. Because of the latter property, electromagnetic waves are transverse waves. In the figure, the electric field $\vec{\mathbf{E}}$ is in the *y*-direction and the magnetic field $\vec{\mathbf{B}}$ is in the *z*-direction. Light propagates in a direction perpendicular to these two fields. That direction is determined by yet another right-hand rule: (1) point the fingers of your right hand in the direction of $\vec{\mathbf{E}}$, (2) curl them in the direction of $\vec{\mathbf{B}}$, (3) the right thumb then points in the direction of propagation of the wave.

Electromagnetic waves travel with the speed of light. In fact, it can be shown that the speed of an electromagnetic wave is related to the permeability and permittivity of the medium through which it travels. Maxwell found this relationship for free space to be

$$c = \frac{1}{\sqrt{\mu_0 \epsilon_0}}$$ **[21.24]**   ◀ Speed of light.

where *c* is the speed of light, $\mu_0 = 4\pi \times 10^{-7} \, \text{N} \cdot \text{s}^2/\text{C}^2$ is the permeability constant of vacuum, and $\epsilon_0 = 8.854 \, 19 \times 10^{-12} \, \text{C}^2/\text{N} \cdot \text{m}^2$ is the permittivity of free space. Substituting these values into Equation 21.24, we find that

$$c = 2.997 \, 92 \times 10^8 \, \text{m/s}$$ **[21.25]**

The fact that electromagnetic waves travel at the same speed as light in vacuum led scientists to conclude (correctly) that **light is an electromagnetic wave**.

Maxwell also proved the following relationship for electromagnetic waves:

$$\frac{E}{B} = c$$ **[21.26]**

which states that the ratio of the magnitude of the electric field to the magnitude of the magnetic field equals the speed of light.

Electromagnetic waves carry energy as they travel through space, and this energy can be transferred to objects placed in their paths. The average rate at which energy passes through an area perpendicular to the direction of travel of a wave, or the average power per unit area, is called the **intensity** *I* of the wave, and is given by

$$I = \frac{E_{max} B_{max}}{2\mu_0}$$ [21.27]

where E_{max} and B_{max} are the *maximum* values of *E* and *B*. The quantity *I* is analogous to the intensity of sound waves introduced in Chapter 14. From Equation 21.26, we see that $E_{max} = cB_{max} = B_{max}/\sqrt{\mu_0\epsilon_0}$. Equation 21.27 can therefore also be expressed as

$$I = \frac{E_{max}^2}{2\mu_0 c} = \frac{c}{2\mu_0} B_{max}^2$$ [21.28]

Note that in these expressions we use the *average* power per unit area. A detailed analysis would show that the energy carried by an electromagnetic wave is shared equally by the electric and magnetic fields.

Light is an electromagnetic wave and transports energy and momentum. ▶

Electromagnetic waves have an average intensity given by Equation 21.28. When the waves strike an area *A* of an object's surface for a given time Δt, energy $U = IA\Delta t$ is transferred to the surface. Momentum is transferred, as well. Hence, pressure is exerted on a surface when an electromagnetic wave impinges on it. In what follows, we assume that the electromagnetic wave transports a total energy *U* to a surface in a time Δt. If the surface absorbs all the incident energy *U* in this time, Maxwell showed that the total momentum $\vec{p}$ delivered to this surface has a magnitude

$$p = \frac{U}{c} \qquad \text{(complete absorption)}$$ [21.29]

If the surface is a perfect reflector, then the momentum transferred in a time Δt for normal incidence is twice that given by Equation 21.29. This is analogous to a molecule of gas bouncing off the wall of a container in a perfectly elastic collision. If the molecule is initially traveling in the positive *x*-direction at velocity *v*, and after the collision is traveling in the negative *x*-direction at velocity $-v$, then its change in momentum is given by $\Delta p = mv - (-mv) = 2mv$. Light bouncing off a perfect reflector is a similar process, so for complete reflection,

$$p = \frac{2U}{c} \qquad \text{(complete reflection)}$$ [21.30]

Although radiation pressures are very small (about $5 \times 10^{-6} \text{ N/m}^2$ for direct sunlight), they have been measured with a device such as the one shown in Figure 21.21. Light is allowed to strike a mirror and a black disk that are connected to each other by a horizontal bar suspended from a fine fiber. Light striking the black disk is completely absorbed, so *all* of the momentum of the light is transferred to the disk. Light striking the mirror head-on is totally reflected; hence, the momentum transfer to the mirror is twice that transmitted to the disk. As a result, the horizontal bar supporting the disks twists counterclockwise as seen from above. The bar comes to equilibrium at some angle under the action of the torques caused by radiation pressure and the twisting of the fiber. The radiation pressure can be determined by measuring the angle at which equilibrium occurs. The apparatus must be placed in a high vacuum to eliminate the effects of air currents. It's interesting that similar experiments demonstrate that electromagnetic waves carry angular momentum, as well.

In summary, electromagnetic waves traveling through free space have the following properties:

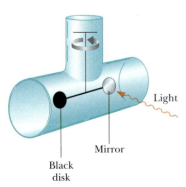

Figure 21.21 An apparatus for measuring the radiation pressure of light. In practice, the system is contained in a high vacuum.

1. Electromagnetic waves travel at the speed of light.
2. Electromagnetic waves are transverse waves, because the electric and magnetic fields are perpendicular to the direction of propagation of the wave and to each other.
3. The ratio of the electric field to the magnetic field in an electromagnetic wave equals the speed of light.
4. Electromagnetic waves carry both energy and momentum, which can be delivered to a surface.

◀ Some properties of electromagnetic waves

Applying Physics 21.2   Solar System Dust

In the interplanetary space in the Solar System, there is a large amount of dust. Although interplanetary dust can in theory have a variety of sizes—from molecular size upward—why are there very few dust particles smaller than about 0.2 μm in the Solar System? [*Hint:* The Solar System originally contained dust particles of all sizes.]

Explanation Dust particles in the Solar System are subject to two forces: the gravitational force toward the Sun and the force from radiation pressure, which is directed away from the Sun. The gravitational force is proportional to the cube of the radius of a spherical dust particle, because it is proportional to the mass (ρV) of the particle. The radiation pressure is proportional to the square of the radius, because it depends on the cross-sectional area of the particle. For large particles, the gravitational force is larger than the force of radiation pressure, and the weak attraction to the Sun causes such particles to move slowly towards it. For small particles, less than about 0.2 μm, the larger force from radiation pressure sweeps them out of the Solar System.

Quick Quiz 21.4

In an apparatus such as that in Figure 21.21, suppose the black disk is replaced by one with half the radius. Which of the following are different after the disk is replaced? (a) radiation pressure on the disk; (b) radiation force on the disk; (c) radiation momentum delivered to the disk in a given time interval.

EXAMPLE 21.8   A Hot Tin Roof (Solar-Powered Homes)

Goal   Calculate some basic properties of light and relate them to thermal radiation.

Problem   Assume that the Sun delivers an average power per unit area of about 1.00×10^3 W/m^2 to Earth's surface. (a) Calculate the total power incident on a flat tin roof 8.00 m by 20.0 m. Assume that the radiation is incident *normal* (perpendicular) to the roof. (b) Calculate the peak electric field of the light. (c) Compute the peak magnetic field of the light. (d) The tin roof reflects some light, and convection, conduction, and radiation transport the rest of the thermal energy away, until some equilibrium temperature is established. If the roof is a perfect blackbody and rids itself of one-half of the incident radiation through thermal radiation, what's its equilibrium temperature? Assume the ambient temperature is 298 K.

(Example 21.8) A solar home in Oregon.

John Neal/Photo Researchers, Inc.

Solution

(a) Calculate the power delivered to the roof.

Multiply the intensity by the area to get the power:

$$\mathscr{P} = IA = (1.00 \times 10^3 \text{ W/m}^2)(8.00 \text{ m} \times 20.0 \text{ m})$$
$$= 1.60 \times 10^5 \text{ W}$$

(b) Calculate the peak electric field of the light.

Solve Equation 21.28 for E_{max}:

$$I = \frac{E_{max}^2}{2\mu_0 c} \quad \rightarrow \quad E_{max} = \sqrt{2\mu_0 cI}$$

$$E_{max} = \sqrt{2(4\pi \times 10^{-7} \, \text{N} \cdot \text{s}^2/\text{C}^2)(3.00 \times 10^8 \, \text{m/s})(1.00 \times 10^3 \, \text{W/m}^2)}$$

$$= \boxed{868 \, \text{V/m}}$$

(c) Compute the peak magnetic field of the light.

Obtain B_{max} using Equation 21.26:

$$B_{max} = \frac{E_{max}}{c} = \frac{868 \, \text{V/m}}{3.00 \times 10^8 \, \text{m/s}} = \boxed{2.89 \times 10^{-6} \, \text{T}}$$

(d) Find the equilibrium temperature of the roof.

Substitute into Stefan's law. Only one-half the incident power should be substituted, and twice the area of the roof (both the top and the underside of the roof count).

$$\mathcal{P} = \sigma e A(T^4 - T_0^4)$$

$$T^4 = T_0^4 + \frac{\mathcal{P}}{\sigma e A}$$

$$= (298 \, \text{K})^4 + \frac{(0.500)(1.60 \times 10^5 \, \text{W/m}^2)}{(5.67 \times 10^{-8} \, \text{W/m}^2 \cdot \text{K}^4)(1)(3.20 \times 10^2 \, \text{m}^2)}$$

$$T = 333 \, \text{K} = \boxed{60.0° \, \text{C}}$$

Remarks   If the incident power could *all* be converted to electric power, it would be more than enough for the average home. Unfortunately, solar energy isn't easily harnessed, and the prospects for large-scale conversion are not as bright as they may appear from this simple calculation. For example, the conversion efficiency from solar to electrical energy is far less than 100%; 10% is typical for photovoltaic cells. Roof systems for using solar energy to raise the temperature of water with efficiencies of around 50% have been built. Other practical problems must be considered, however, such as overcast days, geographic location, and energy storage.

Exercise 21.8
A spherical satellite orbiting Earth is lighted on one side by the Sun, with intensity 1 340 W/m². (a) If the radius of the satellite is 1.00 m, what power is incident upon it? [*Note:* The satellite effectively intercepts radiation only over a cross section—an area equal to that of a disk, πr^2.) (b) Calculate the peak electric field. (c) Calculate the peak magnetic field.

Answer   (a) $4.21 \times 10^3 \, \text{W}$ (b) $1.01 \times 10^3 \, \text{V/m}$ (c) $3.35 \times 10^{-6} \, \text{T}$

EXAMPLE 21.9   Clipper Ships of Space

Goal   Relate the intensity of light to its mechanical effect on matter.

Problem   Aluminized mylar film is a highly reflective, lightweight material that could be used to make sails for spacecraft driven by the light of the sun. Suppose a sail with area 1.00 km² is orbiting the Sun at a distance of 1.50×10^{11} m. The sail has a mass of 5.00×10^3 kg and is tethered to a payload of mass 2.00×10^4 kg. **(a)** If the intensity of sunlight is 1.34×10^3 W and the sail is oriented perpendicular to the incident light, what radial force is exerted on the sail? **(b)** About how long would it take to change the radial speed of the sail by 1.00 km/s? Assume that the sail is perfectly reflecting.

Strategy   Equation 21.30 gives the momentum imparted when light strikes an object and is totally reflected. The change in this momentum with time is a force. For part (b), use Newton's second law to obtain the acceleration. The velocity kinematics equation then yields the necessary time to achieve the desired change in speed.

Solution
(a) Find the force exerted on the sail.

Write Equation 21.30, and substitute $U = \mathcal{P}\,\Delta t = IA\,\Delta t$ for the energy delivered to the sail:

$$\Delta p = \frac{2U}{c} = \frac{2\mathcal{P}\,\Delta t}{c} = \frac{2IA\,\Delta t}{c}$$

Divide both sides by Δt, obtaining the force $\Delta p/\Delta t$ exerted by the light on the sail:

$$F = \frac{\Delta p}{\Delta t} = \frac{2IA}{c} = \frac{2(1340\ \text{W/m}^2)(1.00 \times 10^6\ \text{m}^2)}{3.00 \times 10^8\ \text{m/s}}$$

$$= \boxed{8.93\ \text{N}}$$

(b) Find the time it takes to change the radial speed by 1.00 km/s.

Substitute the force into Newton's second law and solve for the acceleration of the sail:

$$a = \frac{F}{m} = \frac{8.93\ \text{N}}{2.50 \times 10^4\ \text{kg}} = 3.57 \times 10^{-4}\ \text{m/s}^2$$

Apply the kinematics velocity equation:

$$v = at + v_0$$

Solve for t:

$$t = \frac{v - v_0}{a} = \frac{1.00 \times 10^3\ \text{m/s}}{3.57 \times 10^{-4}\ \text{m/s}^2} = \boxed{2.80 \times 10^6\ \text{s}}$$

Remarks   The answer is a little over a month. While the acceleration is very low, there are no fuel costs, and within a few months the velocity can change sufficiently to allow the spacecraft to reach any planet in the solar system. Such spacecraft may be useful for certain purposes and are highly economical, but require a considerable amount of patience.

Exercise 21.9
A laser has a power of 22.0 W and a beam radius of 0.500 mm. (a) Find the intensity of the laser. (b) Suppose you were floating in space and pointed the laser beam away from you. What would your acceleration be? Assume your total mass, including equipment is 72.0 kg and that the force is directed through your center of mass. (*Hint:* The change in momentum is the same as in the nonreflective case.) (c) Compare the acceleration found in part (b) with the acceleration of gravity of a space station of mass 1.00×10^6 kg, if the station's center of mass is 100.0 m away.

Answers   (a) 2.80×10^7 W/m^2 (b) 1.02×10^{-9} m/s^2 (c) 6.67×10^{-9} m/s^2. If you were planning to use your laser welding torch as a thruster to get you back to the station, don't bother—the force of gravity is stronger. Better yet, get somebody to toss you a line.

21.12   THE SPECTRUM OF ELECTROMAGNETIC WAVES

All electromagnetic waves travel in a vacuum with the speed of light, c. These waves transport energy and momentum from some source to a receiver. In 1887, Hertz successfully generated and detected the radio-frequency electromagnetic waves predicted by Maxwell. Maxwell himself had recognized as electromagnetic waves both visible light and the infrared radiation discovered in 1800 by William Herschel. It is now known that other forms of electromagnetic waves exist that are distinguished by their frequencies and wavelengths.

Because all electromagnetic waves travel through free space with a speed c, their frequency f and wavelength λ are related by the important expression

$$c = f\lambda \qquad \text{[21.31]}$$

The various types of electromagnetic waves are presented in Figure 21.22 (page 716). Note the wide and overlapping range of frequencies and wavelengths. For instance, an AM radio wave with a frequency of 5.00 MHz (a typical value) has a wavelength of

$$\lambda = \frac{c}{f} = \frac{3.00 \times 10^8\ \text{m/s}}{5.00 \times 10^6\ \text{s}^{-1}} = 60.0\ \text{m}$$

The following abbreviations are often used to designate short wavelengths and distances:

Wearing sunglasses lacking ultraviolet (UV) protection is worse for your eyes than wearing no sunglasses at all. Sunglasses without protection absorb some visible light, causing the pupils to dilate. This allows more UV light to enter the eye, increasing the damage to the lens of the eye over time. Without the sunglasses, the pupils constrict, reducing both visible and dangerous UV radiation. Be cool: wear sunglasses with UV protection.

Figure 21.22 The electromagnetic spectrum. Note the overlap between adjacent types of waves. The expanded view to the right shows details of the visible spectrum.

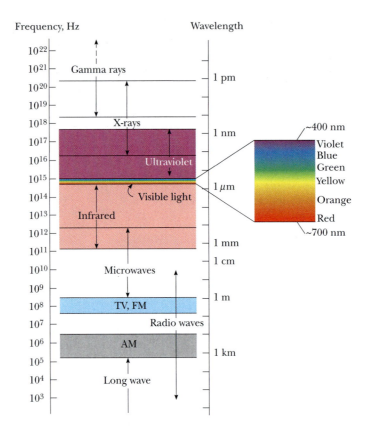

$$1 \text{ micrometer } (\mu\text{m}) = 10^{-6} \text{ m}$$

$$1 \text{ nanometer } (\text{nm}) = 10^{-9} \text{ m}$$

$$1 \text{ angstrom } (\text{Å}) = 10^{-10} \text{ m}$$

The wavelengths of visible light, for example, range from 0.4 μm to 0.7 μm, or 400 nm to 700 nm, or 4 000 Å to 7 000 Å.

Quick Quiz 21.5

Which of the following statements are true about light waves? (a) The higher the frequency, the longer the wavelength. (b) The lower the frequency, the longer the wavelength. (c) Higher frequency light travels faster than lower frequency light. (d) The shorter the wavelength, the higher the frequency. (e) The lower the frequency, the shorter the wavelength.

Brief descriptions of the wave types follow, in order of decreasing wavelength. There is no sharp division between one kind of electromagnetic wave and the next. All forms of electromagnetic radiation are produced by accelerating charges.

Radio waves, which were discussed in Section 21.10, are the result of charges accelerating through conducting wires. They are, of course, used in radio and television communication systems.

Microwaves (short-wavelength radio waves) have wavelengths ranging between about 1 mm and 30 cm and are generated by electronic devices. Their short wavelengths make them well suited for the radar systems used in aircraft navigation and for the study of atomic and molecular properties of matter. Microwave ovens are an interesting domestic application of these waves. It has been suggested that solar energy might be harnessed by beaming microwaves to Earth from a solar collector in space.

Infrared waves (sometimes incorrectly called "heat waves"), produced by hot objects and molecules, have wavelengths ranging from about 1 mm to the longest wavelength of visible light, 7×10^{-7} m. They are readily absorbed by most materials. The infrared energy absorbed by a substance causes it to get warmer because the energy agitates the atoms of the object, increasing their vibrational or translational

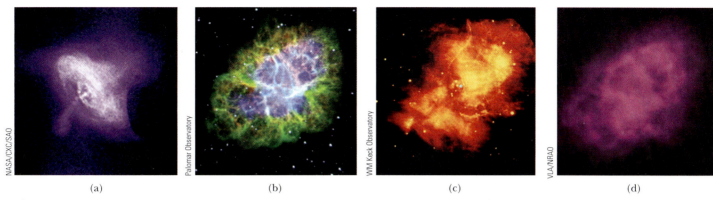

NASA/CXC/SAO   (a)   Palomar Observatory   (b)   WM Keck Observatory   (c)   VLA/NRAO   (d)

Figure 21.23   Observations in different parts of the electromagnetic spectrum show different features of the Crab Nebula. (a) X-ray image. (b) Optical image. (c) Infrared image. (d) Radio image.

motion. The result is a rise in temperature. Infrared radiation has many practical and scientific applications, including physical therapy, infrared photography, and the study of the vibrations of atoms.

Visible light, the most familiar form of electromagnetic waves, may be defined as the part of the spectrum that is detected by the human eye. Light is produced by the rearrangement of electrons in atoms and molecules. The wavelengths of visible light are classified as colors ranging from violet ($\lambda \approx 4 \times 10^{-7}$ m) to red ($\lambda \approx 7 \times 10^{-7}$ m). The eye's sensitivity is a function of wavelength and is greatest at a wavelength of about 5.6×10^{-7} m (yellow green).

Ultraviolet (UV) light covers wavelengths ranging from about 4×10^{-7} m (400 nm) down to 6×10^{-10} m (0.6 nm). The Sun is an important source of ultraviolet light (which is the main cause of suntans). Most of the ultraviolet light from the Sun is absorbed by atoms in the upper atmosphere, or stratosphere. This is fortunate, because UV light in large quantities has harmful effects on humans. One important constituent of the stratosphere is ozone (O_3), produced from reactions of oxygen with ultraviolet radiation. The resulting ozone shield causes lethal high-energy ultraviolet radiation to warm the stratosphere.

X-rays are electromagnetic waves with wavelengths from about 10^{-8} m (10 nm) down to 10^{-13} m (10^{-4} nm). The most common source of x-rays is the acceleration of high-energy electrons bombarding a metal target. X-rays are used as a diagnostic tool in medicine and as a treatment for certain forms of cancer. Because x-rays easily penetrate and damage or destroy living tissues and organisms, care must be taken to avoid unnecessary exposure and overexposure.

Gamma rays—electromagnetic waves emitted by radioactive nuclei—have wavelengths ranging from about 10^{-10} m to less than 10^{-14} m. They are highly penetrating and cause serious damage when absorbed by living tissues. Accordingly, those working near such radiation must be protected by garments containing heavily absorbing materials, such as layers of lead.

When astronomers observe the same celestial object using detectors sensitive to different regions of the electromagnetic spectrum, striking variations in the object's features can be seen. Figure 21.23 shows images of the Crab Nebula made in four different wavelength ranges. The Crab Nebula is the remnant of a supernova explosion that was seen on the Earth in 1054 A.D. (Compare with Fig. 8.28).

Applying Physics 21.3   The Sun and the Evolution of the Eye

The center of sensitivity of our eyes coincides with the center of the wavelength distribution of the Sun. Is this an amazing coincidence?

Explanation   This is not a coincidence; rather it's the result of biological evolution. Humans have evolved with vision most sensitive to wavelengths that are strongest from the Sun. If aliens from another planet ever arrived at Earth, their eyes would have the center of sensitivity at wavelengths different from ours. If their sun were a red dwarf, for example, they'd be most sensitive to red light.

21.13 THE DOPPLER EFFECT FOR ELECTROMAGNETIC WAVES

As we saw in Section 14.6, sound waves exhibit the Doppler effect when the observer, the source, or both are moving relative to the medium of propagation. Recall that in the Doppler effect, the observed frequency of the wave is larger or smaller than the frequency emitted by the source of the wave.

A Doppler effect also occurs for electromagnetic waves, but it differs from the Doppler effect for sound waves in two ways. First, in the Doppler effect for sound waves, motion relative to the medium is most important, because sound waves require a medium in which to propagate. In contrast, the medium of propagation plays no role in the Doppler effect for electromagnetic waves, because the waves require no medium in which to propagate. Second, the speed of sound that appears in the equation for the Doppler effect for sound depends on the reference frame in which it is measured. In contrast, as we shall see in Chapter 26, the speed of electromagnetic waves has the same value in all coordinate systems that are either at rest or moving at constant velocity with respect to one another.

The single equation that describes the Doppler effect for electromagnetic waves is given by the approximate expression

$$f_O \approx f_S \left(1 \pm \frac{u}{c}\right) \qquad \text{if } u \ll c \qquad \text{[21.32]}$$

where f_O is the observed frequency, f_S is the frequency emitted by the source, c is the speed of light in a vacuum, and u is the *relative* speed of the observer and source. Note that Equation 21.32 is valid only if u is much smaller than c. Further, it can also be used for sound as long as the relative velocity of the source and observer is much less than the velocity of sound. The positive sign in the equation must be used when the source and observer are moving toward one another, while the negative sign must be used when they are moving away from each other. Thus, we anticipate an increase in the observed frequency if the source and observer are approaching each other and a decrease if the source and observer recede from each other.

Astronomers have made important discoveries using Doppler observations on light reaching Earth from distant stars and galaxies. Such measurements have shown that most distant galaxies are moving away from the Earth. Thus, the Universe is expanding. This Doppler shift is called a *red shift* because the observed wavelengths are shifted towards the red portion (longer wavelengths) of the visible spectrum. Further, measurements show that the speed of a galaxy increases with increasing distance from the Earth. More recent Doppler effect measurements made with the Hubble Space Telescope have shown that a galaxy labeled M87 is rotating, with one edge moving toward us and the other moving away. Its measured speed of rotation was used to identify a supermassive black hole located at its center.

SUMMARY

Physics⊗Now™ Take a practice test by logging into Physics-Now at **www.cp7e.com** and clicking on the Pre-Test link for this chapter.

21.1 Resistors in an AC Circuit

If an AC circuit consists of a generator and a resistor, the current in the circuit is in phase with the voltage, which means the current and voltage reach their maximum values at the same time.

In discussions of voltages and currents in AC circuits, **rms values** of voltages are usually used. One reason is that

AC ammeters and voltmeters are designed to read rms values. The rms values of currents and voltage (I_{rms} and ΔV_{rms}), are related to the maximum values of these quantities (I_{max} and ΔV_{max}) as follows:

$$I_{rms} = \frac{I_{max}}{\sqrt{2}} \qquad \text{and} \qquad \Delta V_{rms} = \frac{\Delta V_{max}}{\sqrt{2}} \qquad \text{[21.2, 21.3]}$$

The rms voltage across a resistor is related to the rms current in the resistor by **Ohm's law**:

$$\Delta V_{R,rms} = I_{rms} R \qquad \text{[21.4]}$$

21.2   Capacitors in an AC Circuit

If an AC circuit consists of a generator and a capacitor, the voltage lags behind the current by 90°. This means that the voltage reaches its maximum value one-quarter of a period after the current reaches its maximum value.

The impeding effect of a capacitor on current in an AC circuit is given by the **capacitive reactance** X_C, defined as

$$X_C \equiv \frac{1}{2\pi f C} \qquad [21.5]$$

where f is the frequency of the AC voltage source.

The rms voltage across and the rms current in a capacitor are related by

$$\Delta V_{C,\text{rms}} = I_{\text{rms}} X_C \qquad [21.6]$$

21.3   Inductors in an AC Circuit

If an AC circuit consists of a generator and an inductor, the voltage leads the current by 90°. This means the voltage reaches its maximum value one-quarter of a period before the current reaches its maximum value.

The effective impedance of a coil in an AC circuit is measured by a quantity called the **inductive reactance** X_L, defined as

$$X_L \equiv 2\pi f L \qquad [21.8]$$

The rms voltage across a coil is related to the rms current in the coil by

$$\Delta V_{L,\text{rms}} = I_{\text{rms}} X_L \qquad [21.9]$$

21.4   The RLC Series Circuit

In an RLC series AC circuit, the maximum applied voltage ΔV is related to the maximum voltages across the resistor (ΔV_R), capacitor (ΔV_C), and inductor (ΔV_L) by

$$\Delta V_{\text{max}} = \sqrt{\Delta V_R^2 + (\Delta V_L - \Delta V_C)^2} \qquad [21.10]$$

If an AC circuit contains a resistor, an inductor, and a capacitor connected in series, the limit they place on the current is given by the **impedance Z** of the circuit, defined as

$$Z \equiv \sqrt{R^2 + (X_L - X_C)^2} \qquad [21.13]$$

The relationship between the maximum voltage supplied to an RLC series AC circuit and the maximum current in the circuit, which is the same in every element, is

$$\Delta V_{\text{max}} = I_{\text{max}} Z \qquad [21.14]$$

In an RLC series AC circuit, the applied rms voltage and current are out of phase. The **phase angle** ϕ between the current and voltage is given by

$$\tan \phi = \frac{X_L - X_C}{R} \qquad [21.15]$$

21.5   Power in an AC Circuit

The **average power** delivered by the voltage source in an RLC series AC circuit is

$$\mathcal{P}_{\text{av}} = I_{\text{rms}} \Delta V_{\text{rms}} \cos \phi \qquad [21.17]$$

where the constant $\cos \phi$ is called the **power factor**.

21.6   Resonance in a Series RLC Circuit

In general, the rms current in a series RLC circuit can be written

$$I_{\text{rms}} = \frac{\Delta V_{\text{rms}}}{Z} = \frac{\Delta V_{\text{rms}}}{\sqrt{R^2 + (X_L - X_C)^2}} \qquad [21.18]$$

The current has its *maximum* value when the impedance has its *minimum* value, corresponding to $X_L = X_C$ and $Z = R$. The frequency f_0 at which this happens is called the **resonance frequency** of the circuit, given by

$$f_0 = \frac{1}{2\pi\sqrt{LC}} \qquad [21.19]$$

21.7   The Transformer

If the primary winding of a transformer has N_1 turns and the secondary winding consists of N_2 turns, then if an input AC voltage ΔV_1 is applied to the primary, the induced voltage in the secondary winding is given by

$$\Delta V_2 = \frac{N_2}{N_1} \Delta V_1 \qquad [21.22]$$

When N_2 is greater than N_1, ΔV_2 exceeds ΔV_1 and the transformer is referred to as a *step-up transformer*. When N_2 is less than N_1, making ΔV_2 less than ΔV_1, we have a *step-down transformer*. In an ideal transformer, the power output equals the power input.

21.8–21.13   Electromagnetic Waves and their Properties

Electromagnetic waves were predicted by James Clerk Maxwell and experimentally confirmed by Heinrich Hertz. These waves are created by accelerating electric charges, and have the following properties:

1. Electromagnetic waves are transverse waves, because the electric and magnetic fields are perpendicular to the direction of propagation of the waves.
2. Electromagnetic waves travel at the speed of light.
3. The ratio of the electric field to the magnetic field at a given point in an electromagnetic wave equals the speed of light:

$$\frac{E}{B} = c \qquad [21.26]$$

4. Electromagnetic waves carry energy as they travel through space. The average power per unit area is the intensity I, given by

$$I = \frac{E_{\text{max}} B_{\text{max}}}{2\mu_0} = \frac{E_{\text{max}}^2}{2\mu_0 c} = \frac{c}{2\mu_0} B_{\text{max}}^2 \qquad [21.27, 21.28]$$

where E_{max} and B_{max} are the maximum values of the electric and magnetic fields.

5. Electromagnetic waves transport linear and angular momentum as well as energy. The momentum p delivered in time Δt at normal incidence to an object that completely absorbs light energy U is given by

$$p = \frac{U}{c} \qquad \text{(complete absorption)} \qquad [21.29]$$

If the surface is a perfect reflector, then the momentum delivered in time Δt at normal incidence is twice that given by Equation 21.29:

$$p = \frac{2U}{c} \qquad \text{(complete reflection)} \qquad [21.30]$$

6. The speed c, frequency f, and wavelength λ of an electromagnetic wave are related by

$$c = f\lambda \qquad [21.31]$$

The **electromagnetic spectrum** includes waves covering a broad range of frequencies and wavelengths. These waves have a variety of applications and characteristics, depend-ing on their frequencies or wavelengths. The frequency of a given wave can be shifted by the relative velocity of observer and source, with the observed frequency f_O given by

$$f_O \approx f_S\left(1 \pm \frac{u}{c}\right) \qquad \text{if } u \ll c \qquad [21.32]$$

where f_S is the frequency of the source, c is the speed of light in a vacuum, and u is the *relative* speed of the observer and source. The positive sign is used when the source and observer approach each other, the negative sign when they recede from each other.

CONCEPTUAL QUESTIONS

1. Before the advent of cable television and satellite dishes, homeowners either mounted a television antenna on the roof or used "rabbit ears" atop their sets. (See Fig. Q21.1.) Certain orientations of the receiving antenna on a television set gave better reception than others. Furthermore, the best orientation varied from station to station. Explain.

Figure Q21.1

2. What is the impedance of an *RLC* circuit at the resonance frequency?

3. When a DC voltage is applied to a transformer, the primary coil sometimes will overheat and burn. Why?

4. Why are the primary and secondary coils of a transformer wrapped on an iron core that passes through both coils?

5. Receiving radio antennas can be in the form of conducting lines or loops. What should the orientation of each of these antennas be relative to a broadcasting antenna that is vertical?

6. If the fundamental source of a sound wave is a vibrating object, what is the fundamental source of an electromagnetic wave?

7. In radio transmission, a radio wave serves as a carrier wave, and the sound signal is superimposed on the carrier wave. In amplitude modulation (AM) radio, the amplitude of the carrier wave varies according to the sound wave. The Navy sometimes uses flashing lights to send Morse code between neighboring ships, a process that has similarities to radio broadcasting. Is this process AM or FM? What is the carrier frequency? What is the signal frequency? What is the broadcasting antenna? What is the receiving antenna?

8. When light (or other electromagnetic radiation) travels across a given region, what is it that oscillates? What is it that is transported?

9. In space sailing, which is a proposed alternative for transport to the planets, a spacecraft carries a very large sail. Sunlight striking the sail exerts a force, accelerating the spacecraft. Should the sail be absorptive or reflective to be most effective?

10. How can the average value of an alternating current be zero, yet the square root of the average squared value not be zero?

11. Suppose a creature from another planet had eyes that were sensitive to infrared radiation. Describe what it would see if it looked around the room that you are now in. That is, what would be bright and what would be dim?

12. Why should an infrared photograph of a person look different from a photograph taken using visible light?

13. Radio stations often advertise "instant news." If what they mean is that you hear the news at the instant they speak it, is their claim true? About how long would it take for a message to travel across the United States by radio waves, assuming that the waves could travel that great distance and still be detected?

14. Would an inductor and a capacitor used together in an AC circuit dissipate any energy?

15. Does a wire connected to a battery emit an electromagnetic wave?

16. If a high-frequency current is passed through a solenoid containing a metallic core, the core becomes warm due to induction. Explain why the temperature of the material rises in this situation.

17. If the resistance in an *RLC* circuit remains the same, but the capacitance and inductance are each doubled, how will the resonance frequency change?

18. Why is the sum of the maximum voltages across each of the elements in a series *RLC* circuit usually greater than the maximum applied voltage? Doesn't this violate Kirchhoff's loop rule?

19. What is the advantage of transmitting power at high voltages?

20. What determines the maximum voltage that can be used on a transmission line?

21. Will a transformer operate if a battery is used for the input voltage across the primary? Explain.

PROBLEMS

1, 2, 3 = straightforward, intermediate, challenging   □ = full solution available in *Student Solutions Manual/Study Guide*
Physics⊗Now™ = coached problem with hints available at **www.cp7e.com**   🔋 = biomedical application

Section 21.1 Resistors in an AC Circuit

1. An rms voltage of 100 V is applied to a purely resistive load of 5.00 Ω. Find (a) the maximum voltage applied, (b) the rms current supplied, (c) the maximum current supplied, and (d) the power dissipated.

2. (a) What is the resistance of a lightbulb that uses an average power of 75.0 W when connected to a 60-Hz power source with an peak voltage of 170 V? (b) What is the resistance of a 100-W bulb?

3. An AC power supply that produces a maximum voltage of $\Delta V_{max} = 100$ V is connected to a 24.0-Ω resistor. The current and the resistor voltage are respectively measured with an ideal AC ammeter and an ideal AC voltmeter, as shown in Figure P21.3. What does each meter read? Note that an ideal ammeter has zero resistance and an ideal voltmeter has infinite resistance.

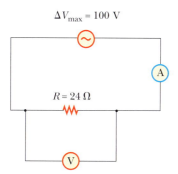

$\Delta V_{max} = 100$ V

$R = 24$ Ω

Figure P21.3

4. Figure P21.4 shows three lamps connected to a 120-V AC (rms) household supply voltage. Lamps 1 and 2 have 150-W bulbs; lamp 3 has a 100-W bulb. Find the rms current and the resistance of each bulb.

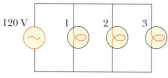

120 V   1   2   3

Figure P21.4

5. An audio amplifier, represented by the AC source and the resistor R in Figure P21.5, delivers alternating voltages at audio frequencies to the speaker. If the source puts out an alternating voltage of 15.0 V (rms), the resistance R is 8.20 Ω, and the speaker is equivalent to a resistance of 10.4 Ω, what is the time-averaged power delivered to the speaker?

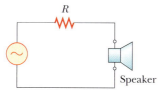

R

Speaker

Figure P21.5

6. An AC voltage source has an output voltage given by $\Delta v = (150$ V$) \sin 377t$. Find (a) the rms voltage output, (b) the frequency of the source, and (c) the voltage at $t = 1/120$ s. (d) Find the maximum current in the circuit when the voltage source is connected to a 50.0-Ω resistor.

Section 21.2 Capacitors in an AC Circuit

7. Show that the SI unit of capacitive reactance X_c is the ohm.

8. What is the maximum current delivered to a circuit containing a 2.20-μF capacitor when it is connected across (a) a North American outlet having $\Delta V_{rms} = 120$ V and $f = 60.0$ Hz and (b) a European outlet having $\Delta V_{rms} = 240$ V and $f = 50.0$ Hz?

9. **Physics⊗Now**™ When a 4.0-μF capacitor is connected to a generator whose rms output is 30 V, the current in the circuit is observed to be 0.30 A. What is the frequency of the source?

10. What maximum current is delivered by an AC generator with a maximum voltage of $\Delta V_{max} = 48.0$ V and a frequency $f = 90.0$ Hz when it is connected across a 3.70-μF capacitor?

11. What must be the capacitance of a capacitor inserted in a 60-Hz circuit in series with a generator of 170 V maximum output voltage to produce an rms current output of 0.75 A?

12. The generator in a purely capacitive AC circuit has an angular frequency of 120π rad/s. If $\Delta V_{max} = 140$ V and $C = 6.00$ μF, what is the rms current in the circuit?

Section 21.3 Inductors in an AC Circuit

13. Show that the inductive reactance X_L has SI units of ohms.

14. The generator in a purely inductive AC circuit has an angular frequency of 120π rad/s. If $V_{max} = 140$ V and $L = 0.100$ H, what is the rms current in the circuit?

15. An inductor has a 54.0-Ω reactance at 60.0 Hz. What will be the *maximum* current if this inductor is connected to a 50.0-Hz source that produces a 100-V rms voltage?

16. An inductor is connected to a 20.0-Hz power supply that produces a 50.0-V rms voltage. What inductance is needed to keep the maximum current in the circuit below 80.0 mA?

17. Determine the maximum magnetic flux through an inductor connected to a standard outlet ($\Delta V_{rms} = 120$ V, $f = 60.0$ Hz).

Section 21.4 The *RLC* Series Circuit

18. An inductor ($L = 400$ mH), a capacitor ($C = 4.43$ μF), and a resistor ($R = 500$ Ω) are connected in series. A 50.0-Hz AC generator connected in series to these elements produces a maximum current of 250 mA in the circuit. (a) Calculate the required maximum voltage ΔV_{max}. (b) Determine the phase angle by which the current leads or lags the applied voltage.

19. A 40.0-μF capacitor is connected to a 50.0-Ω resistor and a generator whose rms output is 30.0 V at 60.0 Hz. Find

(a) the rms current in the circuit, (b) the rms voltage drop across the resistor, (c) the rms voltage drop across the capacitor, and (d) the phase angle for the circuit.

20. A 50.0-Ω resistor, a 0.100-H inductor, and a 10.0-μF capacitor are connected in series to a 60.0-Hz source. The rms current in the circuit is 2.75 A. Find the rms voltages across (a) the resistor, (b) the inductor, (c) the capacitor, and (d) the RLC combination. (e) Sketch the phasor diagram for this circuit.

21. A resistor ($R = 900\ \Omega$), a capacitor ($C = 0.25\ \mu$F), and an inductor ($L = 2.5$ H) are connected in series across a 240-Hz AC source for which $\Delta V_{max} = 140$ V. Calculate (a) the impedance of the circuit, (b) the maximum current delivered by the source, and (c) the phase angle between the current and voltage. (d) Is the current leading or lagging the voltage?

22. An AC source operating at 60 Hz with a maximum voltage of 170 V is connected in series with a resistor ($R = 1.2$ kΩ) and a capacitor ($C = 2.5\ \mu$F). (a) What is the maximum value of the current in the circuit? (b) What are the maximum values of the potential difference across the resistor and the capacitor? (c) When the current is zero, what are the magnitudes of the potential difference across the resistor, the capacitor, and the AC source? How much charge is on the capacitor at this instant? (d) When the current is at a maximum, what are the magnitudes of the potential differences across the resistor, the capacitor, and the AC source? How much charge is on the capacitor at this instant?

23. A 60.0-Ω resistor, a 3.00-μF capacitor, and a 0.400-H inductor are connected in series to a 90.0-V (rms), 60.0-Hz source. Find (a) the voltage drop across the LC combination and (b) the voltage drop across the RC combination.

24. An AC source operating at 60 Hz with a maximum voltage of 170 V is connected in series with a resistor ($R = 1.2$ kΩ) and an inductor ($L = 2.8$ H). (a) What is the maximum value of the current in the circuit? (b) What are the maximum values of the potential difference across the resistor and the inductor? (c) When the current is at a maximum, what are the magnitudes of the potential differences across the resistor, the inductor, and the AC source? (d) When the current is zero, what are the magnitudes of the potential difference across the resistor, the inductor, and the AC source?

25. A person is working near the secondary of a transformer, as shown in Figure P21.25. The primary voltage is 120 V (rms) at 60.0 Hz. The capacitance C_s, which is the stray capacitance between the hand and the secondary winding, is 20.0 pF. Assuming that the person has a body resistance to ground of $R_b = 50.0$ kΩ, determine the rms voltage across the body. (*Hint:* Redraw the circuit with the secondary of the transformer as a simple AC source.)

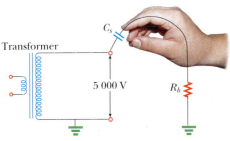

Figure P21.25

26. A coil of resistance 35.0 Ω and inductance 20.5 H is in series with a capacitor and a 200-V (rms), 100-Hz source. The rms current in the circuit is 4.00 A. (a) Calculate the capacitance in the circuit. (b) What is ΔV_{rms} across the coil?

27. **Physics⊗Now™** An AC source with a maximum voltage of 150 V and $f = 50.0$ Hz is connected between points a and d in Figure P21.27. Calculate the rms voltages between points (a) a and b, (b) b and c, (c) c and d, and (d) b and d.

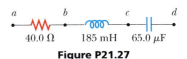

Figure P21.27

Section 21.5 Power in an AC Circuit

28. A 50.0-Ω resistor is connected to a 30.0-μF capacitor and to a 60.0-Hz, 100-V (rms) source. (a) Find the power factor and the average power delivered to the circuit. (b) Repeat part (a) when the capacitor is replaced with a 0.300-H inductor.

29. A multimeter in an RL circuit records an rms current of 0.500 A and a 60.0-Hz rms generator voltage of 104 V. A wattmeter shows that the average power delivered to the resistor is 10.0 W. Determine (a) the impedance in the circuit, (b) the resistance R, and (c) the inductance L.

30. An AC voltage with an amplitude of 100 V is applied to a series combination of a 200-μF capacitor, a 100-mH inductor, and a 20.0-Ω resistor. Calculate the power dissipation and the power factor for frequencies of (a) 60.0 Hz and (b) 50.0 Hz.

31. An inductor and a resistor are connected in series. When connected to a 60-Hz, 90-V (rms) source, the voltage drop across the resistor is found to be 50 V (rms) and the power delivered to the circuit is 14 W. Find (a) the value of the resistance and (b) the value of the inductance.

32. Consider a series RLC circuit with $R = 25\ \Omega$, $L = 6.0$ mH, and $C = 25\ \mu$F. The circuit is connected to a 10-V (rms), 600-Hz AC source. (a) Is the sum of the voltage drops across R, L, and C equal to 10 V (rms)? (b) Which is greatest, the power delivered to the resistor, to the capacitor, or to the inductor? (c) Find the average power delivered to the circuit.

Section 21.6 Resonance in a Series RLC circuit

33. An RLC circuit is used to tune a radio to an FM station broadcasting at 88.9 MHz. The resistance in the circuit is 12.0 Ω and the capacitance is 1.40 pF. What inductance should be present in the circuit?

34. Consider a series RLC circuit with $R = 15\ \Omega$, $L = 200$ mH, $C = 75\ \mu$F, and a maximum voltage of 150 V. (a) What is the impedance of the circuit at resonance? (b) What is the resonance frequency of the circuit? (c) When will the current be greatest—at resonance, at ten percent below the resonant frequency, or at ten percent above the resonant frequency? (d) What is the rms current in the circuit at a frequency of 60 Hz?

35. The AM band extends from approximately 500 kHz to 1 600 kHz. If a 2.0-μH inductor is used in a tuning circuit for a radio, what are the extremes that a capacitor

must reach in order to cover the complete band of frequencies?

36. A series circuit contains a 3.00-H inductor, a 3.00-μF capacitor, and a 30.0-Ω resistor connected to a 120-V (rms) source of variable frequency. Find the power delivered to the circuit when the frequency of the source is (a) the resonance frequency, (b) one-half the resonance frequency, (c) one-fourth the resonance frequency, (d) two times the resonance frequency, and (e) four times the resonance frequency. From your calculations, can you draw a conclusion about the frequency at which the maximum power is delivered to the circuit?

37. **Physics⊗Now™** A 10.0-Ω resistor, a 10.0-mH inductor, and a 100-μF capacitor are connected in series to a 50.0-V (rms) source having variable frequency. Find the energy delivered to the circuit during one period if the operating frequency is twice the resonance frequency.

Section 21.7 The Transformer

38. An AC adapter for a telephone-answering unit uses a transformer to reduce the line voltage of 120 V (rms) to a voltage of 9.0 V. The rms current delivered to the answering system is 400 mA. (a) If the primary (input) coil in the transformer in the adapter has 240 turns, how many turns are there on the secondary (output) coil? (b) What is the rms power delivered to the transformer? Assume an ideal transformer.

39. An AC power generator produces 50 A (rms) at 3 600 V. The voltage is stepped up to 100 000 V by an ideal transformer, and the energy is transmitted through a long-distance power line that has a resistance of 100 Ω. What percentage of the power delivered by the generator is dissipated as heat in the power line?

40. A transformer is to be used to provide power for a computer disk drive that needs 6.0 V (rms) instead of the 120 V (rms) from the wall outlet. The number of turns in the primary is 400, and it delivers 500 mA (the secondary current) at an output voltage of 6.0 V (rms). (a) Should the transformer have more turns in the secondary compared with the primary, or fewer turns? (b) Find the current in the primary. (c) Find the number of turns in the secondary.

41. A transformer on a pole near a factory steps the voltage down from 3 600 V (rms) to 120 V (rms). The transformer is to deliver 1 000 kW to the factory at 90% efficiency. Find (a) the power delivered to the primary, (b) the current in the primary, and (c) the current in the secondary.

42. A transmission line that has a resistance per unit length of 4.50×10^{-4} Ω/m is to be used to transmit 5.00 MW over 400 miles (6.44×10^5 m). The output voltage of the generator is 4.50 kV (rms). (a) What is the line loss if a transformer is used to step up the voltage to 500 kV (rms)? (b) What fraction of the input power is lost to the line under these circumstances? (c) What difficulties would be encountered on attempting to transmit the 5.00 MW at the generator voltage of 4.50 kV (rms)?

Section 21.10 Production of Electromagnetic Waves by an Antenna

Section 21.11 Properties of Electromagnetic Waves

43. The U.S. Navy has long proposed the construction of extremely low frequency (ELF waves) communications

systems; such waves could penetrate the oceans to reach distant submarines. Calculate the length of a quarter-wavelength antenna for a transmitter generating ELF waves of frequency 75 Hz. How practical is this antenna?

44. Experimenters at the National Institute of Standards and Technology have made precise measurements of the speed of light using the fact that, in vacuum, the speed of electromagnetic waves is $c = 1/\sqrt{\mu_0\epsilon_0}$, where the constants $\mu_0 = 4\pi \times 10^{-7}$ N $\cdot$ s^2/C^2 and $\epsilon_0 = 8.854 \times 10^{-12}$ C^2/N $\cdot$ m^2. What value (to four significant figures) does this formula give for the speed of light in vacuum?

45. Oxygenated hemoglobin absorbs weakly in the red (hence its red color) and strongly in the near infrared, while deoxygenated hemoglobin has the opposite absorption. This fact is used in a "pulse oximeter" to measure oxygen saturation in arterial blood. The device clips onto the end of a person's finger and has two light-emitting diodes [a red (660 nm) and an infrared (940 nm)] and a photocell that detects the amount of light transmitted through the finger at each wavelength. (a) Determine the frequency of each of these light sources. (b) If 67% of the energy of the red source is absorbed in the blood, by what factor does the amplitude of the electromagnetic wave change? [*Hint:* The intensity of the wave is equal to the average power per unit area as given by Equation 21.28.]

46. **Operation of the pulse oximeter (see previous problem).** The transmission of light energy as it passes through a solution of light-absorbing molecules is described by the Beer–Lambert law

$$I = I_0 10^{-\epsilon CL} \quad \text{or} \quad \log_{10}\left(\frac{I}{I_0}\right) = -\epsilon CL$$

which gives the decrease in intensity I in terms of the distance L the light has traveled through a fluid with a concentration C of the light-absorbing molecule. The quantity ϵ is called the extinction coefficient, and its value depends on the frequency of the light. (It has units of m^2/mol.) Assume that the extinction coefficient for 660-nm light passing through a solution of oxygenated hemoglobin is identical to the coefficient for 940-nm light passing through deoxygenated hemoglobin. Assume also that 940-nm light has zero absorption ($\epsilon = 0$) in oxygenated hemoglobin and 660-nm light has zero absorption in deoxygenated hemoglobin. If 33% of the energy of the red source and 76% of the infrared energy is transmitted through the blood, determine the fraction of hemoglobin that is oxygenated.

47. A microwave oven is powered by an electron tube called a magnetron that generates electromagnetic waves of frequency 2.45 GHz. The microwaves enter the oven and are reflected by the walls. The standing-wave pattern produced in the oven can cook food unevenly, with hot spots in the food at antinodes and cool spots at nodes, so a turntable is often used to rotate the food and distribute the energy. If a microwave oven is used with a cooking dish in a fixed position, the antinodes can appear as burn marks on foods such as carrot strips or cheese. The separation distance between the burns is measured to be 6.00 cm. Calculate the speed of the microwaves from these data.

48. Assume that the solar radiation incident on Earth is 1 340 W/m^2 (at the top of Earth's atmosphere). Calculate

the total power radiated by the Sun, taking the average separation between Earth and the Sun to be 1.49×10^{11} m.

49. Physics⊗Now™ The Sun delivers an average power of $1\,340$ W/m² to the top of Earth's atmosphere. Find the magnitudes of $\vec{E}_{max}$ and $\vec{B}_{max}$ for the electromagnetic waves at the top of the atmosphere.

Section 21.12 The Spectrum of Electromagnetic Waves

50. A diathermy machine, used in physiotherapy, generates electromagnetic radiation that gives the effect of "deep heat" when absorbed in tissue. One assigned frequency for diathermy is 27.33 MHz. What is the wavelength of this radiation?

51. What are the wavelength ranges in (a) the AM radio band (540–$1\,600$ kHz) and (b) the FM radio band (88–108 MHz)?

52. An important news announcement is transmitted by radio waves to people who are 100 km away, sitting next to their radios, and by sound waves to people sitting across the newsroom, 3.0 m from the newscaster. Who receives the news first? Explain. Take the speed of sound in air to be 343 m/s.

53. Infrared spectra are used by chemists to help identify an unknown substance. Atoms in a molecule that are bound together by a particular bond vibrate at a predictable frequency, and light at that frequency is absorbed strongly by the atom. In the case of the C=O double bond, for example, the oxygen atom is bound to the carbon by a bond that has an effective spring constant of 2 800 N/m. If we assume that the carbon atom remains stationary (it is attached to other atoms in the molecule), determine the resonant frequency of this bond and the wavelength of light that matches that frequency. Verify that this wavelength lies in the infrared region of the spectrum. (The mass of an oxygen atom is 2.66×10^{-26} kg.)

21.13 The Doppler Effect for Electromagnetic Waves

54. A spaceship is approaching a space station at a speed of 1.8×10^5 m/s. The space station has a beacon that emits green light with a frequency of 6.0×10^{14} Hz. What is the frequency of the beacon observed on the spaceship? What is the change in frequency? (Carry five digits in these calculations.)

55. While driving at a constant speed of 80 km/h, you are passed by a car traveling at 120 km/h. If the frequency of light emitted by the taillights of the car that passes you is 4.3×10^{14} Hz, what frequency will you observe? What is the change in frequency?

56. A speeder tries to explain to the police that the yellow warning lights on the side of the road looked green to her because of the Doppler shift. How fast would she have been traveling if yellow light of wavelength 580 nm had been shifted to green with a wavelength of 560 nm? (Note that, for speeds less than $0.03c$, Equation 21.32 will lead to a value for the change of frequency accurate to approximately two significant digits.)

ADDITIONAL PROBLEMS

57. As a way of determining the inductance of a coil used in a research project, a student first connects the coil to a 12.0-V battery and measures a current of 0.630 A. The student

then connects the coil to a 24.0-V (rms), 60.0-Hz generator and measures an rms current of 0.570 A. What is the inductance?

58. The intensity of solar radiation at the top of Earth's atmosphere is $1\,340$ W/m³. Assuming that 60% of the incoming solar energy reaches Earth's surface, and assuming that you absorb 50% of the incident energy, make an order-of-magnitude estimate of the amount of solar energy you absorb in a 60-minute sunbath.

59. A 200-Ω resistor is connected in series with a 5.0-μF capacitor and a 60-Hz, 120-V rms line. If electrical energy costs \$0.080/kWh, how much does it cost to leave this circuit connected for 24 h?

60. A series *RLC* circuit has a resonance frequency of $2\,000/\pi$ Hz. When it is operating at a frequency of $\omega > \omega_0$, $X_L = 12$ Ω and $X_C = 8.0$ Ω. Calculate the values of *L* and *C* for the circuit.

61. Two connections allow contact with two circuit elements in series inside a box, but it is not known whether the circuit elements are *R*, *L*, or *C*. In an attempt to find what is inside the box, you make some measurements, with the following results: when a 3.0-V DC power supply is connected across the terminals, a maximum direct current of 300 mA is measured in the circuit after a suitably long time. When a 60-Hz source with maximum voltage of 3.0 V is connected instead, the maximum current is measured as 200 mA. (a) What are the two elements in the box? (b) What are their values of *R*, *L*, or *C*?

62. (a) What capacitance will resonate with a one-turn loop of inductance 400 pH to give a radar wave of wavelength 3.0 cm? (b) If the capacitor has square parallel plates separated by 1.0 mm of air, what should the edge length of the plates be? (c) What is the common reactance of the loop and capacitor at resonance?

63. A dish antenna with a diameter of 20.0 m receives (at normal incidence) a radio signal from a distant source, as shown in Figure P21.63. The radio signal is a continuous sinusoidal wave with amplitude $E_{max} = 0.20$ μV/m. Assume that the antenna absorbs all the radiation that falls on the dish. (a) What is the amplitude of the magnetic field in this wave? (b) What is the intensity of the radiation received by the antenna? (c) What is the power received by the antenna?

Figure P21.63

64. A particular inductor has appreciable resistance. When the inductor is connected to a 12-V battery, the current in the inductor is 3.0 A. When it is connected to an AC source with an rms output of 12 V and a frequency of 60 Hz, the current drops to 2.0 A. What are (a) the impedance at 60 Hz and (b) the inductance of the inductor?

65. One possible means of achieving space flight is to place a perfectly reflecting aluminized sheet into Earth's orbit and to use the light from the Sun to push this solar sail. Suppose such a sail, of area 6.00×10^4 m^2 and mass 6 000 kg, is placed in orbit facing the Sun. (a) What force is exerted on the sail? (b) What is the sail's acceleration? (c) How long does it take for this sail to reach the Moon, 3.84×10^8 m away? Ignore all gravitational effects, and assume a solar intensity of 1 340 W/m^2. [*Hint:* The radiation pressure by a reflected wave is given by 2(average power per unit area)$/c$.]

66. Suppose you wish to use a transformer as an impedance-matching device between an audio amplifier that has an output impedance of 8.0 kΩ and a speaker that has an input impedance of 8.0 Ω. What should be the ratio of primary to secondary turns on the transformer?

67. Compute the average energy content of a liter of sunlight as it reaches the top of Earth's atmosphere, where its intensity is 1 340 W/m^2.

68. In an *RLC* series circuit that includes a source of alternating current operating at fixed frequency and voltage, the resistance R is equal to the inductive reactance. If the plate separation of the capacitor is reduced to one-half of its original value, the current in the circuit doubles. Find the initial capacitive reactance in terms of R.

ACTIVITIES

1. For this observation, you will need some items that can be found at many electronics stores. You will need a bicolored light-emitting diode (LED), a resistor of about 100 Ω, 2 m of flexible wire, and a step-down transformer with an output of 3 to 6 V. Use the wire to connect the LED and the resistor in series with the transformer. A bicolored LED is designed such that it emits a red color when the current in the LED is in one direction and green when the current reverses. When connected to an AC source, the LED is yellow. Why?

Hold the wires and whirl the LED in a circular path. In a darkened room, you will see red and green bars at equally spaced intervals along the path of the LED. Why?

As you continue to whirl the LED in a circular path, have your partner count the number of green bars in the circle, then measure the time it takes for the LED to travel ten times around the circular path. Based on this information, determine the time it takes for the color of the LED to change from green to red to green. You should obtain an answer of (1/60) s. Why?

2. Rotate a portable radio (with a telescoping antenna) about a horizontal axis while it is tuned to a weak station. Such an antenna detects the varying electric field produced by the station. What can you determine about the direction of the electric field produced by the transmitter?

Now turn on your radio to a nearby station and experiment with shielding the radio from incoming waves. Is the reception affected by surrounding the radio by aluminum foil? By plastic wrap? Use any other material you have available. What kinds of material block the signal? Why?

Light is bent (refracted) as it passes through water, with different wavelengths bending by different amounts (which is called dispersion). Together with reflection, these physical phenomena lead to the creation of a rainbow when light passes through small, suspended droplets of water.

Peter Bowater/Photo Researchers, Inc.

Reflection and Refraction of Light

Light has a dual nature. In some experiments it acts like a particle, while in others it acts like a wave. In this part of the book, we concentrate on the aspects of light that are best understood through the wave model. First we discuss the reflection of light at the boundary between two media and the refraction (bending) of light as it travels from one medium into another. We use these ideas to study the refraction of light as it passes through lenses and the reflection of light from mirrored surfaces. Finally, we describe how lenses and mirrors can be used to view objects with telescopes and microscopes and how lenses are used in photography. The ability to manipulate light has greatly enhanced our capacity to investigate and understand the nature of the universe.

22.1 THE NATURE OF LIGHT

Until the beginning of the 19th century, light was modeled as a stream of particles emitted by a source that stimulated the sense of sight on entering the eye. The chief architect of the particle theory of light was Newton. With this theory, he provided simple explanations of some known experimental facts concerning the nature of light—namely, the laws of reflection and refraction.

Most scientists accepted Newton's particle theory of light. During Newton's lifetime, however, another theory was proposed. In 1678, the Dutch physicist and astronomer Christian Huygens (1629–1695) showed that a wave theory of light could also explain the laws of reflection and refraction.

The wave theory didn't receive immediate acceptance, for several reasons. First, all the waves known at the time (sound, water, and so on) traveled through some sort of medium, but light from the Sun could travel to Earth through empty space.

Further, it was argued that if light were some form of wave, it would bend around obstacles; hence, we should be able to see around corners. It is now known that light does indeed bend around the edges of objects. This phenomenon, known as *diffraction*, is difficult to observe because light waves have such short wavelengths. Even though experimental evidence for the diffraction of light was discovered by Francesco Grimaldi (1618–1663) around 1660, for more than a century most scientists rejected the wave theory and adhered to Newton's particle theory, probably due to Newton's great reputation as a scientist.

The first clear demonstration of the wave nature of light was provided in 1801 by Thomas Young (1773–1829), who showed that under appropriate conditions, light exhibits interference behavior. Light waves emitted by a single source and traveling along two different paths can arrive at some point and combine and cancel each other by destructive interference. Such behavior couldn't be explained at that time by a particle model, because scientists couldn't imagine how two or more particles could come together and cancel each other.

The most important development in the theory of light was the work of Maxwell, who predicted in 1865 that light was a form of high-frequency electromagnetic wave (Chapter 21). His theory also predicted that these waves should have a speed of 3×10^8 m/s, in agreement with the measured value.

Although the classical theory of electricity and magnetism explained most known properties of light, some subsequent experiments couldn't be explained by the assumption that light was a wave. The most striking of these was the *photoelectric effect* (which we will examine more closely in Chapter 27), discovered by Hertz. Hertz found that clean metal surfaces emit charges when exposed to ultraviolet light.

In 1905, Einstein published a paper that formulated the theory of light quanta ("particles") and explained the photoelectric effect. He reached the conclusion that light was composed of corpuscles, or discontinuous quanta of energy. These corpuscles or quanta are now called *photons* to emphasize their particlelike nature. According to Einstein's theory, the energy of a photon is proportional to the frequency of the electromagnetic wave associated with it, or:

$$E = hf \qquad\qquad [22.1]$$

◀ Energy of a photon

where $h = 6.63 \times 10^{-34}$ J · s is *Planck's constant*. This theory retains some features of both the wave and particle theories of light. As we will discuss later, the photoelectric effect is the result of energy transfer from a single photon to an electron in the metal. This means the electron interacts with one photon of light as if the electron had been struck by a particle. Yet the photon has wavelike characteristics, as implied by the fact that a frequency is used in its definition.

In view of these developments, light must be regarded as having a *dual nature*: **In some experiments light acts as a wave and in others it acts as a particle**. Classical electromagnetic wave theory provides adequate explanations of light propagation and of the effects of interference, whereas the photoelectric effect and other experiments involving the interaction of light with matter are best explained by assuming that light is a particle.

So in the final analysis, is light a wave or a particle? The answer is neither and both: light has a number of physical properties, some associated with waves and others with particles.

22.2 REFLECTION AND REFRACTION

When light traveling in one medium encounters a boundary leading into a second medium, the processes of reflection and refraction can occur. In **reflection**, part of the light encountering the second medium bounces off that medium. In **refraction**, the light passing into the second medium bends through an angle with respect to the normal to the boundary. Often, both processes occur at the same time, with part of the light being reflected and part refracted. To study reflection and refraction we need a way of thinking about beams of light, and this is given by the ray approximation.

CHRISTIAN HUYGENS
(1629–1695), Dutch Physicist and Astronomer

Huygens is best known for his contributions to the fields of optics and dynamics. To Huygens, light was a vibratory motion in the ether, spreading out and producing the sensation of light when impinging on the eye. On the basis of this theory, he deduced the laws of reflection and refraction and explained the phenomenon of double refraction.

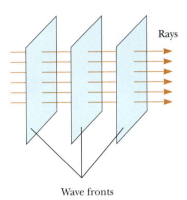

Figure 22.1 A plane wave traveling to the right. Note that the rays, corresponding to the direction of wave motion, are straight lines perpendicular to the wave fronts.

The Ray Approximation in Geometric Optics

An important property of light that can be understood based on common experience is the following: **light travels in a straight-line path in a homogeneous medium, until it encounters a boundary between two different materials**. When light strikes a boundary, it is either reflected from that boundary, passes into the material on the other side of the boundary, or partially does both.

The preceding observation leads us to use what is called the **ray approximation** to represent beams of light. As shown in Figure 22.1, a ray of light is an imaginary line drawn along the direction of travel of the light beam. For example, a beam of sunlight passing through a darkened room traces out the path of a light ray. We will also make use of the concept of wave fronts of light. A **wave front** is a surface passing through the points of a wave that have the same phase and amplitude. For instance, the wave fronts in Figure 22.1 could be surfaces passing through the crests of waves. The rays, corresponding to the direction of wave motion, are straight lines perpendicular to the wave fronts. When light rays travel in parallel paths, the wave fronts are planes perpendicular to the rays.

Reflection of Light

When a light ray traveling in a transparent medium encounters a boundary leading into a second medium, part of the incident ray is reflected back into the first medium. Figure 22.2a shows several rays of a beam of light incident on a smooth, mirrorlike reflecting surface. The reflected rays are parallel to each other, as indicated in the figure. The reflection of light from such a smooth surface is called **specular reflection**. On the other hand, if the reflecting surface is rough, as in Figure 22.2b, the surface reflects the rays in a variety of directions. Reflection from any rough surface is known as **diffuse reflection**. A surface behaves as a smooth surface as long as its variations are small compared with the wavelength of the incident light. Figures 22.2c and 22.2d are photographs of specular and diffuse reflection of laser light, respectively.

APPLICATION

Seeing the Road on a Rainy Night

As an example, consider the two types of reflection from a road surface that a driver might observe while driving at night. When the road is dry, light from oncoming vehicles is scattered off the road in different directions (diffuse reflection) and the road is clearly visible. On a rainy night when the road is wet,

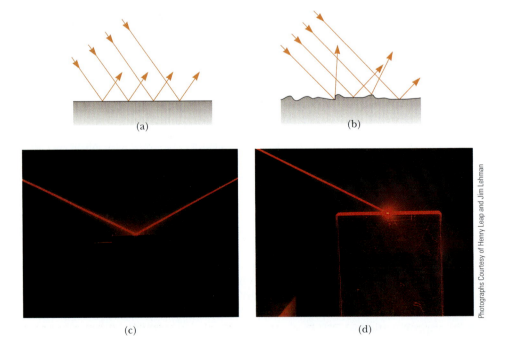

(a)   (b)

(c)   (d)

Photographs Courtesy of Henry Leap and Jim Lehman

Figure 22.2 A schematic representation of (a) specular reflection, where the reflected rays are all parallel to each other, and (b) diffuse reflection, where the reflected rays travel in random directions. (c, d) Photographs of specular and diffuse reflection, made with laser light.

road's irregularities are filled with water. Because the wet surface is smooth, the light undergoes specular reflection. This means that the light is reflected straight ahead, and the driver of a car sees only what is directly in front of him. Light from the side never reaches his eye. In this book we concern ourselves only with specular reflection, and we use the term *reflection* to mean specular reflection.

Quick Quiz 22.1

Which part of Figure 22.3, (a) or (b), better shows specular reflection of light from the roadway?

(a)   (b)

Figure 22.3 (Quick Quiz 22.1)

Consider a light ray traveling in air and incident at some angle on a flat, smooth surface, as in Active Figure 22.4. The incident and reflected rays make angles θ_1 and θ_1', respectively, **with a line perpendicular to the surface** at the point where the incident ray strikes the surface. We call this line the *normal* to the surface. Experiments show that **the angle of reflection equals the angle of incidence**;

$$\theta_1' = \theta_1 \qquad [22.2]$$

You may have noticed a common occurrence in photographs of individuals: their eyes appear to be glowing red. This occurs when a photographic flash device is used and the flash unit is close to the camera lens. Light from the flash unit enters the eye and is reflected back along its original path from the retina. This type of reflection back along the original direction is called *retroreflection*. If the flash unit and lens are close together, retroreflected light can enter the lens. Most of the light reflected from the retina is red, due to the blood vessels at the back of the eye, giving the red-eye effect in the photograph.

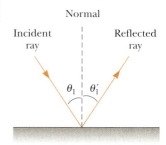

ACTIVE FIGURE 22.4
According to the law of reflection, $\theta_1 = \theta_1'$.

Physics Now™
Log into PhysicsNow at **www.cp7e.com** and go to Active Figure 22.4, where you can vary the incident angle and see the effect on the reflected ray.

A P P L I C A T I O N
Red Eyes in Flash Photographs

Applying Physics 22.1 The Colors of Water Ripples at Sunset

An observer on the west-facing beach of a large lake is watching the beginning of a sunset. The water is very smooth except for some areas with small ripples. The observer notices that some areas of the water are blue and some are pink. Why does the water appear to be different colors in different areas?

Explanation The different colors arise from specular and diffuse reflection. The smooth areas of the water will specularly reflect the light from the west, which is the pink light from the sunset. The areas with small ripples will reflect the light diffusely, so light from all parts of the sky will be reflected into the observer's eyes. Because most of the sky is still blue at the beginning of the sunset, these areas will appear to be blue.

Applying Physics 22.2 Double Images

When looking outdoors through a glass window at night, why do you sometimes see a double image of yourself?

Explanation Reflection occurs whenever there is an interface between two different media. For the glass in the window, there are two such surfaces. The first is the inner surface of the glass, and the second is the outer surface. Each of these interfaces results in an image.

INTERACTIVE EXAMPLE 22.1 The Double-Reflecting Light Ray

Goal Calculate a resultant angle from two reflections.

Problem Two mirrors make an angle of 120° with each other, as in Figure 22.5. A ray is incident on mirror M_1 at an angle of 65° to the normal. Find the angle the ray makes with the normal to M_2 after it is reflected from both mirrors.

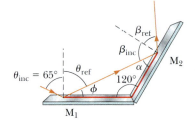

Figure 22.5 (Example 22.1) Mirrors M_1 and M_2 make an angle of 120° with each other.

Strategy Apply the law of reflection twice. Given the incident ray at angle θ_{inc}, find the final resultant angle, β_{ref}.

Solution

Apply the law of reflection to M_1 to find the angle of reflection, θ_{ref}:

$$\theta_{ref} = \theta_{inc} = 65°$$

Find the angle ϕ that is the complement of the angle θ_{ref}:

$$\phi = 90° - \theta_{ref} = 90° - 65° = 25°$$

Find the unknown angle α in the triangle of M_1, M_2, and the ray traveling from M_1 to M_2, using the fact that the three angles sum to 180°:

$$180° = 25° + 120° + \alpha \quad \rightarrow \quad \alpha = 35°$$

The angle α is complementary to the angle of incidence, β_{inc}, for M_2:

$$\alpha + \beta_{inc} = 90° \quad \rightarrow \quad \beta_{inc} = 90° - 35° = 55°$$

Apply the law of reflection a second time, obtaining β_{ref}:

$$\beta_{ref} = \beta_{inc} = \boxed{55°}$$

Remarks Notice the heavy reliance on elementary geometry and trigonometry in these reflection problems.

Exercise 22.1

Repeat the problem if the angle of incidence is 55° and the second mirror makes an angle of 100° with the first mirror.

Answer 45°

Physics⊗Now™ Investigate reflection for various mirror angles by logging into PhysicsNow at **www.cp7e.com** and going to Interactive Example 22.1.

Refraction of Light

When a ray of light traveling through a transparent medium encounters a boundary leading into another transparent medium, as in Active Figure 22.6a, part of the ray is reflected and part enters the second medium. The ray that enters the second medium is bent at the boundary and is said to be *refracted*. The incident ray, the reflected ray, the refracted ray, and the normal at the point of incidence all lie

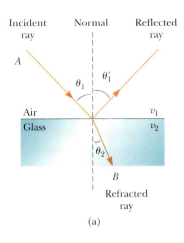

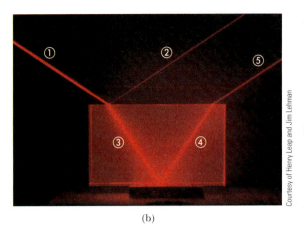

Courtesy of Henry Leap and Jim Lehman

(a)   (b)

ACTIVE FIGURE 22.6
(a) A ray obliquely incident on an air–glass interface. The refracted ray is bent toward the normal because $v_2 < v_1$. (b) Light incident on the Lucite block bends both when it enters the block and when it leaves the block.

Physics⊗Now™
Log into PhysicsNow at **www.cp7e.com** and go to Active Figure 22.6, where you can vary the incident angle and see the effect on the reflected and refracted rays.

in the same plane. The **angle of refraction**, θ_2, in Active Figure 22.6a depends on the properties of the two media and on the angle of incidence, through the relationship

$$\frac{\sin\theta_2}{\sin\theta_1} = \frac{v_2}{v_1} = \text{constant} \qquad \textbf{[22.3]}$$

where v_1 is the speed of light in medium 1 and v_2 is the speed of light in medium 2. Note that the angle of refraction is also measured with respect to the normal. In Section 22.7 we will derive the laws of reflection and refraction using Huygens' principle.

Experiment shows that **the path of a light ray through a refracting surface is reversible**. For example, the ray in Active Figure 22.6a travels from point A to point B. If the ray originated at B, it would follow the same path to reach point A, but the reflected ray would be in the glass.

Quick Quiz 22.2

If beam 1 is the incoming beam in Active Figure 22.6b, which of the other four beams are due to reflection? Which are due to refraction?

When light moves from a material in which its speed is high to a material in which its speed is lower, the angle of refraction θ_2 is less than the angle of incidence. The refracted ray therefore bends toward the normal, as shown in Active Figure 22.7a. If the ray moves from a material in which it travels slowly to a material in which it travels more rapidly, θ_2 is greater than θ_1, so the ray bends away from the normal, as shown in Active Figure 22.7b.

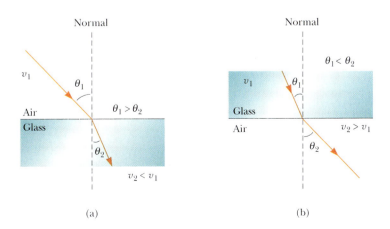

(a)   (b)

ACTIVE FIGURE 22.7
(a) When the light beam moves from air into glass, its path is bent toward the normal. (b) When the beam moves from glass into air, its path is bent away from the normal.

Physics⊗Now™
Log into PhysicsNow at **www.cp7e.com** and go to Active Figure 22.7. Light passes through three layers of material. You can vary the incident angle and see the effect on the refracted rays for a variety of values of the index of refraction (Table 22.1, page 732) of the three materials.

22.3   THE LAW OF REFRACTION

When light passes from one transparent medium to another, it's refracted because the speed of light is different in the two media[1]. The **index of refraction**, n, of a medium is defined as the ratio c/v;

Index of refraction ▶

$$n \equiv \frac{\text{speed of light in vacuum}}{\text{speed of light in a medium}} = \frac{c}{v} \qquad \text{[22.4]}$$

From this definition, we see that the index of refraction is a dimensionless number that is greater than or equal to one because v is always less than c. Further, n is equal to one for vacuum. Table 22.1 lists the indices of refraction for various substances.

As light travels from one medium to another, its frequency doesn't change. To see why, consider Figure 22.8. Wave fronts pass an observer at point A in medium 1 with a certain frequency and are incident on the boundary between medium 1 and medium 2. The frequency at which the wave fronts pass an observer at point B in medium 2 must equal the frequency at which they arrive at point A. If this were not the case, the wave fronts would either pile up at the boundary or be destroyed or created at the boundary. Because neither of these events occurs, the frequency must remain the same as a light ray passes from one medium into another.

Therefore, because the relation $v = f\lambda$ must be valid in both media, and because $f_1 = f_2 = f$, we see that

$$v_1 = f\lambda_1 \qquad \text{and} \qquad v_2 = f\lambda_2$$

Because $v_1 \neq v_2$, it follows that $\lambda_1 \neq \lambda_2$. A relationship between the index of refraction and the wavelength can be obtained by dividing these two equations and making use of the definition of the index of refraction given by Equation 22.4:

$$\frac{\lambda_1}{\lambda_2} = \frac{v_1}{v_2} = \frac{c/n_1}{c/n_2} = \frac{n_2}{n_1} \qquad \text{[22.5]}$$

which gives

$$\lambda_1 n_1 = \lambda_2 n_2 \qquad \text{[22.6]}$$

TIP 22.1   An Inverse Relationship

The index of refraction is *inversely* proportional to the wave speed. Therefore, as the wave speed v decreases, the index of refraction n increases.

TIP 22.2   The Frequency Remains the Same

The *frequency* of a wave does *not* change as the wave passes from one medium to another. Both the wave speed and the wavelength *do* change, but the frequency remains the same.

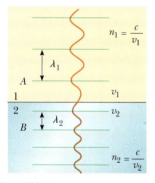

Figure 22.8 As the wave moves from medium 1 to medium 2, its wavelength changes, but its frequency remains constant.

TABLE 22.1

Indices of Refraction for Various Substances, Measured with Light of Vacuum Wavelength $\lambda_0 = 589$ mn

Substance	Index of Refraction	Substance	Index of Refraction
Solids at 20°C		**Liquids at 20°C**	
Diamond (C)	2.419	Benzene	1.501
Fluorite (CaF_2)	1.434	Carbon disulfide	1.628
Fused quartz (SiO_2)	1.458	Carbon tetrachloride	1.461
Glass, crown	1.52	Ethyl alcohol	1.361
Glass, flint	1.66	Glycerine	1.473
Ice (H_2O) (at 0°C)	1.309	Water	1.333
Polystyrene	1.49		
Sodium chloride (NaCl)	1.544	**Gases at 0°C, 1 atm**	
Zircon	1.923	Air	1.000 293
		Carbon dioxide	1.000 45

[1]The speed of light varies between media because the time lags caused by the absorption and reemission of light as it travels from atom to atom depend on the particular electronic structure of the atoms constituting each material.

Let medium 1 be the vacuum, so that $n_1 = 1$. It follows from Equation 22.6 that the index of refraction of any medium can be expressed as the ratio

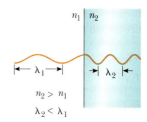

$$n = \frac{\lambda_0}{\lambda_n} \qquad [22.7]$$

where λ_0 is the wavelength of light in vacuum and λ_n is the wavelength in the medium having index of refraction n. Figure 22.9 is a schematic representation of this reduction in wavelength when light passes from a vacuum into a transparent medium.

We are now in a position to express Equation 22.3 in an alternate form. If we substitute Equation 22.5 into Equation 22.3, we get

Figure 22.9 A schematic diagram of the *reduction* in wavelength when light travels from a medium with a low index of refraction to one with a higher index of refraction.

$$n_1 \sin \theta_1 = n_2 \sin \theta_2 \qquad [22.8]$$

◀ Snell's law of refraction

The experimental discovery of this relationship is usually credited to Willebord Snell (1591–1627) and is therefore known as **Snell's law of refraction**.

Quick Quiz 22.3

A material has an index of refraction that increases continuously from top to bottom. Of the three paths shown in Figure 22.10, which path will a light ray follow as it passes through the material?

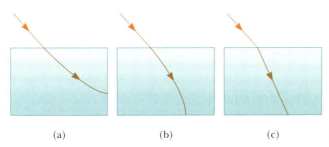

      (a)   (b)   (c)   **Figure 22.10** (Quick Quiz 22.3)

Quick Quiz 22.4

As light travels from a vacuum ($n = 1$) to a medium such as glass ($n > 1$), which of the following properties remains the same? (a) wavelength, (b) wave speed, or (c) frequency?

EXAMPLE 22.2   Angle of Refraction for Glass

Goal   Apply Snell's law to a slab of glass.

Problem   A light ray of wavelength 589 nm (produced by a sodium lamp) traveling through air is incident on a smooth, flat slab of crown glass at an angle of 30.0° to the normal, as sketched in Figure 22.11. Find the angle of refraction, θ_2.

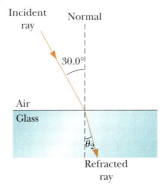

Figure 22.11 (Example 22.2) Refraction of light by glass.

Strategy   Substitute quantities into Snell's law and solve for the unknown angle of refraction, θ_2.

Solution

Solve Snell's law (Eq. 22.8) for $\sin \theta_2$:

$$(1) \qquad \sin \theta_2 = \frac{n_1}{n_2} \sin \theta_1$$

From Table 22.1, find $n_1 = 1.00$ for air and $n_2 = 1.52$ for crown glass. Substitute these values into (1) and take the inverse sine of both sides:

$$\sin \theta_2 = \left(\frac{1.00}{1.52} \right)(\sin 30.0°) = 0.329$$

$$\theta_2 = \sin^{-1}(0.329) = \boxed{19.2°}$$

Remarks   Notice the light ray bends toward the normal when it enters a material of a higher index of refraction. If the ray left the material following the same path in reverse, it would bend away from the normal.

Exercise 22.2

If the light ray moves from inside the glass toward the glass–air interface at an angle of 30.0° to the normal, determine the angle of refraction.

Answer   The ray bends 49.5° *away* from the normal, as expected.

EXAMPLE 22.3   Light in Fused Quartz

Goal   Use the index of refraction to determine the effect of a medium on light's speed and wavelength.

Problem   Light of wavelength 589 nm in vacuum passes through a piece of fused quartz of index of refraction $n = 1.458$. (a) Find the speed of light in fused quartz. (b) What is the wavelength of this light in fused quartz? (c) What is the frequency of the light in fused quartz?

Strategy   Substitute values into Equations 22.4 and 22.7.

Solution
(a) Find the speed of light in fused quartz.

Obtain the speed from Equation 22.4:

$$v = \frac{c}{n} = \frac{3.00 \times 10^8 \text{ m/s}}{1.458} = \boxed{2.06 \times 10^8 \text{ m/s}}$$

(b) What is the wavelength of this light in fused quartz?

Use Eq. 22.7 to calculate the wavelength:

$$\lambda_n = \frac{\lambda_0}{n} = \frac{589 \text{ nm}}{1.458} = \boxed{404 \text{ nm}}$$

(c) What is the frequency of the light in fused quartz?

The frequency in quartz is the same as in vacuum. Solve $c = f\lambda$ for the frequency:

$$f = \frac{c}{\lambda} = \frac{3.00 \times 10^8 \text{ m/s}}{589 \times 10^{-9} \text{ m}} = \boxed{5.09 \times 10^{14} \text{ Hz}}$$

Remarks   It's interesting to note that the speed of light in vacuum, 3.00×10^8 m/s, is an upper limit for the speed of material objects. In our treatment of relativity in Chapter 26, we will find that this upper limit is consistent with experimental observations. However, it's possible for a particle moving in a medium to have a speed that exceeds the speed of light in that medium. For example, it's theoretically possible for a particle to travel through fused quartz at a speed greater than 2.06×10^8 m/s, but it must still have a speed less than 3.00×10^8 m/s.

Exercise 22.3

Light with wavelength 589 nm passes through crystalline sodium chloride. Find (a) the speed of light in this medium, (b) the wavelength, and (c) the frequency of the light.

Answer   (a) 1.94×10^8 m/s   (b) 381 nm   (c) 5.09×10^{14} Hz

---

INTERACTIVE EXAMPLE 22.4   Light Passing through a Slab

Goal   Apply Snell's law when a ray passes into and out of another medium.

Problem   A light beam traveling through a transparent medium of index of refraction n_1 passes through a thick transparent slab with parallel faces and index of refraction n_2 (Fig. 22.12). Show that the emerging beam is parallel to the incident beam.

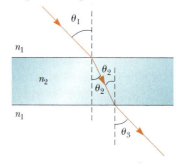

Figure 22.12   (Example 22.4) When light passes through a flat slab of material, the emerging beam is parallel to the incident beam, and therefore $\theta_1 = \theta_3$.

Strategy   Apply Snell's law twice, once at the upper surface and once at the lower surface. The two equations will be related because the angle of refraction at the upper surface equals the angle of incidence at the lower surface. The ray passing through the slab makes equal angles with the normal at the entry and exit points. This procedure will enable us to compare angles θ_1 and θ_3.

Solution

Apply Snell's law to the upper surface:

$$(1) \qquad \sin\theta_2 = \frac{n_1}{n_2}\sin\theta_1$$

Apply Snell's law to the lower surface:

$$(2) \qquad \sin\theta_3 = \frac{n_2}{n_1}\sin\theta_2$$

Substitute Equation 1 into Equation 2:

$$\sin\theta_3 = \frac{n_2}{n_1}\left(\frac{n_1}{n_2}\sin\theta_1\right) = \sin\theta_1$$

Take the inverse sine of both sides, noting that the angles are positive and less than 90°:

$$\theta_3 = \theta_1$$

Remarks   The preceding result proves that the slab doesn't alter the direction of the beam. It does, however, produce a lateral displacement of the beam, as shown in Figure 22.12.

Exercise 22.4
Suppose the ray, in air with $n = 1.00$, enters a slab with $n = 2.50$ at a $45.0°$ angle with respect to the normal, then exits the bottom of the slab into water, with $n = 1.33$. At what angle to the normal does the ray leave the slab?

Answer $32.1°$

Physics⊗Now™   Explore refraction through slabs of various thickness by logging into PhysicsNow at **www.cp7e.com** and going to Interactive Example 22.4.

EXAMPLE 22.5   Refraction of Laser Light in a Digital Video Disk (DVD)

Goal   Apply Snell's law together with geometric constraints.

Problem   A DVD is a video recording consisting of a spiral track about $1.0\ \mu m$ wide with digital information. (See Fig. 22.13a.) The digital information consists of a series of pits that are "read" by a laser beam sharply focused on a track in the information layer. If the width a of the beam at the information layer must equal $1.0\ \mu m$ to distinguish individual tracks, and the width w of the beam as it enters the plastic is 0.7000 mm, find the angle θ_1 at which the conical beam should enter the plastic. (See Fig. 22.13b.) Assume the plastic has a thickness $t = 1.20$ mm and an index of refraction $n = 1.55$. Note that this system is relatively immune to small dust particles degrading the video

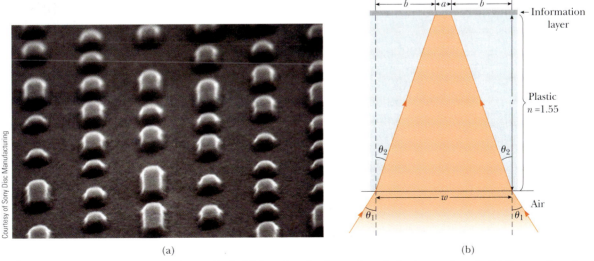

Courtesy of Sony Disc Manufacturing

(a)   (b)

Figure 22.13   (Example 22.5) A micrograph of a DVD surface showing tracks and pits along each track. (b) Cross section of a cone-shaped laser beam used to read a DVD.

quality, because particles would have to be as large as 0.700 mm to obscure the beam at the point where it enters the plastic.

Strategy   Use right-triangle trigonometry to determine the angle θ_2, and then apply Snell's law to obtain the angle θ_1.

Solution

From the top and bottom of Figure 22.13b, obtain an equation relating w, b, and a:

$$w = 2b + a$$

Solve this equation for b and substitute given values:

$$b = \frac{w - a}{2} = \frac{700.0 \times 10^{-6}\,\text{m} - 1.0 \times 10^{-6}\,\text{m}}{2} = 349.5\,\mu\text{m}$$

Now use the tangent function to find θ_2:

$$\tan\theta_2 = \frac{b}{t} = \frac{349.5\,\mu\text{m}}{1.20 \times 10^3\,\mu\text{m}} \quad \rightarrow \quad \theta_2 = 16.2°$$

Finally, use Snell's law to find θ_1:

$$n_1 \sin\theta_1 = n_2 \sin\theta_2$$

$$\sin\theta_1 = \frac{n_2 \sin\theta_2}{n_1} = \frac{1.55 \sin 16.2°}{1.00} = 0.433$$

$$\theta_1 = \sin^{-1}(0.433) = \boxed{25.7°}$$

Remarks   Despite its apparent complexity, the problem isn't that different from Example 22.2.

Exercise 22.5

Suppose you wish to redesign the system to decrease the initial width of the beam from 0.700 0 mm to 0.600 0 mm, but leave the incident angle θ_1 and all other parameters the same as before, except the index of refraction for the plastic material (n_2) and the angle θ_2. What index of refraction should the plastic have?

Answer   1.79

22.4   DISPERSION AND PRISMS

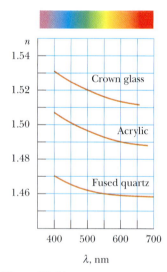

Figure 22.14   Variations of index of refraction in the visible spectrum with respect to vacuum wavelength for three materials.

In Table 22.1, we presented values for the index of refraction of various materials. If we make careful measurements, however, we find that the index of refraction in anything but vacuum depends on the wavelength of light. The dependence of the index of refraction on wavelength is called **dispersion**. Figure 22.14 is a graphical representation of this variation in the index of refraction with wavelength. Because n is a function of wavelength, Snell's law indicates that **the angle of refraction made when light enters a material depends on the wavelength of the light**. As seen in the figure, the index of refraction for a material usually decreases with increasing wavelength. This means that violet light ($\lambda \cong 400$ nm) refracts more than red light ($\lambda \cong 650$ nm) when passing from air into a material.

To understand the effects of dispersion on light, consider what happens when light strikes a prism, as in Figure 22.15a. A ray of light of a single wavelength that is incident on the prism from the left emerges bent away from its original direction of travel by an angle δ, called the **angle of deviation**. Now suppose a beam of white light (a combination of all visible wavelengths) is incident on a prism. Because of dispersion, the different colors refract through different angles of deviation, and the rays that emerge from the second face of the prism spread out in a series of colors known as a visible **spectrum**, as shown in Figure 22.16. These colors, in order of decreasing wavelength, are red, orange, yellow, green, blue, and violet. Violet light deviates the most, red light the least, and the remaining colors in the visible spectrum fall between these extremes.

Prisms are often used in an instrument known as a **prism spectrometer**, the essential elements of which are shown in Figure 22.17a (page 738). This instrument

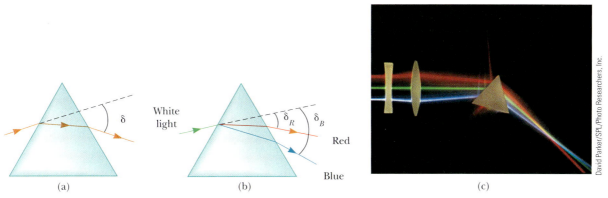

Figure 22.15 (a) A prism refracts a light ray and deviates the light through the angle δ. (b) When light is incident on a prism, the blue light is bent more than the red. (c) Light of different colors passes through a prism and two lenses. Note that as the light passes through the prism, different wavelengths are refracted at different angles.

is commonly used to study the wavelengths emitted by a light source, such as a sodium vapor lamp. Light from the source is sent through a narrow, adjustable slit and lens to produce a parallel, or collimated, beam. The light then passes through the prism and is dispersed into a spectrum. The refracted light is observed through a telescope. The experimenter sees different colored images of the slit through the eyepiece of the telescope. The telescope can be moved or the prism can be rotated in order to view the various wavelengths, which have different angles of deviation. Figure 22.17b (page 738) shows one type of prism spectrometer used in undergraduate laboratories.

All hot, low-pressure gases emit their own characteristic spectra. Thus, one use of a prism spectrometer is to identify gases. For example, sodium emits only two wavelengths in the visible spectrum: two closely spaced yellow lines. (The bright linelike images of the slit seen in a spectroscope are called *spectral lines.*) A gas emitting these, and only these, colors can thus be identified as sodium. Likewise, mercury vapor has its own characteristic spectrum, consisting of four prominent wavelengths—orange, green, blue, and violet lines—along with some wavelengths of lower intensity. The particular wavelengths emitted by a gas serve as "fingerprints" of that gas. Spectral analysis, which is the measurement of the wavelengths emitted or absorbed by a substance, is a powerful general tool in many scientific areas. As examples, chemists and biologists use infrared spectroscopy to identify molecules, astronomers use visible-light spectroscopy to identify elements on distant stars, and geologists use spectral analysis to identify minerals.

APPLICATION

Identifying Gases with a Spectrometer

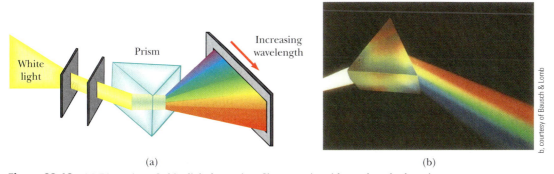

Figure 22.16 (a) Dispersion of white light by a prism. Since *n* varies with wavelength, the prism disperses the white light into its various spectral components. (b) Different colors of light that pass through a prism are refracted at different angles because the index of refraction of the glass depends on wavelength. Violet light bends the most, red light the least.

(a)   (b)

Figure 22.17   (a) A diagram of a prism spectrometer. The colors in the spectrum are viewed through a telescope. (b) A prism spectrometer with interchangeable components.

Applying Physics 22.3   Dispersion

When a beam of light enters a glass prism, which has nonparallel sides, the rainbow of color exiting the prism is a testimonial to the dispersion occurring in the glass. Suppose a beam of light enters a slab of material with parallel sides. When the beam exits the other side, traveling in the same direction as the original beam, is there any evidence of dispersion?

Explanation Due to dispersion, light at the violet end of the spectrum exhibits a larger angle of refraction on entering the glass than light at the red end. All colors of light return to their original direction of propagation as they refract back out into the air. As a result, the outgoing beam is white. But the net shift in the position of the violet light along the edge of the slab is larger than the shift of the red light, so one edge of the outgoing beam has a bluish tinge to it (it appears blue rather than violet, because the eye is not very sensitive to violet light), whereas the other edge has a reddish tinge. This effect is indicated in

Figure 22.18. The colored edges of the outgoing beam of white light are evidence of dispersion.

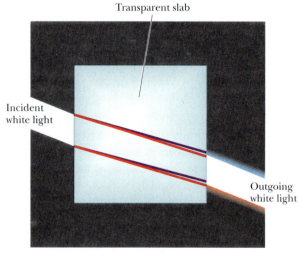

Figure 22.18   (Applying Physics 22.3)

22.5   THE RAINBOW

The dispersion of light into a spectrum is demonstrated most vividly in nature through the formation of a rainbow, often seen by an observer positioned between the Sun and a rain shower. To understand how a rainbow is formed, consider Active Figure 22.19. A ray of light passing overhead strikes a drop of water in the atmosphere and is refracted and reflected as follows: It is first refracted at the front surface of the drop, with the violet light deviating the most and the red light the least. At the back surface of the drop, the light is reflected and returns to the front surface, where it again undergoes refraction as it moves from water into air. The rays leave the drop so that the angle between the incident white light and the returning violet ray is 40° and the angle between the white light and the returning red ray is 42°. This small angular difference between the returning rays causes us to see the bow, as explained in the next paragraph.

Now consider an observer viewing a rainbow, as in Figure 22.20a. If a raindrop high in the sky is being observed, the red light returning from the drop can reach the observer because it is deviated the most, but the violet light passes over the observer because it is deviated the least. Hence, the observer sees this drop as being red. Similarly, a drop lower in the sky would direct violet light toward the observer and appear to be violet. (The red light from this drop would strike the ground and not be seen.) The remaining colors of the spectrum would reach the observer from raindrops lying between these two extreme positions. Figure 22.20b shows a beautiful rainbow and a secondary rainbow with its colors reversed.

22.6 HUYGENS' PRINCIPLE

The laws of reflection and refraction can be deduced using a geometric method proposed by Huygens in 1678. Huygens assumed that light is a form of wave motion rather than a stream of particles. He had no knowledge of the nature of light or of its electromagnetic character. Nevertheless, his simplified wave model is adequate for understanding many practical aspects of the propagation of light.

Huygens' principle is a geometric construction for determining at some instant the position of a new wave front from knowledge of the wave front that preceded it. (A wave front is a surface passing through those points of a wave which have the same phase and amplitude. For instance, a wave front could be a surface passing

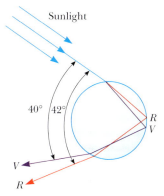

ACTIVE FIGURE 22.19
Refraction of sunlight by a spherical raindrop.

Physics⊗Now™
Log into PhysicsNow at **www.cp7e.com** and go to Active Figure 22.19, where you can vary the point at which the sunlight enters the raindrop and verify that the angles shown are maximum angles.

(a)

(b)

Figure 22.20 (a) The formation of a rainbow. (b) This photograph of a rainbow shows a distinct secondary rainbow with the colors reversed.

Figure 22.21 Huygens' constructions for (a) a plane wave propagating to the right and (b) a spherical wave.

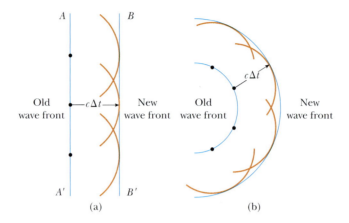

through the crests of waves.) In Huygens' construction, **all points on a given wave front are taken as point sources for the production of spherical secondary waves, called wavelets, which propagate in the forward direction with speeds characteristic of waves in that medium. After some time has elapsed, the new position of the wave front is the surface tangent to the wavelets**.

◄ Huygens' principle

Figure 22.21 illustrates two simple examples of Huygens' construction. First, consider a plane wave moving through free space, as in Figure 22.21a. At $t = 0$, the wave front is indicated by the plane labeled AA'. In Huygens' construction, each point on this wave front is considered a point source. For clarity, only a few points on AA' are shown. With these points as sources for the wavelets, we draw circles of radius $c\Delta t$, where c is the speed of light in vacuum and Δt is the period of propagation from one wave front to the next. The surface drawn tangent to these wavelets is the plane BB', which is parallel to AA'. In a similar manner, Figure 22.21b shows Huygens' construction for an outgoing spherical wave.

Figure 22.22 shows a convincing demonstration of Huygens' principle. Plane waves coming from far off shore emerge from the openings between the barriers as two-dimensional circular waves propagating outward.

Huygens' Principle Applied to Reflection and Refraction

The laws of reflection and refraction were stated earlier in the chapter without proof. We now derive these laws using Huygens' principle. Figure 22.23a illustrates the law of reflection. The line AA' represents a wave front of the incident light. As ray 3 travels from A' to C, ray 1 reflects from A and produces a spherical wavelet of radius AD. (Recall that the radius of a Huygens wavelet is $v\Delta t$.) Because the two

Figure 22.22 This photograph of the beach at Tel Aviv, Israel, shows Huygens wavelets radiating from each opening between breakwalls. Note how the beach has been shaped by the wave action.

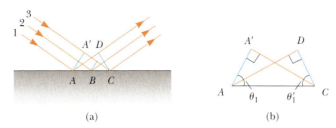

Figure 22.23 (a) Huygens' construction for proving the law of reflection. (b) Triangle ADC is congruent to triangle $AA'C$.

wavelets having radii $A'C$ and AD are in the same medium, they have the same speed v, so $AD = A'C$. Meanwhile, the spherical wavelet centered at B has spread only half as far as the one centered at A, because ray 2 strikes the surface later than ray 1.

From Huygens' principle, we find that the reflected wave front is CD, a line tangent to all the outgoing spherical wavelets. The remainder of our analysis depends on geometry, as summarized in Figure 22.23b. Note that the right triangles ADC and $AA'C$ are congruent because they have the same hypotenuse, AC, and because $AD = A'C$. From the figure, we have

$$\sin\theta_1 = \frac{A'C}{AC} \quad \text{and} \quad \sin\theta_1' = \frac{AD}{AC}$$

The right-hand sides are equal, so $\sin\theta = \sin\theta_1'$, and it follows that $\theta_1 = \theta_1'$, which is the law of reflection.

Huygens' principle and Figure 22.24a can be used to derive Snell's law of refraction. In the time interval Δt, ray 1 moves from A to B and ray 2 moves from A' to C. The radius of the outgoing spherical wavelet centered at A is equal to $v_2 \Delta t$. The distance $A'C$ is equal to $v_1 \Delta t$. Geometric considerations show that angle $A'AC$ equals θ_1 and angle ACB equals θ_2. From triangles $AA'C$ and ACB, we find that

$$\sin\theta_1 = \frac{v_1 \Delta t}{AC} \quad \text{and} \quad \sin\theta_2 = \frac{v_2 \Delta t}{AC}$$

If we divide the first equation by the second, we get

$$\frac{\sin\theta_1}{\sin\theta_2} = \frac{v_1}{v_2}$$

But from Equation 22.4 we know that $v_1 = c/n_1$ and $v_2 = c/n_2$. Therefore,

$$\frac{\sin\theta_1}{\sin\theta_2} = \frac{c/n_1}{c/n_2} = \frac{n_2}{n_1}$$

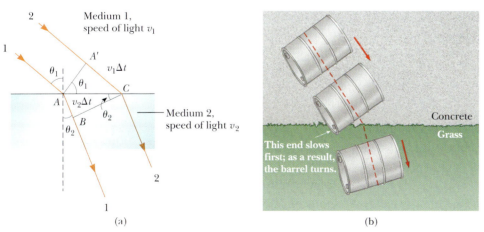

(a)   (b)

Figure 22.24 (a) Huygens' construction for proving the law of refraction. (b) Overhead view of a barrel rolling from concrete onto grass.

and it follows that

$$n_1 \sin \theta_1 = n_2 \sin \theta_2$$

which is the law of refraction.

A mechanical analog of refraction is shown in Figure 22.24b. When the left end of the rolling barrel reaches the grass, it slows down, while the right end remains on the concrete and moves at its original speed. This difference in speeds causes the barrel to pivot, changing its direction of its motion.

22.7   TOTAL INTERNAL REFLECTION

An interesting effect called *total internal reflection* can occur when light encounters the boundary between a medium with a *higher* index of refraction and one with a *lower* index of refraction. Consider a light beam traveling in medium 1 and meeting the boundary between medium 1 and medium 2, where n_1 is greater than n_2 (Active Fig. 22.25). Possible directions of the beam are indicated by rays 1 through 5. Note that the refracted rays are bent away from the normal because n_1 is greater than n_2. At some particular angle of incidence θ_c, called the **critical angle**, the refracted light ray moves parallel to the boundary, so that $\theta_2 = 90°$ (Active Fig. 22.25b). *For angles of incidence greater than θ_c, the beam is entirely reflected at the boundary, as is ray 5 in Active Figure 22.25a. This ray is reflected as though it had struck a perfectly reflecting surface. It and all rays like it obey the law of reflection: the angle of incidence equals the angle of reflection.

We can use Snell's law to find the critical angle. When $\theta_1 = \theta_c$, $\theta_2 = 90°$, Snell's law (Eq. 22.8) gives

$$n_1 \sin \theta_c = n_2 \sin 90° = n_2$$

$$\sin \theta_c = \frac{n_2}{n_1} \qquad \text{for} \qquad n_1 > n_2 \qquad [22.9]$$

Equation 22.9 can be used only when n_1 is greater than n_2, because **total internal reflection occurs only when light attempts to move from a medium of higher index of refraction to a medium of lower index of refraction**. If n_1 were less than n_2, Equation 22.9 would give $\sin \theta_c > 1$, which is an absurd result because the sine of an angle can never be greater than one.

When medium 2 is air, the critical angle is small for substances with large indices of refraction, such as diamond, where $n = 2.42$ and $\theta_c = 24.0°$. By comparison, for crown glass, $n = 1.52$ and $\theta_c = 41.0°$. This property, combined with proper faceting, causes a diamond to sparkle brilliantly.

This photograph shows nonparallel light rays entering a glass prism. The bottom two rays undergo total internal reflection at the longest side of the prism. The top three rays are refracted at the longest side as they leave the prism.

Courtesy of Henry Leap and Jim Lehman

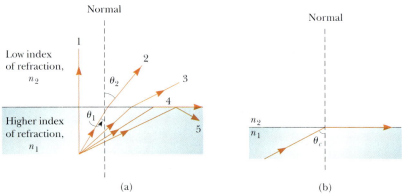

ACTIVE FIGURE 22.25

(a) Rays from a medium with index of refraction n_1 travel to a medium with index of refraction n_2, where $n_1 > n_2$. As the angle of incidence increases, the angle of refraction θ_2 increases until θ_2 is 90° (ray 4). For even larger angles of incidence, total internal reflection occurs (ray 5). (b) The angle of incidence producing a 90° angle of refraction is often called the *critical angle* θ_c.

Physics ⊗ Now™

Log into PhysicsNow at **www.cp7e.com** and go to Active Figure 22.25, where you can vary the incident angle and see how the refracted ray undergoes total internal reflection when the incident angle exceeds the critical angle.

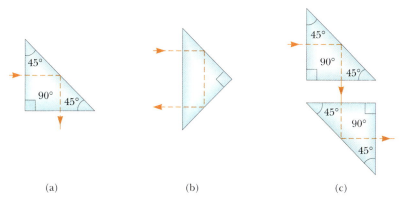

Figure 22.26   Internal reflection in a prism. (a) The ray is deviated by 90°. (b) The direction of the ray is reversed. (c) Two prisms used as a periscope.

(a)   (b)   (c)

A prism and the phenomenon of total internal reflection can alter the direction of travel of a light beam. Figure 22.26 illustrates two such possibilities. In one case the light beam is deflected by 90.0° (Fig. 22.26a), and in the second case the path of the beam is reversed (Fig. 22.26b). A common application of total internal reflection is a submarine periscope. In this device, two prisms are arranged as in Figure 22.26c, so that an incident beam of light follows the path shown and the user can "see around corners."

APPLICATION

Submarine Periscopes

Applying Physics 22.4   Total Internal Reflection and Dispersion

A beam of white light is incident on the curved edge of a semicircular piece of glass, as shown in Figure 22.27. The light enters the curved surface along the normal, so it shows no refraction. It encounters the straight side of the glass at the center of curvature of the curved side and refracts into the air. The incoming beam is moved clockwise (so that the angle θ increases) such that the beam always enters along the normal to the curved side and encounters the straight side at the center of curvature of the curved side. Why does the refracted beam become redder as it approaches a direction parallel to the straight side?

Explanation   When the outgoing beam approaches the direction parallel to the straight side, the incident angle is approaching the critical angle for total internal reflection. Dispersion occurs as the light passes out of the glass. The index of refraction for light at the violet end of the visible spectrum is larger than at the red end. As a result, as the outgoing beam approaches the straight side, the violet light undergoes total internal reflection, followed by the other colors. The red light is the last to undergo total internal reflection, so just before the outgoing light disappears, it's composed of light from the red end of the visible spectrum.

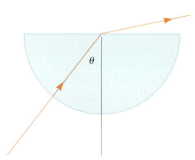

Figure 22.27   (Applying Physics 22.4)

EXAMPLE 22.6   A View from the Fish's Eye

Goal   Apply the concept of total internal reflection. **(a)** Find the critical angle for a water–air boundary if the index of refraction of water is 1.33. **(b)** Use the result of part (a) to predict what a fish will see (Fig. 22.28) if it looks up toward the water surface at angles of 40.0°, 48.8°, and 60.0°.

Strategy   After finding the critical angle by substitution, use the fact that the path of a light ray is reversible: at a given angle, wherever a light beam can go is also where a beam of light can come from, along the same path.

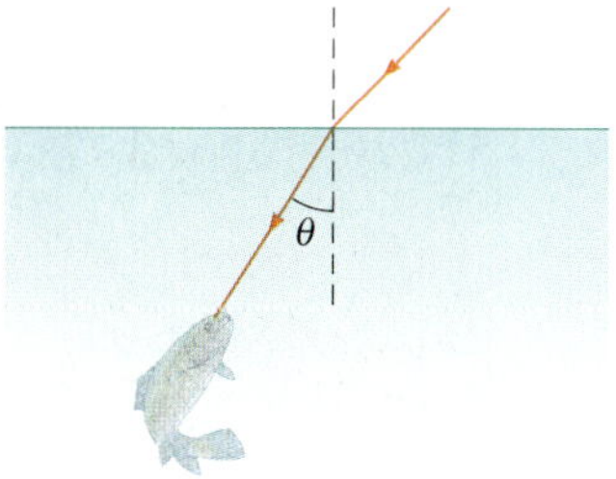

Figure 22.28   (Example 22.6) A fish looks upward toward the water's surface.

Solution

(a) Find the critical angle for a water–air boundary.

Substitute into Equation 22.9 to find the critical angle: $\sin\theta_c = \dfrac{n_2}{n_1} = \dfrac{1.00}{1.33} = 0.752$

$$\theta_c = \boxed{48.8°}$$

(b) Predict what a fish will see if it looks up toward the water surface at angles of 40.0°, 48.8°, and 60.0°.

At an angle of 40.0°, a beam of light from underwater will be refracted at the surface and enter the air above. Because the path of a light ray is reversible (Snell's law works both going and coming), light from above can follow the same path and be perceived by the fish. At an angle of 48.8°, the critical angle for water, light from underwater is bent so that it travels along the surface. This means that light following the same path in reverse can reach the fish only by skimming along the water surface before being refracted towards the fish's eye. At angles greater than the critical angle of 48.8°, a beam of light shot toward the surface will be completely reflected down toward the bottom of the pool. Reversing the path, the fish sees a reflection of some object on the bottom.

Exercise 22.6
Suppose a layer of oil with $n = 1.50$ coats the surface of the water. What is the critical angle for total internal reflection for light traveling in the oil layer and encountering the oil-water boundary?

Answer   62.7°

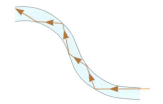

Figure 22.29 Light travels in a curved transparent rod by multiple internal reflections.

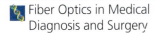

APPLICATION

Fiber Optics in Medical Diagnosis and Surgery

Fiber Optics

Another interesting application of total internal reflection is the use of solid glass or transparent plastic rods to "pipe" light from one place to another. As indicated in Figure 22.29, light is confined to traveling within the rods, even around gentle curves, as a result of successive internal reflections. Such a light pipe can be quite flexible if thin fibers are used rather than thick rods. If a bundle of parallel fibers is used to construct an optical transmission line, images can be transferred from one point to another.

Very little light intensity is lost in these fibers as a result of reflections on the sides. Any loss of intensity is due essentially to reflections from the two ends and absorption by the fiber material. Fiber-optic devices are particularly useful for viewing images produced at inaccessible locations. Physicians often use fiber-optic cables to aid in the diagnosis and correction of certain medical problems without the intrusion of major surgery. For example, a fiber-optic cable can be threaded through the esophagus and into the stomach to look for ulcers. In this application, the cable consists of two fiber-optic lines: one to transmit a beam of light into the stomach for illumination and the other to allow the light to be transmitted out of the stomach. The resulting image can, in some cases, be viewed directly by the physician, but more often is displayed on a television monitor or captured on film.

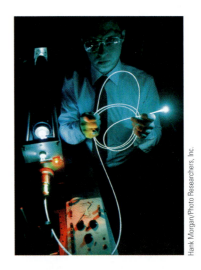

(*Left*) Strands of glass optical fibers are used to carry voice, video, and data signals in telecommunication networks. (*Right*) A bundle of optical fibers is illuminated by a laser.

Dennis O'Clair/Tony Stone Images/Getty Images

Hank Morgan/Photo Researchers, Inc.

In a similar way, fiber-optic cables can be used to examine the colon or to help physicians perform surgery without the need for large incisions.

The field of fiber optics has revolutionized the entire communications industry. Billions of kilometers of optical fiber have been installed in the United States to carry high-speed internet traffic, radio and television signals, and telephone calls. The fibers can carry much higher volumes of telephone calls and other forms of communication than electrical wires because of the higher frequency of the infrared light used to carry the information on optical fibers. Optical fibers are also preferable to copper wires because they are insulators and don't pick up stray electric and magnetic fields or electronic "noise."

Applying Physics 22.5   Design of an Optical Fiber

An optical fiber consists of a transparent core surrounded by cladding, which is a material with a lower index of refraction than the core (Fig. 22.30). A cone of angles, called the acceptance cone, is at the entrance to the fiber. Incoming light at angles within this cone will be transmitted through the fiber, whereas light entering the core from angles outside the cone will not be transmitted. The figure shows a light ray entering the fiber just within the acceptance cone and undergoing total internal reflection at the interface between the core and the cladding. If it is technologically difficult to produce light so that it enters the fiber from a small range of angles, how could you adjust the indices of refraction of the core and cladding to increase the size of the acceptance cone—would you design the indices to be farther apart or closer together?

Explanation The acceptance cone would become larger if the critical angle (θ_c in the figure) could be

made smaller. This can be done by making the index of refraction of the cladding material smaller, so that the indices of refraction of the core and cladding material would be farther apart.

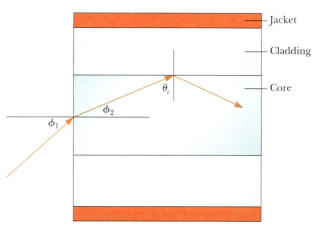

Figure 22.30   (Applying Physics 22.5)

SUMMARY

Physics⊗Now™ Take a practice test by logging into Physics-Now at **www.cp7e.com** and clicking on the Pre-Test link for this chapter.

22.1   The Nature of Light

Light has a dual nature. In some experiments it acts like a wave, in others like a particle, called a photon by Einstein. The energy of a photon is proportional to its frequency,

$$E = hf \qquad [22.1]$$

where $h = 6.63 \times 10^{-34}$ J·s is *Planck's constant.*

22.2   Reflection and Refraction

In the reflection of light off a flat, smooth surface, the angle of incidence, θ_1, with respect to a line perpendicular to the surface is equal to the angle of reflection, θ_1':

$$\theta_1' = \theta_1 \qquad [22.2]$$

Light that passes into a transparent medium is bent at the boundary and is said to be *refracted*. The angle of refraction is the angle the ray makes with respect to a line

perpendicular to the surface after it has entered the new medium.

22.3   The Law of Refraction

The **index of refraction** of a material, n, is defined as

$$n \equiv \frac{c}{v} \qquad [22.4]$$

where c is the speed of light in a vacuum and v is the speed of light in the material. The index of refraction of a material is also

$$n = \frac{\lambda_0}{\lambda_n} \qquad [22.7]$$

where λ_0 is the wavelength of the light in vacuum and λ_n is its wavelength in the material.

The **law of refraction**, or **Snell's law**, states that

$$n_1 \sin \theta_1 = n_2 \sin \theta_2 \qquad [22.8]$$

where n_1 and n_2 are the indices of refraction in the two media. The incident ray, the reflected ray, the refracted ray, and the normal to the surface all lie in the same plane.

22.4 Dispersion and Prisms & 22.5 The Rainbow

The index of refraction of a material depends on the wavelength of the incident light, an effect called *dispersion*. Light at the violet end of the spectrum exhibits a larger angle of refraction on entering glass than light at the red end. Rainbows are a consequence of dispersion.

22.6 Huygens' Principle

Huygens' principle states that all points on a wave front are point sources for the production of spherical secondary waves called wavelets. These wavelets propagate forward at a speed characteristic of waves in a particular medium. After some time has elapsed, the new position of the wave front is the surface tangent to the wavelets. This principle can be used to deduce the laws of reflection and refraction.

22.7 Total Internal Reflection

Total internal reflection can occur when light, traveling in a medium with higher index of refraction, is incident on the boundary of a material with a lower index of refraction. The *maximum angle of incidence* θ_c for which light can move from a medium with index n_1 into a medium with index n_2, where n_1 is greater than n_2, is called the **critical angle** and is given by

$$\sin\theta_c = \frac{n_2}{n_1} \qquad \text{for} \qquad n_1 > n_2 \qquad [22.9]$$

Total internal reflection is used in the optical fibers that carry data at high speed around the world.

CONCEPTUAL QUESTIONS

1. Under certain circumstances, sound can be heard from extremely far away. This frequently happens over a body of water, where the air near the water surface is cooler than the air at higher altitudes. Explain how the refraction of sound waves could increase the distance over which sound can be heard.

2. What are some reasons that most ceilings are made of white textured material?

3. The color of an object is said to depend on the wavelengths the object reflects. So, if you view colored objects under water, in which the wavelength of the light will be different, does the color change?

4. How is it possible that a complete circle of a rainbow can sometimes be seen from an airplane?

5. A ray of light is moving from a material having a high index of refraction into a material with a lower index of refraction. (a) Is the ray bent toward the normal or away from it? (b) If the wavelength is 600 nm in the material with the high index of refraction, is it greater, smaller, or the same in the material with the lower index of refraction? (c) How does the frequency change as the light moves between the two materials? Does it increase, decrease, or remain the same?

6. Why does the arc of a rainbow appear with red on top and violet on the bottom?

7. A scientific supply catalog advertises a material having an index of refraction of 0.85. Is this a good product to buy? Why or why not?

8. Under what conditions is a mirage formed? On a hot day, what are we seeing when we observe a mirage of a water puddle on the road?

9. In dispersive materials, the angle of refraction for a light ray depends on the wavelength of the light. Does the angle of reflection from the surface of the material depend on the wavelength? Why or why not?

10. A type of mirage called a *pingo* is often observed in Alaska. Pingos occur when the light from a small hill passes to an observer by a path that takes the light over a body of water warmer than the air. What is seen is the hill and an inverted image directly below it. Explain how these mirages are formed.

11. Explain why a diamond loses most of its sparkle when submerged in carbon disulfide.

12. Suppose you are told that only two colors of light (X and Y) are sent through a glass prism and that X is bent more than Y. Which color travels more slowly in the prism?

13. The level of water in a clear, colorless glass can easily be observed with the naked eye. The level of liquid helium in a clear glass vessel is extremely difficult to see with the naked eye. Explain. [*Hint:* The index of refraction of liquid helium is close to that of air.]

14. Is it possible to have total internal reflection for light incident from air on water? Explain.

15. Why does a diamond show flashes of color when observed under white light?

16. Explain why an oar partially submearged in water appears to be bent.

17. Why do astronomers looking at distant galaxies talk about looking backward in time?

18. If a beam of light with a given cross section enters a new medium, the cross section of the refracted beam is different from that of the incident beam. Is it larger or smaller, or is there no definite direction to the change?

PROBLEMS

1, 2, 3 = straightforward, intermediate, challenging   ☐ = full solution available in *Student Solutions Manual/Study Guide*

Physics⊗Now™ = coached solution with hints available at **www.cp7e.com**   🧬 = biomedical application

Section 22.1 The Nature of Light

1. During the Apollo XI Moon landing, a retroreflecting panel was erected on the Moon's surface. The speed of light can be found by measuring the time it takes a laser beam to travel from Earth, reflect from the panel, and return to Earth. If this interval is found to be 2.51 s, what is the measured speed of light? Take the center-to-center distance from Earth to Moon to be 3.84×10^8 m. Assume that the Moon is directly overhead and do not neglect the sizes of the Earth and Moon.

2. Figure P22.2 shows the apparatus used by Armand H. L. Fizeau (1819–1896) to measure the speed of light. The basic idea is to measure the total time it takes light to travel from some point to a distant mirror and back. If d is the distance between the light source and the mirror, and if the transit time for one round-trip is t, then the speed of light is $c = 2d/t$. To measure the transit time, Fizeau used a rotating toothed wheel, which converts an otherwise continuous beam of light to a series of light pulses. The rotation of the wheel controls what an observer at the light source sees. For example, assume that the toothed wheel of the Fizeau experiment has 360 teeth and is rotating at a speed of 27.5 rev/s when the light from the source is extinguished—that is, when a burst of light passing through opening A in Figure P22.2 is blocked by tooth B on return. If the distance to the mirror is 7 500 m, find the speed of light.

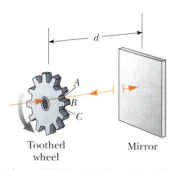

Figure P22.2   (Problems 2 and 3)

3. In an experiment designed to measure the speed of light using the apparatus of Fizeau described in the preceding problem, the distance between light source and mirror was 11.45 km and the wheel had 720 notches. The experimentally determined value of c was 2.998×10^8 m/s. Calculate the minimum angular speed of the wheel for this experiment.

4. Albert A. Michelson very carefully measured the speed of light using an alternative version of the technique developed by Fizeau. (See Problem 22.2.) Figure P22.4 shows the approach Michelson used. Light was reflected from one face of a rotating eight-sided mirror towards a stationary mirror 35.0 km away. At certain rates of rotation, the returning beam of light was directed toward the eye of an observer as shown. (a) What minimum angular speed must the rotating mirror have in order that side A will

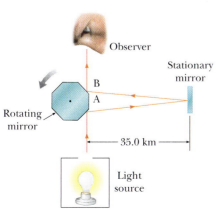

Figure P22.4

have rotated to position B, causing the light to be reflected to the eye? (b) What is the next-higher angular velocity that will enable the source of light to be seen?

5. Figure P22.5 shows an apparatus used to measure the distribution of the speeds of gas molecules. The device consists of two slotted rotating disks separated by a distance d, with the slots displaced by the angle θ. Suppose the speed of light is measured by sending a light beam toward the left disk of this apparatus. (a) Show that a light beam will be seen in the detector (that is, will make it through both slots) only if its speed is given by $c = \omega d/\theta$, where ω is the angular speed of the disks and θ is measured in radians. (b) What is the measured speed of light if the distance between the two slotted rotating disks is 2.500 m, the slot in the second disk is displaced 1/60 of 1 degree from the slot in the first disk, and the disks are rotating at 5 555 rev/s?

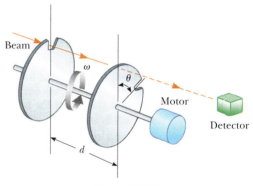

Figure P22.5

Section 22.2 Reflection and Refraction
Section 22.3 The Law of Refraction

6. The two mirrors in Figure P22.6 meet at a right angle. The beam of light in the vertical plane P strikes mirror 1 as shown. (a) Determine the distance the reflected light beam travels before striking mirror 2. (b) In what direction does the light beam travel after being reflected from mirror 2?

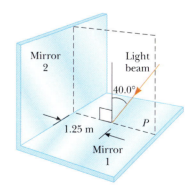

Figure P22.6

7. An underwater scuba diver sees the Sun at an apparent angle of 45.0° from the vertical. What is actual direction of the Sun?

8. Light is incident normal to a 1.00-cm layer of water that lies on top of a flat Lucite® plate with a thickness of 0.500 cm. How much more time is required for light to pass through this double layer than is required to traverse the same distance in air ($n_{\text{Lucite}} = 1.59$)?

9. **Physics⊗Now™** A laser beam is incident at an angle of 30.0° to the vertical onto a solution of corn syrup in water. If the beam is refracted to 19.24° to the vertical, (a) what is the index of refraction of the syrup solution? Suppose the light is red, with wavelength 632.8 nm in a vacuum. Find its (b) wavelength, (c) frequency, and (d) speed in the solution.

10. Light containing wavelengths of 400 nm, 500 nm, and 650 nm is incident from air on a block of crown glass at an angle of 25.0°. (a) Are all colors refracted alike, or is one color bent more than the others? (b) Calculate the angle of refraction in each case to verify your answer.

11. Light of wavelength λ_0 in a vacuum has a wavelength of 438 nm in water and a wavelength of 390 nm in benzene. (a) What is the wavelength λ_0? (b) Using only the given wavelengths, determine the ratio of the index of refraction of benzene to that of water.

12. Light of wavelength 436 nm in air enters a fishbowl filled with water, then exits through the crown-glass wall of the container. Find the wavelengths of the light (a) in the water and (b) in the glass.

13. A ray of light is incident on the surface of a block of clear ice at an angle of 40.0° with the normal. Part of the light is reflected and part is refracted. Find the angle between the reflected and refracted light.

14. The laws of refraction and reflection are the same for sound as for light. The speed of sound is 340 m/s in air and 1 510 m/s in water. If a sound wave traveling in air approaches a plane water surface at an angle of incidence of 12.0°, what is the angle of refraction?

15. The light emitted by a helium–neon laser has a wavelength of 632.8 nm in air. As the light travels from air into zircon, find (a) its speed in zircon, (b) its wavelength in zircon, and (c) its frequency.

16. A flashlight on the bottom of a 4.00-m-deep swimming pool sends a ray upward and at an angle so that the ray strikes the surface of the water 2.00 m from the point directly above the flashlight. What angle (in air) does the emerging ray make with the water's surface?

17. How many times will the incident beam shown in Figure P22.17 be reflected by each of the parallel mirrors?

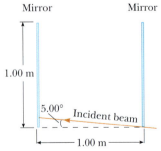

Figure P22.17

18. A ray of light strikes a flat 2.00-cm-thick block of glass ($n = 1.50$) at an angle of 30.0° with the normal (Fig. P22.18). Trace the light beam through the glass, and find the angles of incidence and refraction at each surface.

19. When the light ray in Problem 18 passes through the glass block, it is shifted laterally by a distance d (Fig. P22.18). Find the value of d.

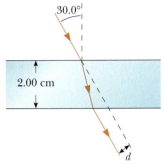

Figure P22.18 (Problems 18, 19, and 20)

20. Find the time required for the light to pass through the glass block described in Problem 19.

21. The light beam shown in Figure P22.21 makes an angle of 20.0° with the normal line NN' in the linseed oil. Determine the angles θ and θ'. (The refractive index for linseed oil is 1.48.)

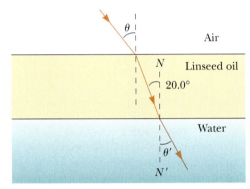

Figure P22.21

22. A submarine is 300 m horizontally out from the shore and 100 m beneath the surface of the water. A laser beam is sent from the sub so that it strikes the surface of the water at a point 210 m from the shore. If the beam just strikes the top of a building standing directly at the water's edge, find the height of the building.

23. Two light pulses are emitted simultaneously from a source. The pulses take parallel paths to a detector 6.20 m away, but one moves through air and the other through a block of ice. Determine the difference in the pulses' times of arrival at the detector.

24. A narrow beam of ultrasonic waves reflects off the liver tumor in Figure P22.24. If the speed of the wave is 10.0% less in the liver than in the surrounding medium, determine the depth of the tumor.

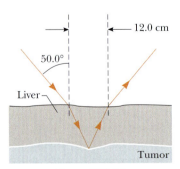

Figure P22.24

25. A beam of light both reflects and refracts at the surface between air and glass, as shown in Figure P22.25. If the index of refraction of the glass is n_g, find the angle of incidence, θ_1, in the air that would result in the reflected ray and the refracted ray being perpendicular to each other. [*Hint*: Remember the identity $\sin(90° - \theta) = \cos \theta$.]

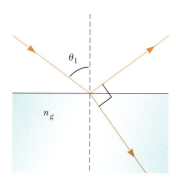

Figure P22.25

26. Three sheets of plastic have unknown indices of refraction. Sheet 1 is placed on top of sheet 2, and a laser beam is directed onto the sheets from above so that it strikes the interface at an angle of 26.5° with the normal. The refracted beam in sheet 2 makes an angle of 31.7° with the normal. The experiment is repeated with sheet 3 on top of sheet 2, and with the same angle of incidence, the refracted beam makes an angle of 36.7° with the normal. If the experiment is repeated again with sheet 1 on top of sheet 3, what is the expected angle of refraction in sheet 3? Assume the same angle of incidence.

27. An opaque cylindrical tank with an open top has a diameter of 3.00 m and is completely filled with water. When the afternoon Sun reaches an angle of 28.0° above the horizon, sunlight ceases to illuminate the bottom of the tank. How deep is the tank?

28. A cylindrical cistern, constructed below ground level, is 3.0 m in diameter and 2.0 m deep and is filled to the brim with a liquid whose index of refraction is 1.5. A small object rests on the bottom of the cistern at its center. How far from the edge of the cistern can a girl whose eyes are 1.2 m from the ground stand and still see the object?

Section 22.4 Dispersion and Prisms

29. The index of refraction for red light in water is 1.331, and that for blue light is 1.340. If a ray of white light enters the water at an angle of incidence of 83.00°, what are the underwater angles of refraction for the blue and red components of the light?

30. A certain kind of glass has an index of refraction of 1.650 for blue light of wavelength 430 nm and an index of 1.615 for red light of wavelength 680 nm. If a beam containing these two colors is incident at an angle of 30.00° on a piece of this glass, what is the angle between the two beams inside the glass?

31. **Physics⊗Now™** A ray of light strikes the midpoint of one face of an equiangular (60°–60°–60°) glass prism ($n = 1.5$) at an angle of incidence of 30°. (a) Trace the path of the light ray through the glass, and find the angles of incidence and refraction at each surface. (b) If a small fraction of light is also reflected at each surface, find the angles of reflection at the surfaces.

32. The index of refraction for violet light in silica flint glass is 1.66, and that for red light is 1.62. What is the angular dispersion of visible light passing through an equilateral prism of apex angle 60.0° if the angle of incidence is 50.0°? (See Fig. P22.32.)

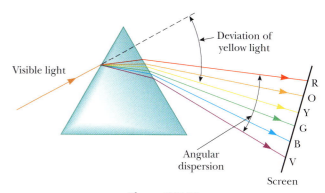

Figure P22.32

Section 22.7 Total Internal Reflection

33. Calculate the critical angles for the following materials when surrounded by air: (a) zircon, (b) fluorite, and (c) ice. Assume that $\lambda = 589$ nm.

34. For light of wavelength 589 nm, calculate the critical angle for the following materials surrounded by air: (a) diamond and (b) flint glass.

35. Repeat Problem 34, but this time suppose that the materials are surrounded by water.

36. **Physics⊗Now™** A beam of light is incident from air on the surface of a liquid. If the angle of incidence is 30.0° and the angle of refraction is 22.0°, find the critical angle for the liquid when surrounded by air.

37. A plastic light pipe has an index of refraction of 1.53. For total internal reflection, what is the minimum angle of incidence if the pipe is in (a) air? (b) water?

38. Determine the maximum angle θ for which the light rays incident on the end of the light pipe in Figure P22.38 are subject to total internal reflection along the walls of the pipe. Assume that the light pipe has an index of refraction of 1.36 and that the outside medium is air.

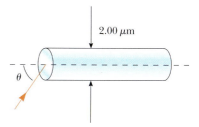

2.00 μm

Figure P22.328

39. Consider a common mirage formed by superheated air just above a roadway. A truck driver whose eyes are 2.00 m above the road, where $n = 1.000\,3$, looks forward. She has the illusion of seeing a patch of water ahead on the road, where her line of sight makes an angle of 1.20° below the horizontal. Find the index of refraction of the air just above the road surface. [*Hint*: Treat this as a problem one involving total internal reflection.]

40. A jewel thief hides a diamond by placing it on the bottom of a public swimming pool. He places a circular raft on the surface of the water directly above and centered over the diamond, as shown in Figure P22.40. If the surface of the water is calm and the pool is 2.00 m deep, find the minimum diameter of the raft that would prevent the diamond from being seen.

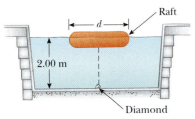

Figure P22.40

41. A room contains air in which the speed of sound is 343 m/s. The walls of the room are made of concrete, in which the speed of sound is 1 850 m/s. (a) Find the critical angle for total internal reflection of sound at the concrete–air boundary. (b) In which medium must the sound be traveling in order to undergo total internal reflection? (c) "A bare concrete wall is a highly efficient mirror for sound." Give evidence for or against this statement.

42. Three adjacent faces (that all share a corner) of a plastic cube of index of refraction n are painted black. A clear spot at the painted corner serves as a source of diverging rays when light comes through it. Show that a ray from this corner to the center of a clear face is totally reflected if $n \geq \sqrt{3}$.

43. The light beam in Figure P22.43 strikes surface 2 at the critical angle. Determine the angle of incidence, θ_i.

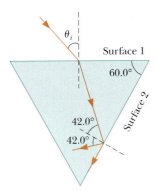

Figure P22.43

ADDITIONAL PROBLEMS

44. (a) Consider a horizontal interface between air above and glass with an index of 1.55 below. Draw a light ray incident from the air at an angle of incidence of 30.0°. Determine the angles of the reflected and refracted rays, and show them on the diagram. (b) Suppose instead that the light ray is incident from the glass at an angle of incidence of 30.0°. Determine the angles of the reflected and refracted rays, and show all three rays on a new diagram. (c) For rays incident from the air onto the air–glass surface, determine and tabulate the angles of reflection and refraction for all the angles of incidence at

10.0° intervals from 0° to 90.0°. (d) Do the same for light rays traveling up to the interface through the glass.

45. A layer of ice having parallel sides floats on water. If light is incident on the upper surface of the ice at an angle of incidence of 30.0°, what is the angle of refraction in the water?

46. A light ray of wavelength 589 nm is incident at an angle θ on the top surface of a block of polystyrene surrounded by air, as shown in Figure P22.46. (a) Find the maximum value of θ for which the refracted ray will undergo total internal reflection at the left vertical face of the block. (b) Repeat the calculation for the case in which the polystyrene block is immersed in water. (c) What happens if the block is immersed in carbon disulfide?

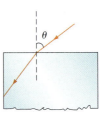

Figure P22.46

47. Figure P22.47 shows the path of a beam of light through several layers with different indices of refraction. (a) If $\theta_1 = 30.0°$, what is the angle θ_2 of the emerging beam? (b) What must the incident angle θ_1 be in order to have total internal reflection at the surface between the medium with $n = 1.20$ and the medium with $n = 1.00$?

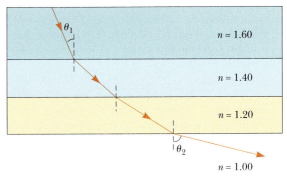

Figure P22.47

48. The walls of a prison cell are perpendicular to the four cardinal compass directions. On the first day of spring, light from the rising Sun enters a rectangular window in the eastern wall. The light traverses 2.37 m horizontally to shine perpendicularly on the wall opposite the window. A prisoner observes the patch of light moving across this western wall and for the first time forms his own understanding of the rotation of the Earth. (a) With what speed does the illuminated rectangle move? (b) The prisoner holds a small square mirror flat against the wall at one corner of the rectangle of light. The mirror reflects light back to a spot on the eastern wall close beside the window. How fast does the smaller square of light move across that wall? (c) Seen from a latitude of 40.0° north, the rising Sun moves through the sky along a line making a 50.0° angle with the southeastern horizon. In what direction does the rectangular patch of light on the western wall of

the prisoner's cell move? (d) In what direction does the smaller square of light on the eastern wall move?

49. As shown in Figure P22.49, a light ray is incident normal to on one face of a 30°–60°–90° block of dense flint glass (a prism) that is immersed in water. (a) Determine the exit angle θ_4 of the ray. (b) A substance is dissolved in the water to increase the index of refraction. At what value of n_2 does total internal reflection cease at point P?

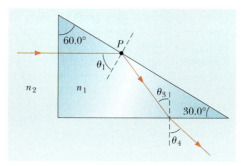

Figure P22.49

50. A narrow beam of light is incident from air onto a glass surface with index of refraction 1.56. Find the angle of incidence for which the corresponding angle of refraction is one-half the angle of incidence. [*Hint:* You might want to use the trigonometric identity $\sin 2\theta = 2 \sin \theta \cos \theta$.]

51. One technique for measuring the angle of a prism is shown in Figure P22.51. A parallel beam of light is directed onto the apex of the prism so that the beam reflects from opposite faces of the prism. Show that the angular separation of the two reflected beams is given by $B = 2A$.

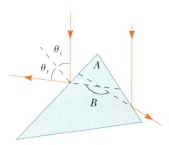

Figure P22.51

52. An optical fiber with index of refraction n and diameter d is surrounded by air. Light is sent into the fiber along its axis, as shown in Figure P22.52. (a) Find the smallest outside radius R permitted for a bend in the fiber if no light is to escape. (b) Does the result for part (a) predict reasonable behavior as d approaches zero? As n increases? As n approaches unity? (c) Evaluate R, assuming that the diameter of the fiber is 100 μm and its index of refraction is 1.40.

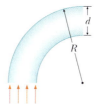

Figure P22.52

53. A piece of wire is bent through an angle θ. The bent wire is partially submerged in benzene (index of refraction = 1.50), so that, to a person looking along the dry part, the wire appears to be straight and makes an angle of 30.0° with the horizontal. Determine the value of θ.

54. A light ray traveling in air is incident on one face of a right-angle prism with index of refraction $n = 1.50$, as shown in Figure P22.54, and the ray follows the path shown in the figure. Assuming that $\theta = 60.0°$ and the base of the prism is mirrored, determine the angle ϕ made by the outgoing ray with the normal to the right face of the prism.

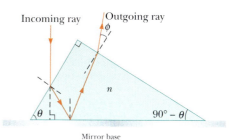

Figure P22.54

55. **Physics⊗Now™** A transparent cylinder of radius $R = 2.00$ m has a mirrored surface on its right half, as shown in Figure P22.55. A light ray traveling in air is incident on the left side of the cylinder. The incident light ray and the exiting light ray are parallel, and $d = 2.00$ m. Determine the index of refraction of the material.

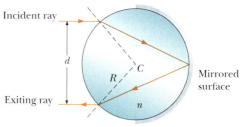

Figure P22.55

56. A laser beam strikes one end of a slab of material, as in Figure P22.56. The index of refraction of the slab is 1.48. Determine the number of internal reflections of the beam before it emerges from the opposite end of the slab.

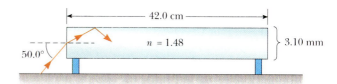

Figure P22.56

57. For this problem, refer to Figure 22.15. For various angles of incidence, it can be shown that the deviation angle δ is a minimum when the ray passes through the glass so that the interior ray is parallel to the base of the prism. A measurement of this minimum angle of deviation enables one to find the index of refraction of the prism material.

Show that n is given by the expression

$$n = \frac{\sin\left[\frac{1}{2}(A + \delta_{min})\right]}{\sin\left(\frac{A}{2}\right)}$$

where A is the apex angle of the prism.

58. A hiker stands on a mountain peak near sunset and observes a rainbow caused by water droplets in the air about 8.00 km away. The valley is 2.00 km below the mountain peak and entirely flat. What fraction of the complete circular arc of the rainbow is visible to the hiker?

59. A light ray incident on a prism is refracted at the first surface, as shown in Figure P22.59. Let ϕ represent the apex angle of the prism and n its index of refraction. Find, in terms of n and ϕ, the smallest allowed value of the angle of incidence at the first surface for which the refracted ray will not undergo total internal reflection at the second surface.

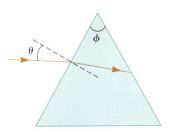

Figure P22.59

60. Students allow a narrow beam of laser light to strike a water surface. They arrange to measure the angle of refraction for selected angles of incidence and record the data shown in the following table:

Angle of Incidence	Angle of Refraction
(degrees)	(degrees)
10.0	7.5
20.0	15.1
30.0	22.3
40.0	28.7
50.0	35.2
60.0	40.3
70.0	45.3
80.0	47.7

Use the data to verify Snell's law of refraction by plotting the sine of the angle of incidence versus the sine of the angle of refraction. From the resulting plot, deduce the index of refraction of water.

61. A light ray enters a rectangular block of plastic at an angle $\theta_1 = 45.0°$ and emerges at an angle $\theta_2 = 76.0°$, as shown in Figure P22.61. (a) Determine the index of refraction of

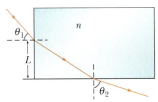

Figure P22.61

the plastic. (b) If the light ray enters the plastic at a point $L = 50.0$ cm from the bottom edge, how long does it take the light ray to travel through the plastic?

62. A. H. Pfund's method for measuring the index of refraction of glass is illustrated in Figure P22.62. One face of a slab of thickness t is painted white, and a small hole scraped clear at point P serves as a source of diverging rays when the slab is illuminated from below. Ray PBB' strikes the clear surface at the critical angle and is totally reflected, as are rays such as PCC'. Rays such as PAA' emerge from the clear surface. On the painted surface there appears a dark circle of diameter d, surrounded by an illuminated region, or halo. (a) Derive an equation for n in terms of the measured quantities d and t. (b) What is the diameter of the dark circle if $n = 1.52$ for a slab 0.600 cm thick? (c) If white light is used, the critical angle depends on color caused by dispersion. Is the inner edge of the white halo tinged with red light or violet light? Explain.

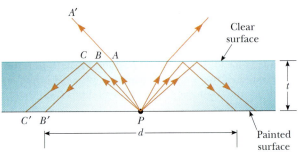

Figure P22.62

ACTIVITIES

1. Tape a coin to the bottom of a large opaque bowl, as shown in Figure A22.1a. Stand over the bowl so that you are looking at the coin, and then move backwards away from the bowl until you can no longer see the coin over the bowl's rim. Remain at that position, and have a friend fill the bowl with water, as shown in Figure A22.1b. You can now see the coin again because the light is refracted at the water–air interface.

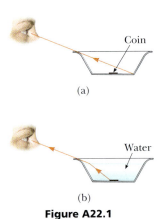

Figure A22.1

2. Tape a piece of black paper to the end of a flashlight and cut a narrow slit in the middle of the paper, as shown in Figure A22.2. Lean a flat mirror against one end of a tray partially filled with water. Shine your flashlight on that

part of the mirror which is under water, and hold a sheet of white paper such that the reflected light shines on the paper. You should observe a spectrum of colors on the paper as the light is dispersed when it travels from air into water and then from water into air. According to your observations, which color is bent the most? Which is bent the least?

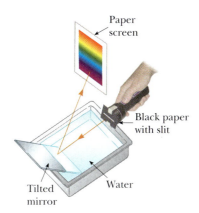

Figure A22.2

3. Create an artificial rainbow by standing with your back to the sun and spraying water into the air with a hose. You should cover the end of the hose slightly with a finger or use a nozzle so that the water is broken up into tiny droplets. Which color is on the outside of the arc? The inside?

If the droplets are close to you in the experiment above, you may be able to see two rainbows, one for each eye. Close one eye, and only one bow is seen.

With the Sun high in the sky, stand on a ladder and spray the hose toward the ground. In this case, you should be able to form a complete circle rainbow.

4. By observing the shadows formed by large and small light sources, you can demonstrate that light rays travel in straight-line paths. Figure A22.4 shows the shadows formed on a screen by a baseball when the light from a lightbulb (a large source) falls on it. Use the figure to explain why the shadow is dark at location A and less dark at B. Replace the light source with a small source, such as a high-intensity lamp with a small, clear bulb. (A substitute for a high-intensity lamp is an ordinary lightbulb with a piece of cardboard close in front of it with a hole about the size of a penny punched in the cardboard. The hole serves as the small source of light.) What kind of shadow is formed in this case? Why?

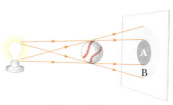

Figure A22.4

This beautiful photograph shows a raindrop suspended from a leaf. The raindrop acts as a lens. It refracts light twice to produce a real, inverted image of the foliage beyond.

Don Hammond/CORBIS

Mirrors and Lenses

The development of the technology of mirrors and lenses led to a revolution in the progress of science. These devices, relatively simple to construct from cheap materials, led to microscopes and telescopes, extending human sight and opening up new pathways to knowledge, from microbes to distant planets.

This chapter covers the formation of images when plane and spherical light waves fall on plane and spherical surfaces. Images can be formed by reflection from mirrors or by refraction through lenses. In our study of mirrors and lenses, we continue to assume that light travels in straight lines (the ray approximation), ignoring diffraction.

23.1 FLAT MIRRORS

We begin by examining the flat mirror. Consider a point source of light placed at O in Figure 23.1, a distance p in front of a flat mirror. The distance p is called the **object distance**. Light rays leave the source and are reflected from the mirror. After reflection, the rays diverge (spread apart), but they appear to the viewer to come from a point I behind the mirror. Point I is called the **image** of the object at O. Regardless of the system under study, **images are formed at the point where rays of light actually intersect or where they appear to originate.** Because the rays in the figure appear to originate at I, which is a distance q behind the mirror, that is the location of the image. The distance q is called the **image distance**.

Images are classified as real or virtual. In the formation of a *real image*, light actually passes through the image point. For a *virtual image*, the light doesn't pass through the image point, but appears to come (diverge) from there. The image formed by the flat mirror in Figure 23.1 is a virtual image. In fact, the images seen in flat mirrors are always virtual (for real objects). Real images can be displayed on a screen (as at a movie), but virtual images cannot.

We will examine some of the properties of the images formed by flat mirrors by using the simple geometric techniques shown in Active Figure 23.2. To find out where an image is formed, it's necessary to follow at least two rays of light as they reflect from the mirror. One of those rays starts at P, follows the horizontal path PQ to the mirror, and reflects back on itself. The second ray follows the oblique path PR and reflects as shown. An observer to the left of the mirror would trace the two reflected rays back to the point from which they appear to have originated: point P'. A continuation of this process for points other than P on the object would result in a virtual image (drawn as a yellow arrow) to the right of the mirror. Because triangles PQR and $P'QR$ are identical, $PQ = P'Q$. Hence, we conclude that **the image formed by an object placed in front of a flat mirror is as far behind the mirror as the object is in front of the mirror**. Geometry also shows that the object height h equals the image height h'. The **lateral magnification** M is defined as

$$M \equiv \frac{\text{image height}}{\text{object height}} = \frac{h'}{h} \qquad [23.1]$$

This is a general definition of the lateral magnification of any type of mirror. For a flat mirror, $M = 1$ because $h' = h$.

In summary, the image formed by a flat mirror has the following properties:

1. The image is as far behind the mirror as the object is in front.
2. The image is unmagnified, virtual, and upright. (By *upright*, we mean that if the object arrow points upward, as in Figure 23.2, so does the image arrow. The opposite of an upright image is an inverted image.)

Finally, note that a flat mirror produces an image having an *apparent* left–right reversal. You can see this reversal standing in front of a mirror and raising your right hand. Your image in the mirror raises his left hand. Likewise, your hair appears to be parted on the opposite side, and a mole on your right cheek appears to be on your image's left cheek.

Quick Quiz 23.1

In the overhead view of Figure 23.3, the image of the stone seen by observer 1 is at C. Where does observer 2 see the image—at A, at B, at C, at E, or not at all?

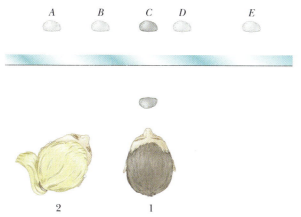

Figure 23.3 (Quick Quiz 23.1)

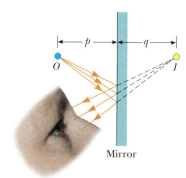

Figure 23.1 An image formed by reflection from a flat mirror. The image point I is behind the mirror at distance q, which is equal in magnitude to the object distance p.

TIP 23.1 Magnification ≠ Enlargement

Note that the word *magnification*, as used in optics, doesn't always mean *enlargement*, because the image could be smaller than the object.

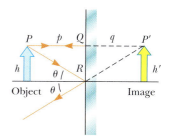

ACTIVE FIGURE 23.2

A geometric construction to locate the image of an object placed in front of a flat mirror. Because the triangles PQR and $P'QR$ are identical, $p = |q|$ and $h = h'$.

Physics Now™

Log into PhysicsNow at **www.cp7e.com** and go to Active Figure 23.2, where you can move the object and see the effect on the image.

EXAMPLE 23.1 "Mirror, Mirror, on the Wall"

Goal Apply the properties of a flat mirror.

Problem A man 1.80 m tall stands in front of a mirror and sees his full height, no more and no less. If his eyes are 0.14 m from the top of his head, what is the minimum height of the mirror?

Strategy   Figure 23.4 shows two rays of light, one from his feet and the other from the top of his head, reflecting off the mirror and entering his eye. The ray from his feet just strikes the bottom of the mirror, so if the mirror were longer, it would be too long; if shorter, the ray would not be reflected. The angle of incidence and the angle of reflection are equal, labeled θ. This means the two triangles, ABD and DBC, are identical because they are right triangles with a common side (DB) and two identical angles θ. Use this key fact and the small isosceles triangle FEC to solve the problem.

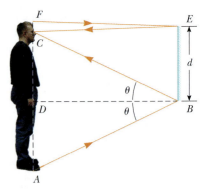

Figure 23.4   (Example 23.1)

Solution

We need to find BE, which equals d. Relate this length to lengths on the man's body:

$$BE = DC + \tfrac{1}{2}CF \qquad (1)$$

We need the lengths DC and CF. Set the sum of sides opposite the identical angles θ equal to AC:

$$AD + DC = AC = (1.80 - 0.14) = 1.66 \text{ m} \qquad (2)$$

$AD = DC$, so substitute into Equation 2 and solve for DC:

$$AD + DC = 2DC = 1.66 \text{ m} \quad \rightarrow \quad DC = 0.83 \text{ m}$$

CF is given as 0.14 m. Substitute this and DC into Equation 1:

$$BE = d = DC + \tfrac{1}{2}CF = 0.83 \text{ m} + \tfrac{1}{2}(0.14 \text{ m}) = \boxed{0.90 \text{ m}}$$

Remarks   The mirror must be exactly equal to half the height of the man in order for him to see only his full height and nothing more or less. Notice that the answer doesn't depend on his distance from the mirror.

Exercise 23.1

How large should the mirror be if he wants to see only the upper third of his body?

Answer   0.30 m

APPLICATION

Day and Night Settings for Rearview Mirrors

Most rearview mirrors in cars have a day setting and a night setting. The night setting greatly diminishes the intensity of the image so that lights from trailing cars will not blind the driver. To understand how such a mirror works, consider Figure 23.5. The mirror is a wedge of glass with a reflecting metallic coating on the back side. When the mirror is in the day setting, as in Figure 23.5a, light from an object behind the car strikes the mirror at point 1. Most of the light enters the wedge, is refracted, and reflects from the back of the mirror to return to the front surface, where it is refracted again as it reenters the air as ray B (for *bright*). In addition, a small portion of the light is reflected at the front surface, as indicated by ray D (for

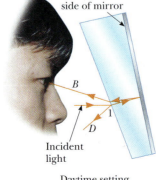

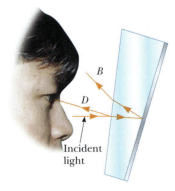

Figure 23.5   Cross-sectional views of a rearview mirror. (a) With the day setting, the silvered back surface of the mirror reflects a bright ray B into the driver's eyes. (b) With the night setting, the glass of the unsilvered front surface of the mirror reflects a dim ray D into the driver's eyes.

Daytime setting

(a)

Nighttime setting

(b)

dim). This dim reflected light is responsible for the image observed when the mirror is in the night setting, as in Figure 23.5b. Now the wedge is rotated so that the path followed by the bright light (ray *B*) doesn't lead to the eye. Instead, the dim light reflected from the front surface travels to the eye, and the brightness of trailing headlights doesn't become a hazard.

Applying Physics 23.1   Illusionist's Trick

The professor in the box shown in Figure 23.6 appears to be balancing himself on a few fingers

Figure 23.6   (Applying Physics 23.1)

with both of his feet elevated from the floor. He can maintain this position for a long time, and appears to defy gravity. How do you suppose this illusion was created?

Explanation This is an example of an optical illusion, used by magicians, that makes use of a mirror. The box that the professor is standing in is a cubical open frame that contains a flat vertical mirror through a diagonal plane. The professor straddles the mirror so that one leg is in front of the mirror and the other leg is behind it, out of view. When he raises his front leg, that leg's reflection rises also, making it appear both his feet are off the ground, creating the illusion that he's floating in the air. In fact, he supports himself with the leg behind the mirror, which remains in contact with the ground.

23.2   IMAGES FORMED BY SPHERICAL MIRRORS

Concave Mirrors

A **spherical mirror**, as its name implies, has the shape of a segment of a sphere. Figure 23.7 shows a spherical mirror with a silvered inner, concave surface; this type of mirror is called a **concave mirror**. The mirror has radius of curvature *R*, and its center of curvature is at point *C*. Point *V* is the center of the spherical segment, and a line drawn from *C* to *V* is called the **principal axis** of the mirror.

Now consider a point source of light placed at point *O* in Figure 23.7b, on the principal axis and outside point *C*. Several diverging rays originating at *O* are shown. After reflecting from the mirror, these rays converge to meet at *I*, called the **image point**. The rays then continue and diverge from *I* as if there were an object there. As a result, a real image is formed. **Whenever reflected light actually passes through a point, the image formed there is real**.

We often assume that all rays that diverge from the object make small angles with the principal axis. All such rays reflect through the image point, as in Figure 23.7b.

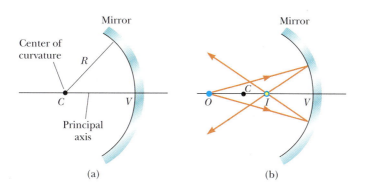

(a)   (b)

Figure 23.7   (a) A concave mirror of radius *R*. The center of curvature, *C*, is located on the principal axis. (b) A point object placed at *O* in front of a concave spherical mirror of radius *R*, where *O* is any point on the principal axis farther than *R* from the surface of the mirror, forms a real image at *I*. If the rays diverge from *O* at small angles, they all reflect through the same image point.

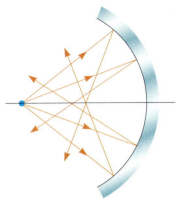

Figure 23.8   Rays at large angles from the horizontal axis reflect from a spherical, concave mirror to intersect the principal axis at different points, resulting in a blurred image. This phenomenon is called *spherical aberration.*

Rays that make a large angle with the principal axis, as in Figure 23.8, converge to other points on the principal axis, producing a blurred image. This effect, called **spherical aberration**, is present to some extent with any spherical mirror and will be discussed in Section 23.7.

We can use the geometry shown in Figure 23.9 to calculate the image distance q from the object distance p and radius of curvature, R. By convention, these distances are measured from point V. The figure shows two rays of light leaving the tip of the object. One ray passes through the center of curvature, C, of the mirror, hitting the mirror head-on (perpendicular to the mirror surface) and reflecting back on itself. The second ray strikes the mirror at point V and reflects as shown, obeying the law of reflection. The image of the tip of the arrow is at the point at which the two rays intersect. From the largest triangle in Figure 23.9 we see that $\tan \theta = h/p$; the light-blue triangle gives $\tan \theta = -h'/q$. The negative sign has been introduced to satisfy our convention that h' is negative when the image is inverted with respect to the object, as it is here. From Equation 23.1 and these results, we find that the magnification of the mirror is

$$M = \frac{h'}{h} = -\frac{q}{p} \qquad \text{[23.2]}$$

From two other triangles in the figure, we get

$$\tan \alpha = \frac{h}{p - R} \qquad \text{and} \qquad \tan \alpha = -\frac{h'}{R - q}$$

from which we find that

$$\frac{h'}{h} = -\frac{R - q}{p - R} \qquad \text{[23.3]}$$

If we compare Equation 23.2 to Equation 23.3, we see that

$$\frac{R - q}{p - R} = \frac{q}{p}$$

Simple algebra reduces this to

Mirror equation ▶

$$\frac{1}{p} + \frac{1}{q} = \frac{2}{R} \qquad \text{[23.4]}$$

This expression is called the **mirror equation**.

If the object is very far from the mirror—if the object distance p is great enough compared with R that p can be said to approach infinity—then $1/p \approx 0$, and we see from Equation 23.4 that $q \approx R/2$. In other words, when the object is very far from the mirror, **the image point is halfway between the center of curvature and the center of the mirror**, as in Figure 23.10a. The incoming rays are essentially parallel in that figure because the source is assumed to be very far from the mirror. In this special case, we call the image point the **focal point** F and the

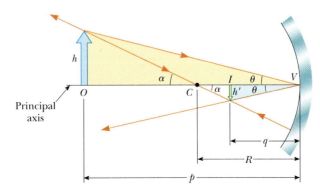

Figure 23.9   The image formed by a spherical concave mirror, where the object at O lies outside the center of curvature, C.

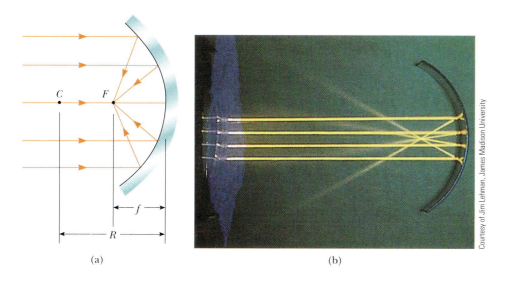

(a)   (b)

Courtesy of Jim Lehman, James Madison University

Figure 23.10 (a) Light rays from a distant object ($p = \infty$) reflect from a concave mirror through the focal point *F*. In this case, the image distance $q = R/2 = f$, where *f* is the focal length of the mirror. (b) A photograph of the reflection of parallel rays from a concave mirror.

image distance the **focal length** *f*, where

$$f = \frac{R}{2}$$ [23.5]   ◀ Focal length

The mirror equation can therefore be expressed in terms of the focal length:

$$\frac{1}{p} + \frac{1}{q} = \frac{1}{f}$$ [23.6]

Note that rays from objects at infinity are always focused at the focal point.

TIP 23.2   Focal Point ≠ *Focus* Point

The focal point is *not* the point at which light rays focus to form an image. The focal point of a mirror is determined *solely* by its curvature—it doesn't depend on the location of any object.

23.3   CONVEX MIRRORS AND SIGN CONVENTIONS

Figure 23.11 shows the formation of an image by a **convex mirror**, which is silvered so that light is reflected from the outer, convex surface. This is sometimes called a **diverging mirror** because the rays from any point on the object diverge after reflection, as though they were coming from some point behind the mirror. The image in Figure 23.11 is virtual rather than real because it lies behind the mirror at the point at which the reflected rays appear to originate. In general, the image formed by a convex mirror is upright, virtual, and smaller than the object.

We won't derive any equations for convex spherical mirrors. If we did, we would find that the equations developed for concave mirrors can be used with convex mirrors if particular sign conventions are used. We call the region in which light rays move the *front side* of the mirror, and the other side, where virtual images are formed, the *back side*. For example, in Figures 23.9 and 23.11, the side to the left of the mirror is the front side and the side to the right is the back side.

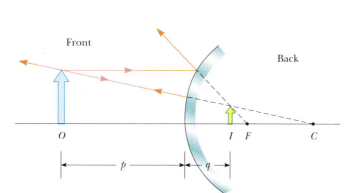

Figure 23.11 Formation of an image by a spherical, convex mirror. Note that the image is virtual and upright.

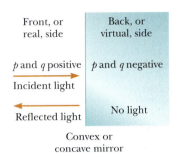

Figure 23.12 A diagram describing the signs of p and q for convex and concave mirrors.

TABLE 23.1

Sign Conventions for Mirrors

Quantity	Symbol	In Front	In Back	Upright Image	Inverted Image
Object location	p	+	−		
Image location	q	+	−		
Focal Length	f	+	−		
Image height	h'			+	−
Magnification	M			+	−

Figure 23.12 is helpful for understanding the rules for object and image distances, and Table 23.1 summarizes the sign conventions for all the necessary quantities. Notice that when the quantities p, q, and f (and R) are located where the light is—in front of the mirror—they are positive, whereas when they are located behind the mirror (where the light isn't), they are negative.

Ray Diagrams for Mirrors

We can conveniently determine the positions and sizes of images formed by mirrors by constructing *ray diagrams* similar to the ones we have been using. This kind of graphical construction tells us the overall nature of the image and can be used to check parameters calculated from the mirror and magnification equations. Making a ray diagram requires knowing the position of the object and the location of the center of curvature. To locate the image, three rays are constructed (rather than just the two we have been constructing so far), as shown by the examples in Active Figure 23.13. All three rays start from the same object point; for these examples, the tip of the arrow was chosen. For the concave mirrors in Active Figure 23.13a and b, the rays are drawn as follows:

1. Ray 1 is drawn parallel to the principal axis and is reflected back through the focal point *F*.
2. Ray 2 is drawn through the focal point and is reflected parallel to the principal axis.
3. Ray 3 is drawn through the center of curvature, *C*, and is reflected back on itself.

Note that rays actually go in all directions from the object; we choose to follow those moving in a direction that simplifies our drawing.

The intersection of any *two* of these rays at a point locates the image. The third ray serves as a check of our construction. The image point obtained in this fashion must always agree with the value of q calculated from the mirror formula.

In the case of a concave mirror, note what happens as the object is moved closer to the mirror. The real, inverted image in Active Figure 23.13a moves to the left as the object approaches the focal point. When the object is at the focal point, the image is infinitely far to the left. However, when the object lies between the focal point and the mirror surface, as in Active Figure 23.13b, the image is virtual and upright.

With the convex mirror shown in Active Figure 23.13c, the image of a real object is always virtual and upright. As the object distance increases, the virtual image shrinks and approaches the focal point as p approaches infinity. You should construct a ray diagram to verify this.

The image-forming characteristics of curved mirrors obviously determine their uses. For example, suppose you want to design a mirror that will help

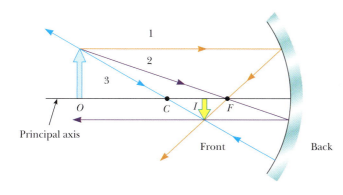

(a)

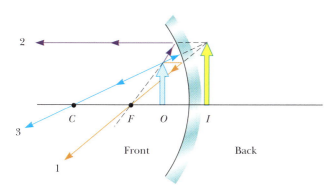
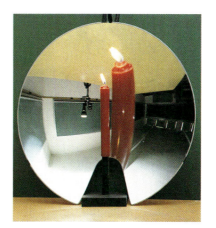

(b)

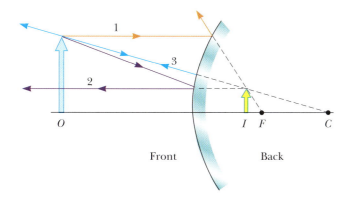
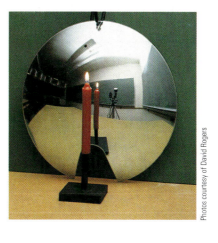

(c)

ACTIVE FIGURE 23.13

Ray diagrams for spherical mirrors and corresponding photographs of the images of candles. (a) When an object is outside the center of curvature of a concave mirror, the image is real, inverted, and reduced in size. (b) When an object is between a concave mirror and the focal point, the image is virtual, upright, and magnified. (c) When an object is in front of a convex mirror, the image is virtual, upright, and reduced in size.

Physics⊗Now™

Log into PhysicsNow at **www.cp7e.com** and go to Active Figure 23.13, where you can move the objects, change the focal lengths of the mirrors, and see the effect on the images.

people shave or apply cosmetics. For this, you need a concave mirror that puts the user inside the focal point, such as the mirror in Active Figure 23.13b. With that mirror, the image is upright and greatly enlarged. In contrast, suppose that the primary purpose of a mirror is to observe a large field of view. In that case,

Figure 23.14 A convex side-view mirror on a vehicle produces an upright image that is smaller than the object. The smaller image means the object is closer than its apparent distance as observed in the mirror.

you need a convex mirror such as the one in Active Figure 23.13c. The diminished size of the image means that a fairly large field of view is seen in the mirror. Mirrors like this one are often placed in stores to help employees watch for shoplifters. A second use of such a mirror is as a side-view mirror on a car (Fig. 23.14). This kind of mirror is usually placed on the passenger side of the car and carries the warning "Objects are closer than they appear." Without such warning, a driver might think she is looking into a flat mirror, which doesn't alter the size of the image. She could be fooled into believing that a truck is far away because it looks small, when it's actually a large semi very close behind her, but diminished in size because of the image formation characteristics of the convex mirror.

Applying Physics 23.2   Concave versus Convex

A virtual image can be anywhere behind a concave mirror. Why is there a maximum distance at which the image can exist behind a *convex* mirror?

Explanation Consider the concave mirror first, and imagine two different light rays leaving a tiny object and striking the mirror. If the object is at the focal point, the light rays reflecting from the mirror will be parallel to the mirror axis. They can be interpreted as forming a virtual image infinitely far away behind the mirror. As the object is brought closer to the mirror, the reflected rays will diverge through larger and larger angles, resulting in their extensions converging closer and closer to the back of the mirror. When the object is brought right up to the mirror, the image is right behind the mirror. When the object is much closer to the mirror than the focal length, the mirrors acts like a flat mirror, and the image is just as far behind the mirror as the object is in front of it. The image can therefore be anywhere from infinitely far away to right at the surface of the mirror. For the convex mirror, an object at infinity produces a virtual image at the focal point. As the object is brought closer, the reflected rays diverge more sharply and the image moves closer to the mirror. As a result, the virtual image is restricted to the region between the mirror and the focal point.

Applying Physics 23.3   Reversible Waves

Large trucks often have a sign on the back saying, "If you can't see my mirror, I can't see you." Explain this sign.

Explanation The trucking companies are making use of the principle of the reversibility of light rays. In order for an image of you to be formed in the driver's mirror, there must be a pathway for rays of light to reach the mirror, allowing the driver to see your image. If you can't see the mirror, then this pathway doesn't exist.

INTERACTIVE EXAMPLE 23.2   Images Formed by a Concave Mirror

Goal Calculate properties of a concave mirror.

Problem Assume that a certain concave spherical mirror has a focal length of 10.0 cm. **(a)** Locate the image and find the magnification for an object distance of 25.0 cm. Determine whether the image is real or virtual, inverted or upright, and larger or smaller. Do the same for object distances of **(b)** 10.0 cm and **(c)** 5.00 cm.

Strategy For each part, substitute into the mirror and magnification equations. Part (b) involves a limiting process, because the answers are infinite.

Solution
(a) Find the image position for an object distance of 25.0 cm.
Calculate the magnification and describe the image.

Use the mirror equation to find the image distance:

(1) $\dfrac{1}{p} + \dfrac{1}{q} = \dfrac{1}{f}$

Substitute and solve for q. According to Table 23.1, p and f are positive.

$\dfrac{1}{25.0 \text{ cm}} + \dfrac{1}{q} = \dfrac{1}{10.0 \text{ cm}}$

$q = \boxed{16.7 \text{ cm}}$

Because q is positive, the image is in front of the mirror and is real. The magnification is given by substituting into Equation 23.2:

$M = -\dfrac{q}{p} = -\dfrac{16.7 \text{ cm}}{25.0 \text{ cm}} = \boxed{-0.668}$

The image is smaller than the object because $|M| < 1$, and inverted because M is negative. (See Fig. 23.13a.)

(b) Locate the image distance when the object distance is 10.0 cm. Calculate the magnification and describe the image.

The object is at the focal point. Substitute $p = 10.0$ cm and $f = 10.0$ cm into the mirror equation:

$\dfrac{1}{10.0 \text{ cm}} + \dfrac{1}{q} = \dfrac{1}{10.0 \text{ cm}}$

$\dfrac{1}{q} = 0 \quad \rightarrow \quad \boxed{q = \infty}$

Since $M = -q/p$, the magnification is infinite, also.

(c) Locate the image distance when the object distance is 5.00 cm. Calculate the magnification and describe the image.

Once again, substitute into the mirror equation:

$\dfrac{1}{5.00 \text{ cm}} + \dfrac{1}{q} = \dfrac{1}{10.0 \text{ cm}}$

$\dfrac{1}{q} = \dfrac{1}{10.0 \text{ cm}} - \dfrac{1}{5.00 \text{ cm}} = -\dfrac{1}{10.0 \text{ cm}}$

$q = \boxed{-10.0 \text{ cm}}$

The image is virtual (behind the mirror) because q is negative. Use Equation 23.2 to calculate the magnification:

$M = -\dfrac{q}{p} = -\left(\dfrac{-10.0 \text{ cm}}{5.00 \text{ cm}} \right) = \boxed{2.00}$

The image is larger (magnified by a factor of 2) because $|M| > 1$, and upright because M is positive. (See Fig. 23.13b.)

Remarks   Note the characteristics of an image formed by a concave, spherical mirror. When the object is outside the focal point, the image is inverted and real; at the focal point, the image is formed at infinity; inside the focal point, the image is upright and virtual.

Exercise 23.2
If the object distance is 20.0 cm, find the image distance and the magnification of the mirror.

Answer $q = 20.0$ cm, $M = -1.00$

Physics⊗Now™ Investigate the image formed for various object positions and mirror focal lengths by logging into PhysicsNow at **www.cp7e.com** and going to Interactive Example 23.2.

EXAMPLE 23.3   Images Formed by a Convex Mirror

Goal   Calculate properties of a convex mirror.

Problem   An object 3.00 cm high is placed 20.0 cm from a convex mirror with a focal length of 8.00 cm. Find **(a)** the position of the image, **(b)** the magnification of the mirror, and **(c)** the height of the image.

Strategy   This problem again requires only substitution into the mirror and magnification equations. Multiplying the object height by the magnification gives the image height.

Solution

(a) Find the position of the image.

Because the mirror is convex, its focal length is negative. Substitute into the mirror equation:

$$\frac{1}{p} + \frac{1}{q} = \frac{1}{f}$$

$$\frac{1}{20.0 \text{ cm}} + \frac{1}{q} = \frac{1}{-8.00 \text{ cm}}$$

Solve for q:

$$q = -5.71 \text{ cm}$$

(b) Find the magnification of the mirror.

Substitute into Equation 23.2:

$$M = -\frac{q}{p} = -\left(\frac{-5.71 \text{ cm}}{20.0 \text{ cm}}\right) = 0.286$$

(c) Find the height of the image.

Multiply the object height by the magnification:

$$h' = hM = (3.00 \text{ cm})(0.286) = 0.858 \text{ cm}$$

Remarks   The negative value of q indicates the image is virtual, or behind the mirror, as in Figure 23.13c. The image is upright because M is positive.

Exercise 23.3

Suppose the object is moved so it is 4.00 cm from the same mirror. Repeat parts (a)–(c).

Answer   (a) -2.67 cm   (b) 0.668   (c) 2.00 cm; the image is upright and virtual.

EXAMPLE 23.4   The Face in the Mirror

Goal   Find a focal length from a magnification and an object distance.

Problem   When a woman stands with her face 40.0 cm from a cosmetic mirror, the upright image is twice as tall as her face. What is the focal length of the mirror?

Strategy   To find f in this example, we must first find q, the image distance. Because the problem states that the image is upright, the magnification must be positive (in this case, $M = +2$), and because $M = -q/p$, we can determine q.

Solution

Obtain q from the magnification equation:

$$M = -\frac{q}{p} = 2$$

$$q = -2p = -2(40.0 \text{ cm}) = -80.0 \text{ cm}$$

Because q is negative, the image is on the opposite side of the mirror and hence is virtual. Substitute q and p into the mirror equation and solve for f:

$$\frac{1}{40.0 \text{ cm}} - \frac{1}{80.0 \text{ cm}} = \frac{1}{f}$$

$$f = 80.0 \text{ cm}$$

Remarks   The positive sign for the focal length tells us the mirror is concave, a fact we already knew because the mirror magnified the object. (A convex mirror would have produced a smaller image.)

Exercise 23.4
Suppose a fun-house mirror makes you appear to have one-third your normal height. If you are 1.20 m away from the mirror, find its focal length. Is the mirror concave or convex?

Answer -0.600 m, convex

23.4   IMAGES FORMED BY REFRACTION

In this section we describe how images are formed by refraction at a spherical surface. Consider two transparent media with indices of refraction n_1 and n_2, where the boundary between the two media is a spherical surface of radius R (Fig. 23.15). We assume that the medium to the right has a higher index of refraction than the one to the left: $n_2 > n_1$. This would be the case for light entering a curved piece of glass from air or for light entering the water in a fishbowl from air. The rays originating at the object location O are refracted at the spherical surface and then converge to the image point I. We can begin with Snell's law of refraction and use simple geometric techniques to show that the object distance, image distance, and radius of curvature are related by the equation

$$\frac{n_1}{p} + \frac{n_2}{q} = \frac{n_2 - n_1}{R} \qquad [23.7]$$

Further, the magnification of a refracting surface is

$$M = \frac{h'}{h} = -\frac{n_1 q}{n_2 p} \qquad [23.8]$$

As with mirrors, certain sign conventions hold, depending on circumstances. First note that real images are formed by refraction on the side of the surface *opposite* the side from which the light comes, in contrast to mirrors, where real images are formed on the *same* side of the reflecting surface. This makes sense, because light reflects off mirrors, so any real images must form on the same side the light comes from. With a transparent medium, the rays pass through and naturally form real images on the opposite side. We define the side of the surface in which light rays originate as the front side. The other side is called the back side. Because of the difference in location of real images, the refraction sign conventions for q and R are the opposite of those for reflection. For example, p, q, and R are all positive in Figure 23.15. The sign conventions for spherical refracting surfaces are summarized in Table 23.2 (page 766).

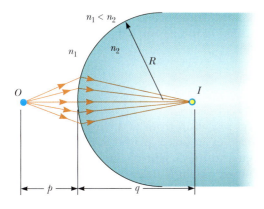

Figure 23.15   An image formed by refraction at a spherical surface. Rays making small angles with the principal axis diverge from a point object at O and pass through the image point I.

TABLE 23.2

Sign Conventions for Refracting Surfaces

Quantity	Symbol	In Front	In Back	Upright Image	Inverted Image
Object location	p	+	−		
Image location	q	−	+		
Radius	R	−	+		
Image height	h'			+	−

Applying Physics 23.4   Underwater Vision

Why does a person with normal vision see a blurry image if the eyes are opened underwater with no goggles or diving mask in use?

Explanation The eye presents a spherical refraction surface. The eye normally functions so that light entering from the air is refracted to form an image in the retina located at the back of the eyeball. The difference in the index of refraction between water and the eye is smaller than the difference in the index of refraction between air and the eye. Consequently, light entering the eye from the water doesn't undergo as much refraction as does light entering from the air, and the image is formed behind the retina. A diving mask or swimming goggles have no optical action of their own; they are simply flat pieces of glass or plastic in a rubber mount. However, they provide a region of air adjacent to the eyes, so that the correct refraction relationship is established and images will be in focus.

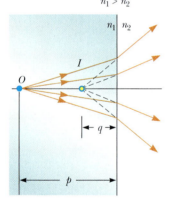

$n_1 > n_2$

ACTIVE FIGURE 23.16
The image formed by a flat refracting surface is virtual and on the same side of the surface as the object. Note that if the light rays are reversed in direction, we have the situation described in Example 22.6.

Physics⊗Now™
Log into PhysicsNow at **www.cp7e.com** and go to Active Figure 23.16, where you can move the object and observe the effect on the location of the image.

Flat Refracting Surfaces

If the refracting surface is flat, then R approaches infinity and Equation 23.7 reduces to

$$\frac{n_1}{p} = -\frac{n_2}{q}$$

$$q = -\frac{n_2}{n_1}p \qquad [23.9]$$

From Equation 23.9 we see that the sign of q is opposite that of p. Consequently, **the image formed by a flat refracting surface is on the same side of the surface as the object**. This is illustrated in Active Figure 23.16 for the situation in which n_1 is greater than n_2, where a virtual image is formed between the object and the surface. Note that the refracted ray bends *away* from the normal in this case, because $n_1 > n_2$.

Quick Quiz 23.2

A person spearfishing from a boat sees a fish located 3 m from the boat at an apparent depth of 1 m. To spear the fish, should the person aim (a) at, (b) above, or (c) below the image of the fish?

Quick Quiz 23.3

True or false? (a) The image of an object placed in front of a concave mirror is always upright. (b) The height of the image of an object placed in front of a concave mirror must be smaller than or equal to the height of the object. (c) The image of an object placed in front of a convex mirror is always upright and smaller than the object.

EXAMPLE 23.5   Gaze into the Crystal Ball

Goal   Calculate the properties of an image created by a spherical lens.

Problem   A coin 2.00 cm in diameter is embedded in a solid glass ball of radius 30.0 cm (Fig. 23.17). The index of refraction of the ball is 1.50, and the coin is 20.0 cm from the surface. **(a)** Find the position of the image of the coin, and **(b)** the height of the coin's image.

Strategy   Because the rays are moving from a medium of high index of refraction (the glass ball) to a medium of lower index of refraction (air), those originating at the coin are refracted away from the normal at the surface and diverge outward. The image is formed in the glass and is virtual. Substitute into Equations 23.7 and 23.8 for the image position and magnification, respectively.

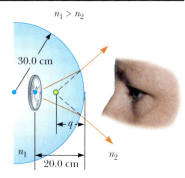

Figure 23.17   (Example 23.5) A coin embedded in a glass ball forms a virtual image between the coin and the surface of the glass.

Solution

Apply Equation 23.7, and take $n_1 = 1.50$, $n_2 = 1.00$, $p = 20.0$ cm, and $R = -30.0$ cm:

$$\frac{n_1}{p} + \frac{n_2}{q} = \frac{n_2 - n_1}{R}$$

$$\frac{1.50}{20.0\ \text{cm}} + \frac{1.00}{q} = \frac{1.00 - 1.50}{-30.0\ \text{cm}}$$

Solve for q:

$$q = -17.1\ \text{cm}$$

To find the image height, we use Equation 23.8 for the magnification:

$$M = -\frac{n_1 q}{n_2 p} = -\frac{1.50(-17.1\ \text{cm})}{1.00(20.0\ \text{cm})} = \frac{h'}{h}$$

$$h' = 1.28h = (1.28)(2.00\ \text{cm}) = 2.56\ \text{cm}$$

Remarks   The negative sign on q indicates that the image is in the same medium as the object (the side of incident light), in agreement with our ray diagram, and therefore must be virtual. The positive value for M means the image is upright.

Exercise 23.5

A coin is embedded 20.0 cm from the surface of a similar ball of transparent substance having radius 30.0 cm and unknown composition. If the coin's image is virtual and located 15.0 cm from the surface, find (a) the index of refraction of the substance and (b) the magnification.

Answers   (a) 2.00   (b) 1.50

EXAMPLE 23.6   The One That Got Away

Goal   Calculate the properties of an image created by a flat refractive surface.

Problem   A small fish is swimming at a depth d below the surface of a pond (Fig. 23.18). **(a)** What is the *apparent depth* of the fish as viewed from directly overhead? **(b)** If the fish is 12 cm long, how long is its image?

Strategy   In this example the refracting surface is flat, so R is infinite. Hence, we can use Equation 23.9 to determine the location of the image, which is the apparent location of the fish.

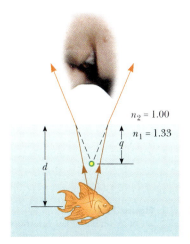

Figure 23.18   (Example 23.6) The apparent depth q of the fish is less than the true depth d.

Solution

(a) Find the apparent depth of the fish.

Substitute $n_1 = 1.33$ for water and $p = d$ into Equation 23.9:

$$q = -\frac{n_2}{n_1}p = -\frac{1}{1.33}d = -0.752d$$

(b) What is the size of the fish's image?

Use Equation 23.9 to eliminate q from the Equation 23.8, the magnification equation:

$$M = \frac{h'}{h} = -\frac{n_1 q}{n_2 p} = -\frac{n_1\left(-\dfrac{n_2}{n_1}p\right)}{n_2 p} = 1$$

$$h' = h = \boxed{12 \text{ cm}}$$

Remarks   Again, because q is negative, the image is virtual, as indicated in Figure 23.18. The apparent depth is three-fourths the actual depth. For instance, if $d = 4.0$ m, then $q = -3.0$ m.

Exercise 23.6
A spear fisherman estimates that a trout is 1.5 m below the water's surface. What is the actual depth of the fish?

Answer   2.0 m

23.5 ATMOSPHERIC REFRACTION

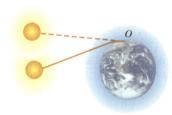

Figure 23.19 Because light is refracted by the Earth's atmosphere, an observer at O sees the Sun even though it has fallen below the horizon.

Images formed by refraction in our atmosphere lead to some interesting phenomena. One such phenomenon that occurs daily is the visibility of the Sun at dusk even though it has passed below the horizon. Figure 23.19 shows why this occurs. Rays of light from the Sun strike the Earth's atmosphere (represented by the shaded area around the planet) and are bent as they pass into a medium that has an index of refraction different from that of the almost empty space in which they have been traveling. The bending in this situation differs somewhat from the bending we have considered previously in that it is gradual and continuous as the light moves through the atmosphere toward an observer at point O. This is because the light moves through layers of air that have a continuously changing index of refraction. When the rays reach the observer, the eye follows them back along the direction from which they appear to have come (indicated by the dashed path in the figure). The end result is that the Sun appears to be above the horizon even after it has fallen below it.

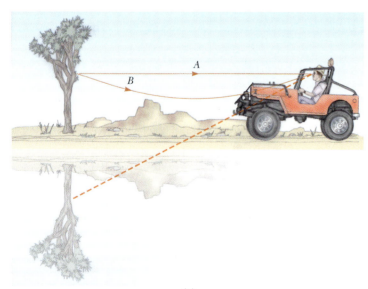

(a)

(b)

John M. Dunay IV, Fundamental Photographs, NYC

Figure 23.20   (a) A mirage is produced by the bending of light rays in the atmosphere when there are large temperature differences between the ground and the air. (b) Notice the reflection of the cars in this photograph of a mirage. The road looks like it's flooded with water, but is actually dry.

The **mirage** is another phenomenon of nature produced by refraction in the atmosphere. A mirage can be observed when the ground is so hot that the air directly above it is warmer than the air at higher elevations. The desert is a region in which such circumstances prevail, but mirages are also seen on heated roadways during the summer. The layers of air at different heights above the Earth have different densities and different refractive indices. The effect this can have is pictured in Figure 23.20a. The observer sees the sky and a tree in two different ways. One group of light rays reaches the observer by the straight-line path *A*, and the eye traces these rays back to see the tree in the normal fashion. In addition, a second group of rays travels along the curved path *B*. These rays are directed toward the ground and are then bent as a result of refraction. As a consequence, the observer also sees an inverted image of the tree and the background of the sky as he traces the rays back to the point at which they appear to have originated. Because an upright image and an inverted image are seen when the image of a tree is observed in a reflecting pool of water, the observer unconsciously calls on this past experience and concludes that the sky is reflected by a pool of water in front of the tree.

23.6   THIN LENSES

A typical **thin lens** consists of a piece of glass or plastic, ground so that each of its two refracting surfaces is a segment of either a sphere or a plane. Lenses are commonly used to form images by refraction in optical instruments, such as cameras, telescopes, and microscopes. The equation that relates object and image distances for a lens is virtually identical to the mirror equation derived earlier, and the method used to derive it is also similar.

Figure 23.21 shows some representative shapes of lenses. Notice that we have placed these lenses in two groups. Those in Figure 23.21a are thicker at the center than at the rim, and those in Figure 23.21b are thinner at the center than at the rim. The lenses in the first group are examples of **converging lenses**, and those in the second group are **diverging lenses**. The reason for these names will become apparent shortly.

As we did for mirrors, it is convenient to define a point called the **focal point** for a lens. For example, in Figure 23.22a (page 770), a group of rays parallel to the axis passes through the focal point *F* after being converged by the lens. The distance from the focal point to the lens is called the **focal length** *f*. **The focal length is the image distance that corresponds to an infinite object distance**. Recall that we are considering the lens to be very thin. As a result, it makes no difference whether we take the focal length to be the distance from the focal point to the surface of the lens or the distance from the focal point to the center of the lens, because the difference between these two lengths is negligible. A thin lens has *two* focal points, as illustrated in Figure 23.22, one on each side of the lens. One focal point corresponds to parallel rays traveling from the left and the other corresponds to parallel rays traveling from the right.

Rays parallel to the axis diverge after passing through a lens of biconcave shape, shown in Figure 23.22b. In this case, the focal point is defined to be the point at which the diverged rays appear to originate, labeled *F* in the figure. Figures 23.22a

| Biconvex | Convex–concave | Plano–convex | Biconcave | Convex–concave | Plano–concave |

(a)   (b)

Figure 23.21   Various lens shapes. (a) Converging lenses have positive focal lengths and are thickest at the middle. (b) Diverging lenses have negative focal lengths and are thickest at the edges.

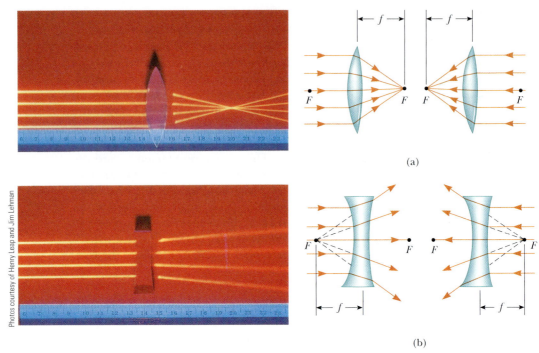

(a)

(b)

Figure 23.22 (*Left*) Photographs of the effects of converging and diverging lenses on parallel rays. (*Right*) The focal points of (a) the biconvex lens and (b) the biconcave lens.

and 23.22b indicate why the names *converging* and *diverging* are applied to these lenses.

Now consider a ray of light passing through the center of a lens. Such a ray is labeled ray 1 in Figure 23.23. For a thin lens, a ray passing through the center is undeflected. Ray 2 in the same figure is parallel to the principal axis of the lens (the horizontal axis passing through O), and as a result it passes through the focal point F after refraction. Rays 1 and 2 intersect at the point that is the tip of the image arrow.

We first note that the tangent of the angle α can be found by using the blue and gold shaded triangles in Figure 23.23:

$$\tan \alpha = \frac{h}{p} \quad \text{or} \quad \tan \alpha = -\frac{h'}{q}$$

From this we find that

$$M = \frac{h'}{h} = -\frac{q}{p} \tag{23.10}$$

The equation for magnification by a lens is the same as the equation for magnification by a mirror. We also note from Figure 23.23 that

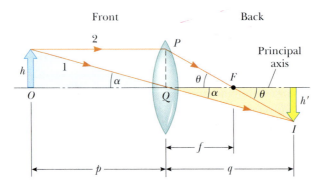

Figure 23.23 A geometric construction for developing the thin-lens equation.

TABLE 23.3

Sign Conventions for Thin Lenses

Quantity	Symbol	In Front	In Back	Convergent	Divergent
Object location	p	+	−		
Image location	q	−	+		
Lens Radii	R_1, R_2	−	+		
Focal Length	f			+	−

$$\tan \theta = \frac{PQ}{f} \quad \text{or} \quad \tan \theta = -\frac{h'}{q-f}$$

However, the height PQ used in the first of these equations is the same as h, the height of the object. Therefore,

$$\frac{h}{f} = -\frac{h'}{q-f}$$

$$\frac{h'}{h} = -\frac{q-f}{f}$$

Using the latter equation in combination with Equation 23.10 gives

$$\frac{q}{p} = \frac{q-f}{f}$$

which reduces to

$$\frac{1}{p} + \frac{1}{q} = \frac{1}{f} \qquad \text{[23.11]}$$

◄ Thin-lens equation

This equation, called the **thin-lens equation**, can be used with both converging and diverging lenses if we adhere to a set of sign conventions. Figure 23.24 is useful for obtaining the signs of p and q, and Table 23.3 gives the complete sign conventions for lenses. Note that **a converging lens has a positive focal length** under this convention and **a diverging lens has a negative focal length**. Hence the names *positive* and *negative* are often given to these lenses.

The focal length for a lens in air is related to the curvatures of its front and back surfaces and to the index of refraction n of the lens material by

$$\frac{1}{f} = (n-1)\left(\frac{1}{R_1} - \frac{1}{R_2}\right) \qquad \text{[23.12]}$$

Figure 23.24 A diagram for obtaining the signs of p and q for a thin lens or a refracting surface.

where R_1 is the radius of curvature of the front surface of the lens and R_2 is the radius of curvature of the back surface. (As with mirrors, we arbitrarily call the side from which the light approaches the *front* of the lens.) Table 23.3 gives the sign conventions for R_1 and R_2. Equation 23.12, called the **lens maker's equation**, enables us to calculate the focal length from the known properties of the lens.

Ray Diagrams for Thin Lenses

Ray diagrams are essential for understanding the overall image formation by a thin lens or a system of lenses. They should also help clarify the sign conventions we have already discussed. Active Figure 23.25 (page 772) illustrates this method for three single-lens situations. To locate the image formed by a converging lens (Active Fig. 23.25a and b), the following three rays are drawn from the top of the object:

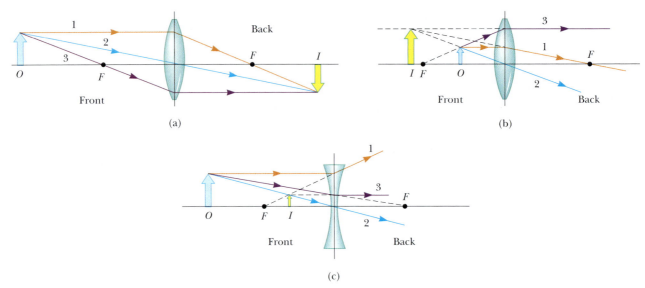

(a)   (b)

(c)

ACTIVE FIGURE 23.25

Ray diagrams for locating the image of an object. (a) The object is outside the focal point of a converging lens. (b) The object is inside the focal point of a converging lens. (c) The object is outside the focal point of a diverging lens.

Physics ⊗ Now™

Log into PhysicsNow at **www.cp7e.com** and go to Active Figure 23.25, where you can move the objects and change the focal lengths of the lenses, observing the effect on the images.

1. The first ray is drawn parallel to the principal axis. After being refracted by the lens, this ray passes through (or appears to come from) one of the focal points.
2. The second ray is drawn through the center of the lens. This ray continues in a straight line.
3. The third ray is drawn through the other focal point and emerges from the lens parallel to the principal axis.

TIP 23.3   We *Choose* Only a Few Rays

Although our ray diagrams in Figure 23.25 only show three rays leaving an object, an infinite number of rays can be drawn between the object and its image.

A similar construction is used to locate the image formed by a diverging lens, as shown in Active Figure 23.25c. The point of intersection of *any two* of the rays in these diagrams can be used to locate the image. The third ray serves as a check on construction.

For the converging lens in Active Figure 23.25a, where the object is *outside* the front focal point ($p > f$), the ray diagram shows the image is real and inverted. When the real object is *inside* the front focal point ($p < f$), as in Active Figure 23.25b, the image is virtual and upright. For the diverging lens of Active Figure 23.25c, the image is virtual and upright.

Quick Quiz 23.4

A clear plastic sandwich bag filled with water can act as a crude converging lens in air. If the bag is filled with air and placed under water, is the effective lens (a) converging or (b) diverging?

Quick Quiz 23.5

In Active Figure 23.25a, the blue object arrow is replaced by one that is much taller than the lens. How many rays from the object will strike the lens?

Quick Quiz 23.6

An object is placed to the left of a converging lens. Which of the following statements are true and which are false? (a) The image is always to the right of the lens. (b) The image can be upright or inverted. (c) The image is always smaller or the same size as the object. Justify your answers with ray diagrams.

Your success in working lens or mirror problems will be determined largely by whether you make sign errors when substituting into the lens or mirror equations. The only way to ensure that you don't make sign errors is to become adept at using the sign conventions. The best way to do this is to work a multitude of problems on your own and to construct confirming ray diagrams. Watching an instructor or reading the example problems is no substitute for practice.

Applying Physics 23.5   Vision and Diving Masks

Diving masks often have a lens built into the glass faceplate for divers who don't have perfect vision. This allows the individual to dive without the necessity of glasses, because the faceplate performs the necessary refraction to produce clear vision. Normal glasses have lenses that are curved on both the front and rear surfaces. The lenses in a diving-mask faceplate often have curved surfaces only on the inside of the glass. Why is this design desirable?

Solution The main reason for curving only the inner surface of the lenses in the diving-mask faceplate is to enable the diver to see clearly while underwater and in the air. If there were curved surfaces on both the front and the back of the diving lens, there would be two refractions. The lens could be designed so that these two refractions would give clear vision while the diver is in air. When the diver went underwater, however, the refraction between the water and the glass at the first interface would differ, because the index of refraction of water is different from that of air. Consequently, the diver's vision wouldn't be clear underwater.

INTERACTIVE EXAMPLE 23.7   Images Formed by a Converging Lens

Goal   Calculate geometric quantities associated with a converging lens.

Problem   A converging lens of focal length 10.0 cm forms images of an object situated at various distances. **(a)** If the object is placed 30.0 cm from the lens, locate the image, state whether it's real or virtual, and find its magnification. **(b)** Repeat the problem when the object is at 10.0 cm and **(c)** again when the object is 5.00 cm from the lens.

Strategy   All three problems require only substitution into the thin-lens equation and the associated magnification equation—Equations 23.10 and 23.11, respectively. The conventions of Table 23.3 must be followed.

Solution
(a) Find the image distance and describe the image when the object is placed at 30.0 cm.

The ray diagram is shown in Figure 23.26a. Substitute into the thin-lens equation to locate the image:

$$\frac{1}{p} + \frac{1}{q} = \frac{1}{f}$$

$$\frac{1}{30.0 \text{ cm}} + \frac{1}{q} = \frac{1}{10.0 \text{ cm}}$$

Solve for q, the image distance. It's positive, so the image is real and on the far side of the lens.

$$q = +15.0 \text{ cm}$$

The magnification of the lens is obtained from Equation 23.10. M is negative and less than one in absolute value, so the image is inverted and smaller than the object:

$$M = -\frac{q}{p} = -\frac{15.0 \text{ cm}}{30.0 \text{ cm}} = -0.500$$

(b) Repeat the problem, when the object is placed at 10.0 cm.

Locate the image by substituting into the thin-lens equation:

$$\frac{1}{10.0 \text{ cm}} + \frac{1}{q} = \frac{1}{10.0 \text{ cm}} \quad \rightarrow \quad \frac{1}{q} = 0$$

This equation is satisfied only in the limit as q becomes infinite.

$$q \rightarrow \infty$$

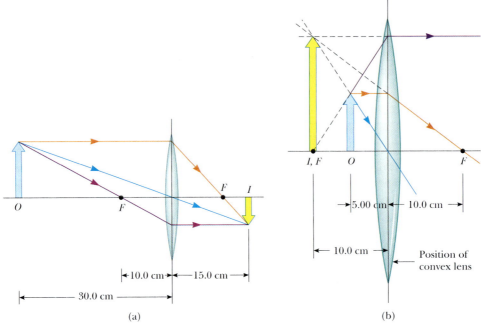

Figure 23.26 (Interactive Example 23.7)

(c) Repeat the problem when the object is placed 5.00 cm from the lens.

See the ray diagram in Figure 23.26b. Substitute into the thin-lens equation to locate the image:

$$\frac{1}{5.00 \text{ cm}} + \frac{1}{q} = \frac{1}{10.0 \text{ cm}}$$

Solve for q, which is negative, meaning the image is on the same side as the object and is virtual:

$$q = \boxed{-10.0 \text{ cm}}$$

Substitute the values of p and q into the magnification equation. M is positive and larger than one, so the image is upright and double the object size:

$$M = -\frac{q}{p} = -\left(\frac{-10.0 \text{ cm}}{5.00 \text{ cm}}\right) = \boxed{+2.00}$$

Remarks   The ability of a lens to magnify objects led to the inventions of reading glasses, microscopes, and telescopes.

Exercise 23.7
Suppose the image of an object is upright and magnified 1.75 times when the object is placed 15.0 cm from a lens. Find the location of the image and the focal length of the lens.

Answers   (a) -26.3 cm (virtual, on the same side as the object)   (b) 34.9 cm

Physics⊗Now™ Investigate the image formed for various object positions and mirror focal lengths by logging into PhysicsNow at **www.cp7e.com** and going to Interactive Example 23.7.

INTERACTIVE EXAMPLE 23.8   The Case of a Diverging Lens

Goal   Calculate geometric quantities associated with a diverging lens.

Problem   Repeat the problem of Example 23.7 for a *diverging* lens of focal length 10.0 cm.

Strategy   Once again, substitution into the thin-lens equation and the associated magnification equation, together with the conventions in Table 23.3, solve the various parts. The only difference is the negative focal length.

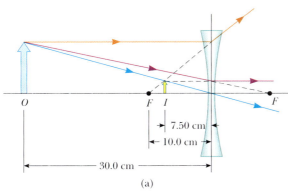

Figure 23.27   (Interactive Example 23.8)

Solution

(a) Locate the image and its magnification if the object is at 30.0 cm.

The ray diagram is given in Figure 23.27a. Apply the thin-lens equation with $p = 30.0$ cm to locate the image:

$$\frac{1}{p} + \frac{1}{q} = \frac{1}{f}$$

$$\frac{1}{30.0 \text{ cm}} + \frac{1}{q} = -\frac{1}{10.0 \text{ cm}}$$

Solve for q, which is negative, hence virtual:

$$q = -7.50 \text{ cm}$$

Substitute into Equation 23.10 to get the magnification. Because M is positive and has absolute value less than one, the image is upright and smaller than the object.

$$M = -\frac{q}{p} = -\left(\frac{-7.50 \text{ cm}}{30.0 \text{ cm}}\right) = +0.250$$

(b) Locate the image and find its magnification if the object is 10.0 cm from the lens.

Apply the thin-lens equation, taking $p = 10.0$ cm:

$$\frac{1}{10.0 \text{ cm}} + \frac{1}{q} = -\frac{1}{10.0 \text{ cm}}$$

Solve for q (once again, the result is negative, so the image is virtual):

$$q = -5.00 \text{ cm}$$

Calculate the magnification. Because M is positive and has absolute value less than 1, the image is upright and smaller than the object.

$$M = -\frac{q}{p} = -\left(\frac{-5.00 \text{ cm}}{10.0 \text{ cm}}\right) = +0.500$$

(c) Locate the image and find its magnification when the object is at 5.00 cm.

The ray diagram is given in Figure 23.27b. Substitute $p = 5.00$ cm into the thin-lens equation to locate the image:

$$\frac{1}{5.00 \text{ cm}} + \frac{1}{q} = -\frac{1}{10.0 \text{ cm}}$$

Solve for q. The answer is negative, so once again the image is virtual.

$$q = -3.33 \text{ cm}$$

Calculate the magnification. Because M is positive and less than one, the image is upright and smaller than the object.

$$M = -\left(\frac{-3.33 \text{ cm}}{5.00 \text{ cm}}\right) = +0.666$$

Remarks   Notice that in every case the image is virtual, hence on the same side of the lens as the object. Further, the image is smaller than the object. For a diverging lens and a real object, this is *always* the case, as can be proven mathematically.

Exercise 23.8

Repeat the calculation, finding the position of the image and the magnification if the object is 20.0 cm from the lens.

Answers $q = -6.67$ cm, $M = 0.334$

Physics⊗Now™ Investigate the image formed for various object positions and mirror focal lengths by logging into PhysicsNow at **www.cp7e.com** and going to Interactive Example 23.8.

Combinations of Thin Lenses

Many useful optical devices require two lenses. Handling problems involving two lenses is not much different from dealing with a single-lens problem twice. First, the image produced by the first lens is calculated as though the second lens were not present. The light then approaches the second lens *as if* it had come from the image formed by the first lens. Hence, **the image formed by the first lens is treated as the object for the second lens**. The image formed by the second lens is the final image of the system. If the image formed by the first lens lies on the back side of the second lens, then the image is treated as a virtual object for the second lens, so p is negative. The same procedure can be extended to a system of three or more lenses. The overall magnification of a system of thin lenses is the *product* of the magnifications of the separate lenses.

INTERACTIVE EXAMPLE 23.9   Two Lenses in a Row

Goal   Calculate geometric quantities for a sequential pair of lenses.

Problem   Two converging lenses are placed 20.0 cm apart, as shown in Figure 23.28a, with an object 30.0 cm in front of lens 1 on the left. **(a)** If lens 1 has a focal length of 10.0 cm, locate the image formed by this lens and determine its

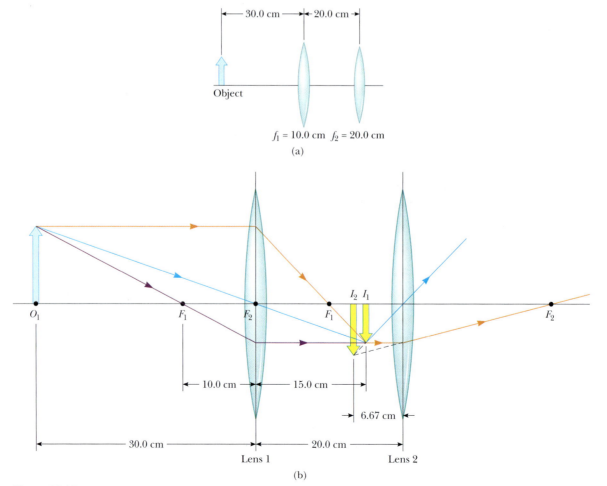

Figure 23.28   (Interactive Example 23.9)

magnification. **(b)** If lens 2 on the right has a focal length of 20.0 cm, locate the final image formed and find the total magnification of the system.

Strategy   We apply the thin-lens equation to each lens. The image formed by lens 1 is treated as the object for lens 2. Also, we use the fact that the total magnification of the system is the product of the magnifications produced by the separate lenses.

Solution

(a) Locate the image and determine the magnification of lens 1.

See the ray diagram, Figure 23.28b. Apply the thin-lens equation to lens 1:

$$\frac{1}{30.0 \text{ cm}} + \frac{1}{q} = \frac{1}{10.0 \text{ cm}}$$

Solve for q, which is positive, hence to the right of the first lens:

$$q = +15.0 \text{ cm}$$

Compute the magnification of lens 1:

$$M_1 = -\frac{q}{p} = -\frac{15.0 \text{ cm}}{30.0 \text{ cm}} = -0.500$$

(b) Locate the final image, and total magnification.

The image formed by lens 1 becomes the object for lens 2. Compute the object distance for lens 2:

$$p = 20.0 \text{ cm} - 15.0 \text{ cm} = 5.00 \text{ cm}$$

Once again apply the thin-lens equation to lens 2 to locate the final image:

$$\frac{1}{5.00 \text{ cm}} + \frac{1}{q} = \frac{1}{20.0 \text{ cm}}$$

$$q = -6.67 \text{ cm}$$

Calculate the magnification of lens 2:

$$M_2 = -\frac{q}{p} = -\frac{(-6.67 \text{ cm})}{5.00 \text{ cm}} = +1.33$$

Multiply the two magnifications to get the overall magnification of the system:

$$M = M_1 M_2 = (-0.500)(1.33) = -0.665$$

Remarks   The negative sign for M indicates that the final image is inverted, and smaller than the object because the absolute value of M is less than one. Because q is negative, the final image is virtual.

Exercise 23.9
If the two lenses in Figure 23.28 are separated by 10.0 cm, locate the final image and find the magnification of the system. [*Hint:* The object for the second lens is virtual!]

Answer   4.00 cm behind the second lens; $M = -0.400$

Physics⊗Now™ Investigate the image formed by a combination of two lenses by logging into PhysicsNow at **www.cp7e.com** and going to Interactive Example 23.9.

23.7   LENS AND MIRROR ABERRATIONS

One of the basic problems of systems containing mirrors and lenses is the imperfect quality of the images, which is largely the result of defects in shape and form. The simple theory of mirrors and lenses assumes that rays make small angles with the principal axis and that all rays reaching the lens or mirror from a point source are focused at a single point, producing a sharp image. This is not always true in the real world. Where the approximations used in this theory do not hold, imperfect images are formed.

If one wishes to analyze image formation precisely, it is necessary to trace each ray, using Snell's law, at each refracting surface. This procedure shows that there is no single point image; instead, the image is blurred. The departures of real (imperfect)

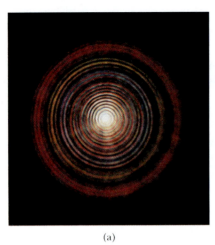 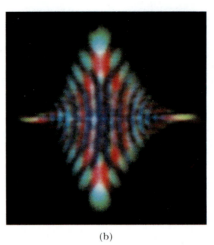 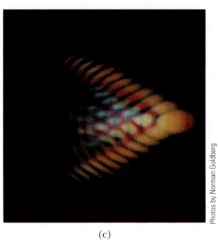

(a)   (b)   (c)

Photos by Norman Goldberg

Figure 23.29 Lenses can produce various forms of aberrations, as shown by these blurred photographic images of a point source. (a) Spherical aberration occurs when light passing through the lens at different distances from the principal axis is focused at different points. (b) Astigmatism is an aberration that occurs when the object is not on the principal axis of the lens. (c) Coma. This aberration occurs when light passing through the lens far from the principal axis focuses at a different part of the focal plane than light passing near the center of the lens.

images from the ideal predicted by the simple theory are called **aberrations**. Two common types of aberrations are spherical aberration and chromatic aberration. Photographs of three forms of lens aberrations are shown in Figure 23.29.

Spherical Aberration

Spherical aberration results from the fact that the focal points of light rays passing far from the principal axis of a spherical lens (or mirror) are different from the focal points of rays with the same wavelength passing near the axis. Figure 23.30 illustrates spherical aberration for parallel rays passing through a converging lens. Rays near the middle of the lens are imaged farther from the lens than rays at the edges. Hence, there is no single focal length for a spherical lens.

Most cameras are equipped with an adjustable aperture to control the light intensity and, when possible, reduce spherical aberration. (An aperture is an opening that controls the amount of light transmitted through the lens.) As the aperture size is reduced, sharper images are produced, because only the central portion of the lens is exposed to the incident light when the aperture is very small. At the same time, however, progressively less light is imaged. To compensate for this loss, a longer exposure time is used. An example of the results obtained with small apertures is the sharp image produced by a pinhole camera, with an aperture size of approximately 0.1 mm.

In the case of mirrors used for very distant objects, one can eliminate, or at least minimize, spherical aberration by employing a parabolic rather than spherical surface. Parabolic surfaces are not used in many applications, however, because they are very expensive to make with high-quality optics. Parallel light rays incident on such a surface focus at a common point. Parabolic reflecting surfaces are used in many astronomical telescopes to enhance the image quality. They are also used in flashlights, in which a nearly parallel light beam is produced from a small lamp placed at the focus of the reflecting surface.

Figure 23.30 Spherical aberration produced by a converging lens. Does a diverging lens produce spherical aberration? (Angles are greatly exaggerated for clarity.)

Chromatic Aberration

The fact that different wavelengths of light refracted by a lens focus at different points gives rise to chromatic aberration. In Chapter 22 we described how the index of refraction of a material varies with wavelength. When white light passes through a lens, for example, violet light rays are refracted more than red light rays (see Fig. 23.31), so the focal length for red light is greater than for violet light.

Figure 23.31 Chromatic aberration produced by a converging lens. Rays of different wavelengths focus at different points. (Angles are greatly exaggerated for clarity.)

Other wavelengths (not shown in the figure) would have intermediate focal points. Chromatic aberration for a diverging lens is opposite that for a converging lens. Chromatic aberration can be greatly reduced by a combination of converging and diverging lenses.

SUMMARY

Physics⊗Now™ Take a practice test by logging into Physics-Now at **www.cp7e.com** and clicking on the Pre-Test link for this chapter.

23.1 Flat Mirrors

Images are formed where rays of light intersect or where they appear to originate. A **real image** is formed when light intersects, or passes through, an image point. In a **virtual image** the light doesn't pass through the image point, but appears to diverge from it.

The image formed by a flat mirror has the following properties:

1. The image is as far behind the mirror as the object is in front.
2. The image is unmagnified, virtual, and upright.

23.2 Images Formed by Spherical Mirrors &
23.3 Convex Mirrors and Sign Conventions

The **magnification** M of a spherical mirror is defined as the ratio of the **image height** h' to the **object height** h, which is the negative of the ratio of the image distance q to the object distance p:

$$M = \frac{h'}{h} = -\frac{q}{p} \qquad [23.2]$$

The **object distance** and **image distance** for a spherical mirror of radius R are related by the **mirror equation**:

$$\frac{1}{p} + \frac{1}{q} = \frac{1}{f} \qquad [23.6]$$

where $f = R/2$ is the **focal length** of the mirror.

Equations 23.2 and 23.6 hold for both concave and convex mirrors, subject to the sign conventions given in Table 23.1.

23.4 Images Formed by Refraction

An image can be formed by refraction at a spherical surface of radius R. The object and image distances for refraction from such a surface are related by

$$\frac{n_1}{p} + \frac{n_2}{q} = \frac{n_2 - n_1}{R} \qquad [23.7]$$

The **magnification of a refracting surface** is

$$M = \frac{h'}{h} = -\frac{n_1 q}{n_2 p} \qquad [23.8]$$

where the object is located in the medium with index of refraction n_1 and the image is formed in the medium with index of refraction n_2. Equations 23.7 and 23.8 are subject to the sign conventions of Table 23.2.

23.6 Thin Lenses

The **magnification of a thin lens** is

$$M = \frac{h'}{h} = -\frac{q}{p} \qquad [23.10]$$

The object and image distances of a thin lens are related by the **thin-lens equation**:

$$\frac{1}{p} + \frac{1}{q} = \frac{1}{f} \qquad [23.11]$$

Equations 23.10 and 23.11 are subject to the sign conventions of Table 23.3.

23.7 Lens and Mirror Aberrations

Aberrations are responsible for the formation of imperfect images by lenses and mirrors. **Spherical aberration** results from the fact that the focal points of light rays far from the principal axis of a spherical lens or mirror are different from those of rays passing through the center. **Chromatic aberration** arises from the fact that light rays of different wavelengths focus at different points when refracted by a lens.

CONCEPTUAL QUESTIONS

1. Tape a picture of yourself on a bathroom mirror. Stand several centimeters away from the mirror. Can you focus your eyes on *both* the picture taped to the mirror *and* your image in the mirror *at the same time*? So where is the image of yourself?

2. One method for determining the position of an image, either real or virtual, is by means of *parallax*. If a finger or another object is placed at the position of the image, as shown in Figure Q23.2, and the finger and the image are viewed simultaneously (the image is viewed through the lens if it is virtual), the finger and image have the same parallax; that is, if the image is viewed from different positions, it will appear to move along with the finger. Use this method to locate the image formed by a lens. Explain why the method works.

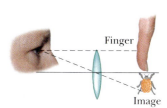

Figure Q23.02

3. A flat mirror creates a virtual image of your face. Suppose the flat mirror is combined with another optical element. Can the mirror form a real image in such a combination?

4. Explain why a mirror cannot give rise to chromatic aberration.

5. You are taking a picture of yourself with a camera that uses an ultrasonic range finder to measure the distance to the object. When you take a picture of yourself in a mirror with this camera, your image is out of focus. Why?

6. A solar furnace can be constructed by using a concave mirror to reflect and focus sunlight into the enclosure of a furnace. What factors in the design of the reflecting mirror will guarantee that very high temperatures can be reached?

7. A virtual image is often described as an image through which light rays don't actually travel, as they do for a real image. Can a virtual image be photographed?

8. What is wrong with the caption of the cartoon shown in Figure Q23.8?

Figure Q23.08 "Most mirrors reverse left and right. This one reverses top and bottom."

9. Suppose you want to use a converging lens to project the image of two trees onto a screen. One tree is a distance x from the lens; the other is at $2x$, as in Figure Q23.9. You adjust the screen so that the near tree is in focus. If you now want the far tree to be in focus, do you move the screen towards or away from the lens?

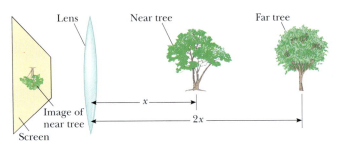

Figure Q23.09

10. Why does a clear stream always appear to be shallower than it actually is?

11. Can a converging lens be made to diverge light if placed in a liquid? How about a converging mirror?

12. A common mirage is formed when the air gets gradually cooler as the height above the ground increases. What might happen if the air grows gradually warmer as the height increases? This often happens over bodies of water or snow-covered ground; the effect is called *looming*.

13. In a Jules Verne novel, a piece of ice is shaped into a magnifying lens to focus sunlight to start a fire. Is this possible?

14. Lenses used in eyeglasses, whether converging or diverging, are always designed such that the middle of the lens curves away from the eye. Why?

15. Why does the focal length of a mirror not depend on the mirror material when the focal length of a lens does depend on the lens material?

16. If a cylinder of solid glass or clear plastic is placed above the words LEAD OXIDE and viewed from the side, as shown in Figure Q23.16, the word LEAD appears inverted, but the word OXIDE does not. Explain.

Figure Q23.16

17. A concave makeup mirror has a focal length of 15 cm. (a) If an object is placed 25 cm in front of the mirror, determine the signs of the focal length, the object distance, and the image distance. (b) Repeat part (a) if the object is placed 5 cm in front of the mirror.

18. Light from an object passes through a lens and forms a visible image on a screen. If the screen is removed, would you be able to see the image if (a) you remained in your present position? (b) you could look at the lens along the its axis, beyond the original position of the screen?

19. An object placed to the left of a converging lens forms a sharp image on a screen to the right of the lens. If the screen is moved towards the lens, the image on the screen (a) gets larger, but remains sharp, (b) gets smaller, but remains sharp, (c) becomes fuzzy and disappears, or (d) remains unchanged.

20. An inverted image of an object is viewed on a screen from the side facing a converging lens. An opaque card is then introduced covering only the upper half of the lens. What happens to the image on the screen? (a) Half the image would disappear. (b) The entire image would appear and remain unchanged. (c) Half the image would disappear and be dimmer. (d) The entire image would appear, but would be dimmer.

PROBLEMS

1, 2, 3 = straightforward, intermediate, challenging   ☐ = full solution available in *Student Solutions Manual/Study Guide*

Physics⊗Now™ = coached solution with hints available at **www.cp7e.com**   🧬 = biomedical application

Section 23.1 Flat Mirrors

1. Does your bathroom mirror show you older or younger than your actual age? Compute an order-of-magnitude estimate for the age difference, based on data that you specify.

2. Use Active Figure 23.2 to give a geometric proof that the virtual image formed by a plane mirror is the same distance behind the mirror as the object is in front of it.

3. A person walks into a room that has, on opposite walls, two plane mirrors producing multiple images. Find the distances from the person to the first three images seen in the left-hand mirror when the person is 5.00 ft from the mirror on the left wall and 10.0 ft from the mirror on the right wall.

4. In a church choir loft, two parallel walls are 5.30 m apart. The singers stand against the north wall. The organist faces the south wall, sitting 0.800 m away from it. So that she can see the choir, a flat mirror 0.600 m wide is mounted on the south wall, straight in front of the organist. What width of the north wall can she see? [*Hint:* Draw a top-view diagram to justify your answer.]

Section 23.2 Images Formed by Spherical Mirrors
Section 23.3 Convex Mirrors and Sign Conventions

In the following problems, algebraic signs are not given. We leave it to you to determine the correct sign to use with each quantity, based on an analysis of the problem and the sign conventions in Table 23.1.

5. **Physics⊗Now™** At an intersection of hospital hallways, a convex mirror is mounted high on a wall to help people avoid collisions. The mirror has a radius of curvature of 0.550 m. Locate and describe the image of a patient 10.0 m from the mirror. Determine the magnification of the image.

6. 🧬 To fit a contact lens to a patient's eye, a *keratometer* can be used to measure the curvature of the cornea—the front surface of the eye. This instrument places an illuminated object of known size at a known distance p from the cornea, which then reflects some light from the object, forming an image of it. The magnification M of the image is measured by using a small viewing telescope that allows a comparison of the image formed by the cornea with a second calibrated image projected into the field of view by a prism arrangement. Determine the radius of curvature of the cornea when $p = 30.0$ cm and $M = 0.013\ 0$.

7. A concave spherical mirror has a radius of curvature of 20.0 cm. Locate the images for object distances of (a) 40.0 cm, (b) 20.0 cm, and (c) 10.0 cm. In each case, state whether the image is real or virtual and upright or inverted, and find the magnification.

8. A dentist uses a mirror to examine a tooth that is 1.00 cm in front of the mirror. The image of the tooth is formed 10.0 cm behind the mirror. Determine (a) the mirror's radius of curvature and (b) the magnification of the image.

9. A large church has a niche in one wall. On the floor plan it appears as a semicircular indentation of radius 2.50 m. A worshiper stands on the centerline of the niche, 2.00 m out from its deepest point, and whispers a prayer. Where is the sound concentrated after it reflects from the back wall of the niche?

10. While looking at her image in a cosmetic mirror, Dina notes that her face is highly magnified when she is close to the mirror, but as she backs away from the mirror, her image first becomes blurry, then disappears when she is about 30 cm from the mirror, and then inverts when she is beyond 30 cm. Based on these observations, what can she conclude about the properties of the mirror?

11. A 2.00-cm-high object is placed 3.00 cm in front of a concave mirror. If the image is 5.00 cm high and virtual, what is the focal length of the mirror?

12. A dedicated sports car enthusiast polishes the inside and outside surfaces of a hubcap that is a section of a sphere. When he looks into one side of the hubcap, he sees an image of his face 30.0 cm in back of it. He then turns the hubcap over, keeping it the same distance from his face. He now sees an image of his face 10.0 cm in back of the hubcap. (a) How far is his face from the hubcap? (b) What is the radius of curvature of the hubcap?

13. A concave makeup mirror is designed so that a person 25 cm in front of it sees an upright image magnified by a factor of two. What is the radius of curvature of the mirror?

14. A certain Christmas tree ornament is a silver sphere having a diameter of 8.50 cm. Determine an object location for which the size of the reflected image is three-fourths the size of the object. Use a principal-ray diagram to arrive at a description of the image.

15. A man standing 1.52 m in front of a shaving mirror produces an inverted image 18.0 cm in front of it. How close to the mirror should he stand if he wants to form an upright image of his chin that is twice the chin's actual size?

16. A convex spherical mirror with a radius of curvature of 10.0 cm produces a virtual image one-third the size of the real object. Where is the object?

17. A child holds a candy bar 10.0 cm in front of a convex mirror and notices that the image is only one-half the size of the candy bar. What is the radius of curvature of the mirror?

18. It is observed that the size of a *real* image formed by a concave mirror is four times the size of the object when the object is 30.0 cm in front of the mirror. What is the radius of curvature of this mirror?

19. A spherical mirror is to be used to form an image, five times as tall as an object, on a screen positioned 5.0 m from the mirror. (a) Describe the type of mirror required. (b) Where should the mirror be positioned relative to the object?

20. A ball is dropped from rest 3.00 m directly above the vertex of a concave mirror having a radius of 1.00 m and lying in a horizontal plane. (a) Describe the motion of the ball's image in the mirror. (b) At what time do the ball and its image coincide?

Section 23.4 Images Formed by Refraction

21. A cubical block of ice 50.0 cm on an edge is placed on a level floor over a speck of dust. Locate the image of the

speck, when viewed from directly above, if the index of refraction of ice is 1.309.

22. The top of a swimming pool is at ground level. If the pool is 2 m deep, how far below ground level does the bottom of the pool appear to be located when (a) the pool is completely filled with water? (b) the pool is filled halfway with water?

23. A paperweight is made of a solid glass hemisphere with index of refraction 1.50. The radius of the circular cross section is 4.0 cm. The hemisphere is placed on its flat surface, with the center directly over a 2.5-mm-long line drawn on a sheet of paper. What length of line is seen by someone looking vertically down on the hemisphere?

24. A flint glass plate ($n = 1.66$) rests on the bottom of an aquarium tank. The plate is 8.00 cm thick (vertical dimension) and covered with water ($n = 1.33$) to a depth of 12.0 cm. Calculate the apparent thickness of the plate as viewed from above the water. (Assume nearly normal incidence of light rays.)

25. A transparent sphere of unknown composition is observed to form an image of the Sun on its surface opposite the Sun. What is the refractive index of the sphere material?

26. A goldfish is swimming at 2.00 cm/s toward the front wall of a rectangular aquarium. What is the apparent speed of the fish as measured by an observer looking in from outside the front wall of the tank? The index of refraction of water is 1.333.

Section 23.6 Thin Lenses

27. A contact lens is made of plastic with an index of refraction of 1.50. The lens has an outer radius of curvature of +2.00 cm and an inner radius of curvature of +2.50 cm. What is the focal length of the lens?

28. The left face of a biconvex lens has a radius of curvature of 12.0 cm, and the right face has a radius of curvature of 18.0 cm. The index of refraction of the glass is 1.44. (a) Calculate the focal length of the lens. (b) Calculate the focal length if the radii of curvature of the two faces are interchanged.

29. **Physics⊗Now™** A converging lens has a focal length of 20.0 cm. Locate the images for object distances of (a) 40.0 cm, (b) 20.0 cm, and (c) 10.0 cm. For each case, state whether the image is real or virtual and upright or inverted, and find the magnification.

30. Where must an object be placed to have unit magnification ($|M| = 1.00$) (a) for a converging lens of focal length 12.0 cm? (b) for a diverging lens of focal length 12.0 cm?

31. A diverging lens has a focal length of 20.0 cm. Locate the images for object distances of (a) 40.0 cm, (b) 20.0 cm, and (c) 10.0 cm. For each case, state whether the image is real or virtual and upright or inverted, and find the magnification.

32. The use of a lens in a certain situation is described by the equation

$$\frac{1}{p} + \frac{1}{-3.50p} = \frac{1}{7.50 \text{ cm}}$$

Determine (a) the object distance and (b) the image distance. (c) Use a ray diagram to obtain a description of the image. (d) Identify a practical device described by the given equation, and write the statement of a problem having a solution that contains the equation.

33. A transparent photographic slide is placed in front of a converging lens with a focal length of 2.44 cm. The lens forms an image of the slide 12.9 cm from it. How far is the lens from the slide if the image is (a) real? (b) virtual?

34. The nickel's image in Figure P23.34 has twice the diameter of the nickel when the lens is 2.84 cm from the nickel. Determine the focal length of the lens.

Figure P23.34

35. A certain LCD projector contains a single thin lens. An object 24.0 mm high is to be projected so that its image fills a screen 1.80 m high. The object-to-screen distance is 3.00 m. (a) Determine the focal length of the projection lens. (b) How far from the object should the lens of the projector be placed in order to form the image on the screen?

36. A person uses a converging lens that has a focal length of 12.5 cm to inspect a gem. The lens forms a virtual image 30.0 cm away. Determine the magnification. Is the image upright or inverted?

37. A diverging lens is to be used to produce a virtual image one-third as tall as the object. Where should the object be placed?

38. An object is 5.00 m to the left of a flat screen. A converging lens for which the focal length is $f = 0.800$ m is placed between object and screen. (a) Show that there are two lens positions that form an image on the screen, and determine how far these positions are from the object. (b) How do the two images differ from each other?

39. A converging lens is placed 30.0 cm to the right of a diverging lens of focal length 10.0 cm. A beam of parallel light enters the diverging lens from the left, and the beam is again parallel when it emerges from the converging lens. Calculate the focal length of the converging lens.

40. An object is placed 20.0 cm to the left of a converging lens of focal length 25.0 cm. A diverging lens of focal length 10.0 cm is 25.0 cm to the right of the converging lens. Find the position and magnification of the final image.

41. Two converging lenses, each of focal length 15.0 cm, are placed 40.0 cm apart, and an object is placed 30.0 cm in front of the first lens. Where is the final image formed, and what is the magnification of the system?

42. Object O_1 is 15.0 cm to the left of a converging lens with a 10.0-cm focal length. A second lens is positioned 10.0 cm to the right of the first lens and is observed to form a final image at the position of the original object O_1. (a) What is the focal length of the second lens? (b) What is the overall magnification of this system? (c) What is the nature (i.e., real or virtual, upright or inverted) of the final image?

43. **Physics⊗Now™** A 1.00-cm-high object is placed 4.00 cm to the left of a converging lens of focal length 8.00 cm. A diverging lens of focal length − 16.00 cm is 6.00 cm to the right of the converging lens. Find the position and height of the final image. Is the image inverted or upright? Real or virtual?

44. Two converging lenses having focal lengths of 10.0 cm and 20.0 cm are placed 50.0 cm apart, as shown in Figure P23.44. The final image is to be located between the lenses, at the position indicated. (a) How far to the left of the first lens should the object be positioned? (b) What is the overall magnification of the system? (c) Is the final image upright or inverted?

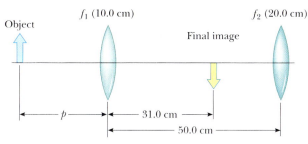

Figure P23.44

45. Lens L_1 in Figure P23.45 has a focal length of 15.0 cm and is located a fixed distance in front of the film plane of a camera. Lens L_2 has a focal length of 13.0 cm, and its distance d from the film plane can be varied from 5.00 cm to 10.0 cm. Determine the range of distances for which objects can be focused on the film.

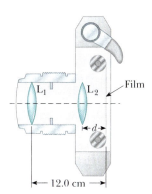

Figure P23.45

46. Consider two thin lenses, one of focal length f_1 and the other of focal length f_2, placed in contact with each other as shown in Figure P23.46. Apply the thin-lens equation to each of these lenses and combine the results to show that this combination of lenses behaves like a thin lens having a focal length f given by $1/f = 1/f_1 + 1/f_2$. Assume that

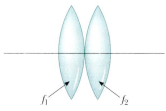

Figure P23.46

the thicknesses of the lenses can be ignored in comparison to the other distances involved.

ADDITIONAL PROBLEMS

47. An object placed 10.0 cm from a concave spherical mirror produces a real image 8.00 cm from the mirror. If the object is moved to a new position 20.0 cm from the mirror, what is the position of the image? Is the final image real or virtual?

48. An object is placed 12 cm to the left of a diverging lens of focal length − 6.0 cm. A converging lens of focal length 12 cm is placed a distance d to the right of the diverging lens. Find the distance d that places the final image at infinity.

49. A convergent lens with a 50.0-mm focal length is used to focus an image of a very distant scene onto a flat screen 35.0 mm wide. What is the angular width α of the scene included in the image on the screen?

50. The object in Figure P23.50 is midway between the lens and the mirror. The mirror's radius of curvature is 20.0 cm, and the lens has a focal length of − 16.7 cm. Considering only the light that leaves the object and travels first towards the mirror, locate the final image formed by this system. Is the image real or virtual? Is it upright or inverted? What is the overall magnification of the image?

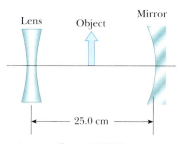

Figure P23.50

51. The lens and mirror in Figure P23.51 are separated by 1.00 m and have focal lengths of + 80.0 cm and − 50.0 cm, respectively. If an object is placed 1.00 m to the left of the lens, locate the final image. State whether the image is upright or inverted, and determine the overall magnification.

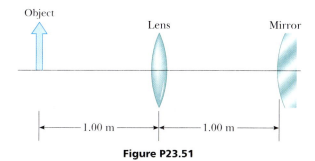

Figure P23.51

52. A diverging lens ($n = 1.50$) is shaped like that in Active Figure 23.25c. The radius of the first surface is 15.0 cm, and that of the second surface is 10.0 cm. (a) Find the focal length of the lens. Determine the positions of the images for object distances of (b) infinity, (c) $3|f|$, (d) $|f|$, and (e) $|f|/2$.

53. A parallel beam of light enters a glass hemisphere perpendicular to the flat face, as shown in Figure P23.53. The radius of the hemisphere is $R = 6.00$ cm, and the index of refraction is $n = 1.560$. Determine the point at which the beam is focused. (Assume paraxial rays—that is, assume that all rays are located close to the principal axis.)

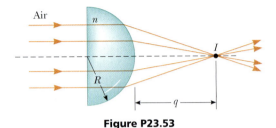

Figure P23.53

54. A converging lens of focal length 20.0 cm is separated by 50.0 cm from a converging lens of focal length 5.00 cm. (a) Find the position of the final image of an object placed 40.0 cm in front of the first lens. (b) If the height of the object is 2.00 cm, what is the height of the final image? Is the image real or virtual? (c) If the two lenses are now placed in contact with each other and the object is 5.00 cm in front of this combination, where will the image be located? (See Problem 46.)

55. **Physics⊗Now™** To work this problem, use the fact that the image formed by the first surface becomes the object for the second surface. Figure P23.55 shows a piece of glass with index of refraction 1.50. The ends are hemispheres with radii 2.00 cm and 4.00 cm, and the centers of the hemispherical ends are separated by a distance of 8.00 cm. A point object is in air, 1.00 cm from the left end of the glass. Locate the image of the object due to refraction at the two spherical surfaces.

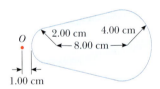

Figure P23.55

56. In a darkened room, a burning candle is placed 1.50 m from a white wall. A lens is placed between candle and wall at a location that causes a larger, inverted image to form on the wall. When the lens is moved 90.0 cm toward the wall, another image of the candle is formed. Find (a) the two object distances that produce the images just described and (b) the focal length of the lens. (c) Characterize the second image.

57. An object 2.00 cm high is placed 40.0 cm to the left of a converging lens having a focal length of 30.0 cm. A diverging lens having a focal length of -20.0 cm is placed 110 cm to the right of the converging lens. (a) Determine the final position and magnification of the final image. (b) Is the image upright or inverted? (c) Repeat parts (a) and (b) for the case where the second lens is a converging lens having a focal length of $+20.0$ cm.

58. A "floating strawberry" illusion can be produced by two parabolic mirrors, each with a focal length of 7.5 cm, facing each other so that their centers are 7.5 cm apart

(Fig. P23.58). If a strawberry is placed on the bottom mirror, an image of the strawberry forms at the small opening at the center of the top mirror. Show that the final image forms at that location, and describe its characteristics. [*Note:* A flashlight beam shone on these *images* has a very startling effect: Even at a glancing angle, the incoming light beam is seemingly reflected off the *images* of the strawberry! Do you understand why?]

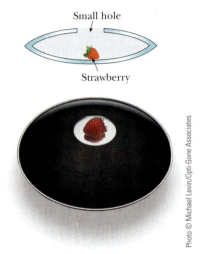

Figure P23.58

59. Figure P23.59 shows a converging lens with radii $R_1 = 9.00$ cm and $R_2 = -11.00$ cm, in front of a concave spherical mirror of radius $R = 8.00$ cm. The focal points (F_1 and F_2) for the thin lens and the center of curvature (C) of the mirror are also shown. (a) If the focal points F_1 and F_2 are 5.00 cm from the vertex of the thin lens, determine the index of refraction of the lens. (b) If the lens and mirror are 20.0 cm apart, and an object is placed 8.00 cm to the left of the lens, determine the position of the final image and its magnification as seen by the eye in the figure. (c) Is the final image inverted or upright? Explain.

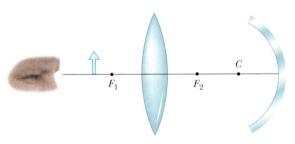

Figure P23.59

60. Find the object distances (in terms of f) for a thin converging lens of focal length f if (a) the image is real and the image distance is four times the focal length; (b) the image is virtual and the image distance is three times the focal length. (c) Calculate the magnification of the lens for cases (a) and (b).

61. The lens maker's equation for a lens with index n_1 immersed in a medium with index n_2 takes the form

$$\frac{1}{f} = \left(\frac{n_1}{n_2} - 1\right)\left(\frac{1}{R_1} - \frac{1}{R_2}\right)$$

A thin diverging glass (index = 1.50) lens with $R_1 = -3.00$ m and $R_2 = -6.00$ m is surrounded by air. An arrow is placed 10.0 m to the left of the lens. (a) Determine the position of the image. Repeat part (a) with the arrow and lens immersed in (b) water (index = 1.33); (c) a medium with an index of refraction of 2.00. (d) How can a lens that is diverging in air be changed into a converging lens.

62. An observer to the right of the mirror–lens combination shown in Figure P23.62 sees two real images that are the same size and in the same location. One image is upright and the other is inverted. Both images are 1.50 times larger than the object. The lens has a focal length of 10.0 cm. The lens and mirror are separated by 40.0 cm. Determine the focal length of the mirror. (Don't assume that the figure is drawn to scale.)

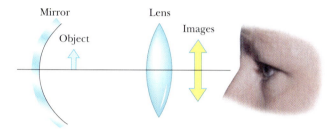

Figure P23.62

63. The lens maker's equation applies to a lens immersed in a liquid if n in the equation is replaced by n_1/n_2. Here n_1 refers to the refractive index of the lens material and n_2 is that of the medium surrounding the lens. (a) A certain lens has focal length of 79.0 cm in air and a refractive index of 1.55. Find its focal length in water. (b) A certain mirror has focal length of 79.0 cm in air. Find its focal length in water.

ACTIVITIES

1. This experiment will enable you to examine some of the properties of images formed by flat mirrors. As shown in Figure A23.1, place a clear plastic sheet between two small candles of the same height. Light one candle placed about 6 inches from the sheet and observe the reflected flame from the front side. Move the unlit candle until it also appears to be lit. The candles should be approximately equidistant from the sheet at this point. Why?

Figure A23.1

When you place your finger on the wick of the unlit candle, you will get the illusion that your finger is burning. Explain your observation.

A similar experiment can be performed with a flat mirror and two pencils. Hold one of the pencils in front of the mirror at a distance of about 10 inches from the mirror and at a position such that about half of the image appears in the mirror. Move the second pencil back and forth behind the mirror until it appears to align perfectly with the image in the mirror. When the two pencils are aligned, measure the distance from the mirror to the location of the second pencil. This pencil is situated at the apparent position of the virtual image. Why? The two pencils should be at equal distances from the mirror. Why?

2. Fill a clear glass tumbler with water and place a pencil or straw into the tumbler as in Figure A23.2. Now observe the pencil from the side at an angle of about 45° to the surface, and note that the line of the portion of the pencil under water is not parallel with the line of the portion in air. That is, the pencil appears to be bent at the point where it enters the water. Use the techniques of Section 23.4 to explain this observation.

Figure A23.2

3. View yourself in a full-length mirror. Stand close to the mirror, and place one piece of tape at the top of the image of your head and another piece at the very bottom of the image of your feet. Now step back a few meters and observe your image. How big is it relative to the original size? How does the distance between the pieces of tape compare with your actual height?

Move to a position in front of the mirror such that you can see a full image of yourself with the top of your head just level with the top of the mirror. Have a friend gradually block off the lower portion of the mirror with a sheet or newspaper page until you can see your complete image—but no more! Measure the length of the mirror and compare this measurement with your height. How do the two compare?

4. Compare the images formed of your face when you look first at the front side and then at the back side of a shiny soupspoon. For each side, observe the change in the image of your face as you move closer to and farther away from the spoon.

5. Draw a sketch of you and your image in a plane mirror when you are walking away from it with velocity $\vec{v}$. Indicate on your diagram the velocity vectors for you and your image. Keep the relative lengths of the two velocities as you believe they will be. Explain your answer. (b) Repeat the procedure in part (a), but now do it as you are walking towards the mirror. Explain why you have the vectors drawn as you do.

The colors in many of a hummingbird's feathers are not due to pigment. The *iridescence* which makes the brilliant colors that often appear on the bird's throat and belly is due to an interference effect caused by structures in the feathers. The colors vary with the viewing angle.

RO-MA/Index Stock Imagery

Wave Optics

Colors swirl on a soap bubble as it drifts through the air on a summer day, and vivid rainbows reflect from the filth of oil films in the puddles of a dirty city street. Beachgoers, covered with thin layers of oil, wear their coated sunglasses that absorb half the incoming light. In laboratories, scientists determine the precise composition of materials by analyzing the light they give off when hot, and in observatories around the world, telescopes gather light from distant galaxies, filtering out individual wavelengths in bands and thereby determining the speed of expansion of the universe.

Understanding how these rainbows are made and how certain scientific instruments can determine wavelengths is the domain of *wave optics*. Light can be viewed as either a particle or a wave. Geometric optics, the subject of the previous chapter, depends on the particle nature of light. Wave optics depends on the wave nature of light. The three primary topics we examine in this chapter are interference, diffraction, and polarization. These phenomena can't be adequately explained with ray optics, but can be understood if light is viewed as a wave.

24.1 CONDITIONS FOR INTERFERENCE

In our discussion of interference of mechanical waves in Chapter 13, we found that two waves could add together either constructively or destructively. In constructive interference, the amplitude of the resultant wave is greater than that of either of the individual waves, whereas in destructive interference, the resultant amplitude is less than that of either individual wave. Light waves also interfere with each other. Fundamentally, all interference associated with light waves arises when the electromagnetic fields that constitute the individual waves combine.

Interference effects in light waves aren't easy to observe because of the short wavelengths involved (about 4×10^{-7} m to about 7×10^{-7} m). For sustained interference between two sources of light to be observed, the following conditions must be met:

Conditions for interference ▶

1. The sources must be **coherent**, which means the waves they emit must maintain a constant phase with respect to each other.
2. The waves must have identical wavelengths.

Two sources (producing two traveling waves) are needed to create interference. To produce a stable interference pattern, the individual waves must maintain a constant phase with one another. When this situation prevails, the sources are said to be coherent. The sound waves emitted by two side-by-side loudspeakers driven by a single amplifier can produce interference because the two speakers respond to the amplifier in the same way at the same time—they are in phase.

If two light sources are placed side by side, however, no interference effects are observed, because the light waves from one source are emitted independently of the waves from the other source; hence, the emissions from the two sources don't maintain a constant phase relationship with each other during the time of observation. An ordinary light source undergoes random changes about once every 10^{-8} s. Therefore, the conditions for constructive interference, destructive interference, and intermediate states have durations on the order of 10^{-8} s. The result is that no interference effects are observed, because the eye can't follow such short-term changes. Ordinary light sources are said to be **incoherent**.

An older method for producing two coherent light sources is to pass light from a single wavelength (monochromatic) source through a narrow slit and then allow the light to fall on a screen containing two other narrow slits. The first slit is needed to create a single wave-front that illuminates both slits coherently. The light emerging from the two slits is coherent because a single source produces the original light beam and the slits serve only to separate the original beam into two parts. Any random change in the light emitted by the source will occur in the two separate beams at the same time, and interference effects can be observed.

Currently it's much more common to use a laser as a coherent source to demonstrate interference. A laser produces an intense, coherent, monochromatic beam over a width of several millimeters. This means that the laser may be used to illuminate multiple slits directly and that interference effects can be easily observed in a fully lighted room. The principles of operation of a laser are explained in Chapter 28.

24.2 YOUNG'S DOUBLE-SLIT EXPERIMENT

Thomas Young first demonstrated interference in light waves from two sources in 1801. Active Figure 24.1a is a schematic diagram of the apparatus used in this experiment. (Young used pinholes rather than slits in his original experiments.)

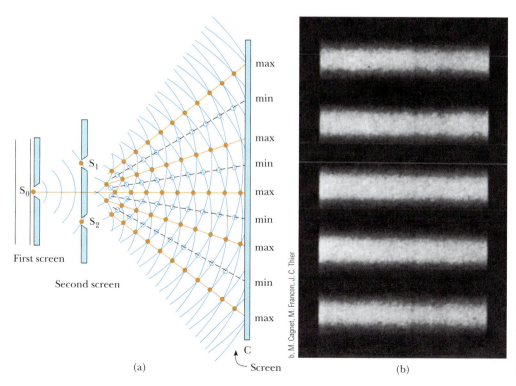

(a)   (b)

b. M. Cagnet, M. Francon, J. C. Thier

ACTIVE FIGURE 24.1
(a) A diagram of Young's double-slit experiment. The narrow slits act as sources of waves. Slits S_1 and S_2 behave as coherent sources that produce an interference pattern on screen C. (The drawing is not to scale.) (b) The fringe pattern formed on screen C could look like this.

Physics⊗Now™
Log into PhysicsNow at **www.cp7e.com** and go to Active Figure 24.1, where you can adjust the slit separation and the wavelength of the light, observing the effect on the interference pattern.

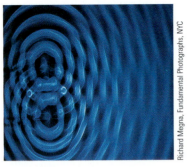

Figure 24.2 An interference pattern involving water waves is produced by two vibrating sources at the water's surface. The pattern is analogous to that observed in Young's double-slit experiment. Note the regions of constructive and destructive interference.

Light is incident on a screen containing a narrow slit S_0. The light waves emerging from this slit arrive at a second screen that contains two narrow, parallel slits S_1 and S_2. These slits serve as a pair of coherent light sources because waves emerging from them originate from the same wave front and therefore are always in phase. The light from the two slits produces a visible pattern on screen C consisting of a series of bright and dark parallel bands called **fringes** (Active Fig. 24.1b). When the light from slits S_1 and S_2 arrives at a point on the screen so that constructive interference occurs at that location, a bright fringe appears. When the light from the two slits combines destructively at any location on the screen, a dark fringe results. Figure 24.2 is a photograph of an interference pattern produced by two coherent vibrating sources in a water tank.

Figure 24.3 is a schematic diagram of some of the ways in which the two waves can combine at screen C of Figure 24.1. In Figure 24.3a, two waves, which leave the two slits in phase, strike the screen at the central point P. Because these waves travel equal distances, they arrive in phase at P, and as a result, constructive interference occurs there and a bright fringe is observed. In Figure 24.3b, the two light waves again start in phase, but the upper wave has to travel one wavelength farther to reach point Q on the screen. Because the upper wave falls behind the lower one by exactly one wavelength, the two waves still arrive in phase at Q, so a second bright fringe appears at that location. Now consider point R, midway between P and Q, in Figure 24.3c. At R, the upper wave has fallen half a wavelength behind the lower wave. This means that the trough of the bottom wave overlaps the crest of the upper wave, giving rise to destructive interference. As a result, a dark fringe can be observed at R.

We can describe Young's experiment quantitatively with the help of Figure 24.4. Consider point P on the viewing screen; the screen is positioned a perpendicular distance L from the screen containing slits S_1 and S_2, which are separated by distance d, and r_1 and r_2 are the distances the secondary waves travel from slit to screen. We assume the waves emerging from S_1 and S_2 have the same constant frequency, have the same amplitude, and start out in phase. The light intensity on the screen at P is the result of light from both slits. A wave from the lower slit, however, travels farther than a wave from the upper slit by the amount $d \sin \theta$. This distance is called the **path difference** δ (lower case Greek delta), where

Path difference ▶

$$\delta = r_2 - r_1 = d \sin \theta \qquad [24.1]$$

Equation 24.1 assumes that the two waves travel in parallel lines, which is approximately true, because L is much greater than d. As noted earlier, the value of this path difference determines whether the two waves are in phase when they arrive at P. If the path difference is either zero or some integral multiple of the wavelength, the two waves are in phase at P and constructive interference results. Therefore, the condition for bright fringes, or **constructive interference**, at P is

Condition for constructive interference (two slits) ▶

$$\delta = d \sin \theta_{\text{bright}} = m\lambda \qquad m = 0, \pm 1, \pm 2, \ldots \qquad [24.2]$$

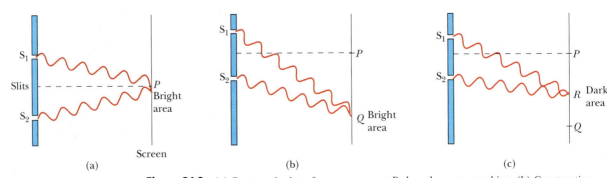

Figure 24.3 (a) Constructive interference occurs at P when the waves combine. (b) Constructive interference also occurs at Q. (c) Destructive interference occurs at R when the wave from the upper slit falls half a wavelength behind the wave from the lower slit. (These figures are not drawn to scale.)

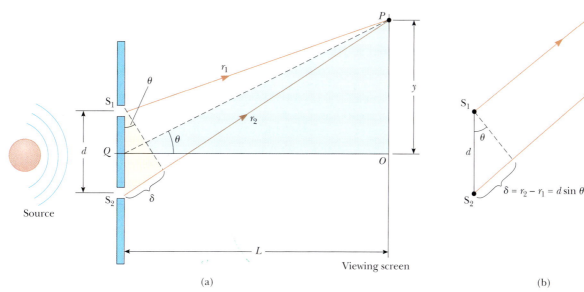

Figure 24.4 A geometric construction that describes Young's double-slit experiment. The path difference between the two rays is $\delta = r_2 - r_1 = d \sin \theta$. (This figure is not drawn to scale.)

The number m is called the **order number**. The central bright fringe at $\theta_{bright} = 0$ ($m = 0$) is called the *zeroth-order maximum*. The first maximum on either side, where $m = \pm 1$, is called the *first-order maximum*, and so forth.

When δ is an odd multiple of $\lambda/2$, the two waves arriving at P are 180° out of phase and give rise to destructive interference. Therefore, the condition for dark fringes, or **destructive interference**, at P is

$$\delta = d \sin \theta_{dark} = (m + \tfrac{1}{2})\lambda \quad m = 0, \pm 1, \pm 2, \dots \qquad \textbf{[24.3]}$$

◀ Condition for destructive interference (two slits)

If $m = 0$ in this equation, the path difference is $\delta = \lambda/2$, which is the condition for the location of the first dark fringe on either side of the central (bright) maximum. Likewise, if $m = 1$, the path difference is $\delta = 3\lambda/2$, which is the condition for the second dark fringe on each side, and so forth.

It's useful to obtain expressions for the positions of the bright and dark fringes measured vertically from O to P. In addition to our assumption that $L \gg d$, we assume that $d \gg \lambda$. These can be valid assumptions because, in practice, L is often on the order of 1 m, d is a fraction of a millimeter, and λ is a fraction of a micrometer for visible light. Under these conditions θ is small, so we can use the approximation $\sin \theta \cong \tan \theta$. Then, from triangle OPQ in Figure 24.4, we see that

$$y = L \tan \theta \approx L \sin \theta \qquad \textbf{[24.4]}$$

Solving Equation 24.2 for $\sin \theta$ and substituting the result into Equation 24.4, we find that the positions of the *bright fringes*, measured from O, are

$$y_{bright} = \frac{\lambda L}{d} m \quad m = 0, \pm 1, \pm 2, \dots \qquad \textbf{[24.5]}$$

Using Equations 24.3 and 24.4, we find that the *dark fringes* are located at

$$y_{dark} = \frac{\lambda L}{d}(m + \tfrac{1}{2}) \quad m = 0, \pm 1, \pm 2, \dots \qquad \textbf{[24.6]}$$

As we will show in Example 24.1, Young's double-slit experiment provides a method for measuring the wavelength of light. In fact, Young used this technique to do just that. In addition, his experiment gave the wave model of light a great deal of credibility. It was inconceivable that particles of light coming through the slits could cancel each other in a way that would explain the dark fringes.

TIP 24.1 Small-Angle Approximation: Size Matters!

The small-angle approximation $\sin \theta \cong \tan \theta$ is true to three-digit precision only for angles less than about 4°.

John S. Shelton

Reflection, interference, and diffraction can be seen in this aerial photograph of waves in the sea.

Applying Physics 24.1   A Smoky Young's Experiment

Consider a double-slit experiment in which a laser beam is passed through a pair of very closely spaced slits and a clear interference pattern is displayed on a distant screen. Now suppose you place smoke particles between the double slit and the screen. With the presence of the smoke particles, will you see the effects of interference in the space between the slits and the screen, or will you see only the effects on the screen?

Explanation You will see the interference pattern both on the screen and in the area filled with smoke between the slits and the screen. There will be bright lines directed toward the bright areas on the screen and dark lines directed toward the dark areas on the screen. This is because Equations 24.5 and 24.6 depend on the distance to the screen, L, which can take any value.

Applying Physics 24.2   Television Signal Interference

Suppose you are watching television by means of an antenna rather than a cable system. If an airplane flies near your location, you may notice wavering ghost images in the television picture. What might cause this phenomenon?

Explanation Your television antenna receives two signals: the direct signal from the transmitting antenna and a signal reflected from the surface of the airplane. As the airplane changes position, there are some times when these two signals are in phase and other times when they are out of phase. As a result, the intensity of the combined signal received at your antenna will vary. The wavering of the ghost images of the picture is evidence of this variation.

Quick Quiz 24.1

In a two-slit interference pattern projected on a screen, the fringes are equally spaced on the screen (a) everywhere (b) only for large angles (c) only for small angles.

INTERACTIVE EXAMPLE 24.1   Measuring the Wavelength of a Light Source

Goal   Show how Young's experiment can be used to measure the wavelength of coherent light.

Problem   A screen is separated from a double-slit source by 1.20 m. The distance between the two slits is 0.030 0 mm. The second-order bright fringe ($m = 2$) is measured to be 4.50 cm from the centerline. Determine (a) the wavelength of the light and (b) the distance between adjacent bright fringes.

Strategy   Equation 24.5 relates the positions of the bright fringes to the other variables, including the wavelength of the light. Substitute into this equation and solve for λ. Taking the difference between y_{m+1} and y_m results in a general expression for the distance between bright fringes.

Solution
(a) Determine the wavelength of the light.

Solve Equation 24.5 for the wavelength and substitute $m = 2$, $y_2 = 4.50 \times 10^{-2}$ m, $L = 1.20$ m, and $d = 3.00 \times 10^{-5}$ m:

$$\lambda = \frac{y_2 d}{mL} = \frac{(4.50 \times 10^{-2}\,\text{m})(3.00 \times 10^{-5}\,\text{m})}{2(1.20\,\text{m})}$$

$$= 5.63 \times 10^{-7}\,\text{m} = \boxed{563\ \text{nm}}$$

(b) Determine the distance between adjacent bright fringes.

Use Equation 24.5 to find the distance between *any* adjacent bright fringes (here, those characterized by m and $m + 1$):

$$\Delta y = y_{m+1} - y_m = \frac{\lambda L}{d}(m + 1) - \frac{\lambda L}{d}m = \frac{\lambda L}{d}$$

$$= \frac{(5.63 \times 10^{-7}\,\text{m})(1.20\,\text{m})}{3.00 \times 10^{-5}\,\text{m}} = \boxed{2.25\ \text{cm}}$$

Remarks   This calculation depends on the angle θ being small, because the small-angle approximation was implicitly used. The measurement of the position of the bright fringes yields the wavelength of light, which in turn is a signature of atomic processes, as will be discussed in the chapters on modern physics. This kind of measurement, therefore, helped open the world of the atom.

Exercise 24.1

Suppose the same experiment is run with a different light source. If the first-order maximum is found at 1.85 cm from the centerline, what is the wavelength of the light?

Answer   463 nm

Physics⊗Now™ Investigate the double-slit interference pattern by logging into PhysicsNow at **www.cp7e.com** and going to Interactive Example 24.1.

24.3   CHANGE OF PHASE DUE TO REFLECTION

Young's method of producing two coherent light sources involves illuminating a pair of slits with a single source. Another simple, yet ingenious, arrangement for producing an interference pattern with a single light source is known as *Lloyd's mirror*. A point source of light is placed at point S, close to a mirror, as illustrated in Figure 24.5. Light waves can reach the viewing point P either by the direct path SP or by the path involving reflection from the mirror. The reflected ray can be treated as a ray originating at the source S' behind the mirror. Source S', which is the image of S, can be considered a virtual source.

At points far from the source, an interference pattern due to waves from S and S' is observed, just as for two real coherent sources. However, the positions of the dark and bright fringes are *reversed* relative to the pattern obtained from two real coherent sources (Young's experiment). This is because the coherent sources S and S' differ in phase by 180°, a phase change produced by reflection.

To illustrate the point further, consider P', the point where the mirror intersects the screen. This point is equidistant from S and S'. If path difference alone were responsible for the phase difference, a bright fringe would be observed at P' (because the path difference is zero for this point), corresponding to the central fringe of the two-slit interference pattern. Instead, we observe a *dark* fringe at P, from which we conclude that a 180° phase change must be produced by reflection from the mirror. In general, **an electromagnetic wave undergoes a phase change of 180° upon reflection from a medium that has an index of refraction higher than the one in which the wave was traveling**.

An analogy can be drawn between reflected light waves and the reflections of a transverse wave on a stretched string when the wave meets a boundary, as in Figure 24.6. The reflected pulse on a string undergoes a phase change of 180°

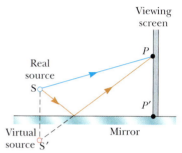

Figure 24.5 Lloyd's mirror. An interference pattern is produced on a screen at P as a result of the combination of the direct ray (blue) and the reflected ray (brown). The reflected ray undergoes a phase change of 180°.

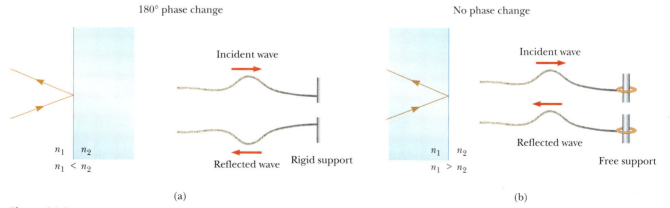

(a)   (b)

Figure 24.6   (a) A ray reflecting from a medium of higher refractive index undergoes a 180° phase change. The right side shows the analogy with a reflected pulse on a string. (b) A ray reflecting from a medium of lower refractive index undergoes no phase change.

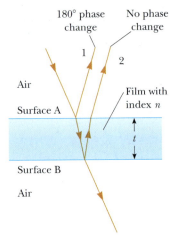

Figure 24.7 Interference observed in light reflected from a thin film is due to a combination of rays reflected from the upper and lower surfaces.

Soap bubbles on water. The colors are due to interference between light rays reflected from the front and back of the thin film of soap making up the bubble. The color depends on the thickness of the film, ranging from black where the film is at its thinnest to magenta where it is thickest.

when it is reflected from the boundary of a denser string or from a rigid barrier and undergoes no phase change when it is reflected from the boundary of a less dense string. Similarly, an electromagnetic wave undergoes a 180° phase change when reflected from the boundary of a medium with index of refraction higher than the one in which it has been traveling. There is no phase change when the wave is reflected from a boundary leading to a medium of lower index of refraction. The transmitted wave that crosses the boundary also undergoes no phase change.

24.4   INTERFERENCE IN THIN FILMS

Interference effects are commonly observed in thin films, such as the thin surface of a soap bubble or thin layers of oil on water. The varied colors observed when incoherent white light is incident on such films result from the interference of waves reflected from the two surfaces of the film.

Consider a film of uniform thickness t and index of refraction n, as in Figure 24.7. Assume that the light rays traveling in air are nearly normal to the two surfaces of the film. To determine whether the reflected rays interfere constructively or destructively, we first note the following facts:

1. An electromagnetic wave traveling from a medium of index of refraction n_1 toward a medium of index of refraction n_2 undergoes a 180° phase change on reflection when $n_2 > n_1$. There is no phase change in the reflected wave if $n_2 < n_1$.
2. The wavelength of light λ_n in a medium with index of refraction n is

$$\lambda_n = \frac{\lambda}{n} \qquad \text{[24.7]}$$

where λ is the wavelength of light in vacuum.

We apply these rules to the film of Figure 24.7. According to the first rule, ray 1, which is reflected from the upper surface A, undergoes a phase change of 180° with respect to the incident wave. Ray 2, which is reflected from the lower surface B, undergoes no phase change with respect to the incident wave. Therefore, ray 1 is 180° out of phase with respect to ray 2, which is equivalent to a path difference of $\lambda_n/2$. However, we must also consider the fact that ray 2 travels an extra distance of $2t$ before the waves recombine in the air above the surface. For example, if $2t = \lambda_n/2$, then rays 1 and 2 recombine in phase, and constructive interference results. In general, the condition for *constructive interference* in thin films is

$$2t = (m + \tfrac{1}{2})\lambda_n \qquad m = 0, 1, 2, \ldots \qquad \text{[24.8]}$$

This condition takes into account two factors: (1) the difference in path length for the two rays (the term $m\lambda_n$) and (2) the 180° phase change upon reflection (the term $\lambda_n/2$). Because $\lambda_n = \lambda/n$, we can write Equation 24.8 in the form

▶ Condition for constructive interference (thin film)

$$2nt = (m + \tfrac{1}{2})\lambda \qquad m = 0, 1, 2, \ldots \qquad \text{[24.9]}$$

If the extra distance $2t$ traveled by ray 2 is a multiple of λ_n, then the two waves combine out of phase and the result is destructive interference. The general equation for *destructive interference* in thin films is

▶ Condition for destructive interference (thin film)

$$2nt = m\lambda \qquad m = 0, 1, 2, \ldots \qquad \text{[24.10]}$$

Equations 24.9 and 24.10 for constructive and destructive interference are valid when there is only one phase reversal. This will occur when the media above and below the thin film both have indices of refraction greater than the film or when both have indices of refraction less than the film. Figure 24.7 is a case in point: the air ($n = 1$) that is both above and below the film has an index of refraction less than that of the film. As a result, there is a phase reversal on reflection off the top layer of the film, but not the bottom, and Equations 24.9 and 24.10 apply. **If the film is placed between two different media, one of lower refractive index than the**

film and one of higher refractive index, Equations 24.9 and 24.10 are reversed: **Equation 24.9 is used for destructive interference, and Equation 24.10 for destructive interference**. In this case, either there is a phase change of 180° for both ray 1 reflecting from surface A and ray 2 reflecting from surface B, as in Figure 24.9 of Example 24.3 (page 794), or there is no phase change for either ray, which would be the case if the incident ray came from underneath the film. Hence, the net change in relative phase due to the reflections is *zero*.

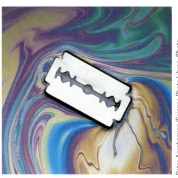

A thin film of oil on water displays interference, evidenced by the pattern of colors when white light is incident on the film. Variations in the film's thickness produce the intersecting color pattern. The razor blade gives you an idea of the size of the colored bands.

Quick Quiz 24.2

Suppose Young's experiment is carried out in air, and then, in a second experiment, the apparatus is immersed in water. In what way does the distance between bright fringes change? (a) They move further apart. (b) They move closer together. (c) There is no change.

Newton's Rings

Another method for observing interference in light waves is to place a planoconvex lens on top of a flat glass surface, as in Figure 24.8a. With this arrangement, the air film between the glass surfaces varies in thickness from zero at the point of contact to some value t at P. If the radius of curvature R of the lens is much greater than the distance r, and if the system is viewed from above light of wavelength λ, a pattern of light and dark rings is observed (Fig. 24.8b). These circular fringes, discovered by Newton, are called **Newton's rings**. The interference is due to the combination of ray 1, reflected from the plate, with ray 2, reflected from the lower surface of the lens. Ray 1 undergoes a phase change of 180° on reflection, because it is reflected from a boundary leading into a medium of higher refractive index, whereas ray 2 undergoes no phase change, because it is reflected from a medium of lower refractive index. Hence, the conditions for constructive and destructive interference are given by Equations 24.9 and 24.10, respectively, with $n = 1$ because the "film" is air. The contact point at O is dark, as seen in Figure 24.8b, because there is no path difference and the total phase change is due only to the 180° phase change upon reflection. Using the geometry shown in Figure 24.8a, we can obtain expressions for the radii of the bright and dark bands in terms of the radius of curvature R, and vacuum wavelength λ. For example, the dark rings have radii of $r \approx \sqrt{m\lambda R/n}$.

One of the important uses of Newton's rings is in the testing of optical lenses. A circular pattern like that in Figure 24.8b is achieved only when the lens is ground to a perfectly spherical curvature. Variations from such symmetry might produce a

TIP 24.2   The Two Tricks of Thin Films

Be sure to include *both* effects—path length and phase change—when you analyze an interference pattern from a thin film.

APPLICATION

Checking for Imperfections in Optical Lenses

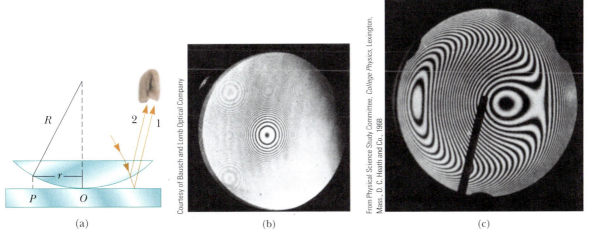

Figure 24.8   (a) The combination of rays reflected from the glass plate and the curved surface of the lens gives rise to an interference pattern known as Newton's rings. (b) A photograph of Newton's rings. (c) This asymmetric interference pattern indicates imperfections in the lens.

Interference in a vertical film of variable thickness. The top of the film appears darkest where the film is thinnest.

pattern like that in Figure 24.8c. These variations give an indication of how the lens must be reground and repolished to remove imperfections.

Problem-Solving Strategy   Thin-Film Interference

The following steps are recommended in addressing thin-film interference problems:

1. Identify the thin film causing the interference, and the indices of refraction in the film and in the media on either side of it.
2. Determine the number of phase reversals: zero, one, or two.
3. Consult the following table, which contains Equations 24.9 and 24.10, and select the correct column for the problem in question:

Equation ($m = 0,1, \ldots$)	1 phase reversal	0 or 2 phase reversals
$2nt = (m + \frac{1}{2})\lambda$ [24.9]	constructive	destructive
$2nt = m\lambda$ [24.10]	destructive	constructive

4. Substitute values in the appropriate equations, as selected in the previous step.

EXAMPLE 24.2   Interference in a Soap-Film

Goal   Calculate interference effects in a thin film when there is one phase reversal.

Problem   Calculate the minimum thickness of a soap-bubble film ($n = 1.33$) that will result in constructive interference in the reflected light if the film is illuminated by light with wavelength 602 nm in free space.

Strategy   There is only one inversion, so the condition for constructive interference is $2nt = (m + \frac{1}{2})\lambda$. The minimum film thickness for constructive interference corresponds to $m = 0$ in this equation.

Remark   The swirling colors in a soap bubble are due to the fact that the thickness of the soap layer varies from one place to another.

Solution
Solve $2nt = \lambda/2$ for the thickness t, and substitute:

$$t = \frac{\lambda}{4n} = \frac{602 \text{ nm}}{4(1.33)} = \boxed{113 \text{ nm}}$$

Exercise 24.2
What other film thicknesses will produce constructive interference?

Answer   339 nm, 566 nm, 792 nm, and so on

INTERACTIVE EXAMPLE 24.3   Nonreflective Coatings for Solar Cells and Optical Lenses

Goal   Calculate interference effects in a thin film when there are two inversions.

Problem   Semiconductors such as silicon are used to fabricate solar cells—devices that generate electric energy when exposed to sunlight. Solar cells are often coated with a transparent thin film, such as silicon monoxide (SiO; $n = 1.45$) to minimize reflective losses (Fig. 24.9). A silicon solar cell ($n = 3.50$) is coated with a thin film of silicon monoxide for this purpose. Assuming normal incidence, determine the minimum thickness of the film that will produce the least reflection at a wavelength of 552 nm.

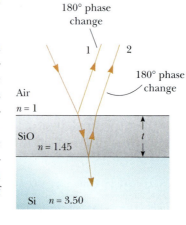

Figure 24.9   (Example 24.3) Reflective losses from a silicon solar cell are minimized by coating it with a thin film of silicon monoxide (SiO).

Strategy   Reflection is least when rays 1 and 2 in Figure 24.9 meet the condition for destructive interference. Note that *both* rays undergo 180° phase changes on reflection. The condition for a reflection *minimum* is therefore $2nt = \lambda/2$.

Solution
Solve $2nt = \lambda/2$ for t, the required thickness:

$$t = \frac{\lambda}{4n} = \frac{552 \text{ nm}}{4(1.45)} = \boxed{95.2 \text{ nm}}$$

Remarks   Typically, such coatings reduce the reflective loss from 30% (with no coating) to 10% (with a coating), thereby increasing the cell's efficiency because more light is available to create charge carriers in the cell. In reality, the coating is never perfectly nonreflecting, because the required thickness is wavelength dependent and the incident light covers a wide range of wavelengths.

Exercise 24.3
Glass lenses used in cameras and other optical instruments are usually coated with one or more transparent thin films, such as magnesium fluoride (MgF_2), to reduce or eliminate unwanted reflection. Carl Zeiss developed this method; his first coating was 1.00×10^2 nm thick. Using $n = 1.38$ for MgF_2, what visible wavelength would be eliminated by destructive interference in the reflected light?

Answer   552 nm

Physics⊗Now™   Investigate the interference for various film properties by logging into PhysicsNow at **www.cp7e.com** and going to Interactive Example 24.3.

EXAMPLE 24.4   Interference in a Wedge-Shaped Film

Goal   Calculate interference effects when the film has variable thickness.

Problem   A pair of glass slides 10.0 cm long and with $n = 1.52$ are separated on one end by a hair, forming a triangular wedge of air as illustrated in Figure 24.10. When coherent light from a helium–neon laser with wavelength 633 nm is incident on the film from above, 15.0 dark fringes per centimeter are observed. How thick is the hair?

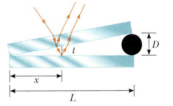

Figure 24.10   (Example 24.4) Interference bands in reflected light can be observed by illuminating a wedge-shaped film with monochromatic light. The dark areas correspond to positions of destructive interference.

Strategy   The interference pattern is created by the thin film of air having variable thickness. The pattern is a series of alternating bright and dark parallel bands. A dark band corresponds to destructive interference, and there is one phase reversal, so $2nt = m\lambda$ should be used. We can also use the similar triangles in Figure 24.10 to obtain the relation $t/x = D/L$. We can find the thickness for any m, and if the position x can also be found, this last equation gives the diameter of the hair, D.

Solution
Solve the destructive-interference equation for the thickness of the film, t, with $n = 1$ for air:

$$t = \frac{m\lambda}{2}$$

If d is the distance from one dark band to the next, then the x-coordinate of the mth band is a multiple of d:

$$x = md$$

By dimensional analysis, d is just the inverse of the number of bands per centimeter.

$$d = \left(15.0 \frac{\text{bands}}{\text{cm}}\right)^{-1} = 6.67 \times 10^{-2} \frac{\text{cm}}{\text{band}}$$

Now use similar triangles, and substitute all the information:

$$\frac{t}{x} = \frac{m\lambda/2}{md} = \frac{\lambda}{2d} = \frac{D}{L}$$

Solve for D and substitute given values:

$$D = \frac{\lambda L}{2d} = \frac{(633 \times 10^{-9} \text{ m})(0.100 \text{ m})}{2(6.67 \times 10^{-4} \text{ m})} = \boxed{4.75 \times 10^{-5} \text{ m}}$$

Remarks Some may be concerned about interference caused by light bouncing off the top and bottom of, say, the upper glass slide. It's unlikely, however, that the thickness of the slide will be half an integer multiple of the wavelength of the helium-neon laser (for some very large value of m). In addition, in contrast to the air wedge, the thickness of the glass doesn't vary.

Exercise 24.4

The air wedge is replaced with water, with $n = 1.33$. Find the distance between dark bands when the helium–neon laser light hits the glass slides.

Answer 5.01×10^{-4} m

24.5 USING INTERFERENCE TO READ CD'S AND DVD'S

APPLICATION

The Physics of CD's and DVD's

Compact disks (CD's) and digital video disks (DVD's) have revolutionized the computer and entertainment industries by providing fast access; high-density storage of text, graphics, and movies; and high-quality sound recordings. The data on these disks are stored digitally as a series of zeros and ones, and these zeros and ones are read by laser light reflected from the disk. Strong reflections (constructive interference) from the disk are chosen to represent zeros and weak reflections (destructive interference) represent ones.

To see in more detail how thin-film interference plays a crucial role in reading CD's, consider Figure 24.11. This shows a photomicrograph of several CD tracks which consist of a sequence of pits (when viewed from the top or label side of the disk) of varying length formed in a reflecting-metal information layer. A cross-sectional view of a CD as shown in Figure 24.12 reveals that the pits appear as bumps to the laser beam, which shines on the metallic layer through a clear plastic coating from below.

As the disk rotates, the laser beam reflects off the sequence of bumps and lower areas into a photodetector, which converts the fluctuating reflected light intensity into an electrical string of zeros and ones. To make the light fluctuations more pronounced and easier to detect, the pit depth t is made equal to one-quarter of a wavelength of the laser light in the plastic. When the beam hits a rising or falling bump edge, part of the beam reflects from the top of the bump and part from the lower adjacent area, ensuring destructive interference and very low intensity when the reflected beams combine at the detector. Bump edges are read as ones, and flat bump tops and intervening flat plains are read as zeros.

In Example 24.5 the pit depth for a standard CD, using an infrared laser of wavelength 780 nm, is calculated. DVDs use shorter wavelength lasers of 635 nm,

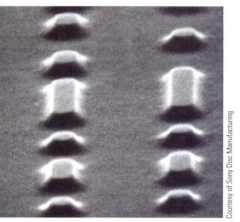

Figure 24.11 A photomicrograph of adjacent tracks on a compact disc (CD). The information encoded in these pits and smooth areas is read by a laser beam.

Courtesy of Sony Disc Manufacturing

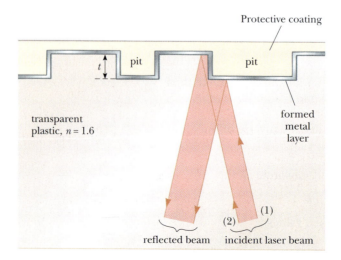

Figure 24.12 Cross section of a CD showing metallic pits of depth t and a laser beam detecting the edge of a pit.

and the track separation, pit depth, and minimum pit length are all smaller. This allows a DVD to store about 30 times more information than a CD.

EXAMPLE 24.5 Pit Depth in a CD

Goal Apply interference principles to a CD.

Problem Find the pit depth in a CD that has a plastic transparent layer with index of refraction of 1.60 and is designed for use in a CD player using a laser with a wavelength of 7.80×10^2 nm in air.

Strategy (See Fig. 24.12.) Rays (1) and (2) both reflect from the metal layer which acts like a mirror, so there is no phase difference due to reflection between those rays. There is, however, the usual phase difference caused by the extra distance $2t$ traveled by ray (2). The wavelength is λ/n, where n is the index of refraction in the substance.

Solution

Use the appropriate condition for destructive interference in a thin film:

$$2t = \frac{\lambda}{2n}$$

Solve for the thickness t and substitute:

$$t = \frac{\lambda}{4n} = \frac{7.80 \times 10^2 \text{ nm}}{(4)(1.60)} = 1.22 \times 10^2 \text{ nm}$$

Remarks Different CD systems have different tolerances for scratches. Anything that changes the reflective properties of the disk can affect the readability of the disk.

Exercise 24.5
Repeat the example for a laser with wavelength 635 nm.

Answer 99.2 nm

24.6 DIFFRACTION

Suppose a light beam is incident on two slits, as in Young's double-slit experiment. If the light truly traveled in straight-line paths after passing through the slits, as in Figure 24.13a, the waves wouldn't overlap and no interference pattern would be seen. Instead, Huygens's principle requires that the waves spread out from the slits, as shown in Figure 24.13b. In other words, the light bends from a straight-line path and enters the region that would otherwise be shadowed. This spreading out of light from its initial line of travel is called **diffraction**.

In general, diffraction occurs when waves pass through small openings, around obstacles, or by sharp edges. For example, when a single narrow slit is placed between a distant light source (or a laser beam) and a screen, the light produces a diffraction pattern like that in Figure 24.14. The pattern consists of a broad, intense central band flanked by a series of narrower, less intense secondary bands (called **secondary maxima**) and a series of dark bands, or **minima**. This phenomenon

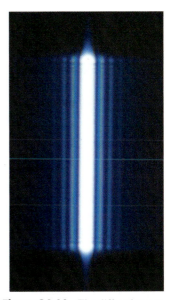

Figure 24.14 The diffraction pattern that appears on a screen when light passes through a narrow vertical slit. The pattern consists of a broad central band and a series of less intense and narrower side bands.

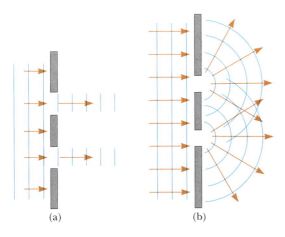

(a)   (b)

Figure 24.13 (a) If light did not spread out after passing through the slits, no interference would occur. (b) The light from the two slits overlaps as it spreads out, filling the expected shadowed regions with light and producing interference fringes.

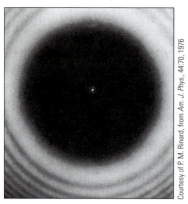

Figure 24.15 The diffraction pattern of a penny placed midway between the screen and the source.

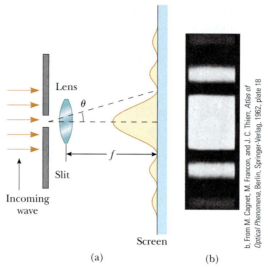

(a)   (b)

ACTIVE FIGURE 24.16

(a) The Fraunhofer diffraction pattern of a single slit. The parallel rays are brought into focus on the screen with a converging lens. The pattern consists of a central bright region flanked by much weaker maxima. (This drawing is not to scale.) (b) A photograph of a single-slit Fraunhofer diffraction pattern.

Physics⊗Now™

Log into PhysicsNow at **www.cp7e.com** and go to Active Figure 24.16, where you can adjust the slit width and the wavelength of the light, observing the effect on the diffraction pattern.

can't be explained within the framework of geometric optics, which says that light rays traveling in straight lines should cast a sharp image of the slit on the screen.

Figure 24.15 shows the diffraction pattern and shadow of a penny. The pattern consists of the shadow, a bright spot at its center, and a series of bright and dark circular bands of light near the edge of the shadow. The bright spot at the center (called the *Fresnel bright spot*) is explained by Augustin Fresnel's wave theory of light, which predicts constructive interference at this point for certain locations of the penny. From the viewpoint of geometric optics, there shouldn't be any bright spot: the center of the pattern would be completely screened by the penny.

One type of diffraction, called **Fraunhofer diffraction**, occurs when the rays leave the diffracting object in parallel directions. Fraunhofer diffraction can be achieved experimentally either by placing the observing screen far from the slit or by using a converging lens to focus the parallel rays on a nearby screen, as in Active Figure 24.16a. A bright fringe is observed along the axis at $\theta = 0$, with alternating dark and bright fringes on each side of the central bright fringe. Active Figure 24.16b is a photograph of a single-slit Fraunhofer diffraction pattern.

24.7 SINGLE-SLIT DIFFRACTION

Until now we have assumed that slits have negligible width, acting as line sources of light. In this section we determine how their nonzero widths are the basis for understanding the nature of the Fraunhofer diffraction pattern produced by a single slit.

We can deduce some important features of this problem by examining waves coming from various portions of the slit, as shown in Figure 24.17. According to Huygens' principle, **each portion of the slit acts as a source of waves. Hence, light from one portion of the slit can interfere with light from another portion**, and the resultant intensity on the screen depends on the direction θ.

To analyze the diffraction pattern, it's convenient to divide the slit into halves, as in Figure 24.17. All the waves that originate at the slit are in phase. Consider waves 1 and 3, which originate at the bottom and center of the slit, respectively. Wave 1 travels farther than wave 3 by an amount equal to the path difference

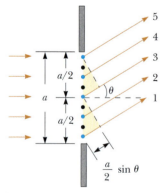

Figure 24.17 Diffraction of light by a narrow slit of width a. Each portion of the slit acts as a point source of waves. The path difference between rays 1 and 3 or between rays 2 and 4 is equal to $(a/2)\sin\theta$. (This drawing is not to scale, and the rays are assumed to converge at a distant point.)

$(a/2) \sin \theta$, where a is the width of the slit. Similarly, the path difference between waves 3 and 5 is $(a/2) \sin \theta$. If this path difference is exactly half of a wavelength (corresponding to a phase difference of 180°), the two waves cancel each other and destructive interference results. This is true, in fact, for any two waves that originate at points separated by half the slit width, because the phase difference between two such points is 180°. Therefore, waves from the upper half of the slit interfere *destructively* with waves from the lower half of the slit when

$$\frac{a}{2} \sin \theta = \frac{\lambda}{2}$$

or when

$$\sin \theta = \frac{\lambda}{a}$$

If we divide the slit into four parts rather than two and use similar reasoning, we find that the screen is also dark when

$$\sin \theta = \frac{2\lambda}{a}$$

Continuing in this way, we can divide the slit into six parts and show that darkness occurs on the screen when

$$\sin \theta = \frac{3\lambda}{a}$$

Therefore, the general condition for **destructive interference** for a single slit of width a is

$$\sin \theta_{dark} = m \frac{\lambda}{a} \quad m = \pm 1, \pm 2, \pm 3, \ldots \quad [24.11]$$

◀ Condition for destructive interference (single slit)

TIP 24.3 The Same, But Different

Although Equations 24.2 and 24.11 have the same form, they have different meanings. Equation 24.2 describes the *bright* regions in a two-slit interference pattern, while Equation 24.11 describes the *dark* regions in a single-slit interference pattern.

Equation 24.11 gives the values of θ for which the diffraction pattern has zero intensity, where a dark fringe forms. However, the equation tells us nothing about the variation in intensity along the screen. The general features of the intensity distribution along the screen are shown in Figure 24.18. A broad central bright fringe is flanked by much weaker bright fringes alternating with dark fringes. The various dark fringes (points of zero intensity) occur at the values of θ that satisfy Equation 24.11. The points of constructive interference lie approximately halfway between the dark fringes. Note that the central bright fringe is twice as wide as the weaker maxima having $m > 1$.

Quick Quiz 24.3

In a single-slit diffraction experiment, as the width of the slit is made smaller, the width of the central maximum of the diffraction pattern (a) becomes smaller, (b) becomes larger, or (c) remains the same.

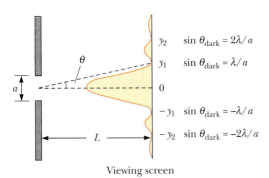

Figure 24.18 Positions of the minima for the Fraunhofer diffraction pattern of a single slit of width a. (This drawing is not to scale.)

Applying Physics 24.3   Diffraction of Sound Waves

If a classroom door is open even just a small amount, you can hear sounds coming from the hallway. Yet you can't see what is going on in the hallway. How can this difference be explained?

Explanation The space between the slightly open door and the wall is acting as a single slit for waves.

Sound waves have wavelengths larger than the width of the slit, so sound is effectively diffracted by the opening, and the central maximum spreads throughout the room. Light wavelengths are much smaller than the slit width, so there is virtually no diffraction for the light. You must have a direct line of sight to detect the light waves.

INTERACTIVE EXAMPLE 24.6   A Single-Slit Experiment

Goal Find the positions of the dark fringes in single-slit diffraction.

Problem Light of wavelength 5.80×10^2 nm is incident on a slit of width 0.300 mm. The observing screen is placed 2.00 m from the slit. Find the positions of the first dark fringes and the width of the central bright fringe.

Strategy This problem requires substitution into Equation 24.11 to find the sines of the angles of the first dark fringes. The positions can then be found with the tangent function, since for small angles $\sin \theta \approx \tan \theta$. The extent of the central maximum is defined by these two dark fringes.

Solution

The first dark fringes that flank the central bright fringe correspond to $m = \pm 1$ in Equation 24.11:

$$\sin \theta = \pm \frac{\lambda}{a} = \pm \frac{5.80 \times 10^{-7}\,\text{m}}{0.300 \times 10^{-3}\,\text{m}} = \pm 1.93 \times 10^{-3}$$

Use the triangle in Figure 24.18 to relate the position of the fringe to the tangent function:

$$\tan \theta = \frac{y_1}{L}$$

Because θ is very small, we can use the approximation $\sin \theta \approx \tan \theta$ and then solve for y_1:

$$\sin \theta \approx \tan \theta \approx \frac{y_1}{L}$$

$$y_1 \approx L \sin \theta = \pm L\,\frac{\lambda}{a} = \boxed{\pm 3.86 \times 10^{-3}\,\text{m}}$$

Compute the distance between the positive and negative first-order maxima, which is the width w of the central maximum:

$$w = +3.86 \times 10^{-3}\,\text{m} - (-3.86 \times 10^{-3}\,\text{m}) = \boxed{7.72 \times 10^{-3}\,\text{m}}$$

Remarks Note that this value of w is much greater than the width of the slit. However, as the width of the slit is *increased*, the diffraction pattern *narrows*, corresponding to smaller values of θ. In fact, for large values of a, the maxima and minima are so closely spaced that the only observable pattern is a large central bright area resembling the geometric image of the slit. Because the width of the geometric image increases as the slit width increases, the narrowest image occurs when the geometric and diffraction widths are equal.

Exercise 24.6

Determine the width of the first-order bright fringe.

Answer 3.86 mm

Physics⊗Now™ Investigate the single-slit diffraction pattern by logging into PhysicsNow at **www.cp7e.com** and going to Interactive Example 24.6.

24.8 THE DIFFRACTION GRATING

The diffraction grating, a useful device for analyzing light sources, consists of a large number of equally spaced parallel slits. A grating can be made by scratching parallel lines on a glass plate with a precision machining technique. The clear panes between

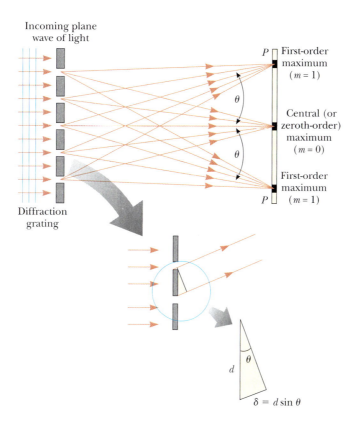

Incoming plane
wave of light

Diffraction
grating

P First-order
maximum
($m = 1$)

Central (or
zeroth-order)
maximum
($m = 0$)

First-order
maximum
P ($m = 1$)

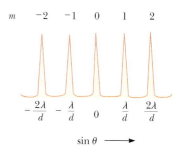

$\delta = d \sin \theta$

Figure 24.19 A side view of a diffraction grating. The slit separation is d, and the path difference between adjacent slits is $d \sin \theta$.

m -2 -1 0 1 2

$-\dfrac{2\lambda}{d}$ $-\dfrac{\lambda}{d}$ 0 $\dfrac{\lambda}{d}$ $\dfrac{2\lambda}{d}$

$\sin \theta \longrightarrow$

ACTIVE FIGURE 24.20

Intensity versus $\sin \theta$ for the diffraction grating. The zeroth-, first-, and second-order principal maxima are shown.

Physics⊗Now™

Log into PhysicsNow at **www.cp7e.com** and go to Active Figure 24.20, where you can choose the number of slits to be illuminated and observe the effect on the diffraction pattern.

scratches act like slits. A typical grating contains several thousand lines per centimeter. For example, a grating ruled with 5 000 lines/cm has a slit spacing d equal to the reciprocal of that number; hence, $d = (1/5\ 000)$ cm $= 2 \times 10^{-4}$ cm.

Figure 24.19 is a schematic diagram of a section of a plane diffraction grating. A plane wave is incident from the left, normal to the plane of the grating. The intensity of the pattern on the screen is the result of the combined effects of interference and diffraction. Each slit causes diffraction, and the diffracted beams in turn interfere with one another to produce the pattern. Moreover, each slit acts as a source of waves, and all waves start in phase at the slits. For some arbitrary direction θ measured from the horizontal, however, the waves must travel *different* path lengths before reaching a particular point P on the screen. From Figure 24.19, note that the path difference between waves from any two adjacent slits is $d \sin \theta$. If this path difference equals one wavelength or some integral multiple of a wavelength, waves from all slits will be in phase at P and a bright line will be observed at that point. Therefore, the condition for **maxima** in the interference pattern at the angle θ is

$$d \sin \theta_{\text{bright}} = m\lambda \qquad m = 0, 1, 2, \ldots \qquad \text{[24.12]}$$

◀ Condition for maxima in the interference pattern of a diffraction grating

Light emerging from a slit at an angle other than that for a maximum interferes nearly completely destructively with light from some other slit on the grating. All such pairs will result in little or no transmission in that direction, as illustrated in Active Figure 24.20.

Equation 24.12 can be used to calculate the wavelength from the grating spacing and the angle of deviation, θ. The integer m is the **order number** of the diffraction pattern. If the incident radiation contains several wavelengths, each wavelength deviates through a specific angle, which can be found from Equation 24.12. All wavelengths are focused at $\theta = 0$, corresponding to $m = 0$. This is called the *zeroth-order maximum*. The *first-order maximum*, corresponding to $m = 1$, is observed at an angle that satisfies the relationship $\sin \theta = \lambda/d$; the *second-order maximum*, corresponding to $m = 2$, is observed at a larger angle θ, and so on. Active Figure 24.20 is a sketch of the intensity distribution for some of the orders

ACTIVE FIGURE 24.21
A diagram of a diffraction grating spectrometer. The collimated beam incident on the grating is diffracted into the various orders at the angles θ that satisfy the equation $d \sin \theta = m\lambda$, where $m = 0, 1, 2, \ldots$

Physics⊗Now™

Log into PhysicsNow at **www.cp7e.com** and go to Active Figure 24.21, where you can use the spectrometer and understand how spectra are measured.

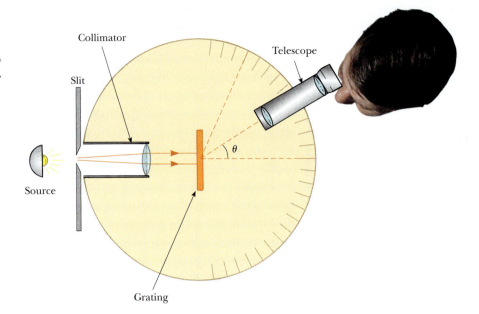

produced by a diffraction grating. Note the sharpness of the principal maxima and the broad range of the dark areas, a pattern in direct contrast to the broad bright fringes characteristic of the two-slit interference pattern.

A simple arrangement that can be used to measure the angles in a diffraction pattern is shown in Active Figure 24.21. This is a form of diffraction-grating spectrometer. The light to be analyzed passes through a slit and is formed into a parallel beam by a lens. The light then strikes the grating at a 90° angle. The diffracted light leaves the grating at angles that satisfy Equation 24.12. A telescope is used to view the image of the slit. The wavelength can be determined by measuring the angles at which the images of the slit appear for the various orders.

Quick Quiz 24.4

If laser light is reflected from a phonograph record or a compact disc, a diffraction pattern appears. The pattern arises because both devices contain parallel tracks of information that act as a reflection diffraction grating. Which device, record or compact disc, results in diffraction maxima that are farther apart?

Applying Physics 24.4   Prism vs. Grating

When white light enters through an opening in an opaque box and exits through an opening on the other side of the box, a spectrum of colors appears on the wall. From this observation, how would you be able to determine whether the box contains a prism or a diffraction grating?

Explanation The determination could be made by noticing the order of the colors in the spectrum relative to the direction of the original beam of white light. For a prism, in which the separation of light is a result of dispersion, the violet light will be refracted more than the red light. Hence, the order of the spectrum from a prism will be from red, closest to the original direction, to violet. For a diffraction grating, the angle of diffraction increases with wavelength, so the spectrum from the diffraction grating will have colors in the order from violet, closest to the original direction, to red. Furthermore, the diffraction grating will produce *two* first-order spectra on either side of the grating, while the prism will produce only a single spectrum.

Applying Physics 24.5   Rainbows from a Compact Disc

White light reflected from the surface of a compact disc has a multicolored appearance, as shown in Figure 24.22. The observation depends on the orientation of the disc relative to the eye and the

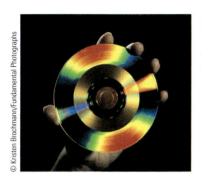

Figure 24.22   (Applying Physics 24.5) Compact discs act as diffraction gratings when observed under white light.

© Kristen Brochmann/Fundamental Photographs

position of the light source. Explain how all this works.

Explanation   The surface of a compact disc has a spiral-shaped track (with a spacing of approximately 1 μm) that acts as a reflection grating. The light scattered by these closely spaced parallel tracks interferes constructively in certain directions that depend on both the wavelength and the direction of the incident light. Any one section of the disc serves as a diffraction grating for white light, sending beams of constructive interference for different colors in different directions. The different colors you see when viewing one section of the disc change as the light source, the disc, or you move to change the angles of incidence or diffraction.

Use of a Diffraction Grating in CD Tracking

APPLICATION

Tracking Information on a CD

If a CD player is to reproduce sound faithfully, the laser beam must follow the spiral track of information perfectly. Sometimes the laser beam can drift off track, however, and without a feedback procedure to let the player know this is happening, the fidelity of the music can be greatly reduced.

Figure 24.23 shows how a diffraction grating is used in a three-beam method to keep the beam on track. The central maximum of the diffraction pattern reads the information on the CD track, and the two first-order maxima steer the beam. The grating is designed so that the first-order maxima fall on the smooth surfaces on either side of the information track. Both of these reflected beams have their own detectors, and because both beams are reflected from smooth surfaces, they should have the same strong intensity when they are detected. If the central beam wanders off the track, however, one of the steering beams will begin to strike bumps on the information track and the amount of light reflected will decrease. This information is then used by electronic circuits to drive the main beam back to its desired location.

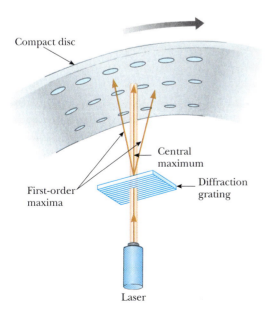

Figure 24.23   The laser beam in a CD player is able to follow the spiral track by using three beams produced with a diffraction grating.

INTERACTIVE EXAMPLE 24.7   A Diffraction Grating

Goal   Calculate different-order principal maxima for a diffraction grating.

Problem   Monochromatic light from a helium–neon laser ($\lambda = 632.8$ nm) is incident normally on a diffraction grating containing 6.00×10^3 lines/cm. Find the angles at which one would observe the first-order maximum, the second-order maximum, and so forth.

Strategy   Find the slit separation by inverting the number of lines per centimeter, then substitute values into Equation 24.12.

Solution

Invert the number of lines per centimeter to obtain the slit separation:

$$d = \frac{1}{6.00 \times 10^3\, \text{cm}^{-1}} = 1.67 \times 10^{-4}\, \text{cm} = 1.67 \times 10^3\, \text{nm}$$

Substitute $m = 1$ into Equation 24.12 to find the sine of the angle corresponding to the first-order maximum:

$$\sin \theta_1 = \frac{\lambda}{d} = \frac{632.8\, \text{nm}}{1.67 \times 10^3\, \text{nm}} = 0.379$$

Take the inverse sine of the preceding result to find θ_1:

$$\theta_1 = \sin^{-1} 0.379 = \boxed{22.3°}$$

Repeat the calculation for $m = 2$:

$$\sin \theta_2 = \frac{2\lambda}{d} = \frac{2(632.8\, \text{nm})}{1.67 \times 10^3\, \text{nm}} = 0.758$$

$$\theta_2 = \boxed{49.3°}$$

Repeat the calculation for $m = 3$:

$$\sin \theta_3 = \frac{3\lambda}{d} = \frac{3(632.8\, \text{nm})}{1.67 \times 10^3\, \text{nm}} = 1.14$$

Because $\sin \theta$ can't exceed one, there is no solution for θ_3.

Remarks   The foregoing calculation shows that there can only be a finite number of principal maxima. In this case, only zeroth-, first-, and second-order maxima would be observed.

Exercise 24.7

Suppose light with wavelength 7.80×10^2 nm is used instead and the diffraction grating has 3.30×10^3 lines per centimeter. Find the angles of all the principal maxima.

Answers $0°, 14.9°, 31.0°, 50.6°$

Physics⊗Now™ Investigate the diffraction pattern from a diffraction grating by logging into PhysicsNow at **www.cp7e.com** and going to Interactive Example 24.7.

24.9   POLARIZATION OF LIGHT WAVES

In Chapter 21, we described the transverse nature of electromagnetic waves. Figure 24.24 shows that the electric and magnetic field vectors associated with an electromagnetic wave are at right angles to each other and also to the direction of wave propagation. The phenomenon of polarization, described in this section, is firm evidence of the transverse nature of electromagnetic waves.

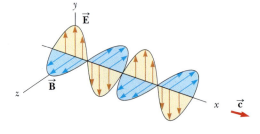

Figure 24.24   A schematic diagram of a polarized electromagnetic wave propagating in the x-direction. The electric field vector $\vec{\mathbf{E}}$ vibrates in the xy-plane, while the magnetic field vector $\vec{\mathbf{B}}$ vibrates in the xz-plane.

An ordinary beam of light consists of a large number of electromagnetic waves emitted by the atoms or molecules of the light source. The vibrating charges associated with the atoms act as tiny antennas. Each atom produces a wave with its own orientation of $\vec{E}$, as in Figure 24.24, corresponding to the direction of atomic vibration. However, because all directions of vibration are possible, the resultant electromagnetic wave is a superposition of waves produced by the individual atomic sources. The result is an **unpolarized** light wave, represented schematically in Figure 24.25a. The direction of wave propagation shown in the figure is perpendicular to the page. Note that *all* directions of the electric field vector are equally probable and lie in a plane (such as the plane of this page) perpendicular to the direction of propagation.

A wave is said to be **linearly polarized** if the resultant electric field $\vec{E}$ vibrates in the same direction *at all times* at a particular point, as in Figure 24.25b. (Sometimes such a wave is described as *plane polarized* or simply *polarized*.) The wave in Figure 24.24 is an example of a wave that is linearly polarized in the *y*-direction. As the wave propagates in the *x*-direction, $\vec{E}$ is always in the *y*-direction. The plane formed by $\vec{E}$ and the direction of propagation is called the *plane of polarization* of the wave. In Figure 24.24, the plane of polarization is the *xy*-plane.

It's possible to obtain a linearly polarized beam from an unpolarized beam by removing all waves from the beam except those with electric field vectors that oscillate in a single plane. We now discuss three processes for doing this: (1) selective absorption, (2) reflection, and (3) scattering.

Polarization by Selective Absorption

The most common technique for polarizing light is to use a material that transmits waves having electric field vectors that vibrate in a plane parallel to a certain direction and absorbs those waves with electric field vectors vibrating in directions perpendicular to that direction.

In 1932, E. H. Land discovered a material, which he called **Polaroid**, that polarizes light through selective absorption by oriented molecules. This material is fabricated in thin sheets of long-chain hydrocarbons, which are stretched during manufacture so that the molecules align. After a sheet is dipped into a solution containing iodine, the molecules become good electrical conductors. However, conduction takes place primarily along the hydrocarbon chains, because the valence electrons of the molecules can move easily only along those chains. (Recall that valence electrons are "free" electrons that can move easily through the conductor.) As a result, the molecules readily *absorb* light having an electric field vector parallel to their lengths and *transmit* light with an electric field vector perpendicular to their lengths. It's common to refer to the direction perpendicular to the molecular chains as the **transmission axis**. In an ideal polarizer, all light with $\vec{E}$ parallel to the transmission axis is transmitted and all light with $\vec{E}$ perpendicular to the transmission axis is absorbed.

Polarizing material reduces the intensity of light passing through it. In Active Figure 24.26, an unpolarized light beam is incident on the first polarizing sheet, called the **polarizer**; the transmission axis is as indicated. The light that passes

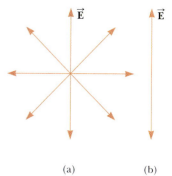

Figure 24.25 (a) An unpolarized light beam viewed along the direction of propagation (perpendicular to the page). The transverse electric field vector can vibrate in any direction with equal probability. (b) A linearly polarized light beam with the electric field vector vibrating in the vertical direction.

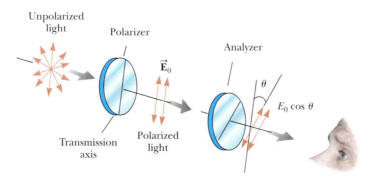

ACTIVE FIGURE 24.26
Two polarizing sheets whose transmission axes make an angle θ with each other. Only a fraction of the polarized light incident on the analyzer is transmitted.

Physics Now™
Log into PhysicsNow at **www.cp7e.com** and go to Active Figure 24.26, where you can rotate the second polarizer and see the effect on the transmitted light.

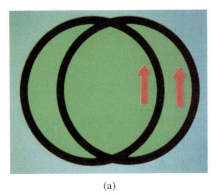
(a)

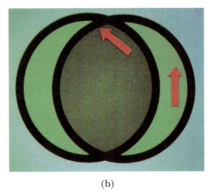
(b)

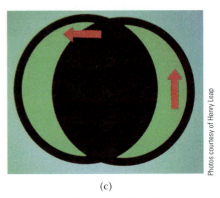
(c)

Photos courtesy of Henry Leap

Figure 24.27   The intensity of light transmitted through two polarizers depends on the relative orientations of their transmission axes. (a) The transmitted light has *maximum* intensity when the transmission axes are *aligned* with each other. (b) The transmitted light intensity diminishes when the transmission axes are at an angle of 45° with each other. (c) The transmitted light intensity is a *minimum* when the transmission axes are at *right angles* to each other.

through this sheet is polarized vertically, and the transmitted electric field vector is $\vec{E}_0$. A second polarizing sheet, called the **analyzer**, intercepts this beam with its transmission axis at an angle of θ to the axis of the polarizer. The component of $\vec{E}_0$ that is perpendicular to the axis of the analyzer is completely absorbed. The component of $\vec{E}_0$ that is parallel to the analyzer axis, $E_0 \cos \theta$, is allowed to pass through the analyzer. Because the intensity of the transmitted beam varies as the *square* of its amplitude E, we conclude that the intensity of the (polarized) beam transmitted through the analyzer varies as

Malus's law ▶

$$I = I_0 \cos^2 \theta \qquad [24.13]$$

where I_0 is the intensity of the polarized wave incident on the analyzer. This expression, known as **Malus's law**, applies to any two polarizing materials having transmission axes at an angle of θ to each other. Note from Equation 24.13 that the transmitted intensity is a maximum when the transmission axes are parallel ($\theta = 0$ or 180°) and is zero (complete absorption by the analyzer) when the transmission axes are perpendicular to each other. This variation in transmitted intensity through a pair of polarizing sheets is illustrated in Figure 24.27.

When unpolarized light of intensity I_0 is sent through a single ideal polarizer, the transmitted linearly polarized light has intensity $I_0/2$. This fact follows from Malus's law, because the average value of $\cos^2 \theta$ is one-half.

Applying Physics 24.6   Polarizing Microwaves

A polarizer for microwaves can be made as a grid of parallel metal wires about a centimeter apart. Is the electric field vector for microwaves transmitted through this polarizer parallel or perpendicular to the metal wires?

Explanation   Electric field vectors parallel to the metal wires cause electrons in the metal to oscillate parallel to the wires. Thus, the energy from the waves with these electric field vectors is transferred to the metal by accelerating the electrons and is eventually transformed to internal energy through the resistance of the metal. Waves with electric field vectors perpendicular to the metal wires are not able to accelerate electrons and pass through the wires. Consequently, the electric field polarization is perpendicular to the metal wires.

EXAMPLE 24.8   Polarizer

Goal   Understand how polarizing materials affect light intensity.

Problem   Unpolarized light is incident upon three polarizers. The first polarizer has a vertical transmission axis, the second has a transmission axis rotated 30.0° with respect to the first, and the third has a transmission axis rotated 75.0° relative to the first. If the initial light intensity of the beam is I_b, calculate the light intensity after the beam passes through (a) the second polarizer and (b) the third polarizer.

Strategy After the beam passes through the first polarizer, it is polarized and its intensity is cut in half. Malus's law can then be applied to the second and third polarizers. The angle used in Malus's law must be relative to the immediately preceding transmission axis.

Solution
(a) Calculate the intensity of the beam after it passes through the second polarizer:

The incident intensity is $I_b/2$. Apply Malus's law to the second polarizer:

$$I_2 = I_0 \cos^2 \theta = \frac{I_b}{2} \cos^2 (30.0^\circ) = \frac{I_b}{2}\left(\frac{\sqrt{3}}{2}\right)^2 = \frac{3}{8}I_b$$

(b) Calculate the intensity of the beam after it passes through the third polarizer.

The incident intensity is now $3I_b/8$. Apply Malus's law to the third polarizer:

$$I_3 = I_2 \cos^2 \theta = \frac{3}{8}I_b \cos^2 (45.0^\circ) = \frac{3}{8}I_b \left(\frac{\sqrt{2}}{2}\right)^2 = \frac{3}{16}I_b$$

Remarks Notice that the angle used in part (b) was not 75.0°, but $75.0^\circ - 30.0^\circ = 45.0^\circ$. The angle is always with respect to the previous polarizer's transmission axis, because the polarizing material physically determines what direction the transmitted electric fields can have.

Exercise 24.8
The polarizers are rotated, so that the second polarizer has a transmission axis of 40.0° with respect to the first polarizer and the third polarizer has an angle of 90.0° with respect to the first. If I_b is the intensity of the original unpolarized light, what is the intensity of the beam after it passes through (a) the second polarizer, and (b) the third polarizer? (c) What is the final transmitted intensity if the second polarizer is removed?

Answers (a) $0.293I_b$ (b) $0.121I_b$ (c) 0

Polarization by Reflection
When an unpolarized light beam is reflected from a surface, the reflected light is completely polarized, partially polarized, or unpolarized, depending on the angle of incidence. If the angle of incidence is either 0° or 90° (a normal or grazing angle), the reflected beam is unpolarized. For angles of incidence between 0° and 90°, however, the reflected light is polarized to some extent. For one particular angle of incidence the reflected beam is completely polarized.

Suppose an unpolarized light beam is incident on a surface, as in Figure 24.28a (page 808). The beam can be described by two electric field components, one parallel to the surface (represented by dots) and the other perpendicular to the first component and to the direction of propagation (represented by brown arrows). It is found that the parallel component reflects more strongly than the other components, and this results in a partially polarized beam. In addition, the refracted beam is also partially polarized.

Now suppose that the angle of incidence, θ_1, is varied until the angle between the reflected and refracted beams is 90° (Fig. 24.28b). At this particular angle of incidence, called the **polarizing angle** θ_p, the reflected beam is completely polarized, with its electric field vector parallel to the surface, while the refracted beam is partially polarized.

An expression relating the polarizing angle to the index of refraction of the reflecting surface can be obtained by the use of Figure 24.28b. From this figure we see that at the polarizing angle, $\theta_p + 90^\circ + \theta_2 = 180^\circ$, so that $\theta_2 = 90^\circ - \theta_p$. Using Snell's law and taking $n_1 = n_{air} = 1.00$ and $n_2 = n$ yields

$$n = \frac{\sin \theta_1}{\sin \theta_2} = \frac{\sin \theta_p}{\sin \theta_2}$$

Because $\sin \theta_2 = \sin(90^\circ - \theta_p) = \cos \theta_p$, the expression for n can be written

Figure 24.28 (a) When unpolarized light is incident on a reflecting surface, the reflected and refracted beams are partially polarized. (b) The reflected beam is completely polarized when the angle of incidence equals the polarizing angle θ_p, satisfying the equation $n = \tan \theta_p$.

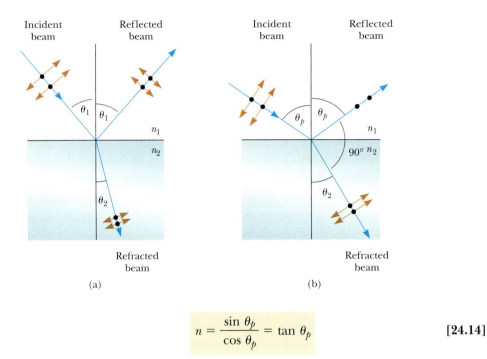

(a)   (b)

Brewster's law ▶

$$n = \frac{\sin \theta_p}{\cos \theta_p} = \tan \theta_p \qquad [24.14]$$

Equation 24.14 is called **Brewster's law**, and the polarizing angle θ_p is sometimes called **Brewster's angle** after its discoverer, Sir David Brewster (1781–1868). For example, Brewster's angle for crown glass (where $n = 1.52$) has the value $\theta_p = \tan^{-1}(1.52) = 56.7°$. Because n varies with wavelength for a given substance, Brewster's angle is also a function of wavelength.

Polarization by reflection is a common phenomenon. Sunlight reflected from water, glass, or snow is partially polarized. If the surface is horizontal, the electric field vector of the reflected light has a strong horizontal component. Sunglasses made of polarizing material reduce the glare, which *is* the reflected light. The transmission axes of the lenses are oriented vertically to absorb the strong horizontal component of the reflected light. Because the reflected light is mostly polarized, most of the glare can be eliminated without removing most of the normal light.

APPLICATION

Polaroid Sunglasses

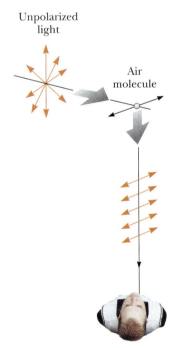

Figure 24.29 The scattering of unpolarized sunlight by air molecules. The light observed at right angles is linearly polarized because the vibrating molecule has a horizontal component of vibration.

Polarization by Scattering

When light is incident on a system of particles, such as a gas, the electrons in the medium can absorb and reradiate part of the light. The absorption and reradiation of light by the medium, called **scattering**, is what causes sunlight reaching an observer on Earth from straight overhead to be polarized. You can observe this effect by looking directly up through a pair of sunglasses made of polarizing glass. Less light passes through at certain orientations of the lenses than at others.

Figure 24.29 illustrates how the sunlight becomes polarized. The left side of the figure shows an incident unpolarized beam of sunlight on the verge of striking an air molecule. When the beam strikes the air molecule, it sets the electrons of the molecule into vibration. These vibrating charges act like those in an antenna except that they vibrate in a complicated pattern. The horizontal part of the electric field vector in the incident wave causes the charges to vibrate horizontally, and the vertical part of the vector simultaneously causes them to vibrate vertically. A horizontally polarized wave is emitted by the electrons as a result of their horizontal motion, and a vertically polarized wave is emitted parallel to the Earth as a result of their vertical motion.

Scientists have found that bees and homing pigeons use the polarization of sunlight as a navigational aid.

Optical Activity

Many important practical applications of polarized light involve the use of certain materials that display the property of **optical activity**. A substance is said to be optically active if it rotates the plane of polarization of transmitted light. Suppose

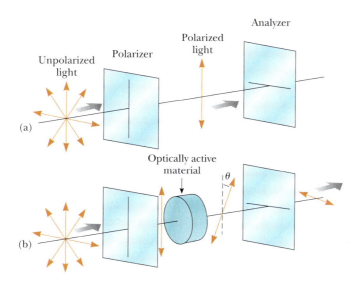

Figure 24.30 (a) When crossed polarizers are used, none of the polarized light can pass through the analyzer. (b) An optically active material rotates the direction of polarization through the angle θ, enabling some of the polarized light to pass through the analyzer.

unpolarized light is incident on a polarizer from the left, as in Figure 24.30a. The transmitted light is polarized vertically, as shown. If this light is then incident on an analyzer with its axis perpendicular to that of the polarizer, no light emerges from it. If an optically active material is placed between the polarizer and analyzer, as in Figure 24.30b, the material causes the direction of the polarized beam to rotate through the angle θ. As a result, some light is able to pass through the analyzer. The angle through which the light is rotated by the material can be found by rotating the polarizer until the light is again extinguished. It is found that the angle of rotation depends on the length of the sample and, if the substance is in solution, on the concentration. One optically active material is a solution of common sugar, dextrose. A standard method for determining the concentration of a sugar solution is to measure the rotation produced by a fixed length of the solution.

Optical activity occurs in a material because of an asymmetry in the shape of its constituent molecules. For example, some proteins are optically active because of their spiral shapes. Other materials, such as glass and plastic, become optically active when placed under stress. If polarized light is passed through an unstressed piece of plastic and then through an analyzer with an axis perpendicular to that of the polarizer, none of the polarized light is transmitted. If the plastic is placed under stress, however, the regions of greatest stress produce the largest angles of rotation of polarized light, and a series of light and dark bands are observed in the transmitted light. Engineers often use this property in the design of structures ranging from bridges to small tools. A plastic model is built and analyzed under different load conditions to determine positions of potential weakness and failure under stress. If the design is poor, patterns of light and dark bands will indicate the points of greatest weakness, and the design can be corrected at an early stage. Figure 24.31 shows examples of stress patterns in plastic.

APPLICATION

Finding the Concentrations of Solutions by Means of Their Optical Activity

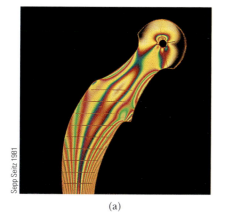

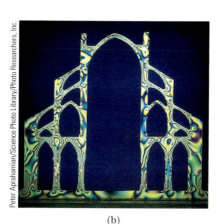

Figure 24.31 (a) Strain distribution in a plastic model of a replacement hip used in a medical research laboratory. The pattern is produced when the model is placed between a polarizer and an analyzer oriented perpendicular to each other. (b) A plastic model of an arch structure under load conditions observed between perpendicular polarizers. Such patterns are useful in the optimum design of architectural components.

APPLICATION

Liquid Crystal Displays (LCD's)

Liquid Crystals

An effect similar to rotation of the plane of polarization is used to create the familiar displays on pocket calculators, wristwatches, notebook computers, and so forth. The properties of a unique substance called a liquid crystal make these displays (called LCD's, for *liquid crystal displays*) possible. As its name implies, a **liquid crystal** is a substance with properties intermediate between those of a crystalline solid and those of a liquid; that is, the molecules of the substance are more orderly than those in a liquid, but less orderly than those in a pure crystalline solid. The forces that hold the molecules together in such a state are just barely strong enough to enable the substance to maintain a definite shape, so it is reasonable to call it a solid. However, small inputs of mechanical or electrical energy can disrupt these weak bonds and make the substance flow, rotate, or twist.

To see how liquid crystals can be used to create a display, consider Figure 24.32a. The liquid crystal is placed between two glass plates in the pattern shown, and electrical contacts, indicated by the thin lines, are made. When a voltage is applied across any segment in the display, that segment turns dark. In this fashion, any number between 0 and 9 can be formed by the pattern, depending on the voltages applied to the seven segments.

To see why a segment can be changed from dark to light by the application of a voltage, consider Figure 24.32b, which shows the basic construction of a portion of

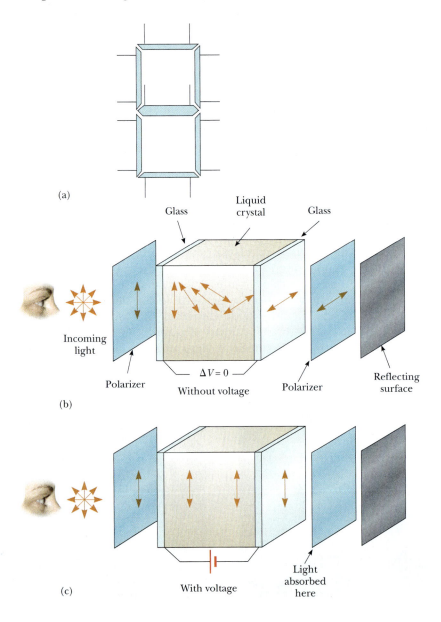

Figure 24.32 (a) The light-segment pattern of a liquid crystal display. (b) Rotation of a polarized light beam by a liquid crystal when the applied voltage is zero. (c) Molecules of the liquid crystal align with the electric field when a voltage is applied.

the display. The liquid crystal is placed between two glass substrates that are packaged between two pieces of Polaroid material with their transmission axes perpendicular. A reflecting surface is placed behind one of the pieces of Polaroid. First consider what happens when light falls on this package and no voltages are applied to the liquid crystal, as shown in Figure 24.32b. Incoming light is polarized by the polarizer on the left and then falls on the liquid crystal. As the light passes through the crystal, its plane of polarization is rotated by 90°, allowing it to pass through the polarizer on the right. It reflects from the reflecting surface and retraces its path through the crystal. Thus, an observer to the left of the crystal sees the segment as being bright. When a voltage is applied as in Figure 24.32c, the molecules of the liquid crystal don't rotate the plane of polarization of the light. In this case, the light is absorbed by the polarizer on the right and none is reflected back to the observer to the left of the crystal. As a result, the observer sees this segment as black. Changing the applied voltage to the crystal in a precise pattern at precise times can make the pattern tick off the seconds on a watch, display a letter on a computer display, and so forth.

SUMMARY

Physics⊗Now™ Take a practice test by logging into Physics-Now at **www.cp7e.com** and clicking on the Pre-Test link for this chapter.

24.1 Conditions for Interference

Interference occurs when two or more light waves overlap at a given point. A sustained interference pattern is observed if (1) the sources are coherent (that is, they maintain a constant phase relationship with one another), (2) the sources have identical wavelengths, and (3) the superposition principle is applicable.

24.2 Young's Double-Slit Experiment

In **Young's double-slit experiment**, two slits separated by distance d are illuminated by a single-wavelength light source. An interference pattern consisting of bright and dark fringes is observed on a screen a distance L from the slits. The condition for **bright fringes** (constructive interference) is

$$d \sin \theta_{bright} = m\lambda \qquad m = 0, \pm 1, \pm 2, \ldots \qquad [24.2]$$

The number m is called the **order number** of the fringe. The condition for **dark fringes** (destructive interference) is

$$d \sin \theta_{dark} = (m + \tfrac{1}{2})\lambda \qquad m = 0, \pm 1, \pm 2, \ldots \qquad [24.3]$$

The position y_m of the bright fringes on the screen can be determined by using the relation $\sin \theta \approx \tan \theta = y_m/L$, which is true for small angles. This can be substituted into Equations 24.2 and 24.3, yielding the location of the bright fringes:

$$y_{bright} = \frac{\lambda L}{d} m \qquad m = 0, \pm 1, \pm 2, \ldots \qquad [24.5]$$

A similar expression can be derived for the dark fringes. This equation can be used either to locate the maxima or to determine the wavelength of light by measuring y_m.

24.3 Change of Phase Due to Reflection &
24.4 Interference in Thin Films

An electromagnetic wave undergoes a phase change of 180° on reflection from a medium with an index of refrac-

tion higher than that of the medium in which the wave is traveling. There is no change when the wave, traveling in a medium with higher index of refraction, reflects from a medium with a lower index of refraction.

The wavelength λ_n of light in a medium with index of refraction n is

$$\lambda_n = \frac{\lambda}{n} \qquad [24.7]$$

where λ is the wavelength of the light in free space. Light encountering a thin film of thickness t will reflect off the top and bottom of the film, each ray undergoing a possible phase change as described above. The two rays recombine, and bright and dark fringes will be observed, with the conditions of interference given by the following table:

Equation ($m = 0, 1, \ldots$)	1 phase reversal	0 or 2 phase reversals
$2nt = (m + \tfrac{1}{2})\lambda$ [24.9]	constructive	destructive
$2nt = m\lambda$ [24.10]	destructive	constructive

24.6 Diffraction &
24.7 Single-Slit Diffraction

Diffraction occurs when waves pass through small openings, around obstacles, or by sharp edges. The **diffraction pattern** produced by a single slit on a distant screen consists of a central bright maximum flanked by less bright fringes alternating with dark regions. The angles θ at which the diffraction pattern has zero intensity (regions of destructive interference) are described by

$$\sin \theta_{dark} = m \frac{\lambda}{a} \qquad m = \pm 1, \pm 2, \pm 3, \ldots \qquad [24.11]$$

where a is the width of the slit and λ is the wavelength of the light incident on the slit.

24.8 The Diffraction Grating

A **diffraction grating** consists of many equally spaced, identical slits. The condition for **maximum intensity** in the interference pattern of a diffraction grating is

$$d \sin \theta_{\text{bright}} = m\lambda \qquad m = 0, 1, 2, \ldots \quad \text{[24.12]}$$

where d is the spacing between adjacent slits and m is the order number of the diffraction pattern. A diffraction grating can be made by putting a large number of evenly spaced scratches on a glass slide. The number of such lines per centimeter is the inverse of the spacing d.

24.9 Polarization of Light Waves

Unpolarized light can be polarized by selective absorption, reflection, or scattering. A material can polarize light if it transmits waves having electric field vectors that vibrate in a plane parallel to a certain direction and absorbs waves with electric field vectors vibrating in directions perpendicular to that direction. When unpolarized light pass through a polarizing sheet, its intensity is reduced by half, and the light becomes polarized. When this light passes through a second polarizing sheet with transmission axis at an angle of θ with respect to the transmission axis of the first sheet, the transmitted intensity is given by

$$I = I_0 \cos^2 \theta \qquad \text{[24.13]}$$

where I_0 is the intensity of the light after passing through the first polarizing sheet.

In general, light reflected from an amorphous material, such as glass, is partially polarized. Reflected light is completely polarized, with its electric field parallel to the surface, when the angle of incidence produces a 90° angle between the reflected and refracted beams. This angle of incidence, called the **polarizing angle** θ_p, satisfies **Brewster's law**, given by

$$n = \tan \theta_p \qquad \text{[24.14]}$$

where n is the index of refraction of the reflecting medium.

CONCEPTUAL QUESTIONS

1. Your automobile has two headlights. What sort of interference pattern do you expect to see from them? Why?

2. Holding your hand at arm's length, you can readily block sunlight from your eyes. Why can you not block sound from your ears this way?

3. Consider a dark fringe in an interference pattern, at which almost no light energy is arriving. Light from both slits is arriving at this point, but the waves cancel. Where does the energy go?

4. If Young's double-slit experiment were performed under water, how would the observed interference pattern be affected?

5. In a laboratory accident, you spill two liquids onto water, neither of which mixes with the water. They both form thin films on the water surface. As the films spread and become very thin, you notice that one film becomes bright and the other black in reflected light. Why might this be?

6. If white light is used in Young's double-slit experiment, rather than monochromatic light, how does the interference pattern change?

7. In our discussion of thin-film interference, we looked at light *reflecting* from a thin film. Consider one light ray, the direct ray, that transmits through the film without reflecting. Then consider a second ray, the reflected ray, that transmits through the first surface, reflects back to the second, reflects again from the first, and then transmits out into the air, parallel to the direct ray. For normal incidence, how thick must the film be, in terms of the wavelength of the light, for the outgoing rays to interfere destructively? Is it the same thickness as for reflected destructive interference?

8. What is the necessary condition on the difference in path length between two waves that interfere (a) constructively and (b) destructively? Assume that the wave sources are coherent.

9. A lens with outer radius of curvature R and index of refraction n rests on a flat glass plate, and the combination is illuminated from white light from above. Is there a dark spot or a light spot at the center of the lens? What does it mean if the observed rings are noncircular?

10. Often, fingerprints left on a piece of glass such as a windowpane show colored spectra like that from a diffraction grating. Why?

11. In everyday experience, why are radio waves polarized, while light is not?

12. Suppose reflected white light is used to observe a thin, transparent coating on glass as the coating material is gradually deposited by evaporation in a vacuum. Describe some color changes that might occur during the process of building up the thickness of the coating.

13. Would it be possible to place a nonreflective coating on an airplane to cancel radar waves of wavelength 3 cm?

14. Certain sunglasses use a polarizing material to reduce the intensity of light reflected from shiny surfaces, such as water or the hood of a car. What orientation of the transmission axis should the material have to be most effective?

15. Why is it so much easier to perform interference experiments with a laser than with an ordinary light source?

16. A simple way of observing an interference pattern is to look at a distant light source through a stretched handkerchief or an open umbrella. Explain how this works.

17. When you receive a chest x-ray at a hospital, the x-rays pass through a series of parallel ribs in your chest. Do the ribs act as a diffraction grating for x-rays?

18. Can a sound wave be polarized? Explain.

19. Astronomers often observe occulations, in which a star passes behind another object, such as the Moon. During an occultation, the intensity of light from the star doesn't suddenly drop to zero as the star passes behind the edge of the Moon. Instead, the intensity fluctuates for a short time before dropping to zero. Why should this happen?

20. In one experiment, light from a laser passes through a double slit and forms an interference pattern on a distant screen. The experiment is repeated after increasing the slit separation by 50%. In which experiment is the distance from the central maximum to the next maximum the greatest?

21. Light in air that is reflected from a water surface is found to be completely polarized at an angle θ. If the light is instead reflected from a glass coffee table, will the new angle for complete polarization be larger or smaller?

22. In one experiment, blue light passes through a diffraction grating and forms an interference pattern on a screen. In a second experiment, red light passes through the same diffraction grating and forms another interference pattern. How do the separations between bright lines in the two experiments compare with each other?

PROBLEMS

1, 2, 3 = straightforward, intermediate, challenging   □ = full solution available in *Student Solutions Manual/Study Guide*

Physics⊗Now™ = coached solution with hints available at **www.cp7e.com**   🧬 = biomedical application

Section 24.2 Young's Double-Slit Experiment

1. A laser beam ($\lambda = 632.8$ nm) is incident on two slits 0.200 mm apart. How far apart are the bright interference fringes on a screen 5.00 m away from the double slits?

2. In a Young's double-slit experiment, a set of parallel slits with a separation of 0.100 mm is illuminated by light having a wavelength of 589 nm, and the interference pattern is observed on a screen 4.00 m from the slits. (a) What is the difference in path lengths from each of the slits to the location of a third-order bright fringe on the screen? (b) What is the difference in path lengths from the two slits to the location of the third dark fringe on the screen, away from the center of the pattern?

3. A pair of narrow, parallel slits separated by 0.250 mm is illuminated by the green component from a mercury vapor lamp ($\lambda = 546.1$ nm). The interference pattern is observed on a screen 1.20 m from the plane of the parallel slits. Calculate the distance (a) from the central maximum to the first bright region on either side of the central maximum and (b) between the first and second dark bands in the interference pattern.

4. Light of wavelength 460 nm falls on two slits spaced 0.300 mm apart. What is the required distance from the slit to a screen if the spacing between the first and second dark fringes is to be 4.00 mm?

5. **Physics⊗Now™** In a location where the speed of sound is 354 m/s, a 2 000-Hz sound wave impinges on two slits 30.0 cm apart. (a) At what angle is the first maximum located? (b) If the sound wave is replaced by 3.00-cm microwaves, what slit separation gives the same angle for the first maximum? (c) If the slit separation is 1.00 μm, what frequency of light gives the same first maximum angle?

6. White light spans the wavelength range between about 400 nm and 700 nm. If white light passes through two slits 0.30 mm apart and falls on a screen 1.5 m from the slits, find the distance between the first-order violet and the first-order red fringes.

7. Two radio antennas separated by 300 m, as shown in Figure P24.7, simultaneously transmit identical signals of the same wavelength. A radio in a car traveling due north receives the signals. (a) If the car is at the position of the second maximum, what is the wavelength of the signals? (b) How much farther must the car travel to encounter the next minimum in reception? [*Hint:* Determine the path difference between the two signals at the two locations of the car.]

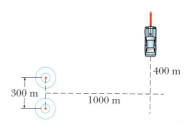

Figure P24.7

8. If the distance between two slits is 0.050 mm and the distance to a screen is 2.50 m, find the spacing between the first- and second-order bright fringes for yellow light of 600-nm wavelength.

9. Waves from a radio station have a wavelength of 300 m. They travel by two paths to a home receiver 20.0 km from the transmitter. One path is a direct path, and the second is by reflection from a mountain directly behind the home receiver. What is the minimum distance from the mountain to the receiver that produces destructive interference at the receiver? (Assume that no phase change occurs on reflection from the mountain.)

10. A pair of slits, separated by 0.150 mm, is illuminated by light having a wavelength of $\lambda = 643$ nm. An interference pattern is observed on a screen 140 cm from the slits. Consider a point on the screen located at $y = 1.80$ cm from the central maximum of this pattern. (a) What is the path difference δ for the two slits at the location y? (b) Express this path difference in terms of the wavelength. (c) Will the interference correspond to a maximum, a minimum, or an intermediate condition?

11. A riverside warehouse has two open doors, as in Figure P24.11. Its interior is lined with a sound-absorbing material. A boat on the river sounds its horn. To person A, the

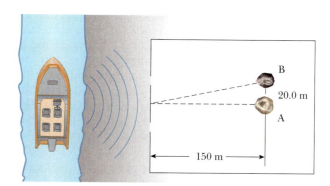

Figure P24.11

sound is loud and clear. To person B, the sound is barely audible. The principal wavelength of the sound waves is 3.00 m. Assuming person B is at the position of the first minimum, determine the distance between the doors, center to center.

12. The waves from a radio station can reach a home receiver by two different paths. One is a straight-line path from the transmitter to the home, a distance of 30.0 km. The second path is by reflection from a storm cloud. Assume that this reflection takes place at a point midway between receiver and transmitter. If the wavelength broadcast by the radio station is 400 m, find the minimum height of the storm cloud that will produce destructive interference between the direct and reflected beams. (Assume no phase changes on reflection.)

13. Radio waves from a star, of wavelength 250 m, reach a radio telescope by two separate paths, as shown in Figure P24.13. One is a direct path to the receiver, which is situated on the edge of a cliff by the ocean. The second is by reflection off the water. The first minimum of destructive interference occurs when the star is 25.0° above the horizon. Find the height of the cliff. (Assume no phase change on reflection.)

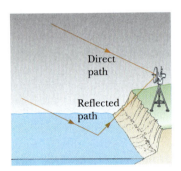

Figure P24.13

Section 24.3 Change of Phase Due to Reflection
Section 24.4 Interference in Thin Films

14. Determine the minimum thickness of a soap film ($n = 1.330$) that will result in constructive interference of (a) the red H_α line ($\lambda = 656.3$ nm); (b) the blue H_γ line ($\lambda = 434.0$ nm).

15. Suppose the film shown in Figure 24.7 has an index of refraction of 1.36 and is surrounded by air on both sides. Find the minimum thickness that will produce constructive interference in the reflected light when the film is illuminated by light of wavelength 500 nm.

16. A thin film of glass ($n = 1.50$) floats on a liquid of $n = 1.35$ and is illuminated by light of $\lambda = 580$ nm incident from air above it. Find the minimum thickness of the glass, other than zero, that will produce destructive interference in the reflected light.

17. A coating is applied to a lens to minimize reflections. The index of refraction of the coating is 1.55, and that of the lens is 1.48. If the coating is 177.4 nm thick, what wavelength is minimally reflected for normal incidence in the lowest order?

18. A transparent oil with index of refraction 1.29 spills on the surface of water (index of refraction 1.33), producing a maximum of reflection with normally incident orange light (wavelength 600 nm in air). Assuming the maximum occurs in the first order, determine the thickness of the oil slick.

19. A possible means for making an airplane invisible to radar is to coat the plane with an antireflective polymer. If radar waves have a wavelength of 3.00 cm and the index of refraction of the polymer is $n = 1.50$, how thick would you make the coating?

20. A beam of light of wavelength 580 nm passes through two closely spaced glass plates, as shown in Figure P24.20. For what minimum non-zero value of the plate separation d will the transmitted light be bright? This arrangement is often used to measure the wavelength of light and is called a Fabry–Perot interferometer.

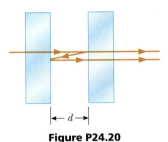

Figure P24.20

21. Astronomers observe the chromosphere of the sun with a filter that passes the red hydrogen spectral line of wavelength 656.3 nm, called the H_α line. The filter consists of a transparent dielectric of thickness d held between two partially aluminized glass plates. The filter is kept at a constant temperature. (a) Find the minimum value of d that will produce maximum transmission of perpendicular H_α light if the dielectric has an index of refraction of 1.378. (b) If the temperature of the filter increases above the normal value increasing its thickness, what happens to the transmitted wavelength? (c) The dielectric will also pass what near-visible wavelength? One of the glass plates is colored red to absorb this light.

22. Two rectangular optically flat plates ($n = 1.52$) are in contact along one end and are separated along the other end by a 2.00-μm-thick spacer (Fig. P24.22). The top plate is illuminated by monochromatic light of wavelength 546.1 nm. Calculate the number of dark parallel bands crossing the top plate (including the dark band at zero thickness along the edge of contact between the plates).

Figure P24.22   (Problems 22 and 23)

23. An air wedge is formed between two glass plates separated at one edge by a very fine wire, as in Figure P24.22. When the wedge is illuminated from above by 600-nm light, 30 dark fringes are observed. Calculate the radius of the wire.

24. A planoconvex lens with radius of curvature $R = 3.0$ m is in contact with a flat plate of glass. A light source and the observer's eye are both close to the normal, as shown in

Figure 24.8a. The radius of the 50th bright Newton's ring is found to be 9.8 mm. What is the wavelength of the light produced by the source?

25. A planoconvex lens rests with its curved side on a flat glass surface and is illuminated from above by light of wavelength 500 nm. (See Fig. 24.8.) A dark spot is observed at the center, surrounded by 19 concentric dark rings (with bright rings in between). How much thicker is the air wedge at the position of the 19th dark ring than at the center?

26. Nonreflective coatings on camera lenses reduce the loss of light at the surfaces of multilens systems and prevent internal reflections that might mar the image. Find the minimum thickness of a layer of magnesium fluoride ($n = 1.38$) on flint glass ($n = 1.66$) that will cause destructive interference of reflected light of wavelength 550 nm near the middle of the visible spectrum.

27. **Physics⊗Now**™ A thin film of MgF$_2$ ($n = 1.38$) with thickness 1.00×10^{-5} cm is used to coat a camera lens. Are any wavelengths in the visible spectrum intensified in the reflected light?

28. A flat piece of glass is supported horizontally above the flat end of a 10.0-cm-long metal rod that has its lower end rigidly fixed. The thin film of air between the rod and the glass is observed to be bright when illuminated by light of wavelength 500 nm. As the temperature is slowly increased by 25.0°C, the film changes from bright to dark and back to bright 200 times. What is the coefficient of linear expansion of the metal?

Section 24.7 Single-Slit Diffraction

29. Helium–neon laser light ($\lambda = 632.8$ nm) is sent through a 0.300-mm-wide single slit. What is the width of the central maximum on a screen 1.00 m from the slit?

30. Light of wavelength 600 nm falls on a 0.40-mm-wide slit and forms a diffraction pattern on a screen 1.5 m away. (a) Find the position of the first dark band on each side of the central maximum. (b) Find the width of the central maximum.

31. Light of wavelength 587.5 nm illuminates a slit of width 0.75 mm. (a) At what distance from the slit should a screen be placed if the first minimum in the diffraction pattern is to be 0.85 mm from the central maximum? (b) Calculate the width of the central maximum.

32. Microwaves of wavelength 5.00 cm enter a long, narrow window in a building that is otherwise essentially opaque to the incoming waves. If the window is 36.0 cm wide, what is the distance from the central maximum to the first-order minimum along a wall 6.50 m from the window?

33. A slit of width 0.50 mm is illuminated with light of wavelength 500 nm, and a screen is placed 120 cm in front of the slit. Find the widths of the first and second maxima on each side of the central maximum.

34. A screen is placed 50.0 cm from a single slit, which is illuminated with light of wavelength 680 nm. If the distance between the first and third minima in the diffraction pattern is 3.00 mm, what is the width of the slit?

Section 24.8 The Diffraction Grating

35. Three discrete spectral lines occur at angles of 10.1°, 13.7°, and 14.8°, respectively, in the first-order spectrum

of a diffraction-grating spectrometer. (a) If the grating has 3 660 slits/cm, what are the wavelengths of the light? (b) At what angles are these lines found in the second-order spectra?

36. Intense white light is incident on a diffraction grating that has 600 lines/mm. (a) What is the highest order in which the complete visible spectrum can be seen with this grating? (b) What is the angular separation between the violet edge (400 nm) and the red edge (700 nm) of the first-order spectrum produced by the grating?

37. The hydrogen spectrum has a red line at 656 nm and a violet line at 434 nm. What angular separation between these two spectral lines obtained with a diffraction grating that has 4 500 lines/cm?

38. A grating with 1 500 slits per centimeter is illuminated with light of wavelength 500 nm. (a) What is the highest-order number that can be observed with this grating? (b) Repeat for a grating of 15 000 slits per centimeter.

39. A light source emits two major spectral lines: an orange line of wavelength 610 nm and a blue-green line of wavelength 480 nm. If the spectrum is resolved by a diffraction grating having 5 000 lines/cm and viewed on a screen 2.00 m from the grating, what is the distance (in centimeters) between the two spectral lines in the second-order spectrum?

40. White light is spread out into its spectral components by a diffraction grating. If the grating has 2 000 lines per centimeter, at what angle does red light of wavelength 640 nm appear in the first-order spectrum?

41. Sunlight is incident on a diffraction grating that has 2 750 lines/cm. The second-order spectrum over the visible range (400–700 nm) is to be limited to 1.75 cm along a screen that is a distance L from the grating. What is the required value of L?

42. Light containing two different wavelengths passes through a diffraction grating with 1 200 slits/cm. On a screen 15.0 cm from the grating, the third-order maximum of the shorter wavelength falls midway between the central maximum and the first side maximum for the longer wavelength. If the neighboring maxima of the longer wavelength are 8.44 mm apart on the screen, what are the wavelengths in the light? [*Hint:* Use the small-angle approximation.]

43. **Physics⊗Now**™ A beam of 541-nm light is incident on a diffraction grating that has 400 lines/mm. (a) Determine the angle of the second-order ray. (b) If the entire apparatus is immersed in water, determine the new second-order angle of diffraction. (c) Show that the two diffracted rays of parts (a) and (b) are related through the law of refraction.

44. Light from a helium–neon laser ($\lambda = 632.8$ nm) is incident on a single slit. What is the maximum width for which no diffraction minima are observed? [*Hint:* Values of $\sin \theta > 1$ are not possible.]

Section 24.9 Polarization of Light Waves

45. The angle of incidence of a light beam in air onto a reflecting surface is continuously variable. The reflected ray is found to be completely polarized when the angle of incidence is 48.0°. (a) What is the index of refraction of the reflecting material? (b) If some of the incident light

(at an angle of 48.0°) passes into the material below the surface, what is the angle of refraction?

46. Unpolarized light passes through two polaroid sheets. The axis of the first is vertical, and that of the second is at 30.0° to the vertical. What fraction of the initial light is transmitted?

47. The index of refraction of a glass plate is 1.52. What is the Brewster's angle when the plate is (a) in air? (b) in water? (See Problem 51.)

48. At what angle above the horizon is the Sun if light from it is completely polarized upon reflection from water?

49. A light beam is incident on heavy flint glass ($n = 1.65$) at the polarizing angle. Calculate the angle of refraction for the transmitted ray.

50. The critical angle for total internal reflection for sapphire surrounded by air is 34.4°. Calculate the Brewster angle for sapphire if the light is incident from the air.

51. Equation 24.14 assumes that the incident light is in air. If the light is incident from a medium of index n_1 onto a medium of index n_2, follow the procedure used to derive Equation 24.14 to show that tan $\theta_p = n_2/n_1$.

52. Plane-polarized light is incident on a single polarizing disk, with the direction of E_0 parallel to the direction of the transmission axis. Through what angle should the disk be rotated so that the intensity in the transmitted beam is reduced by a factor of (a) 3.00, (b) 5.00, (c) 10.0?

53. **Physics⊗Now™** Three polarizing plates whose planes are parallel are centered on a common axis. The directions of the transmission axes relative to the common vertical direction are shown in Figure P24.53. A linearly polarized beam of light with plane of polarization parallel to the vertical reference direction is incident from the left onto the first disk with intensity $I_i = 10.0$ units (arbitrary). Calculate the transmitted intensity I_f when $\theta_1 = 20.0°$, $\theta_2 = 40.0°$, and $\theta_3 = 60.0°$. [*Hint:* Make repeated use of Malus's law.]

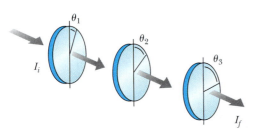

Figure P24.53   (Problems 53 and 62)

54. Light of intensity I_0 and polarized parallel to the transmission axis of a polarizer, is incident on an analyzer. (a) If the transmission axis of the analyzer makes an angle of 45° with the axis of the polarizer, what is the intensity of the transmitted light? (b) What should the angle between the transmission axes be to make $I/I_0 = 1/3$?

55. Light with a wavelength in vacuum of 546.1 nm falls perpendicularly on a biological specimen that is 1.000 μm thick. The light splits into two beams polarized at right angles, for which the indices of refraction are 1.320 and 1.333, respectively. (a) Calculate the wavelength of each component of the light while it is traversing the specimen. (b) Calculate the phase difference between the two beams when they emerge from the specimen.

ADDITIONAL PROBLEMS

56. A beam containing light of wavelengths λ_1 and λ_2 is incident on a set of parallel slits. In the interference pattern, the fourth bright line of the λ_1 light occurs at the same position as the fifth bright line of the λ_2 light. If λ_1 is known to be 540 nm, what is the value of λ_2?

57. Light of wavelength 546 nm (the intense green line from a mercury source) produces a Young's interference pattern in which the second minimum from the central maximum is along a direction that makes an angle of 18.0 min of arc with the axis through the central maximum. What is the distance between the parallel slits?

58. The two speakers are placed 35.0 cm apart. A single oscillator makes the speakers vibrate in phase at a frequency of 2.00 kHz. At what angles, measured from the perpendicular bisector of the line joining the speakers, would a distant observer hear maximum sound intensity? Minimum sound intensity? (Take the speed of sound to be 340 m/s.)

59. Interference effects are produced at point P on a screen as a result of direct rays from a 500-nm source and reflected rays off a mirror, as in Figure P24.59. If the source is 100 m to the left of the screen and 1.00 cm above the mirror, find the distance y (in millimeters) to the first dark band above the mirror.

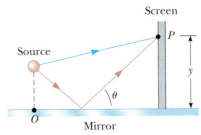

Figure P24.59

60. Many cells are transparent and colorless. Structures of great interest in biology and medicine can be practically invisible to ordinary microscopy. An *interference microscope* reveals a difference in refractive index as a shift in interference fringes, to indicate the size and shape of cell structures. The idea is exemplified in the following problem: An air wedge is formed between two glass plates in contact along one edge and slightly separated at the opposite edge. When the plates are illuminated with monochromatic light from above, the reflected light has 85 dark fringes. Calculate the number of dark fringes that appear if water ($n = 1.33$) replaces the air between the plates.

61. A thin layer of oil ($n = 1.25$) is floating on water. How thick is the oil in the region that strongly reflects green light ($\lambda = 525$ nm)?

62. Three polarizers, centered on a common axis and with their planes parallel to each other, have transmission axes oriented at angles of θ_1, θ_2, and θ_3 from the vertical, as shown in Figure P24.53. Light of intensity I_i, polarized with its plane of polarization oriented vertically, is incident from the left onto the first polarizer. What is the ratio I_f/I_i of the final transmitted intensity to the incident intensity if (a) $\theta_1 = 45°$, $\theta_2 = 90°$, and $\theta_3 = 0°$? (b) $\theta_1 = 0°$, $\theta_2 = 45°$, and $\theta_3 = 90°$?

63. Figure P24.63 shows a radio-wave transmitter and a receiver, both $h = 50.0$ m above the ground and $d = 600$ m apart. The receiver can receive signals directly from the transmitter, and indirectly, from signals that bounce off the ground. If the ground is level between the transmitter and receiver and a $\lambda/2$ phase shift occurs upon reflection, determine the longest wavelengths that interfere (a) constructively and (b) destructively.

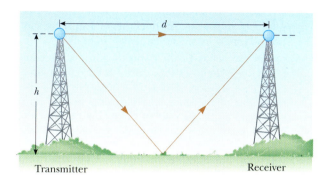

Figure P24.63

64. A planoconvex lens (flat on one side, convex on the other) with index of refraction n rests with its curved side (radius of curvature R) on a flat glass surface of the same index of refraction with a film of index n_{film} between them. The lens is illuminated from above by light of wavelength λ. Show that the dark Newton rings which appear have radii of

$$r \approx \sqrt{m\lambda R/n_{film}}$$

where m is an integer.

65. The transmitting antenna on a submarine is 5.00 m above the water when the ship surfaces. The captain wishes to transmit a message to a receiver on a 90.0-m-tall cliff at the ocean shore. If the signal is to be completely polarized by reflection off the ocean surface, how far must the ship be from the shore?

66. (a) If light is incident at an angle θ from a medium of index n_1 on a medium of index n_2 so that the angle between the reflected ray and refracted ray is β, show that

$$\tan \theta = \frac{n_2 \sin \beta}{n_1 - n_2 \cos \beta}$$

Hint: Use the trigonometric identity:

$$\sin(A + B) = \sin A \cos B + \cos A \sin B$$

(b) Show that the foregoing equation for $\tan \theta$ reduces to Brewster's law when $\beta = 90°$, $n_1 = 1$, and $n_2 = n$.

67. A diffraction pattern is produced on a screen 140 cm from a single slit, using monochromatic light of wavelength 500 nm, The distance from the center of the central maximum to the first-order maximum is 3.00 mm. Calculate the slit width. [*Hint:* Assume that the first-order maximum is halfway between the first- and second-order minima.]

68. A glass plate ($n = 1.61$) is covered with a thin, uniform layer of oil ($n = 1.20$). A light beam of variable wavelength is normally incident from air onto the oil surface. Observation of the reflected beam shows destructive interference at 500 nm and constructive interference at 750 nm. From this information, calculate the thickness of the oil film.

69. The condition for constructive interference by reflection from a thin film in air, as developed in Section 24.4, assumes nearly normal incidence. (a) Show that for large angles of incidence, the condition for constructive interference of light reflecting from a thin film of thickness t, with index of refraction n, and surrounded by air may be written as

$$2nt \cos \theta_2 = \left(m + \frac{1}{2} \right) \lambda$$

where θ_2 is the angle of refraction. (b) Calculate the minimum thickness for constructive interference if sodium light ($\lambda = 590$ nm) is incident at an angle of 30.0° on a film with an index of refraction of 1.38.

70. Figure P24.70 illustrates the formation of an interference pattern by the Lloyd's mirror method. Light from source S reaches the screen via two different pathways. One is a direct path, and the second is by reflection from a horizontal mirror. The effect is as if light from two different sources S and S' had interfered as in the Young's double-slit arrangement. Assume that the actual source S and the virtual source S' are in a plane 25 cm to the left of the mirror and the screen is a distance $L = 120$ cm to the right of that plane. Source S is a distance $h = 2.5$ mm above the top surface of the mirror, and the light is monochromatic with $\lambda = 620$ nm. Determine the distance of the first bright fringe above the surface of the mirror.

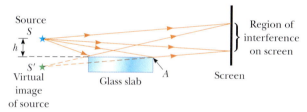

Figure P24.70

71. A piece of transparent material having an index of refraction n is cut into the shape of a wedge as shown in Figure P24.71. The angle of the wedge is small. Monochromatic light of wavelength λ is normally incident from above, and viewed from above. Let h represent the height of the wedge and ℓ its width. Show that bright fringes occur at the positions $x = \lambda\ell(m + \frac{1}{2})/2hn$ and dark fringes occur at the positions $x = \lambda\ell m/2hn$, where $m = 0, 1, 2, \ldots$ and x is measured as shown.

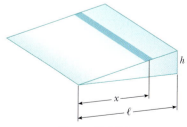

Figure P24.71

ACTIVITIES

1. Place a clear dish or plate on a black surface, such as a sheet of black construction paper. Now add a thin layer of water to the glass, and place a few drops of kerosene or light machine oil on the water. Darken the room and shine a flashlight from an angle, as in Figure A24.1. Note the interference pattern of various colors you observe under the white light. How does the pattern change if you cover the flashlight with a sheet of red, blue, or green cellophane, which acts as a filter?

As an extension of the preceding experiment, observe the colors appearing to swirl on the surface of a soap bubble. What color do you see just before a bubble bursts?

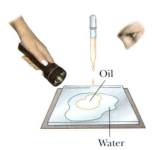

Oil

Water

Figure A24.1

2. Stand a couple of meters from a lightbulb. Facing away from the light, hold a compact disc about 10 cm from your eye and tilt it until the reflection of the bulb is located in the hole at the disc's center. You should see spectra radiating out from the center, with violet on the inside and red on the outside. Now move the disc away from your eye until the violet band is at the outer edge. Carefully measure the distance from your eye to the center of the disc,

and also determine the radius of the disc. Use this information to find the angle θ to the first-order maximum for violet light. Now use the relationship $d \sin \theta = m\lambda$ to determine the spacing between the grooves of the disc. The industry standard is 1.6 μm. How close did you come? While you are observing the spectrum from a CD, note that the color of the light from a given point changes with the viewing angle. Explain this effect in terms of changes in the path length. It is of interest that the blues and blue-greens in hummingbird feathers and butterflies are caused by diffraction off finely aligned structures in feathers and wings. (See chapter opener photo.)

3. (a) Devise a way to use a protractor, a desk lamp, and polarizing sunglasses to measure Brewster's angle for the glass in a window. From this, determine the index of refraction of the glass. (b) Put on a pair of polarizing sunglasses and close one eye. Hold up a lens of a second pair of polarizing glasses in front of your open eye so that light must pass through a lens of each pair before entering your eye. Now rotate the second pair of glasses around. You will note that the light reaching your eye is considerably reduced at some orientations and will pass freely at others. (c) On a sunny day, rotate your polarizing sunglasses in front of your eye and observe how light reflects from a window or the surface of water. Note the change in the amount of light entering your eye for various orientations of the glasses. (d) For a final observation concerning polarized light, rotate a pair of polarizing sunglasses while looking at various areas of the sky. From what direction do you find the light to be most highly polarized?

The Hubble Space Telescope does its viewing above the atmosphere and doesn't suffer from the atmospheric blurring, caused by air turbulence, that plagues ground-based telescopes. Despite this advantage, it does have limitations due to diffraction effects. In this chapter, we show how the wave nature of light limits the ability of any optical system to distinguish between closely spaced objects.

Optical Instruments

We use devices made from lenses, mirrors, or other optical components every time we put on a pair of eyeglasses or contact lenses, take a photograph, look at the sky through a telescope, and so on. In this chapter we examine how these and other optical instruments work. For the most part, our analyses will involve the laws of reflection and refraction and the procedures of geometric optics. To explain certain phenomena, however, we must use the wave nature of light.

25.1 THE CAMERA

The single-lens photographic **camera** is a simple optical instrument having the features shown in Figure 25.1 (page 820). It consists of an opaque box, a converging lens that produces a real image, and a film behind the lens to receive the image. Focusing is accomplished by varying the distance between lens and film—with an adjustable bellows in antique cameras and with some other mechanical arrangements in contemporary models. For proper focusing, which leads to sharp images, the lens-to-film distance depends on the object distance as well as on the focal length of the lens. The shutter, located behind the lens, is a mechanical device that is opened for selected time intervals. With this arrangement, moving objects can be photographed by using short exposure times, dark scenes (with low light levels) by using long exposure times. If this adjustment were not available, it would be impossible to take stop-action photographs. A rapidly moving vehicle, for example, could move far enough while the shutter was open to produce a blurred image. Another major cause of blurred images is movement of the *camera* while the shutter is open. To prevent such movement, you should mount the camera on a tripod or use short exposure times. Typical shutter speeds (that is, exposure times) are 1/30, 1/60, 1/125, and 1/250 s. Stationary objects are often shot with a shutter speed of 1/60 s.

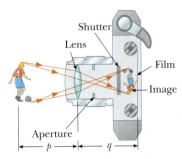

Figure 25.1 A cross-sectional view of a simple camera.

Most cameras also have an aperture of adjustable diameter to further control the intensity of the light reaching the film. When an aperture of small diameter is used, only light from the central portion of the lens reaches the film, so spherical aberration is reduced.

The intensity I of the light reaching the film is proportional to the area of the lens. Because this area in turn is proportional the square of the lens diameter D, the intensity is also proportional to D^2. Light intensity is a measure of the rate at which energy is received by the film per unit area of the image. Because the area of the image is proportional to q^2 in Figure 25.1, and $q \approx f$ (when $p \gg f$, so that p can be approximated as infinite), we conclude that the intensity is also proportional to $1/f^2$, so that $I \propto D^2/f^2$. The brightness of the image formed on the film depends on the light intensity, so we see that it ultimately depends on both the focal length f and diameter D of the lens. The ratio f/D is called the **f-number** (or focal ratio) of a lens:

$$f\text{-number} \equiv \frac{f}{D} \qquad [25.1]$$

The f-number is often given as a description of the lens "speed." A lens with a low f-number is a "fast" lens. Extremely fast lenses, which have an f-number as low as approximately 1.2, are expensive because of the difficulty of keeping aberrations acceptably small with light rays passing through a large area of the lens. Camera lenses are often marked with a range of f-numbers, such as 1.4, 2, 2.8, 4, 5.6, 8, 11, ... Any one of these settings can be selected by adjusting the aperture, which changes the value of D. Increasing the setting from one f-number to the next-higher value (for example, from 2.8 to 4) decreases the area of the aperture by a factor of two. The lowest f-number setting on a camera corresponds to a wide open aperture, and the use of the maximum possible lens area.

Simple cameras usually have a fixed focal length and fixed aperture size, with an f-number of about 11. This high value for the f-number allows for a large **depth of field**. This means that objects at a wide range of distances from the lens form reasonably sharp images on the film. In other words, the camera doesn't have to be focused. Most cameras with variable f-numbers adjust them automatically.

25.2 THE EYE

Like a camera, a normal eye focuses light and produces a sharp image. However, the mechanisms by which the eye controls the amount of light admitted and adjusts to produce correctly focused images are far more complex, intricate, and effective than those in even the most sophisticated camera. In all respects, the eye is a physiological wonder.

Figure 25.2a shows the essential parts of the eye. Light entering the eye passes through a transparent structure called the *cornea*, behind which are a clear liquid (the *aqueous humor*), a variable aperture (the *pupil*, which is an opening in the *iris*), and the *crystalline lens*. Most of the refraction occurs at the outer surface of the eye, at which the cornea is covered with a film of tears. Relatively little refraction occurs in the crystalline lens, because the aqueous humor in contact with the lens has an average index of refraction close to that of the lens. The iris, which is the colored portion of the eye, is a muscular diaphragm that controls pupil size. The iris regulates the amount of light entering the eye by dilating the pupil in low-light conditions and contracting the pupil under conditions of bright light. The f-number range of the eye is from about 2.8 to 16.

The cornea–lens system focuses light onto the back surface of the eye—the *retina*—which consists of millions of sensitive receptors called *rods* and *cones*. When stimulated by light, these structures send impulses to the brain via the optic nerve, converting them into our conscious view of the world. The process by which the brain performs this conversion is not well understood and is the subject of much speculation and research. Unlike film in a camera, the rods and cones chemically

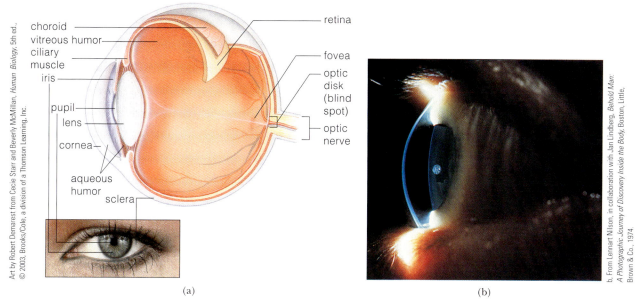

Figure 25.2   (a) Essential parts of the eye. Can you correlate the essential parts of the eye with those of the simple camera in Figure 25.1? (b) Close-up photograph of the human cornea.

adjust their sensitivity according to the prevailing light conditions. This adjustment, which takes about 15 minutes, is responsible for the experience of "getting used to the dark" in such places as movie theaters. Iris aperture control, which takes less than one second, helps protect the retina from overload in the adjustment process.

The eye focuses on an object by varying the shape of the pliable crystalline lens through an amazing process called **accommodation**. An important component in accommodation is the *ciliary muscle*, which is situated in a circle around the rim of the lens. Thin filaments, called *zonules*, run from this muscle to the edge of the lens. When the eye is focused on a distant object, the ciliary muscle is relaxed, tightening the zonules that attach the ciliary muscle to the edge of the lens. The force of the zonules causes the lens to flatten, increasing its focal length. For an object distance of infinity, the focal length of the eye is equal to the fixed distance between lens and retina, about 1.7 cm. The eye focuses on nearby objects by tensing the ciliary muscle, which relaxes the zonules. This action allows the lens to bulge a bit and its focal length decreases, resulting in the image being focused on the retina. All these lens adjustments take place so swiftly that we are not even aware of the change. In this respect, even the finest electronic camera is a toy compared with the eye.

There is a limit to accommodation because objects that are very close to the eye produce blurred images. The **near point** is the closest distance for which the lens can accommodate to focus light on the retina. This distance usually increases with age and has an average value of 25 cm. Typically, at age 10 the near point of the eye is about 18 cm. This increases to about 25 cm at age 20, 50 cm at age 40, and 500 cm or greater at age 60. The **far point** of the eye represents the farthest distance for which the lens of the relaxed eye can focus light on the retina. A person with normal vision is able to see very distant objects, such as the Moon, and so has a far point at infinity.

Conditions of the Eye

When the eye suffers a mismatch between the focusing power of the lens–cornea system and the length of the eye so that light rays reach the retina before they converge to form an image, as in Figure 25.3a, the condition is known as **farsightedness** (or *hyperopia*). A farsighted person can usually see faraway objects clearly but

Figure 25.3 (a) A farsighted eye is slightly shorter than normal; hence, the image of a nearby object focuses *behind* the retina. (b) The condition can be corrected with a converging lens. (The object is assumed to be very small in these figures.)

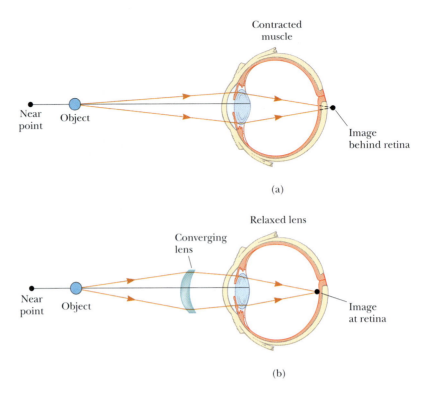

not nearby objects. Although the near point of a normal eye is approximately 25 cm, the near point of a farsighted person is much farther than that. The eye of a farsighted person tries to focus by accommodation, by shortening its focal length. This works for distant objects, but because the focal length of the farsighted eye is longer than normal, the light from nearby objects can't be brought to a sharp focus before it reaches the retina, causing a blurred image. The condition can be corrected by placing a converging lens in front of the eye, as in Figure 25.3b. The lens refracts the incoming rays more toward the principal axis before entering the eye, allowing them to converge and focus on the retina.

Nearsightedness (or *myopia*) is another mismatch condition in which a person is able to focus on nearby objects, but not faraway objects. In the case of *axial myopia*, nearsightedness is caused by the lens being too far from the retina. It is also possible to have *refractive myopia*, in which the lens–cornea system is too powerful for the normal length of the eye. The far point of the nearsighted eye is not at infinity and may be less than a meter. The maximum focal length of the nearsighted eye is insufficient to produce a sharp image on the retina, and rays from a distant object converge to a focus in front of the retina. They then continue past that point, diverging before they finally reach the retina and produce a blurred image (Fig. 25.4a).

Nearsightedness can be corrected with a diverging lens, as shown in Figure 25.4b. The lens refracts the rays away from the principal axis before they enter the eye, allowing them to focus on the retina.

Beginning with middle age, most people lose some of their accommodation ability as the ciliary muscle weakens and the lens hardens. Unlike farsightedness, which is a mismatch of focusing power and eye length, **presbyopia** (literally, "old-age vision") is due to a reduction in accommodation ability. The cornea and lens aren't able to bring nearby objects into focus on the retina. The symptoms are the same as with farsightedness, and the condition can be corrected with converging lenses.

In the eye defect known as **astigmatism**, light from a point source produces a line image on the retina. This condition arises when either the cornea or the lens (or both) are not perfectly symmetric. Astigmatism can be corrected with lenses having different curvatures in two mutually perpendicular directions.

Optometrists and ophthalmologists usually prescribe lenses measured in **diopters**:

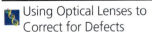

APPLICATION

Using Optical Lenses to Correct for Defects

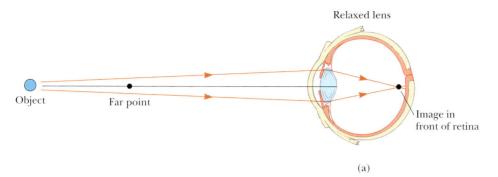

Relaxed lens

Object   Far point

Image in
front of retina

(a)

Figure 25.4   (a) A nearsighted eye
is slightly longer than normal; hence,
the image of a distant object focuses
in front of the retina. (b) The
condition can be corrected with a
diverging lens. (The object is
assumed to be very small in these
figures.)

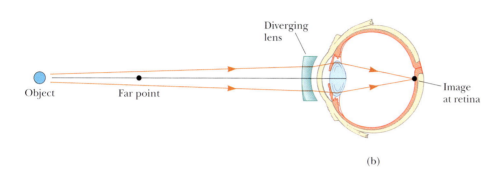

Diverging
lens

Object   Far point

Image
at retina

(b)

> The **power** $\mathcal{P}$ of a lens in diopters equals the inverse of the focal length in
> meters: $\mathcal{P} = 1/f$.

For example, a converging lens with a focal length of $+20$ cm has a power of $+5.0$
diopters, and a diverging lens with a focal length of -40 cm has a power of -2.5
diopters. (Although the symbol is the same as for mechanical power, there is no relationship between the two concepts.)

The position of the lens relative the eye causes differences in power, but this
usually amounts to less than a quarter diopter, which isn't noticeable to most patients. As a result, practicing optometrists deal in increments of a quarter diopter.
Neglecting the eye–lens distance is equivalent to doing the calculation for a contact lens, which rests directly on the eye.

EXAMPLE 25.1   Prescribing a Corrective Lens for a Farsighted Patient

Goal   Apply geometric optics to correct farsightedness.

Problem   The near point of a patient's eye is 50.0 cm. **(a)** What focal length must a corrective lens have to enable
the eye to see clearly an object 25.0 cm away? Neglect the eye–lens distance. **(b)** What is the power of this lens?
(c) Repeat the problem, taking into account the fact that, for typical eyeglasses, the corrective lens is 2.00 cm in front
of the eye.

Strategy   This problem requires substitution into the thin-lens equation (Eq. 23.11) and then using the definition
of lens power in terms of diopters. The object is at 25.0 cm, but the lens must form an image at the patient's near
point, 50.0 cm, the closest point at which the patient's eye can see clearly. In part (c), 2.00 cm must be subtracted
from both the object distance and the image distance to account for the position of the lens.

Solution
(a) Find the focal length of the corrective lens, neglecting its distance from the eye.

Apply the thin-lens equation:
$$\frac{1}{p} + \frac{1}{q} = \frac{1}{f}$$

Substitute $p = 25.0$ cm and $q = -50.0$ cm (the latter is negative because the image must be virtual) on the same side of the lens as the object:

$$\frac{1}{25.0 \text{ cm}} + \frac{1}{-50.0 \text{ cm}} = \frac{1}{f}$$

Solve for f. The focal length is positive, corresponding to a converging lens.

$$f = \boxed{50.0 \text{ cm}}$$

(b) What is the power of this lens?

The power is the reciprocal of the focal length in meters:

$$\mathcal{P} = \frac{1}{f} = \frac{1}{0.500 \text{ m}} = \boxed{+2.00 \text{ diopters}}$$

(c) Repeat the problem, noting that the corrective lens is actually 2.00 cm in front of the eye.

Substitute the corrected values of p and q into the thin-lens equation:

$$\frac{1}{p} + \frac{1}{q} = \frac{1}{23.0 \text{ cm}} + \frac{1}{(-48.0 \text{ cm})} = \frac{1}{f}$$

$$f = 44.2 \text{ cm}$$

Compute the power:

$$\mathcal{P} = \frac{1}{f} = \frac{1}{0.442 \text{ m}} = \boxed{+2.26 \text{ diopters}}$$

Remarks Notice that the calculation in part (c), which doesn't neglect the eye–lens distance, results in a difference of 0.26 diopter.

Exercise 25.1
Suppose a lens is placed in a device that determines its power as 2.75 diopters. Find (a) the focal length of the lens and (b) the minimum distance at which a patient will be able to focus on an object if the patient's near point is 60.0 cm. Neglect the eye–lens distance.

Answers (a) 36.4 cm (b) 22.7 cm

EXAMPLE 25.2 A Corrective Lens for Nearsightedness

Goal Apply geometric optics to correct nearsightedness.

Problem A particular nearsighted patient can't see objects clearly when they are beyond 25 cm (the far point of the eye). **(a)** What focal length should the prescribed contact lens have to correct this problem? **(b)** Find the power of the lens, in diopters. Neglect the distance between the eye and the corrective lens.

Strategy The purpose of the lens in this instance is to take objects at infinity and create an image of them at the patient's far point. Apply the thin-lens equation.

Solution
(a) Find the focal length of the corrective lens.

Apply the thin-lens equation for an object at infinity and image at 25.0 cm:

$$\frac{1}{p} + \frac{1}{q} = \frac{1}{\infty} + \frac{1}{(-25.0 \text{ cm})} = \frac{1}{f}$$

$$f = \boxed{-25.0 \text{ cm}}$$

(b) Find the power of the lens in diopters:

$$\mathcal{P} = \frac{1}{f} = \frac{1}{-0.250 \text{ m}} = -4.00 \text{ diopters}$$

Remarks The focal length is negative, consistent with a diverging lens. Notice that the power is also negative and has the same numeric value as the sum on the left side of the thin-lens equation.

Exercise 25.2
(a) What power lens would you prescribe for a patient with a far point of 35.0 cm? Neglect the eye–lens distance.
(b) Repeat, assuming an eye-corrective lens distance of 2.00 cm.

Answer   (a) -2.86 diopters   (b) -3.03 diopters

Applying Physics 25.1   Vision of the Invisible Man

A classic science fiction story, *The Invisible Man* by H.G. Wells, tells of a person who becomes invisible by changing the index of refraction of his body to that of air. Students who know how the eye works have criticized this story; they claim the invisible man would be unable to see. On the basis of your knowledge of the eye, would he be able to see ?

Explanation   He wouldn't be able to see. In order for the eye to see an object, incoming light must be refracted at the cornea and lens to form an image on the retina. If the cornea and lens have the same index of refraction as air, refraction can't occur, and an image wouldn't be formed.

Quick Quiz 25.1

Two campers wish to start a fire during the day. One camper is nearsighted and one is farsighted. Whose glasses should be used to focus the Sun's rays onto some paper to start the fire? (a) either camper (b) the nearsighted camper (c) the farsighted camper.

25.3   THE SIMPLE MAGNIFIER

The **simple magnifier** is one of the most basic of all optical instruments because it consists only of a single converging lens. As the name implies, this device is used to increase the apparent size of an object. Suppose an object is viewed at some distance p from the eye, as in Figure 25.5. Clearly, the size of the image formed at the retina depends on the angle θ subtended by the object at the eye. As the object moves closer to the eye, θ increases and a larger image is observed. However, a normal eye can't focus on an object closer than about 25 cm, the near point (Fig. 25.6a, page 826). (Try it!) Therefore, θ is a maximum at the near point.

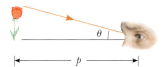

Figure 25.5   The size of the image formed on the retina depends on the angle θ subtended at the eye.

To further increase the apparent angular size of an object, a converging lens can be placed in front of the eye with the object positioned at point O, just inside the focal point of the lens, as in Figure 25.6b. At this location, the lens forms a virtual, upright, and enlarged image, as shown. The lens allows the object to be viewed closer to the eye than is possible without the lens. We define the **angular magnification** m as the ratio of the angle subtended by the object when the lens is in use (angle θ in Fig. 25.6b) to the angle subtended by the object placed at the near point with no lens in use (angle θ_0 in Fig. 25.6a):

$$m \equiv \frac{\theta}{\theta_0} \qquad \text{[25.2]}$$

◀ Angular magnification with the object at the near point

For the case where the lens is held close to the eye, the angular magnification is a maximum when the image formed by the lens is at the near point of the eye, which corresponds to $q = -25$ cm (see Fig. 25.6b). The object distance corresponding to this image distance can be calculated from the thin-lens equation:

$$\frac{1}{p} + \frac{1}{-25 \text{ cm}} = \frac{1}{f} \qquad \text{[25.3]}$$

$$p = \frac{25f}{25 + f}$$

Figure 25.6 (a) An object placed at the near point ($p = 25$ cm) subtends an angle of $\theta_0 \approx h/25$ at the eye. (b) An object placed near the focal point of a converging lens produces a magnified image, which subtends an angle of $\theta \approx h'/25$ at the eye. Note that, in this situation, $q = -25$ cm.

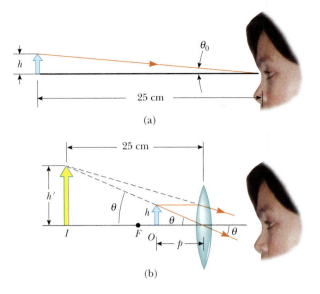

Here, f is the focal length of the magnifier in centimeters. From Figures 25.6a and 25.6b, the small-angle approximation gives

$$\tan \theta_0 \approx \theta_0 \approx \frac{h}{25} \quad \text{and} \quad \tan \theta \approx \theta \approx \frac{h}{p} \qquad [25.4]$$

Equation 25.2 therefore becomes

$$m_{max} = \frac{\theta}{\theta_0} = \frac{h/p}{h/25} = \frac{25}{p} = \frac{25}{25f/(25+f)}$$

so that

$$m_{max} = 1 + \frac{25 \text{ cm}}{f} \qquad [25.5]$$

The maximum angular magnification given by Equation 25.5 is the ratio of the angular size seen with the lens to the angular size seen without the lens, with the object at the near point of the eye. Although the normal eye can focus on an image formed anywhere between the near point and infinity, it's most relaxed when the image is at infinity (Sec. 25.2). For the image formed by the magnifying lens to appear at infinity, the object must be placed at the focal point of the lens, so that $p = f$. In this case, Equation 25.4 becomes

$$\theta_0 \approx \frac{h}{25} \quad \text{and} \quad \theta \approx \frac{h}{f}$$

and the angular magnification is

$$m = \frac{\theta}{\theta_0} = \frac{25 \text{ cm}}{f} \qquad [25.6]$$

With a single lens, it's possible to achieve angular magnifications up to about 4 without serious aberrations. Magnifications up to about 20 can be achieved by using one or two additional lenses to correct for aberrations.

EXAMPLE 25.3 Magnification of a Lens

Goal Compute magnifications of a lens when the image is at the near point and when it's at infinity.

Problem (a) What is the maximum angular magnification of a lens with a focal length of 10.0 cm? (b) What is the angular magnification of this lens when the eye is relaxed? Assume an eye–lens distance of zero.

Strategy The maximum angular magnification occurs when the image formed by the lens is at the near point of the eye. Under these circumstances, Equation 25.5 gives us the maximum angular magnification. In part (b), the eye is relaxed only if the image is at infinity, so Equation 25.6 applies.

Solution

(a) Find the maximum angular magnification of the lens.

Substitute into Equation 25.5:

$$m_{max} = 1 + \frac{25 \text{ cm}}{f} = 1 + \frac{25 \text{ cm}}{10.0 \text{ cm}} = \boxed{3.5}$$

(b) Find the magnification of the lens when the eye is relaxed.

When the eye is relaxed, the image is at infinity, so substitute into Equation 25.6:

$$m = \frac{25 \text{ cm}}{f} = \frac{25 \text{ cm}}{10.0 \text{ cm}} = \boxed{2.5}$$

Exercise 25.3

What focal length would be necessary if the lens were to have a maximum angular magnification of 4.00?

Answer 8.3 cm

25.4 THE COMPOUND MICROSCOPE

A simple magnifier provides only limited assistance with inspection of the minute details of an object. Greater magnification can be achieved by combining two lenses in a device called a compound microscope, a schematic diagram of which is shown in Active Figure 25.7a. The instrument consists of two lenses: an objective with a very short focal length f_o (where $f_o < 1$ cm), and an ocular lens, or eyepiece, with a focal length f_e of a few centimeters. The two lenses are separated by distance L that is much greater than either f_o or f_e.

The basic approach used to analyze the image formation properties of a microscope is that of two lenses in a row: the image formed by the first becomes the object for the second. The object O placed just outside the focal length of the objective forms a real, inverted image at I_1 that is at or just inside the focal point of the eyepiece. This image is much enlarged. (For clarity, the enlargement of I_1 is not shown in Active Fig. 25.7a.) The eyepiece, which serves as a simple magnifier, uses the image at I_1 as its object and produces an image at I_2. The image seen by the eye at I_2 is virtual, inverted, and very much enlarged.

The lateral magnification M_1 of the first image is $-q_1/p_1$. Note that q_1 is approximately equal to L, because the object is placed close to the focal point of the

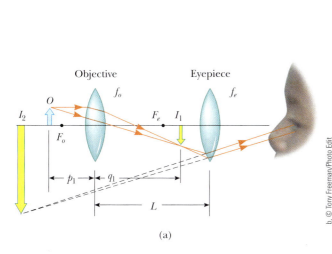

(a)

(b)

b, © Tony Freeman/Photo Edit

ACTIVE FIGURE 25.7
(a) A diagram of a compound microscope, which consists of an objective and an eyepiece, or ocular lens. (b) A compound microscope. The three-objective turret allows the user to switch to several different powers of magnification. Combinations of eyepieces with different focal lengths and different objectives can produce a wide range of magnifications.

Physics⊗Now™

Log into PhysicsNow at **www.cp7e.com** and go to Active Figure 25.7, where you can adjust the focal lengths of the objective and eyepiece lenses, and see the effect on the final image.

objective lens, which ensures that the image formed will be far from the objective lens. Further, because the object is very close to the focal point of the objective lens, $p_1 \approx f_o$. Therefore, the lateral magnification of the objective is

$$M_1 = -\frac{q_1}{p_1} \approx -\frac{L}{f_o}$$

From Equation 25.6, the angular magnification of the eyepiece for an object (corresponding to the image at I_1) placed at the focal point is found to be

$$m_e = \frac{25 \text{ cm}}{f_e}$$

The overall magnification of the compound microscope is defined as the product of the lateral and angular magnifications:

◄ Magnification of a microscope

$$m = M_1 m_e = -\frac{L}{f_o}\left(\frac{25 \text{ cm}}{f_e}\right) \qquad [25.7]$$

The negative sign indicates that the image is inverted with respect to the object.

The microscope has extended our vision into the previously unknown realm of incredibly small objects, and the capabilities of this instrument have increased steadily with improved techniques in precision grinding of lenses. A natural question is whether there is any limit to how powerful a microscope could be. For example, could a microscope be made powerful enough to allow us to see an atom? The answer to this question is no, as long as visible light is used to illuminate the object. In order to be seen, the object under a microscope must be at least as large as a wavelength of light. An atom is many times smaller than the wavelength of visible light, so its mysteries must be probed via other techniques.

The wavelength dependence of the "seeing" ability of a wave can be illustrated by water waves set up in a bathtub in the following way: Imagine that you vibrate your hand in the water until waves with a wavelength of about 6 in. are moving along the surface. If you fix a small object, such as a toothpick, in the path of the waves, you will find that the waves are not appreciably disturbed by the toothpick, but continue along their path. Now suppose you fix a larger object, such as a toy sailboat, in the path of the waves. In this case, the waves are considerably disturbed by the object. The toothpick was much smaller than the wavelength of the waves, and as a result, the waves didn't "see" it. The toy sailboat, however, is about the same size as the wavelength of the waves and hence creates a disturbance. Light waves behave in this same general way. The ability of an optical microscope to view an object depends on the size of the object relative to the wavelength of the light used to observe it. Hence, it will never be possible to observe atoms or molecules with such a microscope, because their dimensions are so small (≈ 0.1 nm) relative to the wavelength of the light (≈ 500 nm).

EXAMPLE 25.4   Microscope Magnifications

Goal   Understand the critical factors involved in determining the magnifying power of a microscope.

Problem   A certain microscope has two interchangeable objectives. One has a focal length of 2.0 cm, and the other has a focal length of 0.20 cm. Also available are two eyepieces of focal lengths 2.5 cm and 5.0 cm. If the length of the microscope is 18 cm, compute the magnifications for the following combinations: the 2.0-cm objective and 5.0-cm eyepiece; the 2.0-cm objective and 2.5-cm eyepiece; the 0.20-cm objective and 5.0-cm eyepiece.

Strategy   The solution consists of substituting into Equation 25.7 for three different combinations of lenses.

Solution

Apply Equation 25.7 and combine the 2.0-cm objective with the 5.0-cm eyepiece:

$$m = -\frac{L}{f_o}\left(\frac{25 \text{ cm}}{f_e}\right) = -\frac{18 \text{ cm}}{2.0 \text{ cm}}\left(\frac{25 \text{ cm}}{5.0 \text{ cm}}\right) = \boxed{-45}$$

Combine the 2.0-cm objective with the 2.5-cm eyepiece: $m = -\dfrac{18\text{ cm}}{2.0\text{ cm}}\left(\dfrac{25\text{ cm}}{2.5\text{ cm}}\right) = -9.0 \times 10^1$

Combine the 0.20-cm objective with the 5.0-cm eyepiece: $m = -\dfrac{18\text{ cm}}{0.20\text{ cm}}\left(\dfrac{25\text{ cm}}{5.0\text{ cm}}\right) = -450$

Remarks   Much higher magnifications can be achieved, but the resolution starts to fall, resulting in fuzzy images that don't convey any details. (See Section 25.6 for further discussion of this point.)

Exercise 25.4
Combine the 0.20-cm objective with the 2.5-cm eyepiece.

Answer 9.0×10^2

25.5   THE TELESCOPE

There are two fundamentally different types of telescope, both designed to help us view distant objects such as the planets in our Solar System. These two types are (1) the **refracting telescope**, which uses a combination of lenses to form an image, and (2) the **reflecting telescope**, which uses a curved mirror and a lens to form an image. Once again, we will be able to analyze the telescope by considering it to be a system of two optical elements in a row. As before, the basic technique followed is that the image formed by the first element becomes the object for the second.

In the refracting telescope, two lenses are arranged so that the objective forms a real, inverted image of the distant object very near the focal point of the eyepiece (Active Fig. 25.8a, page 830). Further, the image at I_1 is formed at the focal point of the objective because the object is essentially at infinity. Hence, the two lenses are separated by the distance $f_o + f_e$, which corresponds to the length of the telescope's tube. Finally, at I_2, the eyepiece forms an enlarged, inverted image of the image at I_1.

The angular magnification of the telescope is given by θ/θ_o, where θ_o is the angle subtended by the object at the objective and θ is the angle subtended by the final image. From the triangles in Active Figure 25.8a, and for small angles, we have

$$\theta \approx \frac{h'}{f_e} \qquad \text{and} \qquad \theta_o \approx \frac{h'}{f_o}$$

Therefore, the angular magnification of the telescope can be expressed as

$$m = \frac{\theta}{\theta_o} = \frac{h'/f_e}{h'/f_o} = \frac{f_o}{f_e} \qquad \text{[25.8]}$$

◄ Angular magnification of a telescope

This equation says that the angular magnification of a telescope equals the ratio of the objective focal length to the eyepiece focal length. Here again, the angular magnification is the ratio of the angular size seen with the telescope to the angular size seen with the unaided eye.

In some applications—for instance, the observation of relatively nearby objects such as the Sun, the Moon, or planets—angular magnification is important. Stars, however, are so far away that they always appear as small points of light regardless of how much angular magnification is used. The large research telescopes used to study very distant objects must have great diameters to gather as much light as possible. It's difficult and expensive to manufacture such large lenses for refracting telescopes. In addition, the heaviness of large lenses leads to sagging, which is another source of aberration.

These problems can be partially overcome by replacing the objective lens with a reflecting, concave mirror, usually having a parabolic shape so as to avoid spherical aberration. Figure 25.9 (page 830) shows the design of a typical reflecting telescope. Incoming light rays pass down the barrel of the telescope and are reflected

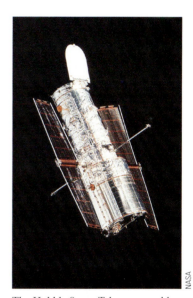

The Hubble Space Telescope enables us to see both further into space and further back in time than ever before.

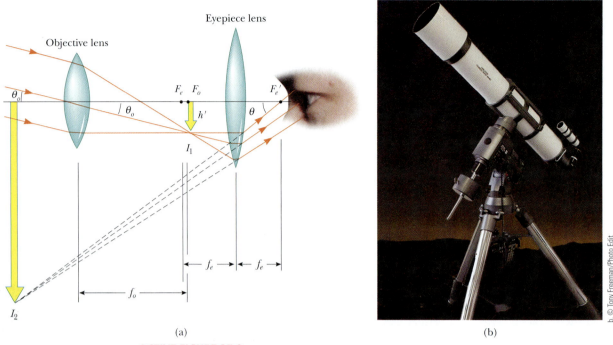

(a)

(b)

ACTIVE FIGURE 25.8

(a) A diagram of a refracting telescope, with the object at infinity. (b) A refracting telescope.

Physics ⊗ Now™

Log into PhysicsNow at **www.cp7e.com** and go to Active Figure 25.8, where you can adjust the focal lengths of the objective and eyepiece lenses, and observe the effect on the final image.

by a parabolic mirror at the base. These rays converge toward point A in the figure, where an image would be formed on a photographic plate or another detector. However, before this image is formed, a small flat mirror at M reflects the light toward an opening in the side of the tube that passes into an eyepiece. This design is said to have a *Newtonian focus*, after its developer. Note that in the reflecting telescope the light never passes through glass (except for the small eyepiece). As a result, problems associated with chromatic aberration are virtually eliminated.

The largest optical telescopes in the world are the two 10-m-diameter Keck reflectors on Mauna Kea in Hawaii. The largest single-mirrored reflecting telescope in the United States is the 5-m-diameter instrument on Mount Palomar in California. (See Fig. 25.10.) In contrast, the largest refracting telescope in the world, at the Yerkes Observatory in Williams Bay, Wisconsin, has a diameter of only 1 m.

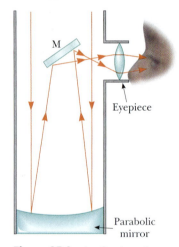

Figure 25.9 A reflecting telescope with a Newtonian focus.

Figure 25.10 The Hale telescope at Mount Palomar Observatory. Just before taking the elevator up to the prime-focus cage, a first-time observer is always told, "Good viewing! And, if you should fall, try to miss the mirror."

EXAMPLE 25.5   Hubble Power

Goal   Understand magnification in telescopes.

Problem   The Hubble telescope is 13.2 m long, but has a secondary mirror that increases its effective focal length to 57.8 m. (See Fig. 25.11.) The telescope doesn't have an eyepiece, because various instruments, not a human eye, record the collected light. However, it can produce images several thousand times larger than they would appear with the unaided human eye. What focal length eyepiece used with the Hubble mirror system would produce a magnification of 8.00×10^3?

Strategy   Equation 25.8 for telescope magnification can be solved for the eyepiece focal length. The equation for finding the angular magnification of a reflector is the same as that for a refractor.

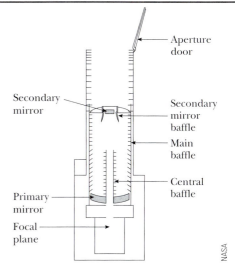

Figure 25.11   A schematic of the Hubble telescope.

Solution

Solve for f_e in Equation 25.8 and substitute values:

$$ m = \frac{f_o}{f_e} \quad \rightarrow \quad f_e = \frac{f_o}{m} = \frac{57.8 \text{ m}}{8.00 \times 10^3} = \boxed{7.23 \times 10^{-3} \text{ m}} $$

Remarks   The result of this magnification is an image with "good" resolution. However, the light-gathering power of a telescope largely determines the resolution of the image, and is far more important than magnification. A high-resolution image can always be magnified so its details can be examined. Such details are often blurred when a low-resolution image is magnified.

Exercise 25.5

The Hale telescope on Mt. Palomar has a focal length of 16.8 m. Find the magnification of the telescope in conjunction with an eyepiece having a focal length of 5.00 mm.

Answer 3.36×10^3

25.6   RESOLUTION OF SINGLE-SLIT AND CIRCULAR APERTURES

The ability of an optical system such as the eye, a microscope, or a telescope to distinguish between closely spaced objects is limited because of the wave nature of light. To understand this difficulty, consider Figure 25.12, which shows two light sources far

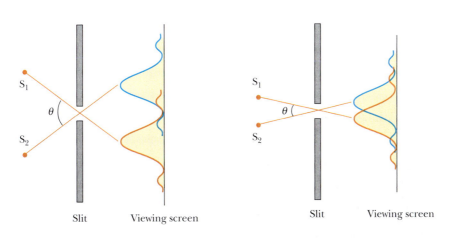

Slit   Viewing screen   Slit   Viewing screen

(a)   (b)

Figure 25.12   Two point sources far from a narrow slit each produce a diffraction pattern. (a) The angle subtended by the sources at the aperture is large enough so that the diffraction patterns are distinguishable. (b) The angle subtended by the sources is so small that the diffraction patterns are not distinguishable. (Note that the angles are greatly exaggerated. The drawing is not to scale.)

from a narrow slit of width a. The sources can be taken as two point sources S_1 and S_2 that are *not* coherent. For example, they could be two distant stars. If no diffraction occurred, two distinct bright spots (or images) would be observed on the screen at the right in the figure. However, because of diffraction, each source is imaged as a bright central region flanked by weaker bright and dark rings. What is observed on the screen is the sum of two diffraction patterns, one from S_1 and the other from S_2.

If the two sources are separated so that their central maxima don't overlap, as in Figure 25.12a, their images can be distinguished and are said to be *resolved*. If the sources are close together, however, as in Figure 25.12b, the two central maxima may overlap and the images are *not resolved*. To decide whether two images are resolved, the following condition is often applied to their diffraction patterns:

Rayleigh's criterion ▶

> When the central maximum of one image falls on the first minimum of another image, the images are said to be just resolved. This limiting condition of resolution is known as **Rayleigh's criterion**.

Figure 25.13 shows diffraction patterns in three situations. The images are just resolved when their angular separation satisfies Rayleigh's criterion (Fig. 25.13a). As the objects are brought closer together, their images are barely resolved (Fig. 25.13b). Finally, when the sources are very close to each other, their images are not resolved (Fig. 25.13c).

From Rayleigh's criterion, we can determine the minimum angular separation θ_{min} subtended by the source at the slit so that the images will be just resolved. In Chapter 24 we found that the first minimum in a single-slit diffraction pattern occurs at the angle that satisfies the relationship

$$\sin\theta = \frac{\lambda}{a}$$

where a is the width of the slit. According to Rayleigh's criterion, this expression gives the smallest angular separation for which the two images can be resolved. Because $\lambda \ll a$ in most situations, $\sin\theta$ is small and we can use the approximation $\sin\theta \approx \theta$. Therefore, the limiting angle of resolution for a slit of width a is

Limiting angle for a slit ▶

$$\theta_{min} \approx \frac{\lambda}{a} \qquad\qquad [25.9]$$

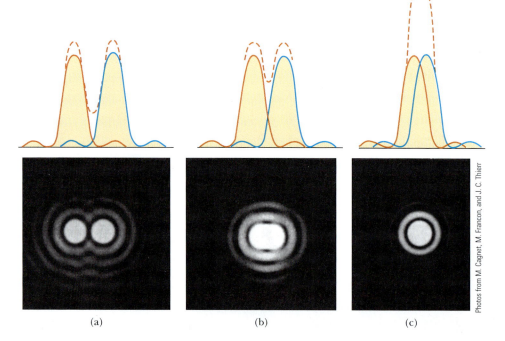

Figure 25.13 The diffraction patterns of two point sources (solid curves) and the resultant pattern (dashed curve) for three angular separations of the sources. (a) The sources are separated such that their patterns are just resolved. (b) The sources are closer together, and their patterns are barely resolved. (c) The sources are so close together that their patterns are not resolved.

(a)   (b)   (c)

Photos from M. Cagnet, M. Francon, and J. C. Thierr

where θ_{min} is in radians. Hence, the angle subtended by the two sources at the slit must be *greater* than λ/a if the images are to be resolved.

Many optical systems use circular apertures rather than slits. The diffraction pattern of a circular aperture (Fig. 25.14) consists of a central circular bright region surrounded by progressively fainter rings. Analysis shows that the limiting angle of resolution of the circular aperture is

$$\theta_{min} = 1.22 \frac{\lambda}{D} \qquad \text{[25.10]}$$

where D is the diameter of the aperture. Note that Equation 25.10 is similar to Equation 25.9, except for the factor 1.22, which arises from a complex mathematical analysis of diffraction from a circular aperture.

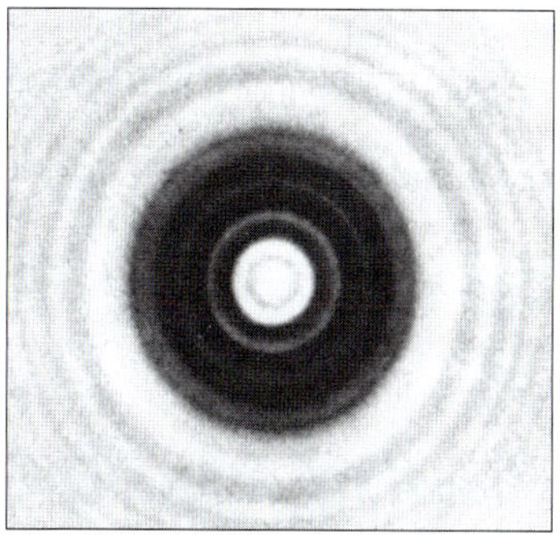

Figure 25.14   The diffraction pattern of a circular aperture consists of a central bright disk surrounded by concentric bright and dark rings.

From M. Cagnet, M. Francon, and J. C. Thier, *Atlas of Optical Phenomena*, Berlin, Springer-Verlag, 1962, plate 34

Quick Quiz 25.2

Suppose you are observing a binary star with a telescope and are having difficulty resolving the two stars. Which color filter will better help resolve the stars? (a) blue (b) red (c) neither—colored filters have no affect on resolution

---

Applying Physics 25.2   Cat's Eyes

Cats' eyes have vertical pupils in dim light. Which would cats be most successful at resolving at night, headlights on a distant car or vertically separated running lights on a distant boat's mast having the same separation as the car's headlights?

Explanation   The effective slit width in the vertical direction of the cat's eye is larger than that in the horizontal direction. Thus, it has more resolving power for lights separated in the vertical direction and would be more effective at resolving the mast lights on the boat.

EXAMPLE 25.6   Resolution of a Microscope

Goal   Study limitations on the resolution of a microscope.

Problem   Sodium light of wavelength 589 nm is used to view an object under a microscope. The aperture of the objective has a diameter of 0.90 cm. **(a)** Find the limiting angle of resolution for this microscope. **(b)** Using visible light of any wavelength you desire, find the maximum limit of resolution for this microscope. **(c)** Water of index of refraction 1.33 now fills the space between the object and the objective. What effect would this have on the resolving power of the microscope, using 589 nm light?

Strategy   Parts (a) and (b) require substitution into Equation 25.10. Because the wavelength appears in the numerator, violet light, with the shortest visible wavelength, gives the maximum resolution. In part (c), the only difference is that the wavelength changes to λ/n, where n is the index of refraction of water.

Solution
(a) Find the limiting angle of resolution for this microscope.

Substitute into Equation 25.10 to obtain the limiting angle of resolution:

$$\theta_{min} = 1.22 \frac{\lambda}{D} = 1.22 \left(\frac{589 \times 10^{-9} \text{ m}}{0.90 \times 10^{-2} \text{ m}} \right)$$
$$= 8.0 \times 10^{-5} \text{ rad}$$

(b) Calculate the microscope's maximum limit of resolution.

To obtain the maximum resolution, substitute the shortest visible wavelength available—violet light, of wavelength 4.0×10^2 nm:

$$\theta_{min} = 1.22 \frac{\lambda}{D} = 1.22 \left(\frac{4.0 \times 10^{-7} \text{ m}}{0.90 \times 10^{-2} \text{ m}} \right)$$
$$= 5.4 \times 10^{-5} \text{ rad}$$

(c) What effect does water between the object and the objective lens have on the resolution, with 589-nm light?

Calculate the wavelength of the sodium light in the water:

$$\lambda_w = \frac{\lambda_a}{n} = \frac{589 \text{ nm}}{1.33} = 443 \text{ nm}$$

Substitute this wavelength into Equation 25.10 to get the resolution:

$$\theta_{min} = 1.22 \left(\frac{443 \times 10^{-9} \text{ m}}{0.90 \times 10^{-2} \text{ m}} \right) = 6.0 \times 10^{-5} \text{ rad}$$

Remarks   In each case, any two points on the object subtending an angle of less than the limiting angle θ_{min} at the objective cannot be distinguished in the image. Consequently, it may be possible to see a cell, but then be unable to clearly see smaller structures within the cell. Obtaining an increase in resolution is the motivation behind placing a drop of oil on the slide for certain objective lenses.

Exercise 25.6
Suppose oil with $n = 1.50$ fills the space between the object and the objective for this microscope. Calculate the limiting angle θ_{min} for sodium light of wavelength 589 nm in air.

Answer 5.3×10^{-5} rad

EXAMPLE 25.7   Resolving Craters on the Moon

Goal   Calculate the resolution of a telescope.

Problem   The Hubble Space Telescope has an aperture of diameter 2.40 m. **(a)** What is its limiting angle of resolution at a wavelength of 6.00×10^2 nm? **(b)** What's the smallest crater it could resolve on the Moon? (The Moon is 3.84×10^8 m from Earth.)

Strategy   After substituting into Equation 25.10 to find the limiting angle, use $s = r\theta$ to compute the minimum size of crater that can be resolved.

Solution
(a) What is the limiting angle of resolution at a wavelength of 6.00×10^2 nm?

Substitute $D = 2.40$ m and $\lambda = 6.00 \times 10^{-7}$ m into Equation 25.10:

$$\theta_{min} = 1.22 \frac{\lambda}{D} = 1.22 \left(\frac{6.00 \times 10^{-7} \text{ m}}{2.40 \text{ m}} \right)$$
$$= 3.05 \times 10^{-7} \text{ rad}$$

(b) What's the smallest lunar crater the Hubble Space Telescope can resolve?

The two opposite sides of the crater must subtend the minimum angle. Use the arc length formula:

$$s = r\theta = (3.84 \times 10^8 \text{ m})(3.05 \times 10^{-7} \text{ rad}) = 117 \text{ m}$$

Remarks   The distance is so great and the angle so small that using the arc length of a circle is justified—the circular arc is very nearly a straight line. The Hubble Space Telescope has produced several gigabytes of data every day for over 20 years.

Exercise 25.7
The Hale telescope on Mount Palomar has a diameter of 5.08 m (200 in.). **(a)** Find the limiting angle of resolution for a wavelength of 6.00×10^2 nm. **(b)** Calculate the smallest crater diameter the telescope can resolve on the Moon. **(c)** The answers appear better than what the Hubble can achieve. Why are the answers misleading?

Answers   **(a)** 1.44×10^{-7} rad **(b)** 55.3 m **(c)** While the numbers are better than Hubble's, the Hale telescope must contend with the effects of atmospheric turbulence, so the smaller space-based telescope actually obtains far better results.

It's interesting to compare the resolution of the Hale telescope with that of a large radio telescope, such as the system at Arecibo, Puerto Rico, which has a diameter of 1 000 ft (305 m). This telescope detects radio waves at a wavelength of 0.75 m. The corresponding minimum angle of resolution can be calculated as 3.0×10^{-3} rad (10 min 19 s of arc), which is more than 10 000 times larger than the calculated minimum angle for the Hale telescope.

With such relatively poor resolution, why is Arecibo considered a valuable astronomical instrument? Unlike its optical counterparts, Arecibo can see through clouds of dust. The center of our Milky Way galaxy is obscured by such dust clouds, which absorb and scatter visible light. Radio waves easily penetrate the clouds, so radio telescopes allow direct observations of the galactic core.

Resolving Power of the Diffraction Grating

The diffraction grating studied in Chapter 24 is most useful for making accurate wavelength measurements. Like the prism, it can be used to disperse a spectrum into its components. Of the two devices, the grating is better suited to distinguishing between two closely spaced wavelengths. We say that the grating spectrometer has a higher *resolution* than the prism spectrometer. If λ_1 and λ_2 are two nearly equal wavelengths between which the spectrometer can just barely distinguish, the **resolving power** of the grating is defined as

$$R \equiv \frac{\lambda}{\lambda_2 - \lambda_1} = \frac{\lambda}{\Delta\lambda} \qquad \textbf{[25.11]}$$

where $\lambda \approx \lambda_1 \approx \lambda_2$ and $\Delta\lambda = \lambda_2 - \lambda_1$. From this equation, it's clear that a grating with a high resolving power can distinguish small differences in wavelength. Further, if N lines of the grating are illuminated, it can be shown that the resolving power in the mth-order diffraction is given by

$$R = Nm \qquad \textbf{[25.12]} \qquad \blacktriangleleft \text{Resolving power of a grating}$$

So the resolving power R increases with the order number m and is large for a grating with a great number of illuminated slits. Note that for $m = 0$, $R = 0$, which signifies that *all wavelengths are indistinguishable* for the zeroth-order maximum. (All wavelengths fall at the same point on the screen.) However, consider the second-order diffraction pattern of a grating that has 5 000 rulings illuminated by the light source. The resolving power of such a grating in second order is $R = 5\,000 \times 2 = 10\,000$. Therefore, the *minimum* wavelength separation between two spectral lines that can be just resolved, assuming a mean wavelength of 600 nm, is calculated from Equation 25.12 to be $\Delta\lambda = \lambda/R = 6 \times 10^{-2}$ nm. For the third-order principal maximum, $R = 15\,000$ and $\Delta\lambda = 4 \times 10^{-2}$ nm, and so on.

EXAMPLE 25.8 Light from Sodium Atoms

Goal Find the necessary resolving power to distinguish spectral lines.

Problem Two bright lines in the spectrum of sodium have wavelengths of 589.00 nm and 589.59 nm, respectively. **(a)** What must the resolving power of a grating be in order to distinguish these wavelengths? **(b)** To resolve these lines in the second-order spectrum, how many lines of the grating must be illuminated?

Strategy This problem requires little more than substituting into Equations 25.11 and 25.12.

Solution
(a) What must the resolving power of a grating be in order to distinguish the given wavelengths?

Substitute into Equation 25.11 to find R:

$$R = \frac{\lambda}{\Delta\lambda} = \frac{589.00 \text{ nm}}{589.59 \text{ nm} - 589.00 \text{ nm}} = \frac{589 \text{ nm}}{0.59 \text{ nm}}$$

$$= 1.0 \times 10^3$$

(b) To resolve these lines in the second-order spectrum, how many lines of the grating must be illuminated?

Solve Equation 25.12 for N and substitute:

$$N = \frac{R}{m} = \frac{1.0 \times 10^3}{2} = 5.0 \times 10^2 \text{ lines}$$

Remarks   The ability to resolve spectral lines is particularly important in experimental atomic physics.

Exercise 25.8

When the lines of a spectrum are examined at high resolution, each line is actually found to be two closely spaced lines called a doublet, due to a phenomenon called electron spin. An example is the doublet in the hydrogen spectrum having wavelengths of 656.272 nm and 656.285 nm. (a) What must be the resolving power of a grating in order to distinguish these wavelengths? (b) How many lines of the grating must be illuminated to resolve these lines in the third-order spectrum?

Answer   (a) 5.0×10^4 (b) 1.7×10^4 lines

25.7   THE MICHELSON INTERFEROMETER

The Michelson interferometer is an optical instrument having great scientific importance. Invented by the American physicist A. A. Michelson (1852–1931), it is an ingenious device that splits a light beam into two parts and then recombines them to form an interference pattern. The interferometer is used to make accurate length measurements.

Active Figure 25.15 is a schematic diagram of an interferometer. A beam of light provided by a monochromatic source is split into two rays by a partially silvered mirror M inclined at an angle of 45° relative to the incident light beam. One ray is reflected vertically upward to mirror M_1, and the other ray is transmitted horizontally through mirror M to mirror M_2. Hence, the two rays travel separate paths, L_1 and L_2. After reflecting from mirrors M_1 and M_2, the two rays eventually recombine to produce an interference pattern, which can be viewed through a telescope. The glass plate P, equal in thickness to mirror M, is placed in the path of the horizontal ray to ensure that the two rays travel the same distance through glass.

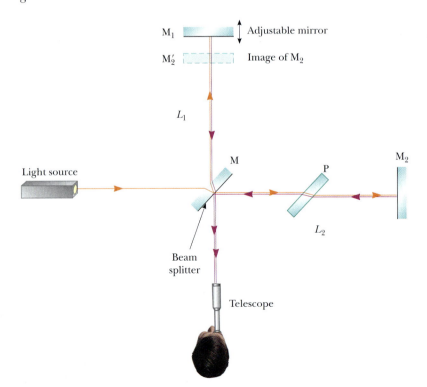

ACTIVE FIGURE 25.15

A diagram of the Michelson interferometer. A single beam is split into two rays by the half-silvered mirror M. The path difference between the two rays is varied with the adjustable mirror M_1.

The interference pattern for the two rays is determined by the difference in their path lengths. When the two rays are viewed as shown, the image of M₂ is at M₂′, parallel to M₁. Hence, the space between M₂′ and M₁ forms the equivalent of a parallel air film. The effective thickness of the air film is varied by using a finely threaded screw to move mirror M₁ in the direction indicated by the arrows in Active Figure 25.15. If one of the mirrors is tipped slightly with respect to the other, the thin film between the two is wedge shaped, and an interference pattern consisting of parallel fringes is set up, as described in Example 24.4. Now suppose we focus on one of the dark lines with the crosshairs of a telescope. As the mirror M₁ is moved to lengthen the path L_1, the thickness of the wedge increases. When the thickness increases by $\lambda/4$, the destructive interference that initially produced the dark fringe has changed to constructive interference, and we now observe a bright fringe at the location of the crosshairs. The term *fringe shift* is used to describe the change in a fringe from dark to light or light to dark. Thus, successive light and dark fringes are formed each time M₁ is moved a distance of $\lambda/4$. The wavelength of light can be measured by counting the number of fringe shifts for a measured displacement of M₁. Conversely, if the wavelength is accurately known (as with a laser beam), the mirror displacement can be determined to within a fraction of the wavelength. Because the interferometer can measure displacements precisely, it is often used to make highly accurate measurements of the dimensions of mechanical components.

If the mirrors are perfectly aligned, rather than tipped with respect to one another, the path difference differs slightly for different angles of view. This arrangement results in an interference pattern that resembles Newton's rings. The pattern can be used in a fashion similar to that for tipped mirrors. An observer pays attention to the center spot in the interference pattern. For example, suppose the spot is initially dark, indicating that destructive interference is occurring. If M₁ is now moved a distance of $\lambda/4$, this central spot changes to a light region, corresponding to a fringe shift.

SUMMARY

Physics⊗Now™ Take a practice test by logging into Physics-Now at **www.cp7e.com** and clicking on the Pre-Test link for this chapter.

25.1 The Camera

The light-concentrating power of a lens of focal length f and diameter D is determined by the *f*-number, defined as

$$f\text{-number} \equiv \frac{f}{D} \qquad [25.1]$$

The smaller the *f*-number of a lens, the brighter is the image formed.

25.2 The Eye

Hyperopia (farsightedness) is a defect of the eye that occurs either when the eyeball is too short or when the ciliary muscle cannot change the shape of the lens enough to form a properly focused image. **Myopia** (nearsightedness) occurs either when the eye is longer than normal or when the maximum focal length of the lens is insufficient to produce a clearly focused image on the retina.

The **power** of a lens in **diopters** is the inverse of the focal length in meters.

25.3 The Simple Magnifier

The **angular magnification of a lens** is defined as

$$m \equiv \frac{\theta}{\theta_0} \qquad [25.2]$$

where θ is the angle subtended by an object at the eye with a lens in use and θ_0 is the angle subtended by the object when it is placed at the near point of the eye and no lens is used. The **maximum angular magnification of a lens** is

$$m_{max} = 1 + \frac{25 \text{ cm}}{f} \qquad [25.5]$$

When the eye is relaxed, the angular magnification is

$$m = \frac{25 \text{ cm}}{f} \qquad [25.6]$$

25.4 The Compound Microscope

The overall **magnification of a compound microscope** of length L is the product of the magnification produced by the objective, of focal length f_o, and the magnification produced by the eyepiece, of focal length f_e:

$$M = -\frac{L}{f_o}\left(\frac{25\text{ cm}}{f_e}\right) \qquad \text{[25.7]}$$

25.5 The Telescope

The **angular magnification of a telescope** is

$$m = \frac{f_o}{f_e} \qquad \text{[25.8]}$$

where f_o is the focal length of the objective and f_e is the focal length of the eyepiece.

25.6 Resolution of Single-Slit and Circular Apertures

Two images are said to be **just resolved** when the central maximum of the diffraction pattern for one image falls on the first minimum of the other image. This limiting condition of resolution is known as **Rayleigh's criterion**. The limiting angle of resolution for a **slit** of width a is

$$\theta_{\min} \approx \frac{\lambda}{a} \qquad \text{[25.9]}$$

The limiting angle of resolution of a **circular aperture** is

$$\theta_{\min} = 1.22\frac{\lambda}{D} \qquad \text{[25.10]}$$

where D is the diameter of the aperture.

If λ_1 and λ_2 are two nearly equal wavelengths between which a grating spectrometer can just barely distinguish, the **resolving power** R of the grating is defined as

$$R \equiv \frac{\lambda}{\lambda_2 - \lambda_1} = \frac{\lambda}{\Delta\lambda} \qquad \text{[25.11]}$$

where $\lambda \approx \lambda_1 \approx \lambda_2$ and $\Delta\lambda = \lambda_2 - \lambda_1$. The **resolving power** of a diffraction grating in the mth order is

$$R = Nm \qquad \text{[25.12]}$$

where N is the number of illuminated rulings on the grating.

CONCEPTUAL QUESTIONS

1. A lens is used to examine an object across a room. Is the lens probably being used as a simple magnifier?

2. Why is it difficult or impossible to focus a microscope on an object across a room?

3. The optic nerve and the brain invert the image formed on the retina. Why don't we see everything upside down?

4. If you want to examine the fine detail of an object with a magnifying glass with a power of + 20.0 diopters, where should the object be placed in order to observe a magnified image of the object?

5. Suppose you are observing the interference pattern formed by a Michelson interferometer in a laboratory and a joking colleague holds a lit match in the light path of one arm of the interferometer. Will this have an effect on the interference pattern?

6. Compare and contrast the eye and a camera. What parts of the camera correspond to the iris, the retina, and the cornea of the eye?

7. Large telescopes are usually reflecting rather than refracting. List some reasons for this choice.

8. If you want to use a converging lens to set fire to a piece of paper, why should the light source be farther from the lens than its focal point?

9. Explain why it is theoretically impossible to see an object as small as an atom regardless of the quality of the light microscope being used.

10. Which is most important in the use of a camera photoflash unit, the intensity of the light (the energy per unit area per unit time) or the product of the intensity and the time of the flash, assuming the time is less than the shutter speed?

11. A patient has a near point of 1.25 m. Is she nearsighted or farsighted? Should the corrective lens be converging or diverging?

12. A lens with a certain power is used as a simple magnifier. If the power of the lens is doubled, does the angular magnification increase or decrease?

PROBLEMS

1, 2, 3 = straightforward, intermediate, challenging   ☐ = full solution available in *Student Solutions Manual/Study Guide*
Physics⊗Now™ = coached problem with hints available at **www.cp7e.com**   🧬 = biomedical application

Section 25.1 The Camera

1. A camera used by a professional photographer to shoot portraits has a focal length of 25.0 cm. The photographer takes a portrait of a person 1.50 m in front of the camera. Where is the image formed, and what is the lateral magnification?

2. The lens of a certain 35 mm camera (35 mm is the width of the film strip) has a focal length of 55 mm and a speed (an *f*-number) of *f*/1.8. Determine the diameter of the lens.

3. A photographic image of a building is 0.092 0 m high. The image was made with a lens with a focal length of 52.0 mm. If the lens was 100 m from the building when the photograph was made, determine the height of the building.

4. The full Moon is photographed using a camera with a 120-mm-focal-length lens. Determine the diameter of the Moon's image on the film. [*Note:* The radius of the Moon is 1.74×10^6 m, and the distance from the Earth to the Moon is 3.84×10^8 m.]

5. A camera is being used with the correct exposure at *f*/4 and a shutter speed of 1/32 s. In order to "stop" a fast-moving subject, the shutter speed is changed to 1/256 s. Find the new *f*-stop that should be used to maintain

satisfactory exposure, assuming no change in lighting conditions.

6. (a) Use conceptual arguments to show that the intensity of light (energy per unit area per unit time) reaching the film in a camera is proportional to the square of the reciprocal of the f-number, as

$$I \propto \frac{1}{(f/D)^2}$$

(b) The correct exposure time for a camera set to $f/1.8$ is $(1/500)$ s. Calculate the correct exposure time if the f-number is changed to $f/4$ under the same lighting conditions.

7. A certain type of film requires an exposure time of 0.010 s with an $f/11$ lens setting. Another type of film requires twice the light energy to produce the same level of exposure. What f-stop does the second type of film need with the 0.010-s exposure time?

8. Assume that the camera in Figure 25.1 has a fixed focal length of 65.0 mm and is adjusted to properly focus the image of a distant object. How far and in what direction must the lens be moved to focus the image of an object that is 2.00 m away?

Section 25.2 The Eye

9. A retired bank president can easily read the fine print of the financial page when the newspaper is held no closer than arm's length, 60.0 cm from the eye. What should be the focal length of an eyeglass lens that will allow her to read at the more comfortable distance of 24.0 cm?

10. A person has far points 84.4 cm from the right eye and 122 cm from the left eye. Write a prescription for the powers of the corrective lenses.

11. The accommodation limits for Nearsighted Nick's eyes are 18.0 cm and 80.0 cm. When he wears his glasses, he is able to see faraway objects clearly. At what minimum distance is he able to see objects clearly?

12. The near point of an eye is 100 cm. A corrective lens is to be used to allow this eye to clearly focus on objects 25.0 in front of it. (a) What should be the focal length of the lens? (b) What is the power of the needed corrective lens?

13. An individual is nearsighted; his near point is 13.0 cm and his far point is 50.0 cm. (a) What lens power is needed to correct his nearsightedness? (b) When the lenses are in use, what is this person's near point?

14. A certain child's near point is 10.0 cm; her far point (with eyes relaxed) is 125 cm. Each eye lens is 2.00 cm from the retina. (a) Between what limits, measured in diopters, does the power of this lens–cornea combination vary? (b) Calculate the power of the eyeglass lens the child should use for relaxed distance vision. Is the lens converging or diverging?

15. An artificial lens is implanted in a person's eye to replace a diseased lens. The distance between the artificial lens and the retina is 2.80 cm. In the absence of the lens, an image of a distant object (formed by refraction at the cornea) falls 2.53 cm behind the retina. The lens is designed to put the image of the distant object on the retina. What is the power of the implanted lens? [*Hint:* Consider the image formed by the cornea to be a virtual object.]

16. **Physics⊗Now™** A person is to be fitted with bifocals. She can see clearly when the object is between 30 cm and 1.5 m from the eye. (a) The upper portions of the bifocals (Fig. P25.16) should be designed to enable her to see distant objects clearly. What power should they have? (b) The lower portions of the bifocals should enable her to see objects comfortably at 25 cm. What power should they have?

Far vision

Near vision

Figure P25.16

Section 25.3 The Simple Magnifier

17. A stamp collector uses a lens with 7.5-cm focal length as a simple magnifier. The virtual image is produced at the normal near point (25 cm). (a) How far from the lens should the stamp be placed? (b) What is the expected angular magnification?

18. A lens having a focal length of 25 cm is used as a simple magnifier. (a) What is the angular magnification obtained when the image is formed at the normal near point ($q = -25$ cm)? (b) What is the angular magnification produced by this lens when the eye is relaxed?

19. A biology student uses a simple magnifier to examine the structural features of the wing of an insect. The wing is held 3.50 cm in front of the lens, and the image is formed 25.0 cm from the eye. (a) What is the focal length of the lens? (b) What angular magnification is achieved?

20. A lens that has a focal length of 5.00 cm is used as a magnifying glass. (a) To obtain maximum magnification, where should the object be placed? (b) What is the magnification?

21. A leaf of length h is positioned 71.0 cm in front of a converging lens with a focal length of 39.0 cm. An observer views the image of the leaf from a position 1.26 m behind the lens, as shown in Figure P25.21. (a) What is the magnitude of the lateral magnification (the ratio of the image size to the object size) produced by the lens? (b) What angular magnification is achieved by viewing the image of the leaf rather than viewing the leaf directly?

Leaf

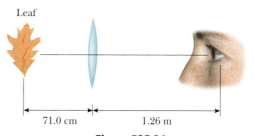

71.0 cm   1.26 m

Figure P25.21

Section 25.4 The Compound Microscope &
Section 25.5 The Telescope

22. The objective lens in a microscope with a 20.0-cm-long tube has a magnification of 50.0, and the eyepiece has a magnification of 20.0. What are the focal lengths of (a) the objective and (b) the eyepiece? (c) What is the overall magnification of the microscope?

23. The desired overall magnification of a compound microscope is 140×. The objective alone produces a lateral magnification of 12×. Determine the required focal length of the eyepiece.

24. A microscope has an objective lens with a focal length of 16.22 mm and an eyepiece with a focal length of 9.50 mm. With the length of the barrel set at 29.0 cm, the diameter of a red blood cell's image subtends an angle of 1.43 mrad with the eye. If the final image distance is 29.0 cm from the eyepiece, what is the actual diameter of the red blood cell?

25. The length of a microscope tube is 15.0 cm. The focal length of the objective is 1.00 cm, and the focal length of the eyepiece is 2.50 cm. What is the magnification of the microscope, assuming it is adjusted so that the eye is relaxed? [*Hint:* To solve this question go back to basics and use the thin lens equation.]

26. A certain telescope has an objective of focal length 1 500 cm. If the Moon is used as an object, a 1.0-cm-long image formed by the objective corresponds to what distance, in miles, on the Moon? Assume 3.8×10^8 m for the Earth–Moon distance.

27. **Physics⊗Now**™ The lenses of an astronomical telescope are 92 cm apart when adjusted for viewing a distant object with minimum eyestrain. The angular magnification produced by the telescope is 45. Compute the focal length of each lens.

28. An elderly sailor is shipwrecked on a desert island, but manages to save his eyeglasses. The lens for one eye has a power of + 1.20 diopters, and the other lens has a power of + 9.00 diopters. (a) What is the magnifying power of the telescope he can construct with these lenses? (b) How far apart are the lenses when the telescope is adjusted for minimum eyestrain?

29. Astronomers often take photographs with the objective lens or mirror of a telescope alone, without an eyepiece. (a) Show that the image size h' for a telescope used in this manner is given by $h' = fh/(f - p)$, where h is the object size, f is the objective focal length, and p is the object distance. (b) Simplify the expression in part (a) if the object distance is much greater than the objective focal length. (c) The "wingspan" of the International Space Station is 108.6 m, the overall width of its solar panel configuration. When it is orbiting at an altitude of 407 km, find the width of the image formed by a telescope objective of focal length 4.00 m.

30. Galileo devised a simple terrestrial telescope that produces an upright image. It consists of a converging objective lens and a diverging eyepiece at opposite ends of the telescope tube. For distant objects, the tube length is the objective focal length less the absolute value of the eyepiece focal length. (a) Does the user of the telescope see a real or virtual image? (b) Where is the final image? (c) If a telescope is to be constructed with a tube of length 10.0 cm and a

magnification of 3.00, what are the focal lengths of the objective and eyepiece?

31. A person decides to use an old pair of eyeglasses to make some optical instruments. He knows that the near point in his left eye is 50.0 cm and the near point in his right eye is 100 cm. (a) What is the maximum angular magnification he can produce in a telescope? (b) If he places the lenses 10.0 cm apart, what is the maximum overall magnification he can produce in a microscope? (Go back to basics and use the thin-lens equation to solve part (b).)

Section 25.6 Resolution of Single-Slit
and Circular Apertures

32. If the distance from the Earth to the Moon is 3.8×10^8 m, what diameter would be required for a telescope objective to resolve a Moon crater 300 m in diameter? Assume a wavelength of 500 nm.

33. A converging lens with a diameter of 30.0 cm forms an image of a satellite passing overhead. The satellite has two green lights (wavelength 500 nm) spaced 1.00 m apart. If the lights can just be resolved according to the Rayleigh criterion, what is the altitude of the satellite?

34. The pupil of a cat's eye narrows to a vertical slit of width 0.500 mm in daylight. What is the angular resolution for a pair of horizontally separated mice? (Use 500-nm light in your calculation.)

35. To increase the resolving power of a microscope, the object and the objective are immersed in oil ($n = 1.5$). If the limiting angle of resolution without the oil is 0.60 μrad, what is the limiting angle of resolution with the oil? [*Hint:* The oil changes the wavelength of the light.]

36. (a) Calculate the limiting angle of resolution for the eye, assuming a pupil diameter of 2.00 mm, a wavelength of 500 nm *in air*, and an index of refraction for the eye of 1.33. (b) What is the maximum distance from the eye at which two points separated by 1.00 cm could be resolved?

37. Two stars in a binary system are 8.0 lightyears away from the observer and can just be resolved by a 20-in. telescope equipped with a filter that allows only light of wavelength 500 nm to pass. What is the distance between the two stars?

38. A spy satellite circles the Earth at an altitude of 200 km and carries out surveillance with a special high-resolution telescopic camera having a lens diameter of 35 cm. If the angular resolution of this camera is limited by diffraction, estimate the separation of two small objects on the Earth's surface that are just resolved in yellow-green light ($\lambda = 550$ nm).

39. Suppose a 5.00-m-diameter telescope were constructed on the Moon, where the absence of atmospheric distortion would permit excellent viewing. If observations were made using 500-nm light, what minimum separation between two objects could just be resolved on Mars at closest approach (when Mars is 8.0×10^7 km from the Moon)?

40. The H_α line in hydrogen has a wavelength of 656.20 nm. This line differs in wavelength from the corresponding spectral line in deuterium (the heavy stable isotope of hydrogen) by 0.18 nm. (a) Determine the minimum number of lines a grating must have to resolve these two wavelengths in the first order. (b) Repeat part (a) for the second order.

41. **Physics⊗Now™** A 15.0-cm-long grating has 6 000 slits per centimeter. Can two lines of wavelengths 600.000 nm and 600.003 nm be separated with this grating? Explain.

Section 25.7 The Michelson Interferometer

42. Light of wavelength 550 nm is used to calibrate a Michelson interferometer. With the use of a micrometer screw, the platform on which one mirror is mounted is moved 0.180 mm. How many fringe shifts are counted?

43. An interferometer is used to measure the length of a bacterium. The wavelength of the light used is 650 nm. As one arm of the interferometer is moved from one end of the cell to the other, 310 fringe shifts are counted. How long is the bacterium?

44. Mirror M_1 in Figure 25.15 is displaced a distance ΔL. During this displacement, 250 fringe shifts are counted. The light being used has a wavelength of 632.8 nm. Calculate the displacement ΔL.

45. A thin sheet of transparent material has an index of refraction of 1.40 and is 15.0 μm thick. When it is inserted in the light path along one arm of an interferometer, how many fringe shifts occur in the pattern? Assume that the wavelength (in a vacuum) of the light used is 600 nm. [*Hint:* The wavelength will change within the material.]

46. The Michelson interferometer can be used to measure the index of refraction of a gas by placing an evacuated transparent tube in the light path along one arm of the device. Fringe shifts occur as the gas is slowly added to the tube. Assume that 600-nm light is used, that the tube is 5.00 cm long, and that 160 fringe shifts occur as the pressure of the gas in the tube increases to atmospheric pressure. What is the index of refraction of the gas? [*Hint:* The fringe shifts occur because the wavelength of the light changes inside the gas-filled tube.]

47. The light path in one arm of a Michelson interferometer includes a transparent cell that is 5.00 cm long. How many fringe shifts would be observed if all the air were evacuated from the cell? The wavelength of the light source is 590 nm and the refractive index of air is 1.000 29. (See the hint in Problem 46.)

ADDITIONAL PROBLEMS

48. A person with a nearsighted eye has near and far points of 16 cm and 25 cm, respectively. (a) Assuming a lens is placed 2.0 cm from the eye, what power must the lens have to correct this condition? (b) Suppose that contact lenses placed directly on the cornea are used to correct the person's eye. What is the power of the lens required in this case, and what is the new near point? [*Hint:* The contact lens and the eyeglass lens require slightly different powers because they are at different distances from the eye.]

49. The near point of an eye is 75.0 cm. (a) What should be the power of a corrective lens prescribed to enable the eye to see an object clearly at 25.0 cm? (b) If, using the corrective lens, the person can see an object clearly at 26.0 cm, but not at 25.0 cm, by how many diopters did the lens grinder miss the prescription?

50. If a typical eyeball is 2.00 cm long and has a pupil opening that can range from about 2.00 mm to 6.00 mm, what

are (a) the focal length of the eye when it is focused on objects 1.00 m away, (b) the smallest *f*-number of the eye when it is focused on objects 1.00 m away, and (c) the largest *f*-number of the eye when it is focused on objects 1.00 m away?

51. **Physics⊗Now™** A cataract-impaired lens in an eye may be surgically removed and replaced by a manufactured lens. The focal length required for the new lens is determined by the lens-to-retina distance, which is measured by a sonarlike device, and by the requirement that the implant provide for correct distance vision. (a) If the distance from lens to retina is 22.4 mm, calculate the power of the implanted lens in diopters. (b) Since there is no accommodation and the implant allows for correct distance vision, a corrective lens for close work or reading must be used. Assume a reading distance of 33.0 cm, and calculate the power of the lens in the reading glasses.

52. Estimate the minimum angle subtended at the eye of a hawk flying at an altitude of 50 m necessary to recognize a mouse on the ground.

53. The wavelengths of the sodium spectrum are $\lambda_1 = 589.00$ nm and $\lambda_2 = 589.59$ nm. Determine the minimum number of lines in a grating that will allow resolution of the sodium spectrum in (a) the first order and (b) the third order.

54. The text discusses the astronomical telescope. Another type is the Galilean telescope, in which an objective lens gathers light (Fig. P25.54) and tends to form an image at point *A*. An eyepiece consisting of a diverging lens intercepts the light before it comes to a focus and forms a virtual image at point *B*. When adjusted for minimum eyestrain, *B* is an infinite distance in front of the lens and parallel rays emerge from the lens, as in Figure P25.54b. An opera glass, which is a Galilean telescope, is used to view a 30.0-cm-tall singer's head that is 40.0 m from the objective lens. The focal length of the objective is + 8.00 cm, and that of the eyepiece is − 2.00 cm. The telescope is adjusted so parallel rays enter the eye. Compute (a) the size of the real image that would have been formed by the objective, (b) the virtual object distance for the diverging lens, (c) the distance between the lenses, and (d) the overall angular magnification.

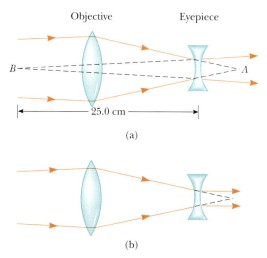

(a)

(b)

Figure P25.54

55. A laboratory (astronomical) telescope is used to view a scale that is 300 cm from the objective, which has a focal length of 20.0 cm; the eyepiece has a focal length of 2.00 cm. Calculate the angular magnification when the telescope is adjusted for minimum eyestrain. [*Note:* The object is not at infinity, so the simple expression $m = f_o/f_e$ is not sufficiently accurate for this problem. Also, assume small angles, so that $\tan \theta \approx \theta$.)

56. If the aqueous humor of the eye has an index of refraction of 1.34 and the distance from the vertex of the cornea to the retina is 2.00 cm, what is the radius of curvature of the cornea for which distant objects will be focused on the retina? (For simplicity, assume that all refraction occurs in the aqueous humor.)

57. A boy scout starts a fire by using a lens from his eyeglasses to focus sunlight on kindling 5.0 cm from the lens. The boy scout has a near point of 15 cm. When the lens is used as a simple magnifier, (a) what is the maximum magnification that can be achieved, and (b) what is the magnification when the eye is relaxed? [*Caution:* The equations derived in the text for a simple magnifier assume a "normal" eye.]

ACTIVITIES

1. (a) Move this book toward your face until the letters just begin to blur. The distance from the book to your eye is your near point. (b) On a sheet of paper, make a dot near the center. Then place an *x* about 3 inches to the left of the dot and another *x* about 3 inches to the right of the dot. With one eye shut and while looking at the dot, move the paper slowly toward your eye. You will notice that a certain distance from your eye, one of the **x**'s will disappear. This is the location of the blind spot of your eye—the point where the optic nerve enters the eye. (c) Stand before a mirror in a darkened room for a few minutes. Then turn on a light in the room and observe your pupils in the mirror as they change size. Such adaptation to the dark also takes place at the rods and cones as they chemically adjust their sensitivity. This adjustment takes 15–30 minutes, as you may have noted whenever you entered a darkened movie theater. Iris aperture control takes less than a second and helps protect the retina from overload.

2. On a sunny day, hold a magnifying glass above a nonflammable surface, such as a sidewalk, so the image of the Sun forms a round spot of light on the surface. Note where the spot formed by the lens is most distinct, or smallest. Use a ruler to measure the distance between the glass and the image. The distance is equal to the focal length of the lens.

3. Hold a pair of prescription glasses about 12 cm from your eye, and look at different objects through the lenses. Try this with different types of glasses, such as those for farsightedness and nearsightedness, and describe what effects the differences have on the image you see. If you have bifocals, how do the images produced by the top and bottom portions of the bifocal lens compare?

4. If you have never experimented with a 35-mm camera with adjustable *f*-numbers and shutter speeds, use up a couple of rolls of film to see what happens. Take several shots of the same object with different settings for these two variables. (You should record your *f*-numbers and shutter speeds for each photograph.) Explain any differences you see in the final images in terms of the settings used.

Courtesy of the Archives, California Institute of Technology

Albert Einstein revolutionized modern physics. He explained the random movements of pollen grains, which proved the existence of atoms, and the photoelectric effect, which showed that light was a particle as well as a wave. His theory of special relativity made clear the foundations of space and time, and his theory of gravitation—general relativity—is the most accurate theory in physics today. He was also deeply concerned with the social impact of scientific discovery.

CHAPTER

26

Relativity

Most of our everyday experiences and observations have to do with objects that move at speeds much less than the speed of light. Newtonian mechanics was formulated to describe the motion of such objects, and its formalism is quite successful in describing a wide range of phenomena that occur at low speeds. It fails, however, when applied to particles having speeds approaching that of light.

This chapter introduces Einstein's theory of special relativity and includes a section on general relativity. The concepts of special relativity often violate our common sense. Moving clocks run slow, and the length of a moving meter stick is contracted. Nonetheless, the theory has been rigorously tested, correctly predicting the results of experiments involving speeds near the speed of light. The theory is verified daily in particle accelerators around the world.

26.1 INTRODUCTION

Experimentally, the predictions of Newtonian theory can be tested at high speeds by accelerating electrons or other charged particles through a large electric potential difference. For example, it's possible to accelerate an electron to a speed of $0.99c$ (where c is the speed of light) by using a potential difference of several million volts. According to Newtonian mechanics, if the potential difference is increased by a factor of 4, the electron's kinetic energy is four times greater and its speed should double to $1.98c$. However, experiments show that the speed of the electron—as well as the speed of any other particle that has mass—always remains *less* than the speed of light, regardless of the size of the accelerating voltage.

The existence of a universal speed limit has far-reaching consequences. It means that the usual concepts of force, momentum, and energy no longer apply for rapidly moving objects. Less obvious consequences include the fact that observers moving at different speeds will measure different time intervals and displacements between the same two events. Newtonian mechanics was contradicted by experimental observations, so it was necessary to replace it with another theory.

In 1905, at the age of 26, Einstein published his special theory of relativity. Regarding the theory, Einstein wrote:

> The relativity theory arose from necessity, from serious and deep contradictions in the old theory from which there seemed no escape. The strength of the new theory lies in the consistency and simplicity with which it solves all these difficulties, using only a few very convincing assumptions.[1]

Although Einstein made many other important contributions to science, his theory of relativity alone represents one of the greatest intellectual achievements of all time. With this theory, experimental observations can be correctly predicted over the range of speeds from $v = 0$ to speeds approaching the speed of light. Newtonian mechanics, which was accepted for more than 200 years, remains valid, but only for speeds much smaller than the speed of light.

At the foundation of special relativity is reconciling the measurements of two observers moving relative to each other. Normally, two such observers will measure different outcomes for the same event. If the measurement is the speed of a car, for example, an observer standing on the road will measure a different speed for the car than an observer in a truck traveling at speed v relative the stationary observer. Special relativity is all about relating two such measurements—and this rather innocuous relating of measurements leads to some of the most bizarre consequences in physics!

26.2 THE PRINCIPLE OF GALILEAN RELATIVITY

In order to describe a physical event, it's necessary to choose a *frame of reference*. For example, when you perform an experiment in a laboratory, you select a coordinate system, or frame of reference, that is at rest with respect to the laboratory. However, suppose an observer in a passing car moving at a constant velocity with respect to the lab were to observe your experiment. Would the observations made by the moving observer differ dramatically from yours? That is, if you found Newton's first law to be valid in your frame of reference, would the moving observer agree with you?

According to the principle of Galilean relativity, **the laws of mechanics must be the same in all inertial frames of reference**. Inertial frames of reference are those reference frames in which Newton's laws are valid. Practically, such frames are those in which objects subjected to no forces move in straight lines at constant speed—thus the name "inertial frame" because objects observed from these frames obey Newton's first law, the law of inertia. For the situation described in the previous paragraph, the laboratory coordinate system and the coordinate system of the moving car are both inertial frames of reference. Consequently, if the laws of mechanics are found to be true in the laboratory, then the person in the car must also observe the same laws.[2]

Consider an airplane in flight, moving with a constant velocity, as in Figure 26.1a. If a passenger in the airplane throws a ball straight up in the air, the passenger observes that the ball moves in a vertical path. The motion of the ball is precisely the same as it would be if the ball were thrown while at rest on Earth. The law of gravity and the equations of motion under constant acceleration are obeyed whether the airplane is at rest or in uniform motion.

Now consider the same experiment when viewed by another observer at rest on Earth. This stationary observer views the path of the ball in the plane to be a parabola, as in Figure 26.1b. Further, according to this observer, the ball has a velocity to the right equal to the velocity of the plane. Although the two observers disagree on the shape of the ball's path, both agree that the motion of the ball obeys the law of gravity and Newton's laws of motion, and even agree on how long

---

[1]A. Einstein and L. Infeld, *The Evolution of Physics* (New York: Simon and Schuster, 1961).
[2]What is an example of a *non*inertial frame? A frame undergoing translational acceleration or a frame rotating with respect to the two inertial frames just mentioned.

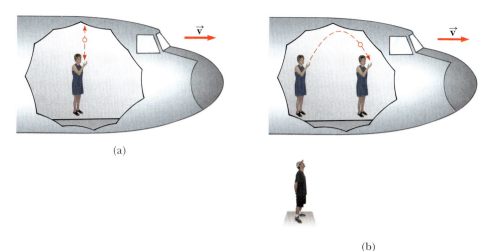

(a)

(b)

Figure 26.1   (a) The observer on the airplane sees the ball move in a vertical path when thrown upward. (b) The observer on Earth views the path of the ball to be a parabola.

the ball is in the air. We draw the following important conclusion: **There is no preferred frame of reference for describing the laws of mechanics**.

26.3   THE SPEED OF LIGHT

It's natural to ask whether the concept of Galilean relativity in mechanics also applies to experiments in electricity, magnetism, optics, and other areas. Experiments indicate the answer is no. For example, if we assume that the laws of electricity and magnetism are the same in all inertial frames, a paradox concerning the speed of light immediately arises. This can be understood by recalling that, according to electromagnetic theory, the speed of light always has the fixed value of $2.997\,924\,58 \times 10^8$ m/s in free space. But this is in direct contradiction to common sense. For example, suppose a light pulse is sent out by an observer in a boxcar moving with a velocity $\vec{v}$ (Fig. 26.2). The light pulse has a velocity $\vec{c}$ relative to observer S' in the boxcar. According to Galilean relativity, the speed of the pulse relative to the stationary observer S outside the boxcar should be $c + v$. This obviously contradicts Einstein's theory, which postulates that the velocity of the light pulse is the same for all observers.

In order to resolve this paradox, we must conclude that either (1) the addition law for velocities is incorrect or (2) the laws of electricity and magnetism are not the same in all inertial frames. Assume that the second conclusion is true; then a preferred reference frame must exist in which the speed of light has the value c, but in any other reference frame the speed of light must have a value that is greater or less than c. It's useful to draw an analogy with sound waves, which propagate through a medium such as air. The speed of sound in air is about 330 m/s when measured in a reference frame in which the air is stationary. However, the speed of sound is greater or less than this value when measured from a reference frame that is moving with respect to the air.

In the case of light signals (electromagnetic waves), recall that electromagnetic theory predicted that such waves must propagate through free space with a speed

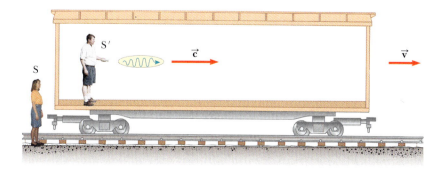

Figure 26.2   A pulse of light is sent out by a person in a moving boxcar. According to Newtonian relativity, the speed of the pulse should be $\vec{c} + \vec{v}$ relative to a stationary observer.

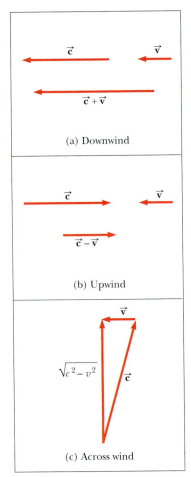

Figure 26.3 If the speed of the ether wind relative to Earth is v, and c is the speed of light relative to the ether, the speed of light relative to Earth is (a) $c + v$ in the downwind direction, (b) $c - v$ in the upwind direction, and (c) $\sqrt{c^2 - v^2}$ in the direction perpendicular to the wind.

equal to the speed of light. However, the theory doesn't require the presence of a medium for wave propagation. This is in contrast to other types of waves, such as water and sound waves, that do require a medium to support the disturbances. In the 19th century, physicists thought that electromagnetic waves also required a medium in order to propagate. They proposed that such a medium existed and gave it the name **luminiferous ether**. The ether was assumed to be present everywhere, even in empty space, and light waves were viewed as ether oscillations. Further, the ether would have to be a massless but rigid medium with no effect on the motion of planets or other objects. These are strange concepts indeed. In addition, it was found that the troublesome laws of electricity and magnetism would take on their simplest forms in a special frame of reference at *rest* with respect to the ether. This frame was called the *absolute frame*. The laws of electricity and magnetism would be valid in this absolute frame, but they would have to be modified in any reference frame moving with respect to the absolute frame.

As a result of the importance attached to the ether and the absolute frame, it became of considerable interest in physics to prove by experiment that they existed. Since it was considered likely that Earth was in motion through the ether, from the view of an experimenter on Earth, there was an "ether wind" blowing through his laboratory. A direct method for detecting the ether wind would use an apparatus fixed to Earth to measure the wind's influence on the speed of light. If v is the speed of the ether relative to Earth, then the speed of light should have its maximum value, $c + v$, when propagating downwind, as shown in Figure 26.3a. Likewise, the speed of light should have its minimum value, $c - v$, when propagating upwind, as in Figure 26.3b, and an intermediate value, $(c^2 - v^2)^{1/2}$, in the direction perpendicular to the ether wind, as in Figure 26.3c. If the Sun were assumed to be at rest in the ether, then the velocity of the ether wind would be equal to the orbital velocity of Earth around the Sun, which has a magnitude of approximately 3×10^4 m/s. Because $c = 3 \times 10^8$ m/s, it should be possible to detect a change in speed of about 1 part in 10^4 for measurements in the upwind or downwind directions. However, as we will see in the next section, all attempts to detect such changes and establish the existence of the ether (and hence the absolute frame) proved futile.

In conclusion, we see that the second hypothesis in our introduction to this section is false—and we now believe that **the laws of electricity and magnetism are the same in all inertial frames**. It is the simple classical addition laws for velocities that are incorrect and must be modified, as shown in Section 26.8.

26.4 THE MICHELSON–MORLEY EXPERIMENT

The most famous experiment designed to detect small changes in the speed of light was first performed in 1881 by Albert A. Michelson (1852–1931) and later repeated under various conditions by Michelson and Edward W. Morley (1838–1923). We state at the outset that the outcome of the experiment contradicted the ether hypothesis.

The experiment was designed to determine the velocity of Earth relative to the hypothetical ether. The experimental tool used was the Michelson interferometer, which was discussed in Section 25.7 and is shown again in Active Figure 26.4. Arm 2 is aligned along the direction of Earth's motion through space. Earth's moving through the ether at speed v is equivalent to the ether's flowing past Earth in the opposite direction with speed v. This ether wind blowing in the direction opposite the direction of Earth's motion should cause the speed of light measured in Earth frame to be $c - v$ as the light approaches mirror M_2 and $c + v$ after reflection, where c is the speed of light in the ether frame.

The two beams reflected from M_1 and M_2 recombine, and an interference pattern consisting of alternating dark and bright fringes is formed. The interference pattern was observed while the interferometer was rotated through an angle of 90°. This rotation supposedly would change the speed of the ether wind along the direction of arm 1. The effect of such rotation should have been to cause the fringe pattern to shift slightly but measurably; however, measurements failed to show any change in the interference pattern! The Michelson–Morley experiment was repeated at different times of the year when the ether wind was expected to change direction, but the results were always the same: **no fringe shift of the magnitude required was ever observed**.

The negative results of the Michelson–Morley experiment not only contradicted the ether hypothesis, but also showed that it was impossible to measure the absolute velocity of Earth with respect to the ether frame. However, as we will see in the next section, Einstein suggested a postulate in the special theory of relativity that places quite a different interpretation on these negative results. In later years, when more was known about the nature of light, the idea of an ether that permeates all of space was relegated to the theoretical graveyard. **Light is now understood to be an electromagnetic wave, which requires no medium for its propagation**. As a result, the idea of an ether in which these waves could travel became unnecessary.

Details of the Michelson–Morley Experiment

As we mentioned earlier, the Michelson–Morley experiment was designed to detect the motion of Earth with respect to the ether. Before we examine the details of this historical experiment, it is instructive to consider a race between two airplanes, as shown in Figure 26.5a. One airplane flies from point O to point A perpendicular to the direction of the wind, and the second airplane flies from point O to point B parallel to the wind. We will assume that they start at O at the same time, travel the same distance L with the same cruising speed c with respect to the wind, and return to O. Which airplane will win the race? In order to answer this question, we calculate the time of flight for both airplanes.

First, consider the airplane that moves along path I parallel to the wind. As it moves to the right, its speed is enhanced by the wind, and its speed with respect to Earth is $c + v$. As it moves to the left on its return journey, it must fly opposite the wind; hence, its speed with respect to Earth is $c - v$. The times of flight to the right and to the left are, respectively,

$$t_R = \frac{L}{c + v} \quad \text{and} \quad t_L = \frac{L}{c - v}$$

and the total time of flight for the airplane moving along path I is

$$t_1 = t_R + t_L = \frac{L}{c + v} + \frac{L}{c - v} = \frac{2Lc}{c^2 - v^2}$$

$$= \frac{2L}{c\left(1 - \dfrac{v^2}{c^2}\right)} \qquad \text{[26.1]}$$

Now consider the airplane flying along path II. If the pilot aims the airplane directly toward point A, it will be blown off course by the wind and won't reach its

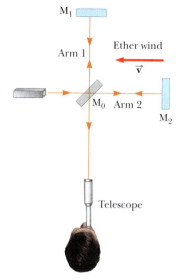

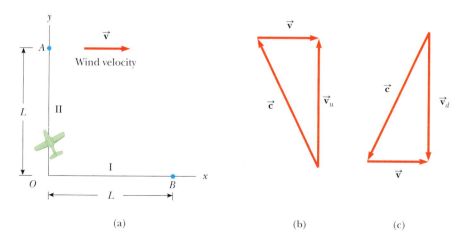

(a)   (b)   (c)

Figure 26.5 (a) If an airplane travels from O to A with a wind blowing to the right, it must head into the wind at some angle. (b) Vector diagram for determining the airplane's direction for the trip from O to A. (c) Vector diagram for determining its direction for the trip from A to O.

destination. To compensate for the wind, the pilot must point the airplane into the wind at some angle, as shown in Figure 26.5a. This angle must be selected so that the vector sum of $\vec{c}$ and $\vec{v}$ leads to a velocity vector pointed directly toward A. The resultant vector diagram is shown in Figure 26.5b, where $\vec{v}_u$ is the velocity of the airplane with respect to the ground as it moves from O to A. From the Pythagorean theorem, the magnitude of the vector $\vec{v}_u$ is

$$v_u = \sqrt{c^2 - v^2} = c\sqrt{1 - \frac{v^2}{c^2}}$$

Likewise, on the return trip from A to O, the pilot must again head into the wind so that the airplane's velocity $\vec{v}_d$ with respect to Earth will be directed toward O, as shown in Figure 26.5c. From this figure, we see that

$$v_d = \sqrt{c^2 - v^2} = c\sqrt{1 - \frac{v^2}{c^2}}$$

The total time of flight for the trip along path II is therefore

$$t_2 = \frac{L}{v_u} + \frac{L}{v_d} = \frac{L}{c\sqrt{1 - \frac{v^2}{c^2}}} + \frac{L}{c\sqrt{1 - \frac{v^2}{c^2}}}$$

$$= \frac{2L}{c\sqrt{1 - \frac{v^2}{c^2}}} \qquad\qquad\qquad \text{[26.2]}$$

Comparing Equations 26.1 and 26.2, we see that the airplane flying along path II wins the race. The difference in flight times is given by

$$\Delta t = t_1 - t_2 = \frac{2L}{c}\left[\frac{1}{\left(1 - \frac{v^2}{c^2}\right)} - \frac{1}{\sqrt{1 - \frac{v^2}{c^2}}}\right]$$

This expression can be simplified by noting that the ratio of wind speed to plane speed, v/c, is usually much smaller than 1, and by using the following binomial expansions in v/c after dropping all terms higher than second order:

$$\left(1 - \frac{v^2}{c^2}\right)^{-1} \approx 1 + \frac{v^2}{c^2}$$

and

$$\left(1 - \frac{v^2}{c^2}\right)^{-1/2} \approx 1 + \frac{1}{2}\frac{v^2}{c^2}$$

The difference in flight times is therefore

$$\Delta t \approx \frac{Lv^2}{c^3} \qquad \text{for} \qquad v/c \ll 1 \qquad\qquad \text{[26.3]}$$

The analogy between this airplane race and the Michelson–Morley experiment is shown in Figure 26.6a. Two beams of light travel along two arms of an interferometer. In this case, the "wind" is the ether blowing across Earth from left to right as Earth moves through the ether from right to left. Because the speed of Earth in its orbital path is approximately 3×10^4 m/s, it is reasonable to use that value for the speed of the ether wind. Notice in this case that $v/c \approx 1 \times 10^{-4} \ll 1$. The two light beams start out in phase and return to form an interference pattern. We assume that the interferometer is adjusted for parallel fringes and that a telescope is focused on one of these fringes. The time difference between the two light beams gives rise to a phase difference between the beams, producing an interference pattern when they combine at the position of the telescope. The difference in the pattern is detected by rotating the interferometer through 90° in a horizontal plane, so that the two beams exchange roles (Fig. 26.6b). This results in a net

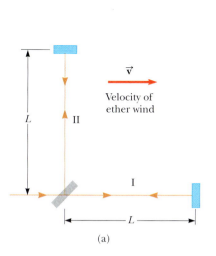

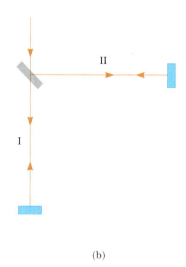

(a)   (b)

Figure 26.6 (a) Top view of the Michelson–Morley interferometer, where $\vec{v}$ is the velocity of the ether and L is the length of each arm. (b) When the interferometer is rotated by 90°, the role of each arm is reversed.

time shift of twice the time difference given by Equation 26.3. The net time difference is therefore

$$\Delta t_{net} = 2\,\Delta t = \frac{2Lv^2}{c^3} \qquad [26.4]$$

The corresponding path difference is

$$\Delta d = c\,\Delta t_{net} = \frac{2Lv^2}{c^2} \qquad [26.5]$$

In the first experiments by Michelson and Morley, each light beam was reflected by the mirrors many times to give an increased effective path length L of about 11 meters. Using this value and taking v to be equal to 3×10^4 m/s gives a path difference of

$$\Delta d = \frac{2(11\ \text{m})(3.0 \times 10^4\ \text{m/s})^2}{(3.0 \times 10^8\ \text{m/s})^2} = 2.2 \times 10^{-7}\ \text{m}$$

This extra travel distance should produce a noticeable shift in the fringe pattern. Specifically, calculations show that if the pattern is viewed while the interferometer is rotated through 90°, a shift of about 0.4 fringe should be observed. The instrument used by Michelson and Morley was capable of detecting a shift in the fringe pattern as small as 0.01 fringe. However, *it detected no shift whatsoever in the fringe pattern.* Since then, the experiment has been repeated many times by different scientists under a wide variety of conditions and no fringe shift has ever been detected. The inescapable conclusion is that motion of Earth with respect to the ether can't be detected.

Many efforts were made to explain the null results of the Michelson–Morley experiment and to save the ether frame concept and the Galilean addition law for the velocity of light. All proposals resulting from these efforts have been shown to be wrong. No experiment in the history of physics has received such valiant efforts to explain the absence of an expected result as was the Michelson–Morley experiment. The stage was set for Einstein, who, at the age of only 26, solved the problem in 1905 with his special theory of relativity.

26.5 EINSTEIN'S PRINCIPLE OF RELATIVITY

In the previous section we noted the serious contradiction between the Galilean addition law for velocities and the fact that the speed of light is the same for all observers. In 1905 Albert Einstein proposed a theory that resolved this contradiction but at the same time completely altered our notions of space and time. He based his special theory of relativity on two postulates:

ALBERT EINSTEIN,
German-American Physicist
(1879–1955)

One of the greatest physicists of all time, Einstein was born in Ulm, Germany. In 1905, at the age of 26, he published four scientific papers that revolutionized physics. Two of these papers were concerned with what is now considered his most important contribution: the special theory of relativity. In 1916, Einstein published his work on the general theory of relativity. The most dramatic prediction of this theory is the degree to which light is deflected by a gravitational field. Measurements made by astronomers on bright stars in the vicinity of the eclipsed Sun in 1919 confirmed Einstein's prediction, and as a result, Einstein became a world celebrity. Einstein was deeply disturbed by the development of quantum mechanics in the 1920s despite his own role as a scientific revolutionary. In particular, he could never accept the probabilistic view of events in nature that is a central feature of quantum theory. The last few decades of his life were devoted to an unsuccessful search for a unified theory that would combine gravitation and electromagnetism.

Postulates of relativity ▶

1. **The principle of relativity**: All the laws of physics are the same in all inertial frames.
2. **The constancy of the speed of light:** The speed of light in a vacuum has the same value, $c = 2.997\ 924\ 58 \times 10^8$ m/s, in all inertial reference frames, regardless of the velocity of the observer or the velocity of the source emitting the light.

The first postulate asserts that *all* the laws of physics are the same in all reference frames moving with constant velocity relative to each other. This postulate is a sweeping generalization of the principle of Galilean relativity, which refers only to the laws of mechanics. From an experimental point of view, Einstein's principle of relativity means that *any* kind of experiment—mechanical, thermal, optical, or electrical—performed in a laboratory at rest, must give the same result when performed in a laboratory moving at a constant speed past the first one. Hence, no preferred inertial reference frame exists, and it is impossible to detect absolute motion.

Although postulate 2 was a brilliant theoretical insight on Einstein's part in 1905, it has since been confirmed experimentally in many ways. Perhaps the most direct demonstration involves measuring the speed of photons emitted by particles traveling at 99.99% of the speed of light. The measured photon speed in this case agrees to five significant figures with the speed of light in empty space.

The null result of the Michelson–Morley experiment can be readily understood within the framework of Einstein's theory. According to his principle of relativity, the premises of the Michelson–Morley experiment were incorrect. In the process of trying to explain the expected results, we stated that when light traveled against the ether wind its speed was $c - v$. However, if the state of motion of the observer or of the source has no influence on the value found for the speed of light, the measured value must always be c. Likewise, the light makes the return trip after reflection from the mirror at a speed of c, not at a speed of $c + v$. Thus, the motion of Earth does not influence the fringe pattern observed in the Michelson–Morley experiment, and a null result should be expected.

If we accept Einstein's theory of relativity, we must conclude that uniform relative motion is unimportant when measuring the speed of light. At the same time, we have to adjust our commonsense notions of space and time and be prepared for some rather bizarre consequences.

26.6   CONSEQUENCES OF SPECIAL RELATIVITY

Almost everyone who has dabbled even superficially in science is aware of some of the startling predictions that arise because of Einstein's approach to relative motion. As we examine some of the consequences of relativity in this section, we'll find that they conflict with some of our basic notions of space and time. We will restrict our discussion to the concepts of length, time, and simultaneity, which are quite different in relativistic mechanics from what they are in Newtonian mechanics. For example, in relativistic mechanics, the distance between two points and the time interval between two events depend on the frame of reference in which they are measured. **In relativistic mechanics, there is no such thing as absolute length or absolute time.** Further, **events at different locations that are observed to occur simultaneously in one frame are not observed to be simultaneous in another frame moving uniformly past the first.**

Absolute length and absolute time intervals are meaningless in relativity. ▶

Simultaneity and the Relativity of Time

A basic premise of Newtonian mechanics is that a universal time scale exists that is the same for all observers. In fact, Newton wrote, "Absolute, true, and mathematical time, of itself, and from its own nature, flows equably without relation to anything external." Newton and his followers simply took simultaneity for granted. In his special theory of relativity, Einstein abandoned that assumption.

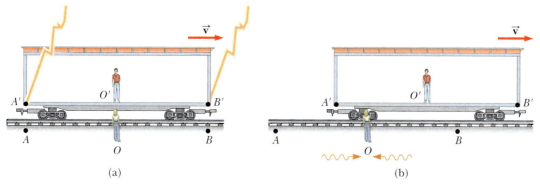

Figure 26.7   Two lightning bolts strike the ends of a moving boxcar. (a) The events appear to be simultaneous to the stationary observer at O, who is midway between A and B. (b) The events don't appear to be simultaneous to the observer at O', who claims that the front of the train is struck *before* the rear.

Einstein devised the following thought experiment to illustrate this point: a boxcar moves with uniform velocity, and two lightning bolts strike its ends, as in Figure 26.7a, leaving marks on the boxcar and the ground. The marks on the boxcar are labeled A' and B', and those on the ground are labeled A and B. An observer at O' moving with the boxcar is midway between A' and B', and an observer on the ground at O is midway between A and B. The events recorded by the observers are the striking of the boxcar by the two lightning bolts.

The light signals recording the instant at which the two bolts struck reach observer O at the same time, as indicated in Figure 26.7b. This observer realizes that the signals have traveled at the same speed over equal distances, and so rightly concludes that the events at A and B occurred simultaneously. Now consider the same events as viewed by observer O'. By the time the signals have reached observer O, observer O' has moved as indicated in Figure 26.7b. Thus, the signal from B' has already swept past O', but the signal from A' has not yet reached O'. In other words, O' sees the signal from B' before seeing the signal from A'. According to Einstein, *the two observers must find that light travels at the same speed.* Therefore, observer O' concludes that the lightning struck the front of the boxcar before it struck the back.

This thought experiment clearly demonstrates that the two events which appear to be simultaneous to observer O do not appear to be simultaneous to observer O'. In other words,

> Two events that are simultaneous in one reference frame are in general not simultaneous in a second frame moving relative to the first. Simultaneity depends on the state of motion of the observer, and is therefore not an absolute concept.

At this point, you might wonder which observer is right concerning the two events. The answer is that *both* are correct, because the principle of relativity states that **there is no preferred inertial frame of reference**. Although the two observers reach different conclusions, both are correct in their own reference frames because the concept of simultaneity is not absolute. In fact, this is the central point of relativity. Any inertial frame of reference can be used to describe events and do physics.

TIP 26.1   Who's Right?

Which person is correct concerning the simultaneity of the two events? Both are correct, because the principle of relativity states that no inertial frame of reference is preferred. Although the two observers may reach different conclusions, both are correct in their own reference frame. Any uniformly moving frame of reference can be used to describe events and do physics.

Time Dilation

We can illustrate the fact that observers in different inertial frames may measure different time intervals between a pair of events by considering a vehicle moving to the right with a speed v as in Active Figure 26.8a (page 852). A mirror is fixed to the ceiling of the vehicle, and an observer O' at rest in this system holds a laser a distance d below the mirror. At some instant, the laser emits a pulse of light

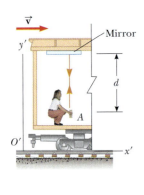

(a)

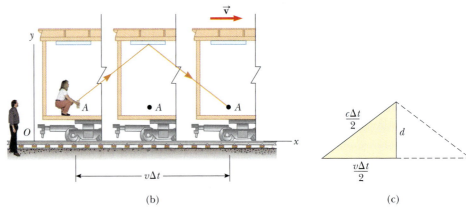

(b)   (c)

ACTIVE FIGURE 26.8

(a) A mirror is fixed to a moving vehicle, and a light pulse leaves O' at rest in the vehicle. (b) Relative to a stationary observer on Earth, the mirror and O' move with a speed v. Note that the distance the pulse travels is greater than $2d$ as measured by the stationary observer. (c) The right triangle for calculating the relationship between Δt and Δt_p.

Physics Now™

Log into PhysicsNow at **www.cp7e.com** and go to Active Figure 26.8, where you can observe the bouncing of the light pulse for various speeds of the train.

directed toward the mirror (event 1), and at some later time after reflecting from the mirror, the pulse arrives back at the laser (event 2). Observer O' carries a clock and uses it to measure the time interval Δt_p between these two events which she views as occurring at the same place. (The subscript p stands for *proper*, as we'll see in a moment.) Because the light pulse has a speed c, the time it takes it to travel from point A to the mirror and back to point A is

$$\Delta t_p = \frac{\text{Distance traveled}}{\text{Speed}} = \frac{2d}{c} \qquad \text{[26.6]}$$

The time interval Δt_p measured by O' requires only a single clock located at the same place as the laser in this frame.

Now consider the same set of events as viewed by O in a second frame, as shown in Active Figure 26.8b. According to this observer, the mirror and laser are moving to the right with a speed v, and as a result, the sequence of events appears different. By the time the light from the laser reaches the mirror, the mirror has moved to the right a distance $v\,\Delta t/2$, where Δt is the time it takes the light pulse to travel from point A to the mirror and back to point A as measured by O. In other words, O concludes that, because of the motion of the vehicle, if the light is to hit the mirror, it must leave the laser at an angle with respect to the vertical direction. Comparing Active Figures 26.8a and 26.8b, we see that the light must travel farther in (b) than in (a). (Note that neither observer "knows" that he or she is moving. Each is at rest in his or her own inertial frame.)

According to the second postulate of the special theory of relativity, both observers must measure c for the speed of light. Because the light travels farther in the frame of O, it follows that the time interval Δt measured by O is longer than the time interval Δt_p measured by O'. To obtain a relationship between these two time intervals, it is convenient to examine the right triangle shown in Active Figure 26.8c. The Pythagorean theorem gives

$$\left(\frac{c\Delta t}{2}\right)^2 = \left(\frac{v\Delta t}{2}\right)^2 + d^2$$

Solving for Δt yields

$$\Delta t = \frac{2d}{\sqrt{c^2 - v^2}} = \frac{2d}{c\sqrt{1 - v^2/c^2}}$$

Because $\Delta t_p = 2d/c$, we can express this result as

$$\Delta t = \frac{\Delta t_p}{\sqrt{1 - v^2/c^2}} = \gamma \, \Delta t_p$$ **[26.7]**   ◀ Time dilation

where

$$\gamma = \frac{1}{\sqrt{1 - v^2/c^2}}$$ **[26.8]**

Because γ is always greater than one, Equation 26.7 says that **the time interval Δt between two events measured by an observer moving with respect to a clock[3] is longer than the time interval Δt_p between the same two events measured by an observer at rest with respect to the clock**. Consequently, $\Delta t > \Delta t_p$, and the proper time interval is expanded or dilated by the factor γ. Hence, this effect is known as **time dilation**.

For example, suppose the observer at rest with respect to the clock measures the time required for the light flash to leave the laser and return. We assume that the measured time interval in this frame of reference, Δt_p, is one second. (This would require a very tall vehicle.) Now we find the time interval as measured by observer O moving with respect to the same clock. If observer O is traveling at half the speed of light ($v = 0.500c$), then $\gamma = 1.15$, and according to Equation 26.7, $\Delta t = \gamma \, \Delta t_p = 1.15(1.00 \text{ s}) = 1.15$ s. Therefore, when observer O' claims that 1.00 s has passed, observer O claims that 1.15 s has passed. Observer O considers the clock of O' to be reading too low a value for the elapsed time between the two events and says that the clock of O' is "running slow." From this phenomenon, we may conclude the following:

> A clock moving past an observer at speed v runs more slowly than an identical clock at rest with respect to the observer by a factor of γ^{-1}.

◀ A clock in motion runs more slowly than an identical stationary clock.

The time interval Δt_p in Equations 26.6 and 26.7 is called the **proper time**. In general, **proper time is the time interval between two events as measured by an observer who sees the events occur at the same position**.

Although you may have realized it by now, it's important to spell out that relativity is a scientific democracy: the view of O' that O is really the one moving with speed v to the left and that O's clock is running more slowly is just as valid as the view of O. The principle of relativity requires that the views of two observers in uniform relative motion be equally valid and capable of being checked experimentally.

We have seen that moving clocks run slow by a factor of γ^{-1}. This is true for ordinary mechanical clocks as well as for the light clock just described. In fact, we can generalize these results by stating that all physical processes, including chemical and biological ones, slow down relative to a clock when those processes occur in a frame moving with respect to the clock. For example, the heartbeat of an astronaut moving through space would keep time with a clock inside the spaceship. Both the astronaut's clock and heartbeat would be slowed down relative to a clock back on Earth (although the astronaut would have no sensation of life slowing down in the spaceship).

Time dilation is a very real phenomenon that has been verified by various experiments involving the ticking of natural clocks. An interesting example of time dilation involves the observation of *muons*—unstable elementary particles that are very similar to electrons, having the same charge, but 207 times the mass. Muons can be produced by the collision of cosmic radiation with atoms high in the atmosphere. These particles have a lifetime of 2.2 μs when measured in a reference frame at rest with respect to them. If we take 2.2 μs as the average lifetime of a muon and assume that their speed is close to the speed of light, we find that

TIP 26.2   Proper Time Interval
You must be able to correctly identify the observer who measures the proper time interval. The proper time interval between two events is the time interval measured by an observer for whom the two events take place at the same position.

---

[3]Actually, Figure 26.8 shows the clock moving and not the observer, but this is equivalent to observer O moving to the left with velocity $\vec{v}$ with respect to the clock.

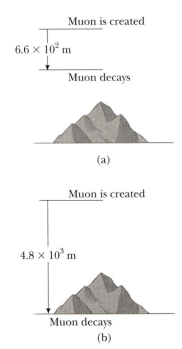

Muon is created

6.6×10^2 m

Muon decays

(a)

Muon is created

4.8×10^3 m

Muon decays

(b)

Figure 26.9   (a) The muons travel only about 6.6×10^2 m as measured in the muons' reference frame, in which their lifetime is about $2.2 \ \mu s$. Because of time dilation, the muons' lifetime is longer as measured by the observer on Earth. (b) Muons traveling with a speed of $0.99c$ travel a distance of about 4.80×10^3 m as measured by an observer on Earth.

these particles can travel only about 600 m before they decay (Fig. 26.9a). Hence, they could never reach Earth from the upper atmosphere where they are produced. However, experiments show that a large number of muons *do* reach Earth, and the phenomenon of time dilation explains how. Relative to an observer on Earth, the muons have a lifetime equal to $\gamma \tau_p$, where $\tau_p = 2.2 \ \mu s$ is the lifetime in a frame of reference traveling with the muons. For example, for $v = 0.99c$, $\gamma \approx 7.1$ and $\gamma \tau_p \approx 16 \ \mu s$. Hence, the average distance muons travel as measured by an observer on Earth is $\gamma v \tau_p \approx 4\,800$ m, as indicated in Figure 26.9b. Consequently, muons can reach Earth's surface.

In 1976 experiments with muons were conducted at the laboratory of the European Council for Nuclear Research (CERN) in Geneva. Muons were injected into a large storage ring, reaching speeds of about $0.9994c$. Electrons produced by the decaying muons were detected by counters around the ring, enabling scientists to measure the decay rate, and hence the lifetime of the muons. The lifetime of the moving muons was measured to be about 30 times as long as that of stationary muons to within two parts in a thousand, in agreement with the prediction of relativity.

Quick Quiz 26.1

Suppose you're an astronaut being paid according to the time you spend traveling in space. You take a long voyage traveling at a speed near that of light. Upon your return to Earth, you're asked how you'd like to be paid: according to the time elapsed on a clock on Earth or according to your ship's clock. Which should you choose in order to maximize your paycheck? (a) the Earth clock (b) the ship's clock (c) Either clock, it doesn't make a difference.

EXAMPLE 26.1   Pendulum Periods

Goal   Apply the concept of time dilation.

Problem   The period of a pendulum is measured to be 3.00 s in the inertial frame of the pendulum. What is the period as measured by an observer moving at a speed of $0.950c$ with respect to the pendulum?

Strategy   Here, we're given the period of the clock as measured by an observer in the rest frame of the clock, so that's a proper time interval Δt_p. We want to know how much time passes as measured by an observer in a frame moving relative to the clock, which is Δt. Substitution into Equation 26.7 then solves the problem.

Solution

Substitute the proper time and relative speed into Equation 26.7:

$$\Delta t = \frac{\Delta t_p}{\sqrt{1 - v^2/c^2}} = \frac{3.00 \text{ s}}{\sqrt{1 - \frac{(0.950c)^2}{c^2}}} = \boxed{9.61 \text{ s}}$$

Remarks   The moving observer considers the *pendulum* to be moving, and moving clocks are observed to run more slowly: while the pendulum oscillates once in 3 s for an observer in the rest frame of the clock, it takes nearly 10 s to oscillate once according the moving observer.

Exercise 26.1

What pendulum period does a third observer moving at $0.900c$ measure?

Answer   6.88 s

The confusion that arises in problems like Example 26.1 lies in the fact that movement is relative: from the point of view of someone in the pendulum's rest frame, the pendulum is standing still (except, of course, for the swinging motion), whereas to someone in a frame that is moving with respect to the pendulum, it's the pendulum that's doing the moving. To keep this straight, always focus on the observer who is doing the measurement, and ask yourself whether the clock being measured is moving with respect to that observer. If the answer is no, then the observer is in the rest frame of the clock and measures the clock's proper time. If the answer is yes, then the time measured by the observer will be dilated — larger than the clock's proper time.

This confusion of perspectives led to the famous "twin paradox."

The Twin Paradox

An intriguing consequence of time dilation is the so-called twin paradox (Fig. 26.10). Consider an experiment involving a set of twins named Speedo and Goslo. When they are 20 years old, Speedo, the more adventuresome of the two, sets out on an epic journey to Planet X, located 20 lightyears from Earth. Further, his spaceship is capable of reaching a speed of $0.95c$ relative to the inertial frame of his twin brother back home. After reaching Planet X, Speedo becomes homesick and immediately returns to Earth at the same speed of $0.95c$. Upon his return, Speedo is shocked to discover that Goslo has aged $2D/v = 2(20 \text{ ly})/(0.95 \text{ ly/y}) = 42$ years and is now 62 years old. Speedo, on the other hand, has aged only 13 years.

Some wrongly consider *this* the paradox; that twins could age at different rates and end up after a period of time having very different ages. While contrary to our common sense, this isn't the paradox at all. The paradox lies in the fact that from Speedo's point of view, *he* was at rest while Goslo (on Earth) sped away from *him* at $0.95c$ and returned later. So Goslo's clock was moving relative to Speedo and hence running slow compared with Speedo's clock. The conclusion: Speedo, not Goslo, should be the older of the twins!

To resolve this apparent paradox, consider a third observer moving at a constant speed of $0.5c$ relative to Goslo. To the third observer, Goslo never changes inertial frames: His speed relative to the third observer is always the same. The third observer notes, however, that Speedo accelerates during his journey, *changing reference frames in the process.* From the third observer's perspective, it's clear that there is something very different about the motion of Goslo when compared to Speedo. The roles played by Goslo and Speedo are not symmetric, so it isn't surprising that time flows differently for each. Further, because Speedo accelerates, he is in a noninertial frame of reference — technically outside the bounds of special relativity (though there are methods for dealing with accelerated motion in relativity). Only Goslo, who is in a single inertial frame, can apply the simple time-dilation formula to Speedo's trip. Goslo finds that instead of aging 42 years,

◀ The space traveler ages more slowly than his twin who remains on Earth.

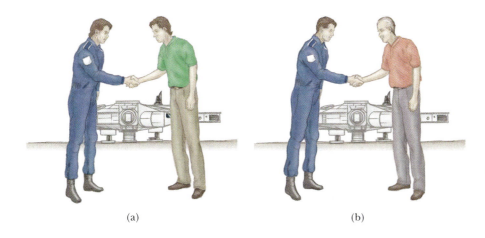

(a)   (b)

Figure 26.10   (a) As the twins depart, they're the same age. (b) When Speedo returns from his journey to Planet X, he's younger than his twin Goslo, who remained on Earth.

Speedo ages only $(1 - v^2/c^2)^{1/2}(42 \text{ years}) = 13$ years. Of these 13 years, Speedo spends 6.5 years traveling to Planet X and 6.5 years returning, for a total travel time of 13 years, in agreement with our earlier statement.

Length Contraction

The measured distance between two points depends on the frame of reference of the observer. The **proper length** L_p of an object is **the length of the object as measured by an observer at rest relative to the object**. The length of an object measured in a reference frame that is moving with respect to the object is always less than the proper length. This effect is known as **length contraction**.

To understand length contraction quantitatively, consider a spaceship traveling with a speed v from one star to another, as seen by two observers, one on Earth and the other in the spaceship. The observer at rest on Earth (and also assumed to be at rest with respect to the two stars) measures the distance between the stars to be L_p. According to this observer, the time it takes the spaceship to complete the voyage is $\Delta t = L_p/v$. Because of time dilation, the space traveler, using his spaceship clock, measures a smaller time of travel: $\Delta t_p = \Delta t/\gamma$. The space traveler claims to be at rest and sees the destination star moving toward the spaceship with speed v. Because the space traveler reaches the star in time Δt_p, he concludes that the distance L between the stars is shorter than L_p. The distance measured by the space traveler is

$$L = v\,\Delta t_p = v\,\frac{\Delta t}{\gamma}$$

Because $L_p = v\,\Delta t$, it follows that

$$L = \frac{L_p}{\gamma} = L_p\sqrt{1 - v^2/c^2} \qquad \text{[26.9]}$$

According to this result, illustrated in Active Figure 26.11, if an observer at rest with respect to an object measures its length to be L_p, an observer moving at a speed v relative to the object will find it to be shorter than its proper length by the factor $\sqrt{1 - v^2/c^2}$. Note that **length contraction takes place only along the direction of motion**.

Time-dilation and length contraction effects have interesting applications for future space travel to distant stars. In order for the star to be reached in a fraction of a human lifetime, the trip must be taken at very high speeds. According to an Earth-bound observer, the time for a spacecraft to reach the destination star will be dilated compared with the time interval measured by travelers. As was discussed in the treatment of the twin paradox, the travelers will be younger than their twins when they return to Earth. Therefore, by the time the travelers reach the star, they will have aged by some number of years, while their partners back on Earth will have aged a larger number of years, the exact ratio depending on the speed of the spacecraft. At a spacecraft speed of $0.94c$, this ratio is about 3:1.

Quick Quiz 26.2

You are packing for a trip to another star, and on your journey you will be traveling at a speed of $0.99c$. Can you sleep in a smaller cabin than usual, because you will be shorter when you lie down? Explain your answer.

Quick Quiz 26.3

You observe a rocket moving away from you. Compared to its length when it was at rest on the ground, you will measure its length to be (a) shorter, (b) longer, or (c) the same. Compared to the passage of time measured by the watch on your wrist, the passage of time on the rocket's clock is (d) faster, (e) slower, or (f) the same. Answer the same questions if the rocket turns around and comes toward you.

TIP 26.3 The Proper Length

You must be able to correctly identify the observer who measures the proper length. The proper length between two points in space is the length measured by an observer at rest with respect to the length. Very often, the proper time interval and the proper length are not measured by the same observer.

Length contraction ▶

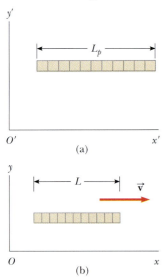

ACTIVE FIGURE 26.11
A meter stick moves to the right with a speed v. (a) The meter stick as viewed by an observer at rest with respect to the meter stick. (b) The meter stick as seen by an observer moving with a speed v with respect to it. The moving meter stick is always measured to be *shorter* than in its own rest frame by a factor of $\sqrt{1 - v^2/c^2}$.

Physics⊗Now™
Log into PhysicsNow at **www.cp7e.com** and go to Active Figure 26.11, where you can view the meter stick from the points of view of two observers and compare the measured lengths of the stick.

EXAMPLE 26.2 Starship Contraction

Goal Apply the concept of length contraction to a moving object.

Problem A starship is measured to be 125 m long while it is at rest with respect to an observer. If this starship now flies past the observer at a speed of $0.99c$, what length will the observer measure for the starship?

Strategy Moving objects are observed to be contracted, or shorter. Substitute into Equation 26.9.

Solution

Substitute into Equation 26.9 to find the length as measured by the observer:

$$L = L_p\sqrt{1 - v^2/c^2} = (125 \text{ m})\sqrt{1 - (0.99\,c)^2/c^2} = \boxed{17.6 \text{ m}}$$

Exercise 26.2

If the ship moves past the observer with a speed of $0.80c$, what length will the observer measure?

Answer 75.0 m

EXAMPLE 26.3 Speedy Plunge

Goal Apply the concept of length contraction to a distance.

Problem (a) An observer on Earth sees a spaceship at an altitude of 4 350 km moving downward toward Earth with a speed of $0.970c$. What is the distance from the spaceship to Earth as measured by the spaceship's captain? (b) After firing his engines, the captain measures her ship's altitude as 267 km, while the observer on Earth measures it to be 625 km. What is the speed of the spaceship at this instant?

Strategy To the captain, the Earth is rushing toward her ship at $0.970c$; hence the distance between her ship and the Earth is contracted. Substitution into Equation 26.9 yields the answer. In part (b) use the same equation, substituting the distances and solving for the speed.

Solution

(a) Find the distance from the ship to Earth as measured by the captain.

Substitute into Equation 26.9, getting the altitude as measured by the captain in the ship.

$$L = L_p\sqrt{1 - v^2/c^2} = (4\,350 \text{ km})\sqrt{1 - (0.970c)^2/c^2}$$
$$= \boxed{1.06 \times 10^3 \text{ km}}$$

(b) What is the subsequent speed of the spaceship if the Earth observer measures the distance from the ship to Earth as 625 km and the captain measures it as 267 km?

Apply the length-contraction equation:

$$L = L_p\sqrt{1 - v^2/c^2}$$

Square both sides of this equation and solve for v:

$$L^2 = L_p^2(1 - v^2/c^2) \quad \rightarrow \quad 1 - v^2/c^2 = \left(\frac{L}{L_p}\right)^2$$

$$v = c\sqrt{1 - (L/L_p)^2} = c\sqrt{1 - (267 \text{ km}/625 \text{ km})^2}$$

$$v = \boxed{0.904c}$$

Remarks The proper length is always the length measured by an observer at rest with respect to that length.

Exercise 26.3

Suppose the observer on the ship measures the distance from Earth as 50.0 km, while the observer on Earth measures the distance as 125 km. At what speed is the spacecraft approaching Earth?

Answer $0.917c$

Length contraction occurs only in the direction of the observer's motion. No contraction occurs perpendicular to that direction. For example, a spaceship at rest relative to an observer may have the shape of an equilateral triangle, but if it passes the observer at relativistic speed in a direction parallel to its base, the base will shorten while the height remains the same. Hence, the craft will be observed to be isosceles. An observer traveling with the ship will still observe it to be equilateral.

26.7 RELATIVISTIC MOMENTUM

Properly describing the motion of particles within the framework of special relativity requires generalizing Newton's laws of motion and the definitions of momentum and energy. These generalized definitions reduce to the classical (nonrelativistic) definitions when v is much less than c.

First, recall that conservation of momentum states that when two objects collide, the total momentum of the system remains constant, assuming that the objects are isolated, reacting only with each other. However, analyzing such collisions from rapidly moving inertial frames, it is found that momentum is not conserved if the classical definition of momentum, $p = mv$, is used. In order to have momentum conservation in all inertial frames—even those moving at an appreciable fraction of c—the definition of momentum must be modified to read

Momentum ▶

$$p \equiv \frac{mv}{\sqrt{1 - v^2/c^2}} = \gamma\, mv \qquad\qquad [26.10]$$

where v is the speed of the particle and m is its mass as measured by an observer at rest with respect to the particle. Note that when v is much less than c, the denominator of Equation 26.10 approaches one, so that p approaches mv. Therefore, the relativistic equation for momentum reduces to the classical expression when v is small compared with c.

EXAMPLE 26.4 The Relativistic Momentum of an Electron

Goal Contrast the classical and relativistic definitions of momentum.

Problem An electron, which has a mass of 9.11×10^{-31} kg, moves with a speed of $0.750c$. Find the classical (nonrelativistic) momentum and compare it to its relativistic counterpart p_{rel}.

Strategy Substitute into the classical definition to get the classical momentum, then multiply by the gamma factor to obtain the relativistic version.

Solution

First, compute the classical (nonrelativistic) momentum with $v = 0.750c$:

$$p = mv = (9.11 \times 10^{-31}\ \text{kg})(0.750 \times 3.00 \times 10^8\ \text{m/s})$$

$$= 2.05 \times 10^{-22}\ \text{kg} \cdot \text{m/s}$$

Multiply this result by γ to obtain the relativistic momentum:

$$p_{rel} = \frac{mv}{\sqrt{1 - v^2/c^2}} = \frac{2.05 \times 10^{-22}\ \text{kg} \cdot \text{m/s}}{\sqrt{1 - (0.750c/c)^2}}$$

$$= 3.10 \times 10^{-22}\ \text{kg} \cdot \text{m/s}$$

Remark The (correct) relativistic result is 50% greater than the classical result. In subsequent calculations, no notational distinction will be made between classical and relativistic momentum. For problems involving relative speeds of $0.2c$, the answer using the classical expression is about 2% below the correct answer.

Exercise 26.4

Repeat the calculation for a proton traveling at $0.600c$.

Answers $p = 3.01 \times 10^{-19}$ kg $\cdot$ m/s, $p_{rel} = 3.76 \times 10^{-19}$ kg $\cdot$ m/s

26.8   RELATIVISTIC ADDITION OF VELOCITIES

Imagine a motorcycle rider moving with a speed of $0.80c$ past a stationary observer, as shown in Figure 26.12. If the rider tosses a ball in the forward direction with a speed of $0.70c$ relative to himself, what is the speed of the ball as seen by the stationary observer at the side of the road? Common sense and the ideas of Newtonian relativity say that the speed should be the sum of the two speeds, or $1.50c$. This answer must be incorrect because it contradicts the assertion that no material object can travel faster than the speed of light.

Einstein resolved this dilemma by deriving an equation for the relativistic addition of velocities. Here, only one dimension of motion will be considered. Let two frames or reference be labeled b and d, and suppose that frame d is moving at velocity v_{db} in the position x-direction relative frame b. If the velocity of an object a as measured in frame d is called v_{ad}, then the velocity of a as measured in frame b, v_{ab}, is given by

$$v_{ab} = \frac{v_{ad} + v_{db}}{1 + \dfrac{v_{ad}v_{db}}{c^2}}$$

[26.11]   ◀ Relativistic velocity addition

The left side of this equation and the numerator on the right are like the equations of Galilean relativity discussed in Chapter 3, and the evaluation of subscripts is applied in the same way as discussed in Section 3.6. The denominator of Equation 26.11 is a correction to Galilean relativity based on length contraction and time dilation.

We apply Equation 26.11 to Figure 26.13, which shows a motorcyclist, his ball, and a stationary observer. We are given

v_{bm} = the velocity of the ball with respect to the motorcycle = $0.70c$

v_{mo} = the velocity of the motorcycle with respect to the stationary observer = $0.80c$,

and we want to find

v_{bo} = the velocity of the ball with respect to the stationary observer.

Thus,

$$v_{bo} = \frac{v_{bm} + v_{mo}}{1 + \dfrac{v_{bm}v_{mo}}{c^2}} = \frac{0.70c + 0.80c}{1 + \dfrac{(0.70c)(0.80c)}{c^2}} = 0.96c$$

The speed of light is the speed limit of the Universe.

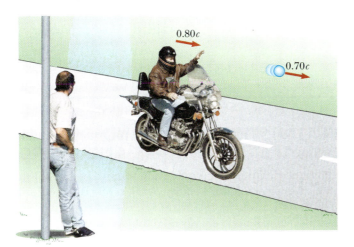

Figure 26.12   A motorcycle moves past a stationary observer with a speed of $0.80c$; the motorcyclist throws a ball in the direction of motion with a speed of $0.70c$ relative to himself.

EXAMPLE 26.5   Urgent Course Correction Needed!

Goal   Apply the concept of the relativistic addition of velocities.

Problem   Suppose that Bob's spacecraft is traveling at $0.600c$ in the positive x-direction, as measured by a nearby observer, while Mike is traveling in his own vehicle directly toward Bob in the negative x-direction at $-0.800c$ relative the nearby observer. What's the velocity of Bob relative to Mike?

Strategy   This problem requires correctly identifying the quantities that go into Equation 26.11, followed by substitution. The measurement of Bob's velocity as determined in the observer's frame O is given, and the measurement of Bob's velocity in Mike's frame is desired.

Solution

Identify the velocity terms in Equation 26.11.

v_{BM} = the velocity of Bob with respect to Mike. This will be substituted for v_{ad} in Equation 26.11.

v_{MO} = the velocity of the Mike with respect to the stationary observer = $-0.800c$. This will be substituted for v_{db} in Equation 26.11.

v_{BO} = the velocity of the Bob with respect to the stationary observer = $0.600c$. This will be substituted for v_{ab} in Equation 26.11.

Substitute the velocity expressions into Equation 26.11. Examining the form of Equation 26.11, we can see intuitively that v_{BM} and v_{MO} belong on the right hand side (the letter M appears in both a first and a second position), so our previous choices are verified.

$$v_{BO} = \frac{v_{BM} + v_{MO}}{1 + \frac{v_{BM}v_{MO}}{c^2}}$$

Substitute given quantities and solve for v_{BM}:

$$0.600c = \frac{v_{BM} - 0.800c}{1 + \frac{v_{BM}(-0.800c)}{c^2}}$$

$$\left(1 - \frac{0.800v_{BM}}{c}\right)0.600c = v_{BM} - 0.800c$$

$$0.600c - 0.480\,v_{BM} = v_{BM} - 0.800c$$

$$v_{BM} = 0.946c$$

Remarks   Notice how much care had to be taken in identifying quantities and their proper signs. Common sense might lead us to believe that Mike would measure Bob's velocity as $1.40c$, but as the calculation shows, Mike measures Bob's velocity as less than that of light.

Exercise 26.5

Suppose Bob shines a laser beam in the direction of his ship's motion. What speed would the nearby observer measure for the beam? Don't guess: do the calculation that proves the answer.

Answer c

26.9   RELATIVISTIC ENERGY AND THE EQUIVALENCE OF MASS AND ENERGY

We have seen that the definition of momentum required generalization to make it compatible with the principle of relativity. Likewise, the definition of kinetic energy requires modification in relativistic mechanics. Einstein found that the correct expression for the **kinetic energy** of an object is

Kinetic energy ▶

$$KE = \gamma mc^2 - mc^2 \qquad\qquad [26.12]$$

The constant term mc^2 in Equation 26.12, which is independent of the speed of the object, is called the **rest energy** of the object, E_R:

$$E_R = mc^2 \qquad \text{[26.13]} \qquad \blacktriangleleft \text{Rest energy}$$

The term γmc^2 in Equation 26.12 depends on the object's speed and is the sum of the kinetic and rest energies. We define γmc^2 to be the **total energy** E, so that

$$\text{total energy} = \text{kinetic energy} + \text{rest energy}$$

or, using Equation 26.12,

$$E = KE + mc^2 = \gamma mc^2 \qquad \text{[26.14]}$$

Because $\gamma = (1 - v^2/c^2)^{-1/2}$, we can also express E as

$$E = \frac{mc^2}{\sqrt{1 - v^2/c^2}} \qquad \text{[26.15]} \qquad \blacktriangleleft \text{Total energy}$$

This is Einstein's famous mass–energy equivalence equation.[4]

The relation $E = \gamma mc^2 = KE + mc^2$ shows the amazing result that **a stationary particle with zero kinetic energy has an energy proportional to its mass**. Further, a small mass corresponds to an enormous amount of energy because the proportionality constant between mass and energy is large: $c^2 = 9 \times 10^{16} \text{ m}^2/\text{s}^2$. The equation $E_R = mc^2$, as Einstein first suggested, indirectly implies that the mass of a particle may be completely convertible to energy and that pure energy—for example, electromagnetic energy—may be converted to particles having mass. This is indeed the case, as has been shown in the laboratory many times. For example, the coming together of a slowly moving electron and its antiparticle, the positron, a particle with the same mass m_e as the electron, but opposite charge, results in the disappearance of both particles and the appearance of a burst of electromagnetic energy in the amount $2m_e c^2$. The reverse process is also fairly easily observed in the laboratory: A high-energy pulse of electromagnetic energy, a gamma ray—disappears near an atom and an electron–positron pair is created with nearly 100% conversion of the gamma ray's energy into mass. Such a pair-production process is shown in the bubble chamber photo of Figure 26.13. We will discuss pair production and annihilation in more detail in Section 26.10.

On a larger scale, nuclear power plants produce energy by the fission of uranium, which involves the conversion of a small amount of the mass of the uranium into energy. The Sun, too, converts mass into energy, and continually loses mass in pouring out a tremendous amount of electromagnetic energy in all directions.

It's extremely interesting that while we have been talking about the interconversion of mass and energy for particles, the expression $E = mc^2$ is universal and applies to all objects, processes, and systems: a hot object has slightly more mass and is slightly more difficult to accelerate than an identical cold object because it has more thermal energy, and a stretched spring has more elastic potential energy and more mass than an identical unstretched spring. A key point, however, is that these changes in mass are often far too small to measure. Our best bet for measuring mass changes is in nuclear transformations, where a measurable fraction of the mass is converted into energy.

Figure 26.13 Bubble-chamber photograph of electron (green) and positron (red) tracks produced by energetic gamma rays. The highly curved tracks at the top are due to the electron and positron in an electron–positron pair bending in opposite directions in the magnetic field.

Lawrence Berkeley Laboratory/Science Photo Library/Photo Researchers, Inc.

EXAMPLE 26.6   Pool Heater

Goal   Combine the concepts of density, rest mass, and heat capacity.

Problem   Suppose some mechanism allowed the conversion of the rest mass of water completely into energy. **(a)** How much rest energy is contained in 0.500 mm³ of water? **(b)** If all this energy is used to heat an Olympic swim-

[4]Although this doesn't look exactly like the famous equation $E = mc^2$, it used to be common to write $m = \gamma m_0$ (Einstein himself wrote it that way), where m is the effective mass of an object moving at speed v and m_0 is the mass of that object as measured by an observer at rest with respect to the object. Then our $E = \gamma mc^2$ becomes the familiar $E = mc^2$. It is currently unfashionable to use $m = \gamma m_0$.

ming pool with dimensions 2.00 m deep, 25.0 m wide, and 50.0 m long, what is the change in temperature of the water?

Strategy   Use the density of water to find the mass in the given volume of water, and multiply by c^2 to get the energy. The heat capacity equation then yields the temperature change.

Solution
(a) How much rest energy is contained in 0.500 mm³ of water?

Use the density to find the mass of this volume of water:

$$\rho = \frac{m}{V} \quad \rightarrow \quad m = \rho V$$

$$m = (1.00 \times 10^3 \,\text{kg/m}^3)(0.500 \,\text{mm}^3)\left(\frac{1.00 \,\text{m}}{1.00 \times 10^3 \,\text{mm}}\right)^3$$

$$= 5.00 \times 10^{-7} \,\text{kg}$$

The energy equivalent of the water is found from Equation 26.13:

$$E_R = mc^2 = (5.00 \times 10^{-7} \,\text{kg})(3.00 \times 10^8 \,\text{m/s})^2$$

$$= 4.50 \times 10^{10} \,\text{J}$$

(b) Find the change in temperature of the pool water.

First find the volume of water in the pool:

$$V = L \times W \times H = (50.0 \,\text{m})(25.0 \,\text{m})(2.00 \,\text{m})$$

$$= 2.50 \times 10^3 \,\text{m}^3$$

Using the definition of density, calculate the mass of the water in the pool:

$$m = \rho V = (1.00 \times 10^3 \,\text{kg/m}^3)(2.50 \times 10^3 \,\text{m}^3)$$

$$= 2.50 \times 10^6 \,\text{kg}$$

Use the heat capacity equation and the result of part (a) to calculate the temperature change of the water in the pool:

$$Q = mc\Delta T$$

$$\Delta T = \frac{Q}{mc} = \frac{4.50 \times 10^{10} \,\text{J}}{(2.50 \times 10^6 \,\text{kg})(4.19 \times 10^3 \,\text{J/kg} \cdot \text{K})}$$

$$= 4.30 \,\text{K}$$

Remarks   Only 12 mm³ of water, completely converted to energy, could raise the water temperature of an Olympic-sized pool by 100 K! However, it's generally impossible to achieve the complete conversion of mass to energy. Nuclear power plants convert only a tiny percentage of the mass of uranium. An exception is the interaction of matter with antimatter.

Exercise 26.6
(a) What mass, when completely converted into energy, would provide the annual energy needs of the entire world (about 4×10^{20} J)? (b) What volume of water contains that much energy?

Answers   (a) 4×10^3 kg   (b) 4 m³

Energy and Relativistic Momentum

Often the momentum or energy of a particle is measured rather than its speed, so it's useful to have an expression relating the total energy E to the relativistic momentum p. This is accomplished by using the expressions $E = \gamma mc^2$ and $p = \gamma mv$. By squaring these equations and subtracting, we can eliminate v. The result, after some algebra, is

$$E^2 = p^2c^2 + (mc^2)^2 \qquad\qquad [26.16]$$

When the particle is at rest, $p = 0$, so $E = E_R = mc^2$. In this special case, the total energy equals the rest energy. For the case of particles that have zero mass, such as photons (massless, chargeless particles of light), we set $m = 0$ in Equation 26.16 and find that

$$E = pc \qquad\qquad\qquad \text{[26.17]}$$

This equation is an exact expression relating energy and momentum for photons, which always travel at the speed of light.

In dealing with subatomic particles, it's convenient to express their energy in electron volts (eV), because the particles are given energy when accelerated through an electrostatic potential difference. The conversion factor is

$$1 \text{ eV} = 1.60 \times 10^{-19} \text{ J}$$

For example, the mass of an electron is 9.11×10^{-31} kg. Hence, the rest energy of the electron is

$$m_e c^2 = (9.11 \times 10^{-31} \text{ kg})(3.00 \times 10^8 \text{ m/s})^2 = 8.20 \times 10^{-14} \text{ J}$$

Converting to eV, we have

$$m_e c^2 = (8.20 \times 10^{-14} \text{ J})(1 \text{ eV}/1.60 \times 10^{-19} \text{ J}) = 0.511 \text{ MeV}$$

Because we frequently use the expression $E = \gamma m c^2$ in nuclear physics, and because m is usually in atomic mass units, u, it is useful to have the conversion factor $1 \text{ u} = 931.494 \text{ MeV}/c^2$. Using this factor makes it easy, for example, to find the rest energy in MeV of the nucleus of a uranium atom with a mass of 235.043 924 u:

$$E_R = mc^2 = (235.043\,924 \text{ u})(931.494 \text{ MeV/u} \cdot c^2)(c^2) = 2.189\,42 \times 10^5 \text{ MeV}$$

Quick Quiz 26.4

A photon is reflected from a mirror. **True or false**: (a) Because a photon has zero mass, it does not exert a force on the mirror. (b) Although the photon has energy, it can't transfer any energy to the surface because it has zero mass. (c) The photon carries momentum, and when it reflects off the mirror, it undergoes a change in momentum and exerts a force on the mirror. (d) Although the photon carries momentum, its change in momentum is zero when it reflects from the mirror, so it can't exert a force on the mirror.

EXAMPLE 26.7   A Speedy Electron

Goal   Compute a total energy and a relativistic kinetic energy.

Problem   An electron moves with a speed $v = 0.850c$. Find its total energy and kinetic energy in mega electron volts (MeV), and compare the latter to the classical kinetic energy (10^6 eV = 1 MeV).

Strategy   Substitute into Equation 26.15 to get the total energy, and subtract the rest mass energy to obtain the kinetic energy.

Solution

Substitute values into Equation 26.15 to obtain the total energy:

$$E = \frac{m_e c^2}{\sqrt{1 - v^2/c^2}} = \frac{(9.11 \times 10^{-31} \text{ kg})(3.00 \times 10^8 \text{ m/s})^2}{\sqrt{1 - (0.850c/c)^2}}$$

$$= 1.56 \times 10^{-13} \text{ J} = (1.56 \times 10^{-13} \text{ J})\left(\frac{1.00 \text{ eV}}{1.60 \times 10^{-19} \text{ J}}\right)$$

$$= \boxed{0.975 \text{ MeV}}$$

The kinetic energy is obtained by subtracting the rest energy from the total energy:

$$KE = E - m_e c^2 = 0.975 \text{ MeV} - 0.511 \text{ MeV} = \boxed{0.464 \text{ MeV}}$$

Calculate the classical kinetic energy:

$$KE_{\text{classical}} = \tfrac{1}{2} m_e v^2$$

$$= \tfrac{1}{2}(9.11 \times 10^{-31} \text{ kg})(0.850 \times 3.00 \times 10^8 \text{ m/s})^2$$

$$= 2.96 \times 10^{-14} \text{ J} = 0.185 \text{ MeV}$$

Remarks   Notice the large discrepancy between the relativistic kinetic energy and the classical kinetic energy.

Exercise 26.7
Calculate the total energy and the kinetic energy in MeV of a proton traveling at 0.600c. (The rest energy of a proton is approximately 938 MeV.)

Answers $E = 1.17 \times 10^3$ MeV, $KE = 232$ MeV

EXAMPLE 26.8   The Conversion of Mass to Kinetic Energy in Uranium Fission

Goal   Understand the production of energy from nuclear sources.

Problem   The fission, or splitting, of uranium was discovered in 1938 by Lise Meitner, who successfully interpreted some curious experimental results found by Otto Hahn as due to fission. (Hahn received the Nobel prize.) The fission of $^{235}_{92}$U begins with the absorption of a slow-moving neutron that produces an unstable nucleus of ^{236}U. The ^{236}U nucleus then quickly decays into two heavy fragments moving at high speed, as well as several neutrons. Most of the kinetic energy released in such a fission is carried off by the two large fragments. **(a)** For the typical fission process

$$^{1}_{0}n + \,^{235}_{92}U \rightarrow \,^{141}_{56}Ba + \,^{92}_{36}Kr + 3^{1}_{0}n$$

calculate the kinetic energy in MeV carried off by the fission fragments. **(b)** What percentage of the initial energy is converted into kinetic energy? The atomic masses involved are given below in atomic mass units.

$$^{1}_{0}n = 1.008\ 665\ u \qquad ^{235}_{92}U = 235.043\ 924\ u \qquad ^{141}_{56}Ba = 140.903\ 496\ u \qquad ^{92}_{36}Kr = 91.907\ 936\ u$$

Strategy   This is an application of the conservation of relativistic energy. Write the conservation law as a sum of kinetic energy and rest energy, and solve for the final kinetic energy. Equation 26.15, solved for v, then yields the speeds.

Solution
(a) Calculate the final kinetic energy for the given process.

Apply the conservation of relativistic energy equation, assuming that $KE_{initial} = 0$:

$$(KE + mc^2)_{initial} = (KE + mc^2)_{final}$$
$$0 + m_n c^2 + m_U c^2 = m_{Ba} c^2 + m_{Kr} c^2 + 3m_n c^2 + KE_{final}$$

Solve for KE_{final} and substitute, converting to MeV in the last step:

$$KE_{final} = [(m_n + m_U) - (m_{Ba} + m_{Kr} + 3m_n)]c^2$$
$$KE_{final} = (1.008\ 665u + 235.043\ 924\ u)c^2$$
$$- [140.903\ 496\ u + 91.907\ 936\ u + 3(1.008\ 665\ u)]c^2$$
$$= (0.215\ 162\ u)(931.494\ \text{MeV/u} \cdot c^2)(c^2)$$
$$= \boxed{200.422\ \text{MeV}}$$

(b) What percentage of the initial energy is converted into kinetic energy?

Compute the total energy, which is the initial energy:

$$E_{initial} = 0 + m_n c^2 + m_U c^2$$
$$= (1.008\ 665u + 235.043\ 924\ u)c^2$$
$$= (236.052\ 59\ u)(931.494\ \text{MeV/u} \cdot c^2)(c^2)$$
$$= 2.198\ 82 \times 10^5\ \text{MeV}$$

Divide the kinetic energy by the total energy and multiply by 100%:

$$\frac{200.422\ \text{MeV}}{2.198\ 82 \times 10^5\ \text{MeV}} \times 100\% = \boxed{9.115 \times 10^{-2}\%}$$

Remarks   This calculation shows that nuclear reactions liberate only about a tenth of one percent of the rest energy of the constituent particles. Some fusion reactions better that number by several times.

Exercise 26.8

In a fusion reaction, light elements combine to form a heavier element. Deuterium, which is also called heavy hydrogen, has an extra neutron in its nucleus. Two such particles can fuse into a heavier form of hydrogen, called tritium, plus an ordinary hydrogen atom. The reaction is

$$^2_1D + ^2_1D \rightarrow ^3_1T + ^1_1H$$

(a) Calculate the energy released in the form of kinetic energy, assuming for simplicity that the initial kinetic energy is zero. (b) What percentage of the rest mass was converted to energy? The atomic masses involved are as follows:

$$^2_1D = 2.014\ 102\ u \qquad ^3_1T = 3.016\ 049\ u \qquad ^1_1H = 1.007\ 825\ u$$

Answers   (a) 4.033 37 MeV   (b) 0.1075%

26.10   PAIR PRODUCTION AND ANNIHILATION

In general, converting mass into energy is a low-yield process. Burning wood or coal, or even the fission or fusion processes presented in Example 26.8, convert only a very small percentage of the available energy. An exception is the reaction of matter with antimatter.

A common process in which a photon creates matter is called **pair production**, illustrated in Figure 26.14. In this process, an electron and a positron are simultaneously produced, while the photon disappears. (Note that the positron is a positively charged particle having the same mass as an electron. The positron is often called the *antiparticle* of the electron.) In order for pair production to occur, energy, momentum, and charge must all be conserved during the process. It's impossible for a photon to produce a single electron because the photon has zero charge and charge would not be conserved in the process.

As we explain in more detail in Chapter 27, the energy of a photon having a frequency f is given by $E = hf$, where h is Planck's constant. The *minimum* energy that a photon must have to produce an electron–positron pair can be found using conservation of energy by equating the photon energy hf_{min} to the total rest energy of the pair. That is,

$$hf_{min} = 2m_e c^2 \qquad \text{[26.18]}$$

Because the energy of an electron is $m_e c^2 = 0.51$ MeV, the minimum energy required for pair production is 1.02 MeV.

Pair production can't occur in a vacuum, but can only take place in the presence of a massive particle such as an atomic nucleus. The massive particle must participate in the interaction in order that energy and momentum be conserved simultaneously.

Pair annihilation is a process in which an electron–positron pair produces two photons—the inverse of pair production. Figure 26.15 is one example of pair annihilation in which an electron and positron initially at rest combine with each other, disappear, and create two photons. Because the initial momentum of the pair is zero, it's impossible to produce a single photon. Momentum can be conserved only if two photons moving in opposite directions, both with the same energy and magnitude of momentum, are produced. We will discuss particles and their antiparticles further in Chapter 30.

26.11   GENERAL RELATIVITY

Special relativity relates observations of inertial observers. Einstein sought a more general theory that would address accelerating systems. His search was motivated in part by the following curious fact: mass determines the inertia of an object and also the strength of the gravitational field. The mass involved in inertia is called inertial mass, m_i, whereas the mass responsible for the gravitational field is called the

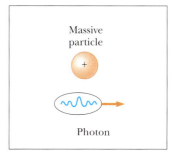
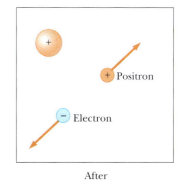

Figure 26.14   Representation of the process of pair production.

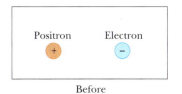
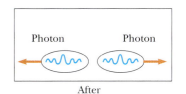

Figure 26.15   Representation of the process of pair annihilation.

gravitational mass, m_g. They appear in Newton's law of gravitation and in the second law of motion:

$$\text{Gravitational property} \qquad F_g = G\,\frac{m_g m_g'}{r^2}$$

$$\text{Inertial property} \qquad F_i = m_i a$$

The value for the gravitational constant G was chosen to make the magnitudes of m_g and m_i numerically equal. Regardless of how G is chosen, however, the strict proportionality of m_g and m_i has been established experimentally to an extremely high degree: a few parts in 10^{12}. It appears that gravitational mass and inertial mass may indeed be exactly equal: $m_i = m_g$.

There is no reason a priori, however, why these two very different quantities should be equal. They seem to involve two entirely different concepts: a force of mutual gravitational attraction between two masses and the resistance of a single mass to being accelerated. This question puzzled Newton and many other physicists over the years and was finally incorporated as a fundamental principle of Einstein's remarkable theory of gravitation, known as *general relativity*, in 1916.

In Einstein's view, the remarkable coincidence that m_g and m_i were exactly equal was evidence for an intimate connection between the two concepts. He pointed out that no mechanical experiment (such as releasing a mass) could distinguish between the two situations illustrated in Figures 26.16a and 26.16b. In each case, a mass released by the observer undergoes a downward acceleration of g relative to the floor.

Einstein carried this idea further and proposed that *no* experiment, mechanical or otherwise, could distinguish between the two cases. This extension to include all phenomena (not just mechanical ones) has interesting consequences. For example, suppose that a light pulse is sent horizontally across the box, as in Figure 26.16c. The trajectory of the light pulse bends downward as the box accelerates upward to meet it. Einstein proposed that a beam of light should also be bent downward by a gravitational field (Fig. 26.16d).

The two postulates of Einstein's **general relativity** are as follows:

1. All the laws of nature have the same form for observers in any frame of reference, accelerated or not.
2. In the vicinity of any given point, a gravitational field is equivalent to an accelerated frame of reference without a gravitational field. (This is the *principle of equivalence*.)

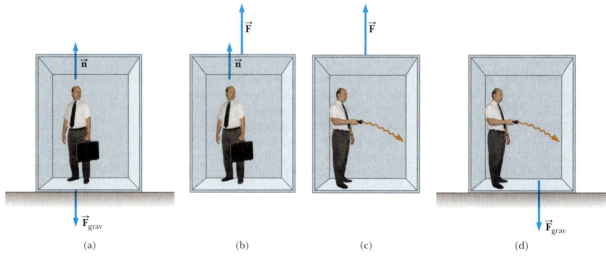

Figure 26.16   (a) The observer in the cubicle is at rest in a uniform gravitational field $\vec{\mathbf{g}}$. He experiences a normal force $\vec{\mathbf{n}}$. (b) Now the observer is in a region where gravity is negligible, but an external force $\vec{\mathbf{F}}$ acts on the frame of reference, producing an acceleration with magnitude g. Again, the man experiences a normal force $\vec{\mathbf{n}}$ that accelerates him along with the cubicle. According to Einstein, the frames of reference in parts (a) and (b) are equivalent in every way. No local experiment could distinguish between them. (c) The observer turns on his pocket flashlight. Because of the acceleration of the cubicle, the beam would appear to bend toward the floor, just as a tossed ball would. (d) Given the equivalence of the frames, the same phenomenon should be observed in the presence of a gravity field.

The second postulate implies that gravitational mass and inertial mass are completely equivalent, not just proportional. What were thought to be two different types of mass are actually identical.

One interesting effect predicted by general relativity is that time scales are altered by gravity. A clock in the presence of gravity runs more slowly than one in which gravity is negligible. As a consequence, light emitted from atoms in a strong gravity field, such as the Sun's, is observed to have a lower frequency than the same light emitted by atoms in the laboratory. This gravitational shift has been detected in spectral lines emitted by atoms in massive stars. It has also been verified on Earth by comparing the frequencies of gamma rays emitted from nuclei separated vertically by about 20 m.

Quick Quiz 26.5

Two identical clocks are in the same house, one upstairs in a bedroom and the other downstairs in the kitchen. Which statement is correct? (a) The clock in the kitchen runs more slowly than the clock in the bedroom. (b) The clock in the bedroom runs more slowly than the clock in the kitchen. (c) Both clocks keep the same time.

The second postulate suggests that a gravitational field may be "transformed away" at any point if we choose an appropriate accelerated frame of reference—a freely falling one. Einstein developed an ingenious method of describing the acceleration necessary to make the gravitational field "disappear." He specified a certain quantity, the *curvature of spacetime*, that describes the gravitational effect at every point. In fact, the curvature of spacetime completely replaces Newton's gravitational theory. According to Einstein, there is no such thing as a gravitational force. Rather, the presence of a mass causes a curvature of spacetime in the vicinity of the mass. Planets going around the Sun follow the natural contours of the spacetime, much as marbles roll around inside a bowl. In 1979, John Wheeler summarized Einstein's general theory of relativity in a single sentence: "Mass one tells spacetime how to curve; curved spacetime tells mass two how to move." The fundamental equation of general relativity can be roughly stated as a proportion as follows:

$$\text{Average curvature of spacetime} \propto \text{energy density}$$

The equation corresponding to this proportion is solved for a mathematical quantity called the *metric*, which can be used to measure the lengths of vectors and to compute trajectories of bodies through space. The metric looks something like a matrix, with different entries at each point of space and time. (There are a few important differences, beyond the level of this course.)

Einstein pursued a new theory of gravity in large part because of a discrepancy in the orbit of Mercury as calculated from Newton's second law. The closest approach of the Mercury to the Sun, called the perihelion, changes position slowly over time. Newton's theory accounted for all but 43 seconds of arc per century; Einstein's general relativity explained the discrepancy.

The most dramatic test of general relativity came shortly after the end of World War I. The theory predicts that a star would bend a light ray by a certain precise amount. Sir Arthur Eddington mounted an expedition to Africa and, during a solar eclipse, confirmed that starlight bent on passing the Sun in an amount matching the prediction of general relativity (Fig. 26.17). When this discovery was announced, Einstein became an international celebrity.

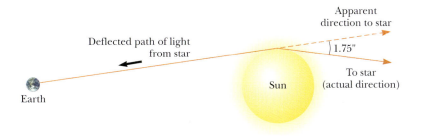

Figure 26.17 Deflection of starlight passing near the Sun. Because of this effect, the Sun and other remote objects can act as a *gravitational lens*. In his general theory of relativity, Einstein calculated that starlight just grazing the Sun's surface should be deflected by an angle of 1.75".

Other tests were proposed and verified long after Einstein's death, including the time delay of radar bounced off Venus, and the gradual lengthening of the period of binary pulsars due to the emission of gravitational radiation. The latter has been verified with such precision that general relativity can lay claim to being the most accurate theory in physics.

General relativity also predicts extreme states of matter created by gravitational collapse. If the concentration of mass becomes very great, as is believed to occur when a large star exhausts its nuclear fuel and collapses to a very small volume, a **black hole** may form. Here the curvature of spacetime is so extreme that all matter and light within a certain radius becomes trapped. This radius, called the *Schwarzschild radius* or *event horizon*, is about 3 km for a black hole with the mass of our Sun. At the black hole's center may lurk a *singularity*—a point of infinite density and curvature where spacetime comes to an end.

There is strong evidence for the existence of a black hole having a mass of millions of Suns at the center of our galaxy.

Applying Physics 26.1 Faster Clocks in a "Mile High City"

Atomic clocks are extremely accurate; in fact, an error of 1 second in 3 million years is typical. This error can be described as about one part in 10^{14}. On the other hand, the atomic clock in Boulder, Colorado, is often 15 ns faster than the one in Washington after only one day. This is an error of about one part in 6×10^{12}, which is about 17 times larger than the typical error. If atomic clocks are so accurate, why does a clock in Boulder not remain synchronous with one in Washington?

Explanation According to the general theory of relativity, the passage of time depends on gravity—clocks run more slowly in strong gravitational fields. Washington is at an elevation very close to sea level, whereas Boulder is about a mile higher in altitude. Hence, the gravitational field at Boulder is weaker than at Washington. As a result, an atomic clock runs more rapidly in Boulder than in Washington. (This effect has been verified by experiment.)

SUMMARY

Physics⊗Now™ Take a practice test by logging into Physics-Now at **www.cp7e.com** and clicking on the Pre-Test link for this chapter.

26.5 Einstein's Principle of Relativity

The two basic postulates of the **special theory of relativity** are as follows:

1. The laws of physics are the same in all inertial frames of reference.
2. The speed of light is the same for all inertial observers, independently of their motion or of the motion of the source of light.

26.6 Consequences of Special Relativity

Some of the consequences of the special theory of relativity are as follows:

1. Clocks in motion relative to an observer slow down, a phenomenon known as **time dilation**. The relationship between time intervals in the moving and at-rest systems is

$$\Delta t = \gamma \Delta t_p \qquad [26.7]$$

where Δt is the time interval measured in the system in relative motion with respect to the clock,

$$\gamma = \frac{1}{\sqrt{1 - v^2/c^2}} \qquad [26.8]$$

and Δt_p is the proper time interval measured in the system moving with the clock.

2. The length of an object in motion is *contracted* in the direction of motion. The equation for **length contraction** is

$$L = L_p\sqrt{1 - v^2/c^2} \qquad [26.9]$$

where L is the length measured by an observer in motion relative to the object and L_p is the proper length measured by an observer for whom the object is at rest.

3. Events that are simultaneous for one observer are not simultaneous for another observer in motion relative to the first.

26.7 Relativistic Momentum

The relativistic expression for the **momentum** of a particle moving with velocity v is

$$p \equiv \frac{mv}{\sqrt{1 - v^2/c^2}} = \gamma mv \qquad [26.10]$$

26.8 Relativistic Addition of Velocities

The relativistic expression for the addition of velocities is

$$v_{ab} = \frac{v_{ad} + v_{db}}{1 + \dfrac{v_{ad}\,v_{db}}{c^2}} \qquad [26.11]$$

where v_{ab} is the velocity of object a with as measured in frame b, v_{ad} is the velocity of object a as measured in frame d, and v_{db} is the velocity of frame d as measured in frame b.

26.9   Relativistic Energy and the Equivalence of Mass and Energy

The relativistic expression for the **kinetic energy** of an object is

$$KE = \gamma mc^2 - mc^2 \qquad [26.12]$$

where mc^2 is the **rest energy** of the object, E_R.

The **total energy** of a particle is

$$E = \frac{mc^2}{\sqrt{1 - v^2/c^2}} \qquad [26.15]$$

This is Einstein's famous mass–energy equivalence equation.

The relativistic momentum is related to the total energy through the equation

$$E^2 = p^2c^2 + (mc^2)^2 \qquad [26.16]$$

26.10   Pair Production and Annihilation

Pair production is a process in which the energy of a photon is converted into mass. In this process, the photon disappears as an electron–positron pair is created. Likewise, the energy of an electron–positron pair can be converted into electromagnetic radiation by the process of **pair annihilation**.

CONCEPTUAL QUESTIONS

1. A spacecraft with the shape of a sphere of diameter D moves past an observer on Earth with a speed $0.5c$. What shape does the observer measure for the spacecraft as it moves past?

2. The equation $E = mc^2$ is often given in popular descriptions of Einstein's theory of relativity. Is this expression strictly correct? For example, does it accurately account for the kinetic energy of a moving mass?

3. You are in a speedboat on a lake. You see ahead of you a wave front, caused by the previous passage of another boat, moving away from you. You accelerate, catch up with, and pass the wave front. Is this scenario possible if you are in a rocket and you detect a wave front of light ahead of you?

4. What two speed measurements will two observers in relative motion always agree upon?

5. The speed of light in water is 2.30×10^8 m/s. Suppose an electron is moving through water at 2.50×10^8 m/s. Does this particle speed violate the principle of relativity?

6. With regard to reference frames, how does general relativity differ from special relativity?

7. Some distant starlike objects, called quasars, are receding from us at half the speed of light (or greater).

What is the speed of the light we receive from these quasars?

8. It is said that Einstein, in his teenage years, asked the question, "What would I see in a mirror if I carried it in my hands and ran at a speed near that of light?" How would you answer this question?

9. List some ways our day-to-day lives would change if the speed of light were only 50 m/s.

10. Two identically constructed clocks are synchronized. One is put into orbit around Earth while the other remains on Earth. Which clock runs more slowly? When the moving clock returns to Earth, will the two clocks still be synchronized.

11. Photons of light have zero mass. How is it possible that they have momentum?

12. Imagine an astronaut on a trip to Sirius, which lies 8 lightyears from Earth. Upon arrival at Sirius, the astronaut finds that the trip lasted 6 years. If the trip was made at a constant speed of $0.8c$, how can the 8-lightyear distance be reconciled with the 6-year duration?

13. Explain why it is necessary, when defining length, to specify that the positions of the ends of a rod are to be measured simultaneously.

PROBLEMS

1, 2, 3 = straightforward, intermediate, challenging   ☐ = full solution available in *Student Solutions Manual/Study Guide*

Physics⊗Now™ = coached problem with hints available at **www.cp7e.com**   = biomedical application

Section 26.4 The Michelson–Morley Experiment

1. Two airplanes fly paths I and II specified in Figure 26.5a. Both planes have air speeds of 100 m/s and fly a distance $L = 200$ km. The wind blows at 20.0 m/s in the direction shown in the figure. Find (a) the time of flight to each city, (b) the time to return, and (c) the difference in total flight times.

2. In one version of the Michelson–Morley experiment, the lengths L in Figure 26.6 were 28 m. Take v to be 3.0×10^4 m/s, and find the time difference caused by rotation of the interferometer and (b) the expected fringe shift, assuming that the light used has a wavelength of 550 nm.

Section 26.6 Consequences of Special Relativity

3. A deep-space probe moves away from Earth with a speed of $0.80c$. An antenna on the probe requires 3.0 s, in probe time, to rotate through 1.0 rev. How much time is required for 1.0 rev according to an observer on Earth?

4. If astronauts could travel at $v = 0.950c$, we on Earth would say it takes $(4.20/0.950) = 4.42$ years to reach Alpha Centauri, 4.20 lightyears away. The astronauts disagree. (a) How much time passes on the astronauts' clocks? (b) What is the distance to Alpha Centauri as measured by the astronauts?

5. A friend in a spaceship travels past you at a high speed. He tells you that his ship is 20 m long and that the identical ship you are sitting in is 19 m long. According to your observations, (a) how long is your ship, (b) how long is his ship, and (c) what is the speed of your friend's ship?

6. An astronaut at rest on Earth has a heartbeat rate of 70 beats/min. When the astronaut is traveling in a spaceship at $0.90c$, what will this rate be as measured by (a) an observer also in the ship and (b) an observer at rest on Earth?

7. The average lifetime of a pi meson in its own frame of reference (i.e., the proper lifetime) is 2.6×10^{-8} s. If the meson moves with a speed of $0.98c$, what is (a) its mean lifetime as measured by an observer on Earth, and (b) the average distance it travels before decaying, as measured by an observer on Earth? (c) What distance would it travel if time dilation did not occur?

8. An astronaut is traveling in a space vehicle that has a speed of $0.500c$ relative to Earth. The astronaut measures his pulse rate at 75.0 per minute. Signals generated by the astronaut's pulse are radioed to Earth when the vehicle is moving perpendicularly to a line that connects the vehicle with an Earth observer. What pulse rate does Earth observer measure? What would be the pulse rate if the speed of the space vehicle were increased to $0.990c$?

9. A muon formed high in Earth's atmosphere travels at speed $v = 0.99c$ for a distance of 4.6 km before it decays into an electron, a neutrino, and an antineutrino ($\mu^- \rightarrow e^- + \nu + \bar{\nu}$). (a) How long does the muon live, as measured in its reference frame? (b) How far does the muon travel, as measured in its frame?

10. A box is cubical with sides of proper lengths $L_1 = L_2 = L_3$, as shown in Figure P26.10, when viewed in its own rest frame. If this block moves parallel to one of its edges with a speed of $0.80c$ past an observer, (a) what shape does it appear to have to this observer, and (b) what is the length of each side as measured by the observer?

Figure P26.10

11. The proper length of one spaceship is three times that of another. The two spaceships are traveling in the same direction and, while both are passing overhead, an Earth observer measures the two spaceships to have the same length. If the slower spaceship is moving with a speed of $0.350c$, determine the speed of the faster spaceship.

12. Observer A measures the length of two rods, one stationary, the other moving with a speed of $0.955c$. She finds that the rods have the same length, L_0. A second observer B travels along with the moving rod. (a) What is the length observer B measures for the rod in observer A's frame? (b) What is the ratio of the length of A's rod to the length of B's rod, according to observer B?

13. A supertrain of proper length 100 m travels at a speed of $0.95c$ as it passes through a tunnel having proper length 50 m. As seen by a trackside observer, is the train ever completely within the tunnel? If so, by how much?

14. An observer moving at a speed of $0.995c$ relative to a rod (Fig. P26.14) measures its length to be 2.00 m and sees its length to be oriented at $30.0°$ with respect to its direction of motion. What is the proper length of the rod? (b) What is the orientation angle in a reference frame moving with the rod?

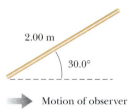

2.00 m

30.0°

Motion of observer

Figure P26.14   View of rod as seen by an observer moving to the right.

15. In 1963, when Mercury astronaut Gordon Cooper orbited Earth 22 times, the press stated that for each orbit, he aged 2 millionths of a second less than that if he remained on Earth. (a) Assuming that he was 160 km above Earth in a circular orbit, determine the time difference between someone on Earth and the orbiting astronaut for the 22 orbits. You will need to use the approximation

$$\frac{1}{\sqrt{1-x}} \approx 1 + \frac{x}{2} \qquad \text{for } x \ll 1$$

(b) Did the press report accurate information? Explain.

16. An interstellar space probe is launched from Earth. After a brief period of acceleration it moves with a constant velocity, 70.0% of the speed of light. Its nuclear-powered batteries supply the energy to keep its data transmitter active continuously. The batteries have a lifetime of 15.0 years as measured in a rest frame. (a) How long do the batteries on the space probe last as measured by mission control on Earth? (b) How far is the probe from Earth when its batteries fail, as measured by mission control? (c) How far is the probe from Earth, as measured by its built-in trip odometer, when its batteries fail? (d) For what total time after launch are data received from the probe by mission control? Note that radio waves travel at the speed of light and fill the space between the probe and Earth at the time the battery fails.

Section 26.7 Relativistic Momentum

17. An electron has a speed $v = 0.90c$. At what speed will a proton have a momentum equal to that of the electron?

18. Calculate the momentum of an electron moving with a speed of (a) $0.010c$, (b) $0.50c$, (c) $0.90c$.

19. **Physics⊗Now™** An unstable particle at rest breaks up into two fragments of *unequal mass*. The mass of the lighter fragment is 2.50×10^{-28} kg, and that of the heavier fragment is 1.67×10^{-27} kg. If the lighter fragment has a speed of $0.893c$ after the breakup, what is the speed of the heavier fragment?

20. The nonrelativistic expression for the momentum of a particle, $p = mv$, can be used if $v \ll c$. For what speed does the use of this formula give an error in the momentum of (a) 1.00% and (b) 10.0%?

Section 26.8 Relativistic Addition of Velocities

21. An electron moves to the right with a speed of $0.90c$ relative to the laboratory frame. A proton moves to the left with a speed of $0.70c$ relative to the electron. Find the speed of the proton relative to the laboratory frame.

22. Spaceship R is moving to the right at a speed of $0.70c$ with respect to Earth. A second spaceship, L, moves to the left at the same speed with respect to Earth. What is the speed of L with respect to R?

23. A Klingon spaceship moves away from Earth at a speed of $0.800c$ (Fig. P26.23). The starship *Enterprise* pursues at a speed of $0.900c$ relative to Earth. Observers on Earth see the *Enterprise* overtaking the Klingon ship at a relative speed of $0.100c$. With what speed is the *Enterprise* overtaking the Klingon ship as seen by the crew of the *Enterprise*?

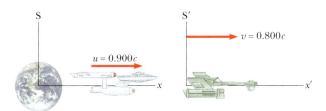

Figure P26.23

24. A spaceship travels at $0.750c$ relative to Earth. If the spaceship fires a small rocket in the forward direction, how fast (relative to the ship) must it be fired for it to travel at $0.950c$ relative to Earth?

25. A rocket moves with a velocity of $0.92c$ to the right with respect to a stationary observer A. An observer B moving relative to observer A finds that the rocket is moving with a velocity of $0.95c$ to the left. What is the velocity of observer B relative to observer A? [*Hint:* Consider observer B's velocity in the frame of reference of the rocket.]

26. A pulsar is a stellar object that emits light in short bursts. Suppose a pulsar with a speed of $0.950c$ approaches Earth and a rocket with a speed of $0.995c$ heads toward the pulsar. (Both speeds are measured in Earth's frame of reference.) If the pulsar emits 10.0 pulses per second in its own frame of reference, at what rate are the pulses emitted in the rocket's frame of reference?

27. Spaceship I, which contains students taking a physics exam, approaches Earth with a speed of $0.60c$, while spaceship II, which contains instructors proctoring the exam, moves away from Earth at $0.28c$, as in Figure P26.27. If the instructors in spaceship II stop the exam after 50 min have passed *on their clock*, how long does the exam last as measured by (a) the students? (b) an observer on Earth?

Figure P26.27

Section 26.9 Relativistic Energy and the Equivalence of Mass and Energy

28. A proton moves with a speed of $0.950c$. Calculate (a) its rest energy, (b) its total energy, and (c) its kinetic energy.

29. What is the speed of a particle whose kinetic energy is equal to its own rest energy?

30. If it takes 3 750 MeV of work to accelerate a proton from rest to a speed of v, determine v.

31. What speed must a particle attain before its kinetic energy is double the value predicted by the nonrelativistic expression $KE = \frac{1}{2}mv^2$?

32. Determine the energy required to accelerate an electron from (a) $0.500c$ to $0.900c$ and (b) $0.900c$ to $0.990c$.

33. **Physics Now™** A cube of steel has a volume of $1.00\ \text{cm}^3$ and a mass of 8.00 g when at rest on Earth. If this cube is now given a speed $v = 0.900c$, what is its density as measured by a stationary observer? Note that relativistic density is E_R/c^2V.

34. An unstable particle with a mass equal to 3.34×10^{-27} kg is initially at rest. The particle decays into two fragments that fly off with velocities of $0.987c$ and $-0.868c$, respectively. Find the masses of the fragments. [*Hint:* Conserve both mass–energy and momentum.]

Section 26.10 Pair Production and Annihilation

35. How much total kinetic energy will an electron–positron pair have if produced by a photon of energy 3.00 MeV?

36. If an electron–positron pair with a total kinetic energy of 2.50 MeV is produced, find (a) the energy of the photon that produced the pair and (b) its frequency.

37. **Physics Now™** Two photons are produced when a proton and an antiproton annihilate each other. What are the minimum frequency and the corresponding wavelength of each photon?

38. An electron moving at a speed of $0.60c$ collides head-on with a positron also moving at $0.60c$. Determine the energy and momentum of each photon produced in the process.

ADDITIONAL PROBLEMS

39. What is the speed of a proton that has been accelerated from rest through a difference of potential of (a) 500 V and (b) 5.00×10^8 V?

40. An electron has a total energy equal to five times its rest energy. (a) What is its momentum? (b) Repeat for a proton.

41. What is the momentum (in units of MeV/c) of an electron with a kinetic energy of 1.00 MeV?

42. An astronomer on Earth observes a meteoroid in the southern sky approaching Earth at a speed of $0.800c$. At the time of its discovery, the meteoroid is 20.0 lightyears from Earth. Calculate (a) the time required for the meteoroid to reach Earth, as measured by Earthbound astronomer, (b) this time as measured by a tourist on the meteoroid, and (c) the distance to Earth as measured by the tourist.

43. Ted and Mary are playing a game of catch in frame S′, which is moving with a speed of $0.60c$; Jim in frame S is watching (Fig. P26.43, page 872). Ted throws the ball to Mary with a speed of $0.80c$ (according to Ted) and their separation (measured in S′) is 1.8×10^{12} m. (a) According to Mary, how fast is the ball moving? (b) According to Mary, how long will it take the ball to reach her? (c) According to Jim, how far apart are Ted and Mary and how fast is the ball moving?

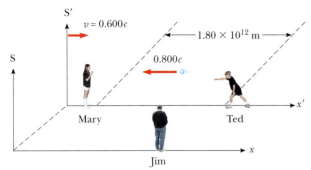

Figure P26.43

44. An alarm clock is set to sound in 10 h. At $t = 0$, the clock is placed in a spaceship moving with a speed of $0.75c$ (relative to Earth). What distance, as determined by an Earth observer, does the spaceship travel before the alarm clock sounds?

45. A radioactive nucleus moves with a speed v relative to a laboratory observer. The nucleus emits an electron in the positive x-direction with a speed of $0.70c$ relative to the decaying nucleus and a speed of $0.85c$ in the $+x$-direction relative to the laboratory observer. What is the value of v?

46. A certain quasar recedes from Earth at $v = 0.870c$. A jet of material ejected from the quasar back toward Earth moves at $0.550c$ relative to the quasar. Find the speed of the ejected material relative to Earth.

47. An astronaut wishes to visit the Andromeda galaxy, making a one-way trip that will take 30.0 years in the spaceship's frame of reference. Assume that the galaxy is 2.00 million light years away and that his speed is constant. (a) How fast must he travel relative to Earth? (b) What will be the kinetic energy of his spacecraft, which has mass is 1.00×10^6 kg? (c) What is the cost of this energy if it is purchased at a typical consumer price for electric energy, 13.0 cents per kWh? The following approximation will prove useful:

$$\frac{1}{\sqrt{1 + x}} \approx 1 - \frac{x}{2} \qquad \text{for } x \ll 1$$

48. The cosmic rays of highest energy are protons that have kinetic energy on the order of 10^{13} MeV. (a) How long would it take a proton of this energy to travel across the Milky Way galaxy, having a diameter $\sim 10^5$ light years, as measured in the proton's frame? (b) From the point of view of the proton, how many kilometers across is the galaxy?

49. A spaceship of proper length 300 m takes 0.75 μs to pass an Earth observer. Determine the speed of this spaceship as measured by the Earth observer.

50. Find the kinetic energy of a 78.0-kg spacecraft launched out of the solar system with speed 106 km/s by using (a) the classical equation $KE = \frac{1}{2}mv^2$ and (b) the relativistic equation. You will need to use the approximation

$$\frac{1}{\sqrt{1 - x}} \approx 1 + \frac{x}{2} \qquad \text{for } x \ll 1$$

51. **Physics⊗Now™** An alien spaceship traveling $0.600c$ toward Earth launches a landing craft with an advance guard of purchasing agents. The lander travels in the same direction with a velocity of $0.800c$ relative to the spaceship. As observed on Earth, the spaceship is 0.200 light year from Earth when the lander is launched. (a) With what velocity is the lander observed to be approaching by observers on Earth? (b) What is the distance to Earth at the time of lander launch, as observed by the aliens on the mother ship? (c) How long does it take the lander to reach Earth, as observed by the aliens on the mother ship? (d) If the lander has a mass of 4.00×10^5 kg, what is its kinetic energy as observed in Earth reference frame?

52. (a) Show that a potential difference of 1.02×10^6 V would be sufficient to give an electron a speed equal to twice the speed of light if Newtonian mechanics remained valid at high speeds. (b) What speed would an electron actually acquire in falling through a potential difference of 1.02×10^6 V?

53. The muon is an unstable particle that spontaneously decays into an electron and two neutrinos. In a reference frame in which the muons are stationary, if the number of muons at $t = 0$ is N_0, the number at time t is given by $N = N_0 e^{-t/\tau}$, where τ is the mean lifetime, equal to 2.2 μs. Suppose that the muons move at a speed of $0.95c$ and that there are 5.0×10^4 muons at $t = 0$. (a) What is the observed lifetime of the muons? (b) How many muons remain after traveling a distance of 3.0 km?

54. An observer in a rocket moves toward a mirror at speed v relative to the reference frame labeled by S in Figure P26.54. The mirror is stationary with respect to S. A light pulse emitted by the rocket travels toward the mirror and is reflected back to the rocket. The front of the rocket is a distance d from the mirror (as measured by observers in S) at the moment the light pulse leaves the rocket. What is the total travel time of the pulse as measured by observers in (a) the S frame and (b) the front of the rocket?

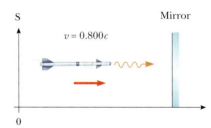

Figure P26.54

55. A physics professor on Earth gives a final exam to her students, who are on a rocket ship traveling at speed v with respect to Earth. The moment the ship passes the professor, she signals the start of the exam. If she wishes her students to have T_0 (rocket time) to complete the exam, show that she should wait a time

$$T = T_0 \sqrt{\frac{1 - v/c}{1 + v/c}}$$

(Earth time) before sending a light signal telling them to stop. [*Hint:* Remember that it takes some time for the second light signal to travel from the professor to the students.]

56. Imagine that the entire Sun collapses to a sphere of radius R_g such that the work required to remove a small mass m from the surface would be equal to its rest energy mc^2. This

radius is called the *gravitational radius* for the Sun. Find R_g. (It is believed that the ultimate fate of very massive stars is to collapse beyond their gravitational radii into black holes.)

57. A rod of length L_0 moves with a speed of v along the horizontal direction. The rod makes an angle of θ_0 with respect to the axis of a coordinate system moving with the rod. (a) Show that the length of the rod, as measured by a stationary observer, is given by

$$L = L_0 \left[1 - \left(\frac{v^2}{c^2} \right) \cos^2 \theta_0 \right]^{1/2}$$

(b) Show that the angle the rod makes with the axis, as seen by the stationary observer, is given by the expression $\tan \theta = \gamma \tan \theta_0$. These results demonstrate that the rod is both contracted and rotated. (Take the lower end of the rod to be at the origin of the moving coordinate system.)

58. The speed limit on a certain roadway is 90.0 km/h. Suppose that speeding fines are made proportional to the amount by which a vehicle's momentum exceeds the momentum the vehicle would have when traveling at the speed limit. The fine for driving at 190 km/h (that is, 100 km/h over the speed limit) is $80.0. What then will be the fine for traveling (a) at 1 090 km/h? (b) at 1 000 000 090 km/h?

59. The identical twins Speedo and Goslo join a migration from the Earth to Planet X. It is 20.0 lightyears away in a reference frame in which both planets are at rest. The twins, of the same age, depart at the same time on different spacecraft. Speedo's craft travels steadily at $0.950c$, Goslo's at $0.750c$. Calculate the age difference between the twins after Goslo's spacecraft lands on Planet X. Which twin is the older?

Color-enhanced scanning electron micrograph of the storage mite *Lepidoglyphus destructor*. These common mites grow to 0.75 mm and feed on molds, flour, and rice. They thrive at 25°C and high humidity and can trigger allergies.

© Eye of Science/Science Source/Photo Researchers, Inc.

CHAPTER

27

Quantum Physics

Although many problems were resolved by the theory of relativity in the early part of the 20th century, many other problems remained unsolved. Attempts to explain the behavior of matter on the atomic level with the laws of classical physics were consistently unsuccessful. Various phenomena, such as the electromagnetic radiation emitted by a heated object (blackbody radiation), the emission of electrons by illuminated metals (the photoelectric effect), and the emission of sharp spectral lines by gas atoms in an electric discharge tube, couldn't be understood within the framework of classical physics. Between 1900 and 1930, however, a modern version of mechanics called *quantum mechanics* or *wave mechanics* was highly successful in explaining the behavior of atoms, molecules, and nuclei.

The earliest ideas of quantum theory were introduced by Planck, and most of the subsequent mathematical developments, interpretations, and improvements were made by a number of distinguished physicists, including Einstein, Bohr, Schrödinger, de Broglie, Heisenberg, Born, and Dirac. In this chapter we introduce the underlying ideas of quantum theory and the wave–particle nature of matter, and discuss some simple applications of quantum theory, including the photoelectric effect, the Compton effect, and x-rays.

27.1 BLACKBODY RADIATION AND PLANCK'S HYPOTHESIS

An object at any temperature emits electromagnetic radiation, called **thermal radiation**. Stefan's law, discussed in Section 11.5, describes the total power radiated. The spectrum of the radiation depends on the temperature and properties of the object. At low temperatures, the wavelengths of the thermal radiation are mainly in the infrared region and hence not observable by the eye. As the temperature of an object increases, the object eventually begins to glow red. At sufficiently high temperatures, it appears to be white, as in the glow of the hot tungsten filament of a lightbulb. A careful study of thermal radiation shows that it consists of a continuous distribution of wavelengths from the infrared, visible, and ultraviolet portions of the spectrum.

From a classical viewpoint, thermal radiation originates from accelerated charged particles near the surface of an object; such charges emit radiation, much as small antennas do. The thermally agitated charges can have a distribution of frequencies, which accounts for the continuous spectrum of radiation emitted by the object. By the end of the 19th century, it had become apparent that the classical theory of thermal radiation was inadequate. The basic problem was in understanding the observed distribution energy as a function of wavelength in the radiation emitted by a blackbody. By definition, a blackbody is an ideal system that absorbs *all* radiation incident on it. A good approximation of a blackbody is a small hole leading to the inside of a hollow object, as shown in Figure 27.1. The nature of the radiation emitted through the small hole leading to the cavity depends *only on the temperature* of the cavity walls, and not at all on the material composition of the object, its shape, or other factors.

Experimental data for the distribution of energy in blackbody radiation at three temperatures are shown in Active Figure 27.2 (page 876). The radiated energy varies with wavelength and temperature. As the temperature of the blackbody increases, the total amount of energy (area under the curve) it emits increases. Also, with increasing temperature, the peak of the distribution shifts to shorter wavelengths. This shift obeys the following relationship, called **Wien's displacement law**,

$$\lambda_{max} T = 0.2898 \times 10^{-2} \text{ m} \cdot \text{K} \qquad [27.1]$$

where λ_{max} is the wavelength at which the curve peaks and T is the absolute temperature of the object emitting the radiation.

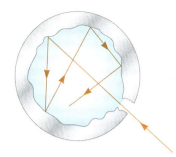

Figure 27.1 An opening in the cavity of a body is a good approximation of a blackbody. As light enters the cavity through the small opening, part is reflected and part is absorbed on each reflection from the interior walls. After many reflections, essentially all of the incident energy is absorbed.

TIP 27.1 Expect to Be Confused

Your life experiences take place in the macroscopic world, where quantum effects are not evident. Quantum effects can be even more bizarre than relativistic effects, but don't despair: confusion is normal and expected. As the Nobel prize-winning physicist Richard Feynman once said, "Nobody understands quantum mechanics."

Applying Physics 27.1 Star Colors

If you look carefully at stars in the night sky, you can distinguish three main colors: red, white, and blue. What causes these particular colors?

Explanation These colors result from the different surface temperatures of stars. A relatively cool star, with a surface temperature of 3 000 K, has a radiation curve like the middle curve in Active Figure 27.2 (page 876). The peak in this curve is above the visible wavelengths, 0.4 μm–0.7 μm, beyond the wavelength of red light, so significantly more radiation is emitted within the visible range at the red end than the blue end of the spectrum. Consequently, the star appears reddish in color, similar to the red glow from the burner of an electric stove.

A hotter star has a radiation curve more like the upper curve in Active Figure 27.2. In this case, the star emits significant radiation throughout the visible range, and the combination of all colors causes the star to look white. This is the case with our own Sun, with a surface temperature of 5 800 K. For very hot stars, the peak can be shifted so far below the visible range that significantly more blue radiation is emitted than red, so the star appears bluish in color.

EXAMPLE 27.1 Thermal Radiation from the Human Body

Goal Apply Wien's law.

Problem The temperature of the skin is approximately 35.0°C. At what wavelength does the radiation emitted from the skin reach its peak?

Strategy This is a matter of substitution into Wien's law, Equation 27.1.

Solution

Apply Wien's displacement law:
$$\lambda_{max} T = 0.289\ 8 \times 10^{-2} \text{ m} \cdot \text{K}$$

Solve for λ_{max}, noting that 35.0°C corresponds to an absolute temperature of 308 K:

$$\lambda_{max} = \frac{0.289\,8 \times 10^{-2}\text{ m}\cdot\text{K}}{308\text{ K}} = 9.41\ \mu m$$

Remark   This radiation is in the infrared region of the spectrum.

Exercise 27.1

(a) Find the wavelength corresponding to the peak of the radiation curve for the heating element of an electric oven at a temperature of 1.20×10^3 K. (Note that although this radiation peak lies in the infrared, there is enough visible radiation at this temperature to give the element a red glow.) (b) The peak in the radiation curve of the Sun is 510 nm. Calculate the temperature of the surface of the Sun.

Answers   (a) 2.42 μm;   (b) 5 700 K

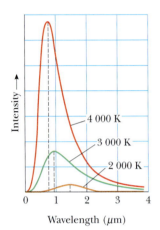

ACTIVE FIGURE 27.2

Intensity of blackbody radiation versus wavelength at three different temperatures. Note that the total radiation emitted (the area under a curve) increases with increasing temperature.

Physics⊗Now™

Log into PhysicsNow at **www.cp7e.com** and go to Active Figure 27.2, where you can adjust the temperature of the blackbody and study the radiation emitted from it.

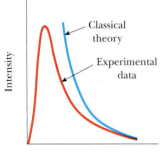

ACTIVE FIGURE 27.3

Comparison of experimental data with the classical theory of blackbody radiation. Planck's theory matches the experimental data perfectly.

Physics⊗Now™

Log into PhysicsNow at **www.cp7e.com** and go to Active Figure 27.3, where you can investigate the discrete energies emitted in the Planck model.

Attempts to use classical ideas to explain the shapes of the curves shown in Active Figure 27.2 failed. Active Figure 27.3 shows an experimental plot of the blackbody radiation spectrum (red curve), together with the theoretical picture of what this curve should look like based on classical theories (blue curve). At long wavelengths, classical theory is in good agreement with the experimental data. At short wavelengths, however, major disagreement exists between classical theory and experiment. As λ approaches zero, classical theory predicts that the amount of energy being radiated should increase. In fact, the theory erroneously predicts that the intensity should be infinite, when the experimental data shows it should approach zero. This contradiction is called the **ultraviolet catastrophe**, because theory and experiment disagree strongly in the short-wavelength, ultraviolet region of the spectrum.

In 1900 Planck developed a formula for blackbody radiation that was in complete agreement with experiments at all wavelengths, leading to a curve shown by the red line in Active Figure 27.3. Planck hypothesized that blackbody radiation was produced by submicroscopic charged oscillators, which he called *resonators*. He assumed that the walls of a glowing cavity were composed of billions of these resonators, although their exact nature was unknown. The resonators were allowed to have only certain discrete energies E_n, given by

$$E_n = nhf \qquad [27.2]$$

where n is a positive integer called a **quantum number**, f is the frequency of vibration of the resonator, and h is a constant known as **Planck's constant**, which has the value

$$h = 6.626 \times 10^{-34}\text{ J}\cdot\text{s} \qquad [27.3]$$

Because the energy of each resonator can have only discrete values given by Equation 27.2, we say the energy is *quantized*. Each discrete energy value represents a different *quantum state*, with each value of n representing a specific quantum state. (When the resonator is in the $n = 1$ quantum state, its energy is hf; when it is in the $n = 2$ quantum state, its energy is $2\,hf$; and so on.)

The key point in Planck's theory is the assumption of quantized energy states. This is a radical departure from classical physics, the "quantum leap" that led to a totally new understanding of nature. It's shocking: it's like saying a pitched baseball can have only a fixed number of different speeds, and no speeds in between those fixed values. When Planck presented his theory, most scientists (including Planck!) didn't consider the quantum concept to be realistic; however, subsequent developments showed that a theory based on the quantum concept (rather than on classical concepts) had to be used to explain a number of other phenomena at the atomic level.

EXAMPLE 27.2   The Quantized Macroscopic Oscillator

Goal   Contrast the classical and quantum oscillator.

Problem   A 2.00-kg mass is attached to a spring having force constant $k = 25.0$ N/m and negligible mass. The spring is stretched 0.400 m from its equilibrium position and released. **(a)** Find the total energy and frequency of oscillation according to classical calculations. **(b)** Assume that Planck's law of energy quantization applies to any oscillator, atomic or large scale, and find the quantum number n for this system. **(c)** How much energy would be carried away in a one-quantum change?

Strategy   We are given the spring constant and the oscillation amplitude, so we can find the total energy with the conservation of mechanical energy, using the point of maximum displacement. Equation 13.10 gives the frequency of a spring system, which can then be used with the quantum hypothesis, Equation 27.2, to obtain the value of the quantum number n. In part (c), a single quantum of energy is always equal to Planck's constant times the frequency.

Solution

(a) Find the energy and the classical frequency of the system.

Substitute into the classical energy when the block is at maximum amplitude:

$$E = \frac{1}{2}kA^2 = \frac{1}{2}(25.0 \text{ N/m})(0.400 \text{ m})^2 = \boxed{2.00 \text{ J}}$$

Compute the frequency of oscillation:

$$f = \frac{1}{2\pi}\sqrt{\frac{k}{m}} = \frac{1}{2\pi}\sqrt{\frac{25.0 \text{ N/m}}{2.00 \text{ kg}}} = \boxed{0.563 \text{ Hz}}$$

(b) Calculate the value of the quantum number n corresponding to the classical energy.

Solve Equation 27.2 for n:

$$E_n = nhf \quad \rightarrow \quad n = \frac{E_n}{hf}$$

$$= \frac{2.00 \text{ J}}{(6.63 \times 10^{-34} \text{ J·s})(0.563 \text{ Hz})} = \boxed{5.36 \times 10^{33}}$$

(c) How much energy would be carried away in a one-quantum change?

Compute the difference between two adjacent energy levels and substitute:

$$\Delta E = E_{n+1} - E_n = hf$$

$$= (6.63 \times 10^{-34} \text{ J·s})(0.563 \text{ Hz}) = \boxed{3.73 \times 10^{-34} \text{ J}}$$

Remarks   The energy carried away by a one-quantum change is such a small fraction of the total energy of the oscillator that we couldn't expect to measure it. Consequently, the energy of an object–spring system decreases by such small quantum transitions that the decrease in energy appears to be continuous. Quantum effects become important and measurable only on the submicroscopic level of atoms and molecules.

Exercise 27.2

A pendulum has a length of 1.50 m. Treating it as a quantum system, calculate (a) its frequency in the presence of Earth's gravitational field and (b) the energy carried away in a change of energy levels from $n = 3$ to $n = 1$.

Answers   (a) 0.407 Hz   (b) 5.39×10^{-34} J

27.2   THE PHOTOELECTRIC EFFECT AND THE PARTICLE THEORY OF LIGHT

In the latter part of the 19th century, experiments showed that light incident on certain metallic surfaces caused the emission of electrons are emitted from the surfaces. This phenomenon is known as the **photoelectric effect**, and the emitted electrons are called **photoelectrons**. The first discovery of this phenomenon was made by Hertz, who was also the first to produce the electromagnetic waves predicted by Maxwell.

MAX PLANCK, German Physicist, (1858–1947)

Planck introduced the concept of a "quantum of action" (Planck's constant *h*) in an attempt to explain the spectral distribution of blackbody radiation, which laid the foundations for quantum theory. In 1918, he was awarded the Nobel Prize for this discovery of the quantized nature of energy.

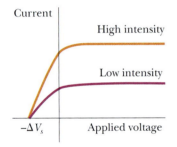

ACTIVE FIGURE 27.5
Photoelectric current versus applied potential difference for two light intensities. The current increases with intensity, but reaches a saturation level for large values of ΔV. At voltages equal to or less than $-\Delta V_s$, where ΔV_s is the stopping potential, the current is zero.

Physics⊗Now™

Log into PhysicsNow at **www.cp7e.com** and go to Active Figure 27.5, where you can sweep through the voltage range and observe the current curve for different intensities of radiation.

ACTIVE FIGURE 27.4
A circuit diagram for studying the photoelectric effect. When light strikes plate E (the emitter), photoelectrons are ejected from the plate. Electrons moving from plate E to plate C (the collector) create a current in the circuit, registered at the ammeter, A.

Physics⊗Now™

Log into PhysicsNow at **www.cp7e.com** and go to Active Figure 27.4, where you can observe the motion of electrons for various frequencies and voltages.

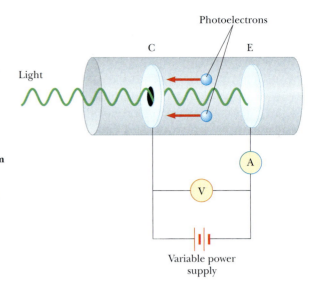

Active Figure 27.4 is a schematic diagram of a photoelectric effect apparatus. An evacuated glass tube known as a photocell contains a metal plate E (the emitter) connected to the negative terminal of a variable power supply. Another metal plate, C (the collector), is maintained at a positive potential by the power supply. When the tube is kept in the dark, the ammeter reads zero, indicating that there is no current in the circuit. However, when plate E is illuminated by light having a wavelength shorter than some particular wavelength that depends on the material used to make plate E, a current is detected by the ammeter, indicating a flow of charges across the gap between E and C. This current arises from photoelectrons emitted from the negative plate E and collected at the positive plate C.

Active Figure 27.5 is a plot of the photoelectric current versus the potential difference ΔV between E and C for two light intensities. At large values of ΔV, the current reaches a maximum value. In addition, the current increases as the incident light intensity increases, as you might expect. Finally, when ΔV is negative—that is, when the power supply in the circuit is reversed to make E positive and C negative—the current drops to a low value because most of the emitted photoelectrons are repelled by the now negative plate C. In this situation, only those electrons having a kinetic energy greater than the magnitude of $e\Delta V$ reach C, where *e* is the charge on the electron.

When ΔV is equal to or more negative than $-\Delta V_s$, the **stopping potential**, no electrons reach C and the current is zero. The stopping potential is *independent* of the radiation intensity. The maximum kinetic energy of the photoelectrons is related to the stopping potential through the relationship

$$KE_{max} = e\Delta V_s \qquad \textbf{[27.4]}$$

Several features of the photoelectric effect can't be explained with classical physics or with the wave theory of light:

- No electrons are emitted if the incident light frequency falls below some **cutoff frequency** f_c, which is characteristic of the material being illuminated. This is inconsistent with the wave theory, which predicts that the photoelectric effect should occur at *any* frequency, provided the light intensity is sufficiently high.
- The maximum kinetic energy of the photoelectrons is independent of light intensity. According to wave theory, light of higher intensity should carry more energy into the metal per unit time and therefore eject photoelectrons having higher kinetic energies.
- The maximum kinetic energy of the photoelectrons increases with increasing light frequency. The wave theory predicts no relationship between photoelectron energy and incident light frequency.
- Electrons are emitted from the surface almost instantaneously (less than 10^{-9} s after the surface is illuminated), even at low light intensities. Classically, we expect the photoelectrons to require some time to absorb the incident radiation before they acquire enough kinetic energy to escape from the metal.

A successful explanation of the photoelectric effect was given by Einstein in 1905, the same year he published his special theory of relativity. As part of a general paper on electromagnetic radiation, for which he received the Nobel Prize in 1921, Einstein extended Planck's concept of quantization to electromagnetic waves. He suggested that a tiny packet of light energy or **photon** would be emitted when a quantized oscillator made a jump from an energy state $E_n = nhf$ to the next lower state $E_{n-1} = (n - 1)hf$. Conservation of energy would require the decrease in oscillator energy, hf, to be equal to the photon's energy E, so that

$$E = hf \qquad [27.5]$$

◀ Energy of a photon

where h is Planck's constant and f is the frequency of the light, which is equal to the frequency of Planck's oscillator.

The key point here is that the light energy lost by the emitter, hf, stays sharply localized in a tiny packet or particle called a photon. In Einstein's model, a photon is so localized that it can give *all* its energy hf to a single electron in the metal. According to Einstein, the maximum kinetic energy for these liberated photoelectrons is

$$KE_{max} = hf - \phi \qquad [27.6]$$

◀ Photoelectric effect equation

where ϕ is called the **work function** of the metal. The work function, which represents the minimum energy with which an electron is bound in the metal, is on the order of a few electron volts. Table 27.1 lists work functions for various metals.

With the photon theory of light, we can explain the previously mentioned features of the photoelectric effect that cannot be understood using concepts of classical physics:

- Photoelectrons are created by absorption of a single photon, so the energy of that photon must be greater than or equal to the work function, else no photoelectrons will be produced. This explains the cutoff frequency.
- From Equation 27.6, KE_{max} depends only on the frequency of the light and the value of the work function. Light intensity is immaterial, because absorption of a single photon is responsible for the electron's change in kinetic energy.
- Equation 27.6 is linear in the frequency, so KE_{max} increases with increasing frequency.
- Electrons are emitted almost instantaneously, regardless of intensity, because the light energy is concentrated in packets rather than spread out in waves. If the frequency is high enough, no time is needed for the electron to gradually acquire sufficient energy to escape the metal.

Experimentally, a linear relationship is observed between f and KE_{max}, as sketched in Figure 27.6. The intercept on the horizontal axis, corresponding to $KE_{max} = 0$, gives the cutoff frequency below which no photoelectrons are emitted, regardless of light intensity. The cutoff *wavelength* λ_c can be derived from Equation 27.6:

$$KE_{max} = hf_c - \phi = 0 \quad \rightarrow \quad h\frac{c}{\lambda_c} - \phi = 0$$

$$\lambda_c = \frac{hc}{\phi} \qquad [27.7]$$

where c is the speed of light. Wavelengths *greater* than λ_c incident on a material with work function ϕ don't result in the emission of photoelectrons.

TABLE 27.1

Work Functions of Selected Metals

Metal	ϕ (eV)
Na	2.46
Al	4.08
Cu	4.70
Zn	4.31
Ag	4.73
Pt	6.35
Pb	4.14
Fe	4.50

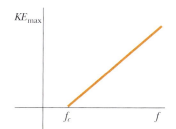

Figure 27.6   A sketch of KE_{max} versus the frequency of incident light for photoelectrons in a typical photoelectric effect experiment. Photons with frequency less than f_c don't have sufficient energy to eject an electron from the metal.

INTERACTIVE EXAMPLE 27.3   Photoelectrons from Sodium

Goal   Understand the quantization of light and its role in the photoelectric effect.

Problem   A sodium surface is illuminated with light of wavelength 0.300 μm. The work function for sodium is 2.46 eV. **(a)** Calculate the energy of each photon in electron volts, **(b)** the maximum kinetic energy of the ejected photoelectrons, and **(c)** the cutoff wavelength for sodium.

Strategy   Parts (a), (b), and (c) require substitution of values into Equations 27.5, 27.6, and 27.7, respectively.

Solution
(a) Calculate the energy of each photon.

Obtain the frequency from the wavelength:

$$c = f\lambda \quad \rightarrow \quad f = \frac{c}{\lambda} = \frac{3.00 \times 10^8 \text{ m/s}}{0.300 \times 10^{-6} \text{ m}}$$

$$f = 1.00 \times 10^{15} \text{ Hz}$$

Use Equation 27.5 to calculate the photon's energy:

$$E = hf = (6.63 \times 10^{-34} \text{ J·s})(1.00 \times 10^{15} \text{ Hz})$$
$$= 6.63 \times 10^{-19} \text{ J}$$

$$= (6.63 \times 10^{-19} \text{ J})\left(\frac{1.00 \text{ eV}}{1.60 \times 10^{-19} \text{ J}}\right) = \boxed{4.14 \text{ eV}}$$

(b) Find the maximum kinetic energy of the photoelectrons.

Substitute into Equation 27.6:

$$KE_{max} = hf - \phi = 4.14 \text{ eV} - 2.46 \text{ eV} = \boxed{1.68 \text{ eV}}$$

(c) Compute the cutoff wavelength.

Convert ϕ from electron volts to joules:

$$\phi = 2.46 \text{ eV} = (2.46 \text{ eV})(1.60 \times 10^{-19} \text{ J/eV})$$
$$= 3.94 \times 10^{-19} \text{ J}$$

Find the cutoff wavelength using Equation 27.7.

$$\lambda_c = \frac{hc}{\phi} = \frac{(6.63 \times 10^{-34} \text{ J·s})(3.00 \times 10^8 \text{ m/s})}{3.94 \times 10^{-19} \text{ J}}$$

$$= 5.05 \times 10^{-7} \text{ m} = \boxed{505 \text{ nm}}$$

Remark   The cutoff wavelength is in the yellow-green region of the visible spectrum.

Exercise 27.3
(a) What minimum-frequency light will eject photoelectrons from a copper surface? (b) If this frequency is tripled, find the maximum kinetic energy (in eV) of the resulting photoelectrons. (Answer in eV.)

Answers   (a) 1.13×10^{15} Hz   (b) 9.40 eV

Physics⊗Now™   Investigate the photoelectric effect for different materials and different wavelengths of light by logging into PhysicsNow at **www.cp7e.com** and going to Interactive Example 27.3.

Photocells

The photoelectric effect has many interesting applications using a device called the *photocell*. The photocell shown in Active Figure 27.4 produces a current in the circuit when light of sufficiently high frequency falls on the cell, but it doesn't allow a current in the dark. This device is used in streetlights: a photoelectric control unit in the base of the light activates a switch that turns off the streetlight when ambient light strikes it. Many garage-door systems and elevators use a light beam and a photocell as a safety feature in their design. When the light beam strikes the photocell, the electric current generated is sufficiently large to maintain a closed circuit. When an object or a person blocks the light beam, the current is interrupted, which signals the door to open.

27.3 X-RAYS

In 1895 at the University of Wurzburg, Wilhelm Roentgen (1845–1923) was studying electrical discharges in low-pressure gases when he noticed that a fluorescent screen glowed even when placed several meters from the gas discharge tube and even when black cardboard was placed between the tube and the screen. He concluded that the effect was caused by a mysterious type of radiation, which he called **x-rays** because of their unknown nature. Subsequent study showed that these rays traveled at or near the speed of light and that they couldn't be deflected by either electric or magnetic fields. This last fact indicated that x-rays did not consist of beams of charged particles, although the possibility that they were beams of uncharged particles remained.

In 1912 Max von Laue (1879–1960) suggested that if x-rays were electromagnetic waves with very short wavelengths, it should be possible to diffract them by using the regular atomic spacings of a crystal lattice as a diffraction grating, just as visible light is diffracted by a ruled grating. Shortly thereafter, researchers demonstrated that such a diffraction pattern could be observed, similar to that shown in Figure 27.7 for NaCl. The wavelengths of the x-rays were then determined from the diffraction data and the known values of the spacing between atoms in the crystal. X-ray diffraction has proved to be an invaluable technique for understanding the structure of matter (as discussed in more detail in the next section).

Typical x-ray wavelengths are about 0.1 nm, which is on the order of the atomic spacing in a solid. We now know that x-rays are a part of the electromagnetic spectrum, characterized by frequencies higher than those of ultraviolet radiation and having the ability to penetrate most materials with relative ease.

X-rays are produced when high-speed electrons are suddenly slowed down— for example, when a metal target is struck by electrons that have been accelerated through a potential difference of several thousand volts. Figure 27.8a shows a schematic diagram of an x-ray tube. A current in the filament causes electrons to be emitted, and these freed electrons are accelerated toward a dense metal target, such as tungsten, which is held at a higher potential than the filament.

Figure 27.9 represents a plot of x-ray intensity versus wavelength for the spectrum of radiation emitted by an x-ray tube. Note that the spectrum has are two distinct components. One component is a continuous broad spectrum that depends on the voltage applied to the tube. Superimposed on this component is a series of sharp, intense lines that depend on the nature of the target material. The accelerating voltage must exceed a certain value, called the **threshold voltage**, in order to observe these sharp lines, which represent radiation emitted by the target atoms as their electrons undergo rearrangements. We will discuss this further in Chapter 28. The continuous radiation is sometimes called **bremsstrahlung**, a German word meaning "braking radiation," because electrons emit radiation when they undergo an acceleration inside the target.

Figure 27.10 (page 882) illustrates how x-rays are produced when an electron passes near a charged target nucleus. As the electron passes close to a positively

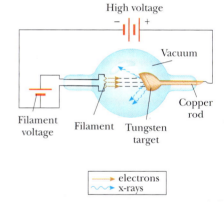

Figure 27.7   X-ray diffraction pattern of NaCl.

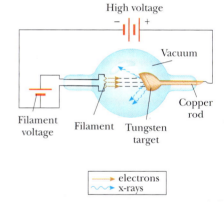

Figure 27.8   (a) Diagram of an x-ray tube. (b) Photograph of an x-ray tube.

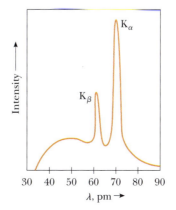

Figure 27.9   The x-ray spectrum of a metal target consists of a broad continuous spectrum plus a number of sharp lines, which are due to *characteristic x-rays*. The data shown were obtained when 35-keV electrons bombarded a molybdenum target. Note that 1 pm = 10^{-12} m = 10^{-3} nm.

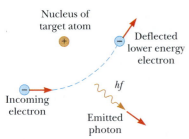

Figure 27.10 An electron passing near a charged target atom experiences an acceleration, and a photon is emitted in the process.

charged nucleus contained in the target material, it is deflected from its path because of its electrical attraction to the nucleus; hence, it undergoes an acceleration. An analysis from classical physics shows that any charged particle will emit electromagnetic radiation when it is accelerated. (An example of this phenomenon is the production of electromagnetic waves by accelerated charges in a radio antenna, as described in Chapter 21.) According to quantum theory, this radiation must appear in the form of photons. Because the radiated photon shown in Figure 27.10 carries energy, the electron must lose kinetic energy because of its encounter with the target nucleus. An extreme example would consist of the electron losing all of its energy in a single collision. In this case, the initial energy of the electron ($e\Delta V$) is transformed completely into the energy of the photon (hf_{max}). In equation form,

$$e\Delta V = hf_{max} = \frac{hc}{\lambda_{min}}$$ [27.8]

where $e\Delta V$ is the energy of the electron after it has been accelerated through a potential difference of ΔV volts and e is the charge on the electron. This says that the shortest wavelength radiation that can be produced is

$$\lambda_{min} = \frac{hc}{e\Delta V}$$ [27.9]

The reason that not all the radiation produced has this particular wavelength is because many of the electrons aren't stopped in a single collision. This results in the production of the continuous spectrum of wavelengths.

Interesting insights into the process of painting and revising a masterpiece are being revealed by x-rays. Long wavelength x-rays are absorbed in varying degrees by some paints, such as those having lead, cadmium, chromium, or cobalt as a base. The x-ray interactions with the paints give contrast, because the different elements in the paints have different electron densities. Also, thicker layers will absorb more than thin layers. To examine a painting by an old master, a film is placed behind it while it is x-rayed from the front. Ghost outlines of earlier paintings and earlier forms of the final masterpiece are sometimes revealed when the film is developed.

APPLICATION

Using X-Rays to Study the Work of Master Painters

EXAMPLE 27.4   An X-Ray Tube

Goal   Calculate the minimum x-ray wavelength due to accelerated electrons.

Problem   Medical x-ray machines typically operate at a potential difference of 1.00×10^5 V. Calculate the minimum wavelength their x-ray tubes produce when electrons are accelerated through this potential difference.

Strategy   The minimum wavelength corresponds to the most energetic photons. Substitute the given potential difference into Equation 27.9.

Solution
Substitute into Equation 27.9:

$$\lambda_{min} = \frac{hc}{e\Delta V}$$

$$= \frac{(6.63 \times 10^{-34}\,\text{J}\cdot\text{s})(3.00 \times 10^8\,\text{m/s})}{(1.60 \times 10^{-19}\,\text{C})(1.00 \times 10^5\,\text{V})}$$

$$= 1.24 \times 10^{-11}\,\text{m}$$

Remarks   X-ray tubes generally operate with half the voltage with respect to Earth, $+50\,000$ V, applied to the anode, and the other half, $-50\,000$ V, applied to the cathode. This lengthens tube lifetime by reducing the probability of voltage breakthroughs.

Exercise 27.4

What potential difference would be necessary to produce gamma rays with wavelength 1.00×10^{-15} m? This wavelength is about the same size as the diameter of a proton.

Solution 1.24×10^9 V

27.4   DIFFRACTION OF X-RAYS BY CRYSTALS

In Chapter 24 we described how a diffraction grating could be used to measure the wavelength of light. In principle, the wavelength of *any* electromagnetic wave can be measured if a grating having a suitable line spacing can be found. The spacing between lines must be approximately equal to the wavelength of the radiation to be measured. X-rays are electromagnetic waves with wavelengths on the order of 0.1 nm. It would be impossible to construct a grating with such a small spacing. However, as noted in the previous section, Max von Laue suggested that the regular array of atoms in a crystal could act as a three-dimensional grating for observing the diffraction of x-rays.

One experimental arrangement for observing x-ray diffraction is shown in Figure 27.11. A narrow beam of x-rays with a continuous wavelength range is incident on a crystal such as sodium chloride. The diffracted radiation is very intense in certain directions, corresponding to constructive interference from waves reflected from layers of atoms in the crystal. The diffracted radiation is detected by a photographic film and forms an array of spots known as a *Laue pattern*. The crystal structure is determined by analyzing the positions and intensities of the various spots in the pattern.

The arrangement of atoms in a crystal of NaCl is shown in Figure 27.12. The smaller red spheres represent Na$^+$ ions, and the larger blue spheres represent Cl$^-$ ions. The spacing between successive Na$^+$ (or Cl$^-$) ions in this cubic structure, denoted by the symbol a in Figure 27.12, is approximately 0.563 nm.

A careful examination of the NaCl structure shows that the ions lie in various planes. The shaded areas in Figure 27.12 represent one example, in which the atoms lie in equally spaced planes. Now suppose an x-ray beam is incident at grazing angle θ on one of the planes, as in Figure 27.13. The beam can be reflected from both the upper and lower plane of atoms. However, the geometric construction in Figure 27.13 shows that the beam reflected from the lower surface travels farther than the beam reflected from the upper surface by a distance of $2d \sin \theta$. The two portions of the reflected beam will combine to produce constructive interference when this path difference equals some integral multiple of the wavelength λ. The condition for constructive interference is given by

$$2d \sin \theta = m\lambda \qquad (m = 1, 2, 3, \ldots) \qquad [27.10]$$

◀ Bragg's law

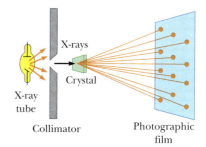

Figure 27.11   Schematic diagram of the technique used to observe the diffraction of x-rays by a single crystal. The array of spots formed on the film by the diffracted beams is called a Laue pattern. (See Fig. 27.7.)

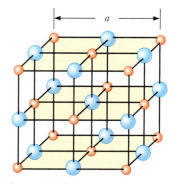

Figure 27.12   A model of the cubic crystalline structure of sodium chloride. The blue spheres represent the Cl$^-$ ions, and the red spheres represent the Na$^+$ ions. The length of the cube edge is $a = 0.563$ nm.

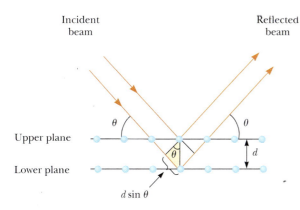

Figure 27.13   A two-dimensional depiction of the reflection of an x-ray beam from two parallel crystalline planes separated by a distance d. The beam reflected from the lower plane travels farther than the one reflected from the upper plane by an amount equal to $2d \sin \theta$.

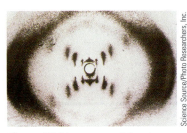

Figure 27.14 An x-ray diffraction photograph of DNA taken by Rosalind Franklin. The cross pattern of spots was a clue that DNA has a helical structure.

|←——— 2 nm ———→|

Figure 27.15 The double-helix structure of DNA.

This condition is known as **Bragg's law**, after W. L. Bragg (1890–1971), who first derived the relationship. If the wavelength and diffraction angle are measured, Equation 27.10 can be used to calculate the spacing between atomic planes.

The method of x-ray diffraction to determine crystalline structures was thoroughly developed in England by W. H. Bragg and his son W. L. Bragg, who shared a Nobel prize in 1915 for their work. Since then, thousands of crystalline structures have been investigated. Most importantly, the technique of x-ray diffraction has been used to determine the atomic arrangement of complex organic molecules such as proteins. Proteins are large molecules containing thousands of atoms that help to regulate and speed up chemical life processes in cells. Some proteins are amazing catalysts, speeding up the slow room temperature reactions in cells by 17 orders of magnitude. In order to understand this incredible biochemical reactivity, it is important to determine the structure of these intricate molecules.

The main technique used to determine the molecular structure of proteins, DNA, and RNA is x-ray diffraction using x-rays of wavelength of about 1.0 Å. This technique allows the experimenter to "see" individual atoms that are separated by about this distance in molecules. Since the biochemical x-ray diffraction sample is prepared in crystal form, the *geometry* (position of the bright spots in space) of the diffraction pattern is determined by the regular three-dimensional crystal lattice arrangement of molecules in the sample. The *intensities* of the bright diffraction spots are determined by the atoms and their electronic distributions in the fundamental building block of the crystal: the unit cell. Using complicated computational techniques, investigators can essentially deduce the molecular structure by matching the observed intensities of diffracted beams with a series of assumed atomic positions that determine the atomic structure and electron density of the molecule. Figure 27.14 shows a classic x-ray diffraction image of DNA made by Rosalind Franklin in 1952.

This and similar x-ray diffraction photos played an important role in the determination of the double-helix structure of DNA by F. H. C. Crick and J. D. Watson in 1953. A model of the famous DNA double helix is shown in Figure 27.15.

EXAMPLE 27.5 X-Ray Diffraction from Calcite

Goal Understand Bragg's law and apply it to a crystal.

Problem If the spacing between certain planes in a crystal of calcite ($CaCO_3$) is 0.314 nm, find the grazing angles at which first- and third-order interference will occur for x-rays of wavelength 0.070 0 nm.

Strategy Solve Bragg's law for $\sin \theta$ and substitute, using the inverse-sine function to obtain the angle.

Solution

Find the grazing angle corresponding to $m = 1$, for first-order interference:

$$\sin \theta = \frac{m\lambda}{2d} = \frac{(0.070\ 0\ \text{nm})}{2(0.314\ \text{nm})} = 0.111$$

$$\theta = \sin^{-1}(0.111) = \boxed{6.37°}$$

Repeat the calculation for third-order interference ($m = 3$):

$$\sin \theta = \frac{m\lambda}{2d} = \frac{3(0.070\ 0\ \text{nm})}{2(0.314\ \text{nm})} = 0.334$$

$$\theta = \sin^{-1}(0.334) = \boxed{19.5°}$$

Remark Notice there is little difference between this kind of problem and a Young's slit experiment.

Exercise 27.5

X-rays of wavelength 0.060 0 nm are scattered from a crystal with a grazing angle of 11.7°. Assume $m = 1$ for this process. Calculate the spacing between the crystal planes.

Answer 0.148 nm

27.5 THE COMPTON EFFECT

Further justification for the photon nature of light came from an experiment conducted by Arthur H. Compton in 1923. In his experiment, Compton directed an x-ray beam of wavelength λ_0 toward a block of graphite. He found that the scattered x-rays had a slightly longer wavelength λ than the incident x-rays, and hence the energies of the scattered rays were lower. The amount of energy reduction depended on the angle at which the x-rays were scattered. The change in wavelength $\Delta\lambda$ between a scattered x-ray and an incident x-ray is called the **Compton shift**.

In order to explain this effect, Compton assumed that if a photon behaves like a particle, its collision with other particles is similar to a collision between two billiard balls. Hence, the x-ray photon carries both measurable *energy* and *momentum*, and these two quantities must be conserved in a collision. If the incident photon collides with an electron initially at rest, as in Figure 27.16, the photon transfers some of its energy and momentum to the electron. As a consequence, the energy and frequency of the scattered photon are lowered and its wavelength increases. Applying relativistic energy and momentum conservation to the collision described in Figure 27.16, the shift in wavelength of the scattered photon is given by

$$\Delta\lambda = \lambda - \lambda_0 = \frac{h}{m_e c}(1 - \cos\theta) \qquad \textbf{[27.11]}$$

where m_e is the mass of the electron and θ is the angle between the directions of the scattered and incident photons. The quantity $h/m_e c$ is called the **Compton wavelength** and has a value of 0.002 43 nm. The Compton wavelength is very small relative to the wavelengths of visible light, so the shift in wavelength would be difficult to detect if visible light were used. Further, note that the Compton shift depends on the scattering angle θ and not on the wavelength. Experimental results for x-rays scattered from various targets obey Equation 27.11 and strongly support the photon concept.

◀ The Compton shift formula

Courtesy of AIP Niels Bohr Library

ARTHUR HOLLY COMPTON,
American Physicist (1892 – 1962)

Compton was born in Wooster, Ohio, and he attended Wooster College and Princeton University. He became director of the laboratory at the University of Chicago, where experimental work concerned with sustained chain reactions was conducted. This work was of central importance to the construction of the first atomic bomb. His discovery of the Compton effect and his work with cosmic rays led to his sharing the 1927 Nobel Prize in physics with Charles Wilson.

Quick Quiz 27.1

An x-ray photon is scattered by an electron. The frequency of the scattered photon relative to that of the incident photon (a) increases, (b) decreases, or (c) remains the same.

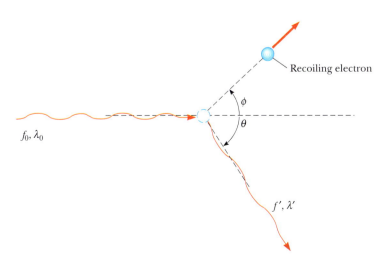

Figure 27.16 Diagram representing Compton scattering of a photon by an electron. The scattered photon has less energy (or a longer wavelength) than the incident photon.

A photon of energy E_0 strikes a free electron, with the scattered photon of energy E moving in the direction opposite that of the incident photon. In this Compton effect interaction, the resulting kinetic energy of the electron is (a) E_0 (b) E (c) $E_0 - E$ (d) $E_0 + E$ (e) None of the above

Applying Physics 27.2   Color Changes through the Compton Effect

The Compton effect involves a change in wavelength as photons are scattered through different angles. Suppose we illuminate a piece of material with a beam of light and then view the material from different angles relative to the beam of light. Will we see a color change corresponding to the change in wavelength of the scattered light?

Explanation There will be a wavelength change for visible light scattered by the material, but the change will be far too small to detect as a color change. The largest possible wavelength change, at 180° scattering, will be twice the Compton wavelength, about 0.005 nm. This represents a change of less than 0.001% of the wavelength of red light. The Compton effect is only detectable for wavelengths that are very short to begin with, so that the Compton wavelength is an appreciable fraction of the incident wavelength. As a result, the usual radiation for observing the Compton effect is in the x-ray range of the electromagnetic spectrum.

INTERACTIVE EXAMPLE 27.6   Scattering X-Rays

Goal   Understand Compton scattering and its effect on the photon's energy.

Problem   X-rays of wavelength $\lambda_0 = 0.200\,000$ nm are scattered from a block of material. The scattered x-rays are observed at an angle of 45.0° to the incident beam. **(a)** Calculate the wavelength of the x-rays scattered at this angle. **(b)** Compute the fractional change in the energy of a photon in the collision.

Solution

(a) Calculate the wavelength of the x-rays.

Substitute into Equation 27.11 to obtain the wavelength shift:

$$\Delta\lambda = \frac{h}{m_e c}(1 - \cos\theta)$$

$$= \frac{6.63 \times 10^{-34}\,\text{J}\cdot\text{s}}{(9.11 \times 10^{-31}\,\text{kg})(3.00 \times 10^8\,\text{m/s})}(1 - \cos 45.0°)$$

$$= 7.11 \times 10^{-13}\,\text{m} = 0.000\,711\,\text{nm}$$

Add this shift to the original wavelength to obtain the wavelength of the scattered photon:

$$\lambda = \Delta\lambda + \lambda_0 = \boxed{0.200\,711\,\text{nm}}$$

(b) Find the fraction of energy lost by the photon in the collision.

Rewrite the energy E in terms of wavelength, using $c = f\lambda$:

$$E = hf = h\frac{c}{\lambda}$$

Compute $\Delta E/E$ using this expression:

$$\frac{\Delta E}{E} = \frac{E_f - E_i}{E_i} = \frac{hc/\lambda_f - hc/\lambda_i}{hc/\lambda_i}$$

Cancel hc and rearrange terms:

$$\frac{\Delta E}{E} = \frac{1/\lambda_f - 1/\lambda_i}{1/\lambda_i} = \frac{\lambda_i}{\lambda_f} - 1 = \frac{\lambda_i - \lambda_f}{\lambda_f} = -\frac{\Delta\lambda}{\lambda_f}$$

Substitute values from part (a):

$$\frac{\Delta E}{E} = -\frac{0.000\,711\,\text{nm}}{0.200\,711\,\text{nm}} = \boxed{-3.54 \times 10^{-3}}$$

Remarks   It is also possible to find this answer by substituting into the energy expression at an earlier stage, but the algebraic derivation is more elegant and instructive.

Exercise 27.6

Repeat the exercise for a photon with wavelength 3.00×10^{-2} nm that scatters at an angle of $60.0°$.

Answers   (a) 3.12×10^{-2} nm   (b) $\Delta E/E = -3.88 \times 10^{-2}$

Physics⊗Now™   Study Compton scattering for different angles by logging into PhysicsNow at **www.cp7e.com** and going to Interactive Example 27.6.

27.6   THE DUAL NATURE OF LIGHT AND MATTER

Light and Electromagnetic Radiation

Phenomena such as the photoelectric effect and the Compton effect offer evidence that when light (or other forms of electromagnetic radiation) and matter interact, the light behaves as if it were composed of particles having energy hf and momentum h/λ. In other contexts, however, light acts like a wave, exhibiting interference and diffraction effects. Is light a wave or a particle?

The answer depends on the phenomenon being observed. Some experiments can be better explained with the photon concept, whereas others are best described with a wave model. The end result is that both models are needed. **Light has a dual nature, exhibiting both wave and particle characteristics**.

To understand why photons are compatible with electromagnetic waves, consider 2.5-MHz radio waves as an example. The energy of a photon having this frequency is only about 10^{-8} eV, too small to allow the photon to be detected. A sensitive radio receiver might require as many as 10^{10} of these photons to produce a detectable signal. Such a large number of photons would appear, on the average, as a continuous wave. With so many photons reaching the detector every second, we wouldn't be able to detect the individual photons striking the antenna.

Now consider what happens as we go to higher frequencies. In the visible region, it's possible to observe both the particle characteristics and the wave characteristics of light. As we mentioned earlier, a light beam shows interference phenomena (thus, it is a wave) and at the same time can produce photoelectrons (thus, it is a particle). At even higher frequencies, the momentum and energy of the photons increase. Consequently, the particle nature of light becomes more evident than its wave nature. For example, the absorption of an x-ray photon is easily detected as a single event, but wave effects are difficult to observe.

LOUIS DE BROGLIE, French Physicist, (1892–1987)

De Broglie was born in Dieppe, France. At the Sorbonne in Paris, he studied history in preparation for what he hoped to be a career in the diplomatic service. The world of science is lucky that he changed his career path to become a theoretical physicist. De Broglie was awarded the Nobel Prize in 1929 for his discovery of the wave nature of electrons.

The Wave Properties of Particles

In his doctoral dissertation in 1924, Louis de Broglie postulated that, **because photons have wave and particle characteristics, perhaps all forms of matter have both properties**. This was a highly revolutionary idea with no experimental confirmation at that time. According to de Broglie, electrons, just like light, have a dual particle–wave nature.

In Chapter 26 we found that the relationship between energy and momentum for a photon, which has a rest energy of zero, is $p = E/c$. We also know from Equation 27.5 that the energy of a photon is

$$E = hf = \frac{hc}{\lambda} \qquad \text{[27.12]}$$

Consequently, the momentum of a photon can be expressed as

$$p = \frac{E}{c} = \frac{hc}{c\lambda} = \frac{h}{\lambda} \qquad \text{[27.13]} \qquad \blacktriangleleft \text{Momentum of a photon}$$

From this equation, we see that the photon wavelength can be specified by its momentum, or $\lambda = h/p$. De Broglie suggested that *all* material particles with momentum p should have a characteristic wavelength $\lambda = h/p$. Because the momentum of a particle of mass m and speed v is $mv = p$, the **de Broglie wavelength** of a particle is

de Broglie's hypothesis ▶

$$\lambda = \frac{h}{p} = \frac{h}{mv} \qquad [27.14]$$

Further, de Broglie postulated that the frequencies of matter waves (waves associated with particles having nonzero rest energy) obey the Einstein relationship for photons, $E = hf$, so that

Frequency of matter waves ▶

$$f = \frac{E}{h} \qquad [27.15]$$

The dual nature of matter is quite apparent in Equations 27.14 and 27.15, because each contains both particle concepts (mv and E) and wave concepts (λ and f). The fact that these relationships had been established experimentally for photons made the de Broglie hypothesis that much easier to accept.

The Davisson–Germer Experiment

De Broglie's proposal in 1923 that matter exhibits both wave and particle properties was first regarded as pure speculation. If particles such as electrons had wave-like properties, then, under the correct conditions, they should exhibit diffraction effects. In 1927, three years after de Broglie published his work, C. J. Davisson (1881–1958) and L. H. Germer (1896–1971) of the United States succeeded in measuring the wavelength of electrons. Their important discovery provided the first experimental confirmation of the matter waves proposed by de Broglie.

The intent of the initial Davisson–Germer experiment was not to confirm the de Broglie hypothesis. In fact, their discovery was made by accident (as is often the case). The experiment involved the scattering of low-energy electrons (about 54 eV) from a nickel target in a vacuum. During one experiment, the nickel surface was badly oxidized because of an accidental break in the vacuum system. After the nickel target was heated in a flowing stream of hydrogen to remove the oxide coating, electrons scattered by it exhibited intensity maxima and minima at specific angles. The experimenters finally realized that the nickel had formed large crystalline regions upon heating and that the regularly spaced planes of atoms in the crystalline regions served as a diffraction grating for electron matter waves. (See Section 27.5.)

Shortly thereafter, Davisson and Germer performed more extensive diffraction measurements on electrons scattered from single-crystal targets. Their results showed conclusively the wave nature of electrons and confirmed the de Broglie relation $\lambda = h/p$. In the same year, G. P. Thomson (1892–1975) of Scotland also observed electron diffraction patterns by passing electrons through very thin gold foils. Diffraction patterns have since been observed for helium atoms, hydrogen atoms, and neutrons. Hence, the universal nature of matter waves has been established in various ways.

Quick Quiz 27.3

A nonrelativistic electron and a nonrelativistic proton are moving and have the same de Broglie wavelength. Which of the following are also the same for the two particles?
(a) speed (b) kinetic energy (c) momentum (d) frequency

Quick Quiz 27.4

We have seen two wavelengths assigned to the electron: the Compton wavelength and the de Broglie wavelength. Which is an actual *physical* wavelength associated with the electron? (a) the Compton wavelength (b) the de Broglie wavelength (c) both wavelengths (d) neither wavelength

EXAMPLE 27.7   The Electron versus the Baseball

Goal   Apply the de Broglie hypothesis to a quantum and a classical object.

Problem   (a) Compare the de Broglie wavelength for an electron ($m_e = 9.11 \times 10^{-31}$ kg) moving at a speed of 1.00×10^7 m/s with that of a baseball of mass 0.145 kg pitched at 45.0 m/s. (b) Compare these wavelengths with that of an electron traveling at $0.999c$.

Strategy   This is a matter of substitution into Equation 27.14 for the de Broglie wavelength. In part (b), the relativistic momentum must be used.

Solution

(a) Compare the de Broglie wavelengths of the electron and the baseball.

Substitute data for the electron into Equation 27.14:

$$\lambda_e = \frac{h}{m_e v} = \frac{6.63 \times 10^{-34} \,\text{J} \cdot \text{s}}{(9.11 \times 10^{-31} \,\text{kg})(1.00 \times 10^7 \,\text{m/s})}$$

$$= 7.28 \times 10^{-11} \,\text{m}$$

Repeat the calculation with the baseball data:

$$\lambda_b = \frac{h}{m_b v} = \frac{6.63 \times 10^{-34} \,\text{J} \cdot \text{s}}{(0.145 \,\text{kg})(45.0 \,\text{m/s})} = 1.02 \times 10^{-34} \,\text{m}$$

(b) Find the wavelength for an electron traveling at $0.999c$.

Replace the momentum in Equation 27.14 with the relativistic momentum:

$$\lambda_e = \frac{h}{m_e v / \sqrt{1 - v^2/c^2}} = \frac{h\sqrt{1 - v^2/c^2}}{m_e v}$$

Substitute:

$$\lambda_e = \frac{(6.63 \times 10^{-34} \,\text{J} \cdot \text{s})\sqrt{1 - (0.999c)^2/c^2}}{(9.11 \times 10^{-31} \,\text{kg})(0.999 \cdot 3.00 \times 10^8 \,\text{m/s})}$$

$$= 1.09 \times 10^{-13} \,\text{m}$$

Remarks   The electron wavelength corresponds to that of x-rays in the electromagnetic spectrum. The baseball, by contrast, has a wavelength much smaller than any aperture through which the baseball could possibly pass, so we couldn't observe any of its diffraction effects. It is generally true that the wave properties of large-scale objects can't be observed. Notice that even at extreme relativistic speeds, the electron wavelength is still far larger than the baseball's.

Exercise 27.7
Find the de Broglie wavelength of a proton ($m_p = 1.67 \times 10^{-27}$ kg) moving with a speed of 1.00×10^7 m/s.

Answer 3.97×10^{-14} m

Application: The Electron Microscope

A practical device that relies on the wave characteristics of electrons is the **electron microscope**. A *transmission* electron microscope, used for viewing flat, thin samples, is shown in Figure 27.17 (page 890). In many respects, it is similar to an optical microscope, but the electron microscope has a much greater resolving power because it can accelerate electrons to very high kinetic energies, giving them very short wavelengths. No microscope can resolve details that are significantly smaller than the wavelength of the radiation used to illuminate the object. Typically, the wavelengths of electrons are about 100 times smaller than those of the visible light used in optical microscopes. (Radiation of the same wavelength as the electrons in an electron microscope is in the x-ray region of the spectrum.)

APPLICATION
Electron Microscopes

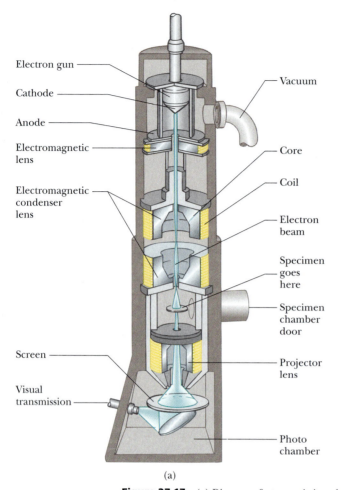

Electron gun

Cathode

Anode

Electromagnetic
lens

Electromagnetic
condenser
lens

Screen

Visual
transmission

Vacuum

Core

Coil

Electron
beam

Specimen
goes
here

Specimen
chamber
door

Projector
lens

Photo
chamber

(a)

(b)

© David Parker/Photo Researchers, Inc.

Figure 27.17 (a) Diagram of a transmission electron microscope for viewing a thin sectioned sample. The "lenses" that control the electron beam are magnetic deflection coils. (b) An electron microscope.

The electron beam in an electron microscope is controlled by electrostatic or magnetic deflection, which acts on the electrons to focus the beam to an image. Rather than examining the image through an eyepiece as in an optical microscope, the viewer looks at an image formed on a fluorescent screen. (The viewing screen must be fluorescent because otherwise the image produced wouldn't be visible.)

Applying Physics 27.3   X-Ray Microscopes?

Electron microscopes (Fig. 27.17) take advantage of the wave nature of particles. Electrons are accelerated to high speeds, giving them a short de Broglie wavelength. Imagine an electron microscope using electrons with a de Broglie wavelength of 0.2 nm. Why don't we design a microscope using 0.2-nm *photons* to do the same thing?

Explanation Because electrons are charged particles, they interact electrically with the sample in the microscope and scatter according to the shape and density of various portions of the sample, providing a means of viewing the sample. Photons of wavelength 0.2 nm are uncharged and in the x-ray region of the spectrum. They tend to simply pass through the thin sample without interacting.

27.7   THE WAVE FUNCTION

De Broglie's revolutionary idea that particles should have a wave nature soon moved out of the realm of skepticism to the point where it was viewed as a necessary concept in understanding the subatomic world. In 1926, the Austrian–German physicist Erwin Schrödinger proposed a wave equation that described how matter waves change in space and time. The Schrödinger wave equation represents a key element in the theory of quantum mechanics. It's as important in quantum mechanics as Newton's laws in classical mechanics. Schrödinger's equation has been successfully applied to the hydrogen atom and to many other microscopic systems.

Solving Schrödinger's equation (beyond the level of this course) determines a quantity Ψ called the **wave function**. Each particle is represented by a wave function Ψ that depends both on position and on time. Once Ψ is found, Ψ^2 gives us information on the **probability** (per unit volume) of finding the particle in any given region. To understand this, we return to Young's experiment involving coherent light passing through a double slit.

First, recall from Chapter 21 that the intensity of a light beam is proportional to the square of the electric field strength E associated with the beam: $I \propto E^2$. According to the wave model of light, there are certain points on the viewing screen where the net electric field is zero as a result of destructive interference of waves from the two slits. Because E is zero at these points, the intensity is also zero, and the screen is dark there. Likewise, at points on the screen at which constructive interference occurs, E is large, as is the intensity; hence, these locations are bright.

Now consider the same experiment when light is viewed as having a particle nature. The number of photons reaching a point on the screen per second increases as the intensity (brightness) increases. Consequently, the number of photons that strike a unit area on the screen each second is proportional to the square of the electric field, or $N \propto E^2$. From a probabilistic point of view, a photon has a high probability of striking the screen at a point at which the intensity (and E^2) is high and a low probability of striking the screen where the intensity is low.

When describing particles rather than photons, Ψ rather than E plays the role of the amplitude. Using an analogy with the description of light, we make the following interpretation of Ψ for particles: If Ψ is a wave function used to describe a single particle, the value of Ψ^2 at some location at a given time is proportional to the probability per unit volume of finding the particle at that location at that time. Adding up all the values of Ψ^2 in a given region gives the probability of finding the particle in that region.

ERWIN SCHRÖDINGER, Austrian Theoretical Physicist (1887–1961)

Schrödinger is best known as the creator of wave mechanics, a less cumbersome theory than the equivalent matrix mechanics developed by Werner Heisenberg. In 1933 Schrödinger left Germany and eventually settled at the Dublin Institute of Advanced Study, where he spent 17 happy, creative years working on problems in general relativity, cosmology, and the application of quantum physics to biology. In 1956, he returned home to Austria and his beloved Tirolean mountains, where he died in 1961.

27.8 THE UNCERTAINTY PRINCIPLE

If you were to measure the position and speed of a particle at any instant, you would always be faced with experimental uncertainties in your measurements. According to classical mechanics, no fundamental barrier to an ultimate refinement of the apparatus or experimental procedures exists. In other words, it's possible, in principle, to make such measurements with arbitrarily small uncertainty. Quantum theory predicts, however, that such a barrier does exist. In 1927, Werner Heisenberg (1901–1976) introduced this notion, which is now known as the **uncertainty principle**:

> If a measurement of the position of a particle is made with precision Δx and a simultaneous measurement of linear momentum is made with precision Δp_x, then the product of the two uncertainties can never be smaller than $h/4\pi$:
>
> $$\Delta x \, \Delta p_x \geq \frac{h}{4\pi} \qquad \text{[27.16]}$$

In other words, **it is physically impossible to measure simultaneously the exact position and exact linear momentum of a particle**. If Δx is very small, then Δp_x is large, and vice versa.

To understand the physical origin of the uncertainty principle, consider the following thought experiment introduced by Heisenberg. Suppose you wish to measure the position and linear momentum of an electron as accurately as possible. You might be able to do this by viewing the electron with a powerful light microscope. For you to see the electron and determine its location, at least one photon of light must bounce off the electron, as shown in Figure 27.18a, and pass through the

WERNER HEISENBERG, German Theoretical Physicist (1901–1976)

Heisenberg obtained his Ph.D. in 1923 at the University of Munich, where he studied under Arnold Sommerfeld. While physicists such as de Broglie and Schrödinger tried to develop physical models of the atom, Heisenberg developed an abstract mathematical model called *matrix mechanics* to explain the wavelengths of spectral lines. Heisenberg made many other significant contributions to physics, including his famous uncertainty principle, for which he received the Nobel Prize in 1932; the prediction of two forms of molecular hydrogen; and theoretical models of the nucleus of an atom.

Figure 27.18 A thought experiment for viewing an electron with a powerful microscope. (a) The electron is viewed before colliding with the photon. (b) The electron recoils (is disturbed) as the result of the collision with the photon.

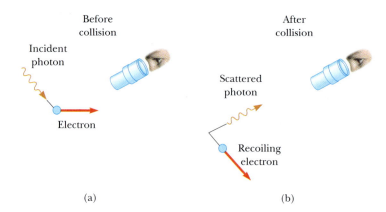

(a)   (b)

microscope into your eye, as shown in Figure 27.18b. When it strikes the electron, however, the photon transfers some unknown amount of its momentum to the electron. Thus, in the process of locating the electron very accurately (that is, by making Δx very small), the light that enables you to succeed in your measurement changes the electron's momentum to some undeterminable extent (making Δp_x very large).

The incoming photon has momentum h/λ. As a result of the collision, the photon transfers part or all of its momentum along the x-axis to the electron. Therefore, the *uncertainty* in the electron's momentum after the collision is as great as the momentum of the incoming photon: $\Delta p_x = h/\lambda$. Further, because the photon also has wave properties, we expect to be able to determine the electron's position to within one wavelength of the light being used to view it, so $\Delta x = \lambda$. Multiplying these two uncertainties gives

$$\Delta x \, \Delta p_x = \lambda \left(\frac{h}{\lambda} \right) = h$$

The value h represents the minimum in the product of the uncertainties. Because the uncertainty can always be greater than this minimum, we have

$$\Delta x \, \Delta p_x \geq h$$

Apart from the numerical factor $1/4\pi$ introduced by Heisenberg's more precise analysis, this inequality agrees with Equation 27.16.

Another form of the uncertainty relationship sets a limit on the accuracy with which the energy E of a system can be measured in a finite time interval Δt:

$$\Delta E \, \Delta t \geq \frac{h}{4\pi} \qquad [27.17]$$

It can be inferred from this relationship that the energy of a particle cannot be measured with complete precision in a very short interval of time. Thus, when an electron is viewed as a particle, the uncertainty principle tells us that (a) its position and velocity cannot both be known precisely at the same time and (b) its energy can be uncertain for a period given by $\Delta t = h/(4\pi \, \Delta E)$.

Applying Physics 27.4   Motion at Absolute Zero

A common, but erroneous, description of the absolute zero of temperature is "that temperature at which all molecular motion ceases." How can the uncertainty principle be used to argue against this description?

Explanation Imagine a particular molecule in a piece of material. The molecule is confined within the material, so there is a fixed uncertainty Δx in its position along one axis, corresponding to the size of that piece of material. If all molecular motion ceased at absolute zero, the given molecule's velocity, in particular, would be exactly zero, so its uncertainty in velocity would be $\Delta v = 0$, meaning its uncertainty in momentum would also be zero, since $p = mv$. The product of zero uncertainty in momentum and a nonzero uncertainty in position is zero, violating the uncertainty principle. So according to the uncertainty principle, there must be some molecular motion even at absolute zero.

EXAMPLE 27.8   Locating an Electron

Goal   Apply Heisenberg's position–momentum uncertainty principle.

Problem   The speed of an electron is measured to be 5.00×10^3 m/s to an accuracy of 0.003 00%. Find the minimum uncertainty in determining the position of this electron.

Strategy   After computing the momentum and its uncertainty, substitute into Heisenberg's uncertainty principle, Equation 27.16.

Solution

Calculate the momentum of the electron:

$$p_x = m_e v = (9.11 \times 10^{-31} \text{ kg})(5.00 \times 10^3 \text{ m/s})$$
$$= 4.56 \times 10^{-27} \text{ kg} \cdot \text{m/s}$$

The uncertainty in p is 0.003 00% of this value:

$$\Delta p_x = 0.000\,030\,0p = (0.000\,030\,0)(4.56 \times 10^{-27} \text{ kg} \cdot \text{m/s})$$
$$= 1.37 \times 10^{-31} \text{ kg} \cdot \text{m/s}$$

Now calculate the uncertainty in position using this value of Δp_x and Equation 27.17:

$$\Delta x \Delta p_x \geq \frac{h}{4\pi} \quad \rightarrow \quad \Delta x \geq \frac{h}{4\pi \Delta p_x}$$

$$\Delta x \geq \frac{6.626 \times 10^{-34} \text{ J} \cdot \text{s}}{4\pi (1.37 \times 10^{-31} \text{ kg} \cdot \text{m/s})} = 0.384 \times 10^{-3} \text{ m}$$

$$= 0.384 \text{ mm}$$

Remarks   Notice that this isn't an exact calculation: the uncertainty in position can take any value, as long as it's greater than or equal to the value given by the uncertainty principle.

Exercise 27.8

Suppose an electron is found somewhere in an atom of diameter 1.25×10^{-10} m. Estimate the uncertainty in the electron's momentum (in one dimension).

Answer $\Delta p \geq 4.22 \times 10^{-25} \text{ kg} \cdot \text{m/s}$

EXAMPLE 27.9   Excited States of Atoms

Goal   Apply the energy–time form of the uncertainty relation.

Problem   As we'll see in the next chapter, electrons in atoms can be found in certain high states of energy called **excited states** for short periods of time. If the average time that an electron exists in one of these states is 1.00×10^{-8} s, what is the minimum uncertainty in energy of the excited state?

Strategy   Substitute values into Equation 27.17, the energy–time form of Heisenberg's uncertainty relation.

Solution

Use Equation 27.17 to obtain the minimum uncertainty in the energy:

$$\Delta E \geq \frac{h}{4\pi \Delta t} = \frac{6.63 \times 10^{-34} \text{ J} \cdot \text{s}}{4\pi (1.00 \times 10^{-8} \text{ s})} = 5.28 \times 10^{-27} \text{ J}$$

$$= 3.30 \times 10^{-8} \text{ eV}$$

Remarks   This is again an imprecise calculation, giving only a lower bound on the uncertainty.

Exercise 27.9

A muon may be considered to be an excited state of an electron, to which it decays in an average of 2.2×10^{-6} s. What's the minimum uncertainty in the muon's (rest) energy, according to the uncertainty principle?

Answer $2.40 \times 10^{-29} \text{ J}$

27.9   THE SCANNING TUNNELING MICROSCOPE[1]

One of the basic phenomena of quantum mechanics—tunneling—is at the heart of a very practical device—the scanning tunneling microscope, or STM—which enables us to get highly detailed images of surfaces with a resolution comparable to the size of a single atom.

Figure 27.19 shows an image of a ring of 48 iron atoms located on a copper surface. Note the high quality of the STM image and the recognizable ring of iron atoms. What makes this image so remarkable is that its resolution—the size of the smallest detail that can be discerned—is about 0.2 nm. For an ordinary microscope, the resolution is limited by the wavelength of the waves used to make the image. An optical microscope has a resolution no better than 200 nm, about half the wavelength of visible light, and so could never show the detail displayed in Figure 27.19. Electron microscopes can have a resolution of 0.2 nm by using electron waves of that wavelength, given by the de Broglie formula $\lambda = h/p$. The electron momentum p required to give this wavelength is $10\,000$ eV/c, corresponding to an electron speed of 2% of the speed of light. Electrons traveling at this speed would penetrate into the interior of the sample in Figure 27.20 and so could not give us information about individual surface atoms.

The STM achieves its very fine resolution by using the basic idea shown in Figure 27.20. A conducting probe with a sharp tip is brought near the surface to be studied. Because it is attracted to the positive ions in the surface, an electron in the surface has a lower total energy than an electron in the empty space between surface and tip. The same thing is true for an electron in the probe tip, which is attracted to the positive ions in the tip. In Newtonian mechanics, this means that electrons cannot move between surface and tip because they lack the energy to escape either material. Because the electrons obey quantum mechanics, however, they can "tunnel" across the barrier of empty space. By applying a voltage between surface and tip, the electrons can be made to tunnel preferentially from surface to tip. In this way, the tip samples the distribution of electrons just above the surface.

Because of the nature of tunneling, the STM is very sensitive to the distance z from tip to surface. The reason is that in the empty space between tip and surface, the electron wave function falls off exponentially with a decay length on the order of 0.1 nm; that is, the wave function decreases by $1/e$ over that distance. For dis-

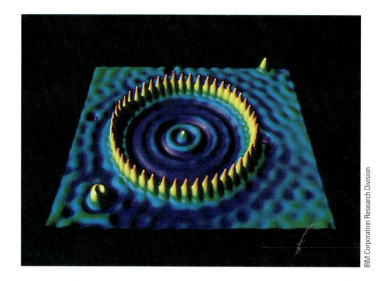

Figure 27.19   This is a photograph of a "quantum corral" consisting of a ring of 48 iron atoms located on a copper surface. The diameter of the ring is 143 nm. The photograph was obtained with a low-temperature scanning tunneling microscope (STM).

IBM Corporation Research Division

[1]This section was written by Roger A. Freedman, University of California, Santa Barbara.

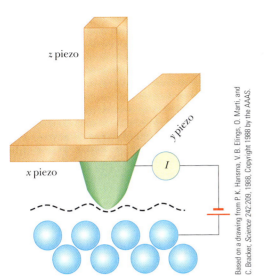

Figure 27.20   A schematic view of an STM. The tip, shown as a rounded cone, is mounted on a piezoelectric x, y, z scanner. A scan of the tip over the sample can reveal contours of the surface down to the atomic level. An STM image is composed of a series of scans displaced laterally from each other.

Based on a drawing from P. K. Hansma, V. B. Elings, O. Marti, and C. Bracker, *Science* 242:209, 1988, Copyright 1988 by the AAAS.

tances z greater than 1 nm (that is, beyond a few atomic diameters), essentially no tunneling takes place. This exponential behavior causes the current of electrons tunneling from surface to tip to depend very strongly on z. This sensitivity is the basis of the operation of the STM: by monitoring the tunneling current as the tip is scanned over the surface, scientists obtain a sensitive measure of the topography of the electron distribution on the surface. The result of this scan is used to make images like that in Figure 27.20. In this way the STM can measure the height of surface features to within 0.001 nm, approximately 1/100 of an atomic diameter!

The STM has, however, one serious limitation: it depends on electrical conductivity of the sample and the tip. Unfortunately, the surfaces of most materials are not electrically conductive. Even metals such as aluminum are covered with nonconductive oxides. A newer microscope—the atomic force microscope, or AFM— overcomes this limitation. It measures the force between a tip and the sample, rather than an electrical current. This force depends strongly on the tip–sample separation just as the electron tunneling current does for the STM. The AFM has comparable sensitivity for measuring topography and has become widely used for technological applications.

Perhaps the most remarkable thing about the STM is that its operation is based on a quantum mechanical phenomenon—tunneling—that was well understood in the 1920s, even though the first STM was not built until the 1980s. What other applications of quantum mechanics may yet be waiting to be discovered?

SUMMARY

27.1   Blackbody Radiation and Planck's Hypothesis

The characteristics of **blackbody radiation** can't be explained with classical concepts. The peak of a blackbody radiation curve is given by **Wien's displacement law**;

$$\lambda_{max} T = 0.289\ 8 \times 10^{-2}\ \text{m} \cdot \text{K} \qquad \textbf{[27.1]}$$

where λ_{max} is the wavelength at which the curve peaks and T is the absolute temperature of the object emitting the radiation.

Planck first introduced the quantum concept when he assumed that the subatomic oscillators responsible for blackbody radiation could have only discrete amounts of energy given by

$$E_n = nhf \qquad \textbf{[27.2]}$$

where n is a positive integer called a **quantum number** and f is the frequency of vibration of the resonator.

27.2   The Photoelectric Effect and the Particle Theory of Light

The **photoelectric effect** is a process whereby electrons are ejected from a metal surface when light is incident on that surface. Einstein provided a successful explanation of this effect by extending Planck's quantum hypothesis to electromagnetic waves. In this model, light is viewed as a stream of particles called photons, each with energy $E = hf$, where f is the light frequency and h is **Planck's constant**. The maximum kinetic energy of the ejected photoelectrons is

$$KE_{max} = hf - \phi \qquad \textbf{[27.6]}$$

where ϕ is the **work function** of the metal.

27.3   X-Rays

27.4   Diffraction of X-Rays by Crystals

X-rays are produced when high-speed electrons are suddenly decelerated. When electrons have been accel-erated through a voltage V, the shortest-wavelength radiation that can be produced is

$$\lambda_{min} = \frac{hc}{eV} \qquad \textbf{[27.9]}$$

The regular array of atoms in a crystal can act as a diffraction grating for x-rays and for electrons. The condition for constructive interference of the diffracted rays is given by **Bragg's law**:

$$2d \sin \theta = m\lambda \qquad (m = 1, 2, 3, \ldots) \quad \textbf{[27.10]}$$

Bragg's law bears a similarity to the equation for the diffraction pattern of a double slit.

27.5   The Compton Effect

X-rays from an incident beam are scattered at various angles by electrons in a target such as carbon. In such a scattering event, a shift in wavelength is observed for the scattered x-rays. This phenomenon is known as the **Compton shift**. Conservation of momentum and energy applied to a photon–electron collision yields the following expression for the shift in wavelength of the scattered x-rays:

$$\Delta\lambda = \lambda - \lambda_0 = \frac{h}{m_e c}(1 - \cos\theta) \qquad \textbf{[27.11]}$$

Here, m_e is the mass of the electron, c is the speed of light, and θ is the scattering angle.

27.6   The Dual Nature of Light and Matter

Light exhibits both a particle and a wave nature. De Broglie proposed that *all* matter has both a particle and a wave nature. The **de Broglie wavelength** of any particle of mass m and speed v is

$$\lambda = \frac{h}{p} = \frac{h}{mv} \qquad \textbf{[27.14]}$$

De Broglie also proposed that the frequencies of the waves associated with particles obey the Einstein relationship $E = hf$.

27.7   The Wave Function

In the theory of **quantum mechanics**, each particle is described by a quantity Ψ called the **wave function**.

The probability per unit volume of finding the particle at a particular point at some instant is proportional to Ψ^2. Quantum mechanics has been highly successful in describing the behavior of atomic and molecular systems.

27.8   The Uncertainty Principle

According to Heisenberg's **uncertainty principle**, it is impossible to measure simultaneously the exact position and exact momentum of a particle. If Δx is the uncertainty in the measured position and Δp_x the uncertainty in the momentum, the product $\Delta x \, \Delta p_x$ is given by

$$\Delta x \, \Delta p_x \geq \frac{h}{4\pi} \qquad \textbf{[27.16]}$$

Also,

$$\Delta E \, \Delta t \geq \frac{h}{4\pi} \qquad \textbf{[27.17]}$$

where ΔE is the uncertainty in the energy of the particle and Δt is the uncertainty in the time it takes to measure the energy.

CONCEPTUAL QUESTIONS

1. If you observe objects inside a very hot kiln why is it difficult to discern the shapes of the objects?

2. Why is an electron microscope more suitable than an optical microscope for "seeing" objects of atomic size?

3. Are blackbodies black?

4. Why is it impossible to simultaneously measure the position and velocity of a particle with infinite accuracy?

5. All objects radiate energy. Why, then, are we not able to see all objects in a dark room?

6. Is light a wave or a particle? Support your answer by citing specific experimental evidence.

7. A student claims that he is going to eject electrons from a piece of metal by placing a radio transmitter antenna adjacent to the metal and sending a strong AM radio signal into the antenna. The work function of a metal is typically a few electron volts. Will this work?

8. Light acts sometimes like a wave and sometimes like a particle. For the following situations, which best describes the behavior of light? Defend your answers. (a) The photoelectric effect. (b) The Compton effect. (c) Young's double-slit experiment.

9. In the photoelectric effect, explain why the stopping potential depends on the frequency of the light but not on the intensity.

10. Which has more energy, a photon of ultraviolet radiation or a photon of yellow light?

11. Why does the existence of a cutoff frequency in the photoelectric effect favor a particle theory of light rather than a wave theory?

12. What effect, if any, would you expect the temperature of a material to have on the ease with which electrons can be ejected from it via the photoelectric effect?

13. The cutoff frequency of a material is f_0. Are electrons emitted from the material when (a) light of frequency greater than f_0 is incident on the material? (b) Less than f_0?

14. The brightest star in the constellation Lyra is the bluish star Vega, whereas the brightest star in Boötes is the reddish star Arcturus. How do you account for the difference in color of the two stars?

15. If the photoelectric effect is observed in one metal, can you conclude that the effect will also be observed in another metal under the same conditions? Explain.

16. A beam of blue light and a beam of red light carry the same total amount of energy. Which beam contains the larger number of photons?

17. An x-ray photon is scattered by an electron which is initially at rest. What happens to the frequency of the scattered photon relative to that of the incident photon?

PROBLEMS

1, 2, 3 = straightforward, intermediate, challenging   ☐ = full solution available in *Student Solutions Manual/Study Guide*
Physics⊗Now™ = coached problem with hints available at **www.cp7e.com**   🧬 = biomedical application

Section 27.1 Blackbody Radiation and Planck's Hypothesis

1. **Physics⊗Now™** What is the surface temperature of (a) Betelgeuse, a red giant star in the constellation of Orion, which radiates with a peak wavelength of about 970 nm? (b) Rigel, a bluish-white star in Orion, radiates with a peak wavelength of 145 nm? Find the temperature of Rigel's surface.

2. (a) Lightning produces a maximum air temperature on the order of 10^4 K, while (b) a nuclear explosion produces a temperature on the order of 10^7 K. Use Wien's displacement law to find the order of magnitude of the wavelength of the thermally produced photons radiated with greatest intensity by each of these sources. Name the part of the electromagnetic spectrum where you would expect each to radiate most strongly.

3. If the surface temperature of the Sun is 5 800 K, find the wavelength that corresponds to the maximum rate of energy emission from the Sun.

4. A beam of blue light and a beam of red light each carry a total energy of 2 500 eV. If the wavelength of the red light is 690 nm and the wavelength of the blue light is 420 nm, find the number of photons in each beam.

5. Calculate the energy in electron volts of a photon having a wavelength (a) in the microwave range, 5.00 cm, (b) in the visible light range, 500 nm, and (c) in the x-ray range, 5.00 nm.

6. A certain light source is found to emit radiation whose peak value has a frequency of 1.00×10^{15} Hz. Find the temperature of the source assuming that it is a blackbody radiator.

7. An FM radio transmitter has a power output of 150 kW and operates at a frequency of 99.7 MHz. How many photons per second does the transmitter emit?

8. 🧬 The threshold of dark-adapted (scotopic) vision is 4.0×10^{-11} W/m^2 at a central wavelength of 500 nm. If light with this intensity and wavelength enters the eye when the pupil is open to its maximum diameter of 8.5 mm, how many photons per second enter the eye?

9. A 1.5-kg mass vibrates at an amplitude of 3.0 cm on the end of a spring of spring constant 20.0 N/m. (a) If the energy of the spring is quantized, find its quantum num-

ber. (b) If n changes by 1, find the fractional change in energy of the spring.

10. A 70.0-kg jungle hero swings at the end of a vine at a frequency of 0.50 Hz at 2.0 m/s as he moves through the lowest point on his arc. (a) Assume the energy is quantized and find the quantum number n for this system. (b) Find the energy carried away in a one-quantum change in the jungle hero's energy.

Section 27.2 The Photoelectric Effect and the Particle Theory of Light

11. **Physics⊗Now™** When light of wavelength 350 nm falls on a potassium surface, electrons having a maximum kinetic energy of 1.31 eV are emitted. Find (a) the work function of potassium, (b) the cutoff wavelength, and (c) the frequency corresponding to the cutoff wavelength.

12. When a certain metal is illuminated with light of frequency 3.0×10^{15} Hz, a stopping potential of 7.0 V is required to stop the most energetic ejected electrons. What is the work function of this metal?

13. What wavelength of light would have to fall on sodium (with a work function of 2.46 eV) if it is to emit electrons with a maximum speed of 1.0×10^6 m/s?

14. Lithium, beryllium, and mercury have work functions of 2.30 eV, 3.90 eV, and 4.50 eV, respectively. If 400-nm light is incident on each of these metals, determine (a) which metals exhibit the photoelectric effect and (b) the maximum kinetic energy of the photoelectrons in each case.

15. From the scattering of sunlight, Thomson calculated that the classical radius of the electron has a value of 2.82×10^{-15} m. If sunlight having an intensity of 500 W/m^2 falls on a disk with this radius, estimate the time required to accumulate 1.00 eV of energy. Assume that light is a classical wave and that the light striking the disk is completely absorbed. How does your estimate compare with the observation that photoelectrons are promptly (within 10^{-9} s) emitted?

16. An isolated copper sphere of radius 5.00 cm, initially uncharged, is illuminated by ultraviolet light of wavelength 200 nm. What charge will the photoelectric effect induce on the sphere? The work function for copper is 4.70 eV.

17. When light of wavelength 254 nm falls on cesium, the required stopping potential is 3.00 V. If light of wavelength 436 nm is used, the stopping potential is 0.900 V. Use this information to plot a graph like that shown in Figure 27.6, and from the graph determine the cutoff frequency for cesium and its work function.

18. Ultraviolet light is incident normally on the surface of a certain substance. The binding energy of the electrons in this substance is 3.44 eV. The incident light has an intensity of 0.055 W/m². The electrons are photoelectrically emitted with a maximum speed of 4.2×10^5 m/s. How many electrons are emitted from a square centimeter of the surface each second? Assume that the absorption of every photon ejects an electron.

Section 27.3 X-Rays

19. The extremes of the x-ray portion of the electromagnetic spectrum range from approximately 1.0×10^{-8} m to 1.0×10^{-13} m. Find the minimum accelerating voltages required to produce wavelengths at these two extremes.

20. Calculate the minimum-wavelength x-ray that can be produced when a target is struck by an electron that has been accelerated through a potential difference of (a) 15.0 kV and (b) 100 kV.

21. What minimum accelerating voltage would be required to produce an x-ray with a wavelength of 0.030 0 nm?

Section 27.4 Diffraction of X-Rays by Crystals

22. A monochromatic x-ray beam is incident on a NaCl crystal surface with $d = 0.353$ nm. The second-order maximum in the reflected beam is found when the angle between the incident beam and the surface is 20.5°. Determine the wavelength of the x-rays.

23. Potassium iodide has an interplanar spacing of $d = 0.296$ nm. A monochromatic x-ray beam shows a first-order diffraction maximum when the grazing angle is 7.6°. Calculate the x-ray wavelength.

24. The spacing between certain planes in a crystal is known to be 0.30 nm. Find the smallest angle of incidence at which constructive interference will occur for wavelength 0.070 nm.

25. X-rays of wavelength 0.140 nm are reflected from a certain crystal, and the first-order maximum occurs at an angle of 14.4°. What value does this give for the interplanar spacing of the crystal?

Section 27.5 The Compton Effect

26. X-rays are scattered from electrons in a carbon target. The measured wavelength shift is 1.50×10^{-3} nm. Calculate the scattering angle.

27. Calculate the energy and momentum of a photon of wavelength 700 nm.

28. A beam of 0.68-nm photons undergoes Compton scattering from free electrons. What are the energy and momentum of the photons that emerge at a 45° angle with respect to the incident beam?

29. A 0.001 6-nm photon scatters from a free electron. For what (photon) scattering angle will the recoiling electron and scattered photon have the same kinetic energy?

30. X-rays with an energy of 300 keV undergo Compton scattering from a target. If the scattered rays are deflected at 37.0° relative to the direction of the incident rays, find (a) the Compton shift at this angle, (b) the energy of the scattered x-ray, and (c) the kinetic energy of the recoiling electron.

31. **Physics⊗Now™** A 0.110-nm photon collides with a stationary electron. After the collision, the electron moves forward and the photon recoils backwards. Find the momentum and kinetic energy of the electron.

32. After a 0.800-nm x-ray photon scatters from a free electron, the electron recoils with a speed equal to 1.40×10^6 m/s. (a) What was the Compton shift in the photon's wavelength? (b) Through what angle was the photon scattered?

33. A 0.45-nm x-ray photon is deflected through a 23° angle after scattering from a free electron. (a) What is the kinetic energy of the recoiling electron? (b) What is its speed?

Section 27.6 The Dual Nature of Light and Matter

34. Calculate the de Broglie wavelength of a proton moving at (a) 2.00×10^4 m/s; (b) 2.00×10^7 m/s.

35. (a) If the wavelength of an electron is 5.00×10^{-7} m, how fast is it moving? (b) If the electron has a speed of 1.00×10^7 m/s, what is its wavelength?

36. A 0.200-kg ball is released from rest at the top of a 50.0-m tall building. Find the de Broglie wavelength of the ball just before it strikes the Earth.

37. The nucleus of an atom is on the order of 10^{-14} m in diameter. For an electron to be confined to a nucleus, its de Broglie wavelength would have to be of that order of magnitude or smaller. (a) What would be the kinetic energy of an electron confined to

this region? (b) On the basis of your result in part (a), would you expect to find an electron in a nucleus? Explain.

38. After learning about de Broglie's hypothesis that particles of momentum p have wave characteristics with wavelength $\lambda = h/p$, an 80.0-kg student has grown concerned about being diffracted when passing through a 75.0-cm-wide doorway. Assume that significant diffraction occurs when the width of the diffraction aperture is less than 10.0 times the wavelength of the wave being diffracted. (a) Determine the maximum speed at which the student can pass through the doorway in order to be significantly diffracted. (b) With that speed, how long will it take the student to pass through the doorway if it is 15.0 cm thick? Compare your result with the currently accepted age of the Universe, which is 4.00×10^{17} s. (c) Should this student worry about being diffracted?

39. De Broglie postulated that the relationship $\lambda = h/p$ is valid for relativistic particles. What is the de Broglie wavelength for a (relativistic) electron whose kinetic energy is 3.00 MeV?

40. A monoenergetic beam of electrons is incident on a single slit of width 0.500 nm. A diffraction pattern is formed on a screen 20.0 cm from the slit. If the distance between successive minima of the diffraction pattern is 2.10 cm, what is the energy of the incident electrons?

41. The resolving power of a microscope is proportional to the wavelength used. A resolution of 1.0×10^{-11} m (0.010 nm) would be required in order to "see" an atom. (a) If electrons were used (electron microscope), what minimum kinetic energy would be required of the electrons? (b) If photons were used, what minimum photon energy would be needed to obtain 1.0×10^{-11} m resolution?

Section 27.7 The Wave Function

Section 27.8 The Uncertainty Principle

42. A 50.0-g ball moves at 30.0 m/s. If its speed is measured to an accuracy of 0.10%, what is the minimum uncertainty in its position?

43. In the ground state of hydrogen, the uncertainty in the position of the electron is roughly 0.10 nm. If the speed of the electron is on the order of the uncertainty in its speed, how fast is the electron moving?

44. Suppose Fuzzy, a quantum mechanical duck, lives in a world in which $h = 2\pi\,\text{J} \cdot \text{s}$. Fuzzy has a mass of 2.00 kg and is initially known to be within a pond 1.00 m wide. (a) What is the minimum uncertainty in his speed? (b) Assuming this uncertainty in speed to prevail for 5.00 s, determine the uncertainty in Fuzzy's position after this time.

45. **Physics⊗Now™** Suppose optical radiation ($\lambda = 5.00 \times 10^{-7}$ m) is used to determine the position of an electron to within the wavelength of the light. What will be the resulting uncertainty in the electron's velocity?

46. (a) Show that the kinetic energy of a nonrelativistic particle can be written in terms of its momentum as $KE = p^2/2m$. (b) Use the results of (a) to find the minimum kinetic energy of a proton confined within a nucleus having a diameter of 1.0×10^{-15} m.

ADDITIONAL PROBLEMS

47. Figure P27.47 shows the spectrum of light emitted by a firefly. Determine the temperature of a blackbody that would emit radiation peaked at the same frequency. Based on your result, would you say firefly radiation is blackbody radiation?

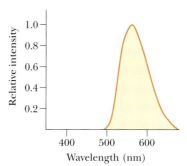

Figure P27.47

48. An x-ray tube is operated at 50 000 V. (a) Find the minimum wavelength of the radiation emitted by this tube. (b) If the radiation is directed at a crystal, the first-order maximum in the reflected radiation occurs when the grazing angle is 2.5°. What is the spacing between reflecting planes in the crystal?

49. The spacing between planes of nickel atoms in a nickel crystal is 0.352 nm. At what angle does a second-order Bragg reflection occur in nickel for 11.3-keV x-rays?

50. Johnny Jumper's favorite trick is to step out of his 16th-story window and fall 50.0 m into a pool. A news reporter takes a picture of 75.0-kg Johnny just before he makes a splash, using an exposure

time of 5.00 ms. Find (a) Johnny's de Broglie wavelength at this moment, (b) the uncertainty of his kinetic energy measurement during such a period of time, and (c) the percent error caused by such an uncertainty.

51. Photons of wavelength 450 nm are incident on a metal. The most energetic electrons ejected from the metal are bent into a circular arc of radius 20.0 cm by a magnetic field with a magnitude of 2.00×10^{-5} T. What is the work function of the metal?

52. A 200-MeV photon is scattered at 40.0° by a free proton that is initially at rest. Find the energy (in MeV) of the scattered photon.

53. A light source of wavelength λ illuminates a metal and ejects photoelectrons with a maximum kinetic energy of 1.00 eV. A second light source of wavelength $\lambda/2$ ejects photoelectrons with a maximum kinetic energy of 4.00 eV. What is the work function of the metal?

54. Red light of wavelength 670 nm produces photoelectrons from a certain photoemissive material. Green light of wavelength 520 nm produces photoelectrons from the same material with 1.50 times the maximum kinetic energy. What is the material's work function?

55. How fast must an electron be moving if all its kinetic energy is lost to a single x-ray photon (a) at the high end of the x-ray electromagnetic spectrum with a wavelength of 1.00×10^{-8} m; (b) at the low end of the x-ray electromagnetic spectrum with a wavelength of 1.00×10^{-13} m?

56. Show that if an electron were confined inside an atomic nucleus of diameter 2.0×10^{-15} m, it would have to be moving relativistically, while a proton confined to the same nucleus can be moving at less than one-tenth the speed of light.

57. A photon strikes a metal with a work function ϕ and produces a photoelectron with a de Broglie wavelength equal to the wavelength of the original photon. (a) Show that the energy of this photon must have been given by

$$E = \frac{\phi(m_e c^2 - \phi/2)}{m_e c^2 - \phi}$$

where m_e is the mass of the electron. [*Hint*: Begin with the conservation of energy, $E + m_e c^2 = \phi + \sqrt{(pc)^2 + (m_e c^2)^2}$.] (b) If one of these photons strikes platinum ($\phi = 6.35$ eV), determine the resulting maximum speed of the photoelectron that is emitted.

58. In a Compton scattering event, the scattered photon has an energy of 120.0 keV and the recoiling electron has a kinetic energy of 40.0 keV. Find (a) the wavelength of the incident photon, (b) the angle θ at which the photon is scattered, and (c) the recoil angle of the electron. [*Hint*: Conserve both mass–energy and relativistic momentum.]

59. A woman on a ladder drops small pellets toward a point target on the floor. (a) Show that, according to the uncertainty principle, the average distance by which she misses the target must be at least

$$\Delta x_f = \left(\frac{2\hbar}{m}\right)^{1/2}\left(\frac{2H}{g}\right)^{1/4}$$

where H is the initial height of each pellet above the floor and m is the mass of each pellet. Assume that the spread in impact points is given by $\Delta x_f = \Delta x_i + (\Delta v_x)t$. (b) If $H = 2.00$ m and $m = 0.500$ g, what is Δx_f?

60. Show that the speed of a particle having de Broglie wavelength λ and Compton wavelength $\lambda_C = h/(mc)$ is

$$v = \frac{c}{\sqrt{1 + (\lambda/\lambda_C)^2}}$$

61. (a) Find the mass of a solid iron sphere 2.00 cm in radius. (b) Assume that it is at 20°C and has emissivity 0.860. Find the power with which it is radiating electromagnetic waves. (c) If this sphere were alone in the Universe, at what rate would its temperature be changing? (d) Assume Wien's law describes the sphere. Find the wavelength λ_{max} of electromagnetic radiation it emits most strongly. Although it emits a spectrum of waves having all different wavelengths, model its whole power output as carried by photons of wavelength λ_{max}. Find (e) the energy of one photon and (f) the number of photons it emits each second. When the sphere is at thermal equilibrium with its surroundings, it emits and also absorbs photons at this rate.

ACTIVITIES

1. Use a black marker or pieces of dark electrical tape to make a very dark area on the outside of a shoebox. Poke a hole in the center of the dark area with a pencil. Now put a lid on the box, and compare the blackness of the hole with the blackness of the surrounding dark area. Based on your observation, explain why the radiation emitted from the hole is like that emitted from a black body.

2. On a clear night, go outdoors far from city lights and find the constellation Orion. Your instructor should be able to furnish you with a star chart to assist you in locating this grouping of stars. Look very carefully at the color of the two stars Betelgeuse and Rigel. (See Fig. A27.2.) Can you tell which star is hotter? Orion is visible only from November through April in the evening sky, so if Orion is not visible when you go out, compare two of the brightest stars you can see, such as Vega in the constellation Lyra and Arcturus in Boötes.

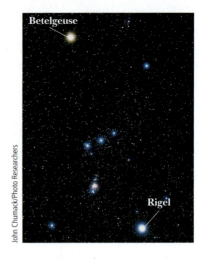

Figure A27.2

"Neon lights," commonly used in advertising signs, consist of thin glass tubes filled with various gases, such as neon and helium. The gas atoms are excited to higher energy levels by electric discharge through the tube. When the electrons in these excited levels return to lower energy levels, the atoms emit light having a wavelength (color) that depends on the type of gas in the tube. For example, a tube filled with neon produces a red-orange color, while helium produces pink.

Atomic Physics

A large portion of this chapter concerns the hydrogen atom. Although the hydrogen atom is the simplest atomic system, it's especially important for several reasons:

- The quantum numbers used to characterize the allowed states of hydrogen can also be used to describe (approximately) the allowed states of more complex atoms. This enables us to understand the periodic table of the elements, one of the greatest triumphs of quantum mechanics.
- The hydrogen atom is an ideal system for performing precise comparisons of theory with experiment and for improving our overall understanding of atomic structure.
- Much of what is learned about the hydrogen atom with its single electron can be extended to such single-electron ions as He^+ and Li^{2+}.

In this chapter we first discuss the Bohr model of hydrogen, which helps us understand many features of that element but fails to explain finer details of atomic structure. Next we examine the hydrogen atom from the viewpoint of quantum mechanics and the quantum numbers used to characterize various atomic states. Quantum numbers aren't mere mathematical abstractions: they have physical significance, such as the role they play in the effect of a magnetic field on certain quantum states. The fact that no two electrons in an atom can have the same set of quantum numbers—the Pauli exclusion principle—is extremely important in understanding the properties of complex atoms and the arrangement of elements in the periodic table. Finally, we apply our knowledge of atomic structure to describe the mechanisms involved in the production of x-rays, the operation of a laser, and the behavior of solid-state devices such as diodes and transistors.

28.1   EARLY MODELS OF THE ATOM

The model of the atom in the days of Newton was a tiny, hard, indestructible sphere. Although this model was a good basis for the kinetic theory of gases, new models had to be devised when later experiments revealed the electronic nature of

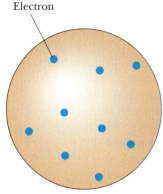

Electron

Figure 28.1 Thomson's model of the atom, with the electrons embedded inside the positive charge like seeds in a watermelon.

Stock Montage, Inc.

SIR JOSEPH JOHN THOMSON,
English Physicist (1856–1940)

Thomson, usually considered the discoverer of the electron, opened up the field of subatomic particle physics with his extensive work on the deflection of cathode rays (electrons) in an electric field. He received the 1906 Nobel prize for his discovery of the electron.

atoms. J. J. Thomson (1856–1940) suggested a model of the atom as a volume of positive charge with electrons embedded throughout the volume, much like the seeds in a watermelon (Fig. 28.1).

In 1911 Ernest Rutherford (1871–1937) and his students Hans Geiger and Ernest Marsden performed a critical experiment showing that Thomson's model couldn't be correct. In this experiment, a beam of positively charged **alpha particles** was projected against a thin metal foil, as in Figure 28.2a. The results of the experiment were astounding. Most of the alpha particles passed through the foil as if it were empty space, but a few particles deflected from their original direction of travel were scattered through large angles. Some particles were even deflected backwards, reversing their direction of travel. When Geiger informed Rutherford of these results, Rutherford wrote, "It was quite the most incredible event that has ever happened to me in my life. It was almost as incredible as if you fired a 15-inch shell at a piece of tissue paper and it came back and hit you."

Such large deflections were not expected on the basis of Thomson's model. According to that model, a positively charged alpha particle would never come close enough to a large positive charge to cause any large-angle deflections. Rutherford explained these astounding results by assuming that the positive charge in an atom was concentrated in a region that was small relative to the size of the atom. He called this concentration of positive charge the **nucleus** of the atom. Any electrons belonging to the atom were assumed to be in the relatively large volume outside the nucleus. In order to explain why electrons in this outer region of the atom were not pulled into the nucleus, Rutherford viewed them as moving in orbits about the positively charged nucleus in the same way that planets orbit the Sun, as shown in Figure 28.2b. Alpha particles themselves were later identified as the nuclei of helium atoms.

There are two basic difficulties with Rutherford's planetary model. First, an atom emits certain discrete characteristic frequencies of electromagnetic radiation and no others; the Rutherford model is unable to explain this phenomenon. Second, the electrons in Rutherford's model undergo a centripetal acceleration. According to Maxwell's theory of electromagnetism, centripetally accelerated charges revolving with frequency f should radiate electromagnetic waves of the same frequency. Unfortunately, this classical model leads to disaster when applied to the atom. As the electron radiates energy, the radius of its orbit steadily decreases and its frequency of revolution increases. This leads to an ever-increasing frequency of emitted radiation and a rapid collapse of the atom as the electron spirals into the nucleus.

28.2 ATOMIC SPECTRA

The hydrogen atom is the simplest atomic system and an especially important one to understand. Much of what we know about the hydrogen atom (which consists of one proton and one electron) can be extended directly to other single-electron

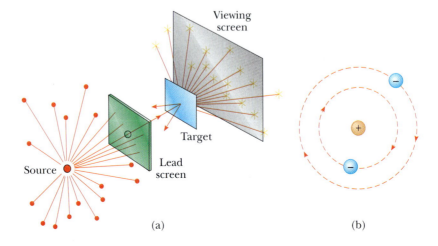

Figure 28.2 (a) Geiger and Marsden's technique for observing the scattering of alpha particles from a thin foil target. The source is a naturally occurring radioactive substance, such as radium. (b) Rutherford's planetary model of the atom.

Viewing screen

Target

Source

Lead screen

(a)

(b)

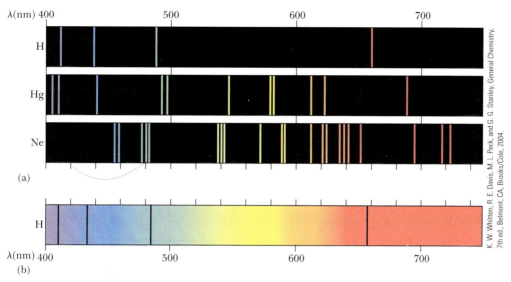

(a)

(b)

Figure 28.3 Visible spectra. (a) Line spectra produced by emission in the visible range for the elements hydrogen, mercury, and neon. (b) The absorption spectrum for hydrogen. The dark absorption lines occur at the same wavelengths as the emission lines for hydrogen shown in (a).

ions such as He^+ and Li^{2+}. Further, a thorough understanding of the physics underlying the hydrogen atom can then be used to describe more complex atoms and the periodic table of the elements.

Suppose an evacuated glass tube is filled with hydrogen (or some other gas) at low pressure. If a voltage applied between metal electrodes in the tube is great enough to produce an electric current in the gas, the tube emits light having a color that depends on the gas inside. (This is how a neon sign works.) When the emitted light is analyzed with a spectrometer, discrete bright lines are observed, each having a different wavelength, or color. Such a series of spectral lines is commonly called an **emission spectrum**. The wavelengths contained in such a spectrum are characteristic of the element emitting the light (Fig. 28.3). Because no two elements emit the same line spectrum, this phenomenon represents a marvelous and reliable technique for identifying elements in a gaseous substance.

The emission spectrum of hydrogen shown in Figure 28.4 includes four prominent lines that occur at wavelengths of 656.3 nm, 486.1 nm, 434.1 nm, and 410.2 nm, respectively. In 1885 Johann Balmer (1825–1898) found that the wavelengths of these and less prominent lines can be described by the simple empirical equation

$$\frac{1}{\lambda} = R_H \left(\frac{1}{2^2} - \frac{1}{n^2} \right) \qquad [28.1] \qquad \blacktriangleleft \text{Balmer series}$$

where n may have integral values of 3, 4, 5, . . . , and R_H is a constant, called the **Rydberg constant**. If the wavelength is in meters, R_H has the value

$$R_H = 1.097\,373\,2 \times 10^7 \text{ m}^{-1} \qquad [28.2] \qquad \blacktriangleleft \text{Rydberg constant}$$

The first line in the Balmer series, at 656.3 nm, corresponds to $n = 3$ in Equation 28.1, the line at 486.1 nm corresponds to $n = 4$, and so on. In addition to the Balmer series of spectral lines, a Lyman series was subsequently discovered in the far ultraviolet, with the radiated wavelengths described by a similar equation.

In addition to emitting light at specific wavelengths, an element can absorb light at specific wavelengths. The spectral lines corresponding to this process form what is known as an **absorption spectrum**. An absorption spectrum can be obtained by passing a continuous radiation spectrum (one containing all wavelengths) through a vapor of the element being analyzed. The absorption spectrum consists of a series of dark lines superimposed on the otherwise bright continuous spectrum. Each line in the absorption spectrum of a given element coincides with a line in the emission spectrum of the element. This means that if hydrogen is the

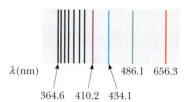

λ(nm)   486.1 656.3

364.6 410.2 434.1

Figure 28.4 The Balmer series of spectral lines for atomic hydrogen, with several lines marked with the wavelength in nanometers. The line labeled 346.6 is the shortest-wavelength line and is in the ultraviolet region of the electromagnetic spectrum. The other labeled lines are in the visible region.

absorbing vapor, dark lines will appear at the visible wavelengths 656.3 nm, 486.1 nm, 434.1 nm, and 410.2 nm, as shown in Figures 28.3b and 28.4.

APPLICATION

Discovery of Helium

The absorption spectrum of an element has many practical applications. For example, the continuous spectrum of radiation emitted by the Sun must pass through the cooler gases of the solar atmosphere before reaching the Earth. The various absorption lines observed in the solar spectrum have been used to identify elements in the solar atmosphere, including one that was previously unknown. When the solar spectrum was first being studied, some lines were found that didn't correspond to any known element. A new element had been discovered! Because the Greek word for Sun is *helios*, the new element was named *helium*. It was later identified in underground gases on Earth. Scientists are able to examine the light from stars other than our Sun in this way, but elements other than those present on Earth have never been detected.

Applying Physics 28.1 Thermal or Spectral?

On observing a yellow candle flame, your laboratory partner claims that the light from the flame originates from excited sodium atoms in the flame. You disagree, stating that because the candle flame is hot, the radiation must be thermal in origin. Before the disagreement leads to fisticuffs, how could you determine who is correct?

Explanation A simple determination could be made by observing the light from the candle flame through

a spectrometer, which is a slit and diffraction grating combination discussed in Chapter 25. If the spectrum of the light is continuous, then it's probably thermal in origin. If the spectrum shows discrete lines, it's atomic in origin. The results of the experiment show that the light is indeed thermal in origin and originates from random molecular motion in the candle flame.

Applying Physics 28.2 Auroras

At extreme northern latitudes, the aurora borealis provides a beautiful and colorful display in the night sky. A similar display occurs near the southern polar region and is called the aurora australis. What's the origin of the various colors seen in the auroras?

Explanation The aurora is due to high speed particles interacting with the Earth's magnetic field and

entering the atmosphere. When these particles collide with molecules in the atmosphere, they excite the molecules in a way similar to the voltage in the spectrum tubes discussed earlier in this section. In response, the molecules emit colors of light according to the characteristic spectrum of their atomic constituents. For our atmosphere, the primary constituents are nitrogen and oxygen, which provide the red, blue, and green colors of the aurora.

28.3 THE BOHR THEORY OF HYDROGEN

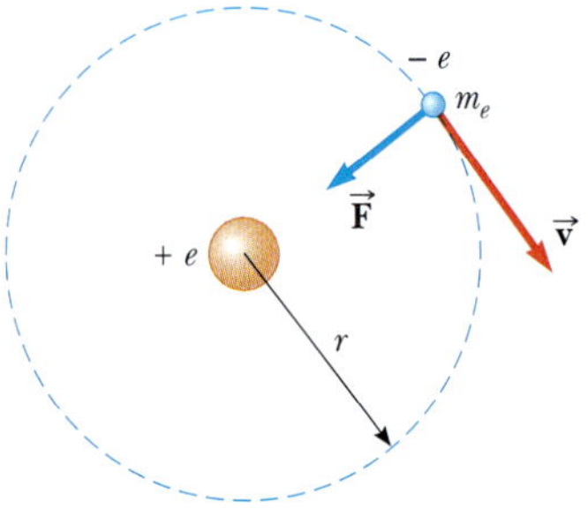

At the beginning of the 20th century, scientists were perplexed by the failure of classical physics to explain the characteristics of spectra. Why did atoms of a given element emit only certain lines? Further, why did the atoms absorb only those wavelengths that they emitted? In 1913 Bohr provided an explanation of atomic spectra that includes some features of the currently accepted theory. Using the simplest atom, hydrogen, Bohr developed a model of what he thought must be the atom's structure in an attempt to explain why the atom was stable. His model of the hydrogen atom contains some classical features, as well as some revolutionary postulates that could not be justified within the framework of classical physics. The basic assumptions of the Bohr theory as it applies to the hydrogen atom are as follows:

Figure 28.5 Diagram representing Bohr's model of the hydrogen atom. The orbiting electron is allowed only in specific orbits of discrete radii.

1. The electron moves in circular orbits about the proton under the influence of the Coulomb force of attraction, as in Figure 28.5. The Coulomb force produces the electron's centripetal acceleration.

2. Only certain electron orbits are stable. These are orbits in which the hydrogen atom doesn't emit energy in the form of electromagnetic radiation. Hence, the total energy of the atom remains constant, and classical mechanics can be used to describe the electron's motion.

3. Radiation is emitted by the hydrogen atom when the electron "jumps" from a more energetic initial state to a less energetic state. The "jump" can't be visualized or treated classically. In particular, the frequency f of the radiation emitted in the jump is related to the change in the atom's energy and is *independent of the frequency of the electron's orbital motion*. The frequency of the emitted radiation is given by

$$E_i - E_f = hf \qquad [28.3]$$

where E_i is the energy of the initial state, E_f is the energy of the final state, h is Planck's constant, and $E_i > E_f$.

4. The size of the allowed electron orbits is determined by a condition imposed on the electron's orbital angular momentum: the allowed orbits are those for which the electron's orbital angular momentum about the nucleus is an integral multiple of $\hbar$ (pronounced "h bar"), where $\hbar = h/2\pi$:

$$m_e v r = n\hbar \quad n = 1, 2, 3, \ldots \qquad [28.4]$$

With these four assumptions, we can calculate the allowed energies and emission wavelengths of the hydrogen atom. We use the model pictured in Figure 28.5, in which the electron travels in a circular orbit of radius r with an orbital speed v.

The electrical potential energy of the atom is

$$PE = k_e \frac{q_1 q_2}{r} = k_e \frac{(-e)(e)}{r} = -k_e \frac{e^2}{r}$$

where k_e is the Coulomb constant. Assuming the nucleus is at rest, the total energy E of the atom is the sum of the kinetic and potential energy:

$$E = KE + PE = \tfrac{1}{2} m_e v^2 - k_e \frac{e^2}{r} \qquad [28.5]$$

We apply Newton's second law to the electron. We know that the electric force of attraction on the electron, $k_e e^2/r^2$, must equal $m_e a_r$, where $a_r = v^2/r$ is the centripetal acceleration of the electron. Thus,

$$k_e \frac{e^2}{r^2} = m_e \frac{v^2}{r} \qquad [28.6]$$

From this equation, we see that the kinetic energy of the electron is

$$\tfrac{1}{2} m_e v^2 = \frac{k_e e^2}{2r} \qquad [28.7]$$

We can combine this result with Equation 28.5 and express the energy of the atom as

$$E = -\frac{k_e e^2}{2r} \qquad [28.8]$$

◀ Energy of the hydrogen atom

where the negative value of the energy indicates that the electron is bound to the proton.

An expression for r is obtained by solving Equations 28.4 and 28.6 for v and equating the results:

$$v^2 = \frac{n^2 \hbar}{m_e^2 r^2} = \frac{k_e e^2}{m_e r}$$

$$r_n = \frac{n^2 \hbar^2}{m_e k_e e^2} \quad n = 1, 2, 3, \ldots \qquad [28.9]$$

◀ The radii of the Bohr orbits are quantized

This equation is based on the assumption that the **electron can exist only in certain allowed orbits determined by the integer n.**

NIELS BOHR, Danish Physicist (1885–1962)

Bohr was an active participant in the early development of quantum mechanics and provided much of its philosophical framework. During the 1920s and 1930s, he headed the Institute for Advanced Studies in Copenhagen. The institute was a magnet for many of the world's best physicists and provided a forum for the exchange of ideas. When Bohr visited the United States in 1939 to attend a scientific conference, he brought news that the fission of uranium had been observed by Hahn and Strassman in Berlin. The results were the foundations of the atomic bomb developed in the United States during World War II. Bohr was awarded the 1922 Nobel Prize for his investigation of the structure of atoms and of the radiation emanating from them.

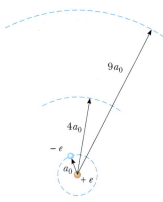

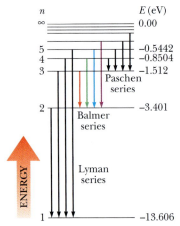

The orbit with the smallest radius, called the **Bohr radius**, a_0, corresponds to $n = 1$ and has the value

$$a_0 = \frac{\hbar^2}{m k_e e^2} = 0.052\ 9\ \text{nm} \qquad [28.10]$$

A general expression for the radius of any orbit in the hydrogen atom is obtained by substituting Equation 28.10 into Equation 28.9:

$$r_n = n^2 a_0 = n^2 (0.052\ 9\ \text{nm}) \qquad [28.11]$$

The first three Bohr orbits for hydrogen are shown in Active Figure 28.6.

Equation 28.9 may be substituted into Equation 28.8 to give the following expression for the energies of the quantum states:

$$E_n = -\frac{m_e k_e^2 e^4}{2\hbar^2}\left(\frac{1}{n^2}\right) \qquad n = 1, 2, 3, \ldots \qquad [28.12]$$

If we insert numerical values into Equation 28.12, we obtain

$$E_n = -\frac{13.6}{n^2}\ \text{eV} \qquad [28.13]$$

The lowest energy state, or **ground state**, corresponds to $n = 1$ and has an energy $E_1 = - m_e k_e^2 e^4/2\hbar^2 = -13.6$ eV. The next state, corresponding to $n = 2$, has an energy $E_2 = E_1/4 = -3.40$ eV, and so on. An energy level diagram showing the energies of these stationary states and the corresponding quantum numbers is given in Active Figure 28.7. The uppermost level shown, corresponding to $E = 0$ and $n \to \infty$, represents the state for which the electron is completely removed from the atom. In this state, the electron's KE and PE are both zero, which means that the electron is at rest infinitely far away from the proton. The minimum energy required to ionize the atom — that is, to completely remove the electron — is called the **ionization energy**. The ionization energy for hydrogen is 13.6 eV.

Equations 28.3 and 28.12 and the third Bohr postulate show that if the electron jumps from one orbit with quantum number n_i to a second orbit with quantum number, n_f, it emits a photon of frequency f given by

$$f = \frac{E_i - E_f}{h} = \frac{m_e k_e^2 e^4}{4\pi\hbar^3}\left(\frac{1}{n_f^2} - \frac{1}{n_i^2}\right) \qquad [28.14]$$

where $n_f < n_i$.

Finally, to compare this result with the empirical formulas for the various spectral series, we use Equation 28.14 and the fact that for light, $\lambda f = c$, to get

$$\frac{1}{\lambda} = \frac{f}{c} = \frac{m_e k_e^2 e^4}{4\pi c\hbar^3}\left(\frac{1}{n_f^2} - \frac{1}{n_i^2}\right) \qquad [28.15]$$

A comparison of this result with Equation 28.1 gives the following expression for the Rydberg constant:

$$R_\text{H} = \frac{m_e k_e^2 e^4}{4\pi c\hbar^3} \qquad [28.16]$$

If we insert the known values of m_e, k_e, e, c, and $\hbar$ into this expression, the resulting theoretical value for R_H is found to be in excellent agreement with the value determined experimentally for the Rydberg constant. When Bohr demonstrated this agreement, it was recognized as a major accomplishment of his theory.

In order to compare Equation 28.15 with spectroscopic data, it is convenient to express it in the form

$$\frac{1}{\lambda} = R_\text{H}\left(\frac{1}{n_f^2} - \frac{1}{n_i^2}\right) \qquad [28.17]$$

We can use this expression to evaluate the wavelengths for the various series in the hydrogen spectrum. For example, in the Balmer series, $n_f = 2$ and $n_i = 3, 4, 5, \ldots$ (Eq. 28.1). For the Lyman series, we take $n_f = 1$ and $n_i = 2, 3, 4, \ldots$. The energy level diagram for hydrogen shown in Active Figure 28.7 indicates the origin of the spectral lines described previously. The transitions between levels are represented by vertical arrows. Note that whenever a transition occurs between a state designated by n_i to one designated by n_f (where $n_i > n_f$), a photon with a frequency $(E_i - E_f)/h$ is emitted. This can be interpreted as follows: the lines in the visible part of the hydrogen spectrum arise when the electron jumps from the third, fourth, or even higher orbit to the second orbit. Likewise, the lines of the Lyman series (in the ultraviolet) arise when the electron jumps from the second, third, or even higher orbit to the innermost ($n_f = 1$) orbit. Hence, the Bohr theory successfully predicts the wavelengths of all the observed spectral lines of hydrogen.

> **TIP 28.1   Energy Depends On *n* Only for Hydrogen**
>
> According to Equation 28.13, the energy depends only on the quantum number *n*. Note that this is only true for the hydrogen atom. For more complicated atoms, the energy levels depend primarily on *n*, but also on other quantum numbers.

INTERACTIVE EXAMPLE 28.1   The Balmer Series for Hydrogen

Goal   Calculate the wavelength, frequency, and energy of a photon emitted during an electron transition in an atom.

Problem   The Balmer series for the hydrogen atom corresponds to electronic transitions that terminate in the state with quantum number $n = 2$, as shown in Figure 28.8. **(a)** Find the longest-wavelength photon emitted in the Balmer series and determine its frequency and energy. **(b)** Find the shortest-wavelength photon emitted in the same series.

Strategy   This is a matter of substituting values into Equation 28.17. The frequency can then be obtained from $c = f\lambda$ and the energy from $E = hf$. The longest wavelength photon corresponds to the one that is emitted when the electron jumps from the $n_i = 3$ state to the $n_f = 2$ state. The shortest wavelength photon corresponds to the one that is emitted when the electron jumps from $n_i = \infty$ to the state $n_f = 2$.

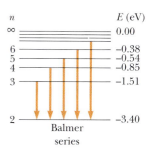

Figure 28.8   (Example 28.1) Transitions responsible for the Balmer series for the hydrogen atom. All transitions terminate at the $n = 2$ level.

Solution

(a) Find the longest wavelength photon emitted in the Balmer series, and determine its energy.

Substitute into Equation 28.17, with $n_i = 3$ and $n_f = 2$:

$$\frac{1}{\lambda} = R_H \left(\frac{1}{n_f^2} - \frac{1}{n_i^2} \right) = R_H \left(\frac{1}{2^2} - \frac{1}{3^2} \right) = \frac{5R_H}{36}$$

Take the reciprocal and substitute, finding the wavelength:

$$\lambda = \frac{36}{5R_H} = \frac{36}{5(1.097 \times 10^7 \text{ m}^{-1})} = 6.563 \times 10^{-7} \text{ m}$$

$$= 656.3 \text{ nm}$$

Now use $c = f\lambda$ to obtain the frequency:

$$f = \frac{c}{\lambda} = \frac{2.998 \times 10^8 \text{ m/s}}{6.563 \times 10^{-7} \text{ m}} = \boxed{4.568 \times 10^{14} \text{ Hz}}$$

Calculate the photon's energy by substituting into Equation 27.5:

$$E = hf = (6.626 \times 10^{-34} \text{ J} \cdot \text{s})(4.568 \times 10^{14} \text{ Hz})$$

$$= 3.027 \times 10^{-19} \text{ J} = \boxed{1.892 \text{ eV}}$$

(b) Find the shortest wavelength photon emitted in the Balmer series.

Substitute into Equation 28.17, with $n_i = \infty$ and $n_f = 2$.

$$\frac{1}{\lambda} = R_H \left(\frac{1}{n_f^2} - \frac{1}{n_i^2} \right) = R_H \left(\frac{1}{2^2} - \frac{1}{\infty} \right) = \frac{R_H}{4}$$

Take the reciprocal and substitute, finding the wavelength:

$$\lambda = \frac{4}{R_H} = \frac{4}{(1.097 \times 10^7 \text{ m}^{-1})} = 3.646 \times 10^{-7} \text{ m}$$

$$= 364.6 \text{ nm}$$

Remarks The first wavelength is in the red region of the visible spectrum. We could also obtain the energy of the photon by using Equation 28.3 in the form $hf = E_3 - E_2$, where E_2 and E_3 are the energy levels of the hydrogen atom, calculated from Equation 28.13. Note that this is the lowest energy photon in the Balmer series, because it involves the smallest energy change. The second photon, the most energetic, is in the ultraviolet region.

Exercise 28.1
(a) Calculate the energy of the shortest wavelength photon emitted in the Balmer series for hydrogen. (b) Calculate the wavelength of a transition from $n = 4$ to $n = 2$.

Answers (a) 3.40 eV (b) 486 nm

Physics⊗Now™ Investigate transitions between various states by logging into PhysicsNow at **www.cp7e.com** and going to Interactive Example 28.1.

Bohr's Correspondence Principle

In our study of relativity in Chapter 26, we found that Newtonian mechanics cannot be used to describe phenomena that occur at speeds approaching the speed of light. Newtonian mechanics is a special case of relativistic mechanics and applies only when v is much smaller than c. Similarly, **quantum mechanics is in agreement with classical physics when the energy differences between quantized levels are very small**. This principle, first set forth by Bohr, is called the **correspondence principle**.

For example, consider the hydrogen atom with $n > 10\ 000$. For such large values of n, the energy differences between adjacent levels approach zero and the levels are nearly continuous, as Equation 28.13 shows. As a consequence, the classical model is reasonably accurate in describing the system for large values of n. According to the classical model, the frequency of the light emitted by the atom is equal to the frequency of revolution of the electron in its orbit about the nucleus. Calculations show that for $n > 10\ 000$, this frequency is different from that predicted by quantum mechanics by less than 0.015%.

28.4 MODIFICATION OF THE BOHR THEORY

The Bohr theory of the hydrogen atom was a tremendous success in certain areas because it explained several features of the hydrogen spectrum that had previously defied explanation. It accounted for the Balmer series and other series; it predicted a value for the Rydberg constant that is in excellent agreement with the experimental value; it gave an expression for the radius of the atom; and it predicted the energy levels of hydrogen. Although these successes were important to scientists, it is perhaps even more significant that the Bohr theory gave us a model of what the atom looks like and how it behaves. Once a basic model is constructed, refinements and modifications can be made to enlarge on the concept and to explain finer details.

The analysis used in the Bohr theory is also successful when applied to *hydrogen-like* atoms. An atom is said to be hydrogen-like when it contains only one electron. Examples are singly ionized helium, doubly ionized lithium, triply ionized beryllium, and so forth. The results of the Bohr theory for hydrogen can be extended to hydrogen-like atoms by substituting Ze^2 for e^2 in the hydrogen equations, where Z is the atomic number of the element. For example, Equations 28.12 and 28.15 become

$$E_n = -\frac{m_e k_e^2 Z^2 e^4}{2\hbar^2}\left(\frac{1}{n^2}\right) \qquad n = 1, 2, 3, \ldots \qquad \text{[28.18]}$$

and

$$\frac{1}{\lambda} = \frac{m_e k_e^2 Z^2 e^4}{4\pi c\hbar^3}\left(\frac{1}{n_f^2} - \frac{1}{n_i^2}\right) \qquad \text{[28.19]}$$

Although many attempts were made to extend the Bohr theory to more complex, multi-electron atoms, the results were unsuccessful. Even today, only approximate methods are available for treating multi-electron atoms.

Quick Quiz 28.1

Consider a hydrogen atom and a singly-ionized helium atom. Which atom has the lower ground state energy? (a) hydrogen (b) helium (c) the ground state energy is the same for both

Quick Quiz 28.2

Consider once again a singly-ionized helium atom. Suppose the remaining electron jumps from a higher to a lower energy level, resulting in the emission of photon, which we'll call photon-He. An electron in a hydrogen atom then jumps between the same two levels, resulting in an emitted photon-H. Which photon has the shorter wavelength? (a) photon-He (b) photon-H (c) The wavelengths are the same.

EXAMPLE 28.2   Singly Ionized Helium

Goal   Apply the modified Bohr theory to a hydrogen-like atom.

Problem   Singly ionized helium, He^+, a hydrogen-like system, has one electron in the $1s$ orbit when the atom is in its ground state. Find **(a)** the energy of the system in the ground state in electron volts, and **(b)** the radius of the ground-state orbit.

Strategy   Part (a) requires substitution into the modified Bohr model, Equation 28.18. In part (b), modify Equation 28.9 for the radius of the Bohr orbits by replacing e^2 by Ze^2, where Z is the number of protons in the nucleus.

Solution

(a) Find the energy of the system in the ground state.

Write Equation 28.18 for the energies of a hydrogen-like system:

$$E_n = -\frac{m_e k_e^2 Z^2 e^4}{2\hbar^2}\left(\frac{1}{n^2}\right)$$

Substitute the constants and convert to electron volts:

$$E_n = -\frac{Z^2(13.6)}{n^2}\ \text{eV}$$

Substitute $Z = 2$ (the atomic number of helium) and $n = 1$ to obtain the ground state energy:

$$E_1 = -4(13.6)\ \text{eV} = \boxed{-54.4\ \text{eV}}$$

(b) Find the radius of the ground state.

Generalize Equation 28.9 to a hydrogen-like atom by substituting Ze^2 for e^2:

$$r_n = \frac{n^2\hbar^2}{m_e k_e Ze^2} = \frac{n^2}{Z}(a_0) = \frac{n^2}{Z}(0.052\ 9\ \text{nm})$$

For our case, $n = 1$ and $Z = 2$:

$$r_1 = \boxed{0.026\ 5\ \text{nm}}$$

Remarks   Notice that for higher Z the energy of a hydrogen-like atom is lower, which means that the electron is more tightly bound than in hydrogen. This results in a smaller atom, as seen in part (b).

Exercise 28.2

Repeat the problem for the first excited state of doubly-ionized lithium ($Z = 3$, $n = 2$).

Answers   (a) $E_2 = -30.6$ eV   (b) $r_2 = 0.070\ 5$ nm

TABLE 28.1

Shell and Subshell Notation

n	Shell Symbol	ℓ	Subshell Symbol
1	K	0	s
2	L	1	p
3	M	2	d
4	N	3	f
5	O	4	g
6	P	5	h
. . .		. . .	

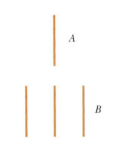

Figure 28.9 A single line (*A*) can split into three separate lines (*B*) in a magnetic field.

Within a few months following the publication of Bohr's paper, Arnold Sommerfeld (1868–1951) extended the Bohr model to include elliptical orbits. We examine his model briefly because much of the nomenclature used in this treatment is still in use today. Bohr's concept of quantization of angular momentum led to the **principal quantum number** n, which determines the energy of the allowed states of hydrogen. Sommerfeld's theory retained n, but also introduced a new quantum number ℓ called the **orbital quantum number**, where the value of ℓ ranges from 0 to $n - 1$ in integer steps. According to this model, an electron in any one of the allowed energy states of a hydrogen atom may move in any one of a number of orbits corresponding to different ℓ values. For each value of n, there are n possible orbits corresponding to different ℓ values. Because $n = 1$ and $\ell = 0$ for the first energy level (ground state), there is only one possible orbit for this state. The second energy level, with $n = 2$, has two possible orbits, corresponding to $\ell = 0$ and $\ell = 1$. The third energy level, with $n = 3$, has three possible orbits, corresponding to $\ell = 0$, $\ell = 1$, and $\ell = 2$.

For historical reasons, **all states with the same principal quantum number n are said to form a shell.** Shells are identified by the letters K, L, M, . . . , which designate the states for which $n = 1, 2, 3, ...$ Likewise, **the states with given values of n and ℓ are said to form a subshell.** The letters $s, p, d, f, g, . . .$ are used to designate the states for which $\ell = 0, 1, 2, 3, 4, ...$ These notations are summarized in Table 28.1.

States that violate the restriction $0 \leq \ell \leq n - 1$, for a given value of n, can't exist. A $2d$ state, for instance, would have $n = 2$ and $\ell = 2$, but can't exist because the highest allowed value of ℓ is $n - 1$, or 1 in this case. For $n = 2$, $2s$ and $2p$ are allowed subshells, but $2d$, $2f$, . . . are not. For $n = 3$, the allowed states are $3s$, $3p$, and $3d$.

Another modification of the Bohr theory arose when it was discovered that the spectral lines of a gas are split into several closely spaced lines when the gas is placed in a strong magnetic field. (This is called the *Zeeman effect*, after its discoverer.) Figure 28.9 shows a single spectral line being split into three closely spaced lines. This indicates that the energy of an electron is slightly modified when the atom is immersed in a magnetic field. In order to explain this observation, a new quantum number, m_ℓ, called the **orbital magnetic quantum number**, was introduced. The theory is in accord with experimental results when m_ℓ is restricted to values ranging from $-\ell$ to $+\ell$ in integer steps. For a given value of ℓ, there are $2\ell + 1$ possible values of m_ℓ.

Finally, very high resolution spectrometers revealed that spectral lines of gases are in fact two very closely spaced lines even in the absence of an external magnetic field. This splitting was referred to as **fine structure**. In 1925 Samuel Goudsmit and George Uhlenbeck introduced the idea of an electron spinning about its own axis to explain the origin of fine structure. The results of their work introduced yet another quantum number, m_s, called the **spin magnetic quantum number**.

For each electron there are two spin states. A subshell corresponding to a given factor of ℓ can contain no more than $2(2\ell + 1)$ electrons. This number comes from the fact that electrons in a subshell must have unique pairs of the quantum numbers (m_ℓ, m_s). There are $2\ell + 1$ different magnetic quantum numbers m_ℓ, and two different spin quantum numbers m_s, making $2(2\ell + 1)$ unique pairs (m_ℓ, m_s). For example, the p subshell ($\ell = 1$) is filled when it contains $2(2 \cdot 1 + 1) = 6$ electrons. This fact can be extended to include all four quantum numbers, as will be important to us later when we discuss the *Pauli exclusion principle*.

All these quantum numbers (addressed in more detail in upcoming sections) were postulated to account for the observed spectra of elements. Only later were comprehensive mathematical theories developed that naturally yielded the same answers as these empirical models.

28.5 DE BROGLIE WAVES AND THE HYDROGEN ATOM

One of the postulates made by Bohr in his theory of the hydrogen atom was that the angular momentum of the electron is quantized in units of $\hbar$, or

$$m_e v r = n\hbar$$

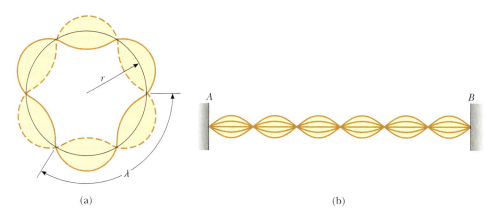

Figure 28.10 (a) Standing-wave pattern for an electron wave in a stable orbit of hydrogen. There are three full wavelengths in this orbit. (b) Standing-wave pattern for a vibrating stretched string fixed at its ends. This pattern also has three full wavelengths.

(a)   (b)

For more than a decade following Bohr's publication, no one was able to explain why the angular momentum of the electron was restricted to these discrete values. Finally, de Broglie gave a direct physical way of interpreting this condition. He assumed that an electron orbit would be stable (allowed) only if it contained an integral number of electron wavelengths. Figure 28.10a demonstrates this point when three complete wavelengths are contained in one circumference of the orbit. Similar patterns can be drawn for orbits containing one wavelength, two wavelengths, four wavelengths, five wavelengths, and so forth. These waves are analogous to standing waves on a string, discussed in Chapter 14. There, we found that strings have preferred (resonant) frequencies of vibration. Figure 28.10b shows a standing-wave pattern containing three wavelengths for a string fixed at each end. Now imagine that the vibrating string is removed from its supports at A and B and bent into a circular shape that brings those points together. The end result is a pattern such as the one shown in Figure 28.10a.

In general, the condition for a de Broglie standing wave in an electron orbit is that the circumference must contain an integral number of electron wavelengths. We can express this condition as

$$2\pi r = n\lambda \qquad n = 1, 2, 3, \ldots$$

Because the de Broglie wavelength of an electron is $\lambda = h/m_e v$, we can write the preceding equation as $2\pi r = nh/m_e v$, or

$$m_e v r = n\hbar$$

This is the same as the quantization of angular momentum condition imposed by Bohr in his original theory of hydrogen.

The electron orbit shown in Figure 28.10a contains three complete wavelengths and corresponds to the case in which the principal quantum number $n = 3$. The orbit with one complete wavelength in its circumference corresponds to the first Bohr orbit, $n = 1$; the orbit with two complete wavelengths corresponds to the second Bohr orbit, $n = 2$; and so forth.

By applying the wave theory of matter to electrons in atoms, de Broglie was able to explain the appearance of integers in the Bohr theory as a natural consequence of standing-wave patterns. This was the first convincing argument that the wave nature of matter was at the heart of the behavior of atomic systems. Although the analysis provided by de Broglie was a promising first step, gigantic strides were made subsequently with the development of Schrödinger's wave equation and its application to atomic systems.

28.6   QUANTUM MECHANICS AND THE HYDROGEN ATOM

One of the first great achievements of quantum mechanics was the solution of the wave equation for the hydrogen atom. The details of the solution are far beyond the level of this course, but we'll describe its properties and implications for atomic structure.

TABLE 28.2

Three Quantum Numbers for the Hydrogen Atom

Quantum Number	Name	Allowed Values	Number of Allowed States
N	Principal quantum number	$1, 2, 3, \ldots$	Any number
ℓ	Orbital quantum number	$0, 1, 2, \ldots, n - 1$	n
m_ℓ	Orbital magnetic quantum number	$-\ell, -\ell + 1, \ldots,$ $0, \ldots, \ell - 1, \ell$	$2\ell + 1$

According to quantum mechanics, the energies of the allowed states are in exact agreement with the values obtained by the Bohr theory (Eq. 28.12) when the allowed energies depend only on the principal quantum number n.

In addition to the principal quantum number, two other quantum numbers emerged from the solution of the wave equation: ℓ and m_ℓ. The quantum number ℓ is called the **orbital quantum number**, and m_ℓ is called the **orbital magnetic quantum number**. As pointed out in Section 28.4, these quantum numbers had already appeared in empirical modifications made to the Bohr theory. The significance of quantum mechanics is that those numbers and the restrictions placed on their values arose directly from mathematics and not from any ad hoc assumptions to make the theory consistent with experimental observation. Because we will need to make use of the various quantum numbers in the sections that follow, the allowed ranges of their values are repeated:

The value of n can range from 1 to ∞ in integer steps.
The value of ℓ can range from 0 to $n - 1$ in integer steps.
The value of m_ℓ can range from $-\ell$ to ℓ in integer steps.

From these rules, it can be seen that for a given value of n, there are n possible values of ℓ, while for a given value of ℓ there are $2\ell + 1$ possible values of m_ℓ. For example, if $n = 1$, there is only 1 value of ℓ, $\ell = 0$. Because $2\ell + 1 = 2 \cdot 0 + 1 = 1$, there is only one value of m_ℓ, which is $m_\ell = 0$. If $n = 2$, the value of ℓ may be 0 or 1; if $\ell = 0$, then $m_\ell = 0$, but if $\ell = 1$, then m_ℓ may be 1, 0, or -1. Table 28.2 summarizes the rules for determining the allowed values of ℓ and m_ℓ for a given value of n.

States that violate the rules given in Table 28.2 cannot exist. For instance, one state that cannot exist is the $2d$ state, which would have $n = 2$ and $\ell = 2$. This state is not allowed because the highest allowed value of ℓ is $n - 1$, or 1 in this case. Thus, for $n = 2$, $2s$ and $2p$ are allowed states, but $2d$, $2f$, $\ldots$ are not. For $n = 3$, the allowed states are $3s$, $3p$, and $3d$.

In general, for a given value of n_1 there are n^2 states with distinct pairs of values of ℓ and m_ℓ.

Quick Quiz 28.3

When the principal quantum number is $n = 5$, how many different values of (a) ℓ and (b) m_ℓ are possible? (c) How many states have distinct pairs of values of ℓ and m_ℓ?

EXAMPLE 28.3 The $n = 2$ Level of Hydrogen

Goal Count states and determine energy based on atomic energy level.

Problem (a) Determine the number of states with a unique set of values for ℓ and m_ℓ in the hydrogen atom for $n = 2$. (b) Calculate the energies of these states.

Strategy This is a matter of counting, following the quantum rules for n, ℓ, and m_ℓ. "Unique" means that no other quantum state has the same pair of numbers for ℓ and m_ℓ the energies are all the same because all states have the same principal quantum number, $n = 2$.

Solution

(a) Determine the number of states with a unique set of values for ℓ and m_ℓ in the hydrogen atom for $n = 2$.

Determine the different possible values of ℓ for $n = 2$:

$0 \le \ell \le n - 1$, so, for $n = 2$, $0 \le \ell \le 1$ and $\ell = 0$ or 1

Find the different possible values of m_ℓ for $\ell = 0$:

$-\ell \le m_\ell \le \ell$, so $-0 \le m_\ell \le 0$ implies $m_\ell = 0$

List the distinct pairs of (ℓ, m_ℓ) for $\ell = 0$:

There is only one: $(\ell, m_\ell) = (0, 0)$.

Find the different possible values of m_ℓ for $\ell = 1$:

$-\ell \le m_\ell \le \ell$, so $-1 \le m_\ell \le 1$ implies $m_\ell = -1, 0$, or 1

List the distinct pairs of (ℓ, m_ℓ) for $\ell = 1$:

There are three: $(\ell, m_\ell) = (1, -1)$, $(1, 0)$, and $(1, 1)$.

Sum the results for $\ell = 0$ and $\ell = 1$:

Number of states $= 1 + 3 = 4$

(b) Calculate the energies of these states.

The common energy of all of the states can be found with Equation 28.13:

$$E_n = -\frac{13.6 \text{ eV}}{n^2} \quad \rightarrow \quad E_2 = -\frac{13.6 \text{ eV}}{2^2} = -3.40 \text{ eV}$$

Remarks   While these states normally have the same energy, applying a magnetic field will result in their taking slightly different energies centered around the energy corresponding to $n = 2$. As seen in the next section, there are in fact twice as many states, corresponding to a new quantum number called *spin*.

Exercise 28.3

(a) Determine the number of states with a unique pair of values for ℓ and m_ℓ in the $n = 3$ level of hydrogen. (b) Determine the energies of those states.

Answers   (a) 9   (b) $E_3 = -1.51$ eV

28.7   THE SPIN MAGNETIC QUANTUM NUMBER

As we'll see in this section, there actually are *eight* states corresponding to $n = 2$ for hydrogen, not four as given in Example 28.3. This happens because another quantum number, m_s, the **spin magnetic quantum number**, has to be introduced to explain the splitting of each level into two.

The need for this new quantum number first came about because of an unusual feature in the spectra of certain gases, such as sodium vapor. Close examination of one of the prominent lines of sodium shows that it is, in fact, two very closely spaced lines. The wavelengths of these lines occur in the yellow region of the spectrum, at 589.0 nm and 589.6 nm. In 1925, when this doublet was first noticed, atomic theory couldn't explain it. To resolve the dilemma, Samuel Goudsmit and George Uhlenbeck, following a suggestion by the Austrian physicist Wolfgang Pauli, proposed the introduction of a fourth quantum number to describe atomic energy levels, called the *spin quantum number*.

In order to describe the spin quantum number, it's convenient (but technically incorrect) to think of the electron as spinning on its axis as it orbits the nucleus, just as the Earth spins on its axis as it orbits the Sun. Strangely, there are only two ways in which the electron can spin as it orbits the nucleus, as shown in Figure 28.11. If the direction of spin is as shown in Figure 28.11a, the electron is said to have "spin up." If the direction of spin is reversed, as in Figure 28.11b, the electron is said to have "spin down." The energy of the electron is slightly different for the two spin directions, and this energy difference accounts for the sodium doublet. The quantum numbers associated with electron spin are $m_s = \frac{1}{2}$ for the spin-up state and $m_s = -\frac{1}{2}$ for the spin-down state. As we'll see in Example 28.5, this new quantum number doubles the number of allowed states specified by the quantum numbers n, ℓ, and m_ℓ.

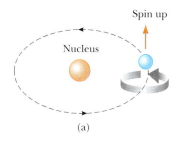

(a)

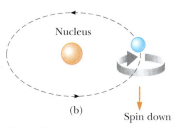

(b)

Figure 28.11   As an electron moves in its orbit about the nucleus, its spin can be either (a) up or (b) down.

TIP 28.2   The Electron Isn't Really Spinning
The electron is *not* physically spinning. Electron spin is a purely quantum effect that gives the electron an angular momentum *as if* it were physically spinning.

Any classical description of electron spin is incorrect because quantum mechanics tells us that since the electron can't be located precisely in space, it cannot be considered to be a spinning solid object, as pictured in Figure 28.11. In spite of this conceptual difficulty, all experimental evidence supports the fact that an electron does have some intrinsic property that can be described by the spin magnetic quantum number.

The spin quantum number didn't come from the original formulation of quantum mechanics by Schrodinger (and independently, by Heisenberg). The English mathematical physicist P. A. M. Dirac developed a relativistic quantum theory in which spin appears naturally.

EXAMPLE 28.4   The Quantum Numbers for the 2p Subshell

Goal   List the distinct quantum states of a subshell by their quantum numbers, including spin.

Problem   List the unique sets of quantum numbers for electrons in the 2p subshell.

Strategy   This is again a matter following the quantum rules for n, ℓ, and m_ℓ, and now m_s as well. The 2p subshell has $n = 2$ (that's the "2" in 2p) and $\ell = 1$ (that's from the p in 2p).

Solution

Because $\ell = 1$, the magnetic quantum number can have the values -1, 0, 1, and the spin quantum number is always $+\frac{1}{2}$ or $-\frac{1}{2}$. Consequently, there are $3 \times 2 = 6$ possible sets of quantum numbers with $n = 2$ and $\ell = 1$, listed in the table at right.

n	ℓ	m_ℓ	m_s
2	1	-1	$-\frac{1}{2}$
2	1	-1	$\frac{1}{2}$
2	1	0	$-\frac{1}{2}$
2	1	0	$\frac{1}{2}$
2	1	1	$-\frac{1}{2}$
2	1	1	$\frac{1}{2}$

Remark   Remember that these quantum states are not just abstractions; they have real physical consequences, such as which electronic transitions can be made within an atom and, consequently, which wavelengths of radiation can be observed.

Exercise 28.4

(a) How many different sets of quantum numbers are there in the 3d subshell? (b) How many sets of quantum numbers are there in a 2d subshell?

Answers   (a) 10   (b) None. A 2d subshell doesn't exist because that would imply a quantum state with $n = 2$ and $\ell = 2$, impossible because $\ell \leq n - 1$.

28.8   ELECTRON CLOUDS

The solution of the wave equation, discussed in Section 27.7, yields a wave function Ψ that depends on the quantum numbers n, ℓ, and m_ℓ. We assume that we have found such a wave function Ψ and see what it may tell us about the hydrogen atom. Let $n = 1$ for the principal quantum number, which corresponds to the lowest energy state for hydrogen. For $n = 1$, the restrictions placed on the remaining quantum numbers are that $\ell = 0$ and $m_\ell = 0$.

The quantity Ψ^2 has great physical significance. If p is a point and V_p a very small volume containing that point, then $\Psi^2 V_p$ is approximately the probability of finding the electron inside the volume V_p. Figure 28.12 gives the probability per unit length of finding the electron at various distances from the nucleus in the 1s state of hydrogen. Some useful and surprising information can be extracted from this curve. First, the curve peaks at a value of $r = 0.052\ 9$ nm, the Bohr radius for the first ($n = 1$) electron orbit in hydrogen. This means that there is a maximum probability of finding the electron in a small interval centered at that distance from the nucleus. However, as the curve indicates, there is also a probability of finding the electron in a small interval centered at any other distance from the nucleus. In other words, the electron is not confined to a particular orbital distance from the nucleus, as assumed in the Bohr model. The electron may be found at various distances from the nucleus, but **the probability of finding it at a distance corresponding to the Bohr radius is a maximum**. Quantum mechanics also predicts that the wave function for the hydrogen atom in the ground state is spherically symmetric; hence the electron can be found in a spherical region surrounding the nucleus. This is in contrast to the Bohr theory, which confines the position of the electron to points in a plane. The quantum mechanical result is often interpreted by viewing the electron as a cloud surrounding the nucleus. An attempt at picturing this cloud-like behavior is shown in Figure 28.13. The densest regions of the cloud represent those locations where the electron is most likely to be found.

If a similar analysis is carried out for the $n = 2$, $\ell = 0$, state of hydrogen, a peak of the probability curve is found at $4a_0$. Likewise, for the $n = 3$, $\ell = 0$ state, the curve peaks at $9a_0$. Thus, quantum mechanics predicts a most probable electron distance to the nucleus that is in agreement with the location predicted by the Bohr theory.

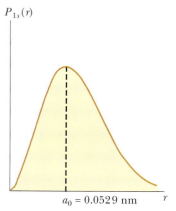

Figure 28.12 The probability per unit length of finding the electron versus distance from the nucleus for the hydrogen atom in the $1s$ (ground) state. Note that the graph has its maximum value when r equals the first Bohr radius, a_0.

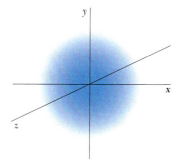

Figure 28.13 The spherical electron cloud for the hydrogen atom in its $1s$ state.

28.9 THE EXCLUSION PRINCIPLE AND THE PERIODIC TABLE

Earlier, we found that the state of an electron in an atom is specified by four quantum numbers: n, ℓ, m_ℓ, and m_s. For example, an electron in the ground state of hydrogen could have quantum numbers of $n = 1$, $\ell = 0$, $m_\ell = 0$, and $m_s = \frac{1}{2}$. As it turns out, the state of an electron in any other atom may also be specified by this same set of quantum numbers. In fact, these four quantum numbers can be used to describe all the electronic states of an atom, regardless of the number of electrons in its structure.

How many electrons in an atom can have a particular set of quantum numbers? This important question was answered by Pauli in 1925 in a powerful statement known as the **Pauli exclusion principle**:

> No two electrons in an atom can ever have the same set of values for the set of quantum numbers n, ℓ, m_ℓ, and m_s.

◄ The Pauli exclusion principle

The Pauli exclusion principle explains the electronic structure of complex atoms as a succession of filled levels with different quantum numbers increasing in energy, where the outermost electrons are primarily responsible for the chemical properties of the element. If this principle weren't valid, every electron would end up in the lowest energy state of the atom and the chemical behavior of the elements would be grossly different. Nature as we know it would not exist—and *we* would not exist to wonder about it!

As a general rule, the order that electrons fill an atom's subshell is as follows: once one subshell is filled, the next electron goes into the vacant subshell that is lowest in energy. If the atom were not in the lowest energy state available to it, it would radiate energy until it reached that state. A subshell is filled when it contains $2(2\ell + 1)$ electrons. This rule is based on the analysis of quantum numbers to be described later. Following the rule, shells and subshells can contain numbers of electrons according to the pattern given in Table 28.3.

The exclusion principle can be illustrated by an examination of the electronic arrangement in a few of the lighter atoms.

Hydrogen has only one electron, which, in its ground state, can be described by either of two sets of quantum numbers: $1, 0, 0, \frac{1}{2}$ or $1, 0, 0, -\frac{1}{2}$. The electronic configuration of this atom is often designated as $1s^1$. The notation $1s$ refers to a

TIP 28.3 The Exclusion Principle is More General

The exclusion principle stated here is a limited form of the more general exclusion principle, which states that no two *fermions* (particles with spin $1/2, 3/2, \ldots$) can be in the same quantum state.

CERN/Courtesy of AIP Emilio Segre Visual Archives

WOLFGANG PAULI (1900–1958)

An extremely talented Austrian theoretical physicist who made important contributions in many areas of modern physics, Pauli gained public recognition at the age of 21 with a masterful review article on relativity that is still considered one of the finest and most comprehensive introductions to the subject. Other major contributions were the discovery of the exclusion principle, the explanation of the connection between particle spin and statistics, and theories of relativistic quantum electrodynamics, the neutrino hypothesis, and the hypothesis of nuclear spin.

TABLE 28.3

Number of Electrons in Filled Subshells and Shells

Shell	Subshell	Number of Electrons in Filled Subshell	Number of Electrons in Filled Shell
K ($n = 1$)	$s(\ell = 0)$	2	2
L ($n = 2$)	$s(\ell = 0)$ $p(\ell = 1)$	2 6	8
M ($n = 3$)	$s(\ell = 0)$ $p(\ell = 1)$ $d(\ell = 2)$	2 6 10	18
N ($n = 4$)	$s(\ell = 0)$ $p(\ell = 1)$ $d(\ell = 2)$ $f(\ell = 3)$	2 6 10 14	32

state for which $n = 1$ and $\ell = 0$, and the superscript indicates that one electron is present in this level.

Neutral *helium* has two electrons. In the ground state, the quantum numbers for these two electrons are 1, 0, 0, $\frac{1}{2}$ and 1, 0, 0, $-\frac{1}{2}$. No other possible combinations of quantum numbers exist for this level, and we say that the K shell is filled. The helium electronic configuration is designated as $1s^2$.

Neutral *lithium* has three electrons. In the ground state, two of these are in the $1s$ subshell and the third is in the $2s$ subshell, because the latter is lower in energy than the $2p$ subshell. Hence, the electronic configuration for lithium is $1s^2 2s^1$.

A list of electronic ground-state configurations for a number of atoms is provided in Table 28.4. In 1871 Dmitri Mendeleev (1834–1907), a Russian chemist, arranged the elements known at that time into a table according to their atomic masses and chemical similarities. The first table Mendeleev proposed contained many blank spaces, and he boldly stated that the gaps were there only because those elements had not yet been discovered. By noting the column in which these missing elements should be located, he was able to make rough predictions about their chemical properties. Within 20 years of this announcement, the elements were indeed discovered.

The elements in our current version of the periodic table are still arranged so that all those in a vertical column have similar chemical properties. For example, consider the elements in the last column: He (helium), Ne (neon), Ar (argon), Kr (krypton), Xe (xenon), and Rn (radon). The outstanding characteristic of these elements is that they don't normally take part in chemical reactions, joining with other atoms to form molecules, and are therefore classified as inert. Because of this "aloofness," they are referred to as the *noble gases*. We can partially understand their behavior by looking at the electronic configurations shown in Table 28.4, page 919. The element helium has the electronic configuration $1s^2$. In other words, one shell is filled. The electrons in this filled shell are considerably separated in energy from the next available level, the $2s$ level.

The electronic configuration for neon is $1s^2 2s^2 2p^6$. Again, the outer shell is filled and there is a large difference in energy between the $2p$ level and the $3s$ level. Argon has the configuration $1s^2 2s^2 2p^6 3s^2 3p^6$. Here, the $3p$ subshell is filled and there is a wide gap in energy between the $3p$ subshell and the $3d$ subshell. Through all the noble gases, the pattern remains the same: a noble gas is formed when either a shell or a subshell is filled, and there is a large gap in energy before the next possible level is encountered.

The elements in the first column of the periodic table are called the *alkali metals* and are highly active chemically. Referring to Table 28.4, we can understand why these elements interact so strongly with other elements. All of these alkali

TABLE 28.4
Electronic Configurations of Some Elements

Z	Symbol	Ground-State Configuration	Ionization Energy (eV)	Z	Symbol	Ground-State Configuration	Ionization Energy (eV)
1	H	$1s^1$	13.595	19	K	[Ar] $4s^1$	4.339
2	He	$1s^2$	24.581	20	Ca	$4s^2$	6.111
				21	Sc	$3d4s^2$	6.54
3	Li	[He] $2s^1$	5.390	22	Ti	$3d^24s^2$	6.83
4	Be	$2s^2$	9.320	23	V	$3d^34s^2$	6.74
5	B	$2s^22p^1$	8.296	24	Cr	$3d^54s^1$	6.76
6	C	$2s^22p^2$	11.256	25	Mn	$3d^54s^2$	7.432
7	N	$2s^22p^3$	14.545	26	Fe	$3d^64s^2$	7.87
8	O	$2s^22p^4$	13.614	27	Co	$3d^74s^2$	7.86
9	F	$2s^22p^5$	17.418	28	Ni	$3d^84s^2$	7.633
10	Ne	$2s^22p^6$	21.559	29	Cu	$3d^{10}4s^1$	7.724
				30	Zn	$3d^{10}4s^2$	9.391
11	Na	[Ne] $3s^1$	5.138	31	Ga	$3d^{10}4s^24p^1$	6.00
12	Mg	$3s^2$	7.644	32	Ge	$3d^{10}4s^24p^2$	7.88
13	Al	$3s^23p^1$	5.984	33	As	$3d^{10}4s^24p^3$	9.81
14	Si	$3s^23p^2$	8.149	34	Se	$3d^{10}4s^24p^4$	9.75
15	P	$3s^23p^3$	10.484	35	Br	$3d^{10}4s^24p^5$	11.84
16	S	$3s^23p^4$	10.357	36	Kr	$3d^{10}4s^24p^6$	13.996
17	Cl	$3s^23p^5$	13.01				
18	Ar	$3s^23p^6$	15.755				

Note: The bracket notation is used as a shorthand method to avoid repetition in indicating inner-shell electrons. Thus, [He] represents $1s^2$, [Ne] represents $1s^22s^22p^6$, [Ar] represents $1s^22s^22p^63s^23p^6$, and so on.

metals have a single outer electron in an *s* subshell. This electron is shielded from the nucleus by all the electrons in the inner shells. Consequently, it's only loosely bound to the atom and can readily be accepted by other atoms that bind it more tightly to form molecules.

The elements in the seventh column of the periodic table are called the *halogens* and are also highly active chemically. All these elements are lacking one electron in a subshell, so they readily accept electrons from other atoms to form molecules.

Quick Quiz 28.4

Krypton (atomic number 36) has how many electrons in its next to outer shell ($n = 3$)?
(a) 2   (b) 4   (c) 8   (d) 18

Applying Physics 28.3   The Periodic Table

Scanning from left to right across one row of the periodic table, the effective size of the atoms first decreases and then increases. What would cause this behavior?

Explanation Starting on the left side of the periodic table and moving toward the middle, the nuclear charge is increasing. As a result, there is an increasing Coulomb attraction between the nucleus and the electrons, and the electrons are pulled into an average position that is closer to the nucleus. From the middle of the row to the right side, the increasing number of electrons being placed in proximity to each other results in a mutual repulsion that increases the average distance from the nucleus and causes the atomic size to grow.

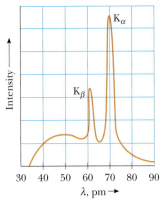

Figure 28.14 The x-ray spectrum of a metal target consists of a broad continuous spectrum (*bremsstrahlung*) plus a number of sharp lines that are due to *characteristic x-rays*. The data shown were obtained when 35-keV electrons bombarded a molybdenum target. Note that 1 pm = 10^{-12} m = 0.001 nm.

28.10 CHARACTERISTIC X-RAYS

X-rays are emitted when a metal target is bombarded with high-energy electrons. The x-ray spectrum typically consists of a broad continuous band and a series of intense sharp lines that are dependent on the type of metal used for the target, as shown in Figure 28.14. These discrete lines, called **characteristic x-rays**, were discovered in 1908, but their origin remained unexplained until the details of atomic structure were developed.

The first step in the production of characteristic x-rays occurs when a bombarding electron collides with an electron in an inner shell of a target atom with sufficient energy to remove the electron from the atom. The vacancy created in the shell is filled when an electron in a higher level drops down into the lower energy level containing the vacancy. The time it takes for this to happen is very short, less than 10^{-9} s. The transition is accompanied by the emission of a photon with energy equaling the difference in energy between the two levels. Typically, the energy of such transitions is greater than 1 000 eV, and the emitted x-ray photons have wavelengths in the range of 0.01 nm to 1 nm.

We assume that the incoming electron has dislodged an atomic electron from the innermost shell, the K shell. If the vacancy is filled by an electron dropping from the next higher shell, the L shell, the photon emitted in the process is referred to as the K_α line on the curve of Figure 28.14. If the vacancy is filled by an electron dropping from the M shell, the line produced is called the K_β line.

Other characteristic x-ray lines are formed when electrons drop from upper levels to vacancies other than those in the K shell. For example, L lines are produced when vacancies in the L shell are filled by electrons dropping from higher shells. An L_α line is produced as an electron drops from the M shell to the L shell, and an L_β line is produced by a transition from the N shell to the L shell.

We can estimate the energy of the emitted x-rays as follows: consider two electrons in the K shell of an atom whose atomic number is Z. Each electron partially shields the other from the charge of the nucleus, Ze, so each is subject to an effective nuclear charge $Z_{\text{eff}} = (Z - 1)e$. We can now use a modified form of Equation 28.18 to estimate the energy of either electron in the K shell (with $n = 1$). We have

$$E_K = -m_e Z_{\text{eff}}^2 \frac{k_e^2 e^4}{2\hbar^2} = -Z_{\text{eff}}^2 E_0$$

where E_0 is the ground-state energy. Substituting $Z_{\text{eff}} = Z - 1$ gives

$$E_K = -(Z - 1)^2 (13.6 \text{ eV}) \qquad \text{[28.20]}$$

As Example 28.5 will show, we can estimate the energy of an electron in an L or an M shell in a similar fashion. Taking the energy difference between these two levels, we can then calculate the energy and wavelength of the emitted photon.

In 1914, Henry G. J. Moseley plotted the Z values for a number of elements against $\sqrt{1/\lambda}$, where λ is the wavelength of the K_α line for each element. He found that such a plot produced a straight line, as in Figure 28.15. This is consistent with our rough calculations of the energy levels based on Equation 28.20. From his plot, Moseley was able to determine the Z values of other elements, providing a periodic chart in excellent agreement with the known chemical properties of the elements.

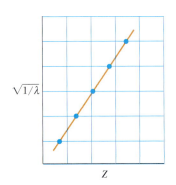

Figure 28.15 A Moseley plot of $\sqrt{1/\lambda}$ versus Z, where λ is the wavelength of the K_α x-ray line of the element of atomic number Z.

EXAMPLE 28.5   Characteristic X-Rays

Goal   Calculate the energy and wavelength of characteristic x-rays.

Problem   Estimate the energy of the characteristic x-ray emitted from a tungsten target when an electron drops from an M shell ($n = 3$ state) to a vacancy in the K shell ($n = 1$ state).

Strategy   Develop two estimates, one for the electron in the K shell ($n = 1$) and one for the electron in the M shell ($n = 3$). For the K-shell estimate, we can use Equation 28.20. For the M shell, we need a new equation. There is one electron in the K shell (because one is missing) and 8 in the L shell, making 9 electrons shielding the nuclear charge. This means $Z_{eff} = 74 - 9$ and $E_M = -Z_{eff}^2 E_3$, where E_3 is the energy of the $n = 3$ level in hydrogen. The difference $E_M - E_K$ is the energy of the photon.

Solution

Use Equation 28.20 to estimate the energy of an electron in the K shell of tungsten, atomic number $Z = 74$:

$$E_K = -(74 - 1)^2(13.6 \text{ eV}) = -72\,500 \text{ eV}$$

Estimate the energy of an electron in the M shell in the same way:

$$E_M = -Z_{eff}^2 E_3 = -(Z - 9)^2\frac{E_0}{3^2} = -(74 - 9)^2\frac{(13.6 \text{ eV})}{9}$$

$$= -6\,380 \text{ eV}$$

Calculate the difference in energy between the M and K shells:

$$E_M - E_K = -6\,380 \text{ eV} - (-72\,500 \text{ eV}) = \boxed{66\,100 \text{ eV}}$$

Find the wavelength of the emitted light:

$$\Delta E = hf = h\frac{c}{\lambda} \quad \rightarrow \quad \lambda = \frac{hc}{\Delta E}$$

$$\lambda = \frac{(6.63 \times 10^{-34} \text{ J·s})(3.00 \times 10^8 \text{ m/s})}{(6.61 \times 10^4 \text{ eV})(1.60 \times 10^{-19} \text{ J/eV})}$$

$$= 1.88 \times 10^{-11} \text{ m} = \boxed{0.018\,8 \text{ nm}}$$

Exercise 28.5

Repeat the problem for a $2p$ electron transiting from the L shell to the K shell. (For technical reasons, the L shell electron must have $\ell = 1$, so a single $1s$ electron and two $2s$ electrons shield the nucleus.)

Answer   (a) 5.54×10^4 eV   (b) $0.022\,4$ nm

28.11   ATOMIC TRANSITIONS

We have seen that an atom will emit radiation only at certain frequencies that correspond to the energy separation between the various allowed states. Consider an atom with many allowed energy states, labeled $E_1, E_2, E_3, \ldots$, as in Figure 28.16. When light is incident on the atom, only those photons whose energy hf matches the energy separation ΔE between two levels can be absorbed by the atom. A schematic diagram representing this **stimulated absorption process** is shown in Active Figure 28.17. At ordinary temperatures, most of the atoms in a sample are in the ground state. If a vessel containing many atoms of a gas is illuminated with a light beam containing all possible photon frequencies (that is, a continuous spectrum), only those photons of energies $E_2 - E_1$, $E_3 - E_1$, $E_4 - E_1$, and so on, can be absorbed. As a result of this absorption, some atoms are raised to various allowed higher energy levels, called **excited states**.

Once an atom is in an excited state, there is a constant probability that it will jump back to a lower level by emitting a photon, as shown in Active Figure 28.18 (page 922).

Figure 28.16   Energy level diagram of an atom with various allowed states. The lowest energy state, E_1, is the ground state. All others are excited states.

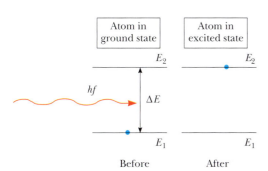

Before   After

ACTIVE FIGURE 28.17

Diagram representing the process of *stimulated absorption* of a photon by an atom. The blue dot represents an electron. The electron is transferred from the ground state to the excited state when the atom absorbs a photon of energy $hf = E_2 - E_1$.

Physics⚛Now™

Log into PhysicsNow at **www.cp7e.com** and go to Active Figure 28.17 to observe stimulated absorption.

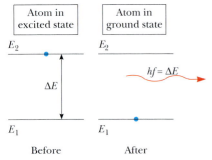

ACTIVE FIGURE 28.18
Diagram representing the process of *spontaneous emission* of a photon by an atom that is initially in the excited state E_2. When the electron falls to the ground state, the atom emits a photon of energy $hf = E_2 - E_1$.

Physics⊗Now™
Log into PhysicsNow at **www.cp7e.com** and go to Active Figure 28.18 to observe spontaneous emission.

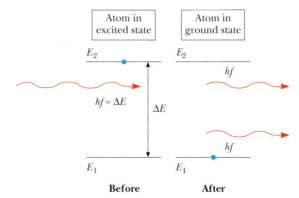

ACTIVE FIGURE 28.19
Diagram representing the process of *stimulated emission* of a photon by an incoming photon of energy *hf*. Initially, the atom is in the excited state. The incoming photon stimulates the atom to emit a second photon of energy $hf = E_2 - E_1$.

Physics⊗Now™
Log into PhysicsNow at **www.cp7e.com** and go to Active Figure 28.19 to observe stimulated emission.

This process is known as **spontaneous emission**. Typically, an atom will remain in an excited state for only about 10^{-8} s.

A third process that is important in lasers, **stimulated emission**, was predicted by Einstein in 1917. Suppose an atom is in the excited state E_2, as in Active Figure 28.19, and a photon with energy $hf = E_2 - E_1$ is incident on it. The incoming photon increases the probability that the excited atom will return to the ground state and thereby emit a second photon having the same energy *hf*. Note that two identical photons result from stimulated emission: the incident photon and the emitted photon. *The emitted photon is exactly in phase with the incident photon.* These photons can stimulate other atoms to emit photons in a chain of similar processes. The many photons produced in this fashion are the source of the intense, coherent (in-phase) light in a laser.

Applying Physics 28.4   Streaking Meteoroids

A physics student is watching a meteor shower in the early morning hours. She notices that the streaks of light from the meteoroids entering the very high regions of the atmosphere last for as long as 2 or 3 seconds before fading. She also notices a lightning storm off in the distance. The streaks of light from the lightning fade away almost immediately after the flash, certainly in much less than 1 second. Both lightning and meteors cause the air to turn into a plasma because of the very high temperatures generated. The light is given off when the stripped electrons in the plasma recombine with the ionized atoms. Why would the light last longer for meteors than for lightning?

Explanation To answer this question, we examine the phrase "the streaks of light from the meteoroids

entering the very high regions of the atmosphere." In the very high regions of the atmosphere, the pressure is very low, so the density is also very low and the atoms of the gas are relatively far apart. Low density means that after the air is ionized by the passing meteoroid, the probability of freed electrons finding an ionized atom with which to recombine is relatively low. As a result, the recombination process occurs over a relatively long time, measured in seconds. Lightning, however, occurs in the lower regions of the atmosphere (the troposphere), where the pressure and density are relatively high. After the ionization by the lightning flash, the electrons and ionized atoms are much closer together than in the upper atmosphere. The probability of a recombination is accordingly much higher, and the time for the recombination to occur is much shorter.

28.12   LASERS AND HOLOGRAPHY

We have described how an incident photon can cause atomic transitions either upward (stimulated absorption) or downward (stimulated emission). The two processes are equally probable. When light is incident on a system of atoms, there is usually a net absorption of energy, because when the system is in thermal equilibrium, there are many more atoms in the ground state than in excited states. However, if the situation can be inverted so that there are more atoms in an excited state than in the ground state, a net emission of photons can result. Such a condition is called **population inversion**. This is the fundamental principle involved in the operation of a laser, an acronym for *light amplification by stimulated emission of radiation*. The amplification corresponds to a buildup of photons in the system as the result of a chain reaction of events. The following three conditions must be satisfied in order to achieve laser action:

1. The system must be in a state of population inversion (that is, more atoms in an excited state than in the ground state).
2. The excited state of the system must be a *metastable state*, which means its lifetime must be long compared with the otherwise usually short lifetimes of excited states. When that is the case, stimulated emission will occur before spontaneous emission.
3. The emitted photons must be confined within the system long enough to allow them to stimulate further emission from other excited atoms. This is achieved by the use of reflecting mirrors at the ends of the system. One end is totally reflecting, and the other is slightly transparent to allow the laser beam to escape.

One device that exhibits stimulated emission of radiation is the helium–neon gas laser. Figure 28.20 is an energy-level diagram for the neon atom in this system. The mixture of helium and neon is confined to a glass tube sealed at the ends by mirrors. A high voltage applied to the tube causes electrons to sweep through it, colliding with the atoms of the gas and raising them into excited states. Neon atoms are excited to state E_3^* through this process and also as a result of collisions with excited helium atoms. When a neon atom makes a transition to state E_2, it stimulates emission by neighboring excited atoms. This results in the production of coherent light at a wavelength of 632.8 nm. Figure 28.21 summarizes the steps in the production of a laser beam.

Figure 28.20 Energy-level diagram for the neon atom in a helium–neon laser. The atom emits 632.8-nm photons through stimulated emission in the transition $E_3^* \rightarrow E_2$. This is the source of coherent light in the laser.

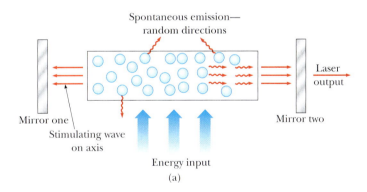

Spontaneous emission—random directions

Mirror one

Stimulating wave on axis

Energy input

Laser output

Mirror two

(a)

(b)

Courtesy of HRL Laboratories LLC, Malibu, CA

Figure 28.21 (a) Steps in the production of a laser beam. The tube contains atoms, which represent the active medium. An external source of energy (optical, electrical, etc.) is needed to "pump" the atoms to excited energy states. The parallel end mirrors provide the feedback of the stimulating wave. (b) Photograph of the first ruby laser, showing the flash lamp surrounding the ruby rod.

(a)   (b)

Figure 28.22 (a) Experimental arrangement for producing a hologram. (b) Photograph of a hologram made with a cylindrical film. Note the detail of the Volkswagen image.

Courtesy of Central Scientific Company

APPLICATION

Laser Technology

Scientist checking the performance of an experimental laser-cutting device mounted on a robot arm. The laser is being used to cut through a metal plate.

Philippe Plailly/Photo Researchers, Inc.

APPLICATION

Holography

Since the development of the first laser in 1960, laser technology has exhibited tremendous growth. Lasers that cover wavelengths in the infrared, visible, and ultraviolet regions of the spectrum are now available. Applications include the surgical "welding" of detached retinas, "lasik" surgery, precision surveying and length measurement, a potential source for inducing nuclear fusion reactions, precision cutting of metals and other materials, and telephone communication along optical fibers. These and other applications are possible because of the unique characteristics of laser light. In addition to being highly monochromatic and coherent, laser light is also highly directional and can be sharply focused to produce regions of extremely intense light energy.

Holography

One interesting application of the laser is holography: the production of three-dimensional images of objects. Figure 28.22a shows how a hologram is made. Light from the laser is split into two parts by a half-silvered mirror at B. One part of the beam reflects off the object to be photographed and strikes an ordinary photographic film. The other half of the beam is diverged by lens L_2, reflects from mirrors M_1 and M_2, and finally strikes the film. The two beams overlap to form an extremely complicated interference pattern on the film, one that can be produced only if the phase relationship of the waves is constant throughout the exposure of the film. This condition is met through the use of light from a laser, because such light is coherent. The hologram records not only the intensity of the light scattered from the object (as in a conventional photograph), but also the phase difference between the reference beam and the beam scattered from the object. Because of this phase difference, an interference pattern is formed that produces an image with full three-dimensional perspective.

A hologram is best viewed by allowing coherent light to pass through the developed film while you look back along the direction from which the beam comes. Figure 28.22b is a photograph of a hologram made using a cylindrical film.

28.13 ENERGY BANDS IN SOLIDS

In this section we trace the changes that occur in the discrete energy levels of isolated atoms when the atoms group together and form a solid. We find that in solids, the discrete levels of isolated atoms broaden into allowed energy bands

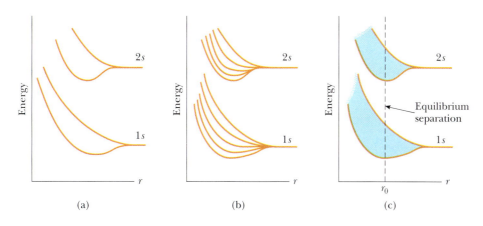

Figure 28.23 (a) Splitting of the 1s and 2s states when two atoms are brought together. (b) Splitting of the 1s and 2s states when five atoms are brought close together. (c) Formation of energy bands when a large number of sodium atoms are assembled to form a solid.

separated by forbidden gaps. The separation and electron population of the highest bands determines whether a given solid is a conductor, an insulator, or a semiconductor.

Consider two identical atoms, initially widely separated, that are brought closer and closer together. If two identical atoms are very far apart, they do not interact, and their electronic energy levels can be considered to be those of isolated atoms. Hence, the energy levels are exactly the same. As the atoms come close together, they essentially become one quantum system, and the Pauli exclusion principle demands that the electrons be in different quantum states for this single system. The exclusion principle manifests itself as a changing or splitting of electron energy levels that were identical in the widely separated atoms, as shown in Figure 28.23a. Figure 28.23b shows that with 5 atoms, each energy level in the isolated atom splits into five different, more closely spaced levels.

If we extend this argument to the large number of atoms found in solids (on the order of 10^{23} atoms/cm^3), we obtain a large number of levels so closely spaced that they may be regarded as a continuous **band** of energy levels, as in Figure 28.23c. An electron can have any energy within an allowed energy band, but cannot have an energy in the **band gap**, or the region between allowed bands. Note that the band gap energy E_g is indicated in Figure 28.23c. In practice we are only interested in the band structure of a solid at some equilibrium separation of its atoms r_0, and so we remove the distance scale on the x-axis and simply plot the allowed energy bands of a solid as a series of horizontal bands, as shown in Figure 28.24 for sodium.

Conductors and Insulators

Figure 28.24 shows that the band structure of a particular solid is quite complicated with individual atomic levels broadening by varying amounts and some levels (3s and 3p) broadening so much that they overlap. Nevertheless, it is possible to gain a qualitative understanding of whether a solid is a conductor, an insulator, or a semiconductor by considering only the structure of the upper or upper two energy bands and whether they are occupied by electrons.

Deciding whether an energy band is empty (unoccupied by electrons), partially filled, or full is carried out in basically the same way as for the energy-level population of atoms: we distribute the total number of electrons from the lowest energy levels up in a way consistent with the exclusion principle. While we omit the details of this process here, one important case is that shown in Figure 28.25a (page 926), where the highest-energy occupied band is only partially full. The other important case, where the highest occupied band is completely full, is shown in Figure 28.25b. Notice that this figure also shows that the highest filled band is called the **valence band** and the next higher empty band is called the **conduction band**. The energy band gap, which varies with the solid, is also indicated as the energy difference E_g between the top of the valence band and the bottom of the conduction band.

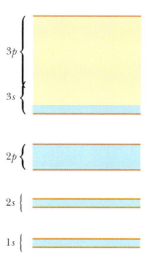

Figure 28.24 Energy bands of sodium. Note the energy gaps (white regions) between the allowed bands; electrons can't occupy states that lie in these forbidden gaps. Blue represents energy bands occupied by the sodium electrons when the atom is in its ground state. Gold represents energy bands that are empty. Note that the 3s and 3p levels broaden so much that they overlap.

Figure 28.25 (a) Half-filled band of a metal, an electrical conductor. (b) An electrical insulator at $T = 0$ K has a filled valence band and an empty conduction band. (c) Band structure of a semiconductor at ordinary temperatures ($T \approx 300$ K). The energy gap is much smaller than in an insulator, and many electrons occupy states in the conduction band.

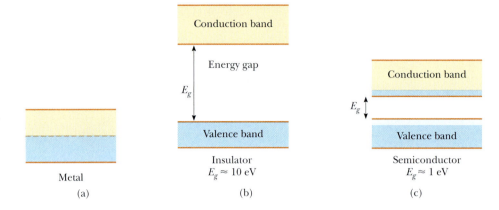

With these ideas and definitions we are now in a position to understand what determines, quantum mechanically, whether a solid will be a conductor or an insulator. When a modest voltage is applied to a good conductor, the electrons accelerate and gain energy. In quantum terms, electron energies increase *if there are higher unoccupied energy levels for electrons to jump to.* For example, electrons near the top of the partially filled band in sodium need to gain very little energy from the applied voltage to reach one of the nearby, closely spaced, empty states. Thus, it is easy for a small voltage to kick electrons into higher energy states, and charge flows easily in sodium, an excellent conductor.

Now consider the case of a material in which the highest occupied band is completely full of electrons and there is a band gap separating this filled valence band from the vacant conduction band, as in Figure 28.25b. A typical case might be diamond (carbon), in which the band gap is about 10 eV. When a voltage is applied, electrons can't easily gain energy, because there are no vacant energy states nearby to which electrons can make transitions. Because the only empty band is the conduction band, an electron must gain an amount of energy at least equal to the band gap in order for it to move through the solid. This large amount of energy can't be supplied by a modest applied voltage, so no charge flows and diamond is a good insulator. In summary then, a conductor has a highest-energy occupied band which is *partially filled*, and in an insulator, has a highest-energy occupied band which is *completely filled* with a large energy gap between the valence and conduction bands.

Semiconductors

To this point, we have completely ignored the influence of temperature on the electronic populations of energy bands. Recalling that the average thermal energy of a particle at temperature T is $3k_BT/2$, we find that an electron at room temperature has an average energy of about 0.04 eV. Because this energy is about 100 times smaller than the band gap in a typical insulator, very few electrons would have enough random thermal energy to jump the energy gap in an insulator and contribute to conduction. However things are different for a semiconductor. As we see in Figure 28.25c, a **semiconductor** is a material with a small band gap of about 1 eV whose conductivity results from appreciable thermal excitation of electrons across the gap into the conduction band at room temperature. The most commonly used semiconductors are silicon and gallium arsenide, with band gaps of 1.14 eV and 1.43 eV, respectively, at 300 K. As you might expect, the resistivity of semiconductors usually decreases with increasing temperature, because k_BT becomes a larger fraction of the band gap energy.

It is interesting that the electrons in the conduction band of a semiconductor don't carry the entire current when a voltage is applied, as Figure 28.26 shows. (It might be said that conduction electrons do not constitute the "whole" story.) The missing electrons in the valence band, shown as a narrow white band in the

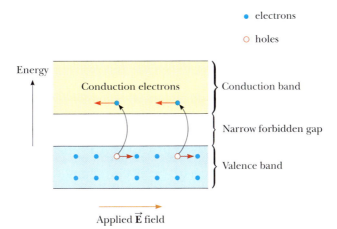

Figure 28.26 Movement of charges (holes and electrons) in a semiconductor. The electrons move in the direction opposite the direction of the external electric field, and the holes move in the direction of the field.

figure, provide a few empty states called **holes** for valence band electrons to fill; so some electrons in the valence band can gain energy and move towards a positive electrode and thus also carry the current. Since the valence band electrons that fill holes leave behind other holes, it is equally valid and more common to view the conduction process in the valence band as a flow of positive holes towards the negative electrode applied to a semiconductor. Thus, a pure semiconductor, such as silicon, can be viewed in a symmetric way: silicon has equal numbers of mobile electrons in the conduction band and holes in the valence band. Furthermore, when an external voltage is applied to the semiconductor, electrons move toward the positive electrode and holes move toward the negative electrode. In the next section we will look at the concepts of an electron and a hole in a simpler, more graphic way as the presence or absence of an outer-shell electron at a particular location in a crystal lattice.

When small amounts of impurities are added to a semiconductor such as silicon (about one impurity atom per 10^7 silicon atoms), both the band structure of the semiconductor and its resistivity are modified. The process of adding impurities, called **doping**, is important in making devices having well-defined regions of different resistivity. For example, when an atom containing five outer-shell electrons, such as arsenic, is added to a semiconductor such as silicon, four of the arsenic electrons form shared bonds with atoms of the semiconductor and one is left over. This extra electron is nearly free of its parent atom and has an energy level that lies in the energy gap, just below the conduction band. Such a pentavalent atom in effect donates an electron to the structure and hence is referred to as a **donor atom**. Because the spacing between the energy level of the electron of the donor atom and the bottom of the conduction band is very small (typically, about 0.05 eV), only a small amount of thermal energy is needed to cause this electron to move into the conduction band. (Recall that the average thermal energy of an electron at room temperature is $3k_B T/2 \approx 0.04$ eV). Semiconductors doped with donor atoms are called **n-type semiconductors**, because the charge carriers are electrons, the charge of which is *negative*.

If a semiconductor is doped with atoms containing three outer-shell electrons, such as aluminum, the three electrons form shared bonds with neighboring semiconductor atoms, leaving an electron deficiency—a hole—where the fourth bond would be if an impurity-atom electron was available to form it. The energy level of this hole lies in the energy gap, just above the valence band. An electron from the valence band has enough energy at room temperature to fill that impurity level, leaving behind a hole in the valence band. Because a trivalent atom, in effect, accepts an electron from the valence band, such impurities are referred to as **acceptor atoms**. A semiconductor doped with acceptor impurities is known as a **p-type semiconductor**, because the majority of charge carriers are *p*ositively charged holes.

28.14 SEMICONDUCTOR DEVICES

The p−n Junction

Now let us consider what happens when a *p*-semiconductor is joined to an *n*-semiconductor to form a *p−n* junction. The junction consists of the three distinct regions shown in Figure 28.27a: a *p*-region, a depletion region, and an *n*-region.

The depletion region, which extends several micrometers to either side of the center of the junction, may be visualized as arising when the two halves of the junction are brought together. Mobile donor electrons from the *n* side nearest the junction (the blue area in Fig. 28.27a) diffuse to the *p* side, leaving behind immobile positive ions. At the same time, holes from the *p* side nearest the junction diffuse to the *n* side and leave behind a region (the red area in Fig. 28.27a) of fixed negative ions. The depletion region is so named because it is depleted of mobile charge carriers.

The depletion region contains an internal electric field (arising from the charges of the fixed ions) on the order of 10^4 to 10^6 V/cm. This field sweeps mobile charge out of the depletion region and keeps it truly depleted. This internal electric field creates an internal potential difference ΔV_0 that prevents further diffusion of holes and electrons across the junction and thereby ensures zero current in the junction when no external potential difference is applied.

Perhaps the most notable feature of the *p−n* junction is its ability to pass current in only one direction. Such *diode* action is easiest to understand in terms of the potential-difference graph shown in Figure 28.27c. If an external voltage ΔV is applied to the junction such that the *p* side is connected to the positive terminal of a voltage source as in Figure 28.27a, the internal potential difference ΔV_0 across the junction is decreased, resulting in a current that increases exponentially with increasing forward voltage, or *forward bias*. In *reverse bias* (where the *n* side of the junction is connected to the positive terminal of a voltage source), the internal potential difference ΔV_0 *increases* with increasing reverse bias. This results in a very small reverse current that quickly reaches a saturation value I_0. The current–voltage relationship for an ideal diode is

$$I = I_0(e^{q\Delta V/k_B T} - 1)$$

[28.21]

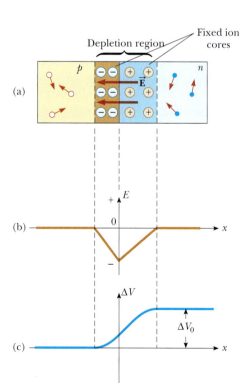

Figure 28.27 (a) Physical arrangement of a *p−n* junction. (b) Internal electric field versus *x* for the *p−n* junction. (c) Internal electric potential ΔV versus *x* for the *p−n* junction. ΔV_0 represents the potential difference across the junction in the absence of an applied electric field.

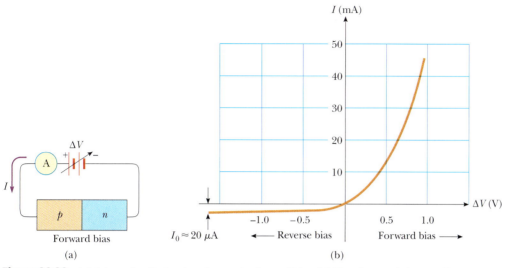

Figure 28.28 (a) Schematic of a $p-n$ junction under forward bias. (b) The characteristic curve for a real $p-n$ junction.

where q is the electron charge, k_B is Boltzmann's constant, and T is the temperature in kelvins. Figure 28.28 shows an $I-\Delta V$ plot characteristic of a real $p-n$ junction, along with a schematic of such a device under forward bias.

The most common use of the semiconductor diode is as a rectifier, a device that changes 120-V AC voltage supplied by the power company to, say the 12-V DC voltage needed by your music keyboard. We can understand how a diode rectifies a current by considering Figure 28.29a, which shows a diode connected in series with a resistor and an AC source. Because appreciable current can pass through the diode in just one direction, the alternating current in the resistor is reduced to the form shown in Figure 28.29b. The diode is said to act as a **half-wave rectifier**, because there is current in the circuit during only half of each cycle.

Figure 28.30a shows a circuit that lowers the AC voltage to 12 V with a step-down transformer and then rectifies both halves of the 12-V AC. Such a rectifier is called a **full-wave rectifier** and when combined with a step-down transformer is the most common DC power supply around the home today. A capacitor added in parallel with the load will yield an even steadier DC voltage.

The Junction Transistor

The invention of the transistor by John Bardeen (1908–1991), Walter Brattain (1902–1987), and William Shockley (1910–1989) in 1948 totally revolutionized the world of electronics. For this work, these three men shared a Nobel prize in 1956. By 1960, the transistor had replaced the vacuum tube in many electronic applications. The advent of the transistor created a multitrillion-dollar industry that produced such popular devices as pocket radios, handheld calculators, computers, television receivers, and electronic games. In this section we explain how a transistor acts as an amplifier to boost the tiny voltages and currents generated in a microphone to the ear-splitting levels required to drive a speaker.

One simple form of the transistor, called the **junction transistor**, consists of a semiconducting material in which a very narrow n region is sandwiched between two p regions. This configuration is called a *pnp* **transistor**. Another configuration is the *npn* **transistor**, which consists of a p region sandwiched between two n regions. Because the operation of the two transistors is essentially the same, we describe only the *pnp* transistor. The structure of the *pnp* transistor, together with its circuit symbol, is shown in Figure 28.31 (page 930). The outer regions are called the **emitter** and **collector**, and the narrow central region is called the **base**. The configuration contains two junctions: the emitter–base interface and the collector–base interface.

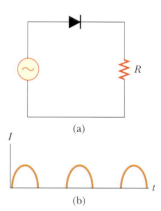

Figure 28.29 (a) A diode in series with a resistor allows current to pass in only one direction. (b) The current versus time for the circuit in (a).

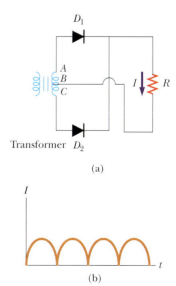

Figure 28.30 (a) A full-wave rectifier circuit. (b) The current versus time in the resistor R.

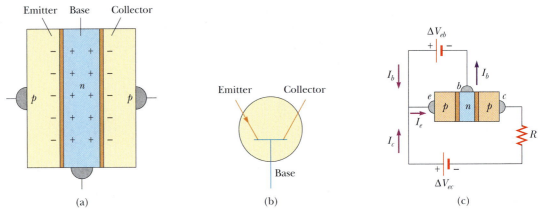

Figure 28.31 (a) The *pnp* transistor consists of an *n* region (base) sandwiched between two *p* regions (emitter and collector). (b) Circuit symbol for the *pnp* transistor. (c) A bias voltage ΔV_{eb} applied to the base as shown produces a small base current I_b that is used to control the collector current I_c in a *pnp* transistor.

Suppose a voltage is applied to the transistor so that the emitter is at a higher electric potential than the collector. (This is accomplished with the battery labeled ΔV_{ec} in Figure 28.31c.) If we think of the transistor as two diodes back to back, we see that the emitter–base junction is forward biased and the base–collector junction is reverse biased. The emitter is heavily doped relative to the base, and as a result, nearly all the current consists of holes moving across the emitter–base junction. Most of these holes do not recombine with electrons in the base because it is very narrow. Instead they are accelerated across the reverse-biased base–collector junction, producing the emitter current I_e in Figure 28.31c.

Although only a small percentage of holes recombine in the base, those that do limit the emitter current to a small value because positive charge carriers accumulating in the base prevent holes from flowing in. In order not to limit the emitter current, some of the positive charge on the base must be drawn off; this is accomplished by connecting the base to the battery labeled ΔV_{eb} in Figure 28.31c. Those positive charges that are not swept across the base–collector junction leave the base through this added pathway. **This base current I_b is very small, but a small change in it can significantly change the collector current I_c.** If the transistor is properly biased, the collector (output) current is directly proportional to the base (input) current and the transistor acts as a current amplifier. This condition may be written

$$I_c = \beta I_b$$

where β, the *current gain* factor, is typically in the range from 10 to 100. Thus, the transistor may be used to amplify a small signal. The small voltage to be amplified is placed in series with the battery V_{eb}. The input signal produces a small variation in the base current, resulting in a large change in the collector current and hence a large change in the voltage across the output resistor.

The Integrated Circuit

Invented independently by Jack Kilby (b. 1923) at Texas Instruments in late 1958 and by Robert Noyce at Fairchild Camera and Instrument in early 1959, the integrated circuit has been justly called "the most remarkable technology ever to hit mankind." Kilby's first device is shown in Figure 28.32a. Integrated circuits have indeed started a "second industrial revolution" and are found at the heart of computers, watches, cameras, automobiles, aircraft, robots, space vehicles, and all sorts of communication and switching networks.

In simplest terms, an **integrated circuit** is a collection of interconnected transistors, diodes, resistors, and capacitors fabricated on a single piece of silicon known as a chip. State-of-the-art chips easily contain several million components in

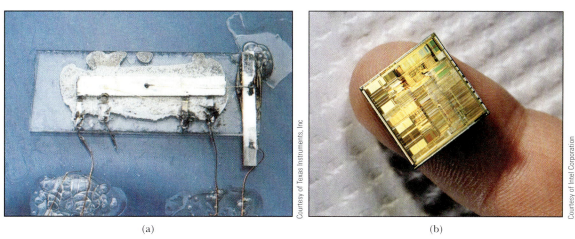

Courtesy of Texas Instruments, Inc

Courtesy of Intel Corporation

(a)   (b)

Figure 28.32   (a) Jack Kilby's first integrated circuit was tested on September 12, 1958. (b) Integrated circuits continue to shrink in size and price while simultaneously growing in capability.

a 1-cm² area, with the number of components per square inch having doubled every year since the integrated circuit was invented.

Integrated circuits were invented partly to solve the interconnection problem spawned by the transistor. In the era of vacuum tubes, power and size considerations of individual components set significant limits on the number of components that could be interconnected in a given circuit. With the advent of the tiny, low-power, highly reliable transistor, design limits on the number of components disappeared and were replaced by the problem of wiring together hundreds of thousands of components. The magnitude of this problem can be appreciated when we consider that second-generation computers (consisting of discrete transistors rather than integrated circuits) contained several hundred thousand components requiring more than a million hand-soldered joints to be made and tested.

In addition to solving the interconnection problem, integrated circuits possess the advantages of miniaturization and fast response, two attributes critical for high-speed computers. The fast response results from the miniaturization and close packing of components, because the response time of a circuit depends on the time it takes for electrical signals traveling at about the speed of light to pass from one component to another. This time is clearly reduced by packing components closely.

SUMMARY

Physics⊗Now™   Take a practice test by logging into PhysicsNow at **www.cp7e.com** and clicking on the Pre-Test link for this chapter.

28.3   The Bohr Theory of Hydrogen &

28.4   Modification of the Bohr Theory

The **Bohr model** of the atom is successful in describing the spectra of atomic hydrogen and hydrogenlike ions. One of the basic assumptions of the model is that the electron can exist only in certain orbits such that its an-

gular momentum mvr is an integral multiple of $\hbar$, where $\hbar$ is Planck's constant divided by 2π. Assuming circular orbits and a Coulomb force of attraction between electron and proton, the energies of the quantum states for hydrogen are

$$E_n = -\frac{m_e k_e^2 e^4}{2\hbar^2}\left(\frac{1}{n^2}\right) \qquad n = 1, 2, 3, \ldots \quad [28.12]$$

where k_e is the Coulomb constant, e is the charge on the electron, and n is an integer called a **quantum number**.

If the electron in the hydrogen atom jumps from an orbit having quantum number n_i to an orbit having quantum number n_f, it emits a photon of frequency f, given by

$$f = \frac{m_e k_e^2 e^4}{4\pi\hbar^3}\left(\frac{1}{n_f^2} - \frac{1}{n_i^2}\right) \qquad \textbf{[28.14]}$$

Bohr's **correspondence principle** states that quantum mechanics is in agreement with classical physics when the quantum numbers for a system are very large.

The Bohr theory can be generalized to hydrogen-like atoms, such as singly ionized helium or doubly ionized lithium. This modification consists of replacing e^2 by Ze^2 wherever it occurs.

28.6 Quantum Mechanics and the Hydrogen Atom &

28.7 The Spin Magnetic Quantum Number

One of the many successes of quantum mechanics is that the quantum numbers n, ℓ, and m_ℓ associated with atomic structure arise directly from the mathematics of the theory. The quantum number n is called the **principal quantum number**, ℓ is the **orbital quantum number**, and m_ℓ is the **orbital magnetic quantum number**. These quantum numbers can take only certain values: $1 \leq n < \infty$ in integer steps, $0 \leq \ell \leq n - 1$, and $-\ell \leq m_\ell \leq \ell$. In addition, a fourth quantum number, called the **spin magnetic quantum number** m_s, is needed to explain a fine doubling of lines in atomic spectra, with $m_s = \pm\frac{1}{2}$.

28.9 The Exclusion Principle and the Periodic Table

An understanding of the periodic table of the elements became possible when Pauli formulated the **exclusion principle**, which states that no two electrons in an atom in the same atom can have the same values for the set of quantum numbers n, ℓ, m_ℓ, and m_s. A particular set of these quantum numbers is called a quantum state. The exclusion priniciple explains how different energy levels in atoms are populated. Once one subshell is filled, the next electron goes into the vacant subshell that is lowest in energy. Atoms with similar configurations in their outermost shell have similar chemical properties and are found in the same column of the periodic table.

28.10 Characteristic X-Rays

Characteristic x-rays are produced when a bombarding electron collides with an electron in an inner shell of an atom with sufficient energy to remove the electron from the atom. The vacancy is filled when an electron from a higher level drops down into the level containing the vacancy, emitting a photon in the x-ray part of the spectrum in the process.

28.11 Atomic Transitions &

28.12 Lasers and Holography

When an atom is irradiated by light of all different wavelengths, it will only absorb only wavelengths equal to the difference in energy of two of its energy levels. This phenomenon, called **stimulated absorption**, places an atom's electrons into **excited states**. Atoms in an excited state have a probability of returning to a lower level of excitation by **spontaneous emission**. The wavelengths that can be emitted are the same as the wavelengths that can be absorbed. If an atom is in an excited state and a photon with energy $hf = E_2 - E_1$ is incident on it, the probability of emission of a second photon of this energy is greatly enhanced. The emitted photon is exactly in phase with the incident photon. This process is called **stimulated emission**. The emitted and original photon can then stimulate more emission, creating an amplifying effect.

Lasers are monochromatic, coherent light sources that work on the principle of **stimulated emission** of radiation from a system of atoms.

CONCEPTUAL QUESTIONS

1. In the hydrogen atom, the quantum number n can increase without limit. Because of this, does the frequency of possible spectral lines from hydrogen also increase without limit?

2. Does the light emitted by a neon sign constitute a continuous spectrum or only a few colors? Defend your answer.

3. In an x-ray tube, if the energy with which the electrons strike the metal target is increased, the wavelengths of the characteristic x-rays do not change. Why not?

4. Must an atom first be ionized before it can emit light? Discuss.

5. Is it possible for a spectrum from an x-ray tube to show the continuous spectrum of x-rays without the presence of the characteristic x-rays?

6. Suppose that the electron in the hydrogen atom obeyed classical mechanics rather than quantum mechanics. Why should such a hypothetical atom emit a continuous spectrum rather than the observed line spectrum?

7. When a hologram is produced, the system (including light source, object, beam splitter, and so on) must be held motionless within a quarter of the light's wavelength. Why?

8. If matter has a wave nature, why is it not observable in our daily experience?

9. Discuss some consequences of the exclusion principle.

10. Can the electron in the ground state of hydrogen absorb a photon of energy less than 13.6 eV? Can it absorb a photon of energy greater than 13.6 eV? Explain.

11. Why do lithium, potassium, and sodium exhibit similar chemical properties?

12. List some ways in which quantum mechanics altered our view of the atom pictured by the Bohr theory.

13. It is easy to understand how two electrons (one with spin up, one with spin down) can fill the 1s shell for a helium atom. How is it possible that eight more electrons can fit into the 2s, 2p level to complete the $1s2s^22p^6$ shell for a neon atom?

14. The ionization energies for Li, Na, K, Rb, and Cs are 5.390, 5.138, 4.339, 4.176, and 3.893 eV, respectively. Explain why these values are to be expected in terms of the atomic structures.

15. Why is stimulated emission so important in the operation of a laser?

16. The Bohr theory of the hydrogen atom is based upon several assumptions. Discuss these assumptions and their significance. Do any of them contradict classical physics?

17. Explain why, in the Bohr model, the total energy of the hydrogen atom is negative.

18. Consider the quantum numbers n, ℓ, m_ℓ, and m_s. (a) Which of these are integers and which are fractional? (b) Which are always positive and which can be negative? (c) If $n = 2$, what is the largest value of ℓ? (d) If $\ell = 1$, what are the possible values of m_ℓ?

19. Photon A is emitted when an electron in a hydrogen atom drops from the $n = 3$ level to the $n = 2$ level. Photon B is emitted when an electron in a hydrogen atom drops from the $n = 4$ level to the $n = 2$ level. (a) In which case is the wavelength of the emitted photon greater? (b) In which case is the energy of the emitted photon greater?

PROBLEMS

1, 2, 3 = straightforward, intermediate, challenging   ☐ = full solution available in *Student Solutions Manual/Study Guide*

Physics⊗Now™ = coached problem with hints available at **www.cp7e.com**   🔬 = biomedical application

Section 28.1 Early Models of the Atom

Section 28.2 Atomic Spectra

1. Use Equation 28.1 to calculate the wavelength of the first three lines in the Balmer series for hydrogen.

2. Show that the wavelengths for the Balmer series satisfy the equation

$$\lambda = \frac{364.5\,n^2}{n^2 - 4}\,\text{nm} \qquad \text{where } n = 3, 4, 5, \ldots$$

3. The "size" of the *atom* in Rutherford's model is about 1.0×10^{-10} m. (a) Determine the attractive electrostatic force between an electron and a proton separated by this distance. (b) Determine (in eV) the electrostatic potential energy of the atom.

4. The "size" of the *nucleus* in Rutherford's model of the atom is about 1.0 fm = 1.0×10^{-15} m. (a) Determine the repulsive electrostatic force between two protons separated by this distance. (b) Determine (in MeV) the electrostatic potential energy of the pair of protons.

5. **Physics⊗Now**™ The "size" of the atom in Rutherford's model is about 1.0×10^{-10} m. (a) Determine the speed of an electron moving about the proton using the attractive electrostatic force between an electron and a proton separated by this distance. (b) Does this speed suggest that Einsteinian relativity must be considered in studying the atom? (c) Compute the de Broglie wavelength of the electron as it moves about the proton. (d) Does this wavelength suggest that wave effects, such as diffraction and interference, must be considered in studying the atom?

6. In a Rutherford scattering experiment, an α-particle (charge $= +2e$) heads directly toward a gold nucleus (charge $= +79e$). The α-particle had a kinetic energy of 5.0 MeV when very far $(r \rightarrow \infty)$ from the nucleus. Assuming the gold nucleus to be fixed in space, determine the distance of closest approach. [*Hint*: Use conservation of energy with $PE = k_e q_1 q_2 / r$.]

Section 28.3 The Bohr Theory of Hydrogen

7. A hydrogen atom is in its first excited state ($n = 2$). Using the Bohr theory of the atom, calculate (a) the radius of the orbit, (b) the linear momentum of the electron, (c) the angular momentum of the electron, (d) the kinetic energy, (e) the potential energy, and (f) the total energy.

8. For a hydrogen atom in its ground state, use the Bohr model to compute (a) the orbital speed of the electron, (b) the kinetic energy of the electron, and (c) the electrical potential energy of the atom.

9. Show that the speed of the electron in the nth Bohr orbit in hydrogen is given by

$$v_n = \frac{k_e e^2}{n\hbar}$$

10. A photon is emitted as a hydrogen atom undergoes a transition from the $n = 6$ state to the $n = 2$ state. Calculate (a) the energy, (b) the wavelength, and (c) the frequency of the emitted photon.

11. A hydrogen atom emits a photon of wavelength 656 nm. From what energy orbit to what lower energy orbit did the electron jump?

12. Following are four possible transitions for a hydrogen atom

   I. $n_i = 2$; $n_f = 5$ II. $n_i = 5$; $n_f = 3$

   III. $n_i = 7$; $n_f = 4$ IV. $n_i = 4$; $n_f = 7$

(a) Which transition will emit the shortest-wavelength photon? (b) For which transition will the atom gain the most energy? (c) For which transition(s) does the atom lose energy?

13. What is the energy of a photon that, when absorbed by a hydrogen atom, could cause (a) an electronic transition from the $n = 3$ state to the $n = 5$ state and (b) an electronic transition from the $n = 5$ state to the $n = 7$ state?

14. A hydrogen atom initially in its ground state ($n = 1$) absorbs a photon and ends up in the state for

which $n = 3$. (a) What is the energy of the absorbed photon? (b) If the atom eventually returns to the ground state, what photon energies could the atom emit?

15. Determine both the longest and the shortest wavelengths in (a) the Lyman series ($n_f = 1$) and (b) the Paschen series ($n_f = 3$) of hydrogen.

16. Show that the speed of the electron in the first (ground-state) Bohr orbit of the hydrogen atom may be expressed as

$$v = (1/137)\,c.$$

17. A monochromatic beam of light is absorbed by a collection of ground-state hydrogen atoms in such a way that six different wavelengths are observed when the hydrogen relaxes back to the ground state. What is the wavelength of the incident beam?

18. A particle of charge q and mass m, moving with a constant speed v, perpendicular to a constant magnetic field, B, follows a circular path. If in this case the angular momentum about the center of this circle is quantized so that $mvr = 2n\hbar$, show that the allowed radii for the particle are

$$r_n = \sqrt{\frac{2\,n\hbar}{qB}}$$

where $n = 1, 2, 3, \ldots$

19. **Physics⊗Now™** (a) If an electron makes a transition from the $n = 4$ Bohr orbit to the $n = 2$ orbit, determine the wavelength of the photon created in the process. (b) Assuming that the atom was initially at rest, determine the recoil speed of the hydrogen atom when this photon is emitted.

20. Consider a large number of hydrogen atoms, with electrons all initially in the $n = 4$ state. (a) How many different wavelengths would be observed in the emission spectrum of these atoms? (b) What is the longest wavelength that could be observed? To which series does it belong?

21. Analyze the Earth–Sun system by following the Bohr model, where the gravitational force between Earth (mass m) and Sun (mass M) replaces the Coulomb force between the electron and proton (so that $F = GMm/r^2$ and $PE = -GMm/r$). Show that (a) the total energy of the Earth in an orbit of radius r is given by (a) $E = -GMm/2r$, (b) the radius of the nth orbit is given by $r_n = r_0 n^2$, where $r_0 = \hbar^2/GMm^2 = 2.32 \times 10^{-138}$ m, and (c) the energy of the nth orbit is given by $E_n = -E_0/n^2$, where $E_0 = G^2 M^2 m^3 / 2\hbar^2 = 1.71 \times 10^{182}$ J. (d) Using the Earth–Sun orbit radius of $r = 1.49 \times 10^{11}$ m, determine the value of the quantum number n. (e) Should you expect to observe quantum effects in the Earth–Sun system?

22. An electron is in the nth Bohr orbit of the hydrogen atom. (a) Show that the period of the electron is $T = t_o n^3$, and determine the numerical value of t_o. (b) On the average, an electron remains in the $n = 2$ orbit for about 10 μs before it jumps down to the $n = 1$ (ground-state) orbit. How many revolutions does the electron make before it jumps to the ground state? (c) If one revolution of the electron is defined as an "electron year" (analogous to an Earth year being one revolution of the Earth around the Sun), does the electron in the $n = 2$ orbit "live" very long? Explain. (d) How does the above calculation support the "electron cloud" concept?

23. Consider a hydrogen atom. (a) Calculate the frequency f of the $n = 2 \rightarrow n = 1$ transition, and compare it with the frequency f_{orb} of the electron orbital motion in the $n = 2$ state. (b) Make the same calculation for the $n = 10\ 000 \rightarrow n = 9\ 999$ transition. Comment on the results.

24. Two hydrogen atoms collide head-on and end up with zero kinetic energy. Each then emits a 121.6-nm photon ($n = 2$ to $n = 1$ transition). At what speed were the atoms moving before the collision?

25. Two hydrogen atoms, both initially in the ground state, undergo a head-on collision. If both atoms are to be excited to the $n = 2$ level in this collision, what is the minimum speed each atom can have before the collision?

26. (a) Calculate the angular momentum of the Moon due to its orbital motion about the Earth. In your calculation, use 3.84×10^8 m as the average Earth–Moon distance and 2.36×10^6 s as the period of the Moon in its orbit. (b) If the angular momentum of the moon obeys Bohr's quantization rule ($L = n\hbar$), determine the value of the quantum number n. (c) By what fraction would the Earth–Moon radius have to be increased to increase the quantum number by 1?

Section 28.4 Modification of the Bohr Theory

Section 28.5 De Broglie Waves and the Hydrogen Atom

27. (a) Find the energy of the electron in the ground state of doubly ionized lithium, which has an atomic number $Z = 3$. (b) Find the radius of its ground-state orbit.

28. (a) Construct an energy level diagram for the He^+ ion, for which $Z = 2$. (b) What is the ionization energy for He^+?

29. The orbital radii of a hydrogen-like atom is given by the equation

$$r = \frac{n^2 \hbar^2}{Z m_e k_e e^2}.$$

What is the radius of the first Bohr orbit in (a) He^+, (b) Li^{2+}, and (c) Be^{3+}?

30. (a) Substitute numerical values into Equation 28.19 to find a value for the Rydberg constant for singly ionized helium, He^+. (b) Use the result of part (a) to find the wavelength associated with a transition from the $n = 2$ state to the $n = 1$ state of He^+. (c) Identify the region of the electromagnetic spectrum associated with this transition.

31. **Physics⊗Now**™ Determine the wavelength of an electron in the third excited orbit of the hydrogen atom, with $n = 4$.

32. Using the concept of standing waves, de Broglie was able to derive Bohr's stationary orbit postulate. He assumed that a confined electron could exist only in states where its de Broglie waves form standing-wave patterns, as in Figure 28.10a. Consider a particle confined in a box of length L to be equivalent to a string of length L and fixed at both ends. Apply de Broglie's concept to show that (a) the linear momentum of this particle is quantized with $p = mv = nh/2L$ and (b) the allowed states correspond to particle energies of $E_n = n^2 E_0$, where $E_0 = h^2/(8mL^2)$.

Section 28.6 Quantum Mechanics and the Hydrogen Atom

Section 28.7 The Spin Magnetic Quantum Number

33. List the possible sets of quantum numbers for electrons in the $3p$ subshell.

34. When the principal quantum number is $n = 4$, how many different values of (a) ℓ and (b) m_ℓ are possible?

35. The ρ-meson has a charge of $-e$, a spin quantum number of 1, and a mass 1 507 times that of the electron. If the electrons in atoms were replaced by ρ-mesons, list the possible sets of quantum numbers for ρ-mesons in the $3d$ subshell.

Section 28.9 The Exclusion Principle and the Periodic Table

36. (a) Write out the electronic configuration of the ground state for oxygen ($Z = 8$). (b) Write out the values for the set of quantum numbers n, ℓ, m_ℓ, and m_s for each of the electrons in oxygen.

37. Two electrons in the same atom have $n = 3$ and $\ell = 1$. (a) List the quantum numbers for the possible states of the atom. (b) How many states would be possible if the exclusion principle did not apply to the atom?

38. How many different sets of quantum numbers are possible for an electron for which (a) $n = 1$, (b) $n = 2$, (c) $n = 3$, (d) $n = 4$, and (e) $n = 5$? Check your results to show that they agree with the general rule that the number of different sets of quantum numbers is equal to $2n^2$.

39. Zirconium ($Z = 40$) has two electrons in an incomplete d subshell. (a) What are the values of n and ℓ for each electron? (b) What are all possible values of m_ℓ and m_s? (c) What is the electron configuration in the ground state of zirconium?

Section 28.10 Characteristic X-Rays

40. The K-shell ionization energy of copper is 8 979 eV. The L-shell ionization energy is 951 eV. Determine the wavelength of the K_α emission line of copper. What must the minimum voltage be on an x-ray tube with a copper target in order to see the K_α line?

41. The K_α x-ray is emitted when an electron undergoes a transition from the L shell ($n = 2$) to the K shell ($n = 1$). Use the method illustrated in Example 28.5 to calculate the wavelength of the K_α x-ray from a nickel target ($Z = 28$).

42. When an electron drops from the M shell ($n = 3$) to a vacancy in the K shell ($n = 1$), the measured wavelength of the emitted x-ray is found to be 0.101 nm. Identify the element.

43. The K series of the discrete spectrum of tungsten contains wavelengths of 0.018 5 nm, 0.020 9 nm, and 0.021 5 nm. The K-shell ionization energy is 69.5 keV. Determine the ionization energies of the L, M, and N shells. Sketch the transitions that produce the above wavelengths.

ADDITIONAL PROBLEMS

44. In a hydrogen atom, what is the principle quantum number of the electron orbit with a radius closest to 1.0 μm?

45. (a) How much energy is required to cause an electron in hydrogen to move from the $n = 1$ state to the $n = 2$ state? (b) If the electrons gain this energy by collision between hydrogen atoms in a high-temperature gas, find the minimum temperature of the heated hydrogen gas. The thermal energy of the heated atoms is given by $3k_B T/2$, where k_B is the Boltzmann constant.

46. A pulsed ruby laser emits light at 694.3 nm. For a 14.0-ps pulse containing 3.00 J of energy, find (a) the physical length of the pulse as it travels through space and (b) the number of photons in it. (c) If the beam has a circular cross section 0.600 cm in diameter, find the number of photons per cubic millimeter.

47. The Lyman series for a (new?) one-electron atom is observed in a distant galaxy. The wavelengths of the first four lines and the short-wavelength limit of this Lyman series are given by the energy-level diagram in Figure P28.47. Based on this information, calculate (a) the energies of the ground state and first four excited states for this one-electron atom and (b) the longest-wavelength (alpha) lines and the short-wavelength series limit in the Balmer series for this atom.

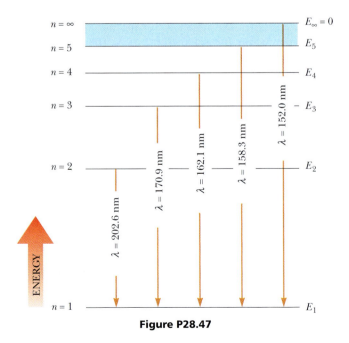

Figure P28.47

48. A dimensionless number that often appears in atomic physics is the fine-structure constant $\alpha = k_e e^2 / \hbar c$, where k_e is the Coulomb constant. (a) Obtain a numerical value for $1/\alpha$. (b) In terms of α, what is the ratio of the Bohr radius a_0 to the Compton wavelength $\lambda_C = h/m_e c$? (d) In terms of α, what is the ratio of the reciprocal of the Rydberg constant $1/R_H$ to the Bohr radius?

49. Mercury's ionization energy is 10.39 eV. The three longest wavelengths of the absorption spectrum of mercury are 253.7 nm, 185.0 nm, and 158.5 nm. (a) Construct an energy-level diagram for mercury. (b) Indicate all emission lines that can occur when an electron is raised to the third level above the ground state. (c) Disregarding recoil of the mercury atom, determine the minimum speed an electron must have in order to make an inelastic collision with a mercury atom in its ground state.

50. Suppose the ionization energy of an atom is 4.100 eV. In this same atom, we observe emission lines that have wavelengths of 310.0 nm, 400.0 nm, and 1 378 nm. Use this information to construct the energy-level diagram with the least number of levels. Assume the higher energy levels are closer together.

51. **Physics⊗Now™** A laser used in eye surgery emits a 3.00-mJ pulse in 1.00 ns, focused to a spot 30.0 μm in diameter on the retina. (a) Find (in SI units) the power per unit area at the retina. (This quantity is called the *irradiance*.) (b) What energy is delivered per pulse to an area of molecular size—say, a circular area 0.600 nm in diameter.

52. An electron has a de Broglie wavelength equal to the diameter of a hydrogen atom in its ground state. (a) What is the kinetic energy of the electron? (b) How does this energy compare with the ground-state energy of the hydrogen atom?

53. Use Bohr's model of the hydrogen atom to show that, when the atom makes a transition from the state n to the state $n - 1$, the frequency of the emitted light is given by

$$f = \frac{2\pi^2 m k_e^2 e^4}{h^3}\left(\frac{2n - 1}{(n - 1)^2 n^2}\right)$$

54. Calculate the classical frequency for the light emitted by an atom. To do so, note that the frequency of revolution is $v/2\pi r$, where r is the Bohr radius. Show that as n approaches infinity in the equation of the preceding problem, the expression given there varies as $1/n^3$ and reduces to the classical frequency. (This is an example of the correspondence principle, which requires that the classical and quantum models agree for large values of n.)

55. A pi meson (π^-) of charge $-e$ and mass 273 times greater than that of the electron is captured by a helium nucleus ($Z = +2$) as shown in Figure P28.55. (a) Draw an energy-level diagram (in units of eV) for this "Bohr-type" atom up to the first six energy levels. (b) When the π-meson makes a transition between two orbits, a photon is emitted that Compton scatters off a free electron initially at rest, producing a scattered photon of wavelength $\lambda' = 0.089\ 929\ 3$ nm at an angle of $\theta = 42.68°$, as shown on the right-hand side of Figure P28.55. Between which two orbits did the π-meson make a transition?

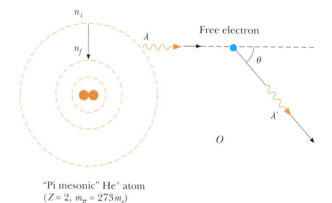

"Pi mesonic" He$^+$ atom
($Z = 2$, $m_\pi = 273 m_e$)

Figure P28.55

56. When a muon with charge $-e$ is captured by a proton, the resulting bound system forms a "muonic atom," which is the same as hydrogen, except with a muon (of mass 207 times the mass of an electron) replacing the electron. For this "muonic atom," determine (a) the Bohr radius and (b) the three lowest energy levels.

57. In this problem, you will estimate the classical lifetime of the hydrogen atom. An accelerating charge loses electromagnetic energy at a rate given by $\mathscr{P} = -2k_e q^2 a^2/(3c^3)$, where k_e is the Coulomb constant, q is the charge of the particle, a is its acceleration, and c is the speed of light in a vacuum. Assume that the electron is one Bohr radius (0.052 9 nm) from the center of the hydrogen atom. (a) Determine its acceleration. (b) Show that $\mathscr{P}$ has units of energy per unit time and determine the rate of energy loss. (c) Calculate the kinetic energy of the electron and determine how long it will take for all of this energy to be converted into electromagnetic waves, assuming that the rate calculated in part (b) remains constant throughout the electron's motion.

58. An electron in a hydrogen atom jumps from some initial Bohr orbit n_i to some final Bohr orbit n_f, as in

Figure P28.58. (a) If the photon emitted in the process is capable of ejecting a photoelectron from tungsten (work function = 4.58 eV), determine n_f. (b) If a minimum stopping potential of $V_0 = 7.51$ volts is required to prevent the photoelectron from hitting the anode, determine the value of n_i.

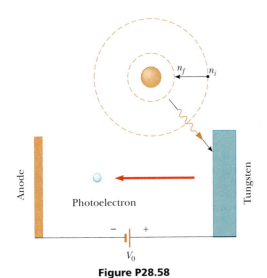

Figure P28.58

ACTIVITIES

1. With your partner not looking, use modeling clay to build one or more mounds on top of a table. Place a piece of cardboard over your mound(s), and assign your partner the task of determining the size, shape, and number of mounds without looking. He is to do this by rolling marbles at the unseen mounds and observing how they emerge. This experiment models the Rutherford scattering experiment.

2. Your instructor can probably lend you a small plastic diffraction grating to enable you to examine the spectrum of different light sources. You can use these gratings to examine a source by holding the grating very close to your eye and noting the spectrum produced by glancing out of the corner of your eye while looking at a light source. You should look at light sources such as sodium vapor lights and mercury vapor lights used in many parking lots, neon lights used in many signs, black lights, ordinary incandescent light bulbs, and so forth.

Aerial view of a nuclear power plant that generates electrical power. Energy is generated in such plants from the process of nuclear fission, in which a heavy nucleus such as ^{235}U splits into smaller particles having a large amount of kinetic energy. This surplus energy can be used to heat water into high pressure steam and drive a turbine.

Nuclear Physics

In 1896, the year that marks the birth of nuclear physics, Henri Becquerel (1852–1908) discovered radioactivity in uranium compounds. A great deal of activity followed this discovery as researchers attempted to understand and characterize the radiation that we now know to be emitted by radioactive nuclei. Pioneering work by Rutherford showed that the radiation was of three types, which he called *alpha, beta,* and *gamma rays*. These types are classified according to the nature of their electric charge and their ability to penetrate matter. Later experiments showed that alpha rays are helium nuclei, beta rays are electrons, and gamma rays are high-energy photons.

In 1911 Rutherford and his students Geiger and Marsden performed a number of important scattering experiments involving alpha particles. These experiments established the idea that the nucleus of an atom can be regarded as essentially a point mass and point charge and that most of the atomic mass is contained in the nucleus. Further, such studies demonstrated a wholly new type of force: the *nuclear force*, which is predominant at distances of less than about 10^{-14} m and drops quickly to zero at greater distances.

Other milestones in the development of nuclear physics include

- the first observations of nuclear reactions by Rutherford and coworkers in 1919, in which naturally occurring α particles bombarded nitrogen nuclei to produce oxygen,
- the first use of artificially accelerated protons to produce nuclear reactions, by Cockcroft and Walton in 1932,
- the discovery of the neutron by Chadwick in 1932,
- the discovery of artificial radioactivity by Joliot and Irene Curie in 1933,
- the discovery of nuclear fission by Hahn, Strassman, Meitner, and Frisch in 1938, and
- the development of the first controlled fission reactor by Fermi and his collaborators in 1942.

In this chapter we discuss the properties and structure of the atomic nucleus. We start by describing the basic properties of nuclei and follow with a discussion of the phenomenon of radioactivity. Finally, we explore nuclear reactions and the various processes by which nuclei decay.

ERNEST RUTHERFORD,
New Zealand Physicist
(1871–1937)

Rutherford was awarded the Nobel Prize in 1908 for discovering that atoms can be broken apart by alpha rays and for studying radioactivity. "On consideration, I realized that this scattering backward must be the result of a single collision, and when I made calculations I saw that it was impossible to get anything of that order of magnitude unless you took a system in which the greater part of the mass of the atom was concentrated in a minute nucleus. It was then that I had the idea of an atom with a minute massive center carrying a charge."

Definition of the unified mass unit u ▶

TIP 29.1   **Mass Number is not the Atomic Mass**

Don't confuse the mass number A with the atomic mass. Mass number is an integer that specifies an isotope and has no units—it's simply equal to the number of nucleons. Atomic mass is an average of the masses of the isotopes of a given element and has units of u.

29.1   SOME PROPERTIES OF NUCLEI

All nuclei are composed of two types of particles: protons and neutrons. The only exception is the ordinary hydrogen nucleus, which is a single proton. In describing some of the properties of nuclei, such as their charge, mass, and radius, we make use of the following quantities:

- the **atomic number** Z, which equals the number of protons in the nucleus,
- the **neutron number** N, which equals the number of neutrons in the nucleus,
- the **mass number** A, which equals the number of nucleons in the nucleus (*nucleon* is a generic term used to refer to either a proton or a neutron).

The symbol we use to represent nuclei is $^A_Z X$, where X represents the chemical symbol for the element. For example, $^{27}_{13} Al$ has the mass number 27 and the atomic number 13; therefore, it contains 13 protons and 14 neutrons. When no confusion is likely to arise, we often omit the subscript Z, because the chemical symbol can always be used to determine Z.

The nuclei of all atoms of a particular element must contain the same number of protons, but they may contain different numbers of neutrons. Nuclei that are related in this way are called **isotopes**. **The isotopes of an element have the same Z value, but different N and A values.** The natural abundances of isotopes can differ substantially. For example, $^{11}_6 C$, $^{12}_6 C$, $^{13}_6 C$, and $^{14}_6 C$ are four isotopes of carbon. The natural abundance of the $^{12}_6 C$ isotope is about 98.9%, whereas that of the $^{13}_6 C$ isotope is only about 1.1%. Some isotopes don't occur naturally, but can be produced in the laboratory through nuclear reactions. Even the simplest element, hydrogen, has isotopes: $^1_1 H$, hydrogen; $^2_1 H$, deuterium; and $^3_1 H$, tritium.

Charge and Mass

The proton carries a single positive charge $+e = 1.602\ 177\ 33 \times 10^{-19}$ C, the electron carries a single negative charge $-e$, and the neutron is electrically neutral. Because the neutron has no charge, it's difficult to detect. The proton is about 1 836 times as massive as the electron, and the masses of the proton and the neutron are almost equal (Table 29.1).

For atomic masses, it is convenient to define the **unified mass unit** u in such a way that the mass of one atom of the isotope ^{12}C is exactly 12 u, where $1\ u = 1.660\ 559 \times 10^{-27}$ kg. The proton and neutron each have a mass of about 1 u, and the electron has a mass that is only a small fraction of an atomic mass unit.

Because the rest energy of a particle is given by $E_R = mc^2$, it is often convenient to express the particle's mass in terms of its energy equivalent. For one atomic mass unit, we have an energy equivalent of

$$E_R = mc^2 = (1.660\ 559 \times 10^{-27}\ \text{kg})(2.997\ 92 \times 10^8\ \text{m/s})^2$$
$$= 1.492\ 431 \times 10^{-10}\ \text{J} = 931.494\ \text{MeV}$$

In calculations, nuclear physicists often express *mass* in terms of the unit MeV/c^2, where

$$1\ u = 931.494\ \text{MeV}/c^2$$

TABLE 29.1

Masses of the Proton, Neutron, and Electron in Various Units

Particle	Mass		
	kg	u	MeV/c^2
Proton	1.6726×10^{-27}	1.007 276	938.28
Neutron	1.6750×10^{-27}	1.008 665	939.57
Electron	9.109×10^{-31}	5.486×10^{-4}	0.511

The Size of Nuclei

The size and structure of nuclei were first investigated in the scattering experiments of Rutherford, discussed in Section 28.1. Using the principle of conservation of energy, Rutherford found an expression for how close an alpha particle moving directly toward the nucleus can come to the nucleus before being turned around by Coulomb repulsion.

In such a head-on collision, the kinetic energy of the incoming alpha particle must be converted completely to electrical potential energy when the particle stops at the point of closest approach and turns around (Active Fig. 29.1). If we equate the initial kinetic energy of the alpha particle to the maximum electrical potential energy of the system (alpha particle plus target nucleus), we have

$$\frac{1}{2}mv^2 = k_e\frac{q_1 q_2}{r} = k_e\frac{(2e)(Ze)}{d}$$

where d is the distance of closest approach. Solving for d, we get

$$d = \frac{4k_e Ze^2}{mv^2}$$

From this expression, Rutherford found that alpha particles approached to within 3.2×10^{-14} m of a nucleus when the foil was made of gold. Thus, the radius of the gold nucleus must be less than this value. For silver atoms, the distance of closest approach was 2×10^{-14} m. From these results, Rutherford concluded that the positive charge in an atom is concentrated in a small sphere, which he called the nucleus, with radius no greater than about 10^{-14} m. Because such small lengths are common in nuclear physics, a convenient unit of length is the *femtometer* (fm), sometimes called the **fermi** and defined as

$$1 \text{ fm} \equiv 10^{-15} \text{ m}$$

Since the time of Rutherford's scattering experiments, a multitude of other experiments have shown that most nuclei are approximately spherical and have an average radius given by

$$r = r_0 A^{1/3} \qquad \text{[29.1]}$$

where A is the total number of nucleons and r_0 is a constant equal to 1.2×10^{-15} m. Because the volume of a sphere is proportional to the cube of its radius, it follows from Equation 29.1 that the volume of a nucleus (assumed to be spherical) is directly proportional to A, the total number of nucleons. This relationship then suggests that **all nuclei have nearly the same density**. Nucleons combine to form a nucleus *as though* they were tightly packed spheres (Fig. 29.2).

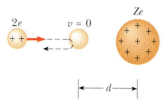

ACTIVE FIGURE 29.1
An alpha particle on a head-on collision course with a nucleus of charge Ze. Because of the Coulomb repulsion between the like charges, the alpha particle will stop instantaneously at a distance d from the nucleus, called the distance of closest approach.

Physics⊗Now™
Log into PhysicsNow at **www.cp7e.com** and go to Active Figure 29.1, where you can adjust the atomic number of the target nucleus and the kinetic energy of the alpha particle. Then observe the approach of the alpha particle toward the nucleus.

Figure 29.2 A nucleus can be visualized as a cluster of tightly packed spheres, each of which is a nucleon.

EXAMPLE 29.1   Sizing a Neutron Star

Goal   Apply the concepts of nuclear size.

Problem   One of the end stages of stellar life is a neutron star, where matter collapses and electrons combine with protons to form neutrons. Some liken neutron stars to a single gigantic nucleus. **(a)** Approximately how many nucleons are in a neutron star with a mass of 3.00×10^{30} kg? (This is the mass number of the star.) **(b)** Calculate the radius of the star, treating it as a giant nucleus. **(c)** Calculate the density of the star, assuming the mass is distributed uniformly.

Strategy   The effective mass number of the neutron star can be found by dividing the star mass in kg by the mass of a neutron. Equation 29.1 then gives an estimate of the radius of the star, which together with the mass determines the density.

Solution
(a) Find the approximate number of nucleons in the star.

Divide the star's mass by the mass of a neutron to find A: $A = \left(\dfrac{3.00 \times 10^{30} \text{ kg}}{1.675 \times 10^{-27} \text{ kg}}\right) = 1.79 \times 10^{57}$

(b) Calculate the radius of the star, treating it as a giant atomic nucleus.

Substitute into Equation 29.1:

$$r = r_0 A^{1/3} = (1.2 \times 10^{-15} \text{ m})(1.79 \times 10^{57})^{1/3}$$
$$= 1.46 \times 10^4 \text{ m}$$

(c) Calculate the density of the star, assuming that its mass is distributed uniformly.

Substitute values into the equation for density and assume the star is a uniform sphere:

$$\rho = \frac{m}{V} = \frac{m}{\frac{4}{3}\pi r^3} = \frac{3.00 \times 10^{30} \text{ kg}}{\frac{4}{3}\pi(1.46 \times 10^4 \text{ m})^3}$$
$$= 2.30 \times 10^{17} \text{ kg/m}^3$$

Remarks   This density is typical of atomic nuclei as well as of neutron stars. A ball of neutron star matter having a radius of only 1 meter would have a powerful gravity field: it could attract objects a kilometer away at an acceleration of over 50 m/s^2!

Exercise 29.1
Estimate the radius of a uranium-235 nucleus.

Answer 7.41×10^{-15} m

MARIA GOEPPERT-MAYER,
German Physicist (1906–1972)

Goeppert-Mayer was born and educated in Germany. She is best known for her development of the shell model of the nucleus, published in 1950. A similar model was simultaneously developed by Hans Jensen, a German scientist. Maria Goeppert-Mayer and Hans Jensen were awarded the Nobel Prize in physics in 1963 for their extraordinary work in understanding the structure of the nucleus.

Courtesy of Louise Barker/AIP Niels Bohr Library

Nuclear Stability

Given that the nucleus consists of a closely packed collection of protons and neutrons, you might be surprised that it can even exist. The very large repulsive electrostatic forces between protons should cause the nucleus to fly apart. However, nuclei are stable because of the presence of another, short-range (about 2 fm) force: the **nuclear force**, an attractive force that acts between all nuclear particles. The protons attract each other via the nuclear force, and at the same time they repel each other through the Coulomb force. The attractive nuclear force also acts between pairs of neutrons and between neutrons and protons.

The nuclear attractive force is stronger than the Coulomb repulsive force within the nucleus (at short ranges). If this were not the case, stable nuclei would not exist. Moreover, the strong nuclear force is nearly independent of charge. In other words, the nuclear forces associated with proton–proton, proton–neutron, and neutron–neutron interactions are approximately the same, apart from the additional repulsive Coulomb force for the proton–proton interaction.

There are about 260 stable nuclei; hundreds of others have been observed, but are unstable. A plot of N versus Z for a number of stable nuclei is given in Figure 29.3. Note that light nuclei are most stable if they contain equal numbers of protons and neutrons, so that $N = Z$, but heavy nuclei are more stable if $N > Z$. This difference can be partially understood by recognizing that as the number of protons increases, the strength of the Coulomb force increases, which tends to break the nucleus apart. As a result, more neutrons are needed to keep the nucleus stable, because neutrons are affected only by the attractive nuclear forces. In effect, the additional neutrons "dilute" the nuclear charge density. Eventually, when $Z = 83$, the repulsive forces between protons cannot be compensated for by the addition of neutrons. Elements that contain more than 83 protons don't have stable nuclei, but decay or disintegrate into other particles in various amounts of time. The masses and some other properties of selected isotopes are provided in Appendix B.

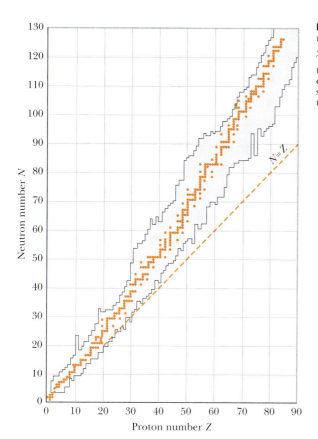

Figure 29.3   A plot of the neutron number N versus the proton number Z for the stable nuclei (solid points). The dashed straight line corresponds to the condition $N = Z$. They are centered on the so-called line of stability. The shaded area shows radioactive (unstable) nuclei.

29.2   BINDING ENERGY

The total mass of a nucleus is always less than the sum of the masses of its nucleons. Also, because mass is another manifestation of energy, **the total energy of the bound system (the nucleus) is less than the combined energy of the separated nucleons**. This difference in energy is called the **binding energy** of the nucleus and can be thought of as the energy that must be added to a nucleus to break it apart into its separated neutrons and protons.

EXAMPLE 29.2   The Binding Energy of the Deuteron

Goal   Calculate the binding energy of a nucleus.

Problem   The nucleus of the deuterium atom, called the deuteron, consists of a proton and a neutron. Calculate the deuteron's binding energy in MeV, given that its atomic mass—that is, *the mass of a deuterium nucleus plus an electron*—is 2.014 102 u.

Strategy   Calculate the sum of the masses of the individual particles and subtract the mass of the combined particle. The masses of the neutral atoms can be used instead of the nuclei because the electron masses cancel. Use the values from Table 29.4 or Table B of the appendix. The mass of an atom given in Appendix B includes the mass of Z electrons, where Z is the atom's atomic number.

Solution

To find the binding energy, first sum the masses of the hydrogen atom and neutron and subtract the mass of the deuteron:

$$\Delta m = (m_p + m_n) - m_d$$
$$= (1.007\ 825\ \text{u} + 1.008\ 665\ \text{u}) - 2.014\ 102\ \text{u}$$
$$= 0.002\ 388\ \text{u}$$

Convert this mass difference to its equivalent in MeV:

$$E_b = (0.002\ 388\ \text{u})\frac{931.5\ \text{MeV}}{1\ \text{u}} = \boxed{2.224\ \text{MeV}}$$

Remarks   This result tells us that to separate a deuteron into a proton and a neutron, it's necessary to add 2.224 MeV of energy to the deuteron to overcome the attractive nuclear force between the proton and the neutron. One way of supplying the deuteron with this energy is by bombarding it with energetic particles.

If the binding energy of a nucleus were zero, the nucleus would separate into its constituent protons and neutrons without the addition of any energy; that is, it would spontaneously break apart.

Exercise 29.2

Calculate the binding energy of $^3_2$He.

Answer   7.718 MeV

It's interesting to examine a plot of binding energy per nucleon, E_b/A, as a function of mass number for various stable nuclei (Fig. 29.4). Except for the lighter nuclei, the average binding energy per nucleon is about 8 MeV. Note that the curve peaks in the vicinity of $A = 60$, which means that nuclei with mass numbers greater or less than 60 are not as strongly bound as those near the middle of the periodic table. As we'll see later, this fact allows energy to be released in fission and fusion reactions. The curve is slowly varying for $A > 40$, which suggests that the nuclear force saturates. In other words, a particular nucleon can interact with only a limited number of other nucleons, which can be viewed as the "nearest neighbors" in the close-packed structure illustrated in Figure 29.2.

Applying Physics 29.1   Binding Nucleons and Electrons

Figure 29.4 shows a graph of the amount of energy required to remove a nucleon from the nucleus. The figure indicates that an approximately constant amount of energy is necessary to remove a nucleon above $A = 40$, whereas we saw in Chapter 28 that widely varying amounts of energy are required to remove an electron from the atom. What accounts for this difference?

Explanation In the case of Figure 29.4, the approximately constant value of the nuclear binding energy is a result of the short-range nature of the nuclear force. A given nucleon interacts only with its few nearest neighbors, rather than with all of the nucleons in the nucleus. Thus, no matter how many nucleons are present in the nucleus, pulling any one nucleon out involves separating it only from its nearest neighbors. The energy to do this, therefore, is approximately independent of how many nucleons are present. For the clearest comparison with the electron, think of averaging the energies required to strip all of the electrons out of a particular atom, from the outermost valence electron to the innermost K-shell electron. This average increases steeply with increasing atomic number. The electrical force binding the electrons to the nucleus in an atom is a long-range force. An electron in an atom interacts with all the protons in the nucleus. When the nuclear charge increases, there is a stronger attraction between the nucleus and the electrons. Therefore, as the nuclear charge increases, more energy is necessary to remove an average electron.

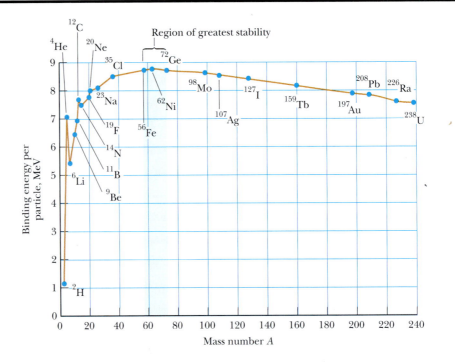

Figure 29.4 Binding energy per nucleon versus the mass number A for nuclei that are along the line of stability shown in Figure 29.3. Some representative nuclei appear as blue dots with labels. (Nuclei to the right of ^{208}Pb are unstable. The curve represents the binding energy for the most stable isotopes.)

29.3 RADIOACTIVITY

In 1896, Becquerel accidentally discovered that uranium salt crystals emit an invisible radiation that can darken a photographic plate even if the plate is covered to exclude light. After several such observations under controlled conditions, he concluded that the radiation emitted by the crystals was of a new type, one requiring no external stimulation. This spontaneous emission of radiation was soon called **radioactivity**. Subsequent experiments by other scientists showed that other substances were also radioactive.

The most significant investigations of this type were conducted by Marie and Pierre Curie. After several years of careful and laborious chemical separation processes on tons of pitchblende, a radioactive ore, the Curies reported the discovery of two previously unknown elements, both of which were radioactive. These were named polonium and radium. Subsequent experiments, including Rutherford's famous work on alpha-particle scattering, suggested that radioactivity was the result of the decay, or disintegration, of unstable nuclei.

Three types of radiation can be emitted by a radioactive substance: alpha (α) particles, in which the emitted particles are $^4_2$He nuclei; beta (β) particles, in which the emitted particles are either electrons or positrons; and gamma (γ) rays, in which the emitted "rays" are high-energy photons. A **positron** is a particle similar to the electron in all respects, except that it has a charge of $+e$. (The positron is said to be the **antiparticle** of the electron.) The symbol e^- is used to designate an electron, and e^+ designates a positron.

It's possible to distinguish these three forms of radiation by using the scheme described in Figure 29.5. The radiation from a radioactive sample is directed into a region with a magnetic field, and the beam splits into three components, two bending in opposite directions and the third not changing direction. From this simple observation it can be concluded that the radiation of the undeflected beam (the gamma ray) carries no charge, the component deflected upward contains positively charged particles (alpha particles), and the component deflected downward contains negatively charged particles (e^-). If the beam includes a positron (e^+), it is deflected upward.

The three types of radiation have quite different penetrating powers. Alpha particles barely penetrate a sheet of paper, beta particles can penetrate a few millimeters of aluminum, and gamma rays can penetrate several centimeters of lead.

The Decay Constant and Half-Life

Observation has shown that if a radioactive sample contains N radioactive nuclei at some instant, then the number of nuclei, ΔN, that decay in a small time interval Δt is proportional to N; mathematically,

$$\frac{\Delta N}{\Delta t} \propto N$$

or

$$\Delta N = -\lambda N \Delta t \qquad \text{[29.2]}$$

where λ is a constant called the **decay constant**. The negative sign signifies that N decreases with time; that is, ΔN is negative. The value of λ for any isotope determines the rate at which that isotope will decay. **The decay rate, or activity R, of a sample is defined as the number of decays per second.** From Equation 29.2, we see that the decay rate is

$$R = \left|\frac{\Delta N}{\Delta t}\right| = \lambda N \qquad \text{[29.3]}$$

◀ Decay rate

Isotopes with a large λ value decay rapidly; those with small λ decay slowly.

MARIE CURIE, Polish Scientist (1867–1934)

In 1903 Marie Curie shared the Nobel Prize in physics with her husband, Pierre, and with Becquerel for their studies of radioactive substances. In 1911 she was awarded a second Nobel Prize in chemistry for the discovery of radium and polonium. Marie Curie died of leukemia caused by years of exposure to radioactive substances. "I persist in believing that the ideas that then guided us are the only ones which can lead to the true social progress. We cannot hope to build a better world without improving the individual. Toward this end, each of us must work toward his own highest development, accepting at the same time his share of responsibility in the general life of humanity."

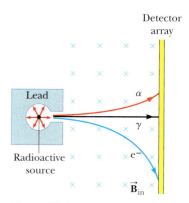

Figure 29.5 The radiation from a radioactive source, such as radium, can be separated into three components using a magnetic field to deflect the charged particles. The detector array at the right records the events. The gamma ray isn't deflected by the magnetic field.

The hands and numbers of this luminous watch contain minute amounts of radium salt. The radioactive decay of radium causes the phosphors to glow in the dark.

A general decay curve for a radioactive sample is shown in Active Figure 29.6. It can be shown from Equation 29.2 (using calculus) that the number of nuclei present varies with time according to the equation

$$N = N_0 e^{-\lambda t} \qquad \text{[29.4a]}$$

where N is the number of radioactive nuclei present at time t, N_0 is the number present at time $t = 0$, and $e = 2.718 \dots$ is Euler's constant. Processes that obey Equation 29.4a are sometimes said to undergo **exponential decay**.[1]

Another parameter that is useful for characterizing radioactive decay is the **half-life $T_{1/2}$**. **The half-life of a radioactive substance is the time it takes for half of a given number of radioactive nuclei to decay.** Using the concept of half-life, it can be shown that Equation 29.4a can also be written as

$$N = N_0 \left(\frac{1}{2}\right)^n \qquad \text{[29.4b]}$$

where n is the number of half-lives. The number n can take any non-negative value and need not be an integer. From the definition, it follows that n is related to time t and the half-life $T_{1/2}$ by

$$n = \frac{t}{T_{1/2}} \qquad \text{[29.4c]}$$

Setting $N = N_0/2$ and $t = T_{1/2}$ in Equation 29.4a gives

$$\frac{N_0}{2} = N_0 e^{-\lambda T_{1/2}}$$

Writing this in the form $e^{\lambda T_{1/2}} = 2$ and taking the natural logarithm of both sides, we get

$$T_{1/2} = \frac{\ln 2}{\lambda} = \frac{0.693}{\lambda} \qquad \text{[29.5]}$$

This is a convenient expression relating the half-life to the decay constant. Note that after an elapsed time of one half-life, $N_0/2$ radioactive nuclei remain (by definition); after two half-lives, half of these will have decayed and $N_0/4$ radioactive nuclei will be left; after three half-lives, $N_0/8$ will be left; and so on.

The unit of activity R is the **curie** (Ci), defined as

$$1 \text{ Ci} \equiv 3.7 \times 10^{10} \text{ decays/s} \qquad \text{[29.6]}$$

This unit was selected as the original activity unit because it is the approximate activity of 1 g of radium. The SI unit of activity is the **becquerel** (Bq):

$$1 \text{ Bq} = 1 \text{ decay/s} \qquad \text{[29.7]}$$

Therefore, 1 Ci = 3.7×10^{10} Bq. The most commonly used units of activity are the millicurie (10^{-3} Ci) and the microcurie (10^{-6} Ci).

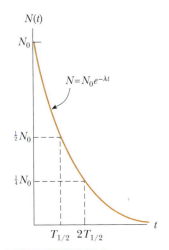

ACTIVE FIGURE 29.6

Plot of the exponential decay law for radioactive nuclei. The vertical axis represents the number of radioactive nuclei present at any time t, and the horizontal axis is time. The parameter $T_{1/2}$ is the half-life of the sample.

Physics⊗Now™

Log into PhysicsNow at **www.cp7e.com** and go to Active Figure 29.6, where you can observe the decay curves for nuclei with varying half-lives.

TIP 29.2   Two Half-Lives Don't Make a Whole-Life

A half-life is the time it takes for half of a given number of nuclei to decay. During a second half-life, half the remaining nuclei decay, so in two half-lives, three-quarters of the original material has decayed, not all of it.

Quick Quiz 29.1

What fraction of a radioactive sample has decayed after two half-lives have elapsed? (a) 1/4   (b) 1/2   (c) 3/4   (d) not enough information to say

Quick Quiz 29.2

Suppose the decay constant of radioactive substance A is twice the decay constant of radioactive substance B. If substance B has a half life of 4 hr, what's the half life of substance A? (a) 8 hr (b) 4 hr (c) 2 hr (d) not enough information to say

[1]Other examples of exponential decay were discussed in Chapter 18 in connection with RC circuits and in Chapter 20 in connection with RL circuits.

INTERACTIVE EXAMPLE 29.3   The Activity of Radium

Goal   Calculate the activity of a radioactive substance at different times.

Problem   The half-life of the radioactive nucleus $^{226}_{88}$Ra is 1.6×10^3 yr. If a sample initially contains 3.00×10^{16} such nuclei, determine **(a)** the initial activity in curies, **(b)** the number of radium nuclei remaining after 4.8×10^3 yr, and **(c)** the activity at this later time.

Strategy   For parts (a) and (c), find the decay constant and multiply it by the number of nuclei. Part (b) requires multiplying the initial number of nuclei by one-half for every elapsed half-life. (Essentially, this is an application of Equation 29.4b.)

Solution

(a) Determine the initial activity in curies.

Convert the half-life to seconds:

$$T_{1/2} = (1.6 \times 10^3 \text{ yr})(3.156 \times 10^7 \text{ s/yr}) = 5.0 \times 10^{10} \text{ s}$$

Substitute this value into Equation 29.5 to get the decay constant:

$$\lambda = \frac{0.693}{T_{1/2}} = \frac{0.693}{5.0 \times 10^{10} \text{ s}} = 1.4 \times 10^{-11} \text{ s}^{-1}$$

Calculate the activity of the sample at $t = 0$, using $R_0 = \lambda N_0$, where R_0 is the decay rate at $t = 0$ and N_0 is the number of radioactive nuclei present at $t = 0$:

$$R_0 = \lambda N_0 = (1.4 \times 10^{-11} \text{ s}^{-1})(3.0 \times 10^{16} \text{ nuclei})$$
$$= 4.2 \times 10^5 \text{ decays/s}$$

Convert to curies to obtain the activity at $t = 0$, using the fact that $1 \text{ Ci} = 3.7 \times 10^{10}$ decays/s:

$$R_0 = (4.2 \times 10^5 \text{ decays/s})\left(\frac{1 \text{ Ci}}{3.7 \times 10^{10} \text{ decays/s}}\right)$$
$$= 1.1 \times 10^{-5} \text{ Ci} = \boxed{11 \ \mu\text{Ci}}$$

(b) How many radium nuclei remain after 4.8×10^3 years?

Calculate the number of half-lives, n:

$$n = \frac{4.8 \times 10^3 \text{ yr}}{1.6 \times 10^3 \text{ yr/half-life}} = 3.0 \text{ half-lives}$$

Multiply the initial number of nuclei by the number of factors of one-half:

$$N = N_0\left(\frac{1}{2}\right)^n \qquad (1)$$

Substitute $N_0 = 3.0 \times 10^{16}$ and $n = 3.0$:

$$N = (3.0 \times 10^{16} \text{ nuclei})\left(\frac{1}{2}\right)^{3.0} = \boxed{3.8 \times 10^{15} \text{ nuclei}}$$

(c) Calculate the activity after 4.8×10^3 yr.

Multiply the number of remaining nuclei by the decay constant to find the activity R:

$$R = \lambda N = (1.4 \times 10^{-11} \text{ s}^{-1})(3.8 \times 10^{15} \text{ nuclei})$$
$$= 5.3 \times 10^4 \text{ decays/s}$$
$$= \boxed{1.4 \ \mu\text{Ci}}$$

Remarks   The activity is reduced by half every half-life, which is naturally the case because activity is proportional to the number of remaining nuclei. The precise number of nuclei at any time is never truly exact, because particles decay according to a probability. The larger the sample, however, the more accurate are the predictions from Equation 29.4.

Exercise 29.3

Find **(a)** the number of remaining radium nuclei after 3.2×10^3 yr and **(b)** the activity at this time.

Answer   **(a)** 7.5×10^{15} nuclei   **(b)** $2.8 \ \mu\text{Ci}$

Physics⊗Now™ Practice evaluating the parameters for the radioactive decay of various isotopes of radium by logging into PhysicsNow at **www.cp7e.com** and going to Interactive Example 29.3.

EXAMPLE 29.4   Radon Gas

Goal   Calculate the number of nuclei after an arbitrary time and the time required for a given number of nuclei to decay.

Problem   Radon $^{222}_{86}$Rn is a radioactive gas that can be trapped in the basements of homes, and its presence in high concentrations is a known health hazard. Radon has a half-life of 3.83 days. A gas sample contains 4.00×10^8 radon atoms initially. **(a)** How many atoms will remain after 14.0 days have passed if no more radon leaks in? **(b)** What is the activity of the radon sample after 14.0 days? **(c)** How long before 99% of the sample has decayed?

Strategy   The activity can be found by substitution into Equation 29.5, as before. Equation 29.4a (or Eq. 29.4b) must be used to find the number of particles remaining after 14.0 days. To obtain the time asked for in part (c), Equation 29.4a must be solved for time.

Solution

(a) How many atoms will remain after 14.0 days have passed?

Determine the decay constant from Equation 29.5:

$$\lambda = \frac{0.693}{T_{1/2}} = \frac{0.693}{3.83 \text{ days}} = 0.181 \text{ day}^{-1}$$

Now use Equation 29.4a, taking $N_0 = 4.0 \times 10^8$, and the value of λ just found to obtain the number N remaining after 14 days:

$$N = N_0 e^{-\lambda t} = (4.00 \times 10^8 \text{ atoms}) e^{-(0.181 \text{ day}^{-1})(14.0 \text{ days})}$$

$$= \boxed{3.17 \times 10^7 \text{ atoms}}$$

(b) What is the activity of the radon sample after 14.0 days?

Express the decay constant in units of s^{-1}:

$$\lambda = (0.181 \text{ day}^{-1})\left(\frac{1 \text{ day}}{8.64 \times 10^4 \text{ s}}\right) = 2.09 \times 10^{-6} \text{ s}^{-1}$$

From Equation 29.3 and this value of λ, compute the activity R:

$$R = \lambda N = (2.09 \times 10^{-6} \text{ s}^{-1})(3.17 \times 10^7 \text{ atoms})$$

$$= 66.3 \text{ decays/s} = \boxed{66.3 \text{ Bq}}$$

(c) How much time must pass before 99% of the sample has decayed?

Solve Equation 29.4a for t, using natural logarithms:

$$\ln(N) = \ln(N_0 e^{-\lambda t}) = \ln(N_0) + \ln(e^{-\lambda t})$$

$$\ln(N) - \ln(N_0) = \ln(e^{-\lambda t}) = -\lambda t$$

$$t = \frac{\ln(N_0) - \ln(N)}{\lambda} = \frac{\ln(N_0/N)}{\lambda}$$

Substitute values:

$$t = \frac{\ln(N_0/0.01 N_0)}{2.09 \times 10^{-6} \text{ s}^{-1}} = 2.20 \times 10^6 \text{ s} = \boxed{25.5 \text{ days}}$$

Remarks   This kind of calculation is useful in determining how long you would have to wait for radioactivity at a given location to fall to safe levels.

Exercise 29.4

(a) Find the activity of the radon sample after 12.0 days have elapsed. **(b)** How long would it take for 85.0% of the sample to decay?

Answer   (a) 95.3 Bq   (b) 9.08×10^5 s = 10.5 days

29.4   THE DECAY PROCESSES

As stated in the previous section, radioactive nuclei decay spontaneously via alpha, beta, and gamma decay. As we'll see in this section, these processes are very different from each other.

Alpha Decay

If a nucleus emits an alpha particle ($^4_2$He), it loses two protons and two neutrons. Therefore, the neutron number N of a single nucleus decreases by 2, Z decreases by 2, and A decreases by 4. The decay can be written symbolically as

$$^A_Z X \quad \rightarrow \quad ^{A-4}_{Z-2} Y + ^4_2 He \qquad [29.8]$$

where X is called the **parent nucleus** and Y is known as the **daughter nucleus**. As examples, ^{238}U and ^{226}Ra are both alpha emitters and decay according to the schemes

$$^{238}_{92} U \quad \rightarrow \quad ^{234}_{90} Th + ^4_2 He \qquad [29.9]$$

and

$$^{226}_{88} Ra \quad \rightarrow \quad ^{222}_{86} Rn + ^4_2 He \qquad [29.10]$$

The half-life for ^{238}U decay is 4.47×10^9 years, and the half-life for ^{226}Ra decay is 1.60×10^3 years. In both cases, note that the A of the daughter nucleus is four less than that of the parent nucleus, while Z is reduced by two. The differences are accounted for in the emitted alpha particle (the ^{4}He nucleus).

The decay of ^{226}Ra is shown in Active Figure 29.7. When one element changes into another, as happens in alpha decay, the process is called **spontaneous decay** or transmutation. As a general rule, (1) the sum of the mass numbers A must be the same on both sides of the equation, and (2) the sum of the atomic numbers Z must be the same on both sides of the equation.

In order for alpha emission to occur, the mass of the parent must be greater than the combined mass of the daughter and the alpha particle. In the decay process, this excess mass is converted into energy of other forms and appears in the form of kinetic energy in the daughter nucleus and the alpha particle. Most of the kinetic energy is carried away by the alpha particle because it is much less massive than the daughter nucleus. This can be understood by first noting that a particle's kinetic energy and momentum p are related as follows:

$$KE = \frac{p^2}{2m}$$

Because momentum is conserved, the two particles emitted in the decay of a nucleus at rest must have equal, but oppositely directed, momenta. As a result, the lighter particle, with the smaller mass in the denominator, has more kinetic energy than the more massive particle.

ACTIVE FIGURE 29.7
The alpha decay of radium-226. The radium nucleus is initially at rest. After the decay, the radon nucleus has kinetic energy KE_{Rn} and momentum $\vec{\mathbf{p}}_{Rn}$, and the alpha particle has kinetic energy KE_α and momentum $\vec{\mathbf{p}}_\alpha$.

Physics⊗Now™
Log into PhysicsNow at **www.cp7e.com** and go to Active Figure 29.7, where you can observe the decay of radium-226.

Quick Quiz 29.3

If a nucleus such as ^{226}Ra that is initially at rest undergoes alpha decay, which of the following statements is true? (a) The alpha particle has more kinetic energy than the daughter nucleus. (b) The daughter nucleus has more kinetic energy than the alpha particle. (c) The daughter nucleus and the alpha particle have the same kinetic energy.

Applying Physics 29.2   Energy and Half-life

In comparing alpha decay energies from a number of radioactive nuclides, why is it found that the half-life of the decay goes down as the energy of the decay goes up?

Explanation It should seem reasonable that the higher the energy of the alpha particle, the more likely it is to escape the confines of the nucleus. The higher probability of escape translates to a faster rate of decay, which appears as a shorter half-life.

EXAMPLE 29.5   Decaying Radium

Goal   Calculate the energy released during an alpha decay.

Problem   We showed that the $^{226}_{88}$Ra nucleus undergoes alpha decay to $^{222}_{86}$Rn (Eq. 29.10). Calculate the amount of energy liberated in this decay. Take the mass of $^{226}_{88}$Ra to be 226.025 402 u, that of $^{222}_{86}$Rn to be 222.017 571 u, and that of $^{4}_{2}$He to be 4.002 602 u, as found in Appendix B.

Strategy   This is a matter of subtracting the neutral masses of the daughter particles from the original mass of the radon nucleus.

Solution

Compute the sum of the mass of the daughter particle, m_d, and the mass of the alpha particle, m_α:

$$m_d + m_\alpha = 222.017\ 571\ u + 4.002\ 602\ u = 226.020\ 173\ u$$

Compute the loss of mass, Δm, during the decay by subtracting the previous result from M_p, the mass of the original particle:

$$\Delta m = M_p - (m_d + m_\alpha) = 226.025\ 402\ u - 226.020\ 173\ u$$
$$= 0.005\ 229\ u$$

Convert the loss of mass Δm to its equivalent energy in MeV:

$$E = (0.005\ 229\ u)(931.494\ \text{MeV/u}) = \boxed{4.871\ \text{MeV}}$$

Remark   The potential barrier is typically higher than this value of the energy, but quantum tunneling permits the event to occur, anyway.

Exercise 29.5

Calculate the energy released when $^{8}_{4}$Be splits into two alpha particles. Beryllium-8 has an atomic mass of 8.005 305 u.

Answer   0.094 1 MeV

Beta Decay

When a radioactive nucleus undergoes beta decay, the daughter nucleus has the same number of nucleons as the parent nucleus, but the atomic number is changed by 1:

$$^{A}_{Z}X \longrightarrow\ ^{A}_{Z+1}Y + e^- \qquad \textbf{[29.11]}$$

$$^{A}_{Z}X \longrightarrow\ ^{A}_{Z-1}Y + e^+ \qquad \textbf{[29.12]}$$

Again, note that the nucleon number and total charge are both conserved in these decays. However, as we will see shortly, these processes are not described completely by these expressions. A typical beta decay event is

$$^{14}_{6}C \longrightarrow\ ^{14}_{7}N + e^- \qquad \textbf{[29.13]}$$

The emission of electrons from a *nucleus* is surprising, because, in all our previous discussions, we stated that the nucleus is composed of protons and neutrons only. This apparent discrepancy can be explained by noting that the emitted electron is created in the nucleus by a process in which a neutron is transformed into a proton. This process can be represented by the equation

$$^{1}_{0}n \longrightarrow\ ^{1}_{1}p + e^- \qquad \textbf{[29.14]}$$

Consider the energy of the system of Equation 29.13 before and after decay. As with alpha decay, energy must be conserved in beta decay. The next example illustrates how to calculate the amount of energy released in the beta decay of $^{14}_{6}C$.

EXAMPLE 29.6   The Beta Decay of Carbon-14

Goal   Calculate the energy released in a beta decay.

Problem   Find the energy liberated in the beta decay of $^{14}_{6}C$ to $^{14}_{7}N$, as represented by Equation 29.13. That equation refers to nuclei, while Appendix B gives the masses of neutral atoms. Adding six electrons to both sides of Equation 29.13 yields

$$^{14}_{6}C\ \text{atom} \longrightarrow\ ^{14}_{7}N\ \text{atom}$$

Strategy   As in preceding problems, finding the released energy involves computing the difference in mass between the resultant particle(s) and the initial particle(s) and converting to MeV.

Solution

Obtain the masses of $^{14}_{6}C$ and $^{14}_{7}N$ from Appendix B and compute the difference between them:

$$\Delta m = m_C - m_N = 14.003\ 242\ u - 14.003\ 074\ u = 0.000\ 168\ u$$

Convert the mass difference to MeV:

$$E = (0.000\ 168\ u)(931.494\ MeV/u) = \boxed{0.156\ MeV}$$

Remarks   The calculated energy is generally more than the energy observed in this process. The discrepancy led to a crisis in physics, because it appeared that energy wasn't conserved. As discussed below, this crisis was resolved by the discovery that another particle was also produced in the reaction.

Exercise 29.6

Calculate the maximum energy liberated in the beta decay of radioactive potassium to calcium: $^{40}_{19}K \rightarrow ^{40}_{20}Ca$.

Answer   1.31 MeV

From Example 29.6, we see that the energy released in the beta decay of ^{14}C is approximately 0.16 MeV. As with alpha decay, we expect the electron to carry away virtually all of this energy as kinetic energy because, apparently, it is the lightest particle produced in the decay. As Figure 29.8 shows, however, only a small number of electrons have this maximum kinetic energy, represented as KE_{max} on the graph; most of the electrons emitted have kinetic energies lower than that predicted value. If the daughter nucleus and the electron aren't carrying away this liberated energy, then where has the energy gone? As an additional complication, further analysis of beta decay shows that the principles of conservation of both angular momentum and linear momentum appear to have been violated!

In 1930 Pauli proposed that a third particle must be present to carry away the "missing" energy and to conserve momentum. Later, Enrico Fermi developed a complete theory of beta decay and named this particle the **neutrino** ("little neutral one") because it had to be electrically neutral and have little or no mass. Although it eluded detection for many years, the neutrino (ν) was finally detected experimentally in 1956. The neutrino has the following properties:

◀ Properties of the neutrino

- Zero electric charge
- A mass much smaller than that of the electron, but probably not zero. (Recent experiments suggest that the neutrino definitely has mass, but the value is uncertain—perhaps less than 1 eV/c^2.)
- A spin of $\frac{1}{2}$
- Very weak interaction with matter, making it difficult to detect

With the introduction of the neutrino, we can now represent the beta decay process of Equation 29.13 in its correct form:

$$^{14}_{6}C \rightarrow ^{14}_{7}N + e^- + \bar{\nu} \qquad [29.15]$$

The bar in the symbol $\bar{\nu}$ indicates an **antineutrino**. To explain what an antineutrino is, we first consider the following decay:

$$^{12}_{7}N \rightarrow ^{12}_{6}C + e^+ + \nu \qquad [29.16]$$

TIP 29.3   Mass Number of the Electron

Another notation that is sometimes used for an electron is $_{-1}^{0}e$. This notation does not imply that the electron has zero rest energy. The mass of the electron is much smaller than that of the lightest nucleon, so we can approximate it as zero when we study nuclear decays and reactions.

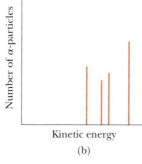

Figure 29.8   (a) Distribution of beta particle energies in a typical beta decay. All energies are observed up to a maximum value. (b) In contrast, the energies of alpha particles from an alpha decay are discrete.

Here, we see that when ^{12}N decays into ^{12}C, a particle is produced which is identical to the electron except that it has a positive charge of $+e$. This particle is called a **positron**. Because it is like the electron in all respects except charge, the positron is said to be the **antiparticle** of the electron. We will discuss antiparticles further in Chapter 30; for now, it suffices to say that, **in beta decay, an electron and an antineutrino are emitted or a positron and a neutrino are emitted**.

Unlike beta decay, which results in a daughter particle with a variety of possible kinetic energies, alpha decays come in discrete amounts, as seen in Figure 29.8b. This is because the two daughter particles have momenta with equal magnitude and opposite direction and are each composed of a fixed number of nucleons.

Gamma Decay

Very often a nucleus that undergoes radioactive decay is left in an excited energy state. The nucleus can then undergo a second decay to a lower energy state—perhaps even to the ground state—by emitting one or more high-energy photons. The process is similar to the emission of light by an atom. An atom emits radiation to release some extra energy when an electron "jumps" from a state of high energy to a state of lower energy. Likewise, the nucleus uses essentially the same method to release any extra energy it may have following a decay or some other nuclear event. In nuclear de-excitation, the "jumps" that release energy are made by protons or neutrons in the nucleus as they move from a higher energy level to a lower level. The photons emitted in the process are called **gamma rays**, which have very high energy relative to the energy of visible light.

A nucleus may reach an excited state as the result of a violent collision with another particle. However, it's more common for a nucleus to be in an excited state as a result of alpha or beta decay. The following sequence of events typifies the gamma decay processes:

$$^{12}_{5}\text{B} \rightarrow ^{12}_{6}\text{C}^{*} + \text{e}^{-} + \overline{\nu} \tag{29.17}$$

$$^{12}_{6}\text{C}^{*} \rightarrow ^{12}_{6}\text{C} + \gamma \tag{29.18}$$

Equation 29.17 represents a beta decay in which ^{12}B decays to ^{12}C*, where the asterisk indicates that the carbon nucleus is left in an excited state following the decay. The excited carbon nucleus then decays to the ground state by emitting a gamma ray, as indicated by Equation 29.18. Note that gamma emission doesn't result in any change in either Z or A.

Practical Uses of Radioactivity

Carbon Dating The beta decay of ^{14}C given by Equation 29.15 is commonly used to date organic samples. Cosmic rays (high-energy particles from outer space) in the upper atmosphere cause nuclear reactions that create ^{14}C from ^{14}N. In fact, the ratio of ^{14}C to ^{12}C (by numbers of nuclei) in the carbon dioxide molecules of our atmosphere has a constant value of about 1.3×10^{-12}, as determined by measuring carbon ratios in tree rings. All living organisms have the same ratio of ^{14}C to ^{12}C because they continuously exchange carbon dioxide with their surroundings. When an organism dies, however, it no longer absorbs ^{14}C from the atmosphere, so the ratio of ^{14}C to ^{12}C decreases as the result of the beta decay of ^{14}C. It's therefore possible to determine the age of a material by measuring its activity per unit mass as a result of the decay of ^{14}C. Through carbon dating, samples of wood, charcoal, bone, and shell have been identified as having lived from 1 000 to 25 000 years ago. This knowledge has helped researchers reconstruct the history of living organism—including human—during that time span.

A particularly interesting example is the dating of the Dead Sea Scrolls. This group of manuscripts was first discovered by a young Bedouin boy in a cave at Qumran near the Dead Sea in 1947. Translation showed the manuscripts to be religious documents, including most of the books of the Old Testament. Because of their historical and religious significance, scholars wanted to know their age. Carbon dating applied to fragments of the scrolls and to the material in which

ENRICO FERMI, Italian Physicist (1901–1954)

Fermi was awarded the Nobel Prize in 1938 for producing the transuranic elements by neutron irradiation and for his discovery of nuclear reactions bought about by slow neutrons. He made many other outstanding contributions to physics, including his theory of beta decay, the free-electron theory of metals, and the development of the world's first fission reactor in 1942. Fermi was truly a gifted theoretical and experimental physicist. He was also well known for his ability to present physics in a clear and exciting manner. "Whatever Nature has in store for mankind, unpleasant as it may be, men must accept, for ignorance is never better than knowledge."

APPLICATION

Carbon Dating of the Dead Sea Scrolls

they were wrapped established that they were about 1950 years old. The scrolls are now stored at the Israel museum in Jerusalem.

Smoke Detectors Smoke detectors are frequently used in homes and industry for fire protection. Most of the common ones are the ionization-type that use radioactive materials. (See Fig. 29.9.) A smoke detector consists of an ionization chamber, a sensitive current detector, and an alarm. A weak radioactive source ionizes the air in the chamber of the detector, which creates charged particles. A voltage is maintained between the plates inside the chamber, setting up a small but detectable current in the external circuit. As long as the current is maintained, the alarm is deactivated. However, if smoke drifts into the chamber, the ions become attached to the smoke particles. These heavier particles do not drift as readily as do the lighter ions, which causes a decrease in the detector current. The external circuit senses this decrease in current and sets off the alarm.

Radon Detection Radioactivity can also affect our daily lives in harmful ways. Soon after the discovery of radium by the Curies, it was found that the air in contact with radium compounds becomes radioactive. It was then shown that this radioactivity came from the radium itself, and the product was therefore called "radium emanation." Rutherford and Soddy succeeded in condensing this "emanation," confirming that it was a real substance: the inert, gaseous element now called **radon** (Rn). Later, it was discovered that the air in uranium mines is radioactive because of the presence of radon gas. The mines must therefore be well ventilated to help protect the miners. Finally, the fear of radon pollution has moved from uranium mines into our own homes. (See Example 29.4.) Because certain types of rock, soil, brick, and concrete contain small quantities of radium, some of the resulting radon gas finds its way into our homes and other buildings. The most serious problems arise from leakage of radon from the ground into the structure. One practical remedy is to exhaust the air through a pipe just above the underlying soil or gravel directly to the outdoors by means of a small fan or blower.

APPLICATION

Smoke Detectors

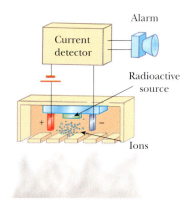

Figure 29.9 An ionization-type smoke detector. Smoke entering the chamber reduces the detected current, causing the alarm to sound.

APPLICATION

Radon Pollution

Applying Physics 29.3   Radioactive Dating of the Iceman

In 1991, a German tourist discovered the well-preserved remains of a man trapped in a glacier in the Italian Alps. (See Fig. 29.10.) Radioactive dating of a sample of bone from this hunter–gatherer, dubbed the "Iceman," revealed an age of 5 300 years. Why did scientists date the sample using the isotope ^{14}C, rather than ^{11}C, a beta emitter with a half-life of 20.4 min?

Explanation ^{14}C has a long half-life of 5 730 years, so the fraction of ^{14}C nuclei remaining after one half-life is high enough to accurately measure changes in the sample's activity. The ^{11}C isotope, which has a very short half-life, is not useful, because its activity decreases to a vanishingly small value over the age of the sample, making it impossible to detect.

If a sample to be dated is not very old — say, about 50 years — then you should select the isotope of some other element with half-life comparable to the age of the sample. For example, if the sample contained hydrogen, you could measure the activity of ^{3}H (tritium), a beta emitter of half-life 12.3 years. As a general rule, the expected age of the sample should be long enough to measure a change in

activity, but not so long that its activity can't be detected.

Hanny Paul/Gamma Liaison

Figure 29.10 (Applying Physics 29.3) The body of an ancient man (dubbed the Iceman) was exposed by a melting glacier in the Alps.

To use radioactive dating techniques, we need to recast some of the equations already introduced. We start by multiplying both sides of Equation 29.4 by λ:

$$\lambda N = \lambda N_0 e^{-\lambda t}$$

From Equation 29.3, we have $\lambda N = R$ and $\lambda N_0 = R_0$. Substitute these expressions into the above equation and divide through by R_0:

$$\frac{R}{R_0} = e^{-\lambda t}$$

R is the present activity and R_0 was the activity when the object in question was part of a living organism. We can solve for time by taking the natural logarithm of both sides of the foregoing equation:

$$\ln\left(\frac{R}{R_0}\right) = \ln(e^{-\lambda t}) = -\lambda t$$

$$t = -\frac{\ln\left(\dfrac{R}{R_0}\right)}{\lambda}$$ [29.19]

EXAMPLE 29.7   Should We Report This to Homicide?

Goal   Apply the technique of carbon-14 dating.

Problem   A 50.0-g sample of carbon is taken from the pelvis bone of a skeleton and is found to have a carbon-14 decay rate of 200.0 decays/min. It is known that carbon from a living organism has a decay rate of 15.0 decays/min · g and that ^{14}C has a half-life of 5 730 yr $= 3.01 \times 10^9$ min. Find the age of the skeleton.

Strategy   Calculate the original activity and the decay constant, and then substitute those numbers and the current activity into Equation 29.19.

Solution

Calculate the original activity R_0 from the decay rate and the mass of the sample:

$$R_0 = \left(15.0 \, \frac{\text{decays}}{\text{min} \cdot \text{g}}\right)(50.0 \text{ g}) = 7.50 \times 10^2 \, \frac{\text{decays}}{\text{min}}$$

Find the decay constant from Equation 29.5:

$$\lambda = \frac{0.693}{T_{1/2}} = \frac{0.693}{3.01 \times 10^9 \text{ min}} = 2.30 \times 10^{-10} \text{ min}^{-1}$$

R is given, so now we substitute all values into Equation 29.19 to find the age of the skeleton:

$$t = -\frac{\ln\left(\dfrac{R}{R_0}\right)}{\lambda} = -\frac{\ln\left(\dfrac{200.0 \text{ decays/min}}{7.50 \times 10^2 \text{ decays/min}}\right)}{2.30 \times 10^{-10} \text{ min}^{-1}}$$

$$= \frac{1.32}{2.30 \times 10^{-10} \text{ min}^{-1}}$$

$$= 5.74 \times 10^9 \text{ min} = \boxed{1.09 \times 10^4 \text{ yr}}$$

Remark   For much longer periods, other radioactive substances with longer half-lives must be used to develop estimates.

Exercise 29.7
A sample of carbon of mass 7.60 g taken from an animal jawbone has an activity of 4.00 decays/min. How old is the jawbone?

Answer 2.77×10^4 yr

APPLICATION

Carbon-14 Dating of the Shroud of Turin

Carbon-14 and the Shroud of Turin

Since the Middle Ages, many people have marveled at a 14-foot-long, yellowing piece of linen found in Turin, Italy, purported to be the burial shroud of Jesus Christ (Fig. 29.11). The cloth bears a remarkable, full-size likeness of a crucified body, with wounds on the head that could have been caused by a crown of thorns and another wound in the side that could have been the cause of death. Skepticism over the authenticity of the shroud has existed since its first public showing in 1354; in fact, a French bishop declared it to be a fraud at the time. Because of its controversial nature, religious bodies have taken a neutral stance on its authenticity.

In 1978 the bishop of Turin allowed the cloth to be subjected to scientific analysis, but notably missing from these tests was carbon-14 dating. The reason for this omission was that, at the time, carbon-dating techniques required a piece of cloth about the size of a handkerchief. In 1988 the process had been refined to the point that pieces as small as one square inch were sufficient, and at that time permission was granted to allow the dating to proceed. Three labs were selected for the testing, and each was given four pieces of material. One of these was a piece of the shroud, and the other three pieces were control pieces similar in appearance to the shroud.

The testing procedure consisted of burning the cloth to produce carbon dioxide, which was then converted chemically to graphite. The graphite sample was subjected to carbon-14 analysis, and in the end all three labs agreed amazingly well on the age of the shroud. The average of their results gave a date for the cloth of A.D. 1 320 ± 60 years, with an assurance that the cloth could not be older than A.D. 1 200. Carbon-14 dating has thus unraveled the most important mystery concerning the shroud, but others remain. For example, investigators have not yet been able to explain how the image was imprinted.

Figure 29.11   The Shroud of Turin as it appears in a photographic negative image.

29.5   NATURAL RADIOACTIVITY

Radioactive nuclei are generally classified into two groups: (1) unstable nuclei found in nature, which give rise to what is called **natural radioactivity**, and (2) nuclei produced in the laboratory through nuclear reactions, which exhibit **artificial radioactivity**.

Three series of naturally occurring radioactive nuclei exist (Table 29.2). Each starts with a specific long-lived radioactive isotope with half-life exceeding that of any of its descendants. The fourth series in Table 29.2 begins with ^{237}Np, a transuranic element (an element having an atomic number greater than that of uranium) not found in nature. This element has a half-life of "only" 2.14×10^6 yr.

The two uranium series are somewhat more complex than the ^{232}Th series (Fig. 29.12). Also, there are several other naturally occurring radioactive isotopes, such as ^{14}C and ^{40}K, that are not part of either decay series.

Natural radioactivity constantly supplies our environment with radioactive elements that would otherwise have disappeared long ago. For example, because the Solar System is about 5×10^9 years old, the supply of ^{226}Ra (with a half-life of only 1 600 yr) would have been depleted by radioactive decay long ago were it not for the decay series that starts with ^{238}U, with a half-life of 4.47×10^9 yr.

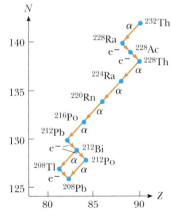

Figure 29.12   Decay series beginning with ^{232}Th.

29.6   NUCLEAR REACTIONS

It is possible to change the structure of nuclei by bombarding them with energetic particles. Such changes are called **nuclear reactions**. Rutherford was the first to observe nuclear reactions, using naturally occurring radioactive sources for the

TABLE 29.2

The Four Radioactive Series

Series	Starting Isotope	Half-life (years)	Stable End Product
Uranium	$^{238}_{92}$U	4.47×10^9	$^{206}_{82}$Pb
Actinium	$^{235}_{92}$U	7.04×10^8	$^{207}_{82}$Pb
Thorium	$^{232}_{90}$Th	1.41×10^{10}	$^{208}_{82}$Pb
Neptunium	$^{237}_{93}$Np	2.14×10^6	$^{209}_{83}$Bi

bombarding particles. He found that protons were released when alpha particles were allowed to collide with nitrogen atoms. The process can be represented symbolically as

$$\ce{^4_2He + ^{14}_7N \rightarrow X + ^1_1H} \qquad [29.20]$$

This equation says that an alpha particle ($^4_2$He) strikes a nitrogen nucleus and produces an unknown product nucleus (X) and a proton ($^1_1$H). Balancing atomic numbers and mass numbers, as we did for radioactive decay, enables us to conclude that the unknown is characterized as $^{17}_8$X. Because the element with atomic number 8 is oxygen, we see that the reaction is

$$\ce{^4_2He + ^{14}_7N \rightarrow ^{17}_8O + ^1_1H} \qquad [29.21]$$

This nuclear reaction starts with two stable isotopes—helium and nitrogen—and produces two different stable isotopes—hydrogen and oxygen.

Since the time of Rutherford, thousands of nuclear reactions have been observed, particularly following the development of charged-particle accelerators in the 1930s. With today's advanced technology in particle accelerators and particle detectors, it is possible to achieve particle energies of at least 1 000 GeV = 1 TeV. These high-energy particles are used to create new particles whose properties are helping to solve the mysteries of the nucleus (and indeed, of the Universe itself).

Quick Quiz 29.4

Which of the following are possible reactions?
(a) $^1_0$n + $^{235}_{92}$U $\rightarrow$ $^{140}_{54}$Xe + $^{94}_{38}$Sr + 2($^1_0$n)
(b) $^1_0$n + $^{235}_{92}$U $\rightarrow$ $^{132}_{50}$Sn + $^{101}_{42}$Mo + 3($^1_0$n)
(c) $^1_0$n + $^{239}_{94}$Pu $\rightarrow$ $^{127}_{53}$I + $^{93}_{41}$Nb + 3($^1_0$n)

EXAMPLE 29.8   The Discovery of the Neutron

Goal   Balance a nuclear reaction to determine an unknown decay product.

Problem   A nuclear reaction of significant note occurred in 1932 when Chadwick, in England, bombarded a beryllium target with alpha particles. Analysis of the experiment indicated that the following reaction occurred:

$$\ce{^4_2He + ^9_4Be \rightarrow ^{12}_6C + ^A_ZX}$$

What is ^A_ZX in this reaction?

Strategy   Balancing mass numbers and atomic numbers yields the answer.

Solution
Write an equation relating the atomic masses on either side: $4 + 9 = 12 + A \ \rightarrow \ A = 1$

Write an equation relating the atomic numbers: $2 + 4 = 6 + Z \ \rightarrow \ Z = 0$

Identify the particle: $^A_ZX = \ ^1_0$n (a neutron)

Remarks   This experiment was the first to provide positive proof of the existence of neutrons.

Exercise 29.8
Identify the unknown particle in this reaction:

$$\ce{^4_2He + ^{14}_7N \rightarrow ^{17}_8O + ^A_ZX}$$

Answer $^A_ZX = \ ^1_1$H (a neutral hydrogen atom)

EXAMPLE 29.9   Synthetic Elements

Goal   Construct equations for a series of radioactive decays.

Problem   **(a)** A beam of neutrons is directed at a target of $^{238}_{92}\text{U}$. The reaction products are a gamma ray and another isotope. What is the isotope? **(b)** The isotope $^{239}_{92}\text{U}$ is radioactive and undergoes beta decay. Write the equation symbolizing this decay and identify the resulting isotope.

Strategy   Balance the mass numbers and atomic numbers on both sides of the equations.

Solution

(a) Identify the isotope produced by the reaction of a neutron with a target of $^{238}_{92}\text{U}$, with production of a gamma ray.

Write an equation for the reaction in terms of the unknown isotope:

$$^{1}_{0}\text{n} + {}^{238}_{92}\text{U} \rightarrow {}^{A}_{Z}\text{X} + \gamma$$

Write and solve equations for the atomic mass and atomic number:

$$A = 1 + 238 = 239; \quad Z = 0 + 92 = 92$$

Identify the isotope:

$$^{A}_{Z}\text{X} = {}^{239}_{92}\text{U}$$

(b) Write the equation for the beta decay of $^{239}_{92}\text{U}$, identifying the resulting isotope.

Write an equation for the decay of $^{239}_{92}\text{U}$ by beta emission in terms of the unknown isotope:

$$^{239}_{92}\text{U} \rightarrow {}^{A}_{Z}\text{Y} + e^{-} + \bar{\nu}$$

Write and solve equations for the atomic mass and charge conservation (the electron counts as -1 on the right):

$$A = 239; \quad 92 = Z - 1 \rightarrow Z = 93$$

Identify the isotope:

$$^{A}_{Z}\text{Y} = {}^{239}_{93}\text{Np (neptunium)}$$

Remarks   The interesting feature of these reactions is the fact that uranium is the element with the greatest number of protons (92) which exists in nature in any appreciable amount. The reactions in parts (a) and (b) do occur occasionally in nature; hence, minute traces of neptunium and plutonium are present. In 1940, however, researchers bombarded uranium with neutrons to produce plutonium and neptunium. These two elements were the first elements made in the laboratory. Since then, the list of synthetic elements has been extended to include those up to atomic number 112. Recently, elements 113 and 115 have been observed, but as of this writing, their existence has not yet been confirmed.

Exercise 29.9
The isotope $^{238}_{93}\text{U}$ is also radioactive and decays by beta emission. What is the end product?

Answer $^{239}_{94}\text{Pu}$

Q Values

We have just examined some nuclear reactions for which mass numbers and atomic numbers must be balanced in the equations. We will now consider the energy involved in these reactions, because energy is another important quantity that must be conserved.
    We illustrate this procedure by analyzing the following nuclear reaction:

$$^{2}_{1}\text{H} + {}^{14}_{7}\text{N} \rightarrow {}^{12}_{6}\text{C} + {}^{4}_{2}\text{He} \qquad [29.22]$$

The total mass on the left side of the equation is the sum of the mass of $^2_1$H (2.014 102 u) and the mass of $^{14}_7$N (14.003 074 u), which equals 16.017 176 u. Similarly, the mass on the right side of the equation is the sum of the mass of $^{12}_6$C (12.000 000 u) plus the mass of $^4_2$He (4.002 602 u), for a total of 16.002 602 u. Thus, the total mass before the reaction is greater than the total mass after the reaction. The mass difference in the reaction is equal to 16.017 176 u − 16.002 602 u = 0.014 574 u. This "lost" mass is converted to the kinetic energy of the nuclei present after the reaction. In energy units, 0.014 574 u is equivalent to 13.576 MeV of kinetic energy carried away by the carbon and helium nuclei.

The energy required to balance the equation is called the Q value of the reaction. In Equation 29.22, the Q value is 13.576 MeV. Nuclear reactions in which there is a release of energy—that is, positive Q values—are said to be **exothermic reactions**.

The energy balance sheet isn't complete, however: We must also consider the kinetic energy of the incident particle before the collision. As an example, assume that the deuteron in Equation 29.22 has a kinetic energy of 5 MeV. Adding this to our Q value, we find that the carbon and helium nuclei have a total kinetic energy of 18.576 MeV following the reaction.

Now consider the reaction

$$^4_2\text{He} + {}^{14}_7\text{N} \rightarrow {}^{17}_8\text{O} + {}^1_1\text{H} \qquad \textbf{[29.23]}$$

Before the reaction, the total mass is the sum of the masses of the alpha particle and the nitrogen nucleus: 4.002 602 u + 14.003 074 u = 18.005 676 u. After the reaction, the total mass is the sum of the masses of the oxygen nucleus and the proton: 16.999 133 u + 1.007 825 u = 18.006 958 u. In this case, the total mass after the reaction is *greater* than the total mass before the reaction. The mass deficit is 0.001 282 u, equivalent to an energy deficit of 1.194 MeV. This deficit is expressed by the negative Q value of the reaction, − 1.194 MeV. Reactions with negative Q values are called **endothermic reactions**. Such reactions won't take place unless the incoming particle has at least enough kinetic energy to overcome the energy deficit.

At first it might appear that the reaction in Equation 29.23 can take place if the incoming alpha particle has a kinetic energy of 1.194 MeV. In practice, however, the alpha particle must have more energy than this. If it has an energy of only 1.194 MeV, energy is conserved but careful analysis shows that momentum isn't. This can be understood by recognizing that the incoming alpha particle has some momentum before the reaction. However, if its kinetic energy is only 1.194 MeV, the products (oxygen and a proton) would be created with zero kinetic energy and thus zero momentum. It can be shown that in order to conserve both energy and momentum, the incoming particle must have a minimum kinetic energy given by

$$KE_{\text{min}} = \left(1 + \frac{m}{M}\right)|Q| \qquad \textbf{[29.24]}$$

where m is the mass of the incident particle, M is the mass of the target, and the absolute value of the Q value is used. For the reaction given by Equation 29.23, we find that

$$KE_{\text{min}} = \left(1 + \frac{4.002\ 602}{14.003\ 074}\right)|-1.194\ \text{MeV}| = 1.535\ \text{MeV}$$

This minimum value of the kinetic energy of the incoming particle is called the **threshold energy**. The nuclear reaction shown in Equation 29.23 won't occur if the incoming alpha particle has a kinetic energy of less than 1.535 MeV, but can occur if its kinetic energy is equal to or greater than 1.535 MeV.

Quick Quiz 29.5

If the Q value of an endothermic reaction is − 2.17 MeV, then the minimum kinetic energy needed in the reactant nuclei if the reaction is to occur must be (a) equal to 2.17 MeV, (b) greater than 2.17 MeV, (c) less than 2.17 MeV, or (d) exactly half of 2.17 MeV.

29.7   MEDICAL APPLICATIONS OF RADIATION

Radiation Damage in Matter

Radiation absorbed by matter can cause severe damage. The degree and kind of damage depend on several factors, including the type and energy of the radiation and the properties of the absorbing material. Radiation damage in biological organisms is due primarily to ionization effects in cells. The normal function of a cell may be disrupted when highly reactive ions or radicals are formed as the result of ionizing radiation. For example, hydrogen and hydroxyl radicals produced from water molecules can induce chemical reactions that may break bonds in proteins and other vital molecules. Large acute doses of radiation are especially dangerous because damage to a great number of molecules in a cell may cause the cell to die. Also, cells that do survive the radiation may become defective, which can lead to cancer.

In biological systems, it is common to separate radiation damage into two categories: somatic damage and genetic damage. **Somatic damage** is radiation damage to any cells except the reproductive cells. Such damage can lead to cancer at high radiation levels or seriously alter the characteristics of specific organisms. **Genetic damage** affects only reproductive cells. Damage to the genes in reproductive cells can lead to defective offspring. Clearly, we must be concerned about the effect of diagnostic treatments, such as x-rays and other forms of exposure to radiation.

Several units are used to quantify radiation exposure and dose. The **roentgen** (R) is defined as **that amount of ionizing radiation which will produce 2.08×10^9 ion pairs in 1 cm^3 of air under standard conditions**. Equivalently, the roentgen is **that amount of radiation which deposits 8.76×10^{-3} J of energy into 1 kg of air**.

For most applications, the roentgen has been replaced by the **rad** (an acronym for *r*adiation *a*bsorbed *d*ose), defined as follows: **One rad is that amount of radiation which deposits 10^{-2} J of energy into 1 kg of absorbing material**.

Although the rad is a perfectly good physical unit, it's not the best unit for measuring the degree of biological damage produced by radiation, because the degree of damage depends not only on the dose, but also on the *type* of radiation. For example, a given dose of alpha particles causes about 10 times more biological damage than an equal dose of x-rays. The **RBE** (*r*elative *b*iological *e*ffectiveness) factor is defined as **the number of rads of x-radiation or gamma radiation that produces the same biological damage as 1 rad of the radiation being used**. The RBE factors for different types of radiation are given in Table 29.3. Note that the values are only approximate because they vary with particle energy and the form of damage.

Finally, the **rem** (*r*oentgen *e*quivalent in *m*an) is defined as the product of the dose in rads and the RBE factor:

$$\text{Dose in rem} = \text{dose in rads} \times \text{RBE}$$

According to this definition, 1 rem of any two kinds of radiation will produce the same amount of biological damage. From Table 29.3, we see that a dose of 1 rad of fast neutrons represents an effective dose of 10 rem and that 1 rad of x-radiation is equivalent to a dose of 1 rem.

TABLE 29.3

RBE Factors for Several Types of Radiation

Radiation	RBE Factor
X-rays and gamma rays	1.0
Beta particles	1.0–1.7
Alpha particles	10–20
Slow neutrons	4–5
Fast neutrons and protons	10
Heavy ions	20

APPLICATION

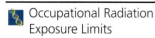 Occupational Radiation
Exposure Limits

Low-level radiation from natural sources, such as cosmic rays and radioactive rocks and soil, delivers a dose of about 0.13 rem/year per person. The upper limit of radiation dose recommended by the U.S. government (apart from background radiation and exposure related to medical procedures) is 0.5 rem/year. Many occupations involve higher levels of radiation exposure, and for individuals in these occupations, an upper limit of 5 rem/year has been set for whole-body exposure. Higher upper limits are permissible for certain parts of the body, such as the hands and forearms. An acute whole-body dose of 400 to 500 rem results in a mortality rate of about 50%. The most dangerous form of exposure is ingestion or inhalation of radioactive isotopes, especially those elements the body retains and concentrates, such as ^{90}Sr. In some cases, a dose of 1000 rem can result from ingesting 1 mCi of radioactive material.

APPLICATION

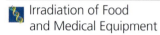 Irradiation of Food
and Medical Equipment

Sterilizing objects by exposing them to radiation has been going on for at least 25 years, but in recent years the methods used have become safer to use and more economical. Most bacteria, worms, and insects are easily destroyed by exposure to gamma radiation from radioactive cobalt. There is no intake of radioactive nuclei by an organism in such sterilizing processes, as there is in the use of radioactive tracers. The process is highly effective in destroying Trichinella worms in pork, salmonella bacteria in chickens, insect eggs in wheat, and surface bacteria on fruits and vegetables that can lead to rapid spoilage. Recently, the procedure has been expanded to include the sterilization of medical equipment while in its protective covering. Surgical gloves, sponges, sutures, and so forth are irradiated while packaged. Also, bone, cartilage, and skin used for grafting is often irradiated to reduce the chance of infection.

Tracing

APPLICATION

 Radioactive Tracers
in Medicine

Radioactive particles can be used to trace chemicals participating in various reactions. One of the most valuable uses of radioactive tracers is in medicine. For example, ^{131}I is an artificially produced isotope of iodine. (The natural, nonradioactive isotope is ^{127}I.) Iodine, a necessary nutrient for our bodies, is obtained largely through the intake of seafood and iodized salt. The thyroid gland plays a major role in the distribution of iodine throughout the body. In order to evaluate the performance of the thyroid, the patient drinks a small amount of radioactive sodium iodide. Two hours later, the amount of iodine in the thyroid gland is determined by measuring the radiation intensity in the neck area.

A medical application of the use of radioactive tracers occurring in emergency situations is that of locating a hemorrhage inside the body. Often the location of the site cannot easily be determined, but radioactive chromium can identify the location with a high degree of precision. Chromium is taken up by red blood cells and carried uniformly throughout the body. However, the blood will be dumped at a hemorrhage site, and the radioactivity of that region will increase markedly.

APPLICATION

Radioactive Tracers in
Agricultural Research

The tracer technique is also useful in agricultural research. Suppose the best method of fertilizing a plant is to be determined. A certain material in the fertilizer, such as nitrogen, can be tagged with one of its radioactive isotopes. The fertilizer is then sprayed onto one group of plants, sprinkled on the ground for a second group, and raked into the soil for a third. A Geiger counter is then used to track the nitrogen through the three types of plants.

Tracing techniques are as wide ranging as human ingenuity can devise. Present applications range from checking the absorption of fluorine by teeth to checking contamination of food-processing equipment by cleansers to monitoring deterioration inside an automobile engine. In the last case, a radioactive material is used in the manufacture of the pistons, and the oil is checked for radioactivity to determine the amount of wear on the pistons.

Computed Axial Tomography (CAT) Scans

The normal x-ray of a human body has two primary disadvantages when used as a source of clinical diagnosis. First, it is difficult to distinguish between various types of tissue in the body because they all have similar x-ray absorption properties.

Second, a conventional x-ray absorption picture is indicative of the average amount of absorption along a particular direction in the body, leading to somewhat obscured pictures. To overcome these problems, an instrument called a CAT scanner was developed in England in 1973; the device is capable of producing pictures of much greater clarity and detail than previously possible.

The operation of a CAT scanner can be understood by considering the following hypothetical experiment: suppose a box consists of four compartments, labeled A, B, C, and D, as in Figure 29.13a. Each compartment has a different amount of absorbing material from any other compartment. What set of experimental procedures will enable us to determine the relative amounts of material in each compartment? The following steps outline one method that will provide this information: first, a beam of x-rays is passed through compartments A and C, as in Figure 29.13b. The intensity of the exiting radiation is reduced by absorption by some number that we assign as 8. (The number 8 could mean, for example, that the intensity of the exiting beam is reduced by eight-tenths of 1% from its initial value.) Because we don't know which of the compartments, A or C, was responsible for this reduction in intensity, half the loss is assigned to each compartment, as in Figure 29.13c. Next, a beam of x-rays is passed through compartments B and D, as in Figure 29.13b. The reduction in intensity for this beam is 10, and again we assign half the loss to each compartment. We now redirect the x-ray source so that it sends one beam through compartments A and B and another through compartments C and D, as in Figure 29.13d. Once more, we measure the absorption. Suppose the absorption through compartments A and B in this experiment is measured to be 7 units. On the basis of our first experiment, we would have guessed it would be 9 units: 4 by compartment A and 5 by compartment B. Thus, we have reduced the guessed absorption for each compartment by 1 unit, so that the sum is 7 rather than 9, giving the numbers shown in Figure 29.13e. Likewise, when the beam is passed through compartments C and D, as in Figure 29.13d, we may find the total absorption to be 11 as compared to our first experiment of 9. In this case, we add 1 unit of absorption to each compartment to give a sum of 11, as in Figure 29.13e. This somewhat crude procedure could be improved by measuring the absorption along other paths. However, these simple measurements are sufficient to enable us to conclude that compartment D contains the most absorbing material and A the least. A visual representation of these results can be obtained by assigning, to each compartment, a shade of gray corresponding to the particular number

CAT Scans

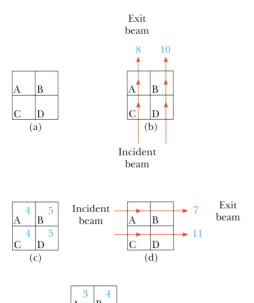

Figure 29.13   An experimental procedure for determining the relative amounts of x-ray absorption by four different compartments in a box.

associated with the absorption. In this example, compartment D would be very dark and compartment A would be very light.

The steps outlined previously are representative of how a CAT scanner produces images of the human body. A thin slice of the body is subdivided into perhaps 10 000 compartments, rather than 4 as in our simple example. The function of the CAT scanner is to determine the relative absorption in each of these 10 000 compartments and to display a picture of its calculations in various shades of gray. Note that "CAT" stands for **computed axial tomography**. The term *axial* is used because the slice of the body to be analyzed corresponds to a plane perpendicular to the head-to-toe axis. *Tomos* is the Greek word for slice and *graph* is the Greek word for picture. In a typical diagnosis, the patient is placed in the position shown in Figure 29.14 and a narrow beam of x-rays is sent through the plane of interest. The emerging x-rays are detected and measured by photomultiplier tubes behind the patient. The x-ray tube is then rotated a few degrees, and the intensity is recorded again. An extensive amount of information is obtained by rotating the beam through 180° at intervals of about 1° per measurement, resulting in a set of numbers assigned to each of the 10 000 "compartments" in the slice. These numbers are then converted by the computer to a photograph in various shades of gray for the segment of the body that is under observation.

A brain scan of a patient can now be made in about 2 s, and a full-body scan requires about 6 s. The final result is a picture containing much greater quantitative information and clarity than a conventional x-ray photograph. Because CAT scanners use x-rays, which are an ionizing form of radiation, the technique presents a modest health risk to the patient being diagnosed.

Magnetic Resonance Imaging (MRI)

At the heart of magnetic resonance imaging (MRI) is the fact that when a nucleus having a magnetic moment is placed in an external magnetic field, its moment precesses about the magnetic field with a frequency that is proportional to the field. For example, a proton, with a spin of 1/2, can occupy one of two energy states when placed in an external magnetic field. The lower energy state corresponds to the case in which the spin is aligned with the field, whereas the higher energy state corresponds to the case in which the spin is opposite the field. Transitions between these two states can be observed with a technique known as **nuclear magnetic resonance**. A DC magnetic field is applied to align the magnetic moments, and a second, weak oscillating magnetic field is applied perpendicular to the DC field. When the frequency of the oscillating field is adjusted to match the precessional frequency of the magnetic moments, the nuclei will "flip" between

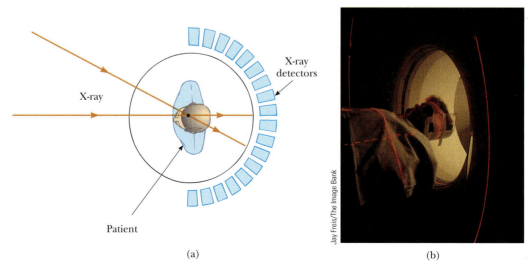

(a)   (b)

Figure 29.14 (a) CAT scanner detector assembly. (b) Photograph of a patient undergoing a CAT scan in a hospital.

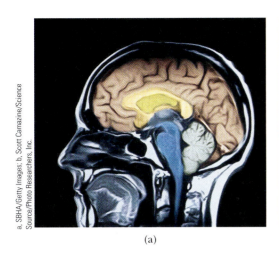
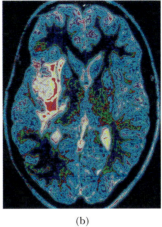

(a)   (b)

Figure 29.15   Computer-enhanced MRI images of (a) a normal human brain and (b) a human brain with a glioma tumor.

the two spin states. These transitions result in a net absorption of energy by the spin system, which can be detected electronically.

In MRI, image reconstruction is obtained using spatially varying magnetic fields and a procedure for encoding each point in the sample being imaged. Some MRI images taken on a human head are shown in Figure 29.15. In practice, a computer-controlled pulse-sequencing technique is used to produce signals that are captured by a suitable processing device. The signals are then subjected to appropriate mathematical manipulations to provide data for the final image. The main advantage of MRI over other imaging techniques in medical diagnostics is that it causes minimal damage to cellular structures. Photons associated with the rf signals used in MRI have energies of only about 10^{-7} eV. Because molecular bond strengths are much larger (on the order of 1 eV), the rf photons cause little cellular damage. In comparison, x-rays or γ-rays have energies ranging from 10^4 to 10^6 eV and can cause considerable cellular damage.

29.8   RADIATION DETECTORS

Although most medical applications of radiation require instruments to make quantitative measurements of radioactive intensity, we have not yet explained how such instruments operate. Various devices have been developed to detect the energetic particles emitted when a radioactive nucleus decays. The **Geiger counter** (Fig. 29.16) is perhaps the most common device used to detect radioactivity. It can be considered the prototype of all counters that use the ionization of a medium as the basic detection process. A Geiger counter consists of a thin wire electrode aligned along the central axis of a cylindrical metallic tube filled with a gas at low pressure. The wire is maintained at a high positive voltage of about 1 000 V relative to the tube. When an energetic charged particle or gamma-ray photon enters the tube through a thin window at one end, some of the gas atoms are ionized. The electrons removed from these atoms are attracted toward the wire electrode,

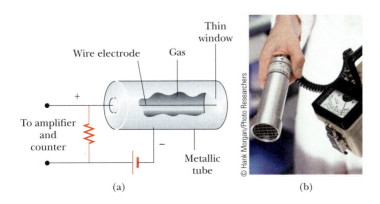

(a)   (b)

Figure 29.16   (a) Diagram of a Geiger counter. The voltage between the wire electrode and the metallic tube is usually about 1 000 V. (b) A Geiger counter.

and in the process they ionize other atoms in their path. This sequential ionization results in an *avalanche* of electrons that produces a current pulse. After the pulse has been amplified, it can either be used to trigger an electronic counter or delivered to a loudspeaker that clicks each time a particle is detected. Although a Geiger counter reliably detects the presence and quantity of radiation, it cannot be used to measure the energy of the detected radiation.

A **semiconductor diode detector** is essentially a reverse biased $p-n$ junction. As an energetic particle passes through the junction, it produces electron–hole pairs that are separated by the internal electric field. This movement of electrons and holes creates a brief pulse of current that is measured with an electronic counter. In a typical device, the duration of the pulse is 10^{-8} s.

A **scintillation counter** usually uses a solid or liquid material having atoms that are easily excited by radiation. The excited atoms then emit photons of visible light when they return to their ground state. Common materials used as scintillators are transparent crystals of sodium iodide and certain plastics. If the scintillator material is attached to one end of a device called a **photomultiplier** (PM) tube, as shown in Figure 29.17, the photons emitted by the scintillator can be converted to an electrical signal. The PM tube consists of numerous electrodes, called *dynodes*, whose electric potentials increase in succession along the length of the tube. Between the top of the tube and the scintillator material is a plate called a *photocathode*. When photons leaving the scintillator hit this plate, electrons are emitted because of the photoelectric effect. As one of these emitted electrons strikes the first dynode, the electron has sufficient kinetic energy to eject several other electrons from the surface of the dynode. When these electrons are accelerated to the second dynode, many more electrons are ejected, and a multiplication process occurs. The end result is 1 million or more electrons striking the last dynode. Hence, one particle striking the scintillator produces a sizable electrical pulse at the PM output, and this pulse is sent to an electronic counter.

Both the scintillator and the semiconductor diode detector are much more sensitive than a Geiger counter, mainly because of the higher mass density of the detecting medium. Both can also be used to measure particle energy from the height of the pulses produced.

Track detectors are various devices used to view the tracks or paths of charged particles directly. High-energy particles produced in particle accelerators may have energies ranging from 10^9 to 10^{12} eV. The energy of such particles can't be measured with the small detectors already mentioned. Instead, their energy and

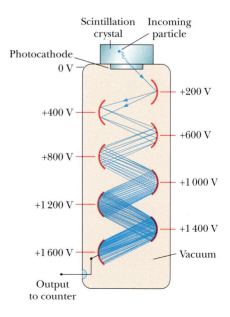

Figure 29.17 Diagram of a scintillation counter connected to a photomultiplier tube.

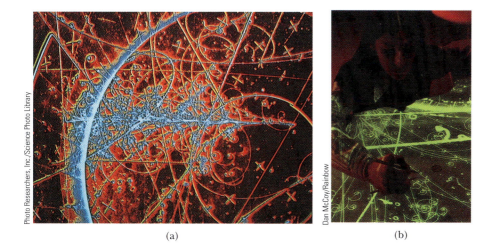

Figure 29.18 (a) Artificially colored photograph showing tracks of particles that have passed through a bubble chamber. (b) This research scientist is studying a photograph of particle tracks made in a bubble chamber at Fermilab.

(a)   (b)

momentum are found from the curvature of their paths in a magnetic field of known magnitude and direction.

A **photographic emulsion** is the simplest example of a track detector. A charged particle ionizes the atoms in an emulsion layer. The path of the particle corresponds to a family of points at which chemical changes have occurred in the emulsion. When the emulsion is developed, the particle's track becomes visible.

A **cloud chamber** contains a gas that has been supercooled to just below its usual condensation point. An energetic charged particle passing through ionizes the gas along its path. The ions serve as centers for condensation of the supercooled gas. The track can be seen with the naked eye and can be photographed. A magnetic field can be applied to determine the charges of the radioactive particles, as well as their momentum and energy.

A device called a **bubble chamber**, invented in 1952 by D. Glaser, uses a liquid (usually liquid hydrogen) maintained near its boiling point. Ions produced by incoming charged particles leave bubblelike tracks, which can be photographed (Fig. 29.18). Because the density of the liquid in a bubble chamber is much higher than the density of the gas in a cloud chamber, the bubble chamber has a much higher sensitivity.

A **wire chamber** consists of thousands of closely spaced parallel wires that collect the electrons created by a passing ionizing particle. A second grid, with wires perpendicular to the first, allows the x,y position of the particle in the plane of the two sets of wires to be determined. Finally several such x,y grids arranged parallel to each other along the z-axis can be used to determine the particle's track in three dimensions. Wire chambers form a part of most detectors used at high-energy accelerator labs and provide electronic readouts to a computer for rapid reconstruction and display of tracks.

SUMMARY

Physics⊗Now™ Take a practice test by logging into PhysicsNow at **www.cp7e.com** and clicking on the Pre-Test link for this chapter.

29.1 Some Properties of Nuclei &
29.2 Binding Energy

Nuclei are represented symbolically as $_Z^A$X, where X represents the chemical symbol for the element. The quantity A is the **mass number**, which equals the total number of nucleons (neutrons plus protons) in the nucleus. The quantity Z is the **atomic number**, which equals the number of protons in the nucleus. Nuclei that contain the same number of protons but different numbers of neutrons are called **isotopes**. In other words, isotopes have the same Z value but different A values.

Most nuclei are approximately spherical, with an average radius given by

$$r = r_0 A^{1/3} \qquad [29.1]$$

where A is the mass number and r_0 is a constant equal to 1.2×10^{-15} m.

The total mass of a nucleus is always less than the sum of the masses of its individual nucleons. This mass difference Δm, multiplied by c^2, gives the **binding energy** of the nucleus.

29.3   Radioactivity

The spontaneous emission of radiation by certain nuclei is called **radioactivity**. There are three processes by which a radioactive substance can decay: alpha (α) decay, in which the emitted particles are ^4_2He nuclei; beta (β) decay, in which the emitted particles are electrons or positrons; and gamma (γ) decay, in which the emitted particles are high-energy photons.

The **decay rate**, or **activity**, R, of a sample is given by

$$R = \left| \frac{\Delta N}{\Delta t} \right| = \lambda N \qquad \textbf{[29.3]}$$

where N is the number of radioactive nuclei at some instant and λ is a constant for a given substance called the **decay constant**.

Nuclei in a radioactive substance decay in such a way that the number of nuclei present varies with time according to the expression

$$N = N_0 e^{-\lambda t} \qquad \textbf{[29.4a]}$$

where N is the number of radioactive nuclei present at time t, N_0 is the number at time $t = 0$, and $e = 2.718 \ldots$ is the base of the natural logarithms.

The **half-life** $T_{1/2}$ of a radioactive substance is the time required for half of a given number of radioactive nuclei to decay. The half-life is related to the decay constant by

$$T_{1/2} = \frac{0.693}{\lambda} \qquad \textbf{[29.5]}$$

29.4   The Decay Processes

If a nucleus decays by alpha emission, it loses two protons and two neutrons. A typical alpha decay is

$$^{238}_{92}\text{U} \quad \rightarrow \quad ^{234}_{90}\text{Th} + ^4_2\text{He} \qquad \textbf{[29.9]}$$

Note that in this decay, as in all radioactive decay processes, the sum of the Z values on the left equals the sum of the Z values on the right; the same is true for the A values.

A typical beta decay is

$$^{14}_6\text{C} \quad \rightarrow \quad ^{14}_7\text{N} + e^- + \overline{\nu} \qquad \textbf{[29.15]}$$

When a nucleus undergoes beta decay, an **antineutrino** is emitted along with an electron, or a **neutrino** along with a positron. A neutrino has zero electric charge and a small mass (which may be zero) and interacts weakly with matter.

Nuclei are often in an excited state following radioactive decay, and they release their extra energy by emitting a high-energy photon called a **gamma ray** (γ). A typical gamma ray emission is

$$^{12}_6\text{C}^* \quad \rightarrow \quad ^{12}_6\text{C} + \gamma \qquad \textbf{[29.18]}$$

where the asterisk indicates that the carbon nucleus was in an excited state before gamma emission.

29.6   Nuclear Reactions

Nuclear reactions can occur when a bombarding particle strikes another nucleus. A typical nuclear reaction is

$$^4_2\text{He} + ^{14}_7\text{N} \quad \rightarrow \quad ^{17}_8\text{O} + ^1_1\text{H} \qquad \textbf{[29.21]}$$

In this reaction, an alpha particle strikes a nitrogen nucleus, producing an oxygen nucleus and a proton. As in radioactive decay, atomic numbers and mass numbers balance on the two sides of the arrow.

Nuclear reactions in which energy is released are said to be **exothermic reactions** and are characterized by positive Q values. Reactions with negative Q values, called **endothermic reactions**, cannot occur unless the incoming particle has at least enough kinetic energy to overcome the energy deficit. In order to conserve both energy and momentum, the incoming particle must have a minimum kinetic energy, called the **threshold energy**, given by

$$KE_{\min} = \left(1 + \frac{m}{M} \right) |Q| \qquad \textbf{[29.24]}$$

where m is the mass of the incident particle and M is the mass of the target atom.

CONCEPTUAL QUESTIONS

1. Isotopes of a given element have different physical properties, such as mass, but the same chemical properties. Why is this?

2. If a heavy nucleus that is initially at rest undergoes alpha decay, which has more kinetic energy after the decay, the alpha particle or the daughter nucleus?

3. A student claims that a heavy form of hydrogen decays by alpha emission. How do you respond?

4. Explain the main differences between alpha, beta, and gamma rays.

5. In beta decay, the energy of the electron or positron emitted from the nucleus lies somewhere in a relatively large range of possibilities. In alpha decay, however, the alpha particle energy can only have discrete values. Why is there this difference?

6. If film is kept in a box, alpha particles from a radioactive source outside the box cannot expose the film, but beta particles can. Explain.

7. In positron decay, a proton in the nucleus becomes a neutron, and the positive charge is carried away by the positron. But a neutron has a larger rest energy than a proton. How is this possible?

8. An alpha particle has twice the charge of a beta particle. Why does the former deflect less than the latter when passing between electrically charged plates, assuming they both have the same speed?

9. Can carbon-14 dating be used to measure the age of a stone?

10. Pick any beta-decay process and show that the neutrino must have zero charge.

11. Why do heavier elements require more neutrons in order to maintain stability?

12. Suppose it could be shown that the intensity of cosmic rays was much greater 10 000 years ago. How would this affect the ages we assign to ancient samples of once-living matter?

13. Compare and contrast a photon and a neutrino.

14. Why is carbon dating unable to provide accurate estimates of very old materials?

15. Two samples of the same radioactive nuclide are prepared. Sample A has twice the intial activity of sample B. How does the half-life of A compare with the half-life of B? After each has passed through five half-lives, what is the ratio of their activities?

16. (a) Describe what happens to the number of protons and neutrons in a nucleus when the nucleus undergoes alpha decay. (b) Repeat for beta decay.

PROBLEMS

1, 2, 3 = straightforward, intermediate, challenging   ☐ = full solution available in *Student Solutions Manual/Study Guide*

Physics⊗Now™ = coached problem with hints available at **www.cp7e.com**   🧬 = biomedical application

Table 29.4 will be useful for many of these problems. A more complete list of atomic masses is given in Appendix B.

Section 29.1 Some Properties of Nuclei

1. Compare the nuclear radii of the following nuclides: ^2_1H, $^{60}_{27}Co$, $^{197}_{79}Au$, $^{239}_{94}Pu$.

2. What is the order of magnitude of the number of protons in your body? Of the number of neutrons? Of the number of electrons?

3. Using the result of Example 29.1, find the radius of a sphere of nuclear matter that would have a mass equal to that of the Earth. The Earth has a mass of 5.98×10^{24} kg and average radius of 6.37×10^6 m.

4. Consider the hydrogen atom to be a sphere of radius equal to the Bohr radius, 0.53×10^{-10} m, and calculate the approximate value of the ratio of the nuclear density to the atomic density.

5. An alpha particle ($Z = 2$, mass 6.64×10^{-27} kg) approaches to within 1.00×10^{-14} m of a carbon nucleus ($Z = 6$). What are (a) the maximum Coulomb

TABLE 29.4

Some Atomic Masses

Element	Atomic Mass (u)	Element	Atomic Mass (u)
$(^0_{-1}e)$	0.000 549	$^{23}_{11}Na$	22.989 770
(^0_1n)	1.008 665	$^{23}_{12}Mg$	22.994 127
^1_1H	1.007 825	$^{27}_{13}Al$	26.981 538
^2_1H	2.014 102	$^{30}_{15}P$	29.978 310
^4_2He	4.002 602	$^{40}_{20}Ca$	39.962 591
^7_3Li	7.016 003	$^{42}_{20}Ca$	41.958 622
^9_4Be	9.012 174	$^{43}_{20}Ca$	42.958 770
$^{10}_5B$	10.012 936	$^{56}_{26}Fe$	55.934 940
$^{12}_6C$	12.000 000	$^{64}_{30}Zn$	63.929 144
$^{13}_6C$	13.003 355	$^{64}_{29}Cu$	63.929 599
$^{14}_7N$	14.003 074	$^{93}_{41}Nb$	92.906 377
$^{15}_7N$	15.000 108	$^{197}_{79}Au$	196.966 543
$^{15}_8O$	15.003 065	$^{202}_{80}Hg$	201.970 617
$^{17}_8O$	16.999 131	$^{216}_{84}Po$	216.001 790
$^{18}_8O$	17.999 160	$^{220}_{86}Rn$	220.011 401
$^{18}_9F$	18.000 937	$^{234}_{90}Th$	234.043 583
$^{20}_{10}Ne$	19.992 435	$^{238}_{92}U$	238.050 784

force on the alpha particle, (b) the acceleration of the alpha particle at this time, and (c) the potential energy of the alpha particle at the same time?

6. Singly ionized carbon atoms are accelerated through 1 000 V and passed into a mass spectrometer to determine the isotopes present. (See Chapter 19.) The magnetic field strength in the spectrometer is 0.200 T. (a) Determine the orbital radii for the ^{12}C and the ^{13}C isotopes as they pass through the field. (b) Show that the ratio of the radii may be written in the form

$$\frac{r_1}{r_2} = \sqrt{\frac{m_1}{m_2}}$$

and verify that your radii in part (a) satisfy this formula.

7. (a) Find the speed an alpha particle requires to come within 3.2×10^{-14} m of a gold nucleus. (b) Find the energy of the alpha particle in MeV.

8. Find the radius of a nucleus of (a) $^4_2$He and (b) $^{238}_{92}$U.

Section 29.2 Binding Energy

9. Calculate the average binding energy per nucleon of $^{93}_{41}$Nb and $^{197}_{79}$Au.

10. Calculate the binding energy per nucleon for (a) ^{2}H, (b) ^{4}He, (c) ^{56}Fe, and (d) ^{238}U.

11. A pair of nuclei for which $Z_1 = N_2$ and $Z_2 = N_1$ are called *mirror isobars*. (The atomic and neutron numbers are interchangeable.) Binding-energy measurements on such pairs can be used to obtain evidence of the charge independence of nuclear forces. Charge independence means that the proton–proton, proton–neutron, and neutron–neutron forces are approximately equal. Calculate the difference in binding energy for the two mirror nuclei $^{15}_8$O and $^{15}_7$N.

12. The peak of the stability curve occurs at ^{56}Fe. This is why iron is prominent in the spectrum of the Sun and stars. Show that ^{56}Fe has a higher binding energy per nucleon than its neighbors ^{55}Mn and ^{59}Co. Compare your results with Figure 29.4.

13. Two nuclei having the same mass number are known as *isobars*. Calculate the difference in binding energy per nucleon for the isobars $^{23}_{11}$Na and $^{23}_{12}$Mg. How do you account for this difference?

14. Calculate the binding energy of the last neutron in the $^{43}_{20}$Ca nucleus. [*Hint*: You should compare the mass of $^{43}_{20}$Ca with the mass of $^{42}_{20}$Ca plus the mass of a neutron. The mass of $^{42}_{20}$Ca = 41.958 622 u.]

Section 29.3 Radioactivity

15. **Physics⊗Now™** The half-life of an isotope of phosphorus is 14 days. If a sample contains 3.0×10^{16} such nuclei, determine its activity. Express your answer in curies.

16. A drug tagged with $^{99}_{43}$Tc (half-life = 6.05 h) is prepared for a patient. If the original activity of the sample was 1.1×10^4 Bq, what is its activity after it has sat on the shelf for 2.0 h?

17. The half-life of ^{131}I is 8.04 days. (a) Calculate the decay constant for this isotope. (b) Find the number of ^{131}I nuclei necessary to produce a sample with an activity of 0.50 μCi.

18. After 2.00 days, the activity of a sample of an unknown type of radioactive material has decreased to 84.2% of the initial activity. (a) What is the half-life of this material? (b) Can you identify it by using the table of isotopes in Appendix B?

19. Suppose that you start with 1.00×10^{-3} g of a pure radioactive substance and 2.0 h later determine that only 0.25×10^{-3} g of the substance remains. What is the half-life of this substance?

20. Radon gas has a half-life of 3.83 days. If 3.00 g of radon gas is present at time $t = 0$, what mass of radon will remain after 1.50 days have passed?

21. Many smoke detectors use small quantities of the isotope ^{241}Am in their operation. The half-life of ^{241}Am is 432 yr. How long will it take for the activity of this material to decrease to 1.00×10^{-3} of the original activity?

22. After a plant or animal dies, its ^{14}C content decreases with a half-life of 5 730 yr. If an archaeologist finds an ancient firepit containing partially consumed firewood, and the ^{14}C content of the wood is only 12.5% that of an equal carbon sample from a present-day tree, what is the age of the ancient site?

23. A freshly prepared sample of a certain radioactive isotope has an activity of 10.0 mCi. After 4.00 h, the activity is 8.00 mCi. (a) Find the decay constant and half-life of the isotope. (b) How many atoms of the isotope were contained in the freshly prepared sample? (c) What is the sample's activity 30 h after it is prepared?

24. A building has become accidentally contaminated with radioactivity. The longest-lived material in the building is strontium-90. (The atomic mass of $^{90}_{38}$Sr is 89.907 7.) If the building initially contained 5.0 kg of this substance, and the safe level is less than 10.0 counts/min, how long will the building be unsafe?

Section 29.4 The Decay Processes

25. Complete the following radioactive decay formulas:

$$^{212}_{83}\text{Bi} \rightarrow ? + {}^4_2\text{He}$$
$$^{95}_{36}\text{Kr} \rightarrow ? + e^-$$
$$? \rightarrow {}^4_2\text{He} + {}^{140}_{58}\text{Ce}$$

26. Complete the following radioactive decay formulas:

$$^{12}_5\text{B} \rightarrow ? + e^-$$
$$^{234}_{90}\text{Th} \rightarrow {}^{230}_{88}\text{Ra} + ?$$
$$? \rightarrow {}^{14}_7\text{N} + e^-$$

27. Complete the following radioactive decay formulas:

$$? \rightarrow e^+ + {}^{40}_{19}\text{K}$$
$$^{94}_{44}\text{Ru} \rightarrow {}^4_2\text{He} + ?$$
$$^{144}_{60}\text{Nd} \rightarrow ? + {}^{140}_{58}\text{Ce}$$

28. Figure P29.28 shows the steps by which $^{235}_{92}$U decays to $^{207}_{82}$Pb. Enter the correct isotope symbol in each square.

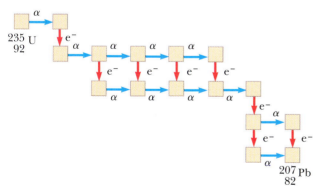

Figure P29.28

29. **Physics⊗Now™** The mass of ^{56}Fe is 55.934 9 u and the mass of ^{56}Co is 55.939 9 u. Which isotope decays into the other and by what process?

30. Find the energy released in the alpha decay of $^{238}_{92}$U. The following mass value will be useful: $^{234}_{90}$Th has a mass of 234.043 583 u.

31. Determine which of the following suggested decays can occur spontaneously:

(a) $^{40}_{20}\text{Ca} \rightarrow e^+ + {}^{40}_{19}\text{K};$ (b) $^{144}_{60}\text{Nd} \rightarrow {}^4_2\text{He} + {}^{140}_{58}\text{Ce}$

32. $^{66}_{28}$Ni (mass = 65.929 1 u) undergoes beta decay to $^{66}_{29}$Cu (mass = 65.9289 u). (a) Write the complete decay formula for this process. (b) Find the maximum kinetic energy of the emerging electrons.

33. An ^{3}H nucleus beta decays into ^{3}He by creating an electron and an antineutrino according to the reaction

$$^3_1\text{H} \rightarrow {}^3_2\text{He} + e^- + \overline{\nu}$$

Use Appendix B to determine the total energy released in this reaction.

34. A piece of charcoal used for cooking is found at the remains of an ancient campsite. A 1.00-kg sample of carbon from the wood has an activity of 2.00×10^3 decays per minute. Find the age of the charcoal. [*Hint:* Living material has an activity of 15.0 decays/minute per gram of carbon present.]

35. A wooden artifact is found in an ancient tomb. Its carbon-14 ($^{14}_6$C) activity is measured to be 60.0% of that in a fresh sample of wood from the same region. Assuming the same amount of ^{14}C was initially present in the wood from which the artifact was made, determine the age of the artifact.

36. A living specimen in equilibrium with the atmosphere contains one atom of ^{14}C (half-life = 5 730 yr) for every 7.70×10^{11} stable carbon atoms. An archeological sample of wood (cellulose, $C_{12}H_{22}O_{11}$) contains 21.0 mg of carbon. When the sample is placed inside a shielded beta counter with 88.0% counting efficiency, 837 counts are accumulated in one week. Assuming that the cosmic-ray flux and the Earth's atmosphere have not changed appreciably since the sample was formed, find the age of the sample.

Section 29.6 Nuclear Reactions

37. The first known reaction in which the product nucleus was radioactive (achieved in 1934) was one in which $^{27}_{13}$Al was bombarded with alpha particles. Produced in the reaction were a neutron and a product nucleus. (a) What was the product nucleus? (b) Find the Q value of the reaction.

38. Complete the following nuclear reactions:

$$? + {}^{14}_7\text{N} \rightarrow {}^1_1\text{H} + {}^{17}_8\text{O}$$
$$^7_3\text{Li} + {}^1_1\text{H} \rightarrow {}^4_2\text{He} + ?$$

39. Identify the unknown particles X and X' in the following nuclear reactions:

$$X + {}_{2}^{4}\text{He} \rightarrow {}_{12}^{24}\text{Mg} + {}_{0}^{1}\text{n}$$

$$ {}_{92}^{235}\text{U} + {}_{0}^{1}\text{n} \rightarrow {}_{38}^{90}\text{Sr} + X + 2{}_{0}^{1}\text{n}$$

$$2{}_{1}^{1}\text{H} \rightarrow {}_{1}^{2}\text{H} + X + X'$$

40. The first nuclear reaction utilizing particle accelerators was performed by Cockcroft and Walton. Accelerated protons were used to bombard lithium nuclei, producing the following reaction:

$$ {}_{1}^{1}\text{H} + {}_{3}^{7}\text{Li} \rightarrow {}_{2}^{4}\text{He} + {}_{2}^{4}\text{He}$$

Since the masses of the particles involved in the reaction were well known, these results were used to obtain an early proof of the Einstein mass–energy relation. Calculate the Q value of the reaction.

41. (a) Suppose ${}^{10}_{5}\text{B}$ is struck by an alpha particle, releasing a proton and a product nucleus in the reaction. What is the product nucleus? (b) An alpha particle and a product nucleus are produced when ${}^{13}_{6}\text{C}$ is struck by a proton. What is the product nucleus?

42. (a) Determine the product of the reaction ${}_{3}^{7}\text{Li} + {}_{2}^{4}\text{He} \rightarrow ? + \text{n}$. (b) What is the Q value of the reaction?

43. Natural gold has only one isotope: ${}_{79}^{197}\text{Au}$. If gold is bombarded with slow neutrons, e^{-} particles are emitted. (a) Write the appropriate reaction equation. (b) Calculate the maximum energy of the emitted beta particles. The mass of ${}_{80}^{198}\text{Hg}$ is 197.966 75 u.

44. Find the threshold energy that an incident neutron must have to produce the reaction

$$ {}_{0}^{1}\text{n} + {}_{2}^{4}\text{He} \rightarrow {}_{1}^{2}\text{H} + {}_{1}^{3}\text{H}$$

45. **Physics⊗Now™** When ${}^{18}\text{O}$ is struck by a proton, ${}^{18}\text{F}$ and another particle are produced. (a) What is the other particle? (b) The reaction has a Q value of -2.453 MeV, and the atomic mass of ${}^{18}\text{O}$ is 17.999 160 u. What is the atomic mass of ${}^{18}\text{F}$?

Section 29.7 Medical Applications of Radiation

46. In terms of biological damage, how many rad of heavy ions is equivalent to 100 rad of x-rays?

47. A person whose mass is 75.0 kg is exposed to a whole-body dose of 25.0 rads. How many joules of energy are deposited in the person's body?

48. A 200-rad dose of radiation is administered to a patient in an effort to combat a cancerous growth. Assuming all of the energy deposited is absorbed by the growth, (a) calculate the amount of energy delivered per unit mass. (b) Assuming the growth has a mass of 0.25 kg and a specific heat equal to that of water, calculate its temperature rise.

49. A "clever" technician decides to heat some water for his coffee with an x-ray machine. If the machine produces 10 rad/s, how long will it take to raise the temperature of a cup of water by 50°C. Ignore heat losses during this time.

50. An x-ray technician works 5 days per week, 50 weeks per year. Assume that the technician takes an average of eight x-rays per day and receives a dose of 5.0 rem/yr as a result. (a) Estimate the dose in rem per x-ray taken. (b) How does this result compare with the amount of low-level background radiation the technician is exposed to?

51. **Physics⊗Now™** A patient swallows a radiopharmaceutical tagged with phosphorus-32 (${}_{15}^{32}\text{P}$), a β^{-} emitter with a half-life of 14.3 days. The average kinetic energy of the emitted electrons is 700 keV. If the initial activity of the sample is 1.31 MBq, determine (a) the number of electrons emitted in a 10-day period, (b) the total energy deposited in the body during the 10 days, and (c) the absorbed dose if the electrons are completely absorbed in 100 g of tissue.

52. A particular radioactive source produces 100 mrad of 2-MeV gamma rays per hour at a distance of 1.0 m. (a) How long could a person stand at this distance before accumulating an intolerable dose of 1 rem? (b) Assuming the gamma radiation is emitted uniformly in all directions, at what distance would a person receive a dose of 10 mrad/h from this source?

ADDITIONAL PROBLEMS

53. A 200.0-mCi sample of a radioactive isotope is purchased by a medical supply house. If the sample has a half-life of 14.0 days, how long will it keep before its activity is reduced to 20.0 mCi?

54. A sample of organic material is found to contain 18 g of carbon. The investigators believe the material to be 20 000 years old, based on samples of pottery found at the site. If so, what is the expected activity of the organic material? Take data from Example 29.7.

55. Deuterons that have been accelerated are used to bombard other deuterium nuclei, resulting in the reaction

$$ {}_{1}^{2}\text{H} + {}_{1}^{2}\text{H} \rightarrow {}_{2}^{3}\text{He} + {}_{0}^{1}\text{n}$$

Does this reaction require a threshold energy? If so, what is its value?

56. Free neutrons have a characteristic half-life of 12 min. What fraction of a group of free neutrons at a thermal energy of 0.040 eV will decay before traveling a distance of 10.0 km?

57. A by-product of some fission reactors is the isotope $^{239}_{94}$Pu, an alpha emitter having a half-life of 24 120 yr. The reaction involved is

$$^{239}_{94}\text{Pu} \rightarrow ^{235}_{92}\text{U} + \alpha$$

Consider a sample of 1.00 kg of pure $^{239}_{94}$Pu at $t = 0$. Calculate (a) the number of $^{239}_{94}$Pu nuclei present at $t = 0$ and (b) the initial activity in the sample. (c) How long does the sample have to be stored if a "safe" activity level is 0.100 Bq?

58. (a) Find the radius of the $^{12}_{6}$C nucleus. (b) Find the force of repulsion between a proton at the surface of a $^{12}_{6}$C nucleus and the remaining five protons. (c) How much work (in MeV) has to be done to overcome this electrostatic repulsion in order to put the last proton into the nucleus? (d) Repeat (a), (b), and (c) for $^{238}_{92}$U.

59. In a piece of rock from the Moon, the ^{87}Rb content is assayed to be 1.82×10^{10} atoms per gram of material and the ^{87}Sr content is found to be 1.07×10^9 atoms per gram. (The relevant decay is ^{87}Rb $\rightarrow$ ^{87}Sr + e$^-$. The half-life of the decay is 4.8×10^{10} yr.) (a) Determine the age of the rock. (b) Could the material in the rock actually be much older? What assumption is implicit in using the radioactive-dating method?

60. Many radioisotopes have important industrial, medical, and research applications. One such radioisotope is ^{60}Co, which has a half-life of 5.2 yr and decays by the emission of a beta particle (energy 0.31 MeV) and two gamma photons (energies 1.17 MeV and 1.33 MeV). A scientist wishes to prepare a ^{60}Co sealed source that will have an activity of at least 10 Ci after 30 months of use. What is the minimum initial mass of ^{60}Co required?

61. A medical laboratory stock solution is prepared with an initial activity due to ^{24}Na of 2.5 mCi/ml, and 10.0 ml of the stock solution is diluted at $t_0 = 0$ to a working solution whose total volume is 250 ml. After 48 h, a 5.0-ml sample of the working solution is monitored with a counter. What is the measured activity? (Note that 1 ml = 1 milliliter.)

62. A theory of nuclear astrophysics is that all the heavy elements such as uranium are formed in supernova explosions of massive stars, which immediately release the elements into space. If we assume that at the time of an explosion there were equal amounts of ^{235}U and ^{238}U, how long ago were the elements that formed our Earth released, given that the present ^{235}U/^{238}U ratio is 0.007? (The half-lives of ^{235}U and ^{238}U are 0.70×10^9 yr and 4.47×10^9 yr, respectively.)

63. A fission reactor is hit by a nuclear weapon, causing 5.0×10^6 Ci of ^{90}Sr ($T_{1/2} = 28.7$ yr) to evaporate into the air. The ^{90}Sr falls out over an area of 10^4 km^2. How long will it take the activity of the ^{90}Sr to reach the agriculturally "safe" level of 2.0 μCi/m^2?

64. After the sudden release of radioactivity from the Chernobyl nuclear reactor accident in 1986, the radioactivity of milk in Poland rose to 2 000 Bq/L due to iodine-131, with a half-life of 8.04 days. Radioactive iodine is particularly hazardous, because the thyroid gland concentrates iodine. The Chernobyl accident caused a measurable increase in thyroid cancers among children in Belarus. (a) For comparison, find the activity of milk due to potassium. Assume that 1 liter of milk contains 2.00 g of potassium, of which 0.011 7% is the isotope ^{40}K, which has a half-life of 1.28×10^9 yr. (b) After what length of time would the activity due to iodine fall below that due to potassium?

65. During the manufacture of a steel engine component, radioactive iron (^{59}Fe) is included in the total mass of 0.20 kg. The component is placed in a test engine when the activity due to the isotope is 20.0 μCi. After a 1 000-h test period, oil is removed from the engine and is found to contain enough ^{59}Fe to produce 800 disintegrations/min per liter of oil. The total volume of oil in the engine is 6.5 L. Calculate the total mass worn from the engine component per hour of operation. (The half-life of ^{59}Fe is 45.1 days.)

66. After determining that the Sun has existed for hundreds of millions of years, but before the discovery of nuclear physics, scientists could not explain why the Sun has continued to burn for such a long time. For example, if it were a coal fire, the Sun would have burned up in about 3 000 yr. Assume that the Sun, whose mass is 1.99×10^{30} kg, originally consisted entirely of hydrogen and that its total power output is 3.76×10^{26} W. (a) If the energy-generating mechanism of the Sun is the transforming of hydrogen into helium via the net reaction

$$4^1_1\text{H} + 2e^- \rightarrow ^4_2\text{He} + 2\nu + \gamma$$

calculate the energy (in joules) given off by this reaction. (b) Determine how many hydrogen atoms constitute the Sun. Take the mass of one hydrogen atom to be 1.67×10^{-27} kg. (c) Assuming that the total power output remains constant, after what time will all the hydrogen be converted into helium, making the Sun die? The actual projected lifetime of the Sun is about 10 billion years, because only the hydrogen in a relatively small core is available as a fuel. (Only in the Sun's core are temperatures and densities high enough for the fusion reaction to be self-sustaining).

ACTIVITIES

1. This experiment will take a little longer to do than most that we have suggested, but the time spent is worthwhile to help you understand the concept of half-life. Obtain a box of sugar cubes and with a pencil make a mark on one side of each of about 200 cubes. Each of these cubes will represent the nucleus of a radioactive substance. Thus, at $t = 0$, you have 200 undecayed nuclei. Now, put the 200 marked cubes in a box and roll them out on a table, just as you would roll dice. Next, count and remove any cubes that have landed marked-side up. These cubes represent nuclei that emitted radiation during the roll. They are no longer radioactive and thus do not participate in the rest of the action. Record the number of undecayed cubes remaining as the number of undecayed nuclei at $t = 1$ roll.

Continue rolling, counting, and removing until you have completed 12 to 15 rolls. By then, you should have only a few cubes remaining. Plot a graph of undecayed cubes versus the roll number and from this determine the "half-roll" of the cubes.

2. Use a nail to punch a hole in the bottom of a large tin can. Hold the can beneath a faucet and adjust the water flow from the faucet to a fine constant stream. Although water flows from the hole at the bottom, you will note that the level of the water in the can rises. As it does so, however, the flow of water leaving the can increases due to increased water pressure caused by the greater depth of water. Unless the flow of water is too great, an equilibrium point will be reached at which the amount of water flowing out of the can each second exactly equals the amount flowing in each second. When this happens, the level of water in the can is constant. As noted in the text, carbon-14 is continually being produced in the atmosphere and is also continually disappearing as it decays into nitrogen. What is the analogy between water entering the can, remaining in the can, and flowing out of the can and the behavior of carbon-14 in the atmosphere?

This photo shows scientist Melissa Douglas and part of the Z machine, an inertial-electrostatic confinement fusion apparatus at Sandia National Laboratories. In the device, giant capacitors discharge through a grid of tungsten wires finer than human hairs, vaporizing them. The tungsten ions implode inward at a million miles an hour. Braking strongly in the grip of a "Z-pinch," they emit powerful x-rays that compress a deuterium pellet, causing collisions between the deuterium atoms that lead to fusion events.

Sandia National Laboratories. Photo by Randy Montoya

Nuclear Energy and Elementary Particles

In this concluding chapter we discuss the two means by which energy can be derived from nuclear reactions: fission, in which a nucleus of large mass number splits into two smaller nuclei, and fusion, in which two light nuclei fuse to form a heavier nucleus. In either case, there is a release of large amounts of energy, which can be used destructively through bombs or constructively through the production of electric power. We end our study of physics by examining the known subatomic particles and the fundamental interactions that govern their behavior. We also discuss the current theory of elementary particles, which states that all matter in nature is constructed from only two families of particles: quarks and leptons. Finally, we describe how such models help us understand the evolution of the Universe.

30.1   NUCLEAR FISSION

Nuclear fission occurs when a heavy nucleus, such as ^{235}U, splits, or fissions, into two smaller nuclei. In such a reaction, **the total mass of the products is less than the original mass of the heavy nucleus**.

Nuclear fission was first observed in 1939 by Otto Hahn and Fritz Strassman, following some basic studies by Fermi. After bombarding uranium ($Z = 92$) with neutrons, Hahn and Strassman discovered two medium-mass elements, barium and lanthanum, among the reaction products. Shortly thereafter, Lise Meitner and Otto Frisch explained what had happened: the uranium nucleus had split into two nearly equal fragments after absorbing a neutron. This was of considerable interest to physicists attempting to understand the nucleus, but it was to have even more far-reaching consequences. Measurements showed that about 200 MeV of energy is released in each fission event, and this fact was to affect the course of human history.

The fission of ^{235}U by slow (low-energy) neutrons can be represented by the sequence of events

$$_0^1\text{n} + {}_{92}^{235}\text{U} \rightarrow {}_{92}^{236}\text{U}^* \rightarrow \text{X} + \text{Y} + \text{neutrons} \qquad [30.1]$$

where $^{236}\text{U}^*$ is an intermediate state that lasts only for about 10^{-12} s before splitting into nuclei X and Y, called **fission fragments**. There are many combinations of X and Y that satisfy the requirements of conservation of energy and charge. In the fission of uranium, about 90 different daughter nuclei can be formed. The process also results in the production of several (typically two or three) neutrons per fission event. On the average, 2.47 neutrons are released per event.

A typical reaction of this type is

$$_0^1\text{n} + {}_{92}^{235}\text{U} \rightarrow {}_{56}^{141}\text{Ba} + {}_{36}^{92}\text{Kr} + 3_0^1\text{n} \qquad [30.2]$$

The fission fragments, barium and krypton, and the released neutrons have a great deal of kinetic energy following the fission event.

The breakup of the uranium nucleus can be compared to what happens to a drop of water when excess energy is added to it. All of the atoms in the drop have energy, but not enough to break up the drop. However, if enough energy is added to set the drop vibrating, it will undergo elongation and compression until the amplitude of vibration becomes large enough to cause the drop to break apart. In the uranium nucleus, a similar process occurs (Fig. 30.1). The sequence of events is as follows:

Sequence of events in a nuclear fission process ▶

1. The ^{235}U nucleus captures a thermal (slow-moving) neutron.
2. The capture results in the formation of $^{236}\text{U}^*$, and the excess energy of this nucleus causes it to undergo violent oscillations.
3. The $^{236}\text{U}^*$ nucleus becomes highly elongated, and the force of repulsion between protons in the two halves of the dumbbell-shaped nucleus tends to increase the distortion.
4. The nucleus splits into two fragments, emitting several neutrons in the process.

The energy released in a typical fission process Q can be estimated. From Figure 29.4, we see that the binding energy per nucleon is about 7.2 MeV for heavy nuclei (those having a mass number of approximately 240) and about 8.2 MeV for nuclei of intermediate mass. This means that the nucleons in the fission fragments are more tightly bound, and therefore have less mass, than the nucleons in the original heavy nucleus. The decrease in mass per nucleon appears as released energy when fission occurs. The amount of energy released is $(8.2 - 7.2)$ MeV per nucleon. Assuming a total of 240 nucleons, we find that the energy released per fission event is

$$Q = 240 \text{ nucleons}/(8.2 \text{ MeV/nucleon} - 7.2 \text{ MeV/nucleon}) = 240 \text{ MeV}$$

This is a large amount of energy relative to the amount released in chemical processes. For example, the energy released in the combustion of one molecule of the octane used in gasoline engines is about one hundred-millionth the energy released in a single fission event!

Figure 30.1 A nuclear fission event as described by the liquid-drop model of the nucleus. (a) A slow neutron approaches a ^{235}U nucleus. (b) The neutron is absorbed by the ^{235}U nucleus, changing it to $^{236}\text{U}^*$, which is a ^{236}U nucleus in an excited state. (c) The nucleus deforms and oscillates like a liquid drop. (d) The nucleus undergoes fission, resulting in two lighter nuclei X and Y, along with several neutrons.

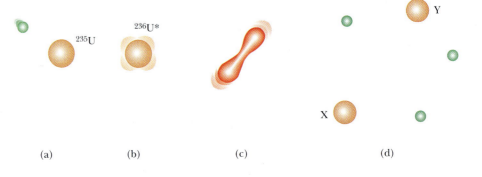

Applying Physics 30.1 Unstable Products

If a heavy nucleus were to fission into just two product nuclei, they would be very unstable. Why is this?

Explanation According to Figure 29.3, the ratio of the number of neutrons to the number of protons increases with Z. As a result, when a heavy nucleus splits in a fission reaction to two lighter nuclei, the lighter nuclei tend to have too many neutrons. This leads to instability, as the nucleus returns to the curve in Figure 29.3 by decay processes that reduce the number of neutrons.

EXAMPLE 30.1 The Fission of Uranium

Goal Balance a nuclear equation to determine details of the fission fragments.

Problem When ^{235}U is struck by a neutron, there are various possible fission fragments. Determine the number of neutrons produced when the fission fragments are ^{140}Xe and ^{94}Sr (isotopes of xenon and strontium).

Strategy This is something like balancing chemical equations: the atomic numbers and mass numbers should balance on either side of the equation.

Solution

Write the equation describing the process, with an unknown number x of neutrons:

$$^{1}_{0}n + {}^{235}_{92}U \rightarrow {}^{140}_{54}Xe + {}^{94}_{38}Sr + x({}^{1}_{0}n)$$

The atomic numbers balance already, as they should. Write an equation relating the mass numbers:

$$1 + 235 = 140 + 94 + x \rightarrow \boxed{x = 2}$$

Remark In this case, the number of protons balanced automatically. If that were not the case, there might be other possible daughter particles, such as protons or helium nuclei (also called alpha particles).

Exercise 30.1

Find the number of neutrons released if the two major fragments are ^{132}Sn and ^{101}Mo.

Answer Three neutrons

Quick Quiz 30.1

In the first atomic bomb, the energy released was equivalent to about 30 kilotons of TNT, where a ton of TNT releases an energy of about 4.0×10^9 J. Estimate the amount of mass converted into energy in this event. (a) 1 μg (b) 1 mg (c) 1 g (d) 1 kg (e) 20 kilotons

EXAMPLE 30.2 A Fission-Powered World

Goal Relate raw material to energy output.

Problem (a) Calculate the total energy released if 1.00 kg of ^{235}U undergoes fission, taking the disintegration energy per event to be $Q = 208$ MeV (a more accurate value than the estimate given previously). (b) How many kilograms of ^{235}U would be needed to satisfy the world's annual energy consumption (about 4×10^{20} J)?

Strategy In part (a), use the concept of a mole and Avogadro's number to obtain the total number of nuclei. Multiplying by the energy per reaction then gives the total energy released. Part (b) requires some light algebra.

Solution

(a) Calculate the total energy released from 1.00 kg of ^{235}U.

Find the total number of nuclei in 1.00 kg of uranium:

$$N = \left(\frac{6.02 \times 10^{23} \text{ nuclei/mol}}{235 \text{ g/mol}} \right)(1.00 \times 10^3 \text{ g})$$

$$= 2.56 \times 10^{24} \text{ nuclei}$$

Multiply N by the energy yield per nucleus, obtaining the total disintegration energy:

$$E = NQ = (2.56 \times 10^{24} \text{ nuclei})\left(208 \frac{\text{MeV}}{\text{nucleus}}\right)$$

$$= 5.32 \times 10^{26} \text{ MeV}$$

(b) How many kilograms would provide for the world's annual energy needs?

Set the energy per kilogram, E_{kg}, times the number of kilograms, N_{kg}, equal to the total annual energy consumption. Solve for N_{kg}:

$$E_{kg} N_{kg} = E_{tot}$$

$$N_{kg} = \frac{E_{tot}}{E_{kg}} = \frac{4 \times 10^{20} \text{ J}}{(5.32 \times 10^{32} \text{ eV/kg})(1.60 \times 10^{-19} \text{ J/eV})}$$

$$= 5 \times 10^{6} \text{ kg}$$

Remark   The calculation implicitly assumes perfect conversion to usable power, which is never the case in real systems. There are enough known uranium deposits to provide the world's current energy requirements for a few hundred years.

Exercise 30.2
How long can one kilogram of U-235 keep a 100-watt lightbulb burning if all its released energy is converted to electrical energy?

Answer   ~30 000 yr

30.2   NUCLEAR REACTORS

We have seen that neutrons are emitted when ^{235}U undergoes fission. These neutrons can in turn trigger other nuclei to undergo fission, with the possibility of a chain reaction (Active Fig. 30.2). Calculations show that if the chain reaction isn't controlled, it will proceed too rapidly and possibly result in the sudden release of an enormous amount of energy (an explosion), even from only 1 g of ^{235}U. If the energy in 1 kg of ^{235}U were released, it would equal that released by the detonation

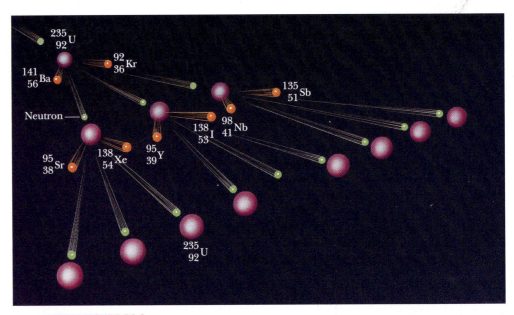

ACTIVE FIGURE 30.2
A nuclear chain reaction initiated by the capture of a neutron.

Physics⊗Now™
Log into PhysicsNow at **www.cp7e.com** and go to Active Figure 30.2 to observe a chain reaction.

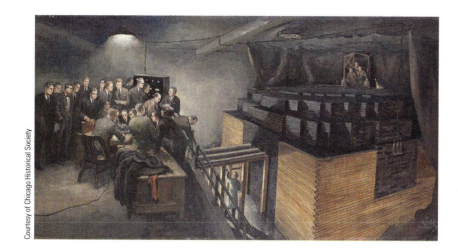

Courtesy of Chicago Historical Society

Painting of the world's first nuclear reactor. Because of wartime secrecy, there are no photographs of the completed reactor, which was composed of layers of graphite interspersed with uranium. A self-sustained chain reaction was first achieved on December 2, 1942. Word of the success was telephoned immediately to Washington with this message: "The Italian navigator has landed in the New World and found the natives very friendly." The historic event took place in an improvised laboratory in the racquet court under the west stands of the University of Chicago's Stagg Field. The Italian navigator was Fermi.

of about 20 000 tons of TNT! An uncontrolled fission reaction, of course, is the principle behind the first nuclear bomb.

A nuclear reactor is a system designed to maintain what is called a **self-sustained chain reaction**. This important process was first achieved in 1942 by a group led by Fermi at the University of Chicago, with natural uranium as the fuel. Most reactors in operation today also use uranium as fuel. Natural uranium contains only about 0.7% of the ^{235}U isotope, with the remaining 99.3% being the ^{238}U isotope. This is important to the operation of a reactor because ^{238}U almost never undergoes fission. Instead, it tends to absorb neutrons, producing neptunium and plutonium. For this reason, reactor fuels must be artificially enriched so that they contain several percent of the ^{235}U isotope.

Earlier we mentioned that an average of about 2.5 neutrons are emitted in each fission event of ^{235}U. In order to achieve a self-sustained chain reaction, one of these neutrons must be captured by another ^{235}U nucleus and cause it to undergo fission. A useful parameter for describing the level of reactor operation is the **reproduction constant K, defined as the average number of neutrons from each fission event that will cause another event**. As we have seen, K can have a maximum value of 2.5 in the fission of uranium. In practice, however, K is less than this because of several factors, which we soon discuss.

A self-sustained chain reaction is achieved when $K = 1$. Under this condition, the reactor is said to be **critical**. When K is less than one, the reactor is subcritical and the reaction dies out. When K is greater than one the reactor is said to be supercritical, and a runaway reaction occurs. In a nuclear reactor used to furnish power to a utility company, it is necessary to maintain a K value close to one.

The basic design of a nuclear reactor is shown in Figure 30.3. The fuel elements consist of enriched uranium. The functions of the remaining parts of the reactor and some aspects of its design are described next.

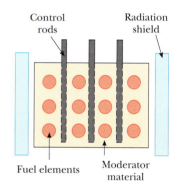

Figure 30.3   Cross section of a reactor core showing the control rods, fuel elements containing enriched fuel, and moderating material, all surrounded by a radiation shield.

Neutron Leakage

In any reactor, a fraction of the neutrons produced in fission will leak out of the core before inducing other fission events. If the fraction leaking out is too large, the reactor will not operate. The percentage lost is large if the reactor is very small because leakage is a function of the ratio of surface area to volume. Therefore, a critical requirement of reactor design is choosing the correct surface-area-to-volume ratio so that a sustained reaction can be achieved.

Regulating Neutron Energies

The neutrons released in fission events are highly energetic, with kinetic energies of about 2 MeV. It is found that slow neutrons are far more likely than fast neutrons to produce fission events in ^{235}U. Further, ^{238}U doesn't absorb slow neutrons. In order for the chain reaction to continue, therefore, the neutrons must be slowed down. This is accomplished by surrounding the fuel with a substance called a **moderator**.

In order to understand how neutrons are slowed down, consider a collision between a light object and a massive one. In such an event, the light object rebounds from the collision with most of its original kinetic energy. However, if the collision is between objects having masses that are nearly the same, the incoming projectile transfers a large percentage of its kinetic energy to the target. In the first nuclear reactor ever constructed, Fermi placed bricks of graphite (carbon) between the fuel elements. Carbon nuclei are about 12 times more massive than neutrons, but after about 100 collisions with carbon nuclei, a neutron is slowed sufficiently to increase its likelihood of fission with ^{235}U. In this design the carbon is the moderator; most modern reactors use heavy water (D_2O) as the moderator.

Neutron Capture

In the process of being slowed down, neutrons may be captured by nuclei that do not undergo fission. The most common event of this type is neutron capture by ^{238}U. The probability of neutron capture by ^{238}U is very high when the neutrons have high kinetic energies and very low when they have low kinetic energies. The slowing down of the neutrons by the moderator serves the dual purpose of making them available for reaction with ^{235}U and decreasing their chances of being captured by ^{238}U.

Control of Power Level

APPLICATION

Nuclear Reactor Design

It is possible for a reactor to reach the critical stage ($K = 1$) after all neutron losses described previously are minimized. However, a method of control is needed to adjust K to a value near one. If K were to rise above this value, the heat produced in the runaway reaction would melt the reactor. To control the power level, control rods are inserted into the reactor core. (See Fig. 30.3.) These rods are made of materials such as cadmium that are highly efficient in absorbing neutrons. By adjusting the number and position of the control rods in the reactor core, the K value can be varied and any power level within the design range of the reactor can be achieved.

A diagram of a pressurized-water reactor is shown in Figure 30.4. This type of reactor is commonly used in electric power plants in the United States. Fission

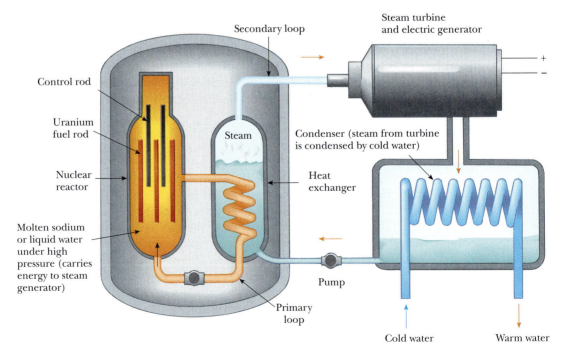

Figure 30.4 Main components of a pressurized-water nuclear reactor.

events in the reactor core supply heat to the water contained in the primary (closed) system, which is maintained at high pressure to keep it from boiling. This water also serves as the moderator. The hot water is pumped through a heat exchanger, and the heat is transferred to the water contained in the secondary system. There the hot water is converted to steam, which drives a turbine–generator to create electric power. Note that the water in the secondary system is isolated from the water in the primary system in order to prevent contamination of the secondary water and steam by radioactive nuclei from the reactor core.

Reactor Safety[1]

The safety aspects of nuclear power reactors are often sensationalized by the media and misunderstood by the public. The 1979 near disaster of Three Mile Island in Pennsylvania and the accident at the Chernobyl reactor in the Ukraine rightfully focused attention on reactor safety. Yet the safety record in the United States is enviable. The records show no fatalities attributed to commercial nuclear power generation in the history of the United States nuclear industry.

Commercial reactors achieve safety through careful design and rigid operating procedures. Radiation exposure and the potential health risks associated with such exposure are controlled by three layers of containment. The fuel and radioactive fission products are contained inside the reactor vessel. Should this vessel rupture, the reactor building acts as a second containment structure to prevent radioactive material from contaminating the environment. Finally, the reactor facilities must be in a remote location to protect the general public from exposure should radiation escape the reactor building.

According to the *Oak Ridge National Laboratory Review*, "The health risk of living within 8 km (5 miles) of a nuclear reactor for 50 years is no greater than the risk of smoking 1.4 cigarettes, drinking 0.5 liters of wine, traveling 240 km by car, flying 9 600 km by jet, or having one chest x-ray in a hospital. Each of these activities is estimated to increase a person's chances of dying in any given year by one in a million."

Another potential danger in nuclear reactor operations is the possibility that the water flow could be interrupted. Even if the nuclear fission chain reaction were stopped immediately, residual heat could build up in the reactor to the point of melting the fuel elements. The molten reactor core would melt its way to the bottom of the reactor vessel and conceivably into the ground below—the so-called China syndrome. Although it might appear that this deep underground burial site would be an ideal safe haven for a radioactive blob, there would be danger of a steam explosion should the molten mass encounter water. This nonnuclear explosion could spread radioactive material to the areas surrounding the power plant. To prevent such an unlikely chain of events, nuclear reactors are designed with emergency core cooling systems, requiring no power, that automatically flood the reactor with water in the event of a loss of coolant. The emergency cooling water moderates heat build-up in the core, which in turn prevents the melting of the reactor vessel.

A continuing concern in nuclear fission reactors is the safe disposal of radioactive material when the reactor core is replaced. This waste material contains long-lived, highly radioactive isotopes and must be stored for long periods of time in such a way that there is no chance of environmental contamination. At present, sealing radioactive wastes in waterproof containers and burying them in deep salt mines seems to be the most promising solution.

Transportation of reactor fuel and reactor wastes poses additional safety risks. However, neither the waste nor the fuel of nuclear power reactors can be used to construct a nuclear bomb.

Accidents during transportation of nuclear fuel could expose the public to harmful levels of radiation. The Department of Energy requires stringent crash

[1]The authors are grateful to Professor Gene Skluzacek of the University of Nebraska at Omaha for rewriting this section.

tests on all containers used to transport nuclear materials. Container manufacturers must demonstrate that their containers will not rupture, even in high-speed collisions.

The safety issues associated with nuclear power reactors are complex and often emotional. All sources of energy have associated risks. Coal, for example, exposes workers to health hazards (including radioactive radon) and produces atmospheric pollution (including greenhouse gases). In each case, the risks must be weighed against the benefits and the availability of the energy source.

30.3 NUCLEAR FUSION

Figure 29.4 shows that the binding energy of light nuclei (those having a mass number lower than 20) is much smaller than the binding energy of heavier nuclei. This suggests a process that is the reverse of fission. **When two light nuclei combine to form a heavier nucleus, the process is called nuclear fusion.** Because the mass of the final nucleus is less than the masses of the original nuclei, there is a loss of mass, accompanied by a release of energy. Although fusion power plants have not yet been developed, a worldwide effort is under way to harness the energy from fusion reactions in the laboratory.

Fusion in the Sun

All stars generate their energy through fusion processes. About 90% of stars, including the Sun, fuse hydrogen, whereas some older stars fuse helium or other heavier elements. Stars are born in regions of space containing vast clouds of dust and gas. Recent mathematical models of these clouds indicate that star formation is triggered by shock waves passing through a cloud. These shock waves are similar to sonic booms and are produced by events such as the explosion of a nearby star, called a *supernova*. The shock wave compresses certain regions of the cloud, causing them to collapse under their own gravity. As the gas falls inward toward the center, the atoms gain speed, which causes the temperature of the gas to rise. Two conditions must be met before fusion reactions in the star can sustain its energy needs: (1) The temperature must be high enough (about 10^7 K for hydrogen) to allow the kinetic energy of the positively charged hydrogen nuclei to overcome their mutual Coulomb repulsion as they collide, and (2) the density of nuclei must be high enough to ensure a high rate of collision.

It's interesting that temperatures inside stars like the Sun are not sufficient to allow colliding protons to overcome Coulomb repulsion. In a certain percentage of collisions, the nuclei pass through the barrier anyway, an example of *quantum tunneling*. So a quantum effect is key in making sunshine.

When fusion reactions occur at the core of a star, the energy that is liberated eventually becomes sufficient to prevent further collapse of the star under its own gravity. The star then continues to live out the remainder of its life under a balance between the inward force of gravity pulling it toward collapse and the outward force due to thermal effects and radiation pressure.

The **proton–proton cycle** is a series of three nuclear reactions that are believed to be the stages in the liberation of energy in the Sun and other stars rich in hydrogen. An overall view of the proton–proton cycle is that four protons combine to form an alpha particle and two positrons, with the release of 25 MeV of energy in the process.

The specific steps in the proton–proton cycle are

$$^1_1\text{H} + {}^1_1\text{H} \rightarrow {}^2_1\text{D} + e^+ + \nu$$

and

$$^1_1\text{H} + {}^2_1\text{D} \rightarrow {}^3_2\text{He} + \gamma \qquad \text{[30.3]}$$

where D stands for deuterium, the isotope of hydrogen having one proton and one neutron in the nucleus. (It can also be written as ^2_1H.) The second reaction is

This photograph of the Sun, taken on December 19, 1973, during the third and final manned Skylab mission, shows one of the most spectacular solar flares ever recorded, spanning more than 588 000 km (365 000 mi) across the solar surface. Several active regions can be seen on the eastern side of the disk. The photograph was taken in the light of ionized helium by the extreme ultraviolet spectro-heliograph instrument of the U.S. Naval Research Laboratory.

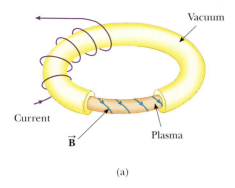

(a)

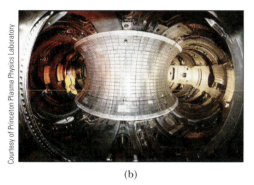

Courtesy of Princeton Plasma Physics Laboratory

(b)

Courtesy of Princeton University

(c)

Figure 30.5 (a) Diagram of a tokamak used in the magnetic confinement scheme. The plasma is trapped within the spiraling magnetic field lines as shown. (b) Interior view of the Tokamak Fusion Test Reactor (TFTR) vacuum vessel located at the Princeton Plasma Physics Laboratory, Princeton University, Princeton, New Jersey. (c) The National Spherical Torus Experiment (NSTX) that began operation in March 1999.

combination of two magnetic fields to confine the plasma inside the doughnut. A strong magnetic field is produced by the current in the windings, and a weaker magnetic field is produced by the current in the toroid. The resulting magnetic field lines are helical, as shown in the figure. In this configuration, the field lines spiral around the plasma and prevent it from touching the walls of the vacuum chamber.

In order for the plasma to reach ignition temperature, some form of auxiliary heating is necessary. A successful and efficient auxiliary heating technique that has been used recently is the injection of a beam of energetic neutral particles into the plasma.

When it was in operation, the Tokamak Fusion Test Reactor (TFTR) at Princeton reported central ion temperatures of 510 million degrees Celsius, more than 30 times hotter than the center of the Sun. TFTR $n\tau$ values for the D–T reaction were well above 10^{13} s/cm^3 and close to the value required by Lawson's criterion. In 1991, reaction rates of 6×10^{17} D–T fusions per second were reached in the JET tokamak at Abington, England.

One of the new generations of fusion experiments is the National Spherical Torus Experiment (NSTX) shown in Figure 30.5c. Rather than generating the donut-shaped plasma of a tokamak, the NSTX produces a spherical plasma that has a hole through its center. The major advantage of the spherical configuration is its ability to confine the plasma at a higher pressure in a given magnetic field. This approach could lead to the development of smaller and more economical fusion reactors.

There are a number of other methods of creating fusion events. In inertial laser confinement fusion, the fuel is put into the form of a small pellet and then collapsed by ultrahigh-power lasers. Fusion can also take place in a device the size of a TV set, and in fact was invented by Philo Farnsworth, one of the fathers of electronic television. In this method, called inertial electrostatic confinement, positively charged particles are rapidly attracted towards a negatively charged grid. Some of the positive particles then collide and fuse.

An international collaborative effort involving four major fusion programs is currently under way to build a fusion reactor called the International Thermonuclear Experimental Reactor (ITER). This facility will address the remaining technological and scientific issues concerning the feasibility of fusion power. The design is completed, and site and construction negotiations are under way. If the planned device works as expected, the Lawson number for ITER will be about six times greater than the current record holder, the JT-60U tokamak in Japan.

30.4   ELEMENTARY PARTICLES

The word "atom" is from the Greek word *atomos*, meaning "indivisible." At one time, atoms were thought to be the indivisible constituents of matter; that is, they were regarded as elementary particles. Discoveries in the early part of the 20th century revealed that the atom is not elementary, but has protons, neutrons, and electrons as its constituents. Until 1932, physicists viewed these three constituent particles as elementary because, with the exception of the free neutron, they are highly stable. The theory soon fell apart, however, and beginning in 1937, many new particles were discovered in experiments involving high-energy collisions between known particles. These new particles are characteristically unstable and have very short half-lives, ranging between 10^{-23} s and 10^{-6} s. So far more than 300 of them have been cataloged.

Until the 1960s, physicists were bewildered by the large number and variety of subatomic particles being discovered. They wondered whether the particles were like animals in a zoo or whether a pattern could emerge that would provide a better understanding of the elaborate structure in the subnuclear world. In the last 30 years, physicists have made tremendous advances in our knowledge of the structure of matter by recognizing that all particles (with the exception of electrons, photons, and a few others) are made of smaller particles called *quarks*. Protons and neutrons, for example, are not truly elementary but are systems of tightly bound quarks. The quark model has reduced the bewildering array of particles to a manageable number and has predicted new quark combinations that were subsequently found in many experiments.

30.5   THE FUNDAMENTAL FORCES OF NATURE

The key to understanding the properties of elementary particles is to be able to describe the forces between them. All particles in nature are subject to four fundamental forces: strong, electromagnetic, weak, and gravitational.

The **strong force** is responsible for the tight binding of quarks to form neutrons and protons and for the nuclear force, a sort of residual strong force, binding neutrons and protons into nuclei. This force represents the "glue" that holds the nucleons together and is the strongest of all the fundamental forces. It is a very short-range force and is negligible for separations greater than about 10^{-15} m (the approximate size of the nucleus). The **electromagnetic force**, which is about 10^{-2} times the strength of the strong force, is responsible for the binding of atoms and molecules. It is a long-range force that decreases in strength as the inverse square of the separation between interacting particles. The **weak force** is a short-range nuclear force that tends to produce instability in certain nuclei. It is responsible for beta decay, and its strength is only about 10^{-6} times that of the strong force. (As we discuss later, scientists now believe that the weak and electromagnetic forces are two manifestations of a single force called the *electroweak* force). Finally, the **gravitational force** is a long-range force with a strength only about 10^{-43} times that of the strong force. Although this familiar interaction is the force that holds the planets, stars, and galaxies together, its effect on elementary

TABLE 30.1

Particle Interactions

Interaction (Force)	Relative Strength[a]	Range of Force	Mediating Field Particle
Strong	1	Short (≈ 1 fm)	Gluon
Electromagnetic	10^{-2}	Long ($\propto 1/r^2$)	Photon
Weak	10^{-6}	Short ($\approx 10^{-3}$ fm)	$W^{\pm}$ and Z bosons
Gravitational	10^{-43}	Long ($\propto 1/r^2$)	Graviton

[a] For two quarks separated by 3×10^{-17} m.

particles is negligible. The gravitational force is by far the weakest of all the fundamental forces.

Modern physics often describes the forces between particles in terms of the actions of field particles or quanta. In the case of the familiar electromagnetic interaction, the field particles are photons. In the language of modern physics, the electromagnetic force is *mediated* (carried) by photons, which are the quanta of the electromagnetic field. Likewise, the strong force is mediated by field particles called *gluons*, the weak force is mediated by particles called the W and Z *bosons*, and the gravitational force is thought to be mediated by quanta of the gravitational field called *gravitons*. All of these field quanta have been detected except for the graviton, which may never be found directly because of the weakness of the gravitational field. These interactions, their ranges, and their relative strengths are summarized in Table 30.1.

30.6 POSITRONS AND OTHER ANTIPARTICLES

In the 1920s, the theoretical physicist Paul Adrien Maurice Dirac (1902–1984) developed a version of quantum mechanics that incorporated special relativity. Dirac's theory successfully explained the origin of the electron's spin and its magnetic moment. But it had one major problem: its relativistic wave equation required solutions corresponding to negative energy states even for free electrons, and if negative energy states existed, we would expect a normal free electron in a state of positive energy to make a rapid transition to one of these lower states, emitting a photon in the process. Normal electrons would not exist and we would be left with a universe of photons and electrons locked up in negative energy states.

Dirac circumvented this difficulty by postulating that all negative energy states are normally filled. The electrons that occupy the negative energy states are said to be in the "Dirac sea" and are not directly observable when all negative energy states are filled. However, if one of these negative energy states is vacant, leaving a hole in the sea of filled states, the hole can react to external forces and therefore can be observed as the electron's positive antiparticle. The general and profound implication of Dirac's theory is that **for every particle, there is an antiparticle with the same mass as the particle, but the opposite charge**. For example, the electron's antiparticle, the *positron*, has a mass of 0.511 MeV/c^2 and a positive charge of 1.6×10^{-19} C. As noted in Chapter 29, we usually designate an antiparticle with a bar over the symbol for the particle. For example, $\overline{p}$ denotes the antiproton and $\overline{\nu}$ the antineutrino. In this book, the notation e^+ is preferred for the positron.

The positron was discovered by Carl Anderson in 1932, and in 1936 he was awarded the Nobel prize for his achievement. Anderson discovered the positron while examining tracks created by electron-like particles of positive charge in a cloud chamber. (These early experiments used cosmic rays—mostly energetic protons passing through interstellar space—to initiate high-energy reactions on the order of several GeV.) In order to discriminate between positive and negative charges, the cloud chamber was placed in a magnetic field, causing moving charges to follow curved paths. Anderson noted that some of the

PAUL ADRIEN MAURICE DIRAC
(1902–1984)

Dirac was instrumental in the understanding of antimatter and in the unification of quantum mechanics and relativity. He made numerous contributions to the development of quantum physics and cosmology, and won the Nobel Prize for physics in 1933.

electronlike tracks deflected in a direction corresponding to a positively charged particle.

Since Anderson's initial discovery, the positron has been observed in a number of experiments. Perhaps the most common process for producing positrons is pair production, introduced in Chapter 26. In this process, a gamma ray with sufficiently high energy collides with a nucleus, creating an electron–positron pair. Because the total rest energy of the pair is $2m_e c^2 = 1.02$ MeV, the gamma ray must have at least this much energy to create such a pair.

Practically every known elementary particle has a distinct antiparticle. Among the exceptions are the photon and the neutral pion (π^0), which are their own antiparticles. Following the construction of high-energy accelerators in the 1950s, many of these antiparticles were discovered. They included the antiproton $\overline{p}$, discovered by Emilio Segrè and Owen Chamberlain in 1955, and the antineutron $\overline{n}$, discovered shortly thereafter.

The process of electron–positron annihilation is used in the medical diagnostic technique of positron emission tomography (PET). The patient is injected with a glucose solution containing a radioactive substance that decays by positron emission. Examples of such substances are oxygen-15, nitrogen-13, carbon-11, and fluorine-18. The radioactive material is carried to the brain. When a decay occurs, the emitted positron annihilates with an electron in the brain tissue, resulting in two gamma ray photons. With the assistance of a computer, an image can be created of the sites in the brain at which the glucose accumulates.

The images from a PET scan can point to a wide variety of disorders in the brain, including Alzheimer's disease. In addition, because glucose metabolizes more rapidly in active areas of the brain, the PET scan can indicate which areas of the brain are involved in various processes such as language, music, and vision.

30.7 MESONS AND THE BEGINNING OF PARTICLE PHYSICS

Physicists in the mid-1930s had a fairly simple view of the structure of matter. The building blocks were the proton, the electron, and the neutron. Three other particles were known or postulated at the time: the photon, the neutrino, and the positron. These six particles were considered the fundamental constituents of matter. Although the accepted picture of the world was marvelously simple, no one was able to provide an answer to the following important question: Because the many protons in proximity in any nucleus should strongly repel each other due to their like charges, what is the nature of the force that holds the nucleus together? Scientists recognized that this mysterious nuclear force must be much stronger than anything encountered up to that time.

The first theory to explain the nature of the nuclear force was proposed in 1935 by the Japanese physicist Hideki Yukawa (1907–1981), an effort that later earned him the Nobel prize. In order to understand Yukawa's theory, it is useful to first note that **two atoms can form a covalent chemical bond by the exchange of electrons**. Similarly, in the modern view of electromagnetic interactions, **charged particles interact by exchanging a photon**. Yukawa used this same idea to explain the nuclear force by proposing a new particle that is exchanged by nucleons in the nucleus to produce the strong force. Further, he demonstrated that the range of the force is inversely proportional to the mass of this particle, and predicted that the mass would be about 200 times the mass of the electron. Because the new particle would have a mass between that of the electron and the proton, it was called a **meson** (from the Greek *meso*, meaning "middle").

In an effort to substantiate Yukawa's predictions, physicists began looking for the meson by studying cosmic rays that enter the Earth's atmosphere. In 1937, Carl Anderson and his collaborators discovered a particle with mass 106 MeV/c^2, about 207 times the mass of the electron. However, subsequent experiments showed that the particle interacted very weakly with matter and hence could not be the carrier of the nuclear force. This puzzling situation inspired several theoreticians

APPLICATION

Positron Emission Tomography

HIDEKI YUKAWA, Japanese Physicist (1907–1981)

Yukawa was awarded the Nobel Prize in 1949 for predicting the existence of mesons. This photograph of Yukawa at work was taken in 1950 in his office at Columbia University.

UPI/Corbis-Bettman

TIP 30.2 The Nuclear Force and the Strong Force

The nuclear force discussed in Chapter 29 was originally called the *strong* force. Once the quark theory was established, however, the phrase *strong force* was reserved for the force between quarks. We will follow this convention: the strong force is between quarks and the nuclear force is between nucleons.

to propose that there are two mesons with slightly different masses, an idea that was confirmed in 1947 with the discovery of the pi meson (π), or simply *pion*, by Cecil Frank Powell (1903–1969) and Guiseppe P. S. Occhialini (1907–1993). The lighter meson discovered earlier by Anderson, now called a *muon* (μ), has only weak and electromagnetic interactions and plays no role in the strong interaction.

The pion comes in three varieties, corresponding to three charge states: π^+, π^-, and π^0. The π^+ and π^- particles have masses of 139.6 MeV/c^2, and the π^0 has a mass of 135.0 MeV/c^2. Pions and muons are highly unstable particles. For example, the π^-, which has a lifetime of about 2.6×10^{-8} s, decays into a muon and an antineutrino. The muon, with a lifetime of 2.2 μs, then decays into an electron, a neutrino, and an antineutrino. The sequence of decays is

$$\pi^- \rightarrow \mu^- + \overline{\nu} \qquad [30.6]$$

$$\mu^- \rightarrow e^- + \nu + \overline{\nu}$$

The interaction between two particles can be understood in general with a simple illustration called a *Feynman diagram*, developed by Richard P. Feynman (1918–1988). Figure 30.6 is a Feynman diagram for the electromagnetic interaction between two electrons. In this simple case, a photon is the field particle that mediates the electromagnetic force between the electrons. The photon transfers energy and momentum from one electron to the other in the interaction. Such a photon, called a *virtual photon*, can never be detected directly because it is absorbed by the second electron very shortly after being emitted by the first electron. The existence of a virtual photon might be expected to violate the law of conservation of energy, but it does not because of the time–energy uncertainty principle. Recall that the uncertainty principle says that the energy is uncertain or not conserved by an amount ΔE for a time Δt such that $\Delta E \, \Delta t \approx \hbar$.

Now consider the pion exchange between a proton and a neutron via the nuclear force (Fig. 30.7). The energy needed to create a pion of mass m_π is given by $\Delta E = m_\pi c^2$. Again, the existence of the pion is allowed in spite of conservation of energy if this energy is surrendered in a short enough time Δt, the time it takes the pion to transfer from one nucleon to the other. From the uncertainty principle, $\Delta E \, \Delta t \approx \hbar$, we get

$$\Delta t \approx \frac{\hbar}{\Delta E} = \frac{\hbar}{m_\pi c^2} \qquad [30.7]$$

Because the pion can't travel faster than the speed of light, the maximum distance d it can travel in a time Δt is $c\Delta t$. Using Equation 30.7 and $d = c\Delta t$, we find this maximum distance to be

$$d \approx \frac{\hbar}{m_\pi c} \qquad [30.8]$$

The measured range of the nuclear force is about 1.5×10^{-15} m. Using this value for d in Equation 30.8, the rest energy of the pion is calculated to be

$$m_\pi c^2 = \frac{\hbar c}{d} = \frac{(1.05 \times 10^{-34}\,\text{J}\cdot\text{s})(3.00 \times 10^8\,\text{m/s})}{1.5 \times 10^{-15}\,\text{m}}$$

$$= 2.1 \times 10^{-11}\,\text{J} \cong 130\,\text{MeV}$$

This corresponds to a mass of 130 MeV/c^2 (about 250 times the mass of the electron), which is in good agreement with the observed mass of the pion.

The concept we have just described is quite revolutionary. In effect, it says that a proton can change into a proton plus a pion, as long as it returns to its original state in a very short time. High-energy physicists often say that a nucleon undergoes "fluctuations" as it emits and absorbs pions. As we have seen, these fluctuations are a consequence of a combination of quantum mechanics (through the uncertainty principle) and special relativity (through Einstein's energy–mass relation $E = mc^2$).

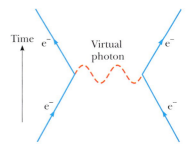

Figure 30.6 Feynman diagram representing a photon mediating the electromagnetic force between two electrons.

RICHARD FEYNMAN, American Physicist (1918–1988)

Feynman, together with Julian S. Schwinger and Shinichiro Tomonaga, won the 1965 Nobel Prize for physics for fundamental work in the principles of quantum electrodynamics. His many important contributions to physics include work on the first atomic bomb in the Manhattan project, the invention of simple diagrams to represent particle interactions graphically, the theory of the weak interaction of subatomic particles, a reformulation of quantum mechanics, and the theory of superfluid helium. Later he served on the commission investigating the *Challenger* tragedy, demonstrating the problem with the O-rings by dipping a scale-model O-ring in his glass of ice water and then shattering it with a hammer. He also contributed to physics education through the magnificent three-volume text *The Feynman Lectures on Physics*.

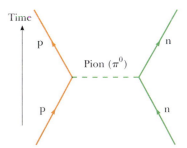

Figure 30.7 Feynman diagram representing a proton interacting with a neutron via the strong force. In this case, the pion mediates the nuclear force.

This section has dealt with the early theory of Yukawa of particles that mediate the nuclear force, pions, and the mediators of the electromagnetic force, photons. Although his model led to the modern view, it has been superseded by the more basic quark–gluon theory, as explained in Sections 30.12 and 30.13.

30.8   CLASSIFICATION OF PARTICLES

Hadrons

All particles other than photons can be classified into two broad categories, hadrons and leptons, according to their interactions. Particles that interact through the strong force are called *hadrons*. There are two classes of hadrons, known as *mesons* and *baryons*, distinguished by their masses and spins. All mesons are known to decay finally into electrons, positrons, neutrinos, and photons. The pion is the lightest of known mesons, with a mass of about 140 MeV/c^2 and a spin of 0. Another is the K meson, with a mass of about 500 MeV/c^2 and spin 0 also.

Baryons have masses equal to or greater than the proton mass (the name *baryon* means "heavy" in Greek), and their spin is always a non-integer value (1/2 or 3/2). Protons and neutrons are baryons, as are many other particles. With the exception of the proton, all baryons decay in such a way that the end products include a proton. For example, the baryon called the Ξ hyperon first decays to a Λ^0 in about 10^{-10} s. The Λ^0 then decays to a proton and a π^- in about 3×10^{-10} s.

Today it is believed that hadrons are composed of quarks. (Later, we will have more to say about the quark model.) Some of the important properties of hadrons are listed in Table 30.2.

TABLE 30.2

Some Particles and Their Properties

Category	Particle Name	Symbol	Anti-particle	Mass (MeV/c^2)	B	L_e	L_μ	L_τ	S	Lifetime(s)	Principal Decay Modes[a]
Leptons	Electron	e^-	e^+	0.511	0	+1	0	0	0	Stable	
	Electron–neutrino	ν_e	$\bar{\nu}_e$	$< 7eV/c^2$	0	+1	0	0	0	Stable	
	Muon	μ^-	μ^+	105.7	0	0	+1	0	0	2.20×10^{-6}	$e^- \bar{\nu}_e \nu_\mu$
	Muon–neutrino	ν_μ	$\bar{\nu}_\mu$	< 0.3	0	0	+1	0	0	Stable	
	Tau	τ^-	τ^+	1 784	0	0	0	+1	0	$< 4 \times 10^{-13}$	$\mu^- \bar{\nu}_\mu \nu_\tau, e^- \bar{\nu}_e \nu_\tau$
	Tau–neutrino	ν_τ	$\bar{\nu}_\tau$	< 30	0	0	0	+1	0	Stable	
Hadrons											
Mesons	Pion	π^+	π^-	139.6	0	0	0	0	0	2.60×10^{-8}	$\mu^+ \nu_\mu$
		π^0	Self	135.0	0	0	0	0	0	0.83×10^{-16}	2γ
	Kaon	K^+	K^-	493.7	0	0	0	0	+1	1.24×10^{-8}	$\mu^+ \nu_\mu, \pi^+ \pi^0$
		K_S^0	$\bar{K}_S^0$	497.7	0	0	0	0	+1	0.89×10^{-10}	$\pi^+ \pi^-, 2\pi^0$
		K_L^0	$\bar{K}_L^0$	497.7	0	0	0	0	+1	5.2×10^{-8}	$\pi^\pm e^\mp \bar{\nu}_e, 3\pi^0$ $\pi^\pm \mu^\mp \bar{\nu}_\mu$
	Eta	η	Self	548.8	0	0	0	0	0	$< 10^{-18}$	$2\gamma, 3\pi$
		η'	Self	958	0	0	0	0	0	2.2×10^{-21}	$\eta \pi^+ \pi^-$
Baryons	Proton	p	$\bar{p}$	938.3	+1	0	0	0	0	Stable	
	Neutron	n	$\bar{n}$	939.6	+1	0	0	0	0	920	$pe^- \bar{\nu}_e$
	Lambda	Λ^0	$\bar{\Lambda}^0$	1 115.6	+1	0	0	0	-1	2.6×10^{-10}	$p\pi^-, n\pi^0$
	Sigma	Σ^+	$\bar{\Sigma}^-$	1 189.4	+1	0	0	0	-1	0.80×10^{-10}	$p\pi^0, n\pi^+$
		Σ^0	$\bar{\Sigma}^0$	1 192.5	+1	0	0	0	-1	6×10^{-20}	$\Lambda^0 \gamma$
		Σ^-	$\bar{\Sigma}^+$	1 197.3	+1	0	0	0	-1	1.5×10^{-10}	$n\pi^-$
	Xi	Ξ^0	$\bar{\Xi}^0$	1 315	+1	0	0	0	-2	2.9×10^{-10}	$\Lambda^0 \pi^0$
		Ξ^-	Ξ^+	1 321	+1	0	0	0	-2	1.64×10^{-10}	$\Lambda^0 \pi^-$
	Omega	Ω^-	Ω^+	1 672	+1	0	0	0	-3	0.82×10^{-10}	$\Xi^0 \pi^0, \Lambda^0 K^-$

[a] Notations in this column, such as $p\pi^-$, $n\pi^0$ mean two possible decay modes. In this case, the two possible decays are $\Lambda^0 \rightarrow p + \pi^-$ and $\Lambda^0 \rightarrow n + \pi^0$.

Leptons

Leptons (from the Greek *leptos*, meaning "small" or "light") are a group of particles that participate in the weak interaction. All leptons have a spin of 1/2. Included in this group are electrons, muons, and neutrinos, which are less massive than the lightest hadron. Although hadrons have size and structure, leptons appear to be truly elementary, with no structure down to the limit of resolution of experiment (about 10^{-19} m).

Unlike hadrons, the number of known leptons is small. Currently, scientists believe there are only six leptons (each having an antiparticle): the electron, the muon, the tau, and a neutrino associated with each:

$$\begin{pmatrix} e^- \\ \nu_e \end{pmatrix} \quad \begin{pmatrix} \mu^- \\ \nu_\mu \end{pmatrix} \quad \begin{pmatrix} \tau^- \\ \nu_\tau \end{pmatrix}$$

The tau lepton, discovered in 1975, has a mass about twice that of the proton.

Although neutrinos have masses of about zero, there is strong indirect evidence that the electron neutrino has a nonzero mass of about 3 eV/c^2, or 1/180 000 of the electron mass. A firm knowledge of the neutrino's mass could have great significance in cosmological models and in our understanding of the future of the Universe.

30.9   CONSERVATION LAWS

A number of conservation laws are important in the study of elementary particles. Although the two described here have no theoretical foundation, they are supported by abundant empirical evidence.

Baryon Number

The law of conservation of baryon number tells us that whenever a baryon is created in a reaction or decay, an antibaryon is also created. This information can be quantified by assigning a baryon number: $B = +1$ for all baryons, $B = -1$ for all antibaryons, and $B = 0$ for all other particles. Thus, the **law of conservation of baryon number** states that whenever a nuclear reaction or decay occurs, the sum of the baryon numbers before the process equals the sum of the baryon numbers after the process.

◀ Conservation of baryon number

Note that if the baryon number is absolutely conserved, the proton must be absolutely stable: if it were not for the law of conservation of baryon number, the proton could decay into a positron and a neutral pion. However, such a decay has never been observed. At present, we can only say that the proton has a half-life of at least 10^{31} years. (The estimated age of the Universe is about 10^{10} years.) In one recent version of a so-called grand unified theory (GUT), physicists have predicted that the proton is actually unstable. According to this theory, the baryon number (sometimes called the *baryonic charge*) is not absolutely conserved, whereas electric charge is always conserved.

EXAMPLE 30.4   Checking Baryon Numbers

Goal   Use conservation of baryon number to determine whether a given reaction can occur.

Problem   Determine whether the following reaction can occur based on the law of conservation of baryon number.

$$p + n \rightarrow p + p + n + \bar{p}$$

Strategy   Count the baryons on both sides of the reaction, recalling that that $B = +1$ for baryons and $B = -1$ for antibaryons.

Solution

Count the baryons on the left: The neutron and proton are both baryons; hence, $1 + 1 = 2$.

Count the baryons on the right: There are three baryons and one antibaryon, so
$$1 + 1 + 1 + (-1) = 2.$$

Remark Baryon number is conserved in this reaction, so it can occur, provided the incoming proton has sufficient energy.

Exercise 30.4
Can the following reaction occur, based on the law of conservation of baryon number?

$$p + n \rightarrow p + p + \bar{p}$$

Answer No. (Show this by computing the baryon number on both sides and finding that they're not equal.)

Lepton Number

Conservation of lepton number ▶

There are three conservation laws involving lepton numbers, one for each variety of lepton. The **law of conservation of electron-lepton number** states that the sum of the electron-lepton numbers before a reaction or decay must equal the sum of the electron-lepton numbers after the reaction or decay. The electron and the electron neutrino are assigned a positive electron-lepton number $L_e = +1$, the antileptons e^+ and $\bar{\nu}_e$ are assigned the electron-lepton number $L_e = -1$, and all other particles have $L_e = 0$. For example, consider neutron decay:

Neutron decay ▶

$$n \rightarrow p^+ + e^- + \bar{\nu}_e$$

Before the decay, the electron-lepton number is $L_e = 0$; after the decay, it is $0 + 1 + (-1) = 0$, so the electron-lepton number is conserved. It's important to recognize that the baryon number must also be conserved. This can easily be seen by noting that before the decay $B = +1$, whereas after the decay $B = +1 + 0 + 0 = +1$.

Similarly, when a decay involves muons, the muon-lepton number L_μ is conserved. The μ^- and the ν_μ are assigned $L_\mu = +1$, the antimuons μ^+ and $\bar{\nu}_\mu$ are assigned $L_\mu = -1$, and all other particles have $L_\mu = 0$. Finally, the tau-lepton number L_τ is conserved, and similar assignments can be made for the τ lepton and its neutrino.

EXAMPLE 30.5   Checking Lepton Numbers

Goal Use conservation of lepton number to determine whether a given process is possible.

Problem Determine which of the following decay schemes can occur on the basis of conservation of lepton number.

$$\mu^- \rightarrow e^- + \bar{\nu}_e + \nu_\mu \tag{1}$$

$$\pi^+ \rightarrow \mu^+ + \nu_\mu + \nu_e \tag{2}$$

Strategy Count the leptons on either side and see if the numbers are equal.

Solution
Because decay 1 involves both a muon and an electron, L_μ and L_e must both be conserved. Before the decay, $L_\mu = +1$ and $L_e = 0$. After the decay, $L_\mu = 0 + 0 + 1 = +1$ and $L_e = +1 - 1 + 0 = 0$. Both lepton numbers are conserved, and on this basis, the decay mode is possible.

Before decay 2 occurs, $L_\mu = 0$ and $L_e = 0$. After the decay, $L_\mu = -1 + 1 + 0 = 0$, but $L_e = +1$. This decay isn't possible because the electron-lepton number is not conserved.

Exercise 30.5
Determine whether the decay $\mu^- \rightarrow e^- + \bar{\nu}_e$ can occur.

Answer No. (Show this by computing muon-lepton numbers on both sides and showing they're not equal.)

Which of the following reactions cannot occur?
(a) $p + p \rightarrow p + p + \bar{p}$ (b) $n \rightarrow p + e^- + \bar{\nu}_e$
(c) $\mu^- \rightarrow e^- + \bar{\nu}_e + \nu_\mu$ (d) $\pi^- \rightarrow \mu^- + \bar{\nu}_\mu$

Which of the following reactions cannot occur?
(a) $p + \bar{p} \rightarrow 2\gamma$ (b) $\gamma + p \rightarrow n + \pi^0$
(c) $\pi^0 + n \rightarrow K^+ + \Sigma^-$ (d) $\pi^+ + p \rightarrow K^+ + \Sigma^+$

Suppose a claim is made that the decay of a neutron is given by $n \rightarrow p^+ + e^-$. Which of the following conservation laws are necessarily violated by this proposed decay scheme? (a) energy (b) linear momentum (c) electric charge (d) lepton number (e) baryon number

30.10   STRANGE PARTICLES AND STRANGENESS

Many particles discovered in the 1950s were produced by the nuclear interaction of pions with protons and neutrons in the atmosphere. A group of these particles, namely the K, Λ, and Σ particles, was found to exhibit unusual properties in their production and decay and hence were called *strange particles*.

One unusual property of strange particles is that they are always produced in pairs. For example, when a pion collides with a proton, two neutral strange particles are produced with high probability (Fig. 30.8) following the reaction

$$\pi^- + p^+ \rightarrow K^0 + \Lambda^0$$

On the other hand, the reaction $\pi^- + p^+ \rightarrow K^0 + n$ has never occurred, even though it violates no known conservation laws and the energy of the pion is sufficient to initiate the reaction.

The second peculiar feature of strange particles is that although they are produced by the strong interaction at a high rate, they do not decay into particles that interact via the strong force at a very high rate. Instead, they decay very slowly, which is characteristic of the weak interaction. Their half-lives are in the range from 10^{-10} s to 10^{-8} s; most other particles that interact via the strong force have lifetimes on the order of 10^{-23} s.

To explain these unusual properties of strange particles, a law called *conservation of strangeness* was introduced, together with a new quantum number S called **strangeness**. The strangeness numbers for some particles are given in Table 30.2. The production of strange particles in pairs is explained by assigning $S = +1$ to one of the particles and $S = -1$ to the other. All nonstrange particles are assigned strangeness $S = 0$. The **law of conservation of strangeness** states that whenever a nuclear reaction or decay occurs, the sum of the strangeness numbers before the process must equal the sum of the strangeness numbers after the process.

The slow decay of strange particles can be explained by assuming that the strong and electromagnetic interactions obey the law of conservation of strangeness, whereas the weak interaction does not. Because the decay reaction involves the loss of one strange particle, it violates strangeness conservation and hence proceeds slowly via the weak interaction.

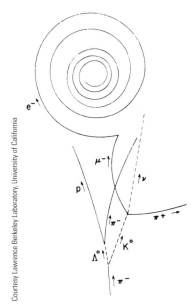

Figure 30.8   This drawing represents tracks of many events obtained by analyzing a bubble-chamber photograph. The strange particles Λ^0 and K^0 are formed (at the bottom) as the π^- interacts with a proton according to the interaction $\pi^- + p \rightarrow \Lambda^0 + K^0$. (Note that the neutral particles leave no tracks, as is indicated by the dashed lines.) The Λ^0 and K^0 then decay according to the interactions $\Lambda^0 \rightarrow \pi + p$ and $K^0 \rightarrow \pi + \mu^- + \nu_\mu$.

◀ Conservation of strangeness number

Applying Physics 30.2 Breaking Conservation Laws

A student claims to have observed a decay of an electron into two neutrinos traveling in opposite directions. What conservation laws would be violated by this decay?

Explanation Several conservation laws are violated. Conservation of electric charge is violated because the negative charge of the electron has disappeared. Conservation of electron lepton number is also violated, because there is one lepton before the decay and two afterward. If both neutrinos were electron-neutrinos, electron lepton number conservation would be violated in the final state. However, if one of the product neutrinos were other than an electron-neutrino, then another lepton conservation law would be violated, because there were no other leptons in the initial state.

Other conservation laws are obeyed by this decay. Energy can be conserved—the rest energy of the electron appears as the kinetic energy (and possibly some small rest energy) of the neutrinos. The opposite directions of the velocities of the two neutrinos allow for the conservation of momentum. Conservation of baryon number and conservation of other lepton numbers are also upheld in this decay.

EXAMPLE 30.6 Is Strangeness Conserved?

Goal Apply conservation of strangeness to determine whether a process can occur.

Problem Determine whether the following reactions can occur on the basis of conservation of strangeness:

$$\pi^0 + n \rightarrow K^+ + \Sigma^- \tag{1}$$

$$\pi^- + p \rightarrow \pi^- + \Sigma^+ \tag{2}$$

Strategy Count strangeness on each side of a given process. If strangeness is conserved, the reaction is possible.

Solution

In the first process, the neutral pion and neutron both have strangeness of zero, so $S_{initial} = 0 + 0 = 0$. Because the strangeness of the K^+ is $S = +1$, and the strangeness of the Σ^- is $S = -1$, the total strangeness of the final state is $S_{final} = +1 - 1 = 0$. Strangeness is conserved and the reaction is allowed.

In the second process, the initial state has strangeness $S_{initial} = 0 + 0 = 0$, but the final state has strangeness $S_{final} = 0 + (-1) = -1$. Strangeness is not conserved and the reaction isn't allowed.

Exercise 30.6

Does the reaction $p^+ + \pi^- \rightarrow K^0 + \Lambda^0$ obey the law of conservation of strangeness? Show why or why not.

Answer Yes. (Show this by computing the strangeness on both sides.)

30.11 THE EIGHTFOLD WAY

Quantities such as spin, baryon number, lepton number, and strangeness are labels we associate with particles. Many classification schemes that group particles into families based on such labels have been proposed. First, consider the first eight baryons listed in Table 30.2, all having a spin of 1/2. The family consists of the proton, the neutron, and six other particles. If we plot their strangeness versus their charge using a sloping coordinate system, as in Figure 30.9a, a fascinating pattern emerges: six of the baryons form a hexagon, and the remaining two are at the hexagon's center. (Particles with spin quantum number 1/2 or 3/2 are called fermions.)

Now consider the family of mesons listed in Table 30.2 with spins of zero. (Particles with spin quantum number 0 or 1 are called bosons.) If we count both particles and antiparticles, there are nine such mesons. Figure 30.9b is a plot of strangeness versus charge for *this* family. Again, a fascinating hexagonal pattern emerges. In this case, the particles on the perimeter of the hexagon lie opposite their antiparticles, and the remaining three (which form their own antiparticles)

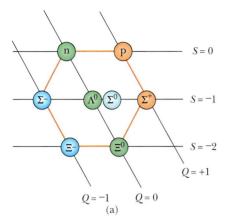
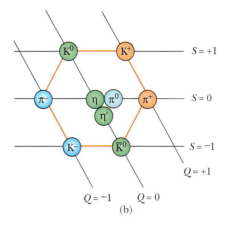

Figure 30.9 (a) The hexagonal eightfold-way pattern for the eight spin $-\frac{1}{2}$ baryons. This strangeness versus charge plot uses a horizontal axis for the strangeness values S, but a sloping axis for the charge number Q. (b) The eightfold-way pattern for the nine spin-zero mesons.

are at its center. These and related symmetric patterns, called the **eightfold way**, were proposed independently in 1961 by Murray Gell-Mann and Yuval Ne'eman.

The groups of baryons and mesons can be displayed in many other symmetric patterns within the framework of the eightfold way. For example, the family of spin-3/2 baryons contains ten particles arranged in a pattern like the tenpins in a bowling alley. After the pattern was proposed, one of the particles was missing—it had yet to be discovered. Gell-Mann predicted that the missing particle, which he called the *omega minus* (Ω^-), should have a spin of 3/2, a charge of -1, a strangeness of -3, and a mass of about $1\,680\ \text{MeV}/c^2$. Shortly thereafter, in 1964, scientists at the Brookhaven National Laboratory found the missing particle through careful analyses of bubble chamber photographs and confirmed all its predicted properties.

The patterns of the eightfold way in the field of particle physics have much in common with the periodic table. Whenever a vacancy (a missing particle or element) occurs in the organized patterns, experimentalists have a guide for their investigations.

30.12   QUARKS

As we have noted, leptons appear to be truly elementary particles because they have no measurable size or internal structure, are limited in number, and do not seem to break down into smaller units. Hadrons, on the other hand, are complex particles with size and structure. Further, we know that hadrons decay into other hadrons and are many in number. Table 30.2 lists only those hadrons that are stable against hadronic decay; hundreds of others have been discovered. These facts strongly suggest that hadrons cannot be truly elementary but have some substructure.

The Original Quark Model

In 1963 Gell-Mann and George Zweig independently proposed that hadrons have an elementary substructure. According to their model, all hadrons are composite systems of two or three fundamental constitutents called **quarks**, which rhymes with "forks" (though some rhyme it with "sharks"). Gell-Mann borrowed the word *quark* from the passage "Three quarks for Muster Mark" in James Joyce's book *Finnegans Wake*. In the original model there were three types of quarks designated by the symbols u, d, and s. These were given the arbitrary names *up*, *down*, and *sideways* (or, now more commonly, *strange*).

An unusual property of quarks is that they have fractional electronic charges, as shown—along with other properties—in Table 30.3 (page 994). Associated with each quark is an antiquark of opposite charge, baryon number, and strangeness. The compositions of all hadrons known when Gell-Mann and Zweig presented their models could be completely specified by three simple rules:

MURRAY GELL-MANN,
American Physicist (1929–)

Gell-Mann was awarded the Nobel Prize in 1969 for his theoretical studies dealing with subatomic particles.

TABLE 30.3

Properties of Quarks and Antiquarks

				Quarks				
Name	Symbol	Spin	Charge	Baryon Number	Strangeness	Charm	Bottomness	Topness
Up	u	$\frac{1}{2}$	$+\frac{2}{3}e$	$\frac{1}{3}$	0	0	0	0
Down	d	$\frac{1}{2}$	$-\frac{1}{3}e$	$\frac{1}{3}$	0	0	0	0
Strange	s	$\frac{1}{2}$	$-\frac{1}{3}e$	$\frac{1}{3}$	−1	0	0	0
Charmed	c	$\frac{1}{2}$	$+\frac{2}{3}e$	$\frac{1}{3}$	0	+1	0	0
Bottom	b	$\frac{1}{2}$	$-\frac{1}{3}e$	$\frac{1}{3}$	0	0	+1	0
Top	t	$\frac{1}{2}$	$+\frac{2}{3}e$	$\frac{1}{3}$	0	0	0	+1

				Antiquarks				
Name	Symbol	Spin	Charge	Baryon Number	Strangeness	Charm	Bottomness	Topness
Anti-up	$\bar{u}$	$\frac{1}{2}$	$-\frac{2}{3}e$	$-\frac{1}{3}$	0	0	0	0
Anti-down	$\bar{d}$	$\frac{1}{2}$	$+\frac{1}{3}e$	$-\frac{1}{3}$	0	0	0	0
Anti-strange	$\bar{s}$	$\frac{1}{2}$	$+\frac{1}{3}e$	$-\frac{1}{3}$	+1	0	0	0
Anti-charmed	$\bar{c}$	$\frac{1}{2}$	$-\frac{2}{3}e$	$-\frac{1}{3}$	0	−1	0	0
Anti-bottom	$\bar{b}$	$\frac{1}{2}$	$+\frac{1}{3}e$	$-\frac{1}{3}$	0	0	−1	0
Anti-top	$\bar{t}$	$\frac{1}{2}$	$-\frac{2}{3}e$	$-\frac{1}{3}$	0	0	0	−1

TABLE 30.4

Quark Composition of Several Hadrons

Particle	Quark Composition
Mesons	
π^+	$\bar{d}u$
π^-	$\bar{u}d$
K^+	$\bar{s}u$
K^-	$\bar{u}s$
K^0	$\bar{s}d$
Baryons	
p	uud
n	udd
Λ^0	uds
Σ^+	uus
Σ^0	uds
Σ^-	dds
Ξ^0	uss
Ξ^-	dss
Ω^-	sss

1. Mesons consist of one quark and one antiquark, giving them a baryon number of 0, as required.
2. Baryons consist of three quarks.
3. Antibaryons consist of three antiquarks.

Table 30.4 lists the quark compositions of several mesons and baryons. Note that just two of the quarks, u and d, are contained in all hadrons encountered in ordinary matter (protons and neutrons). The third quark, s, is needed only to construct strange particles with a strangeness of either $+1$ or -1. Active Figure 30.10 is a pictorial representation of the quark compositions of several particles.

Applying Physics 30.3   Conservation of Meson Number?

We have seen a law of conservation of lepton number and a law of conservation of baryon number. Why isn't there a law of conservation of meson number?

Explanation   We can argue this from the point of view of creating particle–antiparticle pairs from available energy. If energy is converted to the rest energy of a lepton–antilepton pair, then there is no net change in lepton number, because the lepton has a lepton number of $+1$ and the antilepton -1. Energy could also be transformed into the rest energy of a

baryon–antibaryon pair. The baryon has baryon number $+1$, the antibaryon -1, and there is no net change in baryon number.

But now suppose energy is transformed into the rest energy of a quark–antiquark pair. By definition in quark theory, a quark–antiquark pair is a meson. There was no meson before, and now there's a meson, so already there is violation of conservation of meson number. With more energy, we can create more mesons, with no restriction from a conservation law other than that of energy.

Charm and Other Recent Developments

Although the original quark model was highly successful in classifying particles into families, there were some discrepancies between predictions of the model and certain experimental decay rates. Consequently, a fourth quark was proposed by several physicists in 1967. The fourth quark, designated by c, was given a property called **charm**. A charmed quark would have the charge $+2e/3$, but its charm would distinguish it from the other three quarks. The new quark would have a charm $C = +1$, its antiquark would have a charm $C = -1$, and all other quarks would have $C = 0$, as indicated in Table 30.3. Charm, like strangeness, would be conserved in strong and electromagnetic interactions but not in weak interactions.

In 1974 a new heavy meson called the J/ψ particle (or simply, ψ) was discovered independently by a group led by Burton Richter at the Stanford Linear Accelerator (SLAC) and another group led by Samuel Ting at the Brookhaven National Laboratory. Richter and Ting were awarded the Nobel Prize in 1976 for this work. The J/ψ particle didn't fit into the three-quark model, but had the properties of a combination of a charmed quark and its antiquark ($c\bar{c}$). It was much heavier than the other known mesons ($\sim 3\ 100$ MeV/c^2) and its lifetime was much longer than those of other particles that decay via the strong force. In 1975, researchers at Stanford University reported strong evidence for the existence of the tau (τ) lepton, with a mass of $1\ 784$ MeV/c^2. Such discoveries led to more elaborate quark models and the proposal of two new quarks, named *top* (t) and *bottom* (b). To distinguish these quarks from the old ones, quantum numbers called *topness* and *bottomness* were assigned to these new particles and are included in Table 30.3. In 1977 researchers at the Fermi National Laboratory, under the direction of Leon Lederman, reported the discovery of a very massive new meson Υ with composition $b\bar{b}$. In March of 1995, researchers at Fermilab announced the discovery of the top quark (supposedly the last of the quarks to be found) having mass 173 GeV/c^2.

You are probably wondering whether such discoveries will ever end. How many "building blocks" of matter really exist? The numbers of different quarks and leptons have implications for the primordial abundance of certain elements, so at present it appears there may be no further fundamental particles. Some properties of quarks and leptons are given in Table 30.5.

Despite extensive experimental efforts, no isolated quark has ever been observed. Physicists now believe that quarks are permanently confined inside ordinary particles because of an exceptionally strong force that prevents them from escaping. This force, called the color force (which will be discussed in Section 30.13), increases with separation distance (similar to the force of a spring). The great strength of the force between quarks has been described by one author as follows:[2]

> Quarks are slaves of their own color charge, . . . bound like prisoners of a chain gang. . . . Any locksmith can break the chain between two prisoners, but no locksmith is expert enough to break the gluon chains between quarks. Quarks remain slaves forever.

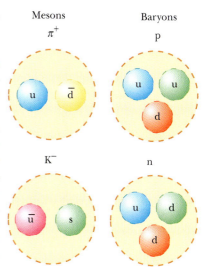

ACTIVE FIGURE 30.10
Quark compositions of two mesons and two baryons. Note that the mesons on the left contain two quarks, and the baryons on the right contain three quarks.

Physics⊗Now™
Log into PhysicsNow at **www.cp7e.com** and go to Active Figure 30.10 to observe the quark compositions for the mesons and baryons.

TABLE 30.5

The Fundamental Particles and Some of Their Properties

Particle	Rest Energy	Charge
Quarks		
u	360 MeV	$+\frac{2}{3}e$
d	360 MeV	$-\frac{1}{3}e$
c	1500 MeV	$+\frac{2}{3}e$
s	540 MeV	$-\frac{1}{3}e$
t	173 GeV	$+\frac{2}{3}e$
b	5 GeV	$-\frac{1}{3}e$
Leptons		
e^-	511 keV	$-e$
μ^-	107 MeV	$-e$
τ^-	1784 MeV	$-e$
ν_e	<30 eV	0
ν_μ	<0.5 MeV	0
ν_τ	<250 MeV	0

[2]Harald Fritzsch, *Quarks: The Stuff of Matter* (London: Allen Lane, 1983).

Computers at Fermilab create a pictorial representation such as this of the paths of particles after a collision.

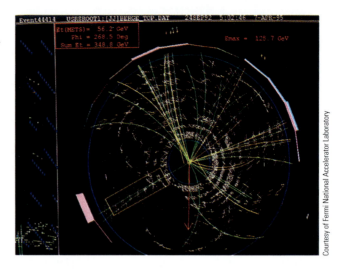

30.13 COLORED QUARKS

Shortly after the theory of quarks was proposed, scientists recognized that certain particles had quark compositions that were in violation of the Pauli exclusion principle. Because all quarks have spins of 1/2, they are expected to follow the exclusion principle. One example of a particle that violates the exclusion principle is the Ω^- (sss) baryon, which contains three s quarks having parallel spins, giving it a total spin of 3/2. Other examples of baryons that have identical quarks with parallel spins are the Δ^{++} (uuu) and the Δ^- (ddd). To resolve this problem, Moo-Young Han and Yoichiro Nambu suggested in 1965 that quarks possess a new property called **color** or **color charge**. This "charge" property is similar in many respects to electric charge, except that it occurs in three varieties, labeled *red*, *green*, and *blue*! (The antiquarks are labeled *anti-red*, *anti-green*, and *anti-blue*.) To satisfy the exclusion principle, all three quarks in a baryon must have different colors. Just as a combination of actual colors of light can produce the neutral color white, a combination of three quarks with different colors is also "white," or colorless. A meson consists of a quark of one color and an antiquark of the corresponding anticolor. The result is that baryons and mesons are always colorless (or white).

Although the concept of color in the quark model was originally conceived to satisfy the exclusion principle, it also provided a better theory for explaining certain experimental results. For example, the modified theory correctly predicts the lifetime of the π^0 meson. The theory of how quarks interact with each other by means of color charge is called **quantum chromodynamics**, or QCD, to parallel quantum electrodynamics (the theory of interactions among electric charges). In QCD, the quark is said to carry a **color charge**, in analogy to electric charge. The strong force between quarks is often called the **color force**. The force is carried by massless particles called **gluons** (which are analogous to photons for the electromagnetic force). According to QCD, there are eight gluons, all with color charge. When a quark emits or absorbs a gluon, its color changes. For example, a blue quark that emits a gluon may become a red quark, and a red quark that absorbs this gluon becomes a blue quark. The color force between quarks is analogous to the electric force between charges: Like colors repel and opposite colors attract. Therefore, two red quarks repel each other, but a red quark will be attracted to an anti-red quark. The attraction between quarks of opposite color to form a meson ($q\bar{q}$) is indicated in Figure 30.11a.

Different-colored quarks also attract each other, but with less intensity than opposite colors of quark and antiquark. For example, a cluster of red, blue, and green quarks all attract each other to form baryons, as indicated in Figure 30.11b. Every baryon contains three quarks of three different colors.

Although the color force between two color-neutral hadrons (such as a proton and a neutron) is negligible at large separations, the strong color force between their constituent quarks does not exactly cancel at small separations of about 1 fm. **This residual strong force is in fact the nuclear force that binds protons and**

TIP 30.3 **Color is Not Really Color**

When we use the word *color* to describe a quark, it has nothing to do with visual sensation from light. It is simply a convenient name for a property analgous to electric charge.

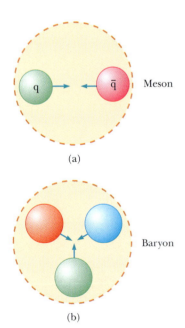

Figure 30.11 (a) A green quark is attracted to an anti-green quark to form a meson with quark structure ($q\bar{q}$). (b) Three different-colored quarks attract each other to form a baryon.

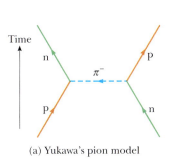

(a) Yukawa's pion model

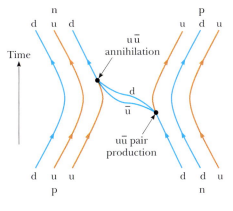

(b) Quark model

Figure 30.12   (a) A nuclear interaction between a proton and a neutron explained in terms of Yukawa's pion exchange model. Because the pion carries charge, the proton and neutron switch identities. (b) The same interaction explained in terms of quarks and gluons. Note that the exchanged $\overline{u}d$ quark pair makes up a π^- meson.

neutrons to form nuclei. It is similar to the residual electromagnetic force that binds neutral atoms into molecules. According to QCD, a more basic explanation of nuclear force can be given in terms of quarks and gluons, as shown in Figure 30.12, which shows contrasting Feynman diagrams of the same process. Each quark within the neutron and proton is continually emitting and absorbing virtual gluons and creating and annihilating virtual $(q\overline{q})$ pairs. When the neutron and proton approach within 1 fm of each other, these virtual gluons and quarks can be exchanged between the two nucleons, and such exchanges produce the nuclear force. Figure 30.12b depicts one likely possibility or contribution to the process shown in Figure 30.12a: a down quark emits a virtual gluon (represented by a wavy line in Fig. 30.12b), which creates a $u\overline{u}$ pair. Both the recoiling d quark and the $\overline{u}$ are transmitted to the proton where the $\overline{u}$ annihilates a proton u quark (with the creation of a gluon) and the d is captured.

30.14   ELECTROWEAK THEORY AND THE STANDARD MODEL

Recall that the weak interaction is an extremely short range force having an interaction distance of approximately 10^{-18} m (Table 30.1). Such a short-range interaction implies that the quantized particles which carry the weak field (the spin one W^+, W^-, and Z^0 bosons) are extremely massive, as is indeed the case. These amazing bosons can be thought of as structureless, pointlike particles as massive as krypton atoms! The weak interaction is responsible for the decay of the c, s, b, and t quarks into lighter, more stable u and d quarks, as well as the decay of the massive μ and τ leptons into (lighter) electrons. **The weak interaction is very important because it governs the stability of the basic particles of matter.**

A mysterious feature of the weak interaction is its lack of symmetry, especially when compared to the high degree of symmetry shown by the strong, electromagnetic, and gravitational interactions. For example, the weak interaction, unlike the strong interaction, is not symmetric under mirror reflection or charge exchange. (*Mirror reflection* means that all the quantities in a given particle reaction are exchanged as in a mirror reflection—left for right, an inward motion toward the mirror for an outward motion, etc. *Charge exchange* means that all the electric charges in a particle reaction are converted to their opposites—all positives to negatives and vice versa.) When we say that the weak interaction is not symmetric, we mean that the reaction with all quantities changed occurs less frequently than the direct reaction. For example, the decay of the K^0, which is governed by the weak interaction, is not symmetric under charge exchange because the reaction $K^0 \rightarrow \pi^- + e^+ + \nu_e$ occurs much more frequently than the reaction $K^0 \rightarrow \pi^+ + e^- + \overline{\nu}_e$.

In 1979, Sheldon Glashow, Abdus Salam, and Steven Weinberg won a Nobel prize for developing a theory called the **electroweak theory** that unified the electromagnetic and weak interactions. This theory postulates that the weak and electromagnetic interactions have the same strength at very high particle energies,

Figure 30.13 The Standard Model of particle physics.

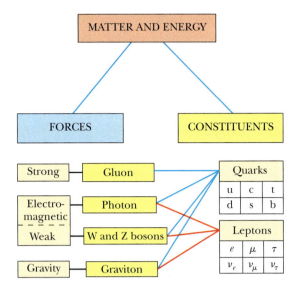

and are different manifestations of a single unifying electroweak interaction. The photon and the three massive bosons ($W^{\pm}$ and Z^{0}) play a key role in the electroweak theory. The theory makes many concrete predictions, but perhaps the most spectacular is the prediction of the masses of the W and Z particles at about 82 GeV/c^2 and 93 GeV/c^2, respectively. A 1984 Nobel Prize was awarded to Carlo Rubbia and Simon van der Meer for their work leading to the discovery of these particles at just those energies at the CERN Laboratory in Geneva, Switzerland.

The combination of the electroweak theory and QCD for the strong interaction form what is referred to in high energy physics as the **Standard Model**. Although the details of the Standard Model are complex, its essential ingredients can be summarized with the help of Figure 30.13. The strong force, mediated by gluons, holds quarks together to form composite particles such as protons, neutrons, and mesons. Leptons participate only in the electromagnetic and weak interactions. The electromagnetic force is mediated by photons, and the weak force is mediated by W and Z bosons. Note that all fundamental forces are mediated by bosons (particles with spin 1) whose properties are given, to a large extent, by symmetries involved in the theories.

However, the Standard Model does not answer all questions. A major question is why the photon has no mass while the W and Z bosons do. Because of this mass difference, the electromagnetic and weak forces are quite distinct at low energies, but become similar in nature at very high energies, where the rest energies of the W and Z bosons are insignificant fractions of their total energies. This behavior during the transition from high to low energies, called **symmetry breaking**, doesn't answer the question of the origin of particle masses. To resolve that problem, a hypothetical particle called the **Higgs boson** has been proposed which provides a mechanism for breaking the electroweak symmetry and bestowing different particle masses on different particles. The Standard Model, including the Higgs mechanism, provides a logically consistent explanation of the massive nature of the W and Z bosons. Unfortunately, the Higgs boson has not yet been found, but physicists know that its mass should be less than 1 TeV/c^2 (10^{12} eV).

In order to determine whether the Higgs boson exists, two quarks of at least 1 TeV of energy must collide, but calculations show that this requires injecting 40 TeV of energy within the volume of a proton. Scientists are convinced that because of the limited energy available in conventional accelerators using fixed targets, it is necessary to build colliding-beam accelerators called **colliders**. The concept of a collider is straightforward. In such a device, particles with equal masses and kinetic energies, traveling in opposite directions in an accelerator ring, collide head-on to produce the required reaction and the formation of new particles. Because the total momentum of the interacting particles is zero, all of their kinetic energy is available for the reaction. The Large Electron–Positron (LEP) collider at CERN, near Geneva, Switzerland, and the Stanford Linear Collider in California collide both electrons and positrons. The Super Proton Synchrotron at CERN accelerates

A view from inside the Large Electron–Positron (LEP) collider tunnel, which is 27 km in circumference.

protons and antiprotons to energies of 270 GeV, and the world's highest-energy proton acclerator, the Tevatron, at the Fermi National Laboratory in Illinois, produces protons at almost 1 000 GeV (or 1 TeV). CERN has started construction of the Large Hadron Collider (LHC), a proton–proton collider that will provide a center-of-mass energy of 14 TeV and allow an exploration of Higgs-boson physics. The accelerator is being constructed in the same 27-km circumference tunnel as CERN's LEP collider, and construction is expected to be completed in 2005.

Following the success of the electroweak theory, scientists attempted to combine it with QCD in a **grand unification theory** (GUT). In this model, the electroweak force was merged with the strong color force to form a grand unified force. One version of the theory considers leptons and quarks as members of the same family that are able to change into each other by exchanging an appropriate particle. Many GUT theories predict that protons are unstable and will decay with a lifetime of about 10^{31} years, a period far greater than the age of the Universe. As yet, proton decays have not been observed.

Applying Physics 30.4   Head-on Collisions

Consider a car making a head-on collision with an identical car moving in the opposite direction at the same speed. Compare that collision to one in which one of the cars collides with a second car that is at rest. In which collision is there a larger transformation of kinetic energy to other forms? How does this idea relate to producing exotic particles in collisions?

Explanation In the head-on collision with both cars moving, conservation of momentum causes most, if not all, of the kinetic energy to be transformed to other forms. In the collision between a moving car and a stationary car, the cars are still moving after the collision in the direction of the moving car, but with reduced speed. Thus, only part of the kinetic energy is transformed to other forms. This suggests the advantage of using colliding beams to produce exotic particles, as opposed to firing a beam into a stationary target. When particles moving in opposite directions collide, all of the kinetic energy is available for transformation into other forms—in this case, the creation of new particles. When a beam is fired into a stationary target, only part of the energy is available for transformation, so particles of higher mass cannot be created.

30.15   THE COSMIC CONNECTION

In this section we describe one of the most fascinating theories in all of science—the Big Bang theory of the creation of the Universe—and the experimental evidence that supports it. This theory of cosmology states that the Universe had a beginning and that this beginning was so cataclysmic that it is impossible to look back beyond it. According to the theory, the Universe erupted from an infinitely dense singularity about 15 to 20 billion years ago. The first few minutes after the Big Bang saw such extremes of energy that it is believed that all four interactions of physics were unified and all matter was contained in an undifferentiated "quark soup."

The evolution of the four fundamental forces from the Big Bang to the present is shown in Figure 30.14. During the first 10^{-43} s (the ultrahot epoch, with

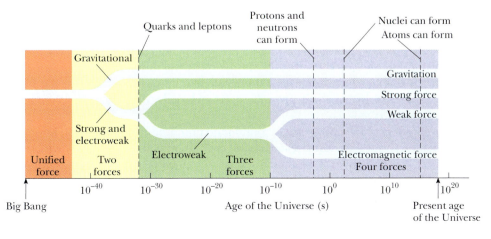

Figure 30.14 A brief history of the Universe from the Big Bang to the present. The four forces became distinguishable during the first microsecond. Following this, all the quarks combined to form particles that interact via the strong force. The leptons remained separate, however, and exist as individually observable particles to this day.

GEORGE GAMOW
(1904–1968)

Gamow and two of his students, Ralph Alpher and Robert Herman, were the first to take the first half hour of the Universe seriously. In a mostly overlooked paper published in 1948, they made truly remarkable cosmological predictions. They correctly calculated the abundances of hydrogen and helium after the first half hour (75% H and 25% He) and predicted that radiation from the Big Bang should still be present and have an apparent temperature of about 5 K.

$T \approx 10^{32}$ K), it is presumed that the strong, electroweak, and gravitational forces were joined to form a completely unified force. In the first 10^{-35} s following the Big Bang (the hot epoch, with $T \approx 10^{29}$ K), gravity broke free of this unification and the strong and electroweak forces remained as one, described by a grand unification theory. This was a period when particle energies were so great ($> 10^{16}$ GeV) that very massive particles as well as quarks, leptons, and their antiparticles, existed. Then, after 10^{-35} s, the Universe rapidly expanded and cooled (the warm epoch, with $T \approx 10^{29}$ to 10^{15} K), the strong and electroweak forces parted company, and the grand unification scheme was broken. As the Universe continued to cool, the electroweak force split into the weak force and the electromagnetic force about 10^{-10} s after the Big Bang.

After a few minutes, protons condensed out of the hot soup. For half an hour the Universe underwent thermonuclear detonation, exploding like a hydrogen bomb and producing most of the helium nuclei now present. The Universe continued to expand, and its temperature dropped. Until about 700 000 years after the Big Bang, the Universe was dominated by radiation. Energetic radiation prevented matter from forming single hydrogen atoms because collisions would instantly ionize any atoms that might form. Photons underwent continuous Compton scattering from the vast number of free electrons, resulting in a Universe that was opaque to radiation. By the time the Universe was about 700 000 years old, it had expanded and cooled to about 3 000 K, and protons could bind to electrons to form neutral hydrogen atoms. Because the energies of the atoms were quantized, far more wavelengths of radiation were not absorbed by atoms than were, and the Universe suddenly became transparent to photons. Radiation no longer dominated the Universe, and clumps of neutral matter grew steadily—first atoms, followed by molecules, gas clouds, stars, and finally galaxies.

Observation of Radiation from the Primordial Fireball

In 1965 Arno A. Penzias (b. 1933) and Robert W. Wilson (b. 1936) of Bell Laboratories made an amazing discovery while testing a sensitive microwave receiver. A pesky signal producing a faint background hiss was interfering with their satellite communications experiments. In spite of their valiant efforts, the signal remained. Ultimately it became clear that they were observing microwave background radiation (at a wavelength of 7.35 cm) representing the leftover "glow" from the Big Bang.

Figure 30.15 Robert W. Wilson (*left*) and Arno A. Penzias (*right*), with Bell Telephone Laboratories' horn-reflector antenna.

The microwave horn that served as their receiving antenna is shown in Figure 30.15. The intensity of the detected signal remained unchanged as the antenna was pointed in different directions. The fact that the radiation had equal strengths in all directions suggested that the entire Universe was the source of this radiation. Evicting a flock of pigeons from the 20-foot horn and cooling the microwave detector both failed to remove the signal. Through a casual conversation, Penzias and Wilson discovered that a group at Princeton had predicted the residual radiation from the Big Bang and were planning an experiment to confirm the theory. The excitement in the scientific community was high when Penzias and Wilson announced that they had already observed an excess microwave background compatible with a 3-K blackbody source.

Because Penzias and Wilson made their measurements at a single wavelength, they did not completely confirm the radiation as 3-K blackbody radiation. Subsequent experiments by other groups added intensity data at different wavelengths, as shown in Figure 30.16. The results confirm that the radiation is that of a blackbody at 2.9 K. This figure is perhaps the most clear-cut evidence for the Big Bang theory. The 1978 Nobel Prize in physics was awarded to Penzias and Wilson for their important discovery.

The discovery of the cosmic background radiation produced a problem, however: the radiation was too uniform. Scientists believed there had to be slight fluctuations in this background in order for such objects as galaxies to form. In 1989, NASA launched a satellite called the Cosmic Background Explorer (COBE, pronounced KOH-bee) to study this radiation in greater detail. In 1992, George Smoot

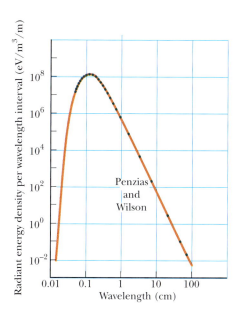

Figure 30.16 Theoretical blackbody (brown curve) and measured radiation spectra (blue points) of the Big Bang. Most of the data were collected from the Cosmic Background Explorer (COBE) satellite. The datum of Wilson and Penzias is indicated.

(b. 1945) at the Lawrence Berkeley Laboratory found that the background was not perfectly uniform, but instead contained irregularities corresponding to temperature variations of 0.000 3 K. It is these small variations that provided nucleation sites for the formation of the galaxies and other objects we now see in the sky.

30.16 PROBLEMS AND PERSPECTIVES

While particle physicists have been exploring the realm of the very small, cosmologists have been exploring cosmic history back to the first microsecond of the Big Bang. Observation of the events that occur when two particles collide in an accelerator is essential in reconstructing the early moments in cosmic history. Perhaps the key to understanding the early Universe is first to understand the world of elementary particles. Cosmologists and particle physicists find that they have many common goals and are joining efforts to study the physical world at its most fundamental level.

Our understanding of physics at short and long distances is far from complete. Particle physics is faced with many questions: why is there so little antimatter in the Universe? Do neutrinos have a small mass, and if so, how much do they contribute to the "dark matter" holding the universe together gravitationally? How can we understand the latest astronomical measurements, which show that the expansion of the universe is accelerating and that there may be a kind of "antigravity force" acting between widely separated galaxies? Is it possible to unify the strong and electroweak theories in a logical and consistent manner? Why do quarks and leptons form three similar but distinct families? Are muons the same as electrons (apart from their different masses), or do they have subtle differences that have not been detected? Why are some particles charged and others neutral? Why do quarks carry a fractional charge? What determines the masses of the fundamental particles? The questions go on and on. Because of the rapid advances and new discoveries in the related fields of particle physics and cosmology, by the time you read this book some of these questions may have been resolved and others may have emerged.

An important question that remains is whether leptons and quarks have a substructure. If they do, one could envision an infinite number of deeper structure levels. However, if leptons and quarks are indeed the ultimate constituents of matter, as physicists today tend to believe, we should be able to construct a final theory of the structure of matter, as Einstein dreamed of doing. In the view of many physicists, the end of the road is in sight, but how long it will take to reach that goal is anyone's guess.

SUMMARY

30.1 Nuclear Fission &
30.2 Nuclear Reactors

In **nuclear fission**, the total mass of the products is always less than the original mass of the reactants. Nuclear fission occurs when a heavy nucleus splits, or fissions, into two smaller nuclei. The lost mass is transformed into energy, electromagnetic radiation, and the kinetic energy of daughter particles.

A **nuclear reactor** is a system designed to maintain a self-sustaining chain reaction. Nuclear reactors using controlled fission events are currently being used to generate electric power. A useful parameter for describing the level of reactor operation is the reproduction constant K, which is the average number of neutrons from each fission event that will cause another event. A self-sustaining reaction is achieved when $K = 1$.

30.3 Nuclear Fusion

In nuclear fusion, two light nuclei combine to form a heavier nucleus. This type of nuclear reaction occurs in the Sun, assisted by a quantum tunneling process that helps particles get through the Coulomb barrier.

Controlled fusion events offer the hope of plentiful supplies of energy in the future. The nuclear fusion reactor is considered by many scientists to be the ultimate energy source because its fuel is water. **Lawson's criterion** states that a fusion reactor will provide a net output power if the product of the plasma ion density n and the plasma confinement time τ satisfies the following relationships:

$$n\tau \geq 10^{14} \text{ s/cm}^3 \quad \text{Deuterium–tritium interaction} \quad \textbf{[30.5]}$$
$$n\tau \geq 10^{16} \text{ s/cm}^3 \quad \text{Deuterium–deuterium interaction}$$

30.5 The Fundamental Forces of Nature

There are four fundamental forces of nature: the **strong** (hadronic), **electromagnetic**, **weak**, and **gravitational** forces. The strong force is the force between nucleons that keeps the nucleus together. The weak force is responsible for beta decay. The electromagnetic and weak forces are now considered to be manifestations of a single force called the **electroweak** force.

Every fundamental interaction is said to be mediated by the exchange of field particles. The electromagnetic interaction is mediated by the photon, the weak interaction by the $W^\pm$ and Z^0 bosons, the gravitational interaction by gravitons, and the strong interaction by gluons.

30.6 Positrons and Other Antiparticles

An antiparticle and a particle have the same mass, but opposite charge, and may also have other properties with opposite values, such as lepton number and baryon number. It is possible to produce particle–antiparticle pairs in nuclear reactions if the available energy is greater than $2mc^2$, where m is the mass of the particle (or antiparticle).

30.8 Classification of Particles

Particles other than photons are classified as hadrons or leptons. **Hadrons** interact primarily through the strong force. They have size and structure and hence are not elementary particles. There are two types of hadrons: *baryons* and *mesons*. Mesons have a baryon number of zero and have either zero or integer spin. Baryons, which generally are the most massive particles, have nonzero baryon numbers and spins of $1/2$ or $3/2$. The neutron and proton are examples of baryons.

Leptons have no known structure, down to the limits of current resolution (about 10^{-19} m). Leptons interact only through the weak and electromagnetic forces. There are six leptons: the electron, e^-; the muon, μ^-; the tau, τ^-; and their associated neutrinos, ν_e, ν_μ, and ν_τ.

30.9 Conservation Laws &
30.10 Strange Particles and Strangeness

In all reactions and decays, quantities such as energy, linear momentum, angular momentum, electric charge, baryon number, and lepton number are strictly conserved. Certain particles have properties called **strangeness** and **charm**. These unusual properties are conserved only in those reactions and decays that occur via the strong force.

30.12 Quarks &
30.13 Colored Quarks

Recent theories postulate that all hadrons are composed of smaller units known as **quarks** which have fractional electric charges and baryon numbers of 1/3 and come in six "flavors": up, down, strange, charmed, top, and bottom. Each baryon contains three quarks, and each meson contains one quark and one antiquark.

According to the theory of **quantum chromodynamics**, quarks have a property called **color**, and the strong force between quarks is referred to as the **color force**. The color force increases as the distance between particles increases, so quarks are confined and are never observed in isolation. When two bound quarks are widely separated, a new quark–antiquark pair forms between them, and the single particle breaks into two new particles, each composed of a quark–antiquark pair.

30.15 The Cosmic Connection

Observation of background microwave radiation by Penzias and Wilson strongly confirmed that the Universe started with a Big Bang about 15 billion years ago and has been expanding ever since. The background radiation is equivalent to that of a blackbody at a temperature of about 3 K.

The cosmic microwave background has very small irregularities, corresponding to temperature variations of 0.000 3 K. Without these irregularities acting as nucleation sites, particles would never have clumped together to form galaxies and stars.

CONCEPTUAL QUESTIONS

1. If high-energy electrons with de Broglie wavelengths smaller than the size of the nucleus are scattered from nuclei, the behavior of the electrons is consistent with scattering from very massive structures much smaller in size than the nucleus, namely, quarks. How is this similar to a classic experiment that detected small structures in an atom?

2. What factors make a fusion reaction difficult to achieve?

3. Doubly charged baryons are known to exist. Why are there no doubly charged mesons?

4. Why would a fusion reactor produce less radioactive waste than a fission reactor?

5. Atoms didn't exist until hundreds of thousands of years after the Big Bang. Why?

6. Particles known as resonances have very short half-lives, on the order of 10^{-23} s. Would you guess they are hadrons or leptons?

7. Describe the quark model of hadrons, including the properties of quarks.

8. In the theory of quantum chromodynamics, quarks come in three colors. How would you justify the statement "All baryons and mesons are colorless?"

9. Describe the properties of baryons and mesons and the important differences between them.

10. Identify the particle decays in Table 30.2 that occur by the electromagnetic interaction. Justify your answer.

11. Kaons all decay into final states that contain no protons or neutrons. What is the baryon number of kaons?

12. When an electron and a positron meet at low speeds in free space, why are *two* 0.511-MeV gamma rays produced, rather than *one* gamma ray with an energy of 1.02 MeV?

13. Two protons in a nucleus interact via the strong interaction. Are they also subject to a weak interaction?

14. Why is a neutron stable inside the nucleus? (In free space, the neutron decays in 900 s.)

15. An antibaryon interacts with a meson. Can a baryon be produced in such an interaction? Explain.

16. Why is water a better shield against neutrons than lead or steel is?

17. How many quarks are there in (a) a baryon, (b) an antibaryon, (c) a meson, and (d) an antimeson? How do you account for the fact that baryons have half-integral spins and mesons have spins of 0 or 1? [*Hint:* quarks have spin $\frac{1}{2}$.]

18. A typical chemical reaction is one in which a water molecule is formed by combining hydrogen and oxygen. In such a reaction, about 2.5 eV of energy is released. Compare this reaction to a nuclear event such as $^{1}_{0}n + ^{235}_{92}U \rightarrow ^{136}_{53}I + ^{98}_{39}Y + 2^{1}_{0}n$. Would you expect the energy released in this nuclear event to be much greater, much less, or about the same as that released in the chemical reaction? Explain.

19. The neutral ρ meson decays by the strong interaction into two pions according to $\rho^0 \rightarrow \pi^+ + \pi^-$, with a half-life of about 10^{-23} s. The neutral K meson also decays into two pions according to $K^0 \rightarrow \pi^+ + \pi^-$, but with a much longer half-life of about 10^{-10} s. How do you explain these observations?

PROBLEMS

1, 2, 3 = straightforward, intermediate, challenging   □ = full solution available in *Student Solutions Manual/Study Guide*
Physics⊗Now™ = coached problem with hints available at **www.cp7e.com**   🧬 = biomedical application

Section 30.1 Nuclear Fission
Section 30.2 Nuclear Reactors

1. If the average energy released in a fission event is 208 MeV, find the total number of fission events required to operate a 100-W lightbulb for 1.0 h.

2. Find the energy released in the fission reaction

$$n + ^{235}_{92}U \rightarrow ^{98}_{40}Zr + ^{135}_{52}Te + 3n$$

The atomic masses of the fission products are 97.912 0 u for $^{98}_{40}Zr$ and 134.908 7 u for $^{135}_{52}$ Te.

3. Find the energy released in the following fission reaction:

$$^{1}_{0}n + ^{235}_{92}U \rightarrow ^{88}_{38}Sr + ^{136}_{54}Xe + 12^{1}_{0}n$$

4. Strontium-90 is a particularly dangerous fission product of ^{235}U because it is radioactive and it substitutes for calcium in bones. What other direct fission products would accompany it in the neutron-induced fission of ^{235}U? [*Note*: This reaction may release two, three, or four free neutrons.]

5. Assume that ordinary soil contains natural uranium in amounts of 1 part per million by mass. (a) How much uranium is in the top 1.00 meter of soil on a 1-acre (43 560-ft^2) plot of ground, assuming the specific gravity of soil is 4.00? (b) How much of the isotope ^{235}U, appropriate for nuclear reactor fuel, is in this soil? [*Hint*: See Appendix B for the percent abundance of $^{235}_{92}U$.]

6. A typical nuclear fission power plant produces about 1.00 GW of electrical power. Assume that the plant has an overall efficiency of 40.0% and that each fission produces 200 MeV of thermal energy. Calculate the mass of ^{235}U consumed each day.

7. Suppose that the water exerts an average frictional drag of 1.0×10^5 N on a nuclear-powered ship. How far can the ship travel per kilogram of fuel if the fuel consists of enriched uranium containing 1.7% of the fissionable isotope ^{235}U and the ship's engine has an efficiency of 20%? (Assume 208 MeV is released per fission event.)

8. It has been estimated that the Earth contains 1.0×10^9 tons of natural uranium that can be mined economically. If all the world's energy needs $(7.0 \times 10^{12}$ J/s) were supplied by ^{235}U fission, how long would this supply last? [*Hint*: See Appendix B for the percent abundance of $^{235}_{92}U$.]

9. Physics⊗Now™ An all-electric home uses approximately 2 000 kWh of electric energy per month. How much uranium-235 would be required to provide this house with its energy needs for 1 year? (Assume 100% conversion efficiency and 208 MeV released per fission.)

Section 30.3 Nuclear Fusion

10. Find the energy released in the fusion reaction

$$^{1}_{1}H + ^{2}_{1}H \rightarrow ^{3}_{2}He + \gamma$$

11. When a star has exhausted its hydrogen fuel, it may fuse other nuclear fuels. At temperatures above 1.0×10^8 K, helium fusion can occur. Write the equations for the following processes: (a) Two alpha particles fuse to produce a nucleus A and a gamma ray. What is nucleus A? (b) Nucleus A absorbs an alpha particle to produce a nucleus B and a gamma ray. What is nucleus B? (c) Find the total energy released in the reactions given in (a) and (b). [*Note*: The mass of $^{8}_{4}$Be $= 8.005\ 305$ u.]

12. Another series of nuclear reactions that can produce energy in the interior of stars is the cycle described below. This cycle is most efficient when the central temperature in a star is above 1.6×10^{7} K. Because the temperature at the center of the Sun is only 1.5×10^{7} K, the following cycle produces less than 10% of the Sun's energy. (a) A high-energy proton is absorbed by ^{12}C. Another nucleus, A, is produced in the reaction, along with a gamma ray. Identify nucleus A. (b) Nucleus A decays through positron emission to form nucleus B. Identify nucleus B. (c) Nucleus B absorbs a proton to produce nucleus C and a gamma ray. Identify nucleus C. (d) Nucleus C absorbs a proton to produce nucleus D and a gamma ray. Identify nucleus D. (e) Nucleus D decays through positron emission to produce nucleus E. Identify nucleus E. (f) Nucleus E absorbs a proton to produce nucleus F plus an alpha particle. What is nucleus F? [*Note*: If nucleus F is not ^{12}C — that is, the nucleus you started with — you have made an error and should review the sequence of events.]

13. If an all-electric home uses approximately 2 000 kWh of electric energy per month, how many fusion events described by the reaction $^{2}_{1}$H $+ {}^{3}_{1}$H $\rightarrow {}^{4}_{2}$He $+ {}^{1}_{0}$n would be required to keep this home running for one year?

14. To understand why plasma containment is necessary, consider the rate at which an unconfined plasma would be lost. (a) Estimate the rms speed of deuterons in a plasma at 4.00×10^{8} K. (b) Estimate the order of magnitude of the time such a plasma would remain in a 10-cm cube if no steps were taken to contain it.

15. The oceans have a volume of 317 million cubic miles and contain 1.32×10^{21} kg of water. Of all the hydrogen nuclei in this water, 0.030 0% of the mass is deuterium. (a) If all of these deuterium nuclei were fused to helium via the first reaction in Equation 30.4, determine the total amount of energy that could be released. (b) The present world electric power consumption is about 7.00×10^{12} W. If consumption were 100 times greater, how many years would the energy supply calculated in part (a) last?

Section 30.6 Positrons and Other Antiparticles

16. Two photons are produced when a proton and an antiproton annihilate each other. What is the minimum frequency and corresponding wavelength of each photon?

17. **Physics ⊗ Now**™ A photon produces a proton–antiproton pair according to the reaction $\gamma \rightarrow$ p $+ \overline{\text{p}}$. What is the minimum possible frequency of the photon? What is its wavelength?

18. A photon with an energy of 2.09 GeV creates a proton–antiproton pair in which the proton has a kinetic energy of 95.0 MeV. What is the kinetic energy of the antiproton?

Section 30.7 Mesons and the Beginning of Particle Physics

19. When a high-energy proton or pion traveling near the speed of light collides with a nucleus, it travels an average distance of 3.0×10^{-15} m before interacting with another particle. From this information, estimate the time for the strong interaction to occur.

20. Calculate the order of magnitude of the range of the force that might be produced by the virtual exchange of a proton.

21. One of the mediators of the weak interaction is the Z^{0} boson, which has a mass of 96 GeV/c^{2}. Use this information to find an approximate value for the range of the weak interaction.

22. If a π^{0} at rest decays into two γ's, what is the energy of each of the γ's?

Section 30.9 Conservation Laws
Section 30.10 Strange Particles and Strangeness

23. Each of the following reactions is forbidden. Determine a conservation law that is violated for each reaction.
    (a) p $+ \overline{\text{p}} \rightarrow \mu^{+} + e^{-}$
    (b) $\pi^{-} +$ p $\rightarrow$ p $+ \pi^{+}$
    (c) p $+$ p $\rightarrow$ p $+ \pi^{+}$
    (d) p $+$ p $\rightarrow$ p $+$ p $+$ n
    (e) $\gamma +$ p $\rightarrow$ n $+ \pi^{0}$

24. For the following two reactions, the first may occur but the second cannot. Explain.

$$K^0 \rightarrow \pi^+ + \pi^- \quad \text{(can occur)}$$
$$\Lambda^0 \rightarrow \pi^+ + \pi^- \quad \text{(cannot occur)}$$

25. **Physics⊗Now**™ Identify the unknown particle on the left side of the reaction

$$? + p \rightarrow n + \mu^+$$

26. Determine the type of neutrino or antineutrino involved in each of the following processes:
    (a) $\pi^+ \rightarrow \pi^0 + e^+ + ?$
    (b) $? + p \rightarrow \mu^- + p + \pi^+$
    (c) $\Lambda^0 \rightarrow p + \mu^- + ?$
    (d) $\tau^+ \rightarrow \mu^+ + ? + ?$

27. The following reactions or decays involve one or more neutrinos. Supply the missing neutrinos.
    (a) $\pi^- \rightarrow \mu^- + ?$
    (b) $K^+ \rightarrow \mu^+ + ?$
    (c) $? + p \rightarrow n + e^+$
    (d) $? + n \rightarrow p + e^-$
    (e) $? + n \rightarrow p + \mu^-$
    (f) $\mu^- \rightarrow e^- + ? + ?$

28. Determine which of the reactions below can occur. For those that cannot occur, determine the conservation law (or laws) that each violates:
    (a) $p \rightarrow \pi^+ + \pi^0$
    (b) $p + p \rightarrow p + p + \pi^0$
    (c) $p + p \rightarrow p + \pi^+$
    (d) $\pi^+ \rightarrow \mu^+ + \nu_\mu$
    (e) $n \rightarrow p + e^- + \bar{\nu}_e$
    (f) $\pi^+ \rightarrow \pi^+ + n$

29. Which of the following processes are allowed by the strong interaction, the electromagnetic interaction, the weak interaction, or no interaction at all?
    (a) $\pi^- + p \rightarrow 2\eta^0$
    (b) $K^- + n \rightarrow \Lambda^0 + \pi^-$
    (c) $K^- \rightarrow \pi^- + \pi^0$
    (d) $\Omega^- \rightarrow \Xi^- + \pi^0$
    (e) $\eta^0 \rightarrow 2\gamma$

30. A K^0 particle at rest decays into a π^+ and a π^-. What will be the speed of each of the pions? The mass of the K^0 is 497.7 MeV/c^2 and the mass of each pion is 139.6 MeV/c^2.

31. Determine whether or not strangeness is conserved in the following decays and reactions:
    (a) $\Lambda^0 \rightarrow p + \pi^-$
    (b) $\pi^- + p \rightarrow \Lambda^0 + K^0$
    (c) $\bar{p} + p \rightarrow \bar{\Lambda}^0 + \Lambda^0$

(d) $\pi^- + p \rightarrow \pi^- + \Sigma^+$
(e) $\Xi^- \rightarrow \Lambda^0 + \pi^-$
(f) $\Xi^0 \rightarrow p + \pi^-$

32. Fill in the missing particle. Assume that (a) occurs via the strong interaction while (b) and (c) involve the weak interaction.
    (a) $K^+ + p \rightarrow \underline{\quad} + p$
    (b) $\Omega^- \rightarrow \underline{\quad} + \pi^-$
    (c) $K^+ \rightarrow \underline{\quad} + \mu^+ + \nu_\mu$

33. Identify the conserved quantities in the following processes:
    (a) $\Xi^- \rightarrow \Lambda^0 + \mu^- + \nu_\mu$
    (b) $K^0 \rightarrow 2\pi^0$
    (c) $K^- + p \rightarrow \Sigma^0 + n$
    (d) $\Sigma^0 \rightarrow \Lambda^0 + \gamma$
    (e) $e^+ + e^- \rightarrow \mu^+ + \mu^-$
    (f) $\bar{p} + n \rightarrow \bar{\Lambda}^0 + \Sigma^-$

Section 30.12 Quarks
Section 30.13 Colored Quarks

34. The quark composition of the proton is uud, while that of the neutron in udd. Show that the charge, baryon number, and strangeness of these particles equal the sums of these numbers for their quark constituents.

35. Find the number of electrons, and of each species of quark, in 1 L of water.

36. The quark compositions of the K^0 and Λ^0 particles are d$\bar{s}$ and uds, respectively. Show that the charge, baryon number, and strangeness of these particles equal the sums of these numbers for the quark constituents.

37. Identify the particles corresponding to the following quark states: (a) suu; (b) $\bar{u}$d; (c) $\bar{s}$d; (d) ssd.

38. What is the electrical charge of the baryons with the quark compositions (a) $\bar{u}\bar{u}$d and (b) $\bar{u}$dd? What are these baryons called?

39. Analyze the first three of the following reactions at the quark level, and show that each conserves the net number of each type of quark; then, in the last reaction, identify the mystery particle:
    (a) $\pi^- + p \rightarrow K^0 + \Lambda^0$
    (b) $\pi^+ + p \rightarrow K^+ + \Sigma^+$
    (c) $K^- + p \rightarrow K^+ + K^0 + \Omega^-$
    (d) $p + p \rightarrow K^0 + p + \pi^+ + ?$

Mathematical Review

A.1 MATHEMATICAL NOTATION

Many mathematical symbols are used throughout this book. You are no doubt familiar with some, such as the symbol $=$ to denote the equality of two quantities.

The symbol $\propto$ denotes a proportionality. For example, $y \propto x^2$ means that y is proportional to the square of x.

The symbol $<$ means *is less than*, and $>$ means *is greater than*. For example, $x > y$ means x is greater than y.

The symbol $\ll$ means *is much less than*, and $\gg$ means *is much greater than*.

The symbol $\approx$ indicates that two quantities are *approximately equal* to each other.

The symbol $\equiv$ means *is defined as*. This is a stronger statement than a simple $=$.

It is convenient to use the notation Δx (read as "delta x") to indicate the *change in the quantity* x. (Note that Δx does not mean "the product of Δ and x.") For example, suppose that a person out for a morning stroll starts measuring her distance away from home when she is 10 m from her doorway. She then moves along a straight-line path and stops strolling 50 m from the door. Her change in position during the walk is $\Delta x = 50$ m $-$ 10 m $= 40$ m or, in symbolic form,

$$\Delta x = x_f - x_i$$

In this equation x_f is the *final position* and x_i is the *initial position*.

We often have occasion to add several quantities. A useful abbreviation for representing such a sum is the Greek letter Σ (capital sigma). Suppose we wish to add a set of five numbers represented by x_1, x_2, x_3, x_4, and x_5. In the abbreviated notation, we would write the sum as

$$x_1 + x_2 + x_3 + x_4 + x_5 = \sum_{i=1}^{5} x_i$$

where the subscript i on x represents any one of the numbers in the set. For example, if there are five masses in a system, m_1, m_2, m_3, m_4, and m_5, the total mass of the system $M = m_1 + m_2 + m_3 + m_4 + m_5$ could be expressed as

$$M = \sum_{i=1}^{5} m_i$$

Finally, the magnitude of a quantity x, written $|x|$, is simply the absolute value of that quantity. The sign of $|x|$ is always positive, regardless of the sign of x. For example, if $x = -5$, $|x| = 5$; if $x = 8$, $|x| = 8$.

A.2 SCIENTIFIC NOTATION

Many quantities that scientists deal with often have very large or very small values. For example, the speed of light is about 300 000 000 m/s and the ink required to make the dot over an i in this textbook has a mass of about 0.000 000 001 kg. Obviously, it is cumbersome to read, write, and keep track of numbers such as these. We avoid this problem by using a method dealing with powers of the number 10:

$$10^0 = 1$$
$$10^1 = 10$$
$$10^2 = 10 \times 10 = 100$$
$$10^3 = 10 \times 10 \times 10 = 1\ 000$$
$$10^4 = 10 \times 10 \times 10 \times 10 = 10\ 000$$
$$10^5 = 10 \times 10 \times 10 \times 10 \times 10 = 100\ 000$$

and so on. The number of zeros corresponds to the power to which 10 is raised, called the **exponent** of 10. For example, the speed of light, 300 000 000 m/s, can be expressed as 3×10^8 m/s.

For numbers less than one, we note the following:

$$10^{-1} = \frac{1}{10} = 0.1$$

$$10^{-2} = \frac{1}{10 \times 10} = 0.01$$

$$10^{-3} = \frac{1}{10 \times 10 \times 10} = 0.001$$

$$10^{-4} = \frac{1}{10 \times 10 \times 10 \times 10} = 0.000\,1$$

$$10^{-5} = \frac{1}{10 \times 10 \times 10 \times 10 \times 10} = 0.000\,01$$

In these cases, the number of places the decimal point is to the left of the digit 1 equals the value of the (negative) exponent. Numbers that are expressed as some power of 10 multiplied by another number between 1 and 10 are said to be in **scientific notation.** For example, the scientific notation for 5 943 000 000 is 5.943×10^9 and that for 0.000 083 2 is 8.32×10^{-5}.

When numbers expressed in scientific notation are being multiplied, the following general rule is very useful:

$$10^n \times 10^m = 10^{n+m} \qquad\qquad \text{[A.1]}$$

where n and m can be *any* numbers (not necessarily integers). For example, $10^2 \times 10^5 = 10^7$. The rule also applies if one of the exponents is negative. For example, $10^3 \times 10^{-8} = 10^{-5}$.

When dividing numbers expressed in scientific notation, note that

$$\frac{10^n}{10^m} = 10^n \times 10^{-m} = 10^{n-m} \qquad\qquad \text{[A.2]}$$

EXERCISES

With help from the above rules, verify the answers to the following:

1. $86\,400 = 8.64 \times 10^4$
2. $9\,816\,762.5 = 9.816\,762\,5 \times 10^6$
3. $0.000\,000\,039\,8 = 3.98 \times 10^{-8}$
4. $(4.0 \times 10^8)(9.0 \times 10^9) = 3.6 \times 10^{18}$
5. $(3.0 \times 10^7)(6.0 \times 10^{-12}) = 1.8 \times 10^{-4}$
6. $\dfrac{75 \times 10^{-11}}{5.0 \times 10^{-3}} = 1.5 \times 10^{-7}$
7. $\dfrac{(3 \times 10^6)(8 \times 10^{-2})}{(2 \times 10^{17})(6 \times 10^5)} = 2 \times 10^{-18}$

A.3 ALGEBRA

A. Some Basic Rules

When algebraic operations are performed, the laws of arithmetic apply. Symbols such as x, y, and z are frequently used to represent quantities that are not specified, what are called the **unknowns.**

First, consider the equation

$$8x = 32$$

If we wish to solve for x, we can divide (or multiply) each side of the equation by the same factor without destroying the equality. In this case, if we divide both sides by 8, we have

$$\frac{8x}{8} = \frac{32}{8}$$

$$x = 4$$

Next consider the equation

$$x + 2 = 8$$

In this type of expression, we can add or subtract the same quantity from each side. If we subtract 2 from each side, we get

$$x + 2 - 2 = 8 - 2$$

$$x = 6$$

In general, if $x + a = b$, then $x = b - a$.

Now consider the equation

$$\frac{x}{5} = 9$$

If we multiply each side by 5, we are left with x on the left by itself and 45 on the right:

$$\left(\frac{x}{5}\right)(5) = 9 \times 5$$

$$x = 45$$

In all cases, **whatever operation is performed on the left side of the equality must also be performed on the right side.**

The following rules for multiplying, dividing, adding, and subtracting fractions should be recalled, where a, b, and c are three numbers:

	Rule	**Example**
Multiplying	$\left(\dfrac{a}{b}\right)\left(\dfrac{c}{d}\right) = \dfrac{ac}{bd}$	$\left(\dfrac{2}{3}\right)\left(\dfrac{4}{5}\right) = \dfrac{8}{15}$
Dividing	$\dfrac{(a/b)}{(c/d)} = \dfrac{ad}{bc}$	$\dfrac{2/3}{4/5} = \dfrac{(2)(5)}{(4)(3)} = \dfrac{10}{12}$
Adding	$\dfrac{a}{b} \pm \dfrac{c}{d} = \dfrac{ad \pm bc}{bd}$	$\dfrac{2}{3} - \dfrac{4}{5} = \dfrac{(2)(5) - (4)(3)}{(3)(5)} = -\dfrac{2}{15}$

EXERCISES

In the following exercises, solve for x:

ANSWERS

1. $a = \dfrac{1}{1 + x}$ $\qquad x = \dfrac{1 - a}{a}$

2. $3x - 5 = 13$ $\qquad x = 6$

3. $ax - 5 = bx + 2$ $\qquad x = \dfrac{7}{a - b}$

4. $\dfrac{5}{2x + 6} = \dfrac{3}{4x + 8}$ $\qquad x = -\dfrac{11}{7}$

B. Powers

When powers of a given quantity x are multiplied, the following rule applies:

$$x^n x^m = x^{n+m} \qquad\qquad \textbf{[A.3]}$$

For example, $x^2 x^4 = x^{2+4} = x^6$.

When dividing the powers of a given quantity, note that

$$\frac{x^n}{x^m} = x^{n-m}$$ [A.4]

For example, $x^8/x^2 = x^{8-2} = x^6$.

A power that is a fraction, such as $\frac{1}{3}$, corresponds to a root as follows:

$$x^{1/n} = \sqrt[n]{x}$$ [A.5]

For example, $4^{1/3} = \sqrt[3]{4} = 1.587\ 4$. (A scientific calculator is useful for such calculations.)

Finally, any quantity x^n that is raised to the mth power is

$$(x^n)^m = x^{nm}$$ [A.6]

Table A.1 summarizes the rules of exponents.

TABLE A.1

Rules of Exponents

$x^0 = 1$
$x^1 = x$
$x^n x^m = x^{n+m}$
$x^n/x^m = x^{n-m}$
$x^{1/n} = \sqrt[n]{x}$
$(x^n)^m = x^{nm}$

EXERCISES

Verify the following:

1. $3^2 \times 3^3 = 243$
2. $x^5 x^{-8} = x^{-3}$
3. $x^{10}/x^{-5} = x^{15}$
4. $5^{1/3} = 1.709\ 975$ (Use your calculator.)
5. $60^{1/4} = 2.783\ 158$ (Use your calculator.)
6. $(x^4)^3 = x^{12}$

C. Factoring

Some useful formulas for factoring an equation are

$$ax + ay + az = a(x + y + z) \qquad \text{common factor}$$
$$a^2 + 2ab + b^2 = (a + b)^2 \qquad \text{perfect square}$$
$$a^2 - b^2 = (a + b)(a - b) \qquad \text{differences of squares}$$

D. Quadratic Equations

The general form of a quadratic equation is

$$ax^2 + bx + c = 0$$ [A.7]

where x is the unknown quantity and a, b, and c are numerical factors referred to as **coefficients** of the equation. This equation has two roots, given by

$$x = \frac{-b \pm \sqrt{b^2 - 4ac}}{2a}$$ [A.8]

If $b^2 \geq 4ac$, the roots will be real.

EXAMPLE

The equation $x^2 + 5x + 4 = 0$ has the following roots corresponding to the two signs of the square root term:

$$x = \frac{-5 \pm \sqrt{5^2 - (4)(1)(4)}}{2(1)} = \frac{-5 \pm \sqrt{9}}{2} = \frac{-5 \pm 3}{2}$$

that is,

$$x_+ = \frac{-5 + 3}{2} = \boxed{-1} \qquad x_- = \frac{-5 - 3}{2} = \boxed{-4}$$

where x_+ refers to the root corresponding to the positive sign and x_- refers to the root corresponding to the negative sign.

EXERCISES

Solve the following quadratic equations:

ANSWERS

1. $x^2 + 2x - 3 = 0$ $x_+ = 1$ $x_- = -3$
2. $2x^2 - 5x + 2 = 0$ $x_+ = 2$ $x_- = 1/2$
3. $2x^2 - 4x - 9 = 0$ $x_+ = 1 + \sqrt{22}/2$ $x_- = 1 - \sqrt{22}/2$

E. Linear Equations

A linear equation has the general form

$$y = ax + b \qquad \text{[A.9]}$$

where a and b are constants. This equation is referred to as being linear because the graph of y versus x is a straight line, as shown in Figure A.1. The constant b, called the **intercept,** represents the value of y at which the straight line intersects the y axis. The constant a is equal to the **slope** of the straight line. If any two points on the straight line are specified by the coordinates (x_1, y_1) and (x_2, y_2), as in Figure A.1, then the slope of the straight line can be expressed

$$\text{Slope} = \frac{y_2 - y_1}{x_2 - x_1} = \frac{\Delta y}{\Delta x} \qquad \text{[A.10]}$$

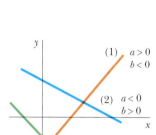

Figure A.1

Note that a and b can have either positive or negative values. If $a > 0$, the straight line has a positive slope, as in Figure A.1. If $a < 0$, the straight line has a *negative* slope. In Figure A.1, both a and b are positive. Three other possible situations are shown in Figure A.2: $a > 0, b < 0$; $a < 0, b > 0$; and $a < 0, b < 0$.

EXERCISES

1. Draw graphs of the following straight lines:
   (a) $y = 5x + 3$ **(b)** $y = -2x + 4$ **(c)** $y = -3x - 6$
2. Find the slopes of the straight lines described in Exercise 1.
   Answers: **(a)** 5   **(b)** -2 **(c)** -3
3. Find the slopes of the straight lines that pass through the following sets of points: **(a)** $(0, -4)$ and $(4, 2)$, **(b)** $(0, 0)$ and $(2, -5)$, and **(c)** $(-5, 2)$ and $(4, -2)$
   Answers: **(a)** $\frac{3}{2}$ **(b)** $-\frac{5}{2}$ **(c)** $-\frac{4}{9}$

Figure A.2

F. Solving Simultaneous Linear Equations

Consider an equation such as $3x + 5y = 15$, which has two unknowns, x and y. Such an equation does not have a unique solution. That is, $(x = 0, y = 3)$, $(x = 5, y = 0)$ and $(x = 2, y = \frac{9}{5})$ are all solutions to this equation.

If a problem has two unknowns, a unique solution is possible only if we have *two* independent equations. In general, if a problem has n unknowns, its solution requires n independent equations. In order to solve two simultaneous equations involving two unknowns, x and y, we solve one of the equations for x in terms of y and substitute this expression into the other equation.

EXAMPLE

Solve the following two simultaneous equations:

$$\textbf{(1)} \quad 5x + y = -8 \qquad \textbf{(2)} \quad 2x - 2y = 4$$

Solution   From (2), we find that $x = y + 2$. Substitution of this into (1) gives

$$5(y + 2) + y = -8$$

$$6y = -18$$

$$\boxed{y = -3}$$

$$x = y + 2 = \boxed{-1}$$

Alternate Solution   Multiply each term in (1) by the factor 2 and add the result to (2):

$$10x + 2y = -16$$

$$\underline{2x - 2y = 4}$$

$$12x = -12$$

$$x = \boxed{-1}$$

$$y = x - 2 = \boxed{-3}$$

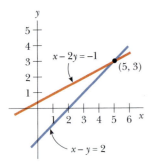

Figure A.3

Two linear equations with two unknowns can also be solved by a graphical method. If the straight lines corresponding to the two equations are plotted in a conventional coordinate system, the intersection of the two lines represents the solution. For example, consider the two equations

$$x - y = 2$$

$$x - 2y = -1$$

These are plotted in Figure A.3. The intersection of the two lines has the coordinates $x = 5$, $y = 3$. This represents the solution to the equations. You should check this solution by the analytical technique discussed above.

EXERCISES

Solve the following pairs of simultaneous equations involving two unknowns:

ANSWERS

1. $x + y = 8$ $x = 5$, $y = 3$
   $x - y = 2$
2. $98 - T = 10a$ $T = 65$, $a = 3.27$
   $T - 49 = 5a$
3. $6x + 2y = 6$ $x = 2$, $y = -3$
   $8x - 4y = 28$

G. Logarithms

Suppose that a quantity x is expressed as a power of some quantity a:

$$x = a^y \qquad \text{[A.11]}$$

The number a is called the **base** number. The **logarithm** of x with respect to the base a is equal to the exponent to which the base must be raised in order to satisfy the expression $x = a^y$:

$$y = \log_a x \qquad \text{[A.12]}$$

Conversely, the **antilogarithm** of y is the number x:

$$x = \text{antilog}_a y \qquad \text{[A.13]}$$

In practice, the two bases most often used are base 10, called the *common* logarithm base, and base $e = 2.718 \ldots$, called the *natural* logarithm base. When common logarithms are used,

$$y = \log_{10} x \qquad (\text{or } x = 10^y) \qquad \text{[A.14]}$$

When natural logarithms are used,

$$y = \ln_e x \qquad (\text{or } x = e^y) \qquad \text{[A.15]}$$

For example, $\log_{10} 52 = 1.716$, so that $\text{antilog}_{10}\ 1.716 = 10^{1.716} = 52$. Likewise, $\ln_e 52 = 3.951$, so $\text{antiln}_e\ 3.951 = e^{3.951} = 52$.

In general, note that you can convert between base 10 and base e with the equality

$$\ln_e x = (2.302\ 585)\log_{10} x \qquad \text{[A.16]}$$

Finally, some useful properties of logarithms are

$$\log (ab) = \log a + \log b \qquad \ln e = 1$$

$$\log (a/b) = \log a - \log b \qquad \ln e^a = a$$

$$\log (a^n) = n \log a \qquad \ln \left(\frac{1}{a}\right) = -\ln a$$

A.4 GEOMETRY

Table A.2 gives the areas and volumes for several geometric shapes used throughout this text:

TABLE A.2

Useful Information for Geometry

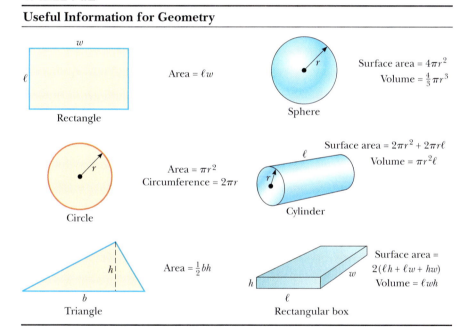

Rectangle	$\text{Area} = \ell w$
Sphere	$\text{Surface area} = 4\pi r^2$ $\text{Volume} = \frac{4}{3}\pi r^3$
Circle	$\text{Area} = \pi r^2$ $\text{Circumference} = 2\pi r$
Cylinder	$\text{Surface area} = 2\pi r^2 + 2\pi r\ell$ $\text{Volume} = \pi r^2\ell$
Triangle	$\text{Area} = \frac{1}{2}bh$
Rectangular box	$\text{Surface area} = 2(\ell h + \ell w + hw)$ $\text{Volume} = \ell wh$

A.5 TRIGONOMETRY

Some of the most basic facts concerning trigonometry are presented in Chapter 1, and we encourage you to study the material presented there if you are having trouble with this branch of mathematics. In addition to the discussion of Chapter 1, certain useful trig identities that can be of value to you follow.

$$\sin^2 \theta + \cos^2 \theta = 1$$

$$\sin \theta = \cos(90° - \theta)$$

$$\cos \theta = \sin(90° - \theta)$$

$$\sin 2\theta = 2 \sin \theta \cos \theta$$

$$\cos 2\theta = \cos^2 \theta - \sin^2 \theta$$

$$\sin(\theta \pm \phi) = \sin \theta \cos \phi \pm \cos \theta \sin \phi$$

$$\cos(\theta \pm \phi) = \cos \theta \cos \phi \mp \sin \theta \sin \phi$$

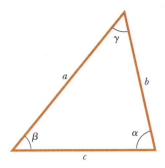

Figure A.4

The following relationships apply to *any* triangle, as shown in Figure A.4:

$$\alpha + \beta + \gamma = 180°$$

$$a^2 = b^2 + c^2 - 2bc \cos \alpha$$

Law of cosines $\quad b^2 = a^2 + c^2 - 2ac \cos \beta$

$$c^2 = a^2 + b^2 - 2ab \cos \gamma$$

Law of sines $\quad \dfrac{a}{\sin \alpha} = \dfrac{b}{\sin \beta} = \dfrac{c}{\sin \gamma}$

An Abbreviated Table of Isotopes

Atomic Number Z	Element	Symbol	Chemical Atomic Mass (u)	Mass Number (* Indicates Radioactive) A	Atomic Mass (u)	Percent Abundance	Half-Life (If Radioactive) $T_{1/2}$
0	(Neutron)	n		1*	1.008 665		10.4 min
1	Hydrogen	H	1.007 94	1	1.007 825	99.988 5	
	Deuterium	D		2	2.014 102	0.011 5	
	Tritium	T		3*	3.016 049		12.33 yr
2	Helium	He	4.002 602	3	3.016 029	0.000 137	
				4	4.002 603	99.999 863	
3	Lithium	Li	6.941	6	6.015 122	7.5	
				7	7.016 004	92.5	
4	Beryllium	Be	9.012 182	7*	7.016 929		53.3 days
				9	9.012 182	100	
5	Boron	B	10.811	10	10.012 937	19.9	
				11	11.009 306	80.1	
6	Carbon	C	12.010 7	10*	10.016 853		19.3 s
				11*	11.011 434		20.4 min
				12	12.000 000	98.93	
				13	13.003 355	1.07	
				14*	14.003 242		5 730 yr
7	Nitrogen	N	14.006 7	13*	13.005 739		9.96 min
				14	14.003 074	99.632	
				15	15.000 109	0.368	
8	Oxygen	O	15.999 4	15*	15.003 065		122 s
				16	15.994 915	99.757	
				18	17.999 160	0.205	
9	Fluorine	F	18.998 403 2	19	18.998 403	100	
10	Neon	Ne	20.179 7	20	19.992 440	90.48	
				22	21.991 385	9.25	
11	Sodium	Na	22.989 77	22*	21.994 437		2.61 yr
				23	22.989 770	100	
				24*	23.990 963		14.96 h
12	Magnesium	Mg	24.305 0	24	23.985 042	78.99	
				25	24.985 837	10.00	
				26	25.982 593	11.01	
13	Aluminum	Al	26.981 538	27	26.981 539	100	
14	Silicon	Si	28.085 5	28	27.976 926	92.229 7	
15	Phosphorus	P	30.973 761	31	30.973 762	100	
				32*	31.973 907		14.26 days
16	Sulfur	S	32.066	32	31.972 071	94.93	
				35*	34.969 032		87.5 days
17	Chlorine	Cl	35.452 7	35	34.968 853	75.78	
				37	36.965 903	24.22	
18	Argon	Ar	39.948	40	39.962 383	99.600 3	
19	Potassium	K	39.098 3	39	38.963 707	93.258 1	
				40*	39.963 999	0.011 7	1.28×10^9 yr

(Continued)

Atomic Number Z	Element	Symbol	Chemical Atomic Mass (u)	Mass Number (* Indicates Radioactive) A	Atomic Mass (u)	Percent Abundance	Half-Life (If Radioactive) $T_{1/2}$
20	Calcium	Ca	40.078	40	39.962 591	96.941	
21	Scandium	Sc	44.955 910	45	44.955 910	100	
22	Titanium	Ti	47.867	48	47.947 947	73.72	
23	Vanadium	V	50.941 5	51	50.943 964	99.750	
24	Chromium	Cr	51.996 1	52	51.940 512	83.789	
25	Manganese	Mn	54.938 049	55	54.938 050	100	
26	Iron	Fe	55.845	56	55.934 942	91.754	
27	Cobalt	Co	58.933 200	59	58.933 200	100	
				60*	59.933 822		5.27 yr
28	Nickel	Ni	58.693 4	58	57.935 348	68.076 9	
				60	59.930 790	26.223 1	
29	Copper	Cu	63.546	63	62.929 601	69.17	
				65	64.927 794	30.83	
30	Zinc	Zn	65.39	64	63.929 147	48.63	
				66	65.926 037	27.90	
				68	67.924 848	18.75	
31	Gallium	Ga	69.723	69	68.925 581	60.108	
				71	70.924 705	39.892	
32	Germanium	Ge	72.61	70	69.924 250	20.84	
				72	71.922 076	27.54	
				74	73.921 178	36.28	
33	Arsenic	As	74.921 60	75	74.921 596	100	
34	Selenium	Se	78.96	78	77.917 310	23.77	
				80	79.916 522	49.61	
35	Bromine	Br	79.904	79	78.918 338	50.69	
				81	80.916 291	49.31	
36	Krypton	Kr	83.80	82	81.913 485	11.58	
				83	82.914 136	11.49	
				84	83.911 507	57.00	
				86	85.910 610	17.30	
37	Rubidium	Rb	85.467 8	85	84.911 789	72.17	
				87*	86.909 184	27.83	4.75×10^{10} yr
38	Strontium	Sr	87.62	86	85.909 262	9.86	
				88	87.905 614	82.58	
				90*	89.907 738		29.1 yr
39	Yttrium	Y	88.905 85	89	88.905 848	100	
40	Zirconium	Zr	91.224	90	89.904 704	51.45	
				91	90.905 645	11.22	
				92	91.905 040	17.15	
				94	93.906 316	17.38	
41	Niobium	Nb	92.906 38	93	92.906 378	100	
42	Molybdenum	Mo	95.94	92	91.906 810	14.84	
				95	94.905 842	15.92	
				96	95.904 679	16.68	
				98	97.905 408	24.13	
43	Technetium	Tc		98*	97.907 216		4.2×10^{6} yr
				99*	98.906 255		2.1×10^{5} yr

Atomic Number Z	Element	Symbol	Chemical Atomic Mass (u)	Mass Number (* Indicates Radioactive) A	Atomic Mass (u)	Percent Abundance	Half-Life (If Radioactive) $T_{1/2}$
44	Ruthenium	Ru	101.07	99	98.905 939	12.76	
				100	99.904 220	12.60	
				101	100.905 582	17.06	
				102	101.904 350	31.55	
				104	103.905 430	18.62	
45	Rhodium	Rh	102.905 50	103	102.905 504	100	
46	Palladium	Pd	106.42	104	103.904 035	11.14	
				105	104.905 084	22.33	
				106	105.903 483	27.33	
				108	107.903 894	26.46	
				110	109.905 152	11.72	
47	Silver	Ag	107.868 2	107	106.905 093	51.839	
				109	108.904 756	48.161	
48	Cadmium	Cd	112.411	110	109.903 006	12.49	
				111	110.904 182	12.80	
				112	111.902 757	24.13	
				113*	112.904 401	12.22	9.3×10^{15} yr
				114	113.903 358	28.73	
49	Indium	In	114.818	115*	114.903 878	95.71	4.4×10^{14} yr
50	Tin	Sn	118.710	116	115.901 744	14.54	
				118	117.901 606	24.22	
				120	119.902 197	32.58	
51	Antimony	Sb	121.760	121	120.903 818	57.21	
				123	122.904 216	42.79	
52	Tellurium	Te	127.60	126	125.903 306	18.84	
				128*	127.904 461	31.74	$>8 \times 10^{24}$ yr
				130*	129.906 223	34.08	$\leq 1.25 \times 10^{21}$ yr
53	Iodine	I	126.904 47	127	126.904 468	100	
				129*	128.904 988		1.6×10^{7} yr
54	Xenon	Xe	131.29	129	128.904 780	26.44	
				131	130.905 082	21.18	
				132	131.904 145	26.89	
				134	133.905 394	10.44	
				136*	135.907 220	8.87	$\geq 2.36 \times 10^{21}$ yr
55	Cesium	Cs	132.905 45	133	132.905 447	100	
56	Barium	Ba	137.327	137	136.905 821	11.232	
				138	137.905 241	71.698	
57	Lanthanum	La	138.905 5	139	138.906 349	99.910	
58	Cerium	Ce	140.116	140	139.905 434	88.450	
				142*	141.909 240	11.114	$>5 \times 10^{16}$ yr
59	Praseodymium	Pr	140.907 65	141	140.907 648	100	
60	Neodymium	Nd	144.24	142	141.907 719	27.2	
				144*	143.910 083	23.8	2.3×10^{15} yr
				146	145.913 112	17.2	
61	Promethium	Pm		145*	144.912 744		17.7 yr
62	Samarium	Sm	150.36	147*	146.914 893	14.99	1.06×10^{11} yr
				149*	148.917 180	13.82	$>2 \times 10^{15}$ yr
				152	151.919 728	26.75	
				154	153.922 205	22.75	

(Continued)

Atomic Number Z	Element	Symbol	Chemical Atomic Mass (u)	Mass Number (* Indicates Radioactive) A	Atomic Mass (u)	Percent Abundance	Half-Life (If Radioactive) $T_{1/2}$
63	Europium	Eu	151.964	151	150.919 846	47.81	
				153	152.921 226	52.19	
64	Gadolinium	Gd	157.25	156	155.922 120	20.47	
				158	157.924 100	24.84	
				160	159.927 051	21.86	
65	Terbium	Tb	158.925 34	159	158.925 343	100	
66	Dysprosium	Dy	162.50	162	161.926 796	25.51	
				163	162.928 728	24.90	
				164	163.929 171	28.18	
67	Holmium	Ho	164.930 32	165	164.930 320	100	
68	Erbium	Er	167.6	166	165.930 290	33.61	
				167	166.932 045	22.93	
				168	167.932 368	26.78	
69	Thulium	Tm	168.934 21	169	168.934 211	100	
70	Ytterbium	Yb	173.04	172	171.936 378	21.83	
				173	172.938 207	16.13	
				174	173.938 858	31.83	
71	Lutecium	Lu	174.967	175	174.940 768	97.41	
72	Hafnium	Hf	178.49	177	176.943 220	18.60	
				178	177.943 698	27.28	
				179	178.945 815	13.62	
				180	179.946 549	35.08	
73	Tantalum	Ta	180.947 9	181	180.947 996	99.988	
74	Tungsten (Wolfram)	W	183.84	182	181.948 206	26.50	
				183	182.950 224	14.31	
				184*	183.950 933	30.64	$>3 \times 10^{17}$ yr
				186	185.954 362	28.43	
75	Rhenium	Re	186.207	185	184.952 956	37.40	
				187*	186.955 751	62.60	4.4×10^{10} yr
76	Osmium	Os	190.23	188	187.955 836	13.24	
				189	188.958 145	16.15	
				190	189.958 445	26.26	
				192	191.961 479	40.78	
77	Iridium	Ir	192.217	191	190.960 591	37.3	
				193	192.962 924	62.7	
78	Platinum	Pt	195.078	194	193.962 664	32.967	
				195	194.964 774	33.832	
				196	195.964 935	25.242	
79	Gold	Au	196.966 55	197	196.966 552	100	
80	Mercury	Hg	200.59	199	198.968 262	16.87	
				200	199.968 309	23.10	
				201	200.970 285	13.18	
				202	201.970 626	29.86	
81	Thallium	Tl	204.383 3	203	202.972 329	29.524	
				205	204.974 412	70.476	
		(Th C″)		208*	207.982 005		3.053 min
		(Ra C″)		210*	209.990 066		1.30 min

Atomic Number Z	Element	Symbol	Chemical Atomic Mass (u)	Mass Number (* Indicates Radioactive) A	Atomic Mass (u)	Percent Abundance	Half-Life (If Radioactive) $T_{1/2}$
82	Lead	Pb	207.2	204*	203.973 029	1.4	$\geq 1.4 \times 10^{17}$ yr
				206	205.974 449	24.1	
				207	206.975 881	22.1	
				208	207.976 636	52.4	
		(Ra D)		210*	209.984 173		22.3 yr
		(Ac B)		211*	210.988 732		36.1 min
		(Th B)		212*	211.991 888		10.64 h
		(Ra B)		214*	213.999 798		26.8 min
83	Bismuth	Bi	208.980 38	209	208.980 383	100	
		(Th C)		211*	210.987 258		2.14 min
84	Polonium	Po					
		(Ra F)		210*	209.982 857		138.38 days
		(Ra C′)		214*	213.995 186		164 μs
85	Astatine	At		218*	218.008 682		1.6 s
86	Radon	Rn		222*	222.017 570		3.823 days
87	Francium	Fr					
		(Ac K)		223*	223.019 731		22 min
88	Radium	Ra		226*	226.025 403		1 600 yr
		(Ms Th$_1$)		228*	228.031 064		5.75 yr
89	Actinium	Ac		227*	227.027 747		21.77 yr
90	Thorium	Th	232.038 1				
		(Rd Th)		228*	228.028 731		1.913 yr
		(Th)		232*	232.038 050	100	1.40×10^{10} yr
91	Protactinium	Pa	231.035 88	231*	231.035 879		32.760 yr
92	Uranium	U	238.028 9	232*	232.037 146		69 yr
				233*	233.039 628		1.59×10^5 yr
		(Ac U)		235*	235.043 923	0.720 0	7.04×10^8 yr
				236*	236.045 562		2.34×10^7 yr
		(UI)		238*	238.050 783	99.274 5	4.47×10^9 yr
93	Neptunium	Np		237*	237.048 167		2.14×10^6 yr
94	Plutonium	Pu		239*	239.052 156		2.412×10^4 yr
				242*	242.058 737		3.73×10^6 yr
				244*	244.064 198		8.1×10^7 yr

[a]Chemical atomic masses are from T. B. Coplen, "Atomic Weights of the Elements 1999," a technical report to the International Union of Pure and Applied Chemistry, and published in *Pure and Applied Chemistry*, 73(4), 667–683, 2001. Atomic masses of the isotopes are from G. Audi and A. H. Wapstra, "The 1995 Update to the Atomic Mass Evaluation," *Nuclear Physics*, A595, vol. 4, 409–480, December 25, 1995. Percent abundance values are from K. J. R. Rosman and P. D. P. Taylor, "Isotopic Compositions of the Elements 1999", a technical report to the International Union of Pure and Applied Chemistry, and published in *Pure and Applied Chemistry*, 70(1), 217–236, 1998.

APPENDIX C
Some Useful Tables

TABLE C.1

Mathematical Symbols Used in the Text and Their Meaning

Symbol	Meaning
$=$	is equal to
$\neq$	is not equal to
$\equiv$	is defined as
$\propto$	is proportional to
$>$	is greater than
$<$	is less than
$\gg$	is much greater than
$\ll$	is much less than
$\approx$	is approximately equal to
$\sim$	is on the order of magnitude of
Δx	change in x or uncertainty in x
Σx_i	sum of all quantities x_i
$\lvert x \rvert$	absolute value of x (always a positive quantity)

TABLE C.2

Standard Symbols for Units

Symbol	Unit	Symbol	Unit
A	ampere	kcal	kilocalorie
Å	angstrom	kg	kilogram
atm	atmosphere	km	kilometer
Bq	bequerel	kmol	kilomole
Btu	British thermal unit	L	liter
C	coulomb	lb	pound
°C	degree Celsius	ly	light year
cal	calorie	m	meter
cm	centimeter	min	minute
Ci	curie	mol	mole
d	day	N	newton
deg	degree (angle)	nm	nanometer
eV	electronvolt	Pa	pascal
°F	degree Fahrenheit	rad	radian
F	farad	rev	revolution
ft	foot	s	second
G	Gauss	T	tesla
g	gram	u	atomic mass unit
H	henry	V	volt
h	hour	W	watt
hp	horsepower	Wb	weber
Hz	hertz	yr	year
in.	inch	μm	micrometer
J	joule	Ω	ohm
K	kelvin		

TABLE C.3

The Greek Alphabet

Alpha	A	α	Nu	N	ν
Beta	B	β	Xi	Ξ	ξ
Gamma	Γ	γ	Omicron	O	o
Delta	Δ	δ	Pi	Π	π
Epsilon	E	ϵ	Rho	P	ρ
Zeta	Z	ζ	Sigma	Σ	σ
Eta	H	η	Tau	T	τ
Theta	Θ	θ	Upsilon	Y	υ
Iota	I	ι	Phi	Φ	ϕ
Kappa	K	κ	Chi	X	χ
Lambda	Λ	λ	Psi	Ψ	ψ
Mu	M	μ	Omega	Ω	ω

TABLE C.4

Physical Data Often Used[a]

Average Earth-Moon distance	3.84×10^8 m
Average Earth-Sun distance	1.496×10^{11} m
Equatorial radius of Earth	6.38×10^6 m
Density of air (20°C and 1 atm)	1.20 kg/m^3
Density of water (20°C and 1 atm)	1.00×10^3 kg/m^3
Free-fall acceleration	9.80 m/s^2
Mass of Earth	5.98×10^{24} kg
Mass of Moon	7.36×10^{22} kg
Mass of Sun	1.99×10^{30} kg
Standard atmospheric pressure	1.013×10^5 Pa

[a] These are the values of the constants as used in the text.

TABLE C.5

Some Fundamental Constants[a]

Quantity	Symbol	Value[b]
Atomic mass unit	u	$1.660\ 540\ 2(10) \times 10^{-27}$ kg $931.494\ 32(28)$ MeV/c^2
Avogadro's number	N_A	$6.022\ 136\ 7(36) \times 10^{23}$ (mol)$^{-1}$
Bohr radius	$a_0 = \dfrac{\hbar^2}{m_e e^2 k_e}$	$0.529\ 177\ 249(24) \times 10^{-10}$ m
Boltzmann's constant	$k_B = R/N_A$	$1.380\ 658(12) \times 10^{-23}$ J/K
Compton wavelength	$\lambda_C = \dfrac{h}{m_e c}$	$2.426\ 310\ 58(22) \times 10^{-12}$ m
Coulomb constant	$k_e = \dfrac{1}{4\pi\epsilon_0}$	$8.987\ 551\ 787 \times 10^9$ N·m^2/C^2 (exact)
Electron mass	m_e	$9.109\ 389\ 7(54) \times 10^{-31}$ kg $5.485\ 799\ 03(13) \times 10^{-4}$ u $0.510\ 999\ 06(15)$ MeV/c^2
Electron volt	eV	$1.602\ 177\ 33(49) \times 10^{-19}$ J
Elementary charge	e	$1.602\ 177\ 33(49) \times 10^{-19}$ C
Gas constant	R	$8.314\ 510(70)$ J/K·mol
Gravitational constant	G	$6.672\ 59(85) \times 10^{-11}$ N·m^2/kg^2
Hydrogen ionization energy	$-E_1 = \dfrac{m_e e^4 k_e^2}{2\hbar^2} = \dfrac{e^2 k_e}{2a_0}$	$13.605\ 698(40)$ eV
Neutron mass	m_n	$1.674\ 928\ 6(10) \times 10^{-27}$ kg $1.008\ 664\ 904(14)$ u $939.565\ 63(28)$ MeV/c^2
Permeability of free space	μ_0	$4\pi \times 10^{-7}$ T·m/A (exact)
Permittivity of free space	$\epsilon_0 = 1/\mu_0 c^2$	$8.854\ 187\ 817 \times 10^{-12}$ C^2/N·m^2 (exact)
Planck's constant	h	$6.626\ 075(40) \times 10^{-34}$ J·s
	$\hbar = h/2\pi$	$1.054\ 572\ 66(63) \times 10^{-34}$ J·s
Proton mass	m_p	$1.672\ 623(10) \times 10^{-27}$ kg $1.007\ 276\ 470(12)$ u $938.272\ 3(28)$ MeV/c^2
Rydberg constant	R_H	$1.097\ 373\ 153\ 4(13) \times 10^7$ m^{-1}
Speed of light in vacuum	c	$2.997\ 924\ 58 \times 10^8$ m/s (exact)

[a] These constants are the values recommended in 1986 by CODATA, based on a least-squares adjustment of data from different measurements. For a more complete list, see Cohen, E. Richard, and Barry N. Taylor, *Rev. Mod. Phys.* **59**:1121, 1987.

[b] The numbers in parentheses for the values below represent the uncertainties in the last two digits.

APPENDIX D
SI Units

TABLE D.1

SI Base Units

Base Quantity	SI Base Unit Name	SI Base Unit Symbol
Length	meter	m
Mass	kilogram	kg
Time	second	s
Electric current	ampere	A
Temperature	kelvin	K
Amount of substance	mole	mol
Luminous intensity	candela	cd

TABLE D.2

Derived SI Units

Quantity	Name	Symbol	Expression in Terms of Base Units	Expression in Terms of Other SI Units
Plane angle	radian	rad	m/m	
Frequency	hertz	Hz	s^{-1}	
Force	newton	N	$kg \cdot m/s^2$	J/m
Pressure	pascal	Pa	$kg/m \cdot s^2$	N/m^2
Energy: work	joule	J	$kg \cdot m^2/s^2$	$N \cdot m$
Power	watt	W	$kg \cdot m^2/s^3$	J/s
Electric charge	coulomb	C	$A \cdot s$	
Electric potential (emf)	volt	V	$kg \cdot m^2/A \cdot s^3$	$W/A, J/C$
Capacitance	farad	F	$A^2 \cdot s^4/kg \cdot m^2$	C/V
Electric resistance	ohm	Ω	$kg \cdot m^2/A^2 \cdot s^3$	V/A
Magnetic flux	weber	Wb	$kg \cdot m^2/A \cdot s^2$	$V \cdot s, T \cdot m^2$
Magnetic field intensity	tesla	T	$kg/A \cdot s^2$	Wb/m^2
Inductance	henry	H	$kg \cdot m^2/A^2 \cdot s^2$	Wb/A

Answers to Quick Quizzes, Odd-Numbered Conceptual Questions and Problems

Chapter 1

CONCEPTUAL QUESTIONS

1. (a) ~ 0.1 m (b) ~ 1 m (c) Between 10 m and 100 m
   (d) ~ 10 m (e) ~ 100 m
3. (a) $\sim 10^6$ beats (b) $\sim 10^9$ beats
5. $\sim 10^9$ s
7. The length of a hand varies from person to person, so it isn't a useful standard of length.
9. The ark has an approximate volume of 10^4 m^3, whereas a home has an approximate volume of 10^3 m^3.
11. A dimensionally correct equation isn't necessarily true. For example, the equation 2 dogs = 5 dogs is dimensionally correct, but isn't true. However, if an equation is not dimensionally correct, it cannot be correct.

PROBLEMS

1. (b) $A_{\text{cylinder}} = \pi R^2$, $A_{\text{rectangular plate}} = \text{length} \times \text{width}$
5. m^3/(kg·s^2)
7. (a) three significant figures (b) four significant figures
   (c) three significant figures (d) two significant figures
9. (a) 797 (b) 1.1 (c) 17.66
11. 115.9 m
13. 228.8 cm
15. 2×10^8 fathoms
17. 1.39×10^3 m^2
19. 1×10^{17} ft
21. 6.71×10^8 mi/h
23. 3.08×10^4 m^3
25. 9.82 cm
27. (a) 1.0×10^3 kg (b) $m_{\text{cell}} = 5.2 \times 10^{-16}$ kg, $m_{\text{kidney}} = 0.27$ kg, $m_{\text{fly}} = 1.3 \times 10^{-5}$ kg
29. 1 800 balls ($\sim 10^3$ balls) (assumes 81 games per season, nine innings per game, an average of ten hitters per inning, and one ball lost for every four hitters)
31. $\sim 10^7$ rev
33. $\sim 10^6$ balls
35. (2.0 m, 1.4 m)
37. $r = 2.2$ m, $\theta = 27°$
39. (a) 6.71 m (b) 0.894 (c) 0.746
41. 3.41 m
43. (a) 3.00 (b) 3.00 (c) 0.800 (d) 0.800 (e) 1.33
45. 5.00/7.00; the angle itself is 35.5°
47. The customer is incorrect. The cost of the pizza should be proportional to the area of the pizza, so the large one should cost more than the small one by a factor of $9^2/6^2 = 81/36 = 2.25$. Therefore, if the small one costs $6, the large one should cost $13.50.
49. The value of k, a dimensionless constant, cannot be found by dimensional analysis.
51. $\sim 10^{-9}$ m
53. $\sim 10^3$ tuners (assumes one tuner per 10 000 residents and a population of 7.5 million)
55. (a) 3.16×10^7 s (b) Between 10^{10} yr and 10^{11} yr

Chapter 2

QUICK QUIZZES

1. (a) 200 yd (b) 0 (c) 0
2. (a) False (b) True (c) True
3. The velocity vs. time graph (a) has a constant slope, indicating a constant acceleration, which is represented by the acceleration vs. time graph (e).

   Graph (b) represents an object with increasing speed, but as time progresses, the lines drawn tangent to the curve have increasing slopes. Since the acceleration is equal to the slope of the tangent line, the acceleration must be increasing, and the acceleration vs. time graph that best indicates this behavior is (d).

   Graph (c) depicts an object which first has a velocity that increases at a constant rate, which means that the object's acceleration is constant. The velocity then stops changing, which means that the acceleration of the object is zero. This behavior is best matched by graph (f).
4. (b)
5. (a) blue graph (b) red graph (c) green graph
6. (e)
7. (c)
8. (a) and (f)

CONCEPTUAL QUESTIONS

1. Yes. If the velocity of the particle is nonzero, the particle is in motion. If the acceleration is zero, the velocity of the particle is unchanging or is constant.
3. Yes. If this occurs, the acceleration of the car is opposite to the direction of motion, and the car will be slowing down.
5. No. They can be used only when the acceleration is constant. Yes. Zero is a constant.
7. Once the objects leave the hand, both are in free fall, and both exhibit the same downward acceleration equal in magnitude to the free-fall acceleration g.
9. Yes. Yes.
11. (a) At the maximum height, the ball is momentarily at rest. (That is, it has zero velocity.) The acceleration remains constant, with magnitude equal to the free-fall acceleration g and directed downward. Thus, even though the velocity is momentarily zero, it continues to change, and the ball will begin to gain speed in the downward direction. (b) The acceleration of the ball remains constant in magnitude and direction throughout the ball's free flight, from the instant it leaves the hand until the instant just before it strikes the ground. The acceleration is directed downward and has a magnitude equal to the free-fall acceleration g.
13. No. If the velocity of A is greater than that of B, car A is moving faster than B at this instant. However, car B may be picking up speed (that is, accelerating) at a greater

rate than A. The driver of B may have stepped very hard on the gas pedal in the recent past, but car B has not yet reached the speed of A.

15. They are the same! You can see this for yourself by solving the kinematic equations for the two cases. The ball that was thrown upward will take longer to reach the ground, but when it arrives, it has the same velocity as the ball that was thrown downward.

17. Ignoring air resistance, in 16 s the pebble would fall a distance $\frac{1}{2}gt^2 = \frac{1}{2}(9.80 \text{ m/s}^2)(16.0 \text{ s})^2 = 1\,250 \text{ m} = 1.25 \text{ km}$. Air resistance is an important force after the first few seconds, when the pebble has gained high speed. Also, part of the 16-s interval must be occupied by the sound returning up the well. Thus, the depth of the well is less than 1.25 km, but on the order of 10^3 m.

19. Above. Your ball has zero initial speed and a smaller average speed during the time of flight to the passing point.

PROBLEMS

1. (a) 52.9 km/h (b) 90.0 km
3. (a) Boat A wins by 60 km (b) 0
5. (a) 180 km (b) 63.4 km/h
7. (a) 4.0 m/s (b) -0.50 m/s (c) -1.0 m/s (d) 0
9. (a) 2.50 m/s (b) -2.27 m/s (c) 0
11. 2.80 h, 218 km
13. 274 km/h
15. (a) 5.00 m/s (b) -2.50 m/s (c) 0 (d) 5.00 m/s
17. (a) 4.0 m/s (b) -4.0 m/s (c) 0 (d) 2.0 m/s
19. (a) 70.0 mi/h·s = 31.3 m/s^2 = 3.19g
    (b) 321 ft = 97.8 m
21. 3.7 s
23. (a) 8.0 m/s^2 (b) about 12 m/s^2
25. 2.74 × 10^5 m/s^2, or 2.79 × 10^4 times g
27. (a) 1.3 m/s^2 (b) 8.0 s
29. (a) 2.32 m/s^2 (b) 14.4 s
31. (a) 8.94 s (b) 89.4 m/s
33. -3.6 m/s^2
35. 200 m
37. (a) 1.5 m/s (b) 32 m
39. (a) 8.2 s (b) 1.3×10^2 m
41. 958 m
43. (a) 31.9 m (b) 2.55 s (c) 2.55 s (d) -25.0 m/s
45. 38.2 m
47. (a) 21.1 m/s (b) 19.6 m (c) 18.1 m/s; 19.6 m
49. (a) 308 m (b) 8.51 s (c) 16.4 s
51. (a) 10.0 m/s upward (b) 4.68 m/s downward
53. 15.0 s
55. (a) -3.50×10^5 m/s^2 (b) 2.86×10^{-4} s
57. 0.60 s
59. (a) 3.00 s (b) -24.5 m/s, -24.5 m/s (c) 23.5 m
61. (a) 5.46 s (b) 73.0 m (c) $v_{\text{Kathy}} = 26.7$ m/s,
    $v_{\text{Stan}} = 22.6$ m/s
63. (a) $t_1 = 5.0$ s, $t_2 = 85$ s (b) 200 ft/s (c) 18 500 ft from starting point (d) 10 s after starting to slow down (total trip time = 100 s)
65. (a) 5.5×10^3 ft (b) 3.7×10^2 ft/s (c) The plane would travel only 0.002 ft in the time it takes the light from the bolt to reach the eye.
67. (a) 7.82 m (b) 0.782 s
69. (a) 3.45 s (b) 10.0 ft
71. 4.2 m/s

Chapter 3

QUICK QUIZZES

1. (c)
2. (b)
3.

Vector	x-component	y-component
$\vec{A}$	$-$	$+$
$\vec{B}$	$+$	$-$
$\vec{A} + \vec{B}$	$-$	$-$

4. (b)
5. (a)
6. (c)
7. (b)

CONCEPTUAL QUESTIONS

1. The components of $\vec{A}$ will both be negative when $\vec{A}$ lies in the third quadrant. The components of $\vec{A}$ will have opposite signs when $\vec{A}$ lies in either the second quadrant or the fourth quadrant.

3. Assuming that the ship sails in a straight line, the wrench will strike the deck at the base of the mast. The wrench has the same horizontal velocity as the ship when it is released, and it will maintain that component of velocity as it falls. Thus, the wrench is always above the same spot on the deck and will strike that spot when it hits the deck. To an observer on the ship, the wrench appears to fall straight downward. To a stationary observer on shore, the wrench is seen to follow a parabolic trajectory.

5. Let v_{0x} and v_{0y} be the initial velocity components for the projectile fired on Earth. Because the projectile fired on the Moon is given the same initial velocity, its initial velocity components have the same values v_{0x} and v_{0y} as those for the projectile on Earth. From $v_y^2 = v_{0y}^2 - 2g\Delta y$, the maximum altitude reached by a projectile is found to be $H = v_{0y}^2/2g$. Because the free-fall acceleration g is smaller on the Moon than on Earth, the maximum altitude will be greater for the projectile fired on the Moon. If they start from the same elevation, the projectile on the Moon will have a longer time of flight and therefore a greater range. The Apollo astronauts made several long golf drives on the Moon.

7. A vector has both magnitude and direction and can be added only to other vectors representing the same physical quantity. A scalar has only a magnitude and can be added only to other scalars representing the same physical quantity.

9. (a) At the top of the projectile's flight, its velocity is horizontal and its acceleration is downward. This is the only point at which the velocity and acceleration vectors are perpendicular. (b) If the projectile is thrown straight up or down, then the velocity and acceleration will be parallel throughout the motion. For any other kind of projectile motion, the velocity and acceleration vectors are never parallel.

11. (a) The acceleration is zero, since both the magnitude and direction of the velocity remain constant. (b) The particle has an acceleration, since the direction of $\vec{v}$ changes.

13. The spacecraft will follow a parabolic path equivalent to that of a projectile thrown off a cliff with a horizontal velocity. As regards the projectile, gravity provides an

acceleration that is always perpendicular to the initial velocity, resulting in a parabolic path. As regards the spacecraft, the initial velocity plays the role of the horizontal velocity of the projectile, and the leaking gas plays the role that gravity plays in the case of the projectile. If the orientation of the spacecraft were to change in response to the gas leak (which is by far the more likely result), then the acceleration would change direction and the motion could become very complicated.

15. For angles $\theta < 45°$, the projectile thrown at angle θ will be in the air for a shorter interval. For the smaller angle, the vertical component of the initial velocity is smaller than that for the larger angle. Thus, the projectile thrown at the smaller angle will not go as high into the air and will spend less time in the air before landing.

17. Yes. The projectile is a freely falling body, because gravity is the only force acting on it. The vertical acceleration will be the local free fall acceleration g; the horizontal acceleration will be zero.

19. If a projectile is to have zero speed at the top of its trajectory, it must be thrown straight upward with zero horizontal velocity. To give a projectile a nonzero speed at the top of its trajectory, the projectile should be thrown at any angle other than 90° to the horizontal.

PROBLEMS

1. Approximately 421 ft at 3° below the horizontal
3. Approximately 15 m at 58° S of E
5. Approximately 310 km at 57° S of W
7. (a) Approximately 5.0 units at $-53°$ (b) Approximately 5.0 units at $+53°$
9. 8.07 m at 42.0° S of E
11. (a) 5.00 blocks at 53.1° N of E (b) 13.0 blocks
13. 47.2 units at 122° from the positive x-axis
15. 157 km
17. 245 km at 21.4° W of N
19. 196 cm at 14.7° below the x-axis
21. (a) (10.0 m, 16.0 m)
23. 6.13 m
25. 12 m/s
27. 48.6 m/s
29. 25 m
31. (a) 32.5 m from the base of the cliff (b) 1.78 s
33. (a) 52.0 m/s horizontally (b) 212 m
35. 390 mi/h at 7.37° N of E
37. 1.88 m
39. 249 ft upstream
41. 18.0 s
43. 196 cm at 14.7° below the positive x-axis
45. (a) 57.7 km/h at 60.0° west of vertical (b) 28.9 km/h downward
47. 68.6 km/h
49. (a) 1.52×10^3 m (b) 36.1 s (c) 4.05×10^3 m
51. $R_{moon} = 18$ m, $R_{Mars} = 7.9$ m
53. (a) +, + (b) −, + (c) (i) must be in either the first or second quadrant
55. (a) 42 m/s (b) 3.8 s (c) $v_x = 34$ m/s, $v_y = -13$ m/s; $v = 37$ m/s
57. 7.5 m/s in the direction the ball was thrown
61. $R/2$
63. 7.5 min
65. 10.8 m

67. (b) $y = Ax^2$ with $A = \dfrac{g}{(2v_i^2)}$ where v_i is the muzzle velocity
(c) 14.5 m/s
69. 227 paces at 165° from the positive x-axis
71. (a) 20.0° above the horizontal (b) 3.05 s
73. (a) 23 m/s (b) 360 m horizontally from the base of cliff

Chapter 4

QUICK QUIZZES

1. (a) True (b) False
2. (a) True (b) True (c) False
3. False
4. (a) (The value of g is smaller on the Moon than on Earth, so, for equal *weights*, the Moon *mass* must be greater.)
5. (c); (d)
6. (c)
7. (c)
8. (b)
9. (b)
10. (b) By exerting an upward force component on the sled, you reduce the normal force on the ground and so reduce the force of kinetic friction.

CONCEPTUAL QUESTIONS

1. (a) Two external forces act on the ball. (i) One is a downward gravitational force exerted by Earth. (ii) The second force on the ball is an upward normal force exerted by the hand. The reactions to these forces are (i) an upward gravitational force exerted by the ball on Earth and (ii) a downward force exerted by the ball on the hand. (b) After the ball leaves the hand, the only external force acting on the ball is the gravitational force exerted by Earth. The reaction is an upward gravitational force exerted by the ball on Earth.

3. No. If the car moves with constant acceleration, then a net force in the direction of the acceleration must be acting on it.

5. The coefficient of static friction is larger than that of kinetic friction. To start the box moving, you must counterbalance the maximum static friction force. This force exceeds the kinetic friction force that you must counterbalance to maintain the constant velocity of the box once it starts moving.

7. The inertia of the suitcase would keep it moving forward as the bus stops. There would be no tendency for the suitcase to be thrown backward toward the passenger. The case should be dismissed.

9. The force causing an automobile to move is the friction between the tires and the roadway as the automobile attempts to push the roadway backward. The force driving a propeller airplane forward is the reaction force exerted by the air on the propeller as the rotating propeller pushes the air backward (the action). In a rowboat, the rower pushes the water backward with the oars (the action). The water pushes forward on the oars and hence the boat (the reaction).

11. When the bus starts moving, Claudette's mass is accelerated by the force exerted by the back of the seat on her body. Clark is standing, however, and the only force acting on him is the friction between his shoes and the floor of the bus. Thus, when the bus starts moving, his feet accelerate forward, but the rest of his body experiences

almost no accelerating force (only that due to his being attached to his accelerating feet!). As a consequence, his body tends to stay almost at rest, according to Newton's first law, relative to the ground. Relative to Claudette, however, he is moving toward her and falls into her lap. Both performers won Academy Awards.

13. The tension in the rope is the maximum force that occurs in *both* directions. In this case, then, since both are pulling with a force of magnitude 200 N, the tension is 200 N. If the rope does not move, then the force on each athlete must equal zero. Therefore, each athlete exerts 200 N against the ground.

15. (a) As the man takes the step, the action is the force his foot exerts on Earth; the reaction is the force exerted by Earth on his foot. (b) Here, the action is the force exerted by the snowball on the girl's back; the reaction is the force exerted by the girl's back on the snowball. (c) This action is the force exerted by the glove on the ball; the reaction is the force exerted by the ball on the glove. (d) This action is the force exerted by the air molecules on the window; the reaction is the force exerted by the window on the air molecules. In each case, we could equally well interchange the terms "action" and "reaction."

17. If the brakes lock, the car will travel farther than it would travel if the wheels continued to roll, because the coefficient of kinetic friction is less than that of static friction. Hence, the force of kinetic friction is less than the maximum force of static friction.

19. (a) The crate accelerates because of a friction force exerted by the floor of the truck on the crate. (b) If the driver slams on the brakes, the crate's inertia tends to keep the crate moving forward.

PROBLEMS

1. (a) 12 N (b) 3.0 m/s^2
3. 2×10^4 N
5. 3.71 N, 58.7 N, 2.27 kg
7. 9.6 N
9. 1.4×10^3 N
11. (a) 0.200 m/s^2 (b) 10.0 m (c) 2.00 m/s
13. 1.1×10^4 N
15. 600 N in vertical cable, 997 N in inclined cable, 796 N in horizontal cable
17. 150 N in vertical cable, 75 N in right-side cable, 130 N in left-side cable
19. 236 N (upper rope), 118 N (lower rope)
21. 613 N
23. 64 N
25. (a) 7.50×10^3 N backward (b) 50.0 m
27. (a) 7.0 m/s^2 horizontal and to the right (b) 21 N (c) 14 N horizontal and to the right
29. 7.90 m/s
31. (a) 14.2 m/s (b) 588 m
33. (a) 2.15×10^3 N forward (b) 645 N forward (c) 645 N rearward (d) 1.02×10^4 N at 74.1° below horizontal and rearward
35. $\mu_s = 0.38$, $\mu_k = 0.31$
37. (a) 0.256 (b) 0.509 m/s^2
39. (a) 14.7 m (b) Neither mass is necessary
41. 32.1 N
43. (a) 33 m/s (b) No. The object will speed up to 33 m/s from any lower speed and will slow down to 33 m/s from any higher speed.

45. $\mu_k = 0.287$
47. (a) 1.78 m/s^2 (b) 0.368 (c) 9.37 N (d) 2.67 m/s
49. 3.30 m/s^2
51. (a) 0.404 (b) 45.8 lb
53. (b) 2-kg block: 5.7 m/s^2 to left; 3-kg block: 5.7 m/s^2 to right; 10-kg block: 5.7 m/s^2 downward (c) 17 N, 41 N
55. (a) 84.9 N upward (b) 84.9 N downward
57. 50 m
59. (a) friction between box and truck (b) 2.94 m/s^2
61. (a) 2.22 m (b) 8.74 m/s down the incline
63. (a) 0.232 m/s^2 (b) 9.68 N
65. (a) 1.7 m/s^2, 17 N (b) 0.69 m/s^2, 17 N
67. 100 N, 204 N
69. (a) $T_1 = 78.0$ N, $T_2 = 35.9$ N (b) 0.656
71. (a) 30.7° (b) 0.843 N
73. 5.5×10^2 N
75. (a) 7.1×10^2 N (b) 8.1×10^2 N (c) 7.1×10^2 N (d) 6.5×10^2 N
77. 72.0 N
79. (a) 0.408 m/s^2 upward (b) 83.3 N
81. (b) 514 N, 558 N, 325 N

Chapter 5

QUICK QUIZZES

1. (C)
2. (d)
3. (c)
4. (c)

CONCEPTUAL QUESTIONS

1. Since no motion is taking place, the rope undergoes no displacement and no work is done on it. For the same reason, no work is being done on the pullers or the ground. Work is being done only within the bodies of the pullers. For example, the heart of each puller is applying forces on the blood to move blood through the body.

3. When the slide is frictionless, changing the length or shape of the slide will not make any difference in the final speed of the child, as long as the difference in the heights of the upper and lower ends of the slide is kept constant. If friction must be considered, the path length along which the friction force does negative work will be greater when the slide is made longer or given bumps. Thus, the child will arrive at the lower end with less kinetic energy (and hence less speed).

5. If we ignore any effects due to rolling friction on the tires of the car, we find that the same amount of work would be done in driving up the switchback and in driving straight up the mountain, since the weight of the car is moved upwards against gravity by the same vertical distance in each case. If we include friction, there is more work done in driving the switchback, since the distance over which the friction force acts is much longer. So why do we use switchbacks? The answer lies in the force required, not the work. The force required from the engine to follow a gentle rise is much smaller than that required to drive straight up the hill. To negotiate roadways running straight uphill, engines would have to be redesigned to enable them to apply much larger forces. (It is for much the same reason that ramps are designed to move heavy objects into trucks, as opposed to lifting the objects vertically.)

7. (a) The tension in the supporting cord does no work, because the motion of the pendulum is always perpendicular to the cord and therefore to the tension force. (b) The air resistance does negative work at all times, because the air resistance is always acting in a direction opposite that of the motion. (c) The force of gravity always acts downwards; therefore, the work done by gravity is positive on the downswing and negative on the upswing.

9. Because the periods are the same for both cars, we need only to compare the work done. Since the sports car is moving twice as fast as the older car at the end of the time interval, it has four times the kinetic energy. Thus, according to the work–energy theorem, four times as much work was done, and the engine must have expended four times the power.

11. During the time that the toe is in contact with the ball, the work done by the toe on the ball is given by

$$W_{toe} = \tfrac{1}{2}m_{ball}v^2 - 0 = \tfrac{1}{2}m_{ball}v^2$$

where v is the speed of the ball as it leaves the toe. After the ball loses contact with the toe, only the gravitational force and the retarding force due to air resistance continue to do work on the ball throughout its flight.

13. Yes, the total mechanical energy of the system is conserved because the only forces acting are conservative: the force of gravity and the spring force. There are two forms of potential energy in this case: gravitational potential energy and elastic potential energy stored in the spring.

15. Let's assume you lift the book slowly. In this case, there are two forces on the book that are almost equal in magnitude: the lifting force and the force of gravity. Thus, the positive work done by you and the negative work done by gravity cancel. There is no net work performed and no net change in the kinetic energy, so the work–energy theorem is satisfied.

17. As the satellite moves in a circular orbit about the Earth, its displacement during any small time interval is perpendicular to the gravitational force, which always acts toward the center of the Earth. Therefore, the work done by the gravitational force during any displacement is zero. (Recall that the work done by a force is defined to be $F\Delta x \cos\theta$, where θ is the angle between the force and the displacement. In this case, the angle is 90°, so the work done is zero.) Since the work–energy theorem says that the net work done on an object during any displacement is equal to the change in its kinetic energy, and the work done in this case is zero, the change in the satellite's kinetic energy is zero: hence, its speed remains constant.

19. If a crate is located on the bed of a truck, and the truck accelerates, the friction force exerted on the crate causes it to undergo the same acceleration as the truck, assuming that the crate doesn't slip. Another example is a car that accelerates because of the frictional forces between the road surface and its tires. This force is in the direction of the motion of the car and produces an increase in the car's kinetic energy.

PROBLEMS

1. 700 J

3. 15.0 MJ

5. (a) 61.3 J (b) −46.3 J (c) 0

7. (a) 79.4 N (b) 1.49 kJ (c) −1.49 kJ

9. (a) 2.00 m/s (b) 200 N

11. 0.265 m/s

13. (a) -5.6×10^2 J (b) 1.2 m

15. (a) 90 J (b) 1.8×10^2 N

17. 1.0 m/s

19. 4.1 m

21. (a) 4.1 m (b) 6.4 m/s

23. $h = 6.94$ m

25. $W_{biceps} = 120$ J, $W_{chin\text{-}up} = 290$ J, additional muscles must be involved

27. 0.459 m

29. 1.53×10^5 N upwards

31. (a) 10.9 m/s (b) 11.6 m/s

33. (a) 544 N/m (b) 19.7 m/s

35. 10.2 m

37. (a) Yes. There are no nonconservative forces acting on the child, so the total mechanical energy is conserved. (b) No. In the expression for conservation of mechanical energy, the mass of the child is included in every term and therefore cancels out. (c) The answer is the same in each case. (d) The expression would have to be modified to include the work done by the force of friction. (e) 15.3 m/s.

39. 2.1 kN

41. 3.8 m/s

43. (a) 5.42 m/s (b) 0.300 (c) 147 J

45. 289 m

47. (a) 24.5 m/s (b) Yes (c) 206 m (d) Unrealistic; the actual retarding force will vary with speed.

49. (a) 1.24 kW (b) 20.9%

51. 8.01 W

53. (a) 2.38×10^4 W = 32.0 hp (b) 4.77×10^4 W = 63.9 hp

55. (a) 24.0 J (b) −3.00 J (c) 21.0 J

57. (a) The graph is a straight line passing through the points (0 m, −16 N), (2 m, 0 N), and (3 m, 8 N). (b) −12.0 J

59. (a) 575 N/m (b) 46.0 J

61. (a) 4.4 m/s (b) 1.5×10^5 N

63. (a) 3.13 m/s (b) 4.43 m/s (c) 1.00 m

65. (a) 0.588 J (b) 0.588 J (c) 2.42 m/s (d) $PE_C = 0.392$ J (e) $KE_C = 0.196$ J

67. (a) 423 mi/gal (b) 776 mi/gal

69. (a) 28.0 m/s (b) 30.0 m (c) 89.0 m beyond the end of the track

71. 1.68 m/s

73. (a) 6.15 m/s (b) 9.87 m/s

75. (a) 101 J (b) 0.410 m (c) 2.84 m/s (d) −9.80 mm (e) 2.85 m/s

77. 914 N/m

79. $W_{net} = 0$, $W_{grav} = -2.0 \times 10^4$ J, $W_{normal} = 0$, $W_{friction} = 2.0 \times 10^4$ J

81. (a) 10.2 kW (b) 10.6 kW (c) 5.82×10^6 J

83. $v = (8gh/15)^{1/2}$

85. between 25.2 km/h and 27.0 km/h

87. (a) 6.75 W/m² (b) 6.64 kW/m² (c) A powerful automobile running on sunlight would have to carry on its roof a solar panel that was huge compared with the size of the car.

89. 4.3 m/s

Chapter 6

QUICK QUIZZES

1. (d)

2. (c)

3. (c)

4. (a)
5. (a) Perfectly inelastic (b) Inelastic (c) Inelastic
6. (a)

CONCEPTUAL QUESTIONS

1. (a) No. It cannot carry more kinetic energy than it possesses. That would violate the law of energy conservation. (b) Yes. By bouncing from the object it strikes, it can deliver more momentum in a collision than it possesses in its flight.

3. If all the kinetic energy disappears, there must be no motion of either of the objects after the collision. If neither is moving, the final momentum of the system is zero, and the initial momentum of the system must also have been zero. A situation in which this could be true would be the head-on collision of two objects having momenta of equal magnitude but opposite direction.

5. Initially, the clay has momentum directed toward the wall. When it collides and sticks to the wall, neither the clay nor the wall appears to have any momentum. Thus, it is tempting to (wrongfully) conclude that momentum is not conserved. However, the "lost" momentum is actually imparted to the wall and Earth, causing both to move. Because of Earth's enormous mass, its recoil speed is too small to detect.

7. Before the step the momentum was zero, so afterward the net momentum must also be zero. Obviously, you have some momentum, so something must have momentum in the opposite direction. That something is Earth, the enormous mass of which ensures that its recoil speed will be too small to detect, but if you want to make Earth move, it is as simple as taking a step.

9. As the water is forced out of the holes in the arm, the arm imparts a horizontal impulse to the water. The water then exerts an equal and opposite impulse on the spray arm, causing the spray arm to rotate in the direction opposite that of the spray.

11. Its speed decreases as its mass increases. There are no external horizontal forces acting on the box, so its momentum cannot change as it moves along the horizontal surface. As the box slowly fills with water, its mass increases with time. Because the product mv must be a constant, and because m is increasing, the speed of the box must decrease.

13. It will be easiest to catch the medicine ball when its speed (and kinetic energy) is lowest. The first option—throwing the medicine ball at the same velocity—will be the most difficult, because the speed will not be reduced at all. The second option, throwing the medicine ball with the same momentum, will reduce the velocity by the ratio of the masses. Since $m_t v_t = m_m v_m$, it follows that

$$v_m = v_t \left(\frac{m_t}{m_m} \right)$$

The third option, throwing the medicine ball with the same kinetic energy, will also reduce the velocity, but only by the square root of the ratio of the masses. Since

$$\frac{1}{2} m_t v_t^2 = \frac{1}{2} m_m v_m^2$$

it follows that

$$v_m = v_t \sqrt{\frac{m_t}{m_m}}$$

Thus, the slowest—and easiest—throw will be made when the momentum is held constant. If you wish to check this answer, try substituting in values of $v_t = 1$ m/s, $m_t = 1$ kg, and $m_m = 100$ kg. Then the same-momentum throw will be caught at 1 cm/s, while the same-energy throw will be caught at 10 cm/s.

15. The follow-through keeps the club in contact with the ball as long as possible, maximizing the impulse. Thus, the ball accrues a larger change in momentum than without the follow-through, and it leaves the club with a higher velocity and travels farther. With a short shot to the green, the primary factor is control, not distance. Hence, there is little or no follow-through, allowing the golfer to have a better feel for how hard he or she is striking the ball.

17. It is the product mv that is the same for both the bullet and the gun. The bullet has a large velocity and a small mass, while the gun has a small velocity and a large mass. Furthermore, the bullet carries much more kinetic energy than the gun.

PROBLEMS

1. 1.39 N·s up
3. (a) 8.35×10^{-21} kg·m/s (b) 4.50 kg·m/s (c) 750 kg·m/s (d) 1.78×10^{29} kg·m/s
5. (a) 31.0 m/s (b) the bullet, 3.38×10^3 J versus 69.7 J
7. $\sim 10^3$ N upward
9. 364 kg·m/s forward, 438 N forward
11. (a) 8.0 N·s (b) 5.3 m/s (c) 3.3 m/s
13. (a) 12 N·s (b) 8.0 N·s (c) 8.0 m/s, 5.3 m/s
15. (a) 9.60×10^{-2} s (b) 3.65×10^5 N (c) $26.6g$
17. 6.7×10^3 N toward the west
19. 65 m/s
21. (a) 1.15 m/s (b) 0.346 m/s directed opposite to girl's motion
23. $KE_E/KE_b \sim 10^{-25}$
25. 1.67 m/s
27. (a) 1.80 m/s (b) 2.16×10^4 J
29. 57 m
31. 15.6 m/s
33. 273 m/s
35. (a) -6.67 cm/s, 13.3 cm/s (b) 0.889
37. 17.1 cm/s (25.0-g object), 22.1 cm/s (10.0-g object)
39. 7.94 cm
41. (a) 2.9 m/s at 32° N of E (b) 7.9×10^2 J converted into internal energy
43. 5.59 m/s north
45. (a) 2.50 m/s at $-60°$ (b) elastic collision
47. (a) 9.0 m/s (b) -15 m/s
49. 1.78×10^3 N on truck driver, 8.89×10^3 N on car driver
51. (a) 8/3 m/s (incident particle), 32/3 m/s (target particle) (b) $-16/3$ m/s (incident particle), 8/3 m/s (target particle) (c) 7.1×10^{-2} J in case (a), and 2.8×10^{-3} J in case (b). The incident particle loses more kinetic energy in case (a), in which the target mass is 1.0 g.
53. 1.1×10^3 N (upward)
55. (a) 1.33 m/s (b) 235 N (c) 0.681 s (d) -160 N·s, 160 N·s (e) 1.82 m (f) 0.454 m (g) -427 J (h) 107 J (i) Equal friction forces act through different distances on the person and the cart to do different amounts of work on them. This is a perfectly inelastic collision in which the total work on both person and cart together is -320 J, which becomes $+320$ J of internal energy.

57. (a) -2.33 m/s, 4.67 m/s (b) 0.277 m (c) 2.98 m
  (d) 1.49 m
59. (a) -0.667 m/s (b) 0.952 m
61. (a) 3.54 m/s (b) 1.77 m (c) 3.54×10^4 N (d) No, the
  normal force exerted by the rail contributes upward momentum to the system.
63. (a) 0.28 or 28% (b) 1.1×10^{-13} J for the neutron,
  4.5×10^{-14} J for carbon
65. $\dfrac{2v_0{}^2}{9\mu_k g} - \dfrac{4d}{9}$
67. (a) No. After colliding, the cars, moving as a unit, would
  travel northeast, so they couldn't damage property
  on the southeast corner. (b) x-component: 16.3 km/h,
  y-component: 9.17 km/h, angle: the final velocity of the
  car is 18.7 km/h at $29.4°$ north of east, consistent with
  part (a).
69. (a) 4.85 m/s (b) 8.41 m
71. 0.312 N to the right
73. (a) 300 m/s, (b) 3.75 m/s, (c) 1.20 m

Chapter 7

QUICK QUIZZES

1. (c)
2. (b)
3. (b)
4. (b)
5. (a)
6. 1. (e) 2. (a) 3. (b)
7. (c)
8. (b), (c)
9. (e)
10. (d)

CONCEPTUAL QUESTIONS

1. (a) The head will tend to lean toward the right shoulder
  (that is, toward the outside of the curve). (b) When there
  is no strap, tension in the neck muscles must produce the
  centripetal acceleration. (c) With a strap, the tension in
  the strap performs this function, allowing the neck muscles to remain relaxed.
3. An object can move in a circle even if the total force on it
  is not perpendicular to its velocity, but then its speed will
  change. Resolve the total force into an inward radial component and a perpendicular tangential component. If the
  tangential force acts in the forward direction, the object
  will speed up, and if the tangential force acts backward,
  the object will slow down.
5. The speedometer will be inaccurate. The speedometer
  measures the number of tire revolutions per second, so its
  readings will be too low.
7. The car cannot round a turn at constant velocity, because
  "constant velocity" means the direction of the velocity is
  not changing. The statement is correct if the word "velocity" is replaced by the word "speed."
9. The gravitational force exerted on the moon by the planet
  produces a centripetal acceleration. Mathematically,
  $$m_{\text{moon}}\left(\frac{v_t{}^2}{r}\right) = \frac{GMm_{\text{moon}}}{r^2}$$

or the mass of the planet is $M = \dfrac{(rv_t{}^2)}{G}$. Both r and v_t can
be determined by observing the motion of the moon, and
the mass of the planet is then easily computed.
11. Consider an individual standing against the inside wall of
  the cylinder with her head pointed toward the axis of the
  cylinder. As the cylinder rotates, the person tends to
  move in a straight-line path tangent to the circular path
  followed by the cylinder wall. As a result, the person
  presses against the wall, and the normal force exerted on
  her provides the radial force required to keep her moving in a circular path. If the rotational speed is adjusted
  such that this normal force is equal in magnitude to her
  weight on Earth, she will not be able to distinguish between the artificial gravity of the colony and ordinary
  gravity.
13. The tendency of the water is to move in a straight-line
  path tangent to the circular path followed by the container. As a result, at the top of the circular path, the water presses against the bottom of the pail, and the normal
  force exerted by the pail on the water provides the radial
  force required to keep the water moving in its circular
  path.
15. Any object that moves such that the *direction* of its velocity
  changes has an acceleration. A car moving in a circular
  path will always have a centripetal acceleration.
17. (b) decrease. The free-fall acceleration near the surface
  of the Earth is given by $g = GM_E/R_E{}^2$, therefore, doubling both the Earth's mass and its radius would reduce
  the value of g by a factor of 2.

PROBLEMS

1. (a) 3.2×10^8 rad (b) 5.0×10^7 rev
3. 1.99×10^{-7} rad/s, $0.986°$/day
5. (a) 821 rad/s^2 (b) 4.21×10^3 rad
7. 3.2 rad
9. Main rotor: 179 m/s $= 0.522v_{\text{sound}}$
  Tail rotor: 221 m/s $= 0.644v_{\text{sound}}$
11. $\sim 10^{-2}$ cm
13. 13.7 rad/s^2
15. (a) 3.37×10^{-2} m/s^2 downward (b) 0
17. (a) 0.35 m/s^2 (b) 1.0 m/s (c) 0.35 m/s^2, 0.94 m/s^2,
  1.0 m/s^2 at $20°$ forward with respect to the direction of a_c
19. (a) 1.10 kN (b) 2.04 times her weight
21. 22.6 m/s
23. (a) 18.0 m/s^2 (b) 900 N (c) 1.84; this large coefficient is
  unrealistic, and she will not be able to stay on the merry-
  go-round.
25. (a) 9.8 N (b) 9.8 N (c) 6.3 m/s
27. (a) 1.58 m/s^2 (b) 455 N upward (c) 329 N upward
  (d) 397 N directed inward and $80.8°$ above horizontal
29. 321 N toward Earth
31. 1.1×10^{-10} N at $72°$ above the $+ x$-axis
33. (a) 2.50×10^{-5} N toward the 500-kg object (b) Between
  the two objects and 0.245 m from the 500-kg object
35. (a) 9.58×10^6 m (b) 5.57 h
37. 1.90×10^{27} kg
39. (a) 1.48 h (b) 7.79×10^3 m/s (c) 6.43×10^9 J
41. (a) 9.40 rev/s (b) 44.1 rev/s^2; $a_r = 2\,590$ m/s^2;
  $a_t = 206$ m/s^2 (c) $F_r = 514$ N; $F_t = 40.7$ N
43. (a) 2.51 m/s (b) 7.90 m/s^2 (c) 4.00 m/s
45. (a) 7.76×10^3 m/s (b) 89.3 min

47. (a) $n = m\left(g - \dfrac{v^2}{r}\right)$ (b) 17.1 m/s

49. (a) $F_{g,\,\text{true}} = F_{g,\,\text{apparent}} + mR_E\omega^2$
    (b) 732 N (equator), 735 N (either pole)

51. 0.131

55. 11.8 km/s

57. 0.75 m

59. (a) $v_0 = \sqrt{g\left(R - \dfrac{2h}{3}\right)}$ (b) $h' = \dfrac{R}{2} + \dfrac{2h}{3}$

61. (a) 15.3 km (b) 1.66×10^{16} kg (c) 1.13×10^4 s

63. 0.835 rev/s = 50.1 rev/min

65. (a) 10.6 kN (b) 14.1 m/s

67. (a) 106 N (b) 0.396

69. (a) 0.605 m/s (b) 17.3 rad/s (c) 5.82 m/s (d) The crank length is unnecessary.

71. (a) 2.0×10^{12} m/s^2 (b) 2.4×10^{11} N (c) 1.4×10^{12} J

Chapter 8

QUICK QUIZZES

1. (d)
2. (b)
3. (b)
4. (a)
5. (c)
6. (c)
7. (a)

CONCEPTUAL QUESTIONS

1. In order for you to remain in equilibrium, your center of gravity must always be over your point of support, the feet. If your heels are against a wall, your center of gravity cannot remain above your feet when you bend forward, so you lose your balance.

3. There are two major differences between torque and work. The primary difference is that the displacement in the expression for work is directed along the force, while the important distance in the torque expression is perpendicular to the force. The second difference involves whether there is motion. In the case of work, work is done only if the force succeeds in causing a displacement of the point of application of the force. By contrast, a force applied at a perpendicular distance from a rotation axis results in a torque regardless of whether there is motion.

    As far as units are concerned, the mathematical expressions for both work and torque are in units that are the product of newtons and meters, but this product is called a joule in the case of work and remains a newton-meter in the case of torque.

5. No. For an object to be in equilibrium, the net external force acting on it must be zero. This is not possible when only one force acts on the object, unless that force should have zero magnitude. In that case, there is really no force acting on the object.

7. As the motorcycle leaves the ground, the friction between the tire and the ground suddenly disappears. If the motorcycle driver keeps the throttle open while leaving the ground, the rear tire will increase its angular speed and, hence, its angular momentum. The airborne motorcycle is now an isolated system, and its angular momentum must be conserved. The increase in angular momentum of the tire directed, say, clockwise must be compensated for by an increase in angular momentum of the entire motorcycle counterclockwise. This rotation results in the nose of the motorcycle rising and the tail dropping.

9. In general, you want the rotational kinetic energy of the system to be as small a fraction of the total energy as possible. That is, you want translation, not rotation. You want the wheels to have as little moment of inertia as possible, so that they present the lowest resistance to changes in rotational motion. Disklike wheels would have lower moments of inertia than hooplike wheels, so disks are preferable. The lower the mass of the wheels, the less is the moment of inertia, so light wheels are preferable. The smaller the radius of the wheels, the less is the moment of inertia, so smaller wheels are preferable—within limits: You want the wheels to be large enough to be able to travel relatively smoothly over irregularities in the road.

11. The angular momentum of the gas cloud is conserved. Thus, the product $I\omega$ remains constant. As the cloud shrinks in size, its moment of inertia decreases, so its angular speed ω must increase.

13. We can assume fairly accurately that the driving motor will run at a constant angular speed and at a constant torque. Therefore, as the radius of the take-up reel increases, the tension in the tape will decrease, in accordance with the equation.

$$T = \tau_{\text{const}} / R_{\text{take-up}} \qquad (1)$$

    As the radius of the source reel decreases, given a decreasing tension, the torque in the source reel will decrease even faster, as the following equation shows:

$$\tau_{\text{source}} = TR_{\text{source}} = \tau_{\text{const}} R_{\text{source}} / R_{\text{take-up}} \qquad (2)$$

    This torque will be partly absorbed by friction in the feed heads (which we assume to be small); some will be absorbed by friction in the source reel. Another small amount of the torque will be absorbed by the increasing angular speed of the source reel. However, in the case of a sudden jerk on the tape, the changing angular speed of the source reel becomes important. If the source reel is full, then the moment of inertia will be large and the tension in the tape will be large. If the source reel is nearly empty, then the angular acceleration will be large instead. Thus, the tape will be more likely to break when the source reel is nearly full. One sees the same effect in the case of paper towels: It is easier to snap a towel free when the roll is new than when it is nearly empty.

15. The initial angular momentum of the system (mouse plus turntable) is zero. As the mouse begins to walk clockwise, its angular momentum increases, so the turntable must rotate in the counterclockwise direction with an angular momentum whose magnitude equals that of the mouse. This conclusion follows from the fact that the final angular momentum of the system must equal the initial angular momentum (zero).

17. When a ladder leans against a wall, both the wall and the floor exert forces of friction on the ladder. If the floor is perfectly smooth, it can exert no frictional force in the horizontal direction to counterbalance the wall's normal

17. 2.171 cm

19. 2.7×10^2 N

21. 1.1 L (0.29 gal)

23. 0.548 gal

25. (a) increases (b) 1.603 cm

27. (a) 627°C (b) 927°C

29. (a) 2.5×10^{19} molecules (b) 4.1×10^{-21} mol

31. 287°C

33. 7.1 m

35. 16.0 cm³

37. 6.21×10^{-21} J

39. 6.64×10^{-27} kg

41. (a) 8.76×10^{-21} J

(b) $v_{\text{rms, He}} = 1.62$ km/s, $v_{\text{rms, Ar}} = 514$ m/s

43. 16 N

45. 0.663 mm at 78.2° below the horizontal

47. 3.55 L

49. 6.57 MPa

51. The expansion of the mercury is almost 20 times that of the flask (assuming Pyrex glass).

53. shorter, by 0.061 mm

55. 2.74 m

57. 1.61 MPa

59. 0.417 L

61. (a) $\theta = \dfrac{(\alpha_2 - \alpha_1)L_0(\Delta T)}{\Delta r}$

(c) The bar bends in the opposite direction.

63. 0.53 kg

Chapter 11

QUICK QUIZZES

1. (a) Water, glass, iron. (b) Iron, glass, water.

2. (b) The slopes are proportional to the reciprocal of the specific heat, so a larger specific heat results in a smaller slope, meaning more energy is required to achieve a given temperature change.

3. (c)

4. (b)

5. (a) 4 (b) 16 (c) 64

CONCEPTUAL QUESTIONS

1. When you rub the surface, you increase the temperature of the rubbed region. With the metal surface, some of this energy is transferred away from the rubbed site by conduction. Consequently, the temperature in the rubbed area is not as high for the metal as it is for the wood, and it feels relatively cooler than the wood.

3. The fruit loses energy into the air by radiation and convection from its surface. Before ice crystals can form inside the fruit to rupture cell walls, all of the liquid water on the skin will have to freeze. The resulting time delay may prevent damage within the fruit throughout a frosty night. Further, a surface film of ice provides some insulation to slow subsequent energy loss by conduction from within the fruit.

5. The operation of an immersion coil depends on the convection of water to maintain a safe temperature. As the water near a coil warms up, the warmed water floats to the top due to Archimedes' principle. The temperature of the coil cannot go higher than the boiling temperature of water, 100°C. If the coil is operated in air, convection is reduced, and the upper limit of 100°C is removed. As a re-

sult, the coil can become hot enough to be damaged. If the coil is used in an attempt to warm a thick liquid like stew, convection cannot occur fast enough to carry energy away from the coil, so that it again may become hot enough to be damaged.

7. One of the ways that objects transfer energy is by radiation. The top of the mailbox is oriented toward the clear sky. Radiation emitted by the top of the mailbox goes upward and into space. There is little radiation coming down from space to the top of the mailbox. Radiation leaving the sides of the mailbox is absorbed by the environment. Radiation from the environment (tree, houses, cars, etc.), however, can enter the sides of the mailbox, keeping them warmer than the top. As a result, the top is the coldest portion and frost forms there first.

9. Tile is a better conductor of energy than carpet, so the tile conducts energy away from your feet more rapidly than does the carpeted floor.

11. The large amount of energy stored in the concrete during the day as the sun falls on it is released at night, resulting in an overall higher average temperature in the city than in the countryside. The heated air in a city rises as it's displaced by cooler air moving in from the countryside, so evening breezes tend to blow from country to city.

13. The fingers are wetted to create a layer of steam between the fingers and the molten lead. The steam acts as an insulator and prevents serious burns. This demonstration is dangerous: we don't recommend it.

15. The increase in the temperature of the ethyl alcohol will be about twice that of the water.

17. (d)

PROBLEMS

1. 10.1°C

3. (a) 1.67×10^{18} J (b) 53.1 yr

5. 1.03×10^3 J

7. 2.85 km

9. 176°C

11. 88 W

13. 185 g

15. 80 g

17. 1.8×10^3 J/kg·°C

19. 0.26 kg

21. 65°C

23. 21 g

25. 2.3 km

27. 16°C

29. $t_{\text{boil}} = 2.8$ min, $t_{\text{evaporate}} = 18$ min

31. (a) all ice melts, $T_f = 40$°C (b) 8.0 g melts, $T_f = 0$°C

33. (a) 0.22 kW (b) 13 mW (c) 56 mW

35. 402 MW

37. 14 ft²·°F·h/Btu

39. 9.0 cm

41. 0.11 kW

43. $\sim 10^3$ W

45. 16:1

47. 12 kW

49. 6.0×10^3 kg

51. 2.3 kg

53. 29°C

55. 30.3 kcal/h

57. 109°C

59. 51.2°C

61. (a) 7 stops (b) Assumes that no energy is lost to the surroundings and that all internal energy generated stays with the brakes.
63. (b) 2.7×10^3 J/kg · °C
65. 1.4 kg
67. 12 h
69. 10.9 g

Chapter 12

QUICK QUIZZES

1. (b)
2. A is isovolumetric, B is adiabatic, C is isothermal, D is isobaric.
3. (c)
4. (b)
5. The number 7 is the most probable outcome. The numbers 2 and 12 are the least probable outcomes.

CONCEPTUAL QUESTIONS

1. First, the efficiency of the automobile engine cannot exceed the Carnot efficiency: it is limited by the temperature of the burning fuel and the temperature of the environment into which the exhaust is dumped. Second, the engine block cannot be allowed to exceed a certain temperature. Third, any practical engine has friction, incomplete burning of fuel, and limits set by timing and energy transfer by heat.

3. If there is no change in internal energy, then, according to the first law of thermodynamics, the heat is equal to the negative of the work done on the gas (and thus equal to the work done *by* the gas). Thus, $Q = -W = W_{\text{by gas}}$.

5. The energy that is leaving the body by work and heat is replaced by means of biological processes that transform chemical energy in the food that the individual ate into internal energy. Thus, the temperature of the body can be maintained.

7. The statement shows a misunderstanding of the concept of heat. Heat is energy in the process of being transferred, not a form of energy that is held or contained. If you wish to speak of energy that is contained, you speak of **internal energy**, not heat. Correct statements would be (1) "Given any two objects in thermal contact, the one with the higher temperature will transfer energy by heat to the other" and (2) "Given any two objects of equal mass, the one with the higher product of absolute temperature and specific heat contains more internal energy."

9. Although no energy is transferred into or out of the system by heat, work is done on the system as the result of the agitation. Consequently, both the temperature and the internal energy of the coffee increase.

11. Practically speaking, it isn't possible to create a heat engine that creates no thermal pollution, because there must be both a hot heat source (energy reservoir) and a cold heat sink (low-temperature energy reservoir). The heat engine will warm the cold heat sink and will cool down the heat source. If either of those two events is undesirable, then there will be thermal pollution.

Under some circumstances, the thermal pollution would be negligible. For example, suppose a satellite in space were to run a heat pump between its sunny side and its dark side. The satellite would intercept some of the energy that gathered on one side and would "dump" it to the dark side. Since neither of those effects would be particularly undesirable, it could be said that such a heat pump produced no thermal pollution.

13. The rest of the Universe must have an entropy change of +8.0 J/K or more.

15. The first law is a statement of conservation of energy that says that we cannot devise a cyclic process that produces more energy than we put into it. If the cyclic process takes in energy by heat and puts out work, we call the device a heat engine. In addition to the first law's limitation, the second law says that, during the operation of a heat engine, some energy must be ejected to the environment by heat. As a result, it is theoretically impossible to construct a heat engine that will work with 100% efficiency.

17. From the point of view of energy principles, the molecules strike the piston and move it through a distance, so the molecules do work on the piston. This work represents a transfer of energy out of the gas. As a result, the internal energy of the gas drops. Because the temperature is related to internal energy, the temperature of the gas drops.

PROBLEMS

3. 1.1×10^4 J
5. (a) -810 J (b) -507 J (c) -203 J
7. (a) -6.1×10^5 J (b) 4.6×10^5 J
9. (a) 1.09×10^3 K (b) -6.81 kJ
11. (a) $Q < 0$, $W = 0$, $\Delta U < 0$ (b) $Q > 0$, $W = 0$, $\Delta U > 0$
13. (a) 567 J (b) 167 J
15. (a) -88.5 J (b) 722 J
17. (a) -180 J (b) $+188$ J
19. (a) -9.12×10^{-3} J (b) -333 J
21. (a) 0.95 J (b) 3.2×10^5 J (c) 3.2×10^5 J
23. 0.489 (or 48.9%)
25. (a) 0.333 (or 33.3%) (b) 2/3
27. (a) 0.672 (or 67.2%) (b) 58.8 kW
29. (a) 0.294 (or 29.4%) (b) 500 J (c) 1.67 kW
31. 1/3
33. 0.49°C
35. (a) -1.2 kJ/K (b) 1.2 kJ/K
37. 57 J/K
39. 3.27 J/K
41. (a)

End Result	Possible Tosses	Total Number of Same Result
All H	HHHH	1
1T, 3H	HHHT, HHTH, HTHH, THHH	4
2T, 2H	HHTT, HTHT, THHT, HTTH, THTH, TTHH	6
3T, 1H	TTTH, TTHT, THTT, HTTT	4
All T	TTTT	1

Most probable result = 2H and 2T.

(b) all H or all T (c) 2H and 2T
43. The maximum efficiency possible with these reservoirs = 50%; the claim is false.
45. (a) $-|Q_h|/T_h$ (b) $|Q_c|/T_c$ (c) $|Q_h|/T_h - |Q_c|/T_c$ (d) 0
47. 2.8°C
49. 18°C
51. (a) 12.2 kJ (b) 4.05 kJ (c) 8.15 kJ

53. (a) 26 J (b) 9.0×10^5 J (c) 9.0×10^5 J
55. (a) 2.49 kJ (b) 1.50 kJ (c) -990 J
57. (a) 0.554 (or 55.4%) (b) The Carnot efficiency is 0.749 (or 74.9%)
59. (a) 2.6×10^3 metric tons/day (b) $\$7.7 \times 10^6$/yr
    (c) 4.1×10^4 kg/s
61. (a) 10 J (b) No work is done; $\Delta U = 42$ J
63. 0.146; 486 kcal

Chapter 13

QUICK QUIZZES

1. (d)
2. (c)
3. (b)
4. (a)
5. (c)
6. (d)
7. (c), (b)
8. (a)
9. (b)

CONCEPTUAL QUESTIONS

1. It will increase. The speed of the wave varies inversely with the mass per unit length of the rope, so it is higher in the lighter rope. The frequency is unchanged as the wave passes from one rope to the other one. The wavelength is $\lambda = v/f$, so with the frequency constant and the speed increasing, the wavelength must increase.

3. No. Because the total energy is $E = \frac{1}{2}kA^2$, changing the mass of the object while keeping A constant has no effect on the total energy. When the object is at a displacement x from equilibrium, the potential energy is $\frac{1}{2}kx^2$, independent of the mass, and the kinetic energy is $KE = E - \frac{1}{2}kx^2$, also independent of the mass.

5. When the spring with two objects on opposite ends is set into oscillation in space, the coil at the exact center of the spring does not move. Thus, we can imagine clamping the center coil in place without affecting the motion. If we do this, we have two separate oscillating systems, one on each side of the clamp. The half-spring on each side of the clamp has twice the spring constant of the full spring, as shown by the following argument: The force exerted by a spring is proportional to the separation of the coils as the spring is extended. Imagine that we extend a spring by a given distance and measure the distance between coils. We then cut the spring in half. If one of the half-springs is now extended by the same distance, the coils will be twice as far apart as they were in the complete spring. Thus, it takes twice as much force to stretch the half-spring, from which we conclude that the half-spring has a spring constant which is twice that of the complete spring. Hence, our clamped system of objects on two half-springs will vibrate with a frequency that is higher than f by a factor of the square root of two.

7. The bouncing ball is not an example of simple harmonic motion. The ball does not follow a sinusoidal function for its position as a function of time. The daily movement of a student is also not simple harmonic motion, since the student stays at a fixed location—school—for a long time. If this motion were sinusoidal, the student would move more and more slowly as she approached her desk, and as soon as she sat down at the desk, she would start to move back toward home again.

9. We assume that the buoyant force acting on the sphere is negligible in comparison to its weight, even when the sphere is empty. We also assume that the bob is small compared with the pendulum length. Then, the frequency of the pendulum is $f = 1/T = (1/2\pi)\sqrt{g/L}$, which is independent of mass. Thus, the frequency will not change as the water leaks out.

11. As the temperature increases, the length of the pendulum will increase due to thermal expansion, and with a greater length, the period of the pendulum increases. Thus, it takes longer to execute each swing, so that each second according to the clock will take longer than an actual second. Consequently, the clock will run slow.

13. A pulse in a long line of people is longitudinal, since the movement of people is parallel to the direction of propagation of the pulse. The speed is determined by the reaction time of the people and the speed with which they can move once a space opens up. There is also a psychological factor, in that people will not want to fill a space that opens up in front of them too quickly, so as not to intimidate the person in front of them. The "wave" at a stadium is transverse, since the fans stand up vertically as the wave sweeps past them horizontally. The speed of this pulse depends on the limits of the fans' abilities to rise and sit rapidly and on psychological factors associated with the anticipation of seeing the pulse approach the observer's location.

15. A wave on a massless string would have an infinite speed of propagation, because the linear mass density of the string is zero.

17. The kinetic energy is proportional to the square of the speed, and the potential energy is proportional to the square of the displacement. Therefore, both must be positive quantities.

19. From $v = \sqrt{F/\mu}$, we see that increasing the tension by a factor of four doubles the wave speed.

PROBLEMS

1. (a) 24 N toward the equilibrium position (b) 60 m/s^2 toward the equilibrium position.
3. (b) 1.81 s (c) No, the force is not of the form of Hooke's law.
5. 0.242 kg
7. (a) 60 J (b) 49 m/s
9. 2.94×10^3 N/m
11. (a) $PE = E/4$ $KE = 3E/4$ (b) $x = A/\sqrt{2}$
13. 0.478 m
15. (a) 0.28 m/s (b) 0.26 m/s (c) 0.26 m/s (d) 3.5 cm
17. 39.2 N
19. (a) You observe uniform circular motion projected on a plane perpendicular to the motion. (b) 0.628 s
21. The horizontal displacement is described by $x(t) = A \cos \omega t$, where A is the distance from the center of the wheel to the crankpin.
23. 0.63 s
25. (a) 1.0 s (b) 0.28 m/s (c) 0.25 m/s
27. (a) 11.0 N toward the left (b) 0.881 oscillations
29. $v = \pm \omega A \sin \omega t$, $a = -\omega^2 A \cos \omega t$
31. 105 complete oscillations
33. (a) slow (b) 9:47 A.M.
35. (a) $L_{Earth} = 25$ cm, $L_{Mars} = 9.4$ cm,
    (b) $m_{Earth} = m_{Mars} = 0.25$ kg

37. (a) 9.00 cm (b) 20.0 cm (c) 40.0 ms (d) 5.00 m/s
39. (a) 11.4 ns (b) 3.41 m
41. 31.9 cm
43. 80.0 N
45. (a) 30.0 N (b) 25.8 m/s
47. 28.5 m/s
49. (a) 0.051 kg/m (b) 20 m/s
51. 13.5 N
53. (a) Constructive interference gives $A = 0.50$ m
     (b) Destructive interference gives $A = 0.10$ m
55. 6.62 cm
57. (a) 588 N/m (b) 0.700 m/s
59. 1.1 m/s
61. (a) 15.8 rad/s (b) 5.23 cm
63. $F_{tangential} = -[(\rho_{air} - \rho_{He})Vg/L]s$, $T = 1.40$ s
67. (a) 0.50 m/s (b) 8.6 cm
69. (a) the longitudinal wave (b) 11.1 min
71. (a) 0.50 m/s (b) 0.39 m/s (c) 5.0 cm

Chapter 14

QUICK QUIZZES

1. (c)
2. (c)
3. (b)
4. (b), (e)
5. (d)
6. (a)
7. (b)

CONCEPTUAL QUESTIONS

1. (a) higher (b) lower
3. The camera is designed to operate at an assumed speed of sound of 345 m/s, the speed of sound at a room temperature of 23°C. If the temperature should decrease to, say, 0°C, the speed of sound will also decrease, and the camera will respond to the fact that it takes longer for the sound to make its round trip. Thus, it will operate as if the object is farther away than it really is.

It is of interest to note that bats use echo sounding like this to locate insects or to avoid obstacles in front of them, something that they must do because of poor eyesight and the high speeds at which they fly. Likewise, blue whales use this technique to help them avoid objects in their path. The need here is obvious, because a typical whale has a mass of 10^5 kg and travels at a relatively fast speed of 20 mi/h, so it takes a long time for it to stop its motion or to change direction.
5. Sophisticated electronic devices break the frequency range of about 60 to 4 000 Hz used in telephone conversations into several frequency bands and then mix them in a predetermined pattern so that they become unintelligible. The descrambler, of course, moves the bands back into their proper order.
7. A rise in temperature will increase the dimensions of the wind instrument much less than it increases the speed of sound in the enclosed air. This effect will raise the resonant frequencies that are produced by the instrument, which will go sharp as the temperature increases and go flat as the temperature decreases.
9. The echo is Doppler shifted, and the shift is like both a moving source and a moving observer. The sound that leaves your horn in the forward direction is Doppler shifted to a higher frequency, because it is coming from a moving source. As the sound reflects back and comes towards you, you are a moving observer, so there is a second Doppler shift to an even higher frequency. If the sound reflects from the spacecraft coming towards you, there is a different moving-source shift to an even higher frequency. The reflecting surface of the spacecraft acts as a moving source.
11. The center of the string is a node for the second harmonic, as well as for every even-numbered harmonic. By placing the finger at the center and plucking, the guitarist is eliminating any harmonic which does not have a node at that point—that is, all the odd harmonics. The even harmonics can vibrate relatively freely with the finger at the center because they exhibit no displacement at that point. The result is a sound with a mixture of frequencies that are integer multiples of the second harmonic, which is one octave higher than the fundamental.
13. The bowstring is pulled away from equilibrium and released, in a manner similar to the way a guitar string is pulled and released when it is plucked. Thus, standing waves will be excited in the bowstring. If the arrow leaves from the exact center of the string, then a series of odd harmonics will be excited. Even harmonics will not be excited, because they have a node at the point where the string exhibits its maximum displacement.
15. At the instant at which there is no displacement of the string, the string is still moving. Thus, the energy is present at that instant entirely as kinetic energy of the string.
17. As the whistle is approaching you, the sound is Doppler shifted to a frequency higher than the natural frequency of the whistle. You will momentarily hear the natural frequency just as the whistle comes parallel with you and is in the act of passing. As soon as the whistle is past, you will hear sound that is Doppler shifted to a frequency lower than the natural frequency of the whistle. As the frequency of the sound steps down from one constant value to a second one, the intensity of the sound varies continuously. The loudness increases smoothly to a maximum and then decreases.
19. The two engines are running at slightly different frequencies, thus producing a beat frequency between them.

PROBLEMS

1. 5.56 km
3. 32°C
5. 516 m
7. 1.99 km
9. (a) 1.00×10^{-2} W/m^2 (b) 105 dB
11. 37 dB
13. 9 additional machines
15. 66.0 dB
17. (a) 1.3×10^2 W (b) 96 dB
21. (a) 75.2-Hz drop (b) 0.953 m
23. 595 Hz
25. 0.391 m/s
27. 19.3 m
29. 48°
31. 800 m
33. (a) 0.240 m (b) 0.855 m
35. (a) Nodes at 0, 2.67 m, 5.33 m, and 8.00 m; antinodes at 1.33 m, 4.00 m, and 6.67 m (b) 18.6 Hz

37. At 0.0891 m, 0.303 m, 0.518 m, 0.732 m, 0.947 m, and 1.16 m from one speaker.
39. (a) 79 N (b) 2.1×10^2 Hz
41. 19.976 kHz
43. 58 Hz
45. 3.0 kHz
47. (a) 0.552 m (b) 317 Hz
49. 5.26 beats/s
51. 3.79 m/s toward the station, 3.88 m/s away from the station
53. (a) 1.98 beats/s (b) 3.40 m/s
55. 1.76 cm
57. 262 kHz
59. 64 dB
61. 439 Hz and 441 Hz
63. 32.9 m/s
65. 3.97 beats/s
67. 1.34×10^4 N
69. 1 204 Hz
71. (a) 617 m (b) 154 m/s

Chapter 15

QUICK QUIZZES

1. (b)
2. (b)
3. (c)
4. (a)
5. (c) and (d)
6. (a)
7. (c)
8. (b)
9. (d)
10. (b) and (d)

CONCEPTUAL QUESTIONS

1. Electrons have been removed from the object.
3. The configuration shown is inherently unstable. The negative charges repel each other. If there is any slight rotation of one of the rods, the repulsion can result in further rotation away from this configuration. There are three conceivable final configurations shown below. Configuration (a) is stable: If the positive upper ends are pushed towards each other, their mutual repulsion will move the system back to the original configuration. Configuration (b) is an equilibrium configuration, but it is unstable: If the lower ends are moved towards each other, their mutual attraction will be larger than that of the upper ends, and the configuration will shift to (c), another possible stable configuration.

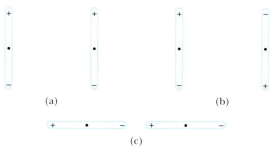

(a)   (b)

(c)

Figure Q15.3

5. Move an object A with a net positive charge so it is near, but not touching, a neutral metallic object B that is insulated from the ground. The presence of A will polarize B, causing an excess negative charge to exist on the side nearest A and an excess positive charge of equal magnitude to exist on the side farthest from A. While A is still near B, touch B with your hand. Additional electrons will then flow from ground, through your body and onto B. With A continuing to be near but not in contact with B, remove your hand from B, thus trapping the excess electrons on B. When A is now removed, B is left with excess electrons, or a net negative charge. By means of mutual repulsion, this negative charge will now spread uniformly over the entire surface of B.

7. An object's mass decreases very slightly (immeasurably) when it is given a positive charge, because it loses electrons. When the object is given a negative charge, its mass increases slightly because it gains electrons.

9. Electric field lines start on positive charges and end on negative charges. Thus, if the fair-weather field is directed into the ground, the ground must have a negative charge.

11. The two charged plates create a region with a uniform electric field between them, directed from the positive toward the negative plate. Once the ball is disturbed so as to touch one plate (say, the negative one), some negative charge will be transferred to the ball and it will be acted upon by an electric force that will accelerate it to the positive plate. Once the ball touches the positive plate, it will release its negative charge, acquire a positive charge, and accelerate back to the negative plate. The ball will continue to move back and forth between the plates until it has transferred all their net charge, thereby making both plates neutral.

13. The electric shielding effect of conductors depends on the fact that there are two kinds of charge: positive and negative. As a result, charges can move within the conductor so that the combination of positive and negative charges establishes an electric field that exactly cancels the external field within the conductor and any cavities inside the conductor. There is only one type of gravitation charge, however, because there is no negative mass. As a result, gravitational shielding is not possible.

15. The electric field patterns of each of these three configurations do not have sufficient symmetry to make the calculations practical. Gauss's law is useful only for calculating the electric fields of highly symmetric charge distributions, such as uniformly charged spheres, cylinders, and sheets.

17. No, the wall is not positively charged. The balloon induces a charge of opposite sign in the wall, causing the balloon and the wall to be attracted to each other. The balloon eventually falls because its charge slowly diminishes as it leaks to ground. Some of the balloon's charge could also be lost due to positive ions in the surrounding atmosphere, which would tend to neutralize the negative charges on the balloon.

19. When the comb is nearby, charges separate on the paper, and the paper is attracted to the comb. After contact, charges from the comb are transferred to the paper, so that it has the same type of charge as the comb. The paper is thus repelled.

21. The attraction between the ball and the object could be an attraction of unlike charges, or it could be an attrac-

tion between a charged object and a neutral object as a result of polarization of the molecules of the neutral object. Two additional experiments could help us determine whether the object is charged. First, a known neutral ball could be brought near the object, and if there is an attraction, the object is negatively charged. Another possibility is to bring a known negatively charged ball near the object. In that case, if there is a repulsion, then the object is negatively charged. If there is an attraction, then the object is neutral.

PROBLEMS

1. 1.1×10^{-8} N (attractive)
3. 91 N (repulsion)
5. (a) 36.8 N   (b) 5.54×10^{27} m/s^2
7. 5.12×10^5 N
9. (a) 2.2×10^{-5} N (attraction)
   (b) 9.0×10^{-7} N (repulsion)
11. 1.38×10^{-5} N at 77.5° below the negative x-axis
13. 0.872 N at 30.0° below the positive x-axis
15. 7.2 nC
17. 1.5×10^{-3} C
19. 7.20×10^5 N/C (downward)
21. 1.2×10^4 N/C
23. (a) 6.12×10^{10} m/s^2 (b) 19.6 μs   (c) 11.8 m
    (d) 1.20×10^{-15} J
25. zero
27. 1.8 m to the left of the -2.5-μC charge
33. (a) 0   (b) 5 μC inside, -5 μC outside   (c) 0 inside,
    -5 μC outside   (d) 0 inside, -5 μC outside
35. 1.3×10^{-3} C
37. (a) 4.8×10^{-15} N   (b) 2.9×10^{12} m/s^2
39. (a) 858 N·m^2/C   (b) 0   (c) 657 N·m^2/C
41. 4.1×10^6 N/C
43. (a) 0   (b) $k_e q/r^2$ outward
47. 57.5 N
49. 24 N/C in the positive x-direction
51. (a) $E = 2k_e qb \, (a^2 + b^2)^{-3/2}$ in the positive x-direction
    (b) $E = k_e Qb(a^2 + b^2)^{-3/2}$ in the positive x-direction
53. (a) 0   (b) 7.99×10^7 N/C (outward)
    (c) 0   (d) 7.34×10^6 N/C (outward)
55. 3.55×10^5 N·m^2/C
57. 4.4×10^5 N/C
59. (a) 10.9 nC   (b) 5.44×10^{-3} N
61. $\sim 10^{-7}$ C
63. (a) 1.00×10^3 N/C   (b) 3.37×10^{-8} s   (c) accelerate at 1.76×10^{14} m/s^2 in the direction opposite that of the electric field

Chapter 16

QUICK QUIZZES

1. (b)
2. (b), (d)
3. (d)
4. (c)
5. (a)
6. (c)
7. (a) C decreases.   (b) Q stays the same.   (c) E stays the same.   (d) ΔV increases.   (e) The energy stored increases.
8. (a) C increases.   (b) Q increases.   (c) E stays the same.   (d) ΔV remains the same.   (e) The energy stored increases.
9. (a)

CONCEPTUAL QUESTIONS

1. (a) The proton moves in a straight line with constant acceleration in the direction of the electric field.   (b) As its velocity increases, its kinetic energy increases and the electric potential energy associated with the proton decreases.

3. The work done in pulling the capacitor plates farther apart is transferred into additional electric energy stored in the capacitor. The charge is constant and the capacitance decreases, but the potential difference between the plates increases, which results in an increase in the stored electric energy.

5. If the power line makes electrical contact with the metal of the car, it will raise the potential of the car to 20 kV. It will also raise the potential of your body to 20 kV, because you are in contact with the car. In itself, this is not a problem. If you step out of the car, however, your body at 20 kV will make contact with the ground, which is at zero volts. As a result, a current will pass through your body and you will likely be injured. Thus, it is best to stay in the car until help arrives.

7. If two points on a conducting object were at different potentials, then free charges in the object would move and we would not have static conditions, in contradiction to the initial assumption. (Free positive charges would migrate from locations of higher to locations of lower potential. Free electrons would rapidly move from locations of lower to locations of higher potential.) All of the charges would continue to move until the potential became equal everywhere in the conductor.

9. The capacitor often remains charged long after the voltage source is disconnected. This residual charge can be lethal. The capacitor can be safely handled after discharging the plates by short-circuiting the device with a conductor, such as a screwdriver with an insulating handle.

11. Field lines represent the direction of the electric force on a positive test charge. If electric field lines were to cross, then, at the point of crossing, there would be an ambiguity regarding the direction of the force on the test charge, because there would be two possible forces there. Thus, electric field lines cannot cross. It is possible for equipotential surfaces to cross. (However, equipotential surfaces at different potentials cannot intersect.) For example, suppose two identical positive charges are at diagonally opposite corners of a square and two negative charges of equal magnitude are at the other two corners. Then the planes perpendicular to the sides of the square at their midpoints are equipotential surfaces. These two planes cross each other at the line perpendicular to the square at its center.

13. You should use a dielectric-filled capacitor whose dielectric constant is very large. Further, you should make the dielectric as thin as possible, keeping in mind that dielectric breakdown must also be considered.

15. (a) ii   (b) i

17. It would make no difference at all. An electron volt is the kinetic energy gained by an electron in being accelerated through a potential difference of 1 V. A proton accelerated through 1 V would have the same kinetic energy, because it carries the same charge as the electron (except for the sign). The proton would be moving in the opposite direction and more slowly after accelerating through 1 V, due to its opposite charge and its larger mass, but it would still gain 1 electron volt, or 1 proton volt, of kinetic energy.

PROBLEMS

1. (a) 6.40×10^{-19} J (b) -6.40×10^{-19} J (c) -4.00 V
3. 1.4×10^{-20} J
5. 1.7×10^{6} N/C
7. (a) 1.13×10^{5} N/C (b) 1.80×10^{-14} N
   (c) 4.38×10^{-17} J
9. (a) 0.500 m (b) 0.250 m
11. (a) 1.44×10^{-7} V (b) -7.19×10^{-8} V
13. (a) 2.67×10^{6} V (b) 2.13×10^{6} V
15. (a) 103 V (b) -3.85×10^{-7} J; positive work must be done to separate the charges.
17. -11.0 kV
19. 2.74×10^{-14} m
21. 0.719 m, 1.44 m, 2.88 m. No. The equipotentials are not uniformly spaced. Instead, the radius of an equipotenial is inversely proportional to the potential.
23. (a) 1.1×10^{-8} F (b) 27 C
25. (a) 11.1 kV/m toward the negative plate (b) 3.74 pF
   (c) 74.7 pC and -74.7 pC
27. (a) 90.4 V (b) 9.04×10^{4} V/m
29. (a) 13.3 μC on each (b) 20.0 μC, 40.0 μC
31. (a) 2.00 μF (b) $Q_3 = 24.0$ μC, $Q_4 = 16.0$ μC,
   $Q_2 = 8.00$ μC, $(\Delta V)_2 = (\Delta V)_4 = 4.00$ V, $(\Delta V)_3 = 8.00$ V
33. (a) 5.96 μF (b) $Q_{20} = 89.5$ μC, $Q_6 = 63.2$ μC,
   $Q_3 = Q_{15} = 26.3$ μC
35. $Q_1 = 16.0$ μC, $Q_5 = 80.0$ μC, $Q_8 = 64.0$ μC,
   $Q_4 = 32.0$ μC
37. (a) $Q_{25} = 1.25$ mC, $Q_{40} = 2.00$ mC (b) $Q'_{25} = 288$ μC,
   $Q'_{40} = 462$ μC, $\Delta V = 11.5$ V
39. $Q'_1 = 3.33$ μC, $Q'_2 = 6.67$ μC
41. 83.6 μC
43. 2.55×10^{-11} J
45. 3.2×10^{10} J
47. $\kappa = 4.0$
49. (a) 8.13 nF (b) 2.40 kV
51. (a) volume 9.09×10^{-16} m^3, area 4.54×10^{-10} m^2
   (b) 2.01×10^{-13} F (c) 2.01×10^{-14} C,
   1.26×10^{5} electronic charges
55. 4.29 μF
57. 6.25 μF
59. 4.47 kV
61. 0.75 mC on C_1, 0.25 mC on C_2
65. 50 N

Chapter 17

QUICK QUIZZES

1. (d)
2. (b)
3. (c), (d)
4. (b)
5. (b)
6. (a)
7. (b)
8. (a)

CONCEPTUAL QUESTIONS

1. Charge. Because an ampere is a unit of current (1 A = 1 C/s) and an hour is a unit of time (1 h = 3 600 s), then 1 A · h = 3 600 C.
3. The gravitational force pulling the electron to the bottom of a piece of metal is much smaller than the electrical repulsion pushing the electrons apart. Thus, free electrons stay distributed throughout the metal. The concept of charges residing on the surface of a metal is true for a metal with an excess charge. The number of free electrons in an electrically neutral piece of metal is the same as the number of positive ions—the metal has zero net charge.
5. A voltage is not something that "surges through" a completed circuit. A voltage is a potential difference that is applied across a device or a circuit. It would be more correct to say "1 ampere of electricity surged through the victim's body." Although this amount of current would have disastrous results on the human body, a value of 1 (ampere) doesn't sound as exciting for a newspaper article as 10 000 (volts). Another possibility is to write "10 000 volts of electricity were applied across the victim's body," which still doesn't sound quite as exciting.
7. We would conclude that the conductor is nonohmic.
9. The shape, dimensions, and the resistivity affect the resistance of a conductor. Because temperature and impurities affect the conductor's resistivity, these factors also affect resistance.
11. The radius of wire B is the square root of three times the radius of wire A. Therefore the cross-sectional area of B three times larger than that of A.
13. The drift velocity might increase steadily as time goes on, because collisions between electrons and atoms in the wire would be essentially nonexistent and the conduction electrons would move with constant acceleration. The current would rise steadily without bound also, because I is proportional to the drift velocity.
15. Once the switch is closed, the line voltage is applied across the bulb. As the voltage is applied across the cold filament when it is first turned on, the resistance of the filament is low, the current is high, and a relatively large amount of power is delivered to the bulb. As the filament warms, its resistance rises and the current decreases. As a result, the power delivered to the bulb decreases. The large current spike at the beginning of the bulb's operation is the reason that lightbulbs often fail just after they are turned on.

PROBLEMS

1. 3.00×10^{20} electrons move past in the direction opposite to the current.
3. 2.00 C
5. 1.05 mA
7. 27 yr
9. (a) n is unaffected (b) v_d is doubled
11. 32 V is 200 times larger than 0.16 V
13. 0.17 mm
15. (a) 30 Ω (b) 4.7×10^{-4} $\Omega \cdot$ m
17. silver ($\rho = 1.59 \times 10^{-8}$ $\Omega \cdot$ m)
19. 256 Ω
21. 1.98 A
23. 26 mA
25. (a) 5.89×10^{-2} Ω (b) 5.45×10^{-2} Ω
27. (a) 3.0 A (b) 2.9 A
29. (a) 1.2 Ω (b) 8.0×10^{-4} (a 0.080% increase)
31. 5.00 A, 24.0 Ω
33. 18 bulbs
35. 11.2 min
37. 34.4 Ω
39. 1.6 cm
41. 295 metric tons/h

43. 26 cents
45. 23 cents
47. $1.2
49. 1.1 km
51. $1.47 \times 10^{-6}\ \Omega \cdot m$; differs by 2.0% from value in Table 17.1
53. (a) $3.06 (b) No. The circuit must be able to handle at least 26 A.
55. (a) 667 A (b) 50.0 km
57. $3.77 \times 10^{28}/m^3$
59. (a) 144 Ω (b) 26 m (c) To fit the required length into a small space. (d) 25 m
61. 37 MΩ
63. 0.48 kg/s
65. (a) $2.6 \times 10^{-5}\ \Omega$ (b) 76 kg
67. (a) 470 W (b) 1.60 mm or more (c) 2.93 mm or more

Chapter 18

QUICK QUIZZES

1. (a), (d)
2. (b)
3. (a)
4. *Parallel*: (a) unchanged (b) unchanged (c) increase (d) decrease
5. *Series*: (a) decrease (b) decrease (c) decrease (d) increase
6. (c)

CONCEPTUAL QUESTIONS

1. No. When a battery serves as a source and supplies current to a circuit, the conventional current flows through the battery from the negative terminal to the positive one. However, when a source having a larger emf than the battery is used to charge the battery, the conventional current is forced to flow through the battery from the positive terminal to the negative one.
3. The total amount of energy delivered by the battery will be less than W. Recall that a battery can be considered an ideal, resistanceless battery in series with the internal resistance. When the battery is being charged, the energy delivered to it includes the energy necessary to charge the ideal battery, plus the energy that goes into raising the temperature of the battery due to I^2r heating in the internal resistance. This latter energy is not available during discharge of the battery, when part of the reduced available energy again transforms into internal energy in the internal resistance, further reducing the available energy below W.
5. The starter in the automobile draws a relatively large current from the battery. This large current causes a significant voltage drop across the internal resistance of the battery. As a result, the terminal voltage of the battery is reduced, and the headlights dim accordingly.
7. An electrical appliance has a given resistance. Thus, when it is attached to a power source with a known potential difference, a definite current will be drawn, and the device can therefore be labeled with both the voltage and the current. Batteries, however, can be applied to a number of devices. Each device will have a different resistance, so the current will vary with the device. As a result, only the voltage of the battery can be specified.

9. Connecting batteries in parallel does not increase the emf. A high-current device connected to two batteries in parallel can draw currents from both batteries. Thus, connecting the batteries in parallel increases the possible current output and, therefore, the possible power output.
11. The lightbulb will glow for a very short while as the capacitor is being charged. Once the capacitor is almost totally charged, the current in the circuit will be nearly zero and the bulb will not glow.
13. The bird is resting on a wire of fixed potential. In order to be electrocuted, a large potential difference is required between the bird's feet. The potential difference between the bird's feet is too small to harm the bird.
15. The junction rule is a statement of conservation of charge. It says that the amount of charge that enters a junction in some time interval must equal the charge that leaves the junction in that time interval. The loop rule is a statement of conservation of energy. It says that the increases and decreases in potential around a closed loop in a circuit must add to zero.
17. A few of the factors involved are as follows: the conductivity of the string (is it wet or dry?); how well you are insulated from ground (are you wearing thick rubber- or leather-soled shoes?); the magnitude of the potential difference between you and the kite; and the type and condition of the soil under your feet.
19. She will not be electrocuted if she holds onto only one high-voltage wire, because she is not completing a circuit. There is no potential difference across her body as long as she clings to only one wire. However, she should release the wire immediately once it breaks, because she will become part of a closed circuit when she reaches the ground or comes into contact with another object.
21. (a) The intensity of each lamp increases because lamp C is short circuited and there is current (which increases) only in lamps A and B. (b) The intensity of lamp C goes to zero because the current in this branch goes to zero. (c) The current in the circuit increases because the total resistance decreases from $3R$ (with the switch open) to $2R$ (after the switch is closed). (d) The voltage drop across lamps A and B increases, while the voltage drop across lamp C becomes zero. (e) The power dissipated increases from $\varepsilon^2/3R$ (with the switch open) to $\varepsilon^2/2R$ (after the switch is closed).
23. The statement is false. The current in each bulb is the same, because they are connected in series. The bulb that glows brightest has the larger resistance and hence dissipates more power

PROBLEMS

1. 4.92 Ω
3. 73.8 W. Your circuit diagram will consist of two 0.800-Ω resistors in series with the 192-Ω resistance of the bulb.
5. (a) 17.1 Ω (b) 1.99 A for 4.00 Ω and 9.00 Ω, 1.17 A for 7.00 Ω, 0.818 A for 10.0 Ω
7. $2.5R$
9. (a) 0.227 A (b) 5.68 V
11. 55 Ω
13. 0.43 A
15. (a) Connect two 50-Ω resistors in parallel, and then connect this combination in series with a 20-Ω resistor.
(b) Connect two 50-Ω resistors in parallel, connect two 20-Ω resistors in parallel, and then connect these two combinations in series with each other.

17. 0.846 A downwards in the 8.00-Ω resistor; 0.462 A downwards in the middle branch; 1.31 A upwards in the right-hand branch
19. (a) 3.00 mA (b) -19.0 V (c) 4.50 V
21. 10.7 V
23. (a) 0.385 mA, 3.08 mA, 2.69 mA
    (b) 69.2 V, with c at the higher potential
25. $I_1 = 3.5$ A, $I_2 = 2.5$ A, $I_3 = 1.0$ A
27. $I_{30} = 0.353$ A, $I_5 = 0.118$ A, $I_{20} = 0.471$ A
29. $\Delta V_2 = 3.05$ V, $\Delta V_3 = 4.57$ V, $\Delta V_4 = 7.38$ V, $\Delta V_5 = 1.62$ V
31. (a) 12 s (b) 1.2×10^{-4} C
33. 1.3×10^{-4} C
35. 0.982 s
37. (a) heater, 10.8 A; toaster, 8.33 A; grill, 12.5 A
    (b) $I_{total} = 31.6$ A, so a 30-A breaker is insufficient.
39. (a) 6.25 A (b) 750 W
41. (a) 1.2×10^{-9} C, 7.3×10^9 K$^+$ ions. Not large, only $1e/290$ A^2
    (b) 1.7×10^{-9} C, 1.0×10^{10} Na$^+$ ions (c) 0.83 μA
    (d) 7.5×10^{-12} J
43. 11 nW
45. 7.5 Ω
47. (a) 15 Ω
    (b) $I_1 = 1.0$ A, $I_2 = I_3 = 0.50$ A, $I_4 = 0.30$ A, and $I_5 = 0.20$ A
    (c) $(\Delta V)_{ac} = 6.0$ V, $(\Delta V)_{ce} = 1.2$ V, $(\Delta V)_{ed} = (\Delta V)_{fd} = 1.8$ V, $(\Delta V)_{cd} = 3.0$ V, $(\Delta V)_{db} = 6.0$ V
    (d) $\mathscr{P}_{ac} = 6.0$ W, $\mathscr{P}_{ce} = 0.60$ W, $\mathscr{P}_{ed} = 0.54$ W, $\mathscr{P}_{fd} = 0.36$ W, $\mathscr{P}_{cd} = 1.5$ W, $\mathscr{P}_{db} = 6.0$ W
49. (a) 12.4 V (b) 9.65 V
51. $I_1 = 0$, $I_2 = I_3 = 0.50$ A,
53. 112 V, 0.200 Ω
55. (a) $R_x = R_2 - \frac{1}{4}R_1$
    (b) $R_x = 2.8$ Ω (inadequate grounding)
59. $\mathscr{P} = \dfrac{(144 \text{ V}^2)R}{(R + 10.0 \text{ Ω})^2}$

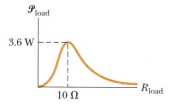

61. (a) 5.68 V (b) 0.227 A
63. 0.395 A; 1.50 V

Chapter 19

QUICK QUIZZES

1. (b)
2. (c)
3. (c)
4. (a)
5. (b)

CONCEPTUAL QUESTIONS

1. The set should be oriented such that the beam is moving either toward the east or toward the west.
3. The proton moves in a circular path upwards on the page. After completing half a circle, it exits the field and moves in a straight-line path back in the direction from whence

it came. An electron will behave similarly, but the direction of traversal of the circle is downward, and the radius of the circular path is smaller.
5. The magnetic force on a moving charged particle is always perpendicular to the particle's direction of motion. There is no magnetic force on the charge when it moves parallel to the direction of the magnetic field. However, the force on a charged particle moving in an electric field is never zero and is always parallel to the direction of the field. Therefore, by projecting the charged particle in different directions, it is possible to determine the nature of the field.
7. The magnetic field produces a magnetic force on the electrons moving toward the screen that produce the image. This magnetic force deflects the electrons to regions on the screen other than the ones to which they are supposed to go. The result is a distorted image.
9. Such levitation could never occur. At the North Pole, where Earth's magnetic field is directed downward, toward the equivalent of a buried south pole, a coffin would be repelled if its south magnetic pole were directed downward. However, equilibrium would be only transitory, as any slight disturbance would upset the balance between the magnetic force and the gravitational force.
11. If you were moving along with the electrons, you would measure a zero current for the electrons, so they would not produce a magnetic field according to your observations. However, the fixed positive charges in the metal would now be moving backwards relative to you, creating a current equivalent to the forward motion of the electrons when you were stationary. Thus, you would measure the same magnetic field as when you were stationary, but it would be due to the positive charges presumed to be moving from your point of view.
13. A compass does not detect currents in wires near light switches, for two reasons. The first is that, because the cable to the light switch contains two wires, one carrying current to the switch and the other carrying it away from the switch, the net magnetic field would be very small and would fall off rapidly with increasing distance. The second reason is that the current is alternating at 60 Hz. As a result, the magnetic field is oscillating at 60 Hz also. This frequency would be too fast for the compass to follow, so the effect on the compass reading would average to zero.
15. The levitating wire is stable with respect to vertical motion: If it is displaced upward, the repulsive force weakens, and the wire drops back down. By contrast, if it drops lower, the repulsive force increases, and it moves back up. The wire is not stable, however, with respect to lateral movement: If it moves away from the vertical position directly over the lower wire, the repulsive force will have a sideways component that will push the wire away.

   In the case of the attracting wires, the hanging wire is not stable with respect to vertical movement. If it rises, the attractive force increases, and the wire moves even closer to the upper wire. If the hanging wire falls, the attractive force weakens, and the wire falls farther. If the wire moves to the right, it moves farther from the upper wire and the attractive force decreases. Although there is a restoring force component pulling it back to the left, the vertical force component is not strong enough to hold the wire up, and it falls.
17. Each coil of the Slinky® will become a magnet, because a coil acts as a current loop. The sense of rotation of the

current is the same in all coils, so each coil becomes a magnet with the same orientation of poles. Thus, all of the coils attract, and the Slinky® will compress.

19. There is no net force on the wires, but there is a torque. To understand this distinction, imagine a fixed vertical wire and a free horizontal wire (see the figure below). The vertical wire carries an upward current and creates a magnetic field that circles the vertical wire, itself. To the right, the magnetic field of the vertical wire points into the page, while on the left side it points out of the page, as indicated. Each segment of the horizontal wire (of length ℓ) carries current that interacts with the magnetic field according to the equation $F = BI\ell \sin \theta$. Apply the right-hand rule on the right side: point the fingers of your right hand in the direction of the horizontal current and curl them into the page in the direction of the magnetic field. Your thumb points downward, the direction of the force on the right side of the wire. Repeating the process on the left side gives a force upward on the left side of the wire. The two forces are equal in magnitude and opposite in direction, so the net force is zero, but they create a net torque around the point where the wires cross.

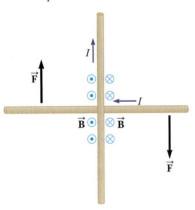

21. (a) The field is into the page. (b) The beam would deflect upwards.

PROBLEMS

1. (a) horizontal and due east (b) horizontal and 30° N of E (c) horizontal and due east (d) zero force
3. (a) into the page (b) toward the right (c) toward the bottom of the page
5. $F_g = 8.93 \times 10^{-30}$ N (downward), $F_e = 1.60 \times 10^{-17}$ N (upward), $F_m = 4.80 \times 10^{-17}$ N (downward)
7. 2.83×10^7 m/s west
9. 0.021 T in the $-y$-direction
11. 8.0×10^{-3} T in the $+z$-direction
13. (a) into the page (b) toward the right (c) toward the bottom of the page
15. 7.50 N
17. 0.131 T (downward)
19. 0.20 T directed out of the page
21. ab: 0, bc: 0.040 0 N in $-x$-direction, cd: 0.040 0 N in the $-z$-direction da: 0.056 6 N parallel to the xz-plane and at 45° to both the $+x$- and the $+z$-directions
23. 9.05×10^{-4} N·m, tending to make the left-hand side of the loop move toward you and the right-hand side move away.
25. (a) 3.97° (b) 3.39×10^{-3} N·m
27. 6.56×10^{-2} T
31. 1.77 cm

33. $r = 3R/4$
35. 20.0 μT
37. 2.4 mm
39. 20.0 μT toward bottom of page
41. 0.167 μT out of the page
43. (a) 4.00 m (b) 7.50 nT (c) 1.26 m (d) zero
45. 4.5 mm
47. 31.8 mA
49. 2.26×10^{-4} N away from the center, zero torque
51. 1.7 N·m
53. (a) 0.500 μT out of the page (b) 3.89 μT parallel to xy-plane and at 59.0° clockwise from $+x$-direction
55. 2.13 cm
57. (a) 1.33 m/s (b) the sign of the emf is independent of the charge
59. 1.41×10^{-6} N
61. 13.0 μT toward the bottom of the page
63. 53 μT toward the bottom of the page, 20 μT toward the bottom of the page, and 0
65. (a) -8.00×10^{-21} kg·m/s (b) 8.90°
67. 1.29 kW
69. (a) 12.0 cm to the left of wire 1 (b) 2.40 A, downward

Chapter 20

QUICK QUIZZES

1. b, c, a
2. (a)
3. (b)
4. (c)
5. (b)

CONCEPTUAL QUESTIONS

1. According to Faraday's law, an emf is induced in a wire loop if the magnetic flux through the loop changes with time. In this situation, an emf can be induced either by rotating the loop around an arbitrary axis or by changing the shape of the loop.
3. As the spacecraft moves through space, it is apparently moving from a region of one magnetic field strength to a region of a different magnetic field strength. The changing magnetic field through the coil induces an emf and a corresponding current in the coil.
5. If the bar were moving to the left, the magnetic force on the negative charges in the bar would be upward, causing an accumulation of negative charge on the top and positive charges at the bottom. Hence, the electric field in the bar would be upward, as well.
7. If, for any reason, the magnetic field should change rapidly, a large emf could be induced in the bracelet. If the bracelet were not a continuous band, this emf would cause high-voltage arcs to occur at any gap in the band. If the bracelet were a continuous band, the induced emf would produce a large induced current and result in resistance heating of the bracelet.
11. As the aluminum plate moves into the field, eddy currents are induced in the metal by the changing magnetic field at the plate. The magnetic field of the electromagnet interacts with this current, producing a retarding force on the plate that slows it down. In a similar fashion, as the plate leaves the magnetic field, a current is induced, and once again there is an upward force to slow the plate.

13. The energy stored in an inductor carrying a current I is equal to $PE_L = (1/2)LI^2$. Therefore, doubling the current will quadruple the energy stored in the inductor.

15. If an external battery is acting to increase the current in the inductor, an emf is induced in a direction to oppose the increase of current. Likewise, if we attempt to reduce the current in the inductor, the emf that is set up tends to support the current. Thus, the induced emf always acts to oppose the change occurring in the circuit, or it acts in the "back" direction to the change.

17. (a) clockwise (b) The net force exerted on the bar must be zero because it moves at constant speed. The component of the gravitational force down the incline is balanced by a component of the magnetic force up the incline.

19. from left to right

PROBLEMS

1. $5.9 \times 10^{-2}\,\text{T}\cdot\text{m}^2$
3. $7.71 \times 10^{-1}\,\text{T}\cdot\text{m}^2$
5. (a) $\Phi_{B,\text{net}} = 0$ (b) 0
7. (a) $3.1 \times 10^{-3}\,\text{T}\cdot\text{m}^2$ (b) $\Phi_{B,\text{net}} = 0$
9. 0.18 T
11. 94 mV
13. 2.7 T/s
15. (a) $4.0 \times 10^{-6}\,\text{T}\cdot\text{m}^2$ (b) 18 μV
17. 10.2 μV
19. 0.763 V
21. (a) toward the east (b) 4.58×10^{-4} V
23. (a) from left to right (b) from right to left
25. (a) $F = N^2B^2w^2v/R$ to the left (b) 0
   (c) $F = N^2B^2w^2v/R$ to the left
27. into the page
29. (a) from right to left (b) from right to left (c) from left to right (d) from left to right
31. 1.9×10^{-11} V
33. (a) 18.1 μV (b) 0
35. (a) 60 V (b) 57 V (c) 0.13 s
37. 20 mV
39. (a) 2.0 mH (b) 38 A/s
43. 12 mH
45. 1.92 Ω
47. 0.140 J
49. (a) 18 J (b) 7.2 J
51. negative ($V_a < V_b$)
53. (a) 20.0 ms (b) 37.9 V (c) 1.52 mV (d) 51.8 mA
55. 1.20 μC
57. (a) 0.500 A (b) 2.00 W (c) 2.00 W
59. 115 kV
61. (a) 0.157 mV (end B is positive) (b) 5.89 mV (end A is positive)
63. (a) 9.00 A (b) 10.8 N (c) b is at the higher potential (d) No
65. $v_t = \dfrac{mgR}{B^2\ell^2}$

Chapter 21

QUICK QUIZZES

1. (c)
2. (b)
3. (b)
4. (b), (c)
5. (b), (d)

CONCEPTUAL QUESTIONS

1. For best reception, the length of the antenna should be parallel to the orientation of the oscillating electric field. Because of atmospheric variations and reflections of the wave before it arrives at your location, the orientation of this field may be in different directions for different stations.

3. The primary coil of the transformer is an inductor. When an AC voltage is applied, the back emf due to the inductance will limit the current in the coil. If DC voltage is applied, there is no back emf, and the current can rise to a higher value. It is possible that this increased current will deliver so much energy to the resistance in the coil that its temperature rises to the point at which insulation on the wire can burn.

5. An antenna that is a conducting line responds to the electric field of the electromagnetic wave—the oscillating electric field causes an electric force on electrons in the wire along its length. The movement of electrons along the wire is detected as a current by the radio and is amplified. Thus, a line antenna must have the same orientation as the broadcast antenna. A loop antenna responds to the magnetic field in the radio wave. The varying magnetic field induces a varying current in the loop (by Faraday's law), and this signal is amplified. The loop should be in the vertical plane containing the line of sight to the broadcast antenna, so the magnetic field lines go through the area of the loop.

7. The flashing of the light according to Morse code is a drastic amplitude modulation—the amplitude is changing from a maximum to zero. In this sense, it is similar to the on-and-off binary code used in computers and compact disks. The carrier frequency is that of the light, on the order of 10^{14} HZ. The frequency of the signal depends on the skill of the signal operator, but it is on the order of a single hertz, as the light is flashed on and off. The broadcasting antenna for this modulated signal is the filament of the lightbulb in the signal source. The receiving antenna is the eye.

9. The sail should be as reflective as possible, so that the maximum momentum is transferred to the sail from the reflection of sunlight.

11. Suppose the extraterrestrial looks around your kitchen. Lightbulbs and the toaster glow brightly in the infrared. Somewhat fainter are the back of the refrigerator and the back of the television set, while the television screen is dark. The pipes under the sink show the same weak glow as the walls, until you turn on the faucets. Then the pipe on the right gets darker and that on the left develops a gleam that quickly runs up along its length. The food on the plates shines, as does human skin, the same color for all races. Clothing is dark as a rule, but your seat and the chair seat glow alike after you stand up. Your face appears lit from within, like a jack-o'-lantern; your nostrils and the openings of your ear canals are bright; brighter still are the pupils of your eyes.

13. Radio waves move at the speed of light. They can travel around the curved surface of the Earth, bouncing between the ground and the ionosphere, which has an altitude that is small compared with the radius of the Earth. The distance across the lower 48 states is approximately 5 000 km, requiring a travel time that is equal to $(5 \times 10^6\,\text{m})/(3 \times 10^8\,\text{m/s}) \sim 10^{-2}$ s. Likewise, radio waves take only 0.07 s to travel halfway around the Earth. In

other words, a speech can be heard on the other side of the world (in the form of radio waves) before it is heard at the back of the room (in the form of sound waves).

15. No. The wire will emit electromagnetic waves only if the current varies in time. The radiation is the result of accelerating charges, which can occur only when the current is not constant.

17. The resonance frequency is determined by the inductance and the capacitance in the circuit. If both L and C are doubled, the resonance frequency is reduced by a factor of two.

19. It is far more economical to transmit power at a high voltage than at a low voltage because the I^2R loss on the transmission line is significantly lower at high voltage. Transmitting power at high voltage permits the use of step-down transformers to make "low" voltages and high currents available to the end user.

21. No. A voltage is induced in the secondary coil only if the flux through the core changes with time.

PROBLEMS

1. (a) 141 V (b) 20.0 A (c) 28.3 A (d) 2.00 kW
3. 70.7 V, 2.95 A
5. 6.76 W
9. 4.0×10^2 Hz
11. 17 μF
15. 3.14 A
17. 0.450 T·m^2
19. (a) 0.361 A (b) 18.1 V (c) 23.9 V (d) $-53.0°$
21. (a) 1.4 kΩ (b) 0.10 A (c) 51° (d) voltage leads current
23. (a) 89.6 V (b) 108 V
25. 1.88 V
27. (a) 103 V (b) 150 V (c) 127 V (d) 23.6 V
29. (a) 208 Ω (b) 40.0 Ω (c) 0.541 H
31. (a) $1.8 \times 10^2 \Omega$ (b) 0.71 H
33. 2.29 μH
35. C_{min} = 4.9 nF, C_{max} = 51 nF
37. 0.242 J
39. 0.18% is lost
41. (a) 1.1×10^3 kW (b) 3.1×10^2 A (c) 8.3×10^3 A
43. 1 000 km; there will always be better use for tax money.
45. $f_{red} = 4.55 \times 10^{14}$ Hz, $f_{IR} = 3.19 \times 10^{14}$ Hz,
    $E_{max,f}/E_{max,i}$ = 0.57
47. 2.94×10^8 m/s
49. $E_{max} = 1.01 \times 10^3$ V/m, $B_{max} = 3.35 \times 10^{-6}$ T
51. (a) 188 m to 556 m (b) 2.78 m to 3.4 m
53. 5.2×10^{13} Hz, 5.8 μm
55. $4.299\ 999\ 84 \times 10^{14}$ Hz; -1.6×10^7 Hz
    (the frequency decreases)
57. 99.6 mH
59. 1.7 cents
61. (a) resistor and inductor (b) R = 10 Ω, L = 30 mH
63. (a) 6.7×10^{-16} T (b) 5.3×10^{-17} W/m^2
    (c) 1.7×10^{-14} W
65. (a) 0.536 N (b) 8.93×10^{-5} m/s^2 (c) 33.9 days
67. 4.47×10^{-9} J

Chapter 22

QUICK QUIZZES

1. (a)
2. Beams 2 and 4 are reflected; beams 3 and 5 are refracted.
3. (b)
4. (c)

CONCEPTUAL QUESTIONS

1. Sound radiated upward at an acute angle with the horizontal is bent back toward Earth by refraction. This means that the sound can reach the listener by this path as well as by a direct path. Thus, the sound is louder.

3. The color will not change, for two reasons. First, despite the popular statement that color depends on wavelength, it actually depends on the *frequency* of the light, which does not change under water. Second, when the light enters the eye, it travels through the fluid within. Thus, even if color did depend on wavelength, the important wavelength is that of the light in the ocular fluid, which does not depend on the medium through which the light traveled to reach the eye.

5. (a) Away from the normal (b) increases (c) remains the same

7. No, the information in the catalog is incorrect. The index of refraction is given by $n = c/v$, where c is the speed of light in a vacuum and v is the speed of light in the material. Because light travels faster in a vacuum than in any other material, it is impossible for the index of refraction of any material to have a value less than 1.

9. There is no dependence of the angle of reflection on wavelength, because the light does not enter deeply into the material during reflection—it reflects from the surface.

11. On the one hand, a ball covered with mirrors sparkles by reflecting light from its surface. On the other hand, a faceted diamond lets in light at the top, reflects it by total internal reflection in the bottom half, and sends the light out through the top again. Because of its high index of refraction, the critical angle for diamond in air for total internal reflection, namely $\theta_c = \sin^{-1}(n_{air}/n_{diamond})$, is small. Thus, light rays enter through a large area and exit through a very small area with a much higher intensity. When a diamond is immersed in carbon disulfide, the critical angle is increased to $\theta_c = \sin^{-1}(n_{carbon\ disulfide}/n_{diamond})$. As a result, the light is emitted from the diamond over a larger area and appears less intense.

13. The index of refraction of water is 1.333, quite different from that of air, which has an index of refraction of about 1. The boundary between the air and water is therefore easy to detect, because of the differing diffraction effects above and below the boundary. (Try looking at a glass half full of water.) The index of refraction of liquid helium, however, happens to be much closer to that of air. Consequently, the defractive differences above and below the helium-air boundary are harder to see.

15. The diamond acts like a prism, dispersing the light into its spectral components. Different colors are observed as a consequence of the manner in which the index of refraction varies with the wavelength.

17. Light travels through a vacuum at a speed of 3×10^8 m/s. Thus, an image we see from a distant star or galaxy must have been generated some time ago. For example, the star Altair is 16 lightyears away; if we look at an image of Altair today, we know only what Altair looked like 16 years ago. This may not initially seem significant; however, astronomers who look at other galaxies can get an idea of what galaxies looked like when they were much younger. Thus, it does make sense to speak of "looking backward in time."

PROBLEMS

1. 3.00×10^8 m/s
3. 114 rad/s for a maximum intensity of returning light
5. (b) 3.000×10^8 m/s
7. 19.5° above the horizontal
9. (a) 1.52 (b) 417 nm (c) 4.74×10^{14} Hz
   (d) 1.98×10^8 m/s
11. (a) 584 nm (b) 1.12
13. 111°
15. (a) 1.559×10^8 m/s (b) 329.1 nm (c) 4.738×10^{14} Hz
17. five times from the right-hand mirror and six times from the left
19. 0.388 cm
21. $\theta = 30.4°$, $\theta' = 22.3°$
23. 6.39 ns
25. $\theta = \tan^{-1}(n_g)$
27. 3.39 m
29. $\theta_{red} = 48.22°$, $\theta_{blue} = 47.79°$
31. (a) $\theta_{1i} = 30°$, $\theta_{1r} = 19°$, $\theta_{2i} = 41°$, $\theta_{2r} = 77°$
    (b) First surface: $\theta_{reflection} = 30°$;
    second surface: $\theta_{reflection} = 41°$
33. (a) 31.3° (b) 44.2° (c) 49.8°
35. (a) 33.4° (b) 53.4°
37. (a) 40.8° (b) 60.6°
39. 1.000 08
41. (a) 10.7° (b) air (c) Sound falling on the wall from most directions is 100% reflected.
43. 27.5°
45. 22.0°
47. (a) 53.1° (b) $\geq 38.7°$
49. (a) 38.5° (b) ≥ 1.44
53. 24.7°
55. 1.93
59. $\theta = \sin^{-1}\left(\sqrt{n^2 - 1}\,\sin\phi - \cos\phi\right)$
61. (a) 1.20 (b) 3.40 ns

Chapter 23

QUICK QUIZZES

1. At C.
2. (c)
3. (a) False (b) False (c) True
4. (b)
5. An infinite number
6. (a) False (b) True (c) False

CONCEPTUAL QUESTIONS

1. You will not be able to focus your eyes on both the picture and your image at the same time. To focus on the picture, you must adjust your eyes so that an object several centimeters away (the picture) is in focus. Thus, you are focusing on the mirror surface. But, your image in the mirror is as far behind the mirror as you are in front of it. Thus, you must focus your eyes beyond the mirror, twice as far away as the picture to bring the image into focus.

3. A single flat mirror forms a virtual image of an object due to two factors. First, the light rays from the object are necessarily diverging from the object, and second, the lack of curvature of the flat mirror cannot convert diverging rays to converging rays. If another optical element is first used to cause light rays to converge, then the flat mirror can be placed in the region in which the converging rays are present, and it will change the direction of the rays so that the real image is formed at a different location. For example, if a real image is formed by a convex lens, and the flat mirror is placed between the lens and the image position, the image formed by the mirror will be real.

5. The ultrasonic range finder sends out a sound wave and measures the time for the echo to return. Using this information, the camera calculates the distance to the subject and sets the camera lens. When the camera is facing a mirror, the ultrasonic signal reflects from the mirror surface and the camera adjusts its focus so that the mirror surface is at the correct focusing distance from the camera. But your image in the mirror is twice this distance from the camera, so it is blurry.

7. Light rays diverge from the position of a virtual image just as they do from an actual object. Thus, a virtual image can be as easily photographed as any object can. Of course, the camera would have to be placed near the axis of the lens or mirror in order to intercept the light rays.

9. We consider the two trees to be two separate objects. The far tree is an object that is farther from the lens than the near tree. Thus, the image of the far tree will be closer to the lens than the image of the near tree. The screen must be moved closer to the lens to put the far tree in focus.

11. If a converging lens is placed in a liquid having an index of refraction larger than that of the lens material, the direction of refractions at the lens surfaces will be reversed, and the lens will diverge light. A mirror depends only on reflection which is independent of the surrounding material, so a converging mirror will be converging in any liquid.

13. This is a possible scenario. When light crosses a boundary between air and ice, it will refract in the same manner as it does when crossing a boundary of the same shape between air and glass. Thus, a converging lens may be made from ice as well as glass. However, ice is such a strong absorber of infrared radiation that it is unlikely you will be able to start a fire with a small ice lens.

15. The focal length for a mirror is determined by the law of reflection from the mirror surface. The law of reflection is independent of the material of which the mirror is made and of the surrounding medium. Thus, the focal length depends only on the radius of curvature and not on the material. The focal length of a lens depends on the indices of refraction of the lens material and surrounding medium. Thus, the focal length of a lens depends on the lens material.

17. (a) all signs are positive (b) f and p are positive, q is negative

19. (c) the image becomes fuzzy and disappears

PROBLEMS

1. on the order of 10^{-9} s younger
3. 10.0 ft, 30.0 ft, 40.0 ft
5. 0.268 m behind the mirror; virtual, upright, and diminished; $M = 0.026\ 8$
7. (a) 13.3 cm in front of mirror, real, inverted, $M = -0.333$
   (b) 20.0 cm in front of mirror, real, inverted, $M = -1.00$
   (c) No image is formed. Parallel rays leave the mirror.
9. Behind the worshipper, 3.33 m from the deepest point in the niche.
11. 5.00 cm

13. 1.0 m
15. 8.05 cm
17. -20.0 cm
19. (a) concave with focal length $f = 0.83$ m
    (b) Object must be 1.0 m in front of the mirror.
21. 38.2 cm below the upper surface of the ice
23. 3.8 mm
25. $n = 2.00$
27. 20.0 cm
29. (a) 40.0 cm beyond the lens, real, inverted, $M = -1.00$
    (b) No image is formed. Parallel rays leave the lens.
    (c) 20.0 cm in front of the lens, virtual, upright, $M = +2.00$
31. (a) 13.3 cm in front of the lens, virtual, upright, $M = +1/3$
    (b) 10.0 cm in front of the lens, virtual, upright, $M = +1/2$
    (c) 6.67 cm in front of the lens, virtual, upright, $M = +2/3$
33. (a) either 9.63 cm or 3.27 cm (b) 2.10 cm
35. (a) 39.0 mm (b) 39.5 mm
37. at distance $2|f|$ in front of lens
39. 40.0 cm
41. 30.0 cm to the left of the second lens, $M = -3.00$
43. 7.47 cm in front of the second lens; 1.07 cm; virtual, upright
45. from 0.224 m to 18.2 m
47. real image, 5.71 cm in front of the mirror
49. 38.6°
51. 160 cm to the left of the lens, inverted, $M = -0.800$
53. $q = 10.7$ cm
55. 32.0 cm to the right of the second surface (real image)
57. (a) 20.0 cm to the right of the second lens; $M = -6.00$
    (b) inverted
    (c) 6.67 cm to the right of the second lens; $M = -2.00$; inverted
59. (a) 1.99
    (b) 10.0 cm to the left of the lens
    (c) inverted
61. (a) 5.45 m to the left of the lens
    (b) 8.24 m to the left of the lens
    (c) 17.1 m to the left of the lens
    (d) by surrounding the lens with a medium having a refractive index greater than that of the lens material.
63. (a) 263 cm (b) 79.0 cm

Chapter 24

QUICK QUIZZES

1. (c)
2. (b)
3. (b)
4. The compact disc

CONCEPTUAL QUESTIONS

1. You will *not* see an interference pattern from the automobile headlights, for two reasons. The first is that the headlights are not coherent sources and are therefore incapable of producing sustained interference. Also, the headlights are so far apart in comparison to the wavelengths emitted that, even if they were made into coherent sources, the interference maxima and minima would be too closely spaced to be observable.

3. The result of the double slit is to redistribute the energy arriving at the screen. Although there is no energy at the location of a dark fringe, there is four times as much energy at the location of a bright fringe as there would be with only a single narrow slit. The total amount of energy arriving at the screen is twice as much as with a single slit, as it must be according to the law of conservation of energy.

5. One of the materials has a higher index of refraction than water, and the other has a lower index. The material with the higher index will appear black as it approaches zero thickness. There will be a 180° phase change for the light reflected from the upper surface, but no such phase change for the light reflected from the lower surface, because the index of refraction for water on the other side is lower than that of the film. Thus, the two reflections will be out of phase and will interfere destructively. The material with index of refraction lower than water will have a phase change for the light reflected from both the upper and the lower surface, so that the reflections from the zero-thickness film will be back in phase and the film will appear bright.

7. For incidence normal to the film, the extra path length followed by the reflected ray is twice the thickness of the film. For destructive interference, this must be a distance of half a wavelength of the light in the material of the film. For a film in air, no 180° phase change will occur in these reflections, so the thickness of the film must be one-quarter wavelength, which is the same as the condition for constructive interference of reflected light. This means that the transmitted light is a minimum when the reflected light is a maximum, and vice versa.

9. Since the light reflecting at the lower surface of the film undergoes a 180° phase change, while light reflecting from the upper surface of the film does not undergo such a change, the central spot (where the film has near zero thickness) will be dark. If the observed rings are not circular, the curved surface of the lens does not have a true spherical shape.

11. For regional communication at the Earth's surface, radio waves are typically broadcast from currents oscillating in tall vertical towers. These waves have vertical planes of polarization. Light originates from the vibrations of atoms or electronic transitions within atoms, which represent oscillations in all possible directions. Thus, light generally is not polarized.

13. Yes. In order to do this, first measure the radar reflectivity of the metal of your airplane. Then choose a light, durable material that has approximately half the radar reflectivity of the metal in your plane. Measure its index of refraction, and place onto the metal a coating equal in thickness to one-quarter of 3 cm, divided by that index. Sell the plane quick, and then you can sell the supposed enemy new radars operating at 1.5 cm, which the coated metal will reflect with extra-high efficiency.

15. If you wish to perform an interference experiment, you need monochromatic coherent light. To obtain it, you must first pass light from an ordinary source through a prism or diffraction grating to disperse different colors into different directions. Using a single narrow slit, select a single color and make that light diffract to cover both slits for a Young's experiment. The procedure is much simpler with a laser because its output is already monochromatic and coherent.

17. Strictly speaking, the ribs do act as a diffraction grating, but the separation distance of the ribs is so much larger than the wavelength of the x-rays that there are no observable effects.

19. As the edge of the Moon cuts across the light from the star, edge diffraction effects occur. Thus, as the edge of

the Moon moves relative to the star, the observed light from the star proceeds through a series of maxima and minima.

21. Larger. From Brewster's law, $n = \tan \theta_p$, we see that the angle increases as n increases.

PROBLEMS

1. 1.58 cm
3. (a) 2.6 mm (b) 2.62 mm
5. (a) 36.2° (b) 5.08 cm (c) 5.08×10^{14} Hz
7. (a) 55.7 m (b) 124 m
9. 75.0 m
11. 11.3 m
13. 148 m
15. 91.9 nm
17. 550 nm
19. 0.500 cm
21. (a) 238 nm (b) λ will increase (c) 328 nm
23. 4.35 μm
25. 4.75 μm
27. No, the wavelengths intensified are 276 nm, 138 nm, 92.0 nm, . . .
29. 4.22 mm
31. (a) 1.1 m (b) 1.7 mm
33. 1.20 mm, 1.20 mm
35. (a) 479 nm, 647 nm, 698 nm (b) 20.5°, 28.3°, 30.7°
37. 5.91° in first order; 13.2° in second order; and 26.5° in third order
39. 44.5 cm
41. 9.13 cm
43. (a) 25.6° (b) 19.0°
45. (a) 1.11 (b) 42.0°
47. (a) 56.7° (b) 48.8°
49. 31.2°
53. 6.89 units
55. (a) 413.7 nm, 409.7 nm (b) 8.6°
57. 0.156 mm
59. 2.50 mm
61. Any positive integral multiple of 210 nm
63. (a) 16.6 m (b) 8.28 m
65. 127 m
67. 0.350 mm
69. 115 nm

Chapter 25

QUICK QUIZZES

1. (c)
2. (a)

CONCEPTUAL QUESTIONS

1. The observer is *not* using the lens as a simple magnifier. For a lens to be used as a simple magnifier, the object distance must be less than the focal length of the lens. Also, a simple magnifier produces a virtual image at the normal near point of the eye, or at an image distance of about $q = -25$ cm. With a large object distance and a relatively short image distance, the magnitude of the magnification by the lens would be considerably less than one. Most likely, the lens in this example is part of a lens combination being used as a telescope.

3. The image formed on the retina by the lens and cornea is already inverted.

5. There will be an effect on the interference pattern—it will be distorted. The high temperature of the flame will change the index of refraction of air for the arm of the interferometer in which the match is held. As the index of refraction varies randomly, the wavelength of the light in that region will also vary randomly. As a result, the effective difference in length between the two arms will fluctuate, resulting in a wildly varying interference pattern.

7. Large lenses are difficult to manufacture and machine with accuracy. Also, their large weight leads to sagging, which produces a distorted image. In reflecting telescopes, light does not pass through glass; hence, problems associated with chromatic aberrations are eliminated. Large-diameter reflecting telescopes are also technically easier to construct. Some designs use a rotating pool of mercury as the reflecting surface.

9. In order for someone to see an object through a microscope, the wavelength of the light in the microscope must be smaller than the size of the object. An atom is much smaller than the wavelength of light in the visible spectrum, so an atom can never be seen with the use of visible light.

11. farsighted; converging

PROBLEMS

1. 30.0 cm beyond the lens, $M = -1/5$
3. 177 m
5. $f/1.4$
7. $f/8.0$.
9. 40.0 cm
11. 23.2 cm
13. (a) -2.00 diopters (b) 17.6 cm
15. $+17.0$ diopters
17. (a) 5.8 cm (b) $m = 4.3$
19. (a) 4.07 cm (b) $m = +7.14$
21. (a) $|M| = 1.22$ (b) $\theta/\theta_0 = 6.08$
23. 2.1 cm
25. $m = -115$
27. $f_o = 90$ cm, $f_e = 2.0$ cm
29. (b) $-fh/p$ (c) -1.07 mm
31. (a) $m = 1.50$ (b) $m = 1.90$
33. 492 km
35. 0.40 μrad
37. 9.1×10^7 km
39. 9.8 km
41. No. A resolving power of 2.0×10^5 is needed, and that available is only 1.8×10^5.
43. 50.4 μm
45. 40
47. 98 fringe shifts
49. (a) $+2.67$ diopters (b) 0.16 diopter too low
51. (a) $+44.6$ diopters (b) 3.03 diopters
53. (a) 1.0×10^3 lines (b) 3.3×10^2 lines
55. $m = 10.7$
57. (a) $m = 4.0$ (b) $m = 3.0$

Chapter 26

QUICK QUIZZES

1. (a)
2. No. From your perspective you're at rest with respect to the cabin, so you will measure yourself as having your normal length, and will require a normal-sized cabin.

3. (a), (e); (a), (e)
4. (a) False (b) False (c) True (d) False
5. (a)

CONCEPTUAL QUESTIONS

1. An ellipsoid. The dimension in the direction of motion would be measured to be less than D.

3. This scenario is not possible with light. Light waves are described by the principles of special relativity. As you detect the light wave ahead of you and moving away from you (which would be a pretty good trick—think about it!), its velocity relative to you is c. Thus, you will not be able to catch up to the light wave.

5. No. The principle of relativity implies that nothing can travel faster than the speed of light in a *vacuum*, which is 3.00×10^8 m/s.

7. The light from the quasar moves at 3.00×10^8 m/s. The speed of light is independent of the motion of the source or the observer.

9. For a wonderful fictional exploration of this question, get a "Mr. Tompkins" book by George Gamow. All of the relativity effects would be obvious in our lives. Time dilation and length contraction would both occur. Driving home in a hurry, you would push on the gas pedal not to increase your speed very much, but to make the blocks shorter. Big Doppler shifts in wave frequencies would make red lights look green as you approached and make car horns and radios useless. High-speed transportation would be both very expensive, requiring huge fuel purchases, as well as dangerous, since a speeding car could knock down a building. When you got home, hungry for lunch, you would find that you had missed dinner; there would be a five-day delay in transit when you watch a live TV program originating in Australia. Finally, we would not be able to see the Milky Way, since the fireball of the Big Bang would surround us at the distance of Rigel or Deneb.

11. A photon transports energy. The relativistic equivalence of mass and energy means that is enough to give it momentum.

13. Your assignment: measure the length of a rod as it slides past you. Mark the position of its front end on the floor and have an assistant mark the position of the back end. Then measure the distance between the two marks. This distance will represent the length of the rod only if the two marks were made simultaneously in your frame of reference.

PROBLEMS

1. (a) $t_{OB} = 1.67 \times 10^3$ s, $t_{OA} = 2.04 \times 10^3$ s
(b) $t_{BO} = 2.50 \times 10^3$ s, $t_{AO} = 2.04 \times 10^3$ s
(c) $\Delta t = 90$ s
3. 5.0 s
5. (a) 20 m (b) 19 m (c) 0.31c
7. (a) 1.3×10^{-7} s (b) 38 m (c) 7.6 m
9. (a) 2.2 μs (b) 0.65 km
11. 0.950c
13. Yes, with 19 m to spare
15. (a) 39.2 μs (b) Accurate to one digit
17. 3.3×10^5 m/s
19. 0.285c
21. 0.54c to the right

23. 0.357c
25. 0.998c toward the right
27. (a) 54 min (b) 52 min
29. $c(\sqrt{3}/2)$
31. 0.786c
33. 18.4 g/cm^3
35. 1.98 MeV
37. 2.27×10^{23} Hz, 1.32 fm for each photon
39. (a) 3.10×10^5 m/s (b) 0.758c
41. 1.42 MeV/c
43. (a) 0.80c (b) 7.5×10^3 s (c) 1.4×10^{12} m, 0.38c
45. 0.37c in the $+ x$-direction
47. (a) $v/c = 1 - 1.12 \times 10^{-10}$ (b) 6.00×10^{27} J
(c) \$$2.17 \times 10^{20}$
49. 0.80c
51. (a) 0.946c (b) 0.160 ly (c) 0.114 yr (d) 7.50×10^{22} J
53. (a) 7.0 μs (c) 1.1×10^4 muons
59. 5.45 yr; Goslo is older.

Chapter 27
QUICK QUIZZES
1. (b)
2. (c)
3. (c)
4. (b)

CONCEPTUAL QUESTIONS

1. The shape of an object is normally determined by observing the light reflecting from its surface. In a kiln, the object will be very hot and will be glowing red. The emitted radiation is far stronger than the reflected radiation, and the thermal radiation emitted is only slightly dependent on the material from which the object is made. Unlike reflected light, the emitted light comes from all surfaces with equal intensity, so contrast is lost and the shape of the object is harder to discern.

3. The "blackness" of a blackbody refers to its ideal property of absorbing all radiation incident on it. If an observed room temperature object in everyday life absorbs all radiation, we describe it as (visibly) black. The black appearance, however, is due to the fact that our eyes are sensitive only to visible light. If we could detect infrared light with our eyes, we would see the object emitting radiation. If the temperature of the blackbody is raised, Wien's law tells us that the emitted radiation will move into the visible range of the spectrum. Thus, the blackbody could appear as red, white, or blue, depending on its temperature.

5. All objects do radiate energy, but at room temperature this energy is primarily in the infrared region of the electromagnetic spectrum, which our eyes cannot detect. (Pit vipers have sensory organs that *are* sensitive to infrared radiation; thus, they can seek out their warm-blooded prey in what we would consider absolute darkness.

7. Most metals have cutoff frequencies corresponding to photons in or near the visible range of the electromagnetic spectrum. AM radio wave photons have far too little energy to eject electrons from the metal.

9. We can picture higher frequency light as a stream of photons of higher energy. In a collision, one photon can give all of its energy to a single electron. The kinetic energy of such an electron is measured by the stopping potential.

The reverse voltage (stopping voltage) required to stop the current is proportional to the frequency of the incoming light. More intense light consists of more photons striking a unit area each second, but atoms are so small that one emitted electron never gets a "kick" from more than one photon. Increasing the intensity of the light will generally increase the size of the current, but will not change the energy of the individual electrons that are ejected. Thus, the stopping potential remains constant.

11. Wave theory predicts that the photoelectric effect should occur at any frequency, provided that the light intensity is high enough. However, as seen in photoelectric experiments, the light must have sufficiently high frequency for the effect to occur.

13. (a) Electrons are emitted only if the photon frequency is greater than the cutoff frequency.

15. No. Suppose that the incident light frequency at which you first observed the photoelectric effect is above the cutoff frequency of the first metal, but less than the cutoff frequency of the second metal. In that case, the photoelectric effect would not be observed at all in the second metal.

17. The frequency of the scattered photon must decrease, because some of its energy is transferred to the electron.

PROBLEMS

1. (a) $\approx 3\,000$ K (b) $\approx 20\,000$ K
3. 500 nm
5. (a) 2.49×10^{-5} eV (b) 2.49 eV (c) 249 eV
7. 2.27×10^{30} photons/s
9. (a) 2.3×10^{31} (b) $\Delta E/E = 4.3 \times 10^{-32}$
11. (a) 2.24 eV (b) 555 nm (c) 5.41×10^{14} Hz
13. 234 nm
15. 148 days, incompatible with observation
17. 4.8×10^{14} Hz, 2.0 eV
19. 1.2×10^{2} V and 1.2×10^{7} V, respectively
21. 41.4 kV
23. 0.078 nm
25. 0.281 nm
27. 1.78 eV, 9.47×10^{-28} kg·m/s
29. 70°
31. 1.18×10^{-23} kg·m/s, 478 eV
33. (a) 1.2 eV (b) 6.5×10^{5} m/s
35. (a) 1.46 km/s (b) 7.28×10^{-11} m
37. (a) $\approx 10^{2}$ MeV (b) No. With kinetic energy much larger than the magnitude of the negative potential energy, the electron would immediately escape.
39. 3.58×10^{-13} m
41. (a) 15 keV (b) 1.2×10^{2} keV
43. $\sim 10^{6}$ m/s
45. 116 m/s
47. $\approx 5\,200$ K; clearly, a firefly is not at that temperature, so this cannot be blackbody radiation.
49. 18.2°
51. 1.36 eV
53. 2.00 eV
55. (a) $0.022\,0c$ (b) $0.999\,2c$
57. (b) 3.72 km/s
59. (b) 5.19×10^{-16} m
61. (a) 0.263 kg (b) 1.81 W
    (c) $-0.015\,3°C/s = -0.919°C/min$ (d) $9.89\ \mu m$
    (e) 2.01×10^{-20} J (f) 8.98×10^{19} photon/s

Chapter 28

QUICK QUIZZES

1. (b)
2. (a)
3. (a) 5 (b) 9 (c) 25
4. (d)

CONCEPTUAL QUESTIONS

1. If the energy of the hydrogen atom were proportional to n (or any power of n), then the energy would become infinite as n grew to infinity. But the energy of the atom is inversely proportional to n^2. Thus, as n grows to infinity, the energy of the atom approaches a value that is above the ground state by a finite amount, namely, the ionization energy 13.6 eV. As the electron falls from one bound state to another, its energy loss is always less than the ionization energy. The energy and frequency of any emitted photon are finite.

3. The characteristic x-rays originate from transitions within the atoms of the target, such as an electron from the L shell making a transition to a vacancy in the K shell. The vacancy is caused when an accelerated electron in the x-ray tube supplies energy to the K shell electron to eject it from the atom. If the energy of the bombarding electrons were to be increased, the K shell electron will be ejected from the atom with more remaining kinetic energy. But the energy difference between the K and L shell has not changed, so the emitted x-ray has exactly the same wavelength.

5. A continuous spectrum without characteristic x-rays is possible. At a low accelerating potential difference for the electron, the electron may not have enough energy to eject an electron from a target atom. As a result, there will be no characteristic x-rays. The change in speed of the electron as it enters the target will result in the continuous spectrum.

7. The hologram is an interference pattern between light scattered from the object and the reference beam. If anything moves by a distance comparable to the wavelength of the light (or more), the pattern will wash out. The effect is just like making the slits vibrate in Young's experiment, to make the interference fringes vibrate wildly so that a photograph of the screen displays only the average intensity everywhere.

9. If the Pauli exclusion principle were not valid, the elements and their chemical behavior would be grossly different, because every electron would end up in the lowest energy level of the atom. All matter would therefore be nearly alike in its chemistry and composition, since the shell structures of each element would be identical. Most materials would have a much higher density, and the spectra of atoms and molecules would be very simple, resulting in the existence of less color in the world.

11. The three elements have similar electronic configurations, with filled inner shells plus a single electron in an s orbital. Since atoms typically interact through their unfilled outer shells, and since the outer shells of these atoms are similar, the chemical interactions of the three atoms are also similar.

13. Each of the eight electrons must have at least one quantum number different from each of the others. They can differ (in m_s) by being spin-up or spin-down. They can differ (in ℓ) in angular momentum and in the general shape of the wave function. Those electrons with $\ell = 1$ can differ (in m_ℓ) in orientation of angular momentum.

15. Stimulated emission is the reason laser light is coherent and tends to travel in a well-defined parallel beam. When a photon passing by an excited atom stimulates that atom to emit a photon, the emitted photon is in phase with the original photon and travels in the same direction. As this process is repeated many times, an intense, parallel beam of coherent light is produced. Without stimulated emission, the excited atoms would return to the ground state by emitting photons at random times and in random directions. The resulting light would not have the useful properties of laser light.

17. The atom is a bound system. The atomic electron does not have enough kinetic energy to escape from its electrical attraction to the nucleus. The electrical potential energy of the atom is negative and is greater than the kinetic energy, so the total energy of the atom is negative.

19. (a) The wavelength of photon A is greater. (b) The energy of photon B is greater.

PROBLEMS

1. 656 nm, 486 nm, and 434 nm
3. (a) 2.3×10^{-8} N (b) -14 eV
5. (a) 1.6×10^6 m/s (b) No, $v/c = 5.3 \times 10^{-3} << 1$
   (c) 0.46 nm (d) Yes. The wavelength is roughly the same size as the atom.
7. (a) 0.212 nm (b) 9.95×10^{-25} kg·m/s
   (c) 2.11×10^{-34} J·s
   (d) 3.40 eV (e) -6.80 eV (f) -3.40 eV
11. $E = -1.51$ eV $(n = 3)$ to $E = -3.40$ eV $(n = 2)$
13. (a) 0.967 eV (b) 0.266 eV
15. (a) 122 nm, 91.1 nm (b) 1.87×10^3 nm, 820 nm
17. 97.2 nm
19. (a) 488 nm (b) 0.814 m/s
21. (d) $n = 2.53 \times 10^{74}$ (e) No. At such large quantum numbers, the allowed energies are essentially continuous.
23. (a) 2.47×10^{14} Hz, $f_{orb} = 8.23 \times 10^{14}$ Hz
   (b) 6.59×10^3 Hz, $f_{orb} = 6.59 \times 10^3$ Hz. For large n, classical theory and quantum theory approach each other in their results.
25. 4.42×10^4 m/s
27. (a) -122 eV (b) 1.76×10^{-11} m
29. (a) 0.026 5 nm (b) 0.017 6 nm (c) 0.013 2 nm
31. 1.33 nm
33. $n = 3, \ell = 1, m_\ell = +1, m_s = \pm 1/2; n = 3, \ell = 1, m_\ell = 0,$
   $m_s = \pm 1/2; n = 3, \ell = 1, m_\ell = -1, m_s = \pm 1/2$
35. Fifteen possible states, as summarized in the following table:

n	3	3	3	3	3	3	3	3	3	3	3	3	3	3	3
ℓ	2	2	2	2	2	2	2	2	2	2	2	2	2	2	2
m_ℓ	+2	+2	+2	+1	+1	+1	0	0	0	−1	−1	−1	−2	−2	−2
m_s	+1	0	−1	+1	0	−1	+1	0	−1	+1	0	−1	+1	0	−1

37. (a) 30 possible states (b) 36
39. (a) $n = 4$ and $\ell = 2$ (b) $m_\ell = (0, \pm 1, \pm 2), m_s = \pm 1/2$
   (c) $1s^2 2s^2 2p^6 3s^2 3p^6 3d^{10} 4s^2 4p^6 4d^2 5s^2 = $ [Kr] $4d^2 5s^2$
41. 0.160 nm
43. L shell: 11.7 keV; M shell: 10.0 keV; N shell: 2.30 keV
45. (a) 10.2 eV (b) 7.88×10^4 K
47. (a) -8.18 eV, -2.04 eV, -0.904 eV, -0.510 eV, -0.325 eV
   (b) 1.09×10^3 nm and 609 nm

49. The four lowest energies are -10.39 eV, -5.502 eV, -3.687 eV, and -2.567 eV (b) The wavelengths of the emission lines are 158.5 nm, 185.0 nm, 253.7 nm, 422.5 nm, 683.2 nm, and 1 107 nm
   (c) 1.31×10^6 m/s
51. (a) 4.24×10^{15} W/m^2 (b) 1.20×10^{-12} J
55. (a) $E_n = (-1.49 \times 10^4 \text{ eV})/n^2$ (b) $n = 4 \rightarrow n = 1$
57. (a) 9.03×10^{22} m/s^2 (b) -4.63×10^{-8} W
   (c) $\sim 10^{-11}$ s

Chapter 29

QUICK QUIZZES
1. (c)
2. (c)
3. (a)
4. (a) and (b)
5. (b)

CONCEPTUAL QUESTIONS
1. Isotopes of a given element correspond to nuclei with different numbers of neutrons. This will result in a variety of different physical properties for the nuclei, including the obvious one of mass. The chemical behavior, however, is governed by the element's electrons. All isotopes of a given element have the same number of electrons and, therefore, the same chemical behavior.

3. An alpha particle contains two protons and two neutrons. Because a hydrogen nucleus contains only one proton, it cannot emit an alpha particle.

5. In alpha decay, there are only two final particles: the alpha particle and the daughter nucleus. There are also two conservation principles: of energy and of momentum. As a result, the alpha particle must be ejected with a discrete energy to satisfy both conservation principles. However, beta decay is a three-particle decay: the beta particle, the neutrino (or antineutron), and the daughter nucleus. As a result, the energy and momentum can be shared in a variety of ways among the three particles while still satisfying the two conservation principles. This allows a continuous range of energies for the beta particle.

7. The larger rest energy of the neutron means that a free proton in space will not spontaneously decay into a neutron and a positron. When the proton is in the nucleus, however, the important question is that of the total rest energy of the nucleus. If it is energetically favorable for the nucleus to have one less proton and one more neutron, then the decay process will occur to achieve this lower energy.

9. Carbon dating cannot generally be used to estimate the age of a stone, because the stone was not alive to take up carbon from the environment. Only the ages of artifacts that were once alive can be estimated with carbon dating.

11. The protons, although held together by the nuclear force, are repelled by the electrostatic force. If enough protons were placed together in a nucleus, the electrostatic force would overcome the nuclear force, which is based on the number of particles, and cause the nucleus to fission.

   The addition of neutrons prevents such fission. The neutron does not increase the electrical force, being electrically neutral, but does contribute to the nuclear force.

13. The photon and the neutrino are similar in that both particles have zero charge and very little mass. (The photon

has zero mass, but recent evidence suggests that certain kinds of neutrinos have a very small mass.) Both must travel at the speed of light and are capable of transferring both energy and momentum. They differ in that the photon has spin (intrinsic angular momentum) $\hbar$ and is involved in electromagnetic interactions, while the neutrino has spin $\hbar/2$, and is closely related to beta decays.

15. Since the two samples are of the same radioactive nuclide, they have the same half-life; the 2:1 difference in activity is due to a 2:1 difference in the mass of each sample. After 5 half lives, each will have decreased in mass by a power of $2^5 = 32$. However, since this simply means that the mass of each is 32 times smaller, the ratio of the masses will still be $(2/32) : (1/32)$, or 2:1. Therefore, the ratio of their activities will *always* be 2:1.

PROBLEMS

1. $A = 2$, $r = 1.5$ fm; $A = 60$, $r = 4.7$ fm; $A = 197$, $r = 7.0$ fm; $A = 239$, $r = 7.4$ fm
3. 1.8×10^2 m
5. (a) 27.6 N (b) 4.16×10^{27} m/s^2 (c) 1.73 MeV
7. (a) 1.9×10^7 m/s (b) 7.1 MeV
9. 8.66 MeV/nucleon for $^{93}_{41}$Nb, 7.92 MeV/nucleon for $^{197}_{79}$Au
11. 3.54 MeV
13. 0.210 MeV/nucleon greater for $^{23}_{11}$Na, attributable to less proton repulsion
15. 0.46 Ci
17. (a) 9.98×10^{-7} s^{-1} (b) 1.9×10^{10} nuclei
19. 1.0 h
21. 4.31×10^3 yr
23. (a) 5.58×10^{-2} h^{-1}, 12.4 h (b) 2.39×10^{13} nuclei (c) 1.9 mCi
25. $^{208}_{81}$Tl, $^{95}_{37}$Rb, $^{144}_{60}$Nd
27. $^{40}_{20}$Ca, $^{94}_{42}$Mo, $^4_2$He
29. e^+ decay, $^{56}_{27}$Co $\rightarrow$ $^{56}_{26}$Fe + e^+ + ν
31. (a) cannot occur spontaneously
    (b) can occur spontaneously
33. 18.6 keV
35. 4.22×10^3 yr
37. (a) $^{30}_{15}$P (b) -2.64 MeV
39. (a) $^{21}_{10}$Ne (b) $^{144}_{54}$Xe (c) X = e^+, X' = ν
41. (a) $^{13}_6$C (b) $^{10}_5$B
43. (a) $^{197}_{79}$Au + n $\rightarrow$ $^{198}_{80}$Hg + e^- + $\overline{\nu}$ (b) 7.88 MeV
45. (a) $^1_0$n (b) Fluoride mass = 18.000 953 u
47. 18.8 J
49. 24 d
51. (a) 8.97×10^{11} electrons (b) 0.100 J (c) 100 rad
53. 46.5 d
55. $Q = 3.27$ MeV > 0, no threshold energy required
57. (a) 2.52×10^{24} (b) 2.29×10^{12} Bq (c) 1.07×10^6 yr
59. (a) 4.0×10^9 yr (b) It could be no older. The rock could be younger if some ^{87}Sr were initially present.
61. 54 μCi
63. 2.3×10^2 yr
65. 4.4×10^{-8} kg/h

Chapter 30

QUICK QUIZZES

1. (c)
2. (a)
3. (b)
4. (d)

CONCEPTUAL QUESTIONS

1. The experiment described is a nice analogy to the Rutherford scattering experiment. In the Rutherford experiment, alpha particles were scattered from atoms and the scattering was consistent with a small structure in the atom containing the positive charge.
3. The largest charge quark is $2e/3$, so a combination of only two particles, a quark and an antiquark forming a meson, could not have an electric charge of $+2e$. Only particles containing three quarks, each with a charge of $2e/3$, can combine to produce a total charge of $2e$.
5. Until about 700 000 years after the Big Bang, the temperature of the Universe was high enough for any atoms that formed to be ionized by ambient radiation. Once the average radiation energy dropped below the hydrogen ionization energy of 13.6 eV, hydrogen atoms could form and remain as neutral atoms for relatively long period of time.
7. In the quark model, all hadrons are composed of smaller units called quarks. Quarks have a fractional electric charge and a baryon number of $\frac{1}{3}$. There are six flavors of quarks: up (u), down (d), strange (s), charmed (c), top (t), and bottom (b). All baryons contain three quarks, and all mesons contain one quark and one antiquark. Section 30.12 has a more detailed discussion of the quark model.
9. Baryons and mesons are hadrons, interacting primarily through the strong force. They are not elementary particles, being composed of either three quarks (baryons) or a quark and an antiquark (mesons). Baryons have a nonzero baryon number with a spin of either $\frac{1}{2}$ or $\frac{3}{2}$. Mesons have a baryon number of zero and a spin of either 0 or 1.
11. All stable particles other than protons and neutrons have baryon number zero. Since the baryon number must be conserved, and the final states of the kaon decay contain no protons or neutrons, the baryon number of all kaons must be *zero*.
13. Yes, but the strong interaction predominates.
15. Unless the particles have enough kinetic energy to produce a baryon–antibaryon pair, the answer is *no*. Antibaryons have a baryon number of -1, baryons have a baryon number of $+1$, and mesons have a baryon number of 0. If such an interaction were to occur and produce a baryon, the baryon number would not be conserved.
17. Baryons and antibaryons contain three quarks, while mesons and antimesons contain two quarks. Quarks have a spin of $1/2$; thus, three quarks in a baryon can only combine to form a net spin that is half-integral. Likewise, two quarks in a meson can only combine to form a net spin of 0 or 1.
19. For the first decay, the half-life is characteristic of the strong interaction, so the ρ^0 must have $S = 0$, and strangeness is conserved. The second decay must occur via the weak interaction.

PROBLEMS

1. 1.1×10^{16} fissions
3. 126 MeV
5. (a) 16.2 kg (b) 117 g
7. 2.9×10^3 km ($\approx 1\ 800$ miles)
9. 1.01 g

11. (a) $^8_4$Be (b) $^{12}_6$C (c) 7.27 MeV
13. 3.07×10^{22} events/yr
15. (a) 3.44×10^{30} J (b) 1.56×10^8 yr
17. (a) 4.53×10^{23} Hz (b) 0.622 fm
19. $\sim 10^{-23}$ s
21. $\sim 10^{-18}$ m
23. (a) conservation of electron-lepton number and conservation of muon-lepton number (b) conservation of charge (c) conservation of baryon number (d) conservation of baryon number (e) conservation of charge
25. $\overline{\nu}_\mu$
27. (a) $\overline{\nu}_\mu$ (b) ν_μ (c) $\overline{\nu}_e$ (d) ν_e (e) ν_μ (f) ν_μ and $\overline{\nu}_e$
29. (a) not allowed; violates conservation of baryon number (b) strong interaction (c) weak interaction (d) weak interaction (e) electromagnetic interaction
31. (a) not conserved (b) conserved (c) conserved (d) not conserved (e) not conserved (f) not conserved
33. (a) charge, baryon number, L_e, L_τ (b) charge, baryon number, L_e, L_μ, L_τ (c) charge, L_e, L_μ, L_τ, strangeness number (d) charge, baryon number, L_e, L_μ, L_τ, strangeness number (e) charge, baryon number, L_e, L_μ, L_τ, strangeness number (f) charge, baryon number, L_e, L_μ, L_τ, strangeness number
35. 3.34×10^{26} electrons, 9.36×10^{26} up quarks, 8.70×10^{26} down quarks
37. (a) Σ^+ (b) π^- (c) K^0 (d) Ξ^-

39.

Reaction	At Quark Level	Net Quarks (before and after)
$\pi^- + p \rightarrow K^0 + \Lambda^0$	$\overline{u}d + uud \rightarrow d\overline{s} + uds$	1 up, 2 down, 0 strange
$\pi^+ + p \rightarrow K^+ + \Sigma^+$	$u\overline{d} + uud \rightarrow u\overline{s} + uus$	3 up, 0 down, 0 strange
$K^- + p \rightarrow$ $K^+ + K^0 + \Omega^-$	$\overline{u}s + uud \rightarrow$ $u\overline{s} + d\overline{s} + sss$	1 up, 1 down, 1 strange

(d) The mystery particle is a Λ^0 or a Σ^0.
41. a neutron, udd
43. 70.45 MeV
45. 18.8 MeV
47. (a) electron-lepton and muon-lepton numbers not conserved (b) electron-lepton number not conserved (c) charge not conserved (d) baryon and electron-lepton numbers not conserved (e) strangeness violated by 2 units
49. (a) 2×10^{24} nuclei (b) ≈ 0.6 kg
51. (a) 1 baryon before and zero baryons after decay. Baryon number is not conserved. (b) 469 MeV, 469 MeV/c (c) 0.999 999 4c
53. (b) 12 days
55. 26 collisions

Credits

Photographs

This page constitutes an extension of the copyright page. We have made every effort to trace the ownership of all copyrighted material and to secure permission from copyright holders. In the event of any question arising as to the use of any material, we will be pleased to make the necessary corrections in future printings. Thanks are due to the following authors, publishers, and agents for permission to use the material indicated.

Chapter 1. **1:** Stone/Getty Images **2:** bottom left, Courtesy of National Institute of Standards and Technology, U.S. Dept. of Commerce; bottom right, Courtesy of Institute of Standards and Technology, U.S. Dept. of Commerce **10:** Billy E. Barnes/Stock Boston **12:** R. Williams (STScI), the HDF-S team, and NASA **19:** top left, Courtesy of George Semple; top right, Courtesy of George Semple **20:** © Sylvain Grandadam/Photo Researchers, Inc. **21:** NASA

Chapter 2. **23:** Caron/Corbis Sygma **39:** Courtesy Amtrak NEC Media Relations **41:** North Wind Archive

Chapter 3. **53:** © Bettmann/Corbis **54:** top, Mack Henley/Visuals Unlimited; bottom, George Semple **62:** HIRB/Index Stock **65:** Mike Powell/Allsport/Getty Images **75:** top right, Joseph Kayne/Dembinsky Photo Associates; bottom left, AP/Wide World Photos

Chapter 4. **81:** top left, © Royalty-Free/Corbis; bottom right, © Lorenzo Ciniglio/Corbis **83:** Roger Viollet, Mill Valley, CA, University Science Books, 1982 **85:** Giraudon/Art Resource **88:** NASA **90:** Jim Gillmoure/corbisstockmarket.com **107:** Guy Sauvage/Photo Researchers, Inc.

Chapter 5. **118:** NASA **121:** top, © Chris Collins/Corbis; bottom, © McCrone Photo/Custom Medical Stock Photo **132:** Wet'n Wild Orlando **141:** top right, Jan Hinsch/Science Photo Library/Photo Researchers, Inc.; bottom right, Gareth Williams, Minor Planet Center **149:** Arthur Tilley/FPG/Getty Images **157:** © Jamie Budge/Corbis **158:** Engraving from *Scientific American,* July 1888.

Chapter 6. **160:** © Reuters/Corbis **162:** © Harold and Esther Edgerton Foundation, 2002, courtesy of Palm Press, Inc. **163:** © Harold and Esther Edgerton Foundation, 2002, Courtesy of Palm Press, Inc. **164:** top right, Tim Wright/Corbis **167:** bottom left, Courtesy of NASA; bottom right, Mike Severns/Stone/Getty Images **171:** Courtesy of Central Scientific Company **181:** Courtesy of Saab **188:** bottom left, Courtesy of CENCO

Chapter 7. **189:** Courtesy NASA **208:** top right, Courtesy of PASCO Scientific

Chapter 8. **226:** © Kevin Fleming/Corbis **230:** David Serway **239:** George Semple **248:** top left, © Benson Krista Hicks/Corbis Sygma; top right, © Benson Krista Hicks/Corbis Sygma; middle left, © Patrick Giardino/Corbis; bottom left, Max Planck Institute for Astronomy and Calar Alto Observatory, K. Meisenheimer and A. Quetz; bottom right, © Smithsonian Institute/Photo Researchers, Inc. **253:** top left, © Eye Ubiquitous/Corbis; top right, Gerard Lacz/NHPA **264:** © Ed Bock/Corbis

Chapter 9. **266:** © Alison Wright/Corbis **267:** Charles D. Winters **271:** NASA **275:** © Royalty-Free/Corbis **276:** Raymond A. Serway **277:** Courtesy of Scientific Company **279:** bottom right, David Frazier **284:** top left, © Royalty-Free/Corbis **285:** top right, © Royalty-Free/Corbis **288:** Andy Sacks/Stone/Getty Images **289:** top right, Kim Vandiver and Harold Edgerton. © Harold and Esther Edgerton Foundation, 2002, courtesy of Palm Press, Inc.; bottom right, George Semple **292:** top left, Corbis-Bettmann; bottom right, Courtesy of Central Scientific Company **299:** Herman Eisenbeiss/Photo Researchers, Inc. **300:** Charles D. Winters **310:** bottom left, Henry Leap and Jim Lehman; bottom right, Pamela Zilly/The Image Bank/Getty Images **318:** The Granger Collection

Chapter 10. **321:** © Jim Sugar/Corbis **328:** top left, George Semple; bottom left, George Semple **329:** AP/Wide World Photos **330:** George Semple **340:** © R. Folwell/Science Photo Library/Photo Researchers, Inc.

Chapter 11. **352:** Steve Bly/Getty Images **353:** By kind permission of the President and Council of the Royal Society **366:** George Semple **371:** Gary Settles/Science Source/Photo Researchers Inc. **373:** bottom left, Daedalus Enterprises, Inc./Peter Arnold, Inc.; bottom right, Photodisc/Getty Images **374:** SPL/Photo Researchers, Inc. **377:** top left, Tom Bean; top right, Tom Bean **380:** Nathan Bilow/Leo de Wys, Inc. **381:** top left, © Jim Bourg/Teuters/Corbis

Chapter 12. **386:** © Tim De Waele/Corbis **404:** Charles D. Winters **405:** J-L. Charmet/SPL/Photo Researchers, Inc. **406:** J-L. Charmet/SPL/Photo Researchers, Inc. **409:** AIP Niels Bohr Library, Lade Collection **413:** top, George Semple; bottom, George Semple **414:** © BSIP/Laurent Science Source/Photo Researchers, Inc.

Chapter 13. **424:** © Rick Doyle/Corbis **429:** Eric Lars Baleke/Black Star **433:** Link/Visuals Unlimited **434:** Telegraph Colour Library/FPG International **448:** Martin

Dohm/SPL/Photo Researchers **451:** Gregg Adams/Stone Getty Images

Chapter 14.   **458:** Patrick Ward/CORBIS **460:** Bernard Benoit/Photo Researchers, Inc. **469:** top right, Courtesy Educational Development Center, Newton, Mass. **472:** top right, © 1973 Kim Vandiver & Harold E. Edgerton/Courtesy of Palm Press, Inc. **473:** © 1994, Comstock **476:** all images, © Richard Megna, Fundamental Photographs, NYC **477:** Charles D. Winters **480:** left, top, Courtesy of Professor Thomas D. Rossing, Northern Illinois University; left, bottom, Ben Rose/The IMAGE Bank Getty Images; bottom left, United Press International Photo; bottom right, United Press International Photo **487:** Royalty Free/Corbis **496:** Sidney Harris

Chapter 15.   **497:** © Keith Kent/Science Photo Library/Photo Researchers, Inc. **498:** © American Philosophical Society/AIP **501:** top, © 1968 Fundamental Photographs; bottom, Photo courtesy of AIP Niels Bohr Library, E. Scott Barr Collection **510:** Photo courtesy of Harold M. Waage, Princeton University **511:** top, Photo courtesy of Harold M. Waage, Princeton University; bottom, Photo courtesy of Harold M. Waage, Princeton University **513:** top, Photo courtesy of Harold M. Waage, Princeton University; middle right, Photo courtesy of Harold M. Waage, Princeton University

Chapter 16.   **531:** Courtesy of Resonance Research Corporation **543:** bottom middle, Riei O'Harra/Black Star/PNI; bottom right, Greig Cranna/Stock, Boston/PNI **546:** Photo courtesy of Harold M. Waage, Princeton University **557:** © Loren Winters/Visuals Unlimited **558:** a. Paul Silverman/Fundamental Photographs; b. George Semple

Chapter 17.   **568:** Telegraph Colour Library/FPG/Getty Images **572:** Michael Dalton, Fundamental Photographs **574:** © Bettmann/CORBIS **575:** Courtesy of Henry Leap and Jim Lehman **578:** Courtesy of Central Scientific Company **580:** Courtesy of IBM Research Laboratory **585:** Courtesy of Medtronic, Inc.

Chapter 18.   **592:** © Lester Lefkowitz/Corbis **593:** George Semple **602:** AIP ESVA, W.F. Meggers Collection **610:** George Semple **613:** Juergen Berger, Max-Planck Institute/Science Photo Library/Photo Researchers, Inc. **616:** middle left, Superstock; middle right, Courtesy of Henry Leap and Jim Lehman

Chapter 19.   **624:** Johnny Johnson/Stone/Getty **625:** Courtesy of Central Scientific Company **626:** a–c. Henry Leap and Jim Lehman **631:** Courtesy of Henry Leap and Jim Lehman **637:** Courtesy of American Honda Motor Co., Inc. **640:** middle left, North Wind Picture Archives; bottom, © Richard Megna, Fundamental Photographs **641:** Leonard de Selva/CORBIS **645:** © Richard Megna, Fundamental Photographs **646:** Courtesy of Henry Leap and Jim Lehman **650:** © Loren Winters/Visuals Unlimited **651:** top right, Courtesy of Central Scientific Company; middle right, Courtesy of Central Scientific Company

Chapter 20.   **660:** PhotoDisc/Getty Images **661:** By kind permission of the President and Council of the Royal Society **666:** Charles D. Winters **667:** Courtesy of PedsLink Pediatric Healthcare Resources, Newport Beach, CA **674:** Luis Castaneda/The Image Bank/Getty Images **678:** North Wind Picture Archives **681:** Courtesy of Saab **688:** bottom left, Courtesy of CENCO

Chapter 21.   **693:** © Bettmann/Corbis **701:** Bettmann/Corbis **704:** Ryan Williams/International Stock Photography **707:** George Semple **708:** top, North Wind Photo Archives; bottom, Bettmann/Corbis **713:** John Neal/Photo Researchers, Inc. **715:** Ron Chapple/Getty Images **717:** a. NASA/CXC/SAO; b. Palomar Observatory; c. WM Keck Observatory; d, VLA/NRAO **720:** George Semple

Chapter 22.   **726:** Peter Bowater/Photo Researchers, Inc. **727:** Courtesy of Rijksmuseum voor de Geschiedenis der Natuurwetenschappen. Courtesy AIP Niels Bohr Library **728:** bottom left and right, Photographs Courtesy of Henry Leap and Jim Lehman **729:** top left and right, Photos by Charles D. Winters **731:** top right, Courtesy of Henry Leap and Jim Lehman **735:** bottom left, Courtesy of Sony Disc Manufacturing **737:** top right, David Parker/SPL/Photo Researchers, Inc.; bottom right, Courtesy of Bausch & Lomb **738:** Courtesy of PASCO Scientific **739:** Mark D. Phillips/Photo Researchers, Inc. **740:** Courtesy of Sabina Zigman/Benjamin Cardozo High School **742:** Courtesy of Henry Leap and Jim Lehman **744:** bottom left, Dennis O'Clair/Tony Stone Images/Getty Images; bottom right, Hank Morgan/Photo Researchers, Inc.

Chapter 23.   **754:** Don Hammond/CORBIS **757:** Courtesy of Henry Leap and Jim Lehman **759:** Courtesy of Jim Lehman, James Madison University **761:** top right, middle right, bottom right, Photos courtesy of David Rogers **762:** © Junebug Clark 1988/Photo Researchers, Inc. **768:** John M. Dunay IV, Fundamental Photographs, NYC **770:** top left, Photos Courtesy of Henry Leap and Jim Lehman **778:** a–c. Photos by Norman Goldberg **780:** middle right, Richard Megna/Fundamental Photographs, NYC **784:** Photo © Michael Levin/Opti-Gone Associates

Chapter 24.   **786:** RO-MA/Index Stock Imagery **787:** M. Cagnet, M. Francon, J. C. Thierr **788:** Richard Megna/Fundamental Photographs, NYC **789:** John S. Shelton **792:** Dr. Jeremy Burgess/Science Photo Library/Photo Researchers, Inc. **793:** top right, Peter Aprahamian/Science Photo Library/Photo Researchers, Inc.; bottom middle, Courtesy of Bausch & Lomb Optical Company; bottom right, From Physical Science Study Committee, *College Physics*, Lexington Mass., D. C. Heath and Co., 1968 **794:** Richard Megna, Fundamental Photographs **796:** Courtesy of Sony Disc Manufacturing **798:** top left, Courtesy of P. M. Rinard, from *Am. J. Phys.*, 44:70, 1976; top right, From M. Cagnet, M. Francon, and J. C. Thierr, *Atlas of Optical Phenomena*, Berlin, Springer-Verlag, 1962, plate 18 **803:** © Kristen Brochmann/Fundamental Photographs **806:** a–c. Photos courtesy of Henry Leap **809:** bottom left, Sepp Seitz 1981; bottom right, Peter Aprahamian/Science Photo Library/Photo Researchers, Inc.

Chapter 25.   **819:** © Denis Scott/CORBIS **821:** top right, From Lennart Nilson, in collaboration with Jan Lindberg, *Behold Man: A Photographic Journey of Discovery Inside the Body*, Boston, Little, Brown & Co., 1974 **827:** bottom right, © Tony Freeman/Photo Edit **829:** NASA **830:** top right, © Tony Freeman/Photo Edit; bottom, Courtesy of Palomar Observatory/California Institute of Technology

832: a–c. Photos from M. Cagnet, M. Francon, and J. C. Thierr 833: From M. Cagnet, M. Francon, and J. C. Thierr, *Atlas of Optical Phenomena*, Berlin, Springer-Verlag, 1962, plate 34

Chapter 26. 843: Courtesy of the Archives, California Institute of Technology 849: AIP Niels Bohr Library 861: Lawrence Berkeley Laboratory/Science Photo Library/Photo Researchers, Inc.

Chapter 27. 874: © Eye of Science/Science Source/Photo Researchers, Inc. 878: © Bettmann/CORBIS 881: Courtesy of GE Medical Systems 884: Science Source/Photo Researchers, Inc. 885: Courtesy of AIP Niels Bohr Library 887: AIP Niels Bohr Library 890: top right, © David Parker/Photo Researchers, Inc. 891: top right, AIP Emilio Segrè Visual Archives; bottom right, Courtesy of the University of Hamburg 894: IBM Corporation Research Division 902: John Chumack/Photo Researchers, Inc.

Chapter 28. 903: Dembinsky Photo Associates 904: Stock Montage, Inc. 907: top right, Princeton University/Courtesy of AIP Emilio Segrè Visual Archives 918: CERN/Courtesy of AIP Emilio Segrè Visual Archives 923: Courtesy of HRL Laboratories LLC, Malibu, CA 924: top right, Courtesy of Central Scientific Company; middle left, Philippe Plailly/Photo Researchers, Inc. 931: top left, Courtesy of Texas Instruments, Inc.; top right, Courtesy of Intel Corporation

Chapter 29. 939: Courtesy of Public Service Electric and Gas Company 940: North Wind Picture Archives 942: Courtesy of Louise Barker/AIP Niels Bohr Library 945: FPG International 946: © Richard Megna/Fundamental Photographs 952: National Accelerator Laboratory 953: Henry Paul/Gamma Liaison 955: Santi Visali/The IMAGE Bank 962: Jay Freis/The Image Bank 963: top left, SBHA/Getty Images; top right, Scott Camazine/Science Source/Photo Researchers, Inc.; bottom right, © Hank Morgan/Photo Researchers 965: top left, Photo Researchers, Inc./Science Photo Library; top right, Dan McCoy/Rainbow

Chapter 30. 973: Sandia National Laboratories/Photo by Randy Montoya 977: Courtesy of Chicago Historical Society

980: NASA 983: top left, Courtesy of Princeton Plasma Physics Laboratory; top right, Courtesy of Princeton University 985: Courtesy AIP Emilio Segrè Visual Archives 986: UPI/Corbis-Bettmann 987: © Shelly Gazin/CORBIS 993: Photo courtesy of Michael R. Dressler 996: Courtesy of Fermi National Accelerator Laboratory 998: Courtesy of CERN 1000: top left, Courtesy of AIP Emilio Segrè Visual Archives; bottom left, AT&T Bell Laboratories

Tables and Illustrations

This page constitutes an extension of the copyright page. We have made every effort to trace the ownership of all copyrighted material and to secure permission from copyright holders. In the event of any question arising as to the use of any material, we will be pleased to make the necessary corrections in future printings. Thanks are due to the following authors, publishers, and agents for permission to use the material indicated.

Chapter 2. 24: Schwadron, Cartoonists & Writers Syndicate/cartoonweb.com 47: By permission of John L. Hart FLP, and Creators Syndicate, Inc.

Chapter 25. 821: Art by Robert Demarest, from Cecie Starr and Beverly McMillan, *Human Biology* 5th ed. Copyright © 2003, Brooks/Cole, a division of Thomson Learning, Inc.

Chapter 27. 894–895: This section was written by Roger A. Freedman, University of California, Santa Barbara 895: Based on a drawing from P. K. Hansma, V. B. Elings, O. Marti, and C. Bracker, *Science* 242:209, 1988, Copyright 1988 by the AAAS

Chapter 28. 905: K. W. Whitten, R. E. Davis, M. L. Peck, and G. G. Stanley, *General Chemistry*, 7th ed., Belmont, CA, Brooks/Cole, 2004

Chapter 30. 1008: © 2005 Sidney Harris

Index

PHYSICAL CONSTANTS

Quantity	Symbol	Value	SI unit
Speed of light in vacuum	c	3.00×10^8	m/s
Permittivity of free space	ϵ_0	8.85×10^{-12}	$C^2/N \cdot m^2$
Coulomb constant, $1/4\pi\epsilon_0$	k_e	8.99×10^9	$N \cdot m^2/C^2$
Permeability of free space	μ_0	1.26×10^{-6} ($4\pi \times 10^{-7}$ exactly)	$T \cdot m/A$
Elementary charge	e	1.60×10^{-19}	C
Planck's constant	h	6.63×10^{-34}	$J \cdot s$
	$\hbar = h/2\pi$	1.05×10^{-34}	$J \cdot s$
Electron mass	m_e	9.11×10^{-31}	kg
		5.49×10^{-4}	u
Proton mass	m_p	$1.672\,65 \times 10^{-27}$	kg
		$1.007\,276$	u
Neutron mass	m_n	$1.674\,95 \times 10^{-27}$	kg
		$1.008\,665$	u
Avogadro's number	N_A	6.02×10^{23}	mol^{-1}
Universal gas constant	R	8.31	$J/mol \cdot K$
Boltzmann's constant	k_B	1.38×10^{-23}	J/K
Stefan-Boltzmann constant	σ	5.67×10^{-8}	$W/m^2 \cdot K^4$
Molar volume of ideal gas at STP	V	22.4	L/mol
		2.24×10^{-2}	m^3/mol
Rydberg constant	R_H	1.10×10^7	m^{-1}
Bohr radius	a_0	5.29×10^{-11}	m
Electron Compton wavelength	$h/m_e c$	2.43×10^{-12}	m
Gravitational constant	G	6.67×10^{-11}	$N \cdot m^2/kg^2$
Standard free-fall acceleration	g	9.80	m/s^2
Radius of Earth (at equator)	R_E	6.38×10^6	m
Mass of Earth	M_E	5.98×10^{24}	kg
Radius of Moon	R_M	1.74×10^6	m
Mass of Moon	M_M	7.36×10^{22}	kg

The values presented in this table are those used in computations in the text. Generally, the physical constants are known to much better precision.